W9-AOT-646

Principles of Fluorescence Spectroscopy

Second Edition

Principles of Fluorescence Spectroscopy

Second Edition

Joseph R. Lakowicz
University of Maryland School of Medicine
Baltimore, Maryland

Kluwer Academic / Plenum Publishers
New York, Boston, Dordrecht, London, Moscow

Library of Congress Cataloging-in-Publication Data

Lakowicz, Joseph R.
Principles of fluorescence spectroscopy / Joseph R. Lakowicz. -- 2nd ed.
p. cm.
Includes bibliographical references and index.
ISBN 0-306-46093-9
1. Fluorescence spectroscopy. I. Title.
QD96.F56L34 1999
543'.08584--dc21 99-30047
CIP

ISBN 0-306-46093-9

233 Spring Street, New York, N.Y. 10013

10 9 8 7 6 5 4 3 2 1

A C.I.P. record for this book is available from the Library of Congress

Printed in the United States of America

To Professor Alexander Jabłoński
on the occasion of his 100th Birthday

Preface

It has been 15 years since publication of the first edition of *Principles of Fluorescence Spectroscopy*. This first volume grew out of a graduate-level course on fluorescence taught at the University of Maryland. The first edition was written during a transition period in the technology and applications of fluorescence spectroscopy. In 1983, time-resolved measurements were performed using methods which are primitive by today's standards. The dominant light sources for time-resolved fluorescence were the nanosecond flashlamps, which provided relatively wide excitation pulses. Detection was accomplished with relatively slow response photomultiplier tubes. In the case of phase-modulation fluorometry, the available instruments operated at one or two fixed light modulation frequencies and thus provided limited information on complex time-resolved decays. Data analysis was also limited because of the lower information content of the experimental data.

Much has changed since 1983. The dominant light sources are now picosecond dye lasers or femtosecond Titanium: sapphire lasers. In the case of phase-modulation fluorometry, frequency-domain instrumentation now operates over a range of light modulation frequencies, allowing resolution of complex decays. The time resolution in both the frequency and the time domain has been increased by the introduction of high-speed microchannel plate photomultiplier tubes. Data analysis has become increasingly sophisticated, not only because of the availability of more powerful computers, but also because of the availability of additional data and the increased resolution available using global analysis. These advanced experimental and analysis capabilities have been extended to provide resolution of complex anisotropy decays, conformational distributions, and complex quenching phenomena.

Another important change since 1983 has been the extensive development of fluorescent probes. Early fluorescent probes were those derived from histochemical staining of cells, a limited number of lipid and conjugatable probes, and, of course, intrinsic fluorescence from proteins. Today the menu of fluorescent probes has expanded manyfold. A wide variety of lipid and protein probes have been developed, and probes have become available with longer excitation and emission wavelengths. There has been extensive development of cation-sensing probes for use in cellular imaging. The nanosecond barrier of dynamic fluorescence information has been broken by the introduction of long-lifetime probes.

Another example of the rapid expansion of fluorescence is DNA sequencing technology. Prior to 1985, most DNA sequencing was performed using radioactive labels. Since that time, sequencing has been accomplished almost exclusively with fluorescent probes. The fluorescence technology for DNA sequencing is advancing rapidly owing to the goal of sequencing the human genome. Finally, who would have expected in 1983 that the gene for the green fluorescent protein could be introduced into cells, with spontaneous folding and formation of the fully fluorescent protein?

Parts of this book were influenced by a course taught at the Center for Fluorescence Spectroscopy, which has been attended by individuals from throughout the world. However, the most important factor stimulating the second edition was the positive comments of individuals who found value in the first edition. Many individuals commented on the value of explaining the basic concepts from their fundamental origins. This has become increasingly important as the number of practitioners of fluorescence spectroscopy has increased, without a significant increase in the number of courses at the undergraduate or graduate level.

In this second edition of *Principles of Fluorescence Spectroscopy*, I have attempted to maintain the emphasis on basics, while updating the examples to include more recent results from the literature. There is a new chapter providing an overview of extrinsic fluorophores. The discussion of time-resolved measurements has been expanded to two chapters. Quenching has also been expanded to two chapters. Energy transfer and anisotropy have each been expanded to three chapters. There is also a new chapter on fluorescence sensing. To enhance the usefulness of this book as a textbook, each chapter is followed by a set of problems. Sections which describe advanced topics are indicated as such, to allow these sections to be skipped in an introductory course. Glossaries of commonly used acronyms and mathematical symbols are provided. For those wanting additional information, Appendix III contains a list of recommended books which expand on various specialized topics.

In closing, I wish to express my appreciation to the many individuals who have assisted me not only in preparation of the book but also in the intellectual developments in my laboratory. My special thanks go to Ms. Mary Rosenfeld for her careful preparation of the text. Mary has cheerfully tolerated the copious typing and numerous revisions of all the chapters. I also thank the many individuals who have proofread various chapters and provided constructive suggestions. These individuals include Felix Castellano, Robert E. Dale, Jonathan Dattelbaum, Maurice Eftink, John Gilchrist, Zygmunt Gryczynski, Petr Herman, Gabor Laczko, Li Li, Harriet Lin, Zakir Murtaza, Leah Tolosa, and Bogumil Zelent. I apologize for any omissions.

I also give my special thanks to Dr. Ignacy Gryczynski and his wife, Krystyna Gryczynska. When I started to write this book, Ignacy said "just go and write, don't worry about the figures." Many of the excellent figures in this book were drawn by Krystyna, with the valuable suggestions of Ignacy. Without their dedicated efforts, the book could not have been completed in any reasonable period of time. I also thank Ms. Suzy Rhinehart for providing a supportive family environment during preparation of this book. Finally, I thank the National Institutes of Health and the National Science Foundation for support of my laboratory.

J. R. Lakowicz
Center for Fluorescence Spectroscopy, Baltimore

Glossary of Acronyms

2,6-ANS 6-Anilinonaphthalene-2-sulfonic acid
ASE Asymptotic standard error
BODIPY Refers to a family of dyes based on 1,3,5,7,8-pentamethylpyrromethene-BF_2, or 4,4-difluoro-4-bora-3a,4a-diaza-s-indacene. BODIPY is a trademark of Molecular Probes, Inc.
CFD Constant fraction discriminator
Dansyl 5-Dimethylaminonaphthalene-1-sulfonic acid
DAPI 4′,6-Diamidino-2-phenylindole
DAS Decay-associated spectra
DNS-Cl Dansyl chloride
DPH 1,6-Diphenyl-1,3,5-hexatriene
EB Ethidium bromide
F Single-letter code for phenylalanine
FAD Flavin adenine dinucleotide
FD Frequency-domain
FISH Fluorescence *in situ* hybridization
FITC Fluorescein-5-isothiocyanate
FMN Flavin mononucleotide
FRET Fluorescence resonance energy transfer
GFP Green fluorescent protein
HIV Human immunodeficiency virus
HSA Human serum albumin
IAEDANS 5-((((2-iodoacetyl)amino)ethyl)amino)-naphthalene-1-sulfonic acid
IAF 5-(Iodoacetamido)fluorescein
ICT Internal charge transfer (state)
IRF Instrument response function
LE Locally excited (state)
MCP Microchannel plate
MLC Metal–ligand complex, usually of a transition metal (Ru, Rh, or Os)
MLCT Metal–ligand charge transfer (state)
MPE Multiphoton excitation
NADH Reduced nicotinamide adenine dinucleotide
NATA *N*-Acetyl-L-tryptophanamide
NATyrA *N*-Acetyl-L-tyrosinamide
NBD 7-Nitrobenz-2-oxa-1,3-diazol-4-yl
NIR Near-infrared
phe Phenylalanine
PC Phosphatidylcholine
PMT Photomultiplier tube
POPOP 1,4-Bis(5-phenyloxazol-2-yl)benzene
PPD 2,5-Diphenyl-1,3,4-oxadazole
PPO 2,5-Diphenyloxazole
Prodan 6-Propionyl-2-(dimethylamino)naphthalene
PSDF Phase-sensitive detection of fluorescence
RET Resonance energy transfer
S_0 Ground electronic state
S_1 First excited singlet state
SPQ 6-Methoxy-*N*-(3-sulfopropyl)quinoline
T_1 First excited triplet state
TAC Time-to-amplitude converter
TCSPC Time-correlated single-photon counting
TD Time-domain
TICT Twisted internal charge-transfer state
TNS 6-(*p*-Toluidinyl)naphthalene-2-sulfonic acid
TRES Time-resolved emission spectra
TRITC Tetramethylrhodamine 5- (and 6-)isothiocyanate
trp Tryptophan
tyr Tyrosine
W Single-letter code for tryptophan
Y Single-letter code for tyrosine

Glossary of Mathematical Terms

A Acceptor or absorption
c Speed of light
C_0 Characteristic acceptor concentration in resonance energy transfer
$C(t)$ Correlation function for spectral relaxation
D Donor, diffusion coefficient, or rotational diffusion coefficient
$D_{\|}$ or $D_{\perp}$ Rate of rotational diffusion around (displacing) the symmetry axis of an ellipsoid of revolution
E Efficiency of energy transfer
F Steady state intensity or fluorescence
F_{χ} Ratio of χ_R^2 values, used to calculate parameter confidence intervals
$F(\lambda)$ Emission spectrum
f_i Fractional steady-state intensities in a multiexponential intensity decay
f_Q Efficiency of collisional quenching
G Correction factor for anisotropy measurements
hw Half-width in a distance or lifetime distribution
$I(t)$ Intensity decay, typically the impulse response function
k_{nr} Nonradiative decay rate
k_S Solvent relaxation rate
k_T Transfer rate in resonance energy transfer
m_{ω} Modulation at a light modulation frequency ω
n Refractive index, when used in consideration of solvent effects
$N(t_k)$ Number of counts per channel, in time-correlated single-photon counting
Q Quantum yield
$P(r)$ Probability function for a distance (r) distribution
pK_a Acid dissociation constant, negative logarithm
r Anisotropy (sometimes distance in a distance distribution)
$\bar{r}$ Average distance in a distance distribution
$r(0)$ Time-zero anisotropy
$r(t)$ Anisotropy decay
r_c Distance of closest approach between donors and acceptors in resonance energy transfer, or fluorophores and quenchers
r_{0i} or $f_0 g_i$ Fractional amplitudes in a multiexponential anisotropy decay
r_0 Fundamental anisotropy in the absence of rotational diffusion
r_{0i} Anisotropy amplitudes in a multiexponential anisotropy decay
r_{∞} Long-time anisotropy in an anisotropy decay
r_{ω} Modulated anisotropy
R_0 Förster distance in resonance energy transfer
α_i Preexponential factors in a multiexponential intensity decay
β Angle between absorption and emission transition moments
Γ Radiative decay rate
γ Inverse of the decay time: $\gamma = 1/\tau$
ε Dielectric constant or extinction coefficient
θ Rotational correlation time
κ^2 Orientation factor in resonance energy transfer
Λ_{ω} Ratio of the modulated amplitudes of the polarized components of the emission
λ Wavelength
λ_{em} Emission wavelength
λ_{em}^{max} Maximum emission wavelength
λ_{ex} Excitation wavelength
λ_{ex}^{max} Maximum excitation or absorption wavelength for the lowest $S_0 \rightarrow S_1$ transition
λ_{max} Emission maximum
μ_E Excited-state dipole moment
μ_G Ground-state dipole moment

$\bar{\nu}$ Wavenumber, in cm^{-1}
$\bar{\nu}_{cg}$ Emission center of gravity
$\nu_{cg}(t)$ Time-resolved emission center of gravity, in cm^{-1}
τ Decay time
$\bar{\tau}$ Average lifetime
τ_ϕ Apparent lifetime calculated from the phase angle at a single frequency
τ_D Donor decay time or solvent dielectric relaxation time
τ_L Solvent longitudinal relaxation time
τ_m Apparent lifetime calculated from the modulation at a single frequency
τ_N Radiative or natural lifetime
τ_S Solvent relaxation time
Δ_ω Differential polarized phase angle, difference in phase between the parallel and perpendicular components of the emission
ϕ_ω Phase angle at a light modulation frequency ω
χ_R^2 Goodness-of-fit parameter, reduced chi-squared
χ^2 Sum of the squared weighted deviations
ω Light modulation frequency in radians per second; 2π times the frequency in cycles per second

Contents

Most sections of this book describe basic aspects of fluorescence spectroscopy, and some sections describe more advanced topics. These sections are marked "Advanced Topics" and can be omitted in an introductory course on fluorescence. The advanced chapters on quenching (Chapter 9), anisotropy (Chapter 12), and energy transfer (Chapters 14 and 15) can be skipped in a first reading. Depending on the interest of the reader, Chapters 18 to 22 can also be skipped.

1. Introduction to Fluorescence

2. Instrumentation for Fluorescence Spectroscopy

3. Fluorophores

4. Time-Domain Lifetime Measurements

5. Frequency-Domain Lifetime Measurements

6. Solvent Effects on Emission Spectra

7. Dynamics of Solvent and Spectral Relaxation

8. Quenching of Fluorescence

9. Advanced Topics in Fluorescence Quenching

10. Fluorescence Anisotropy

11. Time-Dependent Anisotropy Decays

12. Advanced Anisotropy Concepts

13. Energy Transfer

14. Time-Resolved Energy Transfer and Conformational Distributions of Biopolymers

15. Energy Transfer to Multiple Acceptors, in One, Two, or Three Dimensions

16. Protein Fluorescence

17. Time-Resolved Protein Fluorescence

18. Excited-State Reactions

19. Fluorescence Sensing

20. Long-Lifetime Metal–Ligand Complexes

21. DNA Technology

22. Phase-Sensitive and Phase-Resolved Emission Spectra

Appendix I. Corrected Emission Spectra

Appendix II. Fluorescent Lifetime Standards

Appendix III. Additional Reading

Principles of Fluorescence Spectroscopy

Second Edition

Introduction to Fluorescence

1

During the past 15 years there has been a remarkable growth in the use of fluorescence in the biological sciences. Just a few years ago, fluorescence spectroscopy and time-resolved fluorescence were primarily research tools in biochemistry and biophysics. This situation has changed so that fluorescence is now used in environmental monitoring, clinical chemistry, DNA sequencing, and genetic analysis by fluorescence *in situ* hybridization (FISH), to name a few areas of application. Additionally, fluorescence is used for cell identification and sorting in flow cytometry, and in cellular imaging to reveal the localization and movement of intracellular substances by means of fluorescence microscopy. Because of the sensitivity of fluorescence detection, and the expense and difficulties of handling radioactive substances, there is a continuing development of medical tests based on the phenomenon of fluorescence. These tests include the widely used enzyme-linked immunoassays (ELISA) and fluorescence polarization immunoassays.

While there is continued growth in the applications of fluorescence, and continued development of instrumentation and fluorescent probe technology, the principles remain the same and need to be understood by the practitioners. Hence, this volume describes the basic phenomenon of fluorescence, so that these principles can be used successfully in basic and applied research. Throughout the book, we have included examples and applications that illustrate the principles of fluorescence.

1.1. PHENOMENON OF FLUORESCENCE

Luminescence is the emission of light from any substance and occurs from electronically excited states. Luminescence is formally divided into two categories, fluorescence and phosphorescence, depending on the nature of the excited state. In excited singlet states, the electron in the excited orbital is paired (of opposite spin) to the second electron in the ground-state orbital. Consequently, return to the ground state is spin-allowed and occurs rapidly by emission of a photon. The emission rates of fluorescence are typically 10^8 s^{-1}, so that a typical fluorescence lifetime is near 10 ns (10×10^{-9} s). As will be described in Chapter 4, the lifetime (τ) of a fluorophore is the average time between its excitation and its return to the ground state. It is valuable to consider a 1-ns lifetime within the context of the speed of light. Light travels 30 cm or about one foot in one nanosecond. Many fluorophores display subnanosecond lifetimes. Because of the short timescale of fluorescence, measurement of the time-resolved emission requires sophisticated optics and electronics. In spite of the experimental difficulties, time-resolved fluorescence is widely practiced because of the increased information available from the data, as compared with stationary or steady-state measurements.

Phosphorescence is emission of light from triplet excited states, in which the electron in the excited orbital has the same spin orientation as the ground-state electron. Transitions to the ground state are forbidden and the emission rates are slow (10^3–10^0 s^{-1}), so that phosphorescence lifetimes are typically milliseconds to seconds. Even longer lifetimes are possible, as is seen from "glow-in-the-dark" toys: following exposures to light, the phosphorescent substances glow for several minutes while the excited phosphors slowly return to the ground state. Phosphorescence is usually not seen in fluid solutions at room temperature. This is because there exist many deactivation processes which compete with emission, such as nonradiative decay and quenching processes. It should be noted that the distinction between fluorescence and phosphorescence is not always clear. Transition-metal–ligand complexes (MLCs), which contain a metal and one or more organic ligands, display mixed singlet–triplet states. These MLCs display intermediate lifetimes of 400 ns to several microseconds. In this book we will concentrate mainly on the more rapid phenomenon of fluorescence. However, because of the importance of these new metal–ligand luminophores, which bridge the gap between fluorescence and phosphorescence, their properties are described in Chapter 20.

Fluorescence typically occurs from aromatic molecules. Some typical fluorescent substances (fluorophores) are shown in Figure 1.1. One widely encountered fluorophore is quinine, which is present in tonic water. If one observes a glass of tonic water which is exposed to sunlight, a faint blue glow is frequently visible at the surface. This glow is most apparent when the glass is observed at a right angle relative to the direction of the sunlight and when the dielectric constant is decreased by adding less polar solvents like alcohols. The quinine present in the tonic is excited by the ultraviolet (UV) light from the sun. Upon return to the ground state, the quinine emits blue light with a wavelength near 450 nm. The dependence of the brightness of quinine fluorescence on solvent polarity provides a noninvasive means of learning more about our neighbors. The first observation of fluorescence from a quinine solution in sunlight was reported by Sir John Frederick William Herschel in 1845.[1] The following is an excerpt from this early report.

On a Case of Superficial Colour presented by a homogeneous liquid internally colourless. By Sir John Frederick William Herschel, Philosophical Transactions of the Royal Society of London (1845) 135:143–145.
Received January 28, 1845, Read February 13, 1845

The sulphate of quinine is well known to be of extremely sparing solubility in water. It is however easily and copiously soluble in tartaric acid. Equal weights of the sulphate and of crystallized tartaric acid, rubbed up together with addition of a very little water, dissolve entirely and immediately. It is this solution, largely diluted, which exhibits the optical phenomenon in question. Though perfectly transparent and colorless when held between the eye and the light, or a white object, it yet exhibits in certain aspects, and under certain incidences of the light, an extremely vivid and beautiful celestial blue colour, which, from the circumstances of its occurrence, would seem to originate in those strata which the light first penetrates in entering the liquid, and which, if not strictly superficial, at least exert their peculiar power of analyzing the incident rays and dispersing those which compose the tint in question, only through a very small depth within the medium.

To see the colour in question to advantage, all that is requisite is to dissolve the two ingredients above mentioned in equal proportions, in about a hundred times their joint weight of water, and having filtered the solution, pour it into a tall narrow cylindrical glass vessel or test tube, which is to be set upright on a dark colored substance before an open window exposed to strong daylight or sunshine, but with no cross lights, or any strong reflected light from behind. If we look down perpendicularly into the vessel so that the visual ray shall graze the internal surface of the glass through a great part of its depth, the whole of that surface of the liquid on which the light first strikes will appear of a lively blue, . . .

If the liquid be poured out into another vessel, the descending stream gleams internally from all its undulating inequalities, with the same lively yet delicate blue colour, . . . thus clearly demonstrating that contact with a denser medium has no share in producing this singular phenomenon.

The thinnest film of the liquid seems quite as effective in producing this superficial colour as a considerable thickness. For instance, if in pouring it from one glass into another, . . . the end of the funnel be made to touch the internal surface of the vessel well moistened, so as to spread the descending stream over an extensive surface, the intensity of the colour is such that it is almost impossible to avoid supposing that we have a highly colored liquid under our view.

As a footnote, Herschel wrote "I write from recollection of an experiment made nearly twenty years ago, and which I cannot repeat for want of a specimen of the wood."

It is evident from this early description that Herschel recognized the presence of an unusual phenomenon which could not be explained by the scientific knowledge of the time. To this day, the fluorescence of quinine remains one of the most used and most beautiful examples of fluorescence. However, it is unlikely that Herschel's experiment, described from memory 20 years later, would be accepted by the Patent Office. Herschel (Figure 1.2) was from a distinguished family of scientists who lived in England but had their roots in Germany.[2] For most of his life, Herschel did research in astronomy, publishing only a few papers on fluorescence.

It is interesting to notice that the first known fluorophore, quinine, was responsible for stimulating the development of the first spectrofluorometers, which appeared in the 1950s. During World War II, the Department of Defense was interested in monitoring antimalaria drugs, including quinine. This early drug assay resulted in a subsequent program at the National Institutes of Health to develop the first practical spectrofluorometer.[3]

Many other fluorophores are encountered in daily life. The green or red-orange glow sometimes seen in antifreeze is due to trace quantities of fluorescein or rhodamine, respectively (Figure 1.1). Polynuclear aromatic hydrocar-

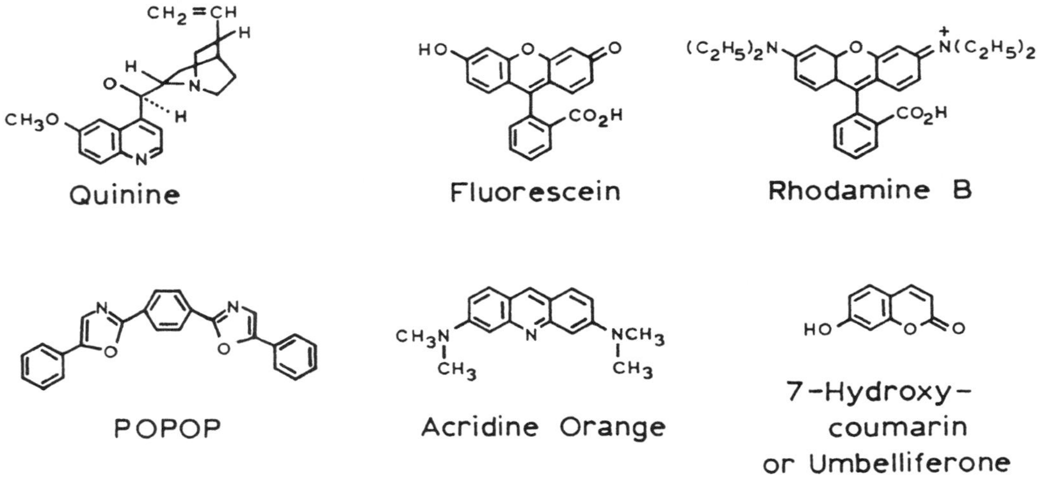

Quinine　Fluorescein　Rhodamine B

POPOP　Acridine Orange　7-Hydroxy-coumarin or Umbelliferone

Figure 1.1. Structures of typical fluorescent substances.

bons, such as anthracene and perylene, are also fluorescent, and the emission from such species is used for environmental monitoring of oil pollution. Some substituted organic compounds are also fluorescent. For example, 1,4-bis(5-phenyloxazol-2-yl)benzene (POPOP) is used in scintillation counting, and Acridine Orange is often used as a DNA stain. Coumarins are also highly fluorescent and are often used as fluorogenic probes in enzyme assays, such as enzyme-linked immunosorbent assays (ELISA). In this case the parent molecule is typically umbelliferyl phosphate, which is nonfluorescent. Hydrolysis of the 7-hydroxyl phosphate by alkaline phosphatase results in a highly fluorescent product.

Numerous additional examples could be presented. Instead of listing them here, examples will appear throughout the book, with reference to the useful properties of the individual fluorophores. In Chapter 3 we summarize the diversity of fluorophores used for research and fluorescence sensing. In contrast to aromatic organic molecules, atoms are generally nonfluorescent in condensed phases. One notable exception is the group of elements commonly known as the lanthanides.[4] The fluorescence from europium and terbium ions results from electronic transitions between *f* orbitals. These orbitals are shielded from the solvent by higher filled orbitals. The lanthanides display long decay times because of this shielding, and they have low emission rates because of their small extinction coefficients.

Fluorescence spectral data are generally presented as emission spectra. A fluorescence emission spectrum is a plot of the fluorescence intensity versus wavelength (nanometers) or wavenumber (cm^{-1}). Two typical fluorescence emission spectra are shown in Figure 1.3. Emission

Figure 1.2. Sir John Frederick William Herschel (March 7, 1792–May 11, 1871). Reproduced courtesy of the Library & Information Centre, Royal Society of Chemistry.

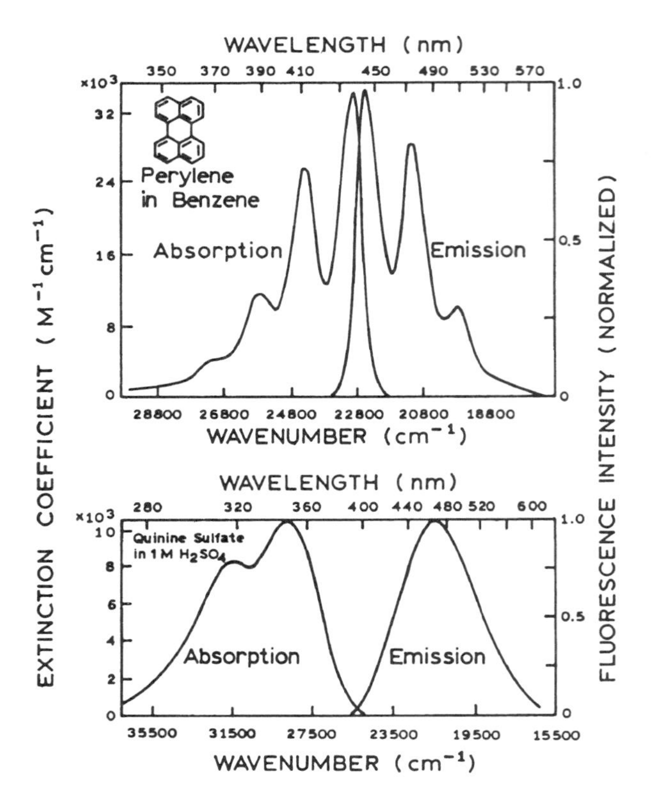

Figure 1.3. Absorption and fluorescence emission spectra of perylene and quinine. Emission spectra cannot be correctly presented on both the wavelength and wavenumber scales. The wavenumber presentation is correct in this instance. Wavelengths are shown for convenience. See Chapter 3. Revised from Ref. 5.

spectra vary widely and are dependent upon the chemical structure of the fluorophore and the solvent in which it is dissolved. The spectra of some compounds, such as perylene, show significant structure due to the individual vibrational energy levels of the ground state and excited states. Other compounds, such as quinine, show spectra which are devoid of vibrational structure.

An important feature of fluorescence is high-sensitivity detection. The sensitivity of fluorescence was used in 1877 to demonstrate that the rivers Danube and Rhine were connected by underground streams.[5] This connection was demonstrated by placing fluorescein (Figure 1.1) into the Danube River. Some 60 hours later, its characteristic green fluorescence appeared in a small river which led to the Rhine. Today fluorescein is still used as an emergency marker for locating individuals at sea, as has been seen on the landing of space capsules in the Atlantic Ocean. Readers interested in the history of fluorescence are referred to the excellent summary by Berlman.[5]

Figure 1.4. Professor Alexander Jabłoński (1898–1980), *circa* 1935. Courtesy of his daughter, Professor Danuta Frąckowiak.

1.2. JABŁOŃSKI DIAGRAM

The processes which occur between the absorption and emission of light are usually illustrated by a Jabłoński[6] diagram. Jabłoński diagrams are often used as the starting point for discussing light absorption and emission. They exist in a variety of forms, to illustrate various molecular processes which can occur in excited states. These diagrams are named after Professor Alexander Jabłoński (Figure 1.4), who is regarded as the father of fluorescence spectroscopy because of his many accomplishments, including his descriptions of concentration depolarization and his definition of the term "anisotropy" to describe the polarized emission from solutions.[7,8]

Brief History of Alexander Jabłoński

Professor Jabłoński was born February 26, 1898, in Voskresenovka, Ukraine. In 1916 he began his study of atomic physics at the University of Kharkov. His study was interrupted by military service first in the Russian Army and later in the newly organized Polish Army during World War I. At the end of 1918, when an independent Poland was recreated after more than

120 years of occupation by neighboring powers, Jabłoński left Kharkov and arrived in Warsaw, where he entered the University of Warsaw to continue his study of physics. His study in Warsaw was again interrupted in 1920 by his military service during the Polish–Bolshevik war.

An enthusiastic musician, Jabłoński played the first violin at the Warsaw Opera from 1921 to 1926 parallel to his studies at the university under Stefan Pienkowski. He received his doctorate in 1930 for work "On the influence of the change of wavelengths of excitation light on the fluorescence spectra." Although Jabłoński left the Warsaw Opera in 1926 and devoted himself entirely to scientific work, music remained his great passion until the last days of his life.

Throughout the 1920s and 1930s the Department of Experimental Physics at the University of Warsaw was an active center for studies on luminescence under S. Pienkowski. During most of this period, Jabłoński worked both theoretically and experimentally on fundamental problems of photoluminescence of liquid solutions as well as on the effects of pressure on atomic spectral lines in gases. The problem that intrigued Jabłoński for many years was the polarization of photoluminescence of solutions. To explain the experimental facts, he distinguished the transition moments in absorption and in emission and analyzed various factors responsible for the depolarization of luminescence.

Jabłoński's work was interrupted once again by World War II. From 1939 to 1945 Jabłoński served in the Polish Army, and he spent time as a prisoner of first the Germany Army and then the Soviet Army. In 1946 he returned to Poland to chair a new Department of Physics in the new Nicholas Copernicus University in Toruń. This beginning occurred in the very difficult postwar years in a country totally destroyed by World War II. Despite all these difficulties, Jabłoński with great energy organized the Department of Physics, which became a scientific center for studies in atomic and molecular physics.

His work continued past his retirement in 1968. Professor Jabłoński created a spectroscopic school of thought which persists even today through his numerous students who now occupy positions at universities in Poland and elsewhere. Professor Jabłoński died on September 9, 1980. More complete accounts of Jabłoński's accomplishments are given in Refs. 7 and 8.

A typical Jabłoński diagram is shown in Figure 1.5. The singlet ground, first, and second electronic states are depicted by S_0, S_1, and S_2, respectively. At each of these electronic energy levels the fluorophores can exist in a number of vibrational energy levels, denoted by 0, 1, 2, etc. In this diagram we have excluded a number of interactions which will be discussed in subsequent chapters, such as quenching, energy transfer, and solvent interactions. The transitions between states are depicted as vertical lines to illustrate the instantaneous nature of light absorption. Transitions occur in about 10^{-15} s, a time too short for significant displacement of nuclei. This is the Franck–Condon principle.

The energy spacing between the various vibrational energy levels is illustrated by the emission spectrum of perylene (Figure 1.3). The individual emission maxima (and hence vibrational energy levels) are about 1500 cm^{-1} apart. At room temperature, thermal energy is not adequate to significantly populate the excited vibrational states. Absorption typically occurs from molecules with the lowest vibrational energy. Of course, the larger energy difference between the S_0 and S_1 excited states is too large for thermal population of S_1, and it is for this reason we use light and not heat to induce fluorescence.

Following light absorption, several processes usually occur. A fluorophore is usually excited to some higher vibrational level of either S_1 or S_2. With a few rare excep-

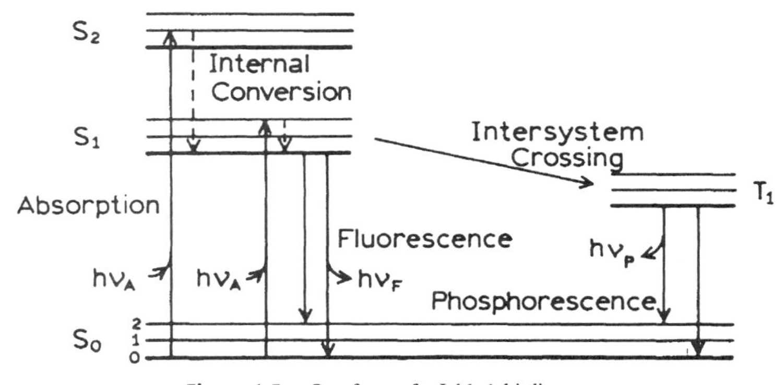

Figure 1.5. One form of a Jabłoński diagram.

tions, molecules in condensed phases rapidly relax to the lowest vibrational level of S_1. This process is called internal conversion and generally occurs in 10^{-12} s or less. Since fluorescence lifetimes are typically near 10^{-8} s, internal conversion is generally complete prior to emission. Hence, fluorescence emission generally results from a thermally equilibrated excited state, that is, the lowest-energy vibrational state of S_1.

Return to the ground state typically occurs to a higher excited vibrational ground-state level, which then quickly (10^{-12} s) reaches thermal equilibrium (Figure 1.5). An interesting consequence of emission to higher vibrational ground states is that the emission spectrum is typically a mirror image of the absorption spectrum of the $S_0 \rightarrow S_1$ transition. This similarity occurs because electronic excitation does not greatly alter the nuclear geometry. Hence, the spacing of the vibrational energy levels of the excited states is similar to that of the ground state. As a result, the vibrational structures seen in the absorption and the emission spectra are similar.

Molecules in the S_1 state can also undergo a spin conversion to the first triplet state, T_1. Emission from T_1 is termed phosphorescence and is generally shifted to longer wavelengths (lower energy) relative to the fluorescence. Conversion of S_1 to T_1 is called intersystem crossing. Transition from T_1 to the singlet ground state is forbidden, and, as a result, rate constants for triplet emission is several orders of magnitude smaller than those for fluorescence. Molecules containing heavy atoms such as bromine and iodine are frequently phosphorescent. The heavy atoms facilitate intersystem crossing and thus enhance phosphorescence quantum yields.

1.3. CHARACTERISTICS OF FLUORESCENCE EMISSION

The phenomenon of fluorescence displays a number of general characteristics. Exceptions are known, but these are infrequent. Generally, if any of the characteristics described in the following sections are not displayed by a given fluorophore, one may infer some special behavior for this compound.

1.3.A. Stokes' Shift

Examination of the Jabłoński diagram (Figure 1.5) reveals that the energy of the emission is typically less than that of absorption. Hence, fluorescence typically occurs at lower energies or longer wavelengths. This phenomenon was first observed by Sir G. G. Stokes in 1852 in Cambridge.[9] These early experiments used relatively simple instrumentation (Figure 1.6). The source of UV excitation was provided by sunlight and a blue glass filter, which was part of a stained glass window. This filter selectively transmitted light below 400 nm, which was absorbed by quinine (Figure 1.3). The exciting light was prevented from reaching the detector (eye) by a yellow glass (of wine) filter. Quinine fluorescence occurs near 450 nm and is therefore easily visible.

It is interesting to read Stokes' description of his observation. The following paragraph is from his report published in 1852.[9]

On the Change of Refrangibility of Light. By G. G. Stokes, M.A., F.R.S., Fellow of Pembroke College, and Lucasian Professor of Mathematics in the University of Cambridge. Philosophical Transactions of the Royal Society of London (1852) 142:463–562.
Received May 11, Read May 27, 1852

The following researches originated in a consideration of the very remarkable phenomenon discovered by Sir John Herschel in a solution of sulphate of quinine, and described by him in two papers printed in the Philosophical Transactions for 1845, entitled "On a Case of Superficial Colour presented by a Homogeneous Liquid internally colourless," and "On the Epipolic Dispersion of Light." The solution of quinine, though it appears to be perfectly transparent and colourless, like

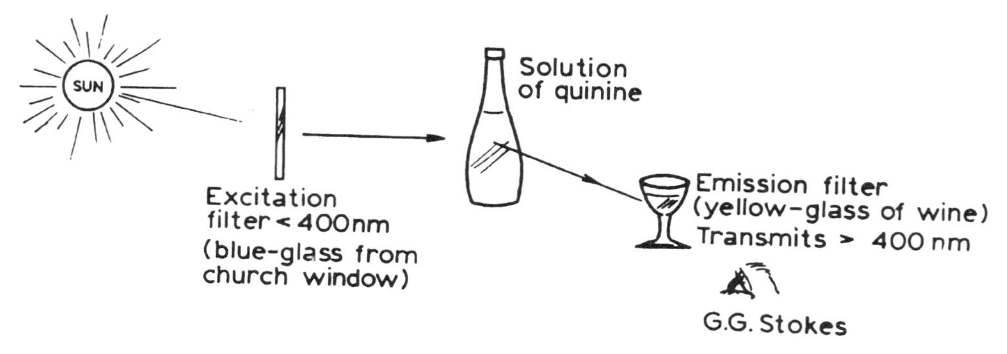

Figure 1.6. Experimental schematic for detection of the Stokes' shift.

> water, when viewed by transmitted light, exhibits nevertheless in certain aspects, and under certain incidences of the light, a beautiful celestial blue colour. It appears from the experiments of Sir John Herschel that the blue colour comes only from a stratum of fluid of small but finite thickness adjacent to the surface by which the light enters. After passing through this stratum, the incident light, though not sensibly enfeebled nor colored, has lost the power of producing the same effect, and therefore may be considered as in some way or other qualitatively different from the original light.

Careful reading of this paragraph reveals several important characteristics of fluorescent solutions. The quinine solution is colorless because it absorbs in the UV, which we cannot see. The blue color comes only from a region near the surface. This is because the quinine solution was relatively concentrated and absorbed all of the UV in the first several millimeters. Hence, Stokes observed the inner filter effect. After passing through the solution, the light was "enfeebled" and no longer capable of causing the blue glow. This occurred because the UV was removed and the "enfeebled" light could no longer excite quinine. However, had Stokes used a second solution of fluorescein, rather than quinine, it would have still been excited because of the longer absorption wavelength of fluorescein.

Energy losses between excitation and emission are observed universally for fluorescent molecules in solution. One common cause of the Stokes' shift is the rapid decay to the lowest vibrational level of S_1. Furthermore, fluorophores generally decay to higher vibrational levels of S_0 (Figure 1.5), resulting in further loss of excitation energy by thermalization of the excess vibrational energy. In addition to these effects, fluorophores can display further Stokes' shifts due to solvent effects, excited-state reactions, complex formation, and/or energy transfer.

Brief History of Sir G. G. Stokes

Professor Stokes (Figure 1.7) was born in Ireland, August 3, 1819. He entered Pembroke College, Cambridge, in 1837 and was elected as a fellow of Pembroke College immediately upon his graduation in 1841. In 1849 Stokes became Lucasian Professor at Cambridge, a chair once held by Newton. Because of a poor endowment for the chair, he also worked in the Government School of Mines.

Stokes was involved with a wide range of scientific problems, including hydrodynamics, elasticity of solids, and diffraction of light. The wave theory of light was already known when he entered Cambridge. In his classic paper on quinine, he understood that light of a higher "refrangibility" or frequency was responsible for the blue glow of lower refrangibility or frequency. Thus, invisible UV rays were absorbed to produce the blue light at the surface. Stokes later suggested the use of optical properties, such as absorption, colored reflection, and fluorescence, to identify organic substances.

Later in life, Stokes was universally honored with degrees and medals. He was knighted in 1889 and became Master of Pembroke College in 1902. After the 1850s, Stokes became involved in administrative matters, and his scientific productivity decreased. Some things never change. Professor Stokes died on February 1, 1903.

1.3.B. Emission Spectra Are Typically Independent of the Excitation Wavelength

Another general property of fluorescence is that the same fluorescence emission spectrum is generally observed irrespective of the excitation wavelength. This is known as

Figure 1.7. Sir George Gabriel Stokes (1819–1903). Lucasian Professor at Cambridge. Reproduced courtesy of the Library & Information Centre, Royal Society of Chemistry.

Kasha's rule,[10] although Vavilov reported in 1926 that quantum yields were generally independent of excitation wavelength.[5] Upon excitation into higher electronic and vibrational levels, the excess energy is quickly dissipated, leaving the fluorophore in the lowest vibrational level of S_1. This relaxation occurs in about 10^{-12} s and is presumably a result of a strong overlap among numerous states of nearly equal energy. Because of this rapid relaxation, emission spectra are usually independent of the excitation wavelength. Exceptions exist, such as fluorophores which exist in two ionization states, each of which displays distinct absorption and emission spectra. Also, some molecules are known to emit from the S_2 level, but such emission is rare and generally not observed in biological molecules.

It is interesting to ask why perylene follows the mirror image rule, but quinine emission exhibits one peak instead of the two peaks seen in its excitation spectrum at 310 and 335 nm (Figure 1.3). In the case of quinine, the shorter-wavelength absorption peak is due to excitation to the second excited state (S_2), which relaxes rapidly to S_1. Hence, emission occurs predominantly from the lowest singlet state (S_1). The emission spectrum of quinine is the mirror image of the $S_0 \rightarrow S_1$ absorption of quinine, not of its total absorption spectrum. This is true for most fluorophores: the emission is the mirror image of the $S_0 \rightarrow S_1$ absorption, not of the total absorption spectrum.

The generally symmetric nature of these spectra is a result of the same transitions being involved in both absorption and emission and the similarities of the vibrational energy levels of S_0 and S_1. In many molecules these energy levels are not significantly altered by the different electronic distributions of S_0 and S_1. According to the Franck–Condon principle, all electronic transitions are vertical, that is, they occur without change in the position of the nuclei. As a result, if a particular transition probability (Franck–Condon factor) between the zeroth and second vibrational levels is largest in absorption, the reciprocal transition is also most probable in emission (Figure 1.8).

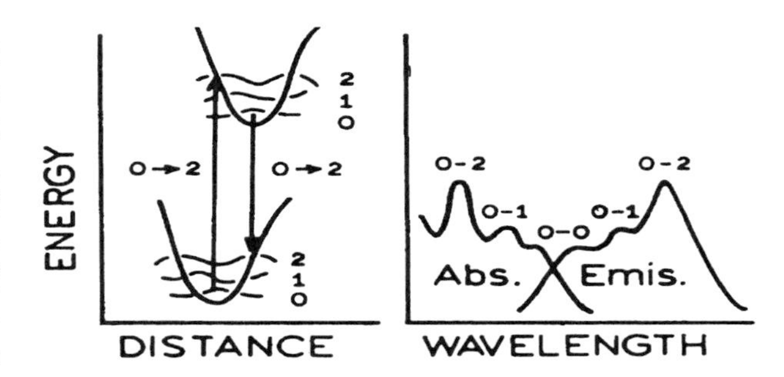

Figure 1.8. Mirror image rule and Franck–Condon factors.

A rigorous test of the mirror image rule requires that the absorption and emission spectra be presented in appropriate units.[11] The closest symmetry should exist between the modified spectra $\varepsilon(\bar{\nu})/\bar{\nu}$ and $F(\bar{\nu})/\bar{\nu}^3$, where $\varepsilon(\bar{\nu})$ is the extinction coefficient at wavenumber ($\bar{\nu}$) and $F(\bar{\nu})$ is the relative photon flux over a wavenumber increment $\Delta\bar{\nu}$. Agreement between these spectra is generally found for polynuclear aromatic hydrocarbons.

1.3.C. Exceptions to the Mirror Image Rule

Although the mirror image rule often holds, many exceptions to this rule occur. This is illustrated for *p*-terphenyl in Figure 1.9. The absorption spectrum of *p*-terphenyl is devoid of structure, but the emission spectrum shows vibrational structure. Such deviations from the mirror image rule usually indicate a different geometric arrangement of nuclei in the excited state as compared to the ground state. Nuclear displacements can occur prior to emission because of the relatively long lifetime of the S_1 state, which allows

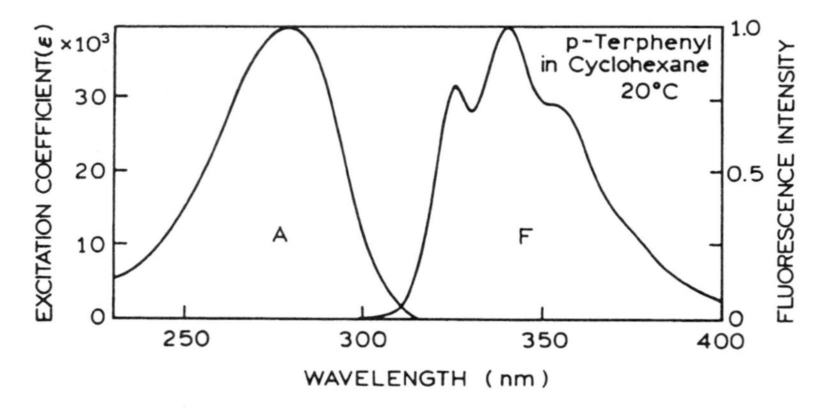

Figure 1.9. Absorption (A) and emission (F) spectra of *p*-terphenyl.

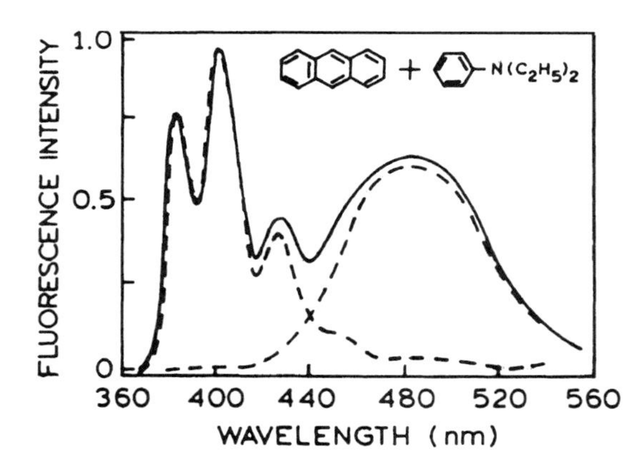

Figure 1.10. Emission spectrum of anthracene in toluene containing 0.2*M* diethylaniline. The dashed curves show the emission spectrum of anthracene and that of its exciplex with diethylaniline. Revised from Ref. 13.

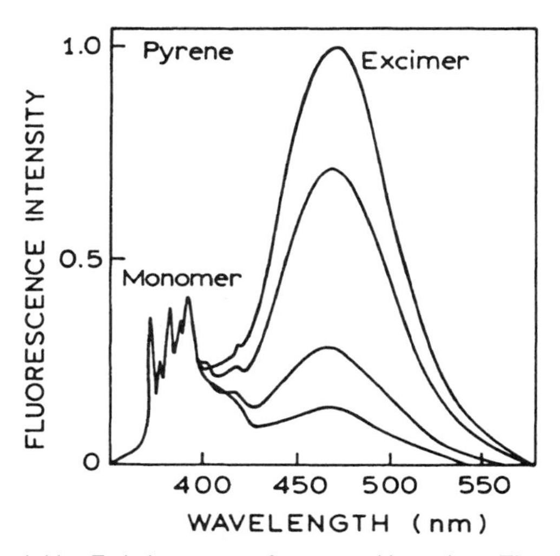

Figure 1.11. Emission spectra of pyrene and its excimer. The relative intensity of the excimer peak (470 nm) decreases as the total concentration of pyrene is decreased from $6 \times 10^{-3} M$ (*top*) to $0.9 \times 10^{-4} M$ (*bottom*). Reproduced with permission from John Wiley and Sons, Inc., from Ref. 11. Birks, J. B., Copyright © 1970, *Photophysics of Aromatic Molecules*, John Wiley & Sons, New York.

time for motion following the instantaneous process of absorption. In the case of *p*-terphenyl, it seems likely that the individual rings become more coplanar in the excited state.[12] As a result, the emission spectrum is more highly structured than the absorption spectrum. In addition to being an exception to the mirror image rule, *p*-terphenyl is unusual in that its emission spectrum shows more vibrational structure than its absorption spectrum. The opposite is generally observed.

Excited-state reactions other than geometric rearrangements can also result in deviations from the mirror symmetry rule. One example is shown in Figure 1.10, which shows the emission spectrum of anthracene in the presence of diethylaniline.[13] The structured emission at shorter wavelengths is a mirror image of the absorption spectrum of anthracene. The unstructured emission at longer wavelengths is due to formation of a charge-transfer complex between the excited state of anthracene and diethylaniline. The unstructured emission is from this complex. Many polynuclear aromatic hydrocarbons, such as pyrene and perylene, also form charge-transfer complexes with amines. These excited-state complexes are referred to as exciplexes.

Some fluorophores can also form complexes with themselves. The best-known example is pyrene. At low concentration, pyrene displays a highly structured emission (Figure 1.11). At higher concentrations, the previously invisible UV emission of pyrene becomes visible at 470 nm. This long-wavelength emission is due to excimer formation, the term excimer being an abbreviation for an excited-state dimer.

Other excited-state processes can occur which shift the emission spectra and may or may not change the spectral profile. Acridine displays two emission spectra which depend on pH (Figure 1.12). The pK_a for dissociation of the proton is 5.45 for acridine in the ground state. However, emission from the acridinium group can be observed at higher pH values. This occurs because the pK_a of the excited state of acridine is 10.7, and thus acridine can bind a proton from the solvent during its excited-state lifetime.[14] Changes in pK_a in the excited state also occur for biochemical fluorophores. For example, phenol and tyrosine each show two emissions, the long-wavelength emission being favored by a high concentration of proton acceptors. The pK_a of the phenolic hydroxyl group decreases from 11 in the ground state to 4 in the excited state. Following excitation, the phenolic proton is lost to proton acceptors in the solution. Depending upon the concentration of these acceptors, either the phenol or the phenolate emission may dominate the emission spectrum.

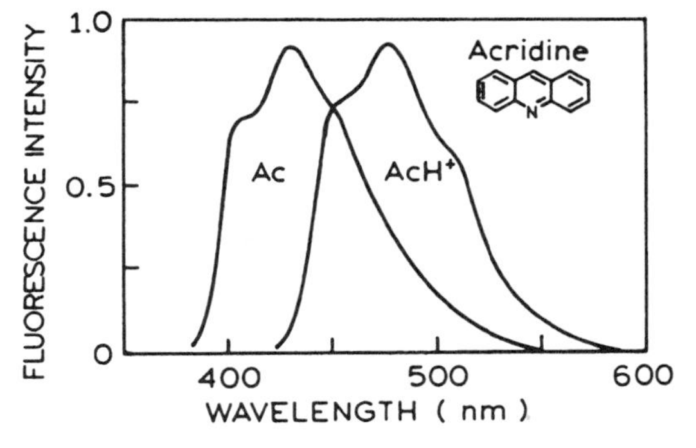

Figure 1.12. Fluorescence emission spectra of the neutral and protonated forms of acridine (Ac). Revised from Ref. 14.

1.4. FLUORESCENCE LIFETIMES AND QUANTUM YIELDS

The fluorescence lifetime and quantum yield are perhaps the most important characteristics of a fluorophore. The quantum yield is the number of emitted photons relative to the number of absorbed photons. Substances with the largest quantum yields, approaching unity, such as rhodamines, display the brightest emission. The lifetime is also important, as the lifetime determines the time available for the fluorophore to interact with or diffuse in its environment, and hence the information available from its emission.

The meaning of the quantum yield and lifetime is best represented by a simplified Jabłoński diagram (Figure 1.13). In this diagram we do not explicitly illustrate the individual relaxation processes leading to the relaxed S_1 state. Instead, we focus attention on those processes responsible for return to the ground state. In particular, we are interested in the emissive rate of the fluorophore (Γ) and its rate of nonradiative decay to S_0 (k_{nr}).

The fluorescence quantum yield is the ratio of the number of photons emitted to the number absorbed. The processes governed by the rate constants Γ and k_{nr} both depopulate the excited state. The fraction of fluorophores which decay through emission, and hence the quantum yield, is given by

$$Q = \frac{\Gamma}{\Gamma + k_{nr}} \qquad [1.1]$$

The quantum yield can be close to unity if the radiationless decay rate is much smaller than the rate of radiative decay, that is, $k_{nr} << \Gamma$. We note that the energy yield of fluorescence is always less than unity because of Stokes' losses. For convenience, we have grouped all possible nonradiative decay processes with the single rate constant k_{nr}.

The lifetime of the excited state is defined by the average time the molecule spends in the excited state prior to return to the ground state. Generally, fluorescence lifetimes are

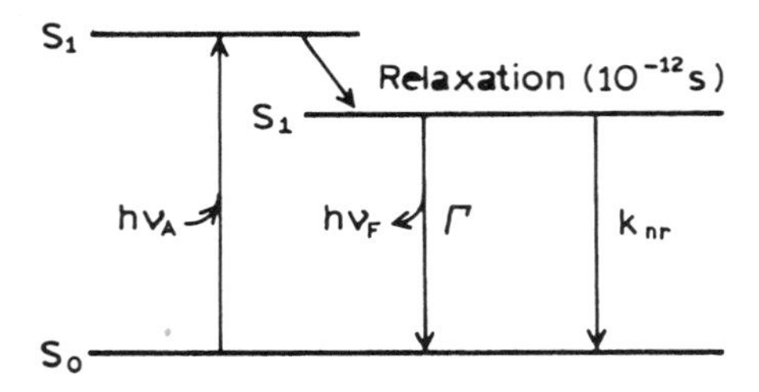

Figure 1.13. A simplified Jabłoński diagram.

near 10 ns. For the fluorophore illustrated in Figure 1.13, the lifetime is

$$\tau = \frac{1}{\Gamma + k_{nr}} \qquad [1.2]$$

One should remember that fluorescence emission is a random process, and few molecules emit their photons at precisely $t = \tau$. The lifetime is an average value of the time spent in the excited state. For a single exponential decay (Eq. [1.13], below), 63% of the molecules have decayed prior to $t = \tau$ and 37% decay at $t > \tau$.

The lifetime of the fluorophore in the absence of nonradiative processes is called the intrinsic or natural lifetime and is given by

$$\tau_n = 1/\Gamma \qquad [1.3]$$

In principle, the natural lifetime τ_n can be calculated from the absorption spectra, extinction coefficient, and emission spectra of the fluorophore. The radiative decay rate Γ can be calculated using[15,16]

$$\Gamma \simeq 2.88 \times 10^{-9}\, n^2 \frac{\int F(\bar{\nu})d\bar{\nu}}{\int F(\bar{\nu})d\bar{\nu}/\bar{\nu}^3} \int \frac{\varepsilon(\bar{\nu})}{\bar{\nu}}\, d\bar{\nu}$$

$$= 2.88 \times 10^{-9} n^2 <\bar{\nu}^{-3}>^{-1} \int \frac{\varepsilon(\bar{\nu})\, d\bar{\nu}}{\bar{\nu}} \qquad [1.4]$$

where $F(\bar{\nu})$ is the emission spectrum plotted on the wavenumber (cm^{-1}) scale, $\varepsilon(\bar{\nu})$ is the absorption spectrum, and n is the refractive index of the medium. The integrals are calculated over the $S_0 \leftrightarrow S_1$ absorption and emission spectra. In many cases this expression works rather well, particularly for solutions of polynuclear aromatic hydrocarbons. For instance, the calculated value[15] of Γ for perylene is 1.8×10^8 s^{-1}, which yields a natural lifetime of 5.5 ns. This value is close to that observed for perylene, which displays a quantum yield near unity. However, there are numerous reasons why Eq. [1.4] can fail. This expression assumes no interaction with the solvent, does not consider changes in refractive index (n) between the absorption and emission wavelength, and assumes no change in excited-state geometry. A more complete form of Eq. [1.4] (not shown) includes a factor $G = g_l/g_u$ on the right-hand side, where g_l and g_u are the degeneracies of the lower and upper states, respectively. For fluorescence transitions, $G = 1$, and for phosphorescence transitions, $G = \frac{1}{3}$.

The natural lifetime can be calculated from the measured lifetime (τ) and quantum yield:

$$\tau_n = \tau/Q \quad [1.5]$$

Equation [1.5] can be derived from Eqs. [1.1]–[1.3]. Many biochemical fluorophores do not behave as predictably as unsubstituted aromatic compounds. Hence, there is often poor agreement between the value of τ_n calculated from Eq. [1.5] and that calculated from the absorption and emission spectra (Eq. [1.4]). These discrepancies occur for a variety of unknown and known reasons, such as a fraction of the fluorophores being located next to quenching groups, which sometimes occurs for tryptophan residues in proteins.

The quantum yield and lifetime can be modified by factors which affect either of the rate constants (Γ or k_{nr}). For example, a molecule may be nonfluorescent as a result of a fast rate of internal conversion or a slow rate of emission. Scintillators are generally chosen for their high quantum yields. These high yields are a result of large Γ values. Hence, the lifetimes are generally short, near 1 ns. The fluorescence emission of aromatic substances containing $-NO_2$ groups is generally weak, primarily as a result of large values for k_{nr}. The quantum yields of phosphorescence are extremely small in fluid solutions at room temperature. The triplet-to-singlet transition is forbidden by symmetry, and the rates of spontaneous emission are about 10^3 s^{-1} or smaller. Since k_{nr} values are near 10^9 s^{-1}, quantum yields of phosphorescence are small at room temperature. From Eq. [1.1] one can predict phosphorescence quantum yields of 10^{-6}.

There are instances where comparison of the natural lifetime, measured lifetime, and quantum yield can be informative. For instance, in the case of the widely used membrane probe 1,6-diphenyl-1,3,5-hexatriene (DPH) the measured lifetime—near 10 ns—is much longer than that calculated from Eq. [1.1], which is near 1.5 ns.[17] In this case the calculation based on the absorption spectrum of DPH is incorrect because the absorption transition is to a state of different electronic symmetry than the emissive state. Such quantum-mechanical effects are rarely seen in more complex fluorophores with heterocyclic atoms.

1.4.A. Fluorescence Quenching

The intensity of fluorescence can be decreased by a wide variety of processes. Such decreases in intensity are called quenching. Quenching can occur by different mechanisms. Collisional quenching occurs when the excited-state fluorophore is deactivated upon contact with some other molecule in solution, which is called the quencher. Collisional quenching is illustrated on the modified Jabłoński diagram in Figure 1.14. In this case the fluorophore is returned to the ground state during a diffusive encounter with the quencher. The molecules are not chemically altered in the process. For collisional quenching, the decrease in intensity is described by the well-known Stern–Volmer equation:

$$\frac{F_0}{F} = 1 + K[Q] = 1 + k_q\tau_0[Q] \quad [1.6]$$

In this expression K is the Stern–Volmer quenching constant, k_q is the bimolecular quenching constant, τ_0 is the unquenched lifetime, and [Q] is the quencher concentration. A wide variety of molecules can act as collisional quenchers. Examples include oxygen, halogens, amines, and electron-deficient molecules like acrylamide. The mechanism of quenching varies with the fluorophore–quencher pair. For instance, quenching of indole by acrylamide is probably due to electron transfer from indole to acrylamide, which does not occur in the ground state. Quenching by halogens and heavy atoms occurs due to spin–orbit coupling and intersystem crossing to the triplet state (Figure 1.5).

Besides collisional quenching, fluorescence quenching can occur by a variety of other processes. Fluorophores can form nonfluorescent complexes with quenchers. This process is referred to as static quenching since it occurs in the ground state and does not rely on diffusion or molecular collisions. Quenching can also occur by a variety of trivial, i.e., nonmolecular, mechanisms, such as attenuation of the incident light by the fluorophore itself or other absorbing species.

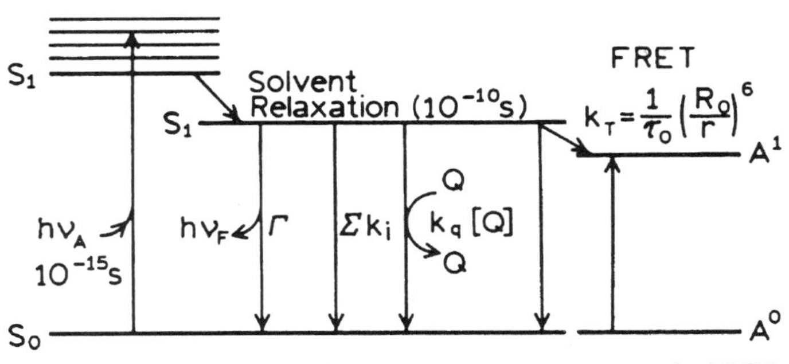

Figure 1.14. Jabłoński diagram with collisional quenching and fluorescence resonance energy transfer (FRET). The term Σk_i is used to represent nonradiative paths to the ground state besides quenching and FRET. Revised from Ref. 20.

1.4.B. Time Scale of Molecular Processes in Solution

The phenomenon of quenching provides a valuable context for understanding the role of the excited-state lifetime in allowing fluorescence measurements to detect dynamic processes in solution or in macromolecules. The basic idea is that absorption is an instantaneous event. According to the Franck–Condon principle, absorption occurs so fast that there is no time for molecular motion during the absorption process. Absorption occurs in the time it takes a photon to travel the length of a photon, in less than 10^{-15} s. Hence, absorption spectroscopy can only yield information on the average ground state of the molecules which absorb light. Only solvent molecules that are immediately adjacent to the absorbing species will affect its absorption spectrum. Absorption spectra are not sensitive to molecular dynamics and can only provide information on the average solvent shell adjacent to the chromophore.

The length of time fluorescent molecules remain in the excited state provides an opportunity for interactions with other molecules in solution. Collisional quenching of fluorescence by molecular oxygen is an excellent example of the expansion of time and distance provided by the fluorescence lifetime. If a fluorophore in the excited state collides with an oxygen molecule, then the fluorophore returns to the ground state without emission of a photon. The diffusion coefficient (D) of oxygen in water at 25 °C is 2.5×10^{-5} cm^2/s. The average distance $(\Delta x^2)^{1/2}$ an oxygen molecule can diffuse in 10^{-8} s or 10 ns is given by the Einstein equation,

$$\Delta x^2 = 2D\tau \qquad [1.7]$$

The distance is about 70 Å, which is comparable to the thickness of a biological membrane or the diameter of a protein. Some fluorophores have lifetimes as long as 400 ns, and hence diffusion of oxygen molecules may be observed over distances of 450 Å. In contrast, absorption measurements are only sensitive to the immediate environment around the fluorophore, and then only sensitive to the instantaneously averaged environment.

Other examples of dynamic processes in solution involve fluorophore–solvent interactions and rotational diffusion. As was observed by Stokes, most fluorophores display emission at lower energies than their absorption. Most fluorophores have larger dipole moments in the excited state than in the ground state. Rotational motions of small solvent molecules in fluid solution are rapid, typically occurring on a timescale of 40 ps or less. The relatively long timescale of fluorescence allows ample time for the solvent molecules to reorient around the excited-state dipole, which lowers its energy and shifts the emission to longer wavelengths. This process is called solvent relaxation and occurs in 10^{-10} s in fluid solution. It is these differences between absorption and emission that result in the high sensitivity of emission spectra to solvent polarity, and the smaller spectral changes seen in absorption spectra. Solvent relaxation can result in substantial Stokes' shifts. In proteins, tryptophan residues absorb light at 280 nm, and their fluorescence emission occurs near 350 nm. Although 10 ns may appear to be a brief time span, it is in fact quite long relative to the motions of small molecules in fluid solution. In the following section we will explain how dynamic rotational diffusion can be studied by measuring the polarization or anisotropy of the emission.

1.5. FLUORESCENCE ANISOTROPY

Anisotropy measurements are commonly used in the biochemical applications of fluorescence. Anisotropy measurements provide information on the size and shape of proteins or the rigidity of various molecular environments. Anisotropy measurements have been used to measure protein–protein associations and fluidity of membranes and for immunoassays of numerous substances.

Anisotropy measurements are based on the principle of photoselective excitation of fluorophores by polarized light. Fluorophores preferentially absorb photons whose electric vectors are aligned parallel to the transition moment of the fluorophore. The transition moment has a defined orientation with respect to the molecular axes. In an isotropic solution, the fluorophores are oriented randomly. Upon excitation with polarized light, one selectively excites those fluorophore molecules whose absorption transition dipole is parallel to the electric vector of the excitation. This selective excitation results in a partially oriented population of fluorophores (photoselection) and in partially polarized fluorescence emission. Emission also occurs with the light polarized along a fixed axis in the fluorophore. The relative angle between these moments determines the maximum measured anisotropy (r_0; see Eq. [10.20]). The fluorescence anisotropy (r) and polarization (P) are defined by

$$r = \frac{I_{\parallel} - I_{\perp}}{I_{\parallel} + 2I_{\perp}} \qquad [1.8]$$

$$P = \frac{I_{\parallel} - I_{\perp}}{I_{\parallel} + I_{\perp}} \qquad [1.9]$$

where $I_{\parallel}$ and $I_{\perp}$ are the fluorescence intensities of the vertically ($\parallel$) and horizontally ($\perp$) polarized emission, when the sample is excited with vertically polarized light. Anisotropy and polarization are both expressions for the same phenomenon, and these values can be interconverted using Eqs. [10.3] and [10.4].

Several phenomena can decrease the measured anisotropy to values lower than the maximum theoretical values. The most common cause is rotational diffusion. Such diffusion occurs during the lifetime of the excited state and displaces the emission dipole of the fluorophore. Measurement of this parameter provides information about the relative angular displacement of the fluorophore between the times of absorption and emission. In fluid solution, most fluorophores rotate extensively in 50–100 ps. Hence, the molecules can rotate many times during the 1- to 10-ns excited-state lifetime, and the orientation of the polarized emission is randomized. For this reason, fluorophores in aqueous nonviscous solution typically display anisotropies near zero. Transfer of excitation between fluorophores also results in decreased anisotropies.

The effects of rotational diffusion can be decreased if the fluorophore is bound to a macromolecule. For instance, it is known that the rotational correlation time for the protein human serum albumin (HSA) is near 50 ns. Suppose HSA is covalently labeled with a fluorophore whose lifetime is 10 ns. Assuming no other processes result in loss of anisotropy, the expected anisotropy is given by the Perrin equation:

$$r = \frac{r_0}{1 + (\tau/\theta)} \qquad [1.10]$$

where r_0 is the anisotropy which would be measured in the absence of rotational diffusion and θ is the rotational correlation time for the diffusion process. In this case, binding of the fluorophore to the protein has slowed the probes' rate of rotational motion. Assuming $r_0 = 0.4$, the anisotropy is expected to be 0.33. Smaller proteins have shorter correlation times and are expected to yield lower anisotropies. The anisotropies of larger proteins can also be low if they are labeled with long-lifetime fluorophores. The essential point is that the rotational correlation times for most proteins are comparable to typical fluorescence lifetimes. As a result, measurements of fluorescence anisotropy will be sensitive to any factor which affects the rate of rotational diffusion. The rotational rates of fluorophores in cell membranes also occur on the nanosecond timescale, and the anisotropy values are thus sensitive to membrane composition. For these reasons, measurements of fluorescence polarization are widely used to study the interactions of biological macromolecules.

1.6. RESONANCE ENERGY TRANSFER

Another important process that occurs in the excited state is resonance energy transfer (RET). This process occurs whenever the emission spectrum of a fluorophore, called the donor, overlaps with the absorption spectrum of another molecule, called the acceptor.[18] Such overlap is illustrated in Figure 1.15. The acceptor does not need to be fluorescent. It is important to understand that RET does not involve emission of light by the donor. RET is not the result of emission from the donor being absorbed by the acceptor. Such reabsorption processes are dependent on the overall concentration of the acceptor, and on nonmolecular factors such as sample size, and are thus of less interest. There is no intermediate photon in RET. The donor and acceptor are coupled by a dipole–dipole interaction. For these reasons, the term RET is preferred to the term fluorescence resonance energy transfer (FRET), which is also in common use.

The extent of energy transfer is determined by the distance between the donor and acceptor and the extent of spectral overlap. For convenience, the spectral overlap (Figure 1.15) is described in terms of the Förster distance (R_0). The rate of energy transfer $k_T(r)$ is given by

$$k_T(r) = \frac{1}{\tau_D}\left(\frac{R_0}{r}\right)^6 \qquad [1.11]$$

where r is the distance between the donor (D) and the acceptor (A), and τ_D is the lifetime of the donor in the absence of energy transfer. The efficiency of energy transfer for a single donor–acceptor pair at a fixed distance is

$$E = \frac{R_0^6}{R_0^6 + r^6} \qquad [1.12]$$

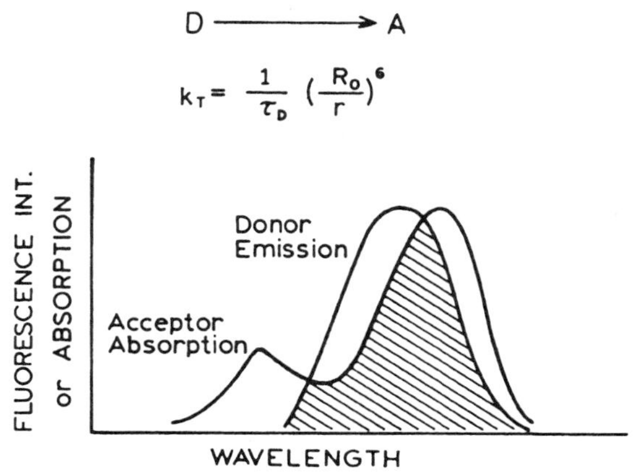

Figure 1.15. Spectral overlap for fluorescence resonance energy transfer (RET).

Hence, the extent of transfer depends on distance (r). Fortunately, the Förster distances are comparable in size to biological macromolecules, 30–60 Å. For this reason, energy transfer has been used as a "spectroscopic ruler" for measurements of distance between sites on proteins.[19] The value of R_0 for energy transfer should not be confused with the fundamental anisotropies (r_0).

The field of RET is large and complex. The theory is different for donors and acceptors that are covalently linked, free in solution, or contained in the restricted geometries of membranes or DNA. Additionally, depending on the donor lifetime, diffusion can increase the extent of energy transfer beyond that predicted by Eq. [1.12].

1.7. STEADY-STATE AND TIME-RESOLVED FLUORESCENCE

Fluorescence measurements can be broadly classified into two types of measurements, steady-state and time-resolved. Steady-state measurements are those performed with constant illumination and observation. This is the most common type of measurement. The sample is illuminated with a continuous beam of light, and the intensity or emission spectrum is recorded (Figure 1.16). Because of the nanosecond timescale of fluorescence, most measurements are steady-state measurements. When the sample is first exposed to light, steady state is reached almost immediately.

The second type of measurements, time-resolved measurements, is used for measuring intensity decays or anisotropy decays. For these measurements, the sample is exposed to a pulse of light, where the pulse width is typically shorter than the decay time of the sample (Figure 1.16). This intensity decay is recorded with a high-speed detection system that permits the intensity or anisotropy to be measured on the nanosecond timescale.

It is important to understand that there exists a rather simple relationship between steady-state and time-resolved measurements. The steady-state observation is simply an average of the time-resolved phenomena over the intensity decay of the sample. For instance, consider a fluorophore which displays a single decay time (τ) and a single rotational correlation time (θ). The intensity and anisotropy decays are given by

$$I(t) = I_0 e^{-t/\tau} \quad [1.13]$$

$$r(t) = r_0 e^{-t/\theta} \quad [1.14]$$

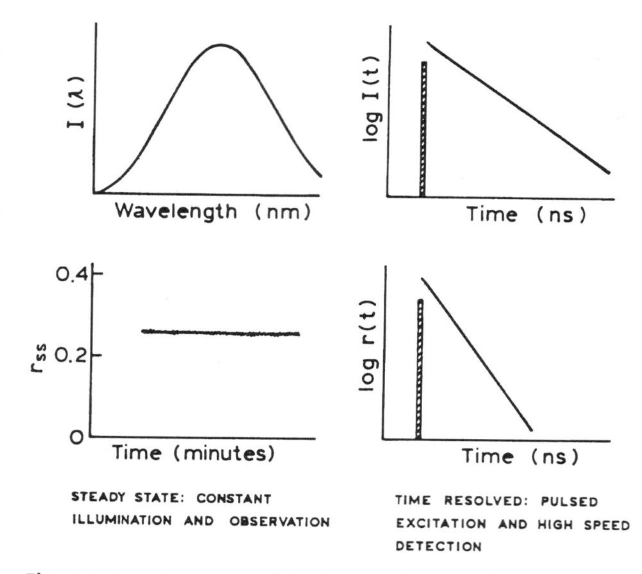

Figure 1.16. Comparison of steady-state and time-resolved fluorescence spectroscopy.

where I_0 and r_0 are, respectively, the intensities and anisotropies at $t = 0$, immediately following the excitation pulse.

Equations [1.13] and [1.14] can be used to illustrate how the decay time determines what can be observed using fluorescence. The steady-state anisotropy (r) is given by the average of $r(t)$ weighted by $I(t)$:

$$r = \frac{\int_0^\infty r(t)I(t)\,dt}{\int_0^\infty I(t)\,dt} \quad [1.15]$$

In this equation the denominator is present to normalize the anisotropy to be independent of total intensity. In the numerator the anisotropy at any time t contributes to the steady-state anisotropy according to the intensity at time t. Substitution of Eqs. [1.13] and [1.14] into Eq. [1.15] yields the Perrin equation (Eq. [1.10]).

Perhaps a simpler example is how the steady-state intensity I_{SS} is related to the decay time. The steady-state intensity is given by

$$I_{SS} = \int_0^\infty I_0 e^{-t/\tau}\,dt = I_0\tau \quad [1.16]$$

The value of I_0 can be considered to be a parameter which depends on the fluorophore concentration and a number of instrumental parameters. Hence, in molecular terms, the steady-state intensity is proportional to the lifetime. This makes sense in consideration of Eqs. [1.1] and [1.2], which showed that the quantum yield was proportional to the lifetime.

1.7.A. Why Time-Resolved Measurements?

Whereas steady-state fluorescence measurements are simple, nanosecond time-resolved measurements typically require complex and expensive instrumentation. Given the relationship between the steady-state and time-resolved measurements, what is the value of these more complex measurements? It turns out that much of the molecular information available from fluorescence is lost during the time-averaging process. For example, anisotropy decays of fluorescent macromolecules are frequently more complex than a single exponential (Eq. [1.14]). The precise shape of the anisotropy decay contains information about the shape of the macromolecule and its flexibility. Unfortunately, this shape information is lost during averaging of the anisotropy over the decay time (Eq. [1.15]). Irrespective of the form of $r(t)$, Eq. [1.15] yields a single steady-state anisotropy. In principle, the value of r still reflects the anisotropy decay and shape of the molecule. In practice, the information from r alone is not sufficient to reveal the form of $r(t)$ or the shape of the molecule.

The intensity decays also contain information that is lost during the averaging process. Frequently, macromolecules can exist in more than a single conformation, and the decay time of a bound probe may depend on conformation. The intensity decay could reveal two decay times, and thus the presence of more than one conformational state. The steady-state intensity will only reveal an average intensity dependent on a weighted average of the two decay times.

There are numerous additional reasons for measuring time-resolved fluorescence. In the presence of energy transfer, the intensity decays reveal how acceptors are distributed in space around the donors. Time-resolved measurements reveal whether quenching is due to diffusion or to complex formation with the ground-state fluorophores. In fluorescence, much of the molecular information content is available only from time-resolved measurements.

1.8. BIOCHEMICAL FLUOROPHORES

Fluorophores are divided into two general classes, intrinsic and extrinsic. Intrinsic fluorophores are those which occur naturally. Extrinsic fluorophores are those which are added to a sample that does not display the desired spectral properties. In proteins, the dominant fluorophore is the indole group of tryptophan (Figure 1.17). Indole absorbs near 280 nm and emits near 340 nm. The emission spectrum of indole is highly sensitive to solvent polarity. The emission of indole may be blue-shifted if the group is buried within a native protein (n), and its emission may shift to longer wavelengths (red shift) when the protein is unfolded (u).

Membranes typically do not display intrinsic fluorescence. For this reason, it is common to label membranes with probes which spontaneously partition into the nonpolar side-chain region of the membranes. One of the most commonly used membrane probes is DPH. Because of its low solubility and quenched emission in water, DPH emission is seen only from membrane-bound DPH. Other lipid probes include fluorophores attached to lipid or fatty acid chains, as shown for rhodamine B in Figure 1.17.

Although DNA contains nitrogenous bases which look like fluorophores, DNA is weakly fluorescent or nonfluorescent. However, a wide variety of dyes bind spontaneously to DNA such as acridines, ethidium bromide, and other planar cationic species. For this reason, staining of cells with dyes that bind to DNA is widely used to visualize and identify chromosomes. There are a few naturally occurring fluorescent bases, such as the Y_t-base, which occurs in the anticodon region of a phenylalanine tRNA (Figure 1.17, bottom).

A wide variety of other substances display significant fluorescence. Among biological molecules, one can observe fluorescence from reduced nicotinamide adenine dinucleotide (NADH), from oxidized flavins (FAD, the adenine dinucleotide, and FMN, the mononucleotide), and from pyridoxal phosphate, as well as from chlorophyll. Occasionally, a species of interest is not fluorescent or is not fluorescent in a convenient region of the UV-visible spectrum. A wide variety of extrinsic probes have been developed for labeling the macromolecules in such cases. Two of the most widely used probes, dansyl chloride (DNS-Cl, which stands for 5-dimethylamino-1-naphthalenesulfonyl chloride) and fluorescein isothiocyanate (FITC), are shown in Figure 1.18. These probes react with the free amino groups of proteins, resulting in proteins which fluoresce at blue (DNS) or green (FITC) wavelengths.

Another approach to obtaining the desired fluorescent signal from the molecule of interest is to synthesize a chemical analog that displays both the chemical properties of the parent molecule and useful fluorescence. For example, adenosine triphosphate (ATP) is essentially nonfluorescent. However, it is possible to create analogs that are fluorescent. The etheno-ATP ($\in$-ATP) analog is fluorescent[21] and might be expected to bind to kinases, but its base-pairing properties are obviously compromised. The *lin*-benzo-AMP derivative is also fluorescent,[22] and one can expect it to display the same base pairing as adenosine

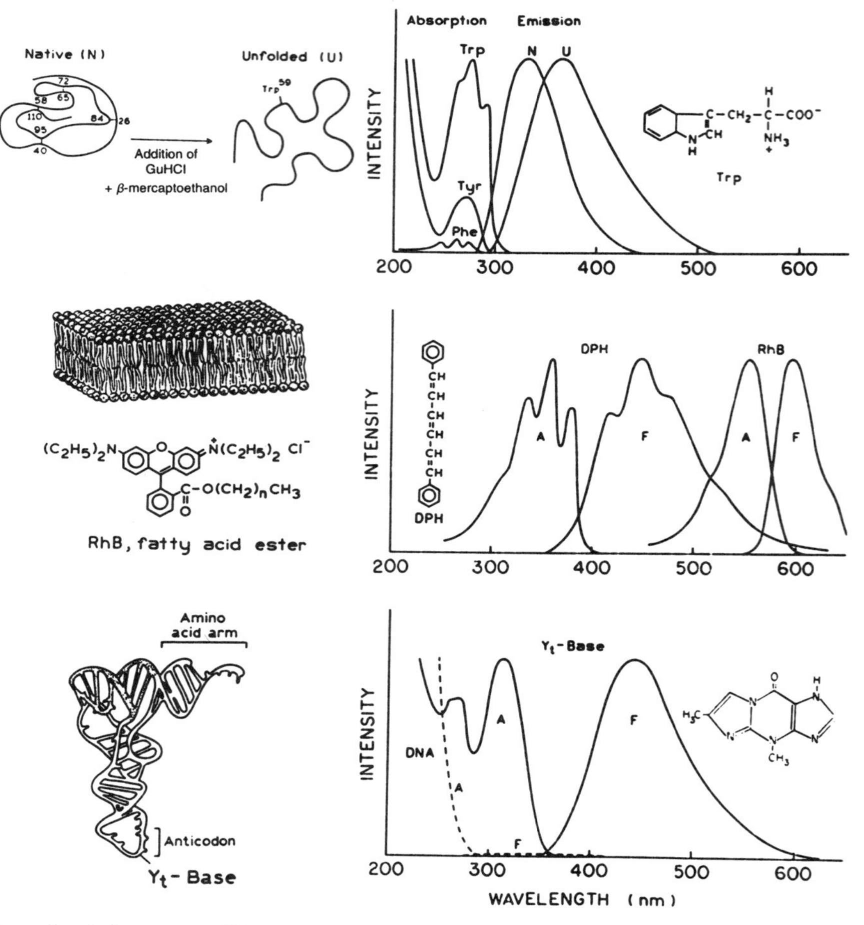

Figure 1.17. Absorption and emission spectra of biomolecules. *Top*: Tryptophan emission from proteins. *Middle*: Spectra of extrinsic membrane probes. *Bottom*: Spectra of the naturally occurring fluorescent base, Y_t-base. DNA itself (— — —) displays very weak emission. Reprinted, with permission from Wiley-VCH, STM, from Ref. 20.

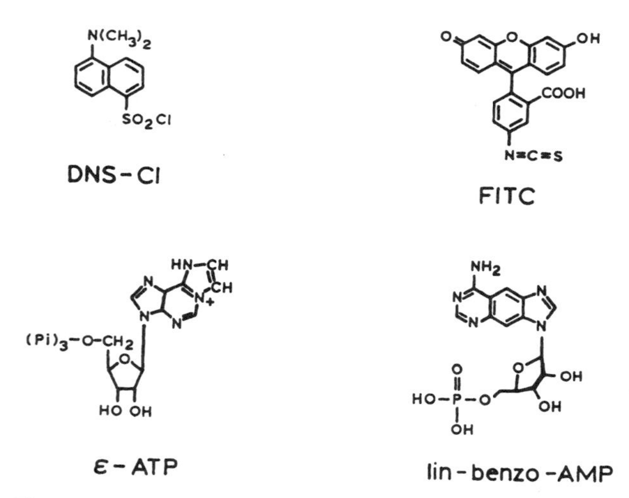

Figure 1.18. Fluorophores for covalent labeling (DNS-Cl and FITC) and fluorescent nucleotide analogs (ε-ATP and *lin*-benzo-AMP).

monophospate (AMP) (Figure 1.18), but it may be too large to fit into some binding sites or in a DNA helix. It turns out that *lin*-benzo-ATP is an excellent substrate for some ATPases but is weakly bound by others.[23] A useful fluorescent probe is one that displays a high intensity, is stable during continued illumination, and does not substantially perturb the biomolecule or process being studied.

1.8.A. Fluorescent Indicators

Another class of fluorophores consists of the fluorescent indicators. These are fluorophores whose spectral properties are sensitive to a substance of interest. One example is PBFI, which is shown in Figure 1.19. This fluorophore contains a central azacrown ether, which binds K^+. Upon K^+ binding, the emission intensity of PBFI increases, allowing the amount of K^+ to be determined. Fluorescent

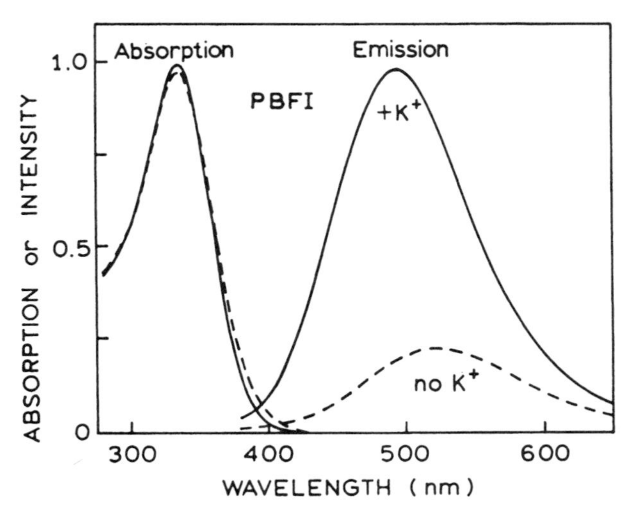

Figure 1.19. Example of a fluorescent indicator, PBFI, which is sensitive to potassium.

indicators are presently available for a wide variety of substances, including Ca^{2+}, Mg^{2+}, Na^{+}, Cl^{-}, and O_2, and for pH. The application of fluorescence to chemical sensing is described in Section 1.10.

1.9. MOLECULAR INFORMATION FROM FLUORESCENCE

1.9.A. Emission Spectra and the Stokes' Shift

The most dramatic aspect of fluorescence is its occurrence at wavelengths longer than those at which absorption occurs. These Stokes' shifts, which are most dramatic for polar fluorophores in polar solvents, are due to interactions between the fluorophore and its immediate environment. The indole group of tryptophan residues in proteins is one such solvent-sensitive fluorophore, and the emission spectra of indole can reveal the location of tryptophan residues in proteins. The emission from an exposed surface residue will occur at longer wavelengths than that from a tryptophan residue in the protein's interior. This phenomenon was illustrated in the top panel of Figure 1.17, which shows a shift in the spectrum of a tryptophan residue upon unfolding of a protein and the subsequent exposure of the tryptophan residue to the aqueous phase. Prior to unfolding, the residue is shielded from the solvent by the folded protein.

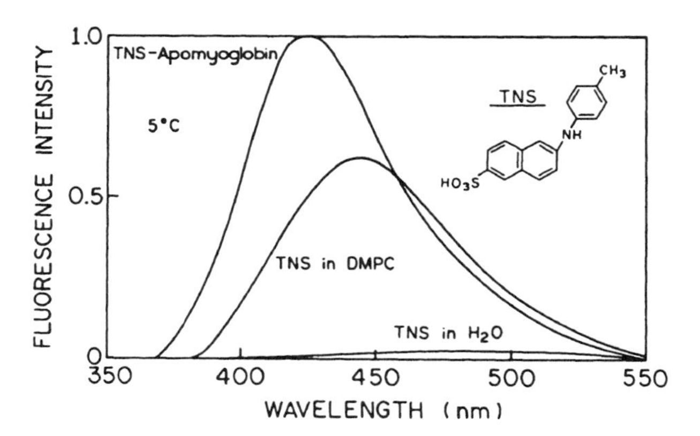

Figure 1.20. Emission spectra of TNS in water, bound to apomyoglobin, and bound to lipid vesicles [dimyristoyl-L-α-phosphatidylcholine (DMPC)]. Reprinted, with permission from Wiley-VCH, STM, from Ref. 20.

Because emission spectra are sensitive to the fluorophore's environment, the spectra of extrinsic probes are often used to determine a probe's location on a macromolecule. For example, one of the widely used probes for such studies is 6-(*p*-toluidinyl)naphthalene-2-sulfonic acid (TNS) (Figure 1.20), which displays the additional favorable property of being very weakly fluorescent in water. Weak fluorescence in water and strong fluorescence when bound to a biomolecule is a convenient property shared by other widely used probes, including the DNA stain ethidium bromide. The protein apomyoglobin contains a hydrophobic pocket which binds the heme group. This pocket can also bind other nonpolar molecules. Upon the addition of apomyoglobin to a solution of TNS, there is a large increase in fluorescence intensity, as well as a shift of the emission spectrum to shorter wavelengths. This increase in TNS fluorescence reflects the nonpolar character of the heme binding site of apomyoglobin. TNS also binds to membranes (Figure 1.20). The emission spectrum of TNS bound to model membranes of dimyristoyl-L-α-phosphatidylcholine (DMPC) is somewhat weaker and at longer wavelengths compared to that of apomyoglobin. This indicates that the TNS binding sites on the surface of the membrane are more polar. From the emission spectrum it appears that TNS binds to the polar head-group region of the membranes, rather than to the nonpolar acyl side-chain region. Hence, the emission spectra of solvent-sensitive fluorophores provide information on the location of the binding sites on the macromolecules.

1.9.B. Quenching of Fluorescence

As described in Section 1.4.A, a wide variety of small molecules or ions can act as quenchers of fluorescence;

that is, they decrease the intensity of the emission. These substances include iodide (I^-), oxygen, and acrylamide. The accessibility of fluorophores to such quenchers can be used to determine the location of probes on macromolecules or the porosity of proteins and membranes to quenchers. This concept is illustrated in Figure 1.21, which shows the emission intensity of a protein- or membrane-bound fluorophore in the presence of the water-soluble quencher iodide, I^-. As shown on the right-hand side of the figure, the emission intensity of a tryptophan on the protein's surface (W_2), or on the surface of a cell membrane (P_2), will be decreased in the presence of a water-soluble quencher. The intensity of a buried tryptophan residue (W_1) or of a probe in the membrane interior (P_1) will be less affected by the dissolved iodide, as seen on the left-hand side of the figure. Alternatively, one can add lipid-soluble quenchers, such as brominated fatty acids, to study the interior acyl side-chain region of membranes through measurements of the extent of quenching by the lipid-soluble quencher.

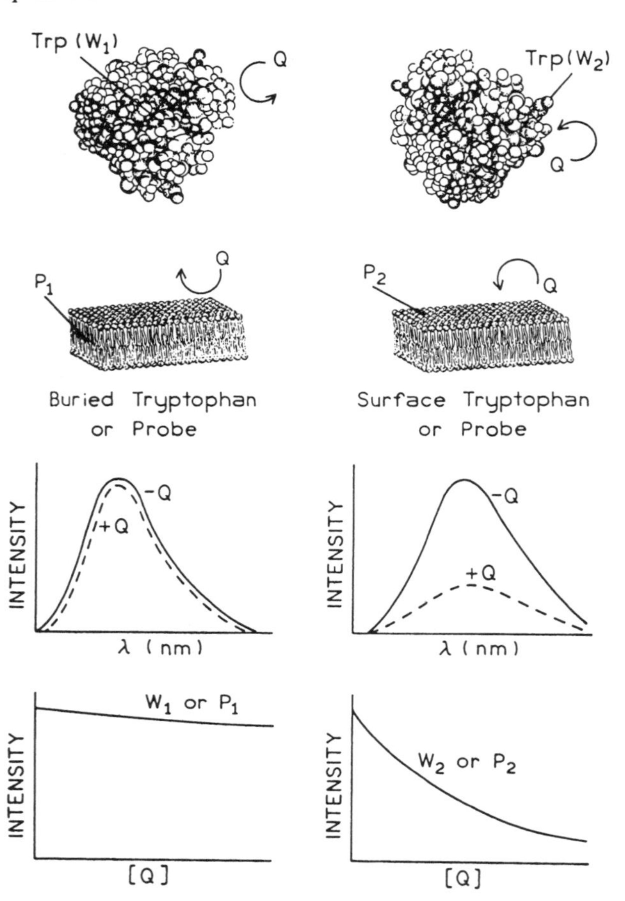

Figure 1.21. Accessibility of fluorophores to the quencher (Q^-). Reprinted, with permission from Wiley-VCH, STM, from Ref. 20.

1.9.C. Fluorescence Polarization or Anisotropy

As described in Section 1.5, fluorophores absorb light along a particular direction with respect to the molecular axes. For example, DPH absorbs only light polarized along its long axis (Figure 1.17). The extent to which a fluorophore rotates during the excited-state lifetime determines its polarization or anisotropy. The phenomenon of fluorescence polarization can be used to measure the apparent volume (or molecular weight) of proteins. This measurement is possible because larger proteins rotate more slowly. Hence, if a protein binds to another protein, the rotational rate decreases, and the anisotropy(s) increases (Figure 1.22). The rotational rate of a molecule is often described by its rotational correlation time θ, which is given by

$$\theta = \frac{\eta V}{RT} \quad [1.17]$$

where η is the viscosity, V is the molecular volume, R is the gas constant, and T is the temperature in Kelvins. Suppose a protein is labeled with DNS-Cl (Figure 1.22, middle). If the protein associates with another protein, the

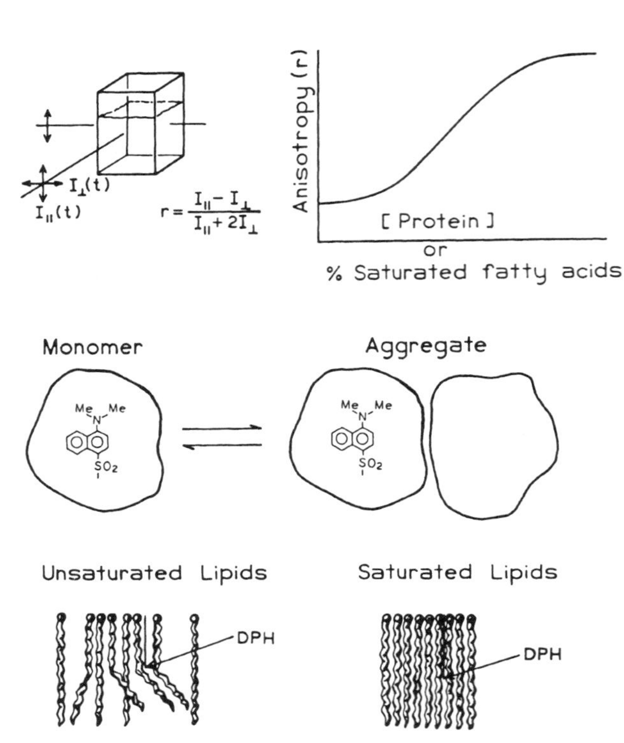

Figure 1.22. Fluorescence polarization, protein association, and membrane microviscosity. From Ref. 20.

volume increases and so does the rotational correlation time. This causes the anisotropy to increase because of the relationship between the steady-state anisotropy *r* and the rotational correlation time θ (Eq. [1.10]).

Fluorescence polarization measurements have also been used to determine the apparent viscosity of the side-chain region (center) of membranes. Such measurements of microviscosity are typically performed using a hydrophobic probe like DPH (Figure 1.22, bottom), which partitions into the membrane. The viscosity of membranes is known to decrease in the presence of unsaturated fatty acid side chains. Hence, an increase in the amount of unsaturated fatty acid is expected to decrease the anisotropy. The apparent microviscosity of the membrane is determined by comparing the polarization of the probe measured in the membrane with that observed in solutions of known viscosity.

Anisotropy measurements are widely used in biochemistry and are even used for clinical immunoassays. One reason for this use is the ease with which these absolute values can be measured and compared between laboratories.

1.9.D. Resonance Energy Transfer

Resonance energy transfer (RET), sometimes called fluorescence resonance energy transfer (FRET), provides an opportunity to measure the distances between sites on macromolecules. Förster distances are typically in the range of 15–60 Å, which is comparable to the diameter of many proteins and to the thickness of membranes. According to Eq. [1.12], the distance between a donor and acceptor can be calculated from the transfer efficiency.

The use of RET to measure protein association and distance is shown in Figure 1.23 for two monomers which associate to form a dimer. Suppose one monomer contains a tryptophan (trp) residue, and the other a dansyl group. The Förster distance is determined by the spectral overlap of the trp donor emission with the dansyl acceptor absorption. Upon association, RET will occur, which decreases the intensity of the donor emission (Figure 1.23). The extent of donor quenching can be used to calculate the donor-to-acceptor distance in the dimer (Eq. [1.12]). It is also important to notice that RET provides a method to measure protein association because it occurs whenever the donor and acceptor are within the Förster distance.

1.10. FLUORESCENCE SENSING

In addition to the use of fluorescence in biochemistry and biophysics, there is a growing interest in its use for analytical and clinical chemistry. The use of fluorescence in clinical diagnostics is part of the continual shift away from the use of radioactive tracers. The use of optical methods eliminates the dangers of handling radioactive materials and the cost of their proper disposal.

Fluorescence is used for a wide variety of biomedical purposes. Fluorescence imaging of gels is used to detect DNA fragments following electrophoretic separation. The newer DNA stains provide high detection sensitivity and can mostly replace the use of ^{32}P and autoradiography. The

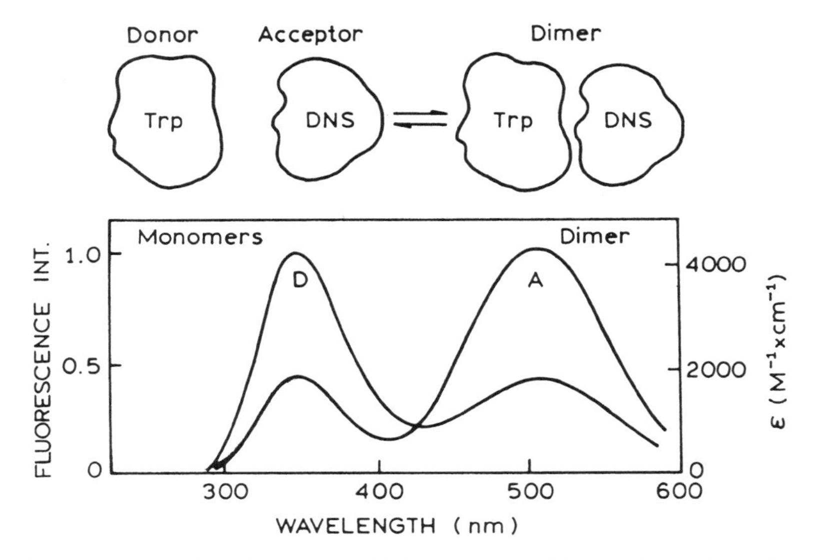

Figure 1.23. Energy transfer between donor (D) and acceptor (A)-labeled monomers, which associate to form a dimer. In this case the donor is tryptophan and the acceptor is a dansyl group (DNS).

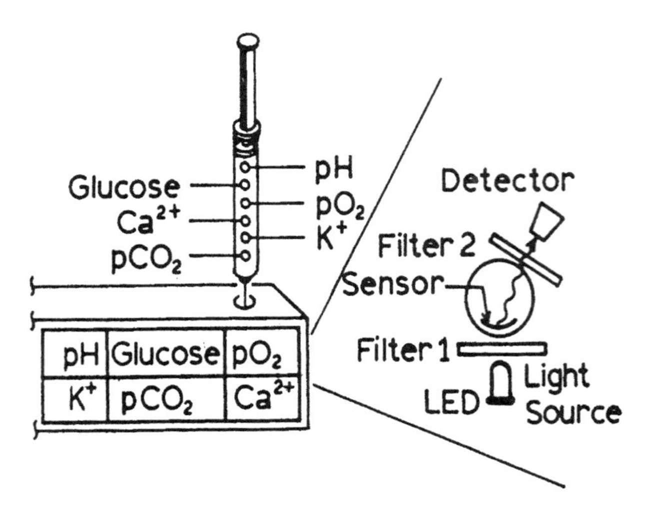

Figure 1.24. Fluorescence sensing in clinical chemistry.

expanding use of DNA sequencing and the Human Genome Project were made practical by the use of fluorescent labels[24,25] in place of radioactive tags.

Another area of active research in the use of fluorescence is clinical chemistry.[26,27] Probes of the type shown in Figure 1.19 are known for a wide variety of analytes, including most of the ionic species in blood. Fluorescence assays have also been described for immunochemical detection of a wide variety of low-molecular-weight drugs and for higher-molecular-weight antigens. Fluorescence assays for blood glucose are also known. While these fluorescence assays exist in numerous formats, the basic idea is as shown in Figure 1.24. The sample may be whole blood. The fluorescent sensing chemistry can be contained within the syringe or capillary tube into which the sample is drawn, or on the side of the container. The sensor is read by a low-cost fluorescence instrument, which may use light-emitting diodes or laser diodes (laser pointers) as the light source. Within the next decade, one can anticipate the introduction of numerous point-of-care fluorescence assays for use at the bedside, in the doctor's office, or for home health care.

1.11. SUMMARY

The essence of any experiment is the existence of an observable quantity and the correlation of the value of this observable with a phenomenon of interest. The time span between the absorption of light and its subsequent reemission allows time for several processes, each of which results in changes of fluorescence spectral observables. These processes include collisions with quenchers, rotational and translational diffusion, formation of complexes with solvents or solutes, and reorientation of the environment surrounding the altered dipole moment of the excited state. These dynamic processes can affect the fluorescence anisotropies, quantum yields, lifetimes, and emission spectra. In addition, resonance energy transfer provides a reliable indicator of molecular proximity on the angstrom size scale. As a result, the spectral characteristics of fluorophores can provide a great deal of information on the solution behavior of macromolecules.

Fluorescence spectroscopy offers the advantages of sensitivity, simplicity, and a wealth of molecular information. Fluorescence measurements are highly sensitive because the observations are made against a dark background. The advantage of a dark background is illustrated by our ability to see stars at night, but not in the daytime when the sky is bright. While instrumentation for time-resolved fluorescence is often complex, instruments for steady-state fluorescence are relatively simple. This is why there is currently a proliferation of instruments for microscopic imaging and macroscopic imaging of gels and multiwell plates. The sensitivity of fluorescence to numerous cations, anions, or nearby acceptors has resulted in the extensive development of assays for clinically relevant species.

And, finally, an important feature of fluorescence is that there is a rather direct connection between the spectral observables and molecular features of the sample. It is easy to visualize how the spectral properties are affected by the local environment, accessibility to quenchers, or the presence of nearby acceptors. It is this connection between what the molecules are doing and how this affects the emitted light which allows the design of experiments that answer our scientific questions.

REFERENCES

1. Herschel, Sir J. F. W., 1845, On a case of superficial colour presented by a homogeneous liquid internally colourless, *Phil. Trans. R. Soc. London* **135:**143–145.
2. Gillispie, C. C. (ed.), 1972, "John Frederick William Herschel," in *Dictionary of Scientific Biography*, Vol. 6, Charles Scribner's Sons, New York, pp. 323–328.
3. Undenfriend, S., 1995, Development of the spectrophotofluorometer and its commercialization, *Protein Sci.* **4:**542–551.
4. Martin, B. R., and Richardson, F., 1979, Lanthanides as probes for calcium in biological systems, *Q. Rev. Biophys.* **12:**181–203.
5. Berlman, I. B. 1971, *Handbook of Fluorescence Spectra of Aromatic Molecules,* 2nd ed., Academic Press, New York.
6. Jabłoński, A., 1935, Über den Mechanisms des Photolumineszenz von Farbstoffphosphoren, *Z. Phys.* **94:**38–46.
7. Szudy, J. (ed.), 1998, *Born 100 Years Ago: Aleksander Jabłoński (1898–1980)*, Uniwersytet Mikolaja Kopernika, Toruń, Poland.
8. Polska Akademia Nauk Instytut Fizyki, 1978, Acta Physica Polonica, *Europhys. J.* **A65**(6).

9. Stokes, G. G., 1852, On the change of refrangibility of light, *Phil. Trans. R. Soc. London* **142:**463–562.
10. Kasha, M., 1950, Characterization of electronic transitions in complex molecules, *Disc. Faraday Soc.* **9:**14–19.
11. Birks, J. B., 1970, *Photophysics of Aromatic Molecules*, John Wiley & Sons, New York.
12. Ref. 11, p. 108.
13. Lakowicz, J. R., and Balter, A., 1982, Analysis of excited state processes by phase-modulation fluorescence spectroscopy, *Biophys. Chem.* **16:**117–132.
14. Gafni, A., and Brand, L., 1978, Excited state proton transfer reactions of acridine studied by nanosecond fluorometry, *Chem. Phys. Lett.* **58:**346–350.
15. Birks, J. B., 1973, *Organic Molecular Photophysics*, John Wiley & Sons, New York, p. 14.
16. Strickler, S. J., and Berg, R. A., 1962, Relationship between absorption intensity and fluorescence lifetime of molecules, *J. Chem. Phys.* **37:**814–822.
17. Ref. 11, p. 120.
18. Föster, Th., 1948, Intermolecular energy migration and fluorescence, *Ann. Phys. (Leipzig)* **2:**55–75. Translated by R. S. Knox.
19. Stryer, L., 1978, Fluorescence energy transfer as a spectroscopic ruler, *Annu. Rev. Biochem.* **47:**819–846.
20. Lakowicz, J. R., 1995, Fluorescence spectroscopy of biomolecules, in *Encyclopedia of Molecular Biology and Molecular Medicine*, R. A. Meyers (ed.), VCH Publishers, New York, pp. 294–306.
21. Secrist, J. A., Barrio, J. R., and Leonard, N. J., 1972, A fluorescent modification of adenosine triphosphate with activity in enzyme synthesis: 1,N^6-ethenoadenosine triphosphate, *Science* **175:**646–647.
22. Scopes, D. I. C., Barrio, J. R., and Leonard, N. J., 1977, Defined dimensional changes in enzyme cofactors: Fluorescent "stretched-out" analogues of adenine nucleotides, *Science* **195:**296–298.
23. Grell, E., Lewitzki, E., Bremer, C., Kramer-Schmitt, S., Weber, J., and Senior, A. E., 1994, *lin*-Boons-ATP and -ADP: Versatile fluorescent probes for spectroscopic and biochemical studies, *J. Fluoresc.* **4**(3):247–250.
24. Smith, L. M., Sanders, J. Z., Kaiser, R. J., Hughes, P., Dodd, C., Connell, C. R., Heiner, C., Kent, S. B. H., and Hood, L. E., 1986, Fluorescence detection in automated DNA sequence analysis, *Nature* **321:**674–679.
25. Prober, J. M., Trainor, G. L., Dam, R. J., Hobbs, F. W., Robertson, C. W., Zagursky, R. J., Cocuzza, A. J., Jensen, M. A., and Baumeister, K., 1987, A system for rapid DNA sequencing with fluorescent chain-terminating dideoxynucleotides, *Science* **238:**330–341.
26. Thompson, R. B. (ed.), 1997, *Advances in Fluorescence Sensing Technology III*, Proceedings of SPIE, Vol. 2980, SPIE, Bellingham, Washington.
27. Lakowicz, J. R. (ed.), 1994, *Topics in Fluorescence Spectroscopy, Volume 4, Probe Design and Chemical Sensing*, Plenum Press, New York.
28. Kasha, M., 1960, Paths of molecular excitation, *Radiat. Res.* **2:**243–275.
29. Lakowicz, J. R., and Weber, G. W., 1973, Quenching of fluorescence by oxygen. A probe for structural fluctuations in macromolecules, *Biochemistry* **12:**4161–4170.
30. Hagag, N., Birnbaum, E. R., and Darnall, D. W., 1983, Resonance energy transfer between cysteine-34, tryptophan-214, and tyrosine-411 of human serum albumin, *Biochemistry* **22:**2420–2427.

PROBLEMS

1.1. *Estimation of Fluorescence and Phosphorescence Quantum Yields*: The quantum yield for fluorescence is determined by the radiative and nonradiative decay rates. The nonradiative rates are typically similar for fluorescence and phosphorescence states, but the emissive rates (Γ) vary greatly. Emission spectra, lifetimes (τ), and quantum yields (θ) for eosin and erythrosin B (Er B) are shown in Figure 1.25.

A. Calculate the natural lifetime (τ_n) and the radiative and nonradiative decay rates for eosin and for Er B. Which rate accounts for the lower quantum yield of Er B?

B. Phosphorescence lifetimes are typically near 1–10 ms. Assume that the natural lifetime for phosphorescence emission of these compounds is 10 ms and that the nonradiative decay rates of the two compounds are the same for the triplet state as for the singlet state. Estimate the phosphorescence quantum yields of eosin and Er B at room temperature.

1.2. *Estimation of Emission from the S_2 State*: If excited to the second singlet state (S_2), fluorophores typically relax to the first singlet state in 10^{-13} s.[28] Using the radiative decay rate calculated for eosin (Problem 1.1), estimate the quantum yield of the S_2 state.

1.3. *Thermal Population of Vibrational Levels*: The emission spectrum of perylene (Figure 1.3) shows equally spaced peaks which are due to various vibrational states as illustrated in Figure 1.5. Use the Boltzmann distribution to estimate the fraction of the ground-state molecules which are in the first vibrationally excited state at room temperature.

1.4. *Anisotropy of a Labeled Protein*: Naphthylamine sulfonic acids are widely used as extrinsic labels of proteins. A number of derivatives are available. One little known but particularly useful derivative is 2-diethylamino-5-

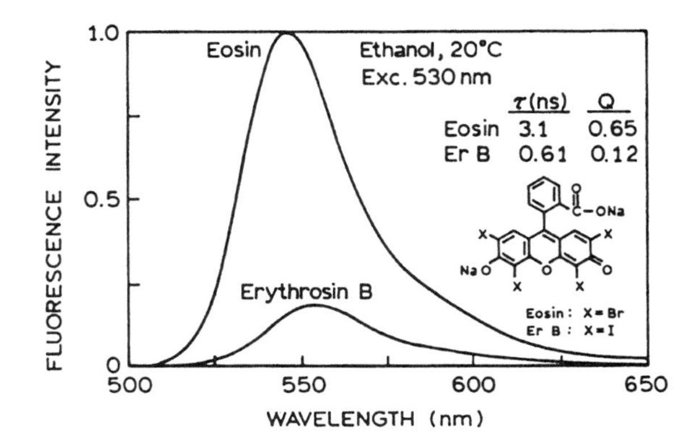

Figure 1.25. Emission spectra of eosin and erythrosin B (Er B).

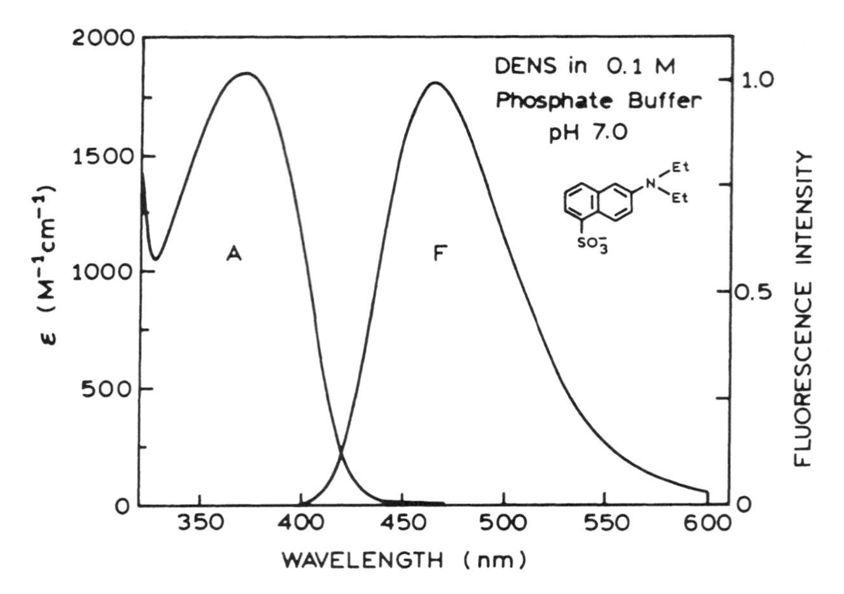

Figure 1.26. Absorption and emission spectra of DENS. The quantum yield relative to that of quinine sulfate is 0.84, and its lifetime is near 30 ns.

naphthalenesulfonicacid(DENS),whichdisplays a lifetime near 30 ns,[29] longer than that of most analogous molecules. Absorption and emission spectra of DENS are shown in Figure 1.26.

A. Suppose the fundamental anisotropy of DENS is 0.30 and that DENS is bound to a protein with a rotational correlation time of 30 ns. What is the anisotropy?
B. Assume now that the protein is bound to an antibody, with a molecular weight of 160,000 and a rotational correlation time of 100 ns. What is the anisotropy of the DENS-labeled protein?

1.5. *Effect of Distance on the Efficiency of FRET*: Assume the presence of a single donor and acceptor and that the distance between them (r) can be varied.

A. Plot the dependence of the energy transfer efficiency on the distance between the donor and the acceptor.
B. What is the transfer efficiency when the donor and the acceptor are separated by $0.5R_0$, R_0, and $2R_0$?

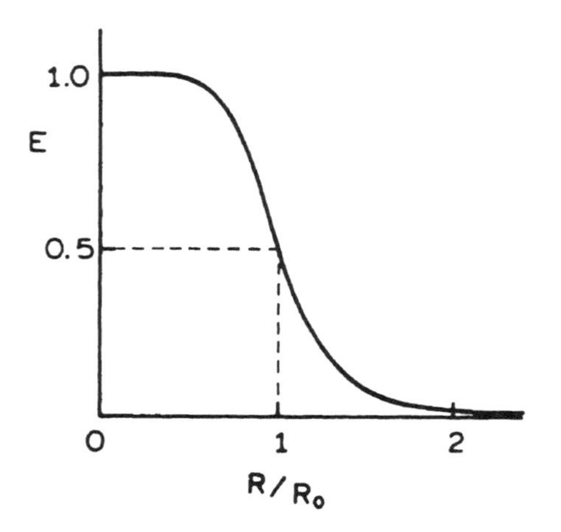

Figure 1.27. Effect of donor-to-acceptor distance on the transfer efficiency.

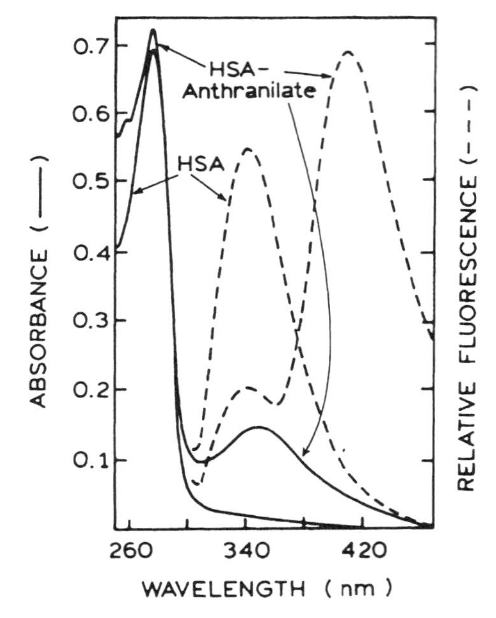

Figure 1.28. Absorption and fluorescence spectra of human serum albumin (HSA) and anthraniloyl-HSA. From Ref. 30.

1.6. *Calculation of a Distance from FRET Data*: The protein human serum albumin (HSA) has a single tryptophan residue at position 214. HSA was labeled with an anthraniloyl group placed covalently on cysteine-34.[30] Emission spectra of the labeled and unlabeled HSA are shown in Figure 1.28. The Förster distance for trp-to-anthraniloyl transfer is 30.3 Å. Use the emission spectra in Figure 1.28 to calculate the trp-to-anthraniloyl distance.

Instrumentation for Fluorescence Spectroscopy

2

The successful application of fluorescence methods requires considerable attention to experimental details and a good understanding of the instrumentation. There are numerous potential artifacts which can distort the data. Fluorescence is a highly sensitive method. The gain or amplification of the instruments can usually be increased to obtain observable signals, even if the sample is nonfluorescent. These signals seen at high amplification may not originate with the fluorophore of interest. Instead, one may observe interference due to background fluorescence from the solvents, light leaks in the instrumentation, stray light passing through the optics, light scattered by turbid solutions, Rayleigh scatter, and/or Raman scatter, to name a few sources of interference. Furthermore, there is no ideal spectrofluorometer, and the available instruments do not yield true excitation or emission spectra. This is because of the nonuniform spectral output of the light sources and the wavelength-dependent efficiency of the monochromators and detectors (photomultiplier tubes). The polarization or anisotropy of the emitted light can also affect the measured fluorescence intensities. To obtain reliable spectral data, one needs to be aware of and control these numerous factors. In this chapter we will discuss the properties of the individual components in a spectrofluorometer and how these properties affect the observed spectral data.

In our discussion of instrumentation factors, we will stress their effects on excitation and emission spectra. However, similar concerns are important in the measurement of fluorescence lifetimes and anisotropies, which will be described in Chapters 4, 5, and 10. Additionally, the optical properties of the samples, such as optical density and turbidity, can also affect the spectra. Specific examples are given to clarify these effects and the means to avoid them.

2.1. EXCITATION AND EMISSION SPECTRA

Generally, one wishes to record both excitation and emission spectra. An emission spectrum is the wavelength distribution of the emission, measured at a single constant excitation wavelength. Conversely, an excitation spectrum is the dependence of emission intensity, measured at a single emission wavelength, upon the excitation wavelength. Such spectra can be presented on either a wavelength scale or a wavenumber scale. Light of a given energy can be described in terms of its wavelength λ, frequency ν, or wavenumber $\bar{\nu}$. The usual units for wavelength are nanometers, and wavenumbers are given in units of reciprocal centimeters (cm^{-1}). Wavelengths and wavenumbers are easily interconverted by taking the reciprocal of each value. For example, 400 nm corresponds to $(400 \times 10^{-7}\ cm)^{-1} = 25{,}000\ cm^{-1}$. The question of whether fluorescence spectra are to be presented on the wavelength or on the wavenumber scale has been the subject of considerable debate. Admittedly, the wavenumber scale is linear in energy. However, most commercially available instrumentation yields spectra on the wavelength scale, and such spectra are more familiar and thus easier to interpret visually. Since corrected spectra are not needed on a routine basis, and since accurately corrected spectra are difficult to obtain, we prefer to use the directly recorded technical spectra on the wavelength scale.

For an ideal instrument, the directly recorded emission spectra would represent the photon emission rate or power emitted at each wavelength, over a wavelength interval determined by the slit widths and dispersion of the emission monochromator. Similarly, the excitation spectrum would represent the relative emission of the fluorophore at each excitation wavelength. For most fluorophores, the quantum yields and emission spectra are independent of excitation wavelength. Hence, the excitation spectrum of a fluorophore is typically superimposable on its absorption spectrum. However, this is rarely observed because the wavelength responses of almost all spectrofluorometers are dependent on wavelength. Even under ideal circumstances, such correspondence requires the presence of only a single type of fluorophore and the absence of other complicating factors, such as a nonlinear response resulting from a high optical density of the sample or the presence of other chromophores in the sample.

Figure 2.1 shows a schematic diagram of a general-purpose spectrofluorometer. This instrument has a xenon lamp as the source of exciting light. Such lamps are generally useful because of their high intensity at all wavelengths ranging upward from 250 nm. The instrument shown is equipped with monochromators to select both the excitation and emission wavelengths. The excitation monochromator in this schematic contains two gratings, which decreases stray light (i.e., light of wavelengths different from the chosen wavelength). In addition, these monochromators use concave gratings, produced by holographic means, which further decrease stray light. In subsequent sections of this chapter we will discuss light sources, detectors, and the importance of spectral purity to minimize interference due to stray light. Both monochromators are motorized to allow automatic scanning of wavelength. The fluorescence is detected with photomultiplier tubes and quantified with the appropriate electronic devices. The output is usually presented in graphical form and is typically stored on the computer disk.

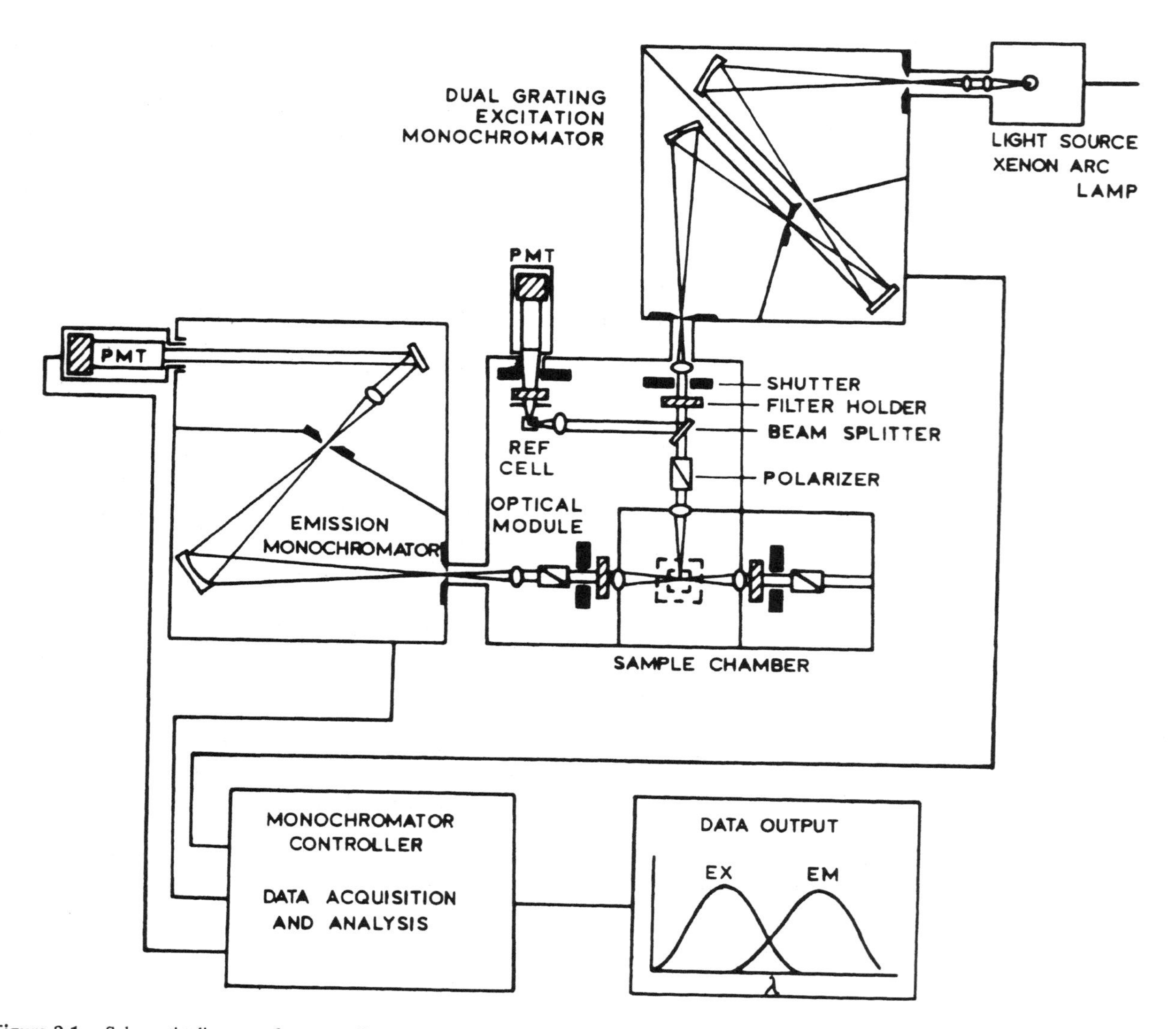

Figure 2.1. Schematic diagram of a spectrofluorometer. PMT, Photomultiplier tube. Revised from commercial literature from SLM Instruments.

The schematic diagram in Figure 2.1 also shows the components of the optical module which surrounds the sample holder. Versatile and stable optical components are indispensable for a research spectrofluorometer. The module shown in Figure 2.1 contains a number of convenient features, which should be considered in the design or selection of such instrumentation. Shutters are provided to eliminate the exciting light or to close off the emission channel. A beam splitter is provided in the excitation light path. This splitter reflects part of the excitation light to a reference cell, which generally contains a stable reference fluorophore. The beam splitter consists of a thin piece of clear quartz, which reflects about 4% of the incident light. This amount is generally adequate for a reference channel, since this channel frequently contains a highly fluorescent quantum counter (Section 2.8). The intensity from the standard solution is typically isolated with a bandpass filter and is proportional to the intensity of the exciting light. Changes in the intensity of the arc lamp may be corrected for by division of the intensity from the sample by that from the reference fluorophore.

Polarizers are present in both the excitation and emission light paths. Generally, the polarizers are removable so that they can be inserted only for measurements of fluorescence anisotropy or when it is necessary to select for particular polarized components of the emission and/or excitation. Accurate measurement of fluorescence anisotropies requires accurate angular positioning of the polarizers. For this reason, the polarizer mounts must be sturdy and accurately indexed to determine the angular orientation. Furthermore, the optical module shown in Figure 2.1 has an additional optical path on the right side of the sample holder. This path allows measurement of fluorescence anisotropy by the T-format method (Chapter 10). There are many occasions when the additional light path is necessary or convenient for experiments.

At present, the trend is toward small, compact spectrofluorometers with all the optical components within a single enclosure. Such instruments are easy to maintain because there is little opportunity to alter the configuration. However, a modular instrument has numerous advantages in a research laboratory. For instance, if the xenon lamp and monochromator are removable, a laser source can be used in place of the arc lamp. On some occasions it is desirable to bypass the emission monochromator and use bandpass filters to collect as much of the emission as possible. Such experiments are only practical if the emission monochromator is removable. If the monochromator cannot be removed, it is useful to remember that gratings act like mirrors at the zero-order diffraction. If the wavelength is set to zero, the monochromator typically transmits all wavelengths. One can then use filters to isolate the desired range of wavelengths.

One should also pay careful attention to the sample holder. If the research involves anisotropy measurements, it will often be necessary to measure the fundamental anisotropy (r_0) in the absence of rotational diffusion. This is accomplished at low temperature, typically –50 °C in glycerol. Low temperature can only be achieved if the sample holder is adequately sized for a high rate of coolant flow, has good thermal contact with the cuvette, and is insulated from the rest of the instrument. Many cuvette holders can maintain a temperature near room temperature but fail if a significantly higher or lower temperature is needed.

Another useful feature is the ability to place optical filters into the excitation or emission light path. Filters are often needed, in addition to a monochromator, to remove unwanted wavelengths in the excitation beam or remove scattered light from the emission channel.

2.1.A. An Ideal Spectrofluorometer

As described above, we wish to record excitation and emission spectra which represent the relative photon intensity per wavelength interval. To obtain such "corrected" emission spectra, the individual components must have the following characteristics:

1. The light source must yield a constant photon output at all wavelengths.
2. The monochromator must pass photons of all wavelengths with equal efficiency.
3. The monochromator efficiency must be independent of polarization.
4. The detector (photomultiplier tube) must detect photons of all wavelengths with equal efficiency.

These characteristics for ideal optical components are illustrated in Figure 2.2. Unfortunately, light sources, monochromators, and photomultiplier tubes with such ideal characteristics are not available. As a result, one is forced to compromise on the selection of components and to correct for the nonideal response of the instrument.

An absorption spectrophotometer contains these same components, and one may wonder why it is possible to record correct absorption spectra. In recording an absorption spectrum, one generally measures the intensity of light transmitted by the sample, relative to that transmitted by the blank. These comparative measurements are performed with the same components, at the same wavelengths. Hence, the nonideal behavior of the components cancels in the comparative measurements. In contrast to absorption

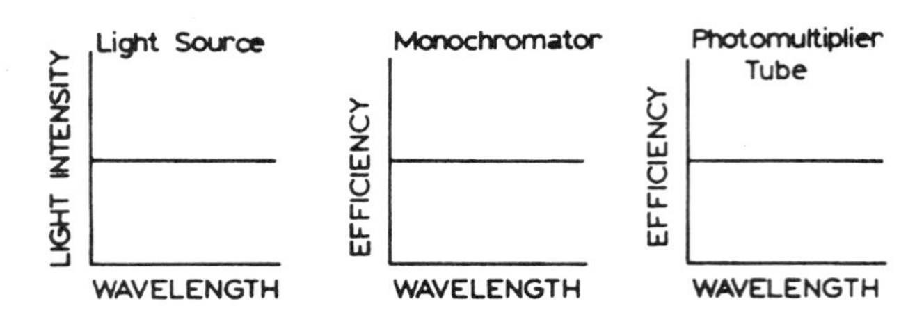

Figure 2.2. Properties of ideal components of a spectrofluorometer.

measurements, fluorescence intensity measurements are absolute, not relative. That is, comparison of the sample fluorescence with a blank is not useful because the blank, in principle, displays no signal. Also, the weak background signal has an unknown spectral distribution and thus cannot be used for correction of the wavelength dependence of the optical components. Hence, the opportunity for internal compensation is limited. As will be described below, a limited number of standard spectra are available for correction purposes. Corrected emission spectra are provided in Appendix I.

2.1.B. Distortions in Excitation and Emission Spectra

To record an excitation spectrum, the emission monochromator is set at the desired wavelength, generally the emission maximum. The excitation monochromator is then scanned through the absorption bands of the fluorophore. The observed signal is distorted for several reasons:

1. The light intensity from the excitation source is a function of wavelength. Even if the intensity of the exciting light is monitored via the beam splitter shown in Figure 2.1, and corrected by division as described earlier in the chapter, the response of the reference solution or detector may be dependent upon wavelength.
2. The transmission efficiency of the excitation monochromators is a function of wavelength.
3. The optical density of the sample may exceed the linear range, which is about 0.1 absorbance units, depending upon sample geometry.

Emission spectra are recorded by choosing an appropriate excitation wavelength and scanning wavelength with the emission monochromator. In addition to the factors discussed above, the emission spectrum is further distorted by the wavelength-dependent efficiency of the emission monochromator and the photomultiplier tubes. The emission spectrum can also be distorted by the absorption of the sample.

2.2. LIGHT SOURCES

We now describe the individual components of a spectrofluorometer. The general characteristics of these components are considered, along with the reason for choosing specific components. Understanding the characteristics of these components allows one to understand the capabilities and limitations of spectrofluorometers. We will first consider light sources.

2.2.A. Arc and Incandescent Lamps

Xenon Lamps. At present, the most versatile light source for a steady-state spectrofluorometer is a high-pressure xenon (Xe) arc lamp. These lamps provide a relatively continuous light output from 250 to 700 nm (Figure 2.3); a number of sharp lines occur near 450 nm and above 800 nm. Xenon arc lamps emit a continuum of light as a result of the recombination of electrons with ionized Xe atoms. These ions are generated by collisions of Xe atoms with the electrons which flow across the arc. Complete separation of the electrons from the atoms yields the continuous emission. Xe atoms which are in excited states but not ionized yield lines rather than broad emission bands. The peaks near 450 nm are due to these excited states. The output intensity drops rapidly below 280 nm. Furthermore, many Xe lamps are classified as being ozone-free, meaning that their operation does not generate ozone in the surrounding air. The quartz envelope used in such ozone-free lamps does not transmit light with wavelengths shorter than 250 nm, and the output of such lamps decreases rapidly with decreasing wavelength.

The wavelength-dependent output of Xe lamps is a major reason for distortion of the excitation spectra of

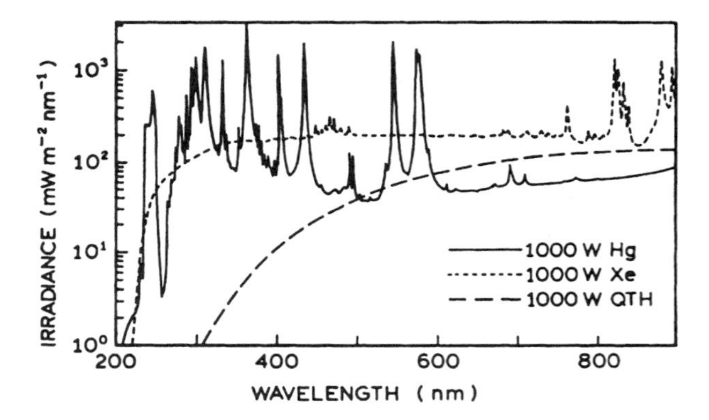

Figure 2.3. Spectral output of high-pressure xenon (Xe) and high-pressure mercury (Hg) arc lamps. Also shown is the output of a quartz–tungsten halogen lamp (QTH). Note that the intensity axis is logarithmic. Revised from technical literature from Oriel Instruments.[1]

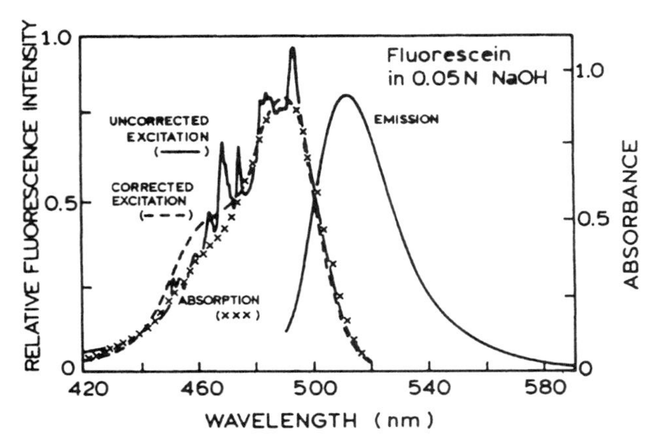

Figure 2.4. Corrected and uncorrected excitation spectra of fluorescein. From Ref. 2.

compounds which absorb in the visible and UV. To illustrate this effect, Figure 2.4 shows corrected and uncorrected excitation spectra of fluorescein. The uncorrected excitation spectrum displays a number of peaks near 450 nm. These peaks are due to the output of the Xe lamp, which also displays peaks near 450 nm (Figure 2.3). Also shown on Figure 2.4 is the excitation spectrum corrected for the wavelength-dependent output of the Xe arc lamp. A quantum counter was used in the reference channel to provide a signal proportional to the lamp intensity, and the intensity of the sample was divided by this reference intensity (Section 2.8). The peaks near 450 nm are no longer apparent in the corrected excitation spectrum, and this spectrum corresponds more closely with the absorption spectrum of fluorescein. The marked difference between the corrected and the uncorrected spectra illustrates the importance of the spectral output of the lamp in the shape of the excitation spectra.

Xenon lamps are usually contained within specially designed housings. The housing which contains the arc lamp serves several important functions (Figure 2.5). The gas in xenon lamps is under high-pressure (about 10 atm), and thus explosion is always a danger. The housing protects the user from the lamp and also from its intense optical output. The housing serves also to direct air over the lamp and remove excess heat and ozone. Of course, a xenon lamp which is on should never be observed directly. The extreme brightness will damage the retina, and the UV light can damage the cornea.

Another important role of the housing is for collecting and collimating the lamp output, which can then be focused into the entrance slit of the monochromator. Some lamp houses have mirrors behind the lamp to direct additional energy toward the output. While one's first impulse is to collect as large an area as possible, most of the light output originates from the small central region between the electrodes, and it is this spot which needs to be focused on the optical entrance slit of the excitation monochromator.

Because of the heat and high intensity of a running xenon lamp, it is not practical to adjust the position of an uncovered lamp. Hence, the lamp housing should have external

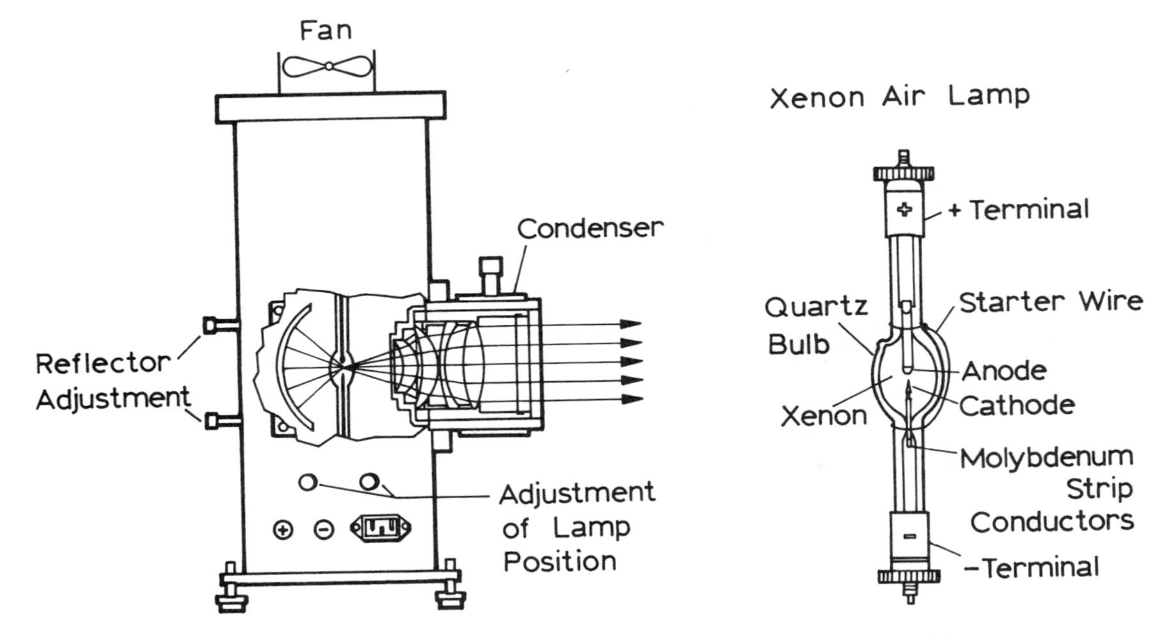

Figure 2.5. Xenon air lamp and a typical housing for the lamp. Revised from Ref. 3.

provisions for position adjustment and focusing. The useful life of a xenon lamp is about 2000 hours. Safety glasses should be worn when these lamps are handled. The quartz envelope should not be touched, and, if touched, it should be cleaned with a solvent such as ethanol. The fingerprint residues will char, resulting in hot spots on the quartz envelope and possibly lamp failure. In disposing of a xenon lamp, one should wrap the lamp in heavy paper and break the quartz envelope to protect the next person handling it. It is important to pay close attention to mounting the lamps in the proper orientation, which can be different for different types of lamps.

The power supplies of arc lamps are also extremely dangerous, generating 25 A at 20 V, for a typical 450-W lamp. Starting the lamps requires high-voltage pulses of 20–40 kV. This voltage can penetrate the skin, and the ensuing high current could be lethal. Starting of xenon lamps is also a frequent cause of damage to nearby electronics, particularly computers. The high-voltage starting pulse can destroy sensitive amplifiers or confuse the computers. If possible, it is preferable to start the lamp first and then turn on other electronic devices.

Xenon arc lamps that have an ellipsoidal reflector as part of the lamp itself are also available (Figure 2.6). The arc is positioned at the focal point of the ellipse, and the output is directed toward the second focal point. Xenon lamps are also available with parabolic reflectors, which collect a large solid angle and provide a collimated output.[4] In addition to having improved light collection efficiency, these lamps are compact, and, as a result, they are found in commercial spectrofluorometers. It is claimed that the same light output is obtained from a parabolic xenon lamp as from a standard lamp housing (Figure 2.5) with twice the wattage.

When a xenon arc lamp is used, it is important to remember that these lamps emit a large amount of infrared radiation, extending beyond the wavelength range of Figure 2.3. Because of the infrared output, the lamp output cannot be passed directly through most optical filters. The filter will heat and/or crack, and the samples will be heated. When the lamp output is passed through a monochromator, the optical components and housing serve as a heat sink. If the xenon lamp output is to be used directly, one should use a heat filter, which is made of heat-resistant glass that absorbs the infrared radiation.

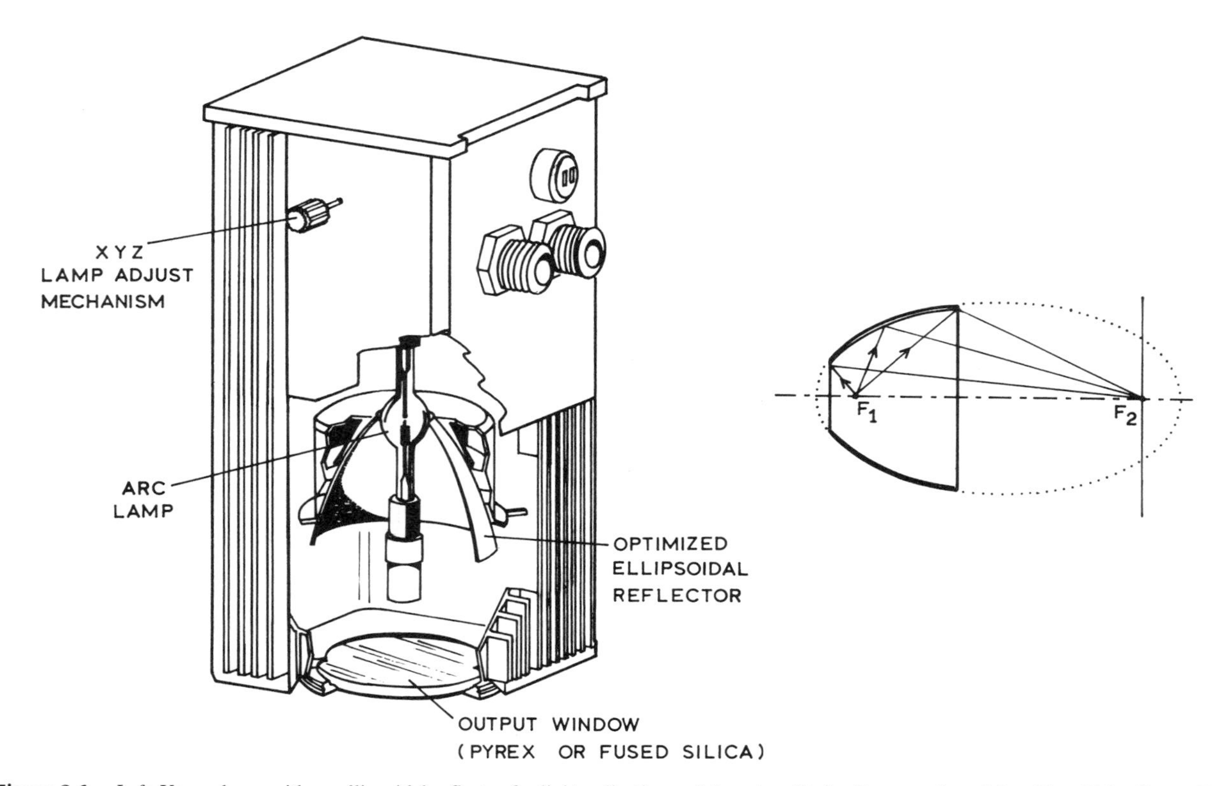

Figure 2.6. *Left*: Xenon lamp with an ellipsoidal reflector for light collection and focusing. *Right*: Cross section of the ellipsoidal reflector. Rays originating at the first focal point, F_1, are focused at F_2. From Ref. 1. Xenon lamps with parabolic reflectors and collimated outputs are also available.[4]

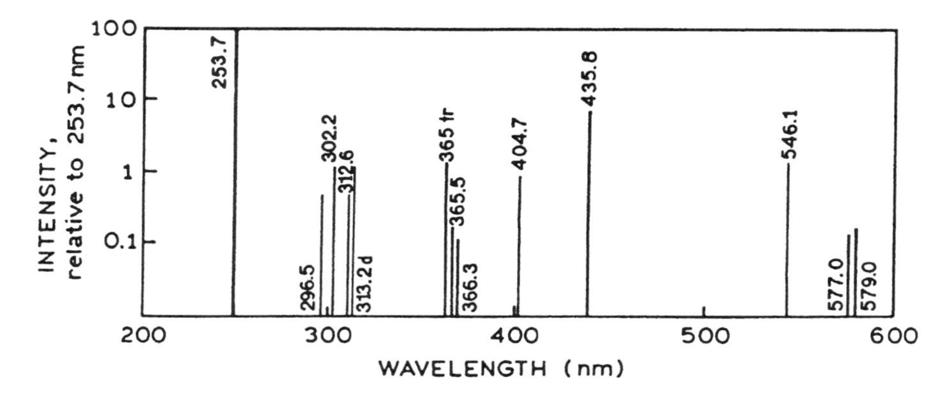

Figure 2.7. Spectral lines from a low-pressure Hg lamp. d, Doublet; tr, triplet. Revised from Ref. 5.

2.2.2. High-Pressure Mercury (Hg) Lamps.

In general, Hg lamps have higher intensities than Xe lamps, but the intensity is concentrated in lines (Figure 2.3). Usually, one requires the ability to choose the excitation wavelengths to suit the fluorophore, rather than vice versa. Hence, these lamps are only useful if the Hg lines are at suitable wavelengths for excitation.

2.2.3. Mercury–Xenon (Hg–Xe) Arc Lamps.

One can also use high-pressure Hg–Xe lamps. These have higher intensities in the UV than Xe lamps, and the presence of Xe tends to broaden the spectral output. The output of these lamps is dominated by the Hg lines and is similar to that shown for the Hg lamps in Figure 2.3, but there is slightly more output between the Hg lines. When first started, the Hg–Xe lamp output is due mostly to Xe. As the lamp reaches operating temperature, all the Hg becomes vaporized, and the Hg output becomes dominant.

2.2.4. Quartz–Tungsten Halogen (QTH) Lamps.

QTH lamps provide continuous output in the visible and infrared (IR) regions of the spectrum (Figure 2.3). Previously, such lamps were not useful for fluorescence because they have low output below 400 nm and are thus not useful for excitation of UV-absorbing fluorophores. However, there is presently increasing interest in fluorophores absorbing in the red and near-infrared (NIR), where the output of a QTH lamp is significant. One can expect an increased use of such light sources as probes are developed for use of longer wavelengths.

2.2.5. Low-Pressure Hg Lamps.

These lamps yield very sharp mercury line spectra which are useful primarily for calibration purposes (Figure 2.7). Details of this procedure are described in Section 2.3.E.

2.2.6. Time-Resolved Measurements.

The light sources described above are used for steady-state or continuous illumination. In general, these cannot be intrinsically pulsed or amplitude-modulated fast enough for use in time-resolved measurements. The arc lamps used for phase-modulation lifetime measurements use external light modulators. Flashlamps are available which provide nanosecond pulses, but the output of these lamps is too weak for steady-state measurements. Light sources for lifetime measurements will be discussed in Chapters 4 and 5.

2.2.B. Solid-State Light Sources

In addition to the lamps described in Section 2.2.A, several other light sources are now being used for fluorescence. These include light-emitting diodes (LEDs) and laser diodes. LEDs are available which provide output over a wide range of wavelengths. One LED cannot cover the entire spectral range, but only a few LEDs are needed to provide light from 430 to 680 nm (Figure 2.8), and LEDs are rather inexpensive. LEDs are solid-state devices which require

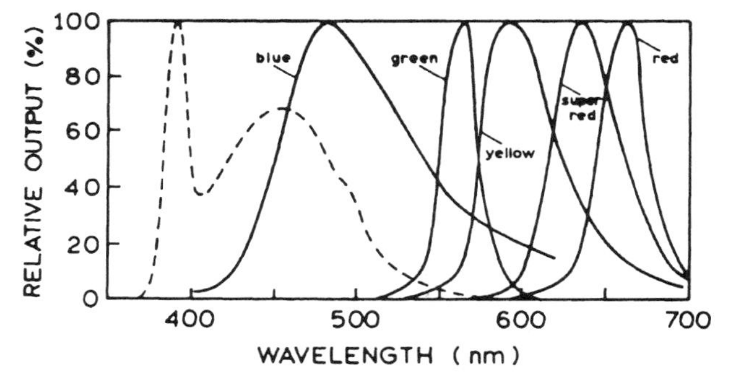

Figure 2.8. Relative spectral output of light-emitting diodes (LEDs). The dashed curve shows the UV and blue output of a LED when driven at high current.[6]

little power and generate little heat. LEDs can be placed close to the samples, and, if needed, the excitation wavelength can be defined better by the use of an excitation filter. Unlike a xenon lamp, the LEDs do not generate significant infrared radiation, so that an additional heat filter is not needed. Remarkably, blue LEDs have recently been shown to provide UV output[6] near 390 nm (dashed curve in Figure 2.8). This occurs when a blue LED is driven at high current. The UV emission can be readily isolated using optical filters. Hence, LEDs are now practical light sources for many fluorophores.

Another light source is the laser diode. In contrast to the LEDs, laser diodes emit monochromatic radiation. At present, laser diodes are available with wavelengths ranging from 630 to 1500 nm. It is likely that laser diodes with shorter wavelengths will soon become available. Laser diodes are convenient because the output is easily focused and manipulated.

In contrast to the output of arc lamps or incandescent lamps, the output of LEDs and laser diodes can be pulsed or modulated. LEDs can be amplitude-modulated up to about 100 MHz, and laser diodes can be modulated to several gigahertz. The use of these light sources for time-resolved measurements will be described in Chapters 4 and 5.

2.3. MONOCHROMATORS

Monochromators are used to disperse polychromatic or white light into the various colors or wavelengths. This dispersion can be accomplished using prisms or diffraction gratings. At present, the monochromators used in most spectrofluorometers employ diffraction gratings rather than prisms. The performance specifications of a monochromator include the dispersion, the efficiency, and the stray light levels. The dispersion is usually given in nanometers per millimeter, where the slit width is expressed in millimeters. In selecting a monochromator for fluorescence spectroscopy, one looks for low stray light levels to avoid problems due to scattered stray light. By stray light we mean light transmitted by the monochromator at wavelengths outside of the chosen wavelength and bandpass. In addition, monochromators are chosen for high efficiency to maximize the ability to detect low light levels. Resolution is usually of secondary importance since emission spectra rarely have peaks with linewidths less than 5 nm. The slit widths are generally variable, and a typical monochromator will have both an entrance and an exit slit. The light intensity which passes through a monochromator is approximately proportional to the square of the slit width. Larger slit widths yield increased signal levels, and therefore higher signal-to-noise ratios. Smaller slit widths yield higher resolution, but at the expense of light intensity. However, if the entrance slit of the excitation monochromator is already wide enough to accept the focused image of the arc, then the intensity will not be increased significantly with wider slit widths. If photobleaching of the sample is a problem, this effect can sometimes be minimized by the use of decreased light levels. Gentle stirring of the sample can also minimize photobleaching. This is because only a fraction of the sample is illuminated and the bleached portion of the sample is continuously replaced by fresh solution.

Monochromators may have planar or concave gratings (Figure 2.9). Planar gratings are usually produced mechanically and may contain imperfections in some of the grooves. Concave gratings are usually produced by holographic and photoresist methods, and imperfections are rare. Imperfections of the gratings are the major source of stray light transmission by the monochromators and of ghost images from the grating. Ghost images can sometimes be seen if one observes the dispersed light within an open monochromator. One may observe diffuse spots of white light on the inside surfaces. Monochromators sometimes contain light blocks to intercept these ghost images.

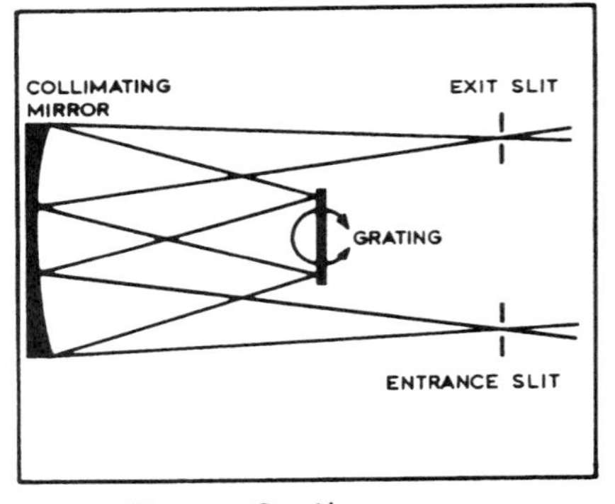

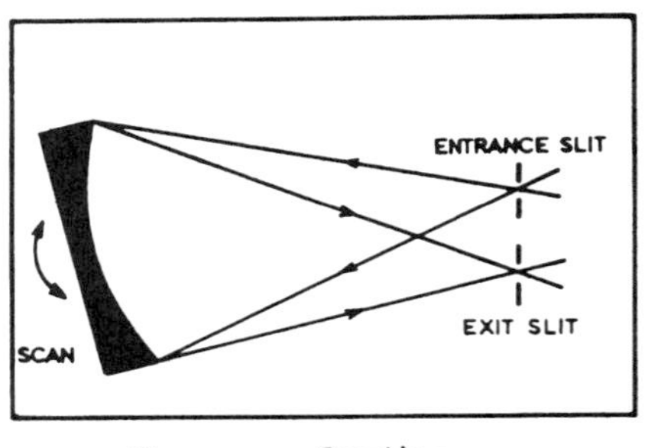

Figure 2.9. Monochromators based on a plane (*top*) or concave (*bottom*) grating. Revised from Ref. 7.

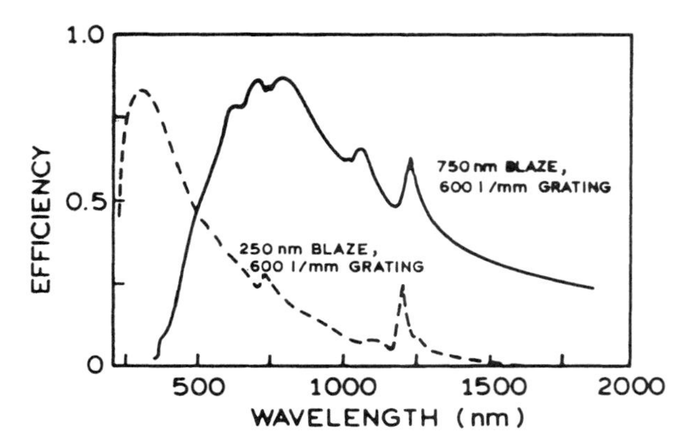

Figure 2.10. Efficiencies of two ruled (600 lines/mm) gratings blazed for different wavelengths. Redrawn from Ref. 3.

Monochromators based on concave gratings can have fewer reflecting surfaces and lower stray light and can be more efficient. A concave grating can serve as both the diffraction and focusing element, resulting in one instead of three reflecting surfaces. For these reasons, the holographic gratings are usually preferable for fluorescence spectroscopy.

It is important to recognize that the transmission efficiency of a grating and monochromator is dependent on wavelength and on the design of the grating. Examples are shown in Figures 2.10 and 2.11. For a mechanically produced plane grating, the efficiency at any given wavelength can be maximized by choice of the blaze angle, which is determined by the shape and angle of the tool used to produce the grating. By choice of this angle, one may obtain maximum diffraction efficiency for a given wavelength region, but the efficiency is less at other wavelengths. For the examples shown in Figure 2.10, the efficiency was optimized for 250 or 750 nm. Generally, one chooses an excitation monochromator with high efficiency in the UV and an emission monochromator with high efficiency at visible wavelengths. Because of the different production methods, holographic gratings are not blazed, but the shape of the grooves can be varied. In general, their peak transmission efficiency is less than for a plane grating, but the efficiency is more widely distributed on the wavelength scale than that of the planar gratings (Figure 2.11).

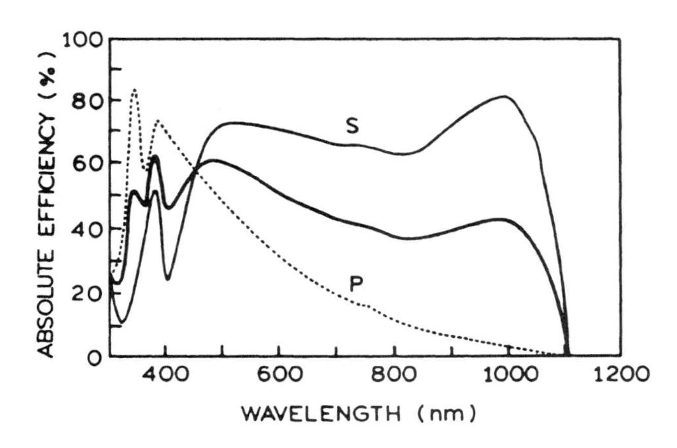

Figure 2.11. Grating efficiency for an 1800-line/mm holographic grating, optimized for the UV. The bold curve is the average, and the other curves are for differently (S and P) polarized light, defined in Figure 2.34. Redrawn from Ref. 7.

2.3.A. Wavelength Resolution and Emission Spectra

Emission spectra of most fluorophores are rather broad and devoid of structure. Hence, the observed emission spectra are typically independent of the spectral resolution. However, for fluorophores which display structured emission, it is important to maintain adequate wavelength resolution, which is adjusted by the slit widths on the monochromator. Emission spectra of *p*-terphenyl are shown in Figure 2.12. The emission spectrum displays vibrational structure when recorded with a resolution of 0.5 nm. However, this structure is nearly lost when the resolution is 10 nm. Although not important for steady-state measurements, it should be recognized that the transit time through a monochromator can depend on wavelength (Section 4.6.E).

2.3.B. Polarization Characteristics of Monochromators

An important characteristic of grating monochromators is that the transmission efficiency depends upon the polarization of the light. This is illustrated in Figure 2.11 for a

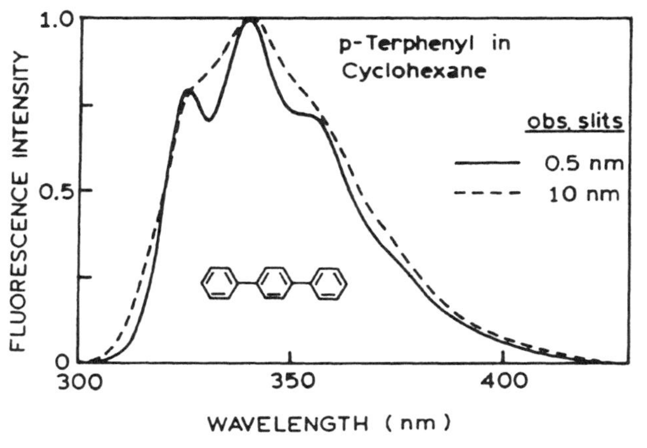

Figure 2.12. Emission spectra of *p*-terphenyl collected with a spectral resolution of 0.5 and 10 nm. From Ref. 8.

concave grating. For this reason, the observed fluorescence intensities can be dependent upon the polarization displayed by the fluorescence emission. In addition, the emission spectrum of the sample can be shifted in wavelength and altered in shape, depending upon the polarization conditions chosen to record the emission spectrum.

For example, consider an emission spectrum recorded with the grating shown in Figure 2.11 and through a polarizer oriented vertically (||) or horizontally (⊥). Assume that the dotted transmission curve corresponds to vertically polarized light. The spectrum recorded through the vertically oriented polarizer would appear shifted to shorter wavelengths relative to that recorded with the polarizer in the horizontal position. This is because the transmission efficiency for vertically polarized light is higher at shorter wavelengths. This spectral shift would be observed irrespective of the sample, and irrespective of whether its emission was polarized or not polarized.

The polarization characteristics of monochromators have important consequences in the measurement of fluorescence anisotropy. Such measurements must be corrected for the varying efficiencies of each optical component; this correction is expressed as the *G*-factor (Section 10.4). However, the extreme properties of the concave gratings (Figure 2.11) can cause difficulties in the measurement of fluorescence polarization. For example, assume that the polarization is to be measured at an excitation wavelength of 450 nm. The excitation intensities will be nearly equal with the excitation polarizers in each orientation, which makes it easier to compare the relative emission intensities. If the emission is unpolarized, the relative intensities of the parallel (||) and perpendicular (⊥) excitation will be nearly equal. However, suppose the excitation is at 340 nm, in which case the intensities of the polarized excitation will be very different. In this case it is more difficult to accurately measure the relative emission intensities because of the larger difference. Measurement of the *G*-factor is generally performed using horizontally polarized light, and the intensity of this component would be low. Hence, anisotropy measurements can be difficult.

The polarization properties of concave gratings can have a dramatic effect on the appearance of emission spectra. This is shown in Figure 2.13 for a probe bound to DNA. The emission spectrum shows a dramatic decrease near 630 nm when the emission polarizer is in the horizontal position. This drop is not seen when the emission polarizer is in the vertical orientation. In addition to this unusual dip in the spectrum, the emission maximum appears to be different for each polarizer orientation. These effects are due to the polarization properties of this particular grating, which displays a minimum in efficiency at 630 nm for horizontally polarized light. Such effects are due to the emission monochromators and are independent of the polarization of the excitation beam.

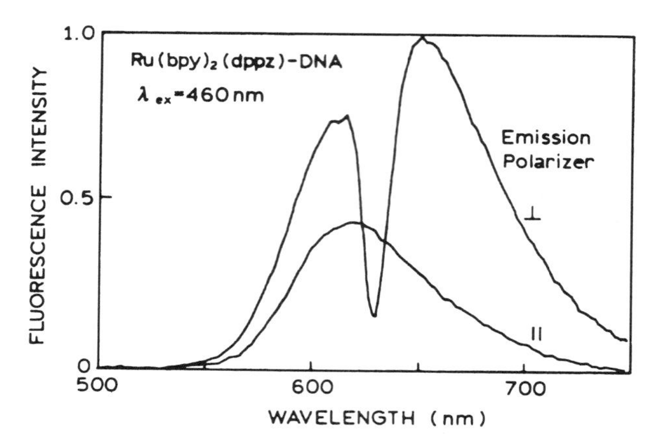

Figure 2.13. Emission spectra of $[Ru(bpy)_2(dppz)]^{2+}$ bound to DNA. Excitation at 460 nm. Except for intensity, the same spectral distribution was observed for vertically as for horizontally polarized excitation. This sample is described in Section 20.3.A. From Ref. 9.

2.3.C. Stray Light in Monochromators

For fluorescence measurements, the stray light level of the monochromator is perhaps the most critical parameter. Stray light is defined as any light which passes through the monochromator besides light of the desired wavelength. Consider first the excitation monochromator. The entire output from the light source enters the monochromator. UV wavelengths are frequently used for excitation, and the UV intensity may be 100-fold less than the visible output of the Xe lamp. Stray light at longer wavelengths can be passed by the excitation monochromator and can easily be as intense as the fluorescence itself. Fluorescence intensities are frequently low, and many biological samples possess significant turbidity. As a result, the incident stray light at the emission wavelength can be scattered and can interfere with measurements of the fluorescence intensity. For these reasons, double-grating monochromators are frequently used, especially for excitation. Stray light levels for such monochromators are frequently 10^{-8}–10^{-12} of the peak intensities. However, double-grating monochromators are less efficient, and sensitivity must be sacrificed.

It is also important to consider the performance of the emission monochromator. Generally, only a low percentage of the exciting light is absorbed by the fluorophores, and fluorescence quantum yields can be small. It is not unusual for the fluorescence signal to be 1000-fold less intense than the exciting light. Now consider a turbid suspension of membranes, from which we wish to observe the fluorescence of membrane-bound proteins. The excitation and emission wavelengths would be near 280 and 340 nm, respectively. Since the emission monochromator is imperfect, some of the scattered light at 280 nm can pass through the emission monochromator set at 340 nm. Assume that the emission monochromator, when set at 340 nm, discriminates against 280 nm by a factor of 10^{-4}. The intensity of scattered light at 280 nm can easily be 10^3-fold more intense than the fluorescence at 340 nm. Hence, 10% of the "fluorescence" may actually be due to scattered exciting light at 280 nm. It is also important to recognize that scattered light is highly polarized, typically 100%. Therefore, stray scattered light can easily invalidate measurements of fluorescence anisotropy.

The stray light rejection of holographic gratings is superior to that of the ruled (i.e., mechanically produced) gratings. It appears that the passage of stray light depends upon imperfections in the gratings, resulting in ghost images which can escape from the monochromators. Fewer such images are present with the holographic gratings because they are produced optically and have fewer imperfections. In addition, monochromators with holographic gratings generally have fewer reflecting surfaces (Figure 2.9). This is because the concave grating can also act as an imaging device, and thus additional concave mirrors are not required for focusing. With fewer reflecting surfaces, there is a decreased probability of stray light escaping from the monochromator.

2.3.D. Second-Order Transmission in Monochromators

Another source of unwanted light is higher-order light diffraction by the monochromator. Light diffraction at the grating can occur as a first-, second-, or higher-order process. These diffraction orders frequently overlap (Figure 2.14). Suppose the excitation monochromator is set at 300 nm. The xenon light source contains output of both 300 and 600 nm. When the monochromator is set at 600 nm, some 300-nm light can be present at the exit slit due to second-order diffraction (Figure 2.14). Hence, when recording an emission spectrum from a turbid solution, one will often observe a peak at twice the excitation wavelength due to second-order transmission through the excitation monochromator. For this reason, we often use bandpass excitation filters to remove unwanted wavelengths from the excitation beam.

Transmission of the second-order diffraction can also result in extraneous light passing through the emission monochromators. Suppose the excitation is at 300 nm, and the emission monochromator is scanned through 600 nm. If the sample is strongly scattering, then some of the scattered light at 300 nm can appear as second-order diffraction when the emission monochromator is set to 600 nm.

2.3.E. Calibration of Monochromators

The wavelength calibration of monochromators should be checked periodically, especially on monochromators where the "calibration" is determined electronically rather than by direct mechanical coupling. For calibration we use a mercury penlight. This low-pressure mercury lamp is shaped like a cylinder, about 5 mm in diameter, and conveniently fits into the cuvette holder. This lamp provides a number of sharp lines which can be used for calibration (Figure 2.7). To hold the lamp stationary, we use a block of metal in which the lamp fits snugly. This holder is the same size as a cuvette. A pinhole on the side of this holder allows a small amount of the light to enter the emission monochromator. A small slit width is used to increase precision of the wavelength determination and to decrease the light intensity. It is important to attenuate the light so that the photomultiplier tube and/or amplifiers are not damaged. Following these precautions, one locates the dominant Hg lines using the emission monochromator. The measured wavelengths are compared with the known values, which are listed in Table 2.1. Since there are multiple Hg lines, it is necessary to observe three or more lines to be certain that the line is assigned the correct wavelength. If the observed wavelengths differ from the known values by a constant amount, one recalibrates the monochromator to obtain coincidence. A more serious problem is encountered if the wavelength scale is nonlinear, that is, the measured wavelengths differ from those in Table 2.1 by an amount which is wavelength-dependent.

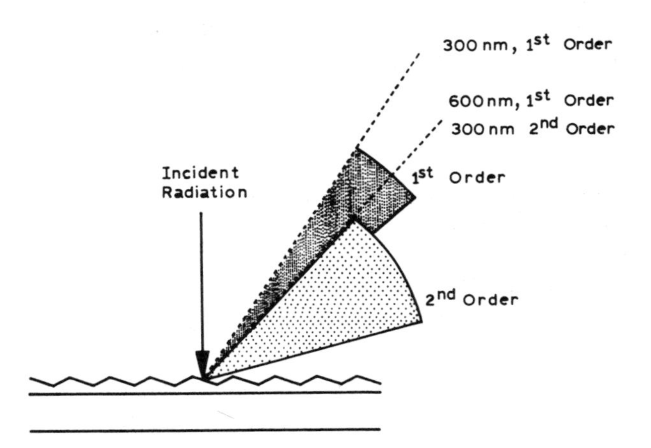

Figure 2.14. First- and second-order diffraction off a diffraction grating. The region where there is no overlap of the first- and second-order diffraction is called the free spectral range.

Table 2.1. Wavelengths and Relative Intensities of Mercury Lines[a]

Wavelength (nm)	Relative intensity
253.7	100.0
296.5	0.6
302.2	1.1
312.6	0.7
313.2	1.1
365.0	0.9
365.5	0.2
366.3	0.1
404.7	0.9
435.8	1.7
546.1	1.2
577.0	0.2
579.0	0.2

[a]Pen-Ray lamp from Ultra-Violet Products, San Gabriel, CA. The values of the relative intensities are only approximate.

In this case the monochromator is generally returned to the manufacturer for realignment.

After calibration of the emission monochromator, the excitation monochromator can be calibrated against this new standard. The slits on both monochromators should be set to the same small value, consistent with the available light intensity. A dilute suspension of glycogen is placed in the cuvette holder to scatter the exciting light. The emission monochromator is set to some arbitrary wavelength. If the excitation monochromator is properly calibrated, then the maximum intensity of the scattered light is seen when the indicated wavelengths are identical. The linearity of the wavelength scale can be determined by setting the emission monochromator at various wavelengths. One advantage of this procedure is that there is no need to remove the light source. One could use the mercury light in place of the xenon lamp to calibrate the excitation monochromator, but then the xenon lamp must be removed.

2.4. OPTICAL FILTERS

2.4.A. Bandpass Filters

Although spectrofluorometers have monochromators for wavelength selection, it is often important to use optical filters in addition to the monochromators. Optical filters are used to compensate for the less than ideal behavior of monochromators. Also, when the spectral properties of a fluorophore are known, maximum sensitivity is often obtained using filters rather than monochromators. The transmission curves of some typical glass filters are shown in Figure 2.15. Filters can be identified which transmit near the emission maximum of proteins, such as the UG-11 and BG-3 (Figure 2.15, top). Other filters transmit near 450 nm, such as the BG-14, which is near the 450-nm emission maximum of many extrinsic fluorophores like DAPI and DPH. The performance of such filters is often nonideal. Many of these bandpass filters also transmit at longer wavelengths, so that one needs to be cautious in their use for eliminating second-order transmission through the monochromator. Long-pass filters, such as those whose transmission curves are shown in the bottom panel of Figure 2.15, are often used to reject scattered light and to transmit the fluorescence.

An important consideration in the use of bandpass filters is the possibility of emission from the filter itself. Many filters are luminescent when illuminated with UV light, which can be scattered from the sample. For this reason, it is usually preferable to locate the filter farther away from the sample, rather than directly against the sample. In spite of the need for these precautions, glass filters of the type shown in Figure 2.15 are highly versatile, effective, and inexpensive, and a wide selection is needed in any fluorescence laboratory. We typically select and use excitation and emission filters for all our experiments, even those using monochromators, to reduce the possibility of undesired wavelengths corrupting the data. When using off-the-shelf filters, one has to be content with the available transmission curves. Custom filters are also available.

2.4.B. Interference Filters

Ten years ago, the selection of interference filters was somewhat limited. Interference filters typically transmit light at a particular wavelength, with a bandpass of 10–20 nm and a peak transmission efficiency of 20%. It is useful to have a variety of interference filters in any laboratory, particularly for wavelengths frequently used for excitation and emission. For intensity or anisotropy decay measurements, we often find it most convenient to isolate the desired wavelengths with an interference filter. The use of an interference filter rather than a bandpass filter for time-resolved measurements is valuable when the emission spectra display time-dependent wavelength shifts. An interference filter defines the emission more closely than a bandpass filter.

At present, a much wider variety of interference filters are available with improved optical properties. This is illustrated by the three filters shown in Figure 2.16. The filter on the top was designed for use with the 488-nm line of an argon-ion laser, and it transmits 80% at the desired wavelength. There is a companion filter which transmits

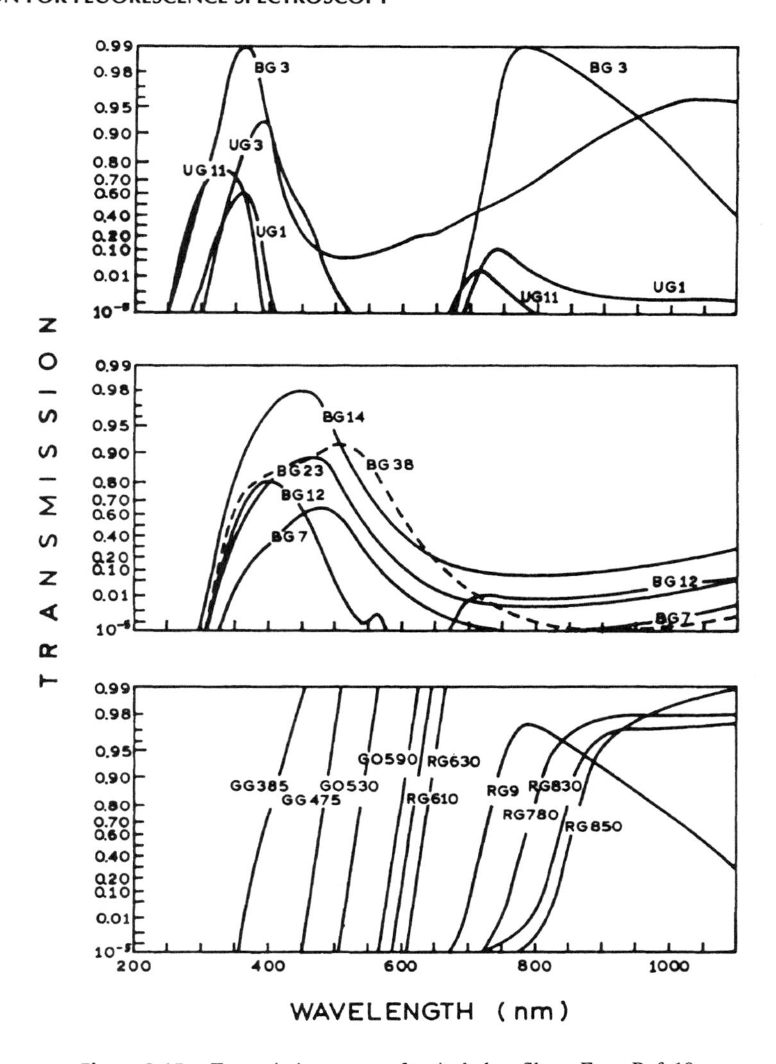

Figure 2.15. Transmission curves of typical glass filters. From Ref. 10.

from 510 to 560 nm, which was probably designed for use with fluorescein (Figure 2.16, middle). These two filters would allow high-sensitivity detection of fluorescein using an air-cooled argon-ion laser, without the need for additional monochromators. More complex transmission profiles are available for specialized applications. This is seen in the bottom panel of Figure 2.16, which shows the transmission curve of an interference filter with two transmission peaks. Interference filters can also be fabricated to serve as excellent short-pass and long-pass filters. Figure 2.17 shows filters which transmit 90% above or below their cut-off or cut-on wavelengths. Basically, with current technology for optical coatings, almost any desired transmission profile can be obtained. In practice, one relies on commonly available filters and uses optimized custom filters only for dedicated and/or high-volume applications.

It is also possible to design filters which reject light at a given wavelength and transmit longer and shorter wavelengths (Figure 2.18). These are known as notch filters, and can be produced by various means.[1,11] Such filters are useful for rejecting scattered light at the excitation wavelength and thus prevent the scattered light from reaching the detector or passing through an imperfect monochromator.

Another type of filter is a heat filter. Most light sources and especially arc lamps generate a good deal of infrared heat. Hence, a xenon lamp cannot be used with only an interference filter to provide monochromatic excitation. Heat filters are tempered glass filters which absorb the infrared heat and transmit the shorter visible and UV wavelengths.

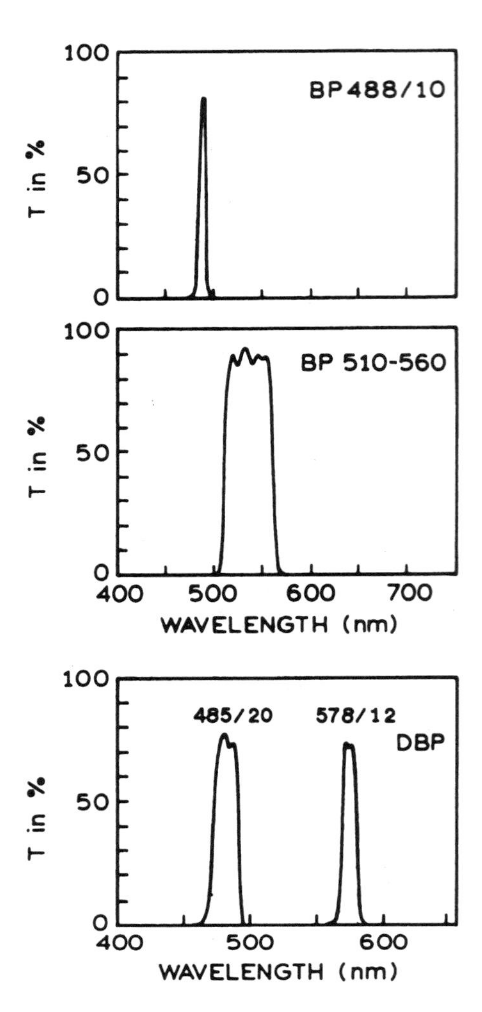

Figure 2.16. Transmission (T) curves of interference filters. Revised from Ref. 10.

2.4.C. Filter Combinations

While one can obtain almost any desired filter with modern coating technology, the design of custom filters for each experiment is usually not practical. If a single filter is not adequate for a given experiment, it is often possible to combine two or more bandpass filters to obtain the desired spectral properties, as illustrated in Figure 2.19. UG-11 and WG 320 filters are often used in this laboratory to isolate protein fluorescence.[12] For probes emitting near 450 nm, we often use a combination of Corning 4-96 and 3-72 filters.[13]. In the example shown in the lower panel of Figure 2.19, the filter combination was selected to reject 702 nm, which was the excitation wavelength for two-photon excitation of Indo-1.[13] This example illustrates an important aspect in selecting filters. Filters should be selected not only for their ability to transmit the desired wavelength, but perhaps more importantly for their ability to reject possible interfering wavelengths.

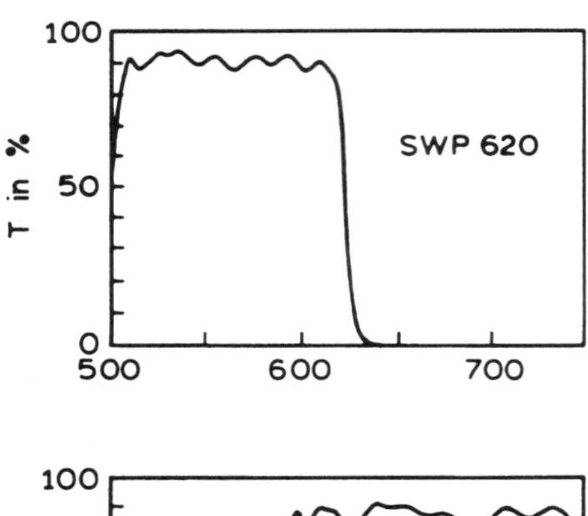

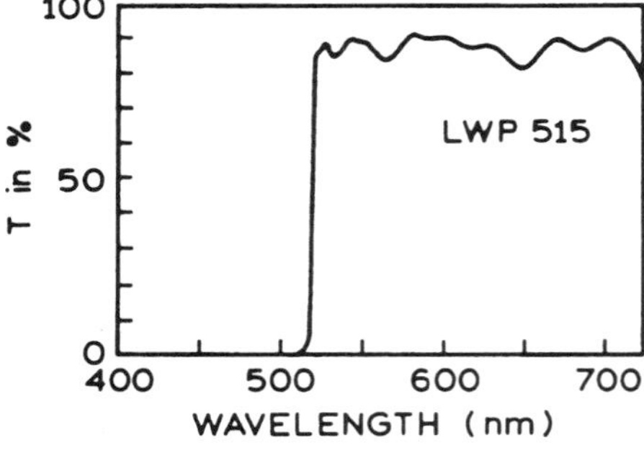

Figure 2.17. Transmission (T) curves of a 620-nm short-wavelength-pass filter and a 515-nm long-wavelength-pass filter. Revised from Ref. 10.

2.4.D. Neutral Density Filters

Neutral density filters are used to attenuate the light equally at all wavelengths. They are typically composed of sheets of glass or quartz coated with a metal to obtain the desired optical density. Quartz transmits in the UV and is preferred unless no work will be done using wavelengths below 360 nm. Neutral density filters are described by their optical density and can typically be obtained in increments

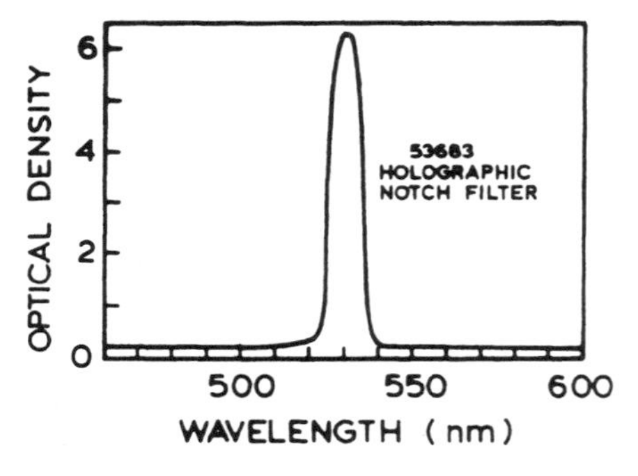

Figure 2.18. Optical density of a 52683 Holographic Notch Filter. Note that the y-axis is optical density. From Ref. 1.

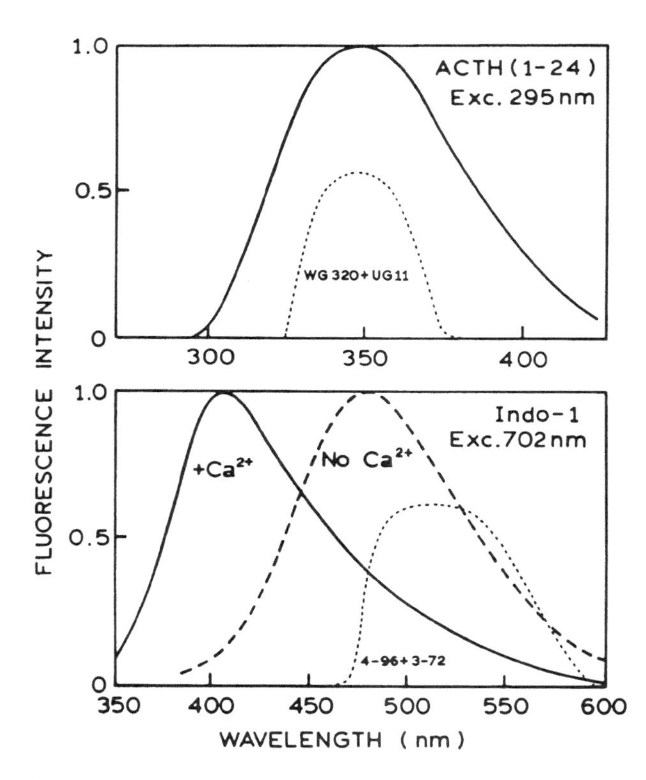

Figure 2.19. Transmission profiles of combinations of Corning and Schott filters used to isolate protein fluorescence (*top*) and Indo-1 fluorescence (*bottom*). Data from Refs. 12 and 13.

of 0.1, up to optical densities of 4. It is often necessary to adjust or match the intensity of two signals, which is conveniently accomplished using neutral density filters.

2.5. OPTICAL FILTERS AND SIGNAL PURITY

A major source of errors in all fluorescence measurements is interference due to scattered light, stray light, or sample impurities. This problem can be minimized by careful selection of the emission filter, by the use of optical filters in addition to the excitation and emission monochromators, and by control experiments designed to reveal the presence of unwanted components. The use of optical filters and control experiments to avoid such artifacts is best illustrated by specific examples.

The bottom panel in Figure 2.20 shows the emission spectrum of a dilute solution of tryptophan in aqueous buffer. The large sharp peak on the left is due to scattered excitation, the broad peak at 360 nm is the tryptophan fluorescence, and the small sharp peak at 310 nm is the Raman scatter. One should be aware of Raman scatter from the solvents. For water, this peak appears at 3600-cm^{-1} lower wavenumber than the exciting light. For excitation at 280 nm, the Raman peak from water occurs at 311 nm. For highly fluorescent samples, the emission spectrum generally overwhelms the Raman peak. However, if the gain of the instrument is increased to compensate for a dilute solution or a low quantum yield, the Raman scatter may become significant and distort the emission spectrum. Since Raman scatter always occurs at a constant wavenumber difference from the incident light, such scatter can be identified by changing the excitation wavelength. Also, the spectral width of the Raman peak will be determined by the resolution of the monochromators.

Frequently, one wishes to observe the fluorescence without regard for its wavelength distribution. Under these circumstances, one may remove the emission monochromator from the system and observe the emission through a filter that removes scattered light. This procedure can

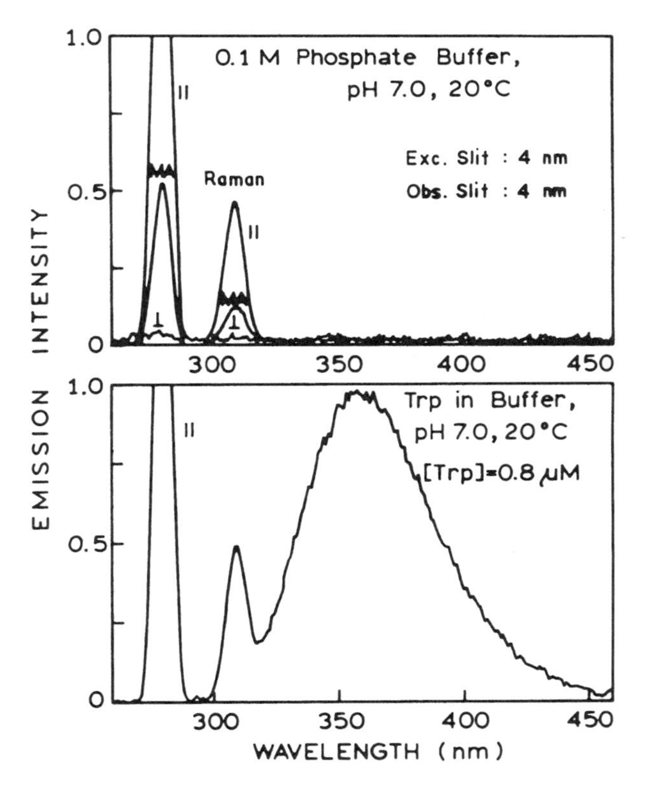

Figure 2.20. *Bottom*: Emission spectrum of a 0.8μ*M* solution of tryptophan in 0.1*M* phosphate buffer, pH 7.0. The observation polarizer was vertically oriented. *Top*: Emission spectrum of a blank buffer solution under the same optical conditions. The emission polarizer was oriented vertically (‖), horizontally (⊥), or at the 54.7° magic-angle (MA) position. From Ref. 8.

result in considerable increases in sensitivity because the bandpass of the observation is increased and the attenuation due to the monochromator is removed. The signal level can often be 50-fold higher when observed through filters rather than a monochromator. Also, it is common practice to perform lifetime measurements using long-pass filters alone to observe the entire emission. Under these conditions, it is important to choose an emission filter that eliminates both the scattered incident light and the Raman scatter.

In any fluorescence experiment, it is essential to examine blank samples, which are otherwise identical to the sample but do not contain the fluorophore. These control samples allow the presence of Rayleigh and Raman scatter to be assessed and controlled. In addition, such samples can reveal the presence of fluorescent impurities. An emission spectrum of the buffer blank for the dilute tryptophan solution is shown in the top panel of Figure 2.20. These control spectra are recorded under the same conditions used to measure the sample, because the only meaningful consideration is whether the blank contributes to the emission under the conditions of a given experiment. In this case the blank spectrum above 320 nm is essentially zero, showing that the peak at 360 nm (Figure 2.20, bottom) is in fact due to the sample. While this may seem like a trivial result, the presence of background fluorescence and/or scattered light is the most common error in fluorescence measurements. Examination of the blank sample also allowed identification of the peak at 310 nm as due to the buffer, and not to the sample.

The most appropriate blank solution is one which is identical to the sample but does not contain the fluorophore. This can be difficult to accomplish with protein or membrane solutions, where the macromolecules themselves are the source of the signal. Such solutions will typically be more strongly scattering than the buffer blanks. In these cases it is useful to add glycogen or colloidal silica (Ludox) to the buffer blank, to mimic the amount of scattering from the sample. This allows one to test whether the chosen filters are adequate to reject scattered light from the sample.

2.5.A. Emission Spectra Taken through Filters

Suppose one wishes to measure the lifetime of the tryptophan sample whose emission spectrum is shown in the bottom panel of Figure 2.20 (bottom), and the measurements will be performed using a bandpass filter to isolate the emission. How can one know that the scattered light has been rejected by the emission filter? This control is performed by collecting an emission spectrum through the filter used to measure the lifetime. In the example shown in the top panel of Figure 2.21, the Schott WG 320 filter rejected both the scattered light and the Raman scatter, as seen by the zero intensities from 280 to 310 nm. If one were concerned about emission from the filter, a similar experiment could be performed using the buffer blank or an equivalent scatterer. The use of an equivalent scattering solution is preferred, as this provides the most rigorous test of the optical arrangement.

It is important to remember that scattered light is usually 100% polarized ($P = r = 1.0$). This is the reason the emission polarizer was vertical in the example shown in the top panel of Figure 2.21, which is the worse-case situation. If the emission spectrum was recorded with the polarizer in the horizontal position, then the scattered light would be rejected by the polarizer (Figure 2.21, bottom). If the lifetimes are then measured with a vertical polarizer, scattered light may be detected. When examining spectra for the presence of scattered light, it is preferable to keep both polarizers in the vertical position and thereby maximize the probability of observing the interfering signal. Conversely, a horizontal emission polarizer can be used to

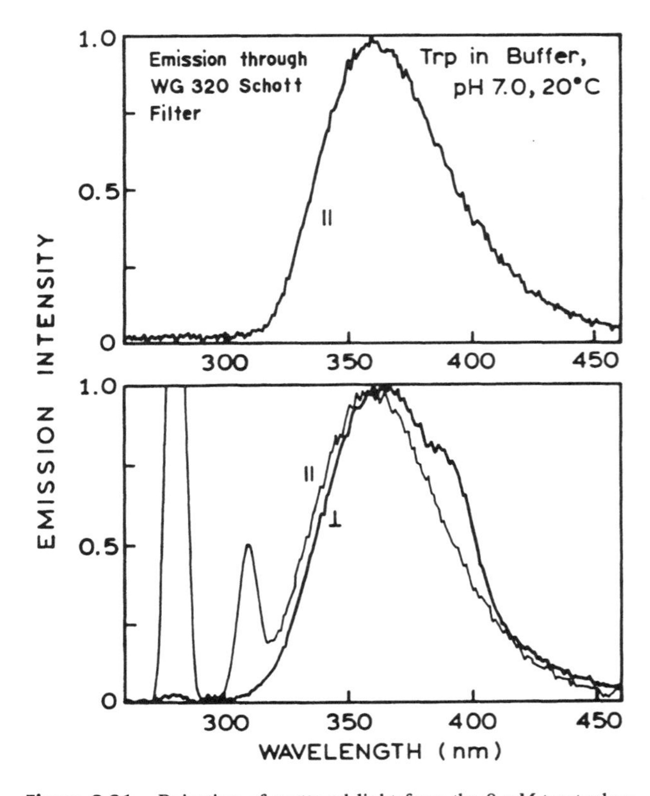

Figure 2.21. Rejection of scattered light from the 8 μM tryptophan solution in Figure 2.20 using a bandpass filter (*top*) or a polarizer but no bandpass filter (*bottom*). The emission polarizer was oriented vertically (‖) or horizontally (⊥). From Ref. 8.

minimize the scattered light if only the emission spectrum needs to be recorded.

Polarization or anisotropy measurements are frequently performed using filters rather than monochromators. Scattered light is 100% polarized (r = 1.0). Hence, a small percentage of scatter can result in serious errors. For example, assume 10% of the observed intensity is due to Raman scatter. Furthermore, assume that the actual anisotropy of a sample, in the absence of scatter, is 0.10. The observed anisotropy is given by

$$r_{\text{obs}} = f_s r_s + f_F r_F \qquad [2.1]$$

where the f_s value represents the fractional contribution of the scattered light, f_F is the fractional contribution of the fluorescence, r_F is the anisotropy of the fluorescence, and r_s is the anisotropy of the scattered light. Substitution into Eq. [2.1] yields r_{obs} = 0.19. Hence, a 10% contribution from scattered light can result in an almost twofold error in the measured anisotropy. The relative error would be still larger if r_F is smaller. These same considerations apply to both Raman and Rayleigh scattering.

The emission spectra in the bottom panel of Figure 2.21 also illustrate how the polarization-dependent transmission properties of the monochromator can distort the emission spectra. For these spectra the excitation was polarized vertically, and the emission spectra were recorded through a polarizer oriented vertically (||) or horizontally (⊥). These spectra are clearly distinct. The spectrum observed through the vertically oriented polarizer is blue-shifted relative to the spectrum observed when the emission polarizer is in the horizontal orientation. The extra shoulder observed at 390 nm is due only to the transmission properties of the monochromator. These results illustrate the need for comparing only those spectra that have been recorded under identical conditions, including the same orientation of the polarizers.

One way to avoid these difficulties is to use a defined orientation of the polarizers when recording emission spectra. One preferred method is to use the so-called "magic angle" conditions—vertically polarized excitation and an emission polarizer oriented 54.7° from the vertical. In effect, the use of these conditions results in a signal proportional to the total fluorescence intensity (I_T), which is given by $I_{\parallel} + 2I_{\perp}$, where $I_{\parallel}$ and $I_{\perp}$ are the intensities of vertically and horizontally polarized emission, respectively. Such precautions are generally taken only when necessary. If the excitation source is unpolarized, then the presence of polarizers in both the excitation and the emission light path results in an approximate fourfold decrease in the signal level. Also, the emission spectra are still distorted by the overall diffraction efficiency of the monochromator.

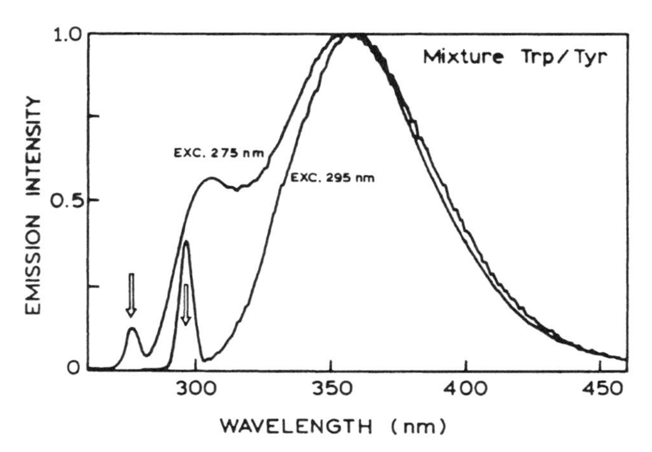

Figure 2.22. Emission spectra of tryptophan with a trace impurity of tyrosine, obtained with excitation at 275 and 295 nm. From Ref. 8.

When measuring tryptophan or protein emission, it is important to recognize that Raman scatter can be mistaken for tyrosine emission. This possibility is shown in Figure 2.22, which shows the emission spectrum of a tryptophan solution that contains a minor amount of tyrosine. Upon excitation at 275 nm, the tyrosine results in a peak near 300 nm. The fact that this peak is due to tyrosine is shown by the spectrum obtained for 295-nm excitation, which shows only the tryptophan emission. If the emission spectrum of tryptophan alone (Figure 2.20, bottom) was recorded at lower resolution, one can readily imagine how the broadened Raman line would become visually similar to tyrosine emission (Figure 2.22).

2.6. PHOTOMULTIPLIER TUBES

Almost all fluorometers use photomultiplier tubes (PMTs) as detectors, and one needs to understand their capabilities and limitations. A PMT is best regarded as a current source, the current being proportional to the light intensity. Although a PMT responds to individual photons, these individual pulses are generally detected as an average signal.

A PMT vacuum tube consists of a photocathode and a series of dynodes, which are the amplification stages (Figure 2.23). The photocathode is a thin film of metal on the inside of the window. Incident photons cause electrons to be ejected from this surface. The generation efficiency of photoelectrons is dependent upon the incident wavelength. The photocathode is held at a high negative potential, typically –1000 to –2000 V. The dynodes are also held at

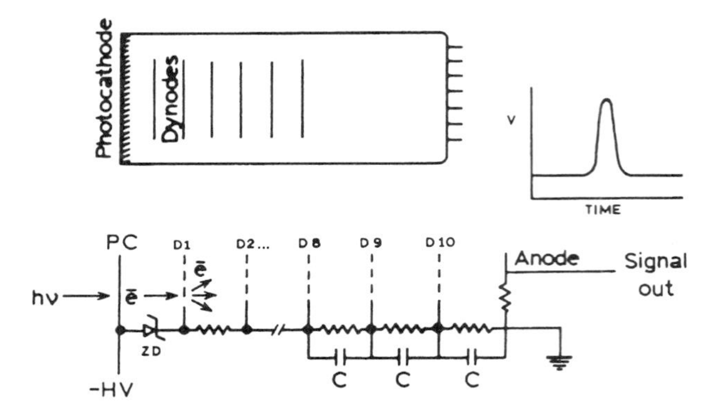

Figure 2.23. Schematic diagram of a PMT and its dynode chain.

negative potentials, but these potentials decrease toward zero along the dynode chain. The potential difference between the photocathode and the first dynode is generally fixed at a constant voltage by a Zener diode, at values ranging from –50 to –200 V. This potential difference cause an ejected photoelectron to be accelerated toward the first dynode. Upon collision with the first dynode, the photoelectron causes 5–20 additional electrons to be ejected, depending on the voltage difference to this dynode. This process continues down the dynode chain until a current pulse arrives at the anode. The size of this pulse depends upon the overall voltage applied to the PMT. Higher voltages result in an increased number of electrons ejected from each dynode, and hence higher amplification. The widespread usefulness of PMTs for low-level light detection is because they are low-noise amplifiers. Little additional noise is created as the electrons pass through the PMT. Amplification outside of the PMT generally results in more noise being added to the signal.

For quantitative measurements, the anode current must be proportional to the light intensity. A nonlinear response can result from an excessive current being drawn from the photocathode. Under high-intensity illumination, the electrical potential of the photocathode can be decreased because of its limited current-carrying capacity. This decreases the voltage difference between the photocathode and the first dynode, which also decreases the gain. In addition, excessive photocurrents can damage the light-sensitive photocathodes, resulting in loss of gain and excessive dark currents. The dark current from a PMT is the current in the absence of incident light.

A linear response also requires that the dynode voltages remain constant, irrespective of the incident light level and anode current. Dynode chains are designed so that the total current through the chain is at least 100-fold greater than the maximum anode current. Consider the 10-stage tube shown in Figure 2.23. Using 100-kΩ resistors, the dynode current would be 1.0 mA when the total voltage across the PMT is 1000 V. Hence, 10 μA should be the maximum anode current. Capacitors are often placed between the higher-numbered dynodes to provide a source of current during a single photoelectron pulse or periods of high illumination. Constant amplification by a PMT requires careful control of the high voltage. A typical PMT will yield a three-fold increase in gain for each 100 volts. Hence, a small change in voltage can result in a significant change in the signal. The high-voltage supply needs to provide a constant, ripple-free voltage that is stable for long periods of time.

PMTs are available in a wide variety of types. They can be classified in various ways, such as according to the design of the dynode chain, size and shape, spectral response, or temporal response.

2.6.A. Spectral Response

Prior to discussing the design features of PMTs, it is valuable to understand their spectral response at various wavelengths. The spectral response is determined by the type of transparent material used for the optically sensitive surface and by the chemical composition of the photocathode. Only light which enters the PMT can generate photocurrent. Hence, the input windows must be transparent to the desired wavelengths. The transmission curves of typical window materials are shown in Figure 2.24. UV-transmitting glass is probably the most commonly used material, transmitting all wavelengths above 200 nm. For detection deeper in the UV, one can choose synthetic quartz. MgF_2 windows are only selected for work in the vacuum UV. Since atmospheric oxygen absorbs strongly below 200 nm, there is little reason for selecting MgF_2

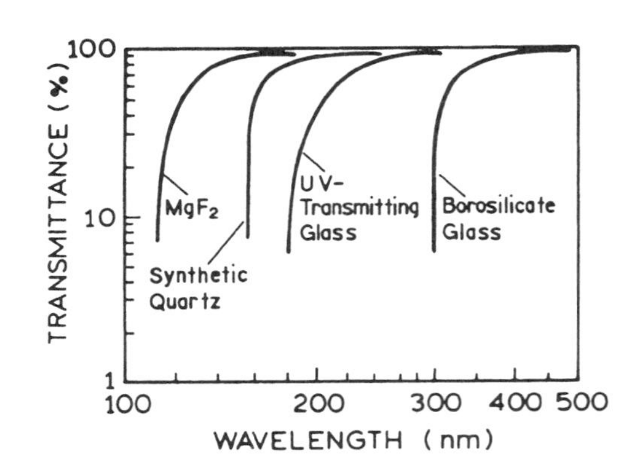

Figure 2.24. Typical transmittance of PMT window materials. Revised from Ref. 14.

unless the apparatus is used in an oxygen-free environment.

The second important factor is the material used for the photocathode. Numerous types of photocathodes are available, and the spectral response curves of just a few are shown in Figure 2.25. Most photocathode materials are sensitive at short wavelengths, so this region of the spectral response is determined by the type of window material. Also, most photocathodes are sensitive in the UV, blue, and green (300–500 nm) regions of the spectrum. The differences in photocathode material are important mostly for wavelengths above 600 nm. One of the most commonly used is the bialkali photocathode (Figure 2.25), which provides high sensitivity and low dark current. One disadvantage of the bialkali photocathode is the rapid decrease in sensitivity above 600 nm. Given the current emphasis on red and NIR fluorescence, the bialkali photocathode is becoming less useful.

The sensitivity above 500 nm has been increased by the introduction of multialkali and extended red multialkali photocathodes, which provide good sensitivity to 700 or 800 nm (Figure 2.25). Red-sensitive PMTs typically have higher dark current, but with the current generation of multialkali photocathodes the dark current is not a problem. Sensitivity to still longer wavelengths can be obtained using the Ag–O–Cs, or S-1, photocathode. However, its quantum efficiency is uniformly poor, and it is often difficult to detect the signal above the dark current with an S-1 PMT. In fact, these PMTs are rarely used without cooling to reduce the dark current.

Note that the quantum efficiency is not constant over any reasonable range of wavelengths. This fact, coupled with the wavelength-dependent efficiencies of monochromators, is the origin of the nonideal wavelength response of spectrofluorometers.

2.6.B. PMT Designs and Dynode Chains

The major types of PMTs and dynode chains used in fluorescence are shown in Figures 2.26 and 2.27, respectively. The most commonly used PMT is the side-window or side-on tube. A large number of variants are available, and all are descendants of one of the earliest PMTs, the 1P-28. These side-on tubes used a circular cage dynode chain, sometimes referred to as a squirrel cage (Figure 2.27). The specifications of one such side-on tube are listed in Table 2.2. Because of the multialkali photocathode, the spectral response is from 185 to 870 nm. This type of circular cage PMT has evolved into the subminiature PMTs (Figure 2.26), which have only recently become

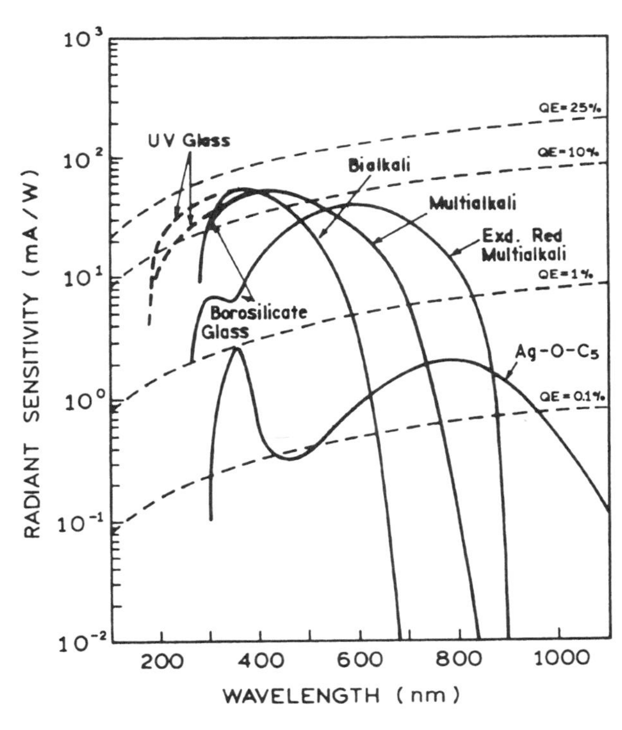

Figure 2.25. Spectral response curves of typical photocathodes. E, Quantum efficiency. From Ref. 14.

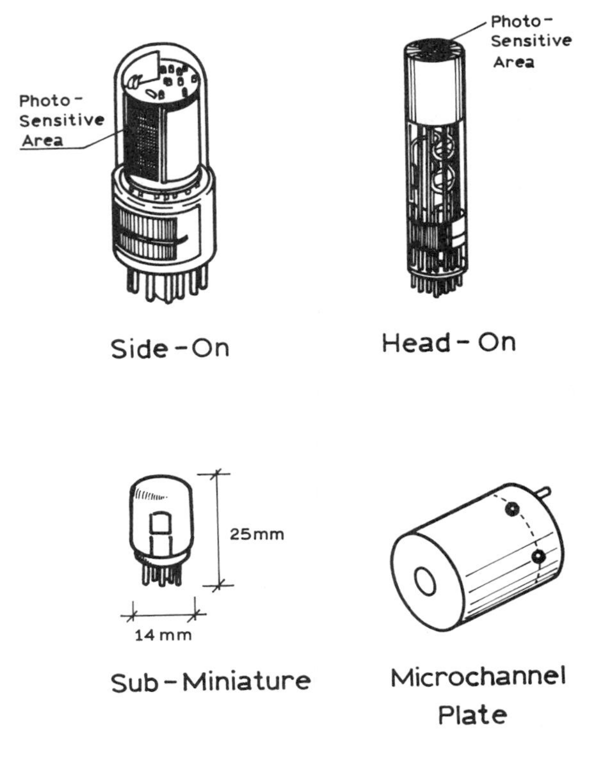

Figure 2.26. Types of PMTs. From Ref. 14.

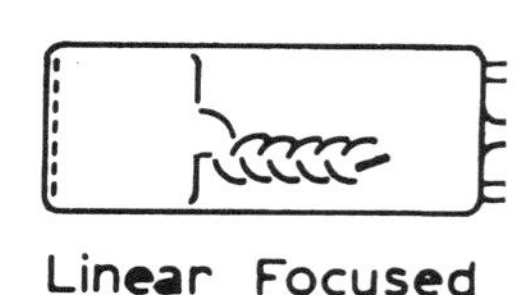

Figure 2.27. Types of PMT dynode chains. From Ref. 14.

available. Because of their compact design, the time response is excellent (Table 2.2). These small PMTs are available complete with a high-voltage supply and dynode chain, all in a compact package. These compact high-sensitivity detectors can be expected to appear in many research and clinical instruments.

Another type of PMT is the head-on design (Figure 2.26). This design is used with various types of dynode chains, such as the box and grid, blind, and mesh designs.[14] For time-resolved fluorescence, the head-on PMTs are most often used with a linear-focused dynode chain (Figure 2.27). The purpose of this design is to minimize the transit time spread and thus improve the time response of the PMT. The use of a head-on design allows the dynode chain to be extended as long as desired, so that the highest amplification is usually available with this type of PMT.

The final type of PMT is the microchannel plate (MCP) PMT. Examination of this PMT (Figure 2.26) does not reveal anything about its unique design. In place of a dynode chain, the MCP PMT has plates which contain numerous small holes (Figure 2.28). The holes in these plates are the microchannels, which are lined with the secondary emissive dynode material. The electrons are amplified as they drop down the voltage gradient across the microchannel plate. Because of the short distances for electron travel, and the restricted range of electron paths, this type of PMT shows the fastest time response and is used in the most demanding time-resolved measurements. MCP PMTs are available with one, two, or three stages of microchannel plates. The amplification is generally lower than for PMTs with discrete dynode chains. Also, the maximum photocurrent is typically 100 nA, as compared with 10–100 μA for a dynode PMT.

2.6.C. Time Response of Photomultiplier Tubes

For steady-state measurements, the time response of a PMT is not important. However, some understanding of the time response of PMTs is essential for an understanding of their use in lifetime measurements. There are three main timing characteristics of PMTs: the transit time, the rise time, and the transit time spread (Figure 2.29). The transit time of a PMT is the time interval between the arrival of a photon at the cathode and the arrival of the amplified pulse at the anode. Typical transit times are near

Table 2.2. Characteristics of Typical Photomultiplier Tubes[a]

	R446	R2560	R3809	R3811
Type	Side window	Head-on	Head-on	Subminiature
Dynode chain	Circular cage	Linear focused	Microchannel Plate	Circular cage
Photocathode	Multialkali	Bialkali	Multialkali S-20	Multialkali
Wavelength range (nm)	185–870	300–650	160–185	185–850
Amplification	5×10^6	6×10^6	2×10^5	1.3×10^6
Rise time (ns)	2.2	2.2	0.15	1.4
Transit time (ns)	22	26	0.55	15
Bandwidth (MHz) (estimate)	200	200	2000	300

[a] The numbers refer to types provided by Hamamatsu, Inc.[14]

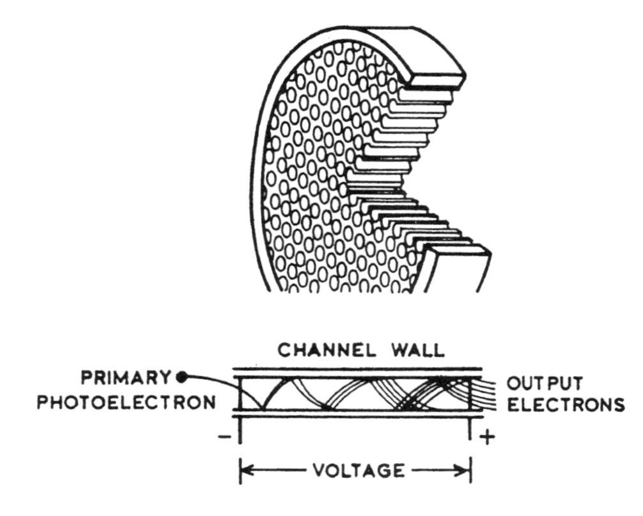

Figure 2.28. Electron amplification in an MCP PMT.

20–50 ns (Figure 2.29). The rise time is the time required for the PMT anode signal to rise from 10% to 90% of its final level. The rise time is determined primarily by the transit time variation in the PMT, that is, the scatter around the average transit time.

The transit time spread is perhaps the most important determinant of the time resolution available with a given

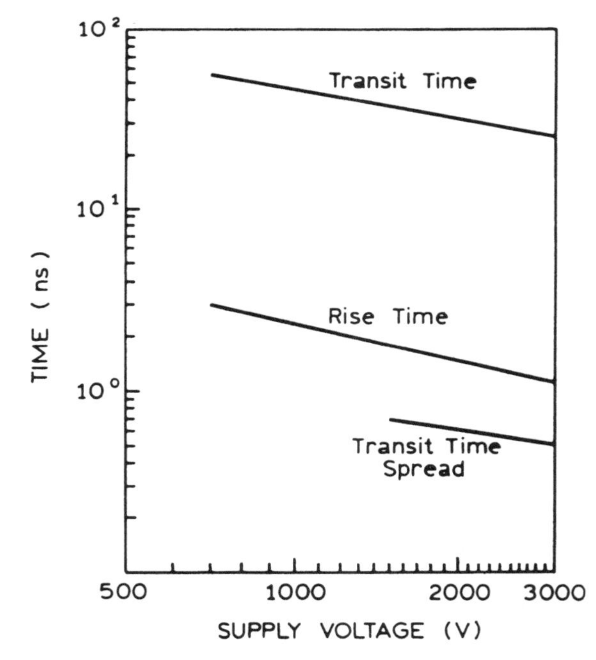

Figure 2.29. Time response of a typical PMT. The data are for an R2059 head-on, 12-stage linear-focused PMT. Revised from Ref. 14.

PMT. These timing variations result from the different geometric paths that the electrons can take from the photocathode to the anode. The photoelectrons can originate from different parts of the photocathode or can have different trajectories from the same region of the photocathode. The electrons subsequently ejected from the dynodes can take slightly different geometric paths through the PMT. This can be seen in Figure 2.23 from the various ejection angles of electrons coming off the first dynode. Transit time spread can be decreased by using photocathode and dynode geometries which minimize the number of different trajectories. This can be accomplished by the use of small illuminated areas or a dynode designed to direct the flight of the electrons along a defined trajectory. Such procedures are generally used where the time response of the PMT is important, such as in the measurement of fluorescence lifetimes.

The most dramatic advance in high-speed PMTs has been the introduction of the MCP PMT. In this case the photoelectrons are proximity-focused into the MCP (Figure 2.27). Once in the MCP, there is very little variation in the electron trajectory. For this reason, MCP PMTs have transit time spreads 10-fold smaller than those of standard PMTs (Table 2.2).

A second source of the time dependence of a PMT is the photocathode itself. Typically, the time response of a PMT is dependent upon the wavelength incident on the photocathode. This property is called the color effect. Apparently, the energy of the ejected electrons is dependent upon the incident wavelength. Color effects are not important for steady-state measurements. However, the color effect can be a significant source of error in the measurement of fluorescence lifetimes (Chapter 4).

2.6.D. Photon Counting versus Analog Detection of Fluorescence

A PMT is capable of detecting individual photons. Each photoelectron results in a burst of 10^5–10^6 electrons, which can be detected as individual pulses at the anode (Figure 2.30). Hence, PMTs can either be operated as photon counters or be used in the analog mode, in which the average photocurrent is measured. Note that we are considering steady-state measurements. Time-correlated photon counting for lifetime measurements will be discussed in Chapter 4.

The use of a PMT in the photon counting mode is shown in Figure 2.30. In the photon counting mode, the individual anode pulses due to each photon are detected and counted. As a result, the detection system is operating at the theoretical limits of sensitivity. Noise or dark current in the PMT frequently results in electrons which do not originate at the photocathode, but from further down the dynode

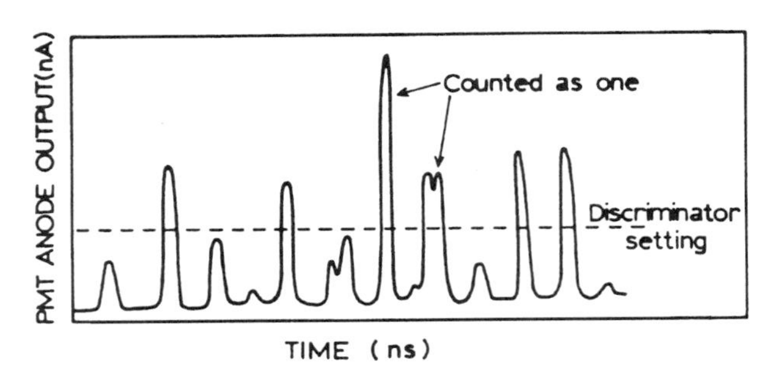

Figure 2.30. Photon counting detection using a PMT. Provided by Dr. R. B. Thompson.

chain. These electron pulses are smaller and can be ignored by setting the detection threshold high enough to count only fully amplified primary photoelectrons. Besides increased sensitivity, the stability of the detection system can be increased. This is a result of the PMT being operated at a constant high voltage. Small drifts in this voltage do not result in significant changes in the efficiency with which each photon is counted. For these reasons, photon counting detection is frequently used when signal levels are low, and when it is necessary to average repetitive wavelength scans to increase the signal-to-noise ratio.

Unfortunately, there are disadvantages of photon counting. The gain of the PMT cannot be varied by changing the amplified voltage. Another disadvantage of photon counting is the limited range of intensity over which the count rate is linear. If two pulses arrive at the anode closely spaced in time, they will be counted as a single pulse. Anode pulses resulting from a single photon are typically 5 ns wide. This limits the frequency response of the PMT to 200 MHz, or 2×10^8 Hz, for a periodic signal. For random events, the count rates need to be about 100-fold less to avoid the simultaneous arrival of two photons (Figure 2.30). Hence, count rates are limited to 2 MHz or less, even in the latest-generation instruments.[15] In practice, the count rates often become sublinear before the theoretical upper limit is reached (Figure 2.31). Higher count rates can be obtained with PMTs which show shorter pulse widths, assuming that the electronic counters are up to the task. Photon counting detection can be inconvenient when signal levels are high. To stay within the linear range, one must adjust the slit widths or the fluorescence intensities (using neutral density filters). Additionally, the signal-to-noise ratio becomes unsatisfactory at count rates below 10,000 photons per second. The linear dynamic range can be as small as 3 log units.[15] This limited intensity range is a drawback of photon counting detection for steady-state measurements of fluorescence, unless the highest sensitivity is required.

In the analog mode the individual pulses are averaged. Since the current from each pulse contributes to the average anode current, the simultaneous arrival of pulses is not a problem. When analog detection is used, the gain of the detection system can be varied by changing either the amplifier gain or the voltage on the PMT. As a result, a wider range of signal levels can be detected without concerns about a nonlinear response. In addition, the precision of the individual measurements seems higher than for the photon counting measurements, presumably because of the higher overall signal levels. However, even the analog measurements have a limited range because all PMTs display saturation above a certain light level if one exceeds the capacity of the photocathode to carry the photocurrent or the ability of the dynode chain to maintain a constant voltage (Figure 2.31).

2.6.E. Symptoms of PMT Failure

PMTs should be handled with care. Their outer surfaces should be free of dust and fingerprints, and these surfaces should not be touched with bare hands. The photocathode is light-sensitive, and it is best to perform all manipulations in dim light. It is convenient to know the common signs of PMT failure. One may observe pulses of current when the applied voltage is high. At lower voltages, this symptom may appear as signal instability. For instance, the gain may change 20% to severalfold over a period of 2–20 seconds. The origin of this behavior is frequently, but not always, a leakage of gas into the tube. The tube cannot be fixed and replacement is necessary. In some instances the tube may perform satisfactorily at lower voltages.

A second symptom is high dark current. This appears as an excessive amount of signal when no light is incident on the PMT. The origin of high dark currents is usually excessive exposure of the tube to light. Such exposure is especially damaging if voltage is applied to the tube at the same time. Again, there is no remedy except replacement or the use of lower voltages.

Signal levels can be unstable for reasons other than a failure of the PMT. If unstable signals are observed, one should determine that there are no light leaks in the instrument and that the high-voltage supplies and amplifiers are functioning properly. In addition, the pins and socket connections of the PMT should be checked, cleaned, and tightened, as necessary. Over a period of years, oxide accumulation can result in decreased electrical contact. Furthermore, photobleaching of a sample may give the appearance of an instrument malfunction. For example, the fluorescence intensity may show a time-dependent decrease due to bleaching and then an increase in intensity

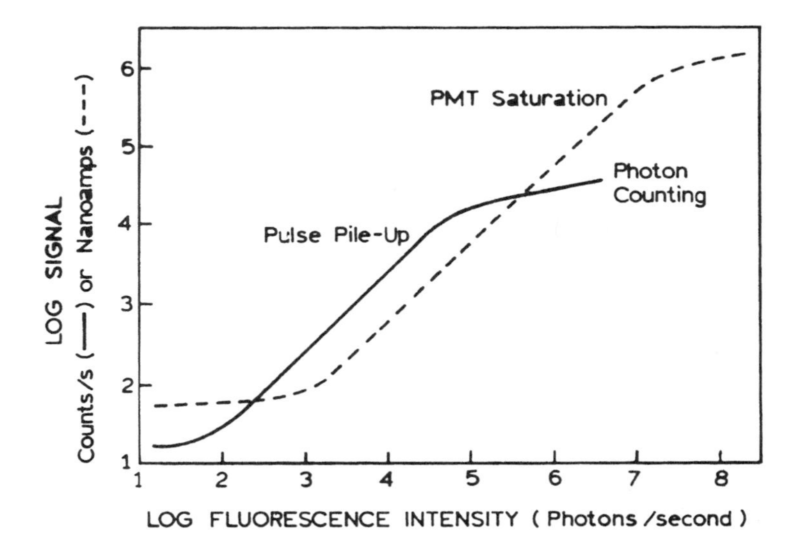

Figure 2.31. Dynamic range available with photon counting and analog detection. Provided by Dr. R. B. Thompson.

due to convection currents which replenish the bleached portion of the sample.

2.6.F. Hybrid Photomultiplier Tubes

For purposes of ruggedness and simplicity, it is desirable to replace PMTs by solid-state detectors. Photodiodes provide a high-speed response and wide spectral sensitivity. Avalanche photodiodes even provide an internal gain of about 100-fold. However, the lack of high amplification and the small active area limit their usefulness for high-sensitivity detection.

A new type of detector is beginning to appear, the hybrid photomultiplier tube (HPMT). These devices have a photocathode like a standard PMT. However, the photoelectrons are amplified in a solid-state structure by an avalanche process, resulting in an electrical pulse which can be further amplified for processing. The gain of the HPMTs is about 2000, with a wide linear response range. HPMTs should become useful for single-photon counting applications.

2.6.G. CCD Detectors

We have focused on PMTs as detectors, but we note here the growing usefulness of charge-coupled devices (CCDs) in fluorescence spectroscopy. CCDs are imaging detectors with remarkable sensitivity and linear dynamic range. CCDs typically contain about 500,000 pixels, but versions with more than 10^6 pixels are available. Each pixel acts as an accumulating detector; that is, charge accumulates in proportion to the total light exposure. The charge at each pixel point can be read out when desired, to obtain a two-dimensional image. CCDs are already in widespread use in fluorescence microscopy.[16,17]

In the future we can expect increased use of CCDs in fluorescence spectroscopy. The charge or signal from rows of pixels can be collected as binned, to improve the signal-to-noise ratio as a two-dimensional detector.[18,19] The combined use of a CCD detector and a concave diffraction grating can result in a simple spectrofluorometer. In the example shown in Figure 2.32, the exit slit of the monochromator was removed. The light of different wavelengths falls on the CCD at different positions, creating a fluorescence spectrograph. Using the instrument depicted in Figure 2.32, it was possible to record an emission spectrum of $10^{-9}M$ anthracene in 10 seconds at high signal-to-noise ratio.[19] CCDs also possess good sensitivity out to 1000 nm. With present CCD technology, the sensitivity can be severalfold better than that of a PMT under low-light conditions.[20] The dark counts in a CCD can be very low. With cooling to –90°C, the dark count can be less than one electron per pixel per day. While CCDs can be useful as steady-state detectors, their time response is too slow for time-resolved measurements.

2.7. POLARIZERS

Prior to discussing polarizers, it is useful to summarize a few conventional definitions. The laboratory z-axis is typi-

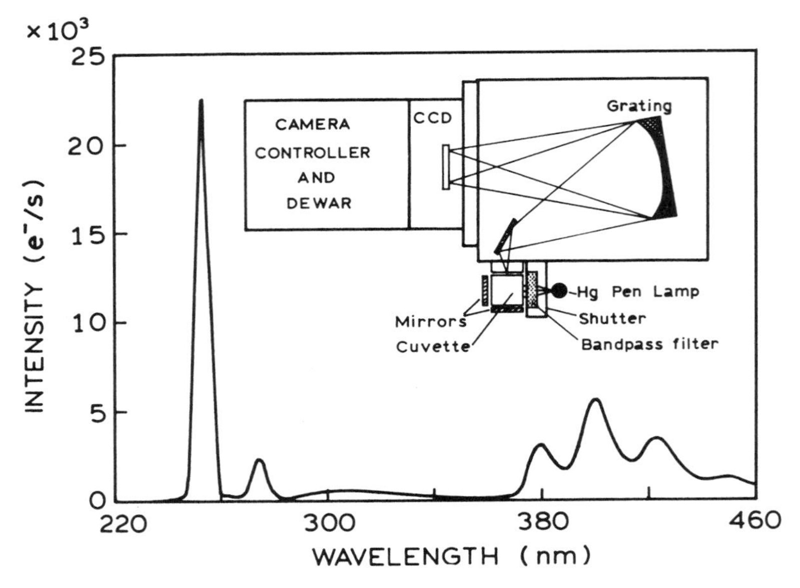

Figure 2.32. Emission spectrum of $10^{-9}M$ anthracene in ethanol using a CCD spectrograph detector. The peaks at 254 and 274 nm are the elastic and Raman scatter, respectively. Reprinted, with permission, from Ref. 20, Copyright © 1989 American Chemical Society.

cally referred to as the vertical axis. Light can be described as having a direction for its electrical component. Unpolarized light, of the type from incandescent or arc lamp sources, has equal amplitudes of the electric vector normal to the direction of light propagation. Polarized light has greater amplitude in one of the directions. Light with its electric vector directed along the z-axis is said to be vertically polarized (Figure 2.33). Light with its electric vector at right angles to the z-axis is said to be horizontally polarized.

In the discussion of polarization, the terms "S" and "P" polarization are often used. These terms are defined relative to the normal to the plane of incidence of the light on the optical interface. The plane of incidence is the plane defined by the light ray and the axis normal to the surface. If the electric vector is in the plane of incidence, the ray is said to be "P" polarized (Figure 2.34). If the electric vector is perpendicular to the plane of incidence, this ray is said to be "S" polarized.

Polarizers transmit light whose electric vector is aligned with the polarization axis and block light which is rotated 90°. These principles are shown in Figure 2.35. The incident light is unpolarized. About 50% of the light is transmitted by the first polarizer, and this beam is polarized along the principal axis of the polarizer. If the second polarizer is oriented in the same direction, then all the light

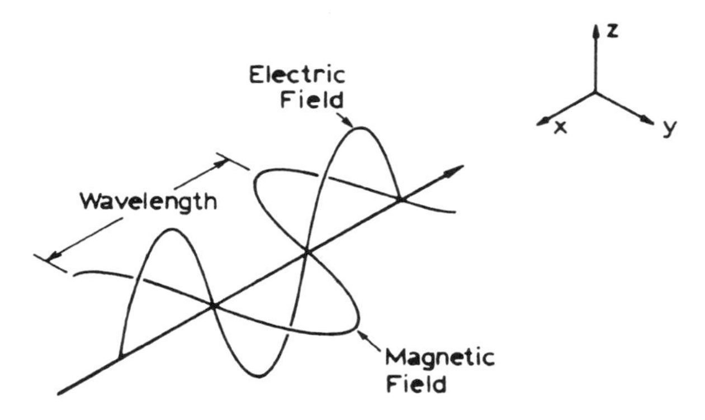

Figure 2.33. Vertically polarized light.

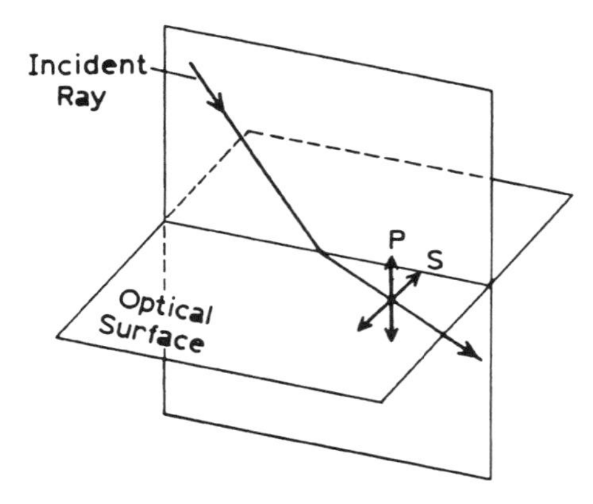

Figure 2.34. Definition of "S" and "P" polarization. P-polarized light has the electric field polarized parallel to the plane of incidence. S-polarized light has the electric field polarized perpendicular to the plane of incidence.

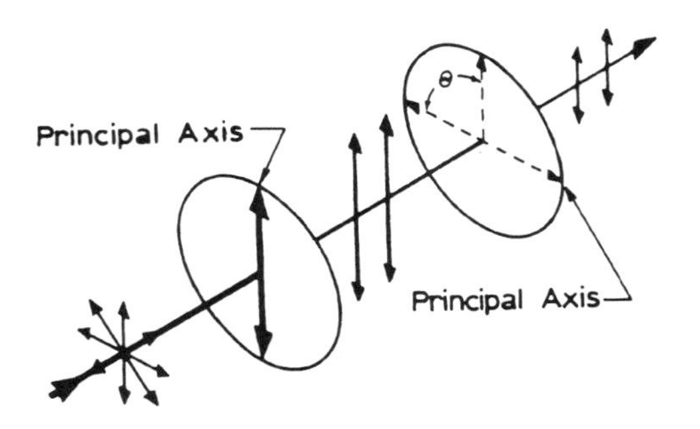

Figure 2.35. Transmission of light through polarizers. The incident beam is unpolarized. Only vertically polarized light passes through the first polarizer. Transmission through the second polarizer is proportional to $\cos^2 \theta$.

is transmitted except for reflection losses. If the second polarizer is rotated through an angle θ, the intensity is given by

$$I = I_{max} \cos^2 \theta \quad [2.2]$$

where I_{max} corresponds to $\theta = 0$. Polarizers are frequently characterized by their extinction ratios. If the first polarizer is illuminated with linearly polarized light along the principal axis, the extinction ratio is the ratio of intensities for parallel ($\theta = 0°$) and crossed polarizers ($\theta = 90°$). A slightly different definition is used if the first polarizer is illuminated with unpolarized light. Extinction ratios range from 10^3 to 10^6.

For general use in fluorescence spectroscopy, a UV-transmitting Glan–Thompson polarizer has the best all-around properties (Figure 2.36). This type of polarizer consists of a calcite prism, which is birefringent, meaning that the refractive index is different along each optical axis of the crystal. The angle of the crystal is cut so that one polarized component undergoes total internal reflection at the interface, and the other continues along its optical path. The reflected beam is absorbed by the black material surrounding the calcite. The purpose of the second prism is to ensure that the desired beam exits the polarizer in the same direction as the entering beam. For high-power laser applications, an exit port is provided to allow the reflected beam to escape.

Calcite transmits well into the UV, and the transmission properties of the polarizer are determined by the material between the two prisms, which can be air or some type of cement. An air space is usually used for UV transmission. The polarizers are typically mounted in 1-in.-diameter cylinders for ease of handling. Glan–Thompson polarizers provide high extinction coefficients—near 10^6—but that is not the reason why they are used in fluorescence. Glan–

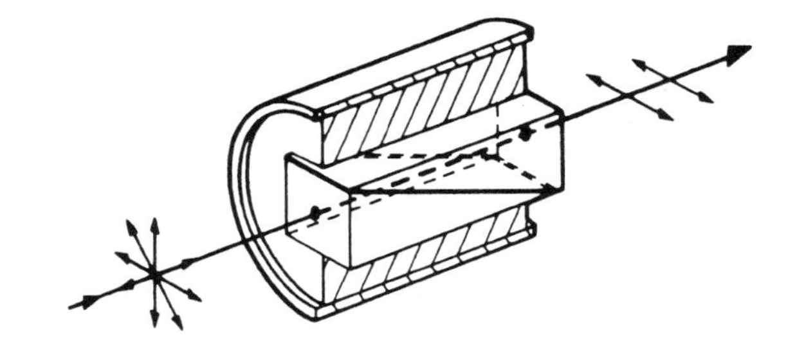

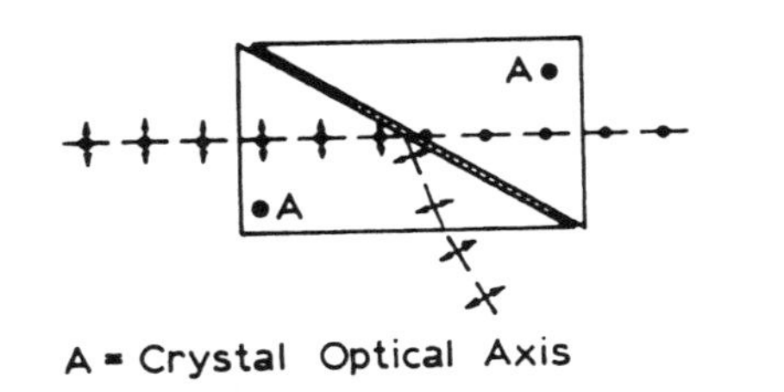

Figure 2.36. Glan–Thompson polarizer (*top*) and light paths (*bottom*).

Thompson polarizers have a high acceptance angle, 10–15°, allowing them to be used where the beams are not well collimated. Another advantage is the low UV and visible absorbance of calcite, providing high transmission efficiency.

Another type of polarizer used in fluorescence is film polarizers, the same type as used in Polaroid glasses. These are thin films of a stretched polymer which transmit the light polarized in one direction and absorb the light polarized in another direction (Figure 2.35). Because the light is absorbed, polarizers of this type are easily damaged by intense laser beams. They have a wide acceptance angle, but overall transmission is poor, especially in the UV.

A wide variety of other polarizers are available. Most of them, such as Wollaston polarizers and Rochon prism polarizers, split the unpolarized light into two beams which must then be spatially selected.

2.8. CORRECTED EXCITATION SPECTRA

The development of methods to obtain excitation and emission spectra corrected for wavelength-dependent effects has been the subject of numerous investigations. Overall, none of these methods is completely satisfactory, especially if the corrected spectra are needed on a regular basis. Prior to correcting spectra, the researcher should determine if such corrections are necessary. Frequently, one only needs to compare emission spectra with other spectra collected on the same instrument. Such comparisons are usually made between the technical (or uncorrected) spectra. Furthermore, the responses of many spectrofluorometers are similar because of the similar components, and comparison with spectra in the literature can frequently be made. Of course, the spectral distributions and emission maximum will differ slightly, but rigorous overlap of spectra obtained in different laboratories is rarely a necessity. And, finally, modern instruments with red-sensitive PMTs, with gratings comparable to that shown in Figure 2.11, can provide spectra which are not very distorted, particularly in the visible to red region of the spectrum. However, corrected spectra are needed for calculation of quantum yields and overlap integrals (Chapter 13), and we will briefly describe the methods we judge to be most useful.

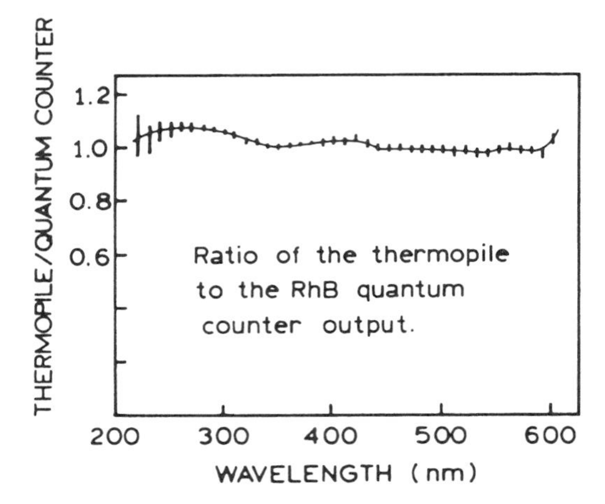

Figure 2.37. Comparison of the thermopile and the rhodamine B (RhB) quantum counter as radiation detectors. Redrawn from Ref. 21.

2.8.A. Use of a Quantum Counter to Obtain Corrected Excitation Spectra

Excitation spectra are distorted primarily by the wavelength dependence of the intensity of the exciting light. This intensity can be converted to a signal proportional to the number of incident photons by the use of a quantum counter. Rhodamine B (RhB) in ethylene glycol (3 g/l) is the best known quantum counter,[21] and to this day, remains the most generally reliable and convenient quantum counter. This concentrated solution absorbs virtually all incident light from 220 to 600 nm. The quantum yield and emission maximum (~630 nm) of rhodamine B are essentially independent of excitation wavelength from 220 to 600 nm.

The principle of a quantum counter is illustrated in Figure 2.37, which shows the ratio of the intensities observed from the RhB quantum counter and a thermopile. It is seen that the ratio remains constant with varying wavelength. Since the emission spectrum of rhodamine B is independent of excitation wavelength, the quantum counter circumvents the wavelength-dependent sensitivity of the reference phototube. Hence, this solution provides a signal of constant emission wavelength, the intensity of which is proportional to the photon flux of the exciting light. Quantum counters can also be made using moderately concentrated solutions of quinine sulfate or fluorescein. Quinine sulfate (4 g/l in $1N$ H_2SO_4) is useful at excitation wavelengths ranging from 220 to 340 nm, and fluorescein (2 g/l in $0.1N$ NaOH) is useful over this same range but is less reliable from 340 to 360 nm[21] where absorption of fluorescein is weaker.

To record corrected excitation spectra, the quantum counter is placed in the reference channel of the spectrofluorometer (Figure 2.1). However, because of the high optical density, the reference cell holder is modified so that the emission is observed from the same surface of the quantum counter as is being illuminated. Alternatively, quantum counters can be used in a transmission mode by observing the fluorescent light exiting the back surface of the illuminated cuvette.[22] In either case, an optical filter is placed between the quantum counter and the PMT which eliminates scattered light but which transmits the fluorescence. With a quantum counter in place, a corrected excitation spectrum may be obtained by scanning the excitation monochromator and measuring the ratio of the fluorescence intensity from the sample to that from the quantum counter. The wavelength-dependent responses of

Table 2.3. Quantum Counters

Solution	Range (nm)	Flatness	Reference
3 g/l rhodamine B in ethylene glycol	220–580	±5%	21
8 g/l rhodamine B in ethylene glycol[a]	250–600	±4%	22
2 g/l fluorescein in $0.1N$ NaOH	240–400[b]	±5%	21
4 g/l quinine sulfate in $1N$ H_2SO_4	220–340	±5%	21
Rhodamine in PVA films[c]	360–600	±3%	23[d]
Coumarins in PVA films	360–480	±3%	23[d]
5 g/l $Ru(bpy)_3^{2+}$ in methanol	360–540	1.1%	24
$Ru(bpy)_3^{+}$ in PVA films	360–530	1%	24[e]
8 g/l HITC[f] in acetonitrile	320–800[g]	±10%	25

[a] In Ref. 22, a higher concentration of rhodamine B is claimed to be preferred for use in the transmission mode.
[b] Response may be 15% low in the range of 340–360 nm.
[c] PVA, Poly (vinyl alcohol).
[d] See Ref. 23 for details on the rhodamines, coumarins, and PVA film preparations.
[e] See Ref. 24 for details.
[f] HITC, 1, 1′, 3, 3, 3′, 3′-Hexamethylindotricarbocyanine.
[g] Deviation up to 20% occurs near 470 nm.

the emission monochromator and phototube are not important because the emission wavelength is unchanged during a scan of the excitation wavelength. This procedure was used to record the corrected excitation spectrum of fluorescein shown in Figure 2.4. In recording such corrected excitation spectra, one needs to determine that the detection system and electronics can record the true ratio over a wide range of signal intensities.

Other quantum counters have been described and are summarized in Table 2.3. The long-wavelength dye HITC extends the range to 800 nm, but its response is not as flat as that of RhB. Unfortunately, there is no perfect quantum counter, and for most applications RhB appears to be the best choice.

2.9. CORRECTED EMISSION SPECTRA

2.9.A. Comparison with Known Emission Spectra

To calculate corrected emission spectra, one needs to know the wavelength-dependent efficiency of the detection system. It is difficult and time-consuming to measure the correction factors directly for any given spectrofluorometer. Even after careful corrections are made, the results are only accurate to ± 10%. For this reason, one generally reports the observed technical spectra. If corrected spectra are necessary, one simple and reliable method of obtaining the necessary correction factors is to compare the observed emission spectrum of a standard substance with the known corrected spectrum for this same substance. Such spectra have been published for a variety of readily available fluorophores including quinine sulfate, β-naphthol, 3-aminophthalimide, 4-dimethylamino-4′-nitrostilbene, and *N,N*-dimethylamino-*m*-nitrobenzene.[26–31] The emission wavelengths of these compounds cover the range from 300 to 800 nm, and the data are presented in graphical and numerical form. More recently, corrected spectra have been published for a series of harmine derivatives, covering the range from 400 to 600 nm.[32] For convenience, some of these corrected spectra are presented in Appendix I.

To obtain correction factors, one records the emission spectrum of a standard compound and compares these data with the standard spectrum. By this simple comparative method, one avoids the difficulties inherent in the more rigorous procedures described below. β-Naphthol should probably not be used as a standard because, under the conditions described, both naphthol and naphtholate emission are observed. The dual emission is a result of an excited-state reaction, the extent of which is difficult to control. The use of quinine sulfate as a standard has been questioned, because its intensity decay may not be a single exponential and its quantum yield may be somewhat dependent on excitation wavelength.[33] However, in spite of these concerns, it seems to be an acceptable standard for most circumstances. One should remember that quinine sulfate is collisionally quenched by chloride,[34] so solutions of quinine used as a quantum yield standard should not contain chloride. A potentially superior standard is β-carboline, whose spectral characteristics are similar to those of quinine sulfate and which displays a single-exponential decay time.[35] The emission spectra of quinine sulfate and β-carboline are similar.

2.9.B. Correction Factors Obtained by Using a Standard Lamp

The correction factors can also be obtained by observing the wavelength-dependent output from a calibrated light source. The wavelength distribution of the light from a tungsten filament lamp can be approximated by that of a blackbody of equivalent temperature. Standard lamps of known color temperature are available from the National Bureau of Standards and other secondary sources. Generally, one uses the spectral output data provided with the lamp [$L(\lambda)$] because the blackbody equation is not strictly valid for a tungsten lamp. The detection system is then calibrated as follows:

1. The intensity of the standard lamp versus wavelength, $I(\lambda)$, is measured using the detection system of the spectrofluorometer.
2. The sensitivity of the detection system, $S(\lambda)$, is calculated using

$$S(\lambda) = \frac{I(\lambda)}{L(\lambda)} \quad [2.3]$$

   where $L(\lambda)$ is the known output of the lamp.
3. The corrected spectra are then obtained by dividing the measured spectra by these sensitivity factors.

It is important to recognize that the operation of a standard lamp requires precise control of the color temperature. In addition, the spectral output of the lamp can vary with the age and usage of the lamp.

2.9.C. Correction Factors Obtained by Using a Quantum Counter and Scatterer

Another method to obtain the correction factors for the emission monochromator and PMT is to calibrate the xenon lamp in the spectrofluorometer for its spectral output.[21] The relative photon output [$L(\lambda)$] can be obtained by placing a quantum counter in the sample compartment. Once this intensity distribution is known, the xenon lamp output is directed onto the detector using a magnesium oxide scatterer. MgO is assumed to scatter all wavelengths with equal efficiency. Correction factors are obtained as follows:

1. The excitation wavelength is scanned with the quantum counter in the sample holder. The output yields the lamp output $L(\lambda)$.
2. The scatterer is placed in the sample compartment, and the excitation and emission monochromators are scanned in unison. This procedure yields the product $L(\lambda) \cdot S(\lambda)$, where $S(\lambda)$ is the sensitivity of the detector system.
3. Division of $S(\lambda) \cdot L(\lambda)$ by $L(\lambda)$ yields the sensitivity factors $S(\lambda)$.

The most critical aspect of this procedure is obtaining a reliable scatterer. The MgO must be freshly prepared and be free of impurities. Although this procedure seems simple, it is difficult to obtain reliable correction factors. It is known that the reflectivity of MgO changes with time and with exposure to UV light, particularly below 400 nm.[36] In addition to changes in reflectivity, it seems probable that the angular distribution of the scattered light and/or collection efficiency changes with wavelength, and one should probably use an integrating sphere at the sample location to avoid spatial effects. Given the complications and difficulties of this procedure, the use of standard spectra is the preferred method when corrected emission spectra are needed.

2.9.D. Conversion between Wavelength and Wavenumber

Occasionally, it is preferable to present spectra on the wavenumber scale ($\bar{\nu}$) rather than on the wavelength scale (λ). Wavelengths are easily converted to wavenumbers simply by taking the reciprocal. However, the bandpass, in wavenumbers, is not constant when the spectrum are recorded with constant wavelength resolution, as is usual with grating monochromators. For example, consider a constant bandpass $\Delta\lambda = \lambda_2 - \lambda_1$, where λ_1 and λ_2 are wavelengths on either side of the transmission maximum (Figure 2.38). At 300 nm a bandpass ($\Delta\lambda$) of 2 nm is equivalent to 222 cm^{-1}. At 600 nm, this same bandpass is equivalent to a resolution ($\Delta\bar{\nu}$) of 55 cm^{-1}. Hence, as the wavelength is increased, the bandpass, in wavenumbers, decreases as the square of the exciting wavelength. From $\bar{\nu} = 1/\lambda$, it follows that $|\Delta\bar{\nu}| = |\Delta\lambda|/\lambda^2$. Therefore, if spectra are obtained in the usual form of intensity per wavelength

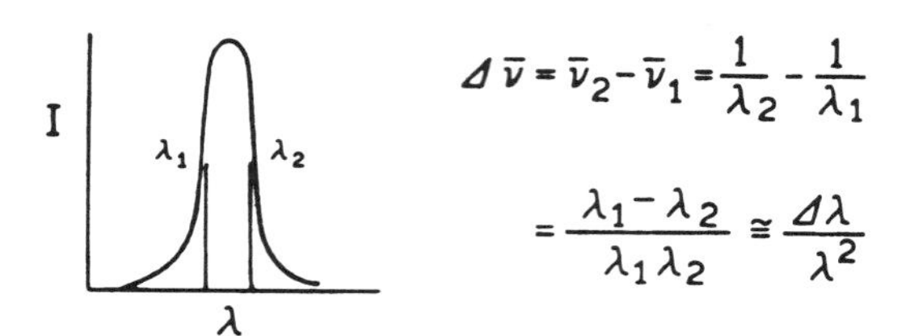

Figure 2.38. Relationship between spectral resolution in wavelength (λ) and in wavenumber ($\bar{\nu}$).

interval, $I(\lambda, \lambda + \Delta\lambda)/\Delta\lambda$, and $I(\bar{\nu}) = I(\bar{\nu}, \bar{\nu} + \Delta\bar{\nu})/\Delta\bar{\nu}$, then conversion to the wavenumber scale requires[37–39] that each intensity be multiplied by λ^2:

$$I(\bar{\nu}) = \lambda^2 I(\lambda) \qquad [2.4]$$

The effect of this wavelength-to-wavenumber conversion is illustrated in Appendix I. Multiplication by λ^2 results in a selective enhancement of the long-wavelength side of the emission, and there is a shift in the apparent emission maximum. It should be noted that even after this correction is performed, the resolution of the spectrum still varies with wavenumber.

2.10. QUANTUM YIELD STANDARDS

The easiest way to estimate the quantum yield of a fluorophore is by comparison with standards of known quantum yield. Some of the most used standards are listed in Table 2.4. The quantum yields of these compounds are mostly independent of excitation wavelength, so the standards can be used wherever they display useful absorption.

Determination of the quantum yield is generally accomplished by comparison of the wavelength-integrated intensity of the unknown to that of the standard. The optical density is kept below 0.05 to avoid inner filter effects, or the optical densities of the sample and reference are matched at the excitation wavelength. The quantum yield of the unknown is calculated using

$$Q = Q_R \frac{I}{I_R} \frac{OD_R}{OD} \frac{n^2}{n_R^2} \qquad [2.5]$$

where Q is the quantum yield, I is the integrated intensity, OD is the optical density, and n is the refractive index. The subscript R refers to the reference fluorophore of known quantum yield. In this expression it is assumed that the sample and reference are excited at the same wavelength, so that it is not necessary to correct for the different excitation intensities of different wavelengths.

Table 2.4. Quantum Yield Standards

Compound	Solvent	λ_{ex} (nm)	T(°C)	Q	Reference(s)
Quinine sulfate	0.1*M* H_2SO_4	350	22	0.577	43
		366	—	0.53 ± 0.023	44
β-Carboline[a]	1*N* H_2SO_4	350	25	0.60	35
Fluorescein	0.1*M* NaOH	496	22	0.95 ± 0.03	45
9,10-DPA[b]	Cyclohexane	—	—	0.95	46
9,10-DPA	Cyclohexone	366	—	1.00 ± 0.05	47, 48
POPOP[c]	Cyclohexane			0.97	46
2-Aminopyridine	0.1*N* H_2SO_4	285	—	0.60 ± 0.05	48,49
Tryptophan	Water	280	—	0.13 ± 0.01	50
Tyrosine	Water	275	23	0.14 ± 0.01	50
Phenylalanine	Water	260	23	0.024	50
Phenol	Water	275	23	0.14 ± 0.01	50
Rhodamine 6G	Ethanol	488	—	0.94	51
Rhodamine 101	Ethanol	450–465	25	1.0	52
Cresyl Violet	Methanol	540–640	22	0.54	53

[a] β-Carboline is 9*H*-pyrido[3,4-*b*]indole.
[b] 9,10-DPA, 9,10-Diphenylanthracene.
[c] POPOP, 1,4-bis(5-phenyloxazol-2-yl)benzene.

This expression is mostly intuitive, except for the use of the ratio of the refractive indices of the sample (n) and reference (n_R). This ratio has its origin in consideration of the intensity observed from a point source in a medium of refractive index n_i by a detector in a medium of refractive index n_o (Figure 2.39). The observed intensity is modified[40,41] by the ratio $(n_i/n_o)^2$. While the derivation was for a point source, the use of the ratio was found to be valid for any detector geometry which did not see past the edge of the beam.[42]

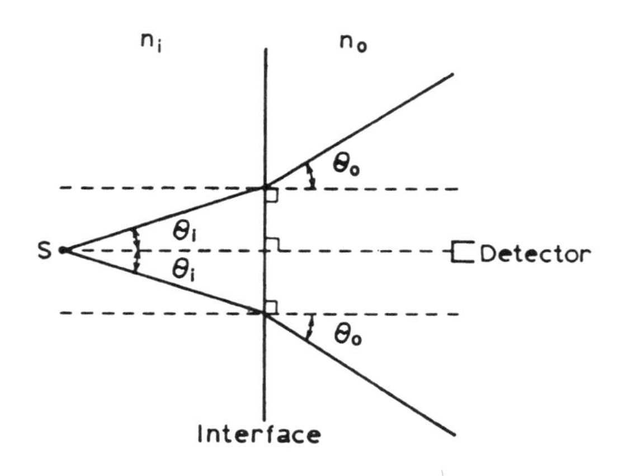

Figure 2.39. Refractive index effects in quantum yield measurements. The point source, S, is in a medium of refractive index n_i, and the detector in a medium of refractive index n_o, with $n_o < n_i$. Reprinted, with permission, from Ref. 40. Copyright © 1971 American Chemical Society.

2.11. EFFECTS OF SAMPLE GEOMETRY

The apparent fluorescence intensity and spectral distribution can be dependent upon the optical density of the sample and the precise geometry of sample illumination. The most common geometry used for fluorescence is right-angle observation of the center of a centrally illuminated cuvette (Figure 2.40, top left). Other geometric arrangements include front-face and off-center illumination. Off-center illumination decreases the path length, which can also be accomplished by using cuvettes with path lengths less than 1 cm. These methods are generally used to decrease the inner filtering effects due to high optical densities or to sample turbidity.

Frequently, front-face illumination is performed using either triangular cuvettes or square cuvettes oriented 45° relative to the incident beam (Figure 2.40). In our opinion, this 45° positioning should be discouraged. A large amount of light is reflected directly into the emission monochromator, increasing the chance that stray light will interfere with the measurements. When front-face illumination is necessary, we prefer to orient the illuminated surface about 30° from the incident beam. This procedure has two advantages. First, less reflected light enters the emission monochromator. Second, the incident light is distributed over a larger surface area, decreasing the sensitivity of the measurement to the precise placement of the cuvette within its holder. One disadvantage of this orientation is a de-

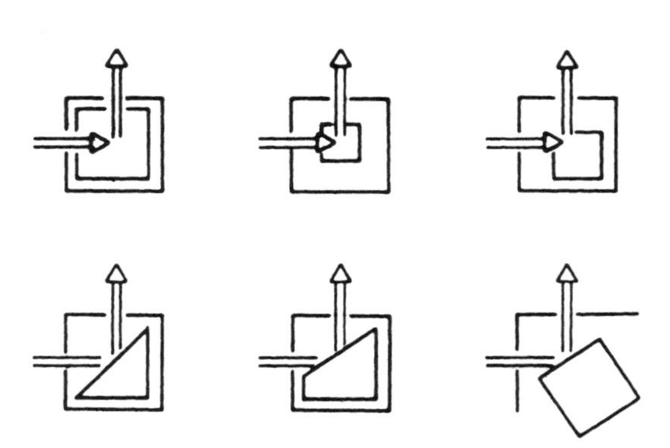

Figure 2.40. Various geometric arrangements for observation of fluorescence.

creased sensitivity because a larger fraction of the incident light is reflected off the surface of the cuvette.

It is important to recognize that fluorescence intensities are proportional to the concentration over only a limited range of optical densities. Consider a 1 × 1 cm cuvette which is illuminated centrally and observed at a right angle (Figure 2.40, top left). Assume further that the optical density at the excitation wavelength is 0.1. Using the definition of optical density ($\log I_0/I$ = OD), the light intensity at the center of the cuvette (I) is $0.88I_0$, where I_0 is the intensity of the light incident upon the cuvette. Since the observed fluorescence intensity is proportional to the intensity of the exciting light, the apparent quantum yield will be about 10% less than that observed for an infinitely dilute solution. This is called an inner filter effect. These effects may decrease the intensity of the excitation at the point of observation or decrease the observed fluorescence by absorption of the fluorescence. The relative importance of each process depends upon the optical densities of the sample at the excitation and emission wavelengths.

To illustrate the effects of optical density on fluorescence intensity, data are presented in Figure 2.41 for quinine sulfate. The measured intensity is proportional to optical density only to an optical density of 0.05. The linear range of the fluorescence intensities could be expanded by using off-center illumination, which reduces the effective light path. These intensities can be approximately corrected for the inner filter effects as follows. Suppose the sample has a significant optical density at both the excitation and the emission wavelength, OD_{ex} and OD_{em}, respectively. These optical densities attenuate the excitation and emission by $10^{-0.5OD_{ex}}$ and $10^{-0.5OD_{em}}$, respectively. Attenuation due to absorption of the incident light and that due to absorption of the emitted light are sometimes called the primary and secondary inner filter effects, respectively.[54,55] The corrected fluorescence intensity is approximately given by

$$F_{corr} = F_{obs} \operatorname{antilog}\left(\frac{OD_{ex} + OD_{em}}{2}\right) \qquad [2.6]$$

The corrected intensities for quinine sulfate are shown on Figure 2.41, and these calculated values are seen to match the initial linear portion of the curve. For precise corrections, it is preferable to prepare calibration curves using the precise compounds and conditions that will be used for the actual experimentation. Empirical corrections are typically used in most procedures to correct for sample absorbance.[54–57]

It is interesting to note that for front-face illumination, the intensities are expected to become independent of concentration at high optical densities.[58,59] Under these conditions, all the incident light is absorbed near the surface of the cuvette. Front-face illumination is also useful for studies of optically dense samples, such as whole blood, or highly scattering solutions. The intensity is expected to be proportional to the ratio of the optical density of the fluorophore to that of the sample.[60] When front-face illumination is used, the total optical density can be very large (20 or larger), but the signal level remains constant. However, high fluorophore concentrations can result in quenching due to a variety of interactions, such as radiative and nonradiative transfer and excimer formation. Fluorophores like fluorescein that have a small Stokes' shift are particularly sensitive to concentration quenching.

High optical densities can distort the emission spectra as well as the apparent intensities. For example, when right-angle observation is used, the short-wavelength emission bands of anthracene are selectively attenuated (Figure 2.42). This occurs because these shorter wavelengths are absorbed by anthracene. Attenuation of the blue edge of the emission is most pronounced for fluorophores that have significant overlap of the absorption and emission spectra. Fluorophores that display a large Stokes' shift are less sensitive to this phenomenon.

Emission spectra can also be distorted at high concentrations when front-face illumination is used. For example, the emission spectrum of 9,10-diphenylanthracene is dependent on excitation wavelength (Figure 2.43). The short-wavelength portion of the emission is attenuated at 365-nm

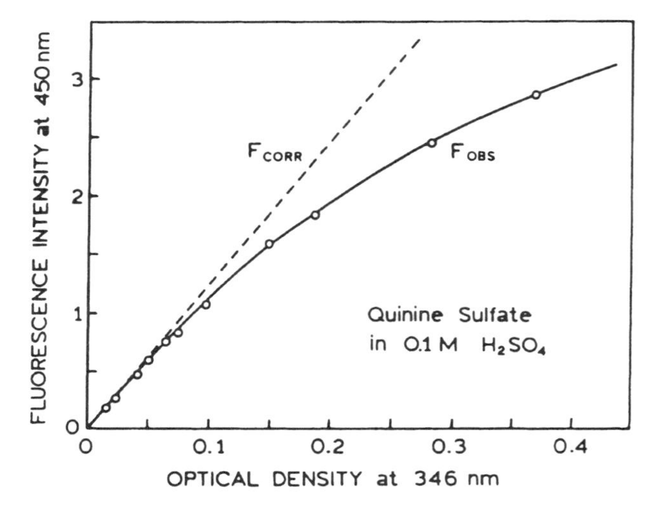

Figure 2.41. Effects of optical density on the fluorescence intensity of quinine sulfate. —, Measured intensities; — — —, intensities corrected according to Eq. [2.6] with OD_{em} = 0. These data were obtained in a 1-cm^2 cuvette which was centrally illuminated.

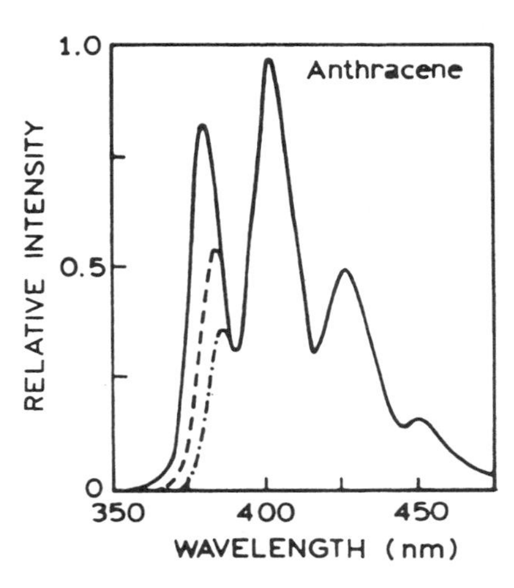

Figure 2.42. Effects of self-absorption of anthracene on its emission spectrum. The concentrations were $10^{-5}M$ (—), $10^{-4}M$ (— — —), and $4 \times 10^{-4}M$ (— • —) in ethanol. A 1-cm^2 cuvette was used with right-angle observation.

excitation, but not at 265-nm excitation. The optical density is less at 365 nm than at 265 nm. As a result, the 365-nm exciting beam penetrates more deeply into the sample, and the probability for reabsorption of the emission is therefore increased. There appears to be no simple and reliable method to precisely quantify and correct for these concentration effects. It is best to avoid these problems by working with dilute solutions.

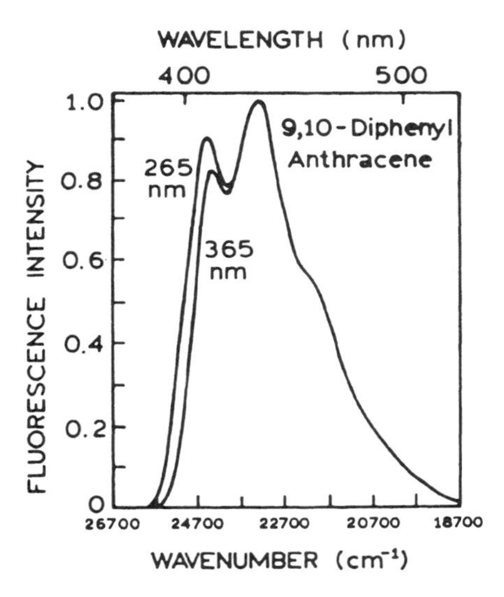

Figure 2.43. Effects of excitation wavelength on the self-absorption of fluorescence by 9,10-diphenylanthracene. Front-face illumination was used at the indicated excitation wavelengths. Revised from Ref. 61.

2.12. COMMON ERRORS IN SAMPLE PREPARATION

It seems valuable to summarize some of the difficulties which can be encountered with any given sample (Figure 2.44). The sample can be too concentrated, in which case all the light is absorbed at the surface facing the light source. In fact, this is one of the more common errors. With highly absorbing solutions and right-angle observations, the signal levels will be very low. Another problem is that the sample contains a fluorescent impurity, or the detected light is contaminated by Rayleigh or Raman scatter. Sometimes one will notice that the signal seems too noisy, given the signal level. Intensity fluctuations can be due to particles which drift through the laser beam and fluoresce or scatter light.

Even if the fluorescence is strong, it is important to consider the possibility of two or more fluorophores, that is, an impure sample. Emission spectra are usually independent of excitation wavelength.[62] Hence, it is useful to determine if the emission spectrum remains the same at different excitation wavelengths.

1. FLUOROPHORE CONCENTRATION TOO HIGH

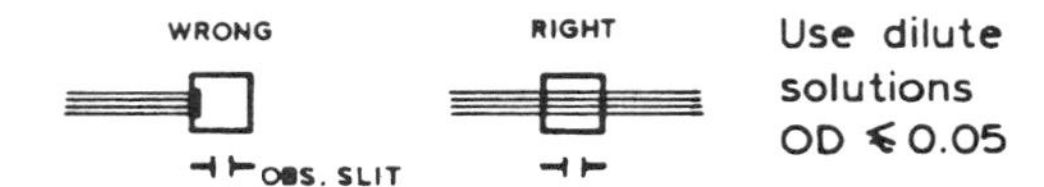

2. CONTAMINATED SOLVENT AND/OR CUVETTE

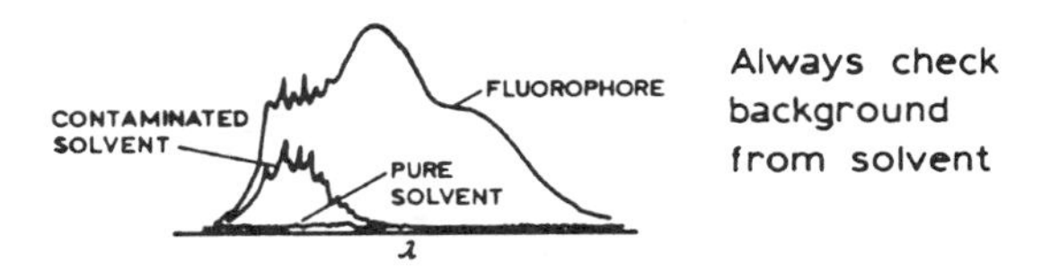

3. SCATTERED LIGHT

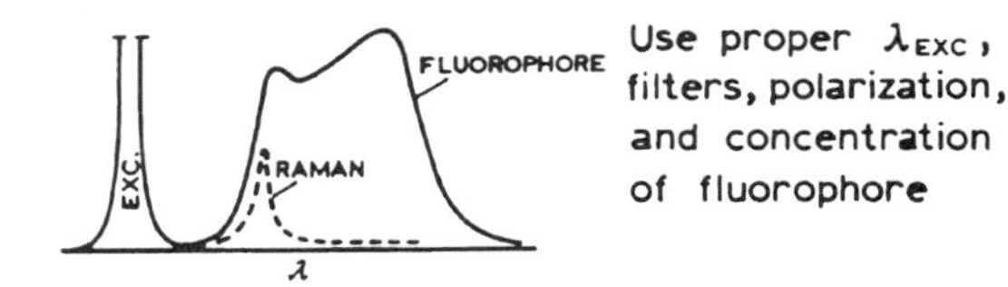

4. PARTICLES IN SOLUTION

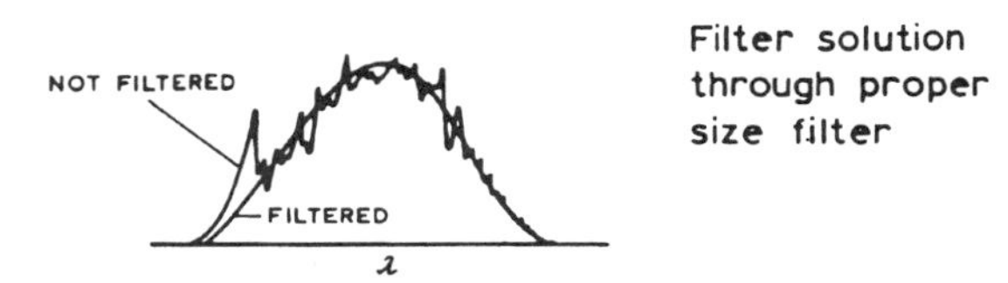

Figure 2.44. Common errors in sample preparation.

One example of a mixture of fluorophores is shown in Figure 2.45, which contains the emission spectrum of a mixture of coumarin 102 (C102) and coumarin 153 (C153). For a pure solution of C102, the same emission spectrum is observed for excitation at 360 and 420 nm (upper panel in Figure 2.45). For a mixture of C102 and C153, one finds an increased intensity above 500 nm for excitation at 420 nm (dashed curve in the lower panel). This is due to C153, as can be seen from its emission spectrum (dotted curve in the lower panel), with an emission maximum at 520 nm. Whenever the emission spectrum changes with excitation wavelength, one should suspect an impurity.

It is interesting to notice the significant difference between the emission spectra of these two coumarin derivatives which results from a small change in structure. The fluorine-substituted coumarin (C153) appears to be more sensitive to solvent polarity. This effect is probably due to an increased charge separation in C153, due to movement of the amino electrons toward the $-CF_3$ group in the excited state. These effects are described in Chapter 6.

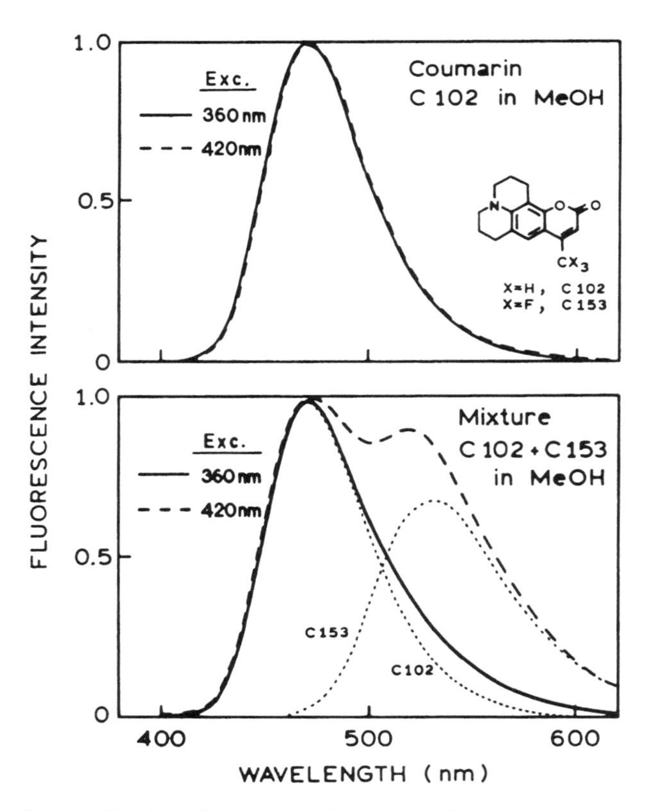

Figure 2.45. Emission spectra of the coumarin derivative C102 (*top*) and a mixture of two coumarin derivatives, C102 and C153 (*bottom*), excited at 360 and 420 nm. From Ref. 63.

2.13. ABSORPTION OF LIGHT AND DEVIATION FROM THE BEER–LAMBERT LAW

A fundamental aspect of fluorescence spectroscopy is the measurement of light absorption. While the theory of light absorption is well known, a number of factors can result in misleading measurements of light absorption. We will first derive the Beer–Lambert law, which allows one to describe the reasons for deviations from this law.

Consider a thin slab of solution of thickness dx which contains n light-absorbing molecules per cubic centimeter (Figure 2.46). Let σ be the effective cross section for absorption, in square centimeters. The light intensity (dI) absorbed per thickness dx is proportional to the intensity of the incident light I and to both σ and n, where n is the number of molecules per cubic centimeter:

$$\frac{dI}{dx} = -I\sigma n \qquad [2.7]$$

Rearrangement and integration, subject to the boundary condition $I = I_0$ at $x = 0$, yield

$$\ln \frac{I_0}{I} = \sigma n d \qquad [2.8]$$

where d is the thickness of the sample. This is the Beer–Lambert equation, which is generally used in an alternative form,

$$\log \frac{I_0}{I} = \varepsilon c d = \text{optical density} \qquad [2.9]$$

where ε is the decadic molar extinction coefficient (in M^{-1} cm^{-1}) and c is the concentration in moles per liter. Combination of Eqs. [2.8] and [2.9] yields the relationship between the extinction coefficient and the cross section for light absorption:

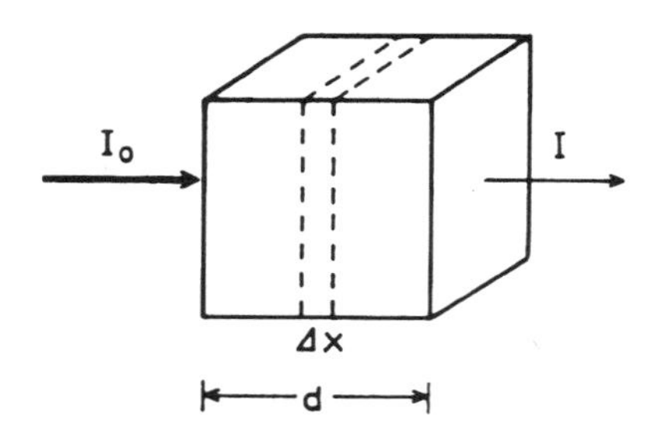

Figure 2.46. Light absorption.

$$\sigma = 2.303 \frac{\varepsilon c}{n} \quad [2.10]$$

Since $n = Nc/10^3$, where N is Avogadro's number, we obtain

$$\sigma = 3.82 \times 10^{-21}\ \varepsilon \ (\text{in cm}^2) \quad [2.11]$$

It is interesting to calculate the absorption cross section for typical aromatic compounds. The extinction coefficients of anthracene are 160,000 and 6300 M^{-1} cm^{-1} at 253 and 375 nm, respectively. These values correspond to cross sections of 6.1 and 0.24 Å^2, respectively. Assuming the molecular cross section of anthracene to be 12 Å^2, we see that anthracene absorbs about 50% of the photons it encounters at 253 nm and 2% of the photons at 375 nm.

Occasionally, one encounters the term oscillator strength. This term represents the strength of an absorption relative to a completely allowed transition. The oscillator strength (f) is related to the integrated absorption of a transition by

$$f = \frac{4.39 \times 10^{-9}}{n} \int \varepsilon(\bar{\nu}) d\bar{\nu} \quad [2.12]$$

where n is the refractive index.

2.13.A. Deviations from Beer's Law

Beer's law predicts that the optical density is directly proportional to the concentration of the absorbing species. Deviations from Beer's law can result from both instrumental and intrinsic causes. Biological samples are frequently turbid because of macromolecules or other large aggregates which scatter light. The optical density resulting from scatter will be proportional to $1/\lambda^4$ (Rayleigh scattering), and may thus be easily recognized as a background absorption which increases rapidly with decreasing wavelength.

If the optical density of the sample is high, and if the absorbing species is fluorescent, the emitted light can reach the detector. This process will yield deviations from Beer's law which are concave toward the concentration axis. However, the fluorescence is omnidirectional whereas the incident light is collimated along an axis. Hence, this effect can be minimized by keeping the detector distant from the sample, thereby decreasing the efficiency with which the fluorescence emission is collected.

If the absorbing species is only partially soluble, it may aggregate in solutions at high concentrations. The absorption spectra of the aggregates may be distinct from the spectrum of the monomers. An example is the common dye bromophenol blue. At concentrations around 10 mg/ml, it appears as a red solution, whereas at lower concentrations it appears blue. Depending upon the wavelength chosen for observation, the deviations from Beer's law may be positive or negative.

The factors described above are due to intrinsic properties of the sample. Instrumental artifacts can also yield optical densities that are nonlinear with concentration. This is particularly true at high optical densities. For example, consider a solution of indole with an optical density of 5 at 280 nm. In order to measure this optical density accurately, the spectrophotometer needs to accurately quantify the intensities of I_0 and I, the latter of which is 10^{-5} times less intense than the incident light I_0. Generally, the stray light passed by the monochromator, at wavelengths where the compound does not absorb, is larger than this value. As a result, one cannot reliably measure such high optical densities unless considerable precautions are taken to minimize stray light.

2.14. TWO-PHOTON AND MULTIPHOTON EXCITATION • Advanced Topic •

The previous discussion of Beer's law leads naturally to the emerging topic of multiphoton excitation. In almost all fluorescence experiments, excitation is due to absorption of a single photon by each fluorophore. Under these conditions, the excitation spectrum of the fluorophore matches its absorption spectrum. However, it is also possible for a fluorophore to absorb two or more long-wavelength photons to reach the same first singlet excited state. This process is illustrated by the Jabłoński diagram for two-photon excitation (Figure 2.47). Two-photon excitation occurs by the simultaneous absorption of two lower-energy photons to reach an excited state.

The nature of two-photon excitation is clarified by an example. Figure 2.48 shows emission spectra of DPH in the nonpolar solvent triacetin. The sample was excited at either the usual one-photon wavelength of 358 nm or the unusual long wavelength of 716 nm. Two photons at 716 nm have the same energy as one photon at 358 nm. Hence, if the molecule can absorb two photons at 716 nm, it can reach the first singlet state.

Two-photon excitation requires special conditions. While the possibility of two-photon excitation was predicted in the 1930s, it was not observed until lasers were invented. This is because the fluorophore must simultaneously absorb two-long wavelength photons. This requires that the two photons be present at the same instant in time,[65] within 10^{-15}–10^{-16} s. This requires high local intensities which can only be obtained from laser sources. In

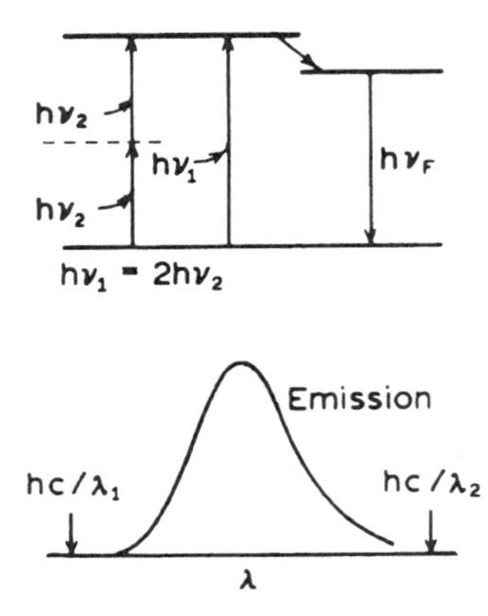

Figure 2.47. Jabłoński diagram for two-photon excitation.

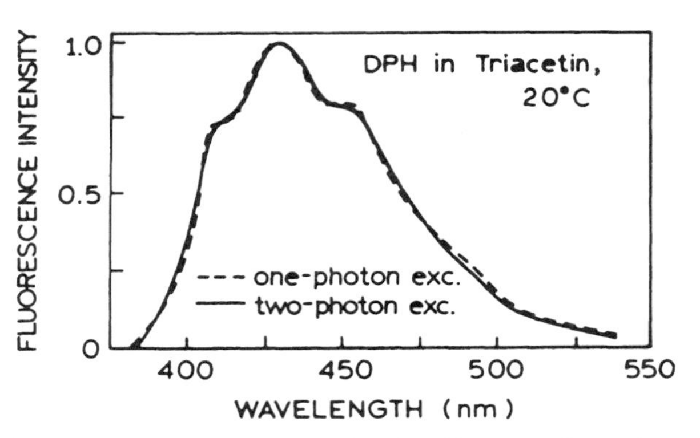

Figure 2.48. Emission spectra of DPH with one- and two-photon excitation, at 358 and 716 nm, respectively. Revised from Ref. 64.

fact, most two-photon and multiphoton experiments are performed with picosecond or femtosecond laser sources, which concentrate the available light into short pulses.

In principle, one-photon and two-photon excitation follow different selection rules.[66] That is, the electronic states which can be reached by one- and two-photon excitation are different. This concept is the same as that which applies to infrared and Raman spectroscopy, where the different selection rules allow observation of different vibrational transitions. In principle, the absorption spectra or excitation spectra are different for one-photon and two-photon excitation. At this time, there is limited information about two-photon absorption spectra, as they are difficult to measure. In some cases, these spectra look like the one-photon spectra at twice the wavelength, and in other cases the spectral shapes are different. However, a large number of emission spectra have been observed with two-photon and even three-photon excitation. In all spectra measured to date, the emission spectra have been the same, irrespective of the nature of the excitation process. One example is DPH, which showed the same emission spectrum for excitation at 358 or 716 nm (Figure 2.48).

Two-photon excitation results in an apparent violation of Beer's law. For one-photon excitation, the amount of light absorbed is proportional to the intensity of the incident light. This results in the emission intensity being directly proportional to the intensity of the excitation. Such behavior is found for DPH for excitation at 358 nm (Figure 2.49). Remarkably different behavior is found for excitation at 716 nm. In this case the emission intensity depends on the square of the incident light intensity. This is a well-known property of two-photon excitation. For three-photon excitation, the emission intensity depends on the third power of the incident intensity.

The fact that two-photon excitation depends quadratically on the incident intensity results in an unusual spatial profile for the excited-state population. For a dilute solution, with one-photon excitation, the total excited-state population is uniform across the sample (Figure 2.50). For two-photon excitation, most of the excitation occurs at the focal point of the excitation, where the local intensity is highest.

For one-photon excitation there is no reason to focus the beam, so collimated excitation beams are usually used. For two-photon excitation, it is essential to focus the excitation to obtain a high local intensity. This property of localized excitation is particularly useful in fluorescence microscopy, where it is possible to excite fluorophores at the focal point of the objective, without photobleaching of the fluorophores above or below the focal plane.

Although multiphoton excitation seems to be an exotic topic, its use is growing rapidly in spectroscopy and imaging. This growth is due to the apparent lack of damaging

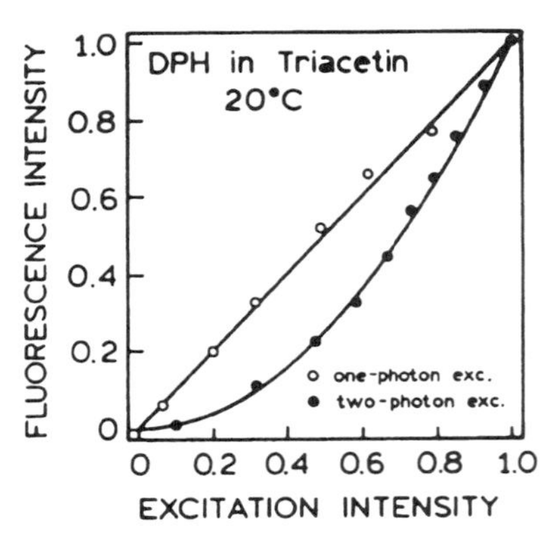

Figure 2.49. Dependence of the emission intensity of DPH on the excitation intensity at 358 (○) and 716 nm (●). The intensities have been normalized to unity at the highest excitation intensities. The solid lines represent linear and quadratic curves. From Ref. 64.

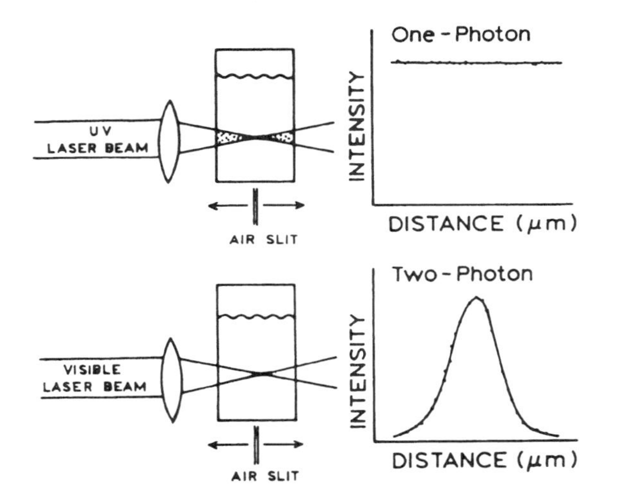

Figure 2.50. Spatial distribution of excited fluorophores for one-photon (*top*) and two-photon excitation (*bottom*).

photochemical processes and the introduction of increasingly simple femtosecond laser sources.

2.15. CONCLUSIONS

At first glance, it seems easy to perform fluorescence experiments. However, there are numerous factors which can compromise the data and invalidate the results. One needs to be constantly aware of the possibility of sample contamination or contamination of the signal by scattered or stray light. Collection of emission spectra, and examination of blank samples, is essential for all experiments. One cannot reliably interpret intensity values, anisotropy, or lifetimes without careful examination of the emission spectra.

In closing this chapter, it seems useful to acknowledge the early report by Udenfriend and co-workers, who reported one of the earliest spectrofluorometers that allowed UV excitation and emission.[67] The availability of such instruments resulted in the remarkable growth of fluorescence spectroscopy over the past 45 years.

REFERENCES

1. Technical Literature, Oriel Instruments, 250 Long Beach Blvd., PO Box 872, Stratford, CT 06497.
2. Laczko, G., and Lakowicz, J. R., unpublished observations.
3. Light Sources, Monochromators & Spectrographs, Detectors & Detection Systems, Fiber Optics, Oriel Corporation, 250 Long Beach Blvd., PO Box 872, Stratford, CT 06497.
4. Cermax Product Specifications for Collimated and Focused Xenon Lamps, ILC Technology, Inc., 399 West Joan Drive, Sunnyvale, CA 94089.
5. Analamp Emission Data Sheet, BHK, Inc. (Subsidiary of Hamamatsu Corporation), 1000 S. Magnolia Ave., Monrovia, CA 91016.
6. Sipior, J., Carter, G. M., Lakowicz, J. R., and Rao, G., 1997, Blue light-emitting diode demonstrated as an ultraviolet excitation source for nanosecond phase-modulation fluorescence lifetime measurements, *Rev. Sci. Instrum.* **68**:2666–2670.
7. 1996 Catalog of Optical Components and Instruments, Optometrics USA, Inc., Nemco Way, Stony Brook Industrial Park, Ayer, MA 01432.
8. Gryczynski, I., and Lakowicz, J. R., unpublished observations.
9. Castellano, P., and Lakowicz, J. R., unpublished observations.
10. Technical Literature, Spindler & Hoyer, Inc., 459 Fortune Blvd., Milford, MA 01757.
11. Flaugh, P. L., O'Donnell, S. E., and Asher, S. A., 1984, Development of a new optical wavelength rejection filter: Demonstration of its utility in Raman spectroscopy, *Appl. Spectrosc.* **386**:847–850.
12. Gryczynski, I., Malak, H., Lakowicz, J. R., Cheung, H. C., Robinson, J., and Umeda, P. K., 1996, Fluorescence spectral properties of troponin C mutant F22W with one-, two- and three-photon excitation, *Biophys. J.* **71**:3448–3453.
13. Szmacinski, H., Gryczynski, I., and Lakowicz, J. R., 1993, Calcium-dependent fluorescence lifetimes of Indo-1 for one- and two-photon excitation of fluorescence, *Photochem. Photobiol.* **58**:341–345.
14. Photomultiplier Tubes, Hamamatsu Photonics K.K., Electron Tube Center, (1994), 314-5, Shimokanzo, Toyooka-village, Iwata-gun, Shizuoka-ken, 438–01 Japan.
15. Leaback, D. H., 1997, Extended theory, and improved practice for the quantitative measurement of fluorescence, *J. Fluoresc.* **7**(1):55S–57S.
16. Hiraoka, Y., Sedat, J. W., and Agard, D. A., 1987, The use of a charge-coupled device for quantitative optical microscopy of biological structures, *Science* **238**:36–41.
17. Aikens, R. S., Agard, D. A., and Sedat, J. W., 1989, Solid-state imagers for microscopy, *Methods Cell Biol.* **29**:291–313.
18. Epperson, P. M., and Denton, M. B., 1989, Binding spectral images in a charge-coupled device, *Anal. Chem.* **61**:1513–1519.
19. Bilhorn, R. B., Sweedler, J. V., Epperson, P. M., and Denton, M. B., 1987, Charge transfer device detectors for analytical optical spectroscopy—operation and characteristics, *Appl. Spectrosc.* **41**:1114–1124.
20. Epperson, P. M., Jalkaian, R. D., and Denton, M. B., 1989, Molecular fluorescence measurements with a charge-coupled device detector, *Anal. Chem.* **61**:282–285.
21. Melhuish, W. H., 1962, Calibration of spectrofluorometers for measuring corrected emission spectra, *J. Opt. Soc. Am.* **52**:1256–1258.
22. Yguerabide, J., 1968, Fast and accurate method for measuring photon flux in the range 2500–6000 Å, *Rev. Sci. Instrum.* **397**:1048–1052.
23. Mandal, K., Pearson, T. D. L., and Demas, J. N., 1980, Luminescent quantum counters based on organic dyes in polymer matrices, *Anal. Chem.* **52**:2184–2189.
24. Mandal, K., Pearson, T. D. L., and Demas, N. J., 1981, New luminescent quantum counter systems based on a transition-metal complex, *Inorg. Chem.* **20**:786–789.
25. Nothnagel, E. A., 1987, Quantum counter for correcting fluorescence excitation spectra at 320- and 800-nm wavelengths, *Anal. Biochem.* **163**:224–237.

26. Lippert, E., Nagelle, W., Siebold-Blakenstein, I., Staiger, U., and Voss, W., 1959, Messung von fluorescenzspektren mit hilfe von spektralphotometern und vergleichsstandards, *Z. Anal. Chem.* **17**:1–18.
27. Schmillen, A., and Legler, R., 1967, *Landolt-Bornstein, Vol. 3, Lumineszenz Organischer Substanzen*, Springer-Verlag, New York, pp. 228–229.
28. Argauer, R. J., and White, C. E., 1964, Fluorescent compounds for calibration of excitation and emission units of spectrofluorometer, *Anal. Chem.* **36**:368–371.
29. Melhuish, W. H., 1960, A standard fluorescence spectrum for calibrating spectrofluorometers, *J. Phys. Chem.* **64**:762–764.
30. Parker, C. A., 1962, Spectrofluorometer calibration in the ultraviolet region, *Anal. Chem.* **34**:502–505.
31. Velapoldi, R. A., 1973, Considerations on organic compounds in solution and inorganic ions in glasses as fluorescent standard reference materials, *Proc. Natl. Bur. Stand.* **378**:231–244.
32. Pardo, A., Reyman, D., Poyato, J. M. L., and Medina, F., 1992, Some β-carboline derivatives as fluorescence standards, *J. Lumin.* **51**:269–274.
33. Chen, R. F., 1967, Some characteristics of the fluorescence of quinine, *Anal. Biochem.* **19**:374–387.
34. Verity, B., and Bigger, S. W., 1996, The dependence of quinine fluorescence quenching on ionic strength, *Int. J. Chem. Kinet.* **2812**:919–923.
35. Ghiggino, K. P., Skilton, P. F., and Thistlethwaite, P. J., 1985, β-Carboline as a fluorescence standard, *J. Photochem.* **31**:113–121.
36. Middleton, W. E. K., and Sanders, C. L., 1951, The absolute spectral diffuse reflectance of magnesium oxide, *J. Opt. Soc. Am.* **41**(6):419–424.
37. Heller, C. A., Henry, R. A., McLaughlin, B. A., and Bliss, D. E., 1974, Fluorescence spectra and quantum yields: Quinine, uranine, 9,10-diphenylanthracene, and 9,10-bis(phenylethynyl)anthracenes, *J. Chem. Eng. Data* **19**(3):214–219.
38. Melhuish, W. H., 1972, Absolute spectrofluorometry, *J. Res. of the National Bureau of Standards* **76A**:547–560.
39. Tazuke, S., and Winnik, M. A., 1986, Fluorescence and phosphorescence spectroscopy in polymer systems: A general introduction, in *Photophysical and Photochemical Tools in Polymer Science*, M. A. Winnik (ed.), D. Reidel, Dordrecht, pp. 15–42.
40. Demas, J. N., and Crosby, G. A., 1971, The measurement of photoluminescence quantum yields. A review, *J. Phys. Chem.* **75**:991–1025.
41. Birks, J. B., 1970, *Photophysics of Aromatic Molecules*, Wiley-Interscience, New York, p. 98.
42. Hermans, J. J., and Levinson, S., 1951, Some geometrical factors in light-scattering apparatus, *J. Opt. Soc. Am.* **41**(7):460–465.
43. Eastman, J. W., 1967, Quantitative spectrofluorimetry—the fluorescence quantum yield of quinine sulfate, *Photochem. Photobiol.* **6**:55–72.
44. Adams, M. J., Highfield, J. G., and Kirkbright, G. F., 1977, Determination of absolute fluorescence quantum efficiency of quinine bisulfate in aqueous medium by optoacoustic spectrometry, *Anal. Chem.* **49**:1850–1852.
45. Brannon, J. H., and Magde, D., 1978, Absolute quantum yield determination by thermal blooming. Fluorescein, *J. Phys. Chem.* **82**:705–709.
46. Mardelli, M., and Olmsted, J., 1977, Calorimetric determination of the 9,10-diphenyl-anthracene fluorescence quantum yield, *J. Photochem.* **7**:277–285.
47. Ware, W. R., and Rothman, W., 1976, Relative fluorescence quantum yields using an integrating sphere. The quantum yield of 9,10-diphenylanthracene in cyclohexane, *Chem. Phys. Lett.* **39**:449–453.
48. Testa, A. C., 1969, Fluorescence quantum yields and standards, *Fluorescence News; Newsletter on Luminescence* **4**(4):1–3.
49. Rusakowicz, R., and Testa, A. C., 1968, 2-Aminopyridine as a standard for low-wavelength spectrofluorometry, *J. Phys. Chem.* **72**:2680–2681.
50. Chen, R. F., 1967, Fluorescence quantum yields of tryptophan and tyrosine, *Anal. Lett.* **1**:35–42.
51. Fischer, M., and Georges, J., 1996, Fluorescence quantum yield of rhodamine 6G in ethanol as a function of concentration using thermal lens spectrometry, *Chem. Phys. Lett.* **260**:115–118.
52. Karstens, T., and Kobe, K., 1980, Rhodamine B and Rhodamine 101 as reference substances for fluorescence quantum yield measurements, *J. Phys. Chem.* **84**:1871–1872.
53. Magde, D., Brannon, J. H., Cremers, T. L., and Olmsted, J., 1979, Absolute luminescence yield of cresyl violet. A standard for the red, *J. Phys. Chem.* **83**:696–699.
54. Kubista, M., Sjöback, R., Eriksson, S., and Albinsson, B., 1994, Experimental correction for the inner-filter effect in fluorescence spectra, *Analyst* **119**:417–419.
55. Yappert, M. C., and Ingle, J. D., 1989, Correction of polychromatic luminescence signals for inner-filter effects, *Appl. Spectrosc.* **43**:759–767.
56. Wiechelman, K. J., 1986, Empirical correction equation for the fluorescence inner filter effect, *Am. Lab.* **18**:49–53.
57. Puchalski, M. M., Morra, M. J., and von Wandruszka, R., 1991, Assessment of inner filter effect corrections in fluorimetry, *Fresenius J. Anal. Chem.* **340**:341–344.
58. Guilbault, G. G. (ed.), 1990, *Practical Fluorescence,* Marcel Dekker, New York, p. 31.
59. Eisinger, J., 1969, A variable temperature, U.V. luminescence spectrograph for small samples, *Photochem. Photobiol.* **9**:247–258.
60. Eisinger, J., and Flores, J., 1979, Front-face fluorometry of liquid samples, *Anal. Biochem.* **94**:15–21.
61. Berlman, I. B., 1971, *Handbook of Fluorescence Spectra of Aromatic Molecules*, 2nd ed., Academic Press, New York.
62. Kasha, M., 1960, Paths of molecular excitation, *Radiat. Res.* **2**:243–275.
63. Gryczynski, I., unpublished observations.
64. Lakowicz, J. R., Gryczynski, I., Kuśba, J., and Danielsen, E., 1992, Two photon induced fluorescence intensity and anisotropy decays of diphenylhexatriene in solvents and lipid bilayers, *J. Fluoresc.* **2**(4):247–258.
65. Xu, C., and Webb, W. W., 1997, Multiphoton excitation of molecular fluorophore and nonlinear laser microscopy, in *Topics in Fluorescence Spectroscopy, Volume 5, Nonlinear and Two-Photon-Induced Fluorescence*, J. R. Lakowicz (ed.), Plenum Press, New York, pp. 471–540.
66. Callis, P. R., 1997, Two-photon induced fluorescence, *Annu. Rev. Phys. Chem.* **48**:271–297.
67. Bowman, R. L., Caulfield, P. A., and Udenfriend, S., 1955, Spectrophotofluorometric assay in the visible and ultraviolet, *Science* **122**:32–33.

PROBLEMS

2.1. *Measurement of High Optical Densities*: Suppose you wish to determine the concentration of a $10^{-4}M$ solution of rhodamine B, which has an extinction coefficient of about 100,000 M^{-1} cm^{-1} at 590 nm. The monochromator in your spectrophotometer is imperfect, and the light presumed to be at 590 nm is 99.99% 590 nm and 0.01% longer wavelengths not absorbed by rhodamine B. What is the true optical density of the solution? What is the apparent optical density measured with your spectrophotometer? Assume that the path length is 1 cm.

2.2. *Calculation of Concentrations by Absorbance*: Suppose that a molecule displays an extinction coefficient of 30,000 M^{-1} cm^{-1} and that you wish to determine its concentration from the absorbance. You have two solutions, with actual optical densities of 0.3 and 0.003 in a 1-cm cuvette. What are the concentrations of the two solutions? Assume the measurement error in percent transmission is 1%. How does the 1% error affect determination of the concentrations?

Fluorophores

3

Fluorescence probes represent the most important area of fluorescence spectroscopy. One can spend a great deal of time describing the instrumentation for fluorescence spectroscopy, including light sources, monochromators, lasers, and detectors. However, in the final analysis, the wavelength and time resolution required of the instruments are determined by the spectral properties of the fluorophores. Furthermore, the information available from the experiments is determined by the properties of the probes. Only probes with nonzero anisotropies can be used to measure rotational diffusion, and the lifetime of the fluorophore must be comparable to the correlation time of interest. Only probes which are sensitive to pH can be used to measure pH. And only probes with reasonably long excitation and emission wavelengths can be used with tissues which display autofluorescence at short excitation wavelengths.

Thousands of fluorescent probes are known, and it is not practical to describe them all. This chapter contains an overview of the various types of fluorophores, their spectral properties, and their applications. Fluorophores can be broadly divided into two main classes, intrinsic and extrinsic. Intrinsic fluorophores are those which occur naturally. These include the aromatic amino acids, NADH, flavins, and derivatives of pyridoxal and chlorophyll. Extrinsic fluorophores are added to the sample to provide fluorescence when none exists or to change the spectral properties of the sample. Extrinsic fluorophores include dansyl chloride, fluorescein, rhodamine, and numerous other substances.

3.1. INTRINSIC OR NATURAL FLUOROPHORES

Intrinsic protein fluorescence originates with the aromatic amino acids,[1,2] tryptophan (trp), tyrosine (tyr), and phenylalanine (phe) (Figure 3.1). The indole groups of tryptophan residues are the dominant source of UV absorbance and emission in proteins. Tyrosine has a quantum yield similar to that of tryptophan (Table 3.1), but its emission spectrum is more narrowly distributed on the wavelength scale (Figure 3.2). This gives the impression of a higher quantum yield for tyrosine. In native proteins the emission of tyrosine is often quenched, which may be due to its interaction with the peptide chain or energy transfer to tryptophan. Denaturation of proteins frequently results in increased tyrosine emission. As in the case of phenol, the pK_a of tyrosine decreases dramatically upon excitation, and excited-state ionization can occur. Emission from phenylalanine is observed only when the sample protein lacks both tyrosine and tryptophan residues, which is a rare occurrence.

The emission of tryptophan is highly sensitive to its local environment, and is thus often used as a reporter group for protein conformational changes. Spectral shifts have been observed as a result of several phenomena, such as binding of ligands and protein–protein association. In addition, the emission maxima of proteins reflect the average exposure of their tryptophan residues to the aqueous phase. Fluorescence lifetimes of tryptophan residues range from 1 to 6 ns. Tryptophan fluorescence is subject to quenching by iodide, acrylamide, and nearby disulfide groups. Tryptophan residues also seem to be quenched by nearby electron-deficient groups like $-NH_3+$, $-CO_2H$, and protonated histidine residues. The presence of multiple tryptophan residues in proteins, each in a different environment, is one reason for the multiexponential intensity decays of proteins.

3.1.A. Fluorescent Enzyme Cofactors

Enzyme cofactors frequently are fluorescent (Figure 3.1). NADH is highly fluorescent, with absorption and emission maxima at 340 and 460 nm, respectively (Figure 3.3). The oxidized form of NADH, NAD^+, is nonfluorescent. The lifetime of NADH in aqueous buffer is near 0.4 ns. The

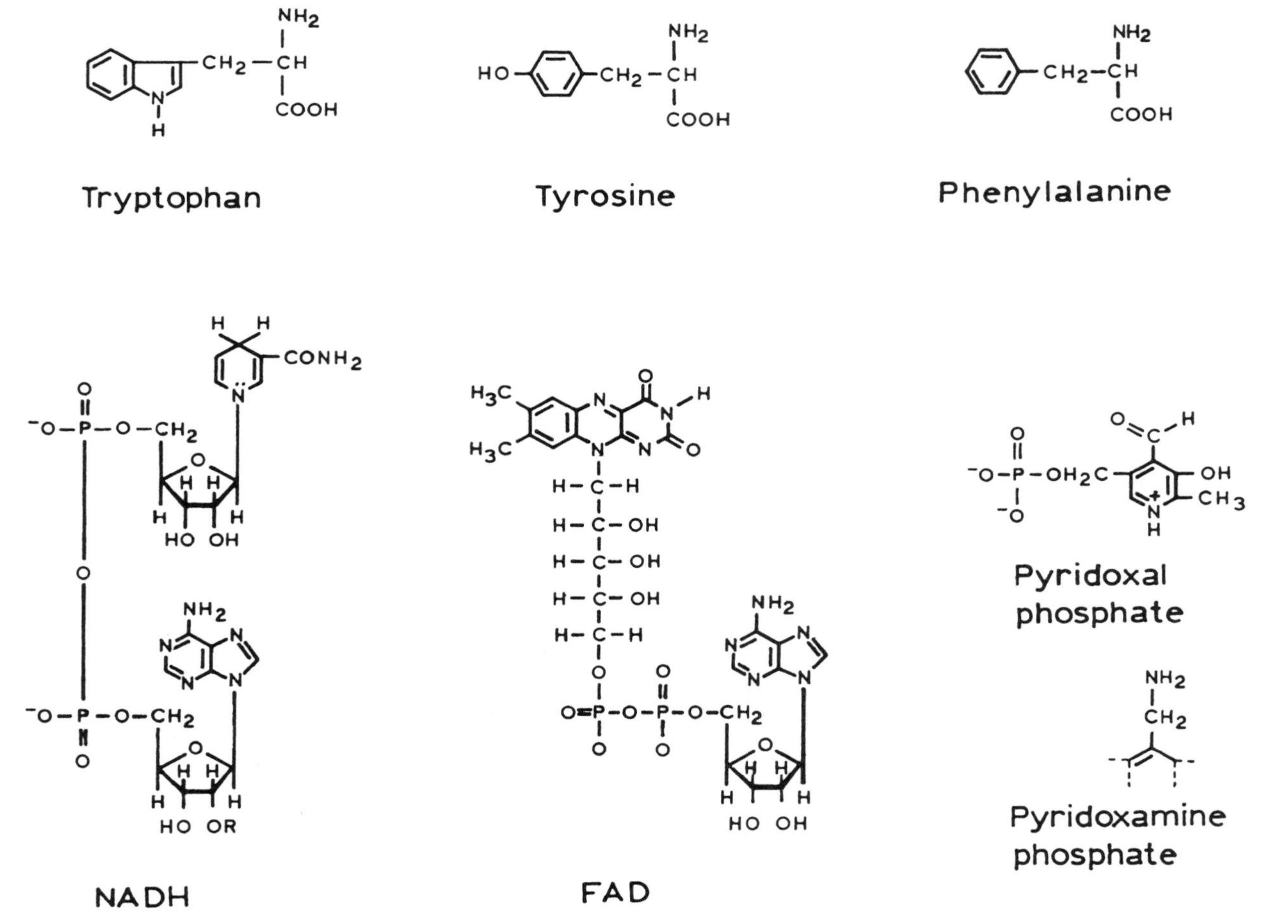

Tryptophan Tyrosine Phenylalanine

NADH FAD Pyridoxal phosphate Pyridoxamine phosphate

Figure 3.1. Intrinsic biochemical fluorophores. In NADH, R is a hydrogen; in NADPH, R is a phosphate group.

fluorescent group is the reduced nicotinamide ring. In solution its fluorescence is partially quenched by collisions or stacking with the adenine moiety. Upon binding of NADH to proteins, the quantum yield of the NADH generally increases fourfold,[3] and the lifetime increases to about 1.2 ns. However, depending on the protein, NADH fluorescence can increase or decrease upon protein binding.[3] The increased quantum yield is generally interpreted as binding of the NADH in an elongated fashion, which prevents contact between adenine and the fluorescent reduced nicotinamide group. Lifetimes as long as 5 ns have been reported for NADH bound to horse liver alcohol dehydrogenase[4] and octopine dehydrogenase.[5] The lifetimes of protein-bound NADH are typically different in the presence and absence of bound enzyme substrate.

The cofactor pyridoxal phosphate is also fluorescent.[6–13] Its absorption and emission spectra are dependent upon its chemical structure in the protein, where pyridoxal groups are often coupled to lysine residues by the aldehyde groups. The emission spectrum of pyridoxamine is at shorter wavelengths than that of pyridoxal phosphate (Figure 3.3). The emission spectrum of pyridoxamine is dependent on pH (not shown), and the emission spectrum of the pyridoxal group depends on its interaction with pro-

Table 3.1. Fluorescence Parameters of Aromatic Amino Acids in Water at Neutral pH[a]

Species	λ_{ex} (nm)	λ_{em} (nm)	Bandwidth (nm)	Quantum yield	Lifetime (ns)
Phenylalanine	260	282	—	0.02	6.8
Tyrosine	275	304	34	0.14	3.6
Tryptophan	295	353	60	0.13	3.1[b]

[a] From Ref. 1.
[b] Mean lifetime.

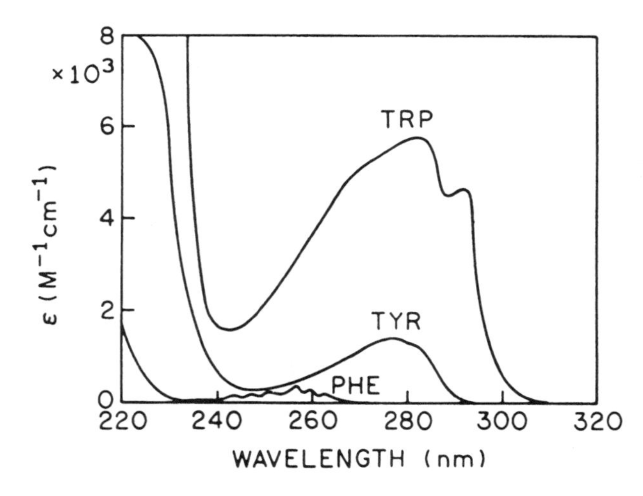

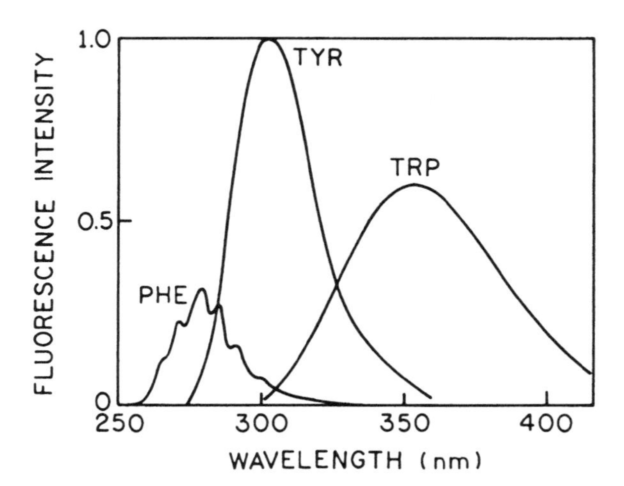

Figure 3.2. Absorption (*top*) and emission spectra (*bottom*) of the fluorescent amino acids in water of pH 7.0.

teins. When bound to phosphorylase, the emission maximum shifts to 535 nm, which seems to be the result of formation of a Schiff base with the enzyme. The spectroscopy of pyridoxal groups is complex, and it seems that this cofactor can exist in a variety of forms.

Riboflavin, FMN, and FAD absorb light in the visible range (at ~450 nm) and emit around 525 nm (Figure 3.3). In contrast to NADH, the oxidized forms of flavins are fluorescent, not the reduced forms. Typical lifetimes for FMN and FAD are 4.7 and 2.3 ns, respectively. As in the case of NADH, the flavin fluorescence is quenched by the adenine. This quenching is due to complex formation between the flavin and the adenine.[15] The latter process is referred to as static quenching. There may also be a dynamic component to the quenching. In contrast to NADH, which is highly fluorescent when bound to proteins, flavoproteins are generally nonfluorescent, but exceptions exist. Intensity decays of protein-bound flavins are typically complex, with multiexponential decay times ranging from 0.1 to 5 ns and mean decay times ranging from 0.3 to 1 ns.[16]

Nucleotides and nucleic acids are generally nonfluorescent. However, some exceptions exist. Yeast $tRNA^{PE}$ contains a highly fluorescent base, known as the Y_t-base, which has an emission maximum near 470 nm and a lifetime of about 6 ns (Figure 1.17). The molecules described above represent the dominant fluorophores in animal tissues. Many additional naturally occurring fluorescent substances are known, and these have been summarized in Ref. 17.

There is presently interest in the use of fluorescence from tissues,[18] either from the intrinsic fluorophores or from extrinsically added probes. Much of the fluorescence from cells and tissues is due to NADH and flavins. While other fluorescent species are seen, their identity is not always known. The best-characterized fluorescence from tissues is from the structural proteins collagen and elastin. When excited below 340 nm, these proteins display emission near 405 nm (Figure 3.3).[14] This fluorescence is thought to be due to cross-links between oxidized lysine residues which ultimately result in hydroxypyridinium groups.[18] However, different emission spectra are observed with longer excitation wavelengths. Intrinsic fluorescence from tissues is not completely understood at this time.

3.1.B. Binding of NADH to a Protein

The above description of intrinsic fluorophores may seem abstract, so it seems useful to consider a specific example. The enzyme 17β-hydroxysteroid dehydrogenase (17β-HSD) catalyzes the last step in the biosynthesis of estradiol from estrogen. The protein consists of two identical subunits, each containing a single tryptophan residue. It binds NADPH as a cofactor.

Emission spectra of 17β-HSD and of NADPH are shown in Figure 3.4. NADPH is identical to NADH (Figure 3.1) except for a phosphate group on the 2'-position of the ribose. For excitation at 295 nm, both the protein and NADPH are excited (Figure 3.4, top). Addition of NADPH to the protein results in a 30% quenching of protein fluorescence and an enhancement of the NADPH fluorescence.[19] The Förster distance for trp → NADPH energy transfer in this system is 23.4 Å. Using Eq. [1.12], one can readily calculate a distance of 26.9 Å between the single tryptophan residue and the NADPH.

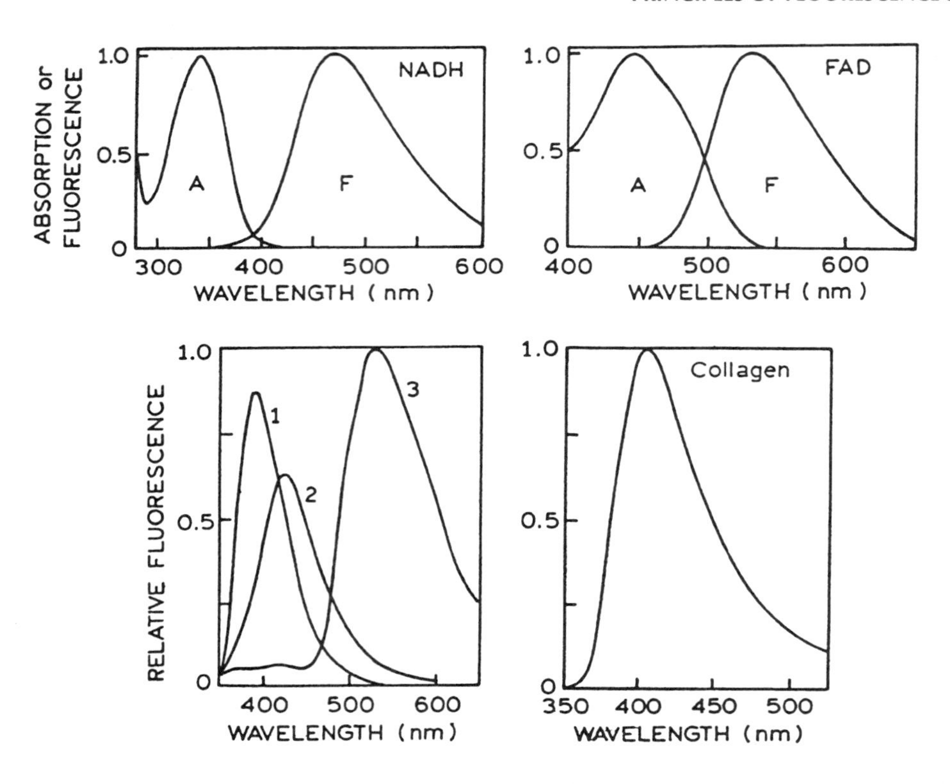

Figure 3.3. *Top*: Absorption and emission spectra of the enzyme cofactors NADH and FAD. *Bottom left*: Emission spectra of pyridoxamine phosphate (1), pyridoxal phosphate (2), and pyridoxal bound to phosphorylase *b* as a Schiff base (3), revised from Ref. 7. *Bottom right*: Emission spectrum of collagen excited at 333 nm. Revised from Ref. 14.

Examination of the emission spectra of NADPH revealed that its emission is more intense in the presence of protein. This can be seen from the increased intensity at 450 nm with 295-nm excitation (Figure 3.4, top). It was not clear from these data whether the quantum yield of NADPH has increased on binding or energy transfer caused the increased NADPH emission. It is possible to excite NADPH alone, and not the protein, using excitation at 340 nm (Figure 3.4, bottom). These spectra show that the emission of NADH is about fourfold higher in the presence of protein. This point can be understood by examination of the titration of protein with NADPH (Figure 3.5). In the absence of protein, the emission increases linearly with NADPH concentration. In the presence of protein, the intensity initially increases more rapidly and then increases as in the absence of protein. The initial increase in the intensity of NADPH is due to binding of NADPH to 17β-HSD, which occurs with a fourfold increase in the quantum yield of NADPH. Once the binding sites on 17β-HSD are saturated, the intensity increases in proportion to the concentration of unbound NADPH.

3.2. EXTRINSIC FLUOROPHORES

There are many instances where the molecule of interest is nonfluorescent, or where the intrinsic fluorescence is not adequate for the desired experiment. For instance, DNA and lipids are essentially devoid of fluorescence (Figure 1.17). In these cases useful fluorescence is obtained by labeling the molecule with extrinsic probes. In the case of proteins, it is frequently desirable to label them with chromophores that have longer excitation and emission wavelengths than the aromatic amino acids. Then the labeled protein can be studied in the presence of other unlabeled proteins.

The number of fluorophores has increased dramatically during the past decade, and no single source summarizes the diverse spectral properties. Useful information on a wide range of fluorophores can be found in the Molecular Probes catalog,[20] and there is a useful summary of fluorophores used for immunoassays.[21] Unfortunately, detailed information on most fluorophores is only available in the primary literature.

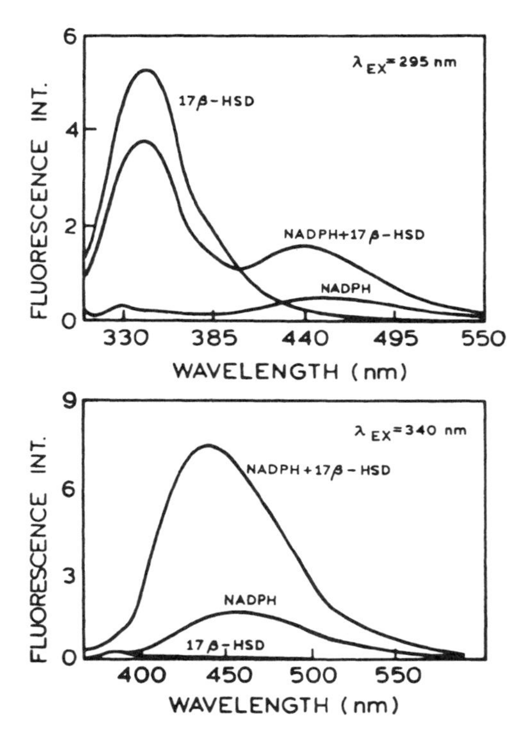

Figure 3.4. Emission spectra of 17β-hydroxysteroid dehydrogenase (17β-HSD) in the presence and absence of NADPH. *Top*: 295-nm excitation; *bottom*: 340-nm excitation. Revised from Ref. 19.

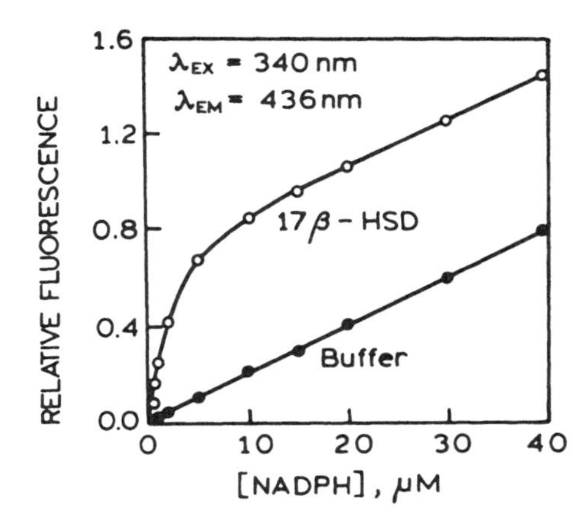

Figure 3.5. Fluorescence intensity of NADPH titrated into buffer (●) or a solution of 17β-HSD (○). Revised from Ref. 19.

3.2.A. Protein-Labeling Reagents

Numerous fluorophores are available for covalent and noncovalent labeling of proteins. The covalent probes can have a variety of reactive groups, for coupling with amine, sulfhydryl, or histidine side chains in proteins. Some of the more widely used probes are shown in Figure 3.6. Dansyl chloride (DNS-Cl) was originally described by Weber,[22] and this early report anticipated the numerous advantages of extrinsic probes in biochemical research. DNS-Cl is widely used to label proteins, especially where polarization measurements are anticipated. This wide use is a result

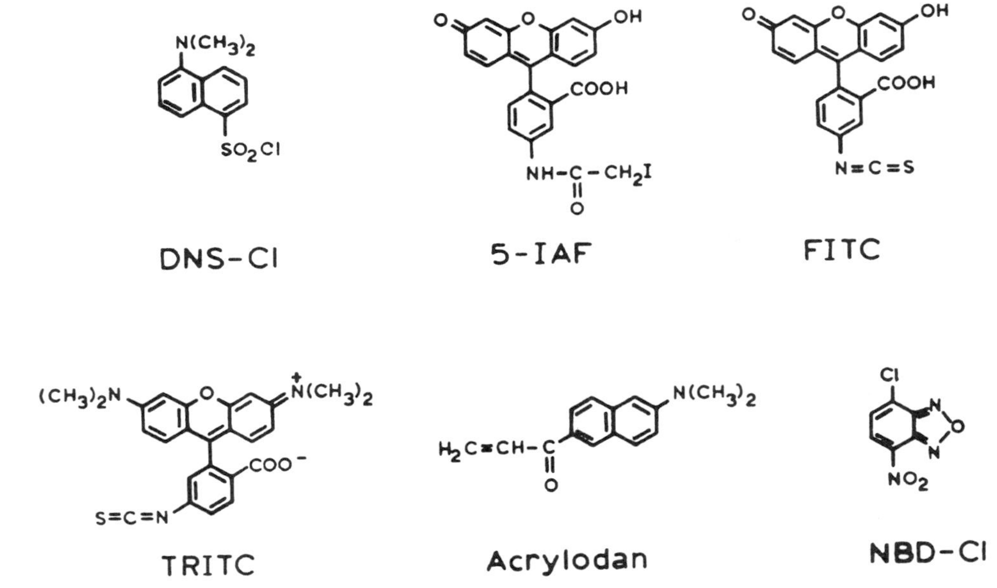

Figure 3.6. Reactive probes for conjugation with macromolecules. DNS-Cl, Dansyl chloride; 5-IAF, 5-(iodoacetamida)fluorescein; FITC, fluorescein 5-isothiocyanate; TRITC, tetramethylrhodamine 5- (and 6-)isothiocyanate; Acrylodan, 6-acryloyl-2-dimethylaminonaphthalene; NBD-Cl, 7-nitrobenz-2-oxa-1,3-diazol-4-yl chloride.

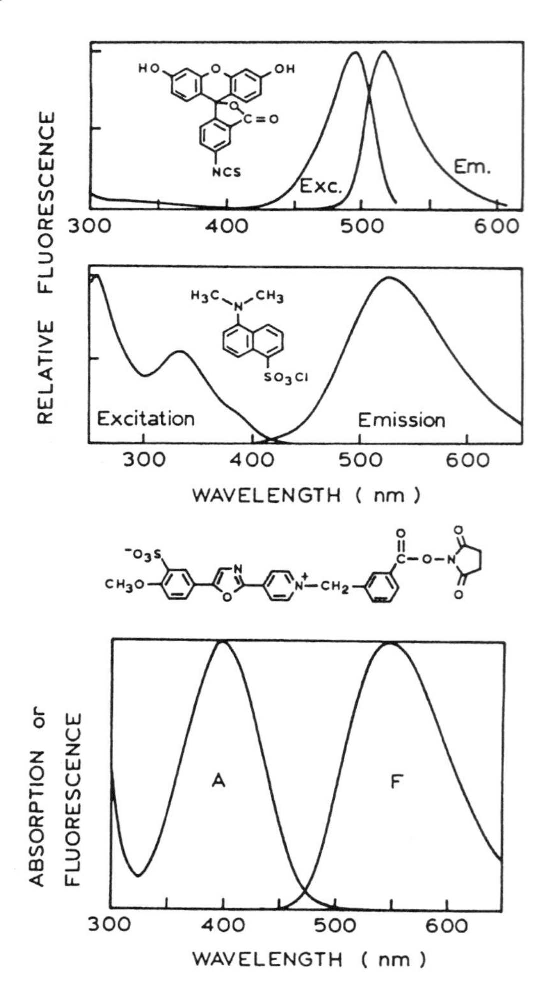

Figure 3.7. Excitation and emission spectra of FITC- (*top*) and DNS-Cl-labeled (*middle*) antibodies. Revised from Ref. 20. Also shown are the excitation and emission spectra of Cascade Yellow in methanol (*bottom*). Revised from Ref. 23.

of its early introduction in the literature and its favorable lifetime (~10 ns). Dansyl groups can be excited at 350 nm, where proteins do not absorb. However, since dansyl groups absorb near 350 nm, they can serve as acceptors of protein fluorescence. The emission spectrum of the dansyl moiety is also highly sensitive to solvent polarity, and the emission maxima are typically near 520 nm (Figure 3.7, middle).

Brief History of Gregorio Weber

In 1997 the fluorescence community lost its pioneer, Professor Gregorio Weber (Figure 3.8). The Professor, as he is referred to by those who knew him, was born in Buenos Aires, Argentina, in 1916. He received an M.D. degree from the University of Buenos Aires in 1942 and went on to graduate studies at Cambridge University. Dr. Weber's extraordinary talents were recognized by Sir Hans Krebs, who recruited him to the University of Sheffield in 1953. During his years at Sheffield, Professor Weber developed the foundations of modern fluorescence spectroscopy. In 1962, Professor Weber joined the University of Illinois at Urbana–Champaign, remaining active until his death in 1997. While at Sheffield, the Professor developed the use of fluorescence polarization for studying macromolecular dynamics. Dr. Weber's laboratory at the University of Illinois was responsible for the first widely used phase-modulation fluorometer, a design that went on to successful commercialization. Perhaps the most important area of Professor Weber's research was the development of probe chemistry and, more importantly, the recognition that fluorescence spectroscopy depends on the probes first and instrumentation second.

While dansyl chloride today seems like a common fluorophore, its introduction by Professor Weber represented a fundamental change in the paradigm of fluorescence spectroscopy. One of Professor Weber's main contributions was the introduction of molecular considerations into fluorescence spectroscopy. The dansyl group is solvent sensitive, and one is thus forced to consider its interactions with its local environment. Professor Weber recognized that proteins could be labeled with fluorophores, which in turn reveal information about the proteins and their interactions with other molecules. The probes which the Professor developed are still in widespread use, including dansyl chloride, ANS, TNS, and Prodan derivatives.

Fluoresceins and rhodamines are also widely used as extrinsic labels. These dyes have favorably long absorption maxima near 480 and 560 nm, respectively. In contrast to the dansyl group, rhodamines and fluoresceins are not sensitive to solvent polarity. An additional reason for their widespread use is their high molar extinction coefficients, near 80,000 M^{-1} cm^{-1} (Table 3.2). A wide variety of reactive derivatives are available, including iodoacetamides, isothiocyanates, and maleimides. Iodoacetamides and maleimides are typically used for labeling sulfhydryl groups whereas isothiocyanates, *N*-hydroxysuccinimide, and sulfonyl chlorides are used for labeling amines.[24] Frequently, commercial labeling reagents are a mixture of isomers.

One common use of fluorescein and rhodamine is for labeling of antibodies. A wide variety of fluorescein- and rhodamine-labeled immunoglobulins are commercially available, and these proteins are frequently used in fluores-

Figure 3.8. Professor Gregorio Weber, examining a high-pressure cell used to study proteins, circa 1972.

Table 3.2. Fluorescence Spectral Properties of Commonly Used Extrinsic Probes[a]

Fluorophore[b]	Absorption (nm)	ε (M^{-1} cm^{-1})	Emission (nm)	Q	τ (ns)
DNS-Cl	340–350	4,300	510–560	0.1–0.3	10–15
FITC	492	72,000	516–525	0.3–0.85	4.5
TRITC	535–545	107,000	570–580	—	2.0
NBD	470	12,900	550	—	
Lucifer Yellow	430	13,000	540	0.2	3.3

[a]From Ref. 21.
[b]DNS-Cl, Dansyl chloride; FITC, fluorescein 5-isothiocyanate; TRITC, tetramethylrhodamine 5- (and 6-)isothiocyanate; NBD, 7-nitrobenz-2-oxa-1,3-diazol-4-yl.

cence microscopy and in immunoassays. The reasons for selecting these probes include high quantum yields and the long wavelengths of absorption and emission, which minimize the problems of background fluorescence from biological samples and eliminate the need for quartz optics. The lifetimes of these dyes are about 4 ns, and their emission spectra are not significantly sensitive to solvent polarity. These dyes are suitable for quantifying the associations of small labeled molecules with proteins via changes in fluorescence polarization.

3.2.B. Role of the Stokes' Shift in Protein Labeling

One problem with fluorescein is its tendency to self-quench. It is well known that the brightness of fluorescein-labeled proteins does not increase linearly with the extent of labeling.[25] In fact, the intensity can decrease as the extent of labeling increases. This effect can be understood by examination of the excitation and emission spectra (Figure 3.7, top). Fluorescein displays a small Stokes' shift. When more than a single fluorescein group is bound to a protein, there can be energy transfer between these groups. This can be understood by realizing that two fluorescein groups attached to the same protein are likely to be within 40 Å of each other, which is within the Förster distance for fluorescein-to-fluorescein transfer. Stated differently, multiple fluorescein groups attached to a protein result in a high local fluorescein concentration.

Dyes which display larger Stokes' shifts are thus desirable for protein labeling. Dansyl chloride is one such probe (Figure 3.7, middle). However, its hydrophobicity becomes a problem at higher degrees of labeling, resulting in protein aggregation. New dyes are being developed which show both a large Stokes' shift and good water solubility. One such dye is Cascade Yellow,[23] which displays excitation and emission maximum near 409 and 558 nm, respectively (Figure 3.7, bottom). The large Stokes' shift minimizes the tendency for homotransfer, and the charges on the aromatic rings aid solubility.

It is convenient to have some representative emission spectra of the commonly used fluorescein and rhodamine derivatives (Figure 3.9, top). The chemical structures of some typical rhodamine derivatives are shown in Figure 3.10. Emission maxima can range from 518 to 621 nm. In the past several years, a new class of dyes has been introduced, under the trademark BODIPY, as replacements for fluorescein and rhodamines.[23] These dyes are based on an unusual boron-containing fluorophore (Figure 3.11). Depending on the structure, a wide range of emission wavelengths can be obtained (Figure 3.9, bottom), from 510 to 675 nm. The BODIPY dyes have the additional advantage of displaying high quantum yields approaching unity, extinction coefficients near 80,000 M^{-1} cm^{-1}, and insensitivity to solvent polarity and pH. The emission spectra are less wide than those of fluorescein and rhodamines, so that more of the light is emitted at the peak wavelength, possibly allowing more individual dyes to be resolved. A disadvantage of the BODIPY dyes is a very small Stokes' shift.[26]

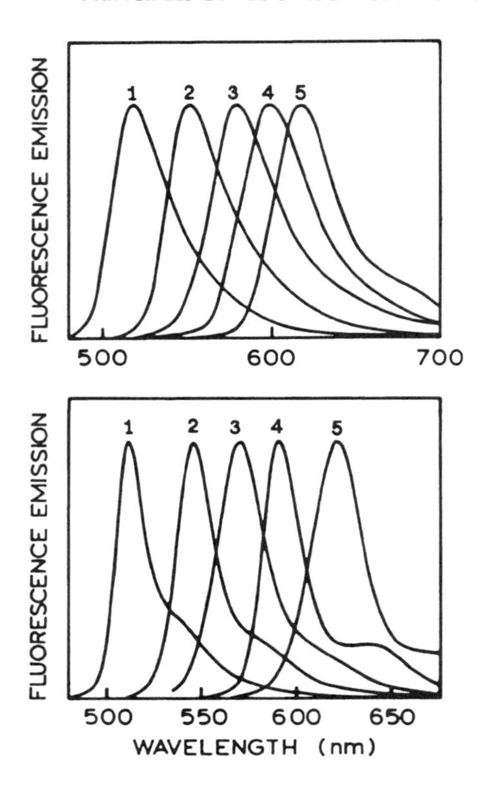

Figure 3.9. *Top*: Normalized fluorescence emission spectra of goat anti-mouse IgG conjugates of commonly used dyes: (1) fluorescein, (2) rhodamine 6G, (3) tetramethylrhodamine, (4) Lissamine rhodamine B, and (5) Texas Red. *Bottom*: Normalized fluorescence emission spectra of BODIPY dyes in methanol: (1) BODIPY-FL, (2) BODIPY-R6G, (3) BODIPY-TMR, (4) BODIPY-581/591, and (5) BODIPY-Texas Red. Structures are shown in Figures 3.10 and 3.11. From Ref. 23, Molecular Probes, Inc.

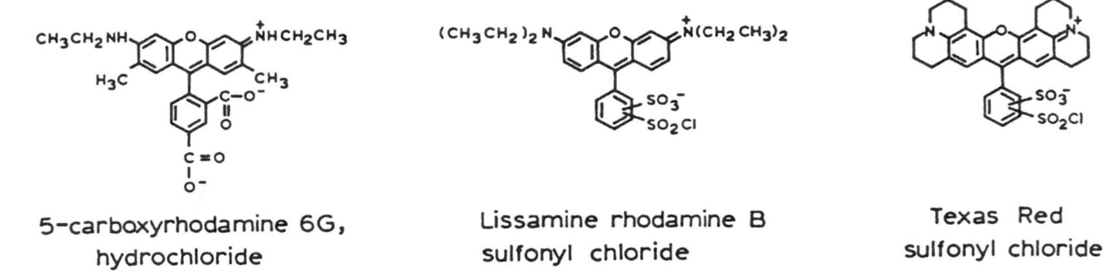

Figure 3.10. Representative structures of rhodamine derivatives. Emission spectra are shown in Figure 3.9. Lissamine and Texas Red are trademarks of Molecular Probes, Inc.

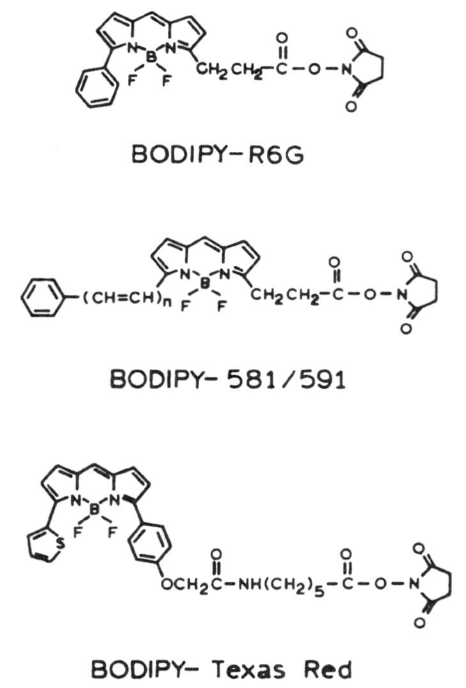

BODIPY-R6G

BODIPY-581/591

BODIPY-Texas Red

Figure 3.11. Representative structures of BODIPY derivatives. Emission spectra are shown in Figure 3.9. BODIPY is a trademark of Molecular Probes, Inc.

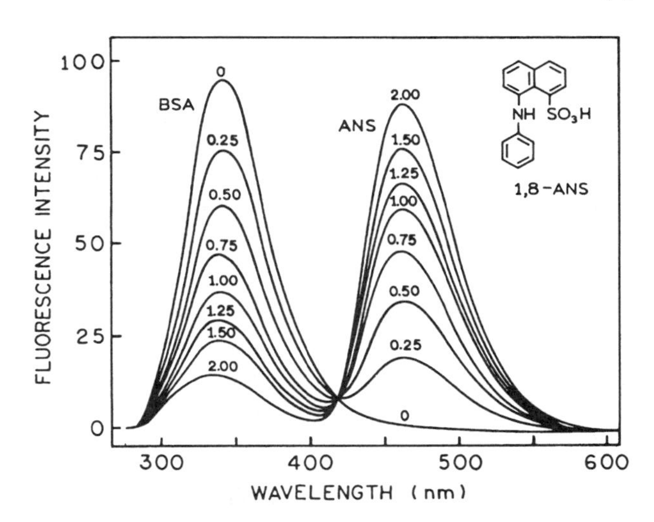

Figure 3.12. Fluorescence emission spectra of bovine serum albumin (BSA) in the presence of increasing concentration of ANS. The numbers indicate the average number of ANS molecules bound per BSA molecule. Excitation at 280 nm. Revised from Ref. 31.

As a result, the dyes transfer to each other with a Förster distance of about 57 Å.

3.2.C. Solvent-Sensitive Probes

In contrast to fluorescein, rhodamines, and BODIPYs, there are fluorophores which display high sensitivity to the polarity of the local environment. One example is Prodan,[27] which is available in the reactive form called acrylodan (Figure 3.6).[28] In the excited state, there is a charge separation from the amino to the carbonyl groups. When bound to membranes, Prodan and its derivatives display large spectral shifts at the membrane phase-transition temperature.[29]

3.2.D. Noncovalent Protein-Labeling Probes

There are also a number of dyes which can be used to label proteins noncovalently. These are typically naphthylamine sulfonic acids, of which 1-anilinonaphthalene-8-sulfonic acid (ANS) and 6-(*p*-toluidinyl)naphthalene-2-sulfonic acid (TNS) are most commonly used.[30] Dyes of this class are frequently weakly fluorescent or nonfluorescent in water but fluoresce strongly when bound to proteins (apomyoglobin) or membranes (DMPC, Figure 1.20). The use of these dyes can be traced to the early studies of Weber and co-workers, who examined their binding to bovine serum albumin (BSA).[31] Figure 3.12 shows the emission spectra of BSA excited at 280 nm as the sample is titrated with ANS. In the absence of BSA, the emission from the ANS dissolved in buffer would be insignificant (not shown). One notices that the tryptophan emission from BSA is quenched upon addition of ANS and that the ANS emission increases as the BSA emission decreases. There is no observable emission from ANS alone, which shows an emission maximum above 500 nm in water. ANS-type dyes are amphipathic, so that the nonpolar region prefers to adsorb onto nonpolar regions of macromolecules. Since the water-phase dye does not contribute to the emission, the observed signal is due to the area of interest, the probe binding site on the macromolecule.

The sensitivity of ANS to its environment was shown by studies of the ANS lifetimes as BSA was progressively saturated with more bound ANS molecules. The greatest enhancements and longest decay times were found for the first two bound ANS molecules. As more ANS was bound, the decay time decreased. These results were interpreted as due to ANS binding to weaker sites which provided a less favorable environment for ANS. In general, a range of emission spectra and decay times are found for ANS bound to different proteins.

Binding of ANS to BSA or human serum albumin (HSA) is one experiment you can do at home. Prepare aqueous solutions of ANS (about 10^{-5} *M*) and BSA (about 10 mg/ml) and observe them under a UV handlamp. Little emission will be seen from either sample. Any emission seen from the ANS solution will be weak and greenish. Then mix the two solutions while illuminating with the UV handlamp. You will notice an immediate increase in fluorescence intensity and a shift of the ANS emission to the blue. We frequently use this demonstration to illustrate fluorescence to students.

3.2.E. Membrane Probes

Labeling of membranes is often accomplished by simple partitioning of water-insoluble probes into the nonpolar regions of membranes. DPH is one of the most commonly used membrane probes. Addition of DPH to a membrane suspension results in complete binding, with no significant emission from DPH in the aqueous phase. All the emission from DPH is then due to DPH in the membrane environment. Within the past few years, the task of labeling membranes has been made easier by the availability of a wide variety of lipid probes. A few examples are shown in Figures 3.13 and 3.14. Lipid probes can be attached to the fatty acid chains or to the phospholipids themselves. The depth of this probe in the bilayer can be adjusted by the length of the various chains, as shown for the anthroyl fatty acid. DPH, which is often used as a partitioning probe, can be localized near the membrane–water interface by attachment of a trimethylammonium group to one of the phenyl rings (TMA-DPH).[32] Pyrene has been attached to lipids (pyrenyl lipid) to estimate diffusive processes in membranes by the extent of excimer formation. Unsaturated fatty acids can also be fluorescent if the double bonds are conjugated as in parinaric acid.[33]

Membranes can also be labeled with more water-soluble probes like fluorescein or rhodamine. In such cases the probes can be forced to localize in the membrane by attachment to long acyl chains or to the phospholipids themselves (Figure 3.14). Depending on chemical structure, the fluorescent group can be positioned either on the fatty acid side chains (fluorenyl-PC) or at the membrane–water interface (Texas Red-PE). In the latter case, the fluorophore Texas Red was selected because of its long-wavelength absorption and high photostability.

It is interesting to consider the spectral properties of two membrane probes that show unusual behavior. The absorption spectrum of α-parinaric acid is highly structured, but the emission is unstructured (Figure 3.15). The explanation for this unusual effect is that the emission originates from a state of different symmetry from that seen in the absorption spectrum. Stated differently, the lowest $S_0 \rightarrow S_1$ transition is not allowed, so the absorption spectrum reflects excitation to a higher level.[33] Emission still occurs as an $S_1 \rightarrow S_0$ transition. This unusual effect is also seen for polyenes like DPH.[34–36] It has the favorable consequence of providing a high extinction coefficient, large Stokes' shift, and long lifetime (near 10 ns for DPH) due to the slower $S_1 \rightarrow S_0$ emission rate. The measured decay times of polyenes are often longer than the natural lifetime, because the natural lifetime is calculated from a different transition.

The pyrenyl-PC probe (Figure 3.14) also displays unusual spectral properties. The emission spectra of pyrenyl-PC liposomes is highly dependent on temperature (Figure 3.16). Of course, the unstructured emission at higher temperatures is due to excimer formation between the pyrene groups.[37] If the pyrenyl-PC is diluted into membranes, the amount of excimer formation can be used to estimate the rate of lateral diffusion of lipids in the membranes.

3.2.F. Membrane Potential Probes

There are also membrane probes which are sensitive to the electrical potential across the membrane. Typical mem-

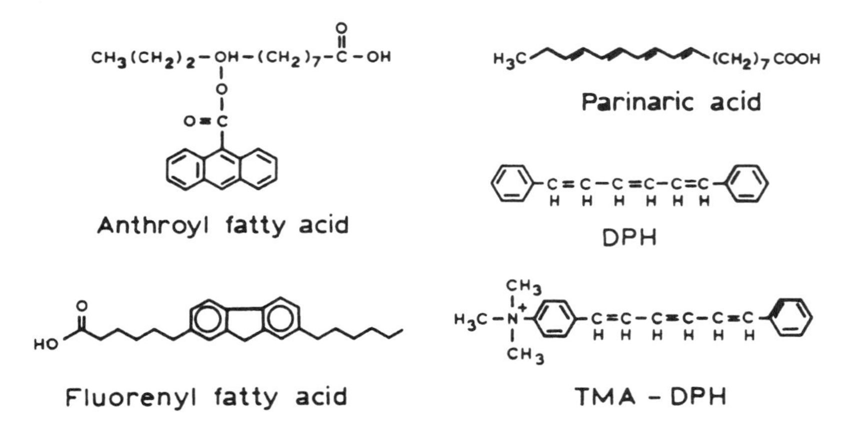

Figure 3.13. Membrane probes which partition into lipid bilayers.

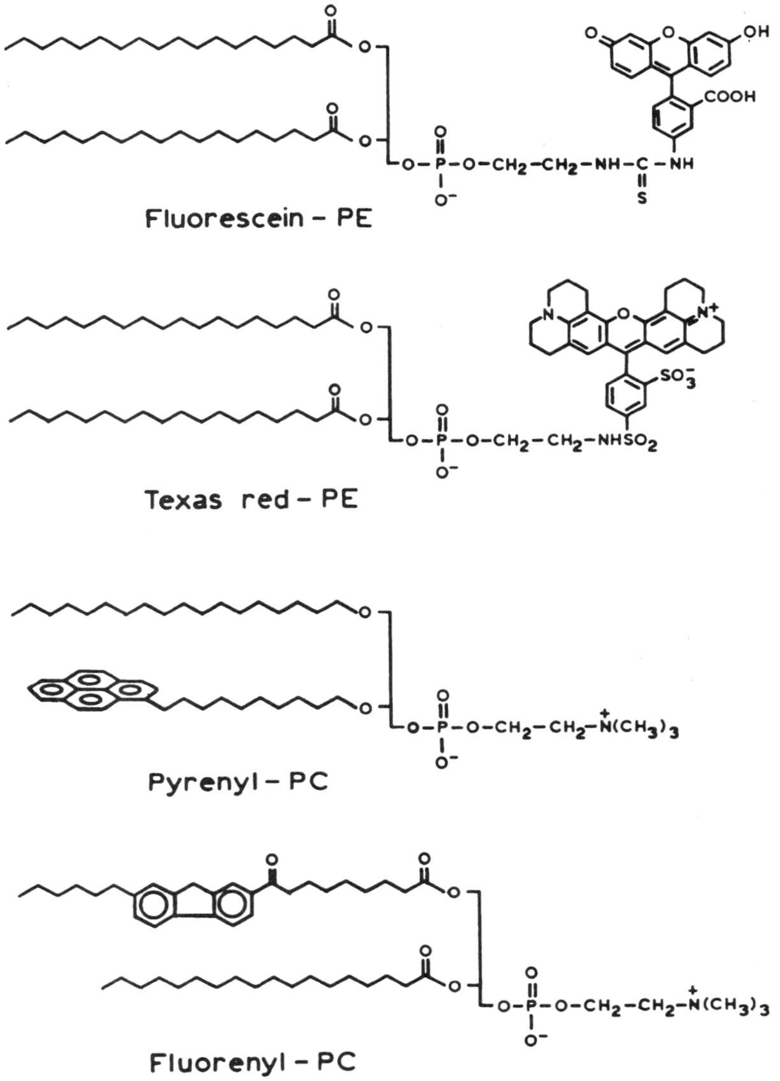

Fluorescein - PE

Texas red - PE

Pyrenyl - PC

Fluorenyl - PC

Figure 3.14. Fluorescent phospholipid analogs. PC, Phosphatidylcholine; PE, phosphatidylethanolamine. Texas Red is a trademark of Molecular Probes, Inc.

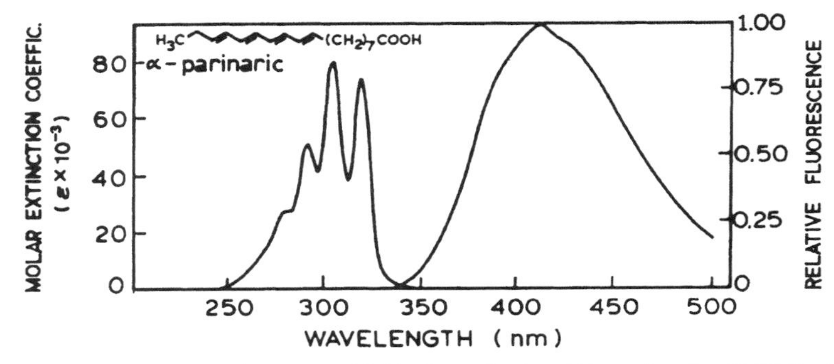

Figure 3.15. Absorption (*left*) and emission spectra (*right*) of α-parinaric acid in methanol at 25 °C. Revised from Ref. 33.

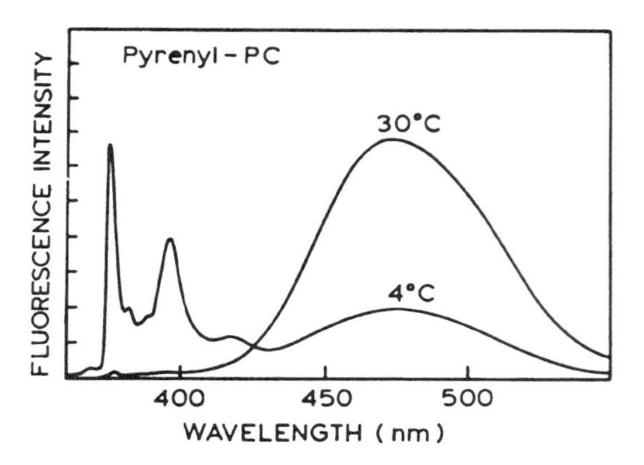

Figure 3.16. Fluorescence emission spectra for pyrenyl-PC measured at 4 and at 30 °C. The lipid probe was dispersed in water. From Ref. 37 and reprinted with permission from Springer-Verlag, Inc.

brane potential probes are shown in Figure 3.17. A number of mechanisms are thought to be responsible, including partitioning of the dye from the water to the membrane phase, reorientation of the dyes in the membrane, aggregation of dyes in the membrane, and the inherent sensitivity of the dyes to the electric field. The carbocyanine dyes typically respond to potential by partitioning and/or aggregation in the membranes,[38,39] whereas the stryryl dyes seem to respond directly to the electric field.[40–43] The merocyanine dyes probably respond to membrane potential by both mechanisms.[44] This field of membrane potential probes is extensive and complex.[45–47]

3.3. RED AND NEAR-INFRARED (NIR) DYES

A discussion of available probes would not be complete without mention of longer-wavelength probes. The cyanine dyes, which were developed for use as membrane potential probes, evolved into some of the more commonly used long-wavelength dyes. Long-wavelength probes are of current interest for several reasons. The sensitivity of fluorescence detection is often limited by the autofluorescence of biological samples. As the excitation wavelength becomes longer, the autofluorescence decreases, and hence detectability over background increases.[48] Long-wavelength dyes can be excited with simple laser sources, such as solid-state laser diodes, such as laser pointers. Long-wavelength dyes are available in the class of cyanine dyes, such as Cy-3, Cy-5, and Cy-7 in Figure 3.18. Such dyes have absorption and emission wavelengths above 650 nm.[49] The cyanine dyes typically display small Stokes' shifts, with the absorption maxima blue-shifted about 30 nm from the emission maxima, as shown for Cy-3 in Figure 3.18. A wide variety of conjugatable cyanine dyes are available. Charged side chains are used for improved water solubility or to prevent self-association, which is a common cause of self-quenching in these dyes.

Additional long-wavelength dyes are shown in Figure 3.19. Some rhodamine derivatives display long-wavelength absorption and emission spectra, as seen for rhodamine 800.[50] The oxazine dyes display surprising long-wavelength absorption and emission maxima, given their small size.[51] Extended conjugated systems result in

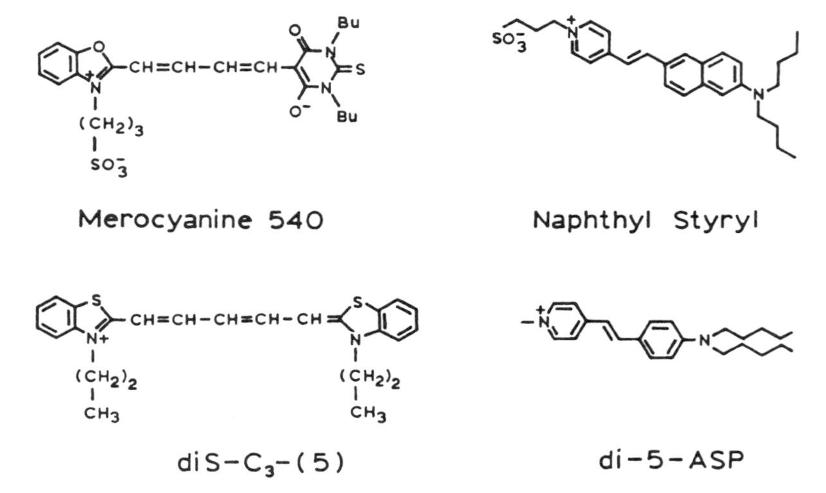

Figure 3.17. Membrane potential probes. Napthyl styryl is di-4-ANEPPS or 1-(3-sulfonatopropyl)-4-[β-[2[(di-*n*-butylamino)-6-naphthyl]vinyl]pyridinium betaine; dis-C_3-(5) is 3,3′-dipropylthiadicarbocyanine; and di-5-ASP is 4-(*p*-dipentyl aminostyryl)-1-methylpyridinium.

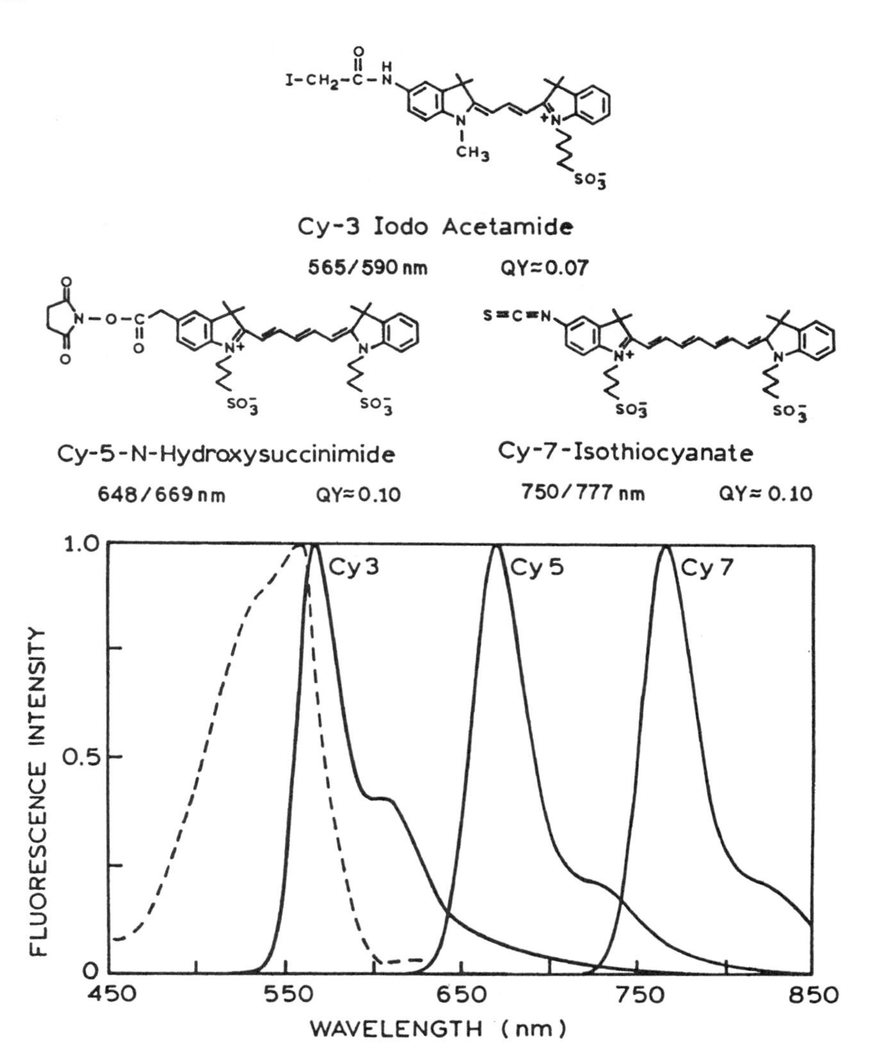

Figure 3.18. Fluorescence emission spectra of chemically reactive cyanine dyes. The dashed curve shows the absorption spectrum of Cy-3.

long absorption and emission wavelengths, as shown for IR-125 and Thiazole Orange. Dyes of this class have been extensively characterized for use as long-wavelength probes and in DNA sequencing.[52] The dye Thiazole Orange can be excited at 735 nm and binds strongly to DNA.[53] Dyes of this type are also used for staining DNA restriction fragments during capillary electrophoresis.

Another class of long-wavelength dyes are the phthalocyanines and naphthalocyanines (Figure 3.19). At present, these dyes are less used in biochemistry owing to their lack of water solubility, their tendency to aggregate, and the lack of conjugatable forms. The spectral properties of the NIR dyes have been summarized in Refs. 48, 54, and 55.

3.3.A. Measurement of Human Serum Albumin with Laser Diode Excitation

The use of an NIR dye for clinical sensing is illustrated in Figure 3.20. The dye AB670 displays properties analogous to those of ANS and TNS, except at longer wavelengths. In the absence of HSA, the AB670 emission is weak, but it increases strongly upon binding to HSA.[56] In the presence of HSA, AB670 is strongly fluorescent. A favorable feature of this probe is that it can be excited at 670 nm; this wavelength is readily available from a laser diode. This dye has been used to develop a clinical assay for HSA. The assay is claimed to be sensitive to 0.7 mg of HSA/l and to provide a low-cost method for measuring HSA.

3.4. DNA PROBES

While very weak intrinsic emission has been observed from unlabeled DNA, this emission is too weak and too deep in the UV for practical applications. Fortunately, there are numerous probes which spontaneously bind to DNA and display enhanced emission.[57–59] Several representative DNA probes are shown in Figure 3.21. One of the most widely used dyes is ethidium bromide (EB). EB is weakly fluorescent in water, and its intensity increases about 30-fold upon binding to DNA. The lifetime of EB is about 1.7 ns in water and increases to about 20 ns upon binding to double-helical DNA. The mode of binding appears to be due to intercalation of the planar aromatic ring between the base pairs of double-helical DNA. Many DNA probes, such as Acridine Orange, also bind by intercalation. Other types of probes bind into the minor groove of DNA, such as 4′, 6-diamidino-2-phenylindole (DAPI) and Hoechst 33342 (Figure 3.21). The fluorescence of DAPI appears to be most enhanced when adjacent to adenine–thymine (AT)-rich regions of DNA.[60] Hoechst 33358 binds with some specificity to certain base-pair sequences.[61–63]

In recent years, improved DNA dyes which bind to DNA with high affinity have been developed. Typical high-affin-

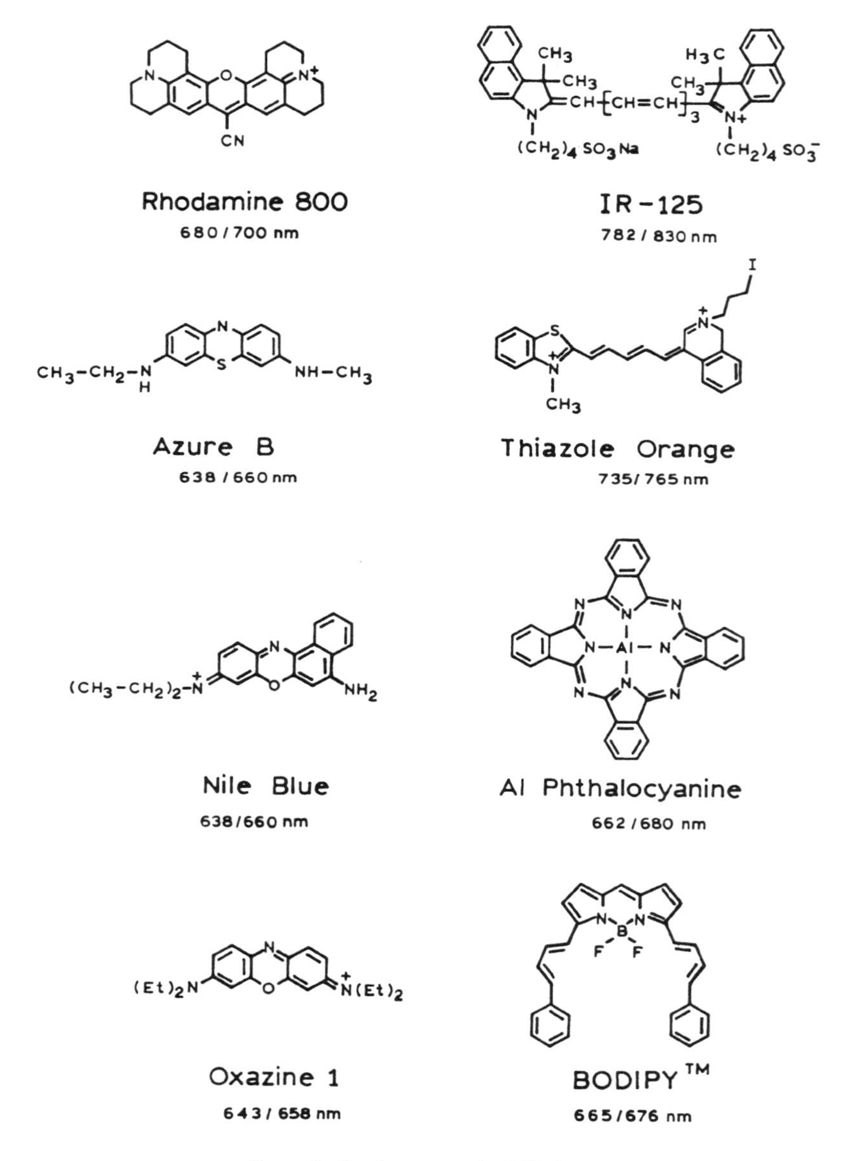

Figure 3.19. Representative NIR dyes.

NC
CN
CN
Cl
Cl
O
N
N(CH3)4
8.0
6.0
4.0
2.0
0
600
650
700
750
WAVELENGTH (nm)
ABSORPTION (——)
or FLUORESCENCE INTENSITY(- - -)
+HSA
-HSA
670nm
Laser Diode

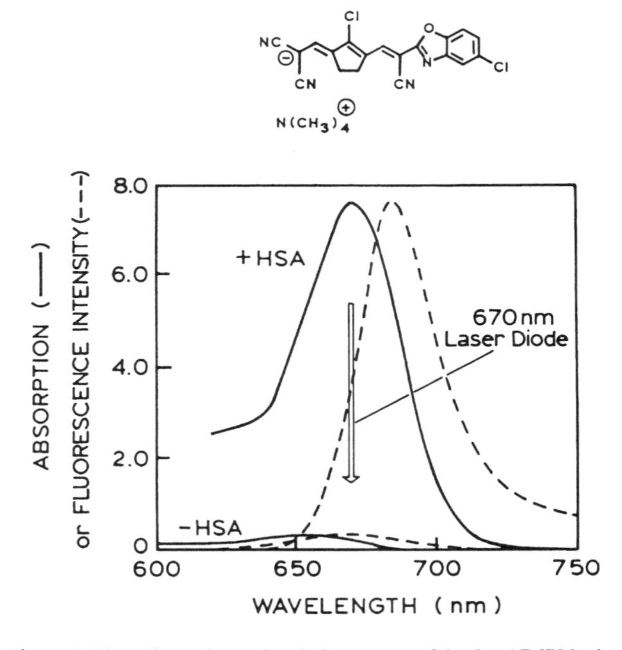

Figure 3.20. Absorption and emission spectra of the dye AB670 in the absence (–HSA) and presence (+HSA) of human serum albumin. From Ref. 56.

ity dyes are dimers of known DNA probes, such as the ethidium homodimer[64] and elongated positively charged dyes like TOTO-1[65] (Figure 3.21). Such dyes remain bound to DNA during gel electrophoresis and allow DNA detection with high sensitivity.

3.4.A. DNA Base Analogs

The native bases of DNA are not useful as fluorescent probes, and thus the use of extrinsic DNA probes is necessary. However, DNA can be made fluorescent by the use of DNA base analogs. 2-Aminopurine (2-AP) is an analog of adenine, and isoxanthopterin (IXP) is an analog of guanine (Figure 3.22).

In solution, 2-AP has a high quantum yield and a single exponential decay time near 10 ns. Upon incorporation into DNA double-strand oligomers, its fluorescence is partially quenched and its decay becomes complex.[66] The sensitivity of 2-AP to its environment makes it a useful probe for studies of DNA conformation and dynamics.[67]

Like 2-AP, IXP is partially quenched when in double-helical DNA (Figure 3.23) but is more fluorescent when

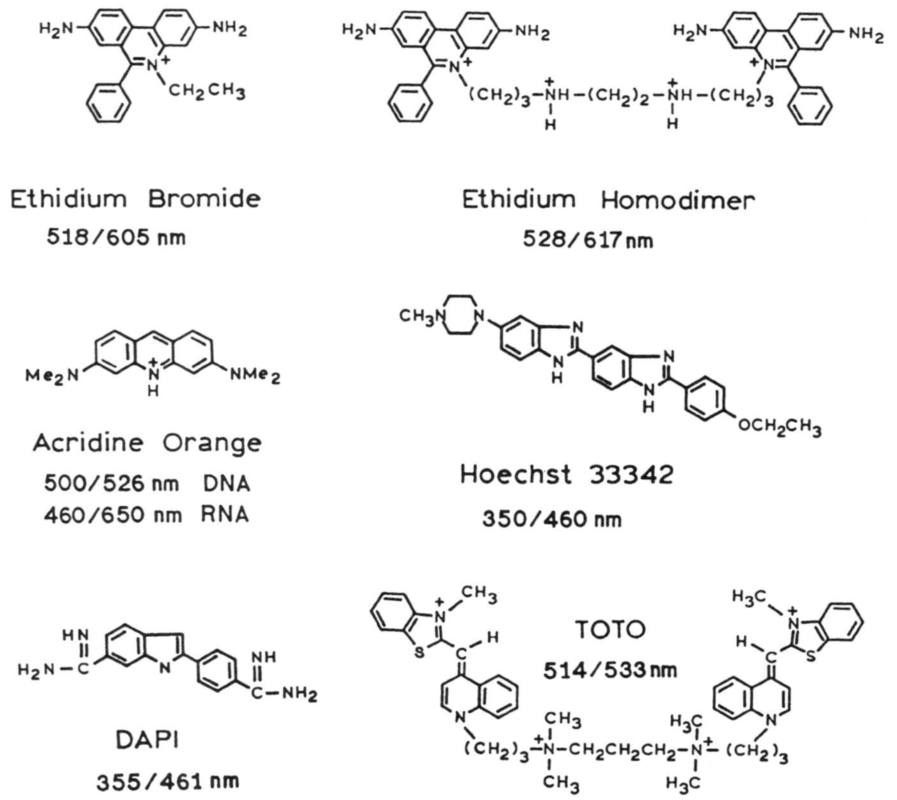

Ethidium Bromide
518/605 nm
Ethidium Homodimer
528/617 nm
Acridine Orange
500/526 nm DNA
460/650 nm RNA
Hoechst 33342
350/460 nm
DAPI
355/461 nm
TOTO
514/533 nm

Figure 3.21. Representative DNA probes. Excitation and emission wavelengths refer to DNA-bound dye.

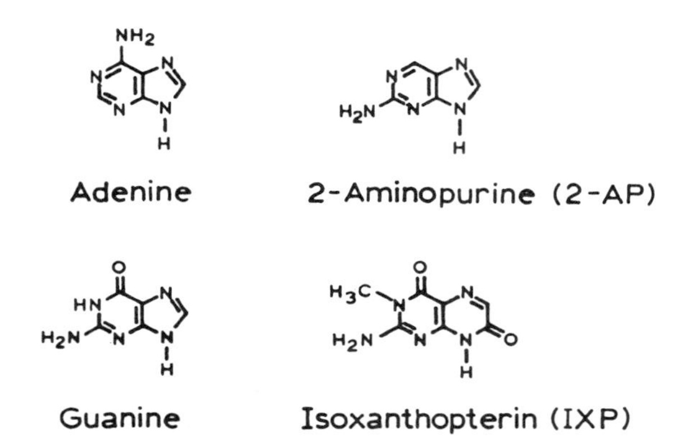

Figure 3.22. DNA purine bases (*left*) and fluorescent base analogs (*right*).

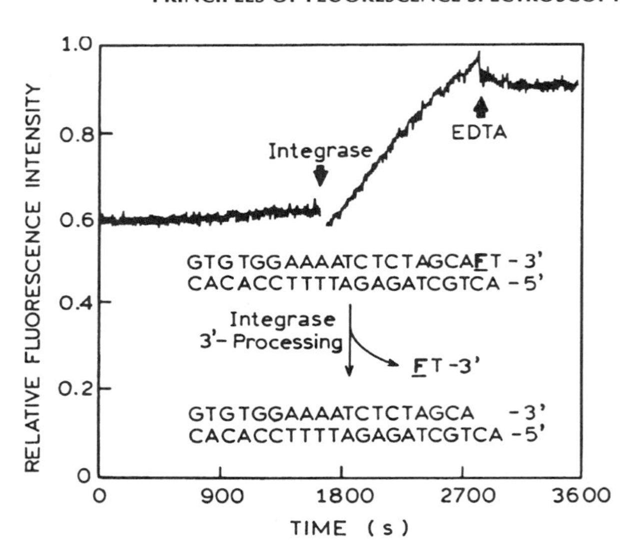

Figure 3.24. HIV integrase assay based on release of the isoxanthopterin-containing nucleotide (F). Revised from Ref. 68, and reprinted with permission of Oxford University Press.

present in a dinucleotide.[68] The dependence of the IXP fluorescence on DNA structure was used for an assay for the human immunodeficiency virus (HIV) integrase protein. This protein is responsible for integration of HIV DNA into the host-cell genome.[69,70] The assay was based on a DNA oligonucleotide which has the sequence specific for HIV integrase (Figure 3.24). The enzyme mechanism involves cleavage of a dinucleotide from the 3′-end of HIV DNA, followed by ligation to the 5′-end of the host DNA. The IXP nucleotide was positioned near the 3′-end of the synthetic substrate. Incubation with HIV integrase resulted in release of the dinucleotide, which was detected by an increase in IXP fluorescence (Figure 3.24).

3.5. CHEMICAL SENSING PROBES

It is often desirable to detect spectroscopically silent substances such as Cl^-, Na^+, or Ca^{2+}. This is possible using

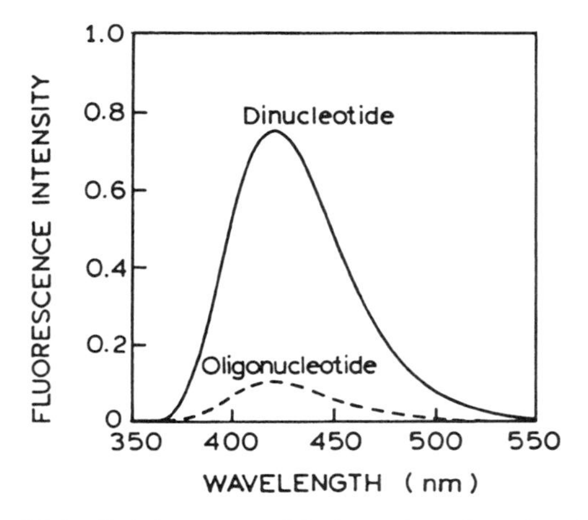

Figure 3.23. Emission spectra of the isoxanthopterin nucleotide in a dinucleotide and an oligonucleotide. From Ref. 68.

sensing probes, some of which are shown in Figure 3.25. The probe N(ethoxycarbonylmethyl)-6-methoxyquinolinium (MQAE) is collisionally quenched by chloride according to the Stern–Volmer equation (Eq. [1.6]), allowing the chloride concentration to be estimated from the extent of quenching.[71] Other probes allow measurement of free Ca^{2+}. Probes such as Fura-2 display Ca^{2+}-dependent spectral shifts. Such probes are called wavelength-ratiometric probes because the analyte (Ca^{2+}) concentration can be determined from a ratio of intensities at different excitation or emission wavelengths. Other probes such as Calcium Green display a Ca^{2+}-dependent increase in intensity but no spectral shift. Wavelength-ratiometric and nonratiometric probes are known for many species, including Na^+, K^+, and Mg^{2+},[72–74] amines, and phosphate,[75] as well as for pH (Figure 3.25). These dyes typically consist of a fluorophore and a region for analyte recognition, such as an azacrown ether for Na^+ or K^+ or a BAPTA group for Ca^{2+}. Such dyes are most often used in fluorescence microscopy and cellular imaging and are trapped in cells either by hydrolysis of cell-permeable esters or by microinjection.

3.6. SPECIAL PROBES

3.6.A. Fluorogenic Probes

Another class of probes is the fluorogenic probes.[76,77] These are dyes which are weakly flourescent or nonfluorescent until some event occurs, such as enzymatic cleav-

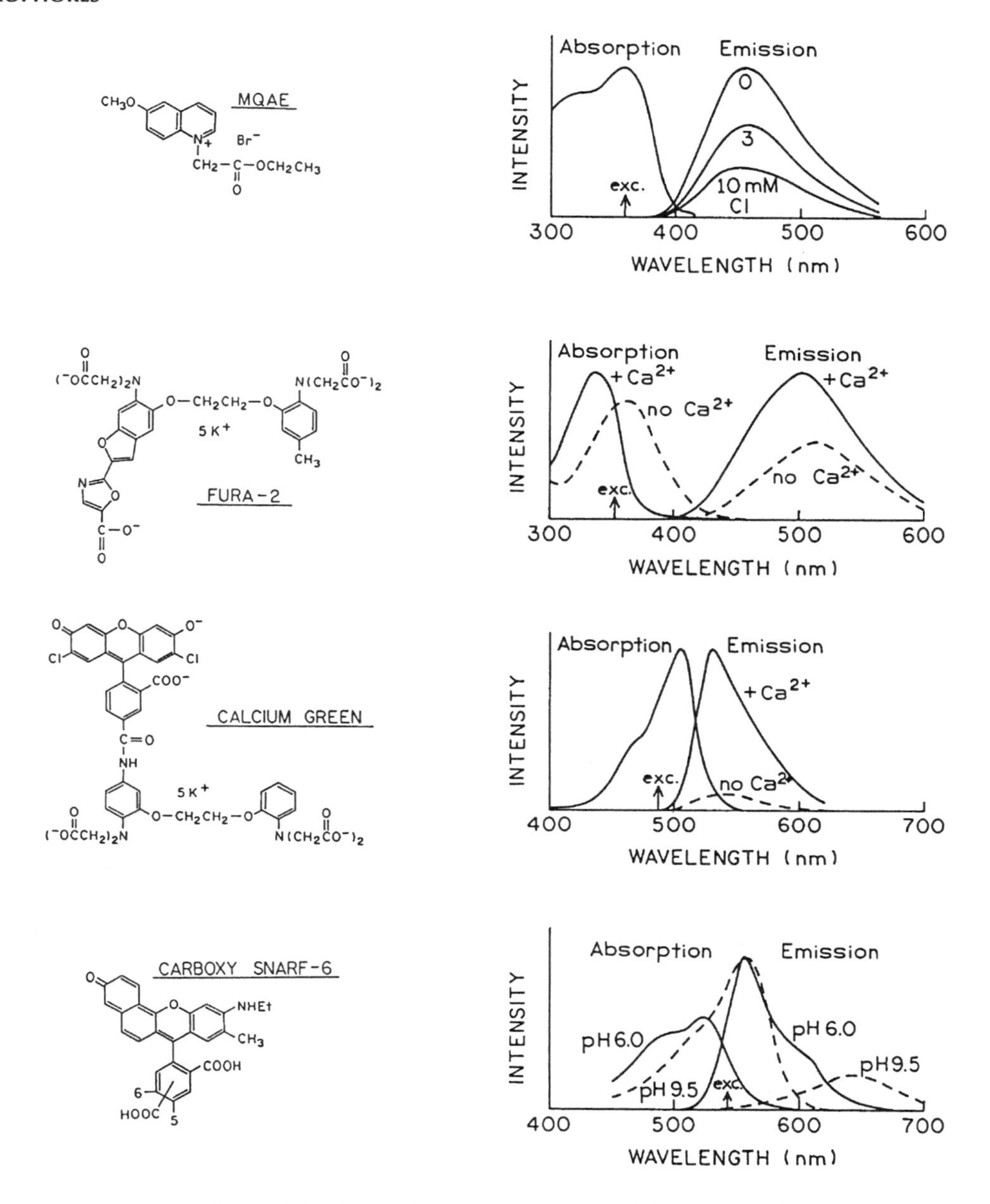

Figure 3.25. Chemical sensing probes (*left*) and their spectra (*right*).

age. Typical fluorogenic probes are shown in Figure 3.26. 7-Umbelliferyl phosphate (7-UmP) is nonfluorescent as the phosphate ester but becomes highly fluorescent upon hydrolysis. 7-UmP is used to measure the activity of alkaline phosphatase. This enzyme is often used as the basis of ELISA assays and is also used in enzyme-amplified DNA assays.

It is often important to measure β-galactosidase activity in cells. This enzyme is often used as a gene marker in cells. Its activity can be detected with the use of a galactoside of umbelliferone or 7-hydroxycoumarin (Figure 3.26, top and middle reactions, respectively). The bottom reaction in Figure 3.26 employs an improved probe. This fluorescein derivative contains a fatty acid chain which serves to retain the probe at the site of hydrolysis. This allows the cells with β-galactosidase activity to be identified under a microscope.

Another class of fluorogenic reagents comprises those which are initially nonfluorescent and become fluorescent upon reacting with amines (Figure 3.27 and Table 3.3). While they have been used for labeling proteins, they are more commonly used in protein sequencing, in the deter-

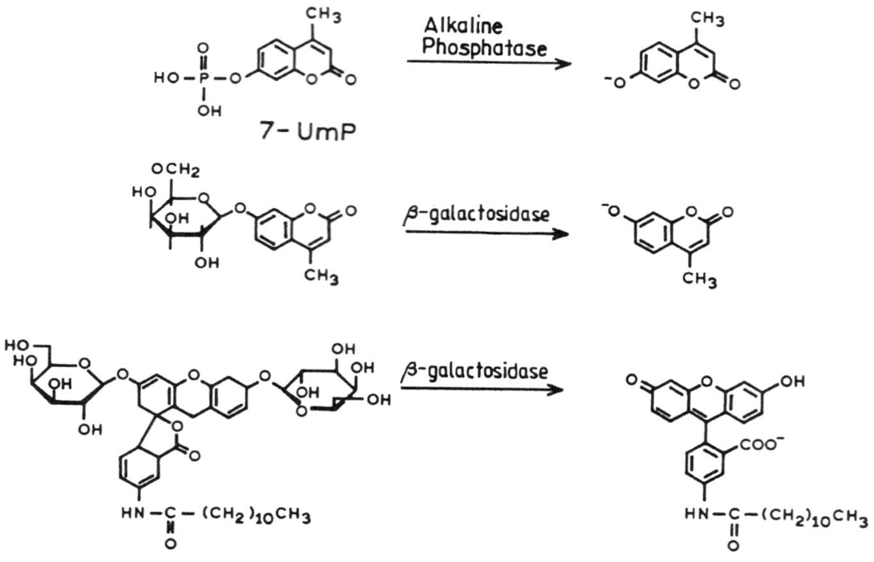

Figure 3.26. Fluorogenic probes.

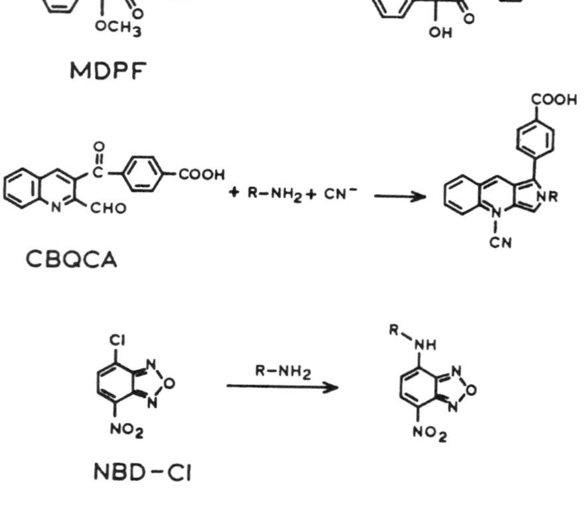

Figure 3.27. Fluorogenic reagents for amines. See Table 3.3.

mination of protein concentration, or for detection of low-molecular-weight amines in chromatography.

Fluorogenic probes can also be based on energy transfer.[78–80] One example is shown in Figure 3.28, in which a peptide is labeled with a donor and an acceptor. The sequence of amino acids was selected to be specific for a protease found in HIV. Cleavage of the peptide results in greater distance between the donor and the acceptor and increased donor intensity. This concept of a decrease in energy transfer upon cleavage has also been applied to lipases which hydrolyze phospholipids. Pyrene-labeled lipids, similar to that shown in Figure 3.14, have been synthesized with an additional trinitrophenyl (TNP) group.[81,82] The pyrene fluorophore is quenched by resonance energy transfer to the nonfluorescent TNP group. Upon cleavage by a lipase, the TNP group is removed and the pyrene becomes fluorescent.

A final type of fluorogenic probe is shown in Figure 3.29. The probe is soluble and nonfluorescent prior to

Table 3.3. Fluorogenic Amino Reagents[a]

Compound[b]	λ_{ex} max (nm)	ε (M^{-1} cm^{-1})	λ_{em} max (nm)
Fluorescamine	395	6,300	480
MDPF	290, 390	6,400	480
CBQCA	465	—	560
NBD	470	12,900	550

[a]From Ref. 21, p. 127 and Ref. 20 p. 42.
[b]MDPF, 2-methoxy-2,4-diphenyl-3 (2H)-furanone; CBQCA, 3-(4-carboxybenzoyl)quinoline-2-carboxaldehyde; NBD, 7-Nitrobenz-2-oxa-1,3-diazol-4-yl.

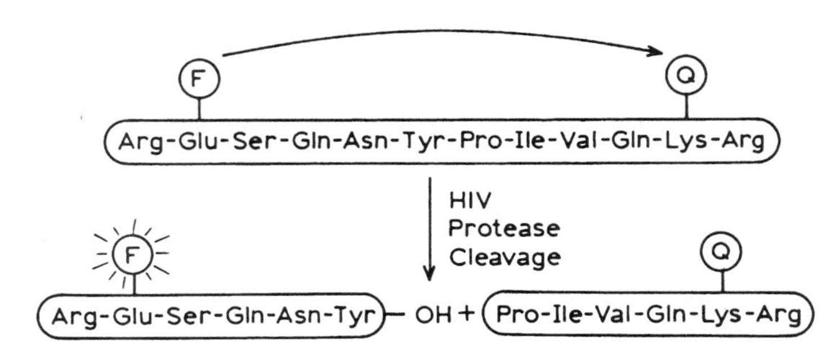

Figure 3.28. Fluorogenic probes for HIV protease. The fluorescence signal is generated when HIV protease releases the fluorophore (F) from the quenching effects of the nearby acceptor chromophore (Q). Revised from Ref. 20, from Molecular Probes, Inc.

enzymatic cleavage of the R group. The enzyme can be a phosphatase or galactosidase. Upon hydrolysis, the probe precipitates where it is formed.[83–85] The precipitates are highly fluorescent and remain at the site of enzymatic activity.

3.6.B. Structural Analogs of Biomolecules

Another approach to designing fluorophores is to make the shape similar to that of the parent biomolecule. This approach was used to make fluorescent analogs of steroids. Two examples are shown in Figure 3.30. Cholesterol is an essential component of cell membranes, and estradiol is important for the expression of female sexual characteristics. Both molecules are nonfluorescent. However, structurally similar molecules have been synthesized which display useful fluorescence. Dehydroergosterol is a fluorescent analog of cholesterol that displays absorption and emission maxima near 325 and 390 nm, respectively. Dehydroergosterol has been used as a probe for the interactions of steroids with membranes.[86–89]

Another type of fluorescent steroid was used as a probe for the estrogen receptor. This receptor plays a role in the growth of hormone-responsive breast cancers, and analysis of receptor levels is an important aspect in the treatment of breast cancer. Estradiol is not fluorescent, so it cannot be used for optical measurement of its receptor. Fluorescent analogs of estradiol have been developed,[90,91] one example being the azatetrahydrochrysene in Figure 3.30. Its absorption and emission spectra are shown in Figure 3.31. This estradiol analog is highly sensitive to solvent polarity, which is an advantage for understanding the local environment of this probe.

The concept of a ligand analog can be extended to other systems. In Figure 1.18 we showed the structures of ε-ATP and *lin*-benzo-AMP, which are fluorescent analogs of ATP and AMP, respectively. Additional substrate analogs include pyrenebutyryl coenzyme A,[92] which is an analog of fatty acid coenzyme A, and formycin, which is a fluorescent analog of adenosine (Figure 3.32).[93]

3.6.C. Viscosity Probes

Although fluorescent quantum yields are often dependent on viscosity, there are relatively few fluorophores characterized as viscosity probes. One such probe is shown in Figure 3.33. This probe displays charge transfer in the

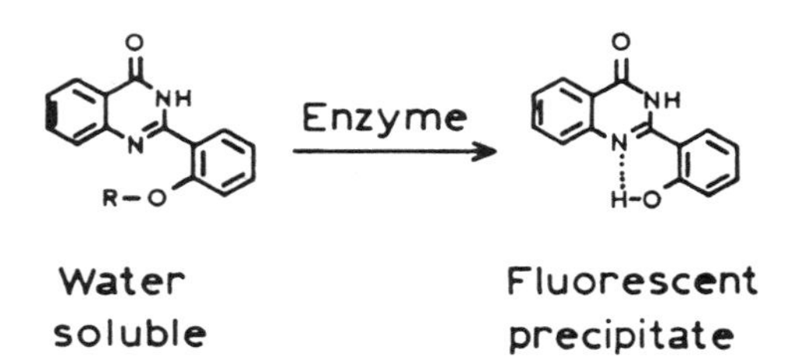

Figure 3.29. A fluorogenic precipitating probe.

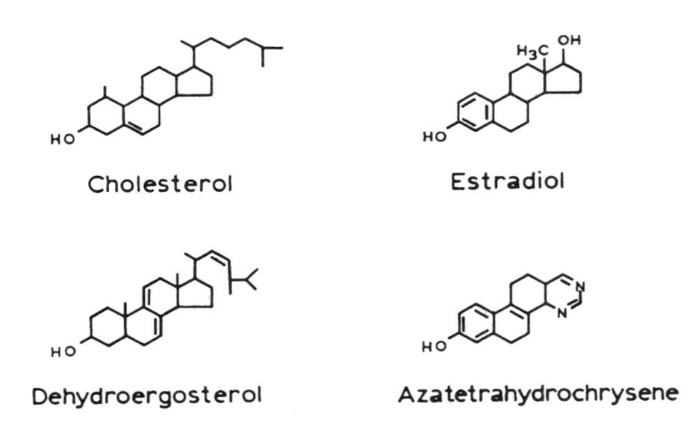

Figure 3.30. Nonfluorescent steroids cholesterol and estradiol and fluorescent analogs dehydroergosterol and 1,3-diaza-9-hydroxy-5,6,11,12-tetrahydrochrysene. Revised from Ref. 86.

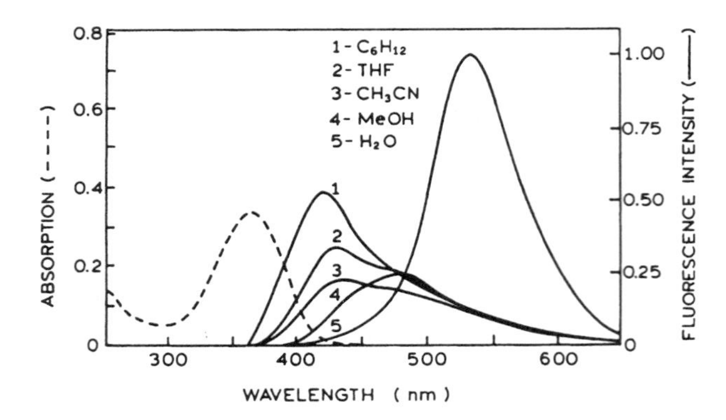

Figure 3.31. Absorption (- - -) and emission spectra (———) of 1,3,-diaza-9-hydroxy-5,6,11,12-tetrahydrochrysene in various solvents. Revised from Ref. 91.

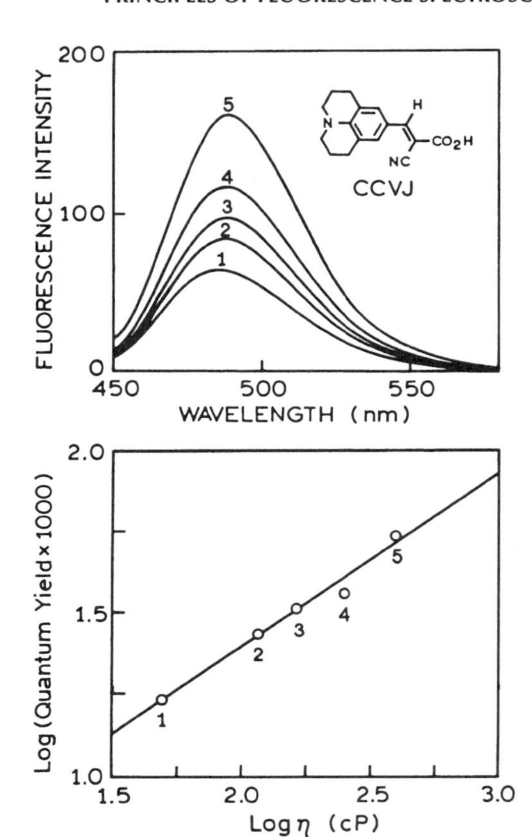

Figure 3.33. Fluorescence emission spectra and relative quantum yields of 9-(2-carboxy-2-cyanovinyl)julolidine (CCVJ) in ethylene glycol/glycol mixtures (1-5) of different viscosities (η). Reprinted, with permission, from Ref. 95, Copyright © 1993, American Chemical Society.

excited state, presumably from the amine to the vinyl group. In a highly viscous environment, the molecule cannot distort as needed for charge transfer, and the decay is radiative.[94,95] In a less viscous environment, the molecule displays internal rotation and charge transfer, which results in radiationless decay. As a result, the quantum yield depends on solvent viscosity. Probes of this type have been used to study the viscosities of membranes[94] and the rigidity of binding sites on proteins. These viscosity probes can be regarded as a subclass of the TICT probes, which are probes that distort in the excited state to form twisted intramolecular charge-transfer states.[96] For probes like DPH, the quantum yield is only weakly dependent on the viscosity. For DPH the viscosity is determined from the anisotropy.

3.7. FLUORESCENT PROTEINS

The fluorescence of proteins is usually due to the aromatic amino acids or to bound prosthetic groups such as NADH, FAD, or pyridoxal phosphate. However, several classes of proteins display intrinsic fluorescence at longer wavelengths.

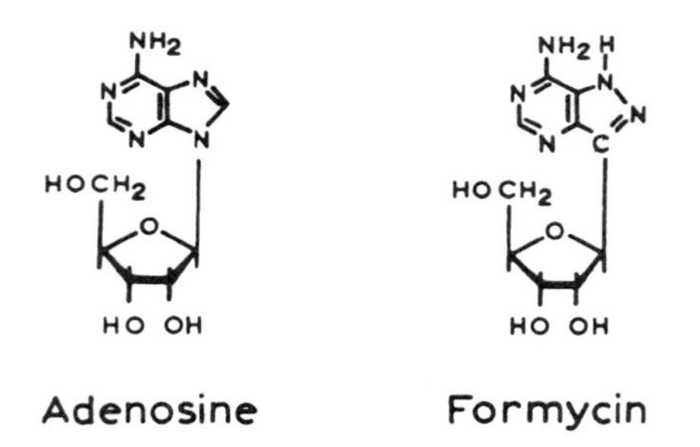

Figure 3.32. Chemical structure of adenosine and the fluorescent analog formycin. Reprinted, with permission, from Ref. 93.

3.7.A. Phycobiliproteins

The phycobiliproteins are intensely fluorescent proteins from blue-green algae and red algae.[97–100] These proteins are contained in phycobilisomes which harvest light that is not absorbed by chlorophyll. The phycobilisomes absorb strongly from 470 to 650 nm (Figure 3.34), in the gap between the blue and far-red absorption of chlorophyll. In intact phycobilisomes the phycobiliproteins are very weakly fluorescent owing to efficient energy transfer to photosynthetic reaction centers. However, upon removal from the phycobilisomes, the phycobiliproteins become highly fluorescent.

The chromophores in the phycobiliproteins are open-chain tetrapyrolle groups called bilins which are covalently

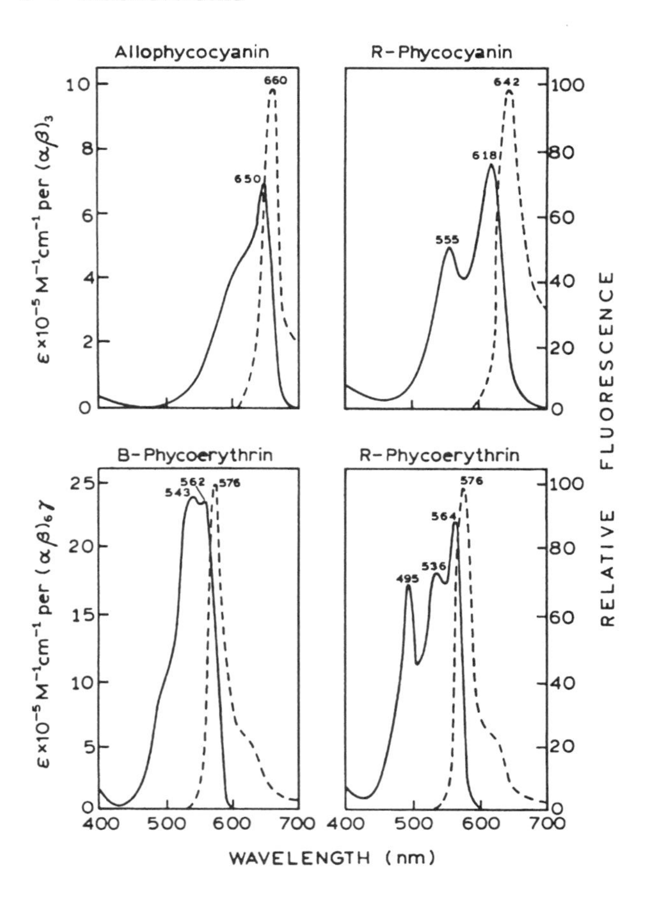

Figure 3.34. Absorption (———) and fluorescence emission spectra (- - -) of four representative phycobiliproteins: Allophycocyanin from the filamentous cyanobacterium *Anabaena variabilis*, R-phycocyanin and B-phycoerythrin from the unicellular red alga *Prophyridium cruentum*, and R-phycoerythrin from the higher red alga *Gastroclonium coulteri*. Revised from Ref. 100, Copyright © 1984, with permission from Elsevier Science.

bound to the phycobiliprotein subunit (Figure 3.35). Each phycobiliprotein displays different spectral properties, depending on the type of bound bilins. There are four main

Phycocyanobilins

Phycoerythrobilins

Phycourobilins

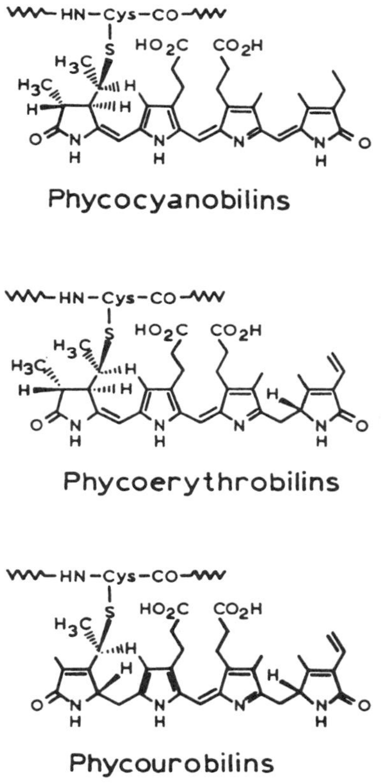

Figure 3.35. Chemical structures of the peptide-linked bilins in phycobiliproteins. Revised from Ref. 100, Copyright © 1984, with permission from Elsevier Science.

types of phycobiliproteins. Representative emission spectra are shown in Figure 3.34. Each phycobiliprotein is made up of a number of subunits, with molecular weights ranging from 100,000 to 240,000 daltons (Table 3.4). A remarkable feature of the phycobiliproteins is the high density of chromophores. The 34 bilins in phycoerythrin correspond to a bilin concentration near 80m*M*. Also, the large number of chromophores results in high extinction

Table 3.4. Properties of Some Major Phycobiliproteins[a]

Protein	Subunit Composition	Approx. mol. wt.	ε (M^{-1} cm^{-1})	Total bilins per protein	λ_{abs}^{max} (nm)	λ_{em}^{max}(nm)
Allophycocyanin	$(\alpha\beta)_3$	100,000	700,000	6	650	660
R-Phycocyanin	$(\alpha\beta)_3$	110,000	1,000,000	9	555, 618	642
B-Phycoerythrin	$(\alpha\beta)_6\gamma$	240,000	2,400,000	34	543, 562	576
R-Phycoerythrin	$(\alpha\beta)_6\gamma$	240,000	2,200,000	34	495, 536, 565	576

[a] From Refs. 98 and 100. C-Phycocyanin (620/642 nm) and C-phycoerythrin (562/576 nm) have a subunit structure $(\alpha\beta)_n$, $n = 1$–6, with molecular weight from 36,500 to 240,000.

coefficients of $2.4 \times 10^6\ M^{-1}\ cm^{-1}$, or about 30 times that of fluorescein.

These spectral properties of the bilin groups result in favorable fluorescence properties of the phycobiliproteins.[100] They display high quantum yields and are up to 20-fold brighter than fluorescein. They are highly water-soluble and stable proteins and can be stored for long periods of time. They are about 10-fold more photostable than fluorescein.[101] They contain a large number of surface lysine groups and are readily conjugatable to other proteins.[102] The long-wavelength absorption and emission make them useful where autofluorescence is a problem. Finally, they display good Stokes' shifts. This is not evident from Figure 3.34 unless one realizes that they can be excited at wavelengths below the excitation maxima.

A minor drawback of the phycobiliproteins is their sensitivity to illumination. This is not due to photobleaching, but to the possibility of exciting more than one chromophore per protein. This results in annihilation of the excited state and a decreased quantum yield and lifetime. The intensity decays are complex, with up to four exponential components ranging from 10 ps to 1.8 ns.[103] In spite of these problems, the phycobiliproteins have been successfully used for immunoassays,[104,105] for marking of cell surface antigens in flow cytometry,[102] and in single-particle detection.[106]

3.7.B. Green Fluorescent Protein

A recent addition to our armada of probes has been provided by the green fluorescent protein (GFP) from the bioluminescent jellyfish *Aequorea victoria*. The biolumi-

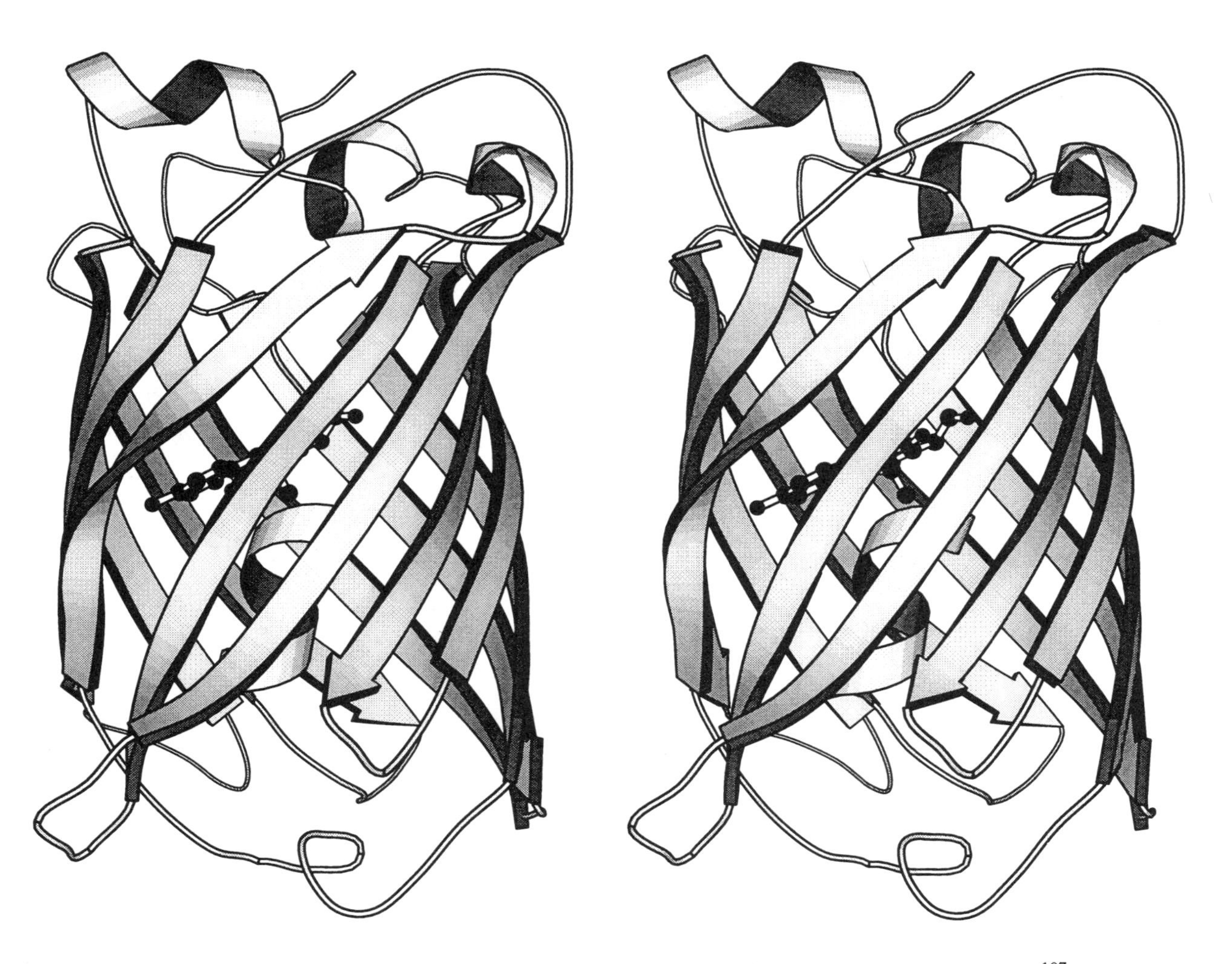

Figure 3.36. Stereo structure of green fluorescent protein. The chromophore is linked to the β-barrel structure by covalent bonds.[107] Figure provided by R. Tsien and S. Remington.

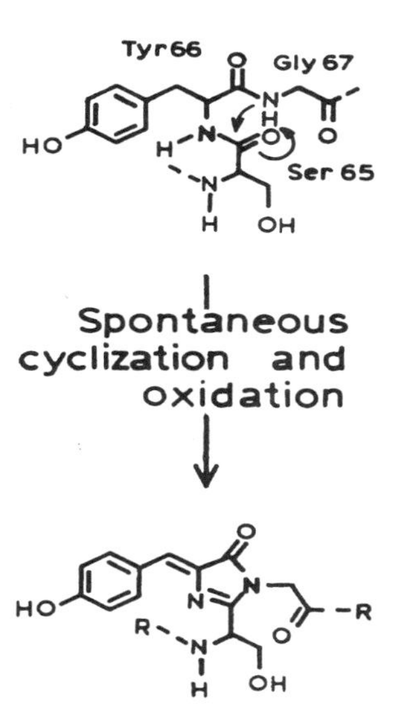

Figure 3.37. Amino acid sequence and chromophore structure in green fluorescent protein. The dashed lines represent the protein backbone. Revised from Ref. 108.

nescence of the primary photoprotein aequorin is blue. The bioluminescence from the jellyfish is green due to a closely associated green fluorescent protein. GFP contains a highly fluorescent group within a highly constrained and protected region of the protein. The chromophore is contained within a barrel of β-sheet protein[107] (Figure 3.36). The remarkable feature of GFP is that the chromophore forms spontaneously upon folding of the polypeptide chain,[108] without the need for enzymatic synthesis (Figure 3.37). As a result, it is possible to insert the gene for GFP into cells and obtain proteins which are synthesized with attached GFP, or even green fluorescent organisms. In fact, green fluorescent mice were shown in a recent publication.[109]

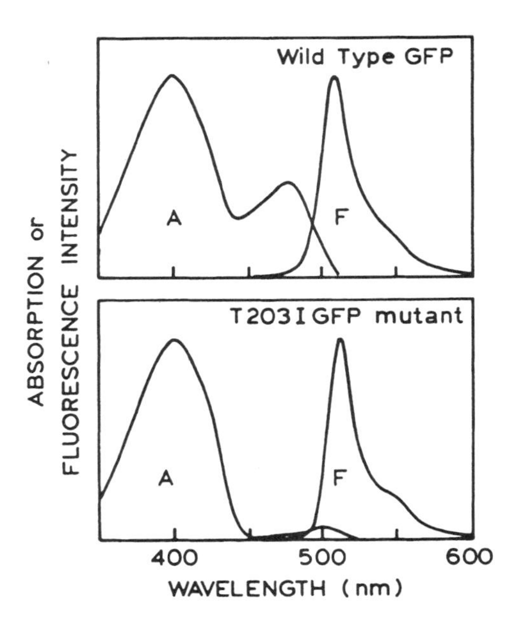

Figure 3.38. Excitation and emission spectra of wild-type GFP (*top*) and a T203I GFP mutant (*bottom*). Revised from Ref. 110, Copyright © 1995, with permission from Elsevier Science.

Absorption and emission spectra of GFP are at favorably long wavelengths (Figure 3.38). The protein displays time-dependent changes in its excitation spectrum upon illumination at 380 nm, but less so for 475 nm.[108] Following these initial changes, GFP is highly photostable. Additionally, it is possible to obtain mutants with different spectral properties. Mutants are known which display longer-wavelength absorption and emission spectra, and which display less dependence on excitation.[110–113]

GFP is one example of a family of proteins which modify the emission of bioluminescent organisms. Another example is the lumazine protein,[114] which contains a highly fluorescent ligand, 6,7-dimethyl-8-(1′-D-ribityl)lumazine, which is similar to flavin (Figure 3.1) but lacks the benzyl ring on the left side. In the case of lumazine protein, the fluorescent group does not form spontaneously as it does in GFP, so that a source of the fluorophore is needed. Yellow fluorescent proteins are also known; these appear to contain flavins as the fluorophore.

3.7.C. Phytofluors—A New Class of Fluorescent Probes

The latest additions to the family of fluorescent probes are the so-called "phytofluors." These fluorescent probes are derived from the phytochromes, which are light-sensitive proteins that are present in photosynthetic organisms. These proteins allow the organisms to adjust to external light conditions and are important in seed germination, flowering, and regulation of plant growth. Phytochromes typically contain a nonfluorescent chromophore which interconverts between two stable forms.

Recently, it was found that some phytochromes spontaneously form covalent adducts[115–117] with phycoerythrobilin (Figure 3.35). Absorption and emission spectra of one phytofluor protein are shown in Figure 3.39. The spectra are at favorably long wavelengths, and the quantum yield is near 0.70. A favorable property of these proteins is the high anisotropy, which occurs because, in contrast to the phycobiliproteins, there is only a single chromophore.

The phytochrome apoproteins are now becoming available as recombinant proteins. Since the fluorescent product is formed spontaneously, these fluorescent phytochromes

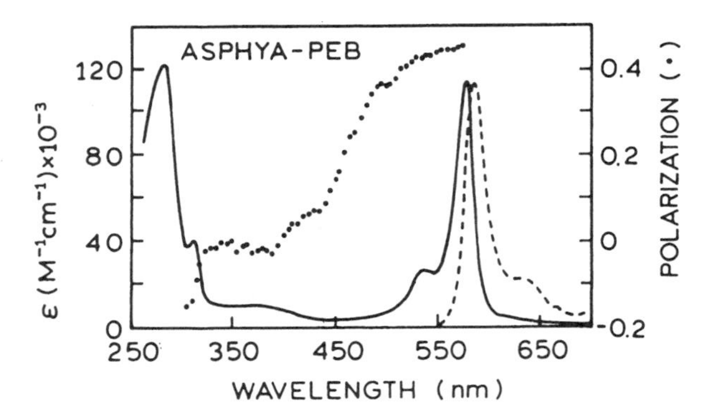

Figure 3.39. Absorption (———), emission (- - -), and excitation polarization (· · ·) spectra of a phytochrome from the plant *Avena sativa*, which contains bound phycoerythrobilin (PEB). Revised and reprinted, with permission, from Ref. 117, Copyright © 1997, Current Biology Ltd.

may become useful as probes for gene expression. However, in contrast to GFP, which spontaneously forms the pigment, it will be necessary to add the phycoerythrobilin pigment, which needs to be transported into the cells.

3.8. LONG-LIFETIME PROBES

The probes described above were organic fluorophores with a wide variety of spectral properties, reactivities, and environmental sensitivities. Although there are numerous organic fluorophores, almost all display lifetimes between 1 and 10 ns, which limit the dynamic information content of fluorescence. There are several exceptions to the short lifetimes of organic fluorophores. Pyrene (Figure 3.40) displays a lifetime near 400 ns in degassed organic solvents. Pyrene has been derivatized by adding fatty acid chains, which typically results in decay times near 100 ns. In labeled macromolecules the intensity decays of pyrene and its derivatives are usually multiexponential. Pyrene seems to display photochemical changes.

Another long-lived organic fluorophore is coronene, which displays a lifetime near 200 ns. In membranes the intensity decay of coronene is multiexponential.[118]

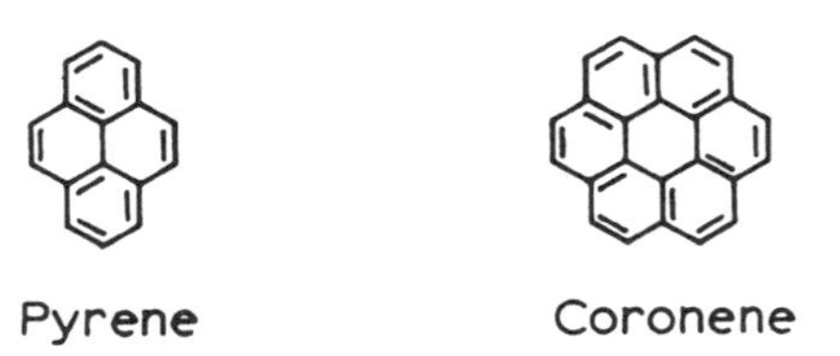

Figure 3.40. Long-lifetime organic fluorophores, pyrene and coronene.

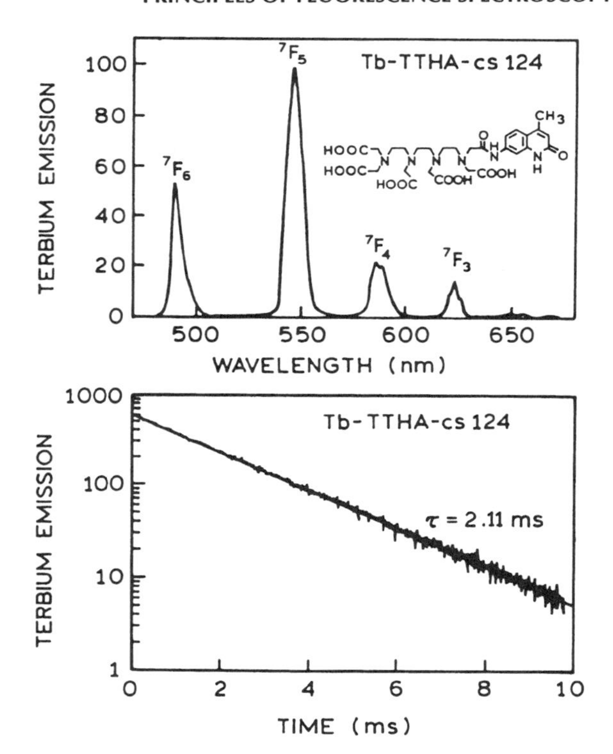

Figure 3.41. Emission spectrum and intensity decay of the lanthanide terbium. TTHA-cs 124 is the conjugate of triethylenetetraamine-hexaacetic acid (TTHA) and 7-amino-4-methyl-2(1*H*)-quinolinone (also called carbostyril 124 or cs 124). Revised and reprinted from Ref. 123, Copyright © 1995, American Chemical Society.

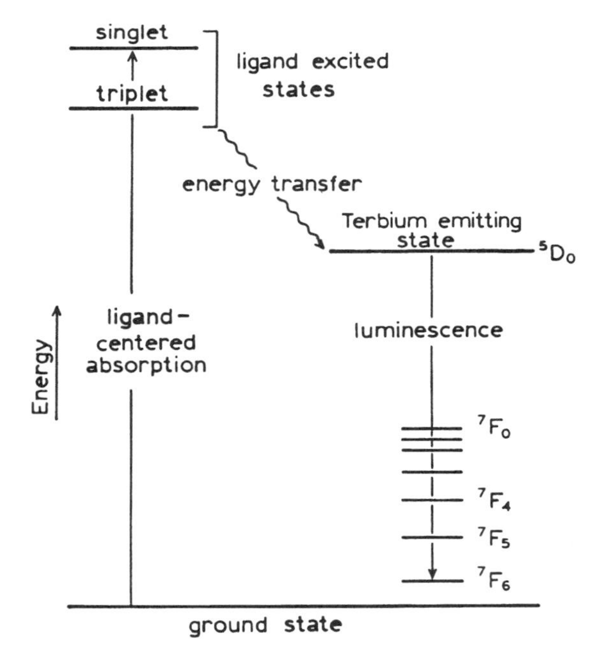

Figure 3.42. Jabłónski diagram for excitation of terbium by energy transfer. Modified and redrawn from Ref. 121, Copyright © 1993, with permission from Elsevier Science.

Coronene has also been conjugated to lipids.[119] Both pyrene and coronene display low initial anisotropies and are only moderately useful for anisotropy experiments. However, as described in the following sections, there are two types of organometallic fluorophores which display long lifetimes and other unique features which allow new types of experiments.

3.8.A. Lanthanides

The lanthanides are uniquely fluorescent metals which display emission in aqueous solution and decay times of 0.5–3 ms.[120–123] Emission results from transitions involving 4*f* orbitals, which are forbidden transitions. As a result, the absorption coefficients are very low, less than 10 M^{-1} cm^{-1}, and the emissive rates are slow, resulting in long lifetimes. The lanthanides behave like atoms and display line spectra[120] (Figure 3.41). Because of the weak absorption, lanthanides are usually not directly excited but rather are excited through chelated organic ligands (Figures 3.41 and 3.42). Hence, the excitation spectrum of the complex shown in Figure 3.41 reflects the absorption spectrum of the ligand and not that of the lanthanide itself.

Lanthanides possess some favorable properties as biochemical probes. They can substitute chemically for calcium in many calcium-dependent proteins.[124–126] A main route of nonradiative decay is by coupling to vibrations of water. For both Eu^{3+} and Tb^{3+}, the lifetime in H_2O and D_2O can be used to calculate the number of bound water molecules (n):

$$n = q\left(\frac{1}{\tau_{H_2O}} - \frac{1}{\tau_{D_2O}}\right) \qquad [3.1]$$

where q is a constant different for each metal.[121] Hence, the decay times of the lanthanides when bound to proteins can be used to calculate the number of bound water molecules in a calcium binding site.

Lanthanides can also be used with proteins which do not have intrinsic binding sites. Reagents have been developed which can be coupled to proteins and which chelate lanthanides.[127,128] Because of their sensitivity to water, lanthanide complexes designed as labels generally have most sites occupied by the ligand.

Lanthanide Immunoassays. Lanthanides have found widespread use in high-sensitivity detection, particularly for immunoassays.[129,130] The basic idea is shown in Figure 3.43. All biological samples display autofluorescence, which is usually the limiting factor in high-sensitivity detection. The autofluorescence usually decays on the nanosecond timescale, as does the fluorescence of most fluorophores. Because of their long decay times, the lanthanides continue to emit following disappearance of the autofluorescence. It is technically simple to turn on the detector after an excitation flash and to integrate the intensity from the lanthanide. It should be noted that the term "time-resolved" is somewhat of a misnomer and does not refer to measurement of the decay time. These time-gated immunoassays are essentially steady-state intensity measurements in which the intensity is measured over a period of time following pulsed excitation.

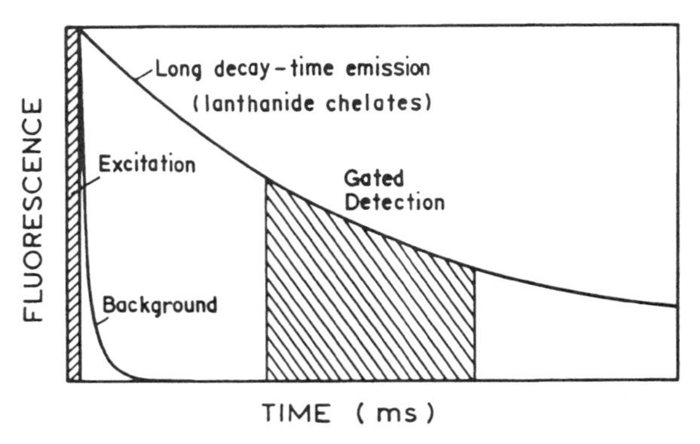

Figure 3.43. Principle of "time-resolved" detection in lanthanide immunoassays. Revised from Ref. 130.

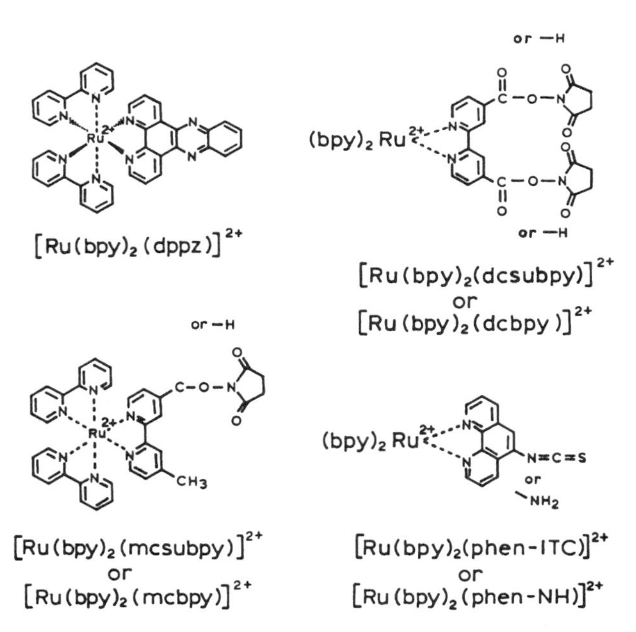

Figure 3.44. Representative ruthenium metal–ligand complexes. These probes have absorption and emission maxima near 450 and 620 nm, respectively, and decay times near 400 ns. The charges (2+) refer to the charge on the metal, not the overall charge of the complex.

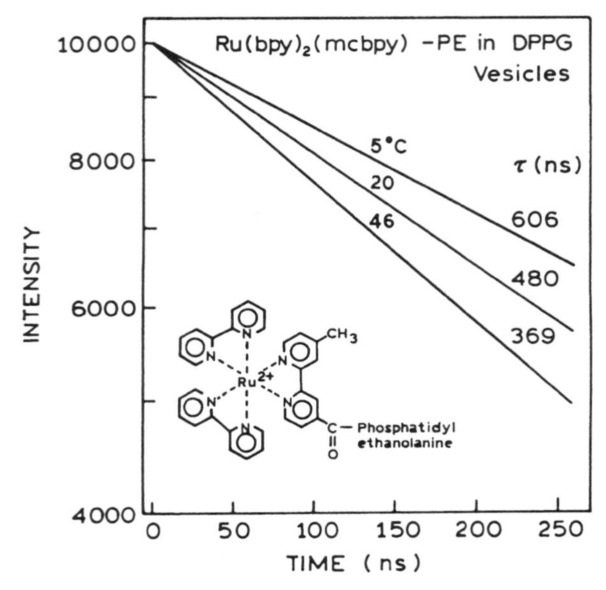

Figure 3.45. Intensity decays of $Ru(bpy)_2(mcbpy)$-PE (PE = phosphatidylethanolamine) in dipalmitoyl-L-α-phosphatidyl glycerol (DPPG) vesicles at 5, 20, and 46%. Modified from Ref. 137.

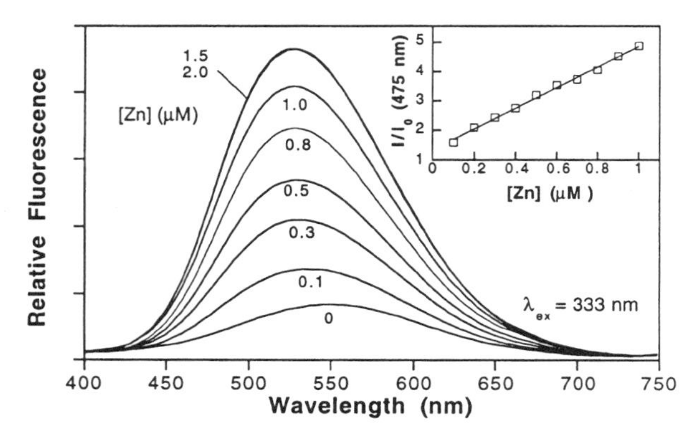

Figure 3.47. Zinc-dependent emission spectra of the dansyl-labeled zinc finger peptide. From Ref. 141.

The lanthanides do suffer several limitations. One is the need to chelate the lanthanide in order to obtain significant emission. This requirement often results in multistep assays. Another difficulty is the absence of polarized emission, so that the lanthanides cannot be used for anisotropy measurements.

3.8.B. Transition-Metal–Ligand Complexes

A final class of probes comprises the long-lifetime transition-metal–ligand complexes (MLCs). These are typically complexes of ruthenium [Ru(II)], rhenium [Re(I)], or osmium [Os(II)] with one or more diimine ligands (Figure 3.44). In contrast to the lanthanides, these compounds display molecular fluorescence from a metal-to-ligand charge-transfer state. The transition is partially forbidden, so that the decay times are long. These complexes are highly stable, like covalent compounds, so there is no significant dissociation of the ligands from the metal.

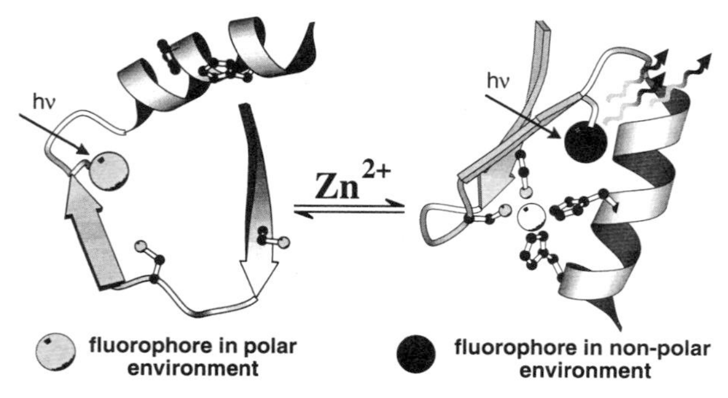

Figure 3.46. A zinc-sensitive peptide based on the dansyl fluorophore. Reprinted, with permission, from Ref. 141, Copyright © 1996, American Chemical Society.

Some representative structures are shown in Figure 3.44. MLCs are available for conjugation with biomolecules.[131–133] There are also intercalating DNA probes (Figure 3.44, top left) which do not fluoresce in water.[134,135] MLCs display lifetimes ranging from 10 ns to 10 μs.[136] For example, $Ru(bpy)_2(mcbpy)$ in Figure 3.45 displayed a decay time near 400 ns when conjugated to proteins[131] and lipids.[137] The MLC probes also are highly photostable and display large Stokes' shifts, so that probe–probe interactions are not expected. These molecules are known to display polarized emission[131,132] and are thus useful for measurement of dynamic processes on the microsecond timescale. The development of MLC probes is in its infancy, and rapid progress in their development can be expected. Additional details can be found in Chapter 20.

3.9. PROTEINS AS SENSORS

The use of fluorescence for chemical sensing requires highly specific probes. One approach to obtaining the needed specificity is to rely on proteins which are known to bind the desired analyte. This approach has been used to develop sensing proteins for calcium,[138] maltose,[139] and zinc.[140,141] One sensor for zinc was based on a zinc finger peptide. Zinc fingers are part of transcription factors. These proteins bind zinc with high affinity and specificity, the metal typically being bound to histidine and/or cysteine residues.

An example of a zinc-sensitive peptide is shown in Figure 3.46. In this case, the zinc finger amino acid sequence was modified to contain a covalently linked dansyl group near the middle of the peptide. Emission spectra of this dansyl zinc finger are shown in Figure 3.47. Upon addition of zinc, the emission intensity increases and the emission maximum shifts to shorter wavelengths. In the absence of zinc, the peptide is unfolded and the dansyl group is exposed to the water. In the presence of zinc, the peptide adopts a folded structure which was anticipated to result in shielding of the dansyl group from water.

3.10. CONCLUSION

A diversity of molecules display fluorescence, and numerous interactions and processes can alter the spectral properties. Fluorophores can be covalently attached to macromolecules, or they can be designed to interact with specific ions. Emission can occur from the UV to the NIR, and probes are available with short (nanosecond) and long (microsecond to millisecond) lifetimes. The technology of probe chemistry is rapidly changing, with new probes allowing previously impossible experiments to be performed. As a result, fluorescence methods are now actively used not only in biophysics, but in cellular imaging, DNA sequencing, flow cytometry, and clinical diagnostics.

REFERENCES

1. Demchenko, A. P., 1981, *Ultraviolet Spectroscopy of Proteins*, Springer-Verlag, New York.
2. Longworth, J. W., 1971, Luminescence of polypeptides and proteins, in *Excited States of Proteins and Nucleic Acids*, R. F. Steiner and I. Welnryb (eds.), Plenum Press, New York, pp. 319–484.
3. Velick, S. F., 1958, Fluorescence spectra and polarization of glyceraldehyde-3-phosphate and lactic dehydrogenase coenzyme complexes, *J. Biol. Chem.* **233:**1455–1467.
4. Gafni, A., and Brand, L., 1976, Fluorescence decay studies of reduced nicotinamide adenine dinucleotide in solution and bound to liver alcohol dehydrogenase, *Biochemistry* **15:**3165–3171.
5. Brochon, J.-C., Wahl, P., Monneuse-Doublet, M.-O., and Olomucki, A., 1977, Pulse fluorimetry study of octopine dehydrogenase-reduced nicotinamide adenine dinucleotide complexes, *Biochemistry* **16:**4594–4599.
6. Churchich, J. E., 1965, Fluorescence properties of pyridoxamine 5-phosphate, *Biochim. Biophys. Acta* **102:**280–288.
7. Honikel, K. O., and Madsen, N. B., 1972, Comparison of the absorbance spectra and fluorescence behavior of phosphorylase *b* with that of model pyridoxal phosphate derivatives in various solvents, *J. Biol. Chem.* **247:**1057–1064.
8. Vaccari, S., Benci, S., Peracchi, A., and Mozzarelli, A., 1997, Time-resolved fluorescence of pyridoxal 5′-phosphate-containing enzymes: Tryptophan synthetase and *O*-acetylserine sulfhydrylase, *J. Fluoresc.* **7:**135S–137S.
9. Kwon, O.-S., Blazquez, M., and Churchich, J. E., 1994, Luminescence spectroscopy of pyridoxic acid and pyridoxic acid bound to proteins, *Eur. J. Biochem.* **219:**807–812.
10. Xiao, G.-S., and Zhou, J.-M., 1996, Conformational changes at the active site of bovine pancreatic RNase A at low concentrations of guanidine hydrochloride probed by pyridoxal 5′-phosphate, *Biochim. Biophys. Acta* **1294:**1–7.
11. Churchich, J. E., 1986, Fluorescence properties of free and bound pyridoxal phosphate and derivatives, in *Pyridoxal Phosphate: Chemical, Biochemical and Medical Aspects, Part A*, D. Dolphin (ed.), Wiley, New York, pp. 545–567.
12. Churchich, J. E., 1976, Fluorescent probe studies of binding sites in proteins and enzymes, in *Modern Fluorescence Spectroscopy*, Vol, 2, E. L. Wehry (ed.), Plenum Press, New York, pp. 217–237.
13. Vaccari, S., Benci, S., Peracchi, A., and Mozzarelli, A., 1996, Time-resolved fluorescence of tryptophan synthase, *Biophys. Chem.* **61:**9–22.
14. Personal communication from Dr. Rebecca Richards-Kortum.
15. Visser, A. J. W. G., 1984, Kinetics of stacking interactions in flavin adenine dinucleotide from time-resolved flavin fluorescence, *Photochem. Photobiol.* **40:**703–706.
16. Leenders, R., Kooijman, M., van Hoek, A., Veeger, C., and Visser, A. J. W. G., 1993, Flavin dynamics in reduced flavodoxins, *Eur. J. Biochem.* **211:**37–45.
17. Wolfbeis, O. S., 1985, The fluorescence of organic natural products, in *Molecular Luminescence Spectroscopy*, S. G. Schulman (ed.), John Wiley & Sons, New York, Part 1, pp. 167–370.
18. Richards-Kortum, R., and Sevick-Muraca, E., 1996, Quantitative optical spectroscopy for tissue diagnosis, *Annu. Rev. Phys. Chem.* **47:**555–606.
19. Li, B., and Lin, S.-X., 1996, Fluorescence-energy transfer in human estradiol 17β-dehydrogenase–NADH complex and studies on the coenzyme binding, *Eur. J. Biochem.* **235:**180–186.
20. Haugland, R. P., 1996, *Handbook of Fluorescent Probes and Research Chemicals*, Molecular Probes, Inc., Eugene, Oregon.
21. Hemmila, I. A., 1991, *Applications of Fluorescence in Immunoassays*, John Wiley & Sons, New York, pp. 107–167.
22. Weber, G., 1951, Polarization of the fluorescence of macromolecules, *Biochem. J.* **51:**155–167.
23. BioProbes 25, 1997, New Products and Applications, Molecular Probes, Inc., Eugene, Oregon.
24. Waggoner, A., 1995, Covalent labeling of proteins and nucleic acids with fluorophores, *Methods Enzymol.* **246:**362–373.
25. Hemmila I. A., 1991, *Applications of Fluorescence in Immunoassays*, John Wiley & Sons, New York, page 67.
26. Johnson, I. D., Kang, H. C., and Haugland, R. P., 1991, Fluorescent membrane probes incorporating dipyrrometheneboron difluoride fluorophores, *Anal. Biochem.* **198:**228–237.
27. Weber, G., and Farris, F. J., 1979, Synthesis and spectral properties of a hydrophobic fluorescent probe: 6-Propionyl-2-(dimethylamino)naphthalene, *Biochemistry* **18:**3075–3078.
28. Prendergast, F. G., Meyer, M., Carlson, G. L., Iida, S., and Potter, J. D., 1983, Synthesis, spectral properties, and use of 6-acryloyl-2-dimethylaminonaphthalene (Acrylodan), *J. Biol. Chem.* **258:**7541–7544.
29. Rottenberg, H., 1992, Probing the interactions of alcohols with biological membranes with the fluorescent probe Prodan, *Biochemistry* **31:**9473–9481.
30. Slavik, J., 1982, Anilinonaphthalene sulfonate as a probe of membrane composition and function, *Biochim. Biophys. Acta* **694:**1–25.
31. Daniel, E., and Weber, G., 1966, Cooperative effects in binding by bovine serum albumin. I. The binding of 1-anilino-8-naphthalenesulfonate. Fluorimetric titrations, in *Cooperative Effects in Binding by Albumin*, **5:**1893–1900.
32. Prendergast, F. G., Haugland, R. P., and Callahan, P. J., 1981, 1-[4-(Trimethylamino)phenyl]-6-phenylhexa-1,3,5 triene: Synthe-

sis, fluorescence properties, and use as a fluorescence probe of lipid bilayers, *Biochemistry* **20:**7333–7338.

33. Sklar, L. A., Hudson, B. S., Petersen, M., and Diamond, J., 1977, Conjugated polyene fatty acids on fluorescent probes: Spectroscopic characterization, *Biochemistry* **16:**813–818.
34. Itoh, T., and Kohler, B. E., 1987, Dual fluorescence of diphenylpolyenes, *J. Phys. Chem.* **91:**1760–1764.
35. Alford, P. C., and Palmer, T. F., 1982, Fluorescence of DPH derivatives, evidence for emission from S_2 and S_1 excited states, *Chem. Phys. Lett.* **86:**248–253.
36. Cundall, R. B., Johnson, I., Jones, M. W., Thomas, E. W., and Munro, I. H., 1979, Photophysical properties of DPH derivatives, *Chem. Phys. Lett.* **64:**39–42.
37. Kinnunen, P. K. J., Koiv, A., and Mustonen, P., 1993, Pyrene-labeled lipids as fluorescent probes in studies on biomembranes and membrane models, in *Fluorescence Spectroscopy: New Methods and Applications*, O. S. Wolfbeis (ed.), Springer-Verlag, New York, pp. 159–171.
38. Smiley, S. T., Reers, M., Mottola-Hartshorn, C., Lin, M., Chen, A., Smith, T. W., Steele, G. D., and Chen, L. B., 1991, Intracellular heterogeneity in mitochondrial membrane potentials revealed by a J-aggregate-forming lipophilic cation JC-1, *Proc. Natl. Acad. Sci. U.S.A.* **88:**3671–3675.
39. Sims, P. J., Waggoner, A. S., Wang, C.-H., and Hoffman, J. F., 1974, Studies on the mechanism by which cyanine dyes measure membrane potential in red blood cells and phosphatidylcholine vesicles, *Biochemistry* **13:**3315–3336.
40. Gross, E., Bedlack, R. S., and Loew, L. M., 1994, Dual-wavelength ratiometric fluorescence measurement of the membrane dipole potential, *Biophys. J.* **67:**208–216.
41. Zhang, J., Davidson, R. M., Wei, M., and Loew, L. M., 1998, Membrane electric properties by combined patch clamp and fluorescence ratio imaging in single neurons, *Biophys. J.* **74:**48–53.
42. Loew, L. M., 1996, Potentiometric dyes: Imaging electrical activity of cell membranes, *Pure Appl. Chem.* **68:**1405–1409.
43. Loew, L. M., 1994, Voltage-sensitive dyes and imaging neuronal activity, *Neuroprotocols* **5:**72–79.
44. Dragsten, P. R., and Webb, W. W., 1978, Mechanism of the membrane potential sensitivity of the fluorescent membrane probe merocyanine 540, *Biochemistry* **17:**5228–5240.
45. Loew, L. M., 1994, Characterization of potentiometric membrane dyes, *Adv. Chem. Ser.* **235:**151–173.
46. Waggoner, A. S., 1979, Dye indicators of membrane potential, *Annu. Rev. Biophys. Bioeng.* **8:**47–68.
47. Loew, L. M., 1982, Design and characterization of electrochromic membrane probes, *J. Biochem. Biophys. Methods* **6:**243–260.
48. Thompson, R. B., 1994, Red and near-infrared fluorometry, in *Topics in Fluorescence Spectroscopy, Volume 4, Probe Design and Chemical Sensing*, J. R. Lakowicz (ed.), Plenum Press, New York, pp. 151–222.
49. Southwick, P. L., Ernst, L. A., Tauriello, E. W., Parker, S. R., Mujumdar, R. B., Mujumdar, S. W., Clever, H. A., and Waggoner, A. S., 1990, Cyanine dye labeling reagents–carboxymethylindocyanine succinimidyl esters, *Cytometry* **11:**418–430.
50. Rahavendran, S. V., and Karnes, H. T., 1996, Application of rhodamine 800 for reversed phase liquid chromatographic detection using visible diode laser induced fluorescence, *Anal. Chem.* **68:**3763–3768.
51. Rahavendran, S. V., and Karnes, H. T., 1996, An oxazine reagent for derivatization of carboxylic acid analytes suitable for liquid chromatographic detection using visible diode laser-induced fluorescence, *J. Pharm. Biomed. Anal.* **15:**83–98.
52. Flanagan, J. H., Romero, S. E., Legendre, B. L., Hammer, R. P., and Soper, A., 1997, Heavy-atom modified near-IR fluorescent dyes for DNA sequencing applications: Synthesis and photophysical characterization, *Proc. SPIE* **2980:**328–337.
53. Owens, C. V., Davidson, Y. Y., Kar, S., and Soper, S. A., 1997, High-resolution separation of DNA restriction fragments using capillary electrophoresis with near-IR, diode-based, laser-induced fluorescence detection, *Anal. Chem.* **69:**1256–1261.
54. Matsuoka, M., 1990, *Infrared Absorbing Dyes*, Plenum Press, New York.
55. Leznoff, C. C., and Lever, A. B. P., 1989, *Phthalocyanines: Properties and Applications*, VCH Publishers, New York.
56. Kessler, M. A., and Wolfbeis, O. S., 1992, Laser-induced fluorometric determination of albumin using longwave absorbing molecular probes, *Anal. Biochem.* **200:**254–259.
57. Steiner, R. F., and Kubota, Y., 1983, Fluorescent dye–nucleic acid complexes, in *Excited States of Biopolymers*, R. F. Steiner (ed.), Plenum Press, New York, pp. 203–254.
58. Georghiou, S., 1977, Interaction of acridine drugs with DNA and nucleotides, *Photochem. Photobiol.* **26:**59–68.
59. Suh, D., and Chaires, J. B., 1995, Criteria for the mode of binding of DNA binding agents, *Bioorg. Med. Chem.* **3:**723–728.
60. Eriksson, S., Kim, S. K., Kubista, M., and Norden, B., 1993, Binding of 4′,6-diamidino-2-phenylindole (DAPI) to AT regions of DNA: Evidence for an allosteric conformational change, *Biochemistry* **32:**2987–2998.
61. Parkinson, J. A., Barber, J., Douglas, K. T., Rosamond, J., and Sharples, D., 1990, Minor-groove recognition of the self-complementary duplex $d(CGCGAATTCGCG)_2$ by Hoechst 33258: A high-field NMR study, *Biochemistry* **29:**10181–10190.
62. Loontiens, F. G., McLaughlin, L. W., Diekmann, S., and Clegg, R. M., 1991, Binding of Hoechst 33258 and 4′,6-diamidino-2-phenylindole to self-complementary decadeoxynucleotides with modified exocyclic base substitutents, *Biochemistry* **30:**182–189.
63. Haq, I., Ladbury, J. E., Chowdhry, B. Z., Jenkins, T. C., and Chaires, J. B., 1997, Specific binding of Hoechst 33258 to the $d(CGCAAATTTGCG)_2$ duplex: Calorimetric and spectroscopic studies, *J. Mol. Biol.* **271:**244–257.
64. Glazer, A. N., Peck, K., and Matheis, R. A., 1990, A stable double-stranded DNA ethidium homodimer complex: Application to picogram fluorescence detection of DNA in agarose gels, *Proc. Natl. Acad. Sci, U.S.A.*, **87:**3851–3855.
65. Rye, H. S., Yue, S., Wemmer, D. E., Quesada, M. A., Haugland, R. P., Mathies, R. A., and Glazer, A. N., 1992, Stable fluorescent complexes of double-stranded DNA with bis-intercalating asymmetric cyanine dyes: Properties and applications, *Nucleic Acids Res.* **20:**2803–2812.
66. Wu, P., Li, H., Nordlund, T. M., and Rigler, R., 1990, Multistate modeling of the time and temperature dependence of fluorescence from 2-aminopurine in a DNA decamer, *Proc. SPIE* **1204:**262–269.
67. Nordlund, T. M., Wu, P., Anderson, S., Nilsson, L., Rigler, R., Graslund, A., McLaughlin, L. W., and Gildea, B., 1990, Structural dynamics of DNA sensed by fluorescence from chemically modified bases, *Proc. SPIE* **1204:**344–353.
68. Hawkins, M. E., Pfleiderer, W., Mazumder, A., Pommier, Y. G., and Balis, F. M., 1995, Incorporation of a fluorescent guanosine analog into oligonucleotides and its application to a real time assay for the HIV-1 integrase 3′-processing reaction, *Nucleic Acids Res.* **23:**2872–2880.
69. Kulkosky, J., and Skalka, A. M., 1990, HIV DNA integration: Observations and inferences, *J. Acquir. Immune Defic. Syndr.* **3:**839–851.

70. Brown, P. O., 1990, Integration of retroviral DNA, in *Current Topics in Microbiology and Immunology, Vol. 157*, Springer-Verlag, Berlin, pp. 19–48.
71. Biwersi, J., Tulk, B., and Verkman, A. S., 1994, Long-wavelength chloride-sensitive fluorescent indicators, *Anal. Biochem.* **219:**139–143.
72. Valeur, B., 1994, Principles of fluorescent probe design for ion recognition, in *Topics in Fluorescence Spectroscopy, Volume 4, Probe Design and Chemical Sensing*, J. R. Lakowicz (ed.), Plenum Press, New York, pp. 21–48.
73. Poenie, M., and Chen, C.-S., 1993, New fluorescent probes for cell biology, in *Optical Microscopy*, B. Herman and J. J. Lemasters (eds.), Academic Press, New York, pp. 1–25.
74. Szmacinski, H., and Lakowicz, J. R., 1994, Lifetime-based sensing, in *Topics in Fluorescence Spectroscopy, Volume 4, Probe Design and Chemical Sensing*, J. R. Lakowicz (ed.), Plenum Press, New York, pp. 295–334.
75. Czarnik, A. W., 1994, Fluorescent chemosensors for cations, anions, and neutral analytes, in *Topics in Fluorescence Spectroscopy, Volume 4, Probe Design and Chemical Sensing*, J. R. Lakowicz (ed.), Plenum Press, New York, pp. 49–70.
76. Haugland, R. P., and Johnson, I. D., 1993, Detecting enzymes in living cells using fluorogenic substrates, *J. Fluoresc.* **3:**119–127.
77. Zhou, M., Upson, R. H., Diwu, Z., and Haugland, R. P., 1996, A fluorogenic substrate for β-glucuronidase: Applications in fluorometric, polyacrylamide gel and histochemical assays, *J. Biochem. Biophys. Methods* **33:**197–205.
78. Gershkovich, A. A., and Kholodovych, V. V., 1996, Fluorogenic substrates for proteases based on intramolecular fluorescence energy transfer (IFETS), *J. Biochem. Biophys. Methods* **33:**135–162.
79. Geoghegan, K. F., 1996, Improved method for converting an unmodified peptide to an energy-transfer substrate for a proteinase, *Bioconjug. Chem.* **7:**385–391.
80. Matayoshi, E. D., Wang, G. T., Krafft, G. A., and Erickson, J., 1990, Novel fluorogenic substrates for assaying retroviral proteases by resonance energy transfer, *Science* **247:**954–957.
81. Zandonella, G., Haalck, L., Spener, F., Faber, K., Paltauf, F., and Hermetter, A., 1995, Inversion of lipase stereospecificity for fluorogenic alkyldiacyl glycerols: Effect of substrate solubilization, *Eur. J. Biochem.* **231:**50–55.
82. Duque, M., Graupner, M., Stütz, H., Wicher, I., Zechner, R., Paltauf, F., and Hermetter, A., 1996, New fluorogenic triacylglycerol analogs as substrates for the determination and chiral discrimination of lipase activities, *J. Lipid Res.* **37:**868–876.
83. Naleway, J. J., Fox, C. M. J., Robinhold, D., Terpetschnig, E., Olson, N. A., and Haugland, R. P., 1994, Synthesis and use of new fluorogenic precipitating substrates, *Tetrahedron Lett.* **35:**8569–8572.
84. Huang, Z., Terpetschnig, E., You, W., and Haugland, R. P., 1992, 2-(2′-Phosphoryloxyphenyl)-4(3*H*)-quinazolinone derivatives as fluorogenic precipitating substrates of phosphatases, *Anal. Biochem.* **207:**32–39.
85. Ziomek, C. A., Lepire, M. L., and Torres, I., 1990, A highly fluorescent simultaneous azo dye technique for demonstration of nonspecific alkaline phosphatase activity, *J. Histochem. Cytochem.* **38:**437–442.
86. Hale, J. E., and Schroeder, F., 1982, Asymmetric transbilayer distribution of sterol across plasma membranes determined by fluorescence quenching of dehydroergosterol, *Eur. J. Biochem.* **122:**649–661.
87. Fischer, R. T., Cowlen, M. S., Dempsey, M. E., and Schroeder, F., 1985, Fluorescence of $\Delta^{5,7,9(11),22}$-ergostatetraen-3β-ol in micelles, sterol carrier protein complexes, and plasma membranes, *Biochemistry* **24:**3322–3331.
88. Schroeder, F., Barenholz, Y., Gratton, E., and Thompson, T. E., 1987, A fluorescence study of dehydroergosterol in phosphatidylcholne bilayer vesicles, *Biochemistry* **26:**2441–2448.
89. Loura, L. M. S., and Prieto, M., 1997, Aggregation state of dehydroergosterol in water and in a model system of membranes, *J. Fluoresc.* **7:**173S–175S.
90. Hwang, K.-J., O'Neil, J. P., and Katzenellenbogen, J. A., 1992, 5,6,11,12-Tetrahydrochrysenes: Synthesis of rigid stilbene systems designed to be fluorescent ligands for the estrogen receptor, *J. Org. Chem.* **57:**1262–1271.
91. Bowen, C. M., and Katzenellenbogen, J. A., 1997, Synthesis and spectroscopic characterization of two aza-tetrahydrochrysenes as potential fluorescent ligands for the estrogen receptor, *J. Org. Chem.* **62:**7650–7657.
92. Wolkowicz, P. E., Pownall, H. J., and McMillin-Wood, J. B., 1982, (1-Pyrenebutyryl)carnitine and 1-pyrenebutyryl coenzyme A: Fluorescent probes for lipid metabolite studies in artificial and natural membranes, *Biochemistry* **21:**2990–2996.
93. Rossomando, E. F., Jahngen, J. H., and Eccleston, J. F., 1981, Formycin 5′-triphosphate, a fluorescent analog of ATP, as a substrate for adenylate cyclase, *Proc. Natl. Acad. Sci. U.S.A.* **78:**2278–2282.
94. Kung, C. E., and Reed, J. K., 1986, Microviscosity measurements of phospholipid bilayers using fluorescent dyes that undergo torsional relaxation, *Biochemistry* **25:**6114–6121. See also *Biochemistry* **28:**6678–6686 (1989).
95. Iwaki, T., Torigoe, C., Noji, M., and Nakanishi, M., 1993, Antibodies for fluorescent molecular rotors, *Biochemistry* **32:**7589–7592.
96. Rettig, W., and Lapouyade, R., 1994, Fluorescence probes based on twisted intramolecular charge transfer (TICT) states and other adiabatic photoreactions, in *Topics in Fluorescence Spectroscopy, Volume 4, Probe Design and Chemical Sensing*, J. R. Lakowicz (ed.), Plenum Press, New York, pp. 109–149.
97. Teale, F. W. J., and Dale, R. E., 1970, Isolation and spectral characterization of phycobiliproteins, *Biochem. J.* **116:**161–169.
98. Glazer, A. N., 1985, Light harvesting by phycobilisomes, *Annu. Rev. Biophys. Biophys. Chem.* **14:**47–77.
99. MacColl, R., and Guard-Friar, D., 1987, *Phycobiliproteins*, CRC Press, Boca Raton, Florida.
100. Glazer. A. N., and Stryer, L., 1984, Phycofluor probes, *Trends Biochem. Soc.* 423–427.
101. White, J. C., and Stryer, L., 1987, Photostability studies of phycobiliprotein fluorescent labels, *Anal. Biochem.* **161:**442–452.
102. Oi, V. T., Glazer, A. N., and Stryer, L., 1982, Fluorescent phycobiliprotein conjugates for analyses of cells and molecules, *J. Cell Biol.* **93:**981–986.
103. Holzwarth, A. R., Wendler, J., and Suter, G. W., 1987, Studies on chromophore coupling in isolated phycobiliproteins, *Biophys. J.* **51:**1–12.
104. Kronick, M. N., and Grossman, P. D., 1983, Immunoassay techniques with fluorescent phycobiliprotein conjugates, *Clin. Chem.* **29:**1582–1586.
105. Kronick, M. N., 1986, The use of phycobiliproteins as fluorescent labels in immunoassays, *J. Immun. Methods* **92:**1–13.
106. Nguyen, D. C., Keller, R. A., Jett, J. H., and Martin, J. C., 1987, Detection of single molecules of phycoerythrin in hydrodynamically focused flows by laser induced fluorescence, *Anal. Chem.* **59:**2158–2161.

107. Ormo, M., Cubitt, A. B., Kallio, K., Gross, L. A., Tsien, R. Y., and Remington, S. J., 1996, Crystal structure of the *Aequorea victoria* green fluorescent protein, *Science* **273:**1392–1395.
108. Chalfie, M., Tu, Y., Euskirchen, G., Ward, W. W., and Prasher, D. C., 1994, Green fluorescent protein as a marker for gene expression, *Science* **263:**802–805.
109. "Jellyfish light up mice," *Science* **277:**41.
110. Ehrig, T., O'Kane, D. J., and Prendergast, F. G., 1995, Green fluorescent protein mutants with altered fluorescence excitation spectra, *FEBS Lett.* **367:**163–166.
111. Delagrave, S., Hawtin, R. E., Silva, C. M., Yang, M. M., and Youvan, D. C., 1995, Red-shifted excitation mutants of the green fluorescent protein, *Bio/Technology* **13:**151–154.
112. Cubitt, A. B., Heim, R., Adams, S. R., Boyd, A. E., Gross, L. A., and Tsien, R. Y., 1995, Understanding, improving and using green fluorescent proteins, *Trends Biochem. Soc.* **20:**448–455.
113. Heim, R., and Tsien, R. Y., 1996, Engineering green fluorescent protein for improved brightness, longer wavelengths and fluorescence resonance energy transfer, *Curr. Biol.* **6:**178–182.
114. Petushkov, V. N., Gibson, B. G., and Lee, J., 1995, Properties of recombinant fluorescent proteins from *Photobacterium leiognathi* and their interaction with luciferase intermediates, *Biochemistry* **34:**3300–3309.
115. Li, L., Murphy, J. T., and Lagarias, J. C., 1995, Continuous fluorescence assay of phytochrome assembly in vitro, *Biochemistry* **34:**7923–7930.
116. Murphy, J. T., and Lagarias, J. C., 1997, Purification and characterization of recombinant affinity peptide-tagged oat phytochrome A, *Photochem. Photobiol.* **65:**750–758.
117. Murphy, J. T., and Lagarias, J. C., 1997, The phytofluors: A new class of fluorescent protein probes, *Curr. Biol.* **7:**870–876.
118. Davenport, L., and Targowski, P., 1996, Submicrosecond phospholipid dynamics using a long lived fluorescence emission anisotropy probe, *Biophys. J.* **71:**1837–1852.
119. Davenport, L., 1994, Fluorescent phospholipid analogs and fatty acid derivatives, U.S. patent 5,332,794, pp. 1–14.
120. Richardson, F. S., 1982, Terbium(III) and europium(III) ions as luminescent probes and stains for biomolecular systems, *Chem. Rev.* **82:**541–552.
121. Sabbatini, N., and Guardigli, M., 1993, Luminescent lanthanide complexes as photochemical supramolecular devices, *Coord. Chem. Rev.* **123:**201–228.
122. Balzani, V., and Ballardini, R., 1990, New trends in the design of luminescent metal complexes, *Photochem. Photobiol.* **52:**409–416.
123. Li, M., and Selvin, P. R., 1995, Luminescent polyaminocarboxylate chelates of terbium and europium: The effect of chelate structure, *J. Am. Chem. Soc.* **117:**8132–8138.
124. Martin, R. B., and Richardson, F. S., 1979, Lanthanides as probes for calcium in biological systems, *Q. Rev. Biophys.* **12:**181–209.
125. Bruno, J., Horrocks, W. DeW., and Zauhar, R. J., 1992, Europium(III) luminescence and tyrosine to terbium(III) energy transfer studies of invertebrate (octopus) calmodulin, *Biochemistry* **31:**7016–7026.
126. Horrocks, W. DeW., and Sudnick, D. R., 1981, Lanthanide ion luminescence probes of the structure of biological macromolecules, *Acc. Chem. Res.* **14:**384–392.
127. Lumture, J. B., and Wensel, T. G., 1993, A novel reagent for labelling macromolecules with intensity luminescent lanthanide complexes, *Tetrahedron Lett.* **34:**4141–4144.
128. Lamture, J. B., and Wensel, T. G., 1995, Intensely luminescent immunoreactive conjugates of proteins and dipicolinate-based polymeric Tb(III) chelates, *Bioconjug. Chem.* **6:**88–92.
129. Lövgren, T., and Pettersson, K., 1990, Time-resolved fluoroimmunoassay, advantages and limitations, in *Luminescence Immunoassay and Molecular Applications*, K. Van Dyke and R. Van Dyke (eds.), CRC Press, Boca Raton, Florida, pp. 233–253.
130. Hemmila, I., 1993, Progress in delayed fluorescence immunoassay, in *Fluorescence Spectroscopy, New Methods and Applications,* O. S. Wolfbeis (ed.), Springer-Verlag, New York, pp. 259–266.
131. Terpetschnig, E., Szmacinski, H., and Lakowicz, J. R., 1997, Long lifetime metal–ligand complexes as probes in biophysics and clinical chemistry, *Methods Enzymol.* **278:**295–321.
132. Szmacinski, H., Terpetschnig, E., and Lakowicz, J. R., 1996, Synthesis and evaluation of Ru-complexes as anisotropy probes for protein hydrodynamics and immunoassays of high molecular-weight antigens, *Biophys. Chem.* **62:**109–120.
133. Guo, X.-Q., Castellano, F. N., Li, L., and Lakowicz, J. R., 1998, Use of a long-lifetime Re(I) complex in fluorescence polarization immunoassays of high-molecular weight analytes, *Anal. Chem.* **70:**632–637.
134. Friedman, A. E., Chambron, J.-C., Sauvage, J.-P., Turro, N. J., and Barton, J. K., 1990, Molecular light switch for DNA $Ru(bpy)_2(dppz)^{2+}$, *J. Am. Chem. Soc.* **112:**4960–4962.
135. Haq, I., Lincoln, P., Suh, D., Norden, B., Chowdhry, B. Z., and Chaires, J. B., 1995, Interaction of Δ- and Λ-$[Ru(phen)_2DPPZ]^{2+}$ with DNA: A calorimetric and equilibrium binding study, *J. Am. Chem. Soc.* **117:**4788–4796.
136. Demas, J. N., and DeGraff, B. A., 1992, Applications of highly luminescent transition metal complexes in polymer systems, *Macromol. Chem. Macromol. Symp.* **59:**35–51.
137. Li, L., Szmacinski, H., and Lakowicz, J. R., 1997, Long-lifetime lipid probe containing a luminescent metal–ligand complex, *Biospectroscopy* **3:**155–159.
138. Giuliano, K. A., Post, P. L., Hahn, K. M., and Taylor, D. L., 1995, Fluorescent protein biosensors: Measurement of molecular dynamics in living cells, *Annu. Rev. Biophys. Biomol. Struct.* **24:**405–434.
139. Marvin, J. S., Corcoran, E. E., Hattangadi, N. A., Zhang, J. V., Gere, S. A., and Hellinga, H. W., 1997, The rational design of allosteric interactions in a monomeric protein and its applications to the construction of biosensors, *Proc. Natl. Acad. Sci. U.S.A.* **94:**4366–4371.
140. Stewart, J. D, Roberts, V. A., Crowder, M. W., Getzoff, E. D., and Benkovic, S. J., 1994, Creation of a novel biosensor for Zn(II), *J. Am. Chem. Soc.* **116:**415–416.
141. Walkup, G. K., and Imperiali, B., 1996, Design and evaluation of a peptidyl fluorescent chemosensor for divalent zinc, *J. Am. Chem. Soc.* **118:**3053–3054.
142. Illsley, N. P., and Verkman, A. S., 1987, Membrane chloride transport measured using a chloride-sensitive fluorescent probe, *Biochemistry* **26:**1215–1219.
143. Kao, J. P. Y., 1994, Practical aspects of measuring $[Ca^{2+}]$ with fluorescent indicators. in *Methods in Cell Biology*, Vol. 40, R. Nuccitelli (ed.), Academic Press, New York, pp. 155–181.

PROBLEMS

3.1. *Binding of Proteins to Membranes or Nucleic Acids*: Suppose you have a protein which displays tryptophan fluorescence, and you wish to determine if the protein binds

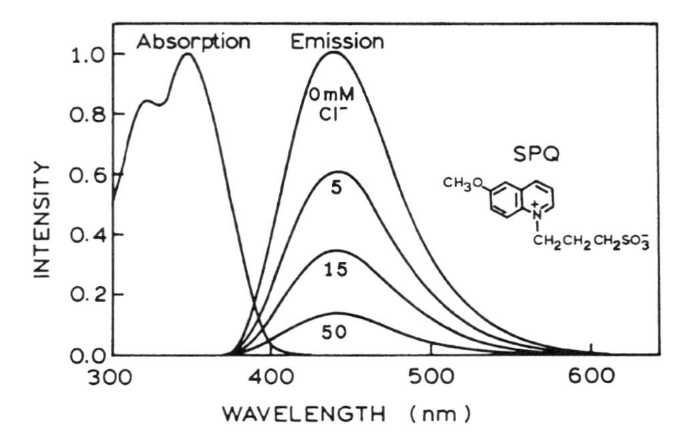

Figure 3.48. Absorption and emission spectra of 6-methoxy-*N*-(3-sulfopropyl)quinolinium (SPQ) in water with increasing amounts of chloride. From Ref. 74.

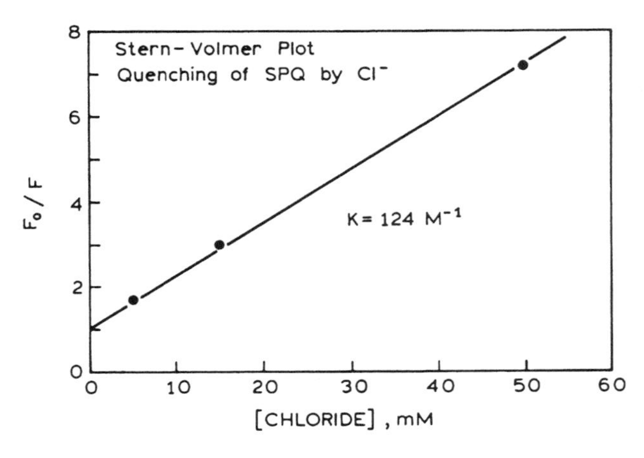

Figure 3.49. Stern–Volmer plot for the quenching of SPQ by chloride.

to DNA or lipid bilayers. Describe how you could use the tryptophan fluorescence to detect binding. Be specific regarding the spectral observables and expected results, including the use of intrinsic fluorescence, anisotropy, and resonance energy transfer.

3.2. *Chloride Quenching of SPQ*: Figure 3.48 shows the absorption and emission spectra of the chloride-sensitive probe 6-methoxy-*N*-(3-sulfopropyl)quinolinium (SPQ) in the presence of increasing amounts of Cl^-. SPQ is collisionally quenched by Cl^-. The unquenched lifetime is 26.3 ns.[142]

A. Use the data in Figure 3.49 to determine the Stern–Volmer quenching constant for chloride.

B. The average concentration of intracellular chloride in blood serum is 103m*M*. What is the lifetime and relative intensity of SPQ in blood serum?

C. Suppose the concentration of Cl^- decreases to 75m*M*. What is the expected lifetime and relative intensity of SPQ?

D. What factors would complicate interpretation of the SPQ intensities and lifetimes as a measure of Cl^- in blood serum?

3.3. *Calcium Concentrations Using Calcium Green and Fura-2*: The probes Calcium Green and Fura-2 display spectral changes in the presence of bound Ca^{2+}. Calcium Green shows changes in fluorescence intensity but not in spectral shape. Fura-2 displays a large shift in its absorption spectrum but little change in the shape of its emission spectrum. Calcium Green and Fura-2 display Ca^{2+} dissociation constants (K_D) near 200n*M*.

A. Derive an expression for the fluorescence intensity of Calcium Green relating its intensity to $[Ca^{2+}]$. For your answer, let F_{min} and F_{max} be the fluorescent intensities of Calcium Green in the absence and presence of saturating Ca^{2+}, respectively, and let K_D be the dissociation constant.

B. Derive an expression for the use of Fura-2 as an excitation wavelength-ratiometric probe of $[Ca^{2+}]$. This is a somewhat difficult problem to solve, and the exact form of the answer depends on how one defines the various terms. Let the subscripts 1 and 2 represent the two excitation wavelengths. Let R_{min} and R_{max} be the ratio of intensities at the two excitation wavelengths for the free (f) and the calcium-bound (b) form of Fura-2, respectively.

Time-Domain Lifetime Measurements

4

Time-resolved measurements are widely used in fluorescence spectroscopy, particularly for studies of biological macromolecules. This is because time-resolved data frequently contain more information than is available from the steady-state data. For instance, consider a protein which contains two tryptophan residues, each with a distinct lifetime. Because of spectral overlap of the absorption and emission, it is not usually possible to resolve the emission from the two residues. However, the time-resolved data may reveal two decay times, which can be used to resolve the emission spectra and relative intensities of the two tryptophan residues. Then one can question how each of the tryptophan residues is affected by the interactions of the protein with its substrate or other macromolecules. Is one of the tryptophan residues close to the binding site? Is a tryptophan residue in a distal domain affected by substrate binding to another domain? Such questions can be answered if one measures the decay times associated with each tryptophan residue.

There are many other examples where the time-resolved data provide information not available from the steady-state data. One can distinguish static and dynamic quenching using lifetime measurements. Formation of static ground-state complexes does not decrease the decay time of the uncomplexed fluorophores because only the unquenched fluorophores are observed in a fluorescence experiment. Dynamic quenching is a rate process acting on the entire excited-state population and thus decreases the mean decay time of the excited-state population. Resonance energy transfer is also best studied using time-resolved measurements. Suppose a protein contains a donor and acceptor, and the steady-state measurements indicate that the donor is 50% quenched by the acceptor. The observation of 50% donor quenching can be due to 100% quenching for one-half of the donors or 50% quenching of all the donors, or some combination of these two limiting possibilities. The steady-state data cannot be used to distinguish between these extreme cases. In contrast, very different donor intensity decays would be observed for each case.

One could list many other cases where the time-resolved data provide more molecular information than steady-state measurements. Additional examples will be seen throughout this book. In this and the following chapters, we present an overview of the two dominant methods for time-resolved measurements, the time-domain (TD) and frequency-domain (FD) methods. Since 1983 there have been remarkable advances in the capabilities of both methods. The discussion of TD measurements in this chapter and of FD measurements in Chapter 5 will not start as a historical survey, but rather with a description of the technology currently in use. We will then describe some of the earlier developments and alternative methods, so that one can understand why specific procedures have been selected. In this chapter we also discuss the important topic of data analysis, which allows use of the copious data provided by modern instruments. We also suggest procedures to avoid misuse of the results by overinterpretation of the data.

4.1. OVERVIEW OF TIME-DOMAIN AND FREQUENCY-DOMAIN MEASUREMENTS

Two methods of measuring time-resolved fluorescence are in widespread use, the time-domain and frequency-domain methods. In time-domain or pulse fluorometry, the sample is excited with a pulse of light (Figure 4.1). The width of the pulse is made as short as possible and is preferably much shorter than the decay time τ of the sample. The time-dependent intensity is measured following the excitation pulse, and the decay time τ is calculated from the slope of a plot of log $I(t)$ versus t, or from the time at which the intensity decreases to $1/e$ of the value at $t = 0$. The intensity decays are often measured through a polarizer oriented at 54.7° from the vertical z-axis. This condition is

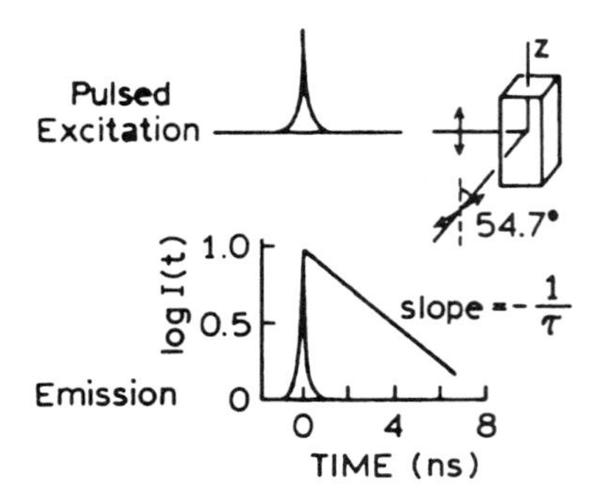

Figure 4.1. Pulse or time-domain lifetime measurements.

used to avoid the effects of rotational diffusion and/or anisotropy on the intensity decay (Chapter 11).

The alternative method of measuring the decay time is the frequency-domain or phase-modulation method. In this case the sample is excited with intensity-modulated light; typically, sine wave modulation is employed (Figure 4.2). The amplitude-modulated excitation should not be confused with the electrical component of an electromagnetic wave. The intensity of the incident light is varied at a high frequency ($\omega = 2\pi$ times the frequency in hertz) comparable to the reciprocal of the decay time τ. When a fluorescent sample is excited in this manner, the emission is forced to respond at the same modulation frequency. Owing to the lifetime of the sample, the emission is delayed in time relative to the excitation. This delay is measured as a phase shift (ϕ) which can be used to calculate the decay time. Magic-angle polarizer conditions are also used in FD measurements.

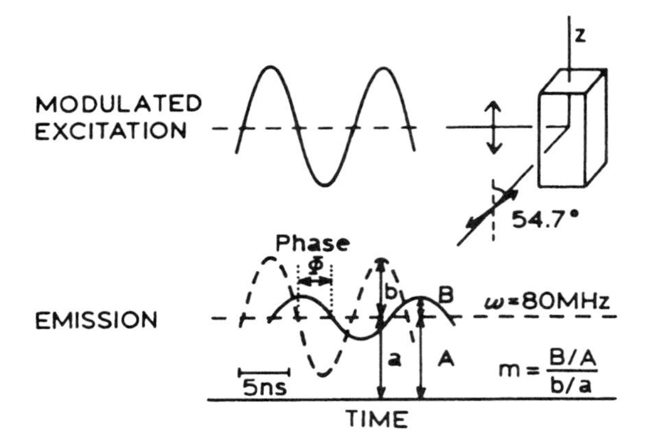

Figure 4.2. Phase-modulation or frequency-domain lifetime measurements. The ratios *B*/*A* and *b*/*a* represent the modulation of the emission and of the excitation, respectively.

Examination of Figure 4.2 reveals another effect of the lifetime, this being a decrease in the peak-to-peak height of the emission relative to that of the modulated excitation. The modulation decreases because some of the fluorophores excited at the peak of the excitation continue to emit when the excitation is at a minimum. The extent to which this occurs depends on the decay time and light modulation frequency. This effect is called demodulation and can also be used to calculate the decay time. At present, pulse and phase-modulation measurements are both in widespread use.

4.1.A. Meaning of the Lifetime or Decay Time

Prior to further discussion of lifetime measurements, it is important to have a firm understanding of the meaning of the lifetime τ. Suppose a sample containing the fluorophore is excited with an infinitely sharp (δ-function) pulse of light. This results in an initial population (n_0) of fluorophores in the excited state. The excited-state population decays with a rate $\Gamma + k_{nr}$ according to

$$\frac{dn(t)}{dt} = -(\Gamma + k_{nr})n(t) \qquad [4.1]$$

where $n(t)$ is the number of excited molecules at time t following excitation, Γ is the emissive rate, and k_{nr} is the nonradiative decay rate. Emission is a random event, and each excited fluorophore has the same probability of emitting in a given period of time. This results in an exponential decay of the excited-state population, $n(t) = n_0 \exp(-t/\tau)$.

In a fluorescence experiment we do not observe the number of excited molecules, but rather a fluorescence intensity, which is proportional to $n(t)$. Hence, Eq. [4.1] can also be written in terms of the time-dependent intensity $I(t)$. Integration of Eq. [4.1] yields the usual expression for a single exponential decay

$$I(t) = I_0 \exp(-t/\tau) \qquad [4.2]$$

where I_0 is the intensity at time zero. The lifetime τ is the inverse of the total decay rate: $\tau = (\Gamma + k_{nr})^{-1}$. In general, the inverse of the lifetime is the sum of the rates which depopulate the excited state. The fluorescence lifetime can be obtained in two ways. This value can be obtained from the time at which the intensity decreases to $1/e$ of its initial value. More commonly, the lifetime is determined from the slope of a plot of log $I(t)$ versus t (Figure 4.1).

The lifetime is also the average amount of time a fluorophore remains in the excited state following excitation. This can be seen by calculating the average time $\langle t \rangle$, which is obtained by averaging t over the intensity decay of the fluorophore:

$$\langle t \rangle = \frac{\int_0^\infty tI(t)dt}{\int_0^\infty I(t)dt} = \frac{\int_0^\infty t\exp(-t/\tau)dt}{\int_0^\infty \exp(-t/\tau)dt} \quad [4.3]$$

The denominator is equal to τ. Following integration by parts, one finds that the numerator is equal to τ^2. Hence, for a single-exponential decay the average time a fluorophore remains in the excited state is equal to the lifetime,

$$\langle t \rangle = \tau \quad [4.4]$$

It is important to note that Eq. [4.4] is not true for more complex decay laws, such as multiexponential or nonexponential decays. Using the observed decay law, an average lifetime can always be calculated using Eq. [4.3.] However, this average lifetime can be a complex function of the parameters describing the actual intensity decay. For this reason, caution is necessary in interpretation of the average lifetime.

Another important concept is that the lifetime is a statistical average, and fluorophores emit randomly throughout the decay. For a large number of fluorophores, some will emit quickly following the excitation, and some will emit at times longer than the lifetime. This time distribution of emitted photons is the intensity decay.

4.1.B. Phase and Modulation Lifetimes

The FD method will be described in more detail in Chapter 5, but it is valuable to understand the basic equations relating lifetimes to phase and modulation. The modulation of the excitation is given by b/a, where a is the average intensity and b is the modulated amplitude of the incident light (Figure 4.2). The modulation of the emission is defined similarly, except using the intensities of the emission, B/A (Figure 4.2). The modulation of the emission is measured relative to the modulation of the excitation, $m = (B/A)/(b/a)$. Although m is actually a demodulation factor, it is usually called the modulation. The other experimental observable is the phase delay, called the phase angle (ϕ), which is usually measured from the zero-crossing times of the modulated components. The phase angle (ϕ) and the modulation (m) can be used to calculate the lifetime from the following relations:

$$\tan\phi = \omega\tau_\phi, \quad \tau_\phi = \omega^{-1}\tan\phi \quad [4.5]$$

$$m = \frac{1}{\sqrt{1+\omega^2\tau_m^2}}, \quad \tau_m = \frac{1}{\omega}\left[\frac{1}{m^2} - 1\right]^{1/2} \quad [4.6]$$

One can use these expressions to calculate the phase (τ_ϕ) and modulation (τ_m) lifetimes for the curves shown in Figure 4.2 (Problem 4.1).

4.1.C. Examples of Time-Domain and Frequency-Domain Lifetimes

It is useful to understand the appearance of the TD and the FD data. TD and FD data are shown for the tryptophan derivative N-acetyl-L-tryptophanamide (NATA) in Figure 4.3. This tryptophan derivative is known to display a single-exponential decay (Chapter 17). In the time domain (Figure 4.3, left), the data are presented as log counts versus time. The data are presented as individual photon counts because most such measurements are performed by single-photon counting. The log intensity of NATA is seen to decay linearly with time, indicating that the decay is a single exponential. The noisy curve in the lower part of this panel [$L(t)$] is the instrument response function (IRF), which shows the shape of the excitation pulse and how this pulse is detected by the instrument. This IRF is clearly not a δ-function, and much of the art of lifetime measurements is in accounting for this nonideal response in analyzing the data.

Analysis of the time domain is accomplished mostly by nonlinear least-squares analysis.[1] In this method one finds the lifetime which results in the best fit between the measured data and data calculated for the assumed lifetime. Although not separately visible in the left-hand panel of Figure 4.3, the calculated intensity decay for $\tau = 5.15$ ns overlaps precisely with the number of photons counted in each channel. The lower left panel of Figure 4.3 shows the deviations between the measured and calculated data, weighted by the standard deviations of the measurements. For a good fit, the deviations are random, indicating that the only source of difference is the random error in the data.

FD data for the same NATA sample are shown in the right-hand panel of Figure 4.3. At present, it is common practice to measure the phase and modulation over a range of light modulation frequencies. As the modulation frequency is increased, the phase angle increases from 0 to 90°, and the modulation decreases from 1 (100%) to 0 (0%). As for the TD data, the FD data are also analyzed by nonlinear least-squares analysis. The filled circles in the upper right panel of Figure 4.3 represent the data, and the solid curves represent the best fit with a single lifetime of 5.09 ns. As for the TD data, the goodness-of-fit is judged by the differences (deviations) between the data and the calculated curves. For the FD data there are two observ-

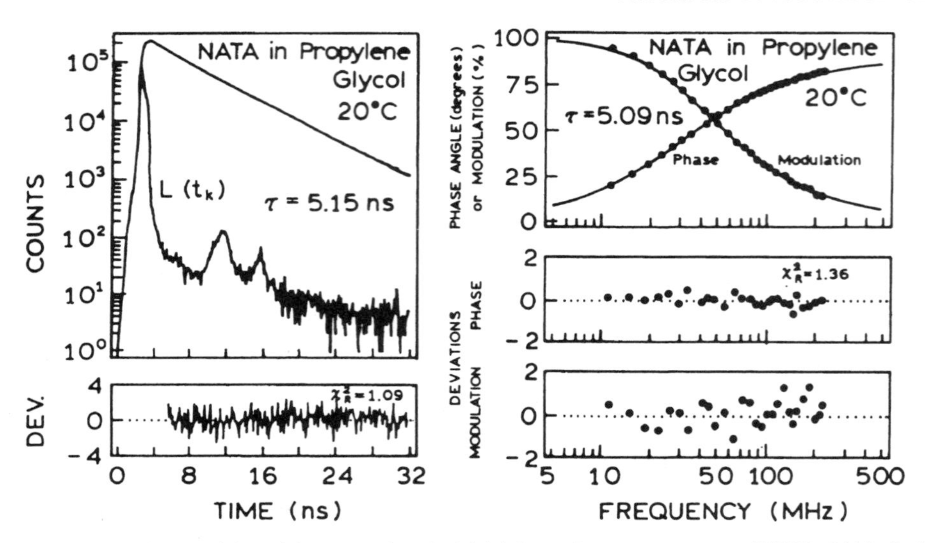

Figure 4.3. Comparison of time-domain (*left*) and frequency-domain (*right*) decay time measurements of NATA. $L(t)$ is the instrument response function.

ables, phase and modulation, so there are two sets of deviations (lower panels). The randomness of the deviations indicates that a single lifetime is adequate to explain the data.

4.2. BIOPOLYMERS DISPLAY MULTIEXPONENTIAL OR HETEROGENEOUS DECAYS

At first glance, the measurement of decay times seems straightforward (Figure 4.3), so one may question why these measurements receive so much attention. Interpretation of the data in Figure 4.3 was relatively simple because the decays were single exponentials. However, most samples display more than one decay time. This situation is illustrated in Figure 4.4 for a protein with two tryptophan residues. Suppose that the two residues display lifetimes of 5 ns each. Then the decay would be a simple single-exponential decay. The decay would be simple to analyze, but one could not distinguish between the two tryptophan residues. Now suppose that one adds a collisional quencher and that only the residue on the surface of the protein is accessible to quenching. Assume that the added quencher reduces the lifetime of the exposed residue to 1 ns. The intensity decay is now a double exponential,

$$I(t) = \alpha_1 e^{-t/5.0} + \alpha_2 e^{-t/1.0} \qquad [4.7]$$

In this expression the α_i values are called the preexponential factors. A fluorophore usually displays the same radiative decay rate in different environments. Thus, for the same fluorophore in different environments, the values of α_i represent the fractional amount of fluorophore in each environment. Hence, for the protein shown in Figure 4.4, one expects $\alpha_1 = \alpha_2 = 0.5$. The presence of two decay times results in curvature in the plot of log $I(t)$ versus time, represented by the dashed line in the upper plot. The goal of the intensity decay measurements is to recover the decay times (τ_i) and amplitudes (α_i) from the $I(t)$ measurements.

The presence of two decay times can also be detected using the FD method. In this case one examines the frequency response of the sample, which consists of a plot of phase and modulation on the log frequency axis (Figure 4.4, bottom). The longer-lifetime tryptophan (τ_1 = 5 ns; solid curve) and the shorter-lifetime tryptophan (τ_2 = 1 ns; dotted curve) each display the curves characteristic of a single decay time. In the presence of both decay times (τ_1 = 5 ns and τ_2 = 1 ns; dashed curve), the frequency response displays a more complex shape which is characteristic of the heterogeneous or multiexponential intensity decay. As for the TD measurements, the FD data are used to recover the individual decay times (τ_i) and amplitudes (α_i) associated with each decay time, typically using the procedure of nonlinear least squares.

Examination of Figure 4.4 shows that the $I(t)$ values start at the same initial value and that the phase and modulation values do not contain any information about the signal intensity. For the protein model shown, one expects the time-zero intensities of each component to be the same, but in general the time-zero intensities of the components in a multiexponential decay are not equal. An important point about lifetime measurements is that the intensity decay, or phase and modulation values, is typically measured without concern about the actual intensity. Intensity decays are typically fit to the multiexponential model

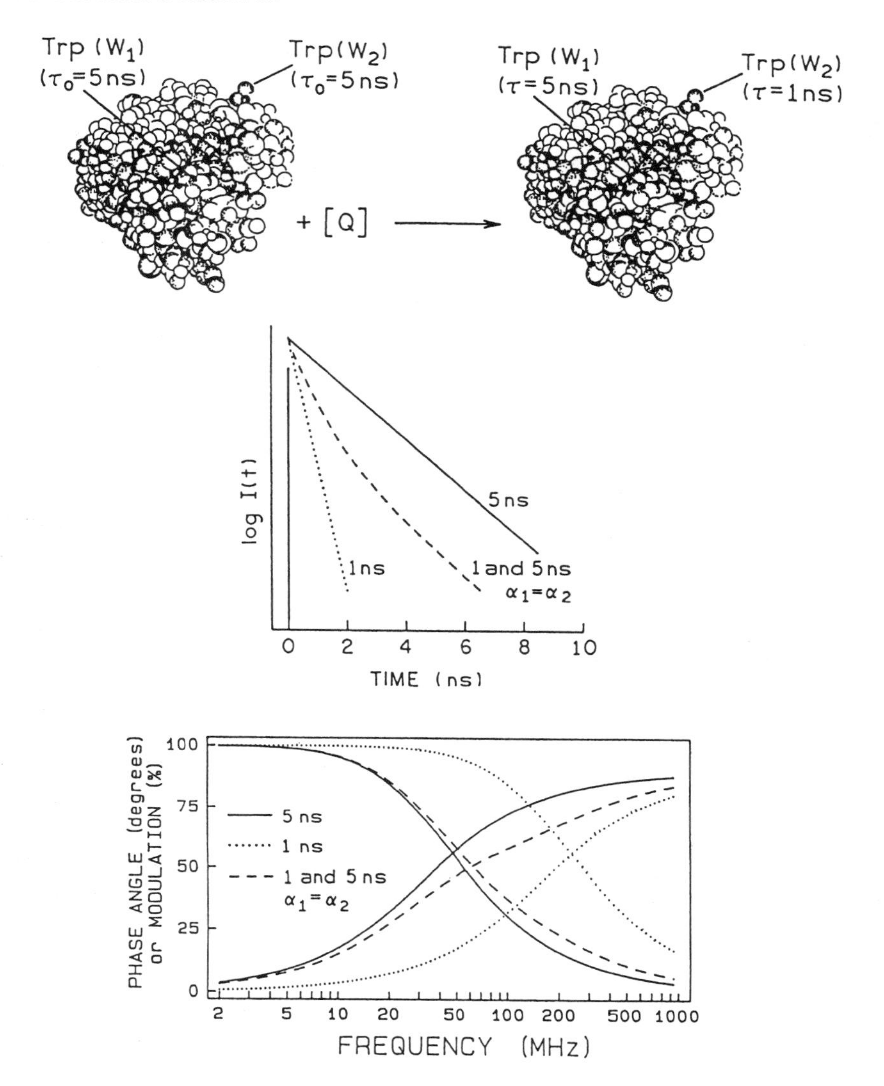

Figure 4.4. Simulated intensity decays of buried (W_1) and exposed (W_2) tryptophan residues in the absence and in the presence of a collisional quencher. From Ref. 2.

$$I(t) = \sum_i \alpha_i \exp(-t/\tau_i) \qquad [4.8]$$

where $\Sigma\alpha_i$ is normalized to unity.

A second example of the need for time-resolved measurements is the complex anisotropy decay displayed by a protein which self-associates into a tetramer (Figure 4.5). Rotational motions can be measured from the decays of anisotropy. For a spherical molecule, one expects a single decay time for the anisotropy, which is called the rotational correlation time (θ),

$$r(t) = r_0 \exp(-t/\theta) \qquad [4.9]$$

In this expression r_0 is the anisotropy at $t = 0$, which is characteristic of the fluorophore. The rotational correlation time θ is the time at which the initial anisotropy has decayed to $1/e$ of its original value and is longer for larger proteins. Suppose that the protein monomers associate to form a tetramer. The rotational correlation time will become longer, and the anisotropy will decay more slowly. However, one often encounters the more complex case where the protein is present in both states. In this case the anisotropy decay will be a double exponential,

$$r(t) = r_0 f_M \exp(-t/\theta_M) + r_0 f_T \exp(-t/\theta_T) \qquad [4.10]$$

where the f_i represent the fraction of the protein present as monomers and tetramers: $f_M + f_T = 1.0$. It is important to recognize that the fractions appearing in Eq. [4.10] actually represent the fractional fluorescence arising from each

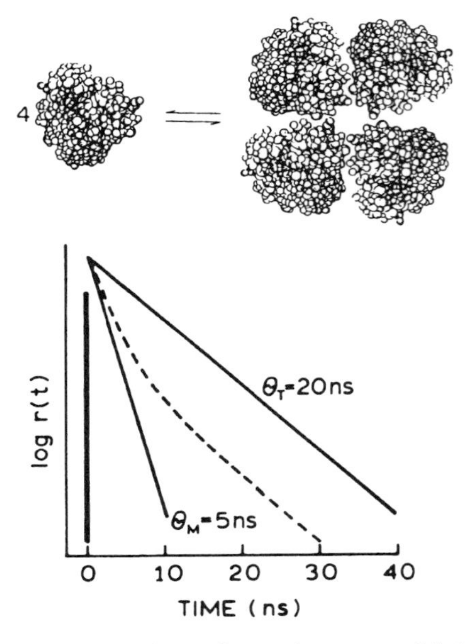

Figure 4.5. Anisotropy decay of a protein monomer (M) that self-associates into a tetramer (T). The dashed line shows the anisotropy decay expected for partially associated monomers.

species. If the quantum yield changes on going from monomer to tetramer, then this change would need to be accounted for in calculating the molecular fraction from the fractional amplitudes in the anisotropy decay.

Anisotropy decays can be more complex than those represented by Eqs. [4.9] and [4.10] and are typically presented as a sum of exponentials

$$r(t) = \sum_j r_{0j} \exp(-t/\theta_j) \qquad [4.11]$$

where the meaning of the amplitudes (r_{0j}) and correlation times (θ_j) depends on the chosen molecular model. The goal of many time-resolved measurements is to determine the form of a complex anisotropy decay. In general, it is more difficult to resolve a multiexponential anisotropy decay (Eq. [4.11]) than a multiexponential intensity decay (Eq. [4.8]).

Although the intensity decays and anisotropy decays have similar mathematical forms, there is no direct linkage between the decay times and rotational correlation times. The decay times are determined by the spectral properties of the fluorophore. The rotational correlation times are determined by the size, shape, and flexibility of the macromolecules. Both the decay times and the rotational correlation times are often on the nanosecond timescale. These conditions result in anisotropies which are sensitive to the size of the protein and its interactions with other macromolecules.

4.2.A. Resolution of Multiexponential Decays Is Difficult

Why is so much attention given to data analysis and obtaining a high signal-to-noise ratio in the time-resolved data? The need for a high signal-to-noise ratio is due to the inherent difficulty in recovering the amplitudes and lifetimes for a multiexponential process. This difficulty was well known to mathematicians and was pointed out to fluorescence spectroscopists in a classic paper by Grinvald and Steinberg.[3] This paper defined the method for analyzing time-resolved fluorescence data, and the same basic approach is still in use today. Another feature of this paper was to illustrate how apparently different multiexponential decays can yield similar $I(t)$ values. Consider the following two double-exponential decays,

$$I_1(t) = 7500\, e^{-t/5.5} + 2500\, e^{-t/8.0} \qquad [4.12]$$

$$I_2(t) = 2500\, e^{-t/4.5} + 7500\, e^{-t/6.7} \qquad [4.13]$$

The preexponential factor sum of 10,000 corresponds to 10,000 photons in the highest intensity channel, which is typical of data for time-correlated single-photon counting (TCSPC). From examination of these equations, one would think that the intensity decays would be distinct. However, a plot of the intensity decays on a linear scale

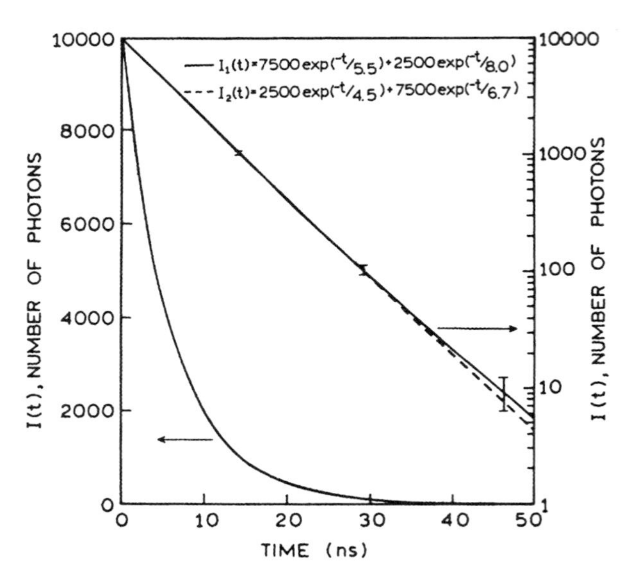

Figure 4.6. Comparison of two intensity decays, on a linear (*left*) and logarithmic scale (*right*). The error bars represent Poisson noise on the photon counts. The decay functions were described in Ref. 3.

shows that they are indistinguishable at all times (Figure 4.6). On a logarithmic scale, one notices some minor differences at 30–50 ns. However, at 50 ns there are only 3–5 photons per time increment (channel), so that the difference between the two decays is just 1–2 photons. In fact, if one adds the Poisson noise that is inevitably present in photon counting data, the difference between the curves is sevenfold less than the uncertainties due to the Poisson noise.[4] This illustrates that it is difficult to distinguish between some multiexponential functions or, conversely, that it is difficult to recover the actual values of α_i and τ_i for a multiexponential decay. We note that a similar result is obtained from simulations of the FD data. The simulated frequency responses are visually indistinguishable for these two decay laws.

Why is it difficult to resolve multiexponential decays? Comparison of $I_1(t)$ and $I_2(t)$ indicates that the lifetimes and amplitudes are different in each decay law. In fact, this is the problem. For a multiexponential decay, one can vary the lifetime to compensate for the amplitude, or vice versa, and obtain similar intensity decays with different values of α_i and τ_i. In mathematical terms, the values of α_i and τ_i are said to be correlated. The problem of correlated parameters has been described within the framework of general least-squares fitting.[5–7] The unfortunate result is that the ability to determine the precise values of α_i and τ_i is greatly hindered by parameter correlation. There is no way to avoid this problem, except by careful experimentation and conservative interpretation of data.

4.3. TIME-CORRELATED SINGLE-PHOTON COUNTING

At present almost all time-domain measurements are performed using TCSPC. Several comprehensive monographs dealing with TCSPC have appeared.[4,8–10] One book is completely devoted to TCSPC and provides numerous valuable details.[8] While somewhat dated, the insightful monograph of Ware[10] clearly describes the concept of TCSPC, and Ware anticipated many of its present applications. Rather than present a history of the method, we will start with current state-of-the-art instrumentation. These instruments use high-repetition-rate picosecond or femtosecond laser light sources and high-speed MCP PMTs. In later sections we will describe other light sources and detectors.

4.3.A. Principles of TCSPC

The principles of TCSPC can be understood by examination of an instrument schematic (Figure 4.7). The experiment starts with the excitation pulse, which excites the sample and starts the time measurement clock. TCSPC is a digital technique, counting photons which are time-correlated in relation to the excitation pulse. The heart of the method is a time-to-amplitude converter (TAC), which can be considered to be analogous to a fast stopwatch.

The sample is repetitively excited using a pulse light source, often from a laser or flashlamp. Each pulse is optically monitored, by a high-speed photodiode or photomultiplier, to produce a start signal which is used to trigger the voltage ramp of the TAC. The voltage ramp is stopped when the first fluorescence photon from the sample is detected. The TAC provides an output pulse whose voltage is proportional to the time between the start and stop signals. A multichannel analyzer (MCA) converts this voltage to a time channel using an analog-to-digital converter (ADC). Summing over many pulses, the MCA builds up a probability histogram of counts versus time channels. The experiment is continued until one has collected more than 10,000 counts in the peak channel. As will be described below in more detail, there can be no more than one photon detected per 100 laser pulses. Under these conditions, the histogram of photon arrival times represents the intensity decay of the sample.

There are many subtleties in TCSPC which are not obvious at first examination. Why is the photon counting rate limited to one photon per 100 laser pulses? Present electronics for TCSPC only allow detection of the first arriving photon. Once the first photon is detected, the dead time in the electronics prevents detection of another photon resulting from the same excitation pulse. Recall that emission is a random event. Following the excitation pulse, more photons are emitted at early times than at late times. If all could be measured, then the histogram of arrival times would represent the intensity decay. However, if many arrive, and only the first is counted, then the intensity decay is distorted to shorter times. This effect is described in more detail in Section 4.5.F.

Another important feature of TCSPC is the use of the rising edge of the photoelectron pulse for timing. This allows phototubes with nanosecond pulse widths to provide subnanosecond resolution. This is possible because the rising edge of the single photon pulses are usually steeper than one would expect from the time response of the PMT. Also, the use of a constant fraction discriminator provides improved time resolution by removing the variability due to the amplitude of each pulse.

4.3.B. Example of TCSPC Data

Prior to examining the electronic components in more detail, it is valuable to examine the actual data. An intensity

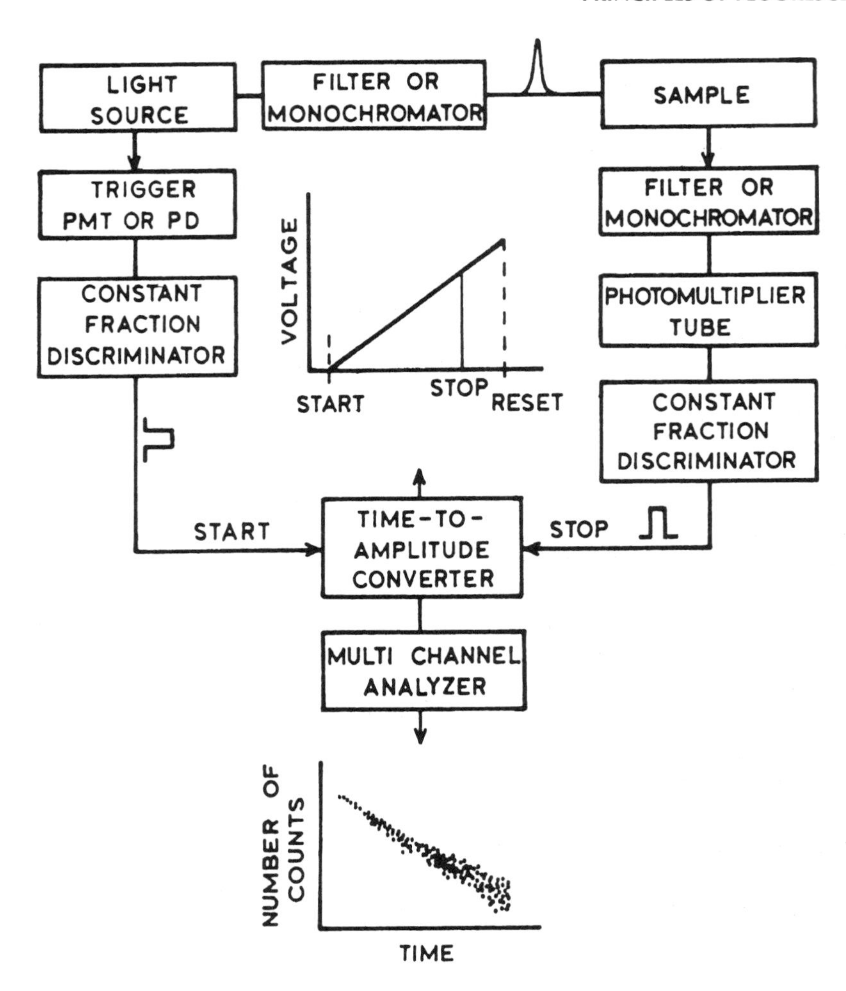

Figure 4.7. Schematic for time-correlated single-photon counting.

decay for the scintillator 2,5-diphenyl-1,3,4-oxadiazole (PPD) is shown in Figure 4.8. These data were obtained with a rhodamine 6G (R6G) dye laser which was cavity-dumped at 1 MHz and frequency-doubled to 300 nm. The detector was an MCP PMT. There are typically three curves associated with an intensity decay. These are the measured data $N(t_k)$, the instrument response function $L(t_k)$, and the calculated decay $N_c(t_k)$. These functions are in terms of discrete times (t_k) because the counted photons are collected into channels, each with a known time (t_k) and width (Δt).

The instrument response function, sometimes called the lamp function, is the response of the instrument to a zero-lifetime sample. This curve is typically collected using a dilute scattering solution such as colloidal silica (Ludox) and no emission filter. This decay represents the shortest time profile which can be measured by the instrument. The lamp function in Figure 4.8 is quite narrow, about 60 ps wide, measured as the full width at half-maximum intensity. The use of a logarithmic intensity scale exaggerates the low-intensity regions of the profile. One notices an afterpulse about 2 ns after the main peak. Such afterpulses are observed with many PMTs. In fact, the instrument response function shown in Figure 4.8 is rather good, and some PMTs give far less ideal profiles. For instance, the profile in Figure 4.3 was measured with an end-on linear-focused PMT, for which the afterpulses and long-time tail are more significant. However, even in this case (Figure 4.3) the number of photons in the peak of the afterpulse is only about 0.05% of the counts in the peak channel.

The next measured curve is the intensity decay of the sample itself $[N(t_k)]$. In Figure 4.8 the peak channel, with the largest number of counts, has recorded approximately 3000 photons. On the log scale the decay is seen to be a straight line, suggesting a single decay time.

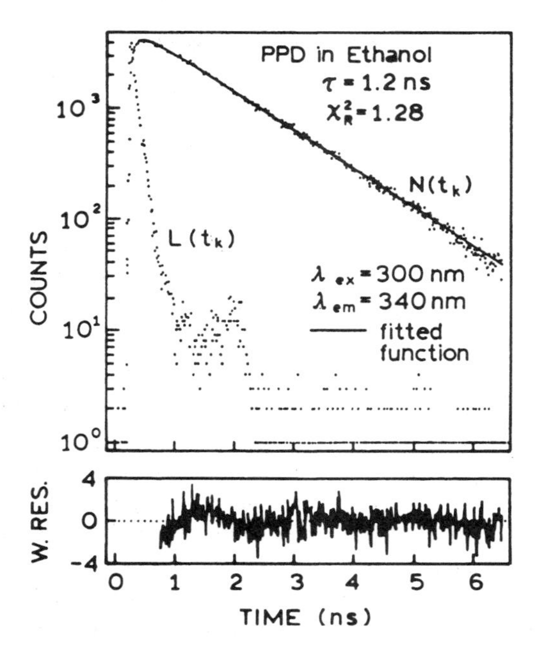

Figure 4.8. TCSPC data for PPD in ethanol. The light source was an R6G dye laser, cavity-dumped at 1 MHz. The detector was an R2809 MCP PMT (Hamamatsu). The left side of the residuals plot (lower panel) shows some minor systematic error. From Ref. 11.

The last curve is the calculated data [$N_c(t_k)$], which is usually called the fitted function. This curve (solid line in Figure 4.8) represents a convolution of the lamp function with the impulse response function, which is also called the intensity decay law. The fitted function is the time profile expected for a given decay law when one considers the form of the instrument response function. The details of calculating the convolution are described in the next section. The decay time is the value of τ which provides the best match between the measured data $N(t_k)$ and the calculated time-dependent intensities $N_c(t_k)$.

4.3.C. Convolution Integral

It is important to understand why the measured intensity decay is a convolution with the lamp function. The impulse response function $I(t)$ is what would be observed with δ-function excitation and a δ-function for the instrument response. Unfortunately, such systems do not exist, and most instrument response functions are several nanoseconds wide. We can consider the excitation pulse to be a series of δ-functions with different amplitudes. Each of these δ-functions excites an impulse response from the sample, with an intensity proportional to the height of the δ-function (Figure 4.9). The measured function [$N(t_k)$, on the right-hand side of Figure 4.9] is the sum of all these exponential decays, starting with different amplitudes and at different times.

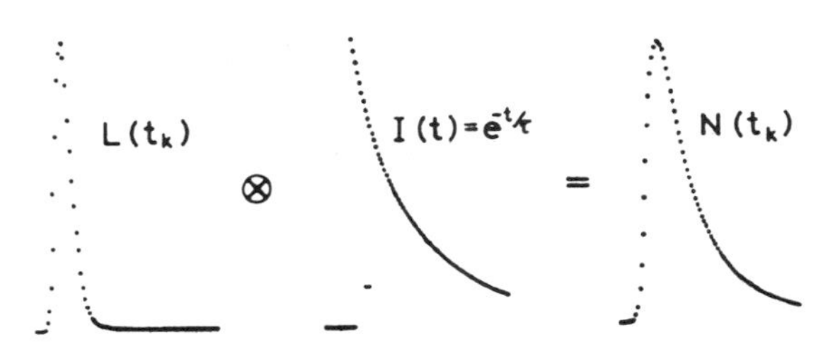

Figure 4.9. Convolution of an impulse response function $I(t)$ with a lamp profile $L(t_k)$ to yield the measured data $N(t_k)$. Revised from Ref. 9.

Mathematically, the concept of convolution can be expressed as follows.[12] Each δ-function excitation is assumed to excite an impulse response at t_k,

$$I_k(t) = L(t_k)\, I(t - t_k)\Delta t \qquad (t > t_k) \tag{4.14}$$

The term $(t - t_k)$ appears because the impulse response is started at $t = t_k$, and it is understood that there is no emission before excitation $(t < t_k)$. The measured decay $N(t_k)$ is the sum of the impulse responses created by all the individual δ-function excitation pulses occurring until t_k,

$$N(t_k) = \sum_{t=0}^{t=t_k} L(t_k) I(t - t_k)\Delta t \tag{4.15}$$

For small values of Δ*t*, this equation can be expressed as an integral:

$$N(t) = \int_0^t L(t') I(t - t')\, dt' \tag{4.16}$$

This expression says that the experimentally measured intensity at time *t* is given by the sum of the intensities expected for all the δ-function excitation pulses which occur until time *t*. It is important to note that new intensity decays are created in the sample as long as there is nonzero intensity in $L(t_k)$. For convenience, the dummy variable of integration is changed using $t' = t - \mu$, so that the convolution integral is expressed as

$$N(t) = \int_0^t L(t - \mu) I(\mu)\, d\mu \tag{4.17}$$

The task is to determine the impulse response function $I(\mu)$ which best matches the experimental data. Since the data are digital, i.e., counts per channel, Eq. [4.15] is perhaps the most convenient form. It is important to notice that time = 0 is not defined in a TCSPC experiment. This is because there is no single δ-function initiating the decay. For convenience, the excitation pulse is positioned close to zero on the time axis so that one can consider the values of $N(t_k)$ relative to the decay time of the sample.

4.4. LIGHT SOURCES FOR TCSPC

The instrumentation for TCSPC is moderately complex, and effective use of this method requires understanding of the various components. Measurement of intensity decays requires a pulsed light source, which at present is often a picosecond dye laser. There is a tendency to use pulsed lasers as "the light source" without developing an appreciation for the theory of operation. One quickly learns that the laser source requires frequent day-to-day adjustments, which are incomprehensible without knowing the operating principles of mode locking and cavity dumping. Similarly, it is not possible to adjust the numerous settings for the TCSPC electronics without understanding their function.

4.4.A. Picosecond Dye Lasers

Given the current advanced technology for TCSPC, it is easy to forget that decay time measurements did not become generally available until the early 1980s. At that time the primary light source for TCSPC was the nanosecond flashlamp (Section 4.4.C). Since that time the preferred light source for TCSPC has become the cavity-dumped dye laser (Figure 4.10). Such lasers provide pulses about 5 ps wide, and the repetition rate is easily chosen from kilohertz rates up to 80 MHz.

A dye laser is a passive device which requires an optical pump source. The primary light source is usually a mode-locked argon-ion laser. Mode-locked neodymium:YAG (Nd:YAG) lasers are also used as the primary source, but they are generally less stable than argon (Ar) lasers. Nd:YAG lasers have their fundamental output at 1064 nm. This fundamental output is frequency-doubled to 532 nm or frequency-tripled to 355 nm in order to pump the dye lasers. The need for frequency doubling is one reason for

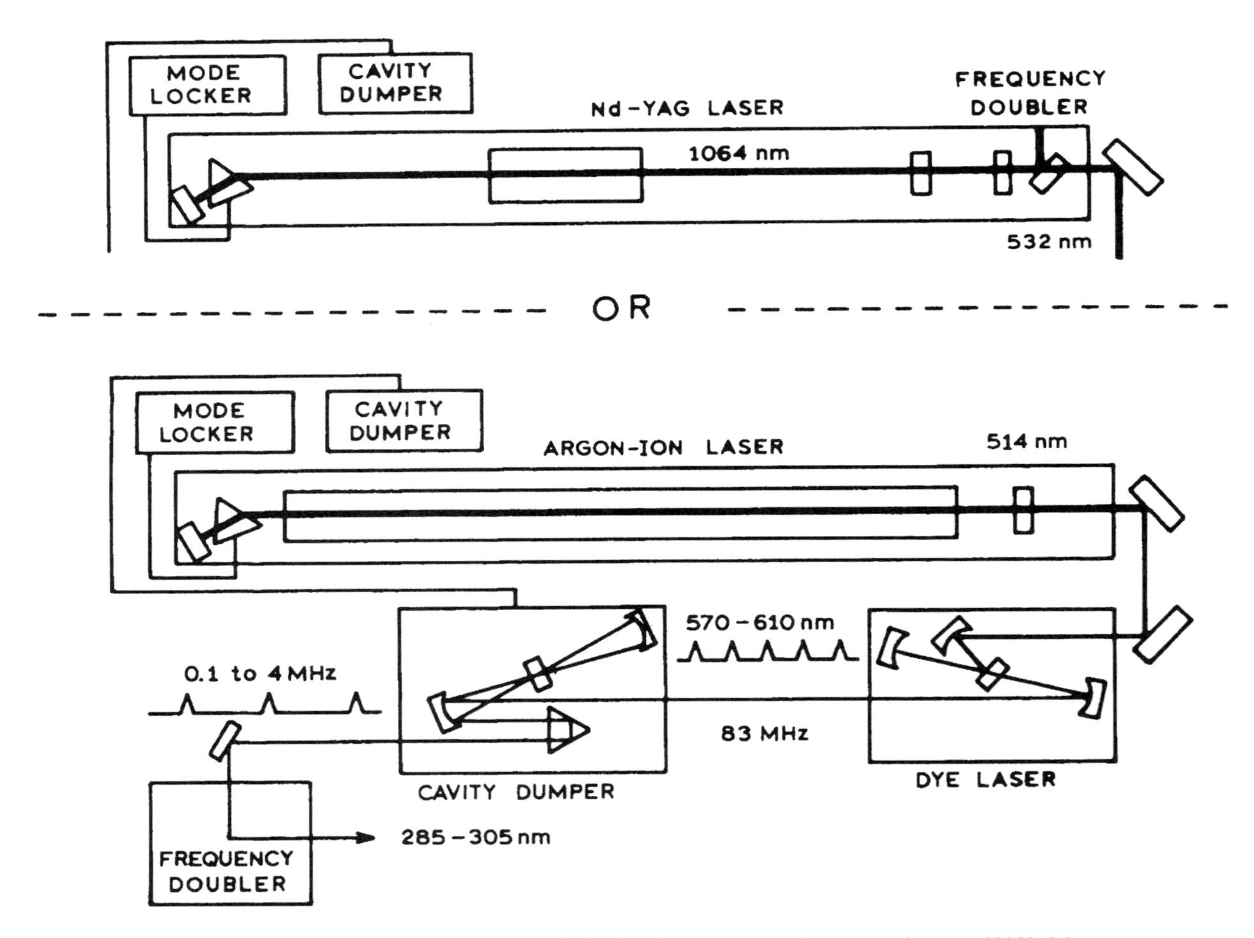

Figure 4.10. Picosecond light sources for TCSPC. The primary pump can be an argon ion or an Nd:YAG laser.

the lower stability of the Nd:YAG lasers, but heating effects in the Nd:YAG laser rod also contribute to the instability. The argon-ion laser has an output at 514 nm, which is used directly for pumping most laser dyes.

It is not practical to provide a complete description of laser physics, and the reader is referred to the many books on this topic[13–15] or to a description of laser principles written for chemists.[16] The argon-ion mode-locked laser provides pulses at 514 nm, about 70 ps wide, with a repetition rate near 80 MHz. Without active mode locking, the argon-ion laser would provide a continuous output. This output is changed to an 80-MHz pulse train by a mode-locking crystal within the laser cavity. This crystal is an acousto-optic device which deflects light out of the cavity based on light diffraction by sound waves in the crystal.[17,18]

In order to obtain mode locking, there is a delicate balance between the length of the argon-ion laser cavity, the resonance frequency at the acousto-optic (AO) crystal, and the driving frequency to the AO crystal. The resonance frequency of the AO crystal and the driving frequency must be matched in order to obtain stable operation. This can be difficult because the resonances of the AO crystal are temperature-dependent. For this reason, the mode-locking crystal is usually thermostated above room temperature. Once the AO crystal is at a stable temperature and resonance, the cavity length of the argon-ion laser must be adjusted so that the round-trip time for photons (near 12 ns for a typical large-frame laser) coincides with the nulls in the AO crystal deflection. Since there are two nulls per cycle, the AO crystal operates near 40 MHz. The laser pulses at 80 MHz exit one end of the laser through a partially transmitting mirror.

The presence of the AO modulator within the laser cavity causes a loss of energy at times other than the round-trip time for the photons and the nulls in the AO crystal. If the cavity length matches the AO crystal frequency, then the photons accumulate in a single bunch and bounce back and forth together within the laser cavity. This is the mode-locked condition.

One may wonder why the argon-ion lasers are usually mode-locked at 514 nm, and not at 488 nm, which is the most intense output. Several reasons have been given for the absence of mode locking at 488 nm. It has been stated that the 488-nm line has more gain, so that one obtains modulated but not mode-locked output. Also, there appear to be two closely spaced lines at 488 nm, which cannot be simultaneously mode-locked. In any event, the 514-nm output is convenient for pumping an R6G dye laser, which is perhaps the most stable and easiest to adjust dye laser for day-to-day use.

Mode-locked argon-ion lasers found use in TCSPC starting in the late 1970s and early 1980s.[19–24] However, this light source was not particularly useful for biochemistry unless one had a fluorophore which could be excited at 514 nm. This problem was solved by the introduction of picosecond dye lasers for TCSPC.[25–29] The mode-locked argon-ion (or Nd:YAG) laser is used as the pumping source for a dye laser, typically R6G. The cavity length of the dye laser is adjusted to be exactly the same as that of the argon-ion laser, so that the round-trip time for the photon bunch in the dye laser is the same as in the argon-ion laser. When the cavity lengths are matched, the incoming 514-nm pulses reinforce a single bunch of photons which oscillates at 80 MHz within the dye laser cavity. To conserve space and to have a stable cavity length, the dye laser cavity is often folded. Because of the wide emission curve of R6G, this dye laser has intrinsically narrow pulses, so that a typical pulse width is near 5 ps. This pulse width is narrower than any available detector response, so that for all practical purposes the dye lasers provide δ-function excitation. When a picosecond dye laser source is used, the width of the instrument response function is due primarily to the detection electronics and PMT.

There are two difficulties with the output of the R6G dye laser. Its wavelength is too long for excitation of most fluorophores, and the repetition rate of 80 MHz is too high for measurement of fluorophores with decay times over 3 ns. At 80 MHz, the pulses occur every 12.5 ns, which is too soon for most intensity decays. One typically measures the intensity decay to about four times the mean decay time, so that decay times longer than 3 ns are too long for a 80-MHz pulse rate. This problem is solved using a cavity dumper, which is an acousto-optic device placed within the dye laser cavity. At the desired time, a burst of radio-frequency (RF) signal (typically near 400 MHz) is put on the AO crystal. This causes the laser beam to be deflected by a small angle, typically 1–3°. The deflected beam is captured by a prism, which deflects the beam out of the laser cavity (Figure 4.10).

For cavity dumping, the AO crystal is pulsed at the desired repetition rate. For instance, for a 1-MHz repetition rate, the RF pulses are sent to the cavity dumping crystal at 1 MHz. The RF pulse width is narrow enough to extract a single optical pulse from the dye laser. The arrival times of the acousto-optic and laser pulses have to be matched. For this reason, the timing of a cavity dumper is typically obtained by dividing the frequency of the mode locker by factors of 2, to obtain progressively lower repetition rates. A somewhat confusing terminology is the use of *continuous wave* (CW) to describe the 80-MHz output of a dye laser. This term refers to continuous operation of the cavity dumper, resulting in a continuous train of pulses at 80 MHz.

A valuable aspect of cavity dumping is that it does not typically decrease the average power from the dye laser, at least within the 1–4 MHz range typical of TCSPC. To be specific, if the 80-MHz output of the dye laser is 100 mW, the output at 4 MHz will also be close to 100 mW. When optical power is not being dumped from the dye laser, the power builds up within the cavity. The individual cavity-dumped pulses become more intense, which turns out to be valuable for frequency-doubling the output of the dye laser.

A final problem with the R6G dye laser output is its long wavelength, from 570 to 610 nm. While shorter-wavelength dyes are available, these will typically require a shorter-wavelength pump laser. Although argon-ion lasers have been mode-locked at shorter wavelengths, this is generally difficult. For instance, there is only one report of the use of a mode-locked argon-ion laser at 351 nm as an excitation source for TCSPC.[30] Even after this was accomplished, the wavelength is too long for excitation of protein fluorescence. Fortunately, there is a relatively easy way to convert the long-wavelength pulses to shorter-wavelength pulses, which is frequency doubling or second-harmonic generation. The cavity-dumped dye laser pulses are quite intense. When they are focused into an appropriate crystal, one obtains photons of twice the energy, or half the wavelength. This process is inefficient, so only a small fraction of the 600-nm light is converted to 300 nm. Hence, careful separation of the long-wavelength fundamental and short-wavelength second harmonic is needed. The important point is that frequency doubling provides picosecond pulses, at any desired repetition rate, with output from 285 to 305 nm when an R6G dye laser is used. These wavelengths are ideal for excitation of intrinsic protein fluorescence.

A convenient feature of dye lasers is the tunable wavelength. The range of useful wavelengths is typically near the emission maximum of the laser dye. Tuning curves of typical dyes are shown in Figure 4.11. Most of these dye

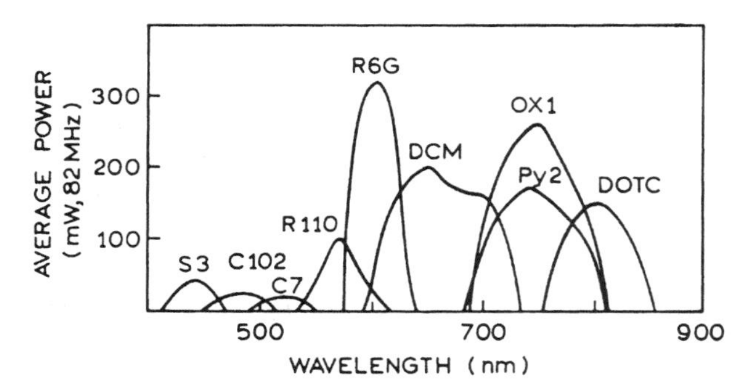

Figure 4.11. Output power of commonly used laser dyes.

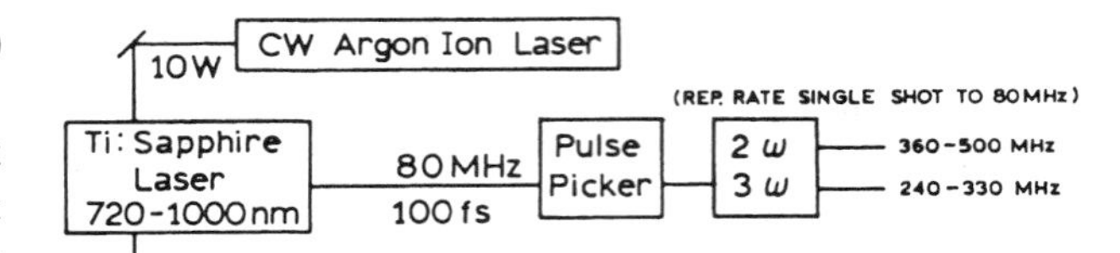

Figure 4.12. Femtosecond Ti:sapphire laser.

lasers are used after frequency doubling. We use R6G for excitation of intrinsic protein fluorescence, and 4-(Dicyanomethylene)-2-methyl-6-(4-dimethylaminostyryl)-4*H*-pyran (DCM) and pyridine-2 (Py2) for excitation of extrinsic probes. Excitation of tyrosine requires output of shorter wavelengths than available from R6G. Rhodamine 560 and rhodamine 575 were found suitable for tyrosine excitation using an argon-ion or an Nd:YAG laser pump source, respectively.[31]

4.4.B. Femtosecond Titanium:Sapphire Lasers

The newest lasers being used for time-resolved fluorescence are the titanium:sapphire lasers (Figure 4.12). These remarkable devices provide pulse widths near 100 fs, and special optics are needed to increase the pulse width to picoseconds. In some ways these lasers are simpler than picosecond dye lasers. The pump source is a continuous, not mode-locked, argon-ion laser. In addition to being simpler, this has the benefit that the continuous output of an argon-ion laser is typically 10- to 15-fold larger than the mode-locked output. Hence, 15 watts or more are available from the argon-ion laser for pumping the Ti:sapphire laser. Typically, Ti:sapphire lasers are pumped with up to 7 W at 514 nm. This allows a Ti:sapphire laser to be pumped with a small-frame argon-ion laser. More recently, Ti:sapphire lasers are being pumped with solid-state diode-pumped lasers, which are similar to Nd:YAG lasers.

A favorable feature of the Ti:sapphire lasers is that they are self-mode-locking. If one taps a Ti:sapphire laser operating in the continuous mode, it will switch to mode-locked operation with 100-fs pulses. This phenomenon is now understood as due to a Kerr lens effect within the Ti:sapphire crystal. The high-intensity pulses create a transient refractive index gradient in the Ti:sapphire crystal which acts like an acousto-optic mode locker. In fact, the phenomenon is referred to as Kerr lens mode-locking. Although the laser can operate in this free-running mode-locked state, an active mode locker can be placed in the cavity to stabilize the mode-locked frequency and to provide synchronization for other parts of the apparatus.

An advantage of the Ti:sapphire laser is that it is a solid-state device. There are no flowing dyes to be replaced, and the Ti:sapphire crystal seems to have an indefinitely long operational life. Self-mode-locking circumvents the need for matching cavity lengths with the pump laser. A disadvantage of the Ti:sapphire laser is the fundamental long-wavelength output, from 720 to 1000 nm. Even after frequency doubling, the wavelengths from 360 to 500 nm are too long for excitation of many fluorophores. This problem can be solved by frequency tripling or third-harmonic generation. However, this is somewhat more complex because one has to overlap the second-harmonic and fundamental beams in a second crystal. The beams need to be overlapped in time and space, which is difficult with femtosecond pulses. A minor disadvantage of the Ti:sapphire lasers for TCSPC is the use of a pulse picker, instead of a cavity dumper, to decrease the repetition rate. Because mode locking occurs within the laser cavity, rather than being accomplished by synchronous pumping, a cavity dumper cannot be used with a Ti:sapphire laser. After the 80-MHz pulses exit the laser, the desired pulses are selected with an AO deflector, called a pulse picker. Since the energy in the other pulses is discarded, there is no increase in peak power as occurs with cavity dumping. In fact, there is a significant decrease in average power when a pulse picker is used. For these reasons, and for lack of general availability, picosecond dye lasers are still the most widely used light source in TCSPC.

Ti:sapphire lasers are being widely used for two-photon and multiphoton excitation. In this case, one can use their intense fundamental output to excite fluorophores by simultaneous absorption of two or more photons. The use of multiphoton excitation has been found to be particularly valuable in microscopy, where localized excitation occurs only at the focal point of the excitation beam.

Another exotic light source is synchrotron radiation. If electrons are circulated at relativistic speeds, they radiate energy over a wide range of wavelengths. These pulses have clean Gaussian shapes and can be very intense. Instruments for TCSPC have been installed at a number of synchrotron sites.[32–37] Unfortunately, it is rather inconvenient to use these light sources. The experimental apparatus must be located at the synchrotron site, and one has to use the beam when it is available. An advantage of the synchrotron source is that a wide range of wavelengths is available, and all wavelengths appear with the same time distribution.

4.4.C. Flashlamps

Prior to the introduction of picosecond lasers, most TCSPC systems used the coaxial flashlamps. A wide range of wavelengths is available, depending on the gas within the flashlamp. These devices typically provide excitation pulses about 2 ns wide, with much less power than is available from a laser source. Flashlamp sources became available in the 1960s,[38–40] but their use in TCSPC did not become widespread until the mid-1970s.[41–44] Because of the lower repetition rate and intensity of the flashlamps, long data acquisition times were necessary. This often resulted in difficulties when fitting the data because the time profile of the lamps changed during data acquisition.[45,46] While these problems still occur, the lamps are now more stable, provide higher repetition rates of up to 50 kHz, and can provide pulse widths near 1 ns.[47–49]

Figure 4.13 shows a typical coaxial flashlamp. Earlier flashlamps were free-running, meaning that the spark occurred whenever the voltage across the electrodes reached the breakdown value. Almost all presently used lamps are gated. The electrodes are charged to high voltage. Both electrodes are charged, as can be seen from the 1-MΩ connecting resistor. At the desired time, the thyratron is gated on to rapidly discharge the top electrode to ground potential. A spark discharge occurs across the electrodes, which results in a flash of light. These pulses are rather weak, and one frequently has to block the room light to see the flash with a naked eye.

Compared to laser sources, flashlamps are simple and inexpensive. Hence, there have been considerable efforts

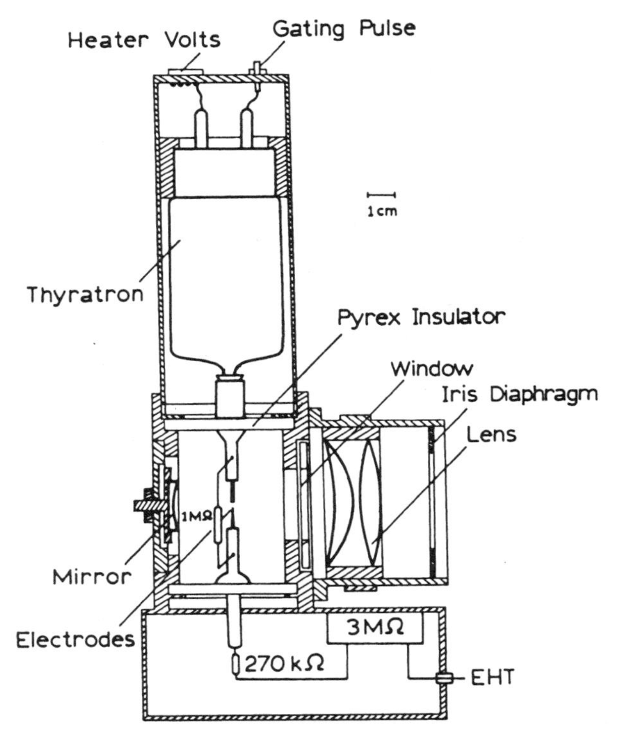

Figure 4.13. Coaxial nanosecond flashlamp. Reprinted with permission from Ref. 47, Copyright © 1981, American Institute of Physics.

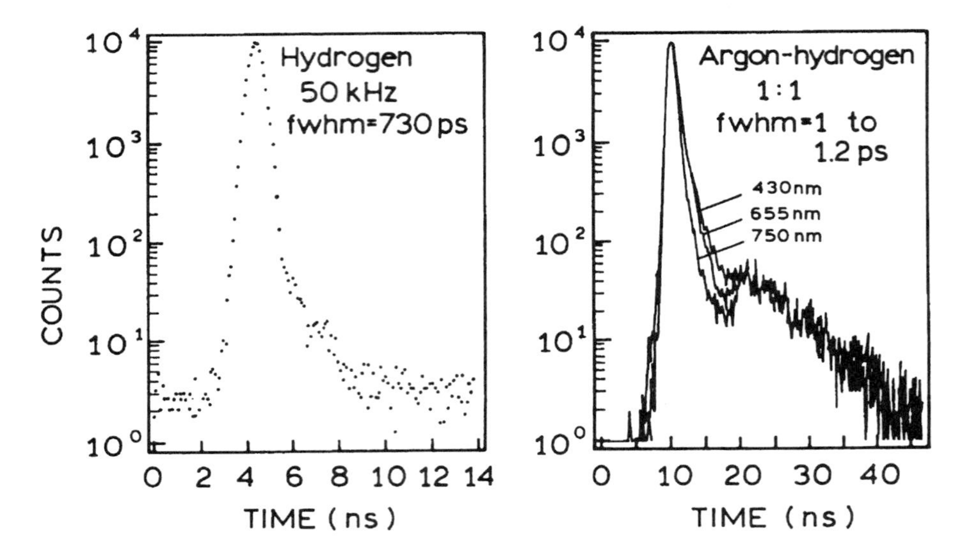

Figure 4.14. Time profiles of coaxial flashlamps. Revised and reprinted from Refs. 9 and 48, Copyright © 1991, American Institute of Physics.

to obtain the shortest pulse widths and highest repetition rates. Despite these efforts, the flashes are much wider than those available with a laser source. One of the shortest time profiles is shown in the left panel of Figure 4.14, where the full width at half-maximum (fwhm) is 730 ps. More typical is the 1.2-ns fwhm for the flashlamp in which the gas is an argon–hydrogen mixture (Figure 4.14, right). Also typical of flashlamps is the long tail that persists after the initial pulse.

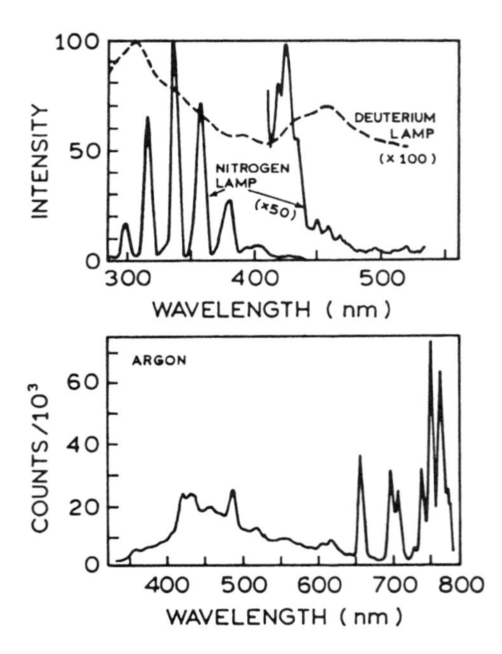

Figure 4.15. Spectral output of flashlamps. The output of the deuterium lamp is 100-fold less than that of the nitrogen lamp. It is possible that the emission below 600 nm for argon is due to impurities. Revised from Refs. 12 and 48.

The spectral output of the pulse lamps depends on the gas, and the pulse width typically depends on both the type of gas and the pressure. Hydrogen or deuterium (Figure 4.15) provides a wide range of wavelengths in the UV, but at low intensity. Nitrogen provides higher intensity at its peak wavelengths, but little output between these wavelengths. In recognition of the growing interest in red and NIR fluorescence, flashlamps have been developed with red and NIR outputs.[48,49] A practical lamp with nanosecond pulse widths was developed using mixtures of argon and hydrogen, providing output to 750 nm. The difficulties of developing a flashlamp are hinted at in the right panel of Figure 4.14, where the temporal lamp output depends on wavelength. Large pulse widths and long tails were found for argon alone. Hydrogen was needed to quench the argon emission and shorten the pulse widths. This is a trial-and-error process, and the gas pressure and electrode spacing must be carefully maintained to obtain similar temporal profiles.

The longer pulse widths obtained using a flashlamp illustrate the need for deconvolution. TCSPC data for the laser dye IR-140 show a decay time of 1.2 ns (Figure 4.16). In this case the width of the excitation pulse is comparable to that of the intensity decay itself. One can see that the intensity values track the long tail of the lamp output. Without the use of deconvolution, these data could not be analyzed.

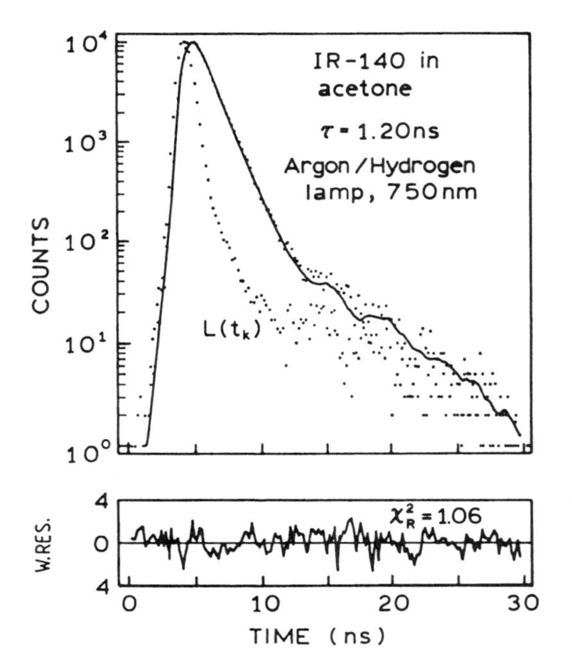

Figure 4.16. Intensity decay of the laser dye IR-140 measured using an Ar–H_2 lamp at 750 nm and an XP-2257B PMT. $L(t)_k$ is the lamp profile. Revised from Ref. 48.

The most significant drawback of using a flashlamp is the low repetition rate. The fastest flashlamps have repetition rates up to 100 kHz, with 20 kHz being more common. Recall that one can only collect about one photon per 50–100 light pulses, so that the maximum photon count rate is near 200 Hz. Hence, a decay curve with 500,000 counts can take up to 40 minutes to accumulate. The data acquisition time can be decreased using a higher repetition rate and a ratio of stop-to-start pulses above 1%. However, even with the high sensitivity of single-photon counting, the low optical output of the flashlamps can limit sensitivity. For these reasons, the higher optical power and faster repetition rates of laser systems make them the preferred light source for TCSPC.

4.4.D. Solid-State Lasers

In the future, we can expect the rather expensive picosecond dye lasers and Ti:sapphire lasers to be replaced by simpler and less expensive devices. A diode-pumped Nd:YAG laser has already been used for time-resolved detection in capillary zone electrophoresis,[50] and one can purchase a streak camera with a pulsed laser diode excitation source.[51] Laser diodes have also been used as the excitation source for FD fluorometry.[52,53] The wavelengths are usually limited to 600–700 nm, but some laser diodes can be frequency-doubled to 410 nm. It is also likely that light-emitting diodes will find use as low-cost pulse light sources, particularly with longer-lived fluorophores. Hence, one can expect the availability of TCSPC instruments to increase as the light sources become less expensive.

4.5. ELECTRONICS FOR TCSPC • Advanced Material •

Like many technologies in current use, the electronics for TCSPC appear to have their origins in defense research. The various components for TCSPC are made to insert into NIM bins, where NIM stands for nuclear instruments modules. The NIM bin supplies electrical power to modules designed for use in high-speed timing and/or TCSPC. In the following sections, we briefly summarize the components most important for TCSPC. Additional details can be found in the monograph by O'Connor and Phillips.[8]

4.5.A. Constant Fraction Discriminators

Among the most important parts of the TCSPC electronics are the constant fraction discriminators (CFDs) (Figure 4.17). The goal is to measure the arrival time of the photoelectron pulse with the highest possible time resolution. This goal is compromised because the pulses have different amplitudes (Figure 4.17). Hence, if one measures the arrival of the pulses by the time when the signal exceeds a threshold, there is a spread Δt in the measured time due to pulse height variations. Although this effect may seem minor, it can be the dominant factor in an instrument

LEADING EDGE DISCRIMINATION

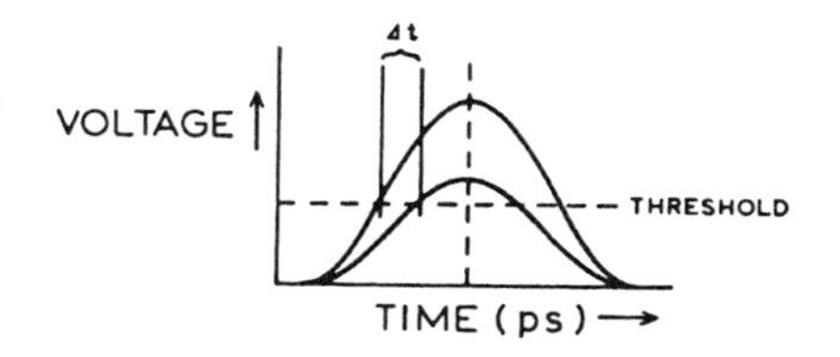

CONSTANT FRACTION DISCRIMINATION

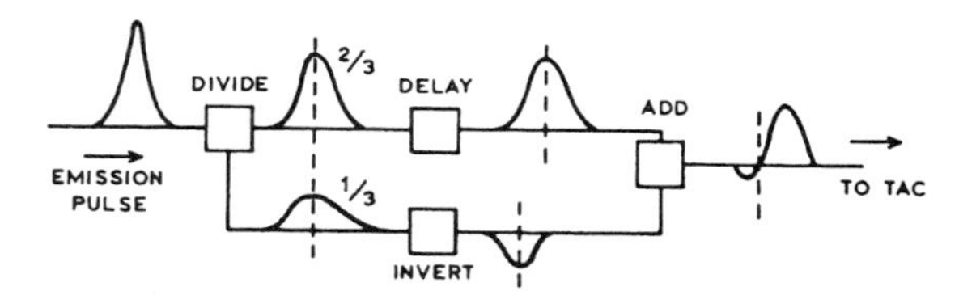

Figure 4.17. Constant fraction discrimination in TCSPC. *Top*: Timing error due to pulse height variations using leading-edge discrimination. *Bottom*: Operation of a constant fraction discriminator.

response function. Leading-edge discriminators can be used for pulses which all have the same height, which may be true for the trigger (start) channel if the laser system is stable.

The contribution of the pulse height distribution can be minimized by the use of CFDs.[54–56] The basic idea of a CFD is to split the signal into two parts, one part of which is delayed by about half of the pulse width. The other part of the signal is inverted. When these two parts are recombined, the zero crossing point is mostly independent of the pulse height. The difference between leading-edge and constant fraction discrimination is remarkable, the timing jitter being 1 ns and 50 ps, respectively.[9] It is important to note that the requirements of a CFD for standard PMTs and MCP PMTs are different. The shorter pulse width from a MCP PMT means that the time delay in the CFD needs to be smaller in order to properly mix the split signals.[57–59]

4.5.B. Amplifiers

Amplifiers have been used after the start and stop detectors in order to obtain adequate signal levels for timing. The present trend is to avoid such amplifiers, as their use can result in additional difficulties. In general, the most noise-free amplification can be obtained within the detector (PMT or photodiode). The cable connecting the detector and the amplifier can act as an antenna, resulting in amplification of the RF noise. The use of amplifiers was necessary when MCP PMTs first appeared because the pulses were too short for the CFDs available at that time. Amplifiers were used to broaden the pulses sent to the CFD.[29] This is no longer necessary with newer CFDs. If amplifiers must be used, they should be positioned as close as possible to the detector.

4.5.C. Time-to-Amplitude Converter (TAC)—Standard and Reversed Configurations

The role of the TAC is to measure the time between the excitation pulse and the first arriving emitted photon, which are the start and stop pulses, respectively. This is accomplished by charging a capacitor during the time interval between the pulses. Typically, the capacitor is charged from 0 to 10 V, over a nanosecond to microsecond time range. For instance, if the chosen range is 50 ns, the capacitor is fully charged at 50 ns. If a stop pulse is received at 25 ns, the charging is stopped at 5 V. If a stop pulse is not received, then the TAC is reset to zero.

In general, the TAC is a rate-limiting component in TCSPC. It takes several microseconds to discharge the capacitor and reset the TAC. This is not a problem with flashlamps: a 50-kHz rate results in start pulses every 20 μs. However, with a high-repetition-rate laser source at 1 MHz, the TAC will be overloaded due to continuous start pulses.

The relatively simple solution to this problem is to operate the TAC in reverse mode.[60–62] In this mode of operation, the first photon detected from the sample serves as the start pulse, and the signal from the excitation pulse is the stop signal. In this way the TAC is activated only if the emitted photon is detected. The only minor disadvantage of reverse TAC operation is that the decay curves appear reversed on the screen of the MCA, but this is easily corrected by software. The reverse mode of TAC operation is not needed with flashlamps because of their lower repetition rates.

An important characteristic of a TAC is its linearity. If the voltage is not linear with time, then the data will contain systematic errors, resulting in difficulties with data analysis. One way to test the linearity of a TAC is to expose the detector to a low level of room light and still use the pulsed light source to trigger the TAC start signal. Since the photons from the room lights are not correlated with the start pulses, the stop pulses should be randomly distributed across the time range, yielding a horizontal line in the MCA.

4.5.D. Multichannel Analyzer (MCA)

The MCA measures the voltage pulses from the TAC and sorts them according to counts at each particular voltage (time). The MCA first performs an analog-to-digital conversion, which typically takes about 5 μs, during which time the MCA is unable to accept another voltage pulse from the TAC. The histogram of the number of counts at each voltage (time) is displayed on the screen of a cathode-ray tube (CRT). This histogram represents the measured intensity decay. MCAs typically have 2048–8192 channels, which can be subdivided into smaller segments. This allows several experiments to be stored in the MCA prior to data transfer and analysis. This ability to store several histograms is particularly important for measurement of anisotropy decays, for which one needs to measure the two polarized intensity decays, as well as one or two lamp profiles.

4.5.E. Delay Lines

Delay lines are incorporated into all TCSPC instruments. The need for delay lines is easily understood by recognizing that there are significant time delays in all components of the instrument. A photoelectron pulse may take 20 ns to exit a PMT. Electrical signals in a cable travel one foot in about 1 ns. It would be difficult to match all these delays in the start and stop detector channels. The need for matching is avoided by the use of calibrated delay lines, which are typically components in the NIM bin. One can also use lengths of coaxial cable as delay lines, but delay lines in factory-built NIM bin housing seem to pick up less RF interference.

Calibrated delay lines are also useful for calibration of the time axis of the MCA. This is accomplished by providing the same input signal to the start and the stop channels of the TAC. The preferred approach is to split an electrical signal, typically from the start detector, and direct this signal to both inputs of the TAC. Since the pulses arrive with a constant time difference, one observes a single peak in the MCA. One then switches the time delay in the start or stop channel by a known amount and finds the peak shift on the MCA display. By repeating this process for several delay times, the TAC and MCA can be calibrated.

4.5.F. Pulse Pileup

In TCSPC one collects only one photon from the sample for every 50–100 excitation pulses. What errors occur if the average number of detected photons is larger? This question cannot be answered directly because the electronics limit the experiment to detecting the first arriving photon. If more than one photon arrives, how does this affect the measured intensity decay? Simulations are shown in Figure 4.18 for a single-exponential decay with larger numbers of arriving photons. The apparent decay time becomes shorter and the decay becomes nonexponential as the number of arriving photons increases. The apparent decay is more rapid because the TAC is stopped by the first arriving photon. Since emission is a random event, the first photon arrives at earlier times for a larger number of arriving photons. Methods to correct for pulse pileup have been proposed,[63,64] but the present consensus is that it is best to avoid pulse pileup by using a low counting rate.

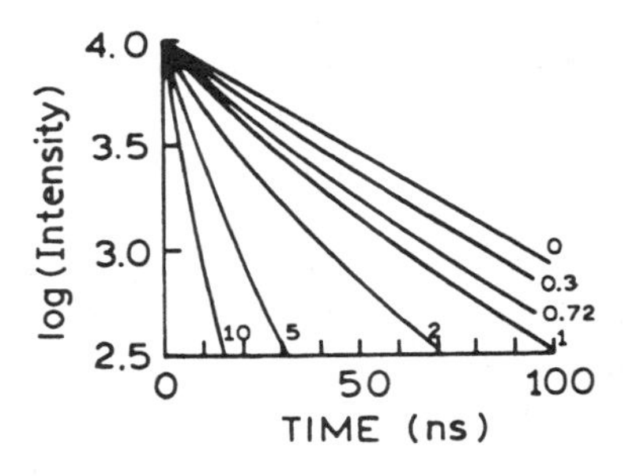

Figure 4.18. Effect of the pulse count rate on a single-exponential decay. The numbers are the number of arriving photons per excitation pulse. For a count rate of 0.01, the curve overlaps the actual decay (curve labeled 0). From Ref. 4.

Methods are being developed to circumvent the 1–2% counting rate typical of TCSPC. One obvious approach is to use multiple sets of detectors and electronics. However, this is too complex and expensive, and it would be difficult to ensure that all components display the same time calibration and impulse response function. The problem is being solved to some extent by the introduction of multianode PMTs and multiplexing methods for detection of multiple photons. MCP PMTs which function as 8–10 separate detectors are available,[65] and devices with up to 96 anodes have been reported.[66] Additionally, arrays of 64 or 96 miniature dynode PMTs in a single package are available, the so-called multichannel PMTs. Each of the anodes in these array detectors can be used for detection of the first observed photon following the excitation pulse.

Dedicated electronic circuits are being developed to allow TCSPC using multianode PMTs. Each detection channel is associated with an individual CFD and a separate memory segment of the MCA.[67–72] For a multiplexed TCSPC instrument with 16 channels, the counting rates were close to 16-fold higher than possible using a single channel. With 64 channels it should be possible to measure photons at up to 38% of the pulse rate, or 380 kHz for a 1-MHz repetition rate. However, one cannot employ a counting rate exceeding the inverse of the dead time of the MCA, typically 2–10 μs, for maximum input rates of 500–100 kHz, respectively. The primary advantage of multiplexed TCSPC seems to be in providing more rapid data acquisition with low-repetition-rate light sources.

The electronics for TCSPC are also becoming less expensive and more compact. Most of the components are now available on PC boards, dispensing with the usual NIM bin electronics.[72] The performance of these systems is reported to be somewhat less than that of full-sized systems, but one can expect further improvements based on application-specific integrated circuits.

4.6. DETECTORS FOR TCSPC • Advanced Material •

4.6.A. MCP PMTs

Perhaps the most critical component for timing is the detector. The timing characteristics of various PMTs have been reviewed in relation to their use in TCSPC.[16,73,74] At present, the detector of choice for TCSPC is the MCP PMT. An MCP PMT provides a 10-fold shorter pulse width than any other PMT and displays lower-intensity afterpulses. Also, the effects of wavelength and spatial location of the light seem to be much smaller with MCP PMTs than with linear-focused or side-window tubes. Although good

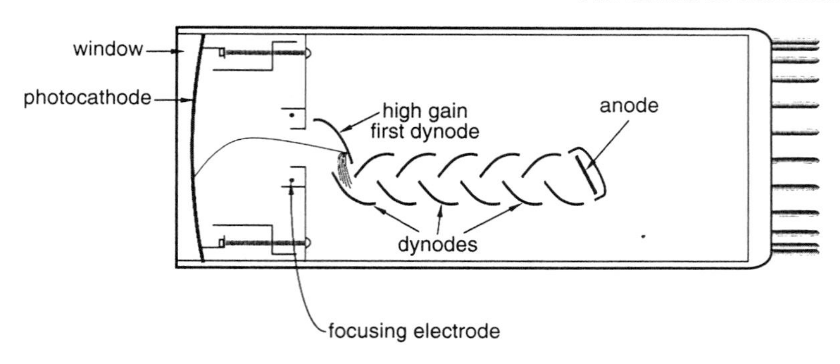

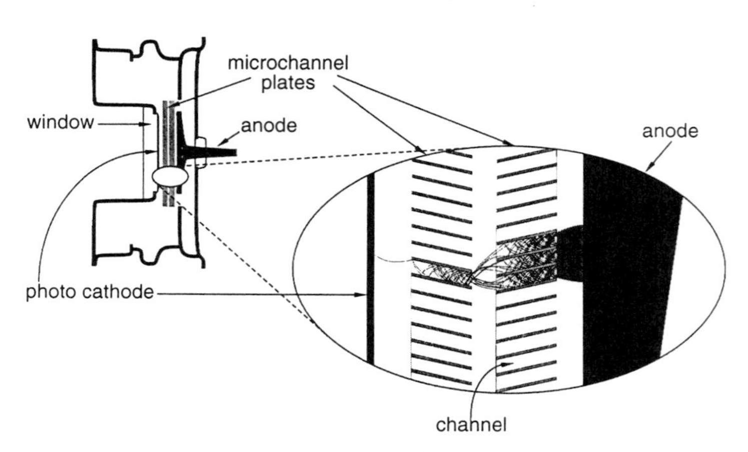

Figure 4.19. Comparison between a traditional dynode chain PMT (RCA 8850 or C31000M) (*top*) and a proximity-focused MCP PMT (*bottom*). From Ref. 16.

time resolution can be obtained with linear-focused and side-window PMTs, the high-speed performance and absence of timing artifacts with MCP PMTs make them the preferred detectors for TCSPC.

Development of MCP PMTs began in the late 1970s,[75–77] with the first useful devices appearing in the early to mid-1980s,[78–82] and their use for TCSPC beginning at the same time.[82–84] The design of an MCP PMT is completely different from that of a dynode chain PMT (Figure 4.19). The factor which limits the time response of a PMT is its transit time spread (TTS), which is the distribution of transit times through the detector. The overall transit time of the electrons through a PMT is not important, as this is just a time delay which is corrected for in the measurements. However, the distribution of transit times, or TTS, is important because this spread limits the time resolution of a PMT. One cannot do timing measurements more accurately than the uncertainty in the time it takes a signal to pass through the detector. In a linear-focused PMT, the TTS is minimized by designing the dynodes so all the electrons tend to travel along the same path. The TTS of most phototubes is near 2 ns, and it can be less than 1 ns with carefully designed PMTs (Table 4.1).

The design of an MCP is completely different because it does not have dynodes. Instead the photoelectrons are amplified along narrow channels lined with the dynode material. Because these channels are very narrow, typically 4–12 μm in diameter, the electrons all travel the same path and hence have the same transit time. Smaller channels result in less TTS. There are a few additional features in the MCP PMT design which provide improved time response. The channels are angled relative to each other, which prevents feedback between the channels and broadening of the time response.[76,77] Also, the first MCP surface is typically covered with aluminum, which prevents secondary electrons emitted from the top of the MCP from entering adjacent channels.

The improvement in time response possible with MCP PMTs is impressive (Figure 4.20). Although difficult to see on the figure, the pulse width for the R1564U (about 60 ps) is 10-fold less than that for the C31000M (0.5 ns), one of the fastest dynode PMTs (Table 4.1). MCP PMTs have been reported to display pulse widths as short as 25 ps (for

Table 4.1. Transit Time Spreads (TTS) of Conventional and Microchannel Plate (MCP) Photomultiplier Tubes (PMTs)[a]

Manufacturer	Photomultiplier	Configuration[b] (upper frequency)	TTS (ns)	Dynode
Hamamatsu	R928	Side-on (300 MHz)	0.9	9 stage
	R1450	Side-on	0.76	10 stage
	R1394	Head-on	0.65	10 stage
	R1828	Head-on	0.55	12 stage
	H5023	Head-on (1 GHz)	0.16	10 stage
RCA	C31000M	Head-on	0.49	12 stage
	8852	Head-on	0.70	12 stage
Philips	XP2020Q	Head-on	0.30	12 stage
Hamamatsu	R1294U	Nonproximity MCP-PMT	0.14	2 MCP
	R1564U	Proximity focused MCP-PMT, 6 micron (1.6–2 GHz)	0.06	2 MCP
	R2809U	Proximity MCP-PMT, 6 micron	0.03[c]	2 MCP
	R3809U	Proximity MCP-PMT, compact size, 6 micron	0.025[c]	2 MCP
	R2566	Proximity MCP-PMT with a grid, 6 micron (5 GHz)[d]	—	2 MCP

[a]Revised from Ref. 81.
[b] The numbers in parentheses are the approximate frequencies where the response is 10% of the low-frequency response. The H5023 has already been used to 1 GHz.
[c] From Ref. 140.
[d] From Ref. 65.

the R3809U). Perhaps the most impressive aspect of the MCP PMT is the absence of afterpulses, which are always present in dynode PMTs. The origin of these afterpulses is not always understood.

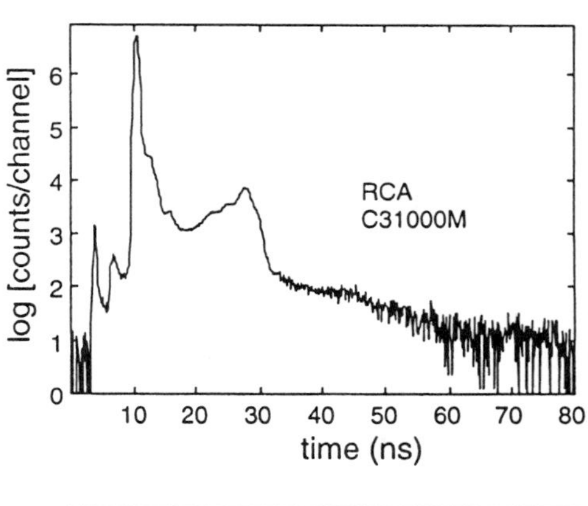

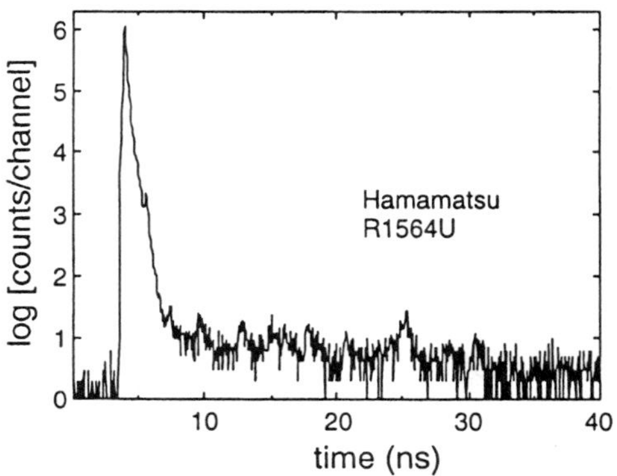

Figure 4.20. Time response of a dynode PMT (RCA C31000M) and an MCP PMT (Hamamatsu R1564U) at 300 nm. From Ref. 16.

MCP PMTs do have some disadvantages. The photocurrent available from an MCP PMT is typically 100 nA (R2908), as compared to 0.1 mA (R928) for a dynode PMT. This means that the MCP PMT responds linearly to light intensity over a smaller range than a dynode PMT. The current-carrying capacity of an MCP PMT is less because the electrical conductance of the MCPs is low. It is known that the pulse widths from an MCP PMT can depend on the count rate,[85] presumably because of voltage changes resulting from the photocurrent. Another disadvantage of the MCP PMTs is their presumed limited useful lifetime, which depends on the photocurrent drawn from the tube. In our hands this has not been a problem, and it is difficult to know whether an MCP PMT has lost gain due to the total current drawn or due to the inevitable overexposure to light which occurs during the lifetime of a PMT. Of course, MCP PMTs are considerably more expensive than dynode PMTs. In general, the expense of an MCP PMT is justified only if it is used with a picosecond light source. Dynode PMTs are adequate for use with flashlamps.

4.6.B. Dynode Chain PMTs

Dynode PMTs cost less than MCP PMTs and are adequate for many TCSPC experiments, especially if the excitation source is a flashlamp. Two types of dynode PMTs are used for TCSPC, the side-window and linear-focused PMTs. Their performance is comparable, but there are minor differences. The side-window tubes are less expensive but can still provide good time resolution. Pulse widths from 112 to 700 ps have been obtained with side-window tubes,[25,86–89] but pulse widths of 1–2 ns are more common. A disadvantage of a side-window PMT is that the time response can depend on the region of the photocathode which is illuminated. Linear-focused PMTs are somewhat more expensive but provide slightly shorter TTSs (Table 4.1) and are less sensitive to which region of the photocathode is illuminated. Linear-focused PMTs are probably still the most widely used detectors in TCSPC, but there is a continual shift toward the MCP PMTs.

In the near future, one can expect improved time resolution at lower costs. Dynode PMTs have been built into centimeter-sized packages (TO-8). Because of the small distances, one can expect low TTSs and performance comparable to that of MCP PMTs. These miniature PMTs may become competitive with MCP PMTs, assuming they are free of timing artifacts.

4.6.C. Photodiodes as Detectors

Photodiodes (PDs) are inexpensive and can respond faster than an MCP PMT. Why are phototubes still the detector of choice? Photodiodes are not usually used for photon counting because of the lack of gain. However, avalanche photodiodes (APDs) have adequate gain and can be as fast as MCP PMTs. The main problem is the small active area. In a PMT or MCP PMT the area of the photocathode is typically 1 cm × 1 cm and is frequently larger. Photons arriving anywhere on the photocathode are detected. In contrast, the active area of an APD is usually less than 1 mm^2, and less than 10 μm × 10 μm for a high-speed APD. It is therefore difficult to focus the fluorescence onto the APD, so the sensitivity is too low for most measurements. Another disadvantage is the relatively long tail following each pulse, the extent of which depends on wavelength. The presence of a wavelength-dependent tail can create problems in data analysis since the instrument response function will depend on wavelength. Methods have been developed to actively quench the tail. Values of the fwhm from 20 to 400 ps have been reported, and APDs have been successfully used in TCSPC.[90–97] One can expect increased use of APDs as time-resolved instruments are designed for specific applications such as DNA sequencing.

Given the low sensitivities of photodiodes, one may question their use as the start detector in the TCSPC instrument schematically represented in Figure 4.7. In this case the light source was a laser, which could be readily focused onto the small active area of a photodiode. Photodiodes are not used as the reference detector with flashlamps because of their low sensitivity. When the light source is a flashlamp, the detector is either a PMT or a wire which acts as an antenna to detect the RF leakage during the lamp pulse.

4.6.D. Color Effects in Detectors

When performing lifetime measurements, one generally compares the response of a fluorescent sample with that of a zero-decay-time scattering sample. Because of the Stokes' shift of the sample, the wavelengths are different when one is measuring the sample and the impulse response function. Unfortunately, the timing characteristics of a PMT can depend on wavelength, and this effect can be substantial.[98–103] The time response of PMTs can also depend on which region of the photocathode is illuminated.

The potential severity of the problem is illustrated in Figures 4.21 and 4.22 for a high-speed PMT (56 DUVP). These traces represent the response to essentially δ-function excitation. One notices that the shape of the pulse is dependent on wavelength (Figure 4.21). To show the magnitude of the color effect more clearly, these responses are plotted on a linear intensity scale in Figure 4.22. The electron pulse arrives earlier for shorter wavelengths, presumably due to a higher velocity of the photoelectrons ejected from the photocathode. There is also a wider spread of transit times at shorter-wavelength illumination, probably due to a range of angles for the ejected electrons and thus a range of path lengths. Such effects can distort a measured lifetime or result in poor fits due to systematic errors. The 56 DUVP is a rather old PMT and is not widely used at the present time; data are shown for this PMT to illustrate the effects possible with high-speed detectors.

A variety of methods have been devised to correct for such color effects. There are two general approaches, one of which is to use a standard with a very short lifetime.[104,105] It is interesting to note that the use of a short-lifetime standard for phase measurements was introduced in 1962.[106] The standard should emit at the wavelength used to measure the sample. Because of the short decay time, one assumes that the measured response is the instru-

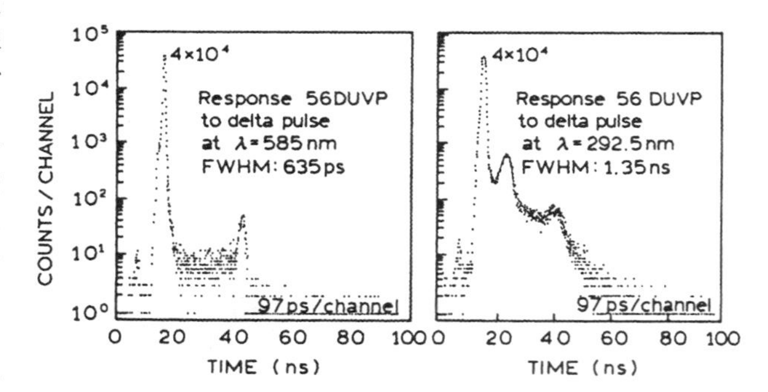

Figure 4.21. Response of the 56 DUVP PMT to a delta pulse. At 585 nm, fwhm = 635 ps, and at 292.5 nm, fwhm = 1.35 ns. Revised and reprinted from Ref. 80. Copyright © 1986, American Institute of Physics.

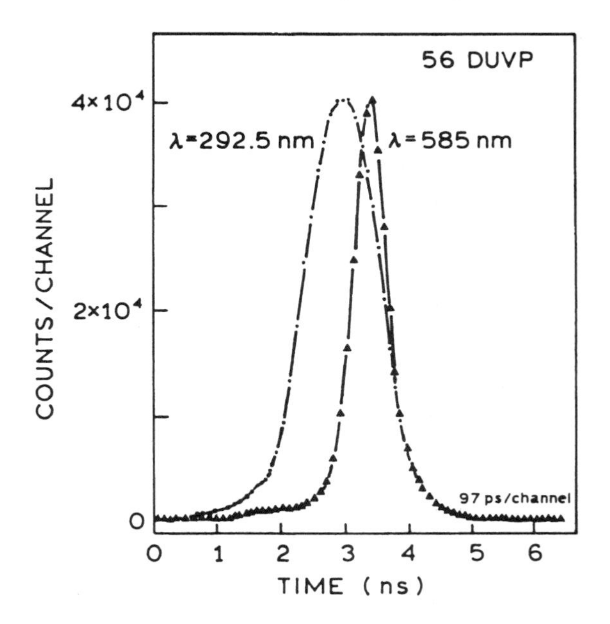

Figure 4.22. Time response of a 56 DUVP PMT on a linear scale. Revised from Ref. 80.

ment response function. Because the wavelengths are matched, one assumes that the color effects are eliminated.

The more common method of correcting for color effects is to use a standard fluorophore which is known to display a single-exponential decay.[107–115] One measures the intensity decay of the sample and the reference fluorophore at the same wavelength. The intensity decays of the reference fluorophore with the known decay time τ_R and of the sample are analyzed simultaneously. In order to correct for the reference lifetime, a different functional form is used for $I(t)$,

$$I(t) = \sum_{i=1}^{n} \alpha_i\left[\delta_0(t) + \left(\frac{1}{\tau_R} - \frac{1}{\tau_i}\right)\exp(-t/\tau_i)\right] \quad [4.18]$$

where $\delta_0(t)$ is the Dirac delta function. In this expression the values of α_i and τ_i have their usual meaning (Eq. [4.8]). This method is best performed when the decay time of the standard is precisely known. However, some groups vary the assumed decay time of the standard to obtain the best fit.

It is valuable to observe how the measurements can be improved using a lifetime reference.[112] NATA is known to display a single-exponential decay in water. Its intensity decay was measured with 295-nm excitation from a cavity-dumped frequency-doubled R6G dye laser.[28] The emission was detected with a Philips PM 2254 PMT, which is a linear-focused dynode PMT comparable to the Philips XP 2020. When NATA was measured relative to scattered light, the fit was fair. However, there were some nonrandom deviations, which are most easily seen in the autocorrelation trace (Figure 4.23, top left). Also, the value of the goodness-of-fit parameter, $\chi_R^2 = 1.2$, is somewhat elevated. The fit was improved when NATA was measured relative to a standard, *p*-terphenyl in ethanol, for which $\tau_R = 1.06$ ns at 20 °C. One can see the contribution of the reference lifetime to the instrument response function as an increased intensity from 5 to 15 ns in the instrument response function (Figure 4.23, right). The use of the *p*-terphenyl reference resulted in more random deviations and a flatter autocorrelation plot (Figure 4.23, top right). Also, the value of χ_R^2 was decreased to 1.1. Comparison of the two sides of Figure 4.23 illustrates the difficulties in judging the quality of the data from any single experiment. It would be difficult to know if the minor deviations seen on the right were due to NATA or to the instrument.

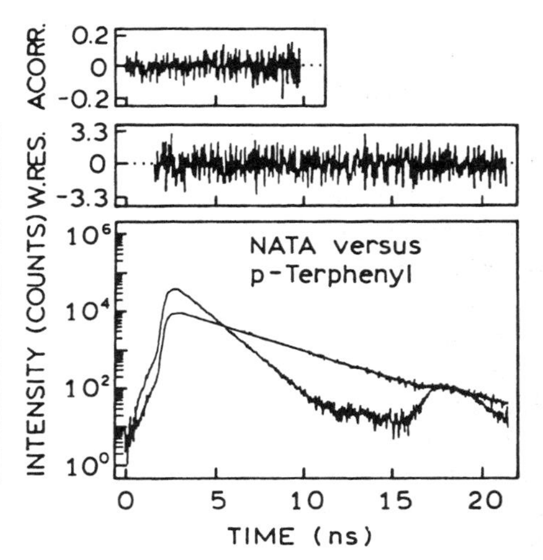

Figure 4.23. Intensity decay of NATA as measured versus scattered light (*left*) and using a *p*-terphenyl lifetime reference (*right*). Revised from Ref. 112.

At present, the need to correct for the wavelength-dependent time response of PMTs has decreased. Side-window tubes such as the R928 show less sensitivity to wavelength than do end-on tubes.[89] Additionally, MCP PMTs are quite insensitive to wavelength or position of illumination. For the R1564 MCP PMT, the instrument response function was found to be the same at 273 and 546 nm (not shown).[80] Moreover, some of the earlier reports overestimated the extent of color effects due to low voltage between the photocathode and first dynode. At present, most PMTs for TCSPC use the highest practical voltage between the photocathode and first dynode to minimize these effects. Also, most programs for analysis of TCSPC allow for a time shift of the lamp function relative to the measured decay. This time shift serves to correct for any residual color effects in the PMT. Although one needs to be aware of the possibility of color effects, the problem is less severe with MCP PMTs. In addition to color effects, TCSPC instruments are prone to systematic errors due to the electronic and optical components; these sources of error have been described in detail in Ref. 116.

Prior to performing any TCSPC measurements, it is desirable to test the performance of the instrument. This is best accomplished using molecules known to display single-exponential decays.[117,118] Assuming the sample is pure and decays are a single exponential, deviations from the expected decay can reveal the presence of systematic errors in the measurements. A number of known single-exponential lifetimes are summarized in Appendix II.

As described in Chapter 2, it is also important to test for background fluorescence from blank solutions. Autofluorescence from the sample can result in errors in the intensity decay, which results in confusion and/or erroneous conclusions. In fact, we have observed a tendency in many researchers to collect the time-resolved data prior to measuring steady-state spectra of the sample and controls. This is particularly dangerous for time-resolved measurements because the data are often collected through filters, without examining the emission spectra. Hence, one can observe an impure sample and obtain a corrupted data set. Subsequent analysis of the data is then unsatisfactory, and one cannot identify the source of error. In our experience, more time is wasted by not having the spectra than the time needed to record blank spectra prior to time-resolved data collection.

Background correction in TCSPC is straightforward. One simply collects data from the background for the same time and conditions as used for the sample. The background can then be subtracted from the sample data. Of course, the source intensity and repetition rate must be the same. Also, inner filter effects in the samples will attenuate its signal. If the control samples have a lower optical density, the measured background can be an overestimation of the actual background. If the number of background counts is small, there is no need to consider the additional Poisson noise in the difference data file. However, if the background level is large, it is necessary to consider the increased noise level in the difference data file.

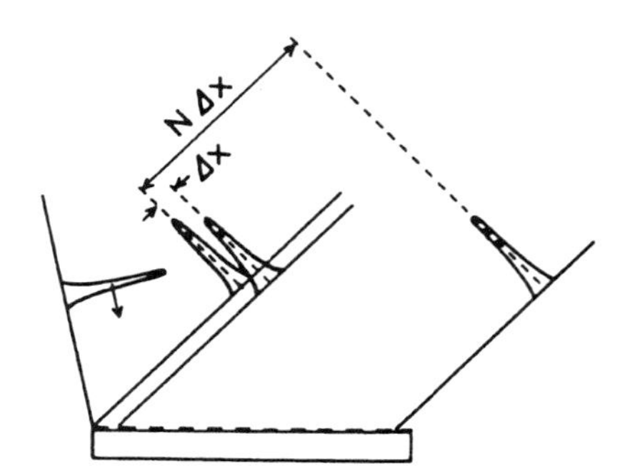

Figure 4.24. Path length difference across a monochromator grating. N is the total number of facets in the grating, and Δx is the path length difference between adjacent reflections. From Ref. 120.

4.6.E. Timing Effects of Monochromators

As the time resolution of the instrumentation increases, one needs to consider the effects of the various optical components. Monochromators can introduce wavelength-dependent time delays and/or broaden the light pulses.[119,120] This effect is illustrated in Figure 4.24, which shows the path length difference for an optical grating with N facets. Monochromators are usually designed in such a manner that the entire grating is illuminated. The maximum time delay is given by[121]

$$t_d = \frac{Nm\lambda}{c} \qquad [4.19]$$

where N is the number of facets, m is the diffraction order (typically 1), λ is the wavelength, and c is the speed of light. A typical grating may have 1200 lines/mm and be 60 mm across. The maximum time delay at 350 nm is thus 84 ps. While picosecond and femtosecond laser pulses are not usually passed through a monochromator, doing so can be expected to broaden the pulse. Alternatively, the apparent intensity decay of a short-lived fluorophore may be broadened by the use of a monochromator to isolate the emission. These effects can be avoided by the use of subtractive dispersion monochromators.[122,123]

4.7. ALTERNATIVE METHODS FOR TIME-RESOLVED MEASUREMENTS • Advanced Material •

4.7.A. Pulse Sampling or Gated Detection

While TCSPC and FD fluorometry are the dominant methods used by biochemists, there are alternative methods for measuring intensity decays. Prior to the introduction of

TCSPC, intensity decays were measured using stroboscopic or pulse sampling methods. The basic idea is to sample the intensity decay repetitively following pulsed excitation (Figure 4.25). The detection gate is displaced across the intensity decay until the entire decay is measured. In fact, the first TD lifetime instruments used gated detection to sample the intensity decay.[39]

Gated detection can be accomplished in two ways. One method is to turn on or gate the gain of the detector for a short period during the intensity decay.[124,125] Surprisingly, this can be accomplished on a timescale adequate for measurement of nanosecond lifetimes. Alternatively, the detector can be on during the entire decay, and the electrical pulse measured with a sampling oscilloscope.[126,127] Such devices can sample electrical signals with a resolution of tens of picoseconds.

While such methods seem direct, they have been mostly abandoned owing to difficulties with systematic errors. The flashlamps generate RF signals which can be picked up by the detection electronics. This difficulty is avoided in TCSPC because low-amplitude noise pulses are rejected, and only the higher-amplitude pulses due to primary photoelectrons are counted. Also, in TCSPC the standard deviation of each channel can be estimated from Poisson statistics, whereas there are no methods to directly estimate the uncertainties with stroboscopic measurements. Hence, TCSPC became the method of choice owing to its high sensitivity and low degree of systematic errors, the latter of which is essential for resolution of complex intensity decays.

In recent years there has been a reintroduction of gated detection methods.[128] The time resolution can be good but is not comparable to that of a laser source and an MCP PMT. Typical instrument response functions are close to 3 ns wide, as seen for the intensity decay of perylene in Figure 4.26. An advantage of this method is that one can detect many photons per lamp pulse, which should provide improved statistics. Gated detection can be used with low-repetition-rate lasers, which are typically less expensive than cavity-dumped dye lasers. A disadvantage of this method is the lack of knowledge of the noise level for each data point, so that one needs to estimate the experimental uncertainties during data analysis.

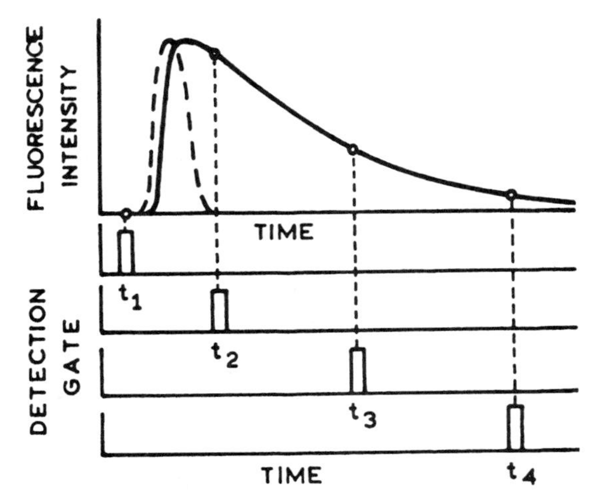

Figure 4.25. Decay time measurements using gated detection in the pulse sampling method. Revised from Ref. 10.

4.7.B. Streak Cameras

Streak cameras can provide time resolution of several picoseconds,[129–139] and some streak cameras have instrument response functions of 400 fs, considerably faster than for TCSPC with an MCP PMT. Streak cameras operate by dispersing the photoelectrons across an imaging screen. This can be accomplished at high speed using deflection plates within the detector (Figure 4.27). In the example shown, the light is dispersed by wavelength in a line across the front of the photocathode. Hence, streak cameras can provide simultaneous measurements of both wavelength and time-resolved decays. Such data are valuable in the study of time-dependent spectral relaxation or samples which contain fluorophores emitting at different wavelengths.

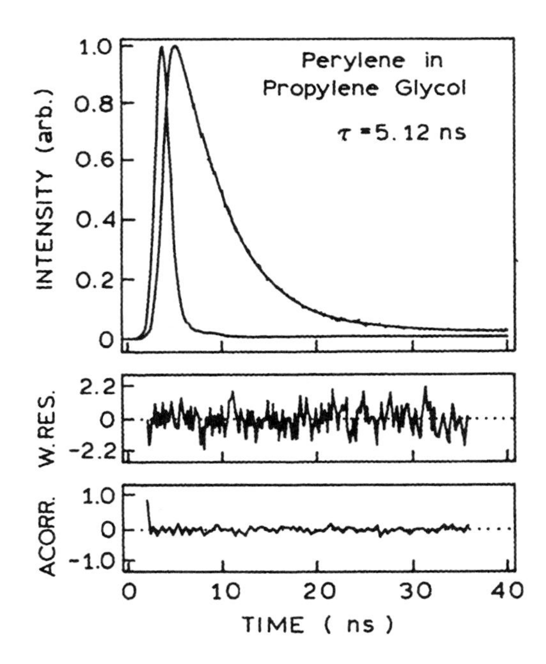

Figure 4.26. Intensity decay of perylene in propylene glycol measured by the pulse sampling method with a gated PMT. The light source is a hydrogen flashlamp with a repetition rate near 20 kHz; fwhm = 1.3 ns. Excitation, 405 nm; emission, 465 nm. Revised and reprinted from Ref. 128. Copyright © 1992, American Institute of Physics.

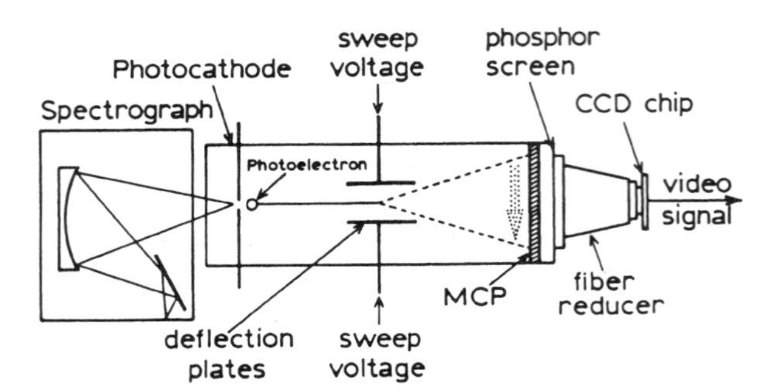

Figure 4.27. Schematic of a streak camera with wavelength resolution. From Ref. 130.

The time resolution obtainable with a streak camera is illustrated in Figure 4.28. The instrument response function is superior to that found with the fastest MCP PMTs. In spite of the high time resolution, streak cameras have not found wide use in biochemical fluorescence studies. This is because the dynamic range of measurable intensities is low, rarely exceeding 1000. Also, the signal-to-noise ratio is typically poor compared to that for TCSPC. These combined factors limit the ability of streak cameras to resolve complex decays. In the past, synchronization was a problem, so that signal averaging was not practical. Modern streak cameras allow averaging with high-repetition-rate lasers, but resolution of complex decays is still inferior to that obtainable with TCSPC.

It is interesting to notice from a comparison of the two MCP PMTs in Figure 4.28 that the performance of MCP PMTs continues to improve. Both MCP PMTs have 6-μm channels. The R3809U is a compact version of the R2809U. The electronics have also improved, so the instrument response function with the R3809U is 25 ps wide.[140]

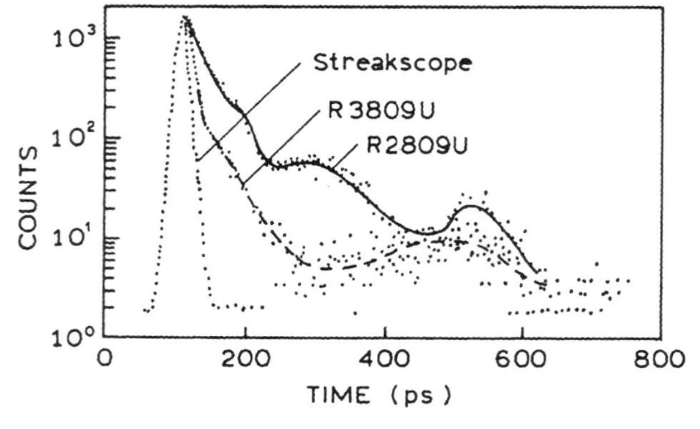

Figure 4.28. Comparison of the instrument response functions of a streak camera and two MCP PMTs (R3809U and R2809U). The 600-nm dye laser pulse width was 2 ps. Revised from Ref. 137.

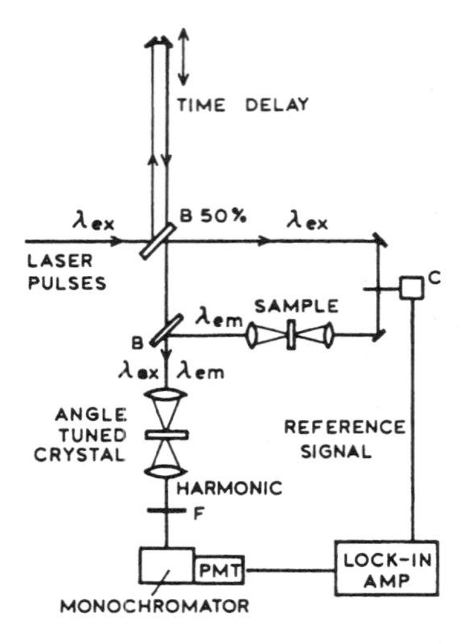

Figure 4.29. Schematic for fluorescence up conversion. B, beamsplitter; C, chopper; F, filter to isolate the harmonic from λ_{ex} and λ_{em}.

4.7.C. Upconversion Methods

The ultimate time resolution is provided by methods which bypass the limited time resolution of the detectors and rely on the picosecond and femtosecond pulse widths available with modern lasers.[141–143] The basic idea is to pass the fluorescence signal through an upconversion crystal and to gate the crystal with another picosecond or femtosecond light pulse (Figure 4.29). One observes the shorter-wavelength harmonic generated by the combined effects of the laser pulse and the emission. The intensity decay is sampled by sweeping the gating pulse with a time delay. The time resolution is determined by the width of the laser pulse. The signals are typically weak, so that an optical chopper and lock-in detector are needed to measure the upconverted signal in the presence of considerable background. The time-resolved decay is obtained by measuring the intensity of the upconverted signal as the delay time is varied.

While upconversion provides impressive time resolution, the instrumentation is too complex for most biochemical experiments. For instance, even a seemingly minor change of the emission wavelength can require a major readjustment of the apparatus since the detector uses an angle-tuned crystal. Also, decay times of more than 1–2 ns are difficult to measure owing to the use of a delay line (~1 ns/ft). Alignment of the delay line can be difficult to maintain as the time delay is altered.

4.8. DATA ANALYSIS

Time-resolved fluorescence data are moderately complex and, in general, cannot be analyzed using graphical methods. Since the mid-1970s many methods have been proposed for analysis of TCSPC data. These include nonlinear least-squares analysis,[3,12,144] the method of moments,[145–147] Laplace transformation[148–151] and the maximum entropy method,[152,153] Prony's method,[154] sine transforms,[155] and phase plane[156] and analytical methods.[157] The method of moments and the Laplace methods are not widely used at the current time. The maximum entropy method is newer and is being used in a number of laboratories. However, we see no reason to question the early reports which indicated that nonlinear least-squares analysis is the most general and reliable method for analysis of time-resolved data.[158]

4.8.A. Assumptions of Nonlinear Least-Squares Analysis

Prior to describing the results of least-squares analysis, it is important to understand its principles, assumptions, and goals. It is often stated that the goal is to fit the data, which is not completely true. The goal is to obtain estimates of parameter values which have the highest probability of being correct. It can be proven mathematically that least-squares analysis provides the best estimates for parameter values if the data satisfy a reasonable set of assumptions, which are as follows[159,160]:

1. All the experimental uncertainty is in the dependent variable (y-axis).
2. The uncertainties in the dependent variable are distributed like a Gaussian, centered on the correct value.
3. There are no systematic errors in either the dependent (y-axis) or independent (x-axis) variables.
4. The assumed fitting function is the correct mathematical description of the system. (Incorrect models yield incorrect parameters.)
5. The data points are all independent observations.
6. There is a sufficient number of data points so that the parameters are overdetermined.

Since these assumptions generally hold for TCSPC, it seems that least squares is the preferred method of analysis. In other fields of science or other areas of biochemistry, one often encounters data in a form that does not satisfy these assumptions, in which case least squares may not be the preferred method of analysis. This can occur when the variables are transformed to yield linear plots, and the error is no longer a Gaussian and/or no longer only in the y-axis.

4.8.B. Overview of Least-Squares Analysis

The data for TCSPC satisfy the assumptions of least-squares analysis, so we can now consider its implementation. In any least-squares analysis one starts with a model which is assumed to describe the data. The goal is then to obtain the parameter values which provide the best match between the data $N(t_k)$ and the calculated decay $N_c(t_k)$ using assumed parameter values. This is accomplished by minimizing the goodness-of-fit parameter χ^2,

$$\chi^2 = \sum_{k=1}^{n} \frac{1}{\sigma_k^2} [N(t_k) - N_c(t_k)]^2 = \sum_{k=1}^{n} \frac{[N(t_k) - N_c(t_k)]^2}{N(t_k)} \quad [4.20]$$

In this expression the sum extends over the number (n) of channels or data points used for a particular analysis, and σ_k is the standard deviation of each data point.

In TCSPC it is straightforward to assign the standard deviations (σ_k). From Poisson statistics the standard deviation is known to be the square root of the number of photon counts, $\sigma_k = [N(t_k)]^{1/2}$. Hence, for a channel with 10,000 counts, $\sigma_k = 100$, and for 10^6 counts, $\sigma_k = 1000$. The relative uncertainty in the data decreases as the number of photons increases. The value of χ^2 is the sum of the squared deviations between the measured values $N(t_k)$ and the expected values $N_c(t_k)$, each divided by squared deviations expected for the number of detected photons.

It is interesting to compare the numerator and denominator in Eq. [4.20]. For instance, assume a channel contains 10^4 counts. Then the expected deviation for this measurement is 100 counts. If the assumed model accounts for the data, then the numerator and denominator of Eq. [4.20] are both $(10^2)^2$, and this data point contributes 1.0 to the value of χ^2. In TCSPC, and also for FD measurements, the number of data points is typically much larger than the number of parameters. Hence, for random errors and the correct model, χ^2 is expected to be approximately equal to the number of data points (channels).

Suppose the data are being analyzed in terms of the multiexponential model (Eq. [4.8]). Then the values of α_i and τ_i are varied until χ^2 is a minimum, which occurs when $N(t_k)$ and $N_c(t_k)$ are most closely matched. A variety of general methods are available for selecting how α_i and τ_i are changed after each iteration. While some procedures work more efficiently than others, all seem to perform adequately.[159,160] These methods include the Gauss–Newton and modified Gauss–Newton algorithms and the Nelder–Mead method. This procedure of matching the calculated curves to the measured data is frequently referred to as deconvolution, which is inaccurate. In reality, an assumed decay law $I(t)$ is convoluted with $L(t_k)$, and the results are compared with $N(t_k)$. This procedure is more correctly called iterative reconvolution.

It is not convenient to interpret the values of χ^2 because χ^2 depends on the number of data points.[1] For this reason one uses the value of reduced χ^2,

$$\chi_R^2 = \frac{\chi^2}{n-p} = \frac{\chi^2}{\nu} \quad [4.21]$$

where n is the number of data points, p is the number of floating parameters, and $\nu = n - p$ is the number of degrees of freedom. If only random errors contribute to χ_R^2, then this value is expected to be near unity. This is because the average χ^2 per data point should be about 1, and typically the number of data points (n) is much larger than the number of parameters. If the model does not fit, the individual values of χ^2 and χ_R^2 are both larger than expected for random errors.

How can one judge the goodness-of-fit from the value of χ_R^2? If one is able to accurately estimate the experimental uncertainties σ_k, then the value of χ_R^2 is expected to be close to unity. This is because each data point is expected to contribute σ_k^2 to χ^2, which is in turn normalized by σ_k^2. If the model does not fit the data, then χ_R^2 will be significantly larger than unity. Although there is an emphasis on statistics in interpretation of TCSPC data, the first step should be a visual comparison of the data and the fitted function as well as a visual examination of the residuals, which are the differences between the measured data and the fitted function. If the data and fitted function are grossly mismatched, there may be a flaw in the program, the program may be trapped in a local minimum far from the correct parameter values, or the model may be incorrect. If the data and fitted functions are closely matched, one has to be careful to avoid using a more complex model when a simple one is adequate. In this laboratory we rely heavily on such visual comparisons. If one cannot visually see that a fit is improved with a more complex model, then we are hesitant to accept the more complex model.

4.8.C. Meaning of the Goodness of Fit, χ_R^2

In analyzing TCSPC data, one is frequently presented with two or more fits to the data, each with a value of χ_R^2. Typically, the magnitude of χ_R^2 decreases for the model with more adjustable parameters. What elevation of χ_R^2 is significant? What decrease in χ_R^2 is adequate to justify accepting the model with more parameters? These questions can be answered in two ways, one based on experience and the other based on mathematics. In mathematical terms, one can predict the probability of obtaining a value of χ_R^2 because of random errors. These values can be found in standard mathematical tables of the χ_R^2 distribution. Selected values are shown in Table 4.2 for various probabilities and numbers of degrees of freedom. Suppose you have in excess of 200 data points and the value of χ_R^2 is 1.25. There is only a 1% chance ($P = 0.01$) that random errors could result in this value, so that the fit can be rejected, assuming that the data are free of systematic errors which can elevate χ_R^2. If the value of χ_R^2 is 1.08, then there is a 20% chance that this value is due to random deviations in the data. While this may seem like a small probability, we feel that it is not advisable to reject a model if the probability exceeds 5%, which corresponds to $\chi_R^2 =$ 1.17 (Table 4.2). In our experience, systematic errors in the data can easily result in a 10–20% elevation of χ_R^2.

While we stated that assumptions 1–6 above were generally true for TCSPC, we are not convinced that assumption 5 is true. After examining many data sets, and the resulting statistics, we have the impression that the data behave as if there are fewer independent observations (degrees of freedom) in the TCSPC data than the number of actual observations (channels). This does not affect our conclusion that least squares is the preferred method of analysis, but it can compromise statistical judgments about the significance of the results. This is another reason for being skeptical about small elevations of χ_R^2.

In our opinion the simple reliance on mathematical tables is dangerous and likely to lead to overinterpretation of the data. The absolute value of χ_R^2 is often of less significance than the relative values of χ_R^2, and systematic errors in the data can easily result in χ_R^2 values in excess of 1.5. In general, we consider twofold or larger decreases in χ_R^2 significant. Smaller changes in χ_R^2 are interpreted with caution, typically based on some prior understanding of the system. In general, we find that for systematic errors, the χ_R^2 value is not significantly decreased using the next more complex model. Hence, if the value of χ_R^2 does not

Table 4.2. χ_R^2 **Distribution**[a]

	Probability (P)					
Degrees of freedom (ν)	0.2	0.1	0.05	0.02	0.01	0.001
10	1.344	1.599	1.831	2.116	2.321	2.959
20	1.252	1.421	1.571	1.751	1.878	2.266
50	1.163	1.263	1.350	1.452	1.523	1.733
100	1.117	1.185	1.243	1.311	1.358	1.494
200	1.083	1.131	1.170	1.216	1.247	1.338

[a] From Ref. 1, Table C-4.

decrease when the data are analyzed with a more complex model, the value of χ_R^2 probably reflects the poor quality of the data.

4.8.D. Autocorrelation Function • Advanced Topic •

Another diagnostic for the goodness-of-fit is the autocorrelation function.[3] For a correct model, and in the absence of systematic errors, one expects the deviations to be randomly distributed around zero. The randomness of the deviations can be judged from the autocorrelation function. Calculation of the correlation function is moderately complex, but the details do not need to be understood to interpret the resulting plots. The autocorrelation function $C(t_j)$ is the extent of correlation between deviations in the kth and the $(k + j)$th channel. The values of $C(t_j)$ are calculated using

$$C(t_j) = \left(\frac{1}{m} \sum_{k=1}^{m} D_k D_{k+j} \right) \Big/ \left(\frac{1}{n} \sum_{k=1}^{n} D_k^2 \right) \qquad [4.22]$$

where D_k is the deviation in the kth data point (see Eq. [4.23]) and D_{k+j} is the deviation in the $(k + j)$th data point. This function measures whether a deviation at one data point (time channel) predicts that the deviation in the jth higher channel will have the same or opposite sign. The calculation is usually extended to test for correlations across half of the data channels ($m = n/2$), so that the autocorrelation plots have half as many data points as the original data set.

One example of an autocorrelation plot was seen in Figure 4.23, where data for NATA were presented for measurements versus scattered light and versus a lifetime reference. This particular instrument showed a minor color effect, which resulted in some systematic deviations between $N_c(t_k)$ and $N(t_k)$. These systematic differences are barely visible in the direct plot of the deviations (Figure 4.23, middle left). The use of the autocorrelation plot (Figure 4.23, upper left) allowed the deviations to be visualized as positive and negative correlations in adjacent or distant channels, respectively. For closely spaced channels, the deviations are likely to both be of the same sign. For more distant channels, the deviations are likely to be of opposite signs. These systematic errors were eliminated by the use of a lifetime reference, as seen by the flat autocorrelation plot (Figure 4.23, upper right).

4.9. ANALYSIS OF MULTIEXPONENTIAL DECAYS

4.9.A. *p*-Terphenyl and Indole—Two Widely Spaced Lifetimes

The best way to understand analysis of the TD data is to consider representative data. For the first example we chose a mixture of *p*-terphenyl and indole, which individually display single-exponential decays of 0.93 and 3.58 ns, respectively. For the TD measurements, a mixture of *p*-terphenyl and indole was observed at 330 nm, where both species emit (Figure 4.30). TCSPC data for this mixture are shown in Figure 4.31. The presence of two decay times is evident from curvature in the plot of log $N(t)$ versus time. The time-dependent data could not be fit to a single decay time, as seen by the mismatch of the calculated convolution integral (dashed line) with the data (dots).

The lower panels show the deviations (D_k) or differences between the measured and calculated data,

$$D_k = \frac{I(t_k) - I_c(t_k)}{\sqrt{I(t_k)}} \qquad [4.23]$$

The weighted residual (W. Res.) or deviation plots are used because it is easier to see the differences between $I(t_k)$ and $I_c(t_k)$ in these plots than in a plot of log $I(t_k)$ versus t_k. Also, the residuals are weighted according to the standard deviation of each data point. For a good fit, these values are expected to be randomly distributed around zero, with a mean value near unity.

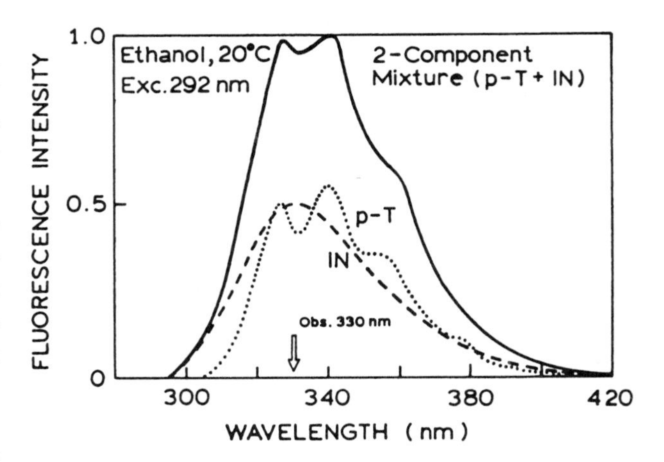

Figure 4.30. Emission spectra of *p*-terphenyl (*p-T*), indole (IN), and their mixture. Excitation was at 292 nm, from a frequency-doubled R6G dye laser. The emission at 330 nm was isolated with a monochromator. From Ref. 161.

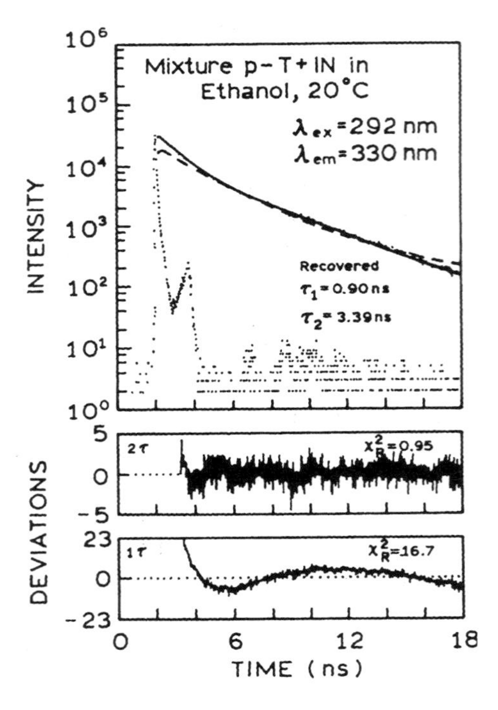

Figure 4.31. Time-domain intensity decay of a two-component mixture of indole (IN) and *p*-terphenyl (*p-T*). The dashed line shows the one-decay-time fit, and the solid line the two-decay-time fit. From Ref. 161.

4.9.B. Comparison of χ_R^2 Values—*F*-Statistic

How can one compare the values of χ_R^2 for two fits? This can be done using the *F*-statistic, which is a ratio of χ_R^2 values. In general, this ratio depends on the number of degrees of freedom (ν) for each fit, and this number will depend on the model. As for χ_R^2, the values of the χ_R^2 ratios that are statistically significant at various levels of probability are available in statistical tables; a few such values are listed in Table 4.3. In practice, there are usually many more data points than parameters, so that ν is almost the same for both fits. For this reason, we did not consider the slightly different numbers of degrees of freedom in the numerator and denominator.

For the mixture of *p*-terphenyl and indole, the single-decay-time fit (Figure 4.31, bottom panel) results in residuals which oscillate across the time axis; this behavior is characteristic of an incorrect model. Also, the value of χ_R^2 = 16.7 is obviously much greater than unity, and according to Table 4.2. there is a less than 0.1% chance that random error could result in such an elevated value of χ_R^2. Additionally, the ratio of χ_R^2 for this model to that for the two-decay time model is 17.6, which is much larger than the values of the *F*-statistic in Table 4.3. Hence, the single-decay-time model is easily rejected for this data set.

Table 4.3. ***F*-Statistic for Comparison of χ_R^2 Values**[a]

Degrees of freedom (ν)	Probability (P)					
	0.32	0.25	0.10	0.05	0.025	0.01
10	1.36	1.55	2.32	2.98	3.96	4.85
15	1.28	1.43	1.97	2.40	2.86	3.52
20	1.24	1.36	1.79	2.12	2.46	2.94
40	1.16	1.24	1.51	1.69	1.88	2.11
60	1.13	1.19	1.40	1.53	1.67	1.84
120	1.09	1.13	1.26	1.35	1.43	1.53
∞	1.00	1.00	1.00	1.00	1.00	1.00

[a] From Ref. 189, Table A-4. In general, the *F*-values are computed for different degrees of freedom for each χ_R^2 value. For TCSPC data and for FD data, the degrees of freedom are usually similar in the numerator and in the denominator. Hence, *F*-values are listed for only one value of ν. Additional *F*-values can be found in Ref. 1, Tables C-6 and C-7.

While the use of the *F*-statistic is common in least-squares analysis, there is a potential problem with the use of an *F*-statistic to compare two χ_R^2 values originating from different mathematical models and the same data set.[162] The use of the *F*-statistic requires that the residuals for each analysis be independent from each other. This is clearly not the case in analysis of TCSPC data, where the residuals result from the same data sets analyzed with different mathematical models. This is most likely to be a problem when the χ_R^2 values are similar.

4.9.C. Parameter Uncertainty—Confidence Intervals

After the analysis is completed, one has a set of α_i and τ_i values. It is important to recognize that there can be considerable uncertainty in these values, particularly for closely spaced decay times. Estimation of the uncertainties in the recovered parameters is an important but often ignored problem. Unfortunately, with nonlinear least-squares analysis there are no general methods for estimating the range of parameter values which are consistent with the data. Uncertainties are reported by almost all the data analysis programs, but these estimates are invariably smaller than the actual uncertainties. In fact, most of us have had the experience of not trusting the results to the degree of precision reported by data analysis programs.

Most software for nonlinear least-squares analysis reports uncertainties which are based on the assumption of no correlation between parameters. These are called the asymptotic standard errors (ASEs).[159] As shown for a mixture having more closely spaced lifetimes in Section 4.9.E, the obvious errors in the recovered parameters often exceed the ASEs. The ASEs usually underestimate the actual uncertainties in the parameter values.

In our opinion the best way to determine the range of parameters consistent with the data is to examine the χ_R^2 surfaces, which is also called a support plane analysis.[160] The basic idea is to change one parameter value from its value where χ_R^2 is a minimum and then rerun the least-squares fit, keeping this parameter value constant at the selected value. By rerunning the fit, the other parameters can adjust to again minimize χ_R^2. If χ_R^2 can be reduced to an acceptable value, then the offset parameter value is said to be consistent with the data. The parameter value is changed again by a larger amount until the χ_R^2 value exceeds an acceptable value, as judged by the F_χ statistic appropriate for p and ν degrees of freedom (see Eq. [4.24]). This procedure is then repeated for the other parameter values.

For determination of the range of parameter values consistent with the data, one typically chooses $P = 0.32$, where P is the probability that the value of F_χ is due to random errors in the data. When the value of P exceeds 0.32, there is less than a 32% chance that the parameter value is consistent with the data. When the value of P is less than 0.32, there is a 68% chance that the parameter value is consistent with the data, which is the usual definition of a standard deviation.

To determine the confidence interval, the value of χ_R^2 with a fixed parameter value, χ_R^2 (par), is compared with the minimum value of χ_R^2 with all parameters variable, χ_R^2 (min).[163,164] The range of parameter values is expanded until χ_R^2 (par) exceeds the F_χ value for the number of parameters (p) and the degrees of freedom (ν) and the chosen probability, typically $P = 0.32$:

$$F_\chi = \frac{\chi_R^2(par)}{\chi_R^2(\text{min})} = 1 + \frac{p}{\nu} F(p, \nu, P) \qquad [4.24]$$

In this expression,[165] $F(p, \nu, P)$ is the F-statistic with p parameters and ν degrees of freedom with a probability of P. The F-statistics needed to calculate F_χ are listed in Table 4.4. It is important to realize that, in general, the uncertainty range will not be symmetrical around the best-fit value of the parameter. For a two-decay-time model ($p = 4$) and 400 degrees of freedom, the χ_R^2 ratio is 1.012 for $P = 0.32$. Calculation of the χ_R^2 surfaces is a time-consuming process and has not yet been automated within most data analysis software. However, these calculations provide a realistic judgment of what one actually knows from the data.

We wish to emphasize that there is no general agreement that the procedure described above represents the correct method to estimate confidence intervals. This is a topic

Table 4.4. F-Statistic for Calculation of Confidence Intervals[a]

Degrees of Freedom (ν)	Probability (P)					
	0.32[b]	0.25	0.10	0.05	0.025	0.01
	One parameter[c]					
10	1.09	1.49	3.29	4.96	6.94	10.0
30	1.02	1.38	2.88	4.17	5.57	7.56
60	1.01	1.35	2.79	4.00	5.29	7.08
120	1.00	1.34	2.75	3.92	5.15	6.85
∞	1.00	1.32	2.71[d]	3.84	5.02	6.63
	Two parameters[c]					
10	1.28	1.60	2.92	4.10	5.46	7.56
30	1.18	1.45	2.49	3.32	4.18	5.39
60	1.16	1.42	2.39	3.15	3.93	4.98
120	1.15	1.40	2.35	3.07	3.80	4.79
∞	1.14	1.39	2.30	3.00	3.69	4.61
	Three parameters[c]					
10	1.33	1.60	2.73	3.71	4.83	6.55
30	1.22	1.44	2.28	2.92	3.59	4.51
60	1.19	1.41	2.18	2.76	3.34	4.18
120	1.18	1.39	2.13	2.68	3.23	3.95
∞	1.17	1.37	2.08	2.60	3.12	3.78
	Five parameters[c]					
10	1.35	1.59	2.52	3.33	4.24	5.64
30	1.23	1.41	2.05	2.53	3.03	3.70
60	1.20	1.37	1.95	2.37	2.79	3.34
120	1.19	1.35	1.90	2.29	2.67	3.17
∞	1.17	1.33	1.85	2.21	2.57	3.02
	Eight parameters[c]					
10	1.36	1.56	2.34	3.07	3.85	5.06
30	1.22	1.37	1.88	2.27	2.65	3.17
60	1.19	1.32	1.77	2.10	2.41	2.82
120	1.17	1.30	1.72	2.02	2.30	2.66
∞	1.16	1.28	1.67	1.94	2.19	2.51
	Ten parameters[c]					
10	1.35	1.55	2.32	2.98	3.72	4.85
30	1.22	1.35	1.82	2.16	2.51	2.98
60	1.18	1.30	1.71	1.99	2.27	2.63
120	1.17	1.28	1.65	1.91	2.16	2.47
∞	1.15	1.25	1.59	1.83	2.05	2.32

[a] From Ref. 189. In the ratio of χ_R^2 values, the degrees of freedom refer to that for the denominator. The degrees of freedom in the numerator are 1, 2, or 3 for one, two, or three additional parameters.

[b] The values for 0.32 were calculated with a program (F-stat) provided by Dr. M. L. Johnson, University of Virginia.

[c] These values refer to the degrees of freedom in the numerator (p).

[d] This value appears to be incorrect in Ref. 189 and was taken from Ref. 1.

which requires further research. Irrespective of whether the F_χ values accurately define the confidence interval, examination of the χ_R^2 surfaces provides valuable insight into the resolution of parameters provided by a given experiment. If the χ_R^2 surfaces do not show well-defined minima, then the data are not adequate to determine the parameters.

There is some disagreement in the statistics literature about the proper form of F_χ for estimating the parameter uncertainty.[162] Some reports[166] argue that since one parameter is being varied, the number of degrees of freedom in the numerator should be 1. In this case F_χ is calculated using

$$F_\chi = \frac{\chi^2(par)}{\chi^2(\min)} = 1 + \frac{1}{\nu} F(p, \nu, P) \qquad [4.25]$$

Since we are varying p parameters to calculate the χ_R^2 surface, we chose to use Eq. [4.24].

An example of a support plane analysis is shown in Figure 4.32 for the mixture of p-terphenyl and indole. The confidence intervals (dashed lines) are given by the intercepts of the χ_R^2 surfaces (solid curves) with the appropriate F_χ values. For comparison, we have also shown the ASEs as solid bars. One notices that the ASEs are about twofold smaller than the confidence intervals. While this is a serious underestimation, this factor of 2 is in fact small compared to what is found for more closely spaced lifetimes. Also, in our opinion, the F_χ value near 1.005 for about 950 degrees of freedom is an underestimation because the time-resolved decay does not have completely independent data points. Suppose the actual number of independent data points was 200. In this case the $F\chi$ value would be near 1.02, which may provide a more realistic range of the uncertainties. For instance, the confidence interval for α_2 would become 0.305 ± 0.015 instead of 0.305 ± 0.005. The uncertainty in the latter value appears to be unrealistically small.

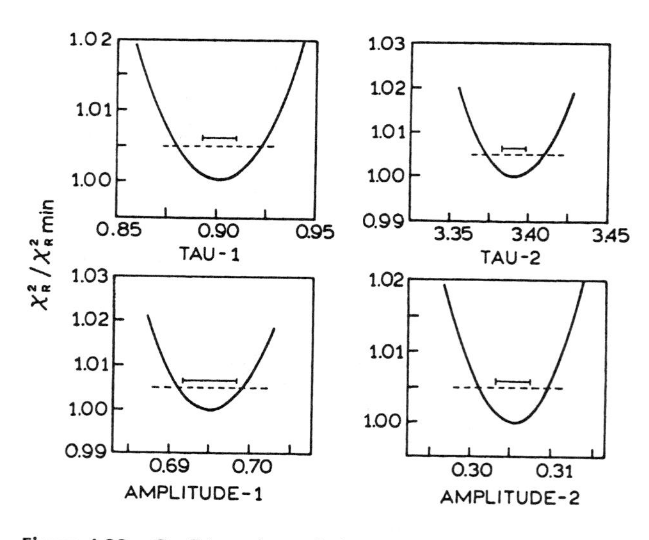

Figure 4.32. Confidence intervals for the τ_i and α_i parameter values from a support plane analysis for the two-component mixture of indole and p-terphenyl in Figure 4.31. The solid bars show the ASEs. From Ref. 161.

Another way to estimate parameter uncertainty is by Monte Carlo simulations.[167] The basic idea is to simulate data using the recovered decay law and the known level of random noise present in a given experiment. Newly generated random noise is added to each simulated data set, which is then analyzed as if it were actual data. New random noise is then added, and the process repeated. This results in a histogram of parameter values. These parameter values are examined to ascertain the range which results from the randomly added noise. It is important that the simulations use a model that correctly describes the system. If not, the results are meaningless. This Monte Carlo method is time-consuming, and thus seldom used. However, with the rapid advances in computer speed, one can expect Monte Carlo simulations to become more widely utilized for estimation of confidence intervals.

4.9.D. Effect of the Number of Photon Counts

For a single-exponential decay, the decay time can usually be determined with adequate accuracy even for a small number of observed photons. However, for multiexponential decays, it is important to measure as many photons as possible to obtain the highest resolution of the parameter values. This is illustrated in Figure 4.33 for the same two-component mixture of p-terphenyl and indole. For these data the number of counts in the peak channel was 10-fold less than in Figure 4.31, 3000 counts versus 30,000 counts. The correct values for the two decay times were still recovered. However, the relative decrease in χ_R^2 for the two-decay-time model was only 1.9-fold, as compared to 17-fold for the higher number of counts. Also, the χ_R^2 surfaces rise more slowly as the lifetimes are varied (Figure 4.33, right), so that the lifetimes are determined with less precision.

4.9.E. Anthranilic Acid and 2-Aminopurine—Two Closely Spaced Lifetimes

The resolution of two decay times becomes more difficult if the decay times are more closely spaced. This is illustrated by a mixture of anthranilic acid (AA) and 2-amino purine (2-AP), which individually display single-exponential decays of 8.53 and 11.27 ns, respectively (shown in Figure

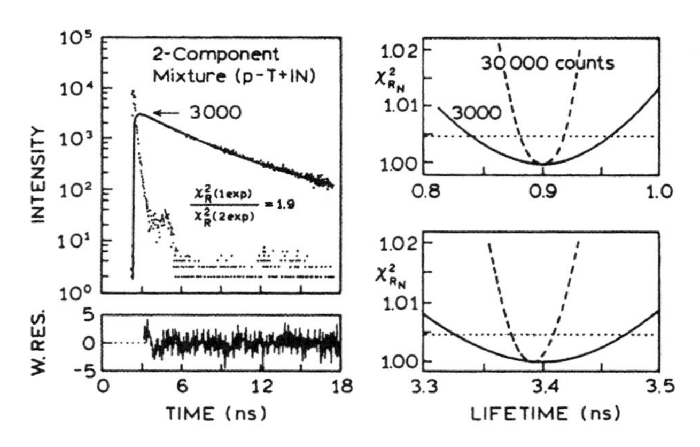

Figure 4.33. Effect of the number of photons counts on the resolution of the lifetimes for the mixture of *p*-terphenyl (*p-T*) and indole (IN) in Figure 4.31. *Left*: Two-decay-time fit of the intensity decay measured with 3000 counts in the peak channel. The corresponding plot for the data obtained with 30,000 counts in the peak channel is shown in Figure 4.31. *Right*: Comparison of the χ_R^2 surfaces for the data obtained with 3000) (—) and 30,000 counts (- - -). From Ref. 161.

4.38 below). Emission spectra for the mixture are shown in the answer to Problem 4.5. The time-dependent data for the mixture of AA and 2-AP are shown in Figure 4.34. At 380 nm, where both fluorophores emit, it is difficult to visually detect the presence of two decay times. The single-decay time model (represented by the dashed line in Figure 4.34) shows only small differences from the data (dots). However, the residual plot shows systematic deviations (Figure 4.34, bottom), which are easier to see on the linear scale used for the deviations than on the logarithmic plot. The 4.5-fold decrease in χ_R^2 for the two-decay-time model is adequate to reject the single-decay-time model.

While the data support acceptance of two decay times, the values of α_i and τ_i are not well determined. This is illustrated in Figure 4.35, which shows that χ_R^2 surfaces for the mixture of AA and 2-AP. This mixture was measured at five emission wavelengths. The data were analyzed individually at each wavelength. Each of the χ_R^2 surfaces shows distinct minima, which leads one to accept the recovered lifetimes. However, one should notice that different lifetimes were recovered at each emission wavelength. This suggests that the actual uncertainties in the recovered lifetimes are larger than expected from the ASEs, and they seem to be even larger than calculated from the χ_R^2 surfaces. Furthermore, the differences in the lifetimes recovered at each emission wavelength seem to be larger than expected even from the χ_R^2 surfaces. This illustrates the difficulties in recovering accurate lifetimes if the values differ by less than twofold.

Another difficulty is that the recovered amplitudes do not follow the emission spectra expected for each compo-

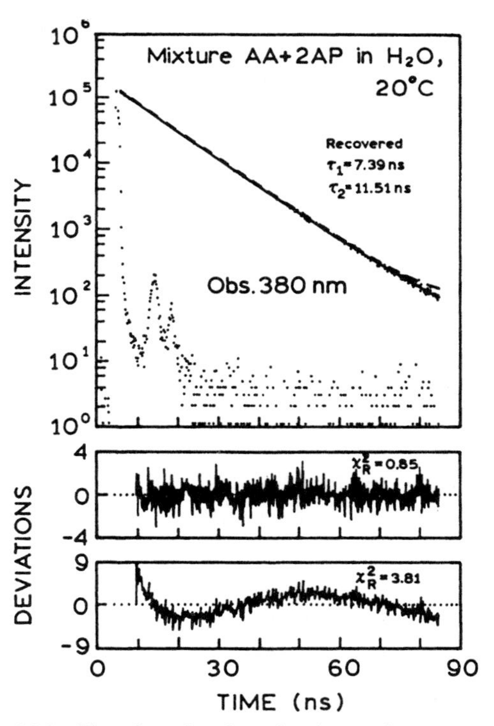

Figure 4.34. Time-dependent intensity decay of a two-component mixture of anthranilic acid (AA) and 2-aminopurine (2AP). The dashed line shows the one-decay-time fit, and the solid line the two-decay-time fit to the data (dots). From Ref. 161.

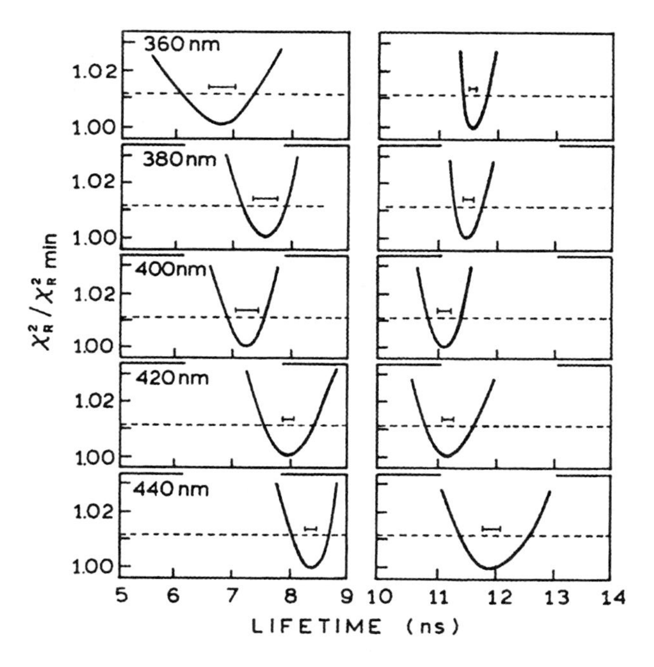

Figure 4.35. Lifetime χ_R^2 surfaces for the two-component mixture of AA and 2AP whose intensity decay is shown in Figure 4.34. The data obtained at each of the five emission wavelengths measured have been analyzed individually. The horizontal bars show the ASEs. From Ref. 161.

Table 4.5. Resolution of a Two-Component Mixture of Anthranilic Acid and 2-Aminopurine, Observed at a Single Wavelength, Using Time-Domain Data

Observation wavelength (nm)	Lifetimes (ns)		Preexponential factors		Fractional intensities		χ_R^2 Number of decay times	
	τ_1	τ_2	α_1	α_2	f_1	f_2	2[a]	1[a]
360	6.72	11.58	0.223	0.777	0.143	0.857	1.01	2.77
380	7.51	11.51	0.404	0.596	0.306	0.694	0.85	3.81
400	7.22	11.08	0.448	0.552	0.347	0.653	0.91	4.51
420	7.99	11.22	0.595	0.405	0.511	0.489	0.93	2.95
440	8.38	11.91	0.763	0.237	0.692	0.308	0.89	2.72

[a] Refers to a two- or one-component fit.

nent (Table 4.5 and Problem 4.5). As the lifetimes become closer together, the parameter values become more highly correlated, and it is difficult to know the true uncertainties. This is not intended to be a criticism of the TD measurements but rather is meant to illustrate the difficulties inherent in the analysis of multiexponential decays.

4.9.F. Global Analysis—Multiwavelength Measurements

One way to improve the resolution of closely spaced lifetimes is to perform measurements at additional wavelengths and to do a global analysis (Section 4.11). The concept of global analysis is based on the assumption that decay times are independent of wavelength. The decay times are global parameters because they are the same in all data sets. The amplitudes are nonglobal because they are different in each data set. Global analysis of the multiwavelength data results in much steeper χ_R^2 surfaces (Figure 4.36), and presumably a higher probability of recovery of the correct lifetimes. The lifetimes are determined with higher certainty from the global analysis because of the steeper χ_R^2 surfaces and the lower value of F_χ with more degrees of freedom (more data). As shown in Problem 4.5, the amplitudes (Table 4.6) recovered from the global analysis more closely reflect the individual emission spectra than the amplitudes recovered from the single-wavelength data.

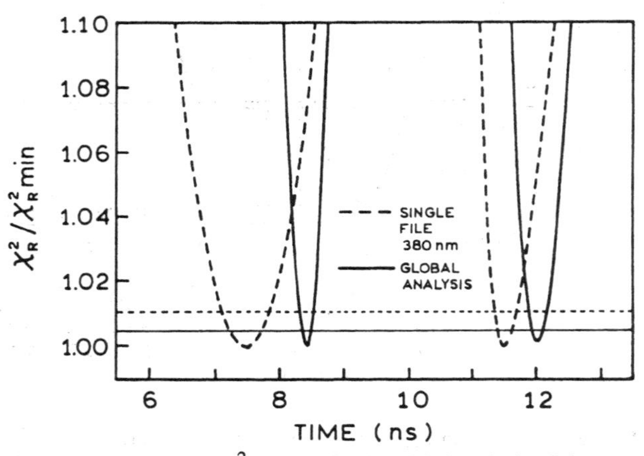

Figure 4.36. Lifetime χ_R^2 surface for the global analysis of the two-component mixture of AA and 2AP (—). Also shown for comparison is the χ_R^2 surface at 380 nm (- - -). The horizontal lines represent the F_χ values. From Ref. 161.

4.9.G. Resolution of Three Closely Spaced Lifetimes

The resolution of multiexponential decays becomes more difficult as the number of decay times increases. To illustrate this case, we chose a mixture of indole (IN), AA, and 2-AP (Figure 4.37). Each individual fluorophore displays a single-exponential decay, with decay times of 4.41, 8.53,

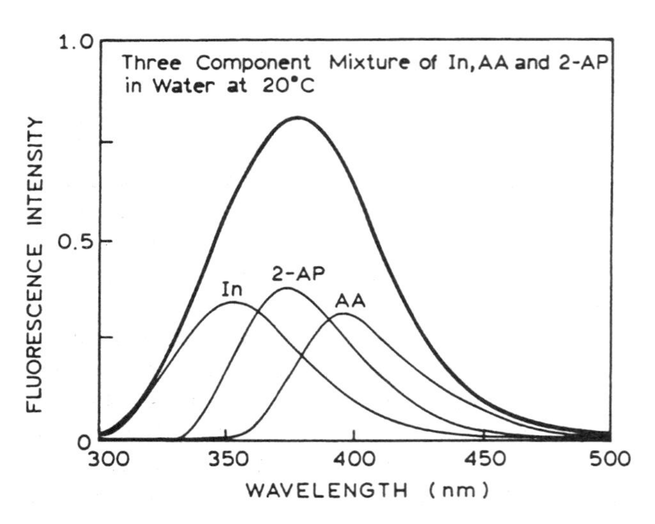

Figure 4.37. Emission spectra of indole (In), 2-aminopurine (2AP), and anthranilic acid (AA) and of the three-component mixture. From Ref. 161.

Table 4.6. Global Analysis of a Two Component Mixture of Anthranilic Acid (M) and 2-Aminopurine (2-AP) Measured at Five Emission Wavelengths: Time-Domain Data[a]

Observation Wavelength (nm)	AA[a] ($\tau_1 = 8.35$ ns)		2-AP ($\tau_2 = 12.16$ ns)	
	α_1	f_1	α_2	f_2
360	0.117	0.089	0.883	0.911
380	0.431	0.357	0.569	0.643
400	0.604	0.528	0.396	0.472
420	0.708	0.640	0.292	0.360
440	0.810	0.758	0.190	0.242

[a]For the two-component fit, $\chi_R^2 = 0.96$; for the one-component fit, $\chi_R^2 = 22.3$.

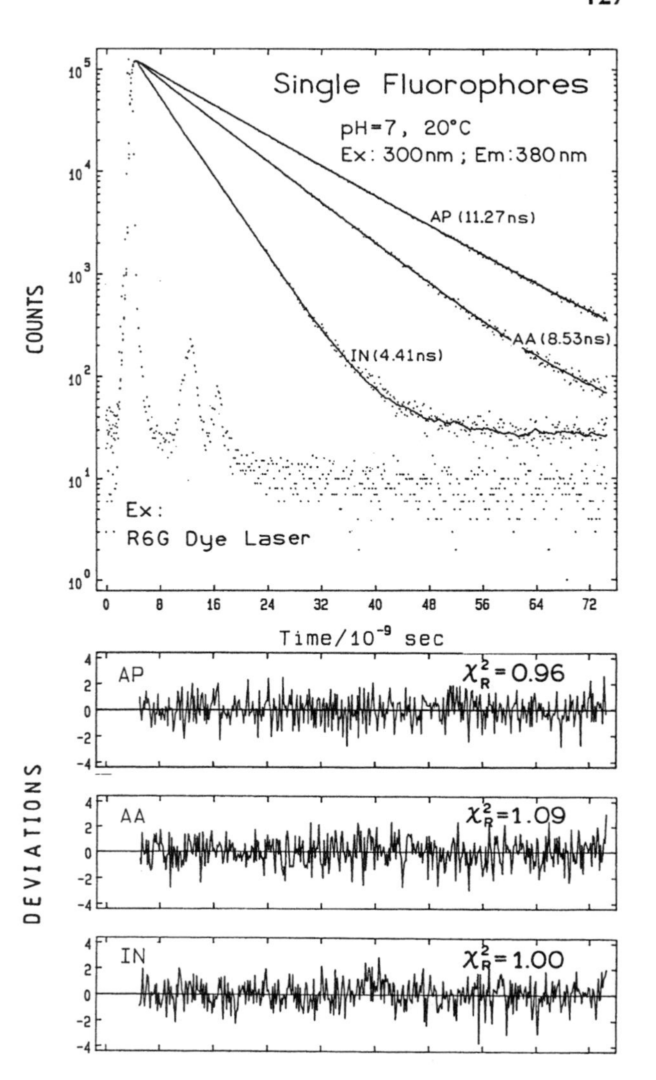

Figure 4.38. TCSPC data for indole (IN), anthranilic acid (AA), and 2-aminopurine (AP). From Ref. 161.

and 11.27 ns, respectively. At 380 nm all three fluorophores emit, and the decay is expected to be a sum of three exponentials.

TCSPC data for the three individual fluorophores are shown in Figure 4.38. Excitation was with the frequency-doubled output of an R6G dye laser, and the emission was detected with an R2809 MCP PMT. The excitation was polarized vertically, and the emission was detected 54.7° from the vertical to avoid the effects of rotational diffusion on the measured intensity decays. The points represent the data $I(t_k)$ i.e., the number of counts measured at each time interval t_k. The solid curves are the fitted functions or calculated data $I_c(t_k)$ using a single decay time. Since $I(t_k)$ and $I_c(t_k)$ are well matched, the decays seem to be single exponentials. As expected for a good fit, the deviations are randomly distributed around zero and the values of χ_R^2 are near unity.

The data for indole illustrate the need to consider the convolution integral when using an MCP PMT, even with a 4.41-ns decay time. At long times the plot of log $I(t_k)$ versus time becomes nonlinear even though there is only a single lifetime. This effect is most visible for indole, which has the shortest lifetime (4.41 ns). This long tail on the intensity decay is due to continued excitation from the tail of the impulse response function. If one did not consider convolution, and calculated the decay times from only the slopes, then one would reach the erroneous conclusion that the indole sample displayed a second long decay time.

Now consider similar data for a mixture of the three fluorophores. Although the decay times range threefold from 4 to 12 ns, this is in fact a difficult resolution. Examination of Figure 4.39 shows that the single-exponential fit (represented by the dashed line) appears to provide a reasonable fit to the data. However, the failure of this model is easily seen in the deviations, which are much larger than unity and are not randomly distributed on the time axis (Figure 4.39, bottom panel). The failure of the single-exponential model can also be seen from the value of $\chi_R^2 = 26.5$, which according to Table 4.2 allows the single-exponential model to be rejected with high certainty. To be more specific, there is a less than 0.1% chance ($P < 0.001$) that this value of χ_R^2 could be the result of random error in the data.

The situation is less clear with the double-exponential fit. In this case the fitted curve overlaps the data (not shown), $\chi_R^2 = 1.22$, and the deviations are nearly random. According to the χ_R^2 table (Table 4.2), there is only a 2%

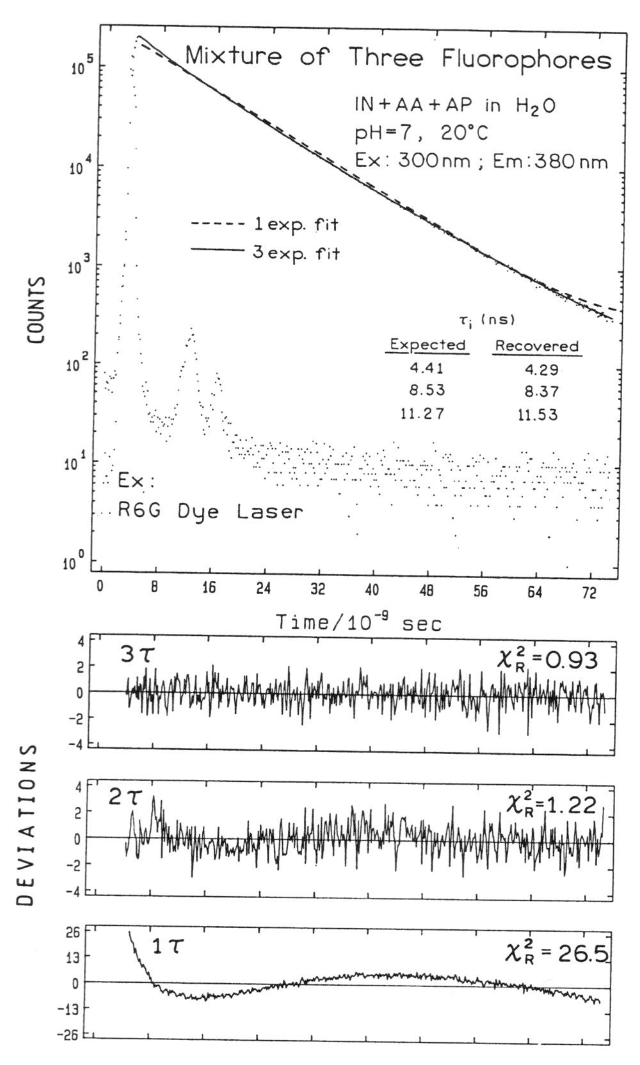

Figure 4.39. TCSPC data for a mixture of indole (IN), anthranilic acid (AA), and 2-aminopurine (AP). From Ref. 161.

chance that $\chi_R^2 = 1.22$ could result from statistical uncertainties in the data. In practice, such values of χ_R^2 are often encountered owing to systematic errors in the data. For instance, the systematic errors in Figure 4.23 resulted in an elevation of χ_R^2 to a similar value. Recall that for an actual experiment we do not know beforehand if the decay is a double-, triple-, or nonexponential decay. One should always accept the simplest model that accounts for the data, so we would be tempted to accept the double-exponential model because of the weak evidence for the third decay time.

An improved fit was obtained with the triple-exponential model, $\chi_R^2 = 0.93$, and the deviations are slightly more random than those for the two-decay-time fit. It is important to understand that such a result indicates that the data are consistent with three decay times but does not prove that the decay is a triple exponential. By least-squares analysis, one cannot exclude other more complex models and can only state that a particular model is adequate to explain the data. In this case, the data are consistent with the triple-exponential model, but the analysis does not exclude the presence of a fourth decay time.

The two- and three-decay-time fits can also be compared by using the ratio of χ_R^2 values. For this mixture the ratio of χ_R^2 values was 1.31. The probability of this ratio occurring due to random deviations in the data is between 5 and 10% (Table 4.3). Hence, there is a relatively low probability of finding this reduction in χ_R^2 (1.22 to 0.93) unless the data actually contain three decay times or are described by some model other than the two-decay-time model. Stated alternatively, there is a 90–95% probability that the two-decay-time model is not an adequate description of the sample.

One may question why there are two tests for the goodness of fit, one based on χ_R^2 itself and the other based on the F-statistic. The values of χ_R^2 are useful when the experimental errors can be accurately estimated, which is usually the case with TCSPC data. In such cases, the value of χ_R^2 provides a test both of the agreement of the measured and calculated $N(t_k)$ values and of whether the only source of noise is Poisson photon statistics. In contrast to χ_R^2, the F-statistic can be used when the experimental uncertainties (σ_k values) are not precisely known. This is usually the case with stroboscopic, gated detection, and streak camera measurements, in which photon counting is not used. This situation also occurs in FD fluorometry, where the uncertainties in the phase and modulation values can only be estimated. The calculated values of χ_R^2 can be very different from unity even for a good fit, because the σ_k^2 values may not equal the values of $[N(t_k) - N_c(t_k)]^2$. This is not a problem as long as the relative values of χ_R^2 are known. In these cases one uses the F-statistic, or relative decrease in χ_R^2, to determine the goodness of fit.

For closely spaced lifetimes, the ASEs will greatly underestimate the uncertainties in the parameters. This underestimation of errors is also illustrated in Table 4.7, which lists the analysis of the three-component mixture when measured at various emission wavelengths. It is clear from these analyses that the recovered lifetimes differ by amounts considerably larger than the ASEs. This is particularly true for the fractional intensities, for which the ASEs are ±0.001. Similar results can be expected for any decay with closely spaced lifetimes.

4.10. INTENSITY DECAY LAWS

So far we have considered methods to measure intensity decays, but we have not considered the forms which are possible. Many examples will be seen in the remainder of this book. A few examples are given here to illustrate the range of possibilities.

4.10.A. Multiexponential Decays

In the multiexponential model the intensity is assumed to decay as the sum of individual single-exponential decays:

$$I(t) = \sum_{i=1}^{n} \alpha_i \exp(-t/\tau_i) \quad [4.26]$$

In this expression the τ_i are the decay times, the α_i represent the amplitudes of the components at $t = 0$, and n is the number of decay times. This is the most commonly used model, but the meaning of the parameters (α_i and τ_i) depends on the system being studied. The most obvious application is to a mixture of fluorophores, each displaying one of the decay times τ_i. In a multi-tryptophan protein the decay times may be assigned to each of the tryptophan residues, although in general this is not possible. Many samples that contain only a single fluorophore display decays more complex than a single exponential. These data are usually interpreted in terms of Eq. [4.26], which then requires explanation of the multiple decay times. If the probe can exist in two environments, such as exposed to and shielded from water, then a decay time can be assigned to each of these states. Hence, single-tryptophan proteins that exist in multiple conformational states may display a decay time for each state. Many papers on protein fluorescence interpret the multiexponential decays in terms of conformational distributions.[168,169]

The meaning of the preexponential factors α_i is different for a mixture of fluorophores and for one fluorophore displaying a complex decay. For the latter case, it is generally safe to assume that the fluorophore has the same radiative decay rate in each environment. In this case the α_i values represent the fraction of the molecules in each conformation at $t = 0$, which corresponds to the ground-state equilibrium. However, the meaning of the α_i values is more complex for a mixture of fluorophores. In this case the relative α_i values depend on the quantum yield and intensity of each fluorophore at the observation wavelength as well as on its concentration.

Irrespective of whether the multiexponential decay originates with a single fluorophore or multiple fluorophores, the values of α_i and τ_i can be used to determine the fractional contribution (f_i) of each decay time to the steady-state intensity. These values are given by

$$f_i = \frac{\alpha_i \tau_i}{\sum_j \alpha_j \tau_j} \quad [4.27]$$

The terms $\alpha_i\tau_i$ are proportional to the area under the decay curve for each decay time. In a steady-state measurement, one measures all the emission irrespective of when the photon is emitted. This is why the intensity is usually weaker for a short decay time: the $\alpha_i\tau_i$ product is smaller. For a mixture of fluorophores, the values of f_i represent the

Table 4.7. Multiexponential Analysis of a Three-Component Mixture of Indole, 2-Aminopurine and Anthranilic Acid

Observation wavelength (nm)	Lifetimes (ns)			Preexponential factors			Fractional intensities[a]			χ_R^2 Number of exponents		
	τ_1	τ_2	τ_3	α_1	α_2	α_3	f_1	f_2	f_3	3	2	1
360	4.79	7.51	11.43	0.314	0.004	0.682	0.161	0.003	0.836	1.10	1.10	17.67
	(0.10)[b]	(1.66)	(0.02)					(0.001)	(0.001)			
380	4.29	8.37	11.53	0.155	0.622	0.223	0.079	0.617	0.304	0.93	1.22	26.45
	(0.33)	(0.05)	(0.02)					(0.001)	(0.001)			
400	4.99	9.50	13.48	0.180	0.722	0.098	0.099	0.755	0.146	0.96	0.97	7.88
	(0.16)	(0.13)	(0.25)					(0.001)	(0.001)			
420	4.32	8.54	11.68	0.072	0.658	0.270	0.034	0.618	0.348	0.93	0.34	4.97
	(0.47)	(0.25)	(0.09)					(0.001)	(0.001)			
440	1.70	7.94	11.07	0.037	0.580	0.383	0.007	0.517	0.476	1.02	1.04	4.14
	(0.61)	(0.24)	(0.06)					(0.001)	(0.001)			

[a] $f_i = \alpha_i\tau_i / \Sigma\alpha_j\tau_j$.
[b] Asymptotic standard errors.

fractional intensity of each fluorophore at each observation wavelength (Tables 4.5 and 4.6). However, the recovered values of f_i may not correlate well with the expected intensities owing to the difficulties of resolving a multiexponential decay.

What are the variable parameters in a multiexponential analysis? Typically, these are the n lifetimes and the n or $n - 1$ amplitudes. In most intensity decay analyses, the total intensity is not measured, and $\Sigma\alpha_i$ is normalized to unity. Also, Σf_i is normalized to unity. Hence, for a three-decay-time fit, there are typically five independently variable parameters—three lifetimes and two amplitudes. However, most programs require that all the amplitudes remain variable during the fitting procedure, and the α_i values are normalized at the end of the analysis. In these cases one is fitting to the total intensity, and there are three variable amplitude parameters.

Finally, it is important to remember that the multiexponential model (Eq. [4.26]) is perhaps the most powerful model. Almost any intensity decay, irrespective of its complexity, can be fit using Eq. [4.26]. This means that one can say that the data are consistent with Eq. [4.26], but the data can also be consistent with many other decay laws.

When the multiexponential decay law is used, it is often useful to determine the average lifetime ($\bar{\tau}$). The average lifetime is given by Eq. [4.3]. For a double-exponential decay, $\bar{\tau}$ is given by

$$\bar{\tau} = \frac{\alpha_1\tau_1^2 + \alpha_2\tau_2^2}{\alpha_1\tau_1 + \alpha_2\tau_2} = f_1\tau_1 + f_2\tau_2 \qquad [4.28]$$

Occasionally, one finds the "average lifetime" given by

$$\langle\tau\rangle = \sum_i \alpha_i\tau_i \qquad [4.29]$$

which is not correct. The value of $\langle\tau\rangle$ is proportional to the area under the decay curve, and for a double-exponential decay it becomes

$$\int_0^\infty I(t)dt = \alpha_1\tau_1 + \alpha_2\tau_2 \qquad [4.30]$$

This value should perhaps be called a lifetime-weighted quantum yield. There are occasions where the value of $\langle\tau\rangle$ is useful. For instance, the efficiency of energy transfer is given by

$$E = 1 - \frac{F_{DA}}{F_D} = 1 - \frac{\int I_{DA}(t)\,dt}{\int I_D(t)\,dt} \qquad [4.31]$$

where $I_{DA}(t)$ and $I_D(t)$ are the intensity decays of the donor in the presence and absence of energy transfer, respectively. The integrals in Eq. [4.31] are proportional to the steady-state intensities in the presence (F_{DA}) and absence (F_D) of acceptor, which are given by Eq. [4.30]. When one is using the results of a multiexponential analysis, the transfer efficiency should be calculated using $<\tau>$ values, since these are proportional to the steady-state intensity.

4.10.B. Lifetime Distributions • Advanced Topic •

There are many situations in which one does not expect a limited number of discrete decay times, but rather a distribution of decay times. Such behavior may be expected for a fluorophore in a mixture of solvents, so that a range of environments exists. One can imagine a fluorophore being surrounded by one, two, three, or more polar molecules, each environment resulting in a different intensity decay. For a single-tryptophan protein, the distribution of decay times may reflect the distribution of protein conformations.[168,169] Another possibility is a protein with many tryptophan residues, so that it is not practical to consider the individual decay times.

In such cases the intensity decays are typically analyzed in terms of a lifetime distribution. In this case the α_i values are replaced by distribution functions $\alpha(\tau)$. The component with each individual τ value is given by

$$I(\tau, t) = \alpha(\tau)e^{-t/\tau} \qquad [4.32]$$

However, one cannot observe these individual components with lifetime τ, but only the entire decay. The total decay law is the sum of the individual decays weighted by the amplitudes,

$$I(t) = \int_{\tau=0}^{\infty} \alpha(\tau)e^{-t/\tau}\,d\tau \qquad [4.33]$$

where $\int \alpha(\tau)\,d\tau = 1.0$.

Lifetime distributions are usually used without a theoretical basis for the $\alpha(\tau)$ distribution. One typically uses arbitrarily selected Gaussian (G) and Lorentzian (L) lifetime distributions. For these functions the $\alpha(\tau)$ values are

$$\alpha_G(\tau) = \frac{1}{\sigma\sqrt{2\pi}} \exp\left[-\frac{1}{2}\left(\frac{\tau - \bar{\tau}}{\sigma}\right)^2\right] \quad [4.34]$$

$$\alpha_L(\tau) = \frac{1}{\pi} \frac{\Gamma/2}{(\tau - \bar{\tau})^2 + (\Gamma/2)^2} \quad [4.35]$$

where $\bar{\tau}$ is the central value of the distribution, σ is the standard deviation of the Gaussian, and Γ is the fwhm for the Lorentzian. For a Gaussian the fwhm is given by 2.345σ. For ease of interpretation, we prefer to describe both distributions by the fwhm. An alternative approach would be to use $\alpha(\tau)$ distributions that are not described by any particular function. This approach may be superior in that it makes no assumptions about the shape of the distribution. However, the use of functional forms for $\alpha(\tau)$ minimizes the number of floating parameters in the fitting algorithms. Without an assumed function form, it may be necessary to place restraints on the adjacent values of $\alpha(\tau)$.

By analogy with the multiexponential model, it is possible that $\alpha(\tau)$ is multimodal. Then

$$\alpha(\tau) = \sum_i g_i \alpha_i^0(\tau) = \sum_i \alpha_i(\tau) \quad [4.36]$$

where i refers to the ith component of the distribution centered at $\bar{\tau}_i$, and g_i represents the amplitude of this component. The g_i values are amplitude factors and the $\alpha_i^0(\tau)$ shape factors describing the distribution. If part of the distribution exists below $\tau = 0$, then the $\alpha_i(\tau)$ values need additional normalization. For any distribution, including those cut off at the origin, the amplitude associated with the ith mode of the distribution is given by

$$\alpha_i = \frac{\int_0^\infty \alpha_i(\tau)\, d\tau}{\int_0^\infty \sum_i \alpha_i(\tau)\, d\tau} \quad [4.37]$$

The fractional contribution of the ith component to the total emission is given by

$$f_i = \frac{\int_0^\infty \alpha_i(\tau)\, \tau d\tau}{\int_0^\infty \sum_i \alpha_i(\tau)\, \tau d\tau} \quad [4.38]$$

In the use of lifetime distributions, each decay time component is associated with three variables, α_i, $\bar{\tau}_i$, and the half-width (σ or Γ). Consequently, one can fit a complex decay with fewer exponential components. For instance, data which can be fit to three discrete decay times can typically be fit to a bimodal distribution model. In general, it is not possible to distinguish between the discrete multiexponential model (Eq. [4.26]) and the lifetime distribution model (Eq. [4.33]), so the model selection must be based on one's knowledge of the system.[170–172]

4.10.C. Stretched Exponentials

A function similar to the lifetime distributions is the stretched exponential

$$I(t) = I_0 \exp[(-t/\tau)^\beta] \quad [4.39]$$

In this expression β is related to the distribution of decay times. The function is not used frequently in biophysics, but it is often found in studies of polymers when one expects a distribution of relaxation times. In a least-squares fit, β and τ would be the variable parameters.

4.10.D. Transient Effects

In many samples the intensity decay can be nonexponential due to phenomena which occur immediately following excitation. This occurs in collisional quenching and in resonance energy transfer. In the presence of a quencher, a fluorophore which displays an unquenched single-exponential lifetime will display a decay of the form

$$I(t) = I_0 \exp[(-t/\tau) - (2bt^{1/2})] \quad [4.40]$$

In this expression, b depends on the quencher concentration and diffusion coefficient. One can fit such decays to the multiexponential model, but one would then erroneously conclude that there are two fluorophore populations. In this case a single fluorophore population gives a nonexponential decay owing to rapid quenching of closely spaced fluorophore–quencher pairs.

Resonance energy transfer (RET) can also result in decays which have various powers of time in the exponent. Depending on whether RET occurs in one, two, or three dimensions, t can appear with powers of $\frac{1}{6}$, $\frac{1}{3}$, or $\frac{1}{2}$, respectively. Hence we see that intensity decays can take a number of forms depending on the underlying molecular phenomenon. In our opinion, it is essential to analyze each decay with the model which correctly describes the samples. Use of an incorrect model, such as use of the multiexponential model to describe transient effects, results in apparent parameter values (α_i and τ_i) which cannot be easily related to the quantities of interest (quencher concentration and diffusion coefficient).

4.11. GLOBAL ANALYSIS

In Section 4.9 we indicated the difficulties of resolving the decay times and amplitudes in a multiexponential decay. In general, the parameters in the various decay functions are correlated and difficult to resolve. The resolution of correlated parameters can be improved by the use of global analysis.[173–178] The basic idea is to combine two or more experiments in which some of the parameters are the same in all measurements, and some are different. This can be illustrated for the emission spectra in Figure 4.37. A nonglobal experiment would be to recover the values of α_i and τ_i from the intensity decay collected at 380 nm, where all three fluorophores emit. A global experiment would be to measure the intensity decays at several wavelengths, say, 360, 380, 400, and 420 nm. The multiple intensity decay curves are then analyzed simultaneously to recover the τ_i values and the $\alpha_i(\lambda)$ values. The τ_i values are assumed to be independent of emission wavelength. In the case of global analysis, the calculation of χ_R^2 extends over several data sets. The global value of χ_R^2 is given by

$$\chi_R^2 = \frac{1}{\nu} \sum_{\lambda} \sum_{k=1}^{n} \frac{[I_c^\lambda(t_k) - I^\lambda(t_k)]^2}{I^\lambda(t_k)} \qquad [4.41]$$

where the additional sum extends over the files measured at each wavelength (λ). For the fitted functions, the α_i values are different at each wavelength [$\alpha_i(\lambda)$] because of the different relative contributions of the three fluorophores. The values of τ_i are assumed to be independent of emission wavelength since each fluorophore is assumed to display a single-exponential decay.

It is easy to see how global analysis can improve resolution. Suppose one of the intensity decays was measured at 320 nm. This decay would be almost completely due to indole (Figure 4.37), which thus would determine its lifetime without contribution from the other fluorophores. Since there is only one decay time, there would be no parameter correlation, and τ_1 would be determined with good certainty. The data at 320 nm will constrain the lifetime of indole in data measured at longer wavelengths. Similarly, even if the choice of wavelengths only partially selects for a given fluorophore, the data serve to determine its decay time and reduce the uncertainty in the remaining parameters.

Global analysis was used to recover the lifetimes across the emission spectrum of the three-component mixture

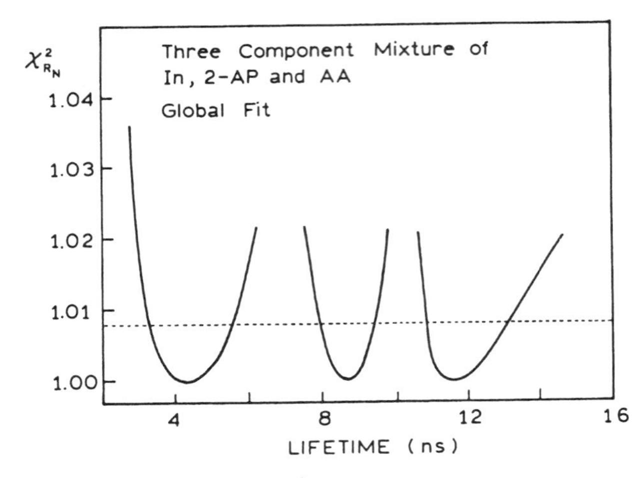

Figure 4.40. Global lifetime χ_R^2 surface for the three-component mixture of IN, AA, and 2AP in Figure 4.39. From Ref. 161.

shown in Figure 4.37, using the decays measured from 360 to 440 nm. The lifetime χ_R^2 surface for the three decay times is shown in Figure 4.40. The expected decay time was recovered for each of the components. However, even with a multiwavelength global analysis the uncertainties are significant. For instance, the value of the 4.1-ns lifetime can range from about 3.2 to 5.5 ns and still be consistent with the data.

4.12. REPRESENTATIVE INTENSITY DECAYS

The concepts described in the preceding sections can be made more understandable by examination of some specific examples.

4.12.A. Intensity Decay for a Single-Tryptophan Protein

The Tet repressor controls the gene in gram-negative bacteria that provides resistance to the antibiotic tetracycline.[179] This protein usually contains two tryptophans, but a mutant protein was engineered which contains a single tryptophan residue at position 43. Intensity decays are shown in Figure 4.41. The light source was a frequency-doubled R6G dye laser at 590 nm, frequency-doubled to 295 nm. The dye laser was cavity-dumped at 80 kHz. The excitation was vertically polarized, and the emission detected through a polarizer set 54.7° from the vertical. The use of magic-angle polarization conditions is essential in this case because the protein can be expected to rotate on a timescale comparable to the intensity decay. A Schott WG 320 filter was used in front of the monochromator to prevent scattered light from entering the monochromator, which was set at 360 nm.

The emission was detected with an XP 2020 PMT. This PMT shows a wavelength-dependent time response, although less than that shown for a related PMT, the 56 DUVP, in Figures 4.21 and 4.22. To avoid color effects, the authors used a short-lifetime reference which shifted the wavelength to the measurement wavelength with minimal time delay.[104] This was accomplished with a solution of *p*-terphenyl highly quenched by CCl_4. The fact that the measurements were performed with a dynode PMT is evident from the width of the impulse response function, which appears to be near 500 ps. Some of this width may be contributed by the short-lifetime standard.

The intensity decay was fit to the single-, double-, and triple-exponential models, resulting in χ_R^2 values of 17, 1.6, and 1.5, respectively. Rejection of the single-exponential model is clearly justified by the data. However, it is less clear that three decay times are needed. The ratio of the χ_R^2 values is 1.07, which is attributable to random error with a probability of over 20% (Table 4.3). In fact, the fractional amplitude of the third component was less than 1%, and the authors accepted the double-exponential fit as descriptive of their protein.

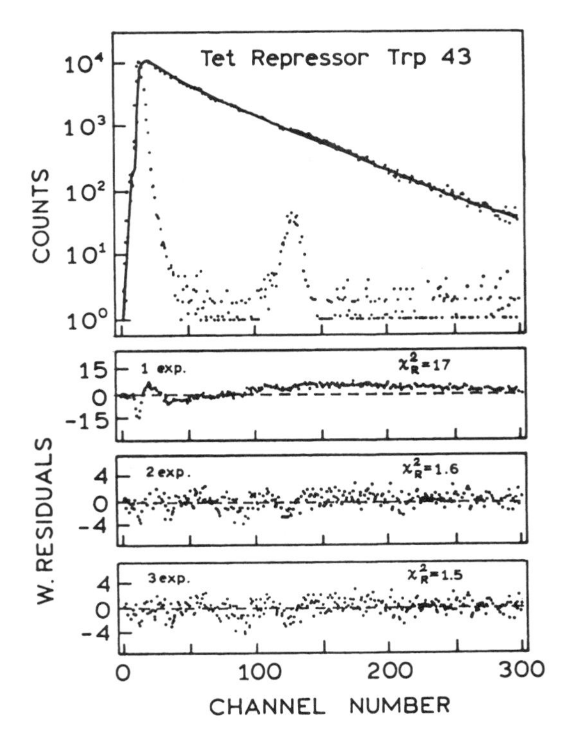

Figure 4.41. Intensity decay of trp-43 in the Tet repressor protein F75 TetR at 360 nm. The calibration is 108 ps/channel. Revised and reprinted with permission from Ref. 179. Copyright © 1992, American Chemical Society.

4.12.B. Green Fluorescent Protein—Systematic Errors in the Data

Green fluorescent protein (GFP) spontaneously becomes fluorescent following synthesis of its amino acid chain. Hence, it is being widely used as a tag to follow gene expression. The intensity decay of GFP was measured with 365-nm excitation, a 1.25-MHz repetition rate, and magic-angle polarizer conditions.[180] The emission was detected above 500 nm, using an MCP PMT.

The intensity decay of GFP could be well fit to a single exponential (Figure 4.42). The value of χ_R^2 is somewhat elevated and not consistent with a single-exponential model. However, the value of χ_R^2 was not decreased by including a second decay time ($\chi_R^2 = 1.18$). Examination of the deviations (Figure 4.42, bottom panel) reveals the presence of a systematic oscillation for which a second decay time does not improve the fit. The failure of χ_R^2 to decrease is typically an indication of systematic error as the origin of the elevated value of χ_R^2.

4.12.C. Erythrosin B—A Picosecond Decay Time

Even with a picosecond laser light source and an MCP PMT, the measurement of short decay times remains challenging. This is illustrated by the intensity decay of erythrosin B in water, shown in Figure 4.43.[181] The width of the instrument response function is seen to be near 100

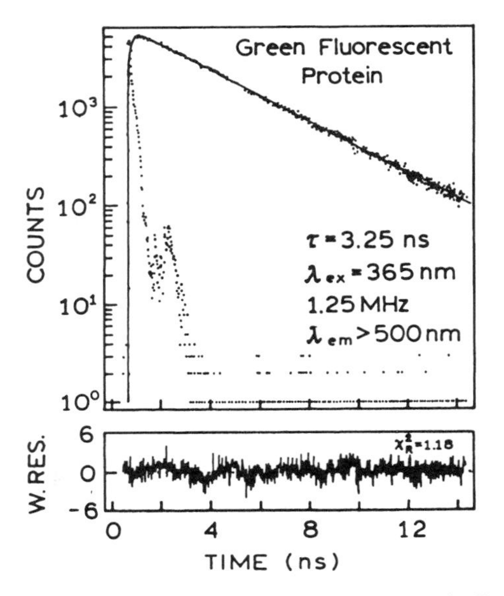

Figure 4.42. Intensity decay of green fluorescent protein. From Ref. 180.

ps. In spite of this fast response, the intensity decay of erythrosin B is still strongly convoluted with the instrument response function. This is because its decay time is only 86 ps. However, the procedure of iterative reconvolution allows the decay time of erythrosin B to be measured in spite of the extensive overlap between the data and instrument response function.

4.12.D. Chlorophyll Aggregates in Hexane

The intensity decay for the tryptophan residues in the Tet repressor was relatively simple. Intensity decays can be much more heterogeneous. One example is the intensity decay of chlorophyll in wet hexane solvents, in which chlorophyll exists in a variety of aggregated states.[182] Data were obtained using a pyridine dye laser at 760 nm, which was cavity-dumped at 1 MHz and frequency-doubled to 380 nm. The emission was detected at 715 nm through an interference filter. The detector was an R2809 MCP PMT, with 6-μm channels. Even though the excitation and emission wavelengths were far apart (380 and 715 nm), color effect corrections did not seem necessary with this MCP PMT. Magic-angle polarizer conditions were used.

The intensity decay of chlorophyll was strongly heterogeneous (Figure 4.44). The decay could not even be approximated by a single decay time. The fit with two decay times was much improved, reducing χ_R^2 from 52.3 to 1.49. A further reduction of 40% in χ_R^2 occurred for the three-decay-time fit. Hence, the two-decay-time model can be rejected because the χ_R^2 ratio of 1.42 would only occur between 1 and 5% of the time due to statistical errors in the data (Table 4.3). Complex intensity decays with up to four lifetimes have been reported for photosynthetic systems.[183,184]

4.12.E. Intensity Decay of FAD

FAD is a cofactor in many enzymatic reactions. The fluorescent moiety is the flavin, which can be quenched on contact with the adenine. In solution, FAD can exist in an open or a stacked configuration. It is known that a significant amount of quenching occurs because cleavage of FAD with phosphodiesterase results in a severalfold increase in fluorescence intensity.

The nature of the flavin quenching by the adenine was studied by TCSPC.[185] Data were obtained using the output of a mode-locked argon-ion laser at 457.9 nm. The detector was an XP 2020 linear-focused PMT, resulting in a relatively wide instrument response function (Figure 4.45). The intensity decay of the flavin alone (FMN) was found to be a single exponential, with a decay time of 4.89 ns. FAD displayed a double-exponential decay with a component of 3.38 ns ($\alpha_1 = 0.46$) and of 0.12 ns ($\alpha_2 = 0.54$). The short-decay-time component was assigned to the stacked forms, allowing calculation of the fractions of FAD present

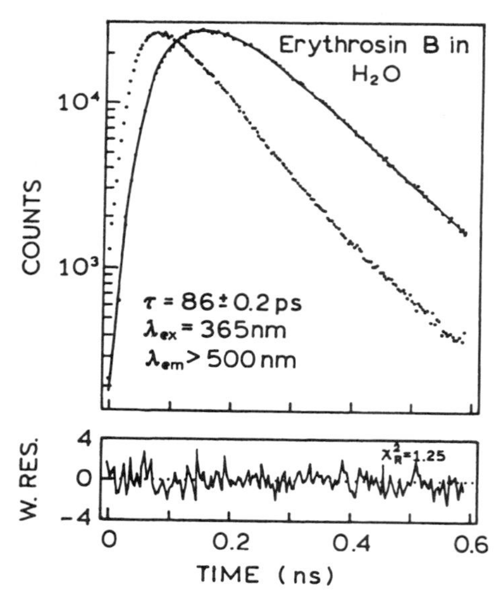

Figure 4.43. Intensity decay of erythrosin B in water. Excitation was obtained from a pyridine 1 dye laser. Measurements were performed with magic-angle polarizer conditions. The solid curve is the best one-decay-time fit. From Ref. 181.

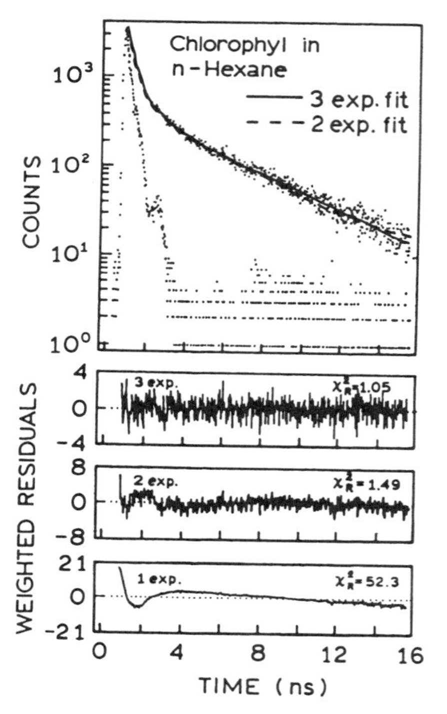

Figure 4.44. Intensity decay of chlorophyll in wet *n*-hexane. Revised from Ref. 182.

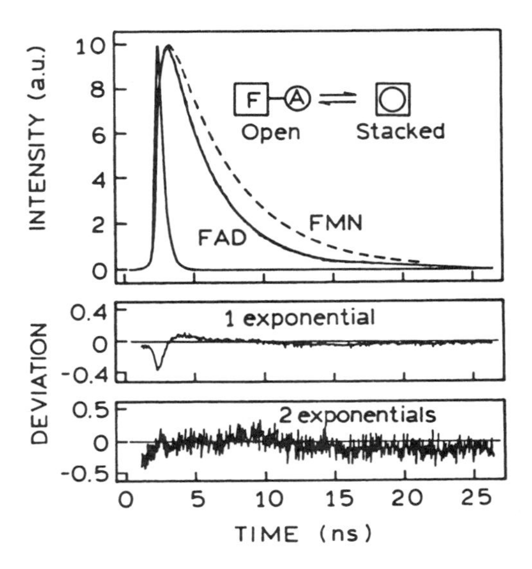

Figure 4.45. Intensity decays of FAD and FMN at pH 7.5, 3 °C. Also shown is the laser pulse profile. The deviations are for fits to the FAD intensity decay. Data from Ref. 185.

in the stacked and open conformations. The lifetime of 3.38 ns is thought to be due to dynamic quenching of the flavin by the adenine moiety.

4.12.F. Microsecond Luminescence Decays

If the decay times become longer, above several hundred nanoseconds, the complexity of TCSPC is no longer necessary. A luminescence intensity decay of a long-lifetime metal–ligand complex in an SiO_2 sol-gel is shown in Figure 4.46.[186] The data were fit to a stretched exponential (Eq. [4.39]) with $\tau = 1.18$ μs and $\beta = 0.843$. These data were obtained by direct digitization of the time-dependent intensity with a digital oscilloscope. The light source was a low-repetition-rate N_2 laser which pumped a coumarin 460 dye laser. The emission was observed continuously with an R928 side-window PMT coupled to a monochromator. The width of the instrument response function was near 14 ns. Because of the long lifetime, relative to the width of the instrument response function, it was not necessary to use deconvolution. The intensity values after the pulse were used for analysis. Because of the high-intensity multiphoton nature of the signal, an average of 100 responses was adequate to quantity the intensity decay.

4.12.G. Subpicosecond Intensity Decays

As the intensity decays become shorter, the measurements become more complex than TCSPC. This may be illustrated by the intensity decays of Nile Blue in aniline.

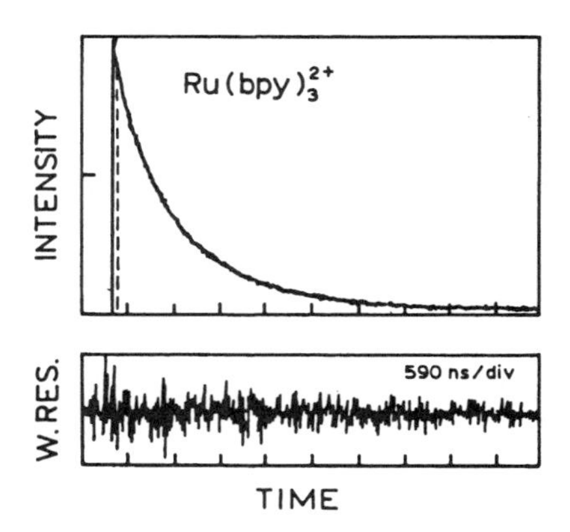

Figure 4.46. Luminescence decay of $Ru(bpy)_3(PF_6)_2$ in an SiO_2 aquogel. Data are fit to a stretched exponential. Revised from Ref. 186.

Dimethylaniline (DMA) quenches Nile Blue by an electron-transfer process, which in neat DMA occurs on a picosecond timescale.[187,188]

Intensity decays were recorded by the upconversion method (Figure 4.47). The excitation source was a special R6G dye laser which provided 70-fs pulses near 600 nm. The width of the instrument response function was less than 1 ps. Emission was detected by mixing the emission with a longer-wavelength dye laser pulse and detecting the harmonics. The intensity decay was obtained by scanning the time delay of the probe pulse. The intensity decay contained components from 0.47 to 5.2 ps. Such time resolution on the subpicosecond timescale is not available from TCSPC.

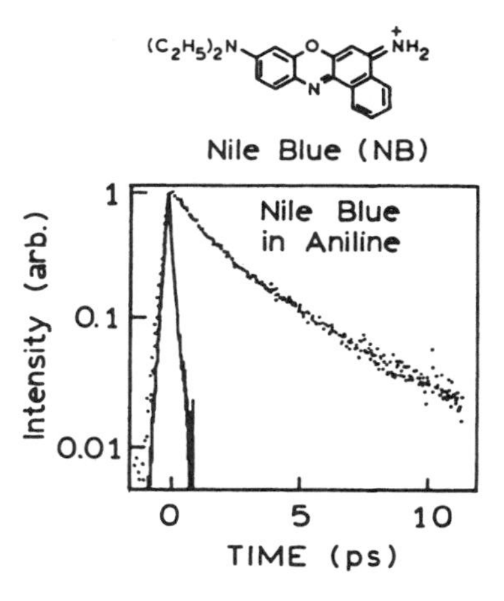

Figure 4.47. Intensity decay of Nile Blue in aniline measured with a streak camera. The decay has components of 470 fs, 1.8 ps, and 5.2 ps. Revised from Ref. 188.

4.13. CLOSING COMMENTS

It may not be evident to the reader, but I have attempted to keep this chapter as brief as possible. There have been numerous developments in the technology since 1983, and I felt these needed to be mentioned for a complete overview of the technology for time-resolved fluorescence. Numerous additional details could have, and perhaps should have, been described. It is hoped that this chapter provides the basics for understanding the extensive literature on time-resolved fluorescence.

REFERENCES

1. Bevington, P. R., 1969, *Data Reduction and Error Analysis for the Physical Sciences*, McGraw-Hill, New York.
2. Lakowicz, J. R., 1996, Fluorescence spectroscopy of biomolecules, in *Encyclopedia of Molecular Biology and Biotechnology*, R. A. Meyers (ed.), VCH Publishers, Weinhein, Germany, pp. 294–306.
3. Grinvald, A., and Steinberg, I. Z., 1974, On the analysis of fluorescence decay kinetics by the method of least-squares, *Anal. Biochem.* **59**:583–593.
4. Demas, J. N., 1983, *Excited State Lifetime Measurements*, Academic Press, New York.
5. Johnson, M. L., 1985, The analysis of ligand binding data with experimental uncertainties in the independent variables, *Anal. Biochem.* **148**:471–478.
6. Bard, J., 1974, *Nonlinear Parameter Estimation*, Academic Press, New York.
7. Johnson, M. L., 1983, Evaluation and propagation of confidence intervals in nonlinear, asymmetrical variance spaces: Analysis of ligand binding data, *Biophys. J.* **44**:101–106.
8. O'Connor, D. V., and Phillips, D., 1984, *Time-Correlated Single Photon Counting*, Academic Press, New York.
9. Birch, D. J. S., and Imhof, R. E., 1991, Time-domain fluorescence spectroscopy using time-correlated single-photon counting, in *Topics in Fluorescence Spectroscopy, Volume 1, Techniques*, J. R. Lakowicz (ed.), Plenum Press, New York, pp. 1–95.
10. Ware, W. R., 1971, Transient luminescence measurements, in *Creation and Detection of the Excited State*, Vol. 1A, A. A. Lamola (ed.), Marcel Dekker, New York, pp. 213–302.
11. Malak, H., unpublished observations.
12. Badea, M. G., and Brand, L., 1971, Time-resolved fluorescence measurements, *Methods in Enzymol.* **61**:378–425.
13. Svelto, O., 1998, *Principles of Lasers*, 4th edition, Translated by David C. Hanna. Plenum Press, New York.
14. Yariv, A., 1989, *Quantum Electronics*, 3rd edition, John Wiley & Sons, New York.
15. Iga, K., 1994, *Fundamentals of Laser Optics*, Plenum Press, New York.
16. Small, E. W., 1991, Laser sources and microchannel plate detectors for pulse fluorometry, in *Topics in Fluorescence Spectroscopy, Volume 1, Techniques*, J. R. Lakowicz (ed.), Plenum Press, New York, pp. 97–182.
17. Wilson, J., and Hawkes, J. F. B., 1983, *Optoelectronics—An Introduction*, Prentice-Hall, Englewood Cliffs, New Jersey.
18. Berg, N. J., and Lee, J. N. (eds.), 1983, *Acousto-Optic Signal Processing*, Marcel Dekker, New York.
19. Visser, A. J. W. G., and Van Hoek, A., 1979, The measurement of subnanosecond fluorescence decay of flavins using time-correlated photon counting and a mode-locked Ar Ion laser, *J. Biochem. Biophys. Methods.* **1**:195–208.
20. Spears, K. G., Cramer, L. E., and Hoffland, L. D., 1978, Subnanosecond time-correlated photon counting with tunable lasers, *Rev. Sci. Instrum.* **49**:255–262.
21. Lytle, E., and Kelsey, M. S., 1974, Cavity-dumped argon-ion laser as an excitable source on time-resolved fluorimetry, *Anal. Chem.* **46**:855–860.
22. Wild, U. P., Holzwarth, A. R., and Good, H. P., 1977, Measurement and analysis of fluorescence decay curves, *Rev. Sci. Instrum.* **48**:1621–1627.
23. Turko, B. T., Nairn, J. A., and Sauer, K., 1983, Single photon timing system for picosecond fluorescence lifetime measurements, *Rev. Sci. Instrum.* **54**:118–120.
24. Alfano, A. J., Fong, F. K., and Lytle, F. E., 1983, High repetition rate subnanosecond gated photon counting, *Rev. Sci. Instrum.* **54**:967–972.
25. Kinoshita, S., Ohta, H., and Kushida, T., 1981, Subnanosecond fluorescence lifetime measuring system using single photon counting method with mode-locked laser excitation, *Rev. Sci. Instrum.* **52**:572–575.
26. Koester, V. J., and Dowben, R. M., 1978, Subnanosecond single photon counting fluorescence spectroscopy using synchronously pumped tunable dye laser excitation, *Rev. Sci. Instrum.* **49**:1186–1191.
27. Zimmerman, H. E., Penn, J. H., and Carpenter, C. W., 1982, Evaluation of single-photon counting measurements of excited-state lifetimes, *Proc. Natl. Acad. Sci. U.S.A.* **79**:2128–2132.
28. van Hoek, A., Vervoort, J., and Visser, A. J. W. G., 1983, A subnanosecond resolving spectrofluorimeter for the analysis of protein fluorescence kinetics, *J. Biochem. Biophys. Methods* **7**:243–254.
29. Small, E. W., Libertini, L. J., and Isenberg, I., 1984, Construction and tuning of a monophoton decay fluorometer with high-resolution capabilities, *Rev. Sci. Instrum.* **55**:879–885.
30. Visser, A. J. W. G., and van Hoek, A., 1981, The fluorescence decay of reduced nicotinamides in aqueous solution after excitation with a UV-mode locked Ar Ion laser, *Photochem. Photobiol.* **33**:35–40.
31. Libertini, L. J., and Small, E. W., 1987, On the choice of laser dyes for use in exciting tyrosine fluorescence decays, *Anal. Biochem.* **163**:500–505.
32. Laws, W. R., and Sutherland, J. C., 1986, The time-resolved photon-counting fluorometer at the national synchrotron light source, *Photochem. Photobiol.* **44**:343–348.
33. Munro, I. H., and Martin, M. M., 1991, Time-resolved fluorescence spectroscopy using synchrotron radiation, in *Topics in Fluorescence Spectroscopy, Volume 1, Techniques*, J. R. Lakowicz (ed.), Plenum Press, New York, pp. 261–291.
34. Munro, I. H., and Schwentner, N., 1983, Time resolved spectroscopy using synchrotron radiation, *Nucl. Instrum. Methods* **208**:819–834.
35. Lopez-Delgado, R., 1978, Comments on the application of synchrotron radiation to time-resolved spectrofluorometry, *Nucl. Instrum. Methods* **152**:247–253.
36. Rehn, V., 1980, Time-resolved spectroscopy in synchrotron radiation, *Nucl. Instrum. Methods* **177**:193–205.
37. van Der Oord, C. J. R., Gerritsen, H. C., Rommerts, F. F. G., Shaw, D. A., Munro, I. H., and Levine, Y. K., 1995, Micro-volume time-resolved fluorescence spectroscopy using a confocal synchrotron radiation microscope, *Appl. Spectrosc.* **49**:1469–1473.
38. Malmberg, J. H., 1957, Millimicrosecond duration of light source, *Rev. Sci. Instrum.* **28**:1027–1029.

39. Bennett, R. G., 1960, Instrument to measure fluorescence lifetimes in the millimicrosecond region, *Rev. Sci. Instrum.* **31:**1275–1279.
40. Yguerabide, J., 1965, Generation and detection of subnanosecond light pulses: Application to luminescence studies, *Rev. Sci. Instrum.* **36:**1734–1742.
41. Birch, D. J. S., and Imhof, R. E., 1977, A single photon counting fluorescence decay-time spectrometer, *J. Phys. E: Sci. Instrum.* **10:**1044–1049.
42. Lewis, C., Ware, W. R., Doemeny, L. J., and Nemzek, T. L., 1973, The measurement of short lived fluorescence decay using the single photon counting method, *Rev. Sci. Instrum.* **44:**107–114.
43. Leskovar, B., Lo, C. C., Hartig, P. R., and Sauer, K., 1976, Photon counting system for subnanosecond fluorescence lifetime measurements, *Rev. Sci. Instrum.* **47:**1113–1121.
44. Bollinger, L. M., and Thomas, G. E., 1961, Measurement of the time dependence of scintillation intensity by a delayed-coincidence method, *Rev. Sci. Instrum.* **32:**1044–1050.
45. Hazan, G., Grinvald, A., Maytal, M., and Steinberg, I. Z., 1974, An improvement of nanosecond fluorimeters to overcome drift problems, *Rev. Sci. Instrum.* **45:**1602–1604.
46. Dreeskamp, H., Salthammer, T., and Laufer, A. G. E., 1989, Time-correlated single-photon counting with alternate recording of excitation and emission, *J. Lumin.* **44:**161–165.
47. Birch, D. J. S., and Imhof, R. E., 1981, Coaxial nanosecond flash-lamp, *Rev. Sci. Instrum.* **52:**1206-1212.
48. Birch, D. J. S., Hungerford, G., and Imhof, R. E., 1991, Near-infrared spark source excitation for fluorescence lifetime measurements, *Rev. Sci. Instrum.* **62:**2405–2408.
49. Birch, D. J. S., Hungerford, G., Nadolski, B., Imhof, R. E., and Dutch, A., 1988, Time-correlated single-photon counting fluorescence decay studies at 930 nm using spark source excitation, *J. Phys. E: Sci. Instrum.* **21:**857–862.
50. Miller, K. J., and Lytle, F. E., 1993, Capillary zone electrophoresis with time-resolved fluorescence detection using a diode-pumped solid-state laser, *J. Chromatogr.* **648:**245–250.
51. Picosecond Fluorescence Lifetime Measurement System, Hamamatsu Literature, Catalog No. SSCS1018E02, Nov/91NB.
52. Thompson, R. B., Frisoli, J. K., and Lakowicz, J. R., 1992, Phase fluorometry using a continuously modulated laser diode, *Anal. Chem.* **64:**2075–2078.
53. Berndt, K. W., Gryczynski, I., and Lakowicz, J. R., 1990, Phase-modulation fluorometry using a frequency-doubled pulsed laser diode light source, *Rev. Sci. Instrum.* **61:**1816–1820.
54. Gedcke, D. A., and McDonald, W. J., 1967, A constant fraction of pulse height trigger for optimum time resolution, *Nucl. Instrum. Methods* **55:**377–380.
55. Gedcke, D. A., and McDonald, W. J., 1966, Design of the constant fraction of pulse height trigger for optimum time resolution, *Nucl. Instrum. Methods* **58:**253–260.
56. Arbel, A., Klein, I., and Yarom, A., 1974, Snap-off constant fraction timing discriminators, *IEEE Trans. Nucl. Sci.* **NS-21:**3–8.
57. Cova, S., Ghioni, M., Zappa, F., and Lacaita, A., 1993, Constant-fraction circuits for picosecond photon timing with microchannel plate photomultipliers, *Rev. Sci. Instrum.* **64:**118–124.
58. Cova, S., Ripamonti, G., and Lacaita, A., 1990, New double constant-fraction trigger circuit for locking on laser pulse trains up to 100 MHz, *Rev. Sci. Instrum.* **61:**1004–1009.
59. Cova, S., and Ripamonti, G., 1990, Improving the performance of ultrafast microchannel plate photomultipliers in time-correlated photon counting by pulse pre-shaping, *Rev. Sci. Instrum.* **61:**1072–1075.
60. Haugen, G. R., Wallin, B. W., and Lytle, F. E., 1979, Optimization of data-acquisition rates in time-correlated single-photon fluorimetry, *Rev. Sci. Instrum.* **50:**64–72.
61. Bowman, L. E., Berglund, K. A., and Nocera, D. G., 1993, A single photon timing instrument that covers a broad temporal range in the reversed timing configuration, *Rev. Sci. Instrum.* **64:**338–341.
62. Baumier, W., Schmalzl, A. X., Göβl, G., and Penzkofer, A., 1992, Fluorescence decay studies applying a cw femtosecond dye laser pumped ungated inverse time-correlated single photon counting system, *Meas. Sci. Technol.* **3:**384–393.
63. Harris, C. M., and Selinger, B. K., 1979, Single-photon decay spectroscopy. II The pileup problem, *Aust. J. Chem.* **32:**2111–2129.
64. Williamson, J. A., Kendall-Tobias, M. W., Buhl, M., and Seibert, M., 1988, Statistical evaluation of dead time effects and pulse pileup in fast photon counting. Introduction of the sequential model, *Anal. Chem.* **60:**2198–2203.
65. Koyama, K., and Fatlowitz, D., 1987, Application of MCP-PMTs to time correlated single photon counting and related procedures, Hamamatsu Technical Information, No. ET-03, pp. 1–18.
66. Howorth, J. R., Ferguson, I., and Wilcox, D., 1995, Developments in microchannel plate photomultipliers, *Proc. SPIE* **2388:**356–362.
67. Beechem, J. M., 1992, Multi-emission wavelength picosecond time-resolved fluorescence decay data obtained on the millisecond scale: Application to protein:DNA interactions and protein folding reactions, *Proc. SPIE* **1640:**676–680.
68. Birch, D. J. S., Holmes, A. S., Imhof, R. E., Nadolski, B. Z., and Cooper, J. C., 1988, Multiplexed time-correlated single photon counting, *Proc. SPIE* **909:**8–14.
69. Birch, D. J. S., McLoskey, D., Sanderson, A., Suhling, K., and Holmes, A. S., 1994, Multiplexed time-correlated single-photon counting, *J. Fluoresc.* **4**(1):91–102.
70. McLoskey, D., Birch, D. J. S., Sanderson, A., Suhling, K., Welch, E., and Hicks, P. J., 1996, Multiplexed single-photon counting. I. A time-correlated fluorescence lifetime camera, *Rev. Sci. Instrum.* **67:**2228–2237.
71. Suhling, K., McLoskey, D., and Birch, D. J. S., 1996, Multiplexed single-photon counting. II. The statistical theory of time-correlated measurements, *Rev. Sci. Instrum.* **67:**2238–2246.
72. Erdmann, R., Becker, W., Ortmann, U., and Enderlein, J., 1995, Simultaneous detection of time-resolved emission spectra using a multianode-PMT and new TCSPC-electronics with 5 MHz count rate, *Proc. SPIE* **2388:**330–334.
73. Candy, B. H., 1985, Photomultiplier characteristics and practice relevant to photon counting, *Rev. Sci. Instrum.* **56:**183–193.
74. Hungerford, G., and Birch, D. J. S., 1996, Single-photon timing detectors for fluorescence lifetime spectroscopy, *Meas. Sci. Technol.* **7:**121–135.
75. Leskovar, B., 1977, Microchannel plates, *Phys. Today* **1977:**42–49.
76. Boutot, J. P., Delmotte, J. C., Miehé, J. A., and Sipp, B., 1977, Impulse response of curved microchannel plate photomultipliers, *Rev. Sci. Instrum.* **48:**1405–1407.
77. Timothy, J. G., and Bybee, R. L., 1977, Preliminary results with microchannel array plates employing curved microchannels to inhibit ion feedback, *Rev. Sci. Instrum.* **48:**292–299.
78. Lo, C. C., and Leskovar, B., 1981, Performance studies of high gain photomultiplier having z-configuration of microchannel plates, *IEEE Trans. Nucl. Sci.* **NS-28:**698–704.
79. Ito, M., Kume, H., and Oba, K., 1984, Computer analysis of the timing properties in micro channel plate photomultiplier tubes, *IEEE Trans. Nucl. Sci.* **NS-31:**408–412.
80. Bebelaar, D., 1986, Time response of various types of photomultipliers and its wavelength dependence in time-correlated single

photon counting with an ultimate resolution of 47 ps FWHM, *Rev. Sci. Instrum.* **57**:1116–1125.

81. Yamazaki, I., Tamai, N., Kume, H., Tsuchiya, H., and Oba, K., 1985, Microchannel plate photomultiplier applicability to the time-correlated photon-counting method, *Rev. Sci. Instrum.* **56**:1187–1194.
82. Uyttenhove, J., Demuynck, J., and Deruytter, A., 1978, Application of a microchannel plate photomultiplier in subnanosecond lifetime measurements, *IEEE Trans. Nucl. Sci.* **NS-25**:566–567.
83. Murao, T., Yamazaki, I., Shindo, Y., and Yoshihara, K., 1982, A subnanosecond time-resolved spectrophotometric system by using synchronously pumped, mode-locked dye laser, *J. Spectrosc. Soc. Jpn.* **1982**:96–103.
84. Murao, T., Yamazaki, I., and Yoshihara, K., 1982, Applicability of a microchannel plate photomultiplier to the time-correlated photon counting technique, *Appl. Opt.* **21**:2297–2298.
85. Boens, N., Tamai, N., Yamazaki, I., and Yamazaki, T., 1990, Picosecond single photon timing measurements with a proximity type microchannel plate photomultiplier and global analysis with reference convolution, *Photochem. Photobiol.* **52**:911–917.
86. Beck, G., 1976, Operation of a 1P28 photomultiplier with subnanosecond response time, *Rev. Sci. Instrum.* **47**:537–541.
87. Kinoshita, S., and Kushida, T., 1982, High-performance, time-correlated single photon counting apparatus using a side-on type photomultiplier, *Rev. Sci. Instrum.* **53**:469–472.
88. Canonica, S., Forrer, J., and Wild, U. P., 1985, Improved timing resolution using small side-on photomultipliers in single photon counting, *Rev. Sci. Instrum.* **56**:1754–1758.
89. Ware, W. R., Pratinidhi, M., and Bauer, R. K., 1983, Performance characteristics of a small side-window photomultiplier in laser single-photon fluorescence decay measurements, *Rev. Sci. Instrum.* **54**:1148–1156.
90. Cova, S., Longoni, A., Andreoni, A., and Cubeddu, R., 1983, A semiconductor detector for measuring ultraweak fluorescence decays with 70ps FWHM resolution, *IEEE J. Quantum Electron.* **QE-19**:630–634.
91. Buller, G. S., Massa, J. S., and Walker, A. C., 1992, All solid-state microscope-based system for picosecond time-resolved photoluminescence measurements on II–VI semiconductors, *Rev. Sci. Instrum.* **63**:2994–2998.
92. Louis, T. A., Ripamonti, G., and Lacaita, A., 1990, Photoluminescence lifetime microscope spectrometer based on time-correlated single-photon counting with an avalanche diode detector, *Rev. Sci. Instrum.* **61**:11–22.
93. Cova, S., Ripamonti, G., and Lacaita, A., 1987, Avalanche semiconductor detector for single optical photons with a time resolution of 60 ps, *Nucl. Instrum. Methods Phys. Res.* **A253**:482–487.
94. Cova, S., Lacaita, A., Ghioni, M., Ripamonti, G., and Louis, T. A., 1989, 20-ps timing resolution with single-photon avalanche diodes, *Rev. Sci. Instrum.* **60**:1104–1110.
95. Cova, S., Longoni, A., and Andreoni, A., 1981, Towards picosecond resolution with single-photon avalanche diodes, *Rev. Sci. Instrum.* **52**:408–412.
96. Louis, T., Schatz, G. H., Klein-Bölting, P., Holzwarth, A. R., Ripamonti, G., and Cova, S., 1988, Performance comparison of a single-photon avalanche diode with a microchannel plate photo-multiplier in time-correlated single-photon counting, *Rev. Sci. Instrum.* **59**:1148–1152.
97. Lacaita, A., Cova, S., and Ghioni, M., 1988, Four-hundred picosecond single-photon timing with commerically available avalanche photodiodes, *Rev. Sci. Instrum.* **59**:1115–1121.
98. Wahl, P., Auchet, J. C., and Donzel, B., 1974, The wavelength dependence of the response of a pulse fluorometer using the single photoelectron counting method, *Rev. Sci. Instrum.* **45**:28–32.
99. Sipp, B., Miehe, J. A., and Lopez-Delgado, R., 1976, Wavelength dependence of the time resolution of high-speed photomultipliers used in single-photon timing experiments, *Opt. Commun.* **16**:202–204.
100. Rayner, D. M., McKinnon, A. F., and Szabo, A. G., 1978, Confidence in fluorescence lifetime determinations: A ratio correction for the photomultiplier time response variation with wavelength, *Can. J. Chem.* **54**:3246–3259.
101. Thompson, R. B., and Gratton, E., 1988, Phase fluorometric method for determination of standard lifetimes, *Anal. Chem.* **60**:670–674.
102. Meister, E. C., Wild, U. P., Klein-Bölting, P., and Holzwarth, A. R., 1988, Time response of small side-on photomultiplier tubes in time-correlated single-photon counting measurements, *Rev. Sci. Instrum.* **59**:499–501.
103. Bauer, R. K., and Balter, A., 1979, A method of avoiding wavelength-dependent errors in decay-time measurements, *Opt. Commun.* **28**:91–96.
104. Kolber, Z. S., and Barkley, M. D., 1986, Comparison of approaches to the instrumental response function in fluorescence decay measurements, *Anal. Biochem.* **152**:6–21.
105. Vecer, J., Kowalczyk, A. A., Davenport, L., and Dale, R. E., 1993, Reconvolution analysis in time-resolved fluorescence experiments—an alternative approach: Reference-to-excitation-to-fluorescence reconvolution, *Rev. Sci. Instrum.* **64**:3413–3424.
106. Kilin, S. F., 1962, The duration of photo- and radioluminescence of organic compounds, *Opt. Spectrosc.* **12**:414–416.
107. Mauzerall, D., Ho, P. P., and Alfano, R. F., 1985, The use of short lived fluorescent dyes to correct for artifacts in the measurements of fluorescence lifetimes, *Photochem. Photobiol.* **42**:183–186.
108. Van Den Zegel, M., Boens, N., Daems, D., and De Schryver, F. C., 1986, Possibilities and limitations of the time-correlated single photon counting technique: A comparative study of correction methods for the wavelength dependence of the instrument response function, *Chem. Phys.* **101**:311–335.
109. James, D. R., Demmer, D. R. M., Verrall, R. E., and Steer, R. P., 1983, Excitation pulse-shape mimic technique for improving picosecond-laser excited time-correlated single-photon counting deconvolutions, *Rev. Sci. Instrum.* **54**:1121–1130.
110. Zuker, M., Szabo, A. G., Bramall, L., Krajcarski, D. T., and Selinger, B., 1985, Delta function convolution method (DFCM) for fluorescence decay experiments, *Rev. Sci. Instrum.* **56**:14–22.
111. Castelli, F., 1985, Determination of correct reference fluorescence lifetimes by self-consitent internal calibration, *Rev. Sci. Instrum.* **56**:538–542.
112. Vos, K., van Hoek, A., and Visser, A. J. W. G., 1987, Application of a reference convolution method to tryptophan fluorescence in proteins, *Eur. J. Biochem.* **165**:55–63.
113. Martinho, J. M. G., Egan, L. S., and Winnik, M. A., 1987, Analysis of the scattered light component in distorted fluorescence decay profiles using a modified delta function convolution method, *Anal. Chem.* **59**:861–864.
114. Ricka, J., 1981, Evaluation of nanosecond pulse-fluorometry measurements—no need for the excitation function, *Rev. Sci. Instrum.* **52**:195–199.
115. Visser, A. J. W. G., Kulinski, T., and van Hoek, A., 1988, Fluorescence lifetime measurements of pseudoazulenes using picosecond-resolved single photon counting, *J. Mol. Struct.* **175**:111–116.

116. Holtom, G. R., 1990, Artifacts and diagnostics in fast fluorescence measurements, *Proc. SPIE* **1204**:2–12.
117. Grinvald, A., 1976, The use of standards in the analysis of fluorescence decay data, *Anal. Biochem.* **75**:260–280.
118. Lampert, R. A., Chewter, L. A., Phillips, D., O'Connor, D. V., Roberts, A. J., and Meech, S. R., 1983, Standards for nanosecond fluorescence decay time measurements, *Anal. Chem.* **55**:68–73.
119. Schiller, N. H., and Alfano, R. R., 1980, Picosecond characteristics of a spectrograph measured by a streak camera/video readout system, *Opt. Commun.* **35**(3):451–454.
120. Rubin, B., and Herman, R. M., 1981, Monochromators as light stretchers, *Am. J. Phys.* **49**:868–871.
121. Imhof, R. E., and Birch, D. J. S., 1982, Distortion of gaussian pulses by a diffraction grating, *Opt. Commun.* **42**(2):83–86.
122. Saari, P., Aaviksoo, J., Freiberg, A., and Timpmann, K., 1981, Elimination of excess pulse broadening at high spectral resolution of picosecond duration light emission, *Opt. Commun.* **39**(1,2):94–98.
123. Bebelaar, D., 1986, Compensator for the time dispersion in a monochromator, *Rev. Sci. Instrum.* **57**:1686–1687.
124. Bhaumik, M. L., Clark, G. L., Snell, J., and Ferder, L., 1965, Stroboscopic time-resolved spectroscopy, *Rev. Sci. Instrum.* **36**:37–40.
125. Barisas, B. G., and Leuther, M. D., 1980, Grid-gated photomultiplier photometer with subnanosecond time response, *Rev. Sci. Instrum.* **51**:74–78.
126. Steingraber, O. J., and Berlman, I. B., 1963, Versatile technique for measuring fluorescence decay times in the nanosecond region, *Rev. Sci. Instrum.* **34**:524–529.
127. Hundley, L., Coburn, T., Garwin, E., and Stryer, L., 1967, Nanosecond fluorimeter, *Rev. Sci. Instrum.* **38**:488–492.
128. James, D. R., Siemiarczuk, A., and Ware, W. R., 1992, Stroboscopic optical boxcar technique for the determination of fluorescence lifetimes, *Rev. Sci. Instrum.* **63**:1710–1716.
129. Nordlund, T. M., 1991, Streak camera for time-domain fluorescence, in *Topics in Fluorescence Spectroscopy, Volume 1, Techniques,* J. R. Lakowicz (ed.), Plenum Press, New York, pp. 183–260.
130. Schiller, N. H., 1984, Picosecond streak camera photonics, in *Semiconductors Probed by Ultrafast Laser Spectroscopy*, Vol. II, Academic Press, pp. 441–458.
131. Campillo, A. J., and Shapiro, S. L., 1983, Picosecond streak camera fluorometry—a review, *IEEE J. Quantum Electron.* **QE-19**:585–603.
132. Knox, W., and Mourou, G., 1981, A simple jitter-free picosecond streak camera, *Opt. Commun.* **37**(3):203–206.
133. Ho, P. P., Katz, A., Alfano, R. R., and Schiller, N. H., 1985, Time response of ultrafast streak camera system using femtosecond laser pulses, *Opt. Commun.* **54**(1):57–62.
134. Bradley, D. J., McInerney, J., Dennis, W. M., and Taylor, J. R., 1983, A new synchroscan streak-camera read-out system for use with CW mode locked lasers, *Opt. Commun.* **44**(5):357–360.
135. Tsuchiya, Y., and Shinoda, Y., 1985, Recent developments of streak cameras, *Proc. SPIE* **533**:110–116.
136. Kinoshita, K., Ito, M., and Suzuki, Y., 1987, Femtosecond streak tube, *Rev. Sci. Instrum.* **58**:932–938.
137. Watanabe, M., Koishi, M., and Roehrenbeck, P. W., 1993, Development and characteristics of a new picosecond fluorescence lifetime system, *Proc. SPIE* **1885**:155–164.
138. Wiessner, A., and Staerk, H., 1993, Optical design considerations and performance of a spectro-streak apparatus for time-resolved fluorescence spectroscopy, *Rev. Sci. Instrum.* **64**:3430–3439.
139. Graf, U., Bühler, C., Betz, M., Zuber, H., and Anliker, M., 1994, Optimized streak-camera system: Wide excitation range and extended time scale for fluorescence lifetime measurement, *Proc. SPIE* **2137**:204–210.
140. Kume, H., Taguchi, T., Nakatsugawa, K., Ozawa, K., Suzuki, S., Samuel, R., Nishimura, Y., and Yamazaki, I., 1992, Compact ultrafast microchannel plate photomultiplier tube, *Proc. SPIE* **1640**:440–447.
141. Porter, G., Reid, E. S., and Tredwell, C. J., 1974, Time resolved fluorescence in the picosecond region, *Chem. Phys. Lett.* **29**:469–472.
142. Beddard, G. S., Doust, T., and Porter, G., 1981, Picosecond fluorescence depolarisation measured by frequency conversion, *Chem. Phys.* **61**:17–23.
143. Kahlow, M. A., Jarzeba, W., DuBruil, T. P., and Barbara, P. F., 1988, Ultrafast emission spectroscopy in the ultraviolet by time-gated upconversion, *Rev. Sci. Instrum.* **59**:1098–1109.
144. Ware, W. R., Doemeny, L. J., and Nemzek, T. L., 1973, Deconvolution of fluorescence and phosphorescence decay curves. A least-squares method, *J. Phys. Chem.* **77**:2038–2048.
145. Isenberg, I., Dyson, R. D., and Hanson, R., 1973, Studies on the analysis of fluorescence decay data by the method of moments, *Biophys. J.* **13**:1090–1115.
146. Small, E. W., and Isenberg, I., 1977, On moment index displacement, *J. Chem. Phys.* **66**:3347–3351.
147. Small, E. W., 1992, Method of moments and treatment of nonrandom error, *Methods Enzymol.* **210**:237–279.
148. Gafni, A., Modlin, R. L., and Brand, L., 1975, Analysis of fluorescence decay curves by means of the Laplace transformation, *Biophys. J.* **15**:263–280.
149. Almgren, M., 1973, Analysis of pulse fluorometry data of complex systems, *Chem. Scri.* **3**:145–148.
150. Ameloot, M., 1992, Laplace deconvolution of fluorescence decay surfaces, *Methods Enzymol.* **210**:237–279.
151. Ameloot, M., and Hendrickx, H., 1983, Extension of the performance of laplace deconvolution in the analysis of fluorescence decay curves, *Biophys. J.* **44**:27–38.
152. Livesey, A. K., and Brochon, J. C., 1987, Analyzing the distribution of decay constants in pulse-fluorimetry using the maximum entropy method, *Biophys. J.* **52**:693–706.
153. Brochon, J.-C., 1994, Maximum entropy method of data analysis in time-resolved spectroscopy, *Methods Enzymol.* **240**:262–311.
154. Zhang, Z., Grattan, K. T. V., Hu, Y., Palmer, A. W., and Meggitt, B. T., 1996, Prony's method for exponential lifetime estimations in fluorescence based thermometers, *Rev. Sci. Instrum.* **67**:2590–2594.
155. López, R. J., González, F., and Moreno, F., 1992, Application of a sine transform method to experiments of single-photon decay spectroscopy: Single exponential decay signals, *Rev. Sci. Instrum.* **63**:3268–3273.
156. Carraway, E. R., Hauenstein, B. L., Demas, J. N., and DeGraff, B. A., 1985, Luminescence lifetime measurements. Elimination of phototube time shifts with the phase plane method, *Anal. Chem.* **57**:2304–2308.
157. Bajzer, Z., Zelic, A., and Prendergast, F. G., 1995, Analytical approach to the recovery of short fluorescence lifetimes from fluorescence decay curves, *Biophys. J.* **69**:1148–1161.
158. O'Connor, D. V. O., Ware, W. R., and Andre, J. C., 1979, Deconvolution of fluorescence decay curves. A critical comparison of techniques, *J. Phys. Chem.* **83**:1333–1343.
159. Johnson, M. L., 1994, Use of least-squares techniques in biochemistry, *Methods Enzymol.* **240**:1–22.

160. Straume, M., Frasier-Cadoret, S. G., and Johnson, M. L., 1991, Least-squares analysis of fluorescence data, in *Topics in Fluorescence Spectroscopy, Volume 2, Principles*, J. R. Lakowicz (ed.), Plenum Press, New York, pp. 177–239.
161. Gryczynski, I., unpublished observations.
162. Johnson, M. L., personal communication.
163. Johnson, M. L., and Faunt, L. M., 1992, Parameter estimation by least-squares methods, *Methods Enzymol.* **210:**1–37.
164. Johnson, M. L., and Frasier, S. G., 1985, Nonlinear least-squares analysis, *Methods Enzymol.* **117:**301–342.
165. Box, G. E. P., 1960, Fitting empirical data, *Ann. N. Y. Acad. Sci.* **86:**792-816.
166. Bates, D. M., and Watts, D. G., 1988, *Nonlinear Regression Analysis and Its Applications*, John Wiley & Sons, New York.
167. Straume, M., and Johnson, M. L., 1992, Monte Carlo method for determining complete confidence probability distributions of estimated model parameters, *Methods Enzymol.* **210:**117–129.
168. Alcala, J. R., 1994, The effect of harmonic conformational trajectories on protein fluorescence and lifetime distributions, *J. Chem. Phys.* **101:**4578–4584.
169. Alcala, J. R., Gratton, E., and Prendergast, F. G., 1987, Fluorescence lifetime distributions in proteins, *Biophys. J.* **51:**597–604.
170. James, D. R., and Ware, W. R., 1985, A fallacy in the interpretation of fluorescence decay parameters, *Chem. Phys. Lett.* **120:**455–459.
171. Vix, A., and Lami, H., 1995, Protein fluorescence decay: Discrete components or distribution of lifetimes? Really no way out of the dilemma?, *Biophys. J.* **68:**1145–1151.
172. Lakowicz, J. R., Cherek, H., Gryczynski, I., Joshi, N., and Johnson, M. L., 1987, Analysis of fluorescence decay kinetics measured in the frequency-domain using distribution of decay times, *Biophys. Chem.* **28:**35–50.
173. Beechem, J. M., Knutson, J. R., Ross, J. B. A., Turner, B. W., and Brand, L., 1983, Global resolution of heterogeneous decay by phase/modulation fluorometry: Mixtures and proteins, *Biochemistry* **22:**6054–6058.
174. Beechem, J. M., Ameloot, M., and Brand, L., 1985, Global analysis of fluorescence decay surfaces: Excited-state reactions, *Chem. Phys. Lett.* **120:**466–472.
175. Knutson, J. R., Beechem, J. M., and Brand, L., 1983, Simultaneous analysis of multiple fluorescence decay curves: A global approach, *Chem. Phys. Lett.* **102:**501–507.
176. Beechem, J. M., 1989, A second generation global analysis program for the recovery of complex inhomogeneous fluorescence decay kinetics, *Chem. Phys. Lipids* **50:**237–251.
177. Beechem, J. M., Gratton, E., Ameloot, M., Knutson, J. R., and Brand, L., 1991, The global analysis of fluorescence intensity and anisotropy decay data: Second-generation theory and programs, in *Topics in Fluorescence Spectroscopy, Volume 2, Principles*, J. R. Lakowicz (ed.), Plenum Press, New York, pp. 241–305.
178. Beechem, J. M., 1992, Global analysis of biochemical and biophysical data, *Methods Enzymol.* **210:**37–55.
179. Chabbert, M., Hillen, W., Hansen, D., Takahashi, M., and Bousquet, J.-A., 1992, Structural analysis of the operator binding domain of Tn10-encoded Tet repressor: A time-resolved fluorescence and anisotropy study, *Biochemistry* **31:**1951–1960.
180. Dattelbaum, J. D., and Castellano, F. N., unpublished observations.
181. Malak, H., and Gryczynski, I., unpublished observations.
182. Frackowiak, D., Zelent, B., Malak, H., Planner, A., Cegielski, R., Munger, G., and Leblanc, R. M., 1994, Fluorescence of aggregated forms of CHl α in various media, *J. Photochem. Photobiol. A: Chem.* **78:**49–55.
183. Werst, M., Jia, Y., Mets, L., and Fleming, G. R., 1992, Energy transfer and trapping in the photosystem I core antenna, *Biophys. J.* **61:**868–878.
184. Gulotty, R. J., Mets, L., Alberte, R. S., and Fleming, G. R., 1986, Picosecond fluorescence studies of excitation dynamics in photosynthetic light-harvesting arrays, in *Applications of Fluorescence in the Biomedical Sciences*, D. L. Taylor, A. S. Waggoner, F. Lanni, R. F. Murphy, and R. R. Birge (eds.), Alan R. Liss, New York, pp. 91–104.
185. Visser, A. J. W. G., 1984, Kinetics of stacking interactions in flavin adenine dinucleotide from time-resolved flavin fluorescence, *Photochem. Photobiol.* **40:**703–706.
186. Castellano, F. N., Heimer, T. A., Tandhasetti, M. T., and Meyer, G. J., 1994, Photophysical properties of ruthenium polypyridyl photonic SiO_2 gels, *Chem. Mater.* **6:**1041–1048.
187. Yoshihara, K., Nagasawa, Y., Yartsev, A., Kumazaki, S., Kandori, H., Johnson, A. E., and Tominaga, K., 1994, Femtosecond intermolecular electron transfer in condensed systems, *J. Photochem. Photobiol. A: Chem.* **80:**169–175.
188. Nagasawa, Y., Yartsev, A. P., Tominaga, K., Johnson, A. E., and Yoshihara, K., 1993, Substituent effects on intermolecular electron transfer: Coumarins in electron-donating solvents, *J. Am. Chem. Soc.* **115:**7922–7923.
189. Montgomery, D. C., and Peck, E. A., 1982, *Introduction to Linear Regression Analysis*, John Wiley & Sons, New York, pp. 466–475.

PROBLEMS

4.1. *Calculation of Lifetimes:* Use the data in Figures 4.1 and 4.2 to estimate the lifetime from the time-domain data and from the phase and modulation.

4.2. *Fractional Intensity of Components in the Tryptophan Intensity Decay:* At pH 7, tryptophan displays a double-exponential intensity decay. At 320 nm the intensity decay law is $I(t) = 0.19 \exp(-t/0.62 \text{ ns}) + 0.81 \exp(-t/3.33 \text{ ns})$. What is the fractional contribution of the 0.62-ns component to the steady-state intensity at 320 nm?

4.3. *Stacking Equilibrium in Flavin Adenine Dinucleotide:* Use the lifetimes obtained from the intensity decays in Figure 4.45 to calculate the collisional rate between the flavin and adenine groups in FAD.

4.4. *Average Lifetime:* Suppose that a protein contains two tryptophan residues with identical lifetimes ($\tau_1 = \tau_2 = 5$ ns) and preexponential factors ($\alpha_1 = \alpha_2 = 0.5$). Now suppose that a quencher is added such that the first tryptophan is quenched 10-fold in both lifetime and steady-state intensity. What is the intensity decay law in the presence of quencher? What is the average lifetime ($\bar{\tau}$) and the lifetime-weighted quantum yield ($\langle\tau\rangle$)? Explain the relative values.

4.5. *Decay Associated Spectra*: Tables 4.5 and 4.6 list the results of the multiexponential analysis of the two-component mixture of anthranilic acid (AA) and 2-aminopurine (2-AP). Use these data to construct the decay-associated spectra (DAS). Explain the results for the DAS recovered from the nonglobal (Table 4.5) and the global (Table 4.6) analysis.

Frequency-Domain Lifetime Measurements

5

In the preceding chapter we described the theory and instrumentation for measuring fluorescence intensity decays using time-domain measurements. In the present chapter we continue this discussion, but we now consider the alternative method called frequency-domain (FD) fluorometry. In this method the sample is excited with light which is intensity-modulated at a high frequency comparable to the reciprocal of the lifetime. When this is done, the emission is also intensity-modulated at the same frequency. However, the emission does not precisely follow the excitation but rather shows time delays and amplitude changes which are determined by the intensity decay law of the sample. To be more precise, the time delay is measured as a phase-angle shift between the excitation and emission, as was shown in Figure 4.2. The peak-to-peak height of the modulated emission is decreased relative to that of the modulated excitation and provides another independent measure of the lifetime.

The interest in time-resolved measurements, whether performed in the time domain or in the frequency domain, originates with the information available from the intensity decay of the sample. Samples with multiple fluorophores typically display multiexponential decays. Even samples with a single fluorophore can display complex intensity decays due to conformational heterogeneity, resonance energy transfer, transient effects in diffusive quenching, or fluorophore–solvent interactions, to name just the most common origins. The goal of the time-resolved measurement is to determine the form of the intensity decay law and to interpret the decay in terms of molecular features of the sample.

It is informative to consider the difference between measuring a single mean decay time and measuring the form of a complex intensity decay. Irrespective of the complexity of the decay, one can always define a mean or apparent decay time. For instance, in the time domain this could be the time when the intensity decays to $1/e$ of its initial value. For a multiexponential decay, the $1/e$ time is typically not equal to any of the decay times. In the frequency domain the mean or apparent decay time could be the apparent lifetime (τ_ϕ) determined from the phase angle (ϕ_ω, Eq. [4.5]) or the apparent lifetime (τ_m) determined from the modulation (m_ω, Eq. [4.6]). In both domains the apparent lifetimes are characteristic of the sample but do not provide a complete description of the complex intensity decay. The values of τ_ϕ and τ_m need not be equal, and each value represents a different weighted average of the decay times displayed by the sample. In general, the apparent lifetime depends on the method of measurement. At present, there are relatively few reports of only the mean decay times. This is because the mean lifetimes represent complex weighted averages of the multiexponential decay. Quantitative interpretation of mean decay times is usually difficult, and the results are often ambiguous.

Prior to 1983, the status of FD fluorometry was mostly to allow determination of mean lifetime, but not to resolve complex intensity decays. This limitation arose because phase-modulation fluorometers only operated at one, two, or three fixed light modulation frequencies. It is well known that resolution of a complex decay requires measurements at a number of modulation frequencies which span the frequency response of the sample. While several variable-frequency instruments were described prior to 1983, these were not generally useful and were limited by systematic errors. The first generally useful variable-frequency instrument was described by Gratton and Limkeman,[1] and shortly thereafter a variable-frequency flourometer was constructed in our laboratory.[2] These instruments allowed phase and modulation measurements from 1 to 200 MHz. This design is the basis of currently available instruments. Since 1983, the use of FD fluorometry has grown rapidly,[3–9] and several companies now offer such instruments. FD fluorometers are now routinely used to study multiexponential intensity decays and nonexponential decays resulting from resonance energy transfer, time-dependent solvent relaxation, and collisional quenching.

In this chapter we describe the instrumentation for FD measurements and the theory used to interpret the experimental data. Additionally, we provide examples which illustrate the many applications of the FD method for the resolution of complex decay kinetics. Later in this chapter, we describe the present state-of-the-art instrumentation, which allows FD data to be obtained to 10 GHz, depending upon the photodetector and associated electronics. Finally, we describe the emergence of simple FD instrumentation based on laser diode or LED excitation.

The objective of both the time- and frequency-domain measurements is to recover the parameters describing the time-dependent decay. Assume that the sample is excited with a δ-function pulse of light. The resulting time-dependent emission is called the impulse response function, which is often represented by the multiexponential model,

$$I(t) = \sum_i \alpha_i e^{-t/\tau_i} \qquad [5.1]$$

In this expression the values of α_i are the preexponential factors, and the τ_i values are the decay times.

It is valuable to understand how a multiexponential decay is related to the steady-state intensity of the sample. The fraction of the intensity observed in the usual steady-state measurement that is due to the ith component in the multiexponential decay is

$$f_i = \frac{\alpha_i \tau_i}{\sum_j \alpha_j \tau_j} \qquad [5.2]$$

Another important concept is to understand whether the values of α_i and τ_i have any physical meaning. The multiexponential model is very powerful and able to account for almost any decay law. Depending upon the actual form of the intensity decay, the values of α_i and τ_i may have direct or indirect molecular significance. For a mixture of two fluorophores, each of which displays a single decay time, the τ_i are the decay times of the two fluorophores, and the f_i are the fractional contributions of each fluorophore to the total emission. In many circumstances there is no obvious linkage between the α_i and τ_i values and the molecular features of the sample. For instance, nonexponential decays occur due to transient effects in quenching, or due to distributions of donor–acceptor distances. These intensity decays can usually be satisfactorily fit by the multiexponential model, but it is difficult to relate the values of α_i and τ_i to the molecular parameters in the actual intensity decay laws which describe transient effects in quenching.

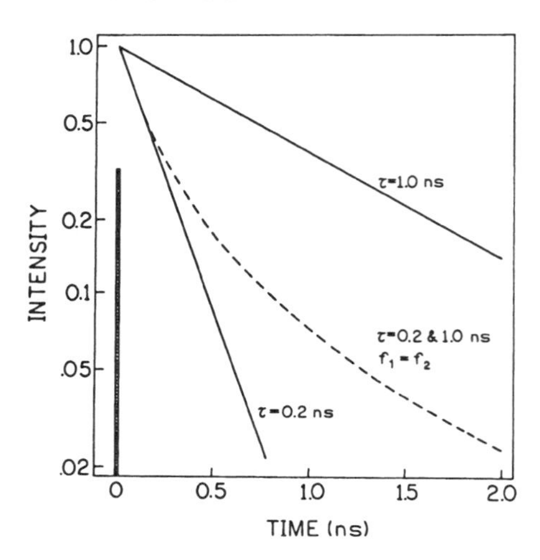

Figure 5.1. Time-domain representation of fluorescence decays. The vertical line shows the δ-function excitation pulse. ——, Intensity decays for single decay times of 0.2 and 1.0 ns; ------, multiexponential decay for $\tau_1 = 0.2$ ns, $\tau_2 = 1.0$ ns, and $f_1 = f_2 = 0.5$. From Ref. 3.

If the emission decays with a single decay time (Figure 5.1, solid lines), it is rather easy to measure the decay time with good accuracy. If the single decay time is long relative to the excitation pulse, then log $I(t)$ decays linearly as a function of time, and the decay time is easily obtained from the slope. The more difficult task is recovery of multiple decay times, which is illustrated for two widely spaced decay times in Figure 5.1 (0.2 and 1.0 ns, dashed curve). In this case, log $I(t)$ does not decay linearly with time. Unfortunately, decay times of the emission from macromolecules are often more closely spaced than the fivefold difference shown in Figure 5.1, and resolution of the decay times becomes increasingly difficult as the decay times become more closely spaced. If the decay times are spaced by 20% (e.g., 1 and 1.2 ns or 5 and 6 ns), it is difficult to distinguish visually a single-exponential decay from a double-exponential decay. In fact, such closely spaced decay times cannot usually be resolved using TD or FD measurements. It is generally difficult to resolve sums of exponentials because the parameters describing the decay are highly correlated. Hence, one requires a high signal-to-noise ratio, or equivalently a large number of photons, to recover the multiple decay times with reasonable confidence.

5.1. THEORY OF FREQUENCY-DOMAIN FLUOROMETRY

In FD fluorometry the excitation source and measurements are rather different than for TD measurements. The pulsed source used in TD measurements is replaced with an intensity-modulated light source.[3] Because of the time lag between absorption and emission, the emission is delayed in time relative to the modulated excitation (Figure 5.2). At each modulation frequency, this delay is described as the phase shift (ϕ_ω), which increases from 0° to 90° with increasing modulation frequency (ω). At first glance, this property of the phase angle is counterintuitive. For a time delay of the type available from an optical delay line, the phase shift can exceed 90° and reach any arbitrary value. For a single-exponential or multiexponential decay, the maximum phase angle is 90°. Hence, the phase angle displayed by any sample is some fraction of 90°, independent of the modulation frequency. Only under special circumstances can the phase angle exceed 90° (Chapter 18).

The finite time response of the sample also results in demodulation of the emission by a factor m_ω. This factor decreases from 1.0 to 0.0 with increasing modulation frequency. At low frequency, the emission closely follows the excitation. Hence, the phase angle is near zero and the modulation is near 1.0. As the modulation frequency is increased, the finite lifetime of the excited state prevents the emission from precisely following the excitation. This results in a phase delay of the emission, and a decrease in the peak-to-peak amplitude of the modulated emission, which is measured relative to the modulated excitation (Figure 5.2). The phase angle and modulation, measured over a wide range of frequencies, constitute the frequency response of the emission. In the presentation of FD data, the modulation frequency on the x-axis (Figure 5.3) is

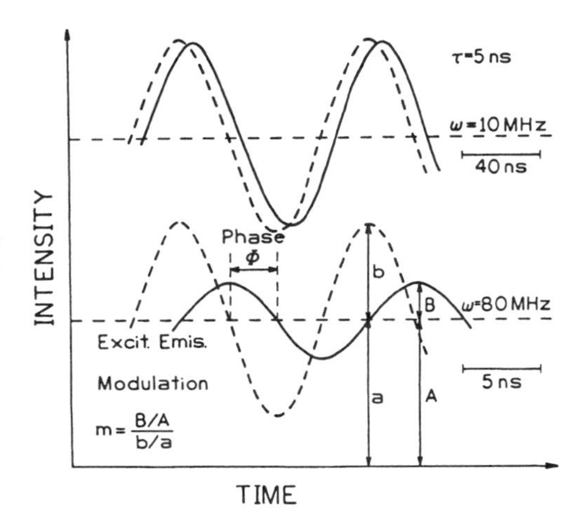

Figure 5.2. Phase and modulation of fluorescence (———) in response to intensity-modulated excitation (– – –). The assumed lifetime is 5 ns, and the modulation frequency is 10 MHz (*top*) or 80 MHz (*bottom*). From Ref. 3.

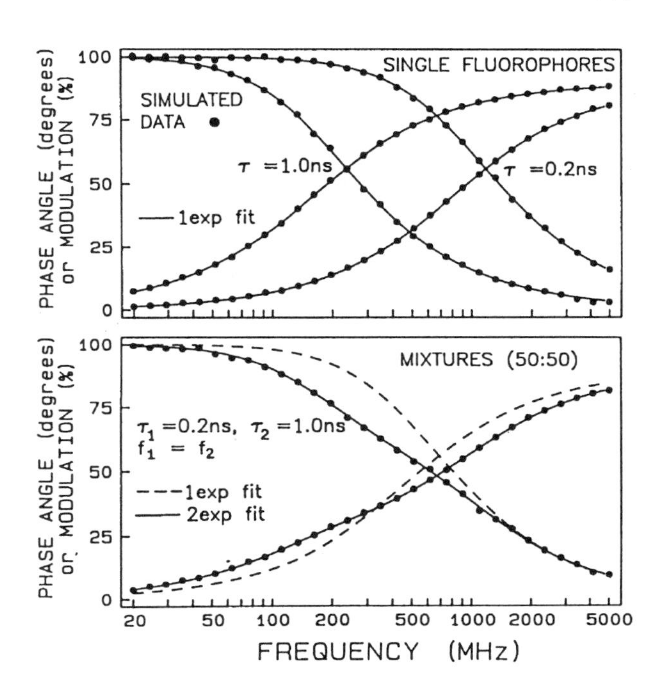

Figure 5.3. Simulated frequency-domain data for single- (*top*) and double- (*bottom*) exponential decays. The phase angle increases and the modulation decreases with increasing modulation frequency. The data points indicate the simulated data. *Top*: The solid curves show the best fits to a single decay time. *Bottom*: The dashed and solid curves show the best single- and double-exponential fits, respectively. From Ref. 3.

usually in units of cycles per second (hertz or megahertz), but for computations one uses the circular modulation frequency ($\omega = 2\pi \times \text{Hz}$) in radians per second.

The characteristic features of the frequency response of a sample are illustrated in Figure 5.3. The shape of the frequency response is determined by the number of decay times displayed by the sample. If the decay is a single exponential (Figure 5.3, top), the frequency response is simple. One can use the phase angle or modulation at any frequency to calculate the lifetime. For a single-exponential decay, the phase and modulation are related to the decay time (τ) by

$$\tan \phi_\omega = \omega \tau \qquad [5.3]$$

and

$$m_\omega = (1 + \omega^2\tau^2)^{-1/2} \qquad [5.4]$$

The derivation of Eqs. [5.3] and [5.4] is given in Section 5.11.A. At higher modulation frequencies, the phase angle of the emission increases and the modulation decreases. This is shown in Figure 5.3 for decay times of 1 ns and 0.2

ns. For the 1-ns decay time, the phase shift at 200 MHz is 51.5°, and the emission is demodulated by a factor of 0.62 relative to the excitation. At a modulation frequency of 1 GHz, the phase angle increases to 81°, and the modulation decreases to 0.16. Most samples of interest display more than one decay time. In this case the lifetimes calculated from the value of ϕ_ω or m_ω, measured at a particular frequency, are only apparent values and are the result of a complex weighting of various components in the emission (Section 5.10). For such samples, it is necessary to measure the phase and modulation values over the widest possible range of modulation frequencies, with the center frequency being comparable to the reciprocal of the mean decay time of the emission.

It is informative to examine the frequency response expected for a multiexponential decay (Figure 5.3, bottom). In this simulation the assumed decay times are 0.2 and 1.0 ns. The objective is to recover the multiple decay times from the experimentally measured frequency response. This is generally accomplished using nonlinear least-squares procedures.[10] The fitting procedure is illustrated by the solid and dashed curves in Figure 5.3. For the single-exponential decays shown in the top half of the figure, it is possible to obtain a good match between the data (represented by the dots) and the curves calculated using the single-exponential model. For a double-exponential decay, as shown in the bottom half of the figure, the data cannot be matched using a single-decay-time fit, represented by the dashed curves. However, the complex frequency response is accounted for by the double-exponential model, represented by the solid curves, with the expected decay times (0.2 and 1 ns) and fractional intensities ($f_1 = f_2 = 0.5$) being recovered from the least-squares analysis.

It is valuable to consider the modulation frequencies needed for various decay times. The useful modulation frequencies are those for which the phase angle is frequency-dependent and there is still measurable modulation (Figure 5.4). Most fluorophores display lifetimes near 10 ns, so that modulation frequencies are typically on the order of 2–200 MHz. If the decay time is near 100 ps, modulation frequencies near 2 GHz are needed. For longer decay times of 1–10 μs, the modulation frequencies can range from 10 kHz to 1 MHz. As the modulation frequency increases, the modulation of the emission decreases. Hence, it becomes more difficult to measure the phase angles as they approach 90°.

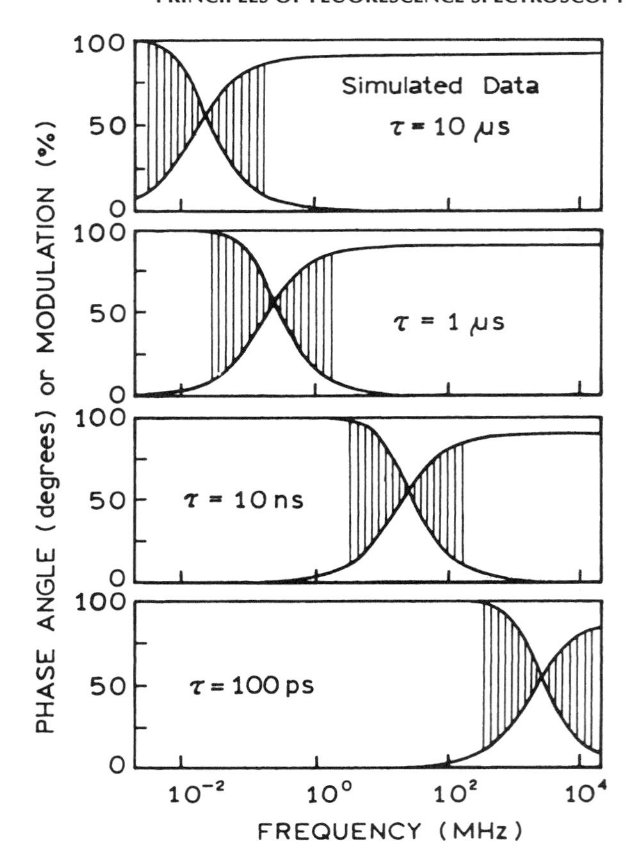

Figure 5.4. Relationship between the decay time and the useful range of light modulation frequencies.

5.1.A. Least-Squares Analysis of Frequency-Domain Intensity Decays

We now describe the analysis used to recover the intensity decay parameters from the frequency response of the emission. For measurement of the intensity decay, it is common practice to use magic-angle conditions, which are vertically polarized excitation and observation 54.7° from the vertical. These conditions eliminate the effects of rotational diffusion on the intensity decay. At present, the FD data are usually analyzed by the method of nonlinear least squares.[10–13] The measured data are compared with values predicted from a model, and the parameters of the model are varied to yield the minimum deviations from the data. It is possible to predict the phase and modulation values for any decay law. The FD data for an intensity decay can be calculated from the sine and cosine transforms of $I(t)$,

$$N_\omega = \frac{\int_0^\infty I(t) \sin \omega t \, dt}{\int_0^\infty I(t) \, dt} \qquad [5.5]$$

$$D_\omega = \frac{\int_0^\infty I(t) \cos \omega t \, dt}{\int_0^\infty I(t) \, dt} \quad [5.6]$$

where ω is the circular modulation frequency (2π times the modulation frequency in hertz). These expressions are sometimes written with $J = \int_0^\infty I(t) \, dt$, which normalizes the expression for the total intensity of the sample. Even if the decay is more complex than a sum of exponentials, it is generally adequate to approximate the decay by such a sum. If needed, nonexponential decay laws can be transformed numerically. For a sum of exponentials the transforms are[12,13]

$$N_\omega = \sum_i \frac{\alpha_i \omega \tau_i^2}{(1 + \omega^2 \tau_i^2)} \Big/ \sum_i \alpha_i \tau_i \quad [5.7]$$

$$D_\omega = \sum_i \frac{\alpha_i \omega \tau_i}{(1 + \omega^2 \tau_i^2)} \Big/ \sum_i \alpha_i \tau_i \quad [5.8]$$

For a multiexponential decay, $J = \Sigma_i \alpha_i \tau_i$, which is proportional to the steady-state intensity of the sample. Because of this normalization factor, one can always fix one of the amplitudes (α_i or f_i) in the analysis of FD data. The calculated frequency-dependent values of the phase angle ($\phi_{c\omega}$) and the demodulation ($m_{c\omega}$) are given by

$$\tan \phi_{c\omega} = N_\omega / D_\omega \quad [5.9]$$

$$m_{c\omega} = (N_\omega^2 + D_\omega^2)^{1/2} \quad [5.10]$$

The parameters (α_i and τ_i) are varied to yield the best fit between the data and the calculated values, as indicated by a minimum value for the goodness-of-fit parameters χ_R^2,

$$\chi_R^2 = \frac{1}{\nu} \sum_\omega \left[\frac{\phi_\omega - \phi_{c\omega}}{\delta \phi} \right]^2 + \frac{1}{\nu} \sum_\omega \left[\frac{m_\omega - m_{c\omega}}{\delta m} \right]^2 \quad [5.11]$$

where ν is the number of degrees of freedom. The value of ν is given by the number of measurements, which is typically twice the number of frequencies minus the number of variable parameters. The subscript c is used to indicate calculated values for assumed values of α_i and τ_i, and $\delta\phi$ and δm are the uncertainties in the phase and modulation values, respectively. Unlike the errors in the photon counting experiments (Chapter 4), these errors cannot be estimated directly from Poisson statistics.

The correctness of a model is judged based on the values of χ_R^2. For an appropriate model and random noise, χ_R^2 is expected to be near unity. If χ_R^2 is greater than unity, then one should consider whether the χ_R^2 value is adequate to reject the model. Rejection is judged from the probability that random noise could be the origin of the value of χ_R^2.[10,11] For instance, a typical FD measurement from this laboratory contains phase and modulation data at 25 frequencies. A double-exponential model contains three floating parameters (two τ_i and one α_i), resulting in 47 degrees of freedom. A value of χ_R^2 equal to 2 is adequate to reject the model with a certainty of 99.9% or higher (see Table 4.2).

In practice, the values of χ_R^2 change depending upon the values of the uncertainties ($\delta\phi$ and δm) used in their calculation. The effects of selecting different values of $\delta\phi$ and δm have been considered in detail.[12,13] The fortunate result is that the recovered parameter values (α_i and τ_i) do not depend strongly on the chosen values of $\delta\phi$ and δm. The parameter values can be expected to be sensitive to $\delta\phi$ and δm if the data are just adequate to determine the parameter values, that is, at the limits of resolution.

For consistency and ease of day-to-day data interpretation, we use constant values of $\delta\phi = 0.2°$ and $\delta m = 0.005$. Although the precise values may vary between experiments, the χ_R^2 values calculated in this way indicate to us the degree of error in a particular data set. For instance, if a particular data set has a poor signal-to-noise ratio, or systematic errors, the value of χ_R^2 is elevated even for the best fit. The use of fixed values of $\delta\phi$ and δm does not introduce any ambiguity in the analysis, as it is the relative values of χ_R^2 which are used in accepting or rejecting a model. We typically compare χ_R^2 for the single-, double-, and triple-exponential fits. If χ_R^2 decreases twofold or more as the model is incremented, then the data probably justify inclusion of the additional decay time. According to Table 4.3, a ratio of χ_R^2 values of 2 is adequate to reject the simpler model with a 99% certainty. It should be remembered that the values of $\delta\phi$ and δm might depend upon frequency, either as a gradual increase in random error with frequency or as higher-than-average uncertainties at discrete frequencies due to interference or other instrumental effects. In most cases the recovered parameter values are independent of the chosen values of $\delta\phi$ and δm. However, caution is needed as one approaches the resolution limits of the measurements. In these cases the values of the recovered parameters might depend upon the values chosen for $\delta\phi$ and δm.

The values of $\delta\phi$ and δm can be adjusted as appropriate for a particular instrument. For instance, the phase data may become more noisy with increasing modulation frequency because the phase angle is being measured from a smaller signal. One can use values of $\delta\phi$ and δm that increase with frequency to account for this effect. In adjusting the values of $\delta\phi$ and δm, we try to give equal weight to the phase and modulation data. This is accomplished by adjusting the relative values of $\delta\phi$ and δm so that the sum of the squared deviations (Eq. [5.11]) are approximately equal for the phase and modulation data.

Another way to estimate the values of $\delta\phi$ and δm is from the data itself. The phase and modulation values at each frequency are typically an average of 10–100 individual measurements. In principle, the values of $\delta\phi$ and δm are given by the standard deviation of the mean of the average value of the phase and the modulation, respectively. In practice, we find that the standard deviation of the mean underestimates the values of $\delta\phi$ and δm. This probably occurs because the individual phase and modulation measurements are not independent of each other. For simplicity and consistency, the use of constant values of $\delta\phi$ and δm is recommended.

In discussing the FD data, we have deliberately avoided the use of phase (τ_ϕ) or modulation (τ_m) lifetimes. These values are the lifetimes calculated from the measured phase and modulation values at a given frequency. The use of phase and modulation lifetimes can be misleading and is best avoided. The characteristics of τ_ϕ and τ_m are discussed in Section 5.10.

5.1.B. Global Analysis of Frequency-Domain Data

As described in Chapter 4, resolution of closely spaced parameters can be improved by global analysis. This applies to the FD data as well as the TD data. The use of global analysis is easiest to visualize for a mixture of fluorophores, each displaying a different emission spectrum. In this case the intensity decay at each wavelength (λ) is given by

$$I(\lambda, t) = \sum_{i=1} \alpha_i(\lambda) e^{-t/\tau_i} \qquad [5.12]$$

where the value of $\alpha_i(\lambda)$ represents the relative contribution of the ith fluorophore at wavelength λ. The frequency response is typically measured at several wavelengths, resulting in wavelength-dependent values of the phase angle $\phi_\omega(\lambda)$ and the modulation $m_\omega(\lambda)$. In this case the values of N_ω^λ and D_ω^λ depend on the observation wavelength and are given by

$$N_\omega^\lambda = \sum_i \frac{\alpha_i(\lambda)\omega\tau_i^2}{1+\omega^2\tau_i^2} \Big/ \sum_i \alpha_i(\lambda)\tau_i \qquad [5.13]$$

$$D_\omega^\lambda = \sum_i \frac{\alpha_i(\lambda)\tau_i}{1+\omega^2\tau_i^2} \Big/ \sum_i \alpha_i(\lambda)\tau_i \qquad [5.14]$$

The wavelength-dependent data sets can be used in a global minimization of χ_R^2,

$$\chi_R^2 = \frac{1}{\nu}\sum_{\lambda,\omega}\left[\frac{\phi_\omega(\lambda) - \phi_{c\omega}(\lambda)}{\delta\phi}\right]^2 + \frac{1}{\nu}\sum_{\lambda,\omega}\left[\frac{m_\omega(\lambda) - m_{c\omega}(\lambda)}{\delta m}\right]^2 \qquad [5.15]$$

where the sum now extends over the frequencies (ω) and wavelengths (λ). Typically, the values of τ_i are assumed to be independent of wavelength and are thus the global parameters. The values of $\alpha_i(\lambda)$ are usually different for each data set; that is, they are nonglobal parameters. The data are normalized at each wavelength, allowing one of the amplitudes at each wavelength to be fixed in the analysis.

5.1.C. Estimation of Parameter Uncertainties

It is important to estimate the range of parameter values which are consistent with the data. As for TCSPC, the ASEs recovered from least-squares analysis do not provide a true estimate of the uncertainty, but provide a significant underestimation of the range of parameter values which are consistent with the data. This effect is due to correlation between the parameters, which is not considered in calculation of the ASEs.

Parameter uncertainty can be determined using the method described in Chapter 4. Algorithms are available to estimate the upper and lower bounds of a parameter based on the extent of correlation.[14–16] These estimates are provided by our own software, but most commercial software provides only the ASEs. If the analysis is at the limits of resolution, we prefer to examine the χ_R^2 surfaces, that is, to perform a support plane analysis. This is accomplished just as for the time-domain data. Each parameter value is varied around its best-fit value, and the value of χ_R^2 is minimized by adjustment of the remaining parameters. The upper and lower limits for a parameter are taken as those which result in an elevation of the F_χ to the value expected for one standard deviation ($P = 0.32$) and the number of degrees of freedom (Section 5.5.A).

5.2. FREQUENCY-DOMAIN INSTRUMENTATION

5.2.A. History of Phase-Modulation Fluorometers

The use of phase-modulation methods for measurements of fluorescence lifetimes has a long history.[17] The first lifetime measurements were performed by Gaviola in 1926 using a phase fluorometer.[18] The first suggestion that phase-angle measurements be used for measuring fluorescence lifetimes appears to have been made even earlier, by Wood in 1921.[19] The use of phase delays to measure short time intervals appears to have been suggested even earlier, in 1899.[20] Hence, the use of phase shifts for timing of rapid processes has been recognized for 100 years. Since the pioneering measurements by Gaviola,[18] a large number of phase-modulation instruments have been described. These include an instrument described by Duschinsky in 1933[21] and an instrument of somewhat more advanced design described by Szymanowski,[22] on which many of Jabłoński's early measurements were performed. Since that time, phase fluorometers have been described by many research groups.[23–40] The first generally useful design appears to be that of Spencer and Weber in 1969.[41] This instrument used a Debye–Sears ultrasonic modulator[42,43] to obtain intensity-modulated light from an arc lamp light source. The Debye–Sears modulator has been replaced by electro-optic modulators in current FD instruments. However, an important feature of Spencer and Weber's instrument[41] is the use of cross-correlation detection (Section 5.11.B). The use of this RF mixing method simplified measurement of the phase angles and modulation values at high frequencies and allowed measurement of the phase angle and modulation with relatively slow timing electronics.

The primary technical factor limiting the development of FD fluorometers was the inability to obtain intensity-modulated light over a range of modulation frequencies. The Debye–Sears modulators are limited to operating at the frequency of the crystal, or multiples thereof. Consequently, the early phase-modulation fluorometers operated at only one to three fixed modulation frequencies. The limited data that they provided were adequate for measuring mean decay times or for detecting the presence of a complex decay. However, the data at a limited number of modulation frequencies were not generally useful for resolution of the parameters describing multi- or nonexponential decays. Another factor limiting the development of the FD instruments was the lack of availability of two stable and closely spaced frequencies, which are needed for cross correlation. In the early instruments this was accomplished by a "phase shifter," which added a small frequency to a higher-megahertz frequency. In present-day instruments the two closely spaced frequencies are available with modern frequency synthesizers.

After the introduction of cross correlation by Spencer and Weber,[41] a number of variable-frequency instruments were described.[17,44–53] These instruments were generally complex and used a CW laser light source modulated by an electro-optic device. Most of these instruments did not use cross-correlation detection and were prone to systematic errors which prevented their use in the analysis of complex decay kinetics. The use of an electro-optic modulator allowed the use of a wide range of frequencies. This feature, combined with cross-correlation detection, resulted in the current generation of variable-frequency instruments.[1,2]

5.2.B. The 200-MHz Frequency-Domain Fluorometer

FD fluorometers are now in widespread use. Most designs are similar to that shown in Figure 5.5, which is in fact not very different from the design of a standard fluorometer. The main differences are the laser light source, the light modulator, and the associated RF electronics. It is difficult to obtain light modulation over a wide range of frequencies. Amplitude modulation of laser sources over a continuous range of frequencies to 200 MHz is now possible with one of several electro-optic (EO) modulators. Light can also be modulated with acousto-optic (AO) modulators. However, AO modulators provide modulation at discrete resonances over a limited range of frequencies.[54,55] Unfortunately, most EO modulators have long, narrow optical apertures, and EO modulators are not easily used with arc lamp sources. Initially, only laser sources seemed practical for use with EO modulators. Surprisingly, it is now possible to use electro-optic devices to modulate arc lamps to 200 MHz, and this is done in the commercial FD instruments. The operational principles of the modulators and the electronic parts needed to construct such an FD instrument are discussed below (Section 5.2.C), along with the rationale for selecting the various components.

The light source for an FD instrument can be almost any CW light source or a high-repetition-rate pulsed laser. The choice of the source is based on the needed wavelengths and power level and the experience and/or budget of the investigators. The He–Cd laser is a convenient CW source, providing CW output at 325 and 442 nm. Unfortunately,

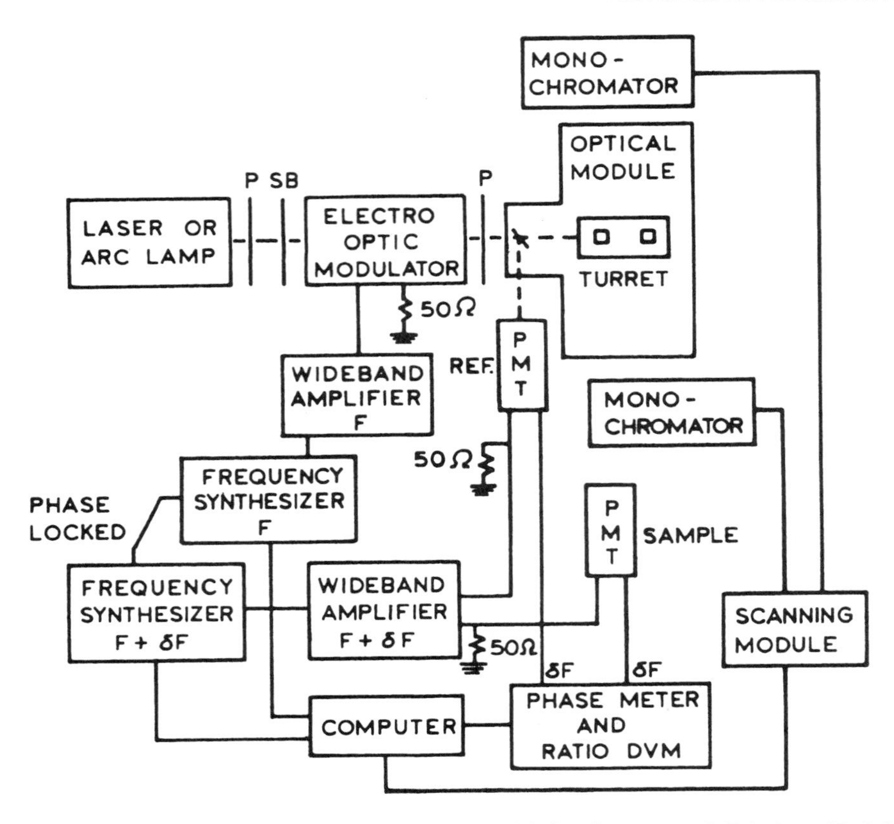

Figure 5.5. Schematic representation of the variable-frequency phase-modulation fluorometer. P, Polarizer; SB, Soleil–Babinet compensator; F, frequency; δF, cross-correlation frequency; PMT, photomultiplier tube; DVM, digital voltmeter. Revised from Ref. 2.

these wavelengths are not suitable for excitation of protein fluorescence. A very versatile source is the argon-ion laser, which can now provide deep UV lines (~275 nm) for intrinsic fluorescent probes. However, only a limited number of UV wavelengths are available. Studies of protein fluorescence usually require 290–300 nm to avoid excitation of tyrosine and to obtain high fundamental anisotropies. However, these wavelengths are not available from an argon-ion laser. The argon-ion laser is an ideal source for pumping dye lasers. The 514-nm line can be mode-locked to synchronously pump a picosecond dye laser system (Section 4.4.A). An Nd:YAG laser can also be the primary source, particularly for pumping dye lasers. In our hands, the Nd:YAG lasers are considerably less stable than an argon-ion laser, with regard to both pulse-to-pulse and long-term stability.

The next apparent difficulty is measurement of the phase angle and modulation at high frequencies. The measurements appear more difficult when one realizes that resolution of multiexponential decays requires accuracy near 0.2° in phase and 0.5% (0.005) in modulation and that this accuracy needs to be maintained from 1 to 200 MHz. In fact, the measurements are surprisingly easy and free of interference because of cross-correlation detection. In cross-correlation detection, the gain of the detector is modulated at a frequency offset ($F + \delta F$) from that of the modulated excitation (F). The difference frequency (δF) is typically in the range of 10–100 Hz. This results in a low-frequency signal at δF which contains the phase and modulation information in the original high-frequency signal (Section 5.11.B). At all modulation frequencies, the phase and modulation can be measured at the same low cross-correlation frequency (δF) with a zero-crossing detector and a ratio digital voltmeter. The use of cross-correlation detection results in the rejection of harmonics and other sources of noise. In newer FD instruments the analog circuits used to measure the phase and modulation are being replaced with signal processing boards which extract the values from the digitized low-frequency signal.

The cross-correlation method is surprisingly robust. The harmonic content (frequency components) of almost any excitation profile can be used if it contains frequency components that are synchronized with the detector. Both pulsed lasers and synchrotron radiation have been used as modulated light sources. Pulsed lasers provide harmonic content to many gigahertz, so the bandwidth of the FD instruments is now limited by the detector and not the light modulator (Section 5.7).

5.2.C. Light Modulators

Use of the FD instruments is facilitated by understanding the operating principles of light modulators. The first step is to obtain intensity-modulated light, which is, in general, a difficult problem. Light can be modulated with high efficiency using AO modulators, which diffract light based on a periodic density gradient caused by sound waves. AO modulators are typically resonant devices, which operate at only certain frequencies. While broadband or variable-frequency AO modulators are known, the active area is usually small, limiting them to use with focused laser sources.

The problem of light modulation was solved by the use of EO modulators (Figure 5.6). EO modulators are constructed of materials which rotate polarized light when the material is exposed to an electric field.[54,55] The modulator is placed between crossed polarizers. In the absence of any voltage, there is no effect on the incident light, and no light is transmitted. If a voltage is applied to the modulator, the electric vector of the light is rotated, and some light passes through the second polarizer. The voltage applied to the modulator is typically an RF signal at the desired modulation frequency.

When an EO modulator is used as just described, without a bias, it provides modulated light at twice the frequency of the electric field applied to the modulator. This occurs because the optical system becomes transmissive whether the voltage is positive or negative (Figure 5.7). However, the amount of light transmitted is rather small. Hence, EO modulators are usually operated with an optical or electrical bias, which results in some light transmission when the voltage applied to the modulator is zero. This can be accomplished by the use of Soleil–Babinet compensator (optical bias) or a DC voltage (electrical bias). This converts the linearly polarized light to circular or elliptically polarized light, which results in partial transmission of the laser beam in the absence of the RF voltage. Application of an RF voltage now results in amplitude modulation of the laser beam at the applied RF frequency (Figure 5.7). The modulated light is polarized according to the orientation of the second polarizer.

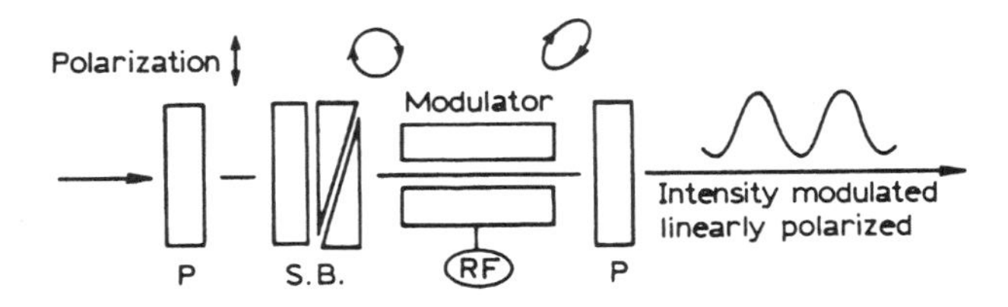

Figure 5.6. Optical arrangement for an electro-optic modulator. P, Polarizer; S.B., Soleil–Babinet compensator; RF, radio-frequency signal. The Soleil–Babinet compensator is replaced by an electrical DC voltage in some commercial FD instruments. Revised from Ref. 2.

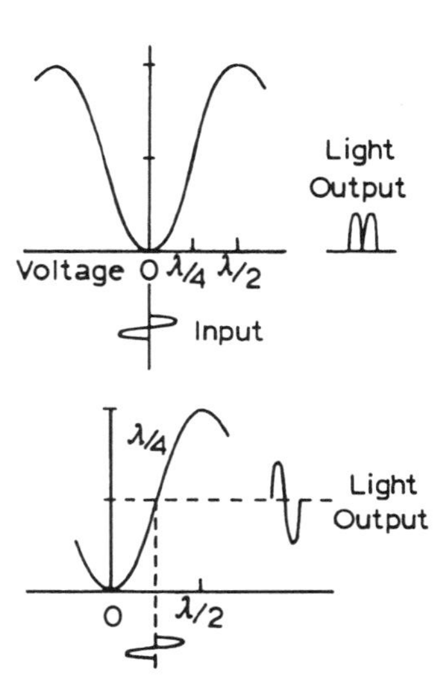

Figure 5.7. Modulated intensity from an electro-optic modulator. The upper drawing shows the modulated intensity without an electrical or optical bias; the lower drawing includes a λ/4 bias.

The usefulness of the EO modulator results from the ability to rapidly change the electric field across the EO crystal. In practice, adequate RF voltage can be applied up to about 200 MHz, which is also the upper frequency limit of most dynode PMTs. The EO modulators are not resonant devices, and they perform equally well at any frequency below the upper frequency limit. However, there are several features of EO modulators which make them not ideal for use in an FD fluorometer. Perhaps the most serious limitation is the high half-wave voltage. The half-wave voltage is the voltage needed to rotate the electric vector by 90° and allow all the light to pass through the second polarizer. Half-wave voltages for commonly used EO modulators are 1–7 kV, whereas the maximum practical RF peak-to-peak voltages are near 100 V. For this reason, one can only obtain a relatively small amount of modulation.

In FD fluorometers the extent of modulation is usually increased by adjusting the optical or electrical bias close to the zero-transmission point. This decreases both the AC and DC intensities, but the DC intensity can be decreased to a greater extent, resulting in a larger AC/DC ratio. Because of the large half-wave voltage, crossed polarizers, and a modulator biased near zero transmission, the overall efficiency of light transmission rarely exceeds 10%.

Another disadvantage of the EO modulators is that they generally require collimated light. Hence, they work much better with lasers than with arc lamp sources. It is possible to obtain useful amounts of modulation with an arc lamp, but the transmission is typically less than 5%. Unfortunately, there does not seem to be any general solution to this problem. As shown in Section 5.7, the limitations of using a light modulator can be avoided by using intrinsically modulated light sources. These include pulsed lasers, laser diodes, and LEDs. However, these sources typically do not provide the UV wavelengths needed for excitation of proteins. LEDs are now available with output near 390 nm, which can be used to excite many extrinsic fluorophores. Longer-wavelength LEDs and laser diodes are widely available.

Modulators are known which have lower half-wave voltages and higher frequency limits, above 1 GHz.[54] These include the longitudinal-field and traveling-wave modulators. These devices typically have long, narrow light paths and are only suitable for use with lasers. Also, they are sensitive to temperature and RF power. This instability limits their use in FD instruments.

5.2.D. Cross-Correlation Detection

The use of cross-correlation detection is an essential feature of the FD measurements. The basic idea is to modulate the gain of the PMT at a frequency offset (δF) from the light modulation frequency. The result is a low-frequency signal from the PMT at frequency δF that contains the phase and modulation information. The phase shift and modulation of the low-frequency signal are the same as one would have observed at high frequency. The phase and modulation of the low-frequency signal are easily measured using either analog or digital methods. Some FD instruments use zero-crossing detectors and a ratio voltmeter (DVM) to measure these values, as was done in the early instruments.[41] In newer FD instruments, the low-frequency signal is digitized and analyzed by a fast Fourier transform.[56–58] It seems that the digital data acquisition of the low-frequency signal decreases the noise in the phase and modulation data by about twofold compared to the analog circuits. The equations describing cross-correlation detection are provided in Section 5.11.B.

Cross-correlation detection provides a significant advantage in addition to allowing low-frequency measurements. The process of cross correlation suppresses harmonic or other frequencies, so that the modulation of the light or PMT gain need not be a pure sine wave. In fact, the excitation waveform can be almost any repetitive waveform, even a laser pulse train. After cross correlation, the phase and modulation values are the values which would have been observed if the modulation were perfectly sinusoidal. This feature of harmonic suppression makes the FD instruments easy to use. One need not be concerned about the shape of the modulated signals, as this will be corrected by cross correlation.

5.2.E. Frequency Synthesizers

The use of cross-correlation detection requires two frequencies which are synchronized but different by a small frequency δF. The cross-correlation frequency δF can be any value and is generally selected to be between 10 and 100 Hz. The synthesizer must provide frequencies to 200 MHz or higher, with 1- or 0.1-Hz resolution. Fortunately, this is not difficult with modern electronics. The requirements for frequency resolution in the synthesizer can be relaxed if one uses higher cross-correlation frequencies,[58] and schemes are being developed which use only one high-stability frequency source.[59] It seems clear that the cost of FD instrumentation will continue to decrease.

5.2.F. Radio-Frequency Amplifiers

The EO modulator requires the highest reasonable voltage, preferably 1500 V peak-to-peak over a wide frequency range. Unfortunately, this is not practical. In order to obtain variable-frequency operation, the circuit is usually terminated with a 50-Ω power resistor. A 25-W amplifier provides only about 100 V into 50 Ω, which is why the overall light transmission is low. One can usually remove the 50-Ω terminating resistor, which results in a twofold increase in voltage. The RF amplifier should be protected from reflected power. It is important to avoid standing waves in the amplifier-to-modulator cable, which should be less than 30 cm long.

In contrast to the high power required by the light modulator, relatively little power is needed to modulate the gain of the PMT. This amplifier is typically near 1 W, and can be less. In fact, we often directly use the output of the frequency synthesizer without amplification for gain modulation of the PMT.

5.2.G. Photomultiplier Tubes

The detector of choice for FD measurements is a PMT. The upper frequency limit of a PMT is determined by its transit time spread (TTS) so that the same detectors that are useful in TCSPC are useful in FD fluorometry. There is a slight difference in that the most important feature for TCSPC is the rise time of the pulse. For FD measurements, the PMTs with the highest frequency limits are those with the most narrow pulses for each photoelectron. While fast rise times usually imply narrow photoelectron pulse widths, some detectors can have fast rise times with long tails.

The approximate upper frequency limits of commonly used PMTs are listed in Table 4.1. These values are estimated based on our experience and product literature. The upper frequency limit of the side window R928 is near 200 MHz. Much higher frequency measurements are possible with MCP PMTs (Section 5.7), but special circuits are needed for cross correlation outside of the PMT.

It is informative to examine the PMT electronics used in an FD fluorometer (Figure 5.8), in this case for an R928 PMT. The circuit starts with a high negative voltage of the photocathode. There is a Zener diode (Z_1) between the photocathode and the first dynode. This diode maintains a constant high 250-V potential. With the use of a constant high potential, the wavelength-dependent and position-dependent time response of the PMT is minimized. The next dynodes are all linked by simple resistors. This allows the gain of the PMT to be varied by changes in applied voltage. Capacitors are included to maintain the voltage difference during periods of transiently high illumination.

Cross correlation is accomplished by injection of a small RF signal at dynode 9 (D_9). The voltage between D_8 and D_9 is held constant by a Zener diode (Z_2). The average voltage between D_9 and ground is adjusted by the bias resistors R_{b1} and R_{b2}. A few volts of RF signal are adequate to obtain nearly 100% gain modulation of the PMT.

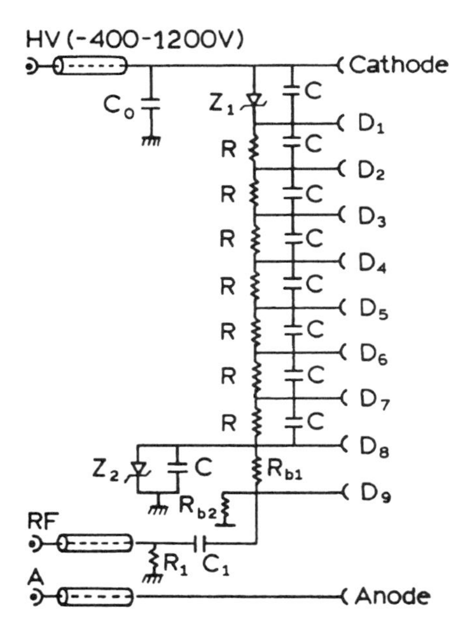

Figure 5.8. Dynode chain for an R928 photomultiplier tube used in a frequency-domain fluorometer. Revised from Ref. 1 and reprinted with permission from the Biophysical Society.

5.2.H. Principle of Frequency-Domain Measurements

When one is performing FD measurements, it is valuable to understand what is being measured. One performs a comparison of the sample emission and scattered light, similar to the comparison in TCSPC. The FD instruments typically contain two detectors, one for the sample and one which serves as a timing reference (Figure 5.5). The reference PMT typically observes reflected light, scattered light, or the emission from a short-decay-time lifetime standard. The sample PMT is exposed alternately to the emission from the sample or to scattered light. The sample and scattering solutions are usually in a rotating sample holder (turret) which precisely positions each solution in the same location. Since everything else in the measurement remains the same, any change in relative phase or modulation is due to the intensity decay of the sample.

Because the scatterer and sample are not observed at the same time, the phase difference and relative modulation cannot be measured at the same time. Instead, all measurements are performed relative to the reference PMT (Figure 5.9). One first measures the phase shift between the scatterer and reference PMT. These signals are shifted by an arbitrary phase angle (ϕ_1) due to the inevitable time delays in the cables and electronic circuits. The second measurement is the phase of the sample relative to the reference PMT. This phase angle (ϕ_2) contains both the arbitrary phase shift ϕ_1 and the value of interest, ϕ_ω. The actual phase shift is then calculated from $\phi_\omega = \phi_2 - \phi_1$.

The modulation is measured in a similar way, except that one typically does not measure the modulation of the reference PMT. The modulation is determined from the AC and DC components of the sample (B/A) and scatterer (b/a), as shown in Figure 5.2. For measurement of the modulation, it does not matter whether one measures the peak-to-peak or root-mean-square (RMS) voltage of the modulated signal, as the method of measurement cancels in the ratio. Although not routinely done, one can use the modulation at the reference PMT as a correction for modulation drifts in the instrument. In this case all measured modulation values are divided by the value at the reference PMT during each particular measurement. This procedure is useful if the extent of modulation is changing during the measurement owing to instabilities in the modulator or light source.

As described in Chapter 4 for TD measurements, it is important to consider the effects of rotational diffusion on the measured intensity decays. Rotational diffusion can also affect the FD measurements (Chapter 11). The use of an excitation and/or emission monochromator results in

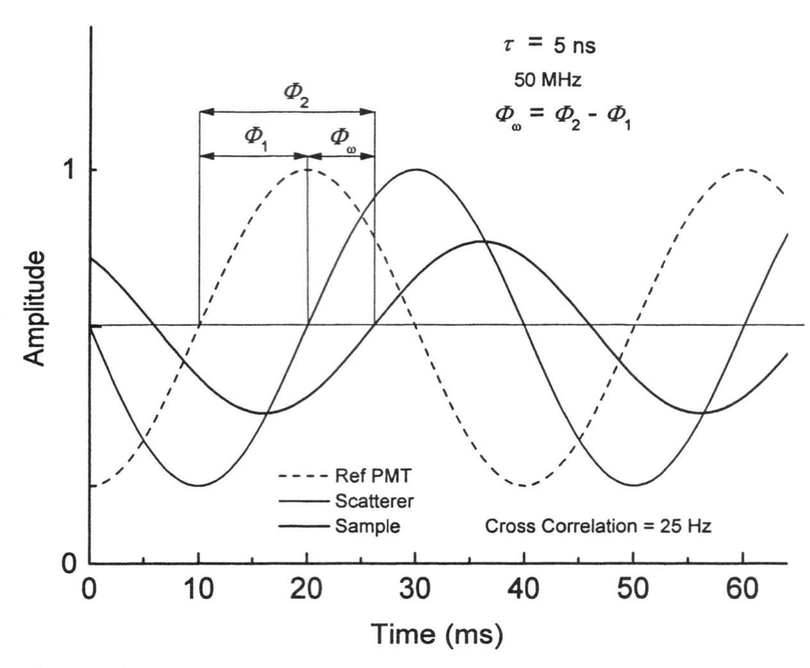

Figure 5.9. Measurement of phase shift and modulation for a modulation frequency of 50 MHz, a lifetime of 5 ns, and a cross-correlation frequency of 25 Hz. The phase angle is 57.5°, and the modulation is 0.54.

either partially polarized excitation or selective observation of one of the polarized components of the emission. If the decay rate is comparable to the rate of rotational diffusion, which is a situation frequently encountered for fluorophores bound to biological macromolecules, then the decay of the individual polarized components of the emission is multiexponential. This effect is unimportant if the rate of rotational diffusion is either much slower or much faster than the decay rate. The effects of rotational diffusion can be canceled by use of vertically polarized excitation and an emission polarizer oriented at 54.7° to the vertical.

5.3. COLOR EFFECTS AND BACKGROUND FLUORESCENCE

5.3.A. Color Effects in Frequency-Domain Measurements

As described in Section 4.6.D, PMTs can display a wavelength-dependent time response. These effects can also be present in the FD measurements.[60–65] In FD measurements the effects are somewhat more difficult to understand. There can be systematic errors in the phase or modulation values, and the direction of the errors is not always intuitively obvious. Fortunately, the color effects are minor with presently used side-window dynode PMTs, and they appear to be negligible with MCP PMTs. Also, systematic errors due to the wavelength-dependent time response are easily corrected using lifetime standards.

In order to correct for a wavelength-dependent response, one uses a lifetime standard in place of the scatterer in the sample turret (Figure 5.5). The standard should display a single-exponential decay of known lifetime τ_R. The lifetime of the standard should be as short as possible, typically near 1 ns, to avoid demodulation, which results in decreased precision of the phase-angle measurements. Another advantage of short-lifetime standards is the minimal effect of dissolved oxygen on the lifetimes. While short lifetimes can be obtained with collisional quenching, this is not recommended because such samples display nonexponential decays. Lifetime standards are summarized in Appendix II. The chosen standard should absorb and emit at wavelengths comparable to those at which the sample absorbs and emits, so that the sample and reference can be observed with the same emission filter and/or monochromator setting. Under these conditions, the PMT observes essentially the same wavelength for both sample and reference measurements, so the color effects are eliminated.

The decay time of the reference results in a phase delay (ϕ_R) and demodulation (m_R) of the reference emission compared to that which would have been observed using a scatterer with a zero lifetime. Of course, ϕ_R and m_R depend on the modulation frequency. The measured values

(ϕ_ω^{obs} and m_ω^{obs}) need to be corrected for this effect. The corrected values are given by

$$\phi_\omega = \phi_\omega^{obs} + \phi_R \quad [5.16]$$

$$m_\omega = m_\omega^{obs} \cdot m_R = m_\omega^{obs} / \sqrt{(1 + \omega^2\tau_R^2)} \quad [5.17]$$

where ϕ_ω^{obs} and m_ω^{obs} are the observed values measured relative to the lifetime standard. These equations can be understood by noting that the observed phase angle is smaller than the actual phase angle owing to the phase angle of the reference ($\phi_\omega^{obs} = \phi_\omega - \phi_R$). Similarly, the observed modulation is the ratio of the modulation of the sample to that of the reference ($m_\omega^{obs} = m_\omega / m_R$). While somewhat different methods have also been proposed,[64,65] we have found this simple approach to be adequate for all samples we have encountered.

5.3.B. Background Correction in Frequency-Domain Measurements

Correcting for background fluorescence from the sample is more difficult than correcting for the color effects.[66,67] In TD measurements, correction for autofluorescence can be accomplished by a relatively straightforward subtraction of the data file measured for the blank from that measured for the sample, with error propagation of Poisson noise if the background level is high. In the frequency domain it is not possible to perform a simple subtraction of the background signal. The background may be due to scattered light with a zero decay time, to impurities with finite lifetimes, or to a combination of scattered light and autofluorescence. One can measure the phase ($\phi_{\omega B}$) and modulation ($m_{\omega B}$) of the background at each light modulation frequency. However, the measured values $\phi_{\omega B}$ and $m_{\omega B}$ cannot be subtracted from the sample data unless the intensities are known and the correction is properly weighted.

Background correction of the FD data is possible, but the procedure is somewhat complex and degrades the resolution of the measurements. Hence, we strive to perform the FD measurements under conditions where background correction is not necessary. If needed, the correction is performed by measuring the frequency response of the background and its fractional contribution (f_B) to the steady-state intensity of the sample. If the background level is low, then the values of $\phi_{\omega B}$ and $m_{\omega B}$ have large uncertainties due to the weak signals. However, this is not usually a problem because if the background is low, its weighted contribution to the sample data is small, so that minimal additional uncertainty is added to the data. If the background is larger, its significance is greater, but it can also be measured with higher precision.

A data file corrected for background is created by the following procedure.[66] Let

$$N_{\omega B} = m_{\omega B} \sin \phi_{\omega B} \quad [5.18]$$

$$D_{\omega B} = m_{\omega B} \cos \phi_{\omega B} \quad [5.19]$$

represent the sine and cosine transforms. In these equations $\phi_{\omega B}$ and $m_{\omega B}$ represent the measured values for the phase and modulation of the background. Alternatively, one can perform a least-squares fit of the phase and modulation data for the background and use the parameter values to calculate $\phi_{\omega B}$ and $m_{\omega B}$. This latter procedure is useful if the background file is not measured at the same modulation frequencies as the sample. In the presence of background, the observed values of N_ω^{obs} and D_ω^{obs} can be represented by

$$N_\omega^{obs} = (1 - f_B)\, m_\omega \sin \phi_\omega + f_B m_{\omega B} \sin \phi_{\omega B} \quad [5.20]$$

$$D_\omega^{obs} = (1 - f_B)\, m_\omega \cos \phi_\omega + f_B m_{\omega B} \sin \phi_{\omega B} \quad [5.21]$$

In these equations f_B is the fraction of the total signal due to the background, and ϕ_ω and m_ω are the correct phase and modulation values in the absence of background. The corrected values of N_ω and D_ω are given by

$$N_\omega = \frac{N_\omega^{obs} - f_B N_{\omega B}}{1 - f_B} \quad [5.22]$$

$$D_\omega = \frac{D_\omega^{obs} - f_B D_{\omega B}}{1 - f_B} \quad [5.23]$$

In using these expressions, the values of $\phi_{\omega B}$ and $m_{\omega B}$ are known from the measured frequency response of the background, and the value of f_B is found from the relative steady-state intensity of the blank measured under the same instrumental conditions and gain as the sample. Except for adjusting the gain and/or intensity, the values of $\phi_{\omega B}$ and $m_{\omega B}$ should be measured under the same conditions as the sample. The corrected values of N_ω and D_ω can be used in Eqs. [5.9] and [5.10] to calculate the corrected phase and modulation values. An important part of this procedure is propagation of errors into the corrected data file. Error propagation is straightforward but complex to describe in detail.[66]

5.4. REPRESENTATIVE FREQUENCY-DOMAIN INTENSITY DECAYS

5.4.A. Exponential Decays

Prior to describing the FD data of biochemical samples, we consider it useful to examine some simple examples.[68] FD intensity decays for anthracene (AN) and 9-cyanoanthracene (9-CA) are shown in Figure 5.10. The samples were in equilibrium with atmospheric oxygen. The emission was observed through a long-pass filter to reject scattered light. Magic-angle polarizer conditions were used, but this is unnecessary for such samples where the emission is completely depolarized. The excitation at 295 nm was obtained from the frequency-doubled output of an R6G dye laser, cavity-dumped at 3.8 MHz. As described in Section 5.7, such light provides intrinsically modulated excitation.

Experimental FD data are represented by the open and filled circles in the upper panel of Figure 5.10. The increasing values are the frequency-dependent phase angles (ϕ_ω), and the decreasing values are the modulation values (m_ω). The solid curves are those calculated for the best single-decay-time fits. In the single-exponential model the decay time is the only variable parameter. The shape of a single-decay-time frequency response is always the same, except that the response is shifted to higher frequencies for shorter decay times.

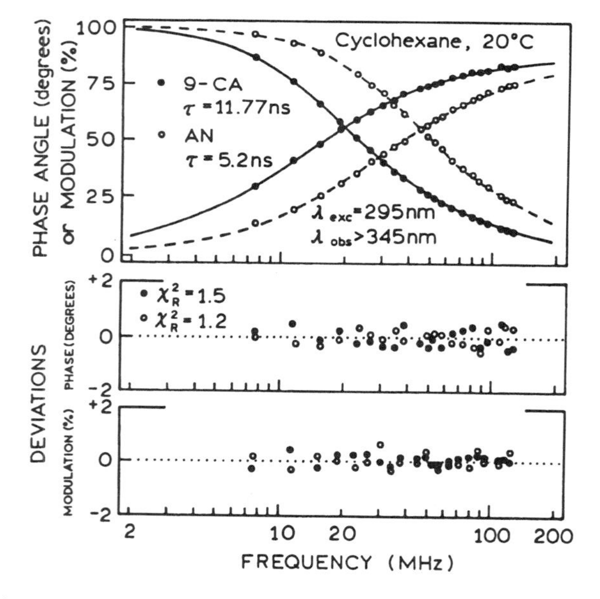

Figure 5.10. Single-exponential decays of anthracene (AN) and 9-cyanoanthracene (9-CA). Samples were equilibrated with atmospheric oxygen. $\delta\phi = 0.2$ and $\delta m = 0.005$. From Ref. 68.

The lower panels show the deviations between the data and the fitted curves. In this case the deviations are presented in units of degrees and percent modulation. For calculation of χ_R^2, the standard errors were taken as $\delta\phi = 0.2$ and $\delta m = 0.005$. The fact that the values of χ_R^2 are close to unity reflects an appropriate choice for the values of $\delta\phi$ and δm. In analysis of the FD data, one cannot rely on the absolute values of χ_R^2 but rather uses the ratio of χ_R^2 values for various models. For AN and 9-CA, the values of χ_R^2 did not decrease for the two-decay-time model, so the single-exponential model was accepted.

5.4.B. Effect of Scattered Light

A critical component of any FD or TD experiment should be collection of emission spectra. One possible artifact is illustrated in Figure 5.11, which shows the emission spectrum of 9,10-diphenylanthracene (DPA) in a solution which also scattered light.[66] 9,10-DPA was dissolved in ethanol which contained a small amount of Ludox scatterer. When the emission was observed without an emission filter (solid curve in Figure 5.11), there was a large peak due to scattered light at the excitation wavelength of 325 nm. The presence of this scattered component would not be recognized without measurement of the emission spectrum and would result in an incorrect intensity decay.

Scattered light is typically rejected from the detector by using emission filters. In this case we used a Corning 3-75 filter, which transmits above 360 nm (dashed curve in Figure 5.11). As a control measurement, one should always record the emission spectrum of the blank sample through the emission filter to ensure that scattered light is rejected. Alternatively, one can measure the intensity of the blank through the filter to determine that the blank contribution is negligible. In such control measurements, it is important

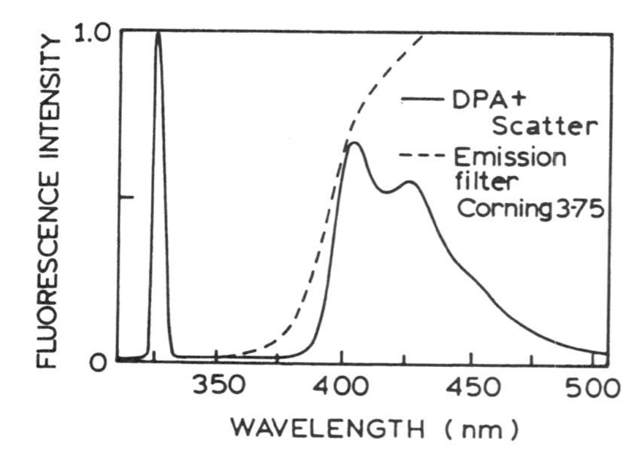

Figure 5.11. Emission spectrum of DPA in ethanol for excitation at 325 nm from a helium–cadmium laser. The solvent contained a small amount of Ludox colloidal silica as the scatterer. The dashed curve is the transmission of the Corning 3-75 filter. Revised from Ref. 66.

that the blank scatters light to the same extent as the sample. Frequently, buffer blanks do not scatter as strongly as the sample containing the macromolecules.

It is useful to understand how scattered light can corrupt the FD data. Frequency responses for the DPA solution are shown in Figure 5.12. For these measurements the excitation source was a helium–cadmium (HeCd) laser at 325 nm. The CW output of this laser was modulated with an EO modulator as shown in Figure 5.6. The effect of scattered light is visually evident in the FD data. When measured without an emission filter, the phase angles (open circles) are considerably smaller than expected for the single-exponential decay of DPA (filled circles). The phase-angle error becomes larger at higher frequencies. It should be noted that the fractional intensity of the background is only 15% ($f_B = 0.15$), so that significant errors in phase angle are expected for even small amounts of scattered light.

It is possible to correct for background from the sample. The filled circles represent the data corrected according to Eqs. [5.18]–[5.23]. The corrected data can be fit to a single decay time with $\tau = 6.01$ ns. An alternative approach is to fit the data with scattered light to include a second component with a lifetime near zero. This also results in a good fit to the data, with a decay time near zero associated with $f_B = 0.15$. However, this procedure is only appropriate if the background is only due to scattered light. In general, autofluorescence will display nonzero lifetimes and nonzero phase angles.

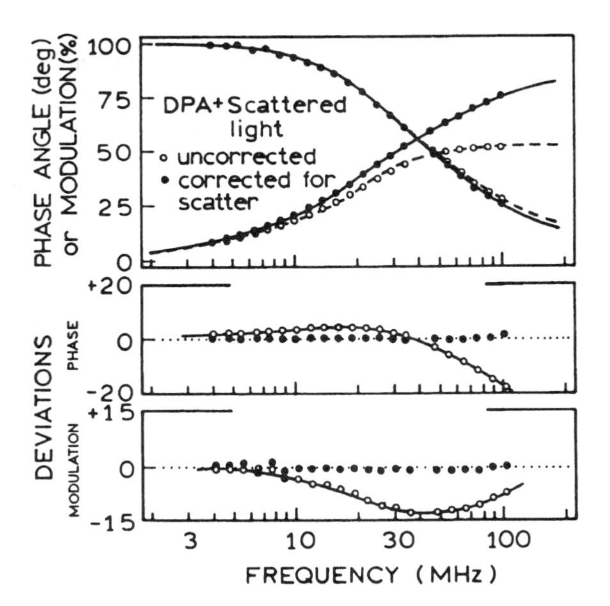

Figure 5.12. Frequency-domain intensity decay of DPA in ethanol with a scatterer. The sample was in equilibrium with dissolved oxygen. Data were measured without an emission filter (○) and then corrected (●) for the scattered light using Eqs. [5.18]–[5.23]. Bottom panels: Deviations from the best single exponential fits for the data with (○), and corrected for (●) scattered light. Revised from Ref. 66.

5.5. ANALYSIS OF MULTIEXPONENTIAL DECAYS

FD data are often analyzed in terms of the multiexponential model. As described in Chapter 4, the amplitudes (α_i) and decay times (τ_i) are usually strongly correlated, so that there can be considerable uncertainty in the values of the recovered parameters. In this section we describe examples of such analyses, with consideration of correlation between the parameter values.

5.5.A. Resolution of Two Widely Spaced Lifetimes

To illustrate the principles of multiexponential analysis, we have chosen the same mixture of *p*-terphenyl and indole as was described in Chapter 4. Emission spectra of this mixture were shown in Figure 4.30. The same 292-nm excitation and 330-nm emission wavelengths were used for the FD measurements as for the TD measurements. The decay times of the individual fluorophores are 0.93 and 3.58 ns for *p*-terphenyl and indole, respectively. For this mixture the decay times are spaced 3.8-fold, making this a moderately easy resolution. FD intensity decay data are shown in Figure 5.13. The presence of more than a single decay time is evident from the shape of the frequency response, which appears to be stretched out along the frequency axis. The fact that the decay is more complex than a single exponential is immediately evident from the attempt to fit the data to a single decay time. The best single-decay-time fit (dashed curves in the top panel of Figure 5.13) is very poor, and the deviations are large and systematic (filled circles, in the lower panels). Also, the value of $\chi_R^2 = 384.6$ is easily rejected as being too large.

Use of a two-decay-time model results in a good fit of the calculated frequency response (solid curves in the top panel of Figure 5.13) to the measured phase and modulation values (filled circles in the top panel). Also, the value of χ_R^2 decreases to 1.34. Use of the three-decay-time model results in a modest reduction of χ_R^2 to 1.24, so that the two-decay-time model is accepted. Once a model is accepted, one examines the recovered parameter values (Table 5.1). For this mixture the recovered decay times of 0.91 and 3.51 ns closely match the lifetimes measured for the individual fluorophores. The recovered amplitudes and fractional intensities suggest that about 64% of the emis-

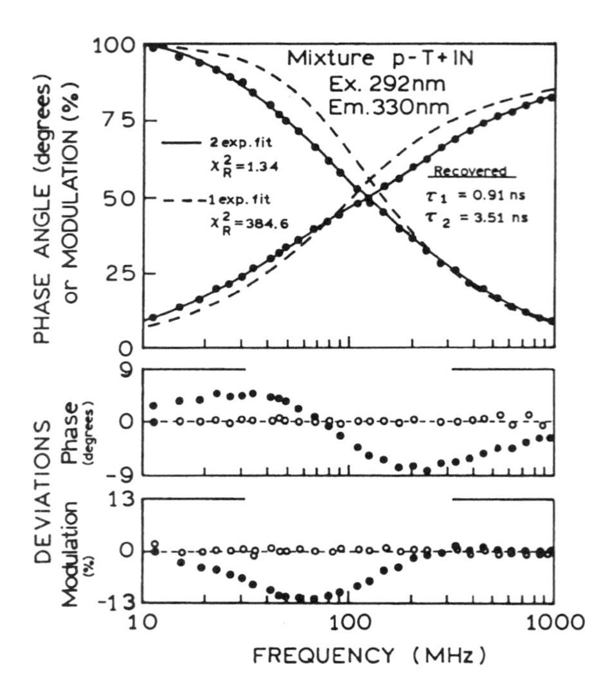

Figure 5.13. Frequency-domain intensity decay data of a two-component mixture of *p*-terphenyl (P-T) and indole (IN) in ethanol, at 20 °C. Magic-angle polarization conditions were used. The sample was in equilibrium with air. The repetition rate was 1.9 MHz from a frequency-doubled R6G dye laser. Emission at 330 nm was isolated with a monochromator. The dashed curves (top panel) and filled circles (lower panels) show the best fit with one decay time; the solid curves (top panel) and open circles (lower panels) show the best fit with two decay times. From Ref. 69.

sion at 330 nm is from the indole, with a decay time of 3.51 ns (Table 5.1).

An important part of data analysis is estimating the confidence intervals for each parameter. Most computer programs report the ASEs, which are the uncertainties calculated under the assumption that the parameters are not correlated. In reality, the range of possible parameters is usually 2- to 10-fold larger than that estimated from the ASEs. For this reason we prefer to examine the χ_R^2 surfaces, that is, to perform a support plane analysis. The upper and lower limits of each parameter are determined from the χ_R^2 ratio as the parameter value is held fixed around the optimal value. The least-squares analysis is then run again to obtain the lowest value of χ_R^2 consistent with the fixed parameter value. One then calculates the ratio of χ_R^2 values,

$$F_\chi = \frac{\chi_R^2(\text{par})}{\chi_R^2(\text{min})} = 1 + \frac{p}{m-p}[F(p, \nu, P)] \qquad [5.24]$$

where $\chi_R^2(\text{par})$ is the value of χ_R^2 with a parameter value held fixed at a value different from that yielding the minimum value of $\chi_R^2(\text{min})$. The upper and lower bounds of each parameter are selected as those where the χ_R^2 ratio matches the F_χ value for one standard deviation ($P = 0.32$) and the number of parameters (p) and degrees of freedom ($\nu = m - p$). It is useful to consider some representative values for the ratio of χ_R^2 values. For the two-component mixture of *p*-terphenyl and indole, there are 29 frequencies and 58 data points. The two-decay-time model has three variable parameters. The *F*-statistic for $p = 3$ and $m = 60$ can be found from Table 4.4 and is 1.19. Hence, the F_χ value used for the confidence interval of each parameter is 1.06. As described in Section 4.9.C, there is some disagreement in the statistics literature as to the exact equation for finding the confidence intervals. We will use Eq. [5.24] but emphasize that its correctness has not yet been proven.

The χ_R^2 surfaces for the two-component mixture are shown in Figure 5.14. The confidence intervals are determined from the intercept of the χ_R^2 surfaces with the χ_R^2 ratio appropriate for the number of parameters and degrees of freedom. Also shown in Figure 5.14, as solid bars, are the ASEs. One notices that the ASEs are about twofold smaller than the confidence intervals. We note that the decay times in this mixture are widely spaced. For more closely spaced decay times, it becomes even more important to consider parameter correlation in the calculation of confidence in-

Table 5.1. Multiexponential Analysis of a Two-Component Mixture of *p*-Terphenyl and Indole[a]

	Lifetimes (ns)		Preexponential factors		Fractional intensities		χ_R^2 Number of decay times	
Sample	τ_1	τ_2	α_1	α_2	f_1	f_2	1[b]	2[b]
p-Terphenyl	0.93		1.0		1.0		1.38	1.42
Indole		3.58		1.0		1.0	1.10	1.13
Mixture	0.91	3.51	0.686	0.314	0.36	0.64	384.6	1.34[c]

[a] $\delta\phi = 0.2$ and $\delta m = 0.005$. From Ref. 69.
[b] Refers to a one- or a two-component fit.
[c] The value of χ_R^2 for the three-decay-time fit for the mixture was 1.24.

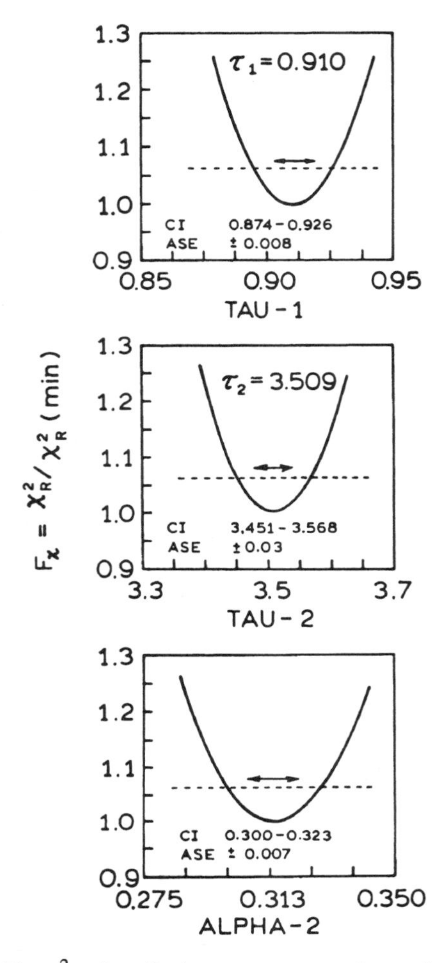

Figure 5.14. χ^2_R surfaces for the two-component mixture of *p*-terphenyl and indole in Figure 5.13. The horizontal arrows show the ASEs, and the dashed lines are at the appropriate F_χ values. From Ref. 69.

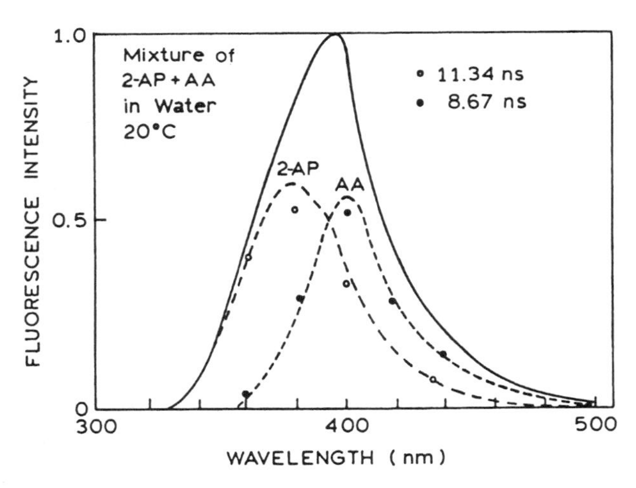

Figure 5.15. Emission spectrum of a two-component mixture of anthranilic acid (AA) and 2-aminopurine (2-AP). Also shown are the amplitudes recovered from the global analysis (Figure 5.18). From Ref. 69.

tervals, and the ASEs can grossly underestimate the true confidence intervals.

5.5.B. Resolution of Two Closely Spaced Lifetimes

The resolution of multiexponential decays becomes more difficult as the decay times become more closely spaced. It was previously noted that two decay times spaced by a factor of 1.4 represents the practical resolution limit for double-exponential decay.[12] It is instructive to examine data for such a mixture because the analysis illustrates the difficulties encountered as one reaches the limits of resolution.

To illustrate a sample with two closely spaced decay times, we have chosen the mixture of anthranilic acid (AA, τ = 8.5 ns) and 2-aminopurine (2-AP, τ = 11.3 ns). While this may seem to be an easy resolution, it is difficult to resolve decay times which are less than twofold different. Emission spectra are shown in Figure 5.15. FD data for the individual fluorophores and for the mixture are shown in Figure 5.16. Each of the single fluorophores displays a frequency response characteristic of a single decay time. The values of χ^2_R for the single-decay-time fits were acceptably low and were not improved by using a two-decay-time model. One notices that the 33% difference in decay time results in only a modest shift on the frequency axis. Also shown in Figure 5.16 (middle panel) are the FD data for the two-component mixture of AA and 2-AP. FD data were measured through a 400-nm interference filter; at this wavelength, both fluorophores contribute almost equally to the measured intensities. For these two closely spaced lifetimes, it is difficult to see a difference between the calculated curves for the one- and two-decay-time fits. The deviations plots (Figure 5.16, lower panels) show larger and systematic deviations for the one-component fit. Also, χ^2_R decreases from 3.31 for the one-decay-time fit to 1.45 for the two-decay-time fit. For these measurements we have approximately 40 degrees of freedom. The *F*-value of 2.3 is seen to be significant at the 1% level (Table 4.3), and there is less than a 0.1% probability that random noise is the origin of the elevated χ^2_R value for the one-decay-time fit (Table 4.2).

In general, we accept the more complex model if χ^2_R decreases by at least 50%, preferably twofold. One often observes a χ^2_R value that is larger than expected but χ^2_R does not decrease for the next more complex model. In these cases the elevated χ^2_R is usually due to systematic errors in the measurements. It is fortunate that in many cases systematic errors cannot be accounted for by another decay time component in the model.

Table 5.2. Resolution of a Two-Component Mixture of Anthranilic Acid and 2-Aminopurine, Observed at a Single Wavelength[a]

Observation wavelength (nm)	Lifetimes (ns)		Preexponential factors		Fractional intensities		χ_R^2	
	τ_1	τ_2	α_1	α_2	f_1	f_2	2[b]	1[b]
360	6.18	11.00	0.037	0.963	0.021	0.979	0.95	1.07
380	8.99	12.26	0.725	0.275	0.659	0.341	1.24	2.06
400	6.82	10.29	0.268	0.732	0.195	0.805	1.54	3.44
420	8.44	12.15	0.832	0.168	0.775	0.225	1.36	2.22
440	7.69	9.73	0.479	0.521	0.421	0.579	1.69	2.01

[a] $\delta\phi = 0.2$ and $\delta m = 0.005$. From Ref. 69.
[b] Refers to a two- or a one-component fit.

While the twofold reduction in χ_R^2 for the two-decay-time model seems reasonable, the recovered parameter values have considerable uncertainty. At 400-nm, the two-component analysis reports a fractional intensity of 80% for the longer lifetime and decay times of 6.82 and 10.29 ns (Table 5.2). These values are considerably different from the expected fractional intensity of about 40% for the longer-decay-time component and the expected decay times of 8.5 and 11.3 ns. In fact, analysis of the intensity decays at a number of wavelengths reveals considerable variability in the recovered lifetimes and fractional intensities. This can be seen in Figure 5.17 from the range of lifetimes found within the ASEs (solid bars). For instance,

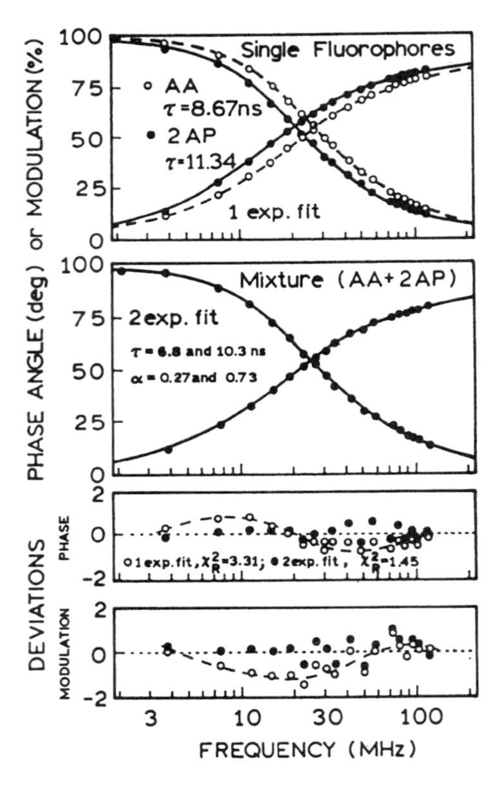

Figure 5.16. Frequency-domain intensity decays for 2-AP and AA. *Top*: Single fluorophore data at 400 nm; *middle*: the two-component mixture measured at 400 nm; *bottom*: deviations for fits to the data for the two-component mixture. The values of χ_R^2 for the one- and two-component fits are 3.31 and 1.45, respectively. From Ref. 69.

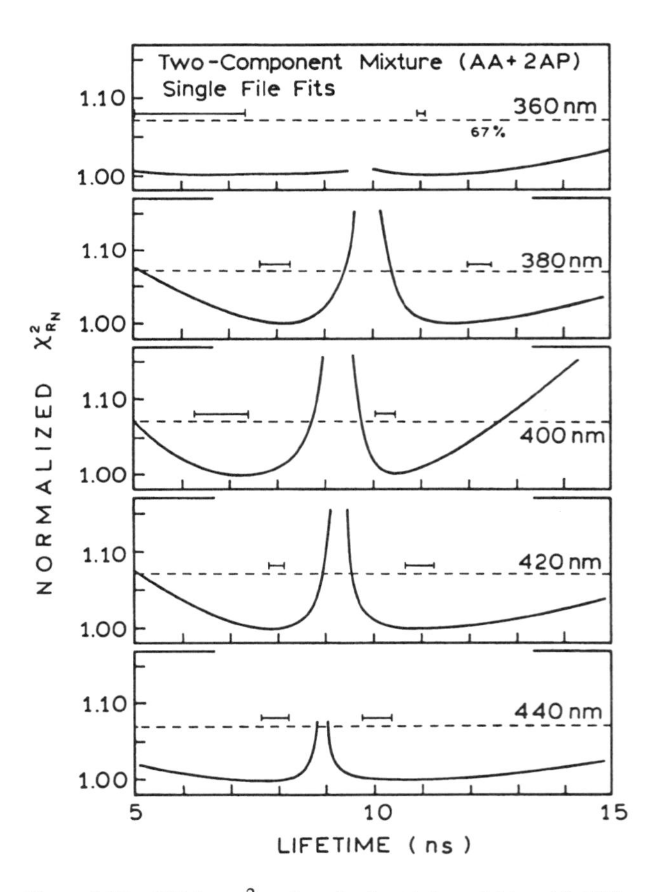

Figure 5.17. Lifetime χ_R^2 surfaces for the mixture of AA and 2-AP for single observation wavelengths. The solid bars show the ASEs. From Ref. 69.

the lifetime of the shorter-decay-time component ranges from 6.18 to 8.99 ns, and the lifetime of the longer component ranges from 9.73 to 12.26 ns (Table 5.2). The amplitudes and fractional intensities recovered from the measurements at different wavelengths do not accurately represent the emission spectra of the two fluorophores. This is seen at 400 nm, where the amplitude of AA is expected to be dominant but a low value was recovered (0.195, Table 5.2). Such variability is typical in the analysis of closely spaced decay times, whether the measurements are performed in the frequency domain or in the time domain.

When the resolution limits are approached, we prefer to estimate the uncertainties from the χ_R^2 surfaces (Figure 5.17). These surfaces for the mixture of AA and 2-AP are moderately wide due to the closely spaced lifetimes and correlation between the lifetimes and amplitudes. The lack of resolution is especially apparent for observation at 380 nm. In this case the emission is due dominantly to just one of the fluorophores (AA). One may wonder why one of the lifetimes can be fixed near 12 ns without an elevation in χ_R^2, when the dominant lifetime is near 8 ns. This probably occurs because the amplitude of the 12-ns component can be decreased as the new χ_R^2 values are calculated.

At observation wavelengths where both AA and 2-AP emit there are modest minima in the χ_R^2 surfaces (Figure 5.17). However, the confidence intervals are large, and in some cases the desired χ_R^2 increase is not reached for any reasonable value of the decay times. It is important to notice that the confidence intervals, which are the ranges of parameter values consistent with the data, are not symmetrical about the best-fit values.

It is easy to be misled by accepting the ASEs from nonlinear least-squares analysis. In Figure 5.17 we showed the ASEs reported for the two-component mixture of *p*-terphenyl and indole. The ASEs are typically much smaller than the confidence intervals estimated from the χ_R^2 values. One also notices that the values of the recovered lifetimes, as seen from the ASEs, are not the same at each wavelength, and in some cases the lifetime ranges from the ASEs do not even overlap. This shows that the variability in the recovered lifetimes is in fact greater than the ASEs, demonstrating that the ASEs are too small. In contrast, the confidence intervals calculated from the χ_R^2 surfaces are larger and overlap at all wavelengths.

5.5.C. Global Analysis of a Two-Component Mixture

The resolution of complex intensity decays can be dramatically enhanced by global analysis, which is the simultaneous analysis of multiple data sets measured under slightly different conditions. For the two-component mixture of AA and 2-AP, data were measured for five emission wavelengths (Figure 5.18), between 360 and 440 nm. The decay time of each fluorophore is expected to be independent of emission wavelength. Hence the global analysis is performed as described in Eqs. [5.13] and [5.15], where the $\alpha_i(\lambda)$ values are assumed to be different at each wavelength but the τ_i values are assumed to be independent of wavelength.

Results of the global analysis are shown in Figure 5.18 and Table 5.3. The value of $\chi_R^2 = 37.4$ for the one-component fit is easily rejected. Use of the two-component model results in a decrease of χ_R^2 to 1.33. For the global analysis, the frequency responses at each emission wavelength are in good agreement with the calculated curves when two wavelength-independent decay times are used. Use of three decay times does not improve χ_R^2, so the two-decay-time model is accepted.

Global analysis results in less uncertainty in the recovered parameters. The lifetime χ_R^2 surfaces from the global analysis are much steeper when calculated using the data at five emission wavelengths (Figure 5.19, solid curves). The elevated values of χ_R^2 are more significant because of the larger number of degrees of freedom. For this global analysis there are approximately 200 data points, and seven variable parameters. Hence, the *F*-statistic is 1.16 (Table 4.4), and the F_χ value is 1.04 (Eq. [5.24]). Global analysis

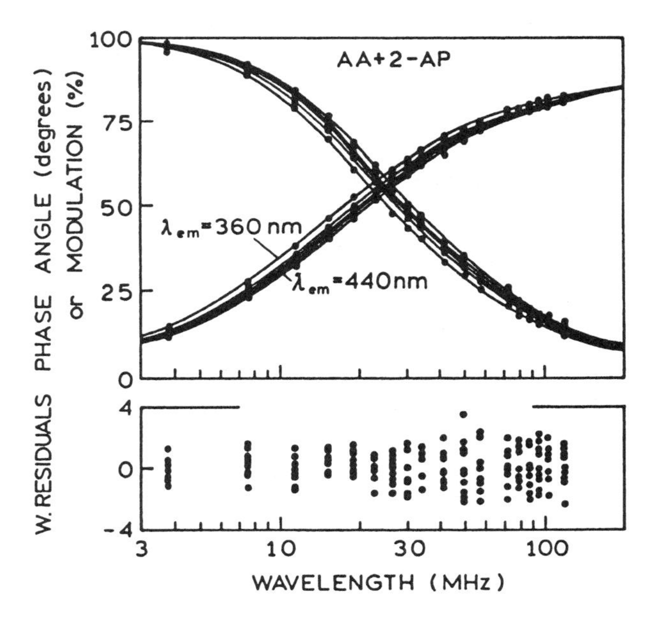

Figure 5.18. Global analysis of the two-component mixture of AA and 2-AP. From left to right, the data are for 360, 380, 400, 420 and 440 nm. The values of χ_R^2 for the global one- and two-component fits are 37.4 and 1.33, respectively. From Ref. 69.

Table 5.3. Global Analysis of a Two-Component Mixture of Anthranilic Acid (AA) and 2-Aminopurine (2-AP) Measured at Five Emission Wavelengths[a,b]

Observation wavelength (nm)	AA (τ_1 = 8.19 ns)[c]		2-AP (τ_2 = 11.18 ns)[c]	
	α_1	f_1	α_2	f_2
360	0.117	0.089	0.883	0.911
380	0.431	0.357	0.569	0.643
400	0.604	0.528	0.396	0.472
420	0.708	0.640	0.292	0.360
440	0.810	0.758	0.190	0.242

[a] $\delta\phi$ = 0.2 and δm = 0.005. From Ref. 69.
[b] For the two-component fit, χ_R^2 = 1.33; for the one-component fit, χ_R^2 = 37.4.
[c] Lifetimes assigned to these fluorophores based on measurements of the individual fluorophores (Figure 5.16).

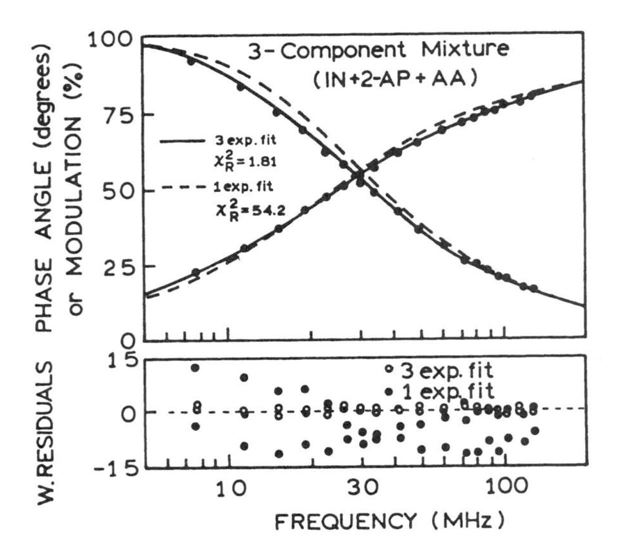

Figure 5.20. Frequency-domain intensity decay for a three-component mixture of indole (IN), 2-aminopurine (2-AP), and anthranilic acid (AA) in water, at 20°C, pH 7, observed at 380 nm. The χ_R^2 values for the one-, two-, and three-decay-time fits are 54.2, 1.71, and 1.81, respectively. From Ref. 69.

also results in improved estimates of the amplitudes. The fractional intensities (f_i) and decay times (τ_i) recovered from the global analysis closely match those expected from the spectral properties of the individual fluorophores (Figure 5.15).

5.5.D. Analysis of a Three-Component Mixture—Limits of Resolution • Advanced Topic •

In our opinion, a three-component mixture with less than a threefold range in lifetime represents the practical limit of resolution for both TD and FD measurements. Analysis of the data from such a sample illustrates important considerations in data analysis at the limits of resolution. FD intensity decay data for the mixture of indole (4.41 ns), anthranilic acid (8.53 ns), and 2-aminopurine (11.27 ns) are shown in Figure 5.20. The data cannot be fit to a single decay time, resulting in χ_R^2 = 54.2, so this model is easily rejected. The situation is less clear for the two- and three-decay-time fits, for which the values of χ_R^2 are 1.71 and 1.81, respectively. At first glance, it seems that χ_R^2 has increased for the triple-exponential fit. However, the increase in χ_R^2 for the three-decay-time model is a result of the larger number of variable parameters and the smaller number of degrees of freedom. In such cases one should examine the value of χ^2, which is the sum of the squared deviations. The values of χ^2 are 2006, 59.7, and 59.6, for the one-, two-, and three-decay-time fits, respectively. Hence, the fit is not worse for the three-decay-time model but is essentially equivalent to that for the two-decay-time model.

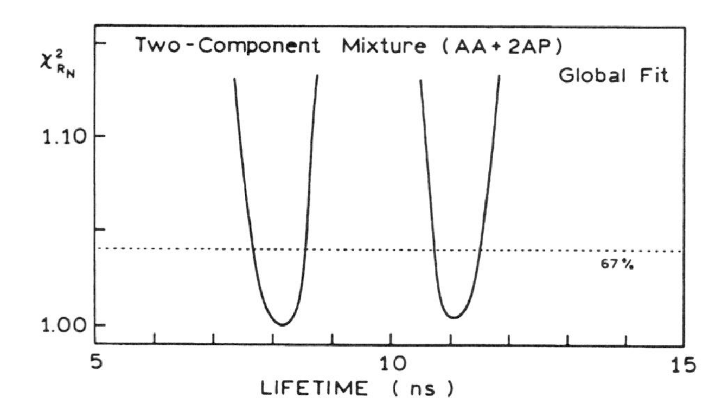

Figure 5.19. Lifetime χ_R^2 surfaces for the mixture of AA and 2-AP. The 67% line refers to the F_χ values for global analysis. From Ref. 69.

Samples such as this three-component mixture are difficult to analyze. In this case we know that there are three decay times, and, in fact, the three decay times are correctly determined by the analysis. However, obtaining the correct values required that the parameter starting values in the least-squares analysis be close to the correct values. Otherwise, the program stopped at incorrect values, apparently trapped in local χ_R^2 minima. Additionally, the χ_R^2 surface is essentially independent of lifetime, as shown in Figure 5.21 for the data measured at 380 nm (dashed line). Hence, without prior knowledge of the presence of three decay times, one would not know whether to accept the two- or three-decay-time fit.

When one reaches this point in an analysis, there is little reason for proceeding further. If the information is not

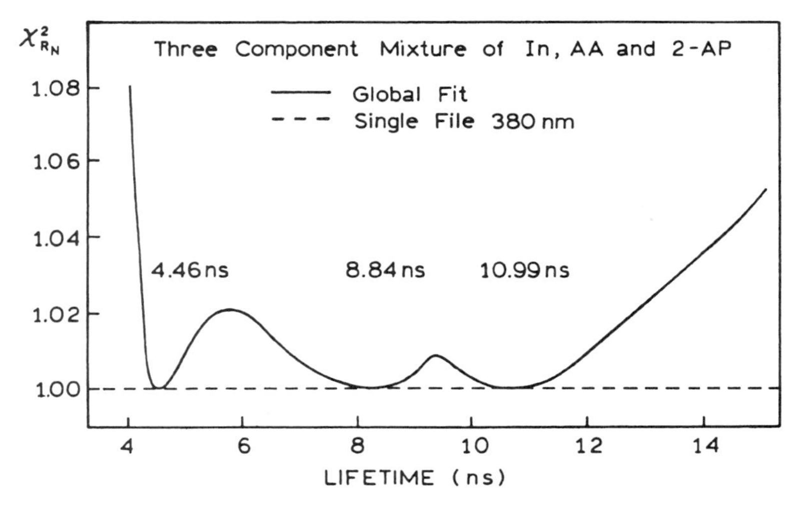

Figure 5.21. Lifetime χ_R^2 surfaces for the three-component mixture in Figure 5.20. – – –, 380 nm; ———, global fit at 360, 380, 400, 420, and 440 nm. From Ref. 69.

present in the data, no amount of analysis will create new information. One can either add new experimental data or add information by restricting parameters based on separate knowledge about the sample. If any of the lifetimes are known, these can be held constant during the least-squares fit. Similarly, one of the amplitudes could be fixed. However, the best approach is to add new data and perform a global analysis.

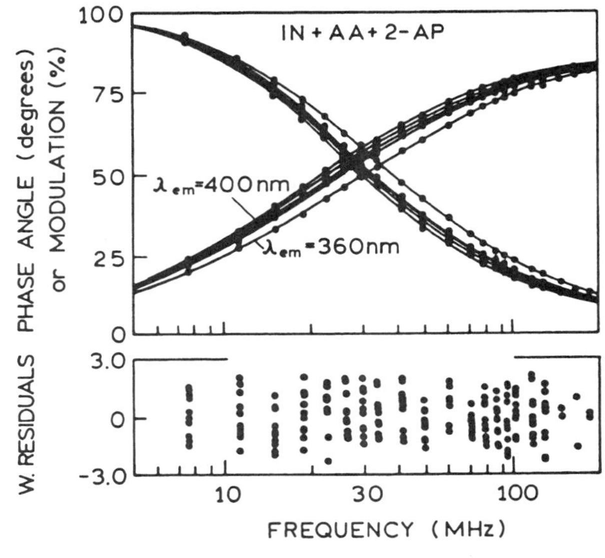

Figure 5.22. Frequency-domain intensity decays of the three-component mixture in Figure 5.20, observed at 360, 380, 400, 420, and 440 nm. The solid curves are for the best fit to three global decay times and nonglobal amplitudes. The values of χ_R^2 for the one-, two-, and three-decay-time fits, are 109.8, 2.3, and 1.1, respectively. From Ref. 69.

The emission from the three-component mixture was measured at five wavelengths, 360, 380, 400, 420, and 440 nm. At each wavelength each fluorophore displays the same decay time but a different fractional amplitude based on its emission spectrum (Figure 5.22). Because of the different amplitudes at each wavelength, the frequency responses are wavelength-dependent (Figure 5.23). One can understand the relative position of these curves by recognizing that indole displays the shortest lifetime (4.41 ns) and emits toward shorter wavelengths. One expects the mean lifetime to be largest near 380 nm, which is the

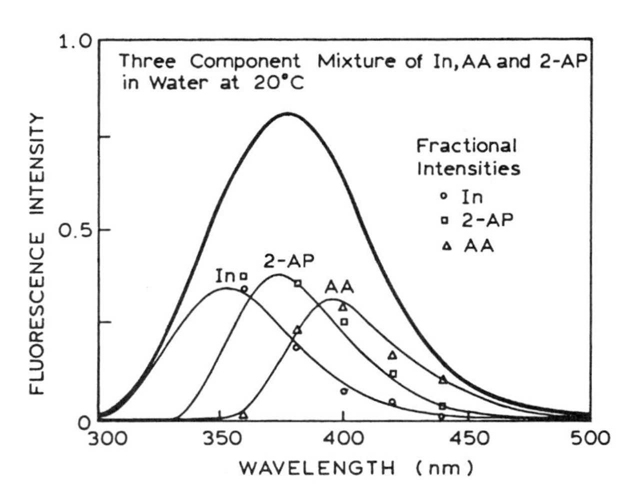

Figure 5.23. Emission spectrum of the three-component mixture from Figure 5.20. Also shown are the fractional intensities recovered from global analysis of the FD data. From Ref. 69.

Table 5.4. Global Analysis of the Frequency-Domain Data for a Three-Component Mixture of Indole, Anthranilic Acid, and 2-Aminopurine[a,b]

Observation wavelength (nm)	$\tau_1 = 4.46$ ns		$\tau_2 = 8.84$ ns		$\tau_3 = 10.99$ ns	
	α_1	f_1	α_2	f_2	α_3	f_3
360	0.700	0.488	0.008	0.011	0.292	0.501
380	0.434	0.254	0.244	0.282	0.322	0.464
400	0.235	0.123	0.429	0.444	0.336	0.433
420	0.306	0.169	0.430	0.471	0.264	0.360
440	0.219	0.121	0.687	0.752	0.094	0.127

[a] $\delta\phi = 0.2$ and $\delta m = 0.005$. From Ref. 69.
[b] $\chi_R^2 = 1.19$, 2.30, and 109.8 for the three-, two-, and one-component fits, respectively.

emission maximum of 2-AP (11.27 ns). The mean lifetime decreases at longer wavelengths as the emission becomes dominated by AA (8.53 ns).

Figure 5.22 and Table 5.4 show the results of global analysis of the wavelength-dependent data. The one- and two-component fits are easily rejected on the basis of the χ_R^2 values of 109.8 and 2.30, respectively, which are both significantly larger than $\chi_R^2 = 1.19$ for the three-decay-time fit. One's knowledge of the parameter values can be tested by examining χ_R^2 surfaces (Figure 5.21). For the single-wavelength data at 380 nm, the value of χ_R^2 was insensitive to fixing of any of the three decay times. When this occurs, recovery of the correct decay times should be regarded as a fortunate coincidence rather than evidence for the resolution obtainable from the data. For the global data, the χ_R^2 surfaces display distinct minima at the correct lifetimes, providing good estimates of the values and range consistent with the data. Also, the recovered amplitudes now closely match those expected from the known emission spectra of the fluorophores (Figure 5.22). By performing additional measurements at different wavelengths and global analysis, a sample that had been unresolvable became a readily resolvable mixture. It is interesting to notice that the magnitude of the χ_R^2 ratio is larger for the more widely spaced decay times. To be specific, the ratio increases to 1.02 between the 4.46- and 8.84-ns lifetimes, and to only 1.01 between the 8.84- and 10.99-ns lifetimes. This effect illustrates why it is more difficult to resolve more closely spaced lifetimes. Finally, it is interesting to consider the F_χ value appropriate for this analysis. There are approximately 200 data points and 13 parameters. The χ_R^2 ratio is 1.15, so $F_\chi = 1.08$. Hence, the confidence intervals overlap for the three lifetimes.

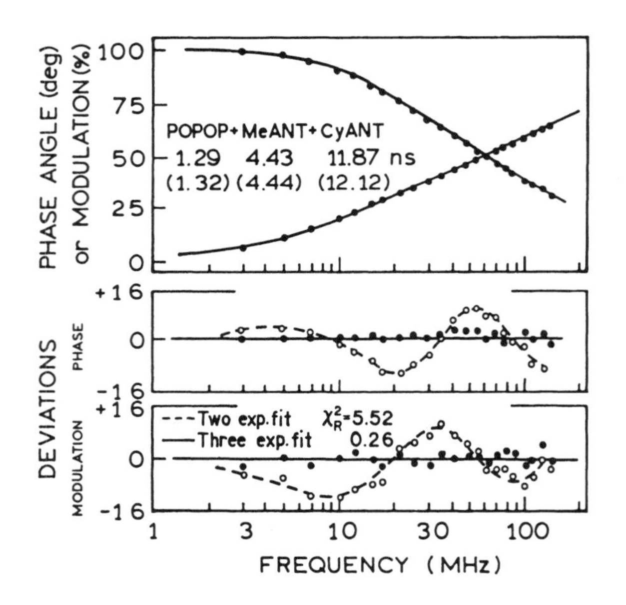

Figure 5.24. Phase and modulation data for a three-component mixture of POPOP, 9-methylanthracene (MeANT), and 9-cyanoanthracene (CyANT) in ethanol at 20 °C, in equilibrium with air. The values shown are recovered from the three-component fit, and the values in parentheses are from separate measurements of the individual pure compounds. The fractional intensities of the three components were equal ($f_1 = f_2 = f_3 = \frac{1}{3}$). The excitation wavelength was 325 nm, and the emission filter was Corning 0-52. The measurements were performed relative to a POPOP reference solution with a reference lifetime of 1.32 ns. $\delta\phi = 0.3°$ and $\delta m = 0.003$. Revised from Ref. 13 and reprinted with permission from the Biophysical Society.

5.5.E. Resolution of a Three-Component Mixture with a 10-Fold Range of Decay Times

The ability to resolve a three-component mixture increases rapidly as the decay times become more widely spaced. A mixture with a 10-fold range of lifetimes is provided by POPOP (1.32 ns), 9-methylanthracene (4.44 ns), and 9-cyanoanthracene (12.12 ns). The relative value of χ_R^2 decreased 20-fold for the three-decay-time fit relative to the two-decay-time fit (Figure 5.24). For this mixture the calculated values for the two-decay-time model (represented by the open circles in Figure 5.24) differ systematically from the data, whereas the deviations from the three-decay-time model (filled circles) are randomly distributed. In this analysis the value of $\chi_R^2 = 0.26$ for the three-decay-time fit seems too small. This is not an error but indicates that the assumed values of $\delta\phi = 0.3$ and $\delta m = 0.003$ are too large and that the actual uncertainties are smaller. These data were measured with an HeCd laser, which provides highly stable illumination. From these results we see that three decay times with a 10-fold range are easily recovered, whereas, as illustrated by the previous example, three lifetimes with a threefold range are difficult to resolve.

5.6. BIOCHEMICAL EXAMPLES OF FREQUENCY-DOMAIN INTENSITY DECAYS

5.6.A. Monellin—A Single-Tryptophan Protein with Three Decay Times

Monellin is a sweet-tasting protein with a single tryptophan residue. The FD intensity decay was measured with 300-nm excitation (Figure 5.25) to avoid excitation of tyrosine residues. The intensity decay is more complex than a single decay-time (not shown). Even the two-decay-time fit results in $\chi_R^2 = 2.83$, which is somewhat too large. The deviations from the two-decay-time fit (open circles in Figure 5.25) show a nonrandom distribution. Analysis with three decay times results in more than a threefold reduction in χ_R^2 to 0.74 and more randomly distributed deviations. The three-decay-time model is thus accepted as describing the sample.

In spite of the improved fit with three decay times, it is important to remember that the results only indicate that the data are consistent with three decay times but do not exclude other models of the same or greater complexity. The actual decay for monellin could have four decay times or be due to a distribution of decay times beyond the resolution limits of the data.

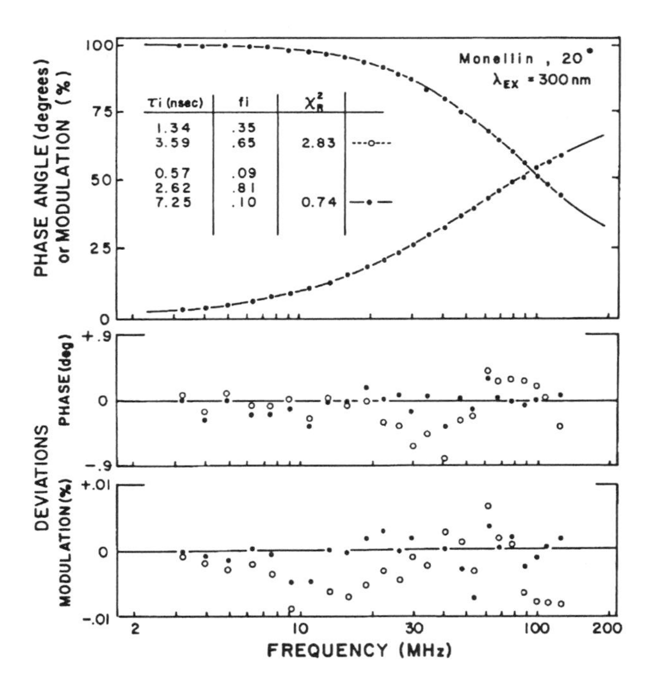

Figure 5.25. Fluorescence intensity decay of the single tryptophan residue in monellin determined by frequency-domain fluorometry. From Ref. 8.

5.6.B. Multiexponential Decays of Staphylococcal Nuclease and Melittin

Most single-tryptophan proteins display multiexponential intensity decays. This is illustrated for two proteins in Figure 5.26.[70] The intensity decay of trp-140 in staphylococcal nuclease is at least a double exponential. The intensity decay of trp-19 in melittin is at least a triple exponential. Under these experimental conditions (1*M* NaCl), melittin is a tetramer, with the monomers each in the α-helical state. The frequency responses and recovered lifetimes in Figures 5.25 and 5.26 are typical of many single- and multitryptophan proteins.

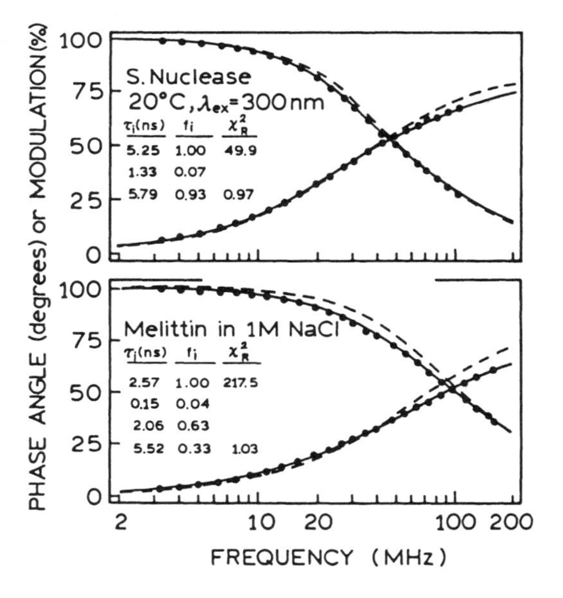

Figure 5.26. Frequency responses of staphylococcal nuclease and melittin tetramer. The data are the solid dots. The best single (– – –) and double- or triple-exponential fits (———) are shown. Data from Ref. 70.

5.6.C. DNA Labeled with DAPI

The probe DAPI is widely used to study DNA.[71–75] DAPI binds to the minor groove of DNA and shows preferential binding to AT-rich regions of DNA. DAPI is only weakly fluorescent in water (Figure 5.27a) and displays an increase in quantum yield upon binding to DNA.[75] The increase in quantum yield is minimal for binding to poly (dG-dC). A more significant enhancement in DAPI fluorescence is found upon binding to DNA containing both GC and AT pairs (circular and linear DNA in Figure 5.27). The largest enhancement of DAPI fluorescence is found for binding of DAPI to poly (dA)-poly (dT) or to poly (dA-dT). The dependence of DAPI fluorescence on the base composition of DNA suggests that DAPI will display complex decay kinetics when bound to DNA because the DAPI will be near both AT and GC base pairs.

FD intensity decays of DAPI are shown in Figure 5.28. Excitation was at 325 nm from an HeCd laser. For the intensity decay measurements, the entire emission was observed using an RG370 long-pass filter. Measurements were performed using POPOP in ethanol as a reference, with a lifetime of 1.35 ns. The frequency responses are visually heterogeneous for DAPI bound to poly (dG-dC) and to linear DNA (Figure 5.28), indicating that DAPI is bound in more than one environment with different decay times. The average lifetime is longest when DAPI is bound to poly (dA)-poly (dT), as seen from the frequency response shifted to the lowest frequencies.

The intensity decays could be fit to the two-decay-time model (Table 5.5). In the absence of DNA, the intensity is dominated by a 0.19-ns component. The decay is nearly unchanged by the presence of poly (dG-dC), except for an increase in the decay time of the short component to 0.6 ns. Substantial changes in the intensity decay are found upon binding to the other DNAs, where the decays are dominated by a 3.8–3.9-ns component. Linear and closed circular (CC) DNA has both AT and GC base pairs. In these cases the intensity decay is heterogeneous due to the presence of DAPI bound to both types of base pairs. Binding of DAPI to a homogeneous DNA, either poly (dA-dT) or poly (dA)-poly (dT), results in a homogeneous decay. These results show how the time-resolved decays can be used to learn about the presence of more than one type of binding site for a fluorophore.

As discussed in Section 5.10, the lifetimes calculated from the phase and modulation at a single frequency are only apparent values. The heterogeneous decay of DAPI in water illustrates this effect.[75] For an observation wavelength of 470 nm and a modulation frequency of 100 MHz,

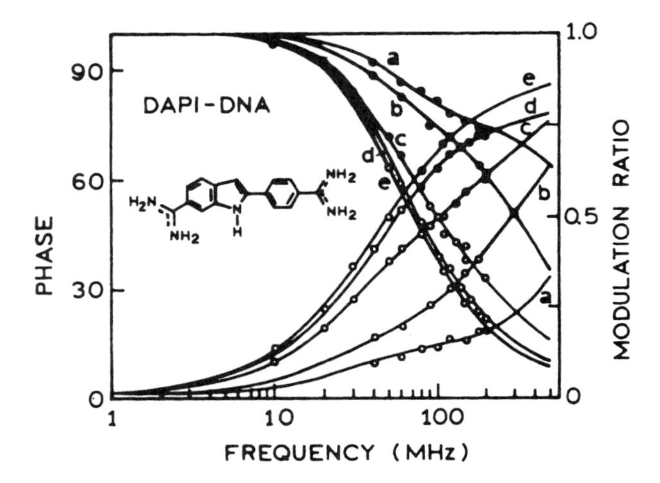

Figure 5.28. Frequency-domain intensity decays of DAPI in water at pH 7.1 (a) and complexed with poly(dG-dC) (b), linear DNA (c), circular DNA (d), and AT polymers (e). The solid curves correspond to the best fits obtained using a double-exponential model. Revised from Ref. 75.

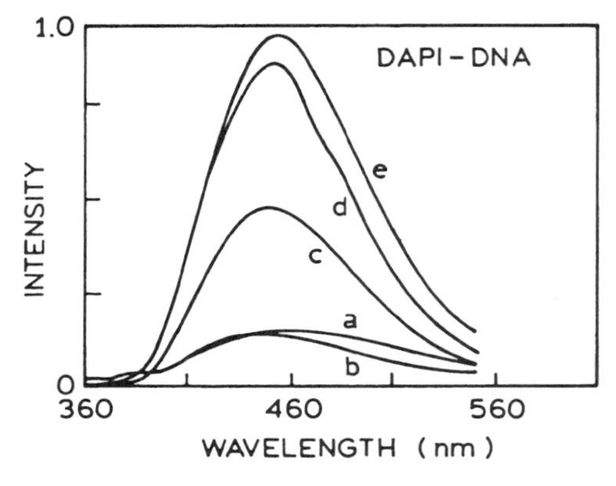

Figure 5.27. Emission spectra of DAPI in water at pH 7.1 (a) and complexed with poly(dG-dC) (b), closed circular DNA (c), linear DNA (d),and poly(dA-dT) (e). Excitation at 340 nm. Revised from Ref. 75.

Table 5.5. Intensity Decays of DAPI Bound to Various DNAs[a]

DNA[b]	τ_1 (ns)	τ_2 (ns)	f_1	f_2
None[c]	0.19	2.80	0.73	0.27
Poly(dG-dC)	0.6	2.9	0.71	0.29
CC-DNA[d]	0.4	3.8	0.05	0.95
Linear DNA	0.5	3.8	0.20	0.80
Poly(dA-dT)	—	3.9	—	1.00
Poly(dA)-poly(dT)	—	3.9	—	1.00

[a]From Ref 75. The DNA phosphate to DAPI ratio was 60.
[b]The measurements with DNA were at 24°C, pH 7.2 in 0.1*M* Tris with 0.1*M* NaCl.
[c]DAPI in water at pH 7.1.
[d]Closed circular DNA-CoEl plasmid DNA.

the apparent phase and modulation lifetimes are τ_ϕ^{app} = 0.47 ns and τ_m^{app} = 1.24 ns. The fact that $\tau_\phi < \tau_m$ indicates that the decay is heterogeneous. In Section 22.2 we show how the wavelength-dependent values of the phase and modulation, or τ_ϕ and τ_m, can be used to recover the emission spectra associated with each decay time.

5.6.D. Quin-2—A Lifetime-Based Sensor for Calcium

Calcium-sensitive fluorophores are widely used in cell biology to measure intracellular calcium concentrations. While most calcium measurements are based on intensity or wavelength-ratiometric measurements, one can also use intensity decay measurements. Lifetime measurements offer the advantage of being independent of the total fluorescence intensity and can thus be used in fluorescence microscopy, where quantitative intensity measurements are difficult.

The use of calcium probes as lifetime-based sensors requires an understanding of how the decay times change in response to Ca^{2+}. One example is shown in Figure 5.29 for Quin-2.[76] The mean decay time of Quin-2 increases from 1.3 ns in the absence of Ca^{2+} to 11.5 ns in the presence of Ca^{2+}. Other laboratories have observed similar lifetime changes for Quin-2.[77,78] This increase in lifetime results in a dramatic shift of the frequency response to lower frequencies. At intermediate Ca^{2+} concentrations, where Quin-2 is partially saturated with Ca^{2+}, one can visually see that the frequency response is heterogeneous, and one can discern the contributions of the short and long decay times of Quin-2.

Data of the type shown in Figure 5.29 are ideal for a global analysis. The two decay times are expected to be constant at all Ca^{2+} concentrations, and the amplitudes are expected to depend on calcium concentration. An alternative approach to improving the resolution is to fix the decay times. In this case the lifetimes in the absence and presence of saturating Ca^{2+} can be used as known values. Analysis of the FD data at one Ca^{2+} concentration can then be used to recover the amplitudes with less uncertainty than if the lifetimes were variable parameters.

5.6.E. SPQ—Collisional Quenching of a Chloride Sensor

In the previous example the solutions contained two fluorescent species, calcium-free and calcium-bound Quin-2. Each form displayed a different decay time, resulting in a complex frequency response in the presence of subsaturating concentrations of Ca^{2+}. Rather different behavior is observed for other types of sensing fluorophores. The probe 6-methoxy-*N*-(3-sulfopropyl)quinoline (SPQ) is used as a chloride probe.[79–81] In contrast to the behavior of Quin-2 in the presence of Ca^{2+}, the emission of SPQ is decreased by increasing amounts of chloride.

Absorption and emission spectra of SPQ were shown in Problem 3.2. Frequency responses of SPQ are shown in Figure 5.30. The frequency responses shift to higher frequencies with increasing amounts of chloride, indicating a decrease in lifetime. One can use the data to calculate the decay times at each chloride concentration (Problem 5.1). These lifetimes are 25.5, 11.3, 5.3, and 2.7 ns, for 0, 10, 30, and 70 m*M* chloride, respectively.

It is interesting to compare the frequency responses of Quin-2 and SPQ. Whereas Quin-2 displayed a complex frequency response, the responses of SPQ appear to be mostly single exponentials. The decays of SPQ are simpler than those of Quin-2 because the chloride is quenching the entire population of SPQ molecules. Hence, the decay is due to a homogeneous population of fluorophores.

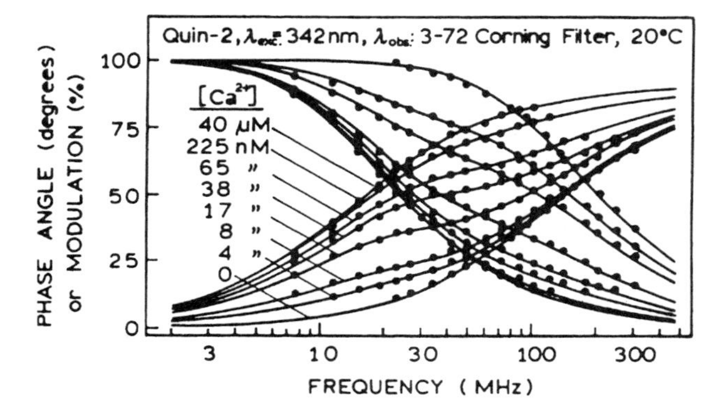

Figure 5.29. Frequency response of the calcium indicator Quin-2 in the presence of increasing amounts of calcium. Revised from Ref. 76.

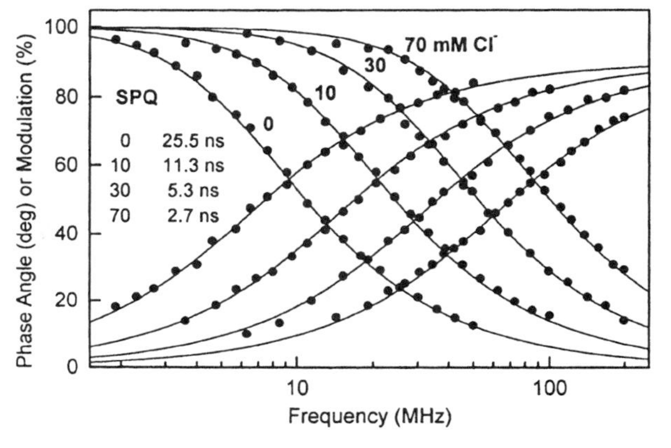

Figure 5.30. Frequency-domain intensity decays of SPQ in the presence of 0, 10, 40, and 70m*M* chloride. From Ref. 82.

5.6.F. Green Fluorescent Protein—One- and Two-Photon Excitation

Green fluorescent protein (GFP) is of wide interest because it can be used to follow gene expression. GFP spontaneously forms a highly fluorescent fluorophore after the amino acid backbone is synthesized. Another area of growing interest is multiphoton excitation.[83] In this process a fluorophore simultaneously absorbs two or more long-wavelength photons to reach the first singlet excited state. Two-photon excitation occurs when there is reasonable probability of two photons being in the same place at the same time, and thus the power density must be rather high. In such experiments, one is often concerned that the fluorophore or protein may be perturbed by the intense illumination. This possibility was investigated by comparing the intensity decays of GFP observed with one-photon and two-photon excitation.

Since GFP is a protein (~28 kDa), the emission is expected to be polarized. For this reason the intensity decays were measured with magic-angle polarizer conditions. Excitation was obtained using femtosecond pulses from a Ti:sapphire laser.[84] A pulse picker was used to reduce the repetition rate to near 4 MHz. For one-photon excitation the laser pulses were frequency-doubled to 400 nm. For two-photon excitation the 800-nm output from the Ti:sapphire laser was used directly to excite the sample. It is essential to select emission filters which reject scattered light at both 400 and 800 nm. The emission was observed through a 510-nm interference filter and a Corning 4-96 filter.

Intensity decays of proteins and labeled macromolecules are typically dependent on the conformation, and any perturbation of the structure is expected to alter the decay times. The intensity decays of GFP were found to be essentially identical for one- and two-photon excitation (Figure 5.31). This indicated that GFP was not perturbed by the intense illumination at 800 nm. The values of χ_R^2 are somewhat high but did not decrease when the two-decay-time model was used. The single-decay-time model was thus accepted for GFP, in agreement with one-photon excitation results from other laboratories.[85]

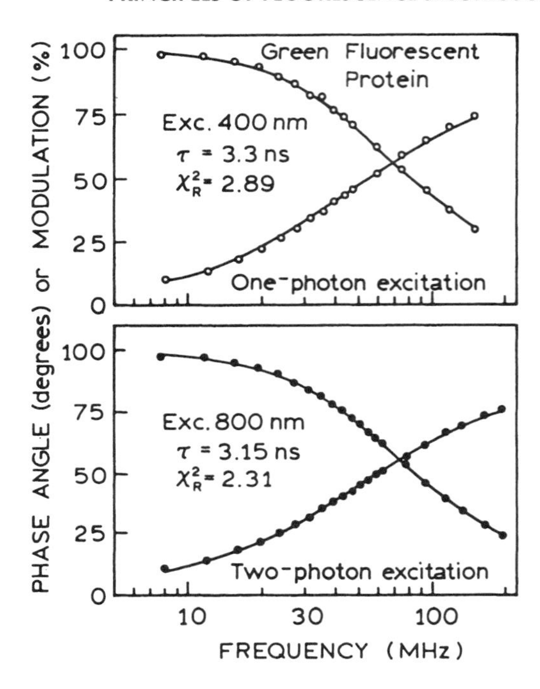

Figure 5.31. Frequency-domain intensity decay data for GFPuv in 0.05*M* phosphate buffer, pH 7, for one-photon excitation at 400 nm and for two-photon excitation at 800 nm. From Ref. 84.

5.6.G. Recovery of Lifetime Distributions from Frequency-Domain Data

In the early days of FD fluorometry, there was skepticism about its ability to determine nonexponential decays, which are those not described by a sum of exponential components. It is now accepted that the FD data, just like the TD data, can be used to resolve any intensity decay within the limits of resolution. For a distribution of lifetimes, the intensity decay is given by

$$I(t) = \int_0^\infty \alpha(\tau) e^{-t/\tau} \qquad [5.25]$$

where the lifetime distribution $\alpha(\tau)$ can be a unimodal or multimodal distribution,

$$\alpha(\tau) = \sum_i \alpha_i(\tau) \qquad [5.26]$$

For a lifetime distribution, the transforms are given by

$$N_\omega J = \int_{t=0}^\infty \int_{\tau=0}^\infty \alpha(\tau) e^{-t/\tau}\, d\tau \sin \omega t\, dt \qquad [5.27]$$

$$D_\omega J = \int_{t=0}^\infty \int_{\tau=0}^\infty \alpha(\tau) e^{-t/\tau}\, d\tau \cos \omega t\, dt \qquad [5.28]$$

where

$$J = \int_{\tau=0}^\infty \alpha(\tau) \tau\, d\tau \qquad [5.29]$$

For such decays, one cannot easily write analytical expressions for the sine and cosine transforms. Hence, the sine and cosine transforms are calculated numerically. This is not a problem with modern computers, which can rapidly do the required numerical integrations. Such procedures have been described in detail for complex decay laws.[86] It is important to recognize that it is difficult to recover all the parameters of a multimodal lifetime distribution and that, in general, a lifetime distribution cannot be distinguished from a multiexponential distribution.[87,88]

5.6.H. Lifetime Distribution of Photosynthetic Components

Phycobilisomes are large complex assemblies of phycobiliproteins designed to collect light and transfer the energy to the photosynthetic reaction centers of cyanobacteria. Phycobiliproteins and phycobilisomes contain several fluorescent species and are known to display complex intensity decays. While these complex decays can be analyzed in terms of the multiexponential model, it is equally probable that these structures display a continuous distribution of decay times.

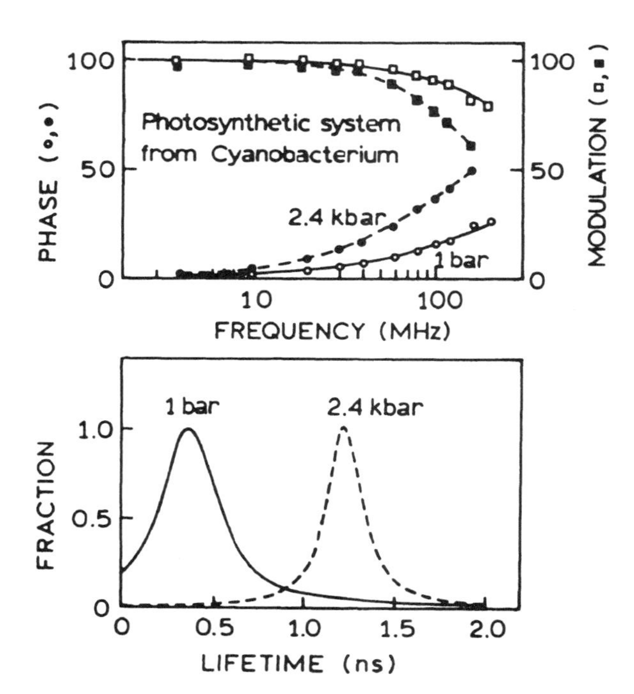

Figure 5.32. Intensity decay kinetics of photosynthetic components from cyanobacterium *Anabaena* sp. *Top*: Frequency response of the emission observed above 620 nm for excitation at 565 nm. Samples were examined at atmospheric pressure (○, □) and 2.4 kbar pressure (●, ■). *Bottom*: Lifetime distribution analysis at atmospheric pressure and 2.4 kbar pressure. (1 bar = 0.987 atm.) Revised from Ref. 89 and reprinted with permission from the Biophysical Society.

FD data were used to study the association of phycobilisomes with photosystem II.[89] Excitation was at 565 nm, from an Nd:YAG-pumped R6G dye laser. The frequency responses were analyzed in terms of a single Lorentzian distribution (Figure 5.32). The samples were examined at atmospheric pressure (open symbols in Figure 5.32) and 2.4 kbar pressure (closed symbols). At the higher pressure, the frequency response shifts to lower frequency (Figure 5.32, top), indicating an increase in the mean lifetime. This increase in lifetime is seen as a shift in the Lorentzian distribution to longer lifetimes (Figure 5.32, bottom). These results were interpreted as due to a pressure-induced suppression of energy transfer from the phycobilisomes to photosystem II. Examination of the data at atmospheric pressure indicates the need for higher modulation frequencies to recover the entire intensity decay, and it is possible that shorter-decay-time components exist at low pressure. Lifetime measurements are particularly useful in studies of pressure-dependent phenomena. It is difficult to measure the intensities using pressure cells. Additionally, anisotropy measurements are difficult because the windows used in these cells typically depolarize the emission. In contrast, the lifetime measurements are mostly independent of these effects.

5.6.I. Lifetime Distributions of the Ca^{2+}-ATPase

Lifetime distributions can also be used to characterize the complex intensity decays of proteins.[90–92] Lifetime distributions can be particularly useful for multitryptophan proteins, where there is no practical possibility of resolving the individual residues. One example is the Ca^{2+}-ATPase. This enzyme contains 13 tryptophan residues, just one of which seems to be sensitive to calcium binding.[93] The FD intensity decay is moderately complex (Figure 5.33) and is consistent with a triple-exponential decay.

Although the decay could be analyzed in terms of the triple-exponential model, there is no reason to believe that the decay of a protein with 13 tryptophan residues is actually a triple exponential. Hence, the FD data were interpreted in terms of a bimodal Lorentzian distribution (Figure 5.34). The analysis does suggest that the distribution of lifetimes in the presence of calcium is different from that in the absence of calcium. The bimodal Lorentzian model contains the same number of variable parameters (five) as the three-decay-time model. The fits are essentially equivalent for these two models, so there is no statistical reason to select the lifetime distribution model over the three-decay-time model. While the data demon-

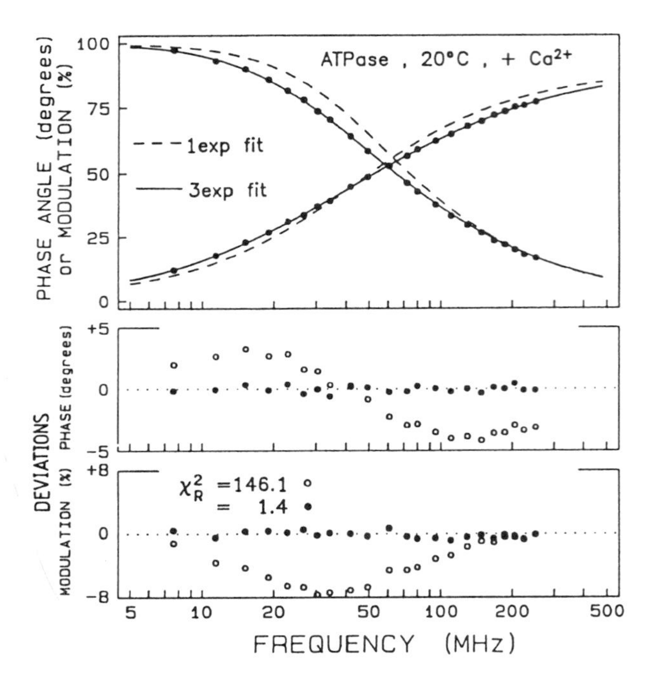

Figure 5.33. Frequency response of the tryptophan emission of the Ca^{2+}-ATPase at 20 °C. The curves are the best single- (– – –) and triple-exponential (———) fits to the data (●). The lower panels show the deviations for the best single- (○) and triple-exponential (●) fits. Revised and reprinted, with permission, from Ref. 93. Copyright © 1989, American Chemical Society.

strate that Ca^{2+} causes a change in the intensity decay of the ATPase, it is difficult to know whether the recovered distributions actually describe the intensity decay or only appear to describe the decay owing to the limited resolution of the data.

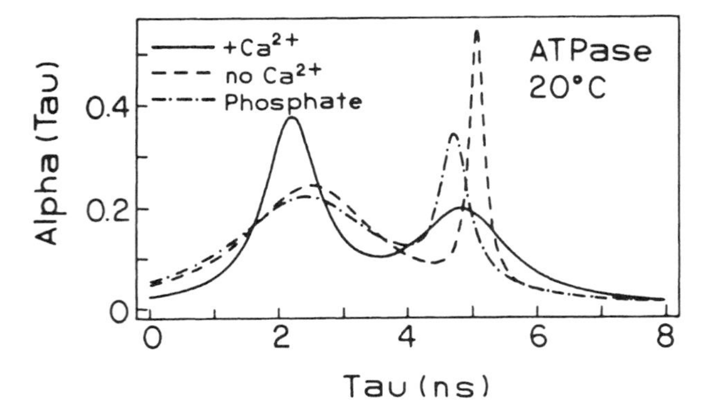

Figure 5.34. Lifetime distributions of the tryptophan emission from ATPase in the presence of (———), in the absence of Ca^{2+} (– – –), and in the presence of phosphate (– · –). Revised and reprinted, with permission, from Ref. 93. Copyright © 1989, American Chemical Society.

5.6.J. Cross Fitting of Models—Lifetime Distributions of Melittin

In the previous section we showed lifetime distributions for the Ca^{2+}-ATPase. When one uses such complex models, it is difficult to know if the results are unique. In such cases we suggest cross fitting of models to determine whether the recovered distributions are statistically different. This procedure is illustrated by the intensity decay of melittin. FD data for the single-tryptophan protein melittin are shown in Figure 5.35. In a mixture of 20% water and 80% methanol, melittin exists as α-helical monomers.The data could be fit to a bimodal Lorentzian,[92] which is shown in the bottom panel of Figure 5.36. The lifetime distributions found for melittin are consistent with the notion that protein structure is an origin of the complex intensity decays of proteins. When dissolved in 6*M* guanidine hydrochloride (GuHCl), which eliminates all structure, the intensity decay becomes equivalent to a double-exponential decay (Figure 5.36, top). In water, melittin is known to have a small amount of residual structure. Under these conditions, one notices that the lifetime distributions become broader (Figure 5.36, middle and bottom panels).

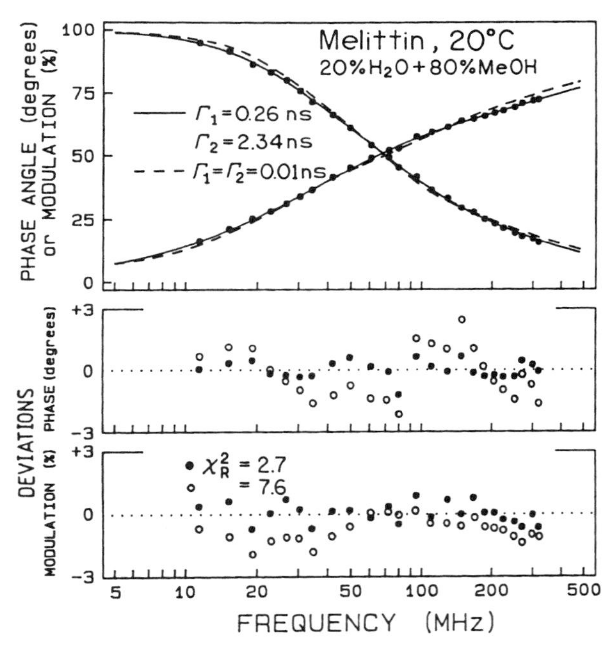

Figure 5.35. Phase and modulation data for synthetic melittin in 20% H_2O + 80% MeOH. The solid curves and the filled circles in the lower panels show the best fit to a bimodal Lorentzian. The dashed curves and the open circles in the lower panels show the best fit when the widths of the distribution Γ_1 and Γ_2 were held constant at the narrow value of 0.01 ns. From Ref. 92.

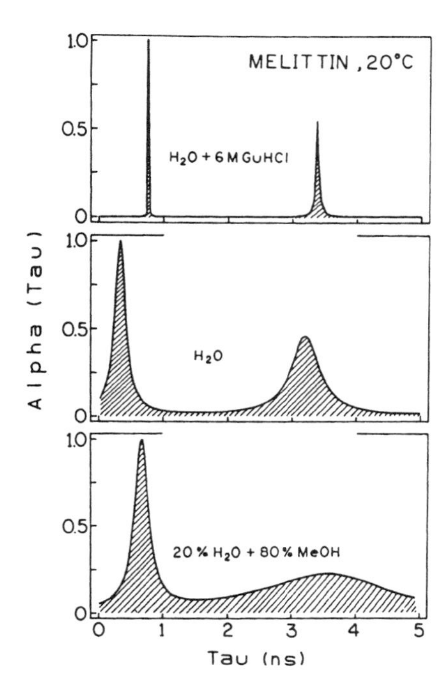

Figure 5.36. Lifetime distributions for synthetic melittin in H_2O with 6*M* GuHCl (*top*), H_2O (*middle*), and a mixture of 20% H_2O + 80% MeOH (*bottom*). From Ref. 92.

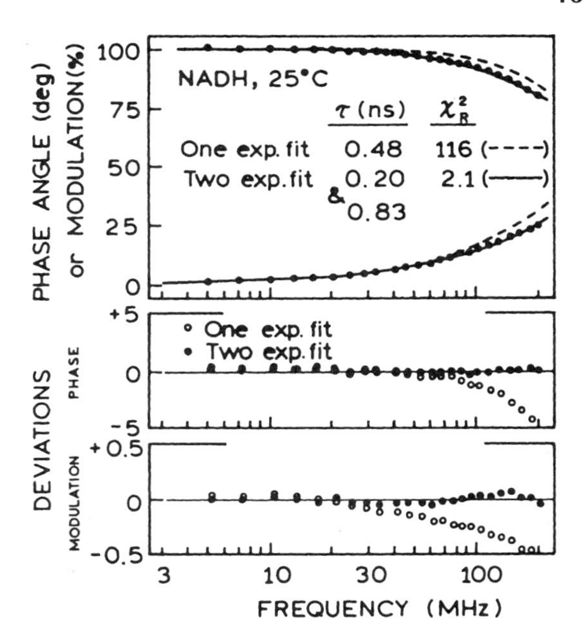

Figure 5.37. Frequency response of NADH dissolved in 0.02*M* Tris (pH 8), at 25 °C. The excitation wavelength was 325 nm from an HeCd laser, which was modulated with an electro-optic modulator. The emission filter was a Corning 0-52 filter. For the two-component analysis, f_1 = 0.57 and f_2 = 0.43. Revised and reprinted from Ref. 2, Copyright © 1985, with permission from Elsevier Science.

While lifetime distributions can be recovered from the FD data, it is important to ask whether the various distributions are distinct within the resolution limits of the data or cannot be distinguished from the data. We answer this question by fitting the data with some of the fixed parameter values. In this case we asked whether the data for melittin in 80% methanol (broad distribution) could be fit with a narrow Lorentzian half-width ($\Gamma_1 = \Gamma_2 = 0.01$ ns). The mean lifetimes and amplitudes of the Lortenzian were still variable parameters. The forced fit resulted in approximately a threefold elevation of χ_R^2, indicating that the intensity decay of melittin in 80% methanol is not consistent with two lifetimes and narrow distributions. A less rigorous test would be to cross fit the data with all the parameters fixed, which results in easier rejection of the alternate decay law. The cross-fitting procedure is recommended whenever one is trying to distinguish between two similar models.

5.6.K. Intensity Decay of NADH

NADH is known to display a subnanosecond decay time near 0.4 ns. Its intensity decay is complex, with decay times near 0.3 and 0.8 ns.[94] FD data for NADH are shown in Figure 5.37. The presence of more than one lifetime is immediately evident from the failure of the single exponential fit (dashed curve) and the systematic deviations (open circles). Use of the two-decay-time model resulted in a 50-fold decrease of χ_R^2. The FD data for NADH illustrate a limitation of the commercially available instruments. An upper frequency of 200 MHz is too low to determine the entire frequency response of NADH or other subnanosecond intensity decays. For this reason, FD instruments were developed to allow measurements at higher modulation frequencies.

5.7. GIGAHERTZ FREQUENCY-DOMAIN FLUOROMETRY

• Advanced Topic •

In FD measurements it is desirable to measure over the widest possible range of frequencies, so as to examine the entire frequency response of the sample. Unfortunately, most FD instruments are limited to an upper frequency near 200 MHz. This limitation arises from two components. First, it is difficult to obtain light modulation above 200 MHz. This limitation is due in part to the large half-wave voltages of most EO modulators. Second, even if light could be modulated at higher frequencies, most dynode PMTs have an upper frequency limit near 200 MHz.

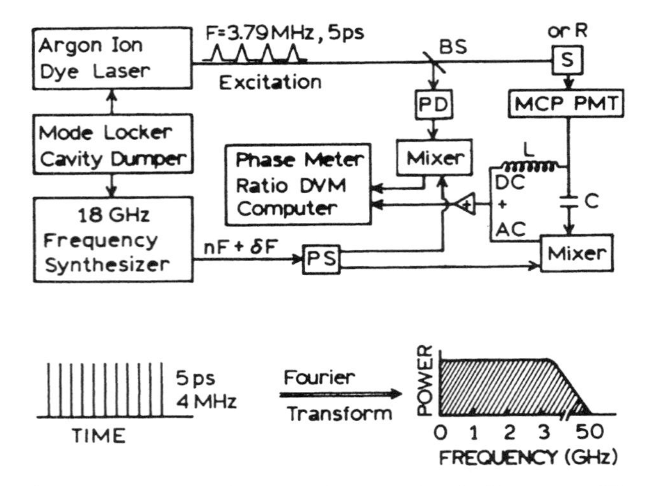

Figure 5.38. Harmonic-content frequency-domain fluorometer. PD, High-speed photodiode; PS, power splitter; DVM, digital voltmeter; MCP PMT, microchannel plate photomultiplier tube; BS, beam splitter; F, pulse repetition rate of the cavity-dumped dye laser; δF, cross-correlation frequency; n, number of the harmonic; S, sample; R, reference or scatterer. The lower panel shows a laser pulse train and its Fourier transform. Revised from Ref. 102.

Fortunately, alternative technologies are available to obtain FD measurements at frequencies above 200 MHz. The need for a light modulator can be eliminated by using the harmonic frequency content of a laser pulse train. Suppose the light source consists of a mode-locked argon-ion laser and a cavity-dumped picosecond dye laser. This source provides 5-ps pulses with a repetition rate near 4 MHz. In the frequency domain, this source is intrinsically modulated to many gigahertz, as shown by the schematic Fourier transform in the lower panel of Figure 5.38. The idea of using the harmonic content of a pulse train was originally proposed for pulsed lasers[95] and later for synchrotron radiation.[96–98] Pulse sources provide intrinsically modulated excitation at each integer multiple of the repetition rate, up to about the reciprocal of the pulse width.[99,100] For a picosecond dye laser, the 4-MHz pulse train can be used for FD measurements at 4, 8, 12, 16 MHz, etc. These harmonics extend to many gigahertz—frequencies that are higher than the upper frequency limit of most detectors.

There are significant advantages in using the pulses from a picosecond laser. The cavity-dumped output of dye lasers is rather easy to frequency-double because of the high peak power. Frequency doubling provides wavelengths for excitation of proteins and other extrinsic probes absorbing in the UV. Importantly, when a picosecond dye laser source is used, it is no longer necessary to use an EO modulator and nearly crossed polarizers, which result in a significant attenuation of the incident light. During the past three years, we have performed numerous measurements with the harmonics extending to 10 GHz. There is no detectable increase in noise up to 10 GHz, suggesting that there is no multiplication of phase noise at the higher harmonics. We no longer see a need for gigahertz light modulators.

The second obstacle to higher-frequency measurements was the detector. The PMT in the 200-MHz instrument (Figure 5.5) was replaced with an MCP PMT. Devices of this type are 10- to 20-fold faster than a standard PMT, and hence a multigigahertz bandwidth was expected. As presently designed, the MCP PMTs do not allow internal cross correlation, which is essential for an adequate signal-to-noise ratio. This problem was circumvented by designing an external mixing circuit,[101,102] which allows cross correlation and preserves both the phase and the modulation data (Figure 5.39). The basic idea is analogous to that represented in Figure 5.8, except that mixing with the low-frequency signal is accomplished after the signal exits the MCP PMT. External cross correlation was found to perform well without any noticeable increase in noise.

What range of frequencies can be expected with a pulsed laser light source and an MCP PMT detector? For Lorentzian-shaped pulses, the harmonic content decreases to one-half of the low-frequency intensity at a frequency $\omega_2 = 2 \ln 2/\Delta t$, where Δt is the pulse width.[100] For 5-ps pulses, the harmonics extend to 280 GHz, higher than the upper frequency limit of any available detector. Hence, for the foreseeable future, the measurements will be limited by the detector.

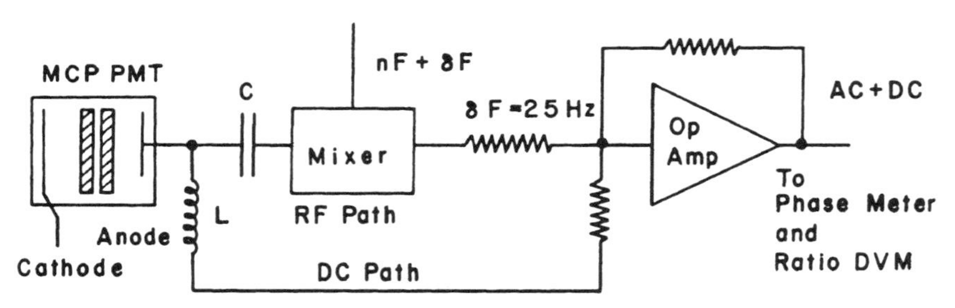

Figure 5.39. External cross correlation of an MCP PMT. From Ref. 101. Reprinted with permission from the American Institute of Physics, Copyright © 1986.

The upper frequency of a detector is limited by the pulse width due to a single photoelectron, or, equivalently, the TTS. Hence, one expects the highest modulation frequencies to be measurable with MCP PMTs that have the smallest TTS (Table 4.1). The relative power at various frequencies can be measured with a spectrum analyzer. This was done for several PMTs using a picosecond pulse train with its high harmonic content as the light source. These results show that the side-window R928 is most useful below 200 MHz (Figure 5.40) and cannot be used for measurements much above about 400 MHz. The R1564U, which is a 6-μm MCP PMT, shows a useful response to 2 GHz. This PMT was used in the first 2-GHz instrument.[101]

To obtain frequencies above 2 GHz, it was necessary to use a specially designed MCP PMT, the R2566. The data in Figure 5.40 are for the 6-μm version of the R2566, which provides measurable power to 10 GHz and allowed construction of a 10-GHz FD instrument.[102] It is interesting to understand why the R2566 provides a higher frequency response. This MCP PMT possesses a grid between the microchannel plates and the anode, which serves to decrease the width of the photoelectron pulses. In the frequency domain, the upper limit of the detector is determined by the reciprocal of the pulse width. In TCSPC the time resolution is determined by the rise time of the PMT, and the overall pulse width is less important.

In Figure 5.40 one notices that the photodiode provides a higher bandwidth than does any of the MCP PMTs. In fact, photodiode detectors were used in several phase fluorometers for measurements at high frequencies.[105–108] Unfortunately, the small active area of photodiodes results in low sensitivity, so that photodiodes are rarely used for fluorescence spectroscopy.

The complete 10-GHz instrument is shown in Figure 5.38. This instrument incorporates a picosecond laser as an intrinsically modulated light source and an MCP PMT as the detector. A photodiode (PD) is adequate as the reference detector because the laser beam can be focused on its small active area. The use of cross correlation allows measurement over the entire frequency range from 1 MHz to 10 GHz without any noticeable increase in noise. Cross correlation allows measurements at any modulation frequency at the same low cross-correlation frequency and avoids the need to measure phase angles and modulation at high frequencies. Another valuable feature of cross correlation is that all the signal appears at the measured frequency. Contrary to intuition, one is not selecting one harmonic component out of many, which would result in low signal levels. The use of cross correlation provides absolute phase and modulation values as if the excitation and detector were both modulated as sine waves. A final favorable feature of this instrument is that the modulation can be higher than 1.0, which is the limit for sine wave modulation. At low frequencies where the detector is fully responsive, the modulation can be as high as 2.0. To understand this unusual result, one needs to examine the Fourier components for a pulse train.

5.7.A. Gigahertz FD Measurements

Several examples of gigahertz FD measurements will illustrate the value of a wide range of frequencies. Short decay times are needed to utilize the high-frequency capabilities. Otherwise, the emission is demodulated prior to reaching the upper frequency limit. A short decay time was obtained using 4-dimethylamino-4′-bromostilbene (DBS) in cyclohexane at 75 °C (Figure 5.41). Because of the short

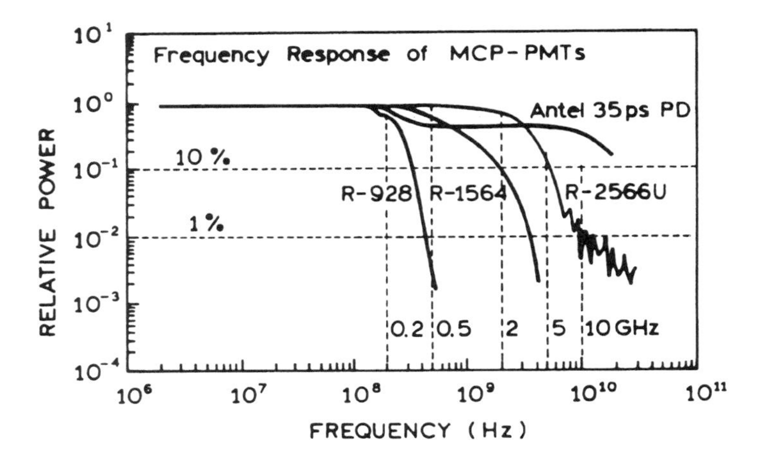

Figure 5.40. Measured frequency responses of several PMTs and a fast photodiode (PD). Data are from Refs. 103 and 104 and technical literature from Hamamatsu, Inc.

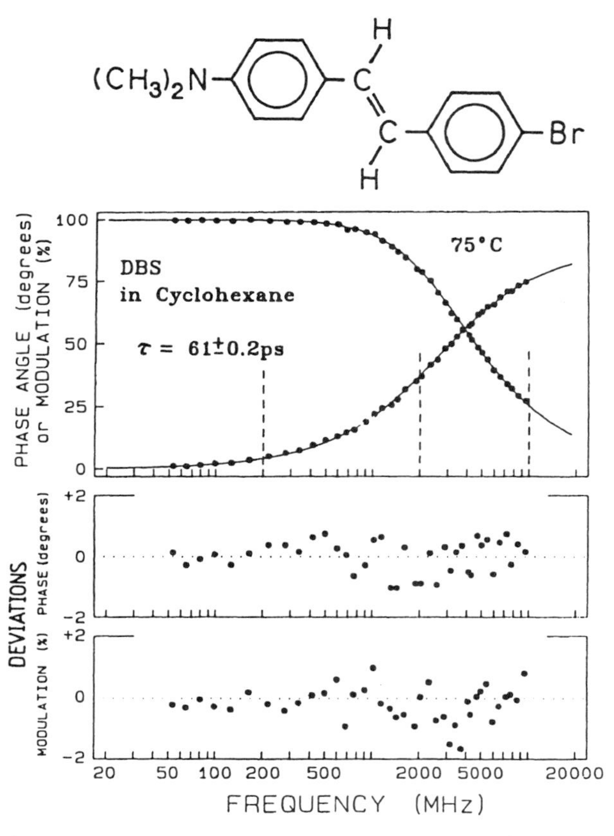

Figure 5.41. Frequency response of 4-dimethylamino-4′-bromostilbene (DBS) up to 10 GHz. The vertical dashed lines are at 200 MHz, 2 GHz, and 10 GHz. From Ref. 3.

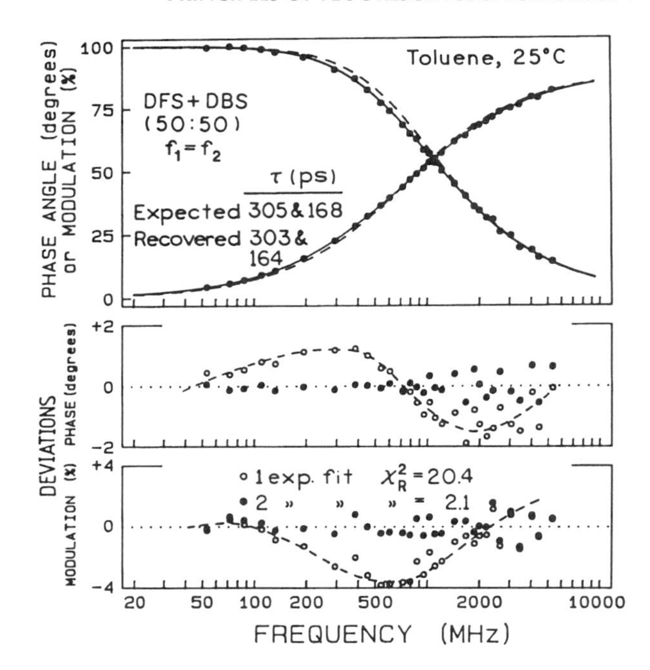

Figure 5.42. Gigahertz frequency-domain data for a 50:50 mixture of 4-dimethylamino-4′-bromostilbene (DBS) and 4-dimethylamino-4′-fluorostilbene (DFS) in toluene at 25 °C. The solid curves show the best two-component fit to the data, and the dashed curves show the best single-exponential fit to the data. The lower panels show the deviations in phase and modulation for double- (●) and single-exponential (○) fits. From Ref. 109.

61-ps lifetime, the phase and modulation data could be measured to 10 GHz. The intensity decay was found to be a single exponential.[3] The vertical dashed lines in Figure 5.41 illustrate how only a fraction of the frequency response could be explored if the upper limit was 200 MHz, or even 2 GHz. It would be difficult to detect additional components in the intensity decay if the data stopped at 200 MHz, which would display a maximum phase angle of 4.4°. An important aspect of these measurements is that no measurable color effect has been observed in the 10-GHz measurements.[102]

If a wide range of frequencies are available, then mixtures of fluorophores with subnanosecond decay times can be resolved.[109] This is illustrated for a mixture of 4-dimethylamino-4′-fluorostilbene (DFS) and DBS in toluene (Figure 5.42). The decay times of the individual fluorophores, 305 and 168 ps, were readily recovered from the measurements to 5 GHz, resulting in a 10-fold decrease of χ_R^2. Resolution of these two decay times would be difficult if data were only available to 200 MHz.

5.7.B. Biochemical Examples of Gigahertz FD Data

While the use of gigahertz measurements may seem exotic, such data are often needed for studies of routine biochemical samples. One example is NADH. At 200 MHz the data only sampled part of the frequency response (Figure 5.37). When data are measured to higher frequencies, one can more dramatically see the difference between the one- and two-decay-time fits (Figure 5.43). The decrease in χ_R^2 for the two-decay-time model is 400-fold. Although we have not performed a support plane analysis on these data, the α_i and τ_i values will be more closely determined using the data extending above 200 MHz.

Another biochemical example is provided by the peptide hormone vasopressin, which acts as an antidiuretic and a vasoconstrictor. Vasopressin is a cyclic polypeptide which contains nine amino acids, including a single tyrosine residue at position 2. Oxytocin has a similar structure but has a distinct physiological activity of stimulating smooth-muscle contraction. Hence, there have been many efforts to use the tyrosine emission to learn about the solution conformation of these peptide hormones.

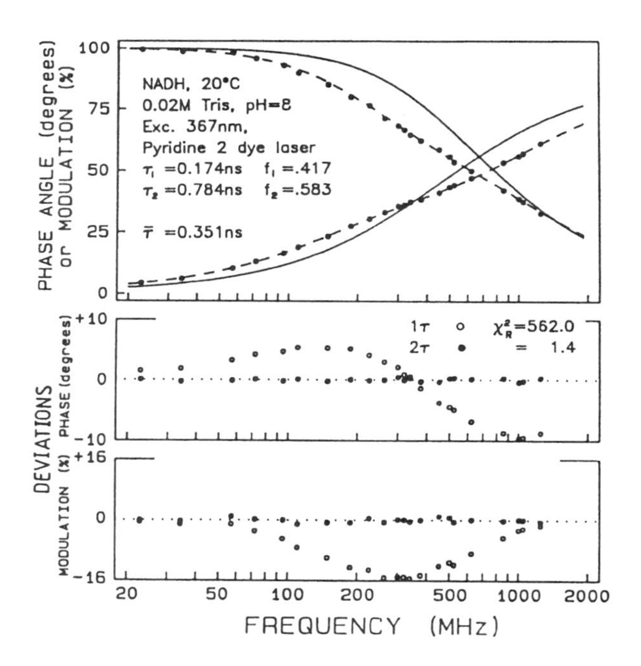

Figure 5.43. Frequency-domain intensity decay of NADH measured up to 2 GHz. From Ref. 3.

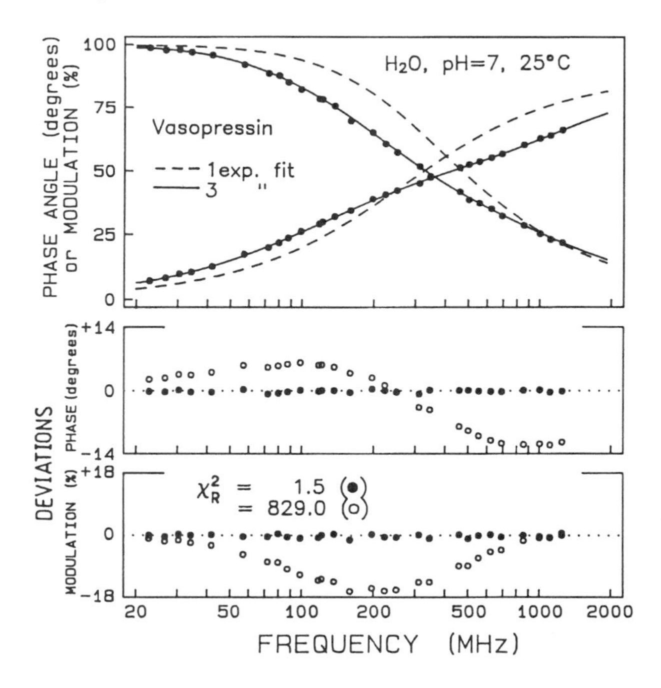

Figure 5.44. Phase and modulation data for the vasopressin tyrosine fluorescence intensity decay. The dashed curves and the open circles in the lower panels show the best single-exponential fit; the solid curves and the filled circles in the lower panels show the best triple-exponential fit. From Ref. 110.

The FD data for vasopressin reveal a complex intensity decay (Figure 5.44).[110] The decay is not even closely approximated by the single-exponential model, represented by the dashed curve. Fitting the data requires three decay times of 0.17, 0.75, and 1.60 ns. These multiple decay times could probably not be recovered if the data were limited to 200 MHz.

MCP PMTs are moderately expensive, and their use in FD measurements requires special circuits for cross correlation. However, the advantages of high-frequency FD data may become available without the use of MCP PMTs. Examination of Figures 5.42 and 5.43 indicates that considerable data could be obtained if the data were available to just 1 GHz. This frequency limit can probably be reached with the new compact PMTs which show short TTSs (Table 4.1). In fact, a dynode PMT (H5023, Table 4.1) has already been used up to 1 GHz.[111] Unfortunately, the use of higher frequencies requires a pulsed laser source to provide the modulated excitation.

5.8. SIMPLE FREQUENCY-DOMAIN INSTRUMENTS

• Advanced Topic •

In the preceding section we described state-of-the-art FD instrumentation. There is presently another state-of-the-art activity focused on simpler, low-cost FD instrumentation. A large fraction of the cost of an FD instrument derives from the light source and/or modulation optics. Under some conditions the light source can be replaced with simple solid-state sources, including laser diodes,[112,113] and LEDs.[114–116] FD measurements have also been accomplished with electroluminescent devices,[117] and even a modulated deuterium lamp.[118]

5.8.A. Laser Diode Excitation

It is well known that the output of laser diodes can be modulated up to several gigahertz. Hence, laser diodes can be used for FD excitation without the use of a modulator. Data are shown in Figure 5.45 for two laser dyes, IR-144 and DOTCI, which were excited with 790- and 670-nm laser diodes, respectively. Frequency-doubled laser diodes have already been used to obtain shorter excitation wavelengths, near 400 nm.[112,119]

5.8.B. LED Excitation

A disadvantage of laser diodes is that they are presently available only with red and NIR wavelengths above 630 nm. Green and blue laser diodes are under development but are not yet available. Fortunately, it is now known that

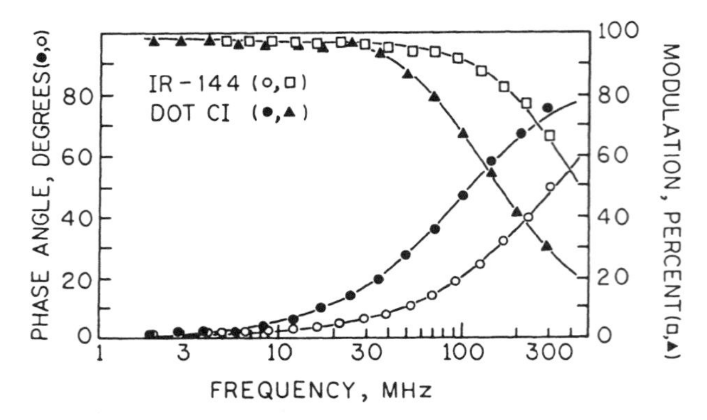

Figure 5.45. Frequency response for IR-144 in ethanol (open symbols) and DOTCI in ethanol (filled symbols). Best one-component fits to each data are indicated by the curves. IR-144 and DOTCI were excited with laser diodes at 790 and 670 nm, respectively. Revised and reprinted, with permission, from Ref. 113, Copyright © 1992, American Chemical Society.

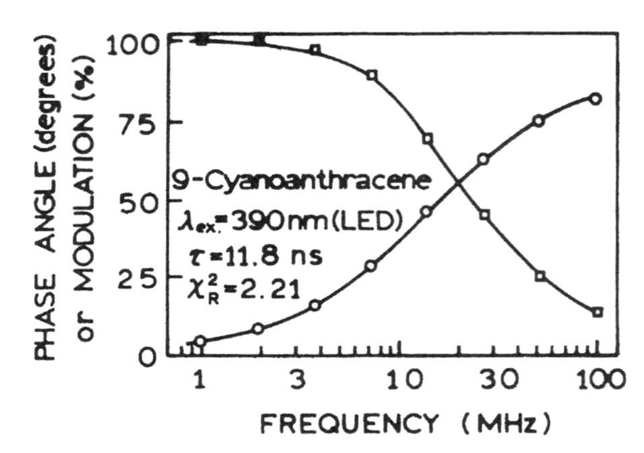

Figure 5.47. Intensity decay of 9-cyanoanthracene in ethanol measured using the 390-nm output of a UV LED as the excitation source. Revised from Ref. 115.

LEDs can be modulated to above 100 MHz.[114–116] Also, when pulsed at high current, some blue LEDs can provide UV output at 390 nm.[115] Hence, LEDs are becoming an alternative to modulated arc lamps.

As shown in Figure 2.8, 390-nm output can be obtained from a blue LED. The UV output of blue LEDs can be modulated up to 100 MHz (Figure 5.46), which is adequate for measurement of decay times over 3 ns. The 390-nm LED output was used to measure the 3.2-ns decay time of GFP (not shown) and the 11.8-ns lifetime of 9-cyanoanthracene[115] (Figure 5.47).

The use of a simple light source such as an LED is likely to find use in analytical chemistry and clinical chemistry. This is illustrated in Figure 5.48, which shows the frequency response of the pH probe SNAFL-2 measured with a blue LED. The decay time of SNAFL-2 changes from about 4.6 ns at pH 5.5 to about 1 ns at pH 10. It is remarkable that such short decay times can be measured with such a simple light source. The LED measurements are in good agreement with those obtained using an argon-ion source, except for a systematic error in the modulation measurement above 20 MHz. As shown in Section 19.6.B, pH measurements can be performed using phase or modulation measurements at a single modulation frequency, and such measurements can be very stable. The possibility of measuring nanosecond decay times using modulated LEDs, and the availability of a wide range of wavelengths to 390 nm, suggests that these light sources will be used for low-cost FD instruments in the near future.

Another application of LEDs will be for excitation of the longer-lived metal–ligand complexes (Chapter 20). The

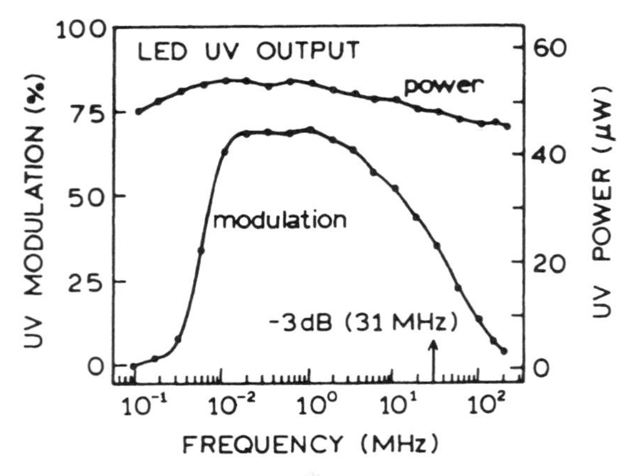

Figure 5.46. Depth of modulation and output power of a UV-emitting blue LED when biased with 60 mA DC and modulated with +13 dBm AC. The -3-dB point for the modulation is 31 MHz. From Ref. 115.

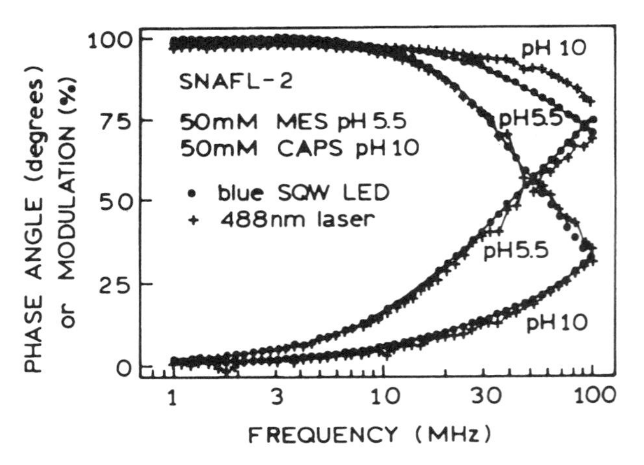

Figure 5.48. Frequency response of the pH probe SNAFL-2 measured with a blue LED (•) or with an externally modulated argon-ion laser at 488 nm (+). From Ref. 115. Reprinted with permission from the American Institute of Physics, Copyright © 1996.

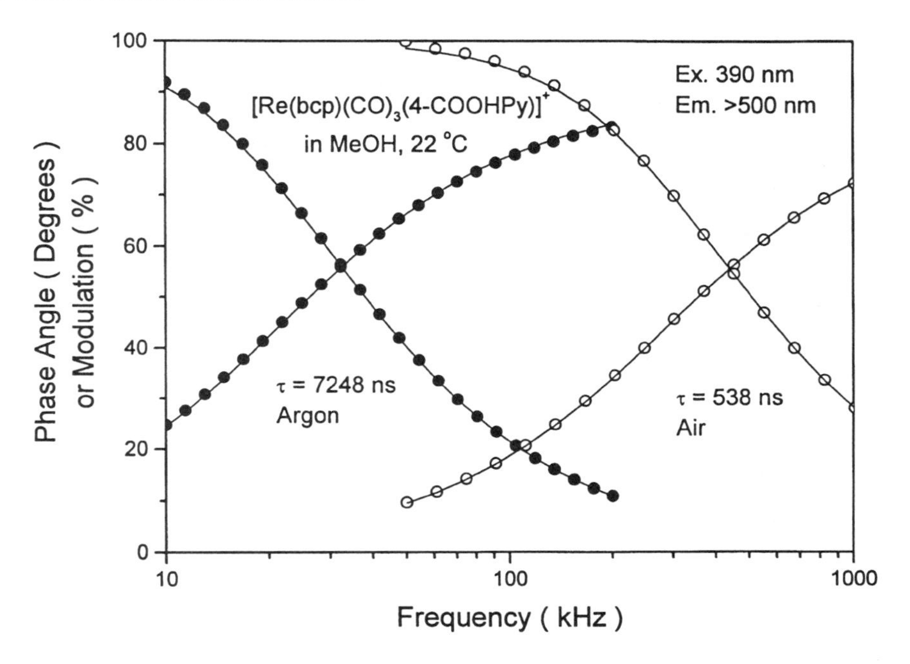

Figure 5.49. Frequency-domain intensity decays of $[Re(bcp)(CO)_3(4\text{-}COOHPy)]^+$ in methanol. Excitation wavelength was 390 nm from a UV LED, and a 500-nm cutoff filter was used to isolate the emission. From Ref. 120.

LEDs are ideal because the 450-nm output is centered on the 450-nm charge-transfer absorption of the ruthenium complexes. The shorter 390-nm output is suitable for excitation of the higher-quantum-yield rhenium complexes. In the example shown in Figure 5.49, the entire frequency response of $[Re(bcp)(CO_3)(4\text{-}COOHPy)]^+$ was measured in the absence (7.25 μs) and in the presence of oxygen (538 ns).[120]

Given the simplicity of such light sources, one can imagine the FD fluorometer being contained in a card in a personal computer (Figure 5.50). The sample may be placed on a detector on the card, or sampling may be done remotely using fiber optics. For example, pH can be measured from lifetime measurements made either in standard cuvettes[121] or through optical fibers.[122] LEDs have been used in phase fluorometric sensors for oxygen[123,124] and carbon dioxide.[125] Phase-modulation lifetimes have been used in high-performance liquid chromotography (HPLC) to assist in the identification and quantitation of polynuclear aromatic hydrocarbons.[126–128] Phase-modulation lifetime measurements have already been used to quantify a wide variety of clinically important analytes.[129] Several companies are already designing simple phase-modulation instruments for use in analytical applications. It is remarkable that the technology of FD fluorometry, once regarded as an exotic methodology, can now be accomplished with components no more complex than consumer electronics.

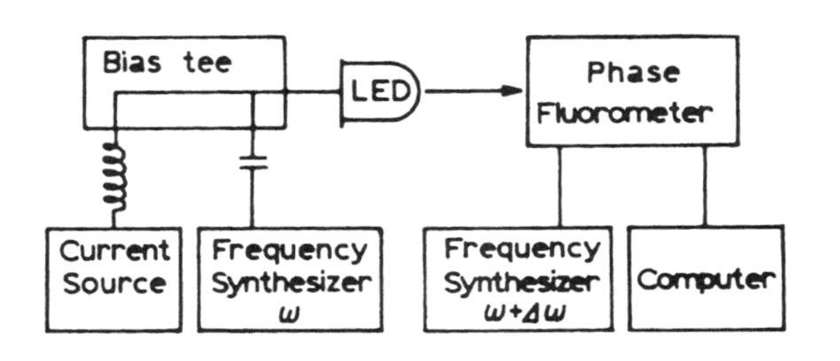

Figure 5.50. Schematic of an LED-based FD instrument. The current source, bias tee, and LED replace the light source, 25-W RF power amplifier, high-voltage bias, Pockels cell, and associated optics found in a standard phase fluorometer. Revised from Ref. 115.

5.9. PHASE ANGLE AND MODULATION SPECTRA
• Advanced Topic •

In all the preceding sections we emphasized FD measurements at a single emission wavelength. When phase-modulation methods are used, it is also possible to record the phase and modulation data as the emission wavelength is scanned. Such data can be referred to as phase-angle or modulation spectra. Given the stability of modern FD instruments, this procedure is quite reliable. For very short decay times, one may need to correct for the wavelength-dependent transit time through the monochromator and/or PMT. For

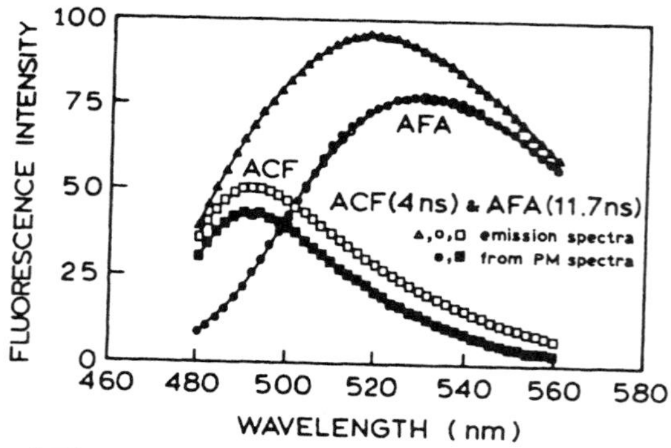

Figure 5.51. Emission spectra and recovered spectra for a mixture of ACF and AFA in propylene glycol at 20 °C. Emission spectra are shown for the mixture (▲) and the individual components (□, ○), together with the spectra recovered from the phase-modulation spectra (■, ●). [ACF] = $5 \times 10^{-7} M$, [AFA] = $2 \times 10^{-5} M$. Revised and reprinted, with permission, from Ref. 130, Copyright © 1990, American Chemical Society.

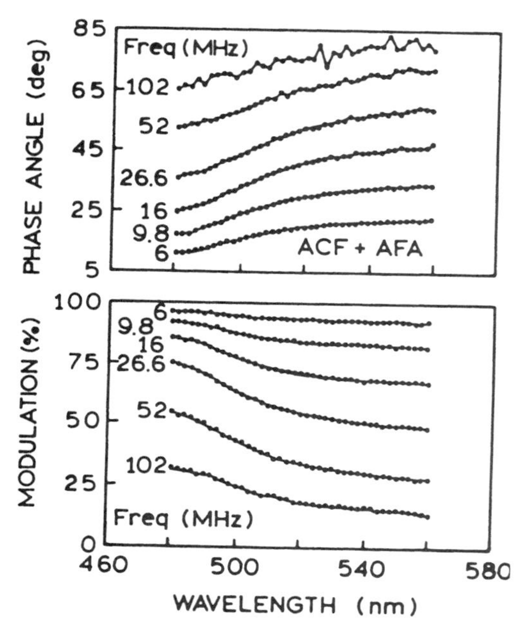

Figure 5.52. Representative phase-angle (*top*) and modulation spectra (*bottom*) for a mixture of ACF and AFA. The complete data set consisted of spectra at 20 modulation frequencies. Revised and reprinted, with permission, from Ref. 130, Copyright © 1990, American Chemical Society.

nanosecond-timescale measurements, such corrections are not necessary.

The use of phase-modulation spectra can be illustrated by the example of a mixture of the fluorophores acriflavine (ACF, 4.0 ns) and 3-aminofluoranthene (AFA, 11.7 ns) (Figure 5.51). Phase-angle and modulation spectra were recorded using the 325-nm output of an HeCd laser modulated by the Lasermetrics 1024 modulator and detected through a monochromator with an R928 PMT. Phase and modulation spectra were collected at various modulation frequencies. As seen in the top and bottom panels, respectively, of Figure 5.52, the phase angles increase and the modulation decreases with wavelength. These effects are due to the increase in mean lifetime as the relative contribution of the 11.7-ns decay time of AFA increases at longer wavelengths.

The phase-modulation spectra can be used to recover the emission spectra and lifetime of each component in the mixture. This is accomplished by a global analysis of the phase-modulation spectra measured at various frequencies. Except for a change in the nature of the data files, the analysis is performed according to Eqs. [5.12]–[5.15]. The emission spectra associated with each fluorophore can be calculated from the recovered values of $\alpha_i(\lambda)$ and the steady-state spectrum of the mixture, $I(\lambda)$. The fractional intensity of each fluorophore at wavelength λ is given by

$$f_i(\lambda) = \frac{\alpha_i(\lambda)\,\tau_i}{\sum_j \alpha_j(\lambda)\,\tau_j} \quad [5.30]$$

and the emission spectrum of each component is given by

$$I_i(\lambda) = f_i(\lambda)\, I(\lambda) \quad [5.31]$$

The fractional contribution of each fluorophore to the total intensity of the sample is given by

$$F_i = \frac{1}{N} \sum_\lambda f_i(\lambda) \quad [5.32]$$

where N is the number of emission wavelengths. The emission spectra recovered from the analysis were in good agreement with those known from the sample preparation (Figure 5.51). For samples such as ACF and AFA, there is little advantage in using phase and modulation spectra, as compared with frequency-swept measurements at a single wavelength followed by changing the wavelength. However, phase and modulation spectra can be more convenient if the fluorophores show highly structured emission spectra. In these cases it may be easier to scan wavelength than to measure at discrete wavelengths adequate to determine the individual emission spectra.

5.9.A. Resolution of the Two Emission Spectra of Tryptophan Using Phase-Modulation Spectra

As described in Chapter 17 on time-resolved protein fluorescence, tryptophan at pH 7 displays a double-exponential

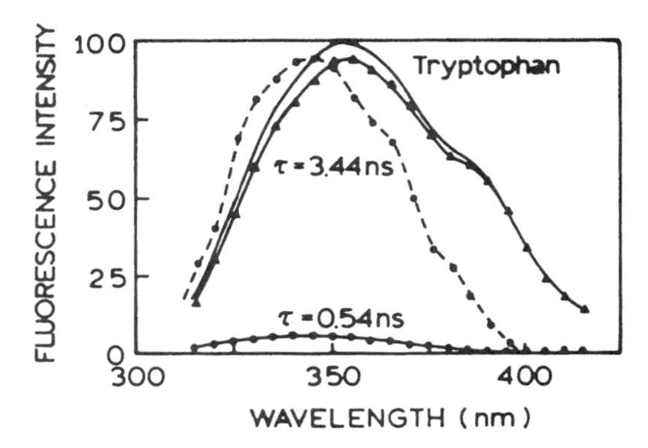

Figure 5.53. Steady-state emission spectrum of tryptophan (———) and emission spectra associated with the two decay times of tryptophan, recovered from the phase and modulation spectra (▲, ●). The dashed curve shows the emission spectrum of the 0.54-ns component normalized to that of the 3.44-ns component. [Trp] = $5 \times 10^{-5} M$. Revised from Ref. 131.

decay with decay times of 3.1 and 0.5 ns. The short-lived component contributes only a small fraction to the total intensity and is thought to display an emission spectrum at shorter wavelengths relative to the total emission of tryptophan (Figure 5.53).

Phase and modulation spectra were used to resolve the emission spectra associated with the two decay times of tryptophan. These spectra display shorter phase angles on the short-wavelength side of the emission (Figure 5.54). For the one-decay-time model, the phase angle and modulation must be independent of wavelength, so this model does not fit the wavelength-dependent data (Figure 5.54). An improved fit was obtained with two decay times, which allow for the phase and modulation to vary with emission wavelength.

The $f_i(\lambda)$ values recovered from the two-decay-time analysis were used to calculate the decay-associated spectra (DAS). The DAS represent the emission spectrum associated with each lifetime. These spectra (Figure 5.53) show a weak component centered at 340 nm, which is in agreement with the results of TCSPC experiments.

5.10. APPARENT PHASE AND MODULATION LIFETIMES

Prior to the availability of variable-frequency instruments, most phase-modulation fluorometers operated at one or a few fixed modulation frequencies. During this time, it became standard practice to report the apparent phase and modulation lifetimes, which are the values calculated from the data at a single modulation frequency. These values are given by

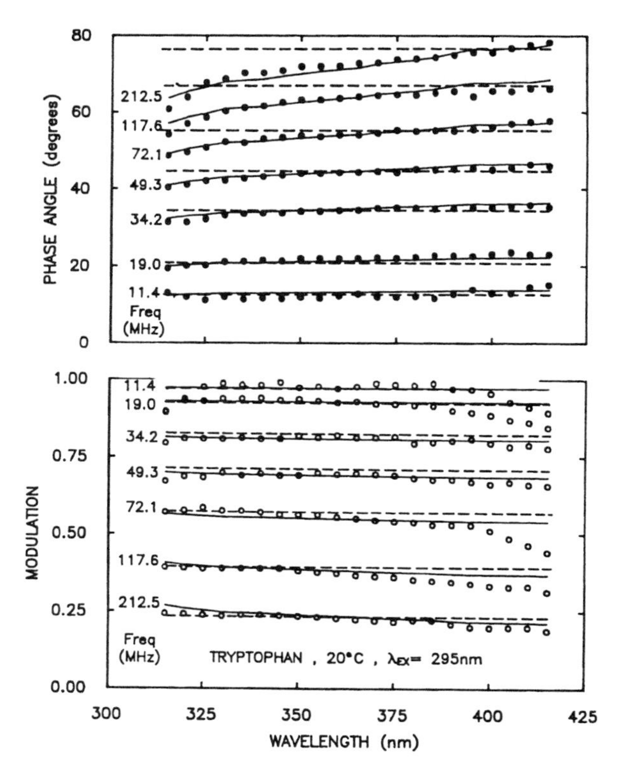

Figure 5.54. Phase (*top*) and modulation spectra (*bottom*) for tryptophan at pH 7.0 and 20 °C. The solid lines represent the best two-component fits, and the dashed lines the best one-component fits. From Ref. 131.

$$\tau_\phi^{app} = \frac{1}{\omega} \tan \phi_\omega \quad [5.33]$$

$$\tau_m^{app} = \frac{1}{\omega}\left[\frac{1}{m_\omega^2} - 1\right]^{1/2} \quad [5.34]$$

where we have dropped the indicator of wavelength for simplicity.

There are several characteristics of the phase and modulation lifetimes which are valuable to know.[41,132] The apparent values are equal only if the intensity decay is a single exponential, for which case

$$\tau_\phi^{app} = \tau_m^{app} = \tau \quad [5.35]$$

For multiexponential or nonexponential decays, the apparent phase lifetimes are shorter than the apparent modulation lifetimes ($\tau_\phi^{app} < \tau_m^{app}$). Also, τ_ϕ^{app} and τ_m^{app} generally decrease at higher modulation frequencies. Hence, their apparent lifetimes depend on the method of measurement and are not true molecular parameters.

The relationship of τ_ϕ^{app} and τ_m^{app} is most easily seen by consideration of a double-exponential decay. Using Eqs. [5.7] and [5.8], one obtains

$$N_\omega(\alpha_1\tau_1 + \alpha_2\tau_2) = \frac{\alpha_1\omega\tau_1^2}{1+\omega^2\tau_1^2} + \frac{\alpha_2\omega\tau_2^2}{1+\omega^2\tau_2^2} \quad [5.36]$$

$$D_\omega(\alpha_1\tau_1 + \alpha_2\tau_2) = \frac{\alpha_1\tau_1}{1+\omega^2\tau_1^2} + \frac{\alpha_2\tau_2}{1+\omega^2\tau_2^2} \quad [5.37]$$

Using Eqs. [5.9] and [5.33], the apparent phase lifetime is given by

$$\tau_\phi^{app} = \frac{\sum_i \alpha_i\tau_i^2/(1+\omega^2\tau_i^2)}{\sum_i \alpha_i\tau_i/(1+\omega^2\tau_i^2)} \quad [5.38]$$

Recall that the average lifetime is given by

$$\bar{\tau} = \frac{\sum_i \alpha_i\tau_i^2}{\sum_j \alpha_j\tau_j} = \sum_i f_i\tau_i = \frac{\alpha_1\tau_1^2 + \alpha_2\tau_2^2}{\alpha_1\tau_1 + \alpha_2\tau_2} \quad [5.39]$$

Comparison of Eqs. [5.38] and [5.39] shows that in τ_ϕ^{app} each decay time is weighted by a factor $\alpha_i\tau_i/(1+\omega^2\tau_i^2)$ rather than a factor $\alpha_i\tau_i = f_i$. For this reason, the components with shorter decay times are weighted more strongly in τ^{app} than in $\bar{\tau}$. Increasing the modulation frequency increases the relative contribution of the short-lived component and hence decreases the value of τ_ϕ^{app}. Using similar reasoning but more complex equations,[41] one can demonstrate that the apparent modulation lifetime is longer than the average lifetime.

An example of the use of apparent phase and modulation lifetimes is given in Figure 5.55, for the mixture of ACF and AFA. This figure shows the phase-angle and modulation spectra in terms of τ_ϕ^{app} and τ_m^{app}. The fact that $\tau_\phi^{app} < \tau_m^{app}$ for a heterogeneous decay is evident by comparison of the upper and lower panels. Also, one immediately notices that the apparent lifetime by phase or modulation depends on modulation frequency and that higher frequencies result in shorter apparent lifetimes. Hence, the apparent lifetimes depend on the method of measurement (phase or modulation) and on the frequency, and it is difficult to interpret these values in terms of molecular features of the sample.

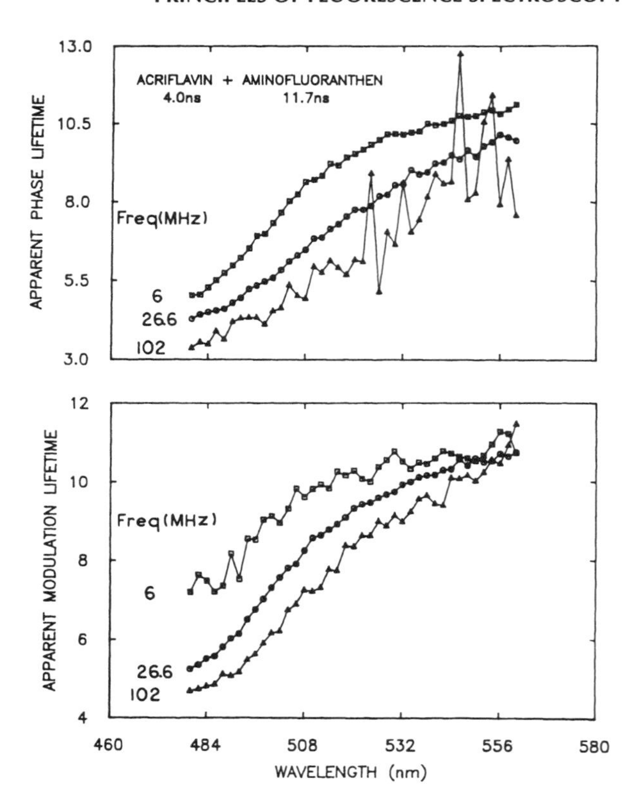

Figure 5.55. Apparent phase (*top*) and modulation lifetimes (*bottom*) for a mixture of ACF and AFA. Revised and reprinted, with permission, from Ref. 130, Copyright © 1990, American Chemical Society.

It is always possible to interpret the phase and modulation values in terms of the apparent lifetimes. However, the use of apparent phase and modulation lifetimes is no longer recommended. These are only apparent values which are the result of a complex weighting of the individual decay times and amplitudes, which depend on the experimental conditions. Also, one does not actually measure apparent lifetimes. These values are interpretations of the measurable quantities, which are the phase and modulation values.

5.11. DERIVATION OF THE EQUATIONS FOR PHASE-MODULATION FLUORESCENCE • SUPPLEMENTAL MATERIAL •

5.11.A. Relationship of the Lifetime to the Phase Angle and Modulation

The equations relating the phase and modulation values to the apparent lifetimes (Eqs. [5.3] and [5.4]) are widely known, but the derivation is rarely given. These expressions have been derived by several routes.[41,133,134] The simplest approach uses the kinetic equations and algebraic manipulation.[41,133] The excitation is assumed to be sinusoidally modulated light,

$$L(t) = a + b \sin \omega t \qquad [5.40]$$

so that $b/a = m_L$ is the modulation of the incident light. The fluorescence emission is forced to respond with the same frequency, but the phase shift and modulation will be different. One can assume that the excited-state population is given by

$$N(t) = A + B \sin(\omega t - \phi) \qquad [5.41]$$

and determine the relationship between fluorescence lifetime and the phase shift (ϕ) and the demodulation (m).

Suppose the intensity decay following δ-function excitation is a single exponential,

$$I(t) = I_0 \exp(-t/\tau) \qquad [5.42]$$

For a single-exponential decay, the differential equation describing the time-dependent intensity is

$$\frac{dI(t)}{dt} = -\frac{1}{\tau} I(t) + L(t) \qquad [5.43]$$

Substitution of Eq. [5.41] into Eq. [5.43] yields

$$\omega B \cos(\omega t - \phi) = -\frac{1}{\tau} [A + B \sin(\omega t - \phi)] + a + b \sin \omega t \qquad [5.44]$$

This equation must be valid for all times. The relationship between the values of a, b, A, and B and the fluorescence lifetime τ can be obtained by expansion of the sine and cosine functions, followed by equating of the constant terms and terms in $\sin \omega t$ and $\cos \omega t$. This yields

$$a - (1/\tau)A = 0 \qquad [5.45]$$

$$\omega \cos \phi - (1/\tau) \sin \phi = 0 \qquad [5.46]$$

$$\omega \sin \phi + (1/\tau) \cos \phi = b/B \qquad [5.47]$$

From Eq. [5.46] one obtains the familiar relationship

$$\frac{\sin \phi}{\cos \phi} = \tan \phi = \omega\, \tau_\phi \qquad [5.48]$$

Squaring Eqs. [5.46] and [5.47], followed by addition, yields

$$\omega^2 + (1/\tau)^2 = (b/B)^2 \qquad [5.49]$$

Recalling that $A = a\tau$ (Eq. [5.45]), one obtains

$$m = \frac{B/A}{b/a} = (1 + \omega^2 \tau_m^2)^{-1/2} \qquad [5.50]$$

which is the usual relationship between the lifetime and the demodulation factor.

An alternative derivation is by the convolution integral.[134] The time-dependent intensity $I(t)$ is given by the convolution of the excitation function (Eq. [5.40]) with the impulse response function (Eq. [5.42]),

$$I(t) = \int_0^\infty L(t') I(t - t')\, dt' \qquad [5.51]$$

Substitution of Eqs. [5.40] and [5.42] yields

$$I(t) = I_0 \int_0^\infty \exp(-t'/\tau)[a + b \cos(\omega t - \omega t')]\, dt' \qquad [5.52]$$

This integral can be calculated by recalling the identities

$$\cos(x - y) = \cos x \cos y + \sin x \sin y \qquad [5.53]$$

$$\int_0^\infty \exp(-kx) \sin mx\, dx = \frac{m}{k^2 + m^2} \qquad [5.54]$$

$$\int_0^\infty \exp(-kx) \cos mx\, dx = \frac{a}{k^2 + m^2} \qquad [5.55]$$

Using these identities yields

$$\int_0^\infty \exp(-t'/\tau) \cos \omega(t - t')\, dt' = \frac{\tau}{\sqrt{1 + \omega^2\tau^2}} \left(\frac{\cos \omega t}{\sqrt{1 + \omega^2\tau^2}} + \frac{\omega t \sin \omega t}{\sqrt{1 + \omega^2\tau^2}} \right) \qquad [5.56]$$

$$= \frac{\tau}{\sqrt{1 + \omega^2\tau^2}} \cos(\omega t - \phi) \qquad [5.57]$$

The last equation (Eq. [5.57]) was obtained using

$$\cos \phi = (1 + \omega^2\tau^2)^{-1/2} \qquad [5.58]$$

and

$$\tan \phi = \frac{\sin \phi}{\cos \phi} \quad [5.59]$$

Hence, the time-dependent intensity is given by

$$I(t) = I_0\tau \left[a + \frac{b}{\sqrt{1 + \omega^2\tau^2}} \cos(\omega t - \phi) \right] \quad [5.60]$$

This expression shows that the emission is demodulated by a factor $(1 + \omega^2 \tau^2)^{-1/2}$ relative to the excitation.

5.11.B. Cross-Correlation Detection

As described in Section 5.2.D, the use of cross-correlation detection transforms the high-frequency emission to a low-frequency signal while preserving the meaning of the phase and modulation values. This can be seen by considering the nature of the signals. The high-frequency time-dependent intensity is given by

$$I(t) = I_0 \,[1 + m \cos(\omega t + \phi)] \quad [5.61]$$

This signal is multiplied by the sinusoidal gain modulation of the detector,[133]

$$G(t) = G_0[1 + m_c \cos(\omega_c t + \phi_c)] \quad [5.62]$$

where G_0 is the average value of the function, and m_c, ω_c, and ϕ_c are, respectively, the modulation, frequency, and phase of the cross-correlation signal. Multiplication of Eqs. [5.62] and [5.63] yields

$$S(t) = N_0G_0[1 + m \cos(\omega t + \phi) + m_c \cos(\omega_c t + \phi_c)$$

$$+ mm_c \cos(\omega t + \phi) \cos(\omega_c t + \omega_c)] \quad [5.63]$$

Using trigonometric identities, the last term becomes

$$\frac{mm_c}{2} [\cos(\Delta\omega t + \Delta\phi) + \cos(\omega_c t + \omega t + \Delta\phi)] \quad [5.64]$$

where $\Delta\omega = \omega_c - \omega$ and $\Delta\phi = \phi_c - \phi$. The frequencies ω_c and ω typically differ by only a small amount. Hence, Eq. [5.63] contains a constant term plus terms with frequencies, ω, ω_c, $\omega + \omega_c$, and $\Delta\omega$. The $\Delta\omega$ term contains the phase and modulation information. In the electronic filtering process, the constant term and terms in ω, ω_c, and $\omega + \omega_c$ all contribute to average intensity, and the term in $\Delta\omega$ determines the phase and amplitude of the low-frequency modulated emission. The presence of the phase and modulation information in the low-frequency signal can also be seen by integration of Eqs. [5.62] and [5.63] over one measurement cycle.[41]

5.12. PERSPECTIVES ON FREQUENCY-DOMAIN FLUOROMETRY

In the 1980s there was considerable debate about the value of phase-modulation lifetime measurements and about whether this method could be useful for the resolution of complex intensity decays. In fact, because of the difficulties in using data at just a few modulation frequencies, phase-modulation fluorometers were the subject of considerable criticism. However, since the introduction of the variable-frequency instruments, phase-modulation fluorometry has undergone a renaissance, and this method is now preferred by many laboratories.

What is the limiting factor in developing lower-cost and simpler variable-frequency instruments? At present, the main limitation appears to be the light source. It is desirable to have available a wide range of wavelengths, so that arc lamps are the preferred light source. However, it is difficult to amplitude-modulate the output of an arc lamp, and useful modulation requires a number of expensive components. Pulsed laser light sources are intrinsically modulated but are expensive and difficult to maintain. Development of an improved modulated light source would allow variable-frequency FD measurements to become simpler and more widely accessible.

REFERENCES

1. Gratton, E., and Limkeman M., 1983, A continuously variable frequency cross-correlation phase fluorometer with picosecond resolution, *Biophys. J.* **44**:315–324.
2. Lakowicz, J. R., and Maliwal, B. P., 1985, Construction and performance of a variable-frequency phase-modulation fluorometer, *Biophys. Chem.* **21**:61–78.
3. Lakowicz, J. R., and Gryczynski, I., 1991, Frequency-domain fluorescence spectroscopy, in *Topics in Fluorescence Spectroscopy, Volume 1, Techniques*, J. R. Lakowicz (ed.), Plenum Press, New York, pp. 293–355.
4. Gratton, E., 1984, The measurement and analysis of heterogeneous emissions by multifrequency phase and modulation fluorometry, *Appl. Spectrosc. Rev.* **20**(1):55–106.
5. Gratton, E., Jameson, D. M., and Hall, R. D., 1984, Multifrequency phase and modulation fluorometry, *Annu. Rev. Biophys. Bioeng.* **13**:105–124.
6. Lakowicz, J. R., 1986, Biochemical applications of frequency-domain fluorometry, in *Applications of Fluorescence in the Biomedical Sciences*, D. Lansing, A. S. Waggoner, F. Lanni, R. F. Murphy, and R. R. Birge (eds.), Alan R. Liss, New York, pp. 225–244.
7. Bright, F. V., Betts, T. A., and Litwiler, K. S., 1990, Advances in multifrequency phase and modulation fluorescence analysis, *Anal. Chem.* **21**:389–405.

8. Lakowicz, J. R., 1985, Frequency-domain fluorometry for resolution of time-dependent fluorescence emission, *Spectroscopy* **1**:28–37.
9. Rabinovich, E. M., O'Brien, M., Srinivasan, B., Elliott, S., Long, X.-C., and Ravinder, K. J., 1998, A compact, LED-based phase fluorimeter-detection system for chemical and biosensor arrays, *Proc. SPIE* **3258**:2–10.
10. Bevington, P. R., 1969, *Data Reduction and Error Analysis for the Physical Sciences*, McGraw-Hill, New York.
11. Taylor, J. R., 1982, *An Introduction to Error Analysis, the Study of Uncertainties in Physical Measurements*, University Science Books, Mill Valley, California.
12. Lakowicz, J. R., Laczko, G., Cherek, H., Gratton, E., and Limkeman, M., 1984, Analysis of fluorescence decay kinetics from variable-frequency phase shift and modulation data, *Biophys. J.* **46**:463–477.
13. Gratton, E., Limkeman, M., Lakowicz, J. R., Maliwal, B., Cherek, H., and Laczko, G., 1984, Resolution of mixtures of fluorophores using variable-frequency phase and modulation data, *Biophys, J.* **46**:479–486.
14. Straume, M., Frasier-Cadoret, S. G., and Johnson, M. L., 1991, Least-squares analysis of fluorescence data, in *Topics in Fluorescence Spectroscopy, Volume 2, Principles*, J. R. Lakowicz (ed.), Plenum Press, New York, pp. 177–239.
15. Johnson, M. L., 1994, Use of least-squares techniques in biochemistry, *Methods Enzymol.* **240**:1–22.
16. Johnson, M. L., and Faunt, L. M., 1992, Parameter estimation by least-squares methods, *Methods Enzymol.* **210**:1–37.
17. Klein, U. K. A., 1984, Picosecond fluorescence decay studied by phase fluorometry and its application to the measurement of rotational diffusion in liquids, *Arabian J. Sci. Eng.* **9**(4):327–344.
18. Gaviola, Z., 1926, Ein Fluorometer, apparat zur messung von fluoreszenzabklingungszeiten, *Z. Phys.* **42**:853–861.
19. Wood, R. W., 1921, The time interval between absorption and emission of light in fluorescence, *Proc. R. Soc. London (A)* **99**:362–371.
20. Abraham and Lemoine, 1899, *C. R. Hebd. Seance Acad. Sci.* **129**:206, as cited in Ref. 23.
21. Duschinsky, V. F., 1933, Der zeitliche intensitatsverlauf von intermittierend angeregter resonanzstrahlung, *Z. Phys.* **81**:7–22.
22. Szymanowski, W., 1935, Verbesserte fluorometermethode zur messung der abklingzeiten der fluoreszensatrahlung, *Z. Phys.* **95**:440–449.
23. Tumerman, L. A., 1941, On the law of decay of luminescence of complex molecules, *J. Phys. (USSR)* **4**:151–166.
24. Maercks, V. O., 1938, Neuartige fluorometer, *Z. Phys.* **109**:685–699.
25. Hupfeld, V. H.-H., 1929, Die nachleuchtdauern der J_2-, K_2, Na_2- und Na-resonanzstrahlung, *Z. Phys.* **54**:484–497.
26. Schmillen, A., 1953, Abklingzeitmessungen an flussigen und festen losungen mit einem neuen fluorometer, *Z. Phys.*. **135**:294–308.
27. Galanin, M. D., 1950, Duration of the excited state of a molecule and the properties of fluorescent solutions, *Tr. Fiz. Inst. Akad. Nauk SSSR* **5**:339–386.
28. Birks, J. B., and Little, W. A., 1953, Photo-fluorescence decay times of organic phosphors, *Proc. Phys. Soc.* **A66**:921–928.
29. Resewitz, V. E.-P., and Lippert, E., 1974, Ein neuartiges phasenfluorometer, *Ber. Bunsenges. Phys. Chem.* **78**:1227–1229.
30. Labhart, V. H., 1964, Eine experimentelle methode zur ermittlung der singulett-triplett-konversionswahrscheinlichkeit und der triplettspektren von gelosten organischen molekeln messungen an 1,2-benzanthracen, *Fasciculus S* **252**:2279–2288.
31. Bailey, E. A., and Rollefson, G. K., 1953, The determination of the fluorescence lifetimes of dissolved substances by a phase shift method, *J. Chem. Phys.* **21**:1315–1326.
32. Bonch-Breuvich, A. M., Kazarin, I. M., Molchanov, V. A., and Shirokov, I. V., 1959, An experimental model of a phase fluorometer, *Instrum. Exp. Tech. (USSR)* **2**:231–236.
33. Bauer, R. K., and Rozwadawski, M., 1959, A new type of fluorometer. Measurements of decay periods of fluorescence of acridine yellow solutions as a function of concentration, *Bull. Acad. Pol. Sci., Ser. Sci. Math. Astron. Phys.* **8**:365–368.
34. Birks, J. B., and Dyson, D. J., 1961, Phase and modulation fluorometer, *J. Sci. Instrum.* **38**:282–285.
35. Muller, A., Lumry, R., and Kokubun, H., 1965, High-performance phase fluorometer constructed from commercial subunits, *Rev. Sci. Instrum.* **36**:1214–1226.
36. Michelbacher, E., 1969, Decay time measurements on pseudo-isocyanine by a phase-fluorometer of 200 Mc modulation frequency, *Z. Naturforsch. A* **24**:790–796.
37. Demtroder, W., 1962, Bestimmung von oszillatorenstarken durch lebensdauermessungen der ersten angeregten niveaus fur die elemente Ga, Al, Mg, Tl und Na, *Z. Phys.*. **42**:42–55.
38. Schlag, E. W., and Wessenhoff, H. V., 1969, Direct timing of the relaxation from selected excited states; beta-naphthylamine, *J. Chem. Phys.* **51**:2508–2514.
39. Venetta, B. D., 1959, Microscope phase fluorometer for determining the fluorescence lifetimes of fluorochromes, *Rev. Sci. Instrum.* **30**:450–457.
40. Schaefer, V. W., 1956, Bestimmung der schwingungsrelaxationszeit in CO/N_2-gasgemischen aus der analyse des frequenzganges eines ultrarot-gasanalysators, *Z. Angew. Phys.* **19**:55–61.
41. Spencer, R. D., and Weber, G., 1969, Measurement of subnansecond fluorescence lifetimes with a cross-correlation phase fluorometer, *Ann. N.Y. Acad. Sci.* **158**:361–376.
42. Debye, P., and Sears, F. W., 1932, On the scattering of light by supersonic waves, *Proc. Natl. Acad. Sci. U.S.A.* **18**:409–414.
43. Lakowicz, J. R., 1983, *Principles of Fluorescence Spectroscopy*, Plenum Press, New York, pp. 76–78.
44. Hauser, M., and Heidt, G., 1975, Phase fluorometer with a continuously variable frequency, *Rev. Sci. Instrum.* **46**:470–471.
45. Salmeen, I., and Rimal, L., 1977, A phase-shift fluorometer using a laser and a transverse electrooptic modulator for subnanosecond lifetime measurements, *Biophys. J.* **20**:335–342.
46. Menzel, E. R., and Popovic, Z. D., 1978, Picosecond-resolution fluorescence lifetime measuring system with a cw laser and a radio, *Rev. Sci. Instrum.* **49**:39–44.
47. Haar, H.-P., and Hauser, M., 1978, Phase fluorometer for measurement of picosecond processes, *Rev. Sci. Instrum.* **49**:632–633.
48. Gugger, H., and Calzaferri, G., 1979, Picosecond time resolution by a continuous wave laser amplitude modulation technique I: A critical investigation, *J. Photochem.* **13**:21–33.
49. Gugger, H., and Calzaferri, G., 1980, Picosecond time resolution by a continuous wave laser amplitude modulation technique II: Experimental basis, *J. Photochem.* **13**:295–307.
50. Gugger, H., and Calzaferri, G., 1981, Picosecond time resolution by a continuous wave laser amplitude modulation technique III: Dual-beam luminescence experiment, *J. Photochem.* **16**:31–41.
51. Baumann, J., and Calzaferri, G., 1983, Development of picosecond time-resolved techniques by continuous-wave laser amplitude modulation IV: Systematic errors, *J. Photochem.* **22**:297–312.
52. Baumann, J., and Calzaferri, G., 1983, Development of picosecond time-resolved techniques by continuous-wave V: Elimination of r.f. interference problems, *J. Photochem.* **23**:387–390.
53. Ide, G., Engelborghs, Y., and Persoons, A., 1983, Fluorescence lifetime resolution with phase fluorometry, *Rev. Sci. Instrum.* **54**:841–844.

54. Kaminov, I. P., 1984, *An Introduction to Electro-Optic Devices*, Academic Press, New York.
55. Wilson, J., and Hawkes, J. F. B., 1983, *Optoelectronics: An Introduction*, Prentice/Hall International, Englewood Cliff, New Jersey, p. 445.
56. Fedderson, B. A., Piston, D. W., and Gratton, E., 1989, Digital parallel acquisition in frequency domain fluorimetry, *Rev. Sci. Instrum.* **60:**2929–2936.
57. Alcala, J. R., 1991, Comment on "Digital parallel acquisition in frequency domain fluorometry," *Rev. Sci. Instrum.* **62:**1672–1673.
58. Barbieri, B., De Piccoli, F., and Gratton, E., 1989, Synthesizers' phase noise in frequency-domain fluorometry, *Rev. Sci. Instrum.* **60:**3201–3206.
59. Levy, R., Guignon, E. F., Cobane, S., St. Louis, E., and Fernandez, S. M., 1997, Compact, rugged and inexpensive frequency-domain fluorometer, *Proc. SPIE* **2980:**81–89.
60. Lakowicz, J. R., Cherek, H., and Balter, A., 1981, Correction of timing errors in photomultiplier tubes used in phase-modulation fluorometry, *J. Biochem. Biophys. Methods* **5:**131–146.
61. Berndt, K., Dürr, H., and Palme, D., 1983, Picosecond phase fluorometry and color delay error, *Opt. Commun.* **47**(5):321–323.
62. Baumann, J., Calzaferri, G., Forss, L., and Hungentobler, Th., 1985, Wavelength-dependent fluorescence decay: An investigation by multiple-frequency picosecond phase fluorometry, *J. Photochem.* **28:**457–473.
63. Pouget, J., Mugnier, J., and Valeur, B., 1989, Correction of systematic phase errors in frequency-domain fluorometry, *J. Phys. E: Sci. Instrum.* **22:**855–862.
64. Barrow, D. A., and Lentz, B. R., 1983, The use of isochronal reference standards in phase and modulation fluorescence lifetime measurements, *J. Biochem. Biophys. Methods* **7:**217–234.
65. Thompson, R. B., and Gratton, E., 1988, Phase fluorometric method for determination of standard lifetimes, *Anal. Chem.* **60:**670–674.
66. Lakowicz, J. R., Jayaweera, R., Joshi, N., and Gryczynski, I., 1987, Correction for contaminant fluorescence in frequency-domain fluorometry, *Anal. Biochem.* **160:**471–479.
67. Reinhart, G. D., Marzola, P., Jameson, D. M., and Gratton, E., 1991, A method for on-line background subtraction in frequency domain fluorometry, *J. Fluoresc.* **1**(3):153–162.
68. Gryczynski, I., and Malak, H., unpublished observations.
69. Gryczynski, I., unpublished observations.
70. Lakowicz, J. R., 1989, Principles of frequency-domain fluorescence spectroscopy and applications to protein fluorescence, in *Cell Structure and Function by Microspectrofluorometry*, E. Kohen and J. G. Hirschberg (eds.), Academic Press, New York, pp. 163–184.
71. Manzini, G., Barcellona, M. L., Avitabile, M., and Quadrifoglio, F., 1983, Interaction of diamidino-2-phenylindole (DAPI) with natural and synthetic nucleic acids, *Nucleic Acids Res.* **11:**8861–8876.
72. Cavatorta, P., Masotti, L., and Szabo, A. G., 1985, A time-resolved florescence study of 4′,6-diamidino-2-phenylindole dihydrochloride binding to polynucleotides, *Biophys. Chem.* **22:**11–16.
73. Tanious, F. A., Veal, J. M., Buczak, H., Ratmeyer, L. S., and Wilson, W. D., 1992, DAPI (4′,6-diamidino-2-phenylindole) binds differently to DNA and RNA: Minor-groove binding at AT sites and intercalation at AU sites, *Biochemistry* **31:**3103–3112.
74. Barcellona, M. L., and Gratton, E., 1989, Fluorescence lifetime distributions of DNA-4′,6-diamidino-2-phenylindole complex, *Biochim. Biophys. Acta* **993:**174–178.
75. Barcellona, M. L., and Gratton, E., 1990, The fluorescence properties of a DNA probe, *Eur. Biophys. J.* **17:**315–323.
76. Lakowicz, J. R., Szmacinski, H., Nowaczyk, N., and Johnson, M. L., 1992, Fluorescence lifetime imaging of calcium using quin-2, *Cell Calcium* **13:**131–147.
77. Miyoshi, N., Hara, K., Kimura, S., Nakanishi, K., and Fukuda, M., 1991, A new method of determining intracellular free Ca^{2+} concentration using Quin-2 fluorescence, *Photochem. Photobiol.* **53:**415–418.
78. Hirshfield, K. M., Toptygin, D., Packard, B. S., and Brand, L., 1993, Dynamic fluorescence measurements of two-state systems: Applications to calcium-chelating probes, *Anal. Biochem.* **209:**209–218.
79. Illsley, N. P., and Verkman, A. S., 1987, Membrane chloride transport measured using a chloride-sensitive fluorescent probe, *Biochemistry* **26:**1215–1219.
80. Verkman, A. S., 1990, Development and biological applications of chloride-sensitive fluorescent indicators, *Am. J. Physiol.* **253:**C375–C388.
81. Verkman, A. S., Sellers, M. C., Chao, A. C., Leung, T., and Ketcham, R., 1989, Synthesis and characterization of improved chloride-sensitive fluorescent indicators for biological applications, *Anal. Biochem.* **178:**355–361.
82. Szmacinski, H., and Lakowicz, J. R., unpublished observations.
83. Lakowicz, J. R. (ed.), 1997, *Topics in Fluorescence Spectroscopy, Volume 5, Nonlinear and Two-Photon Induced Fluorescence*, Plenum Press, New York, 544 pp.
84. Dattelbaum, J. D., Castellano, F. N., Gryczynski, I., and Lakowicz, J. R., 1998, Two-photon spectroscopic properties of a mutant green fluorescent protein, Biphysical Society Meeting, March, 1998, Kansas City, Missouri.
85. Swaminathan, R., Hoang, C. P., and Verkman, A. S., 1997, Photobleaching recovery and anisotropy decay of green fluorescent protein GFP-S65T in solution and cells: Cytoplasmic viscosity probed by green fluorescent protein translational and rotational diffusion, *Biophys. J.* **72:**1900–1907.
86. Kusba, J., and Lakowicz, J. R., 1994, Diffusion-modulated energy transfer and quenching: Analysis by numerical integration of diffusion equation in Laplace space, *Methods Enzymol.* **240:**216–262.
87. Lakowicz, J. R., Cherek, H., Gryczynski, I., Joshi, N., and Johnson, M. L., 1987, Analysis of fluorescence decay kinetics measured in the frequency domain using distributions of decay times, *Biophys. Chem.* **28:**35–50.
88. Alcala, J. R., Gratton, E., and Prendergast, F. G., 1987, Resolvability of fluorescence lifetime distributions using phase fluorometry, *Biophys. J.* **51:**587–596.
89. Foguel, D., Chaloub, R. M., Silva, J. L., Crofts, A. R., and Weber, G., 1992, Pressure and low temperature effects on the fluorescence emission spectra and lifetimes of the photosynthetic components of cyanobacteria, *Biochem. J.* **63:**1613–1622.
90. Alcala, J. R., Gratton, E., and Prendergast, F. G., 1987, Fluorescence lifetime distributions in proteins, *Biophys. J.* **51:**597–604.
91. Alcala, J. R., Gratton, E., and Prendergast, F. G., 1987, Interpretation of fluorescence decays in proteins using continuous lifetime distributions, *Biophys. J.* **51:**925–936.
92. Lakowicz, J. R., Gryczynski, I., Wiczk, W., and Johnson, M. L., 1994, Distributions of fluorescence decay times for synthetic melittin in water–methanol mixtures and complexed with calmodulin, troponin C, and phospholipids, *J. Fluoresc.* **4**(2):169–177.
93. Gryczynski, I., Wiczk, W., Inesi, G., Squier, T., and Lakowicz, J. R., 1989, Characterization of the tryptophan fluorescence from sarcoplasmic reticulum adenosinetriphosphatase by frequency-domain fluorescence spectroscopy, *Biochemistry* **28:**3490–3498.

94. Visser, A. J. W. G., and van Hoek, A., 1981, The fluorescence decay of reduced nicotinamides in aqueous solution after excitation with a uv-mode locked Ar ion laser, *Photochem. Photobiol.* **33**:35–40.
95. Merkelo, H. S., Hartman, S. R., Mar, T., Singhal, G. S., and Govindjee, 1969, Mode-locked lasers: Measurements of very fast radiative decay in fluorescent systems, *Science* **164**:301–303.
96. Gratton, E., and Lopez-Delgado, R., 1980, Measuring fluorescence decay times by phase-shift and modulation techniques using the high harmonic content of pulsed light sources, *Nuovo Cimento* **B56**:110–124.
97. Gratton, E., Jameson, D. M., Rosato, N., and Weber, G., 1984, Multifrequency cross-correlation phase fluorometer using synchrotron radiation, *Rev. Sci. Instrum.* **55**:486–494.
98. Gratton, E., and Delgado, R. L., 1979, Use of synchrotron radiation for the measurement of fluorescence lifetimes with subpicosecond resolution, *Rev. Sci. Instrum.* **50**:789–790.
99. Berndt, K., Duerr, H., and Palme, D., 1982, Picosecond phase fluorometry by mode-locked CW lasers, *Opt. Commun.* **42**:419–422.
100. Gratton, E., and Barbieri, B., 1986, Multifrequency phase fluorometry using pulsed sources: Theory and applications, *Spectroscopy* **1**(6):28–36.
101. Lakowicz, J. R., Laczko, G., and Gryczynski, I., 1986, 2-GHz frequency-domain fluorometer, *Rev. Sci. Instrum.* **57**:2499–2506.
102. Laczko, G., Gryczynski, I., Gryczynski, Z., Wiczk, W., Malak, H., and Lakowicz, J. R., 1990, A 10-GHz frequency-domain fluorometer, *Rev. Sci. Instrum.* **61**:2331–2337.
103. Lakowicz, J. R., Laczko, G., Gryczynski, I., Szmacinski, H., and Wiczk, W., 1989, Frequency-domain fluorescence spectroscopy: Principles, biochemical applications and future developments, *Ber. Bunsenges. Phys. Chem.* **93**:316–327.
104. Lakowicz, J. R., Laczko, G., Gryczynski, I., Szmacinski, H., and Wiczk, W., 1988, Gigahertz frequency domain fluorometry: Resolution of complex decays, picosecond processes and future developments, *J. Photochem. Photobiol. B: Biol.* **2**:295–311.
105. Berndt, K., Durr, H., and Palme, D., 1985, Picosecond fluorescence lifetime detector, *Opt. Commun.* **55**(4):271–276.
106. Berndt, K., 1987, Opto-electronic high-frequency cross-correlation using avalanche photodiodes, *Measurement* **5**(4):159–166.
107. Berndt, K., Klose, E., Schwarz, P., Feller, K.-H., and Faβler, D., 1984, Time resolved fluorescence spectroscopy of cyanine dyes, *Z. Phys. Chem.* **265**:1079–1086.
108. Berndt, K., Durr, H., and Feller, K.-H., 1987, Time resolved fluorescence spectroscopy of cyanine dyes, *Z. Phys. Chem.* **268**:250–256.
109. Lakowicz, J. R., Gryczynski, I., Laczko, G., and Gloyna, D., 1991, Picosecond fluorescence lifetime standards for frequency- and time-domain fluorescence, *J. Fluoresc.* **1**(2):87–93.
110. Gryczynski, I., Szmacinski, H., Laczko, G., Wiczk, W., Johnson, M. L., Kusba, J., and Lakowicz, J. R., 1991, Conformational differences of oxytocin and vasopressin as observed by fluorescence anisotropy decays and transient effects in collisional quenching of tyrosine fluorescence, *J. Fluoresc.* **1**(3):163–176.
111. Vos, R., Strobbe, R., and Engelborghs, Y., 1997, Gigahertz phase fluorometry using a fast high-gain photomultiplier, *J. Fluoresc.* **7**(1):33S–35S.
112. Berndt, K. W., Gryczynski, I., and Lakowicz, J. R., 1990, Phase-modulation fluorometry using a frequency-doubled pulsed laser diode light source, *Rev. Sci. Instrum.* **61**:1816–1820.
113. Thompson, R. B., Frisoli, J. K., and Lakowicz, J. R., 1992, Phase fluorometry using a continuously modulated laser diode, *Anal. Chem.* **64**:2075–2078.
114. Sipior, J., Carter, G. M., Lakowicz, J. R., and Rao, G., 1996, Single quantum well light emitting diodes demonstrated as excitation sources for nanosecond phase-modulation fluorescence lifetime measurements, *Rev. Sci. Instrum.* **67**:3795–3798.
115. Sipior, J., Carter, G. M., Lakowicz, J. R., and Rao, G., 1997, Blue light emitting diode demonstrated as an ultraviolet excitation source for nanosecond phase-modulation fluorescence lifetime measurements, *Rev. Sci. Instrum.* **68**:2666–2670.
116. Fantini, S., Franceschini, M. A., Fishkin, J. B., Barbieri, B., and Gratton, E., 1994, Quantitative determination of the absorption spectra of chromophores in strongly scattering media: A light-emitting diode based technique, *Appl. Opt.* **33**:5204–5213.
117. Berndt, K. W., and Lakowicz, J. R., 1992, Electroluminescent lamp-based phase fluorometer and oxygen sensor, *Anal. Biochem.* **201**:319–325.
118. Morgan, C. G., Hua, Y., Mitchell, A. C., Murray, J. G., and Boardman, A. D., 1996, A compact frequency domain fluorometer with a directly modulated deuterium light source, *Rev. Sci. Instrum.* **67**:41–47.
119. Holavanahali, R., Romauld, M., Carter, G. M., Rao, G., Sipior, J., Lakowicz, J. R., and Bierlein, J. D., 1996, Directly modulated diode laser frequency doubled in a KTP waveguide as an excitation source for CO_2 and O_2 phase fluorometric sensors, *J. Biomed. Opt.* **1**:124–130.
120. Guo, X.-Q., Castellano, F. N., Li, L., Szmacinski, H., and Lakowicz, J. R., 1997, A long-lived, highly luminescent Re(I) metal–ligand complex as a biomolecular probe, *Anal. Biochem.* **254**:179–186.
121. Szmacinski, H., and Lakowicz, J. R., 1993, Optical measurements of pH using fluorescence lifetimes and phase-modulation fluorometry, *Anal. Chem.* **65**:1668–1674.
122. Thompson, R. B., and Lakowicz, J. R., 1993, Fiber optic pH sensor based on phase fluorescence lifetimes, *Anal. Chem.* **65**:853–856.
123. O'Keefe, G., MacCraith, B. D., McEvoy, A. K., McDonagh, C. M., and McGilp, J. F., 1995, Development of a LED-based phase fluorimetric oxygen sensor using evanescent wave excitation of a sol-gel immobilized gel, *Sensors Actuat.* **29**:226–230.
124. Lippitsch, M. E., Pasterhofer, J., Leiner, M. J. P., and Wolfbeis, O. S., 1988, Fibre-optic oxygen sensor with the fluorescence decay time as the information carrier, *Anal. Chim. Acta* **205**:1–6.
125. Sipior, J., Randers-Eichhorn, L., Lakowicz, J. R., Carter, G. M., and Rao, G., 1996, Phase fluorometric optical carbon dioxide gas sensor for fermentation off-gas monitoring, *Biotechnol. Prog.* **12**:266–271.
126. Cobb, W. T., and McGown, L. B., 1987, Phase-modulation fluorometry for on-line liquid chromatographic detection and analysis of mixtures of benzo(*k*)fluoranthene and benzo(*b*)fluoranthene, *Appl. Spectrosc.* **41**:1275–1279.
127. Cobb, W. T., and McGown, L. B., 1989, Multifrequency phase-modulation fluorescence lifetime determinations on-the-fly in HPLC, *Appl. Spectrosc.* **43**:1363–1367.
128. Cobb, W. T., Nithipatikom, K., and McGown, L. B., 1988, Multicomponent detection and determination of polycyclic aromatic hydrocarbons using HPLC and a phase-modulation spectrofluorometer, Special Technical Publication, American Society for Testing and Materials, Vol. 1009, pp. 12–25.
129. Szmacinski, H., and Lakowicz, J. R., 1994, Lifetime-based sensing, in *Topics in Fluorescence Spectroscopy, Vol. 4, Probe Design and Chemical Sensing*, J. R. Lakowicz (ed.), Plenum Press, New York, pp. 295–334.
130. Lakowicz, J. R., Jayaweera, R., Szmacinski, H., and Wiczk, W., 1990, Resolution of multicomponent fluorescence emission using

frequency-dependent phase angle and modulation spectra, *Anal. Chem.* **62**:2005–2012.

131. Lakowicz, J. R., Jayaweera, R., Szmacinski, H., and Wiczk, W., 1989, Resolution of two emission spectra for tryptophan using frequency-domain phase-modulation spectra, *Photochem. Photobiol.* **50**:541–546.
132. Kilin, S. F., 1962, The duration of photo- and radioluminescence of organic compounds, *Opt. Spectrosc.* **12**:414–416.
133. Jameson, D. M., Gratton, E., and Hall, R. D., 1984, The measurement and analysis of heterogeneous emissions by multifrequency phase and modulation fluorometry, *Appl. Spectrosc. Rev.* **20**(1):55–106.
134. Ware, W. R., 1971, Transient luminescence measurements, in *Creation and Detection of the Excited State*, A. A. Lamola (ed.), Marcel Dekker, New York, pp. 213–302.

PROBLEMS

5.1. *Calculation of the Decay Time of SPQ from Phase and Modulation Data*: Use the data in Figure 5.30 to calculate the decay times of SPQ at each chloride concentration. For convenience, selected phase and modulation values are listed in Table 5.6. Data can also be read off Figure 5.30.

5.2. *Determination of Chloride Concentrations with SPQ*: Chloride quenches the fluorescence of SPQ, and this intensity can be used to measure chloride concentrations. Suppose that one is measuring SPQ fluorescence in a fluorescence microscope and that the SPQ concentration is not known. Under these conditions, it is difficult to use the intensity values to measure the chloride concentration. Suggest how the phase or modulation data of SPQ (Figure 5.30) could be used to measure chloride concentrations.

Table 5.6. Selected Phase and Modulation Values for the Chloride Probe SPQ[a]

Chloride concentration	Frequency (MHz)	Phase angle (degrees)	Modulation
0	10	57.4	0.538
	100	86.3	0.060
10m*M*	10	35.1	0.821
	100	82.2	0.141
30m*M*	10	18.0	0.954
	100	73.1	0.286
70m*M*	10	9.4	0.988
	100	59.1	0.505

[a] The listed values were interpolated using the measured frequency responses (Figure 5.30).

Assume that the uncertainties in the phase and modulation values are ±0.2° and ±0.5%, respectively. What is the expected accuracy in the measured chloride concentrations?

5.3. *Effect of Heterogeneity on Apparent Phase and Modulation Lifetimes*: Suppose that you have samples which display a double-exponential decay law, with lifetimes of 0.5 and 5.0 ns. For one sample the preexponential factors are equal ($\alpha_1 = \alpha_2 = 0.5$), and for the other sample the fractional intensities are equal ($f_1 = f_2 = 0.5$). Calculate the apparent phase and modulation lifetimes for these two decay laws at modulation frequencies of 50 and 100 MHz. Explain the relative values of the apparent lifetimes.

Solvent Effects on Emission Spectra

6

Solvent polarity and the local environment have profound effects on the emission spectra of polar fluorophores. These effects are the origin of the Stokes' shift, which is one of the earliest observations in fluorescence. Emission spectra are easily measured, and as a result, there are numerous publications on emission spectra of fluorophores in different solvents and when bound to proteins, membranes, and nucleic acids. One common use of solvent effects is to determine the polarity of the probe binding site on the macromolecule. This is accomplished by comparison of the emission spectra and/or quantum yields of the fluorophore when it is bound to the macromolecule and when it is dissolved in solvents of different polarity. However, there are many additional instances where solvent effects are used. Suppose a fluorescent ligand binds to a protein. Binding is usually accompanied by a spectral shift due to the different environment for the bound ligand. Alternatively, the ligand may induce a spectral shift in the intrinsic or extrinsic protein fluorescence. Additionally, fluorophores often display spectral shifts when they bind to membranes.

The effects of solvent and environment on fluorescence spectra are complex, and there is no single theory which accounts for all these effects. Spectral shifts result from the general effect of solvent polarity whereby the energy of the excited state decreases with increasing solvent polarity. This effect can be accounted for by the Lippert equation (see below). However, spectral shifts also occur due to specific fluorophore–solvent interactions and due to charge separation in the excited state. In the following sections we describe the theory for general solvent effects and provide examples where the spectral properties indicate more complex behavior.

6.1. OVERVIEW OF SOLVENT EFFECTS

Emission from fluorophores generally occurs at wavelengths which are longer than those at which absorption occurs. This loss of energy is due to a variety of dynamic processes which occur following light absorption (Figure 6.1). The fluorophore is typically excited to the first singlet state (S_1), usually to an excited vibrational level within S_1. The excess vibrational energy is rapidly lost to the solvent. If the fluorophore is excited to the second singlet state (S_2), it rapidly decays to the S_1 state in 10^{-12} s due to internal conversion. Solvent effects shift the emission to still lower energy owing to stabilization of the excited state by the polar solvent molecules. Typically, the fluorophore has a larger dipole moment in the excited state (μ_E) than in the ground state (μ_G). Following excitation, the solvent dipoles can reorient or relax around μ_E, which lowers the energy of the excited state. As the solvent polarity is increased, this effect becomes larger, resulting in emission at lower energies or longer wavelengths. In general, only fluorophores which are themselves polar display a large sensitivity to solvent polarity. Nonpolar molecules, such as unsubstituted aromatic hydrocarbons, are much less sensitive to solvent polarity.

Fluorescence lifetimes (1–10 ns) are usually much longer than the time required for solvent relaxation. For fluid solvents at room temperature, solvent relaxation oc-

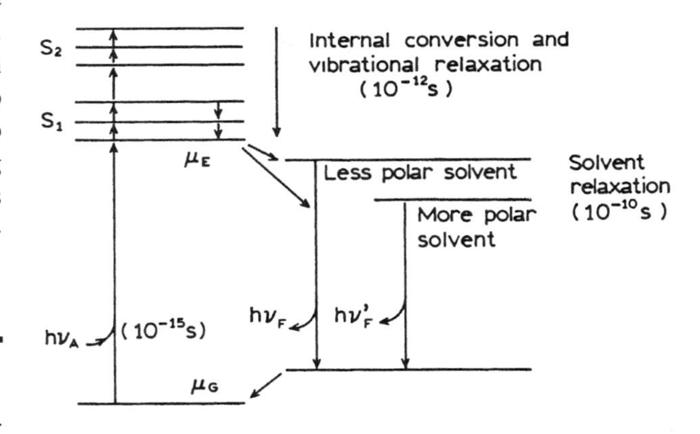

Figure 6.1. Jabłoński diagram for fluorescence with solvent relaxation.

curs in 10–100 ps. For this reason, the emission spectra of fluorophores are representative of the solvent relaxed state. Examination of Figure 6.1 reveals why absorption spectra are less sensitive to solvent polarity than emission spectra. Absorption of light occurs in about 10^{-15} s, a time which is too short for motion of the fluorophore or solvent. Absorption spectra are not affected by the decrease in the excited-state energy which occurs after absorption has occurred.

6.1.A. Polarity Surrounding a Membrane-Bound Fluorophore

Prior to describing the complexities of solvent effects, it is helpful to see an example. Emission spectra of *trans*-4-dimethylamino-4′-(1-oxybutyl)stilbene (DOS) are shown in Figure 6.2.[1] The emission spectra are seen to shift dramatically to longer wavelengths as the solvent polarity is increased from cyclohexane to butanol. In this case the goal was to determine the polarity of the DOS binding site on a model membrane, which was composed of dipalmitoyl-L-α-phosphatidylcholine (DPPC). The emission spectrum of DOS bound to DPPC vesicles was found to be similar to that of DOS in ethyl acetate, which has a dielectric constant (ε) near 5.8. The DOS binding site on DPPC vesicles is obviously more polar than hexane (ε = 1.9) and less polar than butanol (ε = 17.8 at 20 °C). Hence, the emission spectra of DOS indicate that it is in a rather nonpolar environment when bound to DPPC vesicles.

Examination of the chemical structure of DOS suggests a possible mechanism for its high sensitivity to solvent. The dimethylamino group is an electron donor, and the carbonyl group is strongly electron-withdrawing, resulting in a significant dipole moment in the ground state. In the excited state it is likely that this charge separation increases, resulting in a larger dipole moment than in the ground state. Such an increase in dipole moment would explain the sensitivity of the emission spectrum of DOS to solvent polarity. Note that the important parameter for solvent effects is the dipole moment, which depends on charge separation in the fluorophore. The dipole moment should not be confused with the transition moment, which is a quantum-mechanical parameter describing the direction of an electronic transition within the fluorophore.

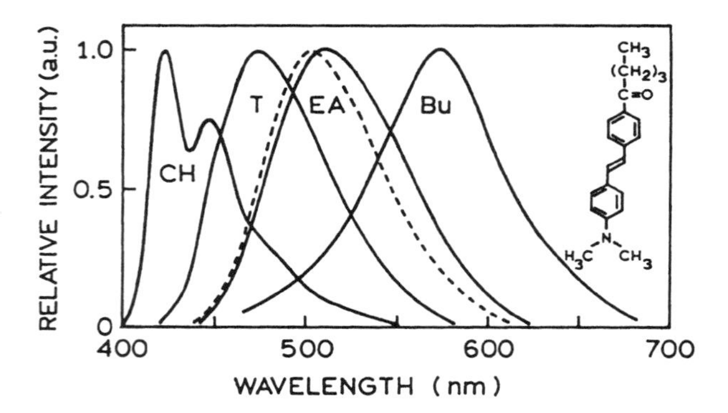

Figure 6.2. Corrected fluorescence emission spectra of DOS in cyclohexane (CH), toluene (T), ethyl acetate (EA), and butanol (Bu). The dashed curve shows the emission of DOS from DPPC vesicles. Revised from Ref. 1.

6.1.B. Mechanisms for Spectral Shifts

While interpretation of solvent-dependent emission spectra appears simple, it is a very complex topic. The complexity is due to the variety of interactions which can result in spectral shifts. At the simplest level, solvent-dependent emission spectra are interpreted in terms of the Lippert equation (Eq. [6.1], Section 6.2), which describes the Stokes' shift in terms of the changes in dipole moment which occur upon excitation, and the energy of a dipole in solvents of various dielectric constants (ε) or refractive indices (n).[2,3] These general solvent effects occur whenever a fluorophore is dissolved in any solvent and are independent of the chemical properties of the fluorophore and the solvent.

Although fluorescence spectral shifts can be interpreted in terms of general solvent effects, this theory is often inadequate for explaining the detailed behavior of fluorophores in a variety of environments. This is because fluorophores often display specific interactions with their local environment, which can shift the spectra by amounts comparable to general solvent effects. For instance, in the nonpolar solvent cyclohexane, indole displays a structured emission (see Figure 16.5) that is a mirror image of its absorption spectrum. Addition of a small amount of ethanol (1–5%) results in a loss of the structured emission. This amount of ethanol is too small to change the solvent polarity significantly and cause a spectral shift due to general solvent effects. The spectral shift seen in the presence of small amounts of ethanol is due to hydrogen bonding of ethanol to the imino nitrogen on the indole ring. Such specific solvent effects occur for many fluorophores and should be considered when interpreting the emission spectra. Hence, the Jabłoński diagram for solvent effects should also reflect the possibility of specific solvent–fluorophore interactions which can lower the energy of the excited state (Figure 6.3). Because of the diversity of solvent–fluorophore interactions, no single theory can be used to describe the solvent-dependent spectra.

In addition to specific solvent–fluorophore interactions, many fluorophores can form an internal charge-transfer (ICT) state or a twisted internal charge-transfer (TICT)

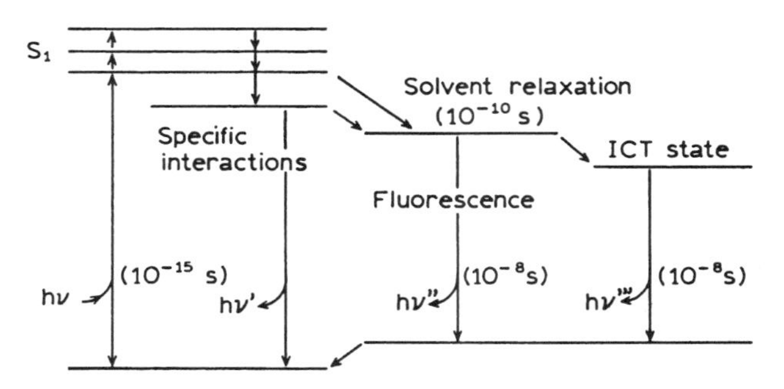

Figure 6.3. Effects of solvent on the energy of the excited state, including specific solvent–fluorophore interactions, general solvent effects, and formation of internal charge-transfer (ICT) states.

state.[4] For instance, suppose that the fluorophore contains both an electron-donating and an electron-accepting group. Such groups could be amino and carbonyl groups, respectively, but numerous other groups are known. Recall Kasha's rule, which states that emission occurs from the lowest-energy excited state. Following excitation, there can be an increase in charge separation within the fluorophore. If the solvent is polar, then a species with charge separation (the ICT state) may become the lowest energy state. In a nonpolar solvent the species without charge separation, the so-called locally excited (LE) state, may have the lowest energy. Hence, the role of solvent polarity is not only to lower the energy of the excited state owing to general solvent effects, but also to govern which state has the lowest energy. In some cases, formation of the ICT state requires rotation of groups on the fluorophore to form the TICT state. Formation of ICT states occurs in many fluorophores, and this effect is not contained within the theory of general solvent effects.

In summary, no single theory can be used for a quantitative interpretation of solvent-dependent emission spectra. The trends observed with solvent polarity follow the theory for general solvent effects, but there are often additional shifts due to specific interactions and to the formation of internal charge-transfer states.

6.2. GENERAL SOLVENT EFFECTS—THE LIPPERT EQUATION

While the theory of general solvent effects cannot explain the detailed behavior of many fluorophores, this theory provides a valuable framework for consideration of solvent-dependent spectral shifts. In the description of general solvent effects, the fluorophore is considered to be a dipole in a continuous medium of uniform dielectric constant (Figure 6.4). Within this framework, specific solvent–fluorophore interactions, or formation of ICT states, can be detected as deviations from the theory. Since this model is devoid of chemical interactions, one cannot expect the theory to explain the effects of specific solvent–fluorophore interactions.

The interactions between the solvent and fluorophore affect the energy difference between the ground and the excited state. To a first approximation, this energy difference (in cm^{-1}) is a property of the refractive index (n) and dielectric constant (ε) of the solvent and is described by the Lippert equation,[2,3]

$$\bar{\nu}_A - \bar{\nu}_F = \frac{2}{hc}\left(\frac{\varepsilon - 1}{2\varepsilon + 1} - \frac{n^2 - 1}{2n^2 + 1}\right)\frac{(\mu_E - \mu_G)^2}{a^3} + \text{constant} \quad [6.1]$$

In this equation, $\bar{\nu}_A$ and $\bar{\nu}_F$ are the wavenumbers (cm^{-1}) of the absorption and emission, respectively. $h = 6.6256 \times 10^{-27}$ erg s is Planck's constant, $c = 2.9979 \times 10^{10}$ cm/s is the speed of light, and a is the radius of the cavity in which the fluorophore resides. Although Eq. [6.1] is only an approximation, there is reasonable correlation between the observed and calculated energy losses in nonprotic solvents. By nonprotic solvents we mean those not having hydroxyl groups or other groups capable of hydrogen bonding. The Lippert equation is an approximation in which higher-order terms are neglected. These terms would account for second-order effects, such as the dipole moments induced in the solvent molecules by the excited fluorophore, and vice versa.

It is instructive to examine the opposite effects of ε and n on the Stokes' shift. An increase in n will decrease this energy loss, whereas an increase in ε results in a larger difference between $\bar{\nu}_A$ and $\bar{\nu}_F$. The refractive index (n) depends on the motion of electrons within the solvent molecules, which is essentially instantaneous and can occur during light absorption. In contrast, the dielectric con-

Figure 6.4. Dipole in a dielectric medium.

stant (ε) is a static property, which depends on both electronic and molecular motions, the latter being solvent reorganization around the excited state. The different effects of ε and n on the Stokes' shift will be explained in detail in Section 6.2.1. Briefly, an increase in refractive index allows both the ground and excited states to be instantaneously stabilized by movements of electrons within the solvent molecules. This electron redistribution results in a decrease in the energy difference between the ground and excited states (Figure 6.5). For this reason, most chromophores display a red shift of the absorption spectrum in solvents relative to the vapor phase.[5–7] An increase in ε will also result in stabilization of the ground and excited states. However, the energy decrease of the excited state due to the dielectric constant occurs only after reorientation of the solvent dipoles. This process requires movement of the entire solvent molecule, not just its electrons. As a result, stabilization of the ground and excited states of the fluorophore depends on the dielectric constant (ε) and is time-dependent. The rate of solvent relaxation depends upon the temperature and viscosity of the solvent (Chapter 7). The excited state shifts to lower energy on a timescale comparable to the solvent reorientation time. In the derivation of the Lippert equation (Section 6.2.A), and throughout this chapter, it is assumed that solvent relaxation is complete prior to emission.

The term inside the large parentheses in Eq. [6.1] is called the orientation polarizability (Δf). The first term, $(\varepsilon - 1)/(2\varepsilon + 1)$, accounts for the spectral shifts due to both the reorientation of the solvent dipoles and the redistribution of the electrons in the solvent molecules. The second term, $(n^2 - 1)/(2n^2 + 1)$, accounts for only the redistribution of electrons. The difference of these two terms accounts for the spectral shifts due to reorientation of the solvent molecules, hence the term orientation polarizability. According to this simple theory, only solvent reorientation is expected to result in substantial Stokes' shifts. The redistribution of electrons occurs instantaneously, and both the ground and excited states are approximately equally stabilized by this process. As a result, the refractive index and electronic redistribution have a comparatively minor effect on the Stokes' shift.

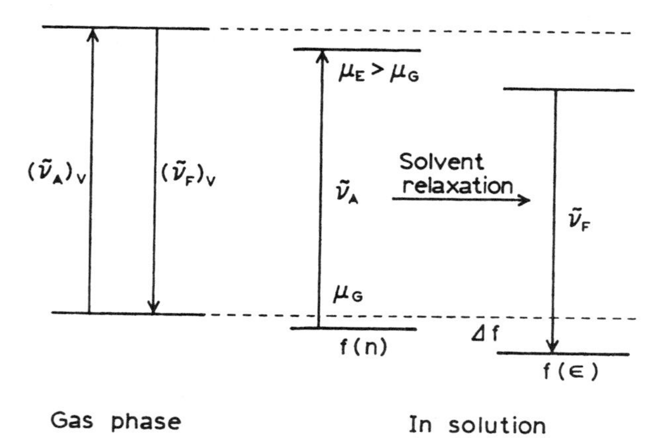

Figure 6.5. Effects of the refractive index (n) and dielectric constant (ε) on the absorption and emission energies.

It is instructive to calculate the magnitude of the spectral shifts which are expected for general solvent effects. Most fluorophores have nonzero dipole moments in the ground and excited states. As an example, we will assume that the ground-state dipole moment is $\mu_G = 6$ D and the excited-state dipole moment is $\mu_E = 20$ D, so that $\mu_E - \mu_G = 14$ D. We will also assume that the cavity radius is 4 Å, which is comparable to the radius of a typical aromatic fluorophore. One debye unit (1 D) is 1.0×10^{-18} esu cm, and 4.8 D is the dipole moment which results from a charge separation of one unit charge (4.8×10^{-10} esu) by 1 Å (10^{-8} cm). An excited-state dipole moment of 20 D is thus comparable to a unit-charge separation of 4.2 Å, which is a distance comparable to the size of a fluorophore. These assumed values of μ_E and μ_G are similar to those observed for fluorophores that are frequently used as polarity probes in biochemical research.

For the model calculation of Stokes' losses, we compare the nonpolar solvent hexane with several polar solvents. Nonpolar solvents such as hexane do not possess a dipole moment. Hence, there are no dipoles to reorient around the excited state of the fluorophore. This physical property of hexane is reflected by $\varepsilon \simeq n^2$ (Table 6.1). From Eq. [6.1] one calculates a small value for the orientation polarizability (Δf), and $\bar{\nu}_A - \bar{\nu}_F$ is expected to be small or zero. For example, if we assume that our model fluorophore absorbs at 350 nm, the emission in hexane is calculated to also be at 350 nm (Table 6.2). However, even in nonpolar solvents, absorption and emission maxima do not coincide this closely. Excitation generally occurs to higher vibrational levels, and this energy is rapidly dissipated in fluid solvents (10^{-12} s). Emission generally occurs to an excited vibrational level of the ground state. As a result, absorption and emission are generally shifted by an amount at least equal to the vibrational energy, or about 1500 cm^{-1}. These energy

Table 6.1. Polarizability Properties of Some Common Solvents

Solvent	ε	n	Δf
Water	78.3	1.33	0.32
Ethanol	24.3	1.35	0.30
Ether	4.35	1.35	0.17
Hexane	1.89	1.37	0.001

Table 6.2. Stokes' Shifts Expected from General Solvent Effects

Solvent	ε	n	Δf^a	$\bar{\nu}_A - \bar{\nu}_F$ $(\text{cm}^{-1})^b$	λ^c_{max}
Hexane	1.874	1.372	-0.0011^d	35	350^d
Chloroform	4.98	1.447	0.1523	4697	418.9
Ethyl acetate	6.09	1.372	0.201	6200	447.0
1-Octanol	10.3	1.427	0.2263	6979	463.1
1-Butanol	17.85	1.399	0.2644	8154	489.8
n-Propanol	21.65	1.385	0.2763	8522	498.8
Methanol	33.1	1.326	0.3098	9554	525.8

[a] $\Delta f = \frac{\varepsilon - 1}{2\varepsilon + 1} - \frac{n^2 - 1}{2n^2 + 1}$.

[b] From Eq. [6.1], assuming $\mu_G = 6$ D, $\mu_E = 20$ D, $\mu_E - \mu_G = 14$ D, and a cavity radius of 4 Å. One debye unit (1.0 D) is 1×10^{-18} esu cm; 4.8 D is the dipole moment which results from a charge separation of 1 unit of charge (4.8×10^{-10} esu) by 1 Å (10^{-8} cm).

[c] Assuming an absorption maximum of 350 nm.

[d] The small negative value of Δf was ignored.

losses are accounted for by the constant term in Eq. [6.1] and would shift emission of our model fluorophore to 370 nm.

In polar solvents such as methanol, substantially larger Stokes' losses are expected. For example, our model fluorophore is expected to emit at 526 nm in this polar solvent (Table 6.2). This shift is due to the larger orientation polarizability of methanol, which is a result of its dipole moment. This sensitivity of the Stokes' shift to solvent polarity is the reason why fluorescence emission spectra are frequently used to estimate the polarity of the environment surrounding the fluorophore.

If fluorophores are sensitive to the reaction field in a solvent, why not use electric fields to measure the excited-state dipole moments? Such studies have been attempted. However, the effects are two orders of magnitude smaller than the effects due to solvent reaction fields, even when one is working near the dielectric breakdown voltage.[8]

6.2.A. Derivation of the Lippert Equation

The interactions responsible for general solvent effects are best understood by derivation of the Lippert equation. This equation can be written as

$$hc\Delta\bar{\nu} = hc(\bar{\nu}_A - \bar{\nu}_F) = \frac{2\Delta f}{a^3}(\mu_E - \mu_G)^2 + \text{constant} \quad [6.2]$$

where $\Delta\bar{\nu}$ is the frequency shift (in cm^{-1}) between absorption and emission, Δf is the orientation polarizability, and μ_E and μ_G are the excited- and ground-state dipole moments, respectively. The Lippert equation is derived by consideration of the interactions of a fluorophore with the solvent and the timescale of these interactions. We need to recall the Franck–Condon principle, which states that nuclei do not move during an electronic transition (10^{-15} s). In contrast, the electrons of the solvent molecules can redistribute around the new excited-state dipole during this time span. In addition, because of the relatively long lifetime of the excited state ($\sim 10^{-8}$ s), the solvent molecules can orientate to their equilibrium positions around the excited state of the fluorophore prior to emission.

The derivation starts with the consideration of a point dipole in a continuous dielectric medium (Figure 6.4). The energy of the dipole in this medium is given by

$$E_{\text{dipole}} = -\mu R \quad [6.3]$$

where R is the reactive field in the dielectric induced by the dipole. The reactive field is parallel and opposite to the direction of the dipole and is proportional to the magnitude of the dipole moment,

$$R = \frac{2\mu}{a^3} f \quad [6.4]$$

In this equation, f is the polarizability of the solvent and a is the cavity radius. The polarizability of the solvent is a result of both the mobility of electrons in the solvent and the dipole moment of the solvent molecules. Each of these components has a different time dependence. Reorientation of the electrons in the solvent is essentially instantaneous. The high-frequency polarizability $f(n)$ is a function of the refractive index,

$$f(n) = \frac{n^2 - 1}{2n^2 + 1} \quad [6.5]$$

The dielectric constant also determines the polarizability of the solvent and includes the effect of molecular orientation of the solvent molecules. Because of the slower timescale of molecular reorientation, this component is called the low-frequency polarizability of the solvent. It is given by

$$f(\varepsilon) = \frac{\varepsilon - 1}{2\varepsilon + 1} \quad [6.6]$$

The difference between these two terms is

$$\Delta f = \frac{\varepsilon - 1}{2\varepsilon + 1} - \frac{n^2 - 1}{2n^2 + 1} \quad [6.7]$$

and is called the orientation polarizability. If the solvent has no permanent dipole moment, $\varepsilon \simeq n^2$ and $\Delta f \simeq 0$. Table

6.1 lists representative values of ε, n, and Δf. From the magnitudes of Δf, one may judge that spectral shifts ($\Delta\bar{\nu}$) will be considerably larger in water than in hexane.

The interactions of a fluorophore with the solvent can be described in terms of its ground-state (μ_G) and excited-state (μ_E) dipole moments and the reactive fields around these dipoles. These fields may be divided into those due to electronic motions (R_{el}^G and R_{el}^E and those due to solvent reorientation (R_{or}^G and R_{or}^E). Assuming equilibrium around the dipole moments of the ground and excited states, these reactive field are

$$R_{el}^G = \frac{2\mu_G}{a^3} f(n) \qquad R_{el}^E = \frac{2\mu_E}{a^3} f(n)$$

$$R_{or}^G = \frac{2\mu_G}{a^3} \Delta f \qquad R_{or}^E = \frac{2\mu_E}{a^3} \Delta f \qquad [6.8]$$

Consider Figure 6.6, which describes these fields during the processes of excitation and emission. For light absorption, the energies of the ground (E^G) and nonequilibrium excited (E^E) states are

$$E^E(\text{absorption}) = E_v^E - \mu_E R_{or}^G - \mu_E R_{el}^E \qquad [6.9]$$

$$E^G(\text{absorption}) = E_v^G - \mu_G R_{or}^G - \mu_G R_{el}^G \qquad [6.10]$$

where E_v^E and E_v^G represent the energy levels of the fluorophore in the vapor state, unperturbed by solvent. The absorption transition energy is decreased by the electronic reaction field induced by the excited-state dipole. This occurs because the electrons in the solvent can follow the rapid change in electron distribution within the fluorophore. In contrast, the orientation of the solvent molecules does not change during the absorption of light. Therefore, the effect of the orientation polarizability, given by $\mu_G R_{or}^G$ and $\mu_E R_{or}^G$, contains only the ground-state orientational reaction field. This separation of effects is due to the Franck–Condon principle. Recalling that energy is related to the wavenumber by $\bar{\nu} = \Delta E/hc$, subtraction of Eq. [6.10] from Eq. [6.9] yields the energy of absorption,

$$hc\bar{\nu}_A = hc(\bar{\nu}_A)_v - (\mu_E - \mu_G)(R_{or}^G) - \mu_E R_{el}^E + \mu_G R_{el}^G \qquad [6.11]$$

where $hc(\bar{\nu}_A)_v$ is the energy difference in a vapor where solvent effects are not present. By a similar consideration, one can obtain the energy of the two electronic levels for emission. These are

$$E^E(\text{emission}) = E_v^E - \mu_E R_{or}^E - \mu_E R_{el}^E \qquad [6.12]$$

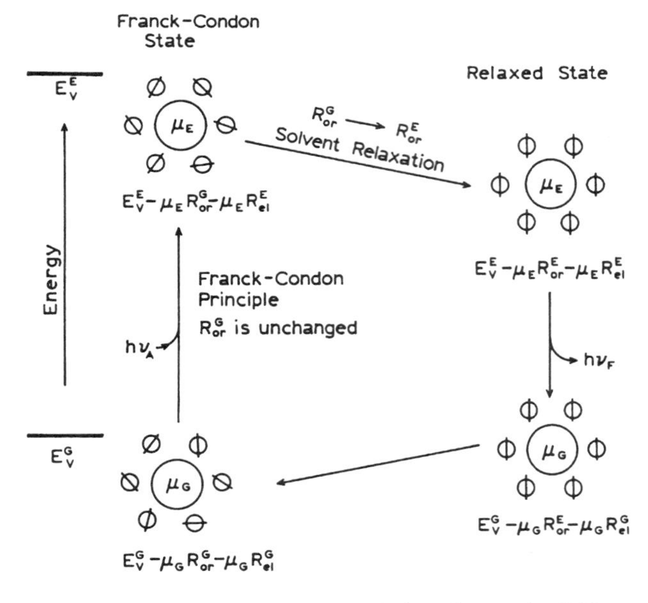

Figure 6.6. Effects of the electronic and orientation reaction fields on the energy of a dipole in a dielectric medium, $\mu_E > \mu_G$. The smaller circles and the lines through them represent the solvent molecules and their dipole moments.

$$E^G(\text{emission}) = E_v^G - \mu_G R_{or}^E - \mu_G R_{el}^G \qquad [6.13]$$

To derive these expressions, we assumed that the solvent relaxed quickly in comparison to the lifetime of the excited state, so that the initial orientation field (R_{or}^G) changed to R_{or}^E prior to emission. The electronic field changed during emission, but the orientation field remained unchanged. The energy of the emission is given by

$$hc\bar{\nu}_F = hc(\bar{\nu}_F)_v - (\mu_E - \mu_G)R_{or}^E - \mu_E R_{el}^E + \mu_G R_{el}^G \qquad [6.14]$$

In the absence of environmental effects, one may expect $\bar{\nu}_A - \bar{\nu}_F$ to be a constant for complex molecules which undergo vibrational relaxation. Hence, subtracting Eq. [6.14] from Eq. [6.11] yields

$$\bar{\nu}_A - \bar{\nu}_F = \frac{-1}{hc}(\mu_E - \mu_G)(R_{or}^E - R_{or}^G) + \text{constant} \qquad [6.15]$$

Substitution from Eq. [6.8] yields the Lippert equation,

$$\bar{\nu}_A - \bar{\nu}_F = \frac{-2}{hca^3}(\mu_E - \mu_G)(\mu_G\Delta f - \mu_E\Delta f) + \text{constant} \qquad [6.16]$$

$$= \frac{2\Delta f}{hca^3}(\mu_E - \mu_G)^2 + \text{constant} \qquad [6.17]$$

It is important to remember that the Lippert equation is only an approximation and contains many assumptions. The fluorophore is assumed to be spherical, and there is no consideration of specific interactions with the solvent. Other more complex equations have been presented, and the interested reader is referred to extensive publications on this subject.[9–14] In all these treatments, the solvent is regarded as a continuum in which the fluorophore is contained. In some cases the fluorophore cavity is described as an ellipse rather than a sphere.[9] Also, we have assumed so far that the ground- and excited-state dipole moments point in the same direction. If these directions are different, the Lippert equation is modified to[13,14]

$$hc\Delta\bar{\nu} = \Delta b\left[\frac{\varepsilon-1}{\varepsilon+2} - \frac{n^2-1}{n^2+2}\right]\frac{(2n^2+1)^2}{(n^2+2)^2} + \text{constant} \qquad [6.18]$$

where

$$\Delta b = \frac{2}{hca^3}(\mu_G^2 - \mu_E^2 - \mu_G\mu_E\cos\alpha) \qquad [6.19]$$

and α is the angle between μ_G and μ_E. It seems reasonable that general solvent effects would depend on the angle between μ_E and μ_G. However, it seems that the directions of the dipole moments are similar in the ground and the excited state for many fluorophores. Given the fact that the spectral shifts due to specific solvent effects and formation of ICT states are often substantial, it seems preferable to use the simplest form of the Lippert equation to interpret the spectral data. Deviations from the predicted behavior can be used to indicate the presence of additional interactions.

6.2.B. Application of the Lippert Equation

The emission spectra of many fluorophores used to label macromolecules are known to be sensitive to solvent polarity. One of the best known examples is the probe TNS. This class of probes has become widespread since its introduction in 1954.[15] TNS and similar molecules are essentially nonfluorescent when in aqueous solution but become highly fluorescent in nonpolar solvents or when bound to proteins and membranes. These probes are highly sensitive to solvent polarity and can potentially reveal the polarity of their immediate environments.[16] For example, the emission maximum of TNS shifts from 416 nm in acetonitrile to about 500 nm in water[17,18] (Figure 6.7), and the position of the emission maximum could be used to estimate the polarity of the binding site of ANS on macromolecules.[19] Another reason for the widespread use of these probes is their low fluorescence in water. For example, the quantum yield of 1,8-ANS is about 0.002 in aqueous buffer, but near 0.4 when bound to serum albumin. This 200-fold enhancement of the quantum yield is useful because the fluorescence of a dye–protein or dye–membrane mixture results almost exclusively from the dye that is bound to the biopolymers, with almost no contribution from the unbound probe. As a result, the spectral properties of the bound fraction may be investigated without interference from free dye.

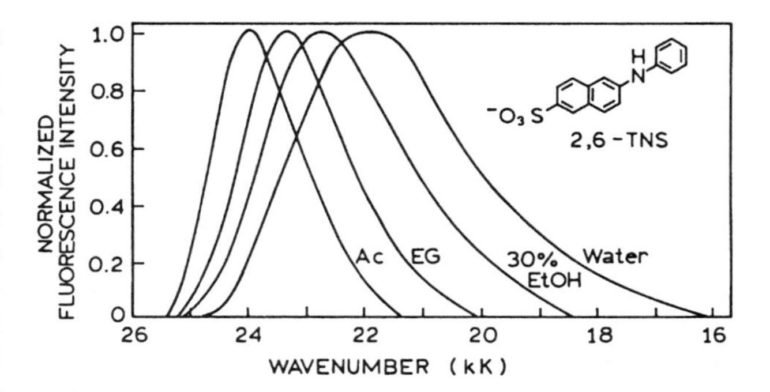

Figure 6.7. Normalized emission spectra for 6-anilinonaphthalene-2-sulfonic acid (ANS). The solvents are acetonitrile (Ac), ethylene glycol (EG), 30% ethanol/70% water (30% EtOH), and water. 1 kK = 1000 cm^{-1}. Revised from Ref. 18.

The solvent sensitivity of a fluorophore can be estimated from a Lippert plot. This is a plot of $(\bar{\nu}_A - \bar{\nu}_F)$ versus the orientation polarizability (Δf). The most sensitive fluorophores are those with the largest changes in dipole moment upon excitation. Representative Lippert plots for two naphthylamine derivatives are shown in Figure 6.8. The

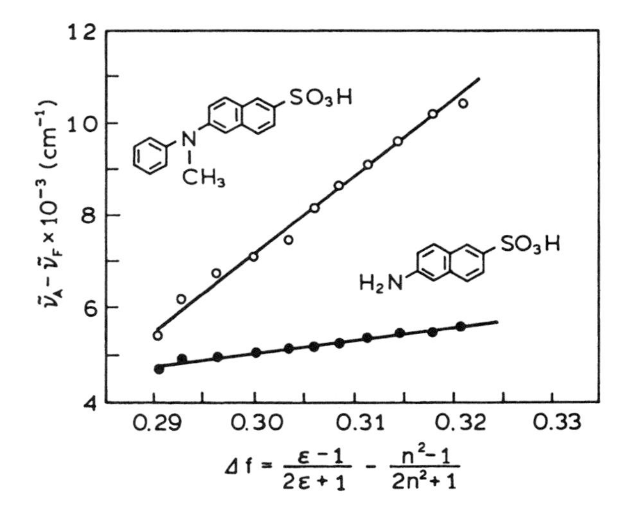

Figure 6.8. Lippert plots for two naphthylamine derivatives in ethanol–water mixtures. Data are shown for *N*-methyl-*N*-phenyl-6-aminonaphthalene-2-sulfonic acid (○) and 6-aminonaphthalene-2-sulfonic acid (•). Revised from Ref. 10.

N-phenyl-*N*-methyl derivative of 6-aminonaphthalene-2-sulfonic acid is more sensitive to solvent polarity than the unsubstituted amino derivatives.[10] The linearity of these plots is often regarded as evidence for the dominant importance of general solvent effects in the spectral shifts. Specific solvent effects lead to nonlinear Lippert plots (Section 6.3). The data in Figure 6.8 are for a limited range of similar solvents. Specifically, these were ethanol–water mixtures, so the same specific effects due to hydrogen bonding were present in all mixtures. In general, the attachment of side chains to the amino group, especially aromatic groups, enhances the sensitivity to solvent polarity. This general trend can be seen from Table 6.3, in which we summarize values of $(\mu_E - \mu_G)$ for a variety of naphthylamine derivatives as determined from the Lippert plots. The aromatic side chains may allow increased charge separation in the excited state, as compared to that for the unsubstituted aminonaphthalene derivatives.

While interpretation of the data for the various aminonaphthalenesulfonic acid derivatives seems clear, the agreement with theory is less satisfactory when the spectral data are examined over a larger range of solvent polarity. Typical data are shown in Figure 6.9 for two ANS derivatives in dioxane–water mixtures.[20] The *y*-axis is the emission energy, which parallels the Stokes' shift since the absorption maxima are relatively insensitive to solvent polarity. The *x*-axis is a different solvent polarity scale [$E_T(30)$], which is described in Section 6.2.C. One notices that there are two regions which are distinguished by their different slopes, suggesting different excited-state dipole moments. This behavior has been interpreted as due to the presence of two different excited states (Figure 6.10).[21–24] Upon excitation, the molecule first reaches an excited state that is localized on the naphthalene ring, which has been called the locally excited (LE) or nonpolar (np) state. If the solvent is nonpolar, the LE state is the low-energy state and the emitting species. In more polar solvents, the charge-transfer state (ct) becomes the low-energy state and emits at longer wavelengths. Electron transfer (et) also serves as

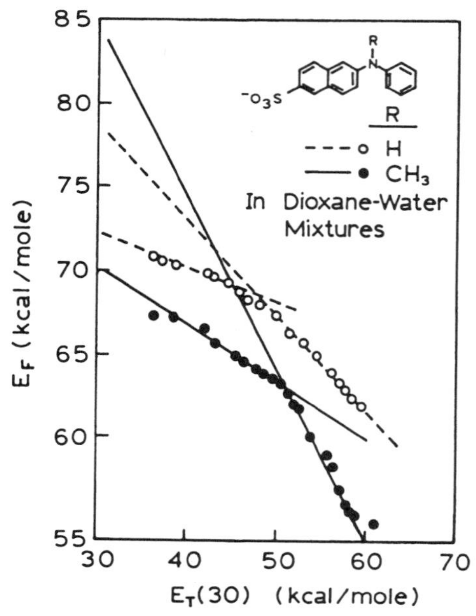

Figure 6.9. Emission energy of ANS derivatives versus the solvent polarity [$E_T(30)$] of dioxane–water mixtures. The probes are *N*-methyl-*N*-phenyl-6-amino-2-naphthalenesulfonate (R = CH_3) and *N*-phenyl-6-amino-2-naphthalenesulfonate (R = H). Revised and reprinted, with permission, from Ref. 20, Copyright © 1978, American Chemical Society.

Table 6.3. Calculated Changes in the Fluorophore Dipole Moment upon Excitation from Data Obtained Using Ethanol–Water Mixtures15

Molecule	Onsager radius (Å)	$\mu_E - \mu_G$ (D)
2-Aminonaphthalene-6-sulfonate	3	9
N,N-Dimethyl-2-aminonaphthalene-6-sulfonate	3	19
N-Phenyl-2-aminonaphthalene	4	20
N-Cyclohexyl-2-aminonaphthalene-6-sulfonate	5	23
N-Phenyl-2-aminonaphthalene-6-sulfonate	5	40
N-Methyl-*N*-phenyl-2-aminonaphthalene-6-sulfonate	5	46
N-(*o*-Toluyl)-2-aminonaphthalene-6-sulfonate	5	49
N-(*m*-Toluyl)-2-aminonaphthalene-6-sulfonate	5	42
N-(*p*-Toluyl)-2-aminonaphthalene-6-sulfonate	5	44

[a] From Ref. 10.

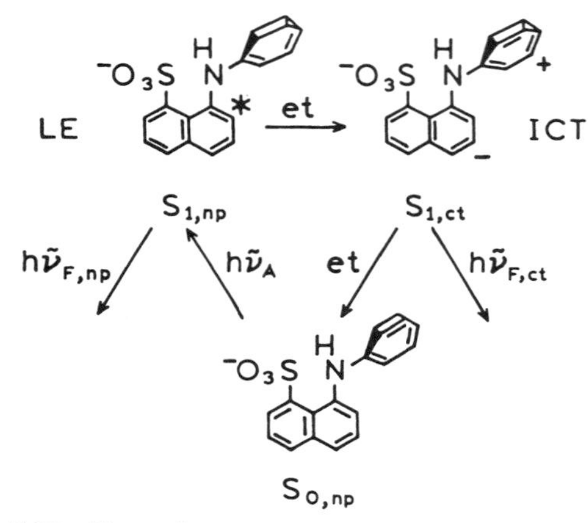

Figure 6.10. Electronic states of 1,8-ANS in the ground and excited states. et, Electron transfer. Emission from the charge-transfer state (ct or ICT) is at longer wavelength than that from the locally excited state (LE). Radiationless decay from $S_{1,ct}$ is typically rapid, resulting in low quantum yields. Revised and reprinted from Ref. 21, Copyright © 1983, American Chemical Society.

an efficient mechanism for radiationess decay, which explains the low fluorescence yield of such compounds in polar solutions.

One may question the difference between the ICT state of a fluorophore and the charge redistribution which exists immediately upon light absorption. The LE state forms immediately upon excitation, without any role of the solvent. In the LE state the electronic distribution is essentially the same as found immediately upon excitation. The ICT state forms in response to solvent polarity and stabilization of the ICT state. In the ICT state the electronic distribution is different from that in the Franck–Condon state, with a greater charge separation delocalized over more atoms. This can occur rapidly if the solvent is already in the proper orientation, or it may occur more slowly if formation of the ICT state requires the solvent to reorient. It should be noted that, following formation of the ICT state, one can expect the solvent to again reorganize around the newly created ICT configuration.

6.2.C. Polarity Scales

So far, we have described solvent polarity in terms of the orientation polarizability Δf. Other polarity scales are also in use.[25–32] These scales have their origin in physical organic chemistry, where solvent polarity was defined in terms of the rate of solvolysis of *t*-butyl chloride (Figure 6.11). This reaction proceeds by an S_N1 mechanism requiring ionization of the *t*-butyl and chloride groups. These ionized species are more stable in a polar solvent, so solvolysis proceeds faster. The *Y* parameter for solvent polarity is given by the log of the rate constant, minus the value in 80% ethanol (Figure 6.11). Given the inconvenience of performing a solvolysis reaction, optical indicators were developed. The first was the iodide salt of an alkylpyridinium derivative (Figure 6.12). This molecule already has separated charges in the ground state. In the excited state the molecule becomes a neutral radical pair, with a smaller dipole moment in the excited state than in the ground state. Since the ground state is stabilized in polar solvents, the charge-transfer absorption band shifts to shorter wavelengths with increasing solvent polarity

$$CH_3-C(CH_3)_2-Cl \rightarrow CH_3-C(CH_3)_2^{\delta+}\cdots Cl^{\delta-}$$

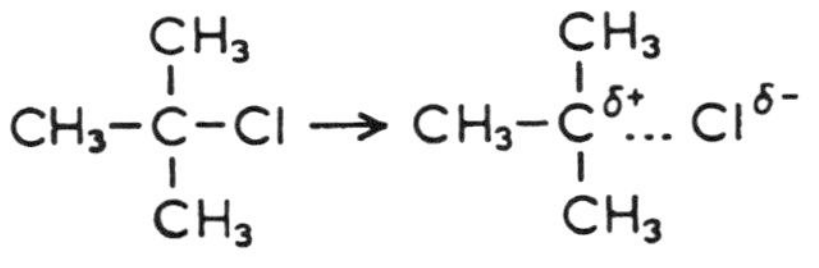

$$Y = \log k_{solvent} - \log k_{80\%\ EtOH}$$

Figure 6.11. Solvolysis reaction used to define the solvent polarity parameter *Y*. Revised from Ref. 25.

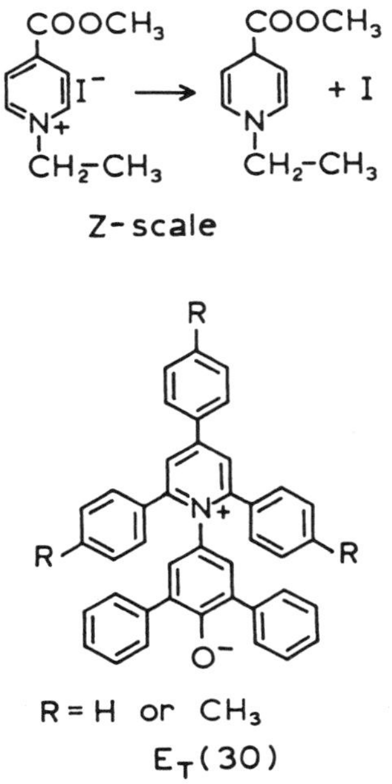

Z-scale

R = H or CH_3

$E_T(30)$

Figure 6.12. Charge-transfer absorbers used to define the Kosower *Z*-scale or the $E_T(30)$ scale. Revised from Ref. 25.

(Figure 6.13). The absorption maxima are transformed to the Kosower *Z* values (kcal/mol) using $Z = 2.859 \times 10^5/\lambda$, where λ is the wavelength of the absorption maximum in angstroms.

Since introduction of the *Z*-scale, a number of other molecules displaying charge-transfer absorption have been proposed as optical indicators of polarity. One of the most widely used indicators is propidium phenol betaine (Figure 6.12).[27,28] The absorption maxima of this indicator are described as the $E_T(30)$ scale. In general, there is good

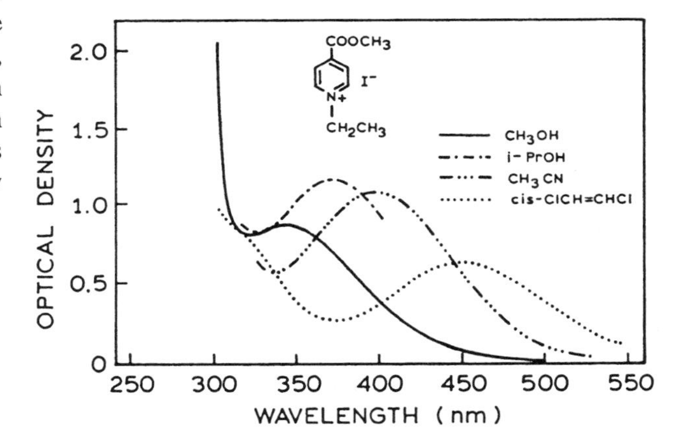

Figure 6.13. The first charge-transfer band of 1-ethyl-4-carbomethoxypyridinium iodide in methanol, isopropyl alcohol, acetonitrile, and *cis*-1,2-dichloroethylene. Revised from Ref. 25.

correlation between the various polarity scales, but the numerical values are different for each scale.

While a number of polarity scales are available, it is important to recognize that there is no such thing as a single polarity parameter.[33] The effects of solvents on fluorescence emission spectra are the result of several interactions which depend not only on ε and *n* but also on specific interactions between the solvent and the fluorophore.

6.3. SPECIFIC SOLVENT EFFECTS

In the preceding sections we described the general interactions between fluorophores and solvents. These interactions are approximately described by the Lippert equation and represent the collective influence of the entire set of surrounding solvent molecules on the fluorophore.[34] These general effects are determined by the electronic polarizability of the solvent (which is described by the refractive index *n*) and the molecular polarizability (which results from reorientation of solvent dipoles). The latter property is a function of the static dielectric constant, ε. In contrast, specific interactions are produced by one or a few neighboring molecules and are determined by the specific chemical properties of both the fluorophore and the solvent.[35,36] Specific effects can be due to hydrogen bonding, acid–base chemistry, or charge-transfer interactions, to name a few of the possible origins. The spectral shifts due to such specific interactions can be substantial and, if not recognized, limit the detailed interpretation of fluorescence emission spectra.

Specific solvent–fluorophore interactions can often be identified by examining emission spectra in a variety of solvents. Typical data for 2-anilinonaphthalene (2-AN) in cyclohexane are shown in Figure 6.14. Addition of low concentrations of ethanol, which are too small to alter the bulk properties of the solvent, result in substantial spectral shifts.[18] Less than 3% ethanol causes a shift in the emission maximum from 372 to 400 nm. Increasing the ethanol concentration from 3 to 100% ethanol caused an additional shift to only 430 nm. Thus, a small percentage of ethanol (3%) caused 50% of the total spectral shift. Upon addition of trace quantities of ethanol, one sees that the intensity of the initial spectrum is decreased, and a new red-shifted spectrum appears. The appearance of a new spectral component is characteristic of specific solvent effects. It is important to recognize that solvent-sensitive fluorophores can yield misleading information on the polarity of their environments if specific interactions occur, or if solvent relaxation is not complete. Because the specific spectral shift occurs at low ethanol concentrations, we believe that the effect is due to hydrogen bonding of ethanol to the

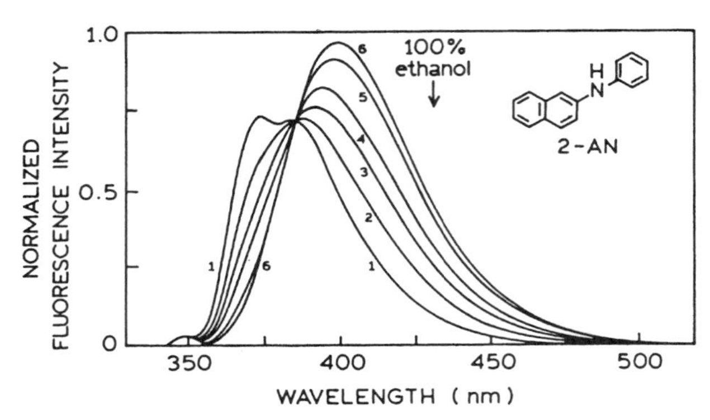

Figure 6.14. Fluorescence emission spectra of 2-anilinonaphthalene in cyclohexane to which ethanol was added. The percentages of ethanol in the solvent were 0% (1), 0.2% (2), 0.4% (3), 0.7% (4), 1.7% (5), and 2.7% (6). The arrow indicates the emission maximum in 100% ethanol. Revised and reprinted, with permission, from Ref. 18, Copyright © 1971, Academic Press, Inc.

amino groups of 2-AN, rather than stabilization of an ICT state by a more polar solvent.

Because of the importance of specific solvent–fluorophore interactions, we present several additional examples. Among the most studied examples are 2-acetylanthracene (2-AA) and its derivatives.[37–39] Emission spectra of 2-AA in hexane containing small amounts of methanol are shown in Figure 6.15. These low concentrations of methanol result in a loss of the structured emission, which is replaced by a longer-wavelength unstructured emission. As the solvent polarity is increased further, the emission spectra continue to shift to longer wavelengths (Figure 6.16). These spectra suggest that the emission of 2-AA is sensitive to both specific solvent effects and general solvent effects in more polar solvents.

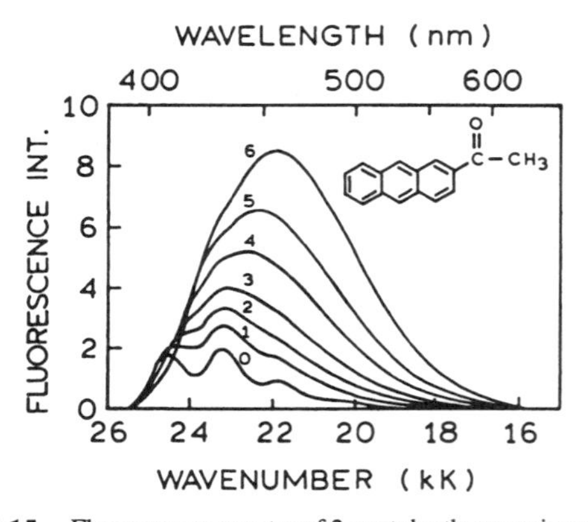

Figure 6.15. Fluorescence spectra of 2-acetylanthracene in methanol–hexane mixtures at 20 °C. Concentrations of methanol in mol dm^{-3}: 0, 0; 1, 0.03; 2, 0.05; 3, 0.075; 4, 0.12; 5, 0.2; 6, 0.34. 1 kK = 1000 cm^{-1}. Revised from Ref. 38.

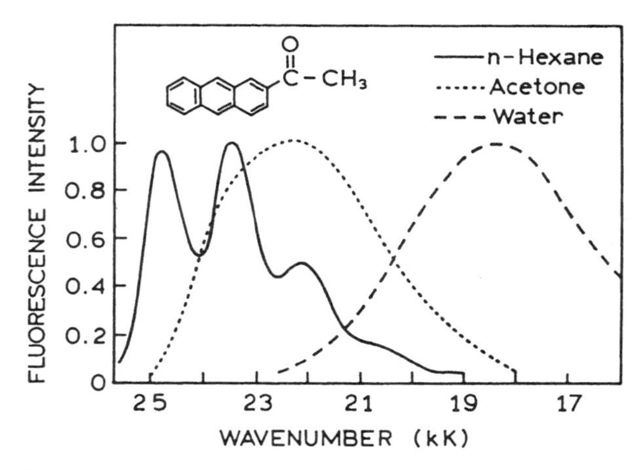

Figure 6.16. Effect of solvent on emission spectra of 2-acetylanthracene. 1 kK = 1000 cm^{-1}. Revised from Ref. 37.

The presence of specific solvent effects can be seen in the dependence of the emission maxima on the percentage of polar solvent (Figure 6.17). In hexane the emission maximum of 2-AA shifted gradually as the percentage of dioxane was increased to 100% (Figure 6.17). These shifts induced by dioxane are probably a result of general solvent effects. In contrast, most of the shift induced by methanol was produced by only about 1–2% methanol. This amount of alcohol is too small to affect the refractive index or dielectric constant of the solvent, and hence this shift is a result of specific solvent effects.

Specific solvent–fluorophore interactions can occur in either the ground state or the excited state. If the interaction only occurs in the excited state, then the polar additive would not affect the absorption spectra. If the interaction occurs in the ground state, then some change in the absorption spectrum is expected. In the case of 2-AA, the absorption spectra showed loss of vibrational structure and a red shift upon addition of methanol to the hexane solvent (Figure 6.18). This suggests that 2-AA and alcohol are already hydrogen-bonded in the ground state. An absence of changes in the absorption spectra would indicate that no ground-state interaction occurs. Alternatively, weak hydrogen bonding may occur in the ground state, and the strength of this interaction may increase following excitation. If specific effects are present for a fluorophore bound to a macromolecule, interpretation of the emission spectrum is complex. For example, a molecule like 2-AA, when bound in a hydrophobic site on a protein, may display an emission spectrum comparable to that seen in water if only a single water molecule is near the carbonyl group.

The presence of specific interactions in the ground state or only in the excited state determines the timescale of these interactions. If the fluorophore and polar solvent are already associated in the ground state, then one expects an immediate spectral shift upon excitation. If the fluorophore and polar solvent only associate in the excited state, then the appearance of the specific solvent effect will depend on the rates of diffusion of the fluorophore and polar solvent. In this case, the dependence on the concentration of polar solvent will be similar to that for quenching reactions.

Another example of specific solvent effects was found for 7-nitrobenz-2-oxa-1,3-diazol-4-yl (NBD). This fluorophore is highly sensitive to solvent polarity and has been conjugated to proteins and lipids in order to probe polarity. Like 2-AN, NBD derivatives also show specific solvent effects.[40] This can be seen in the shift of the emission spectra of NBD derivatives in *n*-hexane upon addition of

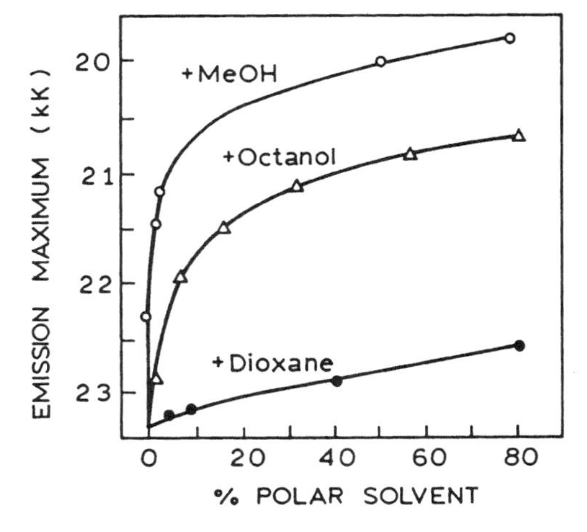

Figure 6.17. Effect of solvent composition on the emission maximum of 2-acethylanthracene. 1kK = 1000 cm^{-1}. Revised from Ref. 37.

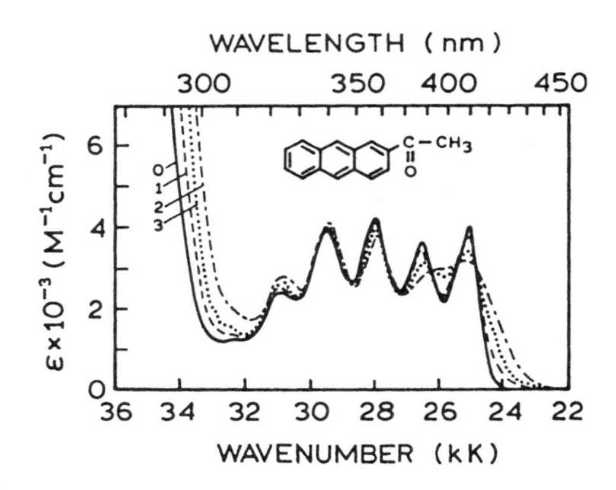

Figure 6.18. Absorption spectra of 2-acetylanthracene in pure hexane (curve 0), a methanol–hexane mixture (curve 1; concentration of methanol = 0.2 mol dm^{-3}), and pure methanol (curve 2). Curve 3 refers to the calculated spectrum of the H-bonded complex. Revised from Ref. 38.

ethanol (Figure 6.19). The magnitude of the specific effects varies between *N*-alkyl derivatives of NBD, and the effects are present in most derivatives.

It is interesting to notice in Figure 6.19 that the specific solvent effects are less dramatic in dioxane than in hexane. This difference can be understood as the result of the somewhat higher polarity of dioxane as compared to hexane. In dioxane, association between ethanol and the fluorophore is probably weaker, resulting in the need for higher alcohol concentrations to observe the solvent- dependent shifts. In fact, the binding in dioxane is so much weaker that the emission spectra shift according to the overall change in solvent polarity. These polarity-dependent properties of NBD have been used to estimate its location in model membranes. Based on the emission spectra and quantum yields of NBD-labeled lipids, it appears that the NBD group is usually located at the lipid–water interface, irrespective of its site of attachment to the lipid.[41]

6.3.A. Specific Solvent Effects and Lippert Plots

Evidence for specific solvent–fluorophore interactions can be seen in the Lippert plots. One example is a long-chain fatty acid derivative of 2-AA. The probe was synthesized for use as a membrane probe.[42,43] One notices from the Lippert plot shown in Figure 6.20 that the Stokes' shift is generally larger in hydrogen-bonding solvents (water, methanol, and ethanol) than in solvents which less readily form hydrogen bonds (Figure 6.20). Such behavior is typical of specific solvent–fluorophore interactions.

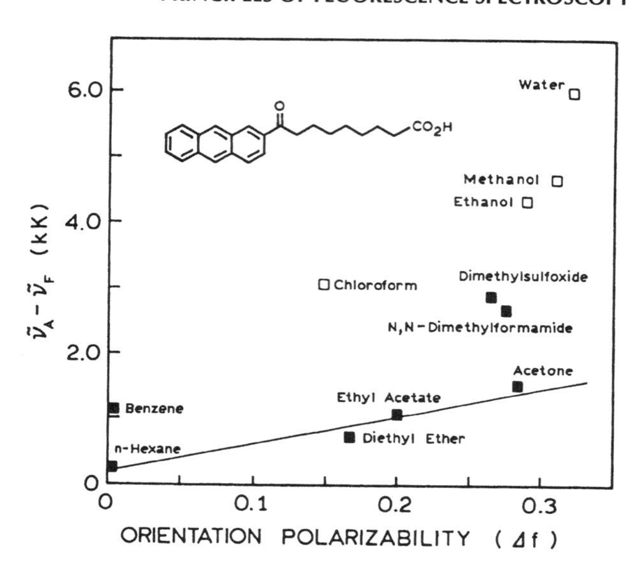

Figure 6.20. Stokes' shifts of methyl 8-(2-anthroyl)octanoate in organic solvents and water. 1 kK = 1000 cm^{-1}. Revised from Ref. 43, Copyright © 1991, with permission from Elsevier Science.

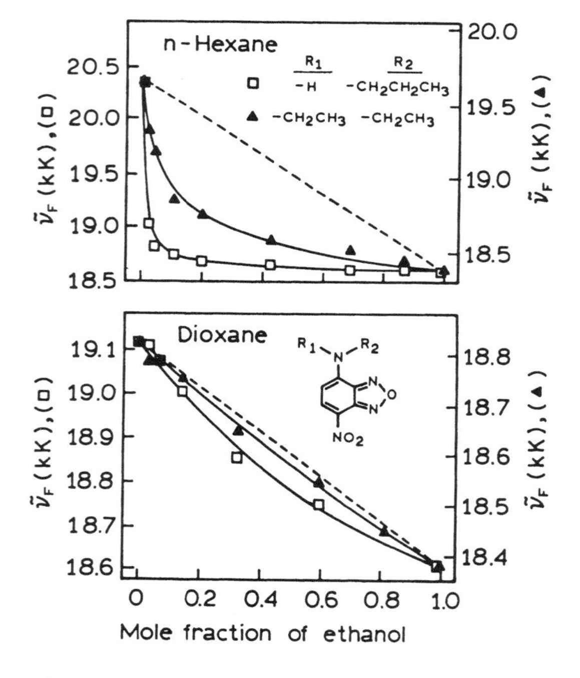

Figure 6.19. Effects of ethanol on NBD derivatives dissolved in *n*-hexane (*top*) or dioxane (*bottom*). Total shifts have been normalized for comparison. In *n*-propylamino-NBD (□), $R_1 = H$ amd $R_2 = CH_2CH_2CH_3$. In diethylamino-NBD (▲), $R_1 = R_1 = CH_2CH_3$. 1kK = 1000 cm^{-1}. Revised from Ref. 40.

Specific solvent effects have also been observed for other commonly used fluorophores. Examples are shown in Figures 6.21 and 6.22 for 1-aminonaphthalene[44] and 9-methyl anthroate,[45,46] respectively. In both instances, the Stokes' shifts are approximately proportional to the orientation polarizability for some solvents, but other solvents produced larger spectral shifts. Larger spectral shifts were found in those solvents capable of forming hydro-

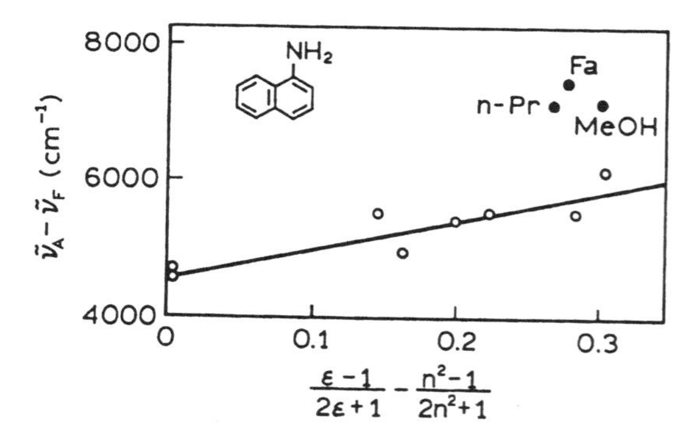

Figure 6.21. Effects of solvent polarity on the Stokes' shift of 1-aminonaphthalene. The filled circles are for *n*-propanol (n-Pr), formamide (Fa), and methanol (MeOH), respectively. The other solvents are nonprotic. Revised from Ref. 44.

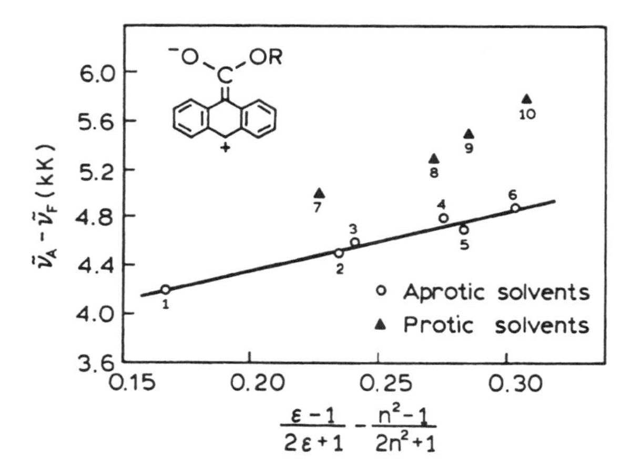

Figure 6.22. Stokes' shift of 9-methyl anthroate in protic and nonprotic solvents. The numbers refer to the following solvents: 1, ethyl ether; 2, benzonitrile; 3, methyl formate; 4, *N,N*-dimethylformamide; 5, acetone; 6, acetonitrile; 7, 80% dioxane–20% water; 8, 60% dioxane–40% water; 9, 40% dioxane—60% water; 10, 20% dioxane–80% water. The structure shows the proposed charge distribution in the excited state. 1 kK = 1000 cm^{-1}. Revised from Ref. 46.

gen bonds. For instance, large Stokes' shifts for 1-aminonaphthalene were observed in *n*-propanol, formamide, and methanol (Figure 6.21). For 9-methyl anthroate, excess Stokes' shifts were seen in all the water–dioxane mixtures, but not in the other solvents. In each case the excess shifts were explained by hydrogen bonding between the polar solvent and the polar group on the fluorophore. For 9-methyl anthroate, the excess shifts were attributed to the localization of a negative charge on the oxygen atom of the ester group (Figure 6.22), and the hydrogen bonding of this group to the protic solvent. Hence, the emission spectra of fluorophores are dependent upon both the orientation polarizability of the solvent and the detailed chemical structures of the fluorophores and the solvents.

In the case of 9-methyl anthroate, it is not clear whether to describe the effect as a specific solvent–fluorophore interaction or as due to formation of a TICT state. In the ground state the carboxyl group is perpendicular to the anthracene ring. In the excited state the carboxyl group and the ring are thought to be coplanar. Hence, 9-methyl anthroate and its derivatives may form a TICT state.[47–50]

The sensitivity of fluorophores to specific interactions with solvents may be regarded as either a problem or a favorable circumstance. It is problematic because these effects can prevent a quantitative interpretation of the emission spectra in terms of the orientation polarizability of the macromolecule to which the probe is bound. The situation is favorable because the specific effects of protic solvents could reveal the accessibility of the macromolecule-bound probe to the aqueous phase. Also, specific solvent effects can cause larger, and hence more easily observed, spectral shifts.

In view of the importance of specific solvent effects, how can one reliably use fluorescence spectral data to indicate the polarity of binding sites? At present, there seems to be no completely reliable method. However, by careful examination of the solvent sensitivity of a fluorophore, reasonable estimates can be made. Consider the following hypothetical experiment. Assume that 2-AA binds strongly to lipid bilayers so that essentially all the 2-AA is bound to the membrane. It seems likely that, irrespective of the location of 2-AA in the bilayer, adequate water will be present, either in the ground state or during the excited-state lifetime, to result in saturation of the specific interactions shown in Figures 6.15 and 6.17. Under these conditions, the solvent sensitivity of this fluorophore will be best represented by that region of Figure 6.17 where the concentration of the protic solvent is greater than 10–20%. Note that essentially equivalent slopes are found within this region for methanol, octanol, and dioxane. These similar slopes indicate a similar mechanism, which is the general solvent effect. Apparently, the specific solvent effects can be saturated or eliminated by an appropriate choice of solvents, as was suggested by the weaker specific effects for NBD in dioxane than in hexane (Figure 6.19). These considerations indicate that a probe will only provide a useful indication of polarity if one considers the relative importance of general and specific solvent effects.

An understanding of specific and general solvent effects can provide a basis for interpreting the emission spectra of fluorophores which are bound to macromolecules. Consider the emission spectra of 2-AN bound to membranes composed of DMPC.[51] The emission spectrum of 2-AN in DPMC is considerably red-shifted relative to the emission in cyclohexane, but it is blue-shifted relative to that in water (Figure 6.23). What is the polarity of the environment of 2-AN in the membranes? Interpretation of the emission maxima of these labeled membranes is also complicated by the time-dependent spectral shifts, a complication that we will ignore for the moment. We noted earlier (Figure 6.14) that 2-AN is highly sensitive to small concentrations of ethanol. It seems likely that cyclohexane containing more than 3% ethanol is the preferable reference solvent for a low-polarity environment. This spectrum is indicated by the dashed line in Figure 6.23. In this solvent the specific effects are saturated. With this adjustment in mind, one may conclude that the environment in which the 2-AN is localized is mostly nonpolar but that this site is accessible to water. Without consideration of specific solvent effects, one might conclude that the 2-AN is in a more polar environment.

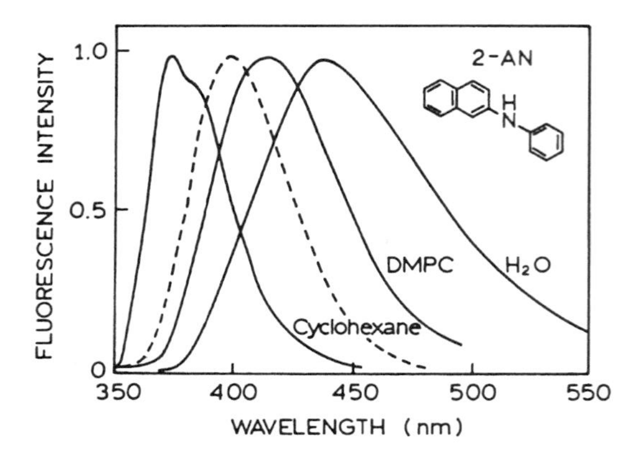

Figure 6.23. Normalized fluorescence emission spectra of 2-anilinonaphthalene in solvents and bound to vesicles of dimyristoyl-L-α-phosphatidylcholine (DMPC). The dashed line indicates the spectrum in cyclohexane containing 3% ethanol. Revised from Ref. 51.

The fact that the hydrogen-bonding interactions of 2-AN are saturated in membranes is supported by time-resolved data. The measurement and interpretation of such data will be described in Chapter 7. Figure 6.24 shows the time-dependent emission maxima of 2-AN bound to DMPC vesicles and in glycerol and cyclohexane. Even at the shortest observable time of one nanosecond, the emission maximum of 2-AN-labeled membranes is similar to that found in the protic solvent glycerol. This initial value for the emission maximum is also comparable to the completely relaxed value found for 2-AN in cyclohexane containing 0.1*M* ethanol. This final value can be regarded as that expected when the specific solvent effects are saturated. This result indicates that in membranes, the specific interactions with water or other polar hydrogen-bonding groups are saturated. These interactions may have occurred in the ground state or on a subnanosecond timescale too rapid to be resolved in this particular experiment. In either event, it seems clear that the emission spectra of 2-AN bound to model membranes can be interpreted more reasonably when compared to their spectra in reference solvents in which the specific solvent effects are saturated.

6.4. TEMPERATURE EFFECTS

In the preceding sections we assumed that solvent relaxation was complete prior to emission, which is true for fluid solvents. At low temperatures the solvent can become more viscous, and the time for solvent reorientation increases. This is illustrated schematically in Figure 6.25, which shows a simplified description of solvent relaxation. Upon excitation, the fluorophore is assumed to be initially in the Franck–Condon state (F). Solvent relaxation proceeds with a rate k_S. If this rate is much slower than the decay rate ($\gamma = 1/\tau$), then one expects to observe the emission spectrum of the unrelaxed F state. If solvent relaxation is much faster than the emission rate ($k_S >> \gamma$), then emission from the relaxed state (R) will be observed. At intermediate temperatures where $k_S \simeq \gamma$, emission and relaxation will occur simultaneously. Under these conditions, an intermediate emission spectrum will be observed. Frequently, this intermediate spectrum (dashed curve in Figure 6.25) is broader on the wavelength scale because of contributions from both the F and R states. Time-dependent spectral relaxation is described in more detail in Chapter 7.

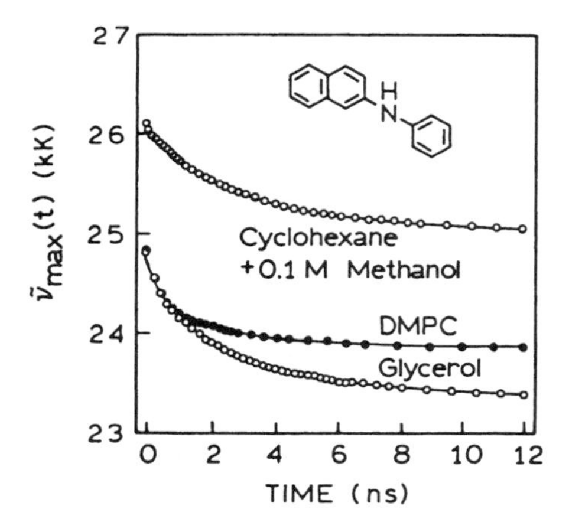

Figure 6.24. Time-resolved emission maxima of 2-anilinonaphthalene in DMPC vesicles and in solvents. 1 kK = 1000 cm^{-1}. Revised from Ref. 51.

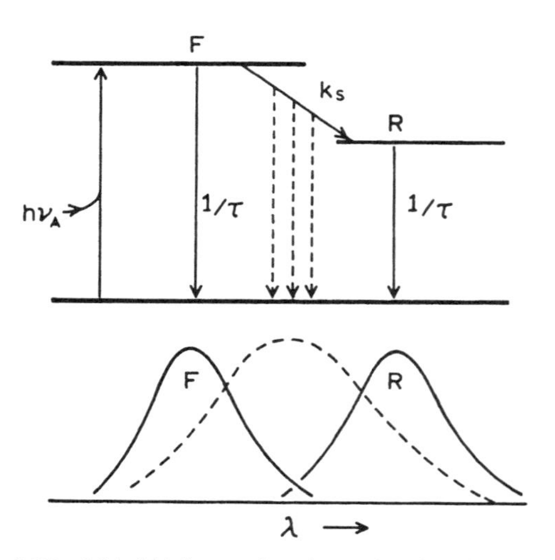

Figure 6.25. Jabłoński diagram for solvent relaxation. F, Franck–Condon state; R, relaxed state.

Examples of temperature-dependent emission spectra are shown in Figure 6.26 for the neutral tryptophan derivative NATA in propylene glycol. Solvents such as propylene glycol, ethylene glycol, and glycerol are frequently used to study fluorescence at low temperature. These solvents are chosen because their viscosity increases gradually with decreasing temperature, and they do not crystallize. Instead, they form a clear highly viscous glass, in which the fluorophores are immobilized. The presence of hydroxyl groups and alkyl chains makes these compounds good solvents for most fluorophores.

As the temperature of NATA in propylene glycol is decreased, the emission spectrum shifts to shorter wavelengths[52] (Figure 6.26). This shift occurs because the decay rate of fluorophores is not very dependent on temperature, but the relaxation rate is strongly dependent on temperature. Hence, at low temperature, emission is observed from the unrelaxed F state. It is important to notice that the structured ^1L_b (see section 16.1.1) emission of NATA is not seen even at the lowest temperature (–68 °C). This is because the hydrogen-bonding properties of propylene glycol persist at low temperature. It is clear that the temperature-dependent spectral shifts for NATA are due to the temperature dependence of the orientation polarizability Δf.

Another example of temperature-dependent spectra is provided by Patman (Figure 6.27).[53] This fluorophore is a lipid-like analog of Prodan, which was developed to be a probe highly sensitive to solvent polarity.[54] The basic idea is that the amino and carbonyl groups serve as the electron donor and acceptor, respectively. In the excited state, one expects a large dipole moment due to charge transfer (Figure 6.28), and thus a high sensitivity to solvent polarity. As will be described in the next section, the behavior of Prodan and its derivatives is actually more complex owing to formation of an ICT state.

As the temperature of Patman in propylene glycol is decreased, the emission spectra shift dramatically to shorter wavelengths (Figure 6.27). The effects of low temperature are similar to those of low-polarity solvents. Because of the decreased rate of solvent motion, emission occurs from the unrelaxed state at low temperature. At intermediate temperature (–30 °C), the emission spectra are broader (Figure 6.27). This is due to emission from both the F and R states. In fact, a plot of the spectral half-width versus temperature shows a maximum width near –30 °C (Figure 6.29). As will be explained in Chapter 7, such behavior indicated a stepwise relaxation process, rather than a continuous shift due to a multitude of solvent–fluorophore interactions.

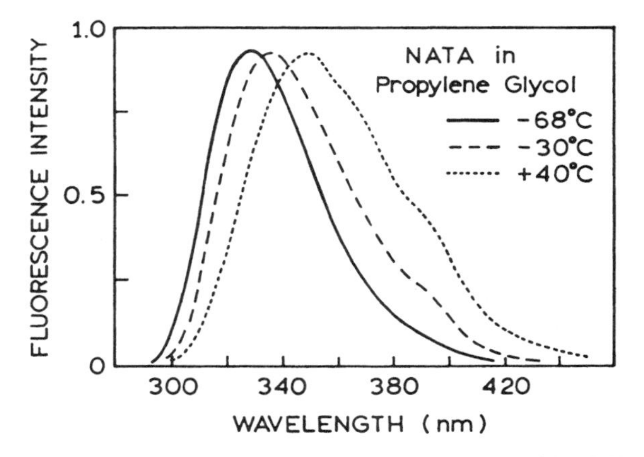

Figure 6.26. Emission spectra of *N*-acetyl-L-tryptophanamide (NATA) in propylene glycol. From Ref. 52.

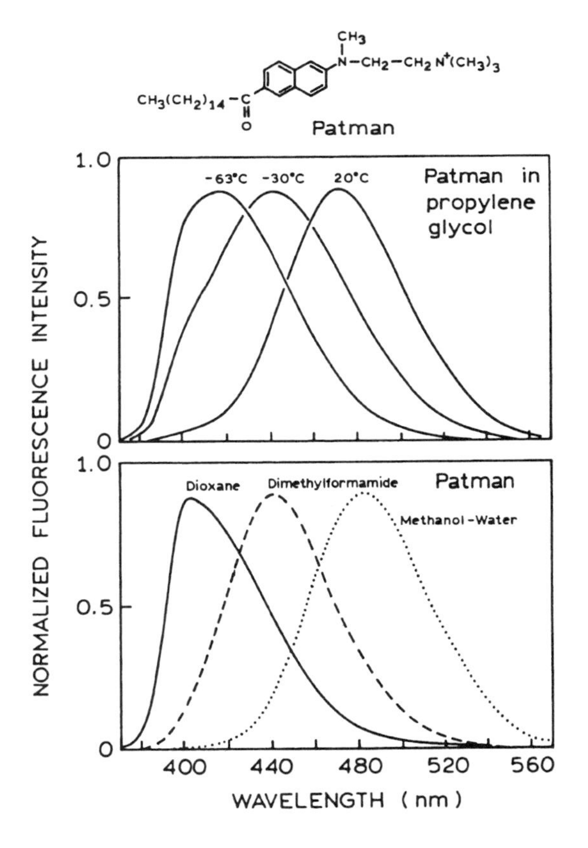

Figure 6.27. Emission spectra of the lipidlike Prodan derivative Patman at various temperatures in propylene glycol (*top*) and in solvents of different polarities (*bottom*). Revised and reprinted from Ref. 53, Copyright © 1983, American Chemical Society.

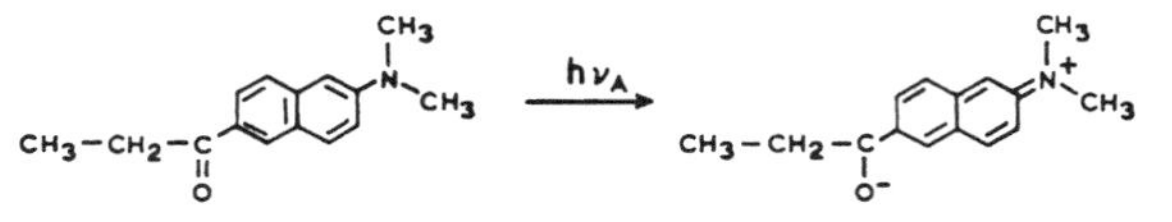

Figure 6.28. Charge separation in the excited state of Prodan [6-propionyl-2-(dimethylamino)naphthalene].

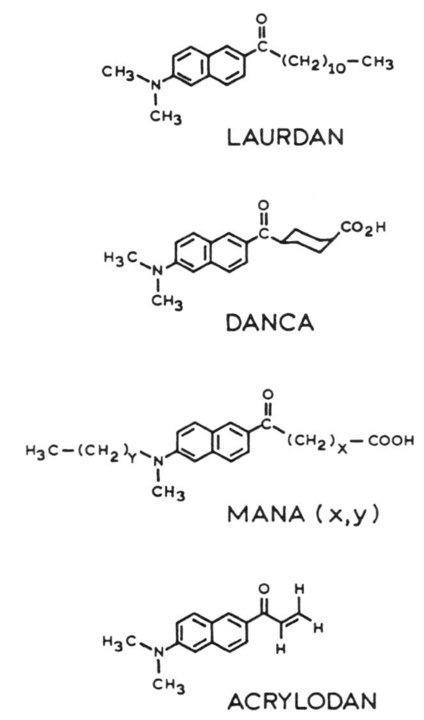

LAURDAN

DANCA

MANA (x,y)

ACRYLODAN

Figure 6.30. Derivatives of Prodan.

6.4.A. LE and ICT States of Prodan

Since its introduction as a solvent-sensitive probe, Prodan and its derivatives (Figure 6.30) have become widely used to label macromolecules.[55–58] Laurdan and Mana (x, y) are usually used to label membranes, and Danca has been used to bind to the hydrophobic regions of proteins. The purpose of the cyclohexanoic acid group in Danca is to make the probe bind in a single orientation. Acrylodan is a conjugatable form of Prodan that preferentially reacts with sulhydryl groups.

This increased use of Prodan-like probes has resulted in further studies to understand their dramatic spectral shifts.[59–65] It now appears that the large spectral shift displayed by Prodan is due to emission from the LE state, which occurs near 400 nm, as well as from an ICT state emitting at longer wavelengths. A hint of this behavior was seen in the top panel of Figure 6.27, where at –30 °C a shoulder appeared on the short-wavelength side of the emission. Such changes in spectral shape often indicate the presence of another state. This new blue-emitting state was more easily seen in ethanol at low temperatures (Figure 6.31). As the temperature of Laurdan in ethanol was decreased, the emission maximum shifted from about 490 nm to 455 nm. As the temperature was lowered to –190 °C, a new emission appeared with a maximum near 420 nm.

The unusual temperature-dependent spectra displayed by Laurdan were explained by the presence of emission from the LE state and from the ICT state. In the LE state, the excitation is localized on the naphthalene ring, so that the molecule is not very polar. In this LE state, the dimethylamino and carbonyl groups are not part of the delocalized electron system. At higher temperature the ICT state forms, with complete charge transfer from the amino group to the carbonyl group. This requires twisting of the dimethylamino group to allow the nitrogen electrons to be in conjugation with the naphthalene ring.[64,65] Hence, the large spectral shift displayed by Prodan-like molecules is somewhat misleading. Part of the shift from 420 to 455 nm is due to formation of the TICT state. The remaining shift from 455 to 490 nm is due to the orientation polarizability (Δf) of the solvent.

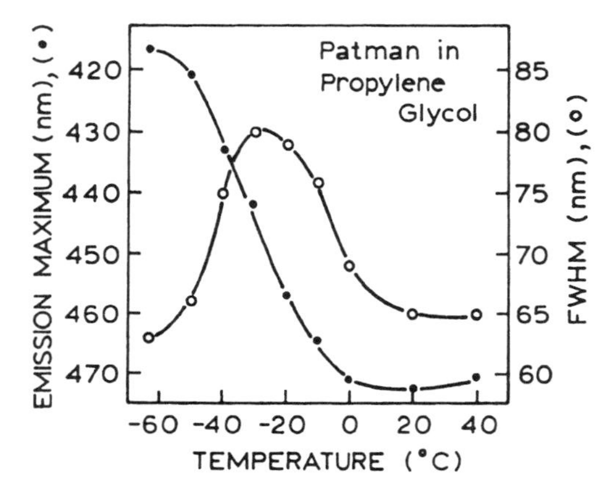

Figure 6.29. Temperature-dependent emission maximum (●) and half-width (○) of Patman in propylene glycol. Revised from Ref. 51.

Prodan is just one example of a large number of molecules which display ICT or TICT emission.[66–73] These molecules typically include an amino group and a cyano group. One example is DMANCN (Figure 6.32). As the temperature is increased, the short-wavelength emission is replaced by the long-wavelength emission of the TICT state, in which the methyl groups are rotated above and below the plane of the naphthalene ring. The long-wavelength emission is anomalous and is called the A-band. The blue emission is called the B-band. At present, the possibility of ICT or TICT emission is being used to design probes with large Stokes' shifts.[66–68]

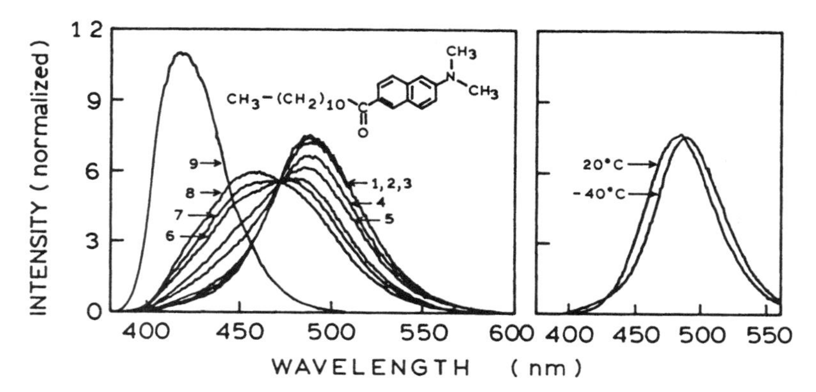

Figure 6.31. *Left:* Emission spectra of Laurdan in ethanol at low temperatures: (1) –50, (2) –60, (3) –70, (4) –80, (5) –85, (6) –90, (7) –100, (8), –110, (9) –190 °C. *Right:* Emission spectra of Laurdan in ethanol at 20 and –40 °C. Revised from Ref. 59.

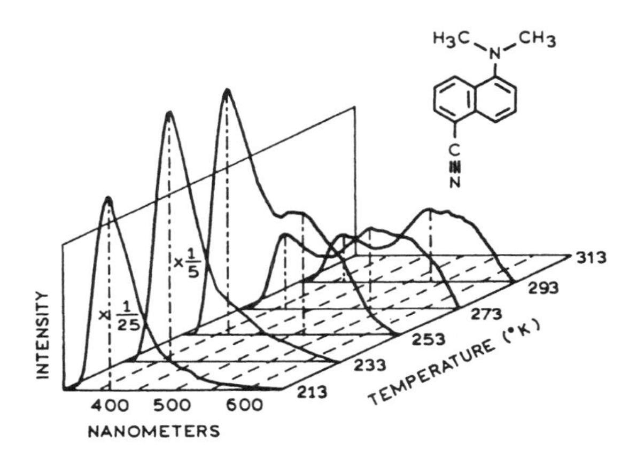

Figure 6.32. Dual fluorescence of DMANCN in propylene glycol as a function of temperature. Revised from Ref. 73.

Although solvent relaxation usually proceeds faster at higher temperatures, high temperature can also prevent the alignment of solvent dipoles. This effect is seen for Laurdan in ethanol at 20 °C (Figure 6.31, right).[59] This emission spectrum is blue-shifted relative to the emission spectrum in ethanol at –40 °C. High-temperature blue shifts have been observed for other fluorophores.[74–78] In general, the most pronounced red shifts occur at temperatures at which the solvent is fluid enough to reorient prior to fluorescence emission but thermal energy is not so great as to disrupt these orientations.

6.5. BIOCHEMICAL EXAMPLES USING PRODAN

The use of solvent-sensitive probes has a long history in biochemical research. We now describe a few examples to illustrate the diversity of applications.

6.5.A. Phase Transition in Membranes

It is well known that model membranes display phase transitions at temperatures which depend on the chemical composition of the membranes. Saturated fatty acids display higher transition temperatures than do unsaturated fatty acids. For example, dipalmitoyl-L-α-phosphatidylcholine (DPPC) displays a transition temperature near 37 °C, whereas dioleoyl-L-α-phosphatidylcholine (DOPC) displays a transition below 0 °C.

These phase transitions can be observed with solvent-sensitive fluorophores. Emission spectra of DPPC and DOPC labeled with Patman are shown in Figure 6.33. In DPPC vesicles at 8 °C the emission of Patman is blue-shifted and representative of the unrelaxed state. At 46 °C, above the DPPC phase transition, the

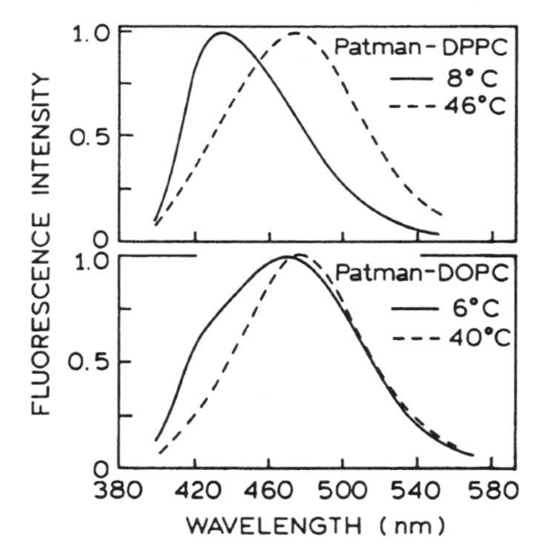

Figure 6.33. Temperature-dependent emission spectra of DPPC (*top*) and DOPC vesicles (*bottom*) labeled with Patman. Revised from Ref. 79, Copyright © 1984, with permission from Elsevier Science.

emission of Patman shifts dramatically to longer wavelengths. In the more fluid DOPC vesicles, the emission of Patman is red-shifted even at low temperature, indicating the higher fluidity of the unsaturated lipids. The emission spectra of Patman-labeled vesicles have been used to study the effects of anesthetics, hydrostatic pressure, or other perturbations on the phase states of membranes.

6.5.B. Protein Association

Actin is a protein which is present in almost all eukaryotic cells and is important in the cytoskeleton and cell motility. Muscle cells are filled with arrays of actin and myosin filaments, which produce motion by ATP hydrolysis and sliding of the filaments. Polymerization of actin (G-actin) to F-actin is widely studied in cell biology.

Protein association reactions are typically studied using fluorescence anisotropy. However, anisotropy measurements are difficult to perform in fluorescence microscopy, which is often used for viewing actin filaments in cells. In contrast, wavelength-ratiometric measurements are frequently used in microscopy. Hence, a method was developed to study actin polymerization using spectral shifts of Prodan. More precisely, a cysteine residue of actin was labeled with acrylodan.[80] The emission spectrum was found to shift to shorter wavelengths upon polymerization of G-actin to F-actin (Figure 6.34). This shift allows the extent of polymerization to be measured by the ratio of fluorescence intensities measured at two emission wavelengths.

6.5.C. Fatty Acid Binding Proteins

Fatty acids are a significant source of energy. In blood, fatty acids are usually transported bound to serum albumin. In cells, fatty acids are transported by fatty acid binding proteins (FABPs), which are a family of small proteins

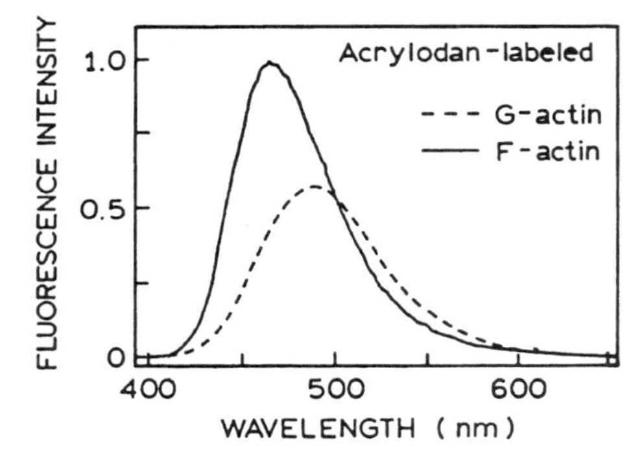

Figure 6.34. Emission spectra of Gactin (- - -) and F-actin (—) labeled with acrylodan. Revised and reprinted, with permission, from Ref. 80, Copyright © 1988, American Chemical Society.

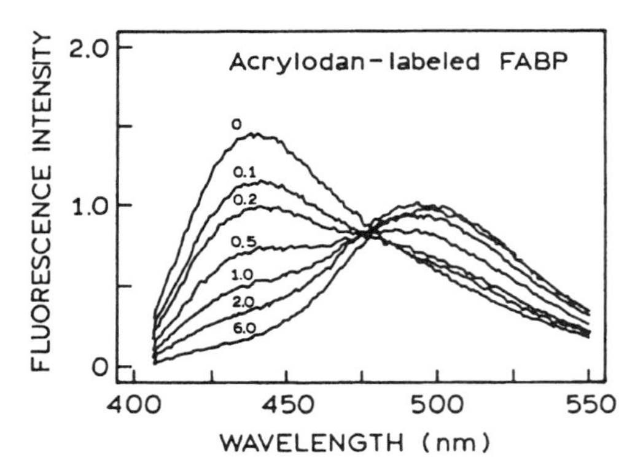

Figure 6.35. Emission spectra of acrylodan-labeled fatty acid binding protein (FABP). The numbers refer to the concentration (*MM*) of fatty acid (oleate) in 0.2 μ*M* FABP. Revised from Ref. 81.

(~15 kDa) with a variety of amino acid sequences. Since fatty acids are spectroscopically silent, their association with FABPs is not easily measured.

The problem of detecting fatty acid binding was solved using FABP labeled with acrylodan.[81–85] Titration of the labeled FABP with fatty acid resulted in a progressive red shift of the emission spectrum (Figure 6.35). Apparently, bound fatty acid displaces this probe from its hydrophobic binding site into the aqueous phase. Displacement of acrylodan from a hydrophobic pocket was supported by a decrease in anisotropy from 0.32 to 0.15 in the presence of bound fatty acid. The spectral shifts were also used to provide a method for the measurement of the concentration of free fatty acids in clinical samples.[86,87]

6.6. BIOCHEMICAL EXAMPLES USING SOLVENT-SENSITIVE PROBES

6.6.A. Exposure of a Hydrophobic Surface on Calmodulin

Calmodulin is involved in calcium signaling and the activation of intracellular enzymes. In the presence of calcium, calmodulin exposes a hydrophobic surface, which can interact with other proteins. This surface can also bind hydrophobic probes. Calmodulin was mixed with one of three different probes, 9-anthroxycholine (9-AC), ANS, or *N*-phenyl-1-naphthylamine (NPN).[88] In aqueous solution, these probes are all weakly fluorescent in the absence of protein (Figure 6.36, —). Addition of calmodulin, without calcium, had only a minor effect on the emission (---). Addition of calcium to calmodulin resulted in a dramatic increase in the intensity of the three probes (———). Since the effect was seen with neutral (NPN), negatively charged (ANS), and partially charged (9-AC) probes, the

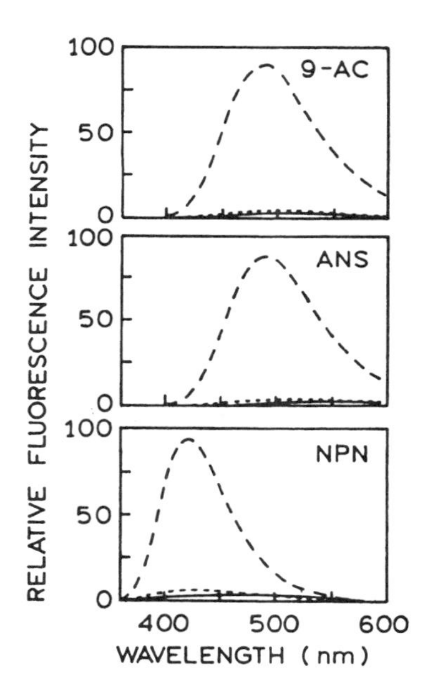

Figure 6.36. Emission spectra of 9-AC, ANS, and NPN in aqueous buffer (—), in the presence of calmodulin (---), and in the presence of calmodulin plus calcium (———). Revised and reprinted, with permission, from Ref. 88, Copyright © 1980, American Chemical Society.

binding was interpreted as due to hydrophobic rather than electrostatic interactions between the probe and calmodulin.

6.6.B. Binding to Cyclodextrins Using a Dansyl Probe

While there are numerous examples of proteins labeled with fluorescent probes, relatively few studies are available

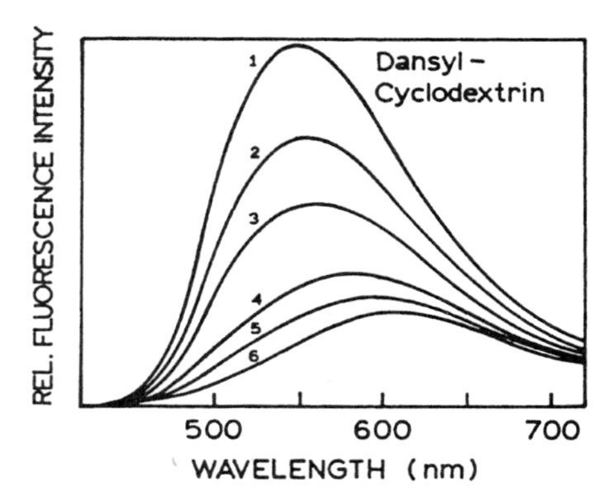

Figure 6.38. Emission spectra of dansyl-cyclodextrin with increasing concentrations of 1-adamantanol. Probe concentration = 10μM; [1-adamantanol]: 1, 0; 2, 20μM; 3, 50μM; 4, 150μM; 5, 300μM; 6, 1mM. Revised from Ref. 89.

using labeled carbohydrates.[89,90] The solvent sensitivity of the dansyl probes have been used to study the binding of organic molecules to cyclodextrins.[89] Cyclodextrins are cyclic sugars. The interior surface of a cyclodextrin is hydrophobic and can bind appropriately sized molecules. Carbohydrates are not fluorescent. In order to obtain a signal, a cyclodextrin was labeled with the dansyl group (Figure 6.37). The labeled cyclodextrin was titrated with 1-adamantanol, which resulted in a decrease in fluorescence intensity and a red shift in the emission spectra (Figure 6.38). In the absence of 1-adamantanol, the dansyl group binds in the cyclodextrin cavity, resulting in a blue-shifted and enhanced emission. Addition of 1-adamantanol, displaces the dansyl group, resulting in greater exposure to the aqueous phase and a red shift in the emission spectra.

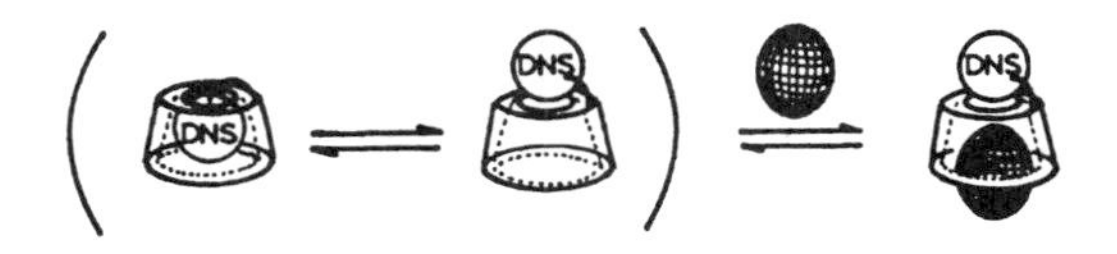

Figure 6.37. Dansyl-labeled cyclodextrin. Also shown is a schematic of the effects of 1-adamantanol, represented by the black ball, on the location of the dansyl (DNS) group. Revised from Ref. 89.

6.6.C. Polarity of a Membrane Binding Site

In Section 6.3 we described specific solvent effects and how these affected 2-AA. This fluorophore was used to synthesize a labeled phospholipid, anthroyl-PC (Figure 6.39), in which the anthroyl group is expected to be buried in the bilayer.[91] Since 2-AA derivatives are highly sensitive to solvent, the emission spectra were used to study probe localization in the bilayer. Emission spectra of anthroyl-PC in various solvents, and when incorporated into egg PC vesicles, are shown in Figure 6.40. In egg PC vesicles the emission is shifted well to the red, to an extent comparable to that observed in dimethylformamide (DMF) or a DMF/water mixture. Also, the absorption spectrum of anthroyl-PC shows the characteristic shape and longer-wavelength tail seen for hydrogen-bonded anthroyl groups (compare Figures 6.40 and 6.18), indicating that the an-

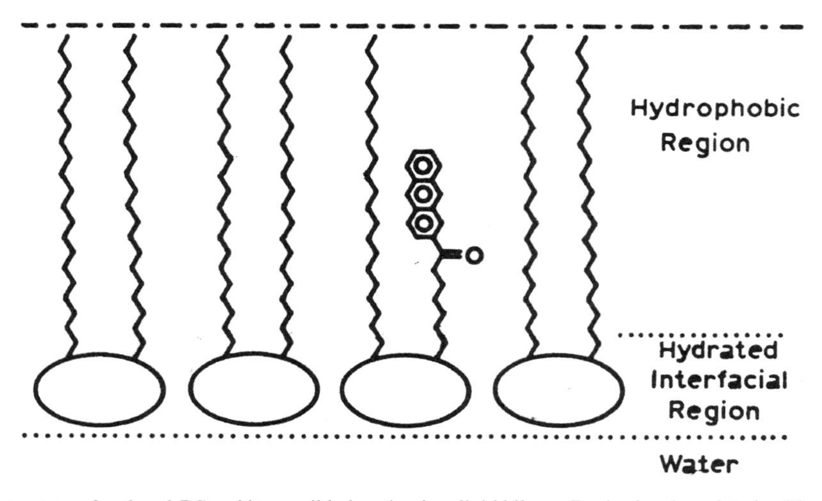

Figure 6.39. Chemical structure of anthroyl-PC and its possible location in a lipid bilayer. Revised and reprinted, with permission, from Ref. 91, Copyright © 1992, American Chemical Society.

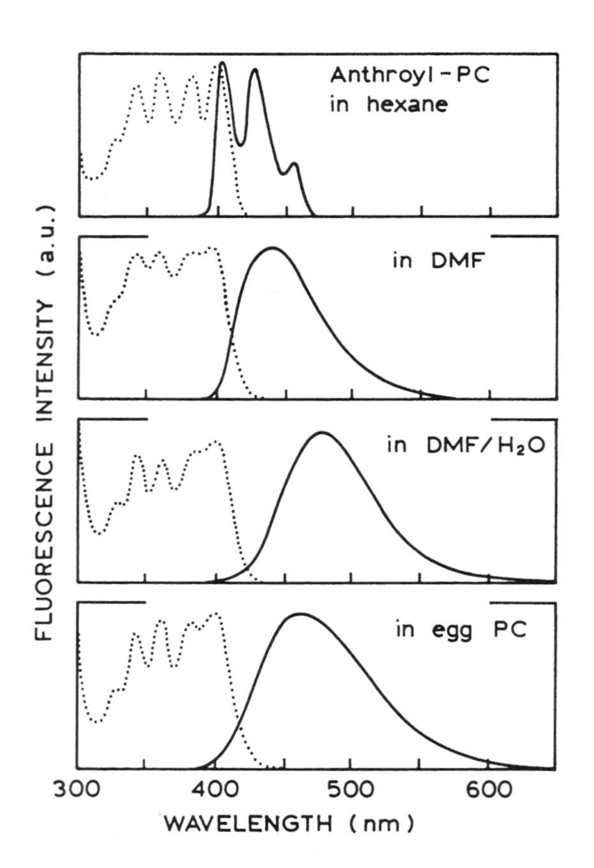

Figure 6.40. Emission spectra of anthroyl-PC in solvents and egg PC vesicles. The dimethylformamide–water mixture was 3.9/6.1 (mole/mole), 20 °C. Revised and reprinted, with permission, from Ref. 91, Copyright © 1992, American Chemical Society.

throyl group is in contact with polar groups in the ground state. In spite of its linkage to the acyl side chain of PC, the probe appears to be in a highly polar region of the membrane.

6.7. DEVELOPMENT OF ADVANCED SOLVENT-SENSITIVE PROBES

The increased understanding of solvent effects and formation of ICT states has resulted in the development of a new class of fluorophores that are highly sensitive to solvent polarity.[92,93] These probes are based on 2,5-diphenyloxazole

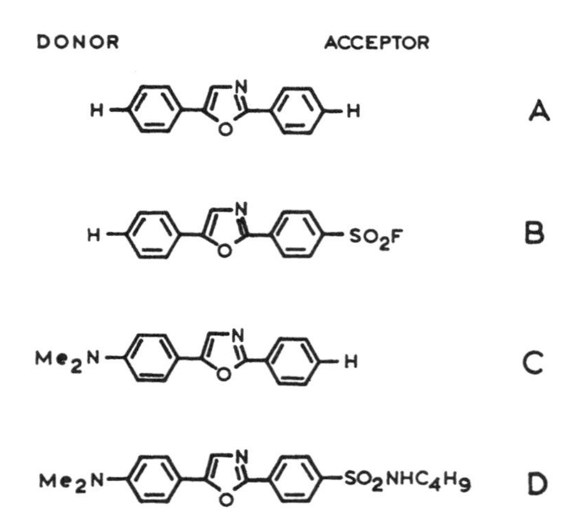

Figure 6.41. Derivatives of DPO (A) with donor and acceptor groups. Revised from Ref. 93.

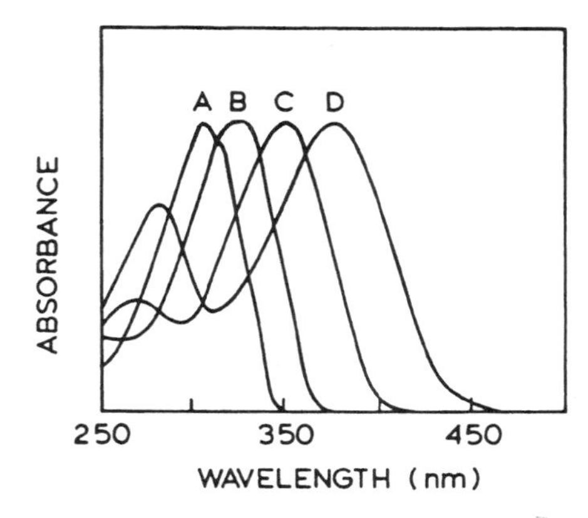

Figure 6.42. Absorption spectra of the DPO derivatives shown in Figure 6.41. The absorption spectra have been normalized. From Ref. 93.

(DPO), which is a well-known scintillator. DPO is soluble mostly in organic solvents, where it displays a high quantum yield. By itself, DPO would not be very sensitive to solvent polarity. Solvent sensitivity was engineered into DPO by the addition of electron donor and acceptor groups. In Figure 6.41, the top compound is DPO; in the lower structures the DPO is modified to contain an electron donor group, acceptor group, or both types of groups.

Absorption and emission spectra of these four DPO derivatives are shown in Figures 6.42 and 6.43. In methanol, DPO displays structured absorption and emission spectra, with only a small Stokes' shift. Addition of the sulfonic acid group alone results in a modest red shift. A larger red shift is observed upon addition of the dimethylamino group. The most dramatic shift is found when both electron-donating and electron-accepting groups are present on the DPO. By addition of both groups, the absorption maximum is shifted to more convenient wavelengths near 380 nm, and the emission is shifted to 600 nm.

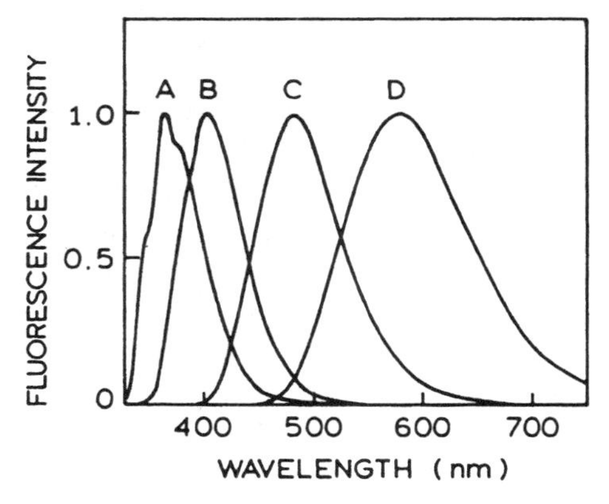

Figure 6.43. Emission spectra of the DPO derivatives shown in Figure 6.41 in methanol. The spectra have been normalized. From Ref. 93.

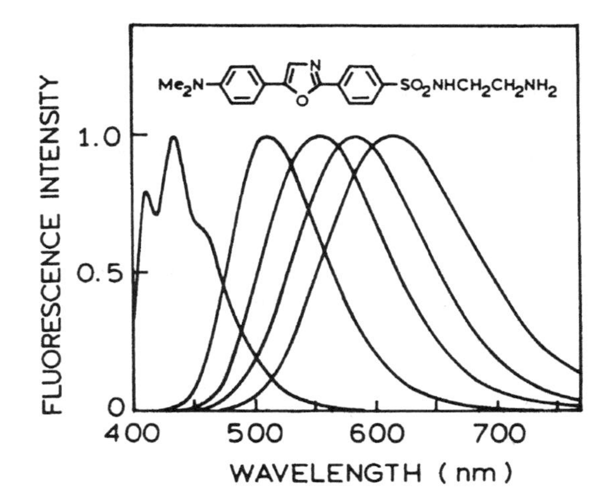

Figure 6.44. Emission spectra of Dapoxyl SEDA (Dapoxyl sulfonyl ethylenediamine) in solvents of increasing polarity: from left to right, hexane, chloroform, ethyl acetate, acetone, acetonitrile, dimethyl sulfoxide, and acetonitrile–water (1:1, vol/vol). From Ref. 93.

Table 6.4. Spectral Properties of Well-Known Solvent-Sensitive Probes[a]

Probe[b]	λ_{abs} (nm) [ε, (M^{-1} cm^{-1})]	λ_F (nm) in $CHCl_3$	λ_F (nm) in MeOH	$\Delta\lambda_F$ (nm)
Dapoxyl SEDA	373 (28,000)	584	509	71
Dansyl EDA	335 (4,600)	526	499	27
ADMAN	360 (15,000)	499	440	59
Prodan	361 (16,000)	498	440	58
1,8-ANS	372 (7,800)	480	490	−10
2,6-ANS	319 (27,000)	422	410	12
7-Ethoxycoumarin	324 (11,000)	399	385	4

[a] From Ref. 93.
[b] Dapoxyl SEDA, Dapoxyl sulfonyl ethylenediamine; Dansyl EDA, dansyl ethylenediamine; ADMAN, 6-acetyl-2-dimethylaminonaphthalene; Prodan, 6-propionyl-2-(dimethylamino)naphthalene; 1,8-ANS, 8-anilinonaphthalene-1-sulfonic acid; 2,6-ANS, 6-anilinonaphthalene-2-sulfonic acid.

The absence of vibronic structure suggests that the long-wavelength emission is due to an ICT state.

Based on these model compounds, reactive forms of dyes called Dapoxyl probes, have been prepared.*[94] These probes are highly sensitive to solvent polarity (Figure 6.44). A structured emission is seen in hexane, which disappears in more polar solvents, suggesting that these dyes form ICT states. These dyes are more sensitive to solvent than other presently used fluorophores (Table 6.4), and they also display usefully high extinction coefficients.

*Dapoxyl is a trademark of Molecular Probes, Inc.

In future years one can expect to see increased use of such dyes designed to display large Stokes' shifts.

6.8. EFFECTS OF SOLVENT MIXTURES

• Advanced Topic •

In Section 6.3 we described the phenomena of specific solvent effects, in which small amounts of a polar solvent result in dramatic shifts of the emission spectra. The presence of such effects suggests the presence of a distribution of fluorophores, each with a somewhat different solvent shell. Such a distribution of environments can result in a complex intensity decay.[94,95]

The effect of a solvent mixture on an intensity decay is illustrated by Y_t-base, which is highly sensitive to solvent polarity. The emission maximum of Y_t-base shifts from 405 nm in benzene to 455 nm in methanol (Figure 6.45). Addition of only 6% methanol, which does not dramatically change the orientation polarizability Δf, results in a large shift of the emission maximum, to 430 nm. A large spectral shift for a small change in the composition of the solvent usually indicates specific solvent effects.

Solvent mixtures provide a natural situation in which one can expect a complex intensity decay, or a distribution

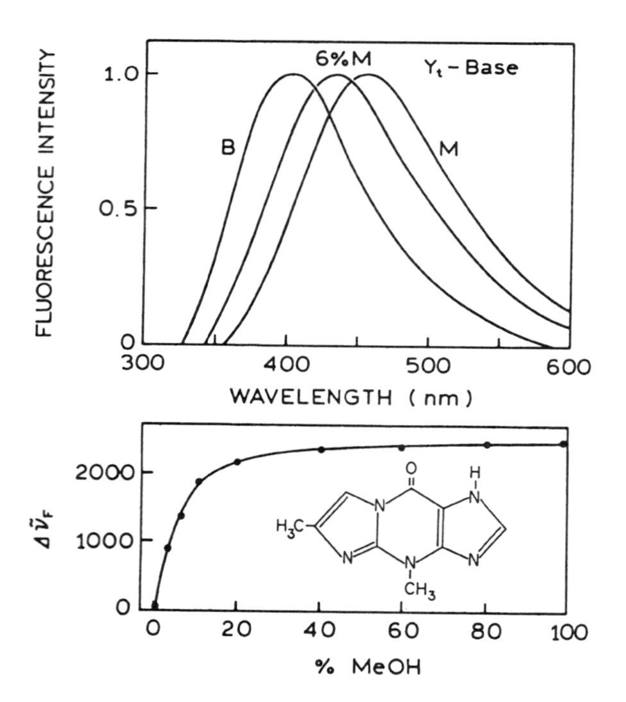

Figure 6.45. *Top:* Emission spectra of Y_t-base in benzene (B), benzene with 6% methanol (6% M), and methanol (M). *Bottom:* Dependence of the spectral shift on the percent methanol in benzene. From Ref. 96.

Table 6.5. Multiexponential Analysis of the Intensity Decays of Y_t-Base in Benzene–Methanol Mixtures[a]

Percent Methanol	Number of decay times	τ_i (ns)	α_i	f_i	χ_R^2
0	1	9.67	1.0	1.0	1.3
4	1	8.80	1.0	1.0	3.2
	2	3.67	0.061	0.026	1.2
		9.10	0.939	0.974	
6	1	8.36	1.0	1.0	8.9
	2	3.50	0.113	0.048	1.2
		8.92	0.887	0.952	
10	1	7.23	1.0	1.0	4.9
	2	2.99	0.082	0.034	1.0
		7.56	0.918	0.966	
100	1	6.25	1.0	1.0	1.9

[a] From Ref. 94.

of lifetimes. The intensity decay of Y_t-base in pure benzene or methanol is mostly a single exponential (Table 6.5). However, the decay times are different in each solvent, so that one can expect a more complex decay in benzene–methanol mixtures. In fact, a more heterogeneous intensity decay was observed in benzene with 6% methanol than in either pure solvent. This can be seen by the elevated value of χ_R^2 for the single-decay-time fit for the intensity decay of Y_t-base in benzene with 6% methanol (Table 6.5). Use of the double-exponential model reduces χ_R^2 from 8.9 to 1.2.

Although the intensity decays of Y_t-base could be fit using a two-decay-time model, it seems unlikely that there would be only two decay times in a solvent mixture. Since the solvent mixture provides a distribution of environments for the fluorophores, one expects the intensity decay to display a distribution of lifetimes. The frequency response of Y_t-base in the solvent mixture could not be fit to a single decay time. However, the data also could be fit to a distribution of lifetimes (Figure 6.46). In 100% benzene or 100% methanol the intensity decays were described by narrow lifetime distributions, which are essentially the same as a single decay time. In benzene–methanol mixtures, wide lifetime distributions were needed to fit the intensity decay (Figure 6.47). Lifetime distributions can be expected for any macromolecule in which there exists a distribution of fluorophore environments.

It is interesting to notice that the χ_R^2 values are the same for the multiexponential and for the lifetime distribution fits for Y_t-base with 6% methanol. This illustrates the frequently encountered situation in which different models

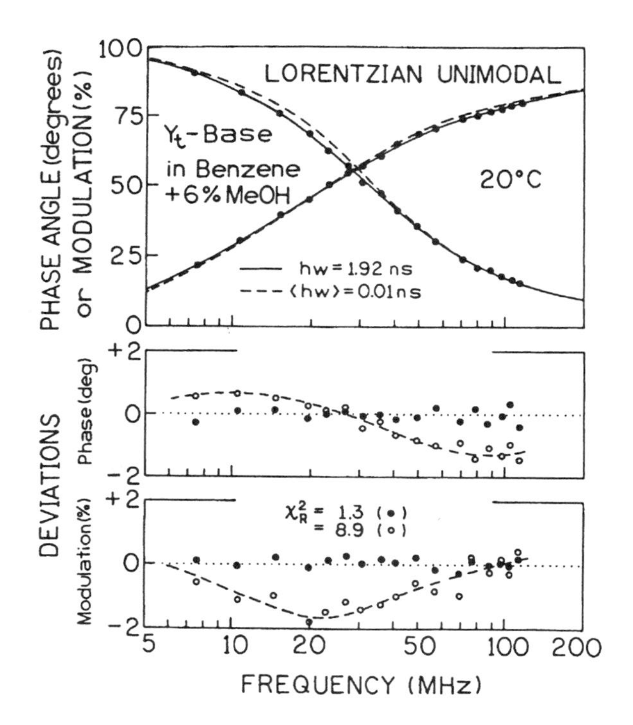

Figure 6.46. Frequency-domain intensity decay of Y_t-base in benzene with 6% methanol. From Ref. 94.

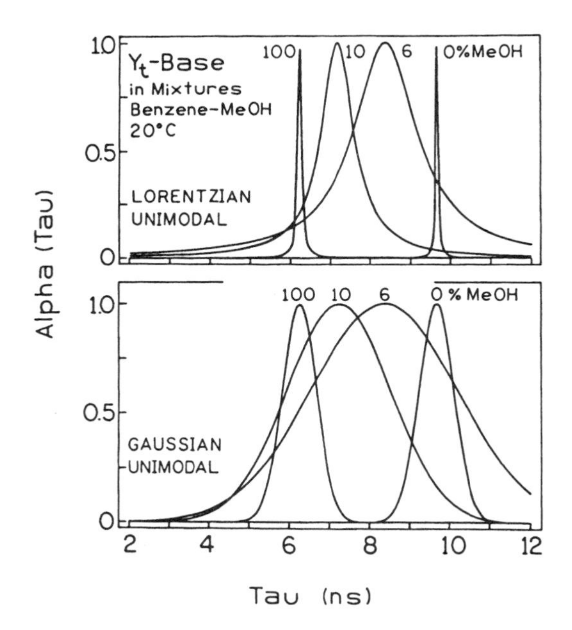

Figure 6.47. Lifetime distribution analysis of the intensity decay of Y_t-base in pure methanol, benzene with 10% methanol, benzene with 6% methanol, and pure benzene. From Ref. 94.

yield equivalent fits. In such cases, one must rely on other information to select the more appropriate model. In this case the lifetime distribution model seems preferable because there is no reason to expect two unique decay times in a solvent mixture.

6.9. SUMMARY OF SOLVENT EFFECTS

A quantitative description of the effects of environment on fluorescence emission spectra is perhaps the most challenging topic in fluorescence spectroscopy. No single theory or type of interactions can be used in all circumstances. There are at least three commonly observed effects:

1. General solvent effects due to the interactions of the dipole of the fluorophore with its environment
2. Specific solvent effects due to fluorophore–solvent interactions
3. Formation of ICT or TICT states depending on the probe structure and the surrounding solvent

Even if only one type of interaction were present, the effects would still be complex and beyond the limits of most models. For instance, the Lippert equation is only an approximation and ignores higher-order terms. Also, this equation only applies to a spherical dipole in a spherical cavity. More complex expressions are needed for nonspherical molecules, but one cannot generally describe the shape of the fluorophore in adequate detail. With regard to specific effects, there is no general theory to predict the shift in emission spectra due to hydrogen-bond formation. Finally, for fluorophores bound to macromolecules or in viscous solvents, spectral relaxation can occur during emission, so that the emission spectrum represents some weighted average of the unrelaxed and the relaxed emission.

Given all these complexities, how can one hope to use the data from solvent-sensitive probes? In our opinion, the best approach is to consider the molecules rather than to rely completely on theory. Observed effects should be considered within the framework of the three interactions listed above. Unusual behavior may be due to the presence of more than one type of interaction. Use the theory as an aid to interpreting plausible molecular interactions, and not as a substitute for careful consideration of probe structure and likely chemical interactions.

REFERENCES

1. Safarzadeh-Amiri, A., Thompson, M., and Krull, U. J., 1989, Trans-4-dimethylamino-4′-(1-oxobutyl)stilbene: A new fluorescent probe

of the bilayer lipid membrane, *J. Photochem. Photobiol. A: Chem.* **47**:299–308.
2. Von Lippert, E., 1957, Spektroskopische bistimmung des dipolmomentes aromatischer verbindungen im ersten angeregten singulettzustand, *Z. Electrochem.* **61**:962–975.
3. Mataga, N., Kaifu, Y., and Koizumi, M., 1956, Solvent effects upon fluorescence spectra and the dipole moments of excited molecules, *Bull. Chem. Soc. Jpn.* **29**:465–470.
4. Rettig, W., 1986, Charge separation in excited states of decoupled systems—TICT compounds and implications regarding the development of new laser dyes and the primary processes of vision and photosynthesis, *Angew. Chem. Int. Ed. Engl.* **25**:971–988.
5. Bayliss, N. S., 1950, The effect of the electrostatic polarization of the solvent on electronic absorption spectra in solution, *J. Chem. Phys.* **18**:292–296.
6. Bayliss, N. S., and McRae, E. G., 1954, Solvent effects in organic spectra: Dipole forces and the Franck–Condon principle, *J. Phys. Chem.* **58**:1002–1006.
7. McRae, E. G., 1956, Theory of solvent effects of molecular electronic spectra. Frequency shifts, *J. Phys. Chem.* **61**:562–572.
8. Baumann, W., and Bischof, H., 1982, Integral electro optical emission measurements. A new method for the determination of dipole moments of molecules in fluorescent states, *J. Mol. Struct.* **84**:181–193.
9. Kawski, A., 1992, Solvent-shift effect of electronic spectra and excited state dipole moments, in *Progress in Photochemistry and Photophysics,* J. F. Rabek (ed.), CRC Press, New York, pp. 1–47.
10. Seliskar, C. J., and Brand, L., 1971, Electronic spectra of 2-aminonaphthalene-6-sulfonate and related molecules. II, Effects of solvent medium on the absorption and fluorescence spectra, *J. Am. Chem. Soc.* **93**:5414–5420.
11. Suppan, P., 1990, Solvatochromic shifts: The influence of the medium on the energy of electronic states, *J. Photochem. Photobiol. A: Chem.* **50**:293–330.
12. Bakhshiev, N. G., 1961, Universal molecular interactions and their effect on the position of the electronic spectra of molecules in two component solutions I. Theory (liquid solutions), *Opt. Spectrosc.* **10**:379–384.
13. Bakhshiev, N. G., 1962, Universal molecular interactions and their effects on the position of electronic spectra of molecules in two-component solutions. IV. Dependence on the magnitude of the Stokes shift in the solvent luminescence spectrum (liquid solutions), *Opt. Spectrosc.* **12**:309–313.
14. Bakhshiev, N. G., 1962, Universal intermolecular interactions and their effect on the position of the electronic spectra of molecules in two-component solutions, *Opt. Spectrosc.* **13**:24–29.
15. Weber, G., and Laurence, D. J. R., 1954, Fluorescent indicators of absorption in aqueous solution and on the solid phase, *Biochem. J.* **56**:xxxi.
16. Slavik, J., 1982, Anilinonaphthalene sulfonate as a probe of membrane composition and function, *Biochim. Biophys. Acta* **694**:1–25.
17. McClure, W. O., and Edelman, G. M., 1954, Fluorescent probes for conformational states of proteins. I. Mechanism of fluorescence of 2-*p*-toluidinylnaphthalene-6-sulfonate, a hydrophobic probe, *Biochemistry* **5**:1908–1919.
18. Brand, L., Seliskar, C. J., and Turner, D. C., 1971, The effects of chemical environment on fluorescence probes, in *Probes of Structure and Function of Macromolecules and Membranes*, B. Chance, C. P. Lee, and J.-K. Blaisie (eds.), Academic Press, New York, pp. 17–39.
19. Turner, D. C., and Brand, L., 1968, Quantitative estimation of protein binding site polarity. Fluorescence of *N*-arylaminonaphthalenesulfonates, *Biochemistry* **7**:3381–3390.
20. Kosower, E. M., and Dodiuk, H., 1978, Intramolecular donor–acceptor systems. 3. A third type of emitting singlet state for *N*-alkyl-6-*N*-arylamino-2-naphthalenesulfonates. Solvent modulation of substituent effects on charge-transfer emissions, *J. Am. Chem. Soc.* **100**:4173–4179.
21. Kosower, E. M., and Kanety, H., 1983, Intramolecular donor–acceptor systems. 10. Multiple fluorescence from 8-(phenylamino)-1-naphthalenesulfonates, *J. Am. Chem. Soc.* **105**:6236–6243.
22. Kosower, E. M., and Huppert, D., 1986, Excited state electron and proton transfers, *Annu. Rev. Phys. Chem.* **37**:127–156.
23. Wilt, J. W., and Chwang, W. K., 1974, Fluorescence of 2-*N*-arylamino-6-naphthalenesulfonates in glycerol, *J. Am. Chem. Soc.* **96**:6195–6196.
24. Huppert, D., Ittah, V., and Kosower, E. M., 1988, New insights into the mechanism of fast intramolecular electron transfer, *Chem. Phys. Lett.* **144**:15–22.
25. Kosower, E. M., 1968, *An Introduction to Physical Organic Chemistry*, John Wiley & Sons, New York, pp. 293–382.
26. Kosower, E. M., 1958, The effect of solvent on spectra. I. A new empirical measure of solvent polarity: Z-values, *J. Am. Chem. Soc.* **80**:3253–3260.
27. Dimroth, K., Reichardt, C., Siepmann, T., and Bohlmann, F., 1963, Über pyridinium-*N*-phenol-betaine und ihre verwendung zur charakterisierung der polarität von lösungsmitteln, *Liebigs Ann. Chem.* **661**:1–37.
28. Reichardt, C., and Harbusch-Görnert, E., 1983, Erweiterung, korrektur und neudefinition der E_T-lösungsmittelpolaritätsskala mit hilfe eines lipophilen penta-tert-butyl-substituierten pyridinium-*N*-phenolat-betainfarbstoffes, *Liebigs Ann. Chem.* **5**:721–743.
29. Kamlet, M. J., Abboud, J. L., and Taft, R. W., 1977, The solvatochromic comparison method. 6. The π^* scale of solvent polarities, *J. Am. Chem. Soc.* **99**:6027–6038.
30. Von Reichardt, C., 1965, Empirische parameter der lösungsmittelpolarität, *Angew. Chem.* **77**:30–40.
31. Buncel, E., and Rajagopal, S., 1990, Solvatochromism and solvent polarity scales, *Acc. Chem. Res.* **23**:226–231.
32. Reichardt, C., 1994, Solvatochromic dyes as solvent polarity indicators, *Chem. Rev.* **94**:2319–2358.
33. Valeur, B., 1993, Fluorescent probes for evaluation of local physical and structural parameters, in *Molecular Luminescence Spectroscopy, Methods and Applications, Part 3*, S. G. Schulman (ed.), John Wiley & Sons, pp. 25–84.
34. Bakhshiev, N. G., 1962, Universal molecular interactions and their effect on the position of electronic spectra of molecules in two-component solutions, *Opt. Spectrosc.* **12**:261–264.
35. Perov, A. N., 1980, Energy of intermediate pair interactions as a characteristic of their nature. Theory of the solvato (fluoro) chromism of three-component solutions, *Opt. Spectrosc.* **49**:371–374.
36. Neporent, B. S., and Bakhshiev, N. G., 1960, On the role of universal and specific intermolecular interactions in the influence of the solvent on the electronic spectra of molecules, *Opt. Spectrosc.* **8**:408–413.
37. Cherkasov, A. S., 1960, Influence of the solvent on the fluorescence spectra of acetylanthracenes, *Akad. Nauk SSSR Bull. Phys. Sci.* **24**:597–601.
38. Tamaki, T., 1982, The photoassociation of 1- and 2-acetylanthracene with methanol, *Bull. Chem. Soc. Jpn.* **55**:1761–1767.
39. Tamaki, T., 1980, Polar fluorescent state of 1- and 2-acylanthracenes. II. The perturbation of protic solvents, *Bull. Chem. Soc. Jpn.* **53**:577–582.

40. Fery-Forgues, S., Fayet, J.-P., and Lopez, A., 1993, Drastic changes in the fluorescence properties of NBD probes with the polarity of the medium: Involvement of a TICT state, *J. Photochem. Photobiol. A: Chem.* **70**:229–243.
41. Mazères, S., Schram, V., Tocanne, J.-F., and Lopez, A., 1996, 7-Nitrobenz-2-oxa-1,3-diazole-4-yl-labeled phospholipids in lipid membranes: Differences in fluorescence behavior, *Biophys. J.* **71**:327–335.
42. Perochon, E., and Tocanne, J.-F., 1991, Synthesis and phase properties of phosphatidylcholine labeled with 8-(2-anthroyl)-octanoic acid, a solvatochromic fluorescent probe, *Chem. Phys. Lipids* **58**:7–17.
43. Perochon, E., Lopez, A., and Tocanne, J. F., 1991, Fluorescence properties of methyl 8-(2-anthroyl) octanoate, a solvatochromic lipophilic probe, *Chem. Phys. Lipids* **59**:17–28.
44. Rosenberg, H. M., and Eimutus, E., 1966, Solvent shifts in electronic spectra I. Stokes' shift in a series of homologous aromatic amines, *Spectrochim. Acta* **22**:1751–1757.
45. Werner, T. C., and Hercules, D. M., 1969, The fluorescence of 9-anthroic acid and its esters. Environmental effects on excited-state behavior, *J. Phys. Chem.* **73**:2005–2011.
46. Werner, T. C., and Hoffman, R. M., 1973, Relation between an excited state geometry change and the solvent dependence of 9-methyl anthroate fluorescence, *J. Phys. Chem.* **77**:1611–1615.
47. Werner, T. C., Matthews, T., and Soller, B., 1976, An investigation of the fluorescence properties of carboxyl substituted anthracenes, *J. Phys. Chem.* **80**:533–541.
48. Garrison, M. D., Doh, L. M., Potts, R. O., and Abraham, W., 1994, Fluorescence spectroscopy of 9-anthroyloxy fatty acids in solvents, *Chem. Phys. Lipids* **70**:155–162.
49. Berberan-Santos, M. N., Prieto, M. J. E., and Szabo, A. G, 1991, Excited-state intramolecular relaxation of the lipophillic probe 12-(9-anthroyloxy)stearic acid, *J. Phys. Chem.* **95**:5471–5475.
50. Thulborn, K. R., and Sawyer, W. H., 1978, Properties and the locations of a set of fluorescent probes sensitive to the fluidity gradient of the lipid bilayer, *Biochim. Biophys. Acta* **511**:125–140.
51. Badea, M. G., De Toma, R. P., and Brand, L., 1978, Nanosecond relaxation processes in liposomes, *Biophys. J.* **24**:197–212.
52. Lakowicz, J. R., and Balter, A., 1982, Direct recording of the initially excited and the solvent relaxed fluorescence emission of a tryptophan derivative in viscous solution by phase sensitive detection of fluorescence, *Photochem. Photobiol.* **36**:125–132.
53. Lakowicz, J. R., Bevan, D. R., Maliwal, B. P., Cherek, H., and Balter, A., 1983, Synthesis and characterization of a fluorescence probe of the phase transition and dynamic properties of membranes, *Biochemistry* **22**:5714–5722.
54. Weber, G., and Farris, F. J., 1979, Synthesis and spectral properties of a hydrophobic fluorescent probe: 6-Propionyl-2-(dimethylamino)naphthalene, *Biochemistry* **18**:3075–3078.
55. Sire, O., Alpert, B., and Royer, C. A., 1996, Probing pH and pressure effects of the apomyoglobin heme pocket with the 2′-(*N,N*-dimethylamino)-6-naphthoyl-4-*trans*-cyclohexanoic acid fluorophore, *Biophys. J.* **70**:2903–2914.
56. Prendergast, F. G., Meyer, M., Carlson, G. L., Iida, S., and Potter, J. D., 1983, Synthesis, spectral properties, and use of 6-acryloyl-2-dimethylaminonaphthalene (acrylodan), *J. Biol. Chem.* **258**:7541–7544.
57. Hendrickson, H. S., Dumdei, E. J., Batchelder, A. G., and Carlson, G. L., 1987, Synthesis of Prodan-phosphatidylcholine, a new fluorescent probe, and its interactions with pancreatic and snake venom phospholipases A_2, *Biochemistry* **26**:3697–3703.
58. Sandez, M. I., Suarez, A., Rios, M. A., Balo, M. C., Fernandez, F., and Lopez, C., 1996, Spectroscopic study of new fluorescent probes, *Photochem. Photobiol.* **64**:486–491.
59. Viard, M., Gallay, J., Vincent, M., Meyer, O., Robert, B., and Paternostre, M., 1997, Laurdan solvatochromism: Solvent dielectric relaxation and intramolecular excited-state reaction, *Biophys. J.* **73**:2221–2234.
60. Zurawsky, W. P., and Scarlata, S. F., 1992, Preferential solvation of 6-propionyl(*N,N*-dimethylamino)naphthalene in binary, polar solvent mixtures, *J. Phys. Chem.* **96**:6012–6016.
61. Ilich, P., and Prendergast, F. G., 1989, Singlet adiabatic states of solvated PRODAN: A semiempirical molecular orbital study, *J. Phys. Chem.* **93**:4441–4447.
62. Catalan, J., Perez, P., Laynez, J., and Blanco, F. G., 1991, Analysis of the solvent effect on the photophysics properties of 6-propionyl-2-(dimethylamino)naphthalene (PRODAN), *J. Fluoresc.* **1**(4):215–223.
63. Parusel, A., Schneider, F. W., and Köhler, G., 1997, An ab initio study on excited and ground state properties of the organic fluorescence probe PRODAN, *J. Mol. Struct. (Theochem)* **398–399**:341–346.
64. Nowak, W., Sygula, A., Adamczak, P., and Balter, A., 1986, On the possibility of fluorescence from twisted intramolecular charge transfer states of 2-dimethylamino-6-acylnaphthalenes. A quantum-chemical study, *J. Mol. Struct.* **139**:13–23.
65. Balter, A., Nowak, W., Pawelkiewicz, W., and Kowalczyk, A., 1988, Some remarks on the interpretation of the spectral properties of Prodan, *Chem. Phys. Lett.* **143**:565–570.
66. Rettig, W., and Lapouyade, R., 1994, Fluorescence probes based on twisted intramolecular charge transfer (TICT) states and other adiabatic photoreactions, in *Topics in Fluorescence Spectroscopy, Volume 4: Probe Design and Chemical Sensing*, J. R. Lakowicz (ed.), Plenum Press, New York, pp. 109–149.
67. Cornelißen, C., and Rettig, W., 1994, Unusual fluorescence red shifts in TICT-forming boranes, *J. Fluoresc.* **4**(1):71–74.
68. Vollmer, F., Rettig, W., and Birckner, E., 1994, Photochemical mechanisms producing large fluorescence Stokes' shifts, *J. Fluoresc.* **4**(1):65–69.
69. Rotkiewicz, K., Grellmann, K. H., and Grabowski, Z. R., 1973, Reinterpretation of the anomalous fluorescence of *p-N,N*-dimethylamino-benzonitrile, *Chem. Phys. Lett.* **19**:315–318.
70. Grabowski, Z. R., Rotkiewicz, K., and Siemiarczuk, A., 1979, Dual fluorescence of donor–acceptor molecules and the twisted intramolecular charge transfer TICT states, *J. Lumin.* **18**:420–424.
71. Belletête, M., and Durocher, G., 1989, Conformational changes upon excitation of dimethylamino para-substituted 3H-indoles. Viscosity and solvent effects, *J. Phys. Chem.* **93**:1793–1799.
72. Jones, G., Jackson, W. R., Choi, C.-Y., and Bergmark, W. R., 1985, Solvent effects on emission yield and lifetime for coumarin laser dyes. Requirements for a rotatory decay mechanism, *J. Phys. Chem.* **89**:294–300.
73. Ayuk, A. A., Rettig, W., and Lippert, E., 1981, Temperature and viscosity effects on an excited state equilibrium as revealed from the dual fluorescence of very dilute solutions of 1-dimethylamino-4-cyanonaphthalene, *Ber. Bunsenges. Phys. Chem.* **85**:553–555.
74. Cherkasov, A. S., and Dragneva, G. I., 1961, Influence of solvent viscosity on the fluorescence spectra of certain organic compounds, *Opt. Spectrosc.* **10**:238–241.
75. Bakhshiev, N. G., and Piterskaya, I. V., 1965, Universal molecular interactions and their effect on the electronic spectra of molecules in two-component solutions X, *Opt. Spectrosc.* **19**:390–395.

76. Macgregor, R. B., and Weber, G., 1981, Fluorophores in polar media. Spectral effects of the Langevin distribution of electrostatic interactions, *Proc. N.Y. Acad. Sci.* **366**:140–154.
77. Piterskaya, I. V., and Bakhshiev, N. G., 1963, Quantitative investigation of the temperature dependence of the absorption and fluorescence spectra of complex molecules, *Bull. Acad. Sci. USSR, Phys. Ser.* **27**:625–629.
78. Kawski, A., 1997, Thermochromic shifts of electronic spectra and excited state dipole moments, *Asian J. Spectrosc.* **1**:27–38.
79. Lakowicz, J. R., Cherek, H., Lazcko, G., and Gratton, E., 1984, Time-resolved fluorescence emission spectra of labeled phospholipid vesicles. As observed using multi-frequency phase-modulation fluorometry, *Biochim. Biophys. Acta* **777**:183–193.
80. Marriott, G., Zechel, K., and Jovin, T. M., 1988, Spectroscopic and functional characterization of an environmentally sensitive fluorescent actin conjugate, *Biochemistry* **27**:6214–6220.
81. Richieri, G. V., Ogata, R. T., and Kleinfeld, A. M., 1992, A fluorescently labeled intestinal fatty acid binding protein, *J. Biol. Chem.* **267**:23495–23501.
82. Richieri, G. V., Anel, A., and Kleinfeld, A. M., 1993, Interactions of long-chain fatty acids and albumin: Determination of free fatty acid levels using the fluorescent probe ADIFAB, *Biochemistry* **32**:7574–7580.
83. Richieri, G. V., Ogata, R. T., and Kleinfeld, A. M., 1996, Kinetics of fatty acid interactions with fatty acid binding proteins from adipocyte, heart, and intestine, *J. Biol. Chem.* **271**:11291–11300.
84. Richieri, G. V., Ogata, R. T., and Kleinfeld, A. M., 1996, Thermodynamic and kinetic properties of fatty acid interactions with rat liver fatty acid-binding protein, *J. Biol. Chem.* **271**:31068–31074.
85. Richieri, G. V., Ogata, R. T., and Kleinfeld, A. M., 1994, Equilibrium constants for the binding of fatty acids with fatty acid-binding proteins from adipocyte, intestine, heart and liver measured with the fluorescent probe ADIFAB, *J. Biol. Chem.* **269**:23918–23930.
86. Richieri, G. V., and Kleinfeld, A. M., 1995, Unbound free fatty acid levels in human serum, *J. Lipid Res.* **36**:229–240.
87. Kleinfeld, A. M., Prothro, D., Brown, D. L., Davis, R. C., Richieri, G. V., and DeMaria, A., 1996, Increases in serum unbound free fatty acid levels following coronary angioplasty, *Am. J. Cardiol.* **78**:1350–1354.
88. LaPorte, D. C., Wierman, B. M., and Storm, D. R., 1980, Calcium-induced exposure of a hydrophobic surface on calmodulin, *Biochemistry* **19**:3814–3819.
89. Wang, Y., Ikeda, T., Ikeda, H., Ueno, A., and Toda, F., 1994, Dansyl-β-cyclodextrins as fluorescent sensors responsive to organic compounds, *Bull. Chem. Soc. Jpn.* **67**:1598–1607.
90. Nagata, K., Furuike, T., and Nishimura, S.-I., 1995, Fluorescence-labeled synthetic glycopolymers: A new type of sugar ligands of lectins, *J. Biochem.* (*Tokyo*) **118**:278–284.
91. Perochon, E., Lopez, A., and Tocanne, J. F., 1992, Polarity of lipid bilayers. A fluorescence investigation, *Biochemistry* **31**:7672–7682.
92. BioProbes 25, New Products and Applications, Molecular Probes, Inc., Eugene, Oregon, May 1997.
93. Diwu, Z., Lu, Y., Zhang, C., Kalubert, D. H., and Haugland, R. P., 1997, Fluorescent molecular probes II. The synthesis, spectral properties and use of fluorescent solvatochromic Dapoxyl™ dyes, *Photochem. Photobiol.* **66**:424–431.
94. Gryczynski, I., Wiczk, W., Lakowicz, J. R., and Johnson, M. L., 1989, Decay time distribution analysis of Y_t-base in benzene–methanol mixtures, *J. Photochem. Photobiol. B: Biol.* **4**:159–170.
95. Gryczynski, I., Wiczk, W., Johnson, M. L., and Lakowicz, J. R., 1988, Lifetime distributions and anisotropy decays of indole fluorescence in cyclohexane/ethanol mixtures by frequency-domain fluorometry, *Biophys. Chem.* **32**:173–185.
96. Gryczynski, I., unpublished observations.

PROBLEMS

6.1. *Calculation of a Stokes' Shift*: Calculate the spectral shift of a fluorophore in methanol with $\mu_E - \mu_G = 14$ D and a cavity radius of 4 Å. That is, confirm the values in Table 6.2. Show that the answer is calculated in wavenumbers (cm^{-1}). Note: 1 esu = 1 $g^{1/2}\, cm^{3/2}/s^{-1}$ and 1 erg = 1 g cm^2/s^{-2}.

6.2. *Calculation of the Excited-State Dipole Moment*: Use the data in Table 6.6 to calculate the change in dipole moment ($\Delta\mu$) upon excitation of Prodan. Calculate the change in dipole moment $\Delta\mu$ for both the locally excited (LE) and the internal charge-transfer (ICT) state. Assume that the cavity radius is 4.2 Å.

Table 6.6. Spectral Properties of Prodan in Various Solvents

No.	Solvent	Absorption maximum (nm)	Emission maximum (nm)	Stokes' shift (cm^{-1})	Δf
1	Cyclohexane	342	401	4302	0.001
2	Benzene	355	421	4416	0.002
3	Triethylamine	343	406	4523	0.102
4	Chlorobenzene	354	430	4992	0.143
5	Chloroform	357	440	5284	0.185
6	Acetone	350	452	6448	0.287
7	Dimethylformamide	355	461	6477	0.276
8	Acetonitrile	350	462	6926	0.304
9	Ethylene glycol	375	515	7249	0.274
10	Propyelene glycol	370	510	7419	0.270
11	Ethanol	360	496	7616	0.298
12	Methanol	362	505	8206	0.308
13	Water	364	531	8646	0.320

Dynamics of Solvent and Spectral Relaxation

7

In the preceding chapter we described the effects of solvents on emission spectra and considered how the solvent-dependent data could be interpreted in terms of the local environment. We assumed that the solvent was in equilibrium around the excited-state dipole prior to emission. Equilibrium around the excited-state dipole is reached in fluid solution because the solvent relaxation times are typically less than 100 ps whereas the decay times are usually 1 ns or longer. However, equilibrium around the excited-state dipole is not reached in more viscous solvents and may not be reached for probes bound to proteins or membranes. In these cases emission occurs during solvent relaxation, and the emission spectrum represents an average of the partially relaxed emission. Under these conditions, the emission spectra display time-dependent changes. These effects are not observed in the steady-state emission spectra but can be seen in the time-resolved data or the intensity decays measured at various emission wavelengths.

As was true for solvent effects, there is no universal theory which provides a quantitative description of all the observed phenomena. Time-dependent shifts occur as the result of general solvent effects, specific solvent–fluorophore interactions, and formation of ICT states. All these time-dependent processes affect the time-resolved data. The effects are often similar and not immediately assignable to a particular molecular event. Hence, it can be challenging to select a model for interpretation of the time-dependent spectra.

Time-dependent spectral shifts are usually studied by measurement of the time-resolved emission spectra (TRES). The TRES are the emission spectra which would be observed if measured at some instant in time following pulsed excitation. Suppose that the rate of solvent or spectral relaxation (k_S) is comparable to the intensity decay rate ($1/\tau$), as shown in Figure 7.1. In this situation, the solvent relaxation time ($\tau_S = k_S^{-1}$) is comparable to the lifetime (τ). At low temperatures in the absence of spectral relaxation, $\tau_S >> \tau$, and one observes emission from the initially excited or Franck–Condon state (F). If relaxation is very rapid, then $\tau_S << \tau$, and emission from the relaxed state (R) is observed. At intermediate temperatures, emission from both states will be observed (dashed curve in Figure 7.1). In general, the decay times of the F(τ_F) and R(τ_R) states can be different. Unless otherwise stated, we will assume that the decay times are not affected by spectral relaxation ($\tau_F = \tau_R = \tau$).

A complex situation is encountered if the rate of solvent relaxation is comparable to the intensity decay rate of the sample. In this case, the intensity decay becomes complex and dependent on the observation wavelength (Figure 7.2). If the emission is observed on the blue side of the emission, the decay [$I_F(t)$] is more rapid than the decay of the total emission [$I_T(t)$]. This is because the short-wavelength emission is decaying by both emission and relaxation. On the long-wavelength side of the emission, one selectively observes those fluorophores which have relaxed prior to

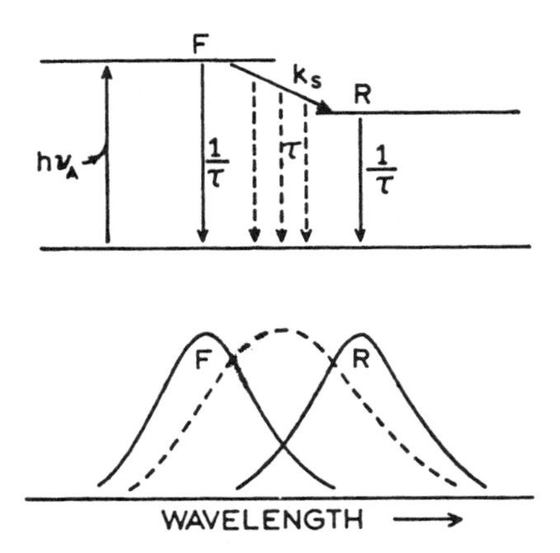

Figure 7.1. Jabłoński diagram for solvent relaxation.

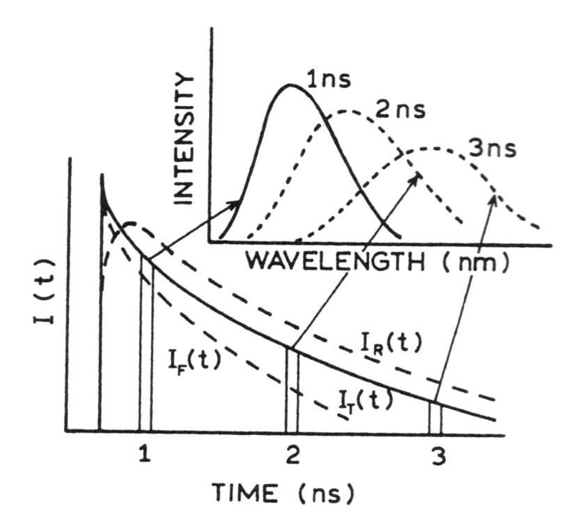

Figure 7.2. Schematic of time-resolved emission spectra. $I_T(t)$ represents the decay of the total emission. $I_F(t)$ and $I_R(t)$ are the intensity decays on the blue and the red side of the emission spectrum, respectively. *Inset*: Emission spectra at t = 1, 2, or 3 ns.

emission. Since these fluorophores are relaxed, they are the longer-lived species. Recall that emission is a random event proceeding at the radiative decay rate. The fluorophores that emit at earlier times tend to have shorter-wavelength emission, and those that emit at later times have longer-wavelength emission. At time zero, there are no relaxed fluorophores. For this reason, one typically observes a rise in the intensity at long wavelengths over a short interval of time, representing formation of the relaxed state. The rise time is frequently associated with a negative preexponential factor in the multiexponential analysis. Following the rise in intensity, the decay $I_R(t)$ typically follows the total emission.

Suppose that emission spectra could be recorded at any desired instant following the excitation pulse (Figure 7.2). If the emission spectrum was observed immediately after excitation (t = 1 ns), then a blue-shifted or unrelaxed emission would be observed. If the time of observation was later, then more of the molecules would have relaxed to longer wavelengths, resulting in a red-shifted emission spectrum. These emission spectra, representing discrete times following excitation, are the TRES. It is technically challenging to determine the TRES, and the molecular interpretation can be equally difficult. In the first part of this chapter, we present an overview of TRES, followed by some biochemical examples. The latter sections describe molecular mechanisms for time-dependent spectral relaxation and show how the wavelength-dependent data can be used to distinguish between models for solvent relaxation.

7.1. CONTINUOUS AND TWO-STATE SPECTRAL RELAXATION

Prior to describing the details of spectral relaxation, we will consider two extreme models, which are called the continuous model and the two-state model. Figure 7.3 shows hypothetical TRES, which have been intensity normalized for easy comparison. In the continuous model, shown in the top panel of the figure, the emission spectrum is assumed to shift progressively to longer wavelengths at longer times.[1–3] The spectral shape is expected to remain the same during relaxation. This type of TRES is expected when there are general solvent effects and/or a multitude of solvent–fluorophore interactions.

An alternative type of relaxation is represented by the two-state model (Figure 7.3, bottom). In this case, the initially excited short-wavelength state (F) is assumed to change to the longer-wavelength relaxed state (R). The F and R states are seen at short and long times, respectively. Each state displays a discrete emission spectrum. This is the same model used to describe excited-state reactions, where the fluorophore undergoes a chemical reaction such as proton loss or excimer formation (Chapter 18). This two-state model also approximates the behavior of fluorophores which display specific solvent effects, as seen for 2-anilinonaphthalene, or formation of an ICT state, as seen for Laurdan (Chapter 6).

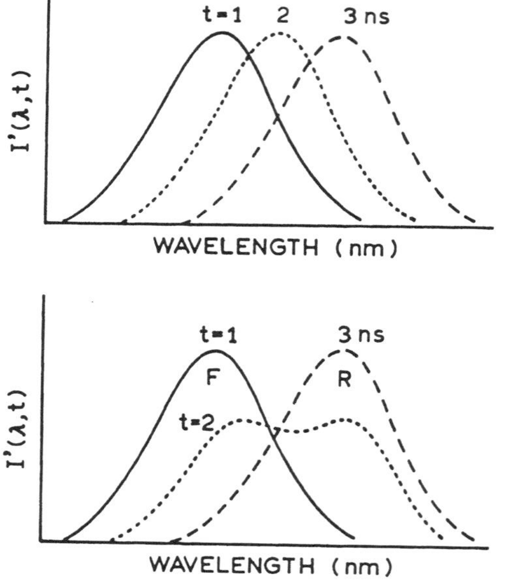

Figure 7.3. Normalized time-resolved emission spectra for continuous relaxation (*top*) and for a two-state process (*bottom*).

Although the continuous and two-state models are only phenomenological models, they provide a useful framework for considering the molecular origin of the time-dependent shifts.

7.2. MEASUREMENT OF TRES

7.2.A. Direct Recording of TRES

Among the methods employed for the measurement of TRES, the most easily understood is direct recording using pulse sampling and time-gated detection (Chapter 4). In fact, the first reported TRES were obtained using this method.[4–6] The sample was 4-aminophthalimide in *n*-propanol at –70 °C. At this temperature the rate of solvent relaxation is comparable to the decay time. The sample was excited with a brief pulse of light, and the detector was gated on for a brief period (typically <0.5 ns) at various times following the excitation pulse. The emission spectrum was then scanned as usual. For short time delays, a blue-shifted emission was observed (Figure 7.4, top). As the delay time was increased, the emission spectrum shifted to longer wavelength. At these longer times, the solvent has relaxed around the excited-state dipole.

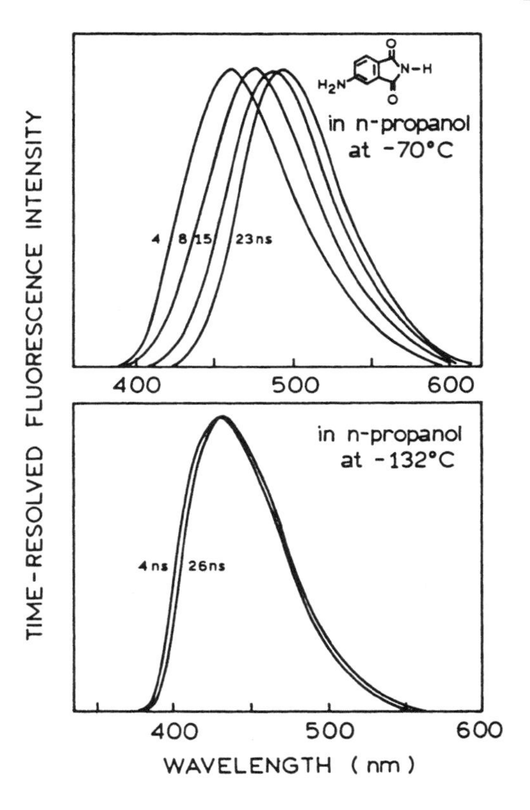

Figure 7.4. Time-resolved fluorescence spectra of 4-aminophthalimide in *n*-propanol at –70 (*top*) and –132 °C (*bottom*) recorded using the pulse sampling method (Chapter 4). The times between excitation and gated sampling of the emission intensity are labeled on the figure. Revised from Ref. 4.

Examination of Figure 7.4 reveals that the shape and/or width of the emission spectrum is unchanged during spectral relaxation. This suggests that relaxation of 4-aminophthalimide in *n*-propanol proceeds as a continuous process. When the temperature was lowered to –132 °C, the emission spectra were similar for time delays of 4 and 26 ns (Figure 7.4, bottom). This is because at lower temperatures the relaxation time of the solvent is longer than the decay time of 4-aminophthalimide, which is near 10 ns.

One can also imagine direct recording of TRES using TCSPC. TRES are occasionally obtained by such direct recording methods. When using TCSPC, one can record photons which arrive within a short time interval. In this instance, one selects the output pulses from the TAC to be within a limited range of voltage values. The range of accepted voltages determines the time window that is observed. The emission spectrum is then recorded with a monochromator. Although intuitively simple, this method is rather inefficient and has not been widely utilized.

Direct recording of TRES is easy to understand, but this procedure is no longer recommended. The directly recorded TRES do not provide for deconvolution of the instrument response function. Hence, the recorded TRES are apparent spectra, which contain distortions due to the instrument response function. One exception is the use of pump-probe or upconversion methods, when the time resolution is fast compared to the rates of spectral relaxation.

7.2.B. TRES from Wavelength-Dependent Decays

At present, TRES are usually obtained indirectly. The procedure starts with measurement of the time-resolved decays at a number of wavelengths across the emission spectrum $[I(\lambda, t)]$.[7–9] The intensity decays are wavelength-dependent, and the emission at short wavelengths decays more rapidly than that at the longer wavelengths. This occurs because the emission at short wavelengths is decaying by both emission and relaxation to longer wavelengths. In contrast, the emission at long wavelengths requires that the fluorophores relax prior to emission and is thus delayed by the relaxation time.

For calculation of the TRES, the intensity decays are usually analyzed in terms of the multiexponential model,

$$I(\lambda, t) = \sum_{i=1}^{n} \alpha_i(\lambda) \exp[-t/\tau_i(\lambda)] \quad [7.1]$$

where the $\alpha_i(\lambda)$ are the preexponential factors and the $\tau_i(\lambda)$ are the decay times, with $\Sigma\alpha_i(\lambda) = 1.0$. In this analysis the decay times can be variables at each wavelength $[\tau_i(\lambda)]$, or they can be assumed to be independent of wavelength (τ_i). Wavelength-dependent decay times are expected for the continuous model, and wavelength-independent lifetimes are expected for the two-state model. However, because of limited resolution and parameter correlation, the data can usually be fit with either wavelength-dependent $[\tau_i(\lambda)]$ or wavelength-independent (τ_i) lifetimes, irrespective of whether relaxation is a continuous or a two-state process. For purposes of calculating the TRES, the choice does not matter. The goal is to obtain a parameterized form of the intensity decays, which are then used to reconstruct the TRES. Typically, no molecular significance is assigned to the intensity decay parameters.

In order to calculate the TRES, one computes a new set of intensity decays, which are normalized so that the time-integrated intensity at each wavelength is equal to the steady-state intensity at that wavelength. Suppose $F(\lambda)$ is the steady-state emission spectrum. One calculates a set of $H(\lambda)$ values using

$$H(\lambda) = \frac{F(\lambda)}{\int_0^\infty I(\lambda, t)\, dt} \quad [7.2]$$

which for the multiexponential analysis becomes

$$H(\lambda) = \frac{F(\lambda)}{\sum_i \alpha_i(\lambda)\tau_i(\lambda)} \quad [7.3]$$

Then, the appropriately normalized intensity decay functions are given by

$$I'(\lambda, t) = H(\lambda)I(\lambda, t) = \sum_i \alpha_i'(\lambda) \exp[-t/\tau_i(\lambda)] \quad [7.4]$$

where $\alpha_i'(\lambda) = H(\lambda)\alpha_i(\lambda)$.

The values of $I'(\lambda, t)$ can be used to calculate the intensity at any wavelength and time, and thus the TRES. The TRES can be shown with the actual intensities, as in Figure 7.2, or peak normalized, as in Figures 7.3 and 7.4. Assuming that the intensity decays have been measured at an adequate number of wavelengths, this procedure (Eqs. [7.1]–[7.4]) yields the actual TRES independent of the nature of the relaxation process. The advantage of this procedure is that the values of $I'(\lambda, t)$ are the actual impulse response functions, corrected for distortions due to convolution with the instrument response function. The time dependence of the spectral shifts and the shape of the TRES are then used to determine the rates of relaxation and the nature of the relaxation process.

The calculation of TRES from the measured intensity decays is best understood by considering a specific example. The data in Figure 7.5 are for the probe 2-anilinonaphthalene (2-AN) in phospholipid vesicles,[9,10] but the procedure is the same for any sample. The time-resolved decays are measured at appropriate wavelengths across the entire emission spectrum. Typical decays at 394, 409, and 435 nm are shown in the upper panels. One may interpret these results as follows. At shorter wavelengths, the intensity decay is faster. This faster decay is evident in both the measured decays $N_k(\lambda, t)$ and the calculated impulse response functions $I(\lambda, t)$. At longer wavelengths, one selectively observes those fluorophores which have relaxed, and hence those which have emitted at later times following excitation. Thus, the overall decay is slower at longer wavelengths.

The time-resolved decays are used to derive the impulse response functions at each emission wavelength, $I(\lambda, t)$. A multiexponential fit is used, but no physical significance is attached to the values of $\alpha_i(\lambda)$ and $\tau_i(\lambda)$. These are simply used to obtain an accurate representation for the observed intensity decays. Note that the x-axis has now been changed to a time axis. This change was made because these are the impulse response functions. For the initial data $N_k(\lambda, t)$, the time is not indicated on the x-axis because the width of the lamp pulse results in an undefined zero time point.

Examination of the upper right panel in Figure 7.5 shows a characteristic feature of solvent relaxation, which is a rise in intensity at long wavelengths (435 nm). This rise occurs because at $t = 0$ no fluorophores are in the relaxed state. The population of the relaxed state increases prior to decreasing due to the total intensity decay. Observation of such a term provides proof that an excited-state process has occurred. If the sample displayed only ground-state heterogeneity, then the decays would be dependent upon wavelength. However, no rise in intensity would be observed and all the preexponential factors would be positive (Chapter 18).

The impulse response functions at each emission wavelength are now adjusted in magnitude so that the integrated intensity corresponds to the steady-state intensity at the same wavelength [$I'(\lambda, t)$; lower left panel in Figure 7.5]. To obtain the TRES, one uses the family of $I'(\lambda, t)$ functions

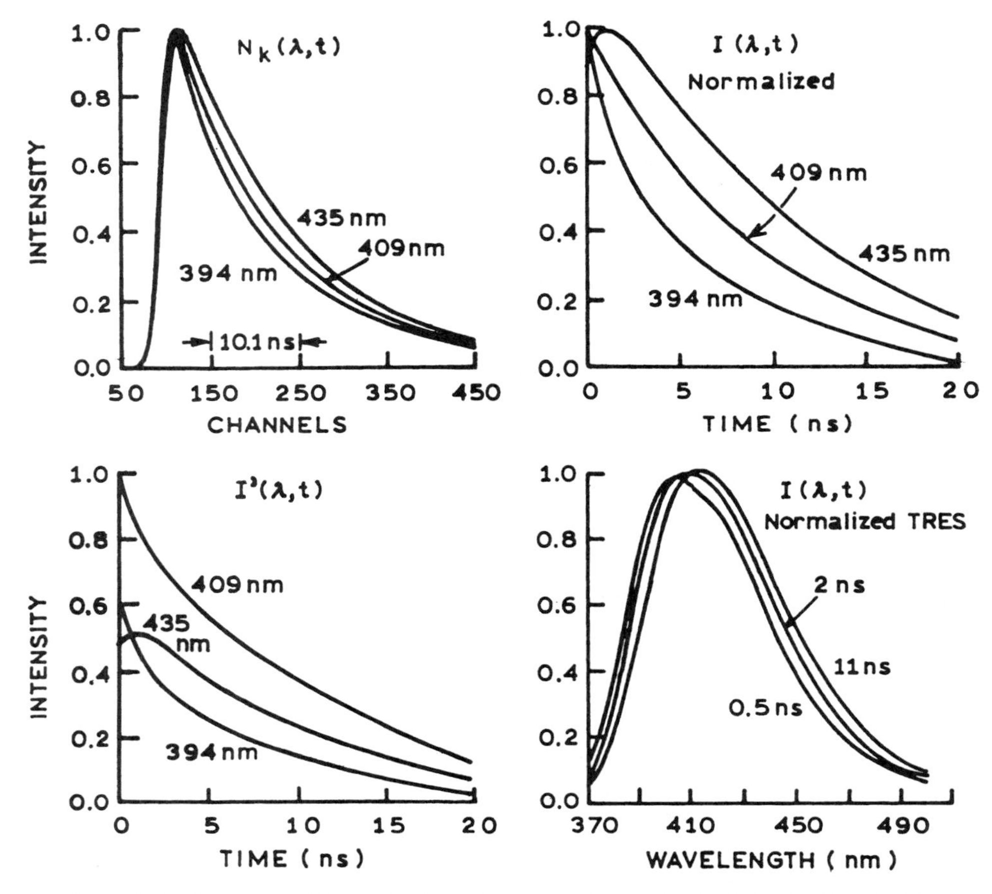

Figure 7.5. Calculation of time-resolved emission spectra. See text for details. Revised from Ref. 9.

to plot the emission spectra for the desired times. The TRES are usually peak normalized to allow easy visualization of the time-dependent spectral shifts. For 2-AN bound to vesicles, one notices that the spectra shift to longer wavelengths at longer times (lower right panel in Figure 7.5).

7.3. BIOCHEMICAL EXAMPLES OF TRES

TRES have been used to study the dynamics of proteins and membranes. The basic idea is that excitation provides an instantaneous perturbation because of the increased dipole moment of the excited state. If the biomolecule is rigid, there will be no relaxation, as was seen above for 4-aminophthalimide in propanol at −132 °C. If the biomolecule is flexible, then the TRES should relax on a timescale characteristic of the macromolecule. Time-dependent spectral shifts have been observed for probes bound to proteins[11–17] and membranes.[18–24] Nanosecond spectral relaxation appears to be typical of labeled biomolecules.

7.3.A. Spectral Relaxation in Apomyoglobin

One example of TRES of a protein is provided by labeled myoglobin. Myoglobin is a muscle protein which binds oxygen from the blood and releases oxygen as needed to the muscles. Myoglobin thus acts as an oxygen reservoir. In myoglobin the oxygen molecule is bound to the heme group, which is near the center of the protein. The heme group can be removed from the protein, leaving a hydrophobic pocket which is known to bind a number of fluorophores.[25,26] Myoglobin without the heme group is called apomyoglobin. The dynamics of myoglobin are of interest because myoglobin cannot bind and release oxygen without undergoing structural fluctuations to allow diffusion of oxygen through the protein. If the protein is flexible on the nanosecond timescale for oxygen penetration, then it seems likely that the protein can be flexible during the nanosecond decay times of bound fluorophores.

In the previous chapter we described Prodan and its derivatives as being highly sensitive to solvent polarity. The dynamics of the heme binding site were studied using the Danca probe (Figure 7.6), which is an analog of Prodan. The carboxycyclohexyl side chain serves to increase the affinity of Danca for apomyoglobin and to ensure that it binds to the protein in a single orientation. A single mode of binding simplifies interpretation of the data by providing a homogeneous probe population. Excitation of Danca results in the instantaneous creation of a new dipole within the apomyoglobin molecule. If myoglobin is flexible on the nanosecond timescale, one expects time-dependent shifts in its emission spectrum as the protein rearranges around the new dipole moment.

Intensity decays were measured at various wavelengths across the emission spectrum (Figure 7.6). The decays are somewhat faster at shorter emission wavelengths. Importantly, there is evidence of a rise time at longer wavelengths, which is characteristic of an excited-state process. A rise time can only be observed if the emission is not directly excited, but rather forms from a previously excited state. In this case the initially excited state does not contribute to the emission at 496 and 528 nm; the emission at these wavelengths is due to relaxation of the initially excited state.

The time-dependent decays were used to construct the TRES (Figure 7.7). These spectra shift progressively to longer wavelength at longer times. Even at the earliest times (20 ps), the TRES are well shifted from the steady-state spectrum observed at 77 K. At this low temperature, solvent relaxation does not occur. As was described in Chapter 6 (Figures 6.28 and 6.31), Prodan-like molecules can emit from LE and ICT states. The strongly blue-shifted

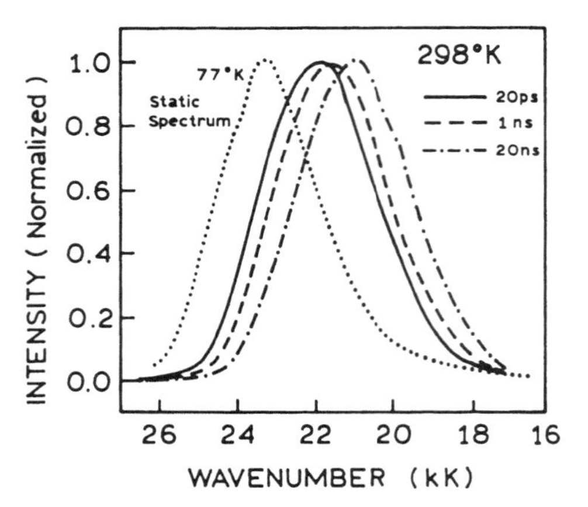

Figure 7.7. TRES of Danca bound to apomyoglobin. Also shown is the steady-state spectrum at 77 K. 1 kK = 1000 cm^{-1}. Revised and reprinted, with permission, from Ref. 15, Copyright © 1992, American Chemical Society.

emission at 77 K is probably due to the LE state. Hence, the emission of Danca-apomyoglobin is from the ICT state, which has undergone nearly complete charge separation.

The TRES can be used to calculate the rates of spectral relaxation. These data are usually presented as the average energy of the emission versus time (Figure 7.8). Alternatively, one can calculate the time-dependent change in the emission maximum. In the case of Danca-apomyoglobin, the decay of the average energy was highly nonexponential, with relaxation times ranging from 20 ps to 20 ns. Spectral relaxation is typically a multi- or nonexponential process for probes in solvents or when bound to macro-

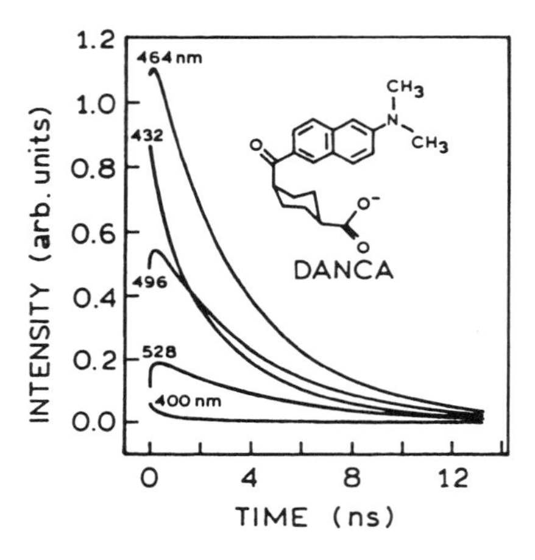

Figure 7.6. Deconvolved fluorescence decays of Danca–apomyoglobin complex from 400 to 528 nm at 298 K. The area under each trace has been scaled to the steady-state intensity at that wavelength. Note that the intensity decays show a rise time at longer wavelengths. Revised and reprinted, with permission, from Ref. 15, Copyright © 1992, American Chemical Society.

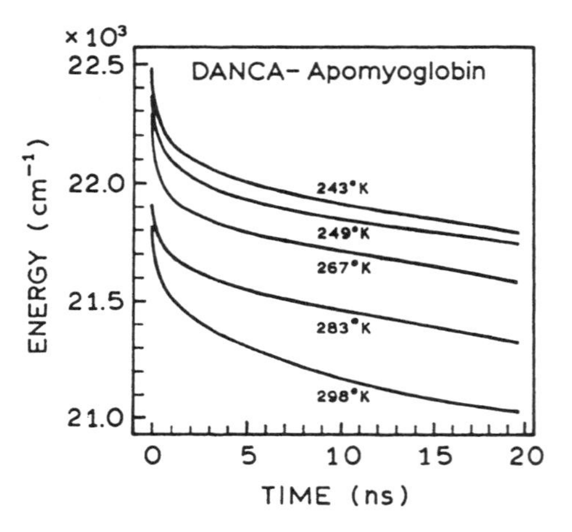

Figure 7.8. Decay of the mean energy of the emission (in cm^{-1}) of Danca-apomyoglobin at various temperatures in water and glycerol–water mixtures. Revised and reprinted, with permission, from Ref. 15, Copyright © 1992, American Chemical Society.

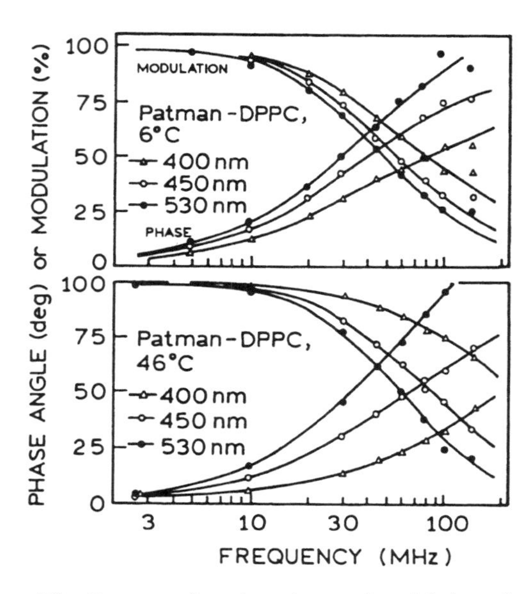

Figure 7.9. Frequency-dependent phase and modulation values for Patman-labeled DPPC vesicles at 6 (*top*) and 46 °C (*bottom*). For both temperatures, data are shown for 400 (△), 450 (○), and 530 nm (●). Revised from Ref. 21, Copyright © 1984, with permission from Elsevier Science.

molecules. Methods used to determine the relaxation times are described in Section 7.3.C.

7.3.B. TRES of Labeled Membranes

TRES have been extensively used to study membrane dynamics.[17–24] While not intuitively obvious, FD data can also be used to determine TRES. This is shown for Patman-labeled DPPC vesicles in Figure 7.9. Patman is also a Prodan derivative, in this case designed to bind to membranes (Chapter 6, Figure 6.27). As for the TD measurements, the FD data are measured for various wavelengths across the emission spectrum. In Figure 7.9, the phase and modulation values are strongly dependent on emission wavelength. The phase angles become larger and the modulation becomes smaller at longer emission wavelengths. These wavelength-dependent changes are characteristic of an excited-state process. In fact, the data prove the presence of an excited-state reaction because the phase angles exceed 90°, which is not possible for a mixture of directly excited fluorophores[27] (Chapter 18). It is interesting to compare the results for Patman-DPPC at low and high temperature (Figure 7.9). On the blue side of the emission, the phase angles are smaller and the modulation values are higher at the higher temperature. This difference reflects the shortened decay time on the blue side of the spectrum due to faster relaxation at the higher temperature. The data at low temperature (6 °C) are also characteristic of an excited-state process, with larger phase angles and lower modulation values on the long-wavelength side of the emission. However, the long-wavelength phase angles do not exceed 90°. The absence of phase angles over 90° does not prove that relaxation has not occurred, only that the conditions were not suited to result in very large phase angles.

The phase and modulation data can be used to derive the wavelength-dependent impulse response functions. In this case the data at all wavelengths could be fit to a set of three wavelength-independent lifetimes (Table 7.1). The three decay times are similar at both 8 and 46 °C. The main distinction between the high- and low-temperature data is the larger contribution of terms with negative preexponential factors at the higher temperature. Such terms are characteristic of emission which is formed by an excited-state process; these are the terms which result in phase angles over 90°. At the longest emission wavelength, these positive and negative terms are nearly equal in magnitude and opposite in sign. This is characteristic of emission from a

Table 7.1. Intensity Decays of Patman-Labeled DPPC Vesicles[a]

	Intensity decay at 6 °C[b]			Intensity decay at 46 °C[b]		
Wavelength	$\alpha_1(\lambda)^a$ ($\tau_1 = 0.98$ ns)	$\alpha_2(\lambda)$ ($\tau_2 = 3.74$ ns)	$\alpha_3(\lambda)$ ($\tau_3 = 5.52$ ns)	$\alpha_1(\lambda)$ ($\tau_1 = 0.95$ ns)	$a_2(\lambda)$ ($\tau_2 = 1.48$ ns)	$a_3(\lambda)$ ($\tau_3 = 3.94$ ns)
400	1.15	0.60	0.25	1.84	—	0.10
430	0.21	1.02	0.77	0.52	1.16	0.32
450	−0.13	0.95	0.92	−0.56	0.91	0.53
470	−0.40	0.73	0.87	−0.77	0.45	0.78
490	−0.50	0.57	0.93	−0.85	0.14	1.01
510	−0.61	0.33	1.06	−0.71	−0.20	1.09
530	−0.74	0.18	1.08	−0.50	−0.42	1.07

[a]From Ref. 21.
[b]The sum of the absolute values of $\alpha_i(\lambda)$ are set to 2.0 at each wavelength.

relaxed state which can be observed without a substantial contribution from the initially excited state.[27]

The recovered impulse response functions (Table 7.1) can be used to calculate the TRES (Figure 7.10). To facilitate comparison of these spectra, we have chosen to display normalized TRES at 0.2, 2, and 20 ns. For DPPC, below its transition temperature near 37 °C, there is only a modest red shift at 20 ns, about 15 nm. At 46 °C, a dramatic spectral shift is seen to occur between 0.2 and 20 ns, about 50 nm. These TRES indicate that the extent and/or rate of spectral relaxation is greater at the higher temperature. At 2 ns and 46 °C, the time-resolved emission spectrum of Patman appears wider than at 0.2 or 20 ns. This is consistent with a two-state model for spectral relaxation. Specifically, at 46 °C and 2 ns, emission is observed from both the unrelaxed and relaxed states of Patman.

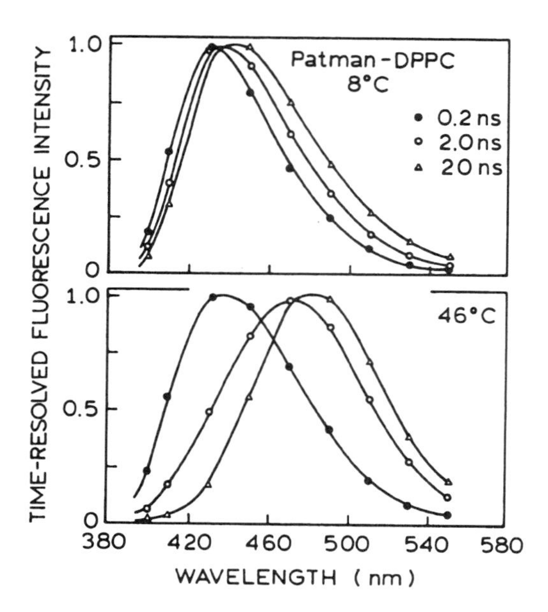

Figure 7.10. TRES of Patman-labeled DPPC vesicles at 8 (*top*) and 46 °C (*bottom*). Time-resolved spectra are shown at 0.2 (●), 2 (○), and 20 ns (△). Revised from Ref. 21, Copyright © 1984, with permission from Elsevier Science.

7.3.C. Analysis of TRES

The time-dependent spectral shifts can be characterized by the time-dependent center of gravity, in wavenumbers or kilokaysers (kK; 1 kK = 10^3 cm^{-1}). The center of gravity is proportional to the average energy of the emission. The center of gravity of the emission is defined by

$$\bar{\nu}_{cg}(t) = \frac{\int_0^\infty I'(\bar{\nu}, t)\bar{\nu}\, d\bar{\nu}}{\int_0^\infty I'(\bar{\nu}, t)\, d\bar{\nu}} \qquad [7.5]$$

where $I'(\bar{\nu},t)$ represents the number of photons per wavenumber interval. These are the intensity decays as normalized in Eq. [7.4], but on the wavenumber scale. The data are typically collected for selected wavelengths, and the center of gravity in kilokaysers is calculated using

$$\bar{\nu}_{cg}(t) = 10{,}000\, \frac{\sum_\lambda I'(\lambda, t)\lambda^{-1}}{\sum_\lambda I'(\lambda, t)} \qquad [7.6]$$

Note that the integral in Eq. [7.5] is over the emission spectrum ($\bar{\nu}$), and not over time. The TRES at any instant in time are used to calculate $\bar{\nu}_{cg}(t)$ at the chosen time. The calculated center of gravity is typically an approximation since the time-resolved data are not collected at all wavelengths. Also, a rigorous calculation of the center of gravity requires use of the corrected emission spectra on the wavenumber scale. Equation [7.6] is simply an expression which uses the available data $[I'(\lambda, t)]$ to obtain an approximate value of $\bar{\nu}_{cg}(t)$.

The time-dependent emission centers of gravity for the Patman-labeled DPPC vesicles considered in the previous section are shown in Figure 7.11. It is apparent that the extent of relaxation is greater at 46 °C than at 8 °C. The rate of relaxation is somewhat faster at the higher temperature. If desired, the values of $\bar{\nu}_{cg}(t)$ versus time can be fit to multiexponential (Eq. [7.15] below) or other nonexponential decay laws for $\bar{\nu}_{cg}(t)$.

The time-dependent spectral half-width $\Delta\bar{\nu}(t)$ (cm^{-1}) can be used to reveal whether the spectral relaxation is best described by a continuous or a two-step model. This half-width can be defined in various ways. One method is to use a function comparable to a standard deviation. In this case $\Delta\bar{\nu}(t)$ can be defined as

$$[\Delta\bar{\nu}(t)]^2 = \frac{\int_0^\infty [\bar{\nu} - \bar{\nu}_{cg}(t)]^2 I'(\bar{\nu}, t)\, d\bar{\nu}}{\int_0^\infty I'(\bar{\nu}, t)\, d\bar{\nu}} \qquad [7.7]$$

For calculation of $\Delta\bar{\nu}(t)$, one uses the TRES calculated for a chosen time t and integrates Eq. [7.7] across the emission

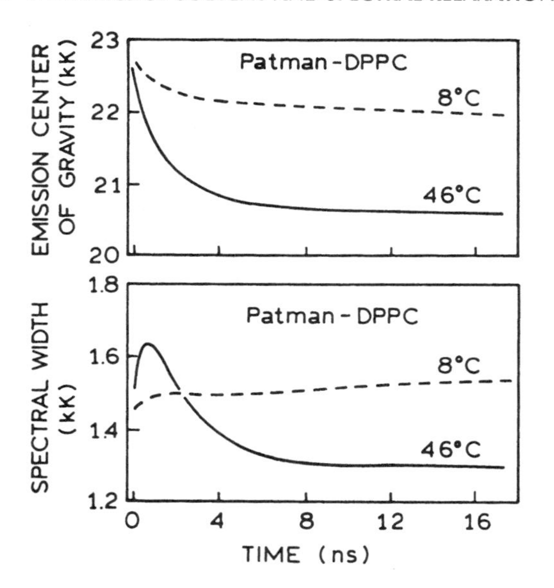

Figure 7.11. Time-resolved emission center of gravity (*top*) and spectral half-width (*bottom*) of Patman-labeled DPPC vesicles at 8 (———) and 46 °C (———) 1 kK = 1000 cm^{-1}. Revised from Ref. 21, Copyright © 1984, with permission from Elsevier Science.

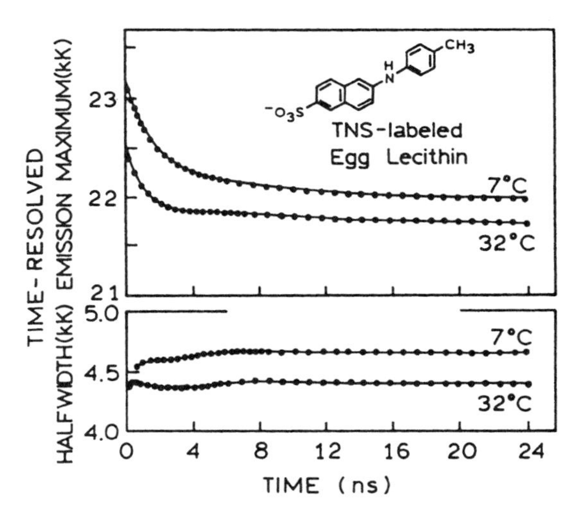

Figure 7.12. Time-resolved emission maxima and spectral half-widths for TNS-labeled egg lecithin. 1 kK = 1000 cm^{-1}. Redrawn from Ref. 17.

spectrum. For data collected at various wavelengths, the value of $\Delta\bar{\nu}(t)$ in kilokaysers is given by

$$\Delta\bar{\nu}(t) = \frac{\sum_{\lambda} [(10{,}000/\lambda) - \bar{\nu}_{cg}(t)]^2 I'(\lambda, t)}{\sum_{\lambda} I'(\lambda, t)} \qquad [7.8]$$

Examination of the time-dependent values of $\Delta\bar{\nu}(t)$ can reveal the nature of the relaxation process. For Patman in DPPC vesicles at 46 °C, there is a dramatic increase in half-width at intermediate times. This suggests that spectral relaxation around Patman is best described as a two-state process. Such behavior is not seen for all probes. For TNS-labeled vesicles,[11,17,21] the half-width remains constant during spectral relaxation (Figure 7.12). This suggests that relaxation around TNS is best described as a continuous process in which an emission spectrum of constant shape slides to longer wavelengths.

The anthroyloxy fatty acids have been extensively used as membrane probes. One advantage of these probes is that the fluorophore can be localized at the desired depth in the membrane by its point of attachment to the fatty acid (Section 9.1). The localized anthroyloxy groups have been used to study the dynamics of spectral relaxation at various depths in membranes. The TRES of these probes are sensitive to the location of the probe in the membrane.[24] This is seen by the larger time-dependent shifts for 2-(9-anthroyloxy)stearic acid (2-AS) than for 16-(9-anthroyloxy)palmitic acid (16-AP). (Figure 7.13). At first glance, this difference seems easy to interpret, with larger spectral shifts for the probe located closer to the polar membrane–water interface. However, closer inspection reveals that the emission spectra of 16-AP are more shifted to the red at early times following excitation. This is evident from the

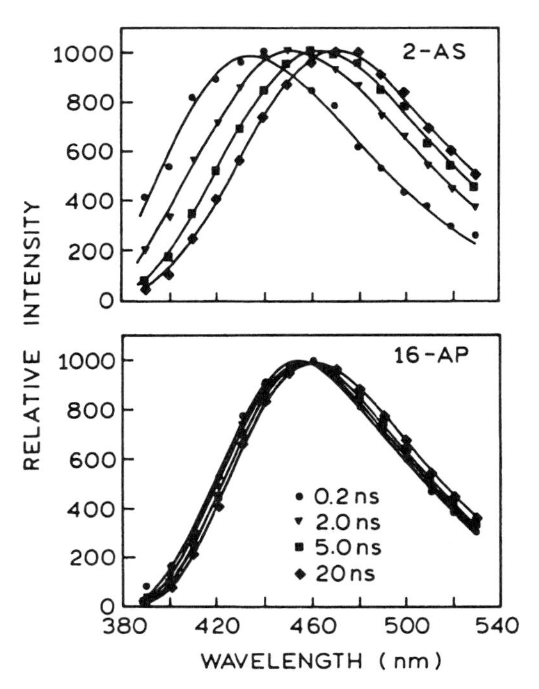

Figure 7.13. TRES of anthroyloxy fatty acids (2-AS and 16-AP) in egg phosphatidylcholine (PC) vesicles. Revised from Ref. 24.

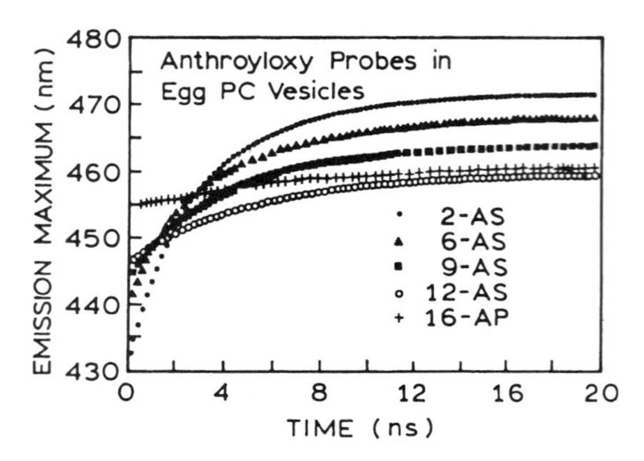

Figure 7.14. Time-dependent emission maxima of anthroyloxy fatty acids in egg PC vesicles. Revised from Ref. 24.

time-dependent emission maxima (Figure 7.14). One would expect that the probes located deeper in the bilayer should display a more blue-shifted emission. The reason for the larger red shift of 16-AP is not understood at this time. However, it is thought that the anthroyloxy probes near the ends of acyl chains can fold back to the lipid–water interface. The data in Figure 7.14 are consistent with the fluorophores in both 12-AS and 16-AP being localized near the lipid–water interface.

The rates of spectral relaxation for the anthroyloxy probes also seem to be unusual. The fluidity of lipid bilayers is thought to increase toward the center of the bilayer. Examination of the time-dependent emission maxima (Figure 7.14) suggests that the relaxation becomes slower for probes localized deeper in the membranes. This is seen more clearly in the correlation functions (Eq. [7.16] below), which display the relaxation behavior on a normalized scale (Figure 7.15). The apparent relaxation times increase from 2.7 to 5.9 ns as the distance between the point of attachment of the anthroyloxy moiety and the carboxy group increases. These results can be understood in terms of the membrane becoming more fluid near the center, if the 12-AS and 16-AP probes are located near the surface. Irrespective of the interpretation, the data demonstrated that membranes relax on the nanosecond timescale, but the details are not completely understood.

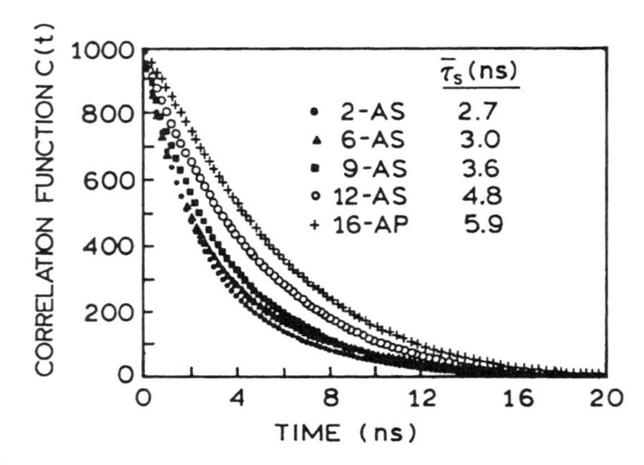

Figure 7.15. Correlation functions for spectral relaxation of the anthroyloxy fatty acids in egg PC vesicles. Revised from Ref. 24.

7.3.D. Spectral Relaxation in Proteins

In contrast to the extensive use of labeled membranes, there are relatively few studies of time-dependent spectral relaxation around tryptophan residues in proteins. The possibility of time-dependent shifts was suggested by the detection of a negative preexponential factor for one protein[28] and by the increase of mean lifetime with emission wavelength.[29] More recently, there have been measurements of TRES of tryptophan in solvents and in proteins that provide good evidence that tryptophan residues can display time-dependent spectral shifts.[30–37]

Time-dependent spectral shifts of tryptophan can be readily observed in solvents at low temperature. Figure 7.16 shows the example of NATA in isobutyl alcohol at

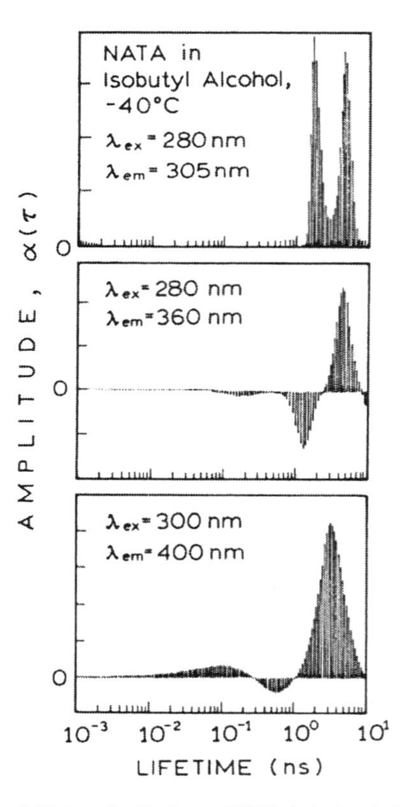

Figure 7.16. Lifetime distributions of NATA in isobutyl alcohol at –40 °C. Revised and reprinted, with permission, from Ref. 36, Copyright © 1995, American Chemical Society.

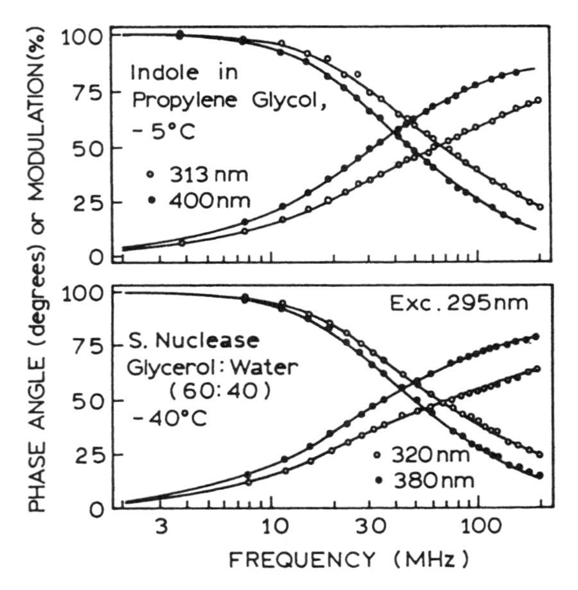

Figure 7.17. Wavelength-dependent phase and modulation data for indole in propylene glycol at –5 °C (*top*) and for staphylococcal nuclease in glycerol/water (60/40) at –40 °C (*bottom*). The frequency responses were measured at the indicated wavelengths on the blue and red sides of the emission spectra. The solid curves represent the best two-component fits (Table 7.2). Revised from Refs. 31 and 33.

–40 °C. In this case, the intensity decays were analyzed in terms of lifetime distributions with the use of the maximum entropy method.[36] The presence of an excited-state reaction is evident from the dependence of the amplitudes on emission wavelength. On the blue side of the emission (305 nm, top panel in Figure 7.16), the amplitudes are all positive. On the red side of the emission (400 nm, bottom panel), there are both positive and negative preexponential factors, proving the presence of time-dependent spectral shifts.

Spectral relaxation of indole has also been observed using the FD method.[31] In this case the frequency responses were dependent on the observation wavelength. The phase angles were larger and the modulations smaller at longer wavelengths (Figure 7.17, top). It is informative to examine the parameters recovered from the multiexponential analysis (Table 7.2). The lifetimes were taken as global parameters. The intensity decay recovered at 400 nm shows a weak negative preexponential factor (α_1 = –0.15). The negative preexponential factor do not need to be equal and opposite the positive preexponential factor to prove that an excited-state reaction is occurring. Spectral overlap of the F and R states can prevent the amplitudes from being equal and opposite. The detection of even a small negative factor proves the existence of an excited-state process. Negative preexponential factors can be masked by spectral overlap or heterogeneity in the intensity decay, so their absence does not exclude an excited-state process. The wavelength-dependent intensity decays were used to determine the TRES for indole. In propylene glycol at –5 °C, one finds time-dependent emission spectra (Figure 7.18, top). The center of gravity was found to shift to longer wavelengths with a relaxation time of 2.8 ns.

TRES have also been observed for proteins.[32–34] The single-tryptophan protein staphylococcal nuclease was examined in glycerol/water at –40 °C. These conditions were used to slow the rate of spectral relaxation so that the frequency responses showed a substantial dependence on emission wavelength (Figure 7.17, bottom). Such data, measured across the emission spectrum, were used to calculate the TRES of staphylococcal nuclease (Figure 7.18, bottom). Under these conditions of low temperature and high viscosity, the relaxation time is near 1.5 ns. In the case of staphylococcal nuclease, a negative preexponential factor was not detected at long wavelength (Table 7.2). This does not indicate the absence of time-dependent relaxation but only suggests that the extent of spectral overlap prevents observation of this component.

Because of the rapid rates of spectral relaxation, it is more difficult to measure protein TRES in aqueous solution at room temperature. However, a few measurements have been performed.[32] Adrenocorticotropin (ACTH) is a small peptide hormone in which the single tryptophan

Table 7.2. Multiexponential Intensity Decay Analysis of Indole and Staphylococcal Nuclease in the Presence of Spectral Relaxation

Sample	λ_{em} (nm)	α_1	τ_1 (ns)	α_2	τ_2 (ns)
Indole in propylene glycol, –5 °C[a]	313	0.61	1.71	0.39	5.56
	360	0.09	"	0.91	"
	400	–0.15	"	0.85	"
Staphylococcal Nuclease, 60% glycerol, –40 °C[b]	313	0.64	1.10	0.36	5.89
	340	0.31	"	0.69	"
	380	0.13	"	0.87	"

[a]Ref. 31.
[b]Ref. 33.

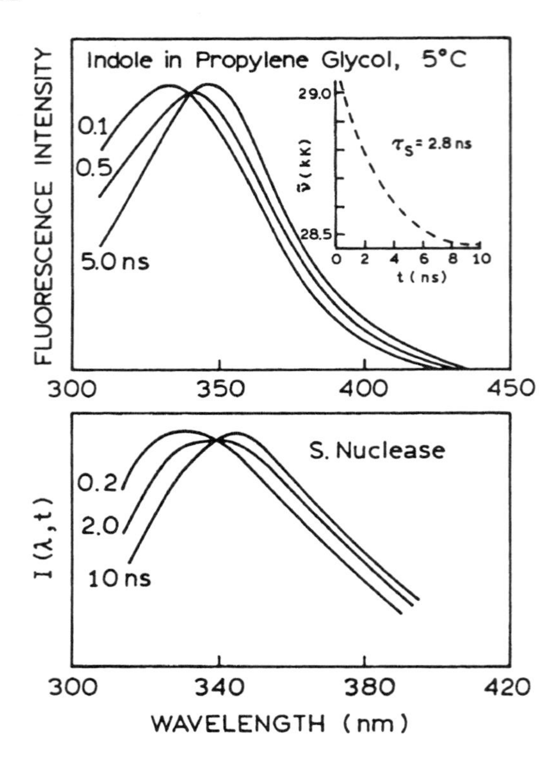

Figure 7.18. TRES of indole in propylene glycol at –5 °C (*top*) and staphylococcal nuclease in 60% glycerol at –40 °C (*bottom*). *Inset*: Time-dependent decrease in the emission center of gravity of indole. 1 kK = 1000 cm^{-1}. Revised from Refs. 31 and 34.

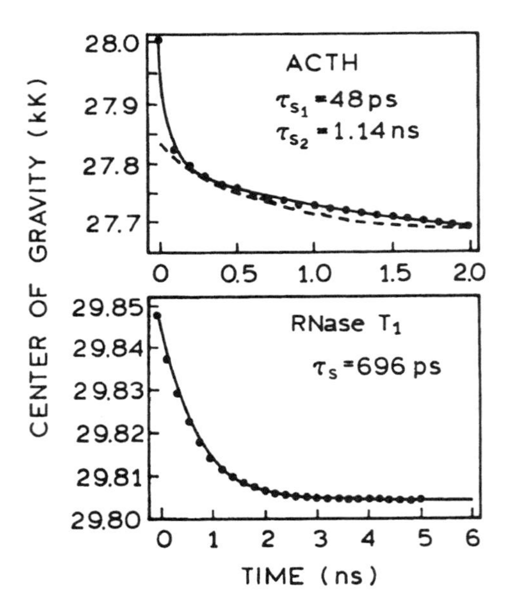

Figure 7.19. Time-resolved center of gravity for the intrinsic tryptophan emission of ACTH (*top*) and ribonuclease T_1 (*bottom*) at 20 °C. The dashed curve in the upper panel shows the best single-relaxation-time fit. 1 kK = 1000 cm^{-1}. From Ref. 32.

residue is exposed to the aqueous phase. Ribonuclease T_1 also contains a single tryptophan residue, but one that is buried in the protein and shielded from water. Both proteins display time-dependent spectral shifts to long wavelength. For ACTH, the emission center of gravity displays two relaxation times (Figure 7.19, top). For ribonuclease T_1, the spectral shift occurs with a single relaxation time (Figure 7.19, bottom). A molecular understanding of such data will depend on a detailed understanding of protein dynamics.

7.4. LIFETIME-RESOLVED EMISSION SPECTRA

• Advanced Topic •

In the preceding sections we described the relatively complex experiments needed to calculate TRES. In some circumstances it is possible to obtain information about the rate of relaxation by using only steady-state measurements. The basic idea is to use collisional quenching to decrease the lifetime of the excited state (τ) to a value comparable to the relaxation time (τ_S). As will be described in Section 7.5, the center of gravity of the emission can be expected to decay exponentially from $\bar{\nu}_0$ at $t = 0$ to $\bar{\nu}_\infty$ at $t = \infty$ according to

$$\bar{\nu}_{cg}(t) = \bar{\nu}_\infty + (\bar{\nu}_0 - \bar{\nu}_\infty)\exp(-t/\tau_S) \qquad [7.9]$$

Suppose the total intensity decays with a single decay time τ. The center of gravity observed in a steady-state emission spectrum is given by the integral average of $\bar{\nu}_{cg}(t)$ over the intensity decay,

$$\bar{\nu}_{cg} = \frac{\int_0^\infty \bar{\nu}_{cg}(t)\exp(-t/\tau)\,dt}{\int_0^\infty \bar{\nu}_{cg}(t)\,dt} \qquad [7.10]$$

Substitution of Eq. [7.9] into Eq. [7.10] yields

$$\bar{\nu}_{cg} = \bar{\nu}_\infty + (\bar{\nu}_0 - \bar{\nu}_\infty)\frac{\tau_S}{\tau_S + \tau} \qquad [7.11]$$

This expression connects the center of gravity observed in a steady-state experiment with the relative values of the decay time τ and the spectral relaxation time τ_S. From this expression one can also understand the spectral shifts observed at low and high temperatures. At low temperature, $\tau_S >> \tau$ and the center of gravity is $\bar{\nu}_0$. At high temperature, $\tau_S << \tau$, and the relaxed emission is observed. The center of gravity and/or emission spectrum is sensitive to temperature or lifetime when $\tau_S \approx \tau$.

Examination of Eq. [7.11] suggests an alternative way to measure spectral relaxation.[38–40] Suppose the lifetime τ can be changed. Then $\bar{\nu}_{cg}$ will vary depending on the relative values of τ and τ_S. One way to vary the lifetime is by collisional quenching. In the presence of a quenching agent, the fluorescence lifetime is decreased according to

$$\tau = \tau_0/(1 + K_D[Q]) \qquad [7.12]$$

where τ_0 is the lifetime in the absence of quenching, K_D is the collisional quenching constant, and [Q] is the concentration of quencher. Strictly speaking, calculation of τ in the presence of a quencher requires the separation of the static and dynamic quenching constants. Hence, the dynamic quenching constant (K_D, Chapter 8) should be used in calculations of the lifetime. For significant quenching, one requires solutions of relatively low viscosity to permit rapid diffusion of the quencher. The quenching procedure is particularly useful for biological macromolecules. This is because it permits the relaxation rates to be measured without variation of the temperature. For proteins and membranes, such temperature changes can themselves alter the relaxation rates. An additional advantage of lifetime-resolved measurements is that only steady-state measurements are required, assuming that τ_0 and K_D are known.

Lifetime-resolved emission spectra of TNS-labeled DMPC/cholesterol vesicles are shown in Figure 7.20. In this case, oxygen was used to decrease the lifetime of TNS. As the average lifetime is decreased from 6.72 to 0.53 ns, the steady-state spectra are seen to shift almost 20 nm to shorter wavelengths. Quenching of fluorescence entails a random collisional encounter between a molecule of TNS and an oxygen molecule. Those fluorophores which remain in the excited state for a longer period of time are more likely to be quenched. The longer-lived fluorophores are also those for which relaxation is more complete. Quenching selectively prevents observation of the emission from these more relaxed fluorophores and thus results in shifts of the average emission to shorter wavelengths. The lack of change in spectral shape is consistent with the continuous relaxation process shown in Figure 7.12.

These lifetime-dependent spectra can be used to measure the spectral relaxation times at the DMPC/cholesterol membranes over a range of temperatures (Figure 7.21). The solid curves through the data points represent the best fit using Eq. [7.11]. It is interesting to note that all these relaxation times fall in a narrow range, from 8.6 to 0.75 ns, and do not show evidence for a phase transition. This is probably because the TNS is localized at the lipid–water interface, where the dynamic properties are similar irrespective of the phase state of the acyl side chains.

Oxygen quenching of fluorescence is particularly useful for lifetime-resolved studies of solvent relaxation. Oxygen quenching can provide a wide range of lifetimes with little change in solvent composition. This is because oxygen diffuses rapidly in most solvents and is a small, nonpolar molecule. In addition, oxygen is highly soluble in organic

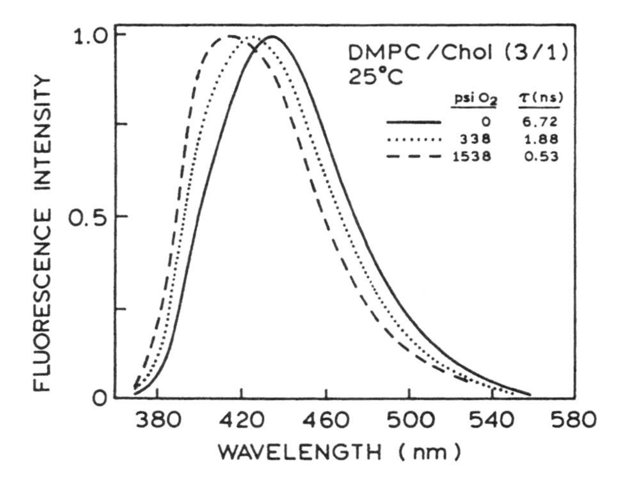

Figure 7.20. Effects of oxygen quenching on the fluorescence emission spectrum of TNS-labeled DMPC/cholesterol (3:1) vesicles. Revised and reprinted, with permission, from Ref. 40, Copyright © 1981, American Chemical Society.

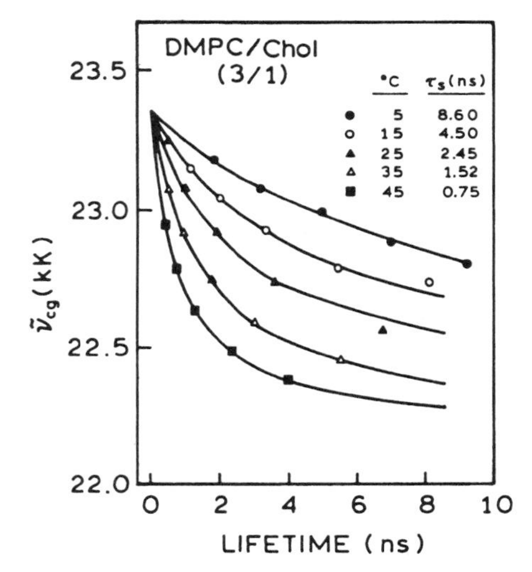

Figure 7.21. Lifetime-resolved centers of gravity for TNS-labeled DMPC/cholesterol vesicles. 1 kK = 1000 cm^{-1}. Revised and reprinted, with permission, from Ref. 40, Copyright © 1981, American Chemical Society.

solvents and is an efficient collisional quencher. These properties result in a large range of accessible lifetimes. However, oxygen quenching requires specialized pressure cells,[38] and, of course, there are dangers associated with the use of high pressures and/or organic solvents.

7.5. PICOSECOND RELAXATION IN SOLVENTS
• Advanced Topic •

In fluid solvents at room temperature, spectral relaxation is usually complete prior to emission and occurs within about 10 ps. This process is too rapid to be resolved with the usual instrumentation for TD or FD fluorescence. However, advances in laser technology and methods for ultrafast spectroscopy have resulted in an increasing interest in picosecond and femtosecond solvent dynamics.[41–48] Because of the rapid timescale, the data on solvent dynamics are usually obtained using fluorescence upconversion. This method is described in Section 4.7.C. Typical data are

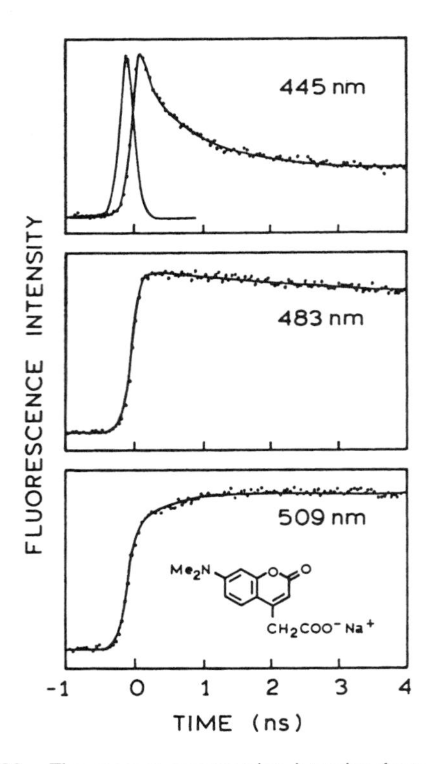

Figure 7.22. Fluorescence upconversion intensity decays of 7-(dimethylamino)coumarin 4-acetate in water. The solid curve through the points is a multiexponential fit to the intensity decay at each wavelength. The peak near zero time in the upper panel is the instrument response function (280-fs fwhm). Revised and reprinted, with permission, from Ref. 48, Copyright © 1988, American Chemical Society.

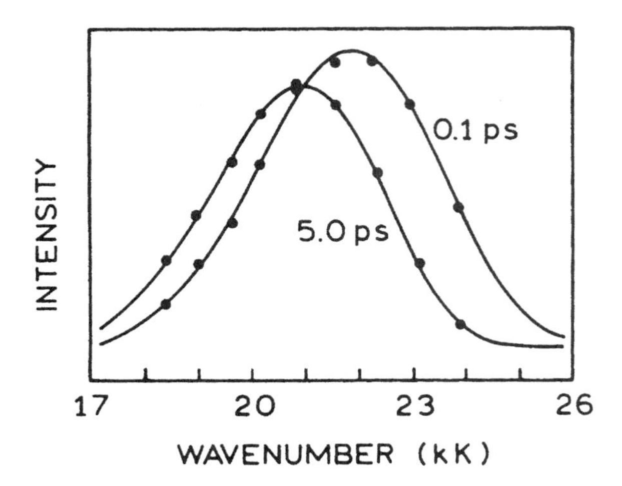

Figure 7.23. TRES of 7-(dimethylamino)coumarin 4-acetate in water at 0.1 and 5.0 ps after excitation. 1 kK = 1000 cm^{-1}. Revised and reprinted, with permission, from Ref. 48, Copyright © 1988, American Chemical Society.

shown in Figure 7.22 for a coumarin derivative. The intensity decays very quickly on the blue side of the emission and displays a rise on the long-wavelength side of the emission. The total intensity decay of this coumarin derivative, with a lifetime of 0.9 ns, is not visible on the picosecond timescale of these measurements, as evidenced by the nearly horizontal line after 1 ps in the lower panel. The wavelength-dependent upconversion data were used to reconstruct the TRES, which are shown in Figure 7.23 at 0.1 and 5.0 ps. By 5 ps, the relaxation is essentially complete, which is why time-dependent effects are usually not considered for fluorophores in fluid solution.

7.5.A. Theory for Time-Dependent Solvent Relaxation

There is no comprehensive theory which can be used to explain all time-dependent spectral shifts. This is because the spectral shifts can have their molecular origin in general solvent effects, specific solvent effects, or other excited-state processes. However, it is possible to predict the time dependence of spectral shifts due to general solvent effects. The basic idea is that the reaction field around the excited molecule relaxes in a manner that is predictable from the dielectric relaxation times (τ_D).[49–51] To be more precise, the spectral relaxation time (τ_S) is expected to be equal to the longitudinal relaxation time (τ_L). The value of τ_L is related to the dielectric relaxation time (τ_D) by

$$\tau_L = \tau_D \frac{2\varepsilon_\infty + \varepsilon_c}{2\varepsilon_0 + \varepsilon_c} \qquad [7.13]$$

Table 7.3. Dielectric Properties of Solvents

Solvent	Temperature (°C)	ε_0	ε_∞	n	τ_D (ps)	τ_L (ps)	Reference(s)
Water	25	78.3	5.2	1.33	8.3	0.4	56
Methanol	19	31.8	2.0	1.33	60	8.2	57
Ethanol	19	26.0	1.9	1.36	90	12.4	57
n-Propanol	19	21.3	1.9	1.38	320	59	57
	−20	27	5.6	—	1300	340	58–60
Glycerol	12	46.8	4.0	—	39		61
	−70	74	4.2	—	110		59

where ε_∞ is the infinite-frequency dielectric constant, ε_0 is the low-frequency dielectric constant, and ε_c is the dielectric constant of the cavity containing the fluorophore and is usually set equal to 1. The values of τ_L are often estimated using various approximations. The value of ε_∞ is often taken as n^2, where n is the refractive index, and the value of ε_0 is taken as the static dielectric constant ε. Expressions used for τ_L include[52–55]

$$\tau_L = \frac{2n^2+1}{2\varepsilon+1}\tau_D \simeq \frac{\varepsilon_\infty}{\varepsilon_0}\tau_D \simeq \frac{n^2}{\varepsilon}\tau_D \qquad [7.14]$$

Values of τ_D and τ_L for typical solvents are given in Table 7.3. For polar solvents, the spectral relaxation times are expected to be 5- to 10-fold smaller than the value of τ_D. Hence, the rate of spectral relaxation is expected to be faster than the rate of dielectric relaxation.

Unfortunately, solvent relaxation even in simple solvents is expected to be complex. This is because most solvents are known to display multiple dielectric relaxation times. For instance, n-propanol is known to display three dielectric relaxation times. Typical values are listed in Table 7.4. The largest value (τ_{D1}) is too large for overall rotation of the alcohol molecules and is attributed to rotation of the alcohol following breakage of a hydrogen bond. The second dielectric relaxation time (τ_{D2}) is assigned to rotational diffusion of the free alcohol molecule. The shortest time (τ_{D3}) is assigned to rotation of the hydroxyl group around the carbon–oxygen bond. Based on this behavior, the center of gravity of the emission spectrum of a fluorophore in n-propanol is expected to decay as

Table 7.4. Dielectric Relaxation Times of Alcohols[a]

	Temperature (°C)	τ_{D1} (ps)	τ_{D2} (ps)	τ_{D3} (ps)
Methanol	20	52	13	1.4
Ethanol	20	191	16	1.6
n-Propanol	20	430	22	2
	40	286	17	2
n-Butanol	20	670	27	2.4
n-Octanol	20	1780	38	3
	40	668	29	3

[a]Refs. 62 and 63.

$$\bar{\nu}_{cg}(t) = \bar{\nu}_\infty + (\bar{\nu}_0 - \bar{\nu}_\infty)\sum_i \beta_i \exp(-t/\tau_{Si}) \qquad [7.15]$$

where the β_i are the amplitudes associated with each relaxation time (τ_{si}). Hence, the time-dependent values of $\bar{\nu}_{cg}(t)$ can show complex multiexponential decays even in simple solvents. Complex decays of $\bar{\nu}_{cg}(t)$ were already seen for a labeled protein in Figure 7.8.

Spectral relaxation data are frequently presented as the correlation function

$$C(t) = \frac{\bar{\nu}_{cg}(t) - \bar{\nu}_\infty}{\bar{\nu}_0 - \bar{\nu}_\infty} \qquad [7.16]$$

This expression has the advantage of normalizing the extent of relaxation to unity, allowing different experiments to be compared. However, when using Eq. [7.16], one loses the information on the emission spectra at $t = 0$ and $t = \infty$. If some portion of the relaxation occurs more rapidly than the time resolution of the measurement, this portion of the energy will be missed and possibly not noticed.

7.5.B. Multiexponential Relaxation in Water

It is of interest to understand the relaxation times expected for probes in water. Such experiments have become possible because of the availability of femtosecond lasers. One such study examined 7-(dimethylamino)coumarin 4-acetate in water (Figure 7.22).[48] The excitation source was a styryl 8 dye laser, with an HITCI saturable absorber, which provided 70-fs pulses at 792 nm. These pulses were frequency-doubled for excitation at 396 nm.

The emission was detected using sum frequency upconversion with a potassium dihydrogen phosphate (KDP) crystal and the residual 792-nm light for gating the crystal. The instrument response function with an fwhm of 280 fs, allowed the intensity decay to be directly recorded (Figure 7.22). These decays showed a rapid component on the short-wavelength side of the emission (top panel), and a nearly single-exponential decay near the middle of the emission spectrum (middle panel), and a rise time on the low-energy side of the emission (bottom panel).

These data allowed construction of the TRES. The time-dependent shifts all occur within 5 ps (Figure 7.23). In this

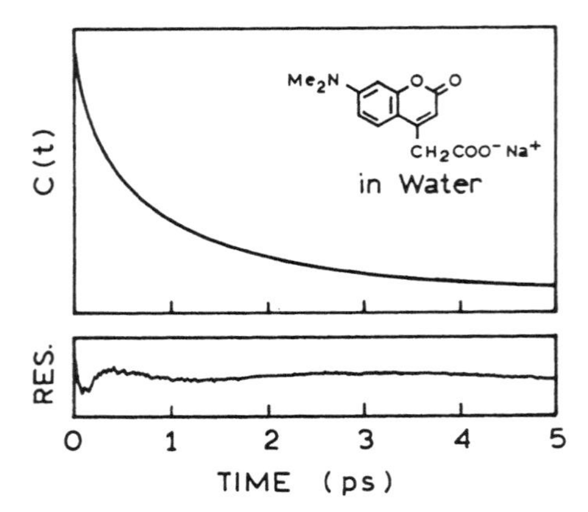

Figure 7.24. Correlation function for spectral relaxation of 7-(dimethylamino)coumarin 4-acetate in water. *Top*: Superposition of an experimentally determined $C(t)$ function for the water and a biexponential function with relaxation times of 0.16 and 1.2 ps. *Bottom*: Residuals for this fit expanded threefold. Revised and reprinted, with permission, from Ref. 48, Copyright © 1988, American Chemical Society.

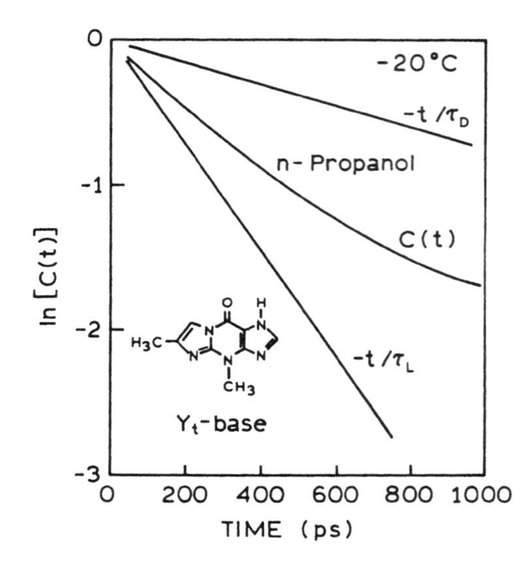

Figure 7.26. Extent of spectral relaxation for Y_t-base [$C(t)$], compared to the dielectric (τ_D) and longitudinal (τ_L) relaxation times of *n*-propanol at –20 °C. Revised from Ref. 64.

case it was necessary to use two correlation times of 0.16 and 1.2 ps to account for the time-dependent shifts (Figure 7.24). Both values are smaller than the dielectric relaxation time of water, which is near 8.5 ps (Table 7.3). Similar results have been observed for other fluorophores in other solvents. In general, spectral relaxation times are shorter than the τ_D values, but one often finds values of τ_S that are smaller or larger than the calculated values of τ_L. This is illustrated by Y_t-base in *n*-propanol at –20 °C.[64] At least two relaxation times are required to explain the time-dependent shifts (Figure 7.25). One can use the correlation

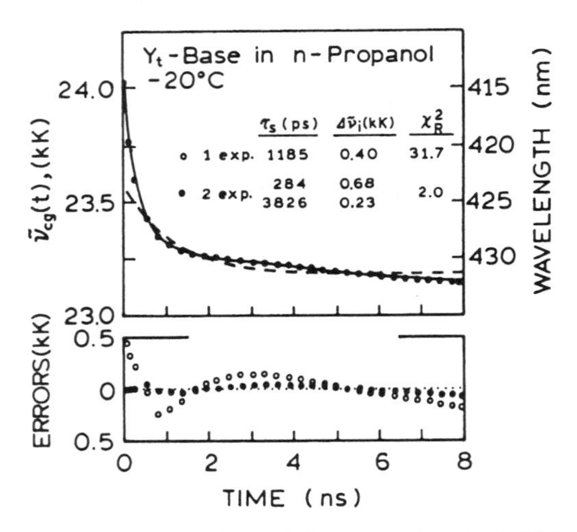

Figure 7.25. Time-dependent emission center of gravity for Y_t-base in *n*-propanol at –20 °C. ———, single-relaxation-time fit; ———, two-relaxation-time fit. 1 kK = 1000 cm^{-1}. From Ref. 64.

functions $C(t)$ to compare the measured and expected relaxation times (Figure 7.26). For Y_t-base in *n*-propanol, the decay of $C(t)$ is faster than predicted from the dielectric relaxation time τ_D, and slower than predicted from the longitudinal relaxation time τ_L. This result is typical of the findings for many fluorophores in polar solvents.[42]

7.6. COMPARISON OF CONTINUOUS AND TWO-STATE RELAXATION • Advanced Topic •

At the beginning of this chapter, we introduced the continuous and two-state models for spectral relaxation. In some instances it is possible to experimentally distinguish these limiting cases. Although this can be accomplished using the TRES, it is often informative to consider the decay times at various wavelengths.

The continuous relaxation model was formalized by Bakhshiev and co-workers.[1–3] In this model the center of gravity of the emission is assumed to decay exponentially according to Eq. [7.9]. The spectral shape of the emission is assumed to remain constant during the time course of the emission, as was illustrated in Figure 7.3. The total emission is assumed to decay exponentially with a lifetime τ. Hence, at any wavenumber, the time-resolved emission $I(\bar{\nu}, t)$ can be described as a product of an exponential decay of total fluorescence intensity and a normalized time-dependent spectral distribution, $A(\bar{\nu}, t)$:

$$I(\bar{\nu}, t) = \frac{1}{\tau} e^{-t/\tau} A(\bar{\nu}, t) \qquad [7.17]$$

The function $A(\bar{\nu}, t)$ describes the time-independent shape of the emission spectrum which shifts in time along the $\bar{\nu}$ axis. The time-resolved decays of intensity are functions of both the emission maximum at time $t[\bar{\nu}_{cg}(t)]$ and the wavenumber that is observed ($\bar{\nu}$). Calculation of the steady-state spectra for this model is moderately difficult. Such calculations require integration over all times:

$$I(\bar{\nu}) = \int_0^\infty I(\bar{\nu}, t)\, dt \qquad [7.18]$$

The presence of a time-dependent spectral shift requires numerical procedures. For the continuous model, the TRES all show the same spectral distribution and half-width, except for the shifting emission maximum.

An alternative model is the discontinuous two-state model. This is formally equivalent to the two-state reaction described in Chapter 18. The emission of the initial state is assumed to be centered at $\bar{\nu}_F$. This state is assumed to change into a relaxed state with a rate constant $k_S = 1/\tau_S$. Both states are assumed to return to the ground state with a rate constant τ^{-1}, but, in general, these rates can be unequal for the F and R states. The emission of the relaxed state is centered at $\bar{\nu}_R$.

The instantaneous or time-resolved emission spectra for the two-state model will be distinctly different from those for the continuous model, at least for the large spectral separations illustrated in Figure 7.3. For the two-state model, the emission spectra of both species will be seen at all times. At short times immediately following excitation, the emission from the F state will dominate. At long times, the emission of the R state will dominate. Unless the decay times of F and R are very different, the emission of both species will be seen at all observable times. Generally, the emission spectra of F and R will be less widely separated than shown in Figure 7.3. In this case individual emission maxima may not be observable. However, the occurrence of relaxation would be revealed by an increase in the half-width of the emission spectrum at the intermediate times.

7.6.A. Experimental Distinction between Continuous and Two-State Relaxation

In principle, the continuous and discontinuous models can be distinguished by two types of experimental data. These are TRES and wavelength-dependent lifetimes. We have already described the TRES and now focus on the wavelength-dependent intensity decays. The two-state model predicts single, constant decay times on the blue- and red-sides of the emission spectra, where a single species (F or R) is being observed. In contrast, the continuous model predicts that these blue- and red-side lifetimes change monotonically in these limiting regions. In particular, the apparent lifetime on the blue edge of the emission approaches zero, and the apparent lifetime on the red edge of the emission increases continually with increasing wavelength. Hence, for continuous relaxation one expects apparent lifetimes which change continuously with emission wavelength. For two-state relaxation, one expects regions of constant lifetime on the blue and red sides of the emission.

In practice, spectral overlap of the initially excited and equilibrium emissions can easily prevent a definitive choice between these models for spectral relaxation. In addition, sensitivity and background considerations decrease the reliability of data obtained in these spectral regions. Nonetheless, these limiting cases serve as a convenient framework within which the observed data can be considered. For instance, a specific solvent effect may result in an essentially instantaneous spectral shift, which is followed by a slower decay of the emission maximum according to the Bakhshiev model. The two-state model can be used to describe the specific interaction, while the continuous model can be used to describe the slower orientational relaxation.

7.6.B. Phase-Modulation Studies of Solvent Relaxation

Historically, phase measurements preceded the use of time-domain measurements for studies of time-dependent spectral shifts. One of the earliest reports appeared in 1965 and described the specific and general solvent effects on 3-amino-*N*-methylphthalimide (3-AP).[65] By a careful analysis of the phase and modulation data measured at various wavelengths, the authors were able to distinguish continuous from two-step relaxation. 3-AP was examined in hexane containing small quantities of pyridine. In the nonpolar solvent hexane, the polar additive forms hydrogen bonds with the amino group on 3-AP. These specific solvent effects lead to significant spectral shifts. Figure 7.27 shows the phase angles observed at various emission wavelengths or wavenumbers. At the time of this study, data were only available at a single modulation frequency, so that TRES were not available. The phase angles observed in pure hexane and in pure pyridine remain relatively constant with wavenumber. In contrast, the phase angles of 3-AP were found to be highly dependent upon wavenumber in hexane solutions containing 0.1 or 0.5% pydridine. The apparent lifetimes are also shown in Figure

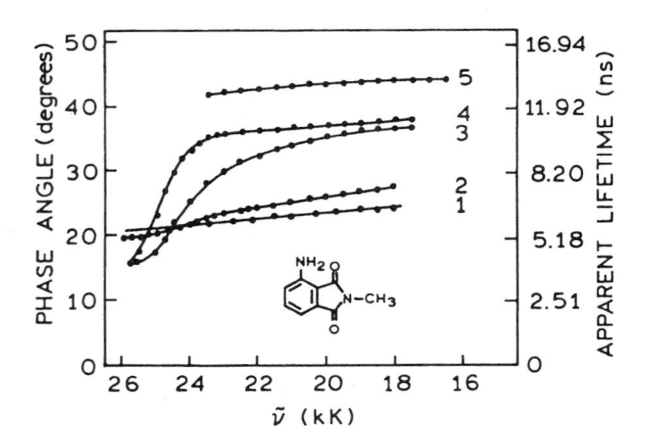

Figure 7.27. Phase angles and apparent lifetimes of 3-amino-*N*-methylphthalimide (3-AP) in pure hexane (1), hexane containing 0.01% (2), 0.1% (3), 0.5% pyridine (4), and pure pyridine (5). The modulation frequency was 11.2 MHz. 1 kK = 1000 cm^{-1}. Revised from Ref. 65.

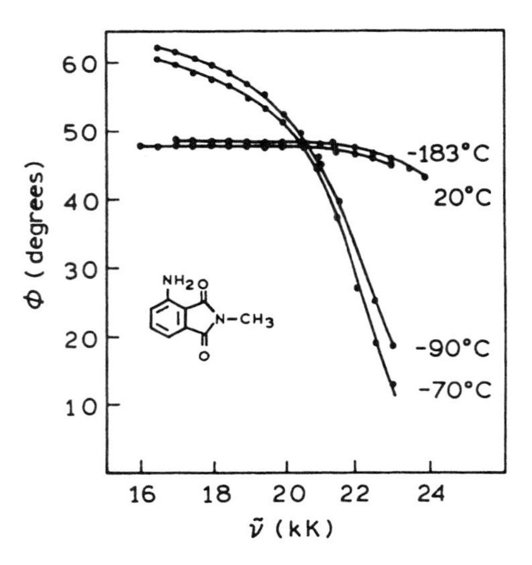

Figure 7.28. Phase angles of 3-amino-*N*-methylphthalimide in *n*-butanol at –183, –90, –70, and 20 °C. 1 kK = 1000 cm^{-1}. Revised from Ref. 65.

7.27 ($\tan \phi = \omega\tau$); one should recall that for an excited-state process, the apparent lifetimes need to be interpreted using the more complex expression described in Chapter 22, assuming that the two-state model is appropriate. In each of the pure solvents, the lifetimes are expected to be independent of emission wavelength, which is consistent with the relatively constant phase angles. In the pyridine-containing hexane solutions, a substantial dependence of phase angle on wavenumber was observed, which was attributed to complex formation between 3-AP and pyridine. The decrease in phase angle (or lifetime) on the blue edge of the spectrum provides strong evidence for an excited-state process. The form of the data, with regions of constant phase angle on the blue and red sides of the emission, is typical of a two-state process.

Contrasting results were obtained for a continuous relaxation process, which was observed for 3-AP in *n*-butanol (Figure 7.28). At low or high temperature, where relaxation either does not occur or is complete, the phase angle is constant across the emission spectrum. At intermediate temperatures (–90 and –70 °C), the phase angle changes continuously on the blue and red sides of the emission spectrum. These data are typical of a continuous relaxation process. The data in Figures 7.27 and 7.28 illustrate how one can use phase angles at various wavelengths to understand the mechanisms of solvent relaxation.

In the discussion of TRES, we noted that negative preexponential factors were proof of an excited-state process. The presence of a negative preexponential factor results in some characteristic features in the phase and modulation data.[66,67] For the product of an excited-state reaction, the phase angle can exceed 90°; in fact, the upper limit is 180°. The apparent phase (τ^{ϕ}) and modulation (τ^{m}) lifetimes also show unusual behavior. For a multiexponential intensity decay, with only positive preexponential factors, τ^{ϕ} is always smaller than τ^{m}. For an excited state reaction, $\tau^{\phi} > \tau^{m}$. This effect is seen in Figure 7.29 for time-dependent relaxation of Patman. Another characteristic is a positive

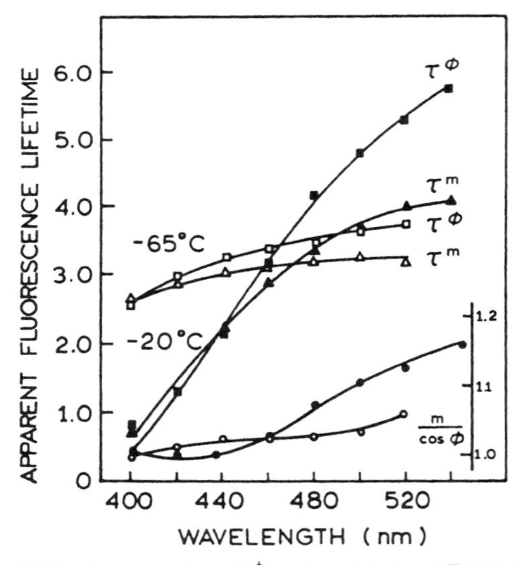

Figure 7.29. Apparent phase (τ^{ϕ}) and modulation (τ^{m}) lifetimes of Patman in propylene glycol at –20 and –65 °C. The modulation frequency was 30 MHz. Also shown are the values of $m/\cos \phi$ at –20 °C (●) and –65 °C (○). Revised and reprinted, with permission, from Ref. 66, Copyright © 1983, American Chemical Society.

value of $m/\cos\phi$, where m is the measured modulation and ϕ is the measured phase angle. For a heterogeneous decay, $m/\cos\phi \le 1.0$. Observation of $m/\cos\phi > 1.0$ is also proof of an excited-state process. These relationships are discussed in more detail in Chapter 18.

7.6.C. Distinction between Solvent Relaxation and Formation of Rotational Isomers

The nature of the TRES can yield insights into the molecular origin of the time-dependent spectral changes.[68] This concept is illustrated by studies of two closely related fluorophores, the hemicyanine dyes BABAPH and BABP (Figure 7.30). These dyes have been used as probes of membrane potential, but an understanding of their excited-state properties was not available. These dyes are known to be solvent dependent, but there was also the possibility that BABP formed rotational isomers in the excited state. In order to separate the solvent-dependent spectral shifts from formation of a polar twisted rotamer, the two closely related molecules BABP and BABAPH were examined. The BABAPH derivative contains an ethylene bridge which prevents rotation of the dibutylamino group in the ground or excited state.

Time-resolved decays were recorded at single wavelengths and analyzed in terms of the multiexponential model. These decays were then used to calculate the TRES. The normalized TRES show that BABP displays a larger time-dependent shift than does the restrained analog BABAPH (Figure 7.31). Importantly, the shape of the TRES was time-dependent for BABP but was independent of time for BABAPH. This result was interpreted in terms of formation of a polar twisted state for BABP. Formation of the twisted state also contributed to the rate of deactivation, resulting in more rapid intensity decays for BABP than for BABAPH (Figure 7.32, lower panel).

The TRES were also used to determine the time-dependent center of gravity $[\bar{\nu}_{cg}(t)]$ and the time-dependent half-width $[\Delta\bar{\nu}(t)]$. The procedure used was different from that described in Section 7.3. In this case the values of the emission maxima and spectral width (fwhm) were determined by nonlinear least-squares fitting of the spectral shape of the TRES. The spectral shape was assumed to follow a lognormal line shape,

$$I(\bar{\nu}) = I_0 \exp\left\{-\left[\ln 2\left(\frac{\ln(\alpha+1)}{b}\right)^2\right]\right\} \tag{7.19}$$

with

$$\alpha = \frac{2b(\bar{\nu} - \bar{\nu}_{max})}{\Delta} > -1$$

where I_0 is amplitude, $\bar{\nu}_{max}$ is the wavenumber of the emission maximum, and the spectral width (fwhm) is given by

BABAPH: $(C_4H_9)_2N$–[2-azaphenanthrene]–N^+–$(CH_2)_4$–SO_3^-

BABP: $(C_4H_9)_2N$–[phenyl]–[pyridinium]–N^+–$(CH_2)_4$–SO_3^-

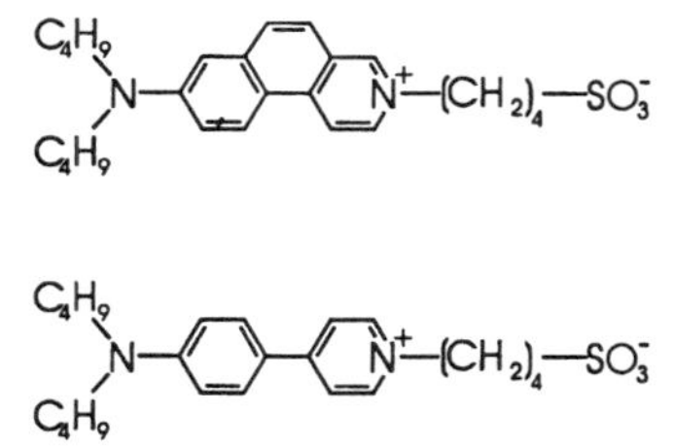

Figure 7.30. Hemicyanine dyes 2-(sulfonatobutyl)-7-(dibutylamino)-2-azaphenanthrene (BABAPH) and 1-(sulfonatobutyl)-4-[4′-(dibutylamino)phenyl]pyridine (BABP). Reprinted, with permission, from Ref. 68, Copyright © 1996, American Chemical Society.

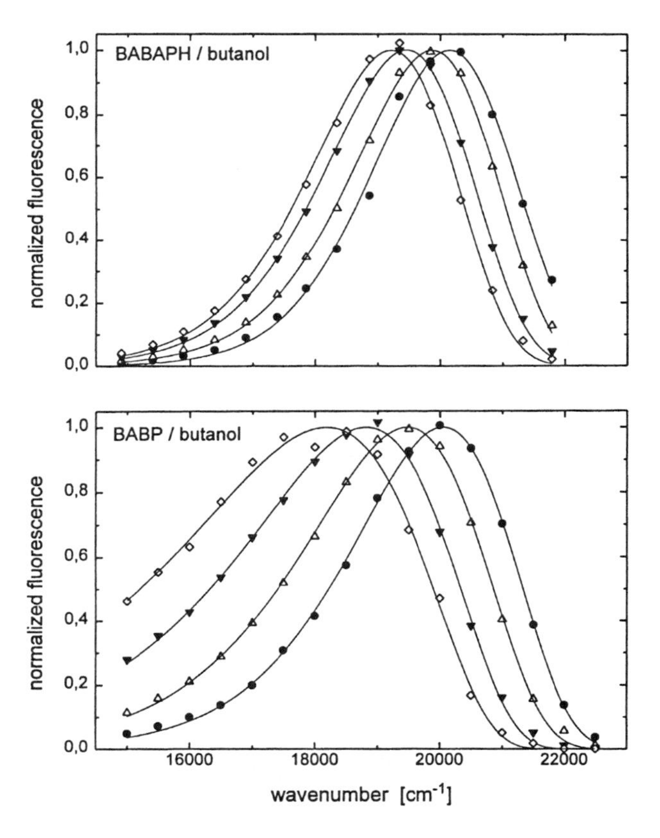

Figure 7.31. Normalized TRES of BABAPH (*top*) and BABP (*bottom*) in butanol, 25 °C, at time delays of 20 (○), 50 (▼), 150 (△), and 2000 ps (●) after excitation. The data are fitted by the lognormal spectral-shape functions. Reprinted, with permission, from Ref. 68, Copyright © 1996, American Chemical Society.

$$\Gamma = \Delta\left(\frac{\sinh(b)}{b}\right) \qquad [7.20]$$

The term b is an asymmetry parameter, and Eq. [7.19] reduces to a Gaussian function for $b = 0$.

The results of least-squares fitting of the TRES line shapes are shown in Figure 7.32. The dramatic difference in spectral shape between BABP and BABAPH is seen in the middle panel. The spectral width of the restrained analog BABAPH is independent of time, whereas BABP shows an increase in spectral width with time. The data are consistent with rotation of the dibutylamino group in BABP to allow a larger charge separation than in BABAPH. The larger charge separation results in a new species emitting at longer wavelengths, and thus the increase in spectral half-width of BABAPH. In this case the TRES provided an explanation for the different spectral shifts displayed by these similar molecules.

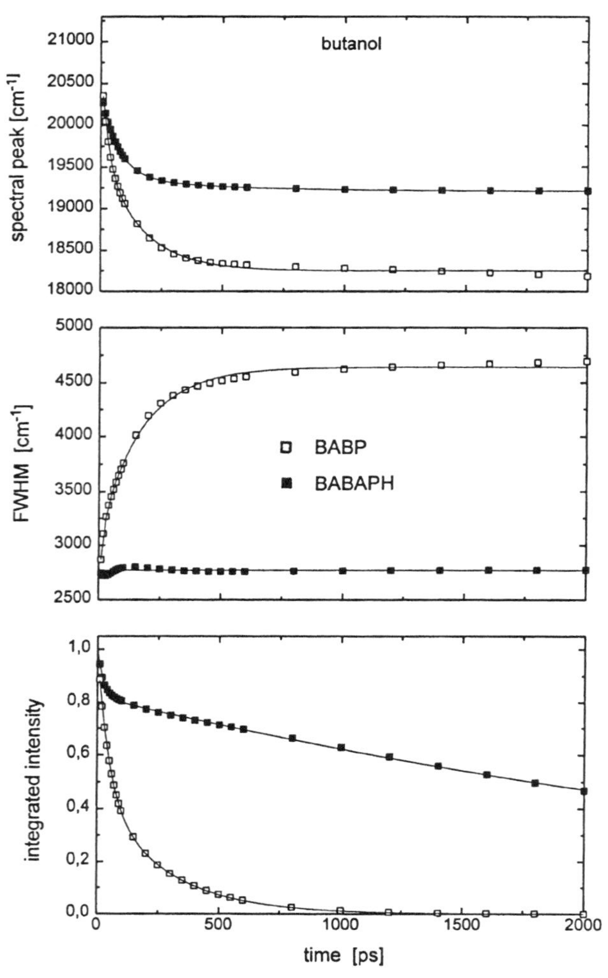

Figure 7.32. Fluorescence dynamics of BABAPH (■) and BABP (□) in butanol at 25 °C. *Top*: Position of spectral maximum. *Middle*: Spectral width (full width at half-maximum). *Bottom*: Integrated intensity normalized to its initial value. Reprinted, with permission, from Ref. 68, Copyright © 1996, American Chemical Society.

7.7. COMPARISON OF TRES AND DAS

In Chapters 4 and 5 we described the use of time-resolved data to calculate the decay-associated spectra (DAS). One may ask how the DAS are related to the TRES. There is no direct relationship between these two types of calculated spectra. This can be seen by examination of the TRES and DAS for a mixture of fluorophores in Figure 7.33. This mixture contained anthracene ($\tau = 4.1$ ns) and 9-cyanoanthracene (9-CA, $\tau = 11.7$ ns).[69] The TRES show contributions from anthracene and 9-CA at all times. At shorter times (approximately 5 ns), the contribution from anthracene is larger than in the steady-state spectrum. At longer times (approximately 20 ns), the relative contribution of the shorter-lived anthracene becomes smaller.

The DAS for this mixture (Figure 7.33, bottom) have a completely different meaning. These calculated spectra

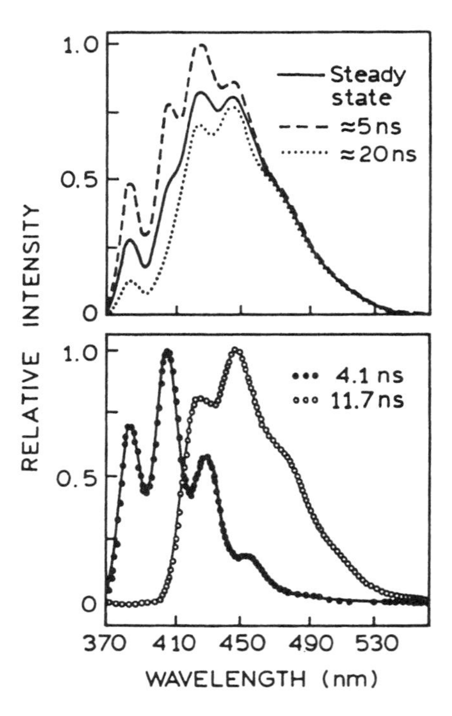

Figure 7.33. *Top*: Steady-state (———) and time-resolved emission spectra of a mixture of anthracene and 9-cyanoanthracene. The TRES are shown for an early time window near 5 ns and a late time window near 20 ns. *Bottom*: Decay-associated spectra for the 4.1-ns (●) and 11.7-ns (○) components. Revised and reprinted, with permission, from Ref. 69, Copyright © 1982, American Chemical Society.

correspond to the emission spectra of the individual components. Also, the DAS are not dependent on time. Interpretation of the DAS is straightforward for a mixture of fluorophores. However, for time-dependent spectral relaxation the DAS can be positive or negative and hence do not correspond to the emission spectra of any particular component. It is possible to calculate species-associated spectra, which reflect the emission spectra of the components with each decay time (Chapter 18).

7.8. RED-EDGE EXCITATION SHIFTS • Advanced Material •

In all the preceding discussions we assumed that the emission spectra were independent of the excitation wavelength. This is a good assumption for fluorophores in fluid solvents. However, this assumption is no longer true in viscous and moderately viscous solvents. For polar fluorophores under conditions where solvent relaxation is not complete, emission spectra shift to longer wavelengths when the excitation is on the long-wavelength edge of the absorption spectrum. This effect can be quite substantial, as is shown for 3-AP in Figure 7.34. This is known as a red-edge excitation shift and has been observed in a number of laboratories for a variety of fluorophores.[70–76]

What is the origin of this unusual behavior? The behavior of polar molecules with red-edge excitation can be understood by considering a Jabłónski diagram which takes into account spectral relaxation (Figure 7.35). Suppose that the fluorophore is in a frozen solvent and that the sample is excited in the center of the last absorption band (λ_C) or on the red edge (λ_R). For excitation at λ_C, the usual emission from the F state is observed. However, excitation at λ_R selects for those fluorophores which have absorption

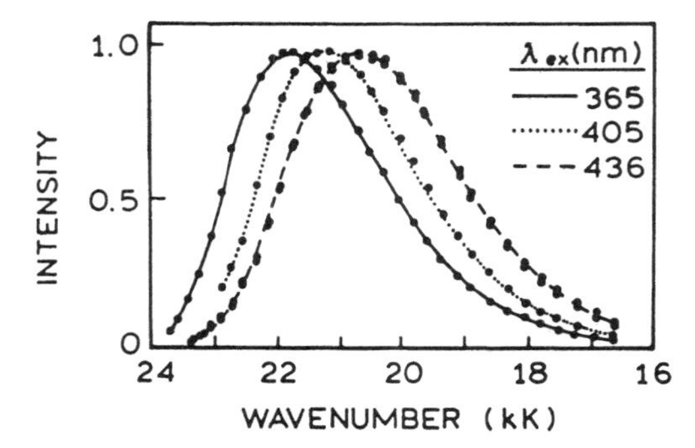

Figure 7.34. Emission spectra of 3-amino-*N*-methylphthalimide in isobutanol at –100 °C. The sample was excited at 365 nm (———), 405 nm (···), and 436 nm (– – –). At 20 °C, all excitation wavelengths yielded the same emission spectrum (– – –). 1 kK = 1000 cm^{-1}. Revised from Ref. 70.

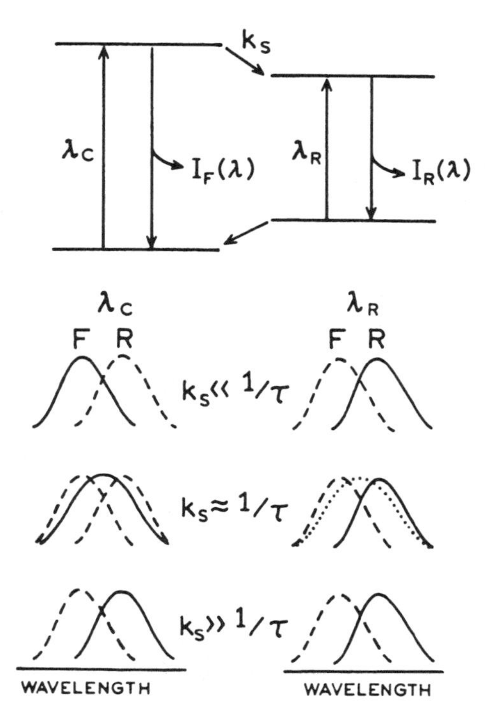

Figure 7.35. Effects of red-edge excitation on emission spectra. The term λ_C indicates excitation in the is center of the last absorption band of the fluorophore. The term λ_R indicates excitation on the red edge of the absorption band. The solid curves represent the observed spectra. To allow comparison, the dashed curves represent the emission spectra of either the F or the R states. The dotted curve (middle panel) represents a possible consequence of reverse relaxation. Revised and reprinted, with permission, from Ref. 77, Copyright © 1984, American Chemical Society.

at lower energy. In any population of molecules in frozen solution, there are some fluorophores which have a solvent configuration which is equivalent to that of the relaxed state. These fluorophores are typically more tightly hydrogen-bonded to the solvent and thus display a red-shifted emission. In frozen solution the fluorophore–solvent configuration persists during the intensity decay, so that the emission is red-shifted.

Knowing the molecular origin of the red-edge shifts, one can easily understand the effects of increasing temperature. It is known that the red-edge shifts disappear in fluid solvents. This is because there is rapid reorientation of the solvent. Hence, even if red-emitting fluorophores are initially excited, the excited fluorophores reach equilibrium with the solvent prior to emission. This implies that there may be some reverse relaxation from the red-shifted emission to the equilibrium condition, as is shown in the middle panel of Figure 7.35 for $k_S \simeq \tau^{-1}$. In fact, relaxation to higher energies has been observed with red-edge excitation.

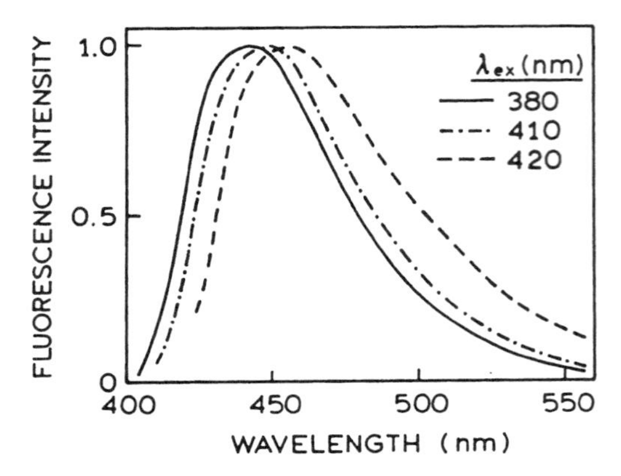

Figure 7.36. Fluorescence emission spectra of 1,8-ANS in 1-propanol at 77 K for various excitation wavelengths. The excitation and the emission bandpass are both 5 nm. Revised from Ref. 74.

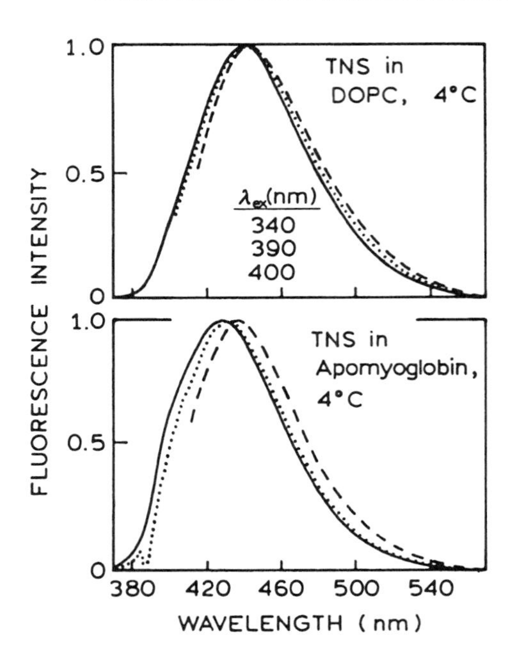

Figure 7.38. Emission spectra of TNS-labeled DOPC vesicles (*top*) and TNS-labeled apomyoglobin (*bottom*) for excitation wavelengths of 340 (——), 390 (· · ·), and 400 nm (-----). Revised and reprinted, with permission, from Ref. 77, Copyright © 1984, American Chemical Society.

Red-edge excitation has been applied to biochemically relevant fluorophores. The widely used probe 1,8-ANS displays a substantial red shift (Figure 7.36). As the excitation wavelength is increased, the emission maxima converge to the value typical of the relaxed emission (Figure 7.37). This suggests that excitation red shifts can be used to estimate biopolymer dynamics.[77–79] Excitation red shifts should only be observed if relaxation is not complete. Also, the magnitude of the red shifts will depend on the rate of spectral relaxation. Excitation red shifts have been observed for labeled proteins and labeled membranes[79–81] and for the intrinsic tryptophan fluorescence of proteins.[33,82] Examples are shown in Figure 7.38 for TNS-labeled vesicles and TNS bound to apomyoglobin. In both cases red-edge shifts are seen, which demonstrate that the probe is not in equilibrium prior to emission. This implies that the equilibrium state starting from the completely red-shifted state cannot be reached during the excited-state lifetime. Of course, we know this from the TRES, but much of the work on red-edge shifts was done when TRES were not readily available.

As described above, red-edge excitation is expected to excite fluorophores that are already in the relaxed state. Hence, there should be smaller time-dependent spectral shifts with red-edge excitation than with excitation in the center-of-gravity absorption band. The time-dependent shifts have been found to be smaller for red-edge excitation.[83–87] This is shown in Figure 7.39 for 1-phenylnaphthylamine (1-AN) in glycerol and in Figure 7.40 for 1-AN bound to egg PC vesicles. For excitation at 337 nm, the TRES display the expected time-dependent shifts. For red-edge excitation (390 or 416 nm), the emission spectra

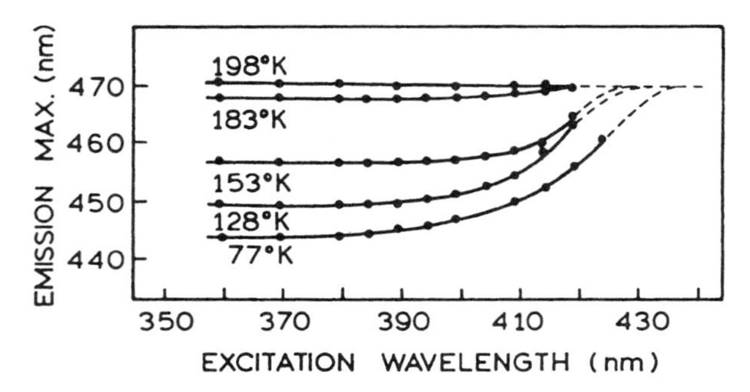

Figure 7.37. Temperature dependence of fluorescence excitation red shift for 1,8-ANS in 1-propanol. Revised from Ref. 74.

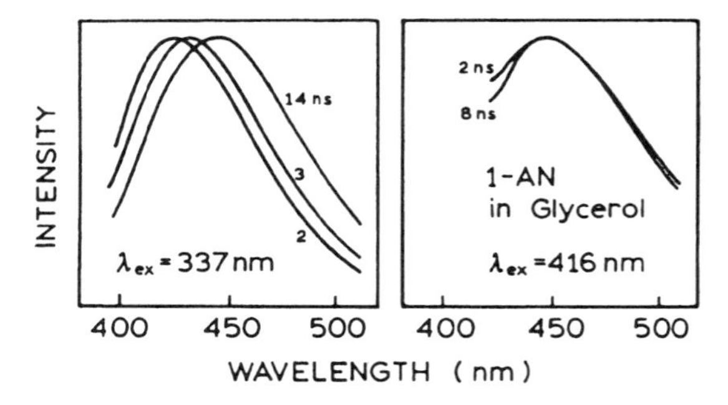

Figure 7.39. TRES of 1-AN in glycerol at different excitation wavelengths. *Left*: 2 ns, 3 ns, and 14 ns after excitation; *right*: 2 ns and 8 ns after excitation. Revised from Ref. 87.

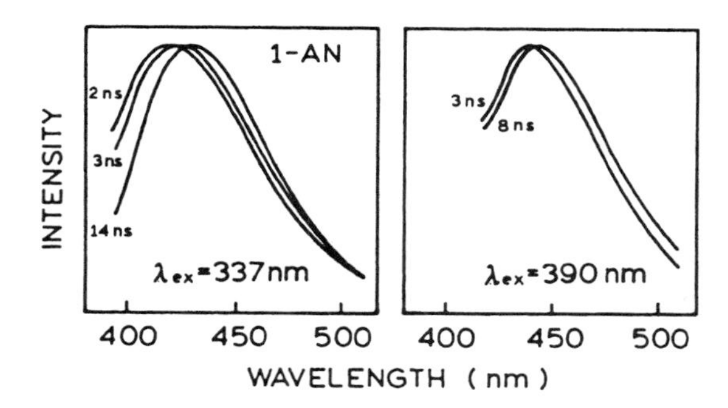

Figure 7.40. TRES of 1-AN in phosphatidylcholine vesicles at different excitation wavelengths. *Left*: 2 ns, 3 ns, and 14 ns after excitation; *right*: 3 ns and 8 ns after excitation. From Ref. 87.

do not shift with time, and the TRES are already red-shifted at the earliest observation times. Similarly, the extent of spectral relaxation of staphylococcal nuclease was smaller with red-edge excitation at 302 nm than with excitation at 295 nm.[33] These results are consistent with the notion that red-edge excitation results in excitation of prerelaxed fluorophores.

Red-edge excitation shifts can explain some unusual observations. There were early reports which suggested the failure of RET on red-edge excitation.[88–90] These studies also reported a decrease in the rate of rotational diffusion upon red-edge excitation. This "new red-edge effect" was explained as being due to excitation of an out-of-plane transition which was not capable of RET. In retrospect, these data are all understandable in terms of the red-edge effect shown in Figure 7.35. The decrease in homotransfer was probably due to the spectral shifts which decreased the overlap integral for homotransfer. The decrease in the rate of rotation can be explained as due to increased hydrogen bonding to the solvent. It is known that the rate of rotational diffusion decreases with increasing numbers of fluorophore–solvent hydrogen bonds.[91] Hence, shifts in the emission spectra can explain the failure of RET with red-edge excitation.

7.9. PERSPECTIVES OF SOLVENT DYNAMICS

In biochemical fluorescence there have been relatively few studies of time-dependent spectral changes. However, it is clear that the TRES contain considerable molecular information about macromolecule dynamics. The TRES are becoming increasingly available from time-resolved instrumentation which simultaneously collects data over a range of wavelengths. The detailed studies of spectral relaxation in solvents are providing increased understanding of the molecular basis of spectral relaxation. Hence, it seems clear that measurements of spectral relaxation will find increased use in studies of biopolymer dynamics.

REFERENCES

1. Bakhshiev, N. G., Mazurenko, Yu. T., and Piterskaya, I. V., 1966, Luminescence decay in different portions of the luminescence spectrum of molecules in viscous solution, *Opt. Spectrosc.* **21**:307–309.
2. Mazurenko, Yu. T., and Bakshiev, N. K., 1970, Effect of orientation dipole relaxation on spectral, time, and polarization characteristics of the luminescence of solutions, *Opt. Spectrosc.* **28**:490–494.
3. Bakhshiev, N. K., Mazurenko, Yu. T., and Piterskaya, I. V., 1969, Relaxation effects in the luminescence characteristics of viscous solutions, *Akad. Nauk SSSR, Bull. Phys. Sci.* **32**:1262–1266.
4. Ware, W. R., Lee, S. K., Brant, G J., and Chow, P. P., 1970, Nanosecond time-resolved emission spectroscopy: Spectral shifts due to solvent–excited solute relaxation, *J. Chem. Phys.* **54**:4729–4737.
5. Ware, W. R., Chow, P., and Lee, S. K., 1968, Time-resolved nanosecond emission spectroscopy: Spectral shifts due to solvent–solute relaxation, *Chem. Phys. Lett.* **2**:356–358.
6. Chakrabarti, S. K., and Ware, W. R., 1971, Nanosecond time-resolved emission spectroscopy of 1-anilino-8-naphthalene sulfonate, *J. Chem. Phys.* **55**:5494–5498.
7. Easter, J. H., DeToma, R. P., and Brand, L., 1976, Nanosecond time-resolved emission spectroscopy of a fluorescence probe adsorbed to L-α-egg lecithin vesicles, *Biophys. J.* **16**:571–583.
8. O'Connor, D. V., and Phillips, D., 1984, *Time-Correlated Single Photon Counting*, Academic Press, New York, pp. 211–251.
9. Badea, M. G., and Brand, L., 1979, Time-resolved fluorescence measurements, *Methods. Enzymol.* **61**:378–425.
10. Badea, M. G., De Toma, R. P., and Brand, L., 1978, Nanosecond relaxation processes in liposomes, *Biophys. J.* **24**:197–212.
11. Brand, L., and Gohlke, J. R., 1971, Nanosecond time-resolved fluorescence spectra of a protein–dye complex, *J. Biol. Chem.* **246**:2317–2324.
12. Gafni, A., De Toma, R. P., Manrow, R. E., and Brand, L., 1977, Nanosecond decay studies of a fluorescence probe bound to apomyoglobin, *Biophys. J.* **17**:155–168.
13. Lakowicz, J. R., and Cherek, H., 1981, Proof of nanosecond timescale relaxation in apomyoglobin by phase fluorometry, *Biochem. Biophys. Res. Commun.* **99**:1173–1178.
14. Lakowicz, J. R., Gratton, E., Cherek, H., Maliwal, B. P., and Laczko, G., 1984, Determination of time-resolved fluorescence emission spectra and anisotropies of a fluorophore–protein complex using frequency-domain phase-modulation fluorometry, *J. Biol. Chem.* **259**:10967–10972.
15. Pierce, D. W., and Boxer, S. G., 1992, Dielectric relaxation in a protein matrix, *J. Phys. Chem.* **96**:5560–5566.
16. Wang, R., Sun, S., Bekos, E. J., and Bright, F. V., 1995, Dynamics surrounding Cys-34 in native, chemically denatured, and silica-adsorbed bovine serum albumin, *Anal. Chem.* **67**:149–159.
17. Demchenko, A. P., Apell, H. -J., Stürmer, W., and Feddersen, B., 1993, Fluorescence spectroscopic studies on equilibrium dipole-relaxational dynamics of Na,K-ATPase, *Biophys. Chem.* **48**:135–147.

18. Easter, J. H., and Brand, L., 1973, Nanosecond time-resolved emission spectroscopy of a fluorescence probe bound to L-α-egg lecithin vesicles, *Biochem. Biophys. Res. Commun.* **52:**1086–1092.
19. Easter, J. H., DeToma, R. P., and Brand, L., 1978, Fluorescence measurements of environmental relaxation at the lipid–water interface region of bilayer membranes, *Biochim. Biophys. Acta* **508:**27–38.
20. DeToma, R. P., Easter, J. H., and Brand, L., 1976, Dynamic interactions of fluorescence probes with the solvent environment, *J. Am. Chem. Soc.* **98:**5001–5007.
21. Lakowicz, J. R., Cherek, H., Laczko, G., and Gratton, E., 1984, Time-resolved fluorescence emission spectra of labeled phospholipid vesicles, as observed using multifrequency phase-modulation fluorometry, *Biochim. Biophys. Acta* **777:**183–193.
22. Parasassi, T., Conti, F., and Gratton, E., 1986, Time-resolved fluorescence emission spectra of laurdan in phospholipid vesicles by multifrequency phase and modulation fluorometry, *Cell. Mol. Biol.* **32:**103–108.
23. Sommer, A., Paltauf, F., and Hermetter, A., 1990, Dipolar solvent relaxation on a nanosecond time scale in ether phospholipid membranes as determined by multifrequency phase and modulation fluorometry, *Biochemistry* **29:**11134–11140.
24. Hutterer, R., Schneider, F. W., Lanig, H., and Hof, M., 1997, Solvent relaxation behaviour of *n*-anthroyloxy fatty acids in PC-vesicles and paraffin oil: A time-resolved emission spectra study, *Biochim. Biophys. Acta* **1323:**195–207.
25. Stryer, L., 1965, The interactions of a naphthalene dye with apomyoglobin and apohemoglobin. A fluorescent probe of non-polar binding sites, *J. Mol. Biol.* **13:**482–495.
26. Murakami, H., and Kushida, T., 1994, Fluorescence properties of Zn-substituted myoglobin, *J. Lumin.* **58:**172–175.
27. Lakowicz, J. R., and Balter, A., 1982, Theory of phase-modulation fluorometry for excited state processes, *Biophys. Chem.* **16:**99–115.
28. Grinvald, A., and Steinberg, I. Z., 1974, Fast relaxation processes in a protein revealed by the decay kinetics of tryptophan fluorescence, *Biochemistry* **13:**5170–5178.
29. Lakowicz, J. R., and Cherek, H., 1980, Dipolar relaxation in proteins on the nanosecond timescale observed by wavelength-resolved phase fluorometry of tryptophan fluorescence, *J. Biol. Chem.* **255:**831–834.
30. Lakowicz, J. R., and Balter, A., 1982, Resolution of initially excited and relaxed states of tryptophan fluorescence by differential-wavelength deconvolution of time-resolved fluorescence decays, *Biophys. Chem.* **15:**353–360.
31. Lakowicz, J. R., Szmacinski, H., and Gryczynski, I., 1988, Picosecond resolution of indole anisotropy decays and spectral relaxation by 2 GHz frequency-domain fluorometry, *Photochem. Photobiol.* **47:**31–41.
32. Szmacinski, H., Lakowicz, J. R., and Gryczynski, I., unpublished observations.
33. Demchenko, A. P., Gryczynski, I., Gryczynski, Z., Wiczk, W., Malak, H., and Fishman, M., 1993, Intramolecular dynamics in the environment of the single tryptophan residue in staphylococcal nuclease, *Biophys. Chem.* **48:**39–48.
34. Demchenko, A. P., 1994, Protein fluorescence, dynamics and function: Exploration of analogy between electronically excited and biocatalytic transition states, *Biochim. Biophys. Acta* **1209:**149–164.
35. Vekshin, N., Vincent, M., and Gallay, J., 1992, Excited-state lifetime distributions of tryptophan fluorescence in polar solutions. Evidence for solvent exciplex formation, *Chem. Phys. Lett.* **199:**459–464.
36. Vincent, M., Gallay, J., and Demchenko, A. P., 1995, Solvent relaxation around the excited state of indole: Analysis of fluorescence lifetime distributions and time-dependence spectral shifts, *J. Phys. Chem.* **99:**14931–14941.
37. Vincent, M., Gallay, J., and Demchenko, A. P., 1997, Dipolar relaxation around indole as evidenced by fluorescence lifetime distributions and time-dependence spectral shifts, *J. Fluoresc.* **7:**107S–110S.
38. Weber, G., and Lakowicz, J. R., 1973, Subnanosecond solvent relaxation studies by oxygen quenching of fluorescence, *Chem. Phys. Lett.* **22:**419–423.
39. Rotkiewicz, K., Grabowski, Z. R., and Jasny, J., 1975, Picosecond isomerization kinetics of excited *p*-dimethylaminobenzonitriles studied by oxygen quenching of fluorescence, *Chem. Phys. Lett.* **34:**55–100.
40. Lakowicz, J. R., and Hogen, D., 1981, Dynamic properties of the lipid–water interface of model membranes as revealed by lifetime-resolved fluorescence emission spectra, *Biochemistry* **20:**1366–1373.
41. Fleming, G. R., and Cho, M., 1996, Chromophore–solvent dynamics, *Annu. Rev. Phys. Chem.* **47:**109–134.
42. Castner, E. W., Maroncelli, M., and Fleming, G. R., 1987, Subpicosecond resolution studies of solvation dynamics in polar aprotic and alcohol solvents, *J. Chem. Phys.* **86:**1090–1097.
43. Su, S.-G., and Simon, J. D., 1987, Solvation dynamics in ethanol, *J. Phys. Chem.* **91:**2693–2696.
44. Simon, J. D., 1988, Time-resolved studies of solvation in polar media, *Acc. Chem. Res.* **21:**128–134.
45. Chapman, C. F., Fee, R. S., and Maroncelli, M., 1995, Measurements of the solute dependence of solvation dynamics in 1-propanol: The role of specific hydrogen-bonding interactions, *J. Phys. Chem.* **99:**4811–4819.
46. Kahlow, M. A., Jarzeba, W., Kang, T. J., and Barbara, P. F., 1989, Femtosecond resolved solvation dynamics in polar solvents, *J. Chem. Phys.* **90:**151–158.
47. Maroncelli, M., and Fleming, G. R., 1987, Picosecond solvation dynamics of coumarin 153: The importance of molecular aspects of solvation, *J. Chem. Phys.* **86:**6221–6239.
48. Jarzeba, W., Walker, G. C., Johnson, A. E., Kahlow, M. A., and Barbara, P. F., 1988, Femtosecond microscopic solvation dynamics of aqueous solutions, *J. Phys. Chem.* **92:**7039–7041.
49. Bagchi, B., Oxtoby, D. W., and Fleming, G. R., 1984, Theory of the time development of the Stokes shift in polar media, *Chem. Phys.* **86:**257–267.
50. Castner, E. W., Fleming, G. R., and Bagchi, B., 1988, Influence of non-Debye relaxation and of molecular shape on the time-dependence of the Stokes shift in polar media, *Chem. Phys. Lett.* **143:**270–276.
51. Castner, E. W., Bagchi, B., Maroncelli, M., Webb, S. P., Ruggiero, A. J., and Fleming, G. R., 1988, The dynamics of polar solvation, *Ber. Bunsenges. Phys. Chem.* **92:**363–372.
52. Bakhshiev, N. G., 1964, Universal intermolecular interactions and their effect on the position of the electronic spectra of molecules in two-component solutions, VII. Theory (general case of an isotropic solution), *Opt. Spectrosc.* **16:**446–451.
53. Castner, E. W., Fleming, G. R., Bagchi, B., and Maroncelli, M., 1988, The dynamics of polar solvation: Inhomogeneous dielectric continuum models, *J. Chem. Phys.* **89:**3519–3534.
54. Piterskaya, I. V., and Bakhshiev, N. G., 1963, Quantitative investigation of the temperature dependence of the absorption and fluorescence spectra of complex molecules, *Akad. Nauk SSSR, Bull. Phys. Sci.* **27:**625–629.

55. Bushuk, B. A., and Rubinov, A. N., 1997, Effect of specific intermolecular interactions on the dynamics of fluorescence spectra of dye solutions, *Opt. Spectrosc.* **82**:530–533.
56. Kaatze, U., and Uhlendorf, V., 1981, The dielectric properties of water at microwave frequencies, *Z. Phys. Chem. N. F.* **126**:151–165.
57. Cole, K. S., and Cole, R. H., 1941, Dispersion and absorption in dielectrics, *J. Chem. Phys.* **9**:341–351.
58. Fellner-Feldegg, H., 1969, The measurement of dielectrics in the time domain, *J. Phys. Chem.* **75**:616–623.
59. Davidson, D. W., and Cole, R. H., 1951, Dielectric relaxation in glycerol, propylene glycol, and *n*-propanol, *J. Chem. Phys.* **19**:1484–1490.
60. Denny, D. J., and Cole, R. H., 1955, Dielectric properties of methanol and methanol–1-propanol solutions, *J. Chem. Phys.* **23**:1767–1772.
61. McDuffie, G. E., and Litovitz, T. A., 1962, Dielectric relaxation in associated liquids, *J. Chem. Phys.* **37**:1699–1705.
62. Gard, S. K., and Smyth, C. P., 1965, Microwave absorption and molecular structure in liquids LXII. The three dielectric dispersion regions of the normal primary alcohols, *J. Phys. Chem.* **69**:1294–1301.
63. Bamford, C. H., and Compton, R. G., 1985, *Chemical Kinetics*, Elsevier, New York, 404 pp.
64. Szmacinski, H., Gryczynski, I., and Lakowicz, J. R., 1996, Resolution of multiexponential spectral relaxation of Y_t-base by global analysis of collisionally quenched samples, *J. Fluoresc.* **6**:177–185.
65. Veselova, T. V., Limareva, L. A., Cherkasov, A. S., and Shirokov, V. I., 1965, Fluorometric study of the effect of solvent on the fluorescence spectrum of 3-amino-*N*-methylphthalimide, *Opt. Spectrosc.* **19**:39–43.
66. Lakowicz, J. R., Bevan, D. R., Maliwal, B. P., Cherek, H., and Balter, A., 1983, Synthesis and characterization of a fluorescence probe of the phase transition and dynamic properties of membranes, *Biochemistry* **22**:5714–5722.
67. Lakowicz, J. R., Cherek, H., and Bevan, D. R., 1980, Demonstration of nanosecond dipolar relaxation in biopolymers by inversion of apparent fluorescence phase shift and demodulation lifetimes, *J. Biol. Chem.* **255**:4403–4406.
68. Röcker, C., Heilemann, A., and Fromherz, P., 1996, Time-resolved fluorescence of a hemicyanine dye: Dynamics of rotamerism and resolvation, *J. Phys. Chem.* **100**:12172–12177.
69. Knutson, J. R., Walbridge, D. G., and Brand, L., 1982, Decay associated fluorescence spectra and the heterogeneous emission of alcohol dehydrogenase, *Biochemistry* **21**:4671–4679.
70. Rudik, K. I., and Pikulik, L. G., 1971, Effect of the exciting light on the fluorescence spectra of phthalimide solutions, *Opt. Spectrosc.* **30**:147–148.
71. Rubinov, A. N., and Tomin, V. I., 1970, Bathochromic luminescence in solutions of organic dyes at low temperatures, *Opt. Spectrosc.* **29**:578–580.
72. Galley, W. C., and Purkey, R. M., 1970, Role of heterogeneity of the solvation site in electronic spectra in solution, *Proc. Natl. Acad. Sci. U.S.A.* **67**:1116–1121.
73. Itoh, K., and Azumi, T., 1973, Shift of emission band upon excitation at the long wavelength absorption edge, *Chem. Phys. Lett.* **22**:395–399.
74. Azumi, T., Itoh, K., and Shiraishi, H., 1976, Shift of emission band upon the excitation at the long wavelength absorption edge. III. Temperature dependence of the shift and correlation with the time dependent spectral shift, *J. Chem. Phys.* **65**:2550–2555.
75. Itoh, K., and Azumi, T., 1975, Shift of the emission band upon excitation at the long wavelength absorption edge. II. Importance of the solute–solvent interaction and the solvent reorientation relaxation process, *J. Chem. Phys.* **62**:3431–3438.
76. Kawski, A., Ston, M., and Janic, I., 1983, On the intensity distribution within photoluminescence bands in rigid and liquid solutions, *Z. Naturforsch. A* **38**:322–324.
77. Lakowicz, J. R., and Keating-Nakamoto, S., 1984, Red-edge excitation of fluorescence and dynamic properties of proteins and membranes, *Biochemistry* **23**:3013–3021.
78. Demchenko, A. P., 1982, On the nanosecond mobility in proteins: Edge excitation fluorescence red shift of protein-bound 2-(*p*-toluidinylnaphthalene)-6-sulfonate, *Biophys. Chem.* **15**:101–109.
79. Shcherbatska, N. V., van Hoek, A., Visser, A. J. W. G., and Koziol, J., 1994, Molecular relaxation spectroscopy of lumichrome, *J. Photochem. Photobiol., A: Chem.* **78**:241–246.
80. Demchenko, A. P., and Shcherbatska, N. V., 1985, Nanosecond dynamics of charged fluorescent probes at the polar interface of a membrane phospholipid bilayer, *Biophys. Chem.* **22**:131–143.
81. Raudino, A., Guerrera, F., Asero, A., and Rizza, V., 1983, Application of red-edge effect on the mobility of membrane lipid polar head groups, *FEBS Lett.* **159**:43–46.
82. Demchenko, A. P., and Ladokhin, A. S., 1988, Temperature-dependent shift of fluorescence spectra without conformational changes in protein: Studies of dipole relaxation in the melittin molecule, *Biochim. Biophys. Acta* **955**:352–360.
83. Conti, C., and Forster, L. S., 1974, Non-exponential decay of indole fluorescence—the red-edge effect, *Biochem. Biophys. Res. Commun.* **57**:1287–1292.
84. Nemkovich, N. A., Rubinov, A. N., and Tomin, V. I., 1981, Kinetics of luminescence spectra of rigid dye solutions due to directed electronic energy transfer, *J. Lumin.* **23**:349–361.
85. Milton, J. G., Purkey, R. M., and Galley, W. C., 1978, The kinetics of solvent reorientation in hydroxylated solvents from the exciting-wavelength dependence of chromophore emission spectra, *J. Chem. Phys.* **68**:5396–5403.
86. Morgenthaler, M. J. E., Meech, S. R., and Yoshihara, K., 1992, The inhomogeneous broadening of the electronic spectra of dyes in glycerol solution, *Chem. Phys. Lett.* **197**:537–541.
87. Gakamsky, D. M., Demchenko, A. P., Nemkovich, N. A., Rubinov, A. N., Tomin, V. I., and Shcherbatska, N. V., 1992, Selective laser spectroscopy of 1-phenylnaphthylamine in phospholipid membranes, *Biophys. Chem.* **42**:49–61.
88. Weber, G., and Shinitzky, M., 1970, Failure of energy transfer between identical aromatic molecules on excitation at the long wave edge of the absorption spectrum, *Proc. Natl. Acad. Sci. U.S.A.* **65**:823–830.
89. Valeur, B., and Weber, G., 1977, Anisotropic rotations in 1-naphthylamine, existence of a red-edge transition moment normal to the ring plane, *Chem. Phys. Lett.* **45**:140–144.
90. Valeur, B., and Weber, G., 1978, A new red-edge effect in aromatic molecules: Anomaly of apparent rotation revealed by fluorescence polarization, *J. Phys. Chem.* **69**:2393–2400.
91. Mantulin, W. W., and Weber, G., 1977, Rotational anisotropy and solvent–fluorophore bonds: An investigation by differential polarized-phase fluorometry, *J. Chem. Phys.* **66**:4092–4099.
92. Brand, L., Seliskar, C. J., and Turner, D. C., 1971, The effects of chemical environment on fluorescence probes, in *Probes of Structure and Function of Macromolecules and Membranes*, B. Chance, C. P. Lee, and J. K. Blaisie (eds.), Academic Press, New York, pp. 17–39.

PROBLEMS

7.1. *Estimation of the Spectral Relaxation Time*: Figure 7.41 shows time-dependent intensity decays of TNS bound to egg lecithin vesicles. The wavelengths of 390, 435, and 530 nm are in the blue, center, and red regions of the emission spectrum. Use the data in Figure 7.41 to calculate the spectral relaxation time for the TNS-labeled residue. Assume that the emission at 390 nm is dominated by the initially excited state (F) and that the emission at 435 nm represents the TNS, unaffected by relaxation.

7.2. *Interpretation of Wavelength-Dependent Lifetimes*: TNS was dissolved in various solvents or bound to vesicles of dioleoyl-L-α-phosphatidylcholine (DOPC). Apparent phase and modulation lifetimes were measured across the emission spectra of these samples (Figure 7.42). Explain the data in Figure 7.42.

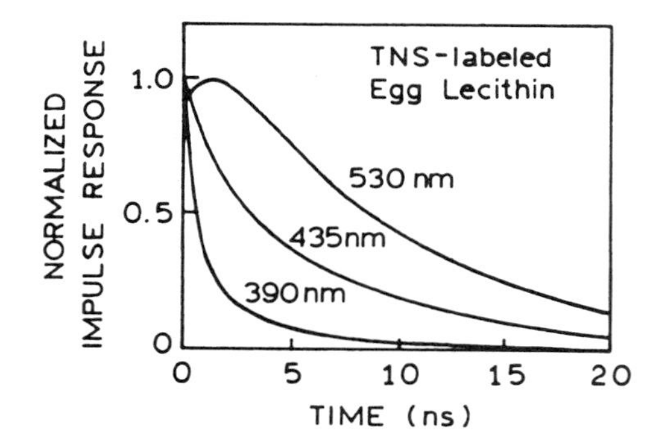

Figure 7.41. Fluorescence impulse response functions of TNS-labeled egg lecithin vesicles. Time-dependent intensities are shown for 390, 435, and 530 nm. From Ref. 92.

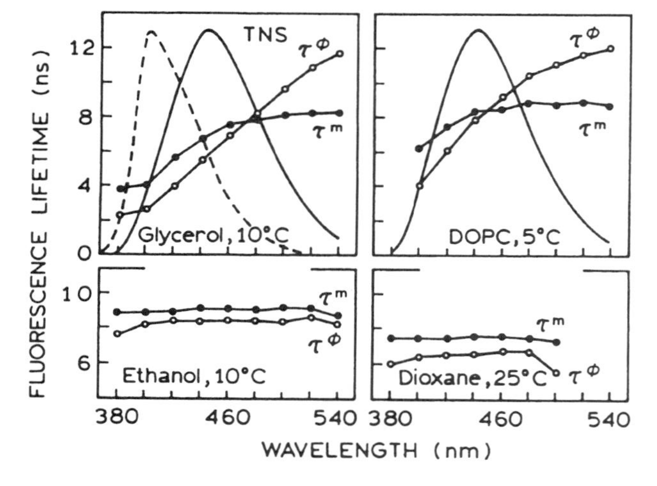

Figure 7.42. Fluorescence lifetimes and spectra of TNS dissolved in various solvents and TNS bound to DOPC vesicles. Apparent phase TNS (○) and modulation (●) lifetimes were measured at 30 MHz. Normalized emission spectra are shown for TNS in glycerol and bound to DOPC vesicles (———) and for TNS in dioxane at 25 °C (– – – –). From Ref. 20.

Quenching of Fluorescence

8

Fluorescence quenching refers to any process which decreases the fluorescence intensity of a sample. A variety of molecular interactions can result in quenching. These include excited-state reactions, molecular rearrangements, energy transfer, ground-state complex formation, and collisional quenching. In this chapter we will be concerned primarily with quenching resulting from collisional encounters between the fluorophore and quencher, which is called collisional or dynamic quenching. Static quenching is a frequent complicating factor in the analysis of quenching data, but it can also be a valuable source of information about binding between the fluorescent sample and the quencher. In addition to the processes described above, apparent quenching can occur due to the optical properties of the sample. For example, high optical densities or turbidity can result in decreased fluorescence intensities. This is a trivial type of quenching which contains little molecular information. Throughout this chapter, we will assume that such trivial effects are not the cause of the observed decreases in fluorescence intensity.

Fluorescence quenching has been widely studied both as a fundamental phenomenon and as a source of information about biochemical systems. The biochemical applications of quenching are due to the intrinsic role of molecular interactions in quenching phenomena. Both static and dynamic quenching require molecular contact between the fluorophore and quencher. In the case of collisional quenching, the quencher must diffuse to the fluorophore during the lifetime of the excited state. Upon contact, the fluorophore returns to the ground state, without emission of a photon. In general, quenching occurs without any permanent change in the molecules, that is, without a photochemical reaction. In the case of static quenching, a complex is formed between the fluorophore and the quencher, and this complex is nonfluorescent. For either static or dynamic quenching to occur, the fluorophore and quencher must be in contact. The requirement of molecular contact results in the numerous applications of quenching. For example, quenching measurements can reveal the accessibility of fluorophores to quenchers. Consider a fluorophore bound either to a protein or to a membrane. If the protein or membrane is impermeable to the quencher, and the fluorophore is located in the interior of the macromolecule, then neither collisional nor static quenching can occur. For this reason, quenching studies can be used to reveal the localization of fluorophores in proteins and membranes and the permeabilities of protons and membranes to quenchers. Additionally, the rate of collisional quenching can be used to determine the diffusion coefficient of the quencher.

It is important to recognize that the phenomenon of collisional quenching results in the expansion of the volume and distance within the solution, which affects the experimental observables. The root-mean-square distance $[\sqrt{(\overline{\Delta x^2})}]$ over which a quencher can diffuse during the lifetime of the excited state (τ) is given by $\sqrt{(\overline{\Delta x^2})} = \sqrt{(2D\tau)}$, where D is the diffusion coefficient. Consider an oxygen molecule in water at 25 °C. Its diffusion coefficient is 2.5×10^{-5} cm^2/s. During a typical fluorescence lifetime of 4 ns, the oxygen molecule can diffuse 44 Å. If the lifetime is longer, diffusion over still larger distances can be observed. For example, for lifetimes of 20 and 100 ns the average distances for oxygen diffusion are 100 Å and 224 Å, respectively. With the introduction of longer-lived probes with microsecond lifetimes (Chapter 20), diffusion over still larger distances can be observed. Hence, fluorescence quenching can reveal the diffusion of quenchers over moderately large distances which are comparable to the sizes of proteins and membranes. This situation is different from that for solvent relaxation. Spectral shifts resulting from reorientation of the solvent molecules are due primarily to the solvent shell immediately adjacent to the fluorophore.

8.1. QUENCHERS OF FLUORESCENCE

A wide variety of substances act as quenchers of fluorescence. One of the best-known collisional quenchers is molecular oxygen,[1] which quenches almost all known fluorophores. Depending upon the sample under investigation, it is frequently necessary to remove dissolved oxygen to obtain reliable measurements of the fluorescence yields or lifetimes. The mechanism by which oxygen quenches fluorescence has been the subject of debate. The most likely mechanism is that the paramagnetic oxygen causes the fluorophore to undergo intersystem crossing to the triplet state. In fluid solutions the long-lived triplets are completely quenched, so that phosphorescence is not observed. Aromatic and aliphatic amines are also efficient quenchers of most unsubstituted aromatic hydrocarbons. For example, anthracene fluorescence is effectively quenched by diethylaniline.[2] In this instance the mechanism of quenching is the formation of an excited charge-transfer complex. The excited-state fluorophore accepts an electron from the amine. In nonpolar solvents, fluorescence from the excited charge-transfer complex (exciplex)

Table 8.1. Quenchers of Fluorescence

Quencher(s)	Typical fluorophore(s)[a]	Reference(s)
Acrylamide	Tryptophan	5–7, 79
Amines	Anthracene, perylene	2, 80–85
Amines	Carbazole	86
Amine anesthetics	Perylene, anthroyloxy probes	87,88
Bromate	—	89
Bromobenzene	Many fluorophores	90
Carboxy groups	Indole	91
Chloride	Quinolinium, SPQ	92–95
Chlorinated compounds	Indoles and carbazoles	96–99
Cobalt (Co^{2+})	NBD, PPO, perylene (energy transfer for some probes)	100–105
Copper (Cu^{2+})	Anthroyloctanoic acid	106
Dimethylformamide	Indole	107
Disulfides	Tyrosine	108
Ethers	9-Arylxanthyl cations	109
Halogens	Anthracene, naphthalene, carbazole	110–125
Halogen anesthetics	Pyrene, tryptophan	126,127
Hydrogen peroxide	Tryptophan	128
Imidazole, histidine	Tryptophan	129
Iodide	Antracene	130–133
Methylmercuric chloride	Carbazole, pyrene	134,135
Nickel (Ni^{2+})	Perylene	136,137
Nitromethane and nitro compounds	Polycyclic aromatic hydrocarbons	138–144
Nitroxides	Naphthalene, Tb^{3+}, anthroyloxy probes	63,145–148
NO (nitric oxide)	Naphthalene, pyrene	149–151
Olefins	Cyanonaphthalene, 2,3-dimethylnaphthalene, pyrene	152–157
Oxygen	Most fluorophores	158–166
Peroxides	Dimethylnaphthalene	167
Picolinium nicotinamide	Tryptophan	168
Pyridine	Carbazole	169
Quinones	Aromatic hydrocarbons, chlorophyll	170,171
Silver (Ag^{+})	Perylene	172
Succinimide	Tryptophan	173,174
Thallium (Tl^{+})	Naphthylaminesulfonic acid	175
Thiocyanate	Anthracene, 5,6-benzoquinoline	176,177
Trifluoroacetamide	Tryptophan	35
Xenon		178

[a]Abbreviations: NBD, 7-Nitrobenz-2-oxa-1,3-diazol-4-yl; PPO, 2,5-diphenyl-1,3,4-oxadiazole; SPQ, 6-methoxy-*N*-(3-sulfopropyl) quinoline.

is frequently observed, and one may regard this process as an excited-state reaction rather than quenching. In polar solvents the exciplex emission is often quenched, so that the fluorophore–amine interaction appears to be that of simple quenching. While it is now known that there is a modest through-space component to almost all quenching reactions, this component is short-range (< 2 Å), so that molecular contact is a requirement for quenching.

Another type of quenching is due to heavy atoms such as iodine and bromine. Halogenated compounds such as trichloroethanol and bromobenzene also act as collisional quenchers. Quenching by the larger halogens such as bromine and iodine may be a result of intersystem crossing to an excited triplet state, promoted by spin–orbit coupling of the excited (singlet) fluorophore and the halogen.[3] Since emission from the triplet state is slow, the triplet emission is highly quenched by other processes. The quenching mechanism is probably different for chlorine-containing substances. Indole, carbazole, and their derivatives are uniquely sensitive to quenching by chlorinated hydrocarbons and by electron scavengers such as protons, histidine, cysteine, NO_3^-, fumarate, Cu^{2+}, Pb^{2+}, Cd^{2+}, and Mn^{2+}.[4] Quenching by these substances probably involves donation of an electron from the fluorophore to the quencher. Additionally, indole and tryptophan and its derivatives are quenched by acrylamide, succinimide, dichloroacetamide, dimethylformamide, pyridinium hydrochloride, imidazolium hydrochloride, methionine, Eu^{3+}, Ag^+, and Cs^+. Quenchers of protein fluorescence have been summarized in several insightful reviews.[5–7] Hence, a variety of quenchers are available for studies of protein fluorescence, especially to determine the surface accessibility of tryptophan residues and the permeation of proteins by the quenchers.

Additional quenchers include purines, pyrimidines, *N*-methylnicotinamide, and *N*-alkylpyridinium and -picolinium salts.[8,9] For example, the fluorescence of both FAD and NADH is quenched by the adenine moiety; flavin fluorescence is quenched by both static and dynamic processes,[10] whereas the quenching of dihydronicotinamide appears to be primarily dynamic.[11] These aromatic substances appear to quench by formation of charge-transfer complexes. Depending upon the precise structure involved, the ground-state complex can be reasonably stable. As a result, both static and dynamic quenching are frequently observed. A variety of other quenchers are known. These are summarized in Table 8.1, which is intended to be an overview and not a complete list. Known collisional quenchers include hydrogen peroxide, nitric oxide (NO), nitroxides, BrO_4–, and even some olefins.

Because of the variety of substances which act as quenchers, one can frequently identify fluorophore–quencher combinations for a desired purpose. It is important to note that not all fluorophores are quenched by all the substances listed above. This fact occasionally allows selective quenching of a given fluorophore. The occurrence of quenching depends upon the mechanism, which in turn depends upon the chemical properties of the individual molecules. Detailed analysis of the mechanism of quenching is complex. In this chapter we will be concerned primarily with the type of quenching, that is, whether quenching is diffusive or static in nature. Later in this chapter, we describe biochemical applications of quenching.

8.2. THEORY OF COLLISIONAL QUENCHING

Collisional quenching of fluorescence is described by the Stern–Volmer equation,

$$\frac{F_0}{F} = 1 + k_q\tau_0[Q] = 1 + K_D[Q] \qquad [8.1]$$

In this equation, F_0 and F are the fluorescence intensities in the absence and presence of quencher, respectively, k_q is the bimolecular quenching constant, τ_0 is the lifetime of the fluorophore in the absence of quencher, and [Q] is the concentration of quencher. The Stern–Volmer quenching constant is given by $k_q\tau_0$. If the quenching is known to be dynamic, the Stern–Volmer constant will be represented by K_D. Otherwise, this constant will be described as K_{SV}.

Quenching data are usually presented as plots of F_0/F versus [Q]. This is because F_0/F is expected to be linearly dependent upon the concentration of quencher. A plot of F_0/F versus [Q] yields an intercept of 1 on the *y*-axis and a slope equal to K_D (Figure 8.1). Intuitively, it is useful to note that K_D^{-1} is the quencher concentration at which F_0/F = 2, or 50% of the intensity is quenched. A linear Stern–Volmer plot is generally indicative of a single class of fluorophores, all equally accessible to quencher. If two fluorophore populations are present, and one class is not accessible to quencher, then the Stern–Volmer plots deviate from linearity toward the *x*-axis. This result is frequently found for the quenching of tryptophan fluorescence in proteins by polar or charged quenchers. These molecules do not readily penetrate the hydrophobic interior of proteins, and only those tryptophan residues on the surface of the protein are quenched.

It is important to recognize that observation of a linear Stern–Volmer plot does not prove that collisional quench-

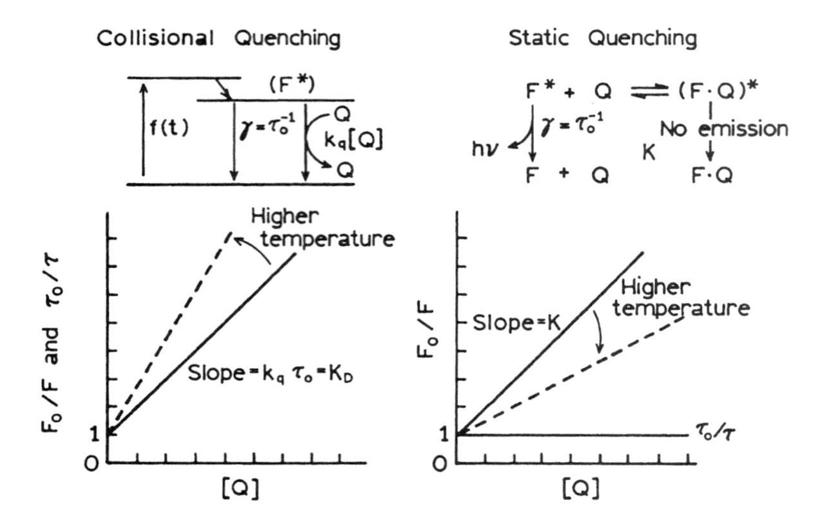

Figure 8.1. Comparison of dynamic (collisional) and static quenching.

ing of fluorescence has occurred. In Section 8.3 we will see that static quenching also results in linear Stern–Volmer plots. Static and dynamic quenching can be distinguished by their differing dependence on temperature and viscosity, or preferably by lifetime measurements. Higher temperatures result in faster diffusion and hence larger amounts of collisional quenching. Higher temperatures will typically result in the dissociation of weakly bound complexes, and hence smaller amounts of static quenching.

8.2.A. Derivation of the Stern–Volmer Equation

The Stern–Volmer equation can be derived by consideration of the fluorescence intensities observed in the absence and presence of quencher. The fluorescence intensity observed for a fluorophore is proportional to its concentration in the excited state [F*]. Under continuous illumination, a constant population of excited fluorophores is established, and therefore $d[F^*]/dt = 0$. In the absence of quencher, the differential equation describing [F*] is

$$\frac{d[\mathrm{F}^*]}{dt} = f(t) - \gamma[\mathrm{F}^*]_0 = 0 \qquad [8.2]$$

and in the presence of quencher,

$$\frac{d[\mathrm{F}^*]}{dt} = f(t) - (\gamma + k_q[\mathrm{Q}])[\mathrm{F}^*] = 0 \qquad [8.3]$$

where $f(t)$ is the constant excitation function, and $\gamma = \tau_0^{-1}$ is the decay rate of the fluorophore in the absence of quencher. In the absence of quenching, the excited-state population decays with a rate $\gamma = (\Gamma + k_{nr})$, where Γ is the radiative decay rate and k_{nr} is the nonradiative decay rate. In the presence of quencher, there is an additional decay rate, k_q[Q]. With continuous excitation, the excited-state population is constant, so the derivative can be easily eliminated from these equations. Division of Eq. [8.3] by Eq. [8.2] yields

$$\frac{F_0}{F} = \frac{\gamma + k_q[\mathrm{Q}]}{\gamma} = 1 + k_q\tau_0[\mathrm{Q}] \qquad [8.4]$$

which is the Stern–Volmer equation.

This equation may also be obtained by considering the fraction of excited fluorophores, relative to the total, which decay by emission. This fraction (F/F_0) is given by the ratio of the decay rate in the absence of quencher (γ) to the total decay rate in the presence of quencher ($\gamma + k_q[\mathrm{Q}]$),

$$\frac{F}{F_0} = \frac{\gamma}{\gamma + k_q[\mathrm{Q}]} = \frac{1}{1 + K_D[\mathrm{Q}]} \qquad [8.5]$$

which is again the Stern–Volmer equation. Since collisional quenching is a rate process which depopulates the excited state, the lifetimes in the absence (τ_0) and presence (τ) of quencher are given by

$$\tau_0 = \gamma^{-1} \qquad [8.6]$$

$$\tau = (\gamma + k_q[\mathrm{Q}])^{-1} \qquad [8.7]$$

and therefore

$$\frac{\tau_0}{\tau} = 1 + k_q\tau_0[\mathrm{Q}] \qquad [8.8]$$

This equation illustrates an important characteristic of collisional quenching, which is an equivalent decrease in fluorescence intensity and in lifetime (Figure 8.1, left); that is, for collisional quenching,

$$\frac{F_0}{F} = \frac{\tau_0}{\tau} \quad [8.9]$$

The decrease in lifetime occurs because quenching is an additional rate process that depopulates the excited state. The decrease in yield occurs because quenching depopulates the excited state without fluorescence emission. Static quenching does not decrease the lifetime because only the fluorescent molecules are observed, and the uncomplexed fluorophores have the unquenched lifetime τ_0.

8.2.B. Interpretation of the Bimolecular Quenching Constant

A frequently encountered value is the bimolecular quenching constant (k_q), which can reflect the efficiency of quenching or the accessibility of the fluorophores to the quencher. As shown below, diffusion-controlled quenching typically results in values of k_q near 1×10^{10} M^{-1} s^{-1}. Smaller values of k_q can result from steric shielding of the fluorophore, and larger apparent values of k_q usually indicate some type of binding interaction.

The meaning of the bimolecular quenching constant can be understood in terms of the frequency of collisions between freely diffusing molecules. The collisional frequency (Z) of a fluorophore with a quencher is given by

$$Z = k_0[Q] \quad [8.10]$$

where k_0 is the diffusion-controlled bimolecular rate constant. This constant may be calculated using the Smoluchowski equation,

$$k_0 = \frac{4\pi RDN}{1000} = \frac{4\pi N}{1000}(R_f + R_q)(D_f + D_q) \quad [8.11]$$

where R is the collision radius, D is the sum of the diffusion coefficients of the fluorophore (D_f) and quencher (D_q), and N is Avogadro's number. The collision radius is generally assumed to be the sum of the molecular radii of the fluorophore (R_f) and quencher (R_q). This equation describes the diffusive flux of a molecule with a diffusion coefficient D through the surface of a sphere of radius R. The factor of 1000 is necessary to keep the units correct when the concentration is expressed in terms of molarity. The term $N/1000$ converts molarity to molecules per cubic centimeter.

The collisional frequency is related to the bimolecular quenching constant by the quenching efficiency f_Q,

$$k_q = f_Q k_0 \quad [8.12]$$

For example, if $f_Q = 0.5$, then 50% of the collisional encounters are effective in quenching, and one expects k_q to be one-half of k_0. Since k_0 can be estimated with moderate precision, the observed value of k_q can be used to judge the efficiency of quenching. Quenchers like oxygen, acrylamide, and I^- generally have efficiencies near unity, but the quenching efficiency of succinimide depends on the solvent. The efficiency is generally less with the lighter halogens. The quenching efficiency of amines depends upon the reduction potential of the fluorophores being quenched, as may be expected for a charge-transfer reaction.

The efficiency of quenching can be calculated from the observed value of k_q, if the diffusion coefficients and molecular radii are known. The radii can be obtained from molecular models, or from the molecular weights and densities of the substances in question. Diffusion coefficients may be obtained from the Stokes–Einstein equation,

$$D = kT/6\pi\eta R \quad [8.13]$$

where k is the Boltzmann constant, η is the solvent viscosity, and R is the molecular radius. Frequently, the Stokes–Einstein equation underestimates the diffusion coefficients of small molecules. For example, quenching efficiencies of 2–3 were calculated for oxygen quenching of fluorophores dissolved in various alcohols.[12] These impossibly large efficiencies were obtained because the diffusion coefficient of oxygen in organic solvents is severalfold larger than predicted by Eq. [8.13]. This equation describes the diffusion of molecules that are larger than the solvent molecules, which is clearly not the case for oxygen in ethanol. As an alternative method, diffusion coefficients can be obtained from nomograms based upon physical properties of the system.[13] Once the diffusion coefficients are known, the bimolecular quenching constant for $f_Q = 1$ can be predicted using the Smoluchowski equation (Eq. [8.11]).

It is instructive to consider typical values for k_q and the concentrations of quencher required for significant quenching. For example, consider the quenching of tryptophan by oxygen.[14] At 25 °C the diffusion coefficient of oxygen in water is 2.5×10^{-5} cm^2/s and that of tryptophan is 0.66×10^{-5} cm^2/s. Assuming a collision radius of 5 Å, substitution into Eq. [8.11] yields $k_0 = 1.2 \times 10^{10}$ M^{-1} s^{-1}. The observed value of the oxygen Stern–Volmer quenching constant was 32.5 M^{-1}. Since the unquenched lifetime of tryptophan is 2.7 ns, $k_q = 1.2 \times 10^{10}\ M^{-1}\ s^{-1}$ which is in excellent agreement with the predicted value. This indicates that essentially every collision of oxygen with tryptophan is effective in quenching, that is, $f_Q = 1.0$. A bimolecular quenching constant near $1 \times 10^{10}\ M^{-1}\ s^{-1}$ may be considered as the largest possible value in aqueous solution. For quenchers other than oxygen, smaller diffusion-limited quenching constants are expected because the diffusion coefficients of the quenchers are smaller. For example, the efficiency of acrylamide quenching of tryptophan fluorescence is also near unity,[15] but $k_q = 5.9 \times 10^9\ M^{-1}\ s^{-1}$. This somewhat smaller value of k_q is a result of the smaller diffusion coefficient of acrylamide relative to that of oxygen. Frequently, data are obtained for fluorophores which are bound to macromolecules. In this case, the fluorophore is not diffusing as rapidly. Also, the quenchers can probably only approach the fluorophores from a particular direction. In such cases, the maximum bimolecular quenching constant is expected to be about 50% of the diffusion-controlled value.[16]

8.3. THEORY OF STATIC QUENCHING

In the previous section we described quenching that resulted from diffusive encounters between the fluorophore and quencher during the lifetime of the excited state. This is a time-dependent process. Quenching can also occur as a result of the formation of a nonfluorescent complex between the fluorophore and quencher. When this complex absorbs light, it immediately returns to the ground state without emission of a photon (Figure 8.1).

The dependence of the fluorescence intensity upon quencher concentration for static quenching is easily derived by consideration of the association constant for complex formation. This constant is given by

$$K_S = \frac{[\mathrm{F-Q}]}{[\mathrm{F}][\mathrm{Q}]} \qquad [8.14]$$

where [F–Q] is the concentration of the complex, [F] is the concentration of uncomplexed fluorophore, and [Q] is the concentration of quencher. If the complexed species is nonfluorescent, then the fraction of the fluorescence that remains, F/F_0, is given by the fraction of the total fluorophores that are not complexed, $f = F/F_0$. Recalling that the total concentration of fluorophore, $[\mathrm{F}]_0$, is given by

$$[\mathrm{F}]_0 = [\mathrm{F}] + [\mathrm{F-Q}] \qquad [8.15]$$

substitution into Eq. [8.14] yields

$$K_S = \frac{[\mathrm{F}]_0 - [\mathrm{F}]}{[\mathrm{F}][\mathrm{Q}]} = \frac{[\mathrm{F}_0]}{[\mathrm{F}][\mathrm{Q}]} - \frac{1}{[\mathrm{Q}]} \qquad [8.16]$$

We can substitute fluorescence intensities for the fluorophore concentrations, and rearrangement of Eq. [8.16] yields

$$\frac{F_0}{F} = 1 + K_S[\mathrm{Q}] \qquad [8.17]$$

Note that the dependence of F_0/F on [Q] is linear and is identical to that observed for dynamic quenching except that the quenching constant is now the association constant. Unless additional information is provided, fluorescence quenching data obtained by intensity measurements alone can be explained by either a dynamic or a static process. As will be shown below, the magnitude of K_S can sometimes be used to demonstrate that dynamic quenching cannot account for the decrease in intensity. The measurement of fluorescence lifetimes is the most definitive method to distinguish static and dynamic quenching. Static quenching removes a fraction of the fluorophores from observation. The complexed fluorophores are nonfluorescent, and the only observed fluorescence is from the uncomplexed fluorophores. The uncomplexed fraction is unperturbed, and hence the lifetime is τ_0. Therefore, for static quenching $\tau_0/\tau = 1$ (Figure 8.1, right). In contrast, for dynamic quenching, $\tau_0/\tau = F_0/F$.

Besides measurement of fluorescence lifetimes, static and dynamic quenching can often be distinguished on the basis of other considerations. Dynamic quenching depends upon diffusion. Since higher temperatures result in larger diffusion coefficients, the bimolecular quenching constants are expected to increase with increasing temperature (Figure 8.1). More specifically, k_q is expected to be proportional to T/η since diffusion coefficients are proportional to this ratio (Eq. [8.13]). In contrast, increased temperature is likely to result in decreased stability of complexes, and thus lower values of the static quenching constants.

One additional method to distinguish static and dynamic quenching is by careful examination of the absorption spectra of the fluorophore. Collisional quenching only affects the excited states of the fluorophores, and thus no changes in the absorption spectra are expected. In contrast, ground-state complex formation will frequently result in perturbation of the absorption spectrum of the fluorophore. In fact, a more complete form of Eq. [8.17] would include the possibility of different extinction coefficients for the free and complexed forms of the fluorophore.

8.4. COMBINED DYNAMIC AND STATIC QUENCHING

In many instances the fluorophore can be quenched both by collisions and by complex formation with the same quencher. The characteristic feature of the Stern–Volmer plots in such circumstances is an upward curvature, concave toward the *y*-axis (Figure 8.2). Then the fractional fluorescence remaining (F/F_0) is given by the product of the fraction not complexed (f) and the fraction not quenched by collisional encounters. Hence,

$$\frac{F}{F_0} = f \frac{\gamma}{\gamma + k_q[\mathrm{Q}]} \qquad [8.18]$$

In the previous section we found that $f^{-1} = 1 + K_S[\mathrm{Q}]$. Inversion of Eq. [8.18] and rearrangement of the last term on the right yields

$$\frac{F_0}{F} = (1 + K_D[\mathrm{Q}])(1 + K_S[\mathrm{Q}]) \qquad [8.19]$$

This modified form of the Stern–Volmer equation is second-order in [Q], which accounts for the upward curvature observed when both static and dynamic quenching occur for the same fluorophore.

The dynamic portion of the observed quenching can be determined by lifetime measurements. That is, $\tau_0/\tau = 1 + K_D[\mathrm{Q}]$. If lifetime measurements are not available, then Eq. [8.19] can be modified to allow a graphical separation of K_S and K_D. Multiplication of the terms in parentheses yields

$$\frac{F_0}{F} = 1 + (K_D + K_S)[\mathrm{Q}] + K_D K_S[\mathrm{Q}]^2 \qquad [8.20]$$

$$= 1 + K_{\mathrm{app}}[\mathrm{Q}] \qquad [8.21]$$

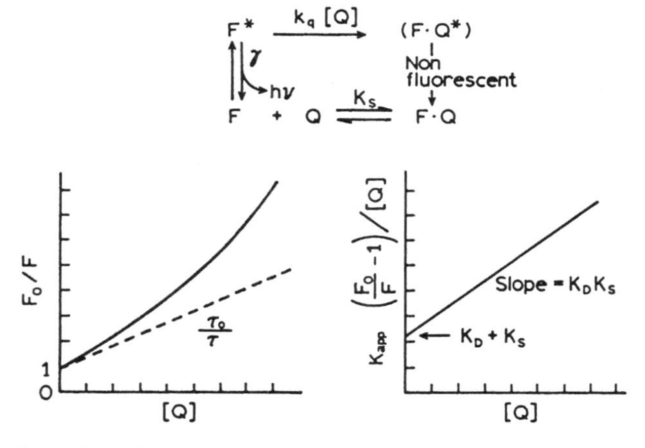

Figure 8.2. Dynamic and static quenching of the same population of fluorophores.

where

$$K_{\mathrm{app}} = \left(\frac{F_0}{F} - 1\right)\frac{1}{[\mathrm{Q}]} = (K_D + K_S) + K_D K_S[\mathrm{Q}] \qquad [8.22]$$

The apparent quenching constant is calculated at each quencher concentration. A plot of K_{app} versus [Q] yields a straight line with an intercept of $K_D + K_S$ and a slope of $K_D K_S$ (Figure 8.2). The individual values can be obtained from the two solutions of the quadratic equation obtained (see Eq. [8.23]). The dynamic component can generally be selected to be the solution comparable in magnitude to the expected diffusion-controlled value. Alternatively, the temperature or viscosity dependence of the values, or other available information, may be used as a basis for assigning the values.

8.5. EXAMPLES OF STATIC AND DYNAMIC QUENCHING

Before proceeding with additional theories and examples of quenching, it seems valuable to present some examples which illustrate both static and dynamic quenching. Data for oxygen quenching of tryptophan are shown in Figure 8.3.[14] The Stern–Volmer plot is linear, which indicates that only one type of quenching occurs. The proportional decrease in the fluorescence lifetime and yields proves that the observed quenching is due to a diffusive process. From the slope of the Stern–Volmer plot, one can calculate that $K_D = 32.5\ M^{-1}$, or that 50% of the fluorescence is quenched at an oxygen concentration of $0.031M$. The value of K_D and the fluorescence lifetime are adequate to calculate the bimolecular quenching constant, $k_q = 1.2 \times 10^{10}\ M^{-1}\ \mathrm{s}^{-1}$. This is the value expected for the diffusion controlled bimolecular rate constant between oxygen and tryptophan (Eq. [8.11]), which indicates efficient quenching by molecular oxygen.

Static quenching is often observed if the fluorophore and quencher can have a stacking interaction. Such interactions often occur between purine and pyrimidine nucleotides and a number of fluorophores.[17–19] One example is quenching of the coumarin derivative C-120 by the nucleosides uridine (U) and deoxycytidine (dC). The intensity Stern–Volmer plot for quenching by U shows clear upward curvature (Figure 8.4). The lifetime Stern–Volmer plot is linear and shows less quenching than the intensity data. It is clear that the intensity of C-120 is being decreased by both complex formation with U as well as collisional quenching by U. Contrasting data were obtained for quenching of C-120 by dC. In this case, the Stern–Volmer plots are linear for both intensities and lifetimes, and F_0/F

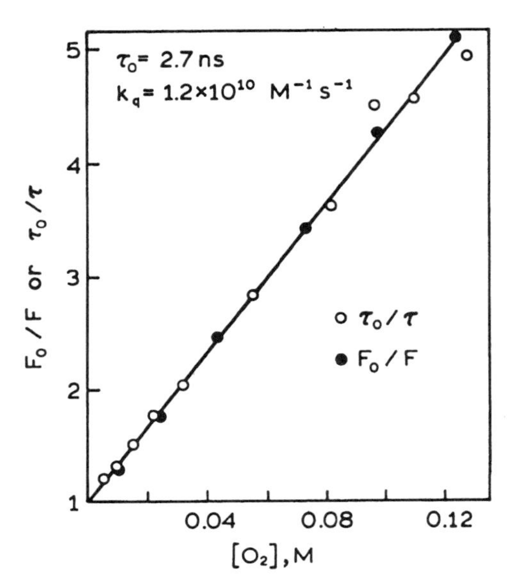

Figure 8.3. Oxygen quenching of tryptophan as observed from fluorescence lifetimes (○) and yields (●). Revised from Ref. 14.

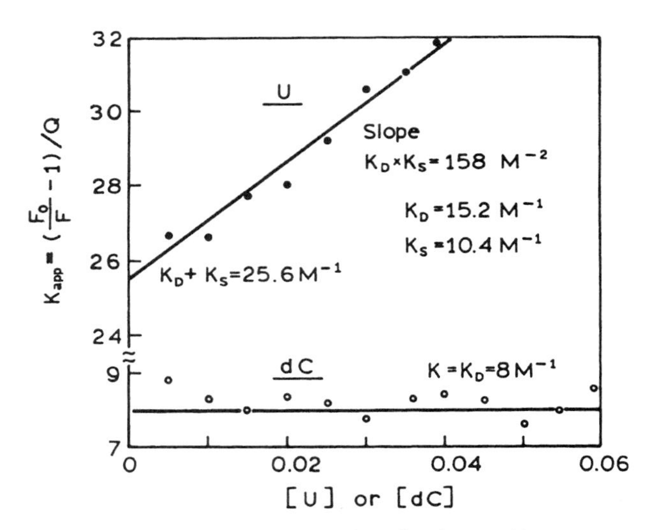

Figure 8.5. Separation of the dynamic and static quenching constants for quenching of C-120 by U or dC. Data from Ref. 19.

$= \tau_0/\tau$. Hence, quenching of C-120 by dC is purely dynamic.

For quenching of C-120 by U, the static and dynamic quenching constants can be determined from a plot of K_{app} versus [U] (Figure 8.5). The slope (S) and intercept (I) were found to be 158 M^{-2} and 25.6 M^{-1}, respectively. Recalling that $I = K_D + K_S$ and $S = K_D K_S$, rearrangement yields

$$K_S^2 - K_S I + S = 0 \qquad [8.23]$$

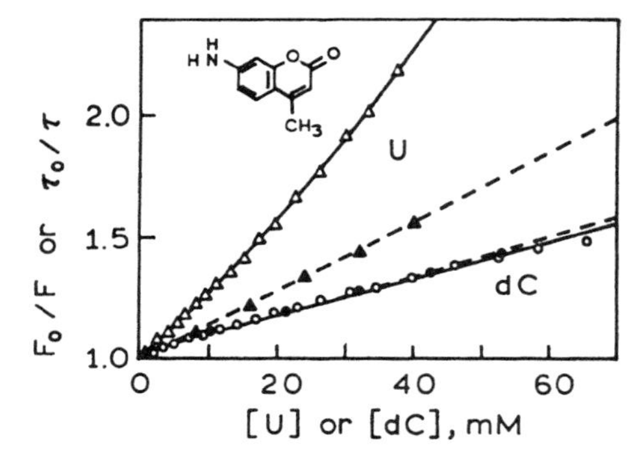

Figure 8.4. Quenching of coumarin C-120 by the nucleosides uridine (Δ, F_0/F; ▲, τ_0/τ) and deoxycytidine (O, F_0/F; ●, τ_0/τ). The sample was excited at the isosbestic point at 360 nm. Revised and reprinted, with permission, from Ref. 19, Copyright © 1996, American Chemical Society.

The solutions for this quadratic equation are $K_S = 15.2$ or 10.4 M^{-1}. From the lifetime data, we know that K_D is near 13.5 M^{-1}. The lower value of 10.4 M^{-1} was assigned as the static quenching constant. At a uridine concentration of 96mM, 50% of the ground state C-120 is complexed and thus nonfluorescent.

It is interesting to mention why the interactions of nucleosides and nucleotides with C-120 were studied. The goal was to develop a method for DNA sequencing using a single electrophoretic lane for all four nucleotides.[19] This would be possible if coumarin derivatives could be identified which display different lifetimes when adjacent to each nucleotide. The DNA sequence could then be determined from the lifetimes observed for each band on the sequencing gel. The use of lifetime measurements in fluorescence sensing is described in Section 19.2.B, and DNA sequencing is described in Section 21.1.

8.6. DEVIATIONS FROM THE STERN–VOLMER EQUATION; QUENCHING SPHERE OF ACTION

Positive deviations from the Stern–Volmer equation are frequently observed when the extent of quenching is large. Two examples of upward-curving Stern–Volmer plots are shown in Figures 8.6 and 8.7 for acrylamide quenching of NATA and of the fluorescent steroid dihydroequilenin (DHE), respectively. The upward-curving Stern–Volmer plots could be analyzed in terms of the static and dynamic quenching constants (Eq. [8.19]). This analysis yields K_S values near 2.8 and 5.2 M^{-1} for acrylamide quenching of

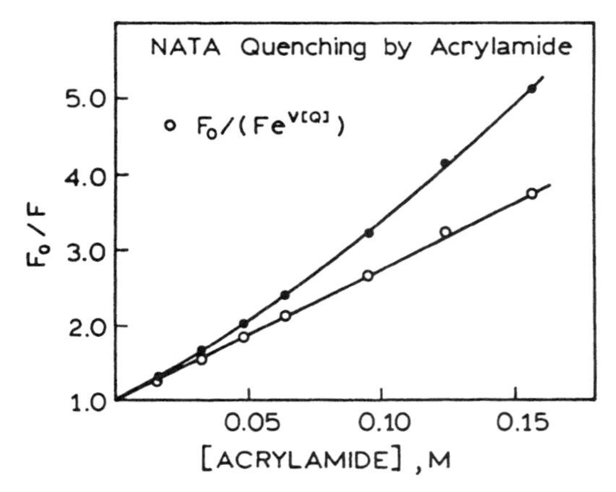

Figure 8.6. Acrylamide quenching of NATA in water. ●, F_0/F; ○, $F_0/(Fe^{V[Q]})$, where $V = 2.0\ M^{-1}$. Revised from Ref. 15.

NATA and DHE, respectively. These values imply that quencher concentrations near $0.3M$ are required to quench one-half of the fluorophores by a static process. Such a weak association suggests that the fluorophores and quenchers do not actually form a ground-state complex. Instead, it seems that the apparent static component is due to the quencher being adjacent to the fluorophore at the moment of excitation. These closely spaced fluorophore–quencher pairs are immediately quenched and thus appear to be dark complexes.

This type of apparent static quenching is usually interpreted in terms of a "sphere of action" within which the probability of quenching is unity. The modified form of the Stern–Volmer equation which describes this situation is

$$\frac{F_0}{F} = (1 + K_D[Q]) \exp([Q]VN/1000) \qquad [8.24]$$

where V is the volume of the sphere.[21] The data in Figures 8.6 and 8.7 are consistent with a sphere radius near 10 Å, which is only slightly larger than the sum of the radii of

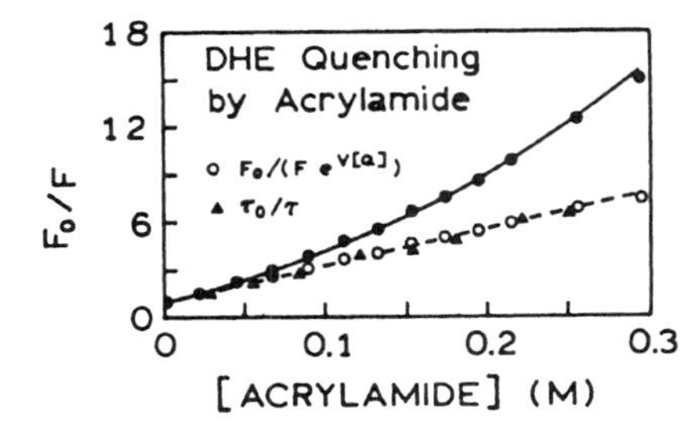

Figure 8.7. Acrylamide quenching of dihydroequilenin (DHE) in buffer containing 10% sucrose at 11 °C. ●, F_0/F; ▲, τ_0/τ; ○, $F_0/(Fe^{V[Q]})$, where $V = 2.4\ M^{-1}$. Revised and reprinted, with permission, from Ref. 20, Copyright © 1990, American Chemical Society.

the fluorophore and quencher. When the fluorophore and quencher are this close, there exists a high probability that quenching will occur before these molecules diffuse apart. As the quencher concentration increases, the probability increases that a quencher is within the first solvent shell of the fluorophore at the moment of excitation.

8.6.A. Derivation of the Quenching Sphere of Action

Assume the existence of a sphere of volume V within which the probability of immediate quenching is unity. Intuitively, if a fluorophore is excited when a quencher is immediately adjacent, then this fluorophore is quenched and is therefore unobservable. The only observable fluorophores are those for which there are no adjacent quenchers. The modified form of the Stern–Volmer equation (Eq. [8.24]) is derived by calculating the fraction of fluorophores which do not contain a quencher within their surrounding sphere of action.[21]

The Poisson probability distribution states that the probability of finding a volume V with n quenchers is

$$P(n) = \frac{\lambda^n}{n!} e^{-\lambda} \qquad [8.25]$$

where λ is the mean number of quenchers per volume V. The average concentration of quenchers (in molecules/cm^3) is given by $[Q]N/1000$, and hence the average number of molecules in the sphere is $\lambda = V[Q]N/1000$. Only those fluorophores without nearby quenchers are fluorescent. The probability that no quenchers are nearby is

$$P(0) = e^{-\lambda} \qquad [8.26]$$

Thus, the existence of the sphere of action reduces the proportion of observable fluorophores by the factor $\exp(-V[Q]N/1000)$, which in turn yields Eq. [8.24]. Division of the values of F_0/F by $\exp(V[Q]N/1000)$ corrects the steady-state intensities for this effect and reveals the dynamic portion of the observed quenching (Figures 8.6 and 8.7). For simplicity, the static term is often expressed in terms of reciprocal concentration.

8.7. EFFECTS OF STERIC SHIELDING AND CHARGE ON QUENCHING

The extent of quenching can be affected by the environment surrounding the fluorophore. One example is the quenching of the steroid DHE by acrylamide. When free in solution, DHE is readily quenched by acrylamide. How-

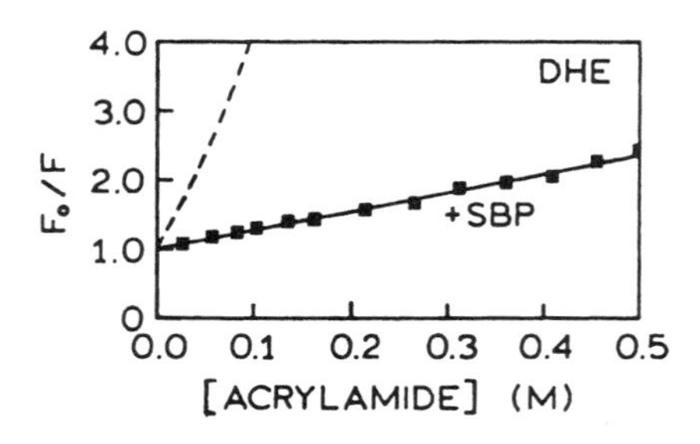

Figure 8.8. Acrylamide quenching of DHE when free in solution (---) and when bound to steroid binding protein (SBP; ■). Revised and reprinted, with permission, from Ref. 20, Copyright © 1990, American Chemical Society.

ever, when bound to a steroid binding protein (SBP), much less quenching occurs (Figure 8.8). In fact, the modest amount of quenching observed was attributed to dissociation of DHE from the protein.[20] Protection from quenching is frequently observed for probes bound to macromolecules[22,23] and even cyclodextrins.[24] In fact, binding of probes to cyclodextrins has been used as a means of obtaining room-temperature phosphorescence.[25] The macromolecules or cyclodextrins provide protection from the solvent but usually not complete protection from diffusing quenchers. Such solutions are usually purged to remove dissolved oxygen in order to observe phosphorescence.

The electronic charge on the quenchers can also have a dramatic effect on the extent of quenching (Figure 8.9). This is illustrated by quenching of 1-ethylpyrene (EP) in micelles, where the detergent molecules have different charges.[26] The quencher was *p-N,N*-dimethylaniline sulfonate (DMAS), which is negatively charged. The micelles were positively charged [dodecyltrimethylammonium chloride (DTAC)], neutral (Brij 35), or negatively charged [sodium dodecyl sulfate (SDS)]. There is extensive quenching of EP in the positively charged DTAC micelles, and essentially no quenching in the negatively charged SDS micelles. In general, one can expect charge effects to be present with charged quenchers such as iodide and to be absent for neutral quenchers like oxygen and acrylamide.

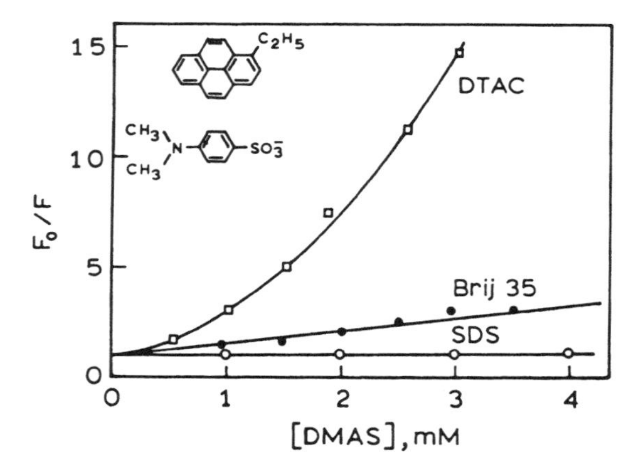

Figure 8.9. Quenching of 1-ethylpyrene (EP) by *p-N,N*-dimethylaniline sulfonate (DMAS), in positively charged micelles of dodecyltrimethylammonium chloride (DTAC), neutral micelles of Brij 35, or negatively charged micelles of sodium dodecyl sulfate (SDS). Revised from Ref. 26.

8.7.A. Accessibility of DNA-Bound Probes to Quenchers

The most dramatic effects of charge and shielding on quenching have been observed for fluorophores bound to DNA. One can expect the extent of quenching to be decreased by intercalation of probes into the DNA double helix. For instance, EB bound to DNA was found to be protected from oxygen quenching by a factor of 30 as compared to EB in solution.[14] Given the high negative charge density of DNA, one can expect the quenching to be sensitive to the charge of the quencher, the ionic strength of the solution, and the rate of quencher diffusion near the DNA helix.[27,28]

Collisional quenching by oxygen was used to study quenching of several DNA-bound probes.[29,30] Oxygen was chosen as the quencher because it is neutral and should thus be unaffected by the charge on the DNA. The probes were selected to have different sizes and different modes of binding to DNA (Figure 8.10). Proflavine intercalates into double-helical DNA and was expected to be protected from quenching. In fact, the bimolecular quenching constant was less than 10% of the diffusion-controlled rate

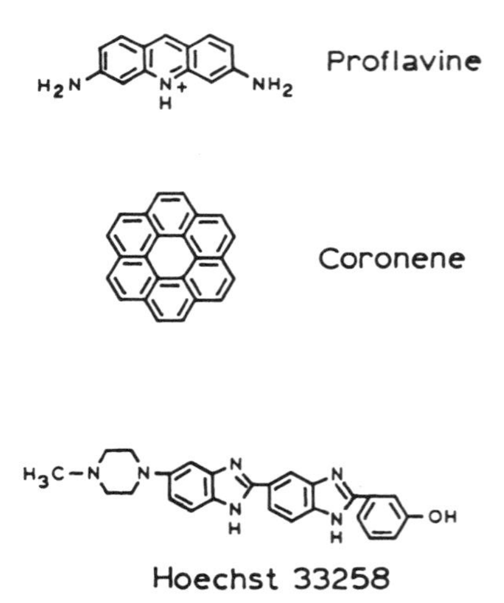

Figure 8.10. Structure of three probes bound to DNA (Table 8.2). From Ref. 30.

Table 8.2. Decay Times and Oxygen Quenching Constants of Probes Bound to DNA[a]

Fluorophore	Type of complex	τ_0 (ns)	k_q (M^{-1} s^{-1})
Proflavine	Intercalation	6.3	$<0.1 \times 10^{10}$
Coronene	Partial intercalation	225	0.17×10^{10}
Hoechst 33258	Minor groove	3.5	1.1×10^{10}

[a]Ref. 30.

(Table 8.2). The k_q value for proflavine may be smaller than shown, as there was little quenching under these experimental conditions. Hoechst 33258 is known to bind to the minor groove of DNA. Surprisingly, the k_q value for Hoechst 33258 bound to DNA was near the diffusion-controlled limit, suggesting complete accessibility by oxygen. The behavior of coronene was intermediate. Coronene is rather large and not able to fully fit into a DNA helix. The intermediate value of k_q, reflecting partial exposure to water, was explained as due to partial intercalation of coronene. These results illustrate how the extent of probe exposure can be correlated with the bimolecular quenching constant. Knowledge of the unquenched fluorescence lifetimes was essential for calculating the values of k_q from the Stern–Volmer quenching constants.

The extent of quenching can also be affected by the charge on the quenchers. This is illustrated by iodide quenching of Hoechst 33258 when free in solution and when bound to DNA (Figure 8.11). Hoechst 33258 is readily quenched by iodide when free in solution but is not quenched when bound to DNA. In the previous paragraph we saw that Hoechst 33258 bound to DNA was completely accessible to the neutral quencher oxygen. Apparently, the negative charges on DNA prevent iodide from coming into contact with Hoechst 33258 when bound to the minor groove of DNA.

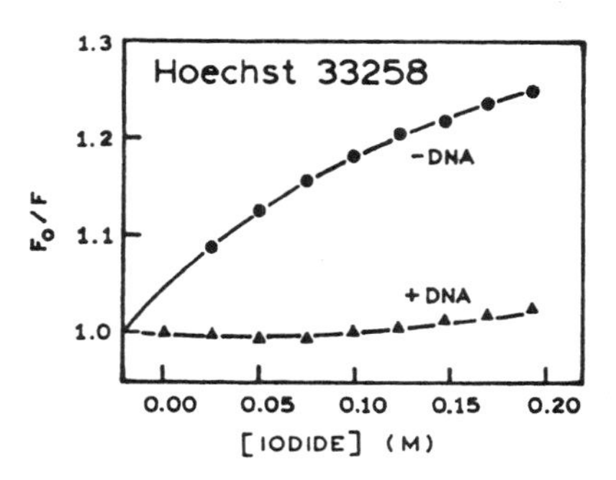

Figure 8.11. Iodide quenching of Hoechst 33258 in the absence (●) and presence (▲) of calf thymus DNA. The ionic strength was kept constant using KCl. Revised from Ref. 31.

8.7.B. Quenching of Ethenoadenine Derivatives

The nucleotide bases of DNA are mostly nonfluorescent. Fluorescent analogs of adenine nucleotides have been created by addition of an etheno bridge, the so-called ε-ATP derivatives (Chapter 3). Depending on the pH and extent of phosphorylation, the charge on the ethenoadenine nucleotides ranges from –3 for ε-ATP to 0 for ethenoadenosine. Hence, one expects the extent of quenching to depend on the charge of the quencher.

Stern–Volmer plots for the various ethenoadenine nucleotides are shown in Figure 8.12. For the neutral quencher acrylamide, there is no effect of charge. For the positively charged quencher Tl^+, the largest Stern–Volmer constant was observed for ε-ATP, with progressively smaller values as the number of negatively charged phosphates decreased. The opposite trend was observed for iodide quenching. Such effects of charge on quenching can be used to determine the local charge around fluorophores on proteins based on quenching by positive, neutral, and negatively charged quenchers.[32–34]

8.8. FRACTIONAL ACCESSIBILITY TO QUENCHERS

Proteins usually contain several tryptophan residues that are in distinct environments. Each residue can be differently accessible to quencher. Hence, one can expect complex Stern–Volmer plots and even spectral shifts due to selective quenching of exposed versus buried tryptophan residues. One example is quenching of lysozyme. This protein from egg white has six tryptophan residues, several of which are known to be near the active site. Lysozyme fluorescence was measured as a function of the concentration of trifluoroacetamide (TFA), which was found to be a collisional quencher of the fluorescence.[35] The Stern–Volmer plot curves downward toward the x-axis (Figure 8.13, left). As will be described below, this is a characteristic feature of two fluorophore populations, one of which is not accessible to the quencher. In the case of lysozyme, the emission spectrum shifts progressively to shorter wavelengths with increasing TFA concentration (Figure 8.13, right). This indicates that those tryptophan residues emitting at longer wavelengths are quenched more readily than the shorter-wavelength tryptophans.

The emission spectrum of the quenched residues can be calculated by taking the difference between the unquenched and quenched emission spectra. This difference spectrum (Figure 8.13, right) shows that the quenched residues display an emission maximum at 348 nm. The

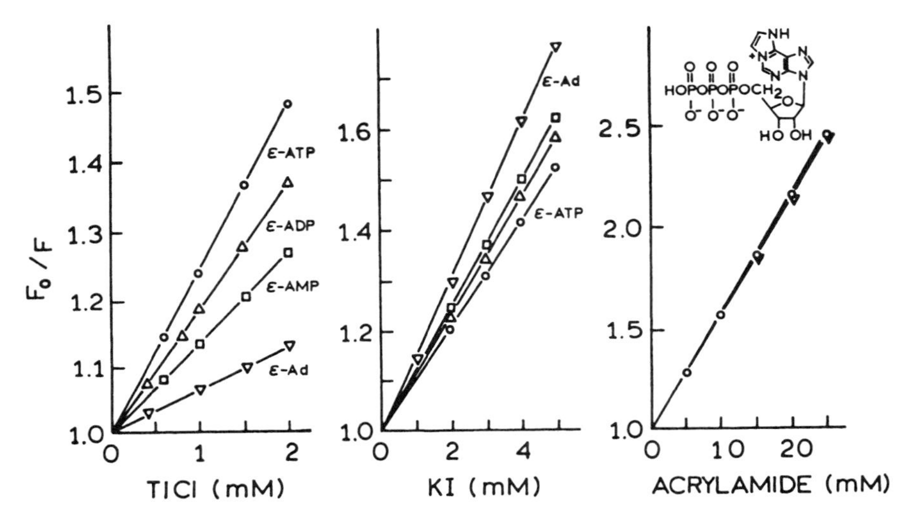

Figure 8.12. Quenching of (ε-ATP (○), ε-ADP (△), ε-AMP (□), and ε-Ad (▽) by Tl^+ (*left*), I^- (*middle*), and acrylamide (*right*) in 10m*M* phosphate buffer, 0.1*M* KCl, 20 °C, pH 7.0. Revised from Ref. 34.

protected residues display an emission maximum at 333 nm.

8.8.A. Modified Stern–Volmer Plots

The differing accessibilities of tryptophan residues in proteins have resulted in the frequent use of quenching to resolve the accessible and inaccessible residues.[36] Suppose that there are two populations of fluorophores, one being accessible (a) to quenchers and the other being inaccessible or buried (b). In this case the Stern–Volmer plot will display downward curvature (Figure 8.13). The total fluorescence in the absence of quencher (F_0) is given by

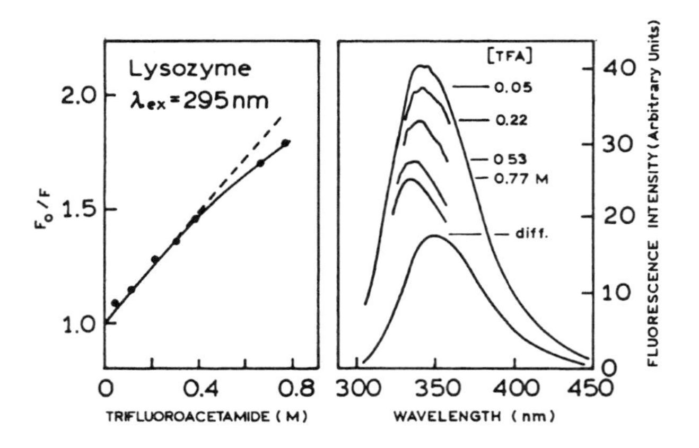

Figure 8.13. Quenching of lysozyme by trifluoroacetamide (TFA). *Left*: Stern–Volmer plot. *Right*: Emission spectra with increasing concentrations of TFA. Also shown is the difference spectrum (diff), 0.0*M* – 0.77*M* TFA. Revised and reprinted from Ref. 35, Copyright © 1984, with permission from Elsevier Science.

$$F_0 = F_{0a} + F_{0b} \quad [8.27]$$

where the 0 subscript once again refers to the fluorescence intensity in the absence of quencher. In the presence of quencher, the intensity of the accessible fraction (f_a) is decreased according to the Stern–Volmer equation, whereas the buried fraction is not quenched. Therefore, the observed intensity is given by

$$F = \frac{F_{0a}}{1 + K_a[Q]} + F_{0b} \quad [8.28]$$

where K_a is the Stern–Volmer quenching constant of the accessible fraction and [Q] is the concentration of quencher. Subtraction of Eq. [8.28] from Eq. [8.27] yields

$$\Delta F = F_0 - F = F_{0a}\left(\frac{K_a[Q]}{1 + K_a[Q]}\right) \quad [8.29]$$

Inversion of Eq. [8.29] followed by division into Eq. [8.27] yields

$$\frac{F_0}{\Delta F} = \frac{1}{f_a K_a[Q]} + \frac{1}{f_a} \quad [8.30]$$

where f_a is the fraction of the initial fluorescence which is accessible to quencher,

$$f_a = \frac{F_{0a}}{F_{0b} + F_{0a}} \quad [8.31]$$

This modified form of the Stern–Volmer equation allows f_a and K_a to be determined graphically (Figure 8.14). A plot of $F_0/\Delta F$ versus $1/[Q]$ yields f_a^{-1} as the intercept and $(f_aK_a)^{-1}$ as the slope. A y-intercept of f_a^{-1} may be understood intuitively. The intercept represents the extrapolation to infinite quencher concentration ($1/[Q] = O$). The value of $F_0/(F_0 - F)$ at this quencher concentration represents the reciprocal of the fluorescence which was quenched. At high quencher concentration, only the inaccessible residues will be fluorescent.

In separation of the accessible and inaccessible fractions of the total fluorescence, it should be realized that there may be more than two classes of tryptophan residues. Also, even the presumed "inaccessible" fraction may be partially accessible to quencher. This possibility is illustrated by the dashed curves in Figure 8.14, which show the expected result if the Stern–Volmer constant for the buried fraction (K_b) is one-tenth of that for the accessible fraction ($K_b = 0.1K_a$). For a limited range of quencher concentrations, the modified Stern–Volmer plot can still appear to be linear. The extrapolated value of f_a would represent an apparent value somewhat larger than seen with $K_b = 0$. Hence, the modified Stern–Volmer plots provide a useful but arbitrary resolution of two assumed classes of tryptophan residues.

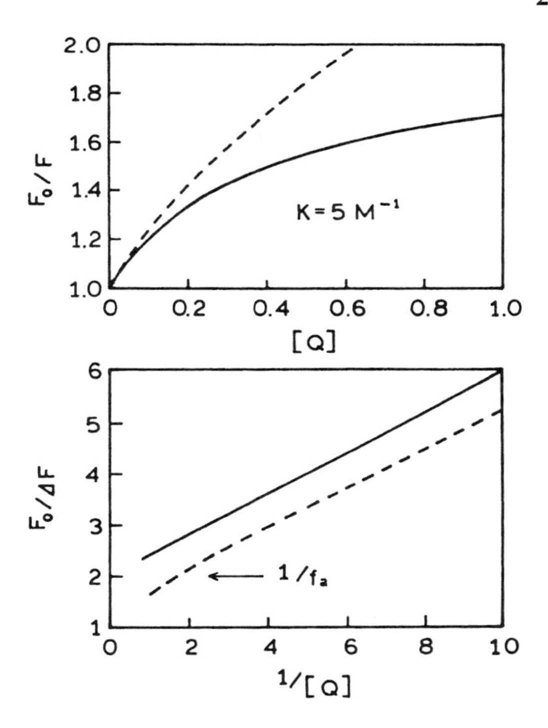

Figure 8.14. Stern–Volmer (*top*) and modified Stern–Volmer plots (*bottom*) for two populations of fluorophores, one of which is inaccessible to quencher. The dashed curves show the effect of the "inaccessible" population being quenched with a K value one-tenth of that for the accessible population.

8.8.B. Experimental Considerations in Quenching

Although quenching experiments are straightforward, there are several potential problems. One should always examine the emission spectra under conditions of maximum quenching. As the intensity is decreased, the contribution from background fluorescence may begin to be significant. Quenchers are often used at high concentrations, and the quenchers themselves may contain fluorescent impurities. Also, the intensity of the Raman and Rayleigh scatter peaks is independent of quencher concentration. Hence, the relative contribution of scattered light always increases with quenching.

It is also important to consider the absorption spectra of the quenchers. Iodide and acrylamide absorb light below 290 nm. The inner filter effect due to their absorption can decrease the apparent fluorescence intensity and thereby distort the quenching data. Regardless of the quencher being used, it is important to determine if the inner filter effects are significant. If inner filter effects are present, the observed fluorescence intensities must be corrected. The lifetime measurements are mostly independent of inner filter effects because these measurements are relatively independent of total intensity.

When iodide or other ionic quenchers are used, it is important to maintain a constant ionic strength. This is usually accomplished by addition of KCl. When iodide is used, it is also necessary to add a reducing agent such as $Na_2S_2O_3$. Otherwise, I_2 is formed, which is reactive and can partition into the nonpolar regions of proteins and membranes.

8.9. APPLICATIONS OF QUENCHING TO PROTEINS

8.9.A. Fractional Accessibility of Tryptophan Residues in Endonuclease III

Since the pioneering study of lysozyme quenching,[36] there have been numerous publications on determining the fraction of protein fluorescence accessible to quenchers.[37–41] In Section 16.5 we show that proteins in the native state often display a fraction of the emission which is not accessible to water-soluble quenchers and that denaturation of the proteins usually results in accessibility of all the tryptophan residues to quenchers. The possibility of buried and exposed residues in a single protein is illustrated by endonuclease III (endo III). Endo III is a DNA repair enzyme which displays both N-glycosylase and apurinine/apyrimidinic lyase activities. The structure of

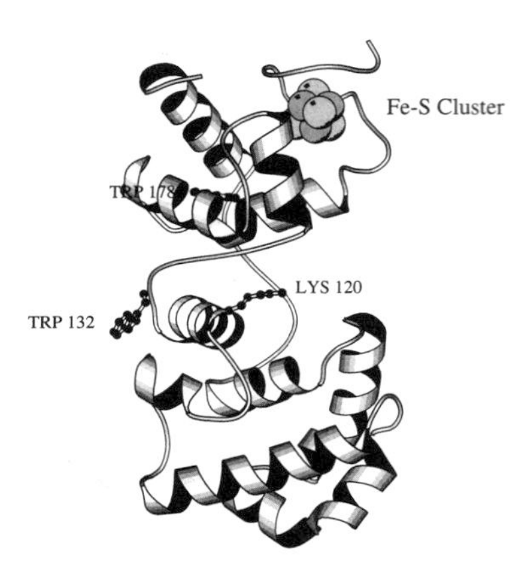

Figure 8.15. Structure of endo III, showing the exposed residue trp-132 and the buried residue trp-178 near the iron–sulfur cluster. Courtesy of Dr. Charles P. Scholes.

endo III shows two domains, with the DNA binding site in the cleft region. Endo III contains two tryptophan residues.[40] Trp-132 is exposed to the solvent, and trp-178 is buried in one of the domains (Figure 8.15). Hence, one expects these residues to be differently accessible to water-soluble quenchers.

The modified Stern–Volmer plot for quenching of endo III by iodide shows clear evidence for a shielded fraction (Figure 8.16). Extrapolation to high iodide concentrations yields an intercept near 2, indicating that only half of the emission can be quenched by iodide. This suggests that both trp residues in endo III are equally fluorescent and that only one residue (trp-132) can be quenched by iodide. Similar results have been obtained for a large number of proteins, and the extent of quenching is known to depend on the size and polarity of the quenchers.[6,7] Quenching of solvent-exposed residues in proteins is now a standard tool in the characterization of proteins.

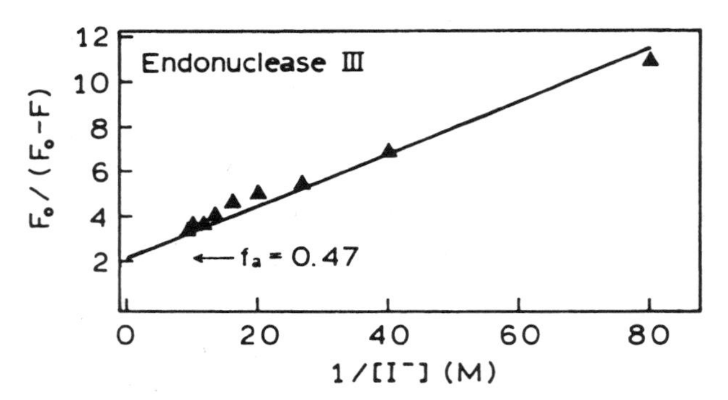

Figure 8.16. Modified Stern–Volmer plot for iodide quenching of endo III, showing evidence for two types of tryptophan residues. The inaccessible fraction is $f_a = 0.47$. Revised from Ref. 40.

8.9.B. Effect of Conformational Changes on Tryptophan Accessibility

The conformational state of a protein can have an influence on the exposure of its tryptophan residues to solvent. This is illustrated by the cyclic AMP receptor protein (CRP) from *Escherichia coli*.[41] This protein regulates the expression of over 20 genes in *E. coli*. CRP consists of two identical polypeptide chains, each containing 209 amino acids. It contains two nonidentical trp residues at positions 13 and 85.

Stern–Volmer plots for the quenching of CRP by acrylamide in the absence and in the presence of bound cyclic AMP (cAMP) are shown in Figure 8.17. In the absence of cAMP, the Stern–Volmer plot shows obvious downward curvature, indicating that one of the trp residues is inaccessible or only slightly accessible to acrylamide. Binding of cAMP results in a dramatic change in the Stern–Volmer plot, which becomes linear. Apparently, binding of cAMP to the CRP causes a dramatic conformational change which results in exposure of the previously shielded trp residue. Changes in accessibility due to conformational changes have been reported for other proteins.[42] Binding of substrates to proteins can also result in shielding of tryptophan, as has been observed for lysozyme[36] and for wheat germ agglutinin.[35]

8.9.C. Quenching of the Multiple Decay Times of Proteins

The intensity decays of proteins are typically multiexponential. Hence, it is natural to follow the individual decay times as the protein is exposed to increasing concentrations of quencher. One example is provided by the investigation of the CRP protein just described.[41] FD data for its intrinsic tryptophan emission yield decay times near 1.5 and 6.8 ns.

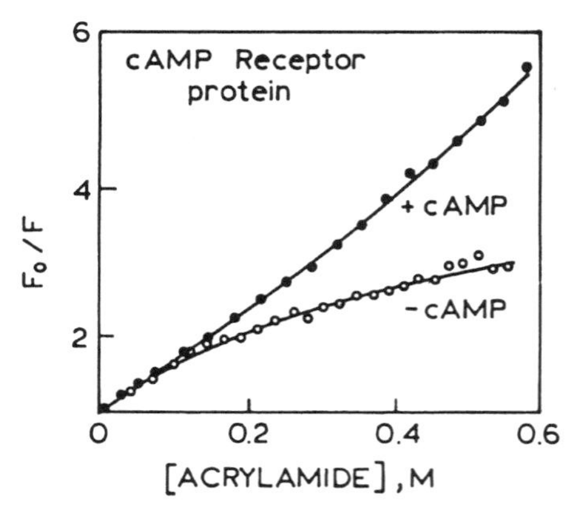

Figure 8.17 Stern–Volmer plot for acrylamide quenching of cAMP receptor protein in the absence (○) and in the presence (●) of cAMP. Revised from Ref. 41.

Similar decay times were observed in the absence and the presence of bound cAMP. For the protein without bound cAMP, the shorter decay time did not change with increasing iodide concentration (Figure 8.18, top), whereas the long lifetime decreased. This indicates that the 6.8-ns component is due to the exposed trp residue. The 1.5-ns component is not quenched by iodide and is assigned to the buried trp residue. In the presence of bound cAMP, both decay times are seen to decrease in the presence of iodide, indicating that both are quenched (Figure 8.18, bottom). These results agree with the linear Stern–Volmer plot found for CRP with bound cAMP (Figure 8.17).

While the results presented in Figure 8.18 show a clear separation of decay times, caution is needed when interpreting decay times in the presence of quencher. For many proteins, the decay times will be closer than 1.5 and 6.8 ns, and the decay time for each trp residue can depend on emission wavelength. Hence, it may not be possible to assign a unique decay time to each tryptophan residue. Additionally, collisional quenching results in nonexponential decays, even if the fluorophore shows a single decay time in the absence of quencher. This change in the intensity decay is due to transient effects in quenching, which are due to the rapid quenching of closely spaced flourophore–quencher pairs, followed by a slower quenching rate due to quencher diffusion. The presence of transient effects results in additional nanosecond decay times. The apparent lifetimes for each residue will be weighted averages that depend on the method of measurement. The assignment of decay times to trp residues in the presence of quenching can be ambiguous. Transient effects in quenching are described in the following chapter.

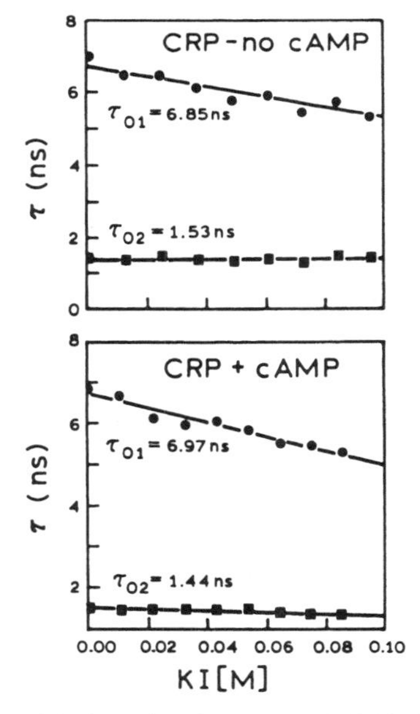

Figure 8.18. Iodide-dependent decay times of CRP in the absence (*top*) and in the presence (*bottom*) of bound cAMP. Revised from Ref. 41.

8.9.D. Effects of Quenchers on Proteins

When one performs quenching experiments, it is important to consider whether the quencher has an adverse effect on the protein. Some quenchers, such as 2,2,2-trichloroethanol, are known to bind to proteins and induce conformational changes.[43] For a time it was thought that acrylamide bound to proteins, but it is now accepted that such binding does not occur except in several specific cases.[44–46] However, even the nonperturbing quencher acrylamide can affect certain proteins, as was found for glyceraldehyde-3-phosphate dehydrogenase (GAPDH).[47] This protein contains three tryptophan residues in each subunit of the tetrameric enzyme. Quenching of the apoenzyme, which lacks NAD^+, by acrylamide yields a Stern–Volmer plot that is highly unusual. The extent of quenching increases rapidly above 0.4*M* acrylamide (Figure 8.19). This effect is not seen for the holoenzyme, which contains bound NAD^+. Acrylamide also caused a slow loss of activity and reduction in the number of thiol groups. Acrylamide appears to bind to GAPDH, reacting with the protein and destroying its activity.

8.9.E. Protein Folding of Colicin E1 • Advanced Topic •

Colicin E1 is a 522-residue polypeptide which is lethal to strains of *E. coli* that do not contain the resistance plasmid. Colicin E1 exerts its toxic effects by forming a channel in the cytoplasmic membrane which depolarizes and deenergizes the cell. The active channel-forming domain consists

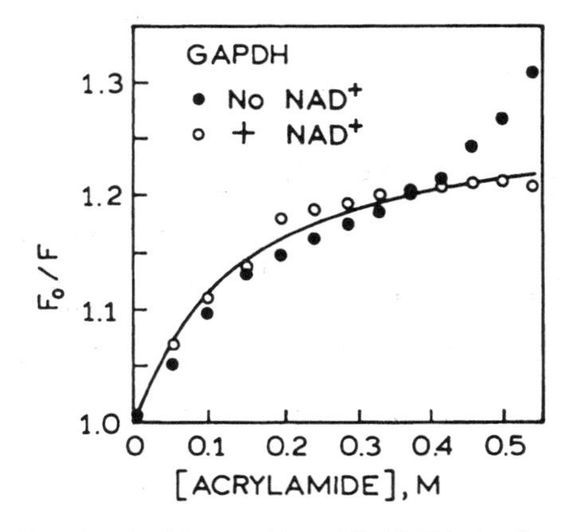

Figure 8.19. Acrylamide quenching of GAPDH in the absence (●) and in the presence (○) of the cofactor NAD^+. Revised from Ref. 47.

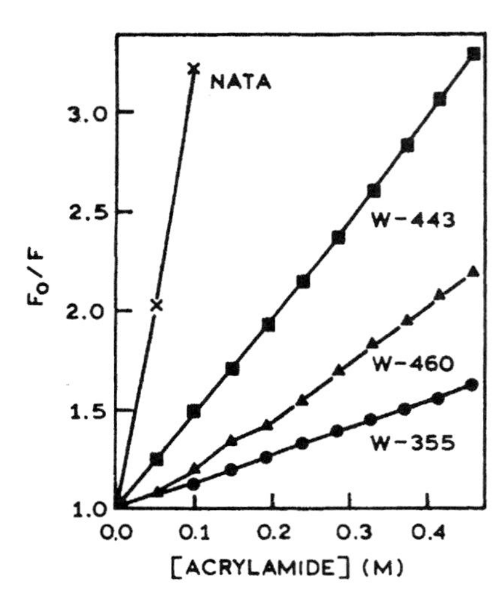

Figure 8.20. Stern–Volmer plots for acrylamide quenching of NATA (x) and three single-tryptophan mutants of the channel-forming peptide of colicin E1, W-355 (●), W-460 (▲), and W-443 (■). Revised from Ref. 48.

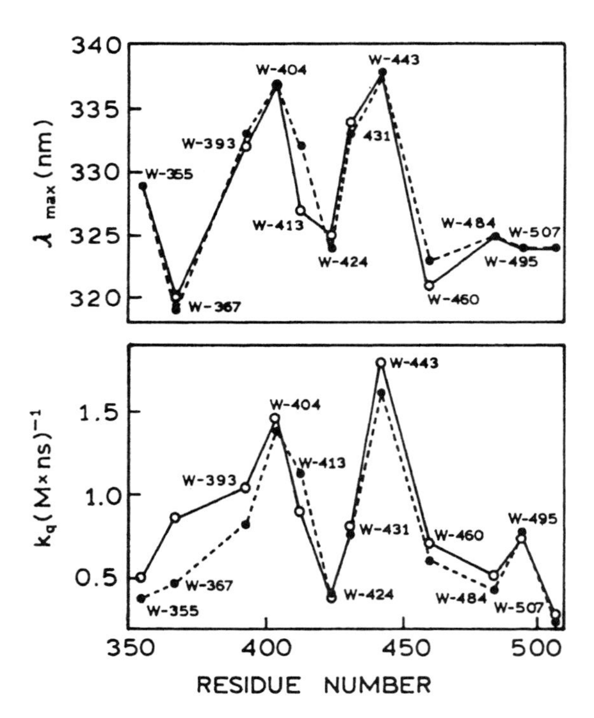

Figure 8.21. Emission maxima (*top*) and bimolecular quenching constants (*bottom*) of 12 single-tryptophan mutants of the channel-forming peptide of colicin E1. Revised and reprinted, with permission, from Ref. 48, Copyright © 1993, American Chemical Society.

of about 200 residues from the carboxy terminus, which form 10 α-helices spanning the membrane.

The conformation of the membrane-bound form of the colicin E1 channel peptide was studied by acrylamide quenching.[48] Twelve single-tryptophan mutants were formed by site-directed mutagenesis. The tryptophan residues were mostly conservative replacements, meaning that the trp residues were placed in positions previously containing phenylalanine or tyrosine. Stern–Volmer plots for acrylamide quenching of three of these mutant proteins and of NATA are shown in Figure 8.20. The accessibility to acrylamide quenching is strongly dependent on the location of the residue, and all residues are shielded relative to NATA. Depending on position, the trp residues also showed different emission maxima (Figure 8.21). The acrylamide bimolecular quenching constants were found to closely follow the emission maxima, with lower values of k_q for the shorter-wavelength tryptophans. Such data can be used to suggest a folding pattern for the channel-forming peptide and to reveal conformational changes that occur upon pH activation of colicin E1.

8.10. QUENCHING-RESOLVED EMISSION SPECTRA

• Advanced Topic •

8.10.A. Fluorophore Mixtures

As shown in Section 8.8, the emission spectra of proteins often shift in the presence of quenching. This effect occurs because the various trp residues are differently sensitive to quenching. Stated alternatively, for a mixture of fluorophores, quenching is expected to be dependent on the observation wavelength. This concept has been extended to calculation of the underlying emission spectra from the wavelength-dependent quenching data.[49–54] This is accomplished by measuring a Stern–Volmer plot for each emission wavelength (λ). For more than one fluorophore, the wavelength-dependent data can be described by

$$\frac{F(\lambda)}{F_0(\lambda)} = \sum_i \frac{f_i(\lambda)}{1 + K_i(\lambda)[Q]} \qquad [8.32]$$

where $f_i(\lambda)$ is the fractional contribution of the ith fluorophore to the steady-state intensity at wavelength λ, and $K_i(\lambda)$ is the Stern–Volmer quencher constant of the ith species at λ. For a single fluorophore, the quenching constant is usually independent of emission wavelength, i.e., $K_i(\lambda) = K_i$.

In order to resolve the individual emission spectra, the data are analyzed by nonlinear least-squares analysis.

Typically, one performs a global analysis in which the K_i values are global, and the $f_i(\lambda)$ values are variable at each wavelength. The result of the analysis is a set of K_i values, one for each component, and the fractional intensities $f_i(\lambda)$ at each wavelength, with $\Sigma f_i(\lambda) = 1.0$. The values of $f_i(\lambda)$ are used to calculate the emission spectrum of each component,

$$F_i(\lambda) = f_i(\lambda)F(\lambda) \qquad [8.33]$$

where $F(\lambda)$ is the steady-state emission spectrum of the sample.

The use of quenching-resolved spectra is illustrated by a sample containing both DPH and 5-((((2-iodoacetyl)amino)ethyl)amino)naphthalene-1-sulfonic acid (IAEDANS). These fluorophores were studied in SDS micelles, where their emission spectra are distinct (Figure 8.22). DPH is not soluble in water, so all the DPH is expected to be dissolved in the SDS micelles. IAEDANS is water-soluble and negatively charged, so it is not expected to bind to the negatively charged SDS micelles. Hence, IAEDANS is expected to be quenched by the water-soluble quencher acrylamide, and DPH is not expected to be accessible to acrylamide quenching.

Stern–Volmer plots for acrylamide quenching of the DPH–IAEDANS mixture as well as the individual fluorophores are shown in Figure 8.23. As predicted from the solubilities of the probes and acrylamide in water, DPH is weakly quenched by acrylamide. In contrast, IAEDANS is strongly quenched. The extent of quenching for the mixture is intermediate between that observed for each probe alone. As expected for a mixture of fluorophores, the

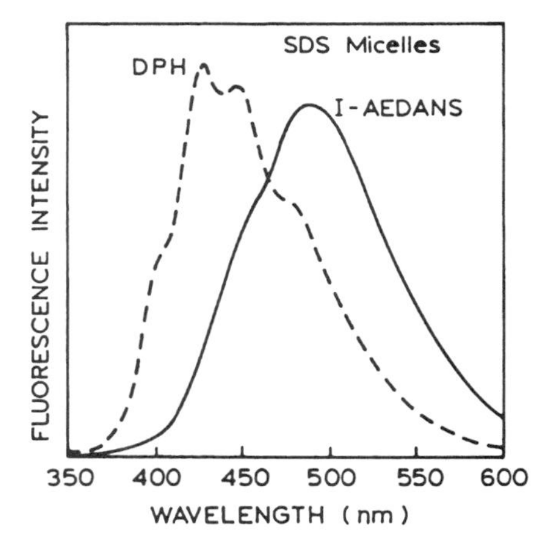

Figure 8.22. Steady-state emission spectra of DPH (---) and IAEDANS (—) in 1.2mM SDS micelles at 23 °C. The concentration of DPH was 2μM, and that of IAEDANS was 170μM. The excitation wavelength was 337 nm. Revised from Ref. 49.

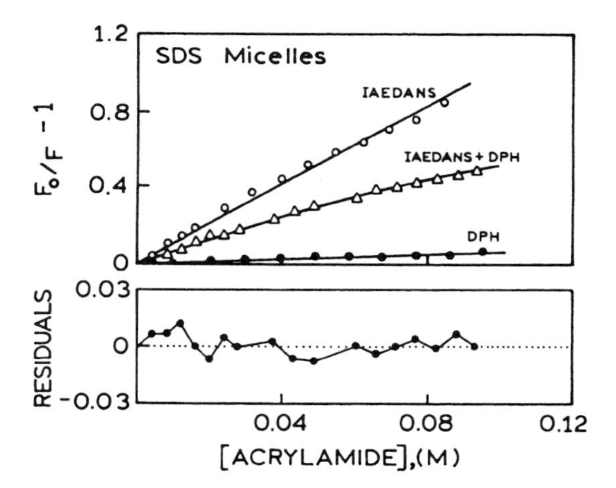

Figure 8.23. Stern–Volmer plots for acrylamide quenching of IAEDANS (○), DPH (●), and IAEDANS and DPH (△) in SDS micelles. For the mixture, the solid line represents the fit with calculated parameters $K_1 = 9.9\ M^{-1}$, $K_2 = 0\ M^{-1}$, $f_1 = 0.69$, and $f_2 = 0.31$ at 473 nm. The lower panel shows the residuals for this fit. Revised from Ref. 49.

Stern–Volmer plot curves downward due to the increasing fractional contribution of the more weakly quenched species at higher quencher concentrations.

The curvature in the Stern–Volmer plots was used to recover the values of $K_i(\lambda)$ and $f_i(\lambda)$ at each wavelength. In this case the $K_i(\lambda)$ values were not used as global parameters, so that $K_1(\lambda)$ and $K_2(\lambda)$ were obtained for each wavelength. At all wavelengths, there were two values near $0.5M^{-1}$ and $9.5M^{-1}$, representing the quenching constants of DPH and IAEDANS, respectively. At the shortest wavelength, below 420 nm, there is only one $K_i(\lambda)$ value because only DPH emits. The recovered values of $f_i(\lambda)$ were used to calculate the individual spectra from the spectrum of the mixture (Figure 8.24). In Chapters 4 and 5 we saw how the component spectra for heterogeneous samples could be resolved using the TD or the FD data. The use of wavelength-dependent quenching provides similar results, without the use of complex instrumentation. Of course, the method depends on the probes being differently sensitive to collisional quenching, which normally occurs if the decay times are different.

8.10.B. Quenching-Resolved Emission Spectra of the *E. coli* Tet Repressor

The Tet repressor from *E. coli* is a DNA-binding protein which controls the expression of genes that confer resistance to tetracycline. This protein contains two tryptophan residues at positions 43 and 75. W43 is thought to be an exposed residue, and W75 is thought to be buried in the protein matrix.[54] Earlier studies of single-tryptophan mu-

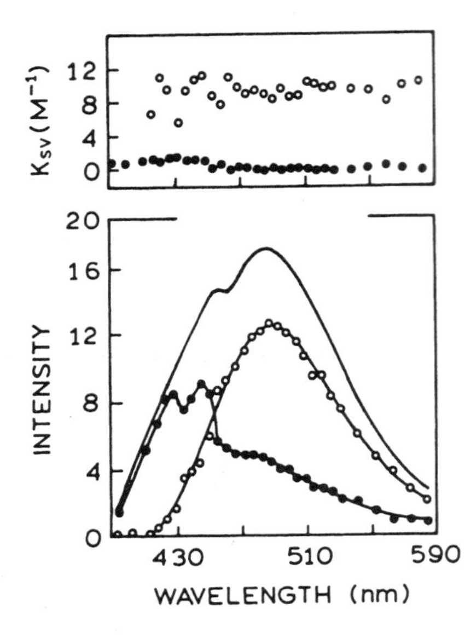

Figure 8.24. *Bottom*: Emission spectrum of the DPH–IAEDANS mixture and quenching-resolved emission spectra of DPH (●) and IAEDANS (○). *Top*: Wavelength dependence of the quenching constants, with average values of $K_1 = 9.6\ M^{-1}$ and $K_2 = 0.47\ M^{-1}$. Revised from Ref. 49.

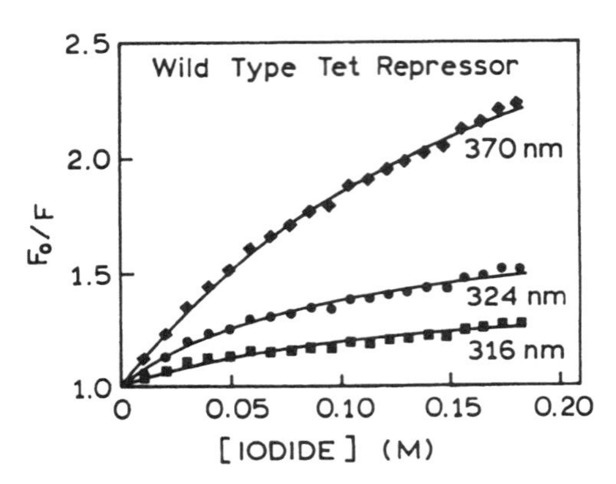

Figure 8.25. Stern–Volmer iodide plots for iodide quenching of the wild-type Tet repressor. The emission wavelengths are indicated on the figure. The solution contained 1 mM sodium thiosulfate to prevent formation of I_3^-. From Ref. 54.

tants of the Tet repressor confirmed the accessibility of W43 to iodide and the shielding of W75 from iodide quenching.[55] Hence, this protein provided an ideal model protein with which to attempt quenching resolution of the individual emission spectra of two tryptophan residues in a protein.

Stern–Volmer plots for iodide quenching of the Tet repressor were measured for various emission wavelengths[54] (Figure 8.25). A larger amount of quenching was observed for longer wavelengths. When the quenching data were analyzed in terms of two components, one of these components was found to be almost inaccessible to iodide. For instance, at 324 nm the recovered values are $f_1 = 0.34$, $K_1 = 16.2\ M^{-1}$, $f_2 = 0.66$, and $K_2 = 0$. The wavelength-dependent data were used to calculate the individual spectra (Figure 8.26). The blue-shifted spectrum with a maximum at 324 nm corresponds to the inaccessible fraction, and the red-shifted spectrum with a maximum at 349 nm is the fraction accessible to iodide quenching. These emission spectra are assigned to W-75 and W-43, respectively.

The results in the top panel of Figure 8.26 illustrate one difficulty often encountered in the determination of quenching-resolved spectra. The quenching constant for a single species can be dependent on emission wavelength. In this case the quenching constant of the accessible tryptophan changed about twofold across its emission spectrum. When this occurs, the values of $K_i(\lambda)$ cannot be treated as global parameters.

The assignments of the quenching-resolved spectra are consistent with the results obtained using single-tryptophan mutants of the Tet repressor[55] (Figure 8.27). Little, if any, quenching was observed for the protein containing only W-75, and W-43 was readily quenched by iodide. Iodide quenching of the wild-type protein is intermediate between that of the two single-tryptophan mutants. While

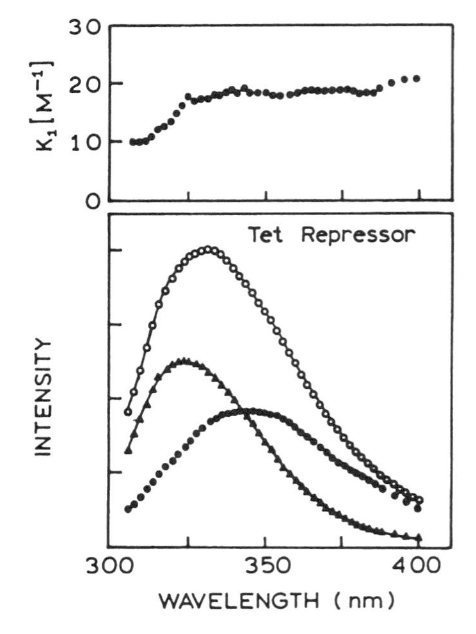

Figure 8.26. *Bottom*: Fluorescence quenching-resolved spectra of wild-type Tet repressor, obtained using potassium iodide as the quencher. *Top*: Wavelength-dependent values of K_1. From Ref. 54.

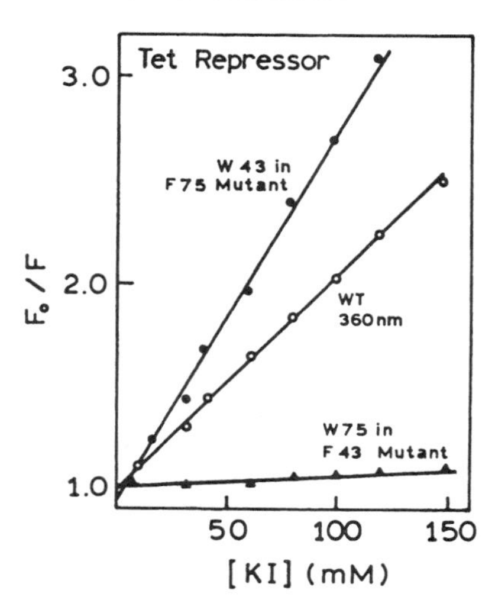

Figure 8.27. Stern–Volmer plots for the iodide quenching of *E. coli* Tet repressor (wild type, WT) and its mutants (F43 and F75). Revised from Ref. 55.

the same information is available from the mutant proteins, the use of quenching provided the resolved spectra with the use of only the wild-type protein.

It is valuable to notice a difference in the method of data analysis for the modified Stern–Volmer plots (Section 8.8.A) and for the quenching-resolved emission spectra. In analyzing a modified Stern–Volmer plot, one assumes that a fraction of the fluorescence is totally inaccessible to quenchers. This may not be completely true because one component can be more weakly quenched, but still quenched to some extent. If possible, it is preferable to analyze the Stern–Volmer plots by nonlinear least-squares analysis when the f_i and K_i values are variable. With this approach, one allows each component to contribute to the data according to its fractional accessibility, instead of forcing one component to be an inaccessible fraction. Of course, such an analysis is more complex, and the data may not be adequate to recover the values of f_i and K_i at each wavelength.

8.11. QUENCHING AND ASSOCIATION REACTIONS

8.11.A. Quenching Due to Specific Binding Interactions

In the preceding sections we considered quenchers that were in solution with the macromolecule but did not display any specific interactions. Such interactions can occur and can appear to be of static quenching.[56–59] One example is provided by serum albumin. Serum albumin consists of a single polypeptide chain, with a molecular weight near 65,000. Albumin is present in high concentrations in blood serum, 35–45 mg/ml, and is important for maintaining osmotic balance and for transport of hydrophobic species. Albumins have hydrophobic sites which are known to bind fatty acids and many fluorescent probes. Human serum albumin (HSA) has a single tryptophan residue, and bovine serum albumin (BSA) has two tryptophan residues.

BSA and HSA can bind halogenated anesthetics. This binding is seen by the effect of chloroform ($CHCl_3$) on the fluorescence intensity of BSA (Figure 8.28). Addition of $CHCl_3$ is seen to result in a progressive decrease in the fluorescence intensity of BSA. In the paper in which these results were reported,[58] there was no mention of photochemical effects, which we found surprising; in our experience, excitation of tryptophan in the presence of $CHCl_3$ results in the formation of blue fluorescent species.

The dependence of the BSA emission on the $CHCl_3$ concentration is shown in Figure 8.29 along with the effects of $CHCl_3$ on the emission of tryptophan and another protein, apomyoglobin. The effect of $CHCl_3$ on BSA is greater than that seen for trp or apomyoglobin, suggesting a specific interaction with BSA. The fact that there is hydrophobic binding of $CHCl_3$ to BSA is shown by the effect of trifluoroethanol (TFE), which disrupts hydrophobic binding. In 50% TFE, $CHCl_3$ no longer quenches BSA to the extent seen in water.

How can one determine whether the quenching seen for BSA in water is due to $CHCl_3$ binding or to collisional quenching? One method is to calculate the apparent bimolecular quenching constant (k_q^{app}). Assume that the decay time of BSA is near 5 ns. The data in Figure 8.29 indicate a K_{SV} value of 400 M^{-1}. The fluorescence is 50% quenched at 2.5mM $CHCl_3$. These values correspond to a bimolecular quenching constant of $k_q^{app} = 8 \times 10^{10}\ M^{-1}\ s^{-1}$, which is about 10-fold larger than the maximum value possible for diffusion-limited quenching

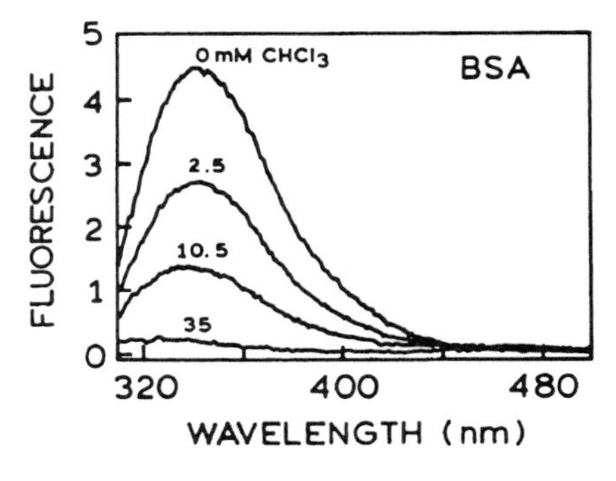

Figure 8.28. Emission spectra of BSA in the presence of various concentrations of chloroform. Revised from Ref. 58.

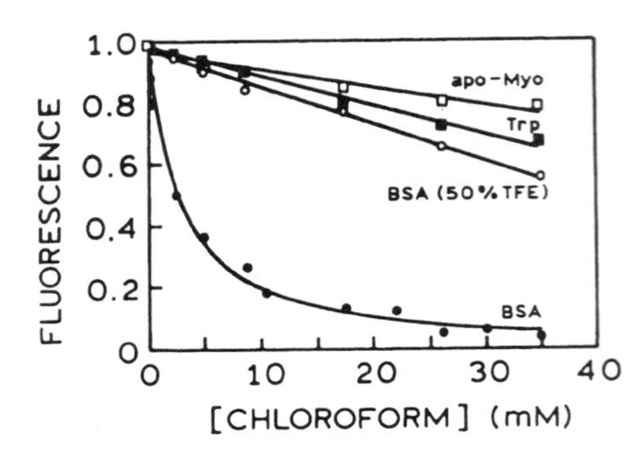

Figure 8.29. Fluorescence intensities of BSA (●), tryptophan (■), and apomyoglobin (□) in the presence of various amounts of dissolved $CHCl_3$. ○, BSA in 50% trifluoroethanol (TFE), which disrupts hydrophobic binding to BSA. Revised from Ref. 58.

in water. Hence, there must be some interaction that increases the local concentrations of $CHCl_3$ around the trp residues in BSA. Since there is no reason to expect ground-state complex formation between trp and $CHCl_3$, the quenching may be dynamic. However, the $CHCl_3$ is probably bound in close proximity to the trp residues, giving the appearance of a dark complex.

Another example of quenching due to a specific binding interaction is shown in Figure 8.30 for binding of caffeine to HSA, an experiment most of us start each morning. The Stern–Volmer plots show a value of $K_{SV} \simeq 7150\ M^{-1}$. This value is obviously too large to be due to collisional quenching, especially for a lifetime near 5 ns. The value of k_q^{app} is $1.4 \times 10^{12}\ s^{-1}$, over 100-fold larger than the maximum diffusion-limited rate. Hence, the caffeine must be bound to the HSA. Caffeine is an electron-deficient molecule and may form ground-state complexes with indole. This possibility could be tested by examination of the absorption spectra of HSA in the absence and presence of caffeine. If ground-state association with indole occurs, then the trp absorption spectrum is expected to change. Another indicator of complex formation is the temperature dependence of the Stern–Volmer plots. For diffusive quenching, one expects more quenching at higher temperatures. In the case of HSA and caffeine, there is less quenching at higher temperatures (Figure 8.30), which suggests that the complex is less stable at higher temperatures.

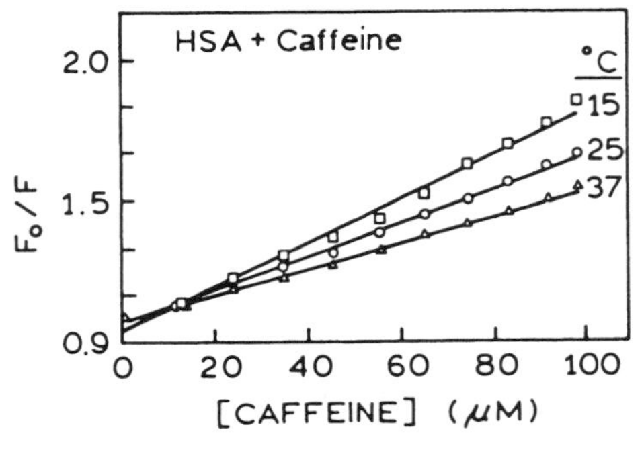

Figure 8.30. Quenching of HSA by caffeine at three temperatures. Revised from Ref. 59.

8.11.B. Binding of Substrates to Ribozymes

The catalytic properties of highly structured RNAs have been the subject of numerous studies since 1982, when it was first reported that certain RNAs, termed ribozymes, could display enzymatic activity. One example is the hairpin ribozyme, which cleaves single-stranded RNA (Figure 8.31). The early observations of quenching by nucleotides[17,18] (Section 8.5) provided the opportunity to study substrate binding to ribozymes. The substrate contained a fluorescein residue covalently linked at the 3′-end. The hairpin ribozyme contains a guanosine residue at the 5′-end. Upon binding of substrate nucleotide to the ribozyme, the fluorescein emission is quenched (Figure 8.32). Quenching also occurs when the fluorescein-labeled substrate binds to the substrate-binding strand (SBS) which contains a 5′-guanosine residue (G-SBS). The guanosine residue is needed for quenching, and the emission of fluorescein is unchanged in the presence of the substrate-binding strand without a 5′-terminal guanosine residue (not shown in Figure 8.32). This example shows how

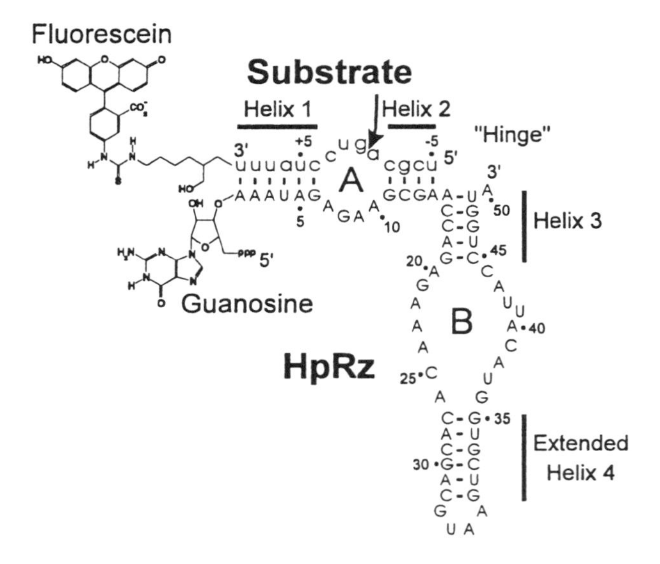

Figure 8.31. Structure of the hairpin ribozyme (HpRz) and the fluorescein-labeled substrate. The substrate-binding strand is the region adjacent to the substrate. Reprinted, with permission, from Ref. 60, Copyright © 1997, Cambridge University Press.

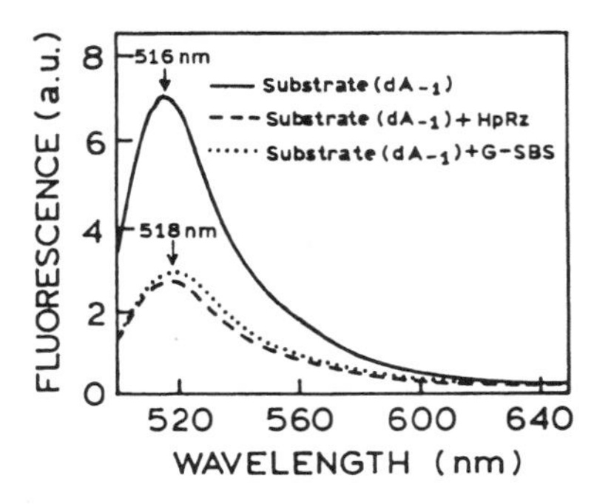

Figure 8.32. Emission spectra of the fluorescein-labeled substrate analog (dA_{-1}) in solution and when bound to the hairpin ribozyme (HpRz) or the guanosine-containing substrate-binding strand (G-SBS). From Ref. 60.

fundamental studies of nucleotide quenching have found useful applications in modern biochemistry.

8.11.C. Association Reactions and Quenching

Fluorescence is often used to measure association reactions. This requires that the fluorescence of one of the reactants changes upon binding. While this often occurs, it is not always the case. One example is the binding of ε-ADP to the DnaB helicase hexamer. The fluorescence of ε-ADP displayed only a small increase upon binding to the protein. Collisional quenching was used to induce a larger change in fluorescence on binding.[61] Acrylamide is an efficient quencher of ε-ADP, which should be quenched more strongly in solution than when bound to the helicase. Hence, if the binding is studied in a solution which contains acrylamide, there should be an increase in ε-ADP fluorescence on binding to the helicase.

Solutions of ε-ADP were titrated with the helicase (Figure 8.33). In the absence of acrylamide, there was little change in the ε-ADP fluorescence. The titrations were performed again, in solutions with increasing amounts of acrylamide. Under these conditions, ε-ADP showed an increase in fluorescence upon binding. This increase occurred because the ε-ADP became shielded from acrylamide upon binding to helicase. The authors also showed that acrylamide had no effect on the affinity of ε-ADP for helicase.[61] In this system the use of a quencher allowed measurement of a binding reaction that would otherwise be difficult to measure.

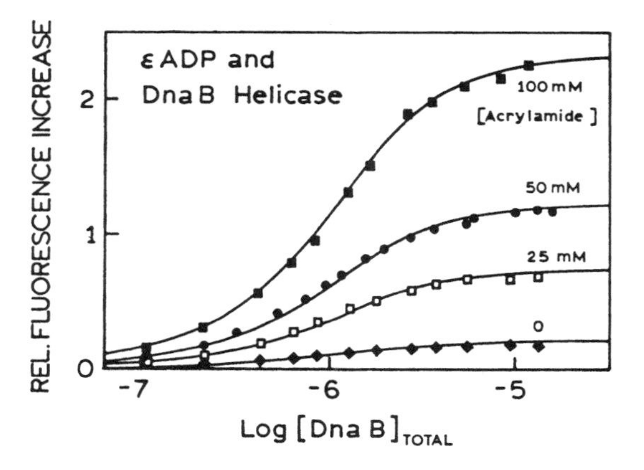

Figure 8.33. Fluorescence titration of ε-ADP, at a constant concentration of nucleotide, with the DnaB helicase in buffer containing different concentrations of acrylamide. Revised and reprinted from Ref. 61, Copyright © 1997, with permission from Elsevier Science.

8.12. INTRAMOLECULAR QUENCHING

Quenching can also occur between covalently linked fluorophore–quencher pairs.[62,63] One common example is the formation of exciplexes by covalently linked aromatic hydrocarbons and amines.[64,65] Another example is the covalent adduct formation by indole and acrylamide. For example, the lifetime of *N*-acetyltryptamine is near 5.1 ns. When the acetyl group is replaced by an acryloyl group, the lifetime is reduced to 31 ps (Figure 8.34). Similarly, covalent attachment of spin labels to a naphthalene derivative reduced its lifetime from 33.7 to 1.1 ns.

The concept of intramolecular quenching can be used to obtain structural information[66] as was demonstrated for a

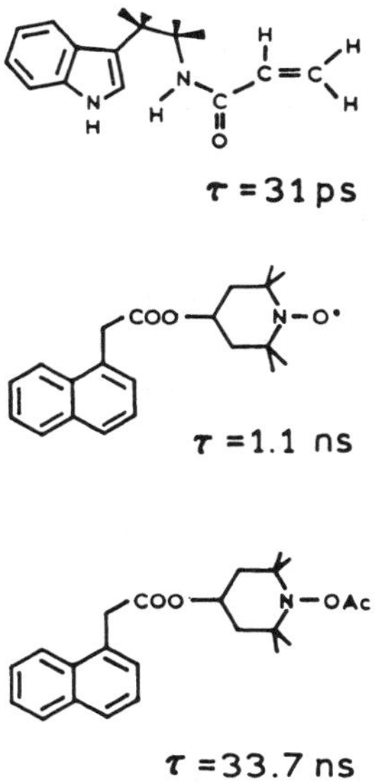

τ = 31 ps

τ = 1.1 ns

τ = 33.7 ns

Figure 8.34. Fluorophore–quencher conjugates which display intramolecular quenching.[62,63]

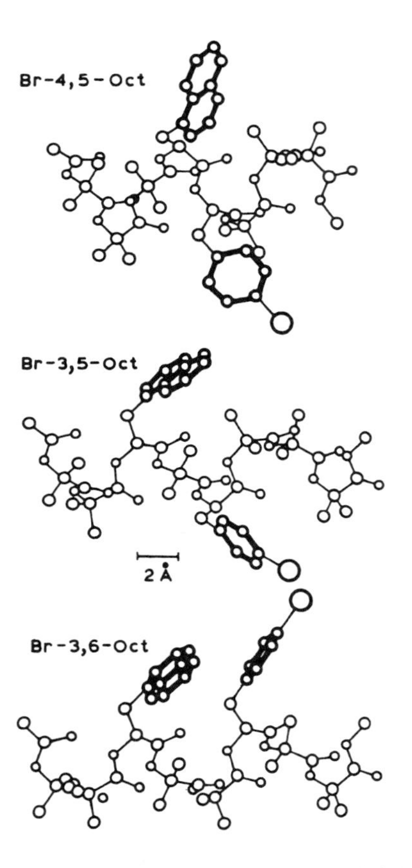

Figure 8.35. Structure of peptides containing naphthylalanine and *p*-bromophenylalanine, separated by 0 (*top*), 1 (*middle*), or 2 (*bottom*) amino acid residues. Revised and reprinted, with permission, from Ref. 66, Copyright © 1993, American Chemical Society.

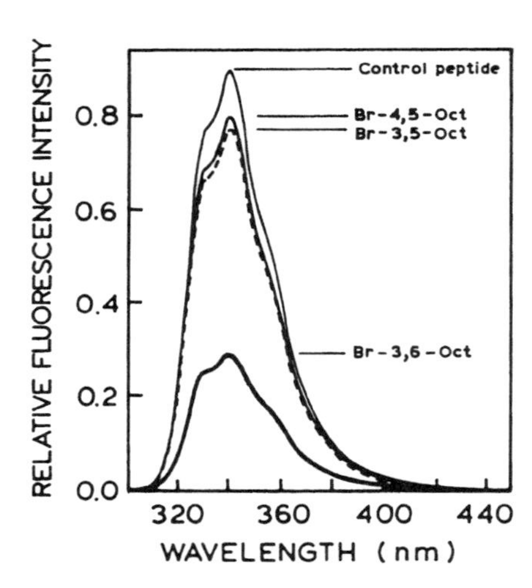

Figure 8.36. Emission spectra of the peptides shown in Figure 8.45. Also shown is the emission spectrum of a control peptide with a phenylalanine group in place of *p*-bromophenylalanine. Revised and reprinted, with permission, from Ref. 66, Copyright © 1993, American Chemical Society.

series of peptides containing a naphthylalanine fluorophore and a *p*-bromophenylalanine quencher (Figure 8.35). The probe and quencher were separated by 0, 1, or 2 amino acid residues. Emission spectra of these peptides show minimal quenching except for separation by 2 amino acid residues (Figure 8.36). In this case the fluorophore and quencher are adjacent, resulting in over 50% quenching of the naphthylalanine. These results suggest that, in solution, this peptide adopts a conformation with the bromo group near the indole ring.

8.13. QUENCHING OF PHOSPHORESCENCE

Phosphorescence is not usually observed in fluid solutions near room temperature. One reason for the absence of phosphorescence is the long phosphorescence lifetimes and the presence of dissolved oxygen and other quenchers. For instance, recent data for tryptophan revealed a phosphorescence lifetime of 1.2 ms at 20 °C.[67] Suppose that the oxygen bimolecular quenching constant is $1 \times 10^{10}\ M^{-1}\ s^{-1}$ and that the aqueous sample is in equilibrium with dissolved oxygen from the air (0.255mM O_2). Using Eq. [8.1], the intensity is expected to be quenched 3000-fold. For this reason, methods have been developed to remove dissolved oxygen from samples used to study phosphorescence.[68,69] In practice, other dissolved quenchers and nonradiative decay rates result in vanishingly small phosphorescence quantum yields in room-temperature solutions. Some exceptions are known, such as when fluorophores are located in highly protected environments within proteins.[70–73] Phosphorescence has also been observed at room temperature for probes bound to cyclodextrins, even in the pres-

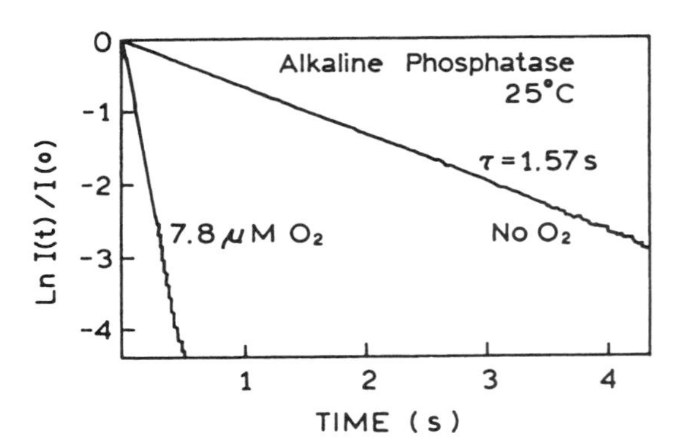

Figure 8.37. Phosphorescence decay of alkaline phosphatase at 25 °C in the absence of oxygen and in the presence of 7.8μM oxygen. Revised and reprinted, with permission, from Ref. 77, Copyright © 1987, Biophysical Society.

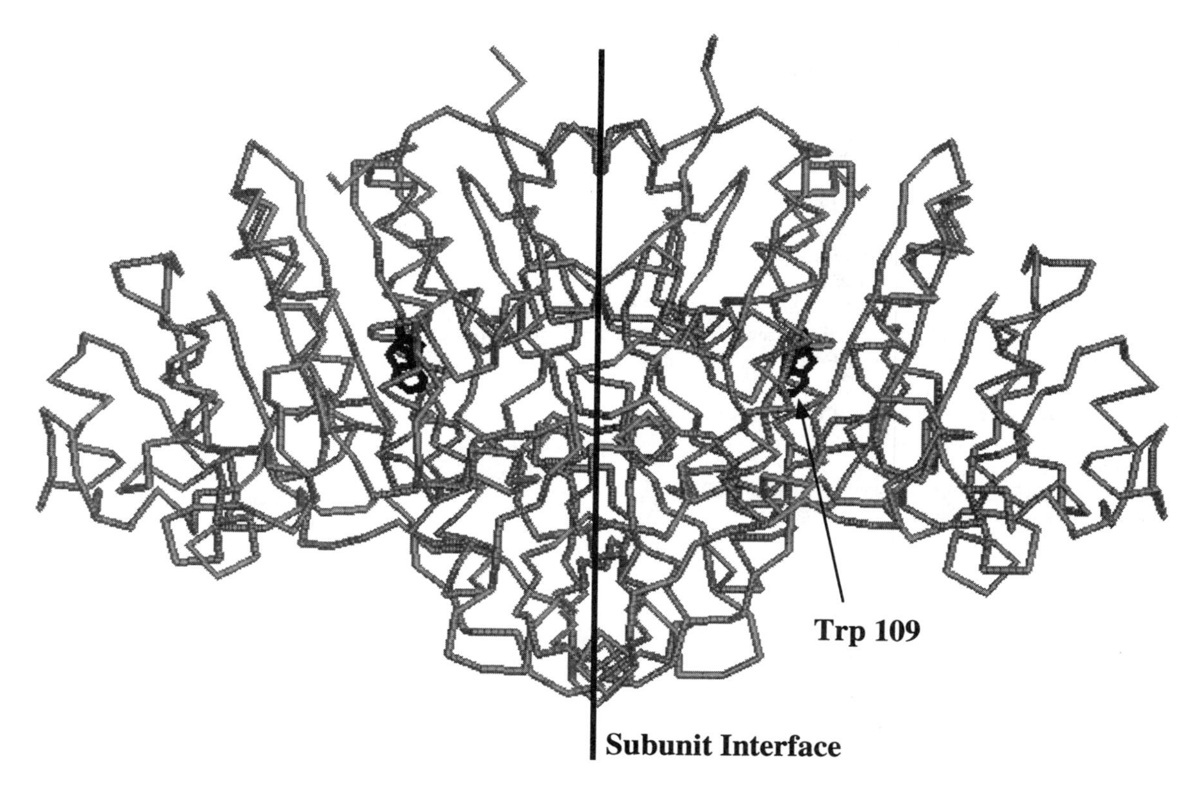

Figure 8.38. Structure of the alkaline phosphatase dimer, showing the phosphorescent residue trp-109. Courtesy of Dr. Ari Gafni, University of Michigan.

ence of oxygen.[74] However, in general, phosphorescence is not commonly observed near room temperature.

Protein phosphorescence can be quenched by a number of small molecules such as amino acids, H_2O, and CS_2,[75,76] in addition to oxygen. An example of the dramatic quenching of protein phosphorescence by even low concentrations of oxygen is shown for alkaline phosphatase phosphorescence[77] in Figure 8.37. It is now known that the phosphorescence from alkaline phosphatase results from one of its three nonidentical tryptophan residues, trp-109. This residue is located in a highly shielded environment near the dimer interface (Figure 8.38). In the presence of 7.8μ*M* oxygen, the phosphorescence lifetime is reduced from 1.57 s to 0.1 s. Note that an oxygen concentration of 7.8μ*M* would have an insignificant effect on nanosecond fluorescence. It is also important to notice that trp-109 in alkaline phosphatase is one of the most shielded residues identified to date in a protein. The lifetimes in the absence and presence of 7.8μ*M* oxygen can be used to calculate a value of $1.2 \times 10^6\ M^{-1}\ s^{-1}$ for the bimolecular quenching constant k_q. If the residue were more typical, with a k_q value of $0.1 \times 10^9\ M^{-1}\ s^{-1}$, one can calculate that the decay time in 7.8μ*M* oxygen would have been reduced to 1.28 ms, and thus quenched by over 1000-fold for a micromolar quencher concentration.

REFERENCES

1. Kautsky, H., 1939, Quenching of luminescence by oxygen, *Trans. Faraday Soc.* **35:**216–219.
2. Knibbe, H., Rehm, D., and Weller, A., 1968, Intermediates and kinetics of fluorescence quenching by electron transfer, *Ber. Bunsenges. Phys. Chem.* **72:**257–263.
3. Kasha, M., 1952, Collisional perturbation of spin–orbital coupling and the mechanism of fluorescence quenching. A visual demonstration of the perturbation, *J. Chem. Phys.* **20:**71–74.
4. Steiner, R. F., and Kirby, E. P., 1969, The interaction of the ground and excited states of indole derivatives with electron scavengers, *J. Phys. Chem.* **73:**4130–4135.
5. Eftink, M. R., and Ghiron, C., 1981, Fluorescence quenching studies with proteins, *Anal. Biochem.* **114:**199–227.
6. Eftink, M. R., 1991, Fluorescence quenching reactions: Probing biological macromolecular structures, in *Biophysical and Biochemical Aspects of Fluorescence Spectroscopy*, T. G. Dewey (ed.), Plenum Press, New York, pp. 1–41.
7. Eftink, M. R., 1991, Fluorescence quenching: Theory and applications, in *Topics in Fluorescence Spectroscopy, Volume 2, Principles*, J. R. Lakowicz (ed.), Plenum Press, New York, pp. 53–126.

8. Davis, G. A., 1973, Quenching of aromatic hydrocarbons by alkylpyridinium halides, *J. Chem. Soc., Chem. Commun.* **1973:**728–729.
9. Shinitzky, M., and Rivnay, B., 1977, Degree of exposure of membrane proteins determined by fluorescence quenching, *Biochemistry* **16:**982–986.
10. Spencer, R. D., and Weber, G., 1972, Thermodynamics and kinetics of the intramolecular complex in flavin-adenine dinucleotide, in *Structure and Function of Oxidation Reduction Enzymes*, A. Akeson and A. Ehrenberg (eds.), Pergamon Press, New York, pp. 393–399.
11. Scott, T. G., Spencer, R. D., Leonard, N. J., and Weber, G., 1970, Emission properties of NADH. Studies of fluorescence lifetimes and quantum efficiencies of NADH, AcPyADH, and simplified synthetic models, *J. Am. Chem. Soc.* **92:**687–695.
12. Ware, W. R., 1962, Oxygen quenching of fluorescence in solution: An experimental study of the diffusion process, *J. Phys. Chem.* **66:**455–458.
13. Othmer, D. F., and Thakar, M. S., 1953, Correlating diffusion coefficients in liquids, *Ind. Eng. Chem.* **45:**589–593.
14. Lakowicz, J. R., and Weber, G., 1973, Quenching of fluorescence by oxygen. A probe for structural fluctuations in macromolecules, *Biochemistry* **12:**4161–4170.
15. Eftink, M. R., and Ghiron, C. A., 1976, Fluorescence quenching of indole and model micelle systems, *J. Phys. Chem.* **80:**486–493.
16. Johnson, D. A., and Yguerabide, J., 1985, Solute accessibility to N′-fluorescein isothiocyanate-lysine-23, cobra α-toxin bound to the acetylcholine receptor, *Biophys. J.* **48:**949–955.
17. Kubota, Y., Nakamura, H., Morishita, M., and Fujisaki, Y., 1978, Interaction of 9-aminoacridine with 7-methylguanosine and $1,N^6$-ethenoadenosine monophosphate, *Photochem. Photobiol.* **27:**479–481.
18. Kubota, Y., Motoda, Y., Shigemune, Y., and Fujisaki, Y., 1979, Fluorescence quenching of 10-methylacridinium chloride by nucleotides, *Photochem. Photobiol.* **29:**1099–1106.
19. Seidel, C. A. M., Schulz, A., and Sauer, M. H. M., 1996, Nucleobase-specific quenching of fluorescent dyes. 1. Nucleobase one-electron redox potentials and their correlation with static and dynamic quenching efficiencies, *J. Phys. Chem.* **100:**5541–5553.
20. Casali, E., Petra, P. H., and Ross, J. B. A., 1990, Fluorescence investigation of the sex steroid binding protein of rabbit serum: Steroid binding and subunit dissociation, *Biochemistry* **29:**9334–9343.
21. Frank, I. M., and Vavilov, S. I., 1931, Über die wirkungssphäre der auslöschuns-vargänge in den flureszierenden flussigHeiten. *Z. Phys.* **69:**100–110.
22. Maniara, G., Vanderkooi, J. M., Bloomgarden, D. C., and Koloczek, H., 1988, Phosphorescence from 2-(*p*-toluidinyl)naphthalene-6-sulfonate and 1-anilinonaphthalene-8-sulfonate, commonly used fluorescence probes of biological structures, *Photochem. Photobiol.* **47:**207–208.
23. Kim, H., Crouch, S. R., and Zabik, M. J., 1989, Room-temperature phosphorescence of compounds in mixed organized media: Synthetic enzyme model-surfactant system, *Anal. Chem.* **61:**2475–2478.
24. Encinas, M. V., Lissi, E. A., and Rufs, A. M., 1993, Inclusion and fluorescence quenching of 2,3-dimethylnaphthalene in β-cyclodextrin cavities, *Photochem. Photobiol.* **57:**603–608.
25. Turro, N. J., Bolt, J. D., Kuroda, Y., and Tabushi, I., 1982, A study of the kinetics of inclusion of halonaphthalenes with β-cyclodextrin via time correlated phosphorescence, *Photochem. Photobiol.* **35:**69–72.
26. Waka, Y., Hamamoto, K., and Mataga, N., 1980, Heteroexcimer systems in aqueous micellar solutions, *Photochem. Photobiol.* **32:**27–35.
27. Atherton, S. J., and Beaumont, P. C., 1986, Quenching of the fluorescence of DNA-intercalated ethidium bromide by some transition metal ions, *J. Phys. Chem.* **90:**2252–2259.
28. Pasternack, R. F., Caccam, M., Keogh, B., Stephenson, T. A., Williams, A. P., and Gibbs, E. J., 1991, Long-range fluorescence quenching of ethidium ion by cationic porphyrins in the presence of DNA, *J. Am. Chem. Soc.* **113:**6835–6840.
29. Poulos, A. T., Kuzmin, V., and Geacintov, N. E., 1982, Probing the microenvironment of benzo[*a*]pyrene diol epoxide–DNA adducts by triplet excited state quenching methods, *J. Biochem. Biophys. Methods* **6:**269–281.
30. Zinger, D., and Geacintov, N. F., 1988, Acrylamide and molecular oxygen fluorescence quenching as a probe of solvent-accessibility of aromatic fluorophores complexed with DNA in relation to their conformations: Coronene-DNA and other complexes, *Photochem. Photobiol.* **47:**181–188.
31. Suh, D., and Chaires, J. B., 1995, Criteria for the mode of binding of DNA binding agents, *Bioorgan. Med. Chem.* **3:**723–728.
32. Ando, T., and Asai, H., 1980, Charge effects on the dynamic quenching of fluorescence of $1,N^6$-ethenoadenosine oligophosphates by iodide, thallium (I) and acrylamide, *J. Biochem.* **88:**255–264.
33. Ando, T., Fujisaki, H., and Asai, H., 1980, Electric potential at regions near the two specific thiols of heavy meromyosin determined by the fluorescence quenching technique, *J. Biochem.* **88:**265–276.
34. Miyata, H., and Asai, H., 1981, Amphoteric charge distribution at the enzymatic site of $1,N^6$-ethenoadenosine triphosphate-binding heavy meromyosin determined by dynamic fluorescence quenching, *J. Biochem.* **90:**133–139.
35. Midoux, P., Wahl, P., Auchet, J.-C., and Monsigny, M., 1984, Fluorescence quenching of tryptophan by trifluoroacetamide, *Biochim. Biophys. Acta* **801:**16–25.
36. Lehrer, S. S., 1971, Solute perturbation of protein fluorescence. The quenching of the tryptophan fluorescence of model compounds and of lysozyme by iodide ion, *Biochemistry* **10:**3254–3263.
37. Eftink, M. R., and Selvidge, L. A., 1982, Fluorescence quenching of liver alcohol dehydrogenase by acrylamide, *Biochemistry* **21:**117–125.
38. Eftink, M., and Hagaman, K. A., 1986, Fluorescence lifetime and anisotropy studies with liver alcohol dehydrogenase and its complexes, *Biochemistry* **25:**6631–6637.
39. Sontag, B., Reboud, A.-M., Divita, G., Di Pietro, A., Guillot, D., and Reboud, J.P., 1993, Intrinsic tryptophan fluorescence of rat liver elongation factor eEF-2 to monitor the interaction with guanylic and adenylic nucleotides and related conformational changes, *Biochemistry* **32:**1976–1980.
40. Xing, D., Dorr, R., Cunningham, R. P., and Scholes, C. P., 1995, Endonuclease III interactions with DNA substrates. 2. The DNA repair enzyme endonuclease III binds differently to intact DNA and to apyrimidinic/apurinic DNA substrates as shown by tryptophan fluorescence quenching, *Biochemistry* **34:**2537–2544.
41. Wasylewski, M., Malecki, J., and Wasylewski, Z., 1995, Fluorescence study of *Escherichia coli* cyclic AMP receptor protein, *J. Protein Chem.* **14:**299–308.
42. Wells, T. A., Nakazawa, M., Manabe, K., and Song, P.-S., 1994, A conformational change associated with the phototransformation of *Pisum* phytochrome A as probed by fluorescence quenching, *Biochemistry* **33:**708–712.

43. Eftink, M. R., Zajicek, J. L., and Ghiron, C. A., 1977, A hydrophobic quencher of protein fluorescence: 2,2,2-Trichloroethanol, *Biochim. Biophys. Acta* **491**:473–481.
44. Blatt, E., Husain, A., and Sawyer, W. H., 1986, The association of acrylamide with proteins. The interpretation of fluorescence quenching experiments, *Biochim. Biophys. Acta* **871**:6–13.
45. Eftink, M. R., and Ghiron, C. A., 1987, Does the fluorescence quencher acrylamide bind to proteins? *Biochim. Biophys. Acta* **916**:343–349.
46. Punyiczki, M., Norman, J. A., and Rosenberg, A., 1993, Interaction of acrylamide with proteins in the concentration range used for fluorescence quenching studies, *Biophys. Chem.* **47**:9–19.
47. Bastyns, K., and Engelborghs, Y., 1992, Acrylamide quenching of the fluorescence of glyceraldehyde-3-phosphate dehydrogenase: Reversible and irreversible effects, *Photochem. Photobiol.* **55**:9–16.
48. Merrill, A. R., Palmer, L. R., and Szabo, A. G., 1993, Acrylamide quenching of the intrinsic fluorescence of tryptophan residues genetically engineered into the soluble colicin E1 channel peptide. Structural characterization of the insertion-competent state, *Biochemistry* **32**:6974–6981.
49. Wasylewski, Z., Kaszycki, P., Guz, A., and Stryjewski, W., 1988, Fluorescence quenching resolved spectra of fluorophores in mixtures and micellar solutions, *Eur. J. Biochem.* **178**:471–476.
50. Wasylewski, Z., Koloczek, H., and Wasniowska, A., 1988, Fluorescence quenching resolved spectroscopy of proteins, *Eur. J. Biochem.* **172**:719–724.
51. Laws, W. R., and Shore, J. D., 1978, The mechanism of quenching of liver alcohol dehydrogenase fluorescence due to ternary complex formation, *J. Biol. Chem.* **23**:8593–8597.
52. Stryjewski, W., and Wasylewski, Z., 1986, The resolution of heterogeneous fluorescence of multitryptophan-containing proteins studied by a fluorescence quenching method, *Eur. J. Biochem.* **158**:547–553.
53. Blicharska, Z., and Wasylewski, Z., 1995, Fluorescence quenching studies of trp repressor using single-tryptophan mutants, *J. Protein Chem.* **14**:739–746.
54. Wasylewski, Z., Kaszycki, P., and Drwiega, M., 1996, A fluorescence study of Tn10-encoded tet repressor, *J. Protein Chem.* **15**:45–52.
55. Hansen, D., Altschmied, L., and Hillen, W., 1987, Engineered tet repressor mutants with single tryptophan residues as fluorescent probes, *J. Biol. Chem.* **29**:14030–14035.
56. Lange, R., Anzenbacher, P., Müller, S., Maurin, L., and Balny, C., 1994, Interaction of tryptophan residues of cytochrome P45scc with a highly specific fluorescence quencher, a substrate analogue, compared to acrylamide and iodide, *Eur. J. Biochem.* **226**:963–970.
57. Johansson, J. S., Eckenhoff, R. G., and Dutton, L., 1995, Binding of halothane to serum albumin demonstrated using tryptophan fluorescence, *Anesthesiology* **83**:316–324.
58. Johansson, J. S., 1997, Binding of the volatile anesthetic chloroform to albumin demonstrated using tryptophan fluorescence quenching, *J. Biol. Chem.* **272**:17961–17965.
59. Gonzalez-Jimenez, J., Frutos, G., and Cayre, I., 1992, Fluorescence quenching of human serum albumin by xanthines, *Biochem. Pharmacol.* **44**:824–826.
60. Walter, N. G., and Burke, J. M., 1997, Real-time monitoring of hairpin ribozyme kinetics through base-specific quenching of fluorescein-labeled substrates, *RNA* **3**:392–404.
61. Jezewska, M. J., and Bujalowski, W., 1997, Quantitative analysis of ligand–macromolecule interactions using differential dynamic quenching of the ligand fluorescence to monitor the binding, *Biophys. Chem.* **64**:253–269.
62. Eftink, M. R., Jia, Y.-W., Graves, D. E., Wiczk, W., Gryczynski, I., and Lakowicz, J. R., 1989, Intramolecular fluorescence quenching in covalent acrylamide–indole adducts, *Photochem. Photobiol.* **49**:725–729.
63. Green, S. A., Simpson, D. J., Zhou, G., Ho, P. S., and Blough, N. V., 1990, Intramolecular quenching of excited singlet states by stable nitroxyl radicals, *J. Am. Chem. Soc.* **112**:7337–7346.
64. Chuang, T. J., Cox, R. J., and Eisenthal, K. B., 1974, Picosecond studies of the excited charge-transfer interactions in anthracene-$(CH_2)_3$-*N,N*-dimethylaniline systems, *J. Am. Chem. Soc.* **96**:6828–6831.
65. Migita, M., Okada, T., Mataga, N., Sakata, Y., Misumi, S., Nakashima, N., and Yoshihara, K., 1981, Picosecond laser spectroscopy of intramolecular heteroexcimer systems. Time-resolved fluorescence studies of *p*-$(CH_3)_2NC_6H_4$-$(CH_2)_n$-(9-anthryl), *p*-$(CH_3)_2NC_6H_4$-$(CH_2)_n$-(1-pyrenyl) systems and 9,9′-bianthryl, *Bull. Chem. Soc. Jpn.* **54**:3304–3311.
66. Basu, G., Anglos, D., and Kuki, A., 1993, Fluorescence quenching in a strongly helical peptide series: The role of noncovalent pathways in modulating electronic interactions, *Biochemistry* **32**:3067–3076.
67. Strambini, G. B., and Gonnelli, M., 1995, Tryptophan phosphorescence in fluid solution, *J. Am. Chem. Soc.* **117**:7646–7651.
68. Englander, S. W., Calhoun, D. B., and Englander, J. J., 1987, Biochemistry without oxygen, *Anal. Biochem.* **161**:300–306.
69. Zhang, H. R., Zhang, J., Wei, Y. S., Jin, E. J., and Liu, C. S., 1997, Study of new facile deoxygenation methods in cyclodextrin induced room temperature phosphorescence, *Anal. Chim. Acta* **357**:119–125.
70. Cioni, P., Puntoni, A., and Strambini, G. B., 1993, Tryptophan phosphorescence as a monitor of the solution structure of phosphoglycerate kinase from yeast, *Biophys. Chem.* **46**:47–55.
71. Gonnelli, M., and Strambini, G. B., 1993, Glycerol effects on protein flexibility: A tryptophan phosphorescence study, *Biophys. J.* **65**:131–137.
72. Strambini, G. B., and Gabellieri, E., 1996, Proteins in frozen solutions: Evidence of ice-induced partial unfolding, *Biophys. J.* **70**:971–976.
73. Vanderkooi, J. M., Calhoun, D. B., and Englander, S. W., 1987, On the prevalence of room-temperature protein phosphorescence, *Science* **236**:568–569.
74. Turro, N. J., Cox, G. S., and Li, X., 1983, Remarkable inhibition of oxygen quenching of phosphorescence by complexation with cyclodextrins, *Photochem. Photobiol.* **37**:149–153.
75. Gonnelli, M., and Strambini, G. B., 1995, Phosphorescence lifetime of tryptophan in proteins, *Biochemistry* **34**:13847–13857.
76. Wright, W. W., Owen, C. S., and Vanderkooi, J. M., 1992, Penetration of analogues of H_2O and CO_2 in proteins studied by room temperature phosphorescence of tryptophan, *Biochemistry* **31**:6538–6544.
77. Strambini, G. B., 1987, Quenching of alkaline phosphatase phosphorescence by O_2 and NO, *Biophys. J.* **52**:23–28.
78. Boaz, H., and Rollefson, G. K., 1950, The quenching of fluorescence. Deviations from the Stern–Volmer law, *J. Am. Chem. Soc.* **72**:3425–3443.
79. Eftink, M. R., and Ghiron, C. A., 1976, Exposure of tryptophanyl residues in proteins. Quantitative determination by fluorescence quenching studies, *Biochemistry* **15**:672–680.
80. Kuzmin, M. G., and Guseva, L. N., 1969, Donor–acceptor complexes of singlet excited states of aromatic hydrocarbons with aliphatic amines, *Chem. Phys. Lett.* **3**:71–72.
81. Rehm, D., and Weller, A., 1970, Kinetics of fluorescence quenching by electron and H-atom transfer, *Israel J. Chem.* **8**:259–271.

82. Obyknovennaya, I. E., Vasileva, I. M., and Cherkasov, A. S., 1986, Quenching of the fluorescence of anthracene by dimethylaniline in aqueous-micellar solvent and formation of luminescent exciplexes, *Opt. Spectrosc.* **60**:169–171.
83. Lewis, F. D., and Bassani, D. M., 1992, Formation and decay of styrene–amine exciplexes, *J. Photochem. Photobiol. A: Chem.* **66**:43–52.
84. Schneider, S., Stammler, W., Bierl, R., and Jager, W., 1994, Ultrafast photoinduced charge separation and recombination in weakly bound complexes between oxazine dyes and *N,N*-dimethylaniline, *Chem. Phys. Lett.* **219**:433–439.
85. Yoshihara, K., Yartsev, A., Nagasawa, Y., Kandori, H., Douhal, A., and Kemnitz, K., 1993, Femtosecond intermolecular electron transfer between dyes and electron-donating solvents, *Pure Appl. Chem.* **65**:1671–1675.
86. Bisht, P. B., and Tripathi, H. B., 1993, Fluorescence quenching of carbazole by triethylamine: Exciplex formation in polar and nonpolar solvents, *J. Lumin.* **55**:153–158.
87. Sikaris, K. A., and Sawyer, W. H., 1982, The interaction of local anaesthetics with synthetic phospholipid bilayers, *Biochem. Pharmacol.* **31**:2625–2631.
88. Fernandez, M. S., and Calderon, E., 1990, The local anaesthetic tetracaine as a quencher of perylene fluorescence in micelles, *J. Photochem. Photobiol. B: Biol.* **7**:75–86.
89. Winkler, M. H., 1969, A fluorescence quenching technique for the investigation of the configurations of binding sites for small molecules, *Biochemistry* **8**:2586–2590.
90. Berlman, I. B., 1973, Empirical study of heavy-atom collisional quenching of the fluorescence state of aromatic compounds in solution, *J. Phys. Chem.* **77**:562–567.
91. James, D. R., and Ware, W. R., 1985, Multiexponential fluorescence decay of indole-3-alkanoic acids, *J. Phys. Chem.* **89**:5450–5458.
92. Illsley, N. P., and Verkman, A. S., 1987, Membrane chloride transport measured using a chloride-sensitive fluorescent probe, *Biochemistry* **26**:1215–1219.
93. Chao, A. C., Dix, J. A., Sellers, M. C., and Verkman, A. S., 1989, Fluorescence measurement of chloride transport in monolayer cultured cells, *Biophys. J.* **56**:1071–1081.
94. Verkman, A. S., 1990, Development and biological applications of chloride-sensitive fluorescent indicators, *Am. J. Phys.* **253**:C375–C388.
95. Martin, A., and Narayanaswamy, R., 1997, Studies on quenching of fluorescence of reagents in aqueous solution leading to an optical chloride-ion sensor, *Sensors Actuators B* **38–39**:330–333.
96. Daems, D., Boens, N., and Schryver, F. C., 1989, Fluorescence quenching with lindane in small unilamellar L,α-dimyristoylphosphatidylcholine vesicles, *Eur. Biophys. J.* **17**:25–36.
97. Namiki, A., Nakashima, N., and Yoshihara, K., 1979, Fluorescence quenching due to the electron transfer. Indole-chloromethanes in rigid ethanol glass, *J. Chem. Phys.* **71**:925–930.
98. Johnson, G. E., 1980, Fluorescence quenching of carbazoles, *J. Phys. Chem.* **84**:2940–2946.
99. Jones, O. T., and Lee, A. G., 1985, Interactions of hexachlorocyclohexanes with lipid bilayers, *Biochim. Biophys. Acta* **812**:731–739.
100. Hariharan, C., Vijaysree, V., and Mishra, A. K., 1997, Quenching of 2,5-diphenyloxazole (PPO) fluorescence by metal ions, *J. Lumin.* **75**:205–211.
101. Morris, S. J., Bradley, D., and Blumenthal, R., 1985, The use of cobalt ions as a collisional quencher to probe surface charge and stability of fluorescently labeled bilayer vesicles, *Biochim. Biophys. Acta* **818**:365–372.
102. Homan, R., and Eisenberg, M., 1985, A fluorescence quenching technique for the measurement of paramagnetic ion concentrations at the membrane/water interface. Intrinsic and X537A-mediated cobalt fluxes across lipid bilayer membranes, *Biochim. Biophys. Acta* **812**:485–492.
103. Salthammer, T., Dreeskamp, H., Birch, D. J. S., and Imhof, R. E., 1990, Fluorescence quenching of perylene by Co^{2+} ions via energy transfer in viscous and non-viscous media, *J. Photochem. Photobiol., A: Chem.* **55**:53–62.
104. Holmes, A. S., Birch, D. J. S., Suhling, K., Imhof, R. E., Salthammer, T., and Dreeskamp, H., 1991, Evidence for donor–donor energy transfer in lipid bilayers: Perylene fluorescence quenching by Co^{2+} ions, *Chem. Phys. Lett.* **186**:189–194.
105. Birch, D. J. S., Suhling, K., Holmes, A. S., Salthammer, T., and Imhof, R. E., 1992, Fluorescence energy transfer to metal ions in lipid bilayers, *Proc. SPIE* **1640**:707–718.
106. Perochon, E., and Tocanne, J.-F., 1991, Synthesis and phase properties of phosphatidylcholine labeled with 8-(2-anthroyl)octanoic acid, a solvatochromic fluorescent probe, *Chem. Phys. Lipids* **58**:7–17.
107. Fucaloro, A. F., Forster, L. S., and Campbell, M. K., 1984, Fluorescence quenching of indole by dimethylformamide, *Photochem. Photobiol.* **39**:503–506.
108. Swadesh, J. K., Mui, P. W., and Scheraga, H. A., 1987, Thermodynamics of the quenching of tyrosyl fluorescence by dithiothreitol, *Biochemistry* **26**:5761–5769.
109. Valentino, M. R., and Boyd, M. K., 1995, Ether quenching of singlet excited 9-arylxanthyl cations, *J. Photochem. Photobiol.* **89**:7–12.
110. Medinger, T., and Wilkinson, F., 1965, Mechanism of fluorescence quenching in solution I. Quenching of bromobenzene, *Trans. Faraday Soc.* **61**:620–630.
111. Ahmad, A., and Durocher, G., 1981, How hydrogen bonding of carbazole to ethanol affects its fluorescence quenching rate by electron acceptor quencher molecules, *Photochem. Photobiol.* **34**:573–578.
112. Bowen, E. J., and Metcalf, W. S., 1951, The quenching of anthracene fluorescence, *Proc. R. Soc. London* **206A**:437–447.
113. Schmidt, R., Janssen, W., and Brauer, H.-D., 1989, Pressure effect on quenching of perylene fluorescence by halonaphthalenes, *J. Phys. Chem.* **93**:466–468.
114. Encinas, M. V., Rubio, M. A., and Lissi, E., 1983, Quenching and photobleaching of excited polycyclic aromatic hydrocarbons by carbon tetrachloride and chloroform in micellar systems, *Photochem. Photobiol.* **37**:125–130.
115. Behera, P. K., Mukherjee, T., and Mishra, A. K., 1995, Quenching of substituted naphthalenes fluorescence by chloromethanes, *J. Lumin.* **65**:137–142.
116. Behera, P. K., and Mishra, A. K., 1993, Static and dynamic model for 1-naphthol fluorescence quenching by carbon tetrachloride in dioxane–acetonitrile mixtures, *J. Photochem. Photobiol., A: Chem.* **71**:115–118.
117. Behera, P. K., Mukherjee, T., and Mishra, A. K., 1995, Simultaneous presence of static and dynamic component in the fluorescence quenching for substituted naphthalene–CCl_4 system, *J. Lumin.* **65**:131–136.
118. Zhang, J., Roek, D. P., Chateauneuf, J. E., and Brennecke, J. F., 1997, A steady-state and time-resolved fluorescence study of quenching reactions of anthracene and 1,2-benzanthracene by carbon tetrabromide and bromoethane in supercritical carbon dioxide, *J. Amer. Chem. Soc.* **119**:9980–9991.
119. Tucker, S. A., Cretella, L. E., Waris, R., Street, K. W., Acree, W. E., and Fetzer, J. C., 1990, Polycyclic aromatic hydrocarbon solute probes. Part VI: Effect of dissolved oxygen and halogenated sol-

vents on the emission spectra of select probe molecules, *Appl. Spectrosc.* **44:**269–273.
120. Wiczk, W. M., and Latowski, T., 1992, The effect of temperature on the fluorescence quenching of perylene by tetrachloromethane in mixtures with cyclohexane and benzene, *Z. Naturforsch. A* **47:**533–535.
121. Wiczk, W. M., and Latowski, T., 1986, Photophysical and photochemical studies of polycyclic aromatic hydrocarbons in solutions containing tetrachloromethane. I. Fluorescence quenching of anthracene by tetrachloromethane and its complexes with benzene, *p*-xylene and mesitylene, *Z. Naturforsch. A* **41:**761–766.
122. Goswami, D., Sarpal, R. S., and Dogra, S. K., 1991, Fluorescence quenching of few aromatic amines by chlorinated methanes, *Bull. Chem. Soc. Jpn.* **64:**3137–3141.
123. Takahashi, T., Kikuchi, K., and Kokubun, H., 1980, Quenching of excited 2,5-diphenyloxazole by CCl_4, *J. Photochem.* **14:**67–76.
124. Alford, P. C., Cureton, C. G., Lampert, R. A., and Phillips, D., 1983, Fluorescence quenching of tertiary amines by halocarbons, *Chem. Phys.* **76:**103–109.
125. Khwaja, H. A., Semeluk, G. P., and Unger, I., 1984, Quenching of the singlet and triplet state of benzene in condensed phase, *Can. J. Chem.* **62:**1487–1491.
126. Washington, K., Sarasua, M. M., Koehler, L. S., Koehler, K. A., Schultz, J. A., Pedersen, L. G., and Hiskey, R. G., 1984, Utilization of heavy-atom effect quenching of pyrene fluorescence to determine the intramembrane distribution of halothane, *Photochem. Photobiol.* **40:**693–701.
127. Lopez, M. M., and Kosk-Kosicka, D., 1998, Spectroscopic analysis of halothane binding to the plasma membrane Ca^{2+}-ATPase, *Biophys. J.* **74:**974–980.
128. Cavatorta, P., Favilla, R., and Mazzini, A., 1979, Fluorescence quenching of tryptophan and related compounds by hydrogen peroxide, *Biochim. Biophys. Acta* **578:**541–546.
129. Vos, R., and Engelborghs, Y., 1994, A fluorescence study of tryptophan–histidine interactions in the peptide alantin and in solution, *Photochem. Photobiol.* **60:**24–32.
130. Mae, M., Wach, A., and Najbar, J., 1991, Solvent effects on the fluorescence quenching of anthracene by iodide ions, *Chem. Phys. Lett.* **176:**167–172.
131. Blatt, E., Ghiggino, K. P., and Sawyer, W. H., 1982, Fluorescence depolarization studies of *n*-(9-anthroyloxy) fatty acids in cetyltrimethylammonium bromide micelles, *J. Phys. Chem.* **86:**4461–4464.
132. Fraiji, L. K., Hayes, D. M., and Werner, T. C., 1992, Static and dynamic fluorescence quenching experiments for the physical chemistry laboratory, *J. Chem. Educ.* **69:**424-428.
133. Zhu, C., Bright, F. V., and Hieftje, G. M., 1990, Simultaneous determination of Br^- and I^- with a multiple fiber-optic fluorescence sensor, *Appl. Spectrosc.* **44:**59–64.
134. Lakowicz, J. R., and Anderson, C. J., 1980, Permeability of lipid bilayers to methylmercuric chloride: Quantification by fluorescence quenching of a carbazole-labeled phospholipid, *Chem.-Biol. Interact.* **30:**309–323.
135. Boudou, A., Desmazes, J. P., and Georgescauld, D., 1982, Fluorescence quenching study of mercury compounds and liposome interactions: Effect of charged lipid and pH, *Ecotoxicol. Environ. Saf.* **6:**379–387.
136. Holmes, A. S., Suhling, K., and Birch, D. J. S., 1993, Fluorescence quenching by metal ions in lipid bilayers, *Biophys. Chem.* **48:**193–204.
137. Birch, D. J. S., Suhling, K., Holmes, A. S., Salthammer, T., and Imhof, R. E., 1993, Metal ion quenching of perylene fluorescence in lipid bilayers, *Pure Appl. Chem.* **65:**1687–1692.
138. Sawicki, E., Stanley, T. W., and Elbert, W. C., 1964, Quenchofluorometric analysis for fluoranthenic hydrocarbons in the presence of other types of aromatic hydrocarbon, *Talanta* **11:**1433–1441.
139. Dreeskamp, H., Koch, E., and Zander, M., 1975, On the fluorescence quenching of polycyclic aromatic hydrocarbons by nitromethane, *Z. Naturforsch. A* **30:**1311–1314.
140. Pandey, S., Fletcher, K. A., Powell, J. R., McHale, M. E. R., Kauppila, A-S. M., Acree, W. E., Fetzer, J. C., Dai, W., and Harvey, R. G., 1997, Spectrochemical investigations of fluorescence quenching agents. Part 5. Effect of surfactants on the ability of nitromethane to selectively quench fluorescence emission of alternant PAHs, *Spectrochim. Acta, Part A* **53:**165–172.
141. Pandey, S., Acree, W. E., Cho, B. P., and Fetzer, J. C., 1997, Spectroscopic properties of polycyclic aromatic compounds. Part 6. The nitromethane selective quenching rule visited in aqueous micellar zwitterionic surfactant solvent media, *Talanta* **44:**413–421.
142. Acree, W. E., Pandey, S., and Tucker, S. A., 1997, Solvent-modulated fluorescence behavior and photophysical properties of polycyclic aromatic hydrocarbons dissolved in fluid solution, *Curr. Top. Solution Chem.* **2:**1–27.
143. Tucker, S. A., Acree, W. E., Tanga, M. J., Tokita, S., Hiruta, K., and Langhals, H., 1992, Spectroscopic properties of polycyclic aromatic compounds: Examination of nitromethane as a selective fluorescence quenching agent for alternant polycyclic aromatic nitrogen hetero-atom derivatives, *Appl. Spectrosc.* **46:**229–235.
144. Tucker, S. A., Acree, W. E., and Upton, C., 1993, Polycyclic aromatic nitrogen heterocycles. Part V: Fluorescence emission behavior of select tetraaza- and diazaarenes in nonelectrolyte solvents, *Appl. Spectrosc.* **47:**201–206.
145. Green, J. A., Singer, L. A., and Parks, J. H., 1973, Fluorescence quenching by the stable free radical di-*t*-butylnitroxide, *J. Chem. Phys.* **58:**2690–2695.
146. Bieri, V. G., and Wallach, D. F. H., 1975, Fluorescence quenching in lecithin:cholesterol liposomes by paramagnetic lipid analogues: Introduction of new probe approach, *Biochim. Biophys. Acta* **389:**413–427.
147. Encinas, M. V., Lissi, E. A., and Alvarez, J., 1994, Fluorescence quenching of pyrene derivatives by nitroxides microheterogeneous systems, *Photochem. Photobiol.* **59:**30–34.
148. Matko, J., Ohki, K., and Edidin, M., 1992, Luminescence quenching by nitroxide spin labels in aqueous solution: Studies on the mechanism of quenching, *Biochemistry* **31:**703–711.
149. Jones, P. F., and Siegel, S., 1971, Quenching of naphthalene luminescence by oxygen and nitric oxide, *J. Chem. Phys.* **54:**3360–3366.
150. Harper, J., and Sailor, M. J., 1996, Detection of nitric oxide and nitrogen dioxide with photoluminescent porous silicon, *Anal. Chem.* **68:**3713–3717.
151. Denicola, A., Souza, J. M., Radi, R., and Lissi, E., 1996, Nitric oxide diffusion in membranes determined by fluorescence quenching, *Arch. Biochem. Biophys.* **328:**208–212.
152. Ware, W. R., Holmes, J. D., and Arnold, D. R., 1974, Exciplex photophysics. II. Fluorescence quenching of substituted anthracenes by substituted 1,1-diphenylethylenes, *J. Am. Chem. Soc.* **96:**7861–7864.
153. Labianca, D. A., Taylor, G. N., and Hammond, G. S., 1972, Structure–reactivity factors in the quenching of fluorescence from naphthalenes by conjugated dienes, *J. Am. Chem. Soc.* **94:**3679–3683.
154. Encinas, M. V., Guzman, E., and Lissi, E. A., 1983, Intramicellar aromatic hydrocarbon fluorescence quenching by olefins, *J. Phys. Chem.* **87:**4770–4772.

155. Abuin, E. B., and Lissi, E. A., 1993, Quenching rate constants in aqueous solution: Influence of the hydrophobic effect, *J. Photochem. Photobiol.* **71**:263–267.
156. Chang, S. L. P., and Schuster, D. I., 1987, Fluorescence quenching of 9,10-dicyanoanthracene by dienes and alkenes, *J. Phys. Chem.* **91**:3644–3649.
157. Eriksen, J., and Foote, C. S., 1978, Electron-transfer fluorescence quenching and exciplexes of cyano-substituted anthracenes, *J. Phys. Chem.* **82**:2659–2662.
158. Fischkoff, S., and Vanderkooi, J. M., 1975, Oxygen diffusion in biological and artificial membranes determined by the fluorochrome pyrene, *J. Gen. Phys.* **65**:663–676.
159. Jameson, D. M., Gratton, E., Weber, G., and Alpert, B., 1984, Oxygen distribution and migration within $MB^{DES\ FE}$ and $HB^{DES\ FE}$, *Biophys. J.* **45**:795–803.
160. Subczynski, W. K., Hyde, J. S., and Kusumi, A., 1989, Oxygen permeability of phosphatidylcholine-cholesterol membranes, *Proc. Natl. Acad. Sci. U.S.A.* **86**:4474–4478.
161. Dumas, D., Muller, S., Gouin, F., Baros, F., Viriot, M.-L., and Stoltz, J.-F., 1997, Membrane fluidity and oxygen diffusion in cholesterol-enriched erythrocyte membrane, *Arch. Biochem. Biophys.* **341**:34–39.
162. Camyshan, S. V., Gritsan, N. P., Korolev, V. V., and Bazhin, N. M., 1990, Quenching of the luminescence of organic compounds by oxygen in glassy matrices, *Chem. Phys.* **142**:59–68.
163. Kikuchi, K., Sato, C., Watabe, M., Ikeda, H., Takahashi, Y., and Miyashi, T., 1993, New aspects on fluorescence quenching by molecular oxygen, *J. Am. Chem. Soc.* **115**:5180–5184.
164. Vaughan, W. M., and Weber, G., 1970, Oxygen quenching of pyrenebutyric acid fluorescence in water. A dynamic probe of the microenvironment, *Biochemistry* **9**:464–473.
165. Abuin, E. B., and Lissi, E. A., 1991, Diffusion and concentration of oxygen in microheterogeneous systems. Evaluation from luminescence quenching data, *Prog. React. Kinet.* **16**:1–33.
166. Parasassi, T., and Gratton, E., 1992, Packing of phospholipid vesicles studied by oxygen quenching of laurdan fluorescence, *J. Fluoresc.* **2**:167–174.
167. Encinas, M. V., and Lissi, E. A., 1983, Intramicellar quenching of 2,3-dimethylnaphthalene fluorescence by peroxides and hydroperoxides, *Photochem. Photobiol.* **37**:251–255.
168. Holmes, L. G., and Robbins, F. M., 1974, Quenching of tryptophyl fluorescence in proteins by *N'*-methylnicotinamide chloride, *Photochem. Photobiol.* **19**:361–366.
169. Martin, M. M., and Ware, W. R., 1978, Fluorescence quenching of carbazole by pyridine and substituted pyridines. Radiationless processes in the carbazole–amine hydrogen bonded complex, *J. Phys. Chem.* **82**:2770–2776.
170. Seely, G. R., 1978, The energetics of electron-transfer reactions of chlorophyll and other compounds, *Photochem. Photobiol.* **27**:639–654.
171. Schroeder, J., and Wilkinson, F., 1979, Quenching of triplet states of aromatic hydrocarbons by quinones due to favourable charge-transfer interactions, *J. Chem. Soc., Faraday Trans. 2* **75**:896–904.
172. Dreeskamp, H., Laufer, A., and Zander, M., 1983, Löschung der perylen-fluoreszenz durch Ag^+-ionen, *Z. Naturforsch. A* **38**:698–700.
173. Badley, R. A., 1975, The location of protein in serum lipoproteins: A fluorescence quenching study, *Biochim. Biophy. Acta* **379**:517–528.
174. Eftink, M. R., and Ghiron, C. A., 1984, Indole fluorescence quenching studies on proteins and model systems: Use of the inefficient quencher succinimide, *Biochemistry* **23**:3891–3899.
175. Moore, H.-P. H., and Raftery, M. A., 1980, Direct spectroscopic studies of cation translocation by *Torpedo* acetylcholine receptor on a time scale of physiological relevance, *Proc. Natl. Acad. Sci. U.S.A.* **77**:4509–4513.
176. Mac, M., Najbar, J., Phillips, D., and Smith, T. A., 1992, Solvent dielectric relaxation properties and the external heavy atom effect in the time-resolved fluorescence quenching of anthracene by potassium iodide and potassium thiocyanate in methanol and ethanol, *J. Chem. Soc., Faraday Trans.* **88**:3001–3005.
177. Carrigan, S., Doucette, S., Jones, C., Marzzacco, C. J., and Halpern, A. M., 1996, The fluorescence quenching of 5,6-benzoquinoline and its conjugate acid by Cl^-, Br^-, SCN^- and I^- ions, *J. Photochem. Photobiol., A: Chem.* **99**:29–35.
178. Horrochs, A. R., Kearvell, A., Tickle, K., and Wilkinson, F., 1966, Mechanism of fluorescence quenching in solution II. Quenching by xenon and intersystem crossing efficiencies, *Trans. Faraday Soc.* **62**:3393–3399.

PROBLEMS

8.1. *Separation of Static and Dynamic Quenching of Acridone by Iodide*: The following data were obtained for quenching of acridone in water at 26 °C.[78] KNO_2 is used to maintain a constant ionic strength and does not quench the fluorescence of arcidone.

[KI] (*M*)	[KNO_2] (*M*)	F_0/F
0.0	1.10	[1.0]
0.04	1.06	4.64
0.10	1.00	10.59
0.20	0.90	23.0
0.30	0.80	37.2
0.50	0.60	68.6
0.80	0.30	137

A. Construct a Stern–Volmer plot.
B. Determine the dynamic (K_D) and static (K_S) quenching constants. Use the quadratic equation to obtain K_D and K_S from the slope and intercept of a plot of K_{app} versus [Q].
C. Calculate the observed bimolecular quenching constant. The unquenched lifetime $\tau_0 = 17.6$ ns.
D. Calculate the diffusion-limited bimolecular quenching constant and the quenching efficiency. The diffusion constant of KI in water is 2.065×10^{-5} cm^2/s for 1*M* KI (*Handbook of Chemistry and Physics*, 55th ed.).
E. Comment on the magnitude of the sphere of action and the static quenching constant, with regard to the nature of the complex.

8.2. *Separation of Static and Dynamic Quenching*: The following table lists the fluorescence lifetimes and relative quantum yields of 10-methylacridinium chloride (MAC) in the presence of adenosine monophosphate (AMP).[18]

[AMP] (mM)	τ (ns)	Intensity
0	32.9	1.0
1.75	26.0	0.714
3.50	21.9	0.556
5.25	18.9	0.426
7.00	17.0	0.333

A. Is the quenching dynamic, static, or both?
B. What is (are) the quenching constant(s)?
C. What is the association constant for the MAC–AMP complex?
D. Comment on the magnitude of the static quenching constant.
E. Assume that the AMP–MAC complex is completely nonfluorescent and that complex formation shifts the absorption spectrum of MAC. Will the corrected excitation spectrum of MAC, in the presence of nonsaturating amounts of AMP, be comparable to the absorption spectrum of MAC or that of the MAC–AMP complex?

8.3. *Effects of Dissolved Oxygen on Fluorescence Intensities and Lifetimes*: Oxygen is known to dissolve in aqueous and organic solutions and is a collisional quencher of fluorescence. Assume that your measurements are accurate to 3%. What are the lifetimes above which dissolved oxygen from the atmosphere will result in changes in the fluorescence intensities or lifetimes that are outside your accuracy limits? Indicate these lifetimes for both aqueous and ethanolic solutions. The oxygen solubility in water is 0.001275M for a partial pressure of 1 atm. Oxygen is fivefold more soluble in ethanol than in water. The following information is needed to answer this question: k_q (in water) $= 1 \times 10^{10}\ M^{-1}\ s^{-1}$; k_q (in ethanol) $= 2 \times 10^{10}\ M^{-1}\ s^{-1}$.

8.4. *Intramolecular Complex Formation by Flavin Adenine Dinucleotide (FAD)*: FAD fluorescence is quenched both by static complex formation between the flavin and adenine rings and by collisions between these two moieties. Flavin mononucleotide (FMN) is similar to FAD except that it lacks the adenine ring. Use the following data for FAD and FMN to calculate the fraction complexed (f) and the collisional deactivation rate (k) of the flavin by the adenine ring. Q is the relative quantum yield. Note that the deactivation rate is in units of s^{-1}.
τ (FMN) = 4.6 ns
τ (FAD) = 2.4 ns
Q(FMN) = 1.0 (assumed unity)
Q(FAD) = 0.09

8.5. *Quenching of Protein Fluorescence; Determination of the Fraction of the Total Fluorescence Accessible to Iodide:* Assume that a protein contains four identical subunits, each containing two tryptophan residues. The following data are obtained in the presence of iodide.

[I$^-$] (M)	Fluorescence intensity
0.00	1.0
0.01	0.926
0.03	0.828
0.05	0.767
0.10	0.682
0.20	0.611
0.40	0.563

A. What fraction of the total tryptophan fluorescence is accessible to quenching?
What property of the Stern–Volmer plots indicates an inaccessible fraction?
B. Assume that all the tryptophans have equal quantum yields and lifetimes (5 ns). How many tryptophan residues are accessible to quenching?
C. What are the bimolecular quenching constants for the accessible and inaccessible residues?
D. Assume that you could selectively excite the accessible tryptophans by excitation at 300 nm. Draw the predicted Stern–Volmer and modified Stern–Volmer plots for the accessible and the inaccessible residues.

8.6. *Quenching of Endonuclease III:* Figure 8.39 shows the effect of a 19-mer of poly(dA-dT) on the intrinsic tryptophan emission of endo nuclease III (Endo III). Explain the data in terms of the structure of endo III (Figure 8.15). Is the quenching collisional or static? Assume that the unquenched decay time is 5 ns. The concentration of endo III is 0.8μM.

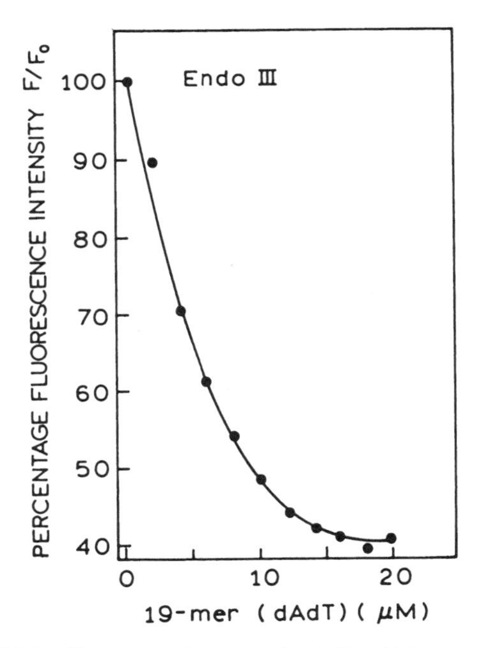

Figure 8.39. Fluorescence intensity of endo III with increasing concentrations of a 19-mer of poly(dA-dT). Revised from Ref. 40.

9

Advanced Topics in Fluorescence Quenching

In the previous chapter we described the basic principles of quenching. A wide variety of quenchers are known. Quenching requires molecular contact between the fluorophore and the quencher. This contact can be due to diffusive encounters (dynamic quenching), or to complex formation (static quenching). Because of the close-range interaction required for quenching, the extent of quenching is sensitive to molecular factors that affect the rate and probability of contact, including steric shielding and charge–charge interactions.

In this chapter we describe more advanced topics in quenching. Quenching in membranes is described in some detail because of the numerous applications. These include localization of membrane-bound probes, estimation of diffusion coefficients in membranes, and the effect of quencher partitioning into membranes. We also describe transient effects in quenching, which result in nonexponential decays whenever diffusive quenching occurs. These effects can complicate the interpretation of the time-resolved data, but they also provide additional information about the diffusion coefficient of the quencher, the interaction radius, and the mechanism of quenching. For those interested in an introduction to fluorescence, the reading of this chapter can be postponed.

9.1. QUENCHING IN MEMBRANES

During the past 15 years, there have been numerous studies of quenching in membranes. These experiments had a number of goals, including determination of the accessibility of probes to water- and lipid-soluble quenchers, determination of quencher partitioning into membranes, localization of probes in membranes relative to the lipid–water interface, and measurement of lateral diffusion coefficients. In the following sections we provide examples of each type of measurement.

9.1.A. Accessibility of Membrane Probes to Water- and Lipid-Soluble Quenchers

Since the early studies of membranes labeled with fluorescent probes, there has been an extensive effort to determine the location of the probes in the membranes. Early membrane probes, such as perylene, 9-vinylanthracene, and DPH,[1–3] were simple hydrophobic molecules that dissolved in the hydrophobic regions of micelles or membranes. Subsequent probes were synthesized and designed to localize the fluorophore at various depths in the membrane. Among the earliest of such probes were the anthroyl-

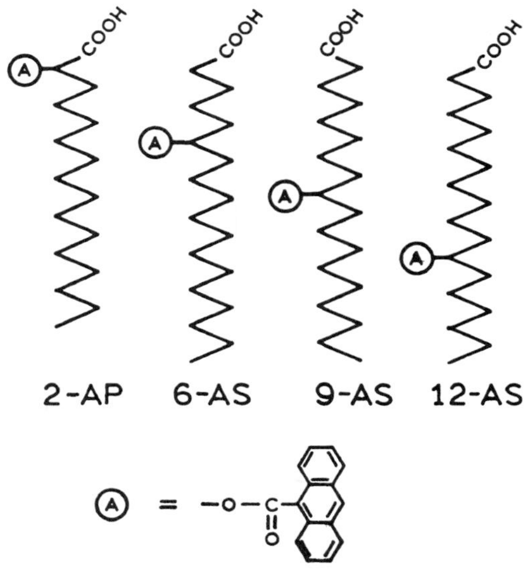

Figure 9.1. Chemical structures of the *n*-(9-anthroyloxy) fatty acids. P is palmitic acid, and S is stearic acid. Revised and reprinted from Ref. 19, Copyright © 1985, with permission from Elsevier Science.

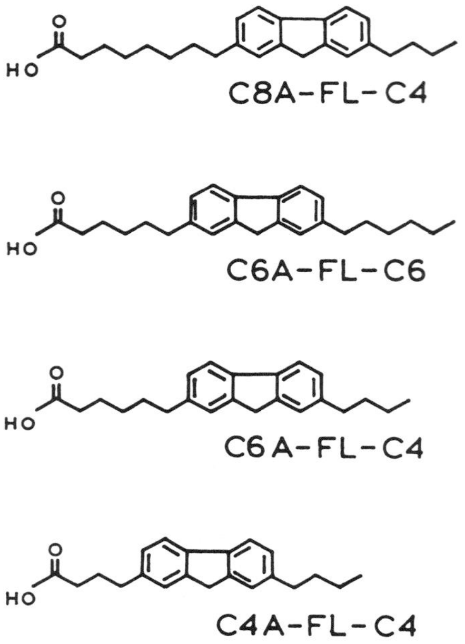

HO C8A-FL-C4

HO C6A-FL-C6

HO C6A-FL-C4

HO C4A-FL-C4

Figure 9.2. Structures of the fluorenyl fatty acids. Revised from Ref. 7.

oxy fatty acids[4–6] and their lipid analogs (Figure 9.1). Subsequently, a wide variety of membrane probes were synthesized, including the fluorenyl fatty acids (Figure 9.2).[7] Other membrane probes include lipids labeled with anthracene,[8–9] pyrene,[10,11] NBD,[12–14] DPH,[15] BODIPY,[16] and carbazole.[17]

The chemical structures of the lipid probes usually suggest their location in a membrane. However, this localization needed to be experimentally confirmed. In fact, this turned out to be important because it is now known that some probes linked to acyl chains can fold back, resulting in fluorophores localized near the surface of the membranes rather than near the center of the bilayer. This occurs with certain membrane probes, but not all probes, and one can expect this effect to depend on the lipid composition of the membrane.

The location of probes in membranes can be determined by fluorescence quenching. One approach is to use quenchers which are either water-soluble or partition into the membranes.[17–19] The basic idea is illustrated in Figure 9.3. Suppose that the fluorophore is attached to a phospholipid. Information about the location of the fluorophore can be obtained from the amount of quenching by water-soluble (Q_w) or membrane-soluble (Q_m) quenchers. If the probe is located deep in the bilayer, one does not expect quenching by the water-soluble quencher. Quenching is expected if the quencher dissolves in the lipid phase. The use of a freely diffusing quencher is illustrated by quenching of 12-(9-anthroyloxy) stearic acid (12-AS) in egg PC bilayers by acrylamide.[20] In 12-AS the anthroyloxy probe is located more than halfway down the acyl chain (Figure 9.1), and it is believed that the continuing alkyl chain keeps 12-AS localized deep in the bilayer. Acrylamide is a water-soluble quencher which is not expected to enter the acyl side-chain region of membranes. The experiments confirmed this prediction, as there was essentially no quenching of 12-AS in lipid bilayers by acrylamide (Figure 9.4). In similar experiments, the carbazole moieties in carbazole-labeled lipids were shown to be inaccessible to

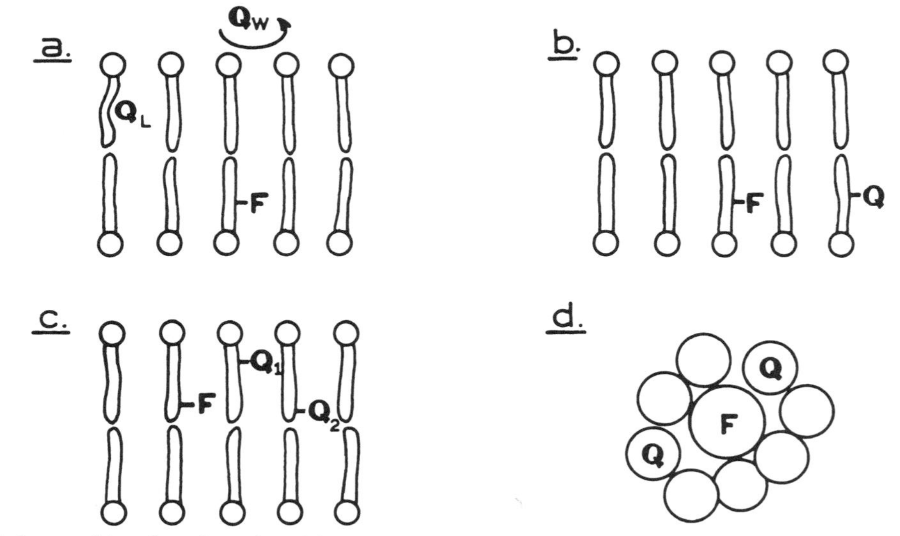

Figure 9.3. Models for quenching of membrane-bound fluorophores. Quenching can be due to freely diffusing quenchers in the water or membrane phase (a) or to quenchers covalently bound to lipids (b). Comparison of quenching by lipid-bound quenchers at different locations can be used to determine probe location in membranes (c). Quenching can also be analyzed in terms of boundary lipid surrounding a probe or protein (d).

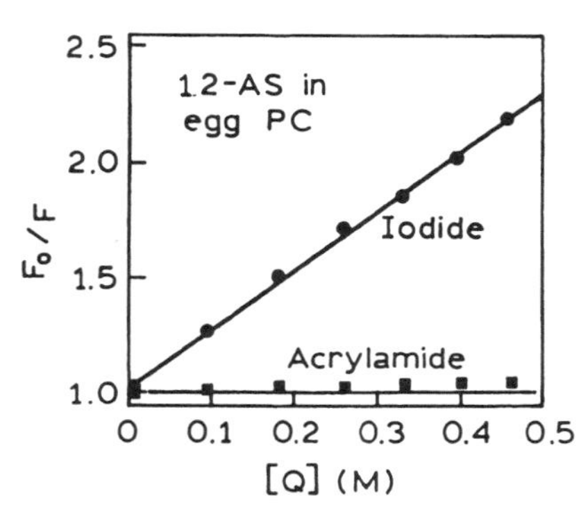

Figure 9.4. Quenching of 12-AS in lipid bilayers by iodide and acrylamide. Revised from Ref. 20.

quenching by the water-soluble quencher *N*-methylnicotinamide when bound to lipid bilayers.[17] However, there is a need for caution in interpreting quenching data for water-soluble quenchers, particularly for iodide quenching. Even though iodide is a water-soluble ion, it readily quenches 12-AS in membranes (Figure 9.4). Iodide quenching of other probes buried in membranes has been reported.[17] This seems to be a characteristic of iodide, which appears to readily enter hydrophobic regions of membranes.

In an effort to determine the location of each of the anthroyloxy groups in the series shown in Figure 9.1, these probes were quenched by the water-soluble quencher Cu^{2+} and the lipid-soluble quencher dimethylaniline (DMA). For quenching by Cu^{2+}, the apparent quenching constants were considerably larger than those possible for a diffusion-controlled reaction (Figure 9.5, left). This is a result of the binding of Cu^{2+} to the lipids and a high local Cu^{2+} concentration, presumably at the lipid–water interface. The unquenched lifetimes of these probes are near 10 ns. The extent of quenching was seen to decrease in the order 2-AP > 6-AS > 9-AS > 12-AS (Figure 9.5, left). These results indicate that the anthroyloxy moieties are localized away from the lipid–water interface, in the same order as predicted from the structures. In these experiments, the lowest extent of quenching was seen for the uncharged methyl ester of 9-anthracenecarboxylic acid (M-9A). Although not a fatty acid derivative, M-9A appears to be localized at the center of the bilayer, farthest away from the lipid–water interface.

To obtain additional information on the location of the anthroyloxy groups, these probes were quenched by DMA, which partitions into the lipid phase. Those fluorophores which are predicted from their structure to be most deeply localized in the membrane (M-9A, 12-AS) showed the largest degree of quenching by the lipid-soluble quencher DMA. Again, the apparent bimolecular quenching constants are larger than those possible for a diffusion-controlled reaction. The discrepancy arises because the concentration axis in Figure 9.5 refers to the total concentration of DMA. This nonpolar compound partitions into the membranes, resulting in an increased concentration of quencher in the region surrounding the fluorophore. In Section 9.1.E, we describe methods to analyze such data in terms of the lipid–water partition coefficients of the quencher. These results illustrate the information obtainable using freely diffusing quenchers that are simply added to the membrane suspension.

Another example of membrane-localized probes is provided by the series of NBD-labeled lipids shown in Figure 9.6. In order to obtain a surface-localized probe, NBD-chloride was reacted with phosphatidylethanolamine to yield NBD-PE. In order to obtain localization at various depths in the membrane, the NBD group was attached to 6-carbon (6-NBD-PA; PA is phosphatidic acid) or 12-carbon (12-NBD-PC) acyl chains. NBD was also attached to cholesterol.

The location of the NBD group in DOPC vesicles was studied using quenching by cobalt (Co^{2+}). NBD-choles-

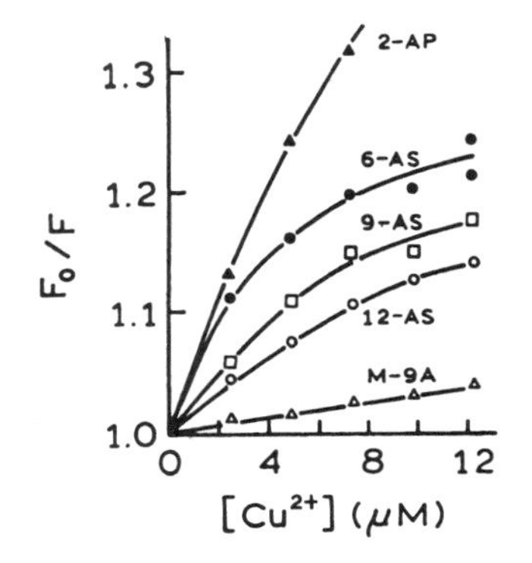

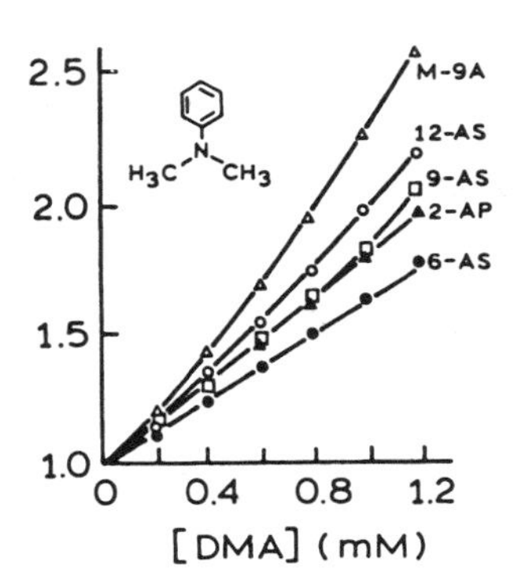

Figure 9.5. Quenching of the anthroyloxy probes by a water-soluble quencher, Cu^{2+}, and by a lipid-soluble quencher, dimethylaniline (DMA). Revised and reprinted from Ref. 4, Copyright © 1978, with permission from Elsevier Science.

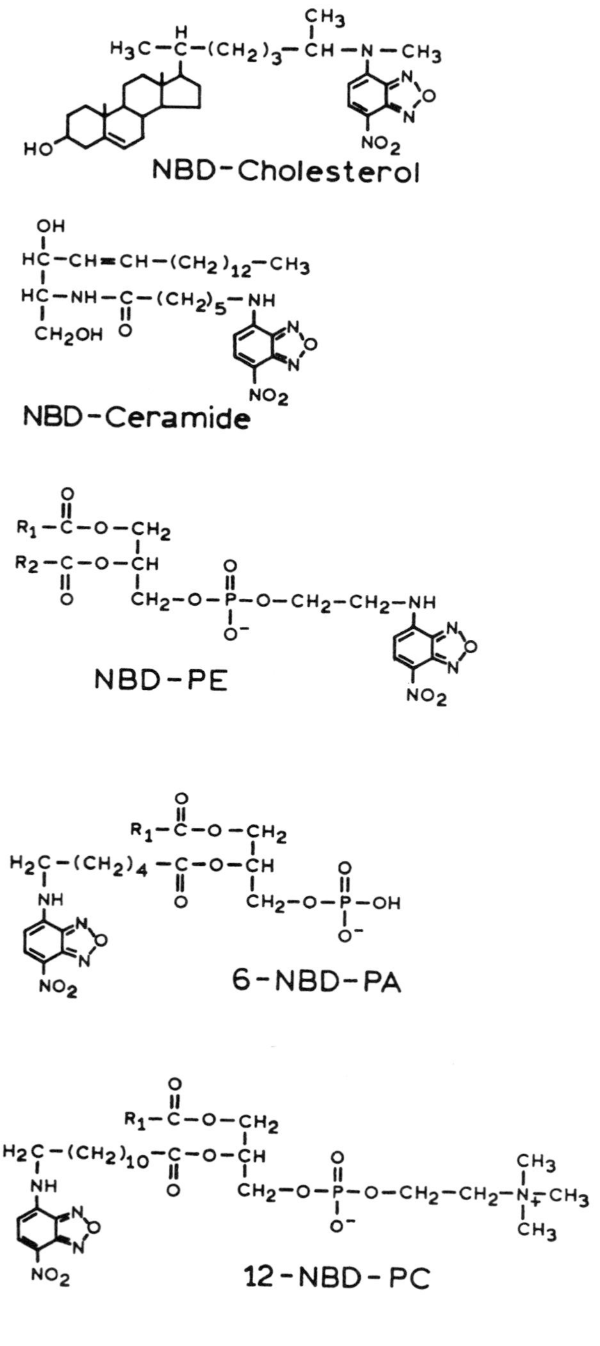

Figure 9.6. Chemical structures of NBD-labeled lipids. Revised and reprinted from Ref. 12, Copyright © 1990, with permission from Elsevier Science.

terol was found to be almost completely inaccessible to quenching by Co^{2+}, suggesting a deep location in the bilayer (Figure 9.7; see also Figure 9.12 below). NBD-PE was readily quenched by Co^{2+}, indicating that the NBD group is near the lipid–water interface, as would be predicted from the structure. Surprising results were obtained for 6-NBD-PC and 12-NBD-PC. The NBD groups in both probes were equally accessible to Co^{2+} quenching, and both were more accessible than NBD-cholesterol (Figure 9.7). This result was interpreted as due to folding of the acyl chains containing the NBD group so that this group became localized near the lipid–water interface (Figure 9.12 below). While the data for Co^{2+} quenching were interpreted as due to collisional quenching, cobalt can also quench NBD emission by RET. The mechanism of quenching depends on the fluorophore–quencher pair[21,22] and can be due to collisional or energy-transfer quenching. The important point is that the location of the lipid-bound fluorophores cannot always be predicted from the chemical structure.

9.1.B. Quenching of Membrane Probes Using Localized Quenchers

The quenchers considered in the previous section diffused freely in the aqueous or membrane phase. A somewhat more advanced approach is to use quenchers which are covalently linked to the phospholipids and thus restricted to particular depths in the lipid bilayers, as illustrated in Figure 9.3b. The most widely used lipid quenchers are the brominated phosphatidylcholines (bromo-PCs) and nitroxide-labeled PCs or fatty acids (Figure 9.8). Such probes have been used to determine the locations of fluorophores in bilayers based on the comparative amounts of quenching when the probe and quencher are at various depths in the membrane.[23–25] One example is quenching

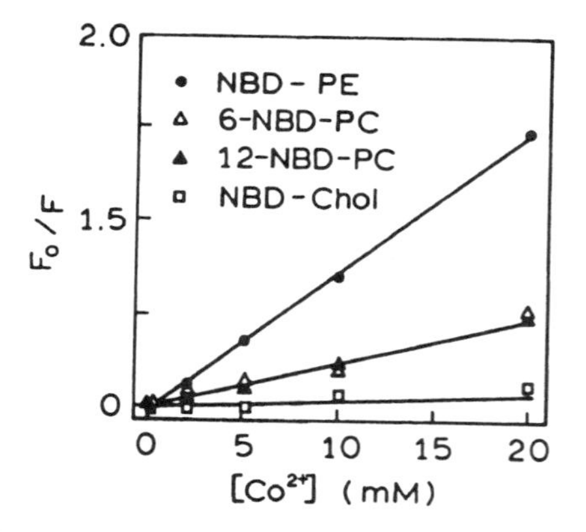

Figure 9.7. Cobalt quenching of NBD-lipid probes in DOPC vesicles. Revised from Ref. 13.

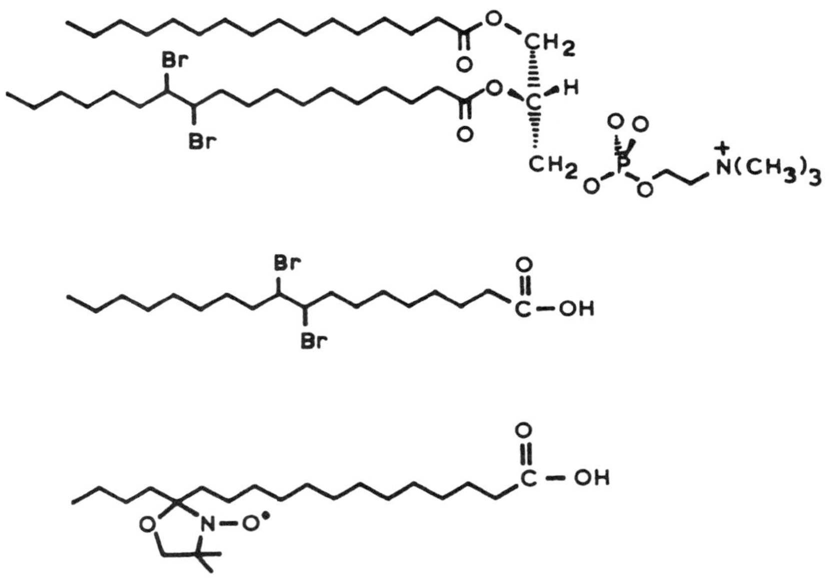

Figure 9.8. Structures of a bromo-PC (*top*), 9,10-dibromostearic acid (*middle*), and a nitroxide labeled fatty acid (*bottom*), which act as localized quenchers in membranes.

of the fluorenyl fatty acids when bound to erythrocyte ghosts. The term ghosts refers to the red blood cell membranes following removal of the hemoglobin by cell lysis. The fluorenyl probes (Figure 9.2) were quenched by 9,10-dibromostearic acid, which partitioned into the membrane. Since the bromine atoms are rather small, they are expected to be localized according to their position on the fatty acid chain. The bromine atoms are located near the center of the fatty acid (Figure 9.8, middle), so one expects maximal quenching for those fluorenyl groups which are located as deep as the bromine atoms. The quenching data reveal larger amounts of quenching as the fluorenyl groups are placed more deeply in the bilayer by a longer methylene chain between the fluorenyl and carboxyl groups (Figure 9.9). Hence, the fluorenyl probes are located as would be expected from their structures. For these fluorenyl fatty acids, the continuing acyl chain beyond the fluorenyl group was important for probe localization.

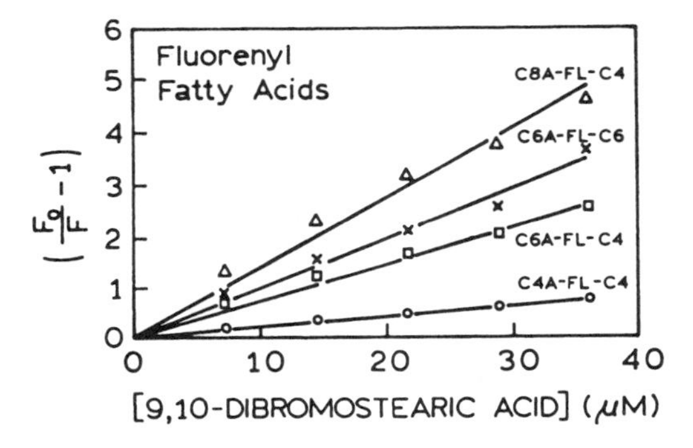

Figure 9.9. Quenching of fluorenyl fatty acids (see Figure 9.2) by 9,10-dibromostearic acid in erythrocyte ghost membranes. Revised and reprinted, with permission, from Ref. 7, Copyright © 1992, American Chemical Society.

Another example of the use of localized quenchers is provided by studies of pyrene-labeled lipids.[24] A number of pyrene-PCs ($Pyr_n PC$, where n is the number of methylene groups between the glycerol backbone and the pyrene) were used. The variation of the number of methylene groups localizes the pyrene groups at different depths in the bilayers (Figure 9.10). Three different bromo-PCs, in which the bromines were at positions 6 and 7, 9 and 10, or 11 and 12, were used as quenchers. The apparent Stern–Volmer quenching constants were measured for each pyrene-PC and each quencher (Figure 9.11). The largest quenching constant was observed for the smallest number of methylene groups in the pyrene-PC (n) when the quencher was $Br_{6,7}PC$, in which the bromine atoms are located just beneath the membrane surface. For large values of n ($n \geq 10$), maximum quenching was observed for $Br_{11,12}PC$, in which the bromine atoms are located more deeply in the membrane. In this case the fluorophores were localized in the bilayer as would be predicted from their structures.

Quenching by brominated PCs has been used in a variety of other experiments. Bromo-PCs are known to quench the fluorescence of membrane-bound proteins[26–28] and have

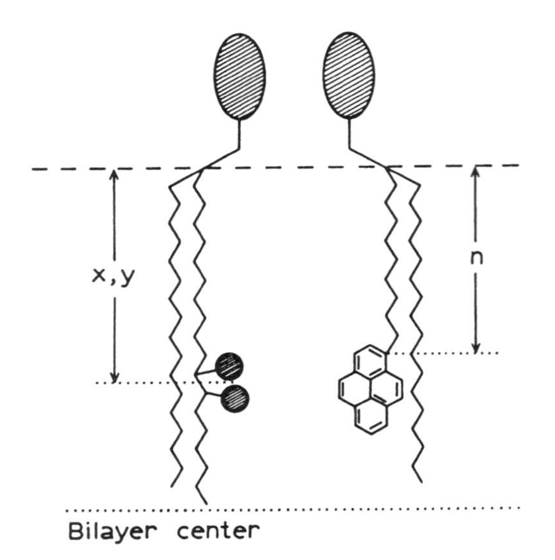

Figure 9.10. Schematic representation of the structures of Pyr_nPC and $Br_{x,y}PC$ species used to study depth-dependent quenching of the pyrene moiety. Revised and reprinted, with permission, from Ref. 24, Copyright © 1995, American Chemical Society.

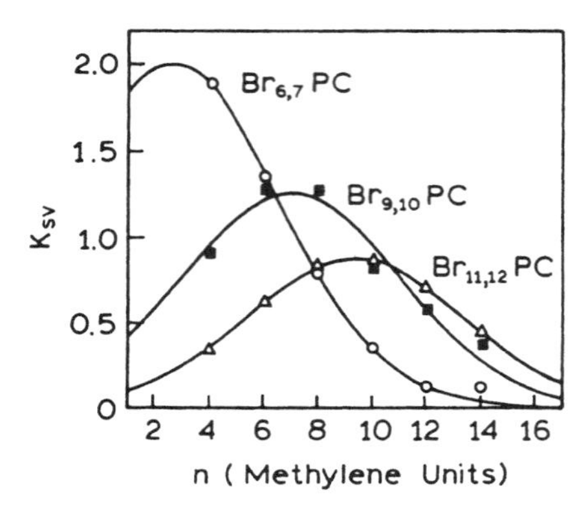

Figure 9.11. Dependence of K_{SV}^{app}, the apparent Stern–Volmer quenching constant, on the number of methylene units (n) in the pyrenylacyl chains of pyrene-PCs in bilayers with different $Br_{x,y}PC$ quenchers. The subscripts x, y indicate the locations of the bromine atoms. The bilayers consisted of 1-palmitoyl-2-oleoylphosphatidylcholine (POPC) with 50 mol % cholesterol. Revised and reprinted, with permission, from Ref. 24, Copyright © 1995, American Chemical Society.

been used to study lateral phase separations in membranes.[29,30]

9.1.C. Parallax Quenching in Membranes • Advanced Topic •

The idea of using localized quenchers in membranes has been extended to allow calculation of the depth of the fluorophores from the center of the bilayer.[31–38] The basic idea is to compare the amount of quenching observed for quenchers which are located at two different depths in the bilayer (Figure 9.3c). The distance of the fluorophore from the center of the bilayer (Z_{CF}) is then calculated from

$$Z_{CF} = L_{C1} + \{[-1n(F_1/F_2)/\pi C] - L_{21}^2\}/2L_{21} \quad [9.1]$$

L_{21} is the distance between the shallow and deep quenchers. The shallow quencher is located at a distance L_{C1} from the center of the bilayer. These distances are shown in Figure 9.12. C is the concentration of quenchers in molecules per unit area. F_1 and F_2 are the relative intensities of the fluorophore in the presence of the shallow and the deep quencher, respectively. This model was derived by assuming that quenching is complete within a critical distance of the quencher and that quenching does not occur across the bilayer. This model also assumes that there is no diffusion in the membrane.

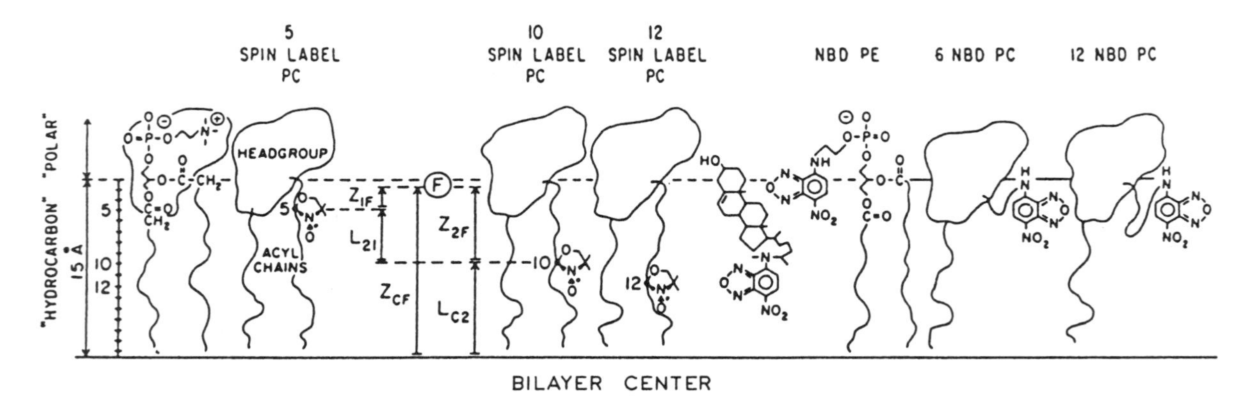

Figure 9.12. Schematic of a lipid bilayer, showing the structures of various probes and quenchers. The solid horizontal line indicates the center of the bilayer. Also shown are the distances used for parallax quenching. Reprinted, with permission, from Ref. 31, Copyright © 1987, American Chemical Society.

This model was tested by examining the quenching of NBD-labeled lipids by nitroxide-labeled (spin-labeled) PCs (Figure 9.12). The relative intensity of each probe was measured as the quencher concentration in the membrane was increased (Figure 9.13). While the intensity differences in Figure 9.13 are not dramatic, one can see that the extent of quenching follows expected trends. For instance, in NBD-PE, the NBD group is attached to the head group of the lipid and localized near the lipid–water interface. NBD-PE is quenched most strongly by the most shallow quencher, 5-SL-PC, and is less quenched by the deeper quencher 12-SL-PC (Figure 9.13, upper left). However, an initially confusing aspect of the data was the similar behavior observed for 6-NBD-PC and 12-NBD-PC, in which the NBD groups were expected to localize deeply in the bilayer. Contrary to expectations, 6-NBD-PC and 12-NBD-PC were more quenched by the shallow quencher 5-SL-PC.

These data were used to calculate the distance of the NBD groups from the center of the bilayer (Figure 9.14). As expected, NBD-PE is most distant from the center, that is, closest to the membrane surface. Surprisingly, 6-NBD-PC and 12-NBD-PC are also rather distant from the bilayer center. These results were interpreted as due to folding of the NBD acyl chain so that the NBD moieties are localized close to the lipid–water interface (Figure 9.12). This result for 6-NBD-PC and 12-NBD-PC is in agreement with the quenching of these probes by cobalt (Figure 9.7) and illustrates the need to determine probe location, rather than

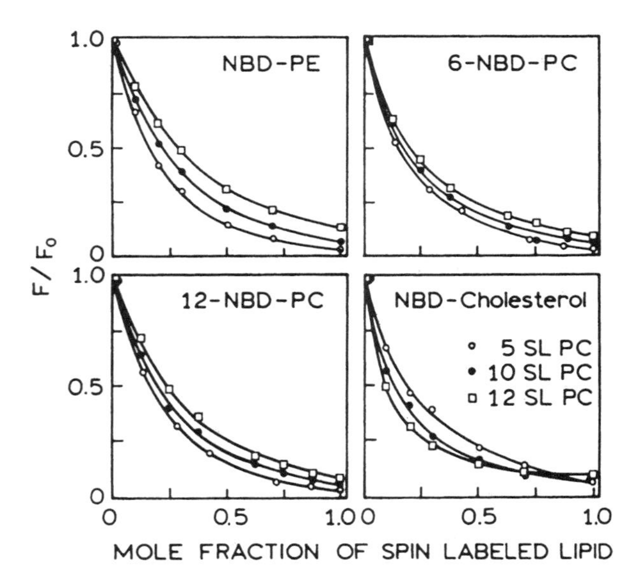

Figure 9.13. Quenching of NBD-labeled lipids in membranes containing one of three spin-labeled (SL) PCs. See Figure 9.6 for the probe structures. From Ref. 31.

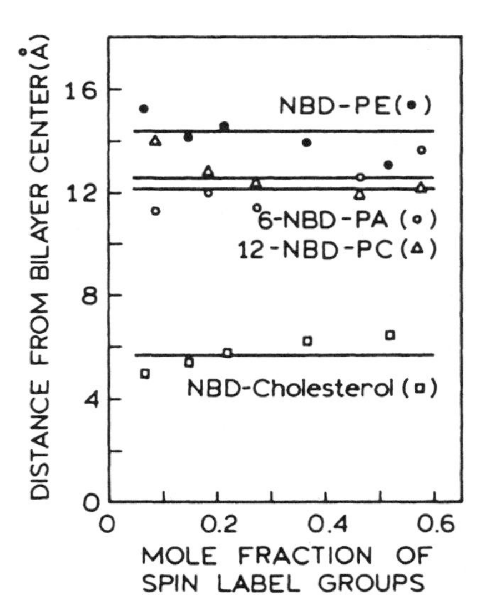

Figure 9.14. Distance of the probe from the bilayer center (Z_{CF}) calculated from the data in Figure 9.13. Revised from Ref. 31.

assuming that the location can be predicted from the structure.

A close examination of Figure 9.13 may cause one to question how NBD-cholesterol was localized. At low quencher concentrations, NBD-cholesterol was quenched most strongly by 12-SL-PC, but this trend did not continue at high quencher concentrations. For calculation of the depth in the membrane, only the data below 40 mol % quencher were used.

The use of spin labels as quenchers requires an understanding of their mechanism of quenching. The mechanism is uncertain because the spin labels are paramagnetic like oxygen but also show visible absorbance suggesting RET. At present, the consensus is that the nitroxide quenching is like oxygen quenching, which is thought to be due to enhanced intersystem crossing caused by the paramagnetic quenchers. The extent of quenching by nitroxide does not correlate with the Förster overlap integrals.[39]

9.1.D. Boundary Lipid Quenching • Advanced Topic •

Quenching in membranes can also be used to study boundary lipids, which are the lipid molecules surrounding a fluorophore or a membrane-bound protein.[40–46] Suppose that a protein is surrounded by a discrete number of lipid molecules and that the tryptophan fluorescence is accessible to quenchers in the membrane phase (Figure 9.3d). Then the number of boundary lipid molecules can be estimated from

$$\frac{F - F_{min}}{F_0 - F_{min}} = (1 - [Q])^n \qquad [9.2]$$

where F_0 is the intensity in the absence of quencher, F_{min} is the intensity when the probe is in pure quencher lipid, and F is the intensity at a given mole fraction of quencher lipid.

This model was tested using small probes in membranes, and for Ca^{2+}-ATPase, which is a large membrane-bound protein containing 11–13 tryptophan residues. For tryptophan octyl ester, the intensity decreased according to $n = 6$, indicating that each tryptophan was surrounded by six lipid molecules (Figure 9.15). For the Ca^{2+}-ATPase, the intensity decreased with $n = 2$. This does not indicate that only two lipid molecules surrounded this protein but rather that only two lipid molecules were in contact with tryptophan residues in the Ca^{2+}-ATPase. Similar results of two boundary lipids were found for Ca^{2+}-ATPase using the quencher 1,2-bis(9,10-dibromooleoyl)phosphatidylcholine.[42]

9.1.E. Effect of Lipid–Water Partitioning on Quenching
• Advanced Topic •

In the preceding examples of quenching in membranes, the quenchers were not soluble in water. Hence, the quencher

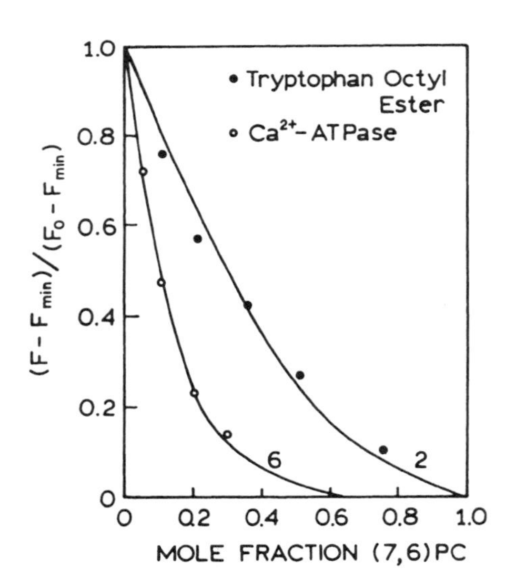

Figure 9.15. Quenching of tryptophan octyl ester (○) and Ca^{2+}-ATPase (●) by a spin-labeled (7,6)PC in egg PC vesicles. (7,6-PC) in a phosphatidyl choline in which the spin label is located on the eighth carbon atom chain of the 2-position fatty acyl group. The structure of the nitroxide spin label is shown in Figure 9.8. The solid curves are for the indicated values of n (Eq. [9.2]). Revised and reprinted, with permission, from Ref. 40, Copyright © 1981, American Chemical Society.

concentrations in the membrane were known from the amount of added quencher. However, there are many instances in which the quencher partitions into the membranes, but some fraction of the quencher remains in the aqueous phase. Consequently, the quencher concentration in the membrane is determined not simply by the amount of quencher added, but also by the total lipid concentration in the sample. In these cases it is necessary to determine the lipid–water partition coefficient in order to interpret the observed quenching.

Consider a quencher which distributes between the membrane and aqueous phases. At nonsaturating concentrations of quencher, the concentrations in the water (w) and membrane (m) phases are related by the partition coefficient

$$P = [Q]_m/[Q]_w \qquad [9.3]$$

The total (T) concentration of quencher added ($[Q]_T$) partitions between the water and membrane phases according to

$$[Q]_T V_T = [Q]_m V_m + [Q]_w V_w \qquad [9.4]$$

where V_m and V_w represent the volumes of the membrane and water phases, respectively. By defining

$$\alpha_m = V_m/V_T \qquad [9.5]$$

to be the volume fraction of membrane phase, one obtains[47]

$$[Q]_m = \frac{P[Q]_T}{P\alpha_m + (1 - \alpha_m)} \qquad [9.6]$$

Substitution of this expression for the membrane concentration of quencher into the Stern–Volmer equation yields

$$\frac{1}{\tau} = \frac{1}{\tau_0} + \frac{k_m P[Q]_T}{P\alpha_m + (1 - \alpha_m)} = \frac{1}{\tau_0} + k_{app}[Q]_T \qquad [9.7]$$

where k_m is the bimolecular quenching constant for the membrane-bound fluorophore. The apparent quenching constant is given by

$$\frac{1}{k_{app}} = \alpha_m\left(\frac{1}{k_m} - \frac{1}{k_m P}\right) + \frac{1}{k_m P} \qquad [9.8]$$

When the fluorophore is present in the membrane phase, the apparent quenching constant is dependent upon P, α_m, and k_m. A plot of k_{app}^{-1} versus α_m allows P and k_m to be determined. Thus, the quenching method allows simultaneous quantitation of both the extent to which a quencher partitions into a bilayer and its diffusion rate (D_q) in this bilayer. The successful determination of the quencher diffusion and partition coefficients requires that the range of lipid concentrations results in a range of fractional partitioning of the quencher. The fraction of the quencher partitioned in the membrane (f_m) is given by

$$f_m = \frac{P\alpha_m}{P\alpha_m + (1 - \alpha_m)} \quad [9.9]$$

To calculate the volume fraction of the lipid, the usual assumption is equal densities for the water and membrane phases. In this case, a 10-mg/ml membrane suspension corresponds to $\alpha_m = 0.01$. The above method of determining the lipid–water partition coefficient only applies when the quencher molecules are present in the bilayer at the moment of excitation. If the diffusional encounters involve molecules in the aqueous phase that diffuse into the lipid phase during the lifetime of the excited state, then no dependence of the apparent quenching on lipid concentration is expected. The situation is more complex when the quenching results from quenchers in both the lipid phase and the water phase.

A number of publications have appeared on the effect of partitioning and quenching.[47–52] One example is a study of quenching of an NBD-labeled lipid by the nitroxide-labeled fatty acid 5-doxylstearate (5-NS). In contrast to the nitroxide-labeled PCs, the fatty acids have a low but significant solubility in water. When NBD-labeled vesicles are titrated with 5-NS, the Stern–Volmer plots are dependent on lipid concentration (Figure 9.16). At lower lipid concentrations, addition of the same total amount of 5-NS results in larger amounts of quenching than observed at higher lipid concentrations. At low lipid concentrations,

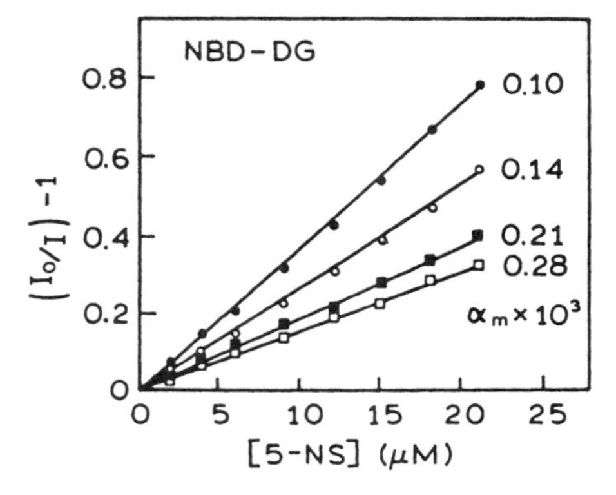

Figure 9.16. Quenching of 1-oleoyl-2-hexanoyl-NBD-glycerol (NBD-DG) by 5-doxylstearate (5-NS). α_m refers to the volume fraction of the egg PC phase. Revised from Ref. 52, Copyright © 1994, with permission from Elsevier Science.

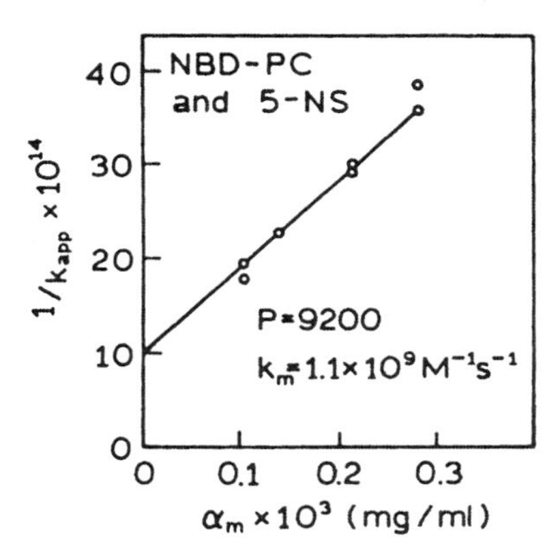

Figure 9.17. Dependence of the apparent quenching constant for the quenching of NBD-DG by 5-NS on lipid concentration (α_m). Revised from Ref. 52.

the added quencher results in a higher quencher concentration in the membrane. This is because there is less lipid into which the quencher can partition. It is this dependence of the apparent quenching constant on lipid concentration which allows the partition coefficient to be determined. This is done by plotting k_{app}^{-1} versus α_m (Figure 9.17). The data indicate that 5-NS partitions almost 10,000-fold into the lipid phase and that the bimolecular quenching constant is $1.1 \times 10^9\ M^{-1}\ s^{-1}$. Use of the Smoluchowski equation yields a mutual diffusion coefficient for the probe and quencher of about 3×10^{-6} cm^2/s. This value is larger than expected for a quencher in a membrane and 100-fold larger than the diffusion coefficient of lipid in membranes determined by fluorescence recovery after photobleaching (FRAP).The interpretation of collisional quenching data in membranes requires additional development.

In some cases the quencher may partition into the membrane phase but the extent of quenching is independent of membrane concentration. This occurs when partitioning is weak and the amount of added quencher is in excess. One example is shown in Figure 9.18 for oxygen quenching of a membrane-bound fluorophore. This figure shows Stern–Volmer plots for 2-methylanthracene in vesicles of DMPC and DPPC, which have phase-transition temperatures (T_c) near 24 and 37 °C, respectively.[53] At the experimental temperature near 31 °C, the DPPC bilayers are below the phase-transition temperature, and the DMPC bilayers are above the phase-transition temperatures. Whereas anisotropy measurements on such bilayers suggest a large change in viscosity at the transition temperature (Chapter 10), the effect on oxygen diffusion is only modest, approximately twofold. This rather surprising result has been confirmed by other fluorescence[54–56] and electron spin resonance experiments.[57] In fact, these experiments sug-

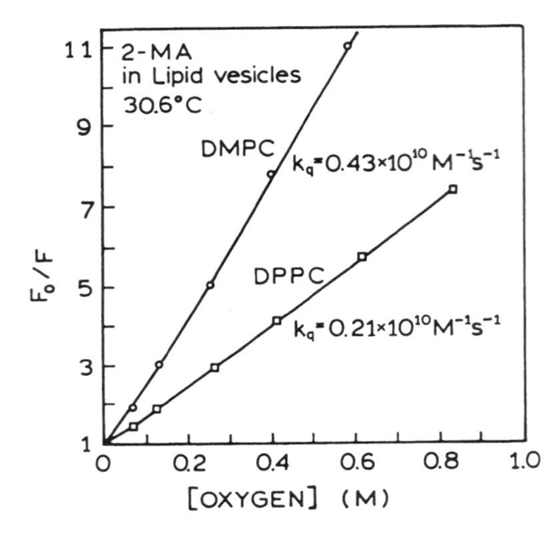

Figure 9.18. Oxygen quenching of 2-methylanthracene (2-MA) at 30.6 °C in DMPC (T_c = 24 °C) and DPPC (T_c = 37 °C) bilayers.

gest that cholesterol, which generally makes membranes more rigid, results in increased rates of oxygen transport. This can be seen by quenching of pyrenedodecanoic acid (PDA) in erythrocyte ghost membranes (Figure 9.19). In this case, the ghost membranes were modified by addition of endogenous cholesterol. Except at the highest cholesterol concentrations, the apparent biomolecular quenching constant increased with increasing amounts of cholesterol (Figure 9.19).

In order to calculate the bimolecular quenching constants, it is necessary to know the oxygen concentration in the membranes. Unfortunately, precise values are not known, particularly for membranes with different lipid compositions. The oxygen solubility in membranes is usually taken as equal to that in nonpolar solvents, and thus approximately fourfold larger than in water. Assuming a lifetime of PDA near 120 ns, one can use the data in Figure 9.19 to calculate $k_q = 0.22 \times 10^{10}\ M^{-1}\ s^{-1}$. The permeability ($P$) of a membrane can be approximated by $P = k_qD/\Delta x$, where D is the oxygen diffusion coefficient and Δx is the membrane thickness. Given that the diffusion coefficient of oxygen is about one-fifth of that in water, and the lipid–water partition coefficient is near 5, this equation suggests that biological membranes do not pose a significant diffusive barrier to oxygen.

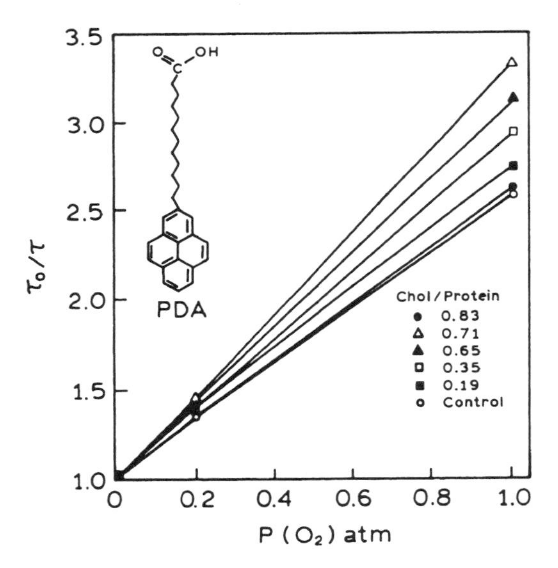

Figure 9.19. Oxygen quenching of pyrenedodecanoic acid in erythrocyte ghost membranes with various cholesterol/protein ratios: ■, 0.19; □, 0.35; ▲, 0.65; △, 0.71; ●, 0.83. ○, Control, no added cholesterol. Revised from Ref. 56.

9.2. DIFFUSION IN MEMBRANES

In the preceding section on localization of probes in membranes, the models did not consider diffusion of the quenchers, except for oxygen quenching and quenching of the NBD-lipid by the nitroxide fatty acid. Diffusion in membranes is a complex topic, and the appropriate model depends on the chemical structure of the quencher. For small, nonpolar quenchers, the membrane can be considered to be a three-dimensional system, with the quencher diffusing freely. The situation changes if the quencher is a lipid, such as the bromo-PCs (Figure 9.10). In this case the quencher will be limited to lateral diffusion within a plane, which requires a different theory.

9.2.A. Quasi-Three-Dimensional Diffusion in Membranes

The concept of quasi-three-dimensional diffusion in membranes is illustrated by pyrene excimer formation.[58] Although excimer formation is not strictly quenching, the monomer emission is decreased by diffusive encounters of excited monomers with ground-state monomers. Hence, the quenching of the monomer emission is described by Stern–Volmer kinetics.

In many instances, it is acceptable to have an observable which reflects the local viscosity of the membrane. The excimer-to-monomer ratios have been widely used for this purpose.[59,60] Emission spectra of pyrene and its excimer were shown in Figure 1.11. The ratio of excimer to monomer emission increases with increasing pyrene concentration and with increasing rates of diffusion. Typical data are shown in Figure 9.20 for pyrene in DMPC vesicles. As the temperature is increased, the excimer-to-monomer ratio increases. The presence of cholesterol results in a lower ratio, reflecting slower pyrene diffusion. By comparison of

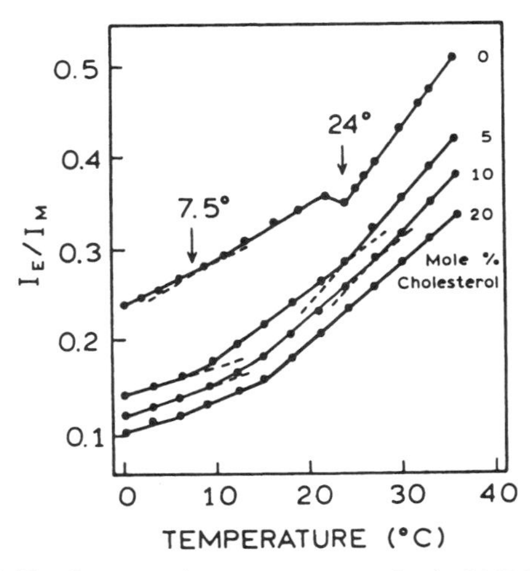

Figure 9.20. Pyrene excimer-to-monomer ratios in DMPC vesicles with added cholesterol as a function of temperature. [DMPC] = 0.3m*M*; 0.5% pyrene by weight of phospholipid. Revised and reprinted, with permission, from Ref. 59, Copyright © 1974, American Chemical Society.

these data with those obtained in solvents of known viscosity, the apparent viscosity of the membrane can be calculated.[60]

The factors which determine the relative amounts of monomer and excimer emission can be understood by considering the kinetic scheme for pyrene in Figure 9.21.[61–66] The fractions of pyrene molecules emitting from the monomer and the excimer state are given by the ratio of the emission rate to the sum of the rates depopulating each state. Hence, the relative intensities F_M and F_E of the monomer and excimer emission, respectively, are

$$F_M = C_M \frac{\Gamma_M}{\Gamma_M + k_{\text{nr}}^M + k_1[\text{P}]} \qquad [9.10]$$

$$F_E = C_E \frac{\Gamma_E}{\Gamma_E + k_{\text{nr}}^E + k_2} \frac{k_1[\text{P}]}{\Gamma_M + k_{\text{nr}}^M + k_1[\text{P}]} \qquad [9.11]$$

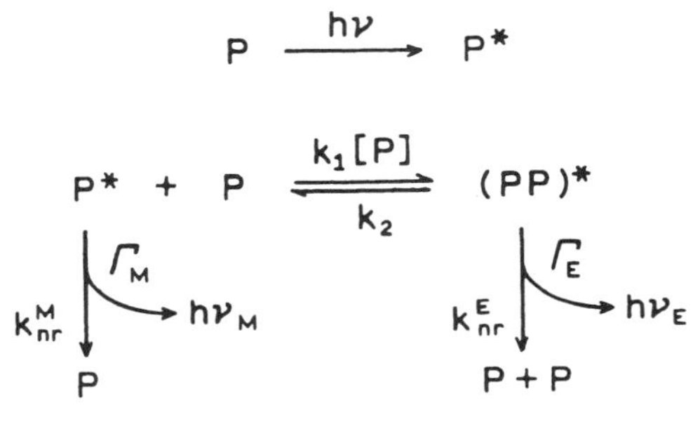

$$\text{P} \xrightarrow{h\nu} \text{P}^*$$

$$\text{P}^* + \text{P} \underset{k_2}{\overset{k_1[\text{P}]}{\rightleftharpoons}} (\text{PP})^*$$

$$\text{P}^* \xrightarrow{k_{nr}^M} \text{P}; \quad \text{P}^* \xrightarrow{\Gamma_M} h\nu_M \qquad (\text{PP})^* \xrightarrow{k_{nr}^E} \text{P} + \text{P}; \quad (\text{PP})^* \xrightarrow{\Gamma_E} h\nu_E$$

Figure 9.21. Kinetic scheme for excimer formation.

In these expressions, [P] is the pyrene concentration, C_M and C_E are constants reflecting the sensitivity of the instrument at each wavelength, Γ_M and Γ_E are the radiative decay rates of the monomer and the excimer, respectively, and k_{nr}^M and k_{nr}^E are the nonradiative decay rates of the monomer and the excimer, respectively. The rate of excimer formation is $k_1[\text{P}]$, and the rate of excimer dissociation is k_2. The last term in Eq. [9.11] is the fraction of the monomers converted to excimers.

The ratio of the excimer to the monomer emission is given by

$$\frac{F_E}{F_M} = C \frac{\Gamma_E}{\Gamma_M} \frac{k_1[\text{P}]}{\Gamma_E + k_{\text{nr}}^E + k_2} = C \frac{\Gamma_E}{\Gamma_M} \tau_E k_1[\text{P}] \qquad [9.12]$$

where C is an instrumental constant for the relative sensitivity at the excimer and monomer wavelengths, and τ_E is the excimer lifetime assuming that this species was directly excited (Chapter 18). It is usually safe to assume that excimer formation is irreversible ($k_2 = 0$), but excimer dissociation is known to occur. The important point is that the ratio of the excimer to the monomer emission is proportional to the pyrene concentration and the forward rate constant k_1. This rate constant is analogous to the bimolecular quenching constant k_q and is thus proportional to the pyrene diffusion coefficient.

If lifetime measurements are available, it is not necessary to use the excimer-to-monomer ratios. This can be seen by considering the monomer lifetime at low concentrations, where excimer formation does not occur (τ_{OM}), and in the presence of excimer formation (τ_M),

$$\tau_{\text{OM}} = \frac{1}{\Gamma_M + k_{\text{nr}}^M} \qquad [9.13]$$

$$\tau_M = \frac{1}{\Gamma_M + k_{\text{nr}}^M + k_1[\text{P}]} \qquad [9.14]$$

Division of Eq. [9.13] by Eq. [9.14] yields

$$\frac{\tau_{\text{OM}}}{\tau_M} = 1 + k_1 \tau_{\text{OM}}[\text{P}] \qquad [9.15]$$

Hence, the pyrene monomer lifetime is decreased according to the Stern–Volmer equation, where the quencher concentration is replaced by the ground-state pyrene monomer concentration.

A disadvantage of the use of pyrene is that the extent of excimer formation depends on the bulk concentration of pyrene in the membranes. For this reason, several groups have synthesized covalently linked pyrenes or pyrene–amine conjugates.[67–72] Examples of such molecules are

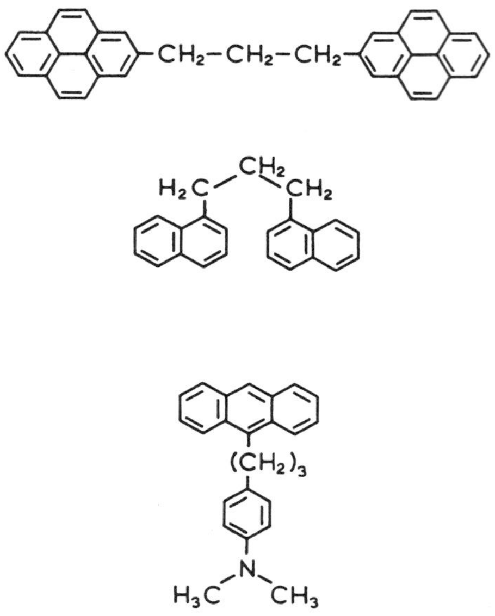

Figure 9.22. Probes which form intramolecular excimers or exciplexes.

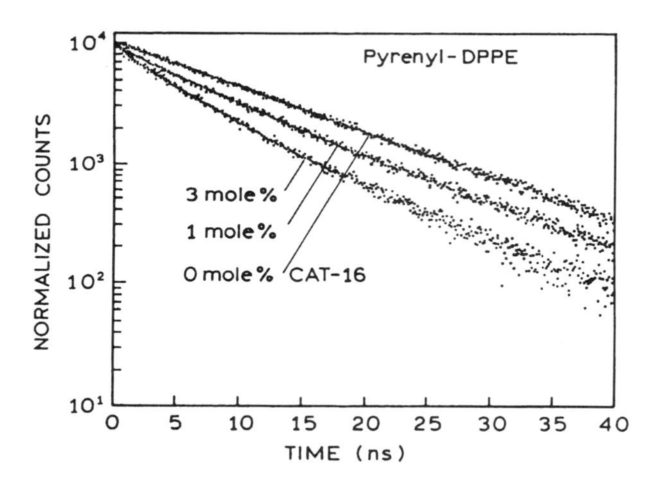

Figure 9.23. Intensity decay of pyrenyl-DPPE [*N*-(1-pyrenylsulfonyl)dipalmitoyl-L-α-phosphatidylethanolamine] in an oleic acid monolayer in the presence of the nitroxide quencher CAT-16 [4-(*N,N*-dimethyl-*N*-hexadecylammonium)-2, 2, 6, 6-tetramethylpiperidine-1-oxyliodide] at 20 °C. Revised and reprinted, with permission, from Ref. 76, Copyright © 1993, American Chemical Society.

shown in Figure 9.22. For such molecules, the extent of excimer or exciplex formation still depends on pyrenyl-to-pyrenyl diffusion, but the excimer-to-monomer ratio is independent of the probe concentration.[72]

9.2.B. Lateral Diffusion in Membranes

Since these earlier studies of excimer formation, there have been advances in the theory for diffusion in two dimensions. The theory is complex, but analytical expressions are now available for the time-dependent decays of fluorophores in membranes, with the quenchers constrained to lateral diffusion in two dimensions.[73–78] The expressions are mathematically complex and are listed below. We first describe one example of quenching of pyrene-DPPE in oleic acid monolayers at an air–water interface. The quencher was a lipid-soluble nitroxide, CAT-16. The pyrene-DPPE decays in the monolayers are shown in Figure 9.23 for 1 and 3 mole % of CAT-16. The decreased pyrene lifetime is visible, and the data allowed recovery of the mutual diffusion coefficient, with a value near 6×10^{-7} cm^2/s. The extent of quenching is modest because the decay time of the pyrene derivative is near 13 ns, not the 200 to 400-ns lifetime normally displayed by pyrene. Another reason for the modest amount of quenching is that lateral diffusion is typically slow in two dimensions. There is usually a rapid component due to quenching of more closely spaced F–Q pairs, following by a longer-lived component due to diffusion of more distant quenchers.[75]

The time-dependent decays expected for Smoluchowski quenching in two dimensions and for the radiation model (Section 9.4) have been reported.[73–76] For the usual assumption of instantaneous quenching on fluorophore–quencher contact, the intensity decay is given by

$$\ln\frac{I(t)}{I_0(t)} = -\gamma t - \frac{1}{2}R^2[\mathrm{Q}]Q(t/\tau_q) \qquad [9.16]$$

where

$$Q(t/\tau_q) = \frac{16}{\pi}\int_0^\infty \frac{1-\exp[-(t/\tau_q)x^2]}{J_0^2(x)+Y_0^2(x)}\frac{dx}{x^3} \qquad [9.17]$$

In these expressions, γ is the reciprocal of the unquenched decay time, [Q] is the quencher concentration in molecules per angstrom squared, R is the interaction radius, and $\tau_q = R^2/D$, where D is the mutual diffusion coefficient. $J_0(x)$ and $Y_0(x)$ are zero-order Bessel functions of the first and second kinds, respectively. Even more complex expressions are needed for the radiation model. Because of the difficulties in evaluating Eqs. [9.16] and [9.17], several approximate analytical expressions have been proposed.[74,75]

9.3. QUENCHING EFFICIENCY

In many instances, such as for membrane-bound fluorophores and quenchers, one wishes to use the observed bimolecular quenching constants to calculate the diffusion coefficient of the quencher. This calculation requires that the quenching efficiency, f_Q, be known and that this efficiency be the same in all samples investigated. For example, the quenching efficiency may be calculated for a given fluorophore–quencher combination by experiments in a solvent of known viscosity. Then quenching of this same fluorophore may be studied when it is bound to a membrane or a protein. In some instances, changes in the solvent or viscosity can result in changes in the quenching efficiency. This dependence of f_Q on viscosity can be visualized as follows. Consider the following reaction scheme:

$$F^* + Q \underset{k_1}{\overset{k_0}{\rightleftharpoons}} (F\text{—}Q)^* \xrightarrow{k_2} F + Q + \text{heat} \quad [9.18]$$

(F—Q)* designates the excited complex of fluorophore and quencher which has not yet been deactivated. For an excited-state reaction, such as exciplex formation with amines, (F—Q)* could be the exciplex. However, we will consider only the case where (F—Q)* is nonfluorescent. If k_2 is much greater than k_1, then the quenching is very efficient, and the quenching constant is given by k_0 (Eq. [8.11]). However, for less efficient quenchers, the dissociation rate k_1 can be comparable to the quenching rate k_2. In this case the quenching constant is

$$k_q = k_0 \frac{k_2}{k_1 + k_2 + \gamma} = k_0 f_Q \quad [9.19]$$

where $\gamma = \tau_0^{-1}$, the reciprocal of the unquenched lifetime. The quenching efficiency is the fraction of the (F—Q)* complexes which are quenched, instead of dissociating or emitting,

$$f_Q = \frac{k_2}{k_1 + k_2 + \gamma} \quad [9.20]$$

For this reason, one might expect the quenching efficiency to increase in more viscous solvents, which slow dissociation of the fluorophore and quencher. The increased duration of each diffusive encounter increases the probability of quenching during this encounter.

For most fluorophore–quencher pairs, quenching is highly efficient, and the quenching efficiency is not known to depend on solvent or viscosity. However, there is a known example where the efficiency is less than unity and dependent on environment, namely, the quenching of indole by succinimide.[79,80] In this case the quenching efficiency is near 0.5 in protic solvents and decreases to less than 0.05 in pure dioxane. The reason for the changes in efficiency is not known. Such changes could compromise interpretation of quenching data for protein- or membrane-bound fluorophores where the local environment is not known in detail.

9.3.A. Steric Shielding Effects in Quenching

The apparent quenching efficiency can also be affected by the extent of fluorophore exposure to the aqueous phase (Figure 9.24). One can readily imagine that the quenching efficiency depends on the solid angle of exposure (θ_0) and the radius of the quencher (R_Q). The bimolecular quenching constant also depends on the depth of the fluorophore in the macromolecule, and even on the rate of rotational diffusion because faster rotational diffusion provides more opportunities for quenching. This theory is moderately complex and not easily summarized.[81–86] An important concept is that the quenching rate for a fluorophore on the surface of a macromolecule can be reduced by a factor of 2 or more due to immobilization of the fluorophore on the surface, even without any additional steric shielding.[85] This is shown in Figure 9.25, where the apparent quenching constant quickly decreases to 50% of the original value as the size of the macromolecule increases. Hence, when the bimolecular quenching constant of a fluorophore on the surface of a macromolecule has a value corresponding to one-half the diffusion-controlled rate, the fluorophore should be regarded as fully accessible to the aqueous phase.

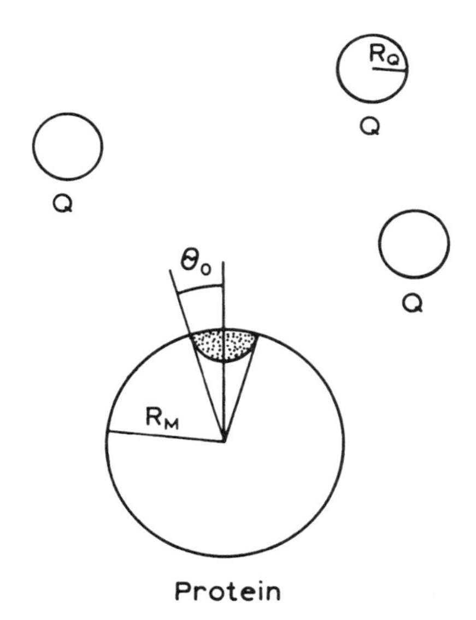

Figure 9.24. Model for the interaction of a uniformly reactive quencher (Q) with a protein that is reactive in a surface area described by cone angle $2\theta_0$. Revised from Ref. 85.

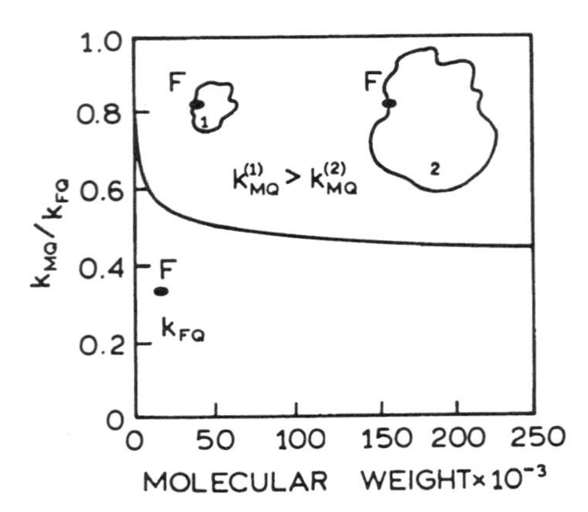

Figure 9.25. Ratio of bimolecular quenching constants for a fluorophore on the surface of a macromolecule (k_{MQ}) and when free in solution (k_{FQ}) as a function of the size of the macromolecule. Revised and reprinted, with permission, from Ref. 85, Copyright © 1985, Biophysical Society.

9.4. TRANSIENT EFFECTS IN QUENCHING • Advanced Topic •

In the discussion of collisional quenching, we assumed that the form of the decay law was not changed by quenching. That is, if the decay was a single exponential before quenching, it remained a single exponential in the presence of quencher, but with a shorter lifetime. In reality, all intensity decays become nonexponential in the presence of collisional quenching.[87] By nonexponential, we mean not described by a sum of exponentials. This effect is due to the so-called transient effects in quenching. For small amounts of quenching in fluid solvents, these effects are barely noticeable. For larger amounts of quenching, especially in moderately viscous solvents, the intensity decays become strongly nonexponential.

The theory for transient effects is complex and has been presented in a monograph on diffusion-controlled reactions.[88] Transient effects were first identified by Smoluchowski, who considered diffusion-controlled reactions between particles in solution.[89] The rate constant for reaction between the particles was shown to be time-dependent:

$$k(t) = 4\pi RN'D[1 + R(\pi Dt)^{-1/2}] \qquad [9.21]$$

where R is the interaction radius [sum of the radius of the fluorophore (F) and that of the quencher (Q)], $N' = 6.02 \times 10^{20}$, and D is the sum of the F and Q diffusion coefficients. The time dependence originates with the random distribution of fluorophores and quenchers at the moment of excitation. Some fluorophore–quencher pairs will be in close proximity, and others will be more widely spaced. The fluorophores with a closely located quencher are extinguished rapidly. With time following excitation, the ensemble of fluorophores evolves, and the fluorophores that remain in the excited state longest are those which were most distant from the closest quencher at the moment of excitation. The time-dependent rate constant decreases from an initially high value to the diffusion-limited value ($k_q = 4\pi RN'D$). The intensity decay in the presence of collisional quenching can be obtained by integration of the differential equation describing $dI(t)/dt$, which includes the time-dependent rate constant. This yields

$$I(t) = I_0 \exp[(-t/\tau) - 2bt^{1/2}] \qquad [9.22]$$

where τ is given by

$$\frac{1}{\tau} = \frac{1}{\tau_0} + k_q[Q] \qquad [9.23]$$

and

$$b = 4R^2N'(\pi D)^{1/2}[Q] \qquad [9.24]$$

While these equations seem complex, the actual situation is still more complex. The Smoluchowski model assumes that the fluorophore is instantaneously deactivated upon contact with the quencher, which results in an infinite quencher concentration gradient around the fluorophore. Also, the Smoluchowski model assumes no quenching in the absence of molecular contact. These assumptions have been modified to allow more realistic modeling of the quenching process.[90,91] Two quenching models are shown schematically in Figure 9.26. The radiation boundary condition (RBC) model assumes that quenching occurs with a finite rate constant κ when F and Q are in contact (Figure 9.26, top). For the Smoluchowski model, $\kappa(r) = \infty$ when F and Q are in contact and $\kappa(r) = 0$ otherwise.

Another model for quenching, the so-called distance-dependent quenching (DDQ) model, assumes that the quenching rate is dependent on the F–Q distance as shown in the lower half of Figure 9.26. Such an exponential dependence on distance is well known for electron-transfer[92] and electron-exchange interactions.[93] For distance-dependent interactions, the rate of quenching at an F–Q distance r is given by

$$k(r) = k_a \exp\left(-\frac{r-a}{r_e}\right) \qquad [9.25]$$

where r_e is the characteristic distance, and k_a is the rate of reaction at a distance a. Typical values of a, the distance of closest approach, are 5–7 Å. These interactions typically decay quickly with distance, with r_e values in the range of 1–2 Å. The time-dependent rate constant is given by

$$k(t) = \frac{4\pi}{C_q^0} \int_a^\infty r^2 k(r) C_q(r, t) dr \quad [9.26]$$

where $C_q(r, t)$ is the concentration of the quencher molecules at distance r from the excited fluorophore at time instant t, and C_q^0 is the bulk quencher concentration.

Unfortunately, the use of these quenching models requires rather complex theory and analysis. For the Smoluchowski, RBC, or DDQ models, the intensity decay is given by

$$I(t) = I_0 \exp(-t/\tau_0) \exp[-[Q] \int_0^t k(t)\, dt] \quad [9.27]$$

where $k(t)$ is the time-dependent rate constant. For the RBC model, $k(t)$ is given by[94–97]

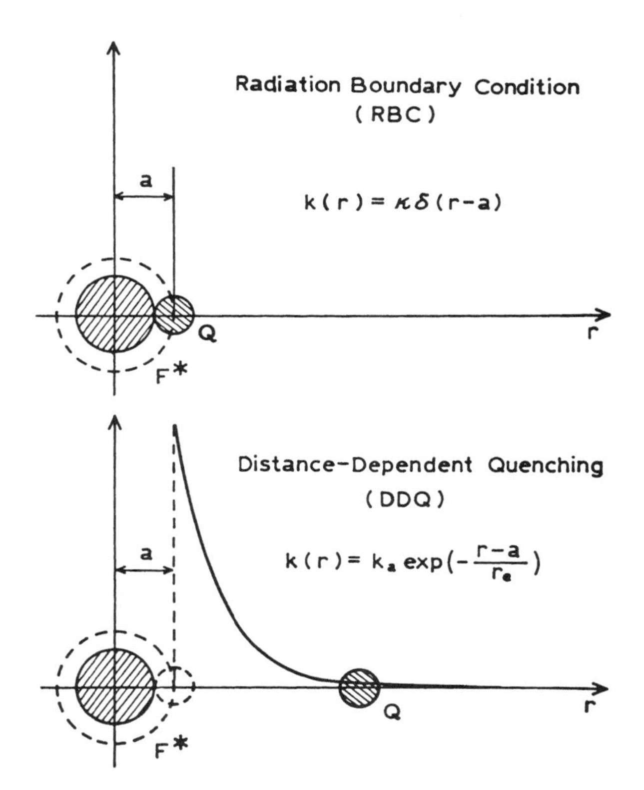

Figure 9.26. Comparison of the RBC and DDQ models for collisional quenching of fluorescence. Revised and reprinted, with permission, from Ref. 100, Copyright © 1996, American Chemical Society.

$$k(t) = \frac{4\pi RDN'}{1 + (D/\kappa R)} \left[1 + \frac{\kappa R}{D} \exp(X^2) \mathrm{erfc}(X) \right] \quad [9.28]$$

$$X = \frac{(Dt)^{1/2}}{R} \left(1 + \frac{\kappa R}{D} \right) \quad [9.29]$$

The function erfc(y) is the complement of the error function,

$$\mathrm{erfc}(y) = \frac{2}{\pi^{1/2}} \int_y^\infty \exp(-z^2)\, dz \quad [9.30]$$

For the DDQ model, analytical expressions for $k(t)$ are not yet known, so numerical procedures are used to calculate the intensity decay $I(t)$.[98–102]

It is useful to visualize how the quenching constant depends on time. The values of $k(t)$ are shown in Figure 9.27 for the Smoluchowski model and for the RBC model with various values of κ. For a diffusion coefficient near 10^{-6} cm^2/s, the transient effects are mostly complete in 10 ps and are no longer present at 1 ns. For these longer times, the value of $k(t)$ becomes equal to the usual bimolecular quenching constant (Eq. [8.11]). At short times, the values of $k(t)$ are initially larger than the calculated value of k_q. This effect is due to the rapid quenching of closely spaced F–Q pairs. In the Smoluchowski model, the value of $k(t)$ diverges to infinity at short times. This difficulty is avoided in the RBC model, where the maximal rate is limited by the value of κ. As κ increases, the RBC model becomes identical with the Smoluchowski model.

9.4.A. Experimental Studies of Transient Effects

Rather than focus on the complex theory of transient effects, we will present several examples. It is important to recognize that detection of transient effects in quenching is difficult. This is illustrated for quenching of NATA by iodide in Figure 9.28. In the absence of iodide, the intensity decay of NATA is a single exponential. The decay remains close to a single exponential even in the presence of 0.35*M* iodide. The deviations from the single-exponential model are difficult to detect even at this highest iodide concentration (Figure 9.28). In fluid solutions, transient effects are often weak and difficult to observe.

It seems that the transient effects are somewhat easier to observe using the FD data. Y_t-base is a naturally fluorescent base found near the anticodon region of the *E. coli* phenylalanine transfer RNA. In the absence of quenching, the FD intensity decay reveals a single-exponential decay (Figure 9.29). In the presence of 0.5*M* acrylamide, the

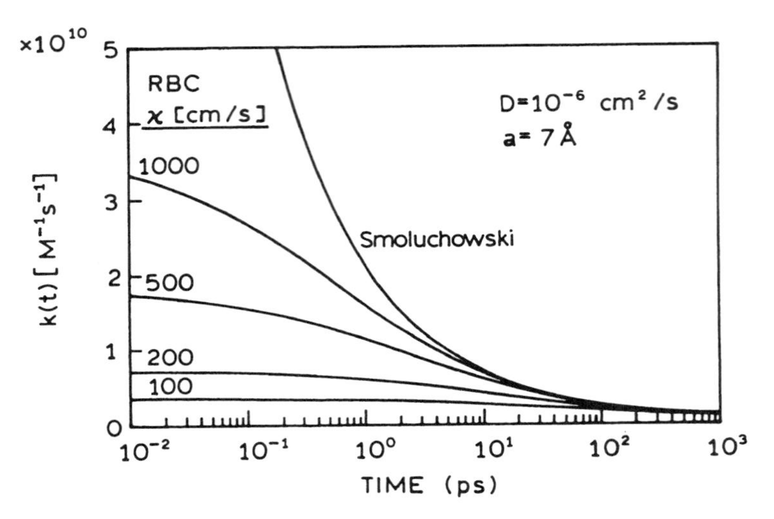

Figure 9.27. Time-dependent rate constants for the Smoluchowski and RBC models. From Ref. 103.

decay becomes strongly heterogeneous and cannot even be approximately fit by a single-exponential decay law ($\chi_R^2 = 413$; Table 9.1). This complexity can be accounted for by using the RBC or DDQ models.[105]

Transient effects can also be observed for proteins.[106,107] The single-tryptophan protein staphylococcal nuclease is known to display a dominant single-exponential decay, as seen by the overlap with the single-decay-time model (Figure 9.30). When quenched by oxygen, the intensity decay is no longer a single exponential (Table 9.1). Oxygen diffusion is rapid, which minimizes the transient part of the rate constant (Eq. [9.21]). Larger effects can be expected

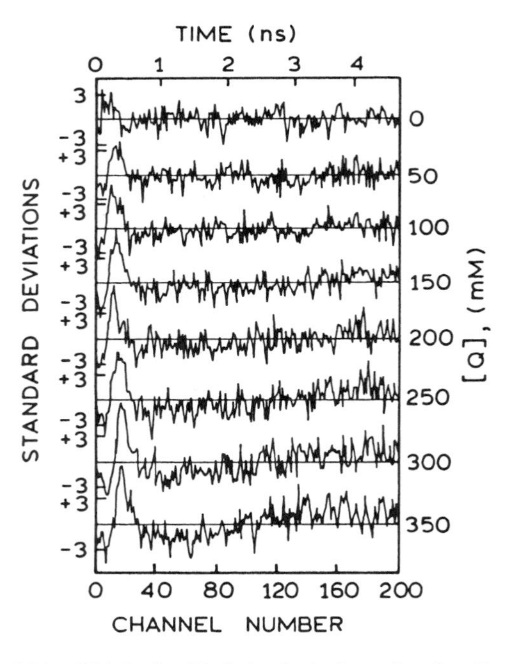

Figure 9.28. Weighted residuals for single-decay-time fits of intensity decays of NATA quenched by iodide at concentrations ranging from 50 to 350 m*M*. pH 7, 22 °C. Revised and reprinted from Ref. 104, Copyright © 1983, with permission from Elsevier Science.

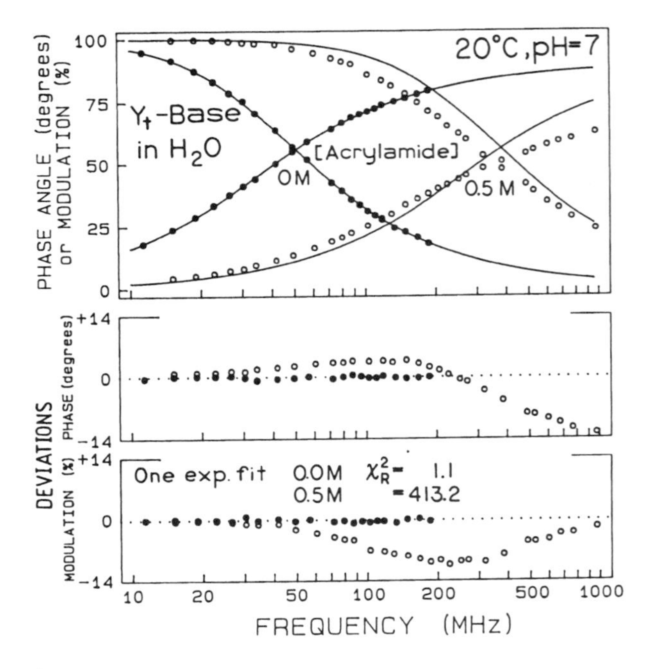

Figure 9.29. Effect of acrylamide quenching on the intensity decay of Y_t-base. The FD intensity decays in the absence (●) and in the presence of 0.5*M* acrylamide are shown. The solid curves represent the best single-decay-time fits to the data. The lower panels show the deviations from the best single-decay-time fits. Revised and reprinted from Ref. 105, Copyright © 1988, with permission from Elsevier Science.

Table 9.1. Intensity Decays of Yt-Base, N-Acetyltryptophanamide (NATA), and Staphylococcal Nuclease in the Absence and in the Presence of Quenching

Sample	Quencher	Single-decay-time fit		Two-decay-time fit			
		τ (ns)	χ_R^2	τ_i (ns)	α_i	f_i	χ_R^2
Y_t-base[a]	—	4.64	1.2				
Y_t-base, water, 20 °C	0.5*M* acrylamide	0.62	413	0.16 0.91	0.56 0.44	0.18 0.82	2.4
NATA[b]	—	2.87	0.8				
NATA, water, 20 °C	0.144*M* oxygen	0.51	175	0.05 0.59	0.37 0.63	0.05 0.95	1.6
Staphylococcal nuclease[b]	—	5.42	5.9	4.89 8.34	0.84 0.16	0.76 0.24	1.1
Staphylococcal nuclease, pH 7.5, 20 °C	0.144*M* oxygen	1.70	26.5	0.29 2.08	0.36 0.64	0.07 0.93	1.6

[a]Ref. 105.
[b]Ref. 106.

if diffusion is slower or if the solution is somewhat viscous. These transient terms can be accounted for by using the multiexponential model (Table 9.1). Transient effects in quenching can result in the appearance of new decay times and can bias the analysis of a multiexponential decay.

What is the value of analyzing quenching data in terms of the various transient models? Perhaps the most important result is that one obtains information on molecular aspects of the quenching process. This is illustrated in

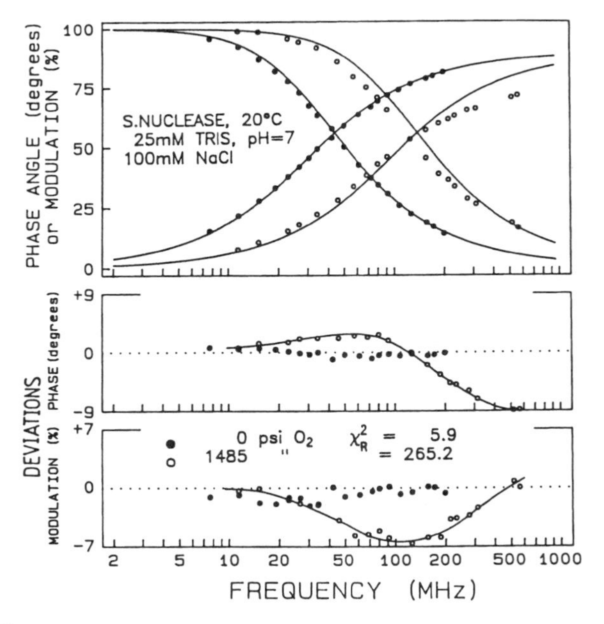

Figure 9.30. Frequency response of staphylococcal nuclease in the absence (●) and presence (○) of 0.144*M* oxygen. The solid curves show the best single-exponential fits to the data. The lower panels show the deviations from the single-exponential fits. From Ref. 106.

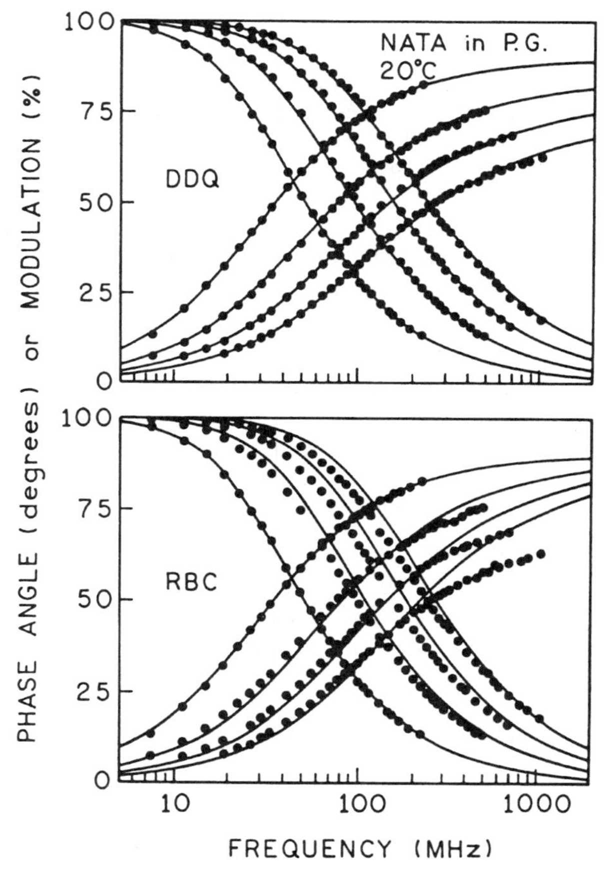

Figure 9.31. FD intensity decays of NATA in propylene glycol at 20 °C in the presence of increasing (from left to right) concentrations of acrylamide (0, 0.5, 1.0, and 1.5*M*). The solid curves in the top panel show the best fit to the DDQ model with a = 5 Å and r_e = 0.319 Å. The solid curves in the bottom panel show the best fit to the RBC model with a = 5 Å. Reprinted, with permission, from Ref. 99, Copyright © 1994, American Society for Photobiology.

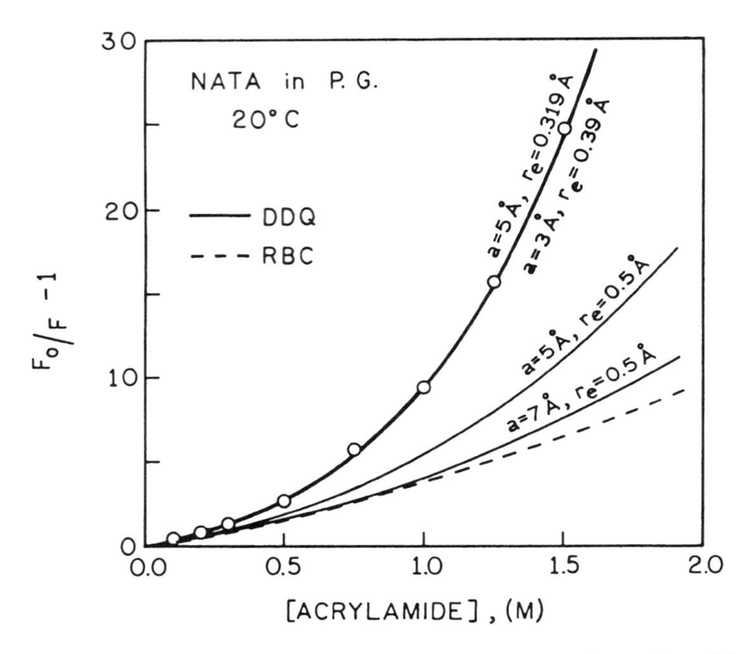

Figure 9.32. Stern–Volmer plots for NATA in propylene glycol at 20 °C quenched by acrylamide (○). The solid curves represent the calculated values of $(F_0/F - 1)$ using the parameter values from the DDQ model for $a = 7$ Å and $r_e = 0.5$ Å, $a = 5$ Å and $r_e = 0.5$ Å, $a = 5$ Å and $r_e = 0.319$ Å, and $a = 3$ Å and $r_e = 0.39$ Å. The dashed curve represents the RBC model with $a = 5$ Å. Reprinted, with permission, from Ref. 99, Copyright © 1994, American Society for Photobiology.

Figure 9.31 for quenching of NATA by acrylamide. The FD data could not be explained by the RBC model but were consistent with the DDQ model. This indicates that the quenching rate displays an exponential dependence on distance and that quenching is not an all-or-none process that occurs only at the contact distance.

Another important result from the transient analysis is an explanation for the upward curvature seen in many Stern–Volmer plots. The Stern–Volmer plot for quenching of NATA by acrylamide shows substantial upward curvature in the moderately viscous solvent propylene glycol (Figure 9.32). Although the curvature could be explained by a weak static quenching constant or by a sphere of action, the upward curvature is a natural consequence of distance-dependent quenching. This explanation of upward deviations in the Stern–Volmer plots is somewhat more satisfying because there is a single molecular interaction, the exponential distance dependence of quenching, which explains the time-dependent and steady-state data.

The transient model can also explain upward-curving Stern–Volmer plots obtained for proteins. This is illustrated by the quenching of staphylococcal nuclease by acrylamide[107] (Figure 9.33). In this case the calculated curve for the intensity data was obtained using the Smoluchowski model (Eq. [9.22]). This equation can be integrated to obtain the Stern–Volmer equation with transient effects[87]:

$$\frac{F_0}{F} = \frac{1 + K_D[Q]}{Y} \qquad [9.31]$$

where $K_D = k_q\tau_0$ and

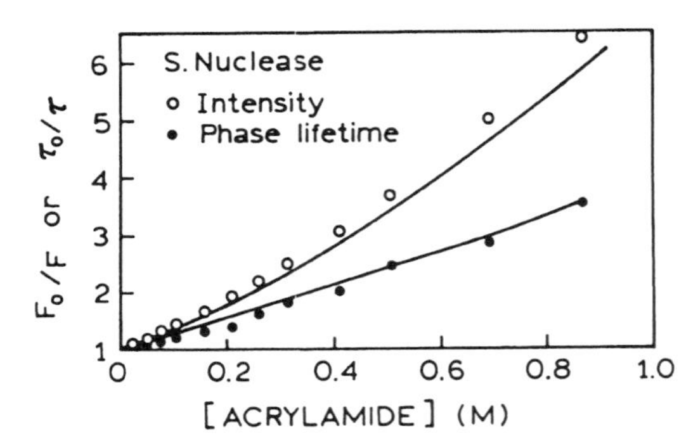

Figure 9.33. Intensity (○) and phase lifetime (●) Stern–Volmer plots for the acrylamide quenching of staphylococcal nuclease A at 20 °C, pH 7. Lifetimes were measured at 50 MHz. Solid curves are a simultaneous fit with R (fixed) = 7 Å, τ_0 = 4.60 ns, and $D = 0.058 \times 10^{-5}$ cm^2/s. Revised from Ref. 107.

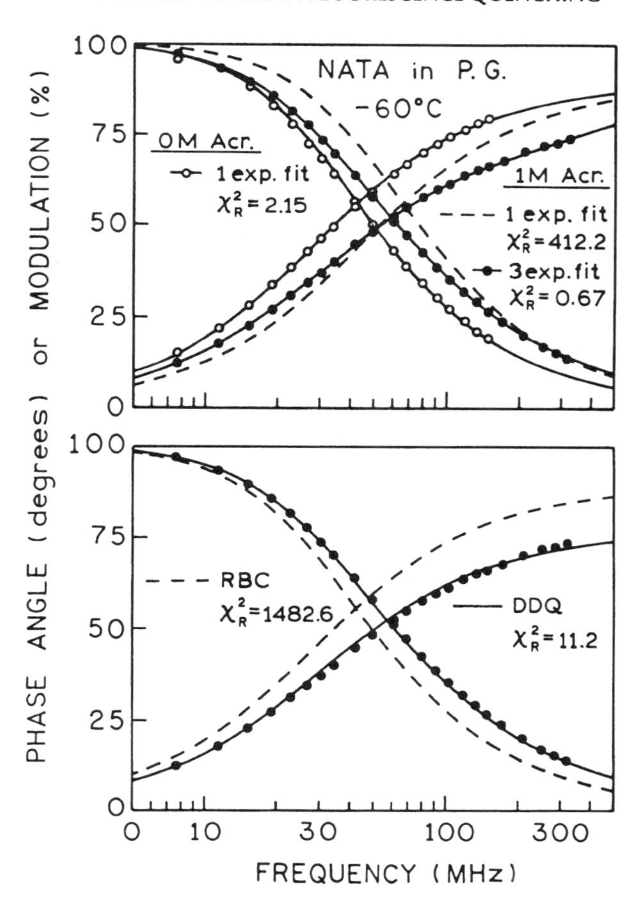

Figure 9.34. Frequency response of the NATA intensity decay in propylene glycol at –60 °C. *Top*: The open circles (○) and solid curve represent the data and best single-exponential fit in the absence of acrylamide. The closed circles (●) and solid curve represent the data and best multiexponential fit with 1.048*M* acrylamide; the dashed curve shows the best single-exponential fit to the data. *Bottom*: The data (●) are for NATA with 1.048*M* acrylamide at –60 °C. The solid curve represents the best DDQ fit. The dashed curve shows the RBC fit with D = 0. Reprinted, with permission, from Ref. 99, Copyright © 1994, American Society for Photobiology.

$$Y = 1 - \frac{b\sqrt{\pi}}{\sqrt{a}} \exp(b^2/a) \operatorname{erfc}(b/\sqrt{a}) \qquad [9.32]$$

with

$$a = \frac{1}{\tau_0} + k_q[\mathrm{Q}] \qquad [9.33]$$

and b as defined in Eq. [9.24].

The validity of distance-dependent quenching is supported by an observation which cannot be explained by other models. In frozen solution, where diffusion does not occur, one still observes quenching of NATA by acrylamide (Figure 9.34). This effect cannot be explained by a static quenching constant or the RBC model. The data show a decrease in lifetime even in the absence of diffusion. This result is not consistent with a sphere of action because the fluorophore–quencher pairs within the sphere are quenched and do not contribute to the intensity decay. Such data (Figure 9.34) can only be explained by a through-space interaction.

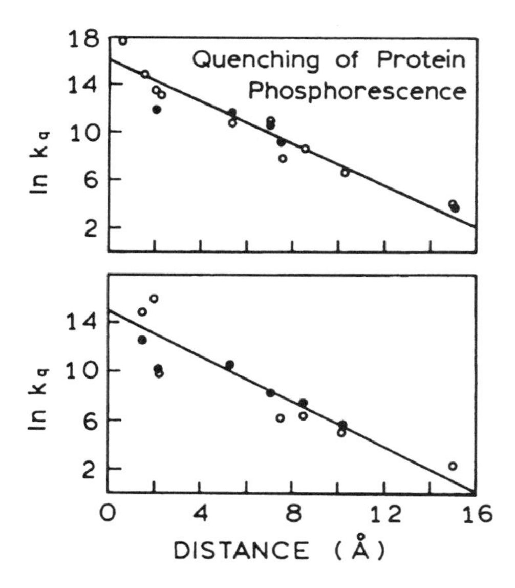

Figure 9.35. Relationship between the bimolecular quenching constants of single-tryptophan proteins and distance of the tryptophan from the protein surface. *Top*: Quenching constants for nitrite (○) and azide (●). *Bottom*: Quenching constants for ethanethiol (○) and nicotinamide (●). From Ref. 108.

9.4.B. Distance-Dependent Quenching in Proteins

The concept of distance-dependent quenching has already been used in biochemistry. Vanderkooi *et al.*[108] examined the quenching of protein phosphorescence by polar molecules (nitrite and azide) which were not expected to penetrate the proteins. A number of single-tryptophan proteins were examined, for which the crystal structures were known. There was a strong correlation between the quenching constant and the distance from the tryptophan to the surface of the protein (Figure 9.35). The quenching constants were not sensitive to solution viscosity, suggesting that the rate of quenching was sensitive to the distance of closest approach. The exponential dependence of quenching constants on distance from the surface indicated that electron transfer or electron exchange governed the quenching process.

REFERENCES

1. Shinitzky, M., Dianoux, A. C., Gitler, C., and Weber, G., 1971, Microviscosity and order in the hydrocarbon region of micelles and membranes determined with fluorescence probes I. Synthetic micelles, *Biochemistry* **10:**2106–2113.
2. Cogen, U., Shinitzky, M., Weber, G., and Nishida, T., 1973, Microviscosity and order in the hydrocarbon region of phospholipid and phospholipid–cholesterol dispersions determined with fluorescent probes, *Biochemistry* **12:**521–528.
3. Shinitzky, M., and Barenholz, Y., 1974, Dynamics of the hydrocarbon layer in liposomes of lecithin and sphingomyelin containing dicetylphosphate, *J. Biol. Chem.* **249:**2652–2657.
4. Thulborn, K. R., and Sawyer, W. H., 1978, Properties and the locations of a set of fluorescent probes sensitive to the fluidity gradient of the lipid bilayer, *Biochim. Biophys. Acta* **511:**125–140.
5. Thulborn, K. R., Tilley, L. M., Sawyer, W. H., and Treloar, F. E., 1979, The use of *n*-(9-anthroyloxy) fatty acids to determine fluidity and polarity gradients in phospholipid bilayers, *Biochim. Biophy. Acta* **558:**166–178.
6. Thulborn, K. R., Treloar, F. E., and Sawyer, W. H., 1978, A microviscosity barrier in the lipid bilayer due to the presence of phospholipids containing unsaturated acyl chains, *Biochem. Biophys. Res. Commun.* **81:**42–49.
7. Lala, A. K., and Koppaka, V., 1992, Fluorenyl fatty acids as fluorescent probes for depth-dependent analysis of artificial and natural membranes, *Biochemistry* **31:**5586–5593.
8. de Bony, J., and Tocanne, J. F., 1984, Photo-induced dimerization of anthracene phospholipids for the study of the lateral distribution of lipids in membranes, *Eur. J. Biochem.* **143:**373–379.
9. de Bony, J., and Tocanne, J. F., 1983, Synthesis and physical properties of phosphatidylcholine labelled with 9-(2-anthryl)nonanoic acid, a new fluorescent probe, *Chem. Phys. Lipids* **32:**105–121.
10. Liu, L., Cheng, K. H., and Somerharju, P., 1993, Frequency-resolved intramolecular excimer fluorescence study of lipid bilayer and non-bilayer phases, *Biophys. J.* **64:**1869–1877.
11. Ollmann, M., Schwarzmann, G., Sandhoff, K., and Galla, H. J., 1987, Pyrene-labeled gangliosides: Micelle formation in aqueous solution, lateral diffusion, and thermotropic behavior in phosphatidylcholine bilayers, *Biochemistry* **26:**5943–5952.
12. Chattopadhyay, A., 1990, Chemistry and biology of *N*-(7-nitrobenz-2-oxa-1,3-diazol-4-yl)-labeled lipids: Fluorescent probes of biological and model membranes, *Chem. Phys. Lipids* **53:**1–15.
13. Chattopadhyay, A., and London, E., 1988, Spectroscopic and ionization properties of *N*-(7-nitrobenz-2-oxa-1,3-diazol-4-yl)-labeled lipids in model membranes, *Biochim. Biophys. Acta* **938:**24–34.
14. Mazeres, S., Schram, V., Tocanne, J.-F., and Lopez, A., 1996, 7-Nitrobenz-2-oxa-1,3-diazole-4-yl-labeled phospholipids in lipid membranes: Differences in fluorescence behavior, *Biophys. J.* **71:**327–335.
15. Kalb, E., Paltauf, F., and Hermetter, A., 1989, Fluorescence lifetime distributions of diphenylhexatriene-labeled phosphatidylcholine as a tool for the study of phospholipid–cholesterol interactions, *Biophys. J.* **56:**1245–1253.
16. Johnson, I. D., Kang, H. C., and Haugland, R. P., 1991, Fluorescent membrane probes incorporating dipyrrometheneboron difluoride fluorophores, *Anal. Biochem.* **198:**228–237.
17. Lakowicz, J. R., and Hogen, D., 1980, Chlorinated hydrocarbon–cell membrane interactions studied by the fluorescence quenching of carbazole-labeled phospholipids: Probe synthesis and characterization of the quenching methodology, *Chem. Phys. Lipids* **26:**1–40.
18. Chalpin, D. B., and Kleinfeld, A. M., 1983, Interaction of fluorescence quenchers with the *n*-(9-anthroyloxy) fatty acid membrane probes, *Biochim. Biophys. Acta* **731:**465–474.
19. Blatt, E., and Sawyer, W. H., 1985, Depth-dependent fluorescent quenching in micelles and membranes, *Biochim. Biophys. Acta* **822:**43–62.
20. Moro, F., Goni, F. M., and Urbaneja, M. A., 1993, Fluorescence quenching at interfaces and the permeation of acrylamide and iodide across phospholipid bilayers, *FEBS Lett.* **330:**129–132.
21. Hariharan, C., Vijaysree, V., and Mishra, A. K., 1997, Quenching of 2,5-diphenyloxazole (PPO) fluorescence by metal ions, *J. Lumin.* **75:**205–211.
22. Salthammer, T., Dreeskamp, H., Birch, D. J. S., and Imhof, R. E., 1990, Fluorescence quenching of perylene by Co^{2+} ions via energy transfer in viscous and non-viscous media, *J. Photochem. Photobiol., A: Chem.* **55:**53–62.
23. Luisetti, J., Mohwald, H., and Galla, H. J., 1977, Paramagnetic fluorescence quenching in chlorophyll A containing vesicles: Evidence for the localization of chlorophyll, *Biochem. Biophys. Res. Commun.* **78:**754–760.
24. Sassaroli, M., Ruonala, M., Virtanen, J., Vauhkonen, M., and Somerharju, P., 1995, Transversal distribution of acyl-linked pyrene moieties in liquid-crystalline phosphatidylcholine bilayers. A fluorescence quenching study, *Biochemistry* **34:**8843–8851.
25. Ladokhin, A. S., 1997, Distribution analysis of depth-dependent fluorescence quenching in membranes: A practical guide, *Methods Enzymol.* **278:**462–473.
26. Lakokhin, A., Wang, L., Steggles, A. W., and Holloway, P. W., 1991, Fluorescence study of a mutant cytochrome b_5 with a single tryptophan in the membrane-binding domain, *Biochemistry* **30:**10200–10206.
27. Everett, J., Zlotnick, A., Tennyson, J., and Holloway, P. W., 1986, Fluorescence quenching of cytochrome b_5 in vesicles with an asymmetric transbilayer distribution of brominated phosphatidylcholine, *J. Biol. Chem.* **261:**6725–6729.
28. Gonzalez-Manas, J. M., Lakey, J. H., and Pattus, F., 1992, Brominated phospholipids as a tool for monitoring the membrane insertion of colicin A, *Biochemistry* **31:**7294–7300.
29. Silvius, J. R., 1990, Calcium-induced lipid phase separations and interactions of phosphatidylcholine/anionic phospholipid vesicles. Fluorescence studies using carbazole-labeled and brominated phospholipids, *Biochemistry* **29:**2930–2938.
30. Silvius, J. R., 1992, Cholesterol modulation of lipid intermixing in phospholipid and glycosphingolipid mixtures. Evaluation using fluorescent lipid probes and brominated lipid quenchers, *Biochemistry* **31:**3398–3403.
31. Chattopadhyay, A., and London, E., 1987, Parallax method for direct measurement of membrane penetration depth utilizing fluorescence quenching by spin-labeled phospholipids, *Biochemistry* **26:**39–45.
32. Abrams, F. S., and London, E., 1992, Calibration of the parallax fluorescence quenching method for determination of membrane penetration depth: Refinement and comparison of quenching by spin-labeled and brominated lipids, *Biochemistry* **31:**5312–5322.
33. Abrams, F. S., Chattopadhyay, A., and London, E., 1992, Determination of the location of fluorescent probes attached to fatty acids using parallax analysis of fluorescence quenching: Effect of carboxyl ionization state and environment on depth, *Biochemistry* **31:**5322–5327.
34. Abrams, F. S., and London, E., 1993, Extension of the parallax analysis of membrane penetration depth to the polar region of model membranes: Use of fluorescence quenching by a spin-label attached

to the phospholipid polar headgroup, *Biochemistry* **32:**10826–10831.

35. Asuncion-Punzalan, E., and London, E., 1995, Control of the depth of molecules within membranes by polar groups: Determination of the location of anthracene-labeled probes in model membranes by parallax analysis of nitroxide-labeled phospholipid induced fluorescence quenching, *Biochemistry* **34:**11460–11466.
36. Kachel, K., Asuncion-Punzalan, E., and London, E., 1995, Anchoring of tryptophan and tyrosine analogs at the hydrocarbon–polar boundary in model membrane vesicles: Parallax analysis of fluorescence quenching induced by nitroxide-labeled phospholipids, *Biochemistry* **34:**15475–15479.
37. Ren, J., Lew, S., Wang, Z., and London, E., 1997, Transmembrane orientation of hydrophobic α-helices is regulated both by the relationship of helix length to bilayer thickness and by the cholesterol concentration, *Biochemistry* **36:**10213–10220.
38. Martin, I., Ruysschaert, J.-M., Sanders, D., and Giffard, C. J., 1996, Interaction of the lantibiotic nisin with membranes revealed by fluorescence quenching of an introduced tryptophan, *Eur. J. Biochem.* **239:**156–164.
39. Matko, J., Ohki, K., and Edidin, M., 1992, Luminescence quenching by nitroxide spin labels in aqueous solution: Studies on the mechanism of quenching, *Biochemistry* **31:**703–711.
40. London, E., and Feigenson, G. W., 1981, Fluorescence quenching in model membranes. 1. Characterization of quenching caused by a spin-labeled phospholipid, *Biochemistry* **20:**1932–1938.
41. London, E., and Feigenson, G. W., 1981, Fluorescence quenching in model membranes. 2. Determination of the local lipid environment of the calcium adenosinetriphosphatase from sarcoplasmic reticulum, *Biochemistry* **20:**1939–1948.
42. East, J. M., and Lee, A. G., 1982, Lipid selectivity of the calcium and magnesium ion dependent adenosinetriphosphatase, studied with fluorescence quenching by a brominated phospholipid, *Biochemistry* **21:**4144–4151.
43. Caffrey, M., and Feigenson, G. W., 1981, Fluorescence quenching in model membranes. 3. Relationship between calcium adenosinetriphosphatase enzyme activity and the affinity of the protein for phosphatidylcholines with different acyl chain characteristics, *Biochemistry* **20:**1949–1961.
44. Markello, T., Zlotnick, A., Everett, J., Tennyson, J., and Holloway, P. W., 1985, Determination of the topography of cytochrome b_5 in lipid vesicles by fluorescence quenching, *Biochemistry* **24:**2895–2901.
45. Froud, R. J., East, J. M., Rooney, E. K., and Lee, A. G., 1986, Binding of long-chain alkyl derivatives to lipid bilayers and to (Ca^{2+}-Mg^{2+})-ATPase, *Biochemistry* **25:**7535–7544.
46. Yeager, M. D., and Feigenson, G. W., 1990, Fluorescence quenching in model membranes: Phospholipid acyl chain distributions around small fluorophores, *Biochemistry* **29:**4380–4392.
47. Lakowicz, J. R., Hogen, D., and Omann, G., 1977, Diffusion and partitioning of a pesticide, lindane, into phosphatidylcholine bilayers: A new fluorescence quenching method to study chlorinated hydrocarbon–membrane interactions, *Biochim. Biophys. Acta* **471:**401–411.
48. Omann, G. M., and Glaser, M., 1985, Dynamic quenchers in fluorescently labeled membranes, *Biophys. J.* **47:**623–627.
49. Fato, R., Battino, M., Esposti, M. D., Castelli, G. P., and Lenaz, G., 1986, Determination of partition of lateral diffusion coefficients of ubiquinones by fluorescence quenching of *n*-(9-anthroyloxy)stearic acids in phospholipid vesicles and mitochondrial membranes, *Biochemistry* **25:**3378–3390.
50. Vermeir, M., and Boens, N., 1992, Partitioning of (±)-5,6-dihydro-6-phenyl-2-*n*-alkylimidazo-[2,1-*b*]thiazoles into large unilamellar liposomes: A steady-state fluorescence quenching study, *Biochim. Biophys. Acta* **1104:**63–72.
51. Lakos, Z., Szarka, A., and Somogyi, B., 1995, Fluorescence quenching in membrane phase, *Biochem. Biophys. Res. Commun.* **208:**111–117.
52. Prieto, M. J. E., Castanho, M., Coutinho, A., Ortiz, A., Aranda, F. J., and Gomez-Fernandez, J. C., 1994, Fluorescence study of a derivatized diacylglycerol incorporated in model membranes, *Chem. Phys. Lipids* **69:**75–85.
53. Lakowicz, J. R., 1980, Fluorescence spectroscopic investigations of the dynamic properties of proteins, membranes, and nucleic acids, *J. Biochem. Biophys. Methods* **2:**90–119.
54. Subczynski, W. K., Hyde, J. S., and Kusumi, A., 1989, Oxygen permeability of phosphatidylcholine-cholesterol membranes, *Proc. Natl. Acad. Sci. U.S.A.* **86:**4474–4478.
55. Fischkoff, S., and Vanderkooi, J. M., 1975, Oxygen diffusion in biological and artificial membranes determined by the fluorochrome pyrene, *J. Gen. Phys.* **65:**663–676.
56. Dumas, D., Muller, S., Gouin, F., Baros, F., Viriot, M.-L., and Stoltz, J.-F., 1997, Membrane fluidity and oxygen diffusion in cholesterol-enriched erythrocyte membrane, *Arch. Biochem. Biophys.* **341:**34–39.
57. Subczynski, W. K., Hyde, J. S., and Kusumi, A., 1991, Effect of alkyl chain unsaturation and cholesterol intercalation on oxygen transport in membranes: A pulse ESR spin labeling study, *Biochemistry* **30:**8578–8590.
58. Vanderkooi, J. M., and Callis, J. B., 1974, Pyrene. A probe of lateral diffusion in the hydrophobic region of membranes, *Biochemistry* **13:**4000–4006.
59. Soutar, A. K., Pownall, H. J., He, A. S., and Smith, L. C., 1974, Phase transitions in bilamellar vesicles. Measurements by pyrene excimer fluorescence and effect on transacylation by lecithin: Cholesterol acyltransferase, *Biochemistry* **13:**2828–2836.
60. Pownall, H. J., and Smith, L. C., 1973, Viscosity of the hydrocarbon region of micelles. Measurement by excimer fluorescence, *J. Am. Chem. Soc.* **95:**3136–3140.
61. Galla, H.-J., Hartmann, W., Theilen, U., and Sackmann, E., 1979, On two-dimensional passive random walk in lipid bilayers and fluid pathways in biomembranes, *J. Membr. Biol.* **48:**215–236.
62. Galla, H.-J., and Hartmann, W., 1980, Excimer-forming lipids in membrane research, *Chem. Phys. Lipids* **27:**199–219.
63. Galla, H.-J., and Sackmann, E., 1974, Lateral diffusion in the hydrophobic region of membranes: Use of pyrene excimers as optical probes, *Biochim. Biophys. Acta* **339:**103–115.
64. Geladé, E., Boens, N., and De Schryver, F. C., 1982, Exciplex formation in dodecylammonium propionate reversed micellar systems, *J. Am. Chem. Soc.* **104:**6288–6292.
65. Anderson, V. C., Craig, B. B., and Weiss, R. G., 1982, Liquid-crystalline solvents as mechanistic probes. 8. Dynamic quenching of pyrene fluorescence by pyrene in the liquid-crystalline and isotropic phases of a cholesteric solvent, *J. Am. Chem. Soc.* **104:**2972–2977.
66. Hresko, R. C., Sugar, I. P., Barenholz, Y., and Thompson, T. E., 1986, Lateral distribution of a pyrene-labeled phosphatidylcholine in phosphatidylcholine bilayers: Fluorescence phase and modulation study, *Biochemistry* **25:**3813–3823.
67. Nutakul, W., Thummel, R. P., and Taggart, A. D., 1979, Intramolecular excimer-forming probes of aqueous micelles, *J. Am. Chem. Soc.* **101:**771–772.
68. Goldenberg, M., Emert, J., and Morawetz, H., 1978, Intramolecular excimer study of rates of conformational transitions. Dependence on

molecular structure and the viscosity of the medium, *J. Am. Chem. Soc.* **100**:7171–7177.

69. Mataga, N., Okada, T., Masuhara, H., Nakashima, N., Sakata, Y., and Misumi, S., 1976, Electronic structure and dynamical behavior of some intramolecular exciplexes, *J. Lumin.* **12/13**:159–168.
70. Turro, N. J., Okubo, T., and Weed, G. C., 1982, Enhancement of intramolecular excimer formation of 1,3-bichromophoric propanes via application of high pressure and via complexation with cyclodextrins. Protection from oxygen quenching, *Photochem. Photobiol.* **35**:325–329.
71. Vauhkonen, M., Sassaroli, M., Somerharju, P., and Eisinger, J., 1990, Dipyrenylphosphatidylcholines as membrane fluidity probes. Relationship between intramolecular and intermolecular excimer formation rates, *Biophys. J.* **57**:291–300.
72. Melnick, R. L., Haspel, H. C., Goldenberg, M., Greenbaum, L. M., and Weinstein, S., 1981, Use of fluorescent probes that form intramolecular excimers to monitor structural changes in model and biological membranes, *Biophys. J.* **34**:499–515.
73. Naqvi, K. R., 1974, Diffusion-controlled reactions in two-dimensional fluids: Discussion of measurements of lateral diffusion of lipids in biological membranes, *Chem. Phys. Lett.* **28**:280–284.
74. Owen, C. S., 1975, Two dimensional diffusion theory: Cylindrical diffusion model applied to fluorescence quenching, *J. Chem. Phys.* **62**:3204–3207.
75. Medhage, B., and Almgren, M., 1992, Diffusion-influenced fluorescence quenching: Dynamics in one to three dimensions, *J. Fluoresc.* **2**:7–21.
76. Caruso, F., Grieser, F., and Thistlethwaite, P. J., 1993, Lateral diffusion of amphiphiles in fatty acid monolayers at the air–water interface: A steady-state and time-resolved fluorescence quenching study, *Langmuir* **9**:3142–3148.
77. Blackwell, M. F., Gounaris, K., Zara, S. J., and Barber, J., 1987, A method for estimating lateral diffusion coefficients in membranes from steady-state fluorescence quenching studies, *Biophys. J.* **51**:735–744.
78. Caruso, F., Grieser, F., Thistlethwaite, P. J., and Almgren, M., 1993, Two-dimensional diffusion of amphiphiles in phospholipid monolayers at the air–water interface, *Biophys. J.* **65**:2493–2503.
79. Eftink, M. R., and Ghiron, C. A., 1984, Indole fluorescence quenching studies on proteins and model systems: Use of the inefficient quencher succinimide, *Biochemistry* **23**:3891–3899.
80. Eftink, M. R., Selva, T. J., and Wasylewski, Z., 1987, Studies of the efficiency and mechanism of fluorescence quenching reactions using acrylamide and succinimide as quenchers, *Photochem. Photobiol.* **46**:23–30.
81. Samson, R., and Deutch, J. M., 1978, Diffusion-controlled reaction rate to a buried active site, *J. Chem. Phys.* **68**:285–290.
82. Shoup, D., Lipari, G., and Szabo, A., 1981, Diffusion-controlled bimolecular reaction rates, *Biophys. J.* **36**:697–714.
83. Solc, K., and Stockmayer, W. H., 1973, Kinetics of diffusion-controlled reaction between chemically asymmetric molecules. II. Approximate steady state solution, *Int. J. Chem. Kinet.* **5**:733–752.
84. Schmitz, K. S., and Schurr, J. M., 1972, The role of orientation constraints and rotational diffusion in bimolecular solution kinetics, *J. Phys. Chem.* **76**:534-545.
85. Johnson, D. A., and Yguerabide, J., 1985, Solute accessibility to *N*-fluorescein isothiocyanate-lysine-23 cobra α-toxin bound to the acetylcholine receptor, *Biophys. J.* **48**:949–955.
86. Somogyi, B., and Lakos, Z., 1993, Protein dynamics and fluorescence quenching, *J. Photochem. Photobiol., B: Biol.* **18**:3–16.
87. Nemzek, T. L., and Ware, W. R., 1975, Kinetics of diffusion-controlled reactions: Transient effects in fluorescence quenching, *J. Chem. Phys.* **62**:477–489.
88. Bamford, C. H., Tipper. C. F. H., and Compton, R. G., 1985, *Chemical Kinetics* Elsevier, New York.
89. Smoluchowski, V. M., 1916, Drei vortage über diffusion, brownsche molekularbewegung und koagulation von kolloidteilchen [Three lectures on diffusion, Brownian molecular motion and coagulation of colloids], *Physik. Z.* **17**:557–571, 585–599. An English translation is available in: Chandrasekhar, S., Kac, M., and Smoluchowski, R., 1986, *Marian Smoluchowski, His Life and Scientific Work*, Polish Scientific Publishers, Warsaw.
90. Collins, F. C., and Kimball, G. E., 1949, Diffusion-controlled reaction rates, *J. Colloid Sci.* **4**:425–437. See also Collins, F. C., 1950, *J. Colloid Sci.* **5**:499–505 for correction.
91. Yguerabide, J., Dillon, M. A., and Burton, M., 1964, Kinetics of diffusion-controlled processes in liquids. Theoretical consideration of luminescent systems: Quenching and excitation transfer in collision, *J. Chem. Phys.* **40**:3040–3052.
92. Marcus, R. A., 1993, Electron transfer reactions in chemistry: Theory and experiment (Nobel lecture), *Angew. Chem. Int. Ed. Engl.* **32**:1111–1121.
93. Turro, N. J., 1978, *Modern Molecular Photochemistry*, Benjamin Cummings Publishing Co., Menlo Park, California.
94. Ware, W. R., and Novros, J. S., 1966, Kinetics of diffusion-controlled reactions. An experimental test of theory as applied to fluorescence quenching, *J. Phys. Chem.* **70**:3246–3253.
95. Ware, W. R., and Andre, J. C., 1980, The influence of diffusion on fluorescence quenching, in *Time-Resolved Fluorescence Spectroscopy in Biochemistry and Biology*, R. B. Cundall and R. E. Dale (eds.), Plenum Press, New York, pp. 363–392.
96. Lakowicz, J. R., Johnson, M. L., Gryczynski, I., Joshi, N., and Laczko, G., 1987, Transient effects in fluorescence quenching measured by 2-GHz frequency-domain fluorometry, *J. Phys. Chem.* **91**:3277–3285.
97. Periasamy, N., Doraiswamy, S., Venkataraman, B., and Fleming, G. R., 1988, Diffusion controlled reactions: Experimental verification of the time-dependent rate equation, *J. Chem. Phys.* **89**:4799–4806.
98. Lakowicz, J. R., Kuśba, J., Szmacinski, H., Johnson, M. L., and Gryczynski, I., 1993, Distance-dependent fluorescence quenching observed by frequency-domain fluorometry, *Chem. Phys. Lett.* **206**:455–463.
99. Lakowicz, J. R., Zelent, B., Gryczynski, I., Kuśba, J., and Johnson, M. L., 1994, Distance-dependent fluorescence quenching of tryptophan by acrylamide, *Photochem. Photobiol.* **60**:205–214.
100. Zelent, B., Kuśba, J., Gryczynski, I., Johnson, M. L., and Lakowicz, J. R., 1996, Distance-dependent fluorescence quenching of *p*-bis[2-(5-phenyloxazolyl)]benzene by various quenchers, *J. Phys. Chem.* **100**:18592–18602.
101. Lakowicz, J. R., Zelent, B., Kuśba, J., and Gryczynski, I., 1996, Distance-dependent quenching of nile blue fluorescence by *N,N*-diethylaniline observed by frequency-domain fluorometry, *J. Fluoresc.* **6**:187–194.
102. Kuśba, J., and Lakowicz, J. R., 1994, Diffusion-modulated energy transfer and quenching: Analysis by numerical integration of diffusion equation in Laplace space, *Methods Enzymol.* **240**:216–262.
103. Kuśba, J., 1998, Personal communication.
104. Winjaendts Van Resandt, R. W., 1983, Picosecond transient effect in the fluorescence quenching of tryptophan, *Chem. Phys. Lett.* **95**:205–208.

105. Gryczynski, I., Johnson, M. L., and Lakowicz, J. R., 1988, Acrylamide quenching of Y_t-base fluorescence in aqueous solution, *Biophys. Chem.* **31**:269–274.
106. Lakowicz, J. R., Joshi, N. B., Johnson, M. L., Szmacinski, H., and Gryczynski, I., 1987, Diffusion coefficients of quenchers in proteins from transient effects in the intensity decays, *J. Biol. Chem.* **262**:10907–10910.
107. Eftink, M. R., 1990, Transient effects in the solute quenching of tryptophan residues in proteins, *Proc. SPIE* **1204**:406–414.
108. Vanderkooi, J. M., Englander, S. W., Papp, S., Wright, W. W., and Owen, C. S., 1990, Long-range electron exchange measured in proteins by quenching of tryptophan phosphorescence, *Proc. Natl. Acad. Sci. U.S.A.* **87**:5099–5103.
109. Nelson, G., and Warner, I. M., 1990, Fluorescence quenching studies of cyclodextrin complexes of pyrene and naphthalene in the presence of alcohols, *J. Phys. Chem.* **94**:576–581.

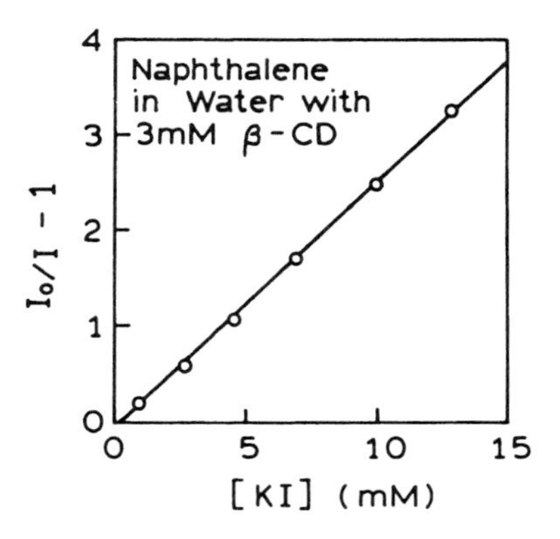

Figure 9.37. Quenching of naphthalene by iodide in the presence of 3m*M* β-cyclodextrin (β-CD). Revised and reprinted, with permission, from Ref. 109, Copyright © 1990, American Chemical Society.

PROBLEMS

9.1. *Interpretation of Apparent Bimolecular Quenching Constants*: Calculate the apparent bimolecular quenching constant of 2-(9-anthroyloxy) palmitate (2-AP) for quenching by copper or dimethylaniline (Figure 9.5). Assume that the unquenched lifetime is 10 ns. Interpret these values with respect to the maximum value of k_q possible in aqueous solution.

9.2. *Oxygen Bimolecular Quenching Constant in a Membrane*: Use the data in Figure 9.36 to calculate k_q for oxygen quenching of pyrene in dimyristoylphosphatidylcholine (DMPC) vesicles.

9.3. *Quenching in the Presence of Cyclodextrins*: Figure 9.37 shows iodide quenching of naphthalene in the presence of 3m*M* β-cyclodextrin (β-CD). Figure 9.38 shows iodide quenching of naphthalene in the presence of 1% (v/v) benzyl alcohol and increasing amounts of β-CD. Explain the results. Assume that the unquenched lifetime of naphthalene is 45 ns.

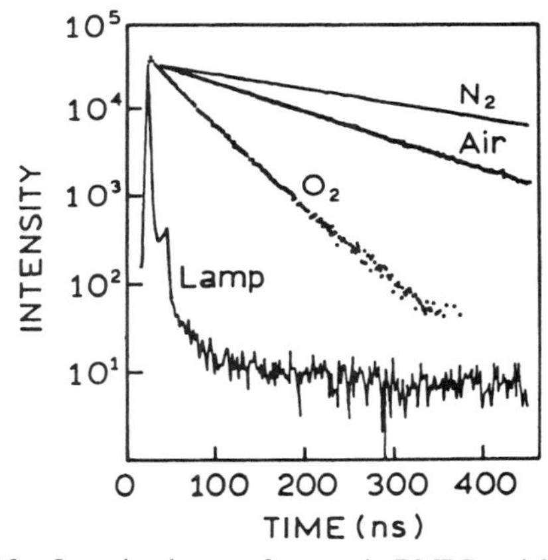

Figure 9.36. Intensity decays of pyrene in DMPC vesicles at 25 °C, equilibrated with various partial pressures of oxygen. Revised and reprinted, with permission, from Ref. 55, Copyright © 1975, Rockefeller University Press.

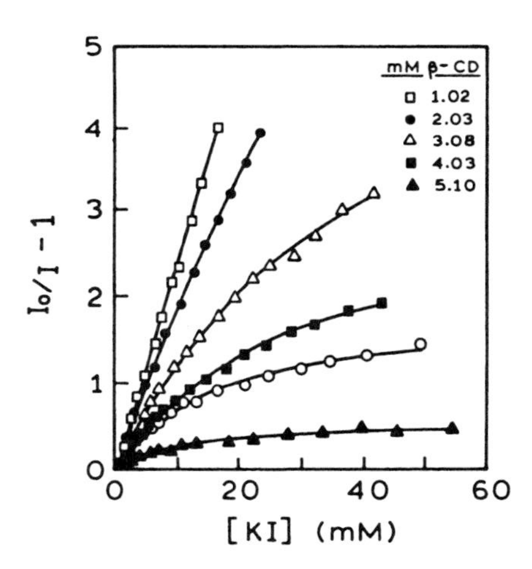

Figure 9.38. Quenching of naphthalene by iodide in the presence of benzyl alcohol (1% v/v) at various concentrations of β-cyclodextrin (β-CD). Revised and reprinted, with permission, from Ref. 109, Copyright © 1990, American Chemical Society.

Fluorescence Anisotropy

10

Upon excitation with polarized light, the emission from many samples is also polarized. The extent of polarization of the emission is described in terms of the anisotropy (r). Samples exhibiting nonzero anisotropies are said to display polarized emission. The origin of these phenomena is based on the existence of transition moments for absorption and emission which lie along specific directions within the fluorophore structure. In homogeneous solution the ground-state fluorophores are all randomly oriented. When exposed to polarized light, those fluorophores which have their absorption transition moments oriented along the electric vector of the incident light are preferentially excited. Hence, the excited-state population is not randomly oriented. Instead, there is a somewhat larger number of excited molecules having their transition moments oriented along the electric vector of the polarized exciting light.

Depolarization of the emission can be caused by a number of phenomena, the relative importance of which depends upon the sample under investigation. Rotational diffusion of fluorophores is one common cause of depolarization. The anisotropy measurements reveal the average angular displacement of the fluorophore that occurs between absorption and subsequent emission of a photon. This angular displacement is dependent upon the rate and extent of rotational diffusion during the lifetime of the excited state. These diffusive motions, depend, in turn, upon the viscosity of the solvent and the size and shape of the rotating molecule. For fluorophores in solution, the rotational rate of the fluorophore is dependent upon the viscous drag imposed by the solvent. As a result, a change in solvent viscosity will result in a change in fluorescence anisotropy. For small fluorophores in solutions of low viscosity, the rate of rotational diffusion is typically faster than the rate of emission. Under these conditions, the emission is depolarized and the anisotropy is close to zero.

The dependence of fluorescence anisotropy upon rotational motion has resulted in numerous applications of fluorescence anisotropy measurements in biochemical research. This is because the timescale of rotational diffusion of biomolecules is comparable to the decay time of many fluorophores. For instance, a protein with a molecular weight of 25 kDa can be expected to have a rotational correlation time near 10 ns. This is comparable to the lifetime of many fluorophores when coupled to proteins. Hence, factors which alter the rotational correlation time will also alter the anisotropy. As examples, fluorescence anisotropy measurements have been used to quantify protein denaturation, protein association with other macromolecules, and the internal dynamics of proteins. In addition, the anisotropies of membrane-bound fluorophores have been used to estimate the internal viscosities of membranes and the effects of lipid composition upon the membrane phase-transition temperature.

In this chapter we describe the fundamental theory for steady-state measurements of fluorescence anisotropy and present selected biochemical applications. In the next chapter we will describe the theory and applications of time-resolved anisotropy measurements.

10.1. DEFINITION OF FLUORESCENCE ANISOTROPY

The measurement of fluorescence anisotropy is illustrated in Figure 10.1. The sample is excited with vertically polarized light. The electric vector of the excitation light is oriented parallel to the vertical or z-axis. One then measures the intensity of the emission through a polarizer. When the emission polarizer is oriented parallel (||) to the direction of the polarized excitation, the observed intensity is called $I_{\parallel}$. Likewise, when the polarizer is perpendicular ($\perp$) to the excitation, the intensity is called $I_{\perp}$. These values are used to calculate the anisotropy[1]:

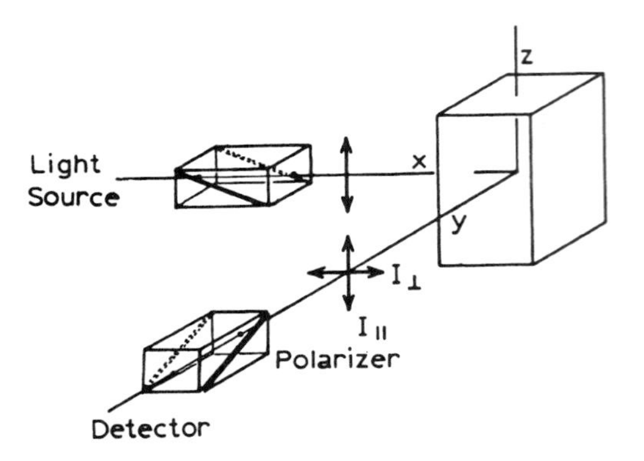

Figure 10.1. Schematic diagram for measurement of fluorescence anisotropies.

$$r = \frac{I_{\parallel} - I_{\perp}}{I_{\parallel} + 2I_{\perp}} \quad [10.1]$$

The anisotropy is a dimensionless quantity which is independent of the total intensity of the sample. This is because the difference ($I_{\parallel} - I_{\perp}$) is normalized by the total intensity, which is $I_T = I_{\parallel} + 2I_{\perp}$.

In earlier publications one frequently encounters the term polarization, which is given by

$$P = \frac{I_{\parallel} - I_{\perp}}{I_{\parallel} + I_{\perp}} \quad [10.2]$$

The polarization and anisotropy values can be interchanged using

$$P = \frac{3r}{2 + r} \quad [10.3]$$

$$r = \frac{2P}{3 - P} \quad [10.4]$$

Although there is nothing incorrect about the notion of polarization, its use should be discouraged. Anisotropy is preferred because most theoretical expressions are considerably simpler when expressed in terms of this parameter, an observation first made by Alexander Jabłoński.[1] As an example of this simplification, consider a mixture of fluorophores, each with polarization P_i and a fractional fluorescence intensity f_i. The polarization of this mixture ($\overline{P}$) is given by[2]

$$\left(\frac{1}{\overline{P}} - \frac{1}{3}\right)^{-1} = \sum_i \frac{f_i}{\left(\frac{1}{P_i} - \frac{1}{3}\right)} \quad [10.5]$$

In contrast, the average anisotropy ($\overline{r}$) is given by

$$\overline{r} = \sum_i f_i r_i \quad [10.6]$$

where the r_i indicate the anisotropies of the individual species. The latter expression is clearly preferable. Furthermore, following pulsed excitation, the decay of fluorescence anisotropy [$r(t)$] of a sphere is given by

$$r(t) = r_0 e^{-t/\theta} \quad [10.7]$$

where r_0 is the anisotropy at $t = 0$, and θ is the rotational correlation time of the sphere. The decay of polarization is not a single exponential, even for a spherical molecule.

Suppose that the light observed through the emission polarizer is completely polarized. Then $I_{\perp} = 0$, and $P = r = 1.0$. This value can be observed for scattered light from an optically dilute scatterer. Completely polarized emission is never observed for fluorescence from homogeneous unoriented samples. The measured values of P or r are smaller due to the angular dependence of photoselection (Section 10.2). Completely polarized emission can be observed for oriented samples.

Now suppose that the emission is completely depolarized. In this case, $I_{\parallel} = I_{\perp}$ and $P = r = 0$. However, it is important to note that P and r are not equal for intermediate values. For the moment, we have assumed that these intensities could be measured without artifacts due to the polarizing properties of the optical components, especially the emission monochromator (Section 2.3.B). In Section 10.4 we will describe methods to correct for such interference.

10.1.A. Origin of the Definitions of Polarization and Anisotropy

One may wonder why two widely used measures exist for the same phenomenon. Both P and r have a rational origin. Consider partially polarized light traveling along the x-axis (Figure 10.2), and assume that one measures the intensities I_z and I_y with the detector and polarizer positioned on the x-axis. The polarization of this light is defined as the fraction of the light that is linearly polarized. Specifically,

$$P = \frac{p}{p + n} \quad [10.8]$$

where p is the intensity of the polarized component, and n is the intensity of the natural component. The intensity of the natural component is given by $n = 2I_y$. The remaining intensity is the polarized component, which is given by $p = I_z - I_y$. For vertically polarized excitation, $I_z = I_{\parallel}$ and $I_y = I_{\perp}$.

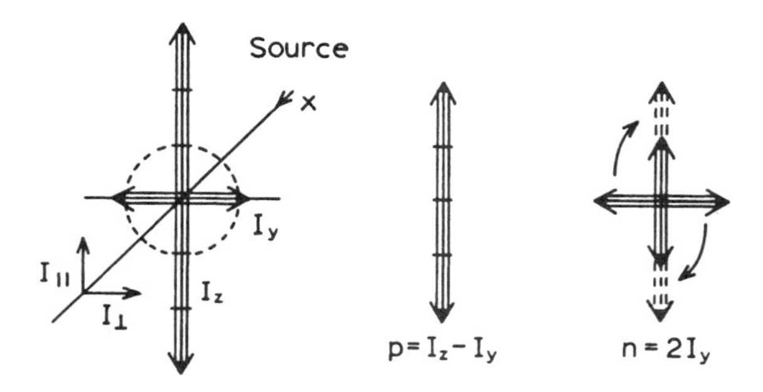

Figure 10.2. Polarization of a ray of light.

Substitution into Eq. [10.8] yields Eq. [10.2], which is the standard definition for polarization.

The anisotropy (r) of a light source is defined as the ratio of the polarized component to the total intensity (I_T),

$$r = \frac{I_z - I_y}{I_x + I_y + I_z} = \frac{I_z - I_y}{I_T} \quad [10.9]$$

When the excitation is polarized along the z-axis, dipolar radiation from the fluorophores is also symmetric around the z-axis. Hence, $I_x = I_y$. Recalling that $I_y = I_{\perp}$ and $I_z = I_{\parallel}$, one obtains Eq. [10.1].

The polarization is an appropriate parameter for describing a light source when a light ray is directed along a particular axis. In this case, $p + n$ is the total intensity, and P is the ratio of the excess intensity along the z-axis divided by the total intensity. In contrast, the radiation emitted by a fluorophore is symmetrically distributed about the z-axis. This distribution of radiated intensity is shown in Figure 10.3 for a dipole oriented along the z-axis. The intensity of the radiated light is proportional to $\cos^2 \zeta$, where ζ is the angle above or below the x–y plane. It is for this reason that, for excitation polarized along the z-axis, the total intensity is not given by $I_{\parallel} + I_{\perp}$, but rather by $I_T = I_{\parallel} + 2 I_{\perp}$ (Section 10.4.F). Hence, the anisotropy is the ratio of the excess intensity that is parallel to the z-axis to the total intensity. It is interesting to notice that a dipole oriented along the z-axis does not radiate along this axis and cannot be observed with a detector on the z-axis.

10.2. THEORY FOR ANISOTROPY

The theory for fluorescence anisotropy can be derived by consideration of a single molecule.[3] Assume for the moment that the absorption and emission transition moments are parallel. This is nearly true for the membrane probe DPH. Assume that this single molecule is oriented with angles θ relative to the z-axis and ϕ relative to the y-axis (Figure 10.4). Of course, the ground-state DPH molecules will be randomly oriented in an isotropic solvent. Our goal is to calculate the anisotropy that would be observed for this oriented molecule in the absence of rotational diffusion. The conditions of parallel dipoles, immobility, and random ground-state orientation simplify the derivation.

It is known that fluorescing fluorophores behave like radiating dipoles.[4] The intensity of light radiated from a

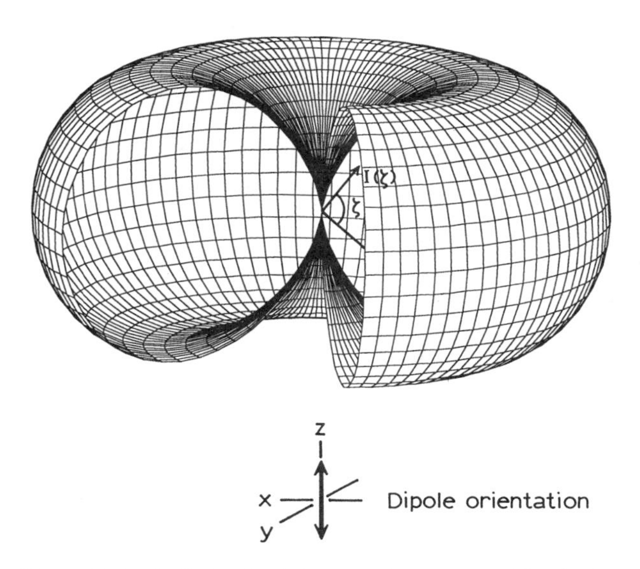

Figure 10.3. Radiating dipole in a coordinate system. The dipole is oriented along the z-axis, and the intensity $I(\zeta)$ of the emission in any direction is proportional to $\cos^2 \zeta$, where ζ is the angle from the x–y plane.

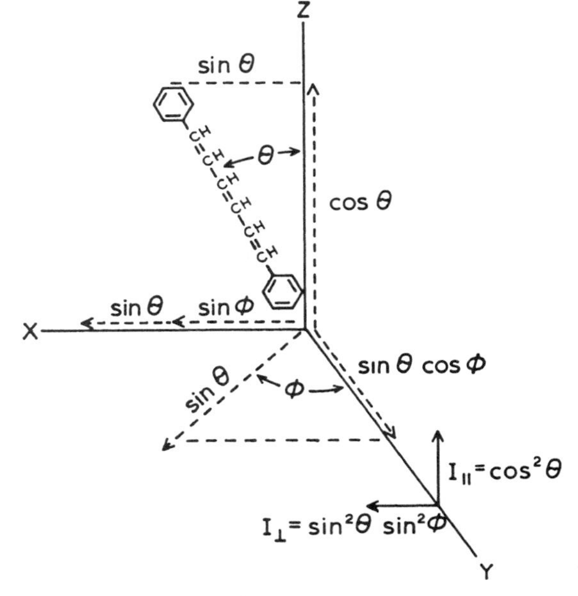

Figure 10.4. Emission intensities for a single fluorophore in a coordinate system.

dipole is proportional to the square of its vector projected onto the axis of observation. One can also reason that the emission is polarized along the transition moment. The intensity observed through a polarizer is proportional to the square of the projection of the electric field of the radiating dipole onto the transmission axis of the polarizer. These projections are given by

$$I_{\|}(\theta, \phi) = \cos^2 \theta \qquad [10.10]$$

$$I_{\perp}(\theta, \phi) = \sin^2 \theta \sin^2 \phi \qquad [10.11]$$

In an actual experiment the solution will contain many fluorophores with a random distribution. The anisotropy is calculated by performing the appropriate average based on excitation photoselection and how the selected molecules contribute to the measured intensity. First, consider excitation polarized along the z-axis. Such excitation must excite all molecules having an angle ϕ with respect to the y-axis with equal probability. That is, the population of excited fluorophores will be symmetrically distributed around the z-axis. Any experimentally accessible population of molecules will be oriented with values of ϕ from 0 to 2π with equal probability. Hence, we can eliminate the ϕ dependence in Eq. [10.11]. The average value of $\sin^2 \phi$ is given by

$$\langle \sin^2 \phi \rangle = \frac{\int_0^{2\pi} \sin^2 \phi \, d\phi}{\int_0^{2\pi} d\phi} = \frac{1}{2} \qquad [10.12]$$

and therefore

$$I_{\|}(\theta) = \cos^2 \theta \qquad [10.13]$$

$$I_{\perp}(\theta) = \frac{1}{2} \sin^2 \theta \qquad [10.14]$$

Now assume that we are observing a collection of fluorophores which are oriented relative to the z-axis with a probability $f(\theta)$. In the following section we will consider the form of $f(\theta)$ expected for excitation photoselection. The measured fluorescence intensities for this collection of molecules are

$$I_{\|} = \int_0^{\pi/2} f(\theta) \cos^2 \theta \, d\theta = k \langle \cos^2 \theta \rangle \qquad [10.15]$$

$$I_{\perp} = \frac{1}{2} \int_0^{\pi/2} f(\theta) \sin^2 \theta \, d\theta \, \frac{k}{2} \langle \sin^2 \theta \rangle \qquad [10.16]$$

where $f(\theta)\, d\theta$ is the probability that a fluorophore is oriented between θ and $\theta + d\theta$, and k is an instrumental constant. Using Eq. [10.11] and the identity $\sin^2 \theta = 1 - \cos^2 \theta$, one finds that

$$r = \frac{3\langle \cos^2 \theta \rangle - 1}{2} \qquad [10.17]$$

Hence, the anisotropy is determined by the average value of $\cos^2 \theta$, where θ is the angle of the emission dipole relative to the z-axis. This is because the observed intensities $I_{\|}$ and $I_{\perp}$ are proportional to the square of the projection of the individual transition moments onto the x- and the z-axis (Figure 10.4).

It is instructive to consider the relationship between r and θ. For a single fluorophore oriented along the z-axis, with collinear transitions, $\theta = 0$ and $r = 1.0$. However, it is not possible to obtain a perfectly oriented excited-state population with optical excitation of homogeneous solutions. Hence, the anisotropies are always less than 1.0. Complete loss of anisotropy is equivalent to $\theta = 54.7°$. This does not mean that each fluorophore is oriented at 54.7° or has rotated through 54.7°. Rather, it means that the average value of $\cos^2 \theta$ is $\frac{1}{3}$, where θ is the angular displacement between the excitation and emission moments. Recall that in the derivation of Eq. [10.17] we assumed that these dipoles were collinear. A slightly more complex expression is necessary for almost all fluorophores because the transition moments are rarely collinear. In addition, we have not yet considered the effects of photoselection on the anisotropy values.

10.2.A. Excitation Photoselection of Fluorophores

Observation of fluorescence requires excitation of the fluorophores. When a sample is illuminated with polarized light, those molecules with their absorption transition moments aligned parallel to the electric vector of the polarized excitation have the highest probability of absorption. The electric dipole of a fluorophore need not be precisely aligned with the z-axis to absorb light polarized along this axis. The probability of absorption is proportional to $\cos^2 \theta$, where θ is the angle the absorption dipole makes with the z-axis.[3] Hence, excitation with polarized light results in a population of excited fluorophores that is symmetrically distributed around the z-axis (Figure 10.5). This phenomenon is called photoselection. Note that the excited-state population is symmetrical around the z-axis. Most of the excited fluorophores are aligned close to the z-axis, and very few fluorophores have their transition moments oriented in the x–y plane. For the random ground-state distribution, which must exist in a disordered solution, the number of molecules at an angle between θ and θ + $d\theta$ is proportional to sin θ $d\theta$. This quantity is proportional to the surface area on a sphere within the angles θ and θ + $d\theta$. Hence, the distribution of molecules excited by vertically polarized light is given by

$$f(\theta)\, d\theta = \cos^2 \theta \sin \theta\, d\theta \qquad [10.18]$$

The probability distribution given by Eq. [10.18] determines the maximum photoselection that can be obtained using one-photon excitation of an isotropic solution. More highly oriented populations can be obtained using multiphoton excitation.[5] Recall that the anisotropy is a simple function of $\langle\cos^2\theta\rangle$ (Eq. [10.17]), so calculation of $\langle\cos^2\theta\rangle$ allows calculation of the anisotropy.

For collinear absorption and emission dipoles, the maximum value of $\langle\cos^2\theta\rangle$ is given by

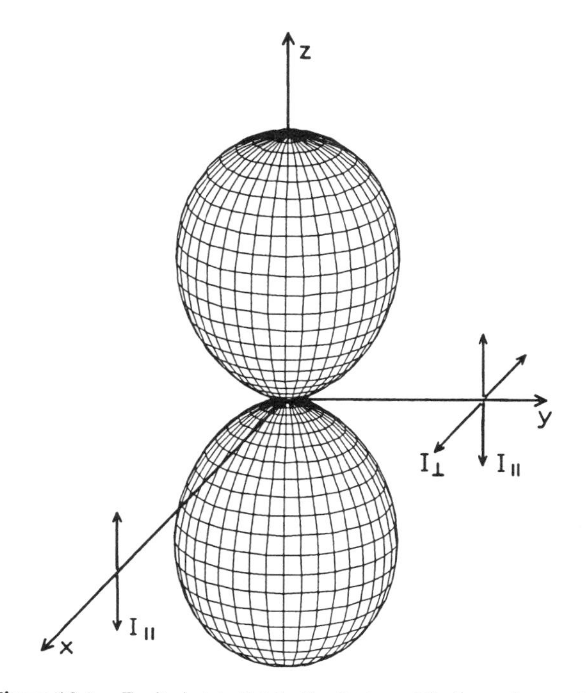

Figure 10.5. Excited-state distribution for immobile fluorophores with $r_0 = 0.4$.

$$\langle\cos^2 \theta\rangle = \frac{\int_0^{\pi/2} \cos^2 \theta\, f(\theta)\, d\theta}{\int_0^{\pi/2} f(\theta)\, d\theta} \qquad [10.19]$$

Substitution of Eq. [10.18] into Eq. [10.19] yields $\langle\cos^2\theta\rangle = \frac{3}{5}$. Recalling Eq. [10.17], one finds a maximum anisotropy of 0.4. This is the value which is observed when the absorption and emission dipoles are collinear, and when there are no processes which result in depolarization. Under these conditions, the excited-state population is preferentially oriented along the z-axis (Figure 10.5), and the value of $I_\perp$ is one-third the value of $I_{\parallel}$ ($I_{\parallel} = 3I_\perp$). We note that this value ($r = 0.4$) is considerably smaller than that possible for a single fluorophore oriented along the z-axis ($r = 1.0$).

It is important to remember that there are other possible origins for polarized light. These include reflections and light scattered by the sample. For a dilute scattering solution, the anisotropy is close to 1.0. Scattered light can interfere with anisotropy measurements. If the measured anisotropy for a randomly oriented sample is greater than 0.4, one can confidently infer the presence of scattered light in addition to fluorescence. The maximum anisotropy of 0.4 for collinear absorption and emission dipoles is a consequence of the $\cos^2\theta$ probability of light absorption. Anisotropy values can exceed 0.4 for multiphoton excitation (Section 10.13).

10.3. EXCITATION ANISOTROPY SPECTRA

In the preceding discussion we assumed that the absorption and emission moments were collinear ($r_0 = 0.4$). Few fluorophores display $r_0 = 0.4$. For most fluorophores, the r_0 values are less than 0.4, and, in general, the anisotropy values depend on the excitation wavelength. This is explained in terms of the transition moments being displaced by an angle β relative to each other. In the previous section (Eqs. [10.10]–[10.17]), we demonstrated that displacement of the emission dipole by an angle θ from the z-axis resulted in a decrease in the anisotropy by a factor of $(3\cos^2\theta - 1)/2$. Similarly, the displacement of the absorption and emission dipoles by an angle β results in a further loss of anisotropy. The observed anisotropy in a vitrified dilute solution is a product of the loss of anisotropy due to photoselection (resulting in a reduction of the anisotropy by a factor of $\frac{2}{5}$) and that due to the angular displacement of the dipoles. Hence, the fundamental anisotropy of a fluorophore is given by

$$r_0 = \frac{2}{5}\left(\frac{3\cos^2\beta - 1}{2}\right) \tag{10.20}$$

The term r_0 is used to refer to the anisotropy observed in the absence of other depolarizing processes such as rotational diffusion or energy transfer. For some molecules, β is close to zero. For example, r_0 values as high as 0.39 have been measured for DPH,[6] although slightly lower values are frequently reported.[7] An anisotropy of 0.39 corresponds to an angle of 7.4° between the dipoles, whereas $r_0 = 0.4$ corresponds to an angle of 0°. It is interesting to note that the fundamental anisotropy value is zero when $\beta = 54.7°$. When β exceeds 54.7°, the anisotropy becomes negative. The maximum negative value (–0.20) is found for $\beta = 90°$. The values of both r_0 and P_0 for $\beta = 0$, 45, 54.7, and 90° are summarized in Table 10.1. One should note that r_0 must be within the range of –0.20–0.40 for an isotropic solution with single-photon excitation.

Measurement of the fundamental anisotropy requires special conditions. In order to avoid rotational diffusion, the probes are usually examined in solvents which form a clear glass at low temperature, such as propylene glycol or glycerol. Additionally, the solutions must be optically dilute to avoid depolarization due to radiative reabsorption and emission or due to RET. One commonly used solvent for measuring fundamental anisotropies is propylene glycol at –60 to –70 °C. Under these conditions, the fluorophores remain immobile during the lifetime of the excited state. The measured anisotropy values (r_0) then provide a measure of the angle between the absorption and emission dipoles (Eq. [10.20]. Since the orientation of the absorption dipole differs for each absorption band, the angle β varies with excitation wavelength.

The changes in the fundamental anisotropy with excitation wavelength can be understood in terms of a rotation of the absorption transition moment. However, a more precise explanation is the changing contributions of two or more electronic transitions, each with a different value of β. As the excitation wavelength changes, so does the fraction of the light absorbed by each transition (see below).

Table 10.1. Relationship between the Angular Displacement of Transition Moments (β) and the Fundamental Anisotropy (r_0) or Polarization (P_0)

β (deg)	r_0	P_0
0	0.40	$0.50 = \frac{1}{2}$
45	0.10	$0.143 = \frac{1}{7}$
54.7	0.00	0.00
90	–0.20	$-0.333 = -\frac{1}{3}$

The anisotropy spectrum is a plot of the anisotropy versus the excitation wavelength for a fluorophore in a dilute vitrified solution. Generally, the anisotropy is independent of the emission wavelength, so only excitation anisotropy spectra are reported. The lack of dependence on emission wavelength is expected because emission is almost always from the lowest singlet state. If emission occurs from more than one state, and if these states show different emission spectra, then the anisotropies can be dependent upon emission wavelength. Such dependence can also be observed in the presence of solvent relaxation (Chapter 7). In this case, anisotropy may decrease with increasing wavelength because the average lifetime is longer for longer wavelengths (Section 10.11). This effect is generally observed when the spectral relaxation time is comparable to the fluorescence lifetime. In completely vitrified solution, where solvent relaxation does not occur, the anisotropy is usually independent of emission wavelength.

Typically, the largest r_0 values are observed for the longest-wavelength absorption band. This is because the lowest singlet state is generally responsible for the observed fluorescence, and this state is also responsible for the longest-wavelength absorption band (Kasha's rule). Hence, absorption and emission involve the same electronic transition and have nearly collinear moments. Larger β values (lower r_0 values) are obtained upon excitation into higher electronic states, which are generally not the states responsible for fluorescence emission. Rather, the fluorophores relax very rapidly to the lowest singlet state. The excitation anisotropy spectrum reveals the angle between the absorption and emission transition moments. However, the directions of these moments within the molecule itself are not revealed. Such a determination requires studies with ordered systems, such as crystals or stretched films.[8–10]

These general features of an anisotropy spectrum are illustrated in Figure 10.6 for DAPI in an isotropic (unstretched) poly(vinyl alcohol) film.[11] This film is very viscous, so the fluorophores cannot rotate during the excited-state lifetime. As is thought to be true for nearly all fluorophores, the absorption and emission dipoles are in the plane of the rings. For excitation wavelengths longer than 330 nm, the r_0 value is relatively constant. The relatively constant r_0 value across the $S_0 > S_1$ transition and a gradual tendency toward higher r_0 values at longer wavelengths are typical of many fluorophores. As the excitation wavelength is decreased, the anisotropy becomes more strongly de-

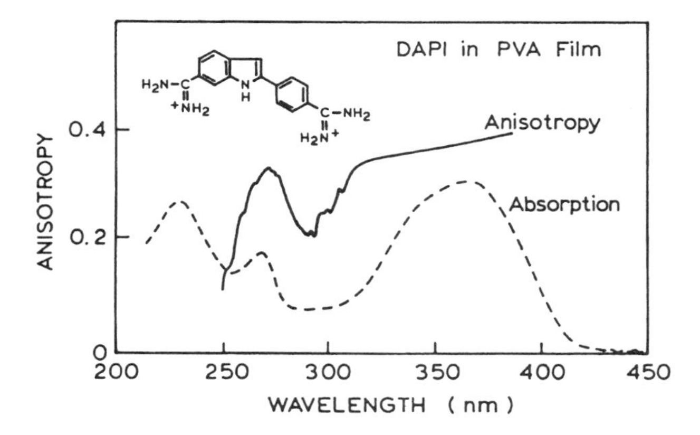

Figure 10.6. Excitation anisotropy (—) and absorption (---) spectra of DAPI in an unstretched poly(vinyl alcohol) (PVA) film. Revised and reprinted, with permission, from Ref. 11, Copyright © 1989, American Chemical Society.

pendent on wavelength. Different anisotropies are expected for the $S_0 \rightarrow S_1$, $S_0 \rightarrow S_2$, and higher transitions.

The excitation anisotropy spectrum of DAPI is rather ideal because r_0 is relatively constant across the long-wavelength absorption band. For many fluorophores, one observes a decrease in r_0 as the excitation wavelength is decreased, as illustrated for 9-(9-anthroyloxy) stearic acid (9-AS) in Figure 10.7. When such probes are used, careful control of the excitation wavelength is needed if the experiments depend on knowledge of the r_0 values.

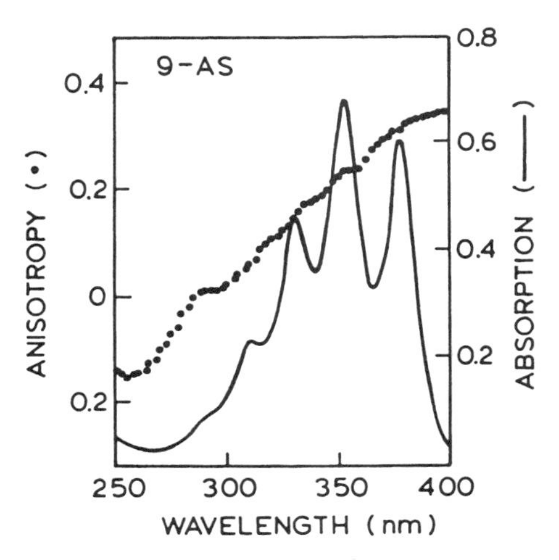

Figure 10.7. Excitation anisotropy spectrum of 9-(9-anthroyloxy) stearic acid (9-AS) in propylene glycol at –52 °C. The same excitation anisotropy spectrum was observed for 2-AS, 7-AS, and 12-AS and 16-(9-anthroyloxy)palmitic acid (16-AP). Revised and reprinted, with permission, from Ref. 12, Copyright © 1982, American Chemical Society.

10.3.A. Resolution of Electronic States from Polarization Spectra

Some fluorophores can display complex anisotropy spectra, even across the apparently longest-wavelength absorption band. One well-known example is indole.[13–15] In this case, the anisotropy varies abruptly with excitation wavelength across the long-wavelength absorption band. This dependence was attributed to the two excited states of indole (^1L_a and ^1L_b) which are responsible for the absorption between 250 and 300 nm. The transition moments are thought to be at an angle of 90° relative to one another.[16,17] Emission of indole occurs mainly from the ^1L_a state.

The complex anisotropy spectrum of indole was used to determine the absorption spectra corresponding to the $S_0 \rightarrow {}^1L_a$ and $S_0 \rightarrow {}^1L_b$ transitions.[14] We present this example because of its didactic value and its importance for a detailed understanding of the fluorescence from tryptophan residues in proteins (Chapter 16). At any excitation wavelength λ, the observed anisotropy is

$$r_0(\lambda) = f_a(\lambda) r_{0a} + f_b(\lambda) r_{0b} \qquad [10.21]$$

where $f_i(\lambda)$ represents the fractional contribution of the ith state to the total absorption at the wavelength λ, and r_{0i} represents the limiting anisotropy of this state. The assumption was made that r_{0a} and r_{0b} are independent of wavelength and that emission occurs only from the ^1L_a state. If excitation and emission occur from the same state, one expects the oscillators to be collinear and r_0 to have a value of 0.4. However, the highest observed value of r_0 was 0.3, and this value was assigned to r_{0a}. That is, the absorption of the ^1L_a state is assumed to be dominant at wavelengths of 300 nm and greater. Selection of a value for r_{0b} is more difficult since there is no obvious wavelength where the absorption of ^1L_b is dominant. Valeur and Weber[14] used the results from another laboratory which indicated that the ^1L_a and ^1L_b oscillators were perpendicular to one another.[16] The anisotropy expected for ^1L_b can be predicted using Eq. [10.20], which describes the loss in anisotropy due to an angular displacement of two oscillators by a known angle. In this equation the maximum value of r_0 (0.4) is replaced by r_{oa}. Hence, the anisotropy of the ^1L_b state is given by

$$r_{0b} = 0.30 \left(\frac{3\cos^2\beta - 1}{2} \right) \qquad [10.22]$$

Using $\beta = 90°$, one obtains $r_{0b} = -0.15$. These values of r_{0a} and r_{0b} are used in Eq. [10.21], along with the restriction that the total fractional absorption is unity $[f_a(\lambda) + f_b(\lambda) = 1]$, to calculate the fractional absorption of each state as a function of wavelength. Rearrangement of Eq. [10.21] yields

$$f_a(\lambda) = \frac{r_0(\lambda) - r_{0b}}{r_{0a} - r_{0b}} = \frac{r_0(\lambda) + 0.15}{0.45} \quad [10.23]$$

$$f_b(\lambda) = \frac{r_{0a} - r_0(\lambda)}{r_{0a} - r_{0b}} = \frac{0.3 - r_0(\lambda)}{0.45} \quad [10.24]$$

Finally, the absorption spectrum of each state is given by

$$A_a(\lambda) = f_a(\lambda)A(\lambda) \quad [10.25]$$

$$A_b(\lambda) = f_b(\lambda)A(\lambda) \quad [10.26]$$

where $A(\lambda)$ is the total absorption spectrum. These resolved spectra are shown in Figure 10.8. The ^1L_b absorption is structured, but this absorption is less intense than the ^1L_a absorption. The peak at 290 nm in the absorption spectrum of indole is seen to be due to the ^1L_b state, and this peak absorption corresponds to a minimum in the r_0 value. Above 295 nm, only the ^1L_a state absorbs, and the anisotropy is relatively constant. This is one reason why

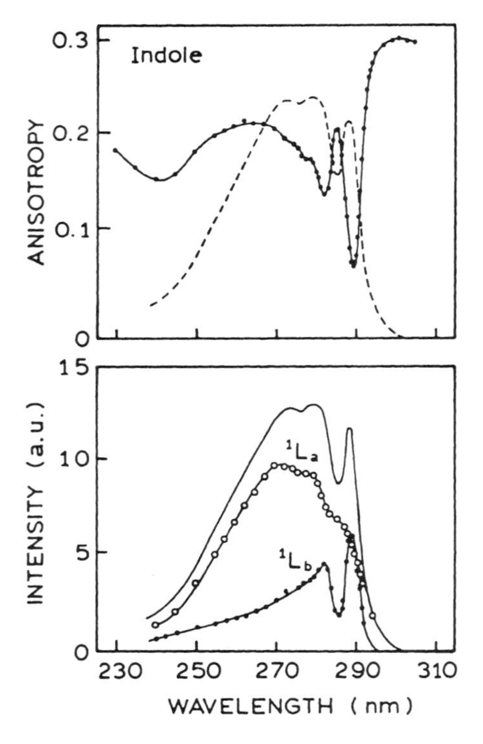

Figure 10.8. *Top*: Excitation anisotropy (—) and absorption (---) spectra of indole in vitrified propylene glycol. *Bottom*: Calculated absorption spectra of the ^1L_a (○) and ^1L_b (●) states. Revised from Ref. 14.

295–300-nm excitation is used when the intrinsic tryptophan emission of proteins is studied. This example illustrates how polarization spectra reveal the electronic properties of fluorophores.

10.4. MEASUREMENT OF FLUORESCENCE ANISOTROPIES

Prior to continuing our discussion of the factors which result in the depolarization of fluorescence, it is useful to understand the methods used to measure the anisotropy or polarization. We now describe steady-state measurements, but similar considerations apply to time-resolved measurements of anisotropies. Two methods are commonly used. These are the L-format method, in which a single emission channel is used, and the T-format method, in which the parallel and perpendicular components are observed simultaneously through separate channels. The procedures described below are intended to correct for the different efficiencies of the instrumentation for detection of the various polarized components of the emission (Chapter 2).

10.4.A. L-Format or Single-Channel Method

The L-format method is used most frequently since most fluorometers have only a single emission channel. Assume that the sample is excited with vertically polarized light, and the emission is observed through a monochromator (Figure 10.9). The monochromator will usually have a different transmission efficiency for vertically and horizontally polarized light. Consequently, rotation of the emission polarizer changes the measured intensities even if the sample emits unpolarized light. Hence, the measured intensities are not the desired parallel and perpendicular intensities. The objective is to measure these actual intensities, $I_{\parallel}$ and $I_{\perp}$, unbiased by the detection system.

We use two subscripts to indicate the orientation of the excitation and emission polarizers, respectively. For example, I_{HV} corresponds to horizontally polarized excitation and vertically polarized emission. This notation is easy to recall since the order of the subscripts represents the order in which the light passes through the two polarizers. Let S_V and S_H be the sensitivities of the emission channel for the vertically and horizontally polarized components, respectively. For vertically polarized excitation, the observed polarized intensities are

$$I_{VV} = kS_V I_{\parallel} \quad [10.27]$$

$$I_{VH} = kS_H I_{\perp} \quad [10.28]$$

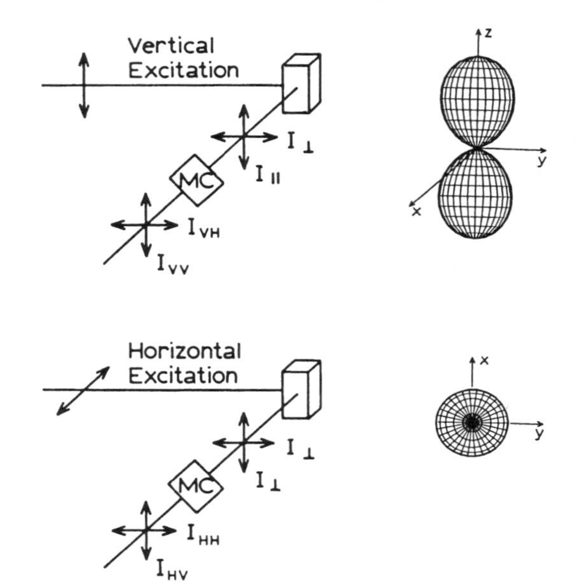

Figure 10.9. Schematic diagram for L-format measurements of fluorescence anisotropy. MC, Monochromator. The shapes at the right are the excited-state distributions.

where k is a proportionality factor to account for the quantum yield of the fluorophore and other instrumental factors besides the polarization-dependent sensitivity. Division of Eq. [10.27] by Eq. [10.28] yields

$$\frac{I_{VV}}{I_{VH}} = \frac{S_V}{S_H}\frac{I_{\parallel}}{I_{\perp}} = G\frac{I_{\parallel}}{I_{\perp}} \qquad [10.29]$$

To calculate the actual intensity ratio ($I_{\parallel}/I_{\perp}$), we need to determine the G-factor, which is the ratio of the sensitivities of the detection system for vertically and horizontally polarized light,

$$G = \frac{S_V}{S_H} \qquad [10.30]$$

For instance, assume $G = 2.0$. This means that the emission monochromator passes vertically polarized light, at the chosen emission wavelength, with twofold greater efficiency than horizontally polarized light. It is important to note that the G-factor is dependent upon the emission wavelength and, to some extent, the bandpass of the monochromator. Frequently, anisotropy measurements are performed using an emission filter rather than a monochromator. Filters generally do not have any significant polarizing effect, and hence one expects $G = 1.0$. Nonetheless, this factor should always be determined since rotation of the emission polarizer can cause the focused image of the fluorescence to change position, altering the effective sensitivity.

The G-factor is easily measured using horizontally polarized excitation. With horizontally polarized excitation, the excited-state distribution is rotated to lie along the observation axis. When this is done, both the horizontally and vertically polarized components are equal and proportional to $I_{\perp}$ (Figure 10.9). This is because both polarizer orientations are perpendicular to the polarization of the excitation. Hence, any measured difference in I_{HV} and I_{HH} must be due to the properties of the detection system. Specifically,

$$\frac{I_{HV}}{I_{HH}} = \frac{S_V I_{\perp}}{S_H I_{\perp}} = \frac{S_V}{S_H} = G \qquad [10.31]$$

Note that it does not matter if the intensity of the excitation is changed when the excitation polarizer is rotated to the horizontal position. This change is a constant factor in the numerator and denominator of Eq. [10.31], and hence cancels. When the G-factor is known, the ratio $I_{\parallel}/I_{\perp}$ can be calculated using

$$\frac{I_{VV}}{I_{VH}}\frac{1}{G} = \frac{I_{VV}}{I_{VH}}\frac{I_{HH}}{I_{HV}} = \frac{I_{\parallel}}{I_{\perp}} \qquad [10.32]$$

The anisotropy is given by

$$r = \frac{(I_{\parallel}/I_{\perp}) - 1}{I_{\parallel}/I_{\perp} + 2} \qquad [10.33]$$

which is frequently used in the alternative formulation

$$R = \frac{I_{VV} - GI_{VH}}{I_{VV} + 2GI_{VH}} \qquad [10.34]$$

10.4.B. T-Format or Two-Channel Anisotropies

In the T-format method, one measures the intensities of the parallel and perpendicular components simultaneously using two separate detection systems (Figure 10.10). The emission polarizers are left unchanged, and hence measurement of the relative sensitivity to each polarization is not necessary. However, one needs to measure the relative sensitivity of the two detection systems, which is accomplished using horizontally polarized excitation. The measurements are performed as follows. The excitation polarizer is first placed in the vertical orientation, and one measures the ratio of the parallel and perpendicular signals (R_V). This ratio is given by

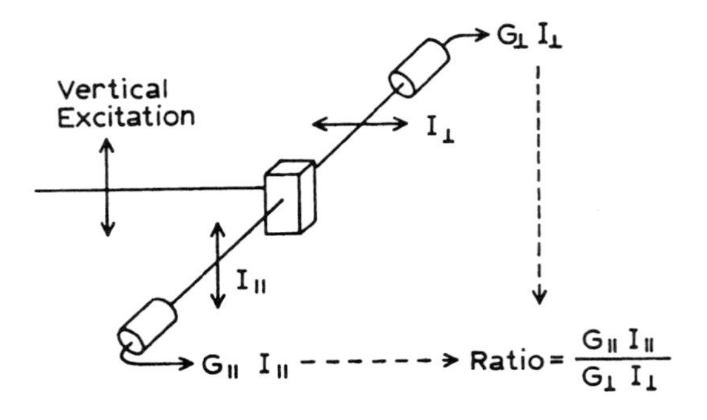

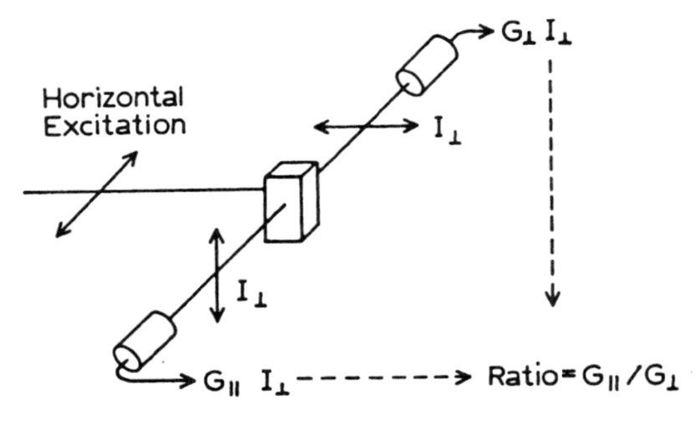

Figure 10.10. Schematic diagram for T-format measurements of fluorescence anisotropy.

$$R_V = \frac{G_\parallel I_\parallel}{G_\perp I_\perp} \quad [10.35]$$

where $G_\parallel$ and $G_\perp$ are the gains of the parallel and perpendicular channels, respectively. This ratio can be adjusted to unity, but in practice it is easier to simply measure the ratio. Because this ratio is measured using horizontally polarized excitation, both emission channels observe $I_\perp$ (Figure 10.10). Hence, the ratio of intensities is given by

$$R_H = \frac{G_\parallel}{G_\perp} \quad [10.36]$$

Division of Eq. [10.35] by Eq. [10.36] yields

$$\frac{R_V}{R_H} = \frac{I_\parallel}{I_\perp} \quad [10.37]$$

which can be used to calculate the anisotropy (Eq. [10.33]). Again, we note that the excitation intensity may vary upon rotation of the excitation polarizer. This factor cancels in taking the ratio. However, difficulties can arise if light from the excitation monochromator is completely polarized in either the vertical or the horizontal direction (Chapter 2). Then either R_H or R_V cannot be measured because of inadequate light intensity. This difficulty can be avoided by using a different excitation wavelength where adequate intensity is available for each polarization. This is acceptable because R_V and R_H are properties of the detection system and are not dependent upon the excitation wavelength or the degree of polarization displayed by the sample. Note that the theoretical maximum for $I_\parallel/I_\perp$ is 3 (r = 0.4). Values of $I_\parallel/I_\perp$ in excess of 3 generally indicate an artifact, typically scattered light. Scattered light is 100% polarized ($I_\parallel/I_\perp = \infty$, $r = 1.0$), and a small percentage of such light can seriously distort the anisotropy values.

10.4.C. Comparison of T-Format and L-Format Measurements

For fluorometers with a single emission channel, one has no choice but to use the L-format method. This procedure requires four individual measurements, I_{VV}, I_{VH}, I_{HV}, and I_{HH}. The latter two define the G-factor, which does not need to be measured each time. The T-format method only requires two measurements, each of which is a ratio. Because of the simultaneous measurement, fluctuations in signal intensity are canceled in T-format measurements. In the early days of fluorescence spectroscopy, T-format measurements were preferred. With modern instruments, there no longer seems to be any significant advantage to T-format measurements, except possibly for a decreased time for data acquisition. L-format measurements are routinely used in our laboratory.

10.4.D. Alignment of Polarizers

Accurate measurement of fluorescence anisotropies requires that the polarizers be precisely positioned in the vertical and horizontal orientations. The alignment can be easily checked and adjusted using a dilute suspension of glycogen or colloidal silica in water. The scattered light is 100% polarized, that is, r = 1.0. In our laboratory we consider the alignment to be adequate when the measured value is 0.97 or larger. It is essential to use dilute suspensions of scatterer. Otherwise, multiple scattering events lead to decreased values of polarization.

Alignment can be accomplished as follows. The excitation polarizer is rotated to the approximate vertical position. Precise vertical alignment is not necessary since the scattered light is vertically polarized. The angular alignment of the emission polarizer is adjusted so that the minimum intensity is observed. This is the horizontal position. It is preferable to use this minimum intensity for alignment; the maximum intensity for the vertical component is less sharply defined. Of course, one should check that the vertical polarizer stop is also properly adjusted. Rotation of the emission polarizer should now yield the maximum and minimum intensities when the polarizer is at the vertical and horizontal stops, respectively. These adjustments should be performed with the emission monochromator removed, or its wavelength chosen for approximately equal transmission efficiencies for vertically and horizontally polarized light. Otherwise, its polarizing properties could interfere with the alignment. The selection of a wavelength at which the transmission efficiencies for vertically and horizontally polarized light are equal can be accomplished using either horizontally polarized excitation, to obtain $I_{\parallel} = I_T$, or a sample whose emission is not polarized. One such solution is 9-cyanoanthracene in the fluid solvent ethanol.

Alignment of the excitation polarizer is performed in a similar manner. The emission polarizer should be set in the vertical position. The minimum and maximum intensities of scattered light should be observed when the excitation polarizer is at its horizontal and vertical stops, respectively. Once again, one should consider the polarization properties of the excitation monochromator.

10.4.E. Magic-Angle Polarizer Conditions

The goal of intensity measurements is usually to measure a signal proportional to the total intensity (I_T), not one proportional to $I_{\parallel}$ or $I_{\perp}$. However, since the transmission efficiency of the emission monochromator depends on polarization, the signal one observes is usually not proportional to $I_{\parallel} + 2\,I_{\perp}$, but rather to some other combination of $I_{\parallel}$ and $I_{\perp}$. With the use of polarizers, the measured intensity can be made proportional to the total intensity $I_T = I_{\parallel} + 2I_{\perp}$, irrespective of the degree of polarization of the sample. To accomplish this, the excitation polarizer is oriented in the vertical position and the emission polarizer is oriented 54.7° from the vertical. Since $\cos^2 54.7°$ is 0.333 and $\sin^2 54.7°$ is 0.667, these polarizer settings result in $I_{\perp}$ being selected twofold over $I_{\parallel}$, forming the correct sum for $I_T = I_{\parallel} + 2I_{\perp}$. The use of these magic-angle conditions is especially important for intensity decay measurements. The intensity decays of the vertically and horizontally polarized components are usually distinct. Hence, if $I_{\parallel}(t)$ and $I_{\perp}(t)$ are not properly weighted, then incorrect decay times are recovered. If the anisotropy is zero, then the correct intensity and intensity decay times are recovered independent of polarizer orientation.

10.4.F. Why Is the Total Intensity Equal to $I_{\parallel} + 2I_{\perp}$?

It is widely known that the total intensity is given by $I_{\parallel} + 2I_{\perp}$, but the origin of this result is less widely understood. This relationship is the result of the transmission properties of polarizers, in particular, the dependence of the intensity on $\cos^2 \alpha$, where α is the angle between the transition moment and the transmitting direction of the polarizer. Consider a collection of fluorophores, each emitting an intensity I_i. The total intensity is given by

$$I_T = \sum_{i=1}^{n} I_i \qquad [10.38]$$

When the intensity is observed through a polarizer (I_P) oriented along an axis p, the intensity is given by

$$I_P = \sum_{i=1}^{n} I_i \cos^2 \alpha_{pi} \qquad [10.39]$$

where α_{pi} is the angle between the direction of the ith emission dipole and the axis of the polarizer. One can choose to measure the intensity along the three Cartesian axes,

$$I_x = \sum_{i=1}^{n} I_i \cos^2 \alpha_{xi} \qquad [10.40]$$

$$I_y = \sum_{i=1}^{n} I_i \cos^2 \alpha_{yi} \qquad [10.41]$$

$$I_z = \sum_{i=1}^{n} I_i \cos^2 \alpha_{zi} \qquad [10.42]$$

where the α_{pi} are the angles between the ith dipole and the representative axis (Figure 10.11). It is easy to see that

$$I_x + I_y + I_z = \sum_{i=1}^{n} I_i(\cos^2 \alpha_{xi} + \cos^2 \alpha_{yi} + \cos^2 \alpha_{zi}) \qquad [10.43]$$

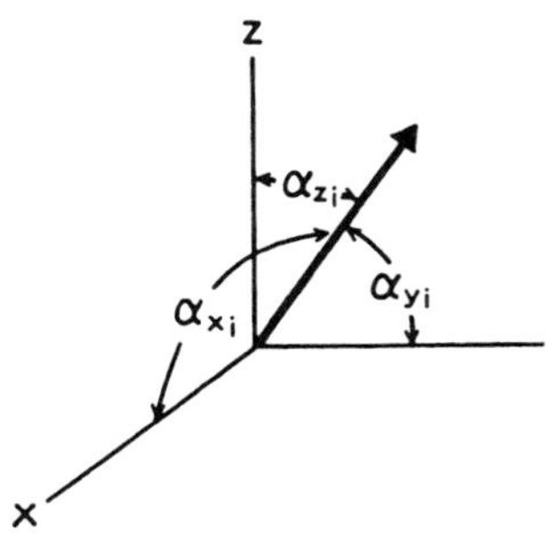

Figure 10.11. Fluorophore at an arbitrary orientation in the Cartesian coordinate system.

$$I_x + I_y + I_z = \sum_{i=1}^{n} I_i = I_T \qquad [10.44]$$

Equation [10.44] is correct because

$$\cos^2 \alpha_{xi} + \cos^2 \alpha_{yi} + \cos^2 \alpha_{zi} = 1.0 \qquad [10.45]$$

is always true. Because $I_x = I_y$ for vertically polarized excitation, $I_T = I_{\parallel} + 2I_{\perp}$. It is interesting to note that this relationship would not be correct if the transmission of polarizers depended on some other function of α_{pi}.

10.4.G. Effect of Radiationless Energy Transfer on the Anisotropy

In Section 10.3 we discussed an intrinsic cause of depolarization, namely, the angular displacement between the absorption and emission moments. The fluorophores were assumed to be immobile, and the excited state was assumed to remain localized on the originally excited fluorophore. The anisotropy can also be decreased by extrinsic factors which act during the lifetime of the excited state.[3] These are rotational diffusion of the fluorophore and the resonance energy transfer (RET) of energy among fluorophores (Chapter 13). Both processes result in additional angular displacement of the emission oscillator and hence lower anisotropies.

The effects of rotational diffusion and energy transfer are easily separated by judicious choice of the experimental conditions. For example, Brownian rotations cause negligible depolarization when the rotational rate is much slower than the rate of fluorescence emission. In contrast, RET occurs only in concentrated solution where the average distance between the fluorophore molecules is comparable to a characteristic distance R_0, which is typically near 40 Å. One may readily calculate that millimolar concentrations are required to obtain this average distance (Chapter 13). Hence, since the usual concentrations required for fluorescence measurements are about $10^{-6}M$, RET is easily avoided by the use of dilute solutions.

The effect of RET on the anisotropy is illustrated by the excitation anisotropy spectra of fluorescein in dilute and in concentrated solution[18] (Figure 10.12). Fluorescein is subject to radiative (emission and reabsorption) and radiationless energy transfer because of the small Stokes' shift. In these experiments, radiative transfer was avoided by using thin samples. In dilute solution, fluorescein displays its characteristic anisotropy spectrum (shown as polarization in Figure 10.12), with high anisotropy for excitation above 380 nm. At high concentration, the anisotropy is decreased, and this effect is due to RET between fluo-

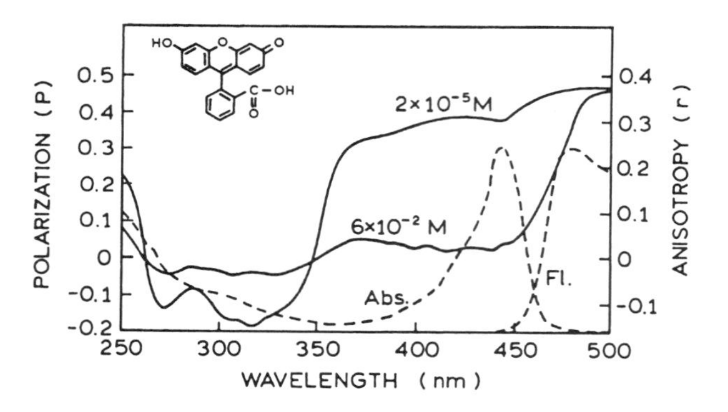

Figure 10.12. Excitation polarization spectra of fluorescein in propylene glycol at −50 °C. Radiative transfer was avoided by using thin samples, 30–50 μm thick. Revised and reprinted, with permission, from Ref. 18.

rescein molecules. In random solution it is known that a single radiationless transfer step reduces the anisotropy to 4% of the initial value.[19–21] Hence, RET is an effective mechanism in depolarization. The presence or absence of RET can usually be predicted from the concentration in the sample and the spectral properties of the probes.

Examination of Figure 10.12 reveals that the anisotropy (polarization) of the concentrated sample increases for excitation wavelengths longer than 460 nm. The anisotropy increases because the red-edge excitation results in a shift of the emission spectrum to longer wavelength, decreased spectral overlap, and less energy transfer (Section 7.8).

10.4.H. Trivial Causes of Depolarization

The measured anisotropies can be lower than the actual values for several trivial reasons, including light scattering, reabsorption, and misalignment of the polarizers. While these are nontrivial experimental problems, they are dependent only on the optical conditions of the experiment and do not provide useful information on the molecular properties of the sample.

Biological samples, such as aqueous suspensions of membranes, are frequently turbid. This turbidity results in scatter of both the incident light and the emitted photons. The scattered incident light can result in excitation, and the emitted photons can be scattered prior to observation. Each scattering event is thought to decrease the anisotropy of the scattered photon to 0.7 of the original value,[22] and the observed anisotropy is expected to decrease linearly with the optical density due to turbidity. The actual proportionality constant depends upon the individual sample under investigation.[23] Therefore, it is advisable to investigate the effect of turbidity for any sample which displays visible turbidity. This can be accomplished by dilution—either by actual dilution of the sample or by using a cuvette with smaller dimensions. If dilution does not result in a change in anisotropy, then there is unlikely to be significant depolarization due to scattering.

Perhaps a more serious effect of scattering is the possibility that scattered light reaches the detector. This is particularly true for dilute solutions where the intensity is low and scattering from the optics and sample can be significant. Since the scattered light will be highly polarized ($r = 1.0$), a small percentage of scattered light can result in significant changes in the anisotropy (Eq. [10.6]; see also Problem 10.2). Hence, the effects of sample turbidity are complex. If the detector system effectively eliminates the scattered light, then the anisotropy will decrease as compared to that measured with more scattered light. If scattered light reaches the detector, then the measured anisotropy is usually increased relative to its true value.

Another trivial cause of depolarization is radiative transfer, which is the reabsorption of emitted photons. It is more difficult to eliminate radiative transfer since it can occur at lower concentrations than RET. In fact, radiative transfer is a frequent cause of low anisotropy values for fluorescein. Because of its large spectral overlap, fluorescein solutions often show the effects of radiative transfer. Surprisingly, radiative transfer is less effective than RET in depolarization. A single radiative transfer step reduces the anisotropy to 28% of its initial value,[24,25] as compared to 4% in a single RET step.

Measured anisotropy values can also be too low due to misalignment or inefficiency of the polarizers. Film polarizers become less ideal at short wavelengths. One can always test for such difficulties by examination of a dilute scattering solution (Section 10.4.D.). Finally, one should always examine a blank sample which scatters light approximately equivalently to the sample. Background signals can be especially problematic for anisotropy measurements because the unwanted signal may be polarized if due to scattered light or unpolarized if due to low-molecular-weight impurities. Hence, background signals can either increase or decrease the anisotropy. In order to correct for background, one should measure the four individual intensities from the blank sample and subtract them from each respective intensity value in Eq. [10.34].

10.4.I. Factors Affecting the Anisotropy

The various factors which can affect the anisotropy are summarized in Ref. 3 in an insightful table which outlines the experimental conditions for various anisotropy measurements (Table 10.2). The fundamental anisotropy can be measured in dilute, highly viscous solutions, where rotational diffusion or RET does not occur. Information about the Förster distance (R_0) for hetero- or homo-RET can be obtained from studies in concentrated, viscous solutions. Perhaps the most interesting condition designated in Table 10.2 is that for dilute, nonviscous solutions. In such solutions, the anisotropy is primarily determined by rotational motion of the fluorophore. For labeled proteins, these motions are dependent upon the size and shape of the protein and its extent of aggregation, among other factors. For membranes, the anisotropy depends upon the chemical composition and phase state of the membranes. As a result, anisotropy measurements are frequently used to study the properties and interactions of biological macromolecules.

Table 10.2. Extrinsic Causes of Depolarization[a]

Condition	Observable	Molecular property
Dilute viscous solution (propylene glycol, −70 °C)	r_0 or cos β (Eq. [10.20])	Angle between absorption and emission dipole as a function of excitation wavelength
Concentrated viscous solution	$0 < \lvert r \rvert < \lvert r_0 \rvert$; anisotropy decreased by energy migration	Distance dependence of radiationless energy transfer
Dilute nonviscous solution (H_2O, EtOH, room temperature)	$0 < \lvert r \rvert < \lvert r_0 \rvert$; anisotropy decreased due to Brownian rotation	Size and shape of fluorophore or macromolecule

[a]From Ref. 3.

10.5. EFFECTS OF ROTATIONAL DIFFUSION ON FLUORESCENCE ANISOTROPIES: THE PERRIN EQUATION

Rotational diffusion of fluorophores is a dominant cause of fluorescence depolarization. This mode of depolarization is described in the simplest case for spherical rotors by the Perrin equation[26–28]

$$\frac{r_0}{r} = 1 + \frac{\tau}{\theta} = 1 + 6D\tau \qquad [10.46]$$

where τ is the fluorescence lifetime, θ is the rotational correlation time, and D is the rotational diffusion coefficient. If the correlation time is much larger than the lifetime ($\theta >> \tau$), then the measured anisotropy (r) is equal to the fundamental anisotropy (r_0). If the correlation time is much shorter than the lifetime ($\theta << \tau$), then the anisotropy is zero.

The Perrin equation has been derived from first principles based on diffusional steps.[3,29] Here we present a straightforward derivation based upon the fact that, following a δ-pulse excitation, the time-resolved decay of anisotropy $r(t)$ for a spherical molecule is a single exponential,

$$r(t) = r_0 e^{-t/\theta} = r_0 e^{-6Dt} \qquad [10.47]$$

In this equation the rotational correlation time of the fluorophore (θ) is given by

$$\theta = \frac{\eta V}{RT} \qquad [10.48]$$

where η is the viscosity, T is the temperature in kelvins, R is the gas constant, and V is the volume of the rotating unit. The rotational correlation time is related to the rotational diffusion coefficient by $\theta = (6D)^{-1}$. Only spherical molecules display a single-exponential anisotropy decay. More complex expressions are predicted for nonsymmetric species or molecules (Chapter 12).

The steady-state anisotropy can be calculated from an average of the anisotropy decay $r(t)$ over the intensity decay $I(t)$,

$$r = \frac{\int_0^\infty I(t) r(t)\, d(t)}{\int_0^\infty I(t)\, dt} \qquad [10.49]$$

For a single-exponential intensity decay, substitution into Eq. [10.49] yields

$$r = \frac{r_0}{1 + (\tau/\theta)} \qquad [10.50]$$

which is a transposed form of the Perrin equation. Using this equation and Eq. [10.48], one can calculate the anisotropy expected for fluorophores in solvents or for labeled macromolecules, assuming that the molecules are spherical. For example, perylene has a lifetime of 6 ns and $r_0 = 0.36$. In ethanol, rotational diffusion is expected to decrease the anisotropy to 0.005 (Problem 10.5).

10.5.A. The Perrin Equation—Rotational Motions of Proteins

Prior to the availability of time-resolved measurements, the apparent molecular volumes of proteins were measured using the Perrin equation. This application can be seen by rearranging Eq. [10.50],

$$\frac{1}{r} = \frac{1}{r_0} + \frac{\tau}{r_0 \theta} \qquad [10.51]$$

For globular proteins, the rotational correlation time is approximately related to the molecular weight (M) of the protein by

$$\theta = \frac{\eta V}{RT} = \frac{\eta M}{RT}(\bar{v} + h) \qquad [10.52]$$

where $\bar{v}$ is the specific volume of the protein and h is the hydration, T is the temperature in kelvins, $R = 8.31 \times 10^7$ erg K^{-1} mol^{-1}, and the viscosity η is in poise (P). Values of $\bar{v}$ for proteins are typically near 0.73 ml/g, and the hydration is near 0.23 g of H_2O per gram of protein. This expression predicts that the correlation time of a hydrated protein is about 30% larger than that expected for an anhydrous sphere. Generally, the observed values of θ are about twice that expected for an anhydrous sphere.[30] For example, for an anhydrous protein sphere with a molecular weight of 50,000, $\bar{v} = 0.73$ ml/g, and $\eta = 0.94$ cP, the calculated rotational correlation time at 25 °C is near 14 ns. The anisotropy decay of an immunoglobulin F_{ab} fragment, labeled at the antigen binding site with dansyl-lysine, yielded a rotational correlation time of 33 ns. This larger correlation time was consistent with a hydration of 0.32 ml/g and an axial ratio near 2. Such a result is typically found for proteins (Table 10.3) and is probably due to the nonspherical shape of most proteins and a larger effective solvent shell for rotational diffusion than for hydration. For convenience, we have listed the calculated rotational correlation times for proteins with different molecular weights and different amounts of hydration in Table 10.4.

The apparent volume of a protein can be determined by measuring the anisotropy at various temperatures and/or

Table 10.3. Rotational Correlation Times for Proteins[a]

Protein	Molecular weight	Observed θ (ns)	$\theta_{obs}/\theta_{calc}$[b]
Apomyoglobin	17,000	8.3	1.9
β-Lactoglobulin (monomer)	18,400	8.5	1.8
Trypsin	25,000	12.9	2.0
Chymotrypsin	25,000	15.1	2.3
Carbonic anhydrase	30,000	11.2	1.4
β-Lactoglobulin (dimer)	36,000	20.3	2.1
Apoperoxidase	40,000	25.2	2.4
Serum albumin	66,000	41.7	2.4

[a] From Ref. 30.
[b] θ_{obs} is the observed rotational correlation time, adjusted to the value of T/η corresponding to water at 25 °C. θ_{calc} is the rotational correlation time calculated for a rigid, unhydrated sphere with the molecular weight of the protein, assuming a partial specific volume of 0.73 ml/g.

Table 10.4. Calculated Rotational Correlation Times for Proteins

T	Molecular weight (kDa)	Correlation time θ^a (ns) $h = 0$	$h = 0.2$	$h = 0.4$
2 °C	10	5.5	6.9	8.4
	25	13.7	17.3	21.1
	50	27.4	34.6	42.0
	100	54.8	69.2	84.0
	500	274.0	346.0	420.0
20 °C	10	3.1	3.9	4.7
	25	7.0	9.7	11.8
	50	15.4	19.5	23.6
	100	30.8	39.0	47.2
	500	154.0	195.0	236.0
37 °C	10	2.0	2.5	3.1
	25	5.0	6.4	7.7
	50	10.0	12.7	15.4
	100	20.1	25.4	30.8
	500	100.5	127.0	154.0

[a] Calculated using $\theta = \eta M(\bar{v} + h)/RT$ with $\bar{v} = 0.75$ ml/g and the indicated degree of hydration (h). The viscosities are η(2 °C) = 1.67 cP, η (20 °C) = 1.00 cP, and η(37 °C) = 0.69 cP.

viscosities. Substitution of Eq. [10.48] into Eq. [10.50] yields a modified form of the Perrin equation:

$$\frac{1}{r} = \frac{1}{r_0} + \frac{\tau RT}{r_0 \eta V} \qquad [10.53]$$

The use of Eq. [10.53] was one of the earliest biochemical applications of fluorescence.[2,31–33] The general approach taken is to label the protein covalently with an extrinsic fluorophore. The fluorophore is chosen primarily on the basis of its fluorescence lifetime. This lifetime should be comparable to the expected rotational correlation time of the protein. In this way the anisotropy will be sensitive to changes in the correlation time. Generally, the fluorescence anisotropies are measured over a range of T/η values. Temperature is varied in the usual manner, and viscosity is generally varied by addition of sucrose or glycerol. For biochemical samples, only a limited range of T/η values are available. At high temperatures the macromolecule may denature, and at low temperatures the solvent may freeze or the macromolecule may not be soluble. The apparent volume of the protein is obtained from a plot of $1/r$ versus T/η (Figure 10.13, upper panel). The intercept of the y-axis represents extrapolation to a very high viscosity and should thus be $1/r_0$, where r_0 is the fundamental anisotropy of the fluorophore. In practice, fluorophores bound to protein often display segmental motions, which are independent of overall rotational diffusion. These motions are often much faster than rotational diffusion and rather insensitive to the macroscopic viscosity. As shown in Section 10.10, such motions can be considered to be an independent factor which depolarizes the emission by a constant factor. The effect is to shift the $1/r$ value to larger values and shift the apparent r_0 value to a smaller value (larger y-intercept). This does not mean that the r_0 value is in fact smaller but rather reflects the inability of the Perrin plot to resolve the faster motion.

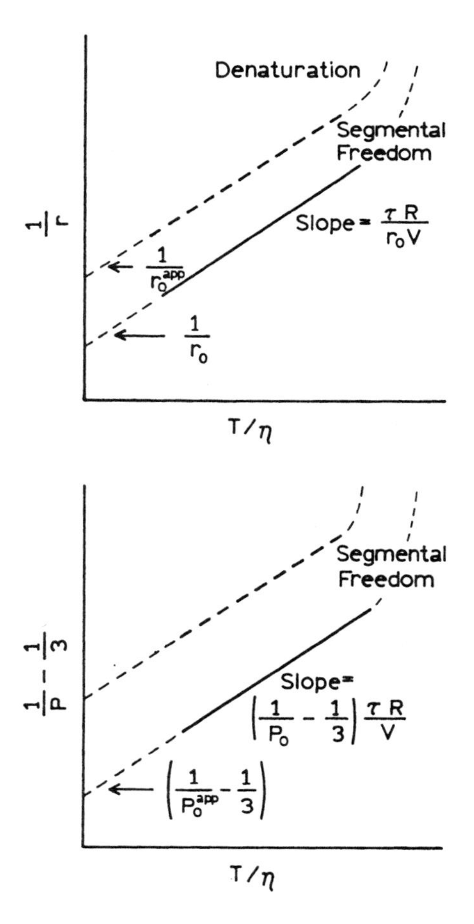

Figure 10.13. Perrin plots for determination of protein volume.

What value of r_0 should be used to calculate the volume of the protein? In general, the extrapolated value of r_0 (r_0^{app}) will yield the best estimate of the volume because this value accounts for segmental motion of the probe, and the effects of this motion on the correlation time θ are at least partially cancelled. For example, assume that the segmental motion of the fluorophore is much more rapid than the rotational diffusion of the protein. Then, the value of r_0 is effectively reduced to

$$r_0^{app} = r_0 \left(\frac{\langle 3\cos^2\alpha \rangle - 1}{2} \right) \qquad [10.54]$$

where α is the angle through which the probe undergoes this segmental motion. The term $1/r_0^{app}$ cancels in the calculation of the molecular volume, and the apparent volume represents overall rotational diffusion of the protein. It is important to note that if the short correlation time for segmental motion (θ_S) is not much faster than the overall correlation time (θ_L), then the apparent volume can be substantially decreased due to contributions from the more rapid motions (Section 10.9). One should not ignore the information available by comparison of the extrapolated and frozen solution values of r_0. If r_0^{app} is much smaller than r_0, one may infer the existence of segmental motions of the probe on the macromolecule, independent motions of domains of the protein, or possibly RET in multiply labeled proteins. Another phenomenon seen in Perrin plots is a more rapid increase in $1/r$ at high temperatures than at lower temperatures. Such an increase is usually caused by denaturation of the macromolecule, resulting in an increase in independent motion of the probe and a decrease in the anisotropy.

In the older literature, one often encounters a different version of the Perrin equation:

$$\left(\frac{1}{P}-\frac{1}{3}\right)=\left(\frac{1}{P_0}-\frac{1}{3}\right)\left(1+\frac{\tau RT}{\eta V}\right) \quad [10.55]$$

$$=\left(\frac{1}{P_0}-\frac{1}{3}\right)\left(1+\frac{3\tau}{\rho}\right) \quad [10.56]$$

This equation is equivalent to Eq. [10.53], except for the use of polarization in place of anisotropy. When using this equation, one plots $(1/P-\frac{1}{3})$ versus T/η to obtain the molecular volume (Figure 10.13, lower panel). The intercept yields an apparent value of $(1/P_0^{app}-\frac{1}{3})$, which can be larger than the true value $(1/P_0-\frac{1}{3})$ if there is segmental motion of the probe. The term ρ is the rotational relaxation time ($\rho = 3\theta$). At present, the use of anisotropy, the rotational correlation time, and Eq. [10.53] is preferred.

10.5.B. Examples of Perrin Plots

It is instructive to examine some representative Perrin plots. Figure 10.14 shows Perrin plots for 9-AS in a viscous paraffin oil, Primol 342.[12] The anisotropies were measured at various excitation wavelengths, corresponding to different r_0 values (Figure 10.7). The y-axis intercepts of the corresponding Perrin plots are different because of the different r_0 values. Also, the slopes are larger for shorter excitation wavelengths because r_0, which has a smaller value at these wavelengths, appears in the denominator of Eq. [10.53]. In spite of the very different values of $1/r$ for

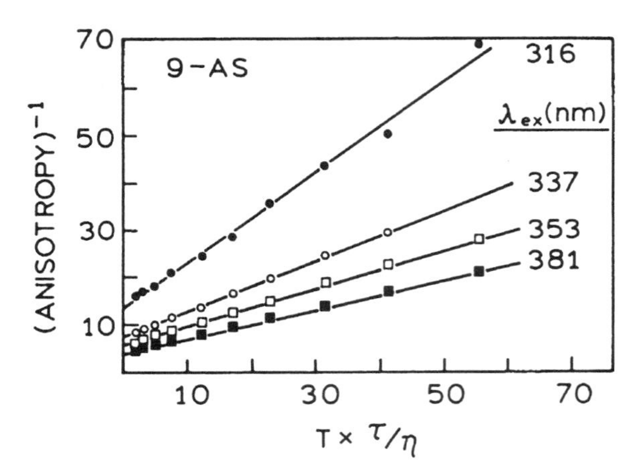

Figure 10.14. Perrin plots of 9-anthroxyloxy stearic acid in Primol 342. The excitation wavelengths were 316 (●), 337 (○), 353 (□) and 381 nm (■). The lifetime, ranging from 10.5 to 9.7 ns, is incorporated into the x-axis. Revised and reprinted with permission from Ref. 12, Copyright © 1982, American Chemical Society.

different excitation wavelengths, the data are all consistent with a correlation time near 15 ns at 25 °C (Table 10.5). For a spherical molecule, the correlation time is independent of r_0. In Chapter 12 we will see that the correlation times can depend on the value of r_0 for nonspherical molecules, an effect which is present to a minor extent for 9-AS.

The y-axis intercepts yield the apparent values of r_0 extrapolated to high viscosity, which can be compared with measured values in a glassy solvent (Table 10.5). One notices that the extrapolated values are about 25% lower than the measured r_0 values. This is a typical result for Perrin plots. Fluorophores are often nonspherical, resulting in multiexponential anisotropy decays. In such cases, the Perrin plots are curved toward the x-axis, resulting in higher apparent intercepts on the y-axis.

Table 10.5. Fundamental Anisotropy (r_0) and Extrapolated Fundamental Anisotropy (r_0^{app}) Values and Rotational Correlation Times for 9-(9-Anthroxyloxy)stearic Acid[a]

Excitation wavelength (nm)	r_0	r_0^{app}	θ (ns)[b]
316	0.090	0.075	13
337	0.175	0.136	14
353	0.231	0.179	14
381	0.323	0.242	18

[a]From Ref. 12.
[b]In Primol 342 at 6 °C.

10.6. PERRIN PLOTS OF PROTEINS

10.6.A. Binding of tRNA to tRNA Synthetase

Perrin plots of labeled macromolecules have been extensively used to determine the apparent hydrodynamic volumes. One example is a study of methionyl-tRNA ($tRNA^{fMet}$) with methionyl-tRNA synthetase (met RS). The 3′-end of the tRNA was labeled with fluorescein (Fl) by periodate oxidation of the tRNA followed by reaction with fluorescein thiosemicarbazide.[34] In this case the Perrin plots were obtained at a single temperature with the viscosity changed by addition of sucrose. As expected, fluorescein free in solution displays a subnanosecond correlation time (Figure 10.15). More surprising is the apparent correlation time for 3′-Fl-tRNA, which is 1.7 ns. Based on the size and shape of tRNA, the rotational correlation time is expected to be near 25 ns. This result indicates that the 3′-Fl label displays significant segmental freedom independent of overall rotational diffusion of the tRNA.

Upon binding of 3′-Fl-tRNA to the synthetase, the anisotropy increases dramatically from 0.062 to 0.197. Also, the small slope of the Perrin plot indicates that the correlation time is now larger than 30 ns. The synthetase consists of two identical subunits, 76 kDa each, so that its rotational correlation time is expected to be 100 ns or longer. Apparently, the 3′-end of the tRNA interacts with the synthetase, immobilizing the fluorescein residue.

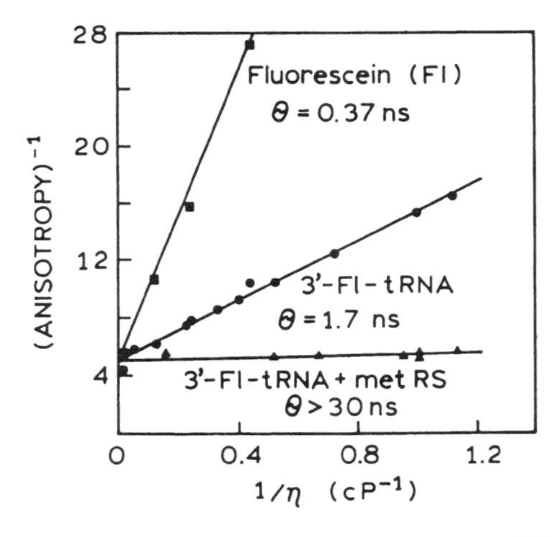

Figure 10.15. Perrin plots for fluorescein (Fl, ■) and $tRNA^{fMet}$ labeled at the 3′-end with fluorescein (3′-Fl-tRNA, ●). Also shown is the plot for the labeled t-RNA when bound to methionyl-tRNA synthetase (met RS, ▲). The experiments were performed at 20 °C. The viscosity η was varied by adding sucrose. Excitation was at 480 nm, and emission was measured at 520 nm. Revised and reprinted, with permission, from Ref. 34, Copyright © 1986, American Chemical Society.

The y-intercepts of the Perrin plots yield the apparent r_0 values, which are near 0.2 for these three fluorescein samples. This value is less than that typical of fluorescein, which is near 0.35. Hence, there appears to be some unresolved motion of the fluorescein even when bound to the synthetase. It is difficult to determine the y-intercept for fluorescein itself, due to the large extent of depolarization.

10.6.B. Molecular Chaperonin cpm 60 (GroEL)

Molecular chaperonins are proteins that assist in protein folding.[35,36] The cpm 60 chaperonin (GroEL) from *E. coli* is a large multisubunit oligomer consisting of 14 identical 60-kDa subunits. These subunits are arranged in two stacked rings, each with 7 subunits. Each subunit consists of three domains. The entire protein of 840 kDa is a cylindrical oligomer, 146 Å tall, 137 Å wide, with a central channel 45 Å in diameter.[37]

The hydrodynamics of cpm 60 were studied using the pyrene-labeled protein.[38] Pyrene was chosen for its long fluorescence lifetime, which can reach 200 ns. In this case the pyrenesulfonyl label displayed a lifetime near 45 ns. The Perrin plot for pyrene-labeled cpm 60 shows significant curvature (Figure 10.16). In this case the data were presented in the older style, using polarization instead of anisotropy. The chaperonin displays considerable flexibility, as can be seen from the dramatic curvature in the Perrin plot. Different apparent correlation times can be obtained for different regions of the curve. At lower viscosities, the

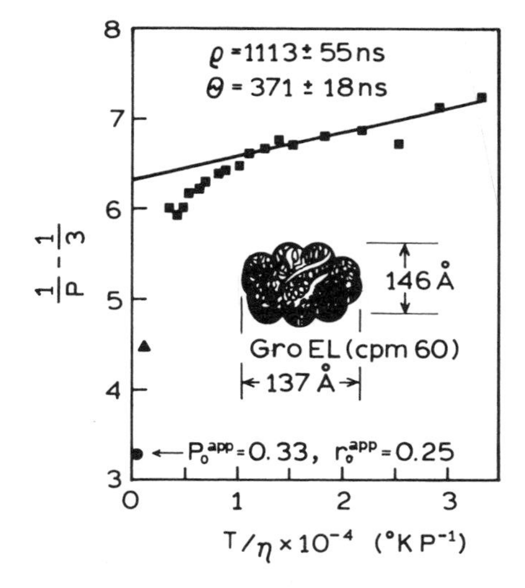

Figure 10.16. Perrin plot for chaperonin cpm 60 labeled with 1-pyrenesulfonyl chloride at 25 °C. Viscosities were varied by adding sucrose. Two points (● and ▲) were measured at 0 °C. The pyrene lifetime was near 45 ns. Revised from Ref. 38.

faster motions are complete, so that the overall rotational correlation time can be estimated from the limiting slope. This calculation results in an apparent correlation time near 370 ns. Calculation of the correlation time using Eq. [10.52], with h = 0.23 g of H_2O/g of protein, yields a correlation time near 347 ns. Hence, the slope from the Perrin plot observed for the lower viscosities provided a reasonable estimate of the overall size. The data at higher viscosity or lower temperature indicate that the domains or subunits of cpm 60 have significant independent mobility.

10.6.C. Perrin Plots of an F_{ab} Immunoglobulin Fragment

A potentially confusing aspect of the Perrin plots is the use of T/η as the x-axis. This implies that the same result should be obtained independent of whether temperature or viscosity is varied. Unfortunately, this is not usually the case. In fact, it is rather common to observe different Perrin plots for variation in temperature as compared to variation in viscosity. This is illustrated in Figure 10.17 by the Perrin plots of F_{ab} fragments labeled with IAEDANS, which show segmental motions that depend on temperature.[39] F_{ab} fragments are the antibody-binding domains of immunoglobulins. Different Perrin plots are observed at 10 and 35 °C (Figure 10.17). At the higher temperature, there is more independent motion of the probe and/or domains of the F_{ab} fragment. Hence, the y-intercepts are different at the two temperatures. Perrin plots which depend differently on temperature and viscosity should not be regarded as incorrect, but rather as reflecting the temperature-dependent dynamics of the protein.

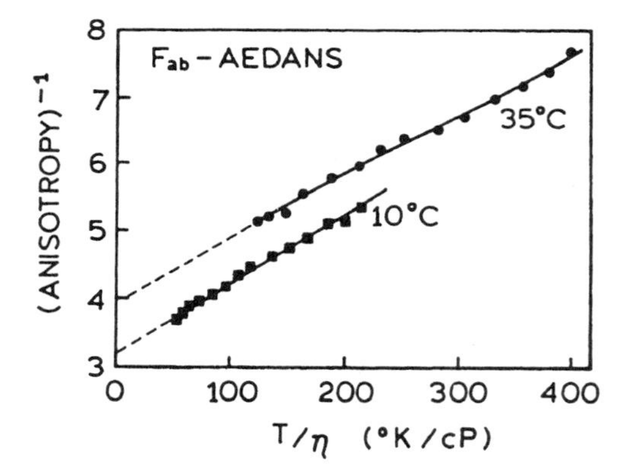

Figure 10.17. Perrin plots for immunoglobulin F_{ab} fragments labeled with IAEDANS at the C-terminus. The viscosity was varied by adding sucrose. Revised and reprinted, with permission, from Ref. 39, Copyright © 1995, American Chemical Society.

10.7. PROTEIN ASSOCIATION REACTIONS

Anisotropy measurements are ideally suited for measuring the association of proteins with other macromolecules. This is because the anisotropy almost always changes in response to a change in correlation time. Also, the experiments are simplified by the independence of the anisotropy from the overall protein concentration.

10.7.A. Peptide Binding to Calmodulin

The use of anisotropy to study protein binding is illustrated by studies of calmodulin. This protein activates a number of intracellular enzymes in response to calcium. One example is myosin light-chain kinase (MLCK).[40] The amino acid sequence which binds to calmodulin contains a single tryptophan residue (Figure 10.18). Since calmodulin contains only tyrosine, the peptide (RS20) can be selectively observed by excitation at 295 nm (Chapter 16). Upon addition of calmodulin, the emission intensity of RS20 increases, and the emission shifts to shorter wavelengths, reflecting a shielded environment for the tryptophan residue. The anisotropy of RS20 increases dramatically on addition of calmodulin (Figure 10.19). These data were used to determine that the stoichiometry of binding is 1:1. The sharp nature of the transition at $10^{-8}M$ calmodulin implies that the binding constant is less than $10^{-9}M$.

How can such data be used to calculate the fraction of the peptide which is free in solution (f_F) and the fraction bound to calmodulin (f_B)? The additivity law for anisotropies (Eq. [10.6]) is appropriate when the f_i are the fractional intensities, not the fractional populations. Near 340 nm there is no change in intensity upon binding. Under

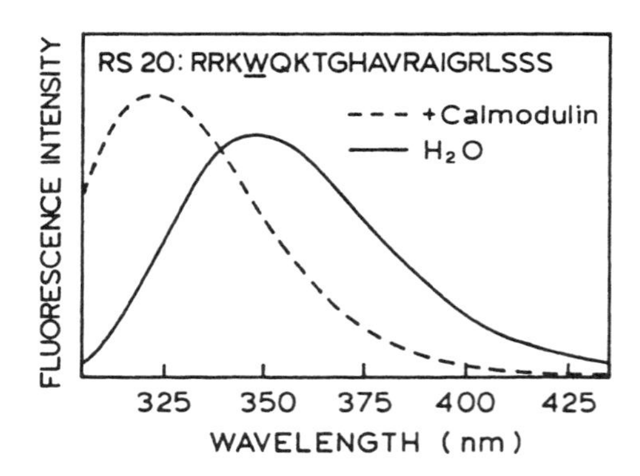

Figure 10.18. Emission spectra of the MLCK peptide RS20 in solution (——) and bound to calmodulin in the presence of calcium (— — —). Excitation at 295 nm. Revised and reprinted, with permission, from Ref. 40, Copyright © 1986, American Chemical Society.

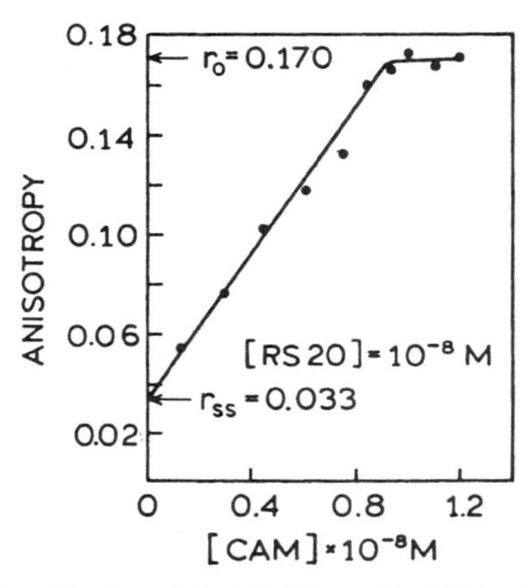

Figure 10.19. Titration of the MLCK peptide RS20 with calmodulin (CAM). Revised and reprinted, with permission, from Ref. 40, Copyright © 1986, American Chemical Society.

these conditions, the fraction of RS20 bound can be calculated using[41]

$$f_B = \frac{r - r_F}{r_B - r_F} \quad [10.57]$$

where r is the measured anisotropy, and r_F and r_B are the anisotropies of the free and bound peptides, respectively. However, at different observation wavelengths the measured anisotropy can be weighted toward the free or the bound form. For instance, at 320 nm the bound peptide contributes more strongly to the measured anisotropy. Under these conditions, one can show that the fraction bound is given by

$$f_B = \frac{r - r_F}{(r - r_F) + R(r_B - r)} \quad [10.58]$$

where $R = F_B/F_F$ is the ratio of intensities of the free and bound forms. Once f_B and f_F are known, the binding constant can be calculated.

10.7.B. Binding of the Trp Repressor to DNA

Another example of anisotropy measurements concerns studies of the binding of the tryptophan repressor protein to DNA.[42] This protein binds to the DNA sequence shown in Figure 10.20. This double-stranded sequence was labeled with fluorescein on one of its 5′-ends. Upon addition of the repressor protein, the fluorescein anisotropy increases due to the decreased rotational rate of the DNA 25-mer when bound to the repressor. The concentration of repressor needed for binding was strongly dependent on the concentration of tryptophan in solution. For instance, the binding to DNA was much stronger in the presence of tryptophan. This is consistent with the known function of

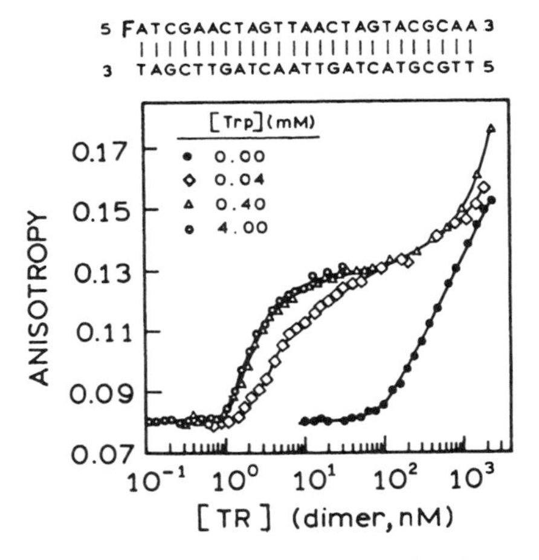

Figure 10.20. Anisotropy of a fluorescein-labeled repressor DNA sequence upon titration with the Trp repressor protein (TR) in solutions containing different concentrations of tryptophan (◇, △, ○) or no tryptophan (●). Tryptophan concentrations are indicated on the figure. The DNA oligomer labeled with fluorescein (F) is shown at the top of the figure. Revised from Ref. 42.

the repressor, which is to turn off the genes responsible for tryptophan synthesis when tryptophan levels are adequate.

The titration curves shown in Figure 10.20 can be understood in terms of a model of Trp repressor binding to DNA (Figure 10.21). Binding of tryptophan to the repressor increases the repressor's affinity for DNA. This model also explains another feature of the titration curves, which is the further increase in anisotropy at higher repressor concentrations (Figure 10.20). Apparently, the DNA 25-mer can bind more than a single repressor dimer, and this additional binding occurs at higher repressor concentrations.

10.7.C. Melittin Association Detected from Homotransfer

In the preceding examples the binding reaction was detected from the increase in anisotropy resulting from the larger correlation time. One can also use the phenomenon of homotransfer to detect binding interactions. This is exemplified by melittin, which self-associates at high salt concentrations (Section 16.6.A). Melittin was randomly labeled with fluorescein isothiocyanate.[43] When the solution contained a small fraction of labeled melittin (1 mole-

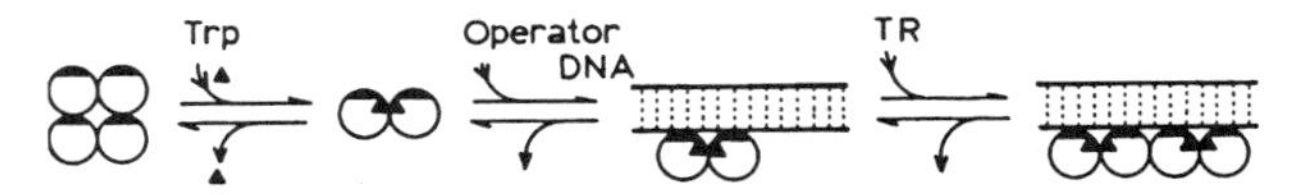

Figure 10.21. Binding of the Trp repressor (TR) to DNA. Revised from Ref. 42.

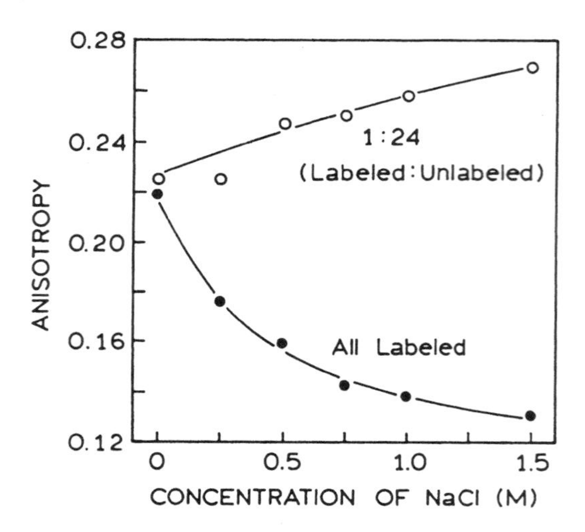

Figure 10.22. Self-association of melittin labeled with fluorescein in a 1:24 mixture with unlabeled melittin (○) and of melittin all labeled with fluorescein (•). Revised and reprinted, with permission, from Ref. 43, Copyright © 1995, Biophysical Society.

cule in 24), the anisotropy increased at 1*M* NaCl (Figure 10.22). When all the melittin molecules were labeled, their anisotropy decreased markedly at high salt concentrations. This decrease is unexpected because melittin forms a tetramer in high-salt solutions, so its correlation time should increase fourfold.

The decrease in anisotropy can be understood from the spectral properties of fluorescein (Figure 3.7). The small Stokes' shift and large overlap result in a Förster distance (R_0) of 53 Å for homotransfer. Hence, the anisotropy decrease in Figure 10.22 is due to homotransfer between fluoresceins in the melittin tetramer.

10.8. ANISOTROPY OF MEMBRANE-BOUND PROBES

Fluorescence anisotropy measurements have been used extensively to study biological membranes. These studies have their origin in the early studies of microviscosity of micelles and membranes.[44–46] The basic idea was to measure the anisotropy of a fluorophore in a reference solvent of known viscosity and then in the membrane. The microviscosity of the membrane was then estimated by comparison with the viscosity calibration curve.

At present, there is less use of the term "microviscosity" because it has been shown that fluorophores in membranes display complex anisotropy decays, which prevents a straightforward comparison of the solvent and membrane data (Chapter 11). However, steady-state and time-resolved anisotropies are still widely used because of the dramatic changes which occur at the membrane phase-

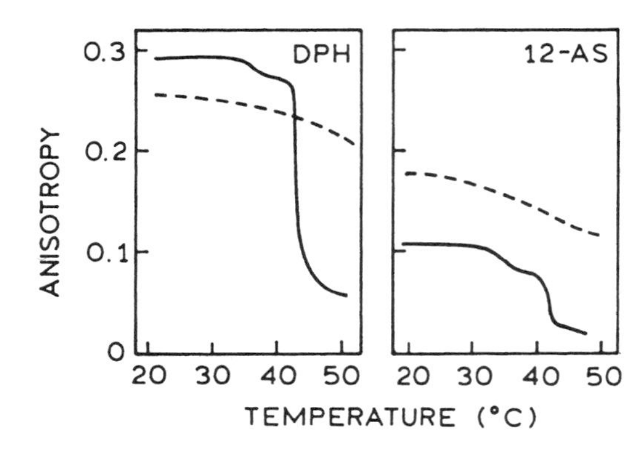

Figure 10.23. Temperature-dependent polarization of DPH and of 12-AS in DPPC vesicles (———) and in DPPC vesicles with 45 mol % cholesterol (— — —). Revised and reprinted from Ref. 47, Copyright © 1982, with permission from Elsevier Science.

transition temperature. This is illustrated by the steady-state polarization of 12-AS and DPH in DPPC vesicles (Figure 10.23). For both 12-AS and DPH, the polarization decreases dramatically at 37 °C, which is where DPPC undergoes its phase transition. The change is more dramatic for DPH than for 12-AS, which is one reason why DPH is so widely used. The polarization values are sensitive to the presence of cholesterol, which tends to make the membranes more rigid and to eliminate the sharp phase transition. In the following chapters we will see that the anisotropy values do not accurately reflect the rate of probe rotation in the membranes, but rather reveal the extent to which the probe motions are restricted by the anisotropic membrane environment.

10.9. LIFETIME-RESOLVED ANISOTROPIES

• Advanced Topic •

In Sections 10.5 and 10.6 we saw how the Perrin equation could be used to estimate the apparent volume of a macromolecule. Examination of Eq. [10.51] suggests an alternative method of estimating the rotational correlation time. Suppose that the lifetime (τ) of the probe could be decreased by collisional quenching. Then a plot of $1/r$ versus τ would have a slope of $(r_0\theta)^{-1}$ and thus would allow measurement of the correlation time. These measurements are called lifetime-resolved anisotropies.[48–52] The first suggestion of lifetime-resolved anisotropies appeared early in the literature as a means to study polymers[53,54] as well as proteins.[55]

There are significant advantages to the use of lifetime-resolved measurements. The lifetime can typically be decreased with only a modest change in solution conditions.

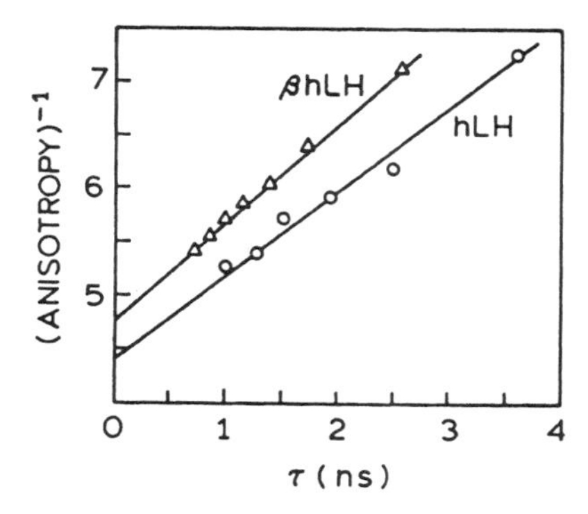

Figure 10.24. Lifetime-resolved anisotropies of human luteinizing hormone (hLH) and its β subunit (βhLH). The lifetimes were varied by oxygen quenching and calculated from the dynamic portion of the observed quenching. Revised and reprinted, with permission, from Ref. 56, Copyright © 1987, American Chemical Society.

This is particularly true for oxygen quenching, since oxygen diffuses rapidly and is an efficient quencher. The use of oxygen quenching is demonstrated in Figure 10.24, which shows the anisotropy values of a peptide hormone when the single tryptophan residue was quenched to various lifetimes.[56] The proteins were human luteinizing hormone (hLH) and its β subunit (βhLH). The intact hormone has a molecular weight of 28,000, and the β-subunit, is 14,000. The apparent correlation times were 6.0 and 4.9 ns, respectively. The calculated correlation times are 10.1 and 5.2 ns, respectively. While the measured and calculated values are reasonably close, the y-intercepts yield apparent r_0 values smaller than the expected value of 0.27. This result is typical of other reports which suggest segmental freedom of tryptophan residues in many proteins.[48–51]

10.9.A. Effect of Segmental Motion on the Perrin Plots

The effect of segmental motion on the Perrin plots can be seen by deriving the Perrin equation for the anisotropy decay expected in the presence of segmental motions. While the results are the same whether one measures anisotropy versus T/η or versus lifetime, these effects are somewhat easier to understand in terms of the lifetime-resolved measurements. In addition, the experiments are easier to interpret because the temperature and solution conditions are not changed, so that changes in protein dynamics and conformation do not complicate the analysis.

Suppose that a fraction α of the total anisotropy is lost due to the segmental motion, with a fast correlation time θ_F, and the remainder decays by overall rotational diffusion of the protein (θ_P). For independent segmental motions and rotational diffusion, the anisotropy is given as the product of the depolarization factors (Section 10.10), and

$$r(t) = r_0[\alpha \exp(-t/\theta_F) + (1-\alpha)] \exp(-t/\theta_P) \quad [10.59]$$

where $f_F + f_P = 1.0$. Substitution of Eq. [10.59] into Eq. [10.49] yields

$$r(\tau) = \frac{\alpha r_0}{1 + \left(\frac{1}{\theta_F} + \frac{1}{\theta_P}\right)\tau} + \frac{(1-\alpha) r_0}{1 + \frac{\tau}{\theta_P}} \quad [10.60]$$

In this expression we have used the notation $r(\tau)$ as a reminder that the steady-state anisotropy depends on the lifetime τ. In many cases, the internal motions are more rapid than overall rotation ($\theta_F << \theta_P$), simplifying Eq. [10.60] to

$$r(\tau) = \frac{\alpha r_0}{1 + \frac{\tau}{\theta_F}} + \frac{r_0(1-\alpha)}{1 + \frac{\tau}{\theta_P}} \quad [10.61]$$

This expression indicates that a fraction of the anisotropy (αr_0) is lost due to the rapid motion, and the remainder $r_0(1-\alpha)$ is lost due to overall rotational diffusion.

The effects of varying degrees of segmental motion are shown in Figure 10.25. The assumed correlation times were $\theta_P = 20$ ns for overall rotational diffusion and $\phi_F = 1.0$ or 0.1 ns for the rapid motion. As the amplitude of the fast motion increases, the y-intercept increases. This shows how segmental motions in proteins result in low apparent r_0 values. Perhaps more important is the observation that the apparent correlation times (θ_A), calculated from the limiting slope of the lifetime Perrin plots, decrease as α increases. This illustrates how segmental freedom in a

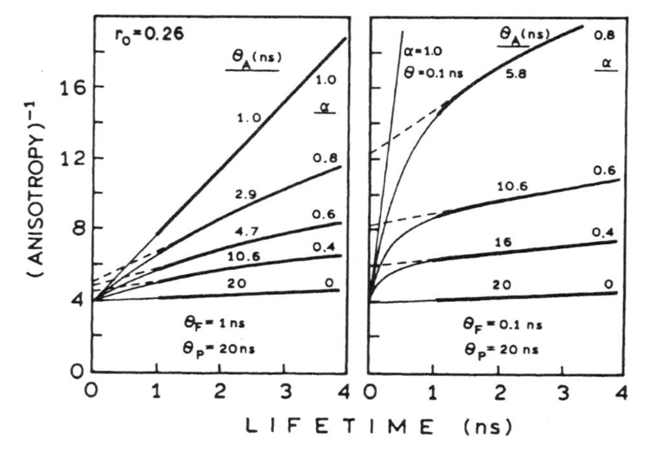

Figure 10.25. Simulated lifetime Perrin plots for a protein with an overall correlation time of $\theta_P = 20$ ns and a fast segmental motion with $\theta_F = 1.0$ (*left*) or 0.1 ns (*right*). The α values represent the fraction of the total anisotropy that is lost due to the segmental motion, and the θ_A values are the apparent correlation times. Revised and reprinted, with permission, from Ref. 48, Copyright © 1983, American Chemical Society.

protein can result in apparent correlation times that are lower than the true values. While these simulated data are for lifetime-resolved measurements, similar effects can be seen in simulated T/η Perrin plots.

10.10. SOLEILLET'S RULE—MULTIPLICATION OF DEPOLARIZATION FACTORS • Advanced Topic •

At the beginning of this chapter, we derived the value of r_0 expected due to photoselection ($r_0 = 0.4$), showed how r_0 was decreased by a factor dependent on the angle β between the absorption and emission transition moments (Eq. [10.20]), and demonstrated that the anisotropy was decreased further due to rotational diffusion (Eq. [10.50]). There is a simple relationship between these different causes of depolarization, which is known as Soleillet's rule.[57–59] The anisotropy is given by the product of the various depolarization factors which occur in a given sample,

$$r = r_0 \prod_i d_i \quad [10.62]$$

where the d_i represent the extent to which the anisotropy is decreased by each depolarization factor. This relationship can be clarified by a specific example. Suppose that a fluorophore has an anisotropy less than 0.4 due to a displacement of the transition moment by an angle β and due to rotational diffusion. The steady-state anisotropy can be written as

$$r = d_1\, d_2\, d_3 = \frac{2}{5}\left(\frac{3\langle\cos^2\beta\rangle - 1}{2}\right)\left(\frac{3\langle\cos^2\omega\rangle - 1}{2}\right) \quad [10.63]$$

In this expression the first depolarization factor (d_1) is 0.4, which accounts for excitation photoselection, $\langle\cos^2\theta\rangle$ (Eq. [10.17]). The second term (d_2) accounts for the angle between the transition moments, $\langle\cos^2\beta\rangle$ (Eq. [10.20]). The third term (d_3) account for the average angular (ω) displacement of the fluorophore during the excited-state lifetime, $\langle\cos^2\omega\rangle$. This factor is also given by $d_3 = [1 + (\tau/\theta)]^{-1}$ (Eq. [10.50]). It is sometimes useful to be aware of Soleillet's rule when attempting to account for the overall loss of anisotropy. The extensive review by Kawski[58] summarizes many advanced concepts about fluorescence anisotropy.

10.11. ANISOTROPIES CAN DEPEND ON EMISSION WAVELENGTH • Advanced Topic •

The anisotropy is generally independent of the emission wavelength. However, the presence of time-dependent spectral relaxation (Chapter 7) can result in a substantial decrease in anisotropy across the emission spectrum.[60–62] A biochemical example is shown in Figure 10.26 for egg PC vesicles labeled with 12-AS. Because of time-dependent reorientation of the local environment around the excited state of 12-AS, the emission spectra display a time-dependent shift to longer wavelengths. While such relaxation is often analyzed in terms of the TRES, one can also determine the mean lifetime at various emission wavelengths (Figure 10.26). The mean lifetime increases with wavelength because the lifetime at short wavelengths is decreased by relaxation, and long-wavelength observation selects for the relaxed species.

Recall that the steady-state anisotropy is determined by $r(t)$ averaged over $I(t)$ (Eq. [10.49]), resulting in the Perrin equation (Eq. [10.50]). Because of the longer average lifetime at long wavelengths, the anisotropy decreases with increasing wavelength (Figure 10.26). If the fluorophores

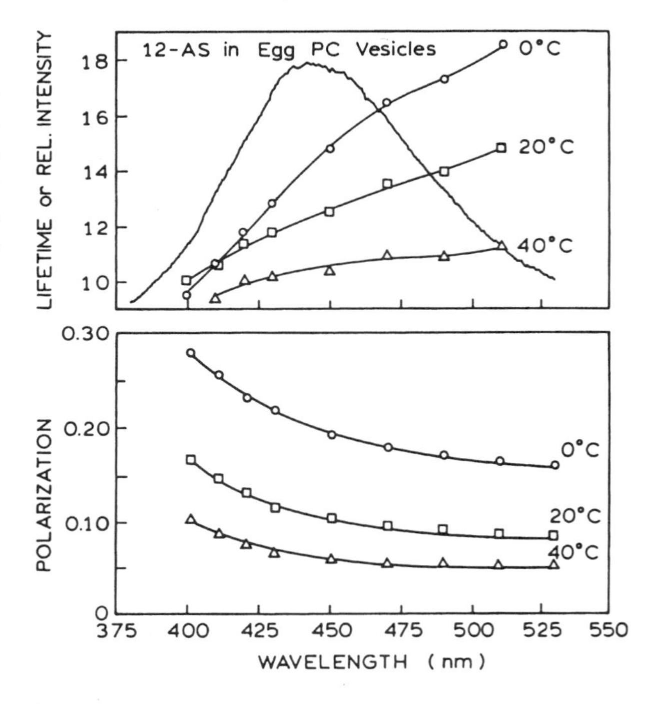

Figure 10.26. Average lifetime of 12-AS in egg PC vesicles at 0, 20, and 40 °C (*top*) and the emission anisotropy (*bottom*) as a function of emission wavelength. The emission spectrum is shown in the upper panel. The average lifetimes were determined from the phase shift at 30 MHz. Revised from Ref. 63.

were in a very fluid or a glassy solvent, the lifetime and anisotropy would be constant across the emission spectrum.

10.12. TRANSITION MOMENTS • Advanced Topic •

Throughout this chapter, we have described the dependence of anisotropy on the direction of the transition moments. The existence of a discrete direction often seems somewhat mysterious because the experiments in isotropic solution do not reveal this direction in the molecule. The direction of the transition moment can be determined if the fluorophores can be oriented. Molecules can be oriented in crystals, liquid crystals, oriented membranes, or stretched polymer films. Stretched films provide the easiest approach to obtaining oriented fluorophores, particularly if the molecules are elongated along one axis, and if uniaxial orientation is adequate.[64–67] The basic procedure is first to dissolve the fluorophore and the polymer in a solvent. The polymer is typically poly(vinyl alcohol) (PVA). After removal of the solvent, the polymer is then stretched about fivefold. For linear molecules like DPH and DAPI, one can obtain nearly complete orientation along the stretching axis. Less asymmetric fluorophores are still aligned, but the orientation function can be more complex.[8]

An example of an oriented system is shown in Figure 10.27 for DPH in a stretched PVA film.[68] Prior to stretching of the film, the absorption spectra are independent of the orientation of the polarizer. After stretching fivefold, the sample absorbs strongly when the polarizer is oriented along the stretching axis, and the absorption is weak when the polarizer is perpendicular to this axis. Since the long axis of DPH orients parallel to the direction of stretching, these spectra provide an experimental demonstration that the transition moment of DPH is oriented along its long axis.

Recall from the theory of anisotropy (Section 10.2) that the highest anisotropy of 0.4 was a consequence of excitation photoselection and that a single fluorophore oriented along the z-axis was predicted to have an anisotropy of 1.0. This prediction can be confirmed from experiments with stretched films (Figure 10.28). As the sample is stretched, the anisotropy of DPH increases well above 0.4, to about 0.82. The value is not 1.0 because the sample is not perfectly aligned, and the value of β for DPH obtained from $\langle \cos^2 \beta \rangle$ seems to be about 8°. Such data have also been used to confirm the cosine-squared dependence of absorption.[69]

Perylene displays regions of high and low anisotropy (Figure 10.29). At wavelengths above 360 nm ($S_0 \rightarrow S_1$), the anisotropy is positive and nearly constant. At wavelengths below 280 nm ($S_0 \rightarrow S_2$), the anisotropy is negative. According to Eq. [10.20], an anisotropy of −0.20 corresponds to an angle of 90° between the absorption and emission moments. This was demonstrated experimentally using the polarized absorption spectra of perylene (Figure 10.30).[8] In this case the spectra are calculated for light polarized parallel or perpendicular to the long axis of perylene. These spectra are calculated from the stretched-film spectra. The long-wavelength absorption ($S_0 \rightarrow S_1$) is thus seen to be oriented along the long axis, and the short-wavelength absorption is oriented along the short

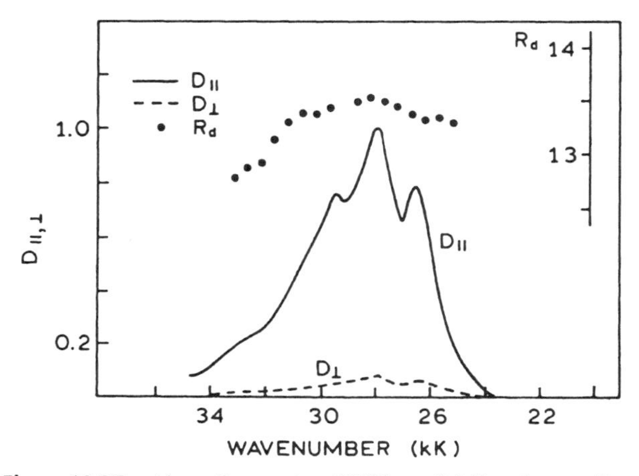

Figure 10.27. Absorption spectra of DPH parallel (∥) and perpendicular (⊥) to the direction of stretching in a PVA film. 1 kK = 1000 cm^{-1}. $R_d = D_\parallel/D_\perp$ is the dichroism ratio. Revised from Ref. 68.

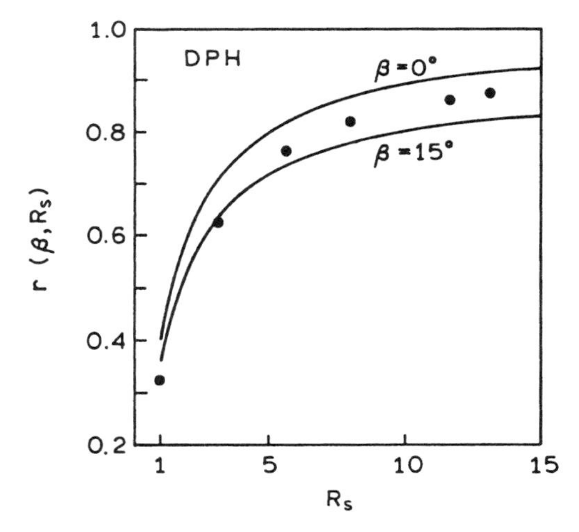

Figure 10.28. Steady-state anisotropy of DPH in stretched PVA films as a function of the stretching ratio (R_s). The solid curves are theoretical curves for β = 0 and 15°. Revised from Ref. 68.

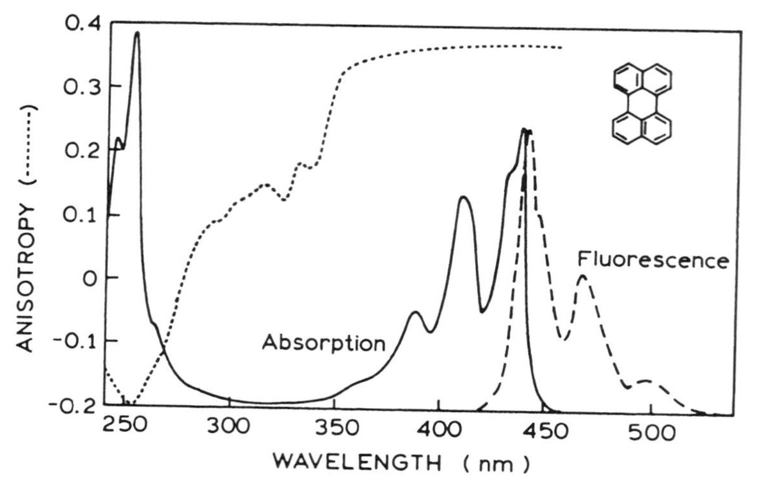

Figure 10.29. Absorption and excitation anisotropy spectra of perylene in propylene glycol.

axis. We will see in the next chapter how the orientation of the transition moments affects the correlation times of nonspherical fluorophores.

As a final example of transition moments, Figure 10.31 shows the polarized absorption spectra of 9-aminoacridinium, again recovered from the experimental spectra for PVA films. In 9-aminoacridinium, the long-wavelength transition displays positive anisotropy, which one tends to associate with transitions along the long axis. However, the long-wavelength ($S_0 \rightarrow S_1$) transition is polarized along the short axis of the molecule. This result is counterintuitive because we generally assign the lowest-energy transition to the longest axis. This result can be understood in terms of movement of the electronic charge between the two nitrogen atoms, each with a different electron density. Absorption spectra in stretched films have been used to determine the direction of transition moments in a number of fluorophores of biochemical interest, including adenine,[71] 2-aminopurine,[72] Y_t-base,[73] and Hoechst 33342.[69]

10.12.A. Anisotropy of Planar Fluorophores with High Symmetry

The fundamental anisotropy can be determined in part by the symmetry of the molecule. This is illustrated by triphenylene, which has threefold symmetry and for which $r_0 = 0.1$ (Figure 10.32).[74] This anisotropy value is found whenever the transition moments are randomized in a plane. The anisotropy value of 0.1 can be calculated from the additivity law of anisotropy (Eq. [10.6]). Following excitation along one of the axes, the emission is randomized among the three identical axes. Since the axes are identical, one-third of the emission originates from each transition, resulting in $r_0 = 0.10$ (Problem 10.8).

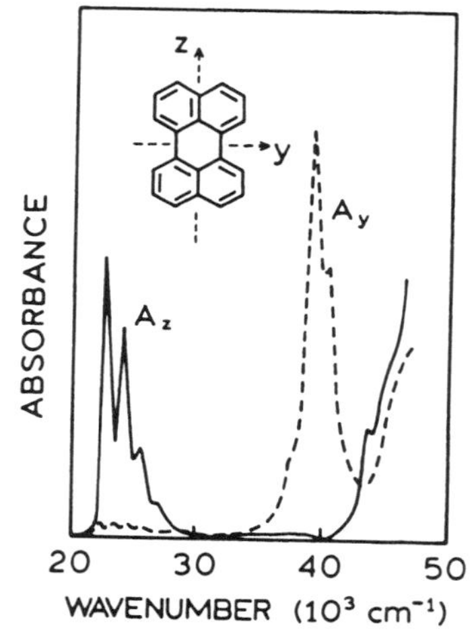

Figure 10.30. Polarized absorption spectra of perylene along the long (z) and short (y) axis. Revised from Ref. 8.

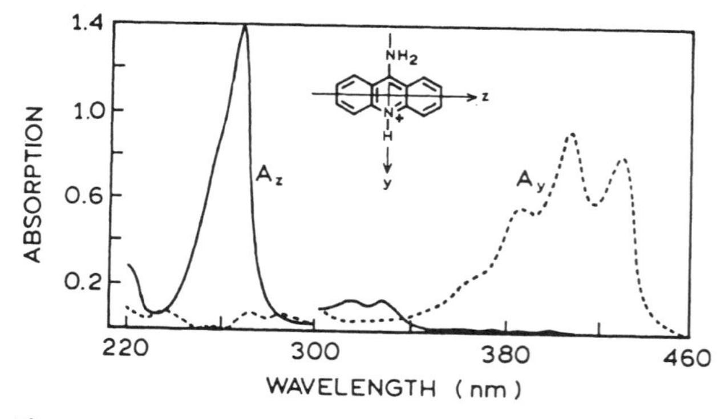

Figure 10.31. Polarized absorption spectra of 9-aminoacridinium in a stretched PVA film. Revised from Ref. 70.

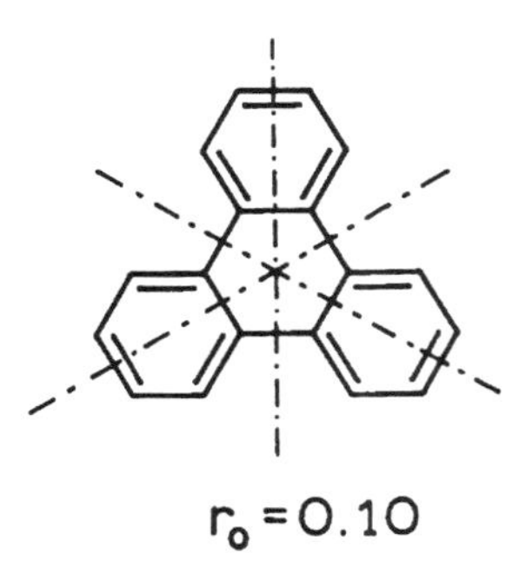

Figure 10.32. Structure and symmetry axes of triphenylene.

10.13. ANISOTROPIES WITH MULTIPHOTON EXCITATION • Advanced Topic •

In Section 2.14 we introduced the topic of two-photon and multiphoton excitation. Two-photon excitation occurs when a fluorophore simultaneously interacts with two longer-wavelength photons, exciting the fluorophore to the first singlet state. Since the fluorophores emit from the same excited state, independent of the mode of excitation, the same emission spectra have been observed for one-, two-, or three-photon excitation.[75] However, the anisotropies can be very different for two-photon as compared to one-photon excitation.[76–78] There are two reasons for the different anisotropies. The first reason is because the excitation photoselection is different depending on the mode of excitation. More specifically, two-photon excitation results in a more strongly aligned population because this process depends on $\cos^4 \theta$ photoselection, rather than $\cos^2 \theta$ photoselection for one-photon excitation. Secondly, the anisotropies can be different because the r_0 values are fundamentally different for multi-photon as compared to one-photon excitation.

10.13.A. Excitation Photoselection for Two-Photon Excitation

As shown in Eq. [10.17], the anisotropy is essentially the average value of $\cos^2 \theta$. The fundamental anisotropy value of 0.4 for one-photon excitation is a consequence of $\cos^2 \theta$ photoselection (Eq. [10.18]). For two-photon excitation, the fluorophore interacts simultaneously with two photons, and each interaction is proportional to $\cos^2 \theta$. Hence, the photoselection function becomes[79]

$$f_2(\theta) = \cos^4 \theta \sin \theta \, d\theta \qquad [10.64]$$

Introduction of this function into the calculation of $\langle \cos^2 \theta \rangle$ (Eq. [10.19]) allows calculation of the anisotropies

Table 10.6. Fundamental Anisotropies for One-, Two-, and Three-Photon Excitation[a]

β (deg)	One-photon r_{01}	Two-photon r_{02}	Three-photon r_{03}
0	0.40	0.57	0.66
4.5	0.10	0.41	0.17
54.7	0.00	0.00	0.00
90	–0.20	–0.29	–0.3

[a]Data from Ref. 79.

expected for collinear transitions (Table 10.6). For $\beta = 0$, the fundamental anisotropy for two-photon excitation is 0.57, rather than 0.4. For three-photon excitation, the fundamental anisotropy can be as large as 0.66 (Figure 10.33).

It is important to recognize the meaning of these anisotropy values. A value of 0.4 for one-photon excitation and a value of 0.57 for two-photon excitation both mean that the absorption and emission transition moments are parallel. There is no new information in the higher anisotropy value for two-photon excitation, except for confirming that the transition moments are still parallel. In some cases the anisotropies do not follow the predictions based on $\cos^2 \theta$, $\cos^4 \theta$ photoselection. For example, tryptophan and indole display lower anisotropies with two-photon excitation (Section 16.10).

While the preceding description of the multiphoton anisotropy seems intuitively direct, we emphasize that it is an oversimplification of the actual situation. An accurate description of the theory requires a description of the transition moments and excited-state distributions in terms of tensors.[76–78]

10.13.B. Two-Photon Anisotropy of DPH

It is instructive to see how the anisotropy depends on the mode of excitation. Figure 10.34 shows the excitation anisotropy spectra of DPH. For one-photon excitation

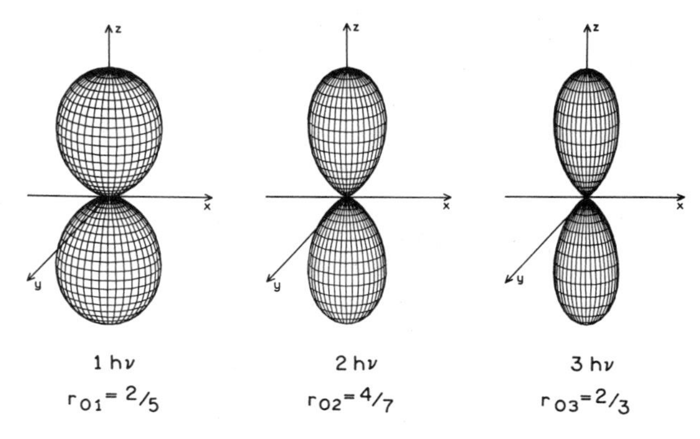

Figure 10.33. Excited-state distributions for $r_0 = 0.40$ with one-, two-, and three-photon excitation.

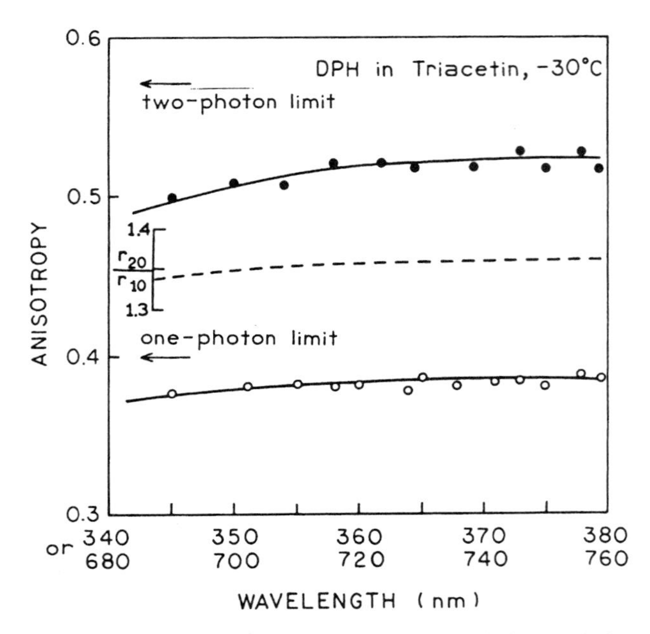

Figure 10.34. Excitation anisotropy spectra of DPH in frozen solution for one-photon and two-photon excitation. From Ref. 80.

(340–380 nm), the anisotropy in frozen solution is always below the one-photon limit of 0.40. For two-photon excitation, the anisotropy is near 0.5, well above the one-photon limit. The larger anisotropy for two-photon excitation is mostly due to $\cos^4 \theta$ photoselection. This can be seen from the ratio of the one- and two-photon anisotropies. This ratio is near 1.35, which is close to the predicted ratio of 1.425 (Table 10.6). These data indicate that

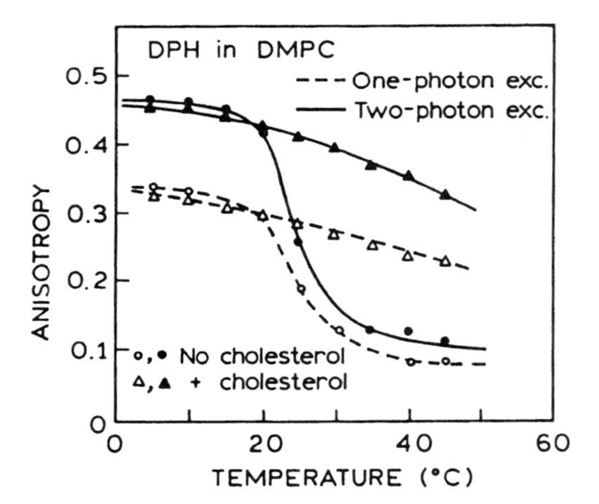

Figure 10.35. Steady-state anisotropies of DPH in vesicles of DMPC, with or without cholesterol, for one-photon (358 nm) or two-photon (716 nm) excitation. Revised from Ref. 80.

the value of β for DPH is nearly the same for one-photon and two-photon excitation.

Multiphoton excitation requires high local light intensities, which are obtained from focused laser beams. Hence, it is natural to ask whether these conditions are compatible with studies of biological molecules. Remarkably, multiphoton excitation is possible without significant heating or damage of many samples. As an example, Figure 10.35 shows the anisotropies of DPH in DMPC vesicles for one-photon (358 nm) and two-photon excitation (716 nm). The same membrane phase transition was observed near 24 °C for one-photon or two-photon excitation. This result indicates that the membranes are not being significantly heated by the locally intense excitation.

REFERENCES

1. Jaboński, A., 1960, On the notion of emission anisotropy, *Bull. Acad. Pol. Sci. Ser. A* **8**:259–264.
2. Weber, G., 1952, Polarization of the fluorescence of macromolecules. I. Theory and experimental method, *Biochem. J.* **51**:145–155.
3. Weber, G., 1966, Polarization of the fluorescence of solutions, in *Fluorescence and Phosphorescence Analysis*, D. M. Hercules (ed.), John Wiley & Sons, New York, pp. 217–240.
4. Selényi, P., 1939, Wide-angle interferences and the nature of the elementary light sources, *Phys. Rev.* **56**:477–479.
5. Lakowicz, J. R., and Gryczynski, I., 1997, Multiphoton excitation of biochemical fluorophores, in *Topics in Fluorescence Spectroscopy, Volume 5, Nonlinear and Two-Photon-Induced Fluorescence*, J. R. Lakowicz (ed.), Plenum Press, New York, pp. 87–144.
6. Weber, G., 1972, Use of fluorescence in biophysics: Some recent developments, *Annu. Rev. Biophys. Bioeng.* **1**:553–570.
7. Shinitzky, M., and Barenholz, Y., 1974, Dynamics of the hydrocarbon layer in liposomes of lecithin and sphingomyelin containing dicetylphosphate, *J. Biol. Chem.* **249**:2652–2657.
8. Michl, J., and Thulstrup, E. W., 1986, *Spectroscopy with Polarized Light*, VCH Publishers, New York.
9. Albinsson, B., Kubista, M., Nordén, B., and Thulstrup, E. W., 1989, Near-ultraviolet electronic transitions of the tryptophan chromophore: Linear dichroism, fluorescence anisotropy, and magnetic circular dichroism spectra of some indole derivatives, *J. Phys. Chem.* **93**:6646–6655.
10. Albinsson, B., Eriksson, S., Lyng, R., and Kubista, M., 1991, The electronically excited states of 2-phenylindole, *Chem. Phys.* **151**:149–157.
11. Kubista, M., Åkerman, B., and Albinsson, B., 1989, Characterization of the electronic structure of 4′,6-diamidino-2-phenylindole, *J. Am. Chem. Soc.* **111**:7031–7035.
12. Vincent, M., de Foresta, B., Gallay, J., and Alfsen, A., 1982, Nanosecond fluorescence anisotropy decays of *n*-(9-anthroyloxy) fatty acids in dipalmitoylphosphatidylcholine vesicles with regard to isotropic solvents, *Biochemistry* **21**:708–716.
13. Weber, G., 1960, Fluorescence-polarization spectrum and electronic-energy transfer in tyrosine, tryptophan and related compounds, *Biochem. J.* **75**:335–345.

14. Valeur, B., and Weber, G., 1977, Resolution of the fluorescence excitation spectrum of indole into the ^1L_a and ^1L_b excitation bands, *Photochem. Photobiol.* **25:**441–444.
15. Eftink, M. R., Selvidge, L. A., Callis, P. R., and Rehms, A. A., 1990, Photophysics of indole derivatives: Experimental resolution of L_a and L_b transitions and comparison with theory, *J. Phys. Chem.* **94:**3469–3479.
16. Yamamoto, Y., and Tanaka, J., 1972, Polarized absorption spectra of crystals of indole and its related compounds, *Bull. Chem. Soc. Jpn.* **65:**1362–1366.
17. Suwaiyan, A., and Zwarich, R., 1987, Absorption spectra of substituted indoles in stretched polyethylene films, *Spectrochim Acta, Part A* **43A:**605–609.
18. Weber, G., and Shinitzky, M., 1970, Failure of energy transfer between identical aromatic molecules on excitation at the long wave edge of the absorption spectrum, *Proc. Natl. Acad. Sci. U.S.A.* **65:**823–830.
19. Kawski, A., 1992, *Fotoluminescencja Roztworów*, Wydawnictwo Naukowe PWN, Warsaw, p. 306.
20. Jaboński, A., 1970, Anisotropy of fluorescence of molecules excited by excitation transfer, *Acta Phys. Pol.* A **38:**453–458.
21. Baumann, J., and Fayer, M. D., 1986, Excitation transfer in disordered two-dimensional and anisotropic three-dimensional systems: Effects of spatial geometry on time-resolved observables, *J. Chem. Phys.* **85:**4087–4107.
22. Teale, F. W. J., 1969, Fluorescence depolarization by light scattering in turbid solutions, *Photochem. Photobiol.* **10:**363–374.
23. Lentz, B. R., 1979, Light scattering effects in the measurement of membrane microviscosity with diphenylhexatriene, *Biophys. J.* **25:**489–494.
24. Berberan-Santos, M. N., Nunes Pereira, E. J., and Martinho, J. M. G., 1995, Stochastic theory of molecular radiative transport, *J. Chem. Phys.* **103:**3022–3028.
25. Nunes Pereira, E. J., Berberan-Santos, M. N., and Martinho, J. M. G., 1996, Molecular radiative transport. II. Monte-Carlo simulation, *J. Chem. Phys.* **104:**8950–8965.
26. Perrin, F., 1929, La fluorescence des solutions. Induction moléculaire. Polarisation et durée d'émission. Photochimie, *Ann. Phys. Ser. 10* **12:**169–275.
27. Perrin, F., 1926, Polarisation de la lumière de fluorescence. Vie moyenne des molécules dans l'état excité, *J. Phys. Radium V, Ser. 6* **7:**390–401.
28. Perrin, F., 1931, Fluorescence. Durée élémentaire d'émission lumineuse, *Conférences d'Actualités Scientifiques et Industrielles* **XXII:**2–41.
29. Weber, G., 1953, Rotational Brownian motion and polarization of the fluorescence of solutions, *Adv. Protein Chem.* **8:**415–459.
30. Yguerabide, J., Epstein, H. F., and Stryer, L., 1970, Segmental flexibility in an antibody molecule, *J. Mol. Biol.* **51:**573–590.
31. Weber, G., 1952, Polarization of the fluorescence of macromolecules. II. Fluorescence conjugates of ovalbumin and bovine serum albumin, *Biochem. J.* **51:**155–167.
32. Laurence, D. J. R., 1952, A study of the absorption of dyes on bovine serum albumin by the method of polarization of fluorescence, *Biochem. J.* **51:**168–180.
33. Gottlieb, Y. Ya., and Wahl, P., 1963, Étude théorique de la polarisation de fluorescence des macromolécules portant un groupe émetteur mobile autour d'un axe de rotation, *J. Chim. Phys.* **60:**849–856.
34. Ferguson, B. Q., and Yang, D. C. H., 1986, Methionyl-tRNA synthetase induced 3′-terminal and delocalized conformational transition in $tRNA^{fMet}$: Steady-state fluorescence of tRNA with a single fluorophore, *Biochemistry* **25:**529–539.
35. Buchner, J., 1996, Supervising the fold: Functional principals of molecular chaperones, *FASEB J.* **10:**10–19.
36. Ellis, R. J., 1996, *The Chaperonins*, Academic Press, New York.
37. Braig, K., Otwinowski, Z., Hegde, R., Bolsvert, D. C., Joachimiak, A., Horwich, A. L., and Sigler, P. B., 1994, The crystal structure of the bacterial chaperonin GroEL at 2.8 Å, *Nature* **371:**578–586.
38. Gorovits, B. M., and Horowitz, P. M., 1995, The molecular chaperonin cpn60 displays local flexibility that is reduced after binding with an unfolded protein, *J. Biol. Chem.* **270:**13057–13062.
39. Lim, K., Jameson, D. M., Gentry, C. A., and Herron, J. N., 1995, Molecular dynamics of the anti-fluorescein 4–4–20 antigen-binding fragment. 2. Time-resolved fluorescence spectroscopy, *Biochemistry* **34:**6975–6984.
40. Lukas, T. J., Burgess, W. H., Prendergast, F. G., Lau, W., and Watterson, D. M., 1986, Calmodulin binding domains: Characterization of a phosphorylation and calmodulin binding site from myosin light chain kinase, *Biochemistry* **25:**1458–1464.
41. Malencik, D. A., and Anderson, S. R., 1984, Peptide binding by calmodulin and its proteolytic fragments and by troponin C, *Biochemistry* **23:**2420–2428.
42. LeTilly, V., and Royer, C. A., 1993, Fluorescence anisotropy assays implicate protein–protein interactions in regulating *trp* repressor DNA binding, *Biochemistry* **32:**7753–7758.
43. Runnels, L. W., and Scarlata, S. F., 1995, Theory and application of fluorescence homotransfer to melittin oligomerization, *Biophys. J.* **69:**1569–1583.
44. Shinitzky, M., Dianoux, A. C., Gitler, C., and Weber, G., 1971, Microviscosity and order in the hydrocarbon region of micelles and membranes determined with fluorescence probes I. Synthetic micelles, *Biochemistry* **10:**2106–2113.
45. Cogen, U., Shinitzky, M., Weber, G., and Nishida, T., 1973, Microviscosity and order in the hydrocarbon region of phospholipid and phospholipid–cholesterol dispersions determined with fluorescent probes, *Biochemistry* **12:**521–528.
46. Thulborn, K. R., Tilley, L. M., Sawyer, W. H., and Treloar, F. E., 1979, The use of *n*-(9-anthroyloxy) fatty acids to determine fluidity and polarity gradients in phospholipid bilayers, *Biochim. Biophys. Acta* **558:**166–178.
47. Thulborn, K. R., and Beddard, G. S., 1982, The effects of cholesterol on the time-resolved emission anisotropy of 12-(9-anthroyloxy)stearic acid in dipalmitoylphosphatidylcholine bilayers, *Biochim. Biophys. Acta* **693:**246–252.
48. Lakowicz, J. R., and Weber, G., 1980, Nanosecond segmental mobilities of tryptophan residues in proteins observed by lifetime-resolved fluorescence anisotropies, *Biophys. J.* **32:**591–601.
49. Lakowicz, J. R., Maliwal, B. P., Cherek, H., and Balter, A., 1983, Rotational freedom of tryptophan residues in proteins and peptides, *Biochemistry* **22:**1741–1752.
50. Lakowicz, J. R., and Maliwal, B. P., 1983, Oxygen quenching and fluorescence depolarization of tyrosine residues in proteins, *J. Biol. Chem.* **258:**4794–4801.
51. Eftink, M., 1983, Quenching-resolved emission anisotropy studies with single and multitryptophan-containing proteins, *Biophys. J.* **43:**323–334.
52. Lakos, Z., Szarka, Á., Koszorús, L., and Somogyi, B., 1995, Quenching-resolved emission anisotropy: A steady state fluorescence method to study protein dynamics, *J. Photochem. Photobiol., B: Biol.* **27:**55–60.
53. Bentz, J. P., Beyl, P., Beinert, G., and Weill, G., 1973, Simultaneous measurements of fluorescence polarization and quenching: A specially designed instrument and an application to the micro-Brownian motion of polymer chains, *Eur. Polym. J.* **11:**711–718.

54. Brown, K., and Soutar, I., 1974, Fluorescence quenching and polarization studies of segmental motion in polystyrene, *Eur. Polym. J.* **10**:433–437.
55. Chen, R. F., 1976, Quenching of the fluorescence of proteins by silver nitrate, *Arch. Biochem. Biophys.* **168**:605–622.
56. Sanyal, G., Charlesworth, M. C., Ryan, R. J., and Prendergast, F. G., 1987, Tryptophan fluorescence studies of subunit interaction and rotational dynamics of human luteinizing hormone, *Biochemistry* **26**:1860–1866.
57. Soleillet, P., 1929, Sur les paramètres caractérisant la polarisation partielle de la lumière dans les phénomènes de fluorescence, *Ann. Phys. Biol. Med.* **12**:23–97.
58. Kawski, A., 1986, Fluorescence anisotropy as a source of information about different photophysical processes, in *Progress and Trends in Applied Optical Spectroscopy*, D. Fassler, K.-H. Feller, and B. Wilhelmi (eds.), Teubner-Texte zur Physik, Vol. 13, Teubner Verlagsgesellschaft, Leipzig, pp. 6–34.
59. Kawski, A., 1993, Fluorescence anisotropy: Theory and applications of rotational depolarization, *Crit. Rev. Anal. Chem.* **23**:459–529.
60. Gurinovich, G. P., Sarzhevskii, A. M., and Sevchenko, A. N., 1963, New data on the dependence of the degree of polarization on the wavelength of fluorescence, *Opt. Spectrosc.* **14**:428–432.
61. Mazurenko, Y. T., and Bakhshiev, N. G., 1970, Effect of orientation dipole relaxation on spectral, time, and polarization characteristics of the luminescence of solutions, *Opt. Spectrosc.* **28**:490–494.
62. Gakamskii, D. M., Nemkovich, N. A., Rubinov, A. N., and Tomin, V. I., 1988, Light-induced rotation of dye molecules in solution, *Opt. Spectrosc.* **64**:406–407.
63. Matayoshi, E. D., and Kleinfeld, A. M., 1981, Emission wavelength-dependent decay of the 9-anthroyloxy-fatty acid membrane probes, *Biophys. J.* **35**:215–235.
64. Thulstrup, E. W., and Michl, J., 1988, Polarized absorption spectroscopy of molecules aligned in stretched polymers, *Spectrochim. Acta, Part A* **44**:767–782.
65. Michl, J., and Thulstrup, E. W., 1987, Ultraviolet and infrared linear dichroism: Polarized light as a probe of molecular and electronic structure, *Acc. Chem. Res.* **20**:192–199.
66. Van Gurp, M., and Levine, Y. K., 1989, Determination of transition moment directions in molecules of low symmetry using polarized fluorescence. I. Theory, *J. Chem. Phys.* **90**:4095–4100.
67. Matsuoka, Y., and Yamaoka, K., 1980, Film dichroism. V. Linear dichroism study of acridine dyes in films with emphasis on the electronic transitions involved in the long-wavelength band of the absorption spectrum, *Bull. Chem. Soc. Jpn* **53**:2146–2151.
68. Kawski, A., and Gryczynski, Z., 1986, On the determination of transition-moment directions from emission anisotropy measurements, *Z. Naturforsch.* A **41**:1195–1199.
69. Kawski, A., Gryczynski, Z., Gryczynski, I., Lakowicz, J. R., and Piszczek, G., 1996, Photoselection of luminescent molecules in anisotropic media in the case of two-photon excitation. Part II. Experimental studies of Hoechst 33342 in stretched poly(vinyl alcohol) films, *Z. Naturforsch. A* **51**:1037–1041.
70. Matsuoka, Y., and Norden, B., 1982, Linear dichroism study of 9-substituted acridines in stretched poly(vinyl alcohol) film, *Chem. Phys. Lett.* **85**:302–306.
71. Holmén, A., Broo, A., Albinsson, B., and B. Nordén, 1997, Assignment of electronic transition moment directions of adenine from linear dichroism measurements, *J. Am. Chem. Soc.* **119**:12240–12250.
72. Holmén, A., Nordén, B., and Albinsson, B., 1997, Electronic transition moments of 2-aminopurine, *J. Am. Chem. Soc.* **119**:3114–3121.
73. Albinsson, B., Kubista, M., Sandros, K., and Nordén, B., 1990, Electronic linear dichroism spectrum and transition moment directions of the hypermodified nucleic acid base wye, *J. Phys. Chem.* **94**:4006–4011.
74. Hall, R. D., Valeur, B., and Weber, G., 1985, Polarization of the fluorescence of triphenylene: A planar molecule with three-fold symmetry, *Chem. Phys. Lett.* **116**(2,3):202–205.
75. Lakowicz, J. R., and Gryczynski, I., 1997, Multiphoton excitation of biochemical fluorophores, in *Topics in Fluorescence Spectroscopy, Volume 5, Nonlinear and Two-Photon-Induced Fluorescence*, J. R. Lakowicz (ed.), Plenum Press, New York, pp. 87–144.
76. Callis, P. R., 1997, The theory of two-photon induced fluorescence anisotropy, in *Topics in Fluorescence Spectroscopy, Volume 5: Nonlinear and Two-Photon-Induced Fluorescence*, J. R. Lakowicz (ed.), Plenum Press, New York, pp. 1–42.
77. Johnson, C. K., and Wan, C., 1997, Anisotropy decays induced by two-photon excitation, in *Topics in Fluorescence Spectroscopy, Volume 5, Nonlinear and Two-Photon-Induced Fluorescence*, J. R. Lakowicz (ed.), Plenum Press, New York, pp. 43–85.
78. Callis, P. R., 1997, Two-photon induced fluorescence, *Annu. Rev. Phys. Chem.* **48**:271-297.
79. Lakowicz, J. R., Gryczynski, I., Gryczynski, Z., Danielson, E., and Wirth, M. J., 1992, Time-resolved fluorescence intensity and anisotropy decays of 2,5-diphenyloxazole by two-photon excitation and frequency-domain fluorometry, *J. Phys. Chem.* **96**:3000–3006.
80. Lakowicz, J. R., Gryczynski, I., Kuśba, J., and Danielsen, E., 1992, Two photon-induced fluorescence intensity and anisotropy decays of diphenylhexatriene in solvents and lipid bilayers, *J. Fluoresc.* **2**(4):247–258.
81. Gryczynski, I., unpublished observations.

PROBLEMS

10.1. *Angles (β) between Absorption and Emission Transition Moments:* Figure 10.29 shows the excitation anisotropy spectrum of perylene in propylene glycol at –60 °C. What is the angle between the transition moments for excitation at 430, 320, and 282 nm?

10.2. *Effect of Scattered Light on the Anisotropy:* Suppose that a membrane-bound fluorophore displays a true anisotropy of 0.30. However, your emission filter allows 20% of the signal to be due to scattered light. What is the measured anisotropy?

10.3. *Calculation of an Anisotropy*: Using a T-format polarization instrument, you obtain the following data for diphenylhexatriene (DPH) in propylene glycol at –60 °C: $I_{HV} = 0.450$; $I_{VV} = 1.330$. The first and second subscripts refer to the orientation of the excitation and emission polarizers, respectively. Calculate r_0 and P_0 for DPH. What is the angle α between the absorption and emission oscillators?

10.4. *Derivation of the Perrin Equation:* Derive the Perrin equation (Eq. [10.50]) for a single-exponential decay of the intensity and anisotropy.

10.5. *Calculation of an Anisotropy in Fluid Solution:* Calculate the expected anisotropy of perylene in ethanol at 20 °C and in propylene glycol at 25 °C. The molecular weight of perylene is 252 g/mol, and the density can be taken as

Table 10.7. Polarization Values of DNS

[BSA] (M)	[DNS] (M)	P
0	1×10^{-7}	0.0149
2×10^{-5}	1×10^{-7}	0.2727
$>>K_d$	1×10^{-7}	0.3913

1.35 g/ml. The viscosity of ethanol at 20 °C is 1.194 cP, and that of propylene glycol at 25 °C is 32 cP. Assume that the lifetime is 6 ns and $r_0 = 0.36$.

10.6. *Binding Reactions by Anisotropy*: Derive Eq. [10.58], which relates the fractional binding to the steady-state anisotropy and the change in intensity on binding.

10.7. *Calculation of a Binding Constant from Anisotropies:* Assume that the probe 5-dimethylaminonaphthalene-1-sulfonic acid (DNS) binds to bovine serum albumin (BSA). The DNS polarization values are given in Table 10.7. The dissociation constant (K_d) for the binding is given by

$$K_d = \frac{[\text{DNS}][\text{BSA}]}{[\text{DNS–BSA}]} \qquad [10.65]$$

A. Calculate the dissociation constant assuming that the quantum yield of the probe is not altered upon binding.

B. Now assume that the quantum yield of DNS increases twofold upon binding to BSA. What is K_d?

C. How would you determine whether the quantum yield of DNS changes upon binding to BSA?

D. Predict the time-resolved anisotropies for each solution, at 20 °C. Assume that $P_0 = 0.3913$ for DNS and that the rotational correlation times of DNS and BSA are well approximated by that predicted for an anhydrous sphere. The molecular weight of BSA is 64,000 g/mol. The viscosity of water at 25 °C is 0.894 cP (P = Ig cm^{-1} s^{-1}). The density of BSA can be taken as 1.2 g/cm^3.

10.8. *Anisotropy of a Planar Oscillator:* Calculate the anisotropy of triphenylene assuming that the emission is randomized among three equivalent axes (Figure 10.32).

10.9. *Correlation Times from Lifetime-Resolved Anisotropies*: Figure 10.36 shows lifetime-resolved anisotropies of melittin. The lifetimes were varied by oxygen quenching. Calculate the correlation times and $r(0)$ values in aqueous buffer without NaCl and 2.4M NaCl, where melittin exists as a monomer and a tetramer, respectively. Assume that the r_0 values for melittin is 0.26 and that the monomeric molecular weight is 3250 g/mol.

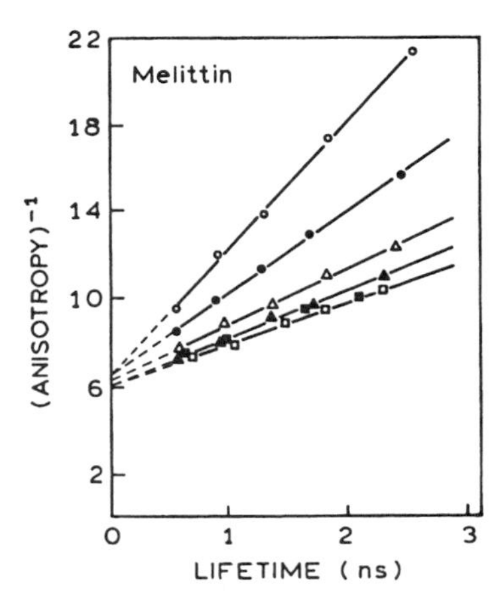

Figure 10.36. Lifetime-resolved anisotropy of melittin in aqueous buffer at 25 °C. The concentrations of NaCl are 0 (○), 0.15 (●), 0.3 (△), 0.6 (▲), 1.5 (■), and 2.4M (□). Revised and reprinted, with permission, from Ref. 49, Copyright © 1983, American Chemical Society.

10.10. *Anisotropies of a Styrene Derivative*: Figure 10.37 shows the steady-state anisotropies of 4-Dimethylamino-ω-diphenylphosphinyl-*trans*-Styrene (DPPS) in n-butanol. The anisotropies are higher for two-photon excitation than for one-photon excitation. In both cases the anisotropies are independent of temperature. Explain both results.

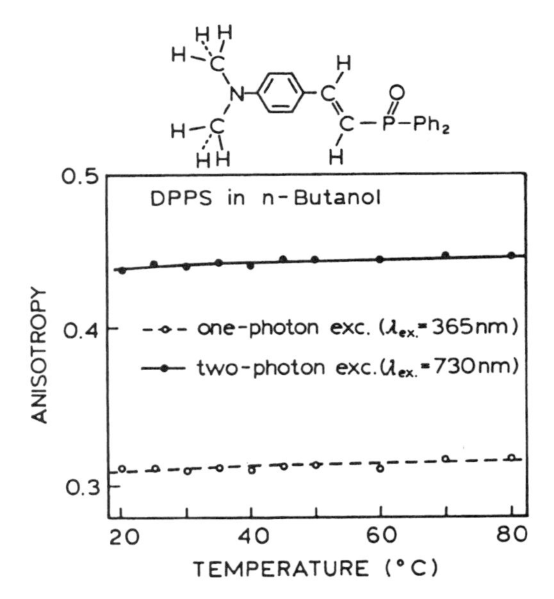

Figure 10.37. Steady-state anisotropies of DPPS for one-photon and two-photon excitation. Revised from Ref. 81.

Time-Dependent Anisotropy Decays

11

In the preceding chapter we described the measurement and interpretation of steady-state fluorescence anisotropies. These values are measured using continuous illumination and represent an average of the anisotropy decay over the intensity decay. Measurement of steady-state anisotropies is simple, but interpretation of the steady-state anisotropies usually depends on an assumed form for the anisotropy decay, which is not directly observed in the experiment. Additional information is available if one measures the time-dependent anisotropy, that is, the values of $r(t)$ following pulsed excitation. The form of the anisotropy decay depends on the size, shape, and flexibility of the labeled molecule, and the data can be compared with the decays calculated from various molecular models. Anisotropy decays can be obtained using the TD or the FD method.

It is valuable to understand the factors which affect the anisotropy decays. For a spherical molecule, the anisotropy is expected to decay with a single rotational correlation time (θ). Perhaps the most frequent interpretation of the correlation time is in terms of the overall rotational correlation time of a protein. The measured values of θ can compare with the value predicted for a hydrated sphere of equivalent molecular weight (Eq. [10.52]). However, numerous factors can result in multiexponential anisotropy decays. Multiexponential anisotropy decays are expected for nonspherical fluorophores or proteins. In this case the correlation times are determined by the rates of rotation about the various molecular axes. Anisotropy decays can be used to estimate the shapes of proteins.

In addition to being affected by shape, anisotropy decays are affected by the segmental flexibility of the macromolecule. For instance, tryptophan anisotropy decays of proteins frequently display correlation times that are too short to be due to rotational diffusion of the whole macromolecule. These short-correlation-time components are often due to independent motions of the tryptophan residue within the protein. Measurements of these components have been widely used to understand the internal dynamics of proteins. Anisotropy decays can also be affected by RET between molecules of the same type of fluorophore, that is, depolarization due to homotransfer.

Anisotropy decays of membrane-bound probes have been particularly informative. Membrane-bound probes often display unusual behavior, whereby the anisotropies do not decay to zero. This behavior occurs because, in contrast to probes in isotropic solutions, probes in membranes do not rotate freely. The extent of rotation is often limited by the anisotropic environment of a membrane. The nonzero anisotropies at long times have been interpreted in terms of the order parameters of the membrane. In this chapter we present examples of simple and complex anisotropy decays to illustrate the wealth of information available from measurements of time-dependent anisotropies. In the following chapter we describe more advanced concepts in anisotropy decay analysis.

11.1. ANALYSIS OF TIME-DOMAIN ANISOTROPY DECAYS

Suppose that a fluorophore is excited with a pulse of vertically polarized light and that it rotates with a single correlation time. The anisotropy decay is determined by measuring the decay of the vertically (‖) and horizontally ($\perp$) polarized components of the emission. If the absorption and emission transition moments are collinear, the time-zero anisotropy is 0.4. In this case the initial intensity of the parallel component is threefold larger than that of the perpendicular component (Figure 11.1, left). Assuming that the fundamental anisotropy is greater than zero ($r_0 > 0$), the vertically polarized excitation pulse results in an initial population of fluorophores that is enriched in the parallel orientation. The decay of the difference between $I_{\parallel}(t)$ and $I_{\perp}(t)$, when properly normalized by the total intensity, is the anisotropy decay (Figure 11.1, right).

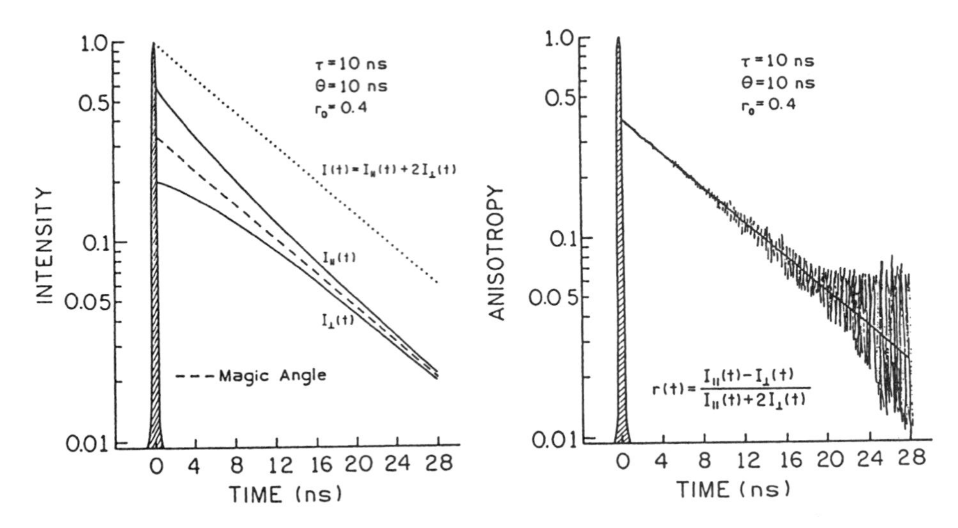

Figure 11.1. Time-dependent polarized decays (*left*) and the calculated anisotropy decay (*right*). From Ref. 1.

Examination of the left panel in Figure 11.1 reveals that the parallel component initially decays more rapidly than the horizontal component. This occurs because the vertically oriented fluorophores are decaying by two processes, the intensity decay with decay time τ and rotation out of the vertical orientation with a correlation time θ. The horizontal component initially decays more slowly because it is repopulated by rotation from the excess vertically oriented population.

Interpretation of anisotropy decays is best understood in terms of the individual components. The decays of the parallel (||) and perpendicular ($\perp$) components of the emission are given by

$$I_{\parallel}(t) = \frac{1}{3} I(t)[1 + 2r(t)] \tag{11.1}$$

$$I_{\perp}(t) = \frac{1}{3} I(t)[1 - r(t)] \tag{11.2}$$

where $r(t)$ is the time-resolved anisotropy. Generally, $r(t)$ can be described as a multiexponential decay,

$$r(t) = r_0 \sum_j g_j \exp(-t/\theta_j) = \sum_j r_{0j} \exp(-t/\theta_j) \tag{11.3}$$

where $r_0 = \Sigma_j r_{0j}$ is the limiting anisotropy in the absence of rotational diffusion, the θ_j are the individual correlation times, and the g_j are the associated fractional amplitudes in the anisotropy decay ($\Sigma g_j = 1.0$). Depending on the circumstances, r_0 may be a known parameter, perhaps from a frozen solution measurement. Alternatively, all the amplitudes (r_{0j}) can be considered to be experimental variables. As shown in the previous chapter, the total intensity for a sample is given by $I_T = I_{\parallel} + 2I_{\perp}$. Similarly, the total (rotation-free) intensity decay is given by

$$I(t) = I_{\parallel}(t) + 2I_{\perp}(t) \tag{11.4}$$

In the time domain, one measures the time-dependent decays of the polarized components of the emission (Eqs. [11.1] and [11.2]). The polarized intensity decays are used to calculate the time-dependent anisotropy,

$$r(t) = \frac{I_{\parallel}(t) - I_{\perp}(t)}{I_{\parallel}(t) + 2I_{\perp}(t)} \tag{11.5}$$

The time-dependent anisotropy decay is then analyzed to determine which model is most consistent with the data.

The experimental procedures and the form of the data are different for FD measurements of the anisotropy decays.[2] The sample is excited with amplitude-modulated light which is vertically polarized (Figure 11.2). As for the TD measurements, the emission is observed through a polarizer, which is rotated between the parallel and the perpendicular orientations. In the frequency domain, there are two observable quantities which characterize the anisotropy decay. These are the phase shift Δ_ω, at the modulation frequency ω, between the perpendicular ($\phi_{\perp}$) and parallel ($\phi_{\parallel}$) components of the emission,

$$\Delta_\omega = \phi_{\perp} - \phi_{\parallel} \tag{11.6}$$

and the ratio of the parallel ($m_{\parallel}$) and the perpendicular ($m_{\perp}$) components of the modulated emission,

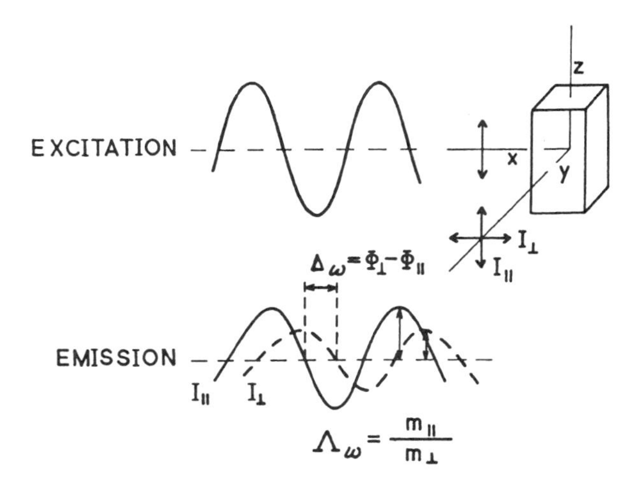

Figure 11.2. FD measurements of anisotropy decays. For simplicity, the average intensity is assumed equal for both polarized components. From Ref 2.

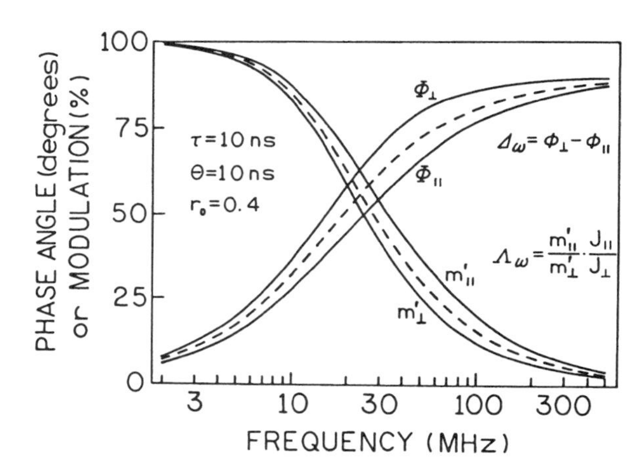

Figure 11.3. Simulated FD data for an anisotropy decay, with $\tau = 10$ ns and $\theta = 10$ ns, showing the phase and actual modulation ($m'_\parallel$ and $m'_\perp$) of the polarized components of the emission, relative to the modulated excitation. The dashed curves show the rotation-free phase and modulation values for the total emission. From Ref. 1.

$$\Lambda_\omega = m_\parallel / m_\perp \qquad [11.7]$$

To avoid confusion, we stress that Λ_ω is the ratio of the modulated amplitudes of the polarized components, not the ratio of the modulation of each polarized component. The ratio Λ_ω is often presented as the frequency-dependent anisotropy (r_ω), which is defined by

$$r_\omega = \frac{\Lambda_\omega - 1}{\Lambda_\omega + 2} \qquad [11.8]$$

The form of the FD anisotropy data is illustrated in Figure 11.3. Analogously to the TD measurements, one could measure the phase and modulation of the polarized component (Figure 11.3, solid curves). One would find that the phase angle of the parallel component ($\phi_\parallel$) would be smaller than the rotation-free phase angle for the total emission and that the modulation would be larger than the rotation-free modulation (Figure 11.3). These effects can be understood as being the result of the shorter mean decay time of the vertically polarized decay (Figure 11.1). Similarly, the phase angle of the perpendicular component is larger, and the modulation smaller, because this component is being repopulated by the excess population in the parallel orientation, resulting in a longer mean decay time for the perpendicular component. While the concept of a mean decay time is useful for understanding the relative behavior of the polarized intensity decays, the use of mean decay times to describe the decay times of the polarized components resulted in some confusion in the early literature.[3–6] For clarity, we note that the modulations shown in Figure 11.3 ($m'_\parallel$ and $m'_\perp$) are the actual modulations of these components [$I_\parallel(t)$ and $I_\perp(t)$]. For calculation of the anisotropy decay, we use the unnormalized amplitudes of the modulated components of the polarized emission.

While one could measure $\phi_\parallel$, $\phi_\perp$, $m'_\parallel$, and $m'_\perp$, to obtain the anisotropy decay, this is not the preferred method. At present, almost all FD anisotropy decays are measured by the differential method. By a simple rotation of the emission polarizer, the differential polarized phase angle (Δ_ω) and modulation ratio (Λ_ω) can be directly measured. It is more accurate to measure the difference and ratio directly, rather than calculating these values from two independently measured values.

The form of the FD anisotropy data measured by the differential method is illustrated in Figure 11.4. The differential phase angles appear to be approximately Lorentzian in shape on the log-frequency scale. The modulated anisotropy increases monotonically with frequency. One can notice that the value of r_ω at low frequency is equal to the steady-state anisotropy:

$$r = \frac{r_0}{1 + \dfrac{\tau}{\theta}} = \frac{r_0}{2} \qquad [11.9]$$

In this case $r = 0.5r_0$ is a result of $\tau = \theta$ for the simulated curves. At high frequency, r_ω approaches r_0. Longer correlation times shift the Δ_ω and r_ω curves to lower modulation frequencies, and shorter correlation times shift these curves to higher frequencies.

It is valuable to understand how changes in the anisotropy decay affect the TD and the FD data. Suppose that the correlation time decreases from 10 to 1.0 ns. In the TD data

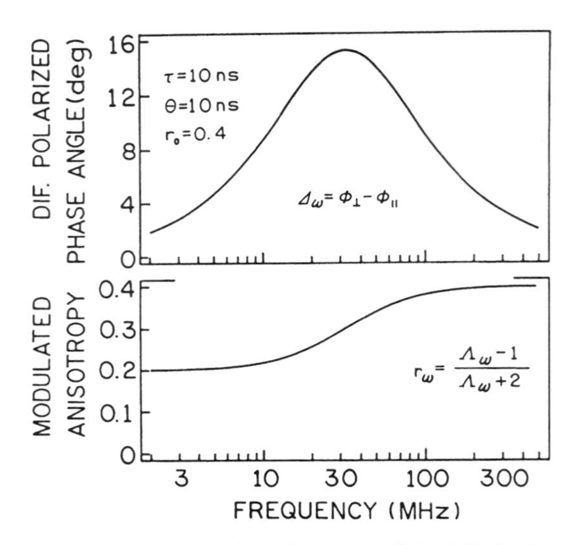

Figure 11.4. Differential phase (Δ_ω) and modulated (Λ_ω) anisotropy for $\tau = 10$ ns, $\theta = 10$ ns, and $r_0 = 0.4$. From Ref. 1.

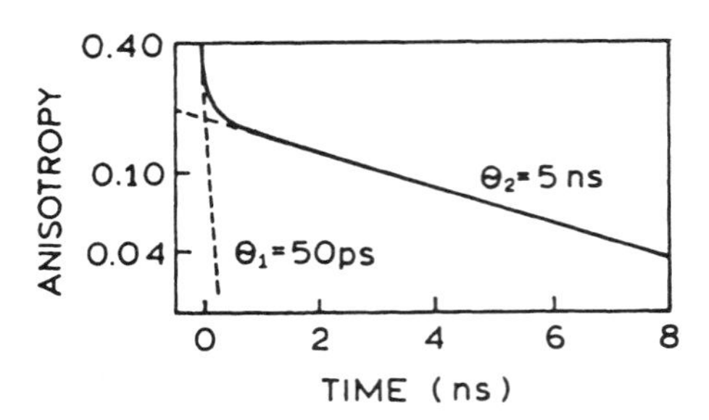

Figure 11.6. Simulated TD anisotropy decay for a double-exponential decay with $r_{01} = r_{02}$, $\theta_1 = 50$ ps, and $\theta_2 = 5$ ns.

the anisotropy decays more rapidly (Figure 11.5). The changes in the FD data are somewhat more complex. The differential phase-angle distribution shifts to higher frequencies with shorter correlation times. Whether the maximum differential phase angle increases or decreases depends on the relative values of τ and θ. A decrease in correlation time results in a decrease in the modulated anisotropy. The limiting value of r_ω at low frequency is the steady-state anisotropy, and at high frequency the limit is still r_0.

Suppose that the anisotropy decays with two correlation times. A typical anisotropy decay for a protein would be characterized by a 5-ns correlation time for overall rotational diffusion and a 50-ps correlation time due to segmental motion of the tryptophan residue (Figure 11.6). The TD anisotropy shows a rapid decay due to the 50-ps component, followed by a slower decay at longer times. Depending upon the resolution of the TD instrument, the fast component may or may not be resolved in the measurements. However, the presence of the fast component can be identified from a time-zero anisotropy [$r(0)$] that is smaller than r_0.

The presence of two decay times also has a dramatic effect on the FD anisotropy data (Figure 11.7). Instead of showing a single Lorentzian distribution for Δ_ω, the differential phase angles show two such distributions, one for each correlation time (dashed curves in Figure 11.7). The presence of the rapid motion is evident from the shift in phase-angle distribution to the higher frequencies. If the amplitude of this rapid motion is increased, the phase angles become smaller at low frequencies and larger at the higher frequencies. If the rapid correlation time is very short, the frequency range of the instrument may not be adequate to detect the presence of this rapid motion. Also, the emission may be highly demodulated owing to the intensity decay, in which case the high-frequency limit of

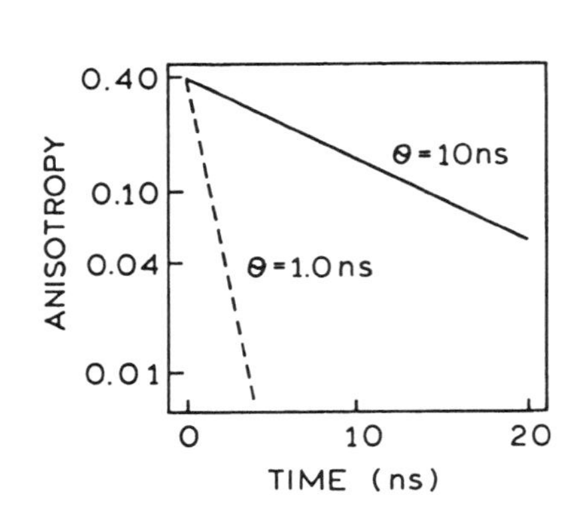

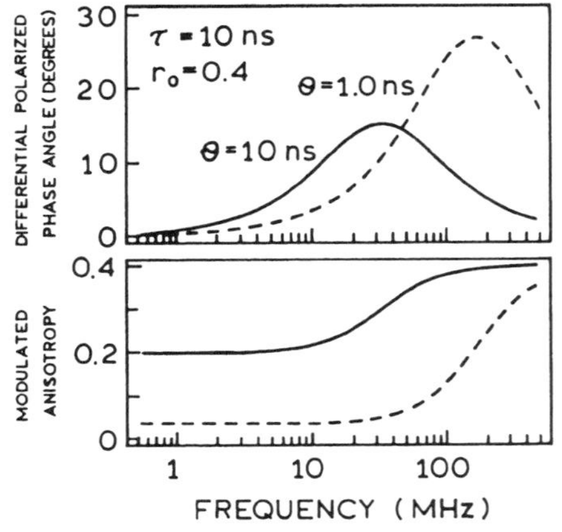

Figure 11.5. Comparison of TD (*left*) and FD anisotropy decays (*right*) with correlation times of 1.0 and 10.0 ns and a lifetime of 10 ns.

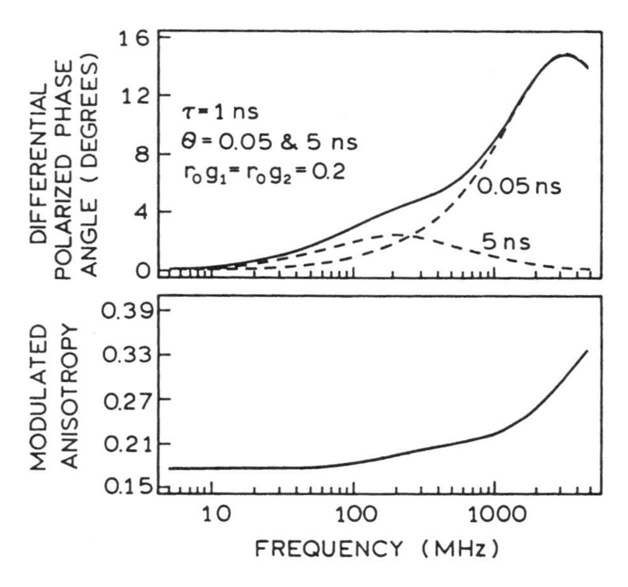

Figure 11.7. FD anisotropy data for a double-exponential anisotropy decay. Revised from Ref. 2.

r_ω cannot be measured. However, one may notice the presence of an unresolved motion by a failure of r_ω to approach the expected limiting value of r_0. This illustrates the advantage of using both the differential phase (Δ_ω) and modulation (r_ω) data in any attempt to resolve a complex and/or rapid anisotropy decay. A failure of r_ω to reach r_0 in the frequency domain is similar to finding $r(0) < r_0$ in the time domain. As for the intensity decays, the parameters describing the anisotropy decay are recovered by comparison of the data with calculated values obtained using various models.

One may notice that the lifetime is included in the FD anisotropy simulations (Figure 11.7) but not in the TD simulations (Figure 11.6). In contrast to the TD anisotropy decay, the FD data depend on the intensity decay time.

11.2. ANISOTROPY DECAY ANALYSIS

11.2.A. Time-Domain Anisotropy Data

The most direct approach to analyzing the anisotropy data is to fit the measured $r(t)$ values to an assumed anisotropy decay law. The measured values are calculated from the polarized intensity decays,

$$r_m(t_k) = \frac{N_{\|}(t_k) - GN_{\perp}(t_k)}{N_{\|}(t_k) + 2GN_{\perp}(t_k)} = \frac{D_m(t_k)}{S_m(t_k)} \qquad [11.10]$$

where $N_{\|}(t_k)$ and $N_{\perp}(t_k)$ are the experimental data convolved with the instrument response function. The measured values $r_m(t_k)$ are then compared with calculated values obtained from the convolution integral

$$r_c(t_k) = \sum_{t=0}^{t=t_k} L(t_k) r(t - t_k)\Delta t \qquad [11.11]$$

where $L(t_k)$ is the instrument response function. The instrument response function is assumed to be independent of the orientation of the emission polarizer. The anisotropy decay is then determined by minimizing χ_R^2,

$$\chi_R^2 = \frac{1}{\nu} \sum_{k=1}^{n} \frac{1}{\sigma_{Rk}^2} [r_m(t_k) - r_c(t_k)]^2 \qquad [11.12]$$

In this equation $r_c(t_k)$ is the anisotropy calculated using Eq. [11.11] and the assumed anisotropy decay law, and ν is the number of degrees of freedom. In using this approach, it is important to remember that the noise in the measurements does not decrease in taking the differences of the intensity values. Instead, the calculated anisotropy values contain all the noise present in the measured polarized decays, and it is important to propagate the noise correctly. In this case the weighting factor is a moderately complex function of the counts in each channel[7,8] and is given by

$$\sigma_{Rk}^2 = r_m^2(t_k)\left(\frac{\sigma_{Sk}^2}{S_m^2(t_k)} + \frac{\sigma_{Dk}^2}{D_m^2(t_k)} - \frac{2\sigma_{SDk}^2}{S_m(t_k)D_m(t_k)}\right) \qquad [11.13]$$

where

$$\sigma_{Dk}^2 = N_{\|}(t_k) + G^2 N_{\perp}(t_k) \qquad [11.14]$$

$$\sigma_{Sk}^2 = N_{\|}(t_k) + 4G^2 N_{\perp}(t_k) \qquad [11.15]$$

$$\sigma_{SDk}^2 = \sigma_{Sk}^2 - 2G^2 \sigma_{Dk}^2 \qquad [11.16]$$

While these expressions seem complex, their origin is simple: it is the propagation of the Poisson noise in each measured intensity into the anisotropy function. The origin of Eq. [11.14] can be seen by considering the Poisson noise in the parallel and perpendicular components, which is given by $[I_{\|}(t_k)]^{1/2}$ and $[I_{\perp}(t_k)]^{1/2}$, respectively. The noise in the sum is given by the sum of the squares of the Poisson noise, with appropriate weighting by the G-factor. The principles of error propagation can be found elsewhere.[9]

Anisotropy decays are sometimes analyzed by separate analysis of the sum and difference decays (Eq. [11.10]). In this method, the sum $S_m(t_k)$ is analyzed first to obtain the parameters describing the intensity decay, which are the α_i and τ_i values when the multiexponential model is used. For this analysis, χ_R^2 is minimized using

$$\chi_R^2 = \frac{1}{\nu} \sum_{k=1}^{n} \frac{1}{\sigma_{Sk}^2} [S_m(t_k) - S_c(t_k)]^2 \qquad [11.17]$$

where σ_{Sk}^2 is given by Eq. [11.15].

The intensity decay also appears in the difference data. The parameters (α_i and τ_i) recovered from the sum analysis are held constant during analysis of $D_m(t_k)$. Once again, χ_R^2 is minimized using[8,10]

$$\chi_R^2 = \frac{1}{\nu} \sum_{k=1}^{n} \frac{1}{\sigma_{Dk}^2} [D_m(t_k) - D_c(t_k)]^2 \qquad [11.18]$$

where σ_{Dk}^2 is given by Eq. [11.14]. When this procedure is used, the values describing $D(t)$ contain both the intensity decay and the anisotropy decay parameters. This can be seen by considering a single-exponential decay of the intensity and the anisotropy. In this case,

$$D(t) = S(t)r(t) = k \exp\left[-t\left(\frac{1}{\tau} + \frac{1}{\theta}\right)\right] \qquad [11.19]$$

where k is a constant. Hence, the difference $D(t)$ decays with an apparent decay time that is shorter than the lifetime (τ) and the correlation time (θ).

Of the two procedures described above, the second one, using separate analyses of $S(t)$ and $D(t)$, is preferable. This is because the use of Eq. [11.10] to calculate the anisotropy is not correct, particularly when the correlation times are close to the width of the instrument response function. The operations of convolution and division do not commute, so that the measured values of $r_m(t_k)$ are not a convolution of the impulse response function $r(t)$ with the lamp function. For instance, even if the anisotropy decay is a single exponential, the calculated values of $r_m(t_k)$ can display unusual shapes, particularly near the rising edge of the curve. Also, the apparent time-zero anisotropies $r(0)$ are often less than the true value of r_0.[11] The use of the calculated sum $[S_m(t_k)]$ and difference $[D_m(t_k)]$ curves is more correct because the operations of addition and subtraction commute with convolution. However, this method still assumes that the lamp function is the same for the parallel and perpendicular components of the emission.

At present, the preferred method of analysis is to directly analyze the polarized intensity decays without calculation of $r_m(t_k)$ or $D_m(t_k)$.[12–15] This is a form of global analysis in which the parallel and perpendicular components (Eqs. [11.1] and [11.2]) are analyzed simultaneously to recover the intensity and anisotropy decay laws. The polarized decays are used with the instrument response function to calculate

$$N_{\parallel}^c(t_k) = \sum_{t=0}^{t=t_k} L_{\parallel}(t_k) I_{\parallel}(t - t_k)\Delta t \qquad [11.20]$$

$$N_{\perp}^c(t_k) = \sum_{t=0}^{t=t_k} L_{\perp}(t_k) I_{\perp}(t - t_k)\Delta t \qquad [11.21]$$

where $L_{\parallel}(t_k)$ and $L_{\perp}(t_k)$ are the instrument response functions for polarized emission. In contrast to the previous method, this method does not assume that the instrument response functions are identical, and they can be rather different without affecting the validity of the procedure.

The calculated polarized intensities are then used to minimize χ_R^2 based on the parameter values in the intensity (α_i and τ_i) and anisotropy decays (r_{0i} and θ_i),

$$\chi_R^2 = \frac{1}{\nu} \sum_{t=0}^{n} \frac{1}{\sigma_{\parallel k}^2} [N_{\parallel}(t_k) - N_{\parallel}^c(t_k)]^2 + \frac{1}{\nu} \sum_{t=0}^{n} \frac{1}{\sigma_{\perp k}^2} [N_{\perp}(t_k) - N_{\perp}^c(t_k)]^2 \qquad [11.22]$$

Since the polarized intensity decays are used directly, the weighting factors are given by

$$\sigma_{\parallel k}^2 = N_{\parallel}(t_k) \qquad [11.23]$$

$$\sigma_{\perp k}^2 = N_{\perp}(t_k) \qquad [11.24]$$

and there is no need to propagate the weighting factors into the sum and difference data. All the parameters are varied simultaneously to obtain the best fit, so that the intensity and anisotropy decay parameters are all optimized to match the actual data, which are the polarized intensity decays.

Sometimes it is necessary to correct the polarized intensity decays for background signals. The counts measured for the blank sample with each polarizer position are subtracted from the measured data for each polarizer position,

$$I_{\|}(t_k) = I_{\|}(t_k)_{\text{sample}} - I_{\|}(t_k)_{\text{background}} \quad [11.25]$$

$$I_{\perp}(t_k) = I_{\perp}(t_k)_{\text{sample}} - I_{\perp}(t_k)_{\text{background}} \quad [11.26]$$

The weighting factor for the corrected data is given by the sum of the weighting factors for the sample and for the background. If the number of background counts is small, this correction to the weighting factor can be ignored. It is important to measure the background for both polarized components, because the background can be different for each component, particularly if scattered light reaches the detector.

It is necessary to know the G-factor in order to calculate the intensity decay. This can be obtained in the usual manner (Chapter 10), in which the intensities are measured with horizontally polarized excitation. For this measurement, one normally measures the polarized intensities for the same period of time, assuming the excitation is constant. Another method is to measure the steady-state anisotropy of the sample, which is then used to constrain the total intensities of the polarized decays. The steady-state anisotropy can be measured on a different instrument. The G-factor can be calculated using[10,12]

$$G = \frac{1-r}{1+2r} \frac{\sum N_{\|}(t_k)}{\sum N_{\perp}(t_k)} \quad [11.27]$$

where the sums are the total number of counts in the polarized intensity decays. If the excitation intensity has changed, or the counting time is different, these values need to be corrected for the different experimental conditions.

11.2.B. Value of r_0

In the anisotropy decay analysis, the value of r_0 can be considered to be a known or an unknown value. If the value of r_0 is known, the anisotropy decay law can be written as

$$r(t) = r_0 \sum_j g_j \exp(-t/\theta_j) \quad [11.28]$$

where the g_j are the fractional amplitudes which decay with the correlation times θ_j. Since $\Sigma g_j = 1.0$, the use of a known r_0 value reduces by one the number of variable parameters. In this type of analysis, the time-zero anisotropy is forced to be equal to r_0.

Alternatively, the total anisotropy can be a variable parameter. In this case,

$$r(t) = \sum_j r_{0j} \exp(-t/\theta_j) \quad [11.29]$$

where the r_{0j} are the fractional anisotropies which decay with the correlation times θ_j. When using this type of analysis, we will refer to the time-zero anisotropy as $r(0)$. If the anisotropy decay contains fast components which are not resolved by the instrument, one usually finds that $\Sigma r_{0j} = r(0)$ is less than the fundamental anisotropy r_0.

11.3. ANALYSIS OF FREQUENCY-DOMAIN ANISOTROPY DECAYS

Analysis of the FD anisotropy data is performed in a manner similar to the intensity decay analysis (Chapter 5). However, there is a somewhat more complex relationship between the data (Δ_ω and Λ_ω) and the transforms. The expected values of Δ_ω ($\Delta_{c\omega}$) and Λ_ω ($\Lambda_{c\omega}$) can be calculated from the sine and cosine transforms of the individual polarized decays,[3,16,17]

$$N_k = \int_0^\infty I_k(t) \sin \omega t \, dt \quad [11.30]$$

$$D_k = \int_0^\infty I_k(t) \cos \omega t \, dt \quad [11.31]$$

where the subscript k indicates the orientation, parallel ($\|$) or perpendicular ($\perp$). The calculated values of Δ_ω and Λ_ω are given by

$$\Delta_{c\omega} = \arctan\left(\frac{D_{\|}N_{\perp} - N_{\|}D_{\perp}}{N_{\|}N_{\perp} + D_{\|}D_{\perp}}\right) \quad [11.32]$$

$$\Lambda_{c\omega} = \left(\frac{N_{\|}^2 + D_{\|}^2}{N_{\perp}^2 + D_{\perp}^2}\right)^{1/2} \quad [11.33]$$

where the N_i and D_i are calculated at each frequency. The parameters describing the anisotropy decay are obtained by minimizing the squared deviations between measured and calculated values, using

$$\chi_R^2 = \frac{1}{\nu} \sum_{\omega} \left(\frac{\Delta_\omega - \Delta_{c\omega}}{\delta\Delta} \right)^2 + \frac{1}{\nu} \sum_{\omega} \left(\frac{\Lambda_\omega - \Lambda_{c\omega}}{\delta\Lambda} \right)^2 \quad [11.34]$$

where $\delta\Delta$ and $\delta\Lambda$ are the uncertainties in the differential phase and modulation ratio, respectively.

In the FD analysis, the rotation-free intensity decay is measured in a separate experiment using magic-angle polarizer conditions. The parameter values, typically α_i and τ_i for the multiexponential model, are held constant during the calculation of χ_R^2 from Eq. [11.34]. In principle, one could measure the phase and modulation of the polarized components and use these data to recover $I(t)$ and $r(t)$. This would be analogous to the method used for TCSPC data. However, it is believed that the anisotropy decay is better determined by direct measurement of the difference (Δ_ω) and ratio (Λ_ω) values.

One may wonder why there is no mention of the G-factor in Eqs. [11.30]–[11.33]. Use of the G-factor is often unnecessary in analysis of the TD data (Section 11.2). This is because TD and FD measurements are typically performed using emission filters rather than a monochromator. The use of an emission monochromator in the steady-state anisotropy measurements is the dominant reason for the G-factor being different from unity. For most time-resolved instruments, especially those using MCP PMT detectors, the detection efficiency is the same for the parallel and perpendicular components, and hence $G = 1.0$.

In the FD measurements, one checks for a sensitivity to polarization by excitation with horizontally polarized light. The measured values of the differential polarized phase angle (Δ_ω) should be zero. Also, the measured value of the modulation ratio (Λ_ω) should be 1.0. If needed, FD anisotropy decays can be measured in a T-format to avoid rotating the emission polarizer.[18]

When the FD anisotropy data are analyzed using Eq. [11.34], the weighting factors are those appropriate for directly measured phase and modulation values. In our analyses we find values of $\delta\Delta_\omega = 0.2$ and $\delta\Lambda_\omega = 0.004$ to be appropriate for measurements. If one separately measured the polarized phase ($\phi_\parallel$ and $\phi_\perp$) and modulation ratios ($m_\parallel'$ and $m_\perp'$), then it would be necessary to propagate the uncertainties into the difference and ratio files, as was done for the TD analyses. A procedure to correct for background fluorescence has been described for the FD anisotropy measurements.[19] This procedure is somewhat more complex than the direct subtraction used for the TD data.

The values of r_0 and $r(0)$ are also treated the same way in the FD analysis as in the TD analysis. The value of r_0 can be a fixed parameter (Eq. [11.28]), or the time-zero anisotropy [$r(0)$] can be a variable in the analysis (Eq. [11.29]). While fixing the value of r_0 avoids the problem of missing a short correlation time present in the sample, the use of an inappropriately large value of r_0 will result in the appearance of a short correlation time in the calculated anisotropy decay.

11.4. ANISOTROPY DECAY LAWS

Depending upon the size and shape of the fluorophore, and its local environment, a wide variety of anisotropy decays are possible. A spherical molecule displays a single rotational correlation time. Anisotropy decays can be more complex if the fluorophore is nonspherical, or if a nonspherical molecule is located in an anisotropic environment. Another origin of complex anisotropy decays is internal flexibility of a fluorophore within a larger macromolecule.

11.4.A. Nonspherical Fluorophores

One origin of multiple correlation times is a nonspherical shape. If a molecule is not spherical, one can imagine different rotational rates around each axis. For instance, perylene is a disklike molecule, and the in-plane rotations are expected to be more rapid than the out-of-plane rotations. The out-of-plane motion requires the displacement of solvent molecules. The in-plane rotations probably require less displacement of solvent and are thus expected to be more rapid. Such a molecule is referred to as an anisotropic rotor. Generally, macromolecules are nonsymmetric, and one expects different rotational diffusion rates about each axis.

The theory for rotational diffusion of anisotropic rotors is complex. This topic is well understood and is described in more detail in Chapter 12. Considerable controversy has surrounded the predicted time-resolved decays for anisotropic molecules.[20–25] It is now agreed[20] that the anisotropy is expected to decay as a sum of exponentials:

$$r(t) = \sum_{j=1}^{5} r_{0j}\, e^{-t/\theta_j} \quad [11.35]$$

There may be as many as five exponential terms for an asymmetric body, but in practice only three correlation times are expected to be distinguishable.[25] For ellipsoids of revolution, which are elongated (prolate) or flattened (oblate) molecules with two equal axes, the anisotropy decay can display only three correlation times. The values of r_{0j} and θ_j are complex functions of the rates of rotation around the molecular axes of the nonsymmetric body and the orientation of the absorption and emission dipoles relative to these axes. In practice, one rarely resolves more than two exponentials, but it is important to recognize that such anisotropic rotations can result in multiexponential decays of anisotropy. For small molecules in solution, the rotational rates around the different axes are rarely different by more than a factor of 10. The resolution of such similar rates is difficult but has been accomplished using TD and FD measurements. The theory of anisotropic rotors is described in Chapter 12, along with examples of the resolution of multiple correlation times.

It is important to recognize that the theory for rotation of nonspherical molecules assumes hydrodynamic behavior, in which the rates of rotation are determined by the viscous drag of the solvent. This theory fails for many small molecules in solution. This failure occurs because small molecules can slip within the solvent, particularly if the motion does not displace solvent or if the molecule is not hydrogen-bonded to the solvent. In these cases one can recover a multiexponential anisotropy decay, but the values of r_{0j} and θ_j may not be understandable using Eq. [11.35] with values or r_{0j} and θ_j appropriate for hydrodynamic rotational diffusion.

When dealing with nonspherical molecules, it is useful to have a definition for the mean correlation time. The most commonly used average is the harmonic mean correlation time θ_H, which is given by

$$\frac{1}{\theta_H} = \frac{\sum_j r_{0j}/\theta_j}{\sum_j r_{0j}} = \frac{1}{r_0} \sum_j \frac{r_{0j}}{\theta_j} \quad [11.36]$$

This expression is sometimes used with $r(0)$ in place of r_0. For a nonspherical molecule, the initial slope of the anisotropy decay is determined by the harmonic mean correlation time.[26]

11.4.B. Hindered Rotors

Decays of fluorescence anisotropy can be complex even for molecules which are isotropic rotors, if these molecules are contained in an anisotropic environment. For example, the emission dipole of DPH is oriented approximately along its long molecular axis. The rotations which displace this dipole are expected to be isotropic (Chapter 12) because the molecule is nearly symmetrical about this axis. Rotation about the long axis of the molecule is expected to be faster than the other rotational rates, but this fast rotation does not displace the emission dipole and hence does not depolarize the emission. Hence, only rotation which displaces the long axis of DPH is expected to depolarize the emission. In solvents, only a single type of rotational motion displaces the emission dipole of DPH, and its anisotropy decay is a single exponential.

However, when DPH is in membranes, one usually finds a complex anisotropy decay.[27–30] The rotational motions of DPH are hindered, and one finds that the anisotropy does not decay to zero. By hindered, we mean that the angular range of the rotational motion is limited. In such cases, a limiting anisotropy (r_∞) is observed at times that are long compared to the fluorescence lifetime. The anisotropy decay is described by

$$r(t) = (r_0 - r_\infty)\exp(-t/\theta) + r_\infty \quad [11.37]$$

In this simple approximate model for a hindered rotor, one assumes that the decay from r_0 to r_∞ still occurs exponentially. More complex expressions may be necessary for a rigorous analysis,[31–33] but the actual data are rarely adequate to resolve more than one correlation time for a hindered rotation. The constant term r_∞ has been interpreted as resulting from an energy barrier that prevents rotational diffusion of the fluorophore beyond a certain angle. Interpretation of the r_∞ values will be discussed in Section 11.5.

11.4.C. Segmental Mobility of a Biopolymer-Bound Fluorophore

Consider a fluorophore which is bound to a macromolecule, and assume that the fluorophore can undergo segmental motions with a fast correlation time θ_F. Let θ_P be the slow correlation time for overall rotation of the macromolecule. A number of theoretical treatments have appeared.[34–36] These more rigorous treatments lead to various expressions for $r(t)$, most of which are well approximated by the simple expressions presented below. Assume that the segmental motions of the fluorophore occur independently of the rotational motion of the macromolecules. Then the anisotropy at any time t is a product of the separate factors responsible for depolarization:

$$r(t) = r_0[\alpha e^{-t/\theta_F} + (1-\alpha)]e^{-t/\theta_P} \quad [11.38]$$

This may be regarded as a slightly more complex case of the hindered rotor in which the anisotropy decays rapidly to $r_\infty = r_0\,(1-\alpha)$ as a result of the segmental motion. However, the anisotropy continues to decay to zero as a result of the overall rotation of the macromolecule. The effect of segmental fluorophore motion within a macromolecule is the appearance of a multiexponential anisotropy decay. This can be understood by multiplying the terms in Eq. [11.38], resulting in two exponentially decaying terms. It is interesting to note that the faster motion must be hindered ($\alpha < 1$) in order for a multiexponential decay of $r(t)$ to be observed. If the segmental motion were completely free, that is, $\alpha = 1$, then the anisotropy would decay with a single apparent correlation time (θ_A). This apparent correlation time would be given by $\theta_A = \theta_P\theta_F/(\theta_P + \theta_F)$. The existence of the segmental motion would only be revealed by the small magnitude of θ_A, relative to the correlation time expected for the macromolecule.

In analyzing time-resolved anisotropy decays, one generally fits the observed values to a sum of exponential decays. Hence, a decay of the form shown by Eq. [11.38] is generally fit to

$$r(t) = r_0(f_S e^{-t/\theta_S} + f_L e^{-t/\theta_L}) \quad [11.39]$$

where the subscripts S and L refer to the short and long correlation times. From comparison of Eqs. [11.38] and [11.39], one can derive the following relationships between the parameters:

$$f_S = \alpha, \qquad f_L = (1 - \alpha) \quad [11.40]$$

$$\frac{1}{\theta_S} = \frac{1}{\theta_F} + \frac{1}{\theta_P}, \qquad \frac{1}{\theta_L} = \frac{1}{\theta_P} \quad [11.41]$$

From Eq. [11.41] one sees that it is acceptable to equate the longer correlation time with that of the overall rotational motion. However, the shorter observed correlation time is not strictly equal to the correlation time of the segmental motion. Only when $\theta_F << \theta_P$ is $\theta_S = \theta_F$, the actual correlation time of the fast motion.

The effects of a fast segmental motion on the TD and FD data were shown in Figures 11.6 and 11.7, respectively. The presence of a 50-ps correlation time results in a rapid initial decrease in $r(t)$. The amplitude of this rapid component depends on the amplitude of the motion, which in Figure 11.6 is assumed to account for half of the total anisotropy ($r_{01} = 0.2$). Following the rapid decrease, the anisotropy decays by the longer correlation time of 5 ns. This is the basis of estimating the overall correlation time of a protein from the correlation time observed at longer times. When the value of θ_F is much less than that of θ_P, and less than the instrumental resolution, the effect of the fast motion is to decrease the apparent value of r_0. The remaining anisotropy decays with a correlation time of θ_P.

The presence of a short correlation time results in a complex appearance of the FD anisotropy data. The single bell-shaped Δ_ω curve is replaced by a more complex curve (Figure 11.7), which is composed of contributions from the two correlation times (dashed curves). Depending on the upper frequency limit of the FD instrument, it may not be possible to measure the maximum values of Δ_ω due to the faster motion. In these cases one observes that the value of Δ_ω increases up to the highest measured frequency. The presence of a rapid correlation time also results in complex behavior for the modulated anisotropy (r_ω). Depending on the upper frequency limit of the measurement, the values of r_ω may not reach the value of r_0.

11.4.D. Correlation Time Distributions • Advanced Material •

As was seen for intensity decays, anisotropy decays may also be analyzed in terms of distributions of correlation times.[37–39] One approach is to describe the correlation time spread in terms of a Gaussian, Lorentzian, or other distribution. The Gaussian (G) and Lorentzian (L) distributions are given by

$$p_G(\theta) = \frac{1}{\sigma\sqrt{2\pi}} \exp\left[-\frac{1}{2}\left(\frac{\theta - \bar{\theta}}{\sigma}\right)^2\right] \quad [11.42]$$

$$p_L(\theta) = \frac{1}{\pi} \frac{\Gamma/2}{(\theta - \bar{\theta})^2 + (\Gamma/2)^2} \quad [11.43]$$

In these expressions $\bar{\theta}$ is the central value, σ is the standard deviation of the Gaussian, and Γ is the full width at half-maximum of the Lorentzian.

Suppose that the anisotropy decay is described by a single-modal distribution, with a single mean value ($\bar{\theta}$). That part of the anisotropy which displays a correlation time θ is given by

$$r(t, \theta) = r_0 p(\theta) \exp(-t/\theta) \quad [11.44]$$

where $p(\theta)$ is the probability of a particular correlation time θ. However, one cannot selectively observe that fraction of the anisotropy which decays with θ, but rather must observe all components. Hence, the observed anisotropy decay is given by the integral equation

$$r(t) = r_0 \int_0^\infty p(\theta) \exp(-t/\theta)\, d\theta \quad [11.45]$$

By analogy with a multimodal lifetime distribution, it is possible to describe the anisotropy decay by a multimodal correlation time distribution. In this case the amplitude which decays with a correlation time θ is given by

$$r(t, \theta) = \sum_j r_{0j} p_j(\theta) \exp(-t/\theta) \quad [11.46]$$

and the observed anisotropy decay is given by

$$r(t) = \sum_j r_{0j} \int_0^\infty p_j(\theta) \exp(-t/\theta)\, d\theta \qquad [11.47]$$

In this formulation the distribution shape factors are normalized so that the integrated probability of each mode of the distribution is equal to unity. Equations [11.45] and [11.47] are properly normalized only if none of the probability occurs below zero.[39] Depending on the values of $\bar{\theta}$, σ, or Γ, part of the probability for the Gaussian or Lorentzian distributions (Eqs. [11.42] and [11.43]) can occur below zero, even if $\bar{\theta}$ is larger than zero. The correlation time distributions can also be obtained using maximum entropy methods, typically without using assumed shapes for the distribution functions.[37,38]

11.4.E. Associated Anisotropy Decays

Multiexponential anisotropy decay can also occur for a mixture of independently rotating fluorophores. As an example, such anisotropy decay can occur for a fluorophore when some of the fluorophore molecules are bound to protein and some are free in solution. The anisotropy from the mixture is an intensity-weighted average of the contribution from the probe in each environment,

$$r(t) = r_1(t)f_1(t) + r_2(t)f_2(t) \qquad [11.48]$$

where $r_1(t)$ and $r_2(t)$ are the anisotropy decays in each environment. The fractional time-dependent intensities for each fluorophore are determined by the decay times in each environment. For single-exponential decays, these fractional contributions are given by

$$f_i(t) = \frac{\alpha_i \exp(-t/\tau_i)}{\alpha_1 \exp(-t/\tau_1) + \alpha_2 \exp(-t/\tau_2)} \qquad [11.49]$$

Such systems can yield unusual anisotropy decays which show minima and increases in anisotropies at long times.[40–43] Associated anisotropy decays are described in more detail in Chapter 12.

11.5. HINDERED ROTATIONAL DIFFUSION IN MEMBRANES

We now consider examples of anisotropy decays which have been extensively used to characterize model and real biological membranes. One of the most widely used probes is DPH, originally proposed as a probe to estimate the microviscosity of cell membranes.[44] The basic idea was to compare the anisotropy observed for the membrane-bound probe with that observed for the probe in solutions of known viscosity. By comparison, the microviscosity of the membrane could be calculated. This procedure assumes that the rotational motions are the same in the reference solvent and in the membranes.

A frequently used reference solvent for DPH is mineral oil. This solvent is used because the decay times of DPH in mineral oil are mostly independent of temperature, whereas the decay times in propylene glycol are dependent on temperature. Polarized intensity decays of DPH are shown in Figure 11.8. In mineral oil the difference between

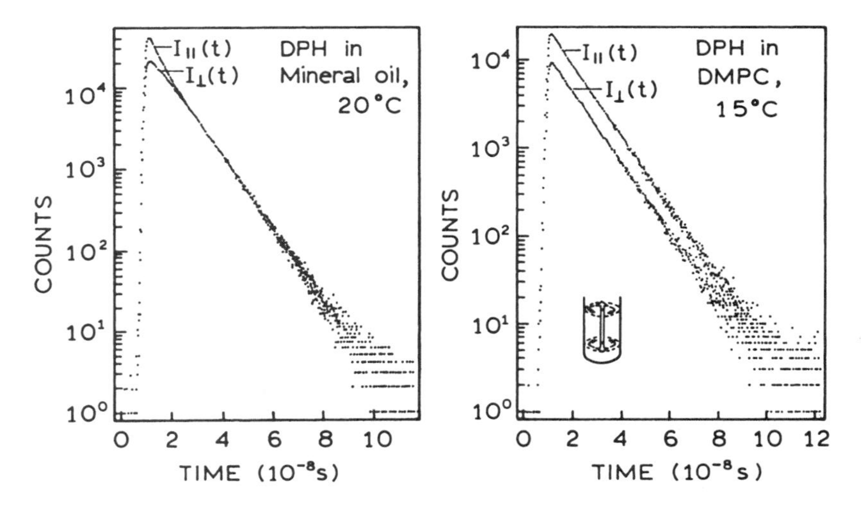

Figure 11.8. Polarized intensity decays of DPH in mineral oil at 20 °C (*left*) and in DMPC vesicles at 15 °C (*right*). Similar data have been reported from several laboratories.[10,27,30]

the polarized intensities decays to zero, indicating that the anisotropy decays to zero. Contrasting results were found for DPH in vesicles of DMPC. In this case the polarized intensities remain different during the entire decay. This result indicates that the anisotropy does not decay to zero at long times (Figure 11.9). At higher temperature, above the membrane phase transition, the long-time anisotropy becomes closer to zero, as shown for DMPC residues at 37.8 °C (Figure 11.9).

A typical result from these studies is that the presence of cholesterol in the membranes results in more hindered rotational diffusion than in the absence of cholesterol. This can be seen in the experimental anisotropy decays of DPH in DPPC vesicles, as the mole fraction of cholesterol is increased (Figure 11.10). Similar behavior has been observed for DPH in a wide range of phospholipids[45–48] and for other membrane-bound probes.[49–52] Such behavior is a general feature of the anisotropy decays of labeled membranes.

Considerable attention has been given to the molecular interpretation of the limiting anisotropies (r_∞). The impetus for the analysis arises from a desire to understand the properties of the membranes that are responsible for the hindered rotation. In one analysis,[31,53] the rodlike DPH molecule is assumed to exist in a square-well potential such that its rotation is unhindered until a certain angle (θ_C) is reached. Rotation beyond this angle is assumed to be

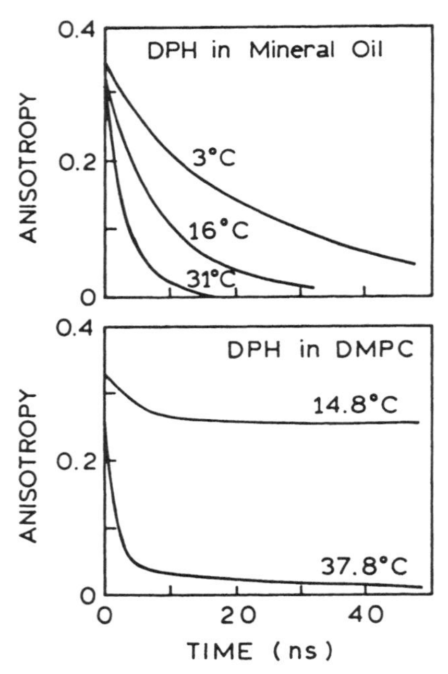

Figure 11.9. Anisotropy decays of DPH in mineral oil (*top*) and in DMPC vesicles (*bottom*). Redrawn from Refs. 10 and 27.

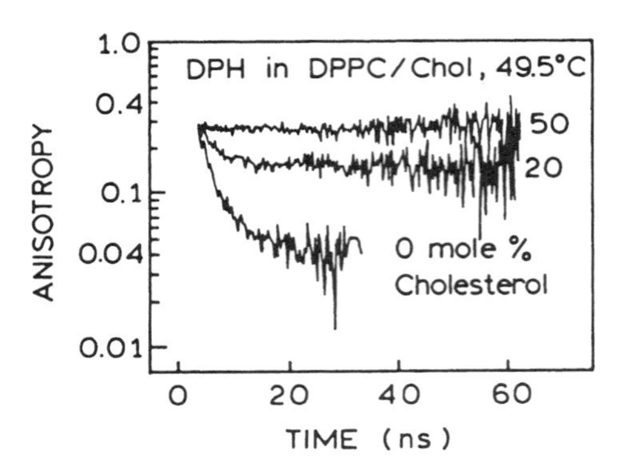

Figure 11.10. Anisotropy decays of DPH in DPPC vesicles at 49.5 °C, containing 0, 20, and 50 mol % cholesterol. At 49.5 °C the DPPC membranes are above their phase-transition temperature, which is near 37 °C. Revised from Ref. 29.

energetically impossible. In this model the limiting anisotropy is related to the cone angle θ_C by

$$\frac{r_\infty}{r_0} = S^2 = \left[\frac{1}{2}(\cos\theta_C)(1+\cos\theta_C)\right]^2 \qquad [11.50]$$

This ratio is also equal to the square of the order parameter (S). Completely unhindered motion is found for $\theta_C = 90°$. The limiting anisotropies of DPH can be interpreted in terms of this model. As the cholesterol content of DPPC vesicles increases, the cone angle decreases (Figure 11.11, left), with the effect being much smaller at higher temperature. If one varies the temperature of the vesicles, the cone angle of DPH in pure DPPC vesicles increases dramatically at the phase-transition temperature near 37 °C, and the presence of cholesterol prevents free rotation of DPH at all temperatures (Figure 11.11, right).

While this interpretation of the r_∞ values is intuitively pleasing, the values of θ_C should be interpreted with caution. One difficulty of this model is that the calculation of θ_C from r_∞/r_0 depends upon the existence of a square-well

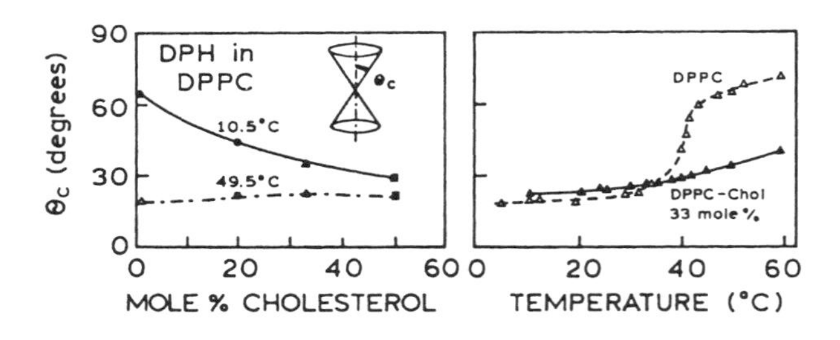

Figure 11.11. Cone angle (Eq. [11.50]) for rotational diffusion of DPH in DPPC vesicles as a function of cholesterol of the vesicles and temperature. Revised from Ref. 29.

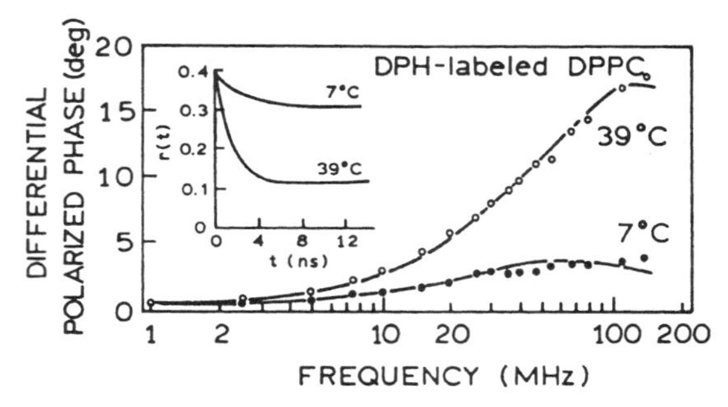

Figure 11.12. Frequency-dependent values of Δ_ω for DPH in DPPC vesicles. The inset shows the recovered anisotropy decay. From Ref. 60.

potential in the membranes. The fact that a nonzero value of r_∞ is observed demonstrates the existence of a barrier to rotation but does not demonstrate that the barrier is as abrupt as a square-well potential. For this reason, caution is advised in the interpretation of θ_C values derived from observed values of r_∞.

Alternatively, it has been shown that r_∞ is related to the order parameter describing the equilibrium orientation distribution of the probe at times much longer than the rotational correlation time.[36,54–57] Specifically,

$$\frac{r_\infty}{r_0} = \left\langle \frac{3\cos^2\theta - 1}{2} \right\rangle^2 \qquad [11.51]$$

where the angle brackets indicate an average over all fluorophores, and θ is the angular rotation of DPH in the membrane. This result is claimed to be independent of any assumed model except for the assumption of cylindrical symmetry. In this case, $r_\infty = 0$ when the average value of θ reaches 54.7°.

The presence of hindered rotations of DPH results in unusual FD data.[58–60] The effect of a nonzero value of r_∞ is a uniform decrease in the differential phase angles (Figure 11.12). Hence, the values of Δ_ω are much smaller below the transition temperature of DPPC vesicles and increase dramatically above the transition temperature, when the DPH molecules can rotate freely. These data can be interpreted using Eqs. [11.30]–[11.34] to recover the time-dependent anisotropy (Figure 11.12, inset).

11.6. TIME-DOMAIN ANISOTROPY DECAYS OF PROTEINS

During the past 15 years, there have been numerous anisotropy decay measurements on proteins, and it is not practical to even cite the many references. Rather, we choose to present examples which illustrate the range of behavior found for proteins.

11.6.A. Intrinsic Tryptophan Anisotropy Decay of Liver Alcohol Dehydrogenase

Liver alcohol dehydrogenase (LADH) is a dimer with two tryptophan residues in each identical subunit and a total molecular weight of 80 kDa. One of the residues is exposed to the solvent (trp-15), and one residue is buried (trp-314). This buried residue can be selectively excited on the red edge of the absorption spectrum at 300 nm.[61]

The anisotropy decay of LADH excited at 300 nm is shown in Figure 11.13. The decay was found to be a single exponential with a correlation time of 33 ns.[61] This single correlation time can be compared with that predicted for a hydrated sphere (0.2 g of H_2O/g of protein) from Eq. [10.52]), which is 31 ns at 20 °C. Hence, this tryptophan residue appears to be rigidly held within the protein matrix.

Even though trp-314 appears to rotate with the protein, the data still suggest the presence of some segmental mobility. This is evident from the apparent time-zero anisotropy, $r(0) = 0.22$, which is less than the fundamental anisotropy of tryptophan at this excitation wavelength. The motions responsible for this loss of anisotropy may be on a timescale faster than the resolution of these measurements. Additionally, the studies have suggested that the apparent correlation times are different with different excitation wavelengths.[15] This cannot occur for a sphere but can occur for nonspherical molecules if the orientation of the transition in the molecule is different for different excitation wavelengths. LADH is thought to be shaped like a prolate ellipsoid with semi-axes of 11 and 6 nm and an axial ratio near 1.8 (Chapter 12).

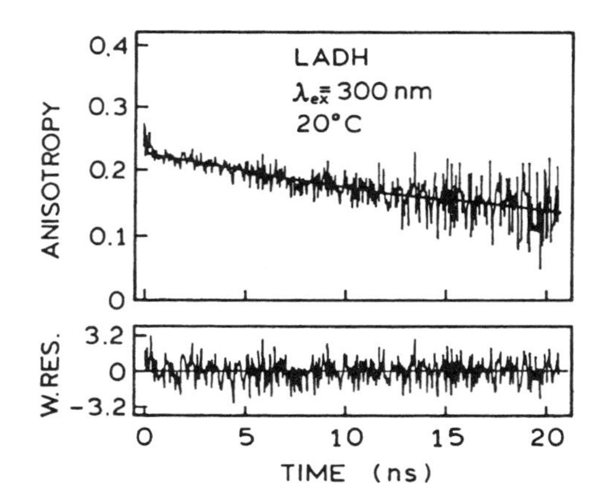

Figure 11.13. Anisotropy decay of liver alcohol dehydrogenase (LADH) excited at 300 nm. The apparent time-zero anisotropy is $r(0) = 0.22$. Revised from Ref. 61.

11.6.B. Phospholipase A_2

An anisotropy decay more typical of proteins is shown by phospholipase A_2. This enzyme catalyzes the hydrolysis of phospholipids and is active mostly when located at a lipid–water interface. Phospholipase A_2 has a single tryptophan residue (trp-3), which serves as the intrinsic probe. The anisotropy decay is clearly more complex than a single exponential.[62] At long times, the correlation time is 6.5 ns, consistent with overall rotational diffusion. However, in comparison with LADH, there is a dramatic decrease in anisotropy at short times (Figure 11.14). The correlation time of the fast component is less than 50 ps, and this motion accounts for one-third of the total anisotropy. Independent tryptophan motions have been observed in a large number of proteins[63–65] and have been predicted by molecular dynamics calculations.[66] Fast components in the anisotropy decay are also observed for labeled proteins.[67,68] Hence, segmental motions of intrinsic and extrinsic fluorophores appear to be a common feature of proteins.

11.6.C. Domain Motions of Immunoglobulins

Anisotropy decay measurements have been used to examine the flexibility of immunoglobulins in solution.[69–73] Early studies of immunoglobulin G (IgG) suggested independent motions of the F_{ab} fragments, independent of overall rotational motion. Many immunoglobulins are Y-shaped proteins. The two tops of the Y are the F_{ab} regions, which bind to the antigen. In the case of IgE (Figure 11.15), the bottom of the Y is the F_c fragment, which binds to a receptor on the plasma membrane.

In order to study IgE dynamics, the antigen dansyl-lysine was bound to the antigen binding sites on the F_{ab} regions. The

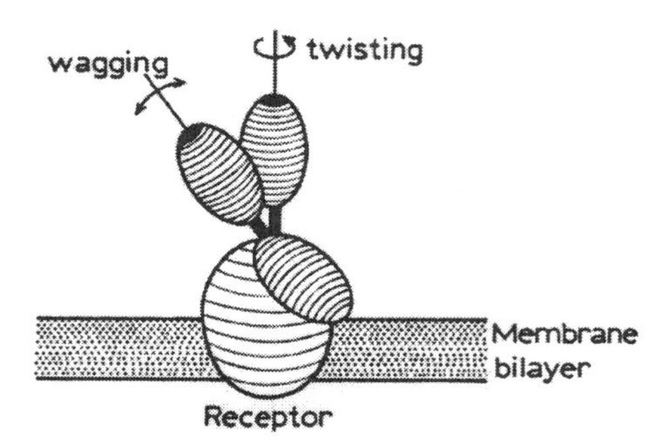

Figure 11.15. Model of IgG complexed with the plasma membrane receptor. The arrows suggest modes of segmental motion of the F_{ab} fragments. Reprinted, with permission, from Ref. 74, Copyright © 1990,

anisotropy decays are dramatically different when the IgE is free in solution and when it is bound to the membrane receptor (Figure 11.16). When it is bound to the receptor, there is a long correlation time of 438 ns (Table 11.1). This correlation time is too long for overall rotational diffusion of the protein and thus reflects the anisotropy decay of the membrane-bound form of IgE. The actual correlation time is probably longer, the reported correlation time being the longest correlation time observable with the 27-ns intensity decay time of the dansyl-lysine.

Domain motions within the IgE molecule are evident from the multiexponential fits to the anisotropy decays (Table 11.1). When free in solution, IgE displays two correlation times of 48 and 125 ns. The larger value is due to overall rotation of IgE, and the shorter value is due to independent motions of the F_{ab} fragments. Assignment of the 48-ns correlation time to F_{ab} motion is supported by the same correlation time being present when the antibody is

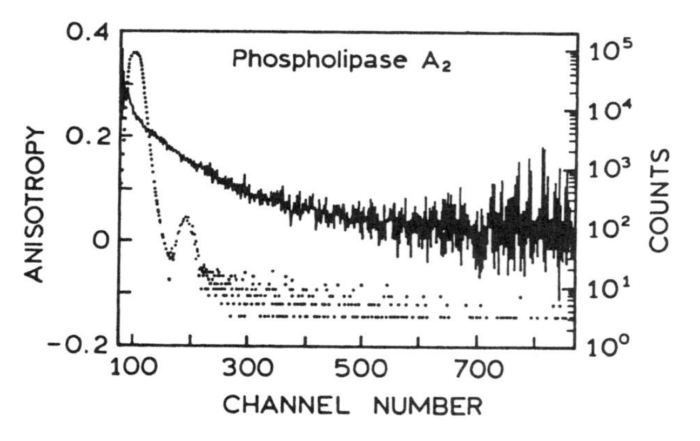

Figure 11.14. Anisotropy decay of trp-3 in phospholipase A_2; 30 ps per channel. The anisotropy decay parameters are $r_{01} = 0.104$, $r_{02} = 0.204$, $\theta_1 < 50$ ps, and $\theta_2 = 6.5$ ns. Revised from Ref. 62.

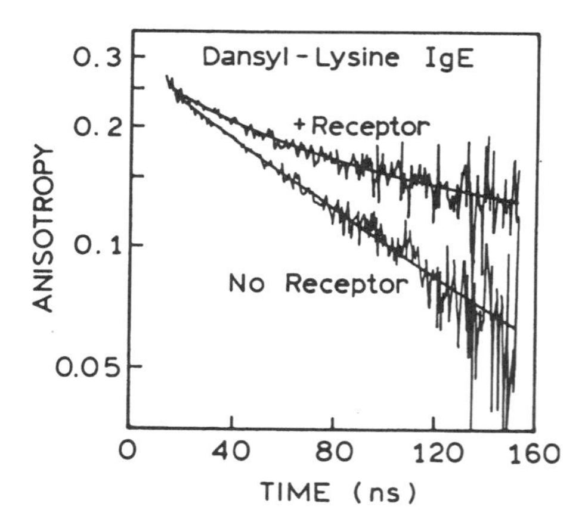

Figure 11.16. Anisotropy decay of dansyl-lysine bound to the antigen binding sites of IgE in the absence and in the presence of the membrane receptor. Revised and reprinted, with permission, from Ref. 74, Copyright © 1990, American Chemical Society.

Table 11.1. Anisotropy Decays of Dansyl-Lysine Bound to IgE[a]

Sample	g_1^b	θ_1 (ns)	g_2^b	θ_2 (ns)
IgE in solution	0.30	48	0.70	125
IgE receptor	0.32	34	0.68	438

[a]From Ref. 74.
[b]The g_j values represent the fraction of the total anisotropy which decays with the correlation time θ_j.

Table 11.2. Flavin Mononucleotide Intensity and Anisotropy Decays in the Presence of Yellow Fluorescent Protein (YFP)[a]

[YFP]	α_1^b	τ_1 (ns)	τ_2 (ns)	r_{01}	θ_1 (ns)	r_{02}	θ_2 (ns)	r^c
5.76μ*M*	0.08	4.4	7.6	0.02	0.15	0.29	14.8	0.157
0.72μ*M*	0.34	4.4	7.6	0.17	0.15	0.19	14.8	0.108
0.18μ*M*	0.69	4.4	7.6	0.26	0.14	0.11	14.8	0.071

[a]From Ref. 75.
[b]$\alpha_1 + \alpha_2 = 1.0$.
[c]Steady-state anisotropy.

bound to the receptor. This suggests that IgE interacts with its receptor through the F_c region and that this interaction does not inhibit motion of the F_{ab} domains (Figure 11.15).

11.6.D. Effects of Free Probe on Anisotropy Decays

Anisotropy decays of intrinsic and extrinsic probes frequently show subnanosecond components, which are usually due to rapid segmental motions. However, such components should be interpreted with caution and can be due to scattered light reaching the detector. Another origin of rapid anisotropy components is the presence of unbound probe in a sample thought to contain only the labeled macromolecule. A free probe will typically display a 50–100-ps correlation time, which can easily be mistakenly attributed to segmental motion.

The effect of free probe is illustrated by anisotropy decays of the yellow fluorescent protein (YFP) from *Vibrio fischeri*. This protein is from a bioluminescent bacterium, and the emitting fluorophore is FMN. The intensity decay time of FMN is 4.4 ns in solution and 7.6 ns when bound to YFP. The binding constant of FMN to YFP is only modest, so that, depending on YFP concentration, some of the FMN can dissociate.

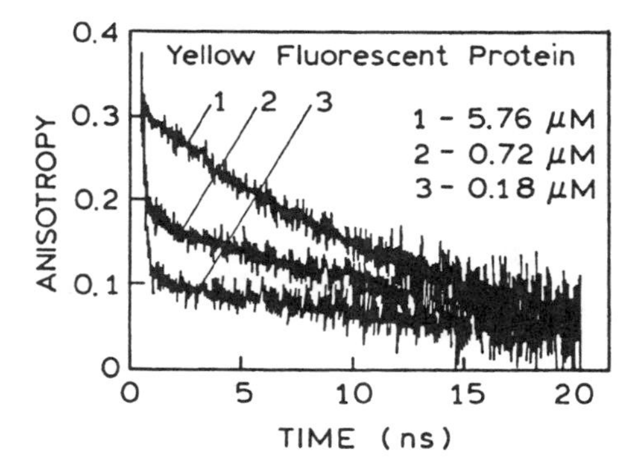

Figure 11.17. Anisotropy decays of yellow fluorescent protein at three concentrations. Revised and reprinted,with permission, from Ref. 75, Copyright © 1997, American Society for Photobiology.

Anisotropy decays of YFP are shown in Figure 11.17. At higher protein concentrations, the decay is predominantly due to a 14.8-ns correlation time assigned to overall protein rotation. This correlation is longer than expected for a protein with a molecular weight of 22.7 kDa (near 9 ns), which suggests an elongated shape for the protein, so it appears that the FMN is rigidly bound to YFP. As the protein is diluted, the anisotropy decay shows a fast component near 0.15 ns, which has been assigned to free FMN (Table 11.2). The fast correlation time is probably beyond the time resolution of the measurements, so that the actual value may be smaller than 0.15 ns. As the protein concentration is decreased, the amplitude of the short-decay-time (4.4 ns) and fast-correlation-time (0.15 ns) component increases because a larger fraction of FMN is in the free form. Such data can be used to calculate the dissociation constant of FMN from the protein.

11.7. FREQUENCY-DOMAIN ANISOTROPY DECAYS OF PROTEINS

The FD method provides good resolution of the complex anisotropy decays displayed by proteins.[76–81] Several examples are provided in Chapter 17, so only a few examples will be presented here.

11.7.A. Apomyoglobin—A Rigid Rotor

Apomyoglobin is known to bind a number of fluorescent probes in its hydrophobic heme binding site. One example is TNS, which is essentially nonfluorescent in aqueous solution and becomes highly fluorescent when bound to apomyoglobin. Differential polarized phase angles for TNS-labeled apomyoglobin are shown in Figure 11.18. These values are consistent with a correlation time of 20.5 ns and an $r(0)$ value of 0.331. Since the r_0 value recovered from the FD data agrees with the frozen solution value ($r_0 = 0.32$), these data indicate that TNS-apomyoglobin

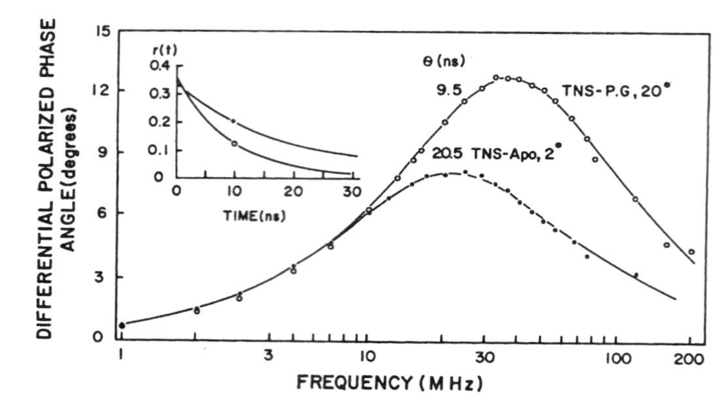

Figure 11.18. FD anisotropy decays of TNS in propylene glycol (PG) and bound to apomyoglobin (Apo). *Inset*: TD anisotropy decays. From Ref. 81.

rotates as a rigid body, without significant free rotation of the TNS group.[81]

Also shown in Figure 11.18 are FD data for rotational diffusion of TNS in propylene glycol at 20 °C. At this temperature the decay time of TNS is propylene glycol is 7.8 ns. The recovered correlation time was 9.5 ns, with $r(0) = 0.351$. As is typical for polar fluorophores in polar solvents, TNS appears to rotate like a spherical molecule.

11.7.B. Melittin Self-Association and Anisotropy Decays

Melittin is a small protein (26 amino acids) which self-associates into tetramers. Melittin was labeled with an anthraniloyl moiety [*N*-methylanthraniloylamide (NMA)] to serve as an extrinsic probe.[78] FD data for the monomeric and tetrameric forms of melittin are shown in Figure 11.19. Also shown are the FD data for the free probe (NMA) not bound to protein. The values of Δ_ω for the free probe are close to zero for all frequencies below 20 MHz and increase to only several degrees near 150 MHz. Also, the modulated anisotropies (r_ω) are near zero at all measurable frequencies. These low values are due to the rapid (73 ps) correlation time of the free probe.

The shapes of the Δ_ω and r_ω plots are rather different for the monomeric (M) and tetrameric (T) forms of melittin. In both cases, the anisotropy decays are complex due to significant segmental mobility of the probe, with a correlation time near 0.2 ns. Upon formation of a tetramer, the shape of the Δ_ω curve shows evidence of overall rotational diffusion with a dominant correlation time near 3.7 ns. In the monomeric state, overall rotational diffusion is not visible, but the data contain a substantial component near 1.6 ns, which is due to monomeric melittin. These changes in the overall rate of rotational diffusion upon tetramer formation can be easily seen in the values of r_ω, which are uniformly large for the tetramer. These dramatic changes in Δ_ω and r_ω show that the FD data are highly sensitive to rotational diffusion and local motions of proteins.

11.7.C. Picosecond Rotational Diffusion of Oxytocin

If the correlation times are very short, these fast motions can be missed in the data. In the time domain, the faster components are resolved by decreasing the width of the instrument response function. In the frequency domain, the resolution of faster components is accomplished by meas-

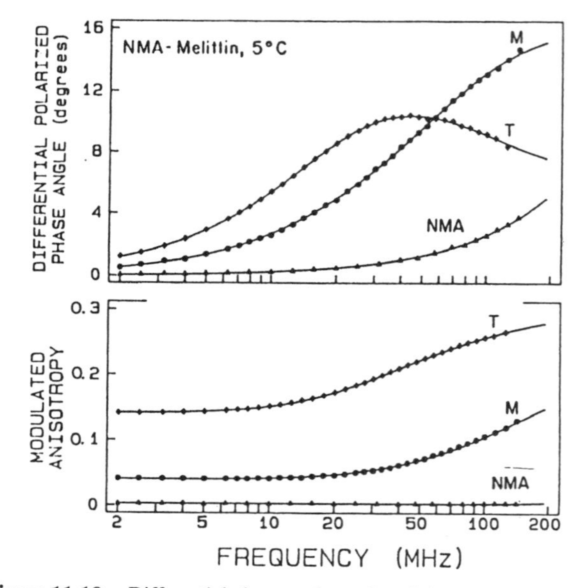

Figure 11.19. Differential phase angles and modulated anisotropies of *N*-methylanthraniloyl-melittin monomer (M) and tetramer (T). Also shown are data for the free probe *N*-methylanthraniloylamide (NMA) at 5 °C. Reprinted from Ref. 78, Copyright © 1986, with permission from Elsevier Science.

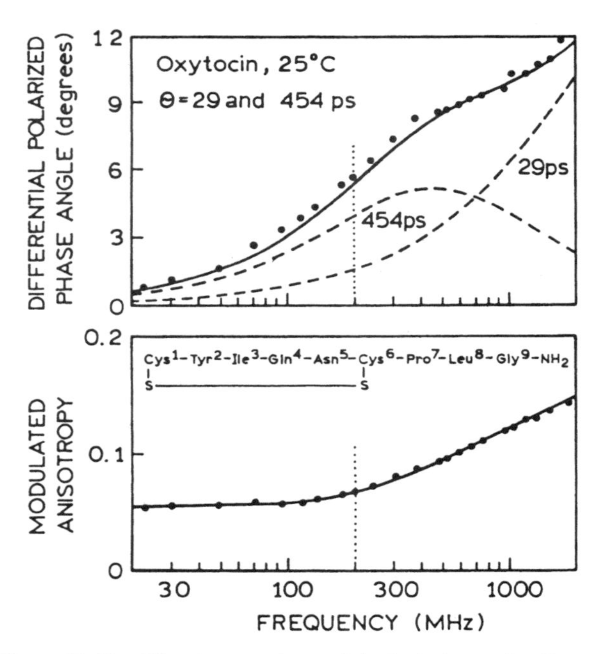

Figure 11.20. FD anisotropy decay of the intrinsic tyrosine fluorescence of oxytocin. The dashed curves show the values expected for $r_{01} = 0.208$, $\theta_1 = 29$ ps, $r_{02} = 0.112$, and $\theta_2 = 454$ ps. Revised and reprinted from Ref. 80, Copyright © 1986, with permission from Elsevier Science.

uring at higher light modulation frequencies (Section 5.7). Measurements to 2 GHz were used to resolve the picosecond anisotropy decay of oxytocin.

Oxytocin is a small cyclic polypeptide which contains nine amino acids and a single tyrosine residue. The FD anisotropy decay of the tyrosine flourescence is shown in Figure 11.20. The FD data to 200 MHz (vertical dotted line) show only increasing values of Δ_ω and little change in the modulated anisotropy. Hence, the data contain incomplete information on the anisotropy decay. This situation was improved by the use of instrumentation which allowed measurements to 2 GHz. In this case there is detectable shape in the values of Δ_ω, and one can see that there are components due to two correlation times, 29 and 454 ps. The 29-ps correlation time is due to segmental motions of the tyrosyl residues, and the 454-ps correlation time is due to overall rotation of the peptide.

11.8. MICROSECOND ANISOTROPY DECAYS

11.8.A. Phosphorescence Anisotropy Decays

The information available from an anisotropy decay is limited to times during which emission occurs. For this

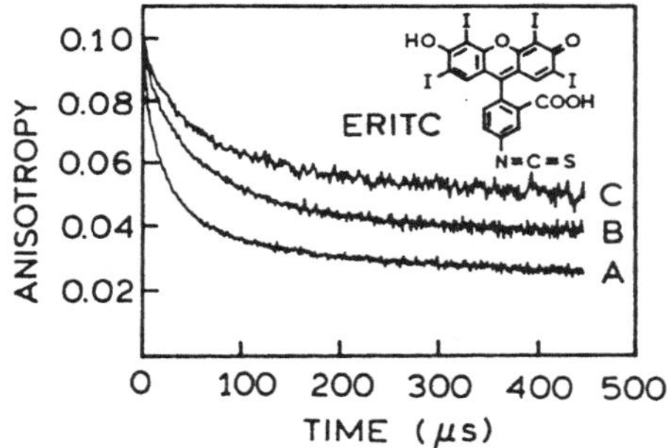

Figure 11.21. Phosphorescence decays of erythrosin-labeled Ca^{2+}-ATPase in sarcoplasmic reticulum vesicles. Anisotropy decays were obtained in the absence of melittin (A) and in the presence of acetylated melittin (B) and native melittin (C). ERITC is erythrosin-5-isothiocynate. Revised and reprinted, with permission, from Ref. 86, Copyright © 1995, American Chemical Society.

reason, it is usually difficult to obtain reliable data at times longer than three intensity decay times. Even for probes having relatively long lifetimes, such as dansyl-lysine (Section 11.6.C), the time window is too small to effectively study membrane-bound proteins.

Rotational motions at longer times can be measured using phosphorescence anisotropy decays.[82–86] These experiments are illustrated by studies of the sarcoplasmic reticulum Ca^{2+}-ATPase, which is a 110-kDa transmembrane protein. Not many phosphorescent probes are available, and one of the most commonly used is erythrosin, which in deoxygenated solution displays a phosphorescence decay time near 100 μs.[83] Typical phosphorescence anisotropy decays are shown in Figure 11.21 for erythrosin-labeled Ca^{2+}-ATPase. In this case the ATPase was cross-linked in the membrane by melittin. The extent of cross-linking was greater for native melittin than for acetylated melittin, which neutralizes the positive charges on melittin and decreases its interactions with the Ca^{2+}-ATPase. Because the extent of cross-linking is less for acetylated melittin, the anisotropy decays more rapidly. There are several disadvantages to the use of erythrosin, eosin, and other phosphorescent probes. The signals are usually weak owing to the low phosphorescence quantum yields; rigorous exclusion of oxygen is needed to prevent quenching; and, finally, fundamental anisotropies are usually low, near 0.1, resulting in decreased resolution of the anisotropy decays.

11.8.B. Long-Lifetime Metal–Ligand Complexes

Another approach to measuring long correlation times is to use luminescent MLCs. These probes, which display lifetimes ranging from 100 ns to 10 μs, have only recently become available[87–90] (Chapter 20). They are typically complexes of transition metals with diimine ligands. A lipid MLC probe was made by covalently linking two phosphatidylethanolamine (PE) lipids to a ruthenium MLC which contained two carboxyl groups. The lipid MLC probe was then incorporated into DPPG vesicles (Figure 11.22). The maximum in the differential phase angle (Δ_ω) is near 1 MHz, suggesting slow rotational diffusion. The lifetime of the Ru-PE_2 probe was near 680 ns at 2 °C, which allowed resolution of the slow correlation time.

The FD anisotropy data in Figure 11.22 were used to resolve a double-exponential anisotropy decay, which showed correlation times of 133 and 1761 ns. It is useful to visualize how these correlation times contribute to the data, which is shown by the dashed lines in Figure 11.23. The correlation time of 1761 ns is consistent with that expected for rotational diffusion of phospholipid vesicles with a diameter of 250 Å. At this time the physical origin of the shorter correlation time is not clear, but presumably this correlation time is due to restricted motion of the probe within the membrane.

It is important to notice that the use of a long-lifetime probe allowed measurement of overall rotation of the phospholipid vesicles. Earlier in this chapter, we saw that DPH displayed nonzero r_∞ values at long times. This occurred because the intensity decay times of DPH are short relative to the correlation times of the vesicles. Hence, the use of nanosecond-decay-time fluorophores provides no information on rotational motions of lipid vesicles.

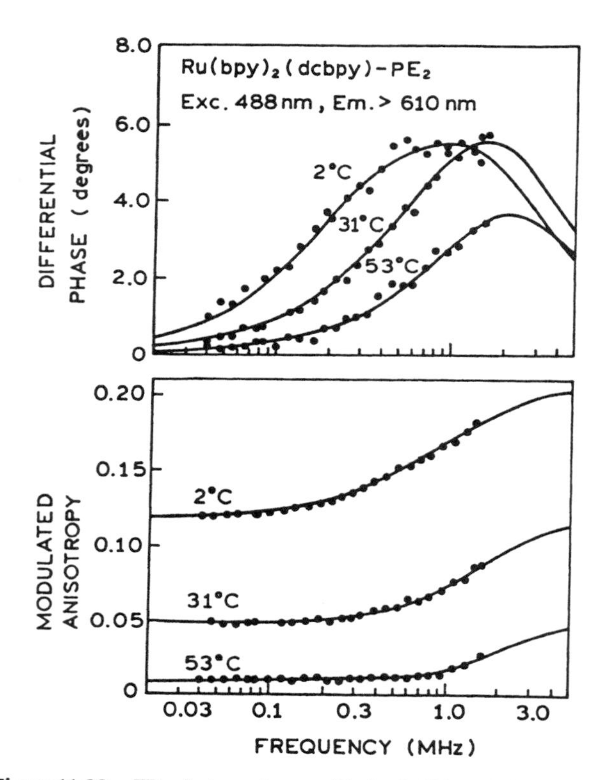

Figure 11.22. FD anisotropy decays of Ru(bpy)$_2$(dcbpy)-PE_2 in DPPG vesicles. Reprinted from Ref. 88, with permission from Academic Press, Inc.

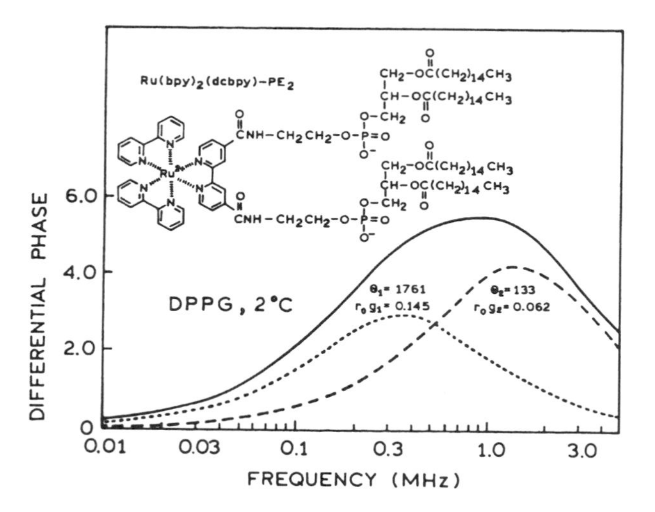

Figure 11.23. Differential phase angles of Ru(bpy)$_2$(dcbpy)-PE_2 in DPPG vesicles at 2 °C. The resolution of the Δ_ω values is based on the correlation times (in nanoseconds) and amplitudes shown on the figure. Data from Ref. 88.

The MLCs have several advantages over the phosphorescent probes. In contrast to phosphorescence, the luminescence from MLCs can be measured in the presence of dissolved oxygen. The MLCs are only partially quenched by ambient oxygen, whereas phosphorescence is usually completely quenched. Additionally, there are relatively few phosphorescent probes, but there are numerous MLCs (Chapter 20).

11.9. ANISOTROPY DECAYS OF NUCLEIC ACIDS

Studies of DNA by fluorescence can be traced to the use of dyes to stain chromatin for fluorescence microscopy. The use of time-resolved fluorescence for DNA dynamics originated with the measurement of anisotropy decays of EB bound to DNA.[91–96] These early studies showed an unusual anisotropy decay, similar to that found for DPH in membranes, in which the anisotropy at long times did not decay to zero (Figure 11.24). At that time, the results were interpreted in terms of the angle through which the EB could rotate within the DNA helix. However, more recent

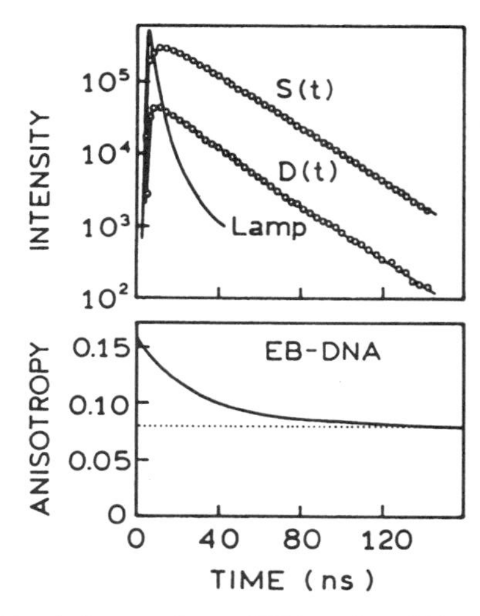

Figure 11.24. Anisotropy decay of EB bound to calf thymus DNA. The excitation was with unpolarized or natural light. Revised from Ref. 91.

studies of EB-DNA anisotropy decays show a continued decrease in anisotropy at longer times, rather than a constant r_∞ value.

One may notice that the anisotropy values in Figure 11.24 are lower than expected for EB. This is because the experiments were performed with natural or unpolarized light for the excitation. When the excitation source is unpolarized, the emission is still polarized. However, the anisotropy values are one-half of those observed with polarized excitation.

Since these early studies, there has been considerable progress in the use of fluorescence to study DNA dynamics.[97–99] Unfortunately, the theory for DNA dynamics is rather complex and not easily summarized. The basic result is that the anisotropy of DNA-bound probes can be depolarized by fast motions of the probes within the DNA helix by bending of DNA about the short axis and by torsional motion of DNA about the long axis. The extent to which these motions contribute to the anisotropy decay depends on the orientation of the transition moments within the DNA helix. The different motions contribute at different times, and the anisotropy decays are expected to be highly nonexponential. Additional information on DNA anisotropy decays is presented in Chapter 12, but one needs to examine the primary literature for a complete description of long-strand DNA dynamics.

11.9.A. Hydrodynamics of DNA Oligomers

It is easier to understand the anisotropy decays of short DNA oligomers. These molecules behave like rigid rods, allowing the data to be interpreted in terms of the rotational correlation times. One example is the anisotropy decay of a DNA 29-base-pair duplex labeled with EB (Figure 11.25).[100] In this case the two polarized components of the anisotropy decay were analyzed separately in terms of Eqs. [11.1] and [11.2], rather than by calculating $r(t)$ from the convolved decay profiles. Examination of these decays shows the initially more rapid decay of the parallel component, and the initially slower decay of the perpendicular component. The anisotropy decays of this and other double-helical DNA oligomers were found to be single exponentials, and the correlation times increased linearly with the number of base pairs (Figure 11.26). Hence, DNA fragments of this size behave like rigid bodies.

It is now known that DNA can adopt a variety of shapes besides linear duplexes. One example is formation of bent helices. In Figure 11.27, the linear structure (AO) is a DNA 50-mer. The bend in the second structure (A5) is due to the insertion of five unpaired adenines. This rather modest change in shape cannot be expected to result in a dramatic change in the anisotropy decay. However, the different rotational properties of these two molecules could be seen in the FD anisotropy data (Figure 11.28). These data were analyzed in terms of a detailed hydrodynamic model and

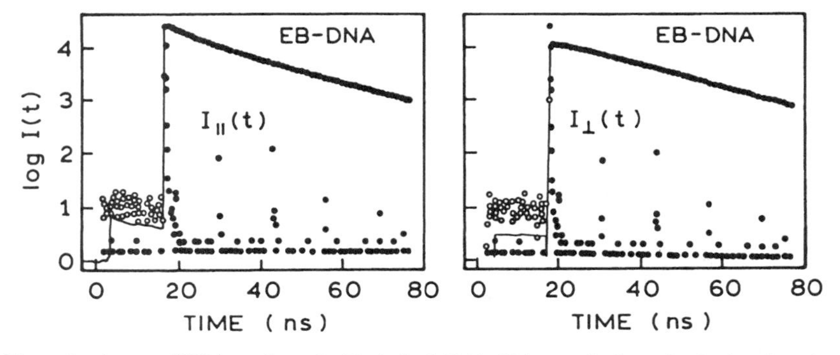

Figure 11.25. Polarized intensity decays of EB bound to a double-helical DNA, 29 base pairs long. Revised and reprinted, with permission, from Ref. 100, Copyright © 1996, American Chemical Society.

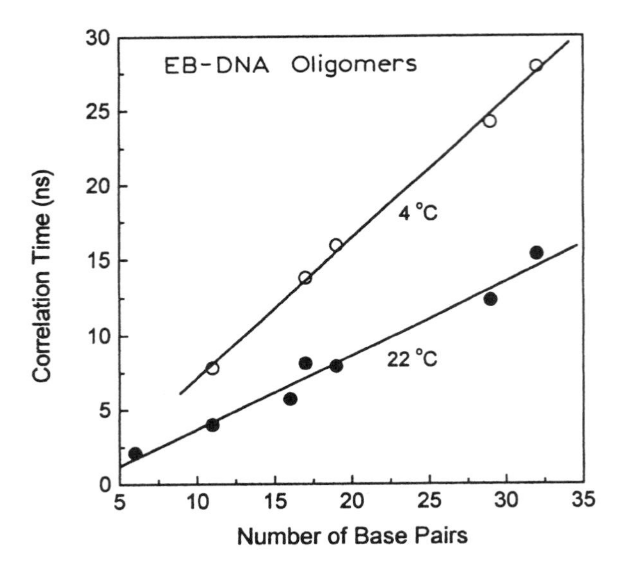

Figure 11.26. Rotational correlation time of EB bound to double-helical DNA oligomers as a function of the size of the DNA fragment. Data from Ref. 100.

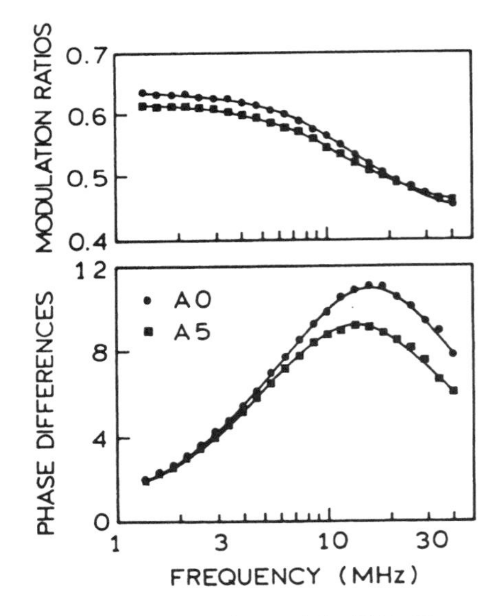

Figure 11.28. FD anisotropy decay of the DNA 50-mers shown in Figure 11.27 (●, AO; ■, A5) labeled with EB. In this figure the modulation data are presented as the ratio of the polarized and modulated intensities, perpendicular divided by parallel. Reprinted from Collini, M., Chirico, G., Baldini, G., and Bianchi, M. E., Conformation of short DNA fragments by modulated fluorescence polarization anisotropy, *Biopolymers* **36**:211–225, Copyright © 1995, by permission of John Wiley & Sons, Inc.

found to be consistent with the known shapes.[101] The different frequency responses can be understood intuitively by recognizing that the straight DNA molecule can rotate more rapidly around the long axis than the bent molecule. This explains the higher frequency of the maximum differential phase angle of AO as compared to A5.

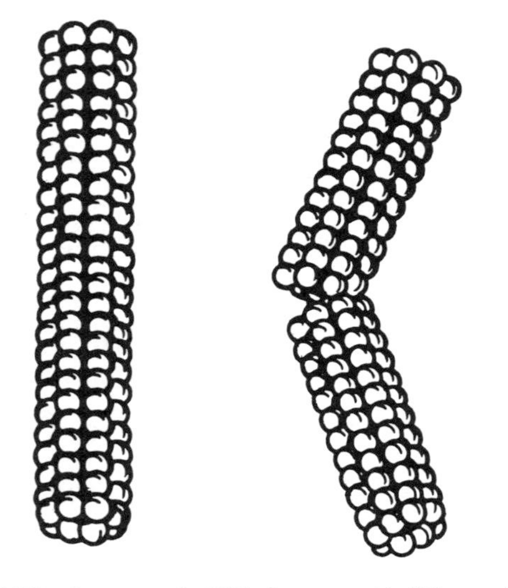

Figure 11.27. Structure of a DNA fragment with 50 base pairs (AO, *left*) and a bent DNA with five unpaired adenines (A5, *right*). Reprinted from Collini, M., Chirico, G., Baldini, G., and Bianchi, M. E., Conformation of short DNA fragments by modulated fluorescence polarization anisotropy, *Biopolymers* **36**:211–225, Copyright © 1995, by permission of John Wiley & Sons, Inc.

The remarkable result is the ability to detect this minor structural difference between the DNA oligomers from the FD data.

11.9.B. Segmental Mobility of DNA

The anisotropy decay of DNA can reflect local motions of the DNA in a manner similar to that found for proteins. However, such motions are more difficult to detect in DNA because the natural bases are virtually nonfluorescent. This problem can be circumvented by the use of modified bases that are fluorescent, one well-known example being 2-aminopurine (2-AP).[102,103] Figure 11.29 shows the anisotropy decay of a DNA 7-mer containing 2-AP as a central residue. The opposite guanine (G) residue does not base-pair with 2-AP, which is thus thought to be pushed out of the helix. The 2-AP displays a 83-ps correlation time which accounts for about 60% of the total anisotropy (Figure 11.29, lower panel).

Such studies of DNA dynamics are the subject of numerous recent publications. The anisotropy decays have been used to detect the effects of ligands on DNA flexibility[104,105] and the effects of cations on the hydrodynamic properties.[106] There is also a recent report in which the intrinsic fluorescence from DNA base pairs was used to study DNA dynamics.[107]

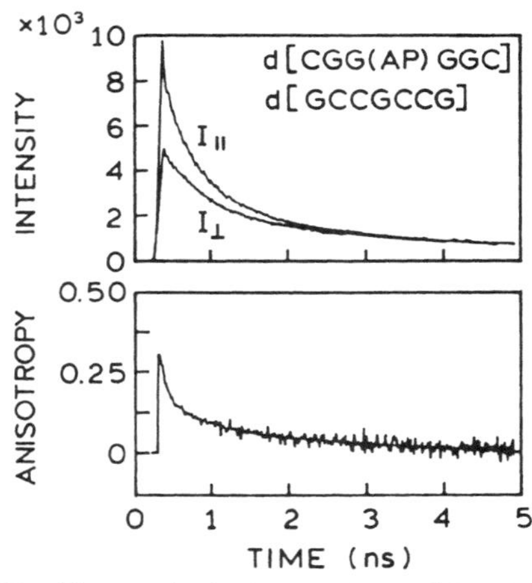

Figure 11.29. Time-resolved anisotropy decay of d[CGG(AP)-GGC]-d[GCCGCCG] in 50m*M* Tris-HCl, pH 7.4, and 0.15*M* NaCl at 4 °C. *Top*: Polarized intensity decays. *Bottom*: Anisotropy decay. Revised and reprinted, with permission, from Ref. 103, Copyright © 1991, American Chemical Society.

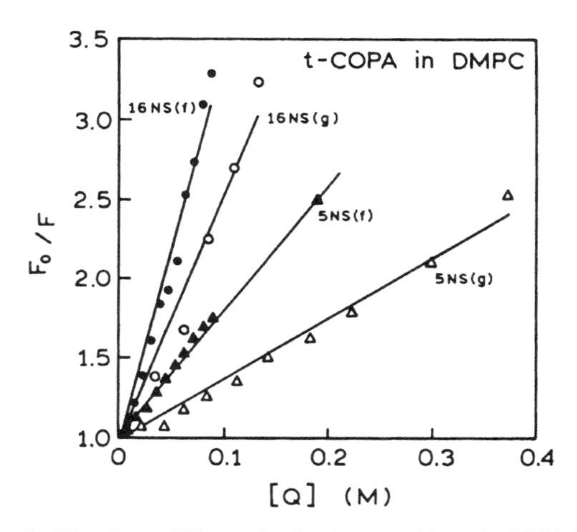

Figure 11.31. Stern–Volmer plot for the quenching of *t*-COPA by the spin probes 5-NS and 16-NS in unilamellar vesicles of DMPC at 17 °C (gel phase, g) and 30 °C (fluid phase, f). The values on the *x*-axis are the quencher concentrations in the membrane phase. Revised and reprinted from Ref. 108, from the Biophysical Society.

11.10. CHARACTERIZATION OF A NEW MEMBRANE PROBE

The concepts described in the preceding sections allow us to readily understand the characteristics of newly developed probes. One recent example is the all-*trans* isomer of 8, 10, 12, 14, 16-octadecapentaenoic acid (*t*-COPA, Figure 11.30). This probe is essentially insoluble and/or nonfluorescent in water, so that the only emission is from *t*-COPA bound to membranes.[108] The absorption spectrum of *t*-COPA is centered at 330 nm, making it an effective acceptor for the intrinsic tryptophan fluorescence of membrane-bound proteins. The effectiveness of *t*-COPA as an acceptor is due in part to its high extinction coefficient, near 105,000 M^{-1} cm^{-1}.

Localization of *t*-COPA in membranes was accomplished by using the spin-labeled fatty acids as quenchers (Section 9.1). DMPC has a phase transition near 24 °C, and quenching is more effective in the fluid phase at 30 °C than in the gel phase at 17 °C (Figure 11.31). The fact that the quenching depends on temperature implies that there is a diffusive component to the quenching, in contrast to the usual assumption of no diffusion in parallax quenching (Section 9.1.C). The larger amounts of quenching by 16-

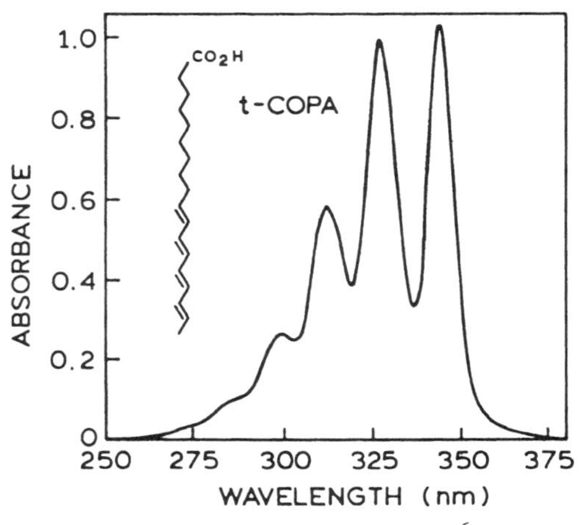

Figure 11.30. Absorption spectrum of *t*-COPA, $10^{-6}M$ in ethanol at 20 °C. Revised and reprinted from Ref. 108, from the Biophysical Society.

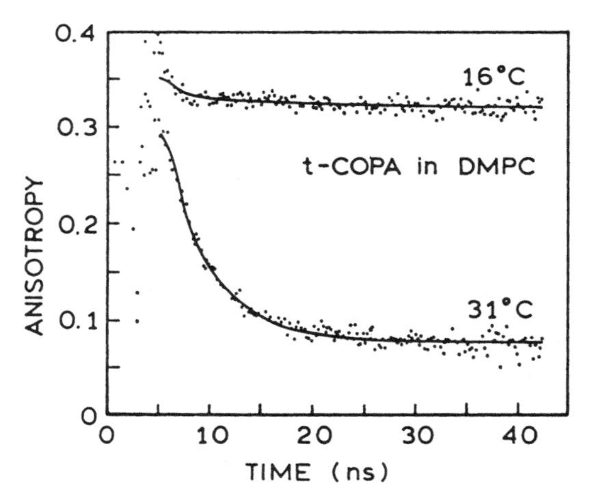

Figure 11.32. Time-resolved fluorescence anisotropy of *t*-COPA in unilamellar vesicles of DMPC in the gel (16 °C) and fluid (31 °C) phases. Revised and reprinted from Ref. 108, from the Biophysical Society.

NS than by 5-NS indicate that the chromophore is buried deeply in the bilayer, away from the lipid–water interface.

The intensity decay of *t*-COPA is multiexponential in solvents and in lipid bilayers. In lipids, the major component in the intensity decay has a lifetime of about 20 ns. This means that the anisotropy decay of *t*-COPA can be measured to longer times than that of DPH, which has a lifetime in membranes typically near 9 ns. The time-zero anisotropy is near 0.385, making *t*-COPA a useful anisotropy probe. The anisotropy decays of *t*-COPA are multiexponential in solvents and in lipid bilayers. In solvents, the anisotropy decays to zero (not shown). In membranes, *t*-COPA behaves like a highly hindered rotor. Below the phase transition, the anisotropy of *t*-COPA displays a high r_∞ value near 0.31. Above the phase transition, the value of r_∞ decreases to 0.07 (Figure 11.32). Because of these dramatic changes in the properties of *t*-COPA, one can expect this fluorophore to become more widely used in studies of membranes.

REFERENCES

1. Lakowicz, J. R., Cherek, H., Kusba, J., Gryczynski, I., and Johnson, M. L., 1993, Review of fluorescence anisotropy decay analysis by frequency-domain fluorescence spectroscopy, *J. Fluoresc.* **3**(2):103–116.
2. Lakowicz, J. R., and Gryczynski, I., 1991, Frequency-domain fluorescence spectroscopy, in *Topics in Fluorescence Spectroscopy, Volume 1, Techniques*, J. R. Lakowicz (ed.), Plenum Press, New York, pp. 293–355.
3. Spencer, R. D., and Weber, G., 1970, Influence of Brownian rotations and energy transfer upon the measurement of fluorescence lifetimes, *J. Chem. Phys.* **52:**1654–1663.
4. Szymanowski, W., 1935, Einfluss der Rotation der Molekule auf die Messungen der Abklingzert des Fluoreszenzstrahlung, *Z. Phys.* **95:**466–473.
5. Kudryashov, P. I., Sveshnikov, B. Y., and Shirokov, V. I., 1960, The kinetics of the concentration depolarization of luminescence and of the intermolecular transfer of excitation energy, *Opt. Spectrosc.* **9:**177–181.
6. Bauer, R. K., 1963, Polarization and decay of fluorescence of solutions, *Z. Naturforsch. A* **18:**718–724.
7. Cross, A. J., and Fleming, G. R., 1984, Analysis of time-resolved fluorescence anisotropy decays, *Biophys. J.* **46:**45–56.
8. Wahl, P. 1979, Analysis of fluorescence anisotropy decays by least square method, *Biophys. Chem.* **10:**91–104.
9. Taylor, J. R., 1982, *An Introduction to Error Analysis*, University Science Books, Mill Valley, California, pp. 173–187.
10. Dale, R. E., Chen, L. A., and Brand, L., 1977, Rotational relaxation of the "microviscosity" probe diphenylhexatriene in paraffin oil and egg lecithin vesicles, *J. Am. Chem. Soc.* **252:**7500–7510.
11. Papenhuijzen, J., and Visser, A. J. W. G., 1983, Simulation of convoluted and exact emission anisotropy decay profiles, *Biophys. Chem.* **17:**57–65.
12. Gilbert, C. W., 1983, A vector method for the non-linear least squares reconvolution-and-fitting analysis of polarized fluorescence decay data, in *Time-Resolved Fluorescence Spectroscopy in Biochemistry and Biology*, R. B. Cundall, and R. E. Dale (eds.), Plenum Press, New York, pp. 605–606.
13. Beechem, J. M., and Brand, L., 1986, Global analysis of fluorescence decay: Applications to some unusual experimental and theoretical studies, *Photochem. Photobiol.* **44:**323–329.
14. Crutzen, M., Ameloot, M., Boens, N., Negri, R. M., and De Schryver, F. C., 1993, Global analysis of unmatched polarized fluorescence decay curves, *J. Phys. Chem.* **97:**8133–8145.
15. Vos, K., van Hoek, A., and Visser, A. J. W. G., 1987, Application of a reference convolution method to tryptophan fluorescence in proteins, *Eur. J. Biochem.* **165:**55–63.
16. Weber, G., 1971, Theory of fluorescence depolarization by anisotropic Brownian rotations. Discontinuous distribution approach, *J. Chem. Phys.* **55:**2399–2407.
17. Merkelo, H., Hammond, J. H., Hartman, S. R., and Derzko, Z. I., 1970, Measurement of the temperature dependence of depolarization time of luminescence, *J. Lumin.* **1,2:**502–512.
18. Lakowicz, J. R., Prendergast, F. G., and Hogen, D., 1979, Differential polarized phase fluorometric investigations of diphenylhexatriene in lipid bilayers: Quantitation of hindered depolarizing rotations, *Biochemistry* **18:**508–519.
19. Reinhart, G. D., Marzola, P., Jameson, D. M., and Gratton, E., 1991, A method for on-line background subtraction in frequency domain fluorometry, *J. Fluoresc.* **1**(3):153–162.
20. Belford, G. G., Belford, R. L., and Weber, G., 1972, Dynamics of fluorescence polarization in macromolecules, *Proc. Natl. Acad. Sci. U.S.A.* **69:**1392–1393.
21. Chuang, T. J., and Eisenthal, K. B., 1972, Theory of fluorescence depolarization by anisotropic rotational diffusion, *J. Chem. Phys.* **57:**5094–5097.
22. Ehrenberg, M., and Rigler, R., 1972, Polarized fluorescence and rotational Brownian diffusion, *Chem. Phys. Lett.* **14:**539–544.
23. Tao, T., 1969, Time-dependent fluorescence depolarization and Brownian rotational diffusion of macromolecules, *Biopolymers* **8:**609–632.
24. Lombardi, J. R., and Dafforn, G. A., 1966, Anisotropic rotational relaxation in rigid media by polarized photoselection, *J. Chem. Phys.* **44:**3882–3887.
25. Small, E. W., and Isenberg, I., 1977, Hydrodynamic properties of a rigid molecule: Rotational and linear diffusion and fluorescence anisotropy, *Biopolymers* **16:**1907–1928.
26. Steiner, R. F., 1991, Fluorescence anisotropy: Theory and applications, in *Topics in Fluorescence Spectroscopy, Volume 2, Principles*, J. R. Lakowicz (ed.), Plenum Press, New York, pp. 1–52.
27. Veatch, W. R., and Stryer, L., 1977, Effect of cholesterol on the rotational mobility of diphenylhexatriene in liposomes: A nanosecond fluorescence anisotropy study, *J. Mol. Biol.* **117:**1109–1113.
28. Chen, L. A., Dale, R. E., Roth, S., and Brand, L., 1977, Nanosecond time-dependent fluorescence depolarization of diphenylhexatriene in dimyristoyllecithin vesicles and the determination of "microviscosity," *J. Biol. Chem.* **252:**2163–2169.
29. Kawato, S., Kinosita, K., and Ikegami, A., 1978, Effect of cholesterol on the molecular motion in the hydrocarbon region of lecithin bilayers studied by nanosecond fluorescence techniques, *Biochemistry* **17:**5026–5031.
30. Hildenbrand, K., and Nicolau, C., 1979, Nanosecond fluorescence anisotropy decays of 1,6-diphenyl-1,3,5-hexatriene in membranes, *Biochim. Biophys. Acta* **553:**365–377.
31. Kinosita, K., Kawato, S., and Ikegami, A., 1977, A theory of fluorescence polarization decay in membranes, *Biophys. J.* **20:**289–305.

32. Kinosita, K., Ikegami, A., and Kawato, S., 1982, On the wobbling-in-cone analysis of fluorescence anisotropy decay, *Biophys. J.* **37**:461–464.
33. Komura, S., Ohta, Y., and Kawato, S., 1990, A theory of optical anisotropy decay in membranes, *J. Phys. Soc. Jpn.* **59**:2584–2595.
34. Wallach, D., 1967, Effects of internal rotation on angular correlation functions, *J. Chem. Phys.* **47**:5258–5268.
35. Gottlieb, Y. Ya., and Wahl, P., 1963, Ètude théorique de la polarisation de fluorescence des macromolecules portant un groupe émetteur mobile autour d'un axe de rotation, *J. Chim. Phys.* **60**:849–856.
36. Lapari, G., and Szabo, A., 1980, Effect of librational motion on fluorescence depolarization and nuclear magnetic resonance relaxation in macromolecules and membranes, *Biophys. J.* **30**:489–506.
37. Vincent, M., and Gallay, J., 1991, The interactions of horse heart apocytochrome *c* with phospholipid vesicles and surfactant micelles: Time-resolved fluorescence study of the single tryptophan residue (Trp-59), *Eur. Biophys. J.* **20**:183–191.
38. Pap, E. H. W., Ter Horst, J. J., Van Hoek, A., and Visser, A. J. W. G., 1994, Fluorescence dynamics of diphenyl-1,3,5-hexatriene-labeled phospholipids in bilayer membranes, *Biophys. Chem.* **48**:337–351.
39. Gryczynski, I., Johnson, M. L., and Lakowicz, J. R., 1994, Analysis of anisotropy decays in terms of correlation time distributions, measured by frequency-domain fluorometry, *Biophys. Chem.* **52**:1–13.
40. Peng, K., Visser, A. J. W. G., van Hoek, A., Wolfs, C. J. A. M., Sanders, J. C., and Hemminga, M. A., 1990, Analysis of time-resolved fluorescence anisotropy in lipid–protein systems. I. Application to the lipid probe octadecyl rhodamine B in interaction with bacteriophage M13 coat protein incorporated in phospholipid bilayers, *Eur. Biophys. J.* **18**:277–283.
41. Visser, A. J. W. G., Van Hoek, A., and Van Paridon, P. A., 1987, Time-resolved fluorescence depolarization studies of parinaroyl phosphatidylcholine in Triton X-100 micelles and rat skeletal muscle membranes, in *Membrane Receptors, Dynamics, and Energetics*, K. W. A. Wirtz (ed.), Plenum Press, New York, pp. 353–361.
42. Brand, L., Knutson, J. R., Davenport, L., Beechem, J. M., Dale, R. E., Walbridge, D. G., and Kowalczyk, A. A., 1985, Time-resolved fluorescence spectroscopy: Some applications of associative behaviour to studies of proteins and membranes, in *Spectroscopy and the Dynamics of Molecular Biological Systems*, P. Bayley and R. E. Dale (Eds.), Academic Press, London, pp. 259–305.
43. Ruggiero, A., and Hudson, B., 1989, Analysis of the anisotropy decay of *trans*-parinaric acid in lipid bilayers, *Biophys. J.* **55**:1125–1135.
44. Shinitzky, M., and Barenholz, Y., 1978, Fluidity parameters of lipid regions determined by fluorescence polarization, *Biochim. Biophys. Acta* **515**:367–394.
45. Kawato, S., Kinosita, K., and Ikegami, A., 1977, Dynamic structure of lipid bilayers studied by nanosecond fluorescence techniques, *Biochemistry* **16**:2319–2324.
46. Stubbs, C. D., and Williams, B. W., 1992, Fluorescence in membranes, in *Topics in Fluorescence Spectroscopy, Volume 3, Biochemical Applications*, J. R. Lakowicz (ed.), Plenum Press, New York, pp. 231–271.
47. Stubbs, C. D., Kouyama, T., Kinosita, K., and Ikegami, A., 1981, Effect of double bonds on the dynamic properties of the hydrocarbon region of lecithin bilayers, *Biochemistry* **20**:4257–4262.
48. Ameloot, M., Hendrickx, H., Herreman, W., Pottel, H., Van Cauwelaert, F., and van der Meer, W., 1984, Effect of orientational order on the decay of fluorescence anisotropy in membrane suspensions, *Biophys. J.* **46**:525–539.
49. Vincent, M., de Foresta, B., Gallay, J., and Alfsen, A., 1982, Nanosecond fluorescence anisotropy decays of *n*-(9-anthroyloxy) fatty acids in dipalmitoylphosphatidylcholine vesicles with regard to isotropic solvents, *Biochemistry* **21**:708–716.
50. Pal, R., Petri, W. A., Ben-Yashar, V., Wagner, R. R., and Barenholz, Y., 1985, Characterization of the fluorophore 4-heptadecyl-7-hydroxycoumarin: A probe for the head-group region of lipid bilayers and biological membranes, *Biochemistry* **24**:573–581.
51. Wolber, P. K., and Hudson, B. S., 1981, Fluorescence lifetime and time-resolved polarization anisotropy studies of acyl chain order and dynamics in lipid bilayers, *Biochemistry* **20**:2800–2810.
52. Davenport, L., and Targowski, P., 1996, Submicrosecond phospholipid dynamics using a long-lived fluorescence emission anisotropy probe, *Biophys. J.* **71**:1837–1852.
53. Kinosita, K., Kawato, S., and Ikegami, A., 1984, Dynamic structure of biological and model membranes: Analysis by optical anisotropy decay measurements, *Adv. Biophys.* **17**:147–203.
54. Jähnig, F., 1979, Structural order of lipids and proteins in membranes: Evaluation of fluorescence anisotropy data, *Proc. Natl. Acad. Sci. U.S.A.* **76**:6361–6365.
55. Heyn, M. P., 1979, Determination of lipid order parameters and rotational correlation times from fluorescence depolarization experiments, *FEBS Lett.* **108**:359–364.
56. Van Blitterswijk, W. J., Van Hoeven, R. P., and Van Der Meer, B. W., 1981, Lipid structural order parameters (reciprocal of fluidity) in biomembranes derived from steady-state fluorescence polarization measurements, *Biochim. Biophys. Acta* **644**:323–332.
57. Best, L., John, E., and Jähnig, F., 1987, Order and fluidity of lipid membranes as determined by fluorescence anisotropy decay, *Eur. Biophys. J.* **15**:87–102.
58. Lakowicz, J. R., Cherek, H., Maliwal, B. P., and Gratton, E., 1985, Time-resolved fluorescence anisotropies of diphenylhexatriene and perylene in solvents and lipid bilayers obtained from multifrequency phase-modulation fluorometry, *Biochemistry* **24**:376–383.
59. Faucon, J. F., and Lakowicz, J. R., 1987, Anisotropy decay of diphenylhexatriene in melittin–phospholipid complexes by multifrequency phase-modulation fluorometry, *Arch. Biochem. Biophys.* **252**:245–258.
60. Lakowicz, J. R., 1985, Frequency-domain fluorometry for resolution of time-dependent fluorescence emission, *Spectroscopy* **1**:28–37.
61. Ross, J. A., Schmidt, C. J., and Brand, L., 1981, Time-resolved fluorescence of the two tryptophans in horse liver alcohol dehydrogenase, *Biochemistry* **20**:4369–4377.
62. Vincent, M., Deveer, A.-M., De Haas, G. H., Verheij, H. M., and Gallay, J., 1993, Stereospecificity of the interaction of porcine pancreatic phospholipase A_2 with micellar and monomeric inhibitors, *Eur. J. Biochem.* **215**:531–539.
63. Bouhss, A., Vincent, M., Munier, H., Gilles, A.-M., Takahashi, M., Barzu, O., Danchin, A., and Gallay, J., 1996, Conformational transitions within the calmodulin-binding site of *Bordetella pertussis* adenylate cyclase studied by time-resolved fluorescence of Trp242 and circular dichroism, *Eur. J. Biochem.* **237**:619–628.
64. Kulinski, T., and Visser, A. J. W. G., 1987, Spectroscopic investigations of the single tryptophan residue and of riboflavin and 7-oxalumazine bound to lumazine apoprotein from *Photobacterium leiognathi*, *Biochemistry* **26**:540–549.
65. Rischel, C., Thyberg, P., Rigler, R., and Poulsen, F. M., 1996, Time-resolved fluorescence studies of the molten globule state of apomyoglobin, *J. Mol. Biol.* **257**:877–885.
66. Axelsen, P. H., Gratton, E., and Prendergast, F. G., 1991, Experimentally verifying molecular dynamics simulations through fluorescence anisotropy measurements, *Biochemistry* **30**:1173–1179.

67. Fa, M., Karolin, J., Aleshkov, S., Strandberg, L., Johansson, L. B.-A., and Ny, T., 1995, Time-resolved polarized fluorescence spectroscopy studies of plasminogen activator inhibitor type I: Conformational changes of the reactive center upon interactions with target proteases, vitronectin and heparin, *Biochemistry* **34:**13833–13840.
68. Broos, J., Visser, A. J. W. G., Engbersen, F. J., Verboom, W., van Hoek, A., and Reinhoudt, D. N., 1995, Flexibility of enzymes suspended in organic solvents probed by time-resolved fluorescence anisotropy. Evidence that enzyme activity and enantioselectivity are directly related to enzyme flexibility, *J. Am. Chem. Soc.* **117:**12657–12663.
69. Yguerabide, J., Epstein, H. F., and Stryer, L., 1970, Segmental flexibility in an antibody molecule, *J. Mol. Biol.* **51:**573–590.
70. Hanson, D. C., Yguerabide, J., and Schumaker, V. N., 1981, Segmental flexibility of immunoglobulin G antibody molecules in solution: A new interpretation, *Biochemistry* **20:**6842–6852.
71. Wahl, P., 1969, Mesure de la décroissance de la fluorescence polarisée de la γ-globuline-1-sulfonyl-5-diméthylaminonaphtalène, *Biochim. Biophys. Acta* **175:**55–64.
72. Wahl, P., Kasai, M., and Changuex, J.-P., 1971, A study of the motion of proteins in excitable membrane fragments by nanosecond fluorescence polarization spectroscopy, *Eur. J. Biochem.* **18:**332–341.
73. Brochon, J.-C., and Wahl, P., 1972, Mesures des déclins de l'anisotropie de fluorescence de la γ-globuline et de ses fragments Fab, Fc et $F(ab)_2$ marqués avec le 1-sulfonyl-5-diméthyl-aminonaphtalène, *Eur. J. Biochem.* **25:**20–32.
74. Holowka, D., Wensel, T., and Baird, B., 1990, A nanosecond fluorescence depolarization study on the segmental flexibility of receptor-bound immunoglobulin E, *Biochemistry* **29:**4607–4612.
75. Visser, A. J. G., van Hoek, A., Visser, N. V., Lee, Y., and Ghisia, S., 1997, Time-resolved fluorescence study of the dissociation of FMN from the yellow fluorescence protein from *Vibrio fischeri*, *Photochem. Photobiol.* **65:**570–575.
76. Lakowicz, J. R., Laczko, G., Gryczynski, I., and Cherek, H., 1986, Measurement of subnanosecond anisotropy decays of protein fluorescence using frequency-domain fluorometry, *J. Biol. Chem.* **261:**2240–2245.
77. Maliwal, B. P., and Lakowicz, J. R., 1986, Resolution of complex anisotropy decays by variable frequency phase-modulation fluorometry: A simulation study, *Biochim. Biophys. Acta* **873:**161–172.
78. Maliwal, B. P., Hermetter, A., and Lakowicz, J. R., 1986, A study of protein dynamics from anisotropy decays obtained by variable frequency phase-modulation fluorometry: Internal motions of *N*-methylanthraniloyl melittin, *Biochim. Biophys. Acta* **873:**173–181.
79. Lakowicz, J. R., Laczko, G., and Gryczynski, I., 1987, Picosecond resolution of tyrosine fluorescence and anisotropy decays by 2-GHz frequency-domain fluorometry, *Biochemistry* **26:**82–90.
80. Lakowicz, J. R., Laczko, G., and Gryczynski, I., 1986, Picosecond resolution of oxytocin tyrosyl fluorescence by 2 GHz frequency-domain fluorometry, *Biophys. Chem.* **24:**97–100.
81. Lakowicz, J. R., Gratton, E., Cherek, H., Maliwal, B. P., and Laczko, G., 1984, Determination of time-resolved fluorescence emission spectra and anisotropies of a fluorophore–protein complex using frequency-domain phase-modulation fluorometry, *J. Biol. Chem.* **259:**10967–10972.
82. Matayoshi, E. D., Sawyer, W. H., and Jovin, T. M., 1991, Rotational diffusion of band 3 in erythrocyte membranes. 2. Binding of cytoplasmic enzymes, *Biochemistry* **30:**3538–3543.
83. Pecht, I., Ortega, E., and Jovin, T. M., 1991, Rotational dynamics of the Fc receptor on mast cells monitored by specific monoclonal antibodies and IgE, *Biochemistry* **30:**3450–3458.
84. Shi, Y., Karon, B. S., Kutchai, H., and Thomas, D. D., 1996, Phospholamban-dependent effects of $C_{12}E_8$ on calcium transport and molecular dynamics in cardiac sarcoplasmic reticulum, *Biochemistry* **35:**13393–13399.
85. Karon, B. S., Geddis, L. M., Kutchai, H., and Thomas, D. D., 1995, Anesthetics alter the physical and functional properties of the Ca-ATPase in cardiac sarcoplasmic reticulum, *Biophys. J.* **68:**936–945.
86. Voss, J. C., Mahaney, J. E., and Thomas, D. D., 1995, Mechanism of Ca-ATPase inhibition by melittin in skeletal sarcoplasmic reticulum, *Biochemistry* **34:**930–939.
87. Terpetschnig, E., Szmacinski, H., and Lakowicz, J. R., 1997, Long-lifetime metal–ligand complexes as probes in biophysics and clinical chemistry, *Methods Enzymol.* **278:**295–321.
88. Li, L., Szmacinski, H., and Lakowicz, J. R., 1997, Synthesis and luminescence spectral characterization of long-lifetime lipid metal–ligand probes, *Anal. Biochem.* **244:**80–85.
89. Guo, X.-Q., Castellano, F. N., Li, L., Szmacinski, H., Lakowicz, J. R., and Sipior, J., 1997, A long-lived, highly luminescent Re(I) metal–ligand complex as a biomolecular probe, *Anal. Biochem.* **254:**179–186.
90. DeGraff, B. A., and Demas, J. N., 1994, Direct measurement of rotational correlation times of luminescent ruthenium (II) molecular probes by differential polarized phase fluorometry, *J. Phys. Chem.* **98:**12478–12480.
91. Wahl, P., Paoletti, J., and Le Pecq, J.-B., 1970, Decay of fluorescence emission anisotropy of the ethidium bromide–DNA complex evidence for an internal motion in DNA, *Proc. Natl. Acad. Sci. U.S.A.* **65:**417–421.
92. Millar, D. P., Robbins, R. J., and Zewail, A. H., 1981, Time-resolved spectroscopy of macromolecules: Effect of helical structure on the torsional dynamics of DNA and RNA, *J. Chem. Phys.* **74:**4200–4201.
93. Ashikawa, I., Kinosita, K., Ikegami, A., Nishimura, Y., Tsuboi, M., Watanabe, K., Iso, K., and Nakano, T., 1983, Internal motion of deoxyribonucleic acid in chromatin. Nanosecond fluorescence studies of intercalated ethidium, *Biochemistry* **22:**6018–6026.
94. Wang, J., Hogan, M., and Austin, R. H., 1982, DNA motions in the nucleosome core particle, *Proc. Natl. Acad. Sci. U.S.A.* **79:**5896–5900.
95. Magde, D., Zappala, M., Knox, W. H., and Nordlund, T. M., 1983, Picosecond fluorescence anisotropy decay in the ethidium/DNA complex, *J. Phys. Chem.* **87:**3286–3288.
96. Genest, D., Wahl, P., Champagne, M. E. M., and Daune, M., 1982, Fluorescence anisotropy decay of ethidium bromide bound to nucleosomal core particles, *Biochimie* **64:**419–427.
97. Schurr, J. M., Fujimoto, B. S., Wu, P., and Song, L., 1992, Fluorescence studies of nucleic acids: Dynamics, rigidities, and structures, in *Topics in Fluorescence Spectroscopy, Volume 3, Biochemical Applications*, J. R. Lakowicz (ed.), Plenum Press, New York, pp. 137–229.
98. Thomas, J. C., Allison, S. A., Appellof, C. J., and Schurr, J. M., 1980, Torsion dynamics and depolarization of fluorescence of linear macromolecules. II. Fluorescence polarization anisotropy measurements on a clean viral 29 DNA, *Biophys. Chem.* **12:**177–188.
99. Barkley, M. D., and Zimm, B. H., 1979, Theory of twisting and bending of chain macromolecules; analysis of the fluorescence depolarization of DNA, *J. Chem. Phys.* **70:**2991–3007.
100. Duhamel, J., Kanyo, J., Dinter-Gottlieb, G., and Lu, P., 1996, Fluorescence emission of ethidium bromide intercalated in defined

DNA duplexes: Evaluation of hydrodynamics components, *Biochemistry* **35:**16687–16697.

101. Collini, M., Chirico, G., Baldini, G., and Bianchi, M. E., 1995, Conformation of short DNA fragments by modulated fluorescence polarization anisotropy, *Biopolymers* **36:**211–225.
102. Wu, P., Li, H., Nordlund, T. M., and Rigler, R., 1990, Multistate modeling of the time and temperature dependence of fluorescence from 2-aminopurine in a DNA decamer, *Proc. SPIE* **1204:**262–269.
103. Guest, C. R., Hochstrasser, R. A., Sowers, L. C., and Millar, D. P., 1991, Dynamics of mismatched base pairs in DNA, *Biochemistry* **30:**3271–3279.
104. Collini, M., Chirico, G., and Baldini, G., 1995, Influence of ligands on the fluorescence polarization anisotropy of ethidium bound to DNA, *Biophys. Chem.* **53:**227–239.
105. Barcellona, M. L., and Gratton, E., 1996, Fluorescence anisotropy of DNA/DAPI complex: Torsional dynamics and geometry of the complex, *Biophys. J.* **70:**2341–2351.
106. Fujimoto, B. S., Miller, J. M., Ribeiro, S., and Schurr, J. M., 1994, Effects of different cations on the hydrodynamic radius of DNA, *Biophys. J.* **67:**304–308.
107. Georghiou, S., Bradrick, T. D., Philippetis, A., and Beechem, J. M., 1996, Large-amplitude picosecond anisotropy decay of the intrinsic fluorescence of double-stranded DNA, *Biophys. J.* **70:**1909–1922.
108. Mateo, C. R., Souto, A. A., Amat-Guerri, F., and Acuña, A. U., 1996, New fluorescent octadecapentaenoic acids as probes of lipid membranes and protein–lipid interactions, *Biophys. J.* **71:**2177–2191.

PROBLEMS

11.1. *Segmental Freedom in Proteins*: Use the data in Figure 11.13 for LADH to calculate the mean angle for the segmental motions of trp-314. Assume the fundamental anisotropy $r_0 = 0.28$.

11.2. *Binding Constant of FMN for YFP*: Use the intensity decay data in Table 11.2 to calculate the dissociation constant (K_D) of FMN for YFP.

Advanced Anisotropy Concepts

12

In the preceding two chapters we described steady-state and time-resolved anisotropy measurements and presented a number of biochemical examples which illustrated the types of information available from these measurements. Throughout these chapters, we stated that anisotropy decay depends on the size and shape of the rotating species. However, the theory which relates the form of the anisotropy decay to the shape of the molecule is complex and was not described in detail. In the present chapter we provide an overview of the rotational properties of nonspherical molecules, as well as representative examples.

Also in this chapter, as the final anisotropy topic, we describe associated anisotropy decays. Such decays occur when the solution contains more than one type of fluorophore or the same fluorophore in different environments. Such systems can result in complex anisotropy decays, even if all the individual species each display a single correlation time.

12.1. ROTATIONAL DIFFUSION OF NONSPHERICAL MOLECULES—AN OVERVIEW

Prior to describing the complex equations for this topic, we present an overview of the results. A rotating fluorophore, or especially a macromolecule, need not be symmetric about any axis. Such a totally unsymmetric shape can only be described by the shape itself, but such a description is not useful for a general theory. Hence, nonspherical molecules are described as being a general ellipsoid or an ellipsoid of revolution (Figure 12.1). An ellipsoid is a shape whose plane sections are all ellipses or circles. A general ellipsoid has three unequal semi-axes $a \neq b \neq c$, and an ellipsoid of revolution has two equal axes and one unique axis. The rotational properties of ellipsoids are described in terms of the rotational rates or rotational diffusion coefficients around each axis (D_1, D_2, and D_3). The theory for rotational diffusion of nonspherical molecules predicts the anisotropy decays for ellipsoids and ellipsoids of revolution. The rotational correlation times in the anisotropy decay are functions of the rotational diffusion coefficients.

Most experiments cannot reveal the shape of a general ellipsoid, so that most data are interpreted in terms of the ellipsoids of revolution. Two cases are possible, the prolate and oblate ellipsoids of revolution. These shapes are often referred to as prolate or oblate ellipsoids. In a prolate ellipsoid, the unique axis is longer than the other two equal axes ($a > b = c$). Prolate ellipsoids are elongated along the symmetry axis. A typical prolate ellipsoid is DPH (Figure 12.2). For an oblate ellipsoid, the unique axis is shorter than the other two equivalent axes ($a < b = c$). Oblate ellipsoids are shaped like flattened spheres. Perylene is an oblate ellipsoid.

Because of the two equivalent axes in ellipsoids of revolution, their hydrodynamics can be described with only two diffusion coefficients (Figure 12.2). Rotation about the unique axis in either a prolate or an oblate

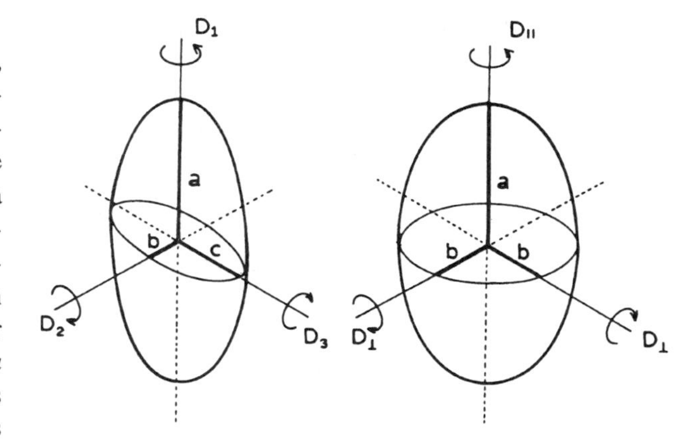

Figure 12.1. *Left*: General ellipsoid with three unequal semi-axes $a \neq b \neq c$; *right*: an ellipsoid of revolution ($a \neq b = c$).

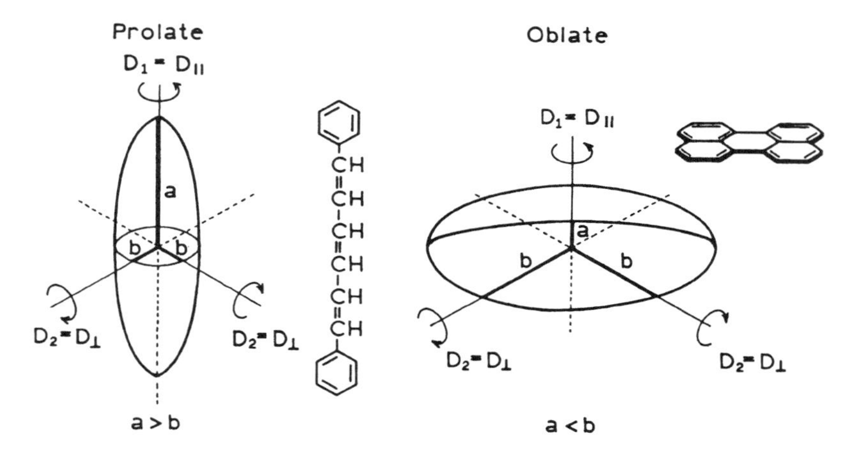

Figure 12.2. Prolate and oblate ellipsoids of revolution.

ellipsoid is called $D_{\parallel}$, and rotation about either of the other two equivalent axes is referred to as $D_{\perp}$. In general, one can expect $D_{\parallel} > D_{\perp}$ because rotation about the unique axis can occur with less displacement of solvent than rotation about the other two axes. However, small molecules do not always rotate as hydrodynamic objects, so that prolate- or oblate-shaped fluorophores may display more than two correlation times.

on the orientation of the transition moments. If one of the transition moments is directed along any of the symmetry axes of the ellipsoid, then the anisotropy decay becomes a double exponential. Hence, one can expect double-exponential anisotropy decays for molecules like DPH or perylene, where the long-wavelength absorption and emission moments are directed along the long axis of the molecules.

12.1.A. Anisotropy Decays of Ellipsoids

The theory for rotational diffusion of ellipsoids, and measurements by fluorescence polarization, can be traced to the classic reports by F. Perrin.[1–4] Since these seminal reports, the theory has been modified to include a description of expected anisotropy decays.[5–10] This theory has been summarized in several reviews.[11–13] For a rigid ellipsoid with three unequal axes, it is now agreed[9] that the anisotropy decays with five correlation times. The correlation times depend on the three rotational diffusion coefficients, and the amplitudes depend on the orientation of the absorption and emission transition moments within the fluorophore and/or ellipsoid. While the theory predicts five correlation times, it is known that two pairs of correlation times will be very close in magnitude, so that in practice only three correlation times are expected for a nonspherical molecule.[10]

The anisotropy decays of nonspherical molecules are usually described in terms of prolate and oblate ellipsoids. The absorption and emission moments can have any arbitrary direction relative to the major (a) and minor (b) axes of the ellipsoid (Figure 12.3). The anisotropy decay of an ellipsoid of revolution can display up to three correlation times, which are functions of the two rotational rates ($D_{\parallel}$ and $D_{\perp}$). The amplitudes of the anisotropy decay depend

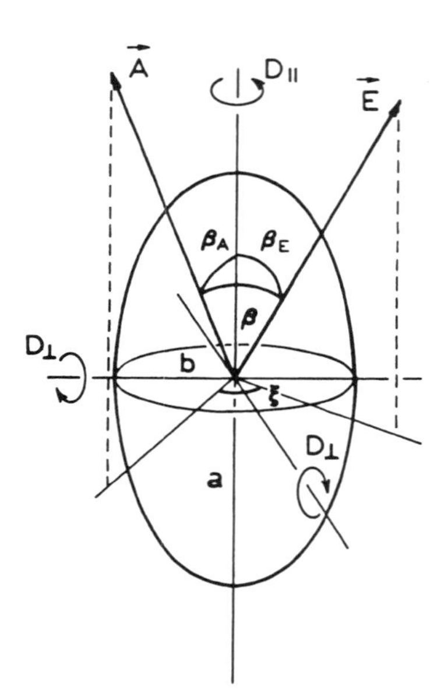

Figure 12.3. Prolate ellipsoid of revolution. $\vec{A}$ and $\vec{E}$ are the directions of the absorption and emission transition moments, respectively. Revised and reprinted, with permission, from Ref. 12, Copyright © 1985, Academic Press, Inc.

12.2. ELLIPSOIDS OF REVOLUTION

We now describe the theory for the anisotropy decays of ellipsoids of revolution. It is important to remember that this theory assumes that the ellipsoid is rigid. In the case of a labeled protein, this assumption implies no independent motion of the fluorophore within the protein. For prolate or oblate ellipsoids, the anisotropy decay is expected to display three correlation times,

$$r(t) = r_1 \exp(-t/\theta_1) + r_2 \exp(-t/\theta_2) + r_3 \exp(-t/\theta_3) \quad [12.1]$$

The amplitudes decaying with each correlation time depend on the angles that the absorption (β_A) and emission (β_E) dipoles make with the symmetry axis. Using the notation defined in Figure 12.3, these amplitudes are

$$r_1 = 0.3 \sin 2\beta_A \sin 2\beta_E \cos \xi \quad [12.2]$$

$$r_2 = 0.3 \sin^2 \beta_A \sin^2 \beta_E \cos 2\xi \quad [12.3]$$

$$r_3 = 0.1(3 \cos^2 \beta_A - 1)(3 \cos^2 \beta_E - 1) \quad [12.4]$$

Occasionally, one finds alternative forms[13,14] for r_1 and r_2,

$$r_1 = 1.2 \sin \beta_A \cos \beta_A \sin \beta_E \cos \beta_E \cos \xi \quad [12.5]$$

$$r_2 = 0.3 \sin^2 \beta_A \sin^2 \beta_E \cos 2\xi \quad [12.6]$$

The fundamental anisotropy is given as usual by

$$r_0 = r_1 + r_2 + r_3 = 0.2(3 \cos^2 \beta - 1) \quad [12.7]$$

where β is the angle between the absorption and emission transition moments. The three correlation times are determined by the two different rotational rates:

$$\theta_1 = (D_{\parallel} + 5D_{\perp})^{-1} \quad [12.8]$$

$$\theta_2 = (4D_{\parallel} + 2D_{\perp})^{-1} \quad [12.9]$$

$$\theta_3 = (6D_{\perp})^{-1} \quad [12.10]$$

Depending upon the shape of the ellipsoid of rotation and the orientation of the transition moments, a variety of complex anisotropy decays can be predicted.[12,13]

In Chapter 10 we saw that one can calculate the rotational correlation time for a spherical molecule (θ_S) based on its volume and the viscosity. Similarly, one can calculate $D_{\parallel}$ and $D_{\perp}$ for ellipsoids of revolution. These values are given by

$$\frac{D_{\parallel}}{D} = \frac{3\rho(\rho - S)}{2(\rho^2 - 1)} \quad [12.11]$$

$$\frac{D_{\perp}}{D} = \frac{3\rho[(2\rho^2 - 1)S - \rho]}{2(\rho^4 - 1)} \quad [12.12]$$

where $D = (6\theta_S)^{-1}$ is the rotational diffusion coefficient of a sphere of equivalent volume, and $\rho = a/b$ is the axial ratio. The value of ρ is greater than 1 for a prolate ellipsoid; for an oblate ellipsoid, $\rho < 1$.

The value of S depends on the type of ellipsoid:

$$S = (\rho^2 - 1)^{-1/2} \ln[\rho + (\rho^2 + 1)^{1/2}] \quad \rho > 1 \quad [12.13]$$

$$S = (1 - \rho^2)^{-1/2} \arctan[(1 - \rho^2)^{1/2}/\rho] \quad \rho < 1 \quad [12.14]$$

These expressions are somewhat complex, but they can be used to predict a variety of anisotropy decays. It is important to recognize that one does not directly measure the rotational diffusion coefficients but rather the correlation times, which are functions of the rotational rates.

12.2.A. Simplified Ellipsoids of Revolution

It is of interest to consider the anisotropy decays expected when the absorption and/or emission transitions are directed along one of the axes. In these cases, Eqs. [12.1]–[12.14] reduce to simpler anisotropy decays, which can often be understood intuitively. Suppose that the fluorophore is shaped like DPH and that both transition moments are aligned with the long axis (Figure 12.4, top). In this case, r_1 and r_2 are zero (Eqs. [12.2] and [12.3]), and r_3 = 0.4. Hence, the anisotropy decays as

$$r(t) = 0.4 \exp(-6D_{\perp}t) \quad [12.15]$$

where the correlation time is $\theta_3 = (6D_{\perp})^{-1}$. This result supports the usual assumption that DPH behaves like an isotropic rotor. The rotation of DPH around its long axis ($D_{\parallel}$) does not displace the collinear transition moments and thus does not decrease the anisotropy. Only the rotation that displaces the transition moments ($D_{\perp}$) results in depolarization.

The results would be rather different if both transition moments were perpendicular to the long axis. In this case the anisotropy decays with two correlation times (Figure 12.4, bottom). The anisotropy decay is more rapid than in the previous case because the faster rotation about the long

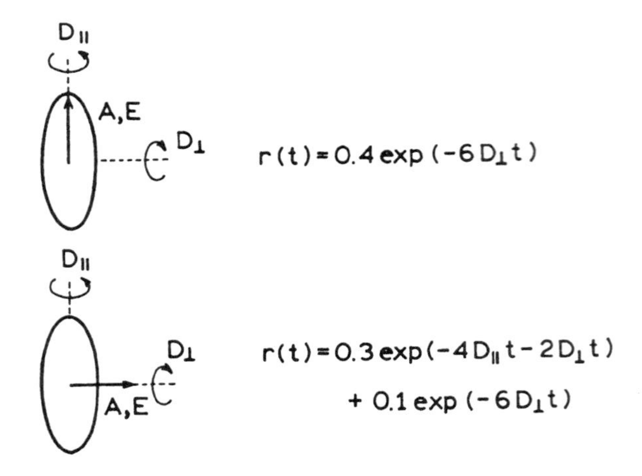

Figure 12.4. Anisotropy decays of prolate ellipsoids of revolution.

axis ($D_{\|}$) now displaces the transition moments. This will result in a rapid randomization about the long axis. At longer times, the slower rotation ($D_{\perp}$) will result in complete depolarization.

One can use similar reasoning to predict the anisotropy decays of disklike oblate ellipsoids (Figure 12.5). For such molecules, the transitions are almost always within the plane of the ring. For collinear transitions, the anisotropy decay has the same form as just described for the prolate ellipsoid with transitions normal to the long axis. The anisotropy decay is initially dominated by the faster in-plane rotation, and at later times by the slower out-of-plane rotation.

It is instructive to consider the anisotropy decays of oblate ellipsoids when the absorption and emission transition moments are not collinear ($\beta \neq 0$). Nonzero angles occur in many fluorophores, including indole and perylene. Suppose that the transition moments are within the plane and displaced by 90° (β = 90°). To be more explicit, $\beta_A = \beta_E$ = 90° and $\beta = \xi$ = 90°. In this case the anisotropy decay displays positive and negative amplitudes,

$$r(t) = -0.3 \exp(-4D_{\|}t - 2D_{\perp}t) + 0.1 \exp(-6D_{\perp}t) \quad [12.16]$$

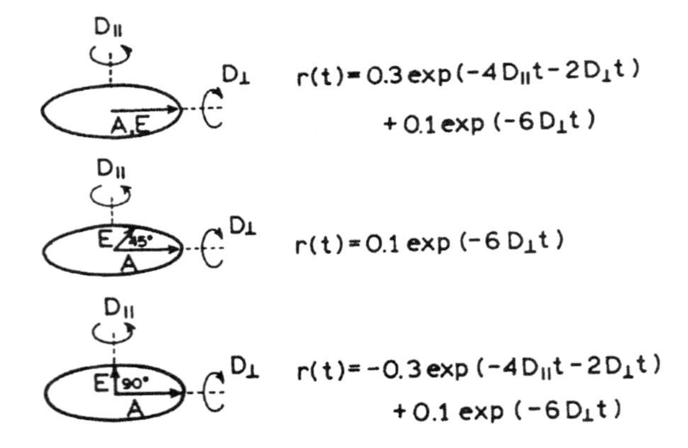

Figure 12.5. Anisotropy decays of oblate ellipsoids of revolution.

From this expression, one can see that the faster in-plane rotation has a larger absolute amplitude and dominates the anisotropy decay at early times. In fact, the anisotropy can become positive prior to decaying to zero. Such behavior has been observed for perylene.

Now consider excitation when β = 45°, with $\beta_A = \beta_E$ = 90°. In this case the anisotropy decay is given by

$$r(t) = 0.1 \exp(-6D_{\perp}t) \quad [12.17]$$

This is the origin of the statement that excitation with r_0 = 0.1 results in measurement of the out-of-plane rotation. When β = 45°, the excitation is randomized in the plane of the fluorophore, so the in-plane rotation has no further effects.

Finally, consider excitation with β = 54.7° and r_0 = 0. For this condition, $r(t)$ remains zero at all times if one of the transitions is aligned with one of the axes. However, if the transitions make nonzero angles with the axes, then $r(t)$ can be nonzero at intermediate times, even with r_0 = 0.

An important aspect of these results (Figures 12.4 and 12.5) is that the correlation times are independent of r_0 but the amplitudes (r_1, r_2, and r_3) depend on the excitation wavelength owing to their dependence on the angles between the transition moments and the axes of the ellipsoid. This provides the opportunity for global analysis based on the anisotropy decays measured for multiple excitation wavelengths. At each excitation wavelength (or r_0 value), one expects different relative contributions of each correlation time to the anisotropy decay. By global analysis of the data, with the correlation times constrained to be the same for all excitation wavelengths, one can expect improved resolution of the multiexponential anisotropy decay.

It is valuable to understand the origin of β values that are not equal to 0° or 90°. Individuals trained in molecular photophysics will note that the electronic transitions of perylene must be directed along the symmetry axes. For this reason, the only expected values of β for perylene are 0° or 90° for the two in-plane transitions, corresponding to r_0 values of 0.40 and –0.20, respectively. Intermediate values of r_0 probably represent a mixture of the two electronic transitions, the relative proportion of which depends on the excitation wavelength. Throughout this book, we are describing such mixed transitions as due to an apparent value of β for a single transition.

12.2.B. Intuitive Description of an Oblate Ellipsoid

The special cases described above can be understood intuitively by considering rotational diffusion of perylene. Assume that perylene is excited in its long-wavelength absorption band where r_0 is near 0.4. The excited-state population will initially be aligned along the vertical or z-axis (Figure 12.6). Rotation of perylene about its symmetry axis (normal to the plane) will displace the transition moments and result in depolarization. Similarly, one of the out-of-plane rotations will displace the transition moments. This agrees with Figure 12.5, which indicates that for an oblate ellipsoid with r_0 = 0.4 both $D_{\parallel}$ and $D_{\perp}$ contribute to the anisotropy decay.

Similarly, consider excitation of perylene with $r_0 = -0.2$. Initially, neither out-of-plane rotation displaces the emission moment, so the only active depolarizing rotation is the in-plane rotation. (Recall that the excited-state population is symmetrical around the z-axis.) Once the molecule has undergone in-plane rotation, then the out-of-plane rotation displaces the emission moment, as seen from Eq. [12.16].

Finally, consider excitation with $\beta = 45°$ or $r_0 = 0.10$. In this case, the excited-state population is immediately randomized around the symmetry axis of the oblate ellipsoid. Hence, only the out-of-plane rotations can further depolarize the emission.

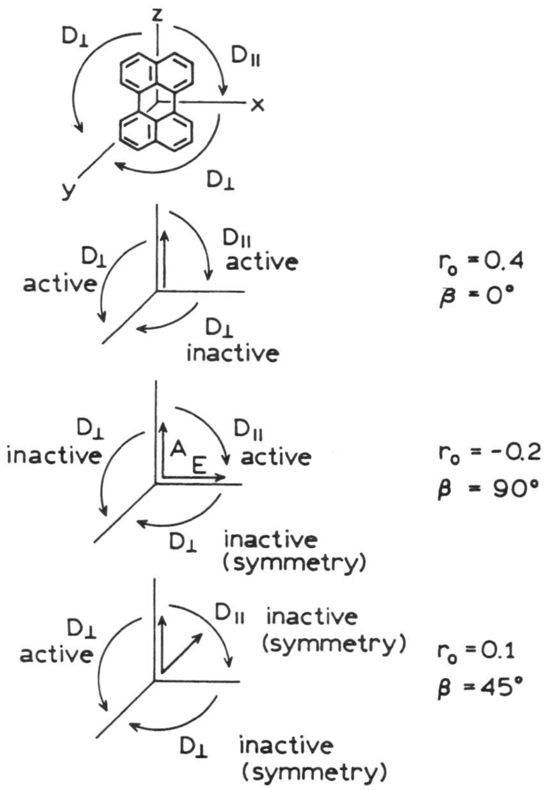

Figure 12.6. Rotational motions of an oblate ellipsoid.

12.2.C. Rotational Correlation Times for Ellipsoids of Revolution

It is interesting to see how the shape of the ellipsoid affects the correlation times. Some representative values of the ratio of the rotational correlation times to that expected for a sphere of equivalent volume are summarized in Table 12.1.[15] The results are quite different for prolate and oblate ellipsoids. For prolate ellipsoids, the three correlation times are all different in magnitude and can be much larger than the correlation time for an equivalent sphere. For the oblate ellipsoids, the three correlation times are all similar in magnitude. Hence, it would be difficult to distinguish an oblate ellipsoid from a sphere, assuming that the oblate ellipsoid displayed hydrodynamic rotation. However, small oblate molecules often display slip diffusion, which results in faster in-plane rotations. For the prolate ellipsoids, one correlation time (θ_3) increases dramatically as the axial ratio increases. This correlation time ($\theta_3 = 6D_{\perp}^{-1}$) is determined by the rotation that displaces the long axis of the prolate ellipsoid. For a prolate ellipsoid, all three correlation times can be distinct for large axial ratios. The ability to detect these correlation times depends on the amplitudes, which depend on the angles with respect to the symmetry axis.

Some fluorophores display r_0 values near 0.4, which indicates that the absorption and emission moments are collinear. In this case, Eqs. [12.2] and [12.3] can be simplified further. Using $\beta_A = \beta_E = \beta_T$ and $\xi = 0$, one finds

$$r_1 = 0.3 \sin^2 2\beta_T \qquad [12.18]$$

Table 12.1. Rotational Correlation Times for Ellipsoids of Revolution[a]

Prolate				Oblate			
ρ	θ_1/θ_S	θ_2/θ_S	θ_3/θ_S	ρ^{-1}	θ_1/θ_S	θ_2/θ_S	θ_3/θ_S
1	1.0	1.0	1.0	1	1.0	1.0	1.0
2	1.32	0.95	1.50	2	1.17	1.49	1.13
5	2.41	0.98	4.64	5	2.30	2.51	2.24
10	3.25	0.99	13.37	10	4.38	4.61	4.30
20	3.73	0.99	41.82	20	8.60	8.85	8.52

[a]From Ref. 15. $\theta_1 = (D_{\parallel} + 5D_{\perp})^{-1}$, $\theta_2 = (4D_{\parallel} + 2D_{\perp})^{-1}$, $\theta_3 = (6D_{\perp})^{-1}$.

$$r_2 = 0.3 \sin^4 \beta_T \quad [12.19]$$

$$r_3 = 0.1(3 \cos^2 \beta_T - 1)^2 \quad [12.20]$$

where β_T is the angle formed by the transition moments with the symmetry axis of the ellipsoid.

In view of the complexity of the equations relating shape to the anisotropy decays, it is useful to have some specific examples. The rotational correlation times for prolate ellipsoids are summarized in Table 12.2. These correlation times are calculated for proteins with a specific volume of $\bar{v} = 0.75$ ml/g and for a hydration of $h = 0$ or $h = 0.2$ ml/g. Values are not listed for oblate ellipsoids since most nonspherical proteins seem to be elongated rather than flattened. Of course, all the correlation times increase with increasing molecular weight. Two of the correlation times (θ_1 and θ_3) increase dramatically as the axial ratio increases. The other correlation time (θ_2) is relatively independent of shape.

It is useful to visualize how the three correlation times depend on the axial ratio. This dependence is shown in Figure 12.7 for a prolate ellipsoid of revolution. The cho-

Table 12.2. Rotational Correlation Times for Prolate Ellipsoids of Revolution[a]

	θ_1 (ns)		θ_2 (ns)		θ_3 (ns)		θ_H (ns)	
ρ	$h = 0$	$h = 0.2$	$h = 0$	$h = 0.2$	$h = 0$	$h = 0.2$	$h = 0$	$h = 0.2$
				MW = 10,000				
1.5	3.46	4.38	2.94	3.72	3.68	4.66	3.27	4.14
2	4.05	5.13	2.93	3.71	4.63	5.87	3.59	4.55
3	5.31	6.72	2.97	3.77	7.18	9.11	4.21	5.34
5	7.40	9.36	3.03	3.83	14.2	18.0	5.00	6.31
10	10.03	12.67	3.06	3.88	41.7	52.1	5.69	7.24
				MW = 25,000				
1.5	8.66	10.96	7.34	9.31	9.21	11.65	8.16	10.37
2	10.11	12.79	7.33	9.29	11.57	14.62	8.96	11.36
3	13.2	16.8	7.43	9.42	17.9	22.8	10.54	13.25
5	18.4	23.4	7.56	9.60	35.4	45.0	12.36	15.96
10	25.1	31.6	7.65	9.71	104.2	128.2	14.23	23.19
				MW = 50,000				
1.5	17.4	21.8	14.7	18.5	18.5	23.1	16.3	20.6
2	20.2	25.5	14.7	18.4	23.1	29.2	17.8	22.4
3	26.7	33.3	15.0	18.7	36.2	45.0	21.0	26.7
5	37.3	46.1	15.2	19.0	72.4	87.7	24.8	31.2
10	50.2	64.1	15.3	19.4	208.3	277.8	28.5	36.2
				MW = 100,000				
1.5	34.7	43.4	29.3	36.7	37.0	46.3	32.6	41.8
2	40.5	50.2	29.4	36.5	46.3	57.4	36.9	44.5
3	53.4	65.3	29.9	37.0	72.4	87.7	42.2	52.1
5	73.0	94.3	30.1	38.1	138.9	185.2	49.5	63.3
10	100.0	128.2	30.4	38.7	416.7	555.5	56.6	72.3
				MW = 500,000				
1.5	173.9	217.9	147.1	185.2	185.2	231.5	163.4	205.8
2	202.4	255.1	147.0	184.5	231.5	292.4	177.9	225.4
3	267.4	333.3	149.7	187.3	362.3	450.5	209.6	267.7
5	373.1	460.8	152.0	190.1	724.6	877.2	248.0	332.0
10	502.5	641.0	153.4	193.8	2083.3	2777.8	285.1	375.4

[a] $\theta_1 = (D_{\parallel} + 5D_{\perp})^{-1}$, $\theta_2 = (4D_{\parallel} + 2D_{\perp})^{-1}$, $\theta_3 = (6D_{\perp})^{-1}$. $D_{\perp}/D$ from Eq. [12.12] and $D_{\parallel}/D$ from Eq. [12.11], S from Eq. [12.13], $\bar{v} = 0.75$ ml/g.

sen parameters (MW = 10,000 $\bar{v}$ = 0.75 ml/g and h = 0.2 ml/g) were selected to simulate a protein. One of the correlation times (θ_3) increases progressively as the axial ratio increases. The other two correlation times (θ_1 and θ_2) reach limiting values at large axial ratios. This occurs because these correlation times contain $D_{\parallel}$, which stays relatively constant as ρ increases. It is interesting to notice that θ_2 actually decreases as ρ increases from 1 to 3. This effect is due to the initial decrease in $D_{\parallel}$ and the large contribution of $D_{\parallel}$ to θ_2.

Also shown in Table 12.2 is the harmonic mean correlation time (θ_H). Calculation of θ_H is somewhat complicated because it depends on the value of r_0 and the orientation of the transition moments relative to the ellipsoid axis. The values in Table 12.2 were calculated using

$$\frac{1}{\theta_H} = \frac{0.4}{\theta_1} + \frac{0.4}{\theta_2} + \frac{0.2}{\theta_3} \qquad [12.21]$$

This definition of θ_H is appropriate when the absorption and emission transition moments are randomly oriented with respect to the helix axis.[11]

The form of the anisotropy decays depends on the axial ratio. Figure 12.8 shows anisotropy decays simulated for various axial ratios using the assumption r_0 = 0.4 (Eqs. [12.18]–[12.20]). For these simulations, we assumed that the transition moments made an angle of 20° with the symmetry axis of the prolate ellipsoid. The anisotropy decays become visually different from a single exponential as the axial ratio exceeds 2. This dependence of the shape of the anisotropy decay on the axial ratio is the basis for determining the shapes of proteins from the time-resolved anisotropies.

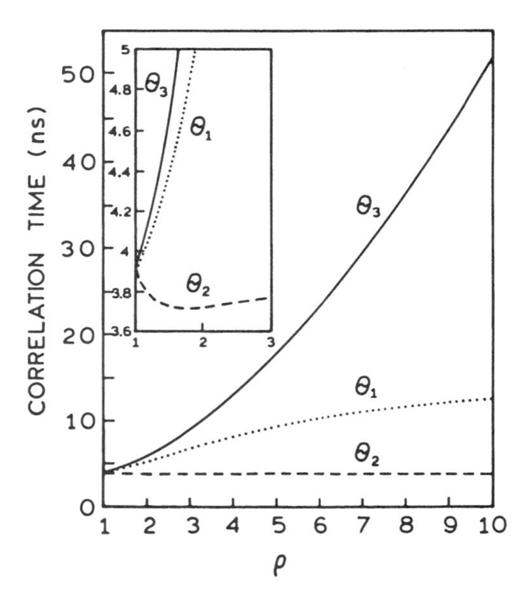

Figure 12.7. Correlation times for a prolate ellipsoid of revolution (MW = 10,000, h = 0.2 mg/g and $\bar{v}$ = 0.75 mg/g) at various axial ratios (ρ). η = 1 cP, $\theta_1 = (D_{\parallel} + 5D_{\perp})^{-1}$, $\theta_2 = (4D_{\parallel} + 2D_{\perp})^{-1}$, $\theta_3 = (6D_{\perp})^{-1}$.

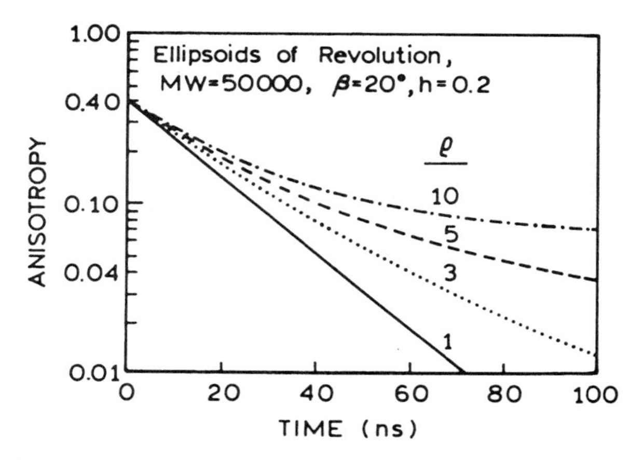

Figure 12.8. Simulated anisotropy decays of a protein (MW = 50,000) shaped like a prolate ellipsoid of revolution, for various axial ratios (ρ). For these simulations, we assumed $\bar{v}$ = 0.75 ml/g, h = 0.2 ml/g, and β = 20°.

12.2.D. Stick versus Slip Rotational Diffusion

The theory described above for rotation of ellipsoids applies only to the "stick" boundary condition. The term "stick" boundary condition refers to rotational diffusion in which the first solvent layer moves with the rotating species, so that rotation is governed by the viscosity of the solvent. Macromolecules and most fluorophores in polar solvents are well described by the stick diffusion. However, small molecules in nonpolar solvents can often display slip rotational diffusion. As an example, perylene in a solvent like hexane can rotate in-plane without significant displacement of solvent. When this occurs, the molecule rotates as if it were in a vacuum, and not affected by solvent viscosity. The theory of slip rotational diffusion is rather complex, and the results are often presented numerically.[16,17] The important point is that the possibility of slip diffusion results in a failure of the theory described above to predict the correlation times of nonspherical molecules. Stated alternatively, one can recover multiple correlation times for small nonspherical molecules, but these values cannot always be interpreted in terms of the correlation times predicted from the hydrodynamic theory (Eqs. [12.1]–[12.14]).

12.3. COMPLETE THEORY FOR ROTATIONAL DIFFUSION OF ELLIPSOIDS

In the preceding sections we described the rotational behavior of ellipsoids with transitions directed along one of the symmetry axes. Although not frequently used, the complete expression for an ellipsoid with three nonequivalent axes (Figure 12.1, left), derived without restricting one of the transitions to be on an axis, is useful to have. In this case the anisotropy decays with five apparent correlation times[9,11,15]:

$$r(t) = \frac{6}{5} \sum_{i=1}^{3} C_i \exp(-t/\theta_i) + [(F+G)/4] \exp[-(6D - 2\Delta)t] + [(F-G)/4] \exp[-(6D + 2\Delta)t] \quad [12.22]$$

In this expression, D is the mean rotational diffusion coefficient,

$$D = \frac{1}{3}(D_1 + D_2 + D_3) \quad [12.23]$$

and

$$\Delta = (D_1^2 + D_2^2 + D_3^2 - D_1D_2 - D_1D_3 - D_2D_3)^{1/2} \quad [12.24]$$

The directions of the transition moments with respect to the axes of the ellipsoid (Figure 12.3) are defined by the cosine of the angles between the absorption transitions and the three axes (α_1, α_2, and α_3) and between the emission transitions and the three axes (ε_1, ε_2, and ε_3). The values of C_i are given by

$$C_i = \alpha_j \, \alpha_k \, \varepsilon_j \, \varepsilon_k \quad [12.25]$$

with ijk = 123, 231, or 312. This other quantities in Eq. [12.22] are defined by

$$\theta_i = (3D + 3D_i)^{-1} \quad [12.26]$$

$$F = \sum_{i=1}^{3} \alpha_i^2 \, \varepsilon_i^2 - 1/3 \quad [12.27]$$

$$G\Delta = \sum_{i=1}^{3} D_i(\alpha_i^2 \, \varepsilon_i^2 + \alpha_j^2 \, \varepsilon_k^2 + \alpha_k^2 \, \varepsilon_j^2) - D \qquad (i \neq j \neq k \neq i) \quad [12.28]$$

In the limiting case of spherical symmetry with $D_1 = D_2 = D_3 = D$, Eq. [12.22] reduces to a single correlation time with $\theta = (6D)^{-1}$. For ellipsoids of revolution, these equations (12.22 to 12.28) reduce to those given in Section 12.2.

12.4. TIME-DOMAIN STUDIES OF ANISOTROPIC ROTATIONAL DIFFUSION

It is surprisingly difficult to detect the presence of anisotropic rotation. However, a few definitive reports have appeared.[18–22] In one such report,[22] the anisotropy decay of 9,10-dimethylanthracene (DMA) in a viscous nonpolar solvent was measured. The sample was excited at 401 nm using the 0.5-ns pulses available from a synchrotron. The solvent was glycerol tripropanoate, which at –38.4 °C had a viscosity of 11 P. Under these conditions, the mean correlation time was 10 ns. The experimental anisotropy decay could not be fit to a single correlation time (Figure 12.9), but the data were well fit using two correlation times (not shown). Although the presence of two correlation times is consistent with DMA being an ellipsoid of revolution, the recovered diffusion coefficients did not agree with those predicted from the hydrodynamic theory for ellipsoids. This failure was attributed to the presence of some slip diffusion. The overall behavior of DMA ap-

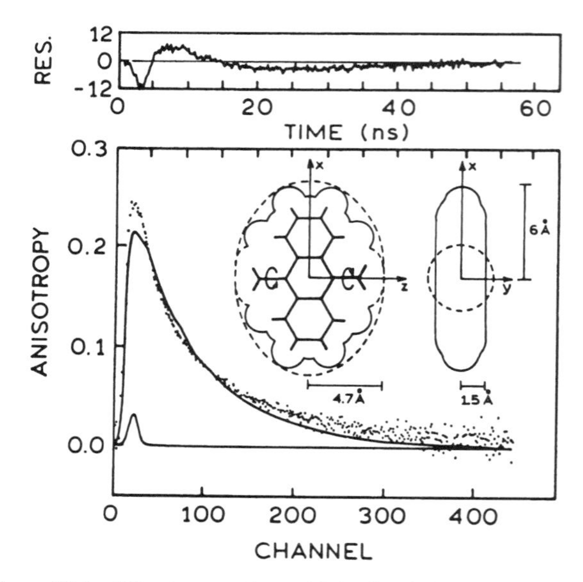

Figure 12.9. TD anisotropy decay of 9,10-dimethylanthracene in glycerol tripropanoate (tripropionin) at –38.4 °C. Excitation at 401 nm was obtained as 0.5-ns pulses from the LURE-ACO synchrotron, Orsay, France. The solid curve shows the single-correlation-time fit, and the upper panel shows the residuals for this fit. Revised and reprinted, with permission, from Ref. 22, Copyright © 1985, American Chemical Society.

peared to be intermediate between the slip and stick diffusion limits.

Perylene has been widely studied as an anisotropic rotor. When perylene is excited in its long-wavelength absorption band, the anisotropy decay is a double exponential, but this is difficult to see in the TD data. However, the use of a shorter excitation wavelength results in a definitive observation.[18] At this wavelength the absorption and emission transition moments are nearly perpendicular, and the time-zero anisotropy is –0.14. For a spherical molecule, the anisotropy is expected to decay from –0.14 to zero with a single correlation time. However, the anisotropy of perylene increases to values above zero and then decays to zero (Figure 12.10). This behavior can be understood in terms of the different rotational motions of perylene.

An intuitive description of the unusual perylene anisotropy decay is shown in Figure 12.11. When perylene is excited at a wavelength with $r_0 = -0.2$, the emission moments are symmetrically distributed in the laboratory x–y plane, and the horizontal intensity is greater than the vertical intensity ($r_0 < 0$). Because the in-plane rotation is more rapid than the out-of-plane rotation, the effect at short times is to rotate the emission dipoles out of the x–y plane toward the vertical axis. This results in the transient anisotropy values above zero (Figure 12.10). At longer times, the slower out-of-plane motions also contribute to depolarization, and the anisotropy decays to zero (Figure 12.11).

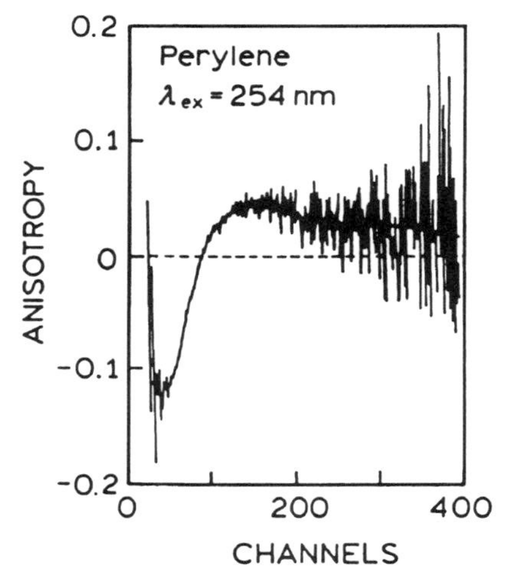

Figure 12.10. Anisotropy decay of perylene in glycerol at 30 °C. Excitation was at 254 nm, where $r_0 = -0.14$. The data were fit as $r(t) = 0.1 \exp(-t/17 \text{ ns}) - 0.24 \exp(-t/2.7 \text{ ns})$. Revised and reprinted, with permission, from Ref. 18, Copyright © 1981, American Institute of Physics.

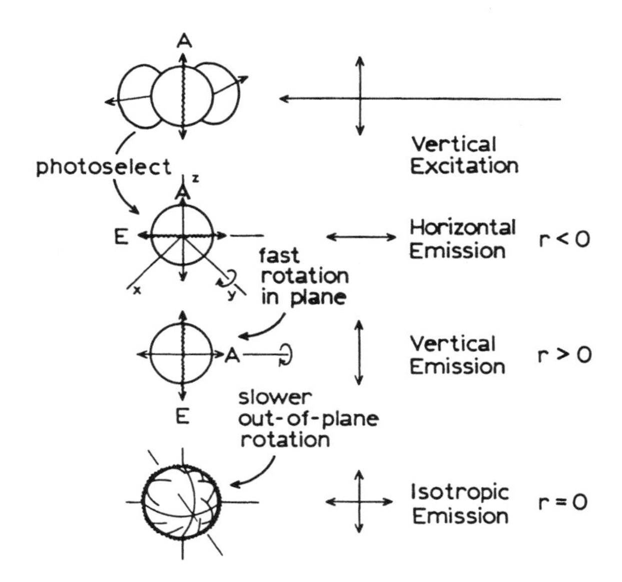

Figure 12.11. Effects of fast in-plane rotation and slower out-of-plane rotation on the anisotropy decay of perylene with $r_0 = -0.2$. Revised and reprinted, with permission, from Ref. 13, Academic Press, Inc.

12.5. FREQUENCY-DOMAIN STUDIES OF ANISOTROPIC ROTATIONAL DIFFUSION

Anisotropic rotational diffusion has been more extensively studied using FD methods. In fact, the earliest reports on the anisotropic rotation of fluorophores concerned experiments performed using fixed-frequency phase-modulation fluorometers.[23–25] At that time the phase-modulation instruments operated at only one or two fixed frequencies. Hence, it was not possible to recover the anisotropy decay law. The experiments were performed by measuring the differential polarized phase angles as the temperature was varied. It is relatively simple to predict the maximum value of Δ_ω for known values of the lifetime and fundamental anisotropy. For an isotropic rotor, the predicted value of Δ_ω is given by

$$\tan \Delta_\omega = \frac{(2D\tau)\omega\tau r_0}{\frac{1}{9}m_0(1+\omega^2\tau^2) + [(2D\tau)/3](2+r_0) + (2D\tau)^2} \quad [12.29]$$

where $m_0 = (1 + 2r_0)(1 - r_0)$. The value of $\tan \Delta_\omega$ is a function of the rotational rate (D), r_0, τ, and ω. The maximum value of $\tan \Delta_\omega$ is given by

$$\tan \Delta_\omega^{\max} = \frac{3\omega\tau r_0}{(2+r_0) + 2[m_0(1+\omega^2\tau^2)]^{1/2}} \quad [12.30]$$

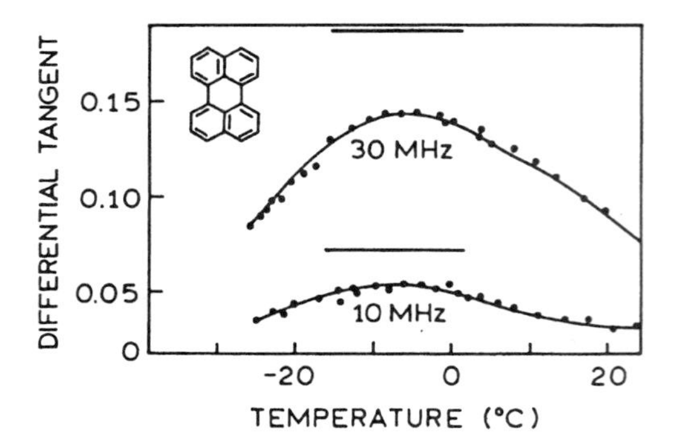

Figure 12.12. Temperature dependence of differential polarized phase angles of perylene in propylene glycol. The horizontal lines show the theoretical maximum values of tan Δ_ω for isotropic rotation. Revised from Ref. 24.

and is thus seen to be independent of the rotational rate. We note that these expressions are only valid for an isotropic rotor with a single fluorescence lifetime. The maximum differential tangent depends only on τ, r_0, and ω. If there is more than one rotational rate, the maximum value of tan Δ is decreased.[23]

Anisotropic rotations of perylene were detected by measurements of the temperature-dependent values of tan Δ, measured at a single modulator frequency.[24] The values of tan Δ are only nonzero in the temperature range where rotational diffusion is on a timescale comparable to the lifetime. At very high or low temperatures, tan Δ = 0. Figure 12.12 shows the temperature-dependent data for perylene in propylene glycol. The maximum values of tan Δ are significantly smaller than the values predicted using Eq. [12.30]. Molecules like DPH, which are not spheres but which display a single rotational rate, display tan Δ_ω^{max} values that agree with the calculated values (Figure 12.13). While the data for perylene did not allow determination of the anisotropy decay parameters, they did provide evidence for the presence of multiexponential anisotropy decays. The presence of anisotropic rotation in perylene was also suggested by early studies with a variable-frequency instrument which indicated that the correlation time for perylene was smaller than predicted based on its size.[27] However, multiple correlation times were not recovered.

At present, the FD measurements provide excellent resolution of complex anisotropy decays.[28] While I do not wish to start a debate, our experience indicates that resolution of complex anisotropy decays is a strength of the FD method. We are not certain why this is so, but one possibility is the ability to measure the difference between the polarized components of the decay directly in the frequency domain. In the time domain, one takes the difference of two independently measured quantities, $I_{\parallel}(t)$ and $I_{\perp}(t)$. Another possible reason is the use of two experimental observables, Δ_ω and Λ_ω, which results in better resolution than the use of either quantity alone.

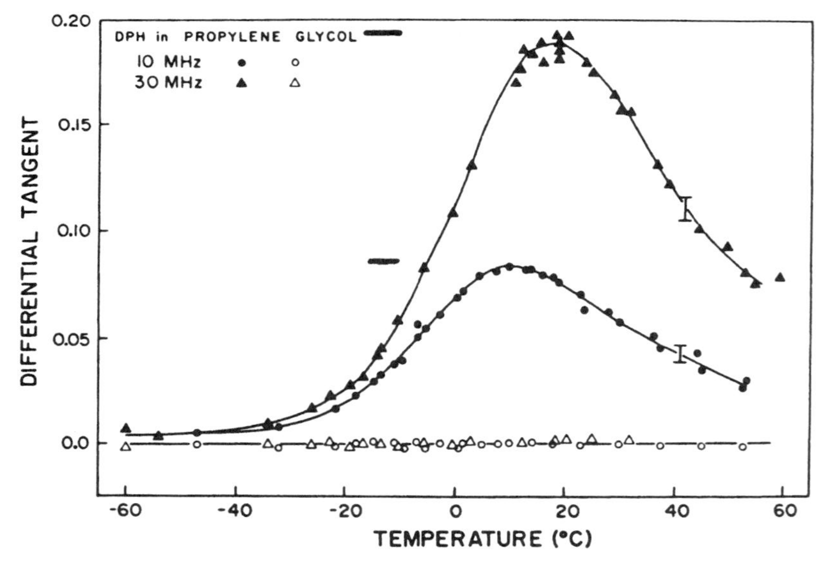

Figure 12.13. Temperature dependence of differential polarized phase angles of DPH in propylene glycol. The solid bars show the maximum values of tan Δ_ω for isotropic rotation. Reprinted, with permission, from Ref. 26, Copyright © 1979, American Chemical Society.

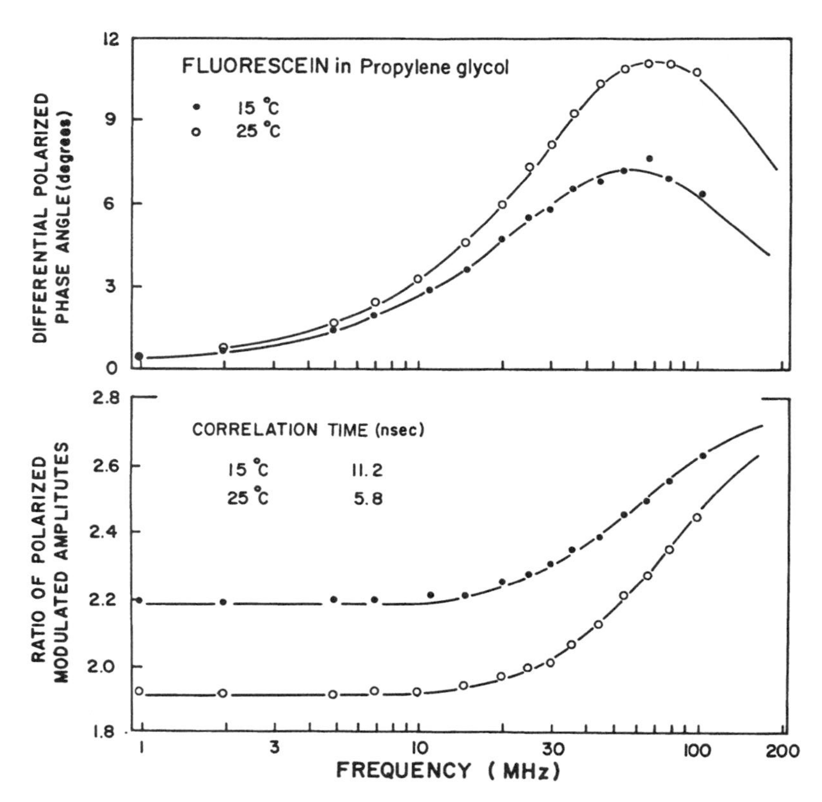

Figure 12.14. FD anisotropy decays of fluorescein in propylene glycol at 15 and 25 °C. The solid curves show the best single-correlation-time fits. The lifetime of fluorescein was 3.7 ns at 15 and 25 °C. Revised and reprinted, with permission, from Ref. 29, Copyright © 1985, American Chemical Society.

Resolution of anisotropic rotational diffusion using FD measurements is illustrated by data for two fluorophores, fluorescein and perylene. FD anisotropy data for fluorescein are shown in Figure 12.14. These data are almost perfectly fit using a single correlation time, indicating that fluorescein behaves like an isotropic rotor.[29] Such behavior is expected for polar fluorophores in polar solvents. In general, it seems that hydrogen-bonding interactions decrease slip diffusion and result in more isotropic rotations.

Rather different results were obtained for perylene (Figure 12.15). In this case the FD data could not be fit to a single correlation time (dashed curve in Figure 12.15), but these data were well fit using two correlation times.[30] The multiexponential anisotropy decay analysis yielded amplitudes of $r_{01} = 0.17$ and $r_{02} = 0.18$, with correlation times of 1.5 and 10.8 ns. This ratio of correlation times is comparable to that recovered from the TD data.[18] The 80-fold decrease in χ_R^2 for the two-correlation-time as compared to the one-correlation-time fit suggests that even more complex anisotropy decays could be recovered from the FD data.

12.6. GLOBAL ANISOTROPY DECAY ANALYSIS WITH MULTIWAVELENGTH EXCITATION

The use of multiple excitation wavelengths provides a useful method for improving the resolution of closely spaced correlation times. For a rigid, nonspherical molecule, different values of r_0 are expected to yield different anisotropy amplitudes, but the same correlation times. Since the contributions of diffusion coefficients are different for each component in the anisotropy decay (Eqs. [12.1]–[12.10]), the use of different excitation wavelengths allows each rotational rate to contribute differently to the data.[31–33]

The concept of using different r_0 values is illustrated by the example of perylene (Figure 12.16). A wide range of r_0 values are available with easily available excitation wavelengths (Figure 10.29). For longer excitation wavelengths (351 and 442 nm), the fundamental anisotropy of perylene is near 0.35. For excitation between 292 and 325 nm, the fundamental anisotropy is near 0.07, and for

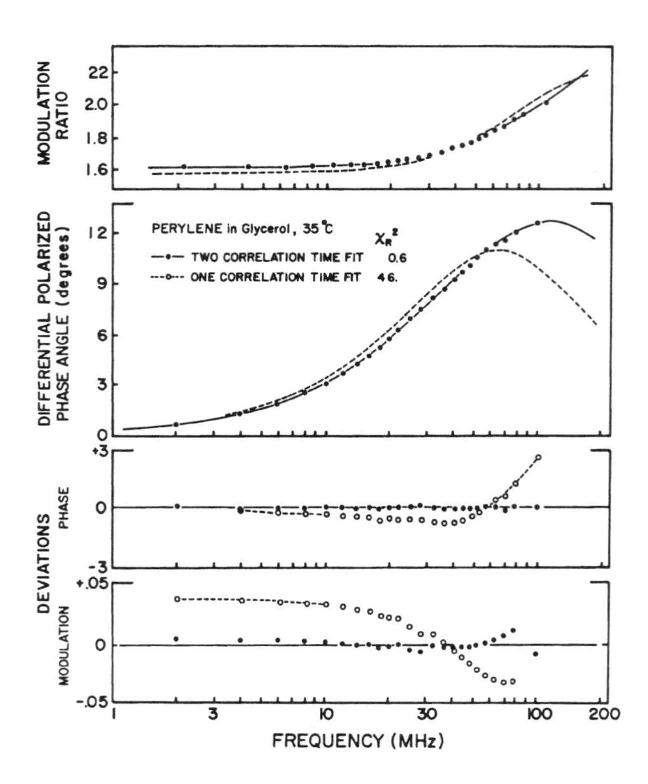

Figure 12.15. FD anisotropy decay of perylene in glycerol at 35 °C. The lifetime was 4.8 ns. Reprinted from Ref. 30, Copyright © 1985, with permission from Elsevier Science.

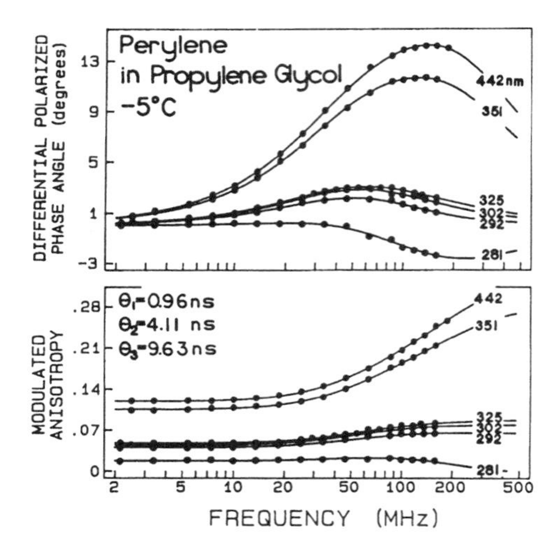

Figure 12.16. FD anisotropy decays of perylene for various excitation wavelengths. From Ref. 34.

excitation at 281 nm, the anisotropy is slightly below zero. Inspection of Figure 12.16 reveals that the frequency of the maximum differential phase angle depends on excitation wavelength. If there were a single rotational correlation time, the curves would be proportional to r_0, but the shape would be unchanged. The shape changes because of the multiple rotational rates and their different contributions at each excitation wavelength.

The wavelength-dependent shifts in the differential polarized phase curves can be understood in terms of the contributions of various rotations to the anisotropy decay. As reasoned in our discussion of Figures 12.5 and 12.6, both the in-plane and out-of-plane rotations are expected to contribute when $r_0 = 0.4$. Hence, the data for 351- and 442-nm excitation represent a weighted average of $D_{\parallel}$ and $D_{\perp}$. For r_0 values near -0.2, one expects the in-plane rotation to be dominant (Figure 12.11). This effect can be seen in the data for 281-nm excitation, for which the maximum value of Δ_{ω} (absolute value) is shifted toward higher modulation frequencies. Finally, for r_0 values near 0.1, one expects the out-of-plane rotation to dominate the anisotropy decay. These slower motions display a maximum value of Δ_{ω} at lower modulation frequencies.

In performing the global analysis, it is assumed that the correlation times are independent of excitation wavelength and that the amplitudes are dependent on wavelength. Under these assumptions, the value of χ_R^2 is calculated using an expanded form of Eq. [11.34], where the sum extends over the excitation wavelengths. The changes in shape of Δ_{ω} provide the opportunity to recover a highly detailed anisotropy decay using global analysis. In fact, the data allowed recovery of three rotational correlation times. The fact that the three correlation times were needed to

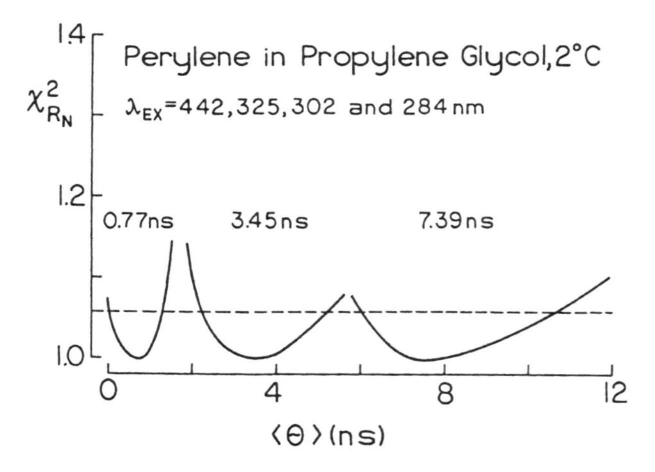

Figure 12.17. The χ_R^2 surfaces for the three correlation times found for perylene. From Ref. 34.

explain the data was demonstrated for the χ^2_R surfaces (Figure 12.17). The presence of three correlation times for perylene is somewhat surprising because the transitions of perylene are along its long and short axes (Figure 10.30), and it is shaped like an oblate ellipsoid. Hence, only two correlation times are expected. Also, the recovered correlation times and amplitudes do not seem to fit any reasonable hydrodynamic shape.[34] It appears that perylene displays partial slip diffusion and thus an anisotropy decay different from that predicted from the hydrodynamic model.

12.7. GLOBAL ANISOTROPY DECAY ANALYSIS WITH COLLISIONAL QUENCHING

As shown in Chapter 10 (Eq. [10.49]), the steady-state anisotropy is the average value of the anisotropy decay, when averaged over the intensity decay of the sample. If the sample displays several correlation times, Eq. [10.49] suggests that their contributions to the anisotropy would be different if the lifetime was changed. It is possible to change the lifetime using energy transfer or collisional quenching.[35–37] Consider a sample with a lifetime of 4 ns and a correlation time of 4.0 or 0.4 ns (Figure 12.18). The information about the anisotropy decay is contained in the difference between the parallel ($I_{\parallel}$) and perpendicular ($I_{\perp}$) components of the emission (Figure 12.18, left). For the 4-ns correlation time (dashed lines), the difference persists throughout the duration of the emission, and all the emitted photons aid in recovering the anisotropy decay. Now suppose that the correlation time is 0.4 ns (solid lines). Then, the polarized components are essentially equal after 1 ns, and the photons detected after this time do not aid in recovering the anisotropy decay. However, the emitted photons all contribute to the emission and to the acquisition time. Because the acquisition times must be finite, and systematic errors or instrumental drifts are always present, it is difficult to collect data adequate to determine the picosecond correlation time.

Now consider the effect of decreasing the decay time to a subnanosecond value of 0.4 ns by quenching (Figure 12.18, right). Because of the shorter lifetime, a larger fraction of the emission contains information on the 0.4-ns correlation time. Hence, information on the faster process is increased by measurement of the quenched samples.

For rapid single-exponential anisotropy decays, analysis of data from the most quenched sample is adequate. However, the usual goal is to improve resolution of the entire anisotropy decay. This can be accomplished by simultaneous analysis of data from the unquenched and several quenched samples. The data from the unquenched sample contribute most to the determination of the longer correlation times, and the data from the quenched samples contribute to resolution of the faster motions.

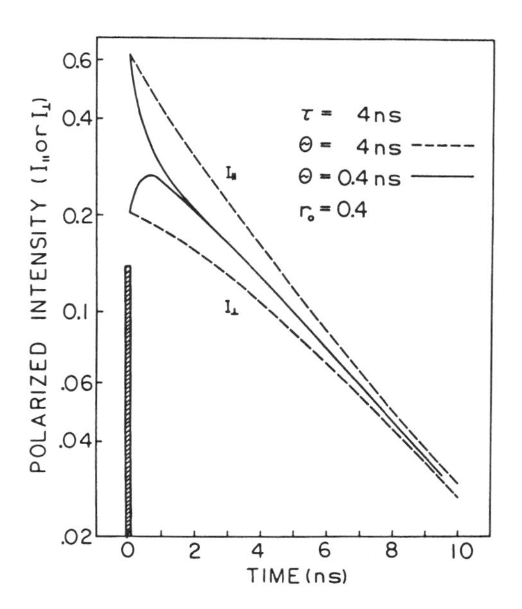

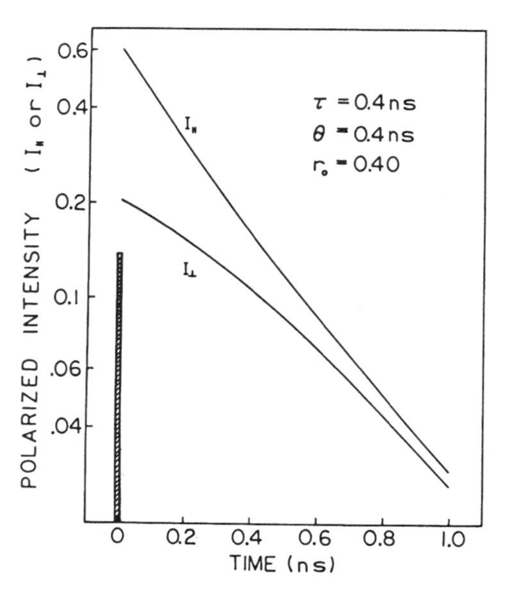

Figure 12.18. Simulated anisotropy decays for unquenched and quenched decays. *Left*: Unquenched decay with τ = 4 ns and θ = 4 (---) or 0.4 ns (———). *Right*: Quenched decay with τ = 0.4 ns and θ = 0.4 ns. For both decays, r_0 = 0.4. Reprinted, with permission, from Ref. 39, Copyright © 1987, Biophysical Society.

12.7.A. Application of Quenching to Protein Anisotropy Decays

While the use of global analysis with quenching is useful for studies of rigid anisotropic rotors, this method is even more useful in studies of complex anisotropy decays from proteins. This is because the correlation times for segmental motions and overall rotations of a protein are usually quite different in magnitude, whereas the correlation times of a rigid rotor are usually similar in magnitude.

The value of quenching can be seen from the simulated FD data presented in Figure 12.19. The unquenched lifetime was again assumed to be 4 ns, but the correlation times were assumed to be 10 ns for overall rotation and 0.1 ns for the segmental motions. To be comparable to the data found for proteins, the fundamental anisotropy (r_0) was 0.30, and the anisotropy amplitude of the segmental motion was taken as 0.07. Assume that the lifetime is reduced to 2.0 or to 0.5 ns by quenching. In practice, the decays become more heterogeneous in the presence of quenching owing to transient effects in diffusion, but this is not important for the simulations.

The existence of two rotational motions is easily visible in the frequency response of the polarized emission. These motions are evident from the two peaks in the plots of differential phase angle versus frequency; these peaks are due to the individual motions with correlation times of 10 and 0.1 ns. The assignment of the 0.1-ns motion to the peak at higher frequencies is demonstrated by the dashed curve in Figure 12.19, which shows the phase angles for the 0.1-ns component alone, with a maximum near 2 GHz. There are two features of the data from the quenched samples which provide increased resolution of the anisotropy decay. First, the data can be obtained to higher frequencies for the samples with shorter decay times. This is because the emission with the shorter lifetimes is less demodulated. The solid regions of the curves show the measurable frequency range where the modulation of the emission is 20% or larger, relative to the modulation of the incident light. Based on this criterion, the upper frequency limit in the absence of quenching ($\tau_0 = 4$ ns) is only 100 MHz. If the lifetime is reduced to 0.5 ns, the upper frequency limit is 2 GHz, which provides more data at frequencies characteristic of the faster motion.

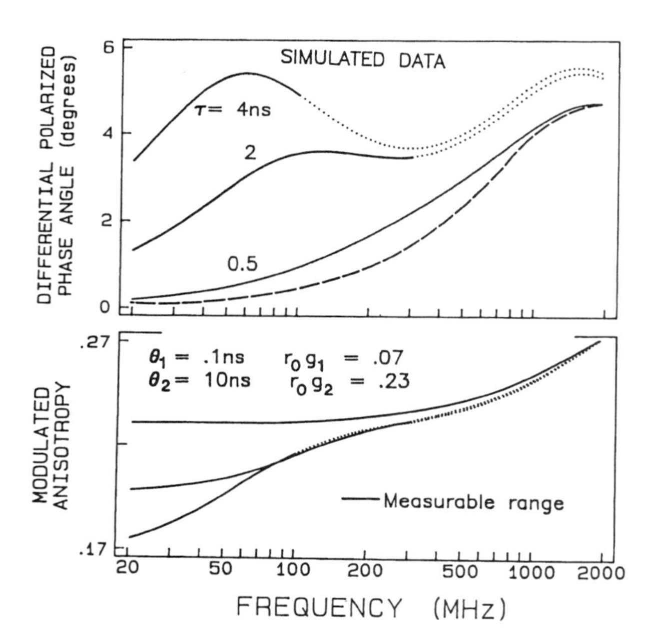

Figure 12.19. Simulated FD anisotropy data. The correlation times were assumed to be 0.10 and 10.0 ns, with amplitudes of 0.07 and 0.23, respectively. Simulated data are shown for decay times of 4, 2, and 0.5 ns. The solid regions of the curves indicate the measurable frequency range, where the modulation is 0.2 or larger for the intensity decay. The dashed curve shows the phase-angle values expected for the 0.1-ns correlation time, with the smaller amplitude (0.07). From Ref. 38.

Quenching also increases the information content of the modulated anisotropies (Figure 12.19, lower panel). At low frequencies, the value of the modulated anisotropy is equal to the steady-state anisotropy, which is increased by quenching. Once again, the measurable range is extended to 2 GHz by collisional quenching. The values of r_ω for the unquenched sample increase over the frequency range from 20 to 100 MHz, which is the portion of r_ω that depends upon overall rotation of the protein. For the quenched sample, the higher-frequency data begin to show a contribution from the faster motion.

An advantage of measuring a series of progressively quenched samples is that the data are differently sensitive to the two motions. The same motions determine the values of Δ_ω and Λ_ω in the quenched and the unquenched samples. However, the proportions that each motion contributes are different. As the decay time is decreased by quenching, the data are shifted toward higher frequencies and the contribution of overall diffusion is decreased (Figure 12.19). Hence, we expect increased resolution by simultaneous analysis of the data from a series of progressively quenched samples.

The value of global anisotropy analysis using quenched samples is illustrated for monellin in Figure 12.20. Monellin is a naturally sweet protein which contains a single tryptophan residue. This residue is rather mobile as seen from the two peaks near 70 and 500 MHz in the plot of differential phase angle versus frequency (Figure 12.20). As the sample was quenched by oxygen, the component due to overall rotation near 70 MHz decreased in amplitude. At the highest amounts of quenching, the FD anisotropy data become dominated by the contribution of the

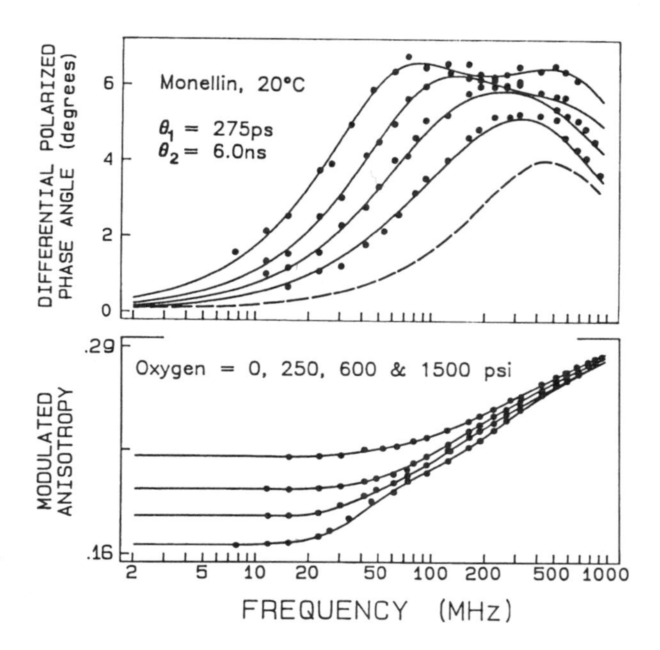

Figure 12.20. FD anisotropy data for monellin with oxygen quenching. Data are shown for oxygen pressures of 0, 250, 600, and 1500 psi. The dashed curve shows the phase-angle values expected for a correlation time of 275 ps, with an amplitude of 0.075. From Ref. 38.

shorter correlation time near 275 ps. Hence, one can visualize how global analysis of quenched and unquenched samples provides improved resolution of the two correlation times.

12.8. DNA

Many fluorophores intercalate into double-helical DNA. Some fluorophores, like EB, become more highly fluorescent when bound to DNA, allowing the emission of the DNA-bound probes to be easily detected. Consequently, there have been numerous studies of the anisotropy decays of DNA-bound probes. At first glance, a DNA-bound probe seems relatively simple. Suppose that the probe is located between the DNA base pairs, which is referred to as being intercalated. The anisotropy decay of such a probe is determined by its freedom within the DNA helix and by the diffusive motions of DNA. These motions are characterized by the torsional motions around the z-axis of DNA and by bending motions about the x-axis (Figure 12.21).

There have been several theoretical reports on the anisotropy decay expected for DNA-bound probes.[40–42] Unfortunately, the theory is complex in its complete form. The contributions of the torsional and bending motions of DNA depend on the orientation of the probe within the DNA helix and the value of r_0. Also, the depolarization due to the torsional motions and that due to the bending motions occur on very different timescales. In fact, the bending motions do not make a significant contribution to the anisotropy decay of EB, with its 23-ns lifetime, and $r(t)$ is dominated by torsional motions around the z-axis.

Because of the complexity of the theory, one often encounters various simplified forms. When the absorption and emission transition moments are collinear, and perpendicular to the helix axis, the anisotropy decay is given by[13]

$$r(t) = \frac{0.75\exp[-\Gamma(t)] + 0.45\exp[-\Delta(t) + \Gamma(t)] + 0.4\exp[-\Delta(t)]}{3 + \exp[-\Delta(t)]} \qquad [12.31]$$

where the twisting decay function is given by

$$\Gamma(t) = 4kT(t/\pi C\rho)^{1/2} \qquad [12.32]$$

and the bending decay function is

$$\Delta(t) = B(t)t^{1/4} \qquad [12.33]$$

In these equations, k is the Boltzmann constant, T is the temperature, C is the torsional stiffness of DNA, and ρ is a frictional coefficient per unit length for rotation about the helix axis. $B(t)$ is a slowly varying function of time which is determined by the bending stiffness of DNA. Equations [12.31]–[12.33] show that the anisotropy decay depends on $t^{1/2}$ due to torsional or twisting motions and on $t^{1/4}$ due to bending motions.

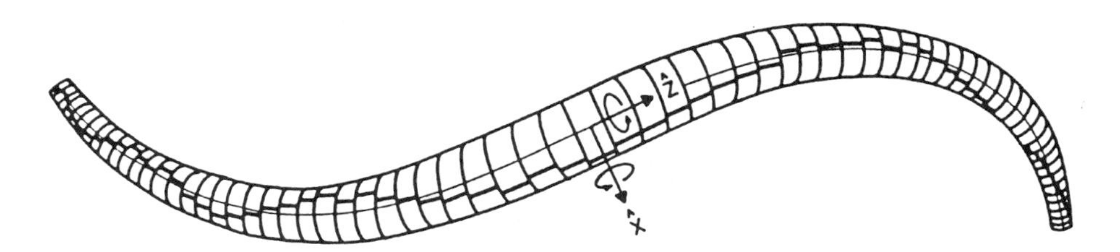

Figure 12.21. Torsional and bending motions of DNA. From Ref. 42.

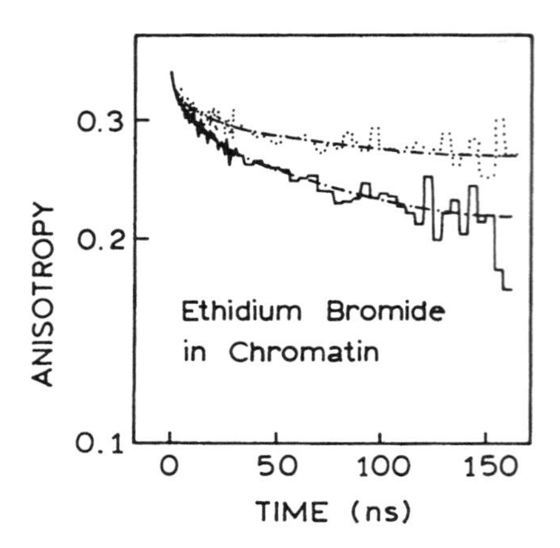

Figure 12.22. Fluorescence anisotropy decay curves of intercalated EB in chromatin DNA. Buffer conditions were 1m*M* Tris and 0.2m*M* EDTA, pH 7.5 (———) and 50m*M* NaCl, 1m*M* Tris, and 0.2m*M* EDTA, pH 7.5 (· · ·). The temperature was 20 °C. Revised and reprinted, with permission, from Ref. 47, Copyright © 1983, American Chemical Society.

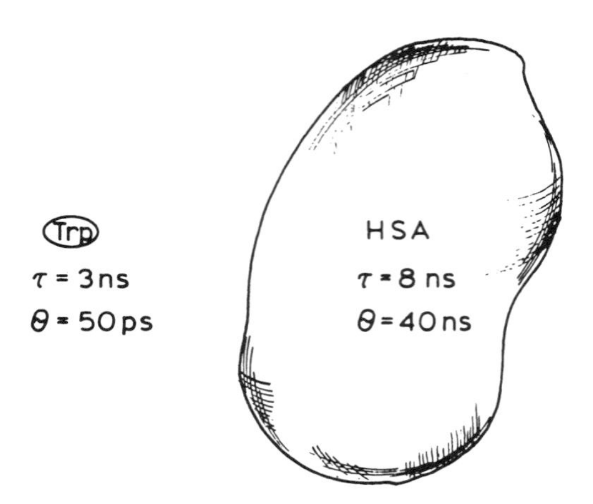

Figure 12.23. Model system for an associated anisotropy decay.

For nanosecond decay times, the bending motions have little effect on the anisotropy decay. Hence, one often encounters these equations is a more simplified form[43–45]:

$$r(t) = r_0[0.75 \exp(-t/\theta_1)^{1/2} + 0.25] \quad [12.34]$$

with

$$\theta_1 = \pi^2 b^2 \eta C / 4k^2 T^2 \quad [12.35]$$

where b is the radius of DNA and η is the viscosity.

While there have been many reports on EB-DNA,[43–47] most of the anisotropy decays are visually similar. The anisotropy decays rapidly at early times and tends toward a constant value at longer times (Figure 12.22). The early portion of the anisotropy decay is due to the torsional motions of DNA, and the constant anisotropy at longer times is due to the bending motions, which are not significant on this timescale.

12.9. ASSOCIATED ANISOTROPY DECAY

In the preceding sections we considered the complex anisotropy decays which result from nonspherical molecules or long flexible molecules like DNA. However, there is another possible origin of complex anisotropy decays, which is a mixture of fluorophores or a single fluorophore in two different environments. This concept is illustrated in Figure 12.23, which considers a mixture of tryptophan and human serum albumin (HSA). Tryptophan has a lifetime near 3 ns and a correlation time near 50 ps. The single tryptophan residue in HSA has a longer lifetime, near 8 ns, and a correlation time near 40 ns. We have assumed that the intensity and anisotropy decays are all single exponentials.

A mixture of fluorophores with correlation times of 40 ns and 50 ps can result in an unusual anisotropy decay. In Section 11.7 we described how rapid segmental motions result in an initial rapid decrease in protein anisotropy, followed by a slower anisotropy decay at longer times due to overall rotational diffusion. Similar correlation times from a mixture can result in a different type of anisotropy decay, in which the anisotropy decreases to a minimum and then increases at longer times (Figure 12.24).

The origin of the unusual behavior can be understood by examining the individual intensity and anisotropy decays

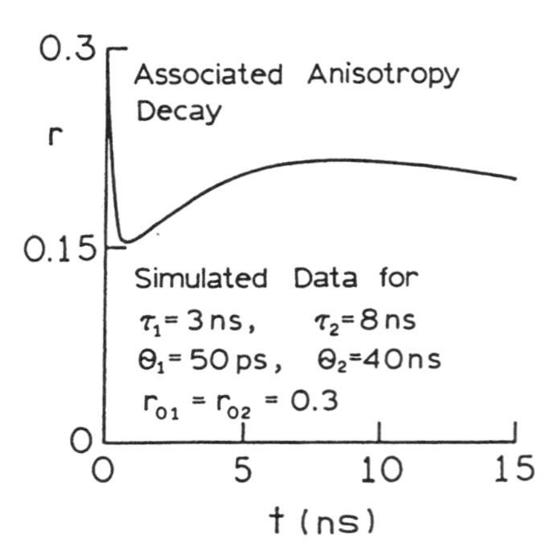

Figure 12.24. Simulated associated anisotropy decay for the mixture of fluorophores in Figure 12.23 (τ_1 = 3 ns, θ_1 = 50 ps, τ_2 = 8 ns, θ_2 = 40 ns, $r_{01} = r_{02}$ = 0.30).

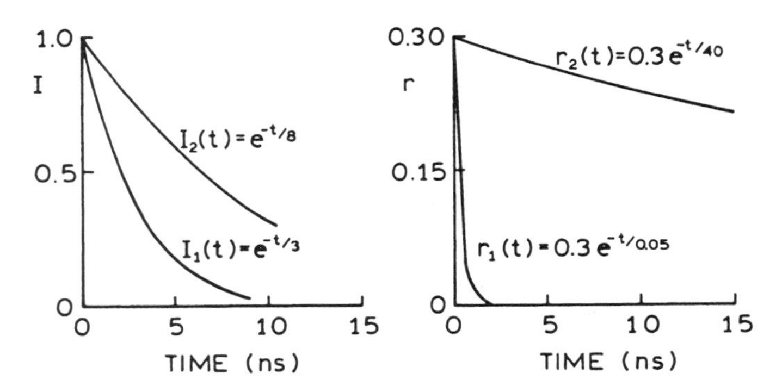

Figure 12.25. Intensity (*left*) and anisotropy decays (*right*) corresponding to the individual species shown in Figure 12.23.

(Figure 12.25). At any time, the anisotropy is given by the additivity law

$$r(t) = f_1(t)r_1(t) + f_2(t)r_2(t) \quad [12.36]$$

where $f_1(t)$ is the fractional intensity of the tryptophan and $f_2(t)$ is the fractional intensity of the HSA. One can readily calculate the fractional intensity of each species using the intensity decays. At $t = 0$, both species contribute equally. Because of the shorter lifetime of the tryptophan, its fractional intensity decreases progressively with time. However, at intermediate times, the low anisotropy of the tryptophan results in a transient decrease in the measured value of $r(t)$. At longer times, the emission becomes dominated by HSA with its larger anisotropy, resulting in an increase in $r(t)$.

12.9.A. Theory for Associated Anisotropy Decays

It is useful to compare the equations describing associated and nonassociated anisotropy decays. For a single fluorophore which displays multiexponential intensity and anisotropy decays, the parallel and perpendicular components of the emission are given by

$$I_\parallel(t) = \frac{1}{3} I(t)[1 + 2r(t)] \quad [12.37]$$

$$I_\perp(t) = \frac{1}{3} I(t)[1 - r(t)] \quad [12.38]$$

These expressions can be easily expanded in terms of the components in $I(t)$ and $r(t)$,

$$I_\parallel(t) = \frac{1}{3} \sum_i \alpha_i \exp(-t/\tau_i)[1 + 2 \sum_j r_{0j} \exp(-t/\theta_j)] \quad [12.39]$$

$$I_\perp(t) = \frac{1}{3} \sum_i \alpha_i \exp(-t/\tau_i)[1 - \sum_j r_{0j} \exp(-t/\theta_j)] \quad [12.40]$$

For the nonassociated decay, a single intensity decay $I(t)$ describes the fluorophore population.

For an associated system, each fluorophore population displays its own intensity and anisotropy decays. The polarized intensity decays are given by

$$I_\parallel(t) = \frac{1}{3} \sum_m I_m(t)[1 + 2r_m(t)] \quad [12.41]$$

$$I_\perp(t) = \frac{1}{3} \sum_m I_m(t)[1 - r_m(t)] \quad [12.42]$$

where m represents each fluorophore population, with an intensity decay $I_m(t)$ and an anisotropy decay $r_m(t)$. The nature of an associated anisotropy decay can be understood by considering Eqs. [12.41] and [12.42] for a two-component mixture and by assuming that each species displays a single lifetime (τ_m) and correlation time (θ_m). The amplitudes at $t = 0$ can be represented by α_m. The anisotropy decay is then given by

$$r(t) = \frac{\sum_m \alpha_m \exp(-t/\tau_m) r_{0m} \exp(-t/\theta_m)}{\sum_m \alpha_m \exp(-t/\tau_m)} \quad [12.43]$$

The fractional intensity of the mth component at any time t is given by

$$f_m(t) = \frac{\alpha_m \exp(-t/\tau_m)}{\sum_m \alpha_m \exp(-t/\tau_m)} \quad [12.44]$$

Hence, the time-dependent anisotropy is given by

$$r(t) = \sum_{m} f_m(t) r_m(t) \quad [12.45]$$

More detailed descriptions of the theory for associated anisotropy decays can be found elsewhere.[12,48–51]

12.9.B. Time-Domain Measurements of Associated Anisotropy Decays

Experimental studies of associated anisotropy decays can be traced to early reports on *cis*- and *trans*-parinaric acid.[52–54] In these studies the anisotropy was observed to increase at long times. An example is shown in Figure 12.26. In this case the probe was *cis*-parinaric acid which was covalently linked to the second position on a phosphatidylcholine molecule (*cis*-parinaroyl-PC).[55,56] The increase in anisotropy at times longer than 10 ns was explained as the result of a population of fluorophores with a longer lifetime, which also displayed a larger order parameter. Associated anisotropy decays have also been observed for labeled oligonucleotides when bound to proteins[57,58] and for rhodamine and proteins in lipid–protein systems.[59,60]

12.9.C. Frequency-Domain Measurements of Associated Anisotropy Decays

Associated anisotropy decays also result in unusual FD data.[61,62] This was illustrated by studies of ANS which was partially bound to apomyoglobin. Each apomyoglobin molecule binds only one ANS molecule, and the extent of binding was limited by using low concentrations of apomyoglobin. FD anisotropy data for the system show differential phase angles that initially increase with frequency and then decrease at higher frequencies (Figure 12.27). Surprisingly, the differential phase angles become negative even though the time-zero anisotropy is positive. This behavior is distinct from that observed for ANS when free in solution or when completely bound to apomyoglobin (Figure 12.28). Data of the type shown in Figure 12.27

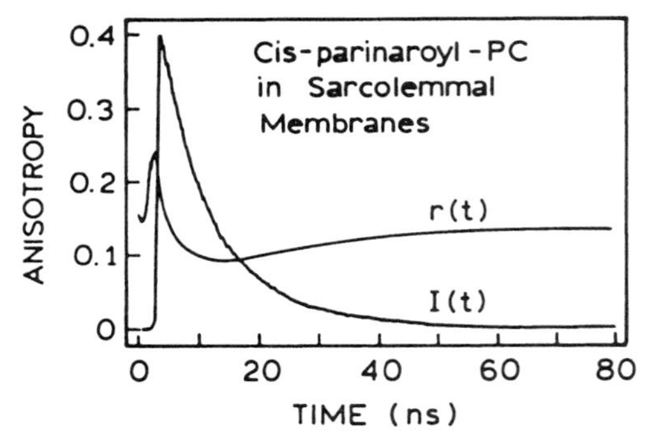

Figure 12.26. Intensity and anisotropy decays of *cis*-parinaroyl-PC in rat skeletal sarcolemmal membranes. From Ref. 55.

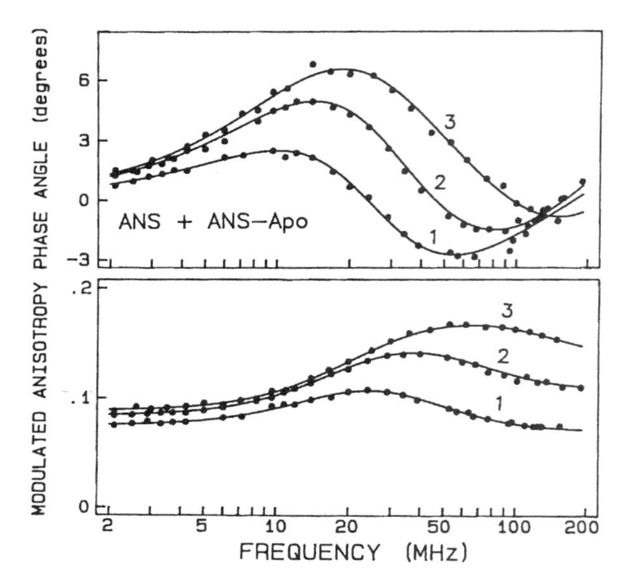

Figure 12.27. FD anisotropy data for samples containing apomyoglobin-ANS and unbound ANS in molar ratios of 0.0087 (1), 0.0158 (2), and 0.079 (3). Reprinted from Ref. 61, Copyright © 1987, with permission from Elsevier Science.

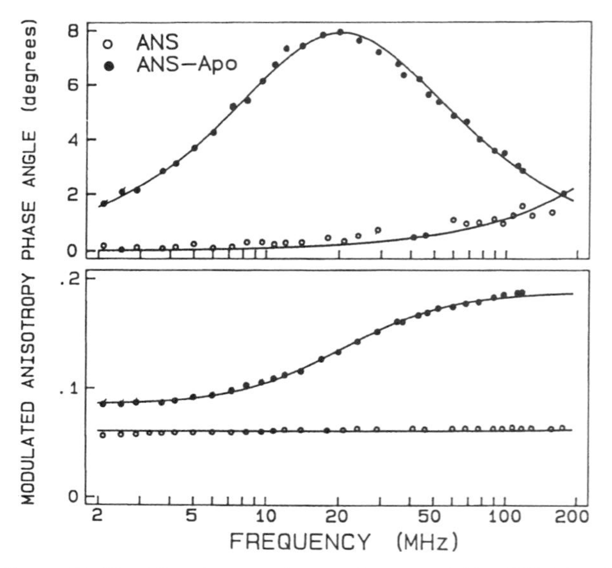

Figure 12.28. FD anisotropy decays of apomyoglobin-ANS (●) and ANS alone (○). The correlation times are 13.2 ns for apomyoglobin-ANS and 97 ps for ANS alone. Reprinted from Ref. 61, Copyright © 1987, with permission from Elsevier Science.

can be analyzed to recover the time-zero anisotropies and correlation times of each fluorophore population.

REFERENCES

1. Perrin, F., 1929, Mouvement brownien d'un ellipsoide (I). Dispersion diélectrique pour des molécules ellipsoidales, *J. Phys. Radium* **10**:497–511.
2. Perrin, F., 1936, Mouvement brownien d'un ellipsoide (II). Rotation libre et dépolarisation des fluorescences. Translation et diffusion de molécules ellipsoidales, *J. Phys. Radium* **1**:1–11.
3. Perrin, F., 1936, Diminution de la polarisation de la fluorescence des solutions résultant du mouvement brownien de rotation, *Acta Phys. Pol.* **5**:335–345.
4. Perrin, F., 1929, La fluorescence des solutions. Induction moléculaire—polarisation et durée d'émission—photochimie, *Ann. Phys.* **12**:169–275.
5. Lombardi, J. R., and Dafforn, G. A., 1966, Anisotropic rotational relaxation in rigid media by polarized photoselection, *J. Chem. Phys.* **44**:3882–3887.
6. Tao, T., 1969, Time-dependent fluorescence depolarization and Brownian rotational diffusion of macromolecules, *Biopolymers* **8**:609–632.
7. Ehrenberg, M., and Rigler, R., 1972, Polarized fluorescence and rotational Brownian motion, *Chem. Phys. Lett.* **14**:539–544.
8. Chuang, T. J., and Eisenthal, K. B., 1972, Theory of fluorescence depolarization by anisotropic rotational diffusion, *J. Chem. Phys.* **57**:5094–5097.
9. Belford, G. G., Belford, R. L., and Weber, G., 1972, Dynamics of fluorescence polarization in macromolecules, *Proc. Natl. Acad. Sci. U.S.A.* **69**:1392–1393.
10. Small, E. W., and Isenberg, I., 1977, Hydrodynamic properties of a rigid molecule: Rotational and linear diffusion and fluorescence anisotropy, *Biopolymers* **16**:1907–1928.
11. Steiner, R. F., 1991, Fluorescence anisotropy: Theory and applications, in *Topics in Fluorescence Spectroscopy, Volume 2, Principles*, J. R. Lakowicz (ed.), Plenum Press, New York, pp. 1–52.
12. Brand, L., Knutson, J. R., Davenport, L., Beechem, J. M., Dale, R. E., Walbridge, D. G., and Kowalczyk, A. A., 1985, Time-resolved fluorescence spectroscopy: Some applications of associative behaviour to studies of proteins and membranes, in *Spectroscopy and the Dynamics of Molecular Biological Systems*, P. Bayley and R. E. Dale (eds.), Academic Press, London, pp. 259–305.
13. Barkley, M. D., Kowalczyk, A. A., and Brand, L., 1981, Fluorescence decay studies of anisotropic rotations: Internal motions in DNA, in *Biomolecular Stereodynamics,* Vol. 1, R. H. Sarma (ed.), Adenine Press, New York, pp. 391–403.
14. Beechem, J. M., Knutson, J. R., and Brand, L., 1986, Global analysis of multiple dye fluorescence anisotropy experiments on proteins, *Biochem. Soc. Trans.* **14**:832–835.
15. Kawski, A., 1993, Fluorescence anisotropy: Theory and applications of rotational depolarization, *Crit. Rev. Anal. Chem.* **23**(6):459–529.
16. Hu, C.-M., and Zwanzig, R., 1974, Rotational friction coefficients for spheroids with the slipping boundary condition, *J. Chem. Phys.* **60**:4354–4357.
17. Youngren, G. K., and Acrivos, A., 1975, Rotational friction coefficients for ellipsoids and chemical molecules with the slip boundary condition, *J. Chem. Phys.* **63**:3846–3848.
18. Barkley, M. D., Kowalczyk, A. A., and Brand, L., 1981, Fluorescence decay studies of anisotropic rotations of small molecules, *J. Chem. Phys.* **75**:3581–3593.
19. Brocklehurst, B., and Young, R. N., 1994, Fluorescence anisotropy decays and viscous behaviour of 2-methyltetrahydrofuran, *J. Chem. Soc., Faraday Trans.* **90**:271–278.
20. Brocklehurst, B., and Young, R. N., 1995, Rotation of perylene in alkanes: Nonhydrodynamic behavior, *J. Phys. Chem.* **99**:40–53.
21. Sasaki, T., Hirota, K., Yamamoto, M., and Y. Nishijima, 1987, Anisotropic rotation of perylene studied by fluorescence depolarization method, *Bull. Chem. Soc. Jpn.* **60**:1165–1167.
22. Viovy, J. L., 1985, Anisotropic rotation of 1,9-dimethylanthracene: A fluorescence anisotropy decay study, *J. Phys. Chem.* **89**:5465–5472.
23. Weber, G., 1977, Theory of differential phase fluorometry: Detection of anisotropic molecular rotations, *J. Chem. Phys.* **66**:4081–4091.
24. Mantulin, W. W., and Weber, G., 1977, Rotational anisotropy and solvent–fluorophore bond: An investigation by differential polarized phase fluorometry, *J. Chem. Phys.* **66**:4092–4099.
25. Weber, G., and Mitchell, G. W., 1976, Demonstration of anisotropic molecular rotations by differential polarized phase fluorometry, in *Excited States of Biological Molecules*, J. B. Birks (ed.), John Wiley & Sons, New York, pp. 72–76.
26. Lakowicz, J. R., Prendergast, F. G., and Hogen, D., 1979, Differential polarized phase fluorometric investigations of diphenylhexatriene in lipid bilayers: Quantitation of hindered depolarizing rotations, *Biochemistry* **18**:508–519.
27. Klein, U. K. A., and Haas, H. P., 1979, Picosecond rotational diffusion of perylene, *Chem. Phys. Lett.* **63**:40–42.
28. Lakowicz, J. R., and Gryczynski, I., 1991, Frequency-domain fluorescence spectroscopy, in *Topics in Fluorescence Spectroscopy, Volume 1, Techniques*, J. R. Lakowicz (ed.), Plenum Press, New York, pp. 293–355.
29. Lakowicz, J. R., Cherek, H., and Maliwal, B. P., 1985, Time-resolved fluorescence anisotropies of diphenylhexatriene and perylene in solvents and lipid bilayers obtained from multifrequency phase-modulation fluorometry, *Biochemistry* **24**:376–383.
30. Lakowicz, J. R., and Maliwal, B. P., 1985, Construction and performance of a variable-frequency phase-modulation fluorometer, *Biophys. Chem.* **21**:61–78.
31. Lakowicz, J. R., Gryczynski, I., Cherek, H., and Laczko, G., 1991, Anisotropy decays of indole, melittin monomer and melittin tetramer by frequency-domain fluorometry and multi-wavelength global analysis, *Biophys. Chem.* **39**:241–251.
32. Gryczynski, I., Cherek, H., Laczko, G., and Lakowicz, J. R., 1987, Enhanced resolution of anisotropic rotational diffusion by multi-wavelength frequency-domain fluorometry and global analysis, *Chem. Phys. Lett.* **135**:193–199.
33. Gryczynski, I., Cherek, H., and Lakowicz, J. R., 1988, Detection of three rotational correlation times for a rigid asymmetric molecule using frequency-domain fluorometry, *Proc. SPIE* **909**:285–292.
34. Gryczynski, I., Danielson, E., and Lakowicz, J. R., unpublished observations.
35. Lakowicz, J. R., Gryczynski, I., and Wiczk, W. M., 1988, Anisotropic rotational diffusion of indole in cyclohexane studied by 2 GHz frequency-domain fluorometry, *Chem. Phys. Lett.* **149**:134–139.
36. Gryczynski, I., Cherek, H., and Lakowicz, J. R., 1988, Detection of three rotational correlation times for a rigid asymmetric molecule using frequency-domain fluorometry, *Biophys. Chem.* **30**:271–277.
37. Gryczynski, I., Wiczk, W., Johnson, M. L., and Lakowicz, J. R., 1988, Lifetime distributions and anisotropy decays of indole fluo-

rescence in cyclohexane/ethanol mixtures by frequency-domain fluorometry, *Biophys. Chem.* **32**:173–185.
38. Lakowicz, J. R., Gryczynski, I., Szmacinski, H., Cherek, H., and Joshi, N., 1991, Anisotropy decays of single tryptophan proteins measured by GHz frequency-domain fluorometry with collisional quenching, *Eur. Biophys. J.* **19**:125–140.
39. Lakowicz, J. R., Cherek, H., Gryczynski, I., Joshi, N., and Johnson, M. L., 1987, Enhanced resolution of fluorescence anisotropy decays by simultaneous analysis of progressively quenched samples, *Biophys. J.* **51**:755–768.
40. Barkley, M. D., and Zimm, B. H., 1979, Theory of twisting and bending of chain macromolecules; analysis of the fluorescence depolarization of DNA, *J. Chem. Phys.* **70**:2991–3007.
41. Thomas, J. C., Allison, S. A., Appellof, C. J., and Schurr, J. M., 1980, Torsion dynamics and depolarization of fluorescence of linear macromolecules II. Fluorescence polarization anisotropy measurements on a clean viral 29 DNA, *Biophys. Chem.* **12**:177–188.
42. Schurr, J. M., Fujimoto, B. S., Wu, P., and Song, L., 1992, Fluorescence studies of nucleic acids: Dynamics, rigidities, and structures, in *Topics in Fluorescence Spectroscopy, Volume 3, Biochemical Applications*, J. R. Lakowicz (ed.), Plenum Press, New York, pp. 137–229.
43. Millar, D. P., Robbins, R. J., and Zewail, A. H., 1981, Time-resolved spectroscopy of macromolecules: Effect of helical structure on the torsional dynamics of DNA and RNA, *J. Chem. Phys.* **74**:4200–4201.
44. Ashikawa, I., Furuno, T., Kinosita, K., Ikegami, A., Takahashi, H., and Akutsu, H., 1983, Internal motion of DNA in bacteriophages, *J. Biol. Chem.* **259**:8338–8344.
45. Ashikawa, I., Kinosita, K., and Ikegami, A., 1984, Dynamics of z-form DNA, *Biochim. Biophys. Acta* **782**:87–93.
46. Genest, D., Wahl, P., Erard, M., Champagne, M., and Daune, M., 1982, Fluorescence anisotropy decay of ethidium bromide bound to nucleosomal core particles, *Biochimie* **64**:419–427.
47. Ashikawa, I., Kinosita, K., Ikegami, A., Nishimura, Y., Tsuboi, M., Watanabe, K., Iso, K., and Nakano, T., 1983, Internal motion of deoxyribonucleic acid in chromatin. Nanosecond fluorescence studies of intercalated ethidium, *Biochemistry* **22**:6018–6026.
48. Fisz, J. J., 1996, Polarized fluorescence decay surface for a mixture of non-interacting species in solution, *Chem. Phys. Lett.* **259**:579–587.
49. Fisz, J. J., 1996, Polarized fluorescence spectroscopy of two-ground and two-excited state systems in solutions, *Chem. Phys. Lett.* **262**:495–506.
50. Fisz, J. J., 1996, Polarized fluorescence decay surface for many-ground- and many-excited-state species in solution, *Chem. Phys. Lett.* **262**:507–518.
51. Bialik, C. N., Wolf, B., Rachofsky, E. L., Ross, J. B. A., and Laws, W. R., 1998, Fluorescence anisotropy decay: Finding the correct physical model, *Proc. SPIE* **3526**:60–67.
52. Wolber, P. K., and Hudson, B. S., 1981, Fluorescence lifetime and time-resolved polarization anisotropy studies of acyl chain order and dynamics in lipid bilayers, *Biochemistry* **20**:2800–2816.
53. Wolber, P. K., and Hudson, B. S., 1982, Bilayer acyl chain dynamics and lipid–protein interaction, *Biophys. J.* **37**:253–262.
54. Ruggiero, A., and Hudson, B., 1989, Analysis of the anisotropy decay of *trans*-parinaric acid in lipid bilayers, *Biophys. J.* **55**:1125–1135.
55. van Paridon, P. A., Shute, J. K., Wirtz, K. W. A., and Visser, A. J. W. G., 1988, A fluorescence decay study of parinaroyl-phosphatidylinositol incorporated into artificial and natural membranes, *Eur. Biophys. J.* **16**:53–63.
56. Visser, A. J. W. G., van Hoek, A., and van Paridon, P. A., 1987, Time-resolved fluorescence depolarization studies of parinaroyl phosphatidylcholine in Triton X-100 micelles and rat skeletal muscle membranes, in *Membrane Receptors, Dynamics, and Energetics*, K. W. A. Wirtz (ed.), Plenum Press, New York, pp. 353–361.
57. Millar, D. P., Allen, D. J., and Benkovic, S. J., 1990, Structure and dynamics of a DNA:polymerase complex by time-resolved fluorescence spectroscopy, *Proc. SPIE* **1204**:392–403.
58. Guest, C. R., Hochstrasser, R. A., Dupuy, C. G., Allen, D. J., Benkovic, S. J., and Millar, D. M., 1991, Interaction of DNA with the Klenow fragment of DNA polymerase I studied by time-resolved fluorescence spectroscopy, *Biochemistry* **30**:8759–8770.
59. Peng, K., Visser, A. J. W. G., van Hoek, A., Wolfs, C. J. A. M., Sanders, J. C., and Hemminga, M. A., 1990, Analysis of time-resolved fluorescence anisotropy in lipid–protein systems, I. Application to the lipid probe octadecyl rhodamine B in interaction with bacteriophage M13 coat protein incorporated in phospholipid bilayers, *Eur. Biopsy. J.* **18**:277–283.
60. Peg, K., Visa, A. J. W. G., van Hoe, A., Wolfs, C. J. A. M., and Hemming, M. A., 1990, Analysis of time-resolved fluorescence an isotropy in lipid–protein systems, II. Application to tryptophan fluorescence of bacteriophage M13 coat protein incorporated in phospholipid bilayers, *Eur. Biophys. J.* **18**:285–293.
61. Szmacinski, H., Jayaweera, R., Cherek, H., and Lakowicz, J. R., 1987, Demonstration of an associated anisotropy decay by frequency-domain fluorometry, *Biophys. Chem.* **27**:233–241.
62. Wang, R., and Bright, F. V., 1993, Rotational reorientation kinetics of dansylated bovine serum albumin, *J. Phys. Chem.* **97**:4231–4238.

PROBLEMS

12.1. *Calculation of an Associated Anisotropy Decay*: Use the intensity and anisotropy decays in Figure 12.25 to calculate the anisotropy at $t = 0$, 1, and 5 ns. Also calculate the anisotropy values at 0, 1, and 5 ns assuming a nonassociated anisotropy decay.

Energy Transfer

13

Fluorescence resonance energy transfer (FRET) is transfer of the excited-state energy from the initially excited donor (D) to an acceptor (A). The donor molecules typically emit at shorter wavelengths which overlap with the absorption spectrum of the acceptor. Energy transfer occurs without the appearance of a photon and is the result of long-range dipole–dipole interactions between the donor and acceptor. The term resonance energy transfer (RET) is preferred because the process does not involve the appearance of a photon. The rate of energy transfer depends upon the extent of spectral overlap of the emission spectrum of the donor with the absorption spectrum of the acceptor, the quantum yield of the donor, the relative orientation of the donor and acceptor transition dipoles, and the distance between the donor and acceptor molecules. The distance dependence of RET has resulted in its widespread use to measure distances between donors and acceptors.

The most common application of RET is to measure the distances between two sites on a macromolecule. Typically, a protein is covalently labeled with a donor and an acceptor (Figure 13.1). The donor is often a tryptophan residue, but extrinsic donors are also used. If there is a single donor and acceptor, and if the D–A distance does not change during the excited-state lifetime, then the D–A distance can be determined from the efficiency of energy transfer. The transfer efficiency can be determined by steady-state measurements of the extent of donor quenching due to the acceptor.

Resonance energy transfer is also used to study macromolecular systems in which a single D–A distance is not present, such as assemblies of proteins and membranes or unfolded proteins where there is a distribution of D–A distances. The extent of energy transfer can also be influenced by the presence of donor-to-acceptor diffusion during the donor lifetime. Although information can be obtained from the steady-state data, such systems are usually studied using time-resolved measurements. These more advanced applications of RET are presented in Chapters 14 and 15.

An important characteristic of energy transfer is that it occurs over distances comparable to the dimensions of biological macromolecules. The distance at which RET is 50% efficient, called the Förster distance,[1] is typically in the range of 20–60 Å. The rate of energy transfer from a donor to an acceptor (k_T) is given by

$$k_T = \frac{1}{\tau_D}\left(\frac{R_0}{r}\right)^6 \qquad [13.1]$$

where τ_D is the decay time of the donor in the absence of acceptor, R_0 is the Förster distance, and r is the donor-to-acceptor (D–A) distance. Hence, the rate of transfer is equal to the decay rate of the donor in the absence of acceptor ($1/\tau_D$) when the D–A distance (r) is equal to the Förster distance (R_0). When the D–A distance is equal to the Förster distance ($r = R_0$), then the transfer efficiency is 50%. At this distance ($r = R_0$), the donor emission would be decreased to one-half of its intensity in the absence of acceptor. The rate of RET depends strongly on distance, being inversely proportional to r^6 (Eq. [13.1]).

Förster distances ranging from 20 to 90 Å are convenient for studies of biological macromolecules. These distances are comparable to the diameter of many proteins, the thickness of biological membranes, and the distance between sites on multisubunit proteins. Any phenomenon which affects the D–A distance will affect the transfer rate, allowing the phenomenon to be quantified. Energy-transfer measurements have been used to estimate the distances between sites on macromolecules and the effects of conformational changes on these distances. In this type of application, one uses the extent of energy transfer between a fixed donor and acceptor to calculate the D–A distance and thus obtain structural information about the macromolecule (Figure 13.1). Such distance measurements have resulted in the description of RET as a "spectroscopic ruler."[2,3] For instance, energy transfer can be used to

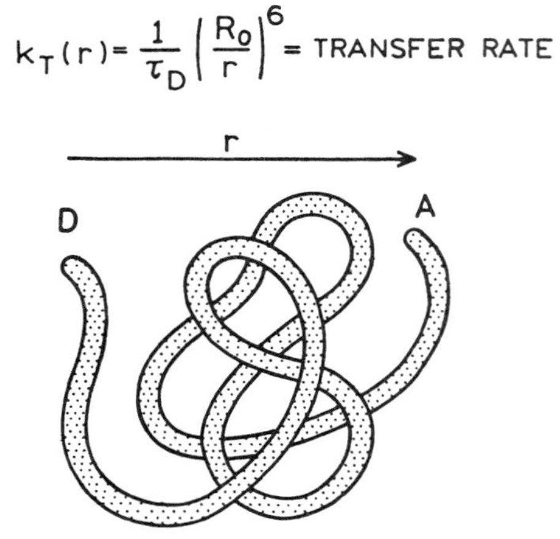

$$k_T(r) = \frac{1}{\tau_D}\left(\frac{R_0}{r}\right)^6 = \text{TRANSFER RATE}$$

r

D A

Figure 13.1. Fluorescence resonance energy transfer for a protein with a single donor (D) and a single acceptor (A).

measure the distance from a tryptophan residue to a ligand binding site when the ligand serves as the acceptor.

In the case of multidomain proteins, RET has been used to measure conformational changes which move the domains closer or further apart. Energy transfer can also be used to measure the distance between a site on a protein and a membrane surface, association between protein subunits, and lateral association of membrane-bound proteins. In the case of macromolecular association reactions, one relies less on determination of a precise D–A distance, and more on the simple fact that energy transfer occurs whenever the donors and acceptors are in close proximity comparable to the Förster distance.

The use of energy transfer as a proximity indicator illustrates an important characteristic of energy transfer. Energy transfer can be reliably assumed to occur whenever the donors and acceptors are within the characteristic Förster distance, and whenever suitable spectral overlap occurs. The value of R_0 can be reliably predicted from the spectral properties of the donors and acceptors. Energy transfer is a through-space interaction which is mostly independent of the intervening solvent and/or macromolecule. In principle, the orientation of the donors and acceptors can prevent energy transfer between a closely spaced D–A pair, but such a result is rare, and possibly nonexistent in biomolecules. Hence, one can assume that RET will occur if the spectral properties are suitable and the D–A distance is comparable to R_0. A wide variety of biochemical interactions result in changes in distance and are thus measurable using RET.

RET is a process which does not involve emission and reabsorption of photons. The theory of energy transfer is based on the concept of a fluorophore as an oscillating dipole, which can exchange energy with another dipole with a similar resonance frequency.[4] Hence, RET is similar to the behavior of coupled oscillators, like two swings on a common supporting beam. In contrast, radiative energy transfer is due to emission and reabsorption of photons and is thus due to inner filter effects. Radiative transfer depends upon less interesting optical properties of the sample, such as the size of the sample container, the path length, the optical densities of the sample at the excitation and emission wavelengths, and the geometric arrangement of the excitation and emission light paths. In contrast, nonradiative energy transfer contains a wealth of structural information concerning the donor–acceptor pair.

RET contains molecular information which is different from that revealed by solvent relaxation, excited-state reactions, fluorescence quenching, or fluorescence anisotropy. These other fluorescence phenomena depend on interactions of the fluorophore with other molecules in the surrounding solvent shell. These nearby interactions are less important for energy transfer, except for their effects on the spectral properties of the donor and acceptor. Nonradiative energy transfer is effective over much longer distances, and the intervening solvent or macromolecule has little effect on the efficiency of energy transfer, which depends primarily on the D–A distance. In this chapter we will describe the basic theory of nonradiative energy transfer. This theory is applicable to a D–A pair separated by a fixed distance, but we also describe examples of RET from membrane-bound proteins to randomly distributed lipid acceptors and the use of RET to study association reactions. More complex formalisms are needed to describe other commonly encountered situations, such as distance distributions (Chapter 14) and the presence of multiple acceptors (Chapter 15).

13.1. THEORY OF ENERGY TRANSFER FOR A DONOR–ACCEPTOR PAIR

The theory for RET is moderately complex, and similar equations have been derived from classical and quantum-mechanical considerations. We will describe only the final equations. Readers interested in the physical basis of RET are referred to the excellent review by Clegg.[4] RET is best understood by considering a single donor and acceptor separated by a distance r. The rate of transfer for a donor and acceptor separated by a distance r is given by

$$k_T(r) = \frac{Q_D\kappa^2}{\tau_D r^6}\left(\frac{9000\,(\ln 10)}{128\,\pi^5 N n^4}\right)\int_0^\infty F_D(\lambda)\varepsilon_A(\lambda)\lambda^4\,d\lambda \qquad [13.2]$$

where Q_D is the quantum yield of the donor in the absence of acceptor; n is the refractive index of the medium, which is typically assumed to be 1.4 for biomolecules in aqueous solution; N is Avogadro's number; τ_D is the lifetime of the donor in the absence of acceptor; $F_D(\lambda)$ is the corrected fluorescence intensity of the donor in the wavelength range λ to $\lambda + \Delta\lambda$, with the total intensity (area under the curve) normalized to unity; $\varepsilon_a(\lambda)$ is the extinction coefficient of the acceptor at λ, which is typically in units of M^{-1} cm^{-1}; κ^2 is a factor describing the relative orientation in space of the transition dipoles of the donor and acceptor and is usually assumed to be equal to $\frac{2}{3}$, which is appropriate for dynamic random averaging of the donor and acceptor (see Section 13.1.A below). In Eq. [13.2] we wrote the transfer rate k_T as a function of r, $k_T(r)$, to emphasize its dependence on distance.

The overlap integral $J(\lambda)$ expresses the degree of spectral overlap between the donor emission and the acceptor absorption,

$$J(\lambda) = \int_0^\infty F_D(\lambda)\varepsilon_A(\lambda)\lambda^4\, d\lambda = \frac{\int_0^\infty F_D(\lambda)\varepsilon_A(\lambda)\lambda^4\, d\lambda}{\int_0^\infty F_D(\lambda)\, d\lambda} \qquad [13.3]$$

$F_D(\lambda)$ is dimensionless. If $\varepsilon_A(\lambda)$ is expressed in units of M^{-1} cm^{-1} and λ is in nanometers, then $J(\lambda)$ is in units of M^{-1} cm^{-1} nm^4. If λ is in centimeters, then $J(\lambda)$ is in units of M^{-1} cm^3. In calculating $J(\lambda)$, one should use the corrected emission spectrum with its area normalized to unity (Eq. [13.3], middle) or normalize the calculated value of $J(\lambda)$ by the area (Eq. [13.3], right). The overlap integral has been defined in several ways, each with different units. In our experience, we find that it is easy to get confused so we recommend the units of nanometers or centimeters for the wavelength and units of M^{-1} cm^{-1} for the extinction coefficient.

Because the transfer rate $k_T(r)$ depends on r, it is inconvenient to use this rate constant in the design of biochemical experiments. For this reason, Eq. [13.2] is written in terms of the Förster distance R_0, at which the transfer rate $k_T(r)$ is equal to the decay rate of the donor in the absence of acceptor (τ_D^{-1}). At this distance, one-half of the donor molecules decay by energy transfer and one-half decay by the usual radiative and nonradiative rates. From Eqs. [13.1] and [13.2] with $k_T(r) = \tau_D^{-1}$, one obtains

$$R_0^6 = \frac{9000(\ln 10)\kappa^2 Q_D}{128\,\pi^5 N n^4} \int_0^\infty F_D(\lambda)\varepsilon_A(\lambda)\lambda^4\, d\lambda \qquad [13.4]$$

This expression allows the Förster distance to be calculated from the spectral properties of the donor and the acceptor and the donor quantum yield. While Eq. [13.4] looks complex, many of the terms are simple physical constants. It is convenient to have simpler expressions for R_0 in terms of the experimentally known values, which is accomplished by combining the constant terms in Eq. [13.4]. If the wavelength is expressed in nanometers, then $J(\lambda)$ is in units of M^{-1} cm^{-1} (nm)^4 and the Förster distance, in angstroms, is given by

$$R_0 = 0.211[\kappa^2 n^{-4} Q_D J(\lambda)]^{1/6} \quad (\text{in Å}) \qquad [13.5]$$

and

$$R_0^6 = 8.79 \times 10^{-5}[\kappa^2 n^{-4} Q_D J(\lambda)] \quad (\text{in Å}^6) \qquad [13.6]$$

If the wavelength is expressed in centimeters and $J(\lambda)$ is in units of M^{-1} cm^3, then the Förster distance is given by

$$R_0^6 = 8.79 \times 10^{-25}[\kappa^2 n^{-4} Q_D J(\lambda)] \quad (\text{in cm}^6) \qquad [13.7]$$

or

$$R_0 = 9.78 \times 10^{3}[\kappa^2 n^{-4} Q_D J(\lambda)]^{1/6} \quad (\text{in Å}) \qquad [13.8]$$

and

$$R_0^6 = 8.79 \times 10^{23}[\kappa^2 n^{-4} Q_D J(\lambda)] \quad (\text{in Å}^6) \qquad [13.9]$$

It is important to recognize that the Förster distances are usually reported for an assumed value of κ^2, typically $\kappa^2 = \frac{2}{3}$. Once the value of R_0 is known, the rate of energy transfer can be easily calculated using

$$k_T(r) = \frac{1}{\tau_D}\left(\frac{R_0}{r}\right)^6 \qquad [13.10]$$

One can then readily determine whether the rate of transfer will be competitive with the decay rate (τ_D^{-1}) of the donor. If the transfer rate is much faster than the decay rate, then energy transfer will be efficient. If the transfer rate is slower than the decay rate, then little transfer will occur during the excited-state lifetime, and RET will be inefficient.

The efficiency of energy transfer (E) is the fraction of photons absorbed by the donor that are transferred to the acceptor. This fraction is given by

$$E = \frac{k_T}{\tau_D^{-1} + k_T} \qquad [13.11]$$

which is the ratio of the transfer rate to the total decay rate of the donor. Recalling that $k_T = \tau_D^{-1}(R_0/r)^6$, one can easily rearrange Eq. [13.11] to yield

$$E = \frac{R_0^6}{R_0^6 + r^6} \quad [13.12]$$

This equation shows that the transfer efficiency is strongly dependent on distance when the D–A distance is near R_0 (Figure 13.2). The efficiency quickly increases to 1.0 as the D–A distance decreases below R_0. For instance, if $r = 0.1R_0$, one can readily calculate that the transfer efficiency is 0.999999, so that the donor emission would not be observable. Conversely, the transfer efficiency quickly decreases to zero if r is greater than R_0. Because E depends so strongly on distance, measurements of the distance (r) are only reliable when r is within a factor of 2 of R_0 (see Problem 13.7). If r is twice the Förster distance ($r = 2R_0$), then the transfer efficiency is 1.56%.

The transfer efficiency is typically measured using the relative fluorescence intensity of the donor, in the absence (F_D) and presence (F_{DA}) of acceptor. The transfer efficiency can also be calculated from the lifetimes under these respective conditions (τ_D and τ_{DA}):

$$E = 1 - \frac{\tau_{DA}}{\tau_D} \quad [13.13]$$

$$E = 1 - \frac{F_{DA}}{F_D} \quad [13.14]$$

It is important to remember the assumptions involved in the derivation of these equations. Equations [13.13] and [13.14] are only applicable to donor–acceptor pairs which are separated by a fixed distance. This situation is frequently encountered for labeled proteins. However, a single fixed donor–acceptor distance is not found for a

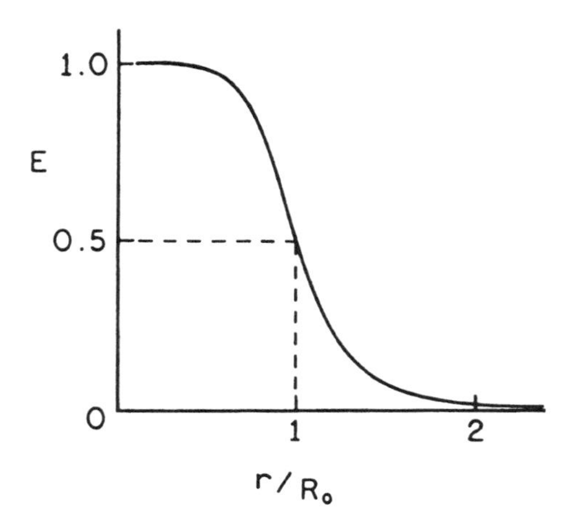

Figure 13.2. Dependence of the energy transfer efficiency (E) on distance. R_0 is the Förster distance.

mixture of donors and acceptors in solution, nor for donors and acceptors dispersed randomly in membranes. More complex expressions are required in these cases, and such expressions are generally derived by averaging the transfer rate over the assumed spatial distribution of donor–acceptor pairs.[5]

The use of lifetimes in Eq. [13.13] has been a source of confusion. In Eq. [13.13] we have assumed that the decay of the donor is a single exponential in the absence (τ_D) and presence (τ_{DA}) of acceptor. Single-exponential decays are rare in biomolecules. If the intensity decays are multiexponential, then it is important to use average decay times in Eq. [13.13] which are proportional to the steady-state intensities. These averages are given by the sum of the $\alpha_i\tau_i$ products. Also, even if the donor decay is a single exponential, the decay rate in the presence of acceptor will only remain a single exponential if there is a single D–A distance. The presence of acceptors at more than one distance can result in more complex decays (Chapter 14).

Assuming that the single-distance model is appropriate, one sees that the rate of energy transfer is dependent upon R_0, which in turn depends upon κ^2, n, Q_D, and $J(\lambda)$. These factors must be known in order to calculate the distance. The refractive index is generally known from the solvent composition or is estimated for the macromolecule. The refractive index is often assumed to be close to that of water (n = 1.33) or small organic molecules (n = 1.39). The quantum yield of the donor, Q_D, is determined by comparison with standard fluorophores. Since Q_D is used as the sixth root in the calculation of R_0, small errors in Q_D do not have a large effect on R_0. The overlap integral must be evaluated for each D–A pair. The greater the overlap of the emission spectrum of the donor with the absorption spectrum of the acceptor, the higher is the value of R_0. Acceptors with larger extinction coefficients result in larger R_0 values. In the equations presented above, it was assumed that the lifetime of the donor was not altered by binding of the acceptor, other than by the rate of energy transfer. For labeled macromolecules, this may not always be the case. Allosteric interactions between the donor and acceptor sites could alter the donor lifetime by enhancement of other decay processes, or by protection from these processes. Under these circumstances, more complex analysis of the apparent transfer efficiency is required, typically by a comparison of the apparent efficiency by donor quenching and enhanced acceptor emission (Section 13.3.D).

The dependence of R_0 on spectral overlap is illustrated in Figure 13.3 for RET from structural isomers of dansyl-labeled phosphatidylethanolamine (DPE) to the eosin-labeled lipid (EPE). Each of the dansyl derivatives of DPE displays a different emission spectrum.[5] As the spectra of the DPE isomers shift to longer wavelengths, the overlap with the absorption spectrum of EPE increases and the R_0 values increase (Table 13.1). One notices that R_0 is not very dependent upon $J(\lambda)$. For instance, for two of the D–A pairs, a 120-fold change in the overlap integral results in a 2.2-fold change in the Förster distance. This is because of the sixth-root dependence in Eq. [13.5]. It should also be noted that the visual impression of overlap is somewhat misleading because the value of $J(\lambda)$ depends on λ^4 (Eq. [13.3]). Comparison of the spectral overlap for 2,5-DPE and 1,5-DPE suggests a larger Förster distance for 1,5-DPE, whereas the calculated value is smaller. The larger Förster distance for 2,5-DPE is due to its larger quantum yield.

Because of the complexity in calculating overlap integrals and Förster distances, it is convenient to have several examples. Values of the overlap integral corresponding to the spectra in Figure 13.3 are summarized in Table 13.1.

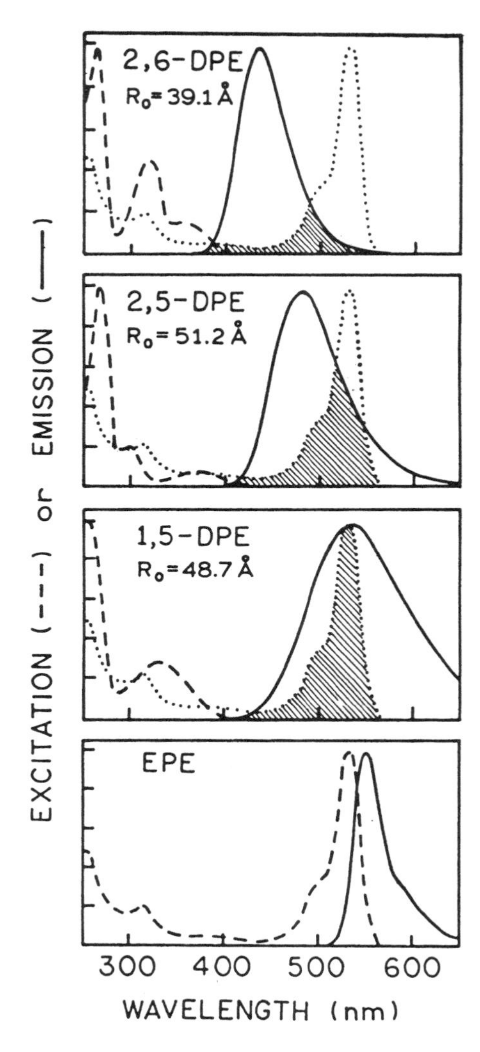

Figure 13.3. Excitation (---) and emission (—) spectra of dansyl-labeled lipids and an eosin-labeled lipid. The eosin- and dansyl-labeled compounds are N-derivatives of phosphatidylethanolamine (PE). In the designations of the dansyl-PE (DPE) derivatives, the numbers refer to the location of the dimethylamino and sulfonyl residues on the naphthalene ring of the dansyl group. The extinction coefficient of eosin-PE (EPE) is 85,000 M^{-1} cm^{-1} at 527 nm. In the top three panels, the long-wavelength absorption spectrum of eosin-PE is shown as a dotted curve. Revised from Ref. 5.

Brief Biographical Sketch of Theodor Förster

The theory for resonance energy transfer was developed by Professor Theodor Förster (Figure 13.4). He was born in Frankfurt, Germany, in 1910. He received a Ph.D. in 1933 for studies of the polarization of reflected electrons. He then became a research assistant in Leipzig, Germany, where he studied light absorption of organic compounds until 1942. In this phase of his work, he applied the newly developed principles of quantum mechanics to chemistry. His most productive period followed World War II. From 1942 to 1945, he held a professorship in Poznan, Poland, where he did not appear to publish any papers, but he did get married. In 1945 he joined the Max-Planck Institute for Physical Chemistry in Göttingen, where he wrote his classic book *Fluoreszenz Organischer Verbindungen*, which has been described as a concentrated "house bible" for the German community of spectroscopists. By 1946 Professor Förster had written his first paper on energy transfer and pointed out the importance of energy transfer in photosynthesis systems. Professor Förster was also among the first scientists to observe excited-state proton transfer, which is now described by the Förster cycle. In 1954 he discovered excimer formation. Professor Förster died of a heart attack in his car on the way to work in 1974. (For additional biographical information, see Ref. 6 and the Introduction about Theodor Förster in Ref. 7.)

13.1.A. Orientation Factor κ^2

A final factor in the analysis of the energy-transfer efficiencies is the orientation factor κ^2, which is given by

$$\kappa^2 = (\cos\theta_T - 3\cos\theta_D\cos\theta_A)^2 \qquad [13.15]$$

$$= (\sin\theta_D\sin\theta_A\cos\phi - 2\cos\theta_D\cos\theta_A)^2 \qquad [13.16]$$

In these equations, θ_T is the angle between the emission transition dipole of the donor and the absorption transition

Table 13.1. Calculated R_0 Values for RET from Structural Isomers of Dansyl-Labeled Phosphatidylethanolamine (DPE) to Eosin-Labeled Ethanolamine (EPE) and from 2,6-DPE to 2,5-DPE[a]

Donor	Acceptor	ϕ_D	J (M^{-1} cm^3)	J (M^{-1} cm^3 (nm)^4)[b]	R_0 (Å)
1,5-DPE	EPE	0.37	2.36×10^{-13}	2.36×10^{15}	48.7
2,5-DPE	EPE	0.76	1.54×10^{-13}	1.54×10^{15}	51.2
2,6-DPE	EPE	0.71	3.31×10^{-14}	3.31×10^{14}	39.1
2,6-DPE	2,5-DPE	0.71	1.3×10^{-15}	1.3×10^{13}	22.8

[a]From Ref. 5. R_0 was calculated using $n = 1.4$ and $\kappa^2 = \frac{2}{3}$.
[b]The factor of 10^{28} between $J(\lambda)$ in M^{-1} cm^3 and M^{-1} cm^3 nm^4 arises from 1 nm = 10^{-7} cm, raised to the fourth power.

dipole of the acceptor, θ_D and θ_A are the angles between these dipoles and the vector joining the donor and the acceptor, and ϕ is the angle between the planes (Figure 13.5). Depending upon the relative orientation of donor and acceptor, κ^2 can range from 0 to 4. For collinear and parallel transition dipoles, $\kappa^2 = 4$, and for parallel dipoles, $\kappa^2 = 1$. Since the sixth root of κ^2 is taken in calculating the distance, variation of κ^2 from 1 to 4 results in only a 26% change in r. With $\kappa^2 = \frac{2}{3}$, as is usually assumed, the calculated distance can be in error by no more than 35%. However, if the dipoles are oriented perpendicular to one another, $\kappa^2 = 0$, which would result in serious errors in the calculated distance. This problem has been discussed in detail.[8–10] By measurements of the fluorescence anisotropy of the donor and the acceptor, one can set limits on κ^2 and thereby minimize uncertainties in the calculated distance.[9–11] An example of calculating the range of possible values of κ^2 is given in Section 13.2.B. In general, variation of κ^2 seems to have not resulted in major errors in the calculated distances.[12,13] Generally, κ^2 is assumed equal to $\frac{2}{3}$, which is the value for donors and acceptors that randomize by rotational diffusion prior to energy transfer. This value is generally assumed for calculation of R_0. Alternatively, one may assume that a range of static donor—acceptor orientations are present and that these orientations do not change during the lifetime of the excited state. In this case, $\kappa^2 = 0.476$.[3] For fluorophores bound to macromolecules, segmental motions of the donor and acceptor tend to randomize the orientations. Further, many

Figure 13.4. Professor Theodor Förster, May 15, 1910–May 20, 1974. Reprinted, with permission, from Ref. 6, Copyright © 1974, Springer-Verlag.

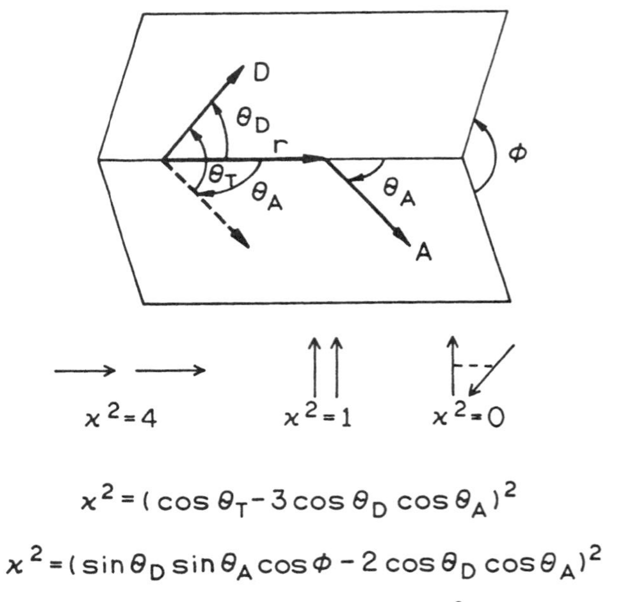

Figure 13.5. Dependence of the orientation factor κ^2 on the directions of the emission dipole of the donor and the absorption dipole of the acceptor.

donors and acceptors display fundamental anisotropies less than 0.4 owing to overlapping electronic transitions. In this case, the range of possible κ^2 values is more limited, and errors in distance are likely to be less than 10%.[14]

13.1.B. Dependence of the Transfer Rate on Distance (*r*), the Overlap Integral (*J*), and κ^2

The theory of Förster predicts that $k_T(r)$ depends on $1/r^6$ (Eq. [13.1]) and linearly on the overlap integral (Eq. [13.2]). Given the complexity and assumptions of RET theory,[4] it was important to demonstrate experimentally that these dependencies are valid. The predicted $1/r^6$ dependence on distance was confirmed experimentally.[15–17] One demonstration used oligomers of poly-L-proline, labeled on opposite ends with a naphthyl (donor) and a dansyl (acceptor) group.[15,16] Poly-L-proline forms a rigid helix of known atomic dimensions, providing fixed distances between the donor and acceptor moieties. By measuring the transfer efficiency with different numbers of proline residues, it was possible to demonstrate that the transfer efficiency in fact decreased as $1/r^6$. These data are described in detail in Problem 13.3.

The linear dependence of k_T on the overlap integral J has also been experimentally proven.[18] This was accomplished using a D–A pair linked by a rigid steroid spacer. The extent of spectral overlap was altered by changing the solvent, which shifted the indole donor emission spectrum and the carbonyl acceptor absorption spectrum. The rate of transfer was found to decrease linearly as the overlap integral decreased. These data are shown in Problem 13.4. To date, there has not been experimental confirmation of the dependence of the transfer rate on κ^2.

Another important characteristic of RET is that the transfer rate is proportional to the decay rate of the fluorophore (Eq. [13.1]). This means that for a D–A pair spaced by the R_0 value, the rate of transfer will be $k_T = \tau_D^{-1}$ whether the decay time is 10 ns or 10 ms. Hence, long-lived lanthanides are expected to display RET over distances comparable to those for the nanosecond-decay-time fluorophores, as demonstrated by transfer from Tb^{3+} to Co^{2+} in thermolysin.[19] This fortunate result occurs because the transfer rate is proportional to the emission rate of the donor. The proportionality to the emissive rate is due to the term Q_D/τ_D in Eq. [13.2]. It is interesting to speculate what would happen if the transfer rate were independent of the decay rate. In this case, a longer-lived donor would allow more time for energy transfer. Then energy transfer would occur over longer distances where the smaller rate of transfer would still be comparable to the donor decay rate.

13.1.C. Homotransfer and Heterotransfer

In the preceding sections we considered only energy transfer between chemically distinct donors and acceptors, which is called heterotransfer. RET can also occur between chemically identical molecules. Such transfer, termed homotransfer, typically occurs for fluorophores which display small Stokes' shifts. One example of homotransfer is provided by the relatively new class of fluorophores referred to as BODIPY* dyes.[20] The absorption and emission spectra of one BODIPY derivative are shown in Figure 13.6. Because of the small Stokes' shift and high extinction coefficient of these probes, the Förster distance for homotransfer is near 57 Å.[20]

At first glance, homotransfer may seem like an unlikely phenomenon, but its occurrence is rather common. For example, it is well known that antibodies labeled with fluorescein do not become more highly fluorescent with higher degrees of labeling.[21] Antibodies are typically brightest with about four fluoresceins per antibody, after which the intensity starts to decrease. This effect is attributed to the small Stokes' shift of fluorescence and homotransfer. In fact, homotransfer among fluorescent molecules was one of the earliest observations in fluores-

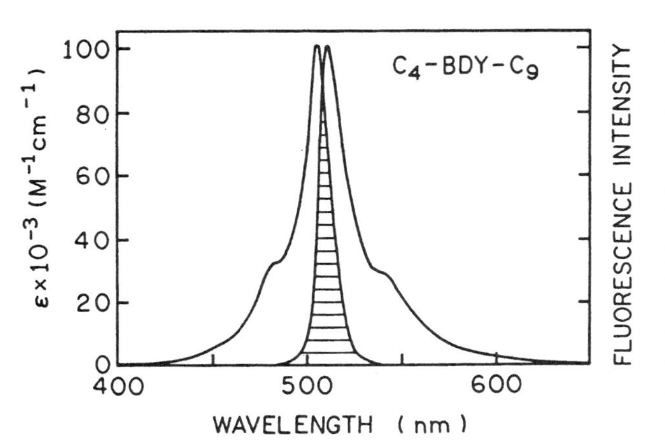

Figure 13.6. Absorption and corrected fluorescence emission (bandpass 2.5 nm) spectra of the BODIPY derivative C_4-BDY-C_9 in methanol, with shaded area representing spectral overlap. Revised and reprinted, with permission, from Ref. 20, Copyright © 1991, Academic Press, Inc.

*BODIPY, 4,4-difluoro-4-bora-3a,4a-diaza-s-indacene. BODIPY is a trademark of Molecular Probes, Inc.

cence and was detected by a decrease in the anisotropy of fluorophores at higher concentrations.[22] The possibility of homotransfer can be readily evaluated by examination of the absorption and emission spectra. For instance, perylene would be expected to display homotransfer, but homotransfer is unlikely for quinine (Figure 1.3).

13.2. DISTANCE MEASUREMENTS USING RET

13.2.A. Distance Measurements in α-Helical Melittin

Because RET can be reliably assumed to depend on $1/r^6$, the transfer efficiency can be used to measure distances between sites in proteins. This use of energy transfer has been recently summarized in useful reviews.[12,23] These articles contain R_0 values for a number of commonly used D–A pairs and offer practical advice. The use of RET in structural biochemistry is illustrated in Figure 13.7 for the peptide melittin.[24] This peptide has 26 amino acids. A single tryptophan residue at position 19 serves as the donor. A single dansyl acceptor was placed on the N-terminal amino group. The spectral properties of this D–A pair are shown in Figure 13.8. These spectral properties result in a Förster distance of 23.6 Å (Problem 13.5).

Depending upon the solvent conditions, melittin can exist in the monomer, tetramer, α-helix, and/or random-coil state.[25–27] In the methanol–water mixture specified on Figure 13.7, melittin is in the rigid α-helical state and exists as a monomer. There is a single dansyl acceptor adjacent to each tryptophan donor, and the helical structure ensures a single D–A distance. Hence, we can use the theory

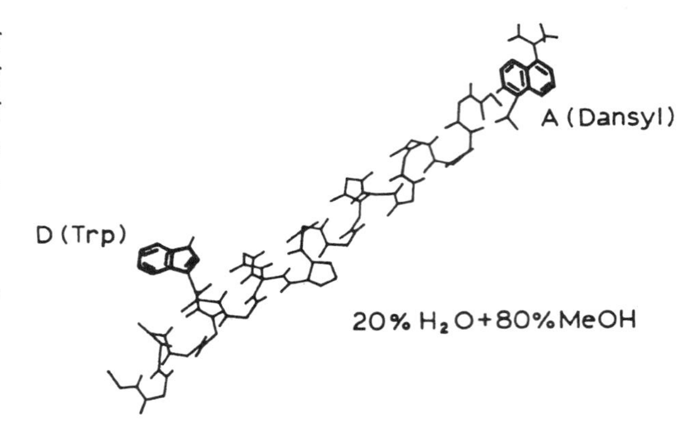

Figure 13.7. Structure of melittin in the α-helical state. The donor is tryptophan-19, and the acceptor is an N-terminal dansyl group. Revised from Ref. 24.

described above, and in particular Eqs. [13.12] and [13.14], to calculate the D–A distance.

In order to calculate the D–A distance, it is necessary to determine the efficiency of energy transfer. This can only be accomplished by comparing the intensity of the donor in the presence of acceptor (F_{DA}) with the donor intensity from a control molecule which lacks the acceptor (F_D). From Figure 13.9 one sees that the value of F_{DA}/F_D is near 0.55, so that the transfer efficiency is less than 50% (E = 0.45). Since E is less than 0.5, we know that the D–A distance must be larger than the R_0 value. Using Eq. [13.12], and the R_0 value of 23.6 Å, one can readily calculate that the tryptophan-to-dansyl distance is 24.4 Å.

It is important to notice the assumptions used in calculating the distance. We assumed that the orientation factor κ^2 was equal to the dynamic average of $\frac{2}{3}$. In the case of melittin, this is a good assumption because both the trp

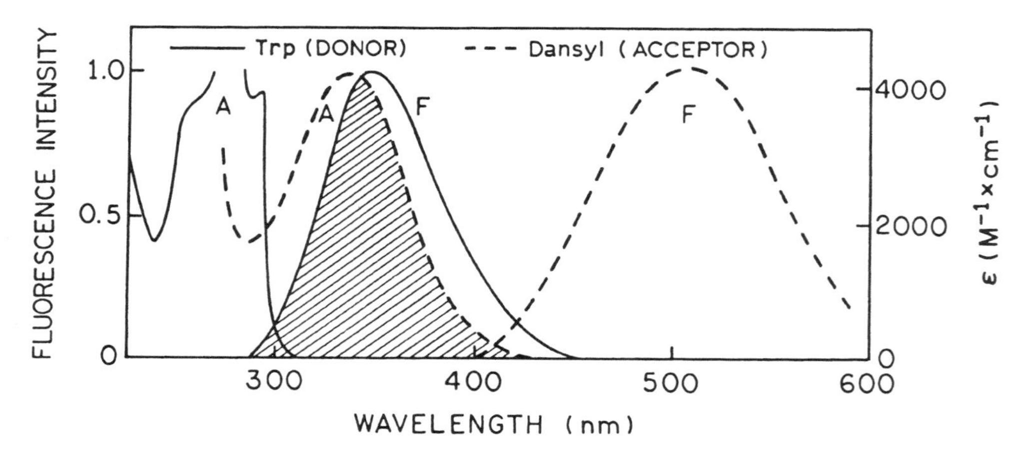

Figure 13.8. Overlap integral (shaded area) for energy transfer from a tryptophan donor to a dansyl acceptor on melittin. R_0 = 23.6 Å. Data from Ref. 24.

donor and dansyl acceptor are fully exposed to the liquid phase, which is highly fluid. The rotational correlation times for such groups are typically near 100 ps, so that the dipoles can randomize during the excited-state lifetime. Perhaps the most dangerous assumption is that the sample is 100% labeled with acceptor. If melittin were incompletely labeled with acceptor, the measured value of F_{DA} would be larger than the true value, and the calculated distance too large. Suppose that the fractional labeling with acceptor is given by f_A, so that the fractional donor population lacking acceptor is given by $1 - f_A$. In this case, Eq. [13.14] becomes[11]

$$E = 1 - \frac{F_{DA} - F_D(1 - f_A)}{F_D f_A} = \left(1 - \frac{F_{DA}}{F_D}\right)\frac{1}{f_A} \qquad [13.17]$$

For a high degree of RET donor quenching ($F_{DA}/F_D << 1$), a small percentage of unlabeled acceptor can result in a large change in the calculated transfer efficiency (Problem 13.9).

Another assumption in calculating the trp to dansyl distance in melittin is that a single conformation exists, i.e., that there is a single D–A distance. This assumption is probably safe for many proteins in the native state, particularly for single-domain proteins. For unfolded peptides or multidomain proteins, a variety of conformations can exist, resulting in a range of D–A distances. In this case, calculation of a single distance using Eq. [13.12] would result in an apparent distance, which would be weighted toward the shorter distances. Such systems are best analyzed in terms of a distance distribution (Chapter 14).

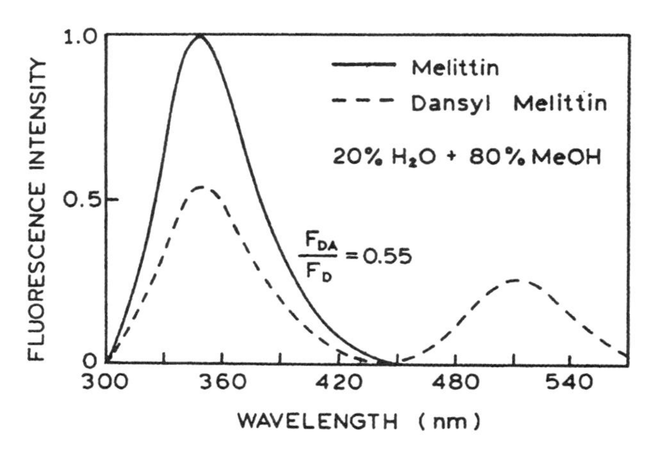

Figure 13.9. Emission spectra of the melittin donor (D) and acceptor-labeled melittin (D–A). Excitation at 282 nm. Revised from Ref. 24.

13.2.B. Effect of κ^2 on the Possible Range of Distances

In distance measurements using RET, there is often concern about the effects of the orientation factor κ^2. At present, there is no way to measure κ^2, short of determination of the X-ray crystal structure, in which case the distance would be known and thus there would be no reason to use energy transfer. However, it is possible to set limits on κ^2, which in turn sets limits on the range of possible D–A distances. These limits are determined from the anisotropies of the donor and acceptor, which reflect the extent of orientational averaging toward the dynamic average of $\kappa^2 = \frac{2}{3}$.

The problem of κ^2 has been discussed in detail by Dale and co-workers[8–10] and summarized by Cheung.[11] The basic idea is that the donor and acceptor move freely within a cone and that energy transfer is rapidly averaged over all available D–A orientations. Interpretation of the formalism described by Dale and co-workers[8–10] is not always straightforward, and we present the method preferred in our laboratory.[28] Although it is not possible to calculate the values of κ^2, it is possible to set upper and lower limits. These values are given by

$$\kappa^2_{min} = \frac{2}{3}\left[1 - \frac{(d^x_D + d^x_A)}{2}\right] \qquad [13.18]$$

$$\kappa^2_{max} = \frac{2}{3}(1 + d^x_D + d^x_A + 3d^x_D d^x_A) \qquad [13.19]$$

where

$$d^x_i = \left(\frac{r_i}{r_0}\right)^{1/2} \qquad [13.20]$$

The value of d^x_i represents the depolarization factor due to segmental motion of the donor (d^x_D) or acceptor (d^x_A), but not the depolarization due to overall rotational diffusion of the protein. Overall rotational diffusion is not important because it does not change the D–A orientation. The values of r_i and r_0 are often taken as the steady-state and fundamental anisotropies, respectively, of the donor or acceptor. If the donor and acceptor do not rotate relative to each other during the excited-state lifetime, then $d^x_D = d^x_A = 1.0$, and $\kappa^2_{min} = 0$ and $\kappa^2_{max} = 4$. If both D and A are independently and rapidly rotating over all space, $\kappa^2_{min} = \kappa^2_{max} = \frac{2}{3}$.

There are several ways to obtain the values of d^x_D and d^x_A . The easiest method is to determine the anisotropy decays of the donor and acceptor, the latter when directly excited. This calculation of a range of κ^2 values is illustrated by the anisotropy decays measured for the tryptophan donor and dansyl acceptor in α-helical melittin (Table 13.2). Both the donor and the acceptor display two correlation times, one near 2 ns due to overall protein rotation, and a shorter correlation time near 0.3 ns due to segmental motions of the donor and the acceptor. It is these faster motions which randomize κ^2. The values of d_D^x and d_A^x are given by the ratio of the long-correlation-time amplitude to the total anisotropy. Hence, for melittin,

$$d_D^x = \left(\frac{0.174}{0.294}\right)^{1/2} = 0.77 \qquad [13.21]$$

$$d_A^x = \left(\frac{0.135}{0.300}\right)^{1/2} = 0.67 \qquad [13.22]$$

Using these values and Eqs. [13.18] and [13.19], one can calculate the limits on κ^2, $\kappa^2_{min} = 0.19$ and $\kappa^2_{max} = 2.66$.

Once the limiting values of κ^2 are known, one may use these values to calculate the maximum and minimum values of the distance that are consistent with the data. In calculating these distances, one must remember that R_0 was calculated with an assumed value of $\kappa^2 = \frac{2}{3}$. Hence, the minimum and maximum distances are given by

$$r_{min} = \left(\frac{\kappa^2_{min}}{2/3}\right)^{1/6} r\left(\frac{2}{3}\right) \qquad [13.23]$$

$$r_{max} = \left(\frac{\kappa^2_{max}}{2/3}\right)^{1/6} r\left(\frac{2}{3}\right) \qquad [13.24]$$

where $r(\frac{2}{3})$ is the distance calculated assuming $\kappa^2 = \frac{2}{3}$. Using the limiting values of κ^2, one finds for the example given above that the distance can be from 0.81 to 1.26 of $r(\frac{2}{3})$, the distance calculated with the assumed value of $\kappa^2 = \frac{2}{3}$. While this range may seem large, it should be remembered that there is an additional depolarization factor due to the transfer process itself, which will further randomize κ^2 toward $\frac{2}{3}$. Equations [13.18]–[13.20] provide a worst-case estimate, which usually overestimates the effects of κ^2 on the calculated distance. For fluorophores with mixed polarization, $r_0 < 0.3$, the error in distance is thought to be below 10%.[14]

There are two other ways to obtain the depolarization factors. One method is to construct a Perrin plot in which the steady-state polarization is measured for various viscosities. Upon extrapolation to the high-viscosity limit, the $1/r_0^{app}$ intercept (Section 10.5) is typically larger than $1/r_0$ in frozen solution. This difference is usually attributed to segmental probe motions and can be used to estimate the depolarization factor, $d_i^x = (r_0^{app}/r_0)^{1/2}$. Another method is to estimate the expected steady-state anisotropy from the lifetime and correlation time of the protein and to use these data to estimate d_D^x and d_A^x (Problem 13.8). The basic idea is to estimate d_A^x and d_D^x by correcting for the decrease in anisotropy resulting from rotational diffusion of the protein. Any loss in anisotropy, beyond that calculated for overall rotation, is assumed to be due to segmental motions of the donor or acceptor.

Table 13.2. Anisotropy Decays for α-Helical Melittin[a]

Fluorophore	r_{0i}	ϕ_i (ns)
Tryptophan-19[b]	0.120	0.23
	0.174	1.77
N-terminal dansyl	0.165	0.28
	0.135	2.18

[a]From Ref. 24.

[b]Determined for donor-only melittin. Similar amplitudes and correlation times were found for trp-19 in dansyl-melittin.

13.2.C. Protein Folding Measured by RET

Energy transfer has also been widely useful for measurements of protein folding. One example is for the protein serine hydroxymethyltransferase. This protein typically has three tryptophan residues, at positions 16, 183, and 385.[29] The acceptor was a pyridoxyl 5′-phosphate (PyP) residue covalently linked to lysine-229. One expects protein folding or unfolding to affect the distance from each of the tryptophan residues to PyP, and thus one expects the extent of RET to be different for the native and folded proteins.

The presence of three tryptophan residues results in emission from three donors, which would be practically impossible to interpret. For this reason, single-tryptophan mutants were prepared which each lacked the other two trp residues. This procedure results in a single trp–PyP pair for each mutant protein. Emission spectra are shown for the trp-183 mutant in Figure 13.10. The protein was initially in 8*M* urea, resulting in a random-coil state; the initial emission spectrum is shown as the dashed curve in Figure 13.10. Upon dilution into buffer without urea, the protein began to refold, as seen by a decrease in the trp-donor emission and an increase in the PyP-acceptor emission. The intensity of the PyP emission increases upon refolding due to increased RET from trp-183. (Enhancement of acceptor emission is described in more detail in Section 13.3.D.) The availability of three single-tryptophan mutants allowed studies of refolding of specific sites on the protein. The PyP acceptor emission at 380 nm was

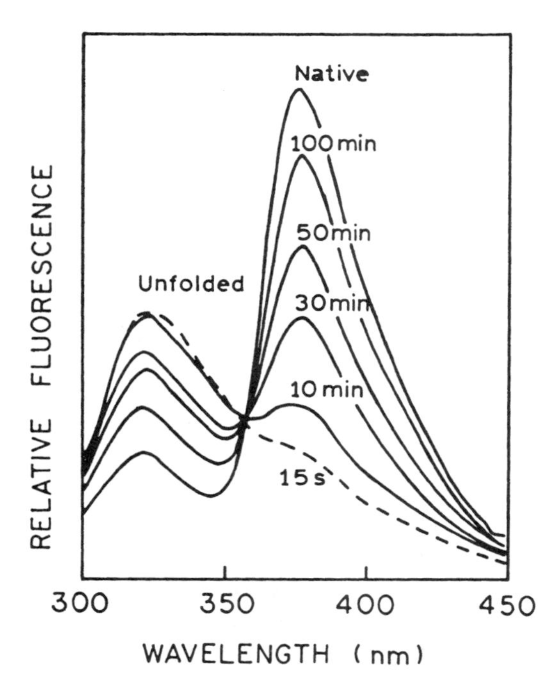

Figure 13.10. Emission spectra of trp-183 in PyP-5′-serine hydroxymethyltransferase during refolding. Excitation at 290 nm. Revised from Ref. 29.

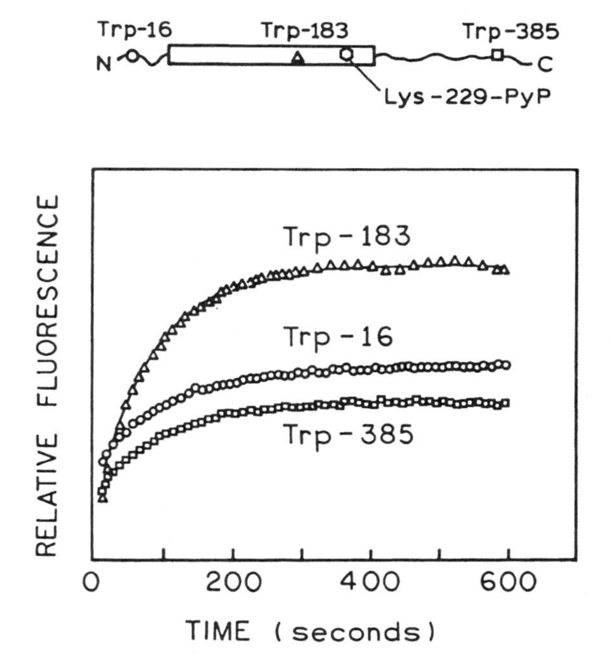

Figure 13.11. Time-dependent intensity of PyP at 380 nm in serine hydroxymethyltransferase during refolding of the single-tryptophan mutants. Excitation at 290 nm. Revised from Ref. 29.

measured during refolding for each of the mutants (trp-183, trp-16, and trp-385 in Figure 13.11), providing information about the folding pathway. For this protein, the greatest acceptor enhancement is seen for the trp-183 mutant, suggesting that trp-183 is closest to PyP in the folded protein. Less transfer is seen from the other two tryptophan residues. The extent of energy transfer from each tryptophan residue allowed determination of the trp-to-PyP distances. By such experiments, one can determine which region of the protein folds first and thus gain an understanding of the folding pathway. These experiments on protein folding represent an important class of experiments based on protein engineering and site-directed mutagenesis. These techniques of molecular biology make it possible to simplify the spectral signal from multitryptophan proteins to obtain structural information.

13.2.D. Orientation of a Protein-Bound Peptide

In the presence of calcium, calmodulin is known to interact with a number of proteins and peptides. One example is binding of a peptide from myosin light-chain kinase (MLCK) to calmodulin.[30] Such peptides are known to bind in the cleft between the two domains of calmodulin (Figure 13.12). When bound to calmodulin, the MLCK peptide

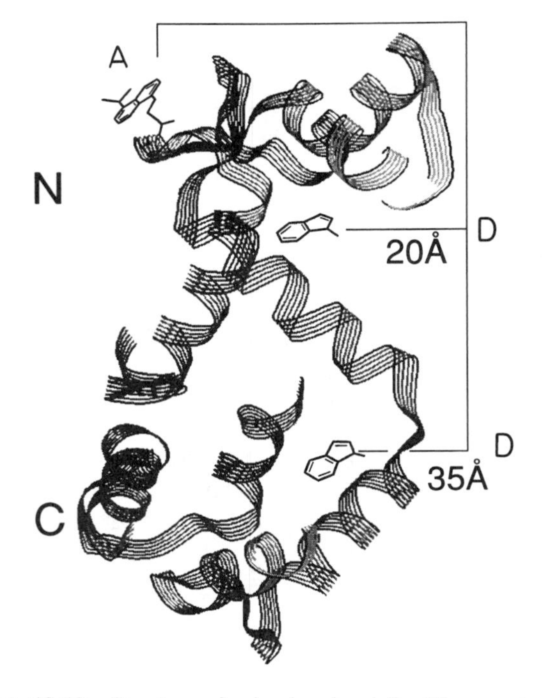

Figure 13.12. Structure of spinach calmodulin. The acceptor is AEDANS on cysteine-26 of calmodulin. The tryptophan donor is on the MLCK peptides shown in Figure 13.13. Revised and reprinted, with permission, from Ref. 30, Copyright © 1992, American Chemical Society.

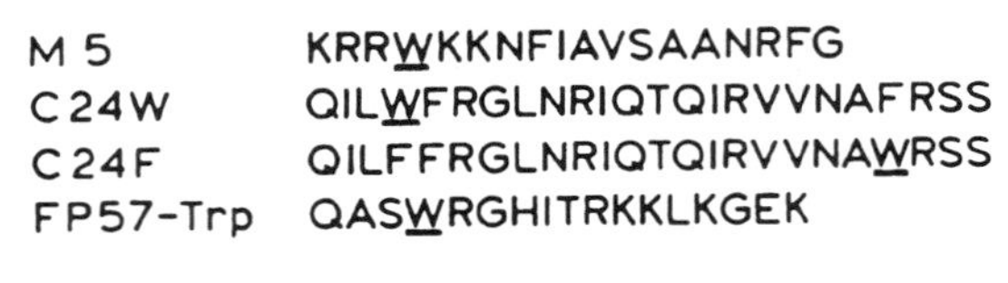

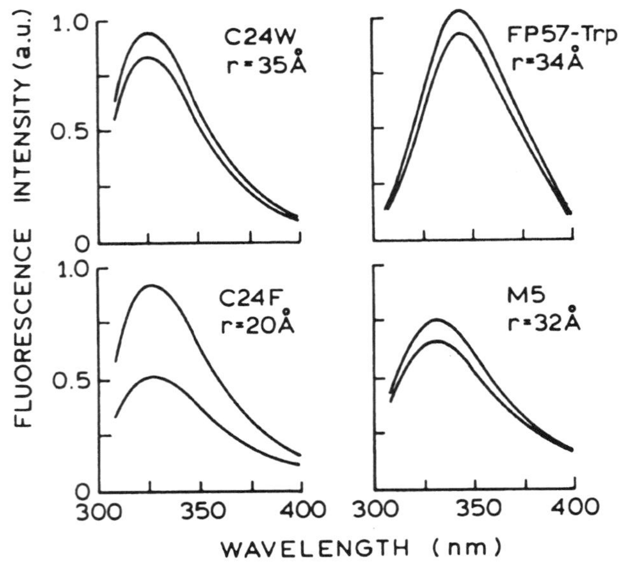

Figure 13.13. Emission spectra of MLCK peptides when free in solution and when bound to AEDANS-calmodulin. The upper and lower spectra in each panel correspond to the peptide emission spectra in the absence and in the presence of AEDANS-calmodulin, respectively. The peptide sequences are shown at the top of the figure, and the calculated distances are shown with the spectra. Excitation was at 295 nm. Revised and reprinted, with permission, from Ref. 30, Copyright © 1992, American Chemical Society.

was known to adopt an α-helical conformation. However, the direction of peptide binding to calmodulin was not known.

Information on the direction of binding was obtained by studying four similar peptides, each of which contained a single tryptophan residue, which served as the donor (Figure 13.13). Calmodulin typically contains only tyrosine residues, so an intrinsic acceptor was not available. This problem was solved by using calmodulin from spinach, which contains a single cysteine residue at position 26. This residue was labeled with 1,5-IAEDANS, which contains a thiol-reactive iodoacetyl group. The Förster distances for trp-to-AEDANS energy transfer ranged from 21 to 24 Å.

Emission spectra of the tryptophan-containing peptides are shown in Figure 13.13. The excitation wavelength was 295 nm to avoid excitation of the tyrosine residues in calmodulin. Upon binding of AEDANS-calmodulin, the tryptophan emission of each peptide was quenched. One of the peptides showed a transfer efficiency of 54%, and the remaining three peptides showed efficiencies ranging from 5 to 16%. These results demonstrated that the N-terminal region of the peptides bound closely to the N-terminal domain of calmodulin and illustrate how structural information can be obtained by comparative studies of analogous structures.

13.3. USE OF RET TO MEASURE MACROMOLECULAR ASSOCIATIONS

Energy transfer is widely useful in biochemistry even apart from its application for the measurement of distances. This is because energy transfer occurs independently of the linker joining the donor and acceptor and depends only on the D–A distance. Hence, any process bringing the donor and acceptor into close proximity will result in energy transfer. This includes biochemical association reactions, as will be illustrated below for protein subunits and DNA oligomers.

13.3.A. Dissociation of the Catalytic and Regulatory Subunits of a Protein Kinase

Cyclic 3′,5′-adenosine monophosphate (cAMP) is an important second messenger in cellular transduction pathways, and it is thus important to develop optical means to detect cAMP. Nucleotides are weakly fluorescent or nonfluorescent, so that it is not practical to use the intrinsic emission of cAMP. One method to detect cAMP is based on the effect of cAMP on the cAMP-dependent protein kinase.[31] The basic idea is to label the protein subunits with donors and acceptors. The presence of cAMP alters the extent of RET, which can be monitored by the donor or acceptor emission.

The cAMP-dependent protein kinase (PK) is composed of four units, two catalytic (C) and two regulatory (R) subunits. These subunits were thought to dissociate in the presence of the substrate cAMP and in the presence of protein kinase inhibitor (PKI). The association of the catalytic (C) and regulatory (R) subunits was examined by covalently labeling the subunits with a fluorescein (Fl-C) and a rhodamine (Rh-R) derivative, respectively.[32] The actual probes used were carboxyfluorescein succinimidyl ester and Texas Red sulfonyl chloride, which is a rhodamine derivative. Absorption and emission spectra are shown in Figure 13.14. The Förster distance for this D–A pair was 51.3 Å and was thus suited for distance measurements between protein subunits. For calculation of R_0, the quantum yield of the donor was taken as 0.5, κ^2 was

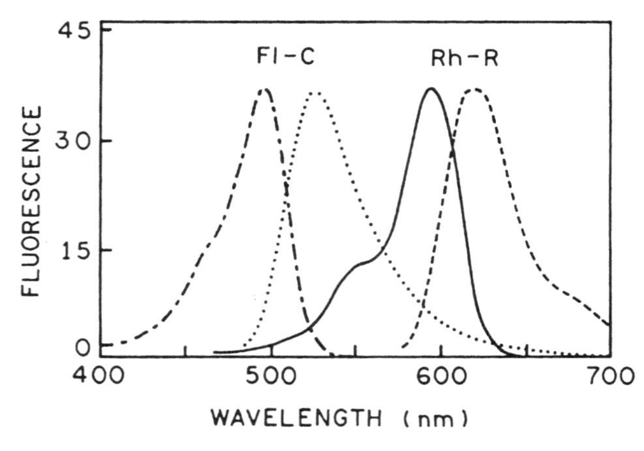

Figure 13.14. Spectral overlap of the fluorescein-labeled catalytic subunit (Fl-C) and the Texas Red-labeled regulator subunit (Rh-R) of a cAMP-dependent protein kinase. —·—, Absorption spectrum of Fl-C; ···, emission spectrum of Fl-C; ———, absorption spectrum of Rh-R; ‒‒, emission spectrum of Rh-R. Revised and reprinted, with permission, from Ref. 32, Copyright © 1993, American Chemical Society.

assumed equal to $\frac{2}{3}$, and the maximum extinction coefficient of the acceptor was set equal to 85,000 M^{-1} cm^{-1}.

Upon addition of cAMP, the extent of energy transfer decreased (Figure 13.15), consistent with a 10-Å increase in the D–A distance. This decrease in RET could be interpreted in terms of a displacement of the donor and acceptor to more distant regions of the protein (Figure 13.16). While the extent of RET allows a distance calculation, this is not necessary. The fact that energy transfer still occurs in the protein kinase with bound cAMP demonstrates that the subunits are still associated in the presence of cAMP. If the subunits were dissociated by cAMP, energy transfer would be eliminated. This is because for unlinked donors and acceptors, the acceptor concentrations need to be in the millimolar range for Förster transfer to occur (Chapter 15).

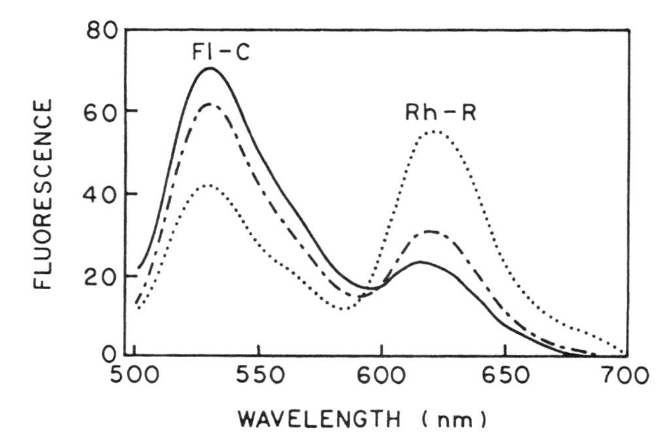

Figure 13.15. Effect of cAMP and protein kinase inhibitor (PKI) on the emission spectrum of a donor- and acceptor-labeled cAMP-dependent protein kinase. The emission spectrum of the holoenzyme without cAMP or PKI (···) and the emission spectra recorded following addition of cAMP (—·—) and PKI (———) are shown. Revised and reprinted, with permission, from Ref. 32, Copyright © 1993, American Chemical Society.

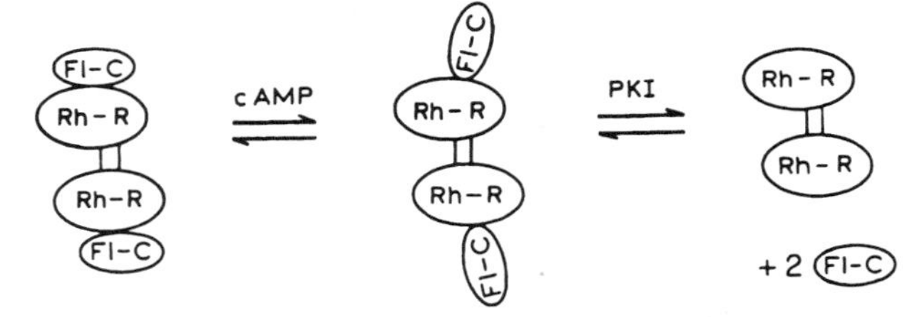

Figure 13.16. Effect of cAMP and protein kinase inhibitor (PKI) on the structure and association of cAMP-dependent protein kinase. The holoenzyme consists of two catalytic and two regulatory subunits. Revised from Refs. 31 and 32.

The protein kinase was examined further by addition of PKI, which resulted in elimination of energy transfer (Figure 13.15). While one could argue about changes in D–A distance versus protein dissociation as the cause of this effect, the important point from these data is that RET can be used as a proximity sensor. For dilute biochemical solutions, energy transfer between nonassociated species can usually be ignored unless the acceptor concentrations are very high (millimolar). Binding will bring donor and acceptor within the Förster distance, resulting in energy transfer. This example shows how RET can be used to measure the extent of association even without knowledge of the Förster distance. Of course, such data could be used to measure the dissociation constant of cAMP from the protein (Problem 13.6). Calculation of the distance is not needed to calculate the dissociation constant. Use of the donor intensity provides a simple method to monitor the rates and extents of any association reaction.

It is important to remember the possibility of inner filter effects, which can distort the measured intensities of the donors or acceptors.[33] For instance, in the previous example suppose that the fluorescence emission was measured at 550 nm, where the acceptor extinction coefficient is near 33,000 M^{-1} cm^{-1}. To maintain the acceptor absorbance below 0.05, the concentration of the acceptor-labeled protein must be below 1.5μM. If higher acceptor concentrations are used, the intensity values must be corrected for the absorption at the excitation and/or emission wavelength.

13.3.B. RET Calcium Indicators

The dependence of RET on proximity has been used to develop indicators for calcium. One example used a pep-

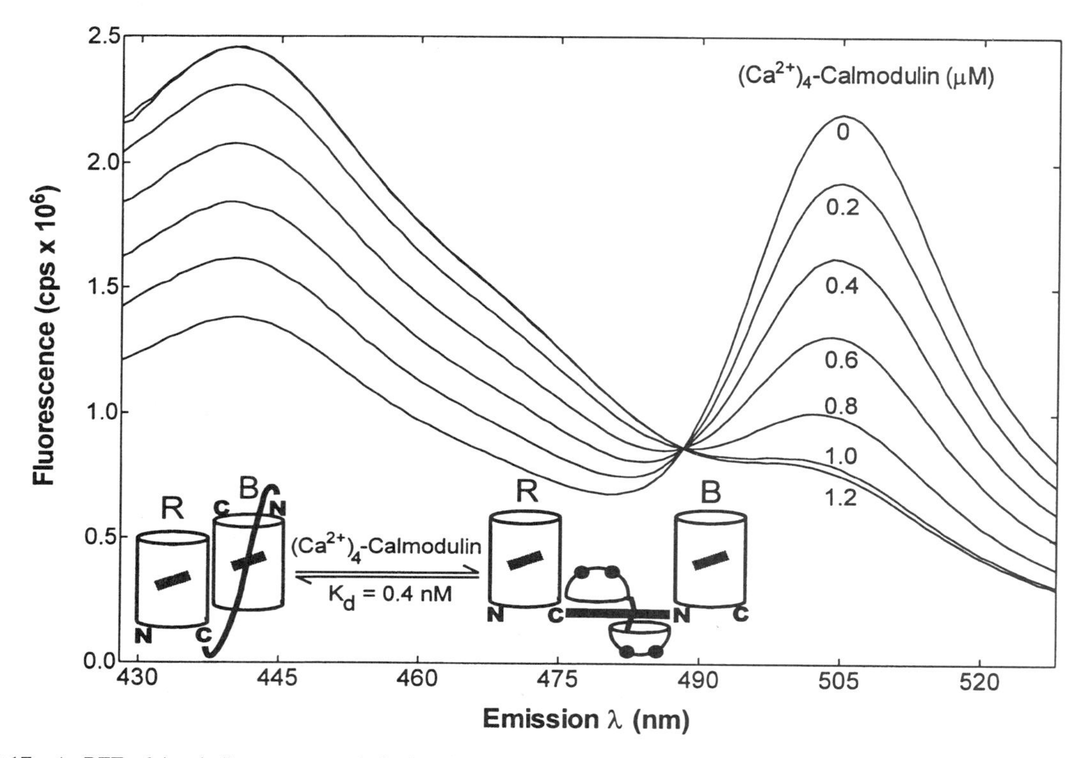

Figure 13.17. An RET calcium indicator composed of calmodulin and two GFP mutants (R and B) linked by the MLCK which binds to calmodulin. From Ref. 34.

tide from MLCK to link two mutants of green fluorescent protein (GFP).[34] By modification of the amino acid sequence of GFP, it is possible to create mutant proteins that emit at shorter or longer wavelengths. Two GFP mutants which would undergo energy transfer, a red (R) and a blue (B) GFP, were selected. These GFPs were placed on opposite ends of the MLCK peptide. In the presence of calcium, calmodulin exposes a hydrophobic surface, which can bind to a number of proteins, including the MLCK peptide. When calmodulin bound the MLCK peptide, the R and B GFPs became more widely separated, and the extent of energy transfer decreased (Figure 13.17). Similar results were obtained for calmodulin covalently linked to the MLCK peptide and the GFPs.[35] An important aspect of this work is that the genes for GFP-labeled proteins can be prepared by molecular biology and expressed in bacteria. This approach bypasses the usual difficulties associated with covalent labeling of specific sites on proteins.

13.3.C. Association Kinetics of DNA Oligomers

The general usefulness of RET in investigations of association reactions is illustrated by studies of donor- and acceptor-labeled DNA oligomers that have a complementary base sequence.[36–38] In such studies, one strand is typically labeled with fluorescein (Fl), and the complementary DNA strand is labeled with rhodamine (Rh) (Figure 13.18). Upon association, the fluorescein donor is quenched.[38] The fluorescence of the donor can be used to measure the rates of association or dissociation of the DNA oligomers (Figure 13.19). In other studies, the extent of RET has been used to detect changes in the melting temperature of DNA oligomers. The melting temperature was found to be sensitive to even a single base pair mismatch.[37]

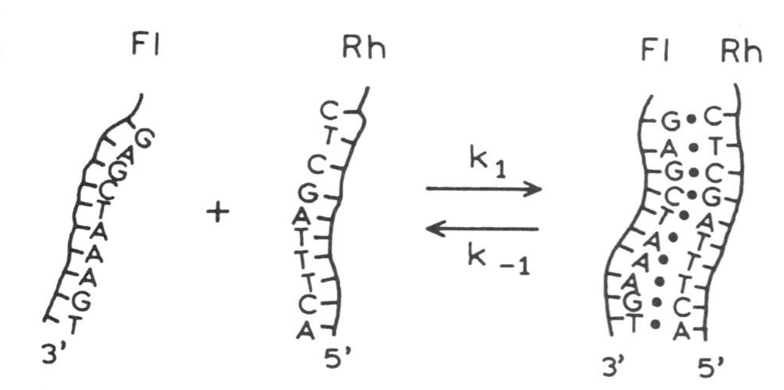

Figure 13.18. DNA association kinetics observed using oligonucleotides labeled with fluorescein (Fl) and rhodamine (Rh). Revised from Ref. 38.

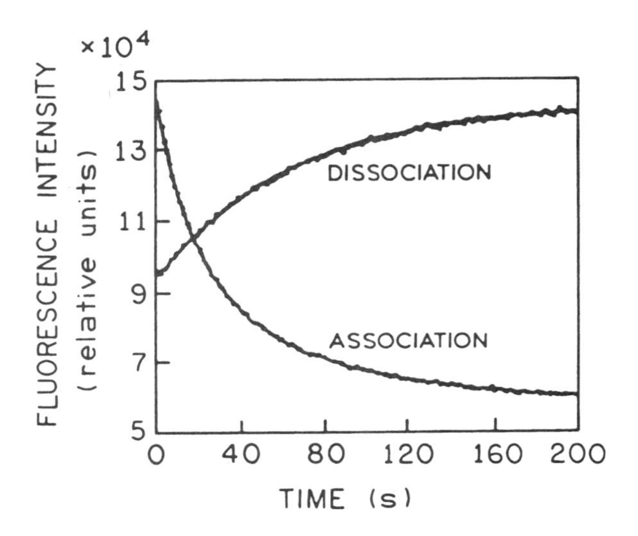

Figure 13.19. Association and dissociation kinetics of complementary donor- and acceptor-labeled oligonucleotides, observed using the donor (fluorescein) emission intensity. Revised and reprinted, with permission, from Ref. 38, Copyright © 1993, American Chemical Society.

RET has also been used to measure the catalytic activity of restriction enzymes, which cleave DNA at specific sites in the sequence. One example of this application is on investigation of the PaeR7 endonuclease.[39] For this restriction enzyme, the recognition site consists of a CTCGAG

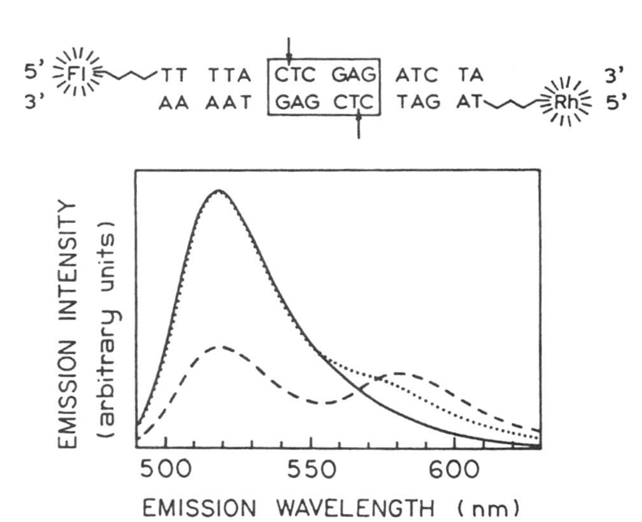

Figure 13.20. Principle of the RET endonuclease assay. The emission spectrum of a 16-mer donor strand (—) is quenched upon addition of its complementary rhodamine-labeled strand (– – –). The fluorescence spectrum taken after addition of PaeR7 endonuclease (· · ·) shows full recovery of the fluorescein emission intensity of the donor strand, indicating complete cleavage of the DNA duplex substrate. The peak at 580 nm in the emission spectra of duplex samples is due to the rhodamine acceptor fluorescence. Revised from Ref. 39.

sequence. Fluorescein-donor- and rhodamine-acceptor-labeled oligonucleotides served as the substrate (Figure 13.20). Binding of the DNA strands resulted in donor quenching and acceptor enhancement. Digestion with the endonuclease resulted in an increase in donor fluorescence equivalent to that in the donor-only control. When the donors and acceptors are on separate DNA fragments, the solution is too dilute for energy transfer.

The concept used in the endonuclease assay was the elimination of RET by enzymatic breakage of a covalently linked D–A pair. This concept has been applied in a variety of other circumstances,[40] including assay of an HIV-1 endonuclease,[41] which is involved in integrating the HIV into the host DNA. RET has also been used to measure the activity of an HIV protease,[42] proteolytic cleavage of linked derivatives of GFP,[43] and the activity of ribozymes in cleaving RNA[44] and as the basis of a general assay for DNA cleavage.[45]

13.3.D. Energy Transfer Efficiency from Enhanced Acceptor Fluorescence

In a RET experiment the acceptor must absorb light at donor emission wavelengths, but the acceptor does not need to be fluorescent. However, fluorescent acceptors are often used. In these cases, light absorbed by the donor and transferred to the acceptor appears as enhanced acceptor emission. This enhanced acceptor emission can be seen in Figure 13.15 at 620 nm for the Rh-labeled subunits of protein kinase when bound to the Fl-labeled subunits and in Figure 13.20 at 580 nm when the donor and acceptor DNA oligomers are associated. By extrapolating the emission spectrum of the donor, one can see that its emission extends to the acceptor wavelengths. Hence, the intensity measured at the acceptor wavelength typically contains some contribution from the donor.

Use of the acceptor intensities is complicated by the need to account for directly excited acceptor emission, which is almost always present.[11,46] In the case of protein kinase, the acceptor emission without RET (shown as the solid curve in Figure 13.15) is about 40% of the intensity with RET. This occurs because the acceptor absorbs at the excitation wavelength used to excite the donor, resulting in acceptor emission without RET.

Calculation of the transfer efficiency from the enhanced acceptor emission requires careful consideration of all the interrelated intensities. Assuming that the donor does not emit at the acceptor wavelength, the efficiency of transfer is given by

$$E = \frac{\varepsilon_A(\lambda_D^{ex})}{\varepsilon_D(\lambda_D^{ex})}\left[\frac{F_{AD}(\lambda_A^{em})}{F_A(\lambda_A^{em})} - 1\right]\left(\frac{1}{f_D}\right) \qquad [13.25]$$

In this expression, $\varepsilon_A(\lambda_D^{ex})$ and $\varepsilon_D(\lambda_D^{ex})$ are the extinction coefficients (single D–A pairs) or absorbance (multiple acceptors) of the acceptor and the donor at the donor excitation wavelength (λ_D^{ex}), and f_D is the fractional labeling with the donor. The acceptor intensities are measured at an acceptor emission wavelength (λ_A^{em}) in the absence $[F_A(\lambda_A^{em})]$ and in the presence $[F_{AD}(\lambda_A^{em})]$ of donor. This expression with $f_D = 1.0$ can be readily obtained by noting that $F_A(\lambda_A^{em})$ is proportional to $\varepsilon_A(\lambda_D^{ex})$ and $F_{AD}(\lambda_A^{em})$ is proportional to $\varepsilon_A(\lambda_D^{ex}) + E\varepsilon_D(\lambda_D^{ex})$. The accuracy of the measured E value is typically less than when using the donor emission (Eq. [13.14]).

It is also important to remember that it may be necessary to correct further for the donor emission at λ_A, which is not accounted for in Eq. [13.25]. The possibility of donor emission at λ_A is easily seen from the donor emission spectra in Figures 13.15 and 13.20. The presence of donor emission at the acceptor wavelength, if not corrected for in measuring the acceptor intensities, will result in an apparent transfer efficiency larger than the actual value (see Problem 13.11). Equation [13.25] can also be applied when multiple acceptors are present, that is, in the case of unlinked donor and acceptor pairs (Chapter 15). In this case, $\varepsilon_D(\lambda_D^{ex})$ and $\varepsilon_A(\lambda_D^{ex})$ are replaced by the optical densities of the donor $[OD_D(\lambda_D^{ex})]$ and of the acceptor $[OD_A(\lambda_D^{ex})]$ at the donor excitation wavelength. The factor $1/f_D$ in Eq. [13.25] is the fractional labeling with the donor. When measuring the acceptor emission, it is important to have complete donor labeling, $f_D = 1.0$.

Occasionally, it is difficult to obtain the transfer efficiency from the sensitized acceptor emission. One difficulty is a precise comparison of the donor-alone and donor–acceptor pair at precisely the same concentration. The need for two samples at the same concentration can be avoided if the donor- and acceptor-labeled sample can be enzymatically digested so as to eliminate energy transfer.[47] Additionally, methods allowing comparison of the donor-alone and donor–acceptor spectra without requiring the concentrations to be the same have been developed. This is accomplished by relying on the shape of the donor emission to subtract its contribution from the emission spectrum of the D–A pair. These methods are best understood by reading the original descriptions.[48,49]

Finally, one should be aware of the possibility that the presence of the acceptor affects the donor fluorescence by a mechanism other than RET. Such effects could occur due to allosteric interactions between the donor and acceptor sites. For example, the acceptor may block diffusion of a quencher to the donor, or it may cause a shift in protein conformation that exposes the donor to solvent. If binding of the acceptor results in quenching of the donor by some other mechanism, then the transfer efficiency determined from the donor will be larger than the true value. In such cases, the transfer efficiency determined from enhanced acceptor emission is thought to be the correct value. The possibility of non-RET donor quenching can be addressed by comparison of the transfer efficiencies observed from donor quenching and acceptor sensitization[50] (see Problem 13.11).

13.4. ENERGY TRANSFER IN MEMBRANES

In the examples of RET described so far, there was a single acceptor attached to each donor molecule. The situation becomes more complex for unlinked donors and acceptors. In this case the bulk concentration of acceptors is important because the acceptor concentration determines the D–A proximity. Also, one needs to consider the presence of more than a single acceptor around each donor. In spite of the complexity, RET has considerable potential for studies of lateral organization in membranes. For example, consider a membrane which contains regions that are in the liquid phase and regions that are in the solid phase. If the donor and acceptor both partition into the same region, one expects the extent of energy transfer to be increased, relative to that expected for a random distribution of donors and acceptors between the phases. Conversely, if donor and acceptor partition into different phases, the extent of energy transfer will decrease relative to a random distribution, an effect which has been observed.[51] Alternatively, consider a membrane-bound protein. If acceptor-labeled lipids cluster around the protein, then the extent of energy transfer will be greater than expected for acceptor randomly dispersed in the membrane. Energy transfer to membrane-localized acceptors can be used to measure the distance of closest approach to a donor site on the protein or the distance from the donor to the membrane surface.

RET in membranes is typically investigated by measuring the transfer efficiency as the membrane acceptor concentration is increased. Quantitative analysis of such data requires knowledge of the extent of energy transfer expected for fluorophores randomly distributed on the surface of a membrane. This is a complex problem which requires one to consider the geometric form of the bilayer (planar or spherical) and transfer between donors and acceptors which are on the same side of the bilayer as well as those on opposite sides. A variety of approaches have been used,[5,52–60] and, in general, numerical simulations and/or computer analyses are necessary. These theories are complex and not easily summarized. The complexity of the problem is illustrated by the fact that an analytical expression for the donor intensity for energy transfer in two dimensions only appeared in 1964[61] and was extended to allow an excluded volume around the donor in 1979.[53] RET to multiple acceptors in one, two, and three dimensions is described in more detail in Chapter 15. Several of these results are presented here to illustrate the general form of the expected data.

A general description of energy transfer on a two-dimensional surface has been given by Fung and Stryer.[5] Assuming no homotransfer between the donors, and no diffusion during the donor excited-state lifetime, the intensity decay of the donor is given by

$$I_D(t) = I_D^0 \exp(-t/\tau_D) \exp[-\sigma S(t)] \qquad [13.26]$$

where

$$S(t) = \int_{r_c}^{\infty} \{1 - \exp[-(t/\tau_D)(R_0/r)^6]\} 2\pi r \, dr \qquad [13.27]$$

In these equations $\exp[-\sigma S(t)]$ describes that portion of the donor decay due to RET, σ is the surface density of the acceptor, and r_c is the distance of closest approach between the donor and acceptors. The energy-transfer efficiency can be calculated by an equation analogous to Eqs. [13.13] and [13.14], except that the intensities or lifetimes are calculated from integrals of the donor intensity decay:

$$E = 1 - \frac{1}{\tau_D} \int \frac{I_D(t)}{I_D^0} dt \qquad [13.28]$$

Equations [13.26]–[13.28] are moderately complex to solve and require use of numerical integration. However, the approach is quite general and can be applied to a wide variety of circumstances by using different expressions for $S(t)$ that correspond to different geometric conditions. Figure 13.21 shows the calculated transfer efficiencies for a case in which the donor and acceptors are constrained to the lipid–water interface region of a bilayer. Several features of these predicted data are worthy of mention. The efficiency of transfer increases with R_0 and is independent of the concentration of donor. The absence of energy transfer between donors is generally a safe assumption unless the donor displays a small Stokes' shift or the donor concentration is high, conditions which favor homotransfer. Only small amounts of acceptor, 0.4 mol %, can result in easily measured quenching. For example, with $R_0 = 40$ Å the transfer efficiency is near 50% for just 0.8 mol % acceptor, or one acceptor per 125 phospholipid molecules.

One may readily visualize how energy quenching data could be used to determine whether the distributions of donor and acceptor are random. Using the calculated value of R_0, one compares the measured extent of donor quenching with the observed efficiency. If the measured quenching efficiency exceeds the calculated value, then a preferential association of donors and acceptors within the membrane is indicated.[62] Less quenching would be observed if the donor and acceptor are localized in different regions of the membrane or if the distance of closest approach is restricted due to steric factors. We note that these calculated values shown in Figure 13.21 are strictly true only for transfer between immobilized donor and acceptor on one side of a planar bilayer. However, this simple model is claimed to be a good approximation for a spherical bilayer.[5] For smaller values of R_0, transfer across the bilayer is not significant.

It is also instructive to examine the time-resolved decays of the donor in the presence of acceptor (Figure 13.22). In the presence of acceptor, the donor decays become significantly nonexponential, especially at higher acceptor concentrations. The origin of this nonexponential decay is the time-dependent distribution of acceptors around the excited donors. At short times following excitation, there exist more donors with nearby acceptors. The donors with nearby acceptors decay more rapidly because of the dis-

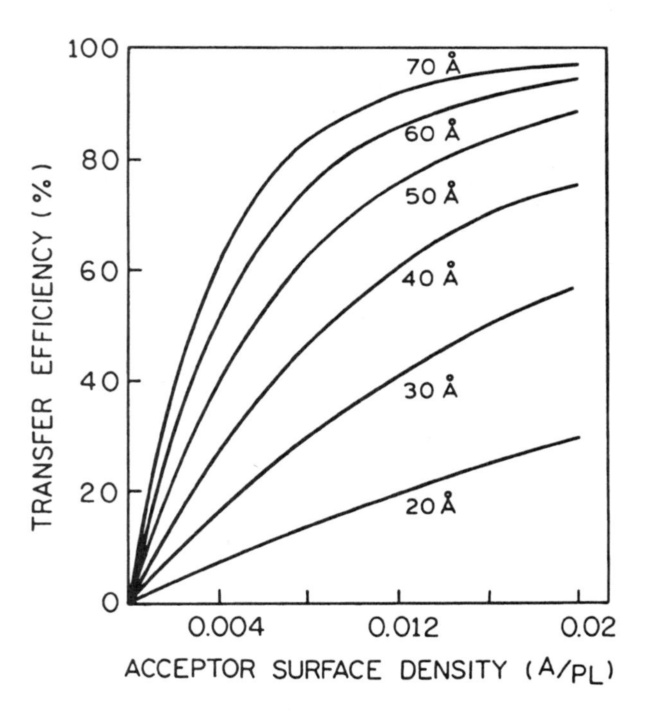

Figure 13.21. Calculated efficiencies of energy transfer for donor–acceptor pairs localized in a membrane. The distances labeled on the curves are the R_0 values for energy transfer, and A/PL is the acceptor-to-phospholipid molar ratio. The area per phospholipid was assumed to be 70 Å^2, so the distance of closest approach was 8.4 Å. Revised from Ref. 5.

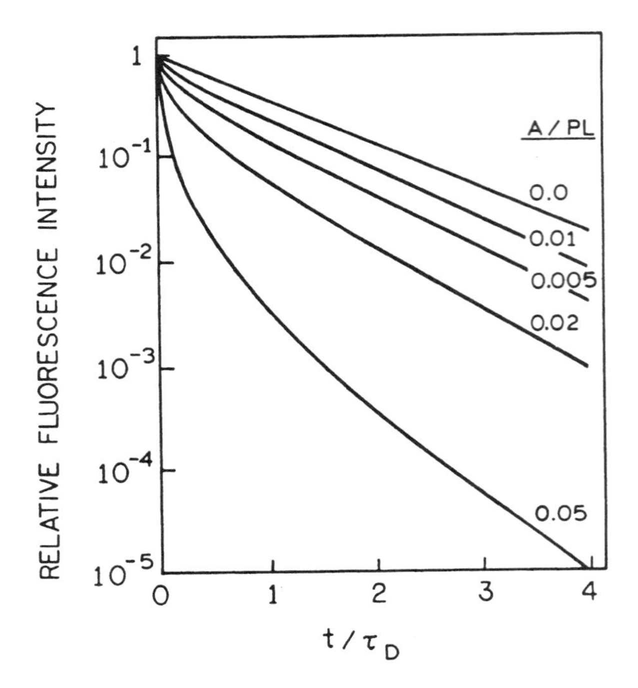

Figure 13.22. Calculated time-resolved decays of donor fluorescence for membrane-bound donors and acceptors. For these calculations R_0 = 40 Å. The values labeled on the curves are the acceptor-to-phospholipid molar ratios (A/PL). Revised from Ref. 5.

tance dependence of energy transfer. At later times, the donors with nearby acceptors have decayed, and the emission results preferentially from donors without nearby acceptors. The decay time of these donors is longer, owing to a slower rate of energy transfer to more distant acceptors.

There is considerable information in these time-resolved decays, and methods to recover this information are described in Chapter 15. In the more general case, the distance between donor and acceptor can vary both as a result of a range of distances and by diffusion. Both factors affect the rates of energy transfer and must be considered in any such analysis.

13.4.A. Lipid Distributions around Gramicidin

Gramicidin is a linear polypeptide antibiotic containing D-and L-amino acids and four tryptophan residues. Its mode of action involves increasing the permeability of membranes to cations and protons. In membranes, this peptide forms a dimer (Figure 13.23)[63] that contains a 4-Å diameter aqueous channel which allows diffusion of cations. The nonpolar amino acids are present on the outside of the helix and are thus expected to be exposed to the acyl side-chain region of the membrane. Hence, gramicidin provides an ideal model with which to examine energy transfer from a membrane-bound protein to membrane-bound acceptors.

It was of interest to determine if membrane-bound gramicidin was surrounded by specific types of phospholipids. This question was addressed by measurement of the transfer efficiencies from the tryptophan donor to dansyl-labeled phosphatidylcholine (PC), 1-acyl-2-[11-[*N*-[5-(dimethylamino)naphthalene-1-sulfonyl]amino]undecanoyl]phosphatidylcholine. The lipid vesicles were composed of PC and phosphatidic acid (PA).[64] Emission spectra of gramicidin bound to PC–PA membranes are

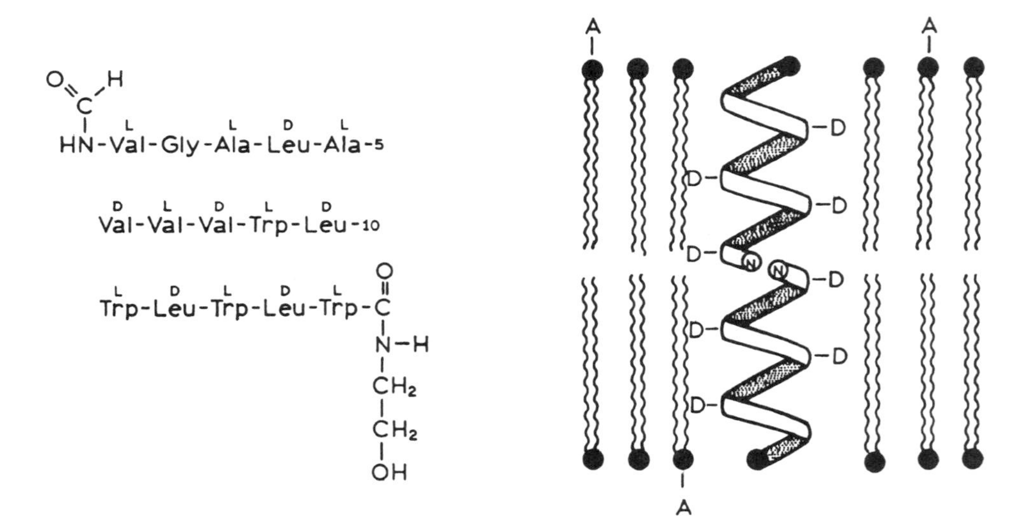

Figure 13.23. Amino acid sequence (*left*) and structure of the membrane-bound dimer (*right*) of gramicidin A. In the amino acid sequence, D and L refer to the optical isomer of the amino acid. Revised from Ref. 63.

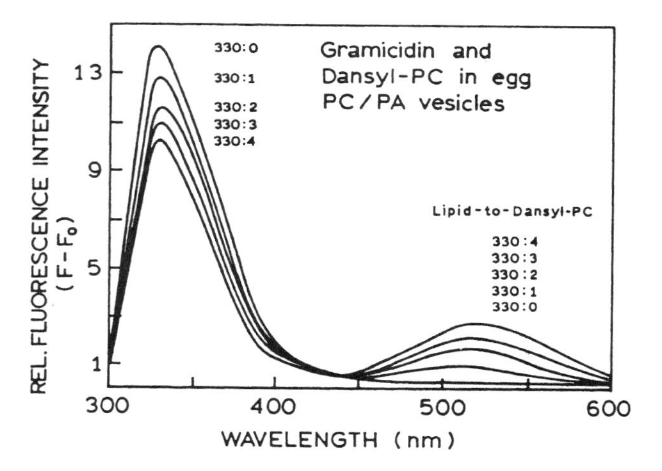

Figure 13.24. Dansyl-PC is 1-acyl-2-[11-[N-[5-(dimethylamino)naphthalene-1-sulfonyl]amino]undecanoyl]phosphatidylcholine. Emission spectra of gramicidin and dansyl-PC in vesicles composed of egg PC and egg PA. λ_{ex} = 282 nm. The lipid-to-dansyl-PC ratios are shown on the figure. Revised and reprinted, with permission, from Ref. 64, Copyright © 1988, American Chemical Society.

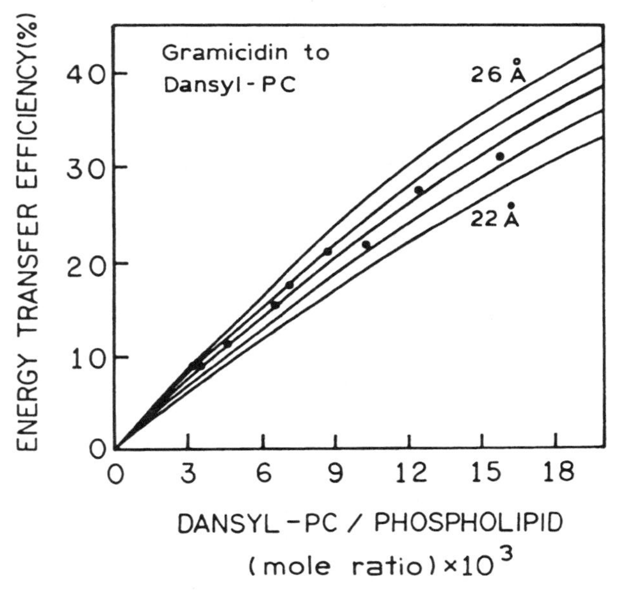

Figure 13.25. Efficiency of energy transfer from gramicidin to dansyl-PC as a function of dansyl-PC/phospholipid ratio. The experimental points (•) were calculated from the tryptophan quenching data, and the solid curves were calculated for a random array of donors and acceptors in two dimensions with R_0 = 22, 23, 24, 25, and 26 Å. A value of R_0 = 24 ± 1 Å gave the best fit to the experimental data. Revised and reprinted with permission from Ref. 64, Copyright © 1988, American Chemical Society.

shown in Figure 13.24. The tryptophan emission is progressively quenched as the dansyl-PC acceptor concentration is increased. The fact that the gramicidin emission is quenched by RET, rather than a collisional process, is supported by the enhanced emission of the dansyl-PC at 520 nm.

The decreasing intensities of the gramicidin emission can be used to determine the tryptophan-to-dansyl-PC transfer efficiencies, as shown in Figure 13.25. The transfer efficiencies are compared with the calculated efficiencies for a random acceptor distribution in two dimensions. These efficiencies were calculated for various values of R_0 using Eqs. [13.26] and [13.28]. The data match the curve calculated for the known R_0 of 24 Å, demonstrating that the distribution of the dansyl-PC acceptors around gramicidin is random. If a particular lipid (PC, PE, or PA) was localized around gramicidin, then RET to the dansyl analog would exceed the calculated values.

13.4.B. Distance of Closest Approach in Membranes

In addition to determining the randomness of lipid distributions, it is also possible to determine the distance of closest approach (r_c) between a membrane-bound protein and an acceptor-labeled lipid. One example is the use of energy transfer to study skeletal protein 4.1.[65] In this case the protein was labeled with IAEDANS, a sulfhydryl-selective dansyl derivative. The acceptor was 3,3′-ditetradecyloxacarbocyanine perchlorate [di-O-C_{14}-(3)], resulting in an R_0 value of 57 Å. Emission spectra are shown in the upper panel of Figure 13.26 for the donor-labeled, acceptor-labeled, and doubly labeled samples as well as the unlabeled sample. The latter spectrum illustrates the need for recording the emission spectra of control samples that do not contain the fluorophores. Depending on the wavelength, the background signal can vary from 20 to nearly 100% of the measured intensity.

In the presence of increasing amounts of acceptor-labeled lipid, the donor-labeled protein is quenched (Figure 13.26, bottom). The fact that the shape of the donor emission spectrum is changing indicates that part of the donor quenching is due to inner filtering by the acceptor absorbance. These data were used to determine the transfer efficiencies to the lipid-bound cyanine dye (Figure 13.27). In this case, the predicted transfer efficiencies were obtained from a simple numerical table[54] which provides a biexponential approximation for various values of r_c/R_0. Use of this table circumvents the need for numerical integration of Eqs. [13.26] and [13.28].

The transfer efficiencies are compared with the calculated curves for various distances of closest approach. As the distance of closest approach becomes larger, the calculated transfer efficiency decreases. The data indicate that the acceptor cannot be closer than 77 Å from the donor-la-

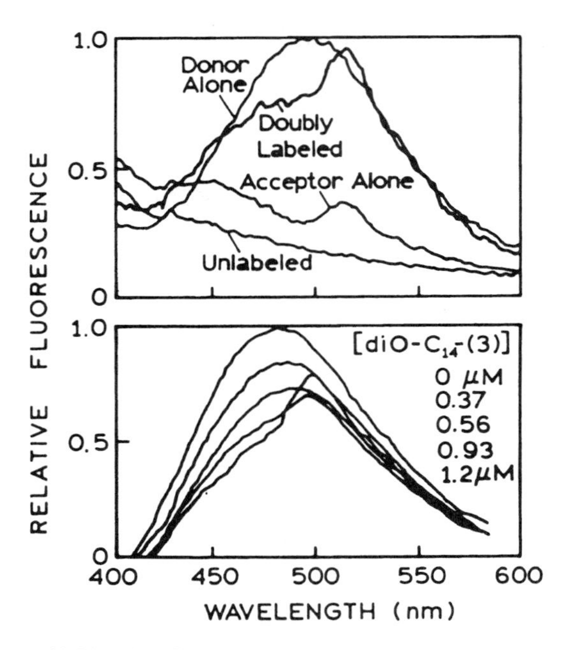

Figure 13.26. *Top:* Emission spectra of skeletal protein 4.1 and of donor-labeled acceptor-labeled, and doubly labeled samples. The donor is IAEDANS and the acceptor is diO-C_{14}-(3). *Bottom:* Emission spectra of IAEDANS-labeled skeletal protein 4.1 in the presence of lipid-bound diO-C_{14}-(3) acceptor. The lipid concentration was 55.4μ*M*, and the acceptor concentrations are shown on the figure. Excitation was at 334 nm. Revised from Ref. 65.

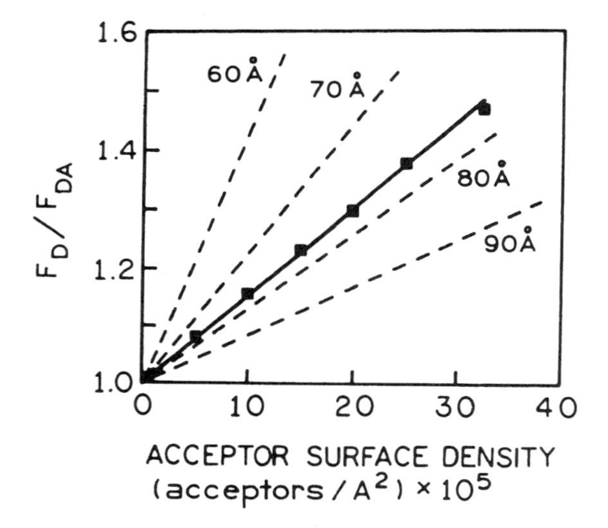

Figure 13.27. RET efficiency from IAEDANS-labeled skeletal protein 4.1 to a lipid-bound cyanine dye. The dashed lines are the predicted transfer efficiencies for various distances of closest approach, calculated according to Ref. 54. Revised from Ref. 65.

beled site and that the donor-labeled site in skeletal protein 4.1 is located deep in the protein. These results suggested that other domains in the protein prevent closer approach of the acceptor.[65] There have been many other reports on energy transfer in membranes, only a few of which can be cited here.[66–69]

13.4.C. Membrane Fusion and Lipid Exchange

Energy transfer has been widely used to study fusion and/or aggregation of membranes. These experiments are shown schematically in Figure 13.28. Suppose that a vesicle contains donor and a surface density of acceptor adequate to quench the donor. As seen from Figure 13.21, the acceptor density does not need to be large. Any process which dilutes the donor and acceptors from the initially labeled vesicles will result in less energy transfer and increased donor emission. For example, if the D–A-labeled residues fuse with an unlabeled vesicle, the acceptor becomes more dilute and the donor intensity increases (Figure 13.28, top). Alternatively, the donor may display a modest water solubility adequate to allow exchange between vesicles (Figure 13.28, left). Then some of the donors will migrate to the acceptor-free vesicles, and again the donor fluorescence will increase. It is also possible to trap a water-soluble fluorophore–quencher pair inside the vesicles. Upon fusion, the quencher can be diluted and/or released. A wide variety of different procedures have been proposed,[70–74] but most rely on these simple proximity considerations.

13.5. ENERGY TRANSFER IN SOLUTION

Energy transfer also occurs for donors and acceptors randomly distributed in three-dimensional solutions. In this case the theory is relatively simple. Following δ-function excitation, the intensity decay of the donor is given by[75–77]

$$I_D(t) = I_D^0 \exp\left[\frac{-t}{\tau_D} - 2\gamma\left(\frac{t}{\tau_D}\right)^{1/2}\right] \qquad [13.29]$$

with $\gamma = A/A_0$, where A is the acceptor concentration. If R_0 is expressed in centimeters, the value of A_0 in moles per liter is given by

$$A_0 = \frac{3000}{2\pi^{3/2} N R_0^3} \qquad [13.30]$$

The relative steady-state quantum yield of the donor is given by

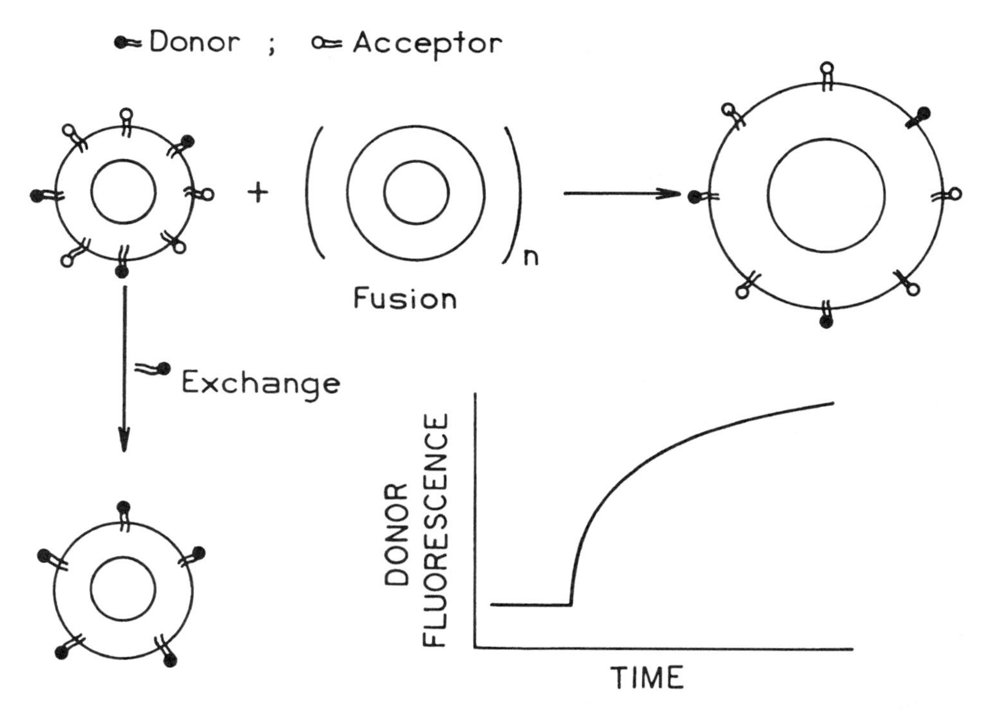

Figure 13.28. Lipid exchange and membrane fusion assays based on energy transfer. Exchange of the donor or acceptor to unlabeled vesicles or fusion of D–A-labeled vesicles with unlabeled vesicles results in dilution of the D–A pairs and increased donor intensities.

$$\frac{F_{DA}}{F_D} = 1 - \sqrt{\pi}\,\gamma \exp(\gamma^2)[1 - \mathrm{erf}(\gamma)] \qquad [13.31]$$

where

$$\mathrm{erf}(\gamma) = \frac{2}{\sqrt{\pi}} \int_0^{\gamma} \exp(-x^2)\, dx \qquad [13.32]$$

These expressions are valid for immobile donors and acceptors for which the orientation factor is randomized by rotational diffusion ($\kappa^2 = \frac{2}{3}$). For randomly distributed acceptors,[3] where rotation is much slower than the donor decay, $\kappa^2 = 0.476$. Still more complex expressions are necessary if the donor and acceptor diffuse during the lifetime of the excited state (Chapters 14 and 15). The complex decay of donor fluorescence reflects the time-dependent population of D–A pairs. Those donors with nearby acceptors decay more rapidly, and donors more distant from acceptors decay more slowly.

The term A_0 is called the critical concentration and represents the acceptor concentration that results in 76% energy transfer. This concentration, in moles per liter (M), can be calculated from Eq. [13.30] or from a simplified expression,[12]

$$A_0 = 447/R_0^3 \qquad [13.33]$$

where R_0 is in units of angstroms. This reveals an important feature of energy transfer between unlinked donors and acceptors, which is that the acceptor concentrations need to be rather high for RET between unlinked donor and acceptors. For instance, if $R_0 = 25$ Å, then $A_0 = 0.029M$ = 29mM. This is why we usually ignore energy transfer between donors and acceptors in different linked D–A pairs. However, it is important to consider inner filter effects when one is comparing intensity values. The high acceptor concentrations needed for RET in solution also make such measurements difficult. The high acceptor concentrations result in high optical densities, requiring front-face observation and careful correction for inner filter effects.

13.5.A. Diffusion-Enhanced Energy Transfer

To this point, we have not considered the effects of diffusion on the extent of energy transfer. This is a complex topic, and typically numerical methods are required for simulations and analysis of these effects. However, one simple case is that in which the donor decay times are very long so that diffusive motions of the donors result in their sampling of the entire available space. This is called the rapid-diffusion limit.[78] With fluorophores displaying nanosecond decay times, this limit is not obtainable. However, lanthanides are known to display much longer decay times—near 0.6–2.5 ms for terbium and europium,[79,80] depending upon the ligands. In relatively fluid solution, the long donor decay times allow the excited donor molecule to diffuse through all available space. The rate of transfer then becomes limited by the distance of closest approach between donors and acceptors. Suppose that an acceptor is buried in a protein or a membrane. One can determine the depth of the acceptor from the extent of energy transfer to the acceptor from terbium or europium chelates in the aqueous phase. Another important feature of energy transfer in the rapid-diffusion limit is that it occurs at much lower acceptor concentrations.[78] For a homogeneous three-dimensional solution, energy transfer is 50% efficient at acceptor concentrations near 1 μ*M*, which is 1000-fold lower than for energy transfer without diffusion.[12] This topic is described in detail in Section 15.4.

13.6. REPRESENTATIVE R_0 VALUES

It is often convenient to have an estimate of an R_0 value prior to performing the complex calculation of the overlap integral or labeling of the macromolecule. Unfortunately, there is no general reference for the Förster distances. A larger number of R_0 values are summarized in a book by Berlman,[81] but most of the listed fluorophores are not used in biochemistry. A large number of Förster distances have been summarized in a recent review,[12] and useful examples with spectra are given in a review by Fairclough and Cantor.[82] Some of these R_0 values are summarized in Table 13.3 for a variety of D–A pairs and in Table 13.4 for tryptophan D–A pairs. In the case of lanthanides in environments where the quantum yield is near unity, and for a multichromophore acceptor, R_0 values as large as 90 Å have been calculated.[82] This is the largest Förster distance reported to this time.

Table 13.3. Representative Förster Distances for Various Donor–Acceptor Pairs[a,b]

Donor	Acceptor	R_0 (Å)	Reference
Naphthalene	Dansyl	22	12
Dansyl	FITC	33–41	82
Dansyl	ODR	43	12
ε-A	NBD	38	12
IAF	TMR	37–50	12
Pyrene	Coumarin	39	12
FITC	TMR	49–54	12
IAEDANS	FITC	49	12
IAEDANS	IAF	46–56	12
IAF	EIA	46	12
BODIPY	BODIPY	57	21
BPE	Cy5	72	12
Terbium	Rhodamine	65	23
Europium	Cy5	70	79
Europium	APC	90	83

[a]Values are from Refs. 12, 20, 23, 79, 82, and 83, which should be consulted for further details.

[b]Abbreviations: Dansyl, 5-dimethylamino-1-naphthalenesulfonic acid; ε-A, 1-N^6-ethenoadenosine; IAF, 5-(iodoacetamido)fluorescein; FITC, fluorescein 5-isothiocyanate; IAEDANS, 5-(((2-iodoacetyl)amino)ethyl)amino) naphthalene-1-sulfonic acid; CF, carboxyfluorescein, succinimidyl ester; BODIPY, 4,4-difluoro-4-bora-3a,4a-diaza-s-indacene; BPE, B-phycoerythrin; ODR, octadecylrhodamine; NBD, 7-nitrobenz-2-oxa-1,3-diazol-4-yl; TMR, tetramethylrhodamine; EIA, 5-(iodoacetetamido)eosin; TR, Texas Red; Cy5, carboxymethylindocyanine-*N*-hydroxysuccinimidyl ester; APC, allophycocyanin.

Table 13.4. Förster Distances for Tryptophan–Acceptor Pairs[a]

Acceptor[b]	R_0 (Å)
Nitrobenzoyl	16
Dansyl	21–24
IAEDANS	22
Anthroyloxy	24
TNB	24
Anthroyl	25
Tyr-NO_2	26
Pyrene	28
Heme	29
NBS	30
DNBS	33
DPH	40

[a]From Ref. 12.

[b]Abbreviations: Dansyl, 5-dimethylamino-1-naphthalenesulfonic acid; IAEDANS, 5-(((2-iodoacetyl)amino)ethyl)amino)naphthalene-1-sulfonic acid; TNB, trinitrophenyl; NBS, nitrobenzenesulfenyl; DNBS, dinitrobenzenesulfonyl; DPH, 1,6-diphenyl-1,3,5-hexatriene.

REFERENCES

1. Förster, Th., 1948, Intermolecular energy migration and fluorescence, *Ann. Phys.* **2**:55–75. Translated by R. S. Knox, Department of Physics and Astronomy, University of Rochester, Rochester, NY 14627.
2. Stryer, L., 1978, Fluorescence energy transfer as a spectroscopic ruler, *Annu. Rev. Biochem.* **47**:819–846.
3. Steinberg, I. Z., 1971, Long-range nonradiative transfer of electronic excitation energy in proteins and polypeptides, *Annu. Rev. Biochem.* **40**:83–114.
4. Clegg, R. M., 1996, Fluorescence resonance energy transfer, in *Fluorescence Imaging Spectroscopy and Microscopy*, X. F. Wang and B. Herman (eds.), John Wiley & Sons, New York, pp. 179–252.
5. Fung, B. K. K., and Stryer, L., 1978, Surface density determination in membranes by fluorescence energy transfer, *Biochemistry* **17**:5241–5248.

6. Weller, A., 1974, Theodor Förster, *Ber. Bunsen-Ges. Phys. Chem.* **78**:969–971.
7. Gordon, M., and Ware, W. R. (eds.), 1975, *The Exciplex*, Academic Press, New York.
8. Dale, R. E., Eisinger, J., and Blumberg, W. E., 1979, The orientational freedom of molecular probes. The orientation factor in intramolecular energy transfer, *Biophys. J.* **26**:161–194. Erratum **30**:365 (1980).
9. Dale, R. E., and Eisinger, J., 1975, Polarized excitation energy transfer, in *Biochemical Fluorescence, Concepts*, Vol. 1. R. F. Chen and H. Edelhoch (eds.), Marcel Dekker, New York, pp. 115–284.
10. Dale, R. E., and Eisinger, J., 1974, Intramolecular distances determined by energy transfer. Dependence on orientational freedom of donor and acceptor, *Biopolymers* **13**:1573–1605.
11. Cheung, H. C., 1991, Resonance energy transfer, in *Topics in Fluorescence Spectroscopy, Vol. 2, Principles*, J. R. Lakowicz (ed.), Plenum Press, New York, pp. 127–176.
12. Wu, P., and Brand, L., 1994, Review—resonance energy transfer: Methods and applications, *Anal. Biochem.* **218**:1–13.
13. Dos Remedios, C. G., and Moens, P. D. J., 1995, Fluorescence resonance energy transfer spectroscopy is a reliable "ruler" for measuring structural changes in proteins, *J. Struct. Biol.* **115**:175–185.
14. Haas, E., Katchalski-Katzir, E., and Steinberg, I. Z., 1978, Effect of the orientation of donor and acceptor on the probability of energy transfer involving electronic transitions of mixed polarizations, *Biochemistry* **17**:5064–5070.
15. Latt, S. A., Cheung, H. T., and Blout, E. R., 1965, Energy transfer. A system with relatively fixed donor–acceptor separation, *J. Am. Chem. Soc.* **877**:996–1003.
16. Stryer, L., and Haugland, R. P., 1967, Energy transfer: A spectroscopic ruler, *Proc. Natl. Acad. Sci. U.S.A.* **58**:719–726.
17. Gabor, G., 1968, Radiationless energy transfer through a polypeptide chain, *Biopolymers* **6**:809–816.
18. Haugland, R. P., Yguerabide, J., and Stryer, L., 1969, Dependence of the kinetics of singlet–singlet energy transfer on spectral overlap, *Proc. Natl. Acad. Sci. U.S.A.* **63**:23–30.
19. Horrocks, W. DeW., Holmquist, B., and Vallee, B. L., 1975, Energy transfer between terbium(III) and cobalt(II) in thermolysin: A new class of metal–metal distance probes, *Proc. Natl. Acad. Sci. U.S.A.* **72**:4764–4768.
20. Johnson, I. D., Kang, H. C., and Haugland, R. P., 1991, Fluorescent membrane probes incorporating dipyrrometheneboron difluoride fluorophores, *Anal. Biochem.* **198**:228–237.
21. Hemmila, I. A. (ed.), 1991, *Applications of Fluorescence in Immunoassays*, John Wiley & Sons, New York. (See p. 113.)
22. Kawski, A., 1983, Excitation energy transfer and its manifestation in isotropic media, *Photochem. Photobiol.* **38**:487–504.
23. Selvin, P. R., 1995, Fluorescence resonance energy transfer, *Methods Enzymol.* **246**:300–334.
24. Lakowicz, J. R., Gryczynski, I., Wiczk, W., Laczko, G., Prendergast, F. C., and Johnson, M. L., 1990, Conformational distributions of melittin in water/methanol mixtures from frequency-domain measurements of nonradiative energy transfer, *Biophys. Chem.* **36**:99–115.
25. Faucon, J. F., Dufourca, J., and Lurson, C., 1979, The self-association of melittin and its binding to lipids, *FEBS Lett.* **102**:187–190.
26. Goto, Y., and Hagihara, Y., 1992, Mechanism of the conformational transition of melittin, *Biochemistry* **31**:732–738.
27. Bazzo, R., Tappin, M. J., Pastore, A., Harvey, T. S., Carver, J. A., and Campbell, I. D., 1988, The structure of melittin: A ^{1}H-NMR study in methanol, *Eur. J. Biochem.* **173**:139–146.
28. Lakowicz, J. R., Gryczynski, I., Cheung, H. C., Wang, C.-K., Johnson, M. L., and Joshi, N., 1988, Distance distributions in proteins recovered by using frequency-domain fluorometry. Applications to troponin I and its complex with troponin C, *Biochemistry* **27**:9149–9160.
29. Cai, K., and Schircht, V., 1996, Structural studies on folding intermediates of serine hydroxymethyltransferase using fluorescence resonance energy transfer, *J. Biol. Chem.* **271**:27311–27320.
30. Chapman, E. R., Alexander, K., Vorherr, T., Carafoli, E., and Storm, D. R., 1992, Fluorescence energy transfer analysis of calmodulin–peptide complexes, *Biochemistry* **31**:12819–12825.
31. Adams, S. R., Bacskai, B. J., Taylor, S. S., and Tsien, R. Y., 1993, Optical probes for cyclic AMP, in *Fluorescent and Luminescent Probes for Biological Activity*, W. T. Mason (ed.), Academic Press, New York, pp. 133–149.
32. Johnson, D. A., Leathers, V. L., Martinez, A.-M., Walsh, D. A., and Fletcher, W. H., 1993, Biomedical example: Use of FRET to measure subunit associations of the regulation (R) and catalytic (C) subunits of a protein kinase, *Biochemistry* **32**:6402–6410.
33. Guptasarma, P., and Raman, B., 1995, Use of tandem cuvettes to determine whether radiative (trivial) energy transfer can contaminate steady-state measurements of fluorescence resonance energy transfer, *Anal. Biochem.* **230**:187–191.
34. Romoser, V. A., Hinkle, P. M., and Persechini, A., 1997, Detection in living cells of Ca^{2+}-dependent changes in the fluorescence emission of an indicator composed of two green fluorescent protein variants linked by a calmodulin-binding sequence, *J. Biol. Chem.* **272**:13270–13274.
35. Miyawaki, A., Llopis, J., Heim, R., McCaffery, J. M., Adams, J. A., Ikura, M., and Tsien, R. Y., 1997, Fluorescent indicators for Ca^{2+} based on green fluorescent proteins and calmodulin, *Nature* **388**:882–887.
36. Cardullo, R. A., Agrawal, S., Flores, C., Zamechnik, P. C., and Wolf, D. E., 1988, Detection of nucleic acid hybridization by nonradiative fluorescence resonance energy transfer, *Proc. Natl. Acad. Sci. U.S.A.* **85**:8790–8794.
37. Parkhurst, K. M., and Parkhurst, L. J., 1996, Detection of point mutations in DNA by fluorescence energy transfer, *J. Biomed. Opt.* **1**:435–441.
38. Morrison, L. E., and Stols, L. M., 1993, Use of FRET to measure association of DNA oligomers, *Biochemistry* **32**:3095–3104.
39. Ghosh, S. S., Eis, P. S., Blumeyer, K., Fearon, K., and Millar, D. P., 1994, Real time kinetics of restriction endonuclease cleavage monitored by fluorescence resonance energy transfer, *Nucleic Acids Res.* **22**:3155–3159.
40. Le Bonniec, B. F., Myles, T., Johnson, T., Knight, C. G., Tapparelli, C., and Stones, S. R., 1996, Characterization of the P'_2 and P'_3 specificities of thrombin using fluorescence-quenched substrates and mapping of the subsites by mutagenesis, *Biochemistry* **35**:7114–7122.
41. Lee, S. P., Censullo, M. L., Kim, H. G., Knutson, J. R., and Han, M. K., 1995, Characterization of endonucleolytic activity of HIV-1 integrase using a fluorogenic substrate, *Anal. Biochem.* **227**:295–301.
42. Mayayoshi, E. D., Wang, G. T., Krafft, G. A., and Erickson, J., 1990, Novel fluorogenic substrates for assaying retroviral proteases by resonance energy transfer, *Science* **247**:954–958.
43. Mitra, R. D., Silva, C. M., and Youvan, D. C., 1996, Fluorescence resonance energy transfer between blue-emitting and red-shifted excitation derivatives of the green fluorescent protein, *Gene* **173**:13–17.

44. Perkins, T. A., Wolf, D. E., and Goodchild, J., 1996, Fluorescence resonance energy transfer analysis of ribozyme kinetics reveals the mode of action of a facilitator oligonucleotide, *Biochemistry* **35:**16370–16377.
45. Lee, P., and Han, M. K., 1997, Fluorescence assays for DNA cleavage, *Methods Enzymol.* **278:**343–361.
46. Bilderback, T., Fulmer, T., Mantulin, W. W., and Glaser, M., 1996, Substrate binding causes movement in the ATP binding domain of *Escherichia coli* adenylate kinase, *Biochemistry* **35:**6100–6106.
47. Epe, B., Steinhauser, K. G., and Woolley, P., 1983, Theory of measurement of Förster-type energy transfer in macromolecules, *Proc. Natl. Acad. Sci. U.S.A.* **80:**2579–2583.
48. Clegg, R. M., 1992, Fluorescence resonance energy transfer and nucleic acids, *Methods Enzymol.* **211:**353–388.
49. Clegg, R. M., Murchie, A. I. H., and Lilley, D. M., 1994, The solution structure of the four-way DNA junction at low-salt conditions: A fluorescence resonance energy transfer analysis, *Biophys. J.* **66:**99–109.
50. Berman, H. A., Yguerabide, J., and Taylor, P., 1980, Fluorescence energy transfer on acetylcholinesterase: Spatial relationships between peripheral site and active center, *Biochemistry* **19:**2226–2235.
51. Pedersen, S., Jorgensen, K., Baekmark, T. R., and Mouritsen, O. G., 1996, Indirect evidence for lipid-domain formation in the transition region of phospholipid bilayers by two-probe fluorescence energy transfer, *Biophys. J.* **71:**554–560.
52. Estep, T. N., and Thomson, T. E., 1979, Energy transfer in lipid bilayers, *Biophys. J.* **26:**195–207.
53. Wolber, P. K., and Hudson, B. S., 1979, An analytical solution to the Förster energy transfer problem in two dimensions, *Biophys. J.* **28:**197–210.
54. Dewey, T. G., and Hammes, G. G., 1980, Calculation of fluorescence resonance energy transfer on surfaces, *Biophys. J.* **32:**1023–1035.
55. Shaklai, N., Yguerabide, J., and Ranney, H. M., 1977, Interaction of hemoglobin with red blood cell membranes as shown by a fluorescent chromophore, *Biochemistry* **16:**5585–5592.
56. Snyder, B., and Freire, E., 1982, Fluorescence energy transfer in two dimensions, *Biophys. J.* **40:**137–148.
57. Bastiaens, P., de Beus, A., Lacker, M., Somerharju, P., Vauhkonen, M., and Eisinger, J., 1990, Resonance energy transfer from a cylindrical distribution of donors to a plane of acceptors, *Biophys. J.* **58:**665–675.
58. Yguerabide, J., 1994, Theory for establishing proximity relations in biological membranes by excitation energy transfer measurements, *Biophys. J.* **66:**683–693.
59. Dewey, T. G., 1991, Fluorescence energy transfer in membrane biochemistry, in *Biophysical and Biochemical Aspects of Fluorescence Spectroscopy*, T. G. Dewey (ed.), Plenum Press, New York, pp. 197–230.
60. Dobretsov, G. E., Kurek, N. K., Machov, V. N., Syrejshehikova, T. I., and Yakimenko, M. N., 1989, Determination of fluorescent probes localization in membranes by nonradiative energy transfer, *J. Biochem. Biophys. Methods* **19:**259–274.
61. Tweet, A. G., Bellamy, W. D., and Gaines, G. L., 1964, Fluorescence quenching and energy transfer in monomolecular films containing chlorophyll, *J. Chem. Phys.* **41:**2068–2077.
62. Loura, L. M. S., Fedorov, A., and Prieto, M., 1996, Resonance energy transfer in a model system of membranes: Application to gel and liquid crystalline phases, *Biophys. J.* **71:**1823–1836.
63. Stryer, L. (ed.), 1995, *Biochemistry*, 4th ed., W. H. Freeman and Company, New York. (See p. 274.)
64. Wang, S., Martin, E., Cimino, J., Omann, G., and Glaser, M., 1988, Distribution of phospholipids around gramicidin and D-β-hydroxybutyrate dehydrogenase as measured by resonance energy transfer, *Biochemistry* **27:**2033–2039.
65. Shahrokh, Z., Verkman, A. S., and Shohet, S. B., 1991, Distance between skeletal protein 4.1 and the erythrocyte membrane bilayer measured by resonance energy transfer, *J. Biol. Chem.* **266:**12082–12089.
66. McCallum, C. D., Su, B., Neuenschwander, P. F., Morrissey, J. H., and Johnson, A. E., 1997, Tissue factor positions and maintains the factor VIIa active site far above the membrane surface even in the absence of the factor VIIa Gla domain, *J. Biol. Chem.* **272:**30160–30166.
67. Davenport, L., Dale, R. E., Bisby, R. H., and Cundall, R. B., 1985, Transverse location of the fluorescent probe 1,6-diphenyl-1,3,5-hexatriene in model lipid bilayer membrane systems by resonance excitation energy transfer, *Biochemistry* **24:**4097–4108.
68. Wolf, D. E., Winiski, A. P., Ting, A. E., Bocian, K. M., and Pagano, R. E., 1992, Determination of the transbilayer distribution of fluorescent lipid analogues by nonradiative fluorescence resonance energy transfer, *Biochemistry* **31:**2865–2873.
69. Isaacs, B. S., Husten, E. J., Esmon, C. T., and Johnson, A. E., 1986, A domain of membrane-bound blood coagulation factor Va is located far from the phospholipid surface. A fluorescence energy transfer measurement, *Biochemistry* **25:**4958–4969.
70. Ladokhin, A. S., Wimley, W. C., Hristova, K., and White, S. H., 1997, Mechanism of leakage of contents of membrane vesicles determined by fluorescence requenching, *Methods Enzymol.* **278:**474–486.
71. Kok, J. W., and Hoekstra, D., 1993, Fluorescent lipid analogues applications in cell and membrane biology, *Fluorescent and Luminescent Probes for Biological Activity*, W. T. Mason (ed.), Academic Press, New York, pp. 101–119.
72. Pyror, C., Bridge, M., and Loew, L. M., 1985, Aggregation, lipid exchange, and metastable phases of dimyristoylphosphatidylethanolamine vesicles, *Biochemistry* **24:**2203–2209.
73. Duzgunes, N., and Bentz, J., 1988, in *Spectroscopic Membrane Probes*, L. D. Loew (ed.), CRC Press, Boca Raton, Florida, pp. 117–159.
74. Silvius, J. R., and Zuckermann, M. J., 1993, Interbilayer transfer of phospholipid-anchored macromolecules via monomer diffusion, *Biochemistry* **32:**3153–3161.
75. Bennett, R. G., 1964, Radiationless intermolecular energy transfer. I. Singlet–singlet transfer, *J. Chem. Phys.* **41:**3037–3041.
76. Eisenthal, K. B., and Siegel, S., 1964, Influence of resonance transfer on luminescence decay, *J. Chem. Phys.* **41:**652–655.
77. Birks, J. B., and Georghiou, S., 1968, Energy transfer in organic systems VII. Effect of diffusion on fluorescence decay, *Proc. R. Soc. (J. Phys. B)* **1:**958–965.
78. Thomas, D. D., Caslsen, W. F., and Stryer, L., 1978, Fluorescence energy transfer in the rapid diffusion limit, *Proc. Natl. Acad. Sci. U.S.A.* **75:**5746–5750.
79. Selvin, P. R., Rana, T. M., and Hearst, J. E., 1994, Luminescence resonance energy transfer, *J. Am. Chem. Soc.* **116:**6029–6030.
80. Li, M., and Selvin, P. R., 1995, Luminescent polyaminocarboxylate chelates of terbium and europium: The effects of chelate structure, *J. Am. Chem. Soc.* **117:**8132–8138.
81. Berlman, I. B., 1973, *Energy Transfer Parameters of Aromatic Compounds*, Academic Press, New York.
82. Fairclough, R. H., and Cantor, C. R., 1978, The use of singlet energy transfer to study macromolecular assemblies, *Methods Enzymol.* **48:**347–379.
83. Mathis, G., 1993, Rare earth cryptates and homogeneous fluoroimmunoassays with human sera, *Clin. Chem.* **39:**1953–1959.

84. Lakowicz, J. R., Johnson, M. L., Wiczk, W., Bhat, A., and Steiner, R. F., 1987, Resolution of a distribution of distances by fluorescence energy transfer and frequency-domain fluorometry, *Chem. Phys. Lett.* **138**:587–593.

PROBLEMS

13.1. *Calculation of a Distance from the Transfer Efficiency*: Use the emission spectra in Figure 13.29 to calculate the distance from the indole donor to the dansyl acceptor. Assume that there is a single D–A distance and that diffusion does not occur during the donor excited-state lifetime. The Förster distance $R_0 = 25.9$ Å, and the donor-alone lifetime is 6.8 ns. What is the D–A distance? What is the donor lifetime in the TUD D–A pair?

13.2. *Measurement of RET Efficiencies (E) from Fluorescence Intensities and Lifetimes*: Use Eq. [13.11] to derive the expressions for E based on intensities (Eq. [13.14]) or lifetimes (Eq. [13.13]).

13.3. *Distance Dependence of Energy Transfer*: The theory of Förster states that the rate of energy transfer depends on $1/r^6$, where r is the donor-to-acceptor distance. This prediction was tested experimentally using naphthyl donors and dansyl acceptors linked by rigid polyprolyl spacers (Figure 13.30). Figure 13.31 shows the excitation spectra for this series of D–A pairs. Assume that each polyprolyl spacer contributes 2.83 Å to the spacing and that the D–A distance ranges from 12 Å ($n = 1$) to 46 Å ($n = 12$).

Use the excitation spectra to demonstrate that k_T depends on $1/r^6$. Note that the dansyl acceptor absorbs maximally at 340 nm and the naphthyl donor has an absorption maximum at 290 nm. Excitation spectra were recorded with the emission monochromator centered on the dansyl emission near 450 nm. What is R_0 for this D–A pair?

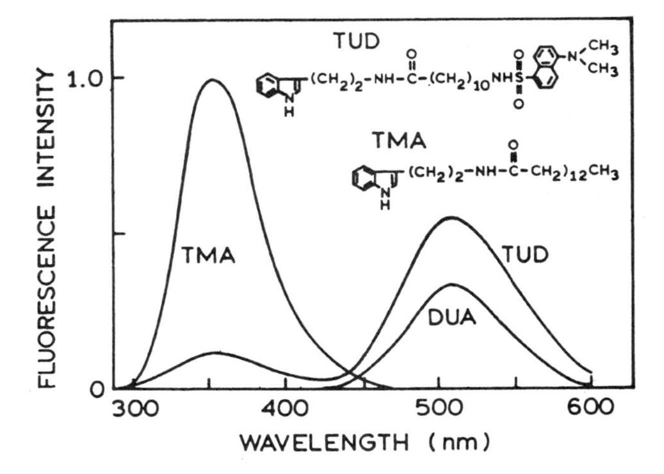

Figure 13.29. Emission spectra of a donor control (TMA) and a donor–acceptor pair (TUD) in propylene glycol at 20 °C. DUA is a dansyl-labeled fatty acid, which is the acceptor-only control sample. Revised from Ref. 84.

n = 1 to 12

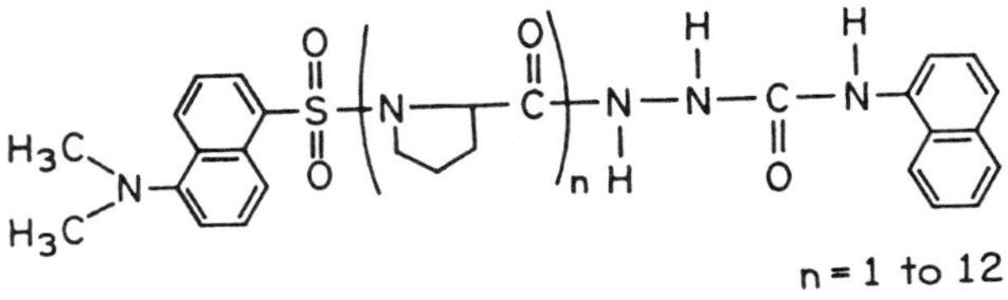

Figure 13.30. Structure of dansyl-(L-prolyl)$_n$-α-naphthyl used for determining the effects of distance on energy transfer. Revised from Ref. 14.

13.4. *Effect of Spectral Overlap on the Rate of Energy Transfer*: Haugland *et al.*[18] investigated the effect of the magnitude of the spectral overlap integral on the rate of fluorescence energy transfer. For this study, they employed the steroids shown in Figure 13.32. They measured the fluorescence lifetimes of compounds I and II. The indole moiety is the donor which transfers energy to the ketone acceptor. Both the absorption spectrum of the ketone and the emission

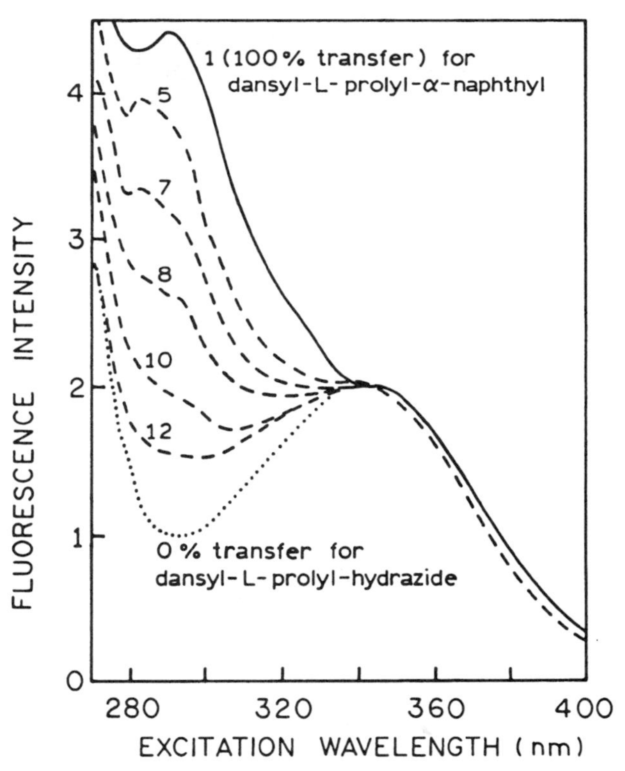

Figure 13.31. Excitation spectra of dansyl-(L-prolyl)$_n$-α-naphthyl molecules. Spectra are shown for dansyl-L-prolyl-hydrazide (···), dansyl-L-prolyl-α-naphthyl (———), and dansyl-(L-prolyl)$_n$-α-naphthyl (— — —), n = 5, 7, 8, 10, and 12. Emission was detected at the dansyl emission maximum near 450 nm. Revised from Ref. 16.

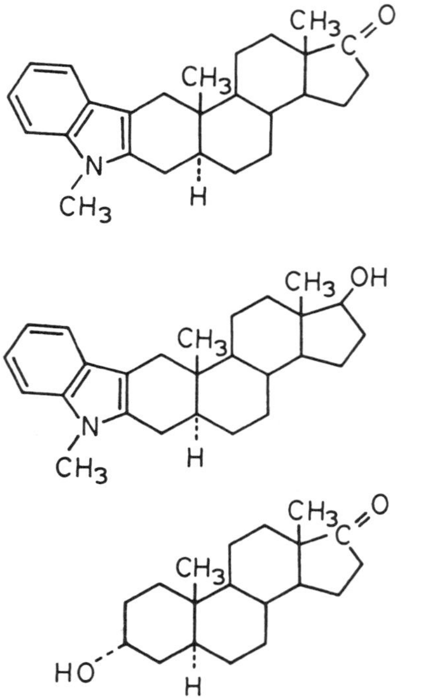

Figure 13.32. Structure of the rigid steroid donor-acceptor pair (I), the steroid containing the donor alone (II), and the steroid containing the acceptor alone (III). Indole is the donor and the carbonyl group is the acceptor. Revised from Ref. 18.

Table 13.5. Fluorescence Spectral Properties of Compounds I and II in a Series of Solvents[a]

Solvent	τ (ns) I	τ (ns) II	n_d	J (cm^6/mmol × 10^{19})
Methanol	5.3	5.6	1.331	1.5
Ethanol	5.6	6.5	1.362	3.0
Dioxane	3.6	5.4	1.423	13.0
Ethyl acetate	3.3	4.7	1.372	12.8
Ethyl ether	2.1	4.5	1.349	30.0
Heptane	1.1	2.8	1.387	60.3

[a]From Ref. 18.

spectrum of the indole are solvent-sensitive. Specifically, the emission spectrum of the indole shifts to shorter wavelengths and the absorption spectrum of the ketone shifts to longer wavelengths as the solvent polarity decreases (Figure 13.33). These shifts result in increasing spectral overlap with decreasing solvent polarity.

Use the data in Table 13.5 to demonstrate that k_T is proportional to the first power of the extent of spectral overlap (J).

13.5. *Calculation of a Förster Distance*: Calculate the Förster distance for the tryptophan-to-dansyl donor–acceptor pair shown in Figure 13.8. The quantum yield of the donor is 0.21.

13.6. *Optical Assay for cAMP*: The effect of cAMP on the donor- and acceptor-labeled protein kinase (Figure 13.15) suggests its use for measuring cAMP. Derive an expres-

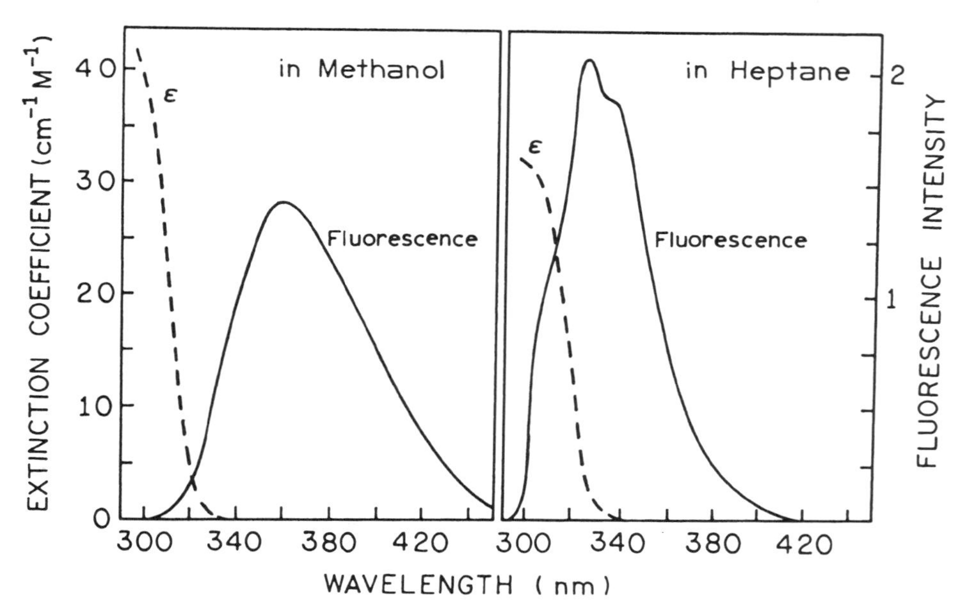

Figure 13.33. Overlap of emission spectrum of the indole donor (II) with the absorption spectrum of the carbonyl acceptor (III). Revised from Ref. 18.

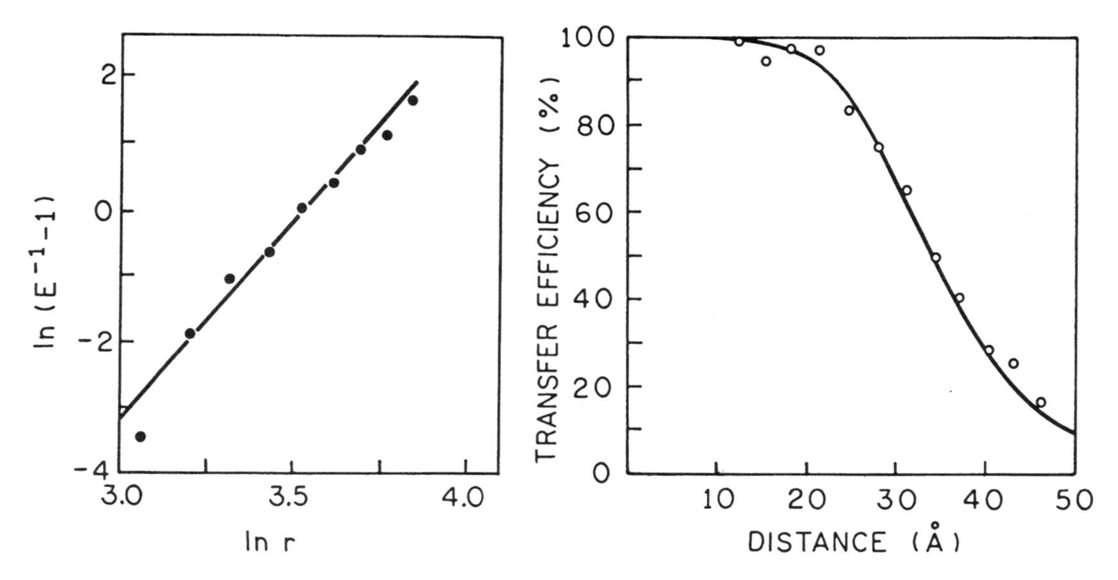

Figure 13.34. Distance dependence of the energy transfer efficiencies in dansyl-(L-propyl)$_n$-α-naphthyl, n = 1–12. Revised from Ref. 16.

sion relating the ratio of donor to acceptor intensities to the protein kinase–cAMP dissociation constant. Assume that the donor and acceptor quantum yields are unchanged upon binding of cAMP, except for the change in energy transfer. Explain any advantages of an assay based on intensity ratios rather than direct intensity measurements.

13.7. *Characteristics of a Closely Spaced D–A Pair.* Assume that you have isolated a protein which contains a single tryptophan residue and binds dinitrophenol (DNP) in the active site. The absorption spectrum of DNP overlaps with the emission spectrum of the tryptophan residue. Assume R_0 = 50 Å and that DNP is not fluorescent. The fluorescence intensities of the tryptophan residue are 20.5 and 4.1 in the absence and presence of DNP, respectively, after correction for the inner filter effects due to the DNP absorption.

A. What is the energy transfer efficiency?
B. Assume that the unquenched lifetime is 5 ns. What is the expected lifetime in the presence of DNP?
C. What is the energy transfer rate?
D. What is the distance between the tryptophan and the DNP?
E. Assume that the solution conditions change so that the distance between the tryptophan and the DNP is 20 Å. What is the expected intensity for the tryptophan fluorescence?
F. For this same solution (r = 20 Å), what would be the effect on the fluorescence intensity of a 1% impurity of a second protein which did not bind DNP? Assume that this second protein has the same lifetime and quantum yield as the protein under investigation.
G. What lifetime would you expect for the sample which contains the impurity? Would this lifetime provide any indication of the presence of an impurity?

13.8. *Effect of κ^2 on the Range of Possible Distances:* Suppose you have a donor- and acceptor-labeled protein which displays the following steady-state anisotropies:

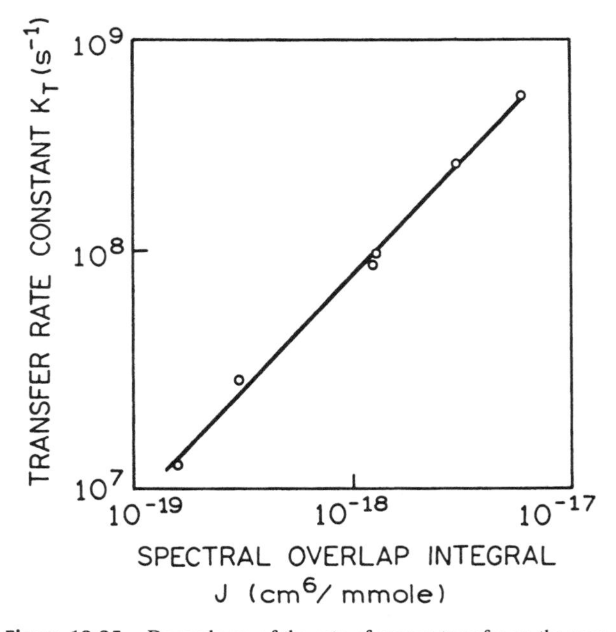

Figure 13.35. Dependence of the rate of energy transfer on the magnitude of the overlap integral. Revised from Ref. 18.

Fluorophore	τ (ns)	r_D or r_A	r_0
Donor-alone control	5	0.1	0.4
Acceptor	15	0.05	0.4

Using an assumed value of $\kappa^2 = \frac{2}{3}$, the D–A distance was calculated to be 25 Å. R_0 is also equal to 25 Å. Assume that the protein displays a rotational correlation time of 5 ns. Use the data provided to calculate the range of distances possible for the D–A pair.

13.9. *Effect of Acceptor Underlabeling on the Calculated Transfer Efficiency*: Suppose that you have a presumed D–A pair. In the absence of acceptor, the donor displays a steady-state intensity $F_D = 1.0$, and in the presence of acceptor, $F_{DA} = 0.5$. Calculate the transfer efficiency assuming that the fractional labeling with acceptor (f_A) is 1.0 or 0.5. How does the change in f_A affect the calculated distance?

13.10. *FRET Efficiency from the Acceptor Intensities*: Derive Eq. [13.25] for the case in which donor labeling is complete; $f_D = 1.0$. Also derive Eq. [13.25] for the case in which donor labeling is incomplete ($f_D < 1$).

13.11. *Correction for Overlapping Donor and Acceptor Emission Spectra*: Equation [13.25] does not consider the possible contribution of the donor emission at the wavelength used to measure acceptor fluorescence (λ_A). Derive an expression for the enhanced acceptor fluorescence when the donor emits at λ_A. Explain how the apparent transfer efficiency, calculated without consideration of the donor contribution, would be related to the true transfer efficiency.

13.12. *Effect of non-RET quenching:* Suppose that you have a protein with a single tryptophan residue. Assume also that the protein noncovalently binds a ligand which serves as a RET acceptor for the tryptophan residue and that the acceptor site is allosterically linked to the donor site such that acceptor binding induces an additional rate of donor quenching, k_q, in addition to k_T.

What is the apparent transfer efficiency upon acceptor binding in terms of τ_D, k_T, and k_q? Is the apparent value (E_D) smaller or larger than the true value (E)?

14

Time-Resolved Energy Transfer and Conformational Distributions of Biopolymers

In the previous chapter we described the principles of resonance energy transfer and how the phenomenon could be used as a "spectroscopic ruler" to measure distances between donor and acceptor sites on macromolecules. Energy transfer was described as a through-space interaction which occurred whenever the emission spectrum of the donor overlapped with the absorption spectrum of the acceptor. For a given donor–acceptor (D–A) pair, the efficiency of energy transfer decreases as r^{-6}, where r is the D–A distance. Each D–A pair has a characteristic distance, the Förster distance (R_0), at which RET is 50% efficient. The extent of energy transfer, as seen from the steady-state data, can be used to measure the distance, to determine the extent of association based on proximity, or to determine the distance of closest approach between the D–A pair.

In this chapter we describe more advanced applications of RET, particularly those which rely on time-resolved measurements of covalently linked D–A pairs. For such pairs, the time-resolved data can be used to recover the conformational distribution or distance distribution between the donor and acceptor. Donor-to-acceptor motions also influence the extent of energy transfer, which can be used to recover the mutual diffusion coefficient. In Chapter 15 we will consider RET to multiple acceptors, in different geometric conditions, such as those which occur in membranes and nucleic acids.

While the phenomenon of energy transfer is the same in all these situations, each application of RET requires a different formalism to interpret the time-resolved data. RET depends on the geometry and dynamics of the system and thus provides a powerful yet complex methodology for studies of donor–acceptor distributions in any dimensionality.

14.1. DISTANCE DISTRIBUTIONS

In the previous chapter we considered energy transfer between a single donor (D) and a single acceptor (A) positioned at a unique distance (r). We now consider the case where a range of D–A distances is possible. This use of RET was pioneered by Haas, Steinberg and co-workers, using flexible polypeptides.[1–3] The concept of a distribution of D–A distances is shown in Figure 14.1. The protein is assumed to be labeled at unique sites by a single donor and a single acceptor. In the native state one expects a single conformation and a single D–A distance. Thus, for the native state, the distance is sharply localized at a particular value of r. This unique distance is expressed as a probability function $P(r)$ which is narrowly distributed along the r axis.

Now suppose that the protein is unfolded to the random-coil state by the addition of denaturant. Since the peptide is in a random state, there exist a range of D–A distances. This range of conformations is described by a wide range of accessible D–A distances, or a wide $P(r)$ distribution (Figure 14.1). For visual clarity, the $P(r)$ distributions

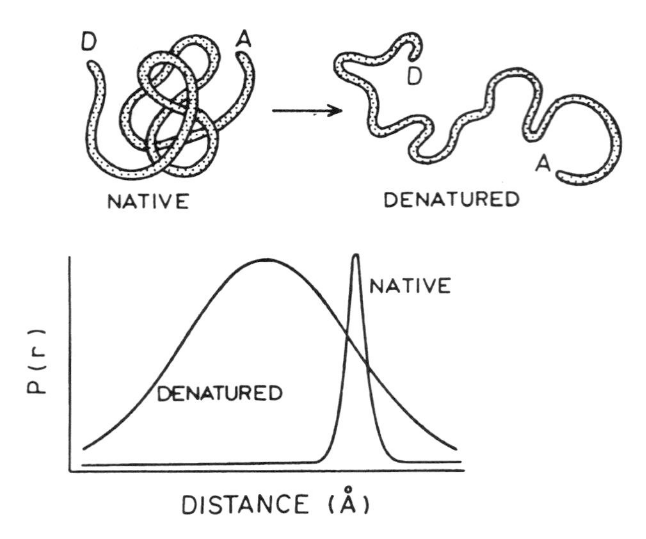

Figure 14.1. Distribution of donor–acceptor distances for a native and a denatured protein. The probability functions $P(r)$ are peak normalized.

shown in Figure 14.1 are peak normalized, but the actual area under each curve should be normalized to unity to correspond to a single acceptor per donor.

The presence of a distribution of distances has a profound impact on the time-resolved decays of the donor. For the native protein, the single D–A distance results in a single transfer rate for all the donors. Hence, the decay time of the donor is shortened, and the donor intensity decay is given by

$$I_{DA}(r, t) = I_D^0 \exp[(-t/\tau_D) - k_T t)] \qquad [14.1]$$

where τ_D is the decay time of the donor in the absence of acceptor, I_D^0 is the donor intensity at $t = 0$, and k_T is the D–A transfer rate, given by

$$k_T(r) = \frac{1}{\tau_D}\left(\frac{R_0}{r}\right)^6 \qquad [14.2]$$

Since there is only one transfer rate, the donor decay remains a single exponential (Figure 14.2, top). It is this assumption of a single distance which allows calculation of the distance from the relative quantum yield of the donor (Chapter 13).

Now assume that there is a range of D–A distances. For these simulated data, we assumed a single-exponential decay time of 5 ns in the absence of acceptor and $R_0 = 25$ Å. Some of the D–A pairs are closely spaced and display shorter decay times, and other D–A pairs are further apart and display longer decay times. The range of distances results in a range of decay times, so that the donor decay becomes more complex than a single exponential (Figure

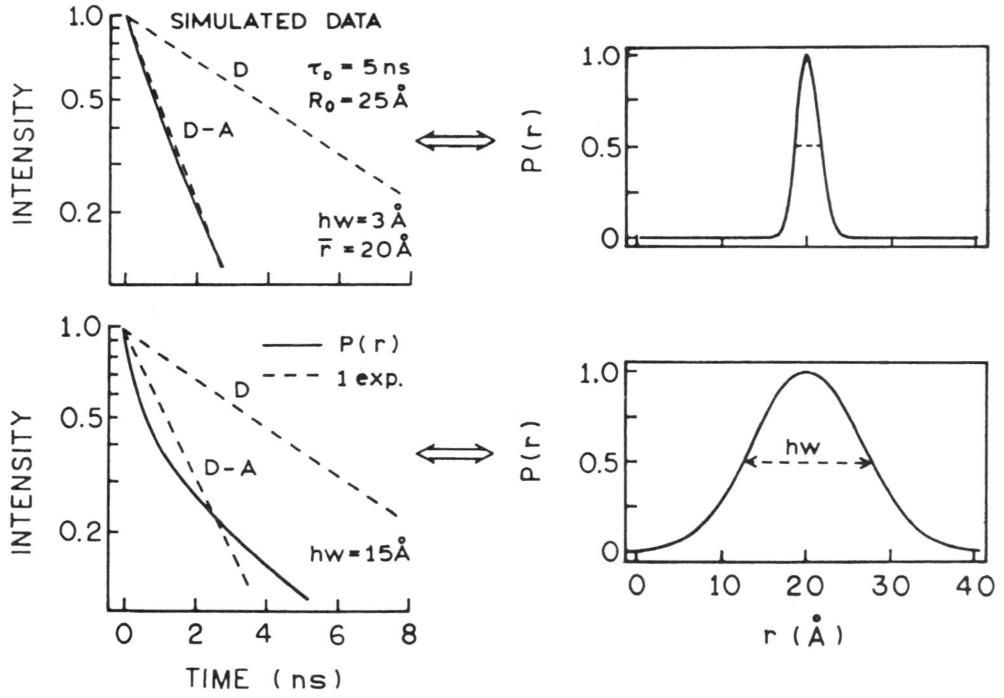

Figure 14.2. Simulated data illustrating the effect of D–A distance distributions of different half-widths (hw) (*right*) on the TD intensity decay of the donor (*left*): ———, Donor intensity decay in the presence of acceptor; – – –, single-exponential fits to the donor intensity decay in the presence (D–A) and absence (D) of acceptor.

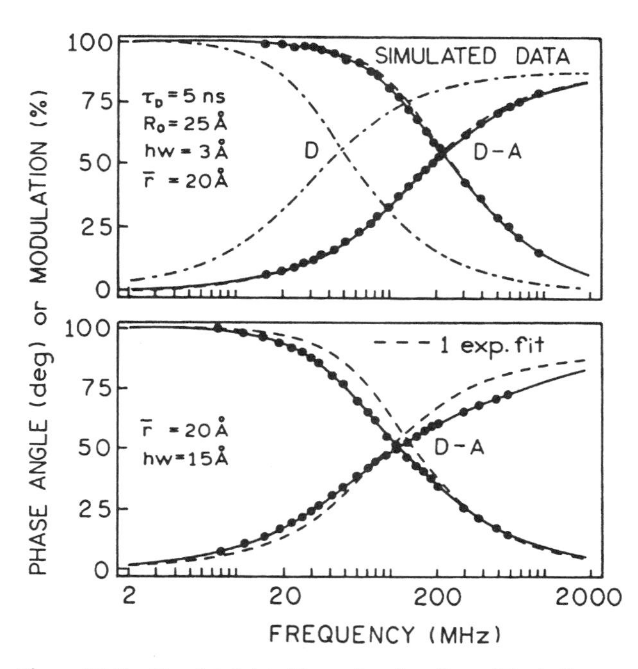

Figure 14.3. Simulated data illustrating the effect of D–A distance distributions of different half-widths (hw) on the FD intensity decay of the donor. ———, Donor intensity decay in the presence of acceptor; ---, best single-exponential fit to the simulated data; -·-, single-exponential fit to the donor intensity decay in the absence of acceptor.

14.2, bottom). Similar results are expected for FD data. If there is a single D–A distance, the FD intensity decay is well approximated by the single-exponential fit (Figure 14.3, top), which is shifted to higher frequencies by the presence of the acceptor. A range of D–A distances results in a frequency response which is spread out along the frequency axis, and one which is no longer a single exponential (Figure 14.3, bottom). The goal of most distance distribution studies is to recover the D–A probability distribution from the nonexponential decays of the donor. It is important to notice the necessity of the time-resolved data. In the presence of a distance distribution, measurement of the relative yield of the donor can be used to calculate an apparent distance, as was done in Problem 13.1. However, the steady-state data cannot be used to determine the distance distribution, and they will not even reveal the presence of a distribution. For the moment, we have ignored donor-to-acceptor motions, which will be discussed in Section 14.5.

Irrespective of whether the donor intensity decays are measured in the time domain or the frequency domain, the information content of the time-resolved data is limited. It is not possible to determine a distance distribution of arbitrary shape. For this reason, it is common practice to describe the distribution using a limited number of parameters. The most appropriate and commonly used distribution is a Gaussian,

$$P(r) = \frac{1}{\sigma\sqrt{2\pi}} \exp\left[-\frac{1}{2}\left(\frac{r-\bar{r}}{\sigma}\right)^2\right] \qquad [14.3]$$

In this equation, $\bar{r}$ is the mean of the Gaussian with a standard deviation of σ. Usually, distance distributions are described by the full width at half-maximum (Figure 14.2, lower right). This half-width (nm) is given by $\text{hw} = 2.354\sigma$. Distance distributions have also been expressed as Lorentzian[4] and other functions,[1,3,5,6] but the principles remain the same. With presently available data, distance distributions can be resolved, but it is difficult to distinguish the precise shape of the distribution.

What equation describes the intensity decay for a distance distribution? Unfortunately, the donor decay can only be described by an integral equation. The intensity decay for those D–A pairs at a distance r is given by Eq. [14.1]. However, these D–A pairs with a unique distance r cannot be individually observed. Rather, one observes a weighted average determined by $P(r)$. The donor intensity decay is a summation of the intensity decays for all accessible distances and is usually written as

$$I_{DA}(t) = \int_{r=0}^{\infty} P(r) I_{DA}(r, t)\, dr$$

$$= I_D^0 \int_{r=0}^{\infty} P(r) \exp\left[\frac{-t}{\tau_D} - \frac{t}{\tau_D}\left(\frac{R_0}{r}\right)^6\right] dr \qquad [14.4]$$

Data analysis is performed by predicting the values of $I_{DA}(t)$ for use with TD or FD measurements and the usual procedures of nonlinear least squares. Typically, the decay time of the donor (τ_D) is known from measurements of the donor in the absence of acceptor. The variable parameters in the analysis are those describing the distance distribution, $\bar{r}$ and hw.

This description of the intensity decay (Eq. [14.4]) illustrates the value of computer programs written in terms of molecular features of the sample. The complex donor decays in the presence of acceptor (Figure 14.2 or 14.3) could be analyzed in terms of the multiexponential model. While it would be possible to fit the data, the value of α_i and τ_i would only indirectly reflect the distance distribution. In contrast, programs based on Eq. [14.4] provide direct information on $P(r)$.

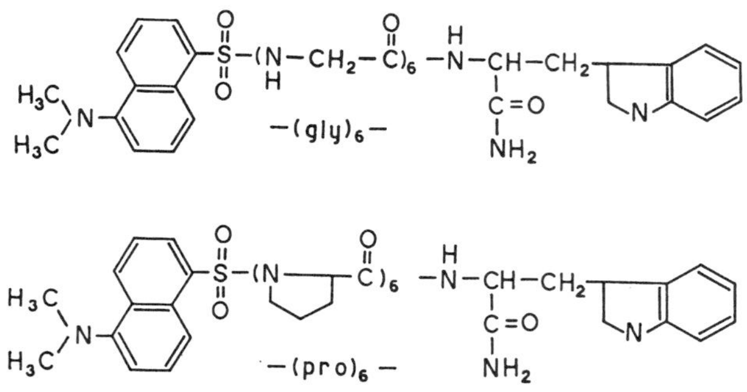

Figure 14.4. Flexible DNS-$(gly)_6$-$TrpNH_2$ and rigid DNS-$(pro)_6$-$TrpNH_2$ D–A pairs. Revised from Ref. 7.

14.2. DISTANCE DISTRIBUTIONS IN PEPTIDES

14.2.A. Comparison for a Rigid and a Flexible Hexapeptide

To illustrate the effects of a distance distribution on the time-resolved decay, we consider two hexapeptides, each containing a tryptophanamide ($TrpNH_2$) donor and a dansyl (DNS) acceptor (Figure 14.4).[7] One of the hexapeptides [DNS-$(pro)_6$-$TrpNH_2$] consists of six rigid proline residues separating Trp and DNS by a single distance. In the other hexapeptide [DNS-$(gly)_6$-$TrpNH_2$], the D–A pair is linked by hexaglycine, which is highly flexible and expected to result in a range of Trp–DNS distances. Emission spectra of these peptides are shown in Figure 14.5. The donor is quenched in both D–A pairs, relative to the donor-alone control, Ac-$(gly)_3$-$TrpNH_2$. The donor is more highly quenched in the flexible $(gly)_6$ peptide relative to the rigid $(pro)_6$ peptide. One could use the relative amounts of donor quenching to calculate the D–A distances. The calculated distance would be correct for the rigid peptide DNS-$(pro)_6$-$TrpNH_2$, but incorrect for the flexible DNS-$(gly)_6$-$TrpNH_2$ peptide. For the latter peptide, one must use time-resolved measurements to interpret the extent of energy transfer.

Information about the distance distribution for flexible molecules can be recovered from the time-resolved decays of the donor. FD data for the two D-A peptides and the donor-alone control molecule are shown in Figure 14.6. In the absence of acceptor, the donor-alone control molecule displays a single-exponential decay with $\tau_D = 5.27$ ns. For the rigid D–A pair (Figure 14.6, top) energy transfer decreases the donor decay time, as seen by the shift of the phase and modulation values to higher frequencies. However, the donor decay remains reasonably close to a single exponential, as seen by the visual similarity of the single- and double-exponential fits. As predicted in Figure 14.2, a single D–A distance results in a single-exponential decay of the donor. Close inspection of the FD intensity decay does reveal some heterogeneity, which is due to a narrow but finite D–A distance distribution.

A remarkably different donor decay is seen for the flexible D–A pair in DNS-$(gly)_6$-$TrpNH_2$. For this case, the donor decay cannot even be approximated by a single

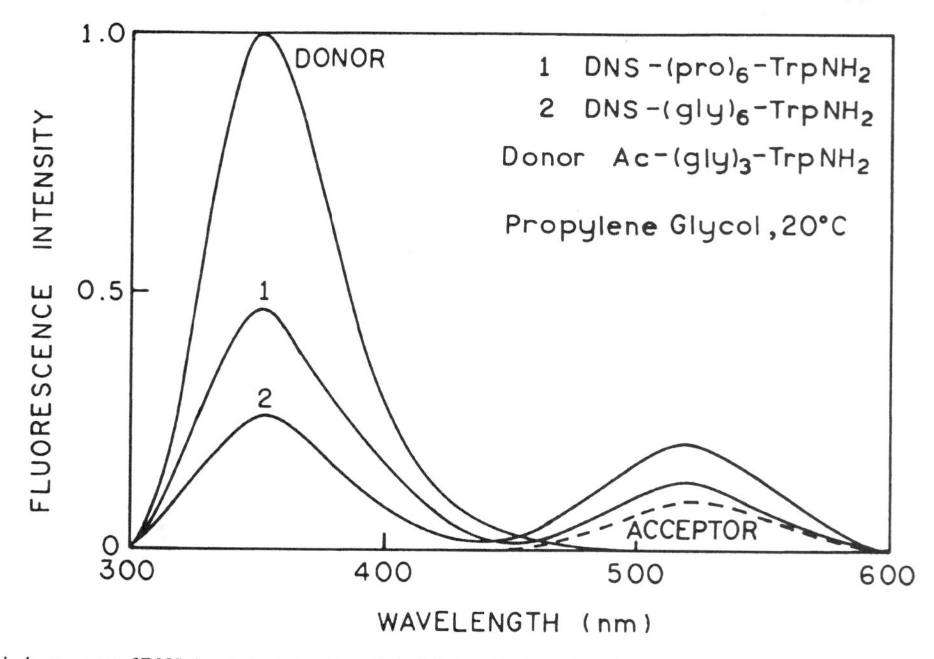

Figure 14.5. Emission spectra of DNS-$(pro)_6$-$TrpNH_2$ (1) and DNS-$(gly)_6$-Trp NH_2 (2) relative to the emission spectrum of the donor-alone control Ac-$(gly)_3$-$TrpNH_2$. Ac refers to an *N*-acetyl group. Revised from Ref. 7.

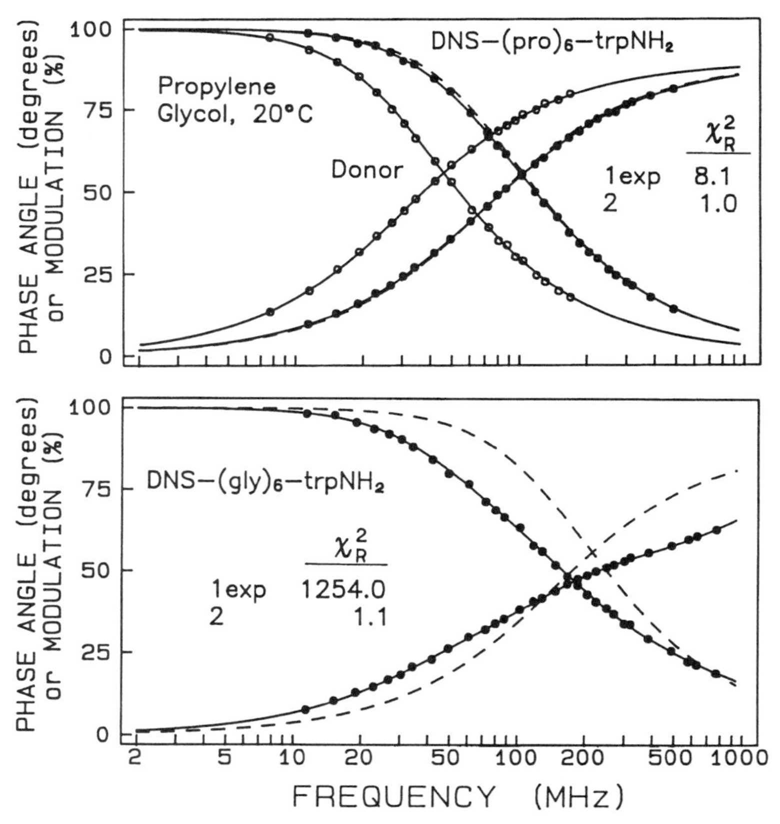

Figure 14.6. FD donor intensity decays of DNS-(pro)$_6$-TrpNH$_2$ (*top*, •), DNS-(gly)$_6$-TrpNH$_2$ (*bottom*), and the donor-alone control Ac-(gly)$_3$-TrpNH$_2$ (*top*, ○).

exponential, as seen from the dashed curve in the lower panel of Figure 14.6, and $\chi^2_R = 1254$ for the single-decay-time fit. A single-decay-time fit is equivalent to fitting the donor decay to a very narrow distance distribution (Eqs. [14.1] and [14.2]). For the flexible peptide, the presence of a range of D–A distances caused the donor decay to become highly nonexponential.

The FD data can be analyzed to recover the distance distribution $P(r)$. This is accomplished by using Eq. [14.4] to predict the phase and modulation values for assumed values of $\bar{r}$ and hw using the general procedures described in Chapter 4 for TD data or in Chapter 5 for FD data. These expressions are shown below (Section 14.3.B) for the case of single- and multiexponential donor decays. Analyses of the FD data in terms of the distance distribution model for both peptides are shown in Figure 14.7. These fits reveal wide and narrow distributions for the flexible and rigid peptides, respectively (Figure 14.8). This result completes the analysis. We have used the complexity of the donor decays to recover the distribution of D–A distances. The complexity of the donor decay caused by the acceptor (Figure 14.6) was used to determine the probability distribution of acceptors around the donor (Figure 14.8).

14.2.B. Cross-fitting Data to Exclude Alternative Models

In addition to obtaining the average distance $\bar{r}$ and half-width hw, it is important to consider whether the data exclude other distance distributions. This can be accomplished by the procedure of cross-fitting the data. In particular, one uses the parameter values from a second competing model to see if they are consistent with the data. If a competing model also fits the data, then we cannot exclude that model from our consideration without additional data. For the flexible peptide DNS-(gly)$_6$-TrpNH$_2$, the alternative model was tested by attempting to fit the data using the half-width of 4 Å found for the rigid peptide. The value of hw = 4 Å was held constant and the least-squares fit run again to minimize χ^2_R using the data for the flexible peptide. The dashed curves in the left panel of Figure 14.7 show that the data for DNS-(gly)$_6$-TrpNH$_2$ are not consistent with a narrow distance distribution. Similarly, the data for DNS-(pro)$_6$-TrpNH$_2$ could not be fit with a wide distribution (dashed curves in the right panel of Figure 14.7). In these cases the attempts to fit the data with different parameter values resulted in obviously incorrect

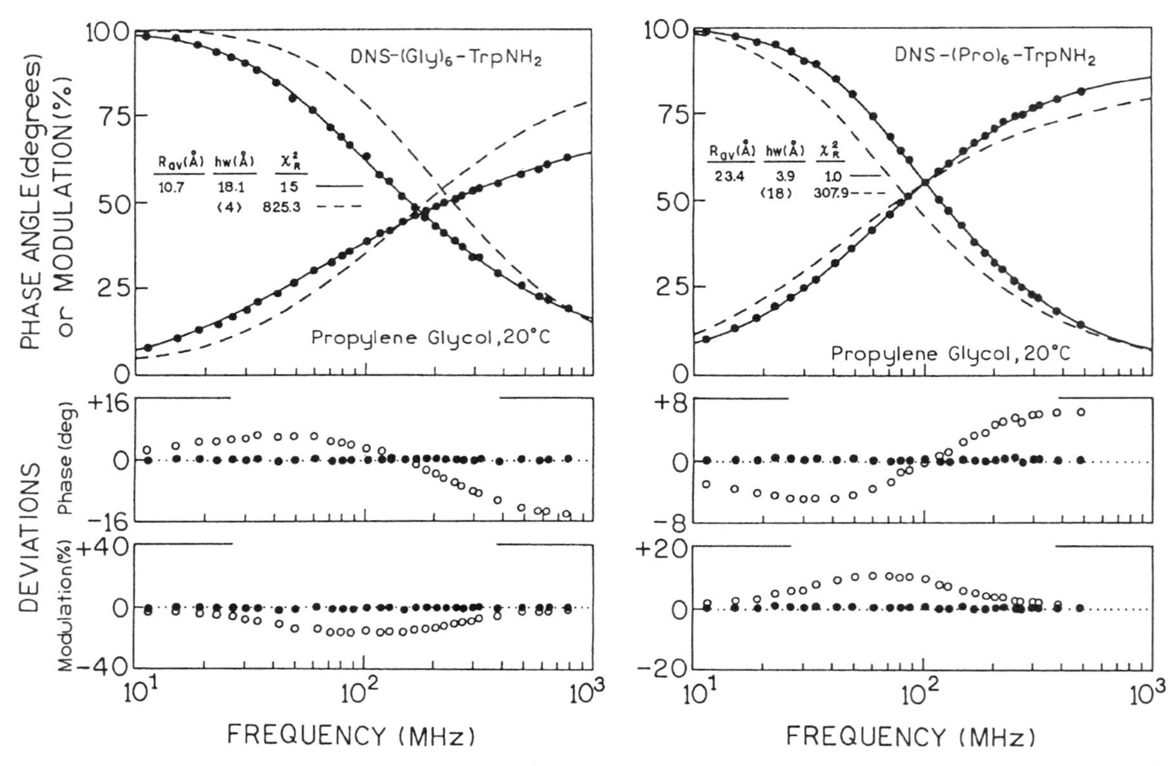

Figure 14.7. Distance distribution fits to the FD donor decays for flexible (*left*) and rigid (*right*) peptides. hw, Half-width of the distance distribution; angular brackets indicate that the value was held constant in the analysis. The lower panels show the residuals with the fixed (○) and variable (●) half widths.

results. However, for conformational changes of proteins, one can expect the distance distributions to be more similar than for these two hexapeptides, so that the ability to reject similar distributions may be more questionable.

14.2.C. Donor Decay without RET

An important aspect of the distance distribution analysis is knowledge of the donor decay time. This is typically

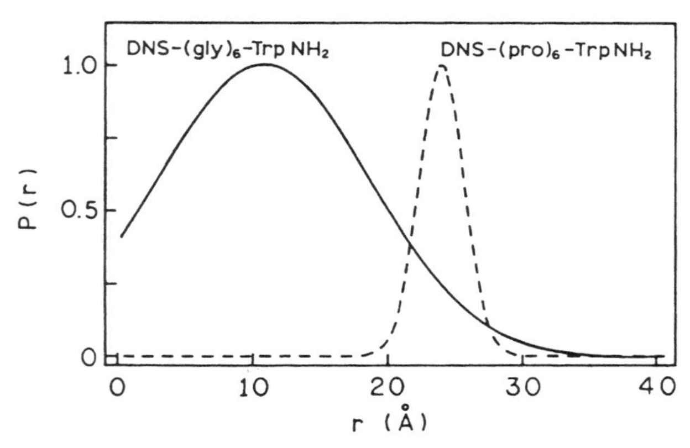

Figure 14.8. Distance distribution fits for DNS-(gly)$_6$-TrpNH$_2$(———) and DNS-(pro)$_6$-TrpNH$_2$ (– – –). Revised from Ref. 7.

measured using a control molecule which is comparable to the D–A pair except that it lacks the acceptor (donor-alone molecule). In the case of the two hexapeptides, the donor control molecule was Ac-(gly)$_3$-TrpNH$_2$, in which the tryptophan donor was attached to a glycine tripeptide. The tripeptide was acetylated at the N-terminus to be more like the D–A pairs. The selection of a suitably designed donor-alone control molecule is a critical step in any energy-transfer experiment. Typically, the donor decay time (or decay times for a multiexponential decay) is measured separately. The data (TD or FD) for the D–A pair are then analyzed with the value of τ_D held fixed. This is necessary because the donor intensity decays of the D–A pair only provide information when compared to the decay that would be observed without energy transfer.

Depending upon the available software, the data for the donor-alone and D–A pairs can also be analyzed simultaneously. In this case the program needs to know that the value of τ_D is determined only by the donor-alone data, as there is no RET in this data set. At first glance, one may think that this simultaneous analysis is identical to fixing τ_D and analyzing the data for the D–A pair. In fact, these two methods are different, and this second analysis is preferred. If the presence of the acceptor has no effect on the donor decay besides that due to RET, both modes of analysis will yield the same value of τ_D. However, if the presence of the acceptor somehow affects τ_D other than by RET, then the second mode of analysis could reveal this effect. Alternatively, since energy transfer decreases the intensity of the donor, an increased contribution of impurity fluorescence to the data for the D–A pair could result. Different values of τ_D would then be obtained for each mode of analysis.

14.2.D. Effect of Concentration of the D–A Pairs

In general, the extent of energy transfer depends on the concentration of the donors and acceptors. As described in the following chapter, for unlinked D–A pairs the acceptor concentration typically needs to be in the millimolar range for significant energy transfer. This is because millimolar concentrations are needed in order for one or more acceptor molecules to be within the Förster distance of the donor. However, for linked D–A pairs, energy transfer is usually dominated by the linked acceptor. Hence, in the preceding sections, we made no mention of the concentration of the linked D–A pairs. Knowledge of the concentration was unnecessary because each donor sees an effective constant concentration of the acceptor determined by the length and flexibility of the linker to which the acceptor is attached. However, the concentration of linked D–A pairs should be low enough to avoid inner filter effects and transfer between donors and acceptors on unlinked D–A pairs. Under these conditions, the extent of energy transfer will be independent of the bulk concentration of the D–A pairs.

14.3. DISTANCE DISTRIBUTIONS IN PROTEINS

The principles described above have been applied to a wide variety of proteins. TD measurements have been used to recover distance distributions in native and unfolded staphylococcal nuclease,[4] ribonuclease,[8] troponin C,[9] and cardiac troponin I.[10] Some of the most elegant reports have been by Haas and co-workers on bovine pancreatic trypsin inhibitor (BPTI).[5,6,11–14] This protein was labeled with a naphthalene donor at the N-terminal amino group and selectively labeled with a coumarin acceptor on each of its four lysine residues. The labeled BPTI was studied during folding and unfolding to determine the folding pathway.[6,11–14] FD measurements have also been used to study proteins, including the ribonuclease S-peptide,[15] myosin subfragment-1,[16] troponin-C[17] and its complexes with troponin C,[18,19] and zinc finger peptides.[20,21] The use of RET with proteins has been reviewed by several authors.[4,22]

14.3.A. Distance Distributions in Melittin

For the purpose of illustrating distance distributions in proteins, we have chosen to present data for melittin.[23,24] Melittin has a single intrinsic tryptophan residue (trp-19) as the donor and was labeled on the N-terminal amino group with a dansyl acceptor. This 26-amino-acid peptide from bee venom was used in Chapter 13 to illustrate measurements of a single D–A distance; in that case, the peptide was in the α-helical state. We now consider melittin in the random-coil state, which folds in a monomeric α-helical state upon addition of methanol (Figure 14.9). Hence, this simple peptide is expected to illustrate how a change in structure affects the D–A distance.

Emission spectra of the donor-alone melittin and dansyl-melittin are shown in Figure 14.10. The efficiency of transfer from tryptophan to the dansyl acceptor can be determined from these donor-normalized spectra. The overall efficiency of RET seems to be higher for α-helical melittin (in 80% methanol) than in aqueous buffer [5m*M* MOPS (morpholinepropanesufonic acid)]. This difference is seen as a weaker donor emission and stronger acceptor emission in 80% methanol. If we calculated a single distance from these data, we would conclude that the trp-to-

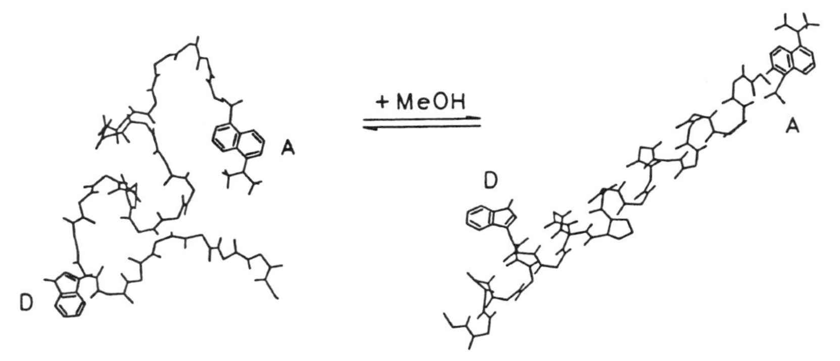

Figure 14.9. Dansyl (acceptor)-labeled melittin in the random-coil (*left*) and the α-helical state (*right*). Revised from Ref. 23, Copyright © 1990, with permission from Elsevier Science.

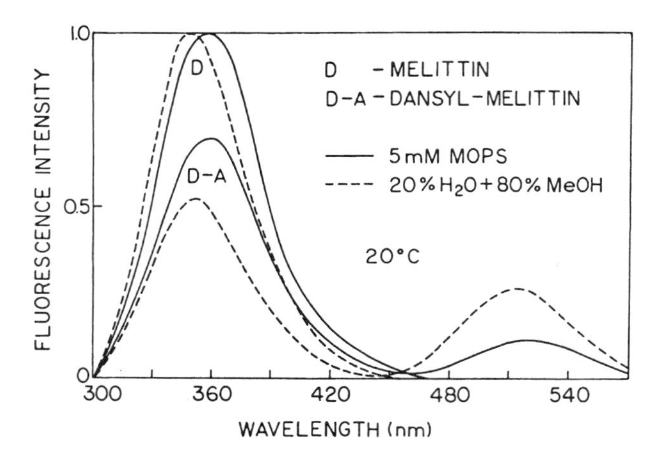

Figure 14.10. Emission spectra of melittin (D) and dansyl-melittin (D–A) in the α-helical (– – –) and the random-coil (———) state. The donor-alone (melittin) intensities are normalized. Reprinted from Ref. 23, Copyright © 1990, with permission from Elsevier Science.

dansyl distance is shorter in the α-helical state than in the random coil state. However, the time-resolved data yield a different result.

FD intensity decays are shown in Figure 14.11 for random-coil melittin, as the donor-alone control, and for dansyl-melittin. The presence of the acceptor shortens the

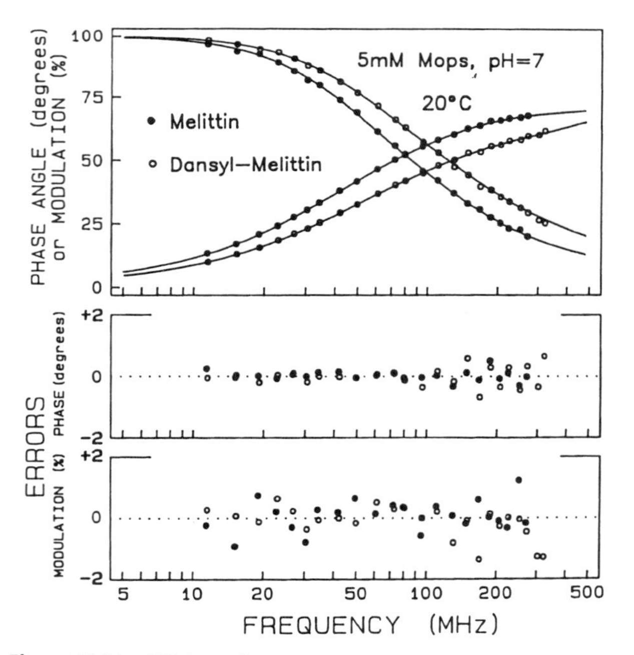

Figure 14.11. FD donor decays of melittin (•) and dansyl-melittin (○) in the random-coil state. The solid lines and residuals (lower panels) are for the best triple exponential fits. Reprinted from Ref. 23, Copyright © 1990, with permission from Elsevier Science.

donor decay time. However, it is difficult to see increased heterogeneity in the tryptophan decay due to the presence of acceptor. The similarity of the donor-alone and donor–acceptor frequency responses in Figure 14.11 illustrates the difficulty in determining distance distributions in proteins. Analysis procedures are needed which not only provide answers, but which test the data for being consistent with or not consistent with competing models.

Another difficulty in measuring distance distributions in proteins is the presence of multiexponential decays even in the absence of acceptor. The intensity decay of melittin itself is a multiexponential which requires at least three decay times for a suitable fit in either solvent (Table 14.1). Also shown in Table 14.1 are the multiexponential analyses for dansyl-melittin in aqueous buffer (random coil) and in 80% methanol (α-helical). The mean decay time ($\bar{\tau}$) of the tryptophan donor is shortened by the presence of acceptor. However, it is difficult, if not impossible, to interpret these values of α_i and τ_i in terms of the melittin conformation. This is why the data are analyzed in terms of distance distributions.

The distance distribution analyses are shown in Figure 14.12, and the resulting distributions are presented in the upper panel of Figure 14.13. In the absence of methanol, the data are consistent with a wide distribution of trp-to-dansyl distances, ranging from 0 to 40 Å. The mean ($\bar{r}$) and half-width (hw) were 17 and 25 Å, respectively. In the presence of methanol, the distance distribution becomes quite sharp, with $\bar{r}$ = 25 Å and hw = 3 Å (Figure 14.13), which is interpreted as being due to the α-helical structure shown in Figure 14.9.

Once again, it is important to test the uniqueness of the result. This was accomplished by cross-fitting the data with alternate parameter values which yielded the dashed curves in Figure 14.12. The data for the random-coil state (Figure 14.12, left) are clearly inconsistent with the α-helical parameters, and the data for the α-helical state (Figure 14.12, right) are clearly inconsistent with the random-coil parameters. The visual differences seen in the cross-fits illustrate an important point. If you cannot visually see the difference, it is probably not there. Trust your eyes, not your computers.

A trp-to-dansyl distance distribution was also obtained for melittin when bound to calmodulin[24] (Figure 14.13, lower panel). In this case also, melittin is expected to be in the α-helical state. It is interesting to notice that the mean distance is somewhat smaller than for the unbound α-helix, suggesting that the hydrophobic patch of calmodulin has imposed some restrictions on the melittin structure. The distance distribution of melittin seems wider when melittin is bound to calmodulin than when it is in solution. There are two possible causes of this result. One is structural

Table 14.1. Intensity Decays of Melittin and Dansyl-Melittin[a]

Sample	%MeOH	$\bar{\tau}$ (ns)[b]	τ_i (ns)	α_i	f_i	χ_R^2 Number of decay times 1	2	3
Melittin	0	3.29	0.32	0.408	0.065			
			2.49	0.401	0.495			
			4.64	0.191	0.400	548	5.7	1.4
Dansyl-melittin	0	2.62	0.14	0.500	0.064			
			1.12	0.287	0.297			
			3.24	0.213	0.639	733	7.4	2.5
Melittin	80	4.19	0.67	0.375	0.092			
			2.96	0.434	0.468			
			6.29	0.192	0.440	551	7.7	2.8
Dansyl-melittin	80	2.96	0.37	0.345	0.083			
			1.73	0.561	0.631			
			4.69	0.094	0.286	368	9.5	1.2

[a]From Ref. 23.
[b]$\bar{\tau}$ is the mean decay time calculated using $\bar{\tau} = f_1\tau_1 + f_2\tau_2 + f_3\tau_3$.

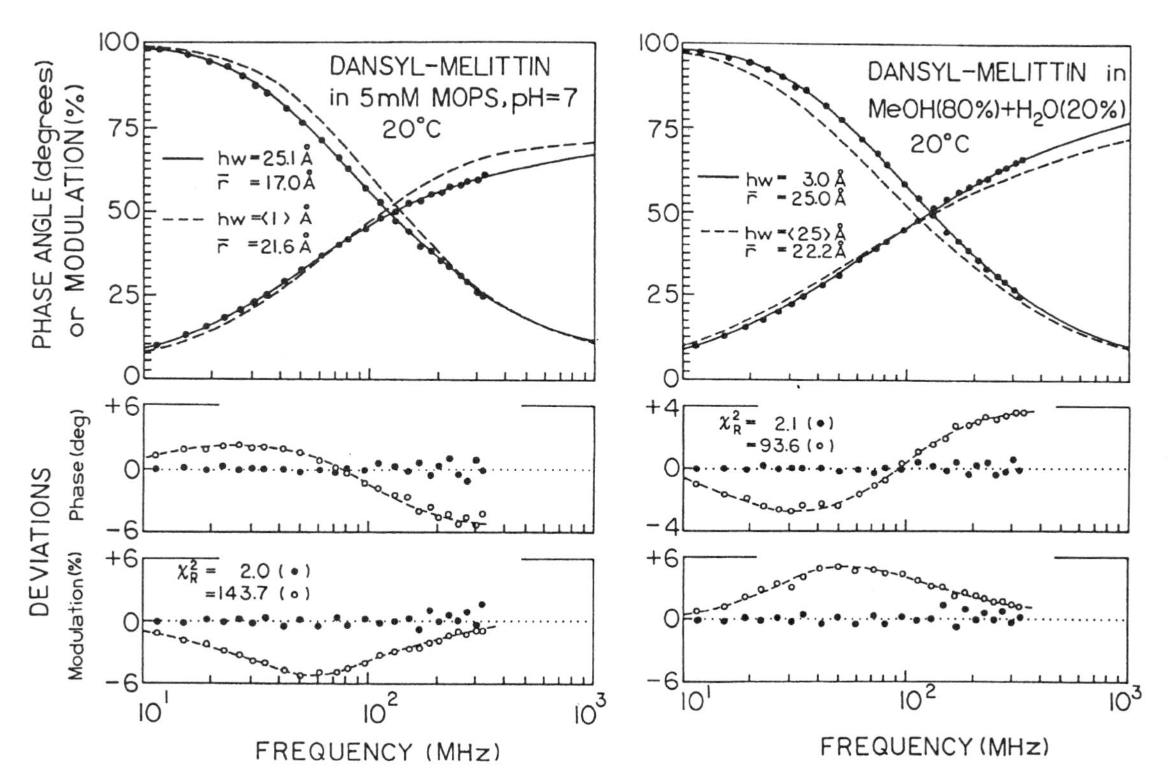

Figure 14.12. FD tryptophan decays of dansyl-melittin in the random-coil (*left*) and the α-helical state (*right*). The solid and dashed lines represent the best distance distribution fits with the half-width variable and fixed, respectively. The lower panels show the residuals with variable (●) and fixed (○) half widths. Revised from Ref. 23, Copyright © 1990, with permission from Elsevier Science.

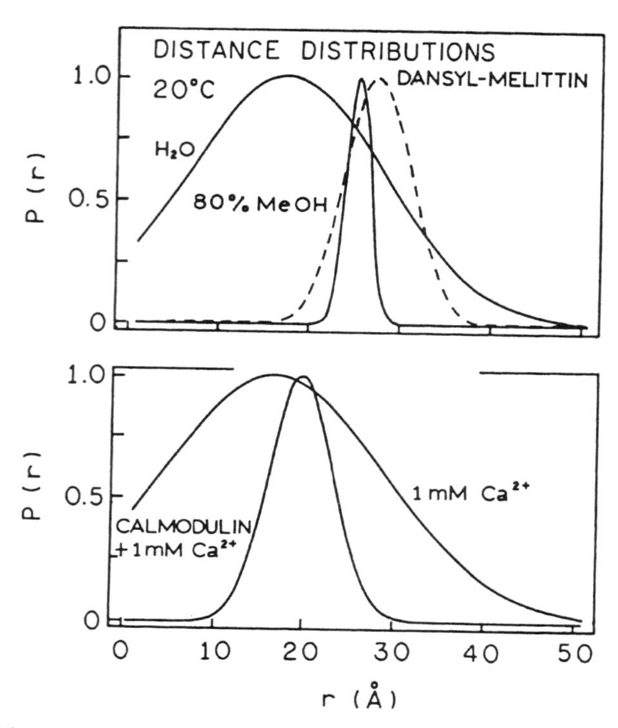

Figure 14.13. Distance distributions of dansyl-melittin in solution and when bound to calmodulin. The dashed curve (*top*) shows the distance distribution obtained with consideration of trp-to-dansyl diffusion; $D = 6.5 \times 10^{-6}$ cm²/s. Revised from Ref. 35.

heterogeneity in the bound state. The second is the use of an incorrect model. In Section 14.5 we describe the effects of donor-to-acceptor diffusion and, in particular, the narrowing of the apparent distribution when data are analyzed without consideration of diffusion (Section 14.5.D), as was done for melittin. In fact, the distribution shown for α-helical melittin in Figure 14.13 is narrowed by diffusion, and consideration of this effect results in a somewhat wider distribution (dashed curve in Figure 14.13).

14.3.B. Distance Distribution Analysis with Frequency-Domain Data

We now consider how the time-resolved data for dansyl-melittin were analyzed to recover the distance distribution. While the description is for the FD data, the same steps are taken for analysis of the TD data. The first step in any distance distribution analysis is determination of the intensity decay of the donor alone, followed by use of this decay law in the $P(r)$ analysis. For any form of the decay law, the phase ($\phi_{c\omega}$) and modulation ($m_{c\omega}$) at a given frequency can be calculated (c) from

$$N_\omega = \frac{\int_0^\infty I(t) \sin \omega t \, dt}{\int_0^\infty I(t)\, dt} \qquad [14.5]$$

$$D_\omega = \frac{\int_0^\infty I(t) \cos \omega t \, dt}{\int_0^\infty I(t)\, dt} \qquad [14.6]$$

Using these transforms, the phase and modulation values can be calculated from

$$\phi_{c\omega} = \arctan (N_\omega / D_\omega) \qquad [14.7]$$

$$m_{c\omega} = (N_\omega^2 + D_\omega^2)^{1/2} \qquad [14.8]$$

If the donor decay is a single exponential, then the decay is given by Eq. [14.4], and the transforms are

$$N_\omega = \frac{1}{J} \int_{r=0}^{\infty} \frac{P(r)\, \omega\, \tau_{DA}^2}{1 + \omega^2 \tau_{DA}^2} dr \qquad [14.9]$$

$$D_\omega = \frac{1}{J} \int_{r=0}^{\infty} \frac{P(r) \tau_{DA}}{1 + \omega^2 \tau_{DA}^2} dr \qquad [14.10]$$

where the normalization factor J is given by

$$J = \left(\int_0^\infty P(r)\, dr \right) \left(\int_0^\infty I_{DA}(t)\, dt \right) \qquad [14.11]$$

In these equations the value of τ_{DA} is given by

$$\frac{1}{\tau_{DA}} = \frac{1}{\tau_D} + \frac{1}{\tau_D} \left(\frac{R_0}{r} \right)^6 \qquad [14.12]$$

Hence, τ_{DA} depends on distance r, as can be seen from Eq. [14.4]. The value of τ_{DA} corresponds to the lifetime of the donor for a particular distance r. As was described above, these specific molecules cannot be observed. Only the entire population can be measured, hence the integrals over r in Eqs. [14.9] and [14.10]. Analytical expressions for

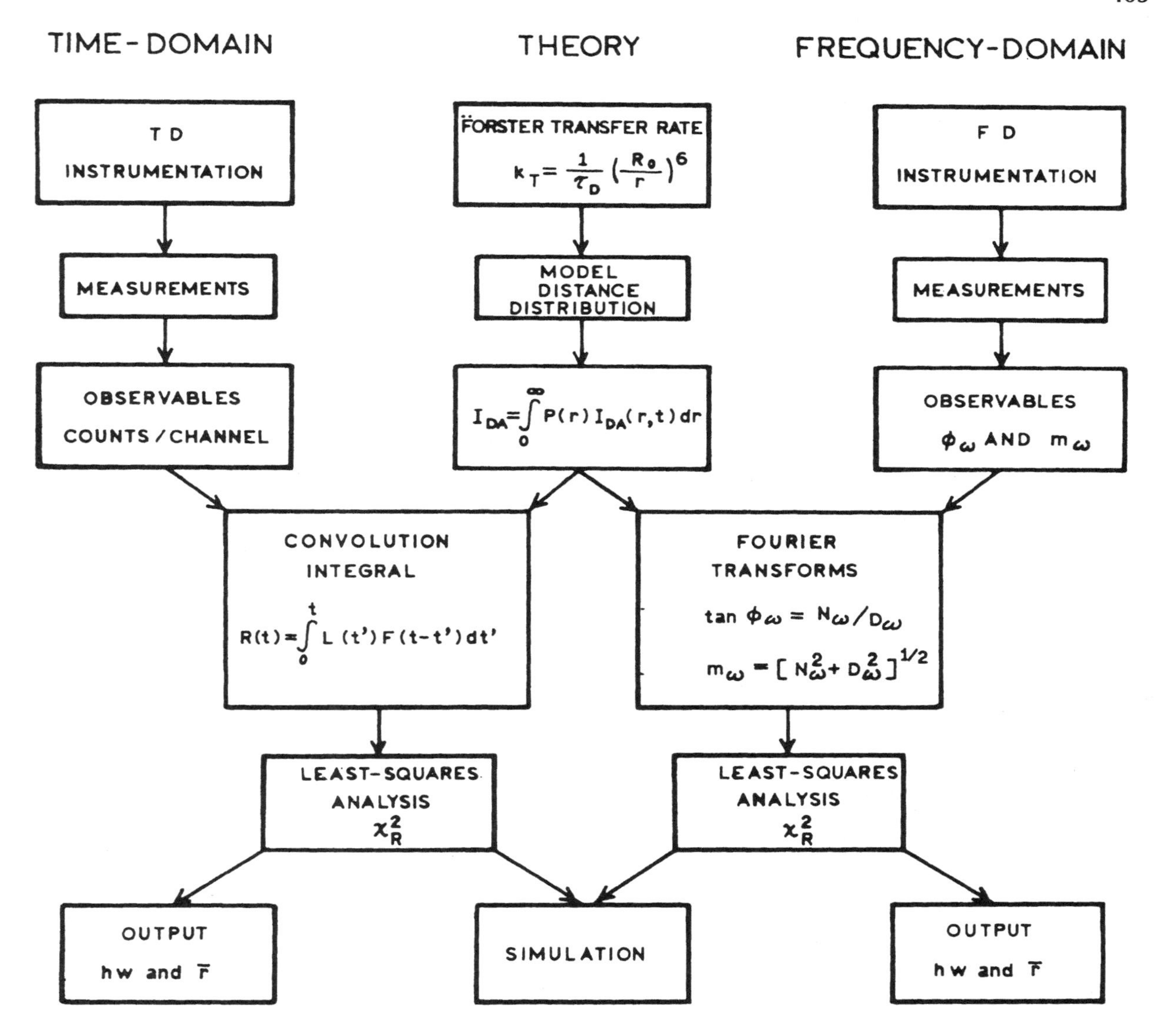

Figure 14.14. Flow chart for distance distribution analysis.

N_ω and D_ω are not available, so these values are calculated numerically. The parameter values describing the distance distribution are recovered from nonlinear least-squares analysis by minimization of χ_R^2,

$$\chi_R^2 = \frac{1}{\nu}\sum_{\omega}\left(\frac{\phi_\omega - \phi_{c\omega}}{\delta\phi}\right)^2 + \frac{1}{\nu}\sum_{\omega}\left(\frac{m_\omega - m_{c\omega}}{\delta m}\right)^2 \qquad [14.13]$$

In this expression, ϕ_ω and m_ω are the experimental values, ν is the number of degrees of freedom, and the subscript c indicates calculated values for assumed values of $\bar{r}$ and hw. The values $\delta\phi$ and δm are the experimental uncertainties in phase and modulation, respectively.

For most proteins or macromolecules, the donor-alone decays are not single exponentials. Hence, we need to consider how to recover the distance distribution for multiexponential-decay donors. In this case the donor-alone decays are described by

$$I_D(t) = \sum_i \alpha_{Di} \exp(-t/\tau_{Di}) \qquad [14.14]$$

where the α_{Di} are the preexponential factors and the τ_{Di} are the decay times for the donor in the absence of any acceptor. At this point we need to make some assumptions. What are the transfer rates and/or Förster distance from each component in the donor decay? In Chapter 13 we showed that the transfer rates were proportional to τ_D^{-1} for each fluorophore. Hence, one can assume that the transfer rate from each decay-time component is given by

$$k_{Ti} = \left(\frac{1}{\tau_{Di}}\right)\left(\frac{R_0}{r}\right)^6 \qquad [14.15]$$

and that the distance-dependent donor decay times are given by

$$\frac{1}{\tau_{DAi}} = \frac{1}{\tau_{Di}} + \left(\frac{1}{\tau_{Di}}\right)\left(\frac{R_0}{r}\right)^6 \qquad [14.16]$$

Other assumptions are possible,[25] but their use does not seem to alter the results. Assuming that Eq. [14.15] correctly describes the transfer rate from each component, the intensity decay of D–A pairs spaced at a distance r is given by

$$I_{DA}(r, t) = \sum_i \alpha_{Di} \exp\left[-\frac{t}{\tau_{Di}} - \frac{t}{\tau_{Di}}\left(\frac{R_0}{r}\right)^6\right] \qquad [14.17]$$

and the intensity decay of the sample is given by

$$I_{DA}(t) = \int_0^\infty P(r) I_{DA}(r, t)\, dr \qquad [14.18]$$

The sine and cosine transforms are

$$N_\omega^{DA} = \frac{1}{J}\int_0^\infty \sum_i \frac{P(r)\alpha_{Di}\omega\tau_{DAi}}{1 + \omega^2\tau_{DAi}^2}\, dr \qquad [14.19]$$

$$D_\omega^{DA} = \frac{1}{J}\int_0^\infty \sum_i \frac{P(r)\alpha_{Di}\tau_{DAi}}{1 + \omega^2\tau_{DAi}^2}\, dr \qquad [14.20]$$

where J is given by Eq. [14.11] using the multiexponential decay law. It is important to notice that a multiexponential decay for the donor does not introduce any additional parameters into the analysis. This is because the intrinsic decays of the donor are measured in a separate experiment, using samples without acceptor. The data from the donor are fit to the multiexponential model, and the parameters (α_{Di} and τ_{Di}) are held constant in Eqs. [14.19] and [14.20] during the least-squares analysis. It should be remembered that τ_{DAi} depends on distance (Eq. [14.16]).

It is apparent from the above description (Eqs. [14.9]–[14.20]) that the distance distribution analysis is moderately complex. A flow chart of the analysis (Figure 14.14) may be easier to understand. The analysis starts with the model (Förster transfer) and measurements (FD instrumentation). Measured and calculated values of ϕ_ω and m_ω are compared, and the parameters are varied to minimize χ_R^2. For an understanding of what values and changes of $\bar{r}$ and hw are resolvable, we strongly recommend the analysis of simulated data. These data should be simulated with parameter values comparable to those expected for your experimental system and should contain the level of random noise found in your data. Analysis of such simulated data allows one to determine whether one can, in fact, resolve the parameters. This is particularly important for similar distributions. We also recommend examination of the χ_R^2 surfaces, which reveal the sensitivity of the data to the described parameters, or the uncertainty range of the parameter values. The use of χ_R^2 surfaces is discussed below in Section 14.5.C. A similar procedure is followed for analysis of the TD data. The only difference is the use of the convolution integral (Section 14.3.D) in place of Eqs. [14.7]–[14.11].

14.3.C. Distance Distributions from Time-Domain Measurements

Distance distributions can also be recovered from the TD data. One example is shown in Figure 14.15. The protein was a mutant of troponin C from turkey skeletal muscle.[9] The structure of troponin C consists of two domains linked by a helical peptide (Figure 14.15). This protein is involved in muscle contraction, and its solution structure is sensitive to calcium.

Troponin C typically lacks tryptophan residues. This is fortunate because it allows insertion of tryptophan donors at any desired location by site-directed mutagenesis. In the mutant shown in Figure 14.15, a single tryptophan residue was placed at position 22 to serve as the donor. As is typical in the creation of mutant proteins, the tryptophan (W) was a conservative replacement for phenylalanine (F). A uniquely reactive site for the acceptor was provided by replacing asparagine-52 with a cysteine residue (N52C). This site was labeled with IAEDANS.

The upper panel in Figure 14.16 shows the time-dependent donor decays of the donor-alone protein and the donor–acceptor protein in the absence of bound calcium. Without acceptor, the intensity decay of trp-22 is close to a single exponential. The intensity decay of trp-22 be-

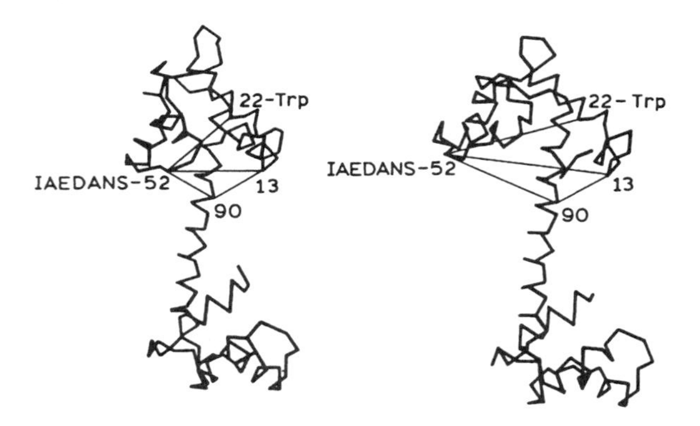

Figure 14.15. The crystal structure of turkey skeletal 2Ca-troponin C (*left*) and the 4Ca-troponin C model (*right*). The distances between positions 13 and 52, 22 and 52, and 52 and 90 are expected to increase significantly in the 4Ca model. From Ref. 9.

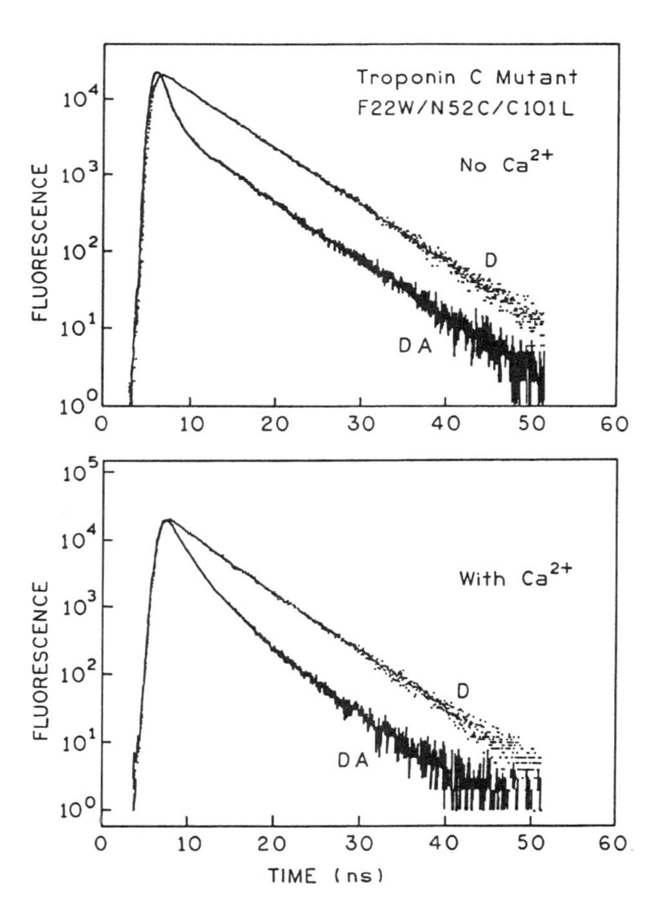

Figure 14.16. Tryptophan-22 intensity decays of troponin C mutant (F22W/N52C/C101L) without (D) and with (DA) IAEDANS on cysteine-52. *Top*: Without Ca^{2+}; *Bottom*: with Ca^{2+}. From Ref. 9.

comes strongly heterogeneous in the presence of IAEDANS acceptor at position 52. This change in the donor decay was used to recover the trp-to-IAEDANS distance distribution, which is seen to be centered near 11 Å and be about 10 Å wide (Figure 14.17). The close D–A distance and relatively wide distribution is one reason for the strongly heterogeneous intensity decay. However, close inspection of Figure 14.16 shows that at longer times the intensity decay of the D–A pair is the same as that for the donor-alone protein. This is due to less than 100% labeling by the acceptor. According to the authors of this study,[9] the extent of acceptor labeling is about 95%. Because the donor is strongly quenched by the nearby acceptor, even 5% donor-alone protein contributes significantly to the measured intensity decay. The distance distribution shown in Figure 14.17 was corrected for acceptor underlabeling (Section 14.7.B). The data in Figure 14.16 illustrate how the presence of only a small amount of donor-alone molecules can distort the data. Complete labeling by acceptor is perhaps the most important step in measuring a distance distribution.

Binding of calcium to the troponin C mutant results in a substantial change in the intensity decay (Figure 14.16, bottom). In this case, the decay of trp-22 appears to be mostly a single exponential in the absence of acceptor. The presence of acceptor results in a decrease in the donor lifetime and some apparent nonexponentiality in the decay. However, the heterogeneity due to the presence of acceptor is far less than observed in the absence of Ca^{2+}. Analysis of the data in terms of a distance distribution yields a calcium-dependent increase in the D–A distance and a restriction to a narrow range of distances (Figure 14.17).

It is informative to examine the multiexponential analyses of the donor decays (Table 14.2). In the absence of acceptor, the tryptophan decays are close to single exponentials, especially in the absence of Ca^{2+}. In the presence of the IAEDANS acceptor, the intensity decay becomes strongly heterogeneous. The extent of heterogeneity can be judged from the χ_R^2 values for the single-decay-time fits. The wider distribution in the absence of Ca^{2+} results in a higher value of χ_R^2 ($\chi_R^2 = 192$) than in the presence of Ca^{2+} ($\chi_R^2 = 32$), where the distribution is narrower. The intensity decays in the absence and presence of Ca^{2+} reveal components with decay times close to that of the donor-alone protein. The α_i values typically represent the fractions of the molecules which display the respective decay times. Thus, the values in Table 14.2 suggest that 6–7% of the troponin C molecules lack a bound acceptor, in agreement with the known extent of labeling.[9]

The fractional intensities of each component (f_i values) represent the fractional contributions to the steady-state intensity or the time-integrated intensity decay. The frac-

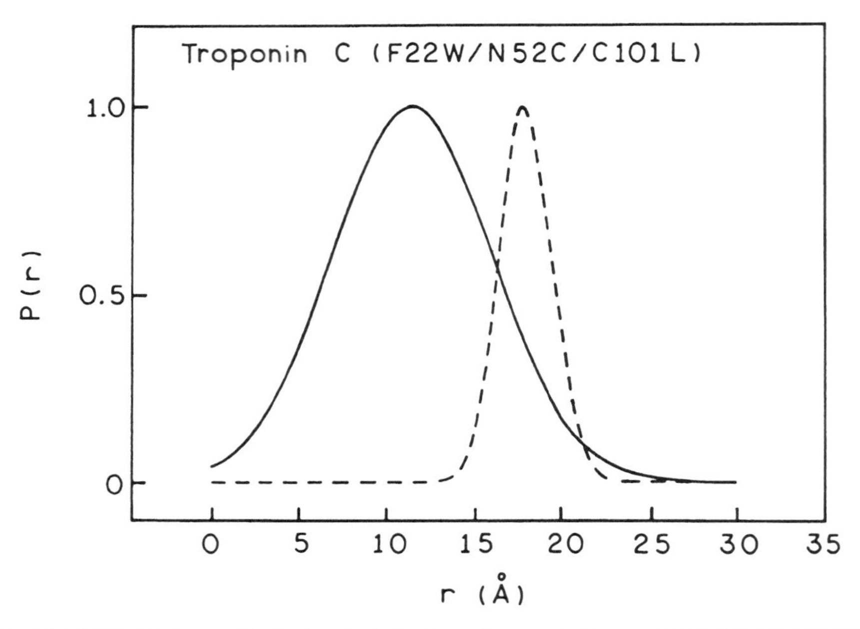

Figure 14.17. Trp-22-to-IAEDANS-52 distance distributions in skeletal muscle troponin C mutant (F22W/N52C/C101L) in the absence (———) and in the presence (– – –) of bound calcium. From Ref. 9.

tional intensity of the component with the donor-alone decay time is nearly 39% in the absence of Ca^{2+} and 17% in the presence of Ca^{2+}. The larger fractional contribution in the absence of Ca^{2+} is due to the larger extent of energy transfer and a resulting increased contribution of the donor-alone molecules. This large contribution (39%) of the donor-alone molecules, when present at a molecular fraction of 7%, illustrates how the data from strongly quenched samples are easily corrupted by small impurity contributions or underlabeling by the acceptor.

The distance distributions recovered for the troponin C mutant can be interpreted in terms of what is known about the structure of troponin C. Figure 14.15 shows the crystal structures of troponin C with two (left) and four (right) bound calciums. Saturation with calcium results in a displacement of the peptide loop containing cys-52 (in the mutant protein) away from the rest of the protein. Apparently, with less bound Ca^{2+}, the loop has considerable flexibility in solution and can exist in a variety of conformations. Saturation with Ca^{2+} results in a more rigid struc-

Table 14.2. Tryptophan Intensity Decays in Troponin C F22W/N52C/C101L[a]

	τ_1 (ns)	α_1 (f_1)[b]	τ_2 (ns)	α_2 (f_2)	τ_3 (ns)	α_3 (f_2)	χ_R^2
Donor alone	5.74	1.00					1.4
	5.27	0.58	6.3	0.42			1.6
Donor–acceptor	1.81	1.0					192.3
(R_0 = 22.1 Å)	5.60	0.08	0.76	0.92			1.5
	5.74	0.07	1.28	0.24	0.50	0.67	1.3
		(0.39)		(0.29)		(0.32)	
Donor alone + Ca^{2+}	4.81	1.0					2.7
	5.39	0.59	3.56	0.41			1.1
Donor–acceptor + Ca^{2+}	2.29	1.0					31.8
(R_0 = 23.9 Å)	4.06	0.18	1.59	0.82			1.9
	5.16	0.060	2.18	0.57	0.79	0.38	1.2
		(0.17)		(0.67)		(0.16)	

[a]Ref. 9.
[b]Values in parentheses are the fractional intensities calculated from $f_i = \alpha_i\tau_i/\Sigma_j\alpha_j\tau_j$.

ture, in which the D–A distribution is narrow because only one conformation is allowed.[9]

These results illustrate the usefulness of time-resolved measurements for detecting conformational heterogeneity in proteins. The short decay times of the tryptophan residues do not allow time for the conformation to be averaged during the excited-state lifetime, so that the individual conformations contribute to a complex donor decay. This is a somewhat unique feature of fluorescence, as most physical methods reveal an average conformation.

As a final point, we wish to stress the importance of time-resolved data for measuring a single distance as well as for measuring distance distributions. If one measures only the steady-state intensities, then one cannot detect the presence of unlabeled or underlabeled molecules. It is easy to be misled by steady-state data that satisfy preconceived notions about the system. In contrast, the time-resolved donor decays contain more information and will often reveal inconsistencies due to incomplete labeling, multiple labeling, or impurity fluorescence. For this reason, measurement of the time-resolved donor decays is recommended for all distance measurements.

14.3.D. Analysis of Distance Distribution from Time-Domain Measurements

The procedure for recovery of the distance distribution from the TD data is analogous to that used for the FD data. The intensity at any time is described by Eq. [14.4]. This decay law is used with the convolution integral and the measured instrument response function, $L(t)$, to obtain calculated values of $I_{DA}(t)$ for any given time t_j,

$$I_{DA}^{c}(t_j) = \sum_{k=1}^{j} L(t_k) I_{DA}(t_j - t_k)\Delta t \qquad [14.21]$$

Equation [14.21] can be understood as the intensity at time t_j being the sum of the intensities due to excitation by a series of δ-functions which, in total, represent the light pulse $L(t)$.

These calculated values are then compared to the measured data $I_{DA}^{m}(t)$ to minimize χ_R^2. The experimental $I_{DA}^{m}(t)$ data are a convolution of the impulse response function $I_{DA}(t)$ with the instrument function $L(t)$. As for the FD measurements, a separate measurement of the donor decay is needed, and this decay law is held constant in the distance distribution analysis.

Irrespective of the type of measurement, TD or FD, it should be noted that the limits of the integral (Eqs. [14.9] and [14.10]) can also be regarded as a variable parameter. Hence, one can evaluate if there exists a distance of closest D–A approach (r_{min}) or a maximum D–A distance (r_{max}) that is allowed by the experimental data. If such r_{min} and r_{max} are used, then one must be careful to normalize the $P(r)$ function to unity.

14.3.E. Domain Motion in Proteins

For troponin C, we showed how energy transfer could be used to measure a conformational distribution within one domain. RET has also been used to measure the range of distances between domains. For many enzymes, it is thought that the substrates cause domains to move relative to each other. One example is phosphoglycerate kinase, which consists of two domains (Figure 14.18). Reactive cysteine residues were introduced at positions 135 and 290 to provide sites for labeling on opposite domains of the proteins. These sites were labeled with an IAEDANS donor and a 5-iodoacetamidofluorescein (IAF) acceptor.[26] The time-resolved donor decays were used to recover the distance distribution (Figure 14.18). In the absence of substrate, a wide D–A distribution was observed. Binding of 3-phosphoglycerate (3-PG) resulted in a displacement of the D–A distance to larger values. Binding of ATP resulted in a still larger distance and a more narrow distribution. This effect is analogous to that observed for tropinin C, where the more extended structure was constrained to a more uniform distribution of conformations.

The donor decays were also used in an attempt to measure diffusion motions between the subunits.[26] As described in Section 14.5.E, longer donor decay times are needed to measure site-to-site domain motions in proteins. Phosphoglycerate kinase has also been used as a model to study protein folding.[27,28] In this case, cysteine residues were introduced throughout the protein to provide probes at different sites in order to study the various steps on the folding pathway.

14.3.F. Distance Distribution Functions

Several different mathematical functions have been used to describe D–A distributions. The Gaussian distribution (Eq. [14.3]) has been most commonly used. This distribution has been used multiplied by different elements, $2\pi r$ and $4\pi r^2$, presumably to account for different volume elements available to the acceptor. Inverted parabolas and other functions have also been used.[1,29] These different functions result in different parameter values, but the overall distant-dependent probabilities look similar.[26,30] In general, there is only enough information to recover an estimate of the distribution, and finer details are not resolved. For most purposes, the simple Gaussian (Eq. [14.3]) is adequate and best suited to describe D–A distributions in macromolecules. In some cases, bimodal Gaus-

0.06
0.04
0.02
0.00
PROBABILITY
0
20
40
60
80
100
DISTANCE (Å)
C 135
C 290

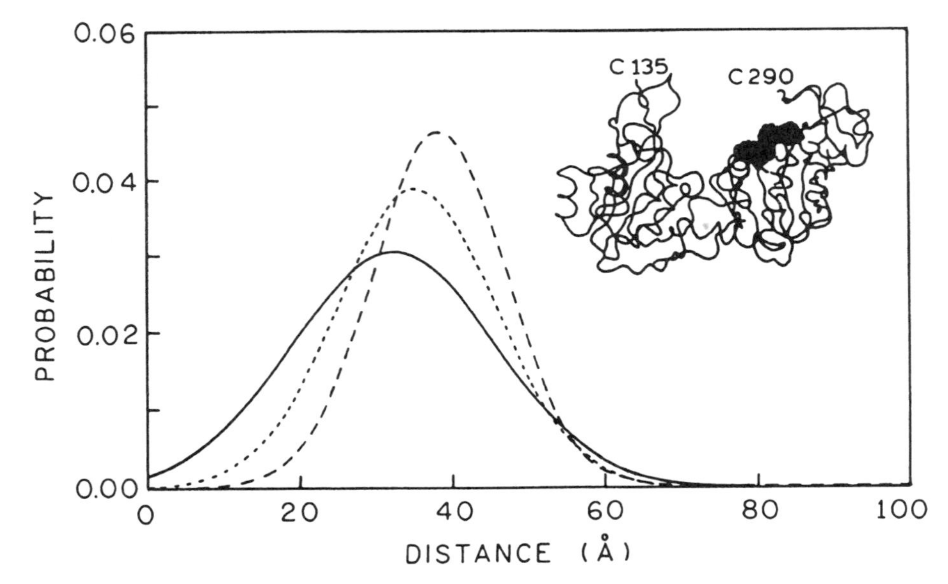

Figure 14.18. Distance distributions between domains in phosphoglycerate kinase, without substrate (———), with bound 3-phosphoglycerate (···), and with bound ATP (- - -). From Ref. 26.

sians have been used to describe a protein in multiple conformational states.[31]

The Lorentzian distribution has also been used,[4] but, in our experience, it results in artificially narrow half-widths. This is because the amplitude of a Lorentzian decreases slowly with distance from the $\bar{r}$ value. As a result, the half-width is determined more by the ability of the time-resolved data to exclude these smaller and larger *r* values than by the actual width of the distribution. For this reason, half-widths determined using a Lorentzian model are invariably smaller than those found using a Gaussian model.

14.4. DISTANCE DISTRIBUTIONS IN A GLYCOPEPTIDE

Whereas fluorescence is widely used to study proteins and membranes, there are relatively few studies of polysaccha-

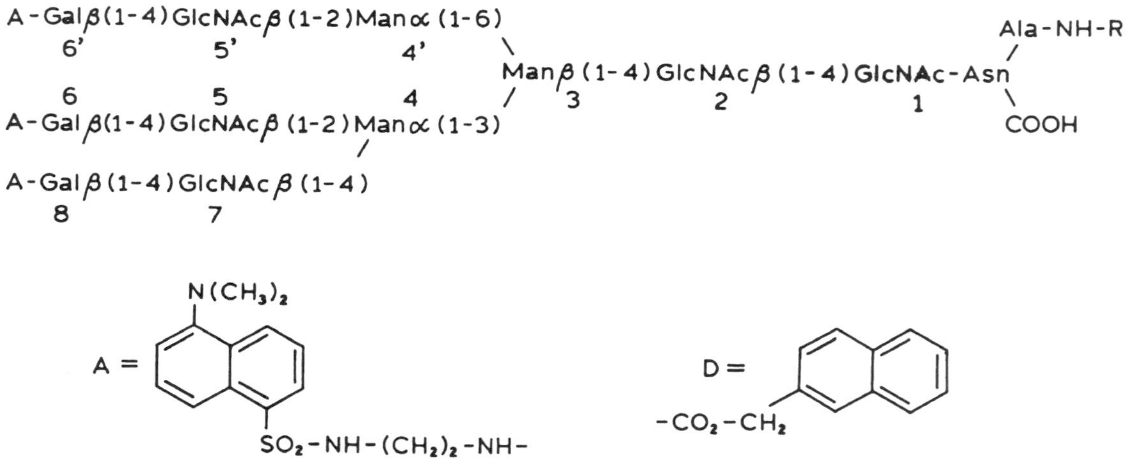

A-Galβ(1-4)GlcNAcβ(1-2)Manα(1-6)
6' 5' 4'
Manβ(1-4)GlcNAcβ(1-4)GlcNAc-Asn
3 2 1
Ala-NH-R
COOH
6 5 4
A-Galβ(1-4)GlcNAcβ(1-2)Manα(1-3)
A-Galβ(1-4)GlcNAcβ(1-4)
8 7
N(CH3)2
A =
SO2-NH-(CH2)2-NH-
D =
-CO2-CH2

Figure 14.19. Structure of triantennary glycopeptide fluorescence energy-transfer probes. D = naphthyl-2-acetyl; A = dansylethylenediamine. GP-A-DanNap glycopeptides all contain D and either A6′, A6, or A8. Revised and reprinted, with permission, from Ref. 32, Copyright © 1991, American Chemical Society.

rides. This is because it is difficult to obtain polysaccharides with a homogeneous structure, and even more difficult to obtain polysaccharides labeled with donor and acceptor fluorophores. Complex oligosaccharides are found in many glycoproteins, including antibodies, hormones, and receptors, yet little is known about their solution conformation.

The elegant experiments by Rice *et al.*[32] illustrate the potential of time-resolved fluorescence to resolve complex conformational distributions. An oligosaccharide was labeled with a naphthyl-2-acetyl donor and a dansylethylenediamine acceptor at one of three locations (Figure 14.19). The authors used both time-resolved and steady-state measurements of RET. The steady-state data provide an important control in all energy-transfer measurements. The transfer efficiency determined from the steady-state and time-resolved data should be the same. If not, something is wrong with the sample or the analysis. Three donor- and acceptor-labeled oligosaccharides were studied, and the data resulted in the remarkable resolution of the multiple conformations existing in aqueous solution (Figure 14.20). The Lorentzian model was used to describe the distance distributions. The distance distributions to two of the acceptor sites were bimodal, suggesting that the oligosaccharide branch could fold back toward the donor site. In contrast, the central branch existed in only one conformation and was unable to fold back toward the donor (Figure 14.19). These results represent the most detailed resolution to date of a conformational distribution in any type of biopolymer.

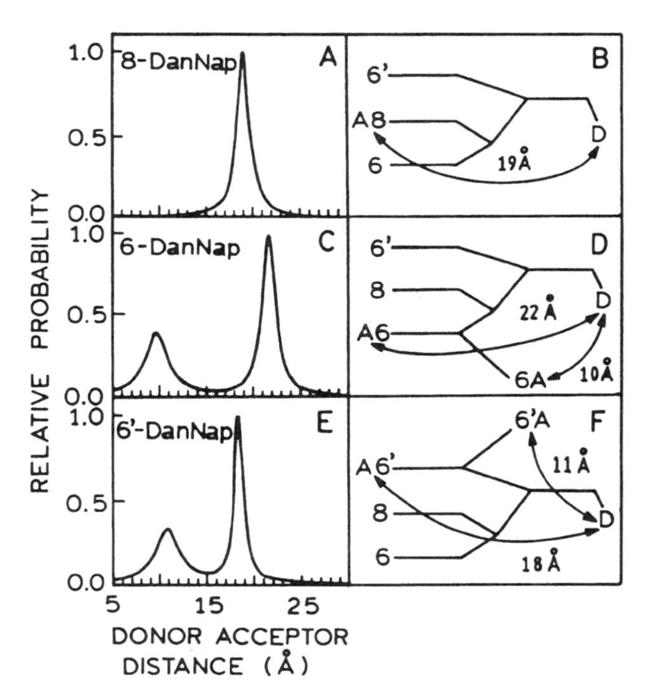

Figure 14.20. Donor/acceptor population distributions. Lorentzian fits of the donor decay curves are shown for GP-8-DanNap (A, B), GP-6-DanNap (C, D), and GP-6′-DanNap (E, F). The plots on the left indicate the probability of a donor/acceptor population occurring at different distances in the triantennary glycopeptide. The ratio of populations for GP-6-DanNap and GP-6′-DanNap were determined to be approximately 60% of the extended conformation and 40% of the folded form. In the illustrations on the right, the double-headed arrows indicate each D–A distance determined from the bimodal Lorentzian fit. Revised and reprinted, with permission, from Ref. 32, Copyright © 1991, American Chemical Society.

14.5. EFFECT OF DIFFUSION FOR LINKED D–A PAIRS

In all the preceding sections we considered the donors and acceptors to be static in space and not to move relative to each other during the excited-state lifetime. However, the nanosecond timescale of the donor lifetime can allow time for donor-to-acceptor diffusion, which is defined by the mutual diffusion coefficient $D = D_D + D_A$ (Figure 14.21). For the lifetimes of 5–20 ns displayed by most fluorophores, there is little translational displacement during the lifetime of the donor, so that the data can be analyzed in terms of a static distance distribution; this was seen in Section 14.2.A for DNS-(gly)$_6$-TrpNH$_2$, in which the Trp donor has a lifetime of 5.3 ns. However, in more fluid solvents, diffusion can influence the extent of energy transfer. The effects of diffusion are complex to calculate, and numerical methods are typically used. We will attempt to illustrate these effects with a minimum of mathematical detail.

14.5.A. Simulations of RET for a Flexible D–A Pair

The effects of donor-to-acceptor diffusion on the donor decay can be seen from the simulated data for a flexible D–A pair. As described by Haas and co-workers,[3,29] the excited-state population is given by

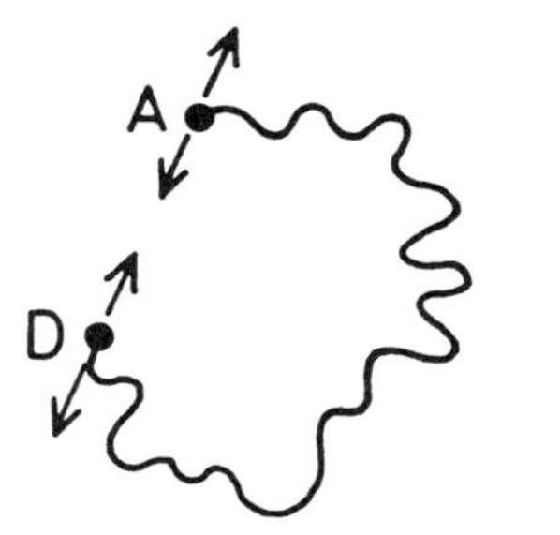

Figure 14.21. Donor–acceptor pair with diffusion.

$$\frac{\partial \overline{N}^*(r,t)}{\partial t} = -\frac{1}{\tau_D}\left[1 + \left(\frac{R_0}{r}\right)^6\right]\overline{N}^*(r,t) + \frac{1}{N_0(r)}\frac{\partial}{\partial r}\left(N_0(r) D \frac{\partial \overline{N}^*(r,t)}{\partial r}\right) \quad [14.22]$$

where $\overline{N}^*(r, t)$ is the distribution of distances for the excited D–A pairs normalized by the time-zero distribution $N_0(r)$. At $t = 0$, the D–A distribution is the equilibrium distribution. However, the distribution changes following excitation owing to more rapid transfer between the more closely spaced D–A pairs. Equation [14.22] is basically the diffusion equation with an added sink term to account for loss of the excited acceptor. The time-dependent intensity decay is given by

$$I_{DA}(t) = I_D^0 \int_{r_{min}}^{r_{max}} P(r)\overline{N}^*(r,t)\, dr \quad [14.23]$$

where r_{min} and r_{max} are chosen to span the accessible distances. Calculation of $I_{DA}(t)$ in the presence of diffusion is moderately complex, and the details can be found elsewhere.[33–35]

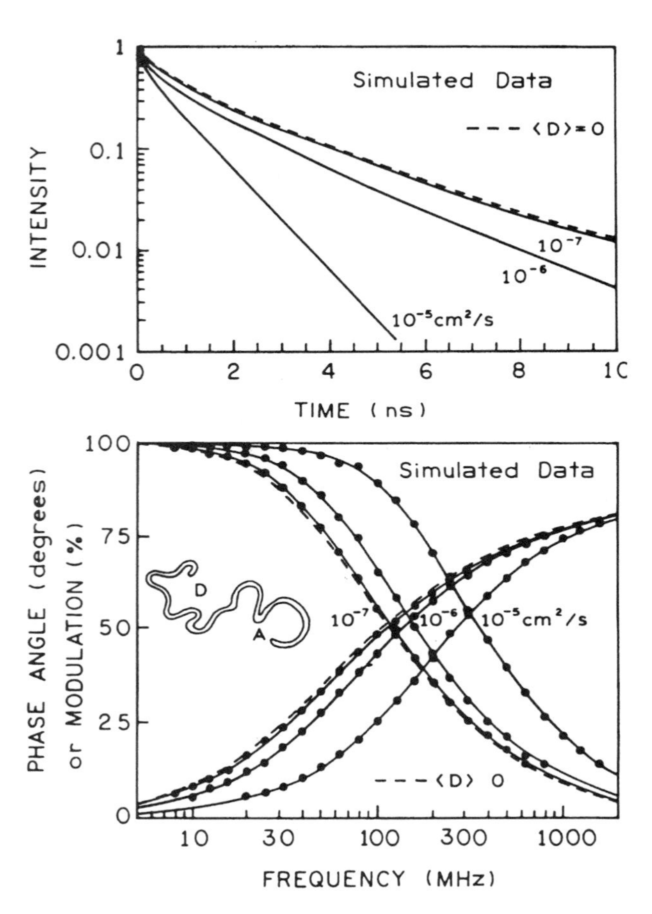

Figure 14.22. Simulated data illustrating the effect of donor-to-acceptor diffusion in a linked D–A pair as seen in the time domain (*top*) and the frequency domain (*bottom*). The assumed values of the diffusion coefficient are labeled on the solid curves; the other assumed parameter values are given in the text. The dashed curves show the time-dependent intensity decay and the frequency response for $D = 0$. Revised from Ref. 34.

To understand the expected effects of diffusion, it is valuable to examine simulated data. Simulated TD and FD data for an assumed donor decay time of 5 ns are shown in Figure 14.22. The distribution of distances at time $t = 0$ was assumed to be a Gaussian with $\bar{r}$ = 20 Å, hw = 15 Å, and a Förster distance of 25 Å. If the value of D is 10^{-7} cm²/s, there is little time for motion during the excited-state lifetime. This can be seen from the near overlap of the simulated data with $D = 0$ and $D = 10^{-7}$ cm²/s. The mean donor-to-acceptor displacement $(\Delta x^2)^{1/2}$ can be calculated to be $(2D\tau_D)^{1/2}$ = 3.2 Å. While significant, this distance is about half the width of a donor or acceptor molecule and is small compared to the assumed $\bar{r}$ of 20 Å. Because of the range of D–A distances, the FD data are spread out along the frequency axis, and the intensity decay is complex. As the diffusion coefficient increases, the intensity decay becomes more rapid, and the frequency response shifts to higher frequency. Both effects are indicative of an increasing amount of energy transfer. It is also important to notice that the intensity decays become more like a single exponential with increasing rates of diffusion. These simulated data show that D–A diffusion increases the extent of energy transfer and alters the shape of the donor intensity decay. Hence, the FD or TD data can, in principle, be used to recover the rate of D–A diffusion as well as the distribution of distances.

14.5.B. Experimental Measurement of D–A Diffusion for a Linked D–A Pair

Understanding of the role of diffusion in energy transfer is best understood by an example. Figure 14.23 shows the emission spectra for a linked indole–dansyl D–A pair TU2D.[34] The donor-alone control molecule is TMA. This flexible D–A pair was found to have the same distance distribution in propylene glycol irrespective of temperature. However, the donor is quenched to a greater extent at higher temperatures, which is due to D–A diffusion during the excited-state lifetime. During the 6-ns donor lifetime, the expected D–A displacement is $(2D\tau)^{1/2}$ = 11 Å if the diffusion coefficient is near 10^{-6} cm²/s. Hence, there is

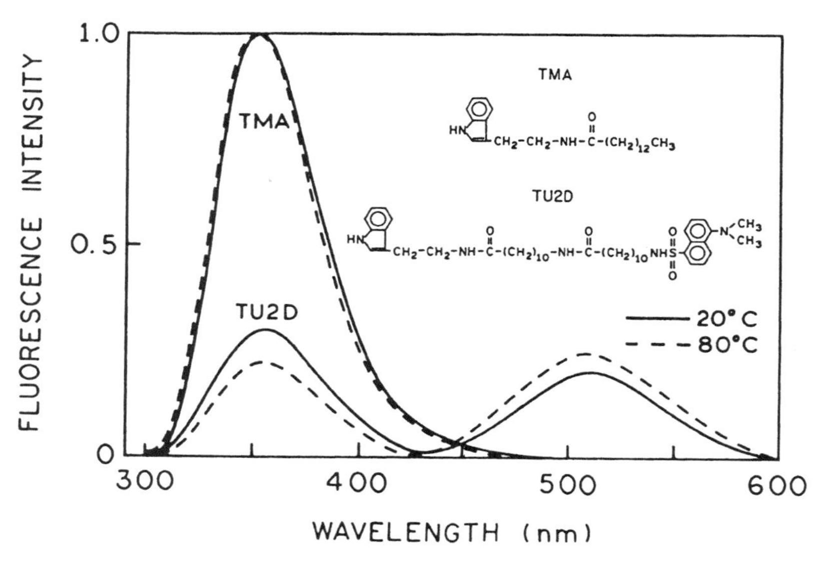

Figure 14.23. Emission spectra of TMA (donor alone) and TU2D (donor–acceptor pair) in propylene glycol at 20 (———) and 80 °C (– – –). The emission spectra of TU2D are normalized to the intensity of TMA at each temperature.

time for donor and acceptor to approach each other while the donor is excited. This results in an increased rate of energy transfer for each D–A pair, and an overall higher transfer efficiency. The increased extent of energy transfer is also seen from the increased acceptor intensity.

FD intensity decays of the donor and the reconstructed TD decays are shown in Figure 14.24. A more rapid decay is seen at higher temperature, and the decay is more like a single exponential. The distance distributions recovered at each temperature are the same (Table 14.3), so that the differences in the intensity decays are due to diffusion. Since the extent of diffusion in propylene glycol is minimal at 20 °C, the differences in the frequency responses at 20 °C and 80 °C can be regarded as the contribution of diffusion. Clearly, D–A diffusion has a significant effect on the FD data.

The effect of diffusion on the shape of the frequency response can be judged by the χ_R^2 values. The data were fit to the model which allows D–A diffusion and to the same model with the diffusion coefficient set equal to zero (Table 14.3). When $D = 0$, the values of χ_R^2 increase at higher temperatures, indicating that the donor decay has become more like a single exponential. In fact, the tendency toward a single exponential can be seen from the recovered values of the half-width. These apparent hw values become smaller at higher temperature. As described in Section 14.5.D, the trend in apparent parameter values can yield useful information about the behavior of complex systems.

Why does D–A diffusion result in increased energy transfer? A more in-depth understanding of diffusion and RET can be obtained by examining the distance distribution in the excited-state. The excited-state distribution evolves in time following pulsed excitation. Such distributions are shown in Figure 14.25 for the simulated data in Figure 14.22 with $D = 10^{-5}$ cm^2/s. The $P(r)$ distributions are shown at $t = 0$ and $t = 1$ ns, in the absence and in the presence of diffusion. In the absence of diffusion, the excited-state population shifts to longer distances at $t = 1$ ns. This occurs because of the distance dependence of k_T, which causes the more closely spaced D–A pairs to transfer energy more rapidly. For rapid D–A diffusion, the population of D–A pairs at shorter distances is replenished by diffusion, resulting in an increase in the closely spaced D–A population, which displays more rapid energy transfer. It is this replenishment of the closely spaced D–A pairs by diffusion which results in increased energy transfer with increasing rates of diffusion.

14.5.C. χ_R^2 Surfaces and Parameter Resolution

The simulations in Figure 14.22 and data for a flexible and diffusing D–A pair in Figure 14.24 provide a good example of how the ability to determine a parameter value depends on the information content of the data. Recall that parameter values are often correlated, so that the values of the goodness-of-fit parameter χ_R^2 depends on the entire set of parameters, not just the value of a single parameter.

Table 14.3. Distance Distribution and End-to-End Diffusion Analysis for TU2D in Propylene Glycol

T (°C)	η (cP)	τ_D (ns)	R_0 (Å)	$\bar{r}$ (Å)	hw (Å)	log D (cm²/s)	χ_R^2
20	48.3	6.59	25.61	20.0	17.2	−6.95	1.8
				20.6	14.9	(−∞)	2.3
40	16.7	6.03	25.16	20.0	16.5	−6.67	2.4
				20.8	12.9	(−∞)	3.5
60	5.8[a]	5.38	24.72	20.5	15.6	−6.32	2.1
				20.9	8.8	(−∞)	5.7
80	2.0[a]	5.00	24.36	19.7	17.5	−5.75	3.1
				20.3	7.1	(−∞)	10.1

[a]Obtained from extrapolation.

More specifically, for highly correlated parameters, a parameter value can be changed with a minimum effect on χ_R^2 if there is a compensating change in the value of the correlated parameter. In the case of distance distributions, there is usually strong correlation between the three parameters $\bar{r}$, hw, and D.

The information content of the data can be judged by examining the χ_R^2 surfaces (Figure 14.26). For the D–A pair TU2D, these surfaces are constructed by holding the value of D fixed at various values surrounding the optimal value. The least-squares analysis is then rerun to again minimize χ_R^2 with different values of $\bar{r}$ and hw. These values of $\bar{r}$ and hw are not shown, but they will typically be different from the values yielding the minimum of χ_R^2. The extent to which χ_R^2 depends on D indicates the certainty with which D is known.

For TU2D in propylene glycol at higher temperatures, diffusion has a significant influence on energy transfer. As a result, the χ_R^2 surface for the diffusion coefficient at 80 °C (Figure 14.26) is a rather steep parabola. At lower

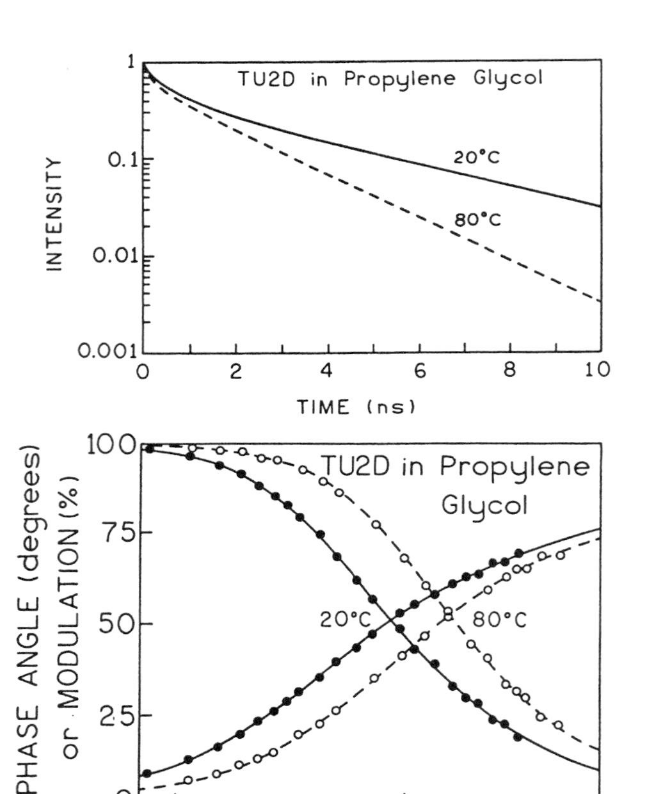

Figure 14.24. *Bottom*: FD donor decays of TU2D in propylene glycol at 20 (———) and 80 °C (– – –). *Top*: Time-dependent decays reconstructed from the FD data. Reprinted, with permission, from Ref. 34, Copyright © 1994, American Society for Photobiology.

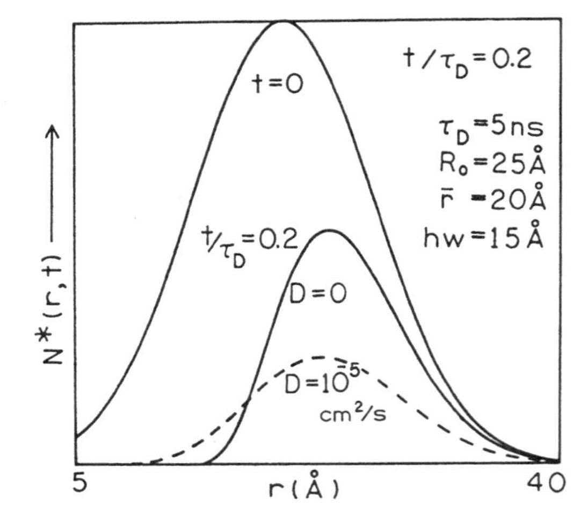

Figure 14.25. Simulated excited-state donor distributions in the absence and in the presence of diffusion. The assumed parameter values are shown on the figure. Reprinted, with permission, from Ref. 34, Copyright © 1994, American Society for Photobiology.

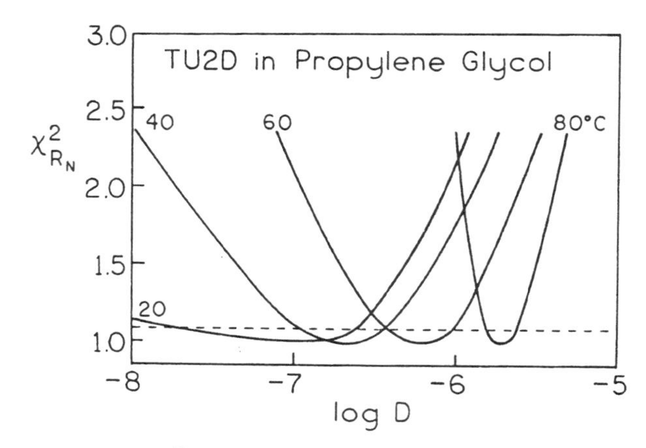

Figure 14.26. χ^2_R surfaces for the donor-to-acceptor diffusion coefficient of TU2D. Reprinted, with permission, from Ref. 34, Copyright © 1994, American Society for Photobiology.

temperatures, D–A diffusion contributes less to energy transfer in TU2D. Hence, the data contain less information about the value of the diffusion coefficient D. This can be seen from the χ^2_R surface (Figure 14.26), which is much flatter in the case of propylene glycol at 20 °C. The broad χ^2_R surface is a consequence of the short decay time of indole ($\tau_D \simeq 6.0$ ns) compared to the rate of diffusion. If the decay time of the donor were longer, then diffusion would have a large influence on RET, D would be better resolved from the data, and the χ^2_R surface would be sharper. In Chapter 4 we described how the uncertainty in a parameter value could be found from the F-statistic, which is shown as the horizontal dashed line in Figure 14.26.

The χ^2_R surfaces for D also illustrate how the uncertainty range of a parameter need not be symmetrical about the optimal value. For the case of TU2D in propylene glycol, the data exclude larger values of D but are less able to exclude smaller values of D. At the limit of no diffusion, energy transfer is not influenced by diffusion and the data contain no information on the value of D. Hence, χ^2_R will not depend on the value of D below a lower limit. This effect is seen by the flatness of the χ^2_R surfaces at lower values of D.

14.5.D. Diffusion and Apparent Distance Distributions

The limited information content of the time-resolved data often allows one to obtain satisfactory fits using an incorrect molecular description of the system. In such cases, the parameter values are only apparent values because of the incorrect molecular description. This is illustrated by the effects of diffusion on the apparent distance distribution that is obtained when the data are analyzed without consideration of diffusion.[36,37] In one such example, distance distributions were recovered for a naphthalene–dansyl D–A pair, NU2D, in propylene glycol (Figure 14.27). In this case, the donor lifetime in the absence of acceptor,was 40–65 ns, depending on temperature. This lifetime allowed time for D–A diffusion. The FD data could be well fit to the Gaussian distance distribution model without consideration of diffusion (not shown).[37] The apparent D–A distribution appears to become sharper and more closely spaced at higher temperatures. (An even sharper apparent distance distribution was found in the less viscous solvent method.) However, the actual distance distribution did not change significantly with temperature.[37] The distribution only appears to become sharper and more closely spaced because diffusion brings the D–A pairs into closer proximity during the excited-state lifetime of the donor.

This example illustrates several important points. First, if not accounted for in the analysis, D–A diffusion will result in apparently sharper and closer D–A distributions. This explains the tendency toward a more single-exponential decay seen in Figure 14.22 at higher rates of diffusion. Second, this example illustrates the importance of considering the reasonableness of a result before accepting the answer. In the case of NU2D, the results in Figure 14.27 should raise the question as to why the distribution would

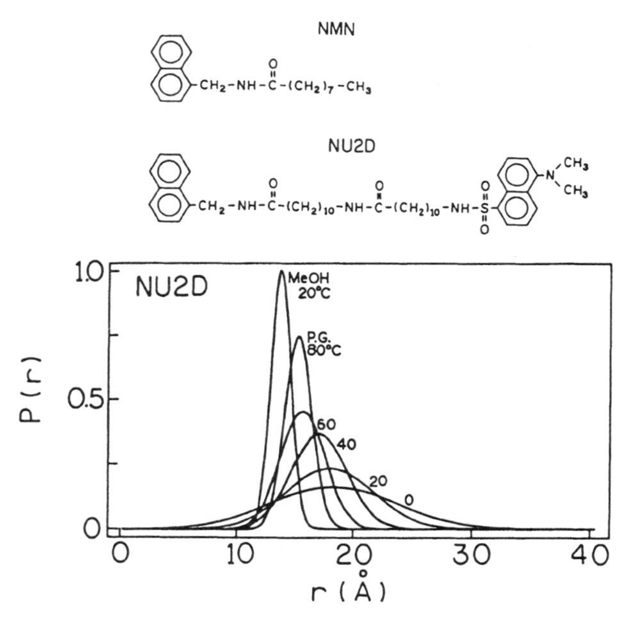

Figure 14.27. Apparent distance distributions for a naphthalene–dansyl D–A pair (NU2D) in the presence of diffusion. NMN is the donor-alone control molecule. Reprinted from Ref. 37, Copyright © 1990, with permission from Elsevier Science.

be closer and sharper at higher temperatures. Since there was no reason for this to occur, it was important to consider what factors (i.e., diffusion) had not been accounted for in the analysis.

14.5.E. RET and Diffusive Motions in Biopolymers

Time-resolved RET has been proposed as a method to measure domain motions in proteins[26,38] and lateral diffusion in membranes. Motions between domains in proteins are known to occur in many proteins and are thought to be essential for protein function.[39] Unfortunately, measurements of domain motions in proteins have been largely unsuccessful owing to the short decay times of the donors, and diffusion in membranes has only been accomplished using pyrene derivatives which display decay times of over 100 ns.[40,41] From the perspective of protein dynamics, the nanosecond decay times of most fluorophores are too short for measurement of site-to-site motions. However, a few measurements have appeared. Ribonuclease A was studied in a partially unfolded state.[42] In this case the carboxy terminus was labeled with a naphthylalanine donor which displayed a donor decay time of 25–44 ns. The acceptor was a coumarin placed on one of three lysine residues. The TD donor decays yielded site-to-site diffusion coefficients of 2 Å^2/ns in the denatured state, but the motions were mostly undetectable in the native state. Similar results were found for a zinc finger peptide, where site-to-site diffusion with D near 10 Å^2/ns was found in the random state, but D decreased to unmeasurable values in the presence of added zinc, which induced folding.[20,21] Haran *et al.*[26] attempted to measure domain motions in phosphoglycerate kinase using RET, but the rate of domain flexing was too slow to allow estimation of the diffusion coefficient with a 20-ns donor lifetime.

Fortunately, this limitation in measuring protein dynamics can be circumvented by using donors with longer decay times. If the decay times are increased from 5 ns (Figure 14.22) to 5 μs, the measurable diffusion coefficients decrease by a factor of 10^3, so that the time-resolved RET becomes sensitive to diffusion coefficients of 10^{-10}–10^{-8} cm^2/s, instead of 10^{-7}–10^{-5} cm^2/s (Figure 14.22). Donors with longer decay times such as the lanthanides europium and terbium display 0.2- to 0.4-ms lifetimes,[43] and the transition-metal–ligand complexes display lifetimes from 100 ns to 10 μs.[44] These complexes are particularly promising because they also display polarized emission and can thus be used to measure rotational motions. The spectral properties of transition-metal–ligand complexes are described in Chapter 20.

14.6. DISTANCE DISTRIBUTIONS IN NUCLEIC ACIDS

14.6.A. Double-Helical DNA

Time-resolved RET measurements have also been applied to nucleic acids.[45–49] Somewhat surprising results were obtained from the complementary oligonucleotides 5′-TGAGGCCTA-3′ and 5′-TAGGCCTCA-3′. Each strand had aminohexylphosphate linkers on the 5′-end (Figure 14.28). One strand was labeled on the 5′-end with fluorescein, and the other with tetramethylrhodamine.[45] One would expect the D–A distance to have a single value, given the rigidity of double-helical DNA.

Intensity decays of the single-strand donor oligonucleotide and of the donor–acceptor duplex are shown in Figure 14.29. Quenching of the donor by RET is clearly seen from the more rapid donor decay in the duplex. However, the data could not be fit to a single distance, as seen from the residuals (A in Figure 14.29) and $\chi_R^2 = 4.87$. Fitting the donor decays required a rather wide distance distribution with a value of $\bar{r}$ near 37 Å and a width near 13 Å (Figure 14.29). This result was explained as due to the aminohexyl linker (Figure 14.28) and illustrates the need for caution in interpreting distance distributions as a property of the labeled molecule. In this case, the distance distribution probably has its origin with the probe labeling procedure. Also, donor-to-acceptor diffusion was not considered in the distance distribution analysis, so it is likely that the actual D–A distribution is wider than shown in Figure 14.29.

These time-resolved data illustrate the experimental difficulty in resolving distance distributions. The fits to a fixed distance and to the Gaussian distribution are barely resolvable within the linewidth of the figure (Figure

D – NH(CH$_2$)$_6$ – TGAGGCCTA – 3′
3′ – ACTCCGGAT – (CH$_2$)$_6$NH – A

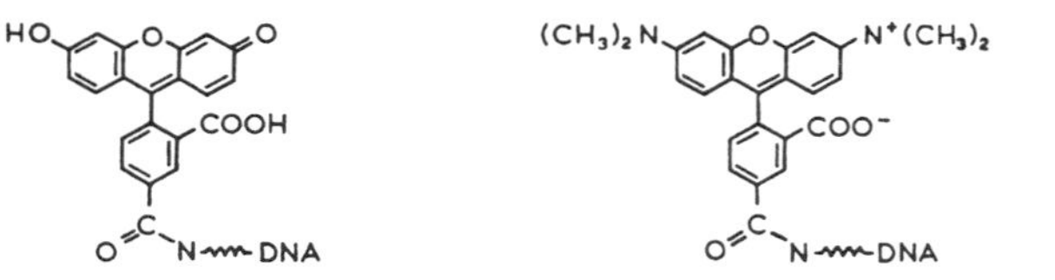

HO O COOH C O N DNA
(CH$_3$)$_2$N N$^+$(CH$_3$)$_2$ COO$^-$ C O N DNA

D : 5-carboxyfluorescein A : 5-carboxytetramethylrhodamine

Figure 14.28. Complementary DNA oligomers labeled with a fluorescent donor and a tetramethylrhodamine acceptor. Revised from Ref. 45.

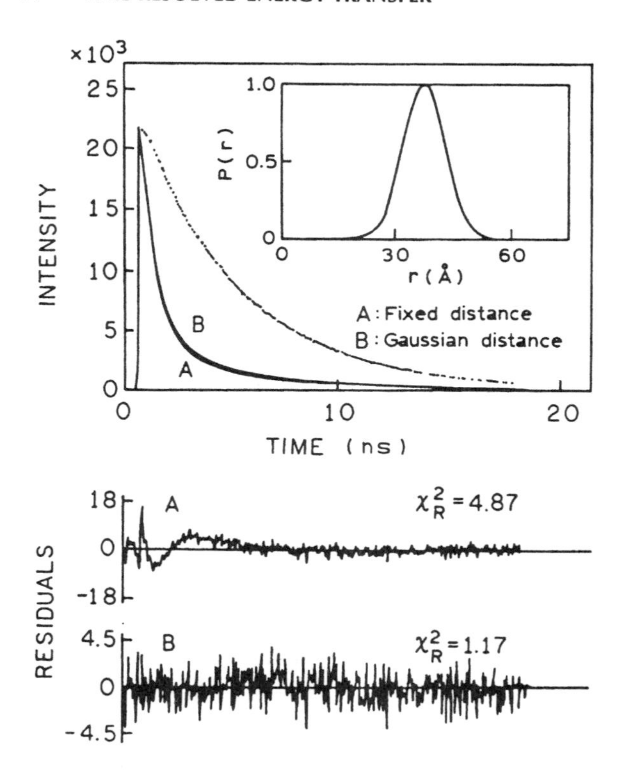

Figure 14.29. Intensity decays of the single-strand donor oligonucleotide and of the donor-acceptor duplex shown in Figure 14.28. The dotted curve shows the intensity decay of the fluorescein donor oligonucleotide alone, and the solid curves show the intensity decay of the donor when bound to the complementary rhodamine-labeled oligonucleotide. A and B represent the fits with a single distance and a Gaussian distribution, respectively. Inset: Recovered distance distribution. Revised from Ref. 45.

14.29). The distribution is determined by the small number of counts present in the residuals to the single-distance fit.

14.6.B. Four-Way Holliday Junction in DNA

Genetic recombination involves exchange of homologous strands between two duplex DNA molecules. This process occurs at a four-way branch point called a Holliday junction (Figure 14.30). This is a mobile structure which can propagate along the DNA strands. These sites are ultimately removed by cellular enzymes, yielding either the parental or recombinant DNA product. Time-resolved energy transfer has been used to study one such junction.[46–49]

The DNA sequence and sites of labeling (I–IV) are shown in Figure 14.30.[47] The same donor and acceptor chemistry was used as for the duplex DNA in Figure 14.28. Examination of the oligonucleotide sequence (Figure 14.30, right) suggests a rather symmetrical structure. However, the time-resolved data reveal different rates of energy transfer when the donor (F) and acceptor (R) are at sites I and II or sites I and IV. The data suggest that sites I and II are further apart than sites I and IV (Figure 14.31). This expectation is confirmed by the recovered distance distributions. The time-resolved data also reveal a difference in the distance distributions between sites I and II and sites I and IV. There is a narrow distance distribution between one pair of ends and a wide distribution between the other pair of ends (Figure 14.31). Results from other laboratories

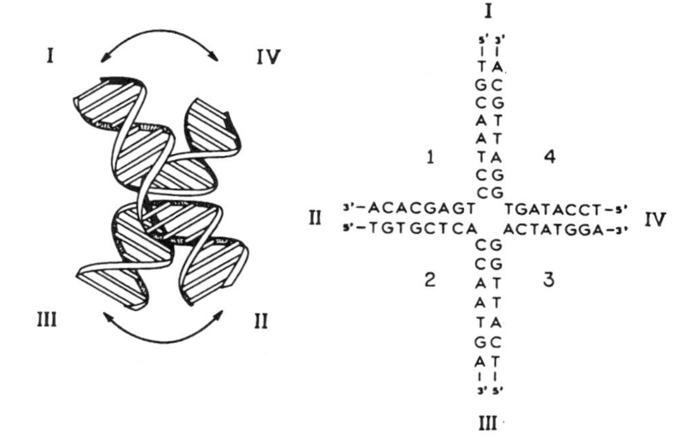

Figure 14.30. DNA Holliday junction. *Right*: Nucleotide sequence; *left*: solution structure. The 5′ ends of each strand contain an aminohexyl moiety for covalent attachment of donor and acceptor dyes. Revised and reprinted, with permission, from Ref. 47, Copyright © 1993, American Chemical Society.

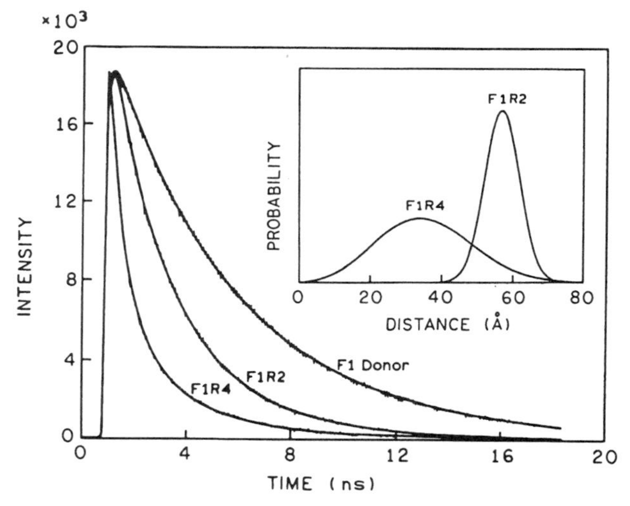

Figure 14.31. Intensity decays and distance distributions in a DNA Holliday junction. F1R2, Donor at site I and acceptor at site II; F1R4, donor at site I and acceptor at site IV (see Figure 14.30). Data from Ref. 47.

have suggested somewhat different structures.[48] Nonetheless, these data for a complex macromolecular assembly illustrate how resolution of distance distributions from the time-resolved energy transfer data has become a widely used tool in structural biology.

14.7. OTHER CONSIDERATIONS

14.7.A. Acceptor Decays

Information about the distance distribution is also contained in the acceptor decays, so that one may ask why we have stressed measurements of the donor decays. Although, in principle, either the donor or the acceptor decay could be used to recover the distance distribution,[50,51] in most circumstances there is less information in the acceptor decay. This can be seen by examining the absorption spectra of tryptophan (indole) and dansyl (Figure 13.8). The acceptors typically absorb light at wavelengths used to excite the donors. As the result of acceptor absorption, the acceptor is excited by two routes, by direct excitation and by RET from the donor. Only the latter portion of the acceptor emission contains information on the distance distribution. Typically, this is the minor fraction of the acceptor intensity, so that most of the acceptor emission is due to the intrinsic decay of the directly excited acceptor. Because of this fact, any analysis of the acceptor decay kinetics must consider directly excited acceptor. Depending on the donor and acceptor emission spectra, it may also be necessary to consider contributions of the donor emission at the wavelength chosen to observe the acceptor. While, in practice, an analysis correcting for donor emission and decay of directly excited acceptor can be performed, such a complete analysis of acceptor decay kinetics has not yet appeared in the literature.

If one can measure the acceptor decay kinetics due only to RET, they will display the unique properties of an excited-state reaction. In the time domain, the characteristics of an excited-state reaction are a rise time in the decay and a negative preexponential factor in the multiexponential analysis. In the frequency domain, the phase angles of the acceptor can exceed 90°. These characteristic features are the result of the acceptor being excited by the donor, rather than directly. These features can be hidden by spectral overlap of the donor and acceptor emission spectra and by decay of directly excited acceptor. Excited-state reactions are described in more detail in Chapter 18.

14.7.B. Effects of Incomplete Labeling

In studies of distance distributions, or in any study of energy transfer, it is critically important to consider the extent of labeling, particularly by the acceptors. Most often, one will measure the donor decay, and paradoxically this means that it is most important to have complete labeling at the acceptor site. The importance of complete acceptor labeling can be understood by considering the effects of incomplete acceptor labeling. If acceptor labeling is incomplete, the donor decay will contain components both from the D–A pairs and from the donors that lack acceptors. The contribution of these donor-alone molecules to the measured emission is greater than would be expected on the basis of their mole fraction. For instance, suppose that the donors are quenched 10-fold by the linked acceptor and that 10% of the donors lack acceptors. The contribution of each donor-alone molecule to the measured emission will be 10-fold greater than that of each D–A pair, so that the donor-alone signal will be equal to that of the D–A pairs. Hence, a small fraction of underlabeling by acceptors can result in severe distortion of the donor decays. This effect was seen in the intensity decay of donor- and acceptor-labeled troponin C (Figure 14.16).

We have found that the presence of a small fraction of donor-alone molecules usually results in a failure of the Gaussian distance distribution analysis to converge. This is a fortunate occurrence, and preferable to a convergent analysis which may be accepted as correct. We have less experience with the properties of the other distributions[1,4,29] in the presence of a significant donor-alone population.

Underlabeling by the acceptor can be accounted for in the least-squares analysis. The donor intensity decay is given by

$$I_{DA}(t) = (1 - f_A) I_D^0 \sum_i \alpha_{0i} \exp\left(\frac{-t}{\tau_{Di}}\right) + f_A I_D^0 \int_0^{\infty} P(r) \sum_i \alpha_{Di} \exp\left[\frac{-t}{\tau_{Di}} - \frac{t}{\tau_{Di}}\left(\frac{R_0}{r}\right)^6\right] \quad [14.24]$$

where f_A is the fraction of the molecules labeled with acceptors.[49,52] While it is possible to correct for acceptor underlabeling, its presence results in decreased information in the donor decay kinetics and quickly results in loss of resolution about the distance distribution. There seems to be strong correlation between f_A, $\bar{r}$, and hw during the least-squares analysis. If the value of f_A is not known, we regard 5% underlabeling by acceptors as the maximum consistent with reasonable confidence in the distance distribution. If f_A is known, and not a variable in the analysis, then a higher level of underlabeling can be tolerated.

Surprisingly, the extent of donor underlabeling is not important in the distance distribution analysis using the donor emission. If molecules lack the donor (i.e., are labeled only with acceptor), these molecules do not contribute to the donor decay. The presence of acceptor-only molecules is not important since the solutions are typically too dilute for transfer to acceptors not covalently linked to donors.

14.7.C. Effect of the Orientation Factor κ^2

As described in the previous chapter, the rate of energy transfer depends on the relative orientation of the donor emission dipole and the acceptor absorption dipole. Fortunately, for studies of distance distributions, the effect of κ^2 is likely to be small, and use of the dynamic average $\kappa^2 = \frac{2}{3}$ is reasonable.[53–55] This is because angular displacements of the donor and acceptor result in a more limited range of κ^2 values which tend toward $\frac{2}{3}$. For a native protein, one can expect the donors and acceptors to be relatively immobile during the excited-state lifetime. This is less likely in the random-coil state, where the donor and acceptor, and the peptide itself, can be expected to rotate and diffuse. For small molecules the size of fluorophores, the rotational correlation times are typically near 100 ps in water or when located on the surface of a macromolecule. The only mechanism by which κ^2 could affect the distance distribution would be if the average D–A orientation were in some way correlated with the D–A distance.[53,54] Given the multitude of conformations accessible to random-coil peptides, this seems unlikely. In general, rotational probe motions occur more rapidly than translational diffusion, so that one can expect rotational averaging even when D–A translational diffusion is minimal. In general, it seems that distance distributions should be largely unaffected by κ^2, especially when the solution is fluid enough to allow rotational motions of the probes.

14.8. DISTANCE DISTRIBUTIONS FROM STEADY-STATE DATA

14.8.A. D–A Pairs with Different R_0 Values

Under special circumstances, it is possible to recover distance distributions without the use of time-resolved data. The use of steady-state data is possible when several measurements of the transfer efficiency are available for different values of R_0. By using different R_0 values, the steady-state energy transfer efficiency samples different regions of the distance distribution. Small R_0 values detect only the closely spaced D–A pairs, and large R_0 values display transfer for all D–A pairs. Different values of R_0 can be obtained by synthesis of analogous D–A pairs, as first suggested by Cantor and Pechukas.[56] This is a rather laborious procedure and has only been accomplished for simple alkyl-chain-linked D–A pairs.[57] These structures and R_0 values are shown in Figure 14.32. The R_0 value is seen to range from 6.3 to 32.7 Å. The steady-state transfer efficiency is related to the distance distribution by

$$E(R_0) = \int_0^\infty \frac{P(r)R_0^6}{R_0^6 + r^6}\, dr \qquad [14.25]$$

As seen in Figure 14.33, the donor emission is progressively quenched as the R_0 value of the acceptor increases. The dependence of $E(R_0)$ on R_0 can be compared with a calculated value E_c by the usual procedures of non-linear least squares using

$$\chi_R^2 = \frac{1}{\nu}\left(\frac{E(R_0) - E_c}{\delta E}\right)^2 \qquad [14.26]$$

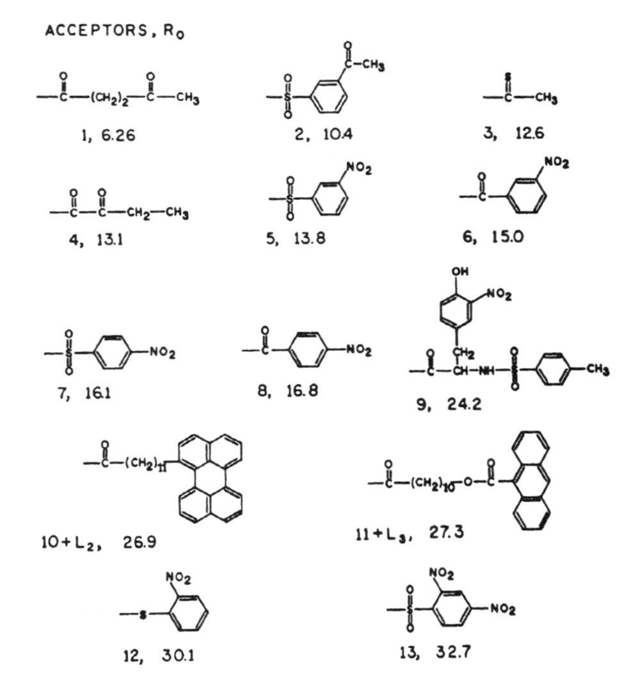

Figure 14.32. Indole–acceptor pairs with Förster distances ranging from 6.3 to 32.7 Å. From Ref. 57.

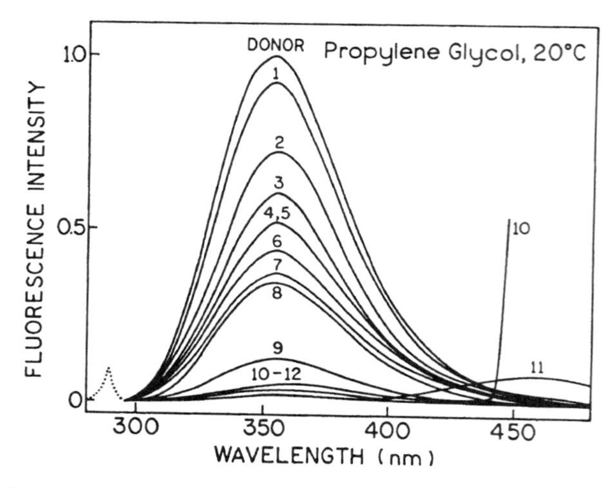

Figure 14.33. Emission spectra for the D–A pairs in Figure 14.32. From Ref. 57.

where ν is the number of degrees of freedom and δE is the uncertainty in the transfer efficiency. In this case with 13 D–A pairs, there are 10 degrees of freedom when using the Gaussian model with two parameters (p), $\nu = n - p - 1$. Such data have been used to obtain a highly resolved distance distribution (Figure 14.34). In this case the distribution was found to be a skewed Gaussian.

14.8.B. Changing R_0 by Quenching

There is an alternative means to obtain different values of R_0. The quantum yield of the donor can be decreased by collisional quenching.[58,59] In this case, the decrease in quantum yield results in a smaller value of R_0, as can be seen from Eq. [13.5]. The new value of R_0, R_0^Q, is given by

$$R_0^Q = R_0 \left(\frac{Q_D^Q}{Q_D} \right)^{1/6} \qquad [14.27]$$

where Q indicates the presence of quencher. Unfortunately, the value of R_0^Q can only be changed to a limited extent for reasonable amounts of quenching. Also, it is important to consider only the dynamic portion of the quenching, since the statically quenched donor does not contribute to the emission.

14.9. APPLICATIONS OF TIME-RESOLVED RET

14.9.A. DNA Hybridization

An instructive example of RET was described by Selvin and co-workers.[60–62] These workers described time-resolved measurements of energy transfer from a lanthanide (terbium) donor on one strand of a DNA 10-mer to a tetramethylrhodamine (TMR) acceptor on the distal end of a complementary DNA strand (Figure 14.35).[60] Because of the high quantum yield of the terbium chelate and the high extinction coefficient of TMR, the D–A pair displayed an R_0 of 65 Å. The objective in showing this example is not to recover a distance distribution, but rather to illustrate the characteristics of the time-resolved RET data with a heterogeneous donor population.

In the absence of the acceptor-DNA strand, the terbium donor-DNA displays a single-exponential decay time of 2.1 ms (Figure 14.36). Upon addition of the acceptor-DNA, the terbium decay becomes clearly multiexponential and shows two decay times of 0.33 and 2.1 ms. The amplitude of the short-decay-time component increases

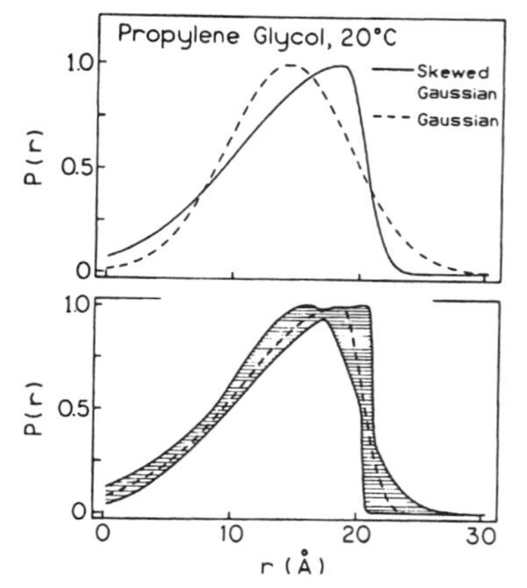

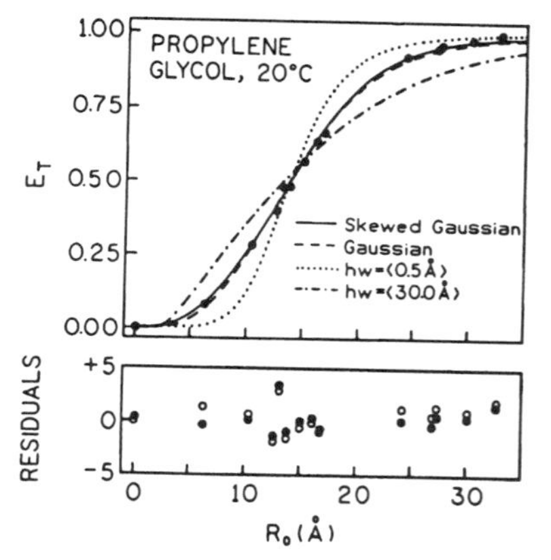

Figure 14.34. Transfer efficiency versus Förster distance (*right*) and distance distribution (*left*) for the alkyl-chain-linked D–A pairs in Figure 14.32. From Ref. 57.

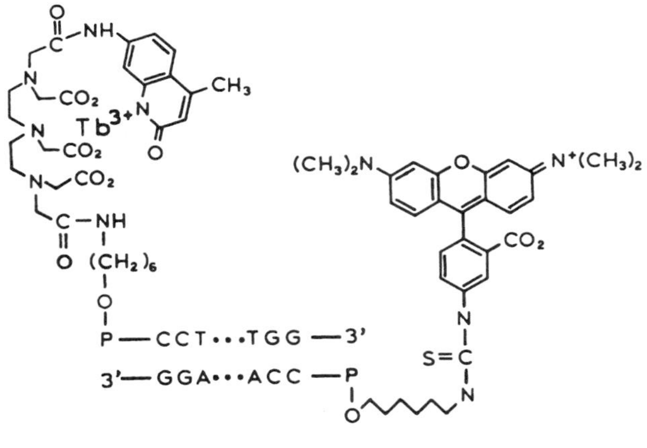

Figure 14.35. Schematic diagram of double-stranded DNA with terbium chelate (donor) at one 5′-end and TMR (acceptor) at other 5′-end. The length of DNA was a 10-mer. Revised from Ref. 60.

with increasing amounts of acceptor DNA. In this case, it is not necessary to measure the extent of donor quenching in the fully bound state (one acceptor per donor) to determine the transfer efficiency. Assuming that the short-decay-time component is due to the D–A pairs, the efficiency of energy transfer can be calculated using Eq. [13.13], yielding $E = 0.84$. Given $R_0 = 65$ Å, this corresponds to a terbium–TMR distance of 0.76 (65 Å) = 49 Å, which is a reasonable distance for the DNA 10-mer. The simple assignment of decay times to the donor-alone and donor–acceptor pairs would not be possible if there was a distribution of distances or multiple D–A distances.

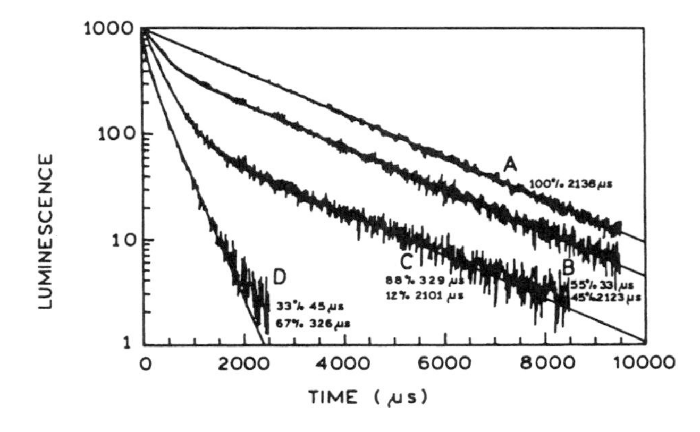

Figure 14.36. Terbium intensity decay of terbium donor DNA at 546 nm in the absence (A) and in the presence of increasing amounts of TMR-DNA acceptors (B and C). Curve D shows the acceptor decay at 570 nm, with the 45-μs component being assigned to an artifact. The intensity decays were as follows, with the percents being the preexponential factors: A, 2138 μs; B, 55% with 331 μs and 45% with 2123 μs; C, 88% with 329 μs and 12% with 2101 μs; D, 33% with 45 μs and 67% with 326 μs. Revised from Ref. 60.

The use of lanthanide donors has a number of advantages, one of which is their symmetrical structure. The emission from lanthanides is unpolarized, and, as a result, the orientation factor can only be important for the acceptor. This limits the range of possible κ^2 values to $\frac{1}{3} < \kappa^2 < \frac{4}{3}$ which, in turn, limits the range of distances to ±11%, as can be calculated by taking the sixth root of the κ^2 values. Additionally, the long decay times of the lanthanides allow time for acceptor motion, which will cause κ^2 to approach $\frac{2}{3}$. Finally, the use of a long-lived donor allows the acceptor emission to be measured without contamination by the directly excited acceptor. This is accomplished by observation of the acceptor emission following the pulsed excitation, possibly using a gated detector. At times longer than the nanosecond acceptor lifetime, the only acceptor emission is that resulting from energy transfer (Figure 14.36). For energy-transfer measurements, this is the emission which contains information on the distance. In this case the long-lived donor, the sensitized acceptor emission decays with the same lifetime as the D–A pairs. However, in contrast to the donor emission, the sensitized acceptor emission is due only to D–A pairs and is thus insensitive to incomplete labeling by donor or acceptor.[56]

14.10. CONCLUSION

In this chapter we have attempted to show the potential of time-resolved measurements for determination of the solution conformations of macromolecules. The determination of conformational distributions appears to be a rather unique property of the time-resolved measurements. Most physical methods reveal only an average conformation. Additionally, the data contain information on the timescale of conformational changes. With the introduction of long-lived probes, conformational dynamics will be measurable on timescales from nanoseconds to tens of microseconds.

As a final comment, we want to emphasize that all the results described in this chapter were for covalently linked donor–acceptor pairs, with a single acceptor per donor. Different and somewhat more complex theory is needed to describe systems with multiple acceptors and unlinked probes. Such systems occur commonly in membranes and nucleic acids and are described in the following chapter.

REFERENCES

1. Haas, E., Wilchek, M., Katchalski-Katzir, E., and Steinberg, I. Z., 1975, Distribution of end-to-end distances of oligopeptides in solu-

tion as estimated by energy transfer, *Proc. Natl. Acad. Sci. U.S.A.* **72:**1807–1811.
2. Grinvald, A., Haas, E., and Steinberg, I. Z., 1972, Evaluation of the distribution of distances between energy donors and acceptors by fluorescence decay, *Proc. Natl. Acad. Sci. U.S.A.* **69:**2273–2277.
3. Haas, E., Katchalski-Katzir, E., and Steinberg, I. Z., 1978, Brownian motion of the ends of oligopeptide chains in solution as estimated by energy transfer between chain ends, *Biopolymers* **17:**11–31.
4. Wu, P., and Brand, L., 1994, Conformational flexibility in a staphylococcal nuclease mutant K45C from time-resolved resonance energy transfer measurements, *Biochemistry* **33:**10457–10462.
5. Amir, D., Krausz, S., and Haas, E., 1992, Detection of local structures in reduced unfolded bovine pancreatic trypsin inhibitor, *Proteins* **13:**162–173.
6. Amir, D., and Haas, E., 1986, Determination of intramolecular distance distributions in a globular protein by nonradiative excitation energy transfer measurements, *Biopolymers* **25:**235–240.
7. Lakowicz, J. R., Wiczk, W., Gryczynski, I., and Johnson, M. L., 1990, Influence of oligopeptide flexibility on donor–acceptor distance distribution by frequency-domain fluorescence spectroscopy, *Proc. SPIE* **1204:**192–205.
8. Haas, E., McWherter, C. A., and Scheraga, H. A., 1988, Conformational unfolding in the N-terminal region of ribonuclease A detected by nonradiative energy transfer: Distribution of interresidue distances in the native, denatured, and reduced-denatured states, *Biopolymers* **27:**1–21.
9. She, M., Xing, J., Dong, W.-J., Umeda, P. K., and Cheung, H., 1998, Calcium binding to the regulatory domain of skeletal muscle troponin C induces a highly constrained open conformation, *J. Mol. Biol.* **281:**445–452 (1998).
10. Dong, W.-J., Chandra, M., Xing, J., She, M., Solaro, R. J., and Cheung, H. C., 1997, Phosphorylation induced distance change in a cardiac muscle troponin I mutant, *Biochemistry* **36:**6754–6761.
11. Amir, D., and Haas, E., 1987, Estimation of intramolecular distance distributions in bovine pancreatic trypsin inhibitor by site-specific labeling and nonradiative excitation energy-transfer measurements, *Biochemistry* **26:**2162–2175.
12. Amir, D., Levy, D. P., Levin, Y., and Haas, E., 1986, Selective fluorescent labeling of amino groups of bovine pancreatic trypsin inhibitor by reductive alkylation, *Biopolymers* **25:**1645–1658.
13. Amir, D., Varshavski, L., and Haas, E., 1985, Selective fluorescent labeling at the α-amino group of bovine pancreatic trypsin inhibitor, *Biopolymers* **24:**623–638.
14. Amir, D., and Haas, E., 1988, Reduced bovine pancreatic trypsin inhibitor has a compact structure, *Biochemistry* **27:**8889–8893.
15. Maliwal, B. P., Lakowicz, J. R., Kupryszewski, G., and Rekowski, P., 1993, Fluorescence study of conformational flexibility of RNase S-peptide: Distance-distribution, end-to-end diffusion, and anisotropy decays, *Biochemistry* **32:**12337–12345.
16. Cheung, H. C., Gryczynski, I., Malak, H., Wiczk, W., Johnson, M. L., and Lakowicz, J. R., 1991, Conformational flexibility of the Cys 697–Cys 707 segment of myosin subfragment 1. Distance distributions by frequency-domain fluorometry, *Biophys. Chem.* **40:**1–17.
17. Cheung, H. C., Wang, C.-K., Gryczynski, I., Wiczk, W., Laczko, G., Johnson, M. L., and Lakowicz, J. R., 1991, Distance distributions and anisotropy decays of troponin C and its complex with troponin I, *Biochemistry* **30:**5238–5247.
18. Lakowicz, J. R., Gryczynski, I., Cheung, H. C., Wang, C.-K, Johnson, M. L., and Joshi, N., 1988, Distance distributions in proteins recovered by using frequency-domain fluorometry. Applications to troponin I and its complex with troponin C, *Biochemistry* **27:**9149–9160.
19. Zhao, X., Kobayashi, T., Malak, H., Gryczynski, I., Lakowicz, J. R., Wade, R., and Collins, J. H., 1995, Calcium-induced troponin flexibility revealed by distance distribution measurements between engineered sites, *J. Biol. Chem.* **270:**15507–15514.
20. Eis, P. S., Kusba, J., Johnson, M. L., and Lakowicz, J. R., 1993, Distance distributions and dynamics of a zinc finger peptide from fluorescence resonance energy transfer measurements, *J. Fluoresc.* 3(1):23–31.
21. Eis, P. S., and Lakowicz, J. R., 1993, Time-resolved energy transfer measurements of donor–acceptor distance distributions and intramolecular flexibility of a CCHH zinc finger peptide, *Biochemistry* **32:**7981–7993.
22. Cheung, H. C., 1991, Resonance energy transfer, in *Topics in Fluorescence Spectroscopy, Volume 2, Principles*, J. R. Lakowicz (ed.), Plenum Press, New York, pp. 127–176.
23. Lakowicz, J. R., Gryczynski, I., Wiczk, W., Laczko, G., Prendergast, F. C., and Johnson, M. L., 1990, Conformational distributions of melittin in water/methanol mixtures from frequency-domain measurements of nonradiative energy transfer, *Biophys. Chem.* **36:**99–115.
24. Lakowicz, J. R., Gryczynski, I., Laczko, G., Wiczk, W., and Johnson, M. L., 1994, Distribution of distances between the tryptophan and the N-terminal residue of melittin in its complex with calmodulin, troponin C, and phospholipids, *Protein Sci.* **3:**628–637.
25. Albaugh, S., and Steiner, R. F., 1989, Determination of distance distribution from time domain fluorometry, *J. Phys. Chem.* **93:**8013–8016.
26. Haran, G., Haas, E., Szpikowska, B. K., and Mas, M. T., 1992, Domain motions in phosphoglycerate kinase: Determination of interdomain distance distributions by site-specific labeling and time-resolved fluorescence energy transfer, *Proc. Natl. Acad. Sci. U.S.A.* **89:**11764–11768.
27. Lillo, M. P., Szpikowska, B. K., Mas, M. T., Sutin, J. D., and Beechem, J. M., 1997, Real-time measurement of multiple intramolecular distances during protein folding reactions: A multisite stopped-flow fluorescence energy-transfer study of yeast phosphoglycerate kinase, *Biochemistry* **36:**11273–11281.
28. Lillo, M. P., Beechem, J. M., Szpikowska, B. K., Sherman, M. A., and Mas, M. T., 1997, Design and characterization of a multisite fluorescence energy-transfer system for protein folding studies: A steady-state and time-resolved study of yeast phosphoglycerate kinase, *Biochemistry* **36:**11261–11272.
29. Katchalski-Katzir, E., Haas, E., and Steinberg, I. A., 1981, Study of conformation and intramolecular motility of polypeptides in solution by a novel fluorescence method, *Ann. N.Y. Acad. Sci.* **36:**44–61.
30. Lakowicz, J. R., Johnson, M. L., Wiczk, W., Bhat, A., and Steiner, R. F., 1987, Resolution of a distribution of distances by fluorescence energy transfer and frequency-domain fluorometry, *Chem. Phys. Lett.* **138:**587–593.
31. Kulinski, T., Wennerberg, A. B. A., Rigler, R., Provencher, S. W., Pooga, M., Langel, U., and Bartfai, T., 1997, Conformational analysis of glanin using end to end distance distribution observed by Förster resonance energy transfer, *Eur. Biophys. J.* **26:**145–154.
32. Rice, K. G., Wu, P., Brand, L., and Lee, Y. C., 1991, Interterminal distance and flexibility of a triantennary glycopeptide as measured by resonance energy transfer, *Biochemistry* **30:**6646–6655.
33. Lakowicz, J. R., Kušba, J., Wiczk, W., Gryczynski, I., and Johnson, M. L., 1990, End-to-end diffusion of a flexible bichromophoric molecule observed by intramolecular energy transfer and frequency domain fluorometry, *Chem. Phys. Lett.* **173:**319–326.
34. Lakowicz, J. R., Gryczynski, I., Kušba, J., Wiczk, W., Szmacinski, H., and Johnson, M. L., 1994, Site-to-site diffusion in proteins as

observed by energy transfer and frequency domain fluorometry, *Photochem. Photobiol.* **59**:16–29.
35. Kušba, J., and Lakowicz, J. R., 1994, Diffusion-modulated energy transfer and quenching: Analysis by numerical integration of diffusion equation in Laplace space, *Methods Enzymol.* **240**:216–262.
36. Lakowicz, J. R., Kušba, J., Wiczk, W., Gryczynski, I., Szmacinski, H., and Johnson, M. L., 1991, Resolution of the conformational distribution and dynamics of a flexible molecule using frequency domain fluorometry, *Biophys. Chem.* **39**:79–84.
37. Lakowicz, J. R., Wiczk, W., Gryczynski, I., Szmacinski, H., and Johnson, M. L., 1990, Influence of end-to-end diffusion on intramolecular energy transfer as observed by frequency-domain fluorometry, *Biophys. Chem.* **38**:99–109.
38. Somogyi, B., Matko, J., Papp, S., Hevessy, J., Welch, G. R., and Damjanovich, S., 1984, Förster-type energy transfer as a probe for changes in local fluctuations of the protein matrix, *Biochemistry* **23**:3404–3411.
39. Gerstein, M., Lesk, A. M., and Chothia, C., 1994, Structural mechanisms for domain movements in proteins, *Biochemistry* **33**:6738–6749.
40. Blackwell, M. F., Gounaris, K., Zara, S. J., and Barber, J., 1987, A method for estimating lateral diffusion coefficients in membranes from steady state fluorescence quenching studies, *J. Biophys. Soc.* **51**:735–744.
41. Ollmann, M., Schwarzmann, G., Sandhoff, K., and Galla, H-J., 1987, Pyrene-labeled gangliosides: Micelle formation in aqueous solution, lateral diffusion, and thermotropic behavior in phosphatidylcholine bilayers, *Biochemistry* **26**:5943–5952.
42. Buckler, D. R., Haas, E., and Scheraga, H. A., 1995, Analysis of the structure of ribonuclease A in native and partially denatured states by time-resolved nonradiative dynamic excitation energy transfer between site-specific extrinsic probes, *Biochemistry* **34**:15965–15978.
43. Sabbatini, N., Guardigli, M., and Lehn, J.-M., 1993, Luminescent lanthanide complexes as photochemical supramolecular devices, *Coordi. Chem. Rev.* **123**:201–228.
44. Demas, J. N., and DeGraff, B. A., 1992, Applications of highly luminescent transition metal complexes in polymer systems, *Macromol. Chem. Macromol. Symp.* **59**:35–51.
45. Hochstrasser, R. A., Chen, S.-M., and Millar, D. P., 1992, Distance distribution in a dye-linked oligonucleotide determined by time-resolved fluorescence energy transfer, *Biophys. Chem.* **45**:133–141.
46. Yang, M., and Millar, D. P., 1997, Fluorescence resonance energy transfer as a probe of DNA structure and function, *Methods Enzymol.* **278**:417–444.
47. Eis, P. S., and Millar, D. P., 1993, Conformational distributions of a four-way DNA junction revealed by time-resolved fluorescence resonance energy transfer, *Biochemistry* **32**:13852–13860.
48. Clegg, R. M., Murchie, A. I. H., and Lilley, D. M., 1994, The solution structure of the four-way DNA junction at low-salt conditions: A fluorescence resonance energy transfer analysis, *Biophys. J.* **66**:99–109.
49. Yang, M., and Millar, D. P., 1996, Conformational flexibility of three-way DNA junctions containing unpaired nucleotides, *Biochemistry* **35**:7959–7967.
50. Beecham, J. M., and Haas, E., 1989, Simulations determination of intramolecular distance distributions and conformational dynamics by global analysis of energy transfer measurements, *Biophys. J.* **55**:1225–1236.
51. Ohmine, I., Silbey, R., and Deutch, J. M., 1997, Energy transfer in labeled polymer chains in semidilute solutions, *Macromolecules* **10**:862–864.
52. Lakowicz, J. R., Gryczynski, I., Wiczk, W., Kušba, J., and Johnson, M. L., 1991, Correction for incomplete labeling in the measurement of distance distributions by frequency-domain fluorometry, *Anal. Biochem.* **195**:243–254.
53. Englert, A., and Leclerc, M., 1978, Intramolecular energy transfer in molecules with a large number of conformations, *Proc. Natl. Acad. Sci. U.S.A.* **75**:1050–1051.
54. Wu, P., and Brand, L., 1992, Orientation factor in steady state and time-resolved resonance energy transfer measurements, *Biochemistry* **31**:7939–7947.
55. Dos Remedios, C. G., and Moens, P. D. J., 1995, Fluorescence resonance energy transfer spectroscopy is a reliable "ruler" for measuring structural changes in proteins, *J. Struct. Biol.* **115**:175–185.
56. Cantor, C. R., and Pechukas, P., 1971, Determination of distance distribution functions by singlet–singlet energy transfer, *Proc. Natl. Acad. Sci. U.S.A.* **68**:2099–2101.
57. Wizk, W., Eis, P. S., Fishman, M. N., Johnson, M. L., and Lakowicz, J. R., 1991, Distance distributions recovered from steady-state fluorescence measurements on thirteen donor–acceptor pairs with different Förster distances, *J. Fluoresc.* **1**(4):273–286.
58. Gryczynski, I., Wiczk, W., Johnson, M. L., and Lakowicz, J. R., 1988, End-to-end distance distributions of flexible molecules from steady state fluorescence energy transfer and quenching induced changes in the Förster distance, *Chem. Phys. Lett.* **145**:439–446.
59. Gryczynski, I., Wiczk, W., Johnson, M. L., Cheung, H. C., Wang, C., and Lakowicz, J. R., 1988, Resolution of the end-to-end distance distribution of flexible molecules using quenching-induced variations of the Förster distance for fluorescence energy transfer, *Biophys. J.* **54**:577–586.
60. Selvin, P. R., and Hearst, J. E., 1994, Luminescence energy transfer using a terbium chelate: Improvements on fluorescence energy transfer, *Proc. Natl. Acad. Sci. U.S.A.* **91**:10024–10028.
61. Selvin, P. R., Rana, T. M., and Hearst, J. E., 1994, Luminescence resonance energy transfer, *J. Am. Chem. Soc.* **116**:6029–6030.
62. Selvin, P. R., 1995, Fluorescence resonance energy transfer, *Methods Enzymol.* **246**:301–334.

PROBLEMS

14.1. *Intensity Decays for Various Donor–Acceptor Pairs*: Assume that you have three different solutions. All contain donors which display the same quantum yield, a single-exponential decay time of 5 ns in the absence of acceptor, and the same Förster distance of 20 Å. One sample (A) consists of a protein with a single donor and three acceptors at distances $r_1 = 15$ Å, $r_2 = 20$ Å, and $r_3 = 25$ Å (Figure 14.37). The second sample (B) contains an equimolar mixture of three proteins, each with a single donor and a single acceptor. However, in each protein there is a different D–A distance, 15, 20, or 25 Å. The third sample (C) is a protein with a single acceptor but with three donors. Each donor displays the same donor-alone lifetime ($\tau_D = 5$ ns) but is at a different distance (15, 20, or 25 Å) from the acceptor.

A. Describe the intensity decays of samples A, B, and C. Which samples display a single-exponential decay law?

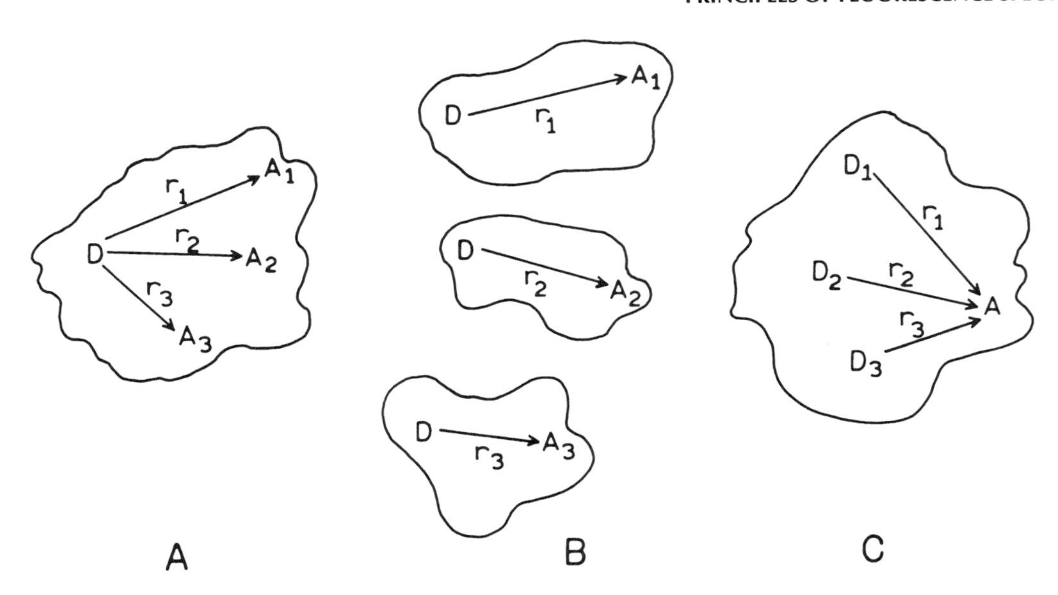

Figure 14.37. Schematic drawings of a protein with one donor and three acceptors (A), three proteins, each with a single donor and acceptor (B), and a protein with three donors and one acceptor (C). The distances are r_1 = 15 Å, r_2 = 20 Å, and r_3 = 25 Å. The Förster distance is 20 Å.

B. Describe the intensity decay law of sample B, including the decay times, preexponential factors (α_i), and fractional intensity (f_i) values.

C. For the three-acceptor protein (sample A), could you detect the presence of three acceptors from the intensity decay?

D. For the three-acceptor protein, what is the apparent D–A distance if you assume the presence of a single acceptor?

14.2. *Acceptor Concentration for RET for Unlinked D–A Pairs*:

A. Suppose that you have a donor and acceptor which are not covalently linked and that the Förster distance is 30 Å. Calculate the acceptor concentration needed to place, on average, one acceptor within a 60-Å cube containing a donor at the center. To a first approximation, this concentration is comparable to that for 50% energy transfer. Compare this acceptor concentration with that calculated using Eq. [13.33].

B. Consider a D–A pair covalently linked by a 30-Å linker. Calculate the effective concentration of acceptors around the donor.

Energy Transfer to Multiple Acceptors, in One, Two, or Three Dimensions

15

In the previous two chapters on energy transfer, we considered primarily covalently linked donor–acceptor pairs, or situations in which there was a single acceptor near each donor. However, there are numerous situations in which there exist multiple acceptors, such as the obvious case of donors and acceptors dissolved in homogeneous solutions. More interesting examples of the multiple-acceptor case occur in membranes and nucleic acids. Suppose that one has a lipid bilayer which contains both donors and acceptors (Figure 15.1, middle). Each donor will be surrounded by acceptors in two dimensions. Since the acceptor distribution is random, each donor sees a different acceptor population. Hence, the intensity decay is an ensemble average and is typically nonexponential. A similar situation exists for donors and acceptors which are intercalated into double-helical DNA (Figure 15.1, right), except that in this case the acceptors are distributed in one dimension along the DNA helix.

The theory for these multiple-acceptor cases is complex, even in three dimensions. For a completely homogeneous and random solution, with no excluded volume, the form of the donor intensity decay and donor quantum yield was described by Förster.[1,2] However, consider a protein with a buried fluorophore which serves as the donor. The exact form of the intensity decay will depend on the acceptor concentration and on the distance of closest approach (r_c) between the donor and acceptor, which could be approximated by the radius of the protein (Figure 15.1, left). The concept of a minimum D–A distance becomes particularly important for membrane-bound proteins, where r_c may reflect the size of the protein, the presence of boundary lipid which excludes the acceptor, or the distance of the donor above the membrane surface. Unfortunately, the theory for RET under these conditions is complex. The theory becomes still more complex if we consider D–A diffusion. Hence, we cannot present a complete description of energy transfer for these diverse conditions. Instead, we will present the best understood cases, with reference to the more complex cases.

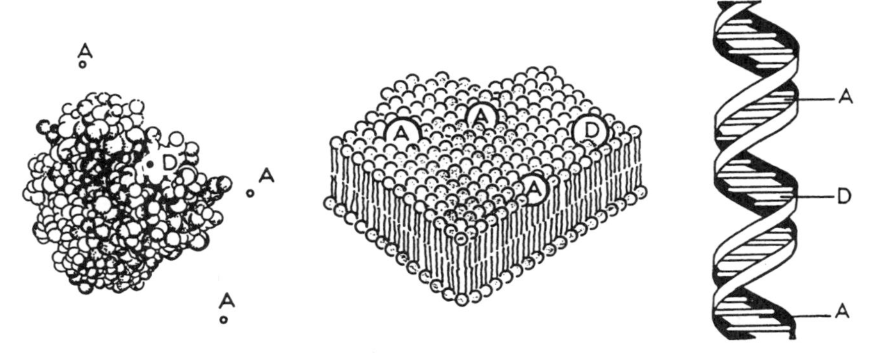

Figure 15.1. RET between unlinked donors (D) and acceptors (A) in a protein (*left*), a lipid bilayer (*middle*), and double-helical DNA (*right*).

15.1. RET IN THREE DIMENSIONS

Consider a homogeneous solution of donors and acceptors, without D–A diffusion. If the concentration of donor is adequately low so that there is no homotransfer between donors, the intensity decay is well known[1–3] and is given by

$$I_{DA}(t) = I_D^0 \exp\left[-\frac{t}{\tau_D} - 2\gamma\left(\frac{t}{\tau_D}\right)^{1/2}\right] \qquad [15.1]$$

where τ_D is the donor decay in the absence of acceptors. The term with the $t^{1/2}$ time dependence is given by

$$\gamma = \frac{\Gamma(1/2)}{2}\frac{C}{C_0} \qquad [15.2]$$

In this expression, $\Gamma(1/2) = \sqrt{\pi} = 1.7724539\ldots$ is the complete gamma function, C is the acceptor concentration, and C_0 is a characteristic acceptor concentration in molecules per cubic centimeter. This concentration is related to the Förster distance by

$$C_0 = \left(\frac{4}{3}\pi R_0^3\right)^{-1} \qquad [15.3]$$

The ratio C/C_0 is thus seen to be the number of acceptor molecules in a sphere of volume $\frac{4}{3}\pi R_0^3$.

For randomly distributed acceptors in three dimensions, the steady-state intensity of the donor is given by

$$\frac{F_{DA}}{F_D} = 1 - \sqrt{\pi}\exp(\gamma^2)[1 - \operatorname{erf}(\gamma)] \qquad [15.4]$$

where the error function erf(γ) was defined in Eq. [13.32]. In contrast to the situation when we were dealing with linked D–A pairs, we have analytical expressions like Eqs. [15.1] and [15.4] for energy transfer between donors and acceptors dissolved in solution. An important consequence of unlinking the donors and acceptors is that the extent of transfer depends on acceptor concentration and that the acceptor concentrations must be quite high for significant energy transfer. The acceptor concentration is often described in terms of C_0, which is the acceptor concentration needed for 72% transfer. The values of C_0 can be calculated from R_0 and are typically in the range of 2–50m*M*. C_0 is the concentration of acceptors needed to statistically place an acceptor within a distance R_0 of the donor. The definition of C_0 is slightly different from that of A_0 in Eq. [13.30], and $C_0 = 0.5\,\pi^{1/2}A_0$. Both C_0 and A_0 are in common use.

The orientation factor κ^2 is not shown explicitly in Eqs. [15.1]–[15.4], as a value of $\frac{2}{3}$ is usually assumed in the calculation of the Förster distance R_0. In general, fluorophores rotate faster than they undergo translational diffusion. As a result, it is often possible to find conditions in which rotational motions result in dynamic averaging of κ^2 to $\frac{2}{3}$. However, if the solution is completely frozen so that rotational motions are slower than the donor decay,[3–5] then for a random three-dimensional solution, $\kappa^2 = 0.476$. Hence, energy transfer is somewhat slower in rotationally frozen solutions compared to that under conditions in which the fluorophores can rotate during the excited-state lifetime. To obtain the same energy transfer as with $\kappa^2 = \frac{2}{3}$ the acceptor concentration needs to be 1.18-fold larger.[5]

There have been numerous studies of energy transfer in three-dimensional solutions. It is generally accepted that Eqs. [15.1]–[15.4] provide an accurate description for randomly distributed donors and acceptors.[6] To illustrate energy transfer in homogeneous solution, we have chosen data for indole as the donor and dansylamide as the acceptor.[7] The donor and acceptor were dissolved in methanol, which is highly fluid, so that diffusion can be expected to increase the extent of energy transfer. In order to obtain significant energy transfer, the dansylamide concentration was 5m*M*. The high acceptor concentrations needed for RET in solution between unlinked D–A pairs make it difficult to measure the donor emission. This is seen from Figure 15.2, where the acceptor emission is much more intense than the donor emission. Of course, the acceptor is excited both by energy transfer and by direct absorption. Careful optical filtering is needed to eliminate the acceptor emission, which is essential for measurement of the donor decay kinetics. Additionally, it is often necessary to use higher than usual concentrations of donor to obtain detectable donor emission in the presence of high acceptor concentrations and high optical densities. For these rea-

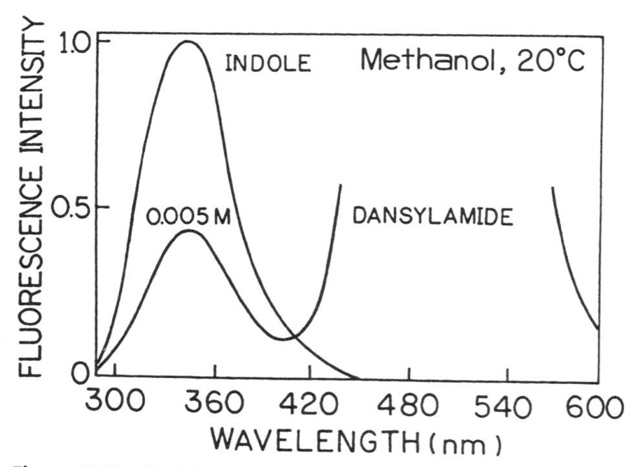

Figure 15.2. Emission spectra of an unlinked indole donor and 5m*M* dansylamide acceptor in methanol at 20 °C. Reprinted, with permission, from Ref. 7, Copyright © 1990, American Chemical Society.

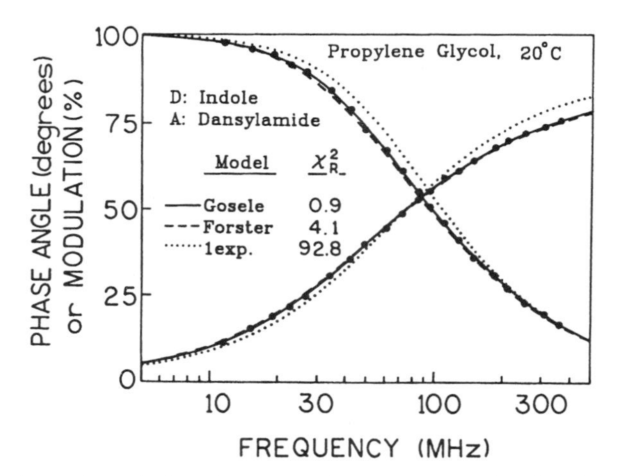

Figure 15.3. FD intensity decays of indole with 12m*M* dansylamide in propylene glycol at 20 °C. The fits to the experimental data (•) are shown for three models. Reprinted, with permission, from Ref. 7, Copyright © 1990, American Chemical Society.

sons, energy-transfer measurements in homogeneous solutions are frequently performed using front-face observation. Even with front-face observation, correction of the donor intensities for the large inner filter effect is difficult and error-prone. For this reason, it is usually more convenient to use time-resolved measurements of the donor, since the intensity decay is independent of the total intensity.

Equation [15.1] predicts that the donor intensity decay becomes nonexponential in the presence of dissolved acceptors. At the moment of excitation, the solution contains donors which, on average, are surrounded by a constant concentration of acceptors at all distances. Some of the donors are located closer to acceptors than other donors. Following pulsed excitation, energy transfer between these more closely spaced D–A pairs occurs rapidly, leading to a rapid component in the intensity decay, which is the $t^{1/2}$ term in Eq. [15.1]. At longer times, the decay becomes dominated by the longer-lived donors, which are, on average, more distant from the acceptors.

In a fluid solvent like methanol, the intensity decay of indole is influenced by translational diffusion, which increases the extent of energy transfer. To understand the effects of diffusion, it is useful to first consider the decay of indole in the relatively viscous solvent propylene glycol, where little diffusion is expected during the 4.23-ns excited-state lifetime of indole. The form of the indole decay was examined using FD measurements (Figure 15.3). In the absence of acceptor, the indole decay was a single exponential (Table 15.1). Because diffusion does not increase the transfer efficiency in this solvent, it was necessary to increase the acceptor concentration to 12m*M*. In the presence of acceptor (12m*M* dansylamide), the decay of indole can no longer be fit to a single decay time (dotted curve in Figure 15.3; $\chi^2_R = 92.8$). The donor intensity decay can be fit using Eq. [15.1], resulting in a much improved value of χ^2_R = 4.1 (dashed curve in Figure 15.3). The variable parameter in this fit is C_0 or R_0. In fact, such measurements are a reliable means of experimentally determining R_0, assuming the acceptor concentration is known. A still better fit of the indole decay was obtained using a model which accounts for D–A diffusion (solid curve in Figure 15.3), because a small amount of D–A diffusion occurs in this solvent.

Table 15.1. Indole Decay Times and Donor-to-Acceptor Diffusion Coefficients in Propylene Glycol and Methanol at 20 °C[a]

Solvent	[A]	Equation no.	τ_D (ns)	R_0 (Å)	10^6D (cm²/s)	χ^2_R
Propylene glycol	0	—[b]	4.23			1.4
	12m*M*	—[b]	2.57			92.8
		15.1	<4.23>[c]	24.9		4.1
		15.1	<4.23>	<24.3>		10.6
		15.8	<4.23>	23.9	1.03	0.9
		15.8	<4.23>	<24.3>	0.62	1.3
Methanol	0	—[b]	4.09			2.4
	5m*M*	—[b]	2.03			30.7
		15.1	<4.09>	37.1		95.8
		15.1	<4.09>	<26.1>		1152.1
		15.8	<4.09>	27.8	26.4	1.8
		15.8	<4.09>	<26.1>	34.0	3.1

[a]From Ref. 7. The acceptor (A) was dansylamide.
[b]The data were fit to a single-exponential decay law.
[c]Angular brackets indicate that the parameter was held fixed during the analysis.

15.1.A. Effect of Diffusion on RET with Unlinked Donors and Acceptors

In Chapter 14 we described how D–A diffusion increases the efficiency of energy transfer between covalently linked D–A pairs. A similar effect occurs for unlinked D–A pairs in homogeneous solution. However, the theory for energy transfer for unlinked donors and acceptors becomes more complex in the presence of diffusion.[8,9] In fact, the theory has not been solved to yield exact analytical expressions to describe the intensity decay. Numerical solutions have been described[8,10]; however, for purposes of least-squares data analysis, it is usually more convenient to use approximate expressions which are available in a closed form. Several approximations are available. For unlinked donors and acceptors in homogeneous solution, the donor decay can be described by[11]

$$I_{DA}(t) = I_D^0 \exp\left[-\frac{t}{\tau_D} - 2B\gamma\left(\frac{t}{\tau_D}\right)^{1/2}\right] \quad [15.5]$$

The parameter B is given by

$$B = \left(\frac{1 + 10.87x + 15.5x^2}{1 + 8.743x}\right)^{3/4} \quad [15.6]$$

where

$$x = D\alpha^{-1/3}t^{2/3}, \qquad \alpha = R_0^6/\tau_D \quad [15.7]$$

A comparable expression is also available[12] where

$$B = \left(\frac{1 + 5.47x + 4.00x^2}{1 + 3.34x}\right)^{3/4} \quad [15.8]$$

It is claimed that this latter approximation is better for longer times.[12,13] If $D = 0$, these expressions become equivalent to Eq. [15.1]. These expressions are thought to be valid for a wide range of diffusion coefficients and acceptor concentrations.[6] The important point is that it is complex to calculate the intensity decay exactly in the presence of D–A diffusion, but the intensity decay can be predicted with good accuracy using the approximate expressions.

The approximate expression for $I_{DA}(t)$ can be used to predict the donor intensity at any time, and thus it can be used for analysis of FD and TD data. Figure 15.4 shows FD data for the indole–dansylamide D–A pairs in methanol, which is less viscous than propylene glycol and allows significant diffusion during the donor's excited-state lifetime. In this case, the Förster equation (Eq. [15.1]) does not fit the data (dashed curve in Figure 15.4) because

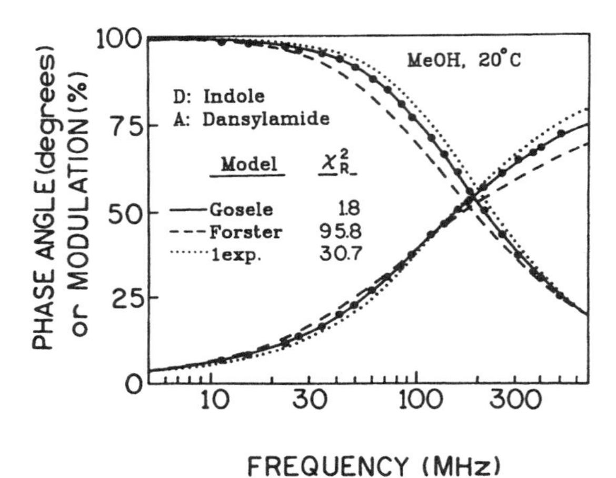

Figure 15.4. FD intensity decays of indole with 5m*M* dansylamide in methanol at 20 °C. The fits to the experimental data (•) are shown for three models. Reprinted, with permission, from Ref. 7, Copyright © 1990, American Chemical Society.

diffusion changes the shape of the intensity decay. Diffusion results in a donor decay which is more like a single exponential, as can be seen by the deviation of the data from the Förster equation toward the single-exponential model (dotted curve in Figure 15.4). The data were well fit by the approximate intensity decay described by Eq. [15.8]. The value of $\chi_R^2 = 95.8$ obtained using the Förster model is artificially low because the fitting procedure increases R_0 to account for diffusion-enhanced energy transfer. If the Förster distance is held constant at the known value of 26.1 Å, then $\chi_R^2 = 1152$ (Table 15.1). This highly elevated χ_R^2 value indicates the significant influence of diffusion in this system. When the correct model is used (Eq. [15.8]), the variable parameters in the analysis are R_0 and the mutual diffusion coefficient D. Least-squares analysis yielded $R_0 = 27.8$ Å and $D = 2.64 \times 10^{-5}$ cm^2/s (Table 15.1), which are reasonably close to the expected values in this solvent.

If the acceptor concentration is known, then the time-resolved data can be used to determine the Förster distance. For the indole–dansylamide D–A pair in propylene glycol, the recovered value of 24.9 Å is in good agreement with the calculated value of 24.3 Å. For the D–A pair in methanol, it is interesting to consider the value of R_0 obtained if diffusion is neglected, which is accomplished by setting $D = 0$ during the least-squares analysis. In this case the apparent value of R_0 is larger, 37.1 Å. This result provides a useful hint. If the time-resolved data yield a larger than expected R_0 value, the cause could be diffusion during the donor excited-state lifetime. It should be noted that we have assumed that the distance of closest approach (r_c) is

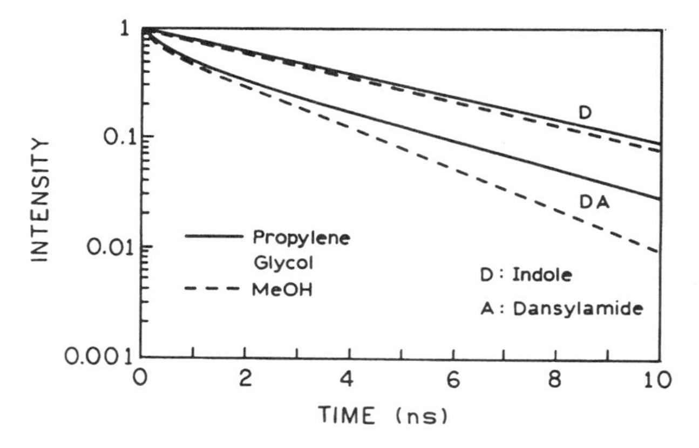

Figure 15.5. Reconstructed TD intensity decays of indole in methanol (– – –) and in propylene glycol (———) without acceptor (D), with 5mM dansylamide in methanol (DA; – – –), and with 12mM dansylamide in propylene glycol (DA; ———).

zero. If r_c is a significant fraction of R_0, then Eqs. [15.1]–[15.8] are not appropriate.

The effect of diffusion on the donor intensity decay can be seen in the TD data. The impulse response functions for the indole–dansylamide mixtures are shown in Figure 15.5. It is visually evident that the presence of the dansylamide acceptor results in a nonexponential indole decay in propylene glycol. In methanol the decay is still heterogeneous but is closer to a single exponential. This trend toward a single exponential donor decay in the presence of diffusion is well known in the literature,[14] and the donor decay becomes a single exponential in the rapid-diffusion limit (Section 15.4).

15.2. EFFECT OF DIMENSIONALITY ON RET

In the previous section we saw that a random distribution of acceptor molecules around a donor resulted in a characteristic donor decay. For randomly distributed acceptors, the form of the donor decay depends on the dimensionality of the acceptor distribution. Different donor decays are expected for acceptor distributions in a volume, in a plane, or along a line. Planar distributions are expected for donors and acceptors in membranes, and linear distributions are expected for dyes intercalated into double-helical DNA (Figure 15.1). Assuming no diffusion, no excluded volume, and a random distribution of donors and acceptors in two dimensions or one dimension, the intensity decays are known in analytical form. For a random two-dimensional distribution,[15,16]

$$I_{DA}(t) = I_D^0 \exp\left[-\frac{t}{\tau_D} - 2\beta\left(\frac{t}{\tau_D}\right)^{1/3}\right] \qquad [15.9]$$

where

$$\beta = \frac{\Gamma(2/3)}{2}\frac{C}{C_0}, \qquad \Gamma(2/3) = 1.354177\ldots \qquad [15.10]$$

and

$$C_0 = (\pi R_0^2)^{-1} \qquad [15.11]$$

Hence, C/C_0 is the number of acceptor molecules in an area equal to πR_0^2, that of a circle with diameter R_0. For a one-dimensional distribution of acceptors, the donor intensity decay is given by

$$I_{DA}(t) = I_D^0 \exp\left[-\frac{t}{\tau_D} - 2\delta\left(\frac{t}{\tau_D}\right)^{1/6}\right] \qquad [15.12]$$

where

$$\delta = \frac{\Gamma(5/6)}{2}\frac{C}{C_0}, \qquad \Gamma(5/6) = 1.128787\ldots \qquad [15.13]$$

and

$$C_0 = \frac{1}{2R_0} \qquad [15.14]$$

In this case, the ratio C/C_0 is the number of acceptor molecules within a linear distance R_0 of the donor. For one-, two-, and three-dimensional distributions of acceptors, there exist components which decay as $t^{1/6}$, $t^{1/3}$, and $t^{1/2}$, respectively. Values of the gamma function can be found from standard mathematical tables using $\Gamma(\alpha + 1) = \alpha\Gamma(\alpha)$. One may be interested in obtaining the relative quantum yields of the donor. Although such expressions are available in one and two dimensions, these are infinite-series and not closed-form expressions. With presently available computers and software, it is equally easy to perform numerical integration of Eqs. [15.1], [15.9], and [15.12] to obtain the donor transfer efficiency. For $C = C_0$, that is, $\gamma = \beta = \delta = 1.0$, the energy transfer efficiencies are 72, 66, and 63% in three, two, and one dimensions, respectively. A graph of the relative donor quantum yields is given in Problem 15.3.

Prior to examining experimental data, it is of interest to examine the forms of the donor decays. The effects of dimensionality on RET are shown using simulated data in Figure 15.6. For these simulations, the acceptor concentrations were taken as equal to the values of C_0, which allows

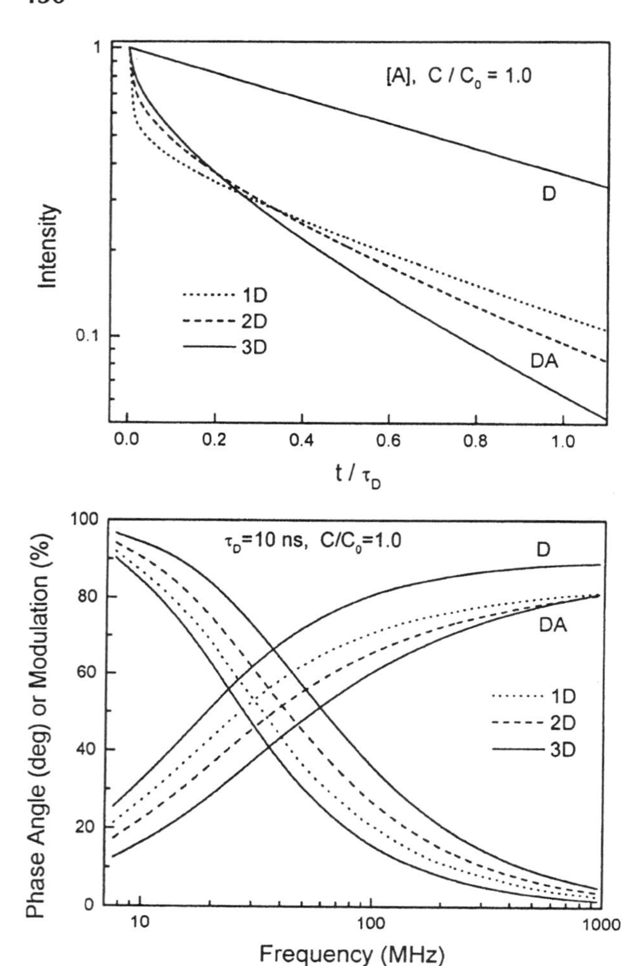

Figure 15.6. Effect of RET in one, two, and three dimensions on the donor intensity decays. Revised from Ref. 17.

the forms of the intensity decays to be compared. As the dimensionality is reduced, the TD intensity decays show an increase in the contribution of the short-time components, which decay faster in one dimension than in three dimensions. The differences in the intensity decays can also be seen in the FD simulations. These simulations suggest that the time-resolved decays can be used to determine the dimensionality of the system. In fact, analysis of the simulated FD data showed that the decay for one-dimensional RET could not be analyzed in terms of RET in two or in three dimensions, and vice versa.[17,18] That is, the form of the intensity decay is unique in each dimension. In such studies, the geometry of the system is often described as a combination of one, two, and three dimensions. Several groups have described the use of RET to determine the fractal dimensions of molecular surfaces.[19–21]

15.2.A. Experimental RET in Two Dimensions

In spite of the considerable interest in membrane organization, there have been relatively few time-resolved studies of RET in two dimensions. We will now describe two reports, one which confirmed the expected decay in two dimensions[22] (Eq. [15.9]) and one which found more complex behavior.[23]

Energy transfer in two dimensions was examined using octadecylrhodamine B (ORB) as the donor and a membrane-bound cyanine dye, $DiIC_1(7)$ as the acceptor (Figure 15.7). These dyes were dispersed in large unilamellar vesicles of DPPC. The intensity decay of ORB became more rapid in vesicles containing increasing concentrations of the acceptor (Figure 15.7). When the temperature was above the lipid phase-transition temperature (~37 °C), the data were adequately fit to the two-dimensional intensity decay law. The data for temperatures both above and below the phase-transition temperature were fit to a general expression,

$$I_{DA}(t) = \exp\left[-\frac{t}{\tau_D} - c\left(\frac{t}{\tau_D}\right)^{d/6}\right] \qquad [15.15]$$

where time-independent c is a constant[21] and d depends on dimensionality. The constant c is related to the surface density of the acceptors and the dimensionality of the system.[24,25]

The time-resolved ORB donor decays (Figure 15.7) were used to recover the value of d in Eq. [15.15]. This was accomplished by least-squares analysis and examination of the χ_R^2 surfaces. For DPPC vesicles at 50 °C, above the phase-transition temperature near 37 °C, the value of d was near 2 and the donor decays were consistent with energy transfer in two dimensions ($t^{1/3}$ dependence). At 25 °C, below the phase transition of the lipids, the time-resolved data were no longer consistent with two-dimensional RET. In this case, the value of d was found to be 0.83–1.0, which is expected for energy transfer in one dimension ($t^{1/6}$ dependence). This effect was explained in terms of co-localization of the donor and acceptor along defect lines in the bilayers. While one may argue with the exact interpretation of the results, it is clear that RET depends on the spatial distribution of donors and acceptors in the lipids. One can use the time-resolved decays to determine d in Eq. [15.15], and thus the fractal dimension of the system. This has been accomplished for lipid bilayers[26] and for dyes adsorbed to silica surfaces[27–29] and to latex spheres.[30]

Returning to ORB adsorbed to DPPC vesicles, an important aspect of the analysis was measurements of the steady-state donor intensities, in addition to the time-

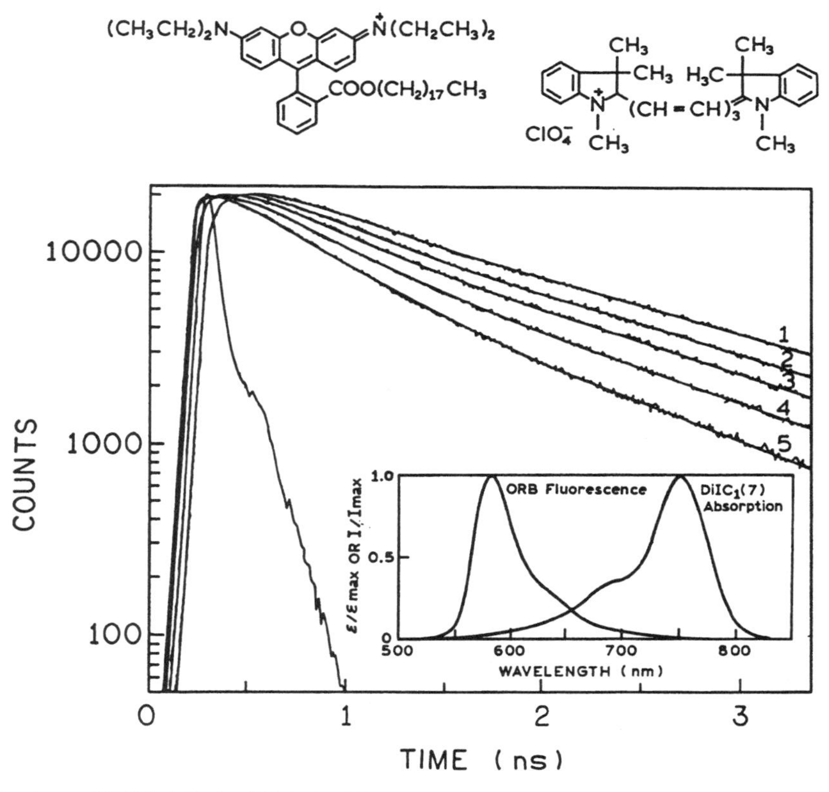

Figure 15.7. TD intensity decay of ORB in DPPC at 50 °C with $DiIC_1(7)$-to-DPPC ratios of 0 (1), 1 to 900 (2), 450 (3), 245 (4), and 160 (5). *Inset*: ORB emission spectrum and $DiIC_1(7)$ absorption spectrum. Revised from Ref. 22.

resolved decays. For the fluid-phase vesicles, the intensities were well fit by theoretical data based on Monte Carlo simulations (Section 15.3). In contrast, for the gel-phase vesicles at 25 °C, the extent of donor quenching was greater than predicted for a random two-dimensional distribution (Figure 15.8). The direction of the deviations was toward the prediction for a one-dimensional system. This is what led the authors[22] to conclude that the donors and acceptors were co-localized around defect lines for the gel-phase lipid. The important point of this comparison is

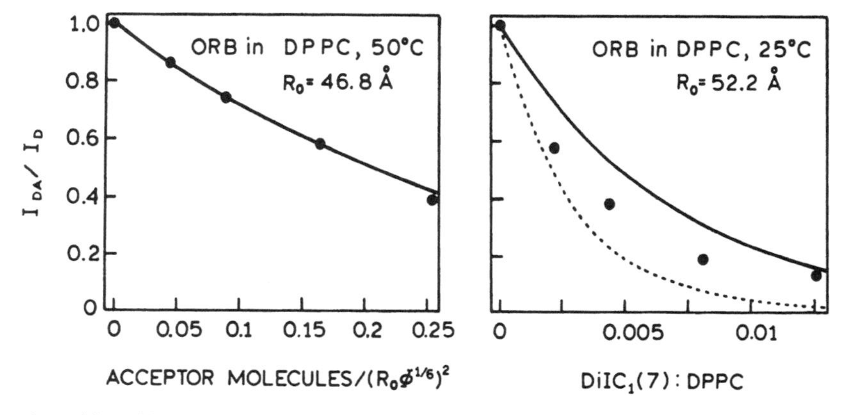

Figure 15.8. ORB donor intensities with increasing concentrations of $DiIC_1(7)$ in DPPC vesicles at 50 °C (*left*) and 25 °C (*right*). *Left*: The curve is for a two-dimensional model[43] with $R_0 = 46.8$ Å. Φ is the donor quantum yield. *Right*: The solid and dashed curves are the theoretical predictions for RET in two and one dimensions, respectively, with $R_0 = 52.2$ Å.[44] From Ref. 22.

that one can gain important insight by comparison of the steady-state and the time-resolved data. Such a comparison is a form of global analysis, that is, inclusion of multiple types of data for comparison with a given molecular model.

For membrane-bound probes, it is not always clear whether energy transfer is occurring in two or in three dimensions. Energy transfer to surface-localized fluorophores on the side of the membrane is likely to be two-dimensional. However, RET may display a different dimensionality for donors on the surface transferring to acceptors dispersed in the acyl side-chain region. Also, acceptors on the opposite surface from the donor may contribute a three-dimensional component to the donor decay. While such systems can be interpreted in terms of fractal dimensions, the concept of fractal dimensions is rather abstract and does not always lead to physical insights. In these cases, it is useful to consider a combination of energy transfer in two and three dimensions.

Energy transfer with a mixed dimensionality was found for the time-resolved donor decays from rhodamine 6G to Malachite Green, when both were adsorbed to vesicles of dihexadecyl phosphate (DHP).[23] The donor decays are shown in Figure 15.9. In this case it was necessary to fit the data to a sum of Eqs. [15.1] and [15.9]. Neither Eq. [15.1] nor Eq. [15.9] alone provided a good fit to these data. While the authors interpreted this effect in terms of a nonrandom acceptor distribution, energy transfer across the bilayer could also have provided a component which appeared to be three-dimensional. Once again, comparison of the steady-state data with predicted donor intensities was essential for selecting between distinct models.[23]

15.2.B. Experimental RET in One Dimension

While there have been numerous studies of the time-resolved fluorescence of dyes bound to DNA,[31] there have been relatively few studies of RET for dyes intercalated into DNA.[18,32] In Figure 15.6 we showed simulations which indicated a rapidly decaying $t^{1/6}$ component for donors and acceptors in one dimension. Such a time-dependent decay has been observed with dyes intercalated into poly d(A-T).[32] The donor was dimethyldiazaperopyrenium (DMPP) and the acceptor was ethidium bromide (EB). Upon binding of the EB acceptor to the poly d(A-T), the DMPP emission was quenched, and the EB emission was enhanced (Figure 15.10). The TD data clearly show a fast component with increasing amplitude as the acceptor concentration is increased (Figure 15.11). Owing to lack of software, the TD data were not analyzed in terms of Eq. [15.12], but the shape of the decays is visually similar to the simulated data for RET in one dimension (Figure 15.6).

As found for energy transfer in membranes, useful information was obtained by examination of the steady-state data. For the D–A pair, the observed transfer efficiency (Figure 15.10, inset) was found to be smaller than that predicted from Monte Carlo simulations. This result was explained in terms of distortion of the DNA by binding of DMPP, which prevented EB from binding to nearby sites.

Energy transfer in one dimension was also studied using the FD method.[18] In this case the donor was Acridine Orange (AO) and the acceptor was a nonfluorescent dye, Nile Blue (NB). Binding of NB to AO-DNA at a low dye/base-pair ratio resulted in significant quenching of the

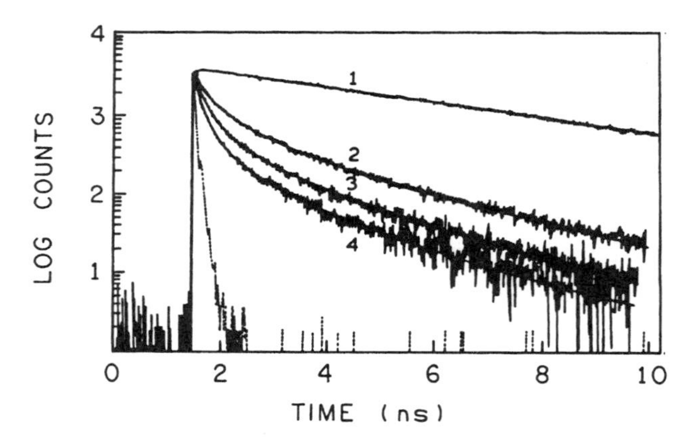

Figure 15.9. Donor decays for energy transfer between donors and acceptors adsorbed to dihexadecyl phosphate vesicles. The donor was rhodamine 6G, and the acceptor was Malachite Green. Concentration of Malachite Green: (1) 0; (2) $3.32 \times 10^{-6} M$; (3) $4.40 \times 10^{-6} M$; (4) $5.27 \times 10^{-6} M$. Revised and reprinted, with permission, from Ref. 23, Copyright © 1987, American Chemical Society.

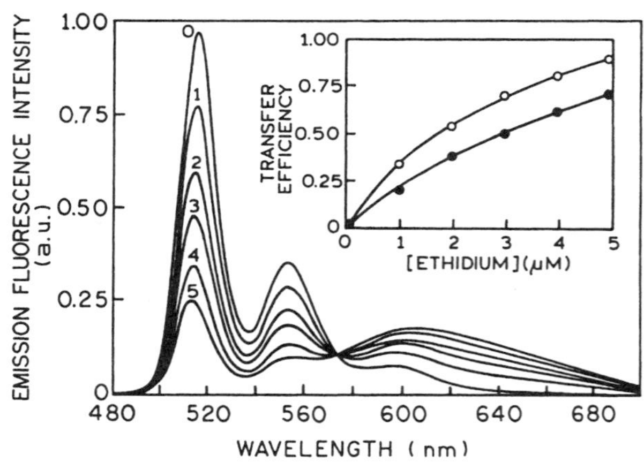

Figure 15.10. Emission spectra of DMPP bound to poly (dA-dT) with different concentrations of EB. The numbers on the spectra are the micromolar concentrations of EB. *Inset*: Comparison of the measured (•) and predicted (○) transfer efficiencies. Revised from Ref. 32.

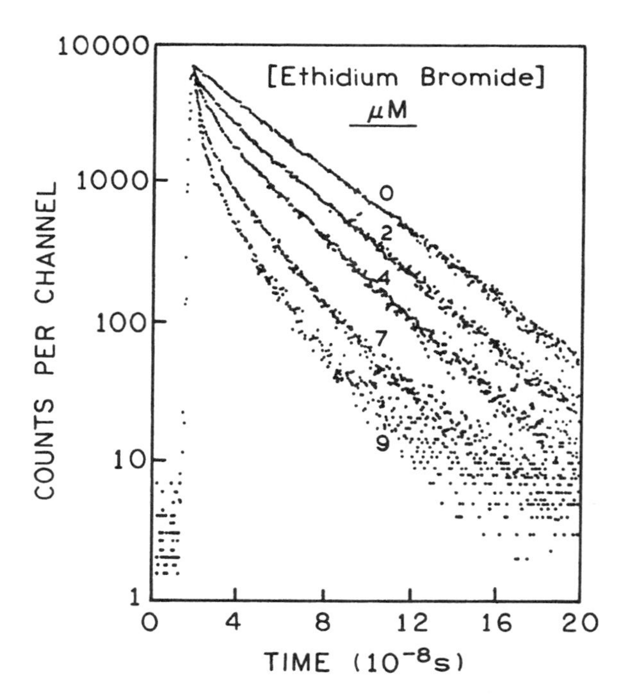

Figure 15.11. Intensity decays of DMPP bound to double-stranded poly (dA-dT) with different concentrations of ethidium bromide. Revised from Ref. 32.

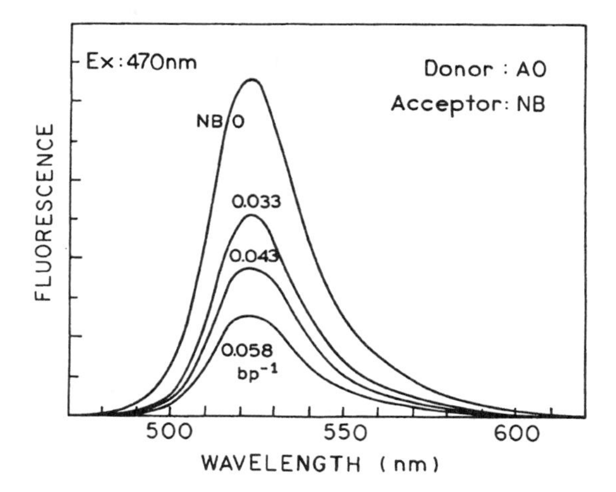

Figure 15.12. Emission spectra of Acridine Orange (AO) bound to DNA with different concentrations of Nile Blue (NB) acceptor per DNA base pair. From Ref. 18, Maliwal, B. P., Kušba, J., and Lakowicz, J. R., Fluorescence energy transfer in one dimension: Frequency-domain fluorescence study of DNA–fluorophore complexes, *Biopolymers* **35**:245–255, Copyright © 1994. Reprinted with permission from John Wiley and Sons, Inc.

AO emission (Figure 15.12). The FD intensity decay data are best fit by the equation for RET in one dimension (Figure 15.13), indicating that energy transfer occurs in one dimension in this system.

While the values of χ_R^2 support the one-dimensional (1D) model, there are only minor visual differences between the fitted functions for 1D, 2D, and 3D RET. However, analysis of the data should also include consideration of the parameter values and the reasonableness of these values. For Eqs. [15.1], [15.9], and [15.12], once R_0 is known, the concentration is the only variable parameter. The concentrations recovered for NB from the analysis in Figure 15.13 are 0.058 acceptors/base pair, 9.9×10^{11} molecules/cm^2, and 1.32mM for the 1D, 2D, and 3D models, respectively. Based on one's understanding of the sample preparation, it should be readily possible to exclude the concentration of 1.32mM recovered from the 3D model. This is the concentration of acceptors in a 3D solution that would result in energy transfer comparable to that observed for AO-DNA. The essential point is that one should always examine the desired parameters for reasonableness based on one's chemical knowledge of the system. In our experience, we have found that the failure of parameter values to follow expected trends is often a sensitive indicator of the lack of validity of the model, often more sensitive than the values of χ_R^2 themselves.

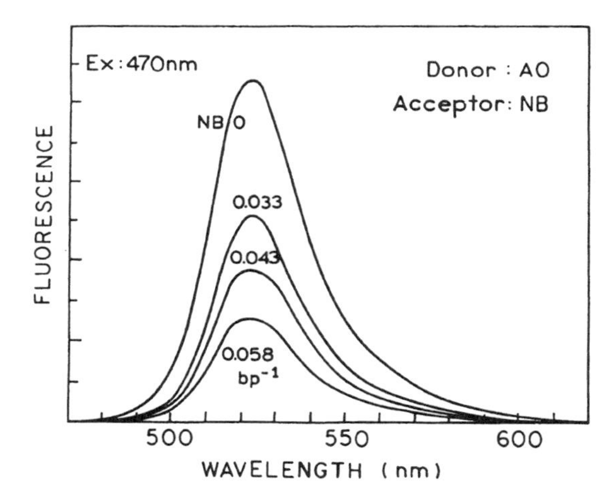

Figure 15.13. FD intensity decay of AO-DNA with 0.058 NB/base pair. The Cower panels show the residuals for the fits of the experimental data using the equations for RET in one (•), two (○), and three dimensions (△). From Ref. 18, Maliwal, B. P., Kušba, J., and Lakowicz, J. R., Fluorescence energy transfer in one dimension: Frequency-domain fluorescence study of DNA–fluorophore complexes, *Biopolymers* **35**:245–255, Copyright © 1994. Reprinted, with permission, from John Wiley and Sons, Inc.

15.3. ENERGY TRANSFER IN RESTRICTED GEOMETRIES

For the systems described above, a random distribution of acceptors, in one, two, or three dimensions, was assumed. One can imagine other situations in which the acceptor distribution is nonrandom. For instance, consider a donor-labeled protein which is bound to a membrane. Depending on the size of the protein, the acceptors may be excluded from a region directly around the donor. In this case the acceptor distribution would be random in two dimensions, but with a minimum distance between the donor and the nearest acceptor (Figure 15.14, inset). Many other nonrandom distributions can be imagined, such as DNA dyes with preferred binding to particular base sequences and distributions of charged acceptors around charged donors.

There have been many attempts to provide analytical expressions for a variety of geometric conditions. These attempts have resulted in complex expressions, which typically are valid under a limited range of conditions. For instance, an analytical solution for 2D RET was given, but this solution only applies when the distance of closest approach (r_c) is much less than R_0.[33] Another approximate solution was presented for the case where $r_c/R_0 > 0.7$.[34] Other solutions have been given,[35–43] and even these complex formulas do not include D–A diffusion. Given the complexity of the equations and the range of possible conditions, there is merit in directly calculating the data for the desired geometry and D–A distribution. This can be accomplished by Monte Carlo simulations[44] or by use of the general expressions.[45]

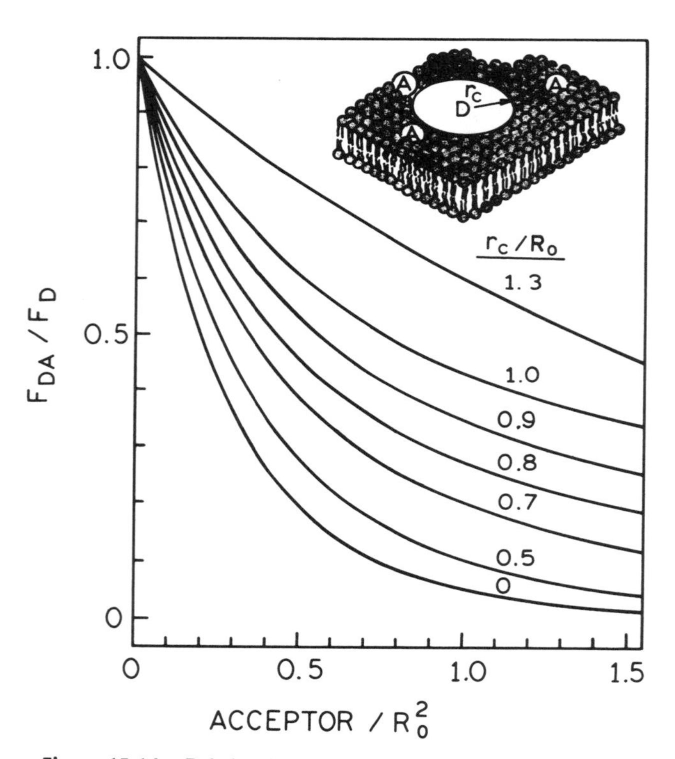

Figure 15.14. Relative donor yields for a random distribution of acceptors in two dimensions. The numbers on the curves are the ratios of the distance of closest approach of the donor and acceptor (r_c) to the Förster distance (R_0). The curves were calculated according to Ref. 33 (Table 15.2).

Monte Carlo simulations provide a rather ideal way to simulate the data for complex systems. The basic idea is to simulate data for one assumed configuration of the system, such as for a given number of acceptor molecules at given distances from the donor. This process is repeated many times, for random configurations generated for the assumed model. As stated in Ref. 44, such simulations can be performed for systems of arbitrary complexity, and to any desired precision. Another positive feature is that one does not need to derive an equation for the donor decay or donor intensity but instead simply writes the differential equation describing the donor. This is clearly an advantage for any but the most mathematically inclined. However, the necessary simulations can be time-consuming even with modern computers, particularly if the Monte Carlo simulations must be done within least-squares fits.

Fortunately, there exists a general solution which can be used for systems of almost any complexity, in the absence of diffusion.[45] The basic idea is to use an integral equation to predict the intensity decay. In two dimensions, the equation is

$$I_D(t) = I_D^0 \exp(-t/\tau_D) \exp[-\sigma S(t)] \quad [15.16]$$

where

$$S(t) = \int_{r_c}^{\infty} \{1 - \exp[-(t/\tau_D)(R_0/r)^6]\}\, 2\pi r\, dr \quad [15.17]$$

The term $\exp[-\sigma S(t)]$ describes that portion of the donor decay due to energy transfer, σ is the surface density of the acceptor, and r_c is the distance of closest approach between the donors and acceptors. The expressions appropriate to one or three dimensions can be written by substituting the surface element $2\pi r$ with 2.0 or $4\pi r^2$ for a line or sphere, respectively. The donor intensity at time t can be calculated for any assumed probability distribution by numerical integration of the appropriate equation. The efficiency of transfer can be calculated from

$$E = 1 - \frac{\int_0^\infty I_{DA}(t)\,dt}{\int_0^\infty I_D(t)\,dt} \qquad [15.18]$$

While the use of these equations will be moderately complex, the approach is general and can be applied to most circumstances. For instance, the existence of an excluded volume around the donor is readily tested by changing the lower limits on the integral over distance. Example calculations using Eqs. [15.16] and [15.17] were shown in Figures 13.21 and 13.22. The transfer efficiency predicted using Eqs. [15.16]–[15.18] can be used as evidence for nonrandom distributions of donors and acceptors in membranes.[46–48] The basic idea is to compare the efficiency of energy transfer (Eq. [15.18]) with that calculated for a random distribution. Larger or smaller transfer efficiencies are taken to indicate co-localization or exclusion of the probes from areas containing the donor. It should be noted that the intensity decays of membrane-bound donors do not always agree with the model for RET in two dimensions.[45]

15.3.A. Effect of an Excluded Area on Energy Transfer in Two Dimensions

Energy transfer from a donor to acceptor is two dimensions in a frequently encountered model in membrane biophysics. Hence, it is valuable to visualize how the donor intensities depend on the surface density of the acceptor and the distance of closest approach (r_c). These values can be obtained from approximation to the exact theory (Table 15.2).[33] The data in Table 15.2 provide a means of calculating the relative quantum yield of the donor using simple numerical equations. These data show that an excluded area results in a significant decrease in the transfer efficiency as soon as r_c exceeds $0.5R_0$. One can use the calculated values in Figure 15.14 for comparison with experimental data to estimate r_c.

15.4. RET IN THE RAPID-DIFFUSION LIMIT

The theory for energy transfer in restricted geometries is obviously complex. Hence, one may question how we can consider the further complications of combining restricted geometries with D–A diffusion. In fact, such theory has not yet appeared in the literature. However, if the donor decay time is very long, then one reaches the rapid-diffusion limit,[49,50] where the theory once again becomes relatively simple.

The basic idea of energy transfer in the rapid-diffusion limit is to use a donor lifetime (τ_D) and acceptor concentration such that the diffusion distance of the donor during τ_D is greater than the mean distance(s) between the donor and acceptor molecules. The rapid-diffusion limit is reached when $D\tau_D/s^2 >> 1$, where $D = D_D + D_A$ is the mutual diffusion coefficient and s is the mean distance between D and A. There are several valuable consequences of being in the rapid-diffusion limit.[49] The concentration of acceptors needed for RET can be 1000-fold less than in the static limit ($D\tau_D/s^2 << 1$), that is, micromolar is, rather than millimolar concentrations (Section 15.1). The donor intensity decays are single exponentials because the acceptor distribution is averaged by diffusion, so that all donors see the same distribution. Finally, the extent of transfer becomes limited by the distance of closest ap-

Table 15.2. Numerical Parameters to Calculate the Relative Donor Quantum Yields in Membranes[a]

r_c/R_0	A_1	k_1	A_2	k_2
0.0	0.6463	4.7497	0.3537	2.0618
0.25	0.6290	4.5752	0.3710	1.9955
0.5	0.6162	4.0026	0.3838	1.4430
0.7	0.6322	3.1871	0.3678	0.7515
0.8	0.6344	2.7239	0.3656	0.4706
0.9	0.6336	2.2144	0.3664	0.2909
1.0	0.6414	1.7400	0.3586	0.1285
1.1	0.6327	1.3686	0.3673	0.04654[b]
1.3	0.6461	0.4899	0.3539	0.005633[b]

[a]From Reference 33. The relative quantum yields are given by $F_{DA}/F_D = A_1 \exp(-k_1 C) + A_2 \exp(-k_2 C)$ where C is concentration of acceptor per R_0^2. The calculated values are accurate to better than 1% for $0 \leq 0.5$.

[b]These values seem to be incorrect in Ref. 33. We decreased the published k_2 values by a factor of 10 and 100, for $r_c/R_0 = 1.1$ and 1.3, respectively. B. Hudson and P. Wolber confirmed that $k_2 = 0.04654$ at $r_c/R_0 = 1.1$.

proach of the donor and acceptor and can thus provide structural information about the investigated system.

What donor decay times are needed to reach the rapid-diffusion limit? This question was addressed by numerical solution of the differential equations describing energy transfer in three dimensions.[8,49] Figure 15.15 shows the calculated transfer efficiencies for donors with various lifetimes when the acceptor concentration is 0.1m*M*.[50] If the decay time is 1 ns, there is little effect of diffusion on energy transfer, even for the highest possible diffusion coefficient of 10^{-5} cm^2/s in aqueous solution at room temperature. This is why D–A diffusion is often neglected in measurements of D–A distances and why RET with nanosecond-lifetime donors has not been used to measure domain dynamics in proteins.

If the donor lifetime is near 1 μs, then the transfer efficiency becomes sensitive to diffusion. Such decay times are available using transition-metal–ligand complexes, as described in Chapter 20. As the donor lifetime becomes longer, 1 ms–1 s, the transfer efficiency reaches an upper limit. This is the rapid-diffusion limit, which can be reached for diffusion coefficients exceeding 10^{-7} cm^2/s if the donor lifetimes are on the millisecond timescale.

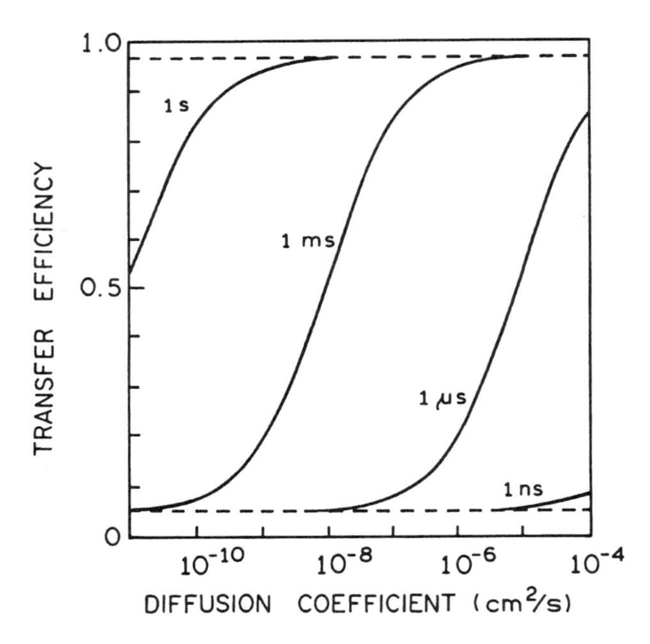

Figure 15.15. Calculated dependence of the transfer efficiency on the diffusion coefficient in three dimensions for donor lifetimes τ_D of 1 ns, 1 μs, 1 ms, and 1 s. *D* is the sum of the diffusion coefficients of the donor and the acceptor. In this calculation, $R_0 = 50$ Å, $r_c = 5$ Å, and the acceptor concentration is 0.1m*M*. Revised and reprinted, with permission, from Ref. 50, Copyright © 1982, Annual Review, Inc.

Such decay times are found in the lanthanides such as europium and terbium, which display lifetimes near 2 ms.

Examination of Figure 15.15 reveals that the transfer efficiency reaches a limiting value of less than 100% for high diffusion coefficients. This rapid-diffusion limit is sensitive to the distance of closest approach of the donor to the acceptors. The transfer efficiency is given by

$$E = \frac{k_T}{\tau_D^{-1} + k_T} \quad [15.19]$$

where k_T is the sum of the transfer rates to all available acceptors. The transfer rate k_T can be calculated from the decay times measured in the absence (τ_D) and presence of acceptors (τ_{DA}),

$$k_T = \frac{1}{\tau_{DA}} - \frac{1}{\tau_D} \quad [15.20]$$

Because of diffusive averaging, it is relatively simple to calculate predicted values of k_T. The precise form of k_T depends on the geometric model. Many such models have been described in detail.[49–53] For spherical donors and acceptors, the diffusion-limited value of k_T is given by[49]

$$k_T = \rho_A \int_{r_c}^{\infty} \frac{1}{\tau_D}\left(\frac{r}{R_0}\right)^6 4\pi r^2\, dr = \frac{4\pi\rho_A R_0^6}{3\tau_D r_c^3} \quad [15.21]$$

where ρ_A is the density of acceptors (molecules/Å^3) and r_c is the distance of closest approach. Equation [15.21] can be understood as the sum of all transfer rates to acceptors randomly distributed in three dimensions starting at r_c. The diffusion-limited transfer rate is thus dependent on r_c^{-3} and the acceptor concentration, as shown in the top panel of Figure 15.16 for various acceptor concentrations. For r_c values less than 10 Å, the transfer efficiencies can exceed 50% for an acceptor concentration of 10μ*M*, considerably lower than in the absence of diffusion (Section 15.1). This model can be employed to measure the distance at which an acceptor is buried in a protein, with the use of a long-lived donor. One such study measured transfer from terbium to the iron binding sites in the protein transferrin.[52] From the transfer rates, the iron binding sites were determined to be deeply buried in the protein, 17 Å below the surface.

Suppose that one had a long-lived lipid derivative which served as the donor and that the membrane also contained acceptors. In two dimensions, the diffusion-limited rate constant is given by

$$k_T = \frac{1}{\tau_D} \int_{r_c}^{\infty} \left(\frac{r}{R_0}\right)^6 \sigma_A 2\pi r\, dr = \frac{\pi \sigma_A R_0^6}{2r_c^4} \qquad [15.22]$$

where σ_A is in units of acceptors/Å^2. Hence, in two dimensions the transfer efficiency is proportional to r_c^{-4}, whereas in three dimensions it is proportional to r_c^{-3}. Given the simplicity of calculating k_T, these expressions have been obtained for a variety of geometric models.[50]

15.4.A. Location of an Acceptor in Lipid Vesicles

Diffusion-limited energy transfer has also been applied to membrane-bound fluorophores, but to a case different from lateral diffusion (Eq. [15.22]). One example, shown in the bottom panel of Figure 15.16, is for donors trapped within a vesicle of radius b. The acceptors are at a known surface density (σ_A in molecules/Å^2) located at a distance r_c below the surface. The diffusion-limited transfer rate is given by

$$k_T = \frac{3\pi b \sigma_A R_0^6}{2\tau_D (b - r_c)^3} \left\{ \frac{1}{2} [(2b - r_c)^{-2} - r_c^{-2}] + \frac{b}{3} [r_c^{-3} - (2b - r_c)^{-3}] \right\} \qquad [15.23]$$

These values are shown in the bottom panel of Figure 15.16 for various surface densities of the acceptor. Energy transfer can be 50% efficient with only 1 acceptor per 5000 phospholipids, if the distance of closest approach is 10 Å.

This model was applied to egg PC vesicles labeled with eosin-PE as the acceptor. The donor was $Tb(DPA)_3$, where DPA is dipicolinic acid. The donor was trapped in the internal aqueous volume of the vesicles. Energy transfer from the $Tb(DPA)_3$ to the eosin acceptor was about 50% efficient at just 1 acceptor per 1000 phospholipid molecules (Figure 15.17). The distance of closest approach of $Tb(DPA)_3$ to eosin was determined by comparison with curves calculated using Eq. [15.23]. The overall diameter of the vesicles was estimated by electron microscopy, allowing b to be fixed at 150 Å. The extent of energy transfer was consistent with a distance of closest approach of 10 Å (Figure 15.17), suggesting that the eosin was localized just under the surface of the membrane. It was not necessary to consider eosin in the outer bilayer since the distance of closest approach determined the transfer efficiency.

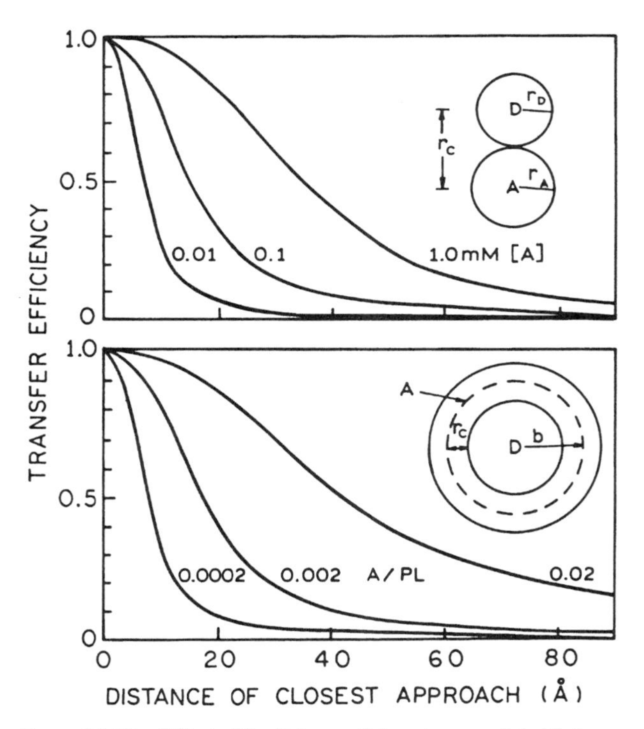

Figure 15.16. Effect of the distance of closest approach (r_c) between the donor and the acceptor on the transfer efficiency in the rapid-diffusion limit. R_0 = 50 Å in this calculation. *Top*: Solution of donors and acceptors. The acceptor concentrations are 1.0, 0.1, and 0.01m*M*. *Bottom*: Solution of donors trapped in the inner aqueous space of a membrane vesicle containing a spherical shell of acceptors at radius b = 150 Å. The surface densities of acceptor are 0.02, 0.002, and 0.0002 per phospholipid (acceptors per 70 Å^2). Revised from Ref. 49.

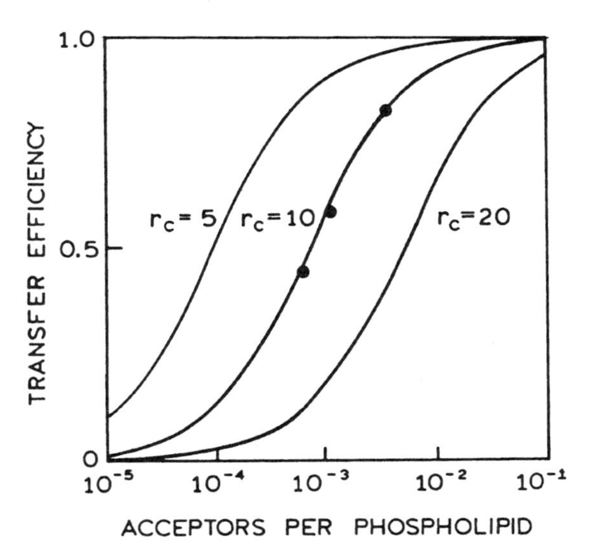

Figure 15.17. Diffusion-limited energy transfer from $Tb(DPA)_3$ trapped in the inner volume of egg PC vesicles labeled with eosin-PE as the acceptor. The solid curves were calculated from Eq. [15.23] assuming R_0 = 45.6 Å, b = 150 Å, and assuming an area of 70 Å^2 per phospholipid. Revised from Ref. 49.

15.4.B. Location of Retinal in Rhodopsin Disk Membranes

The model of a long-lived donor trapped in vesicles was applied to rhodopsin, the photoreceptor protein in retinal rods. This membrane-bound protein contains a retinal chromophore, which serves as the acceptor (Figure 15.18).[51,54] It was possible to prepare vesicles containing rhodopsin which had the same sidedness as native disk membranes.

The absorption spectrum of the acceptor is shown in the bottom panel of Figure 15.19. A favorable aspect of retinal is that it can be photobleached, thus decreasing the acceptor concentration in the disk membranes. Prior to photobleaching, the absorption spectrum of retinal overlaps with that of the Tb^{3+} donor. The emission spectrum of terbium is typical of the lanthanides. The emission is from *f* orbitals of the atom, and the emission spectra are typically highly structured line spectra. Also, it is common to use lanthanide chelates, rather than lanthanides alone. This is because the lanthanides are extremely weak absorbers, with extinction coefficients near 0.1 M^{-1} cm^{-1}. Fortunately, lanthanides can be chelated with aromatic absorbers such as dipicolinate (Figure 15.19). Light absorbed by the ligand is efficiently transferred to the lanthanide, resulting in effective absorption coefficients orders of magnitude larger than those of the uncomplexed lanthanides.

In order to localize the retinal in membranes, Tb^{3+} was added to the inside or the outside of vesicles. Intensity decays for Tb^{3+} trapped inside the disk vesicles are shown in Figure 15.20. The intensity decay is a single exponential in the absence and in the presence of rhodopsin, as predicted for diffusive averaging. The Tb^{3+} decays as a single exponential in the presence of acceptors because all of the

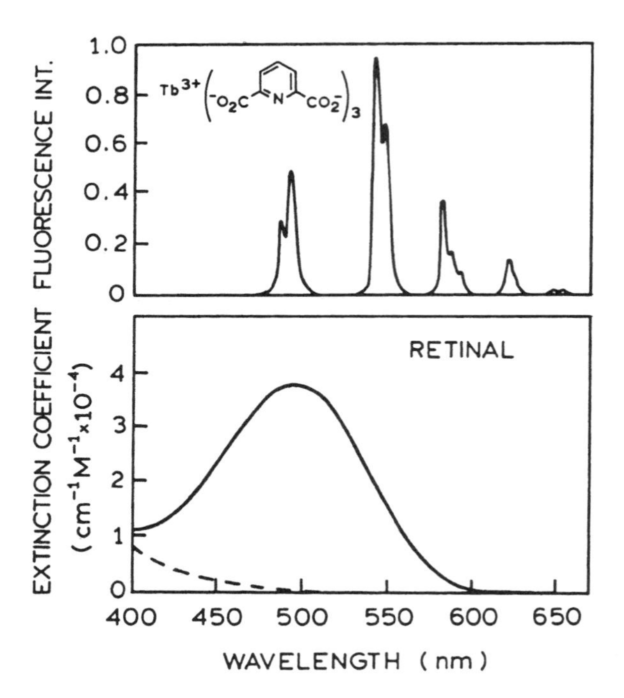

Figure 15.19. *Top*: Emission spectrum of the donor terbium dipicolinate. *Bottom*: Absorption spectrum of the retinal acceptor, before (——) and after (– – –) photobleaching. Reprinted, with permission, from Ref. 51, Copyright © 1982, Academic Press, Inc.

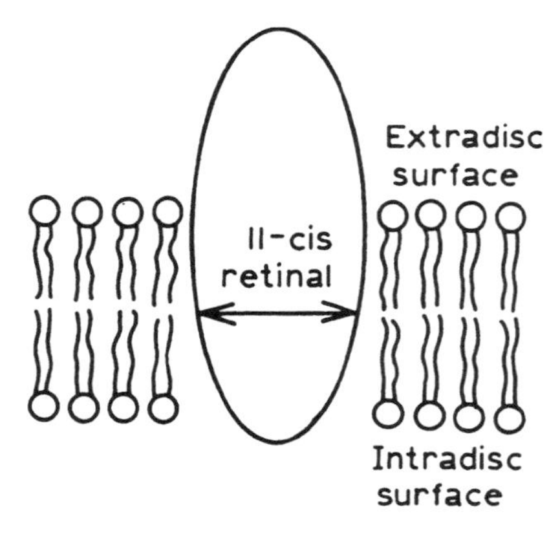

Figure 15.18. Schematic diagram of rhodopsin in a membrane. The single retinal moiety is placed midway between the membrane surfaces, roughly consistent with the energy-transfer data. Reprinted, with permission, from Ref. 51, Copyright © 1982, Academic Press, Inc.

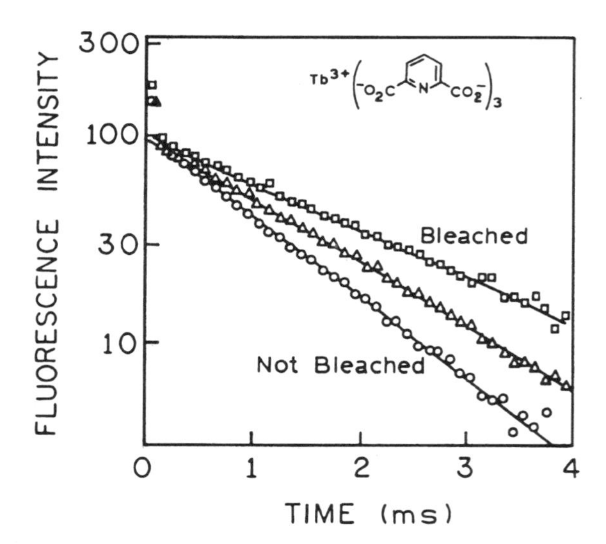

Figure 15.20. Emission kinetics of terbium dipicolinate trapped inside disk membrane vesicles. The fluorescence intensity (plotted on a logarithmic scale) is shown as a function of time after a 1-μs exciting light pulse. ○, Unbleached membranes (fastest decay); △, partially bleached; □, completely bleached. Revised from Ref. 51.

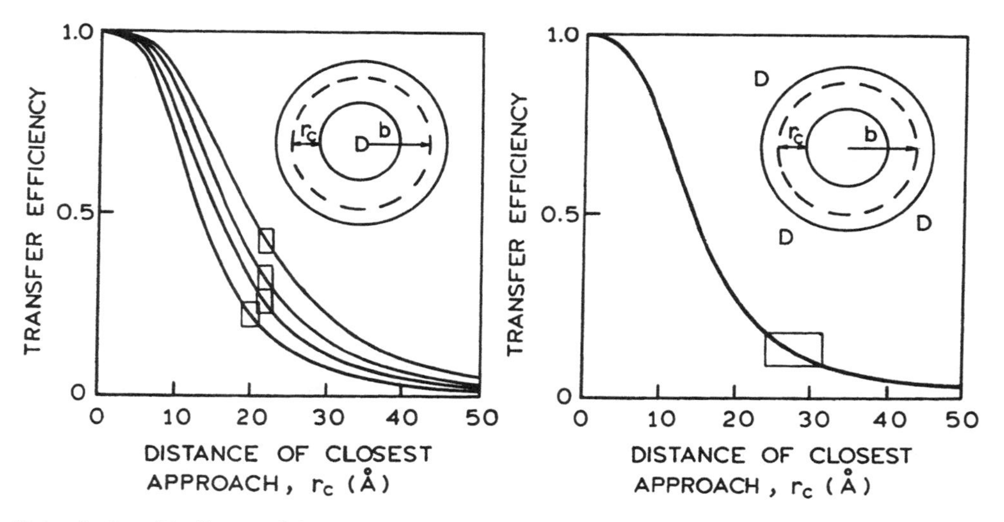

Figure 15.21. Determination of the distance of closest approach of terbium dipicolinate from the inner membrane surface (*left*) and outer membrane surface (*right*) to 11-*cis*-retinal in rhodopsin. The observed transfer efficiencies for different degrees of bleaching are shown as rectangles. The curves show the calculated dependence of the transfer efficiency on r_c for different densities of acceptors (corresponding to different degrees of bleaching). For external donors (*right*), the rhodopsin concentration was 0.4m*M*. The sizes of the rectangles indicate estimated uncertainties in E and r_c. Revised from Ref. 51.

Tb^{3+} donors experience the same diffusion-limited rate of transfer. Photobleaching of the retinal decreases its effective concentration, and the Tb^{3+} lifetime increases.

The decay times found for Tb^{3+} within the vesicles can be used to determine the transfer efficiency. Since the intensity decays are single exponentials, the efficiency can be calculated from

$$E = 1 - \frac{\tau_{DA}}{\tau_D} \quad [15.24]$$

where τ_{DA} and τ_D are the Tb^{3+} decay times in the presence and in the absence of energy transfer. The transfer efficiency is then compared with that predicted using Eq. [15.23] to determine the distance of retinal below the inner membrane surface (Figure 15.21, left). The distance of retinal from the inner membrane surface was estimated to be about 22 Å.

Diffusion-limited RET was also used to determine the distance of the retinal from the outer disk membrane. In this case the diffusion-limited transfer rate is given by

$$k_T = \frac{4\pi \rho_A R_0^6}{\tau_D^3 r_c^3} \left[2 - \frac{r_c}{r_c + b} \right]^{-3} \quad [15.25]$$

where ρ_A is the concentration of acceptor in units of molecules/$Å^3$. In this case the transfer rate depended on the bulk concentration of the rhodopsin. By comparison of the measured efficiency for a given bulk concentration with the corresponding theoretical curve (Figure 15.21, right), it was possible to localize retinal from the outer disk surface. From these combined experiments, it was concluded that the retinal chromophore was located about 28 Å from the outer surface.

15.4.C. Förster Transfer versus Exchange Interactions

While the interpretation of diffusion-enhanced energy transfer seems straightforward, there are some complications. In these examples the distances of closest approach were typically greater than 15 Å. However, there have been difficulties in interpreting the transfer rate constants for smaller molecules in solution where r_c was less than 15 Å.[53,55] At short distances, energy transfer can occur by the exchange mechanism. This mechanism becomes dominant at shorter distances (Figure 15.22) but becomes less important at distances beyond 12 Å. For this reason, diffusion-limited energy transfer cannot be used to calculate distances of closest approach if these distances are within the range of the exchange interaction. The presence of an exchange interaction is typically ascertained from transfer rates that are larger than calculated for Förster transfer.

For comparison of the observed and calculated transfer rates, it is useful to write k_T in terms of a bimolecular (b) rate constant. For spherical donors and acceptors, the rate is given by[53]

$$k_T^b = \frac{2.523 R_0^6}{\tau_D r_c^3} \quad (\text{in } M^{-1}\,\text{s}^{-1}) \quad [15.26]$$

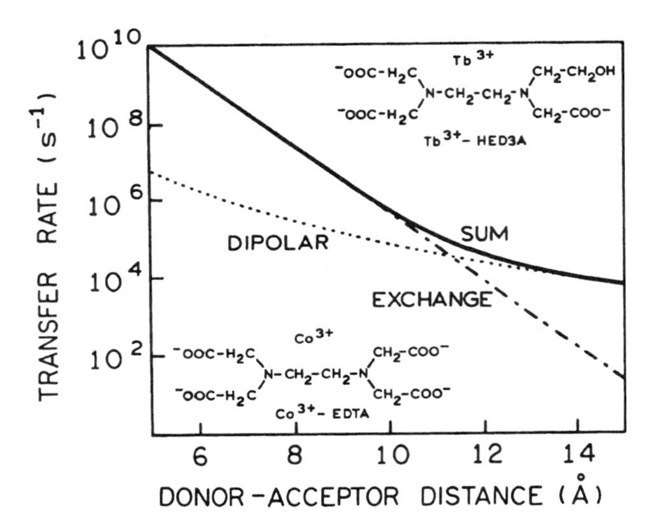

Figure 15.22. Contributions of dipole–dipole interaction and exchange interaction to the rate of energy transfer from Tb^{3+}-HED3A to Co^{3+}-EDTA as a function of donor–acceptor distance. The terbium complex has a zero net charge. Revised from Ref. 51.

This bimolecular rate constant has the usual units of M^{-1} s^{-1} when all distances are in nanometers and time is in seconds. As an example, suppose the lifetime is 2 ms, the Förster distance is 30 Å = 3.0 nm, and r_c = 12 Å = 1.2 nm. Then the maximum value of k_T consistent with Förster transfer is 5.3×10^5 M^{-1} s^{-1}. Measured values of k_T^b larger than the calculated value indicate the presence of an exchange contribution to the transfer rate and/or a distance of closest approach less than 12 Å.

For those familiar with collisional quenching and interpretation of the bimolecular rate constants for quenching (k_q), Eq. [15.26] can be confusing. Why are the calculated values of k_T^b on the order of 10^6 M^{-1} s^{-1}, whereas diffusion-limited quenching results in k_q values on the order of 10^{10} M^{-1} s^{-1}. The difference is due to the unique dependence of energy transfer on the emissive rate of the donor, which appears as the factor τ_D^{-1} in Eq. [15.26]. The rate of energy transfer depends on the decay time of the donor, whereas the rates of collisional quenching are typically independent of the decay time of the fluorophore.

One may question why the exchange interaction is important in quenching of lanthanides, but not in energy transfer from more usual fluorophores with nanosecond lifetimes. For nanosecond-lifetime fluorophores, the rates of transfer are much faster due to the dependence on τ_D^{-1}. Hence, energy transfer occurs as soon as the donors and acceptors are within the Förster distance. The longer lifetimes of the lanthanides result in slower transfer rates and thus the opportunity for the donors and acceptors to approach closely without Förster transfer.

15.5. CONCLUSIONS

The phenomenon of Förster transfer is simultaneously simple and complex. While the theory describing the mechanism of dipolar transfer is complex, the result is dependable and robust. Förster distances can be predicted with good accuracy from the spectral properties of the donor and acceptor. There are no known exceptions to Förster transfer, so that RET can be reliably assumed to occur whenever the donors and acceptors are in close proximity. Hence, RET is a reliable method for the study of the proximity and geometric distributions of donor–acceptor pairs.

The complexity of energy transfer arises from the occurrence of distance distributions, nonrandom distributions, and donor-to-acceptor diffusion. These phenomena result in complex theory, not because of Förster transfer, but because of the need to average the distance dependence over various geometries and timescales.

REFERENCES

1. Förster, V. T., 1949, Experimentelle und theoretische Untersuchung des zwischenmolekularen Ubergangs von Elektronenanregungsenergie, *Z. Naturforsch. A* **4**:321–327.
2. Förster, Th., 1959, 10th Spiers Memorial Lecture, Transfer mechanisms of electronic excitation, *Disc. Faraday Soc.* **27**:7–17.
3. Bojarski, C., and Sienicki, K., 1990, Energy transfer and migration in fluorescent solutions, in *Photochemistry and Photophysics*, Vol. I, J. F. Rabek (ed.), CRC Press, Boca Raton, Florida, pp. 1–57.
4. Galanin, M. D., 1955, The influence of concentration on luminescence in solutions, *Sov. Phys. JETP* **1**:317–325.
5. Maksimov, M. A., and Rozman, I. M., 1962, On the energy transfer in rigid solutions, *Opt. Spectrosc.* **12**:337–338.
6. Millar, D. P., Robbins, R. J., and Zewail, A. H., 1981, Picosecond dynamics of electronic energy transfer in condensed phases, *J. Chem. Phys.* **75**:3649–3659.
7. Lakowicz, J. R., Szmacinski, H., Gryczynski, I., Wiczk, W., and Johnson, M. L., 1990, Influence of diffusion on excitation energy transfer in solutions by gigahertz harmonic content frequency-domain fluorometry, *J. Phys. Chem.* **94**:8413–8416.
8. Steinberg, I. Z., and Katchalski, E., 1968, Theoretical analysis of the role of diffusion in chemical reactions, fluorescence quenching, and nonradiative energy transfer, *J. Chem. Phys.* **48**:2404–2410.
9. Elkana, Y., Feitelson, J., and Katchalski, E., 1968, Effect of diffusion on transfer of electronic excitation energy, *J. Chem. Phys.* **48**:2399–2404.
10. Kušba, J., 1987, Long-range energy transfer in the case of material diffusion, *J. Lumin.* **37**:287–291.
11. Yokota, M., and Tanimato, O., 1967, Effects of diffusion on energy transfer by resonance, *J. Phys. Soc. Jpn.* **22**:779–784.

12. Gösele, U., Hauser, M., Klein, U. K. A., and Frey, R., 1975, Diffusion and long-range energy transfer, *Chem. Phys. Lett.* **34:**519–522.
13. Faulkner, L. R., 1976, Effects of diffusion on resonance energy transfer. Comparisons of theory and experiment, *Chem. Phys. Lett.* **43:**552–556.
14. Birks, J. B., and Georghiou, S., 1968, Energy transfer in organic systems VII. Effect of diffusion on fluorescence decay, *J. Phys. B* **1:**958–965.
15. Tweet, A. O., Bellamy, W. D., and Gains, G. L., 1964, Fluorescence quenching and energy transfer in monomolecular films containing chlorophyll. *J. Chem. Phys.* **41:**2068–2077.
16. Koppel, D. E., Fleming, P. J., and Strittmatter, P., 1979, Intramembrane positions of membrane-bound chromophores determined by excitation energy transfer, *Biochemistry* **24:**5450–5457.
17. Szmacinski, H., 1998, personal communication.
18. Maliwal, B. P., Kušba, J., and Lakowicz, J. R., 1994, Fluorescence energy transfer in one dimension: Frequency-domain fluorescence study of DNA–fluorophore complexes, *Biopolymers* **35:**245–255.
19. Drake, J. M., Klafter, J., and Levitz, P., 1991, Chemical and biological microstructures as probed by dynamic processes, *Science* **251:**1574–1579.
20. Dewey, T. G., 1991, Excitation energy transport in fractal aggregates, *Chem. Phys.* **150:**445–451.
21. Lianos, P., and Duportail, G., 1993, Time-resolved fluorescence fractal analysis in lipid aggregates, *Biophys. Chem.* **48:**293–299.
22. Loura, L. M. M., Fedorov, A., and Prieto, M., 1996, Resonance energy transfer in a model system of membranes: Applications to gel and liquid crystalline phases, *Biophys. J.* **71:**1823–1836.
23. Tamai, N., Yamazaki, T., Yamazaki, I., Mizuma, A., and Mataga, N., 1987, Excitation energy transfer between dye molecules adsorbed on a vesicle surface, *J. Phys. Chem.* **91:**3503–3508.
24. Levitz, P., Drake, J. M., and Klafter, J., 1988, Critical evaluation of the application of direct energy transfer in probing the morphology of porous solids, *J. Chem. Phys.* **89:**5224–5236.
25. Drake, J. M., Levitz, P., Sinha, S. K., and Klafter, J., 1988, Relaxation of excitations in porous silica gels, *Chem. Phys.* **128:**199–207.
26. Dewey, T. G., and Datta, M. M., 1989, Determination of the fractal dimension of membrane protein aggregates using fluorescence energy transfer, *Biophys. J.* **56:**415–420.
27. Drake, J. M., and Kafter, J., 1990, Dynamics of confined molecular systems, *Phys. Today* **1990** (May) 46–55.
28. Pines, D., and Huppert, D., 1987, Time-resolved fluorescence depolarization measurements in mesoporous silicas. The fractal approach, *J. Phys. Chem.* **91:**6569–6572.
29. Pines, D., Huppert, D., and Avnir, D., 1988, The fractal nature of the surfaces of porous silicas as revealed in electronic energy transfer between adsorbates: Comparison of three donor/acceptor pairs, *J. Chem. Phys.* **89:**1177–1180.
30. Nakashima, K., Duhamel, J., and Winnik, M. A., 1993, Photophysical processes on a latex surface: Electronic energy transfer from rhodamine dyes to malachite green, *J. Phys. Chem.* **97:**10702–10707.
31. Schurr, J. M., Fujimoto, B. S., Wu, P., and Song, L., 1992, Fluorescence studies of nucleic acids: Dynamics, rigidities and structures, in *Topics in Fluorescence Spectroscopy, Volume 3, Biochemical Applications*, J. R. Lakowicz (ed.), Plenum Press, New York, pp. 137–229.
32. Mergny, J.-L., Slama-Schwok, A., Montenay-Garestier, T., Rougee, M., and Helene, C., 1991, Fluorescence energy transfer between dimethyldiazaperopyrenium dication and ethidium intercalated in poly d(A-T), *Photochem. Photobiol.* **53:**555–558.
33. Wolber, P. K., and Hudson, B. S., 1979, An analytic solution to the problem of Förster energy transfer problem in two dimensions, *Biophys. J.* **28:**197–210.
34. Dewey, T. G., and Hammes, G. G., 1980, Calculation of fluorescence resonance energy transfer on surfaces, *Biophys. J.* **32:**1023–1036.
35. Hauser, M., Klein, U. K. A., and Gosele, U., 1976, Extension of Förster's theory for long range energy transfer to donor–acceptor pairs in systems of molecular dimensions, *Z. Phys. Chem.* **101:**255–266.
36. Estep, T. N., and Thompson, T. E., 1979, Energy transfer in lipid bilayers, *Biophys. J.* **26:**195–208.
37. Dobretsov, G. E., Kurek, N. K., Machov, V. N., Syrejshchikova, T. I., and Yakimenko, M. N., 1989, Determination of fluorescent probes localization in membranes by nonradiative energy transfer, *J. Biochem. Biophys. Methods* **19:**259–274.
38. Blumen, A., Klafter, J., and Zumofen, G., 1986, Influence of restricted geometries on the direct energy transfer, *J. Chem. Phys.* **84:**1307–1401.
39. Kellerer, H., and Blumen, A., 1984, Anisotropic excitation transfer to acceptors randomly distributed on surfaces, *Biophys. J.* **46:**1–8.
40. Yguerabide, Y., 1994, Theory of establishing proximity relationships in biological membranes by excitation energy transfer measurements, *Biophys. J.* **66:**683–693.
41. Bastiaens, P., de Beun, A., Lackea, M., Somerharja, P., Vauhkomer, M., and Eisinger, J., 1990, Resonance energy transfer from a cylindrical distribution of donors to a plane of acceptors; location of apo-B100 protein on the human low-density lipoprotein particle, *Biophys. J.* **58:**665–675.
42. Baumann, J., and Fayer, M. D., 1986, Excitation transfer in disordered two-dimensional and anisotropic three-dimensional systems: Effects of spatial geometry on time-resolved observables, *J. Chem. Phys.* **85:**4087–4107.
43. Zimet, D. B., Thevenin, B. J.-M., Verkman, A. S., Shohet, S. B., and Abney, J. R., 1995, Calculation of resonance energy transfer in crowded biological membranes, *Biophys. J.* **68:**1592–1603.
44. Snyder, B., and Frieri, E., 1982, Fluorescence energy transfer in two dimensions, *Biophys. J.* **40:**137–148.
45. Fung, B., and Stryer, L., 1978, Surface density measurements in membranes by fluorescence resonance energy transfer, *Biochemistry* **17:**5241–5248.
46. Pedersen, S., Jorgensen, K., Baekmark, T. R., and Mouritsen, O. G., 1996, Indirect evidence for lipid-domain formation in the transition region of phospholipid bilayers by two-probe fluorescence energy transfer, *Biophys. J.* **71:**554–560.
47. Wolf, D. E., Winiski, A. P., Ting, A. E., Bocian, K. M., and Pagano, R. E., 1992, Determination of the transbilayer distribution of fluorescent lipid analogues by nonradiative fluorescence resonance energy transfer, *Biochemistry* **31:**2865–2873.
48. Shaklai, N., Yguerabide, J., and Ranney, H. M., 1977, Interaction of hemoglobin with red blood cell membranes as shown by a fluorescent chromophore, *Biochemistry* **16:**5585–5592.
49. Thomas, D. D., Caslsen, W. F., and Stryer, L., 1978, Fluorescence energy transfer in the rapid diffusion limit, *Proc. Natl. Acad. Sci. U.S.A.* **75:**5746–5750.
50. Stryer, L., Thomas, D. D., and Meares, C. F., 1982, Diffusion-enhanced fluorescence energy transfer, *Annu. Rev. Biophys. Bioeng.* **11:**203–222.
51. Thomas, D. D., and Stryer, L., 1982, Transverse location of the retinal chromophore of rhodopsin in rod outer segment disc membranes, *J. Mol. Biol.* **154:**145–157.

52. Yeh, S. M., and Meares, C. F., 1980, Characterization of transferrin metal-binding sites by diffusion-enhanced energy transfer, *Biochemistry* **19**:5057–5062.
53. Wensel, T. G., Chang, C.-H., and Meares, C. F., 1985, Diffusion-enhanced lanthanide energy-transfer study of DNA-bound cobalt(III) bleomycins: Comparisons of accessibility and electrostatic potential with DNA complexes of ethidium and acridine orange, *Biochemistry* **24**:3060–3069.
54. Stryer, L., Thomas, D. D., and Carlsen, W. F., 1982, Fluorescence energy transfer measurements of distances in rhodopsin and the purple membrane protein, *Methods Enzymol.* **81**:668–678.
55. Cronce, D. T., and Horrocks, W. DeW., 1992, Probing the metal-binding sites of cod paravalbumin using europium(III) ion luminescence and diffusion-enhanced energy transfer, *Biochemistry* **31**:7963–7969.
56. Duportail, G., Merola, F., and Lianos, P., 1995, Fluorescence energy transfer in lipid vesicles. A time-resolved analysis using stretched exponentials, *J. Photochem. Photobiol., A: Chem.* **89**:135–140.

PROBLEMS

15.1. *Estimation of the Distance of Closest Approach (r_c) for Donor and Acceptor Lipids in Vesicles*: Egg yolk phosphatidylethanolamine (PE) vesicles were prepared containing donors and acceptors. The donor was *N*-(7-nitrobenz-2-oxa-1,3-diazol-4-yl)-phosphatidylethanolamine (NBD-PE). The acceptor was *N*-(lissaminerhodamine-β-sulfonyl)-phosphatidylethanolamine (Rh-PE).[56] Emission spectra are shown in Figure 15.23.

Estimate the distance of closest approach of NBD-PE and Rh-PE. Assume that a phospholipid molecule (PE) occupies 70 Å and that the Förster distance is 50 Å.

15.2. *Calculation of the Maximum Transfer Rate for Diffusion-Limited Quenching:* Wensel *et al.*[53] examined quenching of a positively charged terbium chelate ($TbBED2A^+$) by ethidium bromide (EB) bound to double-helical DNA. In the presence of DNA alone, the decay time was 0.844 ms, and in the presence of DNA-EB, the decay time was 0.334 ms (Figure 15.24).

For a donor in an infinite cylinder like DNA, the diffusion-limited transfer rate is given by a modified form of Eq. [15.26],

$$k_T^b = \frac{1.672 R_0^6}{\tau_D r_c^3} \quad (\text{in } M^{-1}\,\text{s}^{-1}) \qquad [15.27]$$

Calculate the observed value of k_T^b and compare it with the maximum theoretical value, assuming $R_0 = 30.2$ Å = 3.02 nm and $r_c = 1.1$ nm due to the combined radii of the donor and DNA. Explain the difference between the measured and theoretical values of k_T^b.

15.3. *Acceptor Concentration for 50% Energy Transfer in One, Two, and Three Dimensions*: The equations describing the intensity decays in one, two, and three dimensions (Eqs.

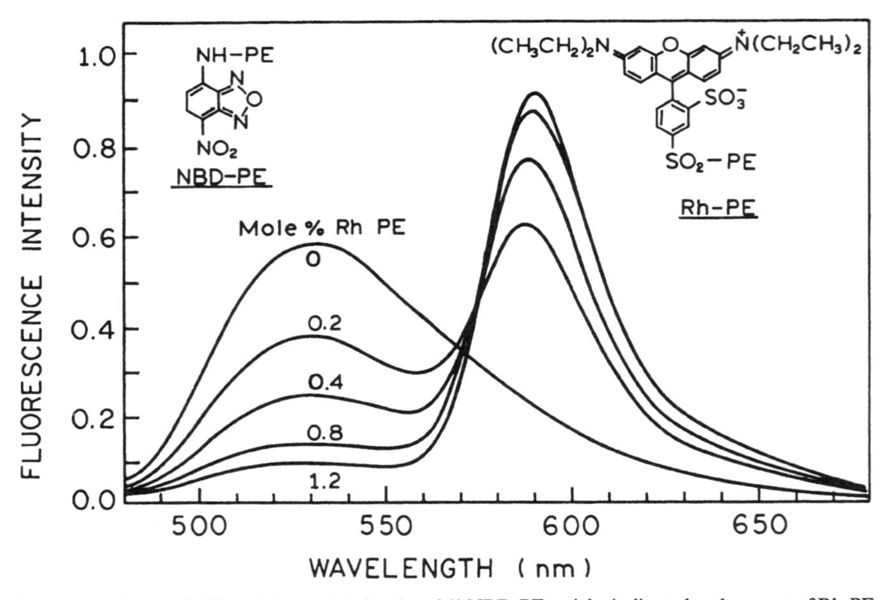

Figure 15.23. Emission spectra of egg yolk PE vesicles containing 1 mol % NBD-PE and the indicated mol percent of Rh-PE. Excitation wavelength was 455 nm. Revised from Ref. 56.

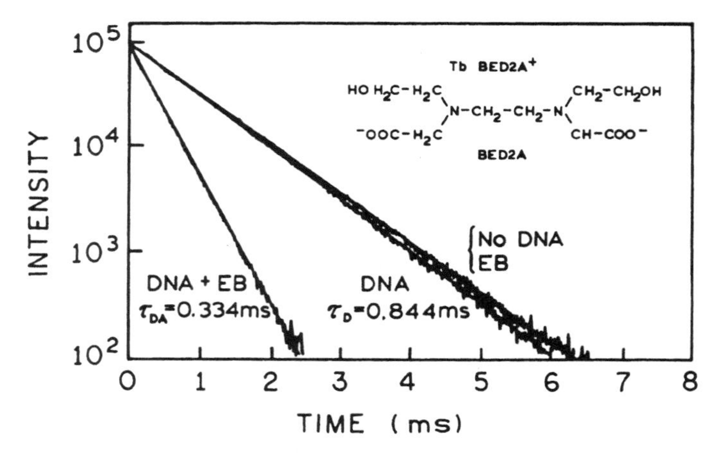

Figure 15.24. Intensity decays of $TbBED2A^+$ in the presence of DNA or of DNA with EB. The concentration of EB was 2.77μ*M*. Revised from Ref. 53.

[15.1], [15.9], and [15.12]) were numerically integrated over time to obtain the relative donor quantum yields versus normalized concentrations (C/C_0). The graph obtained is presented in Figure 15.25. Using this graph, calculate the acceptor concentration corresponding to 50% transfer for a homogeneous solution (m*M*), a membrane (acceptor per phospholipid), and DNA (acceptor per base pair). For the membrane, assume 70 $Å^2$ per phospholipid, and for DNA, assume 3.4 Å per base pair. Use R_0 = 50 Å. Are the acceptor concentrations practical for proteins, membranes, or nucleic acids?

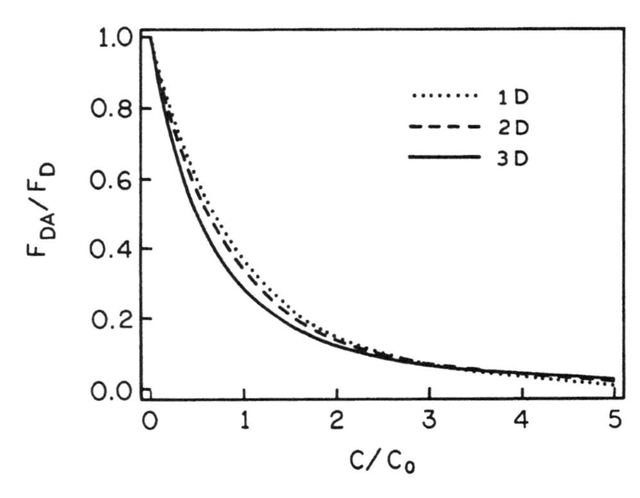

Figure 15.25. Relative donor quantum yield versus C/C_0 in one (1D), two (2D), and three (3D) dimensions. From Ref. 17.

Protein Fluorescence

16

Discussions of biochemical fluorescence frequently start with the subject of protein fluorescence. This is because, among biopolymers, proteins are unique in displaying useful intrinsic fluorescence. Lipids, membranes, and saccharides are essentially nonfluorescent, and the intrinsic fluorescence of DNA is too weak to be useful. In proteins, the three aromatic amino acids—phenylalanine, tyrosine, and tryptophan—are all fluorescent. A favorable feature of protein structure is that these three amino acids are relatively rare in proteins. Tryptophan, which is the dominant intrinsic fluorophore, is generally present at about 1 mol % in proteins. A protein may possess just one or a few tryptophan residues, which facilitates interpretation of the spectral data. If all 20 amino acids were fluorescent, it is probable that protein emission would be too complex to interpret.

A valuable feature of intrinsic protein fluorescence is the high sensitivity of tryptophan to its local environment. One can frequently observe changes in emission spectra of tryptophan in response to protein conformational transitions, subunit association, substrate binding, or denaturation, all of which can affect the local environment surrounding the indole ring. Anisotropy measurements are also informative because tyrosine and tryptophan display high anisotropy, the extent of which is sensitive to protein conformation and the extent of motion during the excited-state lifetime. Also, tryptophan appears to be uniquely sensitive to collisional quenching, apparently due to a tendency of indole to donate electrons while in the excited state. Tryptophan can be quenched by externally added quenchers or by nearby groups in the protein. Hence, one finds numerous reports on the emission spectra, anisotropy, and quenching of tryptophan residues in proteins.

A complicating factor in the interpretation of protein fluorescence is the presence of multiple fluorescent amino acids in most proteins. Since the environment of each residue is distinct, the spectral properties of each residue are generally different. The emission spectra of tryptophan residues in proteins overlap at most usable wavelengths, and one cannot easily separate the spectral contributions of each tryptophan in a multi-tryptophan protein. In addition, complex time-resolved intensity decays are found even for tryptophan itself, as well as for proteins which contain a single tryptophan residue. Most single-tryptophan proteins display multiexponential intensity decays. For this reason, one cannot simply interpret a multiexponential time-resolved decay in terms of the individual tryptophan residues in a protein.

As a further complicating factor, tryptophan displays complex spectral properties due to the presence of two nearly isoenergetic excited states, ^1L_a and ^1L_b. The electronic transitions display distinct absorption, emission, and anisotropy spectra and are differently sensitive to solvent polarity. The complexity of indole photophysics has stimulated detailed studies of protein fluorescence but has also inhibited interpretation of the data.

While the preceding paragraphs present a pessimistic viewpoint of the complexity of protein fluorescence, considerable progress has been made in the past decade. The origin of the multiexponential decay of tryptophan in water is now largely understood as due to the presence of rotational conformational isomers (rotamers). Characterization of single-tryptophan proteins has provided a basis for interpreting the spectral properties of tryptophan in unique environments. Importantly, the introduction of site-directed mutagenesis has allowed creation of mutants which contain one instead of several tryptophan residues, insertion of tryptophan at desired locations in a protein, or modification of the environment around a tryptophan residue. Examination of single-tryptophan proteins and engineered proteins has provided a more detailed understanding of how the local environment determines the spectral properties of tryptophan. It is also possible to insert tryptophan analogs into proteins. These analogs display unique spectral features and are observable in the presence of other tryptophan residues.

In summary, a detailed understanding of indole photophysics, together with the ability to place the tryptophan residues at desired locations, has resulted in an increased ability to use protein fluorescence for a detailed understanding of protein function.

16.1. SPECTRAL PROPERTIES OF THE AROMATIC AMINO ACIDS

Proteins contain three amino acid residues that contribute to their ultraviolet fluorescence: tyrosine (tyr, Y), tryptophan (trp, W), and phenylalanine (phe, F). Several useful reviews and monographs have summarized their spectral properties.[1–4] Absorption and emission spectra of these amino acids are shown in Figure 16.1. The emission of proteins is dominated by tryptophan, which absorbs at the longest wavelength and displays the largest extinction coefficient. Also, because of its long wavelength, energy absorbed by phenylalanine and tyrosine residues is often transferred to the tryptophan residues in the same protein.

Phenylalanine displays the shortest absorption and emission wavelengths. It displays a structured emission with a maximum near 282 nm. The emission of tyrosine in water occurs at 303 nm and is relatively insensitive to solvent polarity. The emission maximum of tryptophan in water occurs near 350 nm and is highly dependent upon polarity and/or local environment. As will be described below (Section 16.1.B), tryptophan is sensitive to general solvent effects and also displays a substantial spectral shift upon

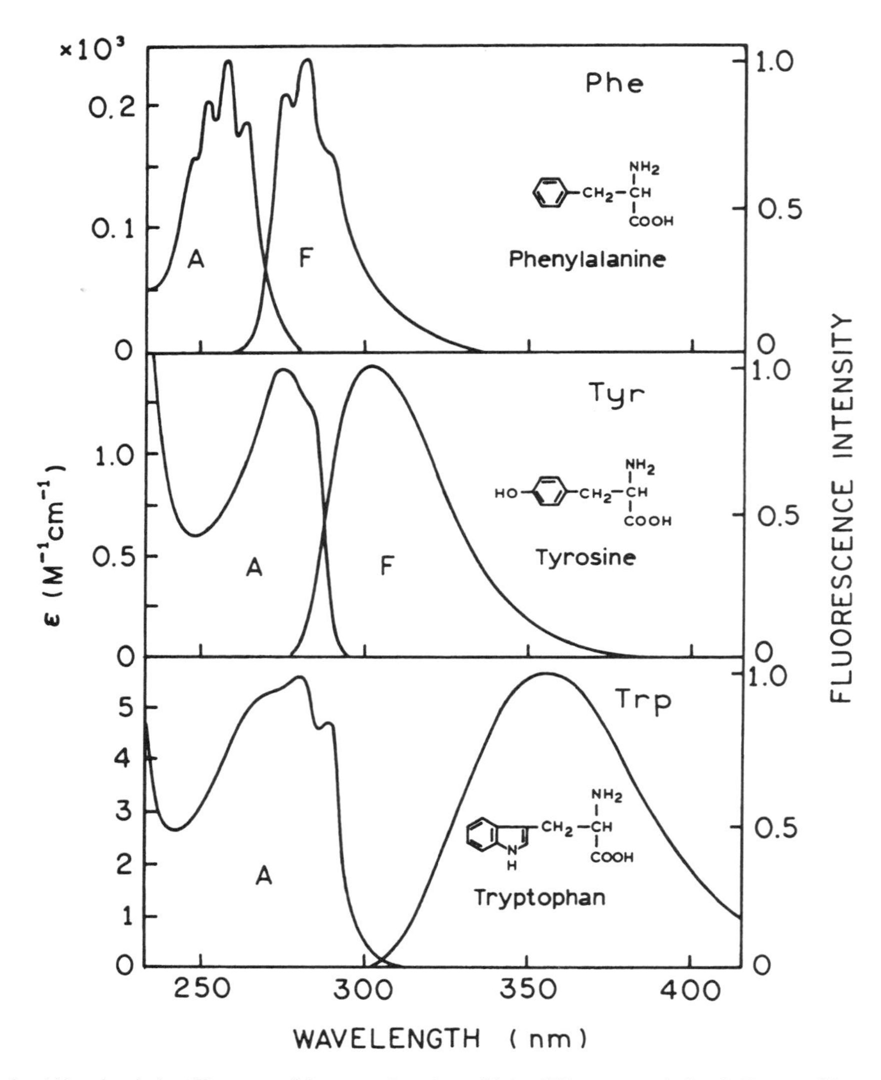

Figure 16.1. Absorption (A) and emission (F) spectra of the aromatic amino acids in pH 7 aqueous solution. I. Gryczynski, unpublished observations.

formation of a hydrogen bond to the imino nitrogen, which can be regarded as a specific solvent effect (Section 6.3). As a result, the emission of each tryptophan residue in a protein depends on the details of its surrounding environment.

Protein fluorescence is generally excited at the absorption maximum near 280 nm or at longer wavelengths. Consequently, phenylalanine is not excited in most experiments. Furthermore, the quantum yield of phenylalanine in proteins is small, typically near 0.03,[1,5] so emission from phenylalanine residues in proteins is rarely observed. The absorption of proteins at 280 nm is due to both tyrosine and tryptophan residues. At 23 °C in neutral aqueous solution, the quantum yields of tyrosine and tryptophan are near 0.14 and 0.13, respectively, with the reported values being somewhat variable.[1–5] At wavelengths longer than 295 nm, the absorption is due primarily to tryptophan. Tryptophan fluorescence can be selectively excited at 295–305 nm. Thus, many papers report the use of 295-nm excitation, which is selected to avoid excitation of tyrosine. An especially comprehensive review of protein fluorescence was published by Longworth.[6] This review describes the spectral properties of the aromatic amino acids, single-tryptophan proteins, and multi-tryptophan proteins.

Tyrosine is often regarded as a rather simple fluorophore. However, under some circumstances, tyrosine can also display complex spectral properties. Tyrosine can undergo an excited-state ionization, resulting in the loss of the proton on the aromatic hydroxyl group. In the ground state, the pK_a of this hydroxyl is about 10. In the excited state, the pK_a decreases to about 4. In neutral solution, the hydroxyl group can dissociate during the lifetime of the excited state, leading to quenching of the tyrosine fluorescence. Tyrosinate is weakly fluorescent at 350 nm, and its emission can be mistaken for tryptophan fluorescence. Although tyrosinate emission is observable in some proteins, it is now thought that excited-state ionization is not a major decay pathway for tyrosine in proteins.

Because of their spectral properties, RET can occur from phenylalanine to tyrosine to tryptophan. Also, blue-shifted tryptophan residues can transfer the excitation to longer-wavelength tryptophan residues. In fact, energy transfer has been repeatedly observed in proteins and is one reason for the minor contribution of phenylalanine and tyrosine to the emission of most proteins. The anisotropy displayed by tyrosine and tryptophan is sensitive to both overall rotational diffusion of proteins and the extent of segmental motion during the excited-state lifetimes. Hence, the intrinsic fluorescence of proteins can provide considerable information about protein structure and dynamics and is often used to study protein folding and association reactions. In this chapter, we present examples of protein fluorescence which illustrate the factors governing the intrinsic fluorescence of proteins. In the next chapter, we describe time-resolved studies of protein fluorescence.

16.1.A. Excitation Polarization Spectra of Tyrosine and Tryptophan

Tryptophan emission is highly sensitive to the local environment, but tyrosine emission is rather insensitive. What is the reason for this distinct behavior? Fluorophores that are sensitive to solvent polarity are typically those that display a large charge separation in the excited state. Although indole is uniquely sensitive to solvent polarity, its spectral properties cannot be completely explained by the change in dipole moment in the excited state. The complex spectral properties of tryptophan and indole are due to the presence of two overlapping transitions in the long-wavelength absorption band, transitions to the ^1L_a and ^1L_b states of indole. For most fluorophores, the long-wavelength absorption represents a single electronic transition to the first singlet state (S_1), which appears to be the case for tyrosine. In the case of tryptophan, indole, and their derivatives, the long-wavelength absorption (240–300 nm) consists of two overlapping transitions to the ^1L_a and ^1L_b excited states. These states have similar energies, and, depending on the environment, either state can have the lower energy. Emission occurs from the lowest-energy state, in accordance with Kasha's rule. The ^1L_a and ^1L_b transitions have different directions in the molecule, so that the anisotropy of indole is complex. Each state has a different dipole moment, so that each state responds differently to solvent polarity.

One way to resolve overlapping electronic transitions is by the use of excitation anisotropy spectra. The anisotropy spectrum is usually measured in frozen solution to prevent rotational diffusion during the excited-state lifetime. For a single electronic transition, the anisotropy is expected to be constant across the absorption band. In the case of tyrosine (Figure 16.2), the anisotropy is relatively constant across the long-wavelength absorption band (260–290 nm). As seen for tyrosine, most fluorophores display some increase in anisotropy as the excitation wavelength increases across the $S_0 \rightarrow S_1$ transition (260–290 nm), but such a monotonic increase is usually regarded as the result of a single transition. It is interesting to note that the data shown in Figure 16.2 are for *N*-acetyl-L-tyrosinamide (NATyrA) instead of tyrosine. Neutral derivatives of the aromatic amino acids are frequently used because the structures of these derivatives more closely resemble those of the residues found in proteins. Also, it is known that the quantum yield and lifetimes can be affected by the ionization state of the amino and carboxyl groups in tyrosine and

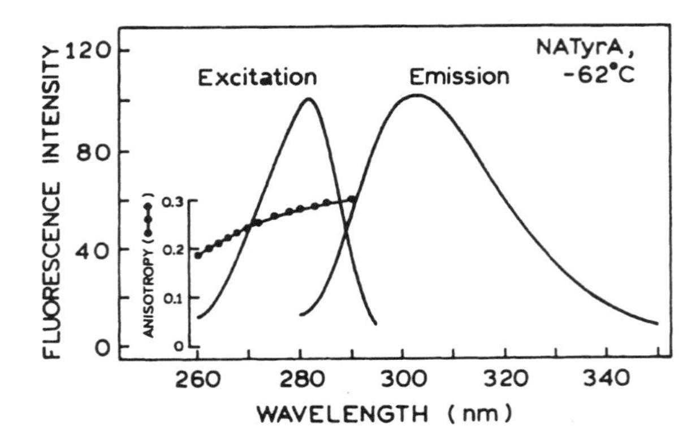

Figure 16.2. Excitation, emission, and excitation anisotropy spectra of *N*-acetyl-L-tyrosinamide (NATyrA). The fluorescence anisotropies (○) were measured in a mixture of 70% propylene glycol with 30% buffer at −62 °C. The emission was observed at 302 nm. Data from Refs. 7–9.

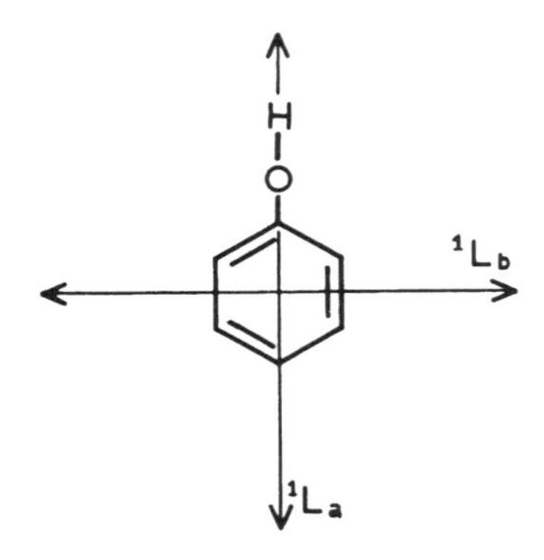

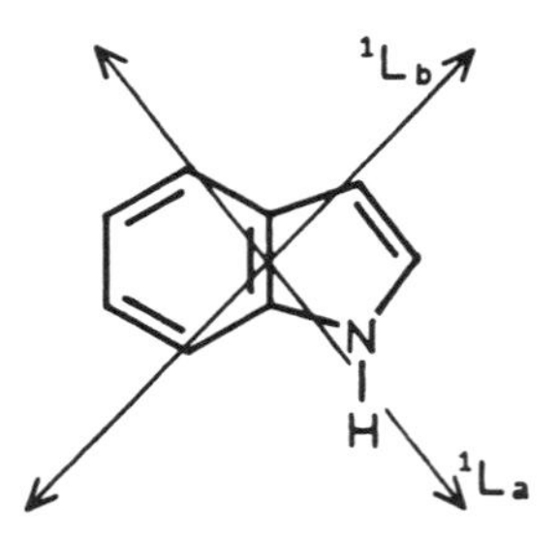

Figure 16.3. Electronic absorption transitions in tyrosine and tryptophan. Data from Refs. 10–11.

tryptophan. These effects can be avoided by use of the neutral analogs.

The lowest electronic transition of tyrosine, with absorption from 260 to 290 nm, is thought to be due to the ^1L_b transition oriented across the phenyl ring[10] (Figure 16.3). The ^1L_a transition is the origin of the more intense absorption below 250 nm. The anisotropy of tyrosine decreases for shorter-wavelength excitation (Figure 16.2) and becomes negative below 240 nm (not shown). This indicates that the transition moment for the ^1L_a absorption is nearly perpendicular to that for the emission from the ^1L_b state. For longer-wavelength excitation, above 260 nm, the anisotropy is positive, indicating that the absorption transition moment is nearly parallel to the transition moment for the emission from the ^1L_b state. Hence, the electronic properties of tyrosine for excitation above 260 nm are relatively simple, with absorption and emission occurring from the same ^1L_b state.

A very different anisotropy spectrum is characteristic of tryptophan, indole, and many of their derivatives. In contrast to tyrosine, tryptophan does not show a constant polarization across the long-wavelength absorption band (Figure 16.4).[11,12] The high r_0 values at 300-nm excitation indicate nearly collinear absorption and emission dipoles. The anisotropy decreases to a minimum at 290-nm excitation and increases at excitation wavelengths from 280 to 250 nm. This complex behavior is attributed to the presence of two electronic transitions to the ^1L_a and ^1L_b states in the last absorption band (Figure 16.3). The transition dipoles for these transitions are oriented nearly perpendicular to each other,[13–17] so that the fundamental anisotropy (r_0) is strongly dependent on the fractional contributions of each state to the absorption transition. For instance, the minimum in the anisotropy spectrum near 290 nm is due to a maximum in the absorption of the ^1L_b state. The emission of tryptophan in solution and of most proteins is unstructured and due to the ^1L_a state. Since the emission is from a state (^1L_a) which is rotated 90° from the absorbing state (^1L_b), the anisotropy is low with 290-nm excitation. This is another reason why protein fluorescence is often excited at 295–300 nm. For these excitation wavelengths, only the ^1L_a state absorbs and the anisotropy is high.

The excitation anisotropy spectra can be used to resolve the absorption spectra of the ^1L_a and ^1L_b states (Figure 16.4).[18] This procedure was described in Chapter 10 and is based on the additivity of anisotropies (Section 10.1). For this resolution, one assumes that the maximal anisotropy is characteristic of absorption to and emission from the same ^1L_a state. A further assumption is that the transitions to the ^1L_a and ^1L_b states are oriented at 90°. Using these assumptions, the ^1L_a transition is found to be unstructured and to extend to longer wavelengths than the structured ^1L_b transition (Figure 16.4). The dominance of the ^1L_a absorption at 300 nm, and emission from the same ^1L_a state, is why proteins display high anisotropy with long-wavelength excitation.

It is well known that the intensity decay of tryptophan at pH 7 is a double exponential, with decay times near 0.5 and 3.1 ns. Time-resolved protein fluorescence will be

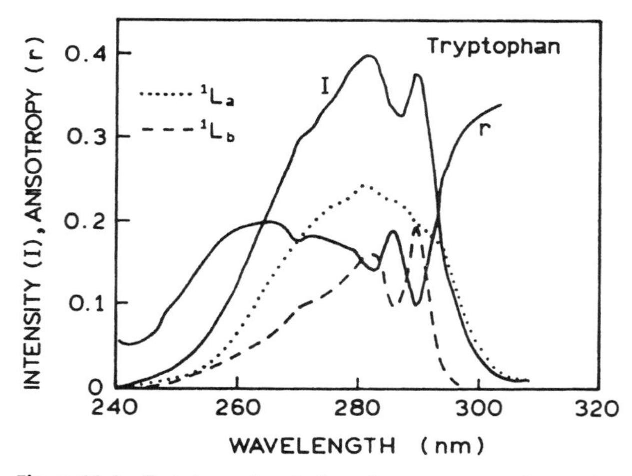

Figure 16.4. Emission and excitation anisotropy spectra of tryptophan in propylene glycol at –50 °C. Also shown are the anisotropy-resolved spectra of the ^1L_a (· · ·) and ^1L_b (— — —) transitions. Data from Ref. 12.

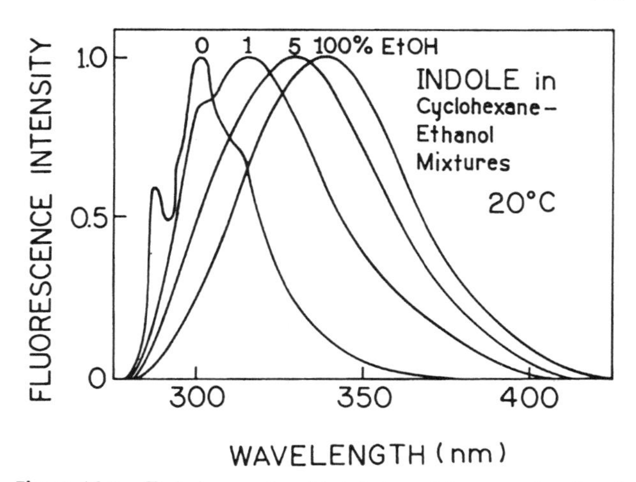

Figure 16.5. Emission spectra of indole in cyclohexane, ethanol, and their mixtures at 20 °C. From Ref. 23.

discussed in more detail in the following chapter. At first, it was thought that the two decay times were due to emission from the ^1L_a and ^1L_b states.[19] However, it is now agreed[20,21] that the two decay times have their origin in the rotamer populations (Chapter 17) and that tryptophan emits only from the ^1L_a state unless the local environment is completely nonpolar. It should be noted that an early report of 2.1 and 5.4 ns for the two decay times of tryptophan[22] was in error owing to photodecomposition of the sample.

16.1.B. Solvent Effects on Tryptophan Emission Spectra

In order to understand protein fluorescence, it is important to understand how each state of indole is affected by the local environment. Valuable hints can be obtained from the solvent-dependent spectra of indole, which can be understood in terms of the ^1L_a and ^1L_b transitions. Emission spectra of indole in cyclohexane are sensitive to trace quantities of hydrogen-bonding solvent (Figure 16.5). These spectra show the presence of specific and general solvent effects.[23] In pure cyclohexane, in the absence of hydrogen bonding, the emission is structured and seems to be a mirror image of the absorption spectrum of the ^1L_b transition (compare Figures 16.4 and 16.5). In the presence of the hydrogen-bonding solvent ethanol, the structured emission is lost and the emission mirrors the ^1L_a transition. Similar results were described in a number of early reports.[24–29] These structured and unstructured emission spectra indicate dual emission from the ^1L_a and ^1L_b states, with the ^1L_a transition being more solvent-sensitive than the ^1L_b transition, and a shifting of the ^1L_a transition to lower energies in polar solvents. A greater solvent sensitivity for the ^1L_a state seems reasonable since the ^1L_a transition more directly involves the polar nitrogen atom of indole (Figure 16.3). It is also known that the excited-state dipole moment is near 6 D for the ^1L_a state but is much smaller for the ^1L_b state.[30] In a completely nonpolar environment, the ^1L_b state can have lower energy and dominate the emission. The electrical field due to the protein, or the solvent reaction field, may also influence the emission spectrum of indole.[31]

The different solvent sensitivities of the ^1L_a and ^1L_b states of indole can also be seen in the absorption spectra of indole in the same cyclohexane–ethanol mixtures (Figure 16.6). As the ethanol concentration is increased, one observes a decrease in the vibrational peak at 288 nm. Similar results have been observed for indole derivatives under a variety of similar conditions.[26,32,33] These spectral changes can be understood in terms of a red shift of the ^1L_a absorption spectrum due to a decrease in energy of the ^1L_a state upon interaction with the polar solvent. This energy shift is due to a hydrogen-bonding interaction with the indole nitrogen. The ^1L_b state is less sensitive to solvent, and its absorption is less affected by polar solvent.

The presence of the ^1L_a and ^1L_b states allows one to understand the complex emission spectra displayed by indole (Figure 16.5). In a completely nonpolar solvent, the ^1L_b state can be the lowest-energy state, resulting in a structured emission. The presence of polar solvent decreases the energy level of the ^1L_a state, so that its unstructured emission dominates, and the structured emission is not observed. At higher ethanol concentrations, these specific solvent interactions are saturated, and indole displays

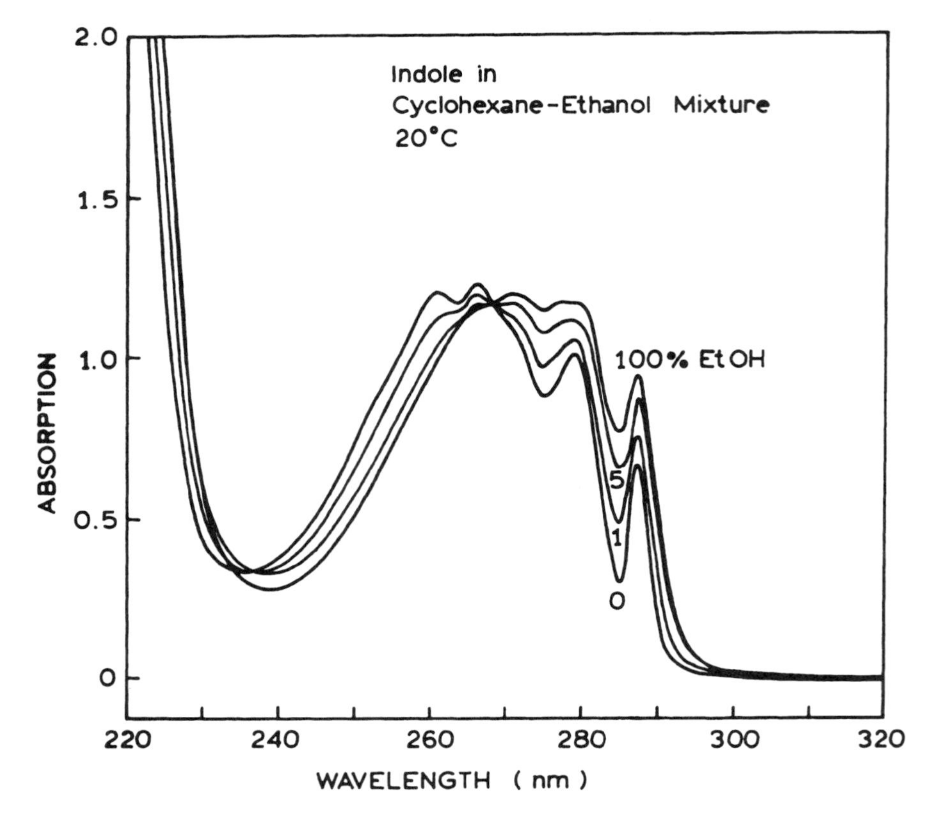

Figure 16.6. Absorption spectra of indole in cyclohexane with increasing amounts of ethanol. Data courtesy of Dr. Ignacy Gryczynski.

a polarity-dependent red shift consistent with general solvent effects.

16.1.C. Excited-State Ionization of Tyrosine

Although tyrosine is usually a simple fluorophore, it is important to recognize the potential for excited-state ionization. Such ionization is expected because the pK_a of the phenolic hydroxyl group decreases from 10.3 in the ground state to about 4 in the excited state. Hence, ionization can occur even at neutral pH. Tyrosinate emission can be most easily observed at high pH, where the phenolic OH group is ionized in the ground state (Figure 16.7). In 0.01*M* NaOH the emission of tyrosine is centered near 345 nm[10,34] and can be mistaken for tryptophan emission. The decay time of tyrosinate at pH 11 has been reported to be 30 ps.[35]

It is important to recognize that tyrosinate emission can also be observed at neutral pH, particularly in the presence of a base which can interact with the excited state. One example is presented in Figure 16.8, which shows the emission spectra of tyrosine at the same pH but with increasing concentrations of acetate buffer. The emission intensity decreases with increasing acetate concentration. This decrease occurs because the weakly basic acetate group can remove the phenolic proton, which has a pK_a of

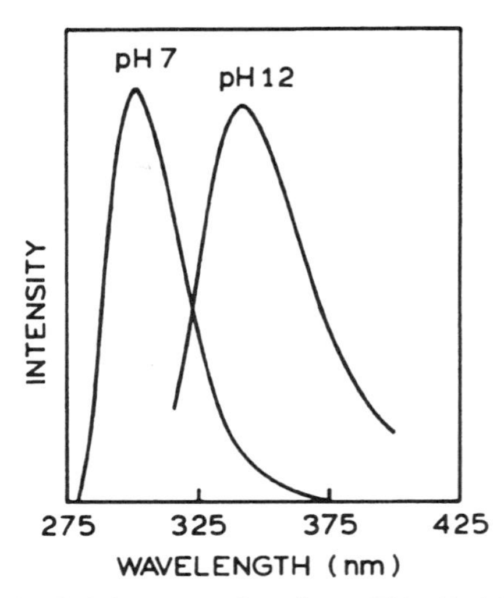

Figure 16.7. Emission spectra of tyrosine at pH 7 and in 0.01*M* NaOH (pH 12). The emission spectra are approximately normalized. Modified from Ref. 10.

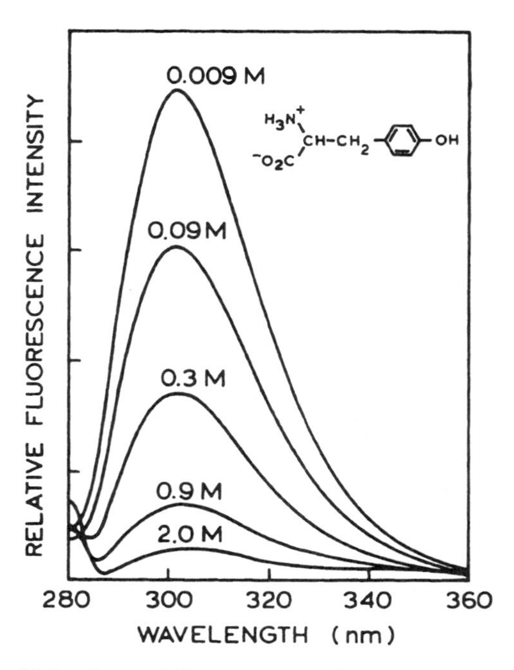

Figure 16.8. Corrected fluorescence emission spectra of tyrosine in acetate/acetic acid buffer solutions of differing ionic strength, pH 6.05. Revised from Ref. 36.

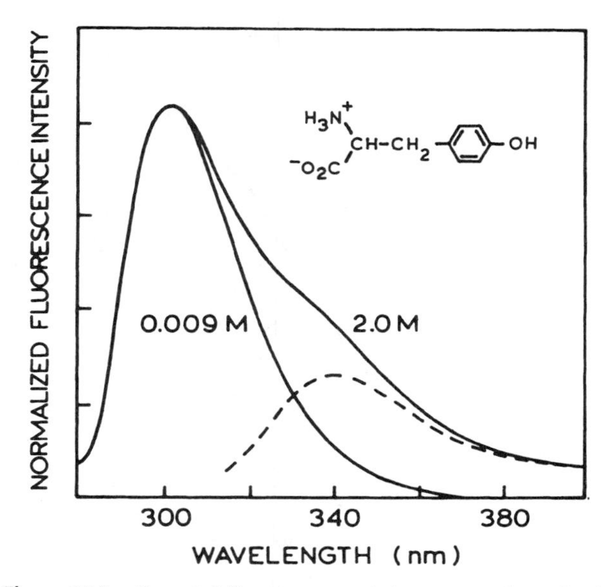

Figure 16.9. Corrected fluorescence emission spectra of tyrosine in 0.009*M* and 2*M* acetate solutions at pH 6.05. The dashed curve shows the difference spectrum due to tyrosinate. Revised from Ref. 36.

4.2–5.3 in the first singlet stage.[34–37] The acetate behaves like a collisional quencher, and the extent of excited-state ionization and quenching depends on the acetate concentration.

Close examination of the emission spectra of tyrosine at high acetate concentrations reveals a spectral shift. This is seen more easily in the peak-normalized emission spectra (Figure 16.9). In 2.0*M* acetate, tyrosine displays additional intensity at 340 nm; this is most clearly seen in the difference emission spectrum (dashed curve in Figure 16.9), which displays a maximum near 345 nm. The essential point is that the phenolic group of tyrosine can ionize even at neutral pH, and the extent to which this occurs depends on the base concentration and exposure of tyrosine to the aqueous phase. However, the extent of excited-state ionization is limited by the short decay time of tyrosine[34] and is slight under most conditions.

16.1.D. Ground-State Complex Formation by Tyrosine

In the example shown above, tyrosinate was formed in the excited state, following excitation of un-ionized tyrosine. More complex behavior is also possible, in which tyrosine forms ground-state complexes.[38,39] In these complexes, the hydroxyl group is not ionized but is prepared to ionize immediately upon excitation. One example is provided by the effects of phosphate on the emission spectrum of tyrosine (Figure 16.10, right).[38] Increasing the phosphate concentration results in a progressive decrease in the tyrosine emission at 305 nm. Concurrently, a new emission is observed at 345 nm, which is due to emission from tyrosinate or tyrosine that is complexed with phosphate. The static mechanism of complex formation between tyrosine and phosphate is seen from a shift of the absorption spectrum to longer wavelengths with increasing phosphate concentration (Figure 16.10, left). If there were no ground-state interaction between tyrosine and phosphate, the absorption spectra would be unchanged. Recall that excitation at 290–300 nm is frequently used to achieve selective excitation of tryptophan residues. From the red-shifted absorption spectrum of the complex (Figure 16.10), one may predict that the 345-nm emission of tyrosinate could be selectively excited at longer wavelengths. Quenching of tyrosine by phosphate and other bases can proceed by both static complex formation and by a collisional Stern–Volmer process.[40,41]

Generally, tyrosinate emission is weak owing to its low quantum yield and, hence, is minor relative to the tryptophan emission. However, the possibility of this emission should be considered in any detailed analysis of the emission spectra and/or fluorescence lifetimes of proteins. In fact, tyrosinate-like emission has been reported for a number of proteins.[42–46] However, it is difficult to be sure whether the proteins in fact displayed tyrosinate emission or the samples were contaminated with tryptophan-con-

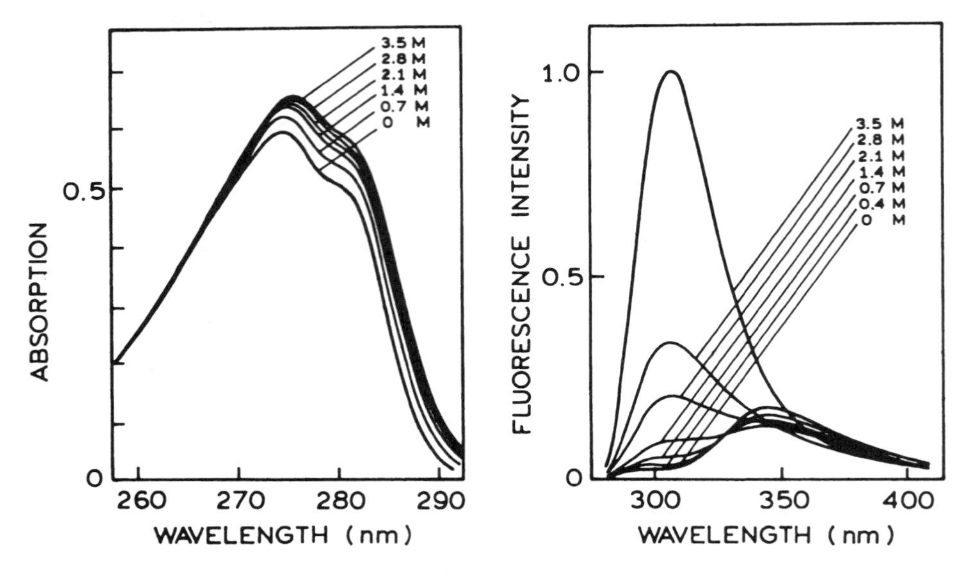

Figure 16.10. Dependence of the absorption and emission spectra of tyrosine on phosphate concentration. The numbers refer to the concentration of K_2HPO_4. The pH was 7.5. Revised from Ref. 38.

taining proteins. Hydrogen-bonding interactions of tyrosine are probably one reason for the low tyrosine emission observed for most native proteins.

16.1.E. Excimer Formation by Phenylalanine

Compared to studies of tyrosine and tryptophan, the number of studies of phenylalanine fluorescence that have been reported is much fewer. One detailed report[47] showed that the quantum yield of phenylalanine in water decreases from 0.04 to 0.01 over the range 0–70 °C, with a value near 0.03 at 20 °C. Over this same temperature range, the decay time decreases from 10 to 2 ns, with a value of 6.8 ns at 20 °C. Like benzene, phenylalanine can also display excimer emission. At concentrations near 0.1*M* in aqueous solution, phenylalanine displays weak unstructured emission on the long-wavelength side of the emission. This emission, centered near 320 nm, is thought to be due to excimers. Excimer emission from phenylalanine residues in proteins has not yet been reported.

16.2. GENERAL FEATURES OF PROTEIN FLUORESCENCE

A number of reviews have summarized the fluorescence spectral properties of proteins.[48–52] Reviews of protein fluorescence frequently begin with the spectral properties of multi-tryptophan proteins, which have been studied since the 1950s. The emission maximum and quantum yield of tryptophan residues in proteins can vary greatly, and these variations are due to the three-dimensional structure of the proteins. Denaturation of proteins results in rather similar emission spectra and quantum yields for the unfolded proteins and an increase in tyrosine emission.[51]

There have been attempts to divide proteins into classes based on their emission spectra.[50] The basic idea is that the tryptophan emission spectrum should reflect the average environment of the tryptophan. For tryptophan in a completely apolar environment, one expects a blue-shifted structured emission characteristic of indole in cyclohexane. As the tryptophan residue moves toward hydrogen-bonding groups and/or becomes exposed to water, the emission shifts to longer wavelengths (Figure 16.11). In fact, individual proteins are known which display this wide range of emission spectra. For example, later in this chapter we will see that azurin displays an emission spectrum characteristic of a completely shielded tryptophan residue. The emission from adrenocorticotropin hormone (ACTH) is characteristic of a fully exposed tryptophan residue.

One might expect that proteins which display a blue-shifted emission spectrum would have higher quantum yields (Q) or lifetimes (τ). Such behavior is expected on the basis of the usual increase in quantum yield when a fluorophore is placed in a less polar solvent. For instance, the lifetime of indole increases from 4.1 ns in ethanol to 7.7 ns in cyclohexane.[23] However, when one examines the values of Q and τ for a number of single-tryptophan proteins, one finds that there is no clear correlation between quantum yield and lifetime (Figure 16.12). Appar-

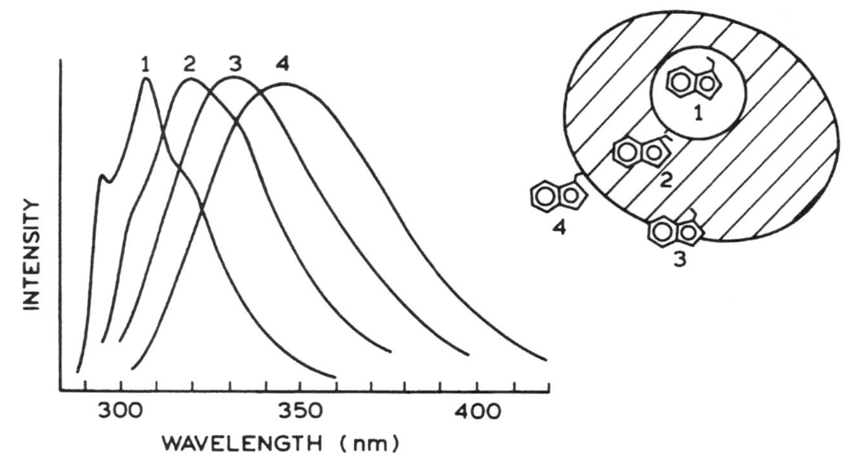

Figure 16.11. Effect of tryptophan environment on the emission spectra of proteins: (1) apoazurin Pfl, (2) ribonuclease T_1, (3) staphylococcal nuclease, (4) glucagon. Revised from Refs. 52 and 53.

ently, the complexity of the protein environment can override the general effects of polarity on the lifetime of tryptophan.

As described in Chapter 1, the natural lifetime (τ_N) of a fluorophore is the reciprocal of the radiative decay rate, which is typically independent of the fluorophore environment. The value of τ_N is calculated from the measured lifetime (τ) and quantum yield (Q), $\tau_N = \tau/Q$. Surprisingly, the apparent values of τ_N also vary considerably among various proteins (Figure 16.12).

Why should the natural lifetimes of proteins be so variable? One explanation is that some tryptophan residues are completely quenched in a multi-tryptophan protein and thus do not contribute to the measured lifetime and the emission maximum. For instance, indole is known to be partially quenched by benzene.[54] Hence, one might expect indole to be quenched by nearby phenylalanine residues in proteins. The nonfluorescent residues still absorb light, and this results in highly variable quantum yields for the different proteins. The lack of correlation between emission spectra and lifetimes of proteins is probably due to the effects of specific protein environments which quench particular tryptophan residues. In the case of single-tryptophan proteins, one can imagine slightly different conformations which bring the residue closer to or farther from a quenching group. Under these conditions, only a fraction of the protein population would be fluorescent, and this fraction would display a longer lifetime than expected from the quantum yield. As will be seen in Chapter 17, there are known examples where particular tryptophan residues are quenched (Section 17.8.B) and examples where normally fluorescent tryptophan residues transfer energy to nonfluorescent tryptophan residues (Section 17.8.C).

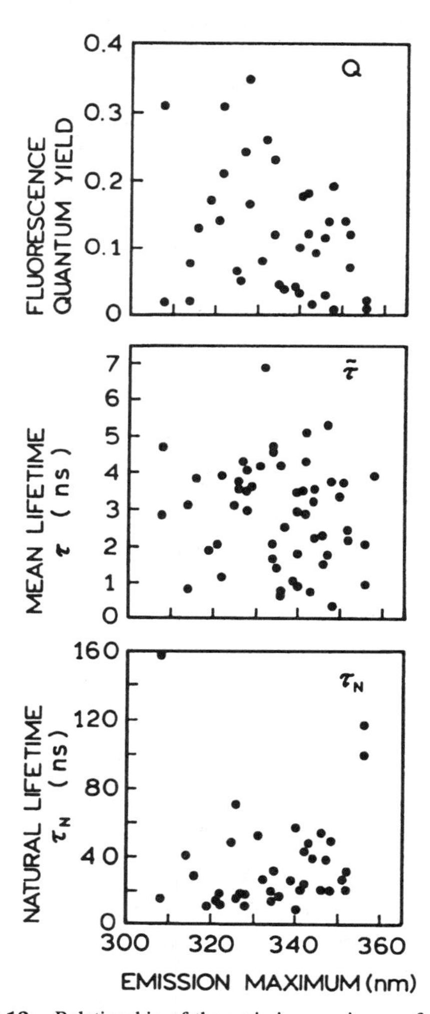

Figure 16.12. Relationship of the emission maximum of proteins to the quantum yield Q (*top*), mean lifetime $\bar{\tau}$ (*middle*), and natural lifetime τ_N (*bottom*). Courtesy of Dr. Maurice Eftink, University of Mississippi.

16.3. TRYPTOPHAN EMISSION IN AN APOLAR PROTEIN ENVIRONMENT

Interpretation of protein fluorescence has been hindered by the presence of multiple tryptophan residues in most proteins. Hence, there have been continuous efforts to identify and characterize single-tryptophan proteins. Fluorescence studies of azurins have been uniquely informative. Azurins are small copper-containing proteins with molecular weights near 15,000. These proteins are involved in the electron-transfer system of denitrifying bacteria. A number of azurins have been characterized, and some contain a single tryptophan residue. A unique feature of the azurins is that some display the most blue-shifted emission observed for a tryptophan residue in proteins.

Emission spectra of the single-tryptophan azurin Pfl from *Pseudomonas fluorescens* is shown in Figure 16.13. In contrast to the emission from most proteins, the emission spectrum shows structure characteristic of the ^1L_b state.[55–57] In fact, as seen in Figure 16.13, the emission spectrum is nearly identical with that of the tryptophan analog 3-methylindole in the nonpolar solvent methylcyclohexane. This suggests that the indole residue is located in a completely nonpolar region of the protein, most probably without a single polar group for a hydrogen bond.

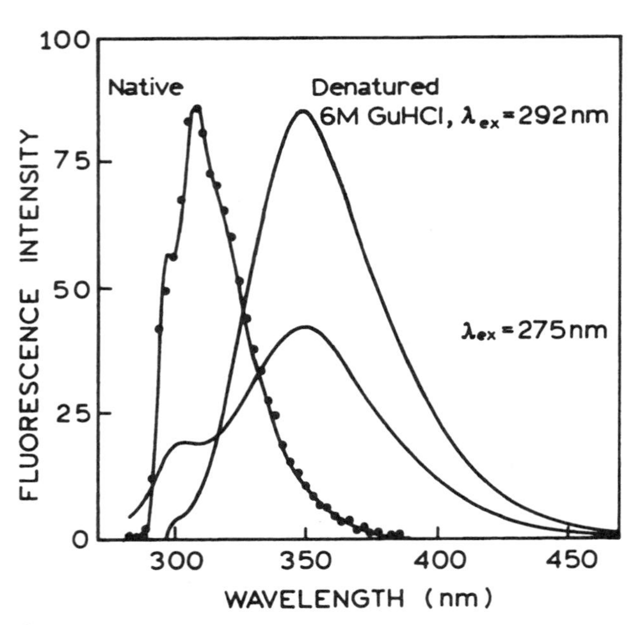

Figure 16.13. Emission spectra of *Pseudomonas fluorescens* (ATCC-13525-2) azurin Pfl. The solid curves show emission spectra for native (0.01*M* cacodylate, pH 5.3) and denatured (6*M* guanidine hydrochloride) azurin Pfl for λ_{ex} = 292 nm. Also shown is the emission spectrum of denatured azurin excited at 275 nm. ●, Emission spectrum of 3-methylindole in methylcyclohexane. Data from Ref. 57.

These results agree with X-ray studies, which show that the indole group is located in the hydrophobic core of the protein.[58,59]

The fact that the structured indole emission of azurin is due to shielding by the protein matrix is shown by denaturation of azurin Pfl. In the presence of 6*M* guanidine hydrochloride, the tryptophan emission loses its structure and shifts to 351 nm, characteristic of a fully exposed tryptophan residue. The emission spectra of denatured azurin (Figure 16.13) illustrate another typical spectral property of proteins. The emission spectra are different for excitation at 275 and 292 nm. For 275-nm excitation, a peak is observed near 300 nm, which is due to the tyrosine residue(s). The occurrence of tyrosine emission indicates that RET from tyrosine to tryptophan is not complete in denatured azurin. Excitation at a wavelength of 292 nm or longer results in excitation of only the tryptophan residue. These spectra demonstrate the wide range of tryptophan spectral properties that can be observed for proteins and suggest that changes in the emission spectra can be used to follow protein unfolding.

16.3.A. Site-Directed Mutagenesis of a Single-Tryptophan Azurin

The tools of molecular biology have provided a powerful means for unraveling the complexities of protein fluorescence. Site-directed mutagenesis and expression of mutant proteins allow the amino acid sequences of proteins to be altered as desired.[60,61] Site-directed mutagenesis was used to determine the effect of amino acid substitutions on the emission spectra of the single tryptophan at position 48 in azurin Pae from *Pseudomonas aeruginosa*.[62] The three-dimensional structure of the protein is known and consists of an α-helix plus eight β-strands which form a hydrophobic core, the so-called β-barrel (Figure 16.14). The single tryptophan residue (W48) is located in the hydrophobic region. The wild-type protein, with no changes in the amino acid sequence, displays the structured emission characteristic of indole in a hydrophobic solvent (Figure 16.15).

It is interesting to observe how small changes in the sequence can affect the structured emission. Can a single amino acid substitution shift the emission spectrum? How many changes are needed to eliminate the structured emission spectrum of tryptophan? This question was answered by mutating two amino acid residues known to be located close to the indole ring in the native protein. In one of these mutant proteins, the nonpolar amino acid isoleucine (ile) in position 7 was replaced with serine (ser). This mutant protein is referred to as Ile7Ser. The amino acid serine contains a hydroxyl group, which might be expected to

Figure 16.14. α-Carbon backbone of wild-type azurin from *Pseudomonas aeruginosa*. The environment around the single tryptophan residue (W48) was varied by mutating isoleucine (I7) or phenylalanine (F110). The shaded sphere is the copper ion. Reprinted, with permission, from Ref. 62, Copyright © 1994, American Chemical Society.

hydrogen bond to indole and thus mimic ethanol in cyclohexane (Figure 16.5). In fact, this single amino acid substitution, Ile7Ser, resulted in a complete loss of the structured emission (Figure 16.15). The emission maximum is still rather blue-shifted (λ_{max} = 313 nm), reflecting the predominantly nonpolar character of the indole environment. The emission of tryptophan-48 was also found to be sensitive to substitution of the phenylalanine residue at position 110 by serine (Phe110Ser, Figure 16.15). These studies demonstrate that just a single hydrogen bond can eliminate the structured emission of indole and that the emission spectra of indole are sensitive to small changes in the local environment.

16.3.B. Emission Spectra of Azurins with One or Two Tryptophan Residues

The azurins also provide examples of single-tryptophan proteins with the tryptophan residues located in different

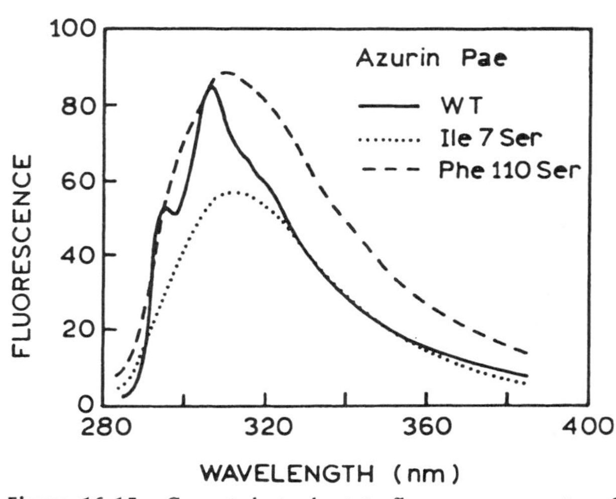

Figure 16.15. Corrected steady-state fluorescence spectra of holoazurin Pae: ——, wild type (WT); · · ·, Ile7Ser (I7S); ---, Phe110Ser (F110S). Data were collected at 298 K in 20m*M* HEPES buffer at pH 8.0. The excitation wavelength was 285 nm. Revised from Ref. 62.

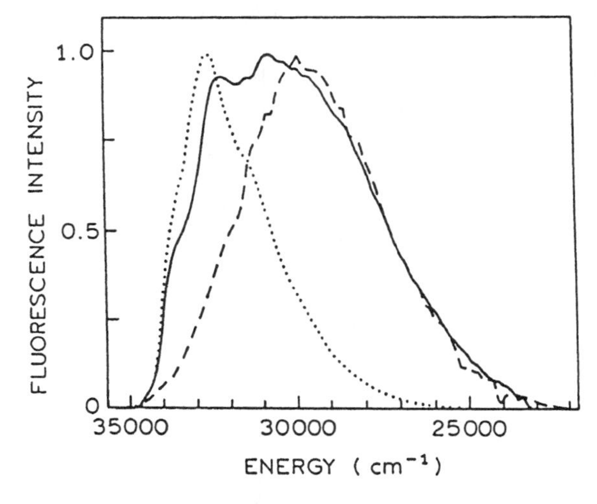

Figure 16.16. Fluorescence spectra of apoazurin Pae (· · ·), apoazurin Ade (———), and apoazurin Afe (— — —). The fluorescence spectra of the holoproteins only differ from those of the apoproteins in having a reduced intensity. Data from Ref. 63.

regions of the protein.[63] Azurins isolated from different microorganisms have somewhat different sequences. As described above, azurin Pae has a single buried tryptophan residue at position 48. Azurin Afe has a single tryptophan on the surface at position 118, and azurin Ade has tryptophan residues at both positions 48 and 118. Each of these azurins displays a distinct emission spectrum (Figure 16.16). The emission of the buried residue W48 is blue-shifted, and the exposed residue W118 displays a red-shifted, featureless spectrum. The emission spectrum of azurin Ade, with both residues, is wider and shows emission from each type of tryptophan residue. Hence, a wide range of environments can exist within a single type of protein.

16.4. ENERGY TRANSFER IN PROTEINS

We now consider the possibility of RET between the aromatic amino acids in proteins. Such transfer is likely to occur because of spectral overlap of the absorption and emission spectra of phe, tyr, and trp (Figure 16.1). Furthermore, the local concentrations of these residues in proteins can be quite large. For instance, consider a typical globular protein with a molecular weight near 50,000. Such a protein contains approximately 450 amino acid residues (about 110 daltons per amino acid residue), about 10 of which will be one of the aromatic amino acids. Using a typical density of proteins of 1.4 g/ml, one can calculate a radius of about 24 Å and a concentration of aromatic amino acids of about $0.28M = 280\text{m}M$.

Typical Förster distances (R_0) and characteristic concentration values (C_0) for energy transfer between the aromatic amino acids are listed in Table 16.1. These values are only meant to illustrate the range of distances and concentrations typical of RET between the fluorescent amino acids. For any given donor–acceptor pair, the actual distances will depend on the quantum yield and emission spectrum of the donor, which can vary greatly for the same residue in different protein environments. Absorption spectra are typically less sensitive to the environment, so the emission spectrum and quantum yield of the donor are the dominant origin of the range of R_0 values. From these values (Table 16.1), it is evident that RET can be expected between the aromatic amino acids in proteins.

It should be noted that the Förster distance for tryptophan-to-tryptophan homotransfer is particularly variable. This is because the extent of spectral overlap is strongly dependent on solvent polarity. In polar solvents the emission spectrum of tryptophan is shifted away from the absorption spectrum, and the Förster distances are smaller. For instance, for the fully exposed tryptophan residues in melittin, the Förster distance was estimated to be just 4 Å.[64,65] At low temperature in a viscous solvent, where the Stokes' shift was smaller, the R_0 value for tryptophan homotransfer[8] was found to be as large as 16 Å. Hence, trp-to-trp transfer can be expected in proteins, particularly if some of the residues display a blue-shifted emission.

Table 16.1. Förster Distances (R_0) and Critical Concentrations (A_0) for RET in Proteins

Donor	Acceptor	R_0 (Å)	A_0 (M)[a]	References
Phe	Tyr	11.5–13.5	0.29–0.18	69–71
Tyr	Tyr	9–16	0.61–0.11	8, 9, 121
Tyr	Trp	9–18	0.61–0.08	8, 67, 71–75
Trp	Trp	4–16	7.0–0.11	8, 64, 76, 154

[a] The critical concentration (A_0) in moles/liter can be calculated from $A_0 = 447/R_0^3$, where R_0 is in angstroms. See Chapter 13.

16.4.A. Tyrosine-to-Tryptophan Energy Transfer in Interferon-γ

The most commonly observed RET in proteins is from tyrosine to tryptophan. This is because most proteins contain both of these amino acids, and both are readily excited at 275 nm. One typical example is human interferon-γ, whose emission spectrum depends on the extent of self-association. Interferon-γ is produced by activated lymphocytes and displays antiviral and immunoregulator activities. Its activity depends on the extent of association to dimers. The intrinsic fluorescence of interferon-γ was used to study its dissociation into monomers.

Interferon-γ is usually a dimer of two identical 17-kDa polypeptides, each containing one tryptophan residue at position 36 and four tyrosine residues (Figure 16.17).[66] The emission spectrum of the interferon-γ dimer (Figure 16.18, top) displays emission from both tyrosine and tryptophan when excited at 270 nm, and from only tryptophan for 295-nm excitation. A difference emission spectrum was obtained by normalizing the spectra at long wavelengths, where only tryptophan emits, followed by subtraction. This difference spectrum is seen to be that expected for tyrosine (dotted curve in the upper panel of Figure

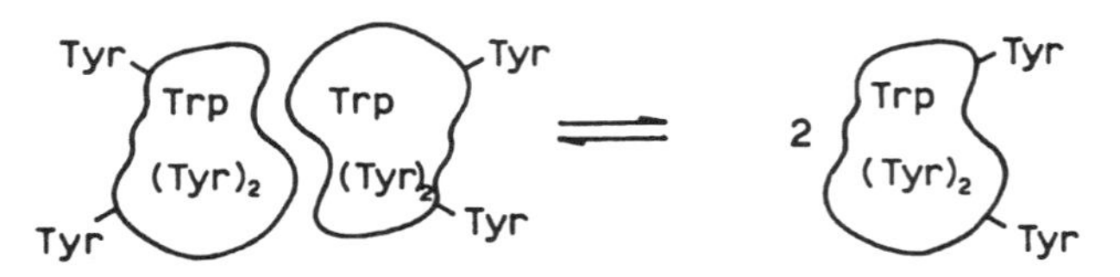

Figure 16.17. Schematic of the interferon-γ dimer. Each monomer contains one tryptophan and four tyrosines.

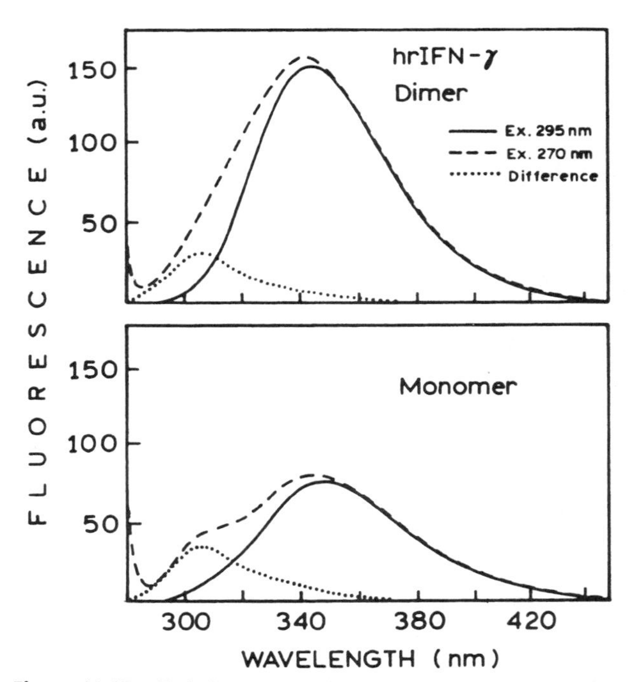

Figure 16.18. Emission spectra of human recombinant interferon-γ (hrIFN-γ) dimer (*top*) and monomer (*bottom*) in 10m*M* Tris buffer at pH 7.7. ———, Excitation at 295 nm; — — —, excitation at 270 nm; ···, difference spectrum. Revised from Ref. 66.

16.18). When incubated under the appropriate conditions, interferon-γ dissociates into monomers. One notices that the relative contribution of the tyrosine increases in the monomeric state (Figure 16.18, lower panel). This increase in tyrosine fluorescence occurs because in the monomeric state the four tyrosines are more distant from the tryptophan acceptors, one of which is on the dissociated subunit. This result also demonstrates that RET can occur between the subunits; otherwise, the extent of tyr-to-trp energy transfer would be independent of association.

16.4.B. Quantitation of RET Efficiencies in Proteins

In the previous section we explained the emission spectra of interferon-γ in terms of energy transfer, but we did not consider the effect of dimer dissociation on the fluorescence intensity of the tryptophan residue. How could we know that the tryptophan quantum yield did not decrease on dissociation of interferon-γ into monomers? In fact, the tryptophan emission intensity decreases approximately twofold on dissociation.[66] Fortunately, there is a means to measure the efficiency of tyr-to-trp energy transfer which is independent of the tryptophan quantum yield.

The efficiency of RET in proteins can be measured by considering the relative absorbance of the aromatic amino acids at various wavelengths[67] (Figure 16.19). These curves can be calculated from the absorption spectra in Figure 16.1. For an equimolar mixture of the three amino acids, tryptophan is the dominant absorber at all wavelengths between 260 and 300 nm. Above 295 nm, only tryptophan absorbs. As the wavelength is decreased, the fractional absorbance of tryptophan decreases, as the fractional absorbance of tyrosine increases. At higher relative concentrations of tyrosine, its fractional absorbance is larger. The extent of tyr-to-trp energy transfer can be found by measuring the relative tryptophan quantum yields for various excitation wavelengths. If energy transfer is 100% efficient, then the tryptophan emits irrespective of whether tyrosine or tryptophan is excited. In this case the relative quantum yield $[Q(\lambda)]$ is independent of excitation wavelength. If there is no tyr-to-trp energy transfer, then the relative tryptophan quantum yield decreases at shorter wavelengths. This decrease occurs because light absorbed by the tyrosine does not result in emission from the tryptophan.

This concept of excitation-dependent quantum yields is shown in Figure 16.19 for an equimolar mixture and a 2:1:1 mixture of tyrosine, tryptophan, and phenylalanine. The upper curves represent the fraction of the total light absorbed by tryptophan. The fractional absorbance due to tryptophan is given by

$$f_{\mathrm{trp}}(\lambda) = \frac{a_{\mathrm{trp}}(\lambda)}{a_{\mathrm{trp}}(\lambda) + a_{\mathrm{tyr}}(\lambda)} \qquad [16.1]$$

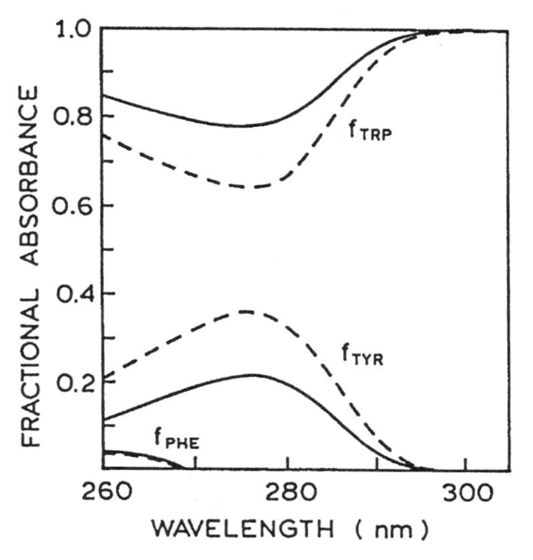

Figure 16.19. Fractional absorbances of tyrosine, tryptophan, and phenylalanine in 1:1:1 (———) and 2:1:1 (— — —) mixtures of the three amino acids. Revised from Ref. 67.

where the $a_i(\lambda)$ are the absorbances of the individual residues at wavelength λ. For simplicity, we have deleted from this expression the minor term due to absorption by phenylalanine. From Figure 16.19, one sees that only tryptophan absorbs at wavelengths longer than 295 nm. This is the basis for using this wavelength to selectively excite tryptophan fluorescence. The relative quantum yield at 295-nm excitation is taken as a reference point at which all the light is absorbed by tryptophan. The fluorescence is monitored at 350 nm or longer to avoid the tyrosine emission. The relative quantum yield at any excitation wavelength is given by

$$Q(\lambda) = \frac{a_{trp}(\lambda) + Ea_{tyr}(\lambda)}{a_{trp}(\lambda) + a_{tyr}(\lambda)} \quad [16.2]$$

where E is the efficiency of energy transfer. Typically, the relative quantum yield is normalized to unity for long-wavelength excitation. If E is 100%, then the relative quantum yield of tryptophan fluorescence is independent of excitation wavelength. If E is zero, then the relative quantum yield is given by the fractional absorbance due to tryptophan.

Representative data for a dipeptide, tyr-trp, and an equimolar mixture of tyrosine and tryptophan are shown in Figure 16.20. In the dipeptide, the donor and acceptor are well within the Förster distance, and the transfer efficiency is expected to be near 100%. This prediction is confirmed by the independence of the tryptophan quantum yield from the excitation wavelength. All the energy absorbed by the solution, irrespective of whether the light is absorbed by tyrosine or tryptophan, appears as tryptophan fluorescence. For the mixture of unlinked tyrosine and tryptophan, the relative quantum yield closely follows the fractional absorbance due to tryptophan ($E = 0$), indicating the absence of significant energy transfer.

From the examination of comparable data for proteins, one can estimate the efficiency of tyr-to-trp energy transfer. The excitation-wavelength-dependent tryptophan quantum yields are shown for interferon-γ in Figure 16.21. These values are measured relative to the quantum yield at 295 nm, where only tryptophan absorbs. In both the monomeric and dimeric states, the relative quantum yield decreases with shorter excitation wavelengths. The quenching efficiency (E) is estimated by comparison with curves calculated for various transfer efficiencies. Using this approach, the transfer efficiency is estimated to be 20% and 60% in the monomer and dimer, respectively. The decrease in relative quantum yield is less in the dimeric state, indicating a greater efficiency of tyr-to-trp energy transfer in the dimer. This implies that the tyrosine residues in one subunit transfer energy to the tryptophan residue in the other subunit. The smaller decrease in relative quantum yield for the dimer at 270 nm can be understood as due to the higher transfer efficiency, so that more of the photons absorbed by tyrosine appear as tryptophan emission.

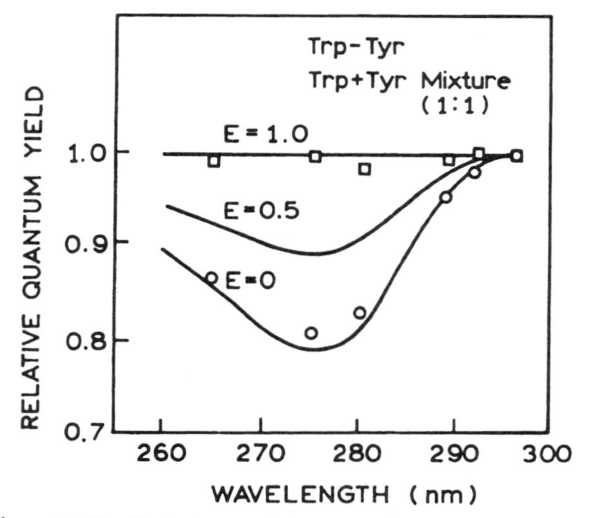

Figure 16.20. Excitation wavelength dependence of the relative quantum yields of the dipeptide tryptophanyltyrosine (□) and an equimolar mixture of tyrosine and tryptophan (○). The curves correspond to energy transfer efficiencies of 0, 50, and 100%. Revised from Ref. 67.

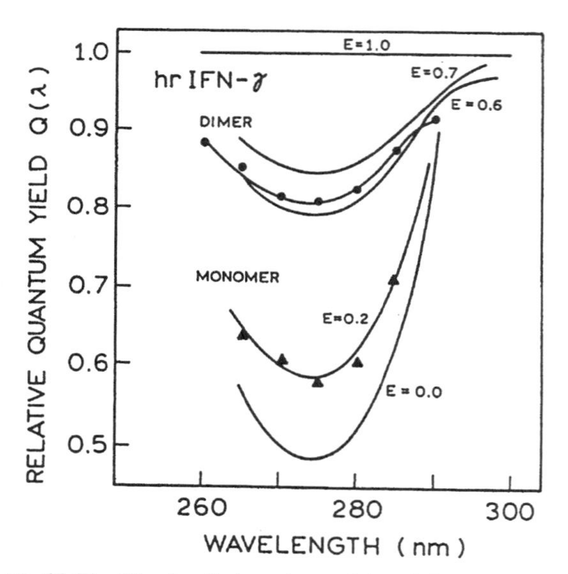

Figure 16.21. Wavelength dependence of the relative quantum yield $Q(\lambda)$ of $1 \mu M$ human recombinant interferon-γ(hrIFN-γ) in the monomer (▲) and dimer (●) states. The solid curves are theoretical curves obtained according to Eisinger[67] for different values of the energy transfer efficiency E. All theoretical and experimental data were normalized to the absorbance at 295 nm. Revised from Ref. 66.

16.4.C. Energy Transfer Detected by Decreases in Anisotropy

Another effect of RET is a decrease in anisotropy. This effect is seen in the excitation anisotropy spectrum of a trp-trp dipeptide shown in Figure 16.22. Compared to tryptophan itself, the dipeptide shows lower anisotropies. This decrease is attributed to RET and the loss of orientation during the transfer process. Although decreases in anisotropy can be used to detect energy transfer, it is difficult to use the data in a quantitative manner. This is because one does not know the extent of depolarization that occurs during transfer. For instance, for randomly distributed donors and acceptors, a single energy-transfer step is expected to decrease the anisotropy to 4% of the initial value. However, the decrease in anisotropy may be less for donors and acceptors that are in partial alignment due to a covalent linkage. Even if the extent of depolarization were known for model compounds, the effect could be different in a protein, dependent on protein conformation, and/or dependent on the relative orientation of the tryptophan residues in the protein.

16.4.D. Phenylalanine-to-Tyrosine Energy Transfer

RET can also occur from phenylalanine to tyrosine. This was found in an unusual histonelike protein, HTa, from a thermophilic archaebacterium *Thermoplasma acidophilum*.[69] HTa associates strongly with DNA to protect it from thermal degradation. This highly unusual protein is a tetramer, with five phenylalanine residues and one tyrosine residue in each monomer, and no tryptophan residues (Figure 16.23). In this class of organisms, the more thermophilic organisms have higher phenylalanine contents in their histonelike proteins. An experimentally useful property of this protein is the ability to remove the tyrosine residues, located at the third position from the carboxy terminus, by digestion with carboxypeptidase A. The digested protein can be used to study the spectral properties of proteins containing only phenylalanines.

Emission spectra of HTa excited at 252 nm are shown in Figure 16.23. The emission appears to be similar to that of tyrosine, with an emission maximum near 300 nm. However, the spectrum also shows a shoulder at 280 nm, which is too blue-shifted to be due to tyrosine. The origin of the dual emission was determined by a comparison with the emission from tyrosine and phenylalanine at equivalent concentrations (Figure 16.23). These spectra suggest that the emission spectrum of HTa is due to the individual contributions of the phe and tyr residues. Assignment of the 280-nm emission to phenylalanine was also accomplished by removal of the tyrosine residue by digestion with carboxypeptidase A, which yielded a spectrum identical to that found for phenylalanine (Figure 16.23, lower panel). Thus, the emission at 280 nm was clearly identified as due to phenylalanine.

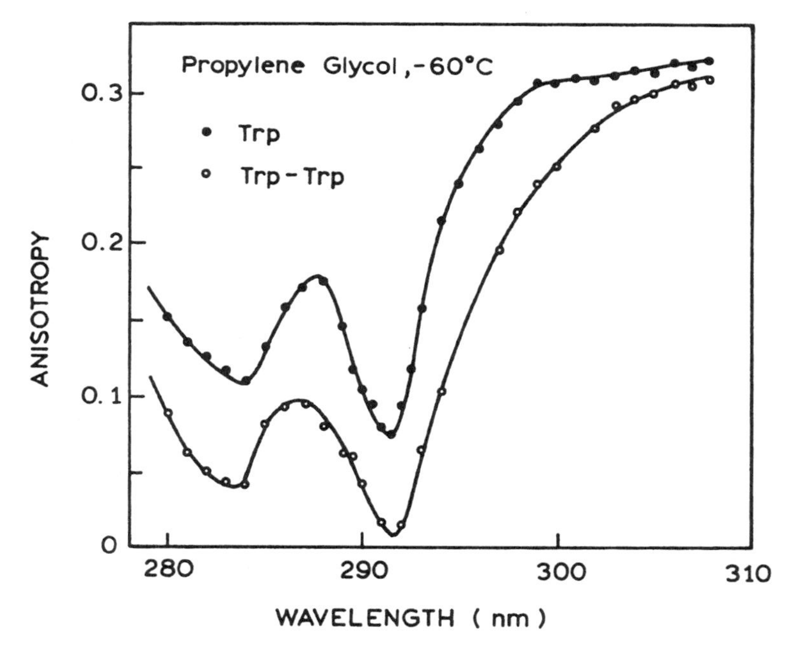

Figure 16.22. Excitation anisotropy spectra of tryptophan (●) and tryptophanyltryptophan (○). From Ref. 68.

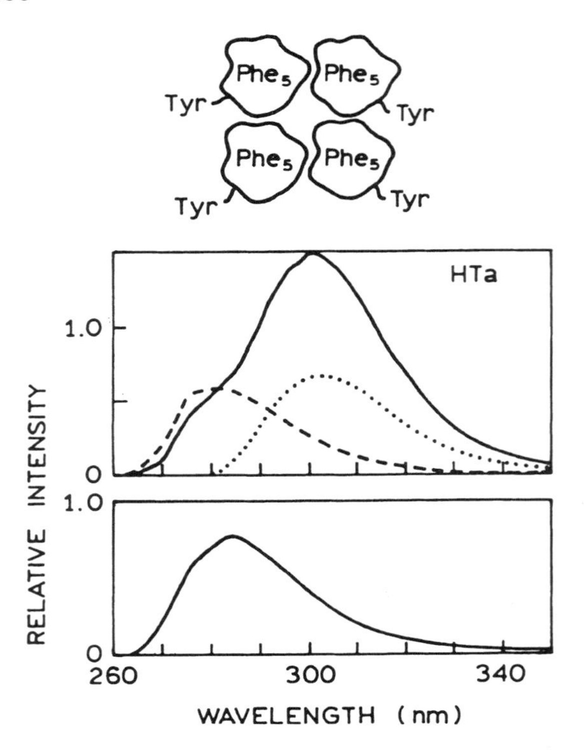

Figure 16.23. *Top*: Emission spectra of archaebacterial histonelike protein (HTa) excited at 252 nm (———) and of equivalent concentrations of phenylalanine (— — —) and tyrosine (· · ·). *Bottom*: Emission spectrum of HTa following removal of the tyrosine by carboxypeptidase A digestion. Revised from Ref. 69.

The presence of phe-to-tyr energy transfer in HTa was detected from the excitation spectra. The reasoning is analogous to that employed in the interpretation of the results in Figure 16.20 for tyr-to-trp transfer. These results are best understood by first examining the absorption spectra (Figure 16.24). The native protein shows the characteristic structured absorption of phenylalanine, as well as absorption at 280 nm (Figure 16.24A). Removal of the tyrosine residues by carboxypeptidase digestion results in loss of absorbance at 280 nm and a structured absorption centered at 260 nm (Figure 16.24C). The excitation spectrum of HTa is shown in Figure 16.25. These data are superimposed on the absorption spectra of tyrosine alone and of a 5:1 mixture of phenylalanine and tyrosine. The excitation spectrum of HTa lies midway between these absorption spectra, indicating a phe-to-tyr transfer efficiency of near 50%. If the transfer efficiency were zero or 100%, the excitation spectrum would be superimposed on the absorption spectrum of tyrosine or of the phe–tyr mixture, respectively. Given R_0 = 13.5 Å for phe-to-tyr energy transfer,[69] and the diameter expected for a protein with a molecular weight of 9934, which is near 28Å, it is not surprising that phenylalanine fluorescence is quenched in HTa. While most proteins are larger, they may also contain multiple phenylalanine and tyrosine residues. Hence phenylalanine-to-tyrosine energy transfer is a common feature of proteins.

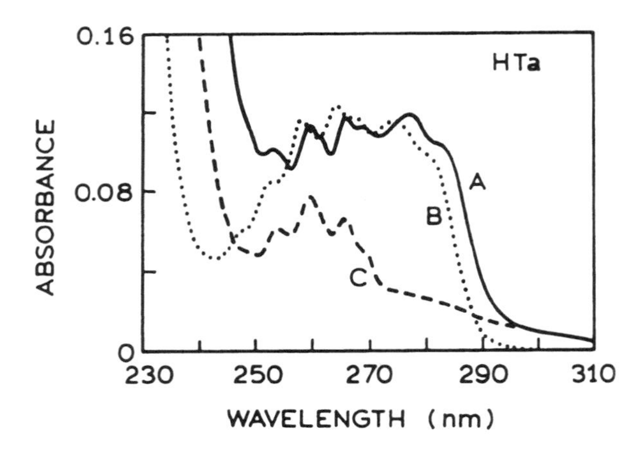

Figure 16.24. Ultraviolet absorption spectra of the histonelike protein HTa. (A) Native protein HTa (80μ*M*); (B) mixture of aromatic amino acids equivalent to that in the protein (80μ*M* tyrosine plus 400μ*M* phenylalanine); (C) HTa (80μ*M*) that has been digested with carboxypeptidase in order to remove the tyrosine residue. Revised from Ref. 69.

Characteristics of Phenylalanine Emission in Proteins. Removal of tyrosine from HTa provides a protein without tyrosine or tryptophan. In this case, the lifetime of phenylalanine was near 22 ns, which is considerably longer than the values of 2–5 ns typical of tyrosine and tryptophan. In the presence of the tyrosine residue, the phenylalanine lifetime is decreased to near 12 ns. This value is consistent with 50% energy transfer from phenylalanine to tyrosine. It is interesting to notice that the 12-ns decay time is observed at all emission wavelengths, even those where only tyrosine is expected to emit.[69] Such long decay times are unlikely to be due to tyrosine itself. In this

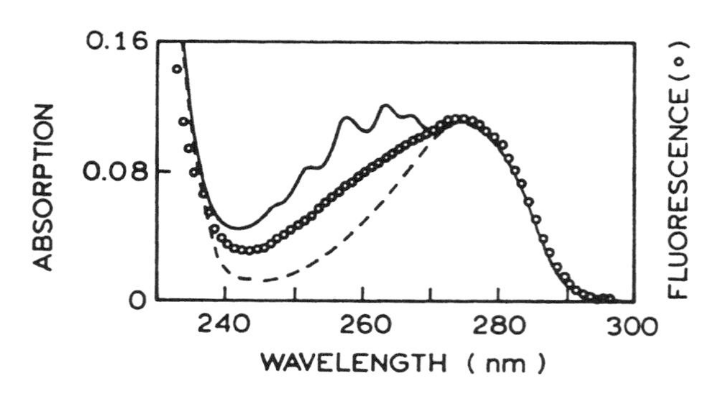

Figure 16.25. Corrected fluorescence excitation spectrum of HTa (○) superimposed upon the ultraviolet absorption spectra of a mixture of tyrosine plus phenylalanine (———) and of tyrosine alone (— — —). Fluorescence emission was measured at 330 nm. Revised from Ref. 69.

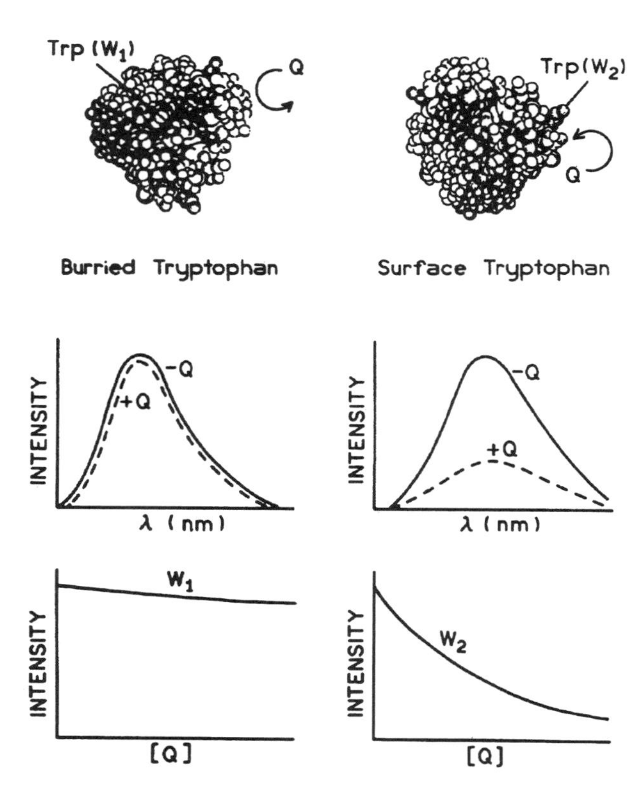

Figure 16.26. Collisional quenching of surface-accessible (W_1) and buried (W_2) tryptophan residues in proteins.

protein the excited phenylalanines continue to transfer energy to tyrosine during their decay, resulting in an apparent 12-ns tyrosine lifetime. This phenomenon will be described more fully in Chapter 18 on excited-state reactions.

16.5. QUENCHING OF TRYPTOPHAN RESIDUES IN PROTEINS

Collisional quenching of proteins has been extensively utilized to determine the extent of tryptophan exposure to the aqueous phase.[77–80] The basic idea is shown in Figure 16.26, which depicts quenching by a water-soluble quencher which does not easily penetrate the protein matrix. Recall that collisional quenching is essentially a contact phenomenon, so that the fluorophore and quencher need to be in molecular contact for quenching to occur. Consequently, if the tryptophan residue is buried inside the protein (W_1), quenching is not expected to occur. If the tryptophan residue is on the protein surface (W_2), then quenching is expected.

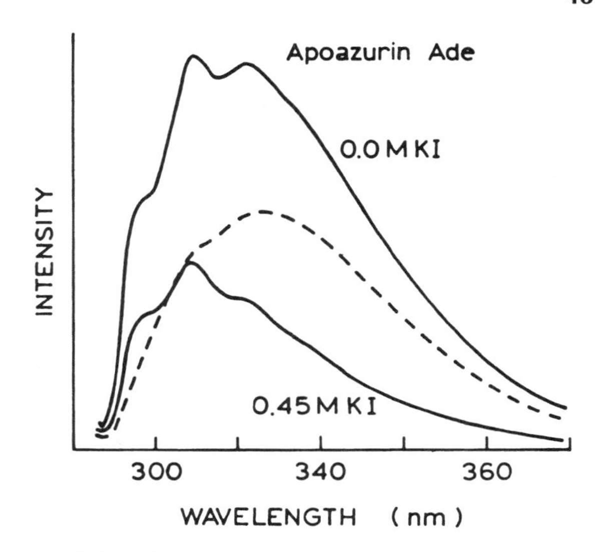

Figure 16.27. Fluorescence emission spectra of apoazurin Ade from *Alcaligenes denitrificans* in the absence (top curve) and presence (bottom curve) of 0.45*M* KI. The dashed curve is the difference spectrum and is due to the easily quenched emission (presumably from Trp-118). The buffer was 0.05*M* sodium acetate, pH 5.0. Revised from Ref. 79.

One example of selective quenching is provided by apoazurin, that is, an azurin lacking copper. Apoazurin Ade has two tryptophan residues, one surface residue and one buried residue.[63] The emission spectrum of this protein, along with those of its single-tryptophan relatives, was shown in Figure 16.16. For our present purposes, the properties of the protein with or without (apo) copper are the same.

Emission spectra were recorded in the absence and presence of 0.45*M* iodide, which is a collisional quencher (Figure 16.27). The intensity decreases in the presence of iodide. More important is the change in the emission spectrum: in the presence of iodide, the spectrum resembles the structured emission seen from the single-tryptophan azurin Pae. The spectrum of the quenched tryptophan residue can be seen from the difference spectrum (dashed curve in Figure 16.27) and is characteristic of an exposed residue in a partially hydrophobic environment. In this favorable case, one residue is quenched and the other is not, providing immediate resolution of the two emission spectra.

16.5.A. Effect of Emission Maximum on Quenching

For water-soluble quenchers, which do not readily penetrate the hydrophobic regions of proteins, there is a strong correlation between the emission maximum and quenching constant.[79] Blue-shifted tryptophan residues are essentially inaccessible to quenching by acrylamide, and

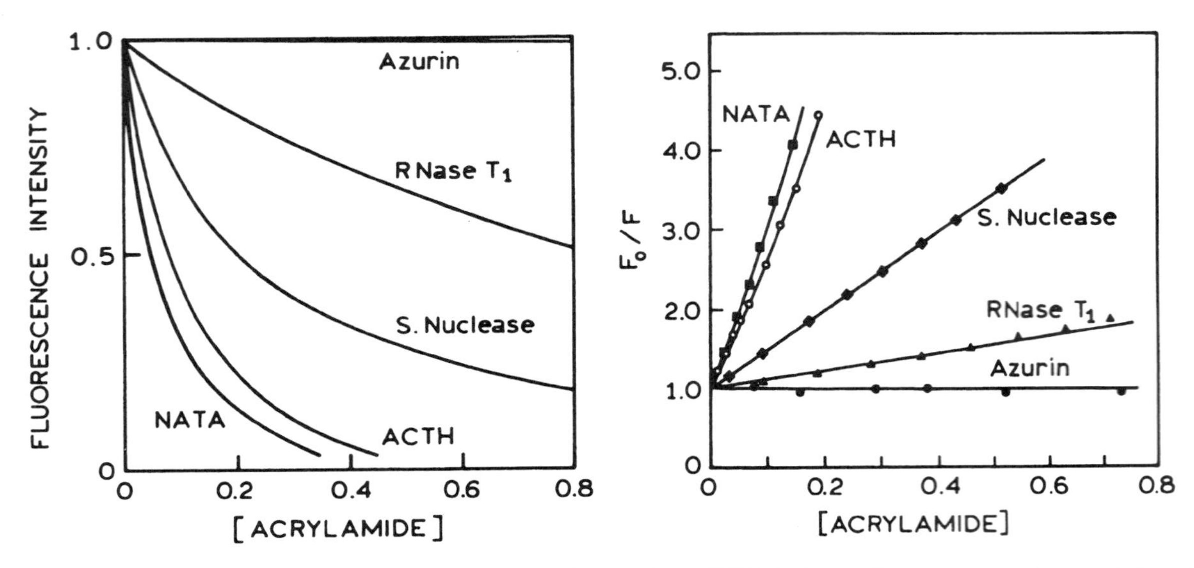

Figure 16.28. Acrylamide quenching of representative single-tryptophan proteins and NATA. *Left*: Intensity versus acrylamide concentration; *right*: Stern–Volmer plots. Data from Refs. 64, 73, and 80.

red-shifted residues are nearly as accessible as tryptophan in water. This can be seen in the dependence of fluorescence intensity on acrylamide concentration. The emission of azurin is essentially unchanged in up to 0.8*M* acrylamide. In contrast, the exposed tryptophan residue in ACTH is almost completely quenched at 0.4*M* acrylamide (Figure 16.28). ACTH activates adenylate cyclase activity in the adrenal cortex, resulting in increased intracellular concentration of cAMP.

Quenching data are typically presented as Stern–Volmer plots, which are shown for several single-tryptophan proteins in the right-hand panel of Figure 16.28. In these plots the larger slopes indicate larger amounts of quenching, and the slopes can be used to calculate the bimolecular quenching constant (k_q). The buried single tryptophan residue in azurin Pae is not affected by acrylamide. In contrast, the fully exposed residue in ACTH is easily quenched by acrylamide. In fact, ACTH is quenched nearly as effectively as NATA. A plot of the bimolecular quenching constant (k_q) for acrylamide quenching versus emission maximum for a group of single-tryptophan proteins is shown in Figure 16.29. These k_q data show that the accessibility of the tryptophan residue to acylamide quenching varies from almost complete inaccessibility for azurin and asparaginase to complete accessibility for glucagon. Glucagon is a relatively small peptide with 29 amino acids which opposes the action of insulin. Glucagon has a random structure in solution.[81–83] Hence, there is little opportunity for the protein to shield the single trp-25 residue from acrylamide quenching. In addition to acrylamide, iodide is another water-soluble quencher for which the bimolecular quenching constant shows a correlation with emission maximum.

The extent of collisional quenching can also be dependent on protein conformation and/or the extent of subunit association. This effect has been illustrated for melittin,

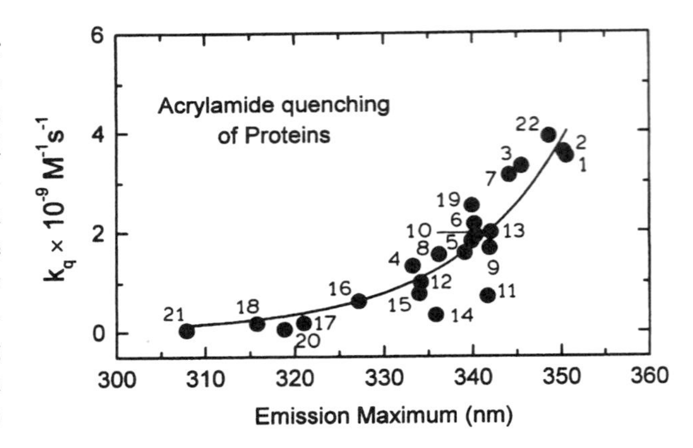

Figure 16.29. Dependence of the apparent acrylamide bimolecular quenching constant (k_q) on the emission maximum for single-tryptophan proteins: 1, glucagon; 2, adrenocorticotropin; 3, melittin monomer; 4, melittin tetramer; 5, gonadotropin; 6, phospholipase A_2; 7, human luteinizing hormone; 8,9, monellin; 10, gonadotropin; 11, human serum albumin, N form; 12, human serum albumin, F form; 13, myelin basic protein; 14, elongation factor Tu-GDP; 15, nuclease; 16, fd phage; 17, ribonuclease T_1; 18, parvalbumin; 19, calcium-depleted parvalbumin; 20, asparaginase; 21, apoazurin Pae; 22, mastoparan X. Revised from Ref. 79.

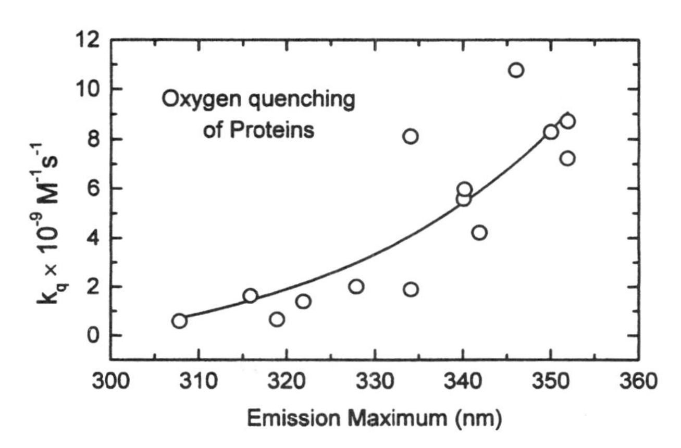

Figure 16.30. Correlation of emission maxima of multi-tryptophan proteins with the oxygen bimolecular quenching constants. Data from Ref. 85.

which is a small peptide of 26 amino acids containing a single tryptophan residue at position 19. Depending on ionic strength, melittin can exist as a monomer or as a tetramer. In the tetramer, the four tryptophan residues, one per monomer, are located in a hydrophobic pocket between the helices.[84] The tryptophan residues in melittin are more easily quenched when melittin is in the monomer state.

Acrylamide and iodide are polar molecules and do not readily penetrate the nonpolar regions of proteins. However, it is known that small nonpolar oxygen molecules readily penetrate all regions of proteins. The oxygen bimolecular quenching constants for a number of proteins show weaker correlation with the emission maxima[85] (Figure 16.30). It is known from the X-ray structures of proteins that the interiors are densely packed, similarly to crystals of organic molecules, so that there is little room for oxygen within the center of most proteins. Surprisingly, the bimolecular quenching constants for oxygen are 20–50% of the value for tryptophan in water (near 1×10^{10} M^{-1} s^{-1}). These results have been taken to indicate that proteins undergo rapid structural fluctuations which allow the oxygen molecules to penetrate during the nanosecond excited-state lifetime. These results were initially surprising to biochemists, and considerable efforts were made to find alternative explanations. However, it is now agreed that oxygen does readily diffuse through most proteins.[86,87] Structural fluctuations on the nanosecond timescale are consistent with the anisotropy decays of most proteins, which typically show nanosecond or subnanosecond components.[88,89] Apparently, acrylamide and iodide are less effective quenchers because it is energetically unfavorable for them to enter nonpolar regions of proteins. Also, acrylamide is somewhat larger than oxygen or iodide, which also plays a role in its more selective quenching.

16.5.B. Fractional Accessibility to Quenching in Multi-Tryptophan Proteins

The fact that polar quenchers are most effective in quenching surface tryptophan residues suggests the use of quenching to resolve the contributions of surface and buried tryptophans to the total fluorescence of a protein. Stellacyanin is a small copper-containing protein which has three tryptophan residues. The emission can be quenched by iodide, but the Stern–Volmer plot for the native protein shows significant downward curvature (Figure 16.31, ○). This suggests that some of the fluorescence is not accessible to iodide quenching.[90] If this protein is unfolded (Fig-

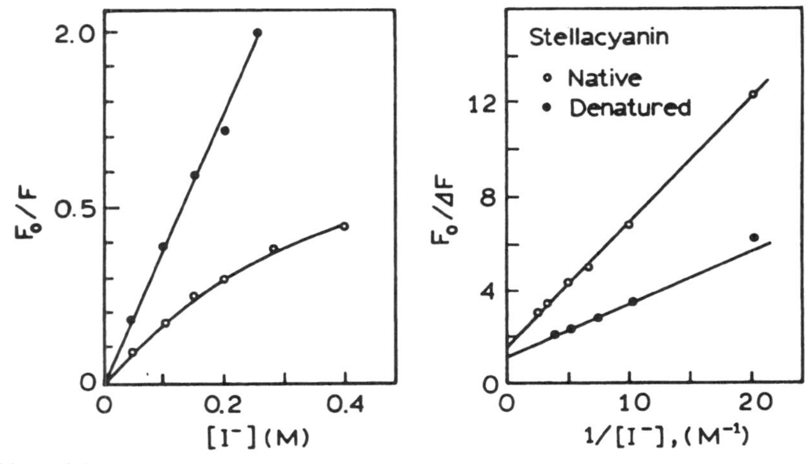

Figure 16.31. Stern–Volmer (*left*) and modified Stern–Volmer plots (*right*) for iodide quenching of native (○) and denatured (●) stellacyanin. Revised from Ref. 90.

ure 16.31, ●), the extent of quenching increases and the Stern–Volmer plot becomes linear.

As described in Section 8.8, the fraction of accessible fluorescence (f_a) can be obtained from a modified Stern–Volmer plot, in which $F_0/\Delta F$ is plotted versus reciprocal quencher concentration. The y-intercept represents extrapolation to infinite quencher concentration. Complete quenching is expected for an accessible fluorophore, so $F_0/\Delta F$ is unity. If a portion of the fluorescence is not accessible, ΔF is smaller than F_0, and the intercept (f_a^{-1}) is greater than unity.

For denatured stellacyanin, the intercept is unity (Figure 16.31, right), so all three tryptophans are accessible to iodide quenching. For native stellacyanin, $f_a = 0.66$, suggesting that two of the three tryptophans are accessible and on the surface of the protein. While a value of $f_a = 0.66$ was easy to interpret for a protein containing three tryptophan residues, it should be recalled that each tryptophan residue need not contribute equally to the total emission. The values of f_a refer to the fraction of fluorescence accessible to quenching, which need not be the same as the fraction of tryptophan residues that is accessible to quenching.

16.5.C. Resolution of Emission Spectra by Quenching

Selective quenching of tryptophan residues in proteins has been extended to allow resolution of the emission spectra of the quenched and unquenched components. The idea is similar to that shown for apoazurin Ade in Figure 16.27. However, for apoazurin Ade, one of the residues was completely inaccessible to quenching. A more general procedure has been developed which can be used to resolve the emission spectra even when both residues are partially accessible to quenching.[91–93] The basic idea is to perform a least-squares fit of the quenching data to recover the quenching constant and fractional intensity at each wavelength (λ),

$$\frac{F(\lambda)}{F_0} = \sum_i \frac{f_i(\lambda)}{1 + K_i(\lambda)[Q]} \qquad [16.3]$$

In this expression the values of $f_i(\lambda)$ represent the fraction of the total emission quenchable with a quenching constant $K_i(\lambda)$ (Section 8.10).

This procedure was applied to a metalloprotease from *Staphylococcus aureus* which contains two tryptophan residues.[94] At any given emission wavelength, the Stern–Volmer plot is curved owing to the different accessibilities of each residue (Figure 16.32). The data were fit by least-squares methods to obtain the values of $K_i(\lambda)$ and $f_i(\lambda)$. Similar data were collected for a range of emission wavelengths. These data can be used to calculate the emission spectrum of each component,

$$F_i(\lambda) = F_0(\lambda) f_i(\lambda) \qquad [16.4]$$

where $F_0(\lambda)$ is the unquenched emission spectrum. For the metalloprotease, this procedure yielded two well-resolved emission spectra (Figure 16.33).

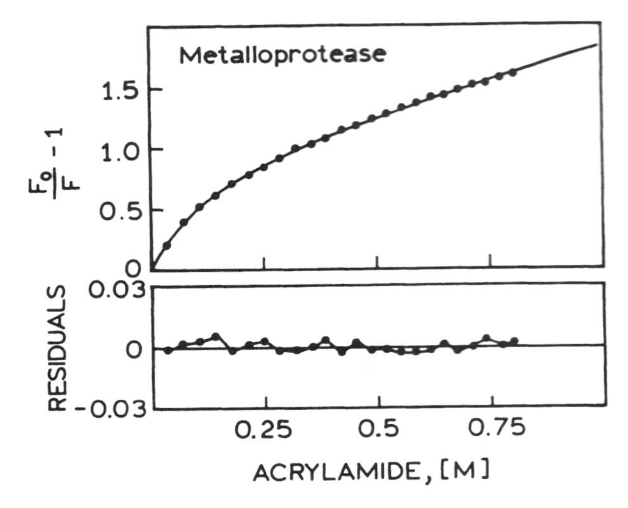

Figure 16.32. Stern–Volmer plot for acrylamide quenching of *S. aureus* metalloprotease. Excitation wavelength was 297 nm, and emission was observed at 336 nm. The solid curve shows the least-squares fit to the data (●) with parameters $K_1 = 14.1\ M^{-1}$, $K_2 = 0.51\ M^{-1}$, $f_1 = 0.52$, and $f_2 = 0.47$. The lower panel shows the residuals. Revised from Ref. 94.

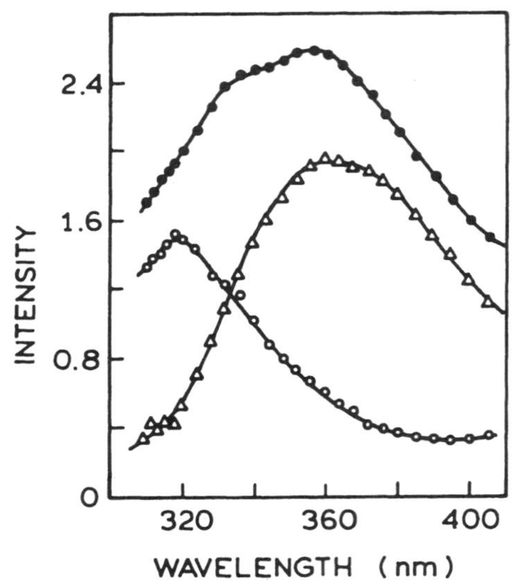

Figure 16.33. Resolution emission spectrum of the two-tryptophan metalloprotease from *S. aureus*: ○, component with $K_1 = 0.513\ M^{-1}$; △, component with $K_2 = 14.13\ M^{-1}$; ●, steady-state spectrum. Revised from Ref. 94.

It should be noted that the use of quenching-resolved spectra may not always be successful. One possible reason for failure would be that one of the tryptophan residues is not in a unique environment. In this case, each tryptophan residue may display more than one emission spectrum, each of which would be quenched to a different extent. In fact, quenching-resolved spectra have already been reported for proteins which contain a single tryptophan residue.[93,95] These results have been interpreted in terms of the protein being present in more than a single conformational state. Also, since a single tryptophan residue can display a multiexponential decay, there is no *a priori* reason to assume that the residue is quenched with the same quenching constant at all emission wavelengths.

16.6. ASSOCIATION REACTIONS OF PROTEINS

16.6.A. Self-Association of Melittin and Binding to Calmodulin

A valuable use of fluorescence is to study the association of proteins with other proteins, lipids, or nucleic acids. Such studies take advantage of the high sensitivity of fluorescence for measurements in dilute solutions. Also, the sensitivity of tryptophan emission to the local environment typically results in spectral changes upon association. Among examples of protein self-association, one of the most studied by fluorescence is the association of melittin monomers to tetramers. When self-association occurs, the random-coil monomers adopt an α-helical conformation. In the α-helical state, one side of the helix is covered with nonpolar amino acid side chains. These nonpolar surfaces come together to form the center of the melittin tetramer. The single tryptophan residue of each monomer is localized in the center of the tetramer, partially shielded from solvent. This results in a blue shift of the emission spectrum (Figure 16.34). Changes in the emission maximum and anisotropy of melittin have allowed a detailed study of solution conditions which favor tetramer formation.[97,98] Melittin is positively charged, and charge repulsion prevents tetramer formation. Increased ionic strength shields these charges and allows the monomers to associate.

The intrinsic tryptophan fluorescence of melittin has also been used to study its binding to calmodulin.[99,100] Calmodulin does not contain tryptophan, so the single tryptophan residue of melittin serves as the intrinsic probe. In the presence of Ca^{2+}, calmodulin undergoes a conformational change and exposes a hydrophobic surface. This surface can interact with the hydrophobic side of the melittin α-helix. In the α-helical state, one side of melittin

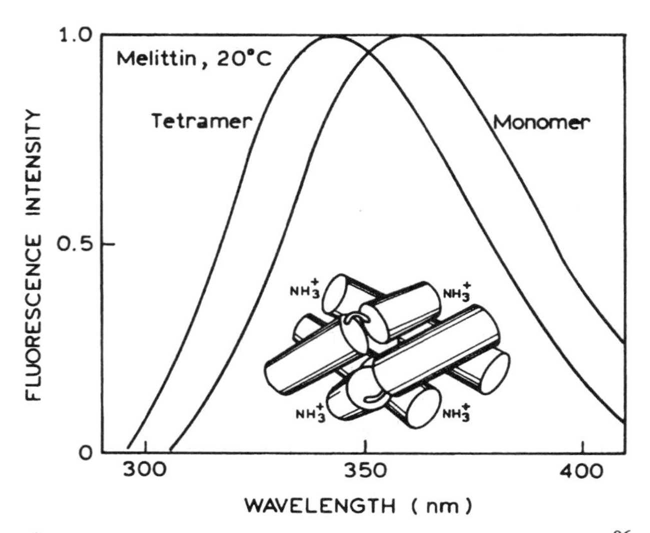

Figure 16.34. Emission spectra of melittin monomer and tetramer.[96] Excitation was at 295 nm. In the schematic structure, the tryptophans are located in the center between the four helices.

is covered with nonpolar amino acids. This side of melittin also contains the tryptophan residue. Binding of melittin to calmodulin results in a blue shift of the melittin emission and an increase in intensity (Figure 16.35). The association of melittin and calmodulin also results in an increase in the anisotropy. From such data, one can determine the binding constants and the stoichiometry of the association. It appears that the increase in anisotropy saturates at a calmodulin/melittin ratio of 1, indicating a 1:1 complex.

16.6.B. Ligand Binding to Proteins

Intrinsic protein fluorescence has also been widely utilized to study binding of ligands to proteins. In many cases the ligands are also fluorescent, and their fluorescence

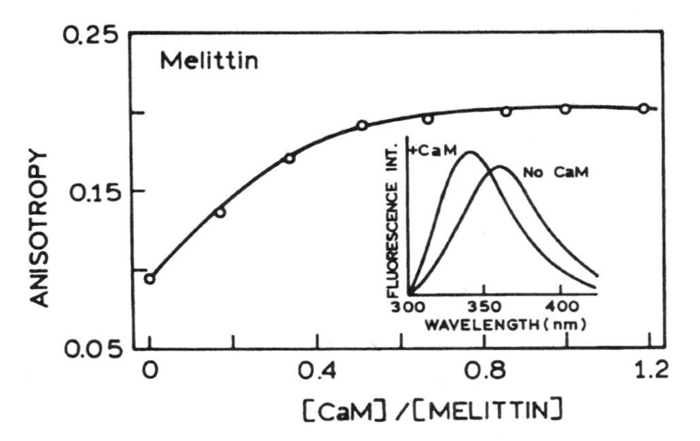

Figure 16.35. Effect of melittin–calmodulin association on the anisotropy of melittin. *Inset*: Emission spectra of melittin in the presence and absence of calmodulin (CaM). Modified from Ref. 100.

changes when they are bound to the protein. For instance, the increased fluorescence of ANS and TNS has been widely used to study their binding to hydrophobic sites on proteins.[101,102] This idea has also been applied to melittin, which binds a molecule of TNS in the tetrameric state, but not in the monomeric state.[103] In aqueous solution TNS is essentially nonfluorescent, and adding TNS to melittin monomer does not result in an enhanced TNS emission. In contrast, adding TNS to tetrameric melittin results in strong emission at 430 nm and quenching of the melittin emission due to RET (Figure 16.36). Apparently, TNS binds to tetrameric melittin because of the formation of a hydrophobic pocket between the four helices. Such data can be used to determine the affinities and stoichiometry of the binding reaction. A similar idea has been widely used to study binding of the coenzyme NADH to dehydrogenases. In this case the increased intensity of the NADH upon binding can be used to determine the extent of binding.[104,105] Alternatively, one can use the quenching of protein fluorescence due to RET to detect coenzyme binding to proteins.[106]

It is important to recognize that ligand binding to proteins can often be detected even if the ligand is spectroscopically silent. This possibility is illustrated by the maltose receptor protein from *E. coli*.[107] Such receptors are important in bacterial chemotaxis, in which they apparently undergo a ligand-induced conformational change. The sensitivity of protein fluorescence to ligand binding is illustrated by the 21% increase in intensity and 6-nm red shift of the emission of the maltose receptor in the presence of maltose (Figure 16.37). This spectral change was specific for maltose and did not occur with similar amounts of glucose. Maltose is a water-soluble compound whose polarity is not very different from that of water itself, so that one does not expect maltose to have a direct effect on tryptophan fluorescence. The occurrence of the spectral shift suggests that the protein conformation is altered by ligand binding. This result illustrates that binding of almost any ligand to a protein can result in a detectable change in the intrinsic protein fluorescence. However, determination of the binding stoichiometries can be difficult, particularly for weak binding.

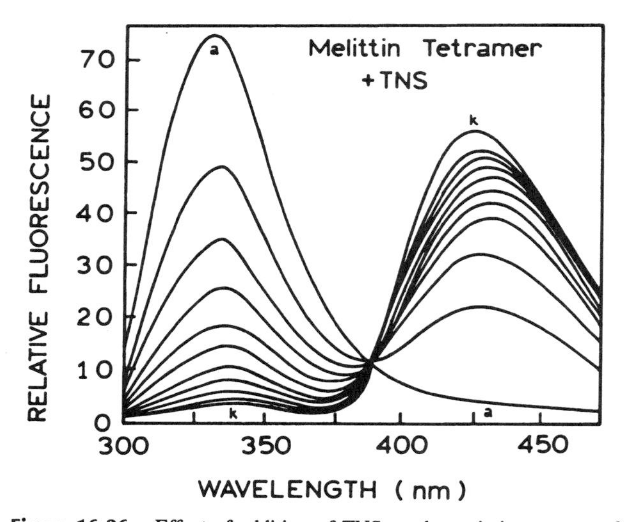

Figure 16.36. Effect of addition of TNS on the emission spectra of tetrameric melittin. The concentration of TNS ranged from 0 (a) to 2 × $10^{-5}M$ (k). Melittin does not have any tyrosine residues, allowing excitation at 280 nm. Data from Ref. 103.

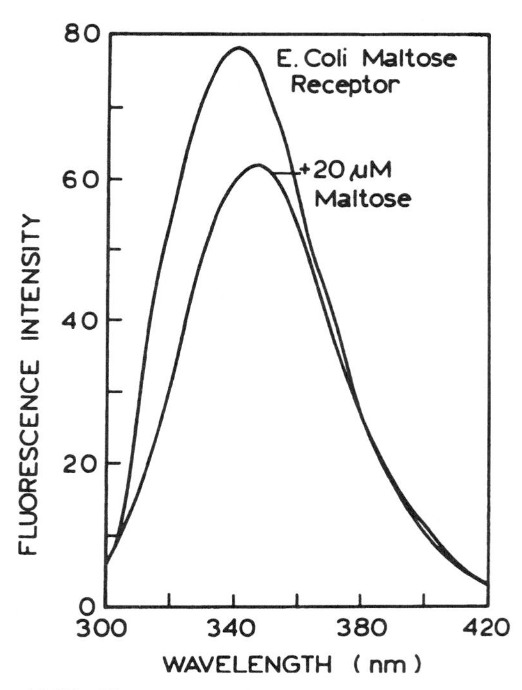

Figure 16.37. Fluorescence emission spectra (uncorrected) of the *E. coli* maltose receptor in the presence and absence of maltose. Excitation was at 288 nm. Revised from Ref. 107.

16.6.C. Correlation of Emission Maxima, Anisotropy, and Quenching Constant for Tryptophan Residues

In the preceding sections we saw how the emission maxima, anisotropy, and accessibility to quenching were all dependent on protein structure and/or ligand binding. These ideas were unified in a single experiment intended to determine the conformation of small peptides bound to calmodulin. Calmodulin is involved in regulation of the activity of intracellular enzymes. This activity is the result of a hydrophobic surface which is expressed in the presence of calcium. This surface interacts with amphipathic α-helices. In fact, calmodulin binds melittin and other α-helical peptides.[108,109] One such peptide is from myosin light-chain kinase (MLCK), which is involved in post-translational phosphorylation of proteins. It was of interest to determine whether peptides bound to calmodulin were in the α-helical form. This could not be determined by circular dichroism (CD) because binding of the peptide changed the conformation of calmodulin, so that changes in the CD spectra were due to both changes in MLCK peptide and calmodulin conformation.

Calmodulin does not contain any tryptophan residues, allowing observation of tryptophan-containing peptides in the presence of calmodulin. Seventeen MLCK-like peptides were synthesized in which a single tryptophan residue was moved along the sequence.[110] If these peptides were bound to calmodulin in the α-helical state, then one expects the single tryptophan residue to be buried between the peptide and calmodulin in some of the peptides, and to be exposed to water in other peptides (Figure 16.38). This dependence is expected to repeat with the periodicity of an α-helix.

The calmodulin-bound state of the peptides was studied by recording the emission spectra, steady-state anisotropies, and acrylamide quenching constants. A periodic behavior was found for all three observables, which oscillated at a rate of about once per 3.4 amino acid residues (Figure 16.39). This period is the same as the number of amino acid residues per turn of an α-helix. The parameter values changed in a manner consistent with periodic exposure and shielding of the tryptophan residue. For instance, at position 3 the emission maximum is at the shortest wavelength. At the same position, the steady-state anisotropy is at its maximum and the acrylamide quenching constant is at a minimum. All three effects are consistent with a shielded tryptophan residue at position 3. At position 8, the emission is at the longest wavelength, the anisotropy is at a minimum, and the tryptophan residue is accessible to quenching. In total, these results demonstrated that the bound peptide was in the α-helical state.

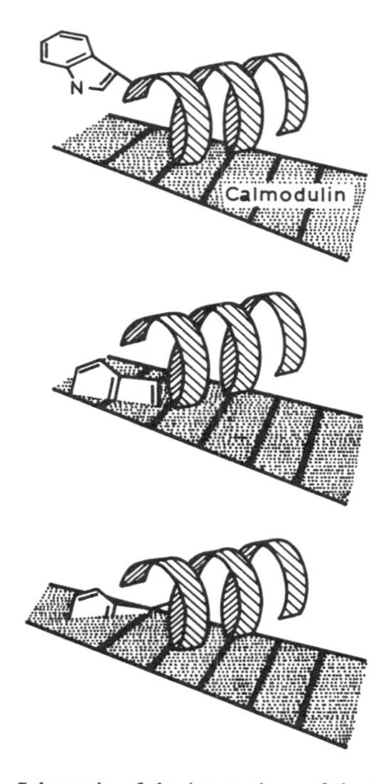

Figure 16.38. Schematic of the interactions of the α-helical MLCK peptide with calmodulin. The position of the single tryptophan residue is moved along the helix in 16 synthetic peptides. Reprinted, with permission, from O'Neil, K. T., Wolfe, H. R., Erickson-Viitanen, S., and DeGrado, W. F., Fluorescence properties of calmodulin-binding peptides reflect alpha-helical periodicity, *Science* **236:**1454–1456, Copyright © 1987, American Association for the Advancement of Science.

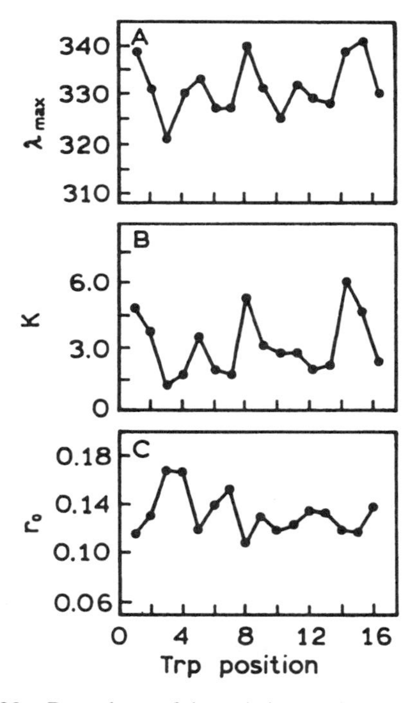

Figure 16.39. Dependence of the emission maxima (A), acrylamide quenching constants (B), and steady-state anisotropies (C) of the MLCK peptides bound to calmodulin on the position of the tryptophan residue. Reprinted, with permission, from O'Neil, K. T., Wolfe, H. R., Erickson-Viitanen, S., and DeGrado, W. F., Fluorescence properties of calmodulin-binding peptides reflect alpha-helical periodicity, *Science* **236:**1454–1456, Copyright © 1987, American Association for the Advancement of Science.

16.6.D. Calmodulin: Resolution of the Four Calcium Binding Sites Using Tryptophan-Containing Mutants

An important tool for modern protein fluorescence is the use of mutant proteins containing single tryptophan residues at desired locations in the protein. This idea was applied to calmodulin and resulted in the ability to detect binding of calcium to each of the four binding sites. Calmodulin is a single polypeptide chain with 145 amino acid residues, with two domains connected by an α-helical linker (Figure 16.40). While calmodulin contains several tyrosine residues, it does not contain tryptophan. This allowed the insertion of tryptophan residues into the sequence to probe various regions of the protein. Calmodulin was known to bind four calcium residues, but little was known about the sequence of binding and possibility of

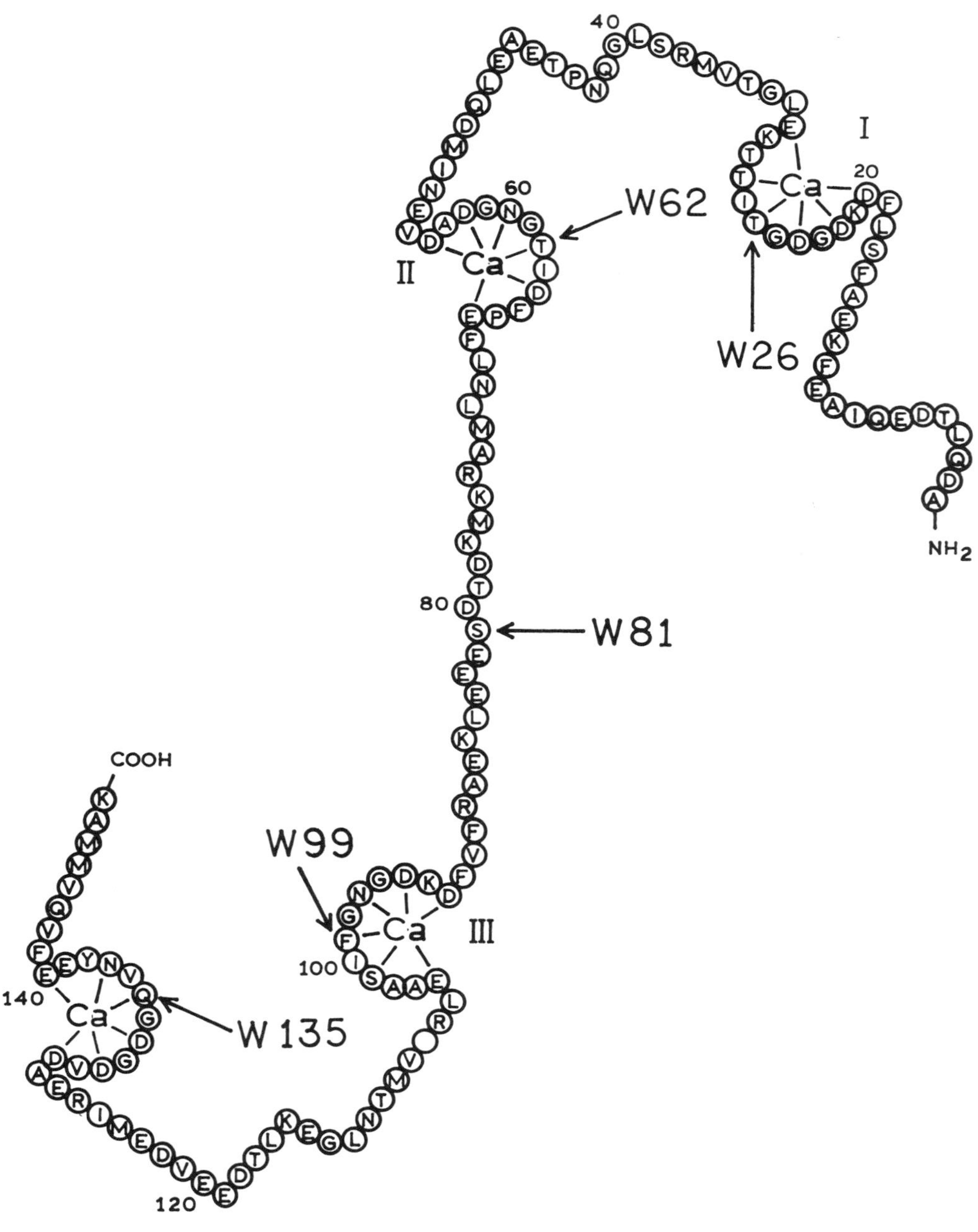

Figure 16.40. Amino acid sequence of calmodulin and five single-tryptophan mutants. Reprinted, with permission, from Ref. 111, Copyright © 1992, American Chemical Society.

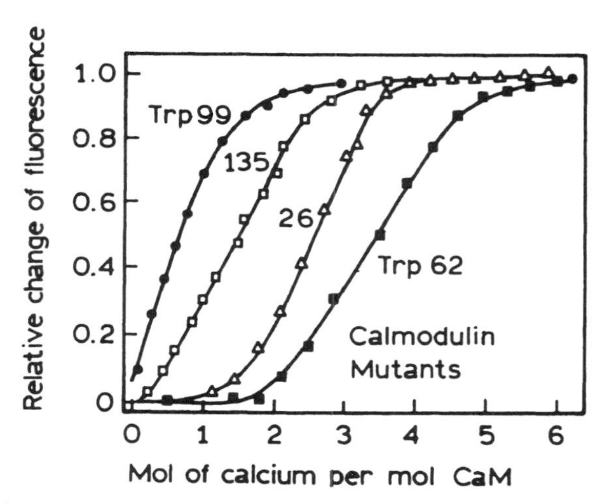

Figure 16.41. Relative changes in the fluorescence intensity of trp-99 (●), trp-135 (□), trp-26 (△), and trp-62 (■) in single-tryptophan mutants of calmodulin (CAM) with changes in the amount of bound calcium. Reprinted, with permission, from Ref. 111, Copyright © 1992, American Chemical Society.

cooperativity between the four binding sites. Extensive studies of calcium binding were not adequate to distinguish between various models for Ca^{2+} binding. Of course, bulk measurements of binding can reveal only the total amount of bound calcium and do not indicate the sequence of calcium binding. This problem was solved by the creation of single-tryptophan mutants.[111] The tryptophan residues were inserted near each of the four Ca^{2+} binding sites, one trp residue per mutant, and in the fifth mutant a trp residue was placed on the α-helical linker.

The authors of the study presented extensive data on the effects of Ca^{2+} on the spectral properties of each single-tryptophan mutant. Their summary results (Figure 16.41) show the intensity changes for each of the four trp binding-site mutants in the presence of various amounts of bound calcium. These measurements yielded the sequence of calcium binding to calmodulin and revealed interactions between the various binding sites. These elegant studies represent the state of the art in combining protein design and fluorescence spectroscopy to provide detailed information on the solution behavior of a protein. Similar tryptophan-containing calmodulin mutants were used to study binding of peptides to calmodulin.[112]

16.7. SPECTRAL PROPERTIES OF GENETICALLY ENGINEERED PROTEINS

The use of molecular biology and site-directed mutagenesis provides an ideal way to gain an understanding of the spectral properties of multi-tryptophan proteins. Suppose that the protein of interest contains two tryptophan residues in the wild-type sequence. Typically, it will be of interest to resolve the contributions of each tryptophan to the total emission. Then, further studies in the presence of substrate or other macromolecules may reveal which region of the protein is involved with its activity. One can attempt to use quenching or time-resolved methods to resolve the emission spectra of the two residues. However, a common result is that the tryptophan residues may interact, so that the wild-type emission is not a simple sum of the contributions of the two residues. In fact, such interactions are difficult to detect using only data from the wild-type two-tryptophan protein.

A powerful approach to resolving the contributions of each tryptophan, and investigating their interactions, is by examination of the two single-tryptophan mutants. Each tryptophan residue is replaced in turn by a similar residue, typically phenylalanine. If the wild-type protein contains three tryptophans, then the three single-tryptophan mutants, each with two tryptophan replacements, are usually created. This approach to resolving protein fluorescence has been used in many laboratories, and it is not practical to summarize all these results in this chapter. Rather, we have attempted to describe a few representative cases which show simple noninteractive tryptophans as well as several more complex cases in which there is trp-to-trp energy transfer.

16.7.A. Protein Tyrosyl Phosphatase—A Simple Two-Tryptophan Protein

A relatively simple example of the use of protein engineering is provided by protein tyrosyl phosphatase (PTP). Tyrosine phosphorylation and dephosphorylation are extensively involved in the regulation of cell growth and transformation. PTP is an 18-kDa single-subunit protein which is involved in the dephosphorylation pathway. The bovine wild-type protein contains two tryptophan residues at positions 39 and 49.

Single-tryptophan mutants of PTP were constructed. In W39F, tryptophan-39 is replaced by phenylalanine (F). In W49F, tryptophan-49 is replaced by phenylalanine. The emission spectra for the two single-tryptophan mutants sum to match the emission spectrum of the wild-type protein (Figure 16.42). Tryptophan-39 is seen to emit at 320 nm (W49F), and tryptophan-49 is seen to be the dominant fluorophore emitting at 350 nm (W39F). The finding that the sum of these two spectra matches the wild-type spectrum indicates that the two tryptophans behave independently and probably do not undergo energy transfer. The use of the two single-tryptophan mutants allowed an unambiguous assignment of the emission spectra to each tryptophan residue.

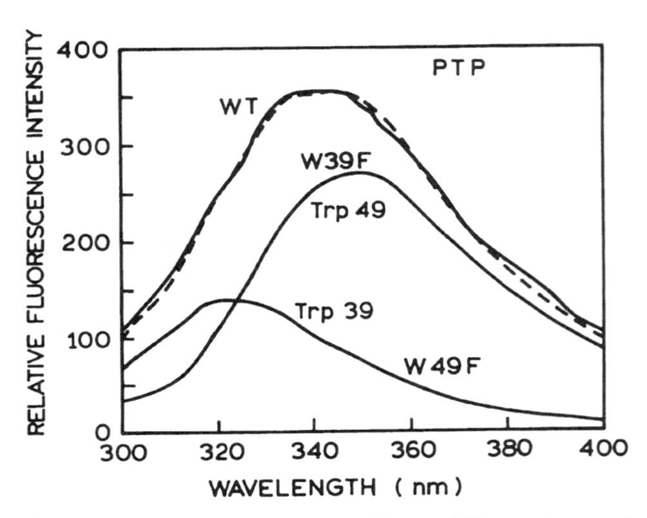

Figure 16.42. Emission spectra of wild-type (WT) protein tyrosyl phosphatase (PTP) and the single-tryptophan mutant proteins W39F and W49F at 20 °C. Excitation was at 295 nm. The composite spectrum from the sum of the contributions due to W39F and W49F is indicated by the dashed curve. Revised from Ref. 113.

16.7.B. Human Tissue Factor Contains Nonfluorescent Tryptophan Residues

In the previous example both tryptophan residues were fluorescent. However, in many proteins one or more of the tryptophans can be nonfluorescent owing to the presence of nearby quenching groups. This situation was found for soluble tissue factor (sTF), which contains four tryptophan residues in the wild-type protein (Figure 16.43). Tissue factors are involved in coagulation following tissue damage. Soluble tissue factor is the cleavage product of tissue factor, containing the first 218 residues. Preparation of mutants lacking one of the four tryptophans yielded a surprising result. Almost all the fluorescence of sTF resulted from just two of the tryptophan residues. This can be seen from the emission spectra of two of the mutants with one of the tryptophans replaced by phenylalanine (W14F or W45F). The emission spectra of these two proteins account for all the emission of the wild-type protein (sTF). This suggests that the other two tryptophan residues, at positions 25 and 158, do not contribute to the emission.

Why are residues 25 and 158 nonfluorescent? This was explained by examination of the crystal structure of sTF. A lysine residue was found close to trp-158, and a disulfide bridge close to trp-25. Protonated amines, disulfides,[115,116] and arginine[117] are known to quench the fluorescence of tryptophan residues in proteins. Hence, individual tryptophan residues in proteins can be nonfluorescent owing to the presence of a nearby quenching group. Unfolding of the protein can result in increased distance from the quencher, and increased fluorescence.

16.7.C. Barnase—A Three-Tryptophan Protein

Barnase is an extracellular ribonuclease which is often used as a model for protein folding. It is a relatively small protein (110 amino acids, 12.4 kDa). Barnase contains three tryptophan residues (Figure 16.44), one of which is located close to histidine-18. Examination of the three single-tryptophan mutants of barnase provided an interesting example of energy transfer between tryptophan residues and the quenching effects of the nearby histidine.[118]

The emission intensity of wild-type barnase, which contains all three tryptophan residues, increases dramatically

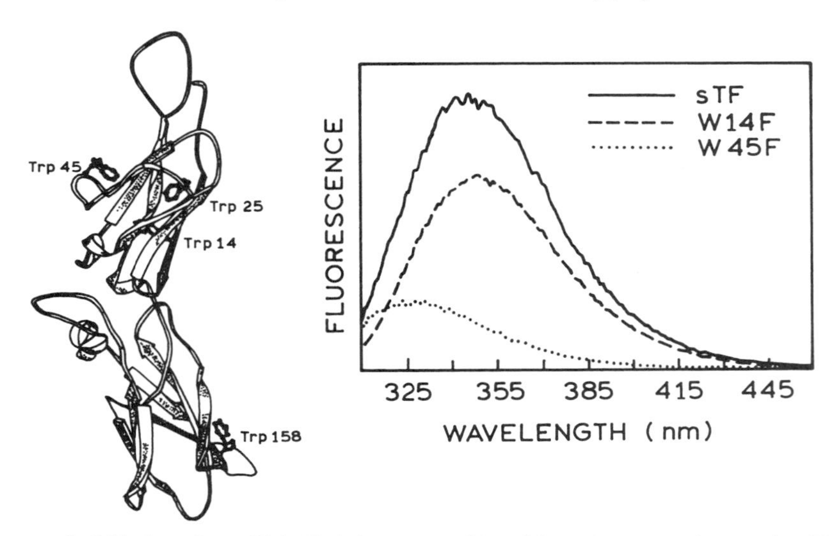

Figure 16.43. *Left*: Structure of soluble tissue factor. *Right*: Emission spectra of the wild-type four-tryptophan protein (sTF) and two of the mutants containing three tryptophan residues (W14F and W45F). From Ref. 114.

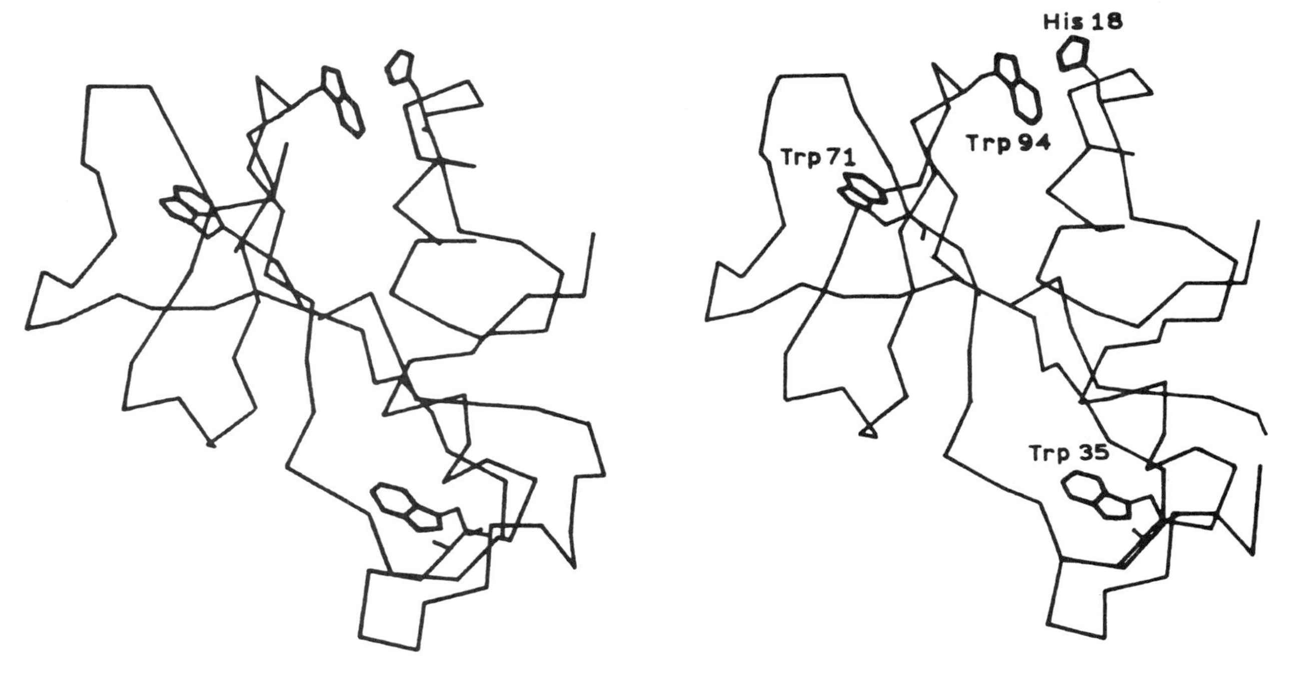

Figure 16.44. Stereoview of the structure of barnase, showing the positions of the three tryptophan residues and histidine-18. Revised and reprinted, with permission, from Ref. 118, Copyright © 1991, American Chemical Society.

as the pH is increased from 7 to 8.5 (Figure 16.45). This increase was attributed to loss of the quenching effect of his-18, which undergoes dissociation of the proton on the imidazole ring at pH 7.5, yielding the neutral side chain. This effect was shown to be due to the presence of trp-94 and the histidine, as only mutants containing both his-18 and trp-94 showed pH-dependent intensities. This result demonstrates that trp residues are quenched by the protonated form of histidine, which exists below pH 7.5. This is not surprising, given the sensitivity of indole to electron-deficient molecules. Quenching of tryptophan by histidine and imidazolium was reported previously.[78,119,120]

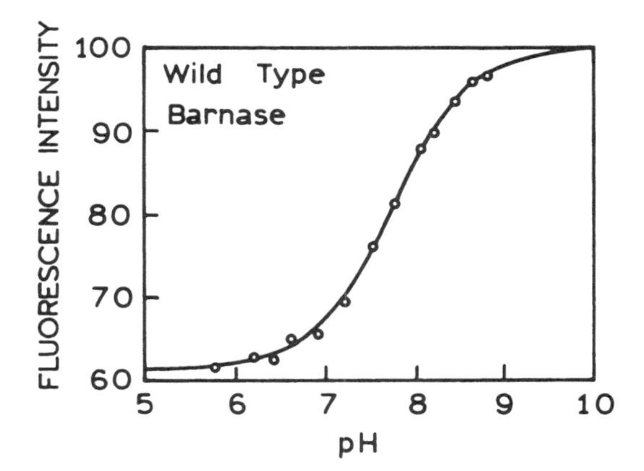

Figure 16.45. Dependence of the intensity of the tryptophan emission of wild-type barnase on pH. Excitation was at 295 nm, and emission was observed at 340 nm. Revised and reprinted, with permission, from Ref. 118, Copyright © 1991, American Chemical Society.

Emission spectra of barnase and its single-tryptophan mutants are shown in Figures 16.46 and 16.47, at pH 5.5 and 9.4, respectively. At first glance, these spectra appear complex. Interpretation of these spectra should be approached by considering the mutant proteins relative to the wild-type protein. First consider the contribution of trp-71. Substitution by tyrosine (W71 $\rightarrow$ Y) has little effect on the spectra. Hence, trp-71 is weakly fluorescent, showing that essentially silent or nonfluorescent trp residues may be a common occurrence in proteins. Such effects may explain the frequent observation of longer than expected lifetimes of proteins which display low quantum yields, and the long apparent natural lifetimes (Figure 16.12).

The effect of his-18 is immediately apparent from the glycine mutant H18 $\rightarrow$ G. The intensity at pH 5.5 increases 2.6-fold over that for the wild-type protein (Figure 16.46), and the intensity at pH 9.4 increases by about 70% (Figure 16.47). This result indicates that both the protonated and the neutral form of histidine quench trp-94, but that quenching by the electron-deficient protonated form present at pH 5.5 is more efficient.

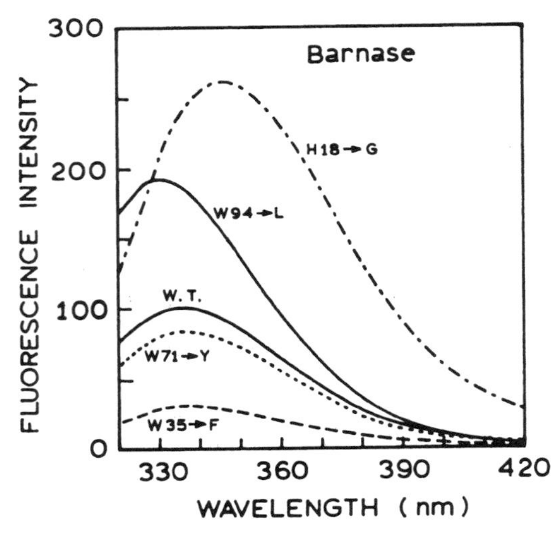

Figure 16.46. Fluorescence emission spectra of wild-type (W.T.) barnase and its three single-tryptophan mutants in buffer at pH 5.5, where histidine-18 (H18) is protonated, and of the H18 → G mutant. Revised and reprinted, with permission, from Ref. 118, Copyright © 1991, American Chemical Society.

Rather remarkable results were found when trp-94 was replaced with leucine (W94 → L). The intensity of the protein emission increased even though the number of tryptophan residues decreased. Other studies showed that these effects were not due to a structural change in the protein.[118] The increase in intensity upon removal of trp-94 can be understood as due to energy transfer from the other trp residues to trp-94. Trp-94 serves as an energy trap for the other trp residues. The emission of trp-94 is red-shifted relative to that of trp-35 and trp-71, as seen in the H18 → G mutants. When trp-94 is removed (W94 → L), the blue-shifted emission from the other residues increases since they no longer transfer energy to the quenched residue trp-94. Examination of the structure (Figure 16.44) suggests that trp-94 is located on the surface of barnase, whereas trp-71 and trp-35 appear to be buried. Trp-94 is red-shifted and serves as an acceptor for the other two tryptophan residues. Trp-94 is also quenched by the presence of a nearby histidine residue. Hence, the overall quantum yield of barnase is strongly influenced by just one tryptophan residue. These results for barnase provide an elegant example of the complex spectral properties of multi-tryptophan proteins and how their behavior can be understood by site-directed mutagenesis.

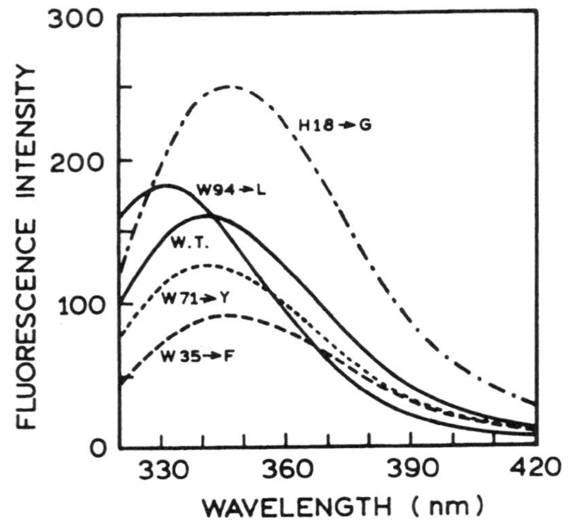

Figure 16.47. Fluorescence emission spectra of wild-type (W.T.) barnase and the barnase mutants in Tris buffer at pH 9.4, where histidine-18 is neutral. From Ref. 118.

16.7.D. Substrate Binding and Site-Directed Mutagenesis

Site-directed mutagenesis and insertion of single tryptophan residues have also been valuable in studies of substrate binding and substrate-induced conformational changes. One example is provided by phosphoglycerate kinase (Figure 16.48), which binds both 3-phosphoglycerate (3-PG) and ATP. The binding sites for the two ligands are in opposite domains of the protein. In this case a series of mutant proteins, each containing a single tryptophan residue, were prepared (Figure 16.48). Binding of ligand (3-PG or ATP) to either domain resulted in changes in the emission from tryptophan residues in the hinge region and on the inner surface of both domains.[121] Emission from W122 and W308, which are distant from the hinge region, were not affected by substrate binding. These results illustrate how the regions of ligand contact and/or conformational change can be mapped by the design of single-tryptophan mutant proteins.

16.7.E. Site-Directed Mutagenesis of Tyrosine Proteins

The concept of engineered proteins has also been applied to tyrosine-only proteins. One example is Δ^5-3-ketosteroid isomerase (KI), which catalyzes the isomerization of steroids. The protein is a dimer, and each subunit contains three tyrosine residues (tyr-14, tyr-55, and tyr-88). The three double-tyrosine deletion mutants were prepared by substitution with phenylalanine.[122] One of the residues (tyr-14) displayed a normal tyrosine emission (Figure 16.49) and a good quantum yield (0.16). The other two mutants (tyr-55 and tyr-88) displayed lower quantum yields (0.06 and 0.03, respectively) and long-wavelength tails on the tyrosine emission. These spectra were interpreted in terms of quenching due to hydrogen-bonding

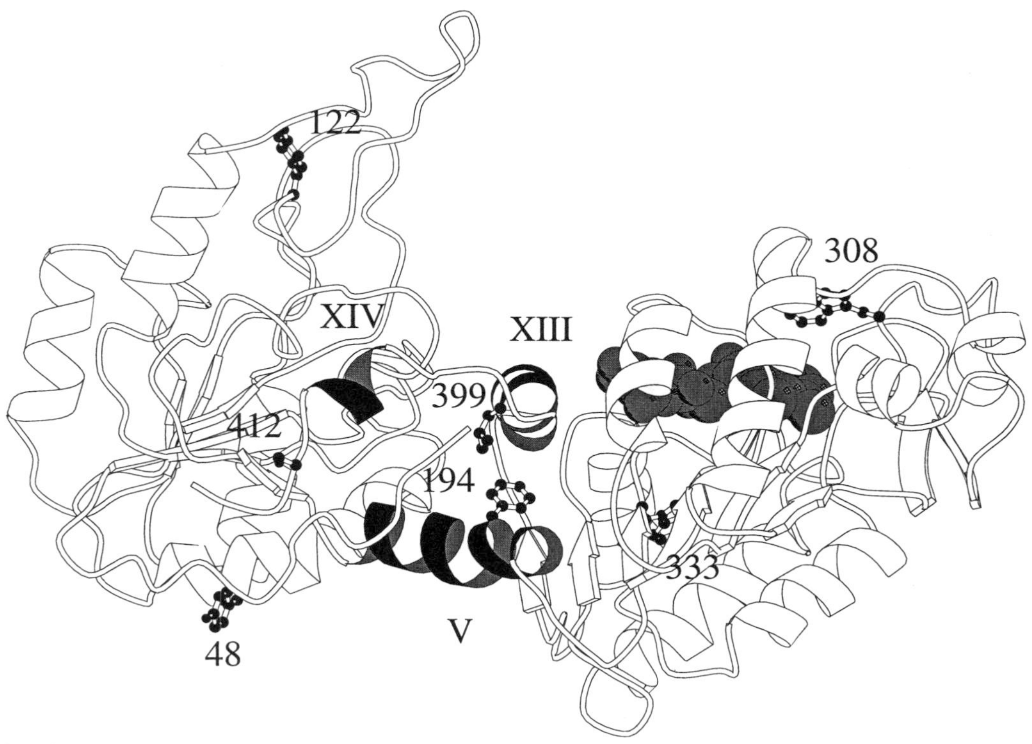

Figure 16.48. Yeast PGK–ATP complex. Numbers indicate positions of single tryptophans used as intrinsic fluorescent probes to detect substrate-induced conformational changes. Three helices situated in the hinge region, V, XIII, and XIV, are shown in black. Reprinted, with permission, from Ref. 121, Copyright © 1996, Cambridge University Press.

interactions with nearby groups, an effect which has also been observed in other proteins.[123] In this protein, the emission is dominated by one of the tyrosine residues.

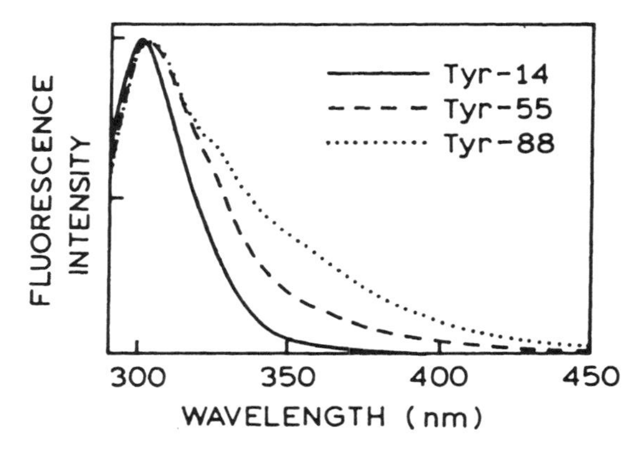

Figure 16.49. Emission spectra of the single-tyrosine mutants of Δ^5-3-ketosteroid isomerase. The wild-type enzyme has three tyrosine residues per subunit. The mutants are Y55F/Y88F (Tyr-14, ———), Y14F/Y88F (Tyr-55, — — —), and Y14F/Y55F (Tyr-88, · · ·). Revised from Ref. 122.

16.8. PROTEIN FOLDING

The intrinsic tryptophan emission of proteins has proven to be particularly valuable in studies of protein folding.[124–133] It is widely recognized that proteins must fold by a particular pathway, as the process is too fast to be explained by a random conformational search. The sensitivity of tryptophan to its local environment usually results in changes in intensity or anisotropy during the folding process. One example is provided by kinetic studies of the refolding of apomyoglobin.[127,128] In this case one of the tryptophan residues was quenched as two helices of myoglobin came into contact. This quenching occurred due to contact of tryptophan-14 on one helix with methionine-131 on another helix. Methionine is known to quench tryptophan fluorescence.[120]

Another example is shown in Figure 16.50 for a mutant nuclease from *S. aureus*. The nuclease contains a single tryptophan residue at position 140 near the carboxy terminus. This residue is mostly on the surface of the protein, but its intensity and anisotropy are sensitive to protein folding. Folding and unfolding reactions were initiated by jumps in pH from 3.2 to 7.0 and from 7.0 to 3.0, respectively. For a pH jump to 7.0, the intensity increased over

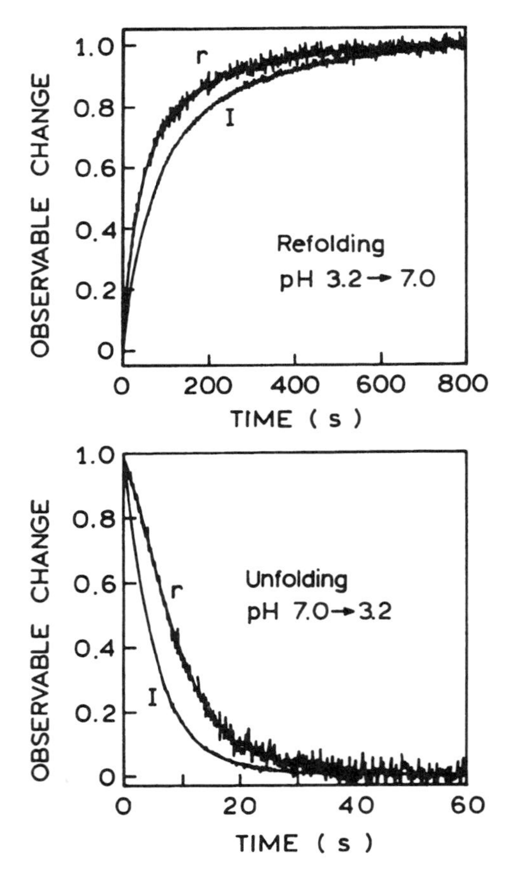

Figure 16.50. Time-dependent intensity (*I*) and anisotropy (*r*) of the single tryptophan in staphylococcal nuclease following a pH jump from 3.2 to 7.0 (*top*) or 7.0 to 3.2 (*bottom*). Revised from Ref. 126.

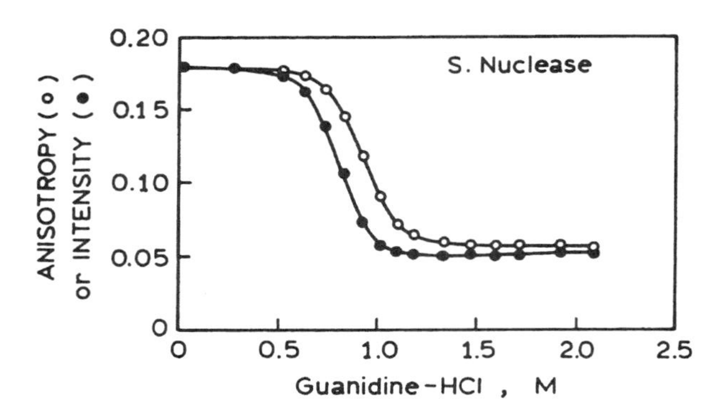

Figure 16.51. Unfolding of staphylococcal nuclease in the presence of various concentrations of denaturant (guanidine hydrochloride) as observed from the steady-state intensity (●) or anisotropy (○). From Ref. 129.

several minutes, indicating that the intensity is larger in the folded state. A time-dependent decrease was observed for unfolding following a pH jump from 7.0 to 3.2.

Protein folding can be followed using fluorescence intensities, anisotropies, or lifetimes. It is important to recognize that the unfolding transitions may not be accurately represented by the lifetimes or anisotropies.[129] For instance, suppose that the intensity and anisotropy both decrease when the protein is unfolded by denaturant. Then the anisotropy data will be more sensitive to the folded state (Figure 16.51). When the intensity decrease indicates that unfolding is one-half complete, the anisotropy change is less than 50% because of the larger contribution of the folded state to the steady-state anisotropy. This is why the folding transition observed from the anisotropy data precedes that observed from the intensities (Figure 16.50, top). When one performs measurements of protein folding, it is important to remember that the aniso-tropy and mean lifetime are intensity-weighted parameters. This means that the measured value depends on the relative fluorescence intensity of each state, as well as the fraction of the protein present in each state.

16.8.A. Protein Engineering of Mutant Ribonuclease for Folding Experiments

Since the classic refolding experiments of Christian Anfinsen on ribonuclease A (RNase A), this protein has been a favorite model for studies of protein folding. RNase A normally contains only tyrosine residues. In order to create a unique probe to study folding, a single tryptophan residue was inserted at position 92.[134] This position was chosen because, in the wild-type protein, the tyrosine residue at position 92 is hydrogen-bonded to aspartate-38. This suggested the possibility that a tryptophan residue inserted at position 92 would be quenched by the nearby carboxyl group in the folded state.

Emission spectra of RNase A with the trp-92 insertion are shown in Figure 16.52. The excitation wavelength was 280 nm, so that both tyrosine and tryptophan are excited. The surprising feature of these spectra is that the tyrosine contribution is highest in the native protein. This is the opposite of what is observed for most proteins. The reason for this unusual result is quenching, in the folded state, of trp-92 by the nearby aspartate residue. Hence, trp-92 in this engineered protein provides a sensitive probe of protein folding, and its intensity increases nearly 100-fold when RNase A is unfolded.

16.8.B. Ribose Binding Protein—Insertion of Tryptophan Residues in Each Domain

Many proteins with a single polypeptide chain have two or more domains in the folded state. The ribose binding protein (RBP) from *E. coli* is one such protein (Figure 16.53). It is of interest to study the folding–unfolding

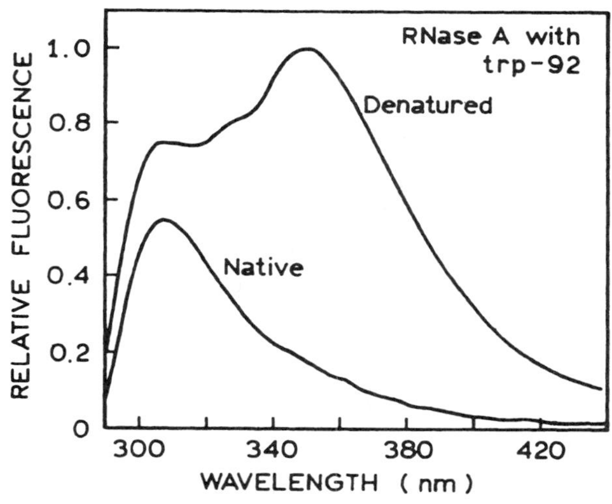

Figure 16.52. Emission spectra of the trp-92 mutant of RNase A, in the native and in the denatured state at pH 5. Excitation was at 280 nm. Revised from Ref. 134.

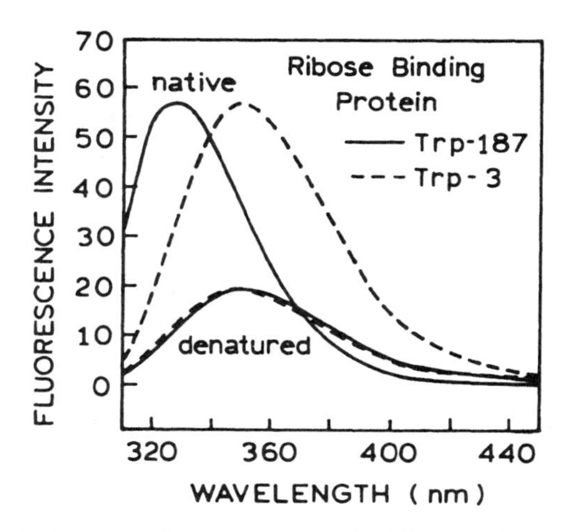

Figure 16.54. Emission spectra of mutants of ribose binding protein in the native and in the denatured state. Revised from Ref. 135.

transition of each domain independently. This can be accomplished by placing a tryptophan residue in each domain.[135] For the RBP, a single tryptophan residue was placed in either the carboxy-terminal (C) domain at position 187 or in the amino-terminal (N) domain at position 3. It was found that the emission of trp-187 was blue-shifted (Figure 16.54), indicating that this residue was shielded from the solvent, whereas trp-3 was exposed to the aqueous phase. Upon unfolding of the protein, both residues displayed the same emission spectrum and quantum yield, demonstrating that the emission of each residue was sensitive to protein structure. In the present case, both the N and C domains displayed similar stability, so both trp-187 and trp-3 showed similar transitions.

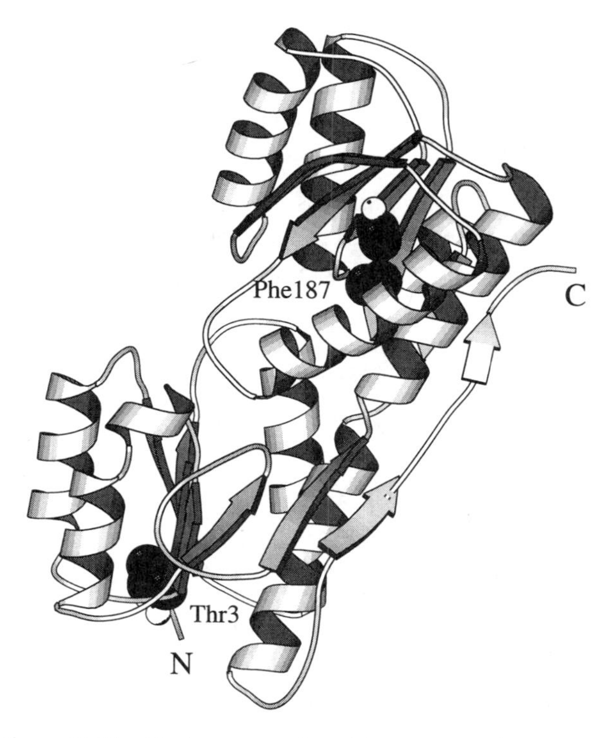

Figure 16.53. The X-ray crystallographic structure of the wild-type ribose binding protein from *E. coli*, showing the locations of trp substitutions for the threonine (thr) residue at position 3 or the phenylalanine (phe) residue at position 187. From Ref. 135.

16.8.C. Emission Spectra of Native and Denatured Proteins

In the previous example of the ribose binding protein, we saw that each tryptophan residue displayed a distinct emission spectrum and that the spectra became nearly identical upon protein unfolding. In fact, this is a general feature of protein fluorescence, which is nicely illustrated by the retinal binding protein CRABPI. This protein has three tryptophan residues at positions 7, 87, and 109 (Figure 16.55). In the native state, the emission spectra are strongly dependent on the excitation wavelength (Figure 16.56, left). This occurs because each of the tryptophan residues is in a different environment, and the absorption and emission spectra of the residues depend on the local environment. When the protein is denatured, the emission spectra become independent of the excitation wavelength (Figure 16.56, right) because the three tryptophan residues are all in a similar environment.

16.8.D. Folding of Lactate Dehydrogenase

In order to determine the folding pathway, it is important to examine multiple positions in a protein.[125,136] This has

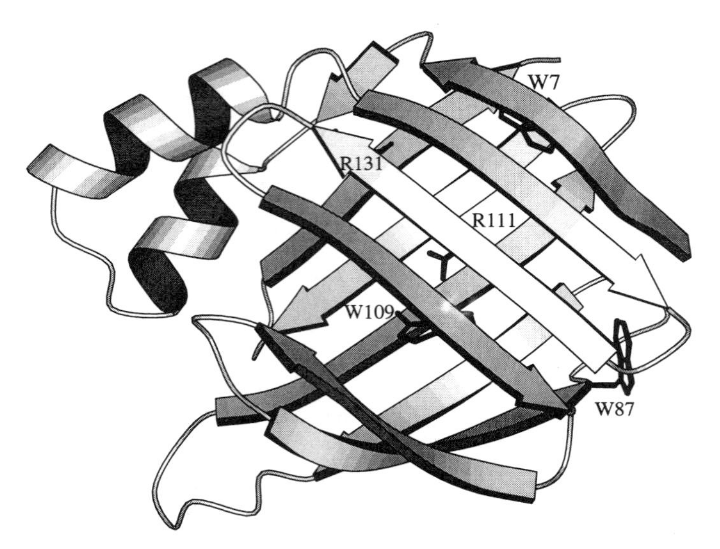

Figure 16.55. Ribbon structure of the native form of the cellular retinal binding protein I (CRABPI). Reprinted, with permission, from Ref. 117, Copyright © 1996, Cambridge University Press.

become feasible as the result of protein engineering. One elegant example is a study of the folding–unfolding transition of lactate dehydrogenase from *Bacillus stearothermophilus*.[125] This protein typically contains three tryptophans (at positions 80, 150, and 203). These were replaced by tyrosines, and nine single-tryptophan mutants were produced, with the trp residues dispersed throughout the protein matrix (Figure 16.57). These mutant proteins were studied in the presence of increasing concentrations of guanidine hydrochloride. Just three of the nine unfolding curves are shown in Figure 16.58. The unfolding transitions occur at different denaturant concentrations for each of the three residues. For instance, the structure surrounding trp-248 persists to higher denaturant concentrations than the structure around residues 279 and 285. From such data, it is possible to elucidate which regions of the protein are more stable and which are first disrupted during protein unfolding.

16.9. TRYPTOPHAN ANALOGS • Advanced Topic •

For calmodulin and other tyrosine-only proteins, a genetically inserted tryptophan residue can serve as a useful probe. However, most proteins contain tryptophan, and in

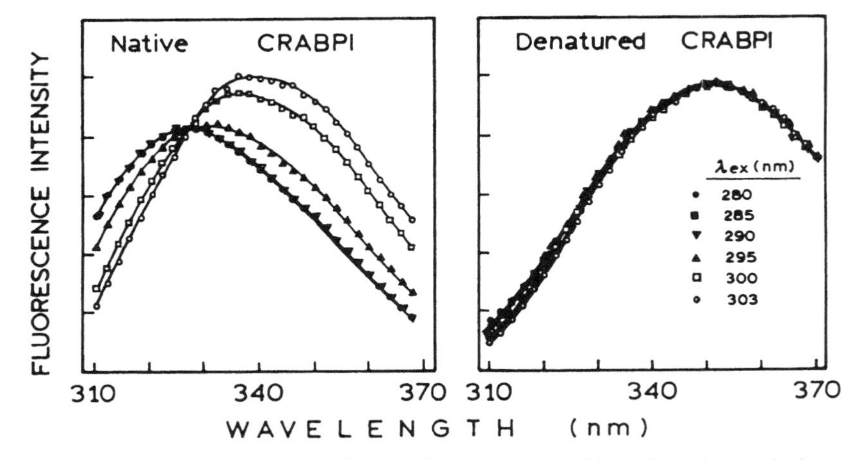

Figure 16.56. Emission spectra of CRABPI, in the native (*left*) and in the denatured state (*right*), for various excitation wavelengths. Revised from Ref. 117.

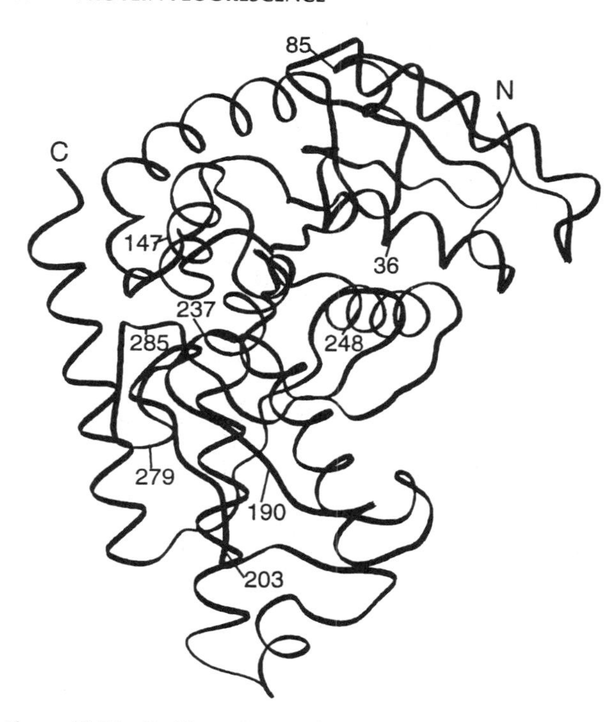

Figure 16.57. Positions of tryptophan probes in the lactate dehydrogenase subunit from *B. stearothermophilus*. The backbone of the protein is shown as a ribbon, and the positions of each single change of tyrosine to tryptophan are indicated by the residue number. Reprinted, with permission, from Ref. 125, Copyright © 1991, American Chemical Society.

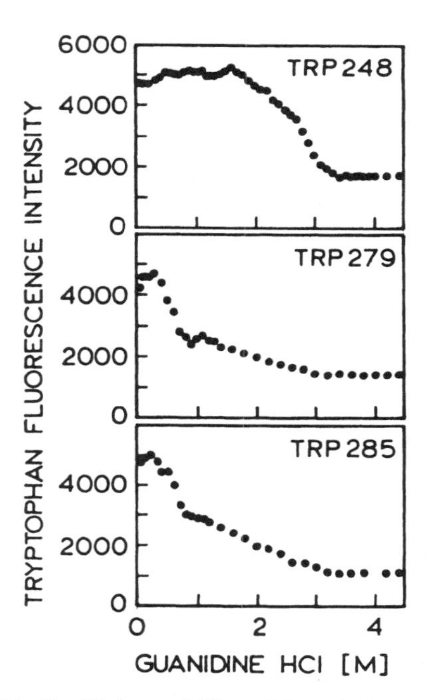

Figure 16.58. Equilibrium unfolding of three single-tryptophan mutants of lactate dehydrogenase monitored by tryptophan fluorescence intensity. Tryptophan fluorescence was excited at 295 nm and measured at 345 nm. Revised from Ref. 125.

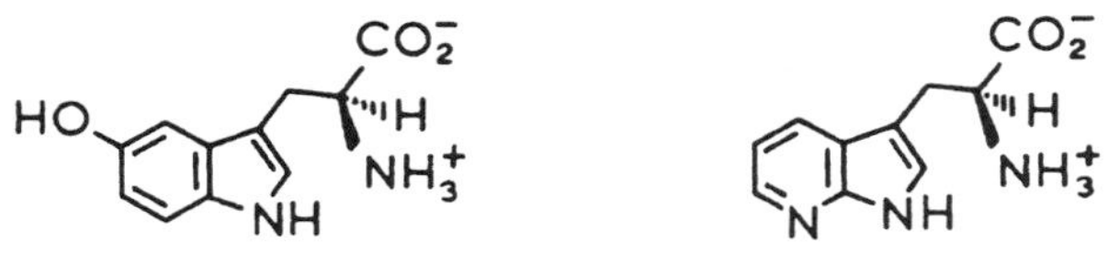

HO CO_2^- H NH_3^+ NH 5 HW

CO_2^- H NH_3^+ N NH 7AW

Figure 16.59. Chemical structures of 5HW and 7AW.

order to study these, or to study a particular protein in the presence of other tryptophan-containing proteins, it would be useful to have tryptophan analogs which could be selectively observed in the presence of tryptophan-containing proteins. This can be accomplished using tryptophan analogs that absorb at longer wavelengths than tryptophan. Two of the most widely used tryptophan analogs are 5-hydroxytryptophan (5HW) and 7-azatryptophan (7AW) (Figure 16.59).[137–141] Both of these analogs display absorption to 320 nm, which is beyond the longest-wavelength absorption of tryptophan (Figure 16.60). These tryptophan analogs are structurally similar to tryptophan and can be incorporated into proteins grown in bacteria that

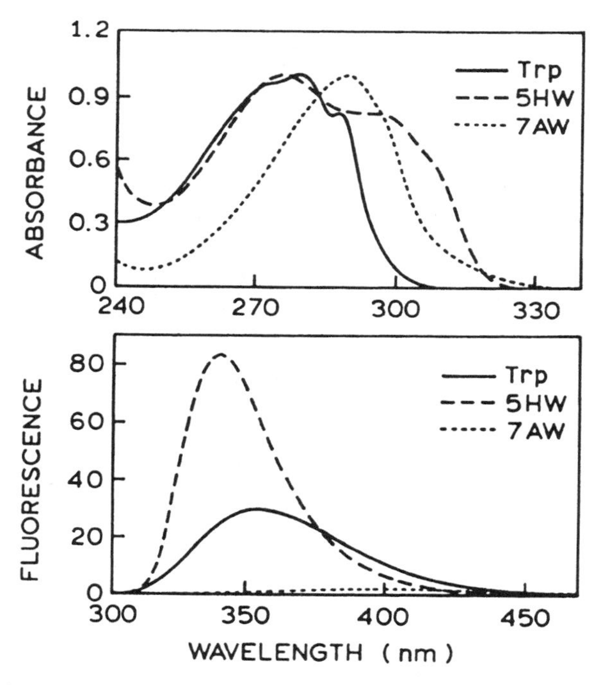

Figure 16.60. *Top*: Absorption spectra of tryptophan (——), 5HW (———), and 7AW (---). The spectra have been normalized to have the same maximum absorbance. *Bottom*: Fluorescence spectra of tryptophan, 5HW, and 7AW. (Aqueous buffer, pH 7.3, room temperature; excitation at 295 nm.) Revised and reprinted, with permission, from Ref. 142, Copyright © 1997, Cambridge University Press.

cannot synthesize their own tryptophan, which are called tryptophan auxotrophs. The resulting proteins are called alloproteins.

5HW and 7AW differ from each other, and from tryptophan, in their spectral properties.[142] In water, 5HW displays a higher quantum yield than tryptophan (0.275 versus 0.14). It is less sensitive to solvent polarity than tryptophan and displays an emission maximum near 339 nm (Figure 16.60). 7AW is highly sensitive to solvent polarity. In water, its quantum yield is low, near 0.017, with an emission maximum near 403 nm.

These two tryptophan analogs have been incorporated into a number of proteins.[137–144] One example is substitution of 5HW and 7AW for the single tryptophan residue in staphylococcal nuclease (Figure 16.61). In this case the emission maxima of tryptophan and 5HW are similar, and the quantum yield of 5HW is lower than that of tryptophan. It is interesting to notice that the emission of 7AW is more intense when located in the protein than in water. This is because the quantum yield of 7AW is highly dependent on solvent polarity and decreases upon contact with water.[145] In fact, this property of 7AW is somewhat problematic. The use of 7AW as an alternative to tryptophan was originally proposed because 7AW displays a simple single-exponential decay.[146,147] Unfortunately, it is now known that 7AW and azaindole display complex decay kinetics owing to the presence of several solvated states.[148–150] A nonexponential decay has also been observed for an octapeptide containing a 7AW residue.[151] 7AW is a useful tryptophan analog, but it is not a simple fluorophore.

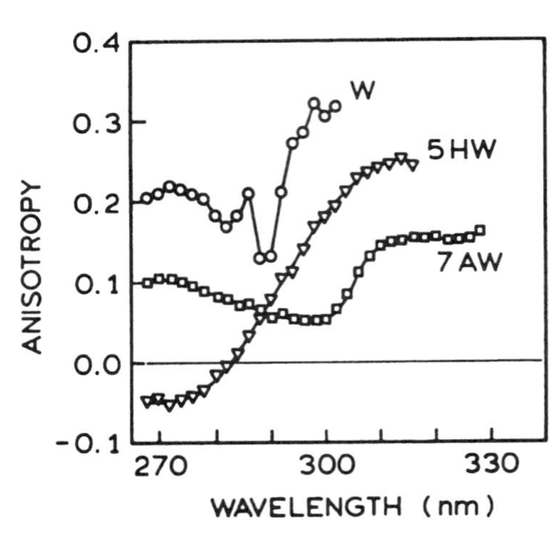

Figure 16.62. Low-temperature excitation anisotropy spectra of tryptophan (○), 5HW (▽), and 7AW (□) in 50% glycerol–phosphate buffer, 77 K. Revised and reprinted, with permission, from Ref. 142, Copyright © 1997, Cambridge University Press.

The advantages of 5HW and 7AW can be seen by their use in studies of more complex biochemical mixtures. 5HW was used to replace the tyrosine residue in insulin. This allowed the fluorescence of 5HW-insulin to be used to study its binding to the insulin receptor.[143] Such studies would not be possible using the tyrosine fluorescence of

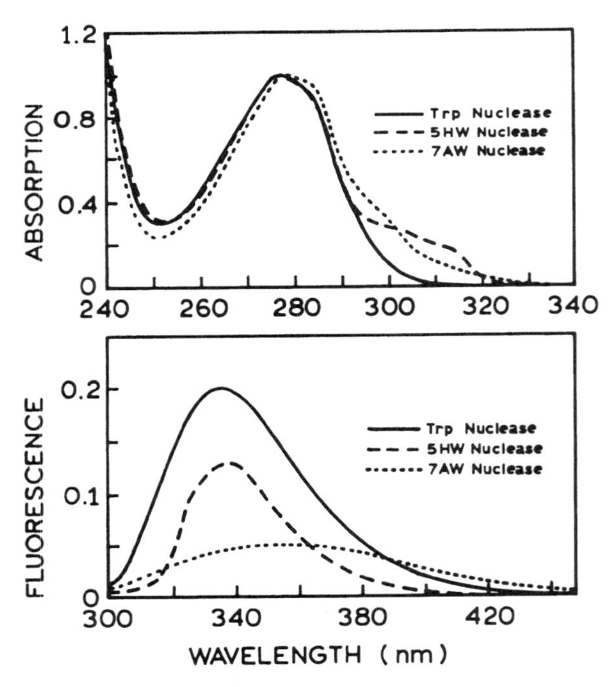

Figure 16.61. *Top*: Absorption spectra of staphylococcal nuclease containing tryptophan (—), 5HW (———), and 7AW (---). *Bottom*: Fluorescence emission spectra of staphylococcal nuclease containing tryptophan (—), 5HW (———), and 7AW (---). (Aqueous buffer, pH 7.3, 20 °C; excitation at 295 nm.) Revised and reprinted, with permission, from Ref. 142, Copyright © 1997, Cambridge University Press.

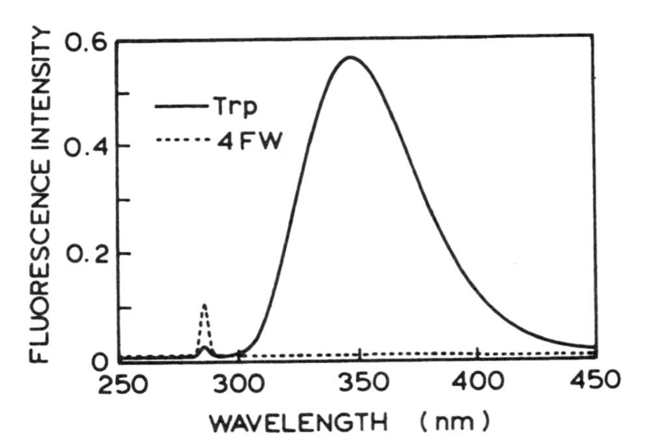

Figure 16.63. Relative fluorescence intensities in aqueous solution of tryptophan (———) and 4FW (— — —) at 25 °C and 285-nm excitation. Revised from Ref. 152.

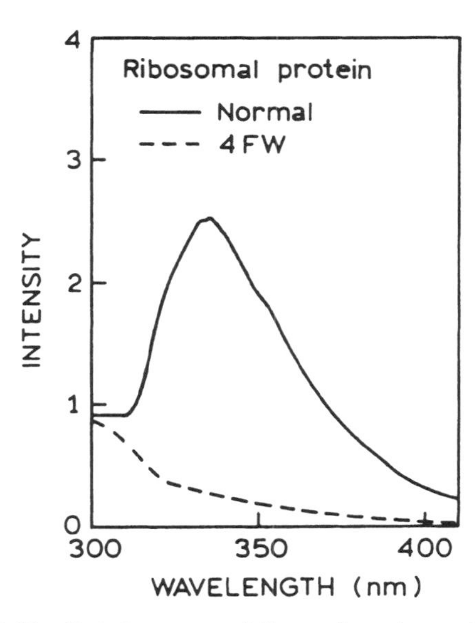

Figure 16.64. Emission spectra of ribosomal protein containing tryptophan (———) or 4FW (— — —). Excitation was at 280 nm. Revised from Ref. 153.

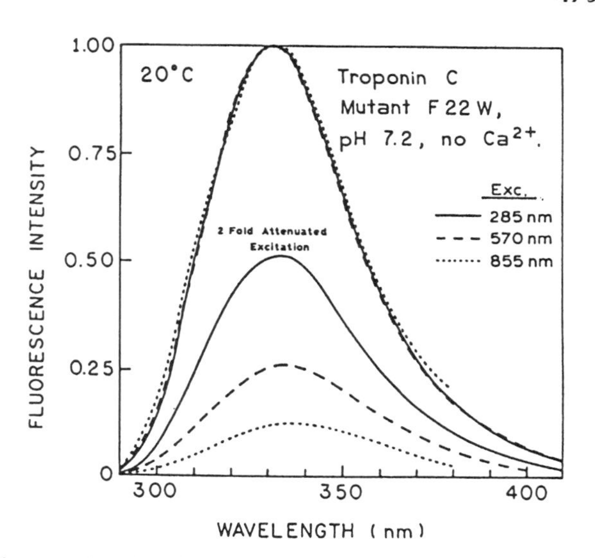

Figure 16.65. Emission spectra of a single-tryptophan mutant of troponin C, F22W, for excitation at 285 (—), 570 (———), and 855 nm (---). The emission spectra are normalized prior to twofold attenuation of the incident light. From Ref. 162.

insulin because this emission is masked by the tryptophan emission of the insulin receptor protein.

These tryptophan analogs are also likely to be useful in studies of protein association reactions, which are often studied using anisotropy measurements. For these purposes, it is valuable to know the excitation anisotropy spectra. It is perhaps unfortunate that the anisotropy of both analogs is lower than that of tryptophan itself (Figure 16.62). The low anisotropy of 7AW is another indication of its complex electronic properties. 5HW has proven valuable in measuring an antigen–antibody association.[137] In this work, 5HW was incorporated into oncomodulin. Binding of this protein to antibodies could be detected even though the antibody possessed numerous tryptophan residues.

Another useful tryptophan analog is 4-fluorotryptophan (4FW). This analog is useful because it is similar in size and shape to tryptophan but is almost nonfluorescent (Figure 16.63).[152] It has been incorporated into ribosomal proteins, resulting in suppression of their fluorescence (Figure 16.64).[153] In these modified proteins the dominant emission is due to the tyrosine residues.

16.10. MULTIPHOTON EXCITATION OF PROTEINS

• Advanced Topic •

During the past several years, there has been an increasing number of reports on multiphoton excitation (MPE).[155] Multiphoton excitation refers to the simultaneous absorption of two or more long-wavelength photons to excite a fluorophore to the first singlet excited state. MPE is typically accomplished using photons of the same wavelength from the same focused laser beam. As may be expected, tryptophan and proteins have already been examined by MPE.[156–159]

MPE of proteins is illustrated in Figure 16.65 for a mutant of troponin C (TnC). This protein from muscle typically contains only tyrosine residues. This TnC mutant contains a single tryptophan residue replacing a phenylalanine residue at position 22 (F22W). Emission spectra are shown for excitation at 285 nm and at the unusual wavelengths of 570 and 855 nm. Remarkably, the same emission spectra were observed for all three excitation wavelengths. Since 570 and 855 nm are much longer than the last absorption band of tryptophan, the emission observed with these excitation wavelengths cannot be due to the usual process of one-photon excitation.

The nature of the excitation process is revealed by the effects of attenuating the intensity of the incident light. At 285 nm a twofold decrease in the incident light results in a twofold decrease in the emission intensity, which is the usual result for one-photon excitation, for which the intensity of the emitted light is directly proportional to the excitation intensity. For excitation at 570 nm, twofold attenuation of the incident light results in a fourfold decrease in emission intensity. At 855 nm the same twofold decrease in incident intensity results in an eightfold decrease in emission intensity. In fact, the emission intensity depends on the square of the incident intensity at 570 nm, and on the cube of the incident intensity at 855 nm (Figure

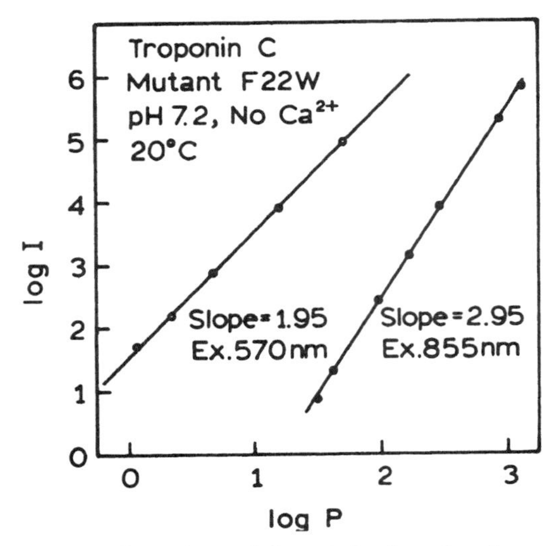

Figure 16.66. Dependence of the emission intensity of troponin C mutant F22W on the logarithm of the incident power (P) at 570 and 855 nm. Data from Ref. 162.

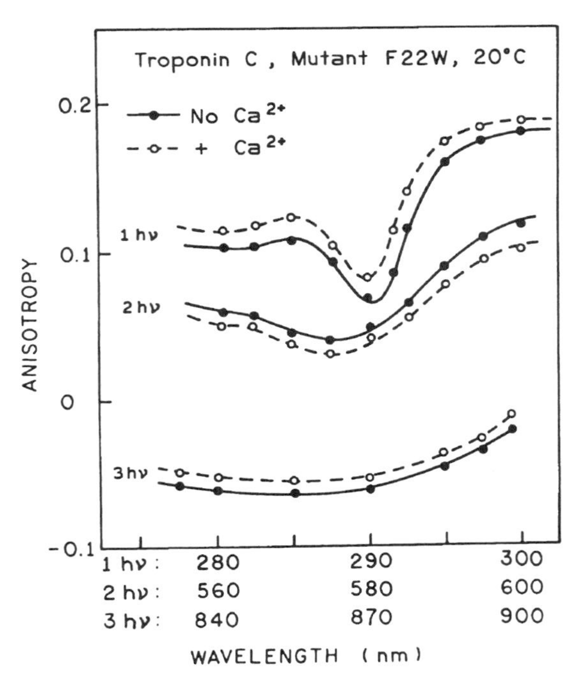

Figure 16.67. Excitation anisotropy spectra of troponin C mutant F22W for one-, two-, and three-photon excitation in the absence (●) and in the presence (○) of Ca^{2+}. From Ref. 162.

16.66). These data indicate that the emission with 570-nm excitation is due to two-photon excitation, and the emission with 855-nm excitation is due to three-photon excitation.

It is of interest to understand how multiphoton excitation is accomplished. The 570-nm excitation was obtained from the cavity-dumped pulses from a rhodamine 6G dye laser. These pulses are about 7 ps wide. Excitation at 855 nm was accomplished using pulses from a Ti:sapphire laser, which are about 70 fs wide. Pulsed excitation is used because it is necessary to have a high instantaneous density of photons in order to have a significant probability of MPE.

What are the advantages of MPE of proteins? One advantage is the possibility of observing intrinsic protein fluorescence in a microscope. The use of intrinsic protein fluorescence in microscopy has not been practical owing to the poor transmission of microscope optics in the ultraviolet. The longer excitation wavelengths employed with MPE are transmitted with high efficiency. Remarkably, the first microscope images of serotonin in neurotransmitter granules[160] were obtained with three-photon excitation near 800 nm, rather than two-photon excitation near 600 nm. This occurred because Ti:sapphire lasers operate only at longer wavelength, beyond the two-photon long-wavelength limit for tryptophan. Ti:sapphire lasers provide higher peak power because the energy is compressed into sub-100-fs pulses, as compared to 5–10-ps pulses from a dye laser.

Another feature of multiphoton excitation is the opportunity for new spectroscopic information. This is illustrated by the unusual anisotropies displayed by tryptophan with two- or three-photon excitation.[157,161,162] Surprisingly, the anisotropies of the tryptophan residue in TnC F22W are lower for two-photon excitation (560–600 nm) than for one-photon excitation (280–300 nm) (Figure 16.67). The anisotropies are still lower for three-photon excitation (840–900 nm). As described in Section 10.13, multiphoton excitation is expected to result in higher anisotropies due to $\cos^4\theta$ or $\cos^6\theta$ photoselection. The lower anisotropies for tryptophan observed with MPE suggest that MPE is occurring primarily to the ^1L_b state of tryptophan, with emission as usual from the ^1L_a state. Apparently, the ^1L_b state displays a higher MPE cross section than the ^1L_a state. Further analysis is needed to confirm or refute this explanation. It seems clear that the increasing availability of picosecond and femtosecond lasers will result in additional studies of MPE of proteins.

16.11. THE CHALLENGE OF PROTEIN FLUORESCENCE

The intrinsic fluorescence of proteins represents a complex spectroscopic challenge. At the initial level, one has to deal with multiple fluorophores with overlapping absorption and emission spectra. The presence of multiple fluorophores is itself a significant challenge. However, the actual situation is still more complex. Tryptophan, the dominant fluorophore, displays complex spectral properties owing to the presence of two overlapping electronic states. Additionally, it is now generally accepted that even for single-tryptophan proteins, the emission often contains multiple spectral contributions due to either multiple conformations or the intrinsic heterogeneity of tryptophan itself. Also, tryptophan is uniquely sensitive to a variety of quenchers, many of which are present in proteins. Small motions of the amino acid side chain can apparently result in changes in tryptophan emission, and some of these motions may occur during the excited-state lifetime. Finally, the fluorescent amino acids in proteins can interact by energy transfer. In some cases a quenched tryptophan residue can serve as a trap for normally fluorescent residues. One can expect such effects to be highly sensitive to the conformation of the proteins and the relative orientation of the fluorescent amino acids.

Given the complexity of tryptophan fluorescence, it is not surprising that the details remained elusive from studies of multi-tryptophan proteins. It was recognized that tryptophan could be quenched by nearby groups and that energy transfer could occur. However, it was not until engineered proteins of known structure became available that we were able to identify convincing examples of these phenomena. These elegant experiments combining sophisticated spectroscopy and protein engineering have provided the first comprehensive understanding of the details governing protein fluorescence.

REFERENCES

1. Permyakov, E. A., 1993, *Luminescent Spectroscopy of Proteins*, CRC Press, Boca Raton, Florida.
2. Demchenko, A. P., 1981, *Ultraviolet Spectroscopy of Proteins*, Springer-Verlag, New York.
3. Konev, S. V., 1967, *Fluorescence and Phosphorescence of Proteins and Nucleic Acids*, Plenum Press, New York.
4. Weinryb, I., and Steiner, R. F., 1971, The luminescence of the aromatic amino acids, in *Excited States of Proteins and Nucleic Acids*, R. F. Steiner and I. Weinryb (eds.), Plenum Press, New York, pp. 277–318.
5. Chen, R. F., 1967, Fluorescence quantum yields of tryptophan and tyrosine, *Anal. Lett.* **1**(1):35–42.
6. Longworth, J. W., 1983, Intrinsic fluorescence of proteins, in *Time-Resolved Fluorescence Spectroscopy in Biochemistry and Biology*, R. B. Cundall and R. E. Dale (eds.), Plenum Press, New York, pp. 651–778.
7. Lakowicz, J. R., and Maliwal, B. P., 1983, Oxygen quenching and fluorescence depolarization of tyrosine residues in proteins, *J. Biol. Chem.* **258**:4794–4801.
8. Weber, G., 1960, Fluorescence polarization spectrum and electronic-energy transfer in tyrosine, tryptophan, and related compounds, *Biochem. J.* **75**:335–345.
9. Weber, G., 1966, Polarization of the fluorescence of solutions, in *Fluorescence and Phosphorescence Analysis*, D. M. Hercules (ed.), Interscience Publishers, New York, pp. 217–240.
10. Ross, J. B. A., Laws, W. R., Rousslang, K. W., and Wyssbrod, H. R., 1992, Tyrosine fluorescence and phosphorescence from proteins and polypeptides, in *Topics in Fluorescence Spectroscopy, Volume 3, Biochemical Applications*, J. R. Lakowicz (ed.), Plenum Press, New York, pp. 1–63.
11. Lakowicz, J. R., Maliwal, B. P., Cherek, H., and Balter, A., 1983, Rotational freedom of tryptophan residues in proteins and peptides, *Biochemistry* **22**:1741–1752.
12. Eftink, M. R., Selvidge, L. A., Callis, P. R., and Rehms, A. A., 1990, Photophysics of indole derivatives: Experimental resolution of ^1L_a and ^1L_b transitions and comparison with theory, *J. Phys. Chem.* **94**:3469–3479.
13. Yamamoto, Y., and Tanaka, J., 1972, Polarized absorption spectra of crystals of indole and its related compounds, *Bull. Chem. Soc. Jpn.* **45**:1362–1366.
14. Song, P.-S., and Kurtin, W. E., 1969, A spectroscopic study of the polarized luminescence of indole, *J. Am. Chem. Soc.* **91**:4892–4906.
15. Albinsson, B., Kubista, M., Norden, B., and Thulstrup, E. W., 1989, Near-ultraviolet electronic transitions of the tryptophan chromophore: Linear dichroism, fluorescence anisotropy, and magnetic circular dichroism spectra of some indole derivatives, *J. Phys. Chem.* **93**:6646–6655.
16. Albinsson, B., and Norden, B., 1992, Excited-state properties of the indole chromophore. Electronic transition moment directions from linear dichroism measurements: Effect of methyl and methoxy substituents, *J. Phys. Chem.* **96**:6204–6212.
17. Callis, P. R., 1997, ^1L_a and ^1L_b transitions of tryptophan: Applications of theory and experimental observations to fluorescence of proteins, *Methods Enzymol.* **278**:113–150.
18. Valeur, B., and Weber, G., 1977, Resolution of the fluorescence excitation spectrum of indole into the ^1L_a and ^1L_b excitation bands, *Photochem. Photobiol.* **25**:441–444.
19. Rayner, D. M., and Szabo, A. G., 1977, Time resolved fluorescence of aqueous tryptophan, *Can. J. Chem.* **56**:743–745.
20. Petrich, J. W., Chang, M. C., McDonald, D. B., and Fleming, G. R., 1983, On the origin of nonexponential fluorescence decay in tryptophan and its derivatives, *J. Am. Chem. Soc.* **105**:3824–3832.
21. Creed, D., 1984, The photophysics and photochemistry of the near-UV absorbing amino acids—I. Tryptophan and its simple derivatives, *Photochem. Photobiol.* **39**:537–562.
22. Fleming, G. R., Morris, J. M., Robbins, R. J., Woolfe, G. J., Thistlewaite, P. J., and Robinson, G. W., 1978, Nonexponential fluorescence decay of aqueous tryptophan and two related peptides by picosecond spectroscopy, *Proc. Natl. Acad. Sci. U.S.A.* **75**:4652–4656.
23. Gryczynski, I., Wiczk, W., Johnson, M. L., and Lakowicz, J. R., 1988, Lifetime distributions and anisotropy decays of indole fluorescence in cyclohexane/ethanol mixtures by frequency-domain fluorometry, *Biophys. Chem.* **32**:173–185.
24. Walker, M. S., Bednar, T. W., and Lumry, R., 1966, Exciplex formation in the excited state, *J. Chem. Phys.* **45**:3455–3456.
25. Hershberger, M. V., Lumry, R., and Verral, R., 1981, The 3-methylindole/*n*-butanol exciplex: Evidence for two exciplex sites in indole compounds, *Photochem. Photobiol.* **33**:609–617.

26. Strickland, E. H., Horwitz, J., and Billups, C., 1970, Near-ultraviolet absorption bands of tryptophan. Studies using indole and 3-methylindole as models, *Biochemistry* **9:**4914–4920.
27. Lasser, N., Feitelson, J., and Lumry, R., 1977, Exciplex formation between indole derivatives and polar solutes, *Isr. J. Chem.* **16:**330–334.
28. Sun, M., and Song, P.-S., 1977, Solvent effects on the fluorescent states of indole derivatives—dipole moments, *Photochem. Photobiol.* **25:**3–9.
29. Lami, H., and Glasser, N., 1986, Indole's solvatochromism revisited, *J. Chem. Phys.* **84:**597–604.
30. Pierce, D. W., and Boxer, S. G., 1995, Stark effect spectroscopy of tryptophan, *Biophys. J.* **68:**1583–1591.
31. Callis, P. R., and Burgess, B. K., 1997, Tryptophan fluorescence shifts in proteins from hybrid simulations: An electrostatic approach, *J. Phys. Chem. B.* **101:**9429–9432.
32. Strickland, E. H., Billups, C., and Kay, E., 1972, Effects of hydrogen bonding and solvents upon the tryptophanyl ^1L_a absorption band. Studies using 2,3-dimethylindole, *Biochemistry* **11:**3657–3662.
33. Van Duuren, B. L., 1961, Solvent effects in the fluorescence of indole and substituted indoles, *J. Org. Chem.* **26:**2954–2960.
34. Willis, K. J., and Szabo, A. G., 1991, Fluorescence decay kinetics of tyrosinate and tyrosine hydrogen-bonded complexes, *J. Phys. Chem.* **95:**1585–1589.
35. Willis, K. J., Szabo, A. G., and Krajcarski, D. T., 1990, The use of Stokes Raman scattering in time correlated single photon counting: Application to the fluorescence lifetime of tyrosinate, *Photochem. Photobiol.* **51:**375–377.
36. Rayner, D. M., Krajcarski, D. T., and Szabo, A. G., 1978, Excited-state acid–base equilibrium of tyrosine, *Can. J. Chem.* **56:**1238–1245.
37. Pal, H., Palit, D. K., Mukherjee, T., and Mittal, J. P., 1990, Some aspects of steady state and time-resolved fluorescence of tyrosine and related compounds, *J. Photochem. Photobiol., A: Chem.* **52:**391–409.
38. Shimizu, O., and Imakuvo, K., 1977, New emission band of tyrosine induced by interaction with phosphate ion, *Photochem. Photobiol.* **26:**541–543.
39. Dietze, E. C., Wang, R. W., Lu, A. Y. H., and Atkins, W. M., 1996, Ligand effects on the fluorescence properties of tyrosine-9 in alpha 1-1 glutathione S-transferase, *Biochemistry* **35:**6745–6753.
40. Behmaarai, T. A., Toulme, J. J., and Helene, C., 1979, Quenching of tyrosine fluorescence by phosphate ions. A model study for protein–nucleic acid complexes, *Photochem. Photobiol.* **30:**533–539.
41. Schnarr, M., and Helene, C., 1982, Effects of excited-state proton transfer on the phosphorescence of tyrosine–phosphate complexes, *Photochem. Photobiol.* **36:**91–93.
42. Szabo, A., Lynn, K. R., Krajcarski, D. T., and Rayner, D. M., 1978, Tyrosinate fluorescence maxima at 345 nm in proteins lacking tryptophan at pH 7, *FEBS Lett.* **94:**249–252.
43. Libertini, L. J., and Small, E. W., 1985, The intrinsic tyrosine fluorescence of histone H1, *Biophys. J.* **47:**765–772.
44. Jordano, J., Barbero, J. L., Montero, F., and Franco, L., 1983, Fluorescence of histones H1, *J. Biol. Chem.* **258:**315–320.
45. Prendergast, F. G., Hampton, P. D., and Jones, B., 1984, Characteristics of tyrosinate fluorescence emission in α- and β-purothionins, *Biochemistry* **23:**6690–6697.
46. Pundak, S., and Roche, R. S., 1984, Tyrosine and tyrosinate fluorescence of bovine testes calmodulin: Calcium and pH dependence, *Biochemistry* **23:**1549–1555.
47. Leroy, E., Lami, H., and Laustriat, G., 1971, Fluorescence lifetime and quantum yield of phenylalanine aqueous solutions. Temperature and concentration effects, *Photochem. Photobiol.* **13:**411–421.
48. Longworth, J. W., 1971, Luminescence of polypeptides and proteins, in *Excited States of Proteins and Nucleic Acids*, R. F. Steiner and I. Weinryb (eds.), Plenum Press, New York, pp. 319–484.
49. Teale, F. W. J., 1960, The ultraviolet fluorescence of proteins in neutral solution, *Biochem. J.* **76:**381–388.
50. Burstein, E. A., Vedenkina, N. S., and Ivkova, M. N., 1974, Fluorescence and the location of tryptophan residues in protein molecules, *Photochem. Photobiol.* **18:**263–279.
51. Kronman, M. J., and Holmes, L. G., 1971, The fluorescence of native, denatured and reduced denatured proteins, *Photochem. Photobiol.* **14:**113–134.
52. Eftink, M. R., 1990, Fluorescence techniques for studying protein structure, *Methods Biochem. Anal.* **35:**117–129.
53. Burstein, E. A., 1976, Luminescence of protein chromophores, in *Model Studies. Science and Technology Results, Volume 6, Biophysics*, VINITI, Moscow.
54. Suwaiyan, A., and Klein, U. K. A., 1989, Picosecond study of solute–solvent interaction of the excited state of indole, *Chem. Phys. Lett.* **159:**244–250.
55. Finazzi-Agro, A., Rotilio, G., Avigliano, L., Guerrieri, P., Boffi, V., and Mondovi, B., 1970, Environment of copper in *Pseudomonas fluorescens* azurin: Fluorometric approach, *Biochemistry* **9:**2009–2014.
56. Burstein, E. A., Permyakov, E. A., Yashin, V. A., Burkhanov, S. A., and Agro, A. F., 1977, The fine structure of luminescence spectra of azurin, *Biochim. Biophys. Acta* **491:**155–159.
57. Szabo, A. G., Stepanik, T. M., Wayner, D. M., and Young, N. M., 1983, Conformational heterogeneity of the copper binding site in azurin, *Biophys. J.* **41:**233–244.
58. Adman, E. T., Stenkamp, R. E., Sieker, L. C., and Jensen, L. H., 1978, A crystallographic model for azurin at 3 Å resolution, *J. Mol. Biol.* **123:**35–47.
59. Adman, E. T., and Jensen, L. H., 1981, Structural features of azurin at 2.7 Å resolution, *Isr. J. Chem.* **21:**8–12.
60. Aueuhel, F. M., Brent, R., Kingston, R. E., Moore, D. D., Seidman, J. G., Smith, J. A., and Struhl, K. (eds.), 1987, *Current Protocols in Molecular Biology*, John Wiley & Sons, New York, Chapter 8.
61. Sambrook, J., Fritsch, E. F., and Maniatis, T., 1989, *Molecular Cloning*, Cold Spring Harbor Laboratory Press, Plainview, New York, USA, Chapter 15.
62. Gilardi, G., Mei, G., Rosato, N., Canters, G. W., and Finazzi-Agro, A., 1994, Unique environment of Trp48 in *Pseudomonas aeruginosa* azurin as probed by site-directed mutagenesis and dynamic fluorescence anisotropy, *Biochemistry* **33:**1425–1432.
63. Petrich, J. W., Longworth, J. W., and Fleming, G. R., 1987, Internal motion and electron transfer in proteins: A picosecond fluorescence study of three homologous azurins, *Biochemistry* **26:**2711–2722.
64. Lakowicz, J. R., Cherek, H., Gryczynski, I., Joshi, N., and Johnson, M. L., 1987, Enhanced resolution of fluorescence anisotropy decays by simultaneous analysis of progressively quenched samples, *Biophys. J.* **51:**755–768.
65. Georghiou, S., Thompson, M., and Mukhopadhyay, A. K., 1982, Melittin–phospholipid interaction studied by employing the single tryptophan residue as an intrinsic fluorescent probe, *Biochim. Biophys. Acta* **688:**441–452.
66. Boteva, R., Zlateva, T., Dorovska-Taran, V., Visser, A. J. W. G., Tsanev, R., and Salvato, B., 1996, Dissociation equilibrium of human recombinant interferon γ, *Biochemistry* **35:**14825–14830.

67. Eisinger, J., 1969, Intramolecular energy transfer in adrenocorticotropin, *Biochemistry* **8**:3902–3908.
68. Gryczynski, I., unpublished observations.
69. Searcy, D. G., Montenay-Garestier, T., and Helene, C., 1989, Phenylalanine-to-tyrosine singlet energy transfer in the archaebacterial histone-like protein HTa, *Biochemistry* **28**:9058–9065.
70. Kupryszewska, M., Gryczynski, I., and Kawski, A., 1982, Intramolecular donor–acceptor separations in methionine- and leucine-enkephalin estimated by long-range radiationless transfer of singlet excitation energy, *Photochem. Photobiol.* **36**:499–502.
71. Gryczynski, I., Kawski, A., Darlak, K., and Grzonka, Z., 1985, Intramolecular electronic excitation energy transfer in dermorphine and its analogues, *J. Photochem.* **30**:371–377.
72. Chiu, H. C., and Bersohn, R., 1977, Electronic energy transfer between tyrosine and tryptophan in the peptides Trp-$(Pro)_n$-Tyr, *Biopolymers* **16**:277–288.
73. Moreno, M. J., and Prieto, M., 1993, Interaction of the peptide hormone adrenocorticotropin, ACTH(1-24), with a membrane model system: A fluorescence study, *Photochem. Photobiol.* **57**:431–437.
74. Pearce, S. F., and Hawrot, E., 1990, Intrinsic fluorescence of binding-site fragments of the nicotinic acetylcholine receptor: Perturbations produced upon binding α-bungarotoxin, *Biochemistry* **29**:10649–10659.
75. Schiller, P. W., 1983, Fluorescence study on the conformation of a cyclic enkephalin analog in aqueous solution, *Biochem. Biophys. Res. Commun.* **114**:268–274.
76. Alfimova, E. Ya., and Likhtenstein, G. I., 1976, Fluorescence study of energy transfer as method of study of protein structure, *Mol. Biol. (Moscow)* **8**(2):127–179.
77. Eftink, M. R., and Ghiron, C. A., 1977, Exposure of tryptophanyl residues and protein dynamics, *Biochemistry* **16**:5546–5551.
78. Eftink, M. R., and Ghiron, C. A., 1981, Fluorescence quenching studies with proteins, *Anal. Biochem.* **114**:199–227.
79. Eftink, M. R., 1991, Fluorescence quenching: Theory and applications, in *Topics in Fluorescence Spectroscopy, Volume 2, Principles*, J. R. Lakowicz (ed.), Plenum Press, New York, pp. 53–126.
80. Eftink, M. R., and Ghiron, C. A., 1976, Exposure of tryptophanyl residues in proteins. Quantitative determination by fluorescence quenching studies, *Biochemistry* **15**:672–680.
81. Wu, C.-S. C., and Yang, J. T., 1980, Helical conformation of glucagon in surfactant solutions, *Biochemistry* **19**:2117–2122.
82. Boesch, C., Bundi, A., Oppliger, M., and Wüthrich, K., 1978, ^{1}H nuclear-magnetic-resonance studies of the molecular conformation of monomeric glucagon in aqueous solution, *Eur. J. Biochem.* **91**:209–214.
83. Edelhoch, H., and Lippoldt, R. E., 1969, Structural studies on polypeptide hormones, *J. Biol. Chem.* **244**:3876–3883.
84. Terwilliger, T. C., and Eisenberg, D., 1982, The structure of melittin, *J. Biol. Chem.* **257**:6016–6022.
85. Lakowicz, J. R., and Weber, G., 1973, Quenching of protein fluorescence by oxygen. Detection of structural fluctuations in proteins on the nanosecond timescale, *Biochemistry* **12**:4171–4179.
86. Calhoun, D. B., Vanderkooi, J. M., and Englander, S. W., 1983, Penetration of small molecules into proteins studied by quenching of phosphorescence and fluorescence, *Biochemistry* **22**:1533–1539.
87. Calhoun, D. B., Englander, S. W., Wright, W. W., and Vanderkooi, J. M., 1988, Quenching of room temperature protein phosphorescence by added small molecules, *Biochemistry* **27**:8466–8474.
88. Kouyama, I., Kinosita, K., and Ikegami, A., 1989, Correlation between internal motion and emission kinetics of tryptophan residues in proteins, *Eur. J. Biochem.* **182**:517–521.
89. Lakowicz, J. R., Gryczynski, I., Cherek, I., Szmacinski, H., and Joshi, N., 1991, Anisotropy decays of single tryptophan proteins measured by GHz frequency-domain fluorometry with collisional quenching, *Eur. Biophys. J.* **19**:125–140.
90. Avigliano, L., Finazzi-Agro, A., and Mondovi, B., 1974, Perturbation studies on some blue proteins, *FEBS Lett.* **38**:205–208.
91. Wasylewski, Z., Kaszycki, P., Guz, A., and Stryjewski, W., 1988, Fluorescence quenching resolved spectra of fluorophores in mixtures and micellar solutions, *Eur. J. Biochem.* **178**:471–476.
92. Stryjewski, W., and Wasylewski, Z., 1986, The resolution of heterogeneous fluorescence of multitryptophan-containing proteins studied by a fluorescence-quenching method, *Eur. J. Biochem.* **158**:547–553.
93. Wasylewski, Z., Kaszycki, P., and Drwiega, M., 1996, A fluorescence study of Tn10-encoded tet repressor, *J. Protein Chem.* **15**:45–52.
94. Wasylewski, Z., Koloczek, H., and Wasniowska, A., 1988, Fluorescence quenching resolved spectroscopy of proteins, *Eur. J. Biochem.* **172**:719–724.
95. Blicharska, Z., and Wasylewski, Z., 1995, Fluorescence quenching studies of Trp repressor using single-tryptophan mutants, *J. Protein Chem.* **14**:739–746.
96. Gryczynski, I., and Lakowicz, J. R., unpublished observations.
97. Talbot, J. C., Dufourcq, J., de Bony, J., Faucon, J. F., and Lussan, C., 1979, Conformational change and self association of monomeric melittin, *FEBS Lett.* **102**:191–193.
98. Georghiou, S., Thompson, M., and Mukhopandhyay, A. K., 1981, Melittin–phospholipid interaction. Evidence for melittin aggregation, *Biochim. Biophys. Acta* **642**:429–432.
99. Steiner, R. F., Marshall, L., and Needleman, D., 1986, The interaction of melittin with calmodulin and its tryptic fragments, *Arch. Biochem. Biophys.* **246**:286–300.
100. Lakowicz, J. R., Gryczynski, I., Laczko, G., Wiczk, W., and Johnson, M. L., 1994, Distribution of distances between the tryptophan and the N-terminal residue of melittin in its complex with calmodulin, troponin C, and phospholipids, *Protein Sci.* **3**:628–637.
101. Daniel, E., and Weber, G., 1966, Cooperative effects in binding by bovine serum albumin, I. The binding of 1-anilino-8-naphthalenesulfonate. *Biochemistry* **5**:1893–1900.
102. Anderson, D. R., and Weber, G., 1969, Fluorescence polarization of the complexes of 1-anilino-8-naphthalenesulfonate with bovine serum albumin. Evidence for preferential orientation of the ligand, *Biochemistry* **8**:371–377.
103. Condie, C. C., and Quay, S. C., 1983, Conformational studies of aqueous melittin, *J. Biol. Chem.* **258**:8231–8234.
104. Ross, J. B. A., Subramanian, S., and Brand, L., 1982, Spectroscopic studies of the pyridine nucleotide coenzymes and their complexes with dehydrogenases, in *The Pyridine Nucleotide Coenzymes*, Academic Press, New York, pp. 19–49.
105. Anderson, S. R., and Weber, G., 1965, Multiplicity of binding by lactate dehydrogenase, *Biochemistry* **4**:1948–1957.
106. Velick, S. F., 1958, Fluorescence spectra and polarization of glyceraldehyde-3-phosphate and lactic dehydrogenase coenzyme complexes, *J. Biol. Chem.* **233**:1455–1467.
107. Zukin, R. S., 1979, Evidence for a conformational change in the *Escherichia coli* maltose receptor by excited state fluorescence lifetime data, *Biochemistry* **18**:2139–2145.
108. Malencik, D. A., and Anderson, S. R., 1984, Peptide binding by calmodulin and its proteolytic fragments and by troponin C, *Biochemistry* **23**:2420–2428.

109. Cox, J. A., Comte, M., Fitton, J. E., and DeGrado, W. F., 1985, The interaction of calmodulin with amphiphilic peptides, *J. Biol. Chem.* **260:**2527–2534.
110. O'Neil, K. T., Wolfe, H. R., Erickson-Viitanen, S., and DeGrado, W. F., 1987, Fluorescence properties of calmodulin-binding peptides reflect alpha-helical periodicity, *Science* **236:**1454–1456.
111. Kilhoffer, M.-C., Kubina, M., Travers, F., and Haiech, J., 1992, Use of engineered proteins with internal tryptophan reporter groups and perturbation techniques to probe the mechanism of ligand–protein interactions: Investigation of the mechanism of calcium binding to calmodulin, *Biochemistry* **31:**8098–8106.
112. Chabbert, M., Lukas, T. J., Watterson, D. M., Axelsen, P. H., and Prendergast, F. G., 1991, Fluorescence analysis of calmodulin mutants containing tryptophan: Conformational changes induced by calmodulin-binding peptides from myosin light chain kinase and protein kinase II, *Biochemistry* **30:**7615–7630.
113. Pokalsky, C., Wick, P., Harms, E., Lytle, F. E., and Van Etten, R. L., 1995, Fluorescence resolution of the intrinsic tryptophan residues of bovine protein tyrosyl phosphatase, *J. Biol. Chem.* **270:**3809–3815.
114. Hasselbacher, C. A., Rusinova, E., Waxman, E., Rusinova, R., Kohanski, R. A., Lam, W., Guha, A., Du, J., Lin, T. C., Polikarpov, I., Boys, C. W. G., Nemerson, Y., Konigsberg, W. H., and Ross, J. B. A., 1995, Environments of the four tryptophans in the extracellular domain of human tissue factor: Comparison of results from absorption and fluorescence difference spectra of tryptophan replacement mutants with the crystal structure of the wild-type protein, *Biophys. J.* **69:**20–29.
115. Sanyal, G., Kim, E., Thompson, F. M., and Brady, E. K., 1989, Static quenching of tryptophan fluorescence by oxidized dithiothreitol, *Biochem. Biophys. Res. Commun.* **165:**772–781.
116. Hennecke, J., Sillen, A., Huber-Wunderlich, M., Engelborghs, Y., and Glockshuber, R., 1997, Quenching of tryptophan fluorescence by the active-site disulfide bridge in the DsbA protein from *Escherichia coli*, *Biochemistry* **36:**6391–6400.
117. Clark, P. L., Liu, Z.-P., Zhang, J., and Gierasch, L. M., 1996, Intrinsic tryptophans of CRABPI as probes of structure and folding, *Protein Sci.* **5:**1108–1117.
118. Loewenthal, R., Sancho, J., and Fersht, A. R., 1991, Fluorescence spectrum of barnase: Contributions of three tryptophan residues and a histidine-related pH dependence, *Biochemistry* **30:**6775–6779.
119. Shopova, M., and Genov, N., 1983, Protonated form of histidine 238 quenches the fluorescence of tryptophan 241 in subtilisin novo, *Int. J. Peptide Res.* **21:**475–478.
120. Eftink, M. R., 1991, Fluorescence quenching reactions, in *Biophysical and Biochemical Aspects of Fluorescence Spectroscopy*, T. Gregory Dewey (ed.), Plenum Press, New York, pp. 1–41.
121. Cheung, C.-W., and Mas, M. T., 1996, Substrate-induced conformational changes in yeast 3-phosphoglycerate kinase monitored by fluorescence of single tryptophan probes, *Protein Sci.* **5:**1144–1149.
122. Wu, P., Li, Y.-K., Talalay, P., and Brand, L., 1994, Characterization of the three tyrosine residues of Δ^5-3-ketosteroid isomerase by time-resolved fluorescence and circular dichroism, *Biochemistry* **33:**7415–7422.
123. Lux, B., Baudier, J., and Gerard, D., 1985, Tyrosyl fluorescence spectra of proteins lacking tryptophan: Effects of intramolecular interactions, *Photochem. Photobiol.* **42:**245–251.
124. Jones, B. E., Beechem, J. M., and Matthews, C. R., 1995, Local and global dynamics during the folding of *Escherichia coli* dihydrofolate reductase by time-resolved fluorescence spectroscopy, *Biochemistry* **34:**1867–1877.
125. Smith, C. J., Clarke, A. R., Chia, W. N., Irons, L. I., Atkinson, T., and Holbrook, J. J., 1991, Detection and characterization of intermediates in the folding of large proteins by the use of genetically inserted tryptophan probes, *Biochemistry* **30:**1028–1036.
126. Otto, M. R., Lillo, M. P., and Beechem, J. M., 1994, Resolution of multiphasic reactions by the combination of fluorescence total-intensity and anisotropy stopped-flow kinetic experiments, *Biophys. J.* **67:**2511–2521.
127. Ballew, R. M., Sabelko, J., and Gruebele, M., 1996, Direct observation of fast protein folding: The initial collapse of apomyoglobin, *Proc. Natl. Acad. Sci. U.S.A.* **93:**5759–5764.
128. Service, R. F., 1996, Folding proteins caught in the act, *Science* **273:**29–30.
129. Eftink, M. R., 1994, The use of fluorescence methods to monitor unfolding transitions in proteins, *Biophys. J.* **66:**482–501.
130. Eftink, M. R., Ionescu, R., Ramsay, G. D., Wong, C.-Y., Wu, J. Q., and Maki, A. H., 1996, Thermodynamics of the unfolding and spectroscopic properties of the V66W mutant of *Staphylococcal* nuclease and its 1-136 fragment, *Biochemistry* **35:**8084–8094.
131. Ropson, I. J., and Dalessio, P. M., 1997, Fluorescence spectral changes during the folding of intestinal fatty acid binding protein, *Biochemistry* **36:**8594–8601.
132. Royer, C. A., Mann, C. J., and Matthews, C. R., 1993, Resolution of the fluorescence equilibrium unfolding profile of *trp* aporepressor using single tryptophan mutants, *Protein Sci.* **2:**1844–1852.
133. Szpikowska, B. K., Beechem, J. M., Sherman, M. A., and Mas, M. T., 1994, Equilibrium unfolding of yeast phosphoglycerate kinase and its mutants lacking one or both native tryptophans: A circular dichroism and steady-state and time-resolved fluorescence study, *Biochemistry* **33:**2217–2225.
134. Sendak, R. A., Rothwarf, D. M., Wedemeyer, W. J., Houry, W. A., and Scheraga, H. A., 1996, Kinetic and thermodynamic studies of the folding/unfolding of a tryptophan-containing mutant of ribonuclease A, *Biochemistry* **35:**12978–12992.
135. Kim, J.-S., and Kim, H., 1996, Stability and folding of a mutant ribose-binding protein of *Escherichia coli*, *J. Protein Chem.* **15:**731–736.
136. Steer, B. A., and Merrill, A. R., 1995, Guanidine hydrochloride-induced denaturation of the colicin E1 channel peptide: Unfolding of local segments using genetically substituted tryptophan residues, *Biochemistry* **34:**7225–7234.
137. Hogue, C. W. V., Rasquinha, I., Szabo, A. G., and MacManus, J. P., 1992, A new intrinsic fluorescent probe for proteins, *FEBS Lett.* **310:**269–272.
138. Ross, J. B. A., Senear, D. F., Waxman, E., Kombo, B. B., Rusinova, E., Huang, Y. T., Laws, W. R., and Hasselbacher, C. A., 1992, Spectral enhancement of proteins: Biological incorporation and fluorescence characterization of 5-hydroxytryptophan in bacteriophage λ CI repressor, *Proc. Natl. Acad. Sci. U.S.A.* **89:**12013–12027.
139. Heyduk, E., and Heyduk, T., 1993, Physical studies on interaction of transcription activator and RNA-polymerase: Fluorescent derivatives of CRP and RNA polymerase, *Cell. Mol. Biol. Res.* **39:**401–407.
140. Laue, T. M., Senear, D. F., Eaton, S., and Ross, J. B. A., 1993, 5-Hydroxytryptophan as a new intrinsic probe for investigating protein–DNA interactions by analytical ultracentrifugation. Study of the effect of DNA on self-assembly of the bacteriophage λ cI repressor, *Biochemistry* **32:**2469–2472.

141. Soumillion, P., Jespers, L., Vervoort, J., and Fastrez, J., 1995, Biosynthetic incorporation of 7-azatryptophan into the phage lambda lysozyme: Estimation of tryptophan accessibility, effect on enzymatic activity and protein stability, *Protein Eng.* **8**:451–456.
142. Wong, C.-Y., and Eftink, M. R., 1997, Biosynthetic incorporation of tryptophan analogues into staphylococcal nuclease: Effect of 5-hydroxytryptophan and 7-azatryptophan on structure and stability, *Protein Sci.* **6**:689–697.
143. Laws, W. R., Schwartz, G. P., Rusinova, E., Burke, G. T., Chu, Y.-C., Katsoyannis, P. G., and Ross, J. B. A., 1995, 5-Hydroxytryptophan: An absorption and fluorescence probe which is a conservative replacement for [A14 tyrosine] in insulin, *J. Protein Chem.* **14**:225–232.
144. Hogue, C. W. V., and Szabo, A. G., 1993, Characterization of aminoacyl-adenylates in *B. subtilis* tryptophanyl-tRNA synthetase, by the fluorescence of tryptophan analogs 5-hydroxytryptophan and 7-azatryptophan, *Biophys. Chem.* **48**:159–169.
145. Guharay, J., and Sengupta, P. K., 1996, Characterization of the fluorescence emission properties of 7-azatryptophan in reverse micellar environments, *Biochem. Biophys. Res. Commun.* **219**:388–392.
146. Negrerie, M., Gai, F., Bellefeuille, S. M., and Petrich, J. W., 1991, Photophysics of a novel optical probe: 7-Azaindole, *J. Phys. Chem.* **95**:8663–8670.
147. Rich, R. L., Negrerie, M., Li, J., Elliott, S., Thornburg, R. W., and Petrich, J. W., 1993, The photophysical probe, 7-azatryptophan, in synthetic peptides, *Photochem. Photobiol.* **58**:28–30.
148. Chen, Y., Gai, F., and Petrich, J. W., 1994, Solvation and excited state proton transfer of 7-azaindole in alcohols, *Chem. Phys. Lett.* **222**:329–334.
149. Chen, Y., Gai, F., and Petrich, J. W., 1994, Single-exponential fluorescence decay of the nonnatural amino acid 7-azatryptophan and the nonexponential fluorescence decay of tryptophan in water, *J. Phys. Chem.* **98**:2203–2209.
150. Chen, Y., Rich, R. L., Gai, F., and Petrich, J. W., 1993, Fluorescent species of 7-azaindole and 7-azatryptophan in water, *J. Phys. Chem.* **97**:1770–1780.
151. English, D. S., Rich, R. L., and Petrich, J. W., 1998, Nonexponential fluorescence decay of 7-azatryptophan induced in a peptide environment, *Photochem. Photobiol.* **67**:76–83.
152. Hott, J. L., and Borkman, R. F., 1989, The non-fluorescence of 4-fluorotryptophan, *Biochem. J.* **264**:297–299.
153. Bronskill, P. M., and Wong, J. T.-F., 1988, Suppression of fluorescence of tryptophan residues in proteins by replacement with 4-fluorotryptophan, *Biochem. J.* **249**:305–308.
154. Willaert, K., Loewenthal, R., Sancho, J., Froeyen, M., Fersht, A., and Engelborghs, Y., 1992, Determination of the excited-state lifetimes of the tryptophan residues in barnase via multifrequency phase fluorometry of tryptophan mutants, *Biochemistry* **31**:711–716.
155. Lakowicz, J. R. (ed.), 1997, *Topics in Fluorescence Spectroscopy, Volume 5, Nonlinear and Two-Photon-Induced Fluorescence*, Plenum Press, New York.
156. Kierdaszuk, B., Gryczynski, I., and Lakowicz, J. R., 1997, Two-photon induced fluorescence of proteins, in *Topics in Fluorescence Spectroscopy, Volume 5, Nonlinear and Two-Photon-Induced Fluorescence*, J. R. Lakowicz (ed.), Plenum Press, New York, pp. 187–209.
157. Lakowicz, J. R., Gryczynski, I., Danielsen, E., and Frisoli, J. K., 1992, Anisotropy spectra of indole and *N*-acetyl-L-tryptophanamide observed for two-photon excitation of fluorescence, *Chem. Phys. Lett.* **194**:282–287.
158. Lakowicz, J. R., and Gryczynski, I., 1993, Tryptophan fluorescence intensity and anisotropy decay of human serum albumin resulting from one-photon and two-photon excitation, *Biophys. Chem.* **45**:1–6.
159. Lakowicz, J. R., Kierdaszuk, B., Gryczynski, I., and Malak, H., 1996, Fluorescence of horse liver alcohol dehydrogenase using one- and two-photon excitation, *J. Fluoresc.* **6**(1):51–59.
160. Maiti, S., Sher, J. B., Williams, R. M., Zipfel, R. W., and Webb, W. W., 1997, Measuring serotonin distribution in live cells with three-photon excitation, *Science* **275**:530–532.
161. Gryczynski, I., Malak, H., and Lakowicz, J. R., 1996, Three-photon excitation of a tryptophan derivative using a fs-Ti:sapphire laser, *Biospectroscopy* **2**:9–15.
162. Gryczynski, I., Malak, H., Lakowicz, J. R., Cheung, H. C., Robinson, J., and Umeda, P. K., 1996, Fluorescence spectral properties of troponin C mutant F22W with one-, two-, and three-photon excitation, *Biophys. J.* **71**:3448–3453.

PROBLEMS

16.1. *Determination of Protein Association and Unfolding by Fluorescence:* Suppose that you have a protein which consists of a single subunit with a molecular weight of 25,000. The protein contains a single tryptophan residue near its central core and several tyrosine residues. The protein also contains a single reactive sulfhydryl residue on the surface.

A. Assume that the monomeric protein can be unfolded by the addition of denaturant. Explain how the fluorescence spectral properties of the unmodified protein could be used to follow the unfolding process.

B. Describe the use of collisional quenchers to probe the accessibility of the tryptophan residue to the solvent.

C. Assume that the unmodified protein self-associates with another subunit to form a dimer. How could fluorescence spectroscopy be used to follow the association process?

D. Describe how you would use fluorescence spectroscopy to measure the distance from the tryptophan residue to the reactive sulfhydryl group. Be specific with regard to the experiments that you would perform and how the data would be interpreted.

E. Describe how you would use energy transfer to measure self-association of the protein, after the protein has been modified on the sulfhydryl group with dansyl chloride.

16.2. *Detection of Protein Dimerization*: Suppose that you have a small protein with a single tryptophan residue which displays $r_0 = 0.30$ and that the protein associates to a dimer. The correlation times of the monomer and dimer are θ_M = 1.25 and θ_D = 2.5 ns, respectively. Upon dimer formation, the lifetime increases from τ_M = 2.5 to τ_D = 5.0 ns, and the relative quantum yield increases twofold. Describe how you would detect dimer formation using the:

A. steady-state intensity

B. intensity decay

C. steady-state anisotropy, or the
D. anisotropy decay.
E. What fraction of the emission is due to the monomers and dimers when 50% of the monomers have formed dimers? What is the steady-state anisotropy? What are the intensity and anisotropy decays?

Time-Resolved Protein Fluorescence

17

In the previous chapter we presented an overview of protein fluorescence. We described the spectral properties of the aromatic amino acids and how these properties depend on protein structure. We now extend this discussion to include time-resolved measurements of intrinsic protein fluorescence. Prior to 1983, most measurements of time-resolved fluorescence were performed using TCSPC. The instruments employed for these measurements typically used a flashlamp excitation source and a standard dynode-chain-type PMT. Such instruments provided instrument response functions with a half-width near 2 ns, which is comparable to the decay time of most proteins. The limited repetition rate of the flashlamps, near 20 kHz, resulted in data of modest statistical accuracy, unless the acquisition times were excessively long. Given the complexity of protein intensity and anisotropy decays, and the inherent difficulty of resolving multiexponential processes, it was difficult to obtain definitive information on the decay kinetics of proteins.

During the past decade, the instrumentation for time-resolved fluorescence of proteins has advanced dramatically. The flashlamp light sources have been replaced by high-repetition-rate (MHz) picosecond dye lasers, which provide both higher excitation intensities and more rapid data acquisition.[1,2] The dynode-chain PMTs have been replaced by MCP detectors, which provide much shorter single-photoelectron pulse widths than a dynode chain PMT. In combination, the new light sources and detectors provide instrument response functions with half-widths near 100 ps, so that picosecond resolution can now be obtained.

Advances in the instrumentation for FD fluorometry have also occurred. Prior to 1983, only fixed-frequency phase-modulation instruments were available. The first generally useful variable-frequency instruments were introduced in the mid-1980s,[3–5] resulting in instruments capable of measurements to 10 GHz.[6] These instruments use the same picosecond dye lasers and MCP PMT detectors as employed for TCSPC, but one measures the phase and modulation of the emission, rather than the arrival time of the individual photons. These instruments provide excellent resolution of the complex intensity and anisotropy decays displayed by proteins. In the past several years, there has been an increased availability of Ti:sapphire lasers, which are more convenient than picosecond dye lasers and which provide femtosecond pulses. However, the Ti:sapphire lasers operate between 700 and 1000 nm and have not been widely used to excite intrinsic protein fluorescence owing to the lack of suitable excitation wavelengths, even after frequency doubling.

The study of time-resolved fluorescence of proteins is challenging because the intensity decays are typically complex owing to the presence of a number of emitting species and the possibility of interactions between the fluorophores. Instruments with high time-resolution and signal-to-noise characteristics are essential for resolving the complex intensity decays from proteins. Once the multiexponential intensity decays were resolved, it was tempting to assign the various components to the individual tryptophan residues in multi-tryptophan proteins. However, it is known that even proteins with a single tryptophan residue typically display two or more decay times.[7,8] Hence, there is no reason to believe that the individual decay times represent individual trp residues, until this fact is experimentally demonstrated. In some cases where the decay times are very different, it has been possible to assign decay times to individual tryptophan residues. However, this is usually difficult unless single-tryptophan mutants are available to support the analysis. An additional complication is that tryptophan itself in solution at neutral pH displays a multiexponential or non-exponential decay. The heterogeneity is moderately weak, and most of the emission occurs with a decay time near 3.1 ns. It is now accepted that a second component exists with a decay time near 0.5 ns. For some time, the origin of this component was unknown. It is now generally agreed that the heterogeneity of the decay is due to the presence of conformational isomers, called rotamers, which display distinct decay times.

Additional factors can also contribute to the complex intensity decays of proteins. Energy transfer can occur from tyrosine to tryptophan, or between tryptophan residues themselves. Following excitation, the tryptophan emission can display time-dependent spectral relaxation. Another source of complexity is the possibility of transient effects in collisional quenching, due to either added quenchers or the presence of nearby quenching groups in the protein. In principle, all these phenomena can be studied using time-resolved measurements. In practice, it is difficult to determine the contributions of each phenomenon to the intensity decay. The use of engineered proteins is critical to resolving these complex interactions.

While the preceding comments are somewhat pessimistic, time-resolved fluorescence offers numerous opportunities for a more detailed understanding of protein structure and function. In favorable cases, it is possible to resolve the emission spectra of individual tryptophan residues based on the time-dependent decays measured at various emission wavelengths. Many time-resolved decays have been interpreted in terms of the location of the tryptophan residues in proteins and the interactions of these residues with nearby amino acid residues in the protein. Conformational changes in proteins often result in changes in the intensity decay due to altered interactions with nearby groups. The time-dependent anisotropy decays are invariably informative about the extent of local protein flexibility and the interactions of a protein with other macromolecules. Also, fluorescence quenching is best studied by time-resolved measurements, which can distinguish between static and dynamic processes. Finally, it is now known that proteins can be phosphorescent at room temperature. The phosphorescence decay times are sensitive to exposure to the aqueous phase as well as the presence of nearby quenchers. In this chapter we present an overview of time-resolved protein fluorescence, with examples which illustrate the range of behavior seen in proteins.

17.1. INTENSITY DECAYS OF TRYPTOPHAN—THE ROTAMER MODEL

One difficulty in interpreting the time-resolved intensity decays of proteins has been a lack of understanding of the intensity decay of tryptophan itself. In neutral aqueous solution, the intensity decay of tryptophan is known to be a double exponential, with decay times near 3.1 and 0.5 ns (Table 17.1).[9–16] To avoid confusion in an already complex topic, we note that incorrect tryptophan decay times have been reported and retracted.[17] One could write many pages describing the numerous results on tryptophan intensity decays. Rather than providing an extensive review, we will describe the final result. The most likely origin of a biexponential decay of tryptophan is the presence of rotational isomers (Figure 17.1). In solution, the side chain of tryptophan can adopt various conformational states, which appear to interchange slowly, on the nanosecond timescale. Hence, a tryptophan solution can be regarded as a mixture of these rotational isomers, called rotamers.

Table 17.1. Intensity Decays of the Aromatic Amino Acids and Related Compounds at 20 °C, pH 7[a]

Compound	τ_1 (ns)	τ_2 (ns)	α_1	α_2	λ_{em} (nm)	Reference(s)
Indole	4.4	—	1.0	—	350	107
Tryptophan	3.1	0.53	0.67	0.33	330	10
NATA	3.0	—	1.0	—	330	10
Phenol	3.16	—	1.0	—	300	34, 36
Tyrosine[b]	3.27	—	1.0	—	300	34, 36
NATyrA	1.66	0.11	0.65	0.35	300	34, 36

[a]See Refs. 30 and 36 for the intensity decays of tryptophan and tyrosine derivatives.
[b]pH6.

In neutral aqueous solution, tryptophan is present in a neutral but zwitterionic form in which the amino group is protonated ($-NH_3^+$) and the carboxy group is ionized ($-CO_2^-$). Recall that indole was found to be uniquely sensitive to quenching, particularly by electron-deficient species. Reported quenchers of indole include acrylamide, imidazolium, ammonium, methionine, tyrosine, disulfides, peptide bonds, trifluoroethanol, and electron scavengers.[18–28] In solution, the emission of tryptophan can be self-quenched by an intramolecular process involving the indole ring and the positively charged ammonium group. Evidence for such a process is found in the well-known dependence of the tryptophan quantum yield on pH (Figure 17.2). As the pH is increased from 8 to 10, the protonated amino group undergoes dissociation to the neutral

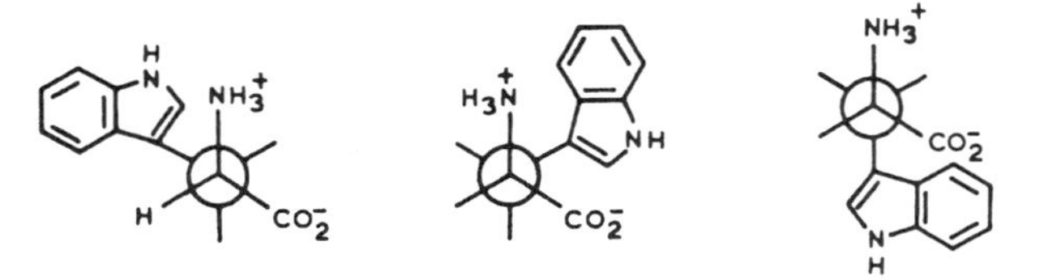

Figure 17.1. Rotational isomers (rotamers) of tryptophan. The rotamer on the left is thought to be responsible for the 0.5-ns decay time. Revised from Ref. 16.

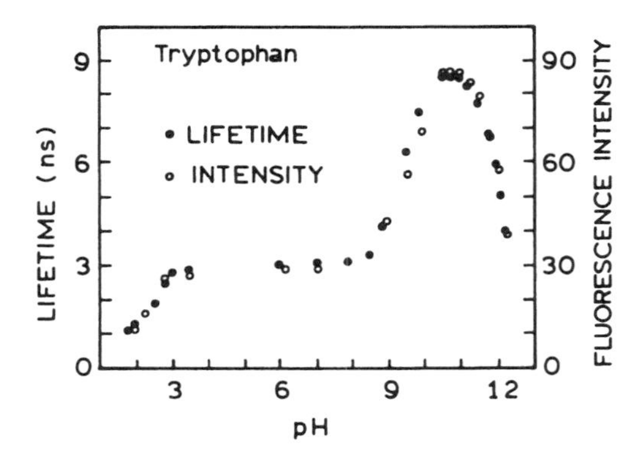

Figure 17.2. Relative fluorescence intensity (○) and mean lifetime (●) of tryptophan as a function of pH. Excitation was at 280 nm, and emission was observed through a Corning 0-52 filter. Mean lifetimes were measured from the phase angle at 10 MHz. Revised from Ref. 29.

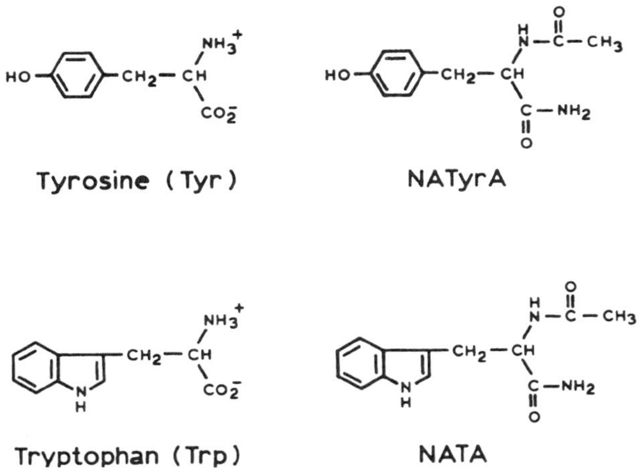

Tyrosine (Tyr) NATyrA

Tryptophan (Trp) NATA

Figure 17.3. Structures of tyrosine and tryptophan and their neutral analogs, *N*-acetyl-L-tyrosinamide (NATyrA) and *N*-acetyl-L-tryptophanamide (NATA).

form, and the quantum yield and mean lifetime increase approximately threefold.[11,29] The quenching effect of a protonated amino group on tryptophan is quite general and occurs for a number of tryptophan-containing peptides as well as tryptophan itself.[30] Furthermore, tryptophan analogs lacking the amino group, such as 3-methylindole,[16] or constrained tryptophan derivatives in which contact of the amino group with the indole ring is prevented,[15] do not show the pH-dependent increase in quantum yield between pH 8 and 10. Complexation of the amino groups to a crown ether, which prevents contact between the charged amino groups and the indole ring, results in a severalfold increase in fluorescence intensity.[31] Examination of the pH-dependent intensities of tryptophan indicates that the intensity decreases further below pH 3 and above pH 11. The decrease in intensity below pH 3 is due to intramolecular quenching of indole by the neutral carboxy group, which serves as an electron acceptor, and provides strong support for the notion of intramolecular quenching of indole by the side chains of aspartate and glutamate residues in proteins. At high pH, indole is quenched by hydroxyl groups, which may be due to collisional quenching by OH^- or excited-state deprotonation of the proton on the indole nitrogen group.

How does quenching by the amino group explain the biexponential decay of tryptophan at pH 7? The basic idea is that this quenching process is most efficient in one of the rotamers and that this species (Figure 17.1, left) displays the shorter 0.5-ns decay time. The complete explanation is probably somewhat more complex. For instance, more than three rotamers are possible, when one also considers the possible orientations of the indole ring.[15] Also, it is known that quenching is usually accompanied by transient effects which appear as short components in the intensity decay.[16] For our purposes, the presence of ground-state rotamers is the dominant origin of the nonexponential decay of tryptophan. Because of the complexity introduced by the amino and carboxyl groups, experiments are frequently performed on uncharged tryptophan analogs (Figure 17.3). In the neutral tryptophan analog NATA, the amino group is acetylated and the carboxyl group is converted to an amide. Uncharged tyrosine derivatives are also used. These forms of the amino acids mimic the structures found when these amino acids are contained in a polypeptide chain.

17.2. TIME-RESOLVED INTENSITY DECAYS OF TRYPTOPHAN AND TYROSINE

Controversy regarding the intensity decay of tryptophan persisted for many years, and one may wonder why the problem took so long to solve. One difficulty was that the measurements pushed the limits of the available instrumentation and methods of data analysis. Many of the early time-resolved decays of tryptophan were measured with flashlamps having nanosecond pulse widths. The low intensities and repetition rates of the flashlamps resulted in data adequate to detect the short-decay-time component, but not to reliably resolve its decay time and amplitude. Also, the emission spectrum of the short-decay-time component is now known to be different from the dominant tryptophan emission, which compromised some of the earlier measurements in which only the long-wavelength emission above 350 nm was observed.

The difficulty in resolving the two intensity decay components is illustrated by the intensity decay of tryptophan at pH 7 (Figure 17.4). The light source was a cavity-dumped rhodamine 6G dye laser, which was frequency-doubled to 295 nm and provided pulses about 7 ps wide. The detector was an MCP PMT detector.[32] This configuration of high-speed components represents the state of the art for TCSPC measurements. The data were fit to the single- and double-exponential models,

$$I(t) = \sum_i \alpha_i \exp(-t/\tau_i) \qquad [17.1]$$

where the α_i are the amplitudes of the components with decay times τ_i. The fitted curves for both models are superimposable on the measured data. Deviations of the data from the single-exponential fit are barely visible in the deviations (Figure 17.4, lower panels), and the decrease in χ_R^2 is only from 2.62 for the single-exponential fit to 1.23 for the double-exponential fit. Hence, even with modern TCSPC instrumentation, it remains difficult to resolve the two decay times of tryptophan. We note that a larger number of counts in the peak channel would have resulted in a larger change in the relative values of χ_R^2.

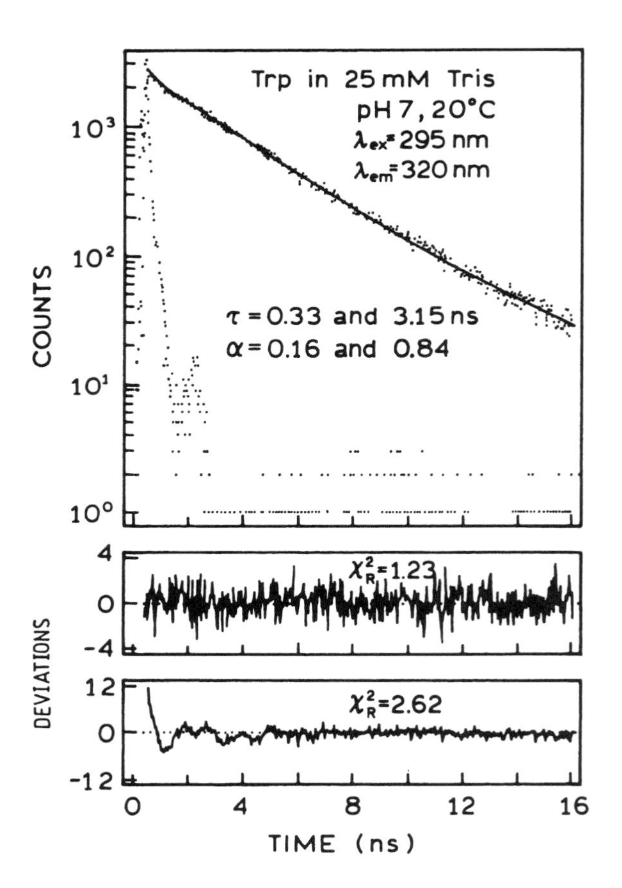

Figure 17.4. TD intensity decay of tryptophan at pH 7. Excitation was at 295 nm, and the emission was observed through a Corning WG 320 filter. The deviations are for the single-exponential (χ_R^2 = 2.62) and the double-exponential (χ_R^2 = 1.23) fits. From Ref. 32.

The multiexponential decay of tryptophan has also been observed using the FD method (Figure 17.5). The data were obtained using the same dye-laser light source and an MCP PMT detector. While we do not wish to initiate a debate about which method is superior, it seems that the 0.6-ns component is easier to detect in the FD data. Visible deviations are seen between the single- and double-exponential fits from 100 to 200 MHz, and the deviations from the single-exponential model are clearly nonrandom (Figure 17.5, lower panels). The 50-fold decrease in χ_R^2 from 58.9 for the single-exponential fit to 0.9 for the double-exponential fit is convincing evidence for the non-single-exponential decay of tryptophan. However, the deviations from a single-exponential decay are modest, so that resolution of the two decay times for tryptophan requires careful experimentation and analysis.

17.2.A. Decay-Associated Emission Spectra of Tryptophan

Another difficulty in recovering the two decay times of tryptophan was the lack of knowledge that the emission spectra were distinct for the 0.5- and 3.1-ns components.

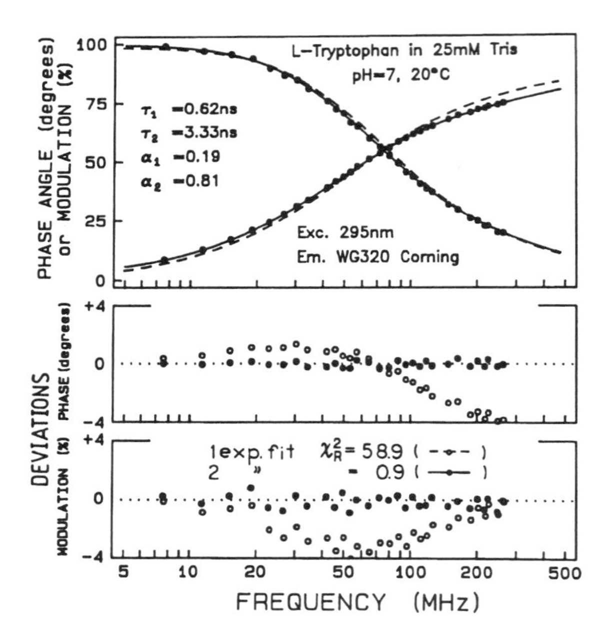

Figure 17.5. FD intensity decay of tryptophan in H_2O at 20 °C and pH 7. Excitation was at 295 nm, and the emission was collected through a long-pass Corning WG 320 filter. From Ref. 5.

It is now known that the 0.5-ns component displays an emission maximum near 320 nm and does not contribute to the emission above 380 nm.[9,10] For this reason, some of the early experiments did not detect the short-decay-time component owing to selection of only the longer-wavelength emission of tryptophan. Long-wavelength filters were often used to avoid possible observation of the ^1L_b emission, which was thought to occur at shorter wavelengths.

Wavelength-dependent data, comparable to those shown in Figure 17.4, were used to reconstruct the emission spectra associated with each decay time.[9,10] To calculate the decay-associated spectra, one has to recall that the fractional contribution (f_i) of each species to the steady-state intensity is proportional to the product $\alpha_i\tau_i$. Hence, the contribution of the short-decay-time component is lower than the relative amplitude (α_i) of this component. Following this procedure, one finds that the short decay-time-component contributes only about 4% to the total emission of tryptophan at neutral pH (Figure 17.6). Emission from the short decay-time-component occurs at slightly shorter wavelengths than for the overall emission and is centered near 335 nm. This blue shift of the emission is consistent with the expected effect of a positive charge on the polar ^1L_a state of indole. Little contribution from the ^1L_b state is expected as conversion of ^1L_b to ^1L_a occurs in less than 2 ps.[33]

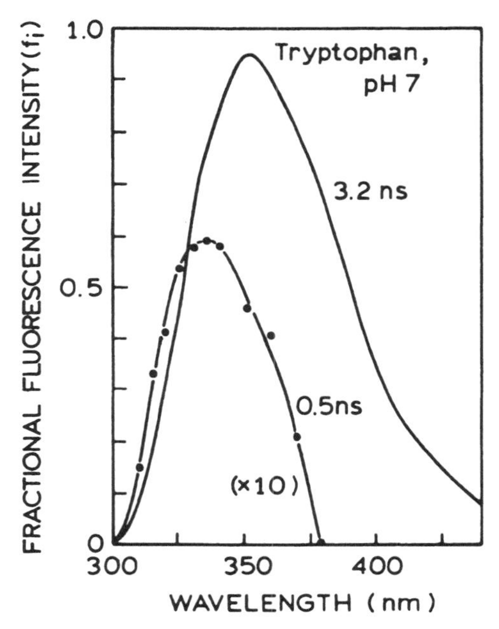

Figure 17.6. Spectral resolution of the 0.5- and 3.2-ns decay components of tryptophan. Revised from Ref. 10.

17.2.B. Intensity Decays of Neutral Tryptophan Derivatives

In proteins, the amino and carboxyl groups of tryptophan are converted into neutral groups by the formation of peptide bonds. Hence, one expects the dominant effect of the $-NH_3^+$ group to be eliminated. In fact, it is known that neutral tryptophan analogs display simpler decay kinetics. The most commonly used analogs is NATA, which has essentially the same structure as a tryptophan residue in proteins (Figure 17.3). The intensity decay of NATA (Figure 17.7) is essentially a single exponential.[10,16,35] Comparable FD data for NATA also revealed a dominant single-exponential decay (Figure 17.8). There is some decrease in χ_R^2 for the double-exponential fit, so that the decay of NATA may display a weak second component. However, for all practical purposes, we can consider NATA to display a single decay time (Table 17.1).

17.2.C. Intensity Decays of Tyrosine and Its Neutral Derivatives

Tyrosine can also display complex decay kinetics, but its properties appear to be opposite to those of tryptophan. FD intensity decays of tyrosine and its neutral analog *N*-acetyl-L-tyrosinamide (NATyrA) are shown in Figure 17.9.

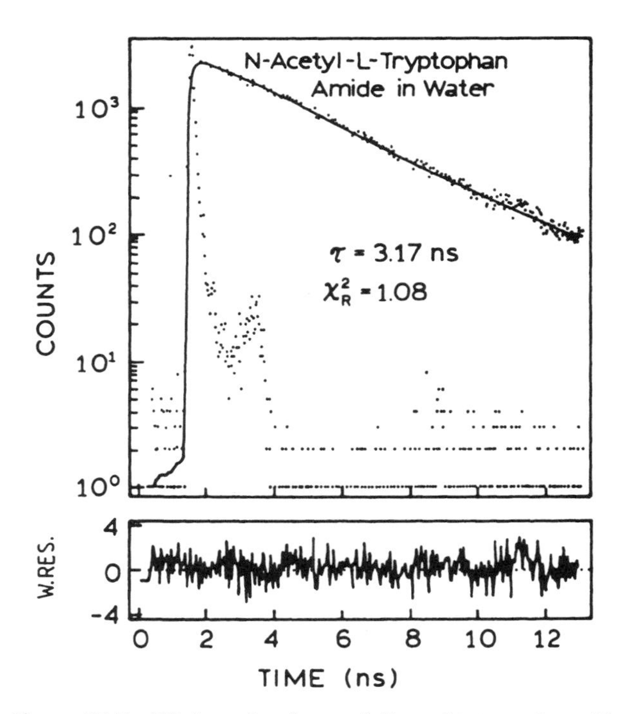

Figure 17.7. TD intensity decay of *N*-acetyl-L-tryptophanamide (NATA) at pH 7. The solid curve and weighted residuals are for the single-decay-time fit. From Ref. 32.

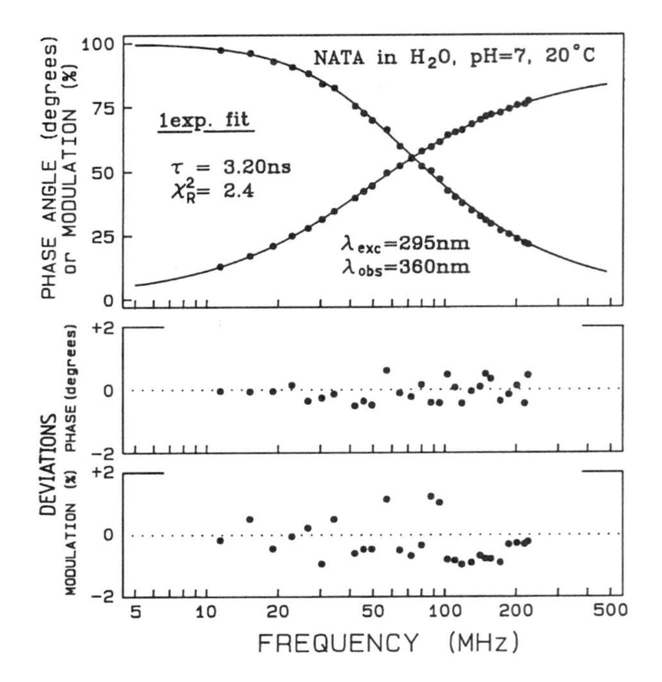

Figure 17.8. FD intensity decay of NATA at pH 7, 20 °C. The solid curves and deviations are for the single-exponential fit ($\tau = 3.20$ ns, $\chi_R^2 = 2.4$). The double-exponential fit yielded $\tau_1 = 1.04$ ns, $\tau_2 = 3.28$ ns, $\alpha_1 = 0.05$, and $\alpha_2 = 0.95$, with $\chi_R^2 = 1.3$. From Ref. 32.

Tyrosine itself displays a single-exponential decay,[36–39] whereas NATyrA displays a biexponential decay (Table 17.1). While the molecular origin of the double-exponential decay of NATyrA is not clear, this phenomenon is accepted to be also due to the presence of ground-state rotamers.[37–39]

Considerably less information is available on the intensity decays of phenylalanine. Some limited decay time information is available,[40,41] but pH-dependent lifetimes have not been reported.

17.3. INTENSITY DECAYS OF PROTEINS

Having a reasonable understanding of the intensity decays of tryptophan, we can now ask how the decays depend on protein structure. Since NATA displays a single decay time, we naturally expect single-tryptophan proteins to display single-exponential decays. However, this is not the case. In a survey of eight single-tryptophan proteins, only one protein (apoazurin) was found to display a single decay time.[7] Most single-tryptophan proteins display double- or triple-exponential decays (Table 17.2), and, of course, multi-tryptophan proteins invariably display multiexponential decays (Table 17.3).[7,8]

What are the general features of protein intensity decays? In general, the variability of the intensity decays seems to be a result of protein structure. The intensity decays and mean decay times seem to be more variable for native proteins than for denatured proteins. The decay times for denatured proteins can be grouped into two classes, with decay times near 1.5 and 4.0 ns, with the latter displaying a longer emission wavelength.[7] This tendency is enhanced in native proteins, and many show lifetimes as long as 7 ns, typically on the red side of the emission. Surprisingly, buried tryptophan residues seem to display shorter lifetimes. There is little correlation between the mean lifetime and emission maximum of proteins (see Figure 16.12). In proteins which contain chromophoric

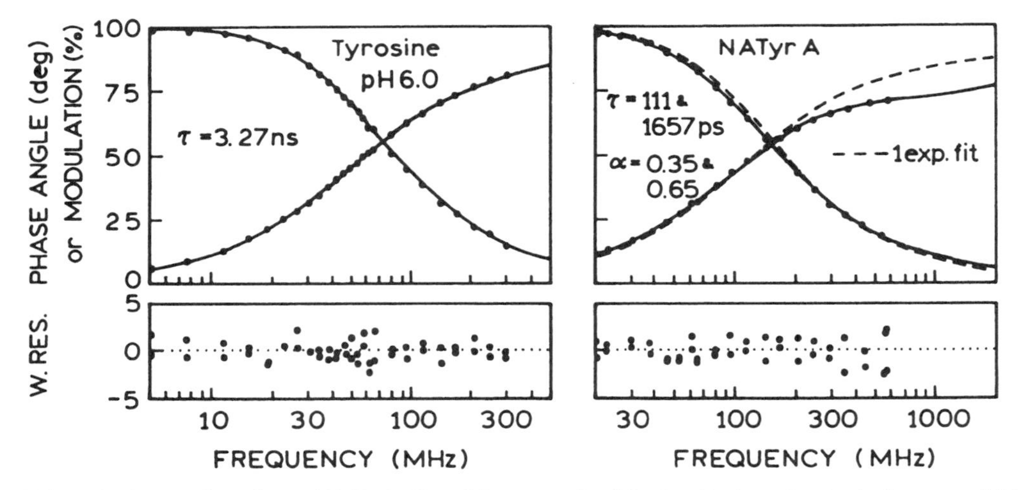

Figure 17.9. FD intensity decays of tyrosine and NATyrA. The solid curves and weighted residuals are for the single-exponential fit for tyrosine and the double-exponential fit for NATyrA. The single-exponential fit for NATyrA is represented by the dashed curve. Data from Ref. 34.

Table 17.2. Intensity Decays of Single-Tryptophan Proteins[a]

Protein	τ_1 (ns)	τ_2 (ns)	τ_3 (ns)	Reference
Azurin	4.8	0.18	—	8
Apoazurin	4.9	—	—	8
Staphylococcal nuclease	5.7	2.0	—	7
RNase T_1 (pH 5.5)	3.87	—	—	
RNase T_1 (pH 7.5)	3.57	0.98	—	
Glucagon	3.6	1.1	—	8
Human serum albumin	7.8	3.3	—	8
Phospholipase A_2	7.2	2.9	0.96	8
Subtilisin	3.82	0.83	0.17	157

[a] Additional intensity decays of single-tryptophan protein can be found in Ref. 158.

groups (ferredoxin and myoglobin), the tryptophan residues are often quenched by RET, resulting in subnanosecond decay times (Table 17.3).

Multiexponential decays are not surprising for multi-tryptophan proteins. However, the origin of multiexponential decays in single-tryptophan proteins is less clear. Given that NATA displays a single decay time, it seems that a single tryptophan residue in a single unique protein environment should result in a single decay time. One obvious origin of multiexponential decays is the existence of multiple protein conformations. Since nearby amino acid residues can act as quenchers, it seems logical that slightly different conformations can result in different decay times for native proteins. However, multiexponential decays can be observed even when the proteins are thought to exist in a single conformation.[42,43] As shown in the following section, unfolding of a protein can change a single-exponential decay into a multiexponential decay, which seems to be consistent with the presence of multiple conformations for a random-coil peptide. It appears that multiexponential decays of single-tryptophan proteins can also be the result of static conformational states of proteins and dynamic processes occurring during the excited-state lifetime.[44,45] These dynamic processes can include nearby motion of quenchers, spectral relaxation, and/or RET.

Table 17.3. Intensity Decays of Multi-tryptophan Proteins[a]

Protein	No. of trp residues	τ_1 (ns)	τ_2 (ns)	τ_3 (ns)
Bovine serum albumin	2	7.1	2.7	—
Liver alcohol dehydrogenase	2	7.0	3.8	—
Ferredoxin	2	6.9	0.5	—
Sperm whale myoglobin	2	2.7	0.1	0.01
Papain	5	7.1	3.7	1.1
Lactate dehydrogenase	6	8.0	4.0	1.0

[a] From Ref. 8.

17.4. EFFECTS OF PROTEIN STRUCTURE ON THE INTENSITY AND ANISOTROPY DECAY OF RIBONUCLEASE T_1

In the previous section we saw that most single-tryptophan proteins display multiexponential decays. However, there are two known exceptions, apoazurin and ribonuclease T_1 (RNase T_1) from *Aspergillus oryzae*. Most ribonucleases do not contain tryptophan. RNase T_1 is unusual in that it contains a single tryptophan residue and, under some conditions, displays a single-exponential decay. RNase T_1 consists of a single polypeptide chain of 104 amino acids. It has four phenylalanines, nine tyrosines, and a single tryptophan residue at position 59. This tryptophan residue is near the active site. The emission maximum is near 323 nm, suggesting that this residue is buried in the protein matrix and not exposed to the aqueous phase. Quenching studies have confirmed that this residue is not easily accessible to quenchers in the aqueous phase.[46–49] RNase T_1 is also unusual in that its intensity decay is a single exponential at pH 5.5.[50–53] The decay becomes a double exponential at pH 7. A single-exponential decay is convenient because changes in the protein structure can be expected to result in more complex decay kinetics. Hence, we will use RNase T_1 as a model to illustrate how the emission spectrum, intensity decay, and anisotropy decay depend on protein structure.

17.4.A. Protein Unfolding Exposes the Tryptophan Residue to Water

Emission spectra of RNase T_1 are shown in Figure 17.10. An excitation wavelength of 295 nm was chosen to avoid excitation of the tyrosine residues. The emission maximum is at 323 nm, which indicates that the indole group is in a nonpolar environment. Although not evident in Figure 17.10, the presence of weak vibrational structure in the emission spectrum has been reported,[50] as also seen for indole in cyclohexane and for azurin. The indole ring of trp-59 is known to be located between the α-helix and β-sheet regions of the protein (Figure 17.11). In this location, the indole ring is expected to be held rigidly by the protein.

Proteins can be unfolded at high temperature or by the addition of denaturants such as urea or guanidine hydrochloride. These conditions result in a dramatic shift of the

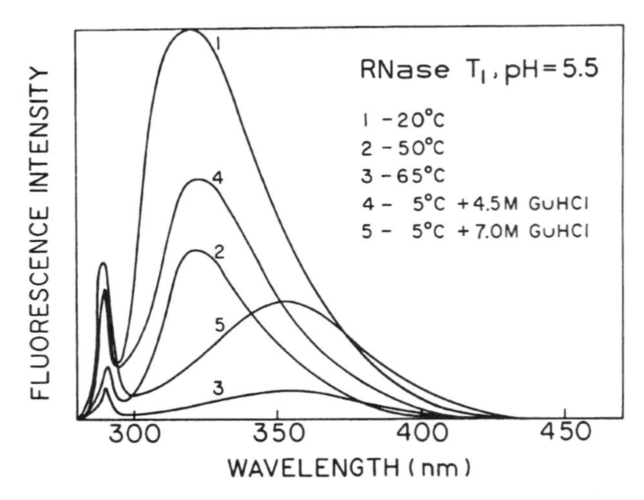

Figure 17.10. Emission spectra of RNase T_1, λ_{ex} = 295 nm. GuHCl, Guanidine hydrochloride. From Ref. 54.

emission spectrum of RNase T_1 to longer wavelengths. In the presence of 7.0*M* guanidine hydrochloride, or at 65 °C, the emission spectrum of RNase T_1 becomes characteristic of a tryptophan residue which is fully exposed to water (Figure 17.10). These results show that protein structure determines the emission maxima of proteins or, conversely, that protein folding can be studied by changes in the fluorescence intensities or emission spectra of proteins.

17.4.B. Conformational Heterogeneity Can Result in Complex Intensity and Anisotropy Decays

If a tryptophan residue exists in a single, unique environment, one expects a single decay time. A single decay time has been observed for RNase T_1 at pH 5.5, as can be seen from the TD (Figure 17.12) or the FD data (Figure 17.13). This is an unusual result, as many single-tryptophan proteins do not display single decay times (Table 17.2). In fact, RNase T_1 only displays a single-exponential decay at pH 5.5. At pH 7.5, the decay becomes more heterogeneous and can be fit to a two-decay-time model.[52]

Another remarkable feature of RNase T_1 is that the anisotropy decay is described by a single correlation time (Figure 17.14), without significant segmental mobility of the tryptophan residue.[50,57,58] This is an unusual result as most tryptophan residues in proteins display significant segmental mobility in addition to overall rotational diffusion of the protein. The correlation time of 9.6 ns at 15.4 °C (Table 17.4) is consistent with the overall rotational diffusion coefficient of the protein. One can use the data in Table 17.4 to show that the rotational rate follows the

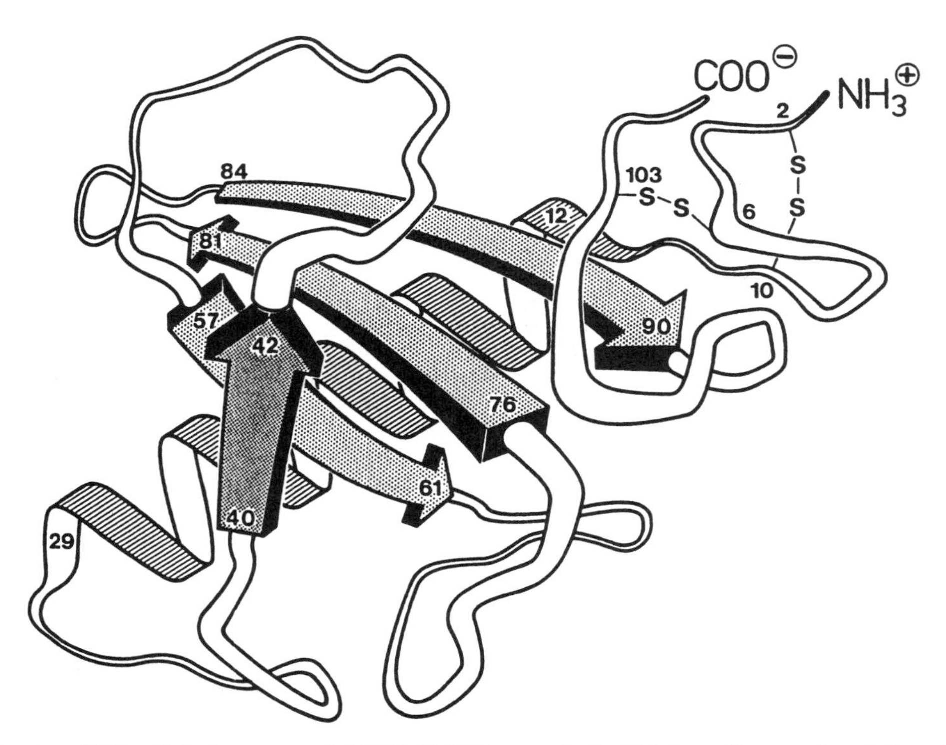

Figure 17.11. Structure of RNase T_1 in the presence of 2′-GMP (2′-GMP removed). Trp-59 is located between the α-helix and β-sheet structure. From Ref. 55.

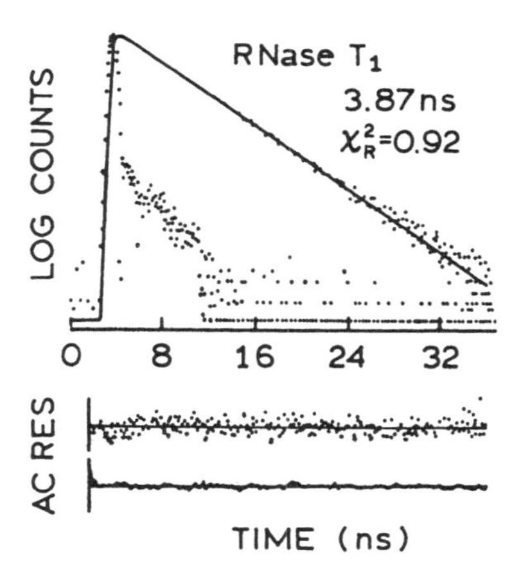

Figure 17.12. Fluorescence decay of RNaseT_1 analyzed by using a single-exponential model. The points represent experimental data, and the solid curve represents the best single-decay-time fit. RES and AC are the residuals and the autocorrelation function, respectively, for the fit. Excitation was at 295 nm, and emission was monitored at 350 nm. The sample was in 0.05*M* acetate buffer, pH 5.5, ionic strength 0.5*M*, 25 °C. Redrawn from Ref. 50.

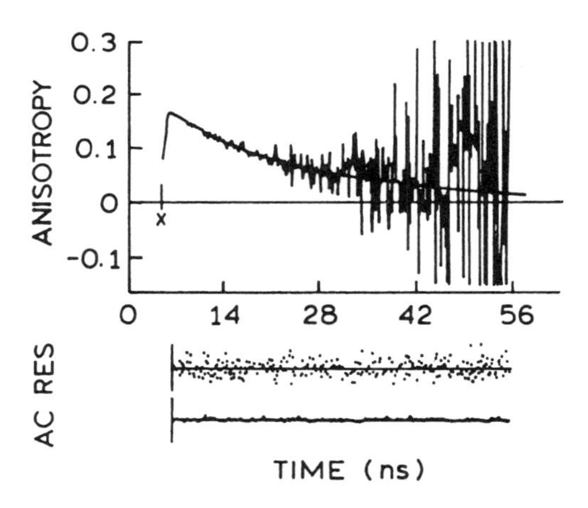

Figure 17.14. Time-resolved fluorescence anisotropy decay of RNase T_1 in buffered aqueous solution at –1.5 °C, pH 5.5. Excitation was at 295 nm. RES and AC are the residuals and the autocorrelation function, respectively, for the single-component fit. Redrawn from Ref. 50.

expected Arrhenius dependence, except at the higher temperatures, where the correlation time is smaller than expected (Problem 17.1). These data[50,57] suggest the absence of segmental motions of the trp residue in the native state of RNase T_1. However, the time-zero anisotropies are less than 0.2 (Table 17.4) and are thus considerably smaller than the expected value of close to 0.3 for excitation at 295 nm. This suggests the presence of some rapid loss of anisotropy which could not be resolved from the data.[56] Since other groups have reported $r(0)$ values near 0.3 for RNase T_1,[51,58] the cause of the low values in Figure 17.14 is not clear.

What is the effect of protein unfolding on the time-resolved decays of RNase T_1? One example is presented in Figure 17.13, which shows the FD intensity decay at 57 °C. In this case the data cannot be even roughly approximated by the single-decay-time fit. Hence, one origin of the multiexponential decays shown by proteins is the conformational heterogeneity that exists in unfolded proteins. The effects of protein unfolding are also evident in the anisotropy decays (Figure 17.15). In the native state (pH

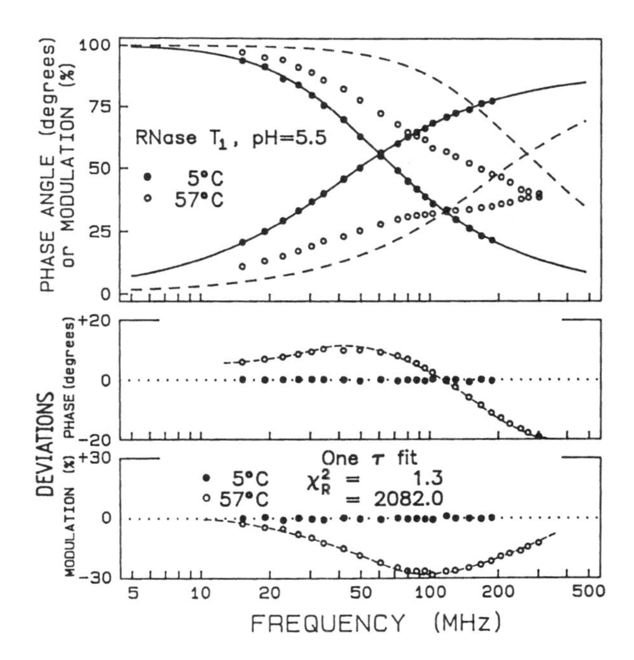

Figure 17.13. Frequency response of RNase T_1 at 5 (●) and 57 °C (○). The solid and dashed curves show the best single-component fits to the data at 5 and 57 °C, respectively. The lower panels show the deviations between the data and the best single-component fits. From Ref. 51.

Table 17.4. Rotational Correlation Times (θ) and Time-Zero Anisotropies [$r(0)$] for RNase T_1 at Various Temperatures[a]

T (°C)	η/T ($\times 10^{-3}$ kg m^{-1} s^{-1} K^{-1})	τ (ns)	$r(0)$	θ (ns)
–1.5	6.63	4.50	0.183 ± 0.007	20.9 ± 1.1
1.5	6.18	4.45	0.176 ± 0.003	15.6 ± 0.8
15.4	3.92	4.19	0.184 ± 0.004	9.6 ± 0.6
29.9	2.63	3.83	0.187 ± 0.004	6.0 ± 0.4
44.4	1.89	3.33	0.197 ± 0.003	3.7 ± 0.2

[a]Also listed are the calculated values of η/T, where η is the solution viscosity. From Ref. 50.

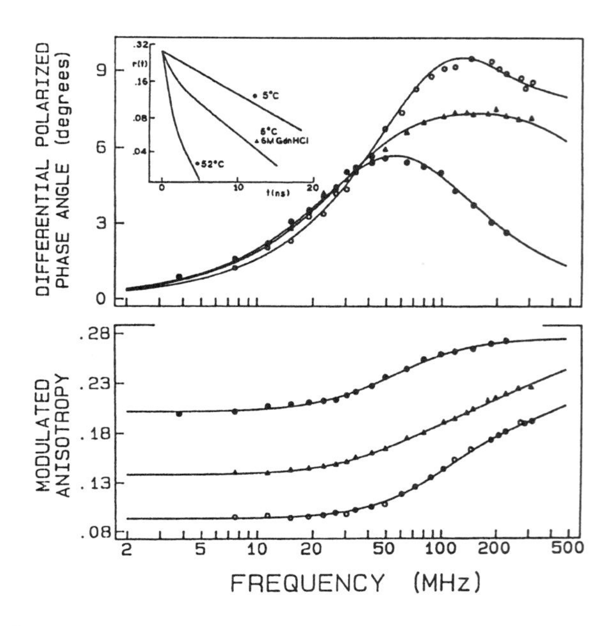

Figure 17.15. FD anisotropy decays of RNase T_1 (pH 5.5) at 5 (●) and 52 °C (○) and at 5 °C with 6*M* guanidine hydrochloride (▲). Inset: Equivalent time-dependent anisotropies reconstructed from the FD data. Similar data were reported in Ref. 58. From Ref. 5.

5.5, 5 °C), RNase T_1 displays FD anisotropy data which are characteristic of a single spherical unit, that is, a single correlation time. If the temperature is increased, or if one adds denaturant, there is an increase in the differential phase angles at high frequency and a decrease in the modulated anisotropy. These effects in the FD anisotropy data are characteristic of rapid motions of the residue in addition to slower overall rotational diffusion (Section 17.5.B). The FD anisotropy data can be used to reconstruct the time-dependent anisotropy decays (Figure 17.15, inset). In the native state, the anisotropy decay is a single exponential. When the protein is unfolded by either guanidine hydrochloride or high temperature, the anisotropy decay displays a fast component due to segmental motions of the tryptophan residue. Hence, rapid components in the anisotropy decay can be expected for random-coil peptides or for tryptophan residues on the surfaces of proteins, which are not constrained by the three-dimensional structure.

17.5. ANISOTROPY DECAYS OF PROTEINS

In the previous section we saw that in the native state RNase T_1 displays a single-exponential anisotropy decay. This was an unusual result, as most proteins show rapid components in the anisotropy decays due to segmental flexibility of the protein (Table 17.5). A typical anisotropy decay is shown in the top panel of Figure 17.16 for the single tryptophan residue in the lumazine protein. The long component near 17 ns is assigned to overall rotational diffusion of the protein. The short component near 0.45 ns is assigned to rapid motions of the tryptophan residue, independent of overall rotational diffusion. Another example is the tryptophan residue in bovine troponin I, which shows a shorter correlation time near 0.90 ns (Figure 17.16, middle). In many cases, the rapid components in the anisotropy decay are too fast to be measured, as shown for a histocompatibility complex protein in the bottom panel of Figure 17.16.

Fast components in the anisotropy decays are characteristic of many proteins[62] and peptides[63] and have been reported in many publications.[64–67] Picosecond-timescale motions were also reported for tyrosine residues in proteins[68,69]; in these studies, a streak camera was used to obtain adequate time resolution. The short correlation time observed in proteins is variable and ranges from 50 to 500 ps, with the values being determined in part by the time resolution of the instrument. The shorter correlation time is approximately equal to that observed for NATA in water or for small peptides (Table 17.5). These shorter correlation times are typically insensitive to protein folding and are not greatly affected by the viscosity of the solution.

The picosecond component in the anisotropy decays of proteins does appear to display some characteristic features. The fraction of the total anisotropy due to this component is typically larger for small unstructured peptides than for tryptophan residues buried in a protein matrix. Unfolding of a protein usually increases the amplitude of the fast motions.

The magnitude of the short correlation time seems to be mostly independent of the size of the protein or peptide. This is illustrated in Figure 17.17, which shows the two

Table 17.5. Anisotropy Decays of NATA and Single-Tryptophan Peptides and Proteins[a]

Species	θ_1 (ps)	θ_2 (ps)	r_{01}	r_{02}	Reference
RNase T_1	6,520	—	0.310	—	73
Staphylococcal nuclease	10,160	91	0.303	0.018	73
Monellin	6,000	360	0.242	0.073	73
ACTH	1,800	200	0.119	0.189	73
Gly-trp-gly	135	39	0.105	0.220	73
NATA	56	—	0.323	—	73
Melittin monomer	1,730	160	0.136	0.187	74
Melittin tetramer	3,400	60	0.208	0.118	74

[a]20 °C, excitation at 300 nm.

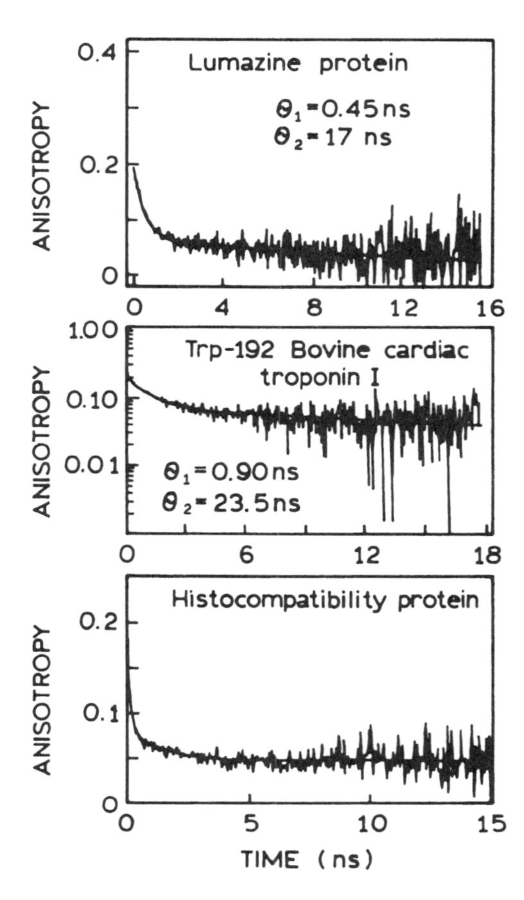

Figure 17.16. Anisotropy decays for the single tryptophan residue in the lumazine protein (*top*), for trp-192 in bovine cardiac troponin I (*middle*), and for a histocompatibility protein (*bottom*). Revised from Refs. 59, 60, and 61.

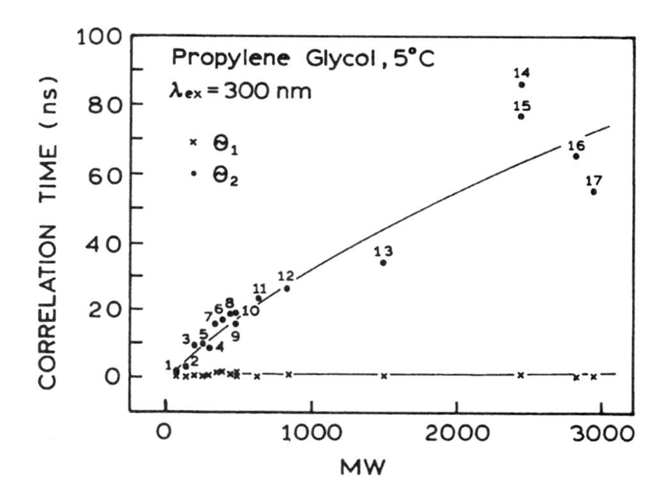

Figure 17.17. Short (×) and long (•) rotational correlation times for indole, tryptophan, and peptides in propylene glycol at 5 °C. 1, Indole; 2, 3-methylindole; 3, trp; 4, NATA; 5, gly-trp; 6, trp-trp; 7, gly-trp-gly; 8, leu-trp-leu; 9, glu-trp-glu; 10, lys-trp-lys; 11, gastrin; 12, pentagastrin; 13, [tyr^4]-bombesin; 14, dynorphin; 15, [asn^{15}]-dynorphin; 16, cosyntropin; 17, melittin. From Ref. 70.

correlation times recovered for molecules spanning a range of sizes, from indole to tryptophan to the tripeptide lys-trp-lys to melittin. The molecules were dissolved in propylene glycol, so the overall correlation times are longer than found in water. Of course, the longer correlation time increases with molecular weight. However, the rapid correlation time is independent of molecular weight and is near 0.5 ns for all peptides.

For proteins it is tempting to assume that tryptophan residues exposed to the aqueous phase will show the most rapid anisotropy decays. However, this is not always the case. As an example, we note the anisotropy decay of staphylococcal nuclease. This protein contains a single tryptophan residue at position 140, which is on the surface of the protein and near the C-terminus. One might expect this region of the protein to rotate freely. In spite of this location in a seemingly unhindered environment, the anisotropy of trp-140 decays predominantly with a correlation time of 8.9 ns,[71,72] which is expected for this size of protein.

17.5.A. Anisotropy Decays of Melittin

The single-tryptophan peptide melittin provides a good example of how an anisotropy decay depends on protein conformation and local environment. As described in Chapter 16, melittin self-associates to form tetramers. During this process, the single tryptophan residue on each chain is buried in the center of the four-helix bundle. Hence, the anisotropy decay is expected to change owing to the larger overall size and restriction of segmental motions. Anisotropy decays for melittin have been reported by several laboratories.[74–78] Typical data for the monomer and tetramer are shown in Figure 17.18. The anisotropy decay was multiexponential in both cases, with a 20–40-ps component in either state. The longer correlation time was 1.4 ns for the monomer and 5.5 ns for the tetramer. These values are consistent with overall rotational diffusion.

Melittin also binds to lipid vesicles because of its nonpolar surface in the α-helical state.[79–81] This binding results in a dramatic change in the anisotropy decay (Figure 17.18), which now shows a correlation time near 12 ns, but there is considerable uncertainty in this value.[75] This result illustrates one of the difficulties with using tryptophan as an anisotropy probe. It is difficult to obtain much information after 10 ns, or three times the mean lifetime, owing to the low remaining intensity. In general, one

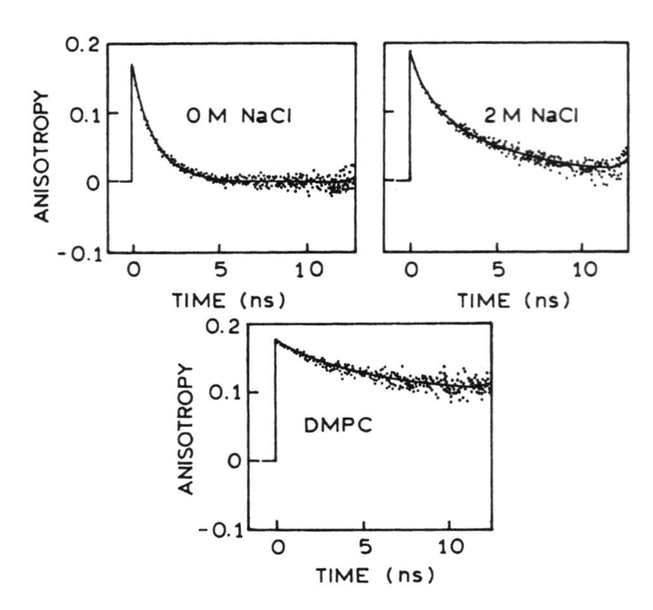

Figure 17.18. Anisotropy decays of monomeric and tetrameric melittin, in water and 2*M* NaCl, respectively, and melittin bound to DMPC lipid vesicles. Excitation was at 300 nm. Data from Ref. 75.

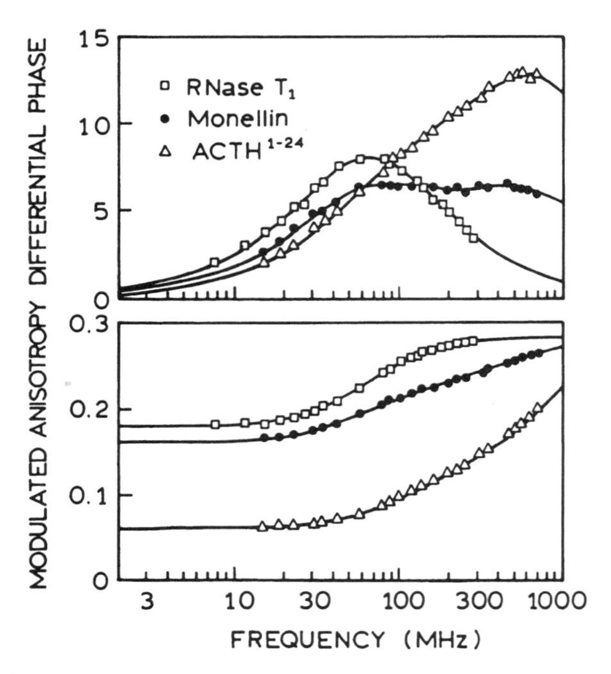

Figure 17.19. FD anisotropy decays of three single-tryptophan proteins: □, RNase T_1; ●, monellin; △, adrenocorticotropin (ACTH (1–24)). Data from Ref. 73.

should attempt to select probes whose decay times are longer than the timescale of the process being measured.

The data in Figure 17.18 illustrate another problem encountered when one is measuring protein anisotropy decays. Examination of the data reveals that the measured time-zero anisotropy [$r(0)$] is less than the fundamental anisotropy ($r_0 = 0.3$) for the 300-nm excitation wavelength. This frequently occurs owing to the limited time resolution of the instrumentation. If the correlation time is too short, the anisotropy decays within the instrument response function, and the apparent time-zero anisotropy is less than the actual value.[82]

17.5.B. Protein Anisotropy Decays as Observed in the Frequency Domain

Anisotropy decays of proteins are often measured using the FD method. Hence, it is valuable to understand how the nature of the anisotropy decay affects the appearance of the FD data. Representative data for three single-tryptophan proteins are shown in Figure 17.19. In one of these proteins, the single tryptophan residue is rigid in the protein (RNase T_1), and in another it is highly mobile (ACTH). The intermediate case of monellin is also shown, in which about 25% of the anisotropy decays by a subnanosecond process (Table 17.5). For the rigid tryptophan residue in RNase T_1, the differential phase angles (Δ_ω) are distributed approximately as a Lorentzian on the log frequency axis (Figure 17.19). As the extent of segmental flexibility of the proteins increases, the differential phase angles increase at higher frequencies. This effect is characteristic of a subnanosecond component in the anisotropy decay.

The extent of segmental mobility also affects the modulated anisotropy (r_ω) data. The values of r_ω at low modulation frequency reflect the steady-state anisotropy and are thus lower for the proteins with higher segmental flexibility. The transition of r_ω toward the high-frequency value of r_0 occurs at higher frequency for proteins with greater flexibility. It is apparent from these data that the FD anisotropy data are highly sensitive to the form of the anisotropy decay and/or the dynamic properties of the proteins.

17.6. TIME-DEPENDENT SPECTRAL RELAXATION IN PROTEINS

Although there have been extensive studies of protein dynamics using the anisotropy decays, less attention has been given to the possibility of time-dependent spectral relaxation in proteins. The occurrence of spectral relaxation seems probable for the solvent-sensitive indole group buried in a semirigid protein matrix. In fact, nanosecond-timescale relaxation of proteins around probes other than tryptophan has been reported by several laboratories.[83,84]

Evidence for time-dependent relaxation of tryptophan residues is pervasive but not specific. One characteristic of spectral relaxation is an increase of the mean decay time with increasing observation wavelength. Since there seems to be no correlation between the emission maxima of proteins and their mean lifetimes (see Figure 16.12), it can be concluded that, in the absence of spectral relaxation, there should be equal numbers of proteins which display increases or decreases in lifetime with increasing wavelength. However, for almost all proteins, the mean lifetime increases with increasing emission wavelength. Even single-tryptophan proteins display lifetimes which increase with increasing wavelength.[85,86] An unambiguous characteristic of an excited-state process is a negative preexponential factor in the intensity decay. Such components have only been rarely observed in proteins, one example being chicken pepsinogen.[87] The absence of negative preexponential does not prove the absence of spectral relaxation because these terms are easily eliminated by spectral overlap.

Why is spectral relaxation rarely observed in proteins? One partial answer is because this effect is usually not considered and the heterogeneity of protein decays is interpreted in terms of conformational states. Also, the time-dependent shifts are small and somewhat difficult to observe. Finally, spectral relaxation probably occurs mostly on a subnanosecond timescale, so that this process is mostly complete prior to emission.

In spite of the difficulties, a limited number of time-resolved emission spectra have been reported for single-tryptophan proteins[88–90] and for model compounds,[91,92] and several reviews have appeared.[93–95] The TRES of melittin bound to lipid vesicles were found to display a modest shift at 5 ns following the excitation pulse (Figure 17.20). In those cases where detailed information is available, the emission center of gravity decays by a multiexponential process, as seen for staphylococcal nuclease (Figure 17.21). One of the spectral relaxation times is typically less than 1 ns,[90,95] so that relaxation may be almost complete prior to emission in most proteins with lifetimes of 2 ns or longer. Nonetheless, it seems that time-dependent spectral shifts contribute to the multiexponential intensity decays of most proteins.

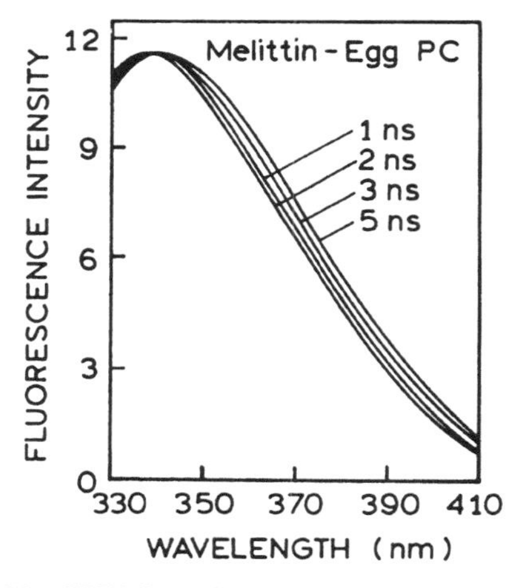

Figure 17.20. TRES for melittin bound to egg PC vesicles. Similar results were found for melittin tetramer. Revised from Ref. 88.

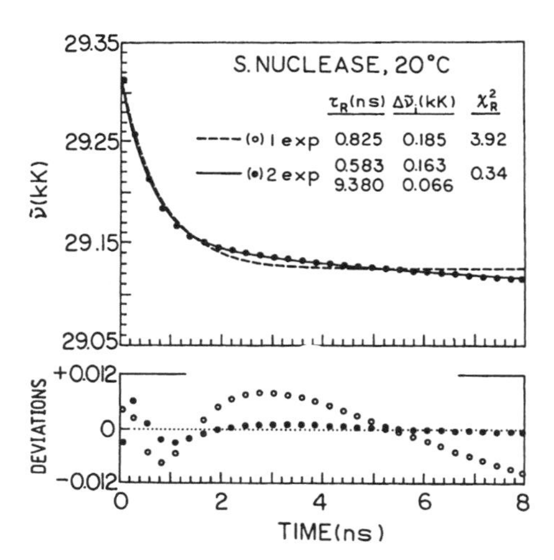

Figure 17.21. Emission center of gravity of staphylococcal nuclease. 1 kK = 1000 cm^{-1}. From Ref. 90.

17.7. DECAY-ASSOCIATED EMISSION SPECTRA OF PROTEINS

An obvious application of time-resolved protein fluorescence is to resolve the contributions of individual tryptophan residues in multi-tryptophan proteins. This is accomplished by measuring the intensity decay across the emission spectrum of the protein, resulting in wavelength-dependent data,

$$I(\lambda, t) = \sum_i \alpha_i(\lambda) \exp(-t/\tau_i) \qquad [17.2]$$

As implied by this equation, a typical, but not necessarily correct, assumption is that the decay times are independent of wavelength. The wavelength-dependent values $\alpha_i(\lambda)$ are then used to construct the emission spectra $[I_i(\lambda)]$ associated with each species,

$$I_i(\lambda) = \frac{\alpha_i(\lambda)\tau_i I(\lambda)}{\sum_j \alpha_j(\lambda)\tau_j} \quad [17.3]$$

In this expression, $I(\lambda)$ is the steady-state emission spectrum. These spectra $I_i(\lambda)$ are called decay-associated spectra (DAS)[96] because $I_i(\lambda)$ represents the emission spectrum of the component emitting with lifetime τ_i.

DAS have been determined for a number of proteins, with an emphasis on proteins which contain two tryptophan residues. In these cases one hopes that each tryptophan will display a single decay time, so that the DAS represent the emission spectra of the individual residues. One example is provided by a study of yeast 3-phosphoglycerate kinase (3-PGK), which has two tryptophan residues. From a number of pH- and wavelength-dependent measurements, the 0.6-ns component in the decay was associated with one residue, and the 3.1- and 7.0-ns components were associated with the second tryptophan residue.[97] The wavelength-dependent intensity decays were then used to resolve the contributions of each residue to the total spectrum of 3-PGK (Figure 17.22). The shorter-lived residue was found to emit at 322 nm, and the longer-lived residue at 338 nm. Similar resolutions have been performed on a number of proteins with two tryptophan residues, including the *E. coli lac* repressor[98] and horse liver alcohol dehydrogenase.[102]

A serious difficulty with the DAS is that each tryptophan residue may not display a unique lifetime. This problem was already noted for 3-PGK, which displayed two decay times for one of its tryptophan residues (Figure 17.22). Another difficulty is that even single-tryptophan proteins often display two or more decay times. These decay times can be used to construct the DAS for a single-tryptophan protein. One example is provided by a study of the single-tryptophan phosphofructokinase from *Bacillus stearothermophilus*.[100] The intensity decay was found to be a double exponential at all emission wavelengths, with lifetimes of 1.6 and 4.4 ns. Since the protein contains a single tryptophan residue, one naturally questions the meaning of the

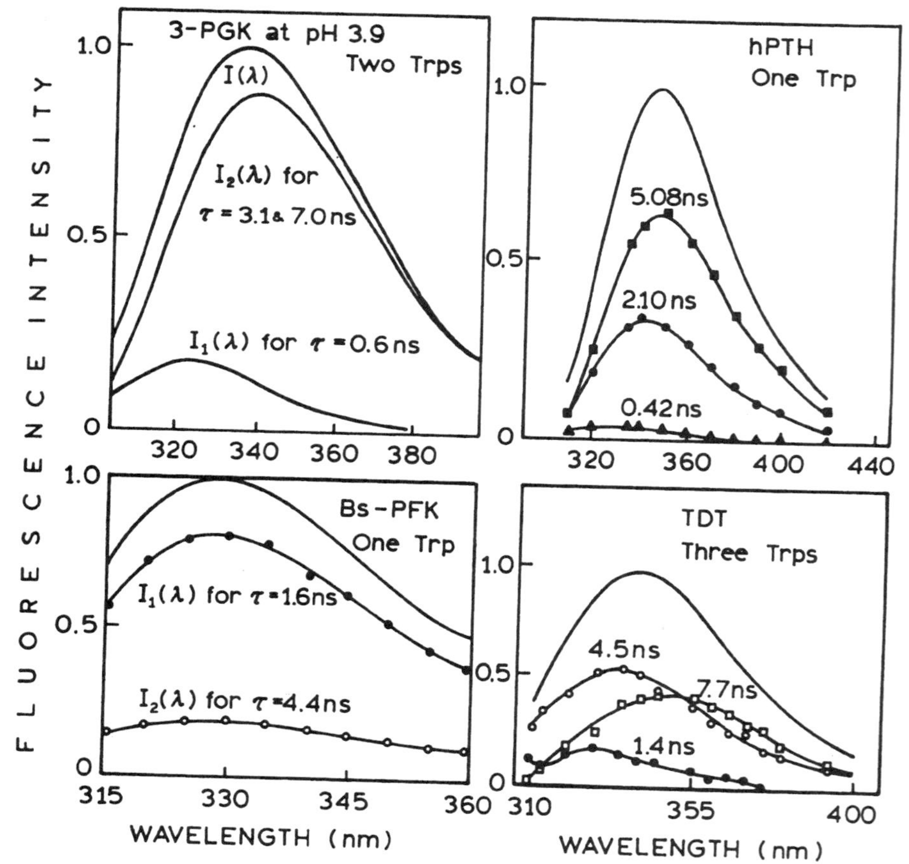

Figure 17.22. DAS of proteins: 3-phosphoglycerate kinase (3-PGK),[98] human parathyroid hormone (hPTH),[99] phosphofructokinase (Bs-PFK),[100] and terminal deoxynucleotide transferase (TDT).[101]

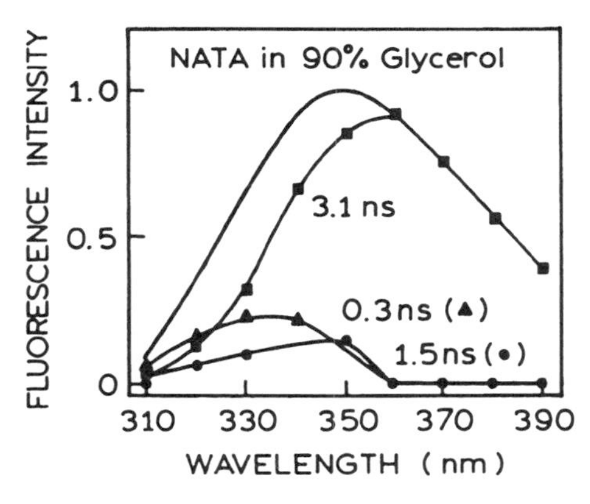

Figure 17.23. DAS of NATA in 90% glycerol. The solid curve without symbols is the steady-state emission spectrum. Revised from Ref. 107.

DAS. In this case the DAS were interpreted as representing the equilibrium between two ground-state conformations of the protein. DAS have also been reported for a number of single-tryptophan proteins and are usually interpreted in terms of multiple conformations.[99,103,104] While the authors of these studies may be correct, it seems surprising that all proteins exist in multiple conformations in solution. Also, DAS based on three decay times imply three conformations. When one examines the DAS for a number of proteins, with one, two, or more tryptophan residues,[105] one finds that the short-lived DAS are almost always at shorter emission wavelengths, and the longer-lived DAS are usually at longer wavelengths. This suggests a common origin for the DAS, which may be time-dependent spectral relaxation. In fact, the DAS often look similar to the DAS spectra of NATA in a viscous solvent (Figure 17.23).[106–108] It seems that DAS can be calculated for almost any protein, whether the protein contains just one or several tryptophan residues. One should exercise caution in assigning physical reality to the DAS until they are supported by additional information beyond the time-dependent decays.

17.8. TIME-RESOLVED STUDIES OF REPRESENTATIVE PROTEINS

In the previous chapter we illustrated the general features of protein fluorescence by a description of specific examples. We now do the same using time-resolved data. Use of the time-resolved data provides considerable information about protein structure and function. For example, the time-resolved intensity decays of proteins can be sensitive to bound ligands, which can displace tryptophan residues to new locations in proteins. The emission from multitryptophan proteins can be resolved through studies of the single-tryptophan mutants. The intensity decays of proteins can be altered by the presence of bound chromophores which serve as RET acceptors. The sensitivity of protein intensity decays to a variety of phenomena has resulted in increased understanding of the functional behavior of proteins.

17.8.A. Annexin V—A Calcium-Sensitive Single-Tryptophan Protein

Annexins are a class of homologous proteins which bind to cell membranes in a calcium-dependent manner. Annexin V possesses a single tryptophan residue, which was found to be highly sensitive to calcium.[109] Addition of calcium results in a shift of the emission maximum from 324 to 348 nm (Figure 17.24). The calcium-dependent shift to long wavelength is accompanied by an increase in the mean decay time (Figure 17.25). A longer lifetime for the solvent-exposed tryptophan is consistent with the earlier observations of long decay times for proteins emitting at longer wavelengths.[7] As expected for a surface-exposed residue, the anisotropy decay also becomes more rapid in the presence of calcium (Figure 17.26). The anisotropy decay was recovered as a distribution of correlation times using the maximum entropy method (MEM) (Figure 17.27). In this case the plot of the correlation time distri-

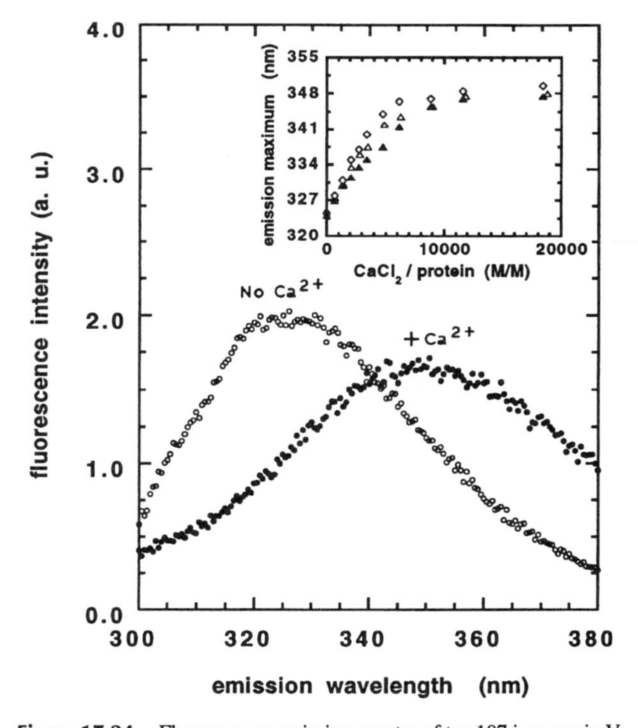

Figure 17.24. Fluorescence emission spectra of trp-187 in annexin V: ○, calcium-free protein; ●, calcium-bound protein. *Inset*: Dependence of the emission maximum of annexin V on the calcium-to-protein molar ratio. Revised and reprinted, with permission, from Ref. 109, Copyright © 1994, American Chemical Society.

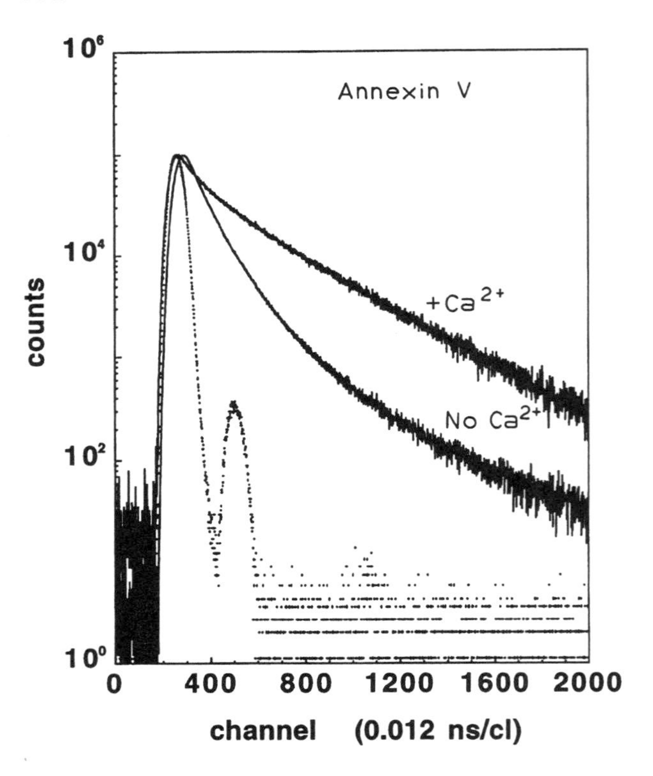

Figure 17.25. Fluorescence intensity decays of trp-187 in annexin V in the absence and in the presence of calcium. Revised and reprinted, with permission, from Ref. 109, Copyright © 1994, American Chemical Society.

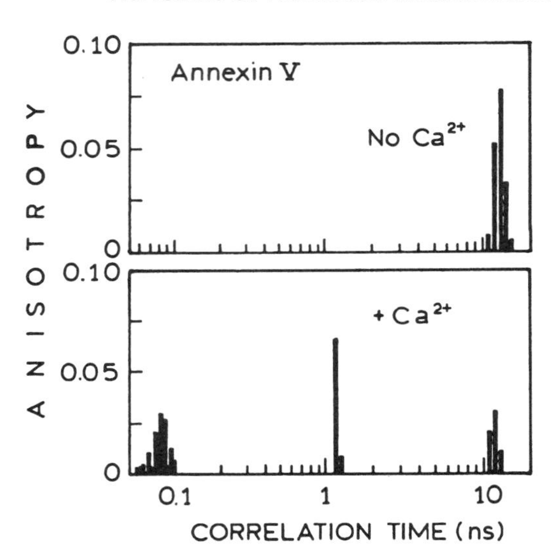

Figure 17.27. MEM-reconstituted correlation time distributions of trp-187 in annexin V in the absence (*top*) and presence (*bottom*) of Ca^{2+}. Revised and reprinted, with permission, from Ref. 109, Copyright © 1994, American Chemical Society.

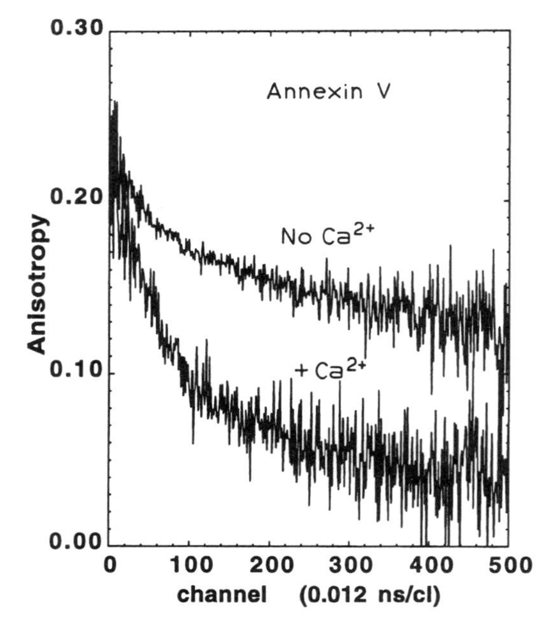

Figure 17.26. Time-resolved anisotropy decays of annexin V in the absence and in the presence of calcium. Data from Ref. 109; figure provided by Dr. Jacques Gallay.

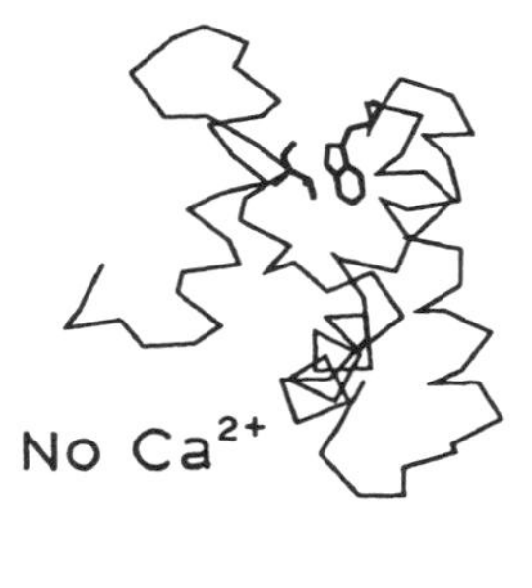

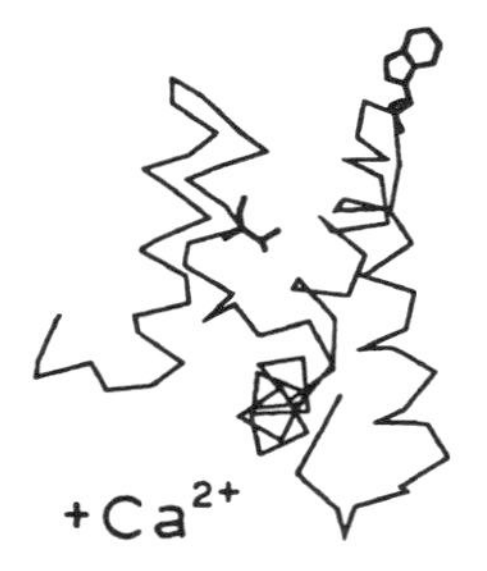

Figure 17.28. Structure of annexin V in the absence (*top*) and presence (*bottom*) of Ca^{2+}. Revised and reprinted, with permission, from Ref. 109, Copyright © 1994, American Chemical Society.

bution is useful for understanding the form of the anisotropy decay. In the absence of Ca^{2+}, annexin V displays a dominant correlation time near 10 ns. In the presence of Ca^{2+}, the anisotropy decay becomes a triple exponential with correlation times near 70 ps, 1 ns, and 10 ns (Figure 17.27). Taken together, these results are consistent with the known structural change induced by Ca^{2+}, which results in displacement of trp-187 from a buried environment to one in which the side chain extends into the aqueous phase (Figure 17.28). It is apparent that the time-resolved intensity and anisotropy decays of proteins are highly sensitive to their solution conformation.

17.8.B. Immunophilin FKBP59—Quenching of Tryptophan Fluorescence by Phenylalanine

Immunophilins are receptors for immunosuppressant drugs such as cyclosporine.[110] The immunophilin domain FKBP59-I contains two tryptophan residues at positions 59 and 89 (Figure 17.29). In general, it would be difficult to separate the contributions of the two tryptophan residues using data for just FKBP59-I. In this case an analogous protein, FKBP12, was available which contained just trp-89. The intensity decays of these two proteins are markedly different, with FKBP12 showing a much lower quantum yield (0.025 versus 0.19 for FKBP59) and a more rapid intensity decay (Figure 17.30). For the protein with two tryptophan residues, one of which is highly fluorescent, the lifetime distribution analysis shows a dominant decay time near 6.26 ns. For the single-tryptophan protein, the intensity decay is dominated by the more weakly emitting residue, which displays a more heterogeneous intensity decay with a dominant decay time near 0.21 ns (Figure 17.30, lower panels). These results revealed an unexpected origin for quenching of trp-89. Apparently, this tryptophan residue is quenched due to a nearby phenylalanine residue. Benzene is known to decrease the decay times of indole.[111] Although not reported, one would expect denaturation of the protein to decrease the extent of quenching and normalize the fluorescence of trp-89.

Another example of tryptophan quenching by nearby amino acid residues is in mutants of HSA.[112] This 64-kDa single-chain protein contains a single tryptophan residue

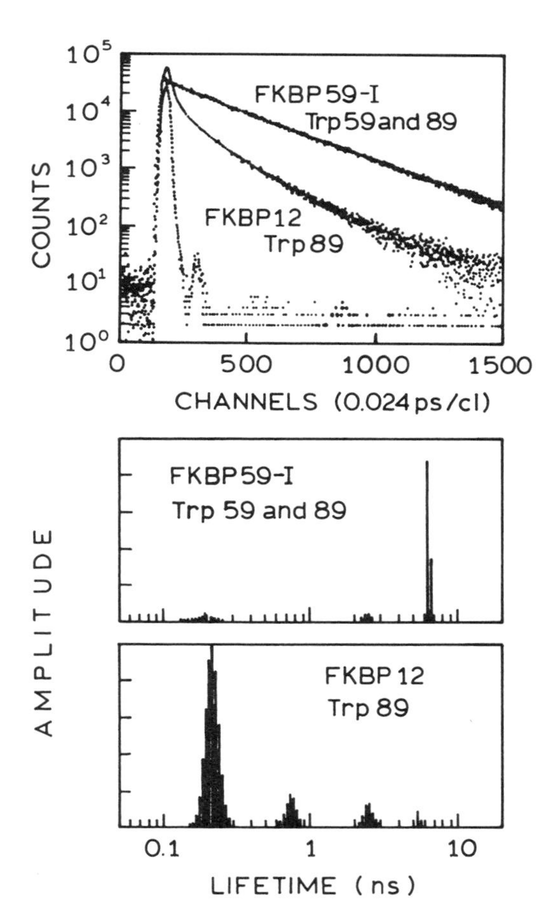

Figure 17.30. *Top*: Time-resolved intensity decays of FKBP59-I and FKBP12. *Bottom*: Lifetime distributions recovered using the maximum entropy method. Excitation at 295 nm; emission at 340 nm; 20 °C. Reprinted, with permission, from Ref. 110, Copyright © 1997, American Chemical Society.

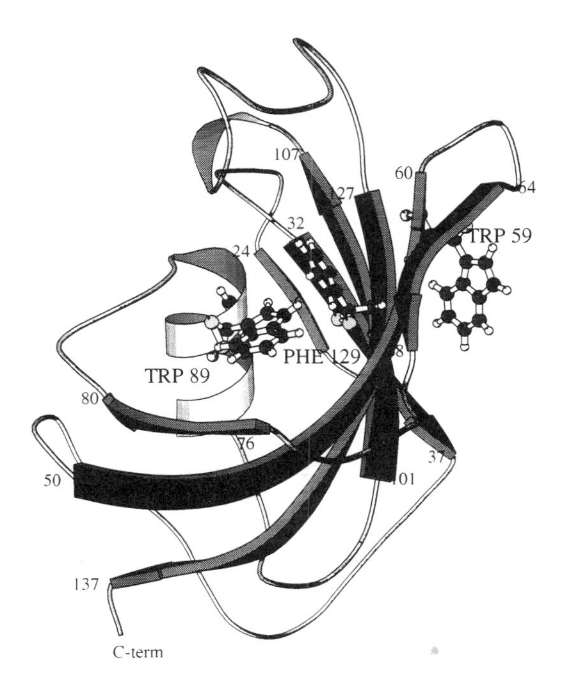

Figure 17.29. Location of the two tryptophan residues in FKBP59-I. FKBP12 has only trp-89. Reprinted, with permission, from Ref. 110, Copyright © 1997, American Chemical Society.

at position 214. There are many genetic varieties of the HSA amino acid sequence, most of which seem to have no physiological significance. However, one mutant displays a high binding affinity for thyroxine (T_4), resulting in elevated levels of serum thyroxine. This mutant HSA has a histidine residue in place of arginine at position 218. The histidine residue quenches trp-214, shortening its mean decay time from 5.6 to 4.2 ns.[112] It is likely that quenching of tryptophan by nearby amino acid residues is a common occurrence in folded proteins.

17.8.C. Trp Repressor—Resolution of the Two Interacting Tryptophans by Site-Directed Mutagenesis

In Section 17.7 we saw the possibility of resolving the emission spectra of individual tryptophan residues using DAS. An alternative and more powerful method is to use site-directed mutagenesis to create mutant proteins, each with a single tryptophan residue. For example, this method has been applied in studies of the Trp repressor from *E. coli*.[113,114] The wild-type protein contains two tryptophans at positions 19 and 99. One mutant, W99F, contains only trp-19, and the other mutant, W19F, contains only trp-99. Emission spectra of the three proteins show that the mutant with trp-19 has a higher quantum yield than the mutant with trp-99, or even the wild-type protein with two tryptophan residues (Figure 17.31).

FD intensity decays of the three repressor proteins show that each displays a distinct average lifetime (Figure 17.32). The longest decay time is shown by the mutant with trp-19, as can be judged by the larger phase angles at lower frequencies. The mean decay time of this protein is even longer than that of the wild-type protein. These results indicate that trp-99 acts as a quencher in the wild-type protein, presumably by RET from trp-19 to trp-99.

The difficulties in calculating the DAS are illustrated by the intensity decays (Table 17.6). Both the wild-type protein and the single-tryptophan mutants display double-exponential decays. Additionally, the decay times overlap for all three proteins. It is difficult to imagine how the emission of the two tryptophan residues could be resolved using data from only the wild-type protein. Also, the interactions between the tryptophan residues imply that the intensity decay of the wild-type protein is not the sum of the intensity decays of the two single-tryptophan mutants.

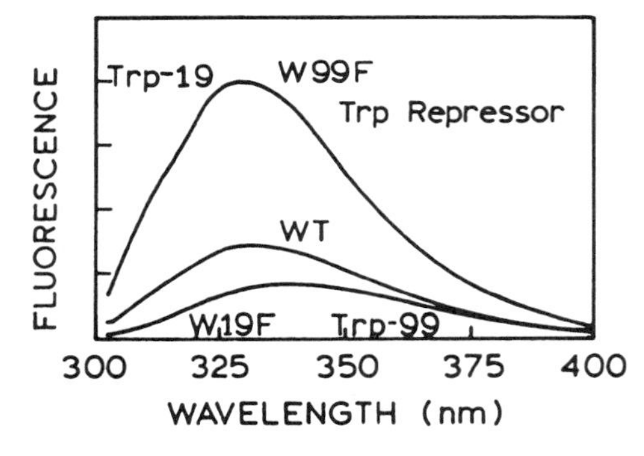

Figure 17.31. Emission spectra of the wild-type *E. coli* Trp repressor (WT) and its two single-tryptophan mutants. Revised from Ref. 113.

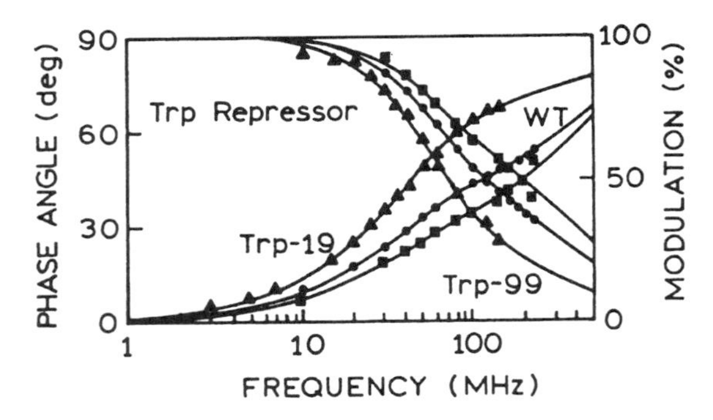

Figure 17.32. FD intensity decays of the wild-type Trp repressor (WT) and its two single-tryptophan mutants. Revised from Ref. 113.

17.8.D. Aspartate Transcarbamylase—Two Noninteracting Tryptophan Residues

Aspartate transcarbamylase (ATCase) is a multisubunit protein with six catalytic (c) and six regulatory (r) subunits and exists as a dodecamer $[(c_3)_2(r_2)_3]$. Each catalytic subunit contains two tryptophan residues at positions 209 and 284, and the regulatory subunit has no tryptophan residues. The intensity decay of the wild-type protein is intermediate between that of the two single-tryptophan mutants (Figure 17.33).[115] This suggests that in ATCase the tryptophans are not interacting. The intensity decays were analyzed in terms of lifetime distributions. Although the results of such an analysis are not always unique, they do allow one to easily visualize the nature of the intensity

Table 17.6. Intensity Decays of the Trp Repressor and Its Single-Tryptophan Mutants[a]

Protein	τ_i (ns)	α_i	f_i	$\bar{\tau}$ (ns)[b]
Wild type	0.57	0.70	0.28	1.41
(trp-19 and trp-99)	3.36	0.30	0.72	
W19F (trp-99)	0.58	0.83	0.47	1.01
	3.14	0.17	0.53	
W99F (trp-19)	0.40	0.28	0.04	2.95
	3.94	0.72	0.96	

[a]From Ref. 113.
[b]Calculated using $\bar{\tau} = \Sigma_i \alpha_i \tau_i$.

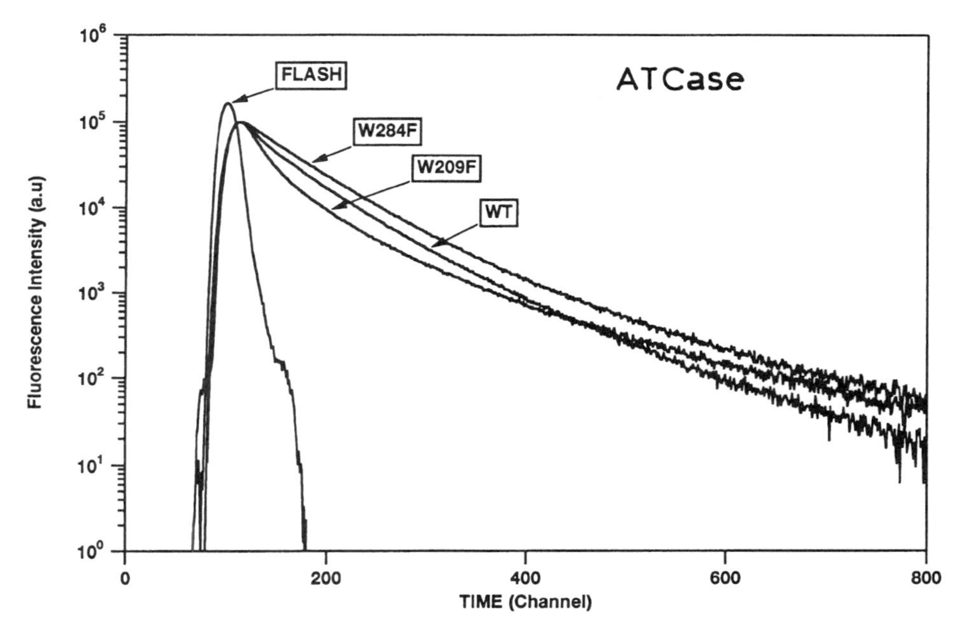

Figure 17.33. Intensity decays of wild-type (WT) aspartate transcarbamylase (ATCase) and its two single-tryptophan mutants. Revised from Ref. 115.

decay. For ATCase, the lifetime distribution of the wild-type protein seems to be a sum of the distributions obtained from each single-tryptophan mutant (Figure 17.34). If the tryptophans interacted by RET, as occurred for the Trp repressor, this simple sum would not occur, as the trp–trp interactions would create new decay-time components in the wild-type protein.

17.8.E. Thermophilic β-Glycosidase—Multi-Tryptophan Protein

In the preceding sections we saw that considerable effort was devoted to obtaining proteins with a limited number of tryptophan residues, each in a known environment. As the number of tryptophan residues increases, it becomes impractical to resolve the individual residues. This is one instance where it is valuable to use lifetime distributions to characterize the intensity decays.

Sulfolobus solfataricus is an extreme thermophile growing at 87 °C. Its β-glycosidase is a 240-kDa tetramer which contains 68 tryptophan residues, 17 per subunit. The high density of tryptophan residues is evident from the three-dimensional structure of one subunit of the protein (Figure 17.35, top). It is obviously impractical to attempt to resolve the individual emission spectra, and equally impractical to prepare single-tryptophan mutants. One can nonetheless characterize the intensity decays of the protein, which in this case were measured using the FD method (Figure 17.35, bottom). The decays are highly heterogeneous and

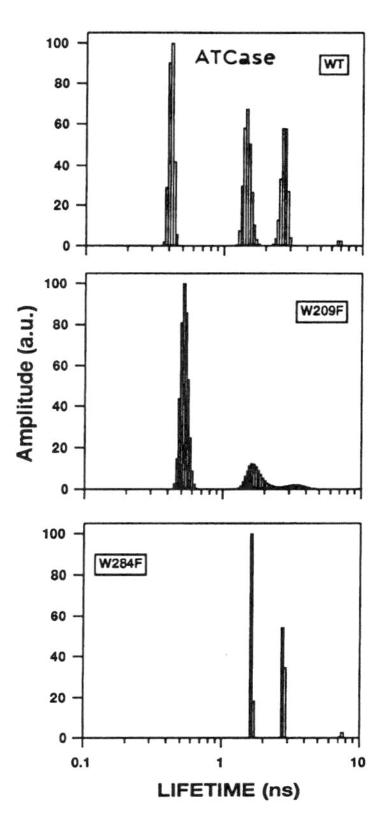

Figure 17.34. Lifetime distributions of wild-type ATCase (WT) and its two single-tryptophan mutants. Revised and reprinted, with permission, from Ref. 115, Copyright © 1992, American Chemical Society.

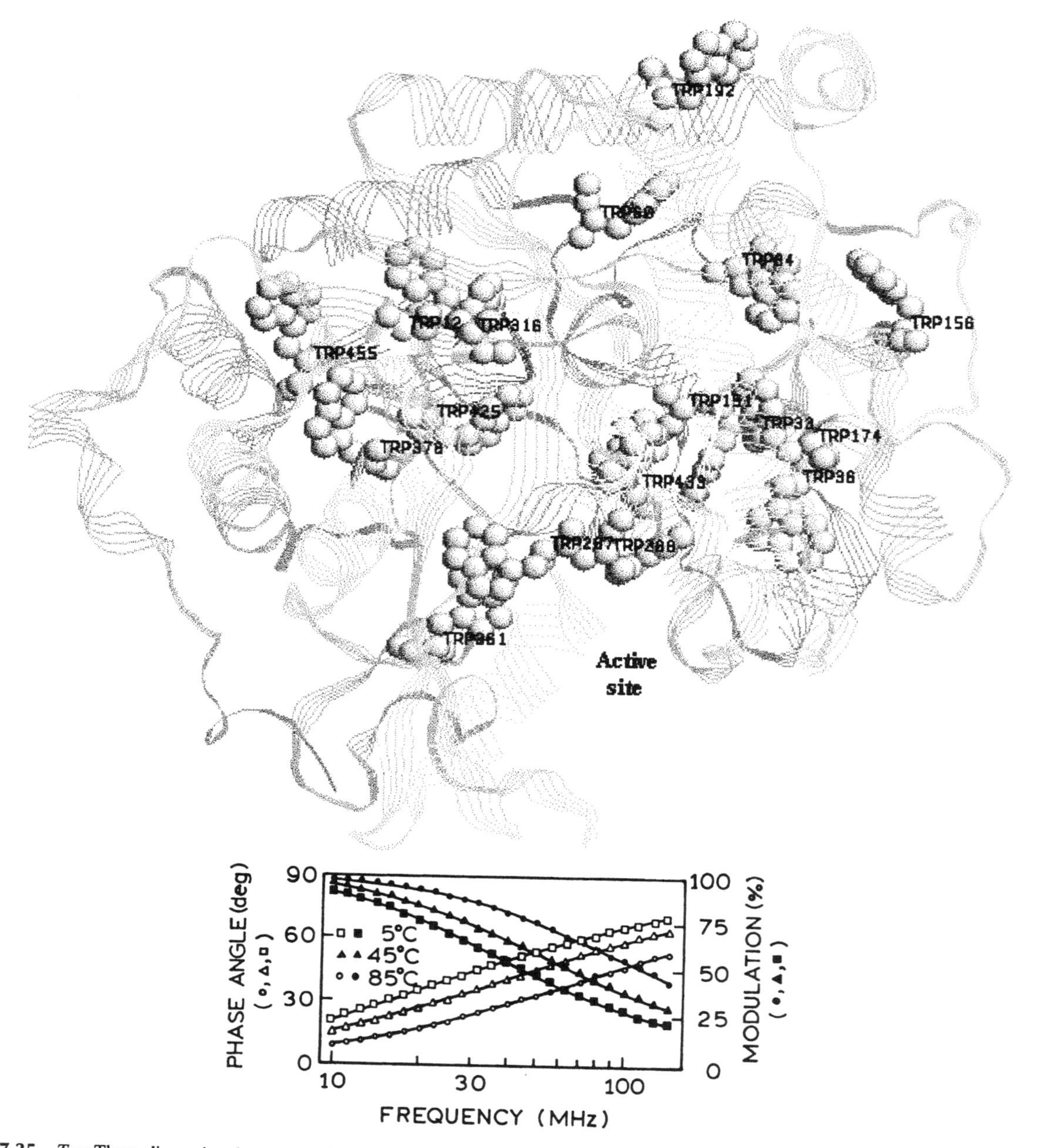

Figure 17.35. *Top*: Three-dimensional structure of one subunit of *S. solfataricus* β-glycosidase, showing the 17 tryptophan residues. Figure provided by Dr. S. D'Auria. *Bottom*: Frequency dependence of the phase shifts and demodulation factors of *S. solfataricus* β-glycosidase fluorescence emission at neutral pH at 5, 45, and 85 °C. Revised from Ref. 116.

were interpreted in terms of a bimodal lifetime distribution (Figure 17.36). As the temperature is increased, the two components shift progressively toward shorter lifetimes. These multiexponential decays are similar to those of native and denatured proteins.[7] However, in this case the β-glycosidase remains native and active to above 85 °C.

17.8.F. Heme Proteins Display Useful Intrinsic Fluorescence

For many years it was assumed that heme proteins were nonfluorescent. This was a reasonable assumption given that the intense Soret absorption band of the heme groups is expected to result in Förster distances (R_0) for trp-to-

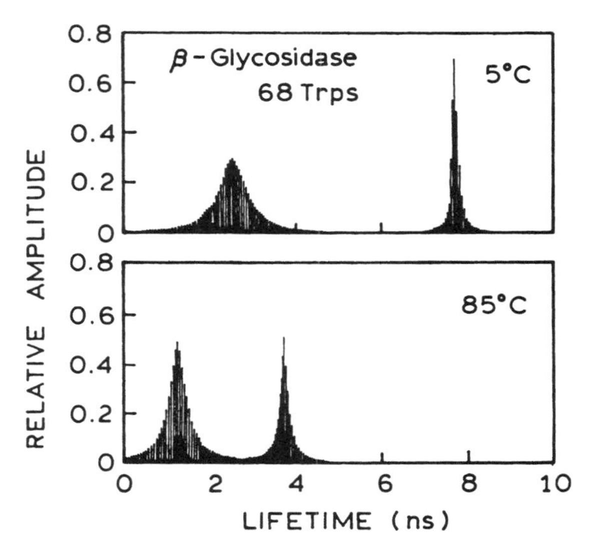

Figure 17.36. Tryptophanyl lifetime distributions of *S. solfataricus* β-glycosidase at 5 (*top*) and 85 °C (*bottom*). Revised from Ref. 116.

Figure 17.37. α-Carbon backbone of *Aplysia* myoglobin, showing the heme and the two-tryptophan residues. From Ref. 125.

heme transfer of about 32–34 Å. Since this distance is larger than the 20-Å radius of a myoglobin molecule, RET is expected to be nearly complete.

Detection of tryptophan emission from heme proteins is experimentally difficult. Suppose that RET quenching is 99% efficient, and suppose that your protein solution contains a 1% impurity of the apoprotein or some nonheme protein. The signal from the impurity will be approximately equal to that from the heme protein. For this reason, it was difficult to prove whether the observed weak emission was due to heme proteins or to impurities. In spite of these formidable problems, some groups were successful in detecting heme-protein emission from the steady-state data.[117–120] Subsequent time-resolved experiments confirmed that heme proteins do display weak intrinsic fluorescence. More importantly, the picosecond decay times found for heme proteins can be correlated with their structure.[121–127]

The structure of myoglobin is shown in Figure 17.37. There are two tryptophan residues at positions 7 and 14. Trp-14 is somewhat closer to the heme (15 Å, as compared to 20 Å for trp-7). Because of the short distance, RET is expected to be rapid. In fact, the decay times are too short to be measured by the usual TCSPC methods, and detection was accomplished with a streak camera. Figure 17.38 shows the time-resolved intensity decays obtained for tuna and sperm whale metmyoglobin. Tuna metmyoglobin has only one tryptophan residue at position 14, which decays with a 31-ps lifetime. Sperm whale metmyoglobin displays two decay times, 30 and 130 ps, assigned to its two residues trp-14 and trp-7, respectively. These decay times are consistent with the known crystal structure of myoglobin and distances from the tryptophan residue(s) to the heme group.[121,123]

The emission of heme proteins has also been studied using the FD method. Representative intensity decays for hemoglobin are shown in Figure 17.39. The FD data revealed both picosecond and nanosecond decay times. Long-lifetime nanosecond components were also found for myoglobin. These long-lived components are thought

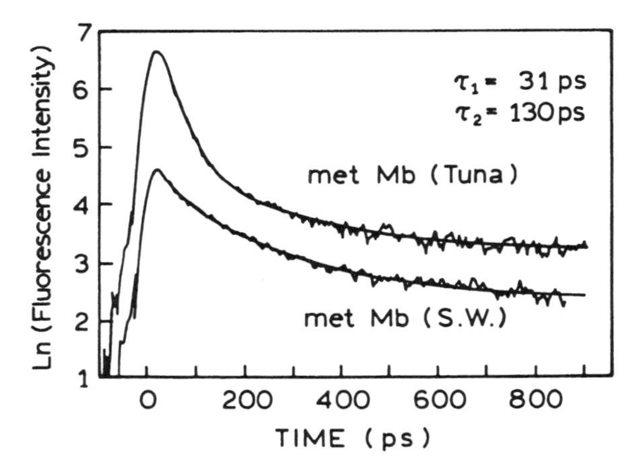

Figure 17.38. Time-resolved intensity decays of trp-7 in tuna metmyoglobin (metMb) and trp-7 and trp-14 in sperm whale (S.W.) metmyoglobin. Revised from Ref. 121.

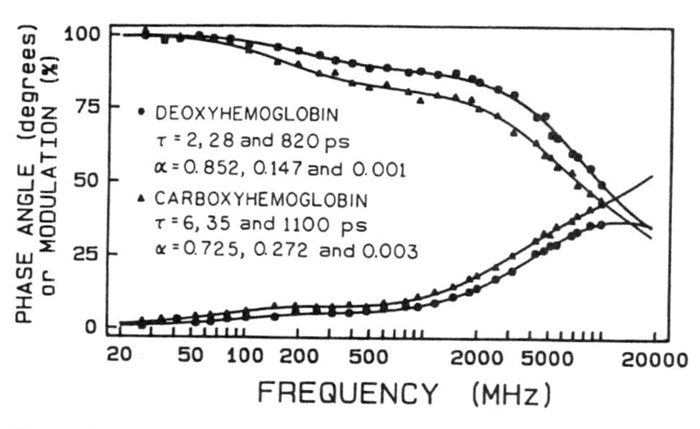

Figure 17.39. FD intensity decays of the intrinsic tryptophan fluorescence from deoxyhemoglobin (●) and carboxyhemoglobin (▲). From Ref. 127.

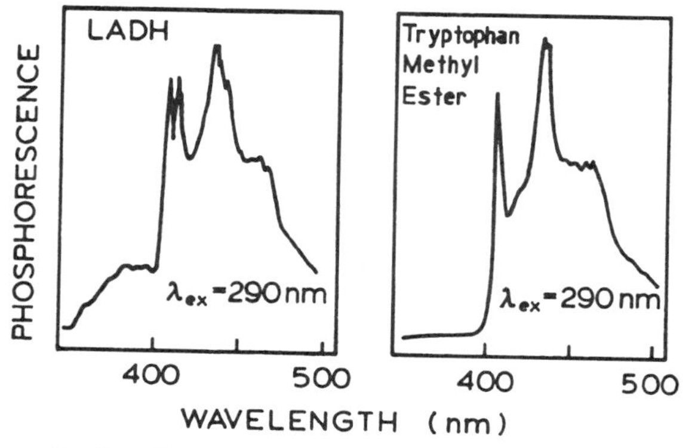

Figure 17.40. Phosphorescence spectra of horse liver alcohol dehydrogenase (LADH; *left*) and tryptophan methyl ester (*right*) in 1:1 ethylene glycol–buffer glasses at 77 K. Revised and reprinted, with permission, from Ref. 129, Copyright © 1970, American Chemical Society.

to be due to conformational isomers of the proteins in which the heme and the tryptophan residue are positioned so that energy transfer cannot occur. The important conclusion from these studies of heme proteins is that the time-resolved data can provide unambiguous information under conditions where the steady-state data are ambiguous.

17.9. PHOSPHORESCENCE OF PROTEINS

For the weakly fluorescent heme proteins, the application of time-resolved measurements allowed clarification of less definitive steady-state data. A similar situation occurred with protein phosphorescence. In this case, there were reports of phosphorescence from proteins at room temperature. However, the signals were weak and variable, so that there was uncertainty as to the existence of the phenomenon and its quantitative interpretation. The use of lifetime measurements allowed a more quantitative interpretation of protein phosphorescence in terms of exposure of the tryptophan residues to the aqueous phase.

Phosphorescence is typically observed at low temperatures in rigid solutions.[128–130] Low temperatures are needed to decrease the possibility of quenching by even very low concentrations of oxygen and other species and to slow the nonradiative decay pathways. It is important to minimize all nonradiative decay pathways because the radiative rates are slow, as the unquenched phosphorescence lifetime of tryptophan is 5.5 s.[130,131] Phosphorescence of tryptophan is structured and occurs at longer wavelengths than fluorescence, as seen at 77 K in a frozen glass (Figure 17.40, right). The sharp peak near 400 nm is the 0–0 vibronic transition. The shoulder below 400 nm in the phosphorescence spectrum of horse liver alcohol dehydrogenase (LADH; Figure 17.40, left) is due to tyrosine residues. A somewhat surprising result is the presence of two peaks near 400 nm in the phosphorescence spectrum of LADH, which contains two tryptophan residues, as compared to one peak in the spectrum of the tryptophan derivative. Recall the Trp repressor, in which trp-19 is buried and trp-99 is more exposed to the aqueous phase.[113] The phosphorescence spectra of the wild-type protein and the two single-tryptophan mutants are shown in Figure 17.41. Each of the single-tryptophan mutants shows a single 0–0 transition, whereas the wild-type protein shows two 0–0 peaks. The presence of two 400-nm peaks has been observed in the phosphorescence spectra of other two-tryptophan proteins.[132–137] It should be noted that the

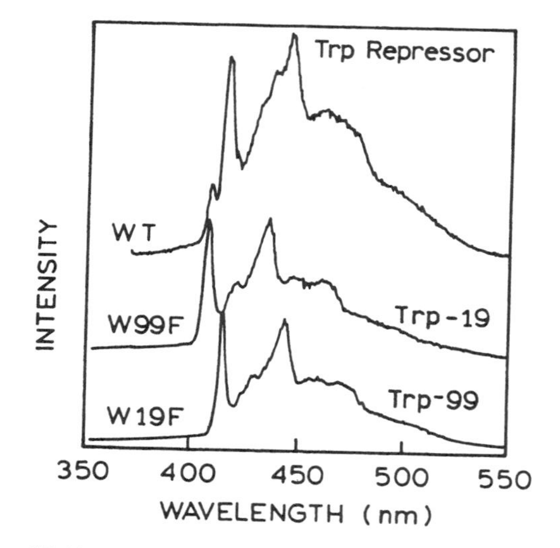

Figure 17.41. Phosphorescence spectra of the wild-type Trp repressor (WT) and its two single-tryptophan mutants at liquid-nitrogen temperature. The spectra are offset for visualization. Revised from Ref. 113.

maximum in the phosphorescence spectrum is expected to be at shorter wavelength for exposed tryptophan residues,[133–135] that is, the opposite trend to that for fluorescence. However, this is not the case for the Trp repressor, where trp-99 is thought to be the more exposed residue. The correlation between solvent exposure and phosphorescence emission maximum is unclear at the present time. It should be recalled that many phosphorescence spectra are recorded at low temperature, where solvent relaxation may not occur. Under these conditions, the exposed residues may display a blue-shifted emission, which could shift to longer wavelengths as the temperature increases.

Gradually, reports appeared about the observation of phosphorescence from proteins at room temperature.[138–140] Room-temperature phosphorescence of LADH was reported even in the presence of oxygen.[141] However, it turned out that the intense illumination of the protein needed to detect the weak phosphorescence apparently caused depletion of oxygen.[142] In spite of the initial confusion, these results pointed toward the possibility of room-temperature phosphorescence from proteins.

Room-temperature protein phosphorescence is now an accepted phenomenon and has been summarized in several reviews.[143–146] Confidence in the observation of room-temperature protein phosphorescence was increased by the use of the phosphorescence decay times, which could be measured more reliably than the intensities. Quantitation of the phosphorescence lifetime required development of a method to continuously remove all traces of dissolved oxygen. Room-temperature phosphorescence decay times were measured for a number of single-tryptophan proteins.[147] The decay times varied with exposure to the aqueous phase (Table 17.7). The longest phosphorescence lifetime found was for azurin, and melittin exhibited one of the shortest lifetimes. The phosphorescence lifetime of parvalbumin increased in the presence of calcium, which results in shielding of its tryptophan residues from water. Subtilisin and staphylococcal nuclease, two proteins with exposed tryptophan residues, were initially reported to be nonphosphorescent.[148] However, a later report indicated that these proteins were phosphorescent, but with phosphorescence lifetimes of less than 1 ms. This value is even less than the 1.2-ms lifetime of tryptophan in water at 20 °C,[149] suggesting some specific quenching interaction.

While this is still an active area of research, it is now accepted that protein phosphorescence is highly sensitive to protein structure. Initially, it was thought that the phosphorescence decays of single-tryptophan proteins were single-exponential decays. It now appears that multiexponential phosphorescence decays are common among many single-tryptophan proteins.[150–154] It is thought that the multiexponential decays reflect the presence of multiple protein conformations that exchange only slowly in solution. A number of amino acids acts as quenchers, including histidine, tyrosine, cysteine, and cystine.[155] The presence of nearby amino acid residues is thus likely to have a dramatic effect on the phosphorescence quantum yield and decay times of the tryptophan residues in proteins. A potentially convenient property of protein phosphorescence is that several of the tryptophan residues may be completely quenched at room temperature, so that the phosphorescence is due to just one or two shielded residues.[156]

Table 17.7. Correlation of Fluorescence Emission Maximum and Phosphorescence Lifetime in Single-Tryptophan Proteins[a]

Protein	Fluorescence maximum (nm)	Phosphorescence lifetime (ms)
Azurin	305	400
Parvalbumin (calcium present)	320	5
Ribonuclease T_1	325	14
Mellitin	340	<0.5
Monellin	345	<0.5
Parvalbumin (no calcium)	350	<0.5

[a]From Ref. 147.

At first glance, it appears that our understanding of protein phosphorescence is progressing faster than past developments in protein fluorescence. However, this is the result of the past efforts which clarified many aspects of protein fluorescence and have provided model proteins with previously characterized tryptophan residues.

17.10. PERSPECTIVES ON PROTEIN FLUORESCENCE

The technology for time-resolved fluorescence now provides picosecond time resolution with excellent signal-to-noise ratio, and one can readily obtain numerous intensity and anisotropy decays. Perhaps the most important goal is to avoid overinterpretation of the data. Subnanosecond motions can be detected in almost all proteins, but that does not demonstrate that the motions are significant for protein function. One should strive for the design of experiments that allow changes in protein dynamics to be linked to structure or function.

Resolution of multiexponential decays is now common, and the data can be readily used to obtain DAS. In most such cases, it is not clear that the assumption of wavelength-independent lifetimes is valid. Also, the apparent DAS do not necessarily correspond to discrete molecular species.

REFERENCES

1. Birch, D. J. S., and Imhof, R. E., 1991, Time-domain fluorescence spectroscopy using time-correlated single-photon counting, in *Topics in Fluorescence Spectroscopy, Volume 1, Techniques*, J. R. Lakowicz (ed.), Plenum Press, New York, pp. 1–95.
2. Small, E. W., 1991, Laser sources and microchannel plate detectors for pulse fluorometry, in *Topics in Fluorescence Spectroscopy, Volume 1, Techniques*, J. R. Lakowicz (ed.), Plenum Press, New York, pp. 97–182.
3. Gratton, E., and Limkeman, M., 1983, A continuously variable frequency cross-correlation phase fluorometer with picosecond resolution, *Biophys. J.* **44:**315–324.
4. Lakowicz, J. R., and Maliwal, B. P., 1985, Construction and performance of a variable-frequency phase-modulation fluorometer, *Biophys. Chem.* **21:**61–78.
5. Lakowicz, J. R., and Gryczynski, I., 1991, Frequency-domain fluorescence spectroscopy, in *Topics in Fluorescence Spectroscopy, Volume 1, Techniques*, J. R. Lakowicz (ed.), Plenum Press, New York, pp. 293–355.
6. Laczko, G., Gryczynski, I., Gryczynski, Z., Wiczk, W., Malak, M., and Lakowicz, J. R., 1990, A 10 GHz frequency-domain fluorometer, *Rev. Sci. Instrum.* **61:**2331–2337.
7. Grinvald, A., and Steinberg, I. Z., 1976, The fluorescence decay of tryptophan residues in native and denatured proteins, *Biochim. Biophys. Acta* **427:**663–678.
8. Beechem, J. M., and Brand, L., 1985, Time-resolved fluorescence of proteins, *Annu. Rev. Biochem.* **54:**43–71.
9. Rayner, D. M., and Szabo, A. G., 1978, Time resolved fluorescence of aqueous tryptophan, *Can. J. Chem.* **56:**743–745.
10. Szabo, A. G., and Rayner, D. M., 1980, Fluorescence decay of tryptophan conformers in aqueous solution, *J. Am. Chem. Soc.* **102:**554–563.
11. Gudgin, E., Lopez-Delgado, R., and Ware, W. R., 1981, The tryptophan fluorescence lifetime puzzle. A study of decay times in aqueous solution as a function of pH and buffer composition, *Can. J. Chem.* **59:**1037–1044.
12. McLaughlin, M. L., and Barkley, M. D., 1997, Time-resolved fluorescence of constrained tryptophan derivatives: Implications for protein fluorescence, *Methods Enzymol.* **278:**190–202.
13. Schiller, P. W., 1985, Application of fluorescence techniques in studies of peptide conformations and interactions, *Peptides* **7:**115–164.
14. Robbins, R. J., Fleming, G. R., Beddard, G. S., Robinson, G. W., Thistlethwaite, P. J., and Woolfe, G. J., 1980, Photophysics of aqueous tryptophan: pH and temperature effects, *J. Am. Chem. Soc.* **102:**6271–6280.
15. Eftink, M. R., Jia, Y., Hu, D., and Ghiron, C. A., 1995, Fluorescence studies with tryptophan analogues: Excited state interactions involving the side chain amino group, *J. Phys. Chem.* **99:**5713–5723.
16. Petrich, J. W., Chang, M. C., McDonald, D. B., and Fleming, G. R., 1983, On the origin of nonexponential fluorescence decay in tryptophan and its derivatives, *J. Am. Chem. Soc.* **105:**3824–3832.
17. Fleming, G. R., Morris, J. M., Robbins, R. J., Woolfe, G. J., Thistlewaite, P. J., and Robinson, G. W., 1978, Nonexponential fluorescence decay of aqueous tryptophan and two related peptides by picosecond spectroscopy, *Proc. Natl. Acad. Sci. U.S.A.* **75:**4652–4656.
18. Chen, Y., Liu, B., Yu, H.-T., and Barkley, M. D., 1996, The peptide bond quenches indole fluorescence, *J. Am. Chem. Soc.* **118:**9271–9278.
19. Chen, Y., Liu, B., and Barkley, M. D., 1995, Trifluoroethanol quenches indole fluorescence by excited-state proton transfer, *J. Am. Chem. Soc.* **117:**5608–5609.
20. Steiner, R. F., and Kirby, E. P., 1969, The interaction of the ground and excited states of indole derivatives with electron scavengers, *J. Phys. Chem.* **15:**4130–4135.
21. Froehlich, P. M., Gantt, D., and Paramasigamani, V., 1977, Fluorescence quenching of indoles by *N,N*-dimethylformamide, *Photochem. Photobiol.* **26:**639–642.
22. Ricci, R. W., and Nesta, J. M., 1976, Inter- and intramolecular quenching of indole fluorescence by carbonyl compounds, *J. Phys. Chem.* **80:**974–980.
23. Shopova, M., and Genov, N., 1983, Protonated form of histidine 238 quenches the fluorescence of tryptophan 241 in subtilisin novo, *Int. J. Peptide Protein Res.* **21:**475–478.
24. Shinitzky, M., and Rivnay, B., 1977, Degree of exposure of membrane proteins determined by fluorescence quenching, *Biochemistry* **16:**982–986.
25. Prutz, W. A., Siebert, F., Butler, J., Land, E. J., Menez, A., and Montenay-Garestier, T., 1982, Intramolecular radical transformations involving methionine, tryptophan and tyrosine, *Biochim. Biophys. Acta* **705:**139–149.
26. Sanyal, G., Kim, E., Thompson, F. M., and Brady, E. K., 1989, Static quenching of tryptophan fluorescence by oxidized dithiothreitol, *Biochem. Biophys. Res. Commun.* **165:**772–781.
27. Butler, J., Land, E. J., Prutz, W. A., and Swallow, A. J., 1982, Charge transfer between tryptophan and tyrosine in proteins, *Biochim. Biophys. Acta* **705:**150–162.
28. Eftink, M. R., 1991, Fluorescence quenching reactions, in *Biophysical and Biochemical Aspects of Fluorescence Spectroscopy*, T. G. Dewey (ed.), Plenum Press, New York, pp. 1–41.
29. Jameson, D. M., and Weber, G., 1981, Resolution of the pH-dependent heterogeneous fluorescence decay of tryptophan by phase and modulation measurements, *J. Phys. Chem.* **85:**953–958.
30. Chen, R. F., Knutson, J. R., Ziffer, H., and Porter, D., 1991, Fluorescence of tryptophan dipeptides: Correlations with the rotamer model, *Biochemistry* **30:**5184–5195.
31. Shizuka, H., Serizawa, M., Shimo, T., Saito, I., and Matsuura, T., 1988, Fluorescence-quenching mechanism of tryptophan. Remarkably efficient internal proton-induced quenching and charge-transfer quenching, *J. Am. Chem. Soc.* **110:**1930–1934.
32. Malak, H., Gryczynski, I., and Lakowicz, J. R., unpublished observations.
33. Ruggiero, A. J., Todd, D. C., and Fleming, G. R., 1990, Subpicosecond fluorescence anisotropy studies of tryptophan in water, *J. Am. Chem. Soc.* **112:**1003–1014.
34. Lakowicz, J. R., Laczko, G., and Gryczynski, I., 1987, Picosecond resolution of tyrosine fluorescence and anisotropy decays by 2-GHz frequency-domain fluorometry, *Biochemistry* **26:**82–90.
35. Döring, K., Konermann, L., Surrey, T., and Jähnig, F., 1995, A long lifetime component in the tryptophan fluorescence of some proteins, *Eur. Biophys. J.* **23:**423–432.
36. Ross, J. B. A., Laws, W. R., Rousslang, K. W., and Wyssbrod, H. R., 1992, Tyrosine fluorescence and phosphorescence from proteins and polypeptides, in *Topics in Fluorescence Spectroscopy, Volume 3, Biochemical Applications*, J. R. Lakowicz (ed.), Plenum Press, New York, pp. 1–63.
37. Laws, W. R., Ross, J. B. A., Wyssbrod, H. R., Beechem, J. M., Brand, L., and Sutherland, J. C., 1986, Time-resolved fluorescence and ^{1}H NMR studies of tyrosine and tyrosine analogues: Correlation of NMR-determined rotamer populations and fluorescence kinetics, *Biochemistry* **25:**599–607.

38. Ross, J. B. A., Laws, W. R., Sutherland, J. C., Buku, A., Katsoyannis, P. G., Schwartz, I. L., and Wyssbrod, H. R., 1986, Linked-function analysis of fluorescence decay functions: Resolution of side-chain rotamer populations of a single aromatic amino acid in small polypeptides, *Photochem. Photobiol.* **44**:365–370.
39. Contino, P. B., and Laws, W. R., 1991, Rotamer-specific fluorescence quenching in tyrosinamide: Dynamic and static interactions, *J. Fluoresc.* **1**(1):5–13.
40. Leroy, E., Lami, H., and Laustriat, G., 1971, Fluorescence lifetime and quantum yield of phenylalanine aqueous solutions. Temperature and concentration effects, *Photochem. Photobiol.* **13**:411–421.
41. Sudhakar, K., Wright, W. W., Williams, S. A., Phillips, C. M., and Vanderkooi, J. M., 1993, Phenylalanine fluorescence and phosphorescence used as a probe of conformation for cod parvalbumin, *J. Fluoresc.* **3**(2):57–64.
42. Eftink, M. R., Ghiron, C. A., Kautz, R. A., and Fox, R. O., 1989, Fluorescence lifetime studies with staphylococcal nuclease and its site-directed mutant: Test of the hypothesis that proline isomerism is the basis for nonexponential decays, *Biophys. J.* **55**:575–579.
43. Dahms, T. E. S., Willis, K. J., and Szabo, A. G., 1995, Conformational heterogeneity of tryptophan in a protein crystal, *J. Am. Chem. Soc.* **117**:2321–2326.
44. Tanaka, F., and Mataga, N., 1987, Fluorescence quenching dynamics of tryptophan in proteins, *Biophys. J.* **51**:487–495.
45. Alcala, J. R., 1994, The effect of harmonic conformational trajectories on protein fluorescence and lifetime distributions, *J. Chem. Phys.* **101**:4578–4584.
46. Eftink, M. R., and Ghiron, C. A., 1975, Dynamics of a protein matrix as revealed by fluorescence quenching, *Proc. Natl. Acad. Sci. U.S.A.* **72**:3290–3294.
47. Eftink, M. R., and Ghiron, C. A., 1977, Exposure of tryptophanyl residues and protein dynamics, *Biochemistry* **16**:5546–5551.
48. Eftink, M. R., and Hagaman, K. A., 1986, The viscosity dependence of the acrylamide quenching of the buried tryptophan residue in parvalbumin and ribonuclease T_1, *Biophys. Chem.* **25**:277–282.
49. Longworth, J. W., 1968, Excited state interactions in macromolecules, *Photochem. Photobiol.* **7**:587–594.
50. James, D. R., Demmer, D. R., Steer, R. P., and Verrall, R. E., 1985, Fluorescence lifetime quenching and anisotropy studies of ribonuclease T1, *Biochemistry* **24**:5517–5526.
51. Gryczynski, I., Eftink, M., and Lakowicz, J. R., 1988, Conformation heterogeneity in proteins as an origin of heterogeneous fluorescence decays, illustrated by native and denatured ribonuclease T_1, *Biochim. Biophys. Acta* **954**:244–252.
52. Eftink, M. R., and Ghiron, C. A., 1987, Frequency domain measurements of the fluorescence lifetime of ribonuclease T_1, *Biophys. J.* **52**:467–473.
53. Chen, L. X.-Q., Longworth, J. W., and Fleming, G. R., 1987, Picosecond time-resolved fluorescence of ribonuclease T_1, *Biophys. J.* **51**:865–873.
54. Gryczynski, I., and Lakowicz, J. R., unpublished observations.
55. Heinemann, U., and Saenger, W., 1982, Specific protein–nucleic acid recognition in ribonuclease T_1–2′-guanylic acid complex: An X-ray study, *Nature* **299**:27–31.
56. MacKerell, A. D., Nilsson, L., Rigler, R., Heinemann, U., and Saenger, W., 1989, Molecular dynamics simulations of ribonuclease T_1: Comparison of the free enzyme and the 2′ GMP–enzyme complex, *Proteins: Structure, Function and Genetics* **6**:20–31.
57. MacKerell, A. D., Rigler, R., Nilsson, L., Hahn, U., and Saenger, W., 1987, A time-resolved fluorescence, energetic and molecular dynamics study of ribonuclease T_1, *Biophys. Chem.* **26**:247–261.
58. Johnson, J. L., and Raushel, F. M., 1996, Influence of primary sequence transpositions on the folding pathways of ribonuclease T_1, *Biochemistry* **35**:10223–10233.
59. Data courtesy of Dr. John Lee, University of Georgia.
60. Liao, R., Wang, C.-K., and Cheung, H. C., 1992, Time-resolved tryptophan emission study of cardiac troponin I, *Biophys. J.* **63**:986–995.
61. Dittes, K., Gakamsky, D. M., Haran, G., Haas, E., Ojcius, D. M., Kourilsky, P., and Pecht, I., 1994, Picosecond fluorescence spectroscopy of a single-chain class I major histocompatibility complex encoded protein in its peptide loaded and unloaded states, *Immunol. Lett.* **40**:125–132.
62. Kouyama, I., Kinosita, K., and Ikegami, A., 1989, Correlation between internal motion and emission kinetics of tryptophan residues in proteins, *Eur. J. Biochem.* **182**:517–521.
63. Chen, L. X.-Q., Petrich, J. W., Fleming, G. R., and Perico, A., 1987, Picosecond fluorescence studies of polypeptide dynamics: Fluorescence anisotropies and lifetimes, *Chem. Phys. Lett.* **139**:55–61.
64. Rischel, C., Thyberg, P., Rigler, R., and Poulsen, F. M., 1996, Time-resolved fluorescence studies of the molten globule state of apomyoglobin, *J. Mol. Biol.* **257**:877–885.
65. Gakamsky, D. M., Haas, E., Robbins, P., Strominger, J. L., and Pecht, I., 1995, Selective steady-state and time-resolved fluorescence spectroscopy of an HLA-A2–peptide complex, *Immunol. Lett.* **44**:195–201.
66. Hansen, J. E., Rosenthal, S. J., and Fleming, G. R., 1992, Subpicosecond fluorescence depolarization studies of tryptophan and tryptophanyl residues of proteins, *J. Phys. Chem.* **96**:3034–3040.
67. Ross, J. B. A., Rousslang, K. W., and Brand, L., 1981, Time-resolved fluorescence and anisotropy decay of the tryptophan in adrenocorticotropin-(1-24), *Biochemistry* **20**:4361–4369.
68. Nordlund, T. M., and Podolski, D. A., 1983, Streak camera measurement of tryptophan and rhodamine motions with picosecond time resolution, *Photochem. Photobiol.* **38**:665–669.
69. Nordlund, T. M., Liu, X.-Y., and Sommer, J. H., 1986, Fluorescence polarization decay of tyrosine in lima bean trypsin inhibitor, *Proc. Natl. Acad. Sci. U.S.A.* **83**:8977–8981.
70. Gryczynski, I., and Lakowicz, J. R., unpublished observations.
71. Brochon, J-C., Wahl, P., and Auchet, J-C., 1974, Fluorescence time-resolved spectroscopy and fluorescence anisotropy decay of the *Staphylococcus aureus* endonuclease, *Eur. J. Biochem.* **41**:577–583.
72. Eftink, M. R., Gryczynski, I., Wiczk, W., Laczko, G., and Lakowicz, J. R., 1991, Effects of temperature on the fluorescence intensity and anisotropy decays of staphylococcal nuclease and the less stable nuclease-ConA-SG28 mutant, *Biochemistry* **30**:8945–8953.
73. Lakowicz, J. R., Gryczynski, I., Szmacinski, H., Cherek, H., and Joshi, N., 1991, Anisotropy decays of single tryptophan proteins measured by GHz frequency-domain fluorometry with collisional quenching, *Eur. Biophys. J.* **19**:125–140.
74. Lakowicz, J. R., Cherek, H., Gryczynski, I., Joshi, N., and Johnson, M. L., 1987, Enhanced resolution of fluorescence anisotropy decays by simultaneous analysis of progressively quenched samples, *Biophys. J.* **51**:755–768.
75. John, E., and Jähnig, F., 1988, Dynamics of melittin in water and membranes as determined by fluorescence anisotropy decay, *Biophys. J.* **54**:817–827.
76. Tran, C. D., and Beddard, G. S., 1985, Studies of the fluorescence from tryptophan in melittin, *Eur. Biophys. J.* **13**:59–64.
77. Lakowicz, J. R., Gryczynski, I., Cherek, H., and Laczko, G., 1991, Anisotropy decays of indole, melittin monomer and melittin tetramer by frequency-domain fluorometry and multi-wavelength global analysis, *Biophys. Chem.* **39**:241–251.

78. Georghiou, S., Thompson, M., and Mukhopadhyay, A. K., 1981, Melittin–phospholipid interaction. Evidence for melittin aggregation, *Biochim. Biophys. Acta* **642**:429–432.
79. Dufourcq, J., and Faucon, J.-F., 1977, Intrinsic fluorescence study of lipid–protein interactions in membrane models, *Biochim. Biophys. Acta* **467**:1–11.
80. Faucon, J. F., Dufourcq, J., and Lussan, C., 1979, The self-association of melittin and its binding to lipids, *FEBS Lett.* **102**:187–190.
81. Kaszycki, P., and Wasylewski, Z., 1990, Fluorescence-quenching-resolved spectra of melittin in lipid bilayers, *Biochim. Biophys. Acta* **1040**:337–345.
82. Papenhuijzen, J., and Visser, A. J. W. G., 1983, Simulation of convoluted and exact emission anisotropy decay profiles, *Biophys. Chem.* **17**:57–65.
83. Lakowicz, J. R., and Cherek, H., 1981, Proof of nanosecond timescale relaxation in apomyoglobin by phase fluorometry, *Biochem. Biophys. Res. Commun.* **99**:1173–1178.
84. Pierce, D. W., and Boxer, S. G., 1992, Dielectric relaxation in a protein matrix, *J. Phys. Chem.* **96**:5560–5566.
85. Lakowicz, J. R., and Cherek, H., 1980, Nanosecond dipolar relaxation in proteins observed by wavelength-resolved lifetimes of tryptophan fluorescence, *J. Biol. Chem.* **255**:831–834.
86. Demchenko, A. P., Gryczynski, I., Gryczynski, Z., Wiczk, W., Malak, H., and Fishman, M., 1993, Intramolecular dynamics in the environment of the single tryptophan residue in staphylococcal nuclease, *Biophys. Chem.* **48**:39–48.
87. Grinvald, A., and Steinberg, I. Z., 1974, Fast relaxation process in a protein revealed by the decay kinetics of tryptophan fluorescence, *Biochemistry* **25**:5170–5178.
88. Georghiou, S., Thompson, M., and Mukhopadhyay, A. K., 1982, Melittin–phospholipid interaction studied by employing the single tryptophan residue as an intrinsic fluorescent probe, *Biochim. Biophys. Acta* **688**:441–452.
89. Kamalov, V. F., Ladokhin, A. S., and Toleutaev, B. N., 1987, Nanosecond intramolecular dynamics of melittin, Translated from *Dokl. Akad. Nauk SSSR* **296**(3):180–182.
90. Szmacinski, H., Lakowicz, J. R., and Johnson, M., 1988, Time-resolved emission spectra of tryptophan and proteins from frequency-domain fluorescence spectroscopy, *Proc. SPIE* **909**:293–298.
91. Lakowicz, J. R., Szmacinski, H., and Gryczynski, I., 1988, Picosecond resolution of indole anisotropy decays and spectral relaxation by 2 GHz frequency-domain fluorometry, *Photochem. Photobiol.* **47**:31–41.
92. Vincent, M., and Gallay, J., 1995, Solvent relaxation around the excited state of indole: Analysis of fluorescence lifetime distributions and time-dependence spectral shifts, *J. Phys. Chem.* **99**:14931–14941.
93. Demchenko, A. P., 1992, Fluorescence and dynamics in proteins, in *Topics in Fluorescence Spectroscopy, Volume 3, Biochemical Applications*, J. R. Lakowicz (ed.), Plenum Press, New York, pp. 65–111.
94. Demchenko, A. P., 1994, Protein fluorescence, dynamics and function: Exploration of analogy between electronically excited and biocatalytic transition states, *Biochim. Biophys. Acta* **1209**:149–164.
95. Demchenko, A. P., Apell, H.-J., Sturmer, W., and Fedderson, B., 1993, Fluorescence spectroscopic studies on equilibrium dipole-relaxation dynamics of Na,K-ATPase, *Biophys. Chem.* **48**:135–147.
96. Knutson, J. R., Walbridge, D. G., and Brand, L., 1982, Decay-associated fluorescence spectra and the heterogeneous emission of alcohol dehydrogenase, *Biochemistry* **21**:4671–4679.
97. Privat, J.-P., Wahl, P., Auchet, J.-C., and Pain, R. H., 1980, Time resolved spectroscopy of tryptophyl fluorescence of yeast 3-phosphoglycerate kinase, *Biophys. Chem.* **11**:239–248.
98. Brochon, J. C., Wahl, P., Charlier, M., Maurizot, J. C., and Helene, C., 1977, Time resolved spectroscopy of the tryptophyl fluorescence of the *E. coli lac* repressor, *Biochem. Biophys. Res. Commun.* **79**:1261–1271.
99. Willis, K. J., and Szabo, A. G., 1992, Conformation of parathyroid hormone: Time-resolved fluorescence studies, *Biochemistry* **31**:8924–8931.
100. Kim, S.-J., Chowdhury, F. N., Stryjewski, W., Younathan, E. S., Russo, P. S., and Barkley, M. D., 1993, Time-resolved fluorescence of the single tryptophan of *Bacillus stearothermophilus* phosphofructokinase, *Biophys. J.* **65**:215–226.
101. Robbins, D. J., Deibel, M. R., and Barkley, M. D., 1985, Tryptophan fluorescence of terminal deoxynucleotidyl transferase: Effects of quenchers on time-resolved emission spectra, *Biochemistry* **24**:7250–7257.
102. Ross, J. A., Schmidt, C. J., and Brand, L., 1981, Time-resolved fluorescence of the two tryptophans in horse liver alcohol dehydrogenase, *Biochemistry* **20**:4369–4377.
103. She, M., Dong, W.-J., Umeda, P. K., and Cheung, H. C., 1997, Time-resolved fluorescence study of the single tryptophans of engineered skeletal muscle troponin C, *Biophys. J.* **73**:1042–1055.
104. Merrill, A. R., Steer, B. A., Prentice, G. A., Weller, M. J., and Szabo, A. G., 1997, Identification of a chameleon-like pH-sensitive segment within the colicin E1 channel domain that may serve as the pH-activated trigger for membrane bilayer association, *Biochemistry* **36**:6874–6884.
105. Hof, M., Fleming, G. R., and Fidler, V., 1996, Time-resolved fluorescence study of a calcium-induced conformational change in prothrombin fragment 1, *Proteins: Structure, Function and Genetics* **24**:485–494.
106. Lakowicz, J. R., and Balter, A., 1982, Resolution of initially excited and relaxed states of tryptophan fluorescence by differential wavelength deconvolution of time-resolved fluorescence decays, *Biophys. Chem.* **15**:353–360.
107. Verkshin, N., Vincent, M., and Gallay, J., 1992, Excited-state lifetime distributions of tryptophan fluorescence in polar solutions. Evidence for solvent exciplex formation, *Chem. Phys. Lett.* **199**:459–464.
108. Vincent, M., Gallay, J., and Demchenko, A. P., 1997, Dipolar relaxation around indole as evidenced by fluorescence lifetime distributions and time-dependence spectral shifts, *J. Fluoresc.* **7**(1):107S–110S.
109. Sopkova, J., Gallay, J., Vincent, M., Pancoska, P., and Lewit-Bentley, A., 1994, The dynamic behavior of annexin V as a function of calcium ion binding: A circular dichroism, UV absorption, and steady state and time-resolved fluorescence study, *Biochemistry* **33**:4490–4499.
110. Rouviere, N., Vincent, M., Craescu, C. T., and Gallay, J., 1997, Immunosuppressor binding to the immunophilin FKBP59 affects the local structural dynamics of a surface β-strand: Time-resolved fluorescence study, *Biochemistry* **36**:7339–7352.
111. Suwaliyan, A., and Klein, U. K. A., 1989, Picosecond study of solute–solvent interaction of the excited state of indole, *Chem. Phys. Lett.* **159**:244–250.
112. Helms, M. K., Petersen, C. E., Bhagavan, N. V., and Jameson, D. M., 1997, Time-resolved fluorescence studies on site-directed mutants of human serum albumin, *FEBS Lett.* **408**:67–70.
113. Eftink, M. R., Ramsay, G. D., Burns, L., Maki, A. H., Mann, C. J., Matthews, C. R., and Ghiron, C. A., 1993, Luminescence studies

of trp repressor and its single-tryptophan mutants, *Biochemistry* **32**:9189–9198.

114. Royer, C. A., 1992, Investigation of the structural determinants of the intrinsic fluorescence emission of the trp repressor using single tryptophan mutants, *Biophys. J.* **63**:741–750.
115. Fetler, L., Tauc, P., Herve, G., Ladjimi, M. M., and Brochon, J.-C., 1992, The tryptophan residues of aspartate transcarbamylase: Site-directed mutagenesis and time-resolved fluorescence spectroscopy, *Biochemistry* **31**:12504–12513.
116. Bismuto, E., Irace, G., D'Auria, S., Rossi, M., and Nucci, R., 1997, Multitryptophan-fluorescence emission decay of β-glycosidase from the extremely thermophilic archaeon *Sulfolobus solfataricus*, *Eur. J. Biochem.* **244**:53–58.
117. Hirsch, R. E., and Nagel, R. L., 1981, Conformational studies of hemoglobins using intrinsic fluorescence measurements, *J. Biol. Chem.* **256**:1080–1083.
118. Hirsch, R. E., and Peisach, J., 1986, A comparison of the intrinsic fluorescence of red kangaroo, horse and sperm whale metmyoglobins, *Biochim. Biophys. Acta* **872**:147–153.
119. Hirsch, R. E., and Noble, R. W., 1987, Intrinsic fluorescence of carp hemoglobin: A study of the R → T transition, *Biochim. Biophys. Acta* **914**:213–219.
120. Sebban, P., Coppey, M., Alpert, B., Lindqvist, L., and Jameson, D. M., 1980, Fluorescence properties of porphyrin-globin from human hemoglobin, *Photochem. Photobiol.* **32**:727–731.
121. Hochstrasser, R. M., and Negus, D. K., 1984, Picosecond fluorescence decay of tryptophans in myoglobin, *Proc. Natl. Acad. Sci. U.S.A.* **81**:4399–4403.
122. Bismuto, E., Irace, G., and Gratton, E., 1989, Multiple conformational states in myoglobin revealed by frequency domain fluorometry, *Biochemistry* **28**:1508–1512.
123. Willis, K. J., Szabo, A. G., Zuker, M., Ridgeway, J. M., and Alpert, B., 1990, Fluorescence decay kinetics of the tryptophyl residues of myoglobin: Effect of heme ligation and evidence for discrete lifetime components, *Biochemistry* **29**:5270–5275.
124. Gryczynski, Z., Lubkowski, J., and Bucci, E., 1995, Heme–protein interactions in horse heart myoglobin at neutral pH and exposed to acid investigated by time-resolved fluorescence in the pico- to nanosecond time range, *J. Biol. Chem.* **270**:19232–19237.
125. Janes, S. M., Holtom, G., Ascenzi, P., Brundri, M., and Hochstrasser, R. M., 1987, Fluorescence and energy transfer of tryptophans in *Aplysia* myoglobin, *Biophys. J.* **51**:653–660.
126. Szabo, A. G., Krajcarski, D., Zuker, M., and Alpert, B., 1984, Conformational heterogeneity in hemoglobin as determined by picosecond fluorescence decay measurements of the tryptophan residues, *Chem. Phys. Lett.* **108**:145–149.
127. Bucci, E., Gryczynski, Z., Fronticelli, C., Gryczynski, I., and Lakowicz, J. R., 1992, Fluorescence intensity and anisotropy decays of the intrinsic tryptophan emission of hemoglobin measured with a 10-GHz fluorometer using front-face geometry on a free liquid surface, *J. Fluoresc.* **2**(1):29–36.
128. Steiner, R. F., and Kolinski, R., 1968, The phosphorescence of oligopeptides containing tryptophan and tyrosine, *Biochemistry* **7**:1014–1018.
129. Purkey, R. M., and Galley, W. C., 1970, Phosphorescence studies of environmental heterogeneity for tryptophyl residues in proteins, *Biochemistry* **9**:3569–3575.
130. Strambini, G. B., and Gonnelli, M., 1985, The indole nucleus triplet-state lifetime and its dependence on solvent microviscosity, *Chem. Phys. Lett.* **115**:196–200.
131. Permyakov, E. A., 1993, *Luminescent Spectroscopy of Proteins*, CRC Press, Boca Raton, Florida.
132. Strambini, G. B., and Gonnelli, M., 1990, Tryptophan luminescence from liver alcohol dehydrogenase in its complexes with coenzyme. A comparative study of protein conformation in solution, *Biochemistry* **29**:196–203.
133. Strambini, G. B., and Gabellieri, E., 1989, Phosphorescence properties and protein structure surrounding tryptophan residues in yeast, pig, and rabbit glyceraldehyde-3-phosphate dehydrogenase, *Biochemistry* **28**:160–166.
134. Cioni, P., Puntoni, A., and Strambini, G. B., 1993, Tryptophan phosphorescence as a monitor of the solution structure of phosphoglycerate kinase from yeast, *Biophys. Chem.* **46**:47–55.
135. Burns, L. E., Maki, A. H., Spotts, R., and Matthews, K. S., 1993, Characterization of the two tryptophan residues of the lactose repressor from *Escherichia coli* by phosphorescence and optical detection of magnetic resonance, *Biochemistry* **32**:12821–12829.
136. Gabellieri, E., Rahuel-Clemont, S., Branlant, G., and Strambini, G. B., 1996, Effects of NAD^+ binding on the luminescence of tryptophans 84 and 310 of glyceraldehyde-3-phosphate dehydrogenase from *Bacillus stearothermophilus*, *Biochemistry* **35**:12549–12559.
137. Strambini, G. B., Cioni, P., and Cook, P. F., 1996, Tryptophan luminescence as a probe of enzyme conformation along the O-acetylserine sulfhydrylase reaction pathway, *Biochemistry* **35**:8392–8400.
138. Kai, Y., and Imakubo, K., 1979, Temperature dependence of the phosphorescence lifetimes of heterogeneous tryptophan residues in globular proteins between 293 and 77 K, *Photochem. Photobiol.* **29**:261–265.
139. Domanus, J., Strambini, G. B., and Galley, W. C., 1980, Heterogeneity in the thermally-induced quenching of the phosphorescence of multi-tryptophan proteins, *Photochem. Photobiol.* **31**:15–21.
140. Barboy, N., and Feitelson, J., 1985, Quenching of tryptophan phosphorescence in alcohol dehydrogenase from horse liver and its temperature dependence, *Photochem. Photobiol.* **41**:9–13.
141. Saviotti, M. L., and Galley, W. C., 1974, Room temperature phosphorescence and the dynamic aspects of protein structure, *Proc. Natl. Acad. Sci. U.S.A.* **71**:4154–4158.
142. Strambini, G. B., 1983, Singular oxygen effects on the room-temperature phosphorescence of alcohol dehydrogenase from horse liver, *Biophys. J.* **43**:127–130.
143. Papp, S., and Vanderkooi, J. M., 1989, Tryptophan phosphorescence at room temperature as a tool to study protein structure and dynamics, *Photochem. Photobiol.* **49**:775–784.
144. Vanderkooi, J. M., and Berger, J. W., 1989, Excited triplet states used to study biological macromolecules at room temperature, *Biochim. Biophys. Acta* **976**:1–27.
145. Subramaniam, V., Gafni, A., and Steel, D. G., 1996, Time-resolved tryptophan phosphorescence spectroscopy: A sensitive probe of protein folding and structure, *IEEE J.* **2**:1107–1114.
146. Schauerte, J. A., Steel, D. G., and Gafni, A., 1997, Time-resolved room temperature tryptophan phosphorescence in proteins, *Methods Enzymol.* **278**:49–71.
147. Vanderkooi, J. M., Calhoun, D. B., and Englander, S. W., 1987, On the prevalence of room-temperature protein phosphorescence, *Science* **230**:568–569.
148. Strambini, G. B., and Gabellieri, E., 1990, Temperature dependence of tryptophan phosphorescence in proteins, *Photochem. Photobiol.* **51**:643–648.
149. Strambini, G. B., and Gonnelli, M., 1995, Tryptophan phosphorescence in fluid solution, *J. Am. Chem. Soc.* **117**:7646–7651.
150. Gonnelli, M., Puntoni, A., and Strambini, G. B., 1992, Tryptophan phosphorescence of ribonuclease T_1 as a probe of protein flexibility, *J. Fluoresc.* **2**(3):157–165.

151. Cioni, P., Gabellieri, E., Gonnelli, M., and Strambini, G. B., 1994, Heterogeneity of protein conformation in solution from the lifetime of tryptophan phosphorescence, *Biophys. Chem.* **52**:25–34.
152. Schlyer, B. D., Schauerte, J. A., Steel, D. G., and Gafni, A., 1994, Time-resolved room temperature protein phosphorescence: Nonexponential decay from single emitting tryptophans, *Biophys. J.* **67**:1192–1202.
153. Hansen, J. E., Steel, D. G., and Gafni, A., 1996, Detection of a pH-dependent conformational change in azurin by time-resolved phosphorescence, *Biophys. J.* **71**:2138–2143.
154. Subramaniam, V., Bergenhem, N. C. H., Gafni, A., and Steel, D. G., 1995, Phosphorescence reveals a continued slow annealing of the protein core following reactivation of *Escherichia coli* alkaline phosphatase, *Biochemistry* **34**:1133–1136.
155. Gonnelli, M., and Strambini, G. B., 1995, Phosphorescence lifetime of tryptophan in proteins, *Biochemistry* **34**:13847–13857.
156. Strambini, G. B., Cioni, P., and Felicioli, R. A., 1987, Characterization of tryptophan environments in glutamate dehydrogenases from temperature-dependent phosphorescence, *Biochemistry* **26**:4968–4975.
157. Tanaka, F., Tamai, N., Mataga, N., Tonomura, B., and Hiromi, K., 1994, Analysis of internal motion of single tryptophan in *Streptomyces* subtilisin inhibitor from its picosecond time-resolved fluorescence, *Biophys. J.* **67**:874–880.
158. Tanaka, F., and Mataga, N., 1992, Non-exponential decay of fluorescence of tryptophan and its motion in proteins, in *Dynamics and Mechanisms of Photoinduced Electron Transfer and Related Phenomena*, N. Mataga, T. Okada, and H. Masuhara (eds.), North-Holland, Amsterdam, pp. 501–512.
159. Eisenberg, D., and Crothers, D., 1979, *Physical Chemistry with Applications to the Life Sciences*, see p. 240.

PROBLEMS

17.1. *Rotational Diffusion of Proteins*: Use the data in Table 17.4 to calculate the activation energy for rotational diffusion of RNase T_1 in aqueous solution. Indicate how the data reveal the presence of protein dynamics at higher temperature. Also, predict the steady-state anisotropy of RNase T_1 at each temperature from the time-resolved data.

17.2. *Rotational Freedom of Tryptophan Residues in Proteins*: Use the data in Table 17.5 to calculate the cone angle for tryptophan motion, independent of overall rotational diffusion. Assume that the fundamental anisotropy is 0.31. Perform the calculation for RNase T_1, staphylococcal nuclease, monellin, melittin monomer, and melittin tetramer.

17.3. *Apparent Time-Zero Anisotropies of Proteins*: The time-zero anisotropies for RNase T_1 are different in Tables 17.4 and 17.5. Describe possible reasons for these differences.

17.4. *Calculation of Decay-Associated Spectra*: The *lac* repressor from *E. coli* is a tetrameric protein with two tryptophan residues per subunit. The intensity decays and emission spectra are shown in Figure 17.42, and the intensity decays are given in Table 17.8.

A. Interpret the intensity decays (Figure 17.42, left) in terms of the mean lifetime at each wavelength.

B. Calculate the DAS and interpret the results.

C. How could you confirm the assignment of the DAS to each tryptophan residue?

Table 17.8. Intensity Decays of the *E. coli lac* repressor[a]

λ_{em} (nm)	α_1 (τ_1 = 3.8 ns)	a_2 (τ_2 = 9.8 ns)
315	0.72	0.28
320	0.64	0.36
330	0.48	0.52
340	0.35	0.65
350	0.28	0.72
360	0.18	0.82
370	0.08	0.92
380	—	1.00

[a]From Ref. 98.

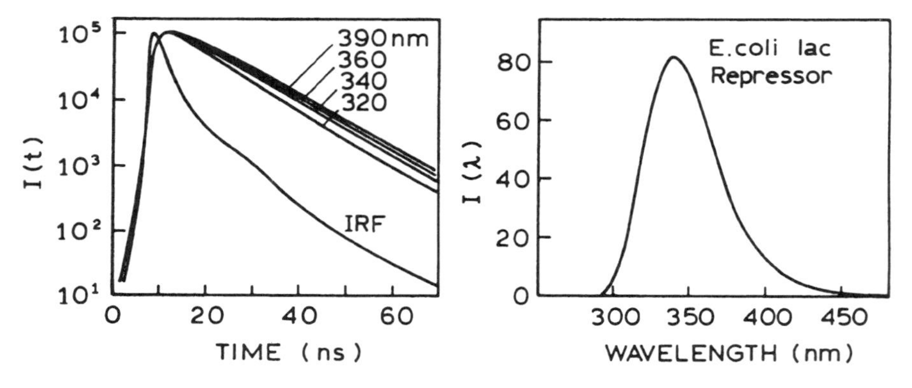

Figure 17.42. Intensity decays at the indicated emission wavelengths (*left*) and emission spectrum (*right*) of the *E. coli lac* repressor. IRF, Instrument response function. Revised from Ref. 98.

Excited-State Reactions

18

In the preceding chapters we saw many examples of excited-state reactions. By an excited-state reaction we mean a molecular process which changes the structure of the excited-state fluorophore, and which occurs subsequent to excitation. Such reactions occur because light absorption frequently changes the electron distribution within a fluorophore, which in turn changes its chemical or physical properties. The best-known example of an excited-state reaction is that of phenol, which in neutral solution can lose the phenolic proton in the excited state. Deprotonation occurs more readily in the excited state because the electrons on the phenolic hydroxyl groups are shifted into the phenol ring, making this hydroxyl group more acidic.

Excited-state reactions are not restricted to ionization. Many dynamic processes that affect fluorescence can be interpreted in terms of excited-state reactions. These processes include spectral relaxation, resonance energy transfer, and excimer formation (Table 18.1). Excited-state reactions occur in biochemical systems, such as the energy transfer between fluorophores in phycobiliproteins and during photosynthesis. In spite of the diversity of phenomena, excited-state processes display characteristic time-dependent decays which can be unambiguously assigned to the presence of an excited-state process. In this chapter we provide an overview of the spectral characteristics of excited-state reactions.

Table 18.1. Phenomena That Display the Characteristics of an Excited-State Reaction

Phenomenon	Examples
Proton loss	Phenols, naphthols
Proton gain	Acridine
Proton tautomerization	7-Azaindole, 3-hydroxyflavone
Resonance energy transfer	Tryptophan-to-dansyl energy transfer
Solvent relaxation	ANS and TNS in viscous solvents
TICT and ICT states	*p*-Dimethylaminobenzonitrile
Excimer formation	Pyrene, 2-phenylindole
Exciplex formation	Anthracene, dimethylaniline

18.1. EXAMPLES OF EXCITED-STATE REACTIONS

While a variety of phenomena can be described as excited-state reactions, in this chapter we will emphasize chemical reactions. Perhaps the predominant type of excited-state reaction is the loss or gain of protons (Table 18.2).[1–6] Whether a fluorophore loses or gains a proton in the excited-state reaction is determined by the direction of the change in pK_a in the excited state. If the pK_a decreases (pK_a^* < pK_a, where the asterisk denotes the excited state), then the fluorophore will tend to lose a proton in the excited state. If the pK_a increases (pK_a^* > pK_a), then the fluorophore may pick up a proton in the excited state. The best-known examples are phenols and acridines. In gen-

Table 18.2. Examples of Excited-State Ionization Reactions

Reaction	Compound	pK_a	pK_a^*	D or P[a]
$ROH \rightleftharpoons RO^- + H^+$	Phenol	10.0	4.1	D
	1-Naphthol	9.2	2.0	D
	2-Naphthol	9.5	2.8	D
	HPTS[b]	7.3	1.0	D
$RNH_2 \rightleftharpoons RNH^- + H^+$	2-Naphthylamine	7.14	12.2	D
$RNH_3^+ \rightleftharpoons RNH_2 + H^+$	2-Naphthylamine (protonated)	4.1	−1.5	D
$ArN + H^+ \rightleftharpoons ArNH^+$	Acridine	5.1	10.6	P
$RCO_2^- + H^+ \rightleftharpoons RCO_2H$	Benzoic acid	4.2	9.5	P
	1-Naphthoic acid	3.7	7.7	P
	Anthracene-9-carboxylic acid	3.7	6.9	P

[a]Deprotonation (D) or protonation (P).
[b]HPTS, 8-Hydroxypyrene-1,3,6,8,10-trisulfonate.

eral, phenols undergo excited-state deprotonation, and acridines undergo excited-state protonation. Electron donors such as –OH, –SH, and $-NH_2$ have a lone pair of electrons, and these electrons tend to become more conjugated to the aromatic ring system in the excited state, resulting in $pK_a^* < pK_a$. Electron acceptors such as $-CO_2^-$ and CO_2H have vacant π orbitals into which electrons can be transferred in the excited state. This increased electron density results in weaker dissociation in the excited state, $pK_a^* > pK_a$. Representative ground-state and excited-state pK_as are given in Table 18.2.

Other examples of excited-state processes include energy transfer, solvent relaxation, and formation of excited-state complexes (Table 18.1). The shared characteristic of all these phenomena is the involvement of an excited-state species which is not excited directly but is formed from the initially excited species. Depending on the magnitude of the spectral shift due to the excited-state process, it may or may not be possible to observe the reaction product without a spectral contribution from the initially excited species.

18.1.A. Excited-State Ionization of Naphthol

Perhaps the most widely studied excited-state reaction is the ionization of aromatic alcohols.[7–19] In the case of 2-naphthol, the pK_a decreases from 9.5 in the ground state to 2.8 in the excited state.[18] In acid solution, the emission is from naphthol, with an emission maximum of 357 nm (Figure 18.1). In basic solution, the emission is from the naphtholate anion and is centered near 409 nm. At intermediate pH values, emission from both species is observed, as shown by the dashed curve in Figure 18.1. The excited-state dissociation of 2-naphthol can be either reversible or irreversible, depending on pH. At pH values near 3 the reaction is reversible, but at pH values above 6 the reaction is irreversible.[14,15] Hence, this system illustrates the characteristics of both reversible and irreversible excited-state reactions.

The spectra shown in Figure 18.1 illustrate another feature of an excited-state process, namely, the appearance of the reaction product under conditions where no product is present in the ground state. At low or high pH, the ground-state fluorophore is present in only one ionization state, and only the emission from this state is observed. The absorption spectra of these low- and high-pH solutions are characteristic of naphthol and naphtholate, respectively (Figure 18.2). However, at pH 3, about 50% of the total emission is from each species. Because the ground state pK_a of 2-naphthol is 9.2, only the un-ionized form is present at pH 3, and the absorption spectrum is that of un-ionized naphthol. Whereas only un-ionized naphthol is present in the ground state at pH 3, the emission spectrum at pH 3 shows emission from both naphthol and naphtholate. At pH 3 the naphtholate emission results from molecules which have undergone dissociation during the duration of the excited state. The invariance of the absorption spectrum under conditions where the emission spectrum of a reacted state is observed is a characteristic feature of an excited-state reaction.

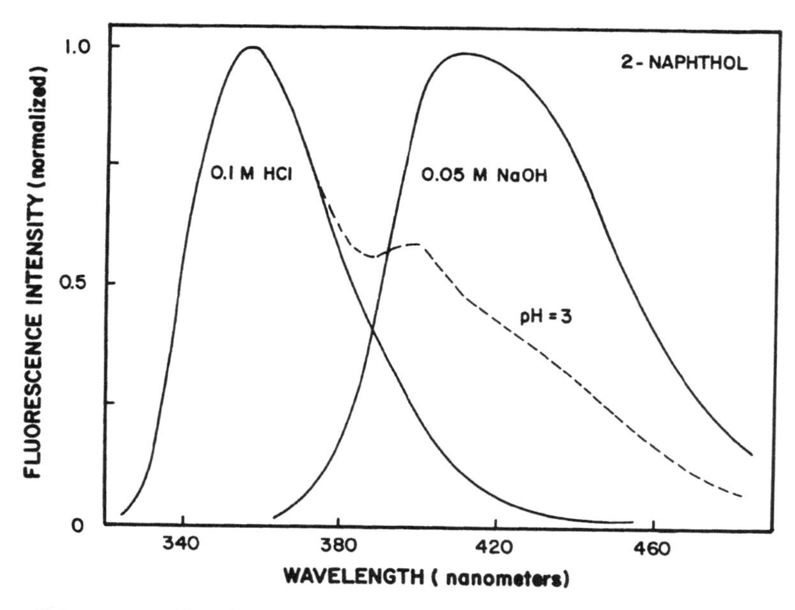

Figure 18.1. Steady-state emission spectra of 2-naphthol at 24 °C in 0.1*M* HCl, 0.05*M* NaOH, and water at pH 3.0 (– – –). Reprinted from Ref. 27, Copyright © 1982, with permission from Elsevier Science.

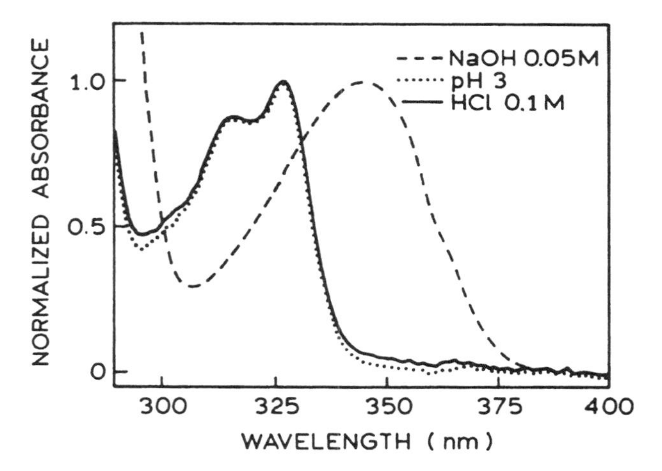

Figure 18.2. Absorption spectra of 2-naphthol in 0.1*M* HCl (——), 0.05*M* NaOH (– – –), and water at pH 3 (· · ·). From Ref. 20.

This characteristic of an excited-state reaction is shown in Figure 18.3. Suppose that one measured the fraction of the signal, absorption or emission, from un-ionized 2-naphthol. The absorption measurement would show a decrease in this fraction at pH 9.2, which is near the ground-state pK_a. The fluorescence measurements would show a change near pH 2, near the pK_a^* value of the excited state. The difference is due to ionization that occurs during the excited-state lifetime. Depending on the fluorophore and solvent, the excited-state reaction can be complete or only partially complete during the excited-state lifetime. It is also important to note that the extent of dissociation depends on the buffer concentration. For instance, suppose that 2-naphthol is dissolved in buffer solutions of the same pH but with increasing concentrations of the buffer. The extent of ionization will increase with increasing buffer concentration owing to reaction of excited 2-naphthol with the weak-base form of the buffer.

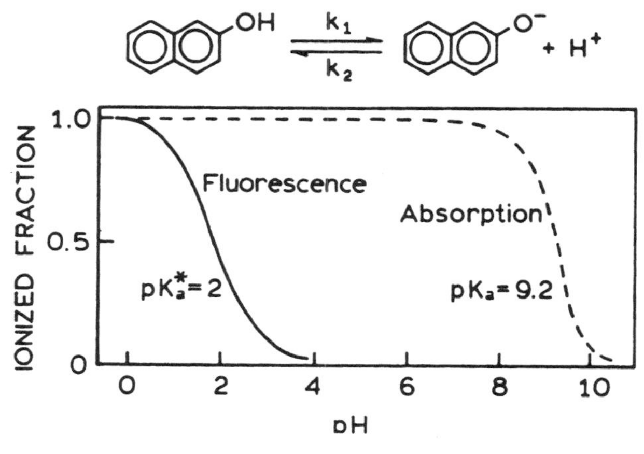

Figure 18.3. Comparison of ground-state (– – –) and excited-state (——) ionization of 2-naphthol.

The spectral shifts that occur upon dissociation in the excited state can be used to calculate the change in pK_a that occurs upon excitation. This is known as the Förster cycle[21–24] (Figure 18.4). The energies of the ground and excited states depend on ionization. Because the dissociation constants are small, the dissociated form of the fluorophore is shown at higher energy. If the pK_a value is lower in the excited state ($pK_a^* < pK_a$), then there is a smaller increase in energy upon dissociation of the excited state ($\Delta H^* < \Delta H$). If one assumes that the entropy change for dissociation is the same for the ground and excited states, then the difference in energy between the ground and excited states of AH and A^- can be related to the change in pK_a values by

$$\Delta pK_a = pK_a - pK_a^* = \frac{E_{HA} - E_{A^-}}{2.3\,RT} \quad [18.1]$$

where R is the gas constant and T is the temperature in kelvins. The energies of the protonated form (E_{HA}) and of the dissociated form (E_{A^-}) are usually estimated from the average of the absorption ($\bar{\nu}_A$) and emission ($\bar{\nu}_F$) maxima of each species,

$$E_i = Nhc\frac{\bar{\nu}_A + \bar{\nu}_F}{2} \quad [18.2]$$

where $\bar{\nu}_A$ and $\bar{\nu}_F$ are in wavenumbers, h is the Planck constant, N is Avogadro's number, and c is the speed of light.

While the Förster cycle calculation is useful for understanding the excited-state ionization of fluorophores, the calculated values of ΔpK_a should be used with caution. This is because at 300 nm an error of 4 nm corresponds to a shift of one pK_a unit. Hence, the Förster cycle is useful

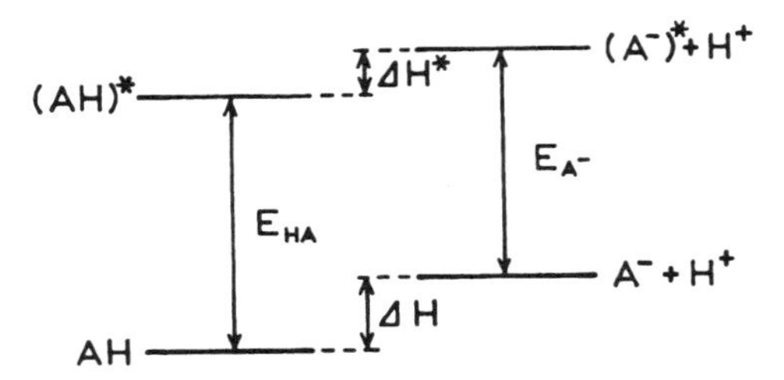

Figure 18.4. Electronic energy levels of an acid AH and its conjugate base A^- in the ground and excited states (the Förster cycle). Reprinted, with permission, from the *Journal of Chemical Education*, Vol. 69, No. 3, 1992, pp. 247–249; Copyright © 1992, Division of Chemical Education, Inc.

for determining the direction of the change in pK_a but is less reliable in estimating the precise value of pK_a^*.

18.2. REVERSIBLE TWO-STATE MODEL

The features of an excited-state reaction are illustrated by the Jabłoński diagram for the reversible two-state model (Figure 18.5). The terms and subscripts F and R refer to the initially excited and the reacted states, respectively. The decay rates of these species are given by $\gamma_F = \Gamma_F + k_1 + k_{nr}^F$ and $\gamma_R = \Gamma_R + k_2 + k_{nr}^R$, where Γ_F and Γ_R are the radiative decay rates and k_{nr}^F and k_{nr}^R are the rates of nonradiative decay. The rate of the forward reaction is given by k_1, and the rate of the reverse reaction by k_2. For simplicity, we have not included the rates of intersystem crossing to the triplet state, nor the rates of solvent relaxation. Depending upon the specific process under consideration, k_1 and k_2 can each be more complex than a simple rate constant. For example, in the case of a reversible loss of a proton, the reverse rate would be $k_2 = k_2'[H^+]$. In the case of exciplex formation, the forward rate would include the concentration of the species (Q) forming the exciplex, $k_1 = k_1'[Q]$. If the reaction is irreversible, $k_2 = 0$. It is simpler to interpret the decay kinetics for irreversible reactions than for reversible reactions.

18.2.A. Steady-State Fluorescence of a Two-State Reaction

Prior to discussing the analysis of excited-state reactions by TD and FD fluorometry, it is useful to describe the spectral properties observed using steady-state methods. Assume that unique wavelengths can be selected where emission occurs only from the unreacted (F) or the reacted

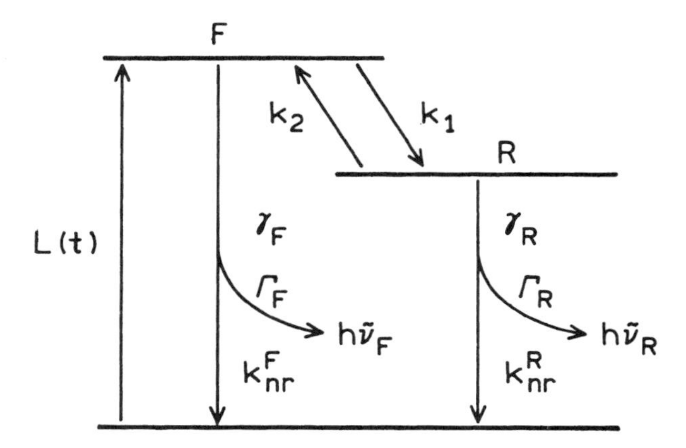

Figure 18.5. Jabłoński diagram for a reversible excited-state reaction. $L(t)$ is the excitation function, which can be a pulsed or amplitude-modulated light beam.

(R) state, and assume initially that the reaction is irreversible, that is, $k_2 = 0$ (Figure 18.5). The relative quantum yield of the F state is given by the ratio of the emissive rate to the total rate of depopulation of the F state. Thus,

$$F_0 = \Gamma_F/(\Gamma_F + k_{nr}^F) \qquad [18.3]$$

$$F = \Gamma_F/(\Gamma_F + k_{nr}^F + k_1) \qquad [18.4]$$

where F_0 and F are the fluorescence intensities observed in the absence and in the presence of the reaction, respectively. Division of Eq. [18.4] by Eq. [18.3] yields

$$\frac{F}{F_0} = \frac{1}{1 + k_1\tau_{0F}} \qquad [18.5]$$

where $\tau_{0F} = \gamma_F^{-1} = (\Gamma_F + k_{nr}^F)^{-1}$ is the lifetime of the F state in the absence of the reaction. Recall that for a biomolecular rate process k_1 would be replaced by k_1 [Q], where Q is the species reacting with the fluorophore. Hence, under some circumstances, the initially excited state can be quenched in accordance with a Stern–Volmer dependence.

The yield of the reaction product R is the fraction of F molecules which have reacted. This yield is given by

$$\frac{R}{R_0} = 1 - \frac{F}{F_0} = \frac{k_1\tau_{0F}}{1 + k_1\tau_{0F}} \qquad [18.6]$$

where R and R_0 are the intensities of the reacted species when the reaction is incomplete and complete, respectively. In the case of excimer formation, R_0 would be observed at high concentrations of the monomer. It is important to note the distinctly different dependences of the F- and R-state intensities on the extent of the reaction. The intensity of the F state decreases monotonically to zero as the reaction rate becomes greater than the reciprocal lifetime. In contrast, the intensity of the R state is zero when the forward rate constant is much less than the decay rate of the F state, $k_1 < \gamma_F$ or $k_1\tau_{0F} << 1$. The intensity of the R state increases to a constant value when $k_1\tau_{0F} >> 1$. The reason for this limiting behavior is that essentially all the F state is converted to the R state. In practice, the fluorescence of the reacted species may not reach a limiting value but may rather decrease as the rate of the forward reaction is increased. This can occur because the conditions needed to drive the reaction to completion may result in other interactions which quench the R state.

18.2.B. Time-Resolved Decays for the Two-State Model

The model presented in Figure 18.5 is described by the following kinetic equations:

$$-\frac{d[\mathrm{F}]}{dt} = \gamma_F[\mathrm{F}] - k_2[\mathrm{R}] - L(t) \qquad [18.7]$$

$$-\frac{d[\mathrm{R}]}{dt} = \gamma_R[\mathrm{R}] - k_1[\mathrm{F}] \qquad [18.8]$$

where [F] and [R] are the concentrations of these states and $L(t)$ is the time profile of the excitation. The solution to these kinetic equations, subject to the boundary conditions $[\mathrm{F}] = [\mathrm{F_0}]$ and $[\mathrm{R}] = 0$ at $t = 0$, has been described previously.[14,15,25,26] Following δ-pulse excitation, the fluorescence decays of the F and R states are given by expressions of the form

$$I_F(t) = \alpha_1 \exp(-\gamma_1 t) + \alpha_2 \exp(-\gamma_2 t) \qquad [18.9]$$

$$I_R(t) = \beta_1 \exp(-\gamma_1 t) - \beta_1 \exp(-\gamma_2 t) \qquad [18.10]$$

where the α_i, β_1, and τ_i values are moderately complex functions of the various rate constants. The values of $\gamma_1 = \tau_1^{-1}$ and $\gamma_2 = \tau_2^{-1}$ are given by

$$\gamma_1, \gamma_2 = \frac{1}{2}\left\{(x+y) \pm [(y-x)^2 + 4k_1k_2]^{1/2}\right\} \qquad [18.11]$$

where

$$x = \Gamma_F + k_{\mathrm{nr}}^F + k_1 = \gamma_F + k_1 \qquad [18.12]$$

$$y = \Gamma_R + k_{\mathrm{nr}}^R + k_2 = \gamma_R + k_2 \qquad [18.13]$$

The important point from these equations is that both species display the same decay times and that the decay of both states is a double-exponential decay. Hence, one consequence of a reversible excited-state reaction is that the F and R states both display a double-exponential decay and that the decay times are the same for both species.

Another important characteristic of an excited-state reaction is seen from the preexponential factors for the R state. If only the R state is observed, then the preexponential factors for the two lifetimes are expected to be equal and opposite (β_1 in Eq. [18.10]). The negative preexponential factor results in a rise in intensity, which is characteristic of excited-state reactions.

For many molecules that display excited-state reactions, there is spectral overlap of the F and R states. If one measures the wavelength-dependent intensity decays, this will be described by

$$I(\lambda, t) = \alpha_1(\lambda)\exp(-t/\tau_1) + \alpha_2(\lambda)\exp(-t/\tau_2) \qquad [18.14]$$

where the same two decay times will be present at all wavelengths. In this case one can perform a global analysis with two wavelength-independent lifetimes. The meaning of the $\alpha_i(\lambda)$ values is complex and has been described in detail.[15,19] As the observation wavelength is increased, one observes an increasing fraction of the R state. The $\alpha_i(\lambda)$ values thus shift from those characteristic of the F state (Eq. [18.9]) to those characteristic of the R state (Eq. [18.10]). On the long-wavelength side of the emission, one expects to observe a negative preexponential factor, unless the spectral shift is small so that the F and R states overlap at all wavelengths. The values of $\alpha_i(\lambda)$ can be used to calculate decay-associated spectra (DAS), which show positive and negative amplitudes. The DAS do not correspond to the emission spectrum of either species. The $I(\lambda, t)$ data can also be used to calculate the species-associated spectra (SAS), which are the spectra of each of the states (Section 18.5).

The theory for an excited-state reaction is considerably simplified if the reaction is irreversible ($k_2 = 0$). In this case, Eq. [18.11] yields $\gamma_1 = \gamma_F + k_1$ and $\gamma_2 = \gamma_R$. Also, one of the preexponential factors in Eq. [18.9] becomes zero so that

$$I_F(t) = \alpha_1 \exp(-\gamma_1 t) \qquad [18.15]$$

$$I_R(t) = \beta_1 \exp(-\gamma_1 t) - \beta_1 \exp(-\gamma_2 t) \qquad [18.16]$$

Hence, the decay of the initially excited state becomes a single exponential. This prediction has been verified in experimental studies of reversible and irreversible reactions. We note that the negative preexponential factor may be associated with either of the decay times (τ_1 or τ_2), depending on the values of the kinetic constants. The negative preexponential factor is always associated with the shorter decay time.[19]

18.3. TIME-DOMAIN STUDIES OF NAPHTHOL DISSOCIATION

The properties of excited-state reactions can be effectively studied using TD methods. Typical data for the excited-state ionization of 2-naphthol are shown in Figure 18.6. The time-dependent data were collected at either 360 or 450 nm. At 360 nm one observes the un-ionized form of 2-naphthol. At 450 nm the emission is due to naphtholate. As predicted by the theory, the naphtholate emission starts at zero and shows a rise time, during which the ionized species is formed.

Time-resolved data collected across the emission spectra can be used to calculate the TRES. These procedures were

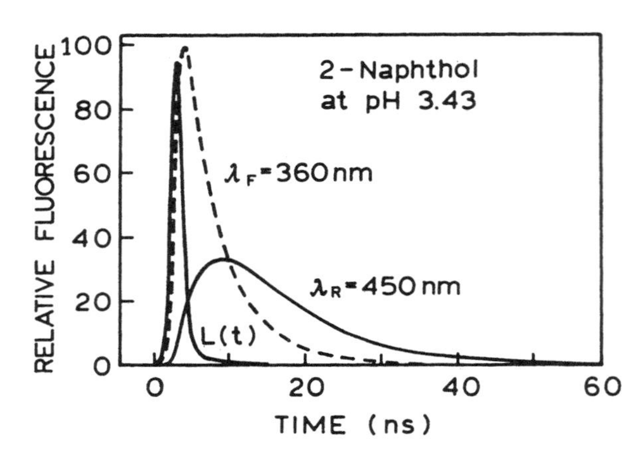

Figure 18.6. Time-dependent intensity decays of 2-naphthol (360 nm) and its ionization product (450 nm). $L(t)$ is the excitation function. Revised from Ref. 15.

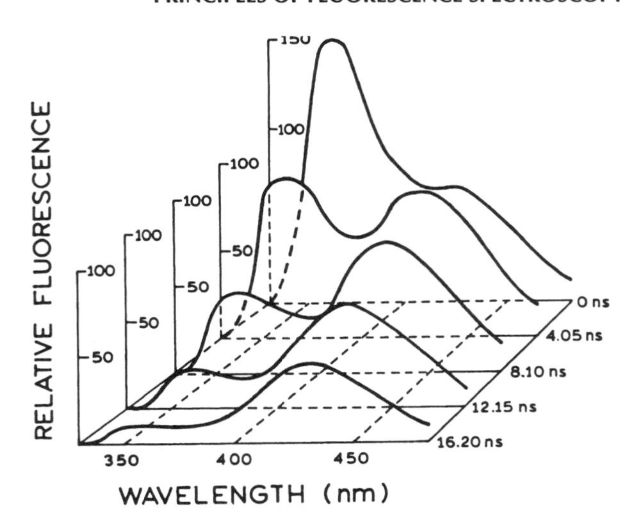

Figure 18.7. TRES of 2,6-naphtholsulfonate. The rate of excited-state ionization was slowed by the addition of 40% ethanol. Revised and reprinted, with permission, from Ref. 14, Copyright © 1972, American Chemical Society.

described in detail in Chapter 7. Such spectra are shown in Figure 18.7 for 2,6-naphtholsulfonate, a sulfonated form of 2-naphthol. In order to obtain TRES within the time resolution available at that time,[14] the reaction was slowed by the addition of ethanol. The TRES shift to longer wavelength following the excitation pulse. At t = 0, the emission is mostly due to the naphthol form emitting near 350 nm. At longer times (16 ns), the emission is almost entirely due to the naphtholate form emitting at longer wavelength. At intermediate times, the emission is due to both forms of the fluorophore.

The theory for a reversible two-state reaction predicts the same two decay times at all emission wavelengths. This is illustrated in the left-hand panel of Figure 18.8 for 2-naphthol at pH 3.0; at this pH, the same decay times were recovered at all emission wavelengths. At higher pH (pH > 6), the ionization of 2-naphthol becomes irreversible, and the F state is predicted to display a single-exponential decay. This is seen from Figure 18.9, where only a single decay time was recovered for the un-ionized species at 350–370 nm.

18.3.A. Differential-Wavelength Methods

In considering excited-state reactions, there is a general principle which clarifies the complex decay kinetics and results in simplified methods of analysis.[27,28] This princi-

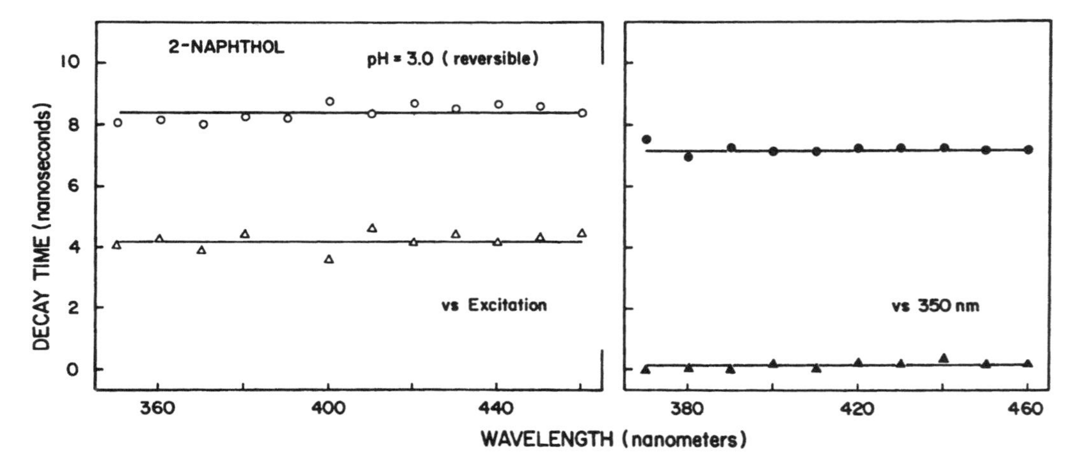

Figure 18.8. Wavelength-dependent decay times of 2-naphthol at pH 3.0. Decay times are shown for deconvolution versus the excitation wavelength (*left*) and versus the emission at 350 nm (*right*). Reprinted from Ref. 27, Copyright © 1982, with permission from Elsevier Science.

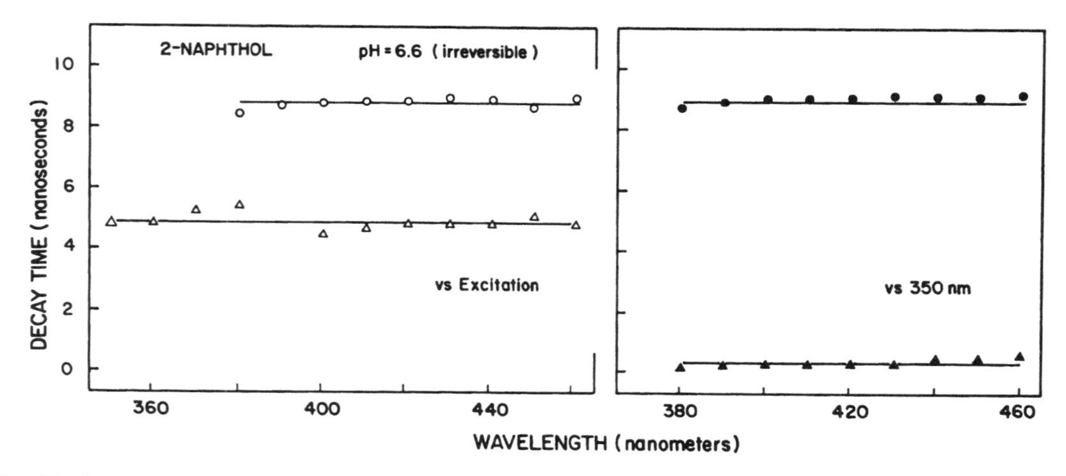

Figure 18.9. Wavelength-dependent decay times of 2-naphthol at pH 6.6. Decay times are shown for deconvolution versus the excitation wavelength (*left*) and versus the emission at 350 nm (*right*). The intensity decay is a single exponential below 380 nm (*left*). Reprinted from Ref. 27, Copyright © 1982, with permission from Elsevier Science.

ple is that the population of the R state can be regarded as a convolution integral with the F state. That is, the F-state population is the excitation pumping function of the R state. Consider a measurement of the R state, made relative to the F state. The F state effectively becomes the lamp function, so that measurements relative to the F state directly reveal the decay kinetics of the R state. The application of this procedure requires a spectral region where the emission from the F state can be observed without overlap from the R state. Since nonoverlap of states is frequently found on the short-wavelength side of the emission, the time profile of the F state can generally be measured directly. Deconvolving the R-state emission with the F-state emission yields the lifetime of the R-state that would be observed if the R state could be directly excited.

The data in the right-hand panels of Figures 18.8 and 18.9 illustrate the simplified interpretation available by deconvolution of the R state versus the F state. For the irreversible case, this procedure yields a zero-decay-time component (Figure 18.9). This component is due to the spectral contribution of the F state at the emission wavelengths used to observe the R state. The 8.7-ns component is the intrinsic lifetime of naphtholate under these experimental conditions. This is the decay time that would be observed if naphtholate could be directly excited under these experimental conditions. For the reversible reaction, a zero-decay-time component is again observed (Figure 18.8). In addition, a new component is observed, with a decay time of 7.2 ns, which is the lifetime of naphtholate under these experimental conditions.

18.4. ANALYSIS OF EXCITED-STATE REACTIONS BY PHASE-MODULATION FLUOROMETRY

The complementary technique of phase-modulation fluorometry provides a number of interesting opportunities for the analysis of excited-state reactions. Of course, one can measure the frequency responses across the emission spectra and use these data to construct the TRES (Chapter 7). However, this procedure does not lead to insights into the meaning of the measured phase and modulation values. Hence, we provide a brief overview of the unique aspects of the phase and modulation values for an excited-state process. Additional details can be found elsewhere.[29–31]

For simplicity, we consider the irreversible reaction ($k_2 = 0$). Assume that by appropriate optical filtering the emission from the F and R states can be individually observed. Then, relative to the phase and modulation of the excitation, the phase and modulation of the F and R state are given by

$$\tan \phi_F = \frac{\omega}{\gamma_F + k_1} = \omega\tau_F \qquad [18.17]$$

$$m_F = \frac{\gamma_F + k_1}{\sqrt{(\gamma_F + k_1)^2 + \omega^2}} = \frac{1}{\sqrt{1 + \omega^2\tau_F^2}} \qquad [18.18]$$

Several aspects of these equations are worthy of mention. Since we have initially assumed that F and R are separately observable, and the reverse reaction does not occur, the decay of F is a single exponential. Hence, for the F state we find the usual expressions for calculation of lifetimes from phase and modulation data. In the absence of any reaction ($k_1 = 0$), the lifetime is $\tau_{0F} = \gamma_F^{-1}$. In the presence of a reaction, the lifetime of F is shortened to $\tau_F = (\gamma_F + k_1)^{-1}$. Thus, for an irreversible reaction, the observed values of ϕ_F and m_F can be used to calculate the true lifetime of the unreacted state.

The phase and modulation of the R state, when measured relative to the excitation, are complex functions of the various kinetic constants. These values are given by

$$\tan \phi_R = \frac{\omega(\gamma_F + \gamma_R + k_1)}{\gamma_R(\gamma_F + k_1) - \omega^2} \quad [18.19]$$

$$m_R = m_F \frac{\gamma_R}{\sqrt{\gamma_R^2 + \omega^2}} = m_F m_{0R} \quad [18.20]$$

In contrast to the values for the F state, the measured values for the R state cannot be directly used to calculate the fluorescence lifetime of the R state. The complexity of the measured values (ϕ_R and m_R) illustrates why use of the apparent phase (τ^{ϕ}) and modulation (τ^m) lifetimes of the relaxed state is not advisable.

Closer examination of Eqs. [18.17]–[18.20] reveals important relationships between the phase and modulation values of the F and R states. For an excited-state process, the phase angles of the F and R states are additive, and the modulations multiply. Once this is understood, the complex expressions (Eqs. [18.19] and [18.20]) become easier to understand. Let ϕ_{0R} be the phase angle of the R state that would be observed if this state could be excited directly. Of course, this is related to the lifetime of the directly excited R state by $\tan \phi_{0R} = \omega\tau_{0R}$. Using Eq. [18.19], the law for the tangent of a sum, and $\tan(\phi_F + \phi_{0R}) = \tan\phi_R$, one finds

$$\phi_R = \phi_F + \phi_{0R} \quad [18.21]$$

Hence, the phase angle of the reacted state, measured relative to the excitation, is the sum of the phase angle of the unreacted state (ϕ_F) and the phase angle of the reacted state that would be observed if this state could be directly excited (ϕ_{0R}). This relationship, represented in Figure 18.10, may be understood intuitively by recognizing that the F state is populating the R state. Of course, these are the same considerations used to describe differential-wavelength deconvolution. For the irreversible reaction, measurement of $\Delta\phi = \phi_R - \phi_F = \phi_{0R}$ reveals directly the intrinsic lifetime of the reacted fluorophore, unaffected by its population through the F state.

$$R:\ \beta_1 e^{-t/\tau_1} - \beta_2 e^{-t/\tau_2}$$

$$F:\ \alpha_1 e^{-t/\tau_1} + \alpha_2 e^{-t/\tau_2}$$

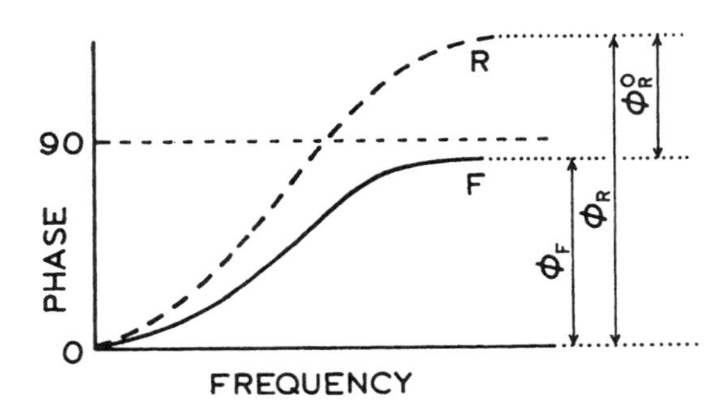

Figure 18.10. Relationship between the phase angles of the F and R states.

The demodulation factors of the two states display similar properties. From Eq. [18.20], one finds that the demodulation of the relaxed state (m_R) is the product of the demodulation of the unrelaxed state (m_F) and that demodulation due to the intrinsic decay of the R state alone (m_{0R}). That is,

$$m_R = m_F \frac{1}{\sqrt{1 + \omega^2\tau_R^2}} = m_F m_{0R} \quad [18.22]$$

A further interesting aspect of the phase difference or the demodulation between F and R is the potential of measuring the reverse reaction rate k_2. This reverse rate can be obtained from the phase-angle difference between the R and F states, or from the modulation of the R state relative to the F state. For a reversible reaction, these expressions are

$$\tan(\phi_R - \phi_F) = \frac{\omega}{(\gamma_R + k_2)} = \omega\tau_R \quad [18.23]$$

$$\frac{m_R}{m_F} = \frac{\gamma_R + k_2}{\sqrt{(\gamma_R + k_2)^2\ \omega^2}} = \frac{1}{\sqrt{1 + \omega^2\tau_R^2}} \quad [18.24]$$

These expressions are similar to the usual expressions for the dependence of phase shift and demodulation on the decay rates of an excited state, except that the decay rate is that of the reacted species ($\gamma_R + k_2$). The initially excited state populates the relaxed state, and the reverse reaction repopulates the F state. Nonetheless, the kinetic constants of the F state do not affect the measurement of $\tan(\phi_R - \phi_F)$ or m_R/m_F. Measurement of either the phase or modulation of the reacted state, relative to that of the unrelaxed state, yields a lifetime of the reacted state. This lifetime is decreased by the reverse reaction rate in a manner analogous to the decrease in the lifetime of the F state by k_1. If the decay rate (γ_R) is known, k_2 may be calculated. If the emission results from a single species which displays one lifetime, and ϕ and m are constant across the emission, then $\phi_R - \phi_F = 0$ and $m_R/m_F = 1$.

Returning to the irreversible model, we note an interesting feature of ϕ_R (Eq. [18.19]). This phase angle can exceed 90°. Specifically, if ω^2 exceeds $\gamma_R(\gamma_F + k_1)$, then $\tan\phi_R < 0$ or $\phi_R > 90°$. In contrast, the phase angle of directly excited species, or the phase angles resulting from a heterogeneous population of fluorophores, cannot exceed 90°. Therefore, observation of a phase angle in excess of 90° constitutes proof of an excited-state reaction.

18.4.A. Effect of an Excited-State Reaction on the Apparent Phase and Modulation Lifetimes

The multiplicative property of the demodulation factors and the additive property of the individual phase angles are the origin of a reversed frequency dependence of the apparent phase-shift and demodulation lifetimes and the inversion of apparent phase and modulation lifetimes when compared to those of a heterogeneous sample. The apparent phase lifetime (τ_R^ϕ) calculated from the measured phase (ϕ_R) of the relaxed state is

$$\tan\phi_R = \omega\tau_R^\phi = \tan(\phi_F + \phi_{0R}) \qquad [18.25]$$

Recalling the law for the tangent of a sum, one obtains

$$\tau_R^\phi = \frac{\tau_F + \tau_{0R}}{1 - \omega^2\tau_F\tau_{0R}} \qquad [18.26]$$

Because of the term $\omega^2\tau_F\tau_{0R}$, an increase in the modulation frequency can result in an increase in the apparent phase lifetime. This result is opposite to that found for a heterogeneous emitting population where the individual species are excited directly. For a heterogeneous sample, an increase in modulation frequency yields a decrease in the apparent phase lifetime.[32] Therefore, the frequency dependence of the apparent phase lifetimes can be used to differentiate a heterogeneous sample from a sample which undergoes an excited-state reactions. Similarly, the apparent modulation lifetime is given by

$$\tau_R^m = \left(\frac{1}{m_R^2} - 1\right)^{-1/2} \qquad [18.27]$$

Recalling Eq. [18.20], one obtains

$$\tau_R^m = (\tau_F^2 + \tau_{0R}^2 + \omega^2\tau_F^2\tau_{0R}^2)^{1/2} \qquad [18.28]$$

Again, increasing ω yields an increased apparent modulation lifetime. This frequency dependence is also opposite to that expected from a heterogeneous sample and is useful in proving that emission results from an excited-state process. In practice, however, the dependence of τ_R^m upon modulation frequency is less dramatic than that of τ_R^ϕ. We again stress that the calculated lifetimes are apparent values and not true lifetimes.

The information derived from phase-modulation fluorometry is best presented in terms of the observed quantities ϕ and m. For a single-exponential decay, and for the unrelaxed state if relaxation is irreversible, $m_F = \cos\phi_F$. Also $m_{0R} = \cos\phi_{0R}$, irrespective of the reversibility of the reaction. A convenient indicator of an excited-state process is the ratio $m/\cos\phi$, where m and ϕ are the experimentally measured demodulation factor and phase angle, respectively. This ratio is unity for a single-exponential decay and is <1 for a heterogeneous emission. In contrast, $m/\cos\phi > 1$ if the emitting species forms subsequent to excitation.[33] Using Eqs. [18.21] and [18.22], and the law for the cosine of a sum, we obtain

$$\frac{m_R}{\cos\phi_R} = \frac{\cos\phi_{0R}\cos\phi_F}{\cos(\phi_{0R} + \phi_F)}$$

$$= \frac{\cos\phi_{0R}\cos\phi_F}{\cos\phi_{0R}\cos\phi_F - \sin\phi_{0R}\sin\phi_F} \qquad [18.29]$$

Dividing the numerator and the denominator by $\cos\phi_{0R}\cos\phi_F$, we obtain

$$\frac{m_R}{\cos\phi_R} = \frac{1}{1 - \tan\phi_{0R}\tan\phi_F} = \frac{1}{1 - \omega^2\tau_{0R}\tau_F} \qquad [18.30]$$

If relaxation is much slower than emission, significant relaxation does not occur. The R state cannot be observed, and Eq. [18.30] cannot be applied. If relaxation is much faster than emission, $\phi_F = 0$ and $m_R/\cos\phi_R = 1$. Importantly, if relaxation and emission occur on comparable timescales, the ratio $m_R/\cos\phi_R$ exceeds unity. Observation of $m/\cos\phi > 1$ proves the occurrence of an excited-state reaction. However, failure to observe $m/\cos\phi > 1$ does not prove that a reaction has not occurred. Heterogeneity, spectral overlap, or the reverse reaction can prevent observation of $m/\cos\phi > 1$.

18.4.B. Wavelength-Dependent Phase and Modulation Values for an Excited-State Reaction

The general features of phase-shift and demodulation data for a sample which undergoes an irreversible excited-state reaction are illustrated by the excited-state protonation of acridine by ammonium ion.[30] In basic solution, an emission maximum of 430 nm is observed (Figure 18.11). Upon acidification of the solution, this spectrum is replaced by a red-shifted spectrum with an emission maximum of 475 nm. The former spectrum is due to neutral acridine (Ac) and the latter is due to the acridinium cation (AcH^+). Since the ground state pK_a of acridine is 5.45, only the neutral species is present at pH 8.3. Nonetheless, increasing concentrations of ammonium ion, at pH 8.3, yield a progressive quenching of the short-wavelength emission, with a concomitant appearance of the emission from the acridinium ion (Figure 18.11). These spectral shifts are the result of protonation of the excited neutral acridine molecules by ammonium ions. In the excited state, the pK_a of neutral acridine increases from the ground-state value of 5.45 to 10.7.

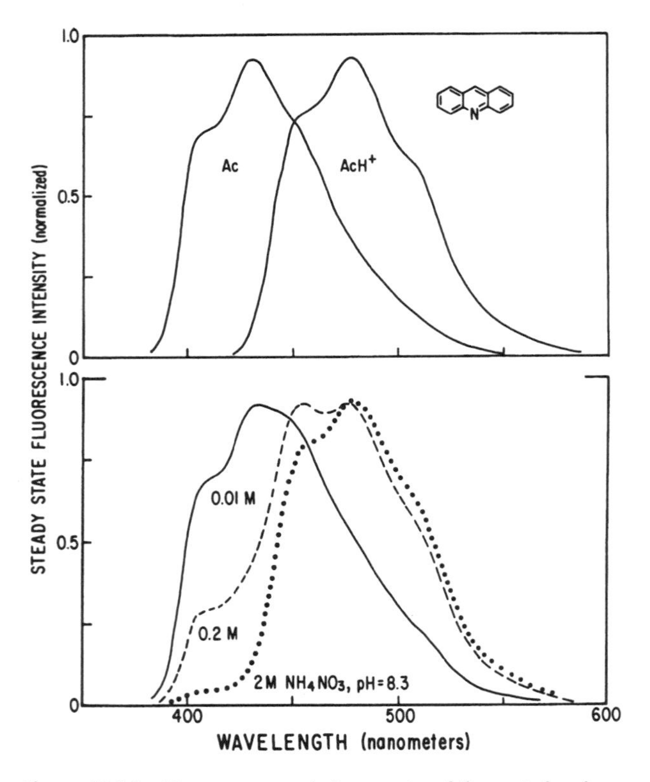

Figure 18.11. Fluorescence emission spectra of the neutral and protonated forms of acridine (*top*) and of acridine at pH 8.3 with various concentrations of ammonium ion (*bottom*). Reprinted from Ref. 30, Copyright © 1982, with permission from Elsevier Science.

This excited-state reaction illustrates the phase-modulation theory just described. It is a two-state process that is essentially irreversible. Time-resolved studies demonstrated that at 410 nm the decay is a single exponential.[6] At 560 nm, the fluorescence decay can be described with two lifetimes, one of which is the same as the decay time observed at 410 nm. The two lifetimes were independent of emission wavelength, but the preexponential factors were dependent upon emission wavelength. These data imply that the lifetime of each species (Ac and AcH^+) is constant across its emission spectrum and that the irreversible two-state theory is appropriate to describe this excited-state process. Examination of the emission spectra (Figure 18.11) reveals a region where only the neutral species emits (390–410 nm), a region of strong overlap (440–500 nm), and a region of moderate overlap where the spectrum is dominated by the AcH^+ emission (> 500 nm).

The apparent phase lifetimes (τ^{ϕ}) for acridine in 0.05*N* NaOH, 0.1*N* H_2SO_4, and 2*M* NH_4NO_3 are shown in Figure 18.12. In acidic or basic solution, one species is present in both the ground and excited states, and the lifetimes (or phase angles) are independent of emission wavelength. In contrast, the lifetimes in 2*M* NH_4NO_3 are highly dependent upon wavelength because of protonation of acridine subsequent to excitation. At short wavelengths (410 nm) where neutral acridine emits, the lifetime is decreased in the presence of ammonium ions. A decreased phase angle (or lifetime) of the initially excited state is a characteristic feature of an excited-state reaction. At longer wavelengths, the apparent lifetime increases with the emission wavelength until a nearly constant value is reached for wavelengths longer than 500 nm. In this wavelength range, the emission is dominated by the acridinium ion. The constant lifetime, or more correctly phase angle, on the red and blue sides of the emission may be regarded as evidence for the two-state model. If the overall emission were shifting to longer wavelengths according to the continuous Bakhshiev model,[34] such regions of constant phase angle would not be expected (Section 7.6). Overall, the phase data for acridine may be regarded as typical for a two-state reaction with moderate spectral overlap. These characteristics include a decrease in lifetime on the blue side of the emission, an increase in apparent lifetime with emission wavelength, and nearly constant lifetimes on the blue and red sides of the emission.

An interesting potential of phase fluorometry is the ability to measure directly the intrinsic lifetime of the reacted species. By "intrinsic" we mean the lifetime that would be observed if this species were formed by direct excitation, rather than by an excited-state reaction. The

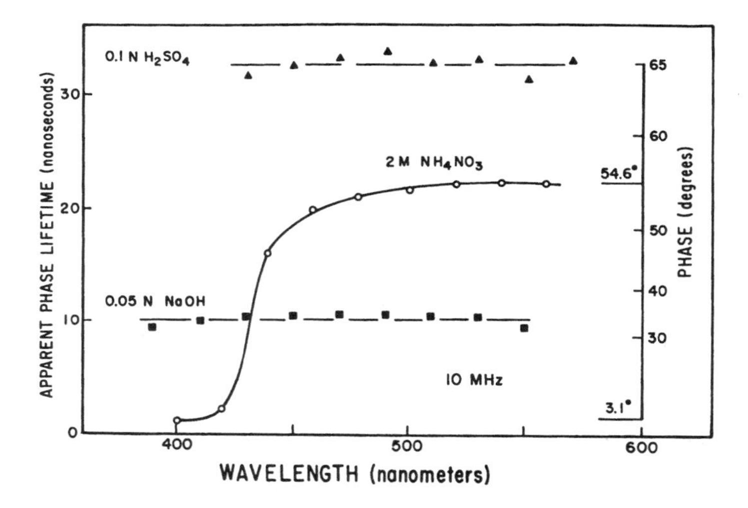

Figure 18.12. Apparent phase lifetimes of acridine in 0.1*N* H_2SO_4 (▲), 0.05*N* NaOH (■), and 2*M* NH_4NO_3 (○). The axis at the right-hand side of the plot indicates the phase angles relative to the exciting light. The phase-angle difference between the red ($\phi_R = 54.6°$) and blue ($\phi_F = 3.1°$) regions of the emission in 2*M* NH_4NO_3 reveals the lifetime of the acridinium cation that would be observed if this species could be directly excited. The phase angles of acridinium, relative to the exciting light, do not yield true lifetimes for the directly excited acridinium cation. Reprinted from Ref. 30, Copyright © 1982, with permission from Elsevier Science.

intrinsic lifetime of the reacted species is revealed by the phase difference ($\Delta\phi$) between the blue and red sides of the emission ($\Delta\phi = \phi_R - \phi_F$). Consider the phase difference between 400 nm and 560 nm shown on Figure 18.12. This phase angle difference ($\Delta\phi = 51.5°$) yields the lifetime of the acridine cation [$\tau(AcH^+)$] according to

$$\tan \Delta\phi = \omega\, \tau(AcH^+) \qquad [18.31]$$

The lifetime is found to be 20 ns in 2*M* NH_4NO_3. Recall that the origin of this simple result is that the excited neutral acridine population is the pumping function for the excited-state acridinium molecules.

Phase measurements alone, at a single modulation frequency, cannot be used to distinguish between ground-state heterogeneity and an excited-state reaction. The increase in phase angle shown in Figure 18.12 could also be attributed to a second directly excited fluorophore with a longer lifetime. Of course, the decrease in the phase lifetime at short wavelengths indicates a quenching process. A rigorous proof of the presence of an excited-state process is obtainable by comparison of the phase shift and demodulation data from the same sample. Figure 18.13 shows apparent phase and modulation lifetimes and the ratio $m/\cos\phi$ for acridine in 0.2*M* NH_4NO_3. On the blue edge, $\tau^\phi \simeq \tau^m$ and $m/\cos\phi \simeq 1$, indicative of a single-exponential decay and therefore an irreversible reaction. In the central overlap region, $\tau^\phi < \tau^m$ and $m/\cos\phi < 1$, which is indicative of emission from more than one state. At longer wavelengths, $\tau^\phi > \tau^m$ and $m/\cos\phi > 1$. These last results are impossible for a multiexponential decay with any degree of heterogeneity. This observation proves that the emission at these longer wavelengths is not a result of direct excitation, but rather is the result of an excited-state reaction.

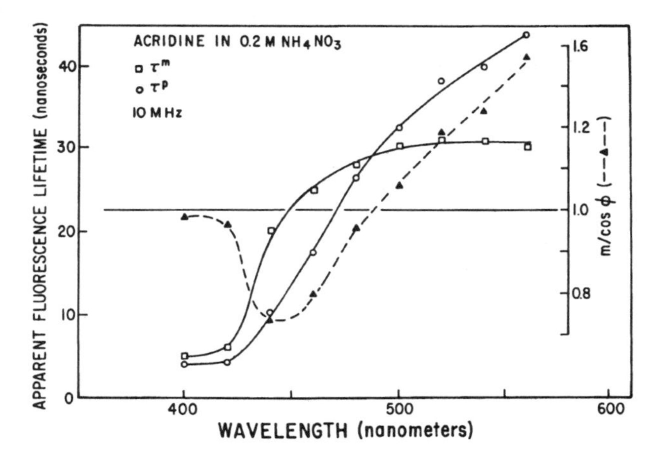

Figure 18.13. Apparent phase (○) and modulation (□) lifetimes of acridine in 0.2*M* NH_4NO_3, pH 8.3. Also shown is the wavelength dependence of $m/\cos\phi$ (▲). Reprinted from Ref. 30, Copyright © 1982, with permission from Elsevier Science.

18.5. FREQUENCY-DOMAIN MEASUREMENT OF EXCIMER FORMATION

In the preceding section we described how the phase and modulation values, when measured at a single modulation frequency, were affected by an excited-state reaction. With the availability of variable-frequency instruments, the intensity decay can be recovered without a detailed interpretation of the individual phase and modulation values. This is illustrated by the example of excimer formation by 2-phenylindole (2-PI).[35] At higher concentrations of 2-PI, there is increased intensity on the long-wavelength side of the emission (Figure 18.14), which is thought to be due to formation of excimers. However, the long-wavelength emission could also be the result of ground-state complexes of 2-PI. These possibilities were distinguished by measurement of the frequency response at various emission wavelengths (indicated by the arrows in Figure 18.14). The data for observation at 360 nm (F) and 480 nm (R) are shown in Figure 18.15. At 360 nm, a typical frequency response is observed. At 480 nm, the phase angles increase rapidly with frequency and exceed 90°,

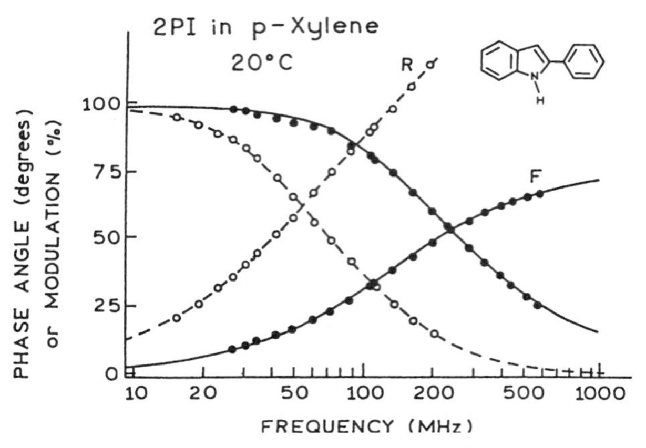

Figure 18.15. FD intensity decays of 2-PI ($6 \times 10^{-3} M$) observed at 360 (F, ●) and 480 nm (R, ○). From Ref. 35.

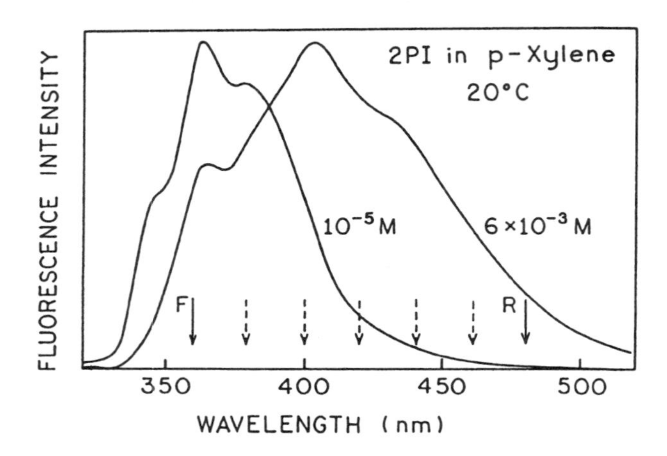

Figure 18.14. Emission spectra of 2-PI in *p*-xylene at concentrations of $10^{-5} M$ and $6 \times 10^{-3} M$. The emission at long wavelengths is due to excimer formation. From Ref. 35.

Table 18.3. Wavelength-Dependent Intensity Decays of 2-Phenylindole ($6 \times 10^{-3} M$) in *p*-Xylene at 20 °C[a]

	$\tau_1 = 0.86$ ns		$\tau_2 = 3.42$ ns	
λ_{em} (nm)	α_1	f_1	α_2	f_2
360	0.967	0.879	0.033	0.121
380	0.951	0.831	0.049	0.069
400	0.731	0.405	0.269	0.595
420	−0.047	−0.012	0.953	0.988
440	−0.326	−0.108	0.674	0.829
460	−0.407	−0.147	0.593	0.853
480	−0.453	−0.172	0.547	0.828

[a] From Ref. 35. The intensity decays measured for each emission wavelength (λ_{em}) were fit with τ_1 and τ_2 as global parameters. The χ_R^2 of this global fit was 3.9.

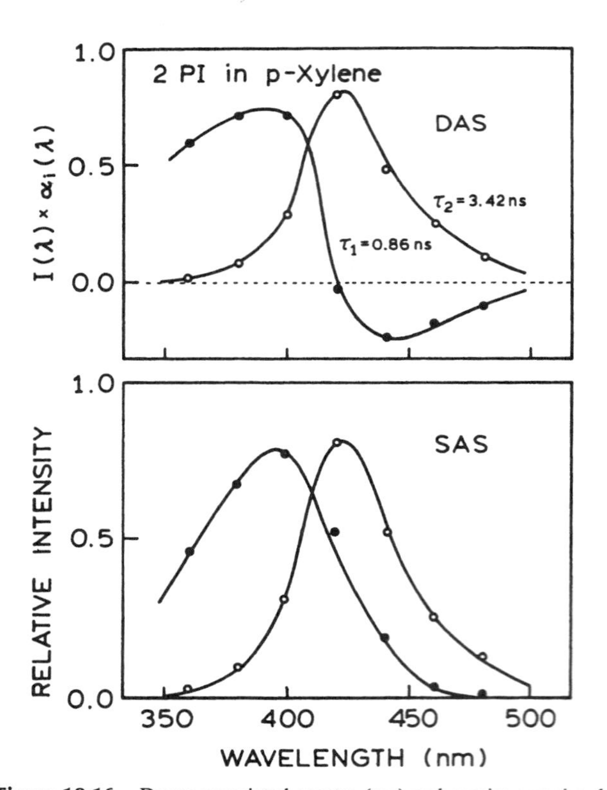

Figure 18.16. Decay-associated spectra (*top*) and species-associated spectra (*bottom*) for 2-PI ($6 \times 10^{-3} M$) in *p*-xylene. From Ref. 35.

which proves that the emission at 480 nm is not directly excited.

The frequency responses at each wavelength were used to recover the intensity decays (Table 18.3). It was possible to fit the data globally with two wavelength-independent lifetimes. As the observation wavelength increases, the amplitude of the 0.86-ns component becomes negative. At 480 nm, the amplitudes of the two components are nearly equal and opposite, suggesting that there is minimal contribution of the F state at this wavelength.

If desired, the decay times and amplitudes in Table 18.3 can be used to construct the TRES, DAS, or SAS. The decay-associated spectrum of the shorter-lifetime component shows regions of positive and negative amplitudes (Figure 18.16). This occurs because there is a negative preexponential factor associated with this decay time at longer wavelengths. The SAS in this case are the emission spectra of the monomer and the excimer. The procedure for calculation of the SAS has been described in detail.[19,36] The SAS indicate that the monomer emission of 2-PI is centered near 390 nm and that the excimer emission is only moderately red-shifted to 420 nm (Figure 18.16).

18.6. BIOCHEMICAL EXAMPLES OF EXCITED-STATE REACTIONS

18.6.A. Excited-State Tautomerization of 7-Azaindole

In the preceding sections we saw examples of excited-state ionization. There are also many examples of intramolecular proton transfer while fluorophores are in the excited state, several of which are shown in Figure 18.17.[37–46] One example is provided by 7-azaindole (7-AI), which has been used as a substitute for indole in tryptophan and in proteins. It was thought that 7-AI may undergo an intramolecular proton-transfer reaction in water (Figure 18.17). However, it was difficult to obtain proof from the steady-state data. An unusual long-wavelength emission from 7-AI was seen in alcohols, but there was only the suggestion of a long-wavelength emission in water (Figure 18.18). Emission from the tautomeric form of 7-AI could be detected from the time-resolved measurements. The intensity decays were distinct at 350 and 550 nm. At the longer wavelength, the decay showed the rise characteristic of an excited-state process (Figure 18.19, bottom). This longer-wavelength component was also evident in the TRES emission spectra (Figure 18.19, top). Intramolecular proton transfer has also been suggested to occur for the DNA probe DAPI and may explain the increase in DAPI fluorescence which occurs upon binding of DAPI to DNA.[47]

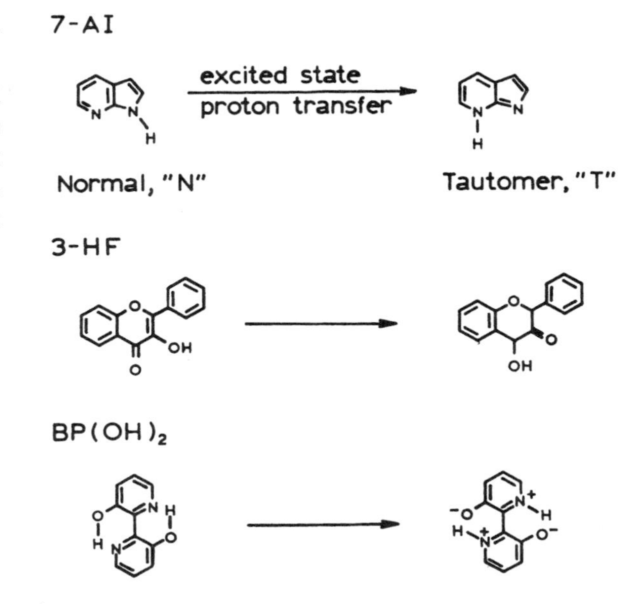

Figure 18.17. Excited-state intramolecular proton transfer in 7-azaindole (7-AI), 3-hydroxyflavone (3-HF), and 2,2′-bipyridyl-3,3′-diol [BP(OH)$_2$].

18.6.B. Exposure of a Membrane-Bound Cholesterol Analog

Excited-state reactions have been used to determine the water accessibility of fluorophores bound to proteins[48,49]

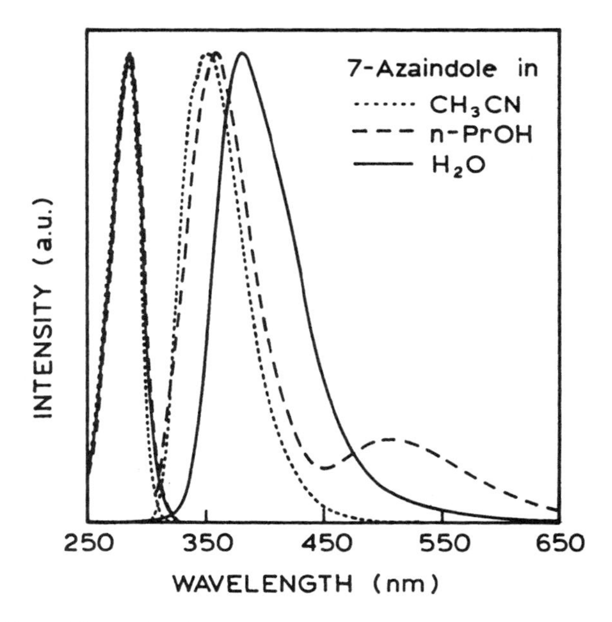

Figure 18.18. Steady-state absorption and emission spectra of 7-AI in acetonitrile (---), *n*-propanol (———), and water (——). Revised and reprinted, with permission, from Ref. 46, Copyright © 1992, American Chemical Society.

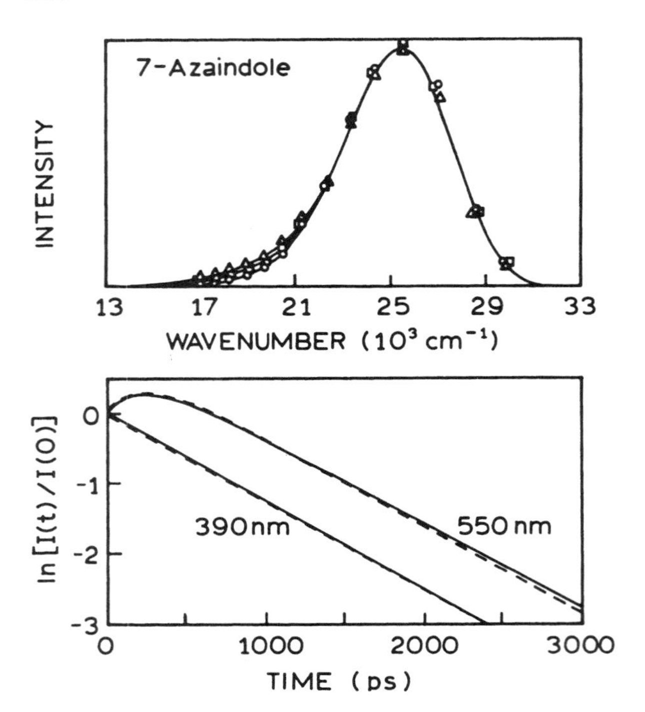

Figure 18.19. *Top*: TRES of 7-AI in water, at t = 0 (○), 500 (□), and 3000 ps (△). *Bottom*: Intensity decays of 7-AI in water at 390 and 550 nm. The solid lines are the experimental data, and the dashed lines are the best fit using an excited state reaction model. Revised and reprinted, with permission, from Ref. 46, Copyright © 1992, American Chemical Society.

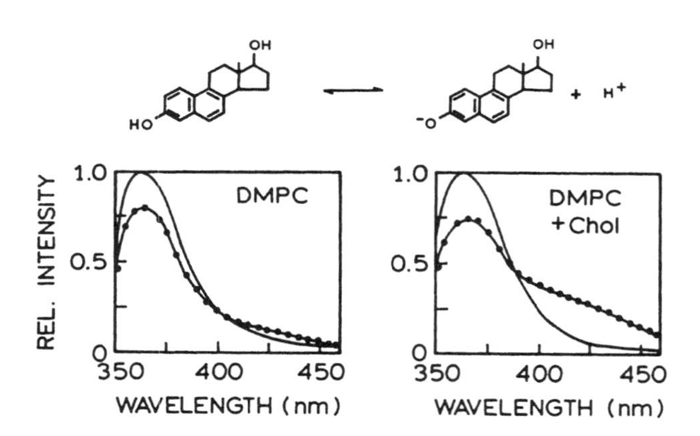

Figure 18.20. *Top*: Excited-state ionization of DHE. *Bottom*: Emission spectra of DHE bound to DMPC vesicles (*left*) or DMPC vesicles containing 10 mol % cholesterol (*right*). Revised and reprinted, with permission, from Ref. 19, Copyright © 1986, American Chemical Society.

or membranes.[7] This is illustrated by the example of the cholesterol analog DHE, which can lose a proton in the excited state (Figure 18.20). The emission spectra of DHE bound to DMPC vesicles without or with 10 mol% cholesterol were examined (Figure 18.20). The long-wavelength emission from the ionized DHE was more intense for the vesicles that also contained cholesterol. This result indicates that cholesterol displaces DHE toward the surface of the bilayers, allowing deprotonation and thus resulting in a more intense long-wavelength emission. The concept of an excited-state reaction was also used to study diffusion of lipids in membranes. Similarly, pyrene-labeled lipids were used to determine the extent to which excimer emission was formed by diffusion or due to preexisting ground-state complexes.[50]

18.7. CONCLUSION

There is an extensive literature on the theoretical and experimental aspects of excited-state reactions. It is known that some coumarins display emission from three states, resulting in more complex data and analysis.[51–54] Also, there have been detailed reports describing methods of identifying excited-state reactions and determining the kinetic parameters.[55–60] In this chapter we could only present a summary to serve as an introduction to this topic.

REFERENCES

1. Ireland, J. F., and Wyatt, P. A. H., 1976, Acid–base properties of electronically excited states of organic molecules, in *Advances in Physical Organic Chemistry*, V. Gold and D. Bethell (eds.), Academic Press, New York, pp. 132–215.
2. Wan, P., and Shukla, D., 1993, Utility of acid–base behavior of excited states of organic molecules, *Chem. Rev.* **93:**571–584.
3. Martynov, I. Y., Demyashkevich, A. B., Uzhinov, B. M., and Kuz'min, M. G., 1977, Proton transfer reactions in the excited electronic states of aromatic molecules, *Usp. Khim. (Russian Chemical Physics)* **46:**3–31.
4. Shizuka, H., 1985, Excited state proton-transfer reactions and proton-induced quenching of aromatic compounds, *Acc. Chem. Res.* **18:**141–147.
5. Schulman, S. G., 1976, Acid–base chemistry of excited singlet states, in *Modern Fluorescence Spectroscopy*, E. L. Wehry (ed.), Plenum Press, New York, pp. 239–275.
6. Gafni, A., and Brand, L., 1978, Excited state proton transfer reactions of acridine studied by nanosecond fluorometry, *Chem. Phys. Lett.* **58:**346–350.
7. Ofran, M., and Feitelson, J., 1973, Time dependence of dissociation in the excited state of β-naphthol, *Chem. Phys. Lett.* **19:**427–431.
8. Tsutsumi, K., and Shizuka, H., 1980, Proton transfer and acidity constant in the excited state of naphthols by dynamic analyses, *Z. Phys. Chem. N. F.* **122:**129–142.

9. Harris, C. M., and Selinger, B. K., 1980, Proton-induced fluorescence quenching of 2-naphthol, *J. Phys. Chem.* **84**:891–898.
10. Harris, C. M., and Sellinger, B. K., 1980, Acid–base properties of 1-naphthol. Proton-induced fluorescence quenching, *J. Phys. Chem.* **84**:1366–1371.
11. Webb, S. P., Yeh, S. W., Philips, L. A., Tolbert, M. A., and Clark, J. H., 1984, Ultrafast excited-state proton transfer in 1-naphthol, *J. Am. Chem. Soc.* **106**:7286–7288.
12. Boyer, R., Deckey, G., Marzzacco, C., Mulvaney, M., Schwab, C., and Halpern, A. M., 1985, The photophysical properties of 2-naphthol, *J. Chem. Educ.* **62**:630–632.
13. Bardez, E., Monnier, E., and Valeur, B., 1985, Dynamics of excited-state reactions in reversed micelles. 2. Proton transfer involving various fluorescent probes according to their sites of solubilization, *J. Phys. Chem.* **89**:5031–5036.
14. Loken, M. R., Hayes, J. W., Gohlke, J. R., and Brand, L., 1972, Excited-state proton transfer as a biological probe. Determination of rate constants by means of nanosecond fluorometry, *Biochemistry* **11**:4779–4786.
15. Laws, W. R., and Brand, L., 1979, Analysis of two-state excited-state reactions. The fluorescence decay of 2-naphthol, *J. Phys. Chem.* **83**:795–802.
16. Htun, M. T., Suwaiyan, A., and Klein, U. K. A., 1995, Time-resolved spectroscopy of 4-hydroxy-1-naphthalenesulphonate in alcohol–water mixtures, *Chem. Phys. Lett.* **243**:506–511.
17. Laws, W. R., Posner, G. H., and Brand, L., 1979, A covalent fluorescence probe based on excited state proton transfer, *Arch. Biochem. Biophys.* **193**:88–100.
18. Marciniak, B., Kozubek, H., and Paszyc, S., 1992, Estimation of pK_a in the first excited single state, *J. Chem. Educ.* **69**:247–249.
19. Davenport, L., Knutson, J. R., and Brand, L., 1986, Excited-state proton transfer of equilenin and dihydroequilenin: Interaction with bilayer vesicles, *Biochemistry* **25**:1186–1195.
20. Lin, H., and Gryczynski, I., unpublished observations.
21. Förster, Th., 1950, Die pH-abhangigkeit der fluoreszenz von naphthalinderivaten, *Z. Electrochem.* **54**:531–553.
22. Grabowski, Z. R., and Grabowska, A., 1976, The Förster cycle reconsidered, *Z. Phys. Chem. N. F.* **104**:197–208.
23. Grabowski, Z. R., 1981, Generalized Förster cycle applied to coordination compounds, *J. Lumin.* **24/25**:559–562.
24. Grabowski, Z. R., and Rubaszewska, W., 1977, Generalised Förster cycle, *J. Chem. Soc., Faraday Trans. I* **73**:11–28.
25. Birks, J. B., 1970, *Photophysics of Aromatic Molecules*, Wiley-Interscience, New York.
26. Brand, L., and Laws, W. R., 1983, Excited-state proton transfer, in *Time-Resolved Fluorescence Spectroscopy in Biochemistry and Biology*, R. D. Cundall and F. E. Dale (eds.), Plenum Press, New York, pp. 319–340.
27. Lakowicz, J. R., and Balter, A., 1982, Differential wavelength deconvolution of time-resolved fluorescence intensities: A new method for the analysis of excited state processes, *Biophys. Chem.* **16**:223–240.
28. Rumbles, G., Smith, T. A., Brown, A. J., Carey, M., and Soutar, I., 1997, Autoreconvolution—an extension to the "reference convolution" procedure for the simultaneous analysis of two fluorescence decays from one sample, *J. Fluoresc.* **7**(3):217–229.
29. Lakowicz, J. R., and Balter, A., 1982, Theory of phase-modulation fluorescence spectroscopy for excited state processes, *Biophys. Chem.* **16**:99–115.
30. Lakowicz, J. R., and Balter, A., 1982, Analysis of excited state processes by phase-modulation fluorescence spectroscopy, *Biophys. Chem.* **16**:117–132.
31. Lakowicz, J. R., and Balter, A., 1982, Detection of the reversibility of an excited state reaction by phase modulation fluorometry, *Chem. Phys. Lett.* **92**:117–121.
32. Spencer, R. D., and Weber, G., 1969, Measurement of subnanosecond fluorescence lifetimes with a cross-correlation phase fluorometer, *Ann. N.Y. Acad. Sci.* **158**:361–376.
33. Veselova, T. V., Limareva, L. A., Cherkasov, A. S., and Shirokov, V. I., 1965, Fluorometric study of the effect of solvent on the fluorescence spectrum of 3-amino-*N*-methylphthalimide, *Opt. Spectrosc.* **19**:39–43.
34. Bakhshiev, N. G., Mazurenko, Yu. T., and Piterskaya, I. V., 1966, Luminescence decay in different portions of the luminescence spectrum of molecules in viscous solutions, *Opt. Spectrosc.* **21**:307–309.
35. Gryczynski, I., unpublished observations.
36. Löfroth, J-E., 1985, Recent developments in the analysis of fluorescence intensity and anisotropy data, in *Analytical Instrumentation*, A. J. W. G. Visser (ed.), Marcel Dekker, New York, pp. 403–431.
37. Itoh, M., Tokumura, K., Tanimoto, Y., Okada, Y., Takeuchi, H., Obi, K., and Tanaka, I., 1982, Time-resolved and steady-state fluorescence studies of the excited state proton transfer in 3-hydroxyflavone and 3-hydroxychromone, *J. Am. Chem. Soc.* **104**:4146–4150.
38. McMorrow, D., and Kasha, M., 1984, Intramolecular excited-state proton transfer in 3-hydroxyflavone. Hydrogen-bonding solvent perturbations, *J. Phys. Chem.* **88**:2235–2243.
39. Bulska, H., 1983, Intramolecular cooperative double proton transfer in [2,2′bipyridyl]-3,3-diol, *Chem. Phys. Lett.* **98**:398–402.
40. Borowicz, P., Grabowska, A., Wortmann, R., and Liptay, W., 1992, Tautomerization in fluorescent states of bipyridyl-diols: A direct confirmation of the intramolecular double proton transfer by electro-optical emission measurements, *J. Lumin.* **52**:265–273.
41. Klöpffer, W., 1977, Intramolecular proton transfer in electronically excited molecules, in *Advances in Photochemistry*, J. N. Pitts, G. S. Hammond, and K. Gollnick (eds.), John Wiley & Sons, New York, pp. 311–358.
42. Waluk, J., Bulska, H., Pakula, B., and Sepiol, J., 1981, Red edge excitation study of cooperative double proton transfer in 7-azaindole, *J. Lumin.* **24/25**:519–522.
43. Kim, Y. T., Yardley, J. T., and Hochstrasser, R. M., 1989, Solvent effects on intramolecular proton transfer, *Chem. Phys.* **136**:311–319.
44. Sytnik, A., Gormin, D., and Kasha, M., 1994, Interplay between excited state intramolecular proton transfer and charge transfer in flavonols and their use as protein-binding site fluorescence probes, *Proc. Natl. Acad. Sci. U.S.A.* **91**:11968–11972.
45. Sekikawa, T., Kobayashi, T., and Inabe, T., 1997, Femtosecond fluorescence study of proton-transfer process in thermochromic crystalline salicylideneanilines, *J. Phys. Chem. B* **101**:10645–10652.
46. Chapman, C. F., and Maroncelli, M., 1992, Excited state tautomerization of 7-azaindole in water, *J. Phys. Chem.* **96**:8430–8441.
47. Szabo, A. G., Krajcarski, D. T., Cavatorta, P., Masotti, L., and Barcellona, M. L., 1986, Excited state pKa behaviour of DAPI. A rationalization of the fluorescence enhancement of DAPI in DAPI–nucleic acid complexes, *Photochem. Photobiol.* **44**:143–150.
48. Gutman, M., Huppert, D., and Nachliel, E., 1982, Kinetic studies of proton transfer in the microenvironment of a binding site, *Eur. J. Biochem.* **121**:637–642.
49. Gutman, M., Nachliel, E., and Huppert, D., 1982, Direct measurement of proton transfer as a probing reaction for the microenvironment of the apomyoglobin heme-binding site, *Eur. J. Biochem.* **125**:175–181.

50. Hresko, R. C., Sugar, I. P., Barenholz, Y., and Thompson, T. E., 1986, Lateral distribution of a pyrene-labeled phosphatidylcholine in phosphatidylcholine bilayers: Fluorescence phase and modulation study, *Biochemistry* **25**:3813–3823.
51. Bauer, R. K., Kowalczyk, A., and Balter, A., 1977, Phase fluorometer study of the excited state reactions of 4-methylumbelliferone, *Z. Naturforsch. A* **32**:560–564.
52. Nemkovich, N. A., Matseiko, V. I., Rubinov, A. N., and Tomin, V. I., 1979, Kinetics of the spontaneous luminescence of β-methylumbelliferone solutions in the nanosecond region, *Opt. Spectrosc.* **47**:490–493.
53. Balter, A., and Rolinski, O., 1984, Excited state reactions of 4-methylumbelliferone studied by nanosecond fluorometry, *Z. Naturforsch. A* **39**:1035–1040.
54. de Melo, J. S., and Macanita, A. L., 1993, Three interconverting excited species: Experimental study and solution of the general photokinetic triangle by time-resolved fluorescence, *Chem. Phys. Lett.* **204**:556–562.
55. Beechem, J. M., Ameloot, M., and Brand, L., 1985, Global analysis of fluorescence decay surfaces: Excited-state reactions, *Chem. Phys. Lett.* **120**:466–472.
56. Ameloot, M., Boens, N., Andriessen, R., Van den Bergh, V., and De Schryver, F. C., 1991, Non a priori analysis of fluorescence decay surfaces of excited state processes. 1. Theory, *J. Phys. Chem.* **95**:2041–2047.
57. Andriessen, R., Boens, N., Ameloot, M., and De Schryver, F. C., 1991, Non a priori analysis of fluorescence decay surfaces of excited-state processes. 2. Intermolecular excimer formation of pyrene, *J. Phys. Chem.* **95**:2047–2058.
58. Boens, N., Andriessen, R., Ameloot, M., Van Dommelen, L., and De Schryver, F. C., 1992, Kinetics and identifiability of intramolecular two-state excited state processes. Global compartmental analysis of the fluorescence decay surface, *J. Phys. Chem.* **96**:6331–6342.
59. Van Dommelen, L., Boens, N., Ameloot, M., De Schryver, F. C., and Kowalczyk, A., 1993, Species-associated spectra and upper and lower bounds on the rate constants of reversible intramolecular two-state excited state processes with added quencher. Global compartmental analysis of the fluorescence decay surface, *J. Phys. Chem.* **97**:11738–11753.
60. Van Dommelen, L., Boens, N., De Schryver, F. C., and Ameloot, M., 1995, Distinction between different competing kinetic models of irreversible intramolecular two-state excited-state processes with added quencher. Global compartmental analysis of the fluorescence decay surface, *J. Phys. Chem.* **99**:8959–8971.

PROBLEMS

18.1. *Lifetime of the R State*: Use the data in Figure 18.15 to calculate the decay time of the 2-PI excimer emission, assuming that this state could be excited directly.

18.2. *Interpretation of Time-Resolved Decays of Acridine*: The excited-state protonation of acridine was examined by time-resolved measurements of the fluorescence decays.[27] Neutral acridine is protonated in the excited state by ammonium ions. The impulse response functions for acridine in 0.2*M* NH_4NO_3, pH 8.3 (Figure 18.11), are listed in Table 18.4.

A. If the absorption spectra of acridine in 0.05*M* NaOH, 0.05*M* H_2SO_4, and 0.2*M* NH_4NO_3 were available, how would you use them to distinguish between ground-state and excited-state protonation of acridine?

B. What characteristics of these data most clearly illustrate that an excited-state reaction is present? Do these data indicate a two-state reaction or some more complex process? Is the reaction reversible or irreversible?

Table 18.4. Time-Resolved Decays of Acridine in 0.2*M* NH_4 NO_3[a,b]

λ (nm)	α_1	τ_1 (ns)	α_2	τ_2 (ns)
400	0.503	3.94	–0.001	33.96
410	0.220	4.00	—	—
420	0.491	3.88	0.002	25.20
450	0.067	3.90	0.028	30.05
500	–0.010	4.91	0.082	29.13
540	–0.036	3.76	0.064	29.83
550	–0.036	3.56	0.058	29.94
560	–0.029	3.86	0.046	29.90

[a]Ref. 27.

[b]The $|\alpha_i\tau_i|$ products are normalized to the steady-state emission intensity at each wavelength.

19

Fluorescence Sensing

Fluorescence sensing of chemical and biochemical analytes is an active area of research.[1–7] This research is being driven by the desire to eliminate radioactive tracers, which are costly to use and dispose of. Additionally, there is a need for rapid and low-cost testing methods for a wide range of clinical, bioprocess, and environmental applications. During the past decade, we have witnessed the introduction of numerous methods based on high-sensitivity fluorescence detection, including DNA sequencing, DNA fragment analysis, fluorescence staining of gels following electrophoretic separation, and a variety of fluorescence immunoassays. Historically, one can trace many of these analytical applications to the classic reports by Undenfriend and co-workers,[8,9] which anticipated many of today's applications of fluorescence. More recent monographs have summarized the numerous analytical applications of fluorescence.[10–14]

Why do we use fluorescence rather than absorption for high-sensitivity detection? The answer lies in the different ways of measuring absorbance and fluorescence. Light absorbance is measured as the difference in intensity between light passing through the reference and the sample which contains the absorbers. In fluorescence, the intensity is measured directly, without comparison with a reference beam. Consider a $10^{-10}M$ solution of a substance with a molar extinction coefficient of $10^5 M^{-1}$ cm^{-1}. The absorbance will be 10^{-5} per centimeter, which is equivalent to a percentage transmission of 99.9977%. Even with exceptional optics and electronics, it will be very difficult to detect the small percentage of absorbed light, 0.0023%. Even if the electronics allow measurement of such a low optical density, the cuvettes will show some variability, which will probably lead to an intensity difference exceeding that due to an absorbance of 10^{-5}. In contrast, fluorescence detection at $10^{-10}M$ is readily accomplished with most fluorometers. This advantage is due to measurement of the fluorescence relative to a dark background, as compared to the bright reference beam in an absorbance measurement. In fact, it is relatively easy to detect low levels of light, and the electronic impulses due to single photons are measurable with most PMTs.

In this chapter we describe the various mechanisms of fluorescence sensing, which include essentially all the phenomena discussed in previous chapters. Fluorescence sensing is described mostly within the framework of its medical applications, but it is clear that fluorescence detection is also widely used in biochemical, chemical, and environmental analysis.

19.1. OPTICAL CLINICAL CHEMISTRY

The value of fluorescence as an analytical method can be seen by imagining future devices which will be used in the doctor's office, at the patient's bedside, or for home health care. Assume that the appropriate fluorescent sensors were available to detect clinically relevant electrolytes (Figure 19.1). Then, a blood sample could be withdrawn directly into the sensing device, which could be a syringe, capillary tube, or self-filling cartridge. The sensor would then be placed into an instrument which would determine the analyte concentration. It is expected that such point-of-care instruments would use solid-state light sources and perform sensing based on fluorescence intensity, polarization, or lifetime. The goal of optical clinical chemistry is to provide information in real time, without handling of blood and without the need to send samples to a central laboratory. While the drawings in Figure 19.1 may seem like science fiction, a capillary sensor for electrolytes has already been reported, and this sensor uses lifetimes to determine the analyte concentrations.[15]

While the devices shown schematically in Figure 19.1 seem simple, one can imagine even simpler devices. For instance, photolithographic methods have been used to create arrays with a different DNA sequence at each position on the array[16,17] and to create patterns of proteins on

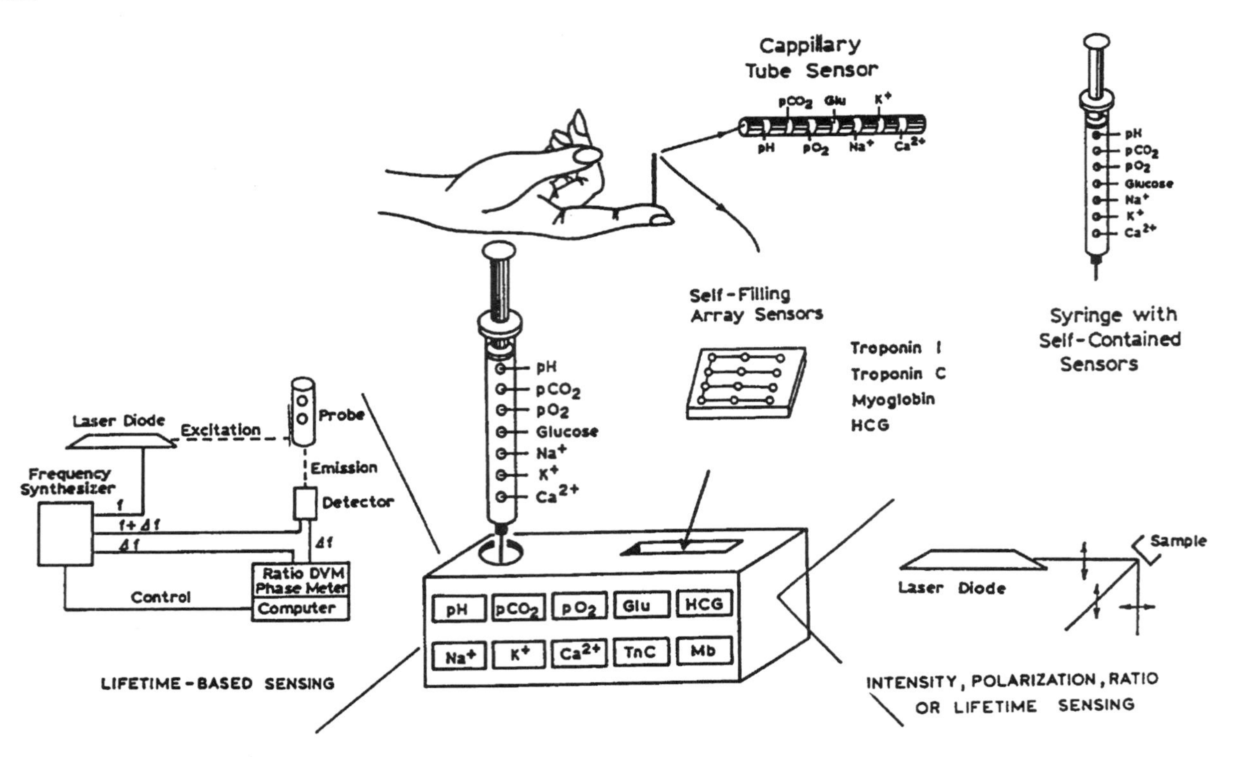

Figure 19.1. Optical clinical chemistry. The sensing fluorophores may be placed on the walls of a syringe or in a self-filling cartridge or capillary tube. HCG, Human chorionic gonadotropin.

two-dimensional surfaces.[18,19] One can imagine these photochemical methods being applied for localization of sensing fluorophores on a two-dimensional sensor (Figure 19.2). Such portable devices would find widespread usefulness, especially in emergency situations. The sensor array could be exposed to blood, and the results would be immediately available. It is this concept of rapid point-of-care clinical chemistry which is driving the rapid development of numerous fluorescence sensing devices. In the following sections, we describe the principles of fluorescence sensing and illustrate how such sensing devices provide analytical data.

19.2. SPECTRAL OBSERVABLES FOR FLUORESCENCE SENSING

In Chapter 1, we described how fluorescein was used to detect an underground connection between two rivers. In this case, sensing was accomplished by detection of the fluorescein in the second river, following its addition to the first river. In the present chapter, sensing is understood in a different context. The goal is to measure the amount of some analyte, not the amount of probe. For example, in the case of blood gas measurements, the analytes are pH, pCO_2, and pO_2. Blood electrolytes—Na^+, K^+, Ca^{2+}, Mg^{2+},

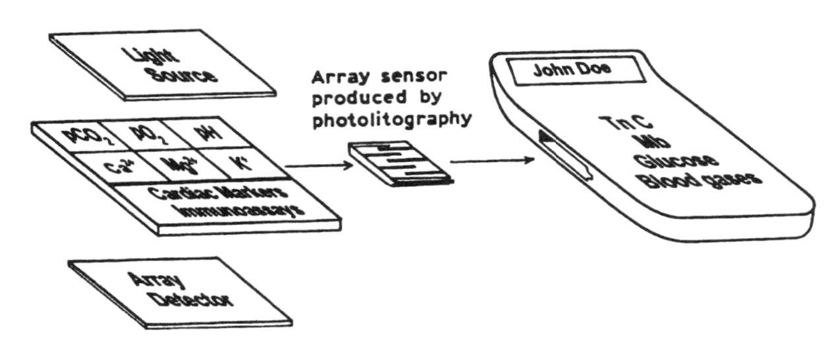

Figure 19.2. Second-generation fluorescence sensor for point-of-care testing. TnC, Troponin C; Mb, myoglobin.

and Cl^-—and many additional analytes are also measured in the clinical laboratory.[20] Fluorescence sensing requires that a fluorescence property change in response to the analyte. Changes can occur in the intensity, excitation spectrum, emission spectrum, anisotropy, or lifetime of the sensing probe. These modes of sensing are shown schematically in Figure 19.3.

The most direct sensing method is based on the fluorescence intensity of the probe changing in response to the analyte. Such changes often occur for fluorophores that are subject to collisional quenching by a relevant species, such as oxygen. While conceptually simple, collisional quenching is only useful with a few clinically relevant analytes. Also, it is often inconvenient or unreliable to use intensity changes, which can occur for a wide variety of reasons. For instance, the use of fiber optics is desirable as a means of locating the sensor at the site of interest and having the light source and detector remotely located.[7] However, it is difficult to perform quantitative intensity measurements through fibers. Intensity measurements are also difficult in fluorescence microscopy. One cannot control the fluorophore concentration at each location in the cell, and the local probe concentration changes continually due to diffusion and/or photobleaching. For this reason, it is important to use measurements that are independent of fluorophore concentration. This has been accomplished using wavelength-ratiometric probes (Figure 19.3), which display shifts in their absorption or emission spectra upon binding of the analyte. The analyte concentration can then be determined from the ratio of fluorescence intensities measured at two excitation or two emission wavelengths. The use of wavelength-ratiometric probes is desirable because these ratios are independent of the probe concentration.

Another ratiometric method is based on the measurement of fluorescence polarization or anisotropy. In this case the analyte causes a change in the polarization of the label. Anisotropy measurements are frequently used in competitive immunoassays, in which the actual analyte displaces labeled analyte that is bound to specific antibody. This results in a decrease in the anisotropy. Anisotropy values are calculated using the ratio of polarized intensity measurements. The use of an intensity ratio makes the anisotropy measurements independent of fluorophore concentration as long as the measurements are not distorted by autofluorescence or poor signal-to-noise ratio. Polarization immunoassays are discussed in Section 19.9.D.

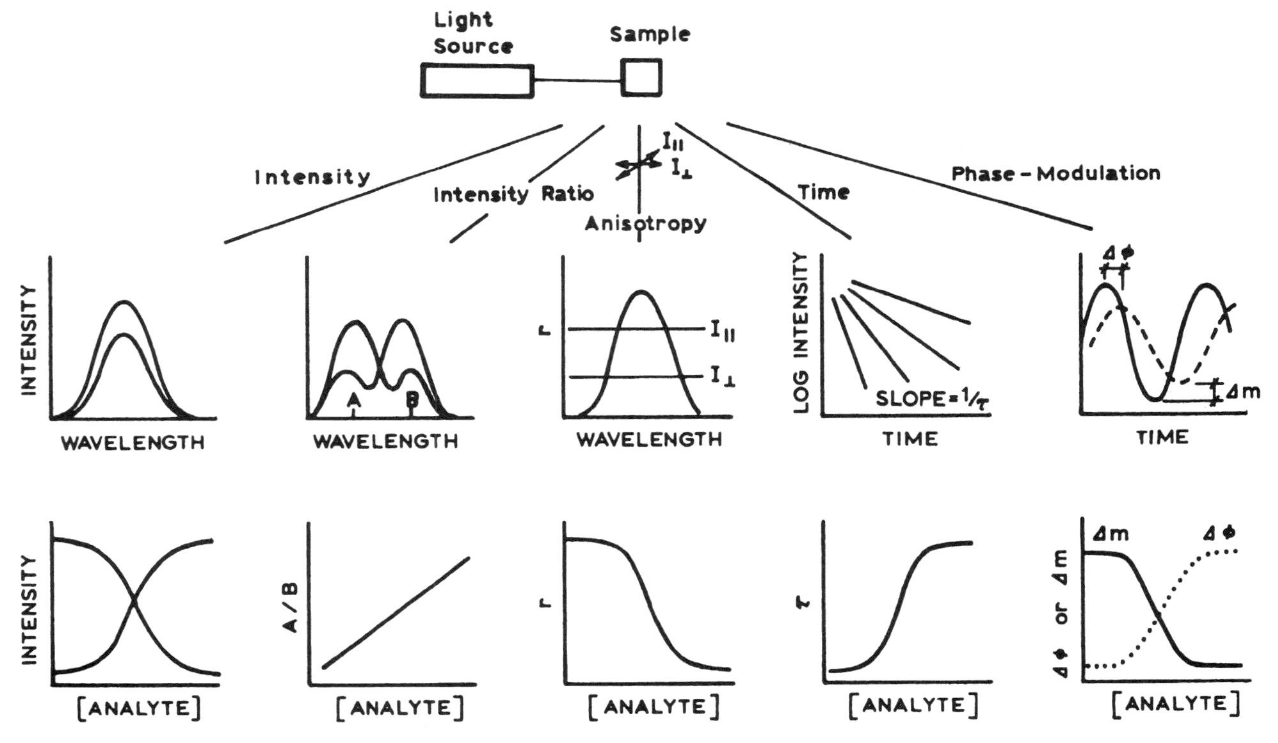

Figure 19.3. Spectral observables for fluorescence sensing. From left to right: Intensity, intensity ratio, anisotropy, time-domain lifetime, and phase-modulation lifetime.

One can also use fluorescence lifetimes for sensing (Figure 19.3). The lifetimes can be measured using either TD or FD methods. A few years ago, lifetime measurements were regarded as too complex for sensing applications. However, advances in electro-optics technology now make it feasible to perform nanosecond-decay-time measurements with small inexpensive instruments. In fact, the use of lifetimes for sensing is presently seen as the next step in making sensors that display the long-term stability needed in real-world applications.[12,21]

19.2.A. Optical Properties of Tissues

The design of fluorescence probes for clinical applications is determined in part by the optical properties of water and tissues. In general, the autofluorescence from tissues or any biological sample is lower for longer excitation wavelengths. Longer wavelengths also allow one to avoid light absorption by hemoglobin and melanin (Figure 19.4). While longer is better, there seem to be few fluorophores that emit at wavelengths longer than 900 nm. This is just as well because water absorption increases above 1000 nm. The region of low absorption from 600 to 1000 nm is sometimes called the therapeutic range.[22] Fortunately, a variety of lasers and solid-state lasers are available for excitation in this range of wavelengths.

19.2.B. Lifetime-Based Sensing

Prior to describing the various mechanisms of sensing, we consider it useful to expand on the use of decay-time measurements for sensing. The advantages of lifetime-based sensing are illustrated in Figure 19.5. In the research laboratory, where clean cuvettes and optical surfaces are easy to maintain, intensity measurements can be accurate and reproducible (top panel in Figure 19.5). However, suppose that the sample is blood, which is contained in a translucent syringe. The intensity will be decreased by the absorbance of the blood and by the scattering properties of the syringe, and it may be difficult to obtain a reliable intensity-based calibration. Now suppose one measures the lifetime instead of the intensity (middle panel in Figure 19.5). Assuming that the intensity is large enough to measure, the intensity decay is independent of attenuation of the signal. Similarly, if the lifetime is measured by phase or modulation (bottom panel in Figure 19.5), the values are expected to be independent of intensity.

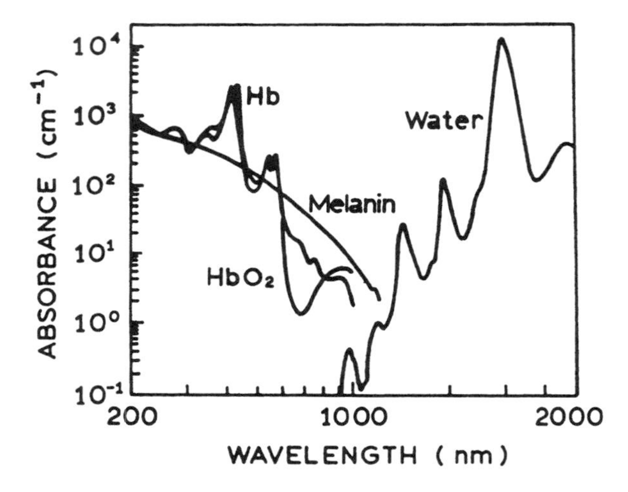

Figure 19.4. Optical absorbance of tissues and water. Hb, Hemoglobin. Revised and reprinted, with permission, from Ref. 22, Copyright © 1996, Annual Reviews.

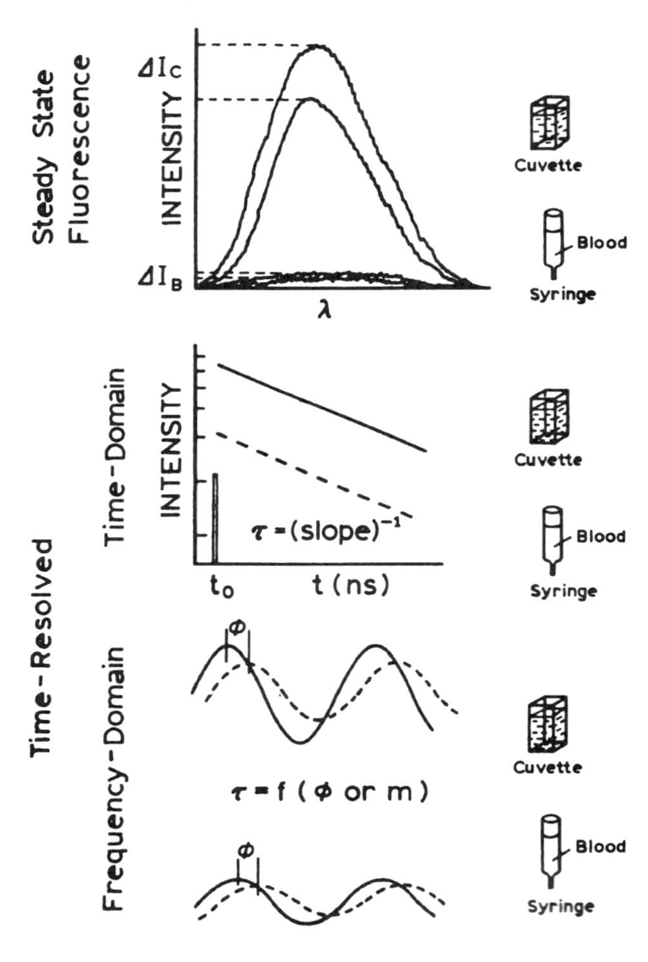

Figure 19.5. Intensity, time-domain, and frequency-domain sensing, as applied in the research laboratory (a cuvette) and in a clinical setting (a blood sample in a syringe). In the top panel, ΔI_C is the intensity change obtained from cuvette measurements, and ΔI_B is that obtained from measurements on blood in a syringe.

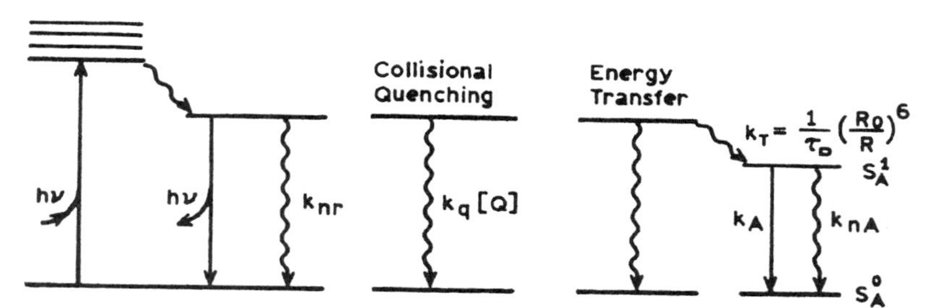

Figure 19.6. Modified Jabłoński diagram for the processes of absorption and fluorescence emission (*left*), dynamic quenching (*middle*), and RET (*right*).

Lifetime-based sensing is presently an active area of research and seems likely to become widely used as a means to avoid the difficulties of quantitative intensity measurements.[12,21] For instance, consider a sensing fluorophore placed on the end of an optical fiber which is used for oceanographic studies. The fiber will be flexing while being towed by the ship, resulting in intensity fluctuations. Additionally, the probe may be leaking from the tip, resulting in decreases in intensity. Quantitative intensity measurements would clearly be difficult under these conditions. Lifetime measurements have already been used for an oceanographic fiber-optic oxygen sensor.[23]

19.3. MECHANISMS OF SENSING

Any phenomenon which results in a change of fluorescence intensity, wavelength, anisotropy, or lifetime can be used for sensing. The simplest mechanism to understand is collisional quenching (Figure 19.6, middle). In this case, one identifies a fluorophore which is quenched by the analyte. Collisional quenching results in a decrease in the intensity or lifetime of the fluorophore, either of which can be used to determine the analyte concentration.

RET is perhaps the most general and valuable phenomenon for fluorescence sensing (Figure 19.6, right). Any process which brings the donor and acceptor into closer proximity will result in a decrease in the donor intensity and/or decay time. Since energy transfer acts over macromolecular distances, it can be used to detect association of proteins, as occurs in immunoassays. However, the applications of RET are not limited to detection of protein association. RET has also been used as the basis for pH and cation sensors. Sensors have been developed which contain acceptors whose absorption spectra are dependent on pH. A change in pH results in a change in absorbance of the acceptor, which in turn alters the donor intensity.

Another mechanism for sensing is available when the fluorophore can exist in two states, depending on the analyte concentration (Figure 19.7). Typically, there is an equilibrium between the fluorophore free in solution and the fluorophore bound to analyte. One form can be nonfluorescent, in which case emission is only seen in the absence or presence of analyte, depending on which form is fluorescent. Probes which act in this manner are not wavelength-ratiometric or lifetime probes. Alternatively, both forms may be fluorescent but differ in quantum yield or emission spectra. This type of behavior is often seen for pH probes, where ionization of the probe results in distinct absorption and/or emission spectra. Spectral shifts are also seen for probes which bind specific cations such as calcium. Such probes allow wavelength-ratiometric measure-

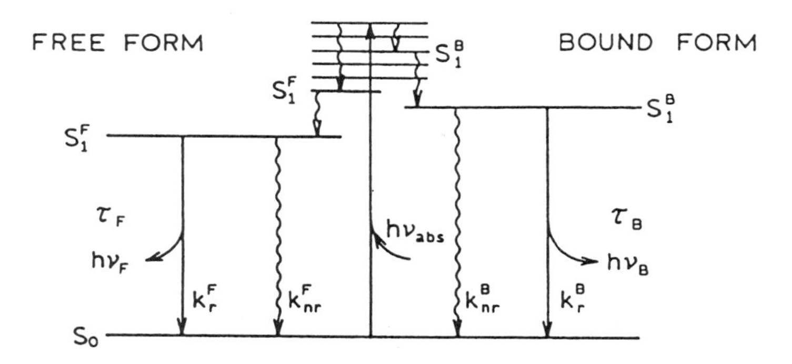

Figure 19.7. Jabłoński diagram for the free (F) and bound (B) forms of a sensing probe. From Ref. 12.

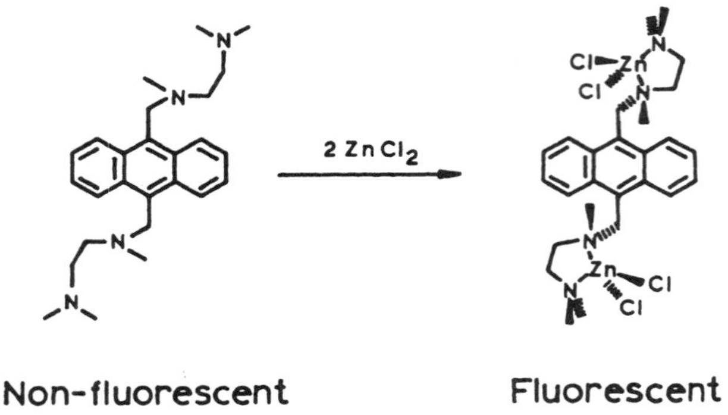

2 $ZnCl_2$

Non-fluorescent Fluorescent

Figure 19.8. A zinc probe based on photoinduced electron transfer. Revised from Ref. 26.

ments. In this case the change in intensity or shift in the emission spectrum is used to determine the analyte concentration. Probes which bind specific analytes are often referred to as probes of analyte recognition.[24]

There are many mechanisms which can be used to design probes that exhibit changes in fluorescence in response to analytes. Fluorescent probes can form TICT states.[25] Another mechanism is photoinduced electron transfer (PET), which has been used to develop sensors for metal ions.[26–28] These sensors often rely on the well-known quenching of polynuclear aromatic hydrocarbons by amines. An example is shown in Figure 19.8 for a zinc sensor. In the absence of zinc, the anthracene is quenched by exciplex formation with the amino groups. Upon binding of zinc, exciplex formation no longer occurs, and the anthracene becomes fluorescent.[26] While the mechanism of this particular sensor is understood, this is not true of all sensors. In many cases, spectral changes are seen for which the mechanism is not certain. In the following sections, we provide examples of each type of sensor (collisional, RET, and analyte recognition).

19.4. SENSING BY COLLISIONAL QUENCHING

19.4.A. Oxygen Sensing

Use of collisional quenching as the sensing mechanism requires the fluorescent probe to be sensitive to quenching by the desired analyte. Collisional quenching results in a decrease in intensity and lifetime, which can be described by the Stern–Volmer equation,

$$\frac{F_0}{F} = \frac{\tau_0}{\tau} = 1 + k_q\tau_0[Q] = 1 + K[Q] \qquad [19.1]$$

In this equation, F_0 (τ_0) and F (τ) are the intensities (lifetimes) in the absence and the presence of the quencher, respectively, K is the Stern–Volmer quenching constant, and k_q is the bimolecular quenching constant. The most obvious application of collisional quenching is oxygen sensing. In order to obtain sensitivity to low concentrations of oxygen, fluorophores are typically chosen which have long lifetimes in the absence of oxygen (τ_0). Long lifetimes are a property of transition-metal complexes[29] (Chapter 20), and such complexes have been frequently used in oxygen sensors.[30,31] For use in an oxygen sensor, the MLC is usually dissolved in silicone, in which oxygen is rather soluble and freely diffusing. Also, the silicone serves as a barrier to other interfering molecules which might interact with the fluorophores and affect the intensity or lifetime.

The high sensitivity of the long-lifetime MLCs to oxygen is shown by the Stern–Volmer plots in Figure 19.9. One notices that $[Ru(Ph_2phen)_3]^{2+}$ is more strongly quenched than $[Ru(phen)_3]^{2+}$. The difference in sensitivity is due to the longer unquenched lifetime (τ_0) of the diphenyl derivative, and thus the larger Stern–Volmer quenching constant (Eq. [19.1]). These long-lifetime probes have been used in real-time oxygen sensors. For

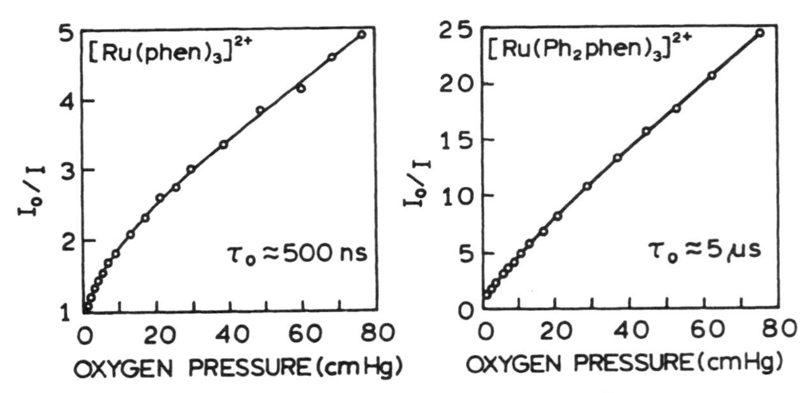

Figure 19.9. Stern–Volmer plots for oxygen quenching of $[Ru(phen)_3]^{2+}$ and $[Ru(Ph_2phen)_3]^{2+}$ in GE RTV 118 silicone. Phen is 1,10-phenanthroline, and Ph_2phen is 4,7-diphenyl-1,10-phenanthroline. Revised from Ref. 29.

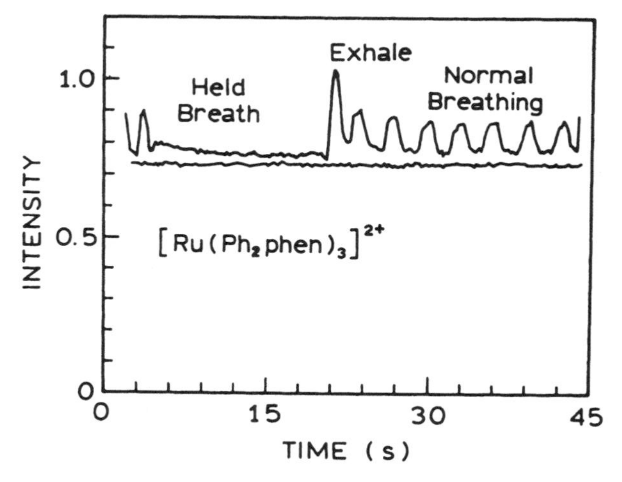

Figure 19.10. Luminescence intensity of an oxygen sensor, with $[Ru(Ph_2phen)_3]^{2+}$ as the probe, when exposed to breathing. Revised from Ref. 29.

example, Figure 19.10 shows the intensity of $[Ru(Ph_2phen)_3]^{2+}$ in silicone while exposed to exhaled air. The intensity increases with each exhale because of the lower O_2 and higher CO_2 content of the exhaled air. The higher intensity on the first exhale, after the breath was held, is due to the lower O_2 content in the air which was retained longer in the lungs. The oxygen sensitivity of the sensor can be adjusted by selecting probes with different lifetimes or by modifying the chemical composition of the supporting media.[32] Hence, a variety of oxygen sensors are available, and one can expect to find increasing use of these simple devices.

19.4.B. Lifetime-Based Sensing of Oxygen

For practical sensing applications, the overall device must be relatively simple and inexpensive, which is possible using the long-lifetime MLCs. The oxygen-sensitive MLCs in Figure 19.9 absorb near 450 nm and are thus easily excited with blue LEDs. One simple oxygen sensor device is shown in Figure 19.11. In fact, such a sensor with an LED light source was described in one of the earliest reports on lifetime-based sensing.[33,34] Because of the long decay times and simple instrumentation, we used oxygen sensors in early studies to demonstrate the stability of phase-angle sensing in the presence of large-amplitude intensity fluctuations.[35] We varied the intensity by waving our fingers in the light path, resulting in fivefold changes in intensity (Figure 19.11). In contrast to the measured intensities, the phase angles remained constant. The stability of phase angles in the presence of intensity changes has resulted in the current enthusiasm for lifetime-based sensing of oxygen, even for imaging applications.[36,37]

It is important to recognize the value of lifetime measurements for probes which are subject to collisional quenching. It is unlikely that such probes will ever display wavelength-ratiometric behavior. The capability for ratiometric measurements can be designed into oxygen sensors by including a nanosecond-lifetime fluorophore in the supporting media as a reference fluorophore that provides a signal mostly insensitive to the oxygen concentration.

19.4.C. Mechanism of Oxygen Selectivity

An important consideration for any sensor is selectivity. For oxygen, the selectivity is provided by a unique combination of the fluorophore and the supporting media. Almost all fluorophores are collisionally quenched by oxygen, so that no fluorophore is completely specific for oxygen. However, the extent of quenching is proportional to the unquenched lifetime τ_0 (Eq. [19.1]). For fluorophores in aqueous solution with decay times under 5 ns, the extent of quenching by dissolved oxygen from the atmosphere is negligible. Hence, a reason for the apparent oxy-

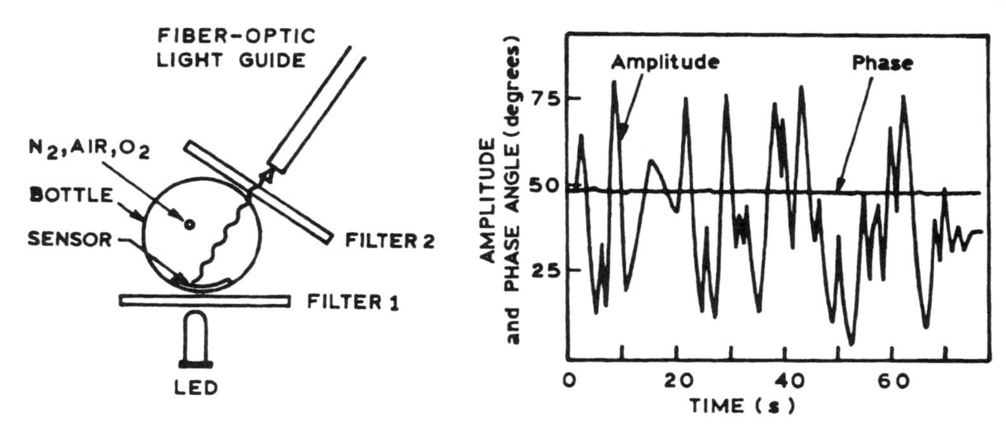

Figure 19.11. Phase-angle stability with intensity fluctuations, measured with an oxygen sensor device. The amplitude of the incident light was varied by waving fingers between the LED and the sensor. Revised from Ref. 35.

gen selectivity of $[Ru(Ph_2phen)_3]^{2+}$ is its long lifetime, near 5 μs, which results in extensive quenching by atmospheric oxygen.

Selectivity of the MLC oxygen sensor is also due to the silicone support. Silicone is impermeable to most polar species, so most possible interferants cannot penetrate the silicone to interact with the probe. Fortunately, oxygen dissolves readily in silicone, so that the support is uniquely permeable to the desired analyte. Finally, there are no other substances in air which act as collisional quenchers. NO is also a quencher but is not usually found in the air. Hence, the sensor is selective for O_2 because of a combination of the long lifetime of the MLC probe and the exclusion of potential interferants from the nonpolar silicone support.

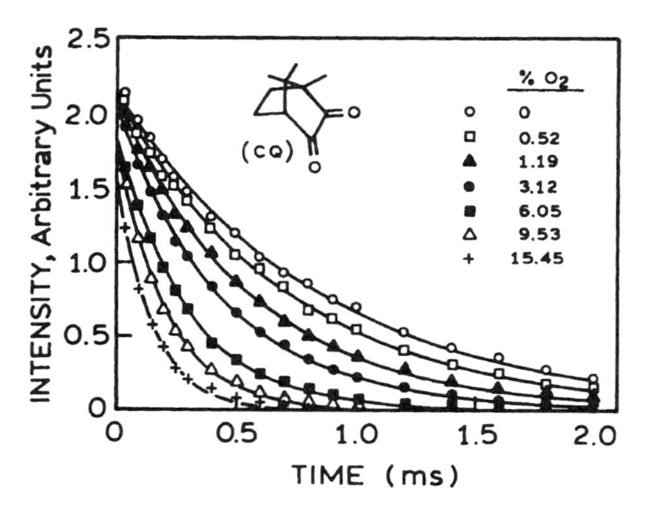

Figure 19.13. Phosphorescence lifetime measurements for camphorquinone in poly(methyl methacrylate) (PMMA), illustrating the influence of oxygen concentration on the triplet decay. The polymeric supports are different in Figures 19.12 and 19.13. Revised and reprinted from Ref. 39, Copyright © 1994, with permission from Elsevier Science.

19.4.D. Other Oxygen Sensors

While $[Ru(Ph_2phen)_3]^{2+}$ is the most commonly used fluorophore in oxygen sensors, other probes are available. Almost any long-lived fluorophore can be used in an oxygen sensor, particularly when dissolved in an organic solvent.[38] Because of the long decay times, phosphorescence has also been frequently used to detect oxygen. One example is presented in Figure 19.12, which shows the excitation and emission spectra of camphorquinone[38] in poly(vinyl chloride) (PVC). The oxygen concentration could be determined from the change of intensity (Figure 19.12) or of the phosphorescence decay times (Figure 19.13). In the case of camphorquinone, the signal was unstable due to photochemical reactions which were occurring in the triplet state. Hence, oxygen measurements based on the intensities were not reliable. However, the decay times consistently yielded the true oxygen concentrations.

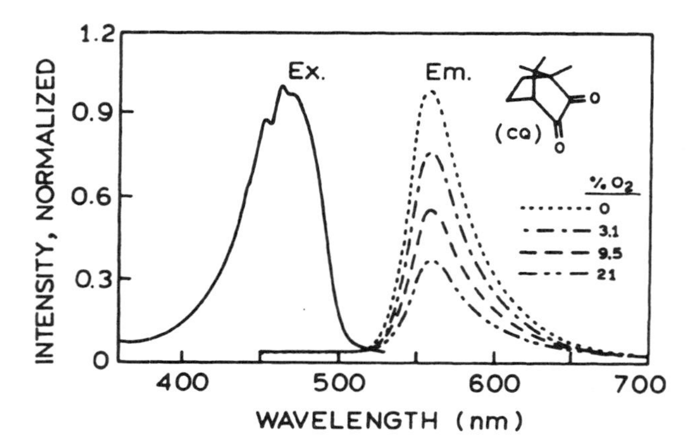

Figure 19.12. Normalized excitation and emission spectra of camphorquinone in poly(vinyl chloride), illustrating the effect of oxygen concentration on phosphorescence intensity. Revised and reprinted from Ref. 39, Copyright © 1994, with permission from Elsevier Science.

For many applications, such as oxygen sensing in blood or through skin, it is useful to have probes which can be excited with red or NIR wavelengths. Several porphyrin derivatives have recently been described which display oxygen-sensitive phosphorescence.[40] One example is platinum(II) octaethylporphyrin ketone (Figure 19.14), which can be excited at 600 nm. This molecule shows a surprisingly large Stokes' shift, with the emission maximum at 758 nm. The lifetime of 61.4 μs results in oxygen-sensitive emission even when the probe is embedded in polystyrene.

Another long-wavelength oxygen probe is $[Os(Ph_2phen)_3]^{2+}$. In contrast to ruthenium MLCs, the osmium MLCs absorb at wavelengths longer than 600 nm. These MLCs can be excited with laser diodes up to 670 nm (Figure 19.15).[41] The unquenched lifetime is near 300 ns, making the Os complex less sensitive than $[Ru(Ph_2phen)_3]^{2+}$ but still adequately sensitive for some applications (Figure 19.16). Such long-wavelength-excitable Os complexes have been used to measure oxygen through skin.[42] Examination of the Stern–Volmer plots for quenching of $[Os(Ph_2phen)_3]^{2+}$ by oxygen reveals that drastically different results were found for the same probe in two different silicones (Figure 19.16). This illustrates an important aspect of sensing: the supporting media is often as important as the probe in determining the response.[32,34,43,44]

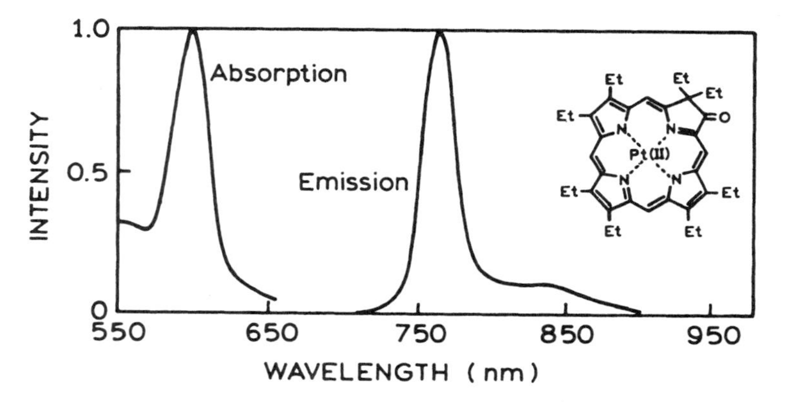

Figure 19.14. Absorption and emission spectra of a phosphorescent porphyrin ketone derivative. Revised from Ref. 40.

19.4.E. Chloride Sensors

It is well known that heavy atoms like bromine and iodine act as collisional quenchers. For sensing applications, chloride is more important because it is prevalent in biological systems. However, chloride is a less effective quencher, and relatively few fluorophores are quenched by chloride. A hint for developing chloride-sensitive probes was available from the knowledge that quinine is quenched by chloride. Quinine contains a quinolinium ring, and other compounds containing this ring were subsequently used to make a variety of chloride-sensitive probes.[45–48] Representative structures are shown in Figure 19.17. It is

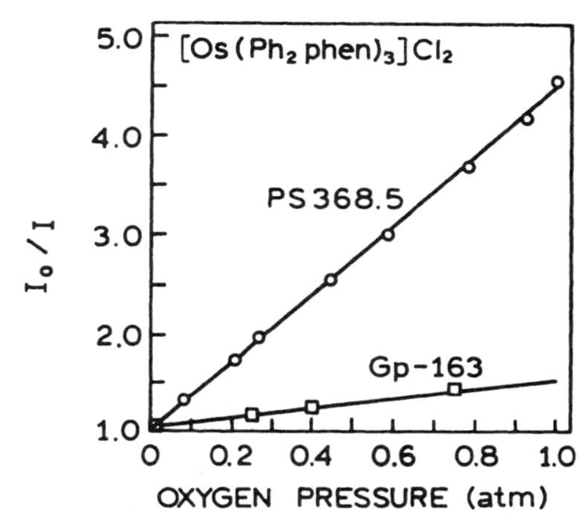

Figure 19.16. Stern–Volmer plots for quenching of $[Os(Ph_2phen)_3]Cl_2$ by oxygen in PS368.5 and Gp-103 silicone. Data from Ref. 41.

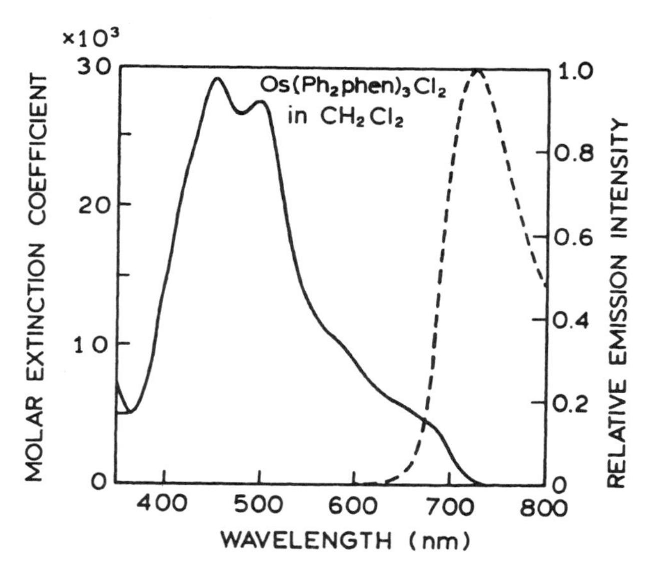

Figure 19.15. Absorption (———) and corrected emission (– – –) spectra of $[Os(Ph_2phen)_3]Cl_2$ in CH_2Cl_2. Revised and reprinted, with permission, from Ref. 41, Copyright © 1996, American Chemical Society.

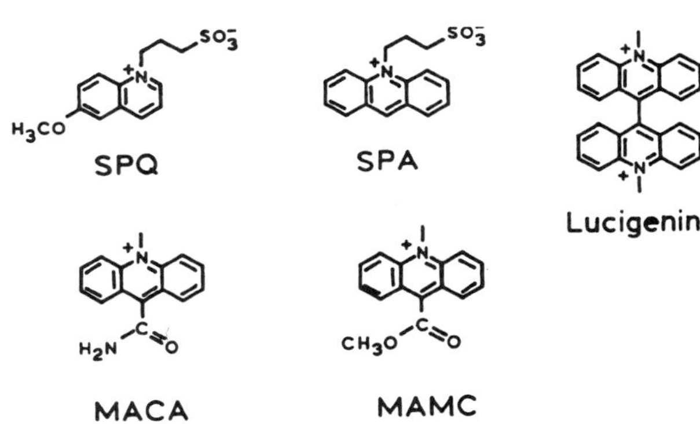

Figure 19.17. Representative chloride probes. These probes are collisionally quenched by chloride (see Table 19.1). Revised and reprinted, with permission, from Ref. 49, Copyright © 1994, Academic Press, Inc.

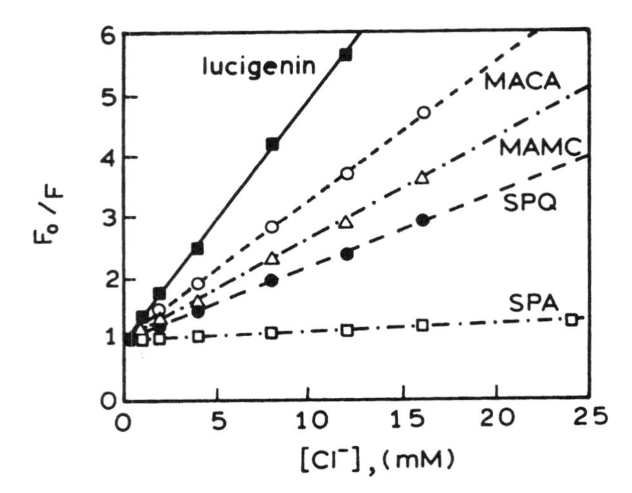

Figure 19.18. Stern–Volmer plots for chloride quenching of SPQ, SPA, lucigenin, MACA, and MAMC. See Table 19.1. Revised and reprinted, with permission, from Ref. 49, Copyright © 1994, Academic Press, Inc.

Table 19.1. Spectral Properties of Representative Chloride Probes[a]

Compound[b]	λ_{Abs} (nm)	λ_{em} (nm)	Quantum yield	$K\ (M^{-1})$
SPQ[c]	318/345	450	0.69	118[d]
SPA	—	—	—	5
Lucigenin	368/955	506	0.67	390
MACA	364/422	500	0.64	225
MAMC	366/424	517	0.24	160

[a]Ref. 49.
[b]Abbreviations: SPQ, 6-methoxy-*N*-(3-sulfoproxyl)quinolinuim; SPA, *N*-sulfopropylacridinium; lucigenin, *N,N'*-dimethyl-9,9'-bisacridinium nitrate; MACA, *N*-methylacridinium-9-carboxamides, MAMC, *N*-methylacridinium-9-methylcarboxylate.
[c]The unquenched lifetime of SPQ is 26 ns.
[d]Stern–Volmer constant in cells is 13 M^{-1} (Ref. 51).

evident from the Stern–Volmer plots in Figure 19.18 that all the quinolines are not equally sensitive to chloride and that the quenching constant depends on the chemical structure (Table 19.1).

These chloride-sensitive probes can be used to measure chloride transport across cell membranes (Figure 19.19). Erythrocyte ghosts are the membranes from red blood cells following removal of the intracellular contents. The ghosts were loaded with SPQ and 100m*M* chloride. The ghosts were then diluted into a solution of 66m*M* K_2SO_4. Sulfate does not quench SPQ. When the ghosts were diluted into sulfate-containing buffer, the intensity of SPQ increased owing to efflux of the chloride. This transport is due to an anion-exchange pathway. Chloride transport could be blocked by dihydro-4,4'-diisothiocyanostilbene-2,2'-disulfonic acid (H_2DIDS), which is an inhibitor of anion transport. Hence, the chloride probes can be used for physiological studies of ion transport.

The chloride probes are subject to interference. They are quenched by bromide, iodide, and thiocyanate.[50] Perhaps more importantly, SPQ is quenched by free amines, which can distort measurements in amine-containing buffers. In fact, the Stern–Volmer quenching constant of SPQ in

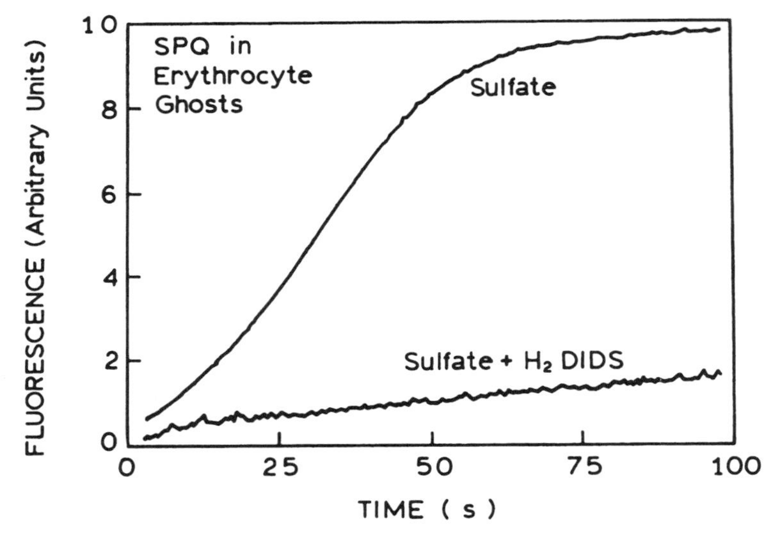

Figure 19.19. Fluorescence intensity of SPQ in erythrocyte ghosts as a function of time, reflecting chloride transport from the erythrocyte ghosts. The two curves are for SPQ-containing erythrocyte ghosts deluted into sulfate-containing buffer in the absence and in the presence of an anion-transport inhibitor (H_2DIDS). Revised and reprinted, with permission, from Ref. 46, Copyright © 1987, American Chemical Society.

aqueous solution is 118 M^{-1}, whereas in cells the quenching constant is near 13 M^{-1}. This decrease has been attributed to quenching of SPQ by nonchloride anions and proteins in cells.[51] Quenching of SPQ by amines was turned into an opportunity, by using the quenching caused by the amine buffers as an indicator of the pH.[50] As the pH increases, more of the buffer is in the free amine form, resulting in a decrease in the intensity of SPQ. Another disadvantage of the chloride probes is that they are not ratiometric probes. Some of the probes leak out of cells, causing the intensity to decrease and preventing accurate measurements of the chloride concentration.

19.4.F. Other Collisional Quenchers

A wide variety of molecules can act as quenchers (Chapter 8), and they permit development of sensors based on collisional quenching. Benzo[*b*]fluoroanthene was found to be highly sensitive to sulfur dioxide.[52] Oxygen interfered with the measurements but was 26-fold less efficient as a quencher than SO_2. Halogenated anesthetics are known to quench protein fluorescence and can be detected by collisional quenching of anthracene and perylene.[53] Carbazole is quenched by a wide variety of chlorinated hydrocarbons.[54] NO, which serves as a signal for blood vessel dilation, is also a collisional quencher.[55,56]

19.5. ENERGY-TRANSFER SENSING

RET offers many opportunities and advantages for fluorescence sensing. Energy transfer occurs whenever the donor and acceptor are within the Förster distance. Changes in energy transfer can occur due to changes in analyte proximity or due to analyte-dependent changes in the absorption spectrum of the acceptor (Figure 19.20). A significant advantage of RET-based sensing is that it simplifies the design of the fluorophore. For collisional quenching, or analyte recognition probes (Section 19.8), the probe must be uniquely sensitive to the analyte. It is frequently difficult to obtain the desired sensitivity and fluorescence spectral properties in the same molecule.

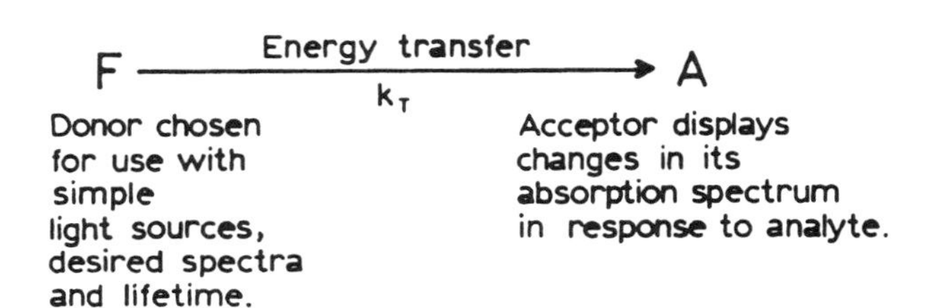

Figure 19.20. Principle of energy-transfer sensing.

However, if RET is used, the donor and acceptor can be separate molecules (Figure 19.20). The donor can be selected for use with the desired light source and need not be intrinsically sensitive to the analyte. The acceptor can be chosen to display a change in absorption in response to the analyte. Alternatively, the sensor can be based on a changing concentration of acceptor around the donor.

19.5.A. pH and pCO_2 Sensing by Energy Transfer

A wide variety of pH indicators are available from analytical chemistry. Since indicators are intended for visual observation, they display pH-dependent absorption spectra, with absorption at visible wavelengths. These indicators have formed the basis for a number of RET pH/pCO_2 sensors. One of the earliest reports used eosin as the donor and Phenol Red as the acceptor.[57] Phenol Red was selected because it displays a pK_a near 7, and the basic form absorbs at 546 nm, where eosin emits. Consequently, the eosin intensity decreased as the pH increased. In the case of this particular sensor, it was not certain whether the decreased intensity was due to RET or to an inner filter effect, but it is clear that RET is a useful mechanism as a basis for designing sensors.

This same basic idea was used to create lifetime-based sensors for pH, pCO_2,[58,59] and ammonia.[60] Spectra for a representative sensor are shown in Figure 19.21. The donor was sulforhodamine 101 (SR 101), whose emission overlapped with the absorption of the acceptor Thymol Blue (TB).[59] The acceptor displayed a pH-dependent change in absorption, which was expected to result in an increase in the extent of energy transfer with an increase in pH. The average lifetime of SR 101 in the sensor was measured by

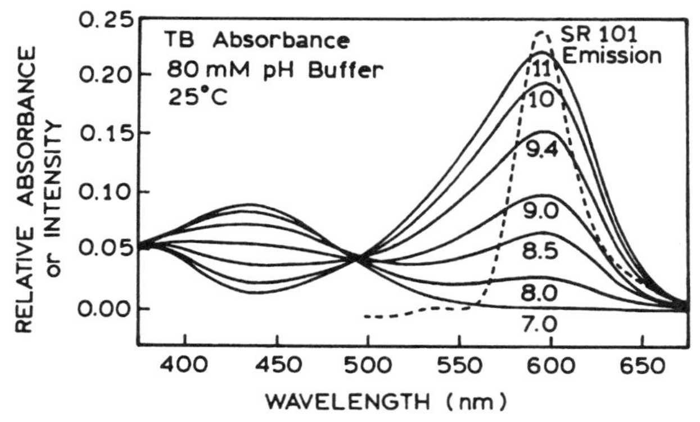

Figure 19.21. Spectral characteristics of the donor and acceptor employed in an RET pCO_2 sensor. The solid curves show the absorption spectra of the Thymol Blue (TB) acceptor at different pH values (labeled on the curves), demonstrating overlap with the emission spectrum of the SR101 donor (---). From Ref. 59.

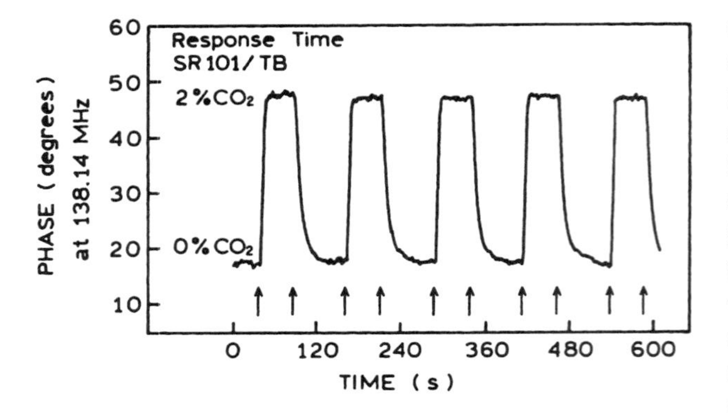

Figure 19.22. Response of an RET pCO_2 sensor employing SR 101 as donor and TB as acceptor to changes in the percentage of CO_2 between 0 and 2%. The phase-angle measurements were made at a light modulation frequency of 138.14 MHz. From Ref. 59.

the FD method. The phase angle was found to be sensitive to the percentage of CO_2 (Figure 19.22). The use of lifetimes confirmed that the change of intensity is due to energy transfer, since inner filter effects are not expected to alter the lifetime. The CO_2 sensor responded in seconds and was highly reproducible. Although not shown in Figure 19.22, the phase angles were demonstrated to be independent of intensity oscillations in the HeCd laser source.[59]

A critical point in sensor design is the support containing the probes. For the pCO_2 sensor, the support consisted of an ethyl cellulose (EC) film coated on a glass slide. The EC contained tetraoctylammonium hydroxide (TOAH), which served as a phase-transfer agent for the CO_2. These details are mentioned to stress the fact that careful consideration of the support and actual use of the sensor is needed to produce a useful device. Depending on the support and analyte, it may be necessary to use phase-transfer agents to facilitate uptake of the analyte into the supporting media.[61] Another important technique for fabricating sensors is the use of sol gels. The basic idea is to hydrolyze tetraethylorthosilicate [TEOS; $Si(OC_2H_5)_4$] in a mixture of water and ethanol. As TEOS is hydrolyzed, it forms silica (SiO_2). Monoliths and glass films of silica can be formed at low temperatures under mild conditions. The porosity of the sol gels can be controlled, and fluorophores or even enzymes can be trapped within the sol gel matrix.[62,63] The result is a solid sensor which contains the trapped molecules.

19.5.B. Glucose Sensing by Energy Transfer

Control of blood glucose is crucial for the long-term health of diabetics. Present methods of measuring glucose require fresh blood, which is obtained by a finger stick. This procedure is painful and inconvenient, making it difficult to determine the blood glucose as frequently as is needed. Erratic blood glucose levels due to diabetes are responsible for adverse long-term problems of blindness and heart disease. These effects are thought to be due to glycosylation of protein in blood vessels.

Because of the medical need, there have been numerous efforts to develop a noninvasive means to measure blood glucose. Unfortunately, such a method is not presently available. There have been continued efforts to develop

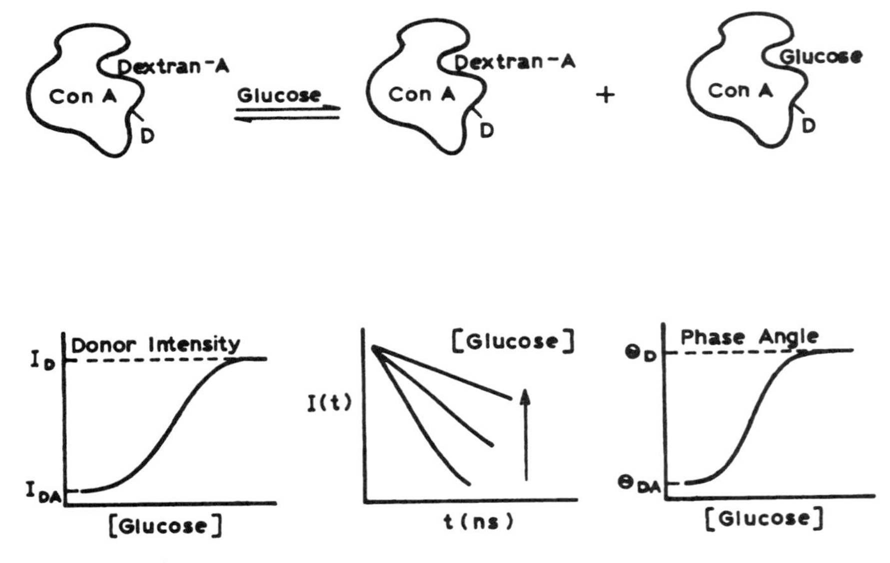

Figure 19.23. Glucose sensing based on RET. Revised from Ref. 64.

fluorescence methods to detect glucose. These have often been based on the glucose binding protein concanavalin A (Con A) and a polysaccharide, typically dextran, which serves as a competitive ligand for glucose (Figure 19.23).[64] Typically, the Con A is labeled with a donor (D) and the dextran with an acceptor (A), but the labels can be reversed. Binding of D-Con A to A-dextran results in a decrease in donor intensity or lifetime. The glucose in the sample competes for the glucose binding sites on D-Con A, releasing D-Con A from the acceptor. The intensity decay time and phase angles of the donor are thus expected to increase with increasing glucose concentration.

These principles were used in the first reports of glucose sensing based on fluorescence intensities.[65–67] Several designs were presented. The first glucose sensor (Figure 19.24, top) used dextran labeled with FITC as the probe and no acceptor. Con A was immobilized on the sides of the sensor. The FITC-dextran bound to Con A on the sides of the sensor was outside of the light path. Hence, in the absence of glucose, the FITC-dextran intensity was low.

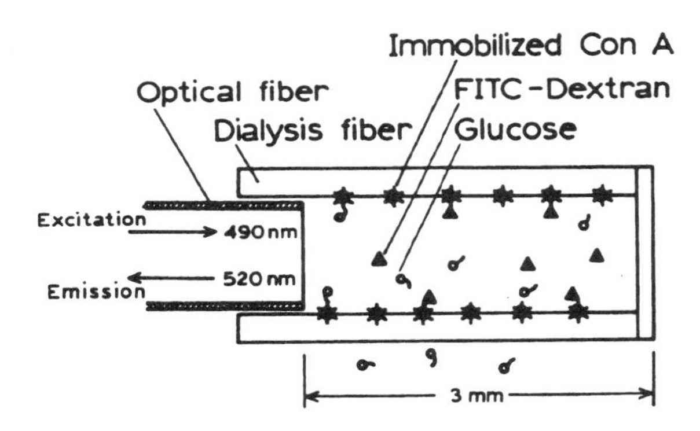

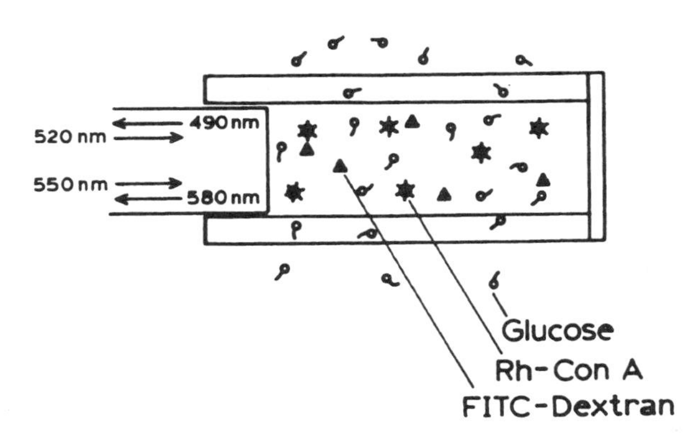

Figure 19.24. RET sensors for glucose. Revised from Ref. 66.

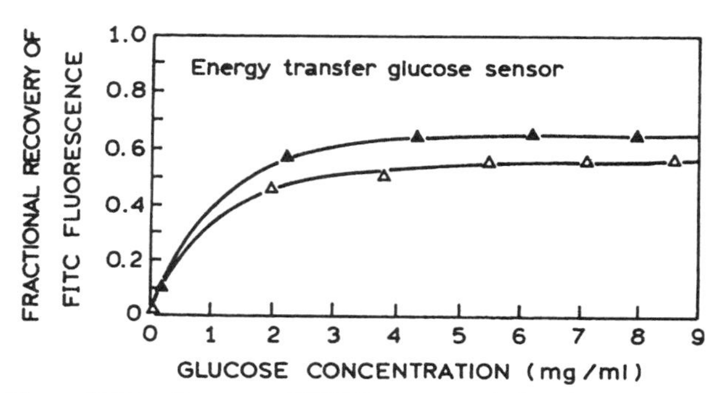

Figure 19.25. Recovery of FITC fluorescence during glucose titration for Rh-Con A/FITC-dextran ratios of 217 (▲) and 870 w/w (△). Revised from Ref. 67.

Addition of glucose released the FITC-dextran to the central region of the sensor, where it could be observed. The sensor design was then changed to a homogeneous format (Figure 19.24, bottom), with FITC-labeled dextran and rhodamine-labeled Con A (Rh-Con A) as the acceptor. The acceptor could be directly excited as a control measurement to determine the amount of Rh-Con A.

The response of this glucose sensor is shown in Figure 19.25. The donor and acceptor were placed on the dextran and Con A, respectively. It is important to notice that the donor fluorescence was not completely recovered at high concentrations of glucose. This lack of complete reversibility is a problem which plagues Con A-based glucose sensors to the present day. It is hoped that these problems can be solved using alternative glucose binding proteins, especially those which have a single glucose binding site and may be less prone to irreversible associations. It seems probable that site-directed mutagenesis will be used to modify the glucose binding proteins to obtain the desired glucose affinity and specificity.

As might be expected, lifetime-based sensing has been applied to glucose. It has been accomplished using nanosecond probes,[68] long-lifetime probes,[69] and laser-diode-excitable probes.[70] The problem of reversibility has been addressed by using sugar-labeled proteins, in an attempt to minimize cross-linking and aggregation of the multivalent Con A. Such glucose sensors are occasionally fully reversible,[70] but there is reluctance to depend on a system in which reversibility is difficult to obtain.

19.5.C. Ion Sensing by Energy Transfer

While wavelength-ratiometric probes are available for Ca^{2+} and Mg^{2+}, the performance of similar probes for Na^+ and K^+ is inadequate (Section 19.8.C). Also, it is difficult to design probes for K^+ and Na^+ with the desired binding constant and selectivity. These difficulties can be understood by considering a sensor for K^+. In blood, the concen-

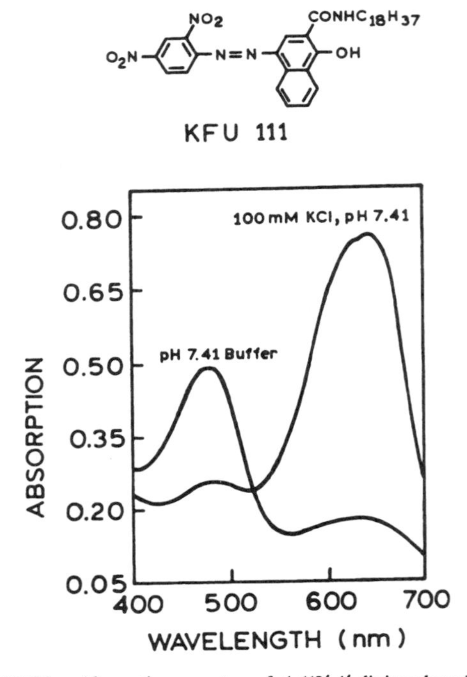

Figure 19.26. Absorption spectra of 4-((2′,4′-dinitrophenyl)azo)-2-((octadecylamino)carbonyl)-1-naphthol (KFU 111) in a plasticized PVC membrane containing potassium tetrakis(4-chlorophenyl)borate (PTCB) and valinomycin, in contact with a 100mM aqueous solution of KCl at pH 7.41 or with potassium-free buffer at pH 7.41. Revised from Ref. 71.

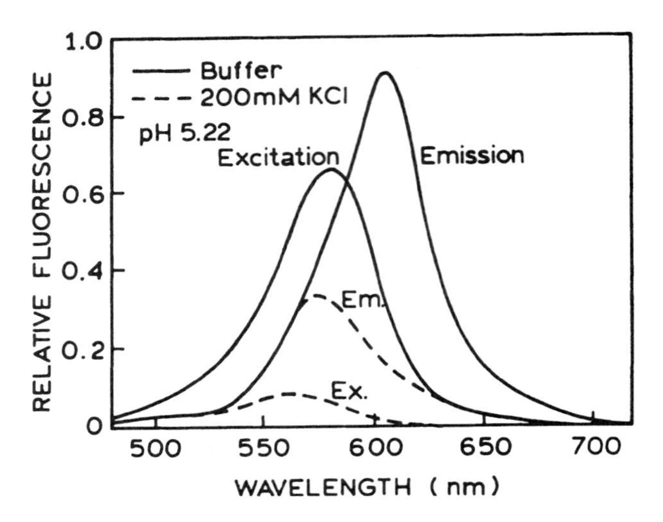

Figure 19.27. Excitation and emission spectra of FluoSphere™ particles contained in a plasticized PVC membrane containing potassium tetrakis(4-chlorophenyl)- borate(PTCB), valinomycin, and KFU 111, in contact with a 200mM KCl solution at pH 5.22 (---) or with plain buffer of pH 5.22 (———). The emission spectrum of the FluoSphere particles in the presence of K^+ is distorted owing to the inner filter effect caused by the blue form of the absorber dye. Revised and reprinted, with permission, from Ref. 71, Copyright © 1993, American Chemical Society.

trations of K^+ and Na^+ are near 4.5mM and 120mM, respectively. A probe for K^+ must be able to bind K^+ selectively and not be saturated by a 25-fold excess of Na^+. Given the similar chemical properties of Na^+ and K^+, such selectivity is difficult to achieve. These problems can be alleviated to some extent by making use of ionophores like valinomycin which display high selectivity for K^+. Valinomycin is a cyclic molecule which unfortunately does not contain any useful chromophoric groups. Consequently, one has to develop a sensor which transduces the binding of K^+ to result in a fluorescence spectral change.

One method that has been employed to develop a K^+ sensor is to use a dye which displays a change in its absorption spectrum on ionization.[71] The sensor was fabricated using pvc and valinomycin, plus fluorescent beads (FluoSphere™).* The acceptor, 4-((2′,4′-dinitrophenyl)-azo)-2-((octadecylamino)carbonyl)-1-naphtho(KFU 111), is shown in Figure 19.26. When K^+ enters the membrane, KFU 111 loses a proton, resulting in an increase in absorbance at 650 nm. This absorbance overlaps with the emission spectrum of the FluoSphere™ particles, resulting in a decreased intensity (Figure 19.27). As the K^+ concentration is increased, the FluoSphere intensity is progressively decreased (Figure 19.28).

Examination of Figure 19.27 reveals that the emission spectrum is distorted at higher concentrations of K^+. This indicates that the mechanism of K^+ sensing is not energy transfer but is an inner filter effect due to KFU 111, as was clearly stated by the authors.[71] In this particular sensor, there was no opportunity for the fluorophore and absorption dye to interact, as the fluorophores were in the beads, and thus distant from the absorber. The important point is that spectral overlap does not imply that the mechanism is energy transfer. Potassium sensors using valinomycin but other donors such as rhodamine and cyanine dyes have also been reported.[72,73] For the cyanine donor 1,1′-dioctadecyl-3,3,3′,3′-tetramethylindodicarbocyanine perchlorate [$DiIC_{18}(5)$], lifetime measurements demonstrated that RET was the dominant mechanism for K^+ sensing.[73]

One difficulty with RET sensing is that the extent of energy transfer depends strongly on acceptor concentration, so that the sensors require frequent calibration. This problem can potentially be circumvented by using covalently linked donors and acceptors. Few such sensors have been reported, primarily owing to the synthetic difficulties.

*FluoSphere is a trademark of Molecular Probes, Inc., Eugene, Oregon.

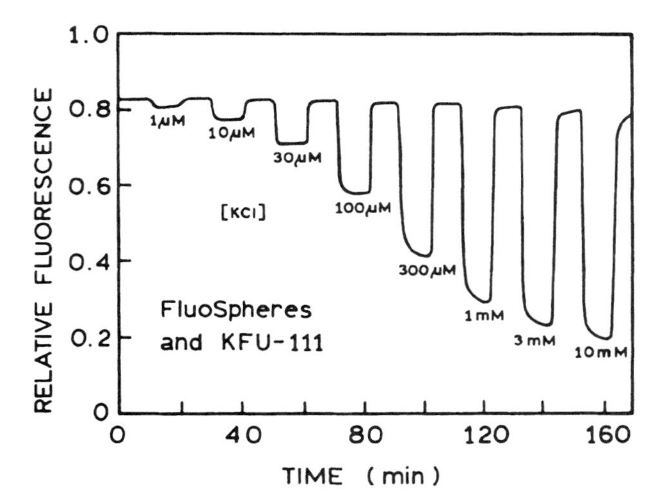

Figure 19.28. Response time, relative signal change, and reversibility of the potassium sensor in the presence of dye KFU 111 in the membrane; pH 5.82; excitation/emission wavelengths set to 560/605 nm. The concentrations of K^+ are indicated on the figure. The sensor did not respond to K^+ without the presence of KFU 111 in the membrane. Revised and reprinted, with permission, from Ref. 71, Copyright © 1993, American Chemical Society.

19.5.D. Theory for Energy-Transfer Sensing

The theory for sensing based on energy transfer is complex and depends on the nature of the sensor. There are two limiting cases: (1) unlinked donors and acceptors distributed randomly in space and (2) covalently linked donor–acceptor pairs. Suppose that the donor–acceptor pair is not linked and that the acceptor can exist in two forms with different absorption spectra and Förster distances (R_{01} and R_{02}). The intensity decay is given by

$$I_{DA}(t) = I_0 \exp\left[-\frac{t}{\tau_D} - 2(\gamma_1 + \gamma_2)\sqrt{\frac{t}{\tau_D}}\right] \qquad [19.2]$$

where τ_D is the donor decay time and γ_1 and γ_2 are functions of the acceptor concentration and are related to R_{01} and R_{02} as described in Chapter 15 (Eqs. [15.2] and [15.3]). Alternatively, the donor and acceptor may be covalently linked. In this case, the intensity decay is given by

$$I_{DA}(t) = I_D\left[g_1 \int_0^\infty P(r) \exp\left(\frac{-t}{\tau_{DA1}}\right) dr + (1-g_1) \int_0^\infty P(r) \exp\left(\frac{-t}{\tau_{DA2}}\right) dr\right] \qquad [19.3]$$

where

$$\frac{1}{\tau_{DAi}} = \frac{1}{\tau_D} + \frac{1}{\tau_D}\left(\frac{R_{0i}}{r}\right)^6 \qquad [19.4]$$

$P(r)$ is the distance distribution, and g_1 and $(1-g_1)$ are the fractional intensities of each form at $t = 0$. The transfer efficiency (E) can be calculated as

$$E = 1 - \frac{F_{DA}}{F_D} = 1 - \int_0^\infty I_{DA}(t)\, dt \Big/ \int_0^\infty I_D(t)\, dt \qquad [19.5]$$

where $I_D(t)$ is the intensity decay of the donor in the absence of acceptor, and F_D and F_{DA} are the relative intensities of the donor in the absence and in the presence of acceptor, respectively. These general expressions can be used to simulate the expected performance of an energy-transfer sensor using known or estimated parameter values.[12]

19.6. TWO-STATE pH SENSORS

19.6.A. Optical Detection of Blood Gases

Prior to considering the spectral properties of fluorescent pH probes, it is useful to understand the clinical need for optical measurements of pH. The phrase "blood gases" refers to the measurement of pH, pCO_2, and pO_2 in arterial blood. Optical detection of blood gases is one of the Holy Grails of optical sensing. Blood gas measurements are frequently performed on patients in the intensive care unit, premature infants, and trauma victims. Since the status of such patients changes rapidly, it is important to obtain the blood gas results as quickly as possible. While the status of optical detection of blood gases is evolving rapidly, many of the currently used methods do not satisfy the needs of intensive care and emergency health care situations. In these unfortunate cases, the blood gases change on the timescale of minutes in response to the patient's physiological status.[74,75]

Determination of blood gases is difficult, time-consuming, and expensive. Measuring a blood gas requires taking a sample of arterial blood, placing it on ice, and transporting it to a central laboratory. At the central laboratory, the pH is measured using an electrode, and O_2 and CO_2 are measured by the Clark and Severinghaus electrodes, respectively. Even for a stat. request, it is difficult to obtain the blood gas report in less than 30 minutes, by which time the patient's status is often quite different. Additionally, handling of blood by health care workers is undesirable with regard to the risk of acquired immunodeficiency syndrome (AIDS) and other infectious diseases. At present, determination of blood gases is estimated to cost at least $400,000,000 per year in the United States.

How can optical sensing of blood gases improve on this situation? One currently available device is shown in Figure 19.29.[75] The basic idea is to place the sensor chemistry in a cassette, which is attached to an arterial line. When needed, blood is drawn into the cassette, and the blood gases are determined by appropriate fluorescent sensors. Oxygen is typically determined using $[Ru(Ph_2phen)_3]^{2+}$, and pH is usually measured using HPTS (see Section 19.6.2).[76] The ability to measure pH also allows pCO_2 to be measured using the bicarbonate couple.[77] This is accomplished by measuring the pH of a bicarbonate solution which is exposed to the CO_2. The concentration of dissolved CO_2 alters the extent of bicarbonate dissociation and hence the pH. With the use of such a device, the blood gas measurements can be made without loss or handling of blood, and the results can be available immediately. Such instruments are currently in use,[78,79] but their performance is sometimes less than desired, and there is a continuing effort to develop simpler and more stable devices. This idea of clinical chemistry using fiber optics originates with the early work of D. W. Lübbers and colleagues.[80]

Although optical detection of blood gases is not routine at present, one can expect point-of-care blood gas measurements to be used more frequently as the technology is perfected. This is because the blood gas data are needed immediately, a fact which may override the efficiencies available in the central clinical laboratory. Catheter-type devices similar to that shown in Figure 19.29 are now specified to provide blood gas measurements for several days. Also, one can imagine blood gas devices similar to that shown in Figure 19.2 being used in ambulances and emergency rooms.

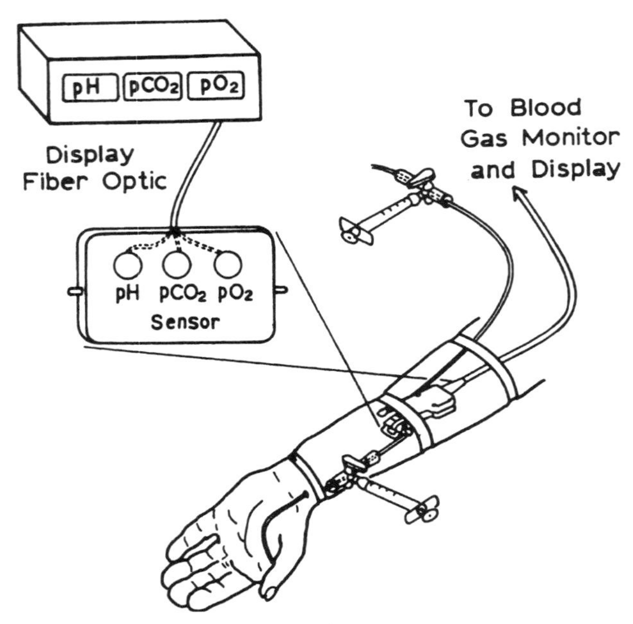

Figure 19.29. Arterial blood gas sensor. Revised from Ref. 75.

19.6.B. pH Sensors

Fluoresceins. Having shown the need for optical pH measurements, we now describe the pH-sensitive probe fluorescein, which was used in one of the earliest pH sensors.[81,82] Fluorescein and other pH-sensitive probes have also been used to measure pCO_2 by the bicarbonate couple.[83,84] One early use of fluorescein was to determine intracellular pH values. However, fluorescein leaks rapidly from cells, so highly charged derivatives are often used, such as 5(6)-carboxyfluorescein or 2′,7′-bis(2-carboxyethyl)-5(6)-carboxyfluorescein (BCECF) (Figure 19.30). Fluorescein displays a complex pH-dependent equilibrium and emission from the various ionic forms[85,86] (Figure 19.31). The lactone form is usually found in organic solvents and is not formed in aqueous solutions above pH 5. Only the two high-pH anionic forms are fluorescent (Figure 19.31).

Fluorescein is a moderately useful excitation wavelength-ratiometric probe. The absorption spectrum shifts to higher wavelengths with a pK_a near 6.5 (Figure 19.32). These absorption and emission spectral changes represent the equilibrium between the two fluorescent forms of fluorescein, the monoanion and dianion forms (Figure 19.31). These spectral changes allow wavelength-ratiometric pH measurements with two excitation wavelengths, near 450 and 495 nm. The intensity ratio increases with increasing pH (Figure 19.33). The data in Figure

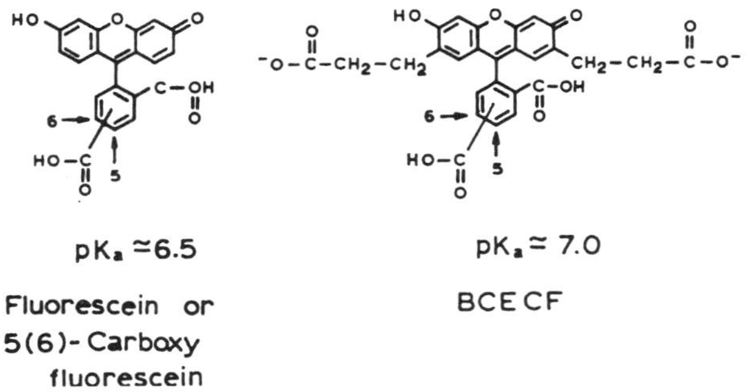

$pK_a \simeq 6.5$

Fluorescein or 5(6)-Carboxy fluorescein

$pK_a \simeq 7.0$

BCECF

Figure 19.30. Fluorescein-type pH probes.

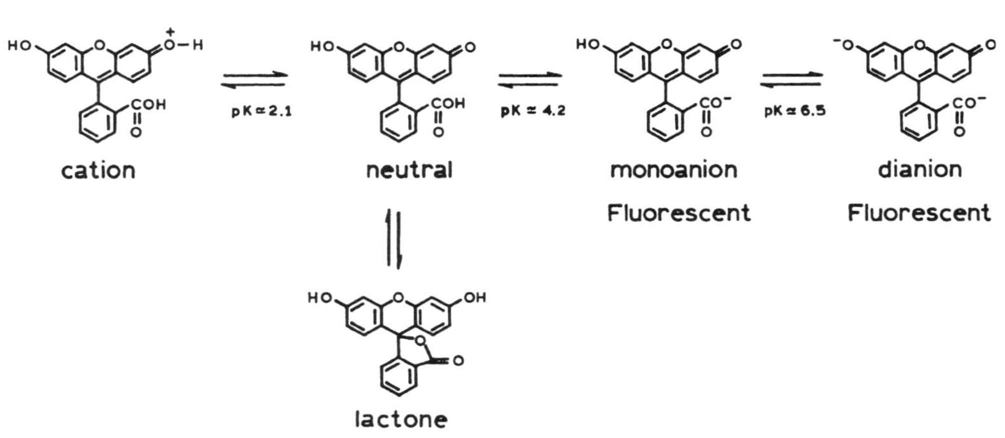

Figure 19.31. The pH-dependent ionization of fluorescein. Only the monoanion and dianion forms of fluorescein are fluorescent.

19.33 are for fluorescein linked to dextran, which was used to prevent the fluorescein from leaking out of the cells.

One disadvantage of fluorescein is that its pK_a is near 6.5, whereas the cytosolic pH of cells is in the range of 6.8–7.4. Hence, it is desirable to have a higher pK_a for accurate pH measurement. BCECF (Figure 19.30) was developed[87] to have spectral properties similar to those of fluorescein but to have a higher pK_a, near 7.0 (Table 19.2). This illustrates an important aspect of all sensing probes: the pK_a value or the analyte dissociation constant must be comparable to the concentration of the analyte to be measured. However, it can be difficult to adjust a pK_a value or dissociation constant. Additionally, the affinities observed in solution may be quite different from the values needed in a sensor, where the sample may be diluted prior to the measurements or the polymeric support may affect the binding constant. For these reasons, significant development is often needed from the point of identifying a sensor to its use in a clinical application.

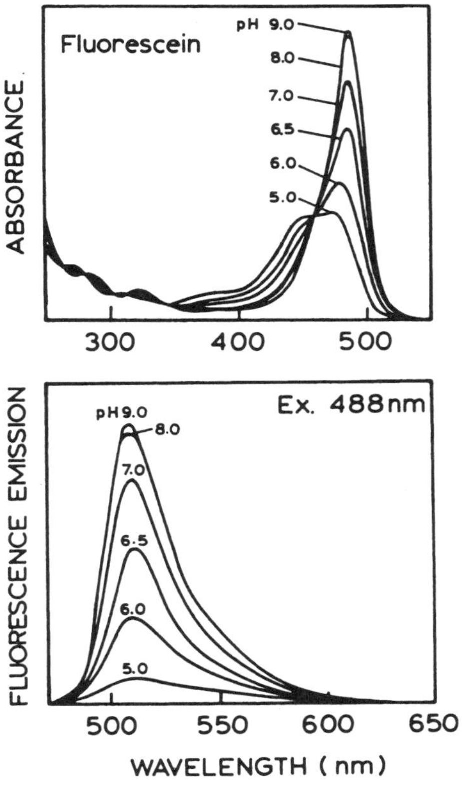

Figure 19.32. Absorption (*top*) and emission spectra (*bottom*) of fluorescein at various pH values. Revised from Ref. 86.

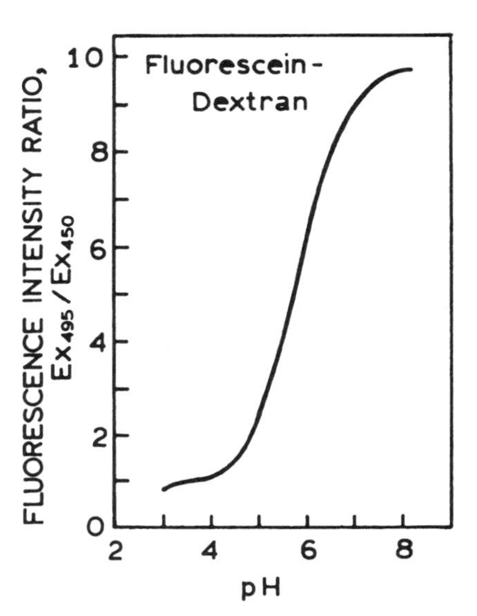

Figure 19.33. Wavelength-ratiometric pH calibration for fluorescein linked to dextran. Revised from Ref. 81.

Table 19.2. Spectral and Lifetime Properties of pH Probes[a]

	Excitation	Emission			
Probe[b]	λ_B (λ_A) (nm)	λ_B (λ_A) (nm)	Q_B (Q_A)	$\bar{\tau}_B(\bar{\tau}_A)$ (ns)	pK_a
BCECF	503 (484)	528 (514)	~0.7	4.49 (3.17)	7.0
SNAFL-1	539 (510)	616 (542)	0.093 (0.33)	1.19 (3.74)	7.7
Carboxy SNAFL-1	540 (508)	623 (543)	0.075 (0.32)	1.11 (3.67)	7.8
Carboxy SNAFL-2	547 (514)	623 (545)	0.054 (0.43)	0.94 (4.60)	7.7
Carboxy SNARF-1	576 (549)	638 (585)	0.091 (0.047)	1.51 (0.52)	7.5
Carboxy SNARF-2	579 (552)	633 (583)	0.110 (0.022)	1.55 (0.33)	7.7
Carboxy SNARF-6	557 (524)	635 (559)	0.053 (0.42)	1.03 (4.51)	7.6
Carboxy SNARF-X	575 (570)	630 (600)	0.160 (0.07)	2.59 (1.79)	7.9
Resorufin	571 (484)	528 (514)	NA[c]	2.92 (0.45)	~5.7
HPTS	454 (403)	511	NA	NA	7.3
$[Ru(deabpy)(bpy)_2]^{2+}$	450 (452)	615 (650)	NA	380 (235)	7.5
Oregon-Green	489 (506)	526	0.65 (0.22)	4.37 (2.47)	1.8
DM-Nerf	497 (510)	527 (536)	0.88 (0.37)	3.98 (2.50)	1.6
Cl-Nerf	504 (514)	540	0.78 (0.19)	4.00 (1.71)	2.3

[a] $\bar{\tau}_A$ and $\bar{\tau}_B$ refer to the mean lifetimes of the acid and base forms, respectively.
[b] Abbreviations: BCECF, 2′,7′-bis(2-carboxyethyl)-5(6)-carboxyfluorescein; HPTS, 8-hydroxypyrene-1,3,6-trisulfonate; deabpy, 4,4′-diethylaminomethyl-2,2′-bipyridine; bpy, 2,2′-bipyridine.
[c] NA, not applicable.

HPTS, a Wavelength-Ratiometric pH Sensor.

A disadvantage of the use of fluorescein in sensors is that fluorescein is difficult to employ as a wavelength-ratiometric probe. This is because the absorption and emission intensity are low for 450-nm excitation (Figure 19.32). The pH probe 8-hydroxypyrene-1,3,6,8-trisulfonate (HPTS)[88–91] displays more favorable properties as a wavelength-ratiometric probe. In HPTS (Figure 19.34), the sulfonate groups provide solubility in water, and the hydroxyl group provides sensitivity to pH.

Excitation and emission spectra of HPTS show a strong dependence on pH (Figure 19.34). As the pH is increased, HPTS shows an increase in absorbance at 450 nm, and a decrease in absorbance below 420 nm. These changes are due to the pH-dependent ionization of the hydroxyl group. The emission spectrum is independent of pH, suggesting that emission occurs only from the ionized form of HPTS. Conveniently, the apparent pK_a of HPTS is near 7.5, making it useful for clinical measurements, which need to be most accurate between pH 7.3 and pH 7.5 (see Table 19.3). HPTS has also been used as a CO_2 sensor when dissolved in the appropriate bicarbonate solution.[92]

One possible disadvantage of HPTS is that it undergoes ionization in the excited state, rather than a ground-state equilibrium. The fact that HPTS undergoes an excited-state reaction can be recognized by noting that the excitation spectra are comparable to the absorption spectra of the

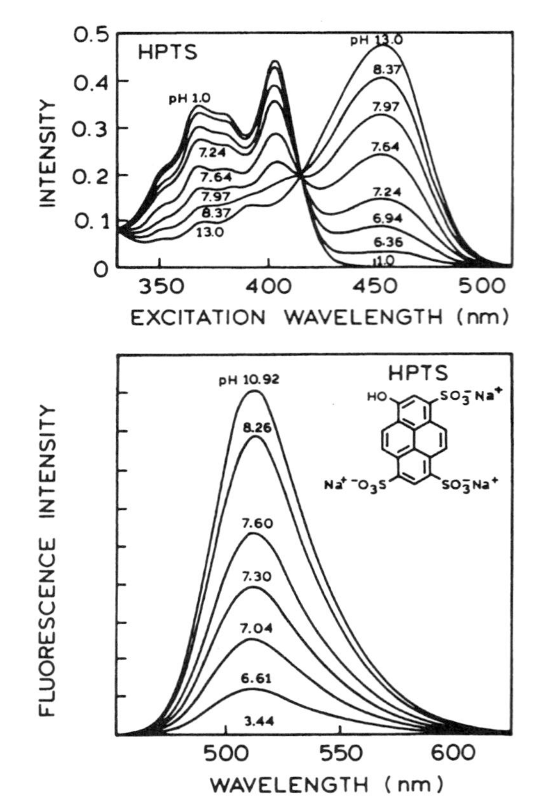

Figure 19.34. *Top*: Excitation spectra of the pH probe 1-hydroxypyrene-3,6,8-trisulfonate (HPTS) in 0.07*M* phosphate buffer at various pHs. *Bottom*: Emission spectra of HPTS when excited at 454 nm. Revised from Ref. 89.

phenol and phenolate forms but that there is only a single long-wavelength emission spectrum (Figure 19.34). The phenol form emits at shorter wavelengths and is only seen in highly acidic media. The presence of excited-state ionization is also indicated by a higher apparent pK_a in pure water than in buffers.[89] It is known that the pK_a values of the hydroxyl group for the ground- and excited-state HPTS are 7.3 and 1.4, respectively,[90] so that HPTS molecules in the protonated state will tend to undergo ionization upon excitation. It seems that any excited-state process will be dependent on the details of the local probe environment. Under most conditions, excited-state ionization of HPTS is complete prior to emission, so that only the phenolate emission is observed. Nonetheless, for sensing purposes we prefer probes that display a ground-state pK_a near 7.5. Another disadvantage of HPTS has been the relatively short excitation wavelength, particularly for the acid form, as it has been difficult to obtain wavelengths below 450 nm without the use of complex light sources. However, the recent discovery of UV emission from blue LEDs (Chapter 2) may result in the increased use of HPTS.

SNAFL and SNARF pH Probes. A family of improved pH probes became available in 1991.[93] These dyes are referred to as seminaphthofluoresceins (SNAFLs) or seminaphthorhodafluors (SNARFs). Representative structures are shown in Figure 19.35. A favorable feature of these probes is that they display shifts in both their absorption and emission spectra with a pK_a from 7.6 to 7.9 (Figure 19.36). Also, the absorption and emission wavelengths are reasonably long, so that both forms of the probes can be excited with visible wavelengths near 540 nm (Table 19.2). The spectral shifts (Figure 19.36) allow the SNAFLs and SNARFs to be used as either excitation or emission wavelength-ratiometric probes.

The fact that both the acid and base forms of these probes are fluorescent suggests their use as lifetime probes. If only one form was fluorescent, then the lifetime would not

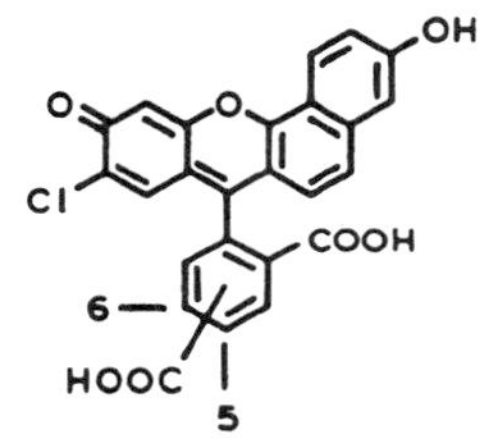

Carboxy SNAFL-2

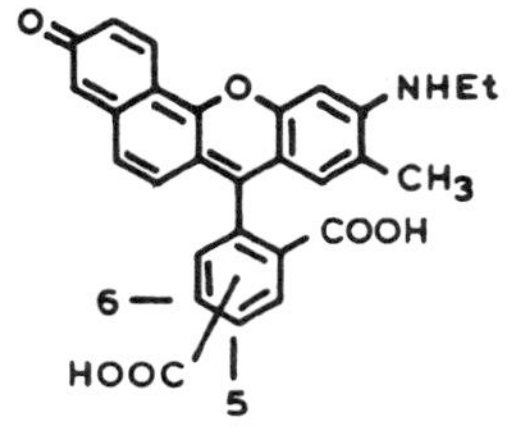

Carboxy SNARF-6

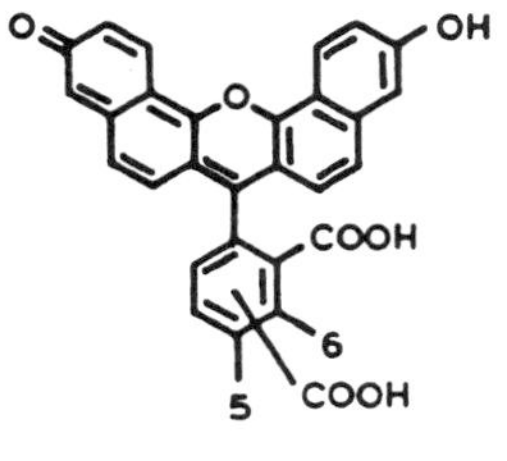

CNF

Figure 19.35. Wavelength-ratiometric pH sensors. Carboxy SNAFL-2 is a seminaphthofluorescein, carboxy SNARF-6 is a seminaphthorhodafluor, and CNF is 5-(and 6-)carboxynaphthofluorescein.

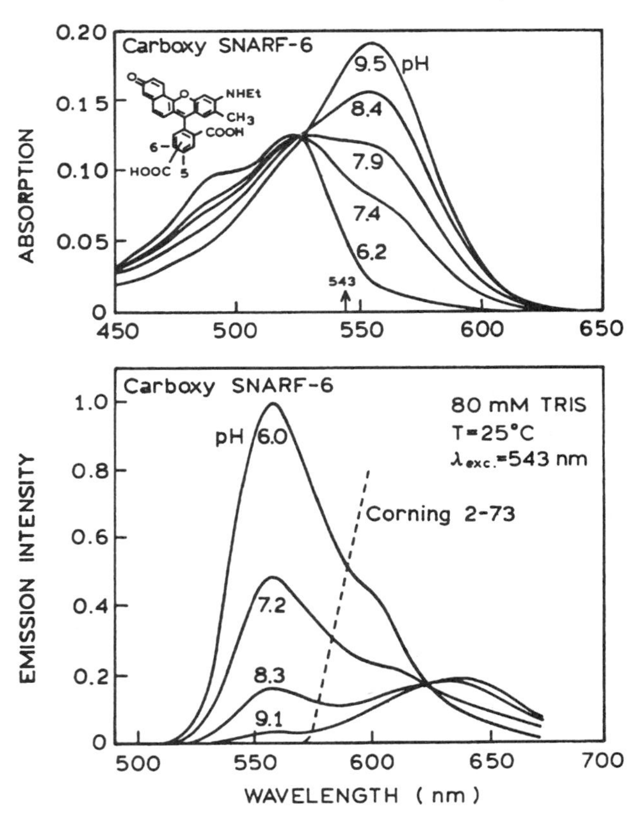

Figure 19.36. pH-dependent absorption (*top*) and emission spectra (*bottom*) of carboxy SNARF-6. The dashed line shows the transmission cutoff of the long-pass filter used for the phase and modulation measurements. Revised from Ref. 94.

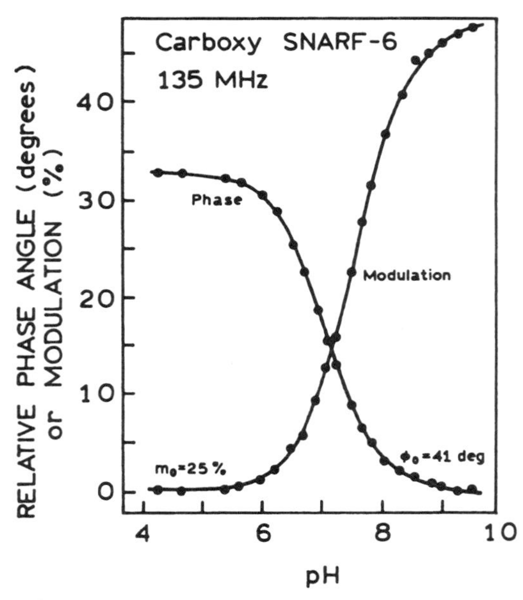

Figure 19.37. pH-dependent phase and modulation of carboxy SNARF-6 when excited at 543 nm with a green HeNe laser. The phase values are relative to the value at high pH, ϕ_0 = 41°. The modulation values are relative to the value at low pH, m_0 = 0.25. Revised and reprinted, with permission, from Ref. 94, Copyright © 1993, American Chemical Society.

change with pH. The pH-dependent phase and modulation data for carboxy SNARF-6 show a strong dependence on pH (Figure 19.37). Apparently, the decay time of the base form is less than that of the acid form. A detailed study revealed the decay times at pH 4.9 and 9.3 to be 4.51 and 0.95 ns, respectively.[94] Lifetime-based pH sensing can provide stable measurements for extended periods of time. The phase angles of carboxy SNARF-6 were found to be constant for a month (Figure 19.38). However, it is important to recognize that lifetime measurements, like intensity ratio measurements, can be affected by interactions of the probes with biological macromolecules. In particular, the intensity decays of carboxy SNARF-1 were found to be sensitive to the presence of serum albumin or intracellular proteins.[95]

For clinical applications, longer wavelengths are usually preferable. This was accomplished with the SNAFL probes by the introduction of an additional benzyl ring into the parent structure (Figure 19.35). This carboxynaphthofluorescein (CNF)[96] also shows shifts in its absorption and emission spectra with pH (Figure 19.39). CNF displays the longest-wavelength absorption and emission available to date in a pH probe, and its pK_a value is near 7.5. A pH probe

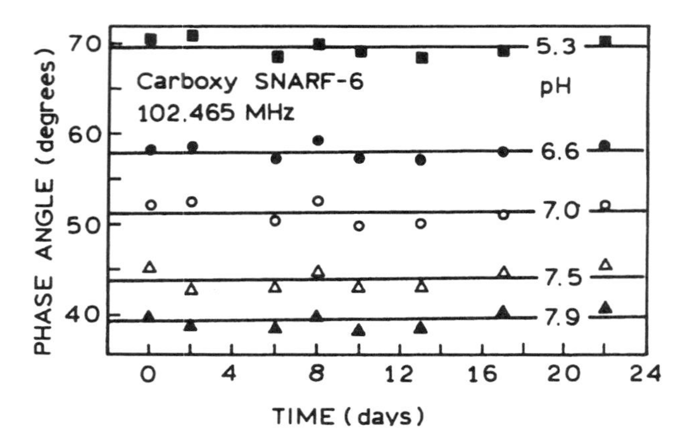

Figure 19.38. Time stability of pH phase-angle sensing by carboxy SNARF-6. The stability was monitored at room temperature, which varied by several degrees Celsius during the monitoring period. Excitation was at 563 nm from an R6G dye laser, and emission was measured at 580 nm using an interference filter. Revised and reprinted, with permission, from Ref. 94, Copyright © 1993, American Chemical Society.

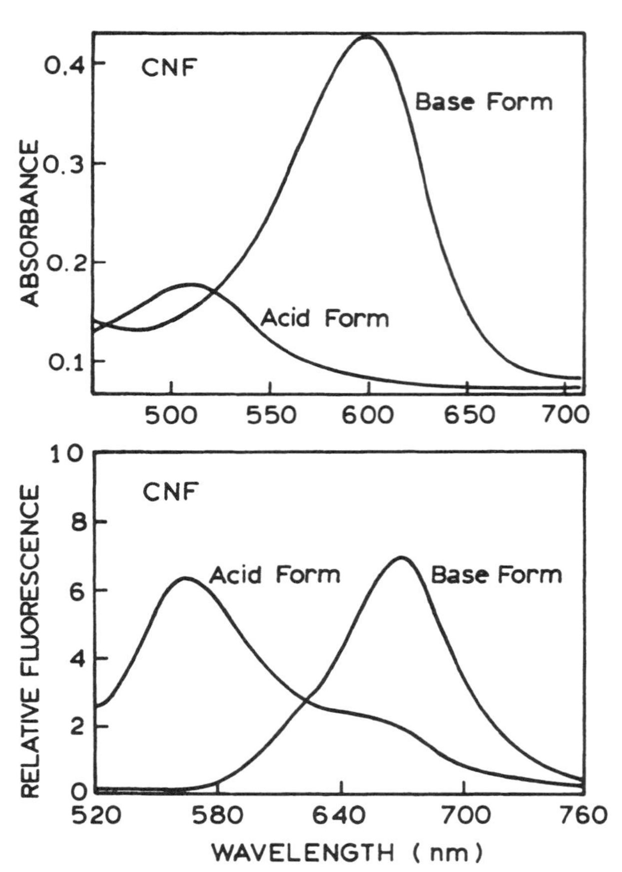

Figure 19.39. Absorption (*top*) and emission spectra (*bottom*) of the acid and base forms of carboxynaphthofluorescein (CNF). Revised from Ref. 96.

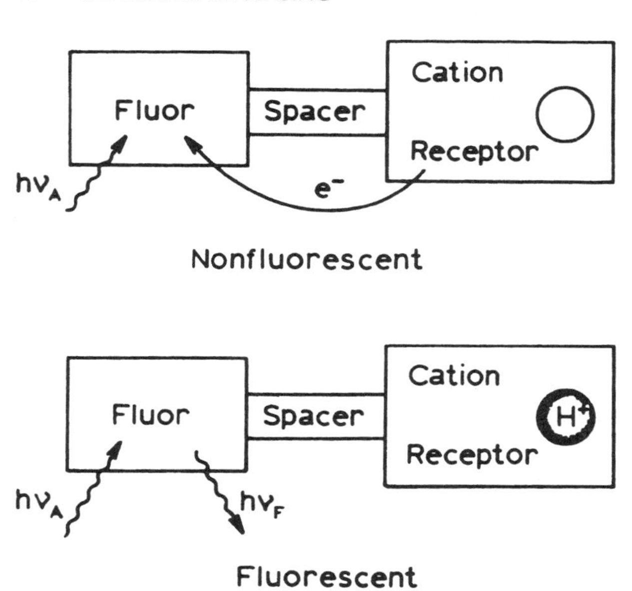

Figure 19.40. Chemical sensing based on photoinduced electron transfer. Revised from Ref. 102.

excitable at 795 nm has been described.[97,98] This carboxy carbocyanine dye shows a decrease in intensity near pH 8.5 but does not display spectral shifts, except at short wavelengths near 435 nm. UV-excitable pH probes with multiple pK_a values from 1.7 to 9.0 have also been described.[99] For clinical applications with simple instruments, it can be valuable to have long-lifetime pH probes. A pH-sensitive ruthenium MLC with a decay time near 300 ns and a pK_a value near 7.5 has been reported.[100] A pH-dependent lanthanide has also been reported.[101] Given the continuing need for pH measurements in analytical and clinical chemistry, one can anticipate additional advances in practical pH sensors.

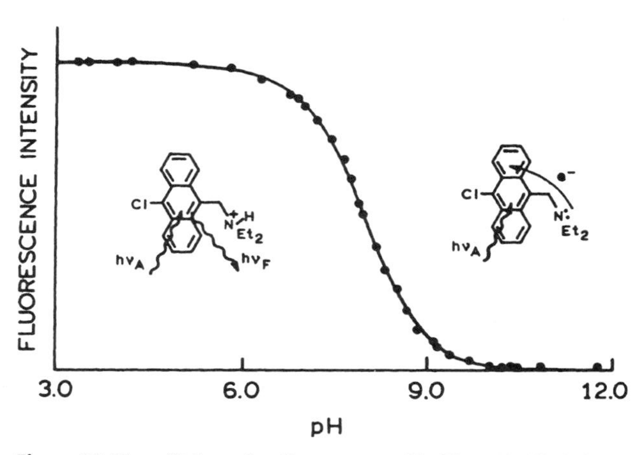

Figure 19.41. pH-dependent fluorescence of 9-chloro-10-(diethylaminomethyl) anthracene. Revised from Ref. 102.

19.7. PHOTOINDUCED ELECTRON-TRANSFER (PET) PROBES FOR METAL IONS AND ANION SENSORS

In the previous section we saw how probes could be designed based on reversible ionization of a group in conjugation with an aromatic ring. Another mechanism for sensing is the quenching interaction of a linked side chain with the fluorophore. The origin of this type of probe can be traced to the early studies of exciplex formation of amines with aromatic hydrocarbons. This phenomenon has been exploited to develop sensors based on quenching of polynuclear aromatic hydrocarbons by amines,[102,103] as illustrated in Figure 19.40. The basic idea is that quenching by amines requires the lone pair of electrons localized on the nitrogen. If this electron pair is bound to a proton, then electron transfer is inhibited, and the fluorophore is not quenched. Such probes are said to undergo photoinduced

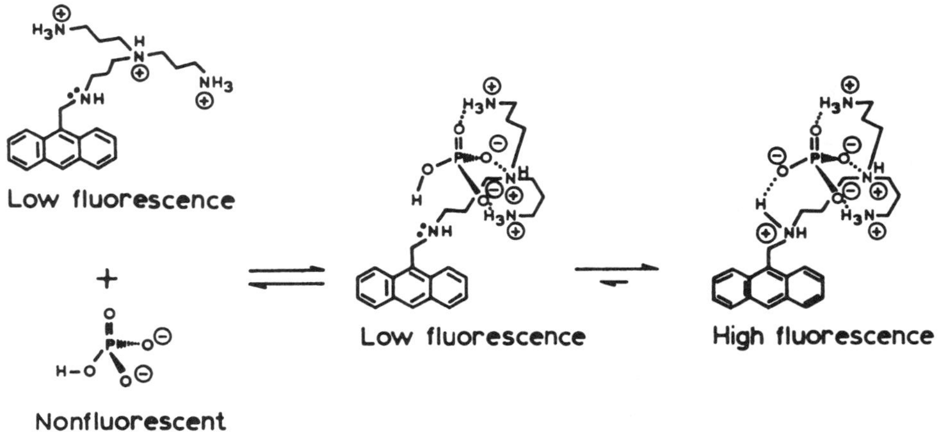

Low fluorescence

Nonfluorescent

Low fluorescence

High fluorescence

Figure 19.42. Phosphate sensing with an alkylaminoanthracene derivative. Revised from Ref. 107.

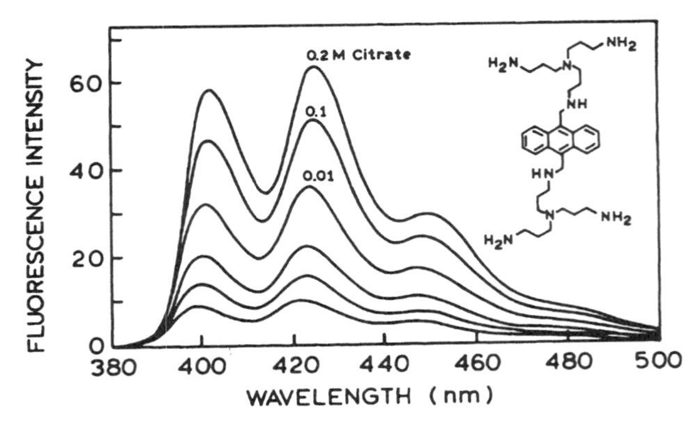

Figure 19.43. Emission spectra of an alkylaminoanthracene derivative in the presence of various amounts of citrate at pH 6. From bottom to top, [citrate] = 0, 0.1m*M*, 1m*M*, 10m*M*, 0.1*M*, and 0.2*M*. Revised from Ref. 107.

electron transfer (PET), which is the light-induced transfer of electrons from the nitrogen into the aromatic ring. A simple example of a PET probe is the alkylaminoanthracene derivative shown in Figure 19.41. At low pH, the amino group is protonated and does not quench the anthracene. As the pH is increased, the amino group becomes unprotonated, and the fluorescence decreases due to PET.

This idea of PET has been extended to create sensors for metal ions[104–106] and for nonmetal anions.[26,107] As an example, Figure 19.42 shows an anthracene derivative with an aminoalkyl side chain which binds phosphate. Hydrogen bonding of phosphate to the amino group results in increased anthracene fluorescence. A similar approach was used to create an anthracene derivative which displays increased fluorescence when bound to citrate (Figure 19.43). The structures of these phosphate and citrate probes illustrate the rational design of fluorophores based on known principles. Unfortunately, the PET mechanism may not be extendable to long-wavelength probes because quenching by amines becomes inefficient at long wavelengths.

19.8. PROBES OF ANALYTE RECOGNITION

The phosphate and citrate probes shown above suggest the design of fluorophores to bind specific analytes. In fact, extensive efforts have been directed toward the design and synthesis of fluorescent probes for cations—Na^+, K^+, Mg^{2+}, and especially Ca^{2+}. These efforts can be traced to the discovery of crown ethers and their ability to form complexes with metal ions[107–111] and subsequent work to create more complex structures to bind a variety of small molecules.[111] The greatest effort has been expended in synthesis of probes for calcium, and entire books have been devoted to calcium probes.[112] Much of this work can be traced to the development of intracellular cation probes by Tsien and colleagues.[113–115] Since these initial publications, many additional cation probes have been developed.[116] It is not possible to completely describe this extensive area of research. Instead, we describe the most commonly used cation sensors and the strengths and weaknesses of existing probes.

Table 19.3. Typical Analyte Concentrations in Blood Serum and in Resting Cells

Analyte	In blood serum (mM)	In resting cells (mM)
H^+	0.034–0.045	0.01–1
(pH)	(7.35–7.46)	(6–9)
Na^+	135–148	4–10
K^+	3.5–5.3	100–140
Li^+	0–2	—
Mg^{2+}	—	0.5–2
Ca^{2+}	4.5–5.5	0.05–0.2
Cl^-	95–110	5–100
HCO_3^-	23–30	—
CO_2	4–7 (% atm)	—
O_2	8–16 (% atm)	—

19.8.A. Specificity of Cation Probes

A survey of the literature reveals that a large number of diverse structures can chelate cations. However, a dominant use of cation probes is imaging of intracellular cations. In this case the indicators have to be sensitive to the intracellular concentrations of cations or anions (Table 19.3). These concentrations define the affinity needed by the chelators for the cation and the degree of discrimination required against other cations. Also, it is desirable to have

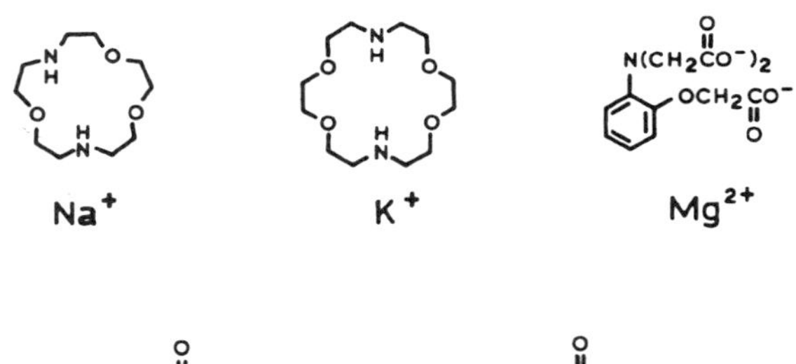

Figure 19.44. Chelating groups for Na^+, K^+, Mg^{2+} and Ca^{2+}.

a means of trapping the probes within cells. Alternatively, when used to measure the concentrations of electrolytes in blood, the probes must have affinity constants comparable to the ion concentrations in blood, which are summarized in Table 19.3. The specificity of these probes must be adequate to discriminate against the other ions present in blood. Once these criteria are taken into account, one arrives at a group of chelators that are useful for intracellular probes. The azacrown ethers have suitable affinity constants for Na^+ and K^+, APTRA is a chelator for Mg^{2+}, and BAPTA is a suitable chelator for Ca^{2+} (Figure 19.44). These are the dominant chelation groups used in intracellular cation probes.

19.8.B. Theory of Analyte Recognition Sensing

Suppose that the probe can exist in two states, free (P_F) and bound (P_B) to the analyte (A). If the binding stoichiometry is 1:1, the dissociation reaction is given by

$$AP_B \rightleftharpoons A + P_F \quad [19.6]$$

and the dissociation constant is defined as

$$K_D = \frac{[P_F]}{[AP_B]}[A] \quad [19.7]$$

The relative concentrations of the free and bound forms of the probe are given by

$$\frac{[P_F]}{[P]} = \frac{K_D}{K_D + [A]} \quad [19.8]$$

$$\frac{[AP_B]}{[P]} = \frac{[A]}{K_D + [A]} \quad [19.9]$$

where [P] is the total concentrations of indicator ($[P] = [P_F] + [AP_B]$).

These equations can be used to calculate the relative amounts of free and bound probes as the analyte concentration is increased (Figure 19.45). The analyte concentrations that can be measured are those for which there exist significant amounts of each form of the probe. Thus, the analyte concentration range over which a probe can be used is determined by the dissociation constant, K_D (Figure 19.45). This is a critical factor in using probes which bind specific analytes. The binding constant of the probe must be comparable to analyte concentration. The useful range of analyte concentrations is typically restricted to $0.1K_D < [A] < 10K_D$. Concentrations lower than $0.1K_D$ and higher than $10K_D$ will produce little change in the observed signal. In the use of fluorescence sensing probes, and Eqs. [19.6]–[19.9], we are assuming that the analyte is present in much greater concentration than the probe. Otherwise, the probe itself becomes a buffer for the analyte and distorts the analyte concentration.

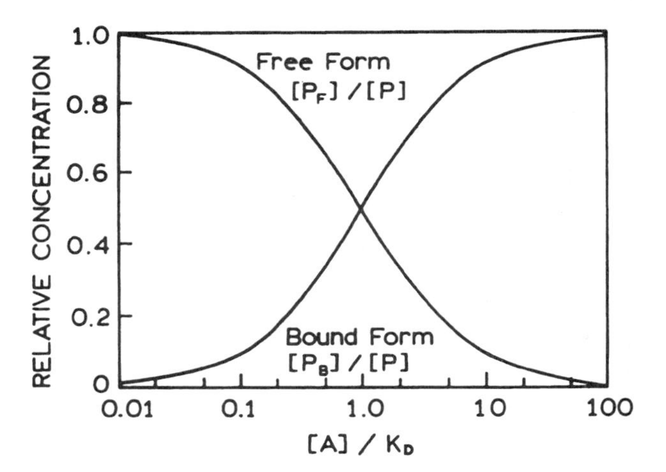

Figure 19.45. Relations between analyte concentration ([A]), dissociation constant (K_D) of the analyte–probe complex, and relative concentrations of the free (P_F) and bound (P_B) forms of the probe. From Ref. 12.

Intensity-Based Sensing. There are a number of probes which display changes in intensity but do not display spectral shifts. Such probes include the calcium probes Calcium Green™, Fluo-3, and Rhod-2. In these cases, the analyte concentration can be obtained from

$$[A] = K_D \frac{F - F_{min}}{F_{max} - F} \quad [19.10]$$

where F_{min} is the fluorescence intensity when the indicator is in the free form, F_{max} is the intensity when the indicator is totally complexed, and F is the intensity when the indicator is partially complexed by analyte. Changes in the fluorescence intensity are typically due to different quantum yields of the free and complexed forms, rather than differences in the absorption spectrum. The changes in quantum yield have been explained as due to formation of TICT states. Binding of the cation raises the energy of the TICT state, so that it is no longer available as a route of radiationless decay.[117]

In order to determine the analyte concentration using Eq. [19.10], all intensities must be determined with the same instrumental configuration, the same optical path length, and the same probe concentration. These requirements are often hard to satisfy, especially in microscopy when observing cells. Measurement of F_{max} and F_{min} requires lysing the cells and titrating the released indicator or using ionophores to saturate the indicator. However, such calibration methods do not compensate for dye loss due to photobleaching or leakage during the experiment and can also alter the spatial distribution of the probe. For this reason, it is desirable to have methods that are independent of probe concentration. This is possible using wavelength-ratiometric probes or lifetime-based sensing.

Wavelength-Ratiometric Probes. Many probes display spectral shifts in their absorption or emission spectra upon binding analytes. In these cases, the analyte concentrations can be determined from a ratio of intensities, independent of the overall probe concentration. For excitation wavelength-ratiometric probes, the analyte concentration can be determined as[115]

$$[A] = K_D\left(\frac{R - R_{min}}{R_{max} - R}\right)\left(\frac{S_F(\lambda_2)}{S_B(\lambda_2)}\right) \quad [19.11]$$

where $R = F(\lambda_1)/F(\lambda_2)$ is the ratio of intensities for the two excitation wavelengths λ_1 and λ_2, and R_{min} and R_{max} are the ratios for the free and the complexed probe, respectively. For an excitation wavelength-ratiometric probe, the values of $S_F(\lambda_2)$ and $S_B(\lambda_2)$ are related to the extinction coefficients (ε) and quantum yields (Φ) of the probe excited at λ_2:

$$\frac{S_F(\lambda_2)}{S_B(\lambda_2)} = \frac{\varepsilon_F}{\varepsilon_B}\frac{\Phi_F}{\Phi_B} \quad [19.12]$$

For an emission wavelength-ratiometric probe, one can use Eq. 19.11 with the values of $S_F(\lambda_2)$ and $S_B(\lambda_2)$; these values are related to the relative intensities of the free and bound forms of the probe:

$$\frac{S_F(\lambda_2)}{S_B(\lambda_2)} = \frac{F_F}{F_B} \quad [19.13]$$

Use of wavelength-ratiometric probes and Eq. [19.11] makes the measurements independent of the probe concentration, unlike intensity-based measurements.

19.8.C. Sodium and Potassium Probes

Typical Na^+ and K^+ probes are shown in Figure 19.46. All of these contain azacrown groups or a closely related structure. The first reported probes[113] were sodium-binding benzofuran isophthalate (SBFI) for Na^+ and potassium-binding benzofuran isophthalate (PBFI) for K^+ (Table 19.4). While designed to be excitation wavelength-ratiometric probes, these probes suffer several disadvantages. They require UV excitation, which results in substantial amounts of autofluorescence from cells. The excitation spectra show only minor changes in shape upon

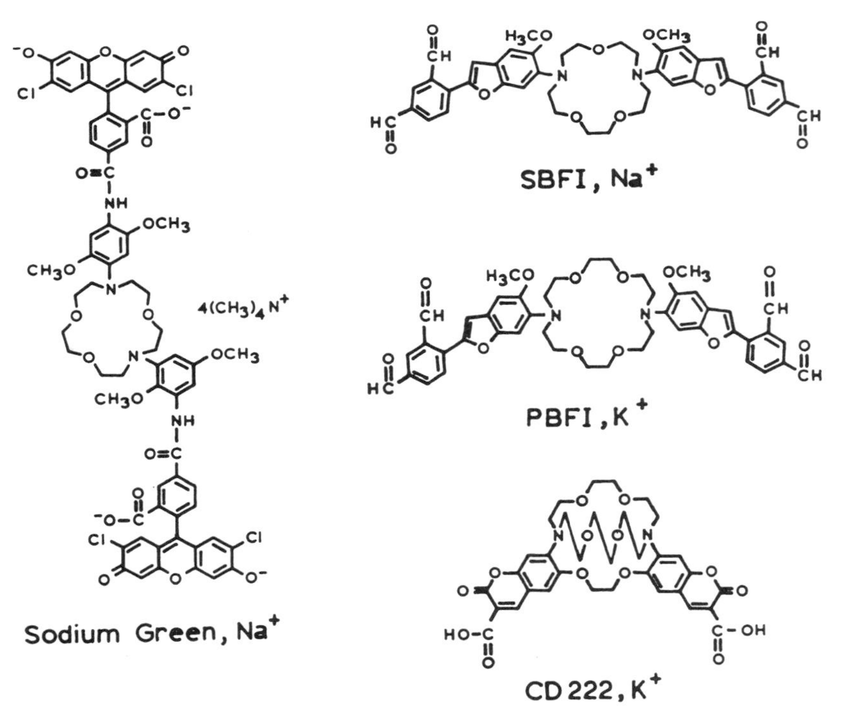

Figure 19.46. Representative Na^+ and K^+ probes. Sodium Green is a trademark of Molecular Probes, Inc.

Table 19.4. Spectral and Lifetime Properties of Mg^{2+}, Na^+, and K^+ Probes[a]

	Excitation	Emission			
Probe[b]	λ_F (λ_B) (nm)	λ_F (λ_B) (nm)	Q_F (Q_B)	$\bar{\tau}_F$ ($\bar{\tau}_B$) (ns)	K_D (m*M*)
Mg^{2+} probes					
Mag-Quin-1	348 (335)	499 (490)	0.0015 (0.009)	0.57 (10.3)	6.7
Mag-Quin-2	353 (337)	487 (493)	0.003 (0.07)	0.84 (8.16)	0.8
Mag-Fura-2	369 (330)	511 (491)	0.24 (0.30)	1.64 (1.72)	1.9
Mag-Fura-5	369 (332)	505 (482)	NA[c]	2.52 (2.39)	2.3
Mag-Indo-1	349 (330)	480 (417)	0.36 (0.59)	1.71 (1.90)	2.7
Mag-Fura-Red	483 (427)	659 (631)	0.012 (0.007)	0.38 (0.35)	2.5
Mg Green	506	532	0.04 (0.42)	0.98 (3.63)	1.0
Mg Orange	550	575	0.13 (0.34)	1.06 (2.15)	3.9
Na^+ probes					
SBFI	348 (335)	499 (490)	0.045 (0.083)	0.27 (0.47)	3.8
SBFO	354 (343)	515 (500)	0.14 (0.44)	1.45 (2.09)	31.0
Na Green	506	535	—[d]	1.14 (2.38)	6.0
K^+ probes					
PBFI	336 (338)	557 (507)	0.24 (0.72)	0.47 (0.72)	5.1
CD 222	396 (363)	480 (467)	3.7-fold	0.17 (0.71)	0.9

[a] *F* and *B* refer to the free and cation-bound forms of the probes, respectively.
[b] Abbreviations: SBFI, sodium-binding benzofuran isophthalate; SBFO, sodium-binding benzofuran oxazole; PBFI, potassium-binding benzofuran isophthalate.
[c] NA: Not applicable.
[d] $Q_B/Q_F = 7$.

binding of Na^+ and K^+ to these probes (Figure 19.47). Apparently, the charge densities of these singly charged cations are not sufficient to result in substantial spectral shifts. In support of this hypothesis, one notices that PBFI, which binds the larger K^+ ion, shows a smaller spectral shift than seen for binding of Na^+ to SBFI.

Recently, a coumarin-based probe for K^+, CD 222, was reported. This probe has a more complex chelating group (Figure 19.46) which is directly connected to the coumarin fluorophore.[118] This probe can be excited at longer wavelengths than SBFI and PBFI and displays larger spectral shifts (Figure 19.47). Given the limited number of probes for K^+, we can expect to see increased use of CD 222. Because the dissociation constant for K^+ binding to CD 222 is near 1m*M* (Table 19.4), this probe is useful for measurements of extracellular K^+; the binding is too strong for measurements of intracellular K^+ (Table 19.3). However, recent studies have shown that the apparent K_D of CD 222 for K^+ is increased in the presence of Na^+. As a result, CD 222 may be useful for measurements of extracellular K^+ in blood in the concentration range of 3–6m*M*.[119] Several other probes with a coumarin fluorophore and a chelator for K^+ have been described.[120,121] Given the availability of blue and UV LEDs, these probes[119–121] may soon find use in simple clinical devices.

In an effort to avoid cellular autofluorescence, several Na^+ and K^+ probes have been developed for longer excitation wavelengths. One of these probes is Sodium Green™, which is a sodium-specific azacrown conjugated on both nitrogens to dichlorofluorescein (Figure 19.46). Sodium Green can be excited at 488 nm and displays increasing intensity in the presence of increasing concentrations of Na^+ (Figure 19.47). Unfortunately, Sodium Green does not display any spectral shifts, so that wavelength-ratiometric measurements are not possible. Furthermore, the analogous probe for potassium has not been reported, so that K^+ probes are limited in number.

The inability to develop wavelength-ratiometric probes for Na^+ and K^+, particularly with long excitation and emission wavelengths, illustrates an advantage of lifetime-based sensing. Sodium Green was found to display a multiexponential decay, with lifetimes of 1.1 and 2.4 ns in the absence and the presence of Na^+, respectively.[122] The phase and modulation values (Figure 19.48) are independent of total intensity, allowing the concentration of Na^+ to be determined even if the probe concentration is unknown. Cation-dependent decay times of SBFI and PBFI have been reported. Unfortunately, SBFI, PBFI, and similar probes display only modest changes in lifetime,[123–125] so that these probes do not seem suitable for lifetime-based sensing of Na^+ or K^+. However, CD 222 was found to display useful changes in lifetime in response to K^+, even in the presence of large amounts of Na^+. In this case, lifetime-based sensing of K^+ at the concentration

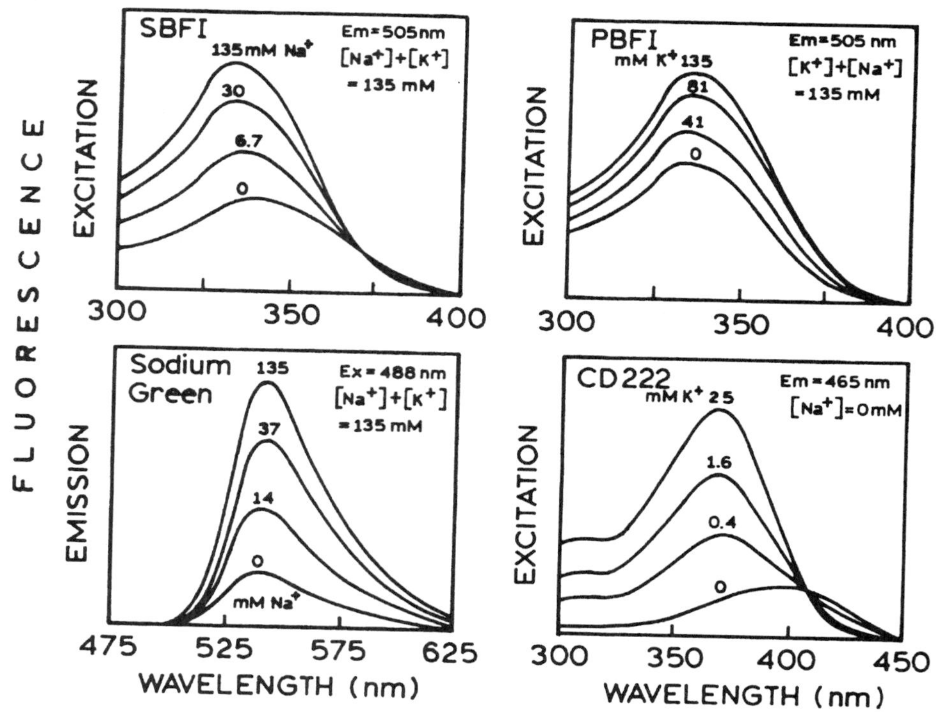

Figure 19.47. Excitation (SBFI, PBFI, and CD 222) and emission (Sodium Green) spectra of Na^+ and K^+ probes in the presence of various concentrations of the ion. Sodium Green is a trademark of Molecular Probes, Inc. Revised from Ref. 116.

present in blood appears to be possible using CD 222. In contrast, CD 222 does not allow wavelength-ratiometric measurements of K^+ in the presence of 100m*M* sodium.[119]

19.8.D. Calcium and Magnesium Probes

Calcium probes (Figure 19.49) are perhaps the most widely used intracellular indicators. All these indicators are based on the BAPTA chelator, which binds Ca^{2+} with affinities near 100 n*M* (Table 19.5). Hence, these probes are suitable for measurements of intracellular Ca^{2+} but bind Ca^{2+} too tightly for measurement of Ca^{2+} in blood or serum, in which the Ca^{2+} concentration is near 5m*M* (Table 19.3). These probes are used most often in fluorescence microscopy, where the local probe concentration is unknown. The salt forms of these dyes (Figure 19.49) do not diffuse across cell membranes, so that the cells need to be labeled by microinjection or electrophoration. BAPTA-based probes are also available with esterified carboxy groups, the so-called acetoxymethyl esters (AM esters). Indo-1 is shown in Figure 19.49 as the AM ester. In this form, the dyes are less polar and passively diffuse across cell membranes. Once inside the cell, the AM esters are cleaved by intracellular esterases, and the negatively charged probe is trapped in the cells.

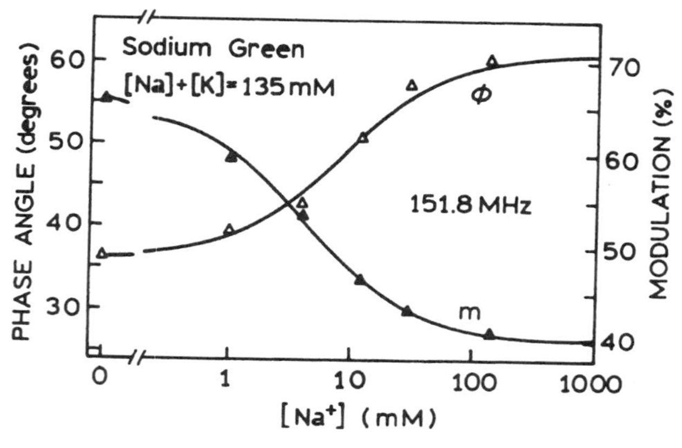

Figure 19.48. Dependence of phase angle (φ) and modulation (*m*) of Sodium Green on Na^+ concentration. Excitation was at 514 nm, and emission above 530 nm was observed. Data from Ref. 122.

Fura-2 and Indo-1 are both wavelength-ratiometric probes. Fura-2 displays a large shift in its absorption spectrum upon binding Ca^{2+} and is thus used with two excitation wavelengths (Figure 19.50).[126] Indo-1 displays a shift in its emission spectrum upon binding Ca^{2+}. Indo-1 can be used with two emission wavelengths and a single excitation wavelength. Thus, Indo-1 is preferred when laser excitation sources are used, such as in a laser scanning

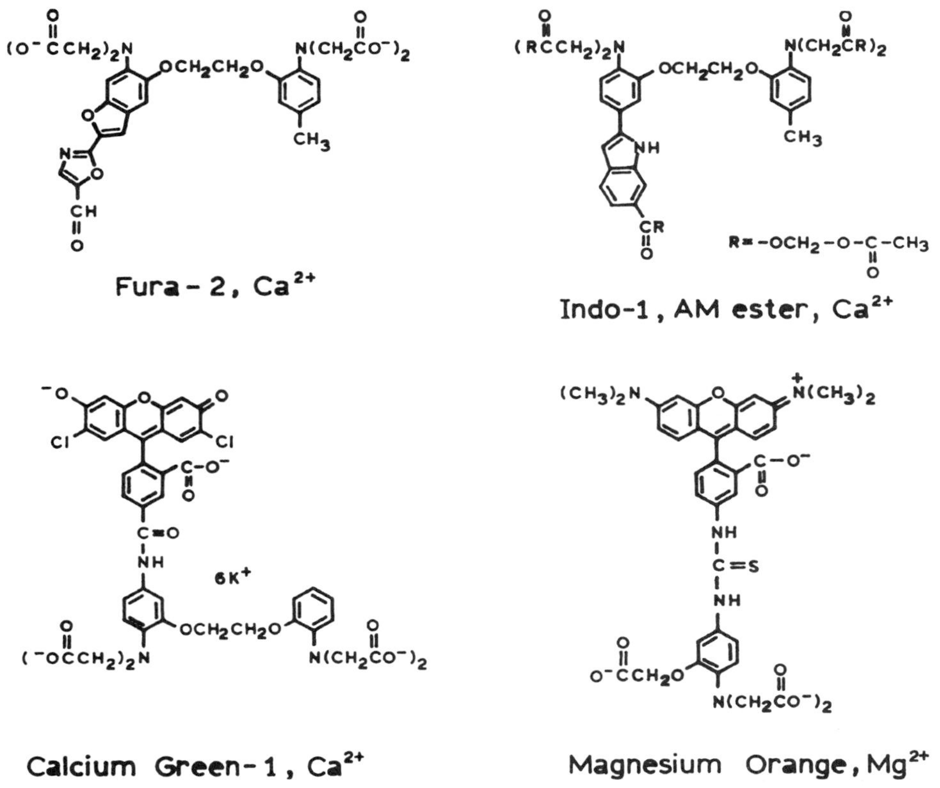

Fura-2, Ca^{2+}

Indo-1, AM ester, Ca^{2+}

Calcium Green-1, Ca^{2+}

Magnesium Orange, Mg^{2+}

Figure 19.49. Representative Ca^{2+} and Mg^{2+} probes.

Table 19.5. Spectral and Lifetime Properties of Ca^{2+} Probes[a]

Probe	Excitation λ_F (λ_B) (nm)	Emission λ_F (λ_B) (nm)	Q_F (Q_B)	$\bar{\tau}_F$ ($\bar{\tau}_B$) (ns)	K_D (nM)
Quin-2	356 (336)	500 (503)	0.03 (0.14)	1.35 (11.6)	60.0
Fura-2	362 (335)	518 (510)	0.23 (0.49)	1.09 (1.68)	145.0
Indo-1	349 (331)	482 (398)	0.38 (0.56)	1.40 (1.66)	230.0
Fura Red	472 (436)	657 (637)	—[b]	0.12 (0.11)	140.0
BTC[c]	464 (401)	531	NA[d]	0.71 (1.38)	
Fluo-3	504	526	40-fold	0.04 (1.28)	390
Rhod-2	550	581	100-fold	NA	570
Ca Green	506	534	0.06 (0.75)	0.92 (3.60)	190
Ca Orange	555	576	0.11 (0.33)	1.20 (2.31)	185
Ca Crimson	588	611	0.18 (0.53)	2.55 (4.11)	185
Ca Green-2	505	536	~100-fold[e]	NA	550
Ca Green-5N	506	536	~30-fold	NA	14,000
Ca Orange-5N	549	582	~5-fold	NA	20,000
Oregon Green					
BAPTA-1	494	523	~14-fold	0.73 (4.0)	170
BAPTA-2	494	523	~35-fold	NA	580
BAPTA-5N	494	521	NA	NA	20,000

[a] F and B refer to the Ca^{2+}-free and Ca^{2+}-bound forms of the probes, respectively.
[b] Low quantum yield.
[c] BTC, coumarin benzothiazole-based indicator.
[d] NA: Not applicable.
[e] The term X-fold refers to the relative increase in fluoresence upon cation binding.

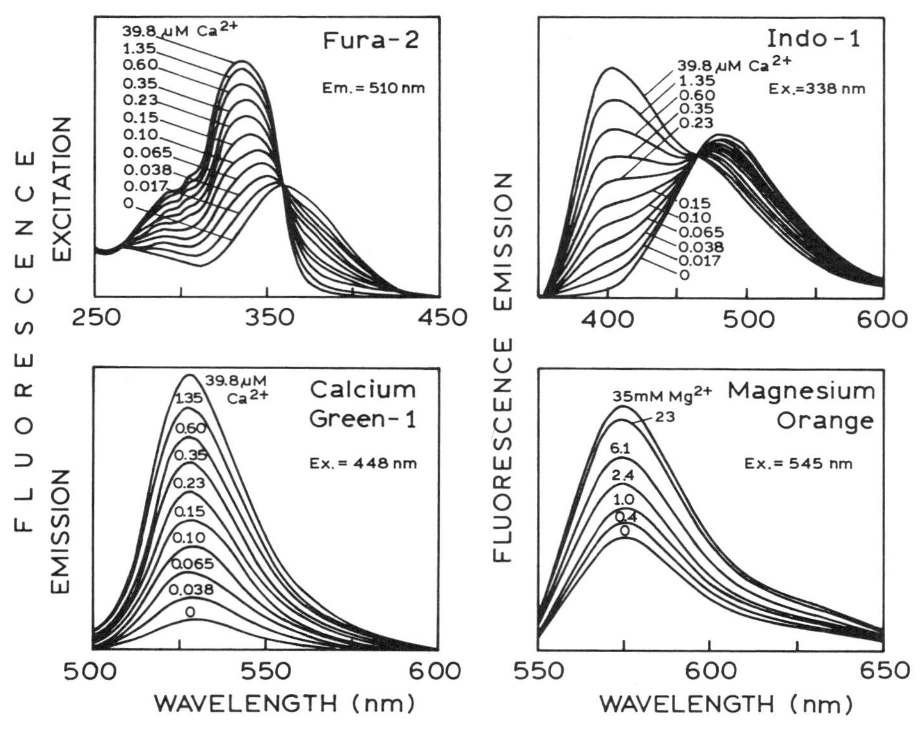

Figure 19.50. Excitation (Fura-2) and emission spectra (Indo-1, Calcium Green-1, and Magnesium Orange) of Ca^{2+} and Mg^{2+} probes in the presence of various concentrations of the ion. Revised from Refs. 114 and 116.

microscope, as it is difficult to get two different excitation wavelengths.

Like the Na^+ and K^+ probes, Fura-2 and Indo-1 absorb in the UV. This is disadvantageous because, in addition to problems of cellular autofluorescence, it is difficult to obtain microscope optics with high UV transmission. Hence, it is desirable to develop calcium probes with longer excitation and emission wavelengths. Coumarin-[127] and styryl-based[128] calcium probes have been developed but have not yet been widely used. Excitation spectra of one such probe are shown in Figure 19.51. These probes allow excitation up to 520 nm, but wavelength-ratiometric measurements require a second excitation wavelength below 430 nm.

Calcium probes based on fluoresceins and rhodamines are also available.[129,130] These probes typically have a BAPTA group linked to the fluorophore, rather than being part of the fluorophore. We refer to such probes as conjugate probes. One example is Calcium Green-1™ (Figure 19.49), which shows approximately an eight-fold increase in fluorescence upon binding calcium (Figure 19.50). Calcium Green-1 is just one member of a series of conjugate probes for Ca^{2+} which display a range of emission wavelengths.[130] Because Calcium Green-1 does not display a spectral shift, it cannot be used for wavelength-radiometric measurements. However, the lifetimes of the Calcium Green series all increase on Ca^{2+} binding, allowing the calcium concentration to be determined from the lifetimes.[131,132] One of the first calcium probes, Quin-2, also displays a 10-fold increase in lifetime when bound to

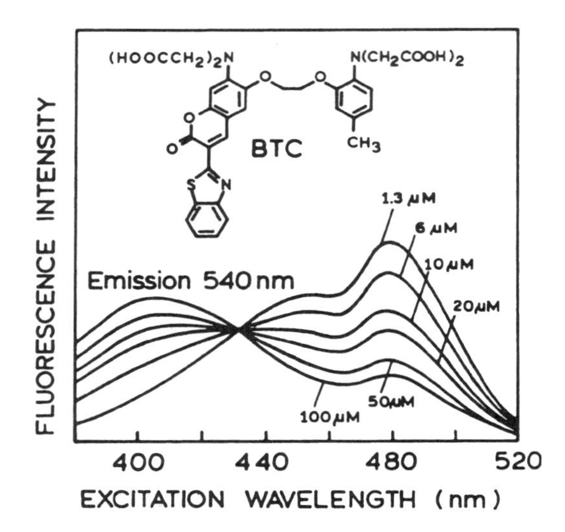

Figure 19.51. Excitation spectra of the coumarin benzothiazole-based indicator (BTC) lithium salt in Ca^{2+} solutions with concentrations ranging from 1.3 to 100μ*M* $CaCl_2$. Revised from Ref. 127.

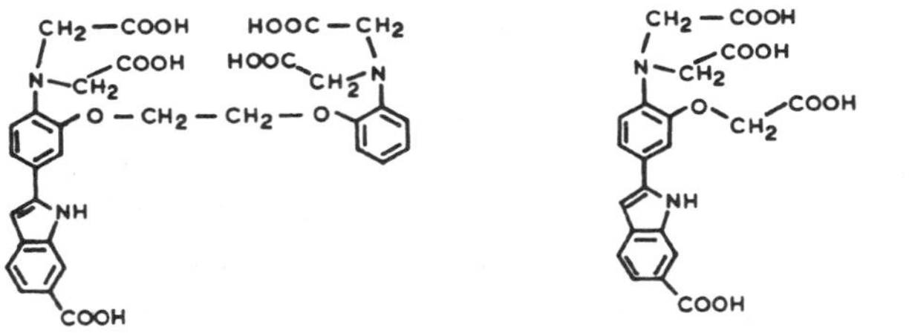

INDO-1 Mag-INDO-1

Figure 19.52. Chemical structures of the calcium probe Indo-1 and the magnesium probe Mag-Indo-1. Revised from Ref. 152.

calcium.[133,134] However, Quin-2 requires UV excitation and displays a low quantum yield, so that it is now used less frequently. Recently, Ca^{2+} probes based on squaraines have been reported, allowing excitation wavelengths as long as 635 nm,[135,136] but these probes have not yet been used for intracellular Ca^{2+} measurements.

Although the use of the calcium probes seems straightforward, calibration is difficult when such probes are located within cells.[137–140] Reference 141 should be consulted for a detailed description of the calibration procedures. When used as intracellular indicators, the calcium probes are typically calibrated in the presence of other intracellular ions at their expected concentrations.[142] Also, it is difficult to maintain nanomolar Ca^{2+} concentrations, and the probes themselves can alter the overall Ca^{2+} concentration. For this reason, calcium buffers have been developed and are commercially available. Finally, the probe may interact with intracellular macromolecules, or by phototransformation during illumination in the microscope, resulting in altered behavior compared to the calibration data.[137,143]

For completeness, we note that Ca^{2+} probes have also been developed using azacrown ethers as the chelator, rather than BAPTA.[135,144–148] However, these probes have been studied mostly in organic solvents and used to study the effects of Ca^{2+} on electron transfer. Such probes have not found use in cell physiology. Magnesium-sensitive probes are also available (Table 19.4),[147–152] and some have been characterized as lifetime probes.[153] These probes typically have the APTRA chelator, rather than BAPTA, as can be seen for the calcium probe Indo-1 and the analogous magnesium probe Mag-Indo-1 (Figure 19.52).

19.8.E. Glucose Probes

How can the principles of analyte recognition be used to develop a fluorophore which is sensitive to glucose? An intrinsic glucose probe would be simpler and more stable than a sensor based on the use of concanavalin A (Section 19.5.B). One approach is to use boronic acid as part of the fluorophore (Figure 19.53).[154] Boron is known to complex

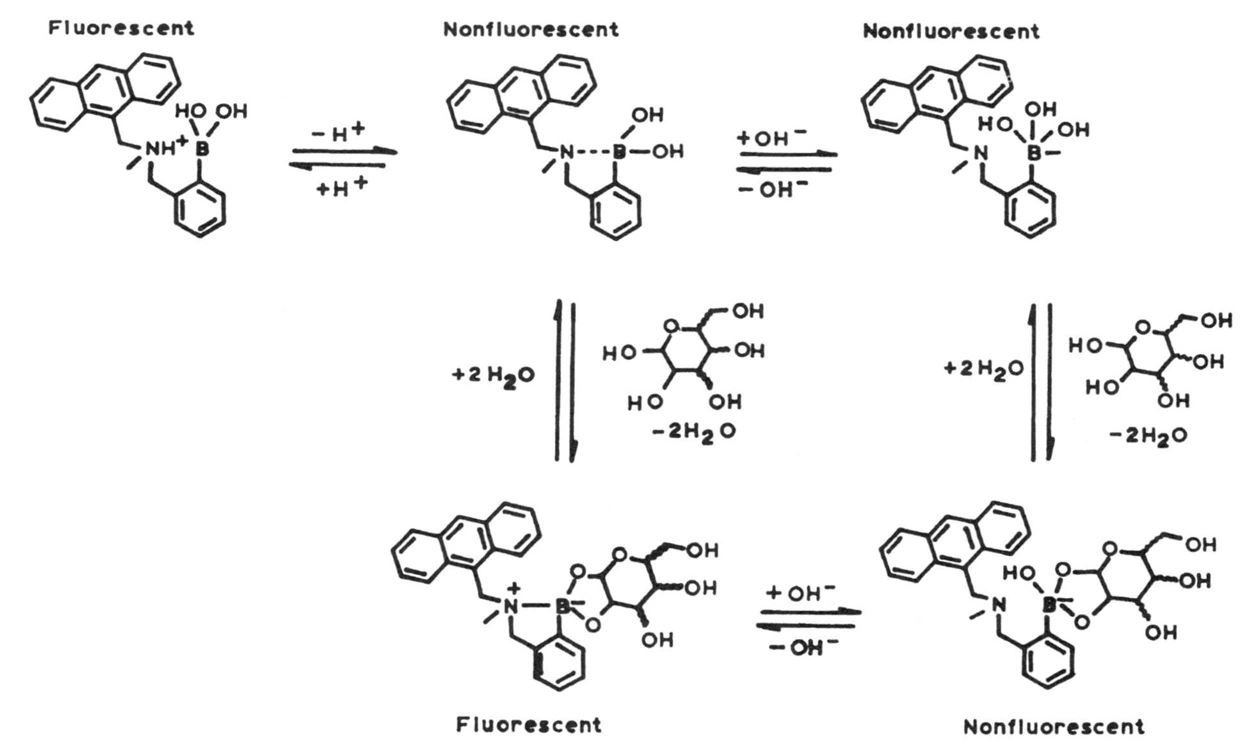

Figure 19.53. Glucose sensor based on photoinduced electron transfer. Revised from Ref. 154.

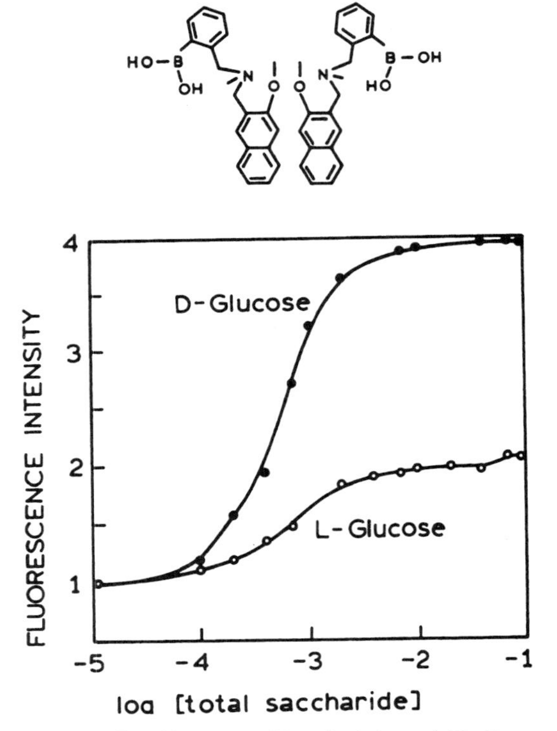

Figure 19.54. *Top*: Glucose-specific probe (R isomer). The lines on the central oxygens and the amino groups represent methyl groups. *Bottom*: Response of the probe to D-glucose (●) and L-glucose (○). Revised from Ref. 156.

diols, and this has been used for decades for oxidation of sugars. Complexation of sugars to boron-containing fluorophores can result in a change in the emission intensity.[154,155] Apparently, sugar complexation to the boron influences the extent of photoinduced electron transfer from the amino group. Depending on pH and sugar concentration, the anthracene group is either fluorescent or nonfluorescent.

A probe reasonably specific for D-glucose has already been reported.[156] In this probe, the boron atoms are located to interact stereospecifically with D-glucose (Figure 19.54). Unfortunately, this glucose probe requires excitation at 289 nm and emits at 358 nm, which is a rather short wavelength for clinical applications. However, given the need for glucose sensors, it seems probable that longer-wavelength glucose probes will be developed.

19.9. IMMUNOASSAYS

Immunoassays constitute a large and diverse family of assays, some of which are based on the principles described above, whereas others are based on unique principles not yet discussed. The basic idea is to couple the association of antibody (Ab) with antigen (Ag) to some other event that yields an observable spectral change. Various mechanisms are possible, including energy transfer, anisotropy, lanthanide trapping and release, or the use of enzymes to amplify the signal from a limited number of antigens. Several reviews are available.[157–162] The use of antibodies as analytical tools can be traced to the development of radioimmunoassays by Berson and Yalow,[163] which resulted in a Nobel Prize. Since then, immunoassays have been widely used, but they are now based mainly on fluorescence detection.

19.9.A. Enzyme-Linked Immunosorbent Assays (ELISA)

The ELISA method is perhaps the most commonly used immunoassay format owing to its high sensitivity and applicability to a wide range of antigens. This method relies on the specific interaction between antigen and antibody. The basic idea is to coat a surface with an antibody specific for the antigen of interest. The sample is incubated with the surface-bound antibody to allow the antibody to capture the antigen (Figure 19.55). The sample is then exposed to a second antibody that is covalently bound to an enzyme, typically alkaline phosphatase (AP), horseradish peroxidase, or β-galactosidase. Hence, the antigen must have more than a single antigenic site, and the second antibody must be different from the first antibody. Following adequate time for binding, the surface is washed to remove unbound enzyme-labeled antibody and the enzyme substrate is added. In the case shown in Figure 19.55, the enzyme alkaline phosphatase cleaves nonfluorescent umbelliferyl phosphate (UmP), yielding the highly fluorescent umbelliferone. A signal is observed only when the antigen is present.

ELISA assays exist in a number of formats. In some cases the reaction product absorbs light, and in other cases the product is strongly fluorescent. The second antibody is not always labeled with enzyme but rather is sometimes detected with yet another antibody that contains the bound enzyme. This procedure eliminates the need to attach probe or enzyme to a specific antibody which may be in short supply.

19.9.B. Time-Resolved Immunoassays

A variant of the ELISA method is the so-called "time-resolved immunoassay."[164–168] This type of assay also uses a polymeric support containing the capture antibody. The second labeled antibody has a covalently bound chelating group which contains a lanthanide such as europium (Figure 19.56). Detection is accomplished by addition of a

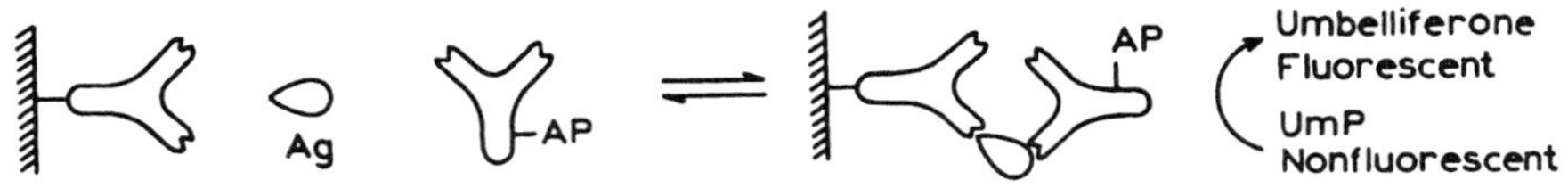

Figure 19.55. Schematic of an enzyme-linked immunosorbent assay (ELISA). AP is alkaline phosphatase, Ag is an antigen, and UmP is umbelliferyl phosphate.

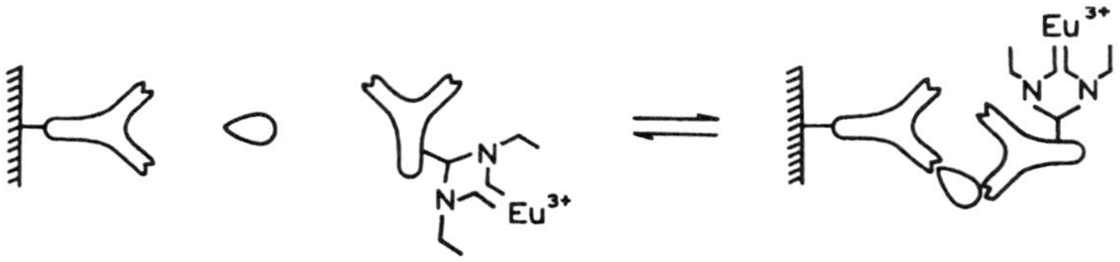

Figure 19.56. Time-resolved immunoassay based on the long-lived emission of europium.

so-called enhancer solution, which chelates the Eu^{3+} and has the necessary chromophore for excitation of the lanthanide by energy transfer (Figure 19.57). The enhancer solution is needed because the lanthanides absorb light very weakly and are rarely excited directly. With the enhancer solution, light is absorbed by chelators, which then transfer their excitation to the lanthanide. The chelating group is typically an EDTA derivative which strongly binds the Eu^{3+}. The Eu^{3+} can be released from the chelator at low pH.

The "time-resolved immunoassays" can be performed by direct detection or in a competitive format. Direct detection is usually used for proteins which contain multiple antigenic sites. One example of direct detection is presented in Figure 19.58, which shows an increase in intensity with increasing amounts of human thyrotropin or thyroid-stimulating hormone (TSH). TSH stimulates synthesis and release of thyroxine (T_4) and triiodothyronine (T_3) from the thyroid. In this case larger amounts of TSH result in larger amounts of bound antibody which is labeled with Eu^{3+}. After washing, the surface-bound Eu^{3+} is released by the enhancer solution, resulting in higher fluorescence at higher TSH concentrations. Competitive assays are also possible and are typically used for low-molecular-weight species. Figure 19.59 shows one such assay for T_4, where the intensity decreases due to release of

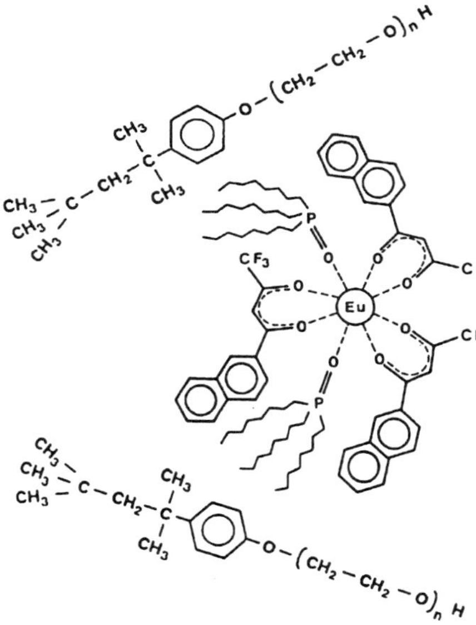

Figure 19.57. Europium in a fluorescent state following addition of enhancer solution. Reprinted, with permission, from Ref. 166, Copyright © 1990, CRC Press.

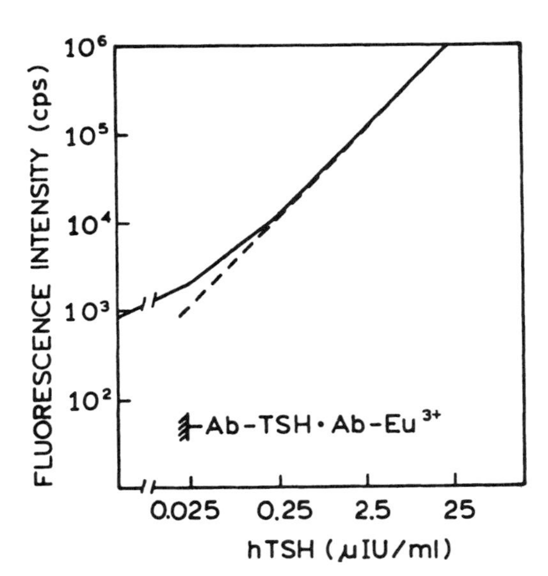

Figure 19.58. Time-resolved direct immunoassay of human thyrotropin (hTSH). The dashed line is the linear response after the background is subtracted. Revised from Ref. 166.

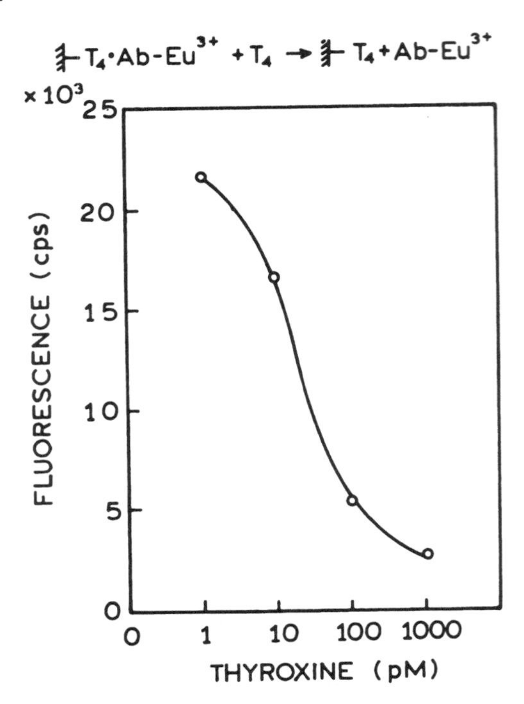

Figure 19.59. Competitive time-resolved immunoassay of thyroxine (T_4). Revised from Ref. 166.

Eu^{2+}-Ab from the surface with increasing concentrations of T_4.

These assays are called "time-resolved" because the sample is excited with a pulse of light, and the detector is gated-on following decay of the prompt autofluorescence. Because the lanthanides display millisecond lifetimes, they continue to emit long after the nanosecond interferences have decayed. The signal is integrated for a period of time, and the assay is based on measurement of integrated intensity, not a decay time. Hence, the phrase "time-resolved" should not be confused with a lifetime measurement. A similar approach was used with pyrene derivatives as long-lifetime donors to phycobiliproteins.[169]

ELISA and time-resolved immunoassays are both described as heterogeneous assays, because the antibodies are bound to a surface and separation steps are required. To minimize sample handling, it is useful to have assays that do not require separation steps. As described in the following sections, such homogeneous assays are possible based on energy transfer or anisotropy.

19.9.C. Energy-Transfer Immunoassays

RET provides an obvious approach to measuring antigen–antibody association and was suggested as a basis for immunoassays in 1976.[170] Such an assay would typically be performed in a competitive format and can be homogeneous. Suppose that the antigen is thyroxine (T_4). Thryoxine is an iodinated hormone produced by the thyroid gland which stimulates metabolism in most tissues. One configuration for such an assay would be for the donor to be labeled with T_4. In the assay shown in Figure 19.60, the donor is a phycobiliprotein, B-phycoerythrin (BPE).[171] The antibody was labeled with the cyanine dye Cy5, which serves as the acceptor (Figure 19.61). In the absence of T_4 from the sample, one expects maximum quenching of the donor. As the T_4 concentration is increased, T_4 competes for binding sites on the antibody, and the donor intensity increases.

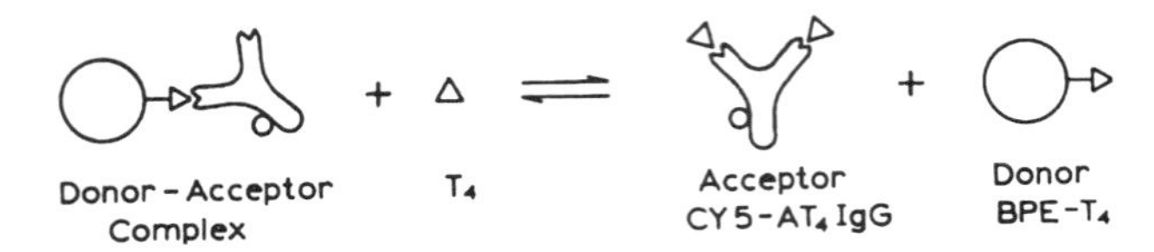

Figure 19.60. Homogeneous energy-transfer immunoassay for thyroxine (T_4). The donor, B-phycoerythrin (BPE), is labeled with T_4, and anti-T_4 antibody (AT_4IgG) is labeled with the cyanine dye Cy5, which serves as the acceptor. Revised from Ref. 171.

To date, the use of energy transfer for immunoassays has been limited. This may be because of the difficulties in measuring intensities. However, energy-transfer immunoassays can be performed using the donor decay times.[171,172] Representative data are shown in Figure 19.62. The decay time of the donor, BPE, is decreased on binding by Cy5-labeled antibody specific for T_4 (C5-AT_4IgG). Addition of free T_4 results in a return to the frequency response seen in the absence of T_4. These phase-angle measurements can be used to quantify T_4 in a homogeneous format (Figure 19.63). We note that phase-angle measurements are preferred to the so-called "phase-re-

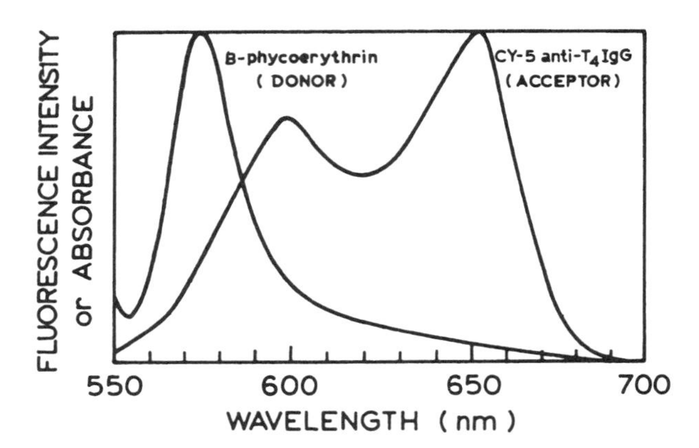

Figure 19.61. Emission spectrum of the fluorescent donor BPE-T_4 and absorption spectrum of the acceptor Cy5-AT_4IgG for the energy-transfer immunoassay in Figure 19.60. From Ref. 171.

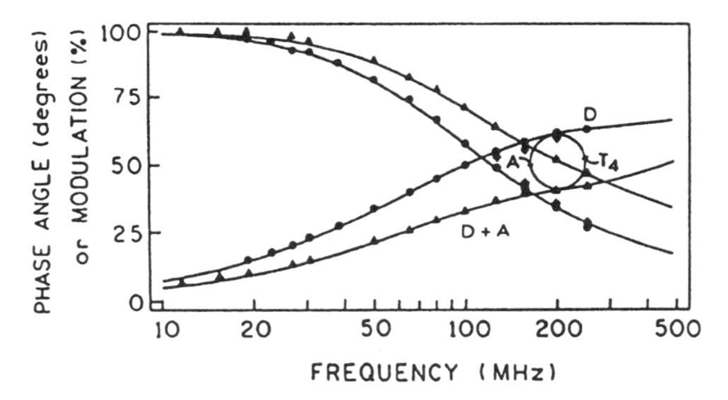

Figure 19.62. Frequency response of 4nM BPE-T_4, D (●), with 50nM Cy5-AT_4IgG, D + A (▲), and in the presence of a near saturating amount of T_4 (◆). From Ref. 171.

solved" immunoassays.[173] The measurements of phase-sensitive intensities in the latter type of assays are still intensity-based measurements.

19.9.D. Fluorescence Polarization Immunoassays

The final type of immunoassay is the fluorescence polarization immunoassay (FPI). Assays of this type are based on anisotropy measurements of labeled antigens.[174,175] The use of the term polarization instead of anisotropy is historical, and now entrenched in the literature. The anisotropy of a mixture (r) is determined by the anisotropies of the free (F) and bound (B) species (r_F and r_B) and their relative fluorescence intensities (f_F and f_B),

$$r = r_F f_F + r_B f_B \tag{19.14}$$

An FPI is a competitive assay, which can be performed in a homogeneous format. Suppose that the antigen is the

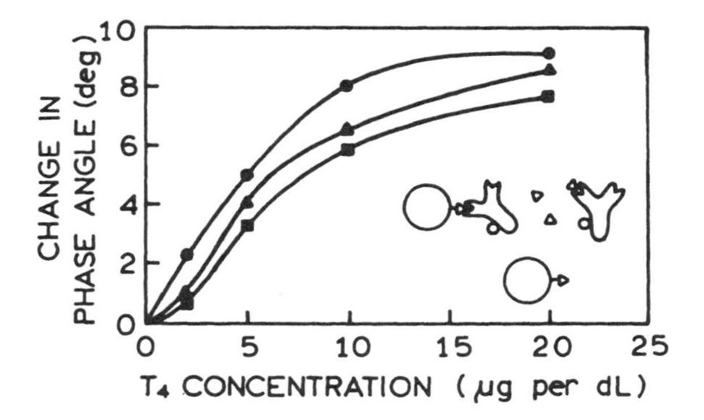

Figure 19.63. Phase fluorescence T_4 immunoassay dose–response curves at selected frequencies: ■, 49.5 MHz; ▲, 64.8 MHz; ●, 99.1 MHz. Experiments were conducted with 2nM BPE-T_4, 10nM Cy5-AT_4IgG, and 0.76mM furosemide. From Ref. 171.

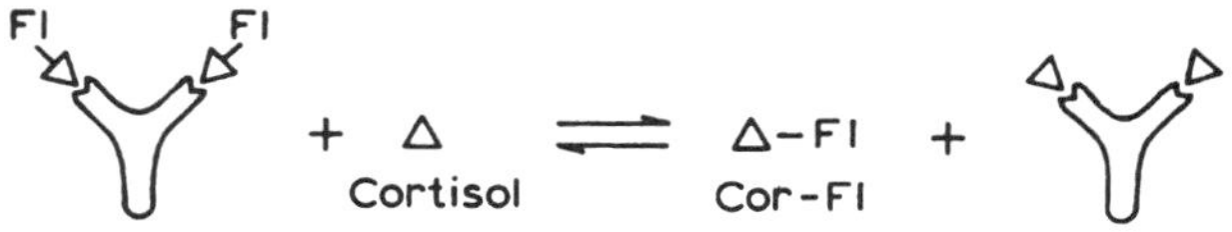

Figure 19.64. Homogeneous fluorescence polarization immunoassay for cortisol (Cor). Fl, Fluorescein.

hormone cortisol (Cor).[176] The assay mixture would contain labeled cortisol, typically labeled with fluorescein (Fl), and antibody specific for cortisol (Figure 19.64). Prior to addition of cortisol from the sample, the anisotropy will be the highest owing to binding of antibody to Cor-Fl. Free cortisol from the sample will displace Cor-Fl from the antibody. The Cor-Fl is now free to rotate, and the anisotropy decreases.

Typical data for a cortisol FPI are shown in Figures 19.65 and 19.66. Cor-Fl was prepared by reaction of cortisol-21-amine with FITC.[176] Upon mixing anti-Cor antibody (Ab) with Cor-Fl, the polarization increased (Figure 19.65), which was presumed to be due to specific binding of Ab to Cor-Fl. The specificity of the reaction was confirmed by adding unlabeled cortisol, which resulted in a decrease in polarization, and also by the absence of a change in polarization due to nonspecific antibody.

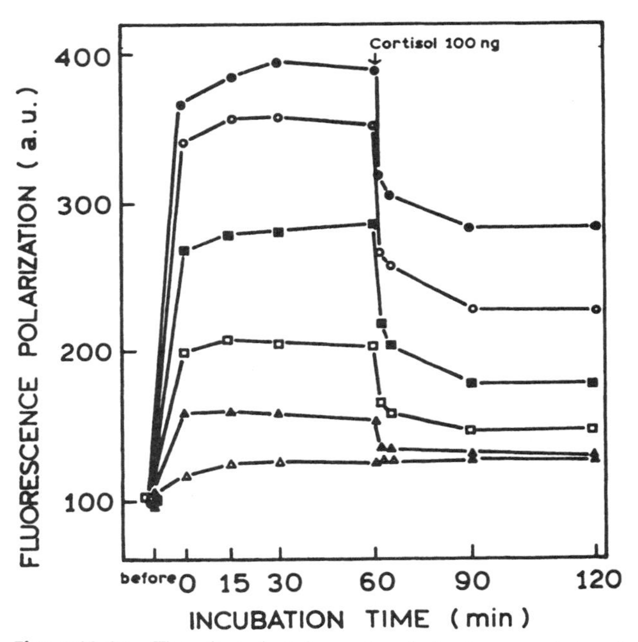

Figure 19.65. Time-dependent changes in polarization upon mixing of antibody to cortisol or nonspecific antibody (△) and fluorescein-labeled cortisol and upon addition, at 60 minutes, of 100 ng of unlabeled cortisol. The antibody is more dilute from top to bottom. Revised from Ref. 176.

To perform the cortisol assay, one uses a mixture of Ab and Cor-Fl, to which is added the serum sample. As the concentration of serum cortisol is increased, the polarization decreases (Figure 19.66). The polarization values are used to determine the cortisol concentration. Similar FPIs have been developed for a wide range of low-molecular-weight analytes, including antibiotics,[177,178] cocaine metabolites,[179,180] therapeutic drugs,[181–184] the immunosuppressant cyclosporine,[185,186] and environmental contaminants,[187,188] to name just a few. Numerous FPIs are routinely performed on automatic clinical analyzers.[189]

With an understanding of the basis of an FPI (Eq. [19.14]), ways to modify the assay for the desired purposes can be envisaged. Suppose that the goal was of the cortisol FPI described above to detect the minimum amount of the antigen. The antibody could be labeled with an energy-transfer acceptor to result in partial quenching of Cor-Fl, decreasing its fractional contribution (f_B). Addition of a small amount of antigen (Cor) would result in release of Cor-Fl, which would display an increase in intensity as well as a lower anisotropy. In this way, one could increase the sensitivity of the assay for release of the labeled antigen.

FPIs have advantages and disadvantages. FPIs do not require multiple antigenic sites, as is needed with heterogeneous capture immunoassays, as detected in section19.9.A. FPIs can be performed in a homogeneous format and may not require separation steps. However,

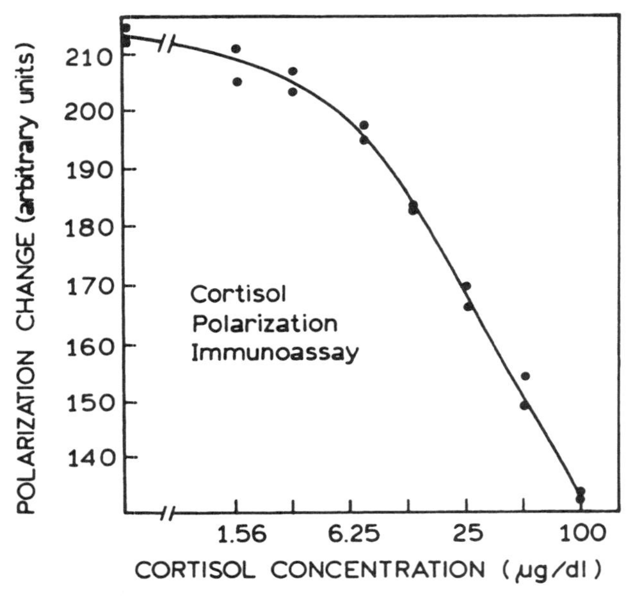

Figure 19.66. Cortisol fluorescence polarization immunoassay. Redrawn from Ref. 176.

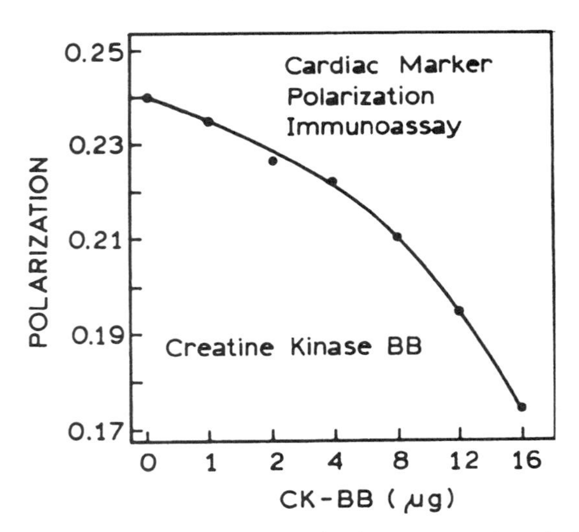

Figure 19.67. Fluorescence polarization immunoassay of creatine kinase BB. Revised from Ref. 192.

because FPIs are usually performed with fluorescein, they are generally limited to low-molecular-weight analytes. This is because the emission must be depolarized in the unbound state, which would not occur for higher-molecular-weight fluorescein-labeled proteins.

The limitation of FPIs to low-molecular-weight analytes is illustrated by the FPI for creatine kinase BB. Creatine kinase (CK) is a dimer, and the subunits can be from muscle (M) or brain (B). CK-MB is used as a marker for cardiac damage, and the presence of CK-BB in the blood may reflect a number of disease states, including brain trauma.[190–192] Figure 19.67 shows an FPI for CK-BB.[192] In this case the protein was labeled with dansylaziridine (DANZA), instead of fluorescein. The immunoassay was also performed with other probes (Table 19.6). One notices that the polarization changes are smaller when the fluorescein derivative IAF or the coumarin derivative CPM is used as the probe. This is because the lifetimes of these

Table 19.6. Fluorescence Polarization Immunoassay of Creatine Kinase (CK) BB[a]

		Polarization	
Fluorophore-CK[b]	τ (ns)	No antibody	With antibody
CPM-CK	~5	0.337	0.342
IAF-CK	~5	0.333	0.339
DNS-CK	~15	0.181	0.224
DANZA-CK	~15	0.170	0.242

[a]Ref. 192.

[b]Abbreviations: CPM, 3-(4-maleimidophenyl)-7-diethylamino-4-methylcoumarin; IAF, 5-(iodoacetamido)fluorescein; DNS, dansyl chloride; DANZA, dansylaziridine.

probes are near 5 ns, whereas the lifetime of DANZA is near 16 ns.[192] The longer lifetime of the dansyl labels (DNS or DANZA) allows more time for the protein to undergo rotational diffusion. This limitation of FPIs to low-molecular-weight substances can be overcome by the use of long-lifetime MLC probes, which is described in the following chapter.

19.10. CONCLUSIONS

Fluorescence sensing is a diverse and rapidly changing field of research. It is not practical to provide a complete description of the numerous results in a single chapter. For example, we have not discussed enzyme-based sensors, which allow determination of a wide range of analytes based on enzymatic activities. One example of such a sensor allows determination of urea based on pH changes induced by urease.[193] Another example of a newly developed probe is the dye Albumin Blue 580, which does not fluoresce in aqueous solution and which becomes highly fluorescent upon binding to albumin.[194] The development of such specific probes and the use of enzyme-coupled sensors suggest that fluorescence-based sensors can be developed for a wide range of clinically relevant analytes.

Another important development is the recent introduction of engineered proteins for use in sensors. This type of sensor is illustrated by the use of a glucose binding protein

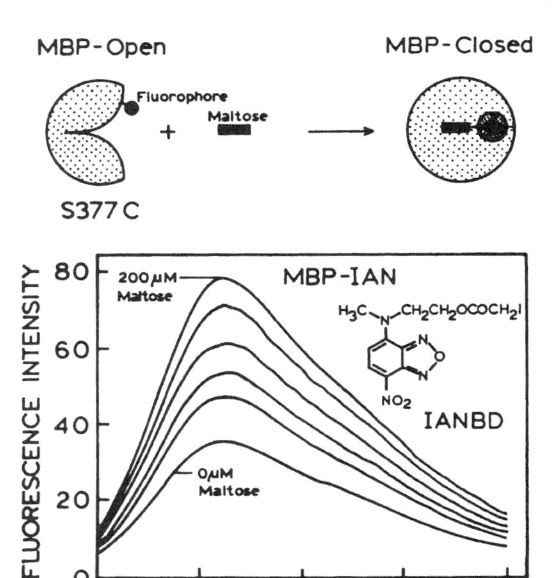

Figure 19.68. *Top*: Binding of maltose to the mutant (S337C) maltose binding protein (MBP) labeled at cysteine-337 with IANBD. *Bottom*: Emission spectra in the presence of various concentrations of maltose. Excitation was at 480 nm. Revised and reprinted, with permission, from Ref. 197, Copyright © 1994, American Chemical Society.

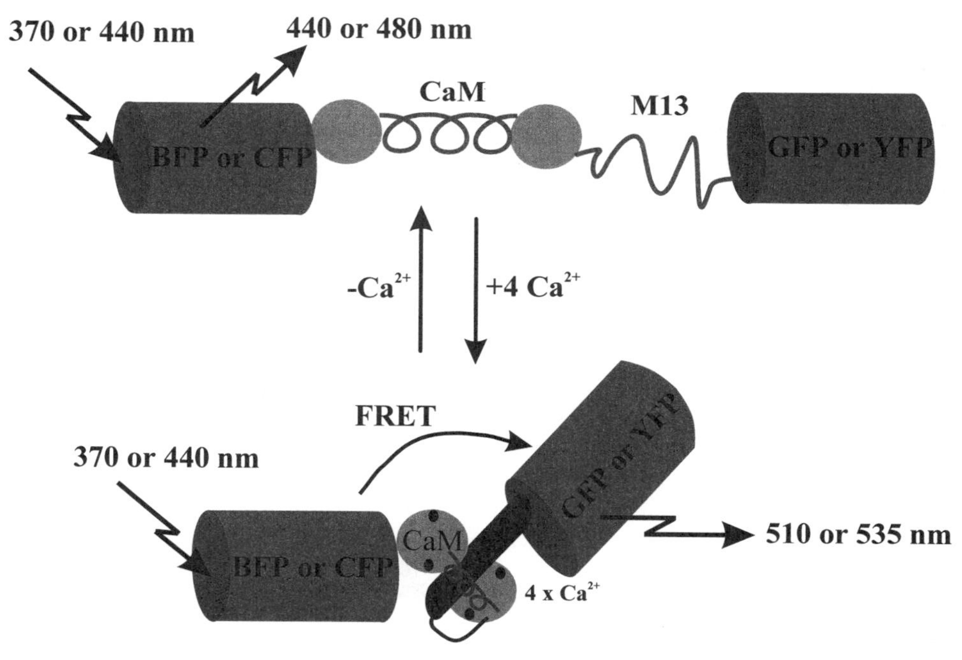

Figure 19.69. Calcium-sensing protein based on RET between two mutant GFPs at opposite ends of calmodulin (CaM). Revised and reprinted, with permission, from Ref. 199, Copyright © 1997, Macmillan Magazines Limited.

in glucose sensors.[195] The amino acid sequence of the glucose–galactose binding protein from *E. coli* was modified to allow labeling by the environmentally sensitive probe Prodan.[195] The emission spectrum of the labeled protein was found to be sensitive to glucose binding. Another example is the maltose binding protein (MBP) from gram-negative bacteria. A mutant MBP which contained a cysteine residue in the binding cleft was prepared,[196,197] (S377C; Figure 19.68). The sulfhydryl group was labeled with a derivative of NBD (IANBD), which is a polarity-sensitive fluorophore. The emission of the labeled protein was found to be highly sensitive to binding of maltose. As a control, the wild-type protein was labeled randomly with IANBD, presumably at methionine residues. The emission of the labeled wild-type protein was not sensitive to binding of maltose. This result illustrates the importance of using the known protein structure and the tools of molecular biology to obtain protein sensors which display useful spectral changes.

The idea of genetically engineered proteins has been extended to create calcium-sensing proteins.[198,199] This has been accomplished by placing mutants of GFP onto each end of calmodulin (Figure 19.69). Mutants of GFP are available which display a range of absorption and emission maxima. Some of these mutants can function as donors and acceptors for RET. A calcium-sensing protein was constructed using molecular biology to contain a donor cyan fluorescent protein (CFP) and an acceptor yellow fluorescent protein (YFP) on each end of calmodulin. This engineered protein also contains a segment of the M13 peptide which binds to calmodulin in the presence of calcium. The presence of this peptide results in a change in energy transfer between the CFP and the YFP upon binding of calcium to calmodulin. Hence, the CFP donor emission is quenched and the YFP acceptor emission is enhanced in the presence of calcium (Figure 19.70). Another example of a calcium-sensing protein was described in Section 13.3.B. Given the specificity of proteins and the power of molecular biology, one can expect to see a wide range of proteins engineered for sensing in the near future.

Finally, we note that GFP can be used as a tag to allow easy detection of gene expression. In these cases the desired genes are linked to the gene for GFP. Protein synthesis can be readily monitored from the amount of GFP formed in the incubator flask.[200,201] Fluorescence sensing is an area of active research, with new developments appearing at a rapid pace. In the present chapter, we have attempted to provide examples of each type of sensing and to illustrate the various principles used in fluorescence sensing. The latest information on fluorescence sensing will continue to be available in monographs and conference proceedings focused on the applications of fluorescence.

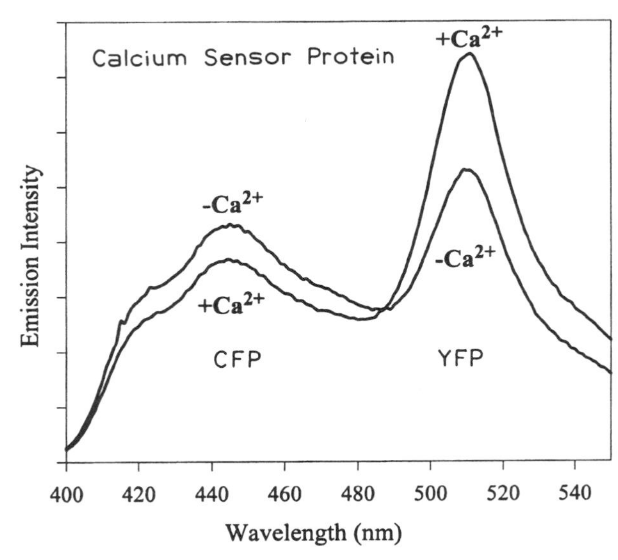

Figure 19.70. Emission spectra, in the presence and in the absence of Ca^{2+}, of a calcium-sensing protein composed of cyan fluorescent protein (CFP), yellow fluorescent protein (YFP), calmodulin, and the M13 peptide. Excitation was at 440 nm. Reprinted, with permission, from Ref. 199, Copyright © 1997, Macmillan Magazines Limited.

REFERENCES

1. Miller, J. N., and Birch, D. J. S. (eds.), 1997, *4th International Conference on Methods and Applications of Fluorescence Spectroscopy, J. Fluoresc.* 7(1):1S–246S.
2. Wolfbeis, O. S. (ed.), 1993, *Proceedings of the 1st European Conference on Optical Chemical Sensors and Biosensors, Europt(R)ode I, Sensors Actuators B* 11. 565 pp.
3. Baldini, F. (ed.), 1995, *Proceedings of the 2nd European Conference on Optical Chemical Sensors and Biosensors, Europt(R)ode II, Sensors Actuators B* 29. 439 pp.
4. Kunz, R. E. (ed.), 1997, *Proceedings of the 3rd European Conference on Optical Chemical Sensors and Biosensors, Part I—Plenary and Parallel Sessions; Part II—Poster Sessions, Europt(R)ode III, Sensors Actuators B* 38. 1–188 and 189–468.
5. Thompson, R. B. (ed.), 1997, *Advances in Fluorescence Sensing Technology III, Proc. SPIE* **2980**.
6. Lakowicz, J. R. (ed.), 1995, *Advances in Fluorescence Sensing Technology II, Proc. SPIE* **2388**.
7. Wolfbeis, O. S. (ed.), 1991, Biomedical applications of fiber optic chemical sensors, in *Fiber Optic Chemical Sensors and Biosensors*, Vol. II, O. S. Wolfbeis (ed.), CRC Press, Boca Raton, Florida, pp. 267–300.
8. Undenfriend, S., 1969, *Fluorescence Assay in Biology and Medicine,* Vol. II, Academic Press, New York. See also Vol. 1, 1962.
9. Duggan, D. E., Bowman, R. L., Brodie, B., and Undenfriend, S., 1957, A spectrophotofluorometric study of compounds of biological interest, *Arch. Biochem. Biophys.* **68**:1–14.

10. Ichinose, N., Schwedt, G., Schnepel, F. M., and Adachi, K., 1987, *Fluorometric Analysis in Biomedical Chemistry*, John Wiley & Sons, New York.
11. Lakowicz, J. R. (ed.), 1994, *Topics in Fluorescence Spectroscopy, Volume 4, Probe Design and Chemical Sensing*, Plenum Press, New York.
12. Szmacinski, H., and Lakowicz, J. R., 1994, Lifetime-based sensing, in *Topics in Fluorescence Spectroscopy, Volume 4, Probe Design and Chemical Sensing*, J. R. Lakowicz (ed.), Plenum Press, New York, pp. 295–334.
13. Schulman, S. G. (ed.), 1993, *Molecular Luminescence Spectroscopy, Part III: Methods and Applications*, John Wiley & Sons, New York. See also Part II, 1988, and Part I, 1985.
14. Czarnik, A. W. (ed.), 1993, *Fluorescent Chemosensors for Ion and Molecule Recognition*, American Chemical Society, Washington, D.C.
15. Kieslinger, D., Draxler, S., Trznadel, K., and Lippitsch, M. E., 1997, Lifetime-based capillary waveguide sensor instrumentation, *Sensors Actuators B* **38–39:**300–304.
16. Pease, A. C., Solas, D., Sullivan, E. J., Cronin, M. T., Holmes, C. P., and Fodor, S. P. A., 1994, Light-generated oligonucleotide arrays for rapid DNA sequence analysis, *Proc. Natl. Acad. Sci. U.S.A.* **91:**5022–5026.
17. Lipshutz, R. J., Morris, D., Chee, M., Hubbell, E., Kozal, M. J., Shah, N., Shen, N., Yang, R., and Fodor, S. P. A., 1995, Using oligonucleotide probe arrays to access genetic diversity, *BioTechniques* **19**(3):442–447.
18. Mooney, J. F., Hunt, A. J., McIntosh, J. R., Liberko, C. A., Walba, D. M., and Rogers, C. T., 1996, Patterning of functional antibodies and other proteins by photolithography of silane monolayers, *Proc. Natl. Acad. Sci. U.S.A.* **93:**12287–12291.
19. Bhatia, S. K., Teixeira, J. L., Anderson, M., Schriver-Lake, L. C., Calvert, J. M., Georger, J. H., Hickman, J. J., Dulcey, C. S., Schoen, P. E., and Ligler, F. S., 1993, Fabrication of surfaces resistant to protein adsorption and application to two-dimensional protein patterning, *Anal. Biochem.* **208:**197–205.
20. Kricka, L. J., Skogerboe, K. J., Hage, D. A., Schoeff, L., Wang, J., Sokol, L. J., Chan, D. W., Ward, K. M., and Davis, K. A., 1997, Clinical chemistry, *Anal. Chem.* **69:**165R–229R.
21. Lippitsch, M. E., Draxler, S., and Kieslinger, D., 1997, Luminescence lifetime-based sensing: New materials, new devices, *Sensors Actuators B* **38–39:**96–102.
22. Richards-Kortum, R., and Sevick-Muraca, E., 1996, Quantitative optical spectroscopy for tissue diagnosis, *Annu. Rev. Phys. Chem.* **47:**555–606.
23. Gouin, J. F., Baros, F., Birot, D., and Andre, J. C., 1997, A fibre-optic oxygen sensor for oceanography, *Sensors Actuators B* **38–39:**401–406.
24. Valeur, B., 1994, Principles of fluorescent probe design for ion recognition, in *Topics in Fluorescence Spectroscopy, Volume 4, Probe Design and Chemical Sensing*, J. R. Lakowicz (ed.), Plenum Press, New York, pp. 21–48.
25. Rettig, W., and Lapouyade, R., 1994, Fluorescence probes based on twisted intramolecular charge transfer (TICT) states and other adiabatic photoreactions, in *Topics in Fluorescence Spectroscopy, Volume 4, Probe Design and Chemical Sensing*, J. R. Lakowicz (ed.), Plenum Press, New York, pp. 109–149.
26. Czarnik, A. W., 1994, Principles of fluorescent probe design for ion recognition, in *Topics in Fluorescence Spectroscopy, Volume 4: Probe Design and Chemical Sensing*, J. R. Lakowicz (ed.), Plenum Press, New York, pp. 49–70.
27. Fabbrizzi, L., and Poggi, A., 1995, Sensors and switches from supramolecular chemistry, *Chem. Soc. Rev.* **24:**197–202.
28. Bryan, A. J., Prasanna de Silva, A., de Silva, S. A., Dayasiri Rupasinghe, A. D., and Samankumara Sandanayake, K. R. A., 1989, Photo-induced electron transfer as a general design logic for fluorescent molecular sensors for cations, *Biosensors* **4:**169–179.
29. Demas, J. N., and DeGraff, B. A., 1994, Design and applications of highly luminescent transition metal complexes, in *Topics in Fluorescence Spectroscopy, Volume 4, Probe Design and Chemical Sensing*, J. R. Lakowicz (ed.), Plenum Press, New York, pp. 71–107.
30. Bacon, J. R., and Demas, J. N., 1987, Determination of oxygen concentrations by luminescence quenching of a polymer immobilized transition metal complex, *Anal. Chem.* **59:**2780–2785.
31. Wolfbeis, O. S., 1991, Oxygen sensors, in *Fiber Optic Chemical Sensors and Biosensors*, Vol. II, O. S. Wolfbeis (ed.), CRC Press, Boca Raton, Florida, pp. 19–53.
32. Mills, A., and Williams, F. C., 1997, Chemical influences on the luminescence of ruthenium diimine complexes and its response to oxygen, *Thin Solid Films* **306:**163–170.
33. Lippitsch, M. E., Pusterhofer, J., Leiner, M. J. P., and Wolfbeis, O. S., 1988, Fibre-optic oxygen sensor with the fluorescence decay time as the information carrier, *Anal. Chim. Acta* **205:**1–6.
34. Draxler, S., Lippitsch, M. E., Klimant, I., Kraus, H., and Wolfbeis, O. S., 1995, Effects of polymer matrices on the time-resolved luminescence of a ruthenium complex quenched by oxygen, *J. Phys. Chem.* **99:**3162–3167.
35. Lakowicz, J. R., Johnson, M. L., Lederer, W. J., Szmacinski, H., Nowaczyk, K., Malak, H., and Berndt, K. W., 1992, Fluorescence lifetime sensing generates cellular images, *Laser Focus World* **28**(5):60–80.
36. Holst, G., Glud, R. N., Kuhl, M., and Klimant, I., 1997, A microoptode array for fine-scale measurement of oxygen distribution, *Sensors Actuators B* **38–39:**122–129.
37. Hartmann, P., Ziegler, W., Holst, G., and Lübbers, D. W., 1997, Oxygen flux fluorescence lifetime imaging, *Sensors Actuators B* **38–39:**110–115.
38. Cox, M. E., and Dunn, B., 1985, Detection of oxygen by fluorescence quenching, *Appl. Opt.* **24:**2114–2120.
39. Charlesworth, J. M., 1994, Optical sensing of oxygen using phosphorescence quenching, *Sensors Actuators B* **22:**1–5.
40. Papkovsky, D. B., Ponomarev, G. V., Trettnak, W., and O'Leary, P., 1995, Phosphorescent complexes of porphyrin ketones: Optical properties and applications to oxygen sensing, *Anal. Chem.* **67:**4112–4117.
41. Xu, W., Kneas, K. A., Demas, J. N., and DeGraff, B. A., 1996, Oxygen sensors based on luminescence quenching of metal complexes: Osmium complexes suitable for laser diode excitation, *Anal. Chem.* **68:**2605–2609.
42. Bambot, S. B., Rao, G., Romauld, M., Carter, G. M., Sipior, J., Terpetschnig, E., and Lakowicz, J. R., 1995, Sensing oxygen through skin using a red diode laser and fluorescence lifetimes, *Biosensors Bioelectron.* **10:**643–652.
43. Xu, W., McDonough, R. C., Langsdorf, B., Demas, J. N., and DeGraff, B. A., 1994, Oxygen sensors based on luminescence quenching. Interactions of metal complexes with the polymer supports, *Anal. Chem.* **66:**4133–4141.
44. Hartmann, P., and Leiner, M. J. P., 1995, Luminescence quenching behavior of an oxygen sensor based on a Ru(II) complex dissolved in polystyrene, *Anal. Chem.* **6:**88–93.
45. Wolfbeis, O. S., and Urbano, E., 1983, Eine fluorimetrische, schwermetallfreie methode zur analyse von chlor, brom und iod in organischen materialien, *Fresenius' Z. Anal. Chem.* **314:**577–581.

46. Illsley, N. P., and Verkman, A. S., 1987, Membrane chloride transport measured using a chloride-sensitive fluorescent probe, *Biochem.* **26**:1215–1219.
47. Verkman, A. S., 1990, Development and biological applications of chloride-sensitive fluorescent indicators, *Am. J. Physiol.* **253**:C375–C388.
48. Verkman, A. S., Sellers, M. C., Chao, A. C., Leung, T., and Ketcham, R., 1989, Synthesis and characterization of improved chloride-sensitive fluorescent indicators for biological applications, *Anal. Biochem.* **178**:355–361.
49. Biwersi, J., Tulk, B., and Verkman, A. S., 1994, Long-wavelength chloride-sensitive fluorescent indicators, *Anal. Biochem.* **219**:139–143.
50. Orosz, D. E., and Carlid, K. D., 1992, A sensitive new fluorescence assay for measuring proton transport across liposomal membranes, *Anal. Biochem.* **210**:7–15.
51. Chao, A. C., Dix, J. A., Sellers, M. C., and Verkman, A. S., 1989, Fluorescence measurement of chloride transport in monolayer cultured cells. Mechanisms of chloride transport in fibroblasts, *Biophys. J.* **56**:1071–1081.
52. Wolfbeis, O. S., and Sharma, A., 1988, Fibre-optic fluorosensor for sulphur dioxide, *Anal. Chim. Acta* **208**:53–58.
53. Sharma, A., Draxler, S., and Lippitsch, M. E., 1992, Time-resolved spectroscopy of the fluorescence quenching of a donor–acceptor pair by halothane, *Appl. Phys. B* **54**:309–312.
54. Omann, G. M., and Lakowicz, J. R., 1982, Interactions of chlorinated hydrocarbon insecticides with membranes, *Biochim. Biophys. Acta* **684**:83–95.
55. Vanderkooi, J. M., Wright, W. W., and Erecinska, M., 1994, Nitric oxide diffusion coefficients in solutions, proteins and membranes determined by phosphorescence, *Biochim. Biophys. Acta* **1207**:249–254.
56. Denicola, A., Souza, J. M., Radi, R., and Lissi, E., 1996, Nitric oxide diffusion in membranes determined by fluorescence quenching, *Arch. Biochem. Biophys.* **328**:208–212.
57. Jordan, D. M., Walt, D. R., and Milanovich, F. P., 1987, Physiological pH fiber-optic chemical sensor based on energy transfer, *Anal. Chem.* **59**:437–439.
58. Lakowicz, J. R., Szmacinski, H., and Karakelle, M., 1993, Optical sensing of pH and pCO_2 using phase-modulation fluorimetry and resonance energy transfer, *Anal. Chim. Acta* **272**:179–186.
59. Sipior, J., Bambot, S., Romauld, M., Carter, G. M., Lakowicz, J. R., and Rao, G., 1995, A lifetime-based optical CO_2 gas sensor with blue or red excitation and Stokes or anti-Stokes detection, *Anal. Biochem.* **227**:309–318.
60. Chang, Q., Sipior, J., Lakowicz, J. R., and Rao, G., 1995, A lifetime-based fluorescence resonance energy transfer sensor for ammonia, *Anal. Biochem.* **232**:92–97.
61. Mills, A., Chang, Q., and McMurray, N., 1992, Equilibrium studies on colorimetric plastic film sensors for carbon dioxide, *Anal. Chem.* **64**:1383–1389.
62. Wolfbeis, O. S., Reisfeld, R., and Oehme, I., 1996, Sol-gels and chemical sensors, *Struct. Bonding* **85**:51–98.
63. Avnir, D., Braun, S., and Ottolenghi, M., 1992, A review of novel photoactive, optical, sensing and bioactive materials. A review. *ACS Symp. Ser.* **499**:384–404.
64. Lakowicz, J. R., 1994, Emerging biomedical applications of time-resolved fluorescence spectroscopy, in *Topics in Fluorescence Spectroscopy, Volume 4, Probe Design and Chemical Sensing*, J. R. Lakowicz (ed.), Plenum Press, New York, pp. 1–9.
65. Schultz, J. S., and Sims, G., 1979, Affinity sensors for individual metabolites, *Biotechnol. Bioeng. Symp.* **9**:65–71.
66. Schultz, J., Mansouri, S., and Goldstein, I. J., 1982, Affinity sensor: A new technique for developing implantable sensors for glucose and other metabolites, *Diabetes Care* **5**(3):245–253.
67. Meadows, D., and Schultz, J. S., 1988, Fiber-optic biosensors based on fluorescence energy transfer, *Talanta* **35**(2):145–150.
68. Lakowicz, J. R., and Maliwal, B. P., 1993, Optical sensing of glucose using phase-modulation fluorometry, *Anal. Chim. Acta* **271**:155–164.
69. Tolosa, L., Szmacinski, H., Rao, G., and Lakowicz, J. R., 1997, Lifetime-based sensing of glucose using energy transfer with a long lifetime donor, *Anal. Biochem.* **250**:102–108.
70. Tolosa, L., Malak, H., Rao, G., and Lakowicz, J. R., 1997, Optical assay for glucose based on the luminescence decay time of the long wavelength dye Cy5™, *Sensors Actuators* **45**:93–99.
71. He, H., Li, H., Mohr, G., Kovacs, B., Werner, T., and Wolfbeis, O. S., 1993, Novel type of ion-selective fluorosensor based on the inner filter effect: An optrode for potassium, *Anal. Chem.* **65**:123–127.
72. Roe, J. N., Szoka, F. C., and Verkman, A. S., 1989, Optical measurement of aqueous potassium concentration by a hydrophobic indicator in lipid vesicles, *Biophys. Chem.* **33**: 295–302.
73. Roe, J. N., Szoka, F. C., and Verkman, A. S., 1990, Fibre optic sensor for the detection of potassium using fluorescence energy transfer, *Analyst* **115**:353–368.
74. Mahutte, C. K., 1994, Continuous intra-arterial blood gas monitoring, *Intensive Care Med.* **20**:85–86.
75. Shapiro, B. A., Mahutte, C. K., Cane, R. D., and Gilmour, I. J., 1993, Clinical performance of a blood gas monitor: A prospective, multicenter trial, *Crit. Care Med.* **21**(4):487–494.
76. Yafuso, M., Arick, S. A., Hansmann, D., Holody, M., Miller, W. W., Yan, C. F., and Mahutte, K., 1989, Optical pH measurements in blood, *Proc. SPIE* **1067**:37–43.
77. Vurek, G. G., Feustel, P. J., and Severinghaus, J. W., 1983, A fiber optic pCO_2 sensor, *Ann. Biomed. Eng.* **11**:499–510.
78. Mahutte, C. K., Holody, M., Maxwell, T. P., Chen, P. A., and Sasse, S. A., 1994, Development of a patient-dedicated, on-demand, blood gas monitor, *Am. J. Respir. Crit. Care Med.* **149**:852–859.
79. Mahutte, C. K., Sasse, S. A., Chen, P. A., and Holody, M., 1994, Performance of a patient-dedicated, on-demand blood gas monitor in medical ICU patients, *Am. J. Respir. Crit. Care Med.* **150**:865–869.
80. Opitz, N., and Lübbers, D. W., 1987, Theory and development of fluorescence-based optochemical oxygen sensors: Oxygen optodes, *Int. Anesthesiol. Clin.* **25**(3):177–197.
81. Ohkuma, S., and Poole, B., 1978, Fluorescence probe measurement of the intralysosomal pH in living cells and the perturbation of pH by various agents, *Proc. Natl. Acad. Sci. U.S.A.* **5**:3327–3331.
82. Thomas, J. A., Buchsbaum, R. N., Zimniak, A., and Racker, E., 1979, Intracellular pH measurements in Ehrlich ascites tumor cells utilizing spectroscopic probes generated in situ, *Biochemistry* **18**:2210–2218.
83. Munkholm, C., Walt, D. R., and Milanovich, F. P., 1988, A fiber-optic sensor for CO_2 measurement, *Talanta* **35**(2):109–112.
84. Kawabata, Y., Kamichika, T., Imasaka, T., and Ishibashi, N., 1989, Fiber-optic sensor for carbon dioxide with a pH indicator dispersed in a poly(ethyleneglycol) membrane, *Anal. Chim. Acta* **219**:223–229.
85. Yguerabide, J., Talavera, E., Alvarez, J. M., and Quintero, B., 1994, Steady-state fluorescence method for evaluating excited-state proton reactions: Application to fluorescein, *Photochem. Photobiol.* **60**:435–441.

86. Haugland, R. P., 1996, Chapter 23, in *Handbook of Fluorescent Probes and Research Chemicals*, Molecular Probes, Eugene, Oregon, pp. 551–561.
87. Rink, T. J., Tsien, R. Y., and Pozzan, T., 1982, Cytoplasmic pH and free Mg^{2+} in lymphocytes, *J. Cell Biol.* **95:**189–196.
88. Clement, N. R., and Gould, J. M., 1981, Pyranine (8-hydroxy-1,3,6-pyrenetrisulfonate) as a probe of internal aqueous hydrogen ion concentration in phospholipid vesicles, *Biochemistry* **20:**1534–1538.
89. Wolfbeis, O. S., Fürlinger, E., Kroneis, H., and Marsoner, H., 1983, Fluorimetric analysis. 1. A study on fluorescent indicators for measuring near neutral ("physiological") pH-values, *Fresenius' Z. Anal. Chem.* **314:**119–124.
90. Schulman, S. G., Chen, S., Bai, F., Leiner, M. J. P., Weis, L., and Wolfbeis, O. S., 1995, Dependence of the fluorescence of immobilized 1-hydroxypyrene-3,6,8-trisulfonate on sodium pH: Extension of the range of applicability of a pH fluorosensor, *Anal. Chim. Acta* **304:**165–170.
91. Zhujun, H., and Seitz, W. R., 1984, A fluorescence sensor for quantifying pH in the range from 6.5 to 8.5, *Anal. Chim. Acta* **160:**47–55.
92. Uttamlal, M., and Walt, D. R., 1995, A fiber-optic carbon dioxide sensor for fermentation monitoring, *BioTechnology* **13:**597–601.
93. Whitaker, J. E., Haugland, R. P., and Prendergast, F. G., 1991, Spectral and photophysical studies of benzo[*c*]xanthene dyes: Dual emission pH sensors, *Anal. Biochem.* **194:**330–344.
94. Szmacinski, H., and Lakowicz, J. R., 1993, Optical measurements of pH using fluorescence lifetimes and phase-modulation fluorometry, *Anal. Chem.* **65:**1668–1674.
95. Srivastava, A., and Krishnamoorthy, G., 1997, Time-resolved fluorescence microscopy could correct for probe binding while estimating intracellular pH, *Anal. Biochem.* **249:**140–146.
96. Wolfbeis, O. S., Rodriguez, N. V., and Werner, T., 1992, LED-compatible fluorosensor for measurement of near-neutral pH values, *Mikrochim. Acta* **108:**133–141.
97. Zen, J-M., and Patonay, G., 1991, Near-infrared fluorescence probe for pH determination, *Anal. Chem.* **63:**2934–2938.
98. Boyer, A. E., Devanathan, S., Hamilton, D., and Patonay, G., 1992, Spectroscopic studies of a near-infrared absorbing pH sensitive carbocyanine dye, *Talanta* **39**(5)**:**505–510.
99. Wolfbeis, O. S., and Marhold, H., 1987, A new group of fluorescent pH-indicators for an extended pH-range, *Anal. Chem.* **327:**347–350.
100. Murtaza, Z., Chang, Q., Rao, G., Lin, H., and Lakowicz, J. R., 1997, Long-lifetime metal–ligand pH probes, *Anal. Biochem.* **247:**216–222.
101. de Silva, A. P., Nimal Gunaratne, H. Q., and Rice, T. E., 1996, Proton-controlled switching of luminescence in lanthanide complexes in aqueous solution: pH sensors based on long-lived emission, *Angew. Chem. Int. Ed. Engl.* **35:**2116–2118.
102. Bryan, A. J., de Silva, P., de Silva, S. A., Rupasinghe, R. A. D. D., and Sandanayake, K. R. A. S., 1989, Photo-induced electron transfer as a general design logic for fluorescent molecular sensors for cations, *Biosensors* **4:**169–179.
103. de Silva, A. P., Gunaratne, H. Q. N., Habib-Jiwan, J.-L, McCoy, C. P., Rice, T. E., and Soumillion, J.-P., 1995, New fluorescent model compounds for the study of photoinduced electron transfer: The influence of a molecular electric field in the excited state, *Angew. Chem. Int. Ed. Engl.* **34:**1728–1731.
104. Akkaya, E. U., Huston, M. E., and Czarnik, A. W., 1990, Chelation-enhanced fluorescence of anthrylazamacrocycle conjugate probes in aqueous solution, *J. Am. Chem. Soc.* **112:**3590–3593.
105. Fages, F., Desvergne, J. P., Bouas-Laurent, H., Marsau, P., Lehn, J.-M., Kotzyba-Hibert, F., Albrecht-Gary, A.-M., and Al-Joubbeh, M., 1989, Anthraceno-cryptands: A new class of cation-complexing macrobicyclic fluorophores, *J. Am. Chem. Soc.* **111:**8672–8680.
106. de Silva, A. P., and de Silva, S. A., 1986, Fluorescent signalling crown ethers; 'switching on' of fluorescence by alkali metal ion recognition and binding in situ, *J. Chem. Soc., Chem. Commun.* **1986:**1709–1710.
107. Huston, M. E., Akkaya, E. U., and Czarnik, A. W., 1989, Chelation enhanced fluorescence detection of non-metal ions, *J. Am. Chem. Soc.* **111:**8735–8737.
108. Pederson, C. J., 1967, Cyclic polyethers and their complexes with metal salts, *J. Am. Chem. Soc.* **89:**7017–7036.
109. Cram, D. J., 1988, The design of molecular hosts, guests, and their complexes, *Science* **240:**760–767.
110. Pederson, C. J., 1988, The discovery of crown ethers, *Science* **241:**536–540.
111. Gokel, G. W., 1991, *Crown Ethers and Cryptands*, The Royal Society of Chemistry, Cambridge.
112. Nuccitelli, R. (ed.), 1994, *Methods in Cell Biology, Vol. 40: A Practical Guide to the Study of Calcium in Living Cells*, Academic Press, New York.
113. Minta, A., and Tsien, R. Y., 1989, Fluorescent indicators for cytosolic sodium, *J. Biol. Chem.* **264:**19449–19457.
114. Grynkiewicz, G., Poenie, M., and Tsien, R. Y., 1985, A new generation of Ca^{2+} indicators with greatly improved fluorescence properties, *J. Biol. Chem.* **260:**3440–3450.
115. Tsien, R. Y., 1989, Fluorescent indicators of ion concentrations, *Methods Cell Biol.* **30:**127–156.
116. Haugland, R. P., 1996, *Handbook of Fluorescent Probes and Research Chemicals*, 6th ed., Molecular Probes, Eugene, Oregon, pp. 503–584.
117. Rettig, W., and Lapouyade, R., 1994, Fluorescence probes based on twisted intramolecular charge transfer (TICT) states and other adiabatic photoreactions, in *Topics in Fluorescence Spectroscopy, Volume 4, Probe Design and Chemical Sensing*, J. R. Lakowicz (ed.), Plenum Press, New York, pp. 109–149.
118. Crossley, R., Goolamali, Z., and Sammes, P. G., 1994, Synthesis and properties of a potential extracellular fluorescent probe for potassium, *J. Chem. Soc., Perkin Trans.* **1994:**1615–1623.
119. Szmacinski, H., and Lakowicz, J. R., 1998, Potassium and sodium measurements in blood using phase-modulation fluorometry, submitted.
120. Crossley, R., Goolamali, Z., Gosper, J. J., and Sammes, P. G., 1994, Synthesis and spectral properties of new fluorescent probes for potassium, *J. Chem. Soc., Perkin Trans. 2* **1994:**513–520.
121. Golchini, K., Mackovic-Basic, M., Gharib, S. A., Masilamani, D., Lucas, M. E., and Kurtz, I., 1990, Synthesis and characterization of a new fluorescent probe for measuring potassium, *Am. J. Physiol.* **258**(*Renal Fluid Electrolyte Physiol.* **27**)**:**F438–F443.
122. Szmacinski, H., and Lakowicz, J. R., 1997, Sodium Green as a probe for intracellular sodium imaging based on fluorescence lifetime, *Anal. Biochem.* **250:**131–138.
123. Lakowicz, J. R., and Szmacinski, H., 1992, Fluorescence lifetime-based sensing of pH, Ca^{2+}, K^+ and glucose, *Sensors Actuators B* **11:**133–143.
124. Meuwis, K., Boens, N., De Schryver, F. C., Gallay, J., and Vincent, M., 1995, Photophysics of the fluorescent K^+ indicator PBFI, *Biophys. J.* **68:**2469–2473.
125. Valeur, B., Bourson, J., and Pouget, J., 1993, Ion recognition detected by changes in photoinduced charge or energy transfer, in

Fluorescent Chemosensors for Ion and Molecule Recognition, A. W. Czarnik (ed.), ACS Symposium Series 538, pp. 25–44.
126. Tsien, R. Y., Rink, T. J., and Poenie, M., 1985, Measurement of cytosolic free Ca^{2+} in individual small cells using fluorescence microscopy with dual excitation wavelengths, *Cell Calcium* **6:**145–157.
127. Iatridou, H., Foukaraki, E., Kuhn, M. A., Marcus, E. M., Haugland, R. P., and Katerinopoulos, H. E., 1994, The development of a new family of intracellular calcium probes, *Cell Calcium* **15:**190–198.
128. Akkaya, E. U., and Lakowicz, J. R., 1993, Styryl-based wavelength ratiometric probes: A new class of fluorescent calcium probes with long wavelength emission and a large Stokes' shift, *Anal. Biochem.* **213:**285–289.
129. Minta, A., Kao, J. P. Y., and Tsien, R. Y., 1989, Fluorescent indicators for cytosolic calcium based on rhodamine and fluorescein chromophores, *J. Biol. Chem.* **264:**8171–8178.
130. Eberhard, M., and Erne, P., 1991, Calcium binding to fluorescent calcium indicators: Calcium green, calcium orange and calcium crimson, *Biochem. Biophy. Res. Comm.* **180:**209–215.
131. Lakowicz, J. R., Szmacinski, H., and Johnson, M. L., 1992, Calcium concentration imaging using fluorescence lifetimes and long-wavelength probes, *J. Fluoresc.* **2**(1):47–62.
132. Hirshfield, K. M., Toptygin, D., Packard, B. S., and Brand, L., 1993, Dynamic fluorescence measurements of two-state systems: Applications to calcium-chelating probes, *Anal. Biochem.* **209:**209–218.
133. Miyoshi, N., Hara, K., Kimura, S., Nakanishi, K., and Fukuda, M., 1991, A new method of determining intracellular free Ca^{2+} concentration using Quin-2 fluorescence, *Photochem. Photobiol.* **53:**415–418.
134. Lakowicz, J. R., Szmacinski, H., Nowaczyk, K., and Johnson, M. L., 1992, Fluorescence lifetime imaging of calcium using Quin-2, *Cell Calcium* **13:**131–147.
135. Oguz, U., and Akkaya, E. U., 1997, One-pot synthesis of a red-fluorescent chemosensor from an azacrown, phloroglucinol and squaric acid: A simple in-solution construction of a functional molecular device, *Tetrahedron Lett.* **38:**4509–4512.
136. Akkaya, E. U., and Turkyilmaz, S., 1997, A squaraine-based near IR fluorescent chemosensor for calcium, *Tetrahedron Lett.* **38:**4513–4516.
137. Wahl, M., Lucherini, M. J., and Gruenstein, E., 1990, Intracellular Ca^{2+} measurement with Indo-1 in substrate-attached cells: Advantages and special considerations, *Cell Calcium* **11:**487–500.
138. Groden, D. L., Guan, Z., and Stokes, B. T., 1991, Determination of Fura-2 dissociation constants following adjustment of the apparent Ca-EGTA association constant for temperature and ionic strength, *Cell Calcium* **12:**279–287.
139. David-Dufilho, M., Montenay-Garestier, T., and Devynck, M.-A., 1989, Fluorescence measurements of free Ca^{2+} concentration in human erythrocytes using the Ca^{2+} indicator Fura-2, *Cell Calcium* **9:**167–179.
140. Hirshfield, K. M., Toptygin, D., Grandhige, G., Kim, H., Packard, B. Z., and Brand, L., 1996, Steady-state and time-resolved fluorescence measurements for studying molecular interactions: Interaction of a calcium-binding probe with proteins, *Biophys. Chem.* **62:**25–38.
141. Kao, J. P. Y., 1994, Practical aspects of measuring [Ca^{2+}] with fluorescent indicators, *Methods Cell Biol.* **40:**155–181.
142. Morris, S. J., Wiegmann, T. B., Welling, L. W., and Chronwall, B. M., 1994, Rapid simultaneous estimation of intracellular calcium and pH, *Methods Cell Biol.* **40:**183–220.
143. Scanlon, M., Williams, D. A., and Fay, F. S., 1987, A Ca^{2+}-insensitive form of fura-2 associated with polymorphonuclear leukocytes, *J. Biol. Chem.* **262:**6308–6312.
144. Bourson, J., Pouget, J., and Valeur, B., 1993, Ion-responsive fluorescent compounds. 4. Effect of cation bonding on the photophysical properties of a coumarin linked to monoaza- and diaza-crown ethers, *J. Phys. Chem.* **97:**4552–4557.
145. Dumon, P., Jonusauskas, G., Dupuy, F., Pee, P., Rulliere, C., Letard, J. F., and Lapouyade, R., 1994, Picosecond dynamics of cation–macrocycle interactions in the excited state of an intrinsic fluorescence probe: The calcium complex of 4-(*N*-monoaza-15-crown-5)-4′-phenylstilbene, *J. Phys. Chem.* **98:**10391–10396.
146. Letard, J.-F., Lapouyade, R., and Rettig, W., 1993, Chemical engineering of fluorescence dyes, *Mol. Cryst. Liq. Cryst.* **236:**41–46.
147. Rurack, K., Bricks, J. L., Kachkovski, A., and Resch, U., 1997, Complexing fluorescence probes consisting of various fluorophores linked to 1-aza-15-crown-5, *J. Fluoresc.* **7**(1):63S–66S.
148. Lohr, H.-G., and Fogtle, F., 1985, Chromo- and fluoroionophores. A new class of dye reagents, *Acc. Chem. Res.* **18:**65–72.
149. de Silva, A. P., Nimal Qunaratne, H. Q., and Maguire, G. E. M., 1994, Off–on fluorescent sensors for physiological levels of magnesium ions based on photoinduced electron transfer (PET), which also behave as photoionic OR logic gates, *J. Chem. Soc., Chem. Commun.* **1994:**1213–1214.
150. Raju, B., Murphy, E., Levy, L. A., Hall, R. D., and London, R. E., 1989, A fluorescent indicator for measuring cytosolic free magnesium, *Am. J. Physiol.* **256:**C540–C548.
151. Illner, H., McGuigan, J. A. S., and Luthi, D., 1992, Evaluation of mag-fura-5, the new fluorescent indicator for free magnesium measurements, *Eur. J. Physiol.,* **422:**179–184.
152. Morelle, B., Salmon, J.-M., Vigo, J., and Viallet, P., 1993, Proton, Mg^{2+} and protein as competing ligands for the fluorescent probe, mag-indo-1: A first step to the quantification of intracellular Mg^{2+} concentration, *Photochem. Photobiol.* **58:**795–802.
153. Szmacinski, H., and Lakowicz, J. R., 1996, Fluorescence lifetime characterization of magnesium probes: Improvement of Mg^{2+} dynamic range and sensitivity using phase-modulation fluorometry, *J. Fluoresc.* **6**(2):83–95.
154. James, T. D., Sandanayake, K. R. A. S., and Shinkai, S., 1994, Novel photoinduced electron-transfer sensor for saccharides based on the interaction of boronic acid and amine, *J. Chem. Soc., Chem. Commun.* **1994:**477–478.
155. Yoon, J., and Czarnik, A. W., 1992, Fluorescent chemosensors of carbohydrates. A means of chemically communicating the binding of polyols in water based on chelation-enhanced quenching, *J. Am. Chem. Soc.* **114:**5874–5875.
156. James, T. D., Sandanayake, K. R. A. S., and Shinkai, S., 1995, Chiral discrimination of monosaccharides using a fluorescent molecular sensor, *Nature* **74:**345–347.
157. Hemmila, I. A., 1992, *Applications of Fluorescence in Immunoassays*, John Wiley & Sons, New York.
158. Van Dyke, K., and Van Dyke, R. (eds.), 1990, *Luminescence Immunoassay and Molecular Applications*, CRC Press, Boca Raton, Florida.
159. Ozinskas, A. J., 1994, Principles of fluorescence immunoassay, in *Topics in Fluorescence Spectroscopy, Volume 4, Probe Design and Chemical Sensing*, J. R. Lakowicz (ed.), Plenum Press, New York, pp. 449–496.
160. Gosling, J. P., 1990, A decade of development in immunoassay methodology, *Clin. Chem.* **36:**1408–1427.

161. Davidson, R. S., and Hilchenbach, M. M., 1990, The use of fluorescent probes in immunochemistry, *Photochem. Photobiol.* **52**:431–438.
162. Vo-Dinh, T., Sepaniak, M. J., Griffin, G. D., and Alarie, J. P., 1993, Immunosensors: Principles and applications, *Immunomethods* **3**:85–92.
163. Berson, S., and Yalow, R., 1959, Quantitative aspects of the reaction between insulin and insulin-binding antibody, *J. Clin. Invest.* **38**:1996–2016.
164. Lovgren, T., Hemmila, I., Pettersson, K., and Halonen, P., 1985, Time-resolved fluorometry in immunoassay, in *Alternative Immunoassays*, W. P. Collins (ed.), John Wiley & Sons, New York, pp. 203–217.
165. Diamandis, E. P., 1988, Immunoassays with time-resolved fluorescence spectroscopy: Principles and applications, *Clin. Biochem.* **21**:139–150.
166. Lövgren, T., and Pettersson, K., 1990, Time-resolved fluoroimmunoassay, advantages and limitations, in *Luminescence Immunoassay and Molecular Applications*, K. Van Dyke and R. Van Dyke (eds.), CRC Press, Boca Raton, Florida, pp. 234–250.
167. Khosravi, M., and Diamandis, E. P., 1987, Immunofluorometry of choriogonadotropin by time-resolved fluorescence spectroscopy, a new europium chelate as label, *Clin. Chem.* **33**:1994–1999.
168. Soini, E., 1984, Pulsed light, time-resolved fluorometric immunoassay, in *Monoclonal Antibodies and New Trends in Immunoassays*, C. A. Bizollon (ed.), Elsevier Science Publishers, New York, pp. 197–208.
169. Morrison, L. E., 1988, Time-resolved detection of energy transfer: Theory and application to immunoassays, *Anal. Biochem.* **174**:101–120.
170. Ullman, E. F., Schwarzberg, M., and Rubenstein, K. E., 1976, Fluorescent excitation transfer immunoassay: A general method for determination of antigens, *J. Biol. Chem.* **251**:4172–4178.
171. Ozinskas, A. J., Malak, H., Joshi, J., Szmacinski, H., Britz, J., Thompson, R. B., Koen, P. A., and Lakowicz, J. R., 1993, Homogeneous model immunoassay of thyroxine by phase-modulation fluorescence spectroscopy, *Anal. Biochem.* **213**:264–270.
172. Lakowicz, J. R., Maliwal, B., Ozinskas, A., and Thompson, R. B., 1993, Fluorescence lifetime energy-transfer immunoassay quantified by phase-modulation fluorometry, *Sensors Actuators B* **12**:65–70.
173. Bright, F. V., and McGown, L. B., 1985, Homogeneous immunoassay of phenobarbital by phase-resolved fluorescence spectroscopy, *Talanta* **32**(1):15–18.
174. Dandliker, W. B., and de Saussure, V. A., 1970, Fluorescence polarization in immunochemistry, *Immunochemistry* **7**:799–828.
175. Spencer, R. D., Toledo, F. B., Williams, B. T., and Yoss, N. L., 1973, Design, construction, and two applications for an automated flow-cell polarization fluorometer with digital read out: Enzyme–inhibitor (antitrypsin) assay and antigen–antibody (insulin–insulin antiserum) assay, *Clin. Chem.* **19**:838–844.
176. Kobayashi, Y., Amitani, K., Watanabe, F., and Miyai, K., 1979, Fluorescence polarization immunoassay for cortisol, *Clin. Chim. Acta* **92**:241–247.
177. Cox, H., Whitby, M., Nimmo, G., and Williams, G., 1993, Evaluation of a novel fluorescence polarization immunoassay for teicoplanin, *Antimicrob. Agents Chemother.* **37**:1924–1926.
178. Mastin, S. H., Buck, R. L., and Mueggler, P. A., 1993, Performance of a fluorescence polarization immunoassay for teicoplanin in serum, *Diagn. Microbiol. Infect. Dis.* **16**:17–24.
179. Ripple, M. G., Goldberger, B. A., Caplan, Y. H., Blitzer, M. G., and Schwartz, S., 1992, Detection of cocaine and its metabolites in human amniotic fluid, *J. Anal. Toxicol.* **16**:328–331.
180. de Kanel, J., Dunlap, L., and Hall, T. D., 1989, Extending the detection limit of the TDx fluorescence polarization immunoassay for benzoylecgonine in urine, *Clin. Chem.* **35**:2110–2112.
181. Uber-Bucek, E., Hamon, M., Huy, C. P., and Dadoun, H., 1992, Determination of thevetin B in serum by fluorescence polarization immunoassay, *J. Pharm. Biomed. Anal.* **10**(6):413–419.
182. Li, P. K., Lee, J. T., Conboy, K. A., and Ellis, E. F., 1986, Fluorescence polarization immunoassay for theophylline modified for use with dried blood spots on filter paper, *Clin. Chem.* **32**:552–555.
183. Haver, V. M., Audino, N., Burris, S., and Nelsnol, M., 1989, Four fluorescence polarization immunoassays for therapeutic drug monitoring evaluated, *Clin. Chem.* **35**:138–140.
184. Klein, C., Batz, H.-G., Draeger, B., Guder, H.-J., Herrmann, R., Josel, H.-P, Nagele, U., Schenk, R., and Bogt, B., 1993, Fluorescence polarization immunoassay, in *Fluorescence Spectroscopy: New Methods and Applications*, O. S. Wolfbeis (ed.), Springer-Verlag, Berlin, pp. 245–258.
185. Wang, P. P., Simpson, E., Meucci, V., Morrison, M., Lunetta, S., Zajac, M., and Boeckx, R., 1991, Cyclosporine monitoring by fluorescence polarization immunoassay, *Clin. Biochem.* **24**:55–58.
186. Winkler, M., Schumann, G., Petersen, D., Oellerich, M., and Wonigeit, K., 1992, Monoclonal fluorescence polarization immunoassay evaluated for monitoring cyclosporine in whole blood after kidney, heart, and liver transplantation, *Clin. Chem.* **38**:123–126.
187. Lukens, H. R., Williams, C. B., Levison, S. A., Dandliker, W. B., Murayama, D., and Baron, R. L., 1977, Fluorescence immunoassay technique for detecting organic environmental contaminants, *Environ. Sci. Technol.* **11**:292–297.
188. Eremin, S. A., 1995, Polarization fluoroimmunoassay for rapid, specific detection of pesticides, in *Immunoanalysis of Agrochemicals*, J. O. Nelson, A. E. Karu, and R. B. Wong (eds.), American Chemical Society, Washington, D.C., pp. 223–234.
189. Fiore, M., Mitchell, J., Doan, T., Nelson, R., Winter, G., Grandone, C., Zeng, K., Haraden, R., Smith, J., Harris, K., Leszczynski, J., Berry, D., Safford, S., Barnes, G., Scholnick, A., and Ludington, K., 1988, The Abbott IMx™ automated benchtop immunochemistry analyzer system, *Clin. Chem.* **34**:1726–1732.
190. Lang, H., and Wurzburg, U., 1982, Creatine kinase, an enzyme of many forms, *Clin. Chem.* **28**:1439–1447.
191. Brayne, C. E. G., Calloway, S. P., and Dow, L., 1982, Blood creatine kinase isoenzymes BB in boxers, *Lancet* **ii**:1308–1309.
192. Grossman, S. H., 1984, Fluorescence polarization immunoassay applied to macromolecules: Creatine kinase-BB, *J. Clin. Immunol.* **7**:96–100.
193. Koncki, R., Mohr, G. J., and Wolfbeis, O. S., 1995, Enzyme biosensor for urea based on a novel pH bulk optode membrane, *Biosensors Bioelectron.* **10**:653–659.
194. Kessler, M. A., Meinitzer, A., and Wolfbeis, O. S., 1997, Albumin blue 580 fluorescence assay for albumin, *Anal. Biochem.* **248**:180–182.
195. Marvin, J. S., and Hellinga, H. W., 1998, Engineering biosensors by introducing fluorescent allosteric signal transducers: Construction of a novel glucose sensor, *J. Am. Chem. Soc.* **120**:7–11.
196. Gilardi, G., Mei, G., Rosato, N., Agró, A. F., and Cass, A. E. G., 1997, Spectroscopic properties of an engineered maltose binding protein, *Protein Eng.* **10**:479–486.
197. Gilardi, G., Zhou, L. Q., Hibbert, L., and Cass, A. E. G., 1994, Engineering the maltose binding protein for reagentless fluorescence sensing, *Anal. Chem.* **66**:3840–3847.

198. Romoser, V. A., Hinkle, P. M., and Persechini, A., 1997, Detection in living cells of Ca^{2+} dependent changes in the fluorescence emission of an indicator composed of two green fluorescent protein variants linked by a calmodulin-binding sequence, A new class of fluorescent indicators *J. Biol. Chem.* **272:**13270–13274.
199. Miyawaki, A., Llopis, J., Heim, R., McCaffery, J. M., Adams, J. A., Ikura, M., and Tsien, R. Y., 1997, Fluorescent indicators for Ca^{2+} based on green fluorescent proteins and calmodulin, *Nature* **388:**882–887.
200. Albano, C. R., Randers-Eichhorn, L., Bentley, W. E., and Rao, G., 1998, Green fluorescent protein as a real time quantitative reporter of heterogeneous protein production, *Biotechnol. Prog.* **14:**351–554.
201. Randers-Richhorn, L., Albano, C. R., Sipior, J., Bentley, W. E., and Rao, G., 1997, On-line green fluorescent protein sensor with LED excitation, *Biotechnol. Bioeng.* **55:**921–926.

PROBLEMS

19.1. *Oxygen Diffusion in a Polymer*: Use the data in Figure 19.13 to calculate the oxygen bimolecular quenching constant of camphorquinone. Compare this value with that expected for a fluorophore in water, which is near 1×10^{10} M^{-1} s^{-1}. Also calculate the diffusion coefficient of oxygen in poly(methyl methacrylate) (PMMA). How does this compare with the value in water, 2.5×10^{-5} cm^2/s? Assume that the solubility of oxygen in PMMA is the same as in water, 0.001275*M*/atm.

19.2. *Lifetimes and Oxygen Quenching*: In Section 19.4.B, we stated that a short-lifetime probe can serve as an intensity reference in an oxygen sensor. Assume that the lifetime of the $[Ru(Ph_2phen)_3]^{2+}$ oxygen sensor is 5 μs and that the lifetime of the reference fluorophore is 5 ns. Describe the relative extents of quenching for each probe.

19.3. *Fluorescence Polarization Immunoassays and Effects of Resonance Energy Transfer*: Suppose that you are performing a fluorescence polarization immunoassay for a small molecule with a rotational correlation time (θ) of 1 ns and that this peptide (P) is labeled with fluorescin (Fl), for which $\tau = 4$ ns and $r_0 = 0.40$.

A. What is the range of possible anisotropy values? Assume that the antibody (Ab) specific for the peptide has a rotational correlation time of 100 ns.
B. Assume that your starting assay contains Fl-P bound to Ab and that upon mixing with the sample, 10% of the Fl-P is displaced. What is the anisotropy?
C. Now assume that the antibody is labeled with rhodamine (Rh) so that Fl-P is 90% quenched by RET. What is the anisotropy when 10% of the Fl-P is displaced?

20 Long-Lifetime Metal–Ligand Complexes

Throughout the previous chapters, we described fluorophores with decay times ranging from 1 to 20 ns. While this timescale is useful for many biophysical measurements, there are numerous instances where longer decay times are desirable. For instance, one may wish to measure rotational motions of large proteins or membrane-bound proteins. In such cases the overall rotational correlation times can easily be longer than 200 ns, and they can exceed 1 μs for larger macromolecular assemblies. Rotational motions on this timescale are not measurable using fluorophores which display nanosecond lifetimes. Processes on the microsecond or even the millisecond timescale have occasionally been measured using phosphorescence.[1–4] However, relatively few probes display useful phosphorescence in room-temperature aqueous solutions. Also, it is usually necessary to perform phosphorescence measurements in the complete absence of oxygen. Hence, there is a clear need for probes which display microsecond lifetimes. In this chapter we describe a family of metal–ligand probes which display decay times ranging from 100 ns to 10 μs. The long lifetimes of the metal–ligand probes allow the use of gated detection, which can be employed to suppress interfering autofluorescence from biological samples and can thus provide increased sensitivity.[5] Finally, the metal–ligand probes display high chemical and photochemical stability. Because of these favorable properties, we expect metal–ligand probes to have numerous uses in biophysical chemistry, clinical chemistry, and DNA diagnostics.

20.1. INTRODUCTION TO METAL–LIGAND PROBES

We use the term metal–ligand complex (MLC) to refer to transition-metal complexes containing one or more diimine ligands. This class of probes is typified by $[Ru(bpy)_3]^{2+}$, where bpy is 2,2′-bipyridine (Figure 20.1, left). This class of compounds was originally developed for use in solar-energy conversion. Upon absorption of light, $[Ru(bpy)_3]^{2+}$ becomes a metal-to-ligand charge-transfer species, in which one of the bpy ligands is reduced and ruthenium is oxidized:

$$[Ru^{II}(bpy)_3]^{2+} \xrightarrow{h\nu} [Ru^{III}(bpy)_2(bpy^{-})]^{2+} \quad [20.1]$$

Ru^{III} is a strong oxidant, and bpy^{-} is a strong reductant. It was hoped that this charge separation could be used to split water to hydrogen and oxygen. Although solar-energy conversion based on such MLCs is not in commercial use, there have been extensive studies of the photophysical properties of the MLCs. A wide variety of luminescent MLCs are now known, some of which are strongly luminescent, and some of which display little or no emission. An extensive review of their spectral properties is available.[6]

Prior to examining the spectral properties of the MLCs, it is useful to have an understanding of their unique elec-

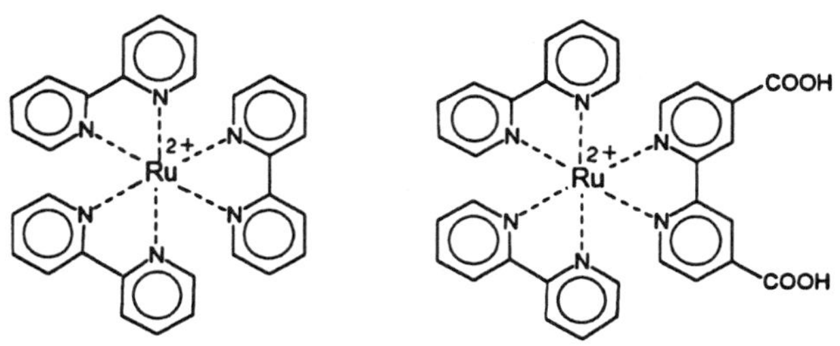

Figure 20.1. Chemical structure of $[Ru(bpy)_3]^{2+}$ and of $[Ru(bpy)_2(dcbpy)]^{2+}$. The latter compound is conjugatable and displays strongly polarized emission.

tronic states. The π orbitals are associated with the organic ligands, and the d orbitals are associated with the metal. All the transition-metal complexes that we will discuss have six d electrons. The presence of ligands splits the d-orbital energy levels into three lower (t) and two higher (e) orbitals (Figure 20.2). The extent of splitting is determined by the crystal-field strength Δ. The three lower-energy d orbitals are filled by the six d electrons. Transitions between the orbitals ($t \rightarrow e$) are formally forbidden. Hence, even if d–d absorption occurs, the radiative rate is low and, like phosphorescence, the emission is quenched. Additionally, electrons in the e orbitals are antibonding with respect to the metal–ligand bonds, so excited d–d states are usually unstable.

The appropriate combination of metal and ligand results in a new transition involving charge transfer between the metal and ligands (Figure 20.3). For the complexes described in this chapter, the electrons are promoted from the metal to the ligand, the so-called metal-to-ligand charge-transfer (MLCT) transition. The MLCT transition is the origin of the intense absorbance of the ruthenium MLCs near 450 nm. Emission from these states is formally phosphorescence. However, these states are somewhat shorter-lived (microseconds) than normal phosphorescent states and thus can emit prior to quenching. The luminescence of MLCs is thought to be short-lived due to spin–orbit coupling with the heavy metal atom, which increases the allowedness of the normally forbidden transition to the ground state.

A Jabłoński diagram for the MLCs is shown in Figure 20.4. Following absorption, the complex undergoes intersystem crossing to the triplet MLCT state. This occurs rapidly,[9] in less than 300 fs, and with high efficiency. Once in the MLCT state, the excited-state complex decays by the usual radiative and nonradiative decay pathways. In

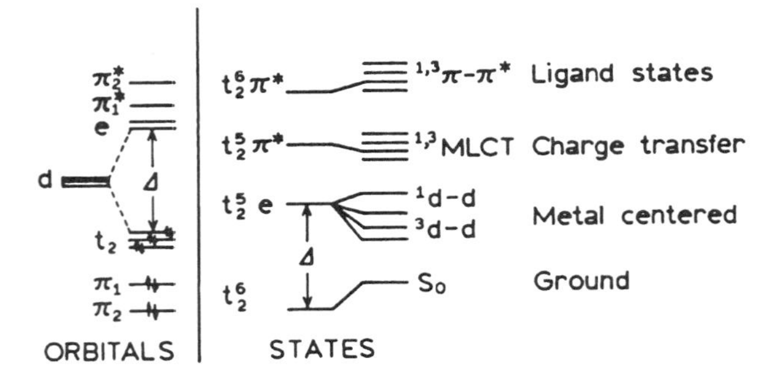

Figure 20.2. Orbital and electronic states of metal–ligand complexes. The d orbitals are associated with the metal and are split by energy Δ due to the crystal field created by the ligands. The π orbitals are associated with the ligands. Revised from Ref. 7.

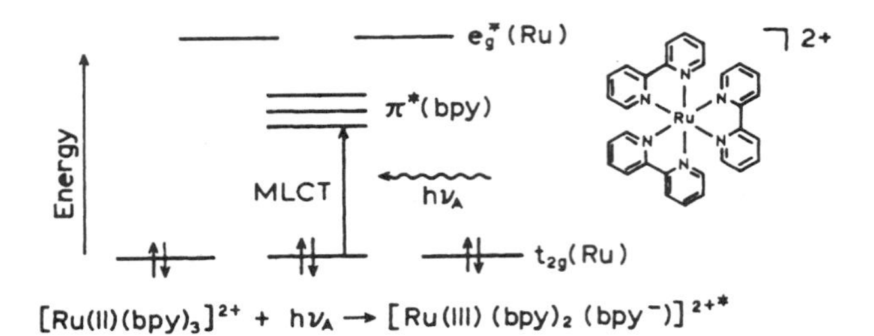

Figure 20.3. Metal-to-ligand charge-transfer (MLCT) transition in $[Ru(bpy)_3]^{2+}$. From Ref. 8.

general, the nonradiative decay rates (k_{nr}) are larger than the radiative decay rates (Γ), and the decay times are determined mostly by the nonradiative decay rates.

For an MLC to be luminescent, several criteria must be satisfied. The crystal field must be strong enough to raise the d–d state above the MLCT state (Figure 20.5). Hence, iron MLCs (FeL_3^{2+}) are nonluminescent owing to the low-lying d–d state. In contrast, RuL_3^{2+} is luminescent because the d–d states are above the MLCT state and do not serve as a major route of radiationless decay. One notices that OsL_3^{2+} has still higher d–d energies. Because the d–d levels are not accessible, osmium complexes are highly photostable. However, osmium MLCs are usually weakly luminescent. This is a result of the energy-gap law (Section 20.2): as the energy of the excited state becomes closer to the ground state, the rate of radiationless decay increases. Osmium MLCs typically have long-wavelength emission, a low-energy MLCT state, and a rapid rate of radiationless decay.

The relative levels of MLCT and d–d states determine the sensitivity of the MLC decay times to temperature. If the d–d levels are close to the MLCT level, then the d–d states are thermally accessible. In such cases, increasing temperature results in decreased lifetimes owing to thermal

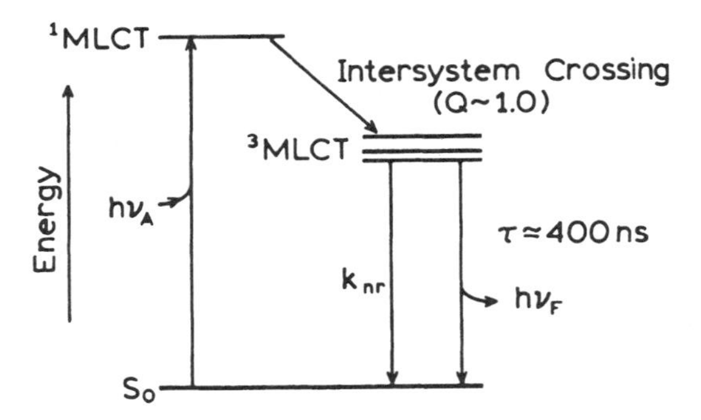

Figure 20.4. Jabłoński diagram for a metal–ligand complex. For $[Ru(bpy)_3]^{2+}$, the decay time is near 400 ns.

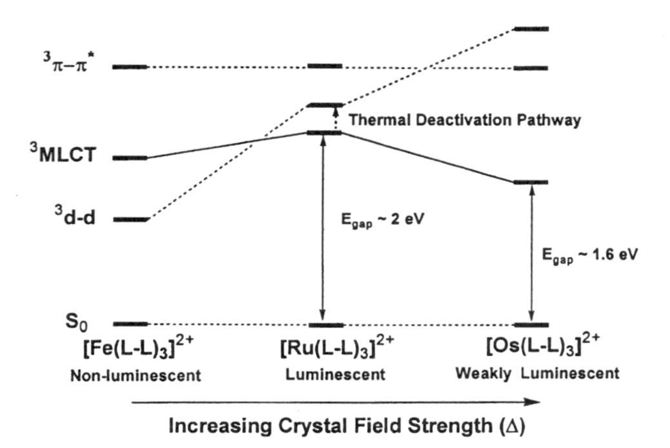

Figure 20.5. Lowest-energy triplet states for metal–ligand complexes with increasing crystal-field strength. Revised from Ref. 10.

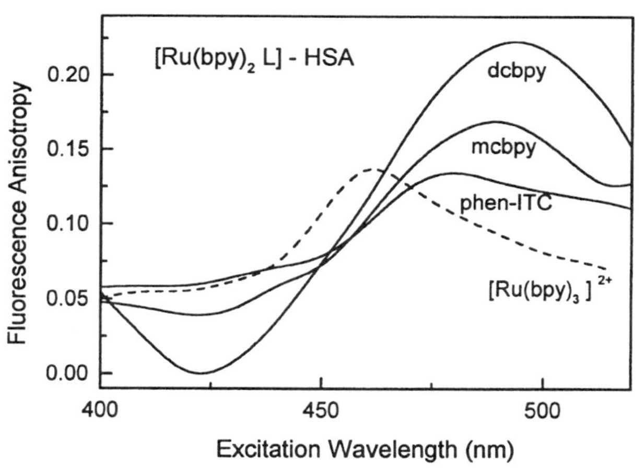

Figure 20.6. Excitation anisotropy spectra of $[Ru(bpy)_2L]$-HSA and $[Ru(bpy)_3]^{2+}$. The anisotropy spectra of $[Ru(bpy)_2(dcbpy)]^{2+}$, $[Ru(bpy)_2(mcbpy)]^{2+}$, and$[Ru(bpy)_2(phen\text{-}ITC)]^{2+}$, conjugated to HSA, in 6:4 (v/v) glycerol:water, $T = -55$ °C, are shown. ---, Anisotropy spectrum of $[Ru(bpy)_3]^{2+}$ in 9:1 (v/v) glycerol:water, $T = -55$ °C.

population of the *d–d* states, followed by rapid radiationless decay. Osmium complexes with thermally inaccessible *d–d* levels are less sensitive to temperature. The spectroscopy of MLCs is described in detail in Refs. 7 and 10–12.

20.1.A. Anisotropy Properties of Metal–Ligand Complexes

The interest in MLCs as biophysical probes was stimulated by the observation that particular MLCs display strongly polarized emission. The structure of $[Ru(bpy)_3]^{2+}$ is highly symmetrical, with three identical ligands. Hence, one does not expect the excited state to be localized on any particular ligand, and the emission is expected to display low or zero anisotropy. This expectation of low anisotropy was supported by the spectral properties of lanthanides, in which the electrons are in a rather symmetrical environment. To the best of our knowledge, the lanthanides have not been shown to display polarized emission. However, it was found that Ru MLCs that contained nonidentical diimine ligands displayed high anisotropy. The first such compound studied[13] was the dicarboxy derivative $[Ru(bpy)_2(dcbpy)]^{2+}$, where dcbpy is 4,4′-dicarboxy-2,2′-bipyridyl (Figure 20.1, right). The excitation anisotropy spectrum of this MLC is shown in Figure 20.6, along with those of $[Ru(bpy)_3]^{2+}$ and two other MLCs. In this case the dicarboxy ligand was conjugated to human serum albumin (HSA) via the carboxy groups. Much higher fundamental anisotropies (r_0) were observed for the dicarboxy derivative than for $[Ru(bpy)_3]^{2+}$. Also, usefully high anisotropies have been observed for a number of other MLCs (Figures 20.6 and 20.7), most of which have been prepared in a form conjugatable with macromolecules.[14–16] It is important to notice that the anisotropy depends strongly on excitation wavelength, which must be carefully selected if one wishes to have the highest value of r_0.

The potential of the MLCs as biophysical probes is illustrated by the range of decay times and quantum yields available with this diverse class of compounds. The chemi-

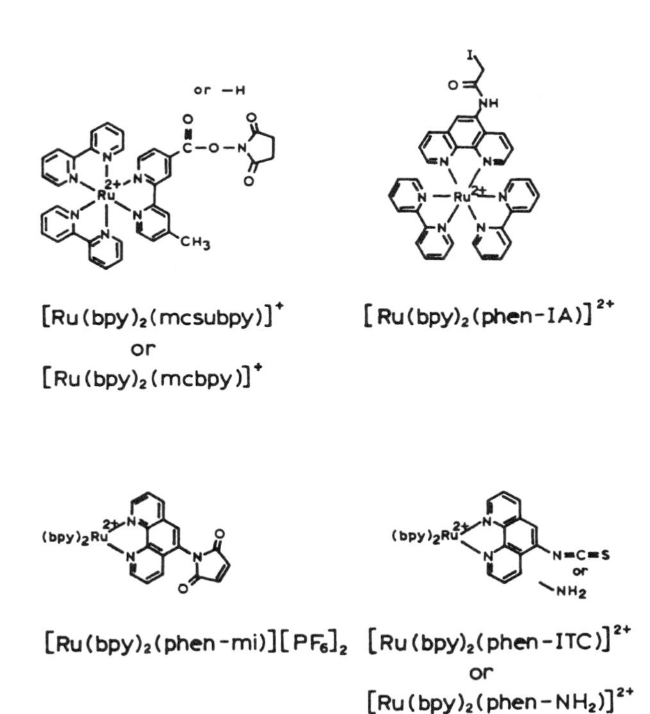

Figure 20.7. Conjugatable metal–ligand complexes.

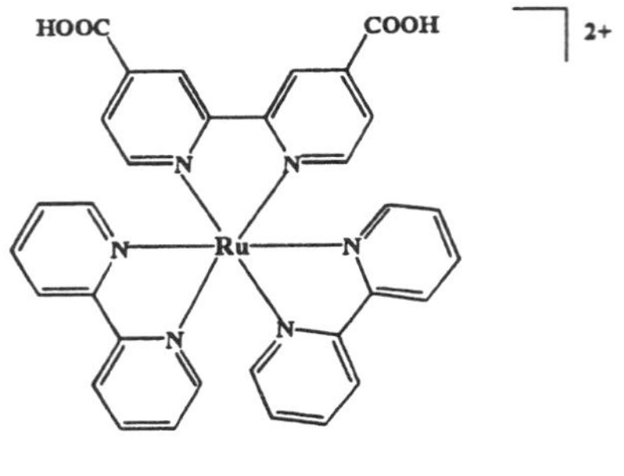

$[Ru(bpy)_2(dcbpy)](PF_6)_2$

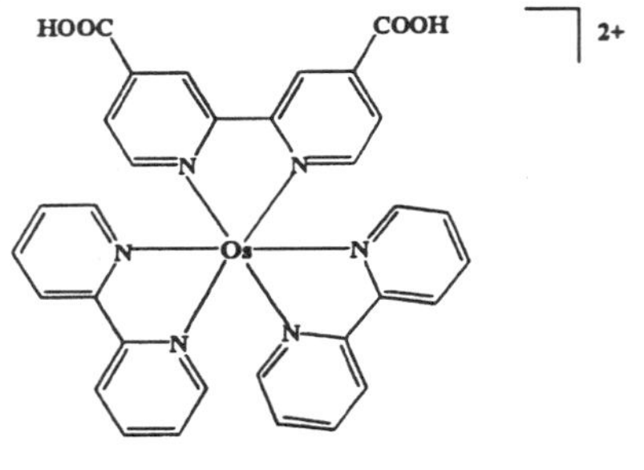

$[Os(bpy)_2(dcbpy)](PF_6)_2$

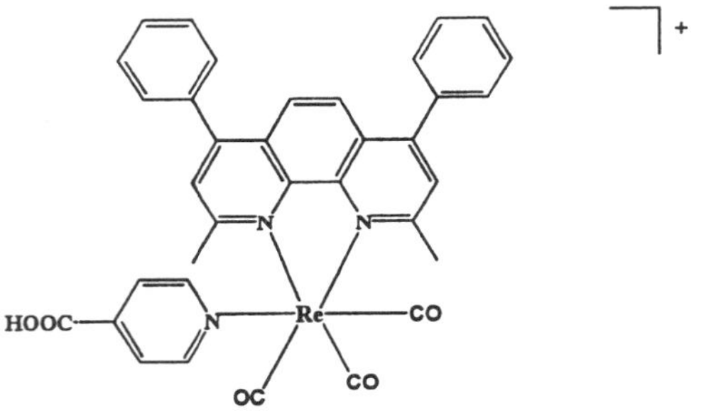

$[Re(bcp)(CO)_3(4\text{-}COOHPy)](PF_6)$

Figure 20.8. Chemical structures of ruthenium(II), osmium(II), and rhenium(I) MLCs. Intensity decays are shown in Figure 20.9.

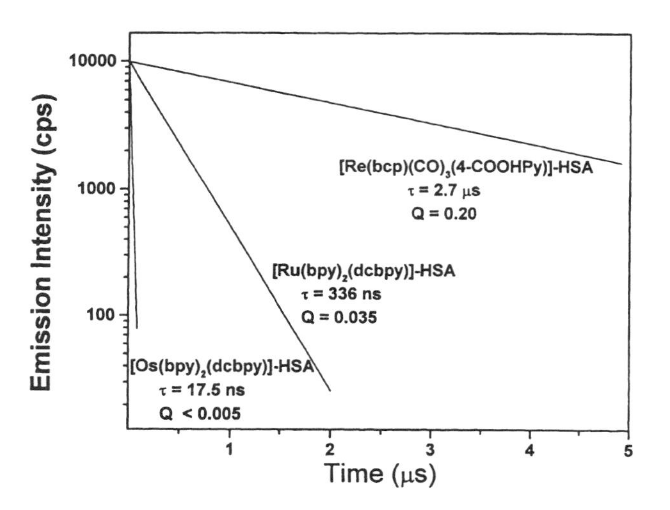

Figure 20.9. Decay times and quantum yields of the Re, Ru, and Os MLCs shown in Figure 20.8 when covalently linked to HSA. The actual intensity decays are multiexponential. The single-exponential intensity decays are shown to illustrate the range of decay times available with MLCs.

cal structures of useful ruthenium (Ru), osmium (Os), and rhenium (Re) complexes are shown in Figure 20.8. When conjugated to HSA, the rhenium MLC displays lifetimes as long as 2.7 μs in oxygenated aqueous solution at room temperature (Figure 20.9), and up to 4 μs in the absence of oxygen.[17] The osmium complex displays a much shorter lifetime and lower quantum yield[18] but can be excited at 680 nm, which is available from a red laser diode. The possibility of long-wavelength excitation and long decay times made us realize that the MLCs had significant potential as biophysical probes,[19] as has been pointed out by others.[7,20]

20.2. SPECTRAL PROPERTIES OF METAL–LIGAND PROBES

In addition to long decay times and polarized emission, the MLCs have favorable absorption and emission spectra. Absorption and emission spectra of three conjugatable Ru MLCs are shown in Figure 20.10. The long-wavelength absorption of the Ru MLCs is not due to absorption by the metal alone or the ligand alone. Ligand absorption, referred to as ligand-centered (LC) absorption, occurs at shorter wavelengths, near 300 nm. Absorption by the metal via *d–d* transitions is forbidden, and the extinction coefficients are very low (1–200 M^{-1} cm^{-1}). The broad absorption band at 450 nm is due to the MLCT transition (Eq. [20.1]). The MLCT transitions display extinction coefficients of 10,000–30,000 M^{-1} cm^{-1}.[7] Although these values are not as large as those of fluorescein or cyanine dyes, these extinction coefficients are comparable to those found for many fluorophores and are thus adequate for most applications.

The emission of the MLCs is also dominated by the MLCT transition, which is centered near 650 nm for the Ru(II) complexes (Figure 20.10). The important point is

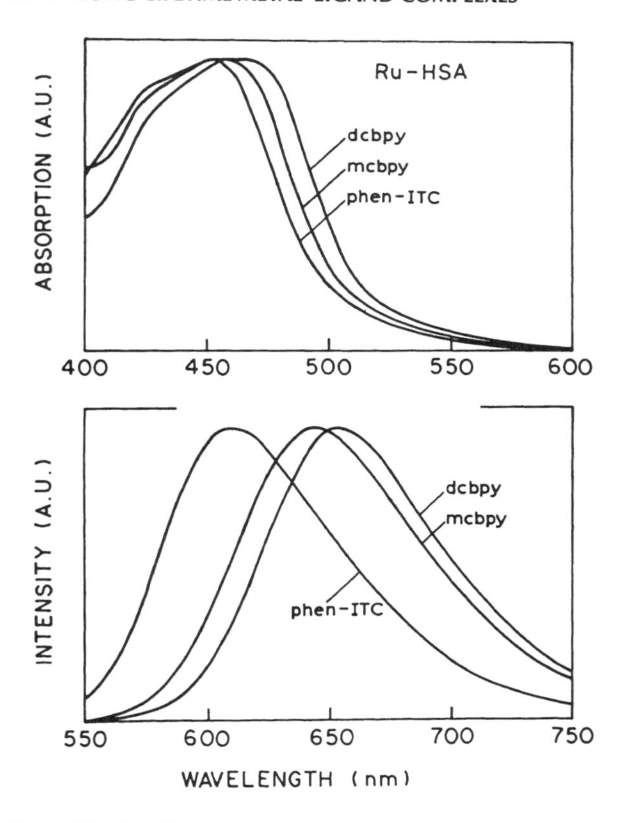

Figure 20.10. Absorption (*top*) and emission spectra (*bottom*) of $[Ru(bpy)_2(dcbpy)]^{2+}$, $[Ru(bpy)_2(mcbpy)]^{+}$, and $[Ru(bpy)_2(phen\text{-}ITC)]^{2+}$ conjugated to HSA. Excitation wavelength, 460 nm; $T = 20$ °C. Structures are shown in Figures 20.1 and 20.7. Revised from Ref. 14, Copyright © 1996, with permission, from Elsevier Science.

that the MLCs behave like a single chromophoric unit. In the case of the MLCs, in contrast to the lanthanides, the absorption and emission are not due to the atom but rather to the entire complex. Also, the metal–ligand bonds are covalent bonds, and the ligands do not dissociate from the metal under any conditions which are remotely physiological.

Examination of Figure 20.10 reveals another favorable spectral property of the MLCs, namely, a large Stokes' shift, which makes it relatively easy to separate the excitation and emission. This large shift results in minimal probe–probe interaction. In contrast to fluorescein, which has a small Stokes' shift, the MLCs do not appear to self-quench when multiple MLCs are attached to a protein molecule.

For the ruthenium complexes, comparison of Figures 20.10 and 20.6 reveals that the MLCs with the longest emission wavelengths display the highest anisotropy. This behavior seems to correlate with the electron-withdrawing properties of the ligand, which are highest for dcbpy and lowest for phenanthroline isothiocyanate (phen-ITC). In general, it seems that having one ligand to preferentially accept the electron in the MLCT transition results in high fundamental anisotropies. This suggests that MLCs with a single chromophoric ligand will have high anisotropies, as has been observed for Re(I) complexes.[17]

20.2.A. The Energy-Gap Law

Another factor which determines the decay times of the MLCs is the energy-gap law. This law states that the nonradiative decay rate of an MLC increases exponentially as the energy gap or emission energy decreases.[21–27] One example of the energy-gap law is shown in Figure 20.11 for Re(I) complexes. In this case the emission maximum is sensitive to the structure of the nonchromophoric ligand. The radiative decay rates (Γ) are relatively independent of the ligand, but the nonradiative decay rate (k_{nr}) is strongly dependent on the ligand and the emission maximum. The dependence of k_{nr} on emission energy for a larger number of Re(I) complexes is shown in Figure 20.12. One sees that k_{nr} increases as the emission energy decreases. Because k_{nr} is much larger than Γ, the decay times of these complexes are determined mostly by the value of k_{nr}.

Another example of the energy-gap law is shown in Figure 20.13 for osmium complexes. In this case the shortest-wavelength (highest-energy) emission was found for the osmium complex with two phosphine (dppene) ligands. When the dppene ligands are replaced with phenanthroline (phen) ligands, the emission maximum occurs at longer wavelengths. When this occurs, the quantum yield (Q) and the lifetime decrease, again illustrating the dependence of the nonradiative decay rate on the energy gap.

Although the energy-gap law is useful in understanding how the quantum yield and lifetime are related to the emission maximum, it cannot be used to compare different types of complexes. That is, the energy-gap law works well within a homologous series of complexes but is less useful when comparing different types of complexes.

20.3. BIOPHYSICAL APPLICATIONS OF METAL–LIGAND PROBES

There are numerous biophysical applications for long-lifetime anisotropy probes. These include studies of all types of macromolecular assemblies. The use of long-lifetime probes in biochemistry is just beginning. We have chosen to show just three representative applications: studies of DNA dynamics, measurement of domain-to-domain mo-

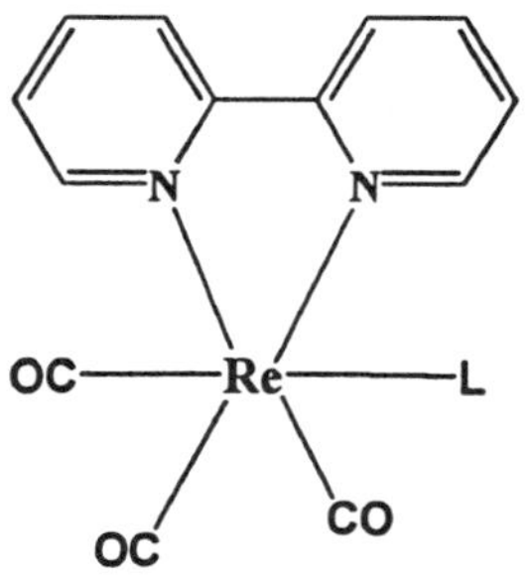

L = Cl^-, 4-NH_2Py, Py, CH_3CN

fac-Re(bpy)$(CO)_3$L

L	λ_{em} (nm)	Q	τ (ns)	Γ (s^{-1})	k_{nr} (s^{-1})
Cl^-	622	0.005	51	9.79×10^4	1.95×10^7
4-NH_2 Py	597	0.052	129	4.06×10^5	7.34×10^6
Py	558	0.16	669	2.36×10^5	1.26×10^6
CH_3CN	536	0.41	1201	3.43×10^5	4.90×10^5

Figure 20.11. Effect of emission maximum on the nonradiative decay rates of Re(I) complexes. The nonradiative decay rate increases exponentially with decreasing emission energy, which is the energy-gap law. Data from Ref. 21.

tions in proteins, and examples of metal–ligand lipid probes.

20.3.A. DNA Dynamics with Metal–Ligand Probes

A spherical molecule will display a single correlation time, and, in general, globular proteins display closely spaced correlation times due to overall rotational diffusion. In contrast, DNA is highly elongated and is expected to display motions on a wide range of timescales, from nanoseconds to microseconds[28,29] (Section 12.8). Most experimental studies of DNA dynamics have been performed using ethidium bromide (EB), which displays a decay time for the DNA-bound state of about 30 ns, or acridine derivatives, which display shorter decay times.[30–32] The short decay times of most DNA-bound dyes are a serious limitation because DNA is expected to display a wide range of relaxation times, and only the nanosecond motions will affect the anisotropy of nanosecond-lifetime probes. In fact, most studies of DNA dynamics report only the torsional motions of DNA, which are detectable on the nanosecond timescale. The slower bending motions of DNA are often ignored when nanosecond-lifetime probes are employed. It seems likely that these slower bending motions are important for packaging of DNA into chromosomes.

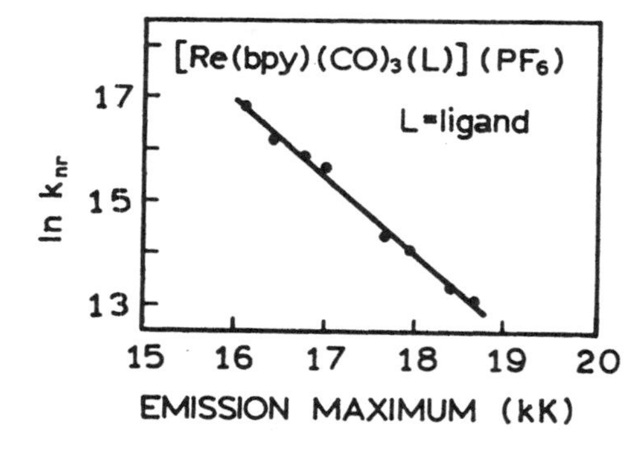

Figure 20.12. Dependence of the nonradiative decay rate of Re(I) complexes on the emission maximum. 1 kK = 1000 cm^{-1}. Revised from Ref. 21.

A MLC probe which intercalates into double-helical DNA, $[Ru(bpy)_2(dppz)]^{2+}$, where dppz is dipyrido-[3,2a:2,'3'-c]phenazine shown in Figure 20.14. This probe is quenched in water and is highly luminescent when bound to DNA. The emission spectrum of $[Ru(bpy)_2(dppz)]^{2+}$ bound to calf thymus DNA is shown

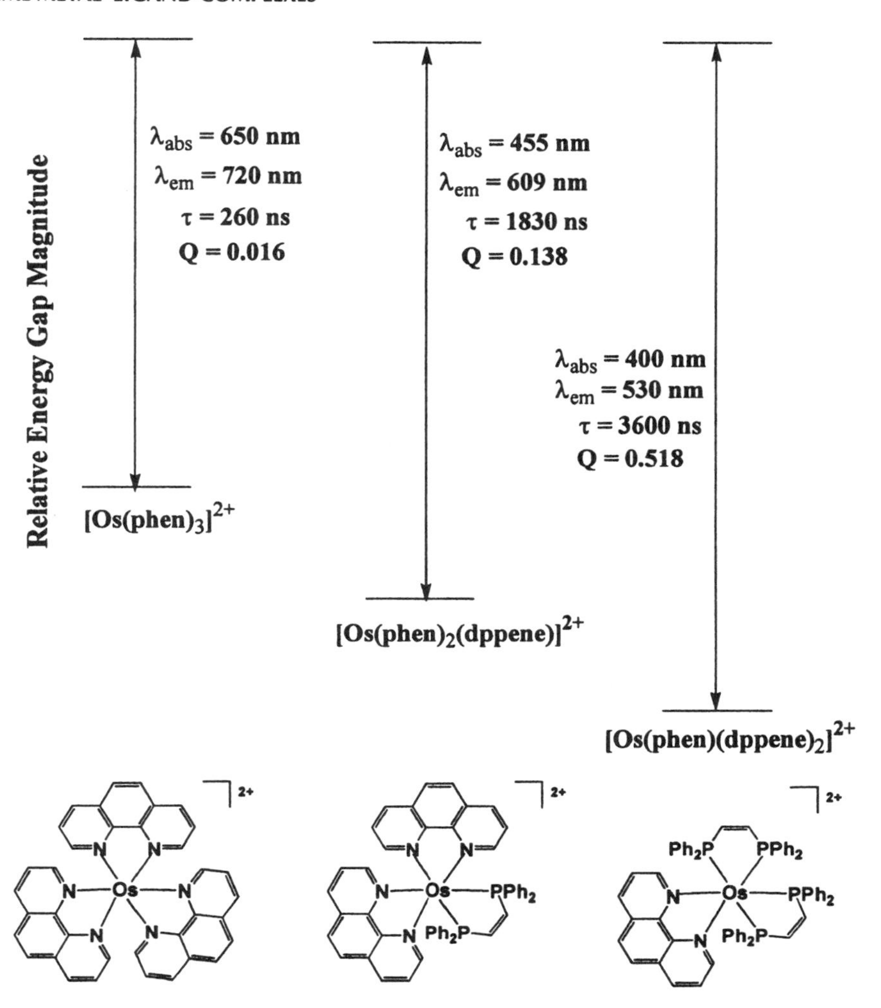

Figure 20.13. Effect of ligands on the emission maximum, quantum yield, and lifetime of osmium MLCs. Data from Ref. 22.

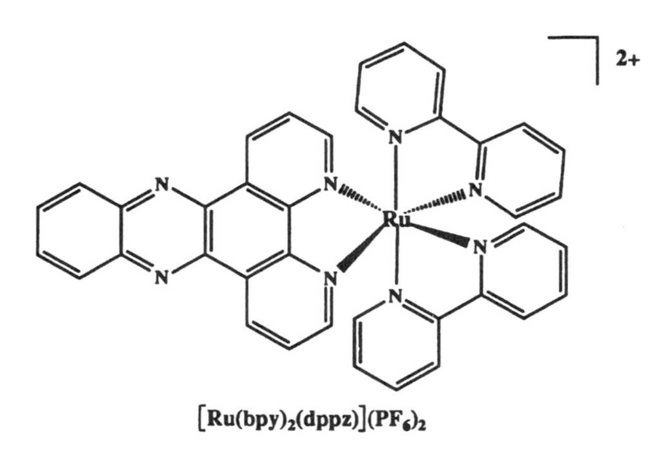

Figure 20.14. Chemical structure of a DNA anisotropy probe, $[Ru(bpy)_2(dppz)]^{2+}$.

in Figure 20.15. In aqueous solution, the probe luminescence is nearly undetectable. In the presence of DNA, the luminescence of $[Ru(bpy)_2dppz]^{2+}$ is remarkably enhanced,[33–35] an effect attributed to intercalation of the dppz ligand into double-helical DNA. This MLC is highly luminescent in aprotic solvents but is dynamically quenched by water or alcohols.[36,37] The increase in fluorescence upon binding to DNA is due to shielding of the nitrogens on the dppz ligand from the solvent. This enhancement of emission upon binding to DNA means that the probe emission is observed from only the DNA-bound forms, without contributions from free probe in solution. In this respect, $[Ru(bpy)_2(dppz)]^{2+}$ is analogous to EB, which also displays significant emission from only the DNA-bound form.

In order to be useful for anisotropy measurements, a probe must display a large fundamental anisotropy (r_0).

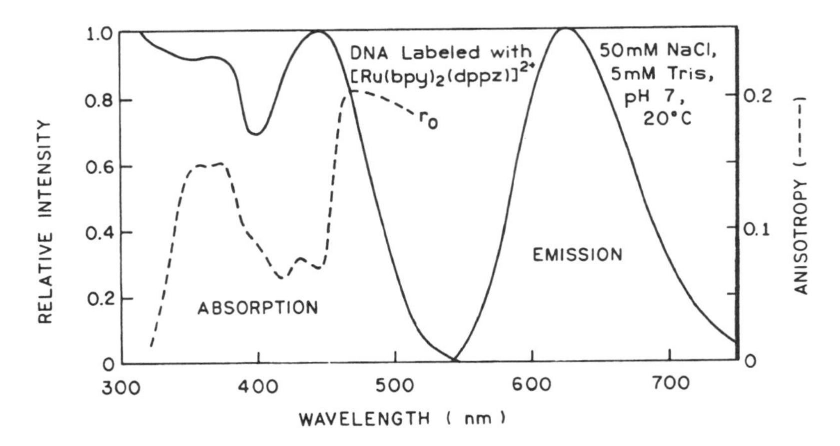

Figure 20.15. Absorption and emission (———) and excitation anisotropy spectra (— — —) of $[Ru(bpy)_2(dppz)]^{2+}$ bound to the calf thymus DNA. The excitation anisotropy spectrum is for a solution in 100% glycerol at –60 °C. From Ref. 38.

The excitation anisotropy spectrum of $[Ru(bpy)_2(dppz)]^{2+}$ in vitrified solution (glycerol, –60 °C) displays maxima at 365 and 490 nm[38,39] (Figure 20.15). The high value of the anisotropy indicates that the excitation is localized on one of the organic ligands, and not randomized among these ligands. It seems reasonable to conclude that the excitation is localized on the dppz ligand because shielding of the dppz ligand results in an increased quantum yield. The time-resolved intensity decay of $[Ru(bpy)_2(dppz)]^{2+}$ bound to calf thymus DNA is shown in Figure 20.16. The intensity decay is best fit by a triple-exponential decay with a mean decay time near 110 ns. Anisotropy decay can typically be measured to about three times the lifetime, suggesting that $[Ru(bpy)_2dppz]^{2+}$ can be used to study DNA dynamics to 300 ns or longer.

The time-resolved anisotropy decay of DNA-bound $[Ru(bpy)_2(dppz)]^{2+}$ is shown in Figure 20.17. The anisotropy decay could be observed to 250 ns, severalfold longer than possible with EB. The anisotropy decay appears to be a triple exponential, with apparent correlation times as long as 189.9 ns. Future intercalative MLC probes may display longer decay times. Studies of DNA-bound MLC probes offer the opportunity to increase the information content of the time-resolved measurements of nucleic acids by extending the observations to the microsecond timescale.

20.3.B. Domain-to-Domain Motions in Proteins

There is presently considerable interest in measuring the rates of domain flexing in multidomain proteins.[40] Domain motions occur in signaling proteins such as calmodulin and sugar receptors. Domain motions are thought to occur in proteins such as hexokinase,[41] creatine kinase,[42,43] protein kinase C, phosphoglycerate kinase,[44] calcium-binding proteins, and immunoglobulins.[45,46] As described in Section 14.5, such motions can be detected by the effects of donor-to-acceptor diffusion on the extent of RET. In spite of this potential, such measurements have not been suc-

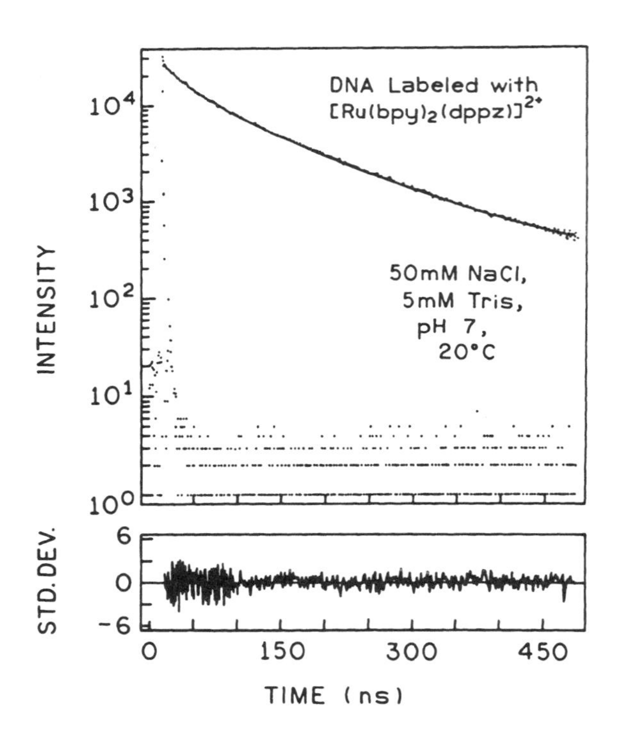

Figure 20.16. Time-dependent intensity decays of DNA labeled with $[Ru(bpy)_2(dppz)]^{2+}$. The data are shown as dots. The solid line and deviations (lower panel) are for the best three-decay-time fit, with decay times of 12.4, 46.6, and 126 ns. From Ref. 38.

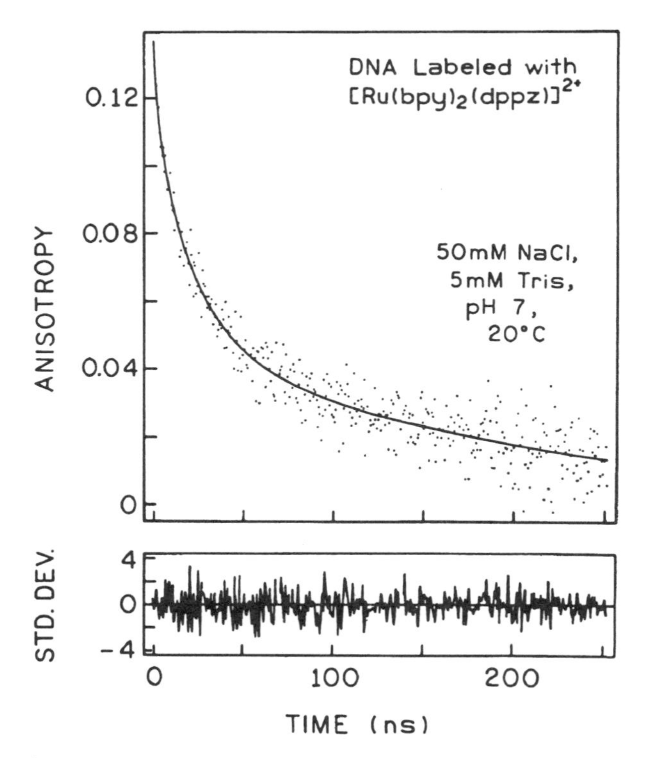

Figure 20.17. Time-dependent anisotropy decay of DNA labeled with $[Ru(bpy)_2(dppz)]^{2+}$. The data are shown as dots. The solid line and deviations (lower panel) are for the best three-correlation-time fit with correlation times of 3.1, 22.2, and 189.9 ns. From Ref. 38.

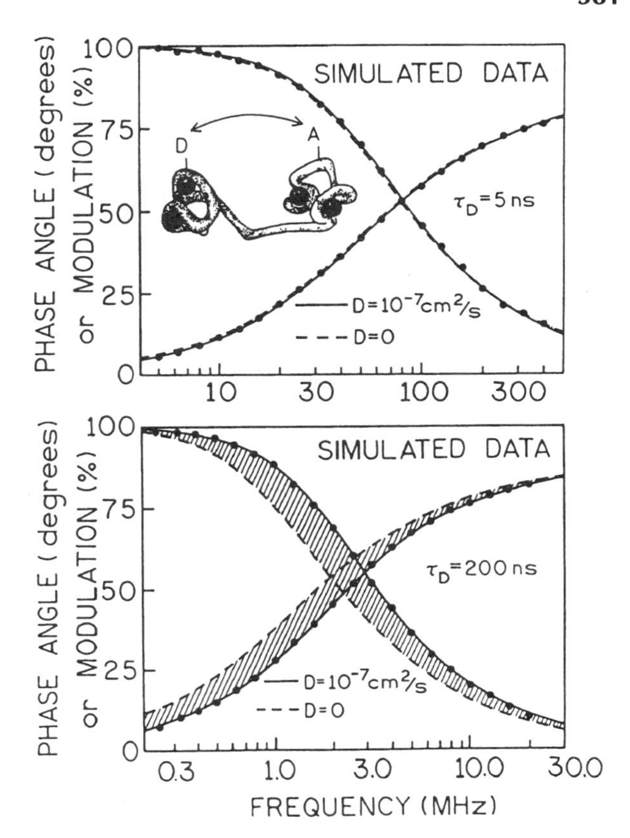

Figure 20.18. Simulated data illustrating the effect of donor lifetime on the contribution of interdomain diffusion to the FD donor decays. For the simulations, we assumed a D–A distance distribution with $R_0 = \bar{r} =$ 30 Å and hw =20 Å, diffusion coefficients of 0 (---) and 10^{-7} cm^2 /s (———), and donor decay times of 5 (*top*) and 200 ns (*bottom*). *Inset*: Schematic representation of donor- (D) and acceptor- (A) labeled domains for calmodulin.

cessful to date, primarily because the decay times of most fluorophores are too short for significant motion to occur during the excited-state lifetime.[44] This fact is illustrated in Figure 20.18, which considers the effect of donor-to-acceptor diffusion on the donor decay, as measured in the frequency domain. For domain-to-domain motions, the mutual diffusion coefficients are expected to be 10^{-7} cm^2/s, or smaller. If the donor decay time is 5 ns, then diffusion has essentially no effect on the extent of RET (Figure 20.18, top). For this reason, the donor decay contains no information on the diffusion coefficient and cannot be used to recover the diffusion coefficient. Suppose now that the donor decay time is increased to 200 ns. Then a diffusion coefficient of 10^{-7} cm^2/s has a significant effect on the extent of RET, as shown by the shaded area in the bottom panel of Figure 20.18.

To date, RET with MLCs has not been used to measure domain flexing in proteins. However, the potential seems clear, especially in light of the long decay times possible with rhenium MLCs. Decay times as long as 4 μs have been found in aqueous solution,[17] suggesting that domain flexing will be measurable with D values smaller than 10^{-8} cm^2/s.

20.3.C. MLC Lipid Probes

Long-lifetime probes are expected to be especially valuable in membrane biophysics. The interest in long-lifetime lipid probes is evident from recent publications on coronene in membranes ($\tau \simeq 200$ ns)[47] and on a 20-ns component from tryptophan in a membrane-bound protein.[48] Long-lifetime MLC probes could be used to study rotational motions of entire lipid vesicles, or to measure diffusion by its effect on RET. Several MLC lipid probes have been described (Figure 20.19), all of which show polarized emission.[17,49,50] The Ru MLC lipid probes display lifetimes near 400 ns, and the Re MLC lipid probe displays a lifetime near 4 μs in dipalmitoyl-L-α-phosphatidylglycerol (DPPG) vesicles, in the presence of dissolved oxygen. Such long-lifetime probes can be used to measure microsecond correlation times in membranes, or even the rotational correlation times of lipid vesicles (Section 11.8.B).

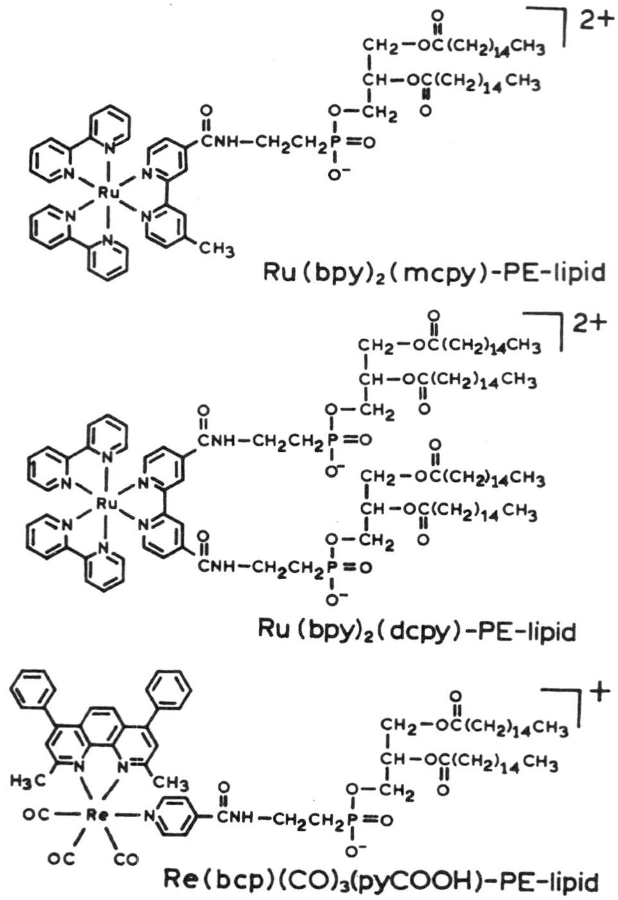

$Ru(bpy)_2(mcpy)$-PE-lipid

$Ru(bpy)_2(dcpy)$-PE-lipid

$Re(bcp)(CO)_3(pyCOOH)$-PE-lipid

Figure 20.19. MLC lipid probes.

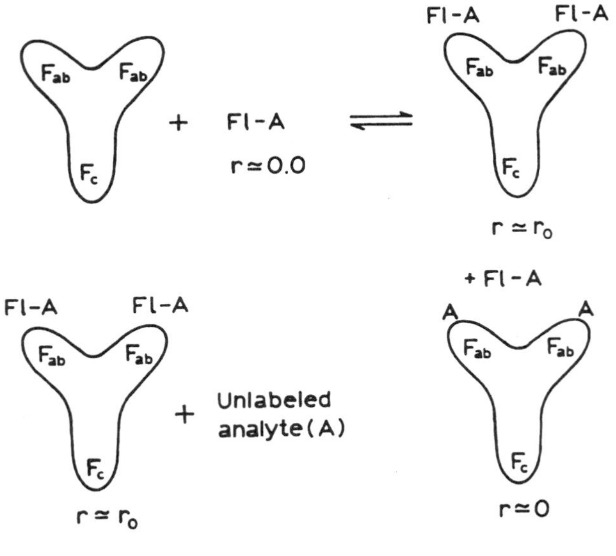

20.4. MLC IMMUNOASSAYS

Fluorescence polarization immunoassays (FPIs) are widely used to measure the amounts of drugs (D) and small molecules in clinical samples (Chapter 19). FPIs are based on the changes in polarization (or anisotropy) which occur when a labeled drug analog (F-D) binds to an antibody specific for that drug (Figure 20.20). The anisotropy of the labeled drug can be estimated from the Perrin equation,

$$r = \frac{r_0}{1 + (\tau/\theta)} \qquad [20.2]$$

where r_0 is the anisotropy observed in the absence of rotational diffusion and θ is the rotational correlation time. Suppose that the fluorophore is fluorescein (Fl), with a lifetime near 4 ns, and the analyte (A) is a small molecule with a rotational correlation time near 100 ps (Figure 20.20, top). The assay is performed using a covalent adduct

Figure 20.20. Schematic of a fluorescence polarization immunoassay.

of fluorescein and the analyte (Fl-A). When Fl-A is free in solution, its anisotropy is expected to be near zero. However, the correlation time of an antibody is near 100 ns. Hence, when Fl-A is bound to immunoglobulin G (IgG), the anisotropy of the Fl-A–IgG complex is expected to be near r_0. FPIs are typically performed in a competitive format. The sample is incubated with a solution containing the labeled drug (Fl-A) and antibody (Figure 20.20, bottom). The larger the amount of unlabeled drug, the more Fl-A is displaced from the antibody, and the lower the polarization. For an FPI to be useful, there needs to be a substantial difference in anisotropy between the free and bound forms of the labeled drug.

The usefulness of MLCs in clinical FPIs is illustrated by consideration of an FPI for a higher-molecular-weight species (Figure 20.21). Suppose that the antigen is HSA, with a molecular weight near 64,000 and a rotational correlation time near 50 ns. This correlation time is already much longer than the lifetime of fluorescein, so that the anisotropy is expected to be near r_0. For this reason, FPIs are typically used to measure only low-molecular-weight substances.

The use of MLC probes can circumvent this limitation of FPIs to low-molecular-weight antigens. The dependence of the anisotropy on the probe lifetime and the molecular weight of the antigen is shown in Figure 20.22. For typical probes with lifetimes near 4 ns (fluorescein or rhodamine), the anisotropy of low-molecular-weight antigens (MW < 1000) can be estimated from Figure 20.22 to be in the vicinity of 0.05. An antibody has a molecular

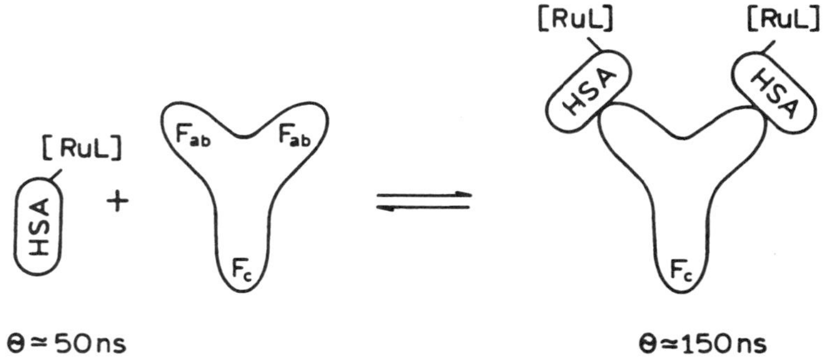

Figure 20.21. Fluorescence polarization immunoassay of a high-molecular-weight species using a Ru(II) MLC.

weight near 160,000, resulting in an anisotropy near 0.29 for the antigen–antibody complex. Hence, a large change in anisotropy is found upon binding of low-molecular-weight species to larger proteins or antibodies. However, if the molecular weight of the antigen is larger, above 20,000, then the anisotropy of the probe changes only slightly upon binding of the labeled antigen to a larger protein. For example, consider an association reaction which changes the molecular weight from 65,000 to 1 million. Such a change could occur for an immunoassay of HSA using polyclonal antibodies, for which the effective molecular weight of the immune complexes could be 1 million or higher. In this case, the anisotropy of a 4-ns probe would change from 0.278 to 0.298, which is too small a change for quantitative purposes. In contrast, with the use of a 400-ns probe, which is near the value found for our MLC, the anisotropy value of the labeled protein with a molecular weight of 65,000 is expected to increase from 0.033 to 0.198 when the molecular weight is increased from 65,000 to 1 million (Figure 20.22).

FPIs of the high-molecular-weight antigen HSA have been performed using the Ru MLC probes.[51,52] These

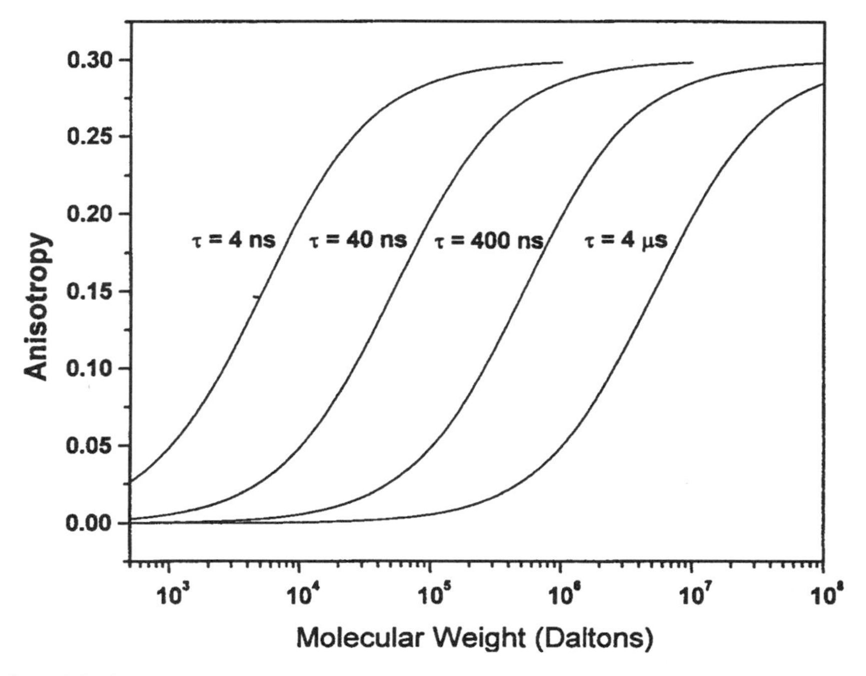

Figure 20.22. Molecular-weight-dependent anisotropies for probe lifetimes from 4 ns to 4 μs. The curves are based on Eq. [20.2], assuming $\overline{v} + h = 1.9$ ml/g for the proteins and $r_0 = 0.30$, in aqueous solution at 20 °C with a viscosity of 1 cP.

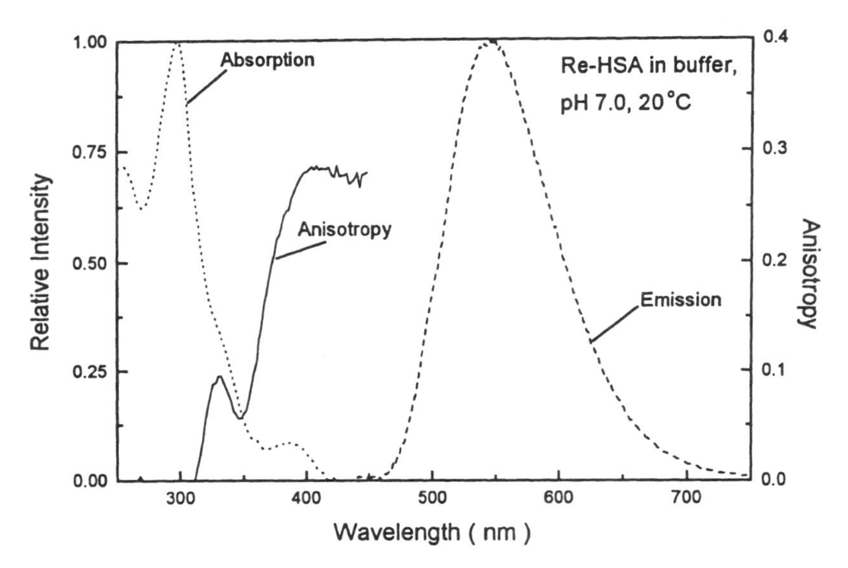

Figure 20.23. Absorption and emission spectra of $[Re(bcp)(CO)_3(4\text{-}COOHPy)]^+$ conjugated to HSA in 0.1*M* phosphate-buffered saline (PBS), pH 7.0. Excitation wavelength was 400 nm. The solid curve shows the excitation anisotropy spectrum in 100% glycerol at –60 °C, with an emission wavelength of 550 nm. Reprinted, with permission, from Ref. 53, Copyright © 1998, American Chemical Society.

first-generation probes displayed relatively low quantum yields, near 0.05. Hence, we show data here for a more recently developed MLC which displays a quantum yield in excess of 0.2 and a lifetime of over 3 μs.[53] The structure of $[Re(bcp)(CO)_3(4\text{-}COOHPy)]^+$ is shown in Figure 20.8. Absorption, emission, and anisotropy spectra of this remarkable probe are shown in Figure 20.23. The Re complex can be excited near 400 nm, which is due in part to the long wavelength absorption of the 4,7-dimethyl-1,10-phenanthroline (bcp) ligand. Although this wavelength may seem short for clinical instruments, wavelengths down to 390 nm can now be obtained from LEDs.[54] The large Stokes' shift, from 350 to 520 nm, makes it easy to isolate the MLC emission. Importantly, the Re MLC displays a high fundamental anisotropy, near 0.3, for excitation at 400 nm.[53,55] This is probably due to the presence of just one chromophoric ligand, so there is no possible randomization of the excitation to other ligands.

Another important feature of this MLC is that it is only moderately quenched by dissolved oxygen in aqueous solution. Modest quenching of the Re complex was also found in the intensity decays (Figure 20.24), where dissolved oxygen was found to decrease the mean decay time from 3.4 to 2.7 μs. These mean lifetimes are 1000-fold longer than those of typical organic fluorophores.

The high quantum-yield rhenium MLC was covalently bound to HSA and used to detect binding of an antibody against HSA (Figure 20.25). The steady-state anisotropy was found to increase nearly fourfold upon binding of IgG specific for HSA, and there was no effect from nonspecific IgG. These results demonstrated that long-lifetime MLCs are useful for immunoassays of high-molecular-weight antigens. It is important to note that the sensitivity of most fluorescence assays is limited not by the ability to detect the emission, but rather by the presence of interfering autofluorescence, which occurs on the 1–10 ns timescale. The availability of probes with longer decay times should also allow increased sensitivity by the use of gated detection following decay of the unwanted autofluorescence.

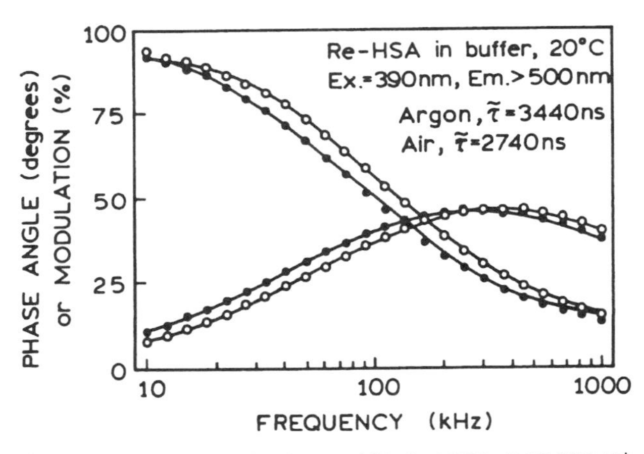

Figure 20.24. FD intensity decays of $[Re(bcp)(CO)_3(4\text{-}COOHPy)]^+$ conjugated to HSA. Excitation wavelength was 390 nm, and a 500-nm cutoff filter was used to isolate the emission. The solid curves show the three-decay-time fits in the absence (●) and presence (○) of dissolved oxygen. Revised and reprinted, with permission, from Ref. 53, Copyright © 1998, American Chemical Society.

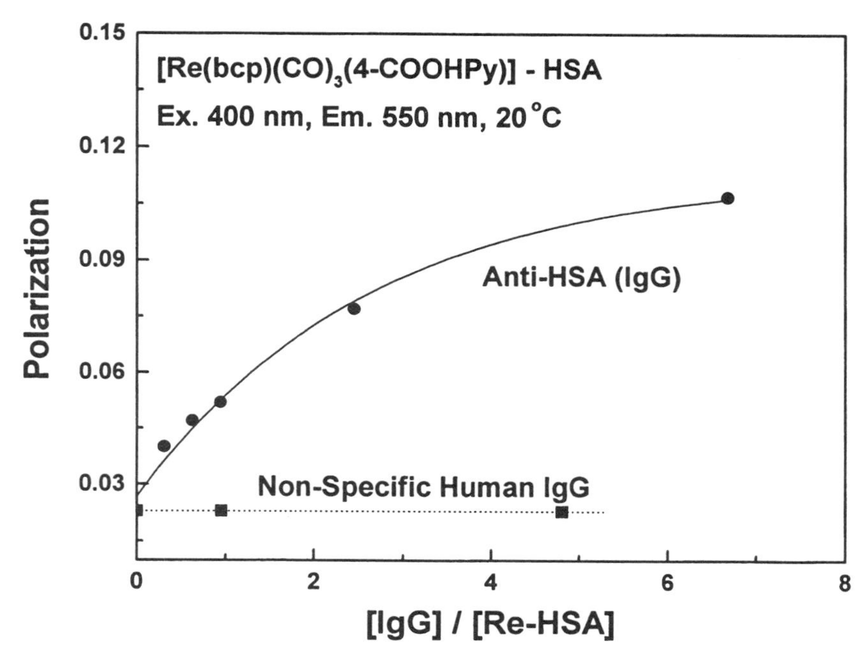

Figure 20.25. Steady-state fluorescence polarization of [Re(bcp)(CO)$_3$(4-COOHPy)]-HSA at various concentrations of IgG specific for HSA (anti-HSA; •) or nonspecific IgG (■). Revised and reprinted, with permission, from Ref. 53, Copyright © 1998, American Chemical Society.

20.4.A. Long-Wavelength Immunoassays

For clinical applications, it is preferable to use the longest possible wavelength for excitation. At longer wavelengths, there is less sample absorbance and less autofluorescence, and the light sources are less expensive. Osmium complexes display longer-wavelength absorption than Re(I) or Ru(II) MLCs. One osmium complex is shown in Figure 20.8, and the absorption and anisotropy spectra of this complex bound to HSA are shown in Figure 20.26. The excitation anisotropy spectrum of [Os(bpy)$_2$(dcbpy)]-HSA was measured in glycerol:water (9:1) at –55 °C. There are two maxima near 0.15 in the anisotropy spectrum, one at 505 and one at 685 nm.[18] The anisotropy maximum at 685 nm of 0.14 is ideal for tracing protein interactions using a 690-nm diode laser. The emission spectrum of this Os complex is centered near 760 nm

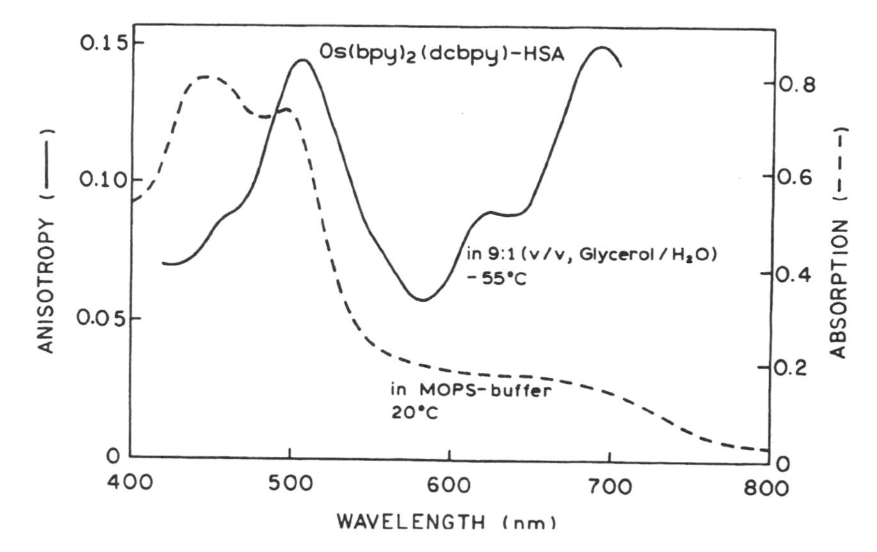

Figure 20.26. Absorption (– – –) and excitation anisotropy spectra (———) of [Os(bpy)$_2$(dcbpy)]-HSA. Revised from Ref. 18.

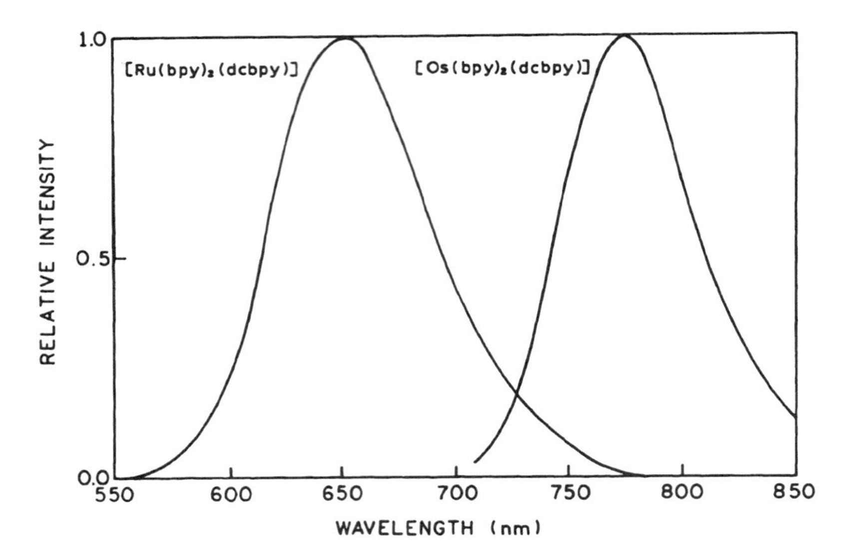

Figure 20.27. Emission spectra of $[Ru(bpy)_2(dcbpy)]$ and $[Os(bpy)_2(dcbpy)]$. From Ref. 18.

(Figure 20.27). Hence the emission can be expected to escape from tissues or even whole blood. The Os MLC was used in the same HSA immunoassay as described above. Though the lifetime of the Os complex is short (14 ns), we observed a nearly twofold increase in steady-state anisotropy upon binding of the antibody (Figure 20.28).

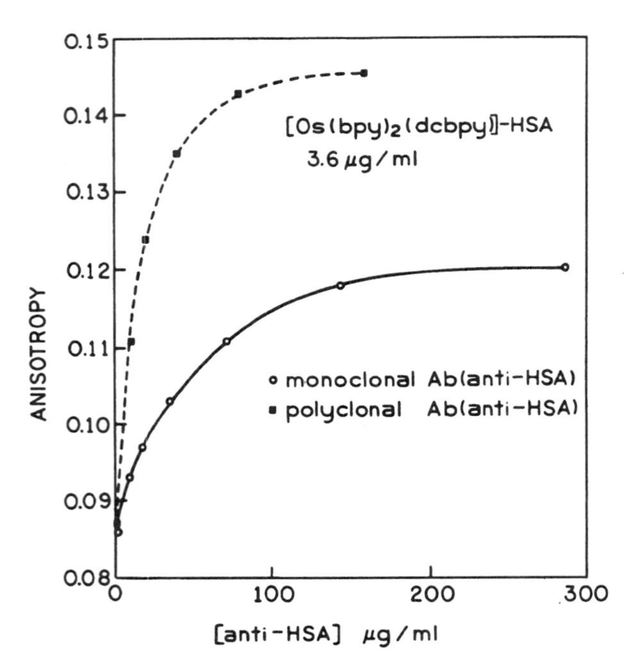

Figure 20.28. Fluorescence polarization immunoassay with $[Os(bpy)_2(dcbpy)]^{2+}$ bound to HSA for excitation near 680 nm using monoclonal (○) and polyclonal (■) antibodies. Data from Ref. 18.

It is presently a challenge to obtain long-wavelength-excitable MLCs which display long lifetimes and high quantum yields. However, some progress has been made. It is known that the decay times of Os MLCs can be increased by the use of tridentate ligands in place of bidentate ligands.[56] Several such compounds have been synthesized[57] (Figure 20.29). These Os complexes display long-wavelength absorption and decay times longer than 100 ns. Another approach to increasing the quantum yield and decay times of Os MLCs is by the use of arsine and phosphine ligands.[22] Typical structures are shown in Figure 20.29. Using this approach, it is possible to obtain high quantum yields. Unfortunately, the absorption spectra shift to shorter wavelengths, and the 450-nm absorption is weak. Nonetheless, work is in progress to improve the spectral properties of the Os MLCs.

20.4.B. Lifetime Immunoassays Based on RET

In Chapter 19, on fluorescence sensing, we described the advantages of lifetime-based sensing. Lifetimes can be mostly independent of the total fluorescence intensity, suggesting that lifetime-based immunoassays can be used in a homogeneous format. Hence, it was of interest to determine if RET could be used to transduce antigen–antibody binding. The donor–acceptor pair used in this study is shown in Figure 20.30. Reactive Blue 4 (RB4) is nonfluorescent, and its absorption spectrum is centered on the emission spectrum of the Ru probe (Figure 20.31). Binding of Ru-labeled HSA to anti-HSA labeled with RB4 resulted in quenching of the Ru emission (Figure 20.32). Binding

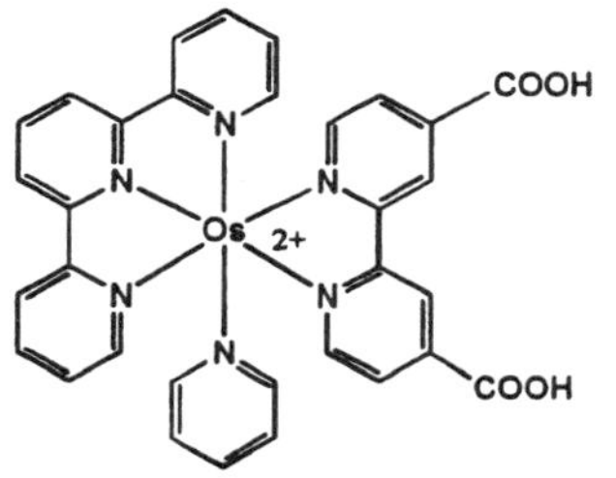

$[Os(tpy)(dcbpy)(py)](PF_6)_2$

λ_{abs} = 550 nm, λ_{em} = 640 nm, τ = 130 ns

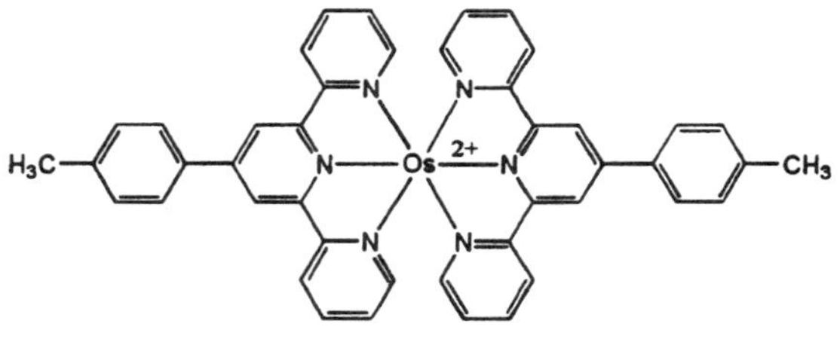

$[Os(ttpy)_2](PF_6)_2$

λ_{abs} = 600 nm, λ_{em} = 735 nm, τ = 220 ns

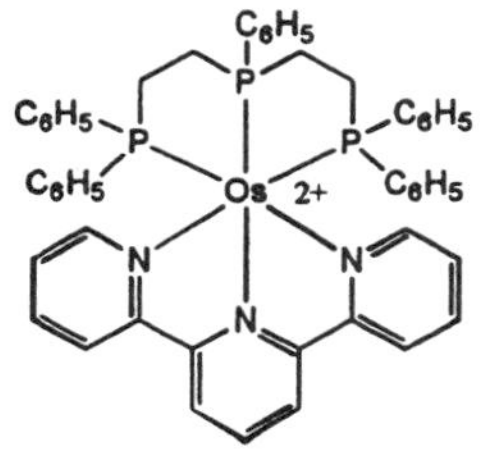

$[Os(tpy)(triphos)](PF_6)_2$

λ_{abs} = 470 nm, λ_{em} = 710 nm, τ = 230 ns

Figure 20.29. Long-wavelength, long-lifetime osmium(II) complexes.

could be detected by a decrease in intensity or lifetime. Addition of unlabeled HSA resulted in increased intensities and lifetimes, due to displacement of Ru-HSA from the acceptor-labeled antibody. As described in Chapter 19, lifetime measurements can be performed when the intensities cannot be accurately measured. Also, a lifetime assay of the type shown in Figure 20.32 could potentially be performed in homogeneous solution, without separation steps.

20.5. CLINICAL CHEMISTRY WITH METAL–LIGAND COMPLEXES

Another area of interest is the use of MLCs in ion sensing for application in either blood chemistry or cellular imaging. At present, ion-sensitive MLCs are just starting to become available. One example is the pH-sensitive MLC $[Ru(bpy)_2(deabpy)]^{2+}$, where deabpy is 4,4′-diethylaminomethyl-2,2′-bipyridine (Figure 20.33).[58] Emission spectra of $[Ru(bpy)_2(deabpy)]^{2+}$ at pH values from 2.23 to 11.75 are shown in Figure 20.34. The emission intensity increases about threefold as the pH increases from 2.52 to 11.75. The pH-dependent intensity changes show a pK_a value near 7.5. This pK_a value is ideally suited for measurements of blood pH, for which the clinically relevant range is from 7.35 to 7.46, with a central value near 7.40. In addition, much cell culture work is performed near pH 7.0–7.2. The changes in emission with changes in pH are believed to be due to deprontonation of the amino groups of $[Ru(bpy)_2(deabpy)]^{2+}$.

The emission spectrum of the MLC pH probe shifts to longer wavelengths as the amino groups are protonated at low pH (Figure 20.34). This suggests the use of $[Ru(bpy)_2(deabpy)]^{2+}$ as a wavelength-ratiometric probe. The emission spectra at various pH values were used to obtain a wavelength-ratiometric calibration curve (Figure

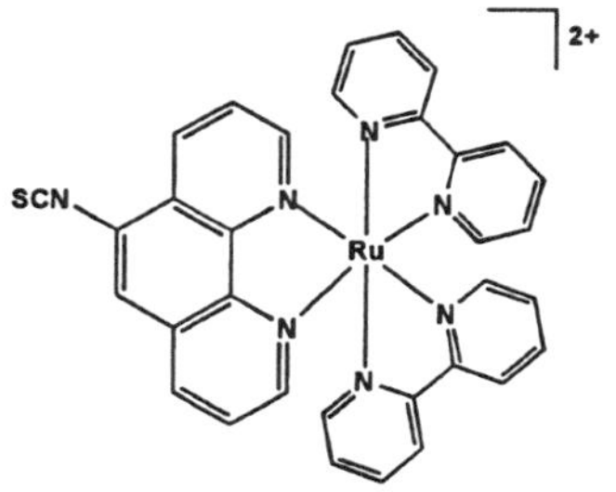

$[Ru(bpy)_2(5\text{-ITCphen})]^{2+}$ Reactive Blue 4 (RB4)

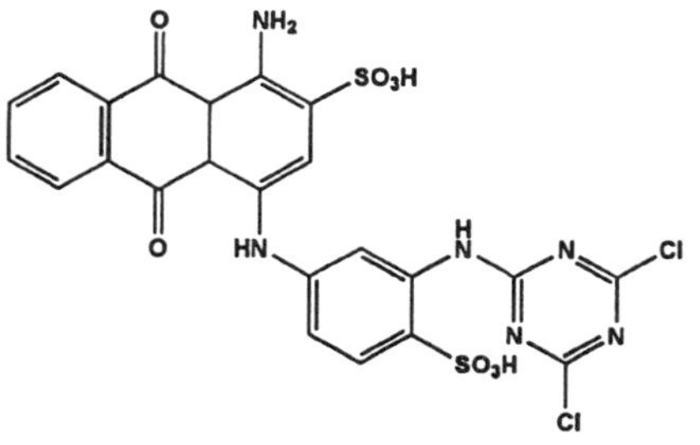

Donor Acceptor

Figure 20.30. Ruthenium MLC donor (*left*) and Reactive Blue 4 acceptor (*right*) used in an energy-transfer immunoassay.

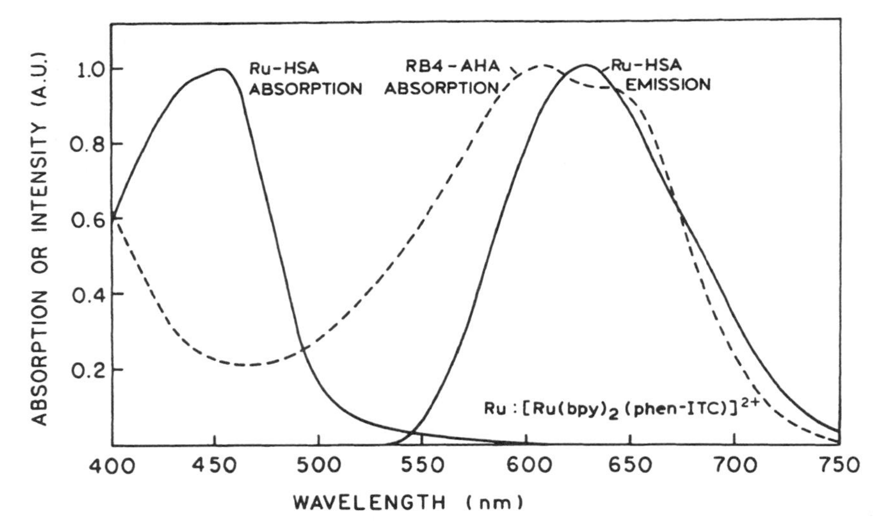

Figure 20.31. Absorption spectra of Ru-HSA (———) and acceptor labeled with anti-HSA (RB4-AHA; — — —) and emission spectrum of Ru-HSA. From Ref. 52.

20.35). Such ratiometric probes are already in widespread use for measurement of Ca^{2+} and pH (Chapter 19), but these are not MLC probes and they display nanosecond decay times. $[Ru(bpy)_2(deabpy)]^{2+}$ is the first MLC probe suitable for use as a ratiometric probe.

The emission shift to longer wavelengths at low pH (Figure 20.34) seems to be generally understandable in terms of the electronic properties of the excited MLCs. The long-wavelength emission is from an MLCT state in which an electron is transferred from Ru to the ligand. The protonated form of deabpy is probably a better electron acceptor; this lowers the energy of the MLCT state, shifting the emission to longer wavelengths and thereby decreasing the lifetime. These results suggest a general approach to designing wavelength-ratiometric MLC probes based on cation-dependent changes in the electron affinity of the ligand. Changes in the emission spectra can be expected to cause changes in lifetime in accordance with the energy-gap law (Section 20.2).

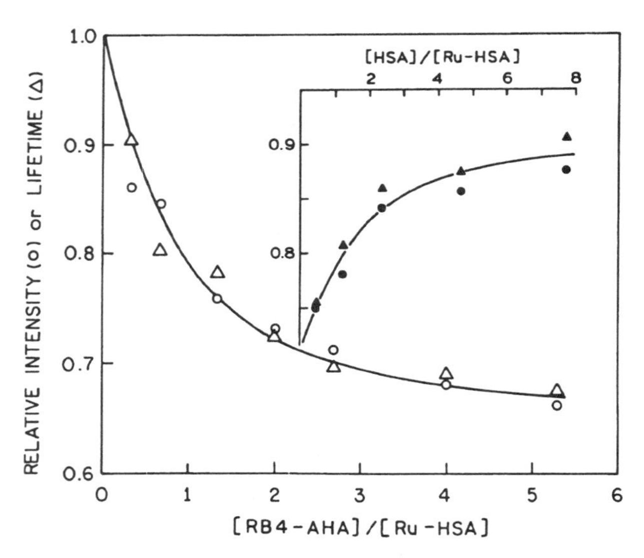

Figure 20.32. Effect of binding of anti-HSA labeled with RB4(RB4-AHA) to Ru-HSA on the relative intensity (○) and lifetime (△) of Ru-HSA. *Inset:* Competitive immunoassay of HSA, showing the increase of fluorescence intensity (●) or lifetime (▲) in the presence of unlabeled HSA. From Ref. 52.

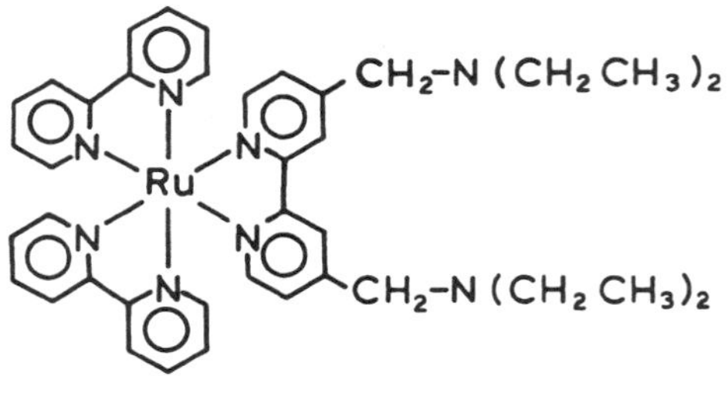

The emission spectra (Figure 20.34) reveal that the probe is luminescent in both the protonated and the unprotonated form. This suggests that it can be useful as a lifetime probe because each form is luminescent and may display distinct decay times. TD or FD measurements could be used to recover the multiexponential decay ex-

$[Ru(bpy)_2(deabpy)](PF_6)_2$

Figure 20.33. Structure of $[Ru(bpy)_2(deabpy)]^{2+}$, a long-lifetime MLC pH probe. From Ref. 58.

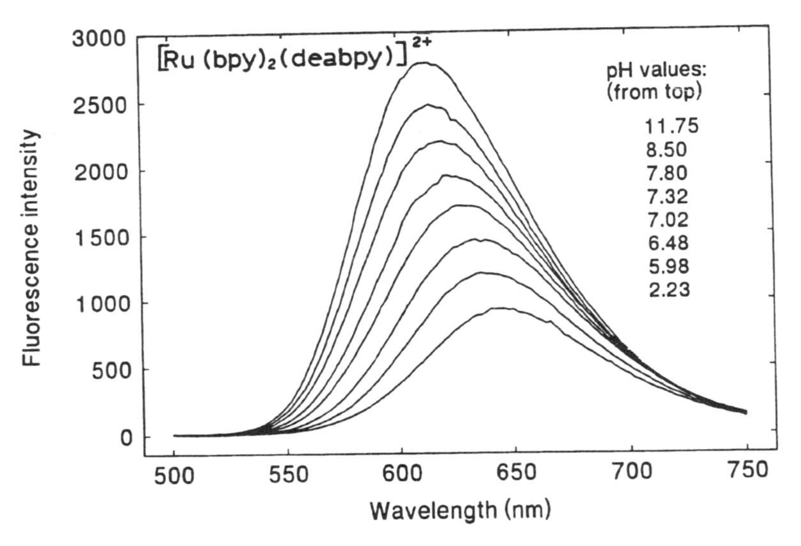

Figure 20.34. pH-dependent emission spectra of $[Ru(bpy)_2(deabpy)(PF_6)_2$. Excitation was at 414 nm. Revised from Ref. 58.

pected in the presence of two or more emitting species. However, this is not necessary for sensing applications, where measurements of the mean lifetime or at a single frequency can be used to recover the analyte concentration.[59] The pH-dependent phase and modulation values of $[Ru(bpy)_2(deabpy)]^{2+}$ are shown in Figure 20.36. As the pH is increased, the phase angles increase from 45 to 57°, and the modulation decreases from 0.69 to 0.52. Similar data were obtained using the blue LED excitation source amplitude-modulated at 823 kHz. Conveniently, these changes occur over the pH range from 6 to 8. With present instrumentation, one can expect the phase angles and modulation to be accurate to 0.1° and 0.005 in modulation.

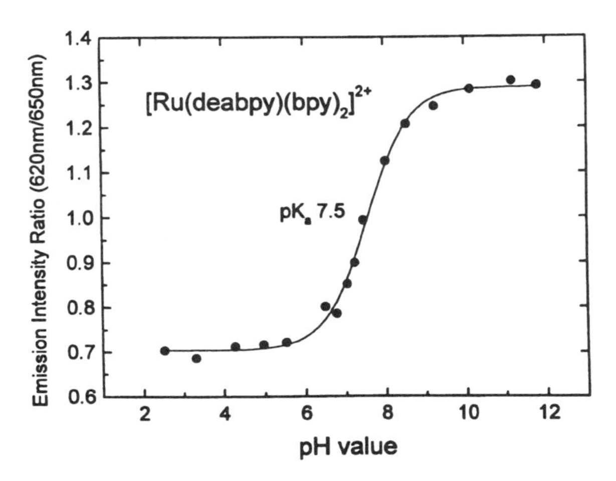

Figure 20.35. Wavelength-ratiometric measurements of pH using the emission intensities of $[Ru(bpy)_2(deabpy)]^{2+}$ at 620 and 650 nm. From Ref. 58.

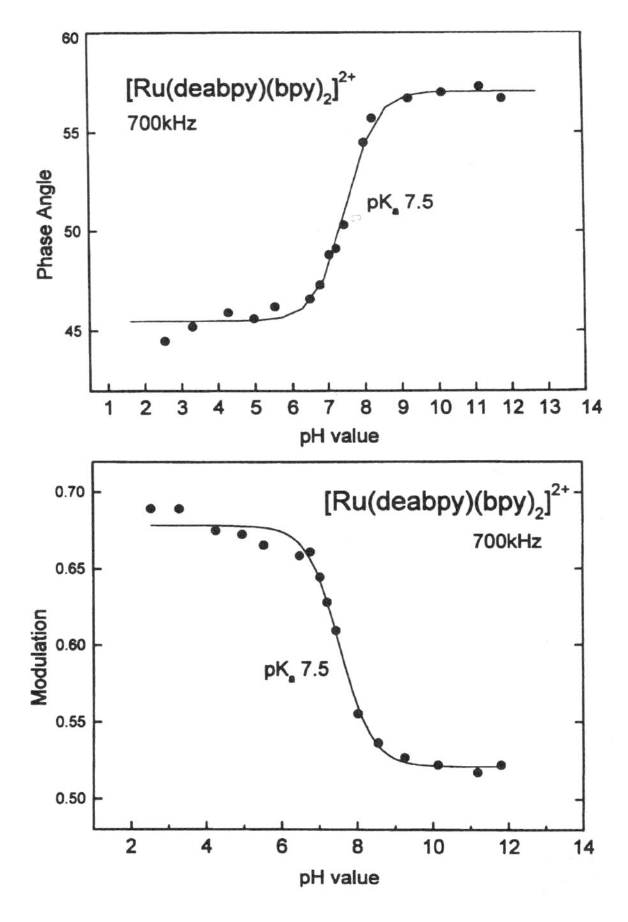

Figure 20.36. pH-dependent phase angles and modulation of $[Ru(bpy)_2(deabpy)]^{2+}$ with a modulation frequency of 700 kHz. From Ref. 58.

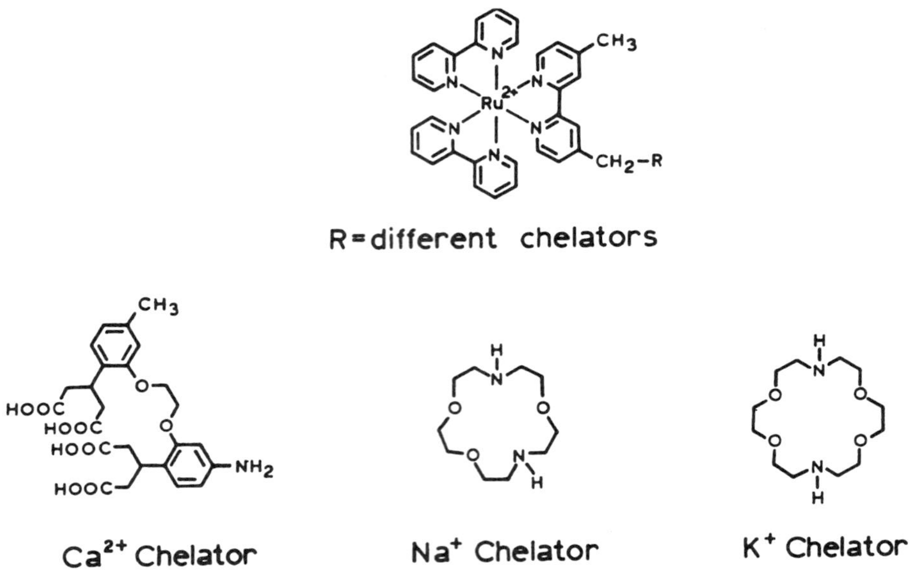

R=different chelators

Ca^{2+} Chelator Na^{+} Chelator K^{+} Chelator

Figure 20.37. Possible structures of cation-sensitive metal–ligand probes.

Hence, the pH can be measured to about 1 part in 50, or ±0.04 from pH 6 to 8. Such resolution is acceptable by medical standards[60]; however, one can expect improved phase and modulation accuracies with dedicated instrumentation and/or the use of multiple light modulation frequencies.

In addition to pH, a variety of other cations and anions are of medical interest. Hence, it seems probable that metal–ligand probes for a wide variety of analytes will be developed.[61,62] For instance, it is now known that nanosecond-lifetime probes with ion-chelating groups display changes in lifetime upon chelation. Such nanosecond-lifetime probes are known for Ca^{2+}, Mg^{2+}, Na^{+}, K^{+}, and Cl^{-}. It seems probable that coupling of the appropriate chelating groups, such as BAPTA or an azacrown ether, to an MLC will result in metal–ligand probes which display ion-sensitive lifetimes (Figure 20.37). One example of such a sensing probe is shown in Figure 20.38 for a Re(I) complex. In this case the emission intensity of the complex is increased upon binding of Pb^{2+} to the covalently attached crown ether. A variety of other cation-sensitive MLCs have been reported,[63–67] as well as a number of pH-sensitive complexes.[68–73] However, in most cases additional development will be needed to obtain complexes which have suitable binding constants, specificity, and solubility for use in cellular or clinical applications.

Once such long-lifetime cation probes are available, one can imagine many applications with simple instrumentation. Because of the long decay time, the light modulation frequencies can be near 1 MHz or lower. Hence, the light source can be an amplitude-modulated LED. The optical output of an LED is easily modulated at kilohertz to low megahertz frequencies.[74–76] If necessary, signal detection can be performed with electronic off-gating of the detector to suppress the nanosecond components due to autofluorescence from the samples. Such probes and simple instrumentation may allow sensing in blood serum or whole blood (Figure 20.39). Immunoassays based on RET with MLCs have already been reported.[52] Hence, the develop-

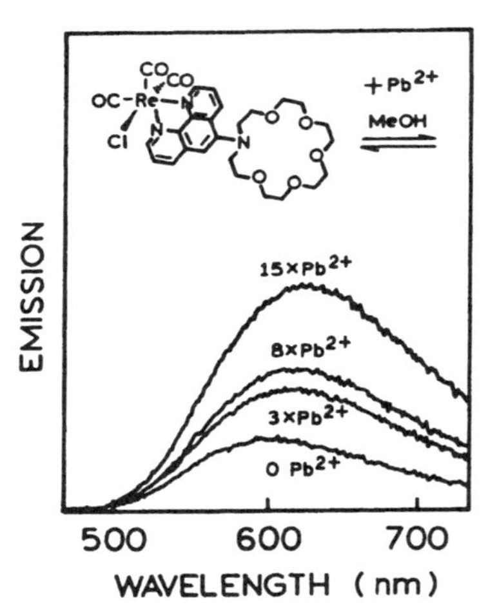

Figure 20.38. Effect of Pb^{2+} on the emission spectrum of a Re(I) complex used as an ion-sensing probe, [Re(5-aza-18-crown-6-1,10-phen)(CO_3)]Cl. The Pb^{2+} concentration is indicated as multiples of the Re(I) complex concentration of 1.4 × 10–4*M* in MeOH. Reprinted, with permission, from Ref. 61, Copyright © 1995, American Chemical Society.

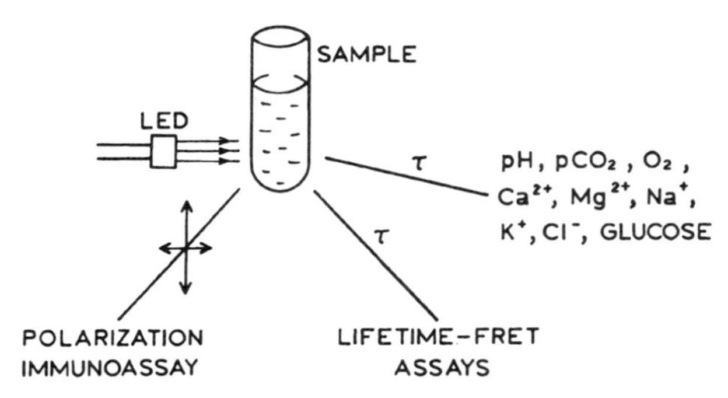

Figure 20.39. Clinical chemistry based on metal–ligand probes with an LED light source.

ment of MLCs can enable the use of simple instrumentation for point-of-care clinical chemistry.

Finally, we wish to point out the high photostability of the MLCs. When $[Ru(bpy)_2(dcbpy)]^{3+}$ and fluorescein were illuminated with the 488-nm output of an argon-ion laser, the MLC was stable for extended periods of time, under conditions where fluorescein was rapidly bleached (Figure 20.40). The initial decrease in the MLC intensity is thought to be due to heating. The long-term photostability of MLCs should make them useful for high-sensitivity detection in fluorescence microscopy, fluorescence *in situ* hybridization, and similar applications.

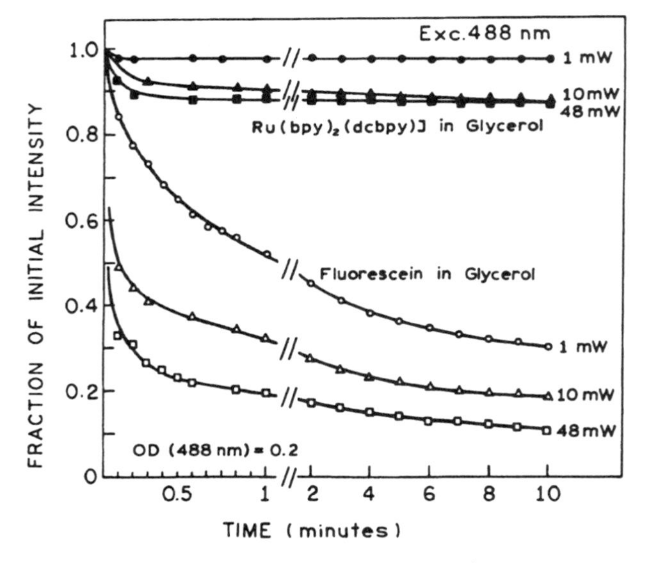

Figure 20.40. Photostability of $[Ru(bpy)_2(dcbpy)]^{2+}$ and fluorescein.

20.6. PERSPECTIVE ON METAL–LIGAND PROBES

In a textbook one expects to find information only on fully matured fields, not on new areas of recent interest. Hence, one may question why there is a chapter on MLCs. This author believes that it is fortunate that fluorescence occurs on the nanosecond timescale, but that applications of fluorescence have also been trapped on this timescale. The development of microsecond MLC probes has resulted in a fundamental change in what is experimentally accessible, particularly detection of processes that are on a timescale slower than nanoseconds. The synthetic and spectral versatility of the MLC probes suggests that they will become a permanent part of the tools used by spectroscopists and biochemists.

REFERENCES

1. Vanderkooi, J. M., 1992, Tryptophan phosphorescence from proteins at room temperature, in *Topics in Fluorescence Spectroscopy. Volume 3, Biochemical Applications*, J. R. Lakowicz (ed.), Plenum Press, New York, pp. 113–136.
2. Hurtubise, R. J., 1990, *Phosphorimetry: Theory, Instrumentation, and Applications*, VCH Publishers, New York.
3. Barthold, M., Barrantes, F. J., and Jovin, T. M., 1981, Rotational molecular dynamics of the membrane-bound acetylcholine receptor revealed by phosphorescence spectroscopy, *Eur. J. Biochem.* **120:**389–397.
4. Che, A., and Cherry, R. J., 1995, Loss of rotational mobility of band 3 proteins in human erythrocyte membranes induced by antibodies to glycophorin A, *Biophys. J.* **68:**1881–1887.
5. Haugen, G. R., and Lytle, F. E., 1981, Quantitation of fluorophores in solution by pulsed laser excitation and time-filtered detection, *Anal. Chem.* **53:**1554–1559.
6. Juris, A., Balzani, V., Barigelletti, F., Campagna, S., Belser, P., and Von Zelewsky, A., 1988, Ru(II) polypyridine complexes: Photophysics, photochemistry, electrochemistry and chemiluminescence, *Coord. Chem. Rev.* **84:**85–277.
7. Demas, J. N., and DeGraff, B. A., 1994, Design and applications of highly luminescent transition metal complexes, in *Topics in Fluorescence Spectroscopy, Volume 4, Probe Design and Chemical Sensing*, J. R. Lakowicz (ed.), Plenum Press, New York, pp. 71–107.
8. F. N. Castellano, personal communication.
9. Damrauer, N. H., Cerullo, G., Yeh, A., Boussie, T. R., Sharnk, C. V., and McCusker, J. K., 1997, Femtosecond dynamics of excited state evolution in $[Ru(bpy)_3]^{2+}$, *Science* **275:**54–57.
10. Demas, J. N., and DeGraff, B. A., 1991, Design and applications of highly luminescent transition metal complexes. *Anal. Chem.* **63:**829A–837A.
11. Kalayanasundaram, K., 1992, *Photochemistry of Polypyridine and Porphyrin Complexes*, Academic Press, New York.
12. Demas, J. N., and DeGraff, B. A., 1997, Applications of luminescent transition metal complexes to sensor technology and molecular probes, *J. Chem. Educ.* **74:**690–695.

13. Terpetschnig, E., Szmacinski, H., Malak., H., and Lakowicz, J. R., 1995, Metal–ligand complexes as a new class of long-lived fluorophores for protein hydrodynamics, *Biophys. J.* **68:**342–350.
14. Szmacinski, H., Terpetschnig, E., and Lakowicz, J.R., 1996, Synthesis and evaluation of Ru-complexes as anisotropy probes for protein hydrodynamics and immunoassays of high-molecular-weight antigens, *Biophys. Chem.* **62:**109–120.
15. Castellano, F. N., Dattelbaum, J. D., and Lakowicz, J. R., 1998, Long-lifetime Ru(II) complexes as labeling reagents for sulfhydryl groups, *Anal. Biochem.* **255:**165–170.
16. Terpetschnig, E., Dattelbaum, J. D., Szmacinski, H., and Lakowicz, J. R., 1997, Synthesis and spectral characterization of a thiol-reactive long-lifetime Ru(II) complex, *Anal. Biochem.* **251:**241–245.
17. Guo X., Li, L., Castellano, F. N., Szmacinski, H., and Lakowicz, J. R., 1997, A long-lived, highly luminescent rhenium (I) metal–ligand complex as a bimolecular probe, *Anal. Biochem.* **254:**179–186.
18. Terpetschnig, E., Szmacinski, H., and Lakowicz, J. R., 1996, Fluorescence polarization immunoassay of a high molecular weight antigen using a long wavelength absorbing and laser diode-excitable metal–ligand complex, *Anal. Biochem.* **240:**54–59.
19. Lakowicz, J. R., Terpetschnig, E., Murtaza, Z., and Szmacinski, H., 1997, Development of long-lifetime metal–ligand probes for biophysics and cellular imaging, *J. Fluoresc.* **7**(1):17–25.
20. Demas, J. N., and DeGraff, B. A., 1992, Applications of highly luminescent transition metal complexes in polymer systems, *Makromol. Chem., Macromol. Symp.* **59:**35–51.
21. Caspar, J., and Meyer, T. J., 1983, Application of the energy gap law to nonradiative, excited-state decay, *J. Phys. Chem.* **87:**952–957.
22. Kober, E. M., Marshall, J. L., Dressick, W. J., Sullivan, B. P., Caspar, J. V., and Meyer, T. J., 1985, Synthetic control of excited states. Nonchromophoric ligand variations in polypyridyl complexes of osmium(II), *Inorg. Chem.* **24:**2755–2763.
23. Kober, E. M., Sullivan, B. P., Dressick, W. J., Caspar, J. V., and Meyer, T. J., 1980, Highly luminescent polypyridyl complexes of osmium(II), *J. Am. Chem. Soc.* **102:**1383–1385.
24. Caspar, J. V., and Meyer, T. J., 1983, Photochemistry of $Ru(bpy)_3^{2+}$—solvent effects, *J. Am. Chem. Soc.* **105:**5583–5590.
25. Caspar, J. V., Kober, E. M., Sullivan, B. P., and Meyer, T. J., 1982, Application of the energy gap law to the decay of charge-transfer excited states, *J. Am. Chem. Soc.* **104:**630–632.
26. Bixon, M., and Jortner, J., 1968, Intramolecular radiationless transitions, *J. Chem. Phys.* **48:**715–726.
27. Freed, K. F., and Jortner, J., 1970, Multiphonon processes in the nonradiative decay of large molecules, *J. Chem. Phys.* **52:**6272–6291.
28. Barkley, M. D., and Zimm, B. H., 1979, Theory of twisting and bending of chain macromolecules: Analysis of the fluorescence depolarization of DNA, *J. Chem. Phys.* **70:**2991–3007.
29. Schurr, J. M., Fujimoto, B. S., Wu, P., and Song, L., 1992, Fluorescence studies of nucleic acids: Dynamics, rigidities and structures, in *Topics in Fluorescence Spectroscopy, Volume 3, Biochemical Applications*, J. R. Lakowicz (ed.), Plenum Press, New York, pp. 137–229.
30. Genest, D., Wahl, P., Erard, M., Champagne, M., and Daune, M., 1982, Fluorescence anisotropy decay of ethidium bromide bound to nucleosomal core particles, *Biochimie* **64:**419–427.
31. Millar, D. P., Robbins, R. J., and Zewail, A. H., 1980, Direct observation of the torsional dynamics of DNA and RNA by picosecond spectroscopy, *Proc. Natl. Acad. Sci. U.S.A.* **77:**5593–5597.
32. Millar, D. P., Robbins, R. J., and Zewail, A. H., 1982, Torsional and bending of nucleic acids studied by subnanosecond time-resolved fluorescence depolarization of intercalated dyes, *J. Chem. Phys.* **76:**2080–2094.
33. Friedman, A. E., Chambron, J.-C., Sauvage, J.-P., Turro, N. J., and Barton, J. K., 1990, Molecular light switch for DNA $Ru(bpy)_2(dppz)^{2+}$, *J. Am. Chem. Soc.* **112:**4960–4962.
34. Jenkins, Y., Friedman, A. E., Turro, N. J., and Barton, J. K., 1992, Characterization of dipyridophenazine complexes of ruthenium(II): The light switch effect as a function of nucleic acid sequence and conformation, *Biochemistry* **31:**10809–10816.
35. Holmlin, R. E., Stemp, E. D. A., and Barton, J. K., 1998, $Ru(phen)_2dppz^{2+}$ luminescence: Dependence on DNA sequences and groove-binding agents, *Inorg. Chem.* **37:**29–34.
36. Chang, Q., Murtaza, Z., Lakowicz, J. R., and Rao, G., 1997, A fluorescence lifetime-based solid sensor for water, *Anal. Chim. Acta* **350:**97–104.
37. Guo, X-Q., Castellano, F. N., Li, L., and Lakowicz, J. R., 1998, A long-lifetime Ru(II) metal–ligand complex as a membrane probe, *Biophys. Chem.* **71:**51–62.
38. Lakowicz, J. R., Malak, H., Gryczynski, I., Castellano, F. N., and Meyer, G. J., 1995, DNA dynamics observed with long lifetime metal–ligand complexes, *Biospectroscopy* **1:**163–168.
39. Malak, H., Gryczynski, I., Lakowicz, J. R., Meyer, G. J., and Castellano, F. N., 1997, Long-lifetime metal–ligand complexes as luminescent probes for DNA, *J. Fluoresc.* **7**(2):107–112.
40. Gerstein, M., Lesk, A. M., and Chothia, C., 1994, Structural mechanisms for domain movements in proteins, *Biochemistry* **33:**6739–6749.
41. Anderson, C. M., Zucker, F. H., and Steitz, T. A., 1979, Space-filling models of kinase clefts and conformation changes, *Science* **204:**375–380.
42. Grossman, S. H. 1990, Resonance energy transfer between the active sites of creatine kinase from rabbit brain, *Biochim. Biophys. Acta* **1040:**276–280.
43. Grossman, S. H. 1989, Resonance energy transfer between the active sites of rabbit muscle creatine kinase: Analysis by steady-state and time-resolved fluorescence, *Biochemistry* **28:**5902–5908.
44. Haran, G., Haas, E., Szpikowska, B. K., and Mas, M. T., 1992, Domain motions in phosphoglycerate kinase: Determination of interdomain distance-distributions by site-specific labeling and time-resolved fluorescence energy transfer, *Proc. Natl. Acad. Sci. U.S.A.* **89:**11764–11768.
45. Holowka, D., Wensel, T., and Baird, B., 1990, A nanosecond fluorescence depolarization study on the segmental flexibility of receptor-bound immunoglobulin E, *Biochemistry* **29:**4607–4612.
46. Zheng, Y., Shopes, B., Holowka, D., and Biard, B., 1991, Conformations of IgE bound to its receptor Fc epsilon RI and in solution, *Biochemistry* **30:**9125–9132.
47. Davenport, L., and Targowski, P., 1996, Submicrosecond phospholipid dynamics using a long-lived fluorescence emission anisotropy probe, *Biophys. J.* **71:**1837–1852.
48. Doring, K., Beck, W., Konermann, L., and Jahnig, F., 1997, The use of a long-lifetime component of tryptophan to detect slow orientational fluctuations of proteins, *Biophys. J.* **72:**326–334.
49. Li, L., Szmacinski, H., and Lakowicz, J. R., 1997, Long-lifetime lipid probe containing a luminescent metal–ligand complex, *Biospectroscopy* **3**(2):155–159.
50. Li, L., Szmacinski, H., and Lakowicz, J.R., 1997, Synthesis and luminescence spectral characterization of long-lifetime lipid metal–ligand probes, *Anal. Biochem.* **244:**80–85.
51. Terpetschnig, E., Szmacinski, H., and Lakowicz, J. R., 1995, Fluorescence polarization immunoassay of a high molecular weight

antigen based on a long-lifetime Ru–ligand complex, *Anal. Biochem.* **227**:140–147.
52. Youn, H. J., Terpetschnig, E., Szmacinski, H., and Lakowicz, J.R., 1995, Fluorescence energy transfer immunoassay based on a long-lifetime luminescence metal–ligand complex, *Anal. Biochem.* **232**:24–30.
53. Guo, X-Q., Castellano, F. N., Li, L., and Lakowicz, J. R., 1998, Use of a long-lifetime Re(I) complex in fluorescence polarization immunoassays of high-molecular weight analytes, *Anal. Chem.* **70**:632–637.
54. Sipior, J., Carter, G. M., Lakowicz, J. R., and Rao, G., 1997, A blue light-emitting diode demonstrated as an ultraviolet excitation source for nanosecond phase-modulation fluorescence lifetime measurements, *Rev. Sci. Instrum.* **68**:2666–2670.
55. Lakowicz, J. R., Murtaza, Z., Jones, W. E., Kim, K., and Szmacinski, H., 1996, Polarized emission from a rhenium metal–ligand complex, *J. Fluoresc.* **6**(4):245–249.
56. Brewer, R. G., Jensen, G. E., and Brewer, K. J., 1994, Long-lived osmium (II) chromophores containing 2,3,5,6-tetrakis(2–pyridyl)pyrazine, *Inorg. Chem.* **33**:124–129.
57. Murtaza, Z., and Lakowicz, J. R., 1999, Long-lifetime and long-wavelength osmium(II) metal compounds containing polypyridine ligands. Excellent red fluorescent dyes for biophysics and sensors, SPIE 3602, in press.
58. Murtaza, Z., Chang, Q., Rao, G., Lin, H., and Lakowicz, J. R., 1997, Long-lifetime metal–ligand pH probe, *Anal. Biochem.* **247**:216–222.
59. Szmacinski, H., and Lakowicz, J. R., 1994, Lifetime-based sensing, *in Topics in Fluorescence Spectroscopy, Volume 4, Probe Design and Chemical Sensing*, J. R. Lakowicz (ed.), Plenum Press, New York, pp. 295–334.
60. Medicare, Medicaid and CLIA programs; Regulations implementing the clinical laboratory improvement amendments of 1988 (CLIA '88), 1992, *Federal Register* **57**:7002–7018.
61. Shen, Y., and Sullivan, B. P., 1995, A versatile preparative route to 5-substituted-1,10-phenanthroline ligands via 1,10-phenanthroline 5,6-epoxide, *Inorg. Chem.* **34**:6235–6236.
62. Shen, Y., and Sullivan, B. P., 1997, Luminescence sensors for cations based on designed transition metal complexes, *J. Chem. Educ.* **74**:685–689.
63. Fujita, E., Milder, S. J., and Brunschwig, B. S., 1992, Photophysical properties of covalently attached $Ru(bpy)_3^{2+}$ and $Mcyclam^{2+}$ (M = Ni, H_2) complexes, *Inorg. Chem.* **31**:2079–2085.
64. Beer, P. D., Kocian, O., Mortimer, R. J., and Ridgway, C., 1991, Syntheses, coordination, spectroscopy and electropolymerisation studies of new alkynyl and vinyl linked benzo- and aza-crown ether-bipyridyl ruthenium(II) complexes. Spectrochemical recognition of group IA/IIA metal cations, *J. Chem. Soc., Chem. Commun.* **1991**:1460–1463.
65. Yoon, D. I., Berg-Brennan, C. A., Lu, H., and Hupp, J. T., 1992, Synthesis and preliminary photophysical studies of intramolecular electron transfer in crown-linked donor–(chromophore–) acceptor complexes, *Inorg. Chem.* **31**:3192–3194.
66. MacQueen, D. B., and Schanze, K. S., 1991, Cation-controlled photophysics in a Re(I) fluoroionophore, *J. Am. Chem. Soc.* **113**:6108–6110.
67. Rawle, S. C., Moore, P., and Alcock, N. W., 1992, Synthesis and coordination chemistry of 1-(2′,2″-bipyridyl-5′-yl-methyl)-1,4,8,11-tetraazacyclotetradecane L^1. Quenching of fluorescence from $[Ru(bipy)_2(L^1)]^{2+}$ by coordination of Ni^{2+} or Cu^{2+} in the cyclam cavity, *J. Chem. Soc., Chem. Commun.* **1992**:684–687.
68. Sun, H., and Hoffman, M. Z., 1993, Protonation of the excited states of ruthenium (II) complexes containing 2,2′-bipyridine, 2,2′-bipyrazine and 2,2′-bipyrimidine ligands in aqueous solution, *J. Phys. Chem.* **97**:5014–5018.
69. Park, J. W., Ahn, J., and Lee, C. 1995, Dependence of the photophysical and photochemical properties of the photosensitizer tris(4,4′-dicarboxy-2,2′-bipyridine)ruthenium (II) on pH, *J. Photochem. Photobiol., A: Chem.* **86**:89–95.
70. de Silva, A. P., Gunaratne, H. Q. N., and Lynch, P. L. M., 1995, Luminescence and charge transfer. Part 4. "On–off" fluorescent PET (photoinduced electron transfer) sensors with pyridine receptors: 1,3-diaryl-5-pyridyl-4,5-dihydropyrazoles, *J. Chem. Soc., Perkin Trans. 2* **1995**:685–690.
71. Walsh, M., Ryan, E. M., O'Kennedy, R., and Vos, J. G., 1996, The pH dependence of the emitting properties of ruthenium polypyridyl complexes bound to poly-L-lysine, *J. Inorg. Biochem.* **63**:215–221.
72. Giordano, P. J., Bock, C. R., and Wrighton, M. S., 1978, Excited state proton transfer of ruthenium(II) complexes of 4,7-dihydroxy-1,10-phenanthroline. Increased acidity in the excited state, *J. Am. Chem. Soc.* **100**:6960–6965.
73. Grigg, R., and Norbert, W. D. J. A., 1992, Luminescent pH sensors based on di(2,2′-bipyridyl) (5,5′-diaminomethyl-2,2′-bipyridyl)-ruthenium(II) complexes, *J. Chem. Soc., Chem. Commun.* **1992**:1300–1302.
74. Fantini, S., Franceschini, M. A., Fishkin, J. B., Barbieri, B., and Gratton, E., 1994, Quantitative determination of the absorption spectra of chromophores in strongly scattering media: A light-emitting diode based technique, *Appl. Opt.* **33**:5204–5213.
75. Lippitsch, M. E., and Wolfbeis, O. S., 1988, Fiber-optics oxygen sensor with the fluorescence decay time as the information carrier, *Anal. Chim. Acta* **205**:1–6.
76. Sipior, J., Carter, G. M., Lakowicz, J. R., and Rao, G., 1996, Single quantum well light emitting diodes demonstrated as excitation sources for nanosecond phase-modulation fluorescence lifetime measurements, *Rev. Sci. Instrum.* **67**:3795–3798.
77. Castellano, F. N., and Lakowicz, J. R., 1998, A water-soluble luminescence oxygen sensor, *Photochem. Photobiol.* **67**:179–183.

PROBLEMS

20.1. *Effect of Off-gating on the Background Level:* Figure 20.41 shows the intensity decay of a long-lifetime sensing probe (τ_2) with an interfering autofluorescence of τ_1 = 7 ns.

A. The decay time of the long-lifetime component is 400 ns. Confirm this by your own calculations.

B. What are the values of α_i in the intensity decay law? That is, describe $I(t)$ in terms of $\alpha_1 e^{-t/\tau_1} + \alpha_2 e^{-t/\tau_2}$.

C. What are the fractional intensities (f_i) of the two components in the steady-state intensity measurements?

D. Suppose that the detector was gated on at 50 ns and that the turn-on time is essentially instantaneous in Figure 20.41. Assume that the intensities are integrated to 5 μs, much longer than τ_2. What are the fractional intensities of each component? Explain the significance of this result for clinical and environmental sensing applications. It is recommended that

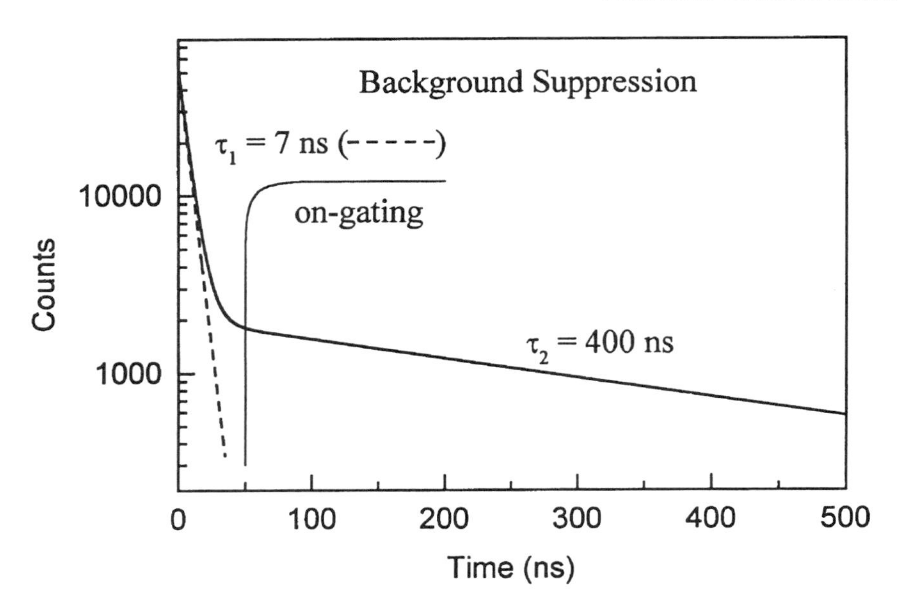

Figure 20.41. Intensity decay of a long-lifetime probe having a decay time of 400 ns, with an interfering autofluorescence of 7 ns.

the integrated fractional intensities be calculated using standard computer programs.

20.2. *Oxygen Bimolecular Quenching Constant for a Metal–Ligand Complex:* The complex of ruthenium with three diphenylphenanthrolines, $[Ru(dpp)_3]^{2+}$, displays a long lifetime of about 5 μs and has found widespread use as an oxygen sensor. Recently, a water-soluble version of the sensor has been synthesized (Figure 20.42).[77] FD intensity data for $[Ru(dpp(SO_3Na)_2)_3]Cl_2$ are shown in Figure 20.43. Calculate the oxygen bimolecular quenching constant for this complex. Assume that the solubility of oxygen in water is 0.001275*M*/atm.

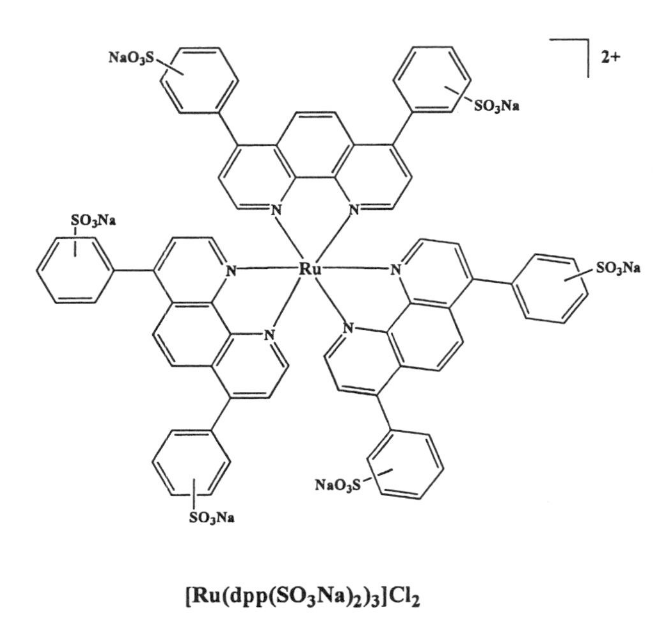

Figure 20.42. Structure of a water-soluble MLC used as an oxygen sensor. From Ref. 77.

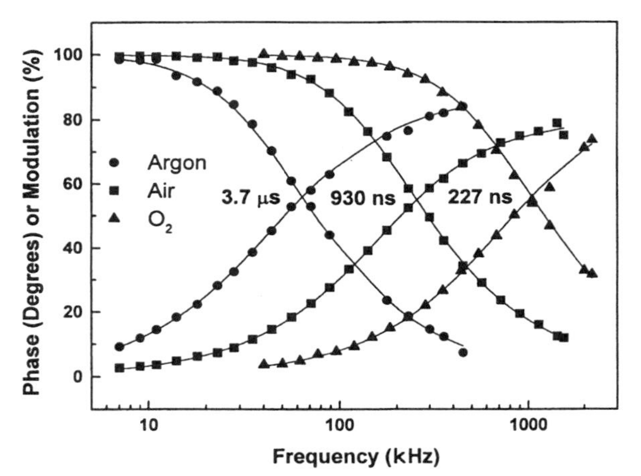

Figure 20.43. Structure of a water-soluble MLC used as an oxygen sensor. From Ref. 77.

DNA Technology

21

During the past 12 years, there have been remarkable advances in the use of fluorescence to study DNA. Fluorescence methods are now used for DNA sequencing, detection of DNA hybridization and restriction enzyme fragments, fluorescence *in situ* hybridization (FISH), and quantitating polymerase chain reaction products. Because of the rapid introduction of new DNA technology, it is surprising to realize that DNA sequencing by fluorescence was first reported just 12 years ago, in 1986. It is not the purpose of this chapter to describe the many specialized methods used in this extensive area of molecular biology and diagnostics. Instead, we give a brief introduction to each topic, followed by a description of the unique fluorophores and principles used for each application.

21.1. DNA SEQUENCING

DNA sequencing first became practical in 1977.[1,2] The original method involved selective chemical degradation of the DNA, followed by chromatography and detection of the fragments by ^{32}P autoradiography. In the same year an improved method based on chain-terminating dideoxynucleotides and ^{32}P also became available.[3] An overview of the history of DNA sequencing methods can be found in the informative text by Watson *et al.*[4]

The basic idea of sequencing using dideoxynucleotide triphosphate (ddNTP) terminators is shown in Figures 21.1 and 21.2. In DNA the nucleotides are linked in a continuous strand via the 5′- and 3′-hydroxyl groups of the pentose sugar. DNA is replicated by adding nucleotides to the 3′-hydroxyl group. This elongation reaction is catalyzed on the unknown sequence by DNA polymerase, starting from a primer location with a known sequence. In the example shown in Figure 21.2, a single fluorescent primer of known sequence is used to initiate the reaction. The reaction mixture is split into four parts for the four bases. The DNA polymerase reaction is randomly terminated along the sequence by the ddNTPs which are added along the growing chain. The absence of a 3′-hydroxyl group on the ddNTPs prevents further elongation and terminates the reaction. This results in a mixture of oligonucleotides of varying length, which are separated by polyacrylamide gel electrophoresis. Remarkably, the numerous fragments differing by just one base pair, among up to several hundred bases can be resolved. Typically, there are four different terminating nucleotides, and each reaction mixture is electrophoresed in a separate lane. The gels separate the DNA fragments according to size, so that the sequence can be determined from the autoradiogram of the separated DNA fragments.

The use of the ddNTP terminators is the preferred sequencing method, but now with the use of fluorescence in place of ^{32}P. The use of radioactive tracers is obviously problematic with regard to cost, safety, and disposal. DNA sequencing using fluorescence became possible in 1986.[5–7]

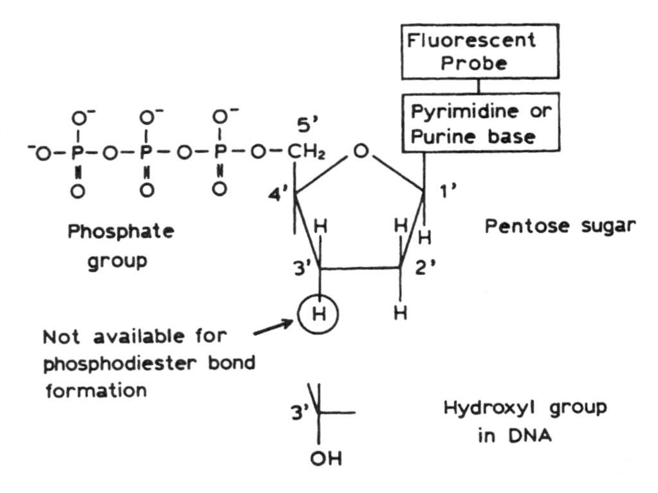

Figure 21.1. Schematic of a dideoxynucleotide triphosphate (ddNTP). Fluorescent and nonfluorescent ddNTPs are used for DNA sequencing, depending on the method. The 2′ group is hydrogen in DNA and is a hydroxyl group in RNA.

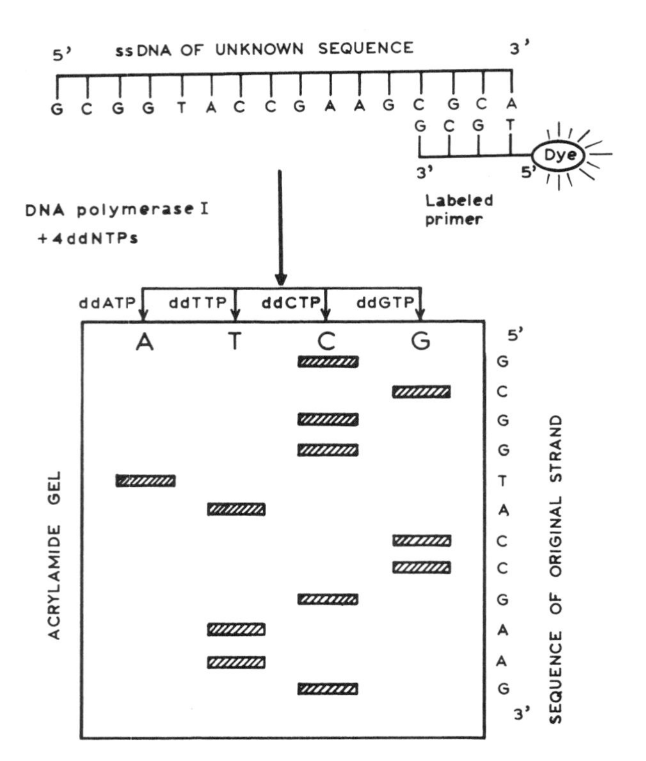

Figure 21.2. Use of ddNTPs and a fluorescent primer for DNA synthesis. Sequencing can also be accomplished in a single lane with the use of four distinctly fluorescent ddNTPs. Revised from Ref. 4.

Several methods were proposed, the first based on the use of four different fluorescent primers and nonfluorescent dideoxynucleotides,[5] and another based on the use of four different fluorescent dideoxynucleotides.[6] DNA sequencing can also be accomplished with a single fluorescent primer and nonfluorescent ddNTPs.[7] Hence, in Figure 21.2, either the primers or the ddNTPs can be fluorescent. If the fluorophores are all distinct, then the DNA can be electrophoresed in a single lane, and the nucleotide sequence identified by the emission spectra. One can also use a single fluorescent primer and nonfluorescent ddNTPs and perform the electrophoresis in four lanes, as shown in Figure 21.2. The nucleotide base present at each position in the sequence is then identified by the presence of the fluorescent primer at each position in the four lanes, each of which was terminated with a different nonfluorescent ddNTP. Several variations are in common use in DNA sequencing.

A variety of fluorophores have been chosen for DNA sequencing, typically a set of four fluorophores, one for each base, A, C, G, and T. The fluorophores are typically selected so that all can be excited using the 488-nm line from an argon-ion laser. The initial set of fluorophores used for four different primers[5] is shown in Figure 21.3. Al-

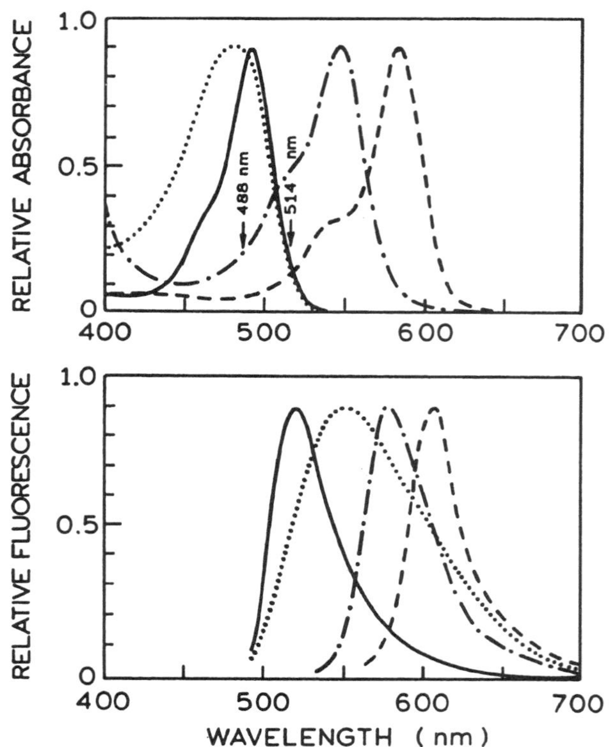

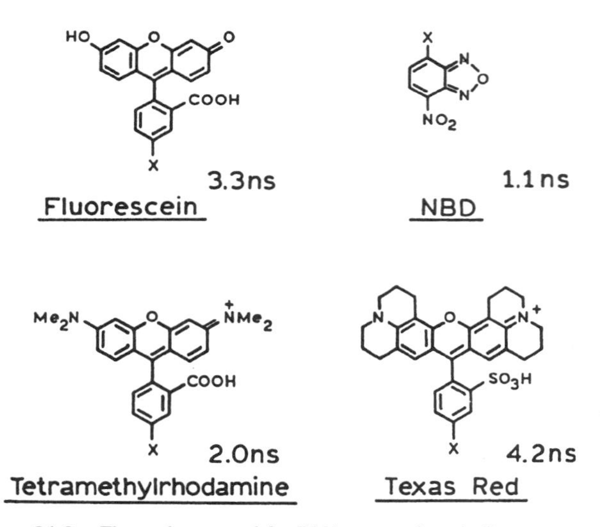

Figure 21.3. Fluorophores used for DNA sequencing via fluorescent primers. *Top*: Absorption spectra; *middle*: emission spectra; *bottom*: probe structures and decay times. The X substituent on the probe was linked to the 5′-end of the DNA using an aliphatic amino group on the 5′-terminus. Revised from Ref. 5. Decay times are from Ref. 28.

though all four dyes could be excited at 488 nm, the absorption of Texas Red and tetramethylrhodamine is obviously weak at 488 nm. For this reason, it was necessary to use excitation at 514 nm to obtain relatively equal intensities for all four probes. Another difficulty with these four dyes is the overlapping of their emission spectra. Therefore, it is necessary to record emission spectra from the gels at more than one excitation and emission wavelength. In spite of these difficulties, the use of four fluorophores allowed use of a single gel column containing the mixture of labeled DNA fragments.

A somewhat improved series of dyes was prepared for use as fluorescent dideoxy terminators. The structures of these dyes are shown in Figure 21.4. One notices the absence of a hydroxyl group on the 3′ portion of the sugar. These nucleotide analogs are unable to elongate the DNA chain. These dyes all display similar extinction coefficients at 488 nm, allowing the use of a single excitation wavelength (Figure 21.5). The fluorescence intensities differ by less than a factor of 2. However, the emission spectra overlap, so that fluorescence detection must be performed through at least two bandpass filters to identify the base pairs.

In the previous paragraphs we mentioned some of the nonideal properties of the dyes. These considerations illustrated spectral features which are important in dyes for DNA sequencing. Useful dyes can be excited with a convenient laser source and yield similar intensities for excitation at a single wavelength. The use of fluorescence has allowed DNA sequencing to become routine in numerous laboratories. Capillary gel electrophoresis is being used in place of electrophoresis on slab gels, providing more rapid separations with improved resolution.[8] Somewhat remarkably, a row of parallel capillary columns can form a waveguide in the direction cutting across the columns. Hence, all the columns can be excited with a single laser beam for use with high-throughput sequencing.[9] Because of the developments in gel and capillary electrophoresis, it is now possible to identify up to 1000 bases in a single separation.[10,11] Some capillary columns have been described as yielding 1000 bases per hour. Other groups have described instruments with up to 100 capillary columns.

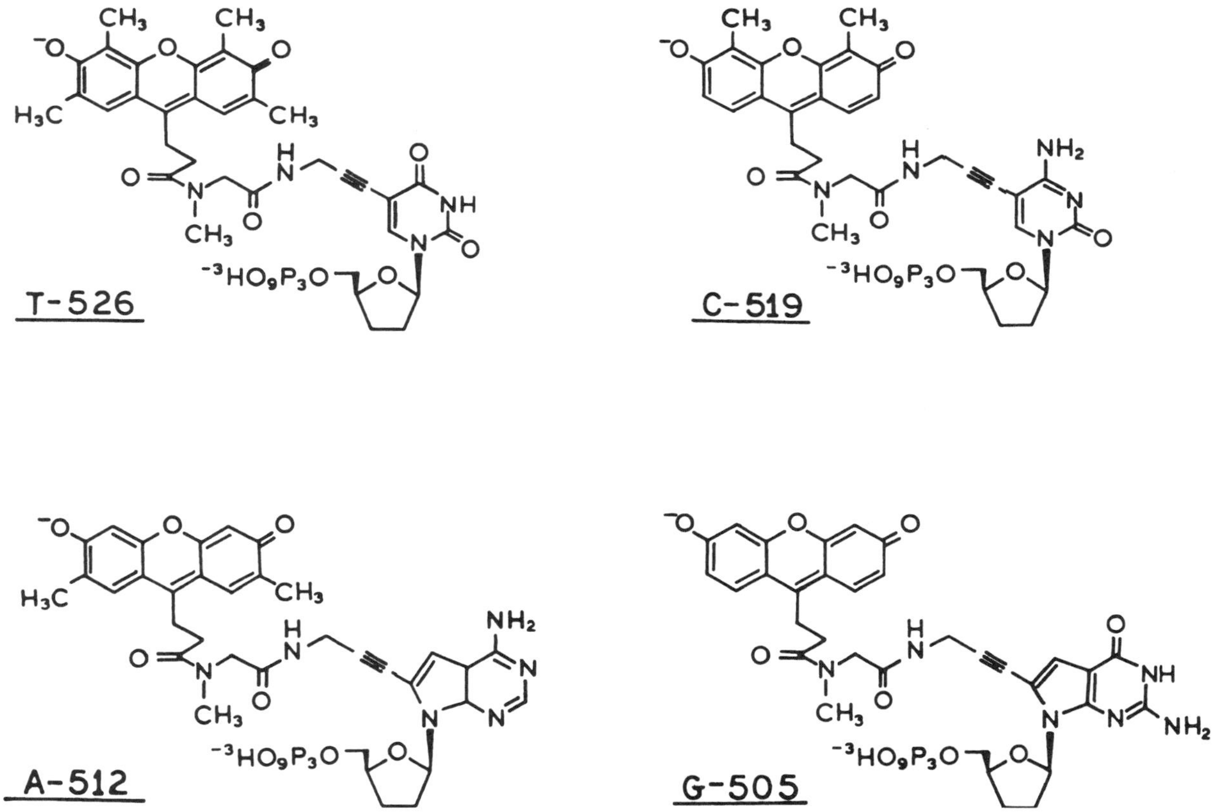

Figure 21.4. Fluorescent chain-terminating dideoxynucleotides. The letters refer to the DNA base, and the numbers refer to the emission maximum. Reprinted, with permission, from Prober, J. M., Trainor, G. L., Dam, R. J., Hobbs, F. W., Robertson, C. W., Zagursky, R. J., Cocuzza, A. J., Jensen, M. A., and Baumeister, K., A system for rapid DNA sequencing with fluorescent chain-terminating dideoxynucleotides, *Science* **238**:336–343, Copyright © 1987, American Association for the Advancement of Science.

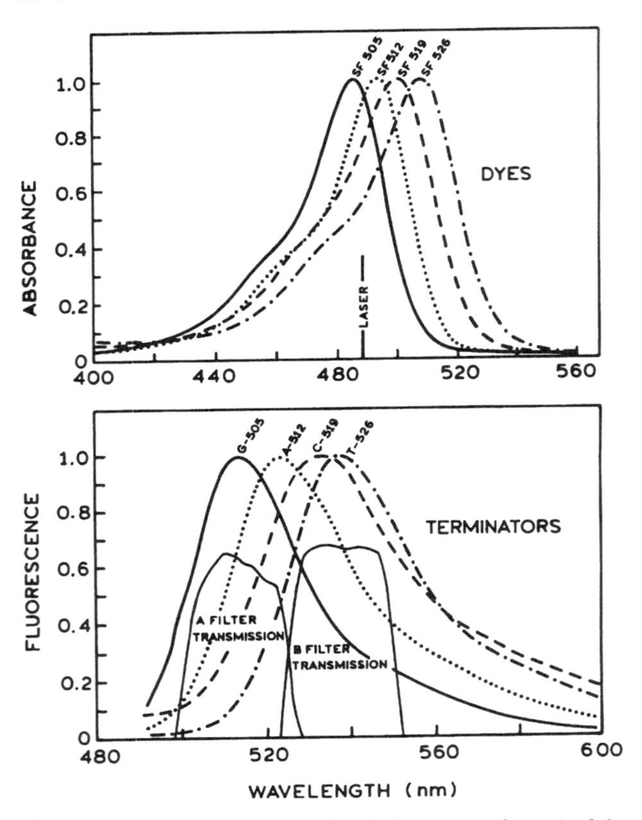

Figure 21.5. Absorption (*top*) and emission spectra (*bottom*) of the fluorescent chain-terminating dideoxynucleotides in Figure 21.4. The absorption spectra are of the succinylfluorescein (SF) dyes, prior to coupling to the amine-reactive ddNTPs. Reprinted, with permission, from Prober, J. M., Trainor, G. L., Dam, R. J., Hobbs, F. W., Robertson, C. W., Zagursky, R. J., Cocuzza, A. J., Jensen, M. A., and Baumeister, K., A system for rapid DNA sequencing with fluorescent chain-terminating dideoxynucleotides, *Science* **238**:336–343, Copyright © 1987, American Association for the Advancement of Science.

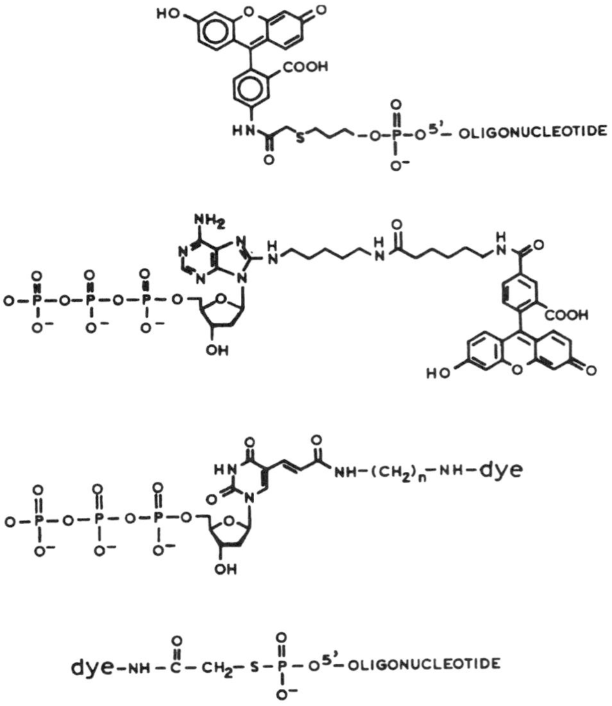

Hence, it seems clear that sequencing technology is poised for further improvements.

21.1.A. Nucleotide Labeling Methods

A wide variety of chemical structures have been used to covalently label DNA.[12,13] One typical linkage was seen in Figure 21.4, which showed acetylene linkages between the flourophores and the nucleotide bases. Other typical structures are shown in Figure 21.6. Probes can be attached to the 5′-end of DNA via a sulfhydryl group linked to the terminal phosphate. Amino groups can also be placed on the terminal phosphate. Alternatively, fluorophores have been linked to the bases themselves, typically opposite to the base-recognition hydrogen-bonding side of the base. The 5′-phosphate can be made reactive with iodoacetamide probes by attaching a terminal $-PO_3S$ residue. This type of attachment is used to label the internal DNA bases.

Figure 21.6. Methods to label DNA. In the top structure, DNA is labeled with a fluorescent primer. The two structures in the middle show labeling of DNA using labeled nucleotide triphosphates. In the bottom structure, DNA is labeled on the 5′-end via a thiophosphate linkage. Revised from Refs. 12 and 13.

21.1.B. Energy-Transfer Dyes for DNA Sequencing

For DNA sequencing, it is desirable to have dyes that display distinct emission spectra and similar intensities with a single excitation wavelength. This is difficult to accomplish using a single fluorophore. Hence, donor–acceptor pairs have been used to accomplish these requirements. One set of donor–acceptor pairs for sequencing was constructed using the fluorophores shown in Figure 21.7. The emission spectra of these four probes are moderately distinct (Figure 21.8), suggesting that they would allow sequencing in a single lane. However, the intensities were found to be rather unequal when excited at a single wavelength of 488 nm. For this reason, the donors and acceptors were covalently linked within the Förster distance (R_0) using reactive oligonucleotides (Figure 21.9, top) or DNA-like sugar polymers without the nucleotide bases (Figure 21.9, bottom).

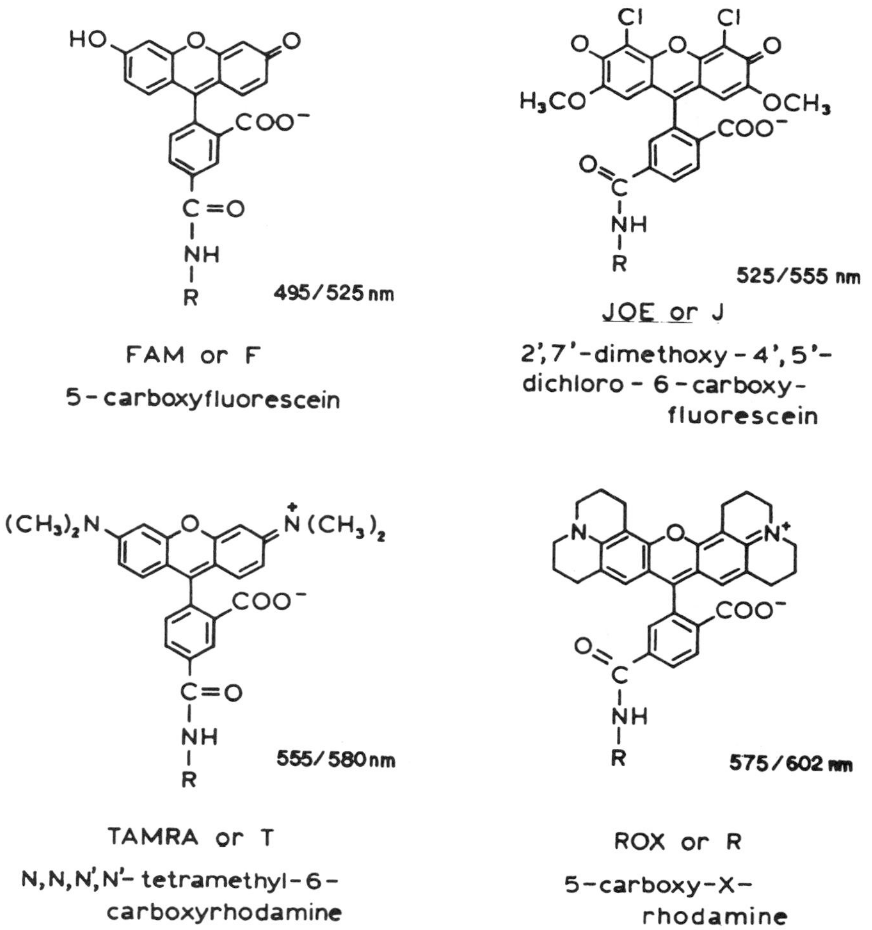

495/525 nm
FAM or F
5-carboxyfluorescein
525/555 nm
JOE or J
2',7'-dimethoxy-4',5'-dichloro-6-carboxy-fluorescein
555/580 nm
TAMRA or T
N,N,N',N'-tetramethyl-6-carboxyrhodamine
575/602 nm
ROX or R
5-carboxy-X-rhodamine

Figure 21.7. Fluorophores used as energy-transfer DNA sequencing probes. The two wavelengths are the excitation and emission maxima. Revised from Ref. 14.

FAM
JOE
TAMRA
ROX
FLUORESCENCE INTENSITY (——)
FILTER TRANSMITTANCE (········) (%)
1.0
0.8
0.6
0.4
0.2
0
100
80
60
40
20
0
450
500
550
600
650
WAVELENGTH (nm)

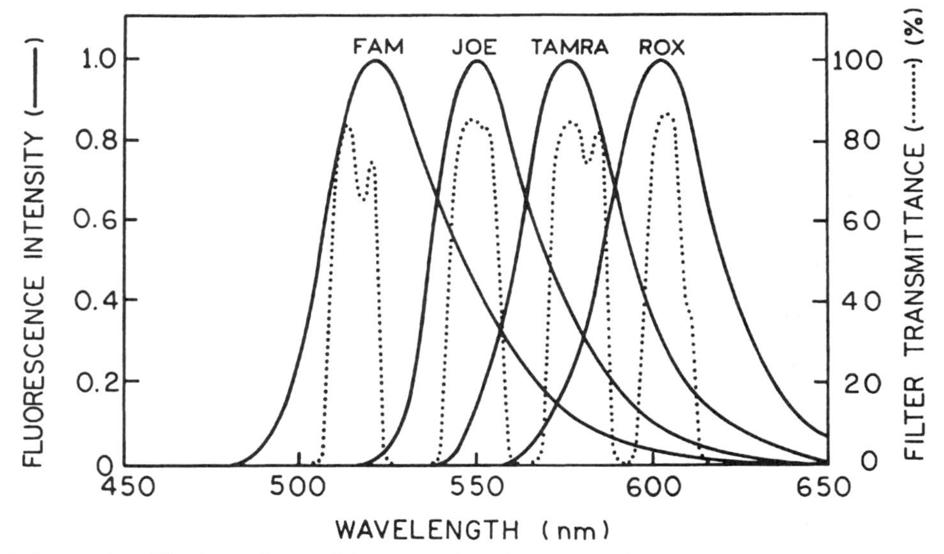

Figure 21.8. Emission spectra of the four probes used for construction of energy-transfer primers. Reprinted, with permission, from Ref. 16, Copyright © 1994, American Chemical Society.

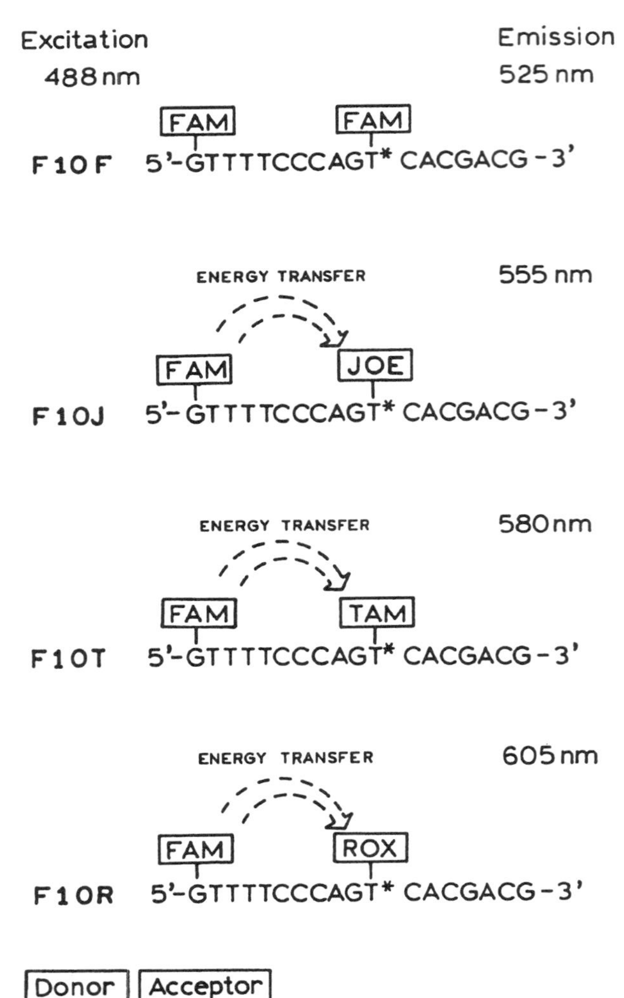

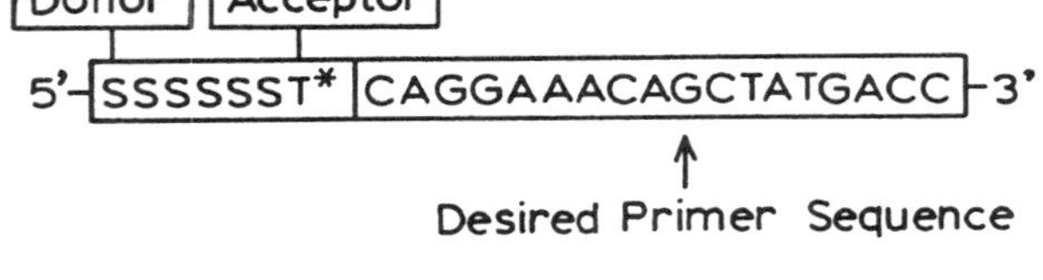

Figure 21.9. Energy-transfer primers used for DNA sequencing. F10F, F10J, F10T, and F10R are energy-transfer primers in which the donor and acceptor (see Figure 21.7 for the structures of F, J, T, and R) have been covalently linked using reactive oligonucleotides. Excitation is at 488 nm, and the emission wavelengths are indicated on the right-hand side of the figure. As shown at the bottom of the figure, donor–acceptor pairs can also be placed on sugar polymers (–SSS–) prior to the primer sequence. Revised from Refs. 14 and 15.

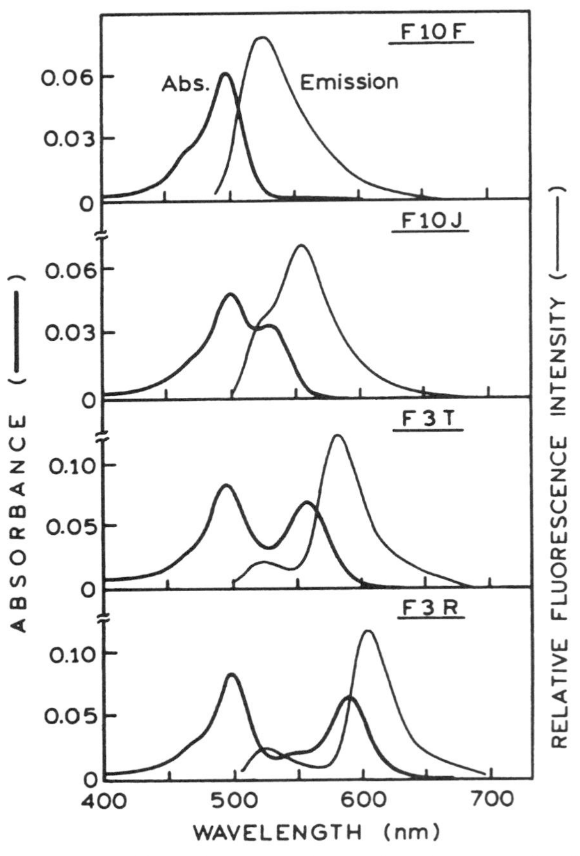

Figure 21.10. Absorption and emission spectra of four energy-transfer primers, showing the relative flourescence intensity for excitation at 488 nm. See Figure 21.8 for the structures of F, J, T, and R. The emission spectrum for each primer pair was determined in solutions having the same absorbance at 260 nm. From Ref. 14.

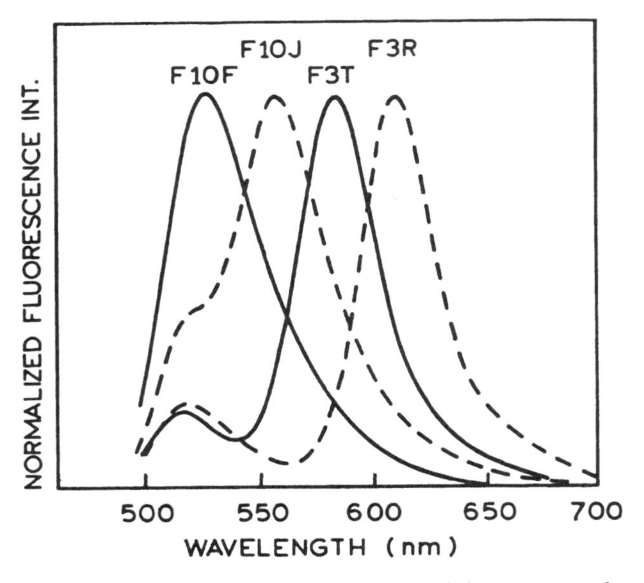

Figure 21.11. Normalized emission spectra of the energy-transfer DNA probes in Figure 21.10. From Ref. 14.

The usefulness of energy transfer for equalization of the intensities is shown in Figure 21.10. All four probes show similar absorbance at 488 nm, and the emission intensities are relatively equal. These probes show emission spectra which are moderately well separated, as is easier to see in

Table 21.1. Comparison of a Laser Diode and an Argon-Ion Laser[a]

	Laser diode	Argon-ion laser
Wavelength	785 nm	488 nm
Life span	100,000 + hr	3000 hr
Optical power output	20 mW	20 mW
Power consumption	0.150 W	1800 W
Replacement cost	$150	<$5000

[a]From Ref. 19.

the normalized emission spectra (Figure 21.11). The bases can be readily identified by measurement at the four emission wavelengths[16] and allow DNA sequencing with capillary electrophoresis using a single 488-nm excitation wavelength.[17] However, close examination of Figure 21.11 shows that the emission spectra overlap and that residual emission from the donors contributes to the intensities at shorter wavelengths. Hence, there is still a need for improved dyes for DNA sequencing. Another set of energy-transfer primers has been reported based on the BODIPY fluorophore.[18] These primers are claimed to show more distinct emission spectra than shown in Figure 21.11. Given the interest in sequencing, one can expect continued development of probes for use with argon-ion laser excitation.

21.1.C. DNA Sequencing with NIR Probes

To satisfy the demands of the human genome project, it is necessary to sequence DNA as rapidly and inexpensively

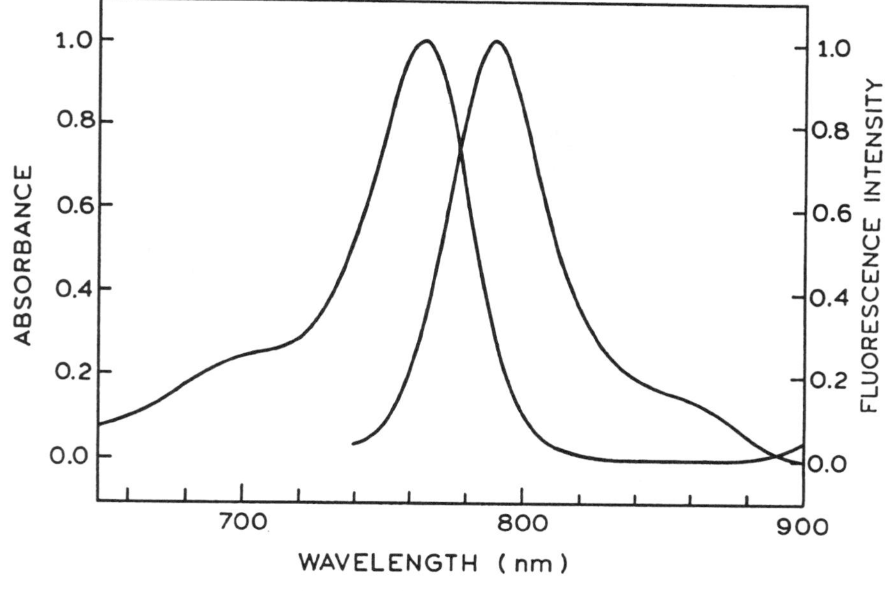

Figure 21.12. Structure and absorption and emission spectra of an NIR DNA primer. Revised from Ref. 20.

as possible. One method to decrease the cost is to use semiconductor laser diodes, which are now available from 630 nm to longer wavelengths. The advantages of a semiconductor laser over a relatively simple air-cooled argon-ion laser are summarized in Table 21.1. These lasers consume little power and can operate for up to 100,000 hours between failures.[19] An additional advantage of red and NIR excitation is the lower autofluorescence from biological samples, gels, solvents, and optical components.

The advantages of NIR wavelengths for DNA sequencing are obvious, and several groups have described NIR dyes for DNA sequencing.[20–24] One such DNA primer is shown in Figure 21.12. Excitation can be accomplished in the NIR at 785 nm, and emission occurs at 810 nm. Such dyes often display small Stokes' shifts,[24] which can result in difficulties in rejecting scattered excitation. The Stokes' shift can be increased using donor–acceptor pairs, as was shown in Section 21.1.B. The quantum yield of the NIR probe shown in Figure 21.12 is low (0.07), and considerably less than that of fluorescein (0.90). Nonetheless, the detection limit was 40-fold lower for this NIR probe, primarily because of the decreased background signal.[20]

In the past, most sequencing instruments have been designed around the spectral properties of available probes. However, there are significant advantages in designing the probes prior to the instrumentation. For instance, the probes shown in Figure 21.3 require an argon ion laser at 488 and 5143 nm for excitation. This was the

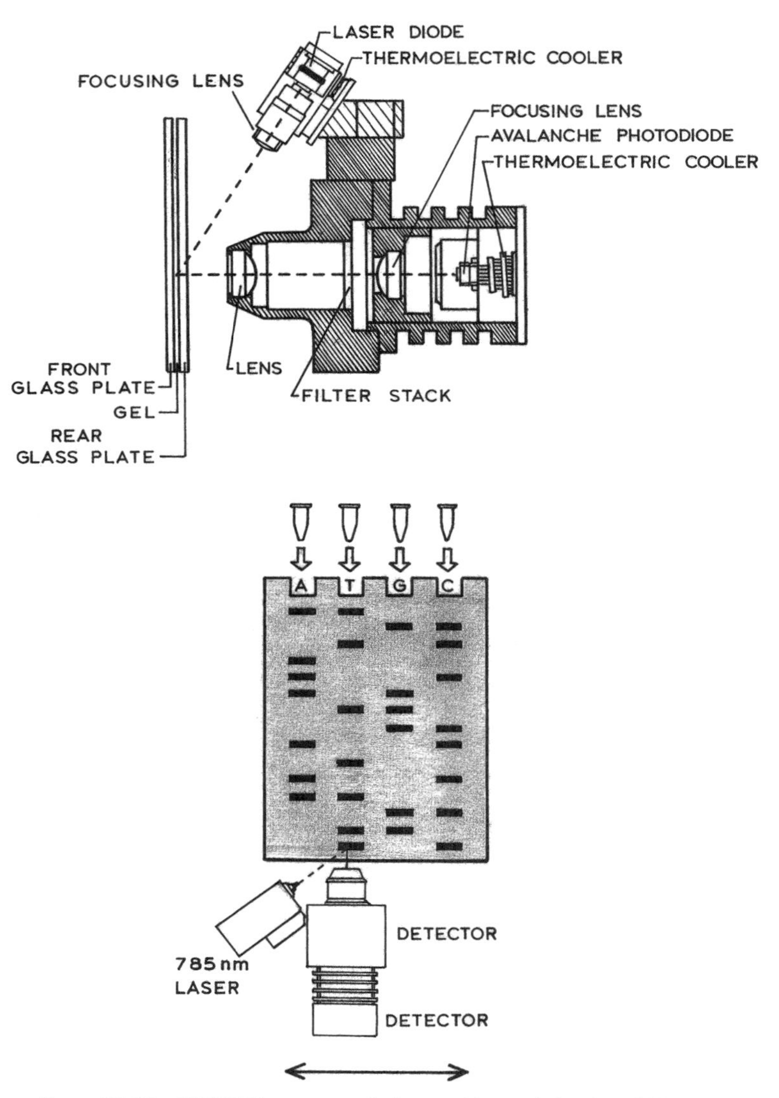

Figure 21.13. NIR DNA sequencer. Redrawn with permission from LiCor, Inc.

philosophy in designing the NIR DNA primer (Figure 21.12), which in turn allowed design of a sensitive and reliable sequencer (Figure 21.13). The long-wavelength absorption maximum allowed use of a 780-nm laser diode as the excitation source.[20] The long-wavelength emission could be efficiently detected with an avalanche photodiode. The excitation beam is incident on the glass plate at the Brewster angle to minimize scattered light. This sequencer illustrates the effectiveness of including probe design as an integral part of the instrument design process.

At present, a set of four NIR probes for single-lane DNA sequencing does not seem to be available, but the spectral

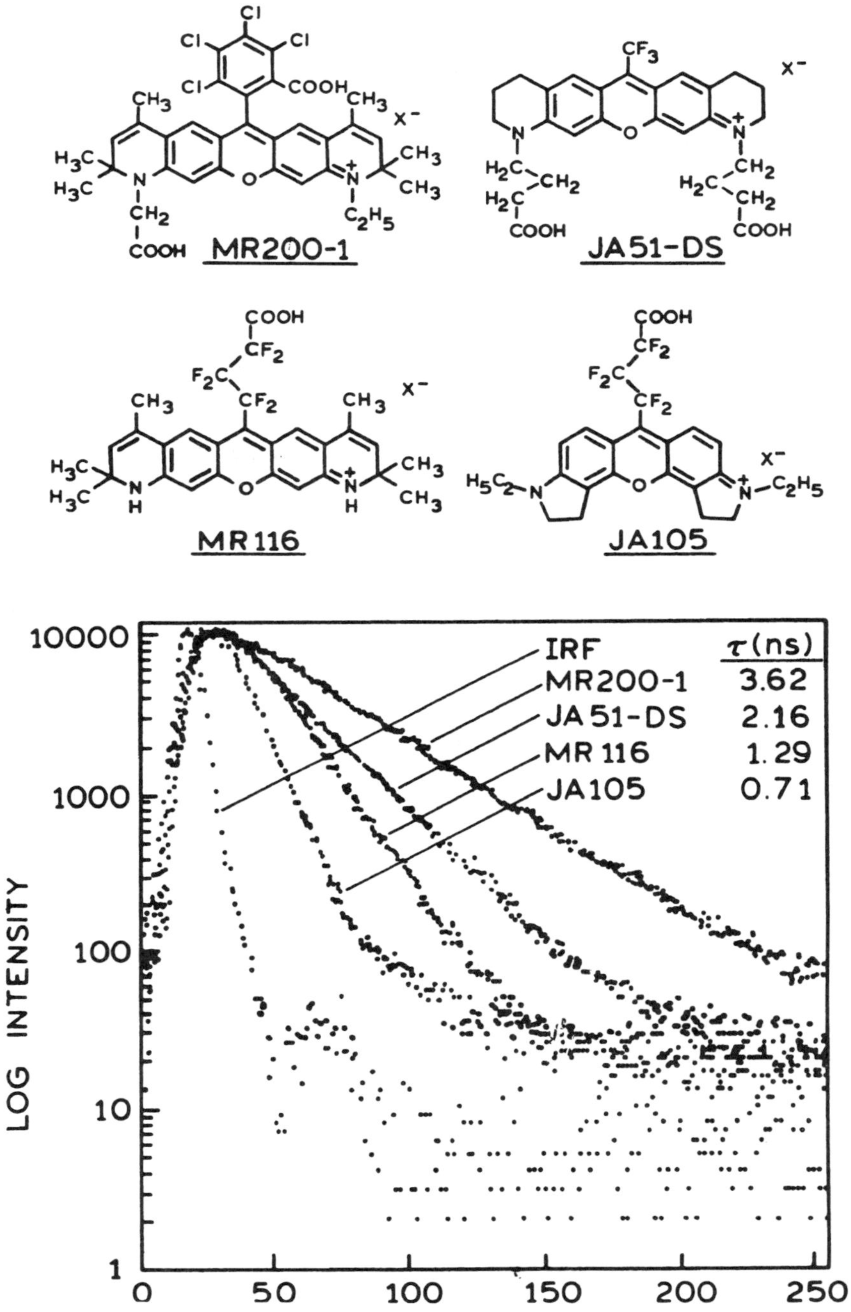

Figure 21.14. Structures and intensity decays of dyes for lifetime-based sequencing. Excitation was with a pulsed laser diode at 636 nm. IRF, Instrument response function. Revised from Ref. 29.

properties of many NIR dyes are known.[25] Most NIR probes are used as primers with nonfluorescent dideoxy terminators. It is difficult and challenging to develop probes for use in DNA sequencing. The dyes must display similar intensities at a single excitation wavelength and must not alter too greatly the electrophoretic mobility of the labeled DNA fragments. Although it seems possible to have four red–NIR dyes for DNA sequencing, each with a distinct emission spectrum, this has not yet been accomplished.

21.1.D. DNA Sequencing Based on Lifetimes

It is difficult to obtain four dyes with similar absorption spectra and different emission spectra. Such dyes would allow determination of all four bases on a single gel column, which is highly desirable for more rapid sequencing. The use of decay times, instead of emission maxima, offers an alternative method of identifying the bases. An additional advantage of lifetime-based sequencing is that the decay times are mostly independent of intensity. If decay times are used to identify the bases, flourophores with overlapping emission spectra can be employed, possibly making it easier to identify suitable fluorophores.

Several groups have made progress toward lifetime-based sequencing.[26–29] The decay times for the initially proposed DNA sequencing dyes (Figure 21.3) have been measured in polyacrylamide gels under sequencing conditions.[28] These decay times are listed on Figure 21.3. Although the decay time is different for each dye, pulsed light sources at 488 and 514 nm are not practical for sequencing. The source would need to be an argon-ion laser, which is pulsed or modulated by an internal (mode-locker) or external (light modulator) device.

A set of lifetime DNA dyes excitable at 636 nm has been proposed (Figure 21.14).[29] The decay times range from 3.6 to 0.7 ns. However, some of these dyes are quenched when bound to oligonucleotides. Nonetheless, the possibility of obtaining different lifetimes suggests that continued research will lead to progress in lifetime-based sequencing. In fact, methods have been described for "on-the-fly" lifetime measurements of labeled DNA primers in capillary electrophoresis.[30,31] Lifetime measurements during capillary electrophoresis are likely to be used in lifetime-based sequencing.

21.2. HIGH-SENSITIVITY DNA STAINS

There are numerous applications which require detection of DNA and DNA fragments. One example is analysis of DNA fragments following digestion with restriction enzymes. Frequently, one wishes to know whether a DNA sample is from a particular individual. This does not require determination of the entire sequence and can be accomplished by examination of the DNA fragments formed by enzymatic degradation of DNA by restriction enzymes. A large number of restriction enzymes are known, each of which is specific for a particular base sequence, but they sometimes recognize more than one sequence. Generally, the enzymes are specific for relatively long sequences of four to nine base pairs, so that a relatively small number of DNA fragments are formed. The basic idea of a restriction fragment analysis is shown in Figure 21.15. The normal DNA has three restriction enzyme sites, and the mutant is missing one of these sites. Following digestion and electrophoresis, the mutant DNA shows one larger DNA fragment, whereas the normal DNA shows two smaller DNA fragments.

Typically, samples of DNA are examined using one or more restriction enzymes. The fragments are different for each individual owing to the sequence polymorphism that occurrs in the population. These different-size fragments are referred to as restriction fragment length polymorphisms (RFLPs). RFLPs do not represent mutations but rather the usual diversity in the gene pool. Following

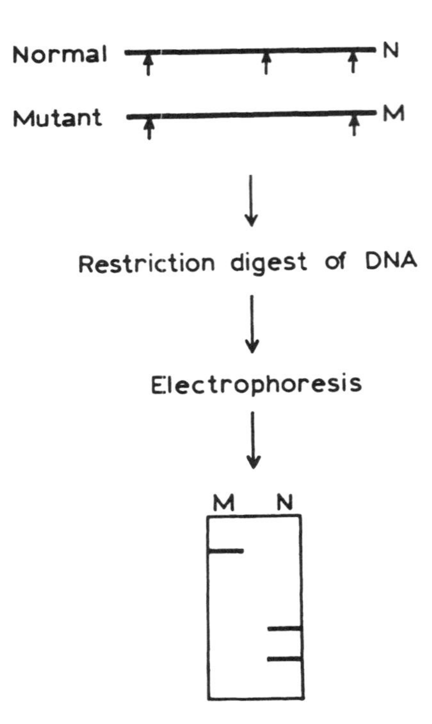

Figure 21.15. Analysis of DNA restriction fragments. The arrows indicate three cleavage sites in normal DNA (N), one of which is missing in the mutant DNA (M).

enzymatic digestion, the fragments are separated on agarose gels. Originally, the DNA was detected using ^{32}P and autoradiography. Today, detection is accomplished mostly by fluorescence.

Detection of DNA using stains has a long history, starting with staining of chromatin with acridine dyes. The situation was improved by the introduction of dyes such as ethidium bromide (EB) and propidium iodide, which fluoresce weakly in water and more strongly when bound to DNA.[32] DNA on gels has been detected by exposing the gels to EB; the gel typically contains micromolar concentrations of EB to ensure that the DNA contains significant amounts of EB. Although such dyes are useful for locating the DNA fragments, sensitivity can be low owing to the background from the free dyes.

21.2.A. High-Affinity Bis DNA Stains

During the past several years, a number of greatly improved dyes have appeared which have high affinity for DNA and exhibit almost no fluorescence in water. Many of these dyes are dimers of acridine or EB.[33,34] The ethidium dimer was found to bind DNA 10^3–10^4 times more strongly than the monomer.[34] Perhaps the most surprising observation was that the homodimer of EB (EthD in Figure 21.16) remained bound to DNA during electrophoresis.[35] This result is surprising because one would expect the positively charged dye to migrate in the opposite direction from the DNA. It was explained as due to slow dissociation of the dyes, such that they do not dissociate on the timescale of electrophoresis. The DNA fragments can thus be stained prior to electrophoresis. As a result, the

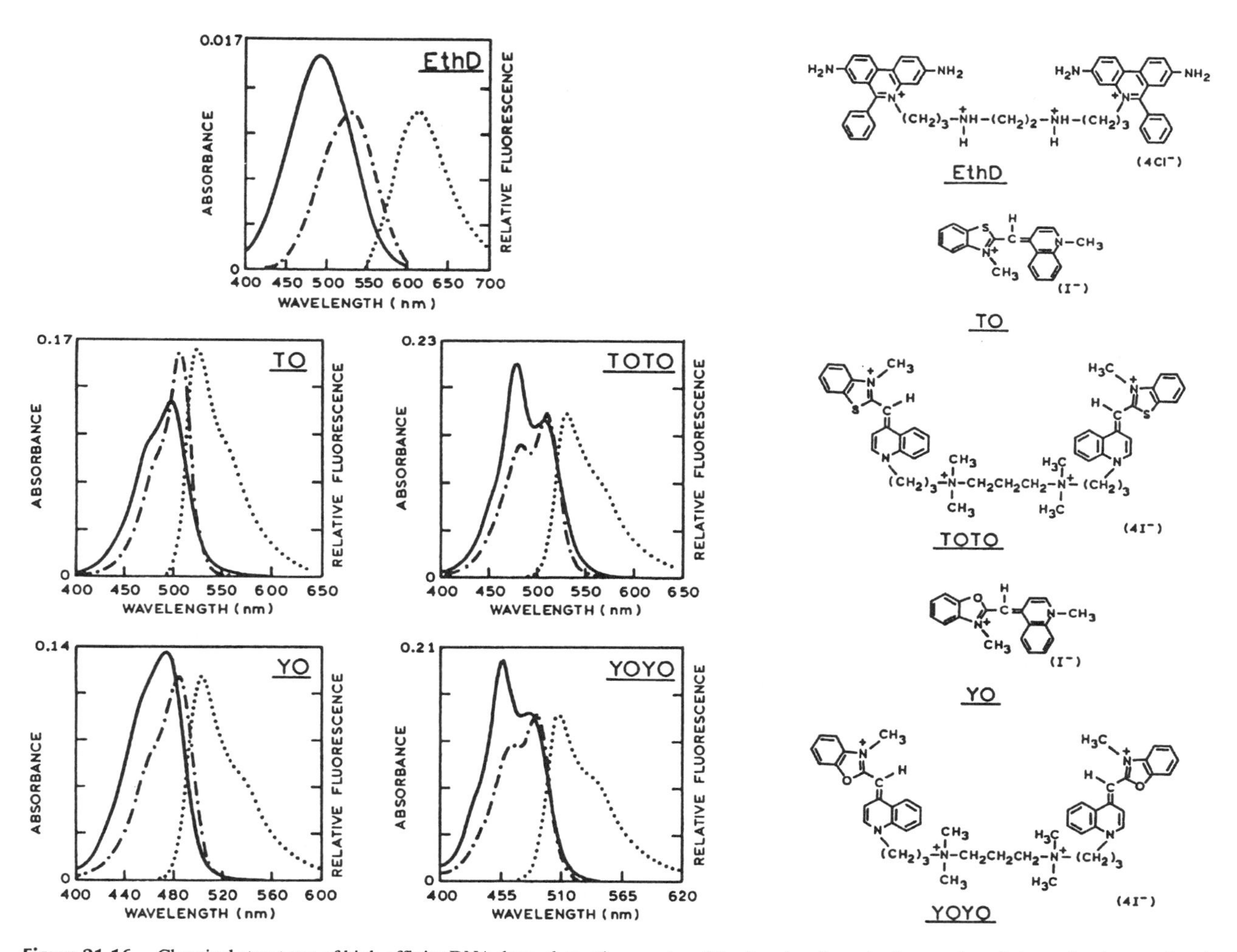

Figure 21.16. Chemical structures of high-affinity DNA dyes, absorption spectra of the free dyes in water (———), and absorption (— • —) and emission spectra (· · ·) of the dyes bound to DNA. The relative enhancements of the flourescence of the dyes on binding to DNA are (top to bottom) 35, 18,900, 1100, 700, and 3200. Revised and reprinted, with permission, from Ref. 36, Copyright © 1992, Oxford University Press.

gels display little background fluorescence, and the DNA fragments can be detected with high sensitivity.

Realization of the usefulness of the EB homodimer resulted in the continued development of DNA dyes with high affinity for DNA.[36–38] Several such structures are shown in Figure 21.16. These dyes are typically positively charged and display large enhancements in fluorescence upon binding to DNA. For example, the EB homodimer displays a 35-fold enhancement, whereas TOTO-1 displays a 1100-fold enhancement. The name TOTO is used to describe thiazole homodimers. The use of such dyes to prestain DNA fragments results in a 500-fold increase in sensitivity as compared to that obtained with gels stained with EB after electrophoresis.[36] These dyes are now widely used as DNA stains, and analogs with slightly longer excitation and emission wavelengths are also available. Additionally, different DNA samples can be stained with different dyes prior to electrophoresis. The dyes do not exchange between the DNA strands, allowing the samples to be identified from the spectral properties. This allows molecular weight standards to be electrophoresed in one lane of the gel with the unknown sample.

21.2.B. Energy-Transfer DNA Stains

While the bis dyes shown in Figure 21.16 display favorable properties, it would be desirable to have dyes excitable at 488 nm that exhibit large Stokes' shifts. Such dyes were created using donor–acceptor pairs.[39,40] One such dye is shown in Figure 21.17. The thiazole dye on the left (TO) serves as the donor for the thiazole-indolenine acceptor (TIN) in TOTIN on the right. Like the previous DNA stains, TOTIN remains bound to DNA during electrophoresis. The half-time for dissociation is 317 min.[40] TOTIN

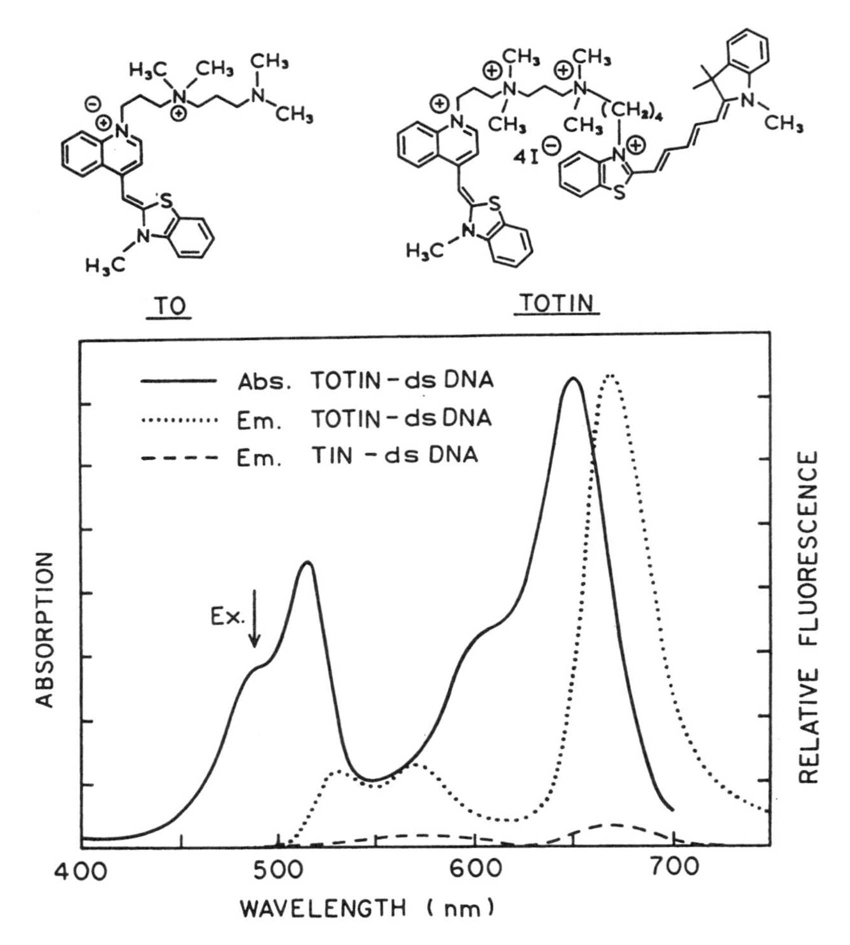

Figure 21.17. *Top*: Structures of Thiazole Orange (TO) and the Thiazole Orange–thiazole-indolenine heterodimer TOTIN, an energy-transfer dye for staining of DNA. *Bottom*: Absorption (———) and emission (· · ·) spectra of TOTIN and emission spectrum of TIN (– – –), the structure on the right side of TOTIN, bound to double-stranded DNA. Excitation was at 488 nm. Reprinted, with permission, from Ref. 40, Copyright © 1995, Academic Press, Inc.

can be excited at 488 nm and displays emission from the TIN moiety near 670 nm (dotted curve in Figure 21.17). For excitation at 488 nm, the emission from TIN alone is much weaker (dashed curve in Figure 21.17). TOTIN also allows red excitation at 630 nm with laser diode or HeNe sources.

21.2.C. DNA Fragment Sizing by Flow Cytometry

The development of high-affinity DNA stains allows a new method of DNA fragment sizing. At present, DNA fragment analysis is performed almost exclusively on slab or capillary gels. These methods are typically limited to fragment sizes up to 50 kilobases (kb) in length, and the size resolution is highly nonlinear.

Flow cytometry is a method in which cells flow one-by-one through an area illuminated by a laser beam. Information about the cells is obtained using fluorescent labels. The development of high-affinity DNA stains allows DNA fragment sizing using flow technology. The basic idea is that the amount of dye bound by the DNA fragments is proportional to the fragment length. Thus, longer DNA fragments bind more dye. The DNA fragments are analyzed in a flow cytometer modified for use with DNA.[41–43] This approach allows measurement of the size and number of DNA fragments, without physical separation of the fragments by chromatography or electrophoresis.

An example of DNA fragment sizing by flow cytometry is shown in Figure 21.18. In this case the DNA was from bacteriophage λ, which was digested with the HindIII restriction enzyme.[42] The DNA was stained with TOTO-1 and excited by an argon-ion laser at 514 nm. As the TOTO-1-stained DNA passes through the laser beam, the instrument records a histogram based on the size of the photon bursts (Figure 21.18, top). For this DNA–enzyme combination, the size of the DNA fragments was known. The photon burst size correlated precisely with fragment size (Figure 21.18, bottom). In other work, the photon burst size was found to be linear with fragment size up to 167 kb.[43] Given the expense and complexity of DNA gels, it seems probable that DNA analysis by flow cytometry will become more widely used in the near future.

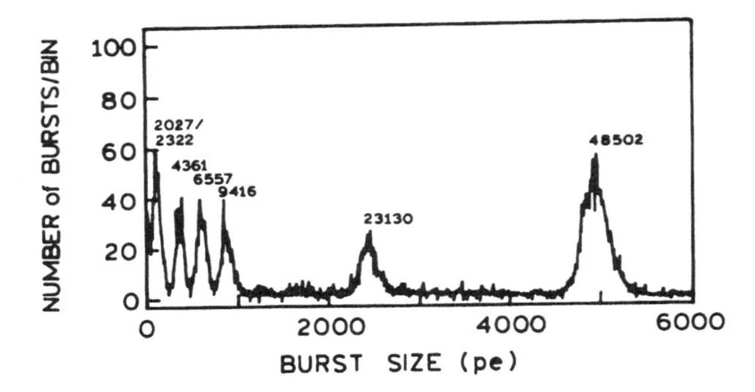

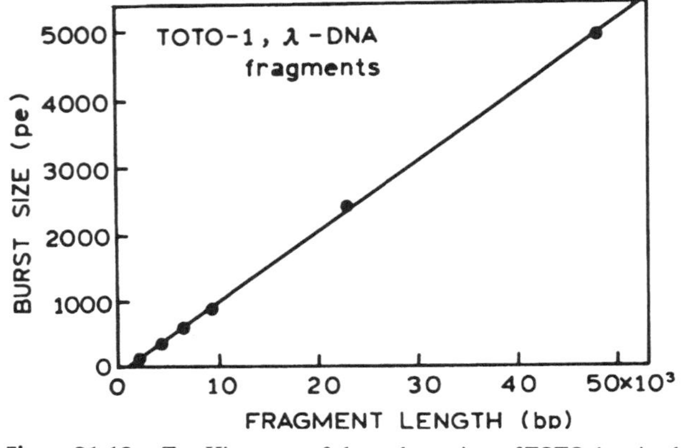

Figure 21.18. *Top*: Histogram of photon burst sizes of TOTO-1-stained DNA from a HindIII digest of λ DNA. *Bottom*: Correlation of the photon burst size with DNA fragment length. Excitation was at 514 nm from an argon-ion laser, and emission was observed through a 550-nm interference filter. Modified From Ref. 42.

21.3. DNA HYBRIDIZATION

Detection of DNA hybridization is widely useful in molecular biology, genetics, and forensics. Hybridization is the basis of the polymerase chain reaction (PCR) and fluorescence *in situ* hybridization. A variety of methods have been used to detect DNA hybridization by fluorescence. Most rely on energy transfer between donor- and acceptor-labeled DNA. Several possible methods are shown schematically in Figure 21.19. The presence of complementary DNA sequences can be detected by increased energy transfer when these sequences are brought into proximity by hybridization.[44–46] This can occur if the complementary strands are labeled with donors and acceptors (Figure 21.19a), as was shown in Figures 13.15 and 13.16. Energy transfer can also occur if the donor- and acceptor-labeled oligonucleotides bind to adjacent regions of a longer DNA sequence (Figure 21.19b). Hybridization can also be detected if a donor intercalates into the double-helical DNA and transfers energy to an acceptor-labeled oligonucleotide (Figure 21.19c). The use of an intercalating dye has been extended to include covalently attached intercalators, whose fluorescence increases in the presence of double-stranded DNA (Figure 21.19d).[47–49] One can also have competitive hybridization,[50] in which the presence of increased amounts of target DNA competes with formation of donor–acceptor pairs (Figure 21.19e). The acceptor can be fluorescent, or it can be nonfluorescent, in which case the donor appears to be quenched.

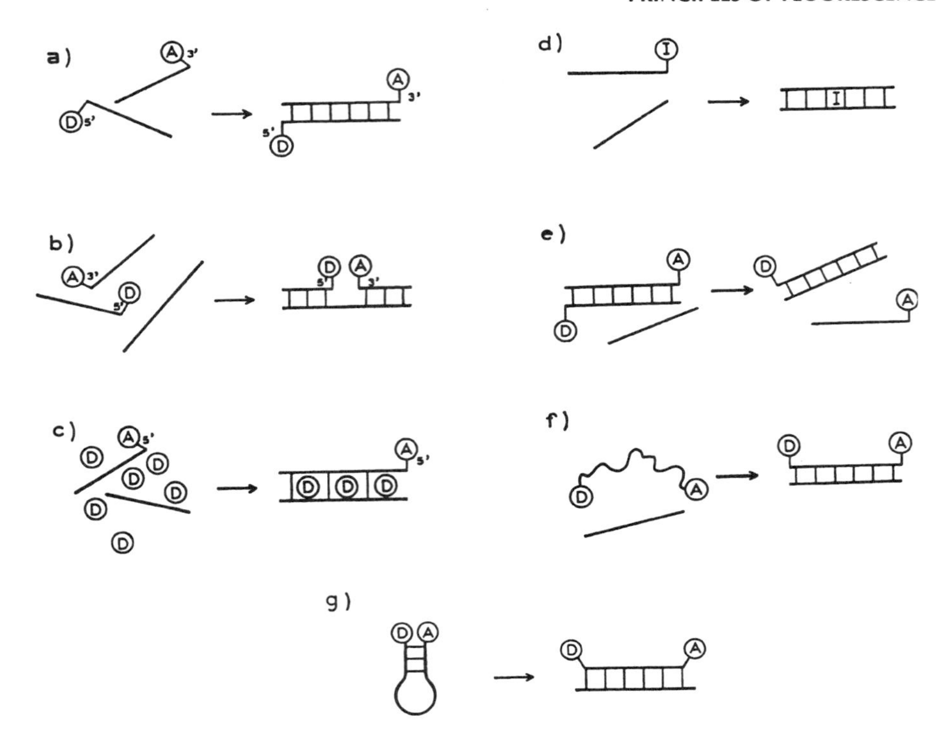

Figure 21.19. Methods to detect DNA hybridization by energy transfer. D, Donor; A, acceptor or quencher; I, intercalating dye. (a)–(e) See Section 21.3; (f) see Section 21.3A; (g) see Section 21.5. From Refs. 20, 45, and 48.

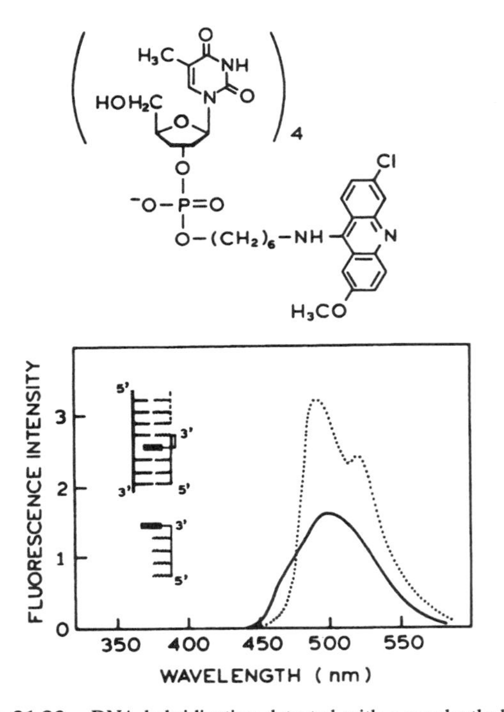

Figure 21.20. DNA hybridization detected with a covalently bound intercalating probe. ———, Flourescence intensity of the probe, an acridine dye covalently linked to the 3′-phosphate of an oligothymidylate; · · ·, flourescence intensity of the probe upon binding to a complementary adenine oligonucleotide. From Ref. 49.

Examples of the various types of hybridization assays can be found throughout the literature. A competitive assay (Figure 21.19e) was performed with complementary DNA strands in which the opposite strands were labeled with fluorescein and rhodamine.[50] Hybridization of the strands resulted in quenching of the donor fluorescence. Increasing amounts of unlabeled DNA, complementary to one of the labeled strands, resulted in displacement of the acceptor and increased donor fluorescence. Such assays can be useful in amplification reactions in which the DNA is thermally denatured during each cycle.

Intercalating DNA dyes have also been used to detect hybridization. One example is shown in Figure 21.20, in which the acridine dye is covalently linked to the 3′-phosphate of an oligothymidylate.[47,48] Upon binding to a complementary adenine oligonucleotide, the acridine fluorescence increases about twofold (Figure 21.20). It seems clear that this method can be extended to dyes which show greater enhancements upon hybridization.

21.3.A. DNA Hybridization Measured with One Donor- and Acceptor-Labeled DNA Probe

Most DNA hybridization methods (Figure 21.19) require two probe DNA molecules, one labeled with donor and the

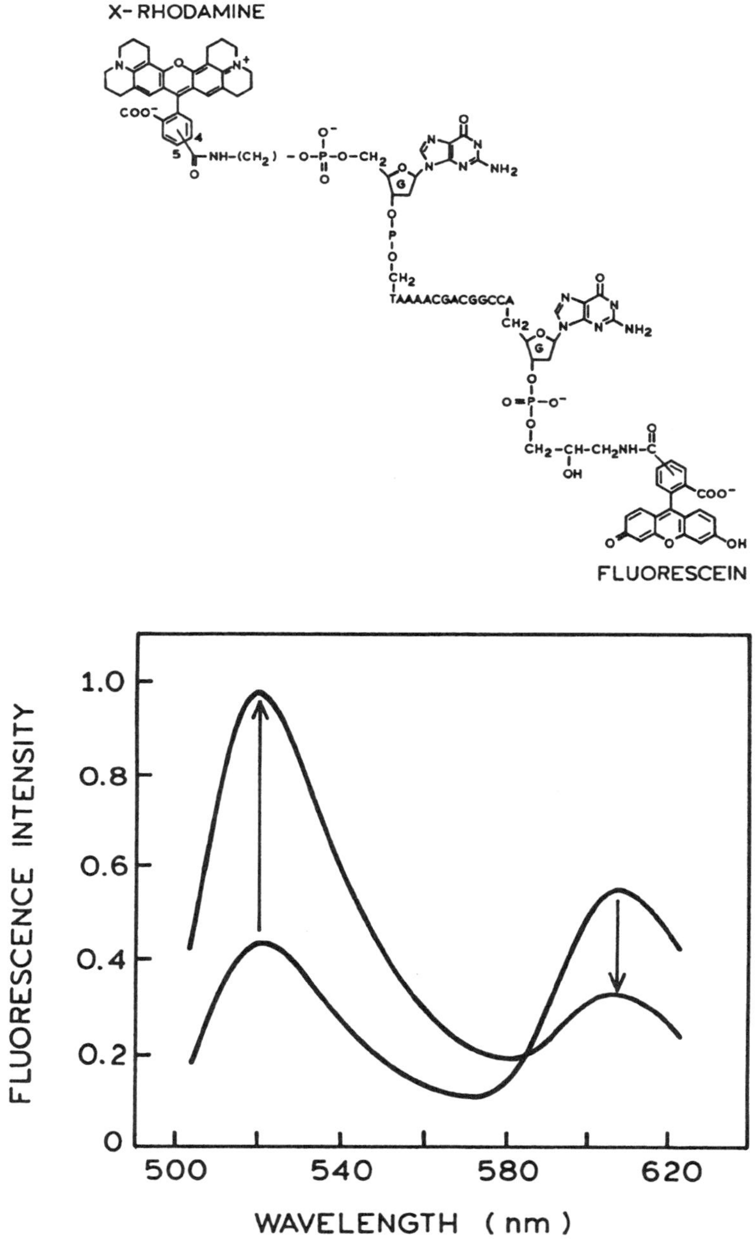

Figure 21.21. Detection of DNA hybridization with a single-donor and acceptor-labeled oligonucleotide. The increase in donor emission and decrease in acceptor emission upon binding of the oligonucleotide to its complementary strand is shown. From Ref. 51.

other with acceptor. It is possible to develop an assay based on a single donor- and acceptor-labeled probe DNA.[51] One example is shown in Figures 21.19f and 21.21, in which single-stranded DNA was labeled on opposite ends with a donor and an acceptor, respectively. In the absence of the complementary strand, the single-stranded probe DNA is flexible. This allows the donor- and acceptor-labeled ends to approach each other closely, resulting in a high RET efficiency. Upon binding of the single-stranded probe DNA to its complementary strand, the donor and acceptor become more distant owing to the greater rigidity of the double-stranded DNA. Hence, hybridization can be detected by an increase in donor emission and a decrease in acceptor emission. One can imagine many circumstances in which a RET assay would be simplified by the use of only a single probe molecule. The donor and acceptor concentrations are forced to remain the same, independent of sample manipulations, because the donor and acceptor are covalently linked. This allows the extent of hybridization to be determined using wavelength-ratiometric measurements.

21.3.B. DNA Hybridization Measured by Excimer Formation

In addition to energy transfer, pyrene excimer formation has also been used to detect DNA hybridization.[52] In this case DNA probes were synthesized with pyrene attached to the 5′ and 3′ ends (Figure 21.22). It is well known that one excited pyrene molecule can form an excited-state complex with another ground-state pyrene. This complex is called an excimer and displays an unstructured emission near 500 nm as compared to the structured emission of pyrene monomer near 400 nm.

The principle of the excimer-forming DNA hybridization assay is shown in Figure 21.23. The assay requires two DNA probes which bind to adjacent sequences on the target DNA. In this case the correct target DNA is a 32-mer oligonucleotide. If both the 5′- and 3′-pyrene probes bind to target DNA, the pyrene monomers are expected to be in close proximity, resulting in excimer emission. Emission spectra of a mixture of the 3′- and 5′-pyrene probes are shown in Figure 21.24. In the absence of target DNA, the emission is near 400 nm and characteristic of pyrene monomer. Titration with increasing amounts of target DNA results in increasing emission from the excimer near 500 nm.

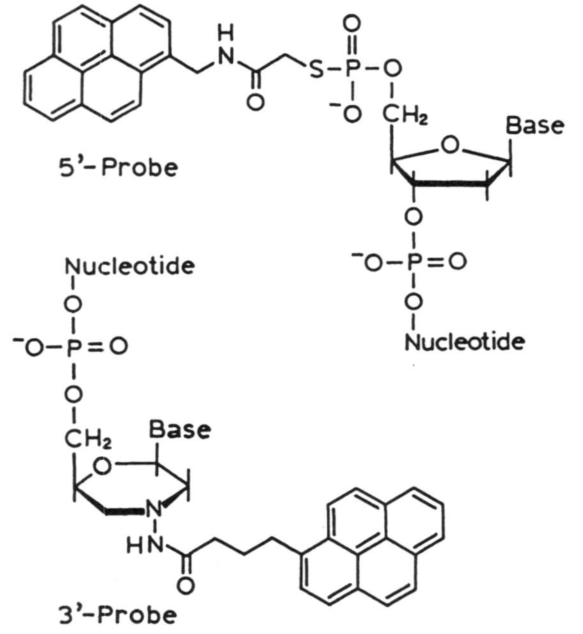

Figure 21.22. Pyrene-labeled oligonucleotide probes. Modified from Ref. 52.

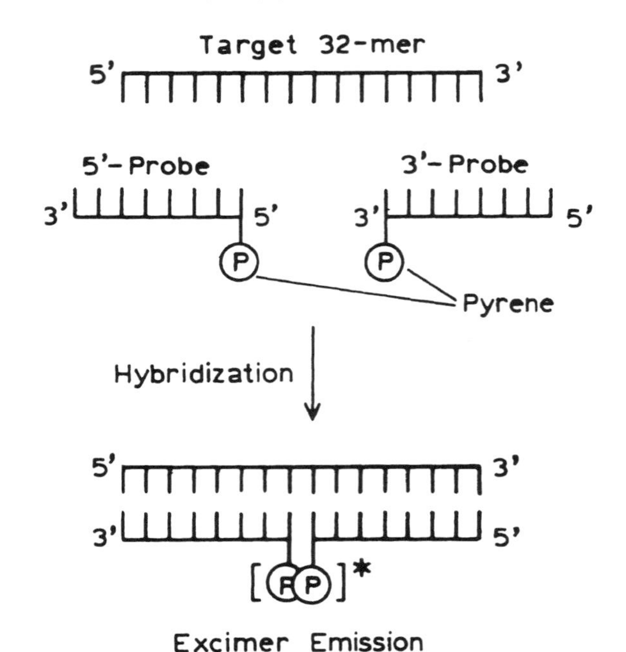

Figure 21.23. Principle of the excimer-forming DNA hybridization assay. Modified from Ref. 52.

This excimer hybridization assay is remarkably sensitive to precise matching of the target sequence with the probe sequence. Just one extra thymine residue in the target DNA, between the pyrene sites on the probe DNA, eliminates most of the excimer emission. This property of the assay distinguishes it from a hybridization assay based on RET. In the case of RET, the donor–acceptor interaction occurs over long distances, so that the additional distance of one base would not abolish RET. In contrast, excimer formation is a short-range interaction which requires molecular contact between the pyrene monomers. For this reason, it is sensitive to small changes in the pyrene-to-pyrene distance.

21.3.C. Polarization Hybridization Assays

DNA hybridization has also been detected using fluorescence anisotropy. The basic idea is to measure the increase in anisotropy when labeled DNA binds to its complementary strand.[53–55] These assays are analogous to the fluorescence polarization immunoassays. Polarization or anisotropy measurements have the favorable property of being independent of the intensity of the signal and dependent on the molecular weight of the labeled molecule.

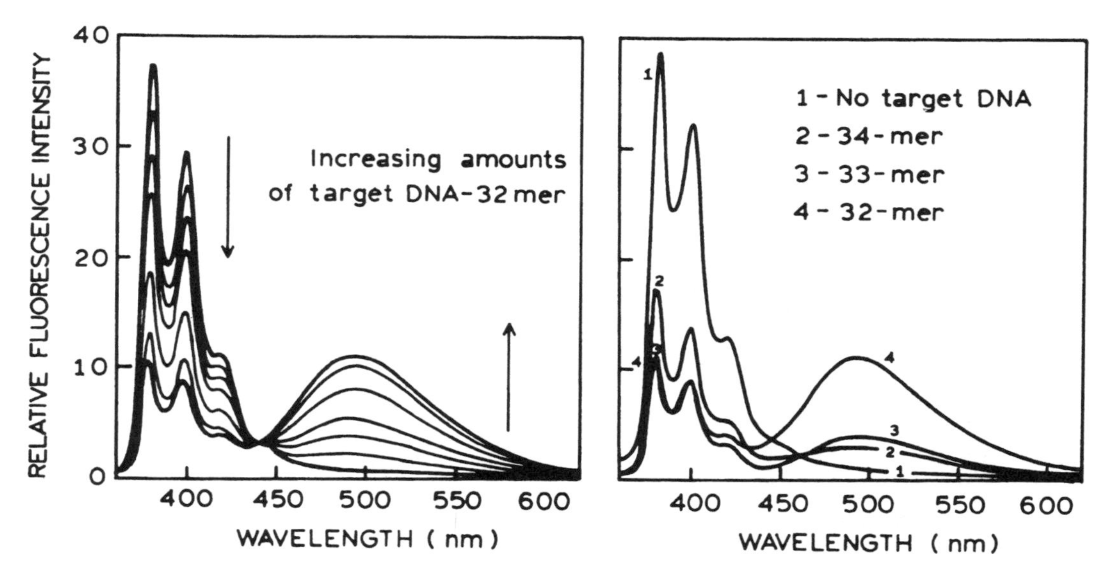

Figure 21.24. Effect of target DNA (32-mer) and mismatched target DNA (33-mer and 34-mer) on the emission from DNA probes labeled with pyrene at the 3′ and 5′ ends. The target DNA 32-mer and the mismatched target DNA have the sequence 5′-AGAGGGCACG-GATACC*GCGAGGTGGAGCGAAT-3′, where the asterisk designates the location of one or two extra thymine residues in the 33-mer and 34-mer, respectively. Modified from Ref. 52.

Also, polarization measurements do not require separation steps.

DNA hybridization assays based on polarization have been reported using the usual fluorescein probes,[54,55] as well as a more novel NIR dye.[53] The structure of the dye, La Jolla Blue™, is shown in Figure 21.25. The central chromophore is a phthalocyanine, which displays the favorable property of absorbing in the NIR and in this case was excited by a pulsed laser diode at 685 nm. The phthalocyanines are poorly soluble in water, and hence the central silicon atom was conjugated to polar groups to increase the water solubility and prevent aggregation.

Polarization values of the La Jolla Blue oligonucleotide are shown in Figure 21.26. The dye-DNA probe was mixed with either complementary or noncomplementary DNA. The polarization increased upon mixing with the complementary strand, but not with the control samples. This

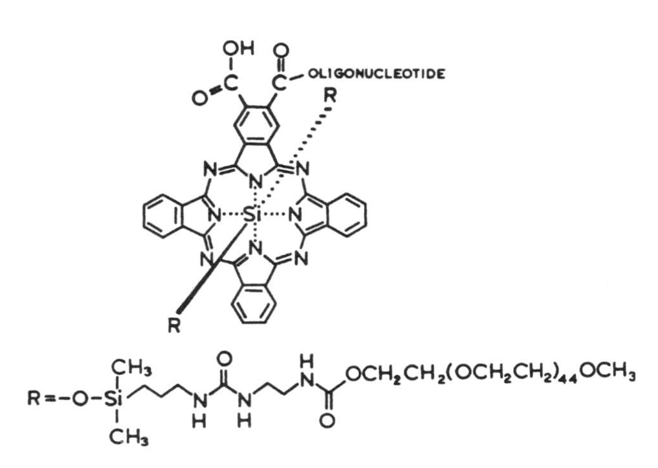

Figure 21.25. Structure of the La Jolla Blue™ oligonucleotide. One mP is equivalent to 0.0001 polarization units. Revised and reprinted, with permission, from Ref. 53, Copyright © 1993, American Association for Clinical Chemistry.

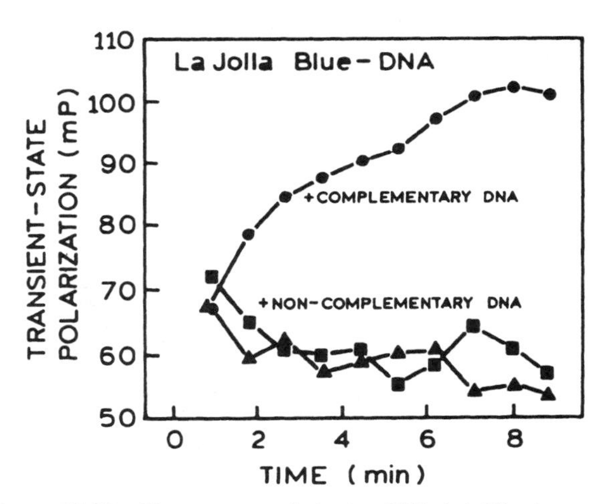

Figure 21.26. Fluorescence polarization DNA hybridization assay. The probe DNA was mixed with complementary (•) or noncomplementary DNA (▲, ■). The excitation source was a pulsed laser diode at 685 nm. The emission at 705 nm was detected after the excitation pulse. Modified from Ref. 53.

result suggests that polarization measurements can be used to monitor the production of complementary DNA by PCR and related amplification methods. A unique aspect of the data in Figure 21.26 is the use of pulsed excitation and gated detection after the excitation pulse. This was done to avoid detection of scattered light and/or background fluorescence from the sample.

21.3.D. Polymerase Chain Reaction

Donor- and acceptor-labeled oligonucleotides, such as that shown in Figure 21.21, have been used to assay the polymerase chain reaction. One such assay is shown in Figure 21.27. The sample initially contains a donor- and acceptor-labeled oligonucleotide, in which the donor fluorescence is quenched.[56,57] During the PCR reaction, this D–A strand is displaced and cleaved by DNA polymerase, which displays some nuclease activity. Upon cleavage of the D–A pair, the donor becomes distant from the acceptor and thus more fluorescent. PCR assays based on fluorogenic D–A pairs are presently used in commercial instruments. The oligonucleotide sequence in the D–A pair is complementary to a portion of the DNA to be amplified. This type of assay can be regarded as an extension of the work on fluorogenic probes described in Chapter 3, wherein the molecule becomes more fluorescent as the result of enzymatic cleavage. Since oligonucleotides are available with NIR dyes excitable at 770 nm,[58] one can expect to see the concept of fluorogenic PCR substrates extended to longer excitation and emission wavelengths. PCR can also be quantified by dyes which bind to double-stranded DNA and by RET.[59]

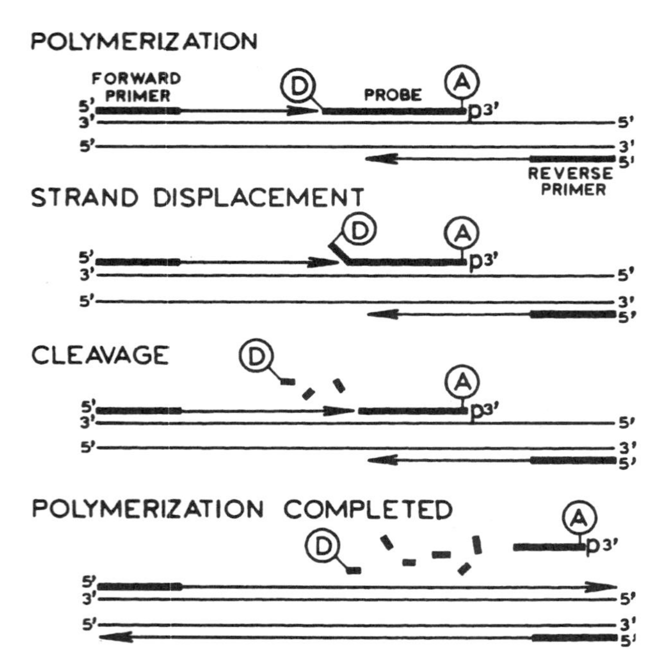

Figure 21.27. Release of donor quenching during polymerase chain reaction. From Ref. 56.

21.3.E. DNA Imagers

The high-sensitivity detection provided by the current generation of DNA stains has resulted in development of a new class of instruments dedicated to imaging the gels containing stained DNA. A schematic of a typical imaging fluorometer is shown in Figure 21.28. A typical imager uses an argon-ion laser operating at 488 or 514 nm. The excitation source and detector are scanned across the gel. The emission is recorded through an emission filter. With the increasing availability of NIR DNA probes,[20,58] one can expect future DNA imagers to be based on laser diode excitation sources.

21.3.F. Light-Generated DNA Probe Arrays

As a final example of DNA hybridization, it is important to mention the new possibility of DNA arrays created by a combination of photochemistry and photolithography. It is now possible to develop DNA "chips" which contain up to 15,000 different DNA sequences.[60–62] This possibility is illustrated in Figure 21.29, which is a schematic of the synthesis of tetranucleotides by photolithographic methods. The surface can be activated by exposure to light, allowing the desired nucleotide to be synthesized at each position on the array. These chips can be incubated with the test DNA which has been labeled with fluorescent

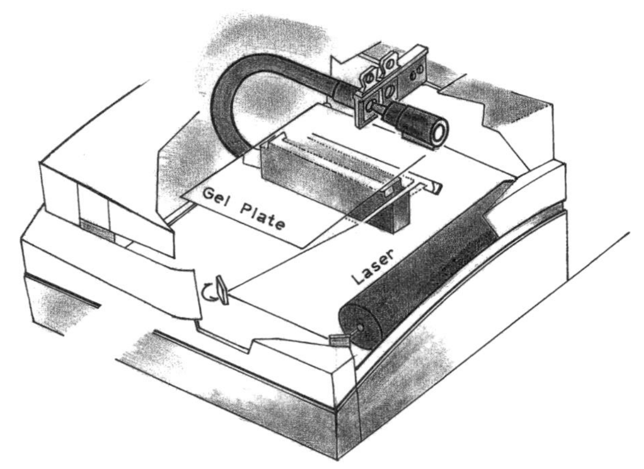

Figure 21.28. Imaging fluorometer for quantitation of stained gels. Redrawn with permission from Molecular Dynamics.

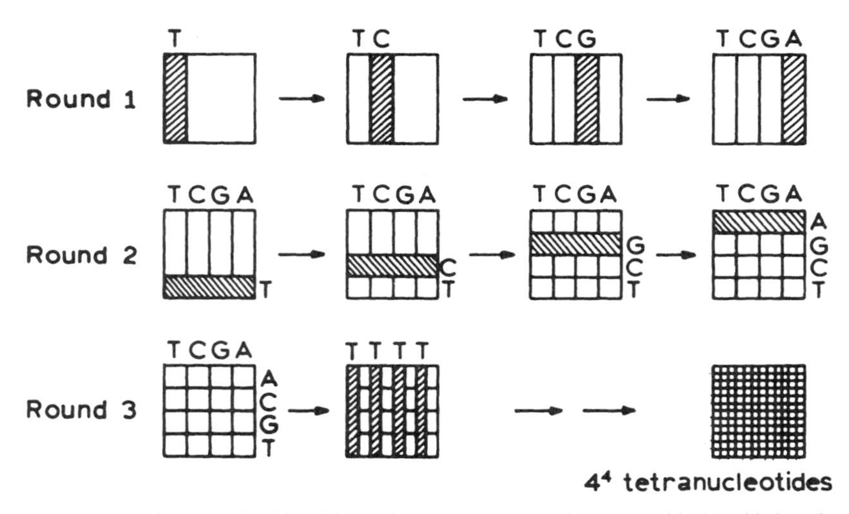

Figure 21.29. Light-generated array of oligonucleotides. The surface is light-activated to react with the added nucleotides, allowing a unique sequence to the synthesized at each position in the array. From Ref. 60.

probes. The presence of certain sequences can be correlated with disease states. For instance, the gene for Duchenne muscular dystrophy is very large, over 250 kilobases long, and more than 500 mutations are known.[61] Hence, it is not practical to sequence the genes or search for each of the possible mutations. Using a DNA chip, one can potentially scan for numerous mutations in a single test. The desired regions of DNA would be amplified by PCR using fluorescent oligonucleotides. Then, if a sequence that is complementary to a sequence on the DNA chip occurs, a bright spot on the array would be detected. These chips could be read with either scanning optics or CCD cameras.

21.4. FLUORESCENCE *IN SITU* HYBRIDIZATION

Fluorescence *in situ* hybridization (FISH) has become widely used in the medical sciences. The basic idea is shown in Figure 21.30. The DNA to be tested, typically metaphase chromosomes, is exposed to probe DNA. The exposure conditions result in denaturation of the chromosomes and hybridization of the chromosomes with the probe DNA. The probe DNA has a base sequence directed toward one or more chromosomes. The probe DNA can be specific for the centromeric or telomeric region of chromosomes, or it can be directed toward the entire chromosome. DNA probes can also be specific for small regions of DNA representing one or several genes.

In situ hybridization has been detected using radioactive tracers and autoradiography. A more complete description can be found in many texts (see, e.g., Ref. 63). For fluorescence *in situ* hybridization, the probe DNA is made as fluorescent as possible by labeling with fluorophores that do not interfere with hybridization. One commonly used fluorescent base is shown in Figure 21.31. The reactive amino group attached to the base can be coupled to any of a wide variety of fluorescent probes. The highest sensitivity has typically been found using the rhodamine dyes.[64] This is because detection of a particular gene sequence requires high-sensitivity detection, and rhodamines are usually more photostable than fluoresceins. DNA oligomers can also be labeled by any of the methods shown in Figure 21.6. A remarkable aspect of the FISH technology is that fluorophores can be incorporated into DNA without disruption of the base pairing. DNA can be synthesized with relatively high levels of fluorescent nucleotides.

The power of FISH is illustrated by its ability to identify all 24 human chromosomes.[65,66] It is possible to develop DNA probes which impart a unique color to each of the human chromosomes. Chromosome-painting probes are developed by PCR amplification of DNA fragments from

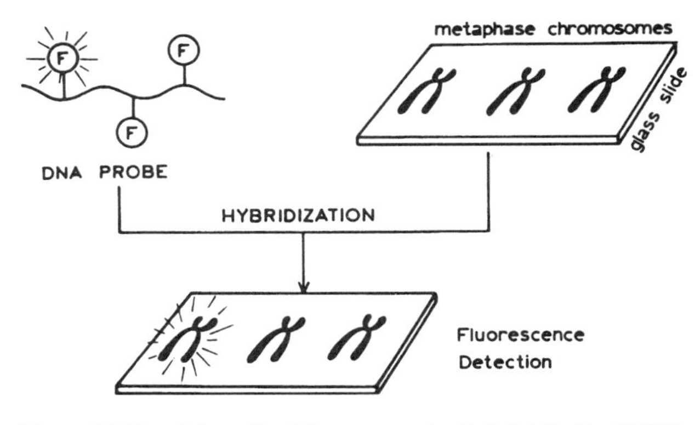

Figure 21.30. Schematic of fluorescence *in situ* hybridization (FISH).

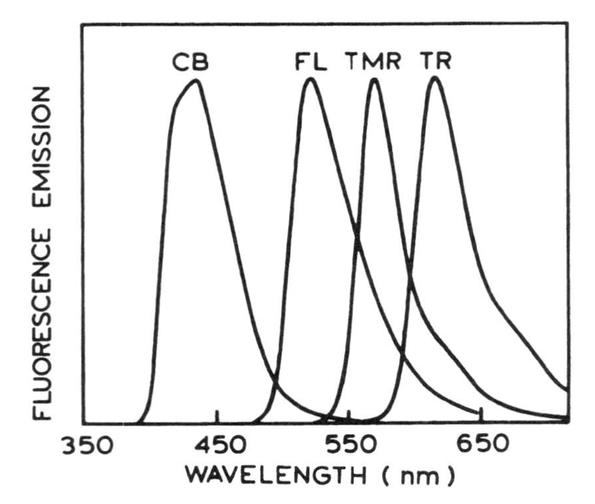

Figure 21.31. Emission spectra of fluorescent deoxynucleotides for incorporation into DNA FISH probes. The fluorophore can be fluorescein (FL) (*top*) or one of a variety of other fluorophores, such as tetramethylrhodamine (TMR), Texas Red (TR), or Cascade Blue (CB). From Molecular Probes literature.

chromosomes that have been previously sorted by flow cytometry. Nonspecific binding is prevented by adding additional DNA which binds to the most common sequences. However, the potential of FISH for chromosome screening can only be realized if all the chromosomes can be simultaneously visualized. Unfortunately, 24 nonoverlapping fluorophores are not available.

A variety of methods have been developed for identifying the chromosomes, two of which are multiplex FISH (M-FISH) and spectral karyotyping (SKY). Most methods include the use of multiple DNA probes, each containing a different fluorophore.[67–69] Differences in the chromosomes are revealed by the different colors or ratio of probes bound to each region of the chromosome. The most spectacular advances in FISH have resulted from characterization of the emission spectra,[67,68] using either an interferometric method[67] or interference filters.[68] By using combinations of just five colors for the chromosome paints, it was possible to identify each of the 24 chromosomes. These were then assigned easily recognized pseudo colors. These advanced FISH methods promise to provide a means of detecting chromosome abnormalities and chromosome rearrangements and of monitoring bone marrow cells following transplantation and cancer chemotherapy.[70–75]

FISH has many other important applications besides identifying chromosomes. With the use of appropriate DNA sequences, FISH can be used to detect the human papillomavirus, considered to be responsible for cervical cancer,[76] and to determine fetal sex from amniotic fluid.[77] The applications of FISH rely on methods for labeling DNA, computerized imaging, and high-sensitivity CCD detection. FISH technology represents a combination of modern optics, molecular biology, and fluorescence spectroscopy and promises to become a central tool in molecular medicine.

21.5. PERSPECTIVES

The use of fluorescence methods in DNA analysis is perhaps the most rapidly growing area of fluorescence spectroscopy. New methods for sequencing, fragment analysis, PCR, and FISH are appearing at a remarkable pace. Also, many additional methods not mentioned in this chapter have been used to amplify, label, and detect DNA.[78–81] As just one example, we note the introduction of "molecular beacons." These "beacons" are nonfluorescent until bound to the complementary strand of DNA.[82,83] Donor quenching occurs by RET to a nonfluorescent acceptor. Prior to annealing with target DNA, the donor and acceptor are closely spaced due to their location on complementary arms of the hairpin stem (Figure 21.19g). Upon binding to the complementary strand of DNA, the donor and acceptor become widely spaced, and the donor fluorescence increases. This type of DNA probe, with either fluorescein or rhodamine donors, is now being used for genotyping of human alleles. No one chapter could summarize these extensive applications, and no chapter on DNA technology can remain up-to-date for long. We have attempted to provide an overview of DNA technology from the viewpoint of the spectroscopist.

REFERENCES

1. Maxam, A. M., and Gilbert, W., 1977, A new method for sequencing DNA, *Proc. Natl. Acad. Sci. U.S.A.* **74**:560–564.
2. Maxam, A. M., and Gilbert, W., 1980, Sequencing end-labeled DNA with base-specific chemical cleavage, *Methods Enzymol.* **65**:499–560.

3. Sanger, F., Nicklen, S., and Coulson, A. R., 1977, DNA sequencing with chain-terminating inhibitors, *Proc. Natl. Acad. Sci. U.S.A.* **74**:5463–5467.
4. Watson, J. D., Gilman, M., Witkowski, J., and Zoller, M., 1992, *Recombinant DNA*, 2nd ed., Scientific American Books, New York.
5. Smith, L. M., Sanders, J. Z., Kaiser, R. J., Hughes, P., Dodd, C., Connell, C. R., Heiner, C., Kent. S. B. H., and Hood, L. E., 1986, Fluorescence detection in automated DNA sequence analysis, *Nature* **321**:674–679.
6. Prober, J. M., Trainor, G. L., Dam, R. J., Hobbs, F. W., Robertson, C. W., Zagursky, R. J., Cocuzza, A. J., Jensen, M. A., and Baumeister, K., 1987, A system for rapid DNA sequencing with fluorescent chain-terminating dideoxynucleotides, *Science* **238**:336–343.
7. Ansorge, W., Sproat, B. S., Stegemann, J., and Schwager, C., 1986, A non-radioactive automated method for DNA sequence determination, *J. Biochem. Biophys. Methods* **13**:315–323.
8. Flick, P. K., 1995, DNA sequencing by nonisotopic methods, in *Nonisotopic Probing, Blotting, and Sequencing*, J. J. Kricka (ed.), Academic Press, New York, pp. 475–492.
9. Dhadwal, H. S., Quesada, M. A., and Studier, F. W., 1997, DNA sequencing by multiple capillaries that form a waveguide, *Proc. SPIE* **2890**:149–162.
10. Hunkapiller, T., Kaiser, R. J., Koop, B. F., and Hood, L., 1991, Large-scale and automated DNA sequence determination, *Science* **254**:59–67.
11. Zimmermann, J., Wiemann, S., Voss, H., Schwager, C., and Ansorge, W., 1994, Improved fluorescent cycle sequencing protocol allows reading nearly 1000 bases, *Biotechniques* **17**(2):302–307.
12. Ansorge, W., Sproat, B., Stegermann, J., Schwager, C., and Zenke, M., 1987, Automated DNA sequencing: Ultrasensitive detection of fluorescent bands during electrophoresis, *Nucleic Acids Res.* **15**:4593–4602.
13. Brumbaugh, J. A., Middendorf, L. R., Grone, D. L., and Ruth, J. L., 1988, Continuous on-line DNA sequencing using oligodeoxynucleotide primers with multiple fluorophores, *Proc. Natl. Acad. Sci U.S.A.* **85**:5610–5614.
14. Ju, J., Ruan, C., Fuller, C. W., Glazer, A. N., and Mathies, R. A., 1995, Fluorescence energy transfer dye-labeled primers for DNA sequencing and analysis, *Proc. Natl. Acad. Sci, U.S.A.* **92**:4347–4351.
15. Ju, J., Glazer, A. N., and Mathies, R. A., 1996, Energy transfer primers: A new fluorescence labeling paradigm for DNA sequencing and analysis, *Nat. Med.* **2**(2):246–249.
16. Takahashi, S., Murakami, K., Anazawa, T., and Kambara, H., 1994, Multiple sheath-flow gel capillary-array electrophoresis for multicolor fluorescent DNA detection, *Anal. Chem.* **66**:1021–1026.
17. Ju, J., Kheterpal, I., Scherer, J. R., Ruan, C., Fuller, C. W., Glazer, A. N., and Mathies, R. A., 1995, Design and synthesis of fluorescence energy transfer dye-labeled primers and their application for DNA sequencing and analysis, *Anal. Biochem.* **231**:131–140.
18. Metzker, M. L., Lu, J., and Gibbs, R. A., 1996, Electrophoretically uniform fluorescent dyes for automated DNA sequencing, *Science* **271**:1420–1422.
19. Middendorf, L. R., Bruce, J. C., Bruce, R. C., Eckles, R. D., Roemer, S. C., and Sloniker, G. D., 1993, A versatile infrared laser scanner/electrophoresis apparatus, *Proc. SPIE* **1885**:423–434.
20. Soper, S. A., Flanagan, J. H., Legendre, B. L., Williams, D. C., and Hammer, R. P., 1996, Near-infrared, laser-induced fluorescence detection for DNA sequencing applications, *IEEE J. Sel. Top. Quantum Electron.* **2**(4):1–11.
21. Shealy, D. B., Lipowska, M., Lipowski, J., Narayanan, N., Sutter, S., Strekowski, L., and Patonay, G., 1995, Synthesis, chromatographic separation, and characterization of near-infrared labeled DNA oligomers for use in DNA sequencing, *Anal. Chem.* **67**:247–251.
22. Williams, D. C., and Soper, S. A., 1995, Ultrasensitive near-IR fluorescence detection for capillary gel electrophoresis and DNA sequencing applications, *Anal. Chem.* **67**:3427–3432.
23. Middendorf, L., Amen, J., Bruce, B., Draney, D., DeGraff, D., Gewecke, J., Grone, D., Humphrey, P., Little, G., Lugade, A., Narayanan, N., Oommen, A., Osterman, H., Peterson, R., Rada, J., Raghavachari, R., and Roemer, S., 1998, Near-infrared fluorescence instrumentation for DNA analysis, S. Daehne et al. (eds.), *Near-Infrared Dyes for High Technology Applications*, 21–54. © 1998 Kluwer Academic Publishers. Printed in the Netherlands.
24. Middendorf, L. R., Bruce, J. C., Bruce, R. C., Eckles, R. D., Grone, D. L., Roemer, S. C., Sloniker, G. D., Steffens, D. L., Sutter, S. L., Brumbaugh, J. A., and Patonay, G., 1992, Continuous, on-line DNA sequencing using a versatile infrared laser scanner/electrophoresis apparatus, *Electrophoresis* **13**:487–494.
25. Middendorf, L., Bruce, R., Brumbaugh, J., Grone, D., Jang, G., Richterich, P., Holtke, H. J., Williams, R. J., and Peralta, J. M., 1995, A two-dimensional infrared fluorescence scanner used for DNA analysis, *Proc. SPIE* **2388**:44–54.
26. Han, K.-T., Sauer, M., Schulz, A., Seeger, S., and Wolfrum, J., 1993, Time-resolved fluorescence studies of labelled nucleosides, *Ber. Bunsenges. Phys. Chem.* **97**:1728–1730.
27. Legendre, B. L., Williams, D. C., Soper, S. A., Erdmann, R., Ortmann, U., and Enderlein, J., 1996, An all solid-state near-infrared time-correlated single photon counting instrument for dynamic lifetime measurements in DNA sequencing applications, *Rev. Sci. Instrum.* **67**:3984–3989.
28. Chang, K., and Force, R. K., 1993, Time-resolved laser-induced fluorescence study on dyes used in DNA sequencing, *Appl. Spectrosc.* **47**:24–29.
29. Sauer, M., Han, K.-T., Ebert, V., Müller, R. Schulz, A., Seeger, S., and Wolfrum, J., 1994, Design of multiplex dyes for the detection of different biomolecules, *Proc. SPIE* **2137**:762–774.
30. Li, L.-C., He, H., Nunnally, B. K., and McGown, L. B., 1997, On-the-fly fluorescence lifetime detection of labeled DNA primers, *J. Chromatogr.* **695**:85–92.
31. Li, L.-C., and McGown, L. B., 1996, On-the-fly frequency-domain fluorescence lifetime detection in capillary electrophoresis, *Anal. Chem.* **68**:2737–2743.
32. Le Pecq, J.-B., and Paoletti, C., 1967, A fluorescent complex between ethidium bromide and nucleic acids, *J. Mol. Biol.* **27**:87–106.
33. Le Pecq, J.-B., Le Bret, M., Barbet, J, and Roques, B., 1975, DNA polyintercalating drugs: DNA binding of diacridine derivatives, *Proc. Natl. Acad. Sci. U.S.A.* **72**:2915–2919.
34. Markovits, J., Roques, B. P., and Le Pecq, J.B., 1979, Ethidium dimer: A new reagent for the fluorimetric determination of nucleic acids, *Anal. Biochem.* **94**:259–264.
35. Glazer, A. N., Peck, K., and Mathies, R. A., 1990, A stable double-stranded DNA ethidium homodimer complex: Application to picogram fluorescence detection of DNA in agarose gels, *Proc. Natl. Acad. Sci. U.S.A.* **87**:3851–3855.
36. Rye, H. S., Yue, S., Wemmer, D. E., Quesada, M. A., Haugland, R. P., Mathies, R. A., and Glazer, A. N., 1992, Stable fluorescent complexes of double-stranded DNA with bis-intercalating asymmetric cyanine dyes: Properties and applications, *Nucleic Acids Res.* **20**:2803–2812.
37. Abramo, K. H., Pitner, J. B., and McGown, L. B., 1997, Spectroscopic studies of single-stranded DNA ligands and oxazole yellow dyes, *Biospectroscopy* **4**:27–35.

38. Nygren, J., Svanvik, N., and Kubista, M., 1998, The interactions between the fluorescent dye thiazole orange and DNA, *Biopolymers* **46**:39–51.
39. Benson, S. C., Mathies, R. A., and Glazer, A. N., 1993, Heterodimeric DNA-binding dyes designed for energy transfer: Stability and applications of the DNA complexes, *Nucleic Acids Res.* **21**:5720–5726.
40. Benson, S. C., Zeng, Z., and Glazer, A. N., 1995, Fluorescence energy-transfer cyanine heterodimers with high affinity for double-stranded DNA, *Anal. Biochem.* **231**:247–255.
41. Goodwin, P. M., Johnson, M. E., Martin, J. C., Ambrose, W. P., Marrone, B. L., Jett, J. H., and Keller, R. A., 1993, Rapid sizing of individual fluorescently stained DNA fragments by flow cytometry, *Nucleic Acids Res.* **21**:803–806.
42. Petty, J. T., Johnson, M. E., Goodwin, P. M., Martin, J. C., Jett, J. H., and Keller, R. A., 1995, Characterization of DNA size determination of small fragments by flow cytometry, *Anal. Chem.* **67**:1755–1761.
43. Huang, Z., Petty, J. T., O'Quinn, B., Longmire, J. L., Brown, N. C., Jett, J. H., and Keller, R. A., 1996, Large DNA fragment sizing by flow cytometry: Application to the characterization of P1 artificial chromosome (PAC) clones, *Nucleic Acids Res.* **24**:4202–4209.
44. Cardullo, R. A., Agrawal, S., Flores, C., Zamecnik, P. C., and Wolf, D. E., 1988, Detection of nucleic acid hybridization by nonradiative fluorescence resonance energy transfer, *Proc. Natl. Acad. Sci. U.S.A.* **85**:8790–8794.
45. Morrison, L. E., and Stols, L. M., 1993, Sensitive fluorescence-based thermodynamic and kinetic measurements of DNA hybridization in solution, *Biochemistry* **32**:3095–3104.
46. Morrison, L. E., 1995, Detection of energy transfer and fluorescence quenching, in *Nonisotopic Probing, Blotting, and Sequencing*, L. J. Kricka (ed.), Academic Press, New York, pp. 429–471.
47. Asseline, U., Toulme, F., Thuong, N. T., Delarue, M., Montenay-Garestier, T., and Helene, C., 1984, Oligodeoxynucleotides covalently linked to intercalating dyes as base sequence-specific ligands. Influence of dye attachment site, *EMBO J.* **3**:795–800.
48. Asseline, U., Delarue, M., Lancelot, G., Toulme, F., Thuong, N. T., Montenay-Garestier, T., and Helene, C., 1984, Nucleic acid-binding molecules with high affinity and base sequence specificity: Intercalating agents covalently linked to oligodeoxynucleotides, *Proc. Natl. Acad. Sci. U.S.A.* **81**:3297–3301.
49. Hélène, C., Montenay-Garestier, T., Saison, T., Takasugi, M., Tolumé, Asseline, U., Lancelot, G., Maurizot, J. C., Tolumé, F., and Thuong, N. T., 1985, Oligodeoxynucleotides covalently linked to intercalating agents: A new class of gene regulatory substances, *Biochime* **67**:777–783.
50. Morrison, L. E., Halder, T. C., and Stols, L. M., 1989, Solution-phase detection of polynucleotides using interacting fluorescent labels and competitive hybridization, *Anal. Biochem.* **188**:231–244.
51. Parkhurst, K. M., and Parkhurst, L. J., 1996, Detection of point mutations in DNA by fluorescence energy transfer, *J. Biomed. Opt.* **1**:435–441.
52. Ebata, K., Masuko, M., Ohtani, H., and Kashiwasake-Jibu, M., 1995, Nucleic acid hybridization accompanied with excimer formation from two pyrene-labeled probes, *Photochem. Photobiol.* **62**:836–839.
53. Devlin, R., Studholme, R. M., Dandliker, W. B., Fahy, E., Blumeyer, K., and Ghosh, S. S., 1993, Homogeneous detection of nucleic acids by transient state polarized fluorescence, *Clin. Chem.* **39**:1939–1943.
54. Murakami, A., Nakaura, M., Nakatsuji, Y., Nagahara, S., Tran-Cong, Q., and Makino, K., 1991, Fluorescent-labeled oligonucleotide probes: Detection of hybrid formation in solution by fluorescence polarization spectroscopy, *Nucleic Acids Res.* **19**:4097–4102.
55. Kumke, M. U., Shu, L., McGown, L. B., Walker, G. T., Pitner, J. B., and Linn, C. P., 1997, Temperature and quenching studies of fluorescence polarization detection of DNA hybridization, *Anal. Chem.* **69**:500–506.
56. Livak, K. J., Flood, S. J. A., Marmaro, J., Giusti, W., and Deetz, K., 1995, Oligonucleotides with fluorescent dyes at opposite ends provide a quenched probe system useful for detecting PCR product and nucleic acid hybridization, *PCR Methods Appli.* **4**:357–362.
57. Gibson, U. E. M., Heid, C. A., and Williams, P. M., 1996, A novel method for real time quantitative RT-PCR, *Genome Res.* **6**:995–1001.
58. Steffens, D. L., Jang, G. Y., Sutter, S. L., Brumbaugh, J. A., Middendorf, L. R., Muhlegger, K., Mardis, E. R., Weinstock, L. A., and Wilson, R. K., 1995, An infrared fluorescent dATP for labeling DNA, *Genome Res.* **5**:393–399.
59. Wittwer, C. T., Herrmann, M. G., Moss, A. A., and Rasmussen, R. P., 1997, Continuous fluorescence monitoring of rapid cycle DNA amplification, *BioTechniques* **22**(1):130–138.
60. Pease, A. C., Solas, D., Sullivan, E. J., Cronin, M. T., Holmes, C. P., and Fodor, S. P. A., 1994, Light-generated oligonucleotide arrays for rapid DNA sequence analysis, *Proc. Natl. Acad. Sci. U.S.A.* **91**:5022–5026.
61. Lipshutz, R. J., Morris, D., Chee, M., Hubbell, E., Kozal, M. J., Shah, N., Shen, N., Yang, R., and Foder, S. P. A., 1995, Using oligonucleotide probe arrays to access genetic diversity, *BioTechniques* **19**:442–447.
62. Cronin, M. T., Fucini, R. V., Kim, S. M., Masino, R. S., Wespi, R. M., and Miyada, C. G., 1996, Cystic fibrosis mutation detection by hybridization to light-generated DNA probe arrays, *Hum. Mutat.* **7**:244–255.
63. Polak, J. M., and McGee, J. O'D., 1990, *In Situ Hybridization, Principles and Practice*, Oxford University Press, New York.
64. Wiegant, J., Wiesmeijer, C. C., Hoovers, J. M. N., Schuuring, E., d'Azzo, A., Vrolijk, J., Tanke, H. J., and Raap, A. K., 1993, Multiple and sensitive fluorescence in situ hybridization with rhodamine-, fluorescein-, and coumarin-labeled DNAs, *Cytogenet. Cell Genet.* **63**:73–76.
65. Schröck, E., du Manoir, S., Veldman, T., Schoell, B., Wienberg, J., Ferguson-Smith, M. A., Ning, Y., Ledbetter, D. H., Bar-Am, I., Soenksen, D., Garini, Y., and Reid, T., 1996, Multicolor spectral karyotyping of human chromosomes, *Science* **273**:494–497.
66. Speicher, M. R., Ballard, S. G., and Ward, D. C., 1996, Karyotyping human chromosomes by combinatorial multi-fluor FISH, *Nat. Genet.* **12**:368–378.
67. Nederlof, P. M., van der Flier, S., Wiegant, J., Raap, A. K., Tanke, H. J., Ploem, J. S., and van der Ploeg, M., 1990, Multiple fluorescence in situ hybridization, *Cytometry* **11**:126–131.
68. Kallioniemi, A., Kallioniemi, O. P., Sudar, D., Rutovitz, D., Gray, J. W., Waldman, F., and Pinkel, D., 1992, Comparative genomic hybridization for molecular cytogenetic analysis of solid tumors, *Science* **258**:818–821.
69. Lewis, R., 1996, Chromosome charting takes a giant step, *Photonics Spectra* **1996**(June):48–49.
70. Le Beau, M. M., 1996, One FISH, two FISH, red FISH, blue FISH, *Nat. Genet.* **12**:341–344.
71. Speicher, M. R., and Ward, D. C., 1996, The coloring of cytogenetics, *Nat. Med.* **2**:1046–1048.
72. Bentz, M., Döhner, H., Cabot, G., and Lichter, P., 1994, Fluorescence in situ hybridization in leukemias: The FISH are spawning, *Leukemia* **8**:1447–1452.

73. Fox, J. L., Hsu, P.-H., Legator, M. S., Morrison, L. E., and Seelig, S. A., 1995, Fluorescence in situ hybridization: Powerful molecular tool for cancer prognosis, *Clin. Chem.* **41**:1554–1559.
74. Popescu, N. C., and Zimonjic, D. B., 1997, Molecular cytogenetic characterization of cancer cell alterations, *Cancer Genet. Cytogenet.* **93**:10–21.
75. Swiger, R. R., and Tucker, J. D., 1996, Fluorescence in situ hybridization, *Environ. Mol. Mutagenesis* **27**:245–254.
76. Siadat-Pajouh, M., Periasamy, A., Ayscue, A. H., Moscicki, A. B., Palefsky, J. M., Walton, L., DeMars, L. R., Power, J. D., Herman, B., and Lockett, S. J., 1994, Detection of human papillomavirus type 16/18 DNA in cervicovaginal cells by fluorescence based in situ hybridization and automated image cytometry, *Cytometry* **15**:245–257.
77. Pandya, P. P., Cardy, D. L. N., Jauniaux, E., Campbell, S., and Nicolaides, K. H., 1994, Rapid determination of fetal sex in coelomic and amniotic fluid by fluorescence in situ hybridization, *Fetal Diagn. Ther.* **10**:66–70.
78. Matthews, J. A., and Kricka, L. J., 1988, Analytical strategies for the use of DNA probes, *Anal. Biochem.* **169**:1–25.
79. Nazarenko, I. A., Bhatnagar, S. K., and Hohman, R. J., 1997, A closed tube format for amplification and detection of DNA based on energy transfer, *Nucleic Acids Res.* **25**:2516–2521.
80. Luehrsen, K. R., Marr, L. L., van der Knaap, E., and Cumberledge, S., 1997, Analysis of differential display RT-PCR products using fluorescent primers and GENESCANTM software, *BioTechniques* **22**:168–174.
81. Alexandre, I., Zammatteo, N., Moris, P., Brancart, F., and Remacle, J., 1997, Comparison of three luminescent assays combined with a sandwich hybridization for the measurement of PCR-amplified human cytomegalovirus DNA, *J. Virol. Methods* **66**:113–122.
82. Kostrikis, L. G., Tyagi, S., Mhlanga, M. M., Ho, D. D., and Kramer, F. R., 1998, Spectral genotyping of human alleles, *Science* **279**:1228–1229.
83. Tyagi, S., Bratu, D. P., and Kramer, F. R., 1998, Multicolor molecular beacons for allele discrimination, *Nat. Biotechnol.* **16**:49–52.

PROBLEMS

21.1. *Spectral Observables Useful for DNA Sequencing*: Intensity, intensity ratio, and lifetime measurements have been used for DNA sequencing. Suggest reasons why anisotropy and collisional quenching have not been used for DNA sequencing.

Phase-Sensitive and Phase-Resolved Emission Spectra

22

In Chapter 5 we described the use of the FD method to measure lifetimes and to resolve complex intensity decays. In FD fluorometers, the sample is excited with intensity-modulated light, and one measures the phase shift and modulation of the emission, both relative to the excitation. The FD method also allows several other types of measurement which can be useful in special circumstances. One method is measurement of phase-sensitive intensities and/or emission spectra. Another method is to use the measured phase and modulation values to resolve the components of species in a mixture based on known decay times.

In phase-sensitive detection of fluorescence (PSDF), the measurements are somewhat different than in FD fluorometers. In PSDF, the emission from the sample is analyzed with a phase-sensitive detector, typically a lock-in amplifier. This measurement procedure selectively attenuates the signal from individual fluorophores on the basis of their fluorescence lifetimes or, more precisely, their phase angles relative to the phase of the detector. PSDF allows the emission from any one species, or, more precisely, the emission with any desired angle, to be suppressed. Phase suppression is accomplished when the phase of the detector is shifted 90° from the phase angle of the emission. Then, the resulting phase-sensitive emission spectrum represents only the emission from the remaining fluorophores. For a two-component mixture, suppression of the emission from one component allows the emission spectrum of the second component to be directly recorded. This procedure is experimentally simple and can be used to record the emission spectra of fluorophores with closely spaced lifetimes.

The second technique is calculation of phase-resolved (PR) emission spectra. In this case one measures the wavelength-dependent phase and/or modulation of the emission. If the lifetimes of the components in the mixture are known, the phase and/or modulation can be used to calculate the emission spectrum of each component in the mixture. The equations for phase-resolved emission spectra have been presented in several equivalent forms, which are summarized in this chapter. These expressions are useful to understand because they clarify one's understanding of what is actually measured in a phase-modulation instrument, and which values are interpretations of the measured values. Also, there are numerous potential applications for these methods, including resolution of heterogeneous fluorescence from proteins and membranes, quantification of macromolecular binding reactions, and analysis of excited-state reactions and solvent dipolar relaxation. All these phenomena are frequently encountered in the application of fluorescence methods to biochemical research. PSDF and PR fluorescence can be profitably used to study these phenomena. Finally, we show how the concept of phase-sensitive detection in fluorescence microscopy has allowed the creation of lifetime images.

22.1. THEORY OF PHASE-SENSITIVE DETECTION OF FLUORESCENCE

In a phase-modulation fluorometer, the sample is illuminated with sinusoidally modulated light which is intensity-modulated with a circular modulation frequency ω (Figure 22.1, upper left). It is known that, irrespective of the complexity of the emission (i.e., single-, multi-, or nonexponential), the emission will also be sinusoidal but will be shifted by a phase angle ϕ and demodulated by a factor m relative to the excitation (Figure 22.1, upper left). For a

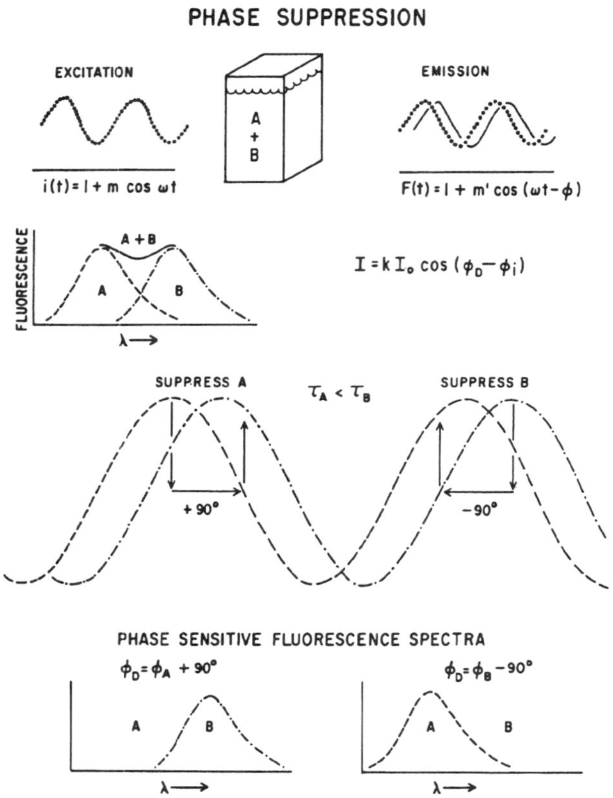

Figure 22.1. Intuitive description of phase-sensitive detection of fluorescence. Reprinted from Ref. 1, Copyright © 1981, with permission from Elsevier Science.

single-exponential decay, as is expected for pure fluorophores in a fluid homogeneous environment, the phase shift and demodulation can be used to calculate the fluorescence lifetime (τ). These relations are

$$\tan \phi_\omega = \omega \tau \qquad [22.1]$$

$$m_\omega = (1 + \omega^2\tau^2)^{-1/2} \qquad [22.2]$$

However, if the sample consists of a mixture of two fluorophores, or a single fluorophore that displays two decay times, then the fluorescence decay is doubly exponential. Multiexponential decays are frequently encountered in biochemical fluorescence, and still more complex decays are found for collisional quenching or in the presence of spectral relaxation. Phase angles and demodulation factors may still be measured, but these values, when interpreted according to Eqs. [22.1] and [22.2], yield only apparent, and not actual, lifetimes.

A phase fluorometer, when coupled with phase-sensitive detection of fluorescence, can be used in a simple manner to resolve heterogeneous fluorescence. Consider a sample containing a single fluorescent species with a lifetime τ. When excited with sinusoidally modulated light, the emission is given by

$$F(t) = 1 + m_L m \sin(\omega t - \phi) \qquad [22.3]$$

where m_L is the modulation of the exciting light. In this equation, m and ϕ are related to the lifetime by Eqs. [22.1] and [22.2]. Because phase-sensitive spectra are typically measured at a single modulation frequency, the subscript ω has been dropped for simplicity. If the sample contains more than one fluorophore, then the modulated emission at each wavelength (λ) is given by

$$F(\lambda, t) = \sum_i I_i(\lambda) f_i m_i \sin(\omega t - \phi_i) \qquad [22.4]$$

In this expression, the $I_i(\lambda)$ are the individual emission spectra; the f_i are the fractional contributions to the total steady-state intensity, $\Sigma f_i = 1.0$; and m_i is the modulation of the ith component and ϕ_i is its phase angle. Depending upon the needs of the experiment, the steady-state spectrum of each species, $I_i(\lambda)$, can be replaced by the steady-state spectrum of the sample, $I(\lambda)$, and the wavelength-dependent fractional intensities:

$$F(\lambda, t) = I(\lambda) \sum_i f_i(\lambda) m_i \sin(\omega t - \phi_i) \qquad [22.5]$$

In Eqs. [22.4] and [22.5] we have assumed that the sample contains discrete lifetimes characterized by m_i and ϕ_i, rather than a nonexponential decay or a lifetime distribution.

Phase-sensitive detection is accomplished by multiplying the emission $F(\lambda, t)$ by a square wave and integrating the result over time to yield a steady-state intensity.[1–9] The square wave is usually regarded as having a value of 0 or 1 depending on the angle within a single period of 2π (Figure 22.2),

$$P(t) = \begin{cases} 0 & \text{from } 0 \text{ to } \phi_D \\ 1 & \text{from } \phi_D \text{ to } \phi_D + \pi \\ 0 & \text{from } \phi_D + \pi \text{ to } \phi_D + 2\pi \end{cases} \qquad [22.6]$$

Typically, the phase angle of the detector (ϕ_D) is varied to integrate the emission over various portions of the 0 to 2π cycle.

The phase-sensitive detector yields a direct current signal proportional to the modulated amplitude and to the

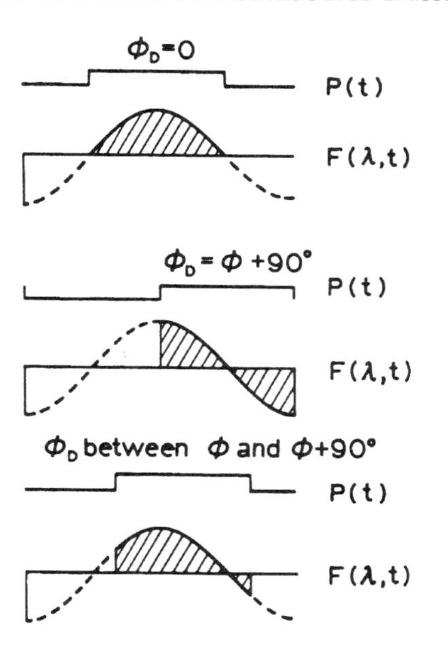

Figure 22.2. Phase-sensitive detection of fluorescence. The detector phase (ϕ_D) can be in-phase with the emission ($\phi_D = \phi$) (*top*), out-of-phase with the emission ($\phi_D = \phi + 90°$) (*middle*), or at some intermediate value (*bottom*). Revised from Ref. 5.

cosine of the phase difference between the detector phase ϕ_D and the phase of the sample. If an emission spectrum of a sample containing a single fluorophore (lifetime) is scanned using phase-sensitive detection, one observes a steady-state spectrum whose amplitude depends on the detector phase angle ϕ_D and the phase angle of the fluorophore ϕ_1:

$$F(\lambda, \phi_D) = kF(\lambda)m \cos(\phi_D - \phi_1) \qquad [22.7]$$

where $F(\lambda)$ is the steady-state emission spectrum, λ is the wavelength, and k is a constant which contains the usual sample and instrumental factors and the constant factor m_L. From Eq. [22.7] one can predict the appearance of the phase-sensitive spectrum of a single-component solution at various detector phase angles. One expects the intensity of the spectra to vary as $\cos(\phi_D - \phi)$, and the spectral distribution to remain unchanged.

The principle and usefulness of phase-sensitive detection are best understood by considering a mixture of two fluorophores, A and B, whose lifetimes (τ_A and τ_B) are each independent of emission wavelength (Figure 22.1). To resolve the spectra of A and B by PSDF, the phase angles (ϕ_A and ϕ_B) or lifetimes must be different. We will assume $\tau_A < \tau_B$. The time-dependent emission is given by

$$F(\lambda, t) = F_A(\lambda)m_A \sin(\omega t - \phi_A) + F_B(\lambda)m_B \sin(\omega t - \phi_B) \qquad [22.8]$$

where $F_A(\lambda)$ and $F_B(\lambda)$ are the intensities of components A and B at wavelength λ in the steady-state spectrum. An important characteristic of the modulated emission is that it is a superimposition of sine waves of the same frequency but differing phases, each resulting from one of the fluorophores (Figure 22.1). The modulated emission can be conveniently examined with a phase-sensitive detector or lock-in amplifier. The resulting unmodulated signal is given by

$$F(\lambda, \phi_D) = F_A(\lambda)m_A \cos(\phi_D - \phi_A) + F_B(\lambda)m_B \cos(\phi_D - \phi_B) \qquad [22.9]$$

For a mixture of two fluorophores, one expects the phase-sensitive spectra to contain contributions from both fluorophores, with fractional contributions dependent on the relative intensities [$F_i(\lambda)$], the modulations (m_i), and, most important, the values of $\phi_D - \phi_i$. The relative contribution of each fluorophore to the phase-sensitive intensity depends on the value of $\cos(\phi_D - \phi_i)$. By selection of $\phi_D - \phi_i = 90°$, the detector can be out-of-phase with one component in the sample. Then, the phase-sensitive spectrum represents the emission spectrum of the other component.

22.1.A. Phase-Sensitive Emission Spectra of a Two-Component Mixture

To illustrate the characteristics of phase-sensitive spectra, it is instructive to consider a mixture of TNS (τ = 12 ns) and Prodan (τ = 4 ns). The emission spectra of TNS and Prodan overlap, but each spectrum can still be visually recognized. TNS dominates the emission up to 440 nm, and TNS and Prodan both contribute to the emission at longer wavelengths (Figure 22.3). The modulated emission of the TNS–Prodan mixture consists of two sine waves, each with a characteristic phase and modulation. When examined with a phase-sensitive detector, the relative intensity of each fluorophore depends on the phase angle of the detector (Figure 22.4). The shape of the phase-sensitive spectra varies with detector phase angle and wavelength. Such changes in shape would not be observed for a single fluorophore or single decay time, and thus they indicate the presence of a heterogeneous decay. As the detector phase angle is increased, the long-wavelength portion of the spectrum decreases more rapidly than the short-wavelength region and crosses zero at smaller phase angles. This is because the phase angle and lifetime of Prodan are smaller than those of TNS. Inspection of the phase-sensitive spectra shows that the emission of Prodan is almost completely suppressed at ϕ_D near 105°, and that of TNS is suppressed at ϕ_D near 125°.

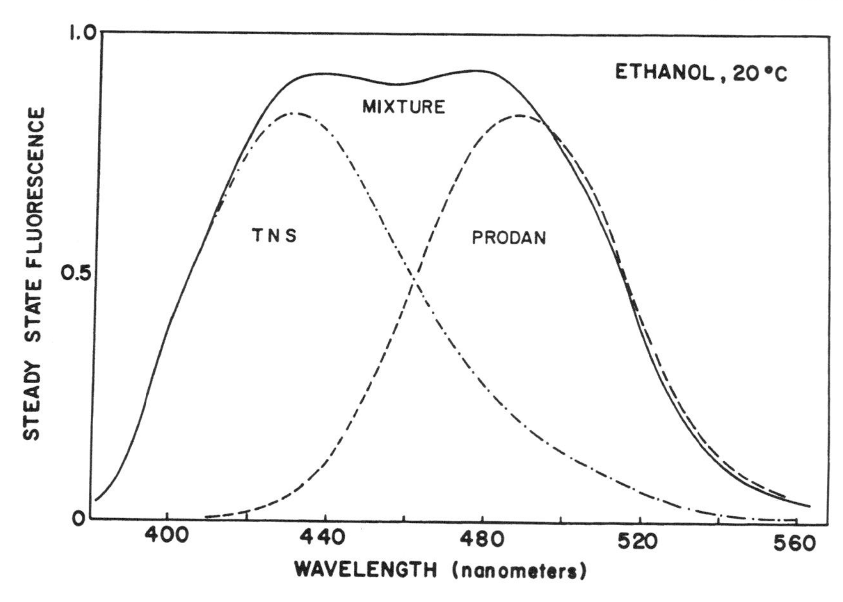

Figure 22.3. Steady-state fluorescence emission spectra of TNS, Prodan, and a mixture of TNS and Prodan. The excitation wavelength was 340 nm. Solutions were purged with inert gas to remove dissolved oxygen. Reprinted from Ref. 1, Copyright © 1981, with permission from Elsevier Science.

It is important to recognize that PSDF, at a single modulation frequency, cannot be used to determine both the spectrum and the lifetime of each component in a two-component mixture. The complete resolution of such a mixture requires that either one of the emission spectra or one of the lifetimes be known from an independent measurement. In spite of this limitation, PSDF can be useful in determining whether the emission from a given sample is heterogeneous, even when the individual spectra or lifetimes are not known and are not resolved. This may be accomplished by observing the phase-sensitive spectra at various phase angles. Generally, the lifetime of each fluorophore is independent of emission wavelength. Of course, exceptions occur in viscous solvents, for excited-state reactions and even for pure compounds. For fluorophores which display a single wavelength-independent lifetime, the phase-sensitive emission spectra are independent of the detector phase angle, except for the variation of intensity according to $\cos(\phi_D - \phi_i)$. This independence from ϕ_D is expected because the emission at each wavelength has the same phase angle, and therefore the entire emission spectrum varies with ϕ_D according to Eq. [22.7]. In contrast, the phase-sensitive spectra are dependent upon the detector phase angles when a mixture is present.

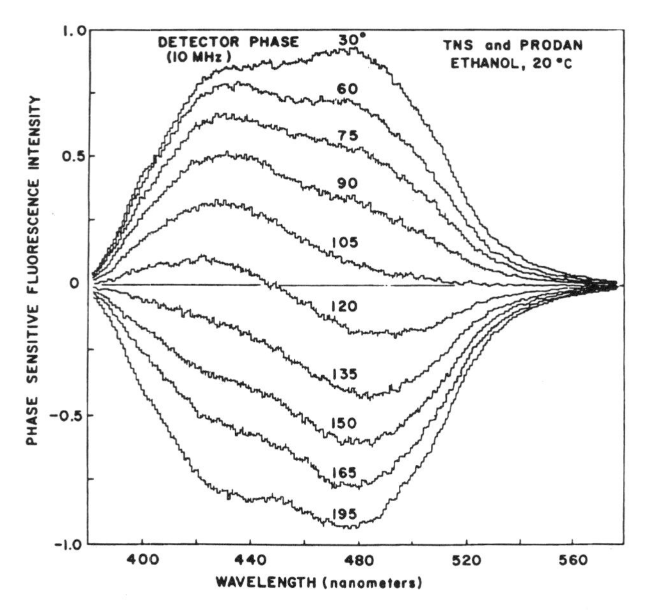

Figure 22.4. Phase-sensitive fluorescence spectra of the TNS–Prodan mixture in Figure 22.3. The detector phase angles are relative to the phase of the excitation, which is considered to be 0°. Reprinted from Ref. 1, Copyright © 1981, with permission from Elsevier Science.

The dependence of the phase-sensitive intensities on the detector phase angle can be used to determine the phase angle of the emission at each wavelength. The phase-sensitive intensities at 410 and 510 nm for the TNS–Prodan mixture are shown in Figure 22.5. These phase-sensitive intensities clearly follow the expected dependencies on $\cos(\phi_D - \phi_i)$. The modulated amplitude at 410 nm is smaller

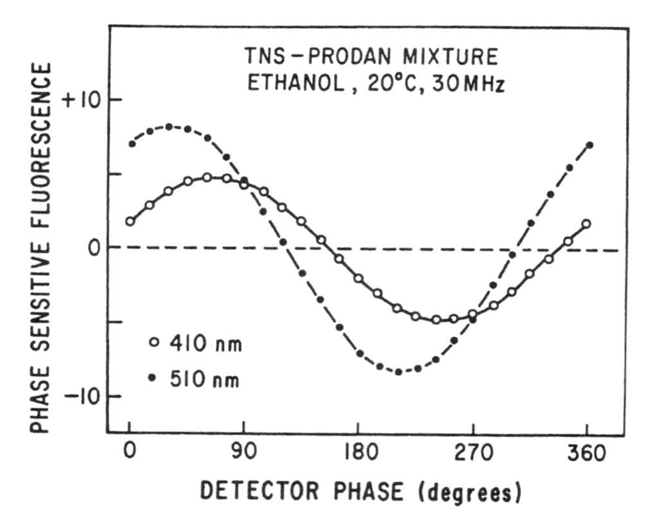

Figure 22.5. Dependence of phase-sensitive fluorescence intensity of the TNS–Prodan mixture in Figure 22.3 on detector phase angle at a modulation frequency of 30 MHz. The 410-nm component is phase-delayed relative to the 510-nm component because of the longer lifetime of TNS (12 ns) relative to that of Prodan (4 ns). The detector phase angles at which the phase-sensitive intensities are zero are in good agreement with those expected from the fluorescence lifetimes, recalling that the spectra overlap at 510 nm. Reprinted from Ref. 1, Copyright © 1981, with permission from Elsevier Science.

than at 510 nm as a result of the longer lifetime of TNS and the smaller value of the modulation of TNS. The zero crossing point occurs at a larger phase angle at 410 nm than at 510 nm, which is due to the longer decay times of TNS as compared to Prodan. It is interesting to notice that measurement of intensity versus phase angle has become used in fluorescence microscopy and is the basis of fluorescence lifetime imaging (Section 22.3).

22.1.B. Phase Suppression

Examination of Figures 22.4 and 22.5 revealed that at certain phase angles the phase-sensitive intensities of TNS or Prodan were near zero. In fact, one can always select a detector phase angle to be out-of-phase with any given component, $|\phi_D - \phi_i| = 90°$. Then, the phase-sensitive emission spectrum contains contributions from only the remaining fluorophores. In the case of a sample which contains two fluorophores, the emission from either component can be suppressed. Then the phase-sensitive emission spectrum reveals the steady-state spectrum of the unsuppressed component. This technique is called phase suppression. For instance, consider species A in Figure 22.1. If $\phi_D = \phi_A + 90°$, then species A does not contribute to the phase-sensitive spectrum. At this detector phase angle, the phase-sensitive spectrum is given by

$$F(\lambda, \phi_A + 90°) = F_B(\lambda) m_B \sin(\phi_B - \phi_A) \quad [22.10]$$

Scanning the wavelength then yields $F_B(\lambda)$, the steady-state spectrum of species B. Conversely, if $\phi_D = \phi_B - 90°$, then

$$F(\lambda, \phi_B - 90°) = F_A(\lambda) m_A \sin(\phi_B - \phi_A) \quad [22.11]$$

The emission from species B is suppressed, and a scan of the emission wavelength yields the steady-state spectrum of species A, $F_A(\lambda)$.

Several aspects of the phase-suppressed spectra are worthy of mention. For a two-component mixture, suppression of one component results in the attenuation of the emission from the other component. Specifically, these spectra are each attenuated to $\sin(\phi_B - \phi_A)$ of the original modulated intensity. Hence, the resolution of components by this method depends on the phase-angle difference between the two emissions, and not on the absolute phase angles or lifetimes. Phase resolution of spectra can be expected to become increasingly difficult as the phase angles or lifetimes of the components become similar. Obviously, if $\phi_B = \phi_A$, suppression of either component results in simultaneous suppression of the other component. One should also recognize that the spectrum of each component is further attenuated by a factor m_A or m_B, where these modulation factors are related to the lifetimes of components A and B by Eq. [22.2]. This indicates that the phase-sensitive intensity from the species with the longer lifetime will be smaller than that from the component with the shorter lifetime. If one wishes to use the phase-sensitive intensities to quantify the amount of each component present in the mixture, these demodulation factors must be considered.

Importantly, for a mixture of fluorophores, the phase angles used for suppression can be used to calculate the lifetime of each species in the mixture. If the steady-state spectrum of one of the species is known, then ϕ_D can be adjusted so that the phase-sensitive spectrum superimposes on this steady-state spectrum. For example, if $F_A(\lambda)$ is known, ϕ_D can be chosen to yield a phase-sensitive spectrum which overlaps with $F_A(\lambda)$. The chosen phase angle will be 90° out-of-phase with the emission from species B and may be used to calculate its lifetime. Of course, this procedure can be reversed to yield the lifetime of species A. For a two-component mixture, phase-sensitive detection of fluorescence allows the lifetimes of both components to be determined for the mixture, if the steady-state spectrum of one component is known. This minimal requirement is satisfied in many experimental situations.

The use of phase suppression is illustrated by recording the individual emission spectra of TNS and Prodan from the mixture. It was possible to identify detector phase angles at which the phase-sensitive emission spectra from the mixture overlapped with the emission spectrum of TNS or Prodan (Figure 22.6). When the detector phase angle was $\phi_D = 12.8° + 90°$, the phase-sensitive emission spectrum resembled the emission spectrum of TNS. The detector phase angle of 12.8° is out-of-phase with the 3.6-ns lifetime of Prodan. Hence, Prodan does not contribute to the phase-sensitive spectrum. Similarly, at a detector phase angle of $\phi_D = 36.5° - 90°$, the phase-sensitive spectrum matched the emission spectrum of Prodan. This result is due to the detector being out-of-phase with the 11.6-ns lifetime of TNS. In contrast to steady-state spectra, phase-sensitive spectra can be positive or negative. Suppression of the shorter-lifetime component (Prodan) and a positive phase-sensitive signal required a +90° phase shift. Suppression of the longer-lived species (TNS) and a positive phase-sensitive signal required a –90° phase shift.

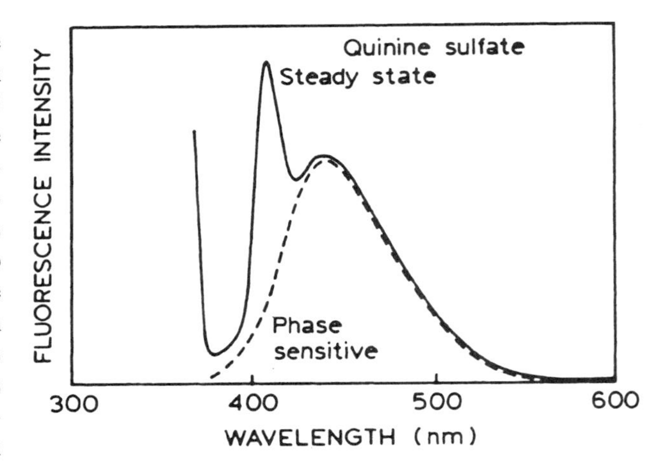

Figure 22.7. Steady state emission spectrum of quinine sulfate (——) and the phase-sensitive spectrum with nulling of the scattered light (– –). Revised from Ref. 6.

22.1.C. Examples of PSDF and Phase Suppression

Since the initial reports on PSDF, this method has been applied to a variety of samples. These applications are illustrated by two examples, suppression of scattered light and resolution of an excited-state reaction.

Scattered light has a zero lifetime and is thus always out-of-phase to some extent with the emission. Several laboratories have suggested the use of phase-sensitive detection to suppress Raman or Rayleigh scattered light.[6,10–13] This application is illustrated in Figure 22.7 for a dilute solution of quinine sulfate excited at 355 nm. There is a large peak due to Rayleigh scatter below 370 nm, and a Raman scatter peak at 410 nm. The scattered light could be suppressed by phase-sensitive detection, allowing the emission spectrum of quinine sulfate to be directly recorded.

PSDF can be used to record the emission spectra of proteins.[2,3] This is illustrated in Figure 22.8 for a mixture of the neutral analogs of tryptophan and tyrosine. At a detector phase angle of 103°, the tyrosine emission is

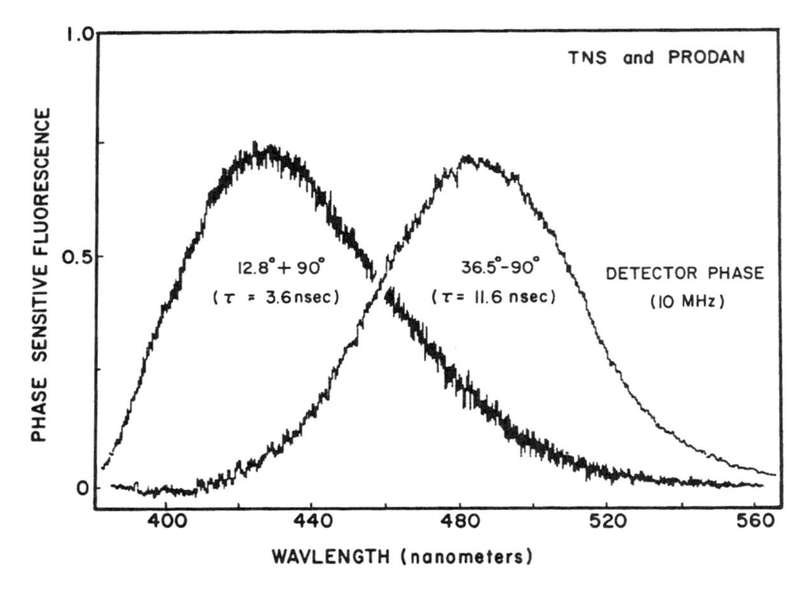

Figure 22.6. Direct recording of the emission spectra of TNS and Prodan from the mixture. When the detector phase angle is out-of-phase with a 11.6-ns component (TNS), only Prodan fluorescence is detected. When the detector phase angle is out-of-phase with a 3.6-ns component (Prodan), only TNS fluorescence is detected. Revised from Ref. 1.

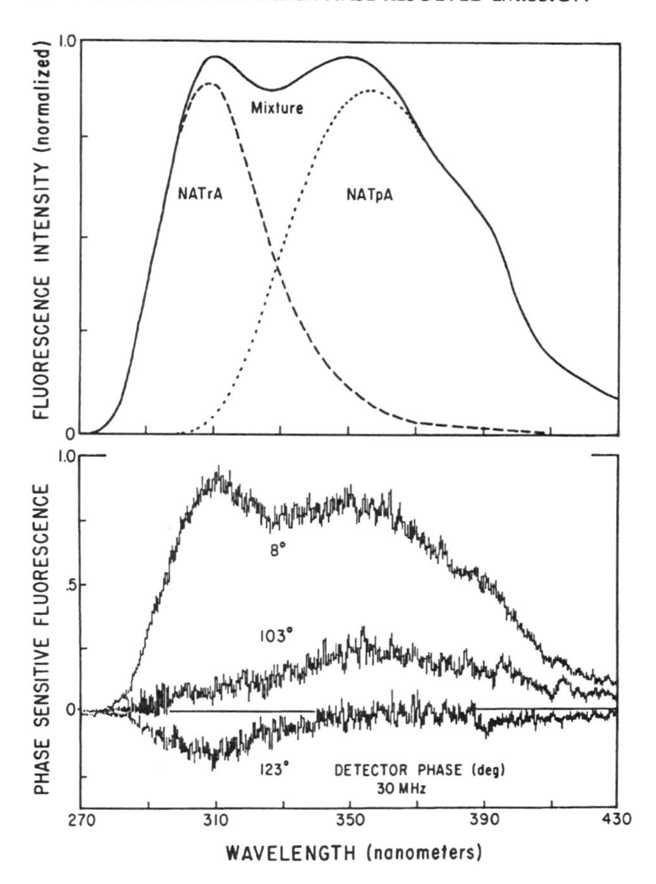

Figure 22.8. Steady-state (*top*) and phase-sensitive emission spectra (*bottom*) of a mixture of *N*-acetly-L-tyrosinamide (NATrA) and *N*-acetyl-L-tryptophanamide (NATpA). Excitation was at 280 nm, and modulation frequency was 30 MHz. From Ref. 2.

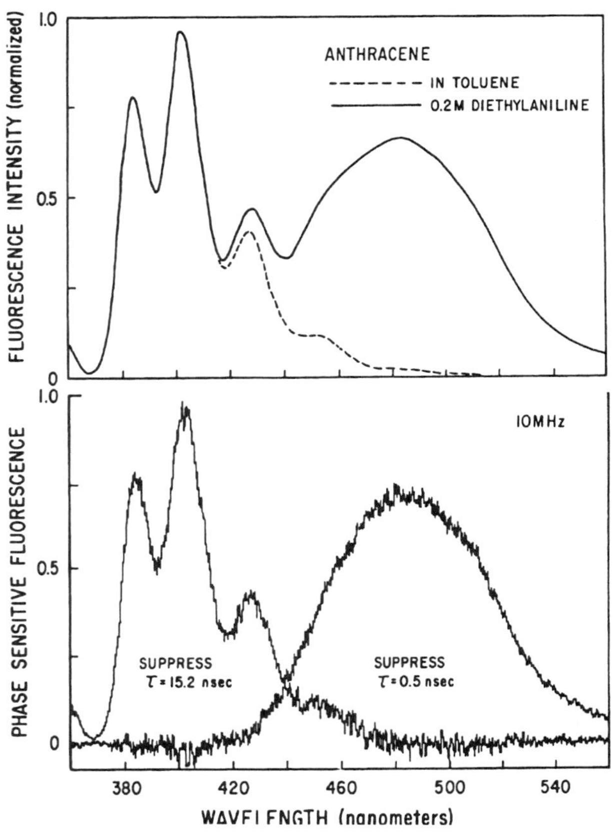

Figure 22.9. Emission spectra of anthracene and its exciplex with diethylaniline. *Top*: Normalized steady-state spectra of anthracene in toluene (– – –) and 0.2*M* diethylaniline (———); *bottom*: phase-sensitive fluorescence spectra of anthracene in the presence of diethylaniline. The excitation wavelength was 357 nm, and the excitation and emission band passes were 8 nm. The solutions were not purged with inert gas. Reprinted from Ref. 1, Copyright © 1981, with permission from Elsevier Science.

nearly suppressed. At a detector phase angle of 123°, the tryptophan emission is nearly suppressed, in this case resulting in a negative phase-sensitive spectrum. The lifetimes of NATrA and NATpA are 1.8 and 2.9 ns, respectively, showing that PSDF is useful for resolving spectra with closely spaced lifetimes.

As a final example of PSDF, we show that it can be used to study excited-state reactions, as well as ground-state multiexponential decays.[1,14–16] This application is illustrated in Figure 22.9 for exciplex formation between anthracene and diethylaniline. In this case, the long-wavelength emission forms subsequent to excitation of anthracene and displays the features of an excited-state reaction. In PSDF it does not matter whether the preexponential factors are positive or negative, or if the phase angle exceeds 90°. Adjustment of the detector phase angle to be out-of-phase with either anthracene or its exciplex allows the emission spectrum of the other species to be recorded (Figure 22.9, lower panel).

The concept of phase suppression can also be applied when the decay is not a sum of exponentials and when the lifetimes of the individual species are not known. This possibility is illustrated by the example of TNS in glycerol. The emission spectrum depends on temperature owing to time-dependent spectral relaxation (Figure 22.10). At intermediate temperatures near –10 °C, the mean lifetime is comparable to the rate of spectral relaxation, and the mean lifetime is expected to increase with increasing observation wavelength.

One can examine such samples using phase-sensitive spectra. The basic idea is to consider the short-wavelength side of the emission as representing the initially excited state, and the red side of the emission as representing the solvent-relaxed state. Suppression of the emission on each side of the emission spectrum allows the apparent relaxed

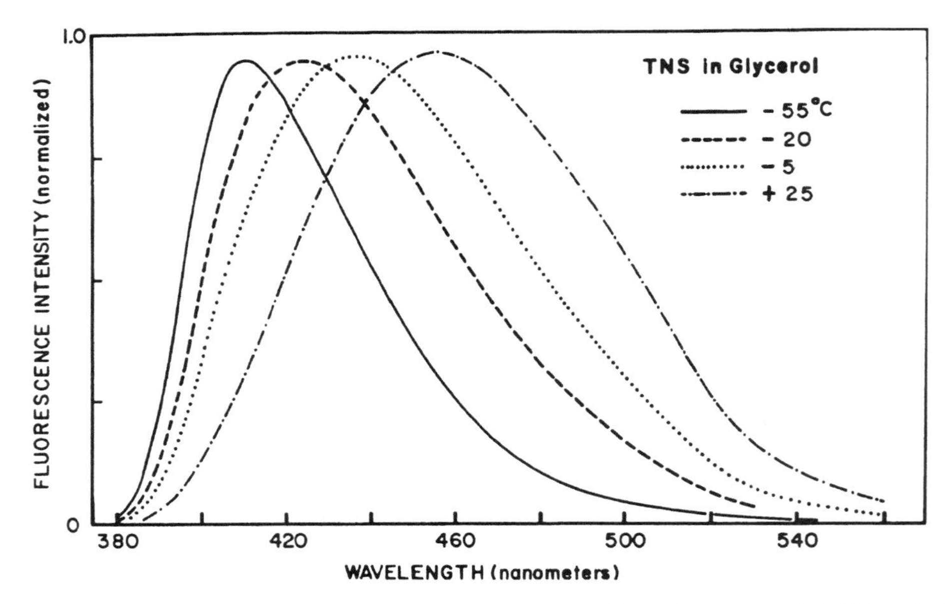

Figure 22.10. Fluorescence emission spectra of TNS in glycerol at –55 (———), –20 (– – –), –5 (· · ·), and +25 °C (- · · -). Reprinted from Ref. 16, Copyright © 1983, with permission from Elsevier Science.

or unrelaxed spectrum to be recorded (Figure 22.11). From these spectra, one can see that the amplitude of the longer-wavelength emission increases as the temperature increases. This result is consistent with faster spectral relaxation at higher temperatures. At high or low temperatures, when spectral relaxation is either complete or very slow, the phase-sensitive spectra do not provide resolution because the emission is dominated by a single state. Phase-sensitive spectra of relaxing systems can be informative. However, it is difficult to interpret the amplitudes of the phase sensitive spectra without additional information about the relaxation rate constants and relative quantum yields of the various states.

In addition to the examples presented here, PSDF has been used to resolve mixtures of four or more fluorophores[17–22] and to suppress background fluorescence.[23] Phase-sensitive detection has also been used to resolve the emission of fluorophores free in solution and bound to macromolecules[24,25] and to study binding between antigens and antibodies.[26–28]

In the preceding examples, one may have noticed that the phase-sensitive emission spectra were somewhat noisy. This is for two reasons. While one can suppress an emission spectrum, the Poisson photon noise remains in the suppressed signal. Secondly, the preceding data were recorded using the older fixed-frequency instruments with a Debye–Sears light modulator. Since the introduction of variable-frequency instruments, there has been less use of PSDF, so that fewer results obtained with the newer instruments have been published.

22.1.D. High-Frequency or Low-Frequency Phase-Sensitive Detection

To this point, we have not described the technical details associated with recording phase-sensitive emission spectra. Are the phase-sensitive spectra recorded using the high-frequency signal prior to cross correlation, or can phase-sensitive detection be performed using the low-frequency cross-correlation signals? In fact, when PSDF was being developed in the early 1980s, it was not known whether the phase resolution shown schematically in Figure 22.1 would work with the low-frequency signal following cross correlation. The first phase-sensitive fluorescence spectra appear to have been recorded by Veselova and co-workers.[29] This group performed phase-sensitive detection directly on the 11.2-MHz PMT output of three phase fluorometers. The use of high-frequency phase-sensitive detection is difficult and prone to artifacts.

Fortunately, it is not necessary to perform phase-sensitive detection at high frequency. It was found that the modulated emission from individual fluorophores could still be resolved using the low-frequency cross-correlation signals. Such low-frequency detection is easy to perform, and one need not be concerned with the possible perturbation of tuned high-frequency circuits. The reference signal for the phase-sensitive detector is provided by the reference phototube, which observes the emission from a ref-

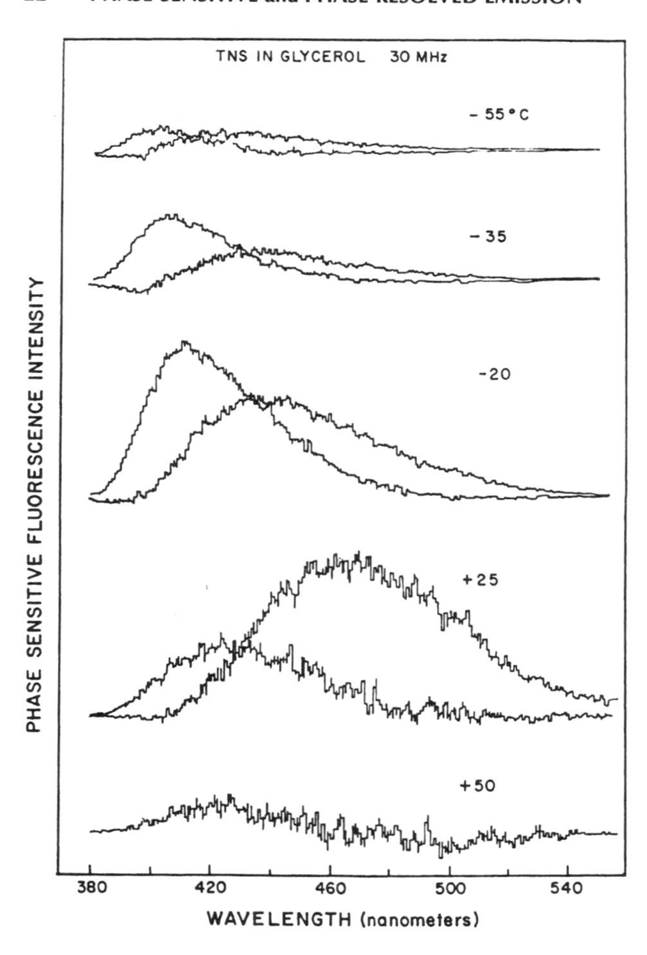

Figure 22.11. Phase-sensitive emission spectra for TNS in glycerol. At each temperature, the emission was suppressed on the blue and red sides of the emission. Revised from Ref. 16, Copyright © 1983, with permission from Elsevier Science.

erence fluorophore or scatterer (Figure 5.5). Following selection of the detector phase angle, the phase-sensitive spectra are collected in the usual manner by scanning the emission monochromator.

22.2. PHASE-MODULATION RESOLUTION OF EMISSION SPECTRA

In recent years, the use of phase-sensitive detection has diminished, probably because of the realization that the phase-sensitive spectra contain less information than the phase-angle and modulation spectra. In the time it takes to record the phase-sensitive spectrum, one can now record the phase angle and/or modulation across the emission spectrum. From these phase-modulation spectra, one can compute the phase-resolved spectra.[30–32] It should be noted that calculation of individual spectra from the phase-angle and modulation spectra has a long history, dating to the early reports by Veselova and co-workers.[33,34] The equation to accomplish these resolutions have been presented in several different forms, which are useful under different circumstances.

22.2.A. Resolution Based on Phase or Modulation Lifetimes

One approach to calculating the phase-resolved spectra is based on use of the apparent phase $[\tau^{\phi}(\lambda)]$ or modulation $[\tau^{m}(\lambda)]$ lifetimes at each wavelength.[30] Suppose that the sample contains two species, with fractional steady-state intensities of $f_1(\lambda)$ and $f_2(\lambda)$, and that the two decay times τ_1 and τ_2 are known and are independent of wavelength. Then the ratio of fractional intensities can be calculated from[30]

$$\frac{f_1(\lambda)}{f_2(\lambda)} = \frac{\tau_2 - \tau_{\phi}(\lambda)}{\tau_{\phi}(\lambda) - \tau_1} \frac{(1 + \omega^2\tau_1^2)}{(1 + \omega^2\tau_2^2)} \qquad [22.12]$$

A similar calculation can be performed using the apparent modulation lifetime:

$$\frac{f_1(\lambda)}{f_2(\lambda)} = \frac{\omega\tau_1 - [(1 + \omega^2\tau_1^2)(1 + \omega^2\tau_2^2)/\{1 + [\omega\tau_m(\lambda)]^2\} - 1]^{1/2}}{[(1 + \omega^2\tau_1^2)(1 + \omega^2\tau_2^2)/\{1 + [\omega\tau_m(\lambda)]^2\} - 1]^{1/2} - \omega\tau_2} \qquad [22.13]$$

An example of two-component resolutions is shown for 6-(diethylamino)naphthalene-1-sulfonate (DENS) and perylene in Figure 22.12. The emission spectra of the individual fluorophores were obtained directly from the phase or modulation values, and similar spectra were obtained using either the phase or the modulation data. An advantage of this direct calculation procedure is that one can change the assumed values of τ_1 and τ_2 to see how these values affect the calculated spectra. Such further calculations are not possible using the phase-sensitive spectra.

22.2.B. Resolution Based on Phase Angles and Modulations

The equations for spectral resolution based on phase and modulation data can be presented in several ways. For simplicity, we will present these equations for a system with two emitting species. For any decay law, the values of $N(\lambda)$ and $D(\lambda)$ are given by[31,32]

$$N(\lambda) = m(\lambda) \sin \phi(\lambda) = f_1(\lambda) m_1 \sin \phi_1 + f_2(\lambda) m_2 \sin \phi_2 \qquad [22.14]$$

$$D(\lambda) = m(\lambda) \cos \phi(\lambda) = f_1(\lambda) m_1 \cos \phi_1 + f_2(\lambda) m_2 \cos \phi_2 \qquad [22.15]$$

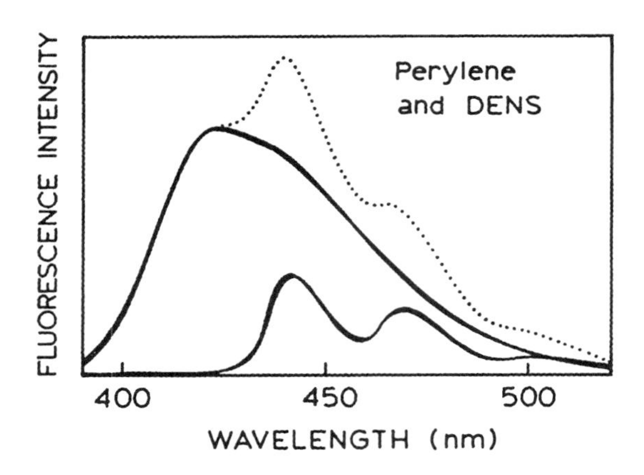

Figure 22.12. Steady-state emission spectrum for a mixture of DENS and perylene in ethanol (· · ·) and the emission spectra calculated from the phase-angle or modulation spectra (———), superimposed on the emission spectra of the individual species. The modulation frequency was 20 MHz. The decay times of perylene and DENS were 4.3 and 12.0 ns, respectively. Revised and reprinted, with permission, from Ref. 23, Copyright © 1984, American Chemical Society.

It is important to understand the meaning of the terms in Eqs. [22.14] and [22.15]. The values of $m(\lambda)$ and $\phi(\lambda)$ are the experimentally determined data; m_i and ϕ_i are constant terms whose values will somehow be known or separately measured. If the intensity decay is due to a mixture of fluorophores each of which displays a single-exponential decay, then $m_1 = \cos\phi_1$ and $m_2 = \cos\phi_2$ and

$$N(\lambda) = f_1(\lambda)\sin\phi_1\cos\phi_1 + f_2(\lambda)\sin\phi_2\cos\phi_2 \quad [22.16]$$

$$D(\lambda) = f_1(\lambda)\cos^2\phi_1 + f_2(\lambda)\cos^2\phi_2 \quad [22.17]$$

However, if the decay is nonexponential, then $m_i \neq \cos\phi_i$.

Application of Cramer's rule to Eqs. [22.14] and [22.15], followed by application of the law for the sine of a difference between two angles, yields

$$f_1(\lambda) = \frac{m(\lambda)\sin[\phi(\lambda) - \phi_2)]}{m_1\sin(\phi_1 - \phi_2)} \quad [22.18]$$

$$f_2(\lambda) = \frac{m(\lambda)\sin[\phi_1 - \phi(\lambda)]}{m_2\sin(\phi_1 - \phi_2)} \quad [22.19]$$

These expressions were first used by Veselova *et al.*[33] to calculate the emission spectra of relaxed and unrelaxed fluorophores during spectral relaxation. Alternative forms of Eqs. [22.18] and [22.19] can be found by noting that $f_1(\lambda) + f_2(\lambda) = 1.0$:

$$f_1(\lambda) = \frac{m(\lambda)\cos\phi(\lambda) - m_2\cos\phi_2}{m_1\cos\phi_1 - m_2\cos\phi_2} \quad [22.20]$$

$$f_2(\lambda) = \frac{m(\lambda)\cos\phi(\lambda) - m_1\cos\phi_1}{m_2\cos\phi_2 - m_1\cos\phi_1} \quad [22.21]$$

Use of either form of these equations requires knowledge of ϕ_1, ϕ_2, m_1, and m_2, or, for a mixture of fluorophores, τ_1 and τ_2.

22.2.C. Resolution of Emission Spectra from Phase and Modulation Spectra

Resolution of emission spectra from the phase and modulation data will be illustrated with the data in Figure 22.13. The dashed curve shows the emission spectrum of anthracene in the presence of diethylaniline (DEA). The structured emission is due to anthracene, and the broad long-wavelength emission is due to the exciplex formed with DEA. The presence of DEA results in a decrease in the phase angle of anthracene seen near 400 nm and increased phase angles at long wavelengths where the exciplex emits. The modulation decreases at wavelengths where the exciplex emission contributes to the intensity.

At long and short wavelengths, the phase and modulation values are constant, allowing assignment of ϕ_1, ϕ_2, m_1, and m_2. The constant phase angles and modulations indicate that the excited-state reaction in a two-state process,

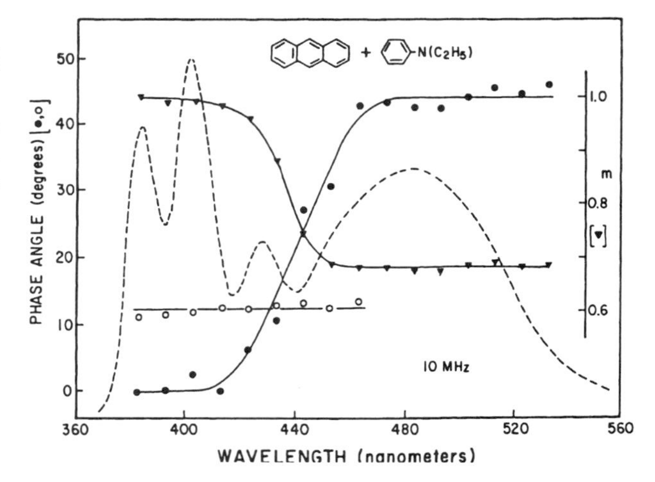

Figure 22.13. Emission spectrum (- - -), phase angles (●), and demodulation factors (▼) for anthracene and its exciplex with diethylaniline (DEA). Anthracene was dissolved in toluene, and the concentration of DEA was 0.2*M*. The excitation was at 357 nm. The solution was not purged with inert gas. ○, Phase angles of anthracene in the absence of DEA. Reprinted from Ref. 32, Copyright © 1982, with permission from Elsevier Science.

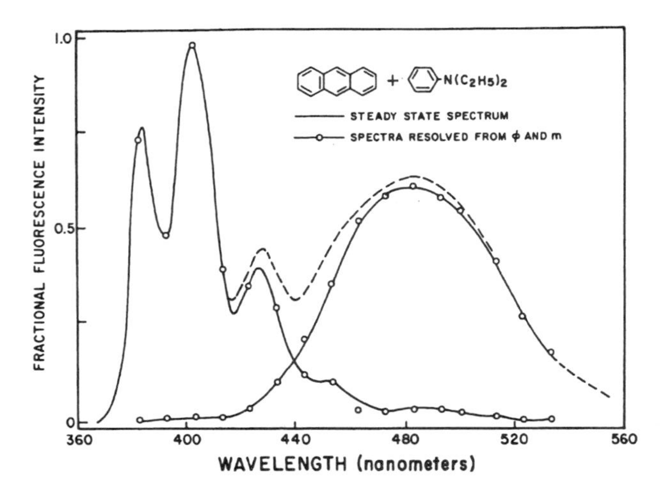

Figure 22.14. Resolution of the monomer and exciplex emission of anthracene. The spectra were calculated using Eqs. [22.20] and [22.21] and the data shown in Figure 22.13. Reprinted from Ref. 32, Copyright © 1982, with permission from Elsevier Science.

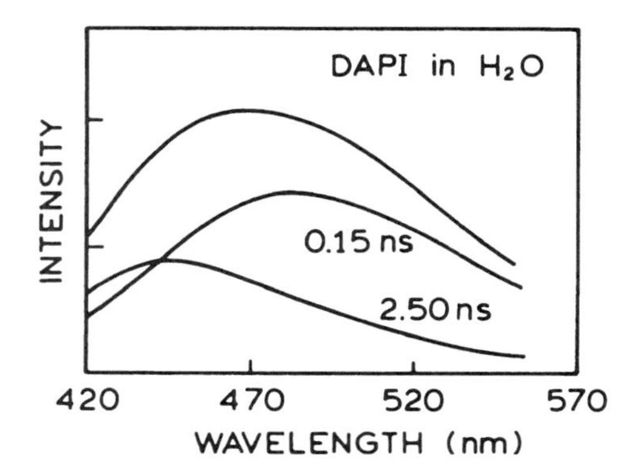

Figure 22.15. Phase and modulation resolved spectra of DAPI at pH 7.1. Excitation was at 325 nm; the modulation frequency was 100 MHz. The spectra for the 0.15- and 2.50-ns components are shown. Revised from Ref. 35 and reprinted with permission from Springer-Verlag.

rather than a continuous process. In the case of an excited-state reaction, it may not be possible to use Eqs. [22.12] and [22.13] because the phase angles can exceed 90° and the value of the modulation is not due to a single modulation lifetime. One can use Eqs. [22.20] and [22.21] to calculate the fractional intensities at each wavelength. By multiplying the steady-state spectrum by the $f_i(\lambda)$ values, one can calculate the emission spectra of anthracene and its exciplex (Figure 22.14).

This procedure can also be used when the phase and modulation values do not display constant values on the blue and red sides of the emission. In this case one obtains apparent spectra, whose molecular significance can only be understood with additional information about the sample.[32]

Another example of the resolution of spectra from the phase-angle and modulation spectra is shown in Figure 22.15 for the DNA probe DAPI dissolved in water at pH 7.1. As described in Section 5.6.C, DAPI displays a multiexponential decay in water with decay times of 0.15 and 2.50 ns.[35] The fractional contribution of each decay time depends on emission wavelength, so that the mean decay time is wavelength-dependent.

The emission spectra associated with each decay time were resolved using the apparent phase lifetimes and Eq. [22.12]. Representative phase and modulation lifetimes are listed in Table 22.1. One notices that $\tau_\phi < \tau_m$ at all wavelengths, which is indicative of a multiexponential decay. If DAPI displayed an excited-state reaction, then one expects $\tau_\phi > \tau_m$ on the long-wavelength side of the emission. The DAS of DAPI (Figure 22.15) show that the shorter-decay-time component displays a longer-wavelength emission.

22.3. FLUORESCENCE LIFETIME IMAGING MICROSCOPY

While phase-sensitive detection of fluorescence is somewhat outdated for the resolution of multicomponent mixtures, this method has found use in fluorescence lifetime imaging microscopy (FLIM).[36–54] The concept of FLIM is illustrated in Figure 22.16. Suppose that the sample contains two regions in which the fluorophore displays different lifetimes, τ_1 and τ_2. The intensity from both regions may be the same, so that the difference in probe environment is not visible in the steady-state image. If the lifetimes

Table 22.1. Apparent Phase and Modulation Lifetimes of DAPI in Water, pH 7.1[a,b]

Emission wavelength (nm)	τ_ϕ^{app} (ns)	τ_m^{app} (ns)	f_1
420	0.71	1.91	0.50
450	0.63	1.61	0.45
500	0.38	0.92	0.24
550	0.31	0.78	0.16

[a]Ref. 35.
[b]Modulation frequency, 100 MHz.

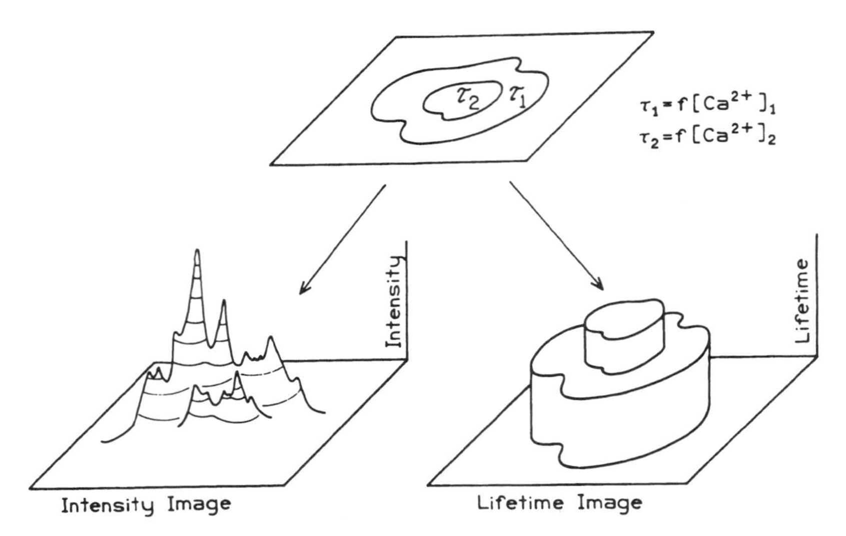

Figure 22.16. Intuitive presentation of the concept of fluorescence lifetime imaging (FLIM). The object is assumed to have two regions that display the same fluorescence intensity ($I_1 = I_2$) but different decay times, $\tau_2 > \tau_1$. From Ref. 37.

were measured in each region of the cell, one would observe regions with the two decay times. These regions can be revealed by color, gray-scale, or contour plots.

The usefulness of FLIM is a result of the difficulties of quantifying fluorescence intensities in fluorescence microscopy. The steady-state fluorescence images can be difficult to interpret and quantify because there is no practical way to determine local concentrations of the probes in various regions of the sample. Moreover, most fluorophores photobleach rapidly, which further complicates the ability to use the intensity images quantitatively. As a result of these difficulties, there have been extensive efforts to develop probes and imaging methods that are independent of the local intensity, such as the wavelength-ratiometric

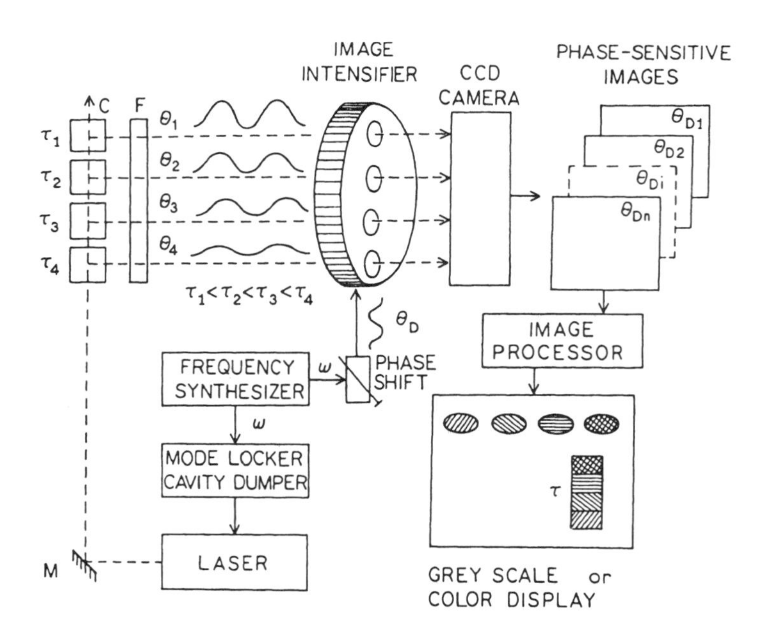

Figure 22.17. Schematic diagram of a FLIM experiment. The "object" consists of a row of four cuvettes and has regions with different decay times, τ_1–τ_4. This object is illuminated with intensity-modulated light. The emission is detected with a phase-sensitive image intensifier and imaged onto a CCD camera. The laser system is a cavity-dumped dye laser, which is synchronously pumped by a mode-locked and frequency-doubled Nd:YAG laser. From Ref. 37.

probes for Ca^{2+}. Lifetime imaging is advantageous because the lifetimes are mostly independent of local probe concentration and photobleaching. Also, lifetime probes are now known for a variety of cations and anions.

Measuring decay times at even a single point in a cuvette can be challenging, and one may wonder how an array of lifetimes, adequate for good spatial resolution, could be obtained. Phase-sensitive detection was found to provide a straightforward means of lifetime imaging.[36,37] The basic idea is to excite the sample with modulated light and to observe the image with an image intensifier. The gain of the image intensifier is modulated at the same frequency as the light modulation (Figure 22.17). Each spot in the image intensifier acts like a phase-sensitive detector, except that the intensities cannot be negative. By varying the phase angle of the image intensifier, one obtains data analogous to those in Figure 22.5, except that one obtains a phase-sensitive image rather than a phase-sensitive intensity.

The principle of FLIM is easier to explain with macroscopic objects than with cellular images. Figure 22.18 shows the phase-sensitive images recorded for a line of four cuvettes, each of which contained a fluorophore with a different lifetime. As the lifetime becomes longer, the phase angles are larger relative to the excitation. Also, the modulation becomes smaller. The phase-sensitive intensities at each detector phase angle can be used to determine the phase angle and modulation at each point in an image.

The results of a calcium-imaging FLIM experiment are shown in Figure 22.19. Kidney cells from a monkey (COS cells) were labeled with the calcium probe Quin-2. The lifetime of Quin-2 increases severalfold in the presence of calcium (Figure 5.29). The intensity of Quin-2 varies dramatically in different regions of the cell (Figure 22.19, left). Without additional information, one does not know whether these variations are due to different probe concentrations or different lifetimes and quantum yields of Quin-2. The phase-angle or calcium concentration images of the Quin-2-labeled cells show a constant phase angle, which indicates that the calcium concentration is the same in all regions of the cell (Figure 22.19, right). FLIM can now be used to image a wide variety of ions, and its use in cell physiology can be expected to increase.

22.3.A. Phase-Suppression Imaging

The concept of phase suppression has also been applied to FLIM. As for PSDF, one can always collect a phase-sensitive image that is out-of-phase with one decay time, thereby suppressing this component in the image. However, the concept of phase suppression can be generalized. Suppose that phase-sensitive images are obtained at two detector phase angles, ϕ_D and $\phi_D + \Delta$, where Δ is the phase-angle difference used in recording the two images.

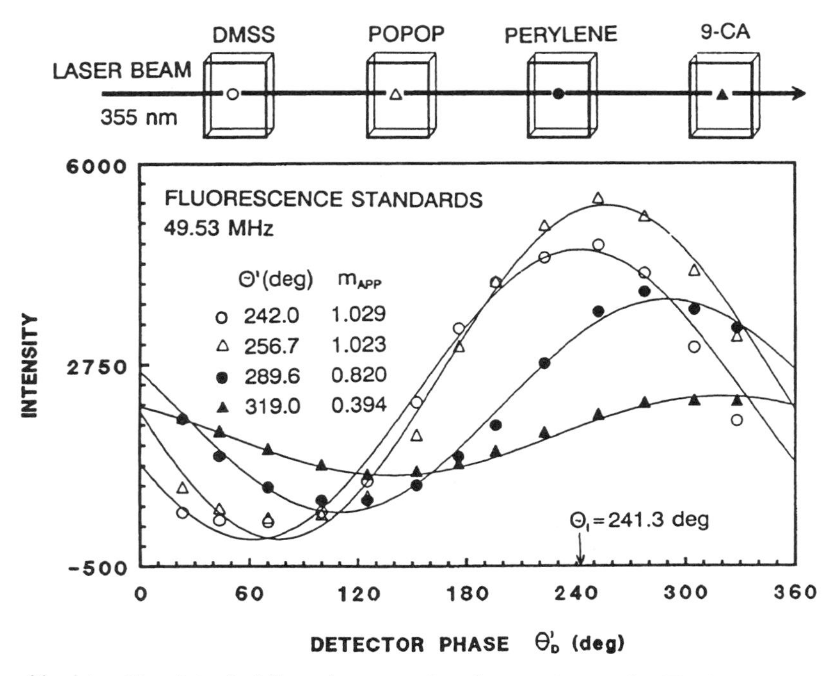

Figure 22.18. Phase-sensitive intensities of standard fluorophores at various detector phase angles. The decay times are 0.04 (○), 1.10 (△), 3.75 (●), and 9.90 ns (▲). The modulation frequency was 49.53 MHz. From Ref. 37.

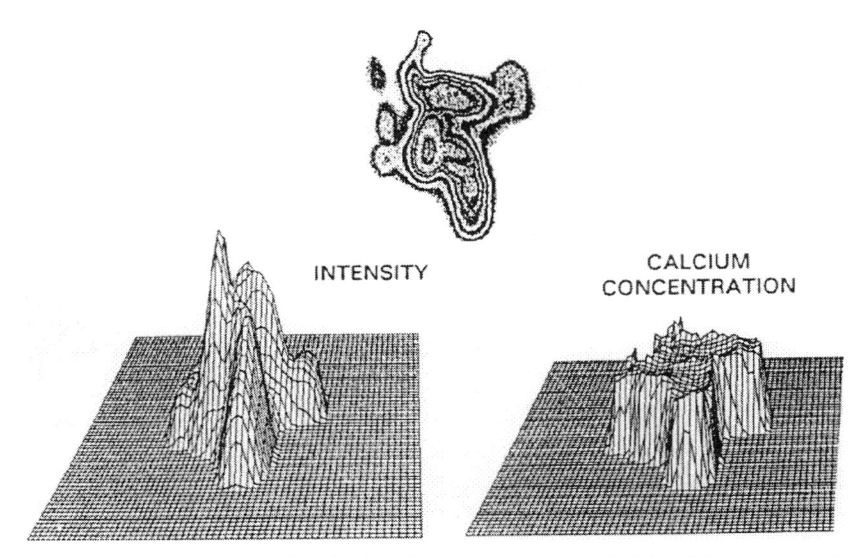

Figure 22.19. Intensity (*left*) and calcium images (*right*) of Quin-2 fluorescence in COS cells. The calcium concentrations were calculated using the calcium-dependent phase angles of Quin-2. Reprinted, with permission, from Ref. 42, Copyright © 1996, John Wiley & Sons, Inc.

One then subtracts the two phase-sensitive images. By a series of trigonometric rearrangements, one can show that the intensity in the difference image is zero for components which display a phase angle of ϕ_S,

$$\phi_S = \phi_D + \Delta/2 \pm n180° \quad [22.22]$$

where n is an integer. The lifetime of the suppressed component is given by

$$\tau_S = \omega^{-1} \tan \theta_S \quad [22.23]$$

The concept of phase suppression is shown schematically in Figure 22.20, where the detector phase-angle difference between two phase-sensitive images is 180°. The suppressed component (τ_S) has equal phase-sensitive intensities at θ_D and $\theta_D + 180°$. If the lifetime (τ_2) is longer than τ_S, then this component displays negative intensity in the difference images. If the lifetime (τ_1) is shorter than τ_S, then this component displays a positive phase-sensitive intensity.

The principle of phase suppression in FLIM can be illustrated by images of NADH in cuvettes. Figure 22.21

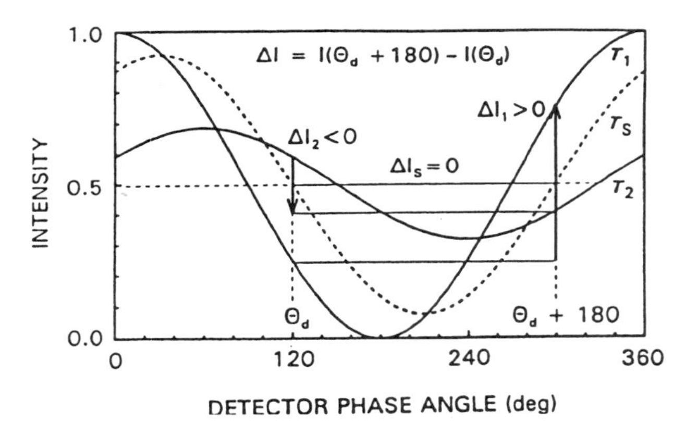

Figure 22.20. Intuitive description of phase suppression. In a difference image with $\Delta I = I(\theta_D + 180°) - I(\theta_D)$, a component with $\tau = \tau_S$ is completely suppressed ($\Delta I_S = 0$). Components with longer lifetimes (τ_2) appear as negative values ($\Delta I_2 < 0$), and those with shorter lifetimes (τ_1) appear to be positive ($\Delta I_1 > 0$). From Ref. 38.

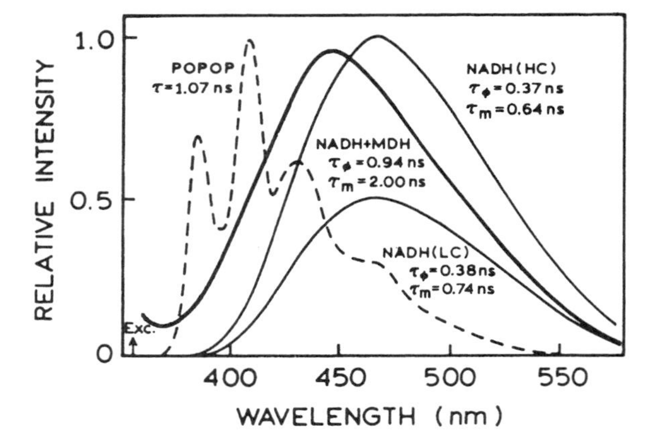

Figure 22.21. Emission spectra of NADH solutions used for phase-suppression imaging. The dashed curve is the emission spectrum of POPOP, which served as a lifetime standard. The concentrations of NADH were 10μ*M* (LC) and 20μ*M* (HC) in the absence of malate dehydrogenase (MDH) and 10μ*M* in the presence of 20μ*M* MDH. Revised from Ref. 36.

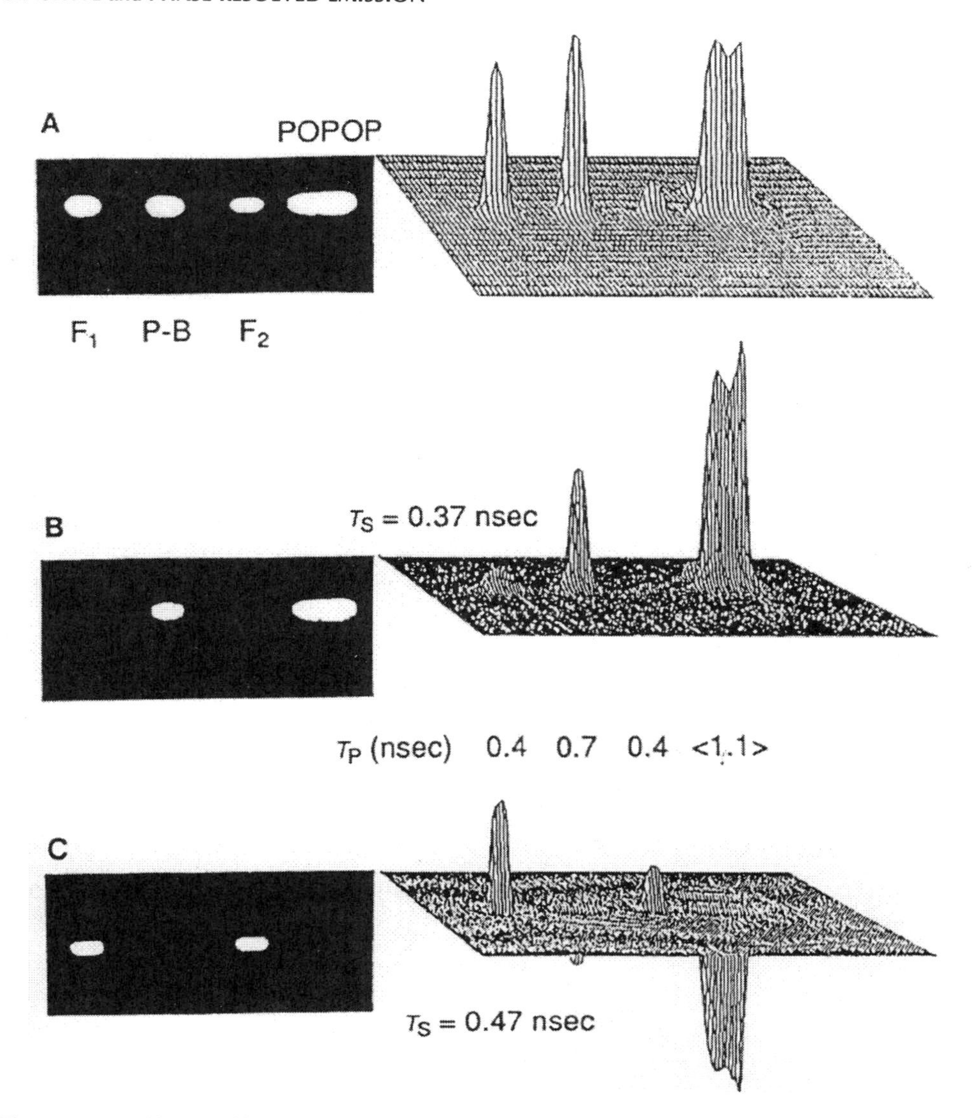

Figure 22.22. Phase-suppressed images of free and protein-bound NADH. (A) Phase-sensitive intensity image of samples (F_1 and F_2, free NADH; P-B, NADH bound to protein; POPOP, standard fluorophore used for lifetime calculation). (B) Difference image with a suppressed lifetime of 0.37 ns. (C) Difference image with a suppressed lifetime of 0.47 ns. From Ref. 39.

shows the emission spectra of NADH free in solution, with a phase lifetime near 0.37 ns, and bound to malate dehydrogenase, with a phase lifetime near 0.94 ns. There are two protein-free samples, each with a different concentration of NADH. Also included is a lifetime standard of POPOP with $\tau_R = 1.07$ ns.

Phase-sensitive images and difference images of these NADH solutions are shown in Figure 22.22. In the nonprocessed image, all the samples display a positive nonzero intensity. Phase-suppressed images are shown in the middle and lower panels, where the suppressed lifetimes are 0.37 and 0.47 ns, respectively. Also shown are the gray-scale images, in which only the positive regions are nonzero. When the suppressed lifetime is 0.37 ns, the emission from both cuvettes with free NADH is suppressed. When the suppressed lifetime is longer (0.47 ns), emission from protein-bound NADH is suppressed, and the difference images show positive signals for the free NADH samples.

REFERENCES

1. Lakowicz, J. R., and Cherek, H., 1981, Phase-sensitive fluorescence spectroscopy. A new method to resolve fluorescence lifetimes or

emission spectra of components in a mixture of fluorophores, *J. Biochem. Biophys. Methods* **5**:19–35.
2. Lakowicz, J. R., and Cherek, H., 1981, Resolution of heterogeneous fluorescence from proteins and aromatic amino acids by phase-sensitive detection of fluorescence, *J. Biol. Chem.* **256**:6348–6353.
3. Lakowicz, J. R., and Cherek, H., 1982, Resolution of heterogeneous fluorescence by phase-sensitive fluorescence spectroscopy, *Biophys. J.* **37**:148–150.
4. Jameson, D. M., Gratton, E., and Hall, R. D., 1984, The measurement and analysis of heterogeneous emissions by multifrequency phase and modulation fluorometry, *Appl. Spectrosc. Rev.* **20**(1):55–106.
5. McGown, L., and Bright, F., 1984, Phase-resolved fluorescence spectroscopy, *Anal. Chem.* **56**:1400–1415.
6. Fugate, R. D., Bartlett, J. D., and Mattheis, J. R., 1984, Phase-resolution in spectrofluorometric measurements: Applications to biochemical systems, *BioTechniques* **1984 (May/June)**:174–180.
7. McGown, L. B., 1985, Determination of fluorescence lifetimes and heterogeneity analysis using time-independent phase-resolved intensity measurements, *Anal. Instrum.* **14**(3&4):251–265.
8. McGown, L. B., and Nithipatikom, K., 1991, Multicomponent determinations of polycyclic aromatic hydrocarbons using synchronous excitation phase-resolved fluorometry, in *Advances in Multidimensional Luminescence*, Vol. 1, I. M. Warner and L. B. McGown (eds.), JAI Press, Greenwich, Connecticut, pp. 97–109.
9. McGown, L. B., and Bright, F. V., 1987, Phase-resolved fluorescence in chemical analysis, *Anal. Chem.* **18**:245–298.
10. Demas, J. N., and Keller, R. A., 1985, Enhancement of luminescence and Raman spectroscopy by phase-resolved background suppression, *Anal. Chem.* **57**:538–545.
11. Nithipatikom, K., and McGown, L. B., 1987, Phase-resolved suppression of scattered light in total luminescence spectra, *Appl. Spectrosc.* **41**:1080–1082.
12. Wirth, M. J., and Chou, S.-H., 1988, Comparison of time and frequency domain methods for rejecting fluorescence from Raman spectra, *Anal. Chem.* **60**:1882–1886.
13. Nithipatikom, K., and McGown, L. B., 1986, Elimination of scatter background in synchronous excitation spectrofluorometry by the use of phase-resolved fluorescence spectroscopy, *Anal. Chem.* **58**:3145–3147.
14. Lakowicz, J. R., and Balter, A., 1982, Direct recording of the initially excited and the solvent relaxed fluorescence emission spectra of tryptophan by phase sensitive detection of fluorescence, *Photochem. Photobiol.* **36**:125–132.
15. Lakowicz, J. R., and Balter, A., 1982, Detection of the reversibility of an excited-state reaction by phase-modulation fluorometry, *Chem. Phys. Lett.* **92**(2):117–121.
16. Lakowicz, J. R., Thompson, R. B., and Cherek, H., 1983, Phase fluorometric studies of spectral relaxation at the lipid–water interface of phospholipid vesicles, *Biochim. Biophys. Acta* **734**:295–308.
17. Nithipatikom, K., and McGown, L. B., 1987, Five- and six-component determinations using phase-resolved fluorescence spectroscopy and synchronous excitation, *Appl. Spectrosc.* **41**:395–398.
18. Bright, F. V., and McGown, L. B., 1986, Three-component determinations using fluorescence anisotropy measurements and wavelength selectivity, *Anal. Chem.* **58**:1424–1427.
19. Bright, F. V., and McGown, L. B., 1985, Phase-resolved fluorometric determinations of four-component systems using two modulation frequencies, *Anal. Chem.* **57**:2877–2880.
20. Bright, F. V., and McGown, L. B., 1985, Four-component determinations using phase-resolved fluorescence spectroscopy, *Anal. Chem.* **57**:55–59.
21. Vitense, K. R., and McGown, L. B., 1987, Simultaneous determination of metals in two-component mixtures with 5-sulfo-8-quinolinol by using phase-resolved fluorimetry, *Anal. Chim. Acta* **193**:119–125.
22. Nithipatikom, K., and McGown, L. B., 1986, Multidimensional data formats for phase-resolved fluorometric multicomponent determinations using synchronous excitation and emission spectra, *Anal. Chem.* **58**:2469–2473.
23. Bright, F. V., and McGown, L. B., 1984, Elimination of bilirubin interference in fluorometric determination of fluorescein by phase-resolved fluorescence spectrometry, *Anal. Chim. Acta* **162**:275–283.
24. Lakowicz, J. R., and Keating, S., 1983, Binding of an indole derivative to micelles as quantified by phase-sensitive detection of fluorescence, *J. Biol. Chem.* **258**:5519–5524.
25. McGown, L. B., 1984, Phase-resolved fluoroimetric determination of two albumin-bound fluorescein species, *Anal. Chim. Acta* **157**:327–332.
26. Nithipatikom, K., and McGown, L. B., 1989, Studies of the homogeneous immunochemical determination of insulin by using a fluorescent label, *Talanta* **36**(1/2):305–309.
27. Nithipatikom, K., and McGown, L. B., 1987, Homogeneous immunochemical technique for determination of human lactoferrin using excitation transfer and phase-resolved fluorometry, *Anal. Chem.* **59**:423–427.
28. Tahboub, Y. R., and McGown, L. B., 1986, Phase-resolved fluoroimmunoassay of human serum albumin, *Anal. Chim. Acta* **182**:185–191.
29. Veselova, T. V., Cherkasov, A. S., and Shirokov, V. I., 1970, Fluorometric method for individual recording of spectra in systems containing two types of luminescent centers, *Opt. Spectrosc.* **29**:617–618.
30. Gratton, E., and Jameson, D. M., 1985, New approach to phase and modulation resolved spectra, *Anal. Chem.* **57**:1694–1697.
31. Lakowicz, J. R., and Balter, A., 1982, Theory of phase-modulation fluorescence spectroscopy for excited state processes, *Biophys. Chem.* **16**:99–115.
32. Lakowicz, J. R., and Balter, A., 1982, Analysis of excited-state processes by phase-modulation fluorescence spectroscopy, *Biophys. Chem.* **16**:117–132.
33. Veselova, T. V., Limareva, L. A., Cherkasov, A. S., and Shirokov, V. I., 1965, Fluorometric study of the effect of solvent on the fluorescence spectrum of 3-amino-*N*-methylphthalimide, *Opt. Spectrosc.* **19**:39–43.
34. Limareva, L. A., Cherkasov, A. S., and Shirokov, V. I., 1968, Evidence of the radiating-centers inhomogeneity of crystalline anthracene in fluorometric phase spectra, *Opt. Spectrosc.* **25**:132–134.
35. Barcellona, M. L., and Gratton, E., 1990, The fluorescence properties of a DNA probe, *Eur. Biophys. J.* **17**:315–323.
36. Lakowicz, J. R., Szmacinski, H., Nowaczyk, K., and Johnson, M. L., 1992, Fluorescence lifetime imaging of free and protein-bound NADH, *Proc. Natl. Acad. Sci. U.S.A.* **89**:1271–1275.
37. Lakowicz, J. R., Szmacinski, H., Nowaczyk, K., Berndt, B. W., and Johnson, M. L., 1992, Fluorescence lifetime imaging, *Anal. Biochem.* **202**:316–330.
38. Lakowicz, J. R., Szmacinski, H., Nowaczyk, K., and Johnson, M. L., 1992, Fluorescence lifetime imaging of calcium using quin-2, *Cell Calcium* **13**:131–147.
39. Szmacinski, H., Lakowicz, J. R., and Johnson, M. L., 1994, Fluorescence lifetime imaging microscopy: Homodyne technique using high-speed gated image intensifier, *Methods Enzymol.* **240**:723–748.

40. Draaijer, A., Sanders, R., and Gerritsen, H. C., 1995, Fluorescence lifetime imaging, a new tool in confocal microscopy, in *Handbook of Biological Confocal Microscopy*, J. B. Pawley (ed.), Plenum Press, New York, pp. 491–505.
41. French, T., Gratton, E., and Maier, J., 1992, Frequency domain imaging of thick tissues using a CCD, *Proc. SPIE* **1640**:254–261.
42. Lakowicz, J. R., and Szmacinski, H., 1996, Imaging applications of time-resolved fluorescence spectroscopy, in *Fluorescence Imaging Spectroscopy and Microscopy*, X. F. Wang and B. Herman (eds.), John Wiley & Sons, New York, pp. 273–311.
43. Wang, X. F., Periasamy, A., Wodnicki, P., Gordon, G. W., and Herman, B., 1996, Time-resolved fluorescence lifetime imaging microscopy: Instrumentation and biomedical applications, in *Fluorescence Imaging Spectroscopy and Microscopy*, X. F. Wang and B. Herman (eds.), John Wiley & Sons, New York, pp. 313–350.
44. Gerritsen, H., and Draaijer, A., 1997, Second International Lifetime Imaging Meeting, Utrecht, The Netherlands, June 14, 1996, *J. Fluoresc.* **7**(1):1–98.
45. Herman, B., Wodnicki, P., Kwon, S., Periasamy, A., Gordon, G. W., Mahajan, N., and Wang, X. F., 1997, Recent developments in monitoring calcium and protein interactions in cells using fluorescence lifetime microscopy, *J. Fluoresc.* **7**(1):85–91.
46. Wagnieres, G., Mizeret, J., Studzinski, A., and van den Bergh, H., 1997, Frequency-domain fluorescence lifetime imaging for endoscopic clinical cancer photodetection: Apparatus design and preliminary results, *J. Fluoresc.* **7**(1):75–83.
47. Hartmann, P., and Ziegler, W., 1996, Lifetime imaging of luminescent oxygen sensors based on all-solid-state technology, *Anal. Chem.* **58**:4512–4514.
48. Cubeddu, R., Canti, G., Pifferi, A., Taroni, P., and Valentini, G., 1997, Fluorescence lifetime imaging of experimental tumors in hematoporphyrin derivative-sensitized mice, *Photochem. Photobiol.* **66**:229–236.
49. Morgan, C. G., and Mitchell, A. C., 1996, Fluorescence lifetime imaging: An emerging technique in fluorescence microscopy, *Chromosome Res.* **4**:261–263.
50. Schneider, P. C., and Clegg, R. M., 1997, Rapid acquisition, analysis, and display of fluorescence lifetime-resolved images for real-time applications, *Rev. Sci. Instrum.* **68**:4107–4119.
51. Dowling, K., Hyde, S. C. W., Dainty, J. C., French, P. M. W., and Hares, J. D., 1997, 2-D fluorescence lifetime imaging using a time-gated image intensifier, *Opt. Commun.* **135**:27–31.
52. Gadella, T. W. J., van Hoek, A., and Visser, A. J. W. G., 1997, Construction and characterization of a frequency-domain fluorescence lifetime imaging microscopy system, *J. Fluoresc.* **7**(1):35–43.
53. vandeVen, M., and Gratton, E., 1993, Time-resolved fluorescence lifetime imaging, in *Optical Microscopy: Emerging Methods and Applications*, B. Herman and J. J. Lemasters (eds.), Academic Press, New York, pp. 373–402.
54. Lakowicz, J. R., Szmacinski, H., Nowaczyk, K., Lederer, W. J., Kirby, M. S., and Johnson, M. L., 1994, Fluorescence lifetime imaging of intracellular calcium in COS cells using Quin-2, *Cell Calcium* **15**:7–27.

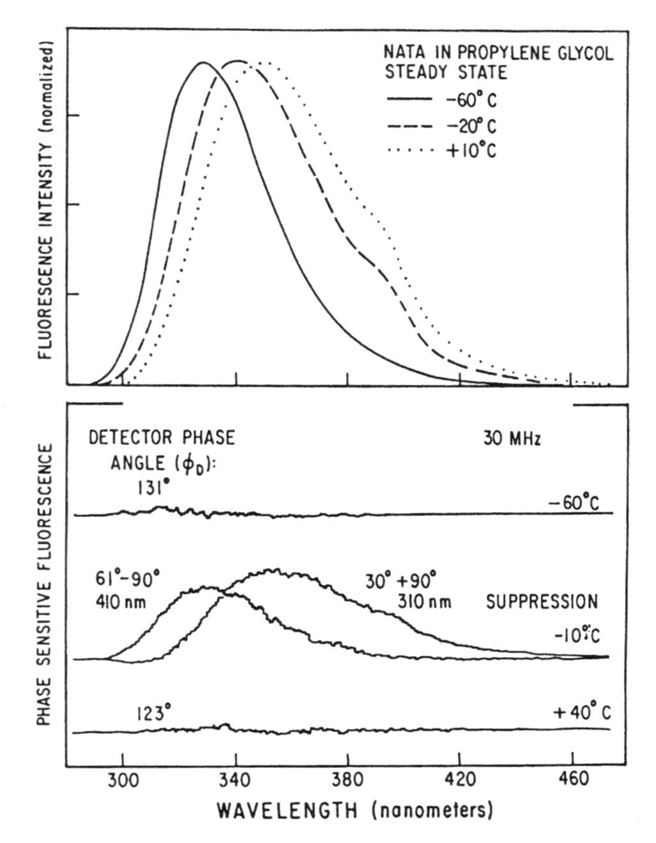

Figure 22.23. Resolution of the initially excited and relaxed states of *N*-acetyl-L-tryptophanamide by phase-sensitive detection of fluorescence. Excitation was at 280 nm. Emission was observed through a monochromator with a bandpass of 8 nm. Reprinted from Ref. 32, Copyright © 1982, with permission from Elsevier Science.

PROBLEMS

22.1. *Determination of the Excitation Wavelength*: The steady-state emission spectrum of quinine sulfate shows a Raman scatter peak at 410 nm (Figure 22.7). What is the excitation wavelength?

22.2. *Attenuation of Phase-Sensitive Spectra*: Assume that the lifetime of quinine sulfate is 20 ns and that the phase-sensitive spectrum in Figure 22.7 was obtained with a light modulation frequency of 10 MHz. What detector phase angle was used to suppress the scattered light? What detector phase angle would yield the highest signal for quinine sulfate? What are the relative values of the quinine sulfate intensity for the maximum intensity and when the light is suppressed?

22.3. Assume that the flourescent probe 5-dimethylaminonaphthalene-1-sulfonic acid (DNS) binds to bovine serum albumin (BSA). Assume further that the yield of DNS increases twofold upon binding and that the lifetime of the free and bound forms are 5 and 10 ns, respectively. Use the following data to calculate the percentage of DNS free in solution and the percentage bound to BSA in the solution containing equimolar concentrations of DNS and BSA. The modulation frequency is 10 MHz. Also, explain the intensity changes between the first two solutions.

Sample	Phase-sensitive intensity at:	
	$\phi_D = 17.44° + 90°$	$\phi_D = 32.1° - 90°$
DNS ($10^{-5}M$)	0	1.0
DNS ($10^{-5}M$) plus excess BSA	1.776	0
DNS ($10^{-5}M$) plus $10^{-5}M$ BSA	0.886	0.50

22.4. *Phase-Sensitive Spectra and Spectral Relaxation*: Phase-sensitive emission spectra were obtained for *N*-acetyl-L-tryptophanamide in propylene glycol at various temperatures (Figure 22.23). These spectra were recorded following adjustment of the detector phase to suppress the emission on the blue or the red side of the emission. Explain the phase-sensitive spectra in Figure 22.23 in terms of the rates of spectral relaxation.

Appendix I Corrected Emission Spectra

A relatively limited number of corrected emission spectra are available. In this appendix, we included those corrected spectra which are well documented. However, it is not possible to state which are the "most correct." Corrected spectra can be interchanged from photons per wavenumber interval $[I(\bar{\nu})]$ to photons per wavelength interval $[I(\lambda)]$ using $I(\lambda) = \lambda^{-2} I(\bar{\nu})$, followed by normalization of the peak intensity to unity.

1. β-CARBOLINE DERIVATIVES AS FLUORESCENCE STANDARDS

Corrected emission spectra of the fluorophores shown in Figure I.1 were recently published.[1] All compounds were measured in 0.1*N* H_2SO_4, except for harmaline, which was measured in 0.01*N* H_2SO_4, at 25 °C. Corrected emission spectra were reported in graphical form (Figure I.2) and in numerical form (Table I.1). Quantum yields and lifetimes were also reported (Table I.2). Quantum yields were determined relative to quinine sulfate in 1.0*N* H_2SO_4, with $Q = 0.546$. Corrected spectra are in relative quanta per wavelength interval. A second corrected emission spectrum of norharmane (β-carboline) was also reported (Table I.3).[3] The spectral properties of β-carboline are similar to those of quinine sulfate.

These compounds satisfy the suggested criteria for emission spectral standards,[3,4] which are as follows:

1. Broad-wavelength emission with no fine structure
2. Chemically stable and easily available and purified
3. High quantum yield
4. Emission spectrum independent of excitation wavelength
5. Completely depolarized emission

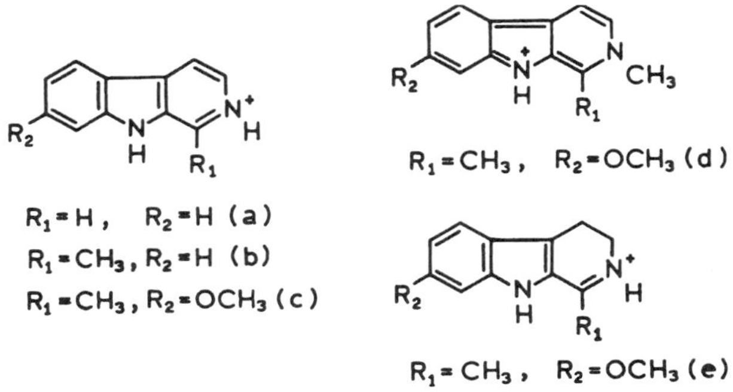

$R_1 = H$, $R_2 = H$ (a)
$R_1 = CH_3$, $R_2 = H$ (b)
$R_1 = CH_3$, $R_2 = OCH_3$ (c)
$R_1 = CH_3$, $R_2 = OCH_3$ (d)
$R_1 = CH_3$, $R_2 = OCH_3$ (e)

Figure I.1. β-Carboline standards. Cationic species structures: (a) β-carboline or norharmane, (b) harmane, (c) harmine, (d) 2-methylharmine, and (e) harmaline. Revised from Ref. 1, Copyright © 1992, with permission from Elsevier Science.

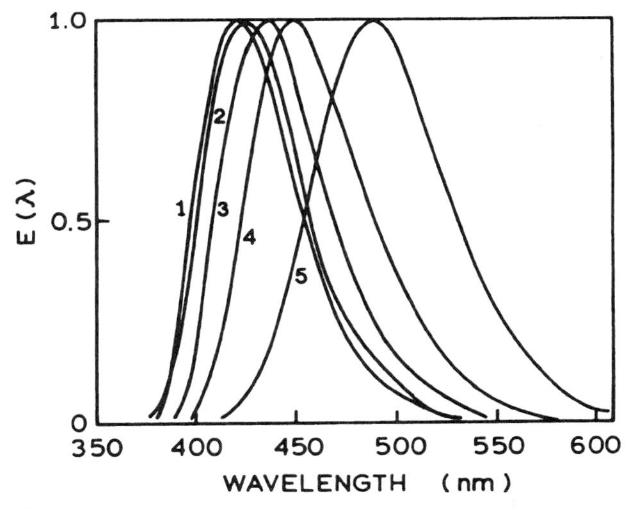

Figure I.2. Corrected and normalized fluorescence spectra for some β-carboline derivatives. (1) Harmine in 0.1*N* H_2SO_4; (2) 2-methylharmine in 0.1*N* H_2SO_4; (3) harmane in 0.1*N* H_2SO_4; (4) norharmane in 0.1*N* H_2SO_4; (5) harmaline in 0.01*N* H_2SO_4. Revised from Ref. 1, Copyright © 1992, with permission from Elsevier Science.

Table I.1. Normalized and Corrected Fluorescence Spectra for Norharmane, Harmane, Harmine, 2-Methylharmine, and Harmaline in H_2SO_4 Aqueous Solutions[a]

Wavelength (nm)	Norharmane or β-carboline[b]	Harmane[b]	Harmine[b]	2-Methylharmine[b]	Harmaline[c]
400			0.63		
405			0.76	0.53	
410			0.87	0.68	
415		0.56	0.93	0.78	
420		0.70	0.98	0.89	
425	0.54	0.81	1.00	0.95	
430	0.67	0.90	0.99	0.98	
433				1.00	
435	0.79	0.97	0.94	0.99	
440	0.90	1.00	0.97[d]	0.97	
445	0.96	0.98	0.80	0.90	
450	0.99	0.94	0.73	0.84	
454	1.00				
455	0.99	0.88	0.64	0.76	
460	0.98	0.82	0.54	0.66	0.48
465	0.96	0.76	0.48	0.58	0.59
470	0.93	0.69		0.52	0.69
475	0.87	0.61		0.44	0.78
480	0.83	0.56			0.86
485	0.78	0.50			0.92
490	0.73				0.96
495	0.67				0.98
498					1.00
500	0.61				0.99
505	0.54				0.97
510					0.96
515					0.92
520					0.89
525					0.83
530					0.77
535					0.74
540					0.71
550					0.59
560					0.49

[a]From Ref. 1.
[b]In 0.1N H_2SO_4.
[c]In 0.01N H_2SO_4.
[d]This number is in question.

Table I.2. Quantum Yields and Lifetimes of the β-Carboline Standards[a,b]

Compound	Quantum yield	Lifetime (ns)
Norharmane	0.58 ± 0.02	21.2
Harmane	0.83 ± 0.03	20.0 ± 0.5
Harmine	0.45 ± 0.03	6.6 ± 0.2
2-Methylharmine	0.45 ± 0.03	6.5 ± 0.2
Harmaline	0.32 ± 0.02	5.3 ± 0.2

[a]From Ref. 1.
[b]Same conditions as in Table I.1.

Table I.3. Corrected Emission Intensities for Norharmane (β-Carboline) in 1.0N H_2SO_4 at 25 °C[a,b]

Wavelength (nm)	Corrected intensity	Wavelength (nm)	Corrected intensity
380	0.001	510	0.417
390	0.010	520	0.327
400	0.068	530	0.255
410	0.243	540	0.193
420	0.509	550	0.143
430	0.795	560	0.107
440	0.971	570	0.082
450	1.000	580	0.059
460	0.977	590	0.044
470	0.912	600	0.034
480	0.810	610	0.025
490	0.687	620	0.019
500	0.540	630	0.011

[a]From Ref. 3.
[b]Excitation at 360 nm.

While the polarization of the β-carboline standards was not measured, these values are likely to be near zero, given that the lifetimes are near 20 ns (Table I.2).

2. CORRECTED EMISSION SPECTRA OF 9,10-DIPHENYLANTHRACENE, QUININE SULFATE, AND FLUORESCEIN

Corrected spectra in quanta per wavelength interval [$I(\lambda)$] were published for 9,10-diphenylanthracene, quinine sulfate, and fluorescein[5] (Figure I.3 and Table I.4). The emission spectrum for quinine sulfate was found to be at somewhat shorter wavelengths than that published by Melhuish.[6]

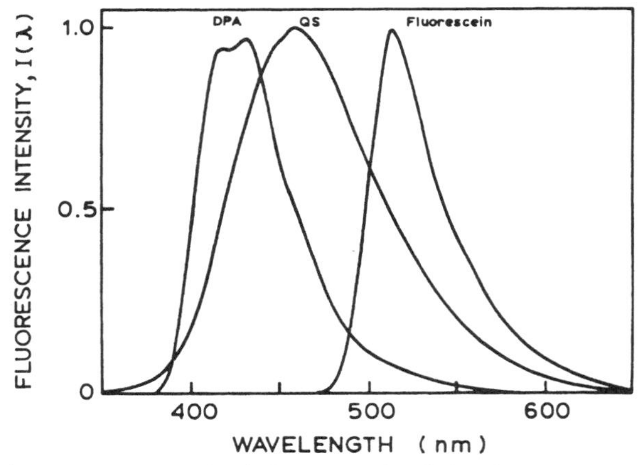

Figure I.3. Corrected emission spectra, in relative photons per wavelength interval [$I(\lambda)$], for 9,10-diphenylanthracene (DPA), quinine sulfate (QS), and fluorescein. From Ref. 5.

Table I.4. Corrected Emission Spectra, in Relative Quanta per Wavelength Interval, for Quinine Sulfate, Fluorescein, and 9,10-Diphenylanthracene (DPA)[a]

Quinine sulfate[b]		Fluorescein[c]		DPA[d]	
λ (nm)	I (λ)	λ (nm)	I (λ)	λ (nm)	I (λ)
310	0	470	0	380	0
350	4	480	7	390	39
380	18	490	151	400	423
400	151	495	360	412	993
410	316	500	567	422	914
420	538	505	795	432	1000
430	735	510	950	440	882
440	888	512	1000	450	607
445	935	515	985	460	489
450	965	520	933	470	346
455	990	525	833	480	222
457.2	1000	530	733	490	150
400	998	540	533	500	103
465	979	550	417	550	4
470	951	560	333	600	0
475	916	570	233		
480	871	580	167		
490	733	600	83		
500	616	620	42		
520	408	640	17		
550	171	650	8		
600	19	670	0		
650	3				
700	0				

[a]From Ref. 5.
[b]In 1.0N H_2SO_4; excitation at 346.5 nm.
[c]In 0.1N NaOH; excitation at 322 nm.
[d]In benzene; excitation at 385 nm.

3. LONG-WAVELENGTH STANDARDS

Corrected emission spectra in relative quanta per wavelength interval were reported in Ref. 4 for quinine sulfate (QS), 3-aminophthalimide (3-AP), and *N,N*-dimethylamino-*m*-nitrobenzene (N,N-DMANB). These standards are useful because they extend the wavelength range to 750 nm (Figure I.4). The data were not reported in numerical form. These corrected emission spectra [$I(\lambda)$] were in agreement with those reported earlier[7] in relative quanta per wavenumber [$I(\bar{\nu})$], following the appropriate transformation.

A more complete set of corrected spectra,[7,8] in relative quanta per wavenumber interval, are presented in Figure I.5 and summarized numerically in Table I.5. These data contain an additional standard, 4-dimethylamino-4′-nitrostilbene (4,4′-DMANS), which extends the wavelength

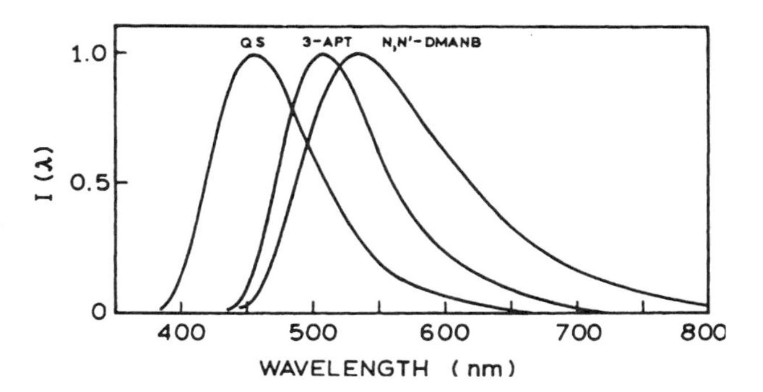

Figure I.4. Corrected emission spectra, in relative quanta per wavelength interval [$I(\lambda)$], for quinine sulfate (QS) in 0.1N H_2SO_4. 3-aminophthalimide (3-AP) in 0.1N H_2SO_4, and *N,N*-dimethylamino-*m*-nitrobenzene (N,N-DMANB) in 30% benzene/70% *n*-hexane. Modified from Ref. 4.

range to 940 nm. These spectra are plotted on the wavenumber scale in Figure I.5. For convenience, the data are transformed to the wavelength scale in Figure I.6 and Table I.6. In summarizing these corrected spectra, we omitted β-naphthol, whose emission spectrum depends on pH and buffer concentration. Because these factors change its spectral shape, β-naphthol is not a good standard.

4. ULTRAVIOLET STANDARDS

2-Aminopyridine has been suggested as a standard from 315 to 480 nm,[9,10] which covers most, but not all, of the wavelengths needed for tryptophan fluorescence (Table I.7

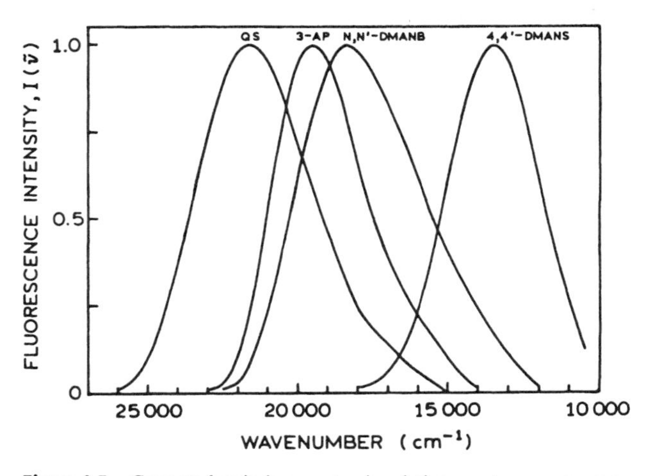

Figure I.5. Corrected emission spectra, in relative quanta per wavenumber interval [$I(\bar{\nu})$], for quinine sulfate (QS), 3-aminophthalimide (3-AP), *N,N*-dimethylamino-*m*-nitrobenzene (N,N′-DMANB), and 4-dimethylamino-4′-nitrostilbene (4,4′ DMANS). See Table I.5 for additional details. Data from Refs. 7 and 8.

Table I.5. Corrected Emission Spectra, in Relative Quanta per Wavenumber Interval,[a] for Quinine Sulfate, 3-Aminophthalimide (3-AP), *N,N*-Dimethylamino-*m*-nitrobenzene (N,N′-DMANB), and 4-Dimethylamino-4′-nitrostilbene (4,4′-DMANS)

Quinine sulfate[b]		3-AP[c]		N,N′-DMANB[d]		4,4′-DMANS[e]	
$\bar{\nu}$ (cm^{-1})	$I(\bar{\nu})$	$\bar{\nu}$ (cm^{-1})	$I(\bar{\nu})$	$\bar{\nu}$ (cm^{-1})	$I(\bar{\nu})$	$\bar{\nu}$ (cm^{-1})	$I(\bar{\nu})$
15.0	0	14.0	1.5	12.0	2.0	10.5	12.5
15.25	1.5	14.25	3.0	12.25	4.0	10.75	18.5
15.5	3.0	14.5	5.0	12.5	6.0	11.0	24.5
15.75	4.5	14.75	7.5	12.75	8.5	11.25	32.5
16.0	6.0	15.0	10.0	13.0	11.0	11.5	41.5
16.25	7.5	15.25	13.0	13.25	13.5	11.75	50.5
16.5	9.5	15.5	16.0	13.5	17.0	12.0	60.0
16.75	11.5	15.75	19.0	13.75	20.0	12.25	70.5
17.0	14.0	16.0	22.0	14.0	23.5	12.5	80.5
17.25	16.0	16.25	25.5	14.25	27.5	12.75	89.0
17.5	18.0	16.5	29.5	14.5	31.0	13.0	95.0
17.75	20.5	16.75	33.5	14.75	35.5	13.25	98.5
18.0	24.0	17.0	38.5	15.0	40.5	13.5	100.0
18.25	28.5	17.25	44.0	15.25	45.0	13.75	98.0
18.5	34.5	17.5	50.0	15.5	50.0	14.0	94.0
18.75	40.5	17.75	56.5	15.75	55.5	14.25	88.0
19.0	46.0	18.0	65.0	16.0	61.5	14.5	81.0
19.25	52.5	18.25	73.0	16.25	68.0	14.75	72.0
19.5	58.5	18.5	82.5	16.5	73.0	15.0	61.5
19.75	65.0	18.75	90.0	16.75	78.0	15.25	51.0
20.0	71.5	19.0	95.0	17.0	82.5	15.5	41.0
20.25	78.5	19.25	98.5	17.25	87.0	15.75	32.0
20.5	84.5	19.5	100.0	17.5	91.5	16.0	24.0
20.75	90.0	19.75	98.5	17.75	95.0	16.25	17.5
21.0	95.0	20.0	94.5	18.0	97.5	16.5	13.0
21.25	98.5	20.25	87.5	18.25	99.5	16.75	9.0
21.5	100.0	20.5	77.5	18.5	99.5	17.0	6.0
21.75	99.5	20.75	66.0	18.75	97.5	17.25	4.0
22.0	98.0	21.0	53.0	19.0	93.5	17.5	2.5
22.25	94.5	21.25	39.5	19.25	87.0	17.75	2.0
22.5	89.0	21.5	28.0	19.5	80.0	18.0	1.5
22.75	82.5	21.75	17.5	19.75	71.5		
23.0	74.0	22.0	11.0	20.0	61.0		
23.25	65.5	22.25	6.0	20.25	51.0		
23.5	55.5	22.5	3.0	20.5	41.5		
23.75	46.0	22.75	1.5	20.75	32.5		
24.0	37.5	23.0	1.0	21.0	23.5		
24.25	29.5			21.25	16.0		
24.5	21.0			21.5	10.5		
24.75	15.0			21.75	6.0		
25.0	10.5			22.0	3.0		
25.25	6.5			22.25	2.0		
25.5	4.0			22.5	1.5		
25.75	2.5						
26.0	1.0						
Max. 21.6	100.0	Max. 19.5	100.0	Max. 18.4	100.0	Max. 13.5	100.0

[a] All listings are in 10^3 cm^{-1}; i.e., 13.5 is 13,500 cm^{-1}.
[b] 10^{-3} *M* in 0.1*N* H_2SO_4, 20 °C.
[c] 5×10^{-4} *M* in 0.1*N* H_2SO_4, 20 °C.
[d] 10^{-4} *M* in 30% benzene/70% *n*-hexane, 20 °C.
[e] In *o*-dichlorobenzene, 20 °C.

Table I.6. Corrected Emission Spectra, in Relative Quanta per Wavelength Interval,[a] for Quinine Sulfate, 3-Aminophthalimide (3-AP), N,N′-Dimethylamino-*m*-nitrobenzene (*N,N*-DMANB), and 4-Dimethylamino-4′-nitrostilbene (4,4′-DMANS)

Quinine sulfate[b]		3-AP[c]		N,N-DMANB[d]		4,4′-DMANS[e]	
λ (nm)	$I(\lambda)$	λ(nm)	$I(\lambda)$	λ(nm)	$I(\lambda)$	λ(nm)	$I(\lambda)$
384.6	1.4	434.8	1.4	444.4	2.2	555.6	2.6
388.3	3.5	439.6	2.0	449.4	2.9	563.4	3.4
392.2	5.5	444.4	4.0	454.5	4.2	571.4	4.1
396.0	8.7	449.4	7.7	459.8	8.3	579.7	6.4
400.0	13.8	454.5	13.9	465.1	14.2	588.2	9.4
404.0	19.4	459.8	21.5	470.6	21.1	597.0	13.6
408.2	26.6	465.1	33.7	476.2	30.2	606.1	19.1
412.4	36.6	470.6	46.4	481.9	40.8	615.4	24.9
416.7	45.5	476.2	60.8	487.8	50.9	625.0	33.2
421.1	54.7	481.9	74.0	493.8	61.0	634.9	42.8
425.5	64.6	487.8	84.8	500.0	71.2	645.2	53.2
430.1	74.6	493.8	93.4	506.3	81.4	655.7	64.0
434.8	82.5	500.0	98.4	512.8	88.7	666.7	74.7
439.6	90.0	506.3	100.0	519.5	94.1	678.0	84.5
444.4	95.0	512.8	99.0	526.3	98.5	689.7	91.9
449.4	98.6	519.5	95.0	533.3	100.0	701.8	96.4
454.5	100.0	526.3	89.2	540.5	99.3	714.3	99.4
459.8	99.2	533.3	82.3	547.9	96.7	727.3	100.0
465.1	97.5	540.5	73.5	555.6	92.2	740.7	98.4
470.6	93.8	547.9	63.3	563.4	87.3	754.7	93.3
476.2	88.3	555.6	54.8	571.4	81.8	769.2	86.7
481.9	81.7	563.4	46.3	579.7	75.5	784.3	78.1
487.8	74.9	571.4	39.9	588.2	69.6	800.0	67.9
493.8	67.9	579.7	34.1	597.0	63.8	816.3	57.1
500.0	60.3	588.2	29.0	606.1	58.0	833.3	46.6
506.3	53.4	597.0	24.5	615.4	52.4	851.1	37.6
512.8	46.9	606.1	20.9	625.0	45.9	869.6	29.6
519.5	41.0	615.4	17.5	634.9	40.2	888.9	22.2
526.3	35.0	625.0	14.7	645.2	35.0	909.1	16.0
533.3	30.0	634.9	12.3	655.7	30.5	930.2	11.5
540.5	24.9	645.2	10.0	666.7	26.6	952.4	7.4
547.9	20.0	655.7	7.9	678.0	22.5		
555.6	16.4	666.7	5.9	689.7	19.0		
563.4	13.6	678.0	4.2	701.8	16.3		
571.4	11.6	689.7	2.7	714.3	13.4		
579.7	10.0	701.8	1.6	727.3	11.0		
588.2	8.5	714.3	0.8	740.7	9.0		
597.0	6.8			754.7	6.9		
606.1	5.5			769.2	5.4		
615.4	4.2			784.3	4.0		
625.0	3.2			800.0	2.7		
634.9	2.4			816.3	1.8		
645.2	1.5			833.3	0.8		
655.7	0.7						
666.7	0.0						

[a]Calculated from the data in Table I.5, using $I(\lambda) = \lambda^{-2} I(\bar{\nu})$, followed by normalization of the peak intensity to 100.

[b]$10^{-3} M$ in 0.1N H_2SO_4, 20 °C.

[c]$5 \times 10^{-4} M$ in 0.1N H_2SO_4, 20 °C.

[d]10^{-4} M in 30% benzene, 70% *n*-hexane, 20 °C.

[e] In *o*-dichlorobenzene, 20 °C.

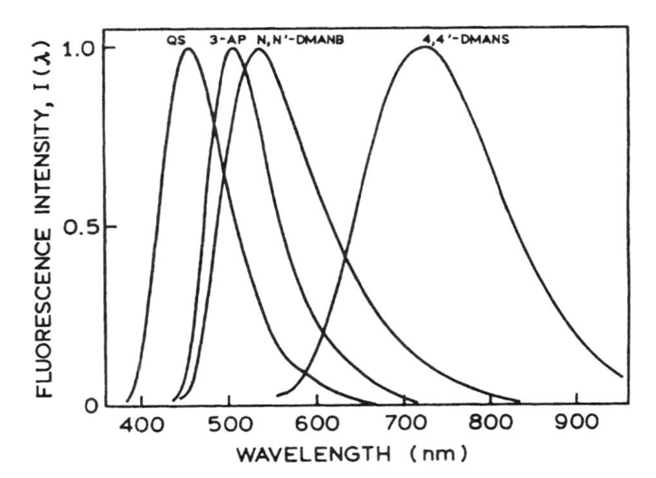

Figure I.6. Corrected emission spectra, in relative quantum per wavelength interval [$I(\lambda)$], for quinine sulfate (QS), 3-aminophthalimide (3-AP), *N,N*-dimethylamino-*m*-nitrobenzene (N,N'-DMANB), and 4-dimethylamino-4'-nitrostilbene (4,4' DMANS). See Table I.6 for additional details. Data from Refs. 7 and 8.

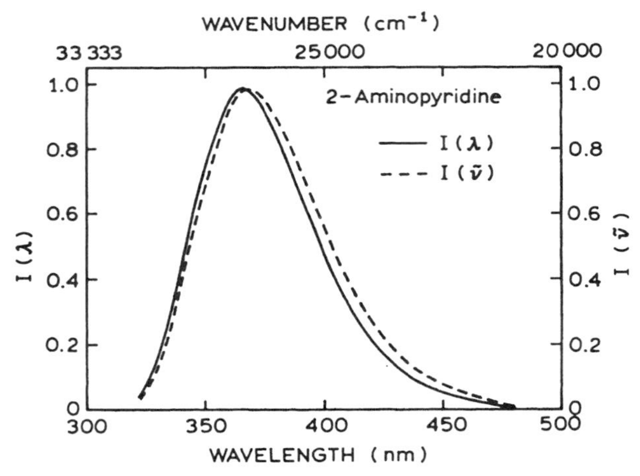

Figure I.7. Corrected emission spectra of 2-aminopyridine, in relative quanta per wavenumber interval [$I(\bar{\nu})$] and per wavelength interval [$I(\lambda)$]. See Table I.7 for additional details. From Refs. 9 and 10.

and Figure I.7). Corrected emission spectra have been reported[11] for phenol and the aromatic amino acids, phenylalanine, tyrosine, and tryptophan (Figure I.8).

5. ADDITIONAL CORRECTED EMISSION SPECTRA

Corrected spectra, as $I(\bar{\nu})$ versus $\bar{\nu}$, can be found in the compendium by Berlman.[12] Included in this volume are a number of UV-emitting species, including indole and phenol, which can be used to obtain corrected emission spectra of proteins. For convenience, Berlman's spectra for phenol and indole are provided in Figure I.9. Cresyl violet in methanol has been proposed as a quantum yield and emission spectral standard for red wavelengths.[13] In methanol at 22 °C, the quantum yield of cresyl violet is reported to be 0.54, with an emission maximum near 614 nm. The use of quinine as a standard has occasionally been questioned.[14–17] Additional discussion about corrected emission spectra can be found in Refs. 18 and 19.

Table I.7. Corrected Emission Spectrum of 2-Aminopyridine[a,b]

$\bar{\nu}$ (cm^{-1})	$I(\bar{\nu})$	λ (nm)	I(λ)[c]
20,800	0.010	480.8	0.006
21,500	0.038	465.1	0.024
22,200	0.073	450.5	0.049
23,000	0.133	434.8	0.095
23,800	0.264	420.2	0.202
24,700	0.450	404.9	0.371
25,600	0.745	390.6	0.660
26,600	0.960	375.9	0.918
27,200	1.00	367.7	1.00
27,800	0.939	359.7	0.981
28,900	0.587	346.0	0.663
30,150	0.121	331.7	0.149
31,000	0.033	322.6	0.049

[a]From Refs. 9 and 10.
[b]$10^{-5}M$ in $0.1N$ H_2SO_4.
[c]Calculated using $I(\lambda) = I(\bar{\lambda})\lambda^{-2}$ followed by normalization of the maximum to unity.

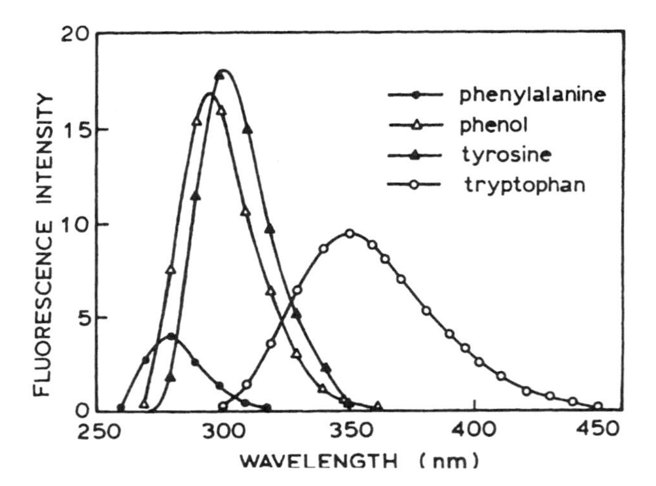

Figure I.8. Corrected emission spectra [$I(\lambda)$] for phenylalanine (●), phenol (△), tyrosine (▲), and tryptophan (○). The areas underneath the curves are proportional to the quantum yields. Data from Ref. 11.

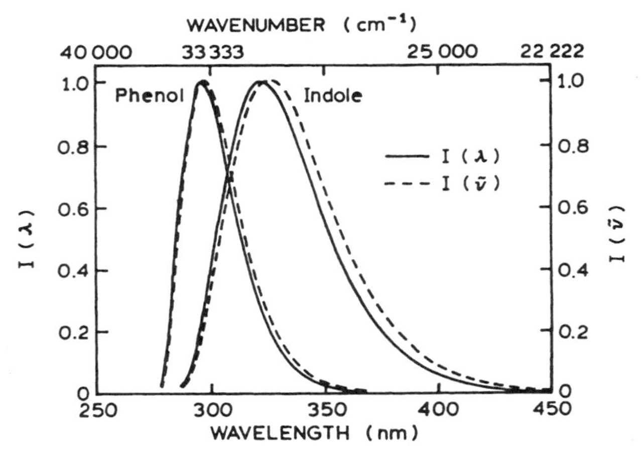

Figure I.9. Corrected emission spectra of phenol in methanol and indole in ethanol. From Ref. 12.

REFERENCES

1. Pardo, A., Reyman, D., Poyato, J. M. L., and Medina, F., 1992, Some β-carboline derivatives as fluorescence standards, *J. Lumin.* **51**:269–274.
2. Melhuish, W. H., 1972, Absolute spectrofluorometry, *J. Res. Natl. Bur. Stand. Sect. A* **A76**:547–560.
3. Ghiggino, K. P., Skilton, P. F., and Thistlethwaite, P. J., 1985, β-Carboline as a fluorescence standard, *J. Photochem.* **31**:113–121.
4. Velapoldi, R. A., 1972, Considerations on organic compounds in solution and inorganic ions in glasses as fluorescent standard reference materials, *Natl. Bur. Stand. Spec. Publ.* **378**:231–244.
5. Heller, C. A., Henry, R. A., McLaughlin, B. A., and Bliss, D. E., 1974, Fluorescence spectra and quantum yields: Quinine, uranine, 9,10-diphenylanthracene, and 9,10-bis(phenylethynyl)anthracenes, *J. Chem. Eng. Data* **19**(3):214–219.
6. Melhuish, W. H., 1960, A standard fluorescence spectrum for calibrating spectrofluorophotometers, *J. Phys. Chem.* **64**:762–764.
7. Lippert, E., Nägele, W., Seibold-Blankenstein, I., Staiger, W., and Voss, W., 1959, Messung von fluorescenzspektren mit hilfe von spektralphotometern und vergleichsstandards, *Z. Anal. Chem.* **170**:1–18.
8. Schmillen, A., and Legler, R., 1967, *Landolt-Börnstein, Volume 3, Lumineszenz Organischer Substanzen,* Springer-Verlag, New York, pp. 228–229.
9. Testa, A. C., 1969, Fluorescence quantum yields and standards, American Instrument Co., Newsletter on Luminescence **4**(4):1–3.
10. Rusakowicz, R., and Testa, A. C., 1968, 2-Aminopyridine as a standard for low-wavelength spectrofluorimetry, *J. Phys. Chem.* **72**:2680–2681.
11. Chen, R. F., 1967, Fluorescence quantum yields of tryptophan and tyrosine, *Anal. Lett.* **1**:35–42.
12. Berlman, I. B., 1971, *Handbook of Fluorescence Spectra of Aromatic Molecules*, 2nd ed., Academic Press, New York.
13. Magde, D., Brannon, J. H., Cremers, T. L., and Olmsted, J., 1979, Absolute luminescence yield of cresyl violet. A standard for the red, *J. Phys. Chem.* **83**:696–699.
14. Chen, R. F., 1967, Some characteristics of the fluorescence of quinine, *Anal. Biochem.* **19**:374–387.
15. Fletcher, A. N., 1968, Fluorescence emission band shift with wavelength of excitation, *J. Phys. Chem.* **72**:2742–2749.
16. Itoh, K., and Azumi, T., 1973, Shift of emission band upon excitation at the long wavelength absorption edge. A preliminary survey for quinine and related compounds, *Chem. Phys. Lett.* **22**:395–399.
17. Gill, J. E., 1969, The fluorescence excitation spectrum of quinine bisulfate, *Photochem. Photobiol.* **9**:313–322.
18. Melhuish, W. H., 1972, Absolute spectrofluorometry, *J. Res. Natl. Bur. Stand., Sect. A* **76**:547–560.
19. Credi, A., and Prodi, L., 1996, Correction of luminescence intensity measurements in solution: A simple method to standardize spectrofluorimeters, *EPA Newsletter* **58**:50–59.

Appendix II Fluorescent Lifetime Standards

It is valuable to have fluorophores of known lifetimes for use as lifetime standards in time-domain or frequency-domain measurements. Perhaps more important than the actual lifetime is knowledge that the fluorophore displays single-exponential decays. Such fluorophores are useful for testing the time-resolved instruments for systematic errors. We have summarized the results on lifetime standards from several laboratories. There is no attempt to compare the values or to evaluate which values are more reliable. Much of the data is from our laboratory because it was readily available with all the experimental details.

1. NANOSECOND LIFETIME STANDARDS

A series of scintillator fluorophores were characterized as standards for correcting timing errors in photomultiplier tubes.[1] Although the decay times were only measured at one or two frequencies, these compounds are thought to display single-exponential decays in ethanol. The decay times were measured for ethanol solutions in equilibrium with air and are not significantly sensitive to temperature (Table II.1). The excitation wavelengths range from 280 to 360 nm, and the emission wavelengths range from 300 to 500 nm (Figure II.1).

One of our most carefully characterized intensity decays is for 2,5-diphenyl-1,3,4-oxadiazole (PPD) in ethanol at 20 °C, in equilibrium with air.[2] The frequency response was measured with a gigahertz frequency-domain instrument. No deviations from a single-exponential decay were detected over the entire range of frequencies (Figure II.2).

Table II.1. Nanosecond Lifetime Reference Fluorophores

Compound[a]	Emission wavelength range (nm)	τ (ns)[b]
p-Terphenyl	310–412	1.05
PPD	310–440	1.20
PPO	330–480	1.40
POPOP	370–540	1.35
$(Me)_2$POPOP	390–560	1.45

[a]Abbreviations: PPD, 2,5-diphenyl-1,3,4-oxadiazole; PPO, 2,5-diphenyloxazole; POPOP, 1,4-bis(5-phenyloxazol-2-yl)]benzene; $(Me)_2$POPOP, 1,4-bis(4-methyl-5-phenyloxazol-2-yl)benzene.

[b]These values are judged to be accurate to ±0.2 ns at 10 and 30 MHz. From Ref. 1.

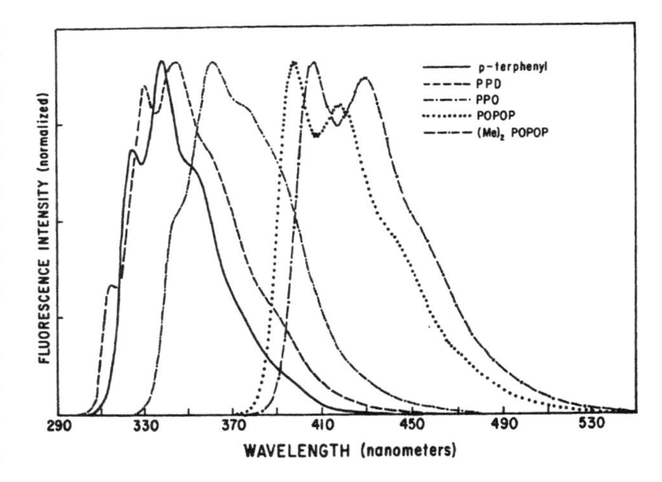

Figure II.1. Emission spectra of nanosecond lifetime reference fluorophores. Reprinted from Ref. 1, Copyright © 1981, with permission from Elsevier Science.

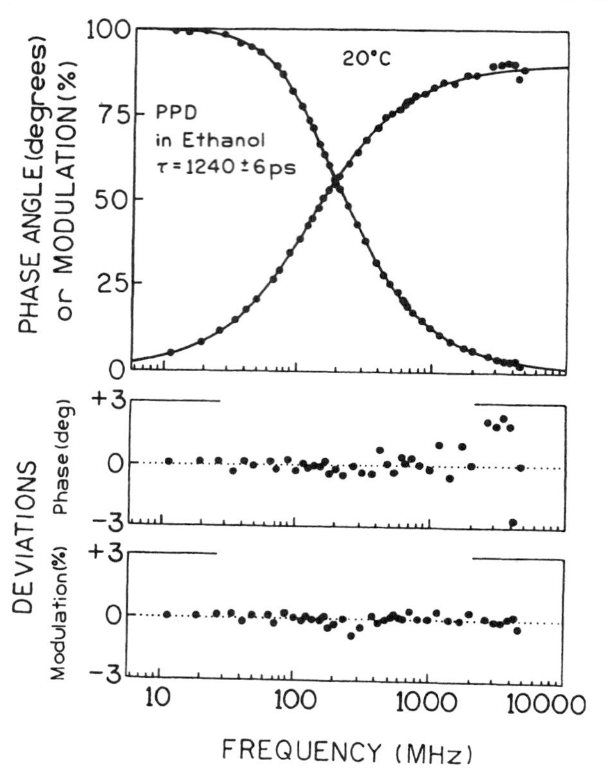

Figure II.2. PPD in ethanol as a single-decay-time standard. The frequency response was measured in ethanol in equilibrium with air. Reprinted, with permission, from Ref. 2, Copyright © 1990, American Institute of Physics.

2. PICOSECOND LIFETIME STANDARDS

Derivatives of dimethylaminostilbene were characterized as lifetime standards with subnanosecond lifetimes[3] (Figure II.3). Excitation wavelengths range up to 420 nm, and emission wavelengths from 340 to over 500 nm (Figure II.4). Representative frequency responses show that the intensity decays are all single exponentials (Figure II.5). The solutions were all in equilibrium with air. The decay times range from 57 to 921 ps (Table II.2 and Figure II.6).

Rose Bengal can serve as a picosecond lifetime standard at longer wavelengths (Figure II.7). Rose Bengal can also be used as a standard for a short rotational correlation time (Figure II.8).

3. REPRESENTATIVE FREQUENCY-DOMAIN INTENSITY DECAYS

It can be useful to have access to the actual lifetime data. Representative frequency-domain data for single-exponential decays are shown in Figure II.9. All samples were in equilibrium with air.[5] Additional frequency-domain data on single-decay-time fluorophores are available in the literature,[6–10] and a cooperative report between several laboratories on lifetime standards is in preparation.[11]

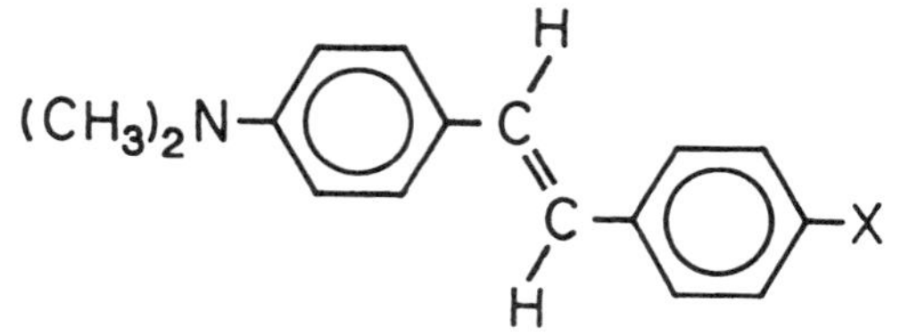

$(CH_3)_2N$–C$_6$H$_4$–CH=CH–C$_6$H$_4$–X

	X
DMS	OCH_3
DFS	F
DBS	Br
DCS	CN

Figure II.3. Picosecond lifetime standard fluorophores: DMS, 4-dimethylamino-4′-methoxystilbene; DFS, 4-dimethylamino-4′-fluorostilbene; DBS, 4-dimethylamino-4′-bromostilbene; DCS, 4-dimethylamino-4′-cyanostilbene. From Ref. 3.

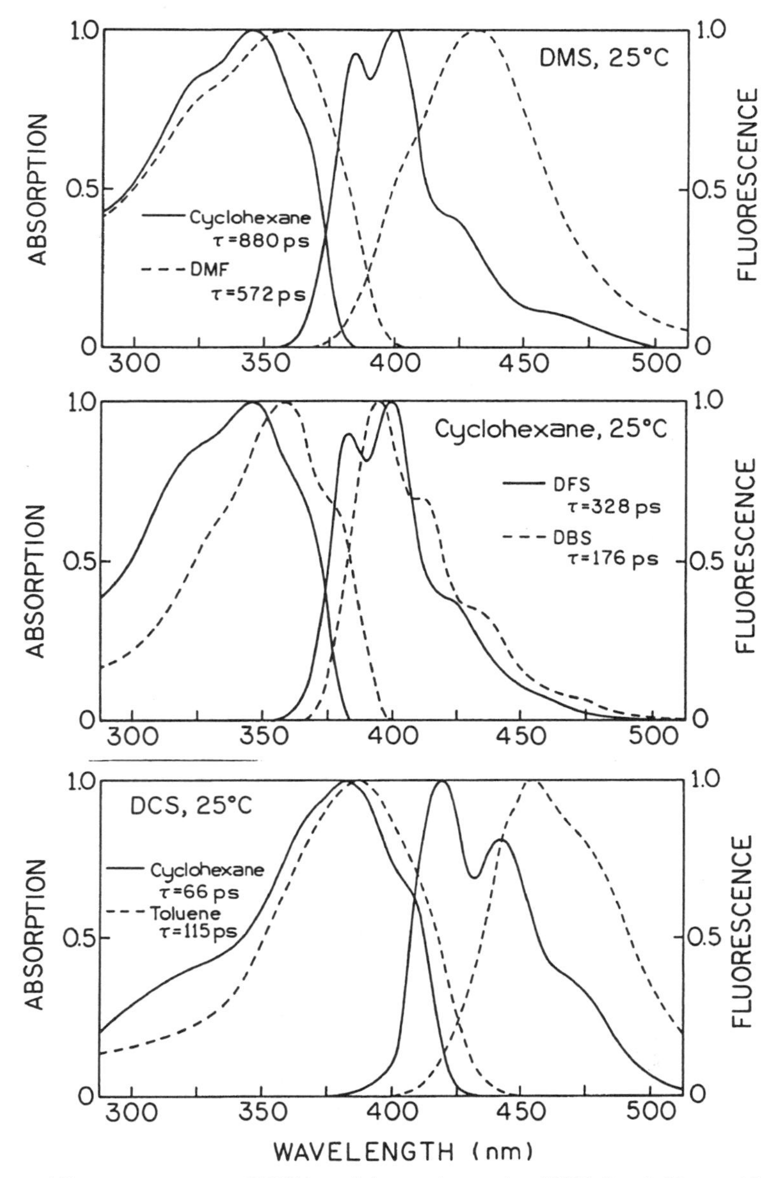

Figure II.4. *Top*: Absorption and fluorescence spectra of DMS in cyclohexane (———) and *N,N'*-dimethylformamide (— — —) at 25 °C. *Middle*: Absorption and fluorescence spectra of DFS (———) and DBS (— — —) in cyclohexane at 25 °C. *Bottom*: Absorption and fluorescence spectra of DCS in cyclohexane (———) and toluene (— — —) at 25 °C. Revised from Ref. 3.

4. TIME-DOMAIN LIFETIME STANDARDS

The need for lifetime standards for time-domain measurements has been recognized for some time.[12] A number of laboratories have suggested samples as single-decay-time standards.[13–17] The data are typically reported only in tables (Tables II.3–II.5), so representative figures are not available. The use of collisional quenching to obtain different lifetimes[17,18] is no longer recommended for lifetime standards, owing to the possibility of transient effects and nonexponential decays. Quinine is not recommended as a lifetime standard owing to the presence of a multiexponential decay.[19,20]

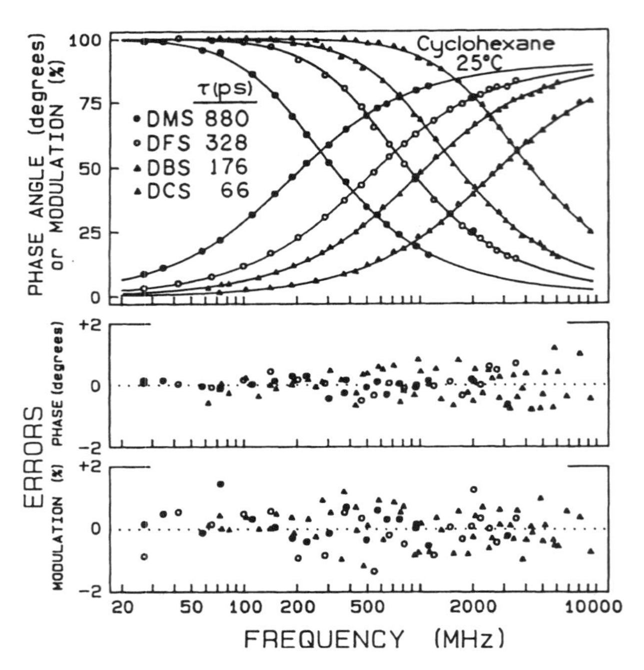

Figure II.5. Representative frequency responses of the picosecond lifetime standards in Figure II.3. From Ref. 3.

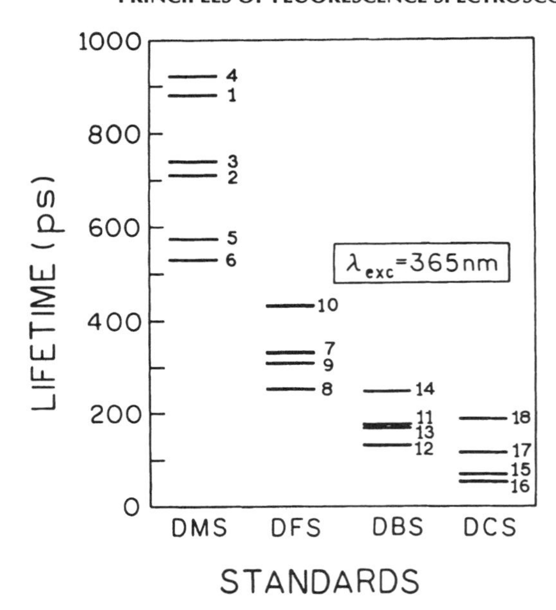

Figure II.6. Lifetimes of the picosecond lifetime standards in Figure II.3 in various solvents at various temperatures. See Table II.2 for the identification of the measurement conditions for 1–18. From Ref. 3.

Table II.2. Picosecond Lifetime Standards[a]

Compound[b]	No.[c]	Solvent[d]	Q[e]	T (°C)	τ (ps)
DMS	1	C	0.59	25	880
	2	C	—	37	771
	3	T	0.32	25	740
	4	T	—	5	921
	5	DMF	0.27	25	572
	6	EA	0.15	25	429
DFS	7	C	—	25	328
	8	C	—	37	252
	9	T	0.16	25	305
	10	T	—	5	433
DBS	11	C	0.11	25	176
	12	C	—	37	133
	13	T	0.12	25	168
	14	T	—	5	248
DCS	15	C	0.06	25	66
	16	C	—	37	57
	17	T	0.06	25	115
	18	T	—	5	186

[a]From Ref. 3. These results were obtained from frequency-domain measurements. The excitation wavelength was 365 nm.
[b]Abbreviations: DMS, 4-dimethylamino-4′-methoxystilbene; DFS, 4-dimethylamino-4′-fluorostilbene; DBS,. 4-dimethylamino-4′-bromostilbene; DCS, 4-dimethylamino-4′-cyanostilbene.
[c]Numbers refer to Figure II.6.
[d]C, Cyclohexane; T, toluene; DMF, dimethylformamide; EA, ethyl acetate.
[e] Quantum yields.

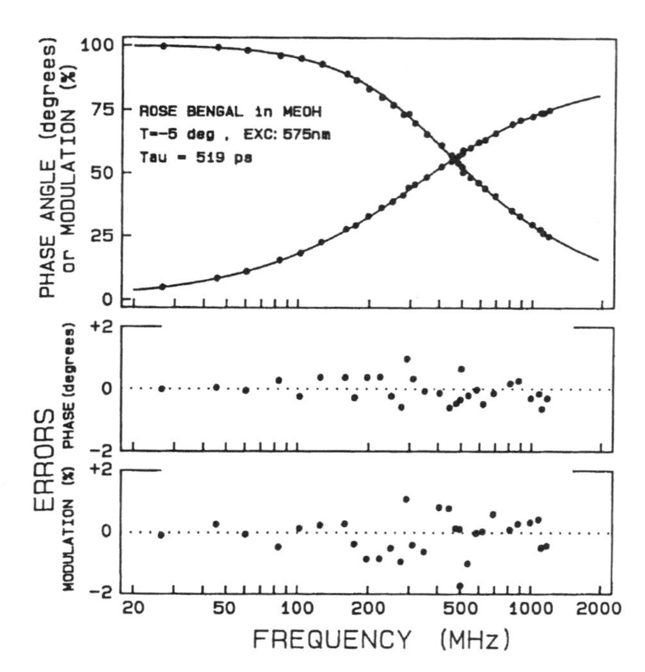

Figure II.7. Rose Bengal as a long-wavelength picosecond lifetime standard. For additional lifetime data on Rose Bengal, see Ref. 4.

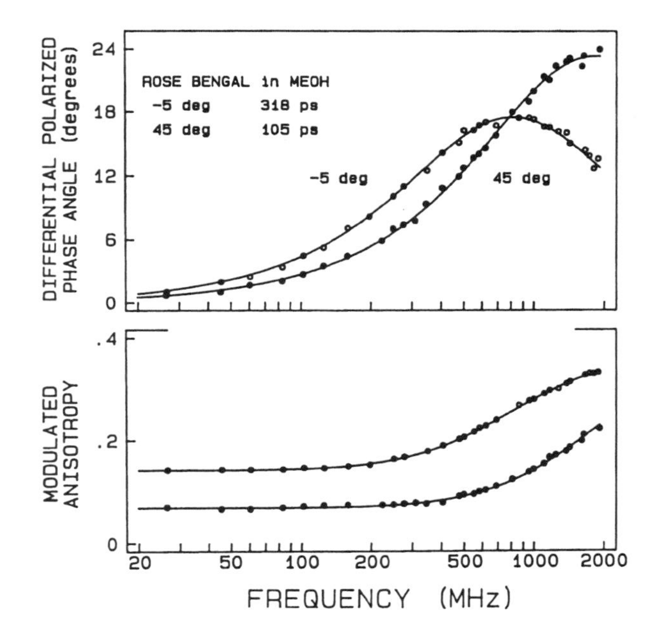

Figure II.8. Rose Bengal as a picosecond correlation- time standard. For additional data on Rose Bengal, see Ref. 4.

Table II.3. Time-Domain Single-Lifetime Standards[a]

Compound	Solvent[b]	λ_{em} (nm)	τ (ns)
PPO[c]	Cyclohexane (degassed)	440	1.42
	Cyclohexane (undegassed)	440	1.28
Anthracene	Cyclohexane (degassed)	405	5.23
	Cyclohexane (undegassed)	405	4.10
1-Cyanonaphthalene	Hexane (degassed)	345	18.23
1-Methylindole	Cyclohexane (degassed)	330	6.24
3-Methylindole	Cyclohexane (degassed)	330	4.36
	Ethanol (degassed)	330	8.17
1,2-Dimethylindole	Ethanol (degassed)	330	5.71

[a] From Ref. 13. The results were obtained using time-correlated single-photon counting.
[b] $T = 20$ °C.
[c] PPO, 2,5-Diphenyloxazole.

Table II.4. Single-Exponential Lifetime Standards[a]

Sample[b]	λ_{em} (nm)	τ (ns)
Anthracene	380	5.47
PPO[c]	400	1.60
POPOP[d]	400	1.38
9-Cyanoanthracene	440	14.76

[a] From Ref. 14. The lifetimes were measured by time-correlated single-photon counting. The paper is unclear on purging, but the values seem consistent with degassed samples.
[b] All samples in ethanol.
[c] PPO, 2,5-Diphenyloxazole.
[d] POPOP, 1,4-Bis(5-phenyloxazol-2-yl)benzene.

Table II.5. Single Exponential Standards[a]

Fluorophore	τ (ns)
POPOP[b] in cyclohexane	1.14 ± 0.01
POPOP in EtOH	1.32 ± 0.01
POPOP in aqueous EtOH	0.87 ± 0.01
Anthracene in EtOH	4.21 ± 0.02
9-Cyanoanthracene in EtOH	11.85 ± 0.03

[a] From Ref. 17. All measurements were at 25 °C, in equilibrium with air, by time-correlated single-photon counting.
[b] POPOP, 1,4-Bis(5-phenyloxazol-2-yl)benzene.

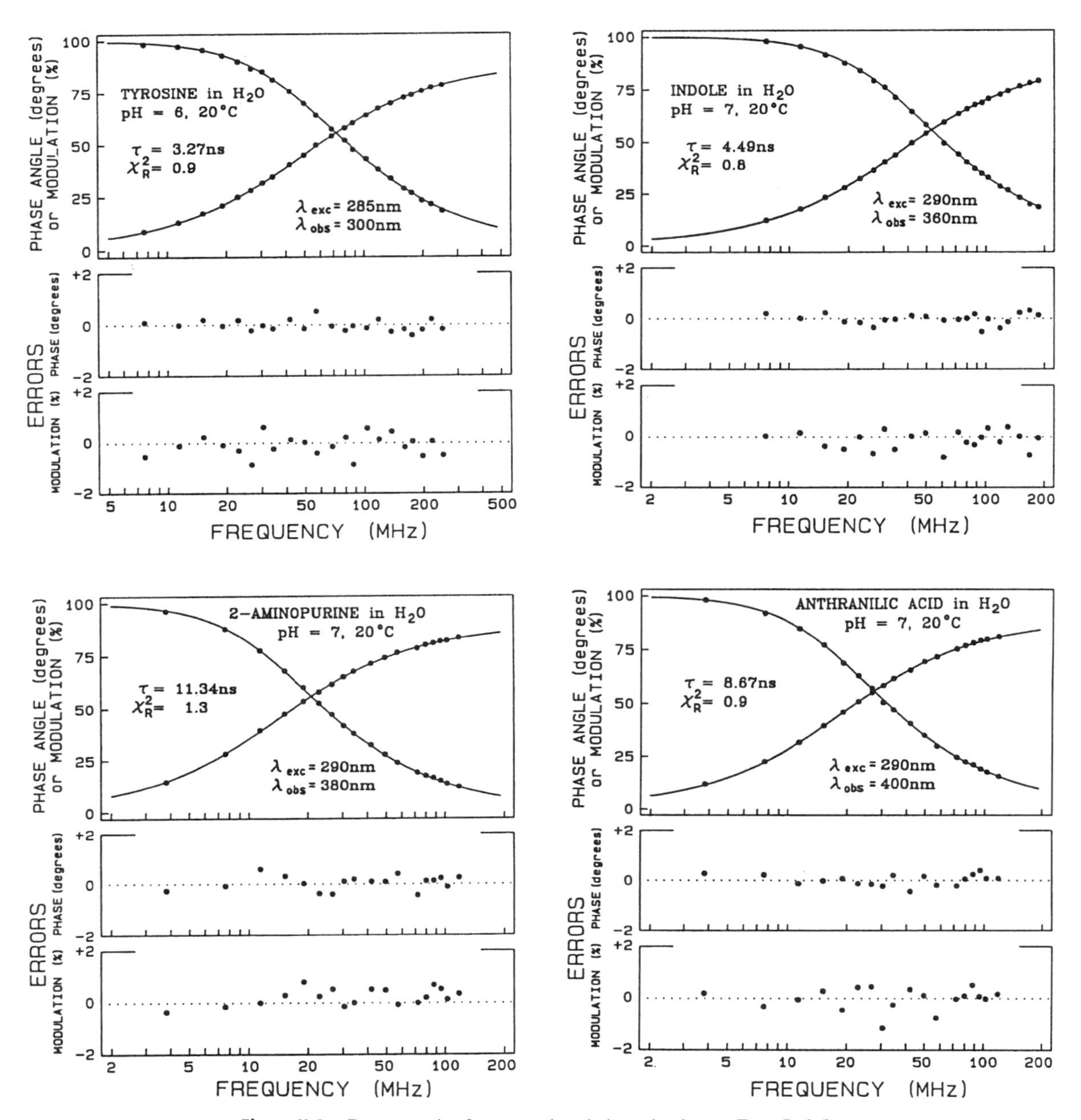

Figure II.9. Representative frequency-domain intensity decays. From Ref. 5.

REFERENCES

1. Lakowicz, J. R., Cherek, H., and Balter, A., 1981, Correction of timing errors in photomultiplier tubes used in phase-modulation fluorometry, *J. Biochem. Biophys. Methods* **5**:131–146.
2. Laczko, G., Gryczynski, I., Gryczynski, Z., Wiczk, W., Malak, H., and Lakowicz, J. R., 1990, A 10-GHz frequency-domain fluorometer, *Rev. Sci. Instrum.* **61**:2331–2337.
3. Lakowicz, J. R., Gryczynski, I., Laczko, G., and Gloyna, D., 1991, Picosecond fluorescence lifetime standards for frequency- and time-domain fluorescence, *J. Fluoresc.* **1**(2):87–93.
4. Lakowicz, J. R., and Gryczynski, I., unpublished observations. For additional lifetime and anisotropy data on Rose Bengal, see; Lakowicz, J. R., Laczko, G., and Gryczynski, I., 1986, 2-GHz frequency-domain fluorometer, *Rev. Sci. Instrum.* **57**:2499–2506.
5. Gryczynski, I., unpublished data.
6. Lakowicz, J. R., and Gryczynski, I., 1993, Characterization of *p*-bis(*O*-methylstyryl)benzene as a lifetime and anisotropy decay standard for two-photon induced fluorescence, *Biophys. Chem.* **47**:1–7.
7. Vos, R., Strobbe, R., and Engelborghs, Y., 1997, Gigahertz phase fluorometry using a fast high-gain photomultiplier, *J. Fluoresc.* **7**(1):33S–35S.

8. Thompson, R. B., and Gratton, E., 1988, Phase fluorometric method for determination of standard lifetimes, *Anal. Chem.* **60**:670–674.
9. Pouget, J., Mugnier, J., and Valeur, B., 1989, Correction of systematic phase errors in frequency-domain fluorometry, *J. Phys. E.: Sci. Instrum.* **22**:855–862.
10. Barrow, D. A., and Lentz, B. R., 1983, The use of isochronal reference standards in phase and modulation fluorescence lifetime measurements, *J. Biochem. Biophys. Methods* **7**:217–234.
11. Boens, N., Hofkens, J., De Schryver, F. C., Ameloot, M., Valeur, B., Pouget, J., Gratton, J., Vandeven, M., Silva, N., Engelborghs, Y., Willaert, K., Rumbles, G., Phillips, D., Visser, A. J. W. G., Van Hoek, A., Lakowicz, J. R., Malak, H., Gryczynski, I., Szabo, A. G., and Krajcarski, D. T., 1998, Fluorescence lifetime standards for time and frequency domain fluorescence spectroscopy, in preparation.
12. Grinvald, A., 1976, The use of standards in the analysis of fluorescence decay data, *Anal. Biochem.* **75**:260–280.
13. Lampert, R. A., Chewter, L. A., Phillips, D., O'Connor, D. V., Roberts, A. J., and Meech, S. R., 1983, Standards for nanosecond fluorescence decay time measurements, *Anal. Chem.* **55**:68–73.
14. Zuker, M., Szabo, A. G., Bramall, L., Krajcarski, D. T., and Selinger, B., 1985, Delta function convolution method (DFCM) for fluorescence decay experiments, *Rev. Sci. Instrum.* **56**:14–22.
15. Rayner, D. M., McKinnon, A. E., and Szabo, A. G., 1977, Correction of instrumental time response variation with wavelength in fluorescence lifetime determinations in the ultraviolet region, *Rev. Sci. Instrum.* **48**:1050–1054.
16. Castelli, F., 1985, Determination of correct reference fluorescence lifetimes by self-consistent internal calibration, *Rev. Sci. Instrum.* **56**:538–542.
17. Kolber, Z. S., and Barkley, M. D., 1986, Comparison of approaches to the instrumental response function in fluorescence decay measurements, *Anal. Biochem.* **152**:6–21.
18. Chen, R. F., 1974, Fluorescence lifetime reference standards for the range 0.189 to 115 nanoseconds, *Anal. Biochem.* **57**:593–604.
19. O'Connor, D. V., Meech, S. R., and Phillips, D., 1982, Complex fluorescence decay of quinine bisulphate in aqueous sulphuric acid solution, *Chem. Phys. Lett.* **88**:22–26.
20. Pant, D., Tripathi, H. B., and Pant, D. D., 1992, Time resolved fluorescence spectroscopy of quinine sulphate, quinidine and 6-methoxyquinoline: pH dependence, *J. Lumin.* **51**:223–230.

Appendix III
Additional Reading

In the preceding chapters we presented an overview of the expanding field of fluorescence spectroscopy and its applications. It is not possible in a single volume to completely describe the molecular photophysics, underlying fluorescence phenomena and the application of these principles to the chemical and biological sciences. The following books are recommended for additional details on specialized topics. This listing is not intended to be inclusive, and the author apologizes for absence of important citations.

1. LIFETIME MEASUREMENTS

Fleming, G. R., 1986, *Chemical Applications of Ultrafast Spectroscopy*, Oxford University Press, New York, 262 pp.

Demas, J. N., 1983, *Excited State Lifetime Measurements*, Academic Press, New York, 267 pp.

O'Connor, D. V., and Phillips, D., 1984, *Time-Correlated Single Photon Counting*, Academic Press, New York, 288 pp.

Lakowicz, J. R. (ed.), 1991, *Topics in Fluorescence Spectroscopy, Volume 1, Techniques*, Plenum Press, New York, 453 pp.

These books provide detailed information on measurements of fluorescence lifetimes. Time-domain methods are described in detail by Demas (1983) and by O'Connor and Phillips (1984). Demas (1983) describes a variety of lifetime methods, and the book by O'Connor and Phillips stresses time-correlated single-photon counting. While the technology for time-resolved fluorescence has advanced rapidly, the principles have remained the same, so these books remain valuable. A more recent summary of time-resolved methods is given in Volume 1 of *Topics in Fluorescence Spectroscopy* (1991). Ultrafast spectroscopy using pump-probe methods has been summarized by Fleming (1986).

2. WHICH MOLECULES ARE FLUORESCENT, REPRESENTATIVE EMISSION SPECTRA, AND PRACTICAL ADVICE

Berlman, I. B., 1973, *Energy Transfer Parameters of Aromatic Compounds*, Academic Press, New York, 379 pp.

Berlman, I. B., 1971, *Handbook of Fluorescence Spectra of Aromatic Molecules*, 2nd ed., Academic Press, New York, 473 pp.

Haugland, R. P., 1996, *Handbook of Fluorescent Probes and Research Chemicals*, 6th ed., M. T. Z. Spence (ed.), Molecular Probes, Inc., Eugene, Oregon, 697 pp.

Landolt-Börnstein, Zahlenwerte und Funktionen aus Naturwissenschaften und Technik, neue Serie, 1967, Springer-Verlag, Berlin, 416 pp.

Schulman, S. G. (ed.), 1985, *Molecular Luminescence Spectroscopy, Methods and Applications: Part I*, John Wiley & Sons, New York, 826 pp.

Schulman, S. G. (ed.), 1988, *Molecular Luminescence Spectroscopy, Methods and Applications: Part 2*, John Wiley & Sons, New York, 526 pp.

Schulman, S. G. (ed.), 1993, *Molecular Luminescence Spectroscopy, Methods and Applications: Part 3*, John Wiley & Sons, New York, 467 pp.

Krasovitskii, B. M., and Bolotin, B. M. 1984, *Organic Luminescent Materials* (translated by V. G. Vopian), VCH Publishers, Weinheim, Germany, 340 pp.

Parker, C. A., 1968, *Photoluminescence of Solutions*, Elsevier, New York, 554 pp.

Guilbault, G. G. (ed.), 1990, *Practical Fluorescence*, 2nd ed., Marcel Dekker, New York, 812 pp.

Prior to experimentation, it is often useful to know which molecules are fluorescent and what is known about their spectral properties. Such information has been gathered in a systematic manner in Part 1 of *Molecular Luminescence Spectroscopy* (Schulman, 1985). Subsequent volumes of this useful series contain information on specialized topics such as luminescence from surfaces, sensors, lanthanides, and Spol'skii' spectroscopy (Schulman, 1988) and chemiluminescence, NIR fluorescence, and biological probes (Shulman, 1993). Useful experimental details are provided in the second edition of *Practical Fluorescence* (Guilbault, 1990), including instrumentation, examples of organic and biochemical fluorophores, environmental analysis, and enzymology.

The book by Krasovitskii and Bolotin (1984) systematically lists classes of organic compounds and their spectral properties, including representative absorption and emission spectra. Examples of absorption and corrected emission spectra of organic molecules can be obtained in the valuable book by Berlman (1971), which unfortunately is no longer in print. Berlman (1973) has also summarized Förster distances for numerous donor–acceptor pairs. Many emission spectra and some corrected standard spectra can be found in Landolt-Börnstein (1967). Data on newly available fluorophores are available in the useful catalogs from Molecular Probes, Inc., and from the Molecular Probes web site, http://www.probes.com.

3. THEORY OF FLUORESCENCE AND PHOTOPHYSICS

Birks, J. B., 1970, *Photophysics of Aromatic Molecules*, Wiley Interscience, New York, 704 pp.

Birks, J. B. (ed.), 1973, *Organic Molecular Photophysics*, Vol. 1, John Wiley & Sons, New York, 600 pp.

Birks, J. B. (ed.), 1975, *Organic Molecular Photophysics*, Vol. 2, John Wiley & Sons, New York, 453 pp.

Kawski, A., 1992, *Fotoluminescencja Roztworow*, Wydawnictwo Naukowe PWN, Warsaw, 369 pp.

As the field of fluorescence spectroscopy is migrating toward applications, the basic principles are described less frequently. The book by Birks (1970) and the two volumes edited by Birks (1973 and 1975) are classic works which summarized organic photophysics. These books start by describing fluorescence from a quantum-mechanical perspective and also contain valuable detailed tables and figures which summarize the spectral properties and photophysical parameters. Unfortunately, these books are no longer in print, but they may be found in libraries.

4. PRINCIPLES OF FLUORESCENCE SPECTROSCOPY

Lakowicz, J. R. (ed.), 1991, *Topics in Fluorescence Spectroscopy, Volume 1, Techniques,* (1991), Plenum Press, New York, 453 pp.

Lakowicz, J. R. (ed.), 1991, *Topics in Fluorescence Spectroscopy, Volume 2, Principles,* Plenum Press, New York, 428 pp.

Lakowicz, J. R. (ed.), 1992, *Topics in Fluorescence Spectroscopy, Volume 3, Biochemical Applications*, Plenum Press, New York, 390 pp.

Hercules, D. M. (ed.), 1966, *Fluorescence and Phosphorescence Analysis*, Interscience Publishers, New York, 258 pp.

Michl, J., and Thulstrup, E. W., 1986, *Spectroscopy with Polarized Light.* VCH Publishers, New York, 573 pp.

Detailed information on specialized areas of fluorescence can be found in the book series *Topics in Fluorescence Spectroscopy*. Each volume emphasizes a different aspect of fluorescence spectroscopy, in which the experts have provided chapters on their specialties. Volume 1 describes instrumentation for time-domain and frequency-domain fluorescence, detectors, streak cameras, correlation methods, microscopy, and flow cytometry. Volume 2 provides information on the biochemical applications of anisotropy, quenching, energy transfer, least-squares analysis, and oriented systems. There are also chapters on solvent interactions and fiber-optic sensors. Volume 3 describes the spectra properties of intrinsic biological fluorophores and labeled macromolecules. There are chapters on proteins, membranes, and nucleic acids, as well as chapters on total internal reflectance fluorescence and microparticle fluorescence.

5. BIOCHEMICAL FLUORESCENCE

Kohen, E., and Hirschberg, J. G. (eds.), 1996, *Analytical Use of Fluorescent Probes in Oncology*, Plenum Press, New York, 448 pp.

Taylor, D. L., Waggoner, A. S., Murphy, R. F., Lanni, F., and Birge, R. R. (eds.), 1986, *Applications of Fluorescence in the Biomedical Sciences*, Alan R. Liss, Inc., New York, 639 pp.

Chen, R. F., and Edelhoch, H. (eds.), 1975, *Biochemical Fluorescence: Concepts*, Vol. 1, Marcel Dekker, New York, 408 pp.

Chen, R. F., and Edelhoch, H. (eds.), 1976, *Biochemical Fluorescence: Concepts*, Vol. 2, Marcel Dekker, New York, pp. 410–944.

Dewey, T. G. (ed.), 1991, *Biophysical and Biochemical Aspects of Fluorescence Spectroscopy*, Plenum Press, New York, 294 pp.

Bach, P. H., Reynolds, C. H., Clark, J. M., Mottley, J., and Poole, P. L. (eds.), 1993, *Biotechnology Applications of Microinjection, Microscopic Imaging, and Fluorescence*, Plenum Press, New York, 255 pp.

Steiner, R. F. (ed.), 1983, *Excited States of Biopolymers*, Plenum Press, New York, 258 pp.

Steiner, R. F., and Weinryb, I. (eds.), 1971, *Excited States of Proteins and Nucleic Acids*, Plenum Press, New York, 487 pp.

Mason, W. T., 1993, *Fluorescent and Luminescent Probes for Biological Activity*, Academic Press, New York, 433 pp.

Jameson, D. M., and Reinhart, G. D. (eds.), 1981, *Fluorescent Biomolecules*, Plenum Press, New York, 461 pp.

Ichinose, N., Schwedt, G., Schnepel, F. M., and Adachi, K., 1987, *Fluorometric Analysis in Biomedical Chemistry*, John Wiley & Sons, New York, 225 pp.

Brand, L., and Johnson, M. L. (eds.), 1997, *Fluorescence Spectroscopy*, Academic Press, New York, 628 pp.

Loew, L. M. (ed.), 1988, *Spectroscopic Membrane Probes,* Vol. I, CRC Press, Boca Raton, Florida, 227 pp.

Loew, L. M. (ed.), 1988, *Spectroscopic Membrane Probes,* Vol. II, CRC Press, Boca Raton, Florida, 206 pp.

Loew, L. M. (ed.), 1988, *Spectroscopic Membrane Probes,* Vol. III, CRC Press, Boca Raton, Florida, 278 pp.

6. PROTEIN FLUORESCENCE

Konev, S.V., 1967, *Fluorescence and Phosphorescence of Proteins and Nucleic Acids* (translated by S. Udenfriend). Plenum Press, New York, 204 pp.

Longworth, J. W., 1971, Luminescence of polypeptides and proteins, in *Excited States of Proteins and Nucleic Acids*, R. F. Steiner and I. Weinryb (eds.), Plenum Press, New York, pp. 319–484.

Permyakov, E. A., 1993, *Luminescent Spectroscopy of Proteins*, CRC Press, Boca Raton, Florida, 164 pp.

Demchenko, A. P., 1981, *Ultraviolet Spectroscopy of Proteins*, Springer-Verlag, New York, 317 pp.

There are numerous publications on intrinsic protein fluorescence, but there are relatively few volumes dedicated to this topic. The book by Demchenko (1981) contains extensive information about the optical spectroscopy of proteins, as does the older but more difficult to find book by Konev (1967). A comprehensive review was presented by Longworth (1971). The more recent book by Permyakov (1993) gives a less detailed overview of protein fluorescence.

7. DATA ANALYSIS AND NONLINEAR LEAST SQUARES

Bevington, P. R. 1969, *Data Reduction and Error Analysis for the Physical Sciences*, McGraw-Hill, New York, 336 pp.

Johnson, M. L., 1983, Evaluation and propagation of confidence intervals in nonlinear, asymmetrical variance spaces: Analysis of ligand binding data, *Biophys. J.* **44:**101–106.

Taylor, J. R., 1982, *An Introduction to Error Analysis*, University Science Books, Mill Valley, California, 270 pp.

Johnson, M. L., and Brand, L. (eds.), 1994, *Methods in Enzymology, Volume 240: Numerical Computer Methods*, Academic Press, New York, 857 pp.

Brand, L., and Johnson, M. L. (eds.), 1992, *Methods in Enzymology, Volume 210: Numerical Computer Methods*, Academic Press, New York, 718 pp.

Mark, H., and Workman, J. 1991, *Statistics in Spectroscopy*, Academic Press, New York, 313 pp.

Magar, M. E., 1972, *Data Analysis in Biochemistry and Biophysics*, Academic Press, New York, 497 pp.

The use of nonlinear least-squares analysis is ubiquitous in the analysis of fluorescence data, particularly time-domain and frequency-domain data. A useful introduction to the principles of least-squares analysis is found in the compact but informative book by Bevington (1969). The applications of these concepts to diverse types of fluorescence data can be found in edited volumes (Brand and Johnson, 1992; Johnson and Brand, 1994). For more basic information about statistics and spectroscopy, one can examine several introductory texts (Taylor, 1982; Mark and Workman, 1991).

8. PHOTOCHEMISTRY

Gilbert, A., and Baggott, J. 1991, *Essentials of Molecular Photochemistry*, CRC Press, Boca Raton, Florida, 538 pp.

Kavarnos, G. J., 1993, *Fundamentals of Photoinduced Electron Transfer*, VCH Publishers, New York, 359 pp.

Turro, N. J. 1978, *Modern Molecular Photochemistry*, Benjamin/Cummings Publishing Co., Menlo Park, California, 678 pp.

Rabek, J. F., 1990, *Photochemistry and Photophysics*, Vol. I, CRC Press, Boca Raton, Florida, 183 pp.

Rabek, J. F., 1990, *Photochemistry and Photophysics*, Vol. II, CRC Press, Boca Raton, Florida, 200 pp.

Rabek, J. F., 1991, *Photochemistry and Photophysics*, Vol. III, CRC Press, Boca Raton, Florida, 202 pp.

Wayne, R. P., 1988, *Principles and Applications of Photochemistry*, Oxford University Press, New York, 268 pp.

For most applications of fluorescence, photochemical effects are to be avoided. However, it can be valuable to understand the chemical reactions that occur in the excited state.

9. FLOW CYTOMETRY

Givan, A. L., 1992, *Flow Cytometry*, Wiley-Liss, New York, 202 pp.

Shapiro, H. M., 1988, *Practical Flow Cytometry*, 2nd ed., Alan R. Liss, New York, 353 pp.

Flow cytometry provides a method for measuring the emission from single cells as they pass through a laser beam. Flow cytometry can be used with DNA stains to measure the DNA content of cells, with antibodies to detect cell-surface antigens, or with sensing probes to measure intracellular concentrations of ions.

10. PHOSPHORESCENCE

Hurtubise, R. J., 1990, *Phosphorimetry: Theory, Instrumentation, and Applications*, VCH Publishers, New York, 370 pp.

11. POLYMER SCIENCE

Winnik, M. A. (ed.), 1986, *Photophysical and Photochemical Tools in Polymer Science: Conformation, Dynamics, Morphology*, D. Reidel, Boston, 647 pp.

12. FLUORESCENCE SENSING

Wolfbeis, O. S. (ed.), 1991, *Fiber Optic Chemical Sensors and Biosensors*, Vol. I, CRC Press, Boca Raton, Florida, 413 pp.

Wolfbeis, O. S. (ed.), 1991, *Fiber Optic Chemical Sensors and Biosensors*, Vol. II, CRC Press, Boca Raton, Florida, 358 pp.

Kunz, R. E. (ed.), 1996, *Proceedings of 3rd European Conference on Optical Chemical Sensors and Biosensors, Europt(R)ode III*, Vol B39. *Sensors Actuators B* **1996**. *Part I: Plenary and Parallel Sessions; Part II: Poster Sessions.*

Lakowicz, J. R. (ed.), 1994, *Topics in Fluorescence Spectroscopy, Volume 4, Probe Design and Chemical Sensing*, Plenum Press, New York, 501 pp.

Lakowicz, J. R. (ed.), 1995, *Advances in Fluorescence Sensing Technology II, Proc. SPIE*, Vol. 2388, 598 pp.

Thompson, R. B. (ed.), 1997, *Advances in Fluorescence Sensing Technology III, Proc. SPIE*, Vol. 2980, 582 pp.

Spichiger-Keller, U. E. 1998, *Chemical Sensors and Biosensors for Medical and Biological Applications*, Wiley-VCH, New York, 413 pp.

Fraser, D. M. (ed.), 1997, *Biosensors in the Body—Continuous in Vivo Monitoring*, Biomaterials Science and Engineering Series, John Wiley & Sons, New York, 268 pp.

Fluorescence spectroscopy is rapidly changing to include analytical and clinical applications. There is also a rapid expansion of the use of fluorescence in DNA analysis and molecular diagnostics. These books describe the design principles of sensing probes, NIR probes, and lifetime-based sensing.

13. IMMUNOASSAYS

Hemmila, I. A. 1991, *Applications of Fluorescence in Immunoassays*, John Wiley & Sons, New York, 343 pp.

Van Dyke, K., and Van Dyke, R., 1990, *Luminescence Immunoassay and Molecular Applications*, CRC Press, Boca Raton, Florida, 341 pp.

14. LATEST APPLICATIONS OF FLUORESCENCE

Kohen, E., and Hirschberg, J. G. (eds.), 1996, *Applications of Optical Engineering to the Study of Cellular Pathology*, Research Signpost, India 182 pp.

Czarnik, A. W. (ed.), 1992, *Fluorescent Chemosensors for Ion and Molecule Recognition*, ACS Symposium Series, Vol. 538, 235 pp.

Slavik, J., 1994, *Fluorescent Probes in Cellular and Molecular Biology*, CRC Press, Boca Raton, Florida, 295 pp.

Slavik, J. (ed.), 1996, *Fluorescence Microscopy and Fluorescent Probes*, Plenum Press, New York, 306 pp.

Slavik, J. (ed.), 1998, *Fluorescence Microscopy and Fluorescent Probes, Volume 2*. Plenum Press, New York, 272 pp.*

Wolfbeis, O. S. (ed.), 1993, *Fluorescence Spectroscopy*, Springer-Verlag, New York, 309 pp.

Baldini, F. (ed.), 1994, *Sensors and Actuators B*, Vol. B29(1–3), 439 pp.

Lakowicz, J. R. (ed.), 1994, *Time-Resolved Spectroscopy in Biochemistry IV, Proc. SPIE*, Vol. 2137, 806 pp.

*Dr. Jan Slavik passed away unexpectedly in January 1999.

Lakowicz, J. R., and Ross, J. B. A., (eds.), 1998, *Advances in Optical Biophysics*, Proc. SPIE, Vol. 3256, 280 pp.
Lakowicz, J. R. (ed.), 1997, *Topics in Fluorescence Spectroscopy, Volume 5, Non-Linear and Two-Photon-Induced Fluorescence*, Plenum Press, New York, 544 pp.
Kwiatkowski, J. S., and Prochorow, J. (eds.), 1999, Proceedings of the Jabłoński Centennial Conference on Luminescence and Photophysic, *Acta Physica Polonica A*, Vol. 95(1), Warsaw, Poland, 196 pp.
Thompson, R. B., and Lakowicz, J. R. (eds.), 1999, Advances in Fluorescence Sensing Technology IV, *SPIE Proceedings*, 3602A in press.

15. INFRARED AND NIR FLUORESCENCE

Matsuoka, M. (ed.), 1990, *Infrared Absorbing Dyes*, Plenum Press, New York, 224 pp.
Leznoff, C. C., and Lever, A. B. P. (eds.), 1989, *Phthalocyanines: Properties and Applications*, VCH Publishers, New York, 436 pp.

All biological samples display interfering autofluorescence, which decreases at longer excitation wavelengths. Because NIR fluorescence can be detected with high sensitivity and inexpensive instrumentation, there is increasing interest in this topic.

16. LASERS

Meyers, R. A., 1991, *Encyclopedia of Lasers and Optical Technology*, Academic Press, New York, 764 pp.
Iga, K., 1994, *Fundamentals of Laser Optics*, R. B. Miles, Tech. Ed., Plenum Press, New York, 285 pp.
Svelto, O. 1998, *Principles of Lasers*, 4th ed., (D. C. Hanna, Editor and Translator), Plenum Press, New York, 594 pp.

Laser technology is evolving rapidly, and most texts are quickly out-of-date. The latest information is typically available at focused meetings and from manufacturers' brochures. The above texts provide relatively simple descriptions of the principles of lasers.

17. FLUORESCENCE MICROSCOPY

Wang, X. F., and Herman, B. 1996, *Fluorescence Imaging Spectroscopy and Microscopy*, John Wiley & Sons, New York, 483 pp.
Wang, Y.-L., and Taylor, D. L. (eds.), 1989, *Fluorescence Microscopy of Living Cells in Culture. Part A: Fluorescent Analogs, Labeling Cells, and Basic Microscopy*, Academic Press, New York, 503 pp.
Taylor, D. L., and Wang, Y.-L. (eds.), 1989, *Fluorescence Microscopy of Living Cells in Culture. Part B: Quantitative Fluorescence Microscopy—Imaging and Spectroscopy*, Academic Press, New York, 503 pp.
Pawley, J. B. (ed.), 1995, *Handbook of Confocal Microscopy*, 2nd ed., Plenum Press, New York, 632 pp. See also the First Edition, 1990.
Matsumoto, B. (ed.), 1993, *Methods in Cell Biology, Volume 38: Cell Biological Applications of Confocal Microscopy*, Academic Press, New York, 380 pp.
Nuccitelli, R. (ed.), 1994, *Methods in Cell Biology, Volume 40: A Practical Guide to the Study of Calcium in Living Cells*, Academic Press, New York, 368 pp.
Herman, B., and Lemasters, J. J., 1993, *Optical Microscopy, Emerging Methods and Applications*, Academic Press, New York, 441 pp.
Herman, B., and Jacobson, K. (eds.), 1990, *Optical Microscopy for Biology*, Wiley-Liss, New York, 658 pp.

A renaissance has occurred in optical microscopy as the result of the availability of cation probes, confocal methods, CCD technology, and multiphoton excitation. These advances have been summarized in numerous recent edited volumes.

18. METAL-LIGAND COMPLEXES AND UNUSUAL LUMINOPHORES

Meyer, G. J. (ed.), 1997, *Molecular Level Artificial Photosynthetic Materials*, John Wiley & Sons, New York, 421 pp.
Kalyanasundaram, K., 1992, *Photochemistry of Polypyridine and Porphyrin Complexes*, Academic Press, New York, 626 pp.
Juris, A., Balzani, V., Barigelletti, F., Campagna, S., Belser P., and Von Zelewsky, A., 1988, Ru(II) polypyridine complexes: Photophysics, photochemistry, electrochemistry, and chemiluminescence, *Coord. Chem. Rev.* **84:**85–277.

The introduction of metal–ligand complexes of the $[Ru(bpy)_3]^{2+}$ class has extended the time range of fluorescence from nanoseconds to microseconds. Most of this information is widely dispersed in the chemical literature, without concern for the biochemical applications. Much of the biochemically relevant information is summarized in Chapter 20. An extensive summary of the spectral properties can be found in the review by Juris *et al.* (1988).

Answers to Problems

CHAPTER 1

1.1. A. The natural lifetimes and radiative decay rates can be calculated from the quantum yields and experimental lifetimes:

$$\tau_N(\text{eosin}) = \frac{\tau}{Q} = \frac{3.1}{0.65} = 4.77 \text{ ns} \qquad [1.18]$$

$$\tau_N(\text{ErB}) = \frac{\tau}{Q} = \frac{0.61}{0.12} = 5.08 \text{ ns} \qquad [1.19]$$

Hence, eosin and erythrosin B have similar natural lifetimes and radiative decay rates (Eq. [1.3]). This is because the two molecules have similar absorption and emission wavelengths and extinction coefficients (Eq. [1.4]).

The nonradiative decay rates can be calculated from Eq. [1.2], which can be rearranged to

$$\frac{1}{\tau} - \frac{1}{\tau_N} = k_{nr} \qquad [1.20]$$

For eosin and erythrosin B, the nonradiative decay rates are 1.1×10^8 s^{-1} and 1.44×10^9 s^{-1}, respectively. The larger nonradiative decay rate of erythrosin B is the reason why it has a shorter lifetime and a lower quantum yield than eosin.

B. The phosphorescence quantum yield (Q_p) can be estimated from an expression analogous to Eq. [1.1]:

$$Q_p = \frac{\Gamma_p}{\Gamma_p + k_{nr}} \qquad [1.21]$$

Using the assumed natural lifetime for phosphorescence emission of 10 ms and $k_{nr} = 1 \times 10^8$ s^{-1}, $Q_p = 10^{-6}$. If k_{nr} is larger, Q_p is still smaller, so that $Q_p \simeq 10^{-7}$ for erythrosin B. This explains why it is difficult to observe phosphorescence at room temperature—most of the molecules that undergo intersystem crossing return to the ground state by nonradiative paths prior to emission.

1.2. The quantum yield (Q_2) of S_2 can be estimated from

$$Q_2 = \frac{\Gamma}{\Gamma + k_{nr}} \qquad [1.22]$$

The value of k_{nr} is given by the rate of internal conversion to S_1, 10^{13} s^{-1}. Using $\Gamma = 2.1 \times 10^8$ (from Problem 1.1), one can estimate $Q_2 = 2 \times 10^{-5}$. Observation of emission from S_2 is unlikely because the molecules relax to S_1 prior to emission from S_2.

1.3. The energy spacing between the various vibrational energy levels is revealed by the emission spectrum of perylene (Figure 1.3). The individual emission maxima (and hence vibrational energy levels) are about 1500 cm^{-1} apart. The Boltzmann distribution describes the relative number of perylene molecules in the 0 and 1 vibrational states. The ratio (R) of the number of molecules in the first vibrationally excited state to that in the ground state is given by

$$R = e^{-\Delta E/kT} \qquad [1.23]$$

where ΔE is the energy difference between the two states, k is the Boltzmann constant, and T is the temperature in kelvins (K). Assuming a room temperature of 300 K, this ratio is about 0.001. Hence, most molecules will be present in the lowest vibrational state, and light absorption results mainly from molecules in this energy level. Because of the larger energy difference between S_0 and S_1, essentially no fluorophores can populate S_1 as a result of thermal energy.

1.4. A. The anisotropy of the DENS-labeled protein is given by Eq. [1.10]. Using $\tau = \theta$, the steady-state anisotropy is expected to be 0.15.

B. If the protein binds to the larger antibody, its rotational correlation time will increase to 100 ns or longer. Hence, the anisotropy will be 0.23 or higher. Such increases in anisotropy upon antigen–antibody binding are the basis of the fluorescence polarization immunoassays, which are used to detect drugs, peptides, and small proteins in clinical samples.

1.5. The dependence of transfer efficiency (E) on distance (r) between a donor and an acceptor can be calculated using

Eq. [1.12] (Figure 1.27). At $r = R_0$, the efficiency is 50%; at $r = 0.5R_0$, $E = 0.98$; and at $r = 2R_0$, $E = 0.02$.

1.6. The distance can be calculated for the relative quantum yield of the donor in the presence and absence of the acceptor. The data in Figure 1.28 reveal a relative tryptophan intensity of 0.25, assuming that the anthraniloyl group does not contribute to the emission at 340 nm. A transfer efficiency of 75% corresponds (Eq. [1.12]) to a distance of $0.83R_0 = 25.2$ Å.

In reality, the actual calculation is more complex, and the tryptophan intensity needs to be corrected for anthraniloyl emission.[30] When this is done, the transfer efficiency is found to be about 63%, and the distance about 31.9 Å.

CHAPTER 2

2.1. The true optical density is 10. Because of stray light, the lowest percent transmission that you can measure is 0.01%. The percent transmission of the rhodamine solution is much less than 0.01%. In fact, $I/I_0 = 10^{-10}$, and % $T = 10^{-8}$%. Hence, your instrument will report an optical density of 4.0. The calculated concentration of rhodamine B would be $4 \times 10^{-5}M$, 2.5-fold less than the true concentration.

2.2. The concentrations of the solutions are $10^{-5}M$ and $10^{-7}M$, respectively. A 1% error in percent transmission means that the OD can be

$$\mathrm{OD} = \log \frac{I_0}{I} = \log \frac{1}{0.51} \text{ or } \log \frac{1}{0.49} \qquad [2.13]$$

Hence, the calculated concentration can be from $0.97 \times 10^{-5}M$ to $1.03 \times 10^{-5}M$.

For the more dilute solution, the 1% error results in a large error in the concentration:

$$\mathrm{OD} = \log \frac{I_0}{I} = \log \frac{1}{1.00} \text{ or } \log \frac{1}{0.98} \qquad [2.14]$$

The measured OD ranges from 0 to 0.009, so the calculated concentration ranges from 0 to $2.9 \times 10^{-7}M$. This shows that it is difficult to determine the concentration from low optical densities. In contrast, it is easy to obtain emission spectra with optical densities near 0.003.

CHAPTER 3

3.1. Binding of the protein to membranes or nucleic acids could be detected by several types of measurements. The most obvious experiment would be to look for changes in the intrinsic tryptophan fluorescence upon mixing with lipid bilayers or nucleic acids. In the case of lipid bilayers, one might expect the tryptophan emission to shift to shorter wavelengths due to shielding of the indole moiety from water. The blue shift of the emission is also likely to be accompanied by an increase in the tryptophan emission intensity. In the case of nucleic acids, tryptophan residues are typically quenched when bound to DNA, so that a decrease in the emission intensity is expected.

Anisotropy measurements of the tryptophan emission could also be used to detect binding. In this case it is difficult to predict the direction of the changes. In general, one expects binding to result in a longer correlation time and higher anisotropy (Eq. [1.10]), and an increase of the tryptophan anisotropy is likely upon binding to proteins. However, the anisotropy increase upon binding to membranes may be smaller than expected if the tryptophan lifetime increases on binding to the membranes (Eq. [1.10]).

In the case of protein binding to nucleic acids, it is difficult to predict the anisotropy change. The tryptophan residues would now be in two states, free (F) and bound (B), and the anisotropy would be given by

$$r = r_F f_F + r_B f_B \qquad [3.2]$$

where f_F and f_B represent the fraction of the total fluorescence from the protein in each state. If the protein is completely quenched on binding to DNA, then the anisotropy will not change because $f_B = 0$. If the protein is partially quenched, the anisotropy will probably increase, but less than expected owing to the small contribution of the DNA-bound protein to the total fluorescence.

Energy transfer can also be used to detect protein binding. Neither DNA nor model membranes possess chromophores which can serve as acceptors for the tryptophan fluorescence. Hence, it is necessary to add extrinsic probes. Suitable acceptors would be probes which absorb near 350 nm, the emission maximum of most proteins. Numerous membranes and nucleic acid probes absorb near 350 nm. The membranes could be labeled with DPH (Figure 3.13), which absorbs near 350 nm. If the protein bound to DPH-labeled membranes, its emission would be quenched by resonance energy transfer to DPH. Similarly, DNA could be labeled with DAPI (Figure 3.21). An advantage of using RET is that the through-space quenching occurs irrespective of the details of the binding interactions. Even if there were no change in the intrinsic tryptophan emission upon binding to lipids or nucleic acids, one still expects quenching of the tryptophan when bound to acceptor-labeled membranes or nucleic acids.

3.2. A. The data in Figure 3.48 can be used to determine the value of F_0/F at each chloride concentration, where $F_0 = 1.0$ is the SPQ fluorescence intensity in the absence of Cl^-, and F is the intensity at each Cl^- concentration. These values are plotted in Figure 3.49. Using the Stern–Volmer equation (Eq. [1.6]), one obtains $K = 124\ M^{-1}$, which is in good agreement with the literature value of 118 M^{-1} (Ref. 142 in Chapter 3).

B. The values of F_0/F and τ_0/τ for 0.103*M* Cl^- can be found from Eq. [1.6]. Using $K = 118\ M^{-1}$, one obtains $F_0/F = \tau_0/\tau = 13.15$. Hence, the intensity of SPQ is $F = 0.076$, relative to the intensity in the absence of Cl^-, $F_0 = 1.0$. The lifetime is expected to be $\tau = 26.3/13.15 = 2.0$ ns.

C. At $[Cl^-] = 0.075M$, $F = 0.102$ and $\tau = 2.67$.

D. The Stern–Volmer quenching constant of SPQ was determined in the absence of macromolecules. It is possible that SPQ binds to proteins or membranes in blood serum. This could change the Stern–Volmer quenching constant by protecting SPQ from collisional quenching. Also, binding to macromolecules could alter τ_0, the unquenched lifetime. Hence, it is necessary to determine whether the quenching constant of SPQ is the same in blood serum as in protein-free solutions.

3.3. A. The dissociation reaction of the probe (P) and analyte (A) is given by

$$P_B \rightleftharpoons A + P_F \quad [3.3]$$

where *B* and *F* refer to the free and bound forms of the probe, respectivley. The ratio of free to bound probe is related to the dissociation constant by

$$K_D = \frac{[P_F]}{[P_B]}[A] \quad [3.4]$$

For the nonratiometric probe Calcium Green, the fluorescence intensity is given by

$$F = k(q_F f_F + q_B f_B) \quad [3.5]$$

where the q_i are the relative quantum yields, f_i is the molecular fraction in each form, and $f_F + f_B = 1.0$. The fluorescence intensity when all the probe is free is $F_{min} = kq_F$; when all the probe is bound, the flourescence intensity is $F_{max} = kq_B$, where *k* is an instrumental constant and *C* is the concentration of the probe.

Equation [3.5] can be used to derive expressions for f_B and f_F in terms of the relative intensities:

$$f_B = \frac{F - F_{min}}{F_{max} - F_{min}} \quad [3.6]$$

$$f_F = \frac{F_{max} - F}{F_{max} - F_{min}} \quad [3.7]$$

The fractions f_B and f_F can be substituted for the probe concentration in Eq. [3.4], yielding

$$[A] = [Ca^{2+}] = K_D \frac{F - F_{min}}{F_{max} - F} \quad [3.8]$$

B. The fluorescence intensity (F_1 or F_2) observed with each excitation wavelength (1 or 2) depends on the intensities and the molecular fractions (f_F and f_B) of the free and the bound forms at each excitation wavelength:

$$F_1 = S_{f1} f_F + S_{b1} f_B \quad [3.9]$$

$$F_2 = S_{f2} C_F + S_{b2} C_B \quad [3.10]$$

The terms S_{fi} and S_{bi} depend on the absorption coefficients and relative quantum yields of the free and bound forms of Fura-2 at each wavelength.

Let $R = F_1/F_2$ be the ratio of intensities at the two excitation wavelengths. In the absence of Ca^{2+},

$$R_{min} = S_{f1}/S_{f2} \quad [3.11]$$

and in the presence of saturating Ca^{2+},

$$R_{max} = S_{b1}/S_{b2} \quad [3.12]$$

Using the definition of the dissociation constant,

$$[Ca^{2+}] = \frac{f_B}{f_F} K_D \quad [3.13]$$

one obtains

$$[Ca^{2+}] = K_D \frac{R - R_{max}}{R_{min} - R} \left(\frac{S_{f2}}{S_{b2}} \right) \quad [3.14]$$

Hence, one can measure $[Ca^{2+}]$ from the ratio of the emission intensities at two excitation wavelengths. However, one needs control measurements, which are the ratio of intensities of the free and bound forms measured at one excitation wavelength as well as measurements of R_{min} and R_{max} (see Ref. 143 in Chapter 3).

CHAPTER 4

4.1. Calculation of the lifetime from the intensity decay is straightforward. The initial intensity decreases to 0.37 (= $1/e$) of the initial value at $t = 5$ ns. Hence, the lifetime is 5 ns.

From Figure 4.2 the phase angle is seen to be about 60°. Using $\omega = 2\pi \cdot 80$ MHz and $\tau_\phi = \omega^{-1} \tan \phi$, one finds $\tau = 3.4$ ns. The modulation of the emission relative to the excitation is near 0.37. Using Eq. [4.6], one finds $\tau_m = 5.0$ ns. Since the phase and modulation lifetimes are not equal, and since $\tau_m > \tau_\phi$, the intensity decay is heterogeneous. Of course, it is difficult to read precise values from Figure 4.2.

4.2. The fractional intensity of the 0.62-ns component can be calculated using Eq. [4.27] and is found to be 0.042 or 4%.

4.3. The short lifetime was assigned to the stacked conformation of FAD (see Section 4.12.E). For the open form, the lifetime of the flavin is reduced from $\tau_0 = 4.89$ ns to $\tau =$

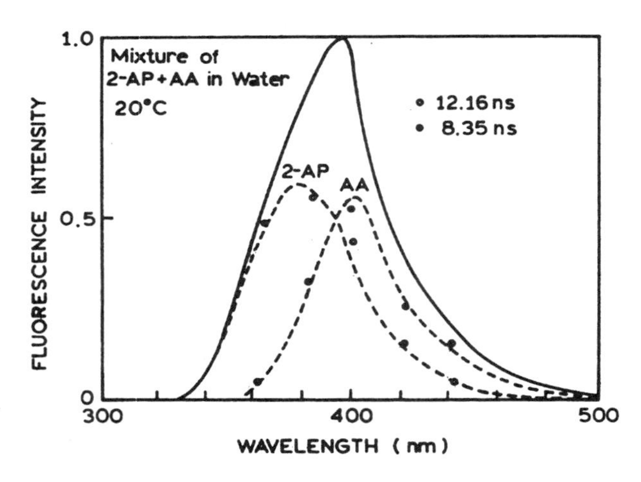

Figure 4.48. Emission spectra of a two-component mixture of anthranilic acid (AA) and 2-aminopurine (2-AP). The data show the fractional amplitudes associated with each decay time recovered from the global analysis. From Ref. 161.

3.38 ns due to collisions with the adenine. The collision frequency is given by $k = \tau^{-1} - \tau_0^{-1} = 9 \times 10^7\ s^{-1}$.

4.4. In the presence of quencher, the intensity decay is given by

$$I(t) = 0.5 \exp(-t/0.5) + 0.5 \exp(-t/5) \quad [4.42]$$

The α_1 and α_2 values remain the same. The fact that the first tryptophan is quenched 10-fold is accounted for by the $\alpha_i\tau_i$ products, $\alpha_1\tau_1 = 0.25$ and $\alpha_2\tau_2 = 2.5$. Using Eqs. [4.28] and [4.29], one can calculate $\bar{\tau} = 4.59$ ns and $\langle\tau\rangle = 2.75$. The average lifetime is close to the unquenched value because the quenched residue ($\tau_1 = 0.5$ ns) contributes only $f_1 = 0.091$ to the steady-state or integrated intensity. If the sample contained two tryptophan residues with equal steady-state intensities and lifetimes of 5.0 and 0.5 ns, then $\bar{\tau} = 0.5\tau_1 + 0.5\tau_2 = 2.75$ ns. The fact that $\langle\tau\rangle$ reflects the relative quantum yield can be seen by noting that $\langle\tau\rangle/\tau_0 = 2.75/5.0 = 0.55$, which is the quantum yield of the quenched sample relative to that of the unquenched sample.

4.5. The DAS can be calculated by multiplying the fractional intensities [f_i (λ)] by the steady-state intensity at each wavelength [$I(\lambda)$]. For the global analysis, these values (Figure 4.48) match the emission spectra of the individual components. However, for the single-wavelength data, the DAS are poorly matched to the individual spectra. This is because the $\alpha_i(\lambda)$ values are not well determined by the data at a single wavelength.

CHAPTER 5

5.1. The decay times can be calculated from either the phase or modulation data at any frequency, using Eqs. [5.3] and [5.4]. These values are listed in Table 5.7. Since the decay times are approximately equal from phase and modulation, the decay is nearly a single exponential. One expects the decay to become nonexponential at high chloride concentrations due to transient effects in quenching. This effect is not yet visible in the FD data for SPQ.

Table 5.7. Apparent Phase and Modulation Lifetimes for the Chloride Probe SPQ

Chloride concentration	Frequency (MHz)	Apparent phase lifetime (τ_ϕ)	Apparent modulation liifetime (τ_m) (ns)
0	10	24.90	24.94
	100	24.62	26.49
10 m*M*	10	11.19	11.07
	100	11.62	11.18
30 m*M*	10	5.17	5.00
	100	5.24	5.36
70 m*M*	10	2.64	2.49
	100	2.66	2.72

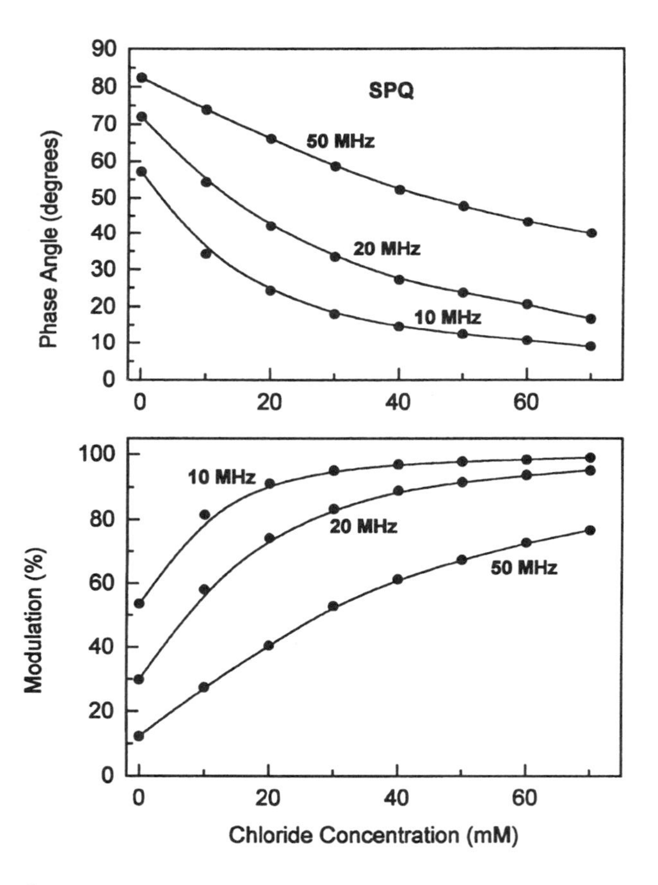

Figure 5.56. Dependence of the phase and modulation of SPQ on chloride concentration.

Table 5.8. Apparent Phase and Modulation Lifetimes for a Double-Exponential Decay

Frequency (MHz)	Decay law[a]	ϕ (deg)	m	τ_ϕ (ns)	τ_m (ns)
50	A	50.5		0.552	
	B	25.6		0.702	
100	A	60.1		0.444	
	B	29.8		0.578	

[a]For decay law A, $\alpha_1 = \alpha_2$; for decay law B, $f_1 = f_2$. In both cases, the lifetimes are 0.5 and 5.0 ns.

5.2. Chloride concentrations can be determined from the phase or modulation values of SPQ at any frequency where these values are sensitive to chloride concentration. Examination of Figure 5.30 indicates that this is a rather wide range, from 5 to 100 MHz. One can prepare calibration curves of phase or modulation of SPQ versus chloride, as shown in Figure 5.56. An uncertainty of ±0.2° in phase or ±0.5% in modulation results in chloride concentrations accurate to approximately ±0.2 and +0.3m*M* respectively, from 0 to 25 m*M*.

5.3. A list of the phase and modulation values for the two decay laws, as well as the apparent phase and modulation lifetimes, is given in Table 5.8. As expected, $\tau_\phi^{app} < \tau_m^{app}$. Both values decrease with higher modulation frequency. The phase angles are smaller, and the modulation is higher, when $f_1 = f_2$ than when $\alpha_1 = \alpha_2$. When $f_1 = f_2$, the α_i values are $\alpha_1 = 0.909$ and $\alpha_2 = 0.091$. The fractional contribution of the short-lifetime component is larger when $f_1 = f_2$.

CHAPTER 6

6.1. The Stokes' shift in wavenumbers can be calculated from the Lippert equation (Eq. [6.17]). Because it is easy to confuse the units, this calculation is shown explicitly:

$$\bar{\nu}_A - \bar{\nu}_F = \frac{2(0.3098)}{(6.6256 \times 10^{-27})(2.9979 \times 10^{10})} \frac{(14 \times 10^{-18})^2}{(4.0 \times 10^{-8})^3} \quad [6.20]$$

The emission maximum in the absence of solvent effects is assumed to be 350 nm, which is 28,571 cm^{-1}. The orientation polarizability of methanol is expected to decrease the excited-state energy by 9554 cm^{-1}, to 19,017 cm^{-1}, which corresponds to 525.8 nm.

The units for $\bar{\nu}_A - \bar{\nu}_F$ are as follows:

$$\frac{(\text{esu cm})^2}{(\text{erg s})(\text{cm/s})(\text{cm}^3)} \quad [6.21]$$

Recalling that 1 erg = 1 g cm^2 s^{-2} and 1 esu = 1 g$^{1/2}$ cm$^{3/2}$/s^{-1}, one obtains $\bar{\nu}_A - \bar{\nu}_F$ in wavenumbers (cm^{-1}).

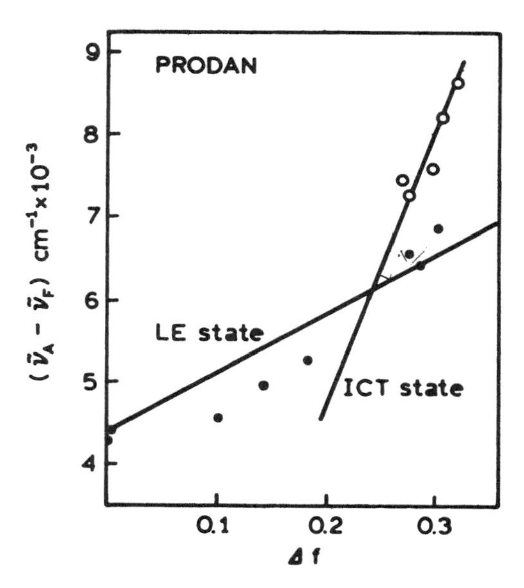

Figure 6.48. Lippert plot of the Stokes' shift of PRODAN. Data from Ref. 54.

6.2. The change in dipole moment can be estimated from the Lippert plot (Figure 6.48). This plot shows biphasic behavior. In low-polarity solvents the emission is probably due to the LE state, and in higher-polarity solvents the emission is due to the ICT state. The slopes for each region of the Lippert plot are

$$\text{slope (LE)} = 7000\,\text{cm}^{-1}$$

$$\text{slope (ICT)} = 33{,}000\,\text{cm}^{-1}$$

The slope is equal to $2(\mu_E - \mu_G)^2/hca^3$. Assuming a radius of 4.2 Å used previously,[54] one obtains for the LE state

$$(\mu_E - \mu_G)^2 = \frac{7000}{2} hca^3$$

$$= \frac{7000}{2}(6.6256 \times 10^{-27})(2.9979 \times 10^{10})$$

$$(4.2 \times 10^{-8})^3 = 5.15 \times 10^{-35} \quad [6.22]$$

The units of $(\mu_E - \mu_G)^2$ are (cm^{-1}) (erg s) (cm/s) (cm^3). Using 1 erg = 1 g cm^2 s^{-2}, one obtains the units g cm^3/s^2 cm^2. Taking the square root yields

$$\frac{g^{1/2}\,\text{cm}^{3/2}}{s}\,\text{cm} \quad [6.23]$$

Since 1 esu = 1 g$^{1/2}$ cm$^{3/2}$ s^{-1}, the result $(\mu_E - \mu_G)$ is in esu cm. This yields $(\mu_E - \mu_G) = 7.1 \times 10^{-18}$ esu cm = 7.1 D. The dipole moment of Prodan is estimated to change by 7.1 debye units upon excitation. An electron separated from a unit positive charge by 1 Å has a dipole moment of 4.8 D. Hence, there is only partial charge separation in the LE state. It should be noted that this value is smaller than initially reported[54] due to a trivial error during the calculations.[65]

For the ICT state, a similar calculation yields $(\mu_E - \mu_G)^2 = 2.42 \times 10^{-34}$ and $\Delta\mu = 15.5 \times 10^{-17}$ esu cm = 15.6 D. This change in dipole moment is equivalent to separation of a unit charge by 3.2 Å, which suggests nearly complete charge separation in the ICT state of Prodan.

CHAPTER 7

7.1. Assume that the decay is a single exponential. Then, the time at which the intensity has decayed to 37% of its original intensity is the fluorescence lifetime. These values are $\tau_F = 1$ ns at 390 nm and $\tau = 5$ ns at 435 nm. The decay time of 5 ns at 435 nm is the decay time that the F state would display in the absence of relaxation. The lifetime of the F state at 390 nm is given by $1/\tau_F = 1/\tau + 1/\tau_S$. This is equivalent to stating that the decay time of the F state (γ_F) is equal to the sum of the rates which depopulate the F state, $\gamma_F = 1/\tau + k_S$. Hence, $\tau_S = 1.25$ ns.

7.2. In the fluid solvents ethanol and dioxane, the apparent lifetimes of TNS are independent of wavelength, indicating that spectral relaxation is complete in these solvents. In glycerol or DOPC vesicles, the apparent lifetimes increase with wavelength, suggesting time-dependent spectral relaxation. The observation of $\tau^{\phi} > \tau^{m}$ at long wavelength is equivalent to observing a negative preexponential factor and proves that relaxation is occurring at a rate comparable to the intensity decay rate.

CHAPTER 8

8.1. Figure 8.40 shows a plot of F_0/F versus $[I^-]$. From the upward curvature of this plot it is apparent that both static and dynamic quenching occur for the same population of fluorophores. The dynamic (K_D) and static (K_S) quenching constants can be obtained from a plot of the apparent quenching constant (K_{app}) versus the concentration of

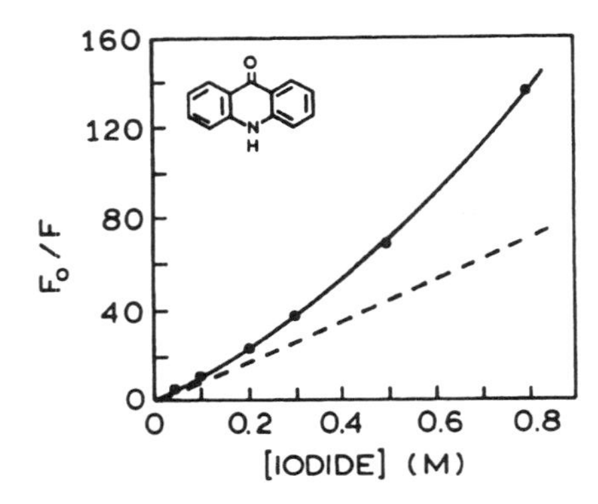

Figure 8.40. Iodide quenching of acridone. Data from Ref. 80.

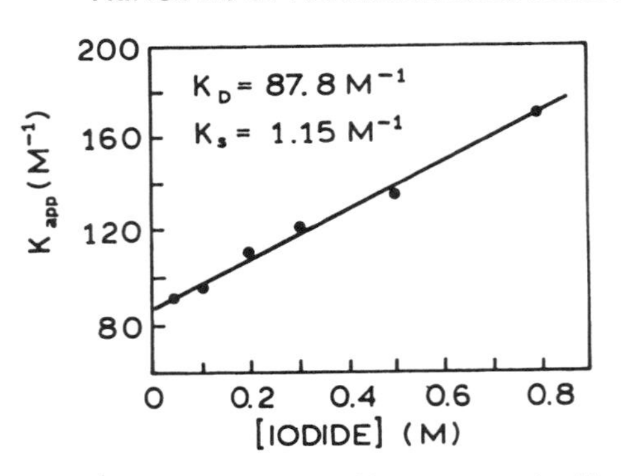

Figure 8.41. Static and dynamic quenching constants of acridone. Data from Ref. 80.

quencher $[I^-]$. The apparent quenching constant is given by $(F_0/F - 1)/[I^-] = K_{app}$.

[K] (M)	K_{app} (M^{-1})
0	—
0.04	91
0.10	96
0.20	110
0.30	121
0.50	135
0.80	170

These results are plotted in Figure 8.41. In the plot the y-intercept is $K_D + K_S = 89\ M^{-1}$, and the slope is $K_D K_S = 101\ M^{-2}$. The quadratic equation can be solved to find K_D and K_S. Assuming that the larger value is K_D, we obtain $K_D = 87.8\ M^{-1}$ and $K_S = 1.15\ M^{-1}$. The bimolecular quenching constant is given by $K_D/\tau_0 = k_q = 4.99 \times 10^9\ M^{-1}\ s^{-1}$. The collisional frequency can be calculated independently from the Smoluchowski equation. Assuming a collision radius of 4 Å, and the diffusion of the quencher to be dominant, one obtains

$$k_0 = 4\pi RDN/10^3$$

$$= \frac{4\pi (4 \times 10^{-8}\ \text{cm})(2.065 \times 10^{-5}\ \text{cm}^2/\text{s})(6.02 \times 10^{+23}\ \text{mol}^{-1})}{10^3\ \text{cm}^3 \text{l}^{-1}}$$

$$= 6.25 \times 10^9\ M^{-1}\ \text{s}^{-1} \quad [8.34]$$

This value describes the quenching constant expected if 100% of the collisional encounters are effective in quenching. Hence, the quenching efficiency $\gamma = k_q/k_0 = 0.80$.

The radius of the sphere of action can be calculated using any of the F_0/F values for which there is excess quenching. The expected values of F_0/F due to dynamic quenching only are indicated by the dashed line on Figure 8.40. At 0.8M iodide,

$$\left(\frac{F_0}{F}\right)_D = 1 + 87.8\ (0.8) = 71.24 \quad [8.35]$$

The observed value is 137. Using

$$\frac{F_0}{F} = (1 + K_D[Q]) \exp([Q]Nv/1000) \qquad [8.36]$$

we obtain exp ([Q]Nv/1000) = 1.92 or [Q]Nv/1000 = 0.654. From these results, one can calculate that the volume of the sphere of action is $v = 1.37 \times 10^{-21}$ cm^3. Using $v = \frac{4}{3}\pi r^3$, where r is the radius, one finds r = 6.9 Å. According to this calculation, whenever an iodide ion is within 6.9 Å of an excited acridone molecule, the probability of quenching is unity.

The static quenching constant is quite small, as is the radius of the sphere of action. It seems that no actual complex is formed in this case. Rather, the static component is due simply to the probability that a fluorophore is adjacent to a quencher at the moment of excitation.

8.2. Using the data in Problem 8.2, one may calculate the following:

[AMP] (mM)	τ_0/τ	F_0/F	$(F_0/F)/(\tau_0/\tau)$
0.0	1.0	1.0	1.0
1.75	1.265	1.40	1.11
3.50	1.502	1.80	1.20
5.25	1.741	2.35	1.35
7.0	1.935	3.00	1.55

The collisional or dynamic quenching constant can be obtained from a plot of τ_0/τ versus [AMP] (Figure 8.42, left). The dynamic quenching constant is 136 M^{-1}. Using the lifetime in the absence of quencher, one finds $k_q = K_D/\tau_0 = 4.1 \times 10^9\ M^{-1}\ s^{-1}$, which is typical for a diffusion-controlled reaction which occurs with high efficiency.

In the previous problem, we obtained K_S and K_D from a plot of the apparent quenching constant versus quencher concentration (Figure 8.41). In this case both the lifetime and yield data are given, and a simpler procedure is possible. We calculated the quantity $(F_0/F)/(\tau_0/\tau)$. From Eqs. [8.8], [8.17], and [8.19], this quantity is seen to reflect only the static component of the quenching,

$$\frac{F_0/F}{\tau_0/\tau} = 1 + K_S\,[\mathrm{AMP}] \qquad [8.37]$$

A plot of $(F_0/F)/(\tau_0/\tau)$ versus [AMP] yields the association constant as the slope (Figure 8.42, right).

In contrast to the apparent association constant for the "acridone–iodide complex," this value (72.9 M^{-1}) is much larger. An actual ground-state complex is likely in this case. This was demonstrated experimentally by examination of the absorption spectrum of MAC, which was found to be changed in the presence of AMP. If the MAC–AMP complex is nonfluorescent, then the only emission observed is that from the uncomplexed MAC. Since these molecules are not complexed, the excitation spectrum of MAC in the presence of AMP will be that of MAC alone.

8.3. The susceptibility of a fluorophore to quenching is proportional to its fluorescence lifetime. Fluorophores with longer lifetimes are more susceptible to quenching. To decide on the upper limit of lifetimes, above which oxygen quenching is significant, we need to consider dissolved oxygen from the air. Based on the assumed accuracy of 3%, we can use $F_0/F = \tau_0/\tau = 1.03$. Since the atmosphere is 20% oxygen, the oxygen concentrations due to atmospheric oxygen are one-fifth the total solubility. For aqueous solutions,

$$\frac{F_0}{F} = \frac{\tau_0}{\tau} = 1.03 = 1 + k_q\tau_0[O_2] \qquad [8.38]$$

Using the information provided for aqueous solutions,

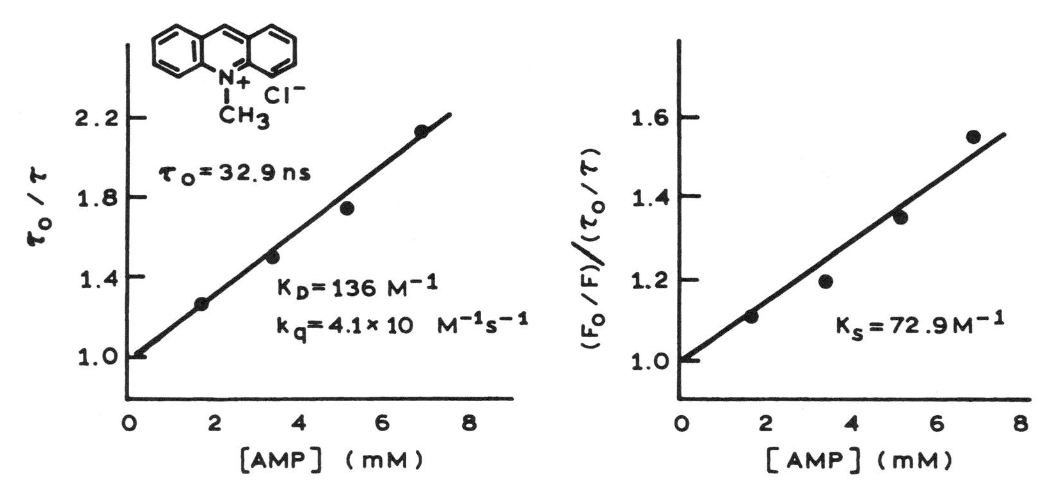

Figure 8.42. Quenching of methyl-acridinium chloride by AMP. From Ref. 18.

$$\tau_0 = \frac{0.03}{k_q[O_2]} = \frac{0.03(5)}{(1 \times 10^{10})(0.001275)} = 11.8 \text{ ns} \quad [8.39]$$

For the ethanol solution,

$$\tau_0 = \frac{0.03(5)}{(2 \times 10^{10})(0.001275)(5)} = 1.2 \text{ ns} \quad [8.40]$$

If the unquenched lifetimes are longer than 1.2 ns in ethanol, or 11.8 ns in water, then dissolved oxygen from the air can result in significant quenching (greater than 3%). If desired, this quenching can be minimized by purging with an inert gas, such as nitrogen or argon.

8.4. The rate of collisional deactivation can be calculated from the decrease in lifetime due to collisions with the adenine ring. The lifetimes in the absence (τ_0) and presence (τ) of the adenine moiety are $\tau_0 = \gamma^{-1}$ and $\tau = (\gamma + k)^{-1}$. Therefore,

$$k = \frac{1}{\tau} - \frac{1}{\tau_0} = 2.0 \times 10^8 \text{ s}^{-1} \quad [8.41]$$

The quantum yield of FAD is decreased by both static and dynamic quenching,

$$\frac{F}{F_0} = \frac{Q(\text{FAD})}{Q(\text{FMN})} = f\frac{\tau}{\tau_0} \quad [8.42]$$

where f is the fraction not complexed. Hence,

$$f = \frac{\tau_0}{\tau}\frac{Q(\text{FAD})}{Q(\text{FMN})} = \frac{(4.6)(0.09)}{(2.4)(1.0)} = 0.17 \quad [8.43]$$

Thus, 83% of the FAD exists as a nonfluorescent complex.

8.5. Using the data provided, one can calculate the following quantities needed for the Stern–Volmer plots.

$[I^-]$ (M)	F_0/F	ΔF	$\Delta_0/\Delta F$	$1/[I^-]$ (M^{-1})
0.0	1.000	0	—	—
0.01	1.080	0.074	13.51	100
0.03	1.208	0.172	5.814	33.3
0.05	1.304	0.233	4.292	20.0
0.10	1.466	0.318	3.145	10.0
0.20	1.637	0.389	2.571	5.0
0.40	1.776	0.437	2.288	2.5

The Stern–Volmer and modified Stern–Volmer plots are shown in Figure 8.43. The downward curvature of the Stern–Volmer plot indicates an inaccessible fraction. From the intercept on the modified Stern–Volmer plot, one finds $f_a = 0.5$. Hence, one tryptophan residue per subunit is accessible to iodide quenching. The slope on the modified Stern–Volmer plot is equal to $(f_a K)^{-1}$. Thus $K = 17.4$ M^{-1}. By assumption, the quenching constant of the inaccessible fraction is zero. Using these results, one can calculate the following quantities in order to predict the quenching plots for each tryptophan residue.

$[I^-]$ (M)	$[I^-]^{-1}$ (M^{-1})	$(F_0/F)_b$	$(F_0/F)_a^a$	$(F_0/\Delta F)_b$	$(F_0/\Delta F)_a^b$
0.0	0	1.0	1.0	∞	—
0.01	100	1.0	1.174	∞	6.747
0.03	33.3	1.0	1.522	∞	2.916
0.05	20.0	1.0	1.870	∞	2.149
0.10	10.0	1.0	2.740	∞	1.575
0.20	5.0	1.0	4.480	∞	1.287
0.40	2.5	1.0	7.960	∞	1.144

[a]Calculated from $F_0/F = 1 + K[Q] = 1 + 17.4[I^-]$.
[b]Calculated from $F_0/\Delta F = (1/K[Q]) + 1$.

For the accessible fraction, the Stern–Volmer plot is linear and the apparent value of f_a is 1 (Figure 8.44). Hence, if the quenching data were obtained using 300-nm

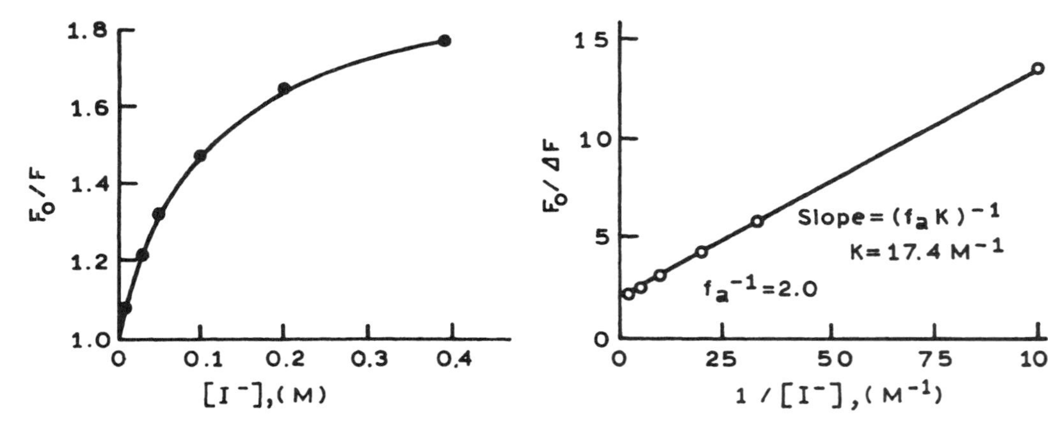

Figure 8.43. Stern–Volmer and modified Stern–Volmer plots for iodide quenching.

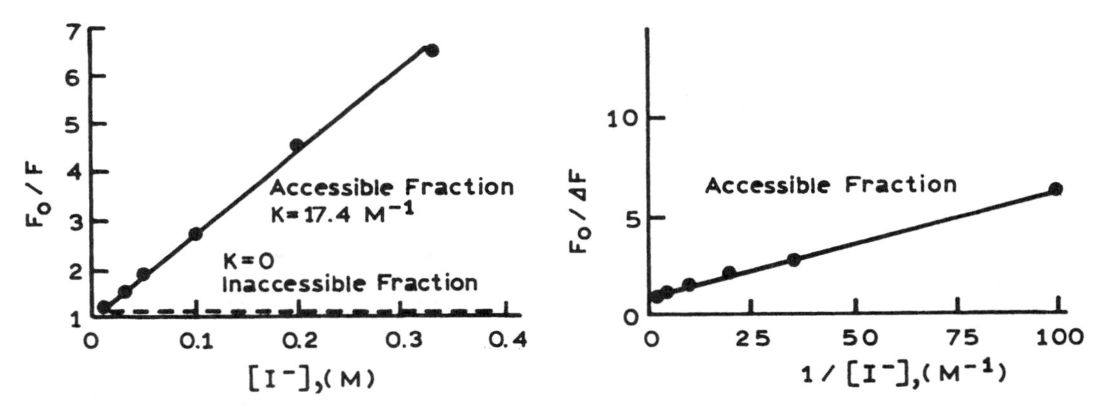

Figure 8.44. Predicted Stern–Volmer plots for the accessible and inaccessible tryptophan residues.

excitation, where only the accessible residue was excited, all the fluorescence would appear to be accessible. Since the inaccessible fraction is not quenched, $F_0/F = 1$ for this fraction. One cannot construct a modified Stern–Volmer plot since $\Delta F = 0$ for this fraction. The bimolecular quenching constant can be calculated using $K = 17.4\ M^{-1}$ and τ = ns, yielding a bimolecular quenching constant $k_a = 0.35 \times 10^{10}\ M^{-1}\ sec^{-1}$

8.6. Quenching of endo III by poly(dA-dT) displays saturation near 20μM poly(dA-dT), which indicates specific binding of poly(dA-dT) to endo III. Assume that the quenching is dynamic. Then K_D is near $10^5\ M^{-1}$, resulting in an apparent value of $k_q = 2 \times 10^{13}\ M^{-1}\ s^{-1}$. This is much larger than the diffusion-controlled limit, so there must be some specific binding.

About 50% of the fluorescence is quenched. In Section 8.9.A we saw that both top residues in Endo III were equally fluorescent. Hence, the titration data (Figure 8.39) suggest that one residue, probably trp-132, is completely quenched when poly(dA-dT) binds to endo III.

CHAPTER 9

9.1. The apparent bimolecular quenching constant for Cu^{2+} can be found by noting that $F_0/F = 1.10$ at $2 \times 10^{-6} M\ Cu^{2+}$. Hence, $K = 50{,}000\ M^{-1}$ and $k_q = 5 \times 10^{12}\ M^{-1}\ s^{-1}$. Similarly, $F_0/F = 1.7$ at 0.001M DMA, yielding $K = 700\ M^{-1}$ and $k_q = 7 \times 10^{10}\ M^{-1}\ s^{-1}$. Both values are larger than the maximum value possible for diffusive quenching in water, $\sim 1 \times 10^{10}\ M^{-1}\ s^{-1}$. This implies some binding or localization of the quenchers near the fluorophores.

9.2. The data in Figure 9.36 can be used to calculate the lifetimes of pyrene, which are 200, 119, and 56 ns in the presence of N_2, air, and O_2, respectively.[55] Assuming that the oxygen solubility in DMPC vesicles is five-fold larger than in water (0.001275M/atm in water), the oxygen bimolecular quenching constant is $k_q = 0.2 \times 10^9\ M^{-1}\ s^{-1}$. This value is about 20% of the value expected for a fluorophore dissolved in water.

9.3. The data in the absence of benzyl alcohol (Figure 9.37) can be used to calculate a bimolecular quenching constant of $0.6 \times 10^9\ M^{-1}\ s^{-1}$. This indicates that the naphthalene is mostly accessible to iodide and probably not bound to the cyclodextrin. This conclusion is supported by the data in the presence of benzyl alcohol. In the presence of benzyl alcohol, the Stern–Volmer plots curve downward in the presence of increasing concentrations of β-CD. This suggests the presence of two naphthalene populations, one of which is less accessible to iodide quenching. In the presence of benzyl alcohol and 5.1mM β-CD, the Stern–Volmer plot is still curved, and the apparent value of k_q decreases, which indicates shielding from iodide quenching. Under these conditions, it seems that naphthalene binds to β-CD only in the presence of benzyl alcohol.

CHAPTER 10

10.1. At 430, 320, and 282 nm, the r_0 values of perylene are 0.35, 0.10, and 0.0, respectively. The angle β between the absorption and emission moments can be calculated using Eq. [10.20]. These calculations yield β = 16.8°, 45°, and 54.7°, respectively.

10.2. If the sample is weakly scattering, the scattered component will be completely polarized ($r = 1.0$). The measured anisotropy can be obtained using Eq. [10.6]:

$$r_{obs} = 0.30(0.80) + 1.0(0.20) = 0.44 \qquad [10.66]$$

An anisotropy value greater than 0.40 should be an immediate warning that the measured value was not due only to fluorescence.

10.3. The corrected ratio, I_{VV}/I_{HV}, is given by 1.33/0.45 = 2.95. Therefore, $r_0 = 0.394$ and $P_0 = 0.494$. The angle between the absorption and emission dipoles can be calculated using Eq. [10.20]. Substitution of $r_0 = 0.394$ yields $\beta = 5.7°$.

10.4. The denominator in Eq. [10.49] is given by

$$\int_0^\infty I(t)\, dt = I_0 \int_0^\infty \exp(-t/\tau)\, dt = I_0\tau \quad [10.67]$$

The numerator in Eq. [10.49] is given by

$$\int_0^\infty I(t)r(t)dt = \int_0^\infty \exp\left[-t\left(\frac{1}{\tau}+\frac{1}{\theta}\right)\right] dt = I_0 r_0 \tau\theta/(\theta+\tau) \quad [10.68]$$

Division of Eq. [10.68] by Eq. [10.67] yields Eq. [10.50].

10.5. The rotational correlation time of perylene in ethanol at 20°C can be calculated using Eq. [10.52]:

$$\theta = \frac{\eta V}{RT} = \frac{(0.01194\text{P})(252\text{ g/mol})(0.74\text{ ml/g})}{(293\text{K})(8.314\times 10^7\text{ erg/mol K})} \quad [10.69]$$

$$= 91\text{ ps} \quad [10.70]$$

The anisotropy can be calculated using Eq. [10.50]:

$$r = \frac{0.36}{1+(6/0.091)} = 0.005 \quad [10.71]$$

A similar calculation for propylene glycol at 25 °C yields $\phi = 2.4$ ns and $r = 0.103$.

10.6. Equation [10.58] can be derived by deducing an expression for the average anisotropy. Suppose that the quantum yields of the free and bound forms are q_F and q_B, respectively. Then the measured anisotropy is

$$r = \frac{f_F q_F r_F + f_B q_B r_B}{f_F q_F + f_B q_B} \quad [10.72]$$

The correctness of this expression can be seen by noting that the numerator is simply a revised form of the additivity law for anisotropies and that the products $f_F q_F$ and $f_B q_B$ represent the intensities of each form of the probe. The denominator normalizes these values to fractional fluorescence intensities. Equation [10.72] can be rearranged to Eq. [10.58] by noting that $f_F + f_B = 1.0$ and $R = q_B/q_F$.

10.7. The observed polarization may be converted into anisotropies using $r = 2P/(3-P)$. The latter are more convenient since

$$\bar{r} = f_F r_F + f_B r_B \quad [10.73]$$

where the subscripts F and B represent the free and bound forms of the fluorophore, respectively, and f_i is the fraction of fluorescence due to each form of the probe. When [BSA] = 0, one observes r_F, and when [BSA] >> K_d, one observes r_B. These considerations are summarized below:

[BSA]	Observable	r
0	r_F	0.010
$2\times 10^{-5}M$	r	0.200
$>>K_d$	r_B	0.300

Using Eq. [10.73], one obtains

$$0.200 = f_F(0.010) + (1-f_F)(0.300) \quad [10.74]$$

and hence $f_F = 0.345$ and $f_B = 0.655$. Since the concentration of DNS is much less than that of BSA, we can assume that the concentration of unliganded BSA is not depleted by the binding of DNS. The ratio of free to bound DNS is given by 0.345/0.655. Hence, from Eq. [10.65],

$$K_d = \frac{(0.345)(2\times 10^5 M)}{(0.655)} = 1.05\times 10^{-5} M \quad [10.75]$$

B. In the use of Eq. [10.73], we assumed that the calculated fractional intensity of each species represented the fraction of the DNS which was bound or free. However, if the relative quantum yield of the bound probe is twofold larger than that of the free probe, then clearly the concentration of the bound form is twofold lower. Therefore,

$$K_d = \frac{(0.345)(2\times 10^5 M)}{0.655/2} = 2.1\times 10^5 M \quad [10.76]$$

C. A change in quantum yield could be readily detected by comparing the intensities of DNS solutions with and without added BSA. Provided that the DNS concentrations are identical, the relative intensities represent the relative quantum yields.

D. Using the data provided, the calculated rotational correlation time of BSA is 20 ns. The anisotropy of free DNS will decay too rapidly for measurement with most currently available instruments. For the solution containing a concentration of BSA adequate to bind all the DNS, one expects

$$r(t) = 0.20e^{-t/20} \quad [10.77]$$

For the $2\times 10^{-5}M$ solution,

$$r(t) = f_B r_0 e^{-t/20} \quad [10.78]$$

$$= 0.131e^{-t/20}$$

10.8. The anisotropy can be calculated using Eqs. [10.6] and [10.20]. The anisotropy from the three transitions can be calculated using $\beta = 0°$ and $\beta = \pm 120°$. Hence, $r = 0.33(0.40) + 0.33(-0.05) + 0.33(-0.05) = 0.10$.

10.9. The apparent $r(0)$ values of melittin are near 0.16, which is considerably less than $r_0 = 0.26$. This indicates that the tryptophan residue in melittin displays fast motions that are not resolved with the available range of lifetimes (0.6–2.4 ns).

The apparent correlation times from melittin can be calculated from the slopes in Figure 10.36. For example, in the absence of NaCl the slope is near 5.8×10^9, which is equal to $[r(0)\theta]^{-1}$. Hence, the apparent correlation time is 1.07 ns.

10.10. The anisotropy of DPPS is higher for two-photon excitation because of $\cos^4\theta$ photoselection. The ratio of the two-photon to one-photon anisotropies is near 1.39, which is close to the predicted values for parallel transitions, 1.425 (Section 10.13).

The anisotropy is independent of temperature because the lifetime decreases with increasing temperature. The decrease in lifetime offsets the decrease in correlation time, resulting in a constant anisotropy.

CHAPTER 11

11.1. The angle can be calculated using Eq. [11.51]. Using the apparent time–zero anisotropy of 0.22 as r_∞, one finds $\langle\cos^2\theta\rangle = 0.924$ and $\theta = 16°$.

11.2. The most direct approach is to use the amplitudes from the intensity decay. The radiative rate of a fluorophore is usually not affected by its environment. Hence, the relative values of α_1 and α_2 represent the fraction of the FMN free or bound to YFP. The dissociation constant is given by

$$K_D = \frac{[\text{FMN}][\text{YFP}]}{[\text{FMN} \cdot \text{YFP}]} \quad [11.52]$$

This equation can be rewritten in terms of the total YFP concentration and the fraction of FMN bound (f_B),

$$K_D = \frac{[\text{YFP}]_T(1-f_B)^2}{f_B} \quad [11.53]$$

where $[\text{YFP}]_T = [\text{YFP}] + [\text{FMN} \cdot \text{YFP}]$ is the total concentration of YFP. This expression can be understood by noticing that the concentrations of free YFP and FMN are both given by $[\text{YFP}]_T(1 - f_B)$ and that the concentration of $[\text{FMN} \cdot \text{YFP}]$ is given by $[\text{YFP}]_T f_B$. At $[\text{YFP}] = 0.18\mu M$, the fraction bound is given by

$$f_B = \frac{\alpha_2}{\alpha_1 + \alpha_2} = 0.31 \quad [11.54]$$

Hence, $K_D = 0.28 \times 10^{-6} M$.

Table 12.3. Associated Anisotropy Decay

t (ns)	$f_1(t)$	$f_2(t)$	$r_1(t)$	$r_2(t)$	$r(t)$
0	0.5	0.5	0.3	0.3	0.30
1	0.45	0.55	0.0	0.29	0.16
5	0.25	0.75	0.0	0.27	0.20

CHAPTER 12

12.1. The anisotropy at any time can be calculated using Eq. [12.36]. The values obtained are listed in Table 12.3. The anisotropy values for the nonassociated decay can be calculated using

$$r(t) = r_0[0.5\exp(-t/0.05) + 0.5\exp(-t/40)] \quad [12.46]$$

For $t = 0$, 1, and 5 ns, these values are 0.30, 0.146, and 0.132, respectively.

CHAPTER 13

13.1. The D–A distance can be calculated using Eq. [13.12]. The transfer efficiency is 90%. Hence, the D–A distance is $r = (0.1)^{1/6}R_0 = 0.68R_0 = 17.6$ Å. The donor lifetime in the D–A pair can be calculated from Eq. [13.13], which can be rearranged to $\tau_{DA} = (1 - E)\tau_D = 0.68$ ns.

13.2. The equations relating the donor intensity to the transfer efficiency can be derived by recalling the expressions for relative quantum yields and lifetimes. The relative intensities and lifetimes are given by

$$F_D = \frac{\Gamma_D}{\Gamma_D + k_{nr}} \qquad \tau_D = \frac{1}{\Gamma_D + k_{nr}} \quad [13.34]$$

$$F_{DA} = \frac{\Gamma_D}{\Gamma_D + k_{nr} + k_T} \qquad \tau_{DA} = \frac{1}{\Gamma_D + k_{nr} + k_T} \quad [13.35]$$

where Γ_D is the emission rate of the donor and k_{nr} is the nonradiative decay rate. The ratio of intensities is given by

$$\frac{F_{DA}}{F_D} = \frac{\Gamma_D + k_{nr}}{\Gamma_D + k_{nr} + k_T} = \frac{\tau_D^{-1}}{\tau_D^{-1} + k_T} \quad [13.36]$$

Hence,

$$1 - \frac{F_{DA}}{F_D} = \frac{k_T}{\tau_D^{-1} + k_T} = E \quad [13.37]$$

One can derive a similar expression for the transfer efficiency E based on lifetimes using the right-hand side of Eqs. [13.34] and [13.35].

It should be noted that k_T was assumed to be a single value, which is equivalent to assuming a single distance. We also assumed that the donor population was homogeneous, so that each donor had a nearby acceptor; that is, labeling by acceptor is 100%.

13.3. The excitation spectra reveal the efficiency of energy transfer by showing the extent to which the excitation of the naphthyl donor at 290 nm results in dansyl emission. The transfer efficiency can be calculated from the emission intensity at 450 nm for 290-nm excitation, which reflects acceptor emission due to excitation of the donor and direct excitation of the acceptor. Dansyl-L-prolyl-hydrazide does not contain a donor, and hence this excitation spectrum defines that expected for 0% transfer. For dansyl-L-prolyl-α-naphthyl, in which the donor and acceptor are closely spaced, energy transfer is 100% efficient. For this D–A pair, the greatest sensitivity of the excitation spectrum to energy transfer is seen near 290 nm, the absorption maximum of the naphthyl donor. For the other derivatives, the intensity is intermediate and dependent upon the length of the spacer. For 290-nm excitation, the transfer efficiency can be calculated from the relative intensity between 0% and 100% transfer. The efficiency of energy transfer decreases as the length of the spacer is increased.

The object of these experiments was to determine the distance dependence of radiationless energy transfer. Hence, we assume that the efficiency of energy transfer depends on distance according to

$$E = \frac{(R_0/r)^j}{(R_0/r)^j + 1} \quad [13.38]$$

where R_0 and r have their usual meanings and j is an exponent to be determined from the observed dependence of E on r. Rearrangement of Eq. [13.38] yields

$$\log(E^{-1} - 1) = j \log r - j \log R_0 \quad [13.39]$$

Hence, a plot of $\log(E^{-1} - 1)$ versus $\log r$ has a slope of j. These data are shown in Figure 13.34. The slope was found to be 5.9 ± 0.3.[16] From this agreement with the predicted value of $j = 6$, the authors of this study concluded that energy transfer followed the predictions of Förster. See Ref. 16 for additional details.

The value of R_0 can be found from the distance at which the transfer efficiency is 50%. From the right-hand side of Figure 13.34, the value of R_0 is seen to be near 33 Å.

13.4. The lifetime of compound I is τ_{DA}, and the lifetime of compound II is τ_D. Compound II serves as a control for the effect of solvent on the lifetime of the indole moiety, in the absence of energy transfer. The rate of energy transfer is given by $k_T = \tau_{DA}^{-1} - \tau_D^{-1}$, and

$$k_T = CJ \quad [13.40]$$

where C is a constant. Hence, a plot of k_T versus J should be linear. The plot of k_T versus J is shown in Figure 13.35. The slope is 1.10. These data confirm the expected dependence of k_T on the overlap integral. See Ref. 16 for additional details.

13.5. If the wavelength (λ) is expressed in nanometers, the overlap integral for Figure 13.8 can be calculated using Eq. [13.3] and is found to be $4.4 \times 10^{13}\ M^{-1}\ \text{cm}^{-1}\ \text{nm}^4$. Using Eq. [13.5], with $n = 1.33$ and $Q_D = 0.21$, one finds $R_0 = 23.6$ Å. If λ is expressed in centimeters, then $J(\lambda) = 4.4 \times 10^{-15}\ M^{-1}\ \text{cm}^3$, and Eq. [13.8] yields $R_0 = 23.6$ Å.

13.6. The dissociation reaction of cAMP from protein kinase (PK) is described by

$$\text{PK} \cdot \text{cAMP} \rightleftharpoons \text{PK} + \text{cAMP}$$

$$\text{B} \rightleftharpoons \text{F} + \text{A} \quad [13.41]$$

where B represents PK with bound cAMP, F represents the PK without bound cAMP, and A represents cAMP. The dissociation constant for cAMP is defined by

$$K_D = \frac{[\text{F}][\text{A}]}{[\text{B}]} \quad [13.42]$$

Using conservation of mass, [F] + [B] = [T], the total protein kinase concentration, one can show that the free and bound fractions of PK are given by

$$f_B = \frac{[\text{B}]}{[\text{T}]} = \frac{[\text{A}]}{K_D + [\text{A}]}, \qquad f_F = \frac{[\text{F}]}{[\text{T}]} = \frac{K_D}{K_D + [\text{A}]} \quad [13.43]$$

Let R_B and R_F represent the ratio of donor to acceptor intensities for each species. At any given cAMP concentration, the observed ratio R is

$$R = f_B R_B + f_F R_F \quad [13.44]$$

Assuming that $f_B + f_F = 1.0$, one obtains

$$[\text{A}] = [\text{cAMP}] = K_D \left(\frac{R - R_B}{R_F - R} \right) \quad [13.45]$$

A ratio of intensities is independent of the total PK concentration and independent of sample-to-sample variations in PK concentration. Hence, intensity ratiometric measurements are convenient and accurate. See Ref. 31 for additional details.

13.7. A. The efficiency of energy transfer can be calculated from Eq. [13.14].

$$E = 1 - \frac{4.1}{20.5} = 0.80 \quad [13.46]$$

B. The expected lifetime in the presence of DNP can be calculated using Eq. [13.13] with $\tau_D = 5.0$ ns and $E = 0.8$:

$$\tau_{DA} = \tau_D (1 - E) = 1 \text{ ns} \qquad [13.47]$$

C. The rate of energy transfer (k_T) can be calculated using Eq. [13.11] with $E = 0.8$ and $\tau_D = 5$ ns:

$$k_T = \frac{E\tau_D^{-1}}{1-E} = 8 \times 10^8 \text{ s}^{-1} \qquad [13.48]$$

D. The distance can be calculated using Eq. [13.12]. Substitution and rearrangement yields $r^6 = 0.25R_0^6$, and therefore $r = 39.7$ Å.

E. The efficiency can be calculated using Eq. [13.12] with $r = 20$ Å. The efficiency is 0.9959. Once the efficiency is known, the intensity can be calculated using Eq. [13.14]. The fluorescence intensity (F_{DA}) is expected to be $0.0041F_D = 0.084$.

F. The 1% impurity would contribute 1% of 20.5, or 0.205, to the total intensity. The contribution from this minor component would be (0.205/0.084) = 2.5 times more intense than the signal from the DNP-binding protein and would invalidate any interpretation of the intensity in the presence of DNP.

G. The lifetime of the sample would be dominated by the impurity and thus would be near 5 ns. Such a result is indicative of an impurity. Specifically, the quantum yield is decreased to 0.0041 of the original value, but the lifetime is relatively unchanged. When this result is found, one should consider the possibility of the presence of a fluorescent impurity.

13.8. In order to calculate the possible effects of κ^2 on distance, we need to determine the depolarization factors due to segmental motion of the donor and the acceptor. Knowledge of the rotational correlation time of the protein (θ) allows us to factor out this component in the steady-state anisotropy. The depolarization factors due to overall rotation (d_{pi}) can be calculated from the Perrin equation. For the donor ($\tau_D = 5$ ns) and acceptor ($\tau_A = 15$ ns), these factors are

$$d_{pD} = \frac{1}{1 + \tau_D/\theta} = 0.5 \qquad [13.49]$$

$$d_{pA} = \frac{1}{1 + \tau_A/\theta} = 0.25 \qquad [13.50]$$

Recall that the overall depolarization is given by Soliellet's rule (Section 10.10), $r = r_0 d_{pi} d_{si}$, when d_{si} is the factor due to segmental motions of the donor or the acceptor. Hence, we can use the steady-state anisotropies to calculate the depolarization factors due to rapid segmental probe motions:

$$d_D^x = \left(\frac{r_D}{r_0 d_{pD}}\right)^{1/2} = 0.71 \qquad [13.51]$$

$$d_A^x = \left(\frac{r_A}{r_0 d_{pA}}\right)^{1/2} = 0.71 \qquad [13.52]$$

Hence, the maximum and minimum values of κ^2 calculated from Eqs. [13.18] and [13.19] are 0.19 and 2.61. Using Eqs. [13.23] and [13.24], the D–A distance can range from $0.81R_0$ to $1.26R_0$, or from 20.3 to 31.5 Å.

13.9. If $f_A = 1.0$, then the transfer efficiency is given by Eq. [13.14],

$$E = 1 - \frac{0.5}{1.0} = 0.5 \qquad [13.53]$$

and the D–A distance is thus equal to R_0. If $f_A = 0.5$, the transfer efficiency is given by Eq. [13.17],

$$E = 1 - \frac{0.5 - 1.0(0.5)}{1.0(0.5)} = 1.0 \qquad [13.54]$$

If $f_A = 0.5$, then the transfer efficiency for the actual D–A pair is 100%, and thus the D–A distance is less than $0.5R_0$. The presence of acceptor underlabeling results in a higher intensity of the donor emission and an overestimation of the true D–A distance.

13.10. Equation [13.25] can be easily derived by writing expressions for the acceptor intensity in the absence (F_A) and in the presence of donor (F_{AD}). These intensities are given by

$$F_A(\lambda_A^{em}) = \varepsilon_A(\lambda_D^{ex})\, C_A\,(\lambda_A^{em}) \qquad [13.55]$$

$$F_{AD}(\lambda_A^{em}) = [\varepsilon_A(\lambda_D^{ex}) + E\varepsilon_D(\lambda_D^{ex})]\, C_A\,(\lambda_A^{em}) \qquad [13.56]$$

where excitation is at λ_D, intensities are measured at λ_A, and E is the transfer efficiency. $C_A(\lambda_A^{em})$ is a constant relating the intensity at λ_A to the acceptor concentration. Dividing Eq. [13.55] by Eq. [13.56], followed by rearrangement, yields Eq. [13.25].

If the extent of donor labeling is less than 1.0, then the acceptor intensities are given by

$$F_A(\lambda_A^{em}) = \varepsilon_A(\lambda_D^{ex})\, C_A\,(\lambda_A^{em}) \qquad [13.57]$$

$$F_{AD}(\lambda_A^{em}) = [\varepsilon_A(\lambda_D^{ex}) + f_D E\varepsilon_D(\lambda_D^{ex})]\, C_A\,(\lambda_A^{em}) \qquad [13.58]$$

where f_D is the fractional labeling with the donor. These expressions can be understood by recognizing that the directly excited acceptor intensity is independent of f_D, but the acceptor intensity due to energy transfer depends on f_D. Rearrangement of Eqs. [13.57] and [13.58] yields Eq. [13.25].

13.11. Let $C_A(\lambda_A)$ and $C_D(\lambda_A)$ be the constants relating the intensities at λ_A to the acceptor and donor concentrations, respectively, when both are excited at λ_D. Since the donor is assumed to emit at λ_A, Eqs. [13.55] and [13.56] become

$$F_A(\lambda_A) = \varepsilon_A(\lambda_D)\, C_A(\lambda_A) \quad [13.59]$$

$$F_{AD}(\lambda_A) = [\varepsilon_A(\lambda_D) + E\varepsilon_D(\lambda_D)]\, C_A(\lambda_A) + \varepsilon_D(\lambda_D) C_{DA}(\lambda_A) \quad [13.60]$$

In Eq. [13.60] we have considered the contribution of the donor in the D–A pair to the intensity at λ_A. In general, $C_{DA}(\lambda_A)$ will be smaller than $C_D(\lambda_A)$ due to RET quenching of the donor. However, $C_D(\lambda_A)$ can be measured with the donor-alone sample. $C_{DA}(\lambda_A)$ can be estimated using the shape of the donor emission to estimate the donor contribution to the emission at λ_A in the doubly labeled sample. Equations [13.59] and [13.60] can be rearranged to

$$\frac{F_{AD}(\lambda_A)}{F_A(\lambda_A)} - 1 = \frac{E\varepsilon_D(\lambda_D)}{\varepsilon_A(\lambda_D)} + \frac{\varepsilon_D(\lambda_D)\, C_{DA}(\lambda_A)}{\varepsilon_A(\lambda_D)\, C_A(\lambda_A)} \quad [13.61]$$

The transfer efficiency as seen from the acceptor emission is given by Eq. 13.25, which assumes that the donor does not emit at the acceptor wavelength. Hence the acceptor emission increases the apparent efficiency to

$$E_{app} = E + \frac{C_{DA}(\lambda_A)}{C_A(\lambda_A)} \quad [13.62]$$

and would thus be larger than the actual efficiency. If the donor does not contribute to the emission at λ_A, then $C_{DA}(\lambda_A) = 0$ and E_{app} becomes the true efficiency. See Ref. 82 for additional details.

13.12. The true transfer efficiency is defined by the proportion of donors which transfer energy to the acceptor and is given by

$$E = \frac{k_T}{\tau_D^{-1} + k_q + k_T} \quad [13.63]$$

The apparent efficiency seen for the donor fluorescence is given by

$$E_D = \frac{k_T + k_q}{\tau_D^{-1} + k_q + k_T} \quad [13.64]$$

The apparent efficiency (E_D) is larger than the true efficiency (E) because the additional quenching pathway decreases the donor emission more than would have occurred by RET alone. See Ref. 50 for additional details.

CHAPTER 14

14.1. A. The intensity decay of A would be a single exponential, but the intensity decays of B and C would be triple exponentials. For sample A, the donors are at a unique distance from acceptors at 15, 20, and 25 Å. The transfer rate is given by

$$k_T = \frac{1}{\tau_D}\left(\frac{20}{15}\right)^6 + \frac{1}{\tau_D}\left(\frac{20}{20}\right)^6 + \frac{1}{\tau_D}\left(\frac{20}{25}\right)^6 \quad [14.28]$$

Calculation of the transfer rate yields $k_T = \tau_D^{-1}(6.88)$. Hence, the decay of sample A is given by Eq. [14.1] and is a single exponential with

$$I_{DA}(t) = I_D^0 \exp\left[-\frac{t}{\tau_D} - \frac{t6.88}{\tau_D}\right] = \exp\left(-\frac{t}{0.63}\right) \quad [14.29]$$

The intensity decay of sample B would be a triple exponential. There would be three different decay times, which can be calculated from the three transfer rates in Eq. [14.28]. The decay times are 0.76 ns, 2.5 ns, and 3.96 ns.

B. Since the unquenched lifetime and quantum yields of the three proteins in sample B are the same, the radiative decay rates are the same, and the relative amplitudes of the three proteins would be the same. Hence, the intensity decay would be given by

$$I(t) = \sum_i \alpha_i e^{-t/\tau_D} \quad [14.30]$$

with $\alpha_1 = \alpha_2 = \alpha_3 = 0.33$ and $\tau_1 = 0.76$, $\tau_2 = 2.5$, and $\tau_3 = 3.96$ ns.

In contrast to the α_i values, the fractional intensities will be very different for each protein in sample B. These values are given by

$$f_i = \frac{\alpha_i \tau_i}{\sum_j \alpha_j \tau_j} \quad [14.31]$$

Hence, the fractional intensities of the three proteins are 0.105, 0.346, and 0.549. The intensity decay of sample C would be the same as sample B, except the α_i values are for the three tryptophan residues.

C. The presence of three acceptors could not be detected in sample A. This is because the only observable would be the decreased donor quantum yield or lifetime. The only way the three acceptors could be detected is from the absorption spectrum, assuming one knows the extinction coefficient for a single acceptor.

D. The apparent distance for an assumed single acceptor can be found from Eq. [14.28]. Numerically, we found $k_T = \tau_D^{-1}(6.88)$, which can be equated to an apparent distance,

$$k_T = \frac{6.88}{\tau_D} = \frac{1}{\tau_D}\left(\frac{R_0}{r_{app}}\right)^6 \quad [14.32]$$

Solving for R_0/r_{app} yields $R_0/r_{app} = 1.38$, so $r_{app} = 14.5$ Å. The extent of energy transfer is thus seen to be dominated by the closest acceptor, at 15 Å. The presence of two more acceptors at 20 and 25 Å only decreases the apparent distance by 0.5 Å.

14.2. A. One acceptor per 60-Å cube corresponds to an acceptor concentration of 8m*M*. Use of Eq. [13.33] with R_0 = 30 Å yields a critical concentration of 17m*M*.
B. A covalently linked acceptor is somewhat equivalent to one acceptor per sphere of 30-Å radius, or 8.84×10^{18} acceptors/cm^3. This is equivalent to an acceptor concentration of 15m*M*. Covalent attachment of an acceptor results in a high effective acceptor concentration.

CHAPTER 15

15.1. Using the data provided in Figure 15.23, one can calculate the following values:

Mol % Rh-PE	Rh-PE/Å^2	Rh-PE/R_0^2	F_{DA}/F_D
0.0	0.0	0.0	1.0
0.2	2.8×10^{-5}	0.071	0.62
0.4	5.7×10^{-5}	0.143	0.40
0.8	11.4×10^{-5}	0.286	0.21
1.2	17.1×10^{-5}	0.429	0.15

The distance of closest approach can be estimated by plotting the last two columns on Figure 15.14. The observed energy-transfer quenching is greater than predicted for no excluded area, so r_c is zero or much less than R_0. This suggests that the donors and acceptors are fully accessible or that perhaps they are somewhat clustered in the PE vesicles. The R_0 value was not reported in Ref. 56.

15.2. The decay times can be used with Eq. [15.20] to obtain the transfer rate, $k_T = 1.85 \times 10^3$ s^{-1}. Dividing by the EB concentration (2.77μ*M*) yields $k_T^b = 6.7 \times 10^8$ M^{-1} s^{-1}.
Using Eq. [15.27] and the values of R_0 and r_c, one finds a maximal bimolecular rate constant of 1.1×10^6 M^{-1} s^{-1}. The measured values could be larger than the theoretical values for two reasons. The positively charged donors may localize around the negatively charged DNA. This would result in a larger apparent concentration of EB. Given the small value of r_c, we cannot exclude the possibility of an exchange contribution to k_T^b.

15.3. To a first approximation, the donor intensity is about 50% quenched when $C/C_0 = 0.5$. This value of C/C_0 can be used to calculate the acceptor concentration in any desired units, as listed in Table 15.3.
Acceptor concentrations near 2m*M* are needed in homogeneous solution. This is generally not practical for proteins because the absorbance due to the acceptor would not allow excitation of the protein. Also, such high concentrations of acceptors are likely to perturb the protein structure.
The situation is much better in lipids and nucleic acids. In this case the acceptors need only to be about 1 per 225 lipids or 1 per 59 base pairs.This favorable situation is the result of a locally high concentration of acceptors due to their localization in the lipid or nucleic acid. The bulk concentration of acceptors can be low and is determined by the bulk concentration of membrane or nucleic acid.

Table 15.3. Approximate Concentrations for 50% Quenching in One, Two, and Three Dimensions

Equation	Concentration for 50% energy transfer
$C_0 = \left(\frac{4}{3}\pi R_0^3\right)^{-1}$	9.55×10^{17} acceptors/cm^3 = 1.59m*M*
$C_0 = (\pi R_0^2)^{-1}$	6.4×10^{11} acceptors/cm^2 = 4.5×10^{-3} acceptors/lipid
$C_0 = (2 R_0)^{-1}$	5×10^5 acceptors/cm = 1.7×10^{-2} acceptors/base pair

CHAPTER 16

16.1. A. Without experimentation, it is not possible to predict how the fluorescence properties of the protein will vary when it is unfolded. In general, one can expect the extent of tyrosine fluorescence to increase when the protein is unfolded. This could be detected by excitation at 280 nm. The tyrosine emission would appear near 308 nm. It is also probable that the fluorescence intensity or the emission maximum of the single tryptophan residue would change as the protein is unfolded. Once the spectral characteristics of the native and unfolded states are determined, the data can be used to quantify the unfolding process. It is important to remember that anisotropy or lifetime measurements may not accurately reflect the fractional populations of the folded and unfolded states. This is particularly true if the quantum yields of the fluorescent residues in the protein change upon unfolding.
B. The extent of exposure to the aqueous phase could be studied by measuring the Stern–Volmer bimolecular quenching constant (k_q) and comparing the measured values with those observed for NATA in the same solvent. One should choose the neutral tryptophan analog to avoid electrostatic effects on the quenching process. The extent of exposure to the aqueous phase can be estimated by comparing the measured quenching constant for the protein with the values found for model compounds.

C. If the protein associates to form a dimer, it is possible that the tryptophan residue becomes shielded from the aqueous phase. In this case one can expect a change in the intensity or emission maximum of the protein. If the tryptophan residue remains exposed to the aqueous phase upon dimer formation, then it is probable that the emission spectrum and intensity will remain the same. In this case the extent of the association should still be detectable from changes in the steady-state anisotropy.

D. In order to measure the distance of the tryptophan to the reactive group, it is necessary to select an appropriate acceptor and to covalently label the protein at the reactive site. It is critical that the protein be completely labeled with acceptor, because any unlabeled fraction will contribute a large amount to the measured intensity, resulting in an underestimation of the distance. Following calculation of the Förster distance R_0 from the spectral properties of the donor and acceptor, the distance can be measured from the decrease in the donor quantum yield due to the presence of acceptor.

E. While not immediately obvious, the extent of energy transfer from a tryptophan donor to an acceptor is expected to change upon association of the acceptor-labeled monomers. This is because, upon dimerization, each tryptophan residue will be brought into proximity of the acceptor on the other subunit. Hence, each tryptophan will transfer to two acceptors, resulting in a higher amount of energy transfer and a lower donor quantum yield in the dimeric state.

16.2. A. Dimerization could be detected from the steady-state intensity or the intensity decay. Upon dimer formation, one will observe a twofold increase in the relative quantum yield. If the dimerization occurs due to a change in protein concentration, then it is necessary to normalize the measured intensities to the same protein concentration. If dimerization occurs as a result of a change in solution conditions, then the intensity change may be observed at a constant protein concentration.

B. Dimer formation could be detected by an increase in the mean lifetime. When dimerization is partially complete, one expects the decay to be a double exponential.

C. Dimerization could not be detected from the steady-state anisotropy. The anisotropy (r) of the monomer and dimer can be calculated using the Perrin equation,

$$r = \frac{r_0}{1 + (\tau/\theta)} \quad [16.5]$$

where τ is the lifetime and θ is the rotational correlation time. The steady-state anisotropy of the monomer (r_M) and that of the dimer (r_D) are both 0.10.

D. Dimerization could be detected by measuring the anisotropy decay, which will display a longer mean correlation time as dimers are formed.

E. When 50% of the monomers have formed dimers, these dimers contribute twice as much as the monomers to the steady-state intensity. Hence, the fractional intensities are $f_M = 0.33$ and $f_D = 0.66$. The steady-state anisotropy is given by

$$r = 0.33 r_M + 0.66 r_D = 0.10 \quad [16.6]$$

and is unchanged during dimerization.

For the intensity decay, we need to calculate the values of α_M and α_D. The relative values are given by $\alpha_M = 0.33/2.5 = 0.13$ and $\alpha_D = 0.66/5.0 = 0.13$. Hence, the intensity decay is given by

$$I(t) = 0.5 \exp(-t/\tau_M) + 0.5 \exp(-t/\tau_D) \quad [16.7]$$

The α_i values are equivalent because we assumed that the intensities and lifetimes both increased by the same amount, meaning that the radiative decay rate stayed the same.

For a mixture of monomers and dimers, the anisotropy decay follows the associated model, where each decay time is associated with one of the correlation times. At any time t, the fractional intensity of the monomer or dimer is given by

$$f_M(t) = \frac{0.5 e^{-t/\tau_M}}{I(t)} \quad [16.8]$$

$$f_D(t) = \frac{0.5 e^{-t/\tau_D}}{I(t)} \quad [16.9]$$

where $I(t)$ is given by Eq. [16.7]. The anisotropy decay is given by

$$r(t) = f_M(t) r_M(t) + f_D(t) r_D(t) \quad [16.10]$$

Hence, this mixture of monomers and dimers displays an associated anisotropy decay.

CHAPTER 17

17.1. The activation energy for any process can be calculated by plotting the logarithm of the rate constant (k) versus the inverse of the temperature in kelvins. For an anisotropy decay, the rotational rate (R) is related to the rotational correlation time (θ) by $\theta = (6R)^{-1}$. The plot of ln(6R) versus (1000/T) is shown in Figure 17.43. The activation energy (E_A) can be calculated from the Arrhenius equation,[159]

$$\ln k = \ln(6R) = -\frac{E_A}{RT} + \ln A \quad [17.4]$$

where A is a constant, and R_g is the gas constant. This equation is simply an expression of the principle that the rate of a process depends on a frequency factor and the energy needed to pass over an energy barrier,

$$k = A\, e^{-E_A/RT} \qquad [17.5]$$

From this analysis (Figure 17.43), one finds $E_A = 6.18$ kcal/mol, which is typical for rotational diffusion of proteins in water. Also, this value is comparable to the activation energy for the temperature-dependent viscosity of water, $E_A = 4.13$ kcal/mol.

The right-hand side of Figure 17.43 represents higher temperatures. At these temperatures, one notices that the rotational rate increases faster than expected from the Arrhenius equation. This is because the protein is beginning to unfold, so that the correlation time is shortened by the additional process of segmental mobility of this tryptophan residue.

The steady-state anisotropy for RNase T_1 at each temperature can be calculated from the Perrin equation,

$$r = \frac{r_0}{1 + (\tau/\theta)} \qquad [17.6]$$

Hence, the values are expected to be 0.151, 0.137, 0.128, 0.114, and 0.104, in order of increasing temperature from −1.5 to 44.4 °C in Table 17.4.

17.2. The cone angle for tryptophan rotational freedom can be calculated from the ratio of the anisotropy amplitude associated with the long correlation time to the total anisotropy. The fractional contribution of the long correlation time (t_L) is given by

$$f_L = \frac{r_{01}}{r_{01} + r_{02}} \qquad [17.7]$$

This fraction can be related to the displacement of the transition dipole according to the definition of anisotropy:

$$\cos^2\beta = \frac{2f_L + 1}{3} \qquad [17.8]$$

Alternatively, this fraction can be related to the angle (θ_c) through which the tryptophan rotates before striking an energy barrier:

$$f_L = \left[\frac{1}{2}\cos\theta_c(1 + \cos\theta_c)\right]^2 \qquad [17.9]$$

Application of these expressions to the data in Table 17.5 yields the results presented in Table 17.9.

17.3. The time-zero anisotropies [$r(0)$] for RNase T_1 in Table 17.4 are from TD data and are lower than those from other reports. One possible origin of the difference is the shorter excitation wavelength (295 nm) for the TD data and the possibility of a small error in the reported excitation wavelength. Another difference is that $r(0)$ was a variable parameter in the TD measurements, whereas $r(0)$ was fixed during analysis of the FD data. It is possible that a short component in the anisotropy decay was missed in the TD data, as suggested by molecular dynamics simulations for RNase T_1.[57]

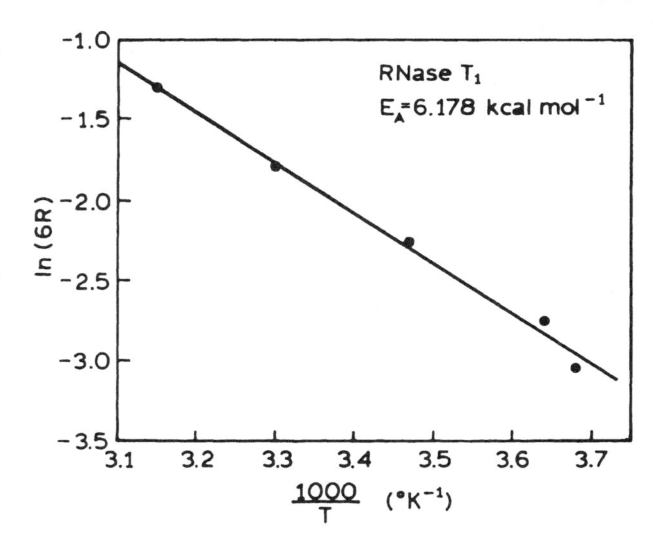

Figure 17.43. Arrhenius plot for the rotational correlation times of RNase T_1. Data from Ref. 50.

17.4. A. The intensity decays more slowly at longer emission wavelengths. This indicates that the mean decay time is increasing. In this case the effect is due to an increasing fractional contribution of the long-lived component (9.8 ns).

B. The DAS are calculated using the data in Table 17.8 and Eq. [17.3], resulting in the DAS shown in Figure 17.44. In order to interpret the DAS, one has to assume that each decay time (3.8 or 9.8 ns) is associated with one of the tryptophan residues. Using this assumption, the 3.8-ns decay time is associated with a blue-shifted emission and a lower quantum yield than the red-shifted 9.8-ns residue.

C. The most rigorous way to confirm assignment of the DAS is to create the single-tryptophan mutants. Each

Table 17.9. Angular Freedom of NATA and Tryptophan Residues in Single-Tryptophan Peptides and Proteins[a]

Species	$r_0 = r_{01} + r_{02}$	f_L	β (deg)	θ_c (deg)
RNase T_1	0.310	1.00	0.0	0.0
Staphylococcal nuclease	0.321	0.944	11.1	11.2
Monellin	0.315	0.768	23.1	23.8
ACTH	0.308	0.386	39.8	43.8
Gly-trp-gly	0.325	0.323	42.2	47.3
NATA	0.323	0.00	54.7	90.0
Melittin monomer	0.323	0.421	38.4	41.9
Melittin tetramer	0.326	0.638	29.4	30.8

[a]At 20 °C.

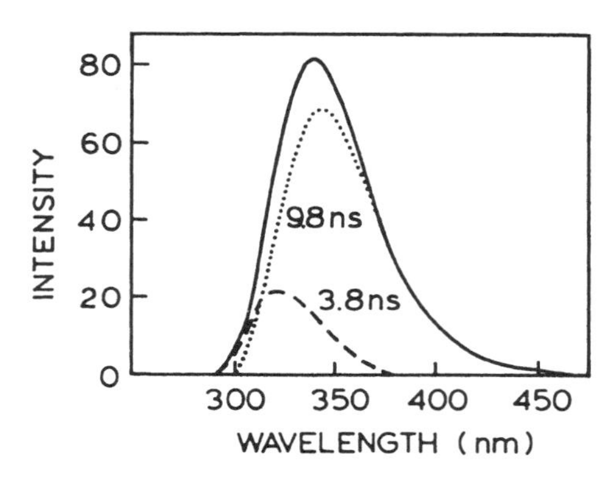

Figure 17.44. Decay-associated spectra calculated from Table 17.8.

mutant should display one of the calculated DAS. One could also use quenching by iodide or acrylamide with the two-tryptophan wild-type protein. In this case one expects the 9.8-ns emission to be more sensitive to quenching, given its longer lifetime and higher exposure to the aqueous phase. The emission spectra could also be resolved by the quenching-resolved method (Section 16.5.C).

CHAPTER 18

18.1. The lifetime of the R state (τ_{OR}) can be calculated from the phase-angle difference $\Delta\phi = \phi_R - \phi_F$. At 100 MHz this difference is 58°, which corresponds to a lifetime of 16 ns.

18.2. A. Acridine and acridinium may be reasonably expected to display distinct absorption spectra. The emission spectrum in 0.2M NH_4NO_3 (Figure 18.11) shows evidence for emission from both acridine and acridinium. Hence, if both species were present in the ground state, the absorption spectrum in 0.2M NH_4NO_3 should be a composite of the absorption spectra in 0.05M H_2SO_4 and in 0.05M NaOH. In contrast, if the acridinium is formed only in the excited state, then the absorption spectrum in 0.2M NH_4NO_3 should be almost identical to that of neutral acridine.

B. Examination of the data in Table 18.4 reveals two decay times which are independent of emission wavelength. This indicates that there are two emitting species and that their decay rates are independent of emission wavelength. On the short-wavelength side of the emission, the decay is a single exponential. This indicates that the reaction is irreversible and that the measured decay times at other wavelengths contain contributions from both acridine and acridinium. Proof of an excited-state reaction is provided by observation of negative preexponential factors. As the observation wavelength is increased, the term having a negative preexponential factor becomes more predominant. At the longest observation wavelengths, one finds that the preexponential factors are nearly equal in magnitude and opposite in sign. This near equality of the preexponential factors indicates that at 560 nm the emission is predominantly from the reacted species. The fact that the magnitude of α_2 is slightly larger than that of α_1 indicates that there is still some emission from neutral acridine at 560 nm.

CHAPTER 19

19.1. The data in Figure 19.13 can be used to determine the lifetimes of camphorquinone in PMMA at various partial pressures of oxygen. These values can be used to construct a lifetime Stern–Volmer plot (Figure 19.71). The oxygen bimolecular quenching constant is near $2.8 \times 10^6\ M^{-1}\ s^{-1}$, which is nearly 10^4 smaller than that for oxygen in water. This suggests a low diffusion coefficient of oxygen in PMMA of about 7×10^{-9} cm^2/s. Of course, the accuracy of these values depends on the assumed oxygen solubility in PMMA.

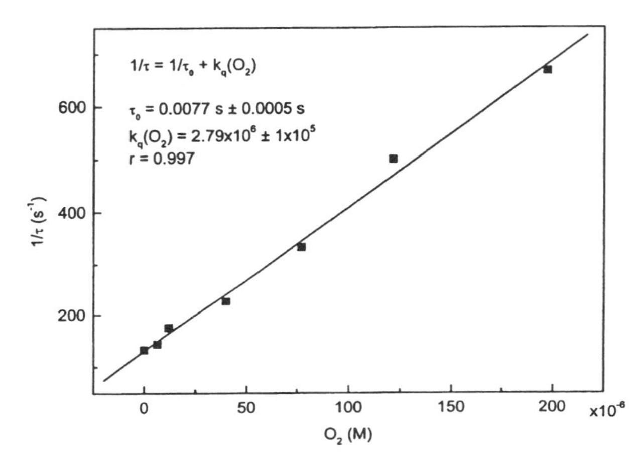

Figure 19.71. Stern–Volmer plot for oxygen quenching of camphorquinone in PMMA. Data from Ref. 39.

19.2. Examination of Figure 19.9 reveals that $[Ru(Ph_2phen)_3]^{2+}$ is quenched 10-fold at 30 torr oxygen. The Stern–Volmer quenching constant is proportional to the lifetime, which is near 5 μs. Hence, for the 5-ns probe, $F_0/F = 1.009$ at this same oxygen pressure. At the highest oxygen pressure of 80 torr, the short-lifetime probe will be quenched less than 3%. This extent of quenching is negligible, so the 5-ns probe can serve as an intensity reference.

19.3. A. The range of anisotropies can be calculated from the Perrin equation,

$$r = \frac{r_0}{1 + (\tau/\theta)} \qquad [19.15]$$

The anisotropy of the free peptide will be 0.080, and the anisotropy of Ab-Fl-P will be 0.385.

B. The anisotropies are additive (Eq. [19.14]). Hence, the anisotropy with 10% free Fl-P is given by

$$r = 0.10(0.08) + 0.90(0.385) = 0.364 \quad [19.16]$$

C. Displacement of Fl-P from Rh-Ab will result in a 10-fold increase in the intensity of Fl-P, due to the elimination of RET. For such cases, the fractional intensities of the free and bound forms are given by

$$f_F = \frac{m_F q_F}{m_F q_F + m_B q_B} \quad [19.17]$$

$$f_B = \frac{m_B q_B}{m_F q_F + m_B q_B} \quad [19.18]$$

where the m_i are the molecular fractions in the free or bound state, and the q_i are the quantum yields. If $q_F = 10q_B$, then

$$f_F = \frac{10m_F}{10m_F + m_B} \quad [19.19]$$

For a molecular fraction of 10%, $f_F = 0.53$. Hence, 10% displacement of Fl-P results in over 50% of the emission from the free peptide. The anisotropy is

$$r = 0.53(0.08) + 0.47(0.40) = 0.230 \quad [19.20]$$

and is seen to decrease more rapidly with displacement of Fl-P. We used 0.40 for the anisotropy of the bound form because RET will decrease the lifetime of the fluorescein to 0.40 ns.

CHAPTER 20

20.1. A. The decay time can be calculated from the slope of the long-lifetime component using any two points. For instance, extrapolating the long decay to zero the intensities of the long delay time component at $t = 0$ and $t = 500$ ns in Figure 20.41 are near 2000 and 600, respectively. For a single-exponential decay, the intensities at two points in time are related by $\ln I(t_1) - \ln I(t_2) = -t_1/\tau + t_2/\tau$. Insertion of the values at $t = 0$ and $t = 500$ ns yields $\tau = 415$ ns.
B. $\alpha_1 = 0.096$, $\alpha_2 = 0.04$. These values are from Figure 20.41, following normalization to $\alpha_1 + \alpha_2 = 1.0$.
C. $f_1 = 0.296, f_2 = 0.704$.
D. $f_1 = 0.0004$, $f_2 = 0.9996$. This result shows that off-gating essentially eliminates the short-lived component, decreasing its fractional contribution from 0.296 to 0.0004. These values were obtained by computer integration of the delay law.

20.2. The oxygen bimolecular quenching constant can be calculated using the decay times in the absence and in the presence of 100% oxygen. The value of $\tau_0/\tau = 16.3 = 1 + k_q\tau_0\,[O_2]$. Using $[O_2] = 0.001275M$ and $\tau_0 = 3.8$ µs, one obtains $k_q = 3.16 \times 10^9\ M^{-1}\ s^{-1}$. This value is reasonably close to the diffusion-controlled limit and indicates that the quenching by oxygen is highly efficient.

CHAPTER 21

21.1. In order to answer this question, we need to design the quenching or anisotropy measurements which could potentially be used for sequencing. One can imagine sequencing with four fluorescent ddNTPs. The Stern–Volmer quenching constants could be different due to either different lifetimes or different accessibilities to the collisional quencher. Then, the sequence could be determined by the quenching constant for each fluorescent band on the gel. Determination of the quenching constant requires a minimum of two intensity measurements, in the absence of quencher and in the presence of a known concentration of quencher. Although such measurements are possible, the use of two samples to measure a single base is too complicated for large-scale sequencing of DNA.

Suppose that the sequencing reaction is performed with a single fluorescent primer. Because anisotropy measurements depend on molecular weight, in principle each oligonucleotide will display a different anisotropy. In practice, the anisotropies for DNA oligomers differing by a single base pair are likely to be too similar in magnitude for useful distinction between oligomers. If the adjacent base pair changes the lifetime of the labeled oligomer, then the anisotropy measurements may be able to identify the base.

One can also imagine the use of four fluorescent ddNTPs, each with a different lifetime. In this case the anisotropy would be different for each base pair, and the anisotropy measurement could be used to identify the base. This approach is more likely to succeed for longer oligomers, for which the anisotropy will become mostly independent of molecular weight. For shorter oligomers, the anisotropy will depend on the fragment length.

CHAPTER 22

22.1. The wavelength of the Raman peak (410 nm) is equivalent to 24,390 cm^{-1}. The Raman peak of water is typically shifted 3600 cm^{-1}. Hence, the excitation wavelength is at 27,990 cm^{-1}, or 357 nm.

22.2. The scattered light has an effective lifetime of zero. Hence, the scattered light can be suppressed with $\phi_D = 90°$. The phase angle of quinine sulfate at 10 MHz can be calculated from $\phi = \mathrm{atan}(\omega\tau) = 51.5°$. The maximum phase-sensitive intensity for quinine sulfate would be observed with ϕ_D = 51.5°. The scattered light is suppressed with ϕ_D = 90°. At this phase angle, the phase-sensitive intensity is attenuated by a factor of $\cos(\phi_D - \phi) = \cos(90 - 51.5) = 0.78$ relative to the phase-sensitive intensity with ϕ_D = 51.5°.

22.3. The detector phases of 17.4° + 90° and 32.1° – 90° are out-of-phase with DNS and DNS-BSA, respectively. This is known because at ϕ_D = 32.1° – 90°, only free DNS is detected. In the equimolar mixture of DNS and BSA, the phase-sensitive intensity of DNS is decreased by 50%. Hence, 50% of the DNS is bound to BSA. Similarly, at ϕ_D = 17.4° + 90°, where only the fluorescence of the DNS–BSA complex is detected, the intensity is 50% of that for the second solution, in which DNS is completely bound. Hence, 50% of the DNS is bound.

The phase-sensitive intensities of the first two solutions may be rationalized as follows. Upon addition of a saturating amount of BSA, all the DNS is bound. Therefore, its contribution to the signal at ϕ_D = 32.1° – 90° is eliminated. The intensity increases twofold and now is observed with ϕ_D = 17.4° + 90°. However, a twofold increase in intensity is not observed because the signal from the bound DNS is more demodulated than that of the free DNS. Specifically, these values are 0.954 and 0.847 for 5 and 10 ns, respectively. Hence, the expected twofold increase in fluorescence intensity is decreased by a factor of 0.847/0.954 = 0.888.

22.4. The viscosity of propylene glycol changes dramatically with temperature, which affects the rate of solvent relaxation. At an intermediate temperature of –10 °C, the relaxation time is comparable to the lifetime. Under these conditions, the emission spectrum contains components of the unrelaxed initially excited state (F) and the relaxed excited state (R). Suppression on the red side of the emission (410 nm) results in recording of the emission spectrum of the F state. Suppression of the blue side of the emission (310 nm) results in recording of the emission spectrum of the R state. Of course, these are only the approximate spectra of these states, but the phase-sensitive spectra appear to show the positions of the unrelaxed and relaxed emission spectra.

At very low temperature (–60 °C), all the emission is from the unrelaxed state, and at high temperature (40 °C) all the emission is from the relaxed state. Since there is only one lifetime across the emission spectra, suppression on either side of the emission suppresses the entire emission spectrum.

Index